LDA	lithium diisopropylamide
LDMAN	lithium 1-(dimethylamino)naphthalenide
LHMDS	=LiHMDS
LICA	lithium isopropylcyclohexylamide
LiHMDS	lithium hexamethyldisilazide
LiTMP	lithium 2,2,6,6-tetramethylpiperidide
LTMP	=LiTMP
LTA	lead tetraacetate
lut	lutidine
m-CPBA	*m*-chloroperbenzoic acid
MA	maleic anhydride
MAD	methylaluminum bis(2,6-di-*t*-butyl-4-methylphenoxide)
MAT	methylaluminum bis(2,4,6-tri-*t*-butylphenoxide)
Me	methyl
MEK	methyl ethyl ketone
MEM	(2-methoxyethoxy)methyl
MIC	methyl isocyanate
MMPP	magnesium monoperoxyphthalate
MOM	methoxymethyl
MoOPH	oxodiperoxomolybdenum(pyridine)-(hexamethylphosphoric triamide)
mp	melting point
MPM	=PMB
Ms	mesyl (methanesulfonyl)
MS	mass spectrometry; molecular sieves
MTEE	methyl *t*-butyl ether
MTM	methylthiomethyl
MVK	methyl vinyl ketone
n	refractive index
NaHDMS	sodium hexamethyldisilazide
Naph	naphthyl
NBA	*N*-bromoacetamide
nbd	norbornadiene (bicyclo[2.2.1]hepta-2,5-diene)
NBS	*N*-bromosuccinimide
NCS	*N*-chlorosuccinimide
NIS	*N*-iodosuccinimide
NMO	*N*-methylmorpholine *N*-oxide
NMP	*N*-methyl-2-pyrrolidinone
NMR	nuclear magnetic resonance
NORPHOS	bis(diphenylphosphino)bicyclo[2.2.1]-hept-5-ene
Np	=Naph
PCC	pyridinium chlorochromate
PDC	pyridinium dichromate
Pent	*n*-pentyl
Ph	phenyl
phen	1,10-phenanthroline
Phth	phthaloyl
Piv	pivaloyl
PMB	*p*-methoxybenzyl
PMDTA	*N,N,N′,N″,N″*-pentamethyldiethylenetriamine
PPA	polyphosphoric acid
PPE	polyphosphate ester
PPTS	pyridinium *p*-toluenesulfonate
Pr	*n*-propyl
PTC	phase transfer catalyst/catalysis
PTSA	*p*-toluenesulfonic acid
py	pyridine
RAMP	(*R*)-1-amino-2-(methoxymethyl)pyrrolidine
rt	room temperature
salen	bis(salicylidene)ethylenediamine
SAMP	(*S*)-1-amino-2-(methoxymethyl)pyrrolidine
SET	single electron transfer
Sia	siamyl (3-methyl-2-butyl)
TASF	tris(diethylamino)sulfonium difluorotrimethylsilicate
TBAB	tetrabutylammonium bromide
TBAF	tetrabutylammonium fluoride
TBAD	=DBAD
TBAI	tetrabutylammonium iodide
TBAP	tetrabutylammonium perruthenate
TBDMS	*t*-butyldimethylsilyl
TBDPS	*t*-butyldiphenylsilyl
TBHP	*t*-butyl hydroperoxide
TBS	=TBDMS
TCNE	tetracyanoethylene
TCNQ	7,7,8,8-tetracyanoquinodimethane
TEA	triethylamine
TEBA	triethylbenzylammonium chloride
TEBAC	=TEBA
TEMPO	2,2,6,6-tetramethylpiperidinoxyl
TES	triethylsilyl
Tf	triflyl (trifluoromethanesulfonyl)
TFA	trifluoroacetic acid
TFAA	trifluoroacetic anhydride
THF	tetrahydrofuran
THP	tetrahydropyran; tetrahydropyranyl
Thx	thexyl (2,3-dimethyl-2-butyl)
TIPS	triisopropylsilyl
TMANO	trimethylamine *N*-oxide
TMEDA	*N,N,N′,N′*-tetramethylethylenediamine
TMG	1,1,3,3-tetramethylguanidine
TMS	trimethylsilyl
Tol	*p*-tolyl
TPAP	tetrapropylammonium perruthenate
TBHP	*t*-butyl hydroperoxide
	…lporphyrin
	…enylmethyl)
	…luenesulfonyl)
	…I) nitrate
	…gen peroxide complex
Z	=Cbz

D1294300

Handbook of Reagents for Organic Synthesis

Activating Agents and Protecting Groups

OTHER TITLES IN THIS COLLECTION

Reagents, Auxiliaries and Catalysts for C–C Bond Formation
Edited by Robert M. Coates and Scott E. Denmark
ISBN 0 471 97924 4

Oxidizing and Reducing Agents
Edited by Steven D. Burke and Rick L. Danheiser
ISBN 0 471 97926 0

Acidic and Basic Reagents
Edited by Hans J. Reich and James H. Rigby
ISBN 0 471 97925 2

Handbook of Reagents for Organic Synthesis

Activating Agents and Protecting Groups

Edited by

Anthony J. Pearson
Case Western Reserve University

and

William R. Roush
University of Michigan

JOHN WILEY & SONS
Chichester • New York • Weinheim • Brisbane • Toronto • Singapore

WITHDRAWN

Copyright © 1999 John Wiley & Sons Ltd
Baffins Lane, Chichester
West Sussex PO19 1UD, UK

National 01243 779777
International (+44) 1243 779777
e-mail (for orders and customer service enquiries): cs-books@wiley.co.uk
Visit our Home Page on
http://www.wiley.co.uk
or http:/www.wiley.com

QD
77
.H37
1999
v.4

Reprinted February 2000

All rights reserved. No part of this publication may be reproduced, stored in a retrieval system, or transmitted, in any form or by any means, electronic, mechanical, photocopying, recording or otherwise, except under the terms of the Copyright Designs and Patents Act 1988 or under the terms of a license issued by the Copyright Licensing Agency, 90 Tottenham Court Road, London, W1P 9HE, UK, without the permission in writing of the publisher.

Other Wiley Editorial Offices

John Wiley & Sons Inc., 605 Third Avenue,
New York, NY 10158-0012, USA

Wiley-VCH Verlag GmbH, Pappelallee 3,
D-69469 Weinheim, Germany

Jacaranda Wiley Ltd, 33 Park Road, Milton,
Queensland 4064, Australia

John Wiley & Sons (Asia) Pte Ltd, 2 Clementi Loop #02-01,
Jin Xing Distripark, Singapore 129809

John Wiley & Sons (Canada) Ltd, 22 Worcester Road,
Rexdale, Ontario M9W 1L1, Canada

Library of Congress Cataloguing-in-Publication Data

Handbook of reagents for organic synthesis.
p. cm.
Includes bibliographical references.
Contents: [1] Reagents, auxiliaries, and catalysts for C–C bond formation / edited by Robert M. Coates and Scott E. Denmark [2] Oxidising and reducing agents / edited by Steven D. Burke and Riek L. Danheiser [3] Acidic and basic reagents / edited by Hans J. Reich and James H. Rigby [4] Activating agents and protecting groups / edited by Anthony J. Pearson and William R. Roush
ISBN 0-471-97924-4 (v. 1). ISBN 0-471-97926-0 (v. 2)
ISBN 0-471-97925-2 (v. 3) ISBN 0-471-97927-9 (v. 4)
1. Chemical tests and reagents. 2. Organic compounds–Synthesis.
QD77.H37 1999
547′.2 dc 21 98-53088
CIP

ADJ-5791

British Library Cataloguing in publication Data

A catalogue record for this book is available from the British Library

ISBN 0 471 97927 9

Typeset by Thomson Press (India) Ltd., New Delhi
Printed and bound in Great Britian by Antony Rowe, Chippenham, Wilts
This book is printed on acid-free paper responsibly manufactured from sustainable forestry, in which at least two trees are planted for each one used in paper production.

Editorial Board

Editor-in-Chief

Leo A. Paquette
The Ohio State University, Columbus, OH, USA

Editors

Steven D. Burke
University of Wisconsin at Madison, WI, USA

Scott E. Denmark
University of Illinois at Urbana-Champaign, IL, USA

Dennis C. Liotta
Emory University Atlanta, CA, USA

Robert M. Coates
University of Illinois at Urbana-Champaign, IL USA

David J. Hart
The Ohio State University Columbus, OH, USA

Anthony J. Pearson
Case Western Reserve University, Cleveland, OH, USA

Rick L. Danheiser
Massachusetts Institute of Technology, Cambridge, MA, USA

Lanny S. Liebeskind
Emory University Atlanta, GA, USA

Hans J. Reich
University of Wisconsin at Madison, WI, USA

James H. Rigby
Wayne State University Detroit, MI, USA

William R. Roush
University of Michigan MI, USA

Assistant Editors

James P. Edwards
Ligand Pharmaceuticals San Diego, CA, USA

Mark Volmer
Emory University Atlanta, GA, USA

International Advisory Board

Leon A. Ghosez
Université Catholique de Louvain, Belgium

Chun-Chen Liao
National Tsing Hua University, Hsinchu, Taiwan

Ryoji Noyori
Nagoya University, Japan

Pierre Potier
CNRS, Gif-sur-Yvette France

Hishashi Yamomoto
Nagoya University, Japan

Jean-Marie Lehn
Université Louis Pasteur Strasbourg, France

Lewis N. Mander
Australian National University, Canberra Australia

Gerald Pattenden
University of Nottingham UK

W. Nico Speckamp
Universiteit van Amsterdam The Netherlands

Steven V. Ley
University of Cambridge UK

Giorgio Modena
Università di Padua Italy

Edward Piers
University of British Columbia, Vancouver Canada

Ekkehard Winterfeldt
Universität Hannover Germany

Managing Editor

Colin J. Drayton
Woking, Surrey, UK

Contents

Preface

As stated in its Preface, the major motivation for our undertaking publication of the *Encyclopedia of Reagents for Organic Synthesis* was "to incorporate into a single work a genuinely authoritative and systematic description of the utility of all reagents used in organic chemistry." By all accounts, this reference compendium has succeeded admirably in attaining this objective. Experts from around the globe contributed many relevant facts that define the various uses characteristic of each reagent. The choice of a masthead format for providing relevant information about each entry, the highlighting of key transformations with illustrative equations, and the incoroporation of detailed indexes serve in tandem to facilitate the retrieval of desired information.

Notwithstanding these accomplishments, the editors have since recognized that the large size of this eight-volume work and its cost of purchase have often served to deter the placement of copies of the *Encyclopedia* in or near laboratories where the need for this type of insight is most critically needed. In an effort to meet this demand in a cost-effective manner, the decision was made to cull from the major work that information having the highest probability for repeated consultation and to incorporate same into a set of handbooks. The latter would also be purchasable on a single unit basis.

The ultimate result of these deliberations is the publication of the *Handbook of Reagents for Organic Synthesis* consisting of the following four volumes:

Reagents, Auxiliaries and Catalysts for C–C Bond Formation
Edited by Robert M. Coates and Scott E. Denmark

Oxidizing and Reducing Agents
Edited by Steven D. Burke and Rick L. Danheiser

Acidic and Basic Reagnets
Edited by Hans J. Reich and James H. Rigby

Activating and Protecting Groups
Edited by Anthony J. Pearson and William R. Roush

Each of the volumes contains a complete compilation of those entries from the original *Encyclopedia* that bear on the specific topic. Ample listings can be found to functionally related reagents contained in the original work. For the sake of current awareness, references to recent reviews and monographs have been included, as have relevant new procedures from *Organic Syntheses*.

The end product of this effort by eight of the original editors of the *Encyclopedia* is an affordable, enlightening set of books that should find their way into the laboratories of all practicing synthetic chemists. Every attempt has been made to be of the broadest synthetic relevance and our expectation is that our colleagues will share this opinion.

Leo A. Paquette
Columbus, Ohio USA

Introduction

The combination of reagents included in this volume reflects the fact that protecting groups and activation procedures are often used in combination, one example being in peptide synthesis, where an amino group of one amino acid component must be blocked before its carboxylic acid is activated for coupling with a second amino acid to form the amide bond. There are many other instances in the synthesis of natural and unnatural products, pharmaceuticals, oligosaccharides, and oligonucleotides, etc., where similar tactics must be employed to prevent undesired activation or reaction of functionality, such as hydroxyl, when more than one such group is present, or to prevent reactive functional groups from entering into unwanted reactions with oxidizing agents, reducing agents, or organometallic reagents commonly employed in organic synthesis. Accordingly, the most important reagents used to protect amines, alcohols, carboxyl, carbonyl and other reactive functional groups are included in this volume.

The selection of activating reagents includes both well known and less traditional ones. Thus, typical peptide coupling reagents that activate carboxylic acids, such as dicyclohexylcarbodiimide, are listed in this volume, in addition to reagents that are not immediately identified as activators. One example of the latter is hexacarbonylchromium, which may be used to activate aromatic substrates toward nucleophilic addition and substitution, via the formation of arenechromium tricarbonyl complexes. Another example is nonacarbonyldiiron, which can serve multiple purposes in activating alkenes and dienes toward nucleophilic attack, or allowing their conversion to cationic allyl- or dienyl- complexes, as well as protecting the same functionality from reactions such as hydroboration, Diels-Alder cycloadditions, etc. Transition metal systems that perform these types of functions could have formed a separate volume if one includes catalytic processes under the heading of activation. To avoid a volume of unmanageable size, the choice of these reagents has been limited to those that are used stoichiometrically, and that are relatively familiar to the organic chemistry community.

Some reagents, such as hexamethylphosphoric triamide (HMPA) and N,N,N',N'-tetramethylethylenediamine (TMEDA) that "activate" enolates and alkyllithium reagents and increase their nucleophilicity, thereby facilitating their reactions, are also included. A number of Lewis acids appear in this volume, including the alkylaluminum halides and the boron halides, as examples of reagents that activate various functional groups by increasing their electrophilicity. The complete entries for all lewis acids and nucleophilic catalysts (e.g. dimethylaminopyridine) also appear in the volume on Acidic and Basic Reagents.

There are many reagents that may be considered as activating in the broadest sense. The phosphorus halides, for example, can be used to activate hydroxyl groups of alcohols or carboxylic acids by converting them to halide leaving groups for nucleophilic substitution or elimination, while the corresponding sulfonate esters activate alcohols in the more traditional manner. Reagents such as N,N'-thiocarbonyldiimidazole and phenyl chlorothionocarbonate, which serve to activate alcohols for subsequent deoxygenation reactions with a trialkyltin hydride reagent, and (methoxycarbonylsulfamoyl)triethylammonium hydroxide, which facilitates the dehydrative elimination of alcohols to alkenes, also qualify as activating reagents in the broadest sense and are included in the present volume. As many examples of activating agents are included in this volume as possible but, again, an effort has been made to produce a work that is not too voluminous in scope.

Finally, there are many reagents that perform functions other than those that are the immediate subject matter of this volume. No attempt has been made to trim the original entries that were prepared for the *Encyclopedia of Reagents for Organic Synthesis*, since we recognize the value of having as much information as possible about each reagent, thus allowing their optimal use in situations where side reactions might be a problem when limited information is at hand.

In preparing this volume we have been aware of the fact that the original *Encyclopedia* entries were written several years ago, and so may not be completely up to date with regard to literature citations. This is inevitable in a work of this kind, but we have tried to ameliorate the problem as much as possible by including references to relevant articles from *Organic Syntheses* Volumes 69–75 that either deal with the preparation of a particular reagent or illustrate its application, as well as recent (since 1993) review articles and monographs that focus on various aspects of the subject matter of this particular volume. For this purpose we have included reviews that may not be directly connected with any particular reagent, but that may be useful to the practicing organic chemist in seeking information concerning use of the various technologies described in the present work. Finally, we have also included expanded lists of "Related Reagents" for each entry which will allow the reader to locate additional information about additional related reagents and methods in the original *Encyclopedia*.

Anthony J. Pearson
Department of Chemistry
Case Western Reserve University
Cleveland, Ohio

William R. Roush
Department of Chemistry
University of Michigan
Ann Arbor, Michigan

Organic Syntheses References

I. Alcohol Activation (Substitution or Elimination Reactions)

Pansare, S. V.; Arnold, L. D.; Vederas, J. C. "Synthesis of *N-tert*-Butyloxycarbonyl-L-serine β-Lactone and the *p*-Toluenesulfonic Acid Salt of (*S*)-3-Amino-2-oxetanone" **OS**, (1991), *70*, 10.

DEAD, Ph_3P (40%)

Dodge, J. A.; Nissen, J. S.; Presnell, M. "A General Procedure for Mitsunobu Inversion of Sterically Hindered Alcohols: Inversion of Menthol. (1*S*, 2*S*, 5*R*)-5-Methyl-2-(1-methylethyl)cyclohexyl 4-Nitrobenzoate" **OS**, (1995), *73*, 110.

$4\text{-}NO_2C_6H_4CO_2H$, PPh_3, EtO_2CNNCO_2Et, THF (85.6%)

Thompson, A. S.; Hartner, F. W., Jr.; Grabowski, E. J. J. "Ethyl (*R*)-2-Azidopropionate" **OS**, (1997), *75*, 31.

$(PhO)_2PON_3$, DBU/THF (57%)

Lynch, K. M.; Dialey, W. P. "3-Chloro-2-(chloromethyl)-1-propene" **OS**, (1997), *75*, 89.

$$C(CH_2OH)_4 \xrightarrow[2.\ SOCl_2]{1.\ \text{pyridine}} HOCH_2C(CH_2Cl)_3 \quad (57\%)$$

Krakowiak, K. E.; Bradshaw, J. S. "4-Benzyl-10,19-diethyl-4,10,19-triaza-1,7,1,16-tetraoxacycloheneicosane (Triaza-21-Crown-7)" **OS**, (1991), *70*, 129.

$SOCl_2$, $CHCl_3$ (73–77%)

Arnold, H.; Overman, L. E.; Sharp, M. J.; Witschel, M. C. "(*E*)-1-Benzyl-3-(1-iodoethylidene)piperidine: Nucleophile-Promoted Alkyne-Iminium Ion Cyclizations" **OS**, (1991), *70*, 111.

MsCl, Et_3N (93–100%)

Behrens, C.; Paquette, L. A. "*N*-Benzyl-2,3-azetidinedione" **OS**, (1997), *75*, 106.

1. MsCl, Et_3N, CH_2Cl_2 2. DBU (80%)

Stille, J. K.; Echavarren, A. M.; Williams, R. M.; Hendrix, J. A. "4-Methoxy-4′-nitrobiphenyl" **OS**, (1992), *71*, 97.

Tf_2O, pyridine (91%)

II. Alcohol Functionalization or Protection:

(a) Acylation

Eberle, M.; Missbach, M.; Seebach, D. "Enantioselective Saponification with Pig Liver Esterase (PLE): (1*S*, 2*S*, 3*R*)-3-Hydroxy-2-nitrocyclohexyl Acetate" **OS**, (1990), *69*, 19.

1. Ac_2O/H^+ 2. H_2O (96%)

Sessler, J. L.; Mozaffari, A.; Johnson, M. R. "3,4-Diethylpyrrole and 2,3,7,8,12,13,17,18-Octaethylporphyrin" **OS**, (1991), *70*, 68.

$+ Ac_2O$, H_2SO_4 (90%)

Sun, R. C.; Okabe, M. "(2*S*, 4*S*)-2,4,5-Trihydroxypentanoic Acid 4,5-Acetonide Methyl Ester" **OS**, (1993), *72*, 48.

1. Br_2, K_2CO_3, H_2O 2. Ac_2O, AcOH, 50°C (72–74%)

Deardorff, D. R.; Windham, C. Q.; Craney, C. L. "Enantioselective Hyrolysis of *cis*-3,5-Diacetoxycyclopentene: (1*R*, 4*S*)-(+)-4-Hydroxy-2-Cyclopentenyl Acetate" **OS**, (1995), *73*, 25.

Ac_2O, imidazole (96–98%)

Braun, M.; Graf, S.; Herzog, S. "(*R*)-(+)-2-Hydroxy-1,2,2-Triphenylethyl Acetate" **OS**, (1993), *72*, 32.

Ph, HO, OH, Ph, Ph — AcCl, Pyridine / CH_2Cl_2 → Ph, AcO, OH, Ph, Ph (92%)

Furuta, K.; Gao, Q.-Z.; Yamamoto, H. "Chiral (Acycloxy)borane Complex-Catalyzed Asymmetric Diels-Alder Reaction: (1*R*)-1,3,4-Trimethyl-3-cyclohexene-1-carboxaldehyde" **OS**, (1993), *72*, 86.

HO, CO_2CH_2Ph, HO, CO_2CH_2Ph + OMe, O, Cl, OMe — Et_3N, DMAP (cat.) / CH_2Cl_2 → OMe, O, CO_2CH_2Ph, O, CO_2CH_2Ph, OH, OMe (78–82%)

(b) Silylation

Tius, M. A.; Kannangara, G. S. K. "Benzoannelation of Ketones: 3,4-Cyclodedeceno-1-methylbenzene" **OS**, (1992), *71*, 158.

O, OH — 1. TMSCl, Et_3N; 2. MgCl, Et_2O → HO, OTMS

Paquette, L. A.; Earle, M. J.; Smith, G. F. "(4*R*)-(+)-*tert*-Butyldimethylsiloxy-2-cyclopenten-1-one" **OS**, (1995), *73*, 36.

O, OH — TBDMSCl / Et_3N, DMAP, CH_2Cl_2 → O, OTBDMS (64%)

Paquette, L. A.; Heidelbaugh, T. M. "(4*S*)-(−)-*tert*-Butyldimethylsiloxy-2-cyclopenten-1-one" **OS**, (1995), *73*, 44.

HO, OAc — TBDMSCl / Et_3N, DMAP, CH_2Cl_2 → TBDMSO, OAc (77–80%)

Mann, J.; Weymouth-Wilson, A. C. "Photoinduced-Addition of Methanol to (5*S*)-(5-*O*-*tert*-Butyldimethylsiloxymethyl)furan-2(5*H*)-one: (4*R*, 5*S*)-4-Hydroxymethyl-(5-*O*-*tert*-Butyldimethylsiloxymethyl)furan-2(5*H*)-one" (**OS**, (1997), *75*, 139.

O, O, CO_2Me — 1. H^+, MeOH; 2. TBDMSCl, imidazole → TBDMSO, O, O (93%)

Overman, L. E.; Rishton, G. M. "3-(*S*)-[(*tert*-Butyldiphenylsilyl)oxy]-2-butanone" **OS**, (1992), *71*, 56.

O, EtO, OH — TBDPSCl / imidazole, THF → O, EtO, OTBDPS (99%)

Wipf, P.; Xu, W. Allylic Alcohols by Alkene Transfer from Zirconium to Zinc: 1-[(*tert*-Butyldiphenylsilyl)oxy]-dec-3-en-5-ol" **OS**, (1996), *74*, 205.

HO — TBDPSCl / imidazole, DMAP, CH_2Cl_2 → TBDPSO (96%)

(c) Ether Formation

Bailey, W. F.; Carson, M. W.; Zarcone, L. M. J. "Selective Protection of 1,3-Diols at the More Hindered Hydroxy Group: 3-(Methoxymethoxy)-1-Butanol" **OS**, (1997), *75*, 177.

O, O, CH_3 — AcCl / cat. $ZnCl_2$, Et_2O → OAc, O, Cl, CH_3 — CH_3OH, (*i*-Pr)$_2$NEt / Et_2O → OAc, O, OCH_3, CH_3 (77–91%)

Tamao, K.; Nakagawa, Y.; Ito, Y. "Regio- and Stereoselective Intramolecular Hydrosilylation of α-Hydroxy Enol Ethers: 2,3-*syn*-2-Methoxymethoxy-1,3-nonanediol" **OS**, (1995), *73*, 94.

Br, OH + $CH_2(OMe)_2$ — P_2O_5 / 0°C to 25°C → Br, OMOM (78%)

Danheiser, R. L.; Romines, K. R.; Koyama, H.; Gee, S. K.; Johnson, C. R.; Medich, J. R. "A Hydroxymethyl Anion Equivalent: Tributyl[(methoxymethoxy)methyl]stannane" **OS**, (1992), *71*, 133.

Bu_3Sn, OH — $CH_2(OMe)_2$, $BF_3{\cdot}OEt_2$ / 4Å molecular sieves, CH_2Cl_2 → Bu_3Sn, OMOM

Dondoni, A.; Merino, P. "Diastereoselective Homologation of D-(*R*)-Glyceraldehyde Acetonide Using 2-(Trimethylsilyl)thiazole: 2-*O*-Benzyl-3,4-isopropylidene-D-erythrose" **OS**, (1993), *72*, 21.

O, O, S, N, OH — NaH / $C_6H_5CH_2Br$ → O, O, S, N, $OCH_2C_6H_5$ (96%)

Bhatia, A. V.; Chaudhary, S. K.; Hernandez, O. "4-Dimethylamino-*N*-triphenylmethylpyridinium Chloride" **OS**, (1997), *75*, 184.

NMe_2, N + Ph_3CCl — 20–25°C / CH_2Cl_2 → NMe_2, N^+, Cl^-, CPh_3 (96%)

(d) Protection of Diols as Ketals

Ley, S. V.; Osborn, H. M. I.; Priepke, H. W. M.; Warriner, S. L.; "(1′*S*, 2′*S*)-Methyl-3*O*,4*O*-(1′, 2′-dimethoxycyclohexane-1′, 2′-diyl)-α-D-mannopyranoside" **OS**, (1997), *75*, 170.

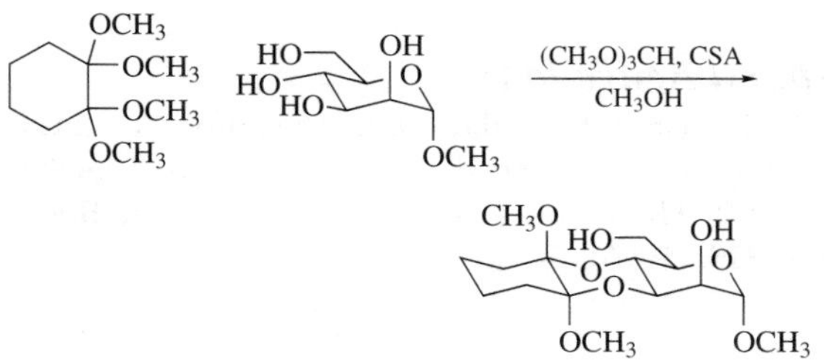

$(CH_3O)_3CH$, CSA / CH_3OH (41–46%)

Schmid, C. R.; Bryant, J. D. "D-(*R*)-Glyceraldehyde Acetonide" **OS**, (1993), *72*, 6.

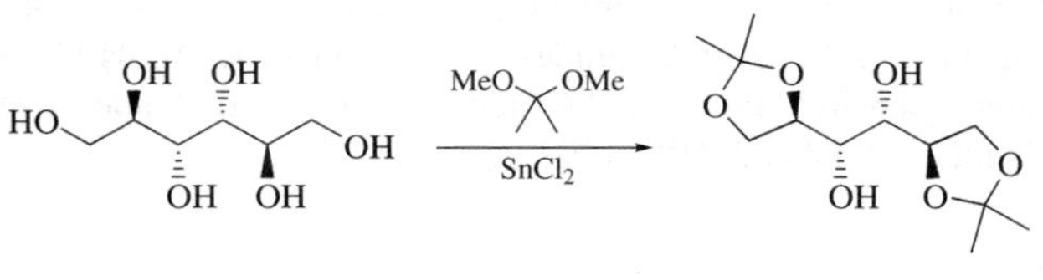

$SnCl_2$ (50–56%)

Sun, R. C.; Okabe, M. "(2*S*, 4*S*)-2,4,5-Trihydroxypentanoic Acid 4,5-Acetonide Methyl Ester" **OS**, (1993), *72*, 48.

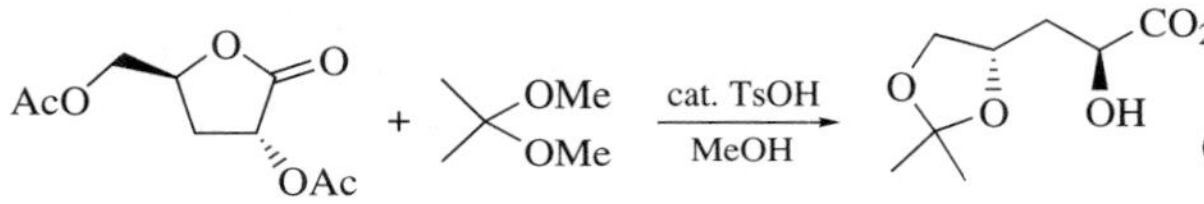

cat. TsOH / MeOH (90%)

III. Amine Protection

Carrasco, M.; Jones, R. J.; Kamel, S.; Rapoport, H. Truong, T. "*N*-(Benzyloxycarbonyl)-L-vinylglycine Methyl Ester" **OS**, (1991), *70*, 29.

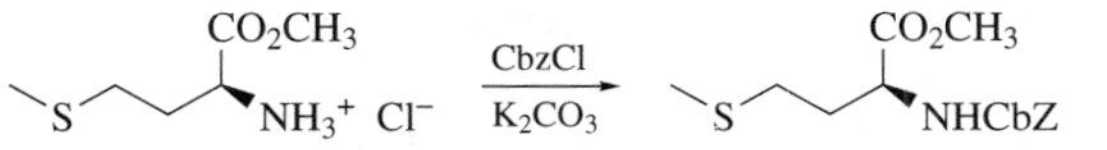

CbzCl / K_2CO_3 (98–99%)

Garner, P.; Park, J. M. "1,1-Dimethylethyl (*S*)- or (*R*)-4-Formyl-2,2-dimethyl-3-oxazolidinecarboxylate: A Useful Serinal Derivative" **OS**, (1991), *70*, 18.

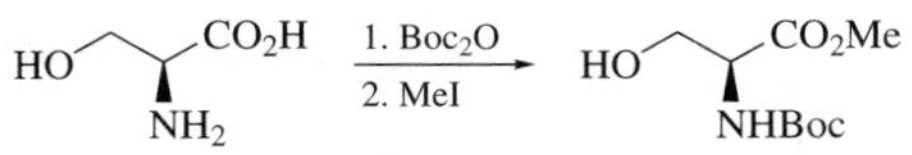

1. Boc_2O 2. MeI (86%)

Lenz, G. R. Lessor, R. A. "Tetrahydro-3-benzazepin-2-ones: Lead Tetraacetate Oxidation of Isoquinoline Enamides" **OS**, (1991), *70*, 139.

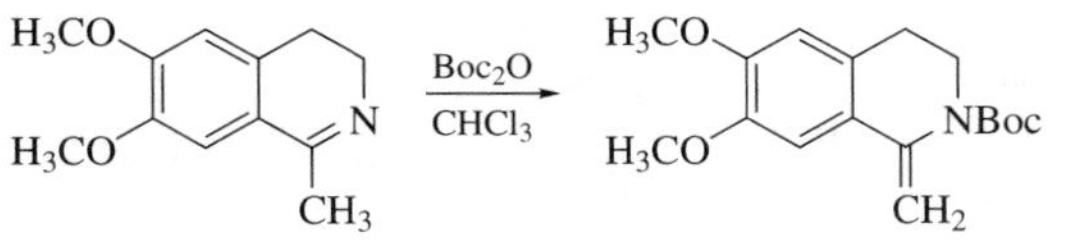

Boc_2O / $CHCl_3$ (96–98%)

Saito, S.; Komada, K.; Moriwake, T. "Diethyl (2*S*, 3*R*)-2-(*N*-*tert*-Butoxycarbonyl)amino-3-hydroxysuccinate" **OS**, (1995), *73*, 184.

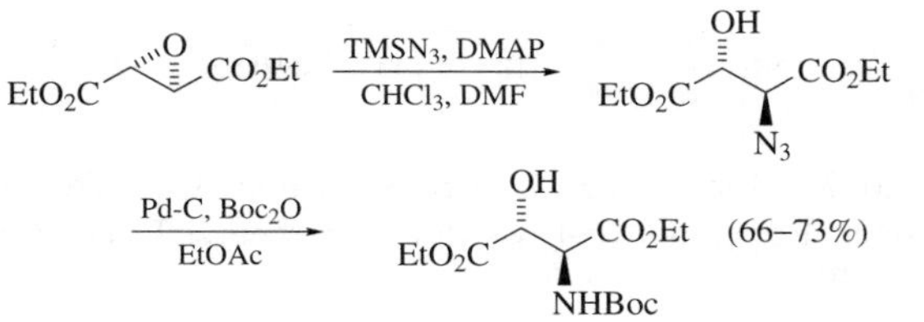

$TMSN_3$, DMAP / $CHCl_3$, DMF (86%); Pd-C, Boc_2O / EtOAc (66–73%)

Nikolic, N. A.; Beak, P. "(*R*)-(+)-2-(Diphenylhydroxymethyl)pyrrolidine" **OS**, (1996), *74*, 23.

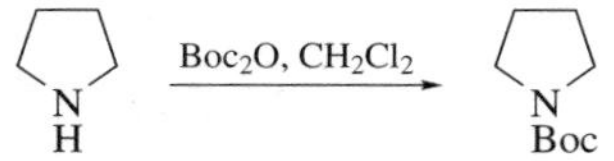

Boc_2O, CH_2Cl_2 (87%)

Chen, W.; Stephenson, E. K.; Cava, M. P.; Jackson, Y. A. "2-Substituted Pyrroles from *N-tert*-Butyloxycarbonyl-2-bromopyrrole: *N-tert*-Butoxy-2-trimethylsilylpyrrole" **OS**, (1991), *70*, 151.

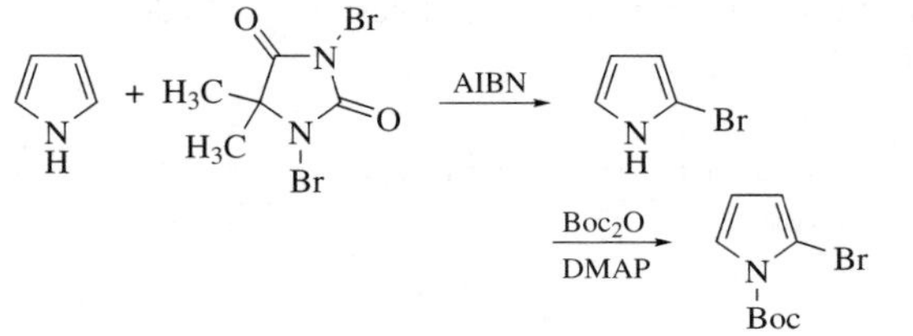

AIBN; Boc_2O / DMAP (82–89%)

Iwao, M.; Kuraishi, T. "Synthesis of 7-Subtituted Indolines via Directed Lithiation of 1-(*tert*-Butoxycarbonyl)indoline: 7-Indoline: 7-Indolinecarboxaldehyde" **OS**, (1995), *73*, 85.

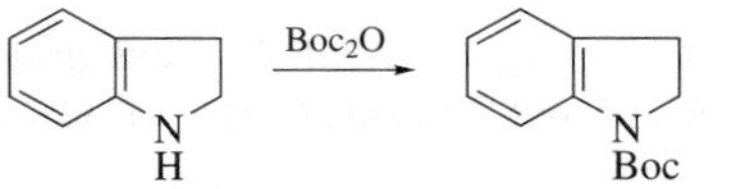

Boc_2O (98%)

Lakner, F. J.; Chu, K. S.; Negrete, G. R.; Konopelski, J. P. "Synthesis of Enantiomerically Pure β-Amino Acids from 2-*tert*-Butyl-1-carbomethoxy-2,3-dihydro-4(1*H*)-pyrimidinone: (*R*)-3-Amino-3-(*p*-methoxyphenyl)propionic Acid" **OS**, (1995), *73*, 201.

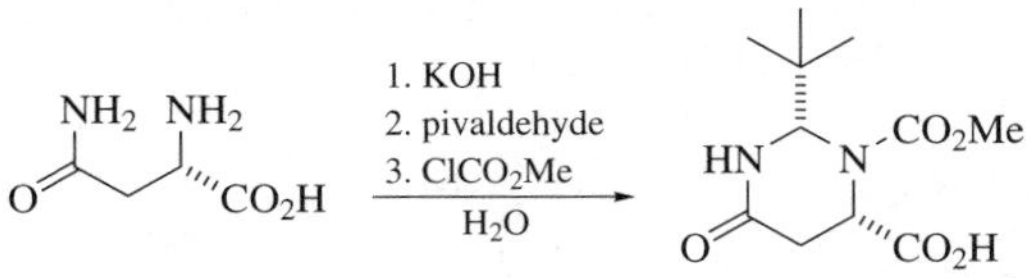

1. KOH 2. pivaldehyde 3. $ClCO_2Me$ / H_2O (72–79%)

Hutchison, D. R.; Khau, V. V.; Martinelli, M. J.; Nayyar, N. K.; Peterson, B. C.; Sullivan, K. A. "Synthesis of *cis*-4a(*S*),8a(*R*)-Perhydro-6(2*H*)-isoquinolinones from Quinine: 4a(*S*),8a(*R*)-2-Benzoyloctahydro-6(2*H*)-isoquinolinone" **OS**, (1997), *75*, 223.

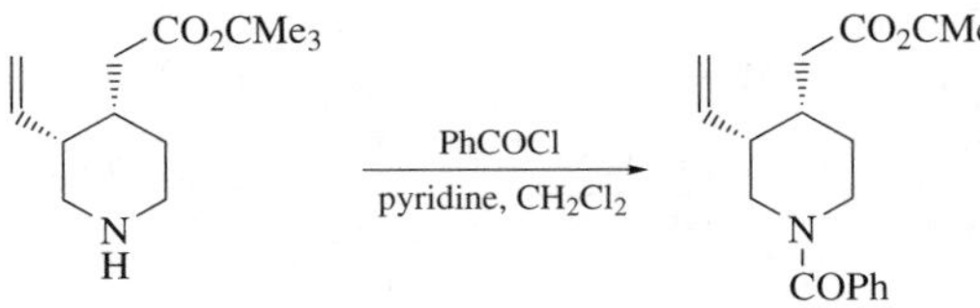

PhCOCl / pyridine, CH_2Cl_2 (100%)

Krakowiak, K. E.; Bradshaw, J. S. "4-Benzyl-10,19-diethyl-4,10,19-triaza-1,7,13,16-tetraoxacycloheneicosane (Triaza-21-Crown-7)" **OS**, (1991), *70*, 129.

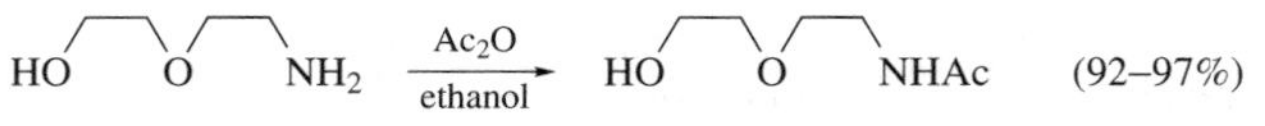

HO–O–NH_2 $\xrightarrow[\text{ethanol}]{Ac_2O}$ HO–O–NHAc (92–97%)

Carrasco, M.; Jones, R. J.; Kamel, S.; Rapoport, H. Truong, T. "*N*-(Benzyloxycarbonyl)-L-vinylglycine Methyl Ester" **OS**, (1991), *70*, 29.

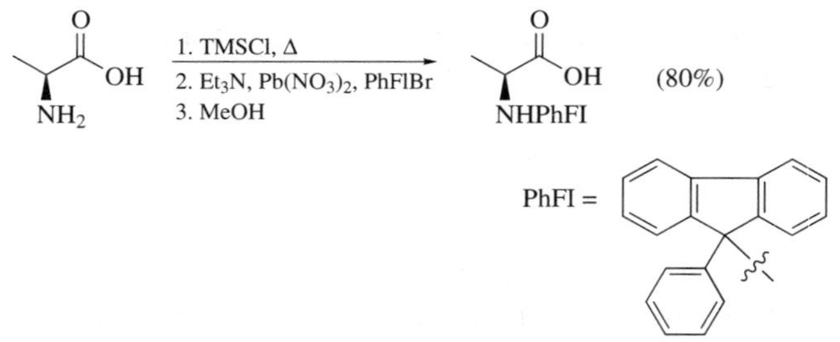

1. TMSCl, Δ; 2. Et_3N, $Pb(NO_3)_2$, PhFlBr; 3. MeOH (80%)

PhFl =

Amat, M.; Hadida, S.; Sathyanarayana, S.; Bosch, J. "Regioselective Synthesis of 3-Substituted Indoles: 3-Ethylindole" **OS**, (1996), *74*, 248.

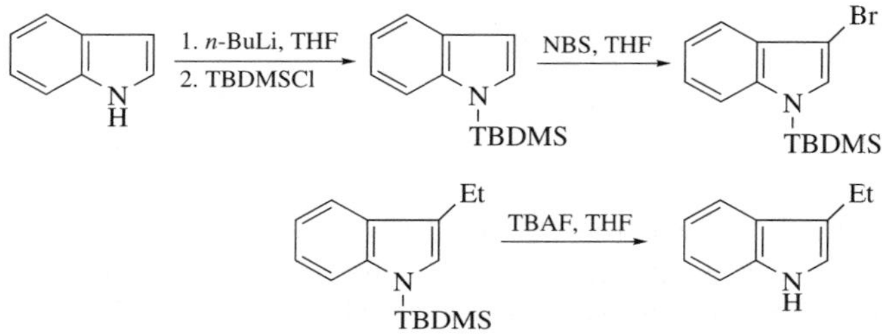

1. *n*-BuLi, THF; 2. TBDMSCl — NBS, THF — TBAF, THF

Weinreb, S. M.; Chase, C. E.; Wipf, P.; Venkatraman, S. "2-Trimethylsilylethanesulfonyl Chloride (SES-Cl)" **OS**, (1997), *75*, 161.

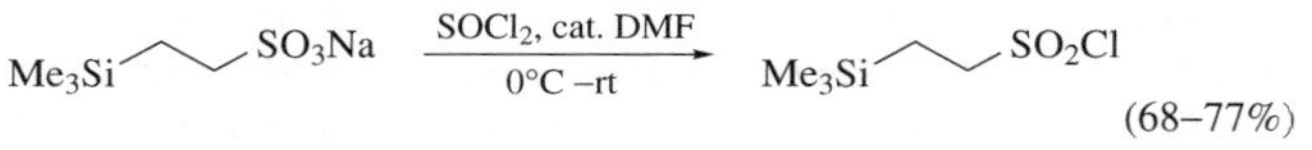

Me_3Si–SO_3Na $\xrightarrow[0°C–rt]{SOCl_2,\ cat.\ DMF}$ Me_3Si–SO_2Cl (68–77%)

Pikal, S.; Corey, E. J. "Enantioselective, Catalytic Diels-Alder Reaction: (1*S*-*endo*)-3-(Bicyclo[2.2.1]hept-5-en-2-ylcarbonyl)-2-oxazolidinone" **OS**, (1992), *71*, 30.

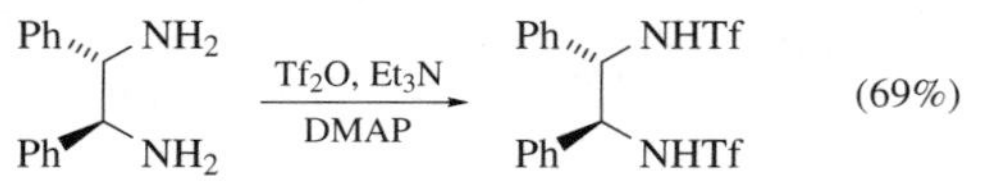

Tf_2O, Et_3N, DMAP (69%)

Schultz, A. G.; Alva. C. W. " Asymmetric Synthesis of *trans*-2-Aminocyclohexanecarboxylic Acid Derivatives from Pyrrolobenzodiazepine-5,11-diones" **OS**, (1995), *73*, 174.

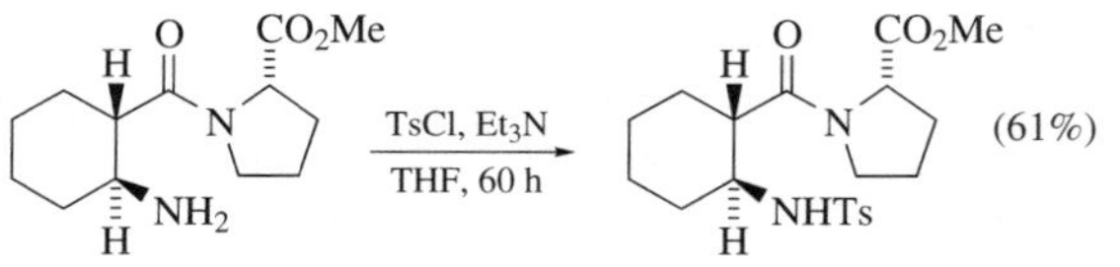

TsCl, Et_3N, THF, 60 h (61%)

Nikolaides, N.; Schipor, I.; Ganem, B. "Conversion of Amines to Phospho Esters: Decyl Diethyl Phosphate" **OS**, (1993), *72*, 246.

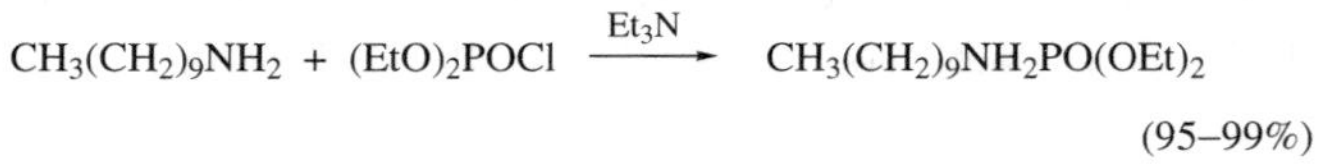

$CH_3(CH_2)_9NH_2 + (EtO)_2POCl \xrightarrow{Et_3N} CH_3(CH_2)_9NH_2PO(OEt)_2$ (95–99%)

IV. Carboxyl Activation:

Xavier, L. C.; Mohan, J. J.; Mathre, D. J.; Thompson, A. S.; Carroll, J. D.; Corley, E. G.; Desmond, R. "(*S*)-Tetrahydro-1-methyl-3,3-diphenyl-1*H*, 3*H*-pyrrolo-[1,2-*c*][1,3,2]oxazaborole-Borane Complex" **OS**, (1996), *74*, 50.

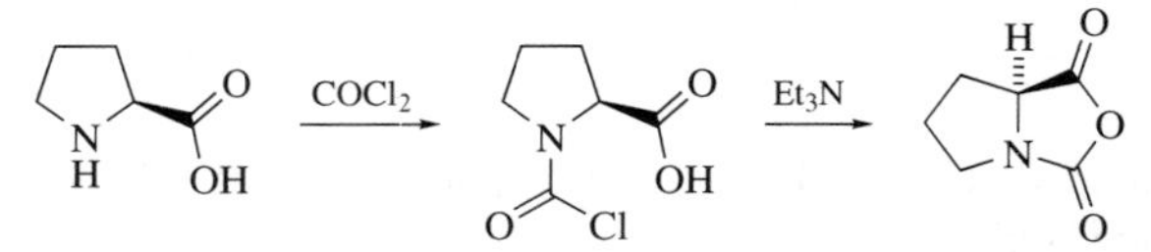

$COCl_2$ — Et_3N

Barton, D. H. R.; MacKinnon, J.; Perchet, R. N.; Tse, C.-L. "Efficient Synthesis of Bromides from Carboxylic Acids Containing a Sensitive Functional Group: Dec-9-enyl Bromide from 10-Undecenoic Acid" **OS**, (1997), *75*, 124.

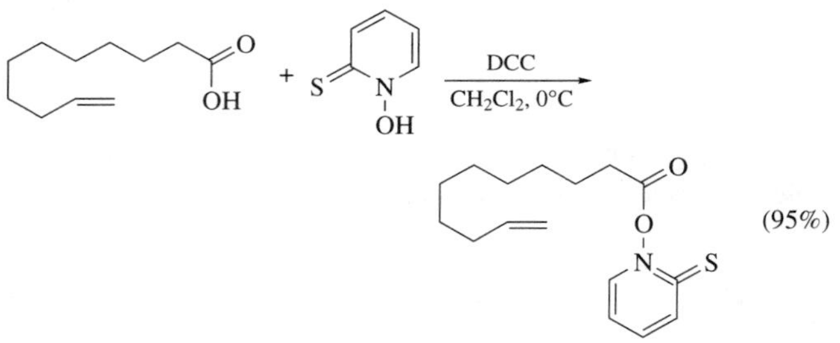

DCC, CH_2Cl_2, 0°C (95%)

Hubschwerlen, C.; Specklin, J.-L. "(3*S*, 4*S*)-3-Amino-1-(3,4-dimethoxybenzyl)-4-[(*R*)-2,2-dimethyl-1,3-dioxolan-4-yl]-2-azetidinone" **OS**, (1993), *72*, 14.

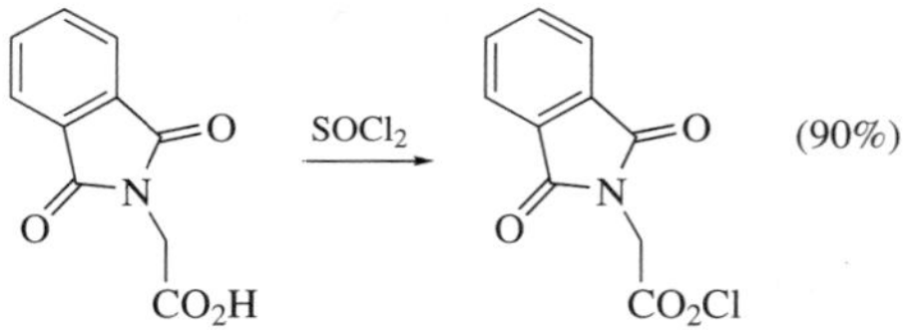

$SOCl_2$ (90%)

Wang, X.; deSilva, S. O.; Reed, J. N.; Billadeau, R.; Griffen, E. J.; Chan, A.; Snieckus, V. "7-Methoxyphthalide" **OS**, (1993), *72*, 163.

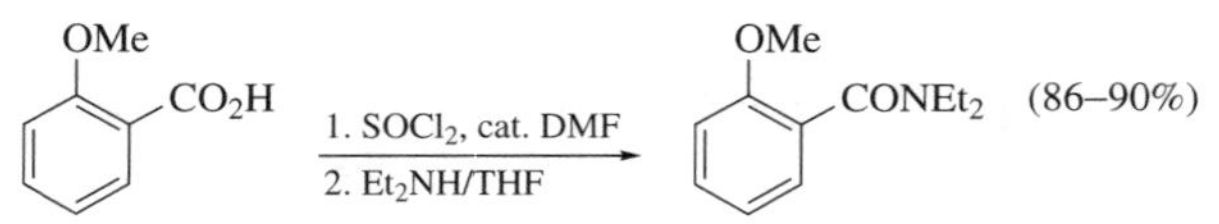

1. $SOCl_2$, cat. DMF; 2. Et_2NH/THF (86–90%)

Meyers, A. I.; Flanagan, M. E. "2,2′-Dimethoxy-6-formylbiphenyl" **OS**, (1992), *71*, 107.

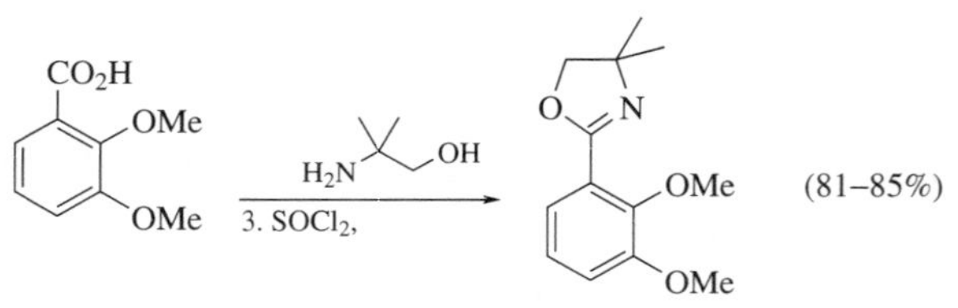

H_2N–C$(CH_3)_2$–CH_2OH; 3. $SOCl_2$, (81–85%)

Gerlach, H.; Kappes, D.; Boeckman, R. K. Jr.; Maw, G. N. "(−)-(1*S*, 4*R*)-Camphanoyl Chloride" **OS**, (1992), *71*, 48.

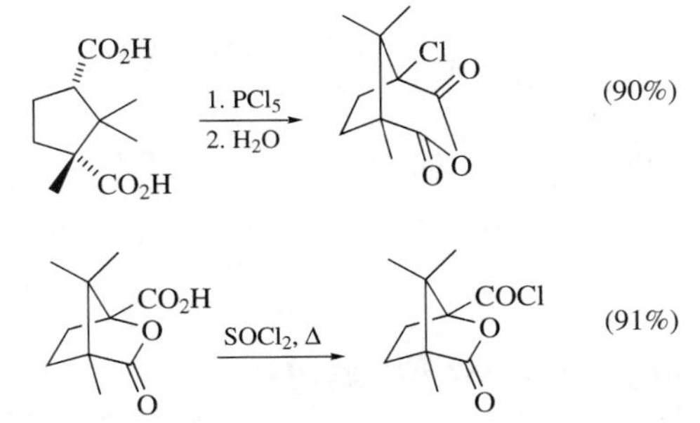

1. PCl_5 2. H_2O (90%)

$SOCl_2$, Δ (91%)

Rosini, G.; Confalonieri, G.; Marotta, E.; Rama, F.; Righi, P. "Preparation of Bicyclo[3.2.0]hept-3-en-6-ones: 1,4-Dimethylbicyclo[3.2.0]hept-3-en-6-one" **OS**, (1996), *74*, 158.

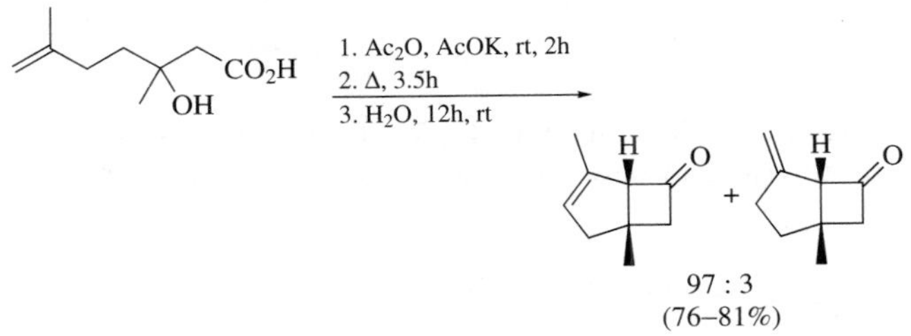

1. Ac_2O, AcOK, rt, 2h 2. Δ, 3.5h 3. H_2O, 12h, rt

97 : 3 (76–81%)

V. Carbonyl Protection:

Ley, S. V.; Osborn, H. M. I.; Priepke, H. W. M.; Warriner, S. L. "(1′*S*, 2′*S*)-Methyl-*O*,4*O*-(1′, 2′-dimethyoxycylohexane-1′, 2′-diyl)-α-D-mannopyranoside" **OS**, (1997), *75*, 170.

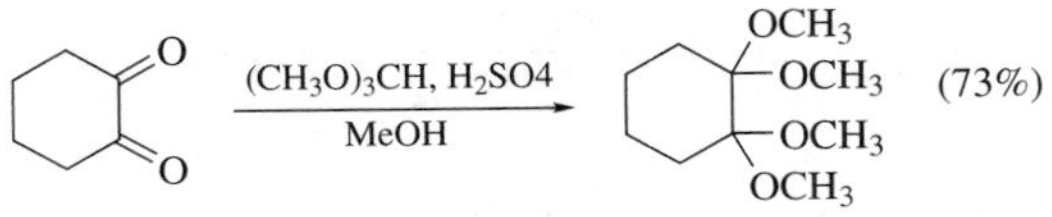

$(CH_3O)_3CH$, H_2SO4, MeOH (73%)

Ni, Z.-J.; Luch, T.-Y. "Nickel-Catalyzed Silylolefination of Allylic Dithioacetals: (*E*, *E*)-Trimethyl(4-phenyl-1,3-butadienyl)silane" **OS**, (1991), *70*, 240.

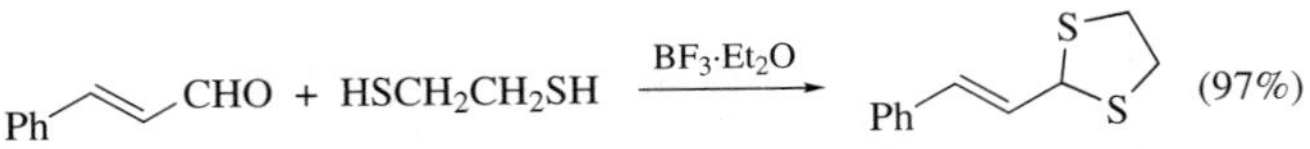

Ph–CH=CH–CHO + $HSCH_2CH_2SH$ $\xrightarrow{BF_3 \cdot Et_2O}$ (97%)

Yuan, T.-M.; Luh, T.-Y. "Nickel-Catalyzed, Geminal Dimethylation of Allylic Dithioacetals; (*E*)-1-Phenyl-3,3-dimethyl-1-butene" **OS**, (1996), *74*, 187.

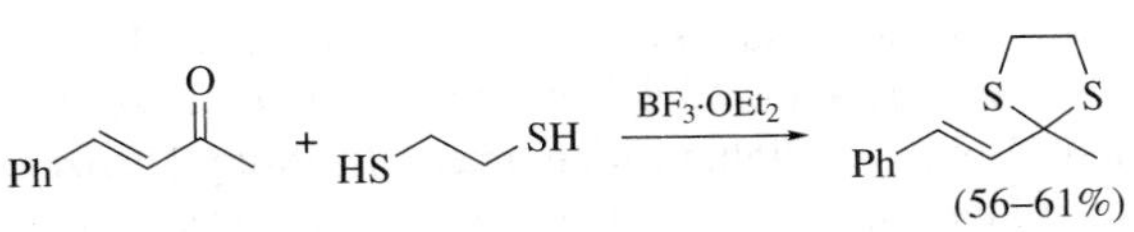

$BF_3 \cdot OEt_2$ (56–61%)

Dahnke, K. R.; Paquette, L. A. "2-Methylene-1,3-dithiolane" **OS**, (1992), *71*, 175.

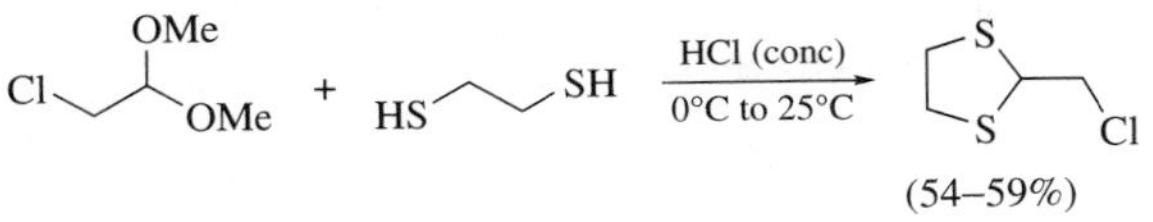

HCl (conc), 0°C to 25°C (54–59%)

VI. Carbonyl Activation or Functionalization:

Schmit, C.; Falmagne, J. B.; Escudero, J.; Vanlierde, H.; Ghosez, L. "A General Synthesis of Cyclobutanones from Olefins and Tertiary Amides: 3-Hexylcyclobutanone" **OS**, (1990), *69*, 199.

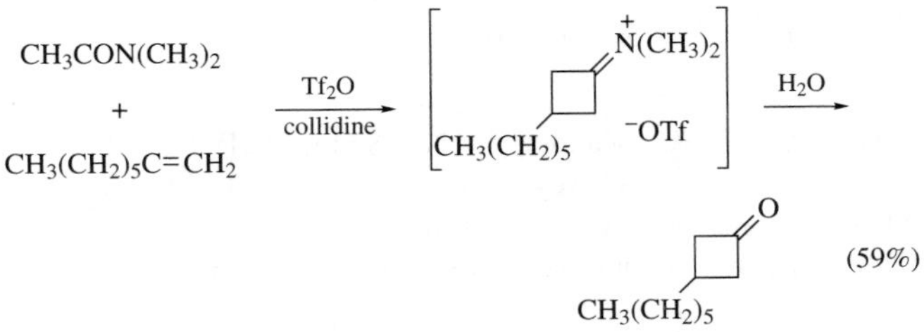

$CH_3CON(CH_3)_2$ + $CH_3(CH_2)_5C{=}CH_2$ $\xrightarrow{Tf_2O,\ collidine}$ [$\overset{+}{N}(CH_3)_2$, ^-OTf, $CH_3(CH_2)_5$] $\xrightarrow{H_2O}$ $CH_3(CH_2)_5$ (59%)

Knapp, S.; Gibson, F. S. "Iodolactamization: 8-*exo*-Iodo-2-azabicyclo[3.3.0]octan-3-one" **OS**, (1991), *70*, 101.

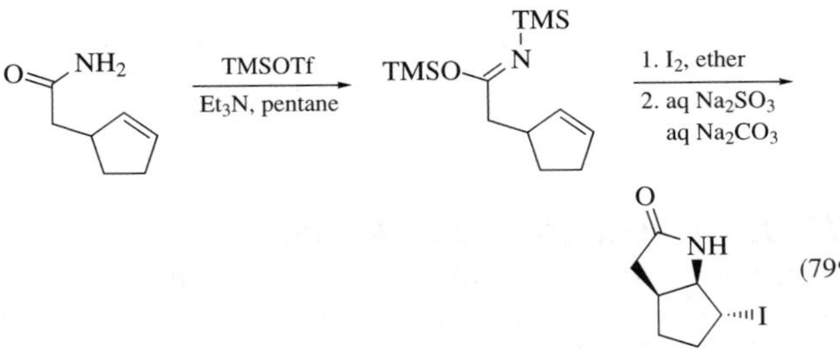

TMSOTf, Et_3N, pentane; 1. I_2, ether 2. aq Na_2SO_3 aq Na_2CO_3 (79%)

Wender, P. A.; White, A. W.; MacDonald, F. E. "Spiroannelation via Organobis(cuprates): 9,9-Dimethylspiro[4.5]decan-7-one" **OS**, (1991), *70*, 204.

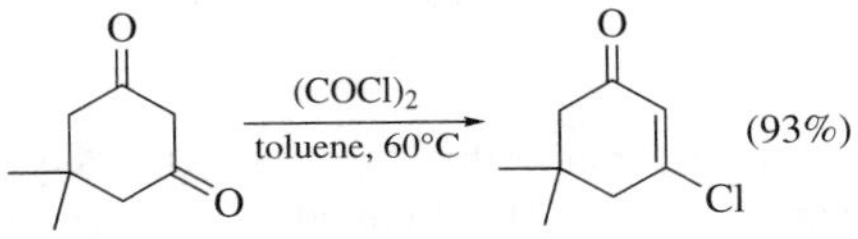

$(COCl)_2$, toluene, 60°C (93%)

Polin, J.; Schottenberger, H. "Conversion of Methyl Ketones into Terminal Acetylenes: Ethynylferrocene" **OS**, (1995), *73*, 262.

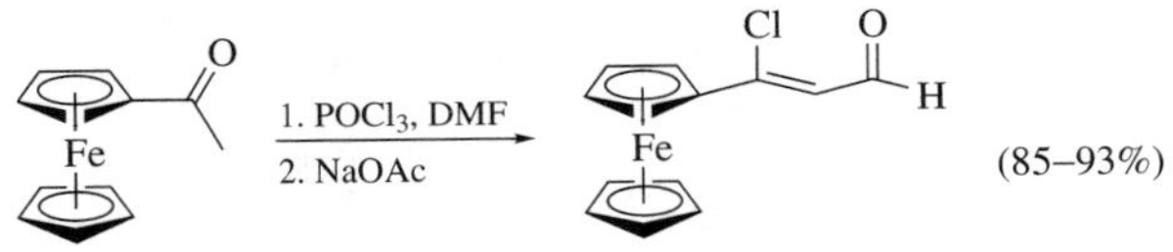

1. $POCl_3$, DMF 2. NaOAc (85–93%)

Myles, D. C.; Bigham, M. H. "Preparation of (*E*, *Z*)-1-Methoxy-2-methyl-3-(trimethylsiloxy)-1,3-pentadiene" **OS**, (1991), *70*, 231.

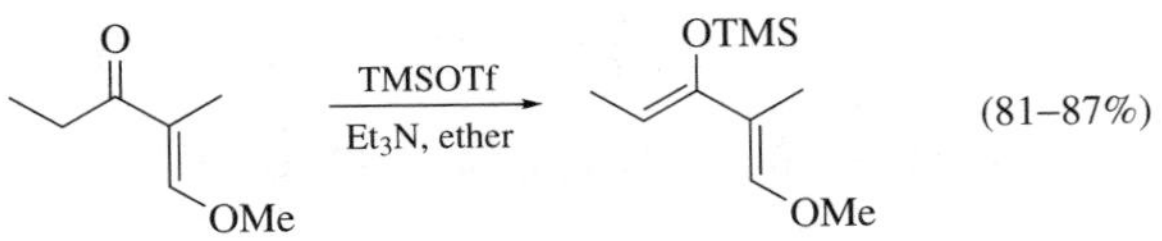

TMSOTf, Et_3N, ether (81–87%)

Umemoto, T.; Tomita, K.; Kawada, K. "*N*-Fluoropyridinium Triflate: An Electrophilic Fluorinating Agent" **OS**, (1990), *69*, 129.

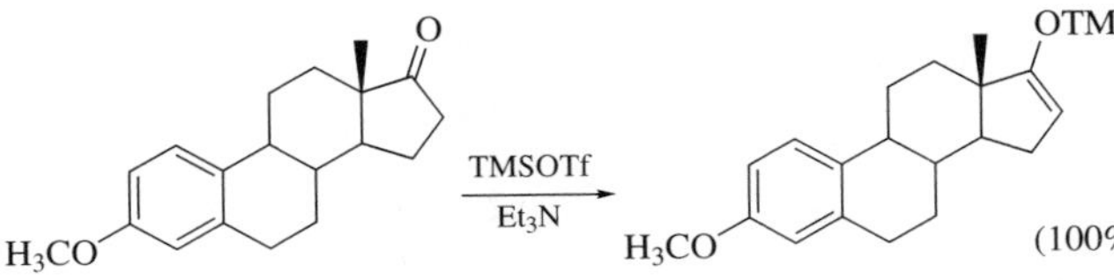

TMSOTf, Et_3N (100%)

Reissig, H.-U.; Reichelt, I.; Kunz, T. "Methoxycarbonylmethylation of Aldehydes via Siloxycyclopropanes: Methyl 3,3-Dimethyl-4-oxobutanoate" **OS**, (1992), *71*, 189.

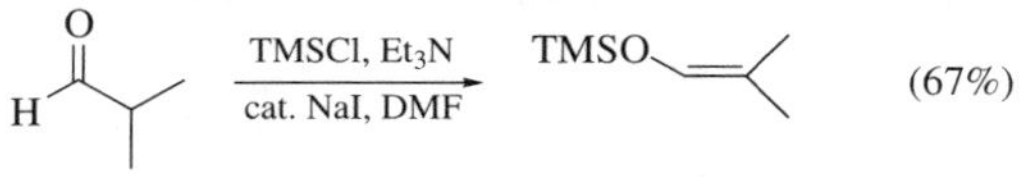

Crabtree S. R.; Mander, L. N.; Sethi, S. P. "Synthesis of β-Keto Esters by C-Acylation of Preformed Enolates with Methyl Cyanoformate: Preparation of Methyl (1α,4aβ,8aα)-2-oxodecahydro-1-naphthoate" **OS**, (1991), *70*, 256.

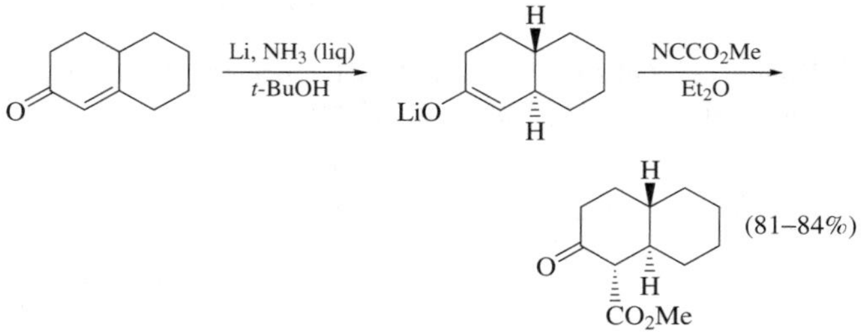

VII. Lewis Acid Promoted Reactions

Overman, L. E.; Rishton, G. M. "Stereocontrolled Preparation of 3-Acyltetrahydrofurans from Acid-Promoted Rearrangements of Allylic Ketals: (2*S*, 3*S*)-3-Acetyl-8-carboethoxy-2,3-dimethyl-1-oxa-8-azaspiro[4.5]decane" **OS**, (1992), *71*, 63.

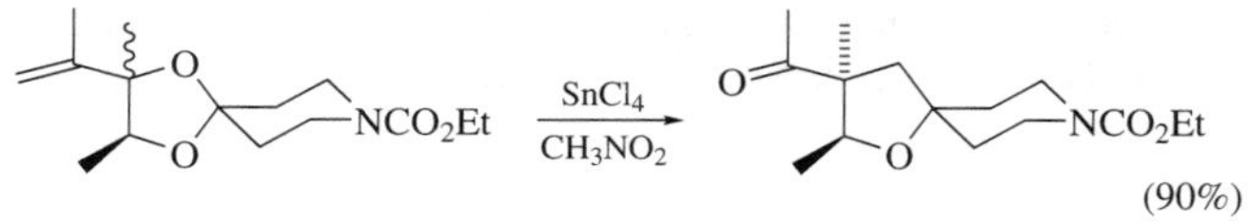

Keck, G. E.; Krishnamurthy, D. "Catalytic Asymmertic Allylation Reactions: (*S*)-1-(Phenylmethoxy)-4-penten-2-ol" **OS**, (1997), *75*, 12.

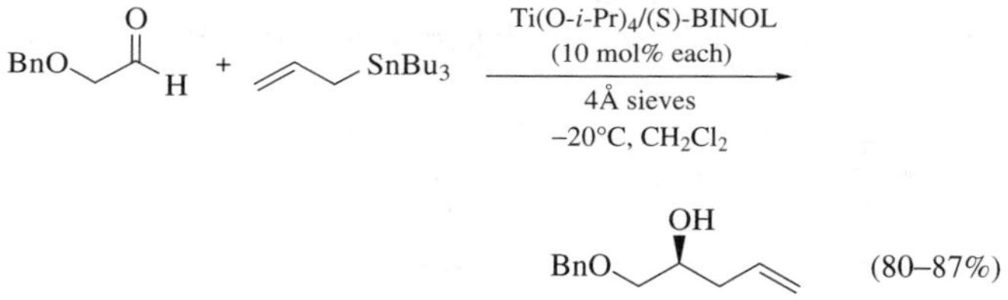

Jager, V.; Poggendorf, P. "Nitroacetaldehyde Diethyl Acetal" **OS**, (1996), *74*, 130.

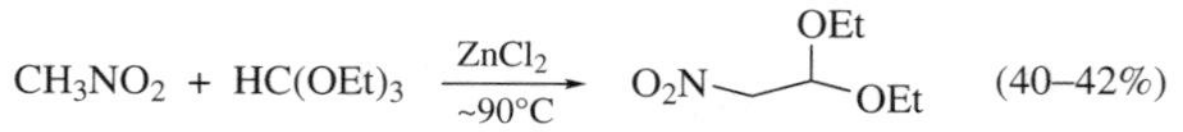

Naruta, Y.; Maruyama, K. "Ubiquinone-1" **OS**, (1992), *71*, 125.

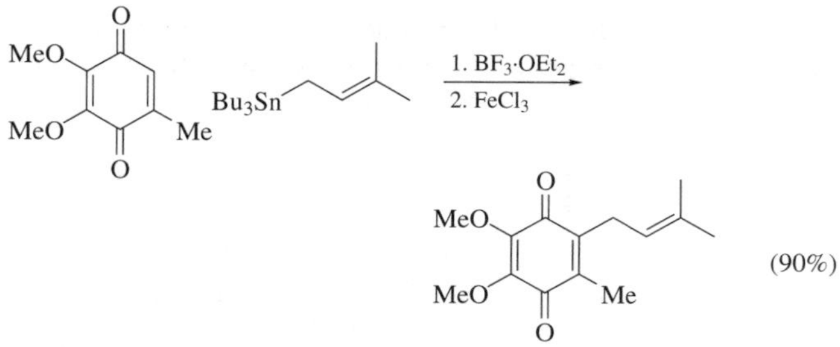

Lee, T. V.; Porter, J. R. "Spiroannelation of Enol Silanes: 2-Oxo-5-methoxyspiro[5.4]decane" **OS**, (1993), *72*, 189.

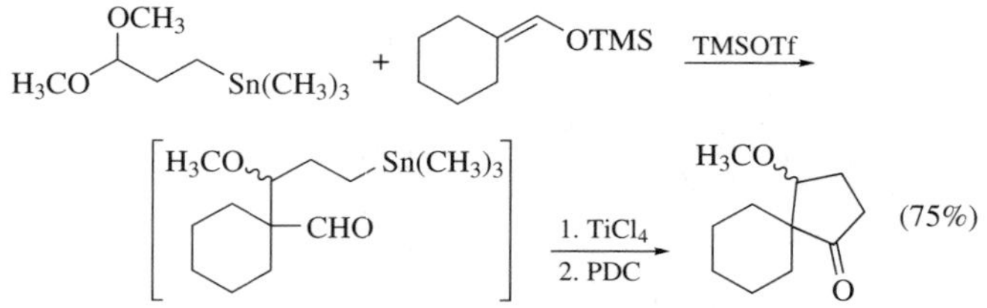

VIII. Sulfonylation Reagents:

Comins, D. L.; Dehghani, A.; Foti, C. J.; Joseph, S. P. "Pyridine-Derived Triflating Reagents: *N*-(2-Pyridyl)triflimide and *N*-(5-Chloro-2-pyridyl)triflimide" **OS**, (1996), *74*, 77.

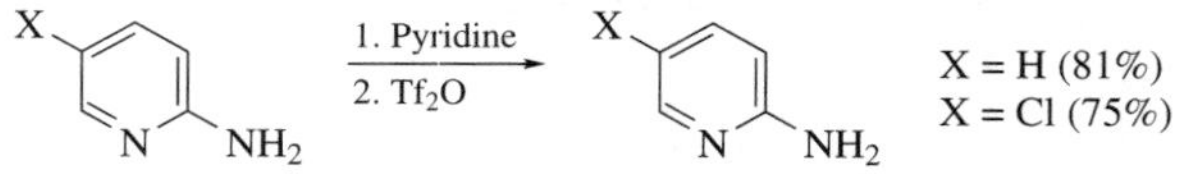

Weinreb, S. M.; Chase, C. E.; Wipf, P.; Venkatraman, S. "2-Trimethylsilylethanesulfonyl Chloride (SES-Cl)" **OS**, (1997), *75*, 161.

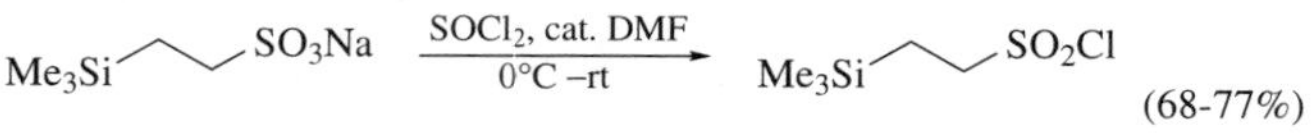

Hazen, G. G.; Billinger, F. W.; Roberts, F. E.; Russ, W. K.; Seman, J. J.; Staskiewicz, S. "4-Dodecylbenzenesulfonyl Azides" **OS**, (1995), *73*, 144.

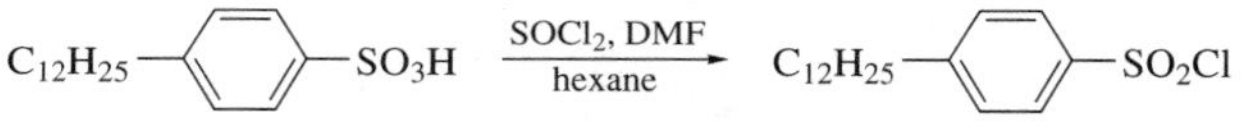

Reid, J. R.; Dufresne, R. F.; Chapman, J. J. "Mesitylenesulfonylhydrazine, and (1α, 2α, 6β)-2,6-Dimethylcyclohexanecarbonitrile and (1α, 2β, 6α)-2,6-Dimethylcyclohexanecarbonitrile as a Racemic Mixture" **OS**, (1996), *74*, 217.

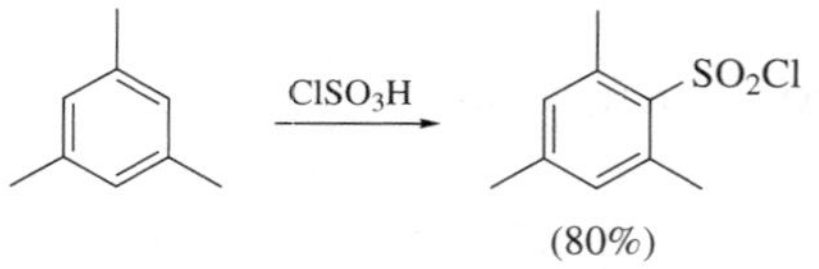

IX. Sulfoxide Activation

McCarthy, J. R.; Matthews, D. P.; Paolini, J. P. "Stereoselective Synthesis of 2,2-Disubstituted 1-Fluoroalkenes: (*E*)-[[Fluoro(2-phenylcyclohexylidene)methyl]sulfonyl] benzene and (*Z*)-[2-(Fluoromethylene)cyclohexyl]benzene" **OS**, (1993), *72*, 209.

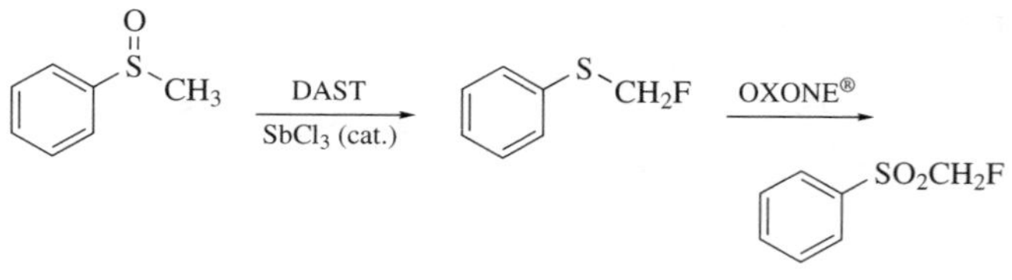

Recent Review Articles and Monographs

General Carboxyl and Hydroxyl Activation

Chatgilialoglu, C.; Ferreri, C. Progress of the Barton-McCombie Methodology: From Tin Hydrides to Silanes. *Res. Chem. Intermed.* **1993**, *19*, 755–775.

Norcross, R. D.; Paterson, I. Total Synthesis of Bioactive Marine Macrolides. *Chem. Rev.* **1995**, *95*, 2041–2114.

Hughes, D. L. Progress in the Mitsunobu Reaction. A Review. *Org. Prep. Proc. Int.*, **1996**, *28*, 127–164.

Ryan, T. A. Phosgene and Related Compounds. Elsevier Science: New York, 1996.

Sherif, S. M.; Erian, A. W. The Chemistry of Trichloroacetonitrile, *Heterocycles* **1996**, *43* (5), 1083–1118.

Cotarca, L.; Delogu, P.; Nardelli, A.; Sunjic, V. Bis(trichloromethyl) Carbonate in Organic Synthesis. *Synthesis* **1996**, 553–576.

Dimon, C.; Hosztafi, S.; Makleit, S. Application of the Mitsunobu Reaction in the Field of Alkaloids. *J. Heterocyclic Chem.* **1997**, *34*, 349–365.

Zard, S. Z. On the Trail of Xanthates: Some New Chemistry from an Old Functional Group. *Angew. Chem. Int. Ed. Engl.* **1997**, *36*, 672–685.

Wisniewski, K.; Koldziejczyk, A. S.; Falkiewicz, B. Applications of the Mitsunobu Reaction in Peptide Chemistry. *J. Peptide Sci.* **1998**, *4*, 1–14.

Peptide Synthesis

Albericio, F.; Carpino, L. A. Coupling reagents and activation. *Method. Enzymol.*, **1997**, *289*, 104–126.

Albericio, F.; Lloyd-Williams, P.; Giralt, E. Convergent solid-phase peptide synthesis. *Method. Enzymol.*, **1997**, *289*, 313–336.

Bodanszky, M. Peptide Chemistry: A Practical Textbook, 2nd ed. Springer-Verlag: Berlin, 1993.

Bodanszky, M. Principles of Peptide Synthesis, 2nd ed. Springer-Verlag: Berlin, 1993.

Bodanszky, M.; Bodanszky, A. The Practice of Peptide Synthesis, 2nd ed. Springer-Verlag: Heidelberg, 1994.

Pennington, M. W.; Dunn, B. M.; Eds. Peptide Synthesis Protocols (In: Methods Mol. Biol. 1994, 35) Humana Press: Totowa, NJ 1994.

Mikhalkin, A. P. The Synthesis, Properties, and Applications of *N*-Acyl-α-aminoacids. *Russ. Chem. Rev.* **1995**, *64* (3), 259–75.

Wipf, P. Synthetic Studies of Biologically Active Marine Cyclopeptides. *Chem. Rev.* **1995**, *95*, 2115–2134.

Gutte, B., Ed. Peptides, Synthesis, Structures, and Applications. Academic Press: San Diego, 1995.

Carpino, L. A; Beyermann, M.; Wenschuh, H.; Bienert, M. Peptide Synthesis via Amino Acid Halides. *Acct. Chem. Res.* **1996**, *29*, 268–274.

Deming, T.J. Polypeptide materials: New synthetic methods and applications. *Adv. Mater.*, **1997**, *9*, 299–311.

Stephanov, V. M. Proteinases as Catalysts in Peptide Synthesis. *Pure Appl. Chem.*, **1996**, *68* (6), 1335–1340.

Ichikawa, J. Fluorine as activator and controller in organic synthesis. *J. Syn. Org. Chem. Jpn.*, **1996**, *54*, 654–664.

North, M. Amines and amides. *Contemp. Org. Synth.*, **1996**, *3*, 323–343.

Humphrey, J. M.; Chamberlin, A. R. Chemical Synthesis of Natural Product Peptides: Coupling Methods for the Incorporation of Noncoded Amino Acids into Peptides. *Chem. Rev.* **1997**, *97*, 2243–2266.

Carbohydrate Activation/Glycosylation

Suzuki, K.; Nagasawa, T. Recent progress in O-glycoside synthesis - methodological aspects. *J. Syn. Org. Chem. Jpn.* **1992**, *50*, 378–390.

Toshima, K.; Tatsuta, K. Recent Progress in *O*-Glycosylation Methods and Its Application to Natural Products Synthesis. *Chem. Rev.*, **1993**, *93*, 1503–1531.

Schmidt, R. R; Kinzy, W. Anomeric-Oxygen Activation for Glycoside Synthesis: The Trichloroacetimidate Method. *Adv. Carbohydrate Chem. Biochem.* **1994**, *50*, 21–123.

Ogawa, T. Haworth Memorial Lecture: Experiments Directed Towards Glycoconjugate Synthesis, CSR **1994**, *23*, 397–407

Wilson, L. J.; Hager, M. W.; El-Kattan, Y.A.; Liotta, D. C. Nitrogen Glycosylation Reactions Involving Pyrimidine and Purine Nucleoside Bases with Furanoside Sugars. Synthesis-Stuttgart **1995**, 1465–1479.

Boons, G.-J. Strategies in Oligosaccharide Synthesis. *Tetrahedron* **1996**, *52*, 1095–1121.

Boons, G.-J. Recent Developments in Chemical Oligosaccharide Synthesis. Contemp. Org. Synth. **1996**, *3*, 173–200.

Danishefsky, S. J.; Bilodeau, M. T. Glycals in Organic Synthesis: The Evolution of Comprehensive Strategies for the Assembly of Oligosaccharides and Glycoconjugates of Biological Consequence. *Angew. Chem. Int. Ed.* **1996**, *35*, 1380–1419.

Voelter, W.; Khan, K. M.; Shekhani, M. S. Anhydro Sugars, Valuable Intermediates in Carbohydrate Syntheses. *Pure Appl. Chem.* **1996**, *68*, 1347–1353.

Khan, S. H.; O'Neill, R. A., Eds. Modern Methods in Carbohydrate Synthesis. Hardwood: Amsterdam, The Netherlands, 1996.

Whitfield, D. M.; Douglas, S. P. Glycosylation reactions: Present status and future directions. *Glycoconjugate J.*, **1996**, *13*, 5–17.

Garegg, P. J. Thioglycosides as Glycosyl Donors in Oligosaccharide Synthesis. *Adv. Carbohydrate Chem. Biochem.* **1997**, *52*, 179–205.

Hanessian, S. Preparative Carbohydrate Chemistry. Marcel Dekker: New York, 1997.

Tsuda, Y. Regioselective manipulation of carbohydrate-hydroxyl groups (selective activation of a hydroxyl group by tin compounds). *J. Synth. Org. Chem. Jpn.* **1997**, 55, 907–919.

Synthesis of Oligonucleotides

Beaucage, S. L.; Iyer, R. P. The Functionalization of Oligonucleotides via Phosphoramidite Derivatives. *Tetrahedron*, **1993**, *49*, 1925–1963.

Beaucage, S. L.; Iyer, R. P. The Synthesis of Modified Oligonucleotides by the Phosphoramidite Approach and Their Applications. *Tetrahedron.* **1993**, *49*, 6123–6194.

Beaucage, S. L.; Iyer, R. P. The Synthesis of Specific Ribonucleotides and Unrelated Phosphorylated Biomolecules by the Phosphoramidite Method. *Tetrahedron.* **1993**, *49*, 10441–10488.

Lesnikowski, Z. J. Stereocontrolled Synthesis of P-Chiral Analogs of Oligonucleotides. *Bioorg. Chem.* **1993**, *21*, 127–155.

Stec, W. J.; Wilk, A. Stereocontrolled Synthesis of Oligo(nucleoside phosphorothioate)s. *Angew. Chem. Int. Ed. Engl.* **1994**, *33*, 709–722

Activation by Lewis Acids

Otera, J. Transesterification. *Chem. Rev.* **1993**, *93*, 1449–1470.

Oh, T.; Reilly. M. Reagent-Controlled Asymmetric Diels-Alder Reactions. A Review. *Org. Prep. Proced. Int.*, **1994**, *26*, 129–148.

Waldmann, H. Asymmetric Hetero Diels-Alder Reactions. *Synthesis*, **1994**, 535–551.

Suzuki, K. Novel Lewis acid catalysis in organic synthesis. *Pure Appl. Chem.* **1994**, *66*, 1557–1564.

Pons, J.-M.; Santelli, M.; Eds. Lewis acids and selectivity in organic synthesis. CRC Press: Boca Raton, FL, 1996.

Hiroi, K. Transition metal or Lewis acid-catalyzed asymmetric reactions with chiral organosulfur functionality. *Rev. Heteroatom Chem.*, **1996**, *14*, 21–57.

Siling, M. I.; Laricheva, T. N. Titanium compounds as catalysts for esterification and transesterification Reactions. *Russ. Chem. Rev.* **1996**, *65*, 279–286.

Engberts, J. B. F. N.; Feringa, B. L.; Keller, E.; Otto, S. Lewis-acid catalysis of carbon carbon bond forming reactions in water. *Recl. Trav. Chim. Pays-Bas* **1996**, *115*, 457–464.

Holloway, C. E.; Melnik, M. Organoaluminium compounds: classification and analysis of crystallographic and structural data. *J. Organomet. Chem.* **1997**, *543*, 1–37.

Activation by Transition Metal Organometallic Systems

Pearson, A. J.; Woodgate, P. D. Aromatic Compounds of the Transition Elements. Second Supplement to the 2nd Edition of Rodd's Chemistry of Carbon Compounds. Vol. $III^B/III^C/III^D$ (partial): Aromatic Compounds. Sainsbury, M., Ed. Elsevier: Amsterdam, The Netherlands, 1995.

Donaldson, W. A. Preparation and Reactivity of Acyclic (Pentadienyl)iron(1+) Cations: Applications to Organic Synthesis. *Aldrichim. Acta* **1997**, *30*, 17–24.

Grée, R.; Lellouche, J. P. Acyclic Diene Tricarbonyliron Complexes in Organic Synthesis. Advances in Metal-Organic Chemistry. Vol. 4. Liebeskind, L. S., Ed. JAI: Greenwich, Connecticut, 1995.

Donohoe, T. J. Stoichiometric Applications of Organotransition Metal Complexes in Organic Synthesis. *Contemp. Org. Synth.* **1996**, *3*, 1–18.

Reviews of Protecting Group Chemistry

Green, T. W.; Wuts, P. G. M. Protective Groups in Organic Chemistry, 2nd Ed; Wiley: New York, 1991.

Reidel, A.; Waldmann, H. Enzymatic Protecting Group Techniques in Bioorganic Synthesis. *J. Prakt. Chem.* **1993**, *335*, 109–127.

Kocienski, P. J. Protecting Groups. Theime Verlag: Stuttgart, 1994.

Sonveaux, E. Protecting Groups in Oligonucleotide Synthesis. in Methods in Molecular Biology, Vol. 26; Agrawal, S., Ed.; Humana Press: Totowa, NJ, 1994, pp. 1–71.

Wong, C.-H.; Whitesides, G. M. Enzymes in Synthetic Organic Chemistry. Pergamon: Oxford, 1994.

Waldmann, H.; Sebastian, D. Enzymatic Protecting Group Techniques. *Chem. Rev.* **1994**, *94*, 911–937.

Jarowicki, K.; Kocienski, P. Protecting groups. *Contemp. Org. Synth.* **1995**, *2*, 315–336; **1996**, *3*, 397–431; and **1997**, *4*, 454–492.

Donaldson, W. A. Transition Metal Alkene, Diene, and Dienyl Complexes: Complexation of Dienes for Protection. in Comprehensive Organometallic Chemistry II, Vol. 12. Hegedus, L. S., Ed. Elsevier: New York, 1995, pp. 623–635.

Schelhaas, M.; Waldmann, H. Protecting Group Strategies in Organic Synthesis. *Angew. Chem., Int. Ed. Engl.*, **1996**, *35*, 2056–2083.

Nelson, T. D.; Crouch, R. D. Selective Deprotection of Silyl Ethers. *Synthesis*, **1996**, 1031–1069.

Ranu, B. C.; Bhar, S. Dealkylation of Ethers. A Review. *Org. Prep. Proced. Int.* **1996**, *28*, 371–409.

Debenham, J. S.; Rodebaugh, R.; Fraser-Reid, B. Recent Advances in *N*-Protection for Amino Sugar Synthesis. *Liebigs Ann./Recueil* **1997**, 791–802.

El Gihani, M. T.; Heaney, H. The Use of Bis(trimethylsilyl)acetamide and Bis(trimethylsilyl)urea for Protection and as Control Reagents in Synthesis. *Synthesis*, **1998**, 357–375.

Acetic Anhydride[1]

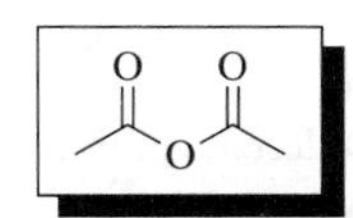

[108-24-7] $C_4H_6O_3$ (MW 102.09)

(useful for the acetylation of alcohols,[2] amines,[3] and thiols,[4] oxidation of alcohols,[5] dehydration,[6] Pummerer[7] reaction, Perkin[8] reaction, Polonovski[9] reaction, *N*-oxide reaction,[10] Thiele[11] reaction, ether cleavage,[12] enol acetate formation,[13] *gem*-diacetate formation[14])

Physical Data: bp 138–140 °C; mp −73 °C; *d* 1.082 g cm^{-3}.

Solubility: sol most organic solvents. Reacts with water rapidly and alcohol solvents slowly.

Form Supplied in: commercially available in 98% and 99+% purities. Acetic anhydride-d_6 is also commercially available.

Analysis of Reagent Purity: IR, NMR.[15]

Preparative Methods: acetic anhydride is prepared industrially by the acylation of ***Acetic Acid*** with ***Ketene***.[1b] A laboratory preparation of acetic anhydride involves the reaction of sodium acetate and ***Acetyl Chloride*** followed by fractional distillation.[16]

Purification: adequate purification is readily achieved by fractional distillation. Acetic acid, if present, can be removed by refluxing with CaC_2 or with coarse magnesium filings at 80–90 °C for 5 days. Drying and acid removal can be achieved by azeotropic distillation with toluene.[17]

Handling, Storage, and Precautions: acetic anhydride is corrosive and a lachrymator and should be handled in a fume hood.

Acetylation. The most notable use of acetic anhydride is for the acetylation reaction of alcohols,[2] amines,[3] and thiols.[4] Acids, Lewis acids, and bases have been reported to catalyze the reactions.

Alcohols. The most common method for acetate introduction is the reaction of an alcohol with acetic anhydride in the presence of pyridine.[2] Often, ***Pyridine*** is used as the solvent and reactions proceed nearly quantitatively (eq 1).

$$\mathrm{ROH} \xrightarrow[\text{pyridine}]{\mathrm{Ac_2O}} \mathrm{ROAc} \qquad (1)$$

If the reaction is run at temperatures lower than 20 °C, primary alcohols can be acetylated over secondary alcohols selectively.[18] Under these conditions, tertiary alcohols are not acylated. Most alcohols, including tertiary alcohols, can be acylated by the addition of DMAP (***4-Dimethylaminopyridine***) and ***Acetyl Chloride*** to the reaction containing acetic anhydride and pyridine. In general, the addition of DMAP increases the rate of acylation by 10^4 (eq 2).[19]

$$\text{(1-methylcyclohexyl)–OH} \xrightarrow[\text{Et}_3\text{N, rt; 86\%}]{\text{DMAP (4-NMe}_2\text{-pyridine), Ac}_2\text{O}} \text{(1-methylcyclohexyl)–OAc} \qquad (2)$$

Recently, Vedejs found that a mixture of ***Tri-n-butylphosphine*** and acetic anhydride acylates alcohols faster than acetic anhydride with DMAP.[20] However, the combination of acetic anhydride with DMAP and ***Triethylamine*** proved superior. It is believed that the Et_3N prevents HOAc from destroying the DMAP catalyst.

Tertiary alcohols have been esterified in good yield using acetic anhydride with calcium hydride or calcium carbide.[21] *t*-Butanol can be esterified to *t*-butyl acetate in 80% yield under these conditions. High pressure (15 kbar) has been used to introduce the acetate group using acetic anhydride in methylene chloride.[22] Yields range from 79–98%. Chemoselectivity is achieved using acetic anhydride and ***Boron Trifluoride Etherate*** in THF at 0 °C. Under these conditions, primary or secondary alcohols are acetylated in the presence of phenols.[23]

α-D-Glucose is peracetylated readily using acetic anhydride in the presence of ***Zinc Chloride*** to give α-D-glucopyranose pentaacetate in 63–72% yield (eq 3).[24]

$$\text{glucose} \xrightarrow[\mathrm{ZnCl_2}]{\mathrm{Ac_2O}} \text{α-D-glucopyranose pentaacetate} \qquad (3)$$

Under basic conditions, α-D-glucose can be converted into β-D-glucopyranose pentaacetate in 56% yield (eq 4).

$$\text{glucose} \xrightarrow[\mathrm{NaOAc}]{\mathrm{Ac_2O}} \text{β-D-glucopyranose pentaacetate} \qquad (4)$$

In the food and drug industry, high-purity acetic anhydride is used in the manufacture of aspirin by the acetylation of salicylic acid (eq 5).[25]

$$\text{salicylic acid} \xrightarrow{\mathrm{Ac_2O}} \text{acetylsalicylic acid} \qquad (5)$$

Amines. The acetylation of amines has been known since 1853 when Gerhardt reported the acetylation of aniline.[3] Acetamides are typically prepared by the reaction of the amine with acetic anhydride (eq 6).

$$\mathrm{PhNH_2} \xrightarrow{\mathrm{Ac_2O}} \mathrm{PhNHCOCH_3} \qquad (6)$$

A unique method for selective acylation of secondary amines in the presence of primary amines involves the use of ***18-Crown-6*** with acetic anhydride and triethylamine.[26] It is believed that the 18-crown-6 complexes primary alkylammonium salts more tightly than the secondary salts, allowing selective acetylation.

In some cases, tertiary amines undergo a displacement reaction with acetic anhydride. A simple example involves the reaction of benzyldimethylamine with acetic anhydride to give dimethylacetamide and benzylacetate (eq 7).[27]

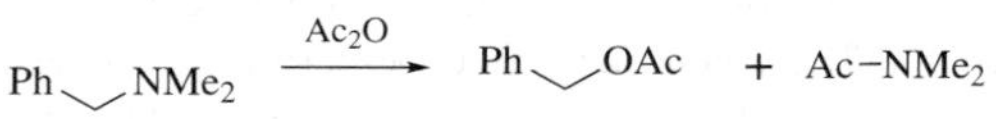

$$\text{PhCH}_2\text{NMe}_2 \xrightarrow{\text{Ac}_2\text{O}} \text{PhCH}_2\text{OAc} + \text{Ac-NMe}_2 \quad (7)$$

Allylic tertiary amines can be displaced by the reaction of acetic anhydride and sodium acetate.[28] The allylic acetate is the major product, as shown in eq 8.

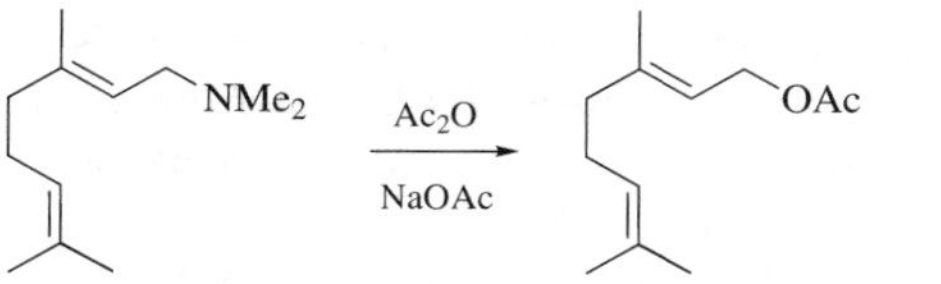

(8)

Cyclic benzylic amines may undergo ring opening upon heating with acetic anhydride (eq 9).[29]

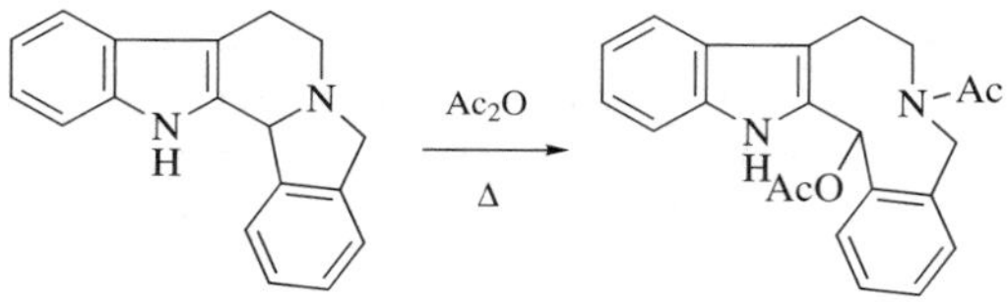

(9)

α-Amino acids react with acetic anhydride in the presence of a base to give 2-acetamido ketones.[30] This reaction is known as the Dakin–West reaction (eq 10) and is believed to go through a oxazolone mechanism. The amine base of choice is 4-dimethylaminopyridine. Under these conditions, the reaction can be carried out at room temperature in 30 min.

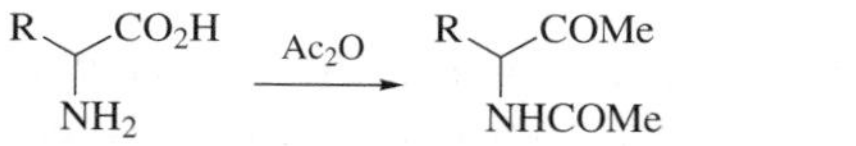

$$\text{RCH(NH}_2\text{)CO}_2\text{H} \xrightarrow{\text{Ac}_2\text{O}} \text{RCH(NHCOMe)COMe} \quad (10)$$

Cyclic β-amino acids rearrange to α-methylene lactams upon treatment with acetic anhydride, as shown in eq 11.[31]

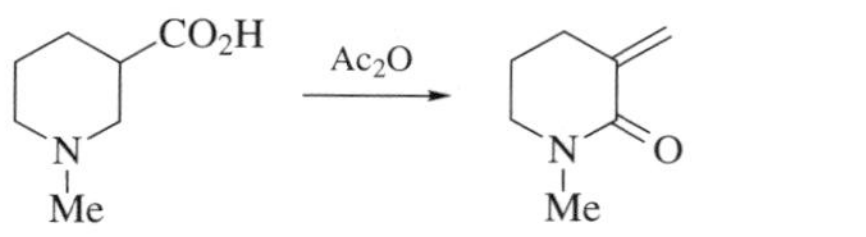

(11)

Thiols. *S*-Acetyl derivatives can be prepared by the reaction of acetic anhydride and a thiol in the presence of potassium bicarbonate.[4] Several disadvantages to the *S*-acetyl group in peptide synthesis include β-elimination upon base-catalyzed hydrolysis. Also, sulfur to nitrogen acyl migration may be problematic.

Oxidation. The oxidation of primary and secondary alcohols to the corresponding carbonyl compounds can be achieved using ***Dimethyl Sulfoxide–Acetic Anhydride***.[5] The reaction proceeds through an acyloxysulfonium salt as the oxidizing agent (eq 12).

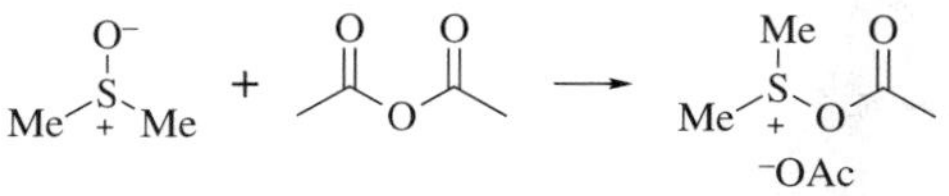

(12)

The oxidations often proceed at room temperature, although long reaction times (18–24 h) are sometimes required. A side product is formation of the thiomethyl ethers obtained from the Pummerer rearrangement.

If the alcohol is unhindered, a mixture of enol acetate (from ketone) and acetate results (eq 13).[32]

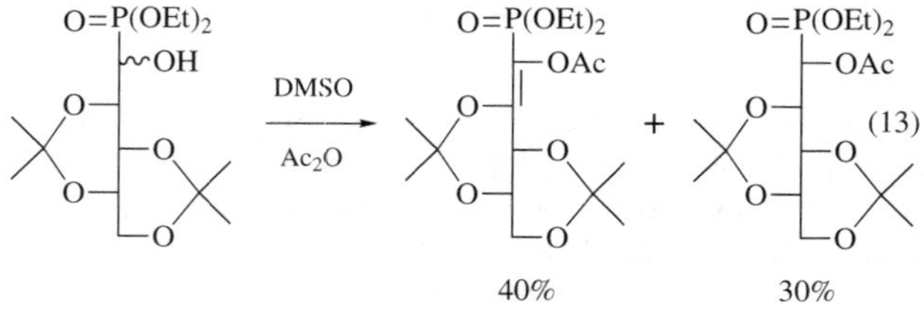

(13)

The oxidation of carbohydrates can be achieved by this method, as Hanessian showed (eq 14).[33]

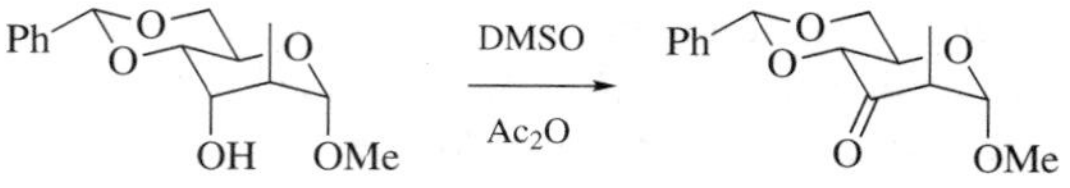

(14)

Aromatic α-diketones can be prepared from the acyloin compounds; however, aliphatic diketones cannot be prepared by this method.[34] The reaction proceeds well in complex systems without epimerization of adjacent stereocenters, as in the yohimbine example (eq 15).[35] This method compares favorably with that of ***Dimethyl Sulfoxide–Dicyclohexylcarbodiimide***.

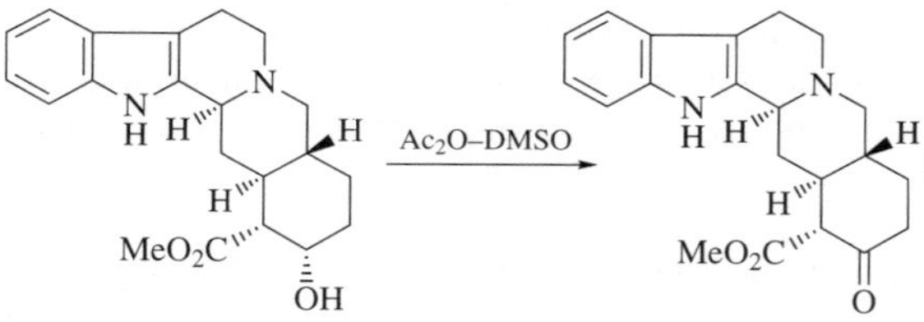

(15)

Dehydration. Many functionalities are readily dehydrated upon reaction with acetic anhydride, the most notable of which is the oxime.[6] Also, dibasic acids give cyclic anhydrides or ketones, depending on ring size.[36]

An aldoxime is readily converted to the nitrile as shown in eq 16.[37]

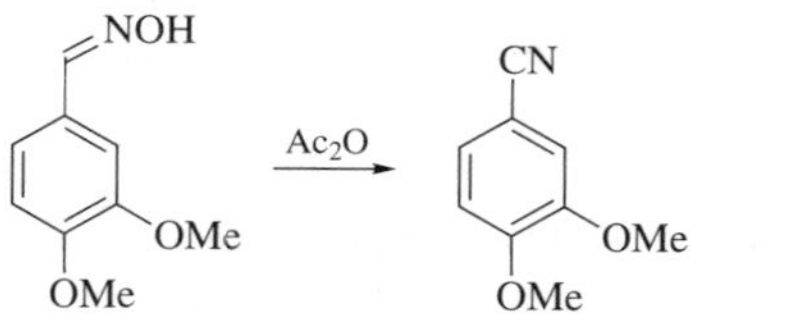

(16)

When oximes of α-tetralones are heated in acetic anhydride in the presence of anhydrous ***Phosphoric Acid***, aromatization occurs as shown in eq 17.[38]

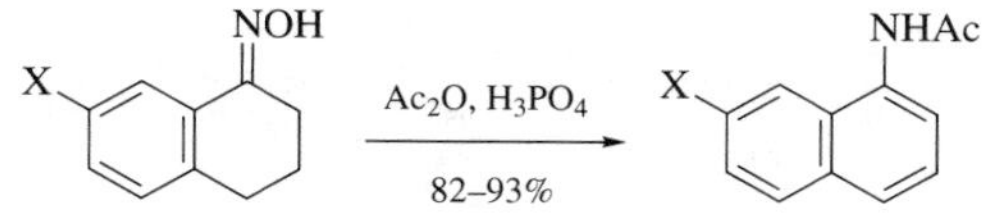

(17)

Oximes of aliphatic ketones lead to enamides upon treatment with acetic anhydride–pyridine, as shown in the steroid example in eq 18.[39]

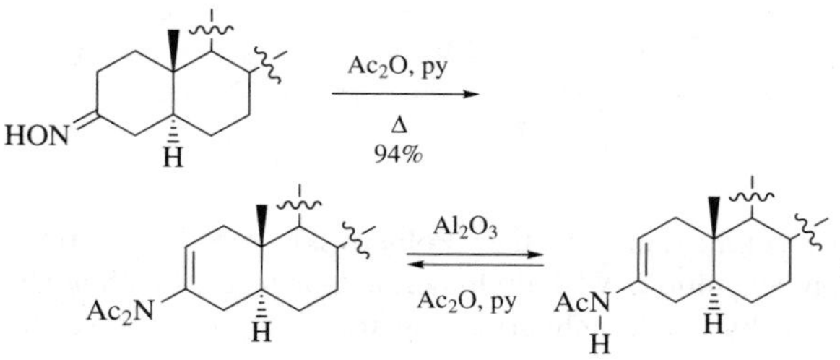

(18)

Upon heating with acetic anhydride, dibasic carboxylic acids lead to cyclic anhydrides of ring size 6 or smaller. Diacids longer than glutaric acid lead to cyclic ketones (eq 19).[36]

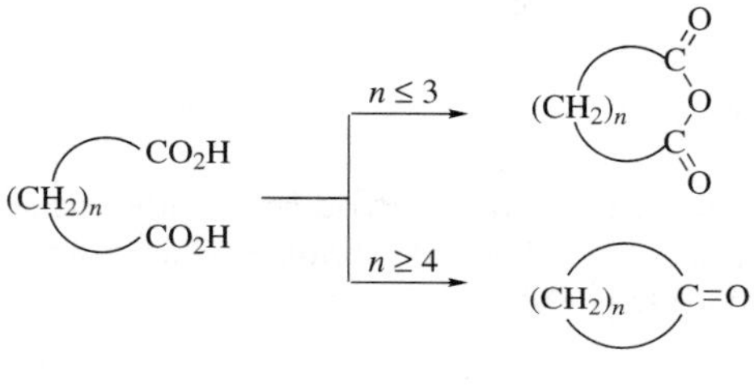

(19)

N-Acylanthranilic acids also cyclize when heated with acetic anhydride (eq 20). The reaction proceeds in 81% yield with slow distillation of the acetic acid formed.[40]

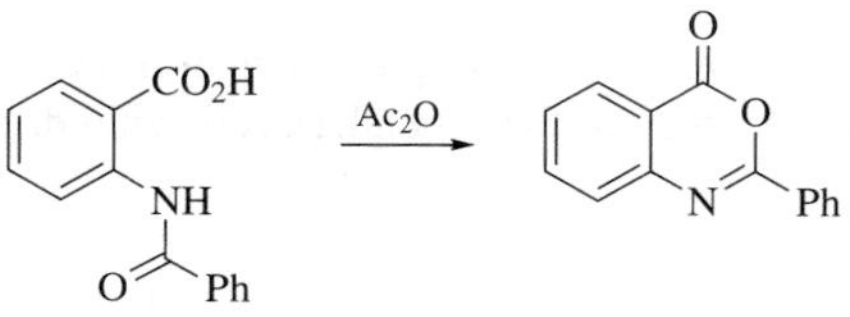

(20)

Pummerer Reaction. In 1910, Pummerer[7] reported that sulfoxides react with acetic anhydride to give 2-acetoxy sulfides (eq 21). The sulfoxide must have one α-hydrogen. Alternative reaction conditions include using ***Trifluoroacetic Anhydride*** and acetic anhydride.

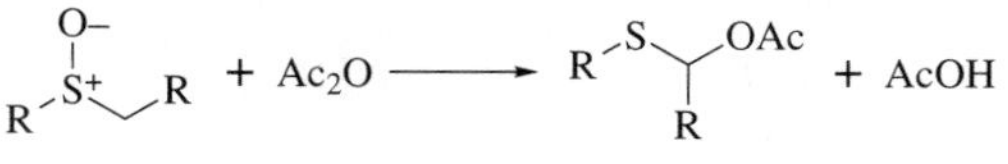

(21)

β-Hydroxy sulfoxides undergo the Pummerer reaction upon addition of sodium acetate and acetic anhydride to give α,β-diacetoxy sulfides. These compounds are easily converted to α-hydroxy aldehydes (eq 22).[41]

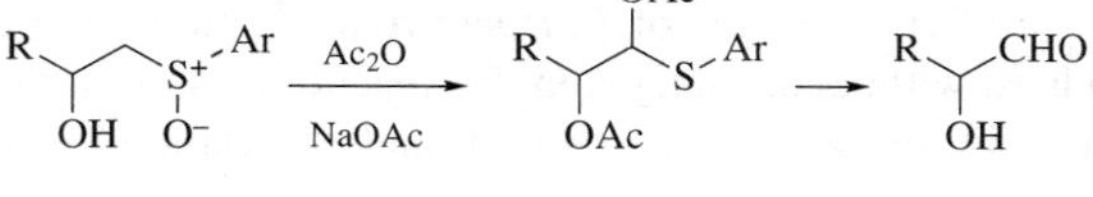

(22)

2-Phenylsulfonyl ketones rearrange in the presence of acetic anhydride–sodium acetate in toluene at reflux to give *S*-aryl thioesters (eq 23).[42] Upon hydrolysis, an α-hydroxy acid is obtained.

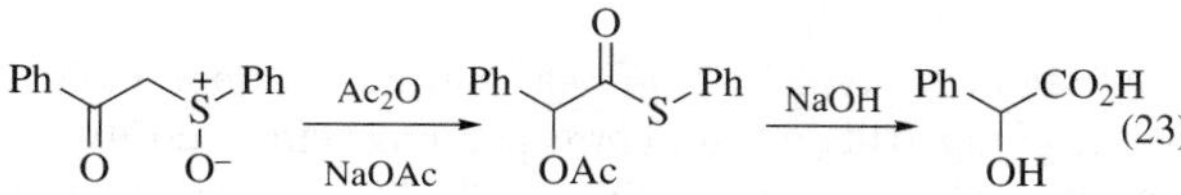

(23)

In the absence of sodium acetate, 2-phenylsulfonyl ketones give the typical Pummerer product. Since β-keto sulfoxides are available by the reaction of esters with the dimsyl anion, this overall process leads to one-carbon homologated α-hydroxy acids from esters (eq 24).[42]

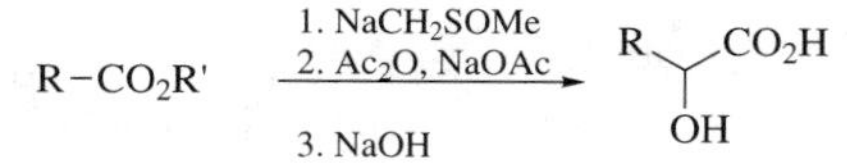

(24)

Also, 2-phenylsulfonyl ketones can be converted to α-phenylthio-α,β-unsaturated ketones via the Pummerer reaction using acetic anhydride and a catalytic amount of ***Methanesulfonic Acid*** (eq 25).[43]

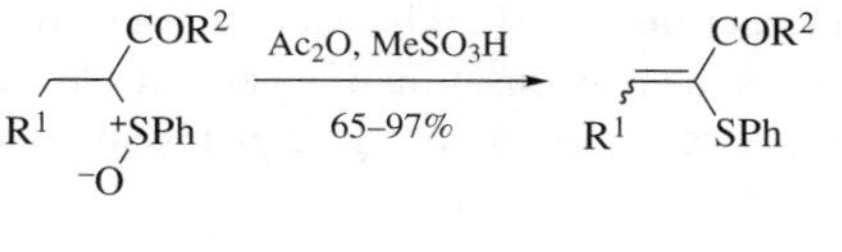

(25)

The Pummerer reaction has been used many times in heterocyclic synthesis as shown in eq 26.

(26)

The Pummerer rearrangement of 4-phenylsulfinylbutyric acid with acetic anhydride in the presence of ***p-Toluenesulfonic Acid*** leads to butanolide formation (eq 27).[44] Oxidation with ***m-Chloroperbenzoic Acid*** followed by thermolysis then leads to an unsaturated compound.

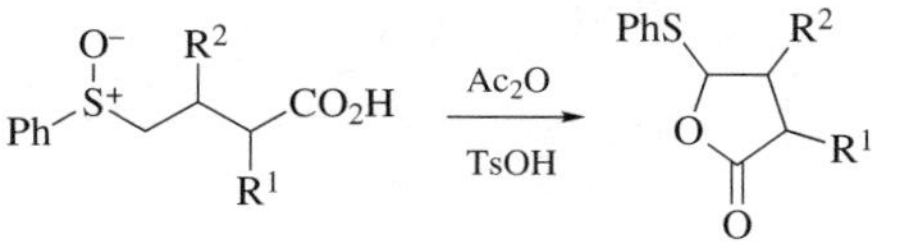

(27)

An unusual Pummerer reaction takes place with penicillin sulfoxide, leading to a ring expansion product as shown in eq 28.[45]

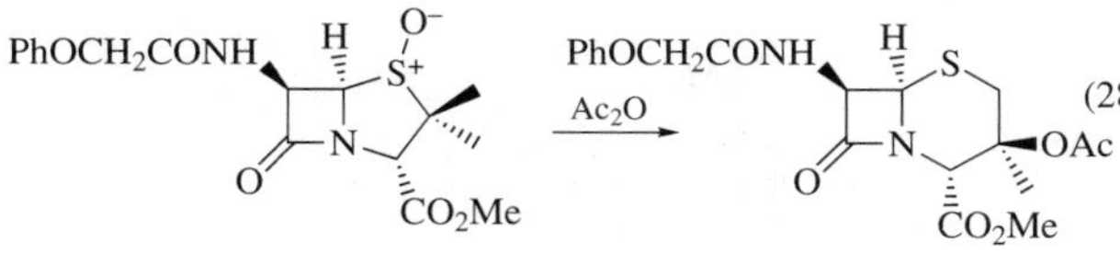

(28)

This led to discovery of the conversion of penicillin V and G to cephalexin,[46] a broad spectrum orally active antibiotic (eq 29).

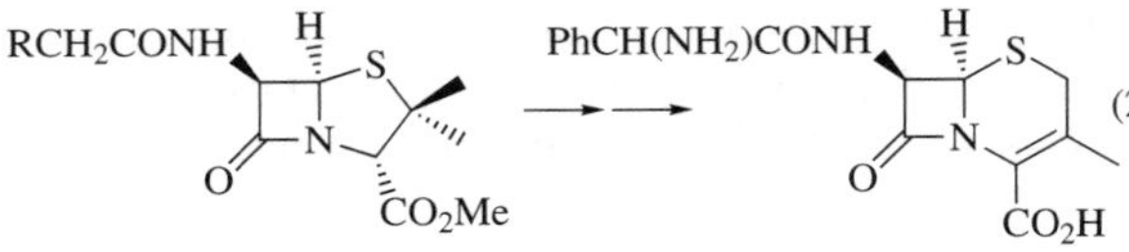

(29)

A Pummerer-type reaction was carried out on a dithiane protecting group to liberate the corresponding ketone (eq 30).[47] These were the only reaction conditions which provided any of the desired ketone.

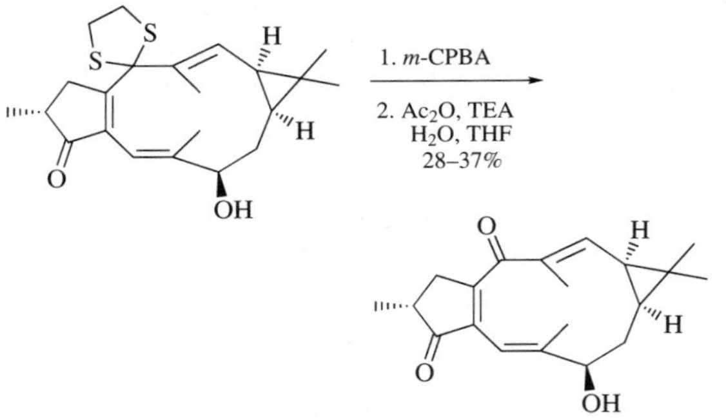

(30)

Perkin Reaction. The Perkin reaction,[8] developed by Perkin in 1868, involves the condensation of an anhydride and an aldehyde in the presence of a weak base to give an unsaturated acid (eq 31).

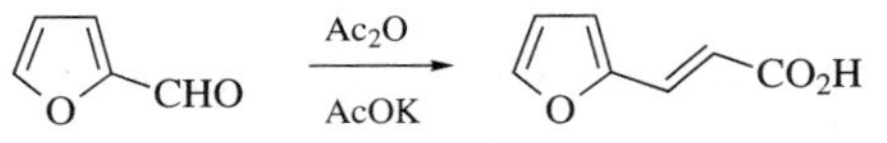

(31)

The reaction is often used for the preparation of cinnamic acids in 74–77% yield (eq 32).

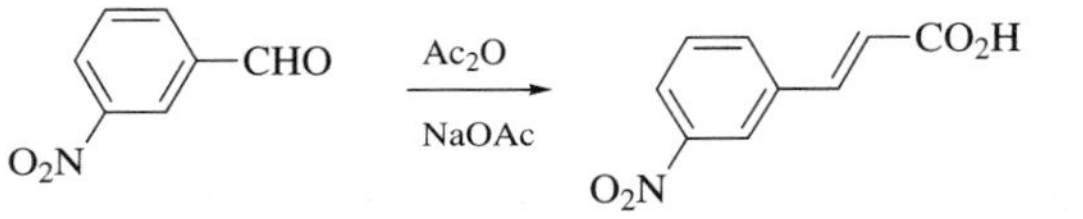

(32)

Aliphatic aldehydes give low or no yields of acid. Coumarin can be prepared by a Perkin reaction of salicylaldehyde and acetic anhydride in the presence of triethylamine (eq 33).

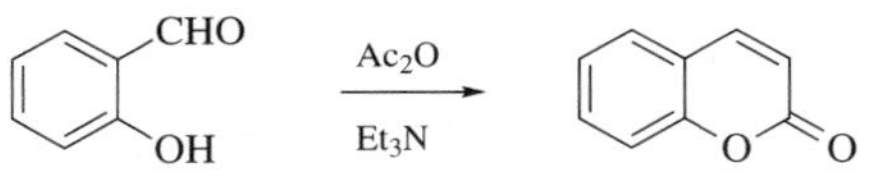

(33)

Polonovski Reaction. In the Polonovski reaction,[9] tertiary amine oxides react with acetic anhydride to give the acetamide of the corresponding secondary amine (eq 34).

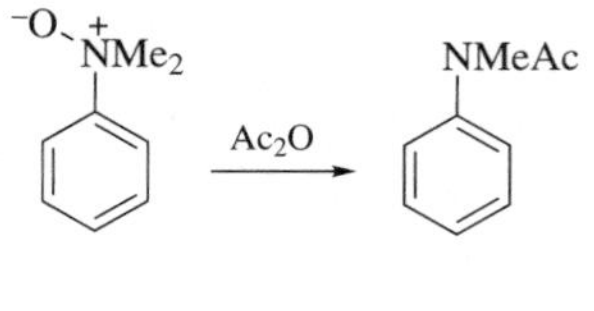

(34)

In nonaromatic cases, the Polonovski reaction gives the *N*-acylated secondary amine as the major product and the deaminated ketone as a minor product (eq 35).[48]

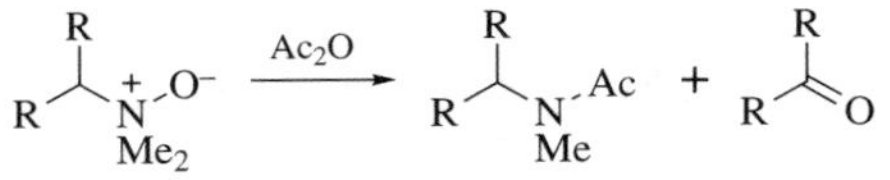

(35)

This reaction has been extended to a synthesis of 2-acetoxybenzodiazepine via an *N*-oxide rearrangement (eq 36).

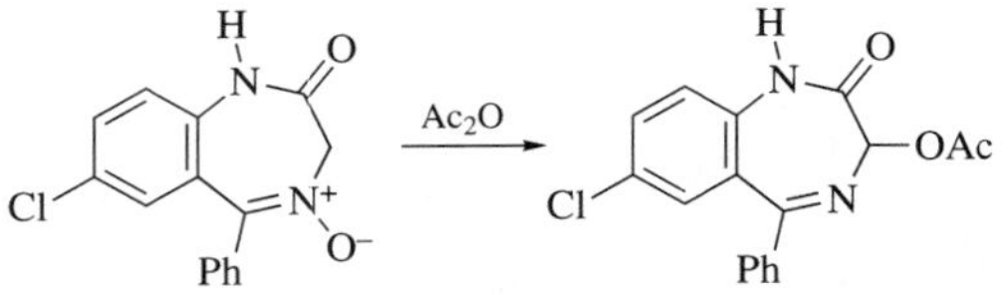

(36)

An application of the Polonovski reaction forms β-carbonylenamines. *N*-Methylpiperidone is reacted with *m*-CPBA followed by acetic anhydride and triethylamine to give the β-carbonyl enamine (eq 37).[49]

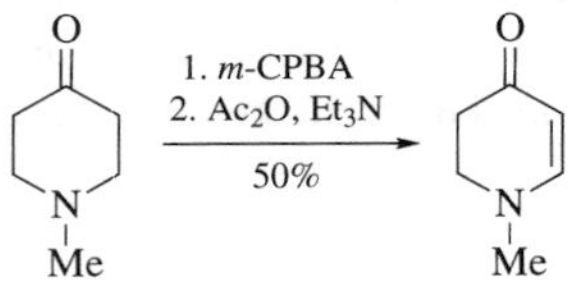

(37)

Reaction with *N*-Oxides. Pyridine 1-oxide reacts with acetic anhydride to produce 2-acetoxypyridine, which can be hydrolyzed to 2-pyridone (eq 38).[10]

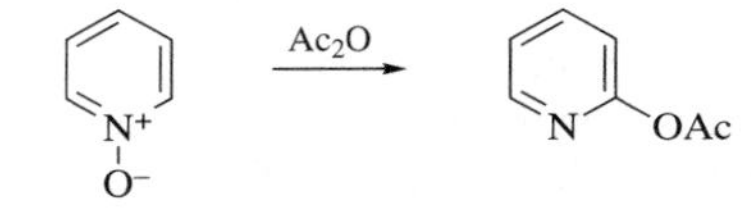

(38)

Open chain *N*-oxides, in particular nitrones, rearrange to amides (almost quantitatively) under acetic anhydride conditions (eq 39).[50]

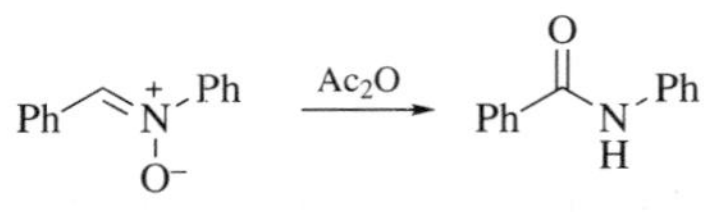

(39)

Heteroaromatic *N*-oxides with a side chain react with acetic anhydride to give side-chain acyloxylation (eq 40).[51]

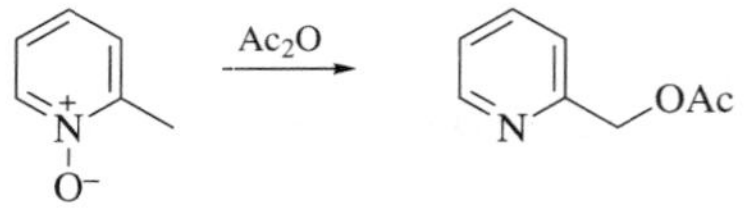

(40)

This reaction has been used in synthetic chemistry as the method of choice to form heterocyclic carbinols or aldehydes.

Thiele Reaction. The Thiele reaction converts 1,4-benzoquinone to 1,2,4-triacetoxybenzene using acetic anhydride and a catalytic amount of ***Sulfuric Acid***.[11] Zinc chloride has been used without advantage. In this reaction, 1,4-addition to the quinone is followed by enolization and acetylation to give the substituted benzene (eq 41).

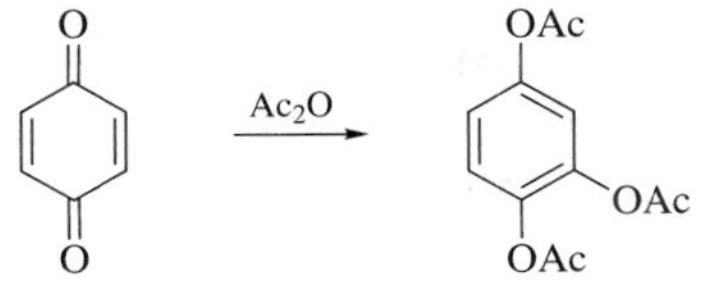

Ac_2O (41)

With unhindered quinones, BF_3 etherate is a more satisfactory catalyst but hindered quinones require the more active sulfuric acid catalyst. 1,2-Naphthoquinones undergo the Thiele reaction with acetic anhydride and sulfuric acid or boron trifluoride etherate (eq 42).

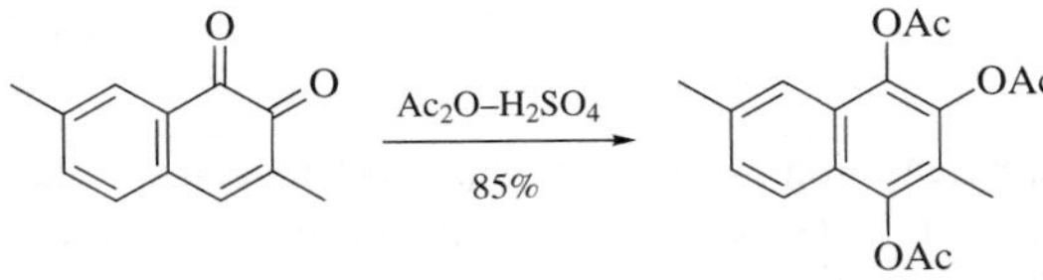

$Ac_2O–H_2SO_4$, 85% (42)

Ether Cleavage. Dialkyl ethers can be cleaved with acetic anhydride in the presence of pyridine hydrochloride or anhydrous ***Iron(III) Chloride***. In both cases, acetate products are produced. As shown in eq 43, the tricyclic ether is cleaved by acetic anhydride and pyridine hydrochloride to give the diacetate in 93% yield.[12]

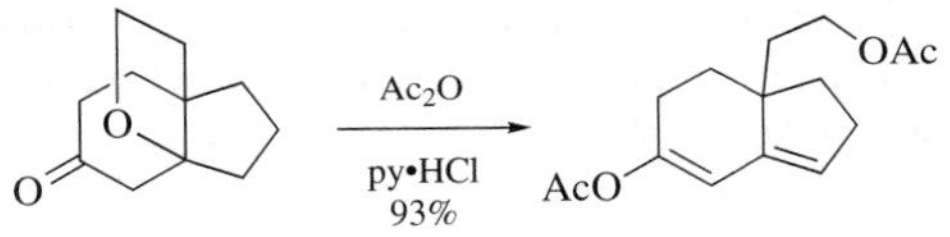

Ac_2O, py•HCl, 93% (43)

Simple dialkyl ethers react with iron(III) chloride and acetic anhydride to produce compounds where both R groups are converted to acetates (eq 44).[52]

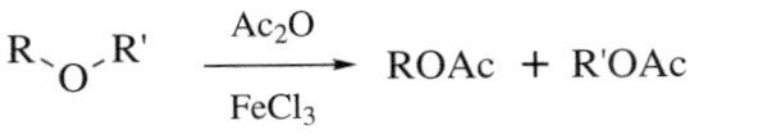

$$\text{R–O–R'} \xrightarrow[\text{FeCl}_3]{\text{Ac}_2\text{O}} \text{ROAc} + \text{R'OAc} \quad (44)$$

Cleavage of allylic ethers can occur using acetic anhydride in the presence of iron(III) chloride (eq 45). The reaction takes place without isomerization of a double bond, but optically active ethers are cleaved with substantial racemization.[53]

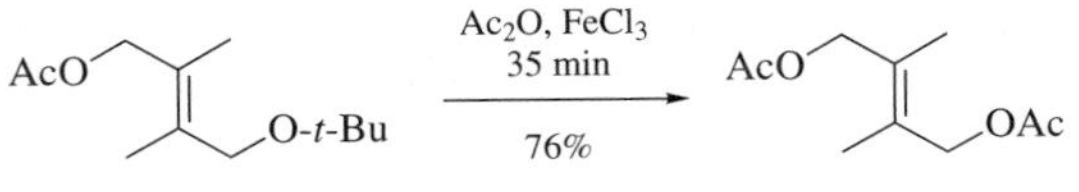

Ac_2O, $FeCl_3$, 35 min, 76% (45)

Cleavage of aliphatic ethers occurs with the reaction of acetic anhydride, boron trifluoride etherate, and ***Lithium Bromide*** (eq 46). The ethers are cleaved to the corresponding acetoxy compounds contaminated with a small amount of unsaturated product.[54] In some cases, the lithium halide may not be necessary.

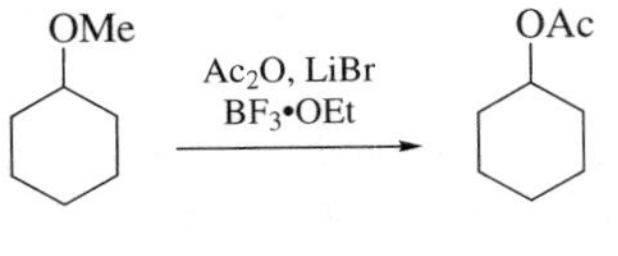

Ac_2O, LiBr, BF_3•OEt (46)

Cyclic ethers are cleaved to ω-bromoacetates using ***Magnesium Bromide*** and acetic anhydride in acetonitrile (eq 47).[55]

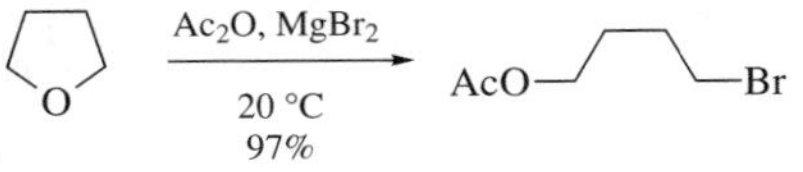

Ac_2O, $MgBr_2$, 20 °C, 97% (47)

The reaction occurs with inversion of configuration, as shown in eq 48.

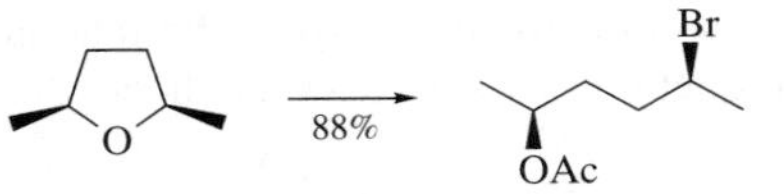

88% (48)

Trimethylsilyl ethers are converted to acetates directly by the action of acetic anhydride–pyridine in the presence of 48% HF or boron trifluoride etherate (eq 49).[56]

$$\text{ROTMS} \xrightarrow[\text{48\% HF or BF}_3\text{•OEt}_2]{\text{Ac}_2\text{O–pyr}} \text{ROAc} \quad (49)$$

Miscellaneous Reactions. Primary allylic alcohols can be prepared readily by the action of *p*-toluenesulfonic acid in acetic anhydride–acetic acid on the corresponding tertiary vinyl carbinol, followed by hydrolysis of the resulting acetate.[57] The vinyl carbinol is readily available from the reaction of a ketone with a vinyl Grignard reagent. Overall yields of allylic alcohols are very good (eq 50).

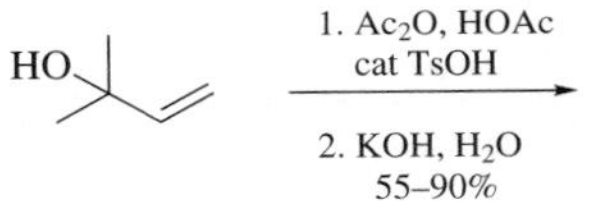

1. Ac_2O, HOAc, cat TsOH; 2. KOH, H_2O, 55–90% (50)

Enol lactonization occurs readily on an α-keto acid using acetic anhydride at elevated temperatures.[58] The reaction shown in eq 51 proceeds in 89% yield. In general, acetic anhydride is superior to acetyl chloride in this reaction.[59]

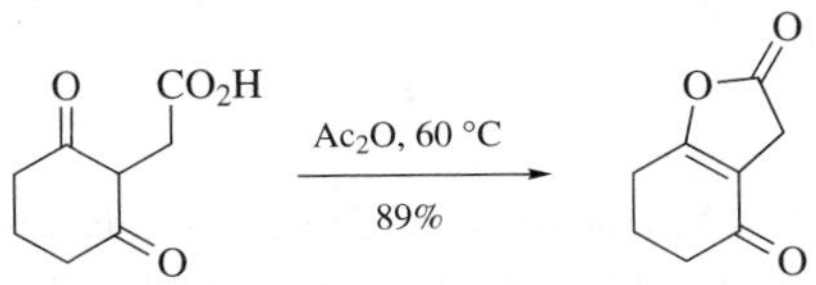

Ac_2O, 60 °C, 89% (51)

Aliphatic aldehydes are easily converted to the enol acetate using acetic anhydride and potassium acetate (eq 52).[13] This reaction only works for aldehydes and is the principal reason for the failure of aldehydes to succeed in the Perkin reaction. Triethylamine and DMAP may also catalyze the reaction.

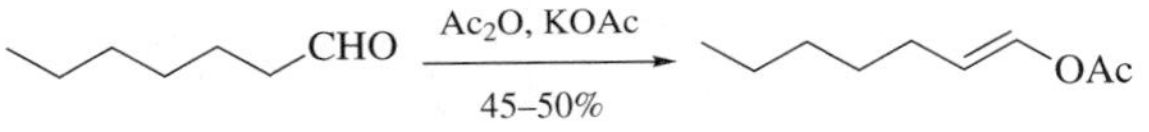

Ac_2O, KOAc, 45–50% (52)

A cyclopropyl ketone is subject to homoconjugate addition using acetic anhydride/boron trifluoride etherate. Upon acetate addition, the enol is trapped as its enol acetate (eq 53).[60]

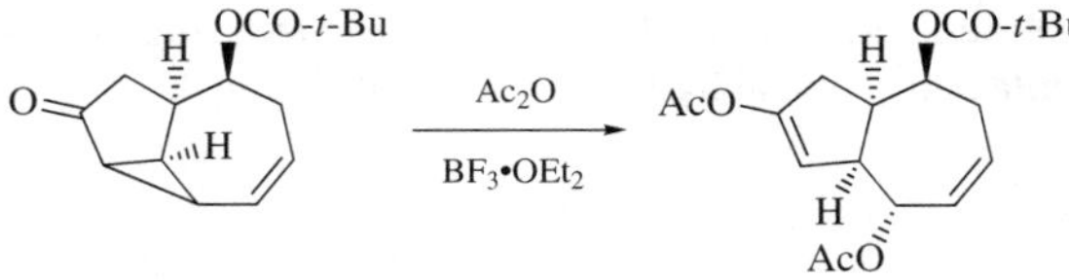

When an aldehyde is treated with acetic anhydride/anhydrous iron(III) chloride, geminal diacetates are formed in good to excellent yields.[14] Aliphatic and unsaturated aldehydes can be used in this reaction as shown in eqs 54–56. Interestingly, if an α-hydrogen is present in an unsaturated aldehyde, elimination of the geminal diacetate product gives a 1-acetoxybutadiene.

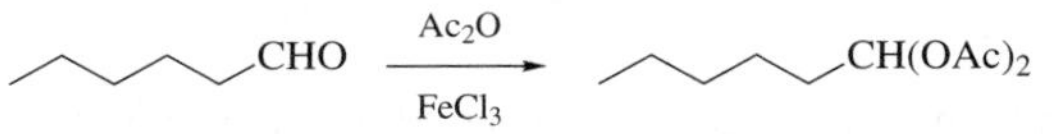

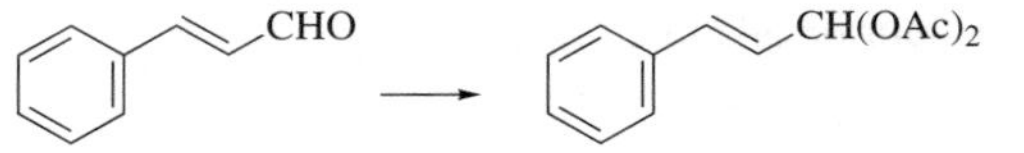

If an aldehyde is treated with acetic anhydride in the presence of a catalytic amount of ***Cobalt(II) Chloride***, a diketone is formed (eq 57).[61]

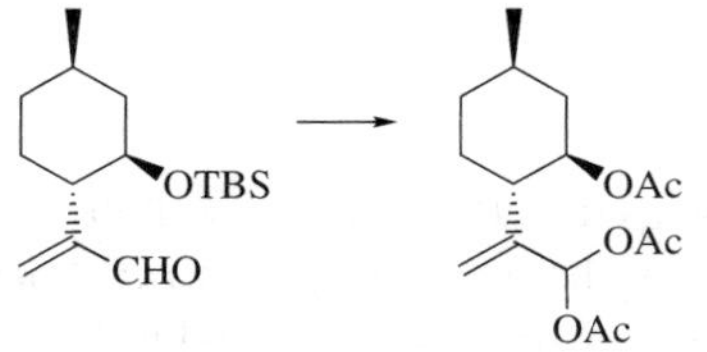

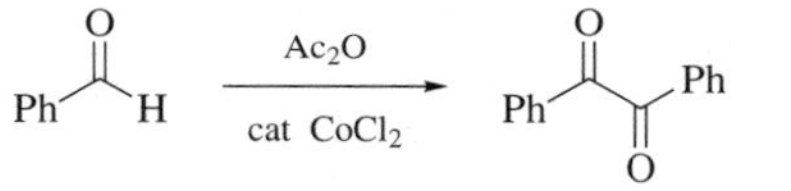

However, if 1.5 equiv of cobalt(II) chloride is added, the geminal diacetate is formed (eq 58).[62]

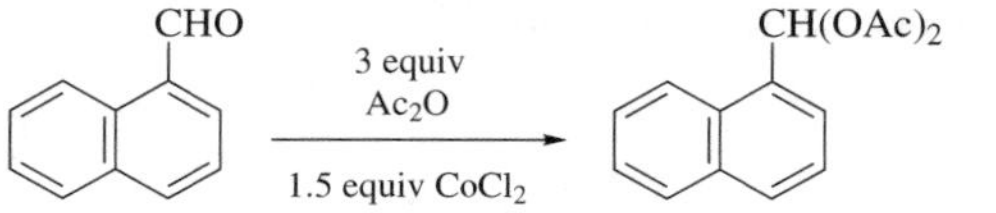

Apparently, the reaction in eq 58 occurs only when the starting material is polyaromatic or with compounds whose carbonyl IR frequency is less than 1685 cm^{-1}.

Acetic anhydride participates in several cyclization reactions. For example, enamines undergo a ring closure when treated with acetic anhydride (eq 59).[63]

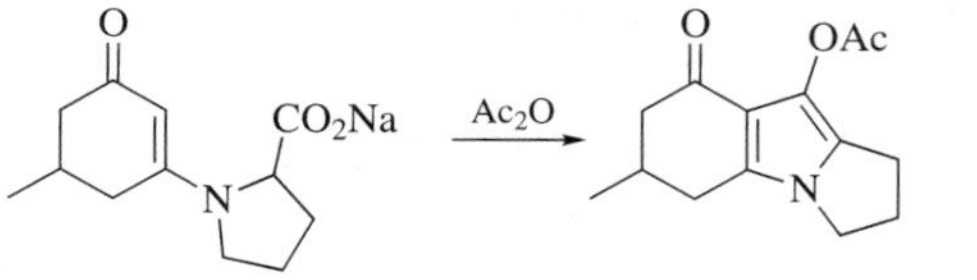

o-Diamine compounds also cyclize when treated with acetic anhydride (eq 60).[64]

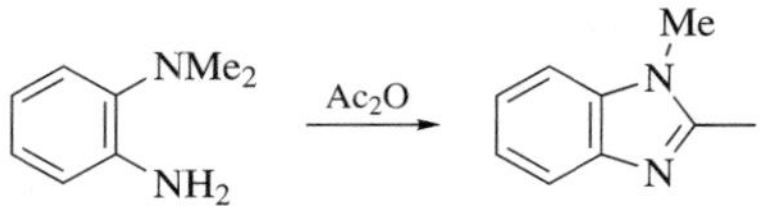

Twistane derivatives were obtained by the reaction of a decalindione with acetic anhydride, acetic acid, and boron trifluoride etherate (eq 61).[65]

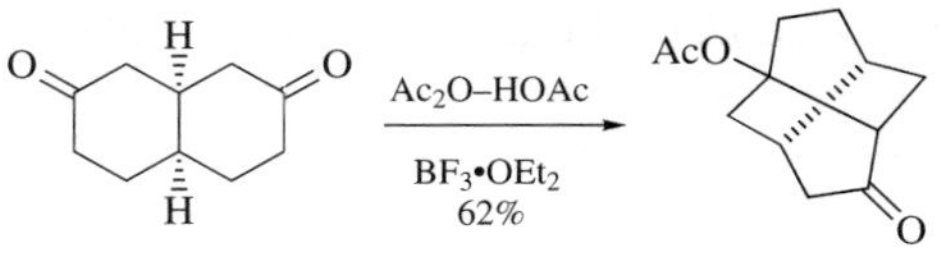

A condensation/cyclization reaction between an alkynyl ketone and a carboxylic acid in the presence of acetic anhydride/triethylamine gives a butenolide (eq 62).[66]

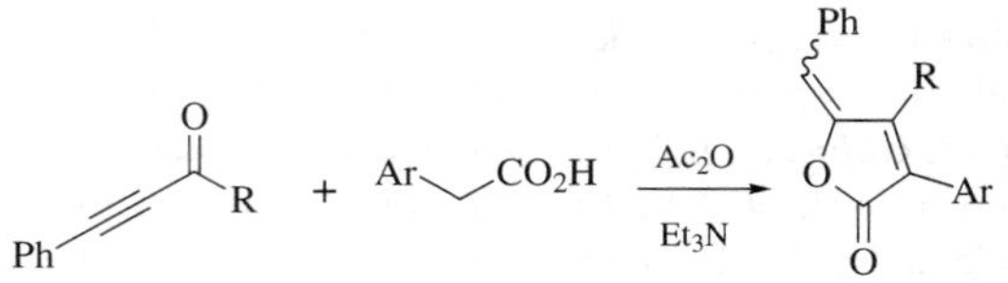

A few rearrangement reactions take place with acetic anhydride. A Claisen rearrangement is involved in the formation of the aromatic acetate in eq 63.[67] The reaction proceeds in 44% yield even after 21 h at 200 °C.

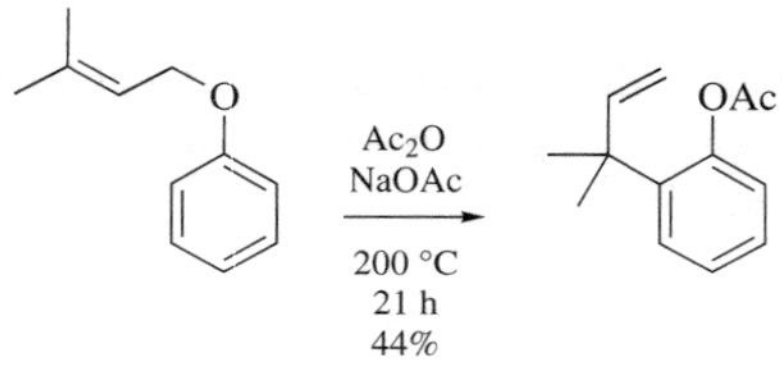

Complex rearrangements have occured using acetic anhydride under basic conditions, as shown in eq 64.[68]

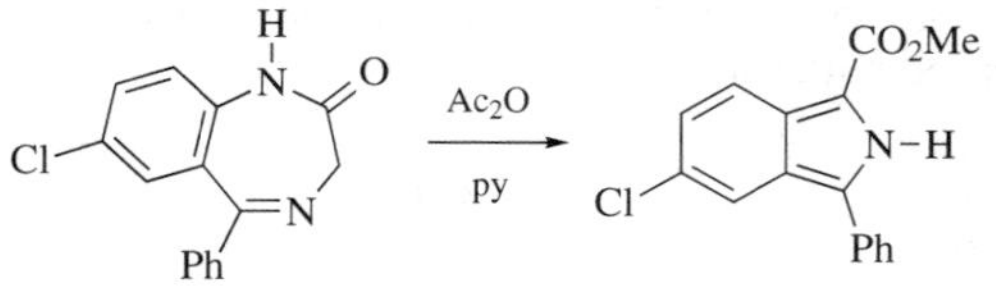

Aromatization occurs readily using acetic anhydride. Aromatization of α-cyclohexanones occurs under acidic conditions to lead to good yields of phenols (eq 65).[69] However, in totally unsubstituted ketones, aldol products are formed.

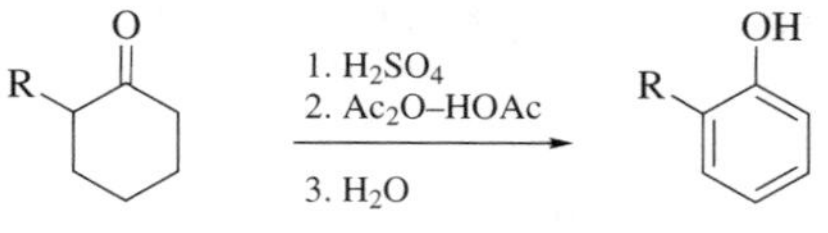

Aromatization of 1,4- and 1,2-cyclohexanediones leads to cresol products (eq 66) in over 90% yield.[70]

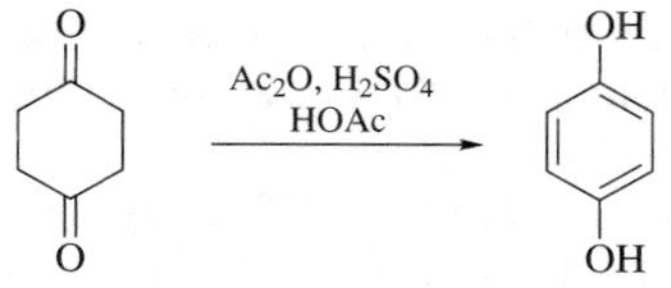

(66)

Alkynes and allenes are formed by the acylation of nitrimines using acetic anhydride/pyridine with DMAP as catalyst (eqs 67 and 68).[71] Nitrimines are prepared by nitration of ketoximes with nitrous acid.

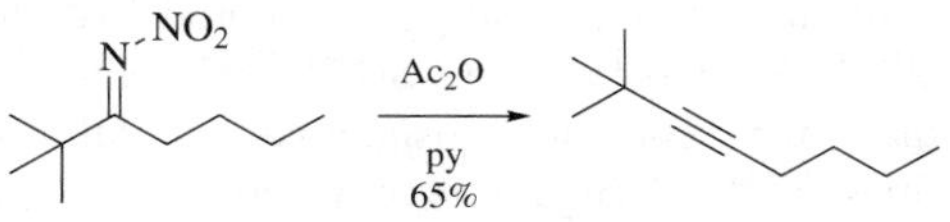

(67)

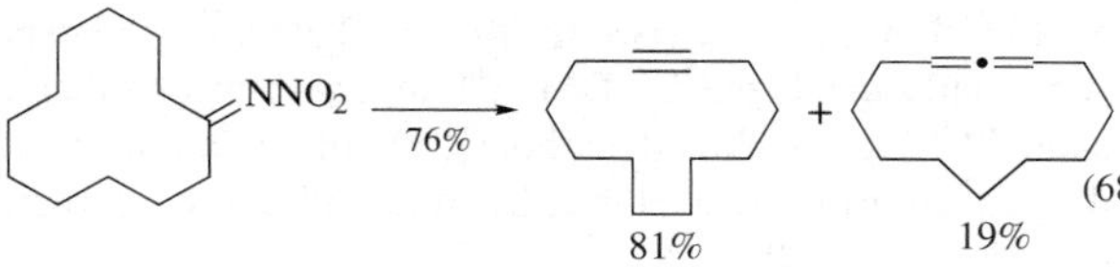

(68)

Lastly, acetic anhydride participates in the Friedel–Crafts reaction.[72] ***Polyphosphoric Acid*** is both reagent and solvent in these reactions (eq 69).

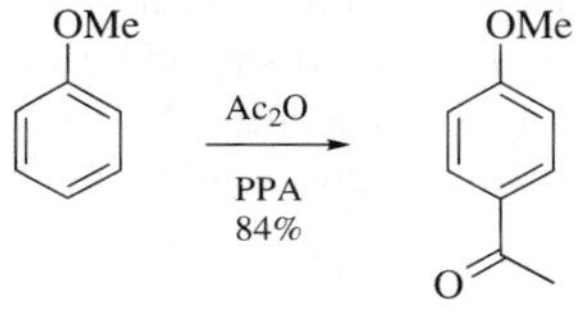

(69)

Related Reagents. Acetyl Chloride; Acetyl Bromide; Isopropenyl Acetate; Vinyl Acetate; Acetyl Cyanide.

1. (a) Kim, D. H. *JHC* **1976**, *13*, 179. (b) Cook, S. L. *Chemical Industries* **1993**, *49*, 145. (c) Joy, E. F.; Barnard, A. J., Jr. *Encyclopedia of Industrial Chemical Analysis*; Interscience: New York, 1967; Vol. 4, p 102.
2. (a) Weber, H.; Khorana, H. G. *J. Mol. Biol.* **1972**, *72*, 219 (b) Zhdanov, R. I.; Zhenodarova, S. M. *S* **1975**, 222.
3. Mariella, R. P.; Brown, K. H. *JOC* **1971**, *36*, 735 and references cited therein.
4. Zervas, L.; Photaki, I.; Ghelis, N. *JACS* **1963**, *85*, 1337.
5. Albright, J. D.; Goldman, L. *JACS* **1967**, *89*, 2416.
6. (a) Buck, J. S.; Ide, W. S. *OSC* **1943**, *2*, 622. (b) White, D. M. *JOC* **1974**, *39*, 1951. (c) Nicolet, B. H.; Pelc, J. J. *JACS* **1922**, *44*, 1145. (d) Bell, M. R.; Johnson, J. R.; Wildi, B. S.; Woodward, R. B. *JACS* **1958**, *80*, 1001.
7. (a) Pummerer, R. *B* **1910**, *43*, 1401. (b) Parham, W. E.; Edwards, L. D. *JOC* **1968**, *33*, 4150. (c) Tanikaga, R.; Yabuki, Y.; Ono, N.; Kaji, A. *TL* **1976**, 2257.
8. (a) Rosen, T. *COS* **1991**, *2*, 395. (b) Merchant, J. R.; Gupta, A. S. *CI(L)* **1978**, 628.
9. Polonovski, M.; Polonovski, M. *BSF* **1927**, *41*, 1190.
10. Katada, M. *J. Pharm. Soc. Jpn.* **1947**, *67*, 51.
11. (a) Thiele, J. *B* **1898**, *31*, 1247. (b) Thiele, J.; Winter, E. *LA* **1899**, *311*, 341.
12. (a) Peet, N. P.; Cargill, R. L. *JOC* **1973**, *38*, 1215. (b) Goldsmith, D. J.; Kennedy, E.; Campbell, R. G. *JOC* **1975**, *40*, 3571.
13. (a) Bedoukin, P. Z. *OSC* **1955**, *3*, 127. (b) Cousineau, T. J.; Cook, S. L.; Secrist, J. A., III *SC* **1979**, *9*, 157.
14. Kochhar, K. S.; Bal, B. S.; Deshpande, R. P.; Rajadhyaksha, S. N.; Pinnick, H. W. *JOC* **1983**, *48*, 1765.
15. For analysis by titration, see Ref. 1c. Spectra available from the Aldrich Library.
16. *Vogel's Textbook of Practical Organic Chemistry*, 4th ed.; Longman: Harlow, 1978; p 499.
17. Perrin, D. D.; Armarego, W. L. F.; Perrin, D. R. *Purification of Laboratory Chemicals*, 2nd ed.; Pergamon: Oxford, 1985; p 77.
18. Stork, G.; Takahashi, T.; Kawamoto, I.; Suzuki, T. *JACS* **1978**, *100*, 8272.
19. Hölfe, G.; Steglich, W.; Vorbrüggen, H. *AG(E)* **1978**, *17*, 569.
20. Vedejs, E.; Diver, S. T. *JACS* **1993**, *115*, 3358.
21. Oppenauer, R. V. *M* **1966**, *97*, 62.
22. Dauben, W. G.; Bunce, R. A.; Gerdes, J. M.; Henegar, K. E.; Cunningham, A. F., Jr.; Ottoboni, T. B. *TL* **1983**, *24*, 5709.
23. Nagao, Y.; Fujita, E.; Kohno, T.; Yagi, M. *CPB* **1981**, *29*, 3202.
24. *Vogel's Textbook of Practical Organic Chemistry*, 4th ed.; Longman: Harlow, 1978; pp 454–455.
25. See Ref. 1(b) and Candoros, F., Rom. Patent 85 726, 1984; *CA* **1985**, *103*, 104 715.
26. Barrett, A. G. M.; Lana, J. C. A. *CC* **1978**, 471.
27. Tiffeneau, M.; Fuhrer, K. *BSF* **1914**, *15*, 162.
28. Fujita, T.; Suga, K.; Watanabe, S. *AJC* **1974**, *27*, 531.
29. Freter, K.; Zeile, K. *CC* **1967**, 416.
30. (a) Dakin, H. D.; West, R. *JBC* **1928**, *78*, 745, 757 (b) Allinger, N. L.; Wang, G. L.; Dewhurst, B. B. *JOC* **1974**, *39*, 1730. (c) Buchanan, G. L. *CS* **1988**, *17*, 91.
31. (a) Ferles, M. *CCC* **1964**, *29*, 2323. (b) Rueppel, M. L.; Rapoport, H. *JACS* **1972**, *94*, 3877.
32. Glebova, Z. I.; Uzlova, L. A.; Zhdanov, Y. A. *ZOB* **1985**, *55*, 1435; *CA* **1986**, *104*, 69 072.
33. Hanessian, S.; Rancourt, G. *CJC* **1977**, *55*, 1111.
34. Newman, M. S.; Davis, C. C. *JOC* **1967**, *32*, 66.
35. (a) Albright, J. D.; Goldman, L. *JACS* **1965**, *87*, 4214. (b) *JOC* **1965**, *30*, 1107.
36. (a) Blanc, H. G. *CR* **1907**, *144*, 1356. (b) Ruzicka, L.; Prelog, V.; Meister, P. *HCA* **1945**, *28*, 1651.
37. Beringer, F. M.; Ugelow, I. *JACS* **1953**, *75*, 2635, and references cited therein.
38. Newnan, M. S.; Hung, W. M. *JOC* **1973**, *38*, 4073.
39. Boar, R. B.; McGhie, J. F.; Robinson, M.; Barton, D. H. R.; Horwell, D. C.; Stick, R. V. *JCS(P1)* **1975**, 1237.
40. Zentmyer, D. T.; Wagner, E. C. *JOC* **1949**, *14*, 967.
41. Iruichijima, S.; Maniwa, K.; Tsuchihashi, G. *JACS* **1974**, *96*, 4280.
42. Iriuchijima, S.; Maniwa, K.; Tsuchihashi, G. *JACS* **1975**, *97*, 596.
43. (a) Monteiro, H. J.; de Souza, J. P. *TL* **1975**, *921*. (b) Monteiro, H. J.; Gemal, A. L. *S* **1975**, 437.
44. Watanabe, M.; Nakamori, S.; Hasegawa, H.; Shirai, K.; Kumamoto, T. *BCJ* **1981**, *54*, 817.
45. Morin, R. B.; Jackson, B. G.; Mueller, R. A.; Lavagnino, E. R.; Scanlon, W. S.; Andrews, S. L. *JACS* **1963**, *85*, 1896.
46. Chauvette, R. R.; Pennington, P. A.; Ryan, C. W.; Cooper, R. D. G.; Jose, F. L.; Wright, I. G.; Van Heyningen, E. M.; Huffman, G. W. *JOC* **1971**, *36*, 1259.
47. Smith, A. B., III; Dorsey, B. D.; Visnick, M.; Maeda, T.; Malamas, M. S. *JACS* **1986**, *108*, 3110.
48. (a) See Ref. 9. (b) Polonovski, M.; Polonovski, M. *BSF* **1926**, *39*, 147. (c) Wenkert, E. *E* **1954**, *10*, 346. (d) Cave, A.; Kan-Fan, C.; Potier, P.; LeMen, J. *T* **1967**, *23*, 4681.
49. Stütz, P.; Stadler, P. A. *TL* **1973**, 5095.
50. Tamagaki, S.; Kozuka, S.; Oae, S. *T* **1970**, *26*, 1795.

51. (a) Kobayashi, G.; Furukawa, S. *CPB* **1953**, *1*, 347. (b) Boekelheide, V.; Lim, W. J. *JACS* **1954**, *76*, 1286. (c) Bullit, O. H., Jr.; Maynard, J. T. *JACS* **1954**, *76*, 1370. (d) Berson, J. A.; Cohen, T. *JACS* **1955**, *77*, 1281.
52. Knoevenagel, E. *LA* **1914**, *402*, 111.
53. Ganem, B.; Small, V. R., Jr. *JOC* **1974**, *39*, 3728.
54. (a) Youssefyeh, R. D.; Mazur, Y. *TL* **1962**, 1287. (b) Narayanan, C. R.; Iyer, K. N. *TL* **1964**, 759.
55. Goldsmith, D. J.; Kennedy, E.; Campbell, R. G. *JOC* **1975**, *40*, 3571.
56. (a) Voaden, D. J.; Waters, R. M. *OPP* **1976**, *8*, 227. (b) For conversion of ROTBS to ROAc using Ac_2O and $FeCl_3$, see Ganem, B.; Small, V. R. *JOC* **1974**, *39*, 3728.
57. Babler, J. H.; Olsen, D. O. *TL* **1974**, 351.
58. Eggette, T. A.; deBoer, J. J. J.; deKoning, H.; Huisman, H. O. *SC* **1978**, *8*, 353.
59. Rosenmund, K. W.; Herzberg, H.; Schutt *CB* **1954**, *87*, 1258.
60. (a) Rigby, J. H.; Senanayake, C. *JACS* **1987**, *109*, 3147. (b) Rigby, J. H.; Senanayake, C. *JOC* **1988**, *53*, 440.
61. Ahmad, S.; Iqbal, J. *CC* **1987**, 692.
62. Fry, A. J.; Rho, A. K.; Sherman, L. R.; Sherwin, C. S. *JOC* **1991**, *56*, 3283.
63. Friary, R. J., Sr.; Gilligan, J. M.; Szajewski, R. P.; Falci, K. J.; Franck, R. W. *JOC* **1973**, *38*, 3487.
64. Meth-Cohn, O.; Suschitzky, H. In *Advances in Heterocyclic Chemistry*; Katritzky, A. R.; Boulton, J., Eds.; Academic Press: New York, 1972; Vol. 14, p 213.
65. Bélanger, A.; Lambert, Y.; Deslongchamps, P. *CJC* **1969**, *47*, 795.
66. Rao, Y. S.; Filler, R. *TL* **1975**, 1457.
67. (a) Rhoads, S. J.; Raulins, N. R. *OR* **1975**, *22*, 1. (b) Karanewsky, D. S.; Kishi, Y.; *JOC* **1976**, *41*, 3026.
68. Gryer, R. I.; Brust, B.; Earley, J. V.; Sternbach, L. H. *JCS(C)* **1967**, 366.
69. Kablaoui, M. S. *JOC* **1974**, *39*, 2126.
70. Kablaoui, M. S. *JOC* **1974**, *39*, 3696.
71. Büchi, G.; Wüest, H. *JOC* **1979**, *44*, 4116.
72. Edwards, J. D.; McGuire, S. E.; Hignite, C. *JOC* **1964**, *29*, 3028.

Regina Zibuck
Wayne State University, Detroit, MI, USA

Acetyl Chloride[1]

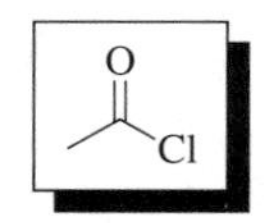

[75-36-5] C_2H_3ClO (MW 78.50)

(useful for electrophilic acetylation of arenes,[2] alkenes,[2a,3] alkynes,[4] saturated alkanes,[3a,5] organometallics, and enolates (on C or O);[6] for cleavage of ethers;[7] for esterification of sterically unhindered[8] or acid-sensitive[9] alcohols; for generation of solutions of anhydrous hydrogen chloride in methanol;[10] as a dehydrating agent; as a solvent for organometallic reactions;[11] for deoxygenation of sulfoxides;[12] as a scavenger for chlorine[13] and bromine;[14] as a source of ketene; and for nucleophilic acetylation[15])

Physical Data: bp 51.8 °C;[1a] mp −112.9 °C;[1a] *d* 1.1051 g cm^{-3};[1a] refractive index 1.38976.[1b] IR (neat) ν 1806.7 cm^{-1};[16] ^{1}H NMR ($CDCl_3$) δ 2.66 ppm; ^{13}C NMR ($CDCl_3$) δ 33.69 ppm (q) and 170.26 ppm (s); the bond angles (determined by electron diffraction[17]) are 127.5° (O–C–C), 120.3° (O–C–Cl), and 112.2° (Cl–C–C).

Analysis of Reagent Purity: a GC assay for potency has been described;[18] to check qualitatively for the presence of HCl, a common impurity, add a few drops of a solution of crystal violet in chloroform;[19] a green or yellow color indicates that HCl is present, while a purple color that persists for at least 10 min indicates that HCl is absent.[1b]

Preparative Methods: treatment of ***Acetic Acid*** or sodium acetate with the standard inorganic chlorodehydrating agents (PCl_3,[1b,23] SO_2Cl_2,[1a,24] or $SOCl_2$[1b,25]) generates material that may contain phosphorus- or sulfur-containing impurities.[1b,23a,26] Inorganic-free material can be prepared by treatment of HOAc with $Cl_2CHCOCl$ (Δ; 70%),[27] PhCOCl (Δ; 88%),[28] $PhCCl_3$ (cat. H_2SO_4, 90 °C; 92.5%),[29] or phosgene[30] (optionally catalyzed by DMF,[30e] magnesium or other metal salts,[30a,b,d] or activated carbon[30b,c]), or by addition of hydrogen chloride to acetic anhydride (85–90 °C; 'practically quantitative').[1a,31]

Purification: HCl-free material can be prepared either by distillation from dimethylaniline[11c,20] or by standard degassing procedures.[20c,21]

Handling, Storage, and Precautions: acetyl chloride should be handled only in a well-ventilated fume hood since it is volatile and toxic via inhalation.[22] It should be stored in a sealed container under an inert atmosphere. Spills should be cleaned up by covering with aq sodium bicarbonate.[1a]

Friedel–Crafts Acetylation. Arenes undergo acetylation to afford aryl methyl ketones on treatment with acetyl chloride (AcCl) together with a Lewis acid, usually ***Aluminum Chloride***$_3$. This reaction, known as the Friedel–Crafts acetylation, is valuable as a preparative method because a single positional isomer is produced from arenes that possess multiple unsubstituted electron-rich positions in many instances.

For example, Friedel–Crafts acetylation of toluene ($AcCl/AlCl_3$, ethylene dichloride, rt) affords *p*-methylacetophenone predominantly (*p*:*m*:*o* = 97.6:1.3:1.2; eq 1).[32]

$AcCl/AlCl_3$, $ClCH_2CH_2Cl$, rt (1)

p:*m*:*o* = 97.6:1.3:1.2

Acetylation of chlorobenzene under the same conditions affords *p*-chloroacetophenone with even higher selectivity (*p*:*m* = 99.5:0.5).[33] Acetylation of bromobenzene[33] and fluorobenzene[33] afford the *para* isomers exclusively. The *para*:*meta*[34] and *para*:*ortho*[32,34] selectivities exhibited by $AcCl/AlCl_3$ are greater than those exhibited by most other Friedel–Crafts electrophiles.

Halogen substituents can be used to control regioselectivity. For example, by introduction of bromine *ortho* to methyl, it is possible to realize '*meta* acetylation of toluene' (eq 2).[35]

Br — $AcCl/AlCl_3$, CH_2Cl_2, rt — Br, O (2)

Regioselectivity is quite sensitive to reaction conditions (e.g. solvent, order of addition of the reactants, concentration, and temperature). For example, acetylation of naphthalene can be directed to produce either a 99:1 mixture of C-1:C-2 acetyl derivatives (by addition of a solution of arene and AcCl in CS_2 to a slurry of $AlCl_3$ in CS_2 at 0 °C) or a 7:93 mixture (by addition of the preformed $AcCl/AlCl_3$ complex in dichloroethane to a dilute solution of the arene in dichloroethane at rt).[36] Similarly, acetylation of 2-methoxynaphthalene can be directed to produce either a 98:2 mixture of C-1:C-6 acetyl derivatives (using the former conditions) or a 4:96 mixture (by addition of the arene to a solution of the preformed $AcCl/AlCl_3$ complex in nitrobenzene).[37] Also, acetylation of 1,2,3-mesitylene can be directed to produce either a 100:0 mixture of C-4:C-5 isomers or a 3:97 mixture.[36c]

Frequently, regioselectivity is compromised by side reactions catalyzed by the HCl byproduct. For example, acetylation of *p*-xylene by treatment with $AlCl_3$ followed by Ac_2O (CS_2, Δ, 1 h) produces a 69:31 mixture of 2,5-dimethylacetophenone and 2,4-dimethylacetophenone, formation of the latter being indicative of competitive acid-catalyzed isomerization of *p*-xylene to *m*-xylene.[38] Also, although acetylation of anthracene affords 9-acetylanthracene regioselectively, if the reaction mixture is allowed to stand for a prolonged time prior to work-up (rt, 20 h) isomerization to a mixture of C-1, C-2, and C-9 acetyl derivatives occurs.[39]

These side reactions can be minimized by proper choice of reaction conditions. Isomerization of the arene can be suppressed by adding the arene to the preformed $AcCl/AlCl_3$ complex. This order of mixing is known as the 'Perrier modification' of the Friedel–Crafts reaction.[40] Acetylation of *p*-xylene using this order of mixing affords 2,5-dimethylacetophenone exclusively.[38] Isomerization of the product aryl methyl ketone can be suppressed by crystallizing the product out of the reaction mixture as it is formed. For example, on acetylation of anthracene in benzene at 5–10 °C, 9-acetylanthracene crystallizes out of the reaction mixture (as its 1/1 $AlCl_3$ complex) in pure form.[39] Higher yields of purer products can also be obtained by substituting ***Zirconium(IV) Chloride***[41] or ***Tin(IV) Chloride***[42] for $AlCl_3$.

AcCl is not well suited for industrial scale Friedel–Crafts acetylations because it is not commercially available in bulk (only by the drum) and therefore must be prepared on site.[1] The combination of ***Acetic Anhydride*** and anhydrous ***Hydrogen Fluoride***, both of which are available by the tank car, is claimed to be more practical.[43] On laboratory scale, $AcCl/AlCl_3$ is more attractive than Ac_2O/HF or $Ac_2O/AlCl_3$. Whereas one equivalent of $AlCl_3$ is sufficient to activate AcCl, 1.5–2 equiv $AlCl_3$ (relative to arene) are required to activate Ac_2O.[36a,37b,38,44] Thus, with Ac_2O, greater amounts of solvent are required and temperature control during the quench is more difficult. Also, slightly lower isolated yields have been reported with Ac_2O than with AcCl in two cases.[36a,45] However, it should be noted that the two reagents generally afford similar ratios of regioisomers.[36a,38,46]

Acetylation of Alkenes. Alkenes, on treatment with $AcCl/AlCl_3$ under standard Friedel–Crafts conditions, are transformed into mixtures of β-chloroalkyl methyl ketones, allyl methyl ketones, and vinyl methyl ketones, but the reaction is not generally preparatively useful because both the products and the starting alkenes are unstable under the hyperacidic reaction conditions. Preparatively useful yields have been reported only with electron poor alkenes such as ethylene (dichloroethane, 5–10 °C; >80% yield of 4-chloro-2-butanone)[47] and ***Allyl Chloride*** (CCl_4, rt; 78% yield of 5-chloro-4-methoxy-2-pentanone after methanolysis),[48] which are relatively immune to the effects of acid.

The acetylated products derived from higher alkenes are susceptible to protonation or solvolysis which produces carbenium ions that undergo Wagner–Meerwein hydride migrations.[49] For example, on subjection of cyclohexene to standard Friedel–Crafts acetylation conditions ($AcCl/AlCl_3$, CS_2 −18 °C), products formed include not only 2-chlorocyclohexyl methyl ketone (in 40% yield)[50] but also 4-chlorocyclohexyl methyl ketone.[2a,51] If benzene is added to the crude acetylation mixture and the temperature is then increased to 40–45 °C for 3 h, 4-phenylcyclohexyl methyl ketone is formed in 45% yield (eq 3).[49a,b]

$AcCl/AlCl_3$, CS_2, −20 °C — [Cl, O; not isolated] — benzene, 40–45 °C, 3 h, 45% — O (3)

Wagner–Meerwein rearrangement also occurs during acetylation of methylcyclohexene, even though the rearrangement is anti-Markovnikov (β-tertiary → γ -secondary; eq 4).[52] Acetylation of *cis*-decalin[53] (see 'Acetylation of Saturated Alkanes' section below) also produces a β-tertiary carbenium ion that undergoes anti-Markovnikov rearrangement. The rearrangement is terminated by intramolecular *O*-alkylation of the acetyl group by the γ-carbenium ion to form a cyclic enol ether in two cases.[49c,53]

$AcCl/AlCl_3$, CS_2, −5 °C → rt — [O, +; O, +] — O, Cl — NaOH, MeOH, rt — O, H (4)

endo:exo = 79:21

Higher alkenes themselves are also susceptible to protonation. The resulting carbenium ions decompose by assorted pathways including capture of chloride (with $SnCl_4$ as the catalyst),[51,54] addition to another alkene to form dimer or polymer,[5b,55] proton loss (resulting in *exo/endo* isomerization), or skeletal rearrangement.[56]

Higher alkenes can be acetylated in synthetically useful yield by treatment with AcCl together with various mild Lewis acids. One that deserves prominent mention is ***Ethylaluminum Dichloride*** (CH_2Cl_2, rt), which is useful for acetylation of all classes of alkenes (monosubstituted, 1,2-disubstituted, and trisubstituted).[57] For example, cyclohexene is converted into an 82/18 mixture of 3-acetylcyclohexene and 2-chlorocyclohexyl methyl ketone in 89% combined yield.

The following Lewis acids are also claimed to be superior to $AlCl_3$: Zn(Cu)/CH_2I_2 (AcCl, CH_2Cl_2, Δ), by which cyclohexene is converted into acetylcyclohexene in 68% yield (after treatment with KOH/MeOH);[58] $ZnCl_2$ (AcCl, Et_2O/CH_2Cl_2, $-75\,^{\circ}C \rightarrow -20\,^{\circ}C$), by which 2-methyl-2-butene is converted into a 15:85 mixture of 3,4-dimethyl-4-penten-2-one and 4-chloro-3,4-dimethyl-2-pentanone in 'quantitative' combined yield;[59] and $SnCl_4$, by which cyclohexene (AcCl, CS_2, $-5\,^{\circ}C \rightarrow$ rt) is converted into acetylcyclohexene in 50% yield (after dehydrochlorination with $PhNEt_2$ at 180 °C),[60] methylcyclohexene (CS_2, rt) is converted into 1-acetyl-2-methylcyclohexene in 48% yield (after dehydrochlorination),[52] and camphene is converted into an acetylated derivative in ≈65% yield.[49c]

Conducting the acetylation in the presence of a nonnucleophilic base or polar solvent is reported to be advantageous. For example, methylenecyclohexane can be converted into 1-cyclohexenylacetone in 73% yield by treatment with $AcSbCl_6$ in the presence of Cy_2NEt (CH_2Cl_2, $-50\,^{\circ}C \rightarrow -25\,^{\circ}C$, 1 h)[61] and cyclohexene can be converted into 3-acetylcyclohexene in 80% yield by treatment with $AcBF_4$ in $MeNO_2$ at $-25\,^{\circ}C$.[62]

Employment of Ac_2O instead of AcCl is also advantageous in some cases. For example, methylcyclohexene can be converted into 3-acetyl-2-methylcyclohexene in 90% yield by treatment with $ZnCl_2$ (neat Ac_2O, rt, 12 h).[63]

Finally, alkenes can be diacetylated to afford pyrylium salts by treatment with excess AcCl/$AlCl_3$,[55b,56,64] albeit in low yield (eq 5).[64a]

(5)

Acetylation of Alkynes. Under Friedel–Crafts conditions (AcCl/$AlCl_3$, CCl_4, 0–5 °C), acetylene undergoes acetylation to afford β-chlorovinyl methyl ketone in 62% yield[4] and under similar conditions ($AcSbF_6$, $MeNO_2$, $-25\,^{\circ}C$) 5-decyne undergoes acetylation to afford 6-acetyl-5-decanone in 73% yield.[65]

Acetylation of Saturated Alkanes. Saturated alkanes, on treatment with a slight excess of AcCl/$AlCl_3$ at elevated temperature, undergo dehydrogenation (by hydride abstraction followed by deprotonation) to alkenes, which undergo acetylation to afford vinyl methyl ketones. The hydride-abstracting species is believed to be either the acetyl cation[66] or $HAlCl_4$,[67] with most evidence favoring the former. Perhaps because the alkenes are generated slowly and consumed rapidly, and therefore are never present in high enough concentration to dimerize, yields are typically higher than those of acetylation of the corresponding alkenes.[53b,68] A similar hypothesis has been offered to explain the phenomenon that the yield from acetylation of tertiary alkyl chlorides is typically higher than the yield from acetylation of the corresponding alkenes.[55b,64a] For example, methylcyclopentane on treatment with AcCl/$AlCl_3$ (CH_2Cl_2, Δ) undergoes acetylation to afford 1-acetyl-2-methylcyclopentene in an impressive 60% yield (eq 6).[53b,66a]

(6)

If the reaction is carried out with excess alkane, a second hydride transfer occurs, resulting in reduction of the enone to the corresponding saturated alkyl methyl ketone.[69,70] For example, stirring AcCl/$AlCl_3$ in excess cyclohexane (30–35 °C, 2.5 h) affords 2-methyl-1-acetylcyclopentane in 50% yield (unpurified; based on AcCl)[55a,69,71] and stirring AcCl/$AlBr_3$ in excess cyclopentane (20 °C, 1 h) affords cyclopentyl methyl ketone in 60% yield (based on AcCl; eq 7).[55c]

(7)

If the reaction is carried out with a substoichiometric amount of alkane, the product is either a 2:1 adduct (if cyclic)[53b,66a] or pyrylium salt (if acyclic).[66b,68b]

Unbranched alkanes also undergo acetylation, but at higher temperature, so yields are generally lower. For example, acetylation of cyclohexane by AcCl/$AlCl_3$ requires refluxing in $CHCl_3$ and affords 1-acetyl-2-methylcyclopentene in only 36% yield.[55c,72]

Despite the modest to low yields, acetylation of alkanes provides a practical method for accessing simple methyl ketones because all the input raw materials are cheap.

Coupling with Organometallic Reagents. Coupling of organometallic reagents with AcCl is a valuable method for preparation of methyl ketones. Generally a catalyst (either a Lewis acid or transition metal salt) is required.

Due to the large number and varied characteristics of the organometallics, comprehensive coverage of the subject would require discussion of each organometallic reagent individually, which is far beyond the scope of this article. Information pertaining to catalyst and condition selection should therefore be accessed from the original literature; some seminal references are given in Table 1.

***C*-Acetylation of Enolates and Enolate Equivalents.** β-Diketones can be synthesized by treatment of metal enolates with AcCl. *O*-Acetylation is often a significant side reaction, but the amount can be minimized by choosing a counterion that is bonded covalently to the enolate[6] such as copper[132] or zinc,[133] and by using AcCl rather than Ac_2O.[6a] Proton transfer from the product β-diketone to the starting enolate is another common side reaction.[134] Alternative procedures for effecting *C*-acetylation that avoid or minimize these side reactions include Lewis acid-catalyzed acetylation of the trimethylsilyl enol ether derivative (AcCl/cat. $ZnCl_2$, CH_2Cl_2 or CH_2Cl_2/Et_2O, rt)[135] and addition of ketene to the morpholine enamine (AcCl/Et_3N, $CHCl_3$, rt).[136]

Table 1 Catalyst Selection Chart[i,ii]

Organo-metallic	R = Alkyl	Vinyl	Aryl	Alkynyl	Allyl[iii]
RLi	N[73 iv]				
R_2Mg	N[74 iv]				
RMgX	Fe[75]				
	Cu[76]				
	Mn[76b,77]				
	D[78]				
	N[79]				
$RCuL^-$	N[80]	N[81]	N[80,82]	N[80,83]	N[84]
				I[83a]	
RZnL	Pd[85,86]	Pd[85]	Pd[87]	Pd[85]	N[88 iv]
	D[89]		D[89]	N[90]	
	N[91]				
RCdL	N[91c,92]		N[91c,92b,d,e]		
RHgL	Al[93]	Al[94]	Al[93]		Al[95]
	Pd[96]	Ti[94 v]	Pd[96,97]		
RBL_3^-	N[98]		N[99,100 vi]	N[101 vi]	
$RAlL_2$	N[102]	N[85,102b]			
	Cu[103]	Pd[85]			
R_4Al^-	Fe[104]	Pd[85]			
	Cu[104]				
$RTlL_2$	N[105]		N[105]		
$RSiL_3$	Al[106,107]	Al[107,108,109 vii]		Al[107,110]	Al[107,111,112 viii 113 ix]
					Ti[114 x]
$RGeL_3$	Al[106b]				
$RSnL_3$	Pd[115]	Pd[115,116]	Pd[117]	Pd[118]	Rh[119 x]
	Al[106b]	Ti[120 iv]	N[117c]		N[121,122 iv]
$RBiL_2$			Pd[123]		
$RTiL_2$					N[124 iv]
$RZrL_3$	N[125]				
RVL_2			N[126]		
RMnL	Cu[127]	Cu[127]	Cu[127]	Cu[127]	Cu[127]
	N[128]	N[128]	N[128]	N[128]	
RRh			N[129]		
RNiL		N[130]			N[130,131]

[i] Codes (in headings): L = unspecified ligand and X = halogen; codes (in entries): N = no catalyst required, Fe = Fe^{III} salt, Cu = Cu^{I} salt, Mn = Mn^{I} salt, D = dipolar aprotic additive, I = LiI, Pd = Pd^0 or Pd^{II} salt, Al = $AlCl_3$, Ti = $TiCl_4$, and Rh = $ClRh(PPh_3)_3$. [ii] Coupling occurs at metal-bearing carbon with retention of configuration to afford RCOMe unless otherwise indicated. [iii] Coupling occurs at γ-carbon unless otherwise indicated. [iv] Product is tertiary alcohol. [v] Coupling occurs with inversion of configuration. [vi] Coupling occurs at β-carbon. [vii] Coupling occurs at δ-carbon. [viii] Substrate is allenic silane and product is furan. [ix] Substrate is propargylsilane. [x] Coupling occurs at α-carbon.

Analogously, esters can be *C*-acetylated by conversion into the corresponding silyl ketene acetal followed by treatment with AcCl. Depending on the coupling conditions (neat AcCl[137] or AcCl/Et_3N[138]), either the *cis*-β-siloxycrotonate ester or the corresponding β,γ-isomer is produced (eqs 8 and 9). The third possible isomer (*trans*-β-siloxycrotonate) is accessible either by silylation of the acetoacetic ester (TMSCl, Et_3N, THF, Δ)[139] or by $HgBr_2/Et_3SiBr$-catalyzed equilibration of the *cis* isomer.[137]

AcCl (neat), rt; 67% (8)

AcCl/Et_3N, THF, rt; 86% (9)

The silyl ketene acetal strategy can also be used to effect γ-acetylation of α,β-unsaturated esters (AcCl/cat. $ZnBr_2$, CH_2Cl_2, rt)[140] and β-ketoesters (AcCl, Et_2O, −78 °C).[141]

Enol Acetylation. Enol acetylation of ketones can be effected by formation of a metal enolate in which the metal is relatively dissociated[6] (such as potassium[142] or magnesium[143]) followed by quenching with AcCl. Alternatively, enol acetates can be synthesized directly from the ketones. For example, 3-keto-$\Delta^{4,6}$-steroids can be converted into $\Delta^{2,4,6}$-trienol acetates by treatment with AcCl/$PhNMe_2$ or into $\Delta^{3,5,7}$-trienol acetates by treatment with AcCl/Ac_2O.[144]

Acetyl Bromide (AcBr) is apparently superior to AcCl as a catalyst for enol acetylation, based on a report that 17β-benzoyloxyestra-4,9(10)-dien-3-one is converted into estradiol 3-acetate-17-benzoate in higher yield at much lower temperature using AcBr rather than AcCl (87.5% yield with 1:2 AcBr:Ac_2O, CH_2Cl_2, rt, 1 h (eq 10) vs. 81.0% yield with 1:2 AcCl:Ac_2O, Δ, 4.5 h).[145]

1/2 AcBr/Ac_2O; CH_2Cl_2, rt, 1 h; 87.5% (10)

β-Keto esters can be converted into either *trans* or *cis* enol acetates. The *trans* isomer is accessible by treatment with AcCl/Et_3N (HMPA, rt)[146] or AcCl/DBU (MeCN, 5 °C → rt; eq 11),[147] while the *cis* isomer is accessible by treatment with isopropenyl acetate/HOTs.[146] Each isomer couples with dialkylcuprates with retention of configuration to afford stereoisomerically enriched α,β-unsaturated esters.[146,148]

DBU; MeCN, 5 °C; then AcCl, rt, 3 h; 53%; *cis*:*trans* = 97:3 (11)

Attempted enol acetylation of β-keto esters by quenching the sodium enolate[146,147] or magnesium chelate[149] with AcCl afforded *C*-acetylated products.

Adducts with Aldehydes and Ketones. AcCl combines with aldehydes[150] (cat. $ZnCl_2$ or $AlCl_3$[150f]) to afford α-chloroalkyl acetates. The reaction is reversible,[151] but at equilibrium the ratio of adduct to aldehyde is usually quite high, and the reaction is otherwise clean (92% yield for acetaldehyde,[150e] 97% yield for benzaldehyde; eq 12[150f]).

AcCl/cat. $ZnCl_2$; 0–10 °C, 1 h (neat) 97% (12)

AcCl also adds to ketones,[150e,151,152] but the adducts are much less thermodynamically stable, so significant amounts of the starting materials are present at equilibrium.[151,152a,b] The equilibrium can be biased in favor of the adduct by employing high concentration, low temperature, a nonpolar solvent, excess AcCl, or AcBr or AcI instead of AcCl.[151,153] For example, the acetone/AcCl adduct can be obtained in good yield (85%) by treatment of acetone with excess (2 equiv) AcCl (cat. $ZnCl_2$, CCl_4, −15 °C).[152c]

Reduction of the aldehyde/AcBr adducts[151,154] with ***Zinc*** or ***Samarium(II) Iodide*** to α-acetoxyalkylzinc[154,155] and samarium[156] compounds, respectively, completes an umpolung of the reactivity of the aldehyde.

Cleavage of Ethers. THF can be opened by treatment with AcCl in combination with either ***Sodium Iodide*** (MeCN, rt, 21 h; 91% yield of 4-iodobutyl acetate)[157] or a Lewis acid such as $ZnCl_2$ (Δ, 1.5 h; 76% yield of 4-chlorobutyl acetate),[158] $SnCl_4$,[159] $CoCl_2$ (rt, MeCN; 90%),[160] $ClPdCH_2Ph(PPh_3)_2/Bu_3SnCl$ (63 °C, 48 h; 95%),[161] $Mo(CO)_6$ (hexane, Δ; 78%),[162] $KPtCl_3(H_2CCH_2)$, and $[ClRh(H_2CCH_2)_2]_2$ (rt; 75% and 83%, respectively).[163] Acyclic dialkyl ethers can also be cleaved efficiently and in many cases regioselectively.[159]

Many of these methods are applicable to deprotection of ether-type protecting groups. For example, benzyl and allyl ethers can be deprotected by treatment with AcCl/cat. $CoCl_2$[160] or AcCl/cat. $ClPdCH_2Ph(PPh_3)_2$/cat. Bu_3SnCl.[161] Dimethyl acetals can be cleaved selectively to aldehydes in the presence of ethylene acetals (AcCl/cat. $ZnCl_2$, Me_2S/THF, 0 °C),[164] or to α-chloro ethers (AcCl/cat. $SOCl_2$, 55 °C).[165] Tetrahydropyranyl (THP) ethers[166] and *t*-butyl ethers[167] can be deprotected by stirring in 1:10 AcCl:HOAc (40–50 °C).

Finally, *t*-alkyl esters can be cleaved to anhydrides and *t*-alkyl chlorides by treatment with AcCl ($MeNO_2$, 70 °C).[168]

Esterification. Although AcCl is intrinsically more reactive than Ac_2O, in combination with various acylation catalysts the reverse reactivity order is exhibited. For example, Ac_2O ***4-Dimethylaminopyridine*** (DMAP) acetylates ethynylcyclohexanol three times faster than AcCl/DMAP ($CDCl_3$, 27 °C).[169] Also, isopropanol does not react with AcCl/Bu_3P (CD_3CN, −8 °C; <5% conversion after 30 min), but after addition of sodium acetate reacts rapidly to form isopropyl acetate (complete in <10 min).[170] As a general rule, therefore, Ac_2O is preferable for acetylation of hindered alcohols while AcCl is preferable for selective monoacetylation of polyols.[171]

Examples of selective acetylations involving AcCl include: acetylation of primary alcohols in the presence of secondary alcohols by AcCl/2,4,6-collidine or *i*-$PrNEt_2$ (CH_2Cl_2, −78 °C);[8,172] acetylation of primary alcohols in the presence of secondary alcohols,[173] and secondary alcohols in the presence of tertiary alcohols,[174] by AcCl/pyridine (CH_2Cl_2, −78 °C); monoacetylation of a 2,4-dihydroxyglucopyranose by AcCl/pyridine/−15 °C (Ac_2O/pyridine/0 °C is less selective);[175] and acetylation of steroidal 5α-hydroxyls (not 5β) by AcCl/$PhNMe_2$ ($CHCl_3$, Δ).[176]

Although Ac_2O/DMAP[177] and Ac_2O/Bu_3P[170] are the preferred reagents for acetylation of most hindered alcohols, satisfactory results can be obtained with AcCl in combination with $PhNMe_2$ ($CHCl_3$, Δ),[178] $PhNEt_2$ ($CHCl_3$, Δ),[179] AgCN (benzene or HMPA, 80 °C),[180] magnesium powder (Et_2O, Δ; 45–55% yield of *t*-BuOAc),[181] and Na_2CO_3 (cat. $PhCH_2NEt_3Cl$, CH_2Cl_2, Δ; 79% yield of *t*-BuOAc).[182] Use of the combination of AcCl/DMAP is not recommended since unidentified byproducts may be generated.[169]

Although acetylations with AcCl/pyridine produce an acidic byproduct (pyridine hydrochloride), it is possible to acetylate highly acid-sensitive alcohols such as 2-(tributylstannylmethyl)allyl alcohol (eq 13)[9b] and 2-(trimethylsilylmethyl)allyl alcohol[9a] with AcCl/pyridine in >90% yield without competing protiodestannylation or protiodesilylation by selecting a solvent (CH_2Cl_2, 0 °C) in which the pyridine hydrochloride is insoluble.

n-Bu_3Sn … HO → AcCl/pyridine; CH_2Cl_2 90% → *n*-Bu_3Sn … AcO (13)

Alternatively, acid-sensitive alcohols may be acetylated by deprotonation with ***n-Butyllithium*** (THF, −78 °C)[183] or ***Ethylmagnesium Bromide*** (Et_2O, rt)[184] followed by quenching with AcCl.

Finally, by using a chiral tertiary amine as the base, it is possible to effect enantioselective acetylations. For example, racemic 1-phenethyl alcohol has been partially resolved by treatment with AcCl in combination with (*S*)-(−)-*N,N*-dimethyl-1-phenethylamine (CH_2Cl_2, −78 °C → rt; ee of acetate 52%, ee of alcohol 59.5%).[185]

Generation of Solutions of Anhydrous Hydrogen Chloride in Methanol. Esterification of alcohols by AcCl proceeds in the absence of HCl scavengers. For example, on addition of AcCl to methanol at rt, a solution of hydrogen chloride and methyl acetate in methanol forms rapidly.[10] This reaction provides a more practical method for access to solutions of HCl in methanol than the apparently simpler method of bubbling anhydrous HCl into methanol because of the difficulty of controlling the amount of anhydrous HCl delivered. Solutions of anhydrous HCl in acetic acid can presumably be prepared analogously by addition of AcCl and an equimolar amount of H_2O to HOAc.

Primary,[186] secondary,[187] and tertiary alcohols[178a,188] also react with AcCl, but the product is the alkyl chloride rather than the ester in most cases. Thus as a preparative esterification method this reaction has limited generality.

AcCl also reacts with anhydrous *p*-toluenesulfonic acid (3–4 equiv AcCl, Δ) to afford acetyl *p*-toluenesulfonate in 97.5% yield along with anhydrous HCl.[189] AcCl does not react with HOAc to generate HCl and Ac_2O, at least in appreciable amounts.[31]

Dehydrating Agent. AcCl reacts with H_2O to afford HCl and HOAc rapidly and quantitatively[31b] and thereby qualifies as a strong dehydrating agent. Examples of reactions in which AcCl functions as a dehydrating agent include: cyclization of dicarboxylic acids to cyclic anhydrides (neat AcCl, Δ);[190] cyclization of keto acids to enol lactones (neat AcCl, Δ);[191] dehydration of nitro compounds into nitrile oxides (by treatment with NaOMe followed by AcCl);[192] and conversion of allylic hydroperoxides into unsaturated ketones (AcCl/pyridine, $CHCl_3$, rt).[193] The dehydrating power of AcCl has been invoked as a possible explanation for its effectiveness for activation of zinc dust.[194]

In Situ Generation of High-Valent Metal Chlorides. Many high-valent metal chlorides are useful as reagents in organic synthesis but are difficult to handle due to their moisture sensitivity. AcCl can be used to generate such reagents in situ from the corresponding metal oxides[11] or acetates.[195] Examples include: α-chlorination of ketones by treatment with AcCl/***Manganese Dioxide*** (HOAc, rt);[196] *cis*-1,2-dichlorination of alkenes by treatment with AcCl/$(Bu_4N)_4Mo_8O_{26}$ (CH_2Cl_2, rt);[197] and dichlorination of alkenes by treatment with AcCl/MnO_2/$MnCl_2$ (DMF, rt).[198] Attempts to dichlorinate alkenes by treatment with AcCl/MnO_2 in THF, however, failed due to cleavage of THF to 4-chlorobutyl acetate.[196,199] A milder reagent that can be used to activate MnO_2 for dichlorination of alkenes in THF is ***Chlorotrimethylsilane***.[199]

Solvent for Organometallic Reactions. Because of its cheapness, volatility, and ability to form moisture-stable solutions of metal chlorides, AcCl is useful as a solvent for reactions involving hygroscopic metal salts.[11] For example, AcCl has been used as a co-solvent for 1,2-chloroacetoxylation of alkenes by ***Chromyl Chloride*** (1:2 AcCl:CH_2Cl_2, −78 °C → rt).[200]

Reaction with Heteroatom Oxides. The key step in a method for α-acetoxylation of aldehydes involves rearrangement of an AcCl–nitrone adduct (eq 14).[201] Analogous methods for α-benzoylation and α-pivaloylation are higher yielding.

t-BuNHOH/HOTs; Na_2SO_4, CH_2Cl_2; 80%

AcCl/Et_3N; Et_2O, 0 °C → rt

HOAc, NaOAc; PhH/H_2O, rt

80% from nitrone (14)

β-Nitrostyrenes cyclize to indolinones on treatment with AcCl ($FeCl_3$, CH_2Cl_2, 0 °C; eq 15).[202]

AcCl/$FeCl_3$; CH_2Cl_2, 0 °C, 5 h; 75% (15)

A high-yielding method for deoxygenation of sulfoxides to sulfides involves treatment with 1.1 equiv ***Tin(II) Chloride*** in the presence of a catalytic amount (0.4 equiv) of AcCl (MeCN/DMF, 0 °C → rt).[12] The mildness of this method is demonstrated by its usability for deoxygenation of a cephalosporin sulfoxide (eq 16).

Another method for deoxygenation of sulfoxides involves treatment with two equiv AcCl (CH_2Cl_2, rt);[203] the oxidized byproduct is claimed to be gaseous chlorine.[203]

0.4 equiv AcCl, 1.1 equiv $SnCl_2$; MeCN/DMF, 0 °C → rt; 98% (16)

Chlorine and Bromine Scavenger. AcCl (cat. H_2SO_4, 40–70 °C) scavenges Cl_2 efficiently (to afford chloroacetyl chloride in 87.1% yield).[13] AcCl also scavenges Br_2 efficiently at 35 °C.[14]

Source of Ketene. AcCl reacts with ***Triethylamine*** at low temperature (−20 °C) to afford acetyltriethylammonium chloride.[204] This salt functions as a source of ketene (or the functional equivalent). For example, it reacts with silyl ketene acetals (THF, rt) to afford silyl enol ethers of acetoacetic esters (eq 9),[138] with α-alkoxycarbonylalkylidenetriphenylphosphoranes (CH_2Cl_2, rt) to afford allenic esters,[205] with enamines (Et_2O, 0 °C) to afford cyclobutanones,[136a] with certain acyl imines (Et_2O, 0 °C) to form formal [4 + 2] ketene cycloadducts,[206] and with certain nonenolizable imines (Et_2O, rt) to afford formal [4 + 2] diketene cycloadducts in up to 55% yield.[207] Also, on refluxing in Et_2O in the absence of a trapping agent, diketene is formed in 50% yield.[208]

AcCl/$AlCl_3$ decomposes to acetylacetone on heating ($CHCl_3$, 54–61 °C, 6 h; 82.5% yield after aqueous work-up).[209] The mechanism presumably involves ketene as an intermediate. However, an attempt to trap the ketene was unsuccessful.[210]

***N*-Acetylation.** Primary and secondary amines can be *N*-acetylated to form acetamides by treatment with AcCl under Schotten–Baumann conditions (aq NaOH),[211] but hydrolysis of AcCl is a significant competing side reaction.[212] Use of Ac_2O (2.5 equiv; Δ, 10–15 min) is therefore recommended.[211]

Tertiary amines react with AcCl to afford acetylammonium salts. Ordinarily, these salts fragment to ketene on warming (see above). However, those that possess a labile alkyl group fragment by loss of the alkyl group (von Braun cleavage). For example, bis(dimethylamino)methane reacts with AcCl (Et_2O, rt) to afford chloromethyldimethylamine,[213] a useful Mannich reagent, and 1,3,5-trimethylhexahydro-*s*-triazine reacts with AcCl ($CHCl_3$, Δ, 1 h) to afford chloromethylmethyl acetamide,[214] a useful amidomethylation reagent.[215] Also, aziridines react with AcCl (PhH, 0 °C) to afford chloroethylacetamides.[216] Allylic amines react with in situ-generated AcI (AcCl/CuI, THF, rt) to afford acetamides.[217]

AcCl also activates pyridines toward nucleophilic addition. For example, phenylmagnesium chloride adds to pyridine in the presence of AcCl (cat. CuI, THF, −20 °C → rt) to afford, after catalytic hydrogenation, *N*-acetyl-4-phenylpiperidine in 65% yield.[218] Also, AcCl catalyzes the reaction between sodium iodide and 2-chloropyridine to afford 2-iodopyridine (MeCN, Δ, 24 h; 55%).[219]

N-Acetylation of enolizable imines to afford enamides can be accomplished by treatment with AcCl/$PhNEt_2$. For example, treatment of crotonaldehyde cyclohexylimine with AcCl followed by $PhNEt_2$ (toluene, rt) affords the enamide in 88% yield.[220]

Primary urethanes can be *N*-acetylated to afford imides by treatment with AcCl (100 °C, 1 h).[221] Alternatively, urethanes can be converted into acetamides by treatment with AcBr (120–130 °C)[221] or in situ-generated AcI[222] (MeCN, 60 °C).[223]

Finally, a convenient method for preparation of ***N-Trimethylsilylacetamide*** (MSA), a useful trimethylsilyl transfer reagent, involves treatment of ***Hexamethyldisilazane*** with AcCl (hexane, Δ; 88%).[224]

***S*-Acetylation.** Both aliphatic and aromatic thiols can be *S*-acetylated by treatment with AcCl (cat. $CoCl_2$, MeCN, rt).[225]

Nucleophilic Acetylation. AcCl together with SmI_2 (MeCN, rt) or $SmCp_2$ delivers the acetyl anion synthon to ketones to afford the corresponding acyloins (eq 17).[15]

O → OH, O
AcCl/SmI_2
MeCN, rt
80%
(17)

Related Reagents. Acetic Anhydride; Acetyl Bromide; Acetyl Fluoride.

---

1. (a) Moretti, T. A. *Kirk-Othmer Encyclopedia of Chemical Technology*, 3rd ed.; Wiley: New York, 1978; Vol. 1, p 162. (b) Wagner, F. S. Jr. *Kirk-Othmer Encyclopedia of Chemical Technology*, 4th ed.; Wiley: New York, 1991; Vol. 1, p 155.
2. (a) Baddeley, G. *QR* **1954**, *8*, 355. (b) Gore, P. H. *Friedel–Crafts and Related Reactions*; Wiley: New York, 1964; Vol. 3, p 1. (c) House, H. O. *Modern Synthetic Reactions*, 2nd ed; Benjamin: Menlo Park, 1972; p 797. (d) Olah, G. A. *Friedel–Crafts Chemistry*; Wiley: New York, 1973; pp 91, 191. (e) Heaney, H. *COC* **1979**, *1*, 241. (f) Olah, G. A.; Meidar, D. *Kirk-Othmer Encyclopedia of Chemical Technology*, 3rd ed.; Wiley: New York, 1980; Vol. 11, p 269. (g) Olah, G. A.; Prakash, G. K. S.; Sommer, J. *Superacids*; Wiley: New York, 1985; p 293.
3. (a) Nenitzescu, C. D.; Balaban, A. T. *Friedel–Crafts and Related Reactions*; Wiley: New York, 1964; Vol. 3, p 1033. (b) Groves, J. K. *CSR* **1972**, *1*, 73. (c) Olah, G. A. *Friedel–Crafts Chemistry*; Wiley: New York, 1973; pp 129, 200.
4. Price, C. C.; Pappalardo, J. A. *JACS* **1950**, *72*, 2613.
5. (a) Olah, G. A. *Friedel–Crafts Chemistry*; Wiley: New York, 1973; p 135. (b) Vol'pin, M.; Akhrem, I.; Orlinkov, A. *NJC* **1989**, *13*, 771.
6. (a) House, H. O.; Auerbach, R. A.; Gall, M.; Peet, N. P. *JOC* **1973**, *38*, 514. (b) Black, T. H. *OPP* **1989**, *21*, 179.
7. (a) Burwell, R. L. Jr. *CRV* **1954**, *54*, 615. (b) Johnson, F. *Friedel–Crafts and Related Reactions*; Wiley: New York, 1965; Vol. 4, p 1. (c) Bhatt, M. V.; Kulkarni, S. U. *S* **1983**, 249.
8. Ishihara, K.; Kurihara, H.; Yamamoto, H. *JOC* **1993**, *58*, 3791.
9. (a) Trost, B. M.; Chan, D. M. T. *JACS* **1983**, *105*, 2315. (b) Trost, B. M.; Bonk, P. J. *JACS* **1985**, *107*, 1778.
10. (a) Freudenberg, K.; Jacob, W. *CB* **1941**, *74*, 1001. (b) Riegel, B.; Moffett, R. B.; McIntosh, A. V. *OS* **1944**, *24*, 41. (c) Fraenkel-Conrat, H.; Olcott, H. S. *JBC* **1945**, *161*, 259. (d) Baker, B. R.; Schaub, R. E.; Querry, M. V.; Williams, J. H. *JOC* **1952**, *17*, 77. (e) De Lombaert, S.; Nemery, I.; Roekens, B.; Carretero, J. C.; Kimmel, T.; Ghosez, L. *TL* **1986**, *27*, 5099. (f) Nashed, E. M.; Glaudemans, C. P. J. *JOC* **1987**, *52*, 5255.
11. (a) Chretien, A.; Oechsel, G. *CR* **1938**, *206*, 254. (b) Paul, R. C.; Sandhu, S. S. *Proc. Chem. Soc.* **1957**, 262. (c) Paul, R. C.; Singh, D.; Sandhu, S. S. *JCS* **1959**, 315. (d) Paul, R. C.; Singh, D.; Sandhu, S. S. *JCS* **1959**, 319. (e) Maunaye, M.; Lang, J. *CR* **1965**, *261*, 3381, 3829.
12. Kaiser, G. V.; Cooper, R. D. G.; Koehler, R. E.; Murphy, C. F.; Webber, J. A.; Wright, I. G.; Van Heyningen, E. M. *JOC* **1970**, *35*, 2430.
13. Scheidmeir, W.; Bressel, U.; Hohenschutz, H. U.S. Patent 3 880 923, 1975.
14. Kharasch, M. S.; Hobbs, L. M. *JOC* **1941**, *6*, 705.
15. (a) Collin, J.; Namy, J.-L.; Dallemer, F.; Kagan, H. B. *JOC* **1991**, *56*, 3118. (b) Ruder, S. M. *TL* **1992**, *33*, 2621.
16. Pouchert, C. J. *The Aldrich Library of FT-IR Spectra*, ed. I; Aldrich: Milwaukee, 1985; Vol. 1, p 723A.
17. Tsuchiya, S.; Kimura, M. *BCJ* **1972**, *45*, 736.
18. *Reagent Chemicals*, 8th ed.; American Chemical Society: Washington, 1993; p 107.
19. Singh, J.; Paul, R. C.; Sandhu, S. S. *JCS* **1959**, 845.
20. (a) Whitmore, F. C. *RTC* **1938**, *57*, 562. (b) Cason, J.; Harman, R. E.; Goodwin, S.; Allen, C. F. *JOC* **1950**, *15*, 860. (c) Perrin, D. D.; Armarego, W. L. F. *Purification of Laboratory Chemicals*, 3rd ed.; Pergamon: Oxford, 1988; p 70.
21. Burton, H.; Praill, P. F. G. *JCS* **1952**, 2546.
22. Sax, N. I. *Dangerous Properties of Industrial Materials*, 6th ed.; Van Nostrand Reinhold: New York, 1984; p 106.
23. (a) Vogel, A. I. *A Text-book of Practical Organic Chemistry*, 3rd ed.; Wiley: New York, 1956; p 367. (b) Damjan, J.; Benczik, J.; Kolonics, Z.; Pelyva, J.; Laborczy, R.; Szabolcs, J.; Soptei, C.; Barcza, I.; Kayos, C. Br. Patent 2 213 144, 1989. (c) Valitova, L. A.; Popova, E. V.; Ibragimov, Sh. N.; Ivanov, B. E. *BAU* **1990**, *39*, 366.
24. Durrans, T. H. U.S. Patent 1 326 040, 1919.
25. Masters, C. L. U.S. Patent 1 819 613, 1931.
26. Montonna, R. E. *JACS* **1927**, *49*, 2114.
27. Mugdan, M.; Wimmer, J. Ger. Patent 549 725, 1931.
28. Brown, H. C. *JACS* **1938**, *60*, 1325.
29. Mills, L. E. U.S. Patent 1 921 767, 1933. U.S. Patent 1 965 556, 1934.
30. (a) Meder, G.; Eggert, E.; Grimm, A. U.S. Patent 2 013 988, 1935. (b) Meder, G.; Geissler, W.; Eggert, E. U.S. Patent 2 013 989, 1935. (c) Meder, G.; Bergheimer, E.; Geisler, W.; Eggert, E. Ger. Patent 638 306, 1936. (d) Eggert, E.; Grimm, A. Ger. Patent 655 683, 1938. (e) Christoph, F. J. Jr.; Parker, S. H.; Seagraves, R. L. U.S. Patent 3 318 950, 1967.
31. (a) Colson, A. *BSF* **1897**, *17*, 55. (b) Inoue, S.; Hayashi, K. *CA* **1954**, *48*, 8255g. (c) Satchell, D. P. N. *JCS* **1960**, 1752. (d) Satchell, D. P. N. *QR* **1963**, *17*, 160.
32. Brown, H. C.; Marino, G.; Stock, L. M. *JACS* **1959**, *81*, 3310.
33. Brown, H. C.; Marino, G. *JACS* **1962**, *84*, 1658.
34. (a) Brown, H. C.; Nelson, K. L. *JACS* **1953**, *75*, 6292. (b) Olah, G. A. *Friedel–Crafts Chemistry*; Wiley: New York, 1973; pp 448, 452.
35. (a) Elwood, T. A.; Flack, W. R.; Inman, K. J.; Rabideau, P. W. *T* **1974**, *30*, 535. (b) Todd, D.; Pickering, M. *J. Chem. Educ.* **1988**, *65*, 1100.
36. (a) Baddeley, G. *JCS* **1949**, S99. (b) Bassilios, H. F.; Makar, S. M.; Salem, A. Y. *BSF* **1954**, *21*, 72. (c) Friedman, L.; Honour, R. J. *JACS* **1969**, *91*, 6344.
37. (a) Girdler, R. B.; Gore, P. H.; Hoskins, J. A. *JCS(C)* **1966**, 181. (b) Arsenijevic, L.; Arsenijevic, V.; Horeau, A.; Jacques, J. *OSC* **1988**, *6*, 34. (c) Magni, A.; Visentin, G. U.S. Patent 4 868 338, 1989.
38. Friedman, L.; Koca, R. *JOC* **1968**, *33*, 1255.
39. (a) Merritt, C. Jr.; Braun, C. E. *OSC* **1963**, *4*, 8. (b) Gore, P. H.; Thadani, C. K. *JCS(C)* **1966**, 1729.
40. (a) Perrier, G. *CB* **1900**, *33*, 815. (b) Perrier, G. *BSF* **1904**, *31*, 859.

41. (a) Heine, H. W.; Cottle, D. L.; Van Mater, H. L. *JACS* **1946**, *68*, 524. (b) Gore, P. H.; Hoskins, J. A. *JCS* **1964**, 5666.
42. Johnson, J. R.; May, G. E. *OSC* **1943**, *2*, 8.
43. (a) Piccolo, O.; Visentin, G.; Blasina, P.; Spreafico, F. U.S. Patent 4 670 603, 1987. (b) Lindley, D. D.; Curtis, T. A.; Ryan, T. R.; de la Garza, E. M.; Hilton, C. B.; Kenesson, T. M. U.S. Patent 5 068 448, 1991.
44. Olah, G. A. *Friedel–Crafts Chemistry*; Wiley: New York, 1973; pp 106, 306.
45. Allen, C. F. H. *OSC* **1943**, *2*, 3.
46. Olah, G. A.; Moffatt, M. E.; Kuhn, S. J.; Hardie, B. A. *JACS* **1964**, *86*, 2198.
47. (a) Sondheimer, F.; Woodward, R. B. *JACS* **1953**, *75*, 5438. (b) Briner, P. H. U.S. Patent 5 124 486, 1992.
48. (a) Kulinkovich, O. G.; Tischenko, I. G.; Sorokin, V. L. *S* **1985**, 1058. (b) Kulinkovich, O. G.; Tischenko, I. G.; Sorokin, V. L. *JOU* **1985**, *21*, 1514. (c) Mamedov, E. I.; Ismailov, A. G.; Zyk, N. V.; Kutateladze, A. G.; Zefirov, N. S. *Sulfur Lett.* **1991**, *12*, 109.
49. (a) Nenitzescu, C. D.; Gavat, I. G. *LA* **1935**, *519*, 260. (b) Johnson, W. S.; Offenhauer, R. D. *JACS* **1945**, *67*, 1045. (c) Crosby, J. A.; Rasburn, J. W. *CI(L)* **1967**, 1365.
50. (a) Darzens, M. G. *CR* **1910**, *150*, 707. (b) Wieland, H.; Bettag, L. *CB* **1922**, *55*, 2246.
51. Royals, E. E.; Hendry, C. M. *JOC* **1950**, *15*, 1147.
52. (a) Turner, R. B.; Voitle, D. M. *JACS* **1951**, *73*, 1403. (b) Dufort, N.; Lafontaine, J. *CJC* **1968**, *46*, 1065.
53. (a) Baddeley, G.; Heaton, B. G.; Rasburn, J. W. *JCS* **1960**, 4713. (b) Morel-Fourrier, C.; Dulcere, J.-P.; Santelli, M. *JACS* **1991**, *113*, 8062.
54. Colonge, J.; Mostafavi, K. *BSF* **1939**, *6*, 335, 342.
55. (a) Nenitzescu, C. D.; Ionescu, C. N. *LA* **1931**, *491*, 189. (b) Baddeley, G.; Khayat, M. A. R. *Proc. Chem. Soc.* **1961**, 382. (c) Akhrem, I. S.; Orlinkov, A. V.; Mysov, E. I.; Vol'pin, M. E. *TL* **1981**, *22*, 3891.
56. Arnaud, M.; Roussel, C.; Metzger, J. *TL* **1979**, 1795.
57. Snider, B. B.; Jackson, A. C. *JOC* **1982**, *47*, 5393.
58. Shono, T.; Nishiguchi, I.; Sasaki, M.; Ikeda, H.; Kurita, M. *JOC* **1983**, *48*, 2503.
59. Baran, J.; Klein, H.; Schade, C.; Will, E.; Koschinsky, R.; Bauml, E.; Mayr, H. *T* **1988**, *44*, 2181.
60. Ruzicka, L.; Koolhaas, D. R.; Wind, A. H. *HCA* **1931**, *14*, 1151.
61. Hoffmann, H. M. R.; Tsushima, T. *JACS* **1977**, *99*, 6008.
62. (a) Smit, W. A.; Semenovsky, A. V.; Kucherov, V. F.; Chernova, T. N.; Krimer, M. Z.; Lubinskaya, O. V. *TL* **1971**, 3101. (b) Smit, V. A.; Semenovskii, A. V.; Lyubinskaya, O. V.; Kucherov, V. F. *DOK* **1972**, *203*, 272.
63. (a) Deno, N. C.; Chafetz, H. *JACS* **1952**, *74*, 3940. (b) Groves, J. K.; Jones, N. *JCS(C)* **1968**, 2215, 2898. (c) Beak, P.; Berger, K. R. *JACS* **1980**, *102*, 3848.
64. (a) Balaban, A. T.; Nenitzescu, C. D. *LA* **1959**, *625*, 74. (b) Balaban, A. T.; Schroth, W.; Fischer, G. *Adv. Heterocycl. Chem.* **1969**, *10*, 241. (c) Erre, C. H.; Roussel, C. *BSF(2)* **1984**, 454.
65. Roitburd, G. V.; Smit, W. A.; Semenovsky, A. V.; Shchegolev, A. A.; Kucherov, V. F.; Chizhov, O. S.; Kadentsev, V. I. *TL* **1972**, 4935.
66. (a) Tabushi, I.; Fujita, K.; Oda, R. *TL* **1968**, 5455. (b) Arnaud, M.; Pedra, A.; Roussel, C.; Metzger, J. *JOC* **1979**, *44*, 2972.
67. (a) Nenitzescu, C. D.; Dragan, A. *CB* **1933**, *66*, 1892. (b) Bloch, H. S.; Pines, H.; Schmerling, L. *JACS* **1946**, *68*, 153.
68. (a) Tabushi, I.; Fujita, K.; Oda, R.; Tsuboi, M. *TL* **1969**, 2581. (b) Arnaud, M.; Pedra, A.; Erre, C.; Roussel, C.; Metzger, J. *H* **1983**, *20*, 761.
69. Nenitzescu, C. D.; Cantuniari, J. P. *LA* **1934**, *510*, 269.
70. Nenitzescu, C. D.; Cioranescu, E. *CB* **1936**, *69*, 1820.
71. Hopff, H. *CB* **1932**, *65*, 482.
72. (a) Tabushi, I.; Fujita, K.; Oda, R. *TL* **1968**, 4247. (b) Harding, K. E.; Clement, K. S.; Gilbert, J. C.; Wiechman, B. *JOC* **1984**, *49*, 2049.
73. Gilman, H.; Van Ess, P. R. *JACS* **1933**, *55*, 1258.
74. Gilman, H.; Schulze, F. *JACS* **1927**, *49*, 2328.
75. (a) Percival, W. C.; Wagner, R. B.; Cook, N. C. *JACS* **1953**, *75*, 3731. (b) Fiandanese, V.; Marchese, G.; Martina, V.; Ronzini, L. *TL* **1984**, *25*, 4805. (c) Babudri, F.; D'Ettole, A.; Fiandanese, V.; Marchese, G.; Naso, F. *JOM* **1991**, *405*, 53.
76. (a) Hosomi, A.; Hayashida, H.; Tominaga, Y. *JOC* **1989**, *54*, 3254. (b) Sproesser, L.; Sperling, K.; Trautmann, W.; Smuda, H. Ger. Patent 3 744 619, 1989.
77. Cahiez, G.; Laboue, B. *TL* **1992**, *33*, 4439.
78. Fauvarque, J.; Ducom, J.; Fauvarque, J.-F. *CR(C)* **1972**, *275*, 511.
79. (a) Gilman, H.; Mayhue, M. L. *RTC* **1932**, *51*, 47. (b) Chan, T. H.; Chang, E.; Vinokur, E. *TL* **1970**, 1137. (c) Stowell, J. C. *JOC* **1976**, *41*, 560. (d) Sato, F.; Inoue, M.; Oguro, K.; Sato, M. *TL* **1979**, 4303.
80. Normant, J. F. *S* **1972**, 63.
81. (a) Marfat, A.; McGuirk, P. R.; Helquist, P. *TL* **1978**, 1363. (b) Fleming, I.; Newton, T. W.; Roessler, F. *JCS(P1)* **1981**, 2527. (c) Corriu, R. J. P.; Moreau, J. J. E.; Vernhet, C. *TL* **1987**, *28*, 2963.
82. (a) Jallabert, C.; Luong-Thi, N.-T.; Riviere, H. *BSF* **1970**, 797. (b) Ebert, G. W.; Rieke, R. D. *JOC* **1984**, *49*, 5280. (c) Ebert, G. W.; Rieke, R. D. *JOC* **1988**, *53*, 4482. (d) Rieke, R. D.; Wehmeyer, R. M.; Wu, T.-C.; Ebert, G. W. *T* **1989**, *45*, 443. (e) Zhu, L.; Wehmeyer, R. M.; Rieke, R. D. *JOC* **1991**, *56*, 1445. (f) Ebert, G. W.; Cheasty, J. W.; Tehrani, S. S.; Aouad, E. *OM* **1992**, *11*, 1560.
83. (a) Bourgain, M.; Normant, J.-F. *BSF* **1973**, 2137. (b) Logue, M. W.; Moore, G. L. *JOC* **1975**, *40*, 131.
84. (a) Corriu, R. J. P.; Guerin, C.; M'Boula, J. *TL* **1981**, *22*, 2985. (b) Fleming, I.; Pulido, F. J. *CC* **1986**, 1010.
85. Negishi, E.; Bagheri, V.; Chatterjee, S.; Luo, F.-T.; Miller, J. A.; Stoll, A. T. *TL* **1983**, *24*, 5181.
86. (a) Tamaru, Y.; Ochiai, H.; Nakamura, T.; Yoshida, Z. *AG(E)* **1987**, *26*, 1157. (b) Jackson, R. F. W.; James, K.; Wythes, M. J.; Wood, A. *CC* **1989**, 644. (c) Harada, T.; Kotani, Y.; Katsuhira, T.; Oku, A. *TL* **1991**, *32*, 1573. (d) Jackson, R. F. W.; Wishart, N.; Wood, A.; James, K.; Wythes, M. J. *JOC* **1992**, *57*, 3397.
87. Grey, R. A. *JOC* **1984**, *49*, 2288.
88. El Alami, N.; Belaud, C.; Villieras, J. *JOM* **1987**, *319*, 303.
89. Grondin, J.; Sebban, M.; Vottero, P.; Blancou, H.; Commeyras, A. *JOM* **1989**, *362*, 237.
90. Verkruijsse, H. D.; Heus-Kloos, Y. A.; Brandsma, L. *JOM* **1988**, *338*, 289.
91. (a) Blaise, E.; Koehler, A. *CR* **1909**, *148*, 489. (b) Jones, R. G. *JACS* **1947**, *69*, 2350. (c) Shirley, D. A. *OR* **1954**, *8*, 28.
92. (a) Gilman, H.; Nelson, J. F. *RTC* **1936**, *55*, 518. (b) Cason, J. *CRV* **1947**, *40*, 15. (c) Kollonitsch, J. *JCS(A)* **1966**, 453. (d) Jones, P. R.; Desio, P. J. *CRV* **1978**, *78*, 491. (e) Burkhardt, E. R.; Rieke, R. D. *JOC* **1985**, *50*, 416.
93. Kurts, A. L.; Beletskaya, I. P.; Savchenko, I. A.; Reutov, O. A. *JOM* **1969**, *17*, P21.
94. (a) Larock, R. C.; Bernhardt, J. C. *TL* **1976**, 3097. (b) Larock, R. C.; Bernhardt, J. C. *JOC* **1978**, *43*, 710.
95. (a) Bundel, Yu. G.; Rozenberg, V. I.; Kurts, A. L.; Antonova, N. D.; Reutov, O. A. *JOM* **1969**, *18*, 209. (b) Larock, R. C.; Lu, Y. *TL* **1988**, *29*, 6761.
96. For Pd-catalyzed coupling of Ph_2Hg and Et_2Hg with acyl bromides, see: Takagi, K.; Okamoto, T.; Sakakibara, Y.; Ohno, A.; Oka, S.; Hayama, N. *CL* **1975**, 951.
97. Bumagin, N. A.; Kalinovskii, I. O.; Beletskaya, I. P. *BAU* **1984**, *33*, 2144.
98. Negishi, E.; Chiu, K.-W.; Yosida, T. *JOC* **1975**, *40*, 1676.
99. Negishi, E.; Abramovitch, A.; Merrill, R. E. *CC* **1975**, 138.

100. Utimoto, K.; Okada, K.; Nozaki, H. *TL* **1975**, 4239.

101. (a) Paetzold, P. I.; Grundke, H. *S* **1973**, 635. (b) Naruse, M.; Tomita, T.; Utimoto, K.; Nozaki, H. *TL* **1973**, 795. (c) Naruse, M.; Tomita, T.; Utimoto, K.; Nozaki, H. *T* **1974**, *30*, 835.

102. (a) Adkins, H.; Scanley, C. *JACS* **1951**, *73*, 2854. (b) Carr, D. B.; Schwartz, J. *JACS* **1977**, *99*, 638. (c) Maruoka, K.; Sano, H.; Shinoda, K.; Nakai, S.; Yamamoto, H. *JACS* **1986**, *108*, 6036.

103. Takai, K.; Oshima, K.; Nozaki, H. *BCJ* **1981**, *54*, 1281.

104. Sato, F.; Kodama, H.; Tomuro, Y.; Sato, M. *CL* **1979**, 623.

105. Marko, I. E.; Southern, J. M. *JOC* **1990**, *55*, 3368.

106. (a) Frainnet, E.; Calas, R.; Gerval, P. *CR* **1965**, *261*, 1329. (b) Sakurai, H.; Tominaga, K.; Watanabe, T.; Kumada, M. *TL* **1966**, 5493. (c) Grignon-Dubois, M.; Dunogues, J.; Calas, R. *S* **1976**, 737. (d) Olah, G. A.; Ho, T.-L.; Prakash, G. K. S.; Gupta, B. G. B. *S* **1977**, 677. (e) Grignon-Dubois, M.; Dunogues, J.; Calas, R. *CJC* **1980**, *58*, 291. (f) Grignon-Dubois, M.; Dunogues, J. *JOM* **1986**, *309*, 35.

107. (a) Chan, T. H.; Fleming, I. *S* **1979**, 761. (b) Parnes, Z. N.; Bolestova, G. I. *S* **1984**, 991.

108. (a) Pillot, J.-P.; Dunogues, J.; Calas, R. *BSF* **1975**, 2143. (b) Fleming, I.; Pearce, A. *JCS(P1)* **1980**, 2485. (c) Babudri, F.; Fiandanese, V.; Marchese, G.; Naso, F. *CC* **1991**, 237.

109. Pillot, J.-P.; Dunogues, J.; Calas, R. *JCR(S)* **1977**, 268.

110. (a) Birkofer, L.; Ritter, A.; Uhlenbrauck, H. *CB* **1963**, *96*, 3280. (b) Walton, D. R. M.; Waugh, F. *JOM* **1972**, *37*, 45.

111. Pillot, J.-P.; Deleris, G.; Dunogues, J.; Calas, R. *JOC* **1979**, *44*, 3397.

112. Danheiser, R. L.; Stoner, E. J.; Koyama, H.; Yamashita, D. S.; Klade, C. A. *JACS* **1989**, *111*, 4407.

113. Pillot, J.-P.; Bennetau, B.; Dunogues, J.; Calas, R. *TL* **1981**, *22*, 3401.

114. (a) Franciotti, M.; Mordini, A.; Taddei, M. *SL* **1992**, 137. (b) Franciotti, M.; Mann, A.; Mordini, A.; Taddei, M. *TL* **1993**, *34*, 1355.

115. Stille, J. K. *AG(E)* **1986**, *25*, 508.

116. (a) Soderquist, J. A.; Leong, W. W.-H. *TL* **1983**, *24*, 2361. (b) Perez, M.; Castano, A. M.; Echavarren, A. M. *JOC* **1992**, *57*, 5047.

117. (a) Milstein, D.; Stille, J. K. *JACS* **1978**, *100*, 3636. (b) Milstein, D.; Stille, J. K. *JOC* **1979**, *44*, 1613. (c) Yamamoto, Y.; Yanagi, A. *H* **1982**, *19*, 41.

118. Logue, M. W.; Teng, K. *JOC* **1982**, *47*, 2549.

119. (a) Kosugi, M.; Shimizu, Y.; Migita, T. *JOM* **1977**, *129*, C36. (b) Andrianome, M.; Delmond, B. *JOC* **1988**, *53*, 542. (c) Andrianome, M.; Haberle, K.; Delmond, B. *T* **1989**, *45*, 1079.

120. Reetz, M. T.; Hois, P. *CC* **1989**, 1081.

121. Gambaro, A.; Peruzzo, V.; Marton, D. *JOM* **1983**, *258*, 291.

122. Yano, K.; Baba, A.; Matsuda, H. *CL* **1991**, 1181.

123. (a) Barton, D. H. R.; Ozbalik, N.; Ramesh, M. *T* **1988**, *44*, 5661. (b) Asthana, A.; Srivastava, R. C. *JOM* **1989**, *366*, 281.

124. Kasatkin, A. N.; Kulak, A. N.; Tolstikov, G. A. *JOM* **1988**, *346*, 23.

125. Hart, D. W.; Schwartz, J. *JACS* **1974**, *96*, 8115.

126. Hirao, T.; Misu, D.; Yao, K.; Agawa, T. *TL* **1986**, *27*, 929.

127. Cahiez, G.; Laboue, B. *TL* **1989**, *30*, 7369.

128. Normant, J.-F.; Cahiez, G. *Modern Synth. Methods* **1983**, *3*, 173.

129. (a) Hegedus, L. S.; Kendall, P. M.; Lo, S. M.; Sheats, J. R. *JACS* **1975**, *97*, 5448. (b) Pittman, C. U. Jr.; Hanes, R. M. *JOC* **1977**, *42*, 1194.

130. Inaba, S.; Rieke, R. D. *JOC* **1985**, *50*, 1373.

131. (a) Baker, R.; Blackett, B. N.; Cookson, R. C.; Cross, R. C.; Madden, D. P. *CC* **1972**, 343. (b) Inaba, S.; Rieke, R. D. *TL* **1983**, *24*, 2451.

132. (a) Tanaka, T.; Kurozumi, S.; Toru, T.; Kobayashi, M.; Miura, S.; Ishimoto, S. *TL* **1975**, 1535. (b) Kurozumi, S.; Toru, T.; Tanaka, T.; Miura, S.; Kobayashi, M.; Ishimoto, S. U.S. Patent 4 009 196, 1977. U.S. Patent 4 139 717, 1979. (c) Lee, S.-H.; Shih, M.-J.; Hulce, M. *TL* **1992**, *33*, 185.

133. Lapkin, I. I.; Saitkulova, F. G. *JOU* **1971**, *7*, 2586.

134. For examples of acetylations of enolates in which proton transfer is not competitive, see Refs. 132a,b, and Evans, D. A.; Ennis, M. D.; Le, T.; Mandel, N.; Mandel, G. *JACS* **1984**, *106*, 1154.

135. (a) Rasmussen, J. K. *S* **1977**, 91. (b) Tirpak, R. E.; Rathke, M. W. *JOC* **1982**, *47*, 5099.

136. (a) Hoch, H.; Hunig, S. *CB* **1972**, *105*, 2660. (b) Nilsson, L. *ACS(B)* **1979**, *33*, 710. (c) Zhang, P.; Li, L. *SC* **1986**, *16*, 957.

137. Burlachenko, G. S.; Mal'tsev, V. V.; Baukov, Yu. I.; Lutsenko, I. F. *JGU* **1973**, *43*, 1708.

138. Rathke, M. W.; Sullivan, D. F. *TL* **1973**, 1297.

139. Chiba, T.; Ishizawa, T.; Sakaki, J.; Kaneko, C. *CPB* **1987**, *35*, 4672.

140. Fleming, I.; Goldhill, J.; Paterson, I. *TL* **1979**, 3209.

141. Brownbridge, P.; Chan, T. H.; Brook, M. A.; Kang, G. J. *CJC* **1983**, *61*, 688.

142. Ladjama, D.; Riehl, J. J. *S* **1979**, 504.

143. (a) Heusler, K.; Kebrle, J.; Meystre, C.; Ueberwasser, H.; Wieland, P.; Anner, G.; Wettstein, A. *HCA* **1959**, *42*, 2043. (b) Ensley, H. E.; Parnell, C. A.; Corey, E. J. *JOC* **1978**, *43*, 1610.

144. Dauben, W. G.; Eastham, J. F.; Micheli, R. A. *JACS* **1951**, *73*, 4496.

145. Snozzi, C.; Goffinet, B.; Joly, R.; Jolly, J. U.S. Patent 3 117 142, 1964.

146. (a) Casey, C. P.; Marten, D. F. *TL* **1974**, 925. (b) Ouannes, C.; Langlois, Y. *TL* **1975**, 3461.

147. Ono, N.; Yoshimura, T.; Saito, T.; Tamura, R.; Tanikaga, R.; Kaji, A. *BCJ* **1979**, *52*, 1716.

148. Casey, C. P.; Marten, D. F.; Boggs, R. A. *TL* **1973**, 2071.

149. (a) Viscontini, M.; Merckling, N. *HCA* **1952**, *35*, 2280. (b) Rathke, M. W.; Cowan, P. J. *JOC* **1985**, *50*, 2622.

150. (a) Adams, R.; Vollweiler, E. H. *JACS* **1918**, *40*, 1732. (b) French, H. E.; Adams, R. *JACS* **1921**, *43*, 651. (c) Ulich, L. H.; Adams, R. *JACS* **1921**, *43*, 660. (d) Euranto, E. K.; Noponen, A.; Kujanpaa, T. *ACS* **1966**, *20*, 1273. (e) Kyburz, R.; Schaltegger, H.; Neuenschwander, M. *HCA* **1971**, *54*, 1037. (f) Neuenschwander, M.; Iseli, R. *HCA* **1977**, *60*, 1061. (g) Neuenschwander, M.; Vogeli, R.; Fahrni, H.-P.; Lehmann, H.; Ruder, J.-P. *HCA* **1977**, *60*, 1073. (h) Bigler, P.; Muhle, H.; Neuenschwander, M. *S* **1978**, 593.

151. Bigler, P.; Neuenschwander, M. *HCA* **1978**, *61*, 2165.

152. (a) Euranto, E.; Kujanpaa, T. *ACS* **1961**, *15*, 1209. (b) Euranto, E.; Leppanen, O. *ACS* **1963**, *17*, 2765. (c) Neuenschwander, M.; Bigler, P.; Christen, K.; Iseli, R.; Kyburz, R.; Muhle, H. *HCA* **1978**, *61*, 2047.

153. (a) Bigler, P.; Schonholzer, S.; Neuenschwander, M. *HCA* **1978**, *61*, 2059. (b) Bigler, P.; Neuenschwander, M. *HCA* **1978**, *61*, 2381.

154. (a) Chou, T.-S.; Knochel, P. *JOC* **1990**, *55*, 4791, 6232. (b) Knochel, P.; Chou, T.-S.; Jubert, C.; Rajagopal, D. *JOC* **1993**, *58*, 588.

155. Knochel, P.; Chou, T.-S.; Chen, H. G.; Yeh, M. C. P.; Rozema, M. J. *JOC* **1989**, *54*, 5202.

156. Enholm, E. J.; Satici, H. *TL* **1991**, *32*, 2433.

157. Oku, A.; Harada, T.; Kita, K. *TL* **1982**, *23*, 681.

158. Cloke, J. B.; Pilgrim, F. J. *JACS* **1939**, *61*, 2667.

159. Duboudin, J.-G.; Valade, J. *BSF* **1974**, 272.

160. (a) Ahmad, S.; Iqbal, J. *CL* **1987**, 953. (b) Iqbal, J.; Srivastava, R. R. *T* **1991**, *47*, 3155.

161. Pri-Bar, I.; Stille, J. K. *JOC* **1982**, *47*, 1215.

162. Alper, H.; Huang, C.-C. *JOC* **1973**, *38*, 64.

163. Fitch, J. W.; Payne, W. G.; Westmoreland, D. *JOC* **1983**, *48*, 751.

164. Chang, C.; Chu, K. C.; Yue, S. *SC* **1992**, *22*, 1217.

165. (a) Straus, F.; Heinze, H. *LA* **1932**, *493*, 191. (b) Quintard, J.-P.; Elissondo, B.; Pereyre, M. *JOM* **1981**, *212*, C31. (c) Quintard, J.-P.; Elissondo, B.; Pereyre, M. *JOC* **1983**, *48*, 1559.

166. (a) Bakos, T.; Vincze, I. *SC* **1989**, *19*, 523. (b) Sabharwal, A.; Vig, R.; Sharma, S.; Singh, J. *IJC(B)* **1990**, *29*, 890.

167. (a) Pop, L.; Oprean, I.; Barabas, A.; Hodosan, F. *JPR* **1986**, *328*, 867. (b) Oprean, I.; Ciupe, H.; Gansca, L.; Hodosan, F. *JPR* **1987**, *329*, 283.

168. Dutka, F.; Marton, A. F. *ZN(B)* **1969**, *24*, 1664.
169. Hofle, G.; Steglich, W.; Vorbruggen, H. *AG(E)* **1978**, *17*, 569.
170. (a) Vedejs, E.; Diver, S. T. *JACS* **1993**, *115*, 3358. (b) Vedejs, E.; Bennett, N. S.; Conn, L. M.; Diver, S. T.; Gingras, M.; Lin, S.; Oliver, P. A.; Peterson, M. J. *JOC* **1993**, *58*, 7286.
171. 1,2-Diols can be selectively monoacetylated by conversion into the cyclic dibutylstannylidene derivative followed by treatment with AcCl: (a) Anchisi, C.; Maccioni, A.; Maccioni, A. M.; Podda, G. *G* **1983**, *113*, 73. (b) Roelens, S. *JCS(P2)* **1988**, 2105. (c) Anderson, W. K.; Coburn, R. A.; Gopalsamy, A.; Howe, T. J. *TL* **1990**, *31*, 169. (d) Getman, D. P.; DeCrescenzo, G. A.; Heintz, R. M. *TL* **1991**, *32*, 5691.
172. γ-Picoline might be unsuitable as a pyridine replacement because it reacts with AcCl to form *N*-acetyl-4-(acetylmethylidene)-1,4-dihydropyridine under standard acetylation conditions (CH_2Cl_2, rt, 8–16 h): Ippolito, R. M.; Vigmond, S. U.S. Patent 4 681 944, 1987.
173. (a) Okamoto, K.; Kondo, T.; Goto, T. *T* **1987**, *43*, 5909. (b) McClure, K. F.; Danishefsky, S. J. *JACS* **1993**, *115*, 6094.
174. (a) Braun, M.; Devant, R. *TL* **1984**, *25*, 5031. (b) Devant, R.; Mahler, U.; Braun, M. *CB* **1988**, *121*, 397.
175. Capek, K.; Steffkova, J.; Jary, J. *CCC* **1966**, *31*, 1854.
176. Bladon, P.; Clayton, R. B.; Greenhalgh, C. W.; Henbest, H. B.; Jones, E. R. H.; Lovell, B. J.; Silverstone, G.; Wood, G. W.; Woods, G. F. *JCS* **1952**, 4883.
177. (a) Litvinenko, L. M.; Kirichenko, A. I. *DOK* **1967**, *176*, 763. (b) Steglich, W.; Hofle, G. *AG(E)* **1969**, *8*, 981. (c) Hofle, G.; Steglich, W. *S* **1972**, 619. (d) Scriven, E. F. V. *CSR* **1983**, *12*, 129.
178. (a) Norris, J. F.; Rigby, G. W. *JACS* **1932**, *54*, 2088. (b) Plattner, Pl. A.; Petrzilka, Th.; Lang, W. *HCA* **1944**, *27*, 513. (c) Hauser, C. R.; Hudson, B. E.; Abramovitch, B.; Shivers, J. C. *OSC* **1955**, *3*, 142. (d) Ohloff, G. *HCA* **1958**, *41*, 845.
179. (a) Plattner, Pl. A.; Furst, A.; Koller, F.; Lang, W. *HCA* **1948**, *31*, 1455. (b) Williams, K. I. H.; Rosenfeld, R. S.; Smulowitz, M.; Fukushima, D. K. *Steroids* **1963**, *1*, 377. (c) Kido, F.; Kitahara, H.; Yoshikoshi, A. *JOC* **1986**, *51*, 1478.
180. (a) Takimoto, S.; Inanaga, J.; Katsuki, T.; Yamaguchi, M. *BCJ* **1976**, *49*, 2335. (b) Amouroux, R.; Chan, T. H. *TL* **1978**, 4453.
181. Spassow, A. *OSC* **1955**, *3*, 144.
182. (a) Illi, V. O. *TL* **1979**, 2431. (b) Szeja, W. *S* **1980**, 402.
183. (a) Perriot, P.; Normant, J. F.; Villieras, J. *CR(C)* **1979**, *289*, 259. (b) Trost, B. M.; Tour, J. M. *JOC* **1989**, *54*, 484. (c) Corey, E. J.; Su, W. *TL* **1990**, *31*, 2089.
184. (a) Evans, D. D.; Evans, D. E.; Lewis, G. S.; Palmer, P. J.; Weyell, D. J. *JCS* **1963**, 3578. (b) Duboudin, J. G.; Ratier, M.; Trouve, B. *JOM* **1987**, *331*, 181.
185. Weidert, P. J.; Geyer, E.; Horner, L. *LA* **1989**, 533.
186. (a) Heyse, M. Ger. Patent 524 435, 1929. (b) Searles, S. Jr.; Pollart, K. A.; Block, F. *JACS* **1957**, *79*, 952. (c) Sharma, K. K.; Torssell, K. B. G. *T* **1984**, *40*, 1085.
187. Kotsuki, H.; Kataoka, M.; Nishizawa, H. *TL* **1993**, *34*, 4031.
188. Bryant, W. M. D.; Smith, D. M. *JACS* **1936**, *58*, 1014.
189. Karger, M. H.; Mazur, Y. *JOC* **1971**, *36*, 528.
190. (a) Lennon, J. J.; Perkin, W. H. Jr. *JCS* **1928**, 1513. (b) Zilkha, A.; Liwschitz, Y. *JCS* **1957**, 4397. (c) Bose, N. K.; Chaudhury, D. N. *T* **1964**, *20*, 49.
191. (a) Turner, R. B. *JACS* **1950**, *72*, 579. (b) Rosenmund, K. W.; Herzberg, H.; Schutt, H. *CB* **1954**, *87*, 1258. (c) Vignau, M.; Bucourt, R.; Tessier, J.; Costerousse, G.; Nedelec, L.; Gasc, J.-C.; Joly, R.; Warnant, J.; Goffinet, B. U.S. Patent 3 453 267, 1969.
192. (a) Harada, K.; Kaji, E.; Zen, S. *CPB* **1980**, *28*, 3296. (b) Fleming, I.; Moses, R. C.; Tercel, M.; Ziv, J. *JCS(P1)* **1991**, 617.
193. Farrissey, W. J. Jr. U.S. Patent 3 291 834, 1966.
194. Stirring zinc dust with AcCl and CuCl (Et_2O, rt $\rightarrow \Delta$) produces an active zinc couple capable of reacting with methylene bromide to form the Simmons–Smith reagent: Friedrich, E. C.; Lewis, E. J. *JOC* **1990**, *55*, 2491.
195. Watt, G. W.; Gentile, P. S.; Helvenston, E. P. *JACS* **1955**, *77*, 2752.
196. Bellesia, F.; Ghelfi, F.; Pagnoni, U. M.; Pinetti, A. *JCR(S)* **1990**, 188.
197. Nugent, W. A. *TL* **1978**, 3427.
198. Bellesia, F.; Ghelfi, F.; Pagnoni, U. M.; Pinetti, A. *SC* **1991**, *21*, 489.
199. Bellesia, F.; Ghelfi, F.; Pagnoni, U. M.; Pinetti, A. *JCR(S)* **1989**, 108.
200. Backvall, J. E.; Young, M. W.; Sharpless, K. B. *TL* **1977**, 3523.
201. Cummins, C. H.; Coates, R. M. *JOC* **1983**, *48*, 2070.
202. (a) Demerseman, P.; Guillaumel, J.; Clavel, J.-M.; Royer, R. *TL* **1978**, 2011. (b) Guillaumel, J.; Demerseman, P.; Clavel, J.-M.; Royer, R.; Platzer, N.; Brevard, C. *T* **1980**, *36*, 2459.
203. Numata, T.; Oae, S. *CI(L)* **1973**, 277.
204. (a) Adkins, H.; Thompson, Q. E. *JACS* **1949**, *71*, 2242. (b) Paukstelis, J. V.; Kim, M. *JOC* **1974**, *39*, 1503.
205. (a) Lang, R. W.; Hansen, H.-J. *HCA* **1980**, *63*, 438. (b) Lang, R. W.; Hansen, H.-J. *OS* **1984**, *62*, 202. (c) Abell, A. D.; Morris, K. B.; Litten, J. C. *JOC* **1990**, *55*, 5217.
206. (a) Burger, K.; Huber, E.; Sewald, N.; Partscht, H. *Chem.-Ztg.* **1986**, *110*, 83. (b) Sewald, N.; Riede, J.; Bissinger, P.; Burger, K. *JCS(P1)* **1992**, 267.
207. Maujean, A.; Chuche, J. *TL* **1976**, 2905.
208. Sauer, J. C. *JACS* **1947**, *69*, 2444.
209. Hunt, C. F. U.S. Patent 2 737 528, 1956.
210. Matoba, K.; Tachi, M.; Itooka, T.; Yamazaki, T. *CPB* **1986**, *34*, 2007.
211. Furniss, B. S.; Hannaford, A. J.; Smith, P. W. G.; Tatchell, A. R. *Vogel's Textbook of Practical Organic Chemistry*, 5th ed.; Longman/Wiley: New York, 1989; pp 916, 1273.
212. (a) Sonntag, N. O. V. *CRV* **1953**, *52*, 237. (b) To minimize the amount of hydrolysis, the acetylation should be run at pH = $(x + 13.25)/2$, where x is the pK_a of the protonated amine: King, J. F.; Rathore, R.; Lam, J. Y. L.; Guo, Z. R.; Klassen, D. F. *JACS* **1992**, *114*, 3028.
213. (a) Bohme, H.; Hartke, K. *CB* **1960**, *93*, 1305. (b) Kinast, G.; Tietze, L.-F. *AG(E)* **1976**, *15*, 239.
214. Kritzler, H.; Wanger, K.; Holtschmidt, H. U.S. Patent 3 242 202, 1966.
215. (a) Ikeda, K.; Morimoto, T.; Sekiya, M. *CPB* **1980**, *28*, 1178. (b) Ikeda, K.; Terao, Y.; Sekiya, M. *CPB* **1981**, *29*, 1156.
216. Okada, I.; Takahama, T.; Sudo, R. *BCJ* **1970**, *43*, 2591.
217. Caubere, P.; Madelmont, J.-C. *CR(C)* **1972**, *275*, 1305.
218. Comins, D. L.; Abdullah, A. H. *JOC* **1982**, *47*, 4315.
219. Corcoran, R. C.; Bang, S. H. *TL* **1990**, *31*, 6757.
220. (a) Oppolzer, W.; Bieber, L.; Francotte, E. *TL* **1979**, 981. (b) Ng, K. S.; Laycock, D. E.; Alper, H. *JOC* **1981**, *46*, 2899.
221. Ben-Ishai, D.; Katchalski, E. *JOC* **1951**, *16*, 1025.
222. Hoffmann, H. M. R.; Haase, K. *S* **1981**, 715.
223. Ihara, M.; Hirabayashi, A.; Taniguchi, N.; Fukumoto, K. *H* **1992**, *33*, 851.
224. (a) Pump, J.; Wannagat, U. *M* **1962**, *93*, 352. (b) Bowser, J. R.; Williams, P. J.; Kurz, K. *JOC* **1983**, *48*, 4111.
225. Ahmad, S.; Iqbal, J. *TL* **1986**, *27*, 3791.

Bruce A. Pearlman
The Upjohn Company, Kalamazoo, MI, USA

Aluminum Chloride[1]

$AlCl_3$

[7446-70-0] $AlCl_3$ (MW 133.34)

(Lewis acid catalyst for Friedel–Crafts, Diels–Alder, [2 + 2] cycloadditions, ene reactions, rearrangements, and other reactions)

Physical Data: mp 190 °C (193–194 °C sealed tube); sublimes at 180 °C; *d* 2.44 g cm^{-3}.
Solubility: sol many organic solvents, e.g. benzene, nitrobenzene, carbon tetrachloride, chloroform, methylene chloride, nitromethane, and 1,2-dichloroethane; insol carbon disulfide.
Form Supplied in: colorless solid when pure, typically a gray or yellow-green solid; also available as a 1.0 M nitrobenzene solution.
Handling, Storage, and Precautions: fumes in air with a strong odor of HCl. $AlCl_3$ reacts violently with H_2O. All containers should be kept tightly closed and protected from moisture.[1c] Use in a fume hood.

Friedel–Crafts Chemistry.[1,2] $AlCl_3$ has traditionally been used in stoichiometric or catalytic[3] amounts to mediate Friedel–Crafts alkylations and acylations of aromatic systems (eq 1).

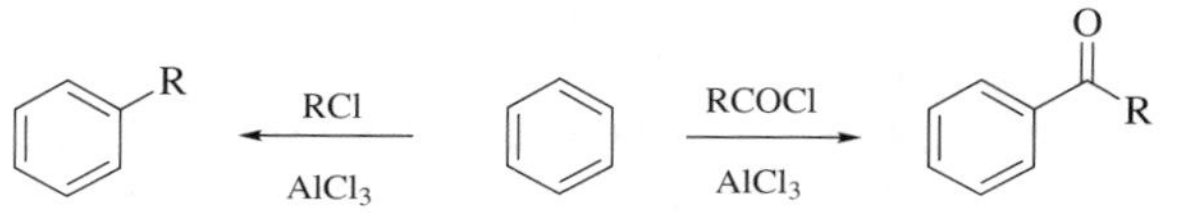

(1)

This is a result of the Lewis acidity of $AlCl_3$ which complexes strongly with carbonyl groups.[4] Adaptations of these basic reactions have been reported.[5] In chiral systems, inter- and intramolecular acylations have been achieved without the loss of optical activity (eq 2).[6]

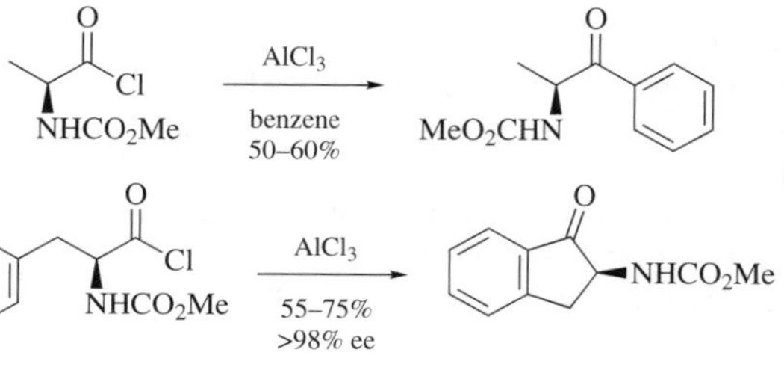

(2)

Friedel–Crafts chemistry at an asymmetric center generally proceeds with racemization, but the use of mesylates or chlorosulfonates as leaving groups has resulted in alkylations with excellent control of stereochemistry.[7] The reactions proceed with inversion of configuration (eq 3). Cyclopropane derivatives have been used as three-carbon units in acylation reactions (eq 4).[8] In conjunction with triethylsilane, a net alkylation is possible under acylation conditions (eq 5).[9] These conditions are compatible with halogen atoms present elsewhere in the molecule. Acylation reactions of phenolic compounds with heteroaromatic systems have also been accomplished (eq 6).[10]

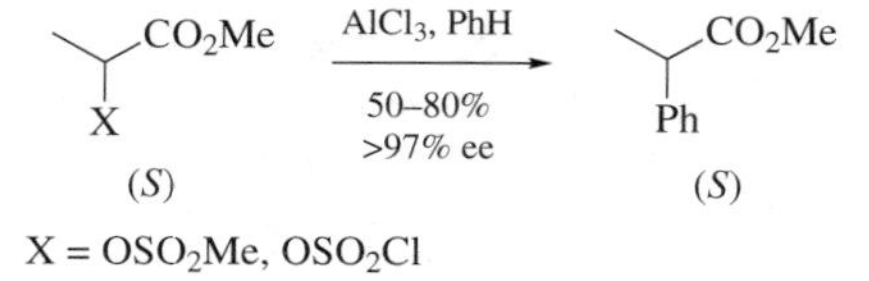

(3)

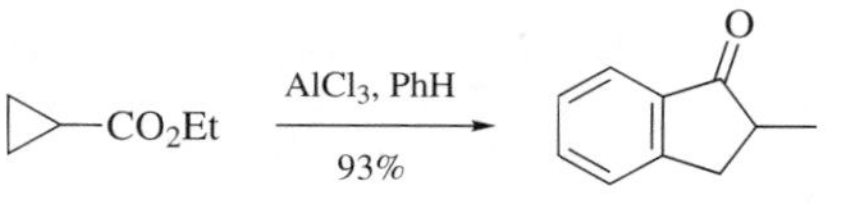

X = OSO_2Me, OSO_2Cl

(4)

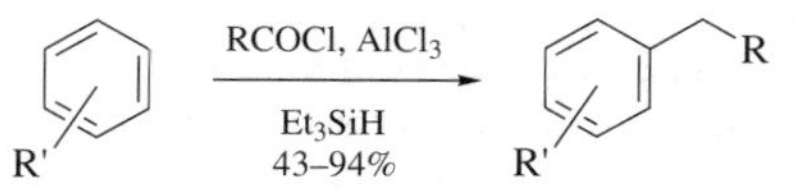

(5)

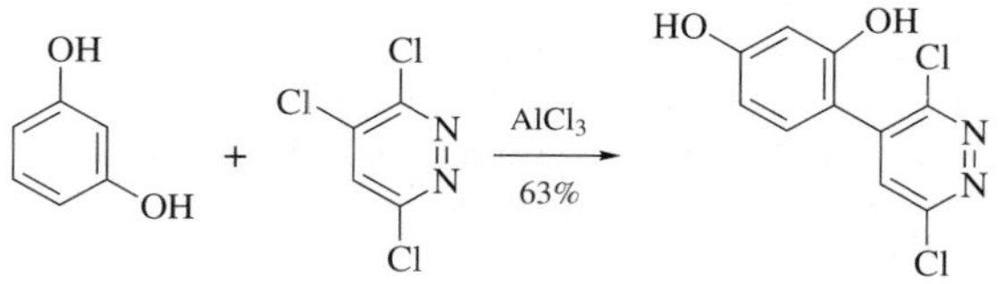

(6)

Treatment of aryl azides with $AlCl_3$ has been reported to give polycyclic aromatic compounds (eq 7),[11] or aziridines when the reactions are run in the presence of alkenes (eq 8).[12]

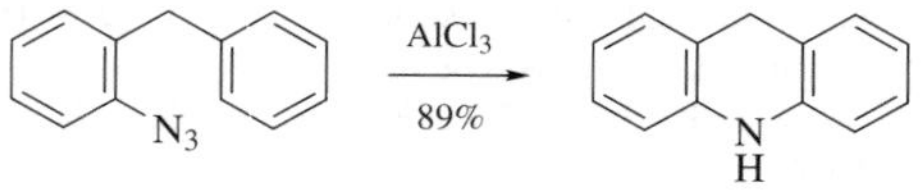

(7)

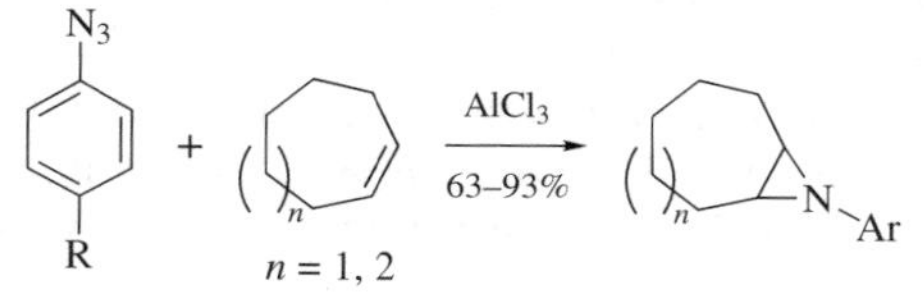

(8)

The scope of Friedel–Crafts chemistry has been expanded beyond aromatic systems to nonaromatic systems, such as alkenes and alkynes and the mechanistic details have been investigated.[13] The Friedel–Crafts alkylation[14] and acylation[15] of alkenes provide access to a variety of organic systems (eq 9). The acylation of alkynes provides access to cyclopentenone derivatives (eq 10).[16] In addition, one can use this chemistry to access indenyl systems[17] and vinyl chlorides.[18] Allylic sulfones can undergo allylation chemistry (eq 11).[19]

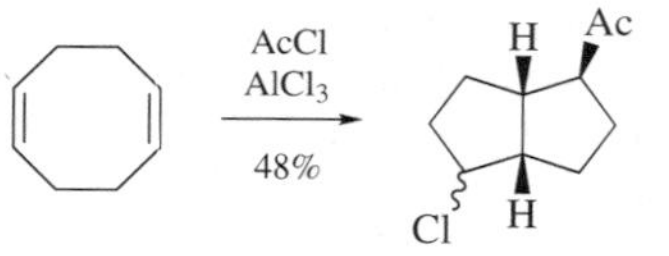

(9)

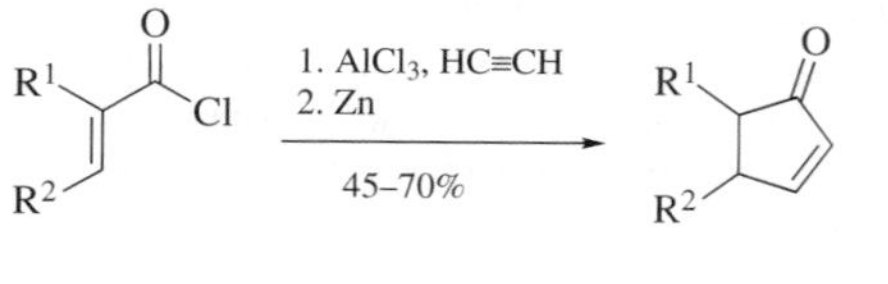

(10)

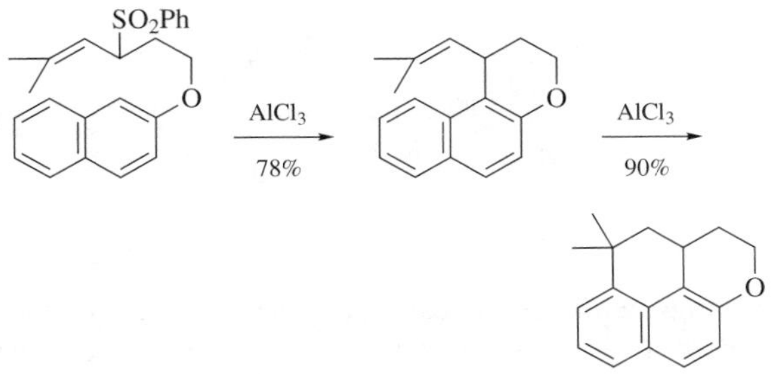

(11)

The use of silyl derivatives in Friedel–Crafts chemistry has not only improved the regioselectivity but extended the scope of these reactions. Substitution at the *ipso* position occurs with aryl silanes (eq 12).[20] The ability of silyl groups to stabilize β-carbenium ions (β-effect) affords acylated products with complete control of regiochemistry (eq 13).[21]

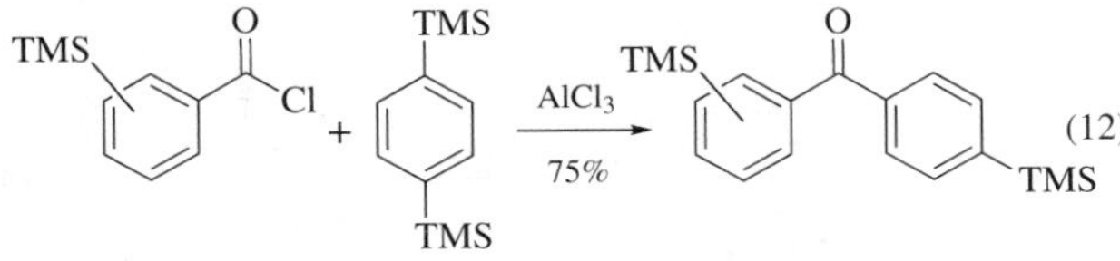

(12)

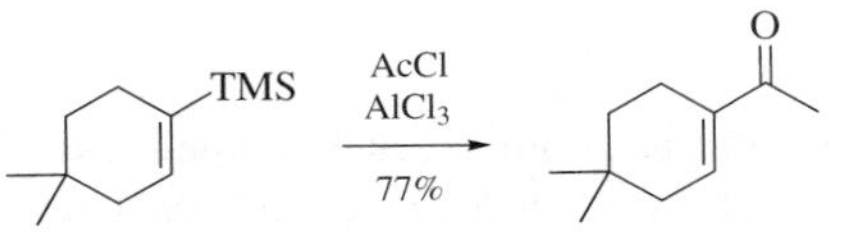

(13)

The use of silylacetylenes gives ynones (eq 14),[22] cyclopentenone derivatives (eq 15),[23] and α-amino acid derivatives (eq 16).[24]

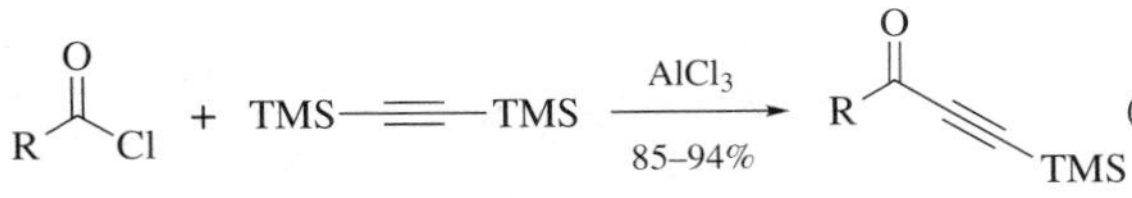

(14)

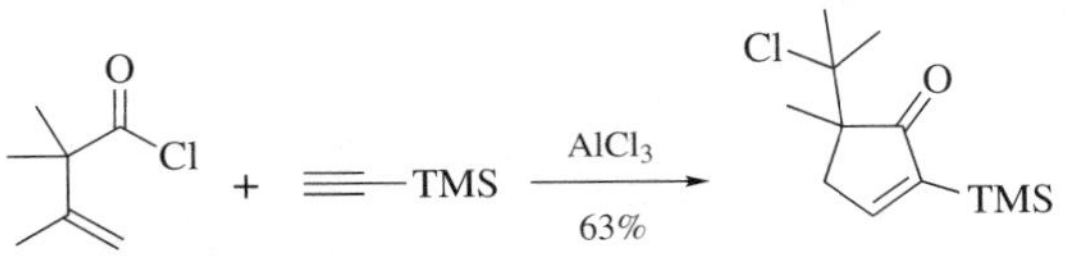

(15)

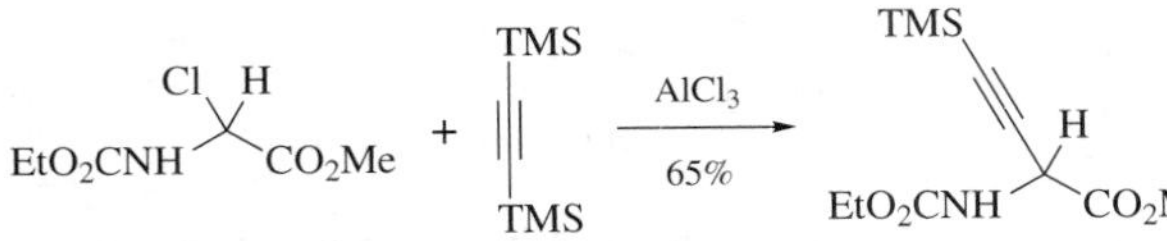

(16)

Propargylic silanes undergo acylation to generate allenyl ketones (eq 17),[25] while alkylsilanes afford cycloalkanones (eq 18).[26]

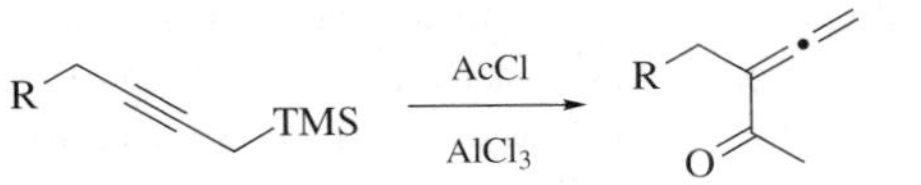

(17)

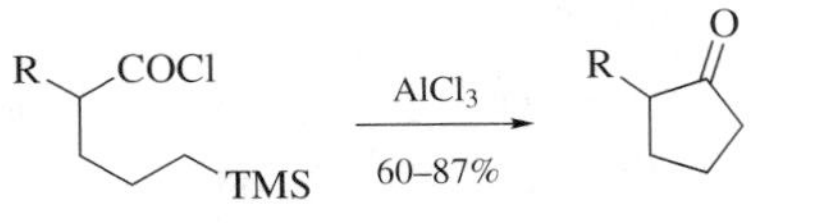

(18)

Several name reactions are promoted by $AlCl_3$. For example, the Darzens–Nenitzescu reaction is simply the acylation of alkenes. The Ferrario reaction generates phenoxathiins from diphenyl ethers (eq 19).[27] The rearrangement of acyloxy aromatic systems is known as the Fries rearrangement (eq 20).[28] Aryl aldehydes are produced by the Gatterman aldehyde synthesis (eq 21).[29] The initial step of the Haworth phenanthrene synthesis makes use of a Friedel–Crafts acylation.[30] The acylation of phenolic compounds is called the Houben–Hoesch reaction (eq 22).[31] The Leuckart amide synthesis generates aryl amides from isocyanates (eq 23).[32]

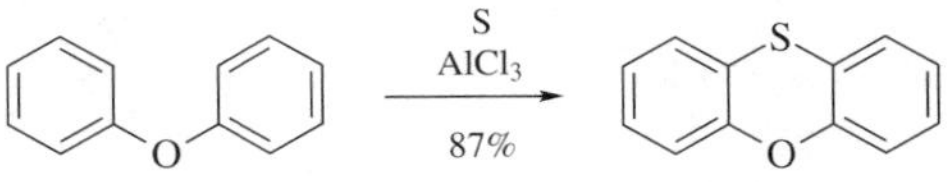

(19)

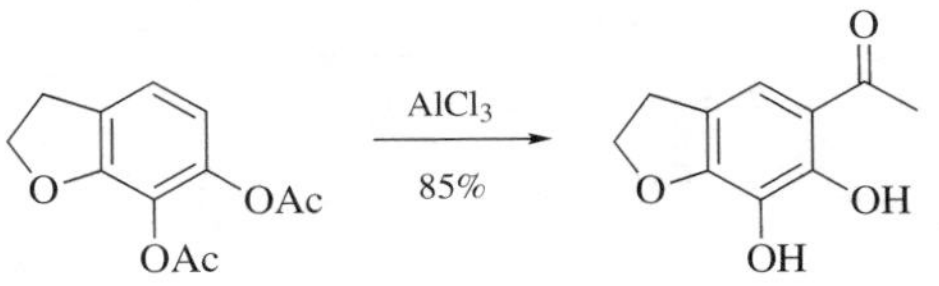

(20)

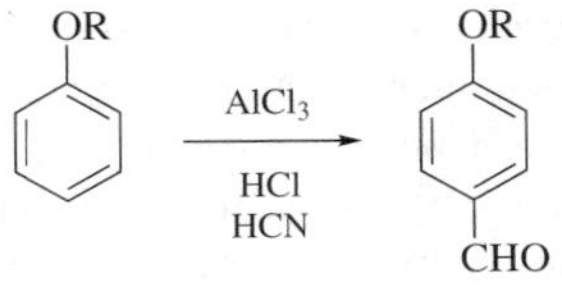

(21)

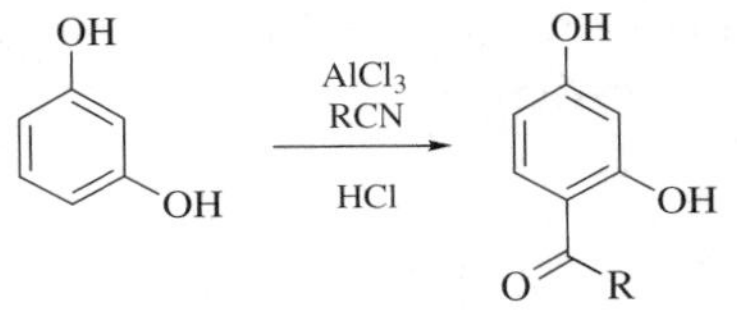

(22)

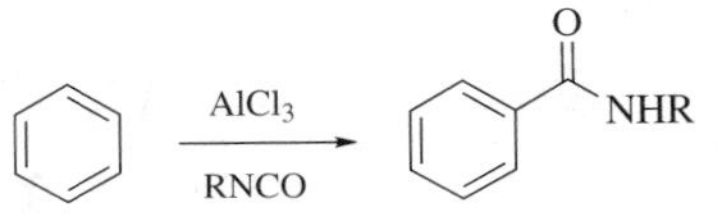

(23)

Amides can also be obtained by $AlCl_3$ catalyzed ester amine exchange which proceeds primarily without racemization of chiral centers (eq 24).[33] The reaction of phenols with β-keto esters is known as the Pechmann condensation (eq 25).[34] Aryl amines are used in the Riehm quinoline synthesis (eq 26).[35] Aromatic systems may be coupled via the Scholl reaction (eq 27)[36] and indole derivatives are prepared in the Stolle synthesis (eq 28).[37] In the Zincke-Suhl reaction, phenols are converted to dienones (eq 29).[38]

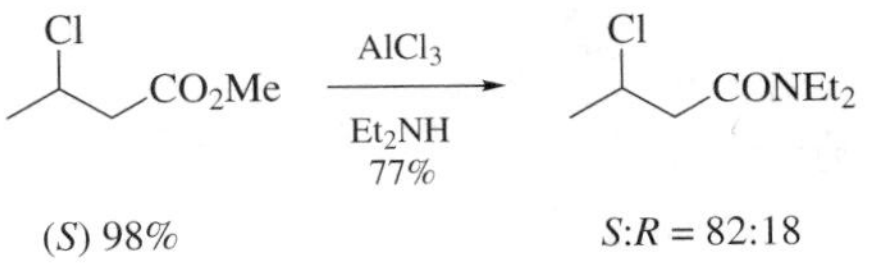

(24)

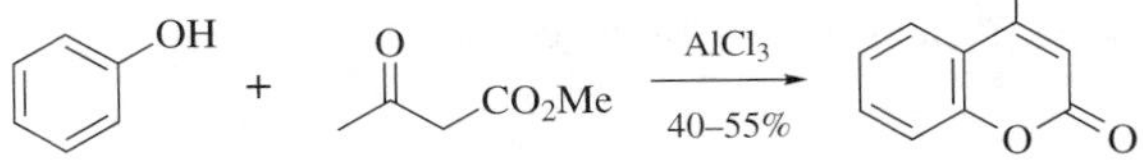

(*S*) 98% *S*:*R* = 82:18

(25)

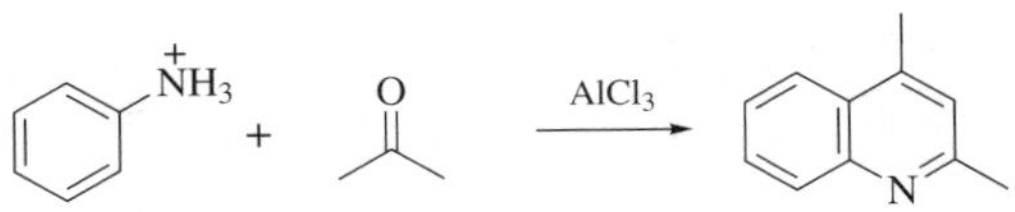

(26)

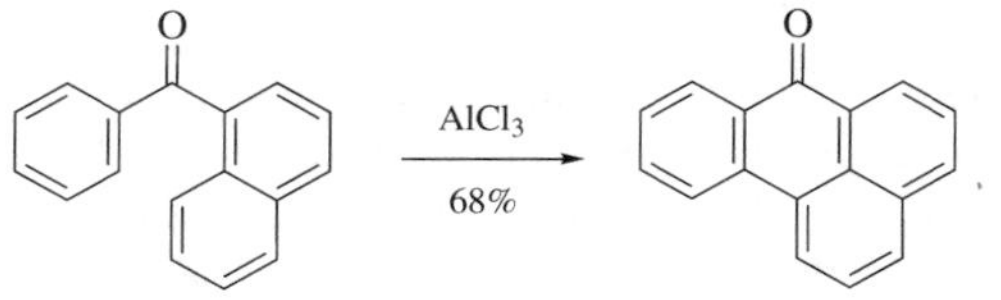

O O $AlCl_3$ 68% (27)

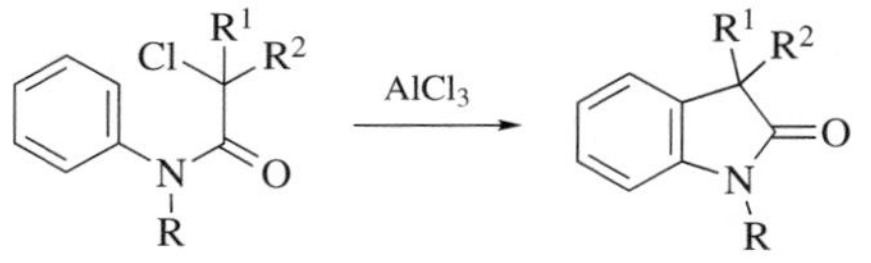

R^1 R^2 Cl N O R $AlCl_3$ R^1 R^2 O N R (28)

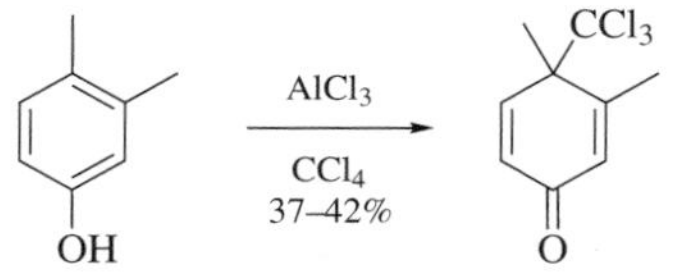

OH $AlCl_3$ CCl_4 37–42% CCl_3 O (29)

Diels–Alder Reactions. There is some evidence that $AlCl_3$ catalysis of Diels–Alder reactions changes the transition state from a synchronous to an asynchronous one.[39] This also enhances asymmetric induction by increasing steric interactions at one end of the dieneophile. There are many examples of $AlCl_3$ promoted Diels–Alder reactions (eq 30).[40] Hetero-Diels–Alder reactions can be used to generate oxygen (eq 31)[41] and nitrogen (eq 32)[42] containing heterocycles.

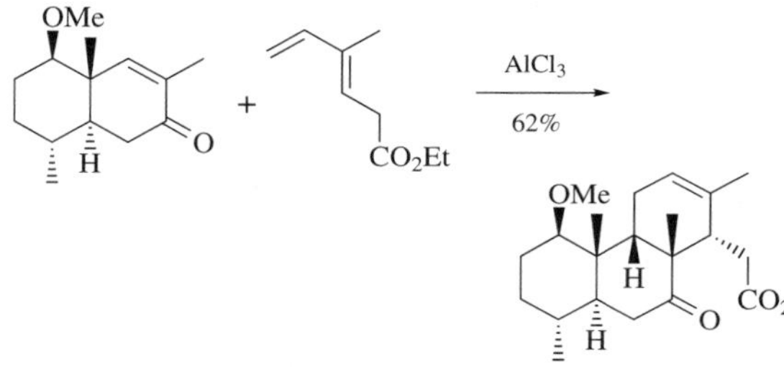

OMe H O + CO_2Et $AlCl_3$ 62% OMe H H O CO_2Et (30)

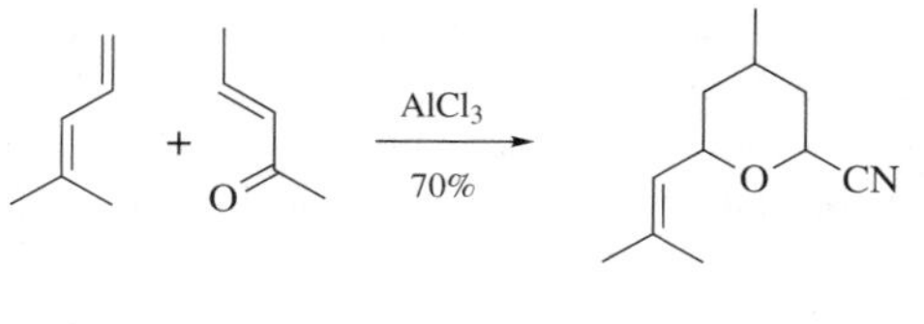

+ O $AlCl_3$ 70% O CN (31)

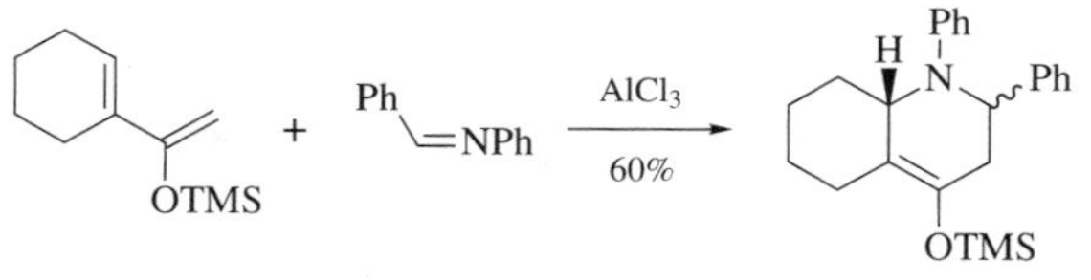

OTMS + Ph NPh $AlCl_3$ 60% H Ph N Ph OTMS (32)

$AlCl_3$ can also be used to catalyze [2 + 2] cycloaddition reactions (eq 33)[43] and ene reactions (eq 34).[44]

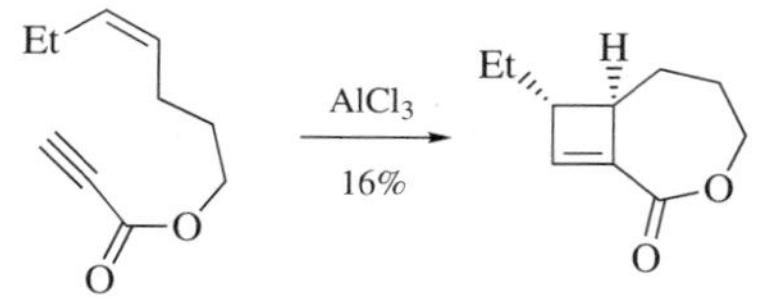

Et O O $AlCl_3$ 16% Et H O O (33)

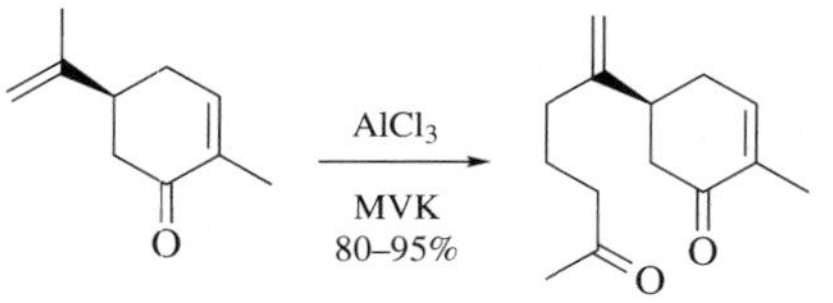

O $AlCl_3$ MVK 80–95% O O (34)

Rearrangements. $AlCl_3$ catalyzed rearrangement of hydrocarbon derivatives to adamantanes has been well documented (eq 35).[45] Other rearrangements have been used in triquinane synthesis (eq 36).[46]

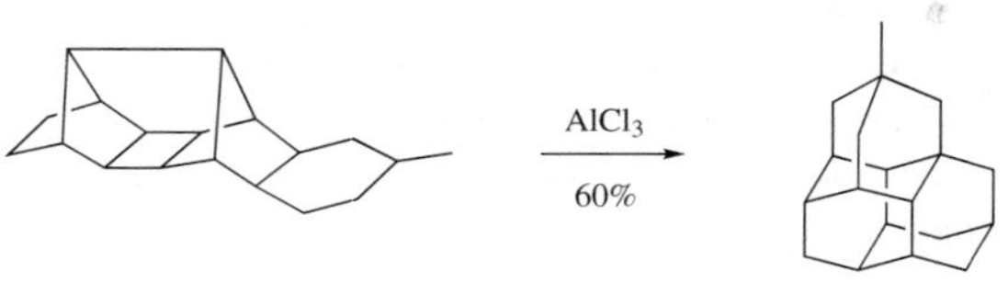

$AlCl_3$ 60% (35)

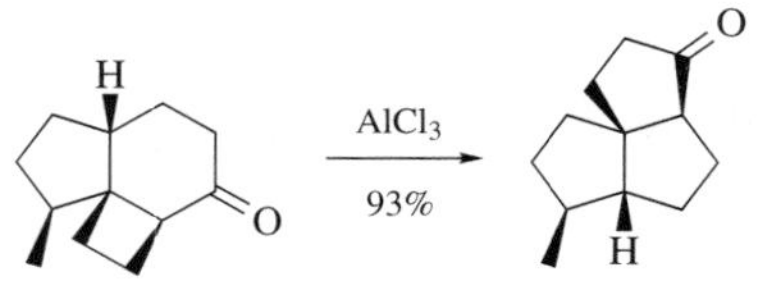

H O $AlCl_3$ 93% O H (36)

Miscellaneous Reactions. $AlCl_3$ has been used to catalyze the addition of allylsilanes to aldehydes and acid chlorides (eq 37).[47] Cyclic ethers (pyrans and oxepins) have been prepared with hydroxyalkenes (eq 38).[48] The course of reactions between aldehydes and allylic Grignard reagents can be completely diverted to α-allylation by $AlCl_3$ (eq 39).[49] The normal course of the reaction gives γ-allylation products.

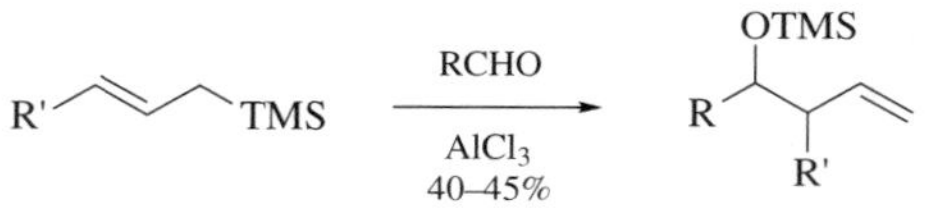

R' TMS RCHO $AlCl_3$ 40–45% OTMS R R' (37)

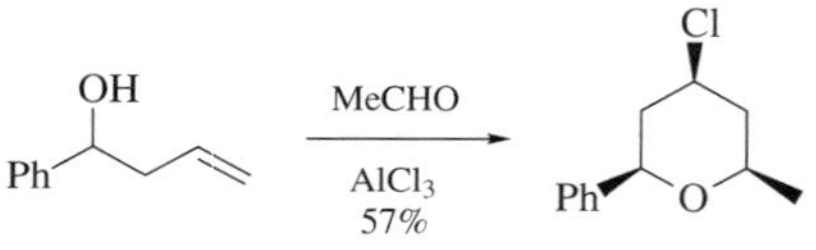

OH Ph MeCHO $AlCl_3$ 57% Cl Ph O (38)

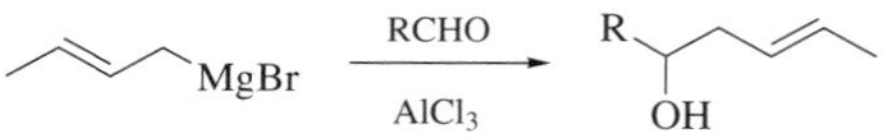

MgBr RCHO $AlCl_3$ R OH (39)

$AlCl_3$ can be used to remove *t*-butyl groups from aromatic rings (eq 40),[50] thereby using this group as a protecting element for a ring position. $AlCl_3$ has also been used to remove *p*-nitrobenzyl (PNB) and benzhydryl protecting groups (eq 41).[51] The combination of $AlCl_3$ and ***Ethanethiol*** has formed the basis of a push–pull mechanism for the cleavage of many types of bonds including C–X,[52] C–NO_2,[53] C=C,[54] and C–O.[55] Furthermore, $AlCl_3$ has been used to catalyze chlorination of aromatic rings,[56] open epoxides,[57] and mediate addition of dichlorophosphoryl groups to alkanes.[58]

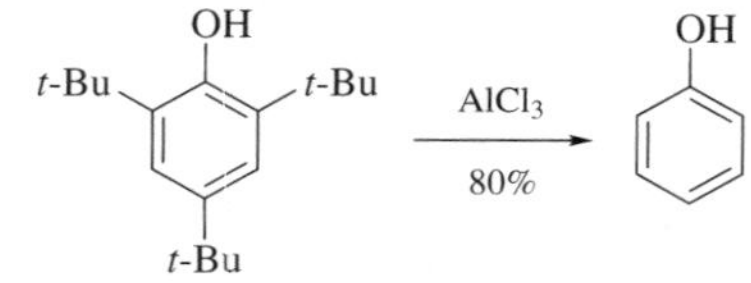

OH *t*-Bu *t*-Bu *t*-Bu $AlCl_3$ 80% OH (40)

R, $NHCO_2PNB$, S, N, O, CO_2PNB —$AlCl_3$, 24–82%→ R, NH_2, S, N, O, CO_2H (41)

Related Reagents. Antimony(V) Fluoride; Boron Trifluoride Etherate; Diethylaluminum Chloride; Dimethylaluminum Chloride; Ethylaluminum Dichloride; Methylaluminum Dichloride; Tin(IV) Chloride; Titanium(IV) Chloride; Triethylaluminum; Trimethylaluminum.

1. (a) Thomas, C. A. *Anhydrous Aluminum Chloride in Organic Chemistry*; ACS Monograph Series; Reinholdt: New York, 1941. (b) Shine, H. J. *Aromatic Rearrangements*; Elsevier: Amsterdam, 1967. (c) *FF* **1967**, *1*, 24. (d) Olah, G. A. *Friedel–Crafts Chemistry*; Wiley: New York, 1973. (e) Roberts, R. M.; Khalaf, A. A. *Friedel–Crafts Alkylation Chemistry*; Marcel Dekker: New York, 1984.
2. Gore, P. H. *CR* **1955**, *55*, 229.
3. Pearson, D. E.; Buehler, C. A. *S* **1972**, 533.
4. (a) Tan, L. K.; Brownstein, S. *JOC* **1982**, *47*, 4737. (b) Tan, L. K.; Brownstein, S. *JOC* **1983**, *48*, 3389.
5. Drago, R. S.; Getty, E. E. *JACS* **1988**, *110*, 3311.
6. McClure, D. E.; Arison, B. H.; Jones, J. H.; Baldwin, J. J. *JOC* **1981**, *46*, 2431.
7. Piccolo, O.; Spreafico, F.; Visentin, G.; Valoti, E. *JOC* **1985**, *50*, 3945.
8. Pinnick, H. W.; Brown, S. P.; McLean, E. A.; Zoller, L. W. *JOC* **1981**, *46*, 3758.
9. Jaxa-Chamiel, A.; Shah, V. P.; Kruse, L. I. *JCS(P1)* **1989**, 1705.
10. (a) Pollak, A.; Stanovnik, B.; Tisler, M. *JOC* **1966**, *31*, 4297. (b) Coates, W. J.; McKillop, A. *JOC* **1990**, *55*, 5418.
11. Takeuchi, H.; Maeda, M.; Mitani, M.; Koyama, K. *CC* **1985**, 287.
12. Takeuchi, H.; Shiobara, Y.; Kawamoto, H.; Koyama, K. *JCS(P1)* **1990**, 321.
13. (a) Puck, R.; Mayr, H.; Rubow, M.; Wilhelm, E. *JACS* **1986**, *108*, 7767. (b) Brownstein, S.; Morrison, A.; Tan, L. K. *JOC* **1985**, *50*, 2796.
14. Mayr, H.; Striepe, W. *JOC* **1983**, *48*, 1159.
15. (a) Ansell, M. F.; Ducker, J. W. *JCS* **1960**, 5219. (b) Cantrell, T. S. *JOC* **1967**, *32*, 1669. (c) Groves, JK. *CSR* **1972**, *1*, 73. (d) House, H. O. *Modern Synthetic Reactions*; Benjamin-Cummings: Menlo Park, CA, 1972; pp 786–816.
16. (a) Martin, G. J.; Rabiller, C.; Mabon, G. *TL* **1970**, 3131. (b) Rizzo, C. J.; Dunlap, N. A.; Smith, A. B. *JOC* **1987**, *52*, 5280.
17. Maroni, R.; Melloni, G.; Modena, G. *JCS(P1)* **1974**, 353.
18. Maroni, R.; Melloni, G.; Modena, G. *JCS(P1)* **1973**, 2491.
19. Trost, B. M.; Ghadiri, M. R. *JACS* **1984**, *106*, 7260.
20. (a) Eaborn, C. *JCS* **1956**, 4858. (b) Habich, D.; Effenberger, F. *S* **1979**, 841.
21. (a) Fleming, I.; Pearce, A. *CC* **1975**, 633. (b) Fristad, W. E.; Dime, D. S.; Bailey, T. R.; Paquette, L. A. *TL* **1979**, 1999.
22. (a) Walton, D. R. M.; Waugh, F. *JOM* **1972**, *37*, 45. (b) Newman, H. *JOC* **1973**, *38*, 2254.
23. Karpf, M. *TL* **1982**, *23*, 4923.
24. Casara, P.; Metcalf, B. W. *TL* **1978**, 1581.
25. Flood, T.; Peterson, P. E. *JOC* **1980**, *45*, 5006.
26. Urabe, H.; Kuwajima, I. *JOC* **1984**, *49*, 1140.
27. Ferrario, E. *BSF* **1911**, *9*, 536.
28. (a) Blatt, A. H. *OR* **1942**, *1*, 342. (b) Gammill, R. B. *TL* **1985**, *26*, 1385.
29. Truce, W. E. *OR* **1957**, *9*, 37.
30. Berliner, E. *OR* **1949**, *5*, 229.
31. Spoerri, P. E.; Dubois, A. S. *OR* **1949**, *5*, 387.
32. Effenberger, F.; Gleiter, R. *CB* **1964**, *97*, 472.
33. Gless, R. D. *SC* **1986**, *16*, 633.
34. Sethna, S.; Phadke, R. *OR* **1953**, *7*, 1.
35. Elderfield, R. C.; McCarthy, J. R. *JACS* **1951**, *73*, 975.
36. Clowes, G. A. *JCS(P1)* **1968**, 2519.
37. Sumpter, W. C. *CR* **1944**, *34*, 393.
38. Newman, M. S.; Wood, L. L. *JACS* **1959**, *81*, 6450.
39. Tolbert, L. M.; Ali, M. B. *JACS* **1984**, *106*, 3806.
40. (a) Cohen, N.; Banner, B. L.; Eichel, W. F. *SC* **1978**, *8*, 427. (b) Fringuelli, F.; Pizzo, F.; Taticchi, A.; Wenkert, E. *SC* **1979**, *9*, 391. (c) Ismail, Z. M.; Hoffmann, H. M. R. *JOC* **1981**, *46*, 3549. (d) Vidari, G.; Ferrino, S.; Grieco, P. A. *JACS* **1984**, *106*, 3539. (e) Angell, E. C.; Fringuelli, F.; Guo, M.; Minuti, L.; Taticchi, A.; Wenkert, E. *JOC* **1988**, *53*, 4325.
41. Ismail, Z. M.; Hoffmann, H. M. R. *AG(E)* **1982**, *21*, 859.
42. LeCoz, L.; Wartski, L.; Seyden-Penne, J.; Chardin, P.; Nierlich, M. *TL* **1989**, *30*, 2795.
43. Jung, M. E.; Haleweg, K. M. *TL* **1981**, *22*, 2735.
44. (a) Snider, B. B.; Rodini, D. J.; Conn, R. S. E.; Sealfon, S. *JACS* **1979**, *101*, 5283. (b) Mehta, G.; Reddy, A. V. *TL* **1979**, 2625. (c) Snider, B. B. *ACR* **1980**, *13*, 426.
45. (a) Bingham, R. C.; Schleyer, P. R. *Top. Curr. Chem.* **1971**, *18*, 1. (b) McKervey, M. A. *CSR* **1974**, *3*, 479. (c) McKervey, M. A. *T* **1980**, *36*, 971.
46. Kakiuchi, K.; Ue, M.; Tsukahara, H.; Shimizu, T.; Miyao, T.; Tobe, Y.; Odaira, Y.; Yasuda, M.; Shima, K. *JACS* **1989**, *111*, 3707.
47. (a) Deleris, G.; Donogues, J.; Calas, R. *TL* **1976**, 2449. (b) Pillot, J.-P.; Donogues, J.; Calas, R. *TL* **1976**, 1871.
48. Coppi, L.; Ricci, A.; Taddai, M. *JOC* **1988**, *53*, 911.
49. Yamamoto, Y.; Maruyama, K. *JOC* **1983**, *48*, 1564.
50. Lewis, N.; Morgan, I. *SC* **1988**, *18*, 1783.
51. Ohtani, M.; Watanabe, F.; Narisada, M. *JOC* **1984**, *49*, 5271.
52. Node, M.; Kawabata, T.; Ohta, K.; Fujimoto, M.; Fujita, E.; Fuji, K. *JOC* **1984**, *49*, 3641.
53. Node, M.; Kawabata, T.; Ueda, M.; Fujimoto, M.; Fuji, K.; Fujita, E. *TL* **1982**, *23*, 4047.
54. Fuji, K.; Kawabata, T.; Node, M.; Fujita, E. *JOC* **1984**, *49*, 3214.
55. Node, M.; Nishide, K.; Ochiai, M.; Fuji, K.; Fujita, E. *JOC* **1981**, *46*, 5163.
56. Watson, W. D. *JOC* **1985**, *50*, 2145.
57. Eisch, J. J.; Liu, Z.-R.; Ma, X.; Zheng, G.-X. *JOC* **1992**, *57*, 5140.
58. Olah, G. A.; Farooq, O.; Wang, Q.; Wu, A.-H. *JOC* **1990**, *55*, 1224.

Paul Galatsis
University of Guelph, Ontario, Canada

Antimony(V) Fluoride[1]

[7783-70-2] F_5Sb (MW 216.74)

(one of the strongest Lewis acids,[1] used as catalyst for Friedel–Crafts reactions, isomerization, and other acid related chemistry;[1] an efficient acid system for preparation of carbocations and onium ions as well as their salts;[1] a fluorinating agent, and a strong oxidant)

Alternate Name: antimony pentafluoride.
Physical Data: bp 149.5 °C.
Solubility: SbF_5 reacts with most organic solvents, forming solids with ether, acetone, carbon disulfide, and petroleum ether. SbF_5 is soluble in SO_2 and SO_2ClF.
Form Supplied in: viscous liquid; commercially available.
Handling, Storage, and Precautions: SbF_5 is extremely corrosive, toxic, and moisture sensitive. It can be purified by distillation. SbF_5 fumes when exposed to atmosphere. It should be stored under anhydrous conditions in a Teflon bottle and handled using proper gloves in a well ventilated hood.

Antimony pentafluoride is one of the strongest Lewis acids reported and is capable of forming stable conjugate superacid systems with HF and FSO_3H (see ***Hydrogen Fluoride-Antimony(V) Fluoride*** and ***Fluorosulfuric Acid–Antimony(V) Fluoride***).[1] Its complex with ***Trifluoromethanesulfonic Acid*** is, however, less stable and cannot be stored for extended periods of time. Nevertheless, CF_3SO_3H/SbF_5 is also a useful acid system when prepared in situ.[2] The most important properties of SbF_5 include its high acidity, strong oxidative ability, and great tendency to form stable anions.[3] The chemistry of SbF_5 is mainly characterized by these properties. The major applications of SbF_5 in organic synthesis include oxidation, fluorination, and as a catalyst for Friedel–Crafts type reactions and other acid related chemistry, and as a medium for preparation of carbocations and onium ions.[1]

Friedel–Crafts and Related Chemistry. The use of SbF_5 as a Friedel–Crafts catalyst has significantly expanded the scope of these reactions.[3] This is made possible through the strong Lewis acidity and oxidative ability of SbF_5. Influenced by the strong acidity, compounds of otherwise very weak nucleophilicity such as perfluorocarbons can react readily with aromatics. Furthermore, the strong oxidative ability of SbF_5 enables unconventional substrates such as methane to be oxidized to form positively charged species, which in turn can be applied as electrophilic reagents. On the other hand, these factors also considerably restrict the use of SbF_5 in organic synthesis since the high reactivity makes it a less selective Lewis acid for many reactions.

Alkylation of arenes proceeds readily under SbF_5 catalysis.[3] In addition to alkyl halides, alkyl esters and haloesters have also been applied to alkylate arenes under the reaction conditions.[4] Perfluoro or perchlorofluoro compounds have been similarly used as alkylating agents in the presence of SbF_5.[5] For example, perfluorotoluene reacts with pentafluorobenzene to form perfluorodiphenylmethane in 68% yield after the reaction mixture is quenched with HF. When H_2O is used in the quenching step, perfluorobenzophenone is obtained in 93% yield (eq 1).[5a]

$C_6F_5CF_3$ → (1. C_6F_5H, SbF_5; 2. HF; 68%) → $C_6F_5CF_2C_6F_5$
$C_6F_5CF_3$ → (1. C_6F_5H, SbF_5; 2. H_2O; 93%) → $C_6F_5COC_6F_5$ (1)

Acylation of pentafluorobenzene to form ketones either with acid halides or with anhydrides has been achieved with an excess of SbF_5.[6] In the case of dicarboxylic acid dichlorides or anhydrides as acylating agents, diketones are obtained as the reaction products (eq 2). The reaction of the anhydrides of dicarboxylic acids with aromatic compounds does not stop at the keto acid stage, as is the case when other Lewis acids are used. Using phosgene in place of acid chlorides results in the formation of pentafluorobenzoic acid in good yield.[3]

$ClCO(CF_2)_nCOCl$ → (C_6F_5H, SbF_5, 20–74%) → $C_6F_5CO(CF_2)_nCOC_6F_5$ (2)

Perfluorobenzenium salt can be conveniently prepared from the reaction of 1,4-perfluorocyclohexadiene with SbF_5.[7a–c] In the presence of SbF_5, this salt is able to react with three equivalents of pentafluorobenzene to yield perfluoro-1,3,5-triphenylbenzene (eq 3).[7a–c] When 2,2′-di-*H*-octafluorobiphenyl is used as the substrate, perfluorotriphenylene is obtained in 50% yield (eq 4). Perfluoronaphthlenenium ion also reacts with polyfluorinated arenes in a similar fashion.[7a–c] The above reaction offers a facile approach for the preparation of these perfluorinated polynuclear aromatic compounds. Oxidation of perchlorobenzene with SbF_5 in the presence of pentafluorobenzene leads to the formation of coupling products.[7d]

$C_6F_7^+ \; Sb_nF_{5n+1}^-$ → (C_6F_5H; SbF_5, H_2O; 70%) → 1,3,5-$(C_6F_5)_3C_6F_3$ (3)

$C_6F_7^+ \; Sb_nF_{5n+1}^-$ + 2,2′-di-H-octafluorobiphenyl → (SbF_5, 50%) → perfluorotriphenylene (4)

Friedel–Crafts sulfonylation of aromatics with alkane- and arenesulfonyl halides and anhydrides has been studied.[8] Good yields of sulfones are generally obtained (eq 5). In the case of pentafluorobenzenesulfonyl fluoride with pentafluorobenzene, decafluorodiphenyl sulfone is formed along with decafluorodiphenyl.[8c] A convenient approach for synthesizing symmetrical aryl sulfones is to react aromatics with ***Fluorosulfuric Acid*** in the presence of SbF_5 (eq 6).[9] Certain phenylacetylenes react with SO_2 and benzene in the presence of SbF_5 to form benzothiophene *S*-oxides[10] (eq 7). In some cases, 1,1-diphenylvinylsulfinic acids were also obtained as side products of the reaction. Sulfinyl fluoride reacts with arenes similarly under the catalysis of SbF_5 to give sulfoxides (eq 8).[8c]

C_6H_5R → ($R'SO_2X$, SbF_5) → $RC_6H_4SO_2R'$ (5)

$X = Cl, F, R'SO_3$

Oxidation of elemental sulfur and selenium with SbF_5 leads to the formation of doubly charged polyatomic cations.[11] These cations are able to react with polyfluorinated arenes to form diaryl sulfides or selenides (eq 9).[12a–c] Under similar treatment, polyfluorodiaryl disulfides or diselenides also react with aromatics to form diaryl sulfides and selenides, respectively (eq 10).[12a,12b]

In the presence of SbF_5, inorganic halides such as NaCl and NaBr can serve as electrophilic halogenating agents.[5b,13] Even deactivated arenes such as 5*H*-nonafluoroindan can be brominated under the reaction conditions (eq 11).[13b]

Functionalization of alkanes has been achieved under SbF_5 catalysis.[3] Halogenation of alkanes is one of the widely studied reactions in this field.[14] A valuable synthetic procedure is the reaction of alkanes with methylene chloride (or bromide) in the presence of SbF_5 (eq 12).[14d] In these reactions, halonium ions are initially formed, which in turn abstract hydride from hydrocarbons. Quenching of the resulting carbocations with halides leads to the desired haloalkanes.

$$\mathrm{RH} \xrightarrow[64\text{–}88\%]{\mathrm{CH_2X_2,\ SbF_5}} \mathrm{RX} \qquad (12)$$

RH = secondary, tertiary alkanes
X = Cl, Br

In the presence of SbF_5, alkyl chlorides are ionized to form carbocations, which can be trapped with CO.[15] Quenching the reaction intermediates with water or alcohols yields the corresponding carboxylic acids or esters. For instance, halogenated trishomobarrelene reacts with CO in SbF_5/SO_2ClF to give, after treatment with alcohol and water, a mixture of acid and ether (eq 13).[15a]

Studies have been carried out on the alkylation of alkenes in the presence of SbF_5, especially fluoroalkenes.[16] For example, treatment of 1,1,1-trifluoroethane with SbF_5 in the presence of tetrafluoroethylene yields 90% of 1,1,1,2,2,3,3-heptafluorobutane.[16c] Perfluoroallyl or -benzoyl compounds with varying structures have also been utilized as alkyating agents (eqs 14 and 15).[16b,e–h] At elevated temperatures, intramolecular alkylation was observed in certain cases with perfluorodienes, forming perfluorocyclopentenes or perfluorocyclobutenes (eq 16).[17]

Acylation of alkenes can also be similarly effected with SbF_5.[18] Reaction of acetyl fluoride with trifluoroethylene produces 1,1,1,2-tetrafluorohexane-3,5-dione in 40% yield.[18a] When antimony pentafluoride is used in excess (more than 6 molar equivalents), 1,1,1,2-tetrafluorobutan-3-one is formed as a side product. Benzoyl fluoride reacts with difluoroethylene to form the expected ketone in 39% yield.[18a] α,β-Unsaturated carboxylic acid fluorides have also been used in reactions with perfluoroalkenes.[18b] In these cases, α,β-unsaturated ketones are obtained (eq 17). Enol acetates of perfluoroisopropyl methyl ketone and perfluoro-*t*-butyl methyl ketone react with acetyl fluoride to provide the corresponding β-diketones (eq 18).[18c]

SbF_5, MeCOF; 32–35%, 32–40% (18)

Oxygenation of dienes with molecular oxygen to form Diels–Alder-type adducts can be effected by Lewis acids and some salts of stable carbenium ions.[19] In the case of ergosteryl acetate, SbF_5 is by far the most active catalyst (eq 19).[19a]

SbF_5, O_2, 97% (19)

Isomerization and Rearrangement. Isomerization of perfluoroalkenes can be realized with SbF_5 catalysis.[20] The terminal carbon–carbon bonds of these alkenes are usually moved to the 2-position under the influence of this catalyst (eq 20). A further inward shift generally occurs only if H or Cl atoms are present in the 4-position of the alkenes. As a rule, the isomerization leads to the predominant formation of the *trans* isomers. Terminal fluorodienes also isomerize exothermally into dienes containing internal double bonds in the presence of SbF_5. With a catalytic amount of SbF_5, perfluoro-1,4-cyclohexadiene disproportionates to hexafluorobenzene and perfluorocyclohexene. The disproportionation proceeds intramolecularly in the case of perfluoro-1,4,5,8-tetrahydronaphthalene. The starting material is completely converted to perfluorotetralin (eq 21).[20f]

$$R_FCF_2FC{=}CF_2 \xrightarrow{SbF_5} R_FFC{=}CFCF_3 \quad (20)$$

SbF_5, SO_2ClF, –90 to –80 °C; 100% (21)

Like most Lewis acids, SbF_5 promotes the rearrangement of epoxides to carbonyl compounds,[21] and SbF_5 is an efficient catalyst for this reaction. However, the migratory aptitude of substitutent groups in the reaction under SbF_5 catalysis is much less selective compared to that promoted by weak Lewis acids such as ***Methylaluminum Bis(4-bromo-2,6-di-t-butylphenoxide)***.[21a] Nevertheless, SbF_5 is well suited for the rearrangement of perfluoroepoxides.[21b–e] Excellent yields of ketones are obtained from the reaction. When diepoxides are used, diketones are obtained as the reaction products (eq 22).[21c]

SbF_5 → $CF_3CO(CF_2)_2COCF_3$ (22)

Fluorination and Transformation of Fluorinated Compounds. SbF_5 is a strong fluorinating agent. However, its use is largely limited to the preparation of perfluoro- or polyfluoroorganic compounds.[22] SbF_5 has also been used for transformation of perfluoro compounds.[22]

When hexachlorobenzene is subjected to SbF_5 treatment, 44% of 1,2-dichloro-3,3,4,4,5,5,6,6-octafluorocyclohex-1-ene is obtained as the major product of the reaction (eq 23).[23] Using $SbCl_5$ as reaction solvent leads to better reaction control, resulting in an increase in the product yield.[23c] In the case of polyfluoronaphthalenes, polyfluorinated tetralins are obtained.[23d] Treatment of hexafluoro-2-trichloromethyl-2-propanol with SbF_5 gives perfluoro-*t*-butyl alcohol in 92% yield (eq 24).[24]

SbF_5, 100 °C, 44% (23)

$$(F_3C)_2C(CCl_3)OH \xrightarrow[92\%]{SbF_5} (F_3C)_3COH \quad (24)$$

Hydrolysis of CF_3 groups to CO_2H can be induced by SbF_5.[25a] In this reaction, the CF_3-bearing compounds are first treated with SbF_5 and the resulting reaction mixtures are subsequently quenched with water to form the desired products. For example, perfluorotoluene was converted to pentafluorobenzoic acid in 86% yield by this procedure. When perfluoroxylenes and perfluoromesitylene go through the same treatment, the corresponding di- or triacids are obtained in high yields. It is also possible to partially hydrolyze perfluoroxylenes and perfluoromesitylene through a stepwise procedure and to isolate intermediate products. The hydrolysis procedure is also applicable to other halogen-containing compounds.[25b]

Under SbF_5 catalysis, ***Trifluoromethanesulfonic Anhydride*** is readily decomposed to trifluoromethyl triflate in high yield (eq 25).[26] This method is the most convenient procedure reported for preparation of the triflate.

$$(CF_3SO_2)_2O \xrightarrow[94\%]{SbF_5,\ 25\ °C} CF_3SO_3CF_3 \quad (25)$$

Preparation of Carbocations, Onium Ions, and Their Salts. SbF_5 is a preferred medium for the preparation of carbocations and onium ions.[1] In fact, the first observation of stable carbocations was achieved in this medium.[1,27] By dissolving *t*-butyl fluoride in an excess of SbF_5, the *t*-butyl cation was obtained (eq 26). Subsequently, many alkyl cations have been obtained in SbF_5 (either neat or diluted with SO_2, SO_2ClF, or SO_2F_2). The 2-norbornyl cation, one of the most controversial ions in the history of physical organic chemistry,[28] was prepared from *exo*-2-fluoronorbornane in SbF_5/SO_2 (or SO_2ClF) solution (eq 27).[1] Bridgehead cations such as 1-adamantyl, 1-trishomobarrelyl, and 1-trishomobullvalyl cations have also been similarly prepared with the use of SbF_5.[15a,29]

SbF_5 → $Sb_nF_{5n+1}^-$ (26)

SbF_5, SO_2ClF → $Sb_nF_{5n+1}^-$ (27)

Carbodications have also been studied.[1] One convenient way of preparing alkyl dications is to ionize dihalides with SbF_5 in

SO_2ClF (eq 28). In these systems, separation of the two cation centers by at least two methylene groups is necessary for the ions to be observable.[1] Aromatic dications are usually prepared by oxidizing the corresponding aromatic compounds with SbF_5 (eq 29).[1,30] In the case of pagodane containing a planar cyclobutane ring, oxidation leads to the formation of cyclobutane dication which was characterized as a frozen two-electron Woodward–Hoffmann transition state model (eq 30).[31]

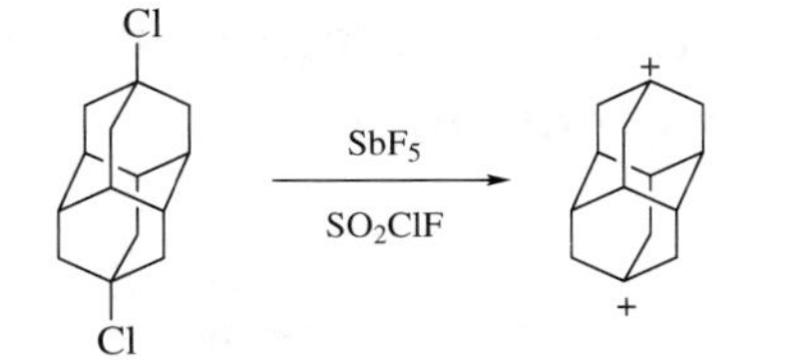

SbF_5, SO_2ClF (28)

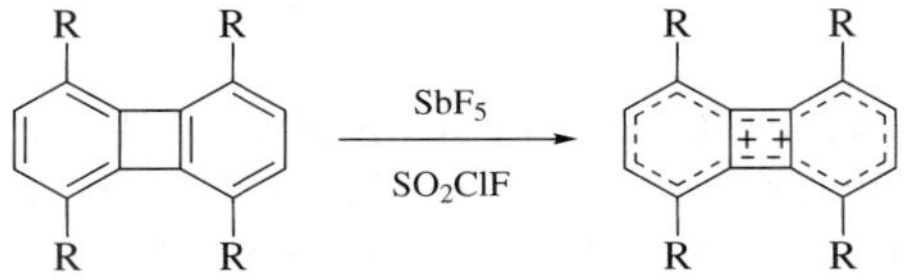

SbF_5, SO_2ClF (29)

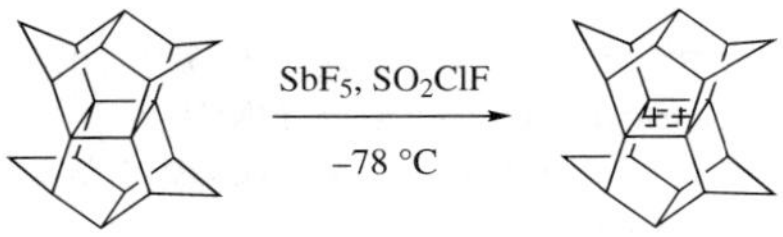

SbF_5, SO_2ClF, −78 °C (30)

SbF_5 has also been found useful in the preparation of homoaromatic cations.[1,32] For example, the simplest 2π monohomoaromatic cations, the homocyclopropenyl cations, can be prepared from corresponding 3-halocyclobutenes in SbF_5 (eq 31).[32a]

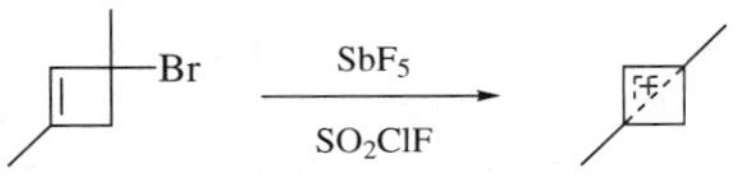

SbF_5, SO_2ClF (31)

Perfluorocarbocations are generally prepared with the aid of SbF_5.[33] Perfluorobenzyl cations and perfluorinated allyl cations are among the most well studied perfluorinated carbocations. It was reported that in the perfluoroallyl cations containing a pentafluorophenyl group at the 1- or 2-position, the phenyl groups were partially removed from the plane of the allyl triad (eq 32).[33b]

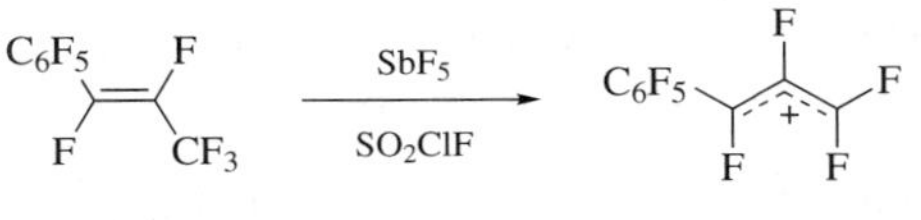

SbF_5, SO_2ClF (32)

Use of SbF_5 as ionizing agent offers a convenient route to acyclic and cyclic halonium ions from alkyl halides.[1] Three-membered ring cyclic halonium ions are important intermediates in the electrophilic halogenations of carbon–carbon double bonds.[1] More recently, a stable 1,4-bridged bicyclic bromonium ion, 7-bromoniabicyclo[2.2.1]heptane, has been prepared through the use of SbF_5 involving an unprecedented transannular participation in a six-membered ring (eq 33).[34]

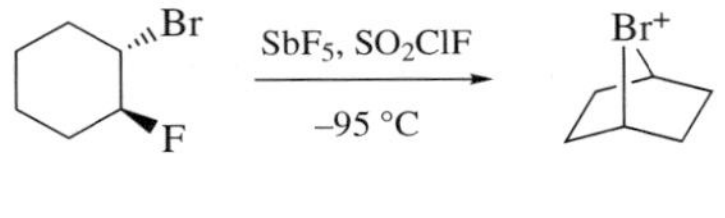

SbF_5, SO_2ClF, −95 °C (33)

Other Reactions. Reaction of α,β-unsaturated carbonyl compounds with diazo compounds generally gives low yields of cyclopropyl compounds. The rapid formation of 1-pyrazolines and their subsequent rearrangement products 2-pyrazolines is the reason for the inefficiency of cyclopropanation of these substrates. However, in the presense of SbF_5, the cyclopropanation of α,β-unsaturated carbonyl compounds with diazocarbonyl compounds proceeds very well to produce the desired products in good yields (eq 34).[35]

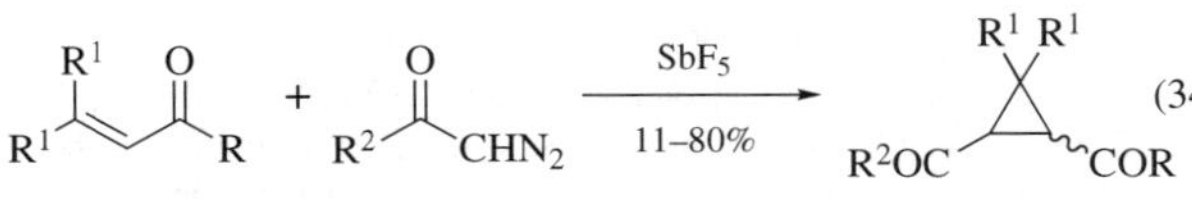

SbF_5, 11–80% (34)

R = H, OMe; R^1 = H, Me; R^2 = Ph, OEt

Related Reagents. Fluorosulfuric Acid–Antimony(V) Fluoride; Hydrogen Fluoride–Antimony(V) Fluoride; Methyl Fluoride–Antimony(V) Fluoride.

1. Olah, G. A.; Prakash, G. K. S.; Sommer, J. *Superacids*; Wiley: New York, 1985.
2. For example, see (a) Ohwada, T.; Yamagata, N.; Shudo, K. *JACS* **1991**, *113*, 1364. (b) Farooq, O.; Farnia, S. M. F.; Stephenson, M.; Olah, G. A. *JOC* **1988**, *53*, 2840. (c) Carre, B.; Devynck, J. *Anal. Chim. Acta.* **1984**, *159*, 149. (d) Choukroun, H.; Germain, A.; Brunel, D.; Commeyras, A. *NJC* **1983**, *7*, 83.
3. Yakobson, G. G.; Furin, G. G. *S* **1980**, 345.
4. (a) Olah, G. A.; Nishimura, J. *JACS* **1974**, *96*, 2214. (b) Booth, B. L.; Haszeldine, R. N.; Laali, K. *JCS(P1)* **1980**, 2887.
5. (a) Pozdnyakovich, Y. V.; Shteingarts, V. D. *JFC* **1974**, *4*, 317. (b) Brovko, V. V.; Sokolenko, V. A.; Yakobson, G. G. *JOU* **1974**, *10*, 300.
6. Furin, G. G.; Yakobson, G. G.; *Izv. Sib. Otd. Akad. Nauk SSSR, Ser. Khim. Nauk* **1974**, 78 (*CA* **1974**, *80*, 120 457c.)
7. (a) Shteingarts, V. D. In *Synthetic Fluorine Chemistry*, Olah, G. A.; Chambers, R. D.; Prakash, G. K. S., Eds.; Wiley: New York, 1992; Chapter 12. (b) Pozdnyakovich, Y. V.; Shteingarts, V. D. *JOU* **1977**, *13*, 1772. (c) Pozdnyakovich, Y. V.; Chuikova, T. V.; Bardin, V. V.; Shteingarts, V. D. *JOU* **1976**, *12*, 687. (d) Bardin, V. V.; Yakobson, G. G. *IZV* **1976**, 2350.
8. Olah, G. A.; Kobayashi, S.; Nishimura, J. *JACS* **1973**, *95*, 564. (b) Olah, G. A.; Lin, H. C. *S* **1974**, 342. (c) Furin, G. G.; Yakobson, G. G.; *Izv. Sib. Otd. Akad. Nauk SSSR, Ser. Khim. Nauk* **1976**, 120 (*CA* **1976**, *85*, 77 801z.)
9. Tanaka, M.; Souma, Y. *JOC* **1992**, *57*, 3738.
10. (a) Fan, R.-L.; Dickstein, J. I.; Miller, S. I. *JOC* **1982**, *47*, 2466. (b) Miller, S. I.; Dickstein, J. I. *ACR* **1976**, *9*, 358.
11. Gillespie, R. J.; Peel, T. E. In *Advances in Physical Organic Chemistry*, Gold, V., Ed.; Academic Press: London, 1971; Vol. 9, p 1.
12. (a) Furin, G. G.; Terent'eva, T. V.; Yakobson, G. G. *JOU* **1973**, *9*, 2221. (b) Yakobson, G. G.; Furrin, G. G.; Terent'eva, T. V. *IZV* **1972,** 2128. (c) Yakobson, G. G.; Furin, G. G.; Terent'eva, T. V. *JOU* **1974**, *10*, 802. (d) Furin, G. G.; Schegoleva, L. N.; Yakobson, G. G. *JOU* **1975**, *11*, 1275.
13. (a) Lukmanov, V. G.; Alekseeva, L. A.; Yagupol'skii, L. M. *JOU* **1977**, *13*, 1979. (b) Furin, G. G.; Malyuta, N. G.; Platonov, V. E.; Yakobson, G. G. *JOU* **1974**, *10*, 832.
14. (a) Olah, G. A.; Mo, Y. K. *JACS* **1972**, *94*, 6864. (b) Olah, G. A.; Renner, R.; Schilling, P.; Mo, Y. K. *JACS* **1973**, *95*, 7686. (c) Halpern, Y. *Isr. J. Chem.* **1975**, *13*, 99. (d) Olah, G. A.; Wu, A.; Farooq, O. *JOC* **1989**, *54*, 1463.
15. (a) de Meijere, A.; Schallner, O. *AG(E)* **1973**, *12*, 399. (b) Farcasiu, D. *CC* **1977**, 394.

16. (a) Belen'kii, G. G.; Savicheva, G. I.; German, L. S. *IZV* **1978**, 1433. (b) Belen'kii, G. G.; Lur'e, E. P.; German, L. S. *IZV* **1976**, 2365. (c) Petrov, V. A.; Belen'kii, G. G.; German, L. S. *IZV* **1980**, 2117. (d) Karpov, V. M.; Mezhenkova, T. V.; Platonov, V. E.; Yakobson, G. G. *JOU* **1984**, *20*, 1220. (e) Petrov, V. A.; Belen'kii, G. G.; German, L. S.; Kurbakova, A. P.; Leites, L. A. *IZV* **1982**, 170. (f) Karpov, V. M.; Mezhenkova, T. V.; Platonov, V. E.; Yakobson, G. G. *IZV* **1985**, 2315. (g) Petrov, V. A.; Belen'kii, G. G.; German, L. S.; Mysov, E. I. *IZV* **1981**, 2098. (h) Petrov, V. A.; Belen'kii, G. G.; German, L. S. *IZV* **1981**, 1920.
17. (a) Petrov, V. A.; Belen'kii, G. G.; German, L. S. *IZV* **1982**, 2411. (b) Petrov, V. A.; Belen'kii, G. G.; German, L. S. *IZV* **1989**, 385. (c) Petrov, V. A.; German, L. S.; Belen'kii, G. G. *IZV* **1989**, 391.
18. (a) Belen'kii, G. G.; German, L. S.; *IZV* **1974**, 942. (b) Chepik, S. D.; Belen'kii, G. G.; Cherstkov, V. F.; Sterlin, S. R.; German, L. S. *IZV* **1991**, 513. (c) Knunyants, I. L.; Igumnov, S. M. *IZV* **1982**, 204.
19. (a) Barton, D. H. R.; Haynes, R. K.; Magnus, P. D.; Menzies, I. D. *CC* **1974**, 511. (b) Barton, D. H. R.; Haynes, R. K.; Leclerc, G.; Magnus, P. D.; Menzies, I. D. *JCS(P1)* **1975**, 2055. (c) Haynes, R. K. *AJC* **1978**, *31*, 131.
20. (a) Filyakova, T. I.; Belen'kii, G. G.; Lur'e, E. P.; Zapevalov, A. Y.; Kolenko, I. P.; German, L. S. *IZV* **1979**, 681. (b) Belen'kii, G. G.; Savicheva, G. I.; Lur'e, E. P.; German, L. S. *IZV* **1978**, 1640. (c) Filyakova, T. I.; Zapevalov, A. Y. *JOU* **1991**, *27*, 1605. (d) Petrov, V. A.; Belen'kii, G. G.; German, L. S. *IZV* **1989**, 385. (e) Chepik, S. D.; Petrov, V. A.; Galakhov, M. V.; Belen'kii, G. G.; Mysov, E. I.; German, L. S. *IZV* **1990**, 1844. (f) Avramenko, A. A.; Bardin, V. V.; Furin, G. G.; Karelin, A. I.; Krasil'nikov, V. A.; Tushin, P. P. *JOU* **1988**, *24*, 1298.
21. (a) Maruoka, K.; Ooi, T.; Yamamoto, H. *T* **1992**, *48*, 3303. (b) Zapevalov, A. Y.; Filyakova, T. I.; Kolenko, I. P.; Kodess, M. I. *JOU* **1986**, *22*, 80. (c) Filyakova, T. I.; Ilatovskii, R. E.; Zapevalov, A. Y. *JOU* **1991**, *27*, 1818. (d) Filyakova, T. I.; Matochkina, E. G.; Peschanskii, N. V.; Kodess, M. I.; Zapevalov, A. Y. *JOU* **1992**, *28*, 20. (e) Filyakova, T. I.; Matochkina, E. G.; Peschanskii, N. V.; Kodess, M. I.; Zapevalov, A. Y. *JOU* **1991**, *27*, 1423.
22. Yakobson, G. G.; Vlasov, V. M. *S* **1976**, 652.
23. (a) Leffler, A. J. *JOC* **1959**, *24*, 1132. (b) McBee, E. T.; Wiseman, P. A.; Bachman, G. B. *Ind. Eng. Chem.* **1947**, *39*, 415. (c) Christe, K. O.; Pavlath, A. E. *JCS* **1963**, 5549. (d) Pozdnyakovich, Y. V.; Shteingarts, V. D. *JOU* **1978**, *14*, 2069.
24. Dear, R. E. A. *S* **1970**, 361.
25. (a) Karpov, V. M.; Panteleev, I. V.; Platonov, V. E. *JOU* **1991**, *27*, 1932. (b) Karpov, V. M.; Platonov, V. E.; Yakoboson, G. G. *IZV* **1979**, 2082.
26. Taylor, S. L.; Martin, J. C. *JOC* **1987**, *52*, 4147.
27. Olah, G. A.; Tolgyesi, W. S.; Kuhn, S. J.; Moffatt, M. E.; Bastien, I. J.; Baker, E. B. *JACS* **1963**, *85*, 1328.
28. For reviews, see (a) Olah, G. A.; Prakash, G. K. S.; Saunders, M. *ACR* **1983**, *16*, 440. (b) Brown, H. C. *ACR* **1983**, *16*, 432. (c) Walling, C. *ACR* **1983**, *16*, 448.
29. (a) Schleyer, P. v. R.; Fort, R. C. Jr.; Watts, W. E.; Comisarow, M. B.; Olah, G. A. *JACS* **1964**, *86*, 4195. (b) Olah, G. A.; Prakash, G. K. S.; Shih, J. G.; Krishnamurthy, V. V.; Mateescu, G. D.; Liang, G.; Sipos, G.; Buss, V.; Gund, T. M.; Schleyer, P. v. R. *JACS* **1985**, *107*, 2764.
30. (a) Mills, N. S. *JOC* **1992**, *57*, 1899. (b) Krusic, P. J.; Wasserman, E. *JACS* **1991**, *113*, 2322. (c) Mullen, K.; Meul, T.; Schade, P.; Schmickler, H.; Vogel, E. *JACS* **1987**, *109*, 4992. (d) Lammertsma, K.; Olah, G. A.; Berke, C. M.; Streitwieser, A. Jr. *JACS* **1979**, *101*, 6658. (e) Olah, G. A.; Lang, G.; *JACS* **1977**, *99*, 6045.
31. Prakash, G. K. S.; Krishnamurthy, V. V.; Herges, R.; Bau, R.; Yuan, H.; Olah, G. A.; Fessner, W-D.; Prinzbach, H. *JACS* **1988**, *110*, 7764.
32. (a) Olah, G. A.; Staral, J. S.; Spear, R. J.; Liang, G. *JACS* **1975**, *97*, 5489. (b) Olah, G. A.; Staral, J. S.; Paquette, L. A. *JACS* **1976**, *98*, 1267.
33. (a) Pozdnyakovich, Y. V.; Shteingarts, V. D. *JFC* **1974**, *4*, 283. (b) Galakhov, M. V.; Petrov, V. A.; Chepik, S. D.; Belen'kii, G. G.; Bakhmutov, V. I.; German, L. S. *IZV* **1989**, 1773. (c) Petrov, V. A.; Belen'kii, G. G.; German, L. S. *IZV* **1984**, 438.
34. Prakash, G. K. S.; Aniszfeld, R.; Hashimoto, T.; Bausch, J. W.; Olah, G. A. *JACS* **1989**, *111*, 8726.
35. Doyle, M. P.; Buhro, W. E.; Dellaria, J. F. Jr. *TL* **1979**, 4429.

George A. Olah, G. K. Surya Prakash, Qi Wang & Xing-ya Li
University of Southern California, Los Angeles, CA, USA

Azidotris(dimethylamino)phosphonium Hexafluorophosphate

$$(Me_2N)_3\overset{+}{P}{-}N_3 \quad PF_6^-$$

[50281-51-1] $C_6H_{18}F_6N_6P_2$ (MW 350.19)

(coupling reagent for amides and peptide synthesis via an acyl azide;[1] in the absence of a nucleophile undergoes a Curtius rearrangement to give the corresponding isocyanate;[1,2] by irradiation gives the iminophosphonium salt;[3,4] diazo transfer reagent[5])

Physical Data: mp >250 °C; IR, $\nu(N_3)$ 2176 cm^{-1}.
Solubility: sol acetone, acetonitrile, DMF.
Form Supplied in: crystalline colorless solid; not available from commercial sources.
Preparative Methods: ***Bromine*** (1 equiv) is added dropwise to ***Hexamethylphosphorous Triamide*** (1 equiv) in dry diethyl ether. After the addition is complete, ***Potassium Hexafluorophosphate*** (1 equiv), dissolved in H_2O is added, and the solid formed is filtered and washed with H_2O. After drying under vacuum, the solid is dissolved in acetone and ***Sodium Azide*** (1.5 equiv) is added. The mixture is stirred overnight at 25 °C, then the sodium bromide formed and the excess of sodium azide is filtered off and the solvent is removed under vacuum.
Purification: by crystallization in a mixture of acetone and diethyl ether.
Handling, Storage, and Precautions: is very stable, not hygroscopic, and can be stored indefinitely under vacuum. The related bromine derivative is an exceptionally safe azide, that does not detonate by shock, friction, rapid heating, or even flame.[5] When used in combination with carboxylic acids, the potential carcinogenic ***Hexamethylphosphoric Triamide*** is a side-product and the reagent must be handled with caution.[6]

Coupling Reagent. Azidotris(dimethylamino)phosphonium hexafluorophosphate has been used as a condensing reagent for the preparation of peptides.[1] Reaction of α-amino protected amino acids or dipeptides with equimolar amounts of α-carboxyl protected amino acids or dipeptides and ***Triethylamine*** (2 equiv) in DMF proceed cleanly overnight at −10 to 0 °C, giving the corresponding peptides in good yields (81–97%) (eq 1). Neither acylation of the hydroxy group of serine nor dehydration of the side chain of asparagine is reported during the coupling.[1]

$$RCO_2H + R'NH_2 \xrightarrow[\substack{Et_3N \\ 81\text{–}97\%}]{(Me_2N)_3\overset{+}{P}\text{–}N_3\ PF_6^-} RCONHR' \quad (1)$$

Racemization from azidotris(dimethylamino)phosphonium hexafluorophosphate-mediated couplings is analyzed by different methods. While no racemization is detected by the Anderson[7] and Weinstein[8] methods, the Young[9] method gave some degree (3%) of racemization.

Although azidotris(dimethylamino)phosphonium hexafluorophosphate has not been used in solid-phase peptide synthesis,[10] other commercial phosphonium salt derivatives such as ***Benzotriazol-1-yloxytris(dimethylamino)phosphonium Hexafluorophosphate*** (BOP),[11] benzotriazol-1-yloxytris(pyrrolidino)phosphonium hexafluorophosphate (PyBOP),[12] or bromotris(pyrrolidino)phosphonium hexafluorophosphate (PyBroP)[13] are commonly used in that strategy.

Isocyanate Formation. The coupling reaction described above occurs via an acyl azide intermediate, which at 50 °C suffers a Curtius rearrangement to give an isocyanate (eq 2),[1] which can further react with amines to give the corresponding ureas.

$$\text{Ar(D)-CO}_2\text{H} \xrightarrow[\text{Et}_3\text{N}]{(\text{Me}_2\text{N})_3\overset{+}{\text{P}}\text{-N}_3\ \text{PF}_6^-} \text{Ar(D)-CON}_3 \xrightarrow{50\ °\text{C}} \text{Ar(D)-NCO}\quad 78\% \qquad (2)$$

Iminophosphonium Salt Formation. Irradiation of azidotris(dimethylamino)phosphonium hexafluorophosphate in acetonitrile at 254 nm using a Rayonnet photochemical reactor for 15 h at 25 °C gives the iminophosphonium salt with release of nitrogen (eq 3). This is the first example reported of a Curtius-type rearrangement involving a charged atom.[3,4]

$$(\text{Me}_2\text{N})_3\overset{+}{\text{P}}\text{-N}_3\ \text{PF}_6^- \xrightarrow[-\text{N}_2]{h\nu} (\text{Me}_2\text{N})_2\overset{+}{\text{P}}\text{=N-NMe}_2\ \text{PF}_6^- \leftrightarrow (\text{Me}_2\text{N})_2\text{P-}\overset{+}{\text{N}}\text{=NMe}_2\ \text{PF}_6^- \qquad (3)$$

Diazo Transfer Reagent. The related compound azidotris(diethylamino)phosphonium bromide has been used for the preparation of diazo products.[5] Active methylene compounds react with azidotris(diethylamino)phosphonium bromide in dry diethyl ether in the presence of only catalytic amounts of base, such as ***Potassium t-Butoxide*** or other alkoxide, yielding the corresponding diazo products in good yields (70–78%) (eq 4).

$$\text{MeCOCH}_2\text{COMe} \xrightarrow[\text{base, 70–78\%}]{(\text{Et}_2\text{N})_3\overset{+}{\text{P}}\text{-N}_3\ \text{Br}^-} \text{MeCOC(N}_2\text{)COMe} \qquad (4)$$

1. Castro, B.; Dormoy J. R. *BSF(2)* **1973**, *12*, 3359.
2. Masson, M. A.; Dormoy J. R. *J. Labelled Compd. Radiopharm.* **1979**, *16*, 785.
3. Majoral, J. P.; Bertrand, G.; Baceiredo, A.; Mulliez, M.; Schmutzler, R. *PS* **1983**, *18*, 221.
4. Mulliez, M.; Majoral, J. P.; Bertrand, G. *JCS(C)* **1984**, 284.
5. McGuiness, M.; Shechter, H. *TL* **1990**, *31*, 4987.
6. Zapp, J. A., Jr. *Science* **1975**, *190*, 422.
7. (a) Anderson, G. W.; Young, R. W. *JACS* **1952**, *74*, 5307. (b) Anderson, G. W.; Zimmermann, J. E.; Callanan, F. M. *JACS* **1967**, *89*, 5012.
8. Weinstein, B.; Pritchard, A. E. *JCS(C)* **1972**, 1015.
9. Williams, M. W.; Young, G. T. *JCS(C)* **1963**, 881.
10. Fields, G. B.; Tian, Z.; Barany, G. In *Synthetic Peptides: A User's Guide*; Grant, G. A., Ed.; Freeman: New York, 1992; p 77.
11. Castro, B.; Dormoy J. R.; Evin, G. *TL* **1975**, 1219.
12. Coste, J.; Le-Nguyen, D.; Castro, B. *TL* **1990**, *31*, 205.
13. Frérot, E.; Coste, J.; Pantaloni, A.; Dufour, M. N.; Jouin, P. *T* **1991**, *47*, 259.

Fernando Albericio & Steven A. Kates
Millipore Corporation, Bedford, MA, USA

Azobisisobutyronitrile

$$\text{Me}_2\text{C(CN)-N=N-C(CN)Me}_2$$

[78-67-1] $C_8H_{12}N_4$ (MW 164.21)

(reagent for initiation of radical reactions)

Alternate Name: AIBN.
Physical Data: mp 103–104 °C.
Solubility: sol benzene, toluene.
Form Supplied in: commercially available as a white powder.
Preparative Method: prepared from the corresponding hydrazine by oxidation with nitrous acid.[1]
Purification: recrystallized from ether.
Handling, Storage, and Precautions: avoid exposure of the reagent to light and heat; store protected from light in a refrigerator or freezer. AIBN is harmful; old or incorrectly stored samples may also present an explosion risk.

Preparation. Azobisisobutyronitrile (AIBN) and analogs can be prepared by a two-step procedure involving initial treatment of a ketone with ***Hydrazine*** in the presence of ***Sodium Cyanide***, followed by oxidation of the so-formed hydrazine (eq 1).[1,2]

$$\text{Me}_2\text{C=O} \xrightarrow[\text{2. [O]}]{\text{1. NH}_2\text{NH}_2\text{, NaCN}} \text{Me}_2\text{C(CN)-N=N-C(CN)Me}_2 \qquad (1)$$

Use in Radical Chemistry. AIBN is the initiator most commonly employed in radical reactions, especially those using ***Tri-n-butylstannane***. At temperatures above about 60 °C, AIBN decomposes with evolution of nitrogen to generate isobutyronitrile radicals, which are able to initiate radical chemistry, usually by abstraction of a hydrogen atom from Bu_3SnH or some other donor. The useful working temperature range of AIBN is 60–120 °C, with the half-life at 80 °C being 1 h. In many applications involving

C–C bond formation, the radical chemistry is designed to proceed by a chain reaction, and so only small amounts (typically ca. 0.1 equiv) of the initiator is required. For best results, especially at more elevated temperatures, AIBN is best added in portions or as a solution by means of a syringe pump. Typical Bu_3SnH-mediated radical reactions, which involve the use of AIBN as initiator, include cyclizations of alkyl and vinyl radicals (eqs 2 and 3).[3,4]

1. $(imidazolyl)_2C(S)$ 2. Bu_3SnH, AIBN (2)

Bu_3SnH AIBN 85% (3)

In eq 2, generation of an alkyl radical is achieved by attack of a tributylstannyl radical on a Barton derivative, whereas in eq 3, the vinyl radical required is generated by addition of the stannyl radical to an alkyne. This type of cyclization has been extended to many types of substrate, including those in which a carbonyl group acts as the acceptor (eq 4),[5] and those in which additional types of radical rearrangement or fragmentation occur (eq 5).[5,6] The latter sequence involves addition of the initially formed radical onto the adjacent carbonyl group, followed by reopening of the so-formed cyclopropanoxy radical in regioselective fashion.[6] The reaction effects a one-carbon ring expansion in good chemical yield.

Bu_3SnH AIBN (1 equiv) Δ or $h\nu$ (4)

Bu_3SnH AIBN 75% (5)

AIBN can also be used in combination with many other types of radical-generating systems other than those employing Bu_3SnH. Examples include cyclizations using ***Tris(trimethylsilyl)silane***,[7] ***Thiophenol***,[8] ***Diphenylphosphine***,[9] and carbonylation of radicals using Ph_3GeH and CO under pressure[10] (eqs 6–9).

$(Me_3Si)_3SiH$ AIBN 63% (6)

PhSH, AIBN TMTHF 52% major product (7)

TMTHF = 2,2,5,5-tetramethyltetrahydrofuran

Ph_2PH, AIBN 66% (8)

AIBN (0.1 equiv) Ph_3GeH (0.1 equiv) $NaBH_3CN$ (2.9 equiv) 1400 psi CO, 105 °C 37% (9)

ACN, 110 °C toluene 72% (10)

For certain types of radical reaction, AIBN may not always give optimal results. Keck and Burnett found in their synthesis of $PGF_{2\alpha}$ that the introduction of the lower side chain, involving radical addition onto a β-stannyl enone, was inefficient when AIBN was used as initiator at 65 °C. By conducting the reaction in refluxing toluene (110 °C), and replacing AIBN with the related ***1,1′-Azobis-1-cyclohexanenitrile*** (ACN), a markedly better yield of the desired adduct was obtained (eq 10).[11]

For reactions which require temperatures above benzene reflux, ACN, which decomposes markedly slower than AIBN, may be a generally useful alternative to AIBN, although this possibility has not been examined in detail. At elevated temperatures, other alternative initiators such as dialkyl peroxides and peresters should also be considered, although the reactivity of the radicals generated from these initiators is quite different to that of the isobutyronitrile radicals generated from AIBN.

Related Reagents. 1,1′-Azobis-1-cyclohexanenitrile (ACN).

1. Thiele, J.; Heuser, K. *LA* **1896**, *290*, 1.
2. Overburger, C. G.; O'Shaughnessy, M. T.; Shalit, H. *JACS* **1949**, *71*, 2661.
3. RajanBabu, T. V. *JACS* **1987**, *109*, 609.
4. Stork, G.; Mook, Jr., R. *JACS* **1987**, *109*, 2829.
5. Nishida, A.; Takahashi, H.; Takeda, H.; Takada, N.; Yonemitsu, O. *JACS* **1990**, *112*, 902.
6. Dowd, P.; Choi, S.-C. *T* **1989**, *45*, 77.
7. Kopping, B.; Chatgilialoglu, C.; Zehnder, M.; Giese, B. *JOC* **1992**, *57*, 3994.
8. Broka, C. A.; Reichert, D. E. C. *TL* **1987**, *28*, 1503.
9. Brumwell, J. E.; Simpkins, N. S.; Terrett, N. K. *TL* **1993**, *34*, 1215.
10. Gupta, V.; Kahne, D. *TL* **1993**, *34*, 591.
11. Keck, G. E.; Burnett, D. A. *JOC* **1987**, *52*, 2958.

Nigel S. Simpkins
University of Nottingham, UK

B

Benzotriazol-1-yloxytris(dimethylamino)phosphonium Hexafluorophosphate[1]

N N N O $\overset{+}{P}(NMe_2)_3$ PF_6^-

[566002-33-6] $C_{12}H_{22}F_6N_6OP_2$ (MW 442.29)

(peptide coupling agent providing high yield and low racemization levels;[1] promotes peptide cyclization[19] and other lactamization,[23] especially β-lactam formation;[24] used to effect various amidification reactions[25–29] and selective esterification;[31] reagent for nucleotidic coupling[32])

Alternate Names: BOP; Castro's reagent.

Physical Data: mp 138 °C (dec.).

Solubility: insol H_2O; sol THF, CH_2Cl_2, MeCN, acetone, DMF, NMP, DMSO.

Form Supplied in: white solid; commercially available. Purity 97–98%.

Analysis of Reagent Purity: ^1H NMR (acetone-d_6): 3.0 (d, 18H, J_{H-P} = 10 Hz, NMe_2), 7.9 (m, 4H, arom.). ^{31}P NMR (CH_2Cl_2): +43.7 (s, P^+), −144.2 (septet, PF_6^-). IR (KBr): 1010 (P–N), 840, 770, 560 (PF_6^-).

Preparative Methods: BOP reagent was initially obtained[2a] by reaction of ***Hexamethylphosphorous Triamide*** (tris(dimethylamino)phosphine, TDAP) with CCl_4 in the presence of HOBt (***1-Hydroxybenzotriazole***) followed by precipitation of the phosphonium salt by anion exchange with aqueous KPF_6 solution. It was then prepared[2b,c] by reaction of cheaper $(Me_2N)_3PO$ (HMPA) with $COCl_2$ (phosgene) to generate the chlorophosphonium chloride intermediate, with treatment of the latter with HOBt and base, and precipitation as below. $COCl_2$ was further replaced by $POCl_3$.[2d,e] The crude final product can, if necessary, be purified by recrystallization from a mixture of acetone–ether or CH_2Cl_2 (DCM)–ether.

Handling, Storage, and Precautions: irritant and harmful. Light sensitive. Incompatibility: strong oxidizing agents. Avoid breathing dust; avoid contact with eyes, skin, and clothing. Keep container closed and store in the dark. Refrigerate. Although BOP is widely used, it is important to stress that the carcinogenic HMPA is a side product and the reagent must be handled with caution. All operations should be carried out in a fume hood.

Peptide Coupling Reagent. First prepared with the aim to associate both the coupling agent (phosphonium salt as dehydrating agent) and the racemization suppressor (HOBt additive) in the same compound, the BOP reagent has since then been widely developed, mainly in peptide synthesis, as well as in other fields. Mechanistic studies on the probable intermediates (acyloxyphosphonium salt, active ester, symmetrical anhydride, oxazolone, eq 1) during the activation step of carboxylic acids with BOP have been made. Early investigations indicating the intermediacy of a benzotriazolyl active ester,[1a,2b,3] as in activation with the DCC/HOBt (see ***1,3-Dicyclohexylcarbodiimide***) reagent system, were later questioned[1c] in the light of better results obtained with BOP in a number of cases. After further studies, the direct formation of the acyloxyphosphonium salt, presumably through the cyclic complex (eq 1), was postulated as likely.[1c]

$RCO_2^-\,HNR_3^+ + BtO\overset{+}{P}Y_3\,PF_6^- \rightleftharpoons RCO_2^-\,BtO\overset{+}{P}Y_3 + HNR_3^+\,PF_6^-$

$\longrightarrow RCO_2\overset{+}{P}Y_3\,BtO^- + HNR_3^+\,PF_6^-$ (acyloxyphosphonium salt) $\longrightarrow$ RCOO....HOBt (oxalone)

acyloxyphosphonium salt $\longrightarrow$ RCO_2Bt active ester + OPY_3

acyloxyphosphonium salt $\rightleftharpoons$ ($RCO_2^-\,HNR_3^+$) $RCO_2\overset{+}{P}Y_3\,RCO_2^- + BtO^-\,HNR_3^+$ $\longrightarrow$ $(RCO)_2O$ + O=PY_3 (anhydride) (1)

acyloxyphosphonium salt $\longrightarrow$ cyclic complex (R, O, O, $\overset{+}{P}Y_3$, NH, R', H, O, Bt)

Y = NMe_2

Unoptimized initial work in peptide coupling with BOP had already shown high yields and short reaction times, even with bulky amino acids (eq 2) and little if no side reactions with some sensitive amino acids bearing unprotected side chain functional groups.[2]

$$\text{Boc-Ile-OH} + \text{HCl, H-Ile-OMe} \xrightarrow[\text{MeCN, rt; 80\%}]{\text{BOP, NEt}_3} \text{Boc-Ile-Ile-OMe} \quad (2)$$

Unlike DCC, BOP reacts exclusively with carboxylate salts, with carboxylic acids remaining unchanged. The trifluoroacetate anion is not activated by BOP; therefore the trifluoroacetylation side reaction is never encountered.

In usual peptide synthesis (e.g. with the Boc/TFA deprotection strategy), a typical coupling step is conducted as follows: BOP is added to the mixture of the Boc-amino acid (or peptide acid segment) and the C-terminal-protected peptide salt (generally the TFA salt). The solvent can be THF, DCM, MeCN, DMF, or NMP, according to the solubility of the reaction partners, but using as high a concentration as possible. No reaction occurs until the tertiary base, *N*-methylmorpholine, ***Triethylamine***, or preferably ***Diisopropylethylamine*** (DIEA), is added. In every case the reaction mixture has to be kept rather basic in order to ensure a fast coupling. Three equiv of base are necessary to neutralize the carboxylic acid, the amine salt, and the acidic HOBt. In such conditions the coupling rate is so high that racemization is negligible in urethane-protected amino acid coupling and fairly low, albeit not negligible, in segment coupling (see eq 3).[4] The excess of acid and BOP is typically 1.1 mol equiv in solution synthesis.

Boc-Val-Val-OH + HCl, H-Leu-OMe —(BOP, NEt_3; MeCN, rt; 95%)→ Boc-Val-Val-Leu-OMe (3)

LL L

LDL/LLL% = 17.2

Racemization in segment coupling has been studied with realistic models[4a] in solution; it appears to be very small when nonhindered amino acids (Ala, Leu, Lys ...) are activated, but coupling of segments terminated with bulky residues (Val, Ile, Thr ...) are to be avoided (eq 3).[4b]

Segment condensation with BOP is generally peculiarly fast compared with conventional use of the classical DCC/HOBt method, as a few hours are enough to achieve coupling of equimolar quantities of both partners, provided the highest concentration is used (eq 4).[5]

Boc-Tyr-Glu(O-*t*-Bu)-Arg-Gly-OH + H-Val-Asp(O-*t*-Bu)-Nle-Ala-Arg(NO_2)-Leu-Gly-OBn

↓ BOP, DIEA, DMF; 84%

Boc-Tyr-Glu(O-*t*-Bu)-Arg-Gly-Val-Asp(O-*t*-Bu)-Nle-Ala-Arg(NO_2)-Leu-Gly-OBn (4)

It is very characteristic that this procedure allows the skipping of a separate neutralization step, either in solution or in SPPS (solid phase peptide synthesis) using the Boc/TFA strategy. This is not only time-saving but also useful to minimize diketopiperazine formation during the third step of a synthesis.

The coupling reaction of Boc-Ile-OH with H-His-Pro-OMe · TFA is generally a very low yield reaction as the neutralization of the dipeptide salt is very readily followed by cyclization to cyclo(His-Pro) (eq 5). Using the BOP procedure allows the efficient 'in situ' trapping of the dipeptide amine by the acylating reagent. This feature also makes strategies of segment coupling at a proline residue possible, which are very often restricted by the necessity of using prolyl *t*-butyl esters in order to avoid the cyclization reaction.

Boc-Ile-OH + H-His-Pro-OMe•TFA —(BOP, DIEA; MeCN; 61%)→ Boc-Ile-His-Pro-OMe (5)

H-His-Pro-OMe•TFA —(base)→ cyclo(His-Pro)

This specific ability of BOP for avoiding diketopiperazine formation is illustrated in liquid phase synthesis[5a] as well as in the SPPS.[5c,d,6,7b]

Another facility offered by BOP is the possibility to couple DCHA salts, due to the high rate ratio of amino acid coupling towards dicyclohexylamine (DCHA). This feature is especially useful in SPPS where the small amount of dicyclohexylamide formed is easily washed out. The main application of that procedure lies in the introduction of histidine as the cheap and easily acessible Boc-His(Boc)-OH both in liquid (eq 6)[5a] and solid[7b–d] phase synthesis.

Boc-His(Boc)-OH, DCHA + H-Leu-Leu-Val-Tyr(Bn)-Ser(Bn)-OBn

—(BOP, DIEA; DMF; 92%)→ Boc-His(Boc)-Leu-Leu-Val-Tyr(Bn)-Ser(Bn)-OBn (6)

Many peptides have been synthesized in solution,[5a,b,8] stepwise or by segment coupling, using BOP as coupling agent.

In solid phase synthesis, BOP exhibits some distinct advantages, including the suppression of a separate neutralization step previously mentioned; indeed, after TFA treatment and washing, it is convenient to introduce the next Boc-amino acid and BOP as solids, with a minimal volume of solvent and, at last, DIEA. Coupling is generally completed after 15–30 min, with the pH maintained at 8 with excess of acylating reagents (1.5–3M). Finally, the water- and DCM-soluble byproducts, HMPA and HOBt, allow very easy washing of the reaction product, in contrast to DCC condensation where elimination of insoluble urea often remains a problem.

The use of BOP in SPPS, initially limited,[7] has since been widely developed, with Boc/TFA,[9,10] Boc/HF,[11] and mainly Fmoc/piperidine[12–15] deprotection systems. In most cases it compares favorably with other reagents such as DCC, DCC/HOBt, and symmetrical anhydride. As already observed in the liquid phase, some amino acids with side chain functional groups such as Tyr, Thr, Ser (hydroxy group),[2a,9b] or poorly soluble pGlu (γ-lactam)[7d,9c] can be used without protecting groups; other functional groups such as $CONH_2$ (Asn) can lead in some cases to partial dehydration (e.g. β-cyanoalanine)[11,14b,c] or cyclization (succinimide)[9a,30] and need to be protected; such classical dehydration reactions, not observed by others,[1a,2a,b] are sometimes dependent on reactions conditions (e.g. repeated coupling cycles) and must be carefully examined.

Peptide Cyclization. Various results were provided[16–20] in peptide chain cyclization with BOP, arising either from the reaction conditions or from the ring size, and from the linear precursor conformation.

Better yields and purities are obtained[16] with less powerful coupling agents such as DPPA (***Diphenyl Phosphorazidate***) and NDPP (norborn-5-ene-2,3-dicarboxamidodiphenyl phosphate) in cyclization of tetra- and pentapeptide sequences, despite the need for longer reaction times than with BOP or HBTU (***O-Benzotriazol-1-yl-N,N,N′,N′-tetramethyluronium Hexafluorophosphate***; 1,1,3,3-tetramethyluronium analog of BOP).

Comparable yields are given for Boc-γ-D-Glu-Tyr-Nle-D-Lys-Trp-OMe cyclizations[17] with BOP ($NaHCO_3$, 4 h, 76%) and MTPA (dimethylphosphinothioyl azide, NEt_3, 36 h, 79%), albeit in different reaction conditions; in this work the reported racemization associated with the use of BOP is not shown.

Cyclization, leading to Gly/Lys-containing peptide macrocycles,[18] has been investigated with a number of coupling reagents. Among them DEPC (***Diethyl Phosphorocyanidate***), DPPA, and BOP are the most effective ones, giving near-quantitative yields.

Macro ring closure of a partly proteinogenic sequence in the synthesis[19] of the cyclodepsipeptide Nordidemnin B was carried out using BOP in heterogeneous conditions (eq 7).

TFA, D-Val-Ψ(CHOH)-Gly-HIP-Leu-Pro-MeTyr(Me)—Z-Thr-OH

$\xrightarrow[\text{DMF, 54\%}]{\text{BOP, NaHCO}_3}$ Z-Thr-D-Val-Ψ(CHOH)-Gly-HIP—MeTyr(Me)-Pro-Leu (cyclic) (7)

In SPPS, side chain to side chain cyclization by the BOP procedure not only proceeds more rapidly, but also gives a purer cyclic product than the DCC/HOBt procedure.[20]

The use of BOP has been reported in the final cyclization of cyclosporin,[21b] even if some racemization is observed. The very difficult coupling of *N*-methyl amino acids is achieved in the same work,[21] as in the preceding syntheses.[19] For the coupling of hindered *N*-methyl[22a] and α,α-dialkyl[22b] amino acids, new reagents[22] such as PyBroP, PyCloP, or PyClU seem to be more promising.

Besides the peptide field, cyclization leading to macrocyclic polyether tetralactams (crown ethers) have been carried out with BOP.[23]

β-Lactam Cyclization. Cyclization of β-amino acids to β-lactams is efficiently effected by treatment with BOP (eq 8).[24a] The present method appears to be limited to the formation of β-lactams from *N*-unsubstituted β-amino acids ($R^1 = R^2 = R^3 = R^5 = H$, $R^4 = Me$, yield 10%).

HO_2C–C(R^1)(R^2)–C(R^3)(R^4)–NHR^5 $\xrightarrow[\text{MeCN, 80 °C, 46–99\%}]{\text{BOP, NEt}_3}$ β-lactam (8)

R^1 = H, Me; R^2 = H, OPh; R^3 = H, Me, Ph;
R^4 = H, Me, *n*-Pr; R^5 = CH_2Ph, *c*-C_6H_{11}, *n*-C_8H_{17}

In the cephem series, cyclization yields of 80% are obtained with BOP, in DCM at rt, with retention of the initial chirality.[24b]

Amidation. A number of other various amidation reactions have been conducted using BOP. Such preparations include *N,O*-dimethyl hydroxamates of amino acids[25a] and peptides[25b] (precursors of chiral peptidyl aldehydes), heterocyclic amide fragments in the synthesis of macrolide[26] and porphyrin models,[27] 'dansylglycine anhydride' as a mixed sulfonic–carboxylic imide by dehydration–cyclization of dansylglycine,[28] and selective monoacylation of heterocyclic diamine in carbohydrate series (eq 9).[29]

CO_2H $\xrightarrow[\text{DCM, rt, >75\%}]{\text{o-}C_6H_4(NH_2)_2\text{, BOP, NEt}_3}$ CONH–C_6H_4–NH_2 (9)

Esterification. Treatment of N-protected amino acids with phenol, NEt_3, and BOP in the appropriate solvent (DCM, MeCN, DMF) affords the corresponding phenyl esters in good yield (eq 10).[30]

$$\text{P-AA-OH + PhOH} \xrightarrow[\text{solvent, rt, 73–97\%}]{\text{BOP, NEt}_3} \text{P-AA-OPh} \quad (10)$$

P = Boc, Z; AA = Ala, Ile, Phe, Trp, Pro, Ser, Met, Cys(SBn), Lys(εZ)

BOP in combination with ***Imidazole*** (ImH) as catalyst provides a very mild system for selective esterification of polyhydroxy compounds such as carbohydrates; thus trehalose is converted into mono-, di-, and tripalmitate mixtures, with the various quantities of each of them depending on the solvent and acid/alcohol ratio. Monoacylated compounds can be obtained selectively, albeit in moderate yield (31%).[31]

Nucleotidic Coupling and Related Reactions. BOP has been used as a condensing agent to promote internucleotidic bond formation in phosphotriester oligodeoxyribonucleotide synthesis.[32] ^{31}P NMR studies[32a] show that a benzotriazolyl phosphate derivative is involved as an active ester intermediate, as with carboxylic acids.[3]

The BOP procedure was applied to the preparation of undecanucleoside decaphosphate.[32b] The same type of activation with BOP is carried out for introducing a phosphonamide bond in dipeptide models[33] with good efficiency (≥60%); coupling reactions by activation of phosphinic acid derivatives[34] with BOP give poor yields.

1. (a) Le Nguyen, D.; Castro, B. *Peptide Chemistry 1987*; Protein Research Foundation: Osaka, 1988; p 231. (b) Kiso, Y.; Kimura, T. *Yuki Gosei Kagaku Kyokai Shi* **1990**, *48*, 1032 (*CA* **1991**, *114*, 164 722k). (c) Coste, J.; Dufour, M. N.; Le Nguyen, D.; Castro, B. In *Peptides, Chem. Struct. Biol.* ESCOM: Leiden, 1990; pp 885–888.
2. (a) Castro, B.; Dormoy, J. R.; Evin, G.; Selve, C. *TL* **1975**, 1219. (b) Castro, B.; Dormoy, J. R.; Evin, G.; Selve, C. *Peptides 1976*; University of Bruxelles: Brussels, 1976; p 79. (c) Castro, B.; Dormoy, J. R.; Dourtoglou, B.; Evin, G.; Selve, C.; Ziegler, J. C. *S* **1976**, 751. (d) Dormoy, J. R.; Castro, B. *TL* **1979**, 3321. (e) Dormoy, J. R.; Castro, B. *T* **1981**, *37*, 3699.
3. Castro, B.; Dormoy, J. R.; Evin, G.; Selve, C. *JCR(S)* **1977**, 182; *JCR(M)* **1977**, 2118.
4. (a) Castro, B.; Dormoy, J. R.; Le Nguyen, D. *TL* **1978**, 4419. (b) Castro, B.; Dormoy, J. R.; Le Nguyen, D. *Peptides 1978*; Wroclaw University Press, Poland, 1979; p 155. (c) Le Nguyen, D.; Dormoy, J. R.; Castro, B.; Prevot, D. *T* **1981**, *37*, 4229.
5. (a) Le Nguyen, D.; Seyer, R.; Heitz, A.; Castro, B. *JCS(P1)* **1985**, 1025. (b) Fehrentz, J. A.; Seyer, R.; Heitz, A.; Fulcrand, P.; Castro, B.; Corvol, P. *Int. J. Pept. Protein Res.* **1986**, *28*, 620. (c) Seyer, R.; Aumelas, A.; Tence, M.; Marie, J.; Bonnafous, J. C.; Jard, S.; Castro, B. *Int. J. Pept. Protein Res.* **1989**, *34*, 235. (d) Seyer, R.; Aumelas, A.; Marie, J.; Bonnafous, J. C.; Jard, S.; Castro, B. *HCA* **1989**, *72*, 678.
6. Gairi, M.; Lloyd-Williams, P.; Albericio, F.; Giralt, E. *T* **1990**, *31*, 7363.
7. (a) Rivaille, P.; Gautron, J. P.; Castro, B.; Milhaud, G. *T* **1980**, *36*, 3413. (b) Le Nguyen, D.; Heitz, A.; Castro, B. *JCS(P1)* **1987**, 1915. (c) Ratman, M.; Le Nguyen, D.; Rivier, J.; Sargent, P. B.; Lindstrom, J. *B* **1986**, *25*, 2633. (d) Seyer, R.; Aumelas, A.; Caraty, A.; Rivaille, P.; Castro, B. *Int. J. Pept. Protein Res.* **1990**, *35*, 465. (e) Evin, G.; Galen, F. X.; Carlson, W. D.; Handschumacher, M.; Novotny, J.; Bouhnik, J.; Menard, J.; Corvol, P.; Haber, E. *B* **1988**, *27*, 156. (f) Bouhnik, J.; Galen, F. X.; Menard, J.; Corvol, P.; Seyer, R.; Fehrentz, J. A.; Le Nguyen, D.; Fulcrand, P.; Castro, B. *JBC* **1987**, *262*, 2913. (g) Fehrentz, J. A.; Heitz, A.; Seyer, R.; Fulcrand, P.; Devilliers, R.; Castro, B.; Heitz, F.; Carelli, C. *B* **1988**, *27*, 4071. (h) Bonnafous, J. C.; Tence, M.; Seyer, R.; Marie, J.; Aumelas,

A.; Jard, S. *BJ* **1988**, *251*, 873. (i) Liu, C. F.; Fehrentz, J. A.; Heitz, A.; Le Nguyen, D.; Castro, B.; Heitz, F.; Carelli, C.; Galen, F. X.; Corvol, P. *T* **1988**, *44*, 675.

8. (a) Eid, M.; Evin, G.; Castro, B.; Menard, J.; Corvol, P. *BJ* **1981**, *197*, 465. (b) Cumin, F.; Evin, G.; Fehrentz, J. A.; Seyer, R.; Castro, B.; Menard, J.; Corvol, P. *JBC* **1985**, *260*, 9154.
9. (a) Fournier, A.; Wang, C. T.; Felix, A. M. *Int. J. Pept. Protein Res.* **1988**, *31*, 86. (b) Fournier, A.; Danho, W.; Felix, A. M. *Int. J. Pept. Protein Res.* **1989**, *33*, 133. (c) Forest, M.; Fournier, A. *Int. J. Pept. Protein Res.* **1990**, *35*, 89.
10. Jarrett, J. T.; Landsbury, Jr, P. T. *TL* **1990**, *31*, 4561.
11. Jezek, J.; Houghten, R. A. *Peptides 1990*; ESCOM: Leiden, 1991; p 74.
12. Rule, W. K.; Shen, J. H.; Tregear, G. W.; Wade, J. D. *Peptides 1988*: de Gruyter: Berlin, 1989; p 238.
13. Hudson, D. *JOC* **1988**, *53*, 617.
14. (a) Frank, R. W.; Gausepohl, H. *Modern Methods in Protein Chemistry*; de Gruyter: Berlin, 1988; Vol. 3. p 41. (b) Gausepohl, H.; Kraft, M.; Frank, R. W. *Int. J. Pept. Protein Res.* **1989**, *34*, 287. (c) Gausepohl, H.; Kraft, M.; Frank, R. W. *Peptides 1988*; de Gruyter: Berlin, 1989; p 241.
15. Waki, M.; Nakahara, T.; Ohno, M. *Peptides Chemistry 1990*; Protein Research Foundation: Osaka, 1991; p 95.
16. Schmidt, R.; Neubert, K. *Int. J. Pept. Protein Res.* **1991**, *37*, 502.
17. Ueki, M.; Kato, T. see ref. 15, p 49.
18. Crusi, E.; Huerta, J. M.; Andreu, D.; Giralt, E. *TL* **1990**, *31*, 4191.
19. Jouin, P.; Poncet, J.; Dufour, M. N.; Pantaloni, A.; Castro, B. *JOC* **1989**, *54*, 617.
20. Felix, A. M.; Wang, C. T.; Heimer, E. P.; Fournier, A. *Int. J. Pept. Protein Res.* **1988**, *31*, 231.
21. (a) Wenger, R. M. *HCA* **1983**, *66*, 2672. (b) Wenger, R. M. *HCA* **1984**, *67*, 502.
22. (a) Coste, J.; Frerot, E.; Jouin, P.; Castro, B. *TL* **1991**, *32*, 1967. (b) Frerot, E.; Coste, J.; Pantaloni, A.; Dufour, M. N.; Jouin, P. *T* **1991**, *47*, 259.
23. Duriez, M. C.; Pigot, T.; Picard, C.; Cazaux, L.; Tisnes, P. *T* **1992**, *48*, 4347.
24. (a) Kim, S.; Lee, T. A. *Bull. Kor. Chem. Soc.* **1988**, *9*, 189. (b) Roze, J. C.; Pradere, J. P.; Duguay, G.; Guevel, A.; Quiniou, H.; Poignant, S. *CJC* **1983**, *61*, 1169.
25. (a) Fehrentz, J. A.; Castro, B. *S* **1983**, 676. (b) Fehrentz, J. A.; Heitz, A.; Castro, B. *Int. J. Pept. Protein Res.* **1985**, *26*, 236.
26. Somers, P. K.; Wandless, T. J.; Schreiber, S. L. *JACS* **1991**, *113*, 8045.
27. Selve, C.; Niedercorn, F.; Nacro, M.; Castro, B.; Gabriel, M. *T* **1981**, *37*, 1903.
28. Weinhold, E. G.; Knowles, J. R. *JACS* **1992**, *114*, 9270.
29. Chapleur, Y.; Castro, B. *JCS(P1)* **1980**, 2683.
30. Castro, B.; Evin, G.; Selve, C.; Seyer, R. *S* **1977**, 413.
31. Chapleur, Y.; Castro, B.; Toubiana, R. *JCS(P1)* **1980**, 1940.
32. (a) Molko, D.; Guy, A.; Teoule, R.; Castro, B.; Dormoy, J. R. *NJC* **1982**, *6*, 277. (b) Molko, D.; Guy, A.; Teoule, R. *Nucleosides Nucleotides* **1982**, *1*, 65.
33. Dumy, P.; Escale, R.; Vidal, J. P.; Girard, J. P.; Parello, J. *CR(C)* **1991**, *312*, 235.
34. Elhaddadi, M.; Jacquier, R.; Petrus, C.; Petrus, F. *PS* **1991**, *63*, 255.

Jean-Robert Dormoy & Bertrand Castro
Sanofi Chimie, Gentilly, France

O-Benzotriazol-1-yl-*N*,*N*,*N*′,*N*′-tetramethyluronium Hexafluorophosphate

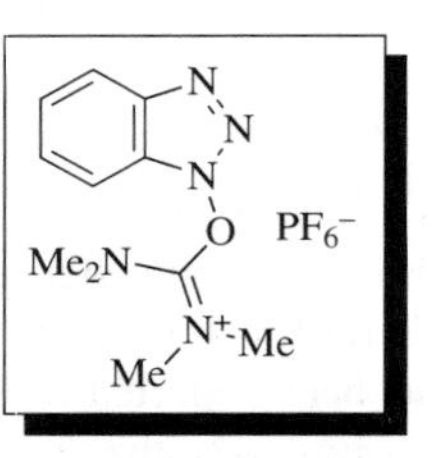

[94790-37-1] $C_{11}H_{16}F_6N_5OP$ (MW 379.29)

(coupling reagent for peptide synthesis[1])

Alternate Name: HBTU.
Physical Data: mp 206–207 °C (dec).
Solubility: sol DMF (0.5 mol L^{-1}). With the tetrafluoroborate counterion the corresponding salt (TBTU) is slightly more soluble (0.6 mol L^{-1}).
Form Supplied in: white solid; widely available.
Purification: by crystallization in a mixture of MeCN and CH_2Cl_2.
Handling, Storage, and Precautions: very stable, not hygroscopic, and can be stored indefinitely. Solutions in DMF (0.45 M) can be stored in an inert atmosphere for weeks. Syntheses of peptides carried out with freshly prepared, 6 and 13 week stored solutions show similar quality of the crude product.[3] Violent decomposition can occur when dried at elevated temperature.[4]

Both *O*-benzotriazol-1-yl-*N,N,N′,N′*-tetramethyluronium hexafluorophosphate (HBTU) and tetrafluoroborate (TBTU) salts have been used as condensing reagents for the preparation of peptides in both solution and solid-phase strategies.[1,4] In solution, reaction of α-amino protected amino acids or dipeptides with equimolar amounts of α-carboxyl protected amino acids or dipeptides and ***Diisopropylethylamine*** (DIEA) (2 equiv) in DMF proceeds cleanly for 15 min at 0 °C, yielding the corresponding peptides in good yields (80–96%) (eq 1).[1,4]

$$R^1CO_2H + R^2NH_2 \xrightarrow[\text{DIEA} \atop 80\text{–}96\%]{\text{HBTU}} R^1CONHR^2 \quad (1)$$

HBTU activation has been adapted for automated stepwise solid-phase peptide synthesis for both *t*-butoxycarbonyl (Boc) and fluorenylmethoxycarbonyl (Fmoc) strategies.[5] For the former, a simple, effective protocol has been developed, which involves simultaneous in situ neutralization with coupling (extra equivalents of base are required). This protocol is particularly suitable for assembling complex peptides, arising from sequence-dependent peptide chain aggregation. Since aggregation occurs when protonated α-ammonium peptide-resin intermediates are neutralized, simultaneous neutralization and acylation can help to overcome this phenomenon.[6] For Fmoc chemistry, efficient protocols are also available for both batch and continuous-flow systems.[3,7]

Various protected peptide segments have been coupled on solid-phase using a combination of TBTU, HOBt, and DIEA for the syntheses of several small proteins. Coupling yields, using three-fold excess of peptide segment and coupling times of ~12 h, are greater than 95%.[8]

Racemization from uronium salt mediated couplings has been determined by analysis of the epimeric products by HPLC using different models. In all cases, racemization is similar or lower than that obtained with carbodiimide and benzotriazole based phosphonium salt methods.[1,2,4,9]

In the absence of the carboxylic component, HBTU reacts with amino groups leading to the formation of a Schiff base (eq 2).[10] Thus in syntheses conducted in solution, the excess of both HBTU and amino component should be avoided. In both solution and solid-phase strategies the sequence of reagent addition is critical. HBTU should be delivered to the carboxylic component for preactivation, prior to the addition of the amine. The Schiff base formation can also occur during slow reactions, such as the preparation of cyclic peptides, where both amino and carboxylic components are in equimolar amounts and an excess of the uronium salt can block the amino group.[11] For the synthesis of cyclic peptides the phosphonium derivatives ***Benzotriazol-1-yloxytris(dimethylamino)phosphonium Hexafluorophosphate*** (BOP)[12] and benzotriazol-1-yloxytris(pyrrolidino)phosphonium hexafluorophosphate (PyBOP),[13] can be more useful.

PF_6^- + H_2N–CH(CH_2Ph)–$CONH_2$ $\xrightarrow{\text{DIEA}}$ Me_2N(NMe$_2$)C=N–CH(CH_2Ph)–$CONH_2$ (2)

Uronium salts have been used for cyclization of linear peptides in both solution and solid-phase modes. Since this reaction is sequence dependent, there are no general conditions. Thus cyclization in solution with uronium salt methods give, for some peptides, better results than the classical reagent ***Diphenyl Phosphorazidate*** (DPPA),[14] but for another case the converse is true.[15] Furthermore, dehydration of *C*-terminal aspartylamide peptides during cyclization with HBTU has been described. This side-reaction can be prevented by the addition of one equivalent of HOBt.[16] Although good results have been obtained in the solid-phase,[17] guanidino formation side-reactions have also been reported.[11]

Recently, other uronium salts, such as *O*-(*N*-succinimidyl)-*N,N,N′,N′*-tetramethyluronium tetrafluoroborate (TSTU),[4,18] *O*-(*N*-5-norbornene-*endo*-2, 3-dicarboximidyl)-*N, N, N′, N′*-tetramethyluronium tetrafluoroborate (TNTU),[4,18] *O*-(2-oxo-1(2*H*)-pyridyl)-*N,N,N′,N′*-bis(pentamethylene)uronium tetrafluoroborate (TOPPipU),[19] ***N-[(Dimethylamino)-1H-1,2,3-triazolo[4,5-b]pyridin-1-ylmethylene]-N-methylmethanaminium Hexafluorophosphate N-Oxide*** (HATU),[20] and *O*-(7-azabenzotriazol-1-yl)-*N,N,N′,N′*-bis(tetramethylene)uronium hexafluorophosphate (HAPyU)[21] have been described and are commerically available. The uronium salts derived from 7-aza-1-hydroxybenzotriazole (HATU and HAPyU) have been shown to be superior to their benzotriazole analogs in terms of coupling efficiency,[20,21] racemization,[22] and cyclization,[23] in both solution and solid-phase strategies.

Finally, an X-ray structure determination of HBTU revealed that the solid-state structure differs considerably from the formulation commonly presented in the literature. The solid-state structure is not the *N,N,N′,N′*-tetramethyluronium salt but rather the guanidinium *N*-oxide (**3**).[24]

(**3**)

Related Reagents. *N*-[(Dimethylamino)-1*H*-1,2,3-tri azolo-[4,5-b]pyridin-1-ylmethylene]-*N*-methylmethanaminium Hexafluorophosphate *N*-Oxide.

1. Dourtoglou, V.; Ziegler, J. C.; Gross, B. *TL* **1978**, 1269.
2. Dourtoglou, V.; Gross, B.; Lambropoulou, V.; Zioudrou, C. *S* **1984**, 572.
3. Fields, C. G.; Lloyd, D. H.; Macdonald, R. L.; Otteson, K. M.; Noble, R. L. *Peptide Res.* **1991**, *4*, 95.
4. Knorr, R.; Trzeciak, A.; Bannwarth, W.; Gillessen, D. *TL* **1989**, *30*, 1927.
5. Fields, G. B.; Tian, Z.; Barany, G. In *Synthetic Peptides: A User's Guide*, Grant, G. A. Ed.; Freeman: New York, 1992; pp 77–183.
6. (a) Schnoelzer, M.; Alewood, P; Jones, A.; Alewood, D.; Kent, S. B. H. *Int. J. Peptide Protein Res.* **1992**, *40*, 180. (b) Reid, G. E.; Simpson, R. J. *Anal. Biochem.* **1992**, *200*, 301. (c) Sueiras-Diaz, J.; Horton, J. *TL* **1992**, *33*, 2721.
7. Poulat, F.; Guichard, G.; Gozé, C.; Heitz, F.; Calas, B.; Berta, P. *FEBS Lett.* **1992**, *309*, 385.
8. Surovoy, A.; Metzger, J.; Jung, G. In *Innovation and Perspectives in Solid-Phase Synthesis: Peptides, Polypeptides and Oligonucleotides*, Epton, R. Ed.; Intercept: Andover, 1992, pp 467–473 (*CA* **1993**, *118*, 234 476s).
9. Benoiton, N. L.; Lee, Y. C.; Steinauer, R.; Chen, F. M. F. *Int. J. Peptide Protein Res.* **1992**, *40*, 559.
10. Gausepohl, H.; Pieles, U.; Frank, R. W. In *Proceedings of the Twelth American Peptide Symposium*; Smith, J. A.; Rivier, J. E. Eds.; ESCOM: Leiden, 1992; pp 523–524. (*CA* **1992**, *117*, 171 981j).
11. Story, S. G.; Aldrich, J. V. *Int. J. Peptide Protein Res.* **1994**, *43*, 292.
12. Castro, B.; Dormoy, J. R.; Evin, G. *TL* **1975**, 1219.
13. Coste, J.; Le-Nguyen, D.; Castro, B. *TL* **1990**, *31*, 205.
14. (a) Hoffmann, E.; Beck-Sickinger, A. G.; Jung G. *LA* **1991**, 585. (b) Zimmer, S.; Hoffmann, E.; Jung, G.; Kessler, H. *LA* **1993**, 497.
15. Schmidt, R.; Neubert, K. *Int. J. Peptide Protein Res.* **1991**, *37*, 502.
16. Rovero, P.; Pegoraro, S.; Bonelli, F.; Triolo, A. *TL* **1993**, *34*, 2199.
17. (a) Trzeciak, A.; Bannwarth, W. *TL* **1992**, *33*, 4557. (b) Neugebauer, W.; Willick, G. In *Proceedings of the Twenty-Second European Peptide Symposium*; Schneider, C. H.; Eberle, A. N. Eds.; ESCOM: Leiden, 1993; pp 393–394.
18. Bannwarth, W.; Schmidt, D.; Stallard, R. L.; Hornung, C.; Knorr, R.; Müller, F. *HCA* **1988**, *71*, 2085.
19. Henklein, P.; Beyermann, M.; Bienert, M.; Knorr, R. In *Proceedings of the Twenty-First European Peptide Symposium*; Giralt, E.; Andreu, D. Eds.; ESCOM: Leiden, 1991; pp 67–68 (*CA* **1991** *115*, 232 811m).
20. Carpino, L. A. *JACS* **1993**, *115*, 4397.
21. Carpino, L. A.; El-Faham, A.; Minor, C. A.; Albericio, F. *JCS (P1)* **1994**, 201.
22. Carpino, L. A.; E.-Faham, A. *JOC* **1994**, *59*, 695.

23. Ehrlich, A.; Rothemund, S.; Brudel, M.; Beyermann, M.; Carpino, L. A.; Bienert, M. *TL* **1993**, *34*, 4781.
24. Abdelmoty, I.; Albericio, F.; Carpino, L.A.; Foxman, B. M.; Kates, S. A. *Lett. Peptide Sci.* **1994**, in press.

Fernando Albericio & Steven A. Kates
Millipore Corporation, Bedford, MA, USA

Benzoyl Chloride[1]

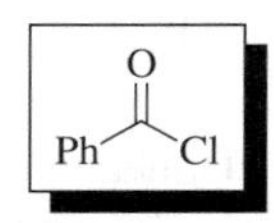

[98-88-4] C_7H_5ClO (MW 140.57)

(useful acylating agent; preparation of ketones from organometallic compounds;[2,3] preparation of 1,3-dicarbonyl compounds from enolates or enols;[4,5] Friedel–Crafts acylation and related reactions[6] with aromatic and heterocyclic compounds,[7] alkenes,[8] alkynes,[9] enoxysilanes,[10] and silicon compounds;[11] acylation of enol ethers, ketene acetals,[12] and enamines;[13] protection of alcohols as benzoates;[14] protection of amines as benzamides[15])

Physical Data: bp 197.2 °C; mp −1 °C; *d* 1.21 g cm^{-3}.
Solubility: slowly decomposed by water and alcohols; sol most organic solvents.
Form Supplied in: colorless liquid; penetrating odor; widely available.
Purification: the good commercial grade of benzoyl chloride can be purified by distillation. The technical material must be purified as follows: a 50% solution of PhCOCl in ether or cyclohexane is washed with cold 5% aqueous sodium bicarbonate; after drying over $CaCl_2$ the solvent is eliminated under vacuo and the PhCOCl is distilled.[16]
Handling, Storage, and Precautions: use in a fume hood; lachrymatory, irritating to skin, eyes and mucous membranes; toxic by inhalation and ingestion.

Acylation of Organometallic Compounds.[2,3] PhCOCl reacts with various organometallic reagents to give the corresponding phenyl ketones in variable yields. The first acylation reactions were performed with organocadmium and organozinc compounds (eqs 1 and 2).[3a,b] In THF, organozinc reagents give poor results; however, the yields are clearly improved in the presence of ***Tetrakis(triphenylphosphine)palladium(0)*** (eq 3).[3c]

$$\text{MeCdCl} + \text{PhCOCl} \xrightarrow[\text{85\%}]{\text{benzene, reflux}} \text{MeCOPh} \quad (1)$$

$$\text{MeZnI} + \text{PhCOCl} \xrightarrow[\text{80\%}]{\text{benzene, 0 °C}} \text{MeCOPh} \quad (2)$$

$$\text{MeZnCl} + \text{PhCOCl} \xrightarrow[\text{25 °C}]{\text{THF}} \text{MeCOPh} \quad (3)$$

no catalyst, 12 h: 10%
5% $Pd(PPh_3)_4$, 30 min: 80%

With the very reactive organomagnesium compounds, the acylation takes place in the presence of $Fe(acac)_3$[3d] or $MnCl_2$[3e] (eqs 4 and 5). It is worthy of note that the 1,2-addition, which is rapid under these reaction conditions (10–20 °C), is completely avoided.

$$\text{PhCOCl} + \text{MeMgCl} \xrightarrow[\text{80\%}]{\text{3\% Fe(acac)}_3\text{, THF, rt}} \text{MeCOPh} \quad (4)$$

$$\text{PhCOCl} + \text{BuMgCl} \xrightarrow[\text{90\%}]{\text{3\% MnCl}_4\text{Li}_2\text{, THF, 0 °C, 45 min}} \text{BuCOPh} \quad (5)$$

Organocopper (60% of PhCOMe from $MeCu{\cdot}PBu_3$)[3f] and cuprate reagents (eq 6),[3g] which are now widely used, give only moderate yields of phenyl ketones. Better yields are obtained at low temperature by using a large excess of organocuprate[3h] or heterocuprate (eq 7).[3i]

$$\text{Me}_2\text{CuLi} + \text{PhCOCl} \xrightarrow[\text{58\%}]{\text{ether, −5 °C}} \text{MeCOPh} \quad (6)$$

$$t\text{-Bu(PhS)CuLi} + \text{PhCOCl} \xrightarrow[\text{84–87\%}]{\text{THF, −78 to −20 °C}} t\text{-BuCOPh} \quad (7)$$

Organomanganese compounds afford excellent yields by the use of a stoichiometric amount of reagents under mild conditions (eq 8).[3j]

$$\text{BuMnCl} + \text{PhCOCl} \xrightarrow[\text{90\%}]{\text{THF, 20 °C}} \text{BuCOPh} \quad (8)$$

Acylation of Enolates, Enols, and Related Reactions.[4,5] According to the reaction conditions, PhCOCl reacts with ketone enolates to lead to enol benzoates (*O*-acylation, kinetic product) or β-diketones (*C*-acylation, thermodynamic product) (eq 9).[5a]

Acetone $\xrightarrow{\text{NaNH}_2\text{, ether}}$ sodium enolate (ONa) $\xrightarrow{\text{PhCOCl}}$ enol benzoate ($PhCO_2$) + 1-phenyl-1,3-butanedione (Ph) (9)

	enol benzoate	β-diketone
2 equiv of PhCOCl:	41%	6%
2 equiv of enolate:	9%	33%

PhCOCl very often gives a mixture of the *O*- and *C*-acylated products. To prepare the enol ester (kinetic conditions) a more reactive acylating agent such as ***Acetyl Chloride*** is generally used. Moreover, carboxylic acid anhydrides are generally preferred to acyl halides. Accordingly, PhCOCl is preferred to prepare 1,3-dicarbonyl compounds. Ketones (eq 10),[5b] esters (eq 11),[5c] and more commonly β-keto esters or related CH acidic compounds (eqs 12–15)[5d–g] can be *C*-acylated via the preformed enolate (eq 11) or directly under basic conditions (eq 13).[5e]

$$\text{PhCOMe} \xrightarrow[\text{79\%}]{\text{1. } t\text{-C}_5\text{H}_{11}\text{ONa, benzene, 0 °C; 2. PhCOCl}} \text{PhCOCH}_2\text{COPh} \quad (10)$$

$$\text{Me}_2\text{CHCO}_2\text{-}t\text{-Bu} \xrightarrow[\text{2. PhCOCl, benzene, 0 °C to rt; 55\%}]{\text{1. LDA, hexane, −78 °C to rt}} \text{PhCO–CMe}_2\text{–CO}_2\text{-}t\text{-Bu} \quad (11)$$

$$\text{CH}_2(\text{CO}_2\text{Et})_2 \xrightarrow[\text{73\%}]{\text{1. NaH, THF, rt; 2. PhCOCl, 0 °C}} \text{PhCOCH}(\text{CO}_2\text{Et})_2 \quad (12)$$

$$\text{PhCOCH}_2\text{CO}_2\text{Et} \xrightarrow[\text{90\%}]{\text{PhCOCl, PhNEt}_2\text{, rt, 24 h}} \text{PhCOCH(COPh)CO}_2\text{Et} \quad (13)$$

$$\text{(2,2,5,5-tetramethyl-3-oxotetrahydrofuran-4-carbaldehyde)} \xrightarrow[\text{90\%}]{\text{1. Na, ether, rt; 2. PhCOCl, }\Delta\text{, 3 h then rt}} \text{(4-COPh, 4-CHO derivative)} \quad (14)$$

$$(\text{EtO})_2\text{P(O)CH}_2\text{CO}_2\text{Et} \xrightarrow[\text{2. PhCOCl, 1 h, }\Delta\text{ then 12 h, rt; 56\%}]{\text{1. (EtO)}_2\text{Mg, EtOH, ether, }\Delta} (\text{EtO})_2\text{P(O)CH(COPh)CO}_2\text{Et} \quad (15)$$

Friedel–Crafts Acylation and Related Reactions.[6,7] Aromatic compounds (eqs 16 and 17)[7a,b] are acylated by PhCOCl in the presence of a Lewis acid such as $AlCl_3$, $TiCl_4$, BF_3, $SnCl_4$, $ZnCl_2$, or $FeCl_3$, or of a strong acid such as polyphosphoric acid or CF_3SO_3H.[6] Metallic Al or Fe and iodine (in situ formation of a Lewis acid) can also act as a catalyst.[6] Various solvents that have been used to perform this reaction are CS_2, CH_2Cl_2, 1,2-dichloroethane, nitrobenzene, and nitromethane.[6] PhCOCl is less reactive than aliphatic carboxylic acid chlorides (with benzene in nitromethane the relative reaction rates are PhCOCl:MeCOCl = 6:100).[7a] As for all electrophilic substitutions, the rate and the regioselectivity of the acylation closely depend on the nature and on the position of the substituents on the aromatic system[6] (eqs 16 and 18[7c]). The nature of the solvent can also exert a strong influence.[6]

$$\text{4-}t\text{-Bu-toluene} + \text{PhCOCl} \xrightarrow[\text{85\%}]{\text{AlCl}_3\text{, PhNO}_2\text{, 10 to 40 °C then rt}} \text{(2-COPh-4-}t\text{-Bu-toluene)} \quad (16)$$

$$\text{thiophene} + \text{PhCOCl} \xrightarrow[\text{88–90\%}]{\text{AlCl}_3\text{, CS}_2\text{, rt, 12 h then }\Delta\text{, 3.5 h}} \text{2-COPh-thiophene} \quad (17)$$

$$\text{2-NHCOPh-naphthalene} \xrightarrow[\text{74\%}]{\text{PhCOCl, SnCl}_4\text{, 100 °C}} \text{1-COPh-2-NHCOPh-naphthalene} \quad (18)$$

The electrophilic acylation of alkenes or alkynes is another example of Friedel–Crafts reactions and is carried out under similar conditions.[6,8,9] With PhCOCl, alkenes can lead to β-chloro alkyl ketones or, more frequently, to the corresponding α,β-ethylenic ketones (eq 19)[8b] according to the reaction conditions.[6,8a] PhCOCl also adds to triple bonds to give a β-chloro vinyl ketone[9] (eq 20).[9b]

$$\text{H}_2\text{C=CH}_2 + \text{PhCOCl} \xrightarrow[\text{70\%}]{\text{AlCl}_3\text{, CHCl}_2\text{CHCl}_2\text{, 5 °C}} \text{H}_2\text{C=CHCOPh} \quad (19)$$

$$\text{HC}\equiv\text{CH} + \text{PhCOCl} \xrightarrow[\text{65–70\%}]{\text{AlCl}_3\text{, ClCH}_2\text{CH}_2\text{Cl, 0 °C}} \text{PhCOCH=CHCl} \quad (20)$$

Under Friedel–Crafts conditions, PhCOCl reacts with various silicon compounds. Thus β-diketones are easily obtained from enoxysilanes[10a,b] (eq 21).[10c] Conversely, vinyl-, aryl-, and allylsilanes lead to the corresponding vinyl, aryl, and allyl ketones in good yields[11a–c] (eq 22[11d] and eq 23[11e]). Vinylsilanes can lead to a mixture of β-chloro alkyl ketones and conjugated enones; a basic treatment is then necessary to obtain only the enone (eq 23).

$$\text{1-Me}_3\text{SiO-cyclohexene} + \text{PhCOCl} \xrightarrow[\text{81\%}]{\text{ZnCl}_2\text{, rt}} \text{2-COPh-cyclohexanone} \quad (21)$$

$$\text{PhCH=CHSiMe}_3 \xrightarrow[\text{2. Et}_3\text{N, ether, }\Delta\text{; 73\%}]{\text{1. PhCOCl, AlCl}_3\text{, CH}_2\text{Cl}_2\text{, 0 °C}} \text{PhCH=CHCOPh} \quad (22)$$

$$\text{H}_2\text{C=CHCH}_2\text{SiMe}_3 \xrightarrow[\text{90\%}]{\text{PhCOCl, AlCl}_3\text{, CH}_2\text{Cl}_2\text{, −30 °C}} \text{H}_2\text{C=CHCH}_2\text{COPh} \quad (23)$$

Acylation of Enol Ethers, Ketene Acetals,[12] and Enamines.[13] Enol ethers,[12a] ketene acetals (eq 24),[12b] and enamines (eq 25)[13g] react with PhCOCl to provide the corresponding β-acylated products (eq 24).[12b] The acylation of enamines has been extensively studied (eq 25);[13a–f] the resulting β-acyl enamine is generally hydrolyzed to give β-diketones (eq 25).

$$(\text{MeO})_2\text{C=CH}_2 + \text{PhCOCl} \xrightarrow[\text{55\%}]{\text{Et}_3\text{N, ether, }\Delta} (\text{MeO})_2\text{C=CHCOPh} \quad (24)$$

$$\text{1-pyrrolidinocyclopentene} \xrightarrow[\text{53\%}]{\text{1. PhCOCl, dioxane, 0 °C to }\Delta\text{; 2. 1N HCl, }\Delta} \text{2-COPh-cyclopentanone} \quad (25)$$

Protection of Alcohols as Benzoates[14] and of Amines as Benzamides.[15] PhCOCl easily reacts with alcohols to give the corresponding benzoates in excellent yields.[14a] The acylation is performed in the presence of an amine, very often pyridine or triethylamine. CH_2Cl_2 or a large excess of amine is generally used as solvent (eq 26).[14d] The reaction has also been performed by phase transfer catalysis (PhCOCl, benzene, Bu_4NCl, 40% NaOH).[14e] Alternatively, the alcohol can be converted to lithium alcoholate (with BuLi), which readily reacts with PhCOCl to give quantitatively the corresponding benzoate (eq 27).[14f] Tributyltin ether prepared by treatment of the alcohol with ***Bis(tri-n-butyltin) Oxide*** has also been used as intermediate.[14g] Benzoates are very often prepared to protect alcohols[14b,c] because they are more stable than acetates and their tendency to migrate to adjacent hydroxyl groups is lower.[14h,i] In most cases the benzoylation of polyhydroxylated molecules is more selective than the acetylation.[14h,i] PhCOCl has also been used to monoprotect diols,[14f] and to acylate a primary alcohol in the presence of a secondary alcohol.[14j] Under similar conditions (pyridine), phenols are converted to aryl benzoates.[14a,k]

HO, HO, OMe —(PhCOCl, pyridine, −45 °C to rt; 90%)→ $PhCO_2$, HO, OMe (26)

$$HO(CH_2)_4OH \xrightarrow[82\%]{\text{1. BuLi, THF, } -78\,^\circ C \quad \text{2. PhCOCl, } 0\,^\circ C} PhCO(CH_2)_4OH \quad (27)$$

PhCOCl reacts also easily with primary or secondary amines in the presence of a base, aqueous alkali (Schotten–Bauman procedure), or tertiary amines (pyridine or Et_3N), to afford the corresponding amides (eq 28).[15b] This reaction is used to protect amines.[15c]

$$PhCH_2CH_2NH_2 \xrightarrow[89\text{–}98\%]{\text{PhCOCl, pyridine, } 0\,^\circ C} PhCH_2CH_2NHCOPh \quad (28)$$

Miscellaneous Reactions. PhCOCN (see ***Acetyl Cyanide***), which is used as acylating agent (for instance, to protect alcohols)[17a] can be prepared by reacting PhCOCl with ***Copper(I) Cyanide*** (60–65%).[17b,c] Diazoalkanes are readily acylated by PhCOCl to give diazo ketones.[18a,b] These compounds are interesting as intermediates[18c–e] to prepare α-halo ketones, α-hydroxy ketones or arylacetic acids (Arndt–Eistert reaction).[18e] PhCOCl has also been used to prepare volatile acyl chlorides (C_2 to C_6) from the corresponding carboxylic acids.[19] On the other hand, it reacts with sulfoxides to lead generally to the corresponding α-chloro sulfides (Pummerer rearrangement).[20]

Related Reagents. Acetic Anhydride; Acetyl Chloride; Acetyl Cyanide; Benzoic Anhydride; Benzoyl Trifluoromethanesulfonate; *p*-Nitrobenzoyl Chloride.

---

1. *The Chemistry of Acyl Halides*; Patai, S., Ed. Wiley: London, 1972. (b) Sonntag, N. O. V. *CRV* **1953**, *52*, 237.
2. (a) March, J. *Advanced Organic Chemistry, Reactions, Mechanisms and Structures*. 4th ed.; Wiley: New York, 1992; pp 487–488. (b) Sheverdina, N. I.; Kocheskov, K. A. *The Organic Compounds of Zinc and Cadmium*; North Holland: Amsterdam, 1967. (c) Shirley, D. A. *OR* **1954**, *8*, 28. (d) Jorgenson, M. J. *OR* **1970**, *18*, 1. (e) Posner, G. H. *OR* **1975**, *22*, 253. (f) Wipf, P. *S* **1993**, 537. (g) Larock, R. C. *Comprehensive Organic Transformations*; VCH: New York, 1989; pp 686–692. (h) See also *n*-Butylmangenese Chloride.
3. (a) Gilman, H.; Nelson, J. F. *RTC* **1936**, *55*, 518. (b) Blaise, E. E. *BSF* **1911**, *9*, I–XXVI. (c) Negishi, E.; Bagheri, V., Chatterjee, S.; Luo, F.; Miller, J. A.; Stoll, A. T. *TL* **1983**, *24*, 5181. (d) Fiandanese, V.; Marchese, G.; Martina, V.; Ronzini, L. *TL* **1984**, *25*, 4805. (e) Cahiez, G.; Laboue, B. *TL* **1992**, *33*, 4439. (f) Whitesides, G. M.; Casey, C. P.; San Filippo Jr, J.; Panek, E. J. *Trans. N.Y. Acad. Sci.* **1967**, *29*, 572. (g) Jallabert, C.; Luong-Thi, N. T.; Rivière, H. *BSF* **1970**, 797. (h) Posner, G. H.; Whitten, C. E. *TL* **1970**, 4647. (i) Posner, G. H.; Whitten, C. E. *OSC* **1988**, *6*, 248. See also also Ref. 2e. (j) Cahiez, G.; Laboue, B. *TL* **1989**, *30*, 7369.
4. (a) House, H. O. *Modern Synthetic Reactions*, 2nd ed; Benjamin: Menlo Park, CA, 1972, pp 734–786. (b) See Ref. 2a, p 490. (c) See Ref. 2g, pp 742 and 764–768. (d) Hauser, C. R.; Swamer, F. W.; Adams, J. T. *OR* **1954**, *8*, 59.
5. (a) Muir, W. M.; Ritchie, P. D.; Lyman, D. J. *JOC* **1966**, *31*, 3790. (b) Vavon, G.; Conia, J. M. *CR* **1951**, *233*, 876. (c) Logue, M. W. *JOC* **1974**, *39*, 3455. (d) Lawesson, S. O.; Busch, T. *ACS* **1959**, *13*, 1717. (e) Wright, P. E.; McEven, W. E. *JACS*, **1954**, *76*, 4540. (f) Korobitsyna, I. K.; Severina, T. A.; Yur'ev, Yu. K. *ZOB* **1959**, *29*, 1960 (*CA* **1960**, *54*, 8772). (g) Gough, S. T. D.; Trippett, S. *JCS* **1962**, 2333.
6. (a) Olah, G. O. *Friedel Crafts and Related Reactions*; Wiley: New York, 1963–1965. Vol. 1–4. (b) See Ref. 4a, pp 786–816. (c) Gore, P. H. *CRV* **1955**, *55*, 229. (d) Gore, P. H. *CI(L)* **1974**, 727. (e) Pearson, D. E.; Buehler, C. A. *S* **1972**, 533. (f) See Ref. 2a, pp 539, 598 and 821. (g) See Ref. 2g, pp 703 and 708.
7. (a) Gore, P. H.; Hoskins, J. A.; Thornburn, S. *JCS(B)* **1970**, 1343. (b) Minnis, W. *OSC*, **1943**, *2*, 520. (c) Orchin, M.; Reggel, L. *JACS* **1951**, *73*, 436.
8. (a) Groves, J. K. *CSR* **1972**, *1*, 73. (b) Matsumoto, T.; Hata, K. *JACS* **1957**, *79*, 5506.
9. (a) Pohland, A. E.; Benson, W. R. *CRV* **1966**, *66*, 161. (b) Kochetkov, N. K.; Khorlin, A. Ya.; Karpeiskii, M. Ya. *ZOB* **1956**, *26*, 595 (*CA* **1956**, *50*, 13 799).
10. (a) Weber, W. P. *Silicon Reagents for Organic Synthesis*; Springer: Berlin, 1983; pp 214. (b) See Ref. 2g, p 753. (c) Fleming, I.; Iqbal, J.; Krebs, E. P. *T* **1983**, *39*, 841.
11. (a) See Ref. 10a, pp 86, 120, and 175. See Ref. 2g, pp 687 and 688. (c) Fleming, I.; Dunoguès, J.; Smithers, R. *OR* **1989**, *37*, 57. (d) Fleming, I.; Pearce, A. *JCS(P1)* **1980**, 2485. (e) Calas, R.; Dunoguès, J.; Pillot, J. P.; Biran, C.; Pisciotti, F.; Arreguy, B. *JOM* **1975**, *85*, 149.
12. (a) Andersson, C.; Hallberg, A. *JOC* **1988**, *53*, 4257. (b) McElvain, S. M.; McKay, Jr., G. R. *JACS* **1956**, *78*, 6086.
13. (a) See Ref. 4a, pp 766–772. (b) See Ref. 2a, pp 601–603 (c) *Enamines: Synthesis, Structure and Reactions*, 2nd ed; Cook, A. G., Ed.; Dekker: New York, 1988. (d) Hünig, S.; Hoch, H. *Fortschr. Chem. Forsch.* **1970**, *14*, 235. (e) Hickmott, P. W. *CI(L)* **1974**, 731. (f) Hickmott, P. W. *T* **1984**, *40*, 2989; *T* **1982**, *38*, 1975 and 3363. (g) Campbell, R. D.; Harmer, W. L. *JOC* **1963**, *28*, 379.
14. (a) See Ref. 2a, p 392. (b) Greene, T. W. *Protective Groups in Organic Synthesis*; Wiley: New York, 1981; p 61. (c) Reese, C. B. *Protective Groups in Organic Chemistry*; McOmie, J. F. W., Ed.; Plenum: London, 1973; p 111. (d) Seymour, F. R. *Carbohydr. Res.* **1974**, *34*, 65. (e) Szeja, W. *S* **1979**, 821. (f) Castellino, A. J.; Rapoport, H. *JOC* **1986**, *51*, 1006. (g) Hanessian, S.; Roy, R. *CJC* **1985**, *63*, 163. (h) Haines, A. H. *Adv. Carbohydr. Chem. Biochem.* **1976**, *33*, 11. (i) Kozikowski, A. P.; Xia, Y.; Rusnak, J. M. *CC* **1988**, 1301. (j) Schlessinger, R. H.; Lopes, A. *JOC* **1981**, *46*, 5252. (k) See Ref. 14b, p 103 and Ref. 14c, pp 171–177.
15. (a) See Ref. 2a, pp 417–418. (b) White, E. *OSC*, **1973**, *5*, 336. (c) See Ref. 14b, pp 261–263 and Ref. 14c, pp 52–53.
16. See Ref. 17b, p 113, note 3.
17. (a) See Ref. 14b, p 61. (b) Oakwood, T. S.; Weisgerber, C. A. *OSC* **1955**, *3*, 112. (c) See Ref. 2a, p 495.
18. (a) Franzen, V. *LA* **1957**, *602*, 199. (b) Bestmann, H. J.; Kolm, H. *CB* **1963**, *96*, 1948. (c) Bridson, J. N.; Hooz, J. *OSC*, **1988**, *6*, 386. (d) Ried, W.; Mengler, H. *Fortschr. Chem. Forsch.* **1965**, *5*, 1. (e) Fridman, A. L.; Ismagilova, G. S.; Zalesov, V. S.; Novikov, S. S. *RCR* **1972**, *41*, 371. (f) See Ref 2a, pp 1083–1085. (g) Meier, H.; Zeller, K.-P. *AG(E)* **1975**, *14*, 32.

19. (a) Finan, P. A.; Fothergill, G. A. *JCS* **1963**, 2723. (b) Brown, H. C. *JACS* **1938**, *60*, 1325.
20. (a) See Ref 2a, p 1236. (b) Mikolajczyk, M.; Zatorski, A.; Grzejszczak, S.; Costisella, B.; Midura, W. *JOC* **1978**, *43*, 2518. (c) De Lucchi, O.; Miotti, U.; Modena, G. *OR* **1991**, *40*, 157.

G. Cahiez
Université Pierre et Marie Curie, Paris, France

Benzyl Bromide

Ph⌒Br

[100-39-0] C_7H_7Br (MW 171.04)

(benzylating agent for a variety of heteroatomic functional groups as well as carbon nucleophiles)

Physical Data: mp −3 to −1 °C; bp 198–199 °C; *d* 1.438 g cm^{-3}.
Solubility: sol ethereal, chlorinated, and dipolar aprotic solvents.
Form Supplied in: 98–99% pure liquid.
Handling, Storage, and Precautions: the reagent is a potent lachrymator and should be handled in a fume hood.

Benzylation of Heteroatomic Functional Groups. Benzylation of various heteroatomic functional groups is readily achieved with this reagent under a variety of conditions and finds widespread application in organic synthesis, primarily as a protecting group.[1]

Alcohols and phenols are benzylated upon treatment with benzyl bromide under basic conditions. For example, treatment of alcohols with ***Sodium Hydride*** or ***Potassium Hydride*** in ethereal solvent[2] or DMF[3] generates alkoxides, which subsequently undergo Williamson reactions with benzyl bromide. Selective benzylation of a primary alcohol in the presence of a secondary alcohol has been accomplished in DMF at low temperature.[4]

Benzylation of alcohols using ***Potassium Fluoride–Alumina*** and benzyl bromide in acetonitrile at room temperature is effective.[5] Silver oxide in DMF is yet another base system.[6] Of particular interest in carbohydrate applications is the reaction of benzyl bromide with carbohydrate derivatives which have been pretreated with tin reagents. Thus it is possible to benzylate an equatorial alcohol in the presence of an axial alcohol (eq 1)[7] and also to selectively benzylate an anomeric hydroxy through ***Di-n-butyltin Oxide***.[8]

n-Bu_2SnO, MeOH; $PhCH_2Br$ (1) 60%

In some instances the sluggish reactivity of sterically hindered alcohols toward benzyl bromide may be overcome through addition of a catalytic iodide source such as ***Tetra-n-butylammonium Iodide***, which generates the more reactive benzyl iodide in situ (see ***Benzyl Iodide***). Benzylation of phenols proceeds well under the conditions described for aliphatic alcohols. Owing to the greater acidity of phenols it is possible to use weaker bases such as ***Potassium Carbonate*** for these reactions.[9]

Benzyl bromide will readily alkylate amino groups. Reactions are normally carried out in the presence of additional base and dibenzylation of primary amines is usually predominant.[10] Selective quaternization of a less hindered tertiary amine in the presence of a more hindered tertiary amine has been described.[11] Amide and lactam nitrogens can be benzylated under basic conditions,[12] as can those of sulfonamides[13] and nitrogen heterocycles.[14]

Thiols,[15] silyl thioethers,[16] and thiosaccharins[17] may be benzylated with benzyl bromide under basic conditions. Thus L-cysteine is *S*-benzylated under basic conditions (eq 2).[18] Benzylation of selenols is likewise possible.[19] A synthesis of benzylic sulfones is possible using ***Benzenesulfonyl Chloride*** and ***Sodium O,O-Diethyl Phosphorotelluroate*** with benzyl bromide.[20]

$PhCH_2Br$, 1N NaOH, rt, 84% (2)

Although the preparation of benzyl carboxylate esters from benzyl bromide and carboxylate anions is not the most common route to these compounds, the reaction is possible when carried out in DMF[20] or using zinc carboxylates.[21]

Nucleophilic attack on benzyl bromide by cyanide and azide anions is feasible with ion-exchange resins or with the corresponding salts.[22]

Reactions with Active Methylene Compounds. Enolates of ketones,[23] esters,[24] enediolates,[25] 1,3-dicarbonyl compounds,[26] amides and lactams,[27] as well as nitrile-stabilized carbanions,[28] can be alkylated with benzyl bromide. Cyclohexanone may be benzylated in 92% ee using a chiral amide base.[29] Amide bases as well as alkoxides have been employed in the case of nitrile alkylations.[28b] Benzylation of metalloenamines may be achieved[30a] and enantioselective reactions are possible using a chiral imine (eq 3).[30b] However, reactions between benzyl bromide and enamines proceed in low yield.[31] The benzylation of a ketone via its enol silyl ether, promoted by fluoride, has been observed.[32]

1. LDA 2. $ZnBr_2$ 3. $PhCH_2Br$ 4. H_3O^+, 84%; 79% ee (3)

Reactions with Metals and Organometallics. Difficulties encountered in the preparation of benzylic metal compounds with active metals are due primarily to the tendency of these compounds to undergo Wurtz coupling (self condensation).[33] Benzylmagnesium bromide may nevertheless be prepared from benzyl bromide and used under standard[34] or Barbier conditions.[35] Benzyllithium cannot be obtained practically from benzyl bromide. Benzylzinc bromide and the cyanocuprate BnCu(CN)ZnBr have both been prepared. The cuprate undergoes 1,2-additions with aldehydes and ketones.[36]

The propensity of benzyl bromide to undergo coupling with organometallic reagents may be used to advantage, as organolithiums,[37] Grignard reagents,[38] organocuprates,[39] organocadmiums,[40] organochromiums,[41] and organoiron reagents[42] are all known to give coupling products. An interesting coupling of benzyl bromide with *N*-methylphthalimide under dissolving metal conditions has been reported (eq 4).[43]

N-methylphthalimide —(2.1 equiv Li, NH_3; $PhCH_2Br$; 96%)→ 3-benzyl-3-hydroxy-2-methylisoindolin-1-one (4)

Related Reagents. Benzyl Chloride; Benzyl Iodide; Benzyl *p*-Toluenesulfonate; Benzyl 2,2,2-Trichloroacetimidate; Benzyl Trifluoromethanesulfonate; 3,4-Dimethoxybenzyl Bromide; *p*-Methoxybenzyl Chloride; 4-Methoxybenzyl 2,2,2-Trichloroacetimidate.

---

1. (a) Greene, T. W.; Wuts, P. G. M. *Protective Groups in Organic Synthesis*, 2nd ed.; Wiley: New York, 1991. (b) *Protective Groups in Organic Chemistry*; McOmie, J. F. W., Ed.; Plenum: New York, 1973.
2. Nicolaou, K. C.; Pavia, M. R.; Seitz, S. P. *JACS* **1981**, *103*, 1224.
3. Hanessian, S.; Liak, T. J.; Dixit, D. M. *Carbohydr. Res.* **1981**, *88*, C14.
4. Fukuzawa, A.; Sato, H.; Masamune, T. *TL* **1987**, *28*, 4303.
5. Ando, T.; Yamawaki, J.; Kawate, T.; Sumi, S.; Hanafusa, T. *BCJ* **1982**, *55*, 2504.
6. Kuhn, R.; Low, I.; Trischmann, H. *CB* **1957**, *90*, 203.
7. (a) Nashed, M. A.; Anderson, L. *TL* **1976**, 3503. (b) Cruzado, C.; Bernabe, M.; Martin-Lomas, M. *JOC* **1989**, *54*, 465.
8. Bliard, C.; Herczegh, P.; Olesker, A.; Lukacs, G. *Carbohydr. Res.* **1989**, *8*, 103.
9. Schmidhammer, H.; Brossi, A. *JOC* **1971**, *93*, 746.
10. (a) Yamazaki, N.; Kibayashi, C. *JACS* **1989**, *111*, 1396. (b) Gray, B. D.; Jeffs, P. W. *CC* **1987**, 1329.
11. (a) Chung, B.-H.; Zymalkowski, F. *AP* **1984**, *317*, 307. (b) Chung, B.-H.; Zymalkowski, F. *AP* **1984**, *317*, 323.
12. (a) Landini, D.; Penso, M. *SC* **1988**, *18*, 791. (b) Staskun, B. *JOC* **1979**, *44*, 875. (c) Sato, R.; Senzaki, T.; Goto, T.; Saito, M. *BCJ* **1986**, *59*, 2950.
13. Bergeron, R. J.; Hoffman, P. G. *JOC* **1979**, *44*, 1835.
14. Chivikas, C. J.; Hodges, J. C. *JOC* **1987**, *52*, 3591.
15. Harpp, D. N.; Kobayashi, M. *TL* **1986**, *27*, 3975.
16. Ando, W.; Furuhata, T.; Tsumaki, H.; Sekiguchi, A. *SC* **1982**, *12*, 627.
17. Yamada, H.; Kinoshita, H.; Inomata, K.; Kotake, H. *BCJ* **1983**, *56*, 949.
18. Dymicky, M.; Byler, D. M. *OPP* **1991**, *23*, 93.
19. Mitchell, R. H. *CC* **1974**, 990.
20. Huang, X.; Pi, J.-H. *SC* **1990**, *20*, 2291.
21. (a) Comber, M. F.; Sargent, M. V.; Skelton, B. W.; White, A. H. *JCS(P1)* **1989**, 441. (b) Shono, T.; Ishige, O.; Uyama, H.; Kashimura, S. *JOC* **1986**, *51*, 546.
22. (a) Gordon, M.; Griffin, C. E. *CI(L)* **1962**, 1019. (b) Hassner, A.; Stern, M. *AG* **1986**, *98*, 479. (c) Bram, G.; Loupy, A.; Pedoussaut, M. *BSF(2)* **1986**, 124. (d) Ravindranath, B.; Srinivas, P. *T* **1984**, *40*, 1623.
23. (a) Gall, M.; House, H. O. *OSC* **1988**, *6*, 121. (b) Sato, T.; Watanabe, T.; Hayata, T.; Tsukui, T. *CC* **1989**, 153.
24. (a) Seebach, D.; Estermann, H. *TL* **1987**, *28*, 3103. (b) Lerner, L. M. *JOC* **1976**, *41*, 2228.
25. Duhamel, L.; Poirier, J.-M. *BSF(2)* **1982**, 297.
26. (a) Berry, N. M.; Darey, M. C. P.; Harwood, L. M. *S* **1986**, 476. (b) Bassetti, M.; Cerichelli, G.; Floris, B. *G* **1986**, *116*, 583. (c) Asaoka, M.; Miyake, K.; Takei, H. *CL* **1975**, 1149. (d) Ogura, K.; Yahata, N.; Minoguchi, M.; Ohtsuki, K.; Takahashi, K.; Iida, H. *JOC* **1986**, *51*, 508.
27. (a) Woodbury, R. P.; Rathke, M. W. *JOC* **1977**, *42*, 1688. (b) Klein, U.; Sucrow, W. *CB* **1977**, *110*, 1611. (c) Meyers, A. I.; Harre, M.; Garland, R. *JACS* **1984**, *106*, 1146.
28. (a) Arseniyadis, S.; Kyler, K. S.; Watt, D. S. *OR* **1984**, *31*, 1. (b) Cope, A. C.; Holmes, H. L.; House, H. O. *OR* **1957**, *9*, 107.
29. Murakata, M.; Nakajima, M.; Koga, K. *CC* **1990**, 1657.
30. (a) Stork, G.; Dowd, S. R. *OSC* **1988**, *6*, 526. (b) Saigo, K.; Kashahara, A.; Ogawa, S.; Nohira, H. *TL* **1983**, *24*, 511.
31. (a) *Enamines: Synthesis, Structure, and Reactions*, 2nd ed.; Cook, A. G., Ed.; Dekker: New York, 1988. (b) Brannock, K. C.; Burpitt, R. D. *JOC* **1961**, *26*, 3576. (c) Opitz, G.; Hellmann, H.; Mildenberger, M.; Suhr, H. *LA* **1961**, *649*, 36.
32. (a) Kuwajima, I.; Nakamura, E. *JACS* **1975**, *97*, 3257. (b) Binkley, E. S.; Heathcock, C. H. *JOC* **1975**, *40*, 2156.
33. Wakefield, B. J. *Organolithium Methods*; Academic: New York, 1988.
34. (a) Kharasch, M. S.; Reinmuth, O. *Grignard Reactions of Nonmetallic Substances*; Constable: London, 1954. (b) Reuvers, A. J. M.; van Bekkum, H.; Wepster, B. M. *T* **1970**, *26*, 2683.
35. Blomberg, C.; Hartog, F. A. *S* **1977**, 18.
36. Berk, S. C.; Knochel, P.; Yeh, M. C. P. *JOC* **1988**, *53*, 5789.
37. (a) Hirai, K.; Matsuda, H.; Kishida, Y. *TL* **1971**, 4359. (b) Hirai, K.; Kishida, Y. *TL* **1972**, 2743. (c) Villieras, J.; Rambaud, M; Kirschleger, B.; Tarhouni, R. *BSF(2)* **1985**, 837.
38. Rahman, M. T.; Nahar, S. K. *JOM* **1987**, *329*, 133.
39. (a) Kobayashi, Y.; Yamamoto, K.; Kumadaki, I. *TL* **1979**, 4071. (b) Furber, M.; Taylor, R. J. K.; Burford, S. C. *TL* **1985**, *26*, 3285.
40. (a) Emptoz, G.; Huet, F. *BSF(2)* **1974**, 1695.
41. Wellmann, J.; Steckhan, E. *S* **1978**, 901.
42. (a) Sawa, Y.; Ryang, M.; Tsutsumi, S. *JOC* **1970**, *35*, 4183. (b) Cookson, R. C.; Farquharson, G. *TL* **1979**, 1255. (c) Sawa, Y.; Ryang, M.; Tsutsumi, S. *TL* **1969**, 5189.
43. Flynn, G. A. *CC* **1980**, 862.

William E. Bauta
Sandoz Research Institute, East Hanover, NJ, USA

Benzyl Chloroformate[1]

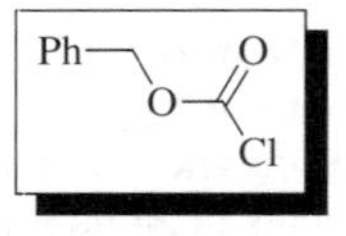

[501-53-1] $C_8H_7ClO_2$ (MW 170.60)

(protecting agent for many functional groups, especially for the N-protection of amino acids;[7] activates pyridine rings toward nucleophilic attack;[53] reagent for the preparation of other benzyloxycarbonylating agents[3])

Alternate Name: benzyloxycarbonyl chloride (CbzCl).
Physical Data: colorless oily liquid, bp 103 °C/20 mmHg (with slow decarboxylation to benzyl chloride at this temperature); bp 85–87 °C/7 mmHg; mp 0 °C; *d* (20 °C) 1.195 g cm^{-3}; n_D^{20} 1.5190.
Form Supplied in: widely available.

Preparative Methods: CbzCl can be readily prepared by the reaction of benzyl alcohol and phosgene, either in toluene solution[4] or neat.[1b,5,6] Purification of such freshly prepared CbzCl is often not necessary; indeed, ca. 25% residual toluene solvent does not interfere with derivatization reactions.[4]

Purification: commercial CbzCl is typically better than 95% pure and is often contaminated by benzyl chloride, benzyl alcohol, toluene, and HCl.[2,3] On storage it can undergo a slow HCl-catalyzed decomposition, even at low temperature.[1a,3] It has been recommended[1b,2] that CbzCl which has been stored for long periods be purified by flushing with a stream of dry air to remove dissolved CO_2 and HCl impurities. Filtration and storage over Na_2SO_4 may be followed by distillation; it is important that low temperatures (oil bath $\leq 85\,^\circ C$)[1b] and high vacuum be used to minimize thermal decomposition.[3]

Handling, Storage, and Precautions: CbzCl is highly toxic and is a cancer suspect agent; it is a lachrymator with an acrid odor and should be handled with caution in a fume hood.

Protection of Amines. The most widespread application of CbzCl is in the protection of primary and secondary amines as the corresponding benzyl carbamates. These reactions are usually performed either under Schotten–Baumann conditions,[7] by using aq $NaHCO_3$ or aq Na_2CO_3 as base in an organic solvent (eq 1),[8] or in the presence of an organic base (typically Et_3N) in CH_2Cl_2.[9] The isolation and/or crystallization of amines is often facilitated by Cbz derivatization of the crude amine generated in another reaction.[10] A wide variety of primary and secondary aliphatic amines, including aziridines,[11] azetidines,[12] and other sensitive[7,9] and electron-deficient[8c] amines, can be selectively protected in the presence of alcohol,[8b,13] phenol,[14] and thiol[15] functionality; however, the selective protection of hindered secondary amines over secondary alcohols has proved difficult.[16]

BnOC(O)Cl, aq $NaHCO_3$, ether, 95% (1)

The N-protection of amino acids during peptide synthesis constitutes a major application of this methodology.[1,8a] References to the N-protection of individual amino acids using CbzCl have been tabulated.[1b,e] Use of the *N*-Cbz protecting group generally affords crystalline amino acid derivatives and suppresses racemization at the α-stereocenter during peptide bond formation. The *N*-Cbz group is stable to conditions used for the formation of peptide bonds. Benzyl carbamates are highly complementary to *t*-butyl (i.e. Boc) carbamates in peptide synthesis (see ***t-Butyl Azidoformate***). Schotten–Baumann conditions are most commonly employed for the preparation of *N*-Cbz amino acids (eq 2);[1b,4] racemization at the α stereocenter is suppressed under these basic conditions by the adjacent negatively charged carboxylate group. The use of a lower reaction temperature ($-20\,^\circ C$) has been shown in one case to afford a higher yield of pure *N*-Cbz derivative.[17] One disadvantage of the standard Schotten–Baumann protocol is the need for either the simultaneous addition of CbzCl and aq NaOH, or the sequential addition of small aliquots of these reagents, throughout the course of the reaction. A modified procedure[1b] using $NaHCO_3$ as the base[8d] avoids this inconvenience. A recent report indicates that *N*-Cbz derivatization of a range of α-amino acids can be conveniently achieved, albeit in only moderate yield, by refluxing with CbzCl in ethyl acetate without added base; interestingly, no racemization was observed.[18]

BnOC(O)Cl, 2M aq NaOH, toluene, rt, 30 min, 80–90% (2)

The selective protection of the ω-amine in α,ω-diamino acids can be achieved by chelation of the α-amino and α-carboxylate functionality by Cu^{II} ion (eq 3).[1b,f,18,19] The Cu^{II} chelate can be cleaved by H_2S,[1f] by thioacetamide,[19] or by use of a chelating ion-exchange resin.[17] Selective protection of the ε-amino group in lysine, the imidazole ring nitrogen in histidine, and the α-amino function in serine, can all be accomplished in homogeneous aqueous medium in the absence of added base using Cbz-imidazolium chloride, a water-soluble benzyloxycarbonylating agent which is prepared in situ from CbzCl and *N*-methylimidazole.[20] Benzyl succinimidyl carbonate, best prepared by reaction of the dicyclohexylammonium salt of *N*-hydroxysuccinimide with CbzCl,[21] is also useful for the *N*-Cbz protection of α-amino acids without the need for added base.

1. $CuCl_2$, aq NaOH; 2. BnOC(O)Cl, pH 9.5, 0 °C, 3.5 h; 3. $MeC(S)NH_2$; 85% (3)

The *N*-Cbz group is stable to a variety of weakly acidic and basic conditions.[22] Benzyl carbamates are most commonly cleaved to the free amine by catalytic hydrogenolysis (H_2/Pd–C)[23] although a plethora of alternate reductive (e.g. transfer hydrogenolysis,[24] Li/liq NH_3[25]), strongly acidic (e.g. HBr/HOAc[26]), and neutral conditions have been employed.[1] The use of transfer hydrogenolysis (cyclohexadiene, 10% Pd–C) allows for the clean deprotection of *N*-Cbz pyrimidines without concomitant reduction of the 5,6-double bond that is observed under standard catalytic hydrogenolysis conditions.[27] The acid stability of the *N*-Cbz group can be manipulated by the introduction of electron-withdrawing or electron-donating substituents onto the phenyl ring. Selective deprotection of an *N*-Cbz group in the presence of an *S*-Cbz moiety can be achieved using HBr/HOAc at rt.[28]

Primary and secondary arylamines can also be *N*-carbamoylated using CbzCl, typically in the presence of aq Na_2CO_3 (eq 4)[29] or pyridine.[30] The protection of electron-deficient anilines can be accomplished using MgO in acetone.[31] Protection of a pyrrole nitrogen can be achieved by initial deprotonation to the potassium salt.[32]

BnOC(O)Cl, aq Na_2CO_3, THF, 0 °C, 30 min, 84% (4)

N-Benzyl carbamates prepared as outlined above can also serve as useful intermediates in the overall *N*-methylation of secondary amines, since $LiAlH_4$ reduction of a benzyl carbamate affords the corresponding tertiary *N*-methylamine (eq 5).[33]

$LiAlH_4$, THF, reflux, 3 h, 90% (5)

Other N-Protection Reactions. The protection of an amide as its *N*-Cbz derivative can be achieved using CbzCl with Et_3N/DMAP,[34] or by initial deprotonation of the amide using BuLi or $(Me_3Si)_2NLi$.[35] The N-protection of nucleosides is usually ineffective with CbzCl but can be selectively and efficiently accomplished using *N*-Cbz imidazolium salts generated in situ from the reaction of CbzCl with *N*-alkylimidazoles (eq 6).[27,36]

1. BnOC(O)Cl, toluene; 2. $Et_3O^+BF_4^-$, CH_2Cl_2, 0 °C; R = TBDMSO…OTBDMS, CH_2Cl_2; 95% (6)

Protection of Alcohols and Phenols. Protection of alcohols and phenols as the corresponding benzyl carbonates is typically achieved using CbzCl in the presence of an organic base (e.g. Et_3N,[37] pyridine[38]) in CH_2Cl_2 or ether at low temperature (−20 °C to rt). The selective protection of one secondary alcohol in the presence of other secondary and tertiary alcohols has been accomplished using DMAP at very low temperatures (−40 to −20 °C) for 3 days (eq 7).[39]

BnOC(O)Cl, DMAP, CH_2Cl_2, −40 to −20 °C, 3 days, 87% (7)

The use of NaH as base allows for the efficient protection of even hindered electron-deficient alcohols (eq 8).[40] A secondary alcohol can be selectively protected over an indole nitrogen.[41] 1-*O*-Methylated 2-deoxyribose sugars can be protected at the 3- and 5-hydroxyl groups[38] while the hemiacetal function is selectively protected over secondary alcohols in 1-unprotected sugars.[42] Both alcohol-[38,43] and phenol-derived[44] benzyl carbonates are efficiently deprotected by catalytic hydrogenolysis; benzyl carbonates are more readily cleaved under these conditions than benzyl ethers.[38] Removal of the *O*-Cbz group can also be achieved by electrolytic reduction.[45]

NaH, BnOC(O)Cl, Bu_4NI, CH_2Cl_2, 92% (8)

Protection of Thiols. The protection of thiols using CbzCl can be achieved using Et_3N as base.[46] The use of aq $NaHCO_3$ as base allows for the selective protection of the SH group in cysteine (eq 9), while *N,S*-bisprotection is observed under Schotten–Baumann conditions (eq 10).[47] Selective protection of the cysteine SH moiety can also be achieved using Cbz-imidazolium chloride, a water-soluble benzyloxycarbonylating agent prepared in situ from CbzCl and *N*-methylimidazole.[20] Both *N*- and *S*-Cbz groups are deprotected using refluxing CF_3CO_2H[48] or by electrolytic reduction.[45] The use of NaOMe/MeOH at rt for 5–10 min allows for the selective cleavage of *S*-Cbz over *N*-Cbz groups.[28]

BnOC(O)Cl, aq $NaHCO_3$, 0 °C, 1 h; 10 °C, 1 h, 67% (9)

BnOC(O)Cl, aq NaOH, 30 min, 71% (10)

Decarboxylative Esterification of Carboxylic Acids. The preparation of benzyl esters from sterically uncongested carboxylic acids,[49] including α-keto acids[37,50] and malonic acid half esters,[51] can be achieved using CbzCl in the presence of an organic base (Et_3N,[37,50,51] pyridine,[52] or Et_3N/DMAP[49] in CH_2Cl_2 or THF). The lack of reactivity of sterically crowded carboxylic acids[49] allows for the selective esterification of the less crowded of two acid moieties (eq 11).[52] A major attraction of the benzyl ester group is its susceptibility to cleavage to the free carboxylic acid by catalytic hydrogenolysis[1d,23] or transfer hydrogenolysis.[1d]

BnOC(O)Cl, pyridine, CH_2Cl_2, rt, overnight, 57% (11)

Activation of Pyridine Ring Towards Nucleophilic Attack. Pyridine and various 4-substituted pyridines can be activated by CbzCl toward regioselective nucleophilic attack by

$NaBH_4$[53] and by alkyl,[54] aryl,[55] and alkynyl[56] Grignard reagents, affording the corresponding 2-substituted 1,2-dihydropyridines (eq 12). Treatment of a 3-substituted pyridine with CbzCl and $NaBH_4$ can afford a mixture of 1,2- and 1,4-reduction products; however, reaction with CbzCl and Red-Al proceeds via regioselective 1,2-reduction at the less crowded α-carbon (eq 13).[57]

1. (ethoxyethoxy)alkyl MgCl (3), BnOC(O)Cl, THF, −20 °C; 2. aq oxalic acid, 85% (12)

BnOC(O)Cl, $NaAlH_2(OCH_2CH_2OMe)_2$, THF, toluene, −70 °C, 30 min, 58% (13)

+ 16% two other regioisomers

Generation of Other Benzyloxycarbonylating Agents. A variety of other benzyloxycarbonylating agents can be prepared from CbzCl.[3] For example, reaction of CbzCl with sodium benzyl carbonate affords dibenzyl carbonate, a promising stable crystalline alternative to CbzCl for achieving clean and efficient *N*-benzyloxycarbonylation.[3] *N*-Cbz-imidazole, prepared from CbzCl and imidazole, has proved a suitable reagent for the selective protection of primary over secondary amines.[58]

Related Reagents. Allyl Chloroformate; 4-Bromobenzyl Chloroformate; *t*-Butyl Chloroformate; Ethyl Chloroformate; 9-Fluorenylmethyl Chloroformate; Isobutyl Chloroformate; Methyl Chloroformate; 2,2,2-Tribromoethyl Chloroformate; 2,2,2-Trichloroethyl Chloroformate; 2,2,2-Trichloro-*t*-butoxycarbonyl Chloride; 2-(Trimethylsilyl)ethyl Chloroformate; Vinyl Chloroformate.

1. (a) Bodanszky, M. *Peptide Chemistry A Practical Textbook*; Springer: Berlin, 1988. (b) Greenstein, J. P.; Winitz, M. *Chemistry of the Amino Acids*; Wiley: New York, 1961; Vol. 2, pp 887–895. (c) Wunsch, E. *MOC* **1974**, *15(1)*, 47–69. (d) Meienhofer, J. In *Chemistry and Biochemistry of the Amino Acids*, Barrett, G. C., Ed.; Chapman & Hall: London, 1985; pp 299–302. (e) Pettit, G. R. *Synthetic Peptides*; Van Nostrand Reinhold: New York, 1970–71; Vols. 1,2, Academic: New York, 1975; Vol. 3. (f) Bailey, P. D. *An Introduction to Peptide Chemistry*; Wiley: Chichester, 1990.
2. Perrin, D. D.; Armerago, W. L. F. *Purification of Laboratory Chemicals*, 3rd ed.; Pergamon: Oxford, 1988; p 97.
3. Wunsch, E.; Graf, W.; Keller, O.; Keller, W.; Wersin, G. *S* **1986**, 958.
4. Carter, H. E.; Frank, R. L.; Johnston, H. W. *OSC* **1955**, *3*, 167.
5. Katchalski, E. *Methods Enzymol.* **1957**, *3*, 541.
6. Farthing, A. C. *JCS* **1950**, 3213.
7. Krow, G. R.; Cannon, K. C.; Carey, J. T.; Ma, H.; Raghavachari, R.; Szczepanski, S. W. *JOC* **1988**, *53*, 2665.
8. (a) Bergmann, M.; Zervas, L. *CB* **1932**, *65*, 1192 (*CA* **1932**, *26*, 5072). (b) Austin, G. N.; Baird, P. D.; Fleet, G. W. J.; Peach, J. M.; Smith, P. W.; Watkin, D. J. *T* **1987**, *43*, 3095. (c) Takahashi, Y.; Yamashita, H.; Kobayashi, S.; Ohno, M. *CPB* **1986**, *34*, 2732. (d) Akhtar, M.; Gani, D. *T* **1987**, *43*, 5341.
9. Meyers, A. I.; Dupre, B. *H* **1987**, *25*, 113.
10. Bashyal, B. P.; Fleet, G. W. J.; Gough, M. J.; Smith, P. W. *T* **1987**, *43*, 3083.
11. Senter, P. D.; Pearce, W. E.; Greenfield, R. S. *JOC* **1990**, *55*, 2975.
12. Baldwin, J. E.; Adlington, R. M.; Jones, R. H.; Schofield, C. J.; Zaracostas, C.; Greengrass, C. W. *T* **1986**, *42*, 4879.
13. Patel, A.; Poller, R. C. *RTC* **1988**, *107*, 182.
14. Van der Eycken, J.; et al. *TL* **1989**, *30*, 3873.
15. Atkinson, J. G.; Girard, Y.; Rokach, J.; Rooney, C. S. *JMC* **1979**, *22*, 99.
16. Iida, H.; Watanabe, Y.; Kibayashi, C. *JACS* **1985**, *107*, 5534.
17. Scott, J. W.; Parker, D.; Parrish, D. R. *SC* **1981**, *11*, 303.
18. Kruse, C. H.; Holden, K. G. *JOC* **1985**, *50*, 2792.
19. Payne, L. S.; Boger, J. *SC* **1985**, *15*, 1277.
20. Guibe-Jampel, E.; Bram, G.; Vilkas, M. *BSF(2)* **1973**, 1021 (*CA* **1973**, *79*, 31 982g).
21. Paquet, A. *CJC* **1982**, *60*, 976.
22. (a) Boisonnas, R. A. *Adv. Org. Chem.* **1963**, *3*, 159. (b) Greene, T. W.; Wuts, P. G. M. *Protective Groups in Organic Synthesis*, 2nd ed.; Wiley: New York, 1991; pp 441–444.
23. Hartung, W. H.; Simonoff, R. *OR* **1953**, *7*, 263.
24. Jackson, A. E.; Johnstone, R. A. W. *S* **1976**, 685.
25. Boutin, R. H.; Rapoport, R. H. *JOC* **1986**, *51*, 5320.
26. Ben-Ishai, D.; Berger, A. *JOC* **1952**, *17*, 1564.
27. Watkins, B. E.; Kiely, J. S.; Rapoport, H. *JACS* **1982**, *104*, 5702.
28. Sokolovsky, M.; Wilchek, M.; Patchornik, A. *JACS* **1964**, *86*, 1202.
29. Jones, G. B.; Moody, C. J. *JCS(P1)* **1989**, 2455.
30. Torrini, I.; Zecchini, G. P.; Agrosi, F.; Paradisi, M. P. *JHC* **1986**, *23*, 1459.
31. Rewcastle, G. W.; Baguley, B. C.; Cain, B. F. *JMC* **1982**, *25*, 1231.
32. Drew, M. G. B.; George, A. V.; Isaacs, N. S.; Rzepa, H. S. *JCS(P1)* **1985**, 1277.
33. Iida, H.; Yamazaki, N.; Kibayashi, C. *JOC* **1987**, *52*, 1956.
34. Fukuyama, T.; Nunes, J. J. *JACS* **1988**, *110*, 5196.
35. Nagasaka, T.; Koseki, Y.; Hamaguchi, F. *TL* **1989**, *30*, 1871.
36. Watkins, B. E.; Rapoport, H. *JOC* **1982**, *47*, 4471.
37. Arnoldi, A.; Merlini, L.; Scaglioni, L. *JHC* **1987**, *24*, 75.
38. Bobek, M.; Bloch, A.; Kuhar, S. *TL* **1973**, 3493.
39. Baker, W. R.; Clark, J. D.; Stephens, R. L.; Kim, K. H. *JOC* **1988**, *53*, 2340.
40. Sutherland, J. K.; Watkins, W. J.; Bailey, J. P.; Chapman, A. K.; Davies, G. M. *CC* **1989**, 1386.
41. Kuehne, M. E.; Podhorez, D. E.; Mulamba, T.; Bornmann, W. G. *JOC* **1987**, *52*, 347.
42. Saito, H.; Nishimura, Y.; Kondo, S.; Umezawa, H. *CL* **1987**, 799.
43. Daubert, B. F.; King, C. G. *JACS* **1939**, *61*, 3328.
44. Kuhn, M.; von Wartburg, A. *HCA* **1969**, *52*, 948.
45. Mairanovsky, V. G. *AG(E)* **1976**, *15*, 281.
46. Barton, D. H. R.; Manly, D. P.; Widdowson, D. A. *JCS(P1)* **1975**, 1568.
47. Berger, A.; Noguchi, J.; Katchalski, E. *JACS* **1956**, *78*, 4483.
48. Zervas, L.; Photaki, I.; Ghelis, N. *JACS* **1963**, *85*, 1337.
49. (a) Kim, S.; Kim, Y. C.; Lee, J. I. *TL* **1983**, *24*, 3365. (b) Kim, S.; Lee, J. I.; Kim, Y. C. *JOC* **1985**, *50*, 560.
50. Domagala, J. M. *TL* **1980**, *21*, 4997.
51. Gutman, A. L.; Boltanski, A. *TL* **1985**, *26*, 1573.
52. Baldwin, J. E.; Otsuka, M.; Wallace, P. M. *T* **1986**, *42*, 3097.
53. Sundberg, R. J.; Amat, M.; Fernando, A. M. *JOC* **1987**, *52*, 3151.
54. (a) Comins, D. L.; Abdullah, A. H.; Smith, R. K. *TL* **1983**, *24*, 2711. (b) Comins, D. L.; Brown, J. D. *TL* **1986**, *27*, 4549.
55. Kurita, J.; Iwata, K.; Tsuchiya, T. *CPB* **1987**, *35*, 3166.
56. Natsume, M.; Ogawa, M. *H* **1983**, *20*, 601.

57. Natsume, M.; Utsonomiya, I.; Yamaguchi, K.; Sakai, S.-I. *T* **1985**, *41*, 2115.

58. Sharma, S. K.; Miller, M. J.; Payne, S. M. *JMC* **1989**, *32*, 357.

Paul Sampson
Kent State University, Kent, OH, USA

Benzyl Chloromethyl Ether[1]

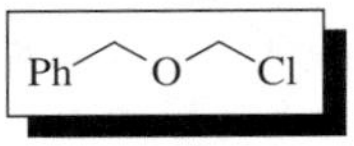

Ph⌒O⌒Cl

[3587-60-8] C_8H_9ClO (MW 156.61)

(protecting group for alcohols[2] and amines;[3] electrophile in alkylation of carbanions;[4] and source of nucleophilic benzyloxymethyl anion[5])

Physical Data: bp 53–56 °C/1.5 mmHg; 96–99 °C/11 mmHg; *d* 1.13 g cm^{-3}; n_D^{20} 1.5268–1.5279.
Solubility: sol THF, ether, dioxane, dichloromethane, chloroform, benzene, toluene, DMF, DMSO, and acetonitrile.
Form Supplied in: liquid.
Analysis of Reagent Purity: NMR, GC.
Preparative Methods: benzyl alcohol plus aqueous formaldehyde and hydrogen chloride gas;[1,6,7] reaction of benzyl methyl ether with boron trichloride;[8] cleavage of benzyloxymethyl methyl sulfide with sulfuryl chloride.[9]
Purification: distillation.
Handling, Storage, and Precautions: stable indefinitely when kept dry but destroyed by water and other nucleophilic reagents.

Alcohol Protecting Group. The foremost use for this reagent is as a protecting group for the alcohol functional group. This can be illustrated in a step utilized in the total synthesis of kijanolide (eq 1).[10] An enantioselective synthesis of the taxol side chain used the benzyloxymethyl (BOM) ether for protection of an alcohol group (eq 2).[11]

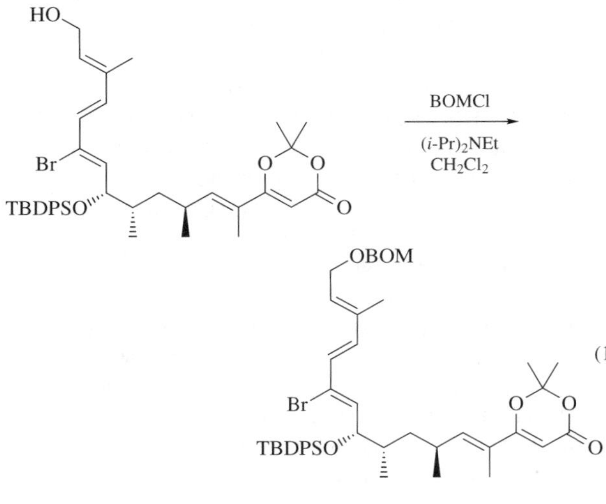

BOMCl, (*i*-Pr)$_2$NEt, CH_2Cl_2 (1)

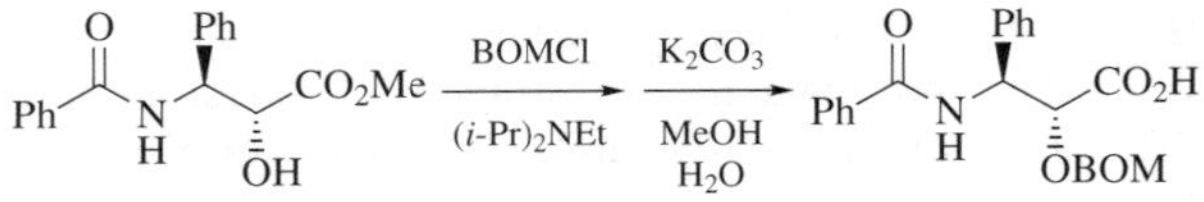

BOMCl, (*i*-Pr)$_2$NEt; K_2CO_3, MeOH, H_2O (2) 82%

Ethyl lactate can be protected as a BOM ether for subsequent nucleophilic additions to the corresponding aldehyde with high diastereoselectivity.[12] A synthesis of zoapatanol used selective protection of a diol (eq 3).[13] Primary alcohols can be protected selectively in the presence of secondary hydroxy groups,[14] and secondary hydronyls react more quickly than tertiary ones with benzyl chloromethyl ether and base.[15] Removal of this protecting group can be carried out by hydrogenolysis over Pd.

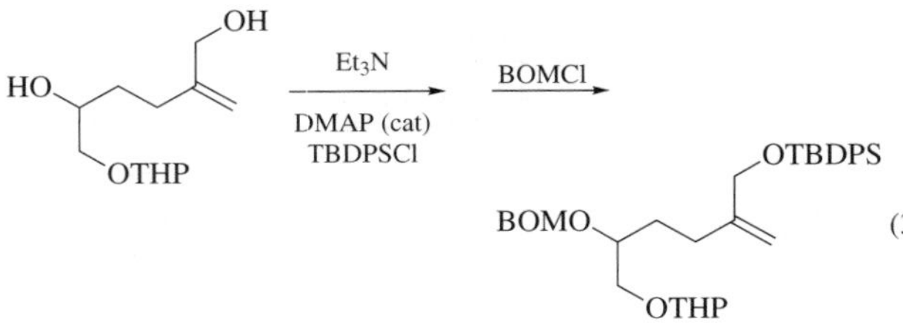

Et_3N, DMAP (cat), TBDPSCl; BOMCl (3) 88%

Amine Protecting Group. 2,3,5-Triiodoimidazole reacts with ***Potassium Carbonate*** in DMF followed by benzyl chloromethyl ether to give the 1-BOM derivative in 83% yield.[3] A substituted purine undergoes protection with benzyl chloromethyl ether and ***Sodium Hydride*** (eq 4).[16] Imide nitrogens can be protected in the presence of lactam nitrogens because of the difference in acidity.[17]

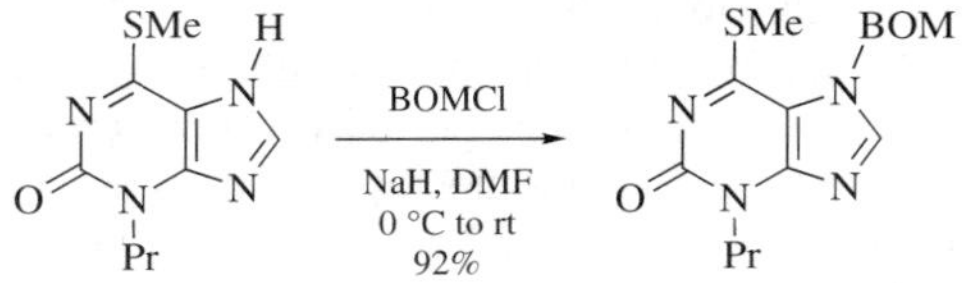

BOMCl, NaH, DMF, 0 °C to rt, 92% (4)

Alkylation of Carbanions. Methyl 2-phenylpropionate can be deprotonated with LDA and alkylated on carbon with benzyl chloromethyl ether.[4] Introduction of the BOM group in both chiral amides (eq 5)[18] and dianions (eq 6)[19] gives good yields. Removal of the benzyl group by hydrogenolysis completes the hydroxymethylation process.

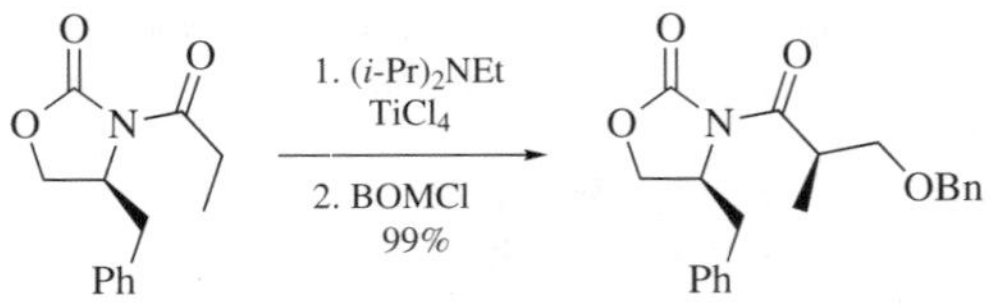

1. (*i*-Pr)$_2$NEt, $TiCl_4$; 2. BOMCl, 99% (5)

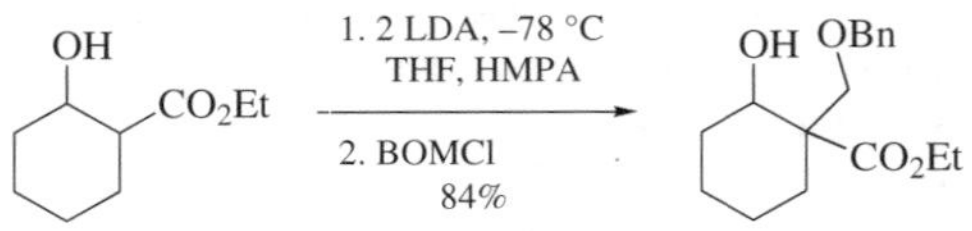

1. 2 LDA, –78 °C, THF, HMPA; 2. BOMCl, 84% (6)

Benzyloxymethyl Anion. Both Grignard and organolithium reagents of the benzyloxymethyl anion can be prepared. The Grignard reagent is formed by reaction with ***Magnesium***.[20] Trapping this species with ***Tri-n-butylchlorostannane*** and reaction of the stannane with ***n-Butyllithium*** gives the organolithium.[20] Addition to electrophiles is predictable (eq 7).[5] Removal of the benzyl group via hydrogenolysis reveals that the ether anion is a masked methanol dianion.

O — BOMLi → BnO— OH (7)

$(Cbz)_2N$ O — $(Cbz)_2N$ O

Related Reagents. Benzyl Chloromethyl Sulfide; *t*-Butyl Chloromethyl Ether; 2-Chloroethyl Chloromethyl Ether; Chloromethyl Methyl Ether; Chloromethyl Methyl Sulfide; (*p*-Methoxybenzyloxy)methyl Chloride; 2-Methoxyethoxymethyl Chloride; 2-(Trimethylsilyl)ethoxymethyl Chloride.

1. Connor, D. S.; Klein, G. W.; Taylor, G. N. *OS* **1972**, *52*, 16.
2. Stork, G.; Isobe, M. *JACS* **1975**, *97*, 6260.
3. Groziak, M. P.; Wei, L. *JOC* **1991**, *56*, 4296.
4. Miyamoto, K.; Tsuchiya, S.; Ohta, H. *JACS* **1992**, *114*, 6256.
5. Wipf, P.; Kim, Y. *JOC* **1993**, *58*, 1649.
6. Hill, A. J.; Keach, D. T. *JACS* **1926**, *48*, 257.
7. Paraformaldehyde can be used in place of formalin: Guedin-Vuong, D.; Yoichi, N. *BSF* **1986**, 245.
8. Goff, D. A.; Harris, R. N., III; Bottaro, J. C.; Bedford, C. D. *JOC* **1986**, *51*, 4711.
9. Benneche, T.; Strande, P.; Undheim, K. *S* **1983**, 762.
10. Roush, W. R.; Brown, B. B. *JOC* **1993**, *58*, 2162.
11. Denis, J. N.; Greene, A. E.; Serra, A. A.; Luche, M. J. *JOC* **1986**, *51*, 46.
12. Banfi, L.; Bernardi, A.; Colombo, L.; Gennari, C.; Scolastico, C. *JOC* **1984**, *49*, 3784.
13. Nicolaou, K. C.; Claremon, D. A.; Barnette, W. E. *JACS* **1980**, *102*, 6611.
14. McCarthy, P. A. *TL* **1982**, 4199.
15. Bender, S. L.; Widlanski, T.; Knowles, J. R.; *B* **1989**, *28*, 7560.
16. Shimada, J.; Suzuki, F. *TL* **1992**, *33*, 3151.
17. Yamagishi, M.; Yamada, Y.; Ozaki, K.; Asao, M.; Shimizu, R.; Suzuki, M.; Matsumoto, M.; Matsuoka, Y.; Matsumoto, K. *JMC* **1992**, *35*, 2085.
18. Evans, D. A.; Urpi, F.; Somers, T. C.; Clark, J. S.; Bilodeau, M. T. *JACS* **1990**, *112*, 8215.
19. Fang, C.; Suganuma, K.; Suemune, H.; Sakai, K. *JCS(P1)*, **1991**, 1549.
20. Still, W. C. *JACS* **1978**, *100*, 1481.

Harold W. Pinnick
Bucknell University, Lewisburg, PA, USA

Benzyl 2,2,2-Trichloroacetimidate

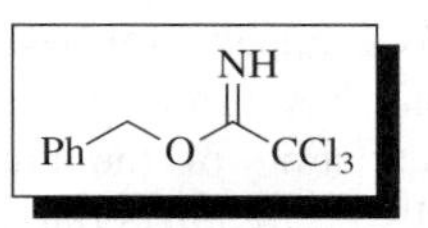

[81927-55-1] $C_9H_8Cl_3NO$ (MW 252.53)

(reagent for protection of hydroxy groups as benzyl ethers under mildly acidic conditions[1])

Alternate Name: BTCA.
Physical Data: bp 106–114 °C/0.5 mmHg; *d* 1.359 g cm^{-3}
Solubility: reactions are most often carried out in a mixture of cyclohexane/dichloromethane;[1] decomposition by rearrangement to *N*-benzyl trichloroacetamide is accelerated in more polar solvents.
Form Supplied in: colorless oil.
Purification: distillation under reduced pressure.
Handling, Storage, and Precautions: moisture sensitive.

Hydroxy Group Protection. The benzyl ether is one of the most versatile and frequently used protecting groups in organic synthesis.[2] The most common way of introducing this group is with a strong base and ***Benzyl Bromide*** using the classical Williamson ether synthesis methodology. Recently, benzyl trichloroacetimidate[3] has found widespread use to protect hydroxy functionalities as their benzyl ethers under mildly acidic conditions,[1,4,5a] when the more conventional strong base-mediated methods either failed or gave lower yields due to side reactions and/or starting material decomposition.

The imidate can be prepared from the sodium alkoxide ion of benzyl alcohol and trichloroacetonitrile according to procedures based on those developed by Cramer[3] in the late 1950s. Substituted benzyl ethers have also been prepared this way, for example using 4-methoxybenzyl trichloroacetimidate,[5] 3,4-dimethoxybenzyl trichloroacetimidate,[6] and 2,6-dichlorobenzyl trichloroacetimidate,[7] and the method has also been applied for the synthesis of allyl[1b] and *t*-butyl ethers,[8] as well as for 2-phenylisopropyl esters[9] for peptide synthesis. The benzyl and 4-methoxybenzyl 2,2,2-trichloroacetimidates are available commercially.

The reaction is performed in nonpolar media to prevent rearrangement of the acetimidate to *N*-benzyl trichloroacetamide, and as an added advantage the solid trichloroacetamide byproduct can simply be removed by filtration. These constraints limit the substrate range to those molecules soluble in this type of solvent; for example, a sugar triol which was insoluble in dichloromethane, when reacted in DMSO yielded very little of the benzylated product desired.[10] It was also noted that differential (primary versus secondary) hydroxy protection in carbohydrates (eq 1) was poor,[10] and the ratio of the 6-OBn to 4-OBn products was at best 2:1.

The reagent has been used widely in carbohydrate[1,10–14] chemistry. The method was found to be useful in cases where *O*-benzoyl group migration had been observed under basic conditions (eq 2),[11] and has also been applied to inositol derivatives,[15] when the same problem[15a] was encountered.

When *N*-acetyl-protected sugar amines were under investigation[1,12] the *N*-acetyl group was observed to undergo transformation to the *O*-benzyl imidate in competition with OH benzylation (eq 3). It appears for best yields at least 1 equiv of reagent per OH and NAc group is needed, and subsequent hydrolysis and reacetylation of the amine are necessary to obtain are the originally desired material. Similar problems were encountered with the morpholine derived lactams (eq 4),[16a] where mixed products were obtained, but with β-lactams[16b] with *N*-amide protection (eq 5), *O*-benzylation with benzyl trichloroacetimidate proceeded smoothly. BTCA has recently been used to convert diketopiperazines into their bis-benzyl imino ethers.[17]

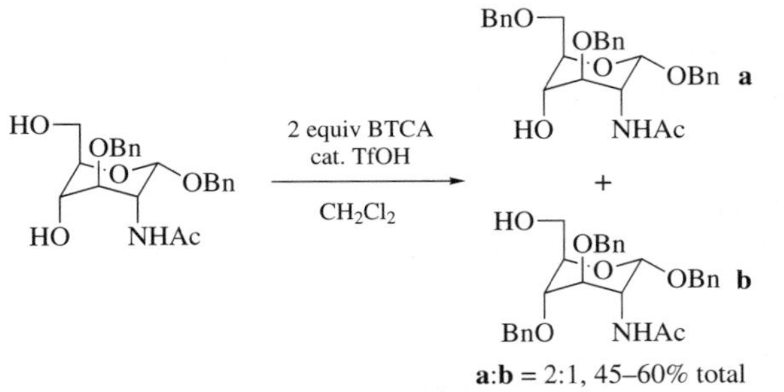

2 equiv BTCA, cat. TfOH, CH_2Cl_2 (1)

a:**b** = 2:1, 45–60% total

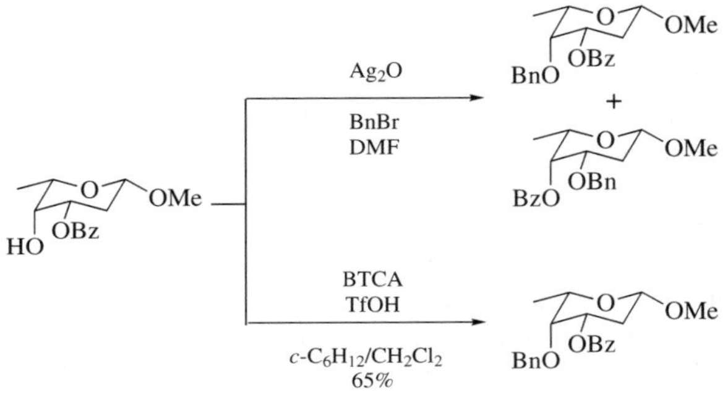

Ag_2O, BnBr, DMF; BTCA, TfOH, c-C_6H_{12}/CH_2Cl_2, 65% (2)

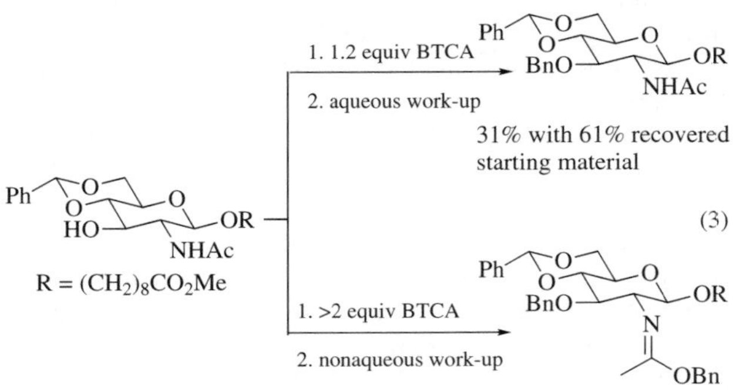

1. 1.2 equiv BTCA 2. aqueous work-up: 31% with 61% recovered starting material; 1. >2 equiv BTCA 2. nonaqueous work-up (3)

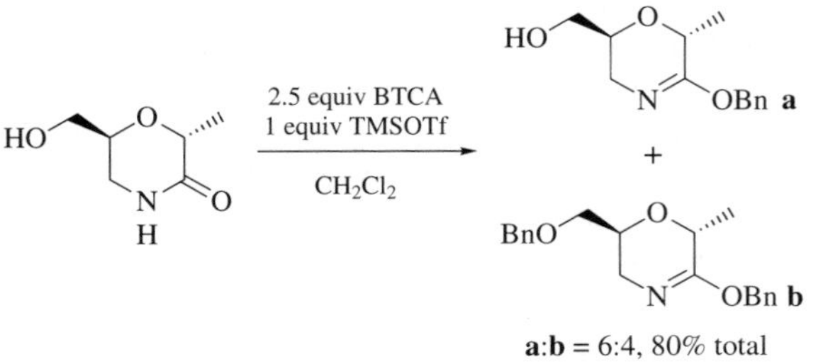

R = $(CH_2)_8CO_2Me$

2.5 equiv BTCA, 1 equiv TMSOTf, CH_2Cl_2 (4)

a:**b** = 6:4, 80% total

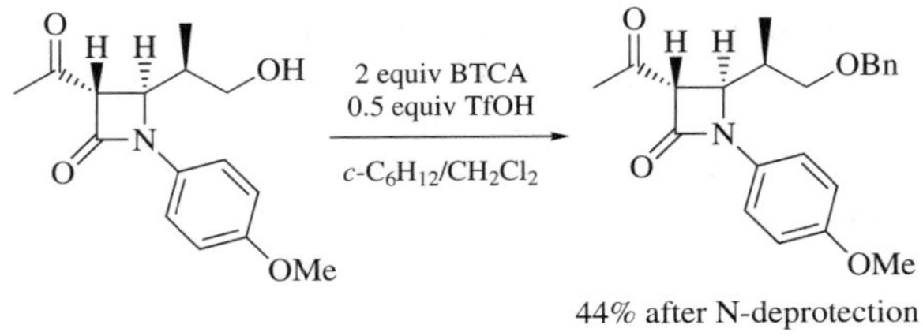

2 equiv BTCA, 0.5 equiv TfOH, c-C_6H_{12}/CH_2Cl_2 (5)

44% after N-deprotection

Other classes of molecules which have been benzylated using this reagent include hydroxy ketones,[4] hydroxy halides[4,7,18] and esters[19–23] such as tartrates,[19] hydroxy propionates or butanoates,[20] long alkyl chain hydroxy esters,[21] and hydroxy lactones.[4] Under basic conditions problems such as enolization, aldol reactions, elimination and epoxide formation can cause problems and when 2-substituted esters, such as ethyl (*S*)-lactate,[22] are required as homochiral starting materials, then protection using this method circumvents the considerable risk of racemization that accompanies the use of bases (eq 6).

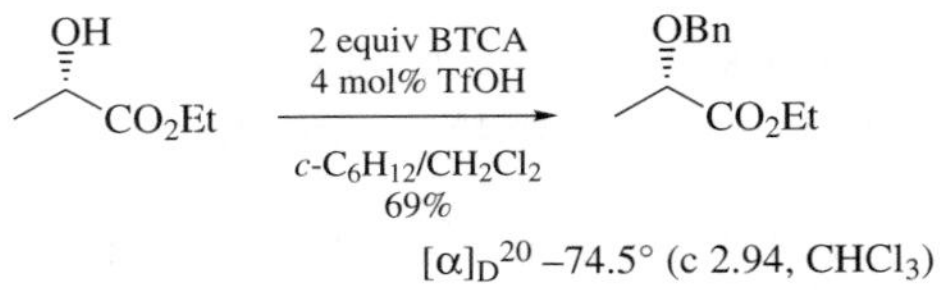

2 equiv BTCA, 4 mol% TfOH, c-C_6H_{12}/CH_2Cl_2, 69% (6)

$[\alpha]_D^{20}$ −74.5° (c 2.94, $CHCl_3$)

Under these benzylation reaction conditions, it has been shown that the *N*-*t*-butoxycarbonyl (N-Boc) (eq 7) and *N*-(2-trimethylsilyl)ethoxycarbonyl (*N*-TEOC) (eq 8) protected amino acid derivatives are converted into their *N*-benzyloxycarbonyl (N-Z) analogs.[24] In most cases yields were moderate, but surprisingly the *t*-butyldimethylsilyloxy and *t*-butyl ester functions are relatively unaffected by the acidic conditions.

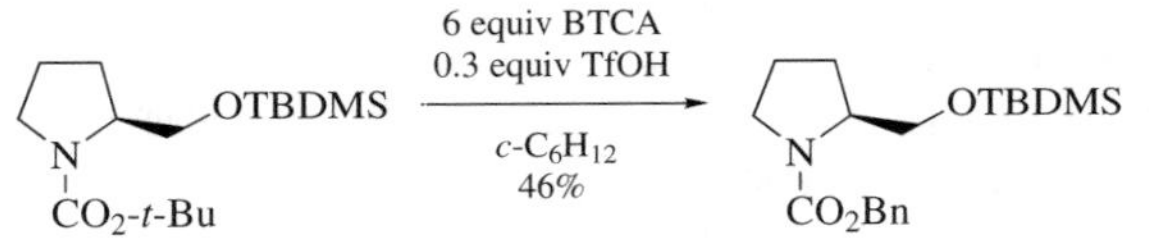

6 equiv BTCA, 0.3 equiv TfOH, c-C_6H_{12}, 46% (7)

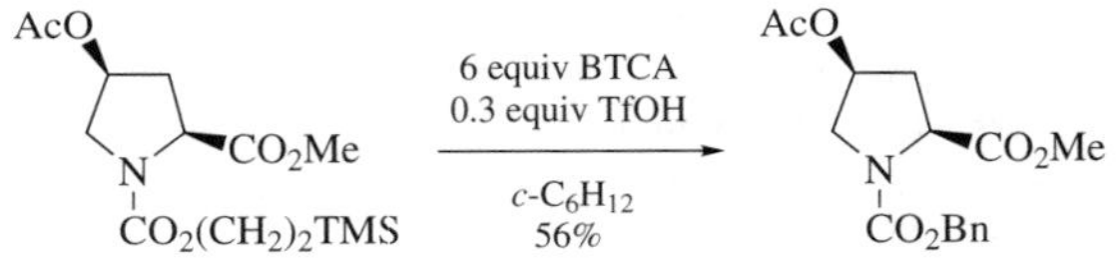

6 equiv BTCA, 0.3 equiv TfOH, c-C_6H_{12}, 56% (8)

The most frequently employed acid catalyst is TfOH (***Trifluoromethanesulfonic Acid***, or triflic acid), but others that have been used include $BF_3\cdot OEt_2$,[4,8] TMSOTf,[4,16] *p*-TsOH,[6] and TFA.[23,16] Camphorsulfonic acid, pyridinium *p*-toluenesulfonate, and trityl perchlorate have also been investigated in conjunction with the more reactive 4-methoxybenzyl trichloroacetimidate.[5a]

Functional groups sometimes sensitive to acidic conditions such as simple esters,[1,19–23] lactones,[4,5d] acetals,[1,5a,15] epoxides,[5a,13] 2-methoxyethyloxymethyl (MEM) groups,[6] *t*-butyldimethylsilyl ethers,[14,24–27] and orthoformates[15b] are generally unaffected by the catalytic amounts of acid used, although some tertiary alcohols[4] and 1-trimethylsilylalkynes[4] have been noted to give poor yields.

Related Reagents. Benzyl Bromide; Benzyl Chloride; Benzyl Iodide; Benzyl Trifluoromethanesulfonate; 3,4-Dimethoxybenzyl

Bromide; *p*-Methoxybenzyl Chloride; 4-Methoxybenzyl 2,2,2-Trichloroacetimidate.

1. (a) Wessel, H.-P.; Iversen, T.; Bundle, D. R. *JCS(P1)* **1985**, 2247. (b) Iversen, T.; Bundle, D. R. *CC* **1981**, 1240.
2. Greene, T. W.; Wuts, P. G. M. *Protective Groups in Organic Synthesis*, 2nd ed.; Wiley: New York, 1991.
3. (a) Cramer, F.; Pawelzik, K.; Baldauf, H. J. *CB* **1958**, *91*, 1049. (b) Cramer, F.; Hennrich, N. *CB* **1961**, *94*, 976.
4. Eckenberg, P.; Groth, U.; Huhn, T.; Richter, N.; Schmeck, C. *T* **1993**, *49*, 1619.
5. (a) Nakajima, N.; Horita, K.; Abe, R.; Yonemitsu, O. *TL* **1988**, *29*, 4139. (b) Takaku, H.; Ueda, S.; Ito, T. *TL* **1983**, *24*, 5363. (c) Romo, D.; Johnson, D. D.; Plamondon, L.; Miwa, T.; Schreiber, S. L. *JOC* **1992**, *57*, 5060. (d) Adams, E.; Hiegemann, M.; Duddeck, H.; Welzel, P. *T* **1990**, *46*, 5975.
6. Takaku, H.; Ito, T.; Imai, K. *CL* **1986**, 1005.
7. Amouroux, R.; Gerin, B.; Chastrette, M. *T* **1985**, *41*, 5321.
8. Armstrong, A.; Brackenridge, I.; Jackson, R. F. W.; Kirk, J. M. *TL* **1988**, *29*, 2483.
9. Yue, C.; Thierry, J.; Potier, P. *TL* **1993**, *34*, 323.
10. Bovin, N. V.; Musina, L. Y.; Khorlin, A. Y. *BAU* **1986**, *35*, 614.
11. Yasumori, T.; Sato, K.; Hashimoto, H.; Yoshimura, J. *BCJ* **1984**, *57*, 2538.
12. (a) Kusumoto, S.; Yamamoto, M.; Shiba, T. *TL* **1984**, *25*, 3727. (b) Imoto, M.; Yoshimura, H.; Yamamoto, M.; Shimamoto, T.; Kusumoto, S.; Shiba, T. *TL* **1984**, *25*, 2667.
13. Schubert, J.; Schwesinger, R.; Knothe, L.; Prinzbach, H. *LA* **1986**, 2009.
14. Takahashi, S.; Terayama, H.; Kuzuhara, H. *TL* **1992**, *33*, 7565.
15. (a) Ozaki, S.; Kondo, Y.; Nakahira, H.; Yamaoka, S.; Watanabe, Y. *TL* **1987**, *28*, 4691. (b) Baudin, G.; Glanzer, B. I.; Swaminathan, K. S.; Vasella, A.; *HCA* **1988**, *71*, 1367. (c) Estevez, V. A.; Prestwich, G. D. *TL* **1991**, *32*, 1623.
16. (a) Danklmaier, J.; Honig, H. *LA* **1989**, 665. (b) Kawabata, T.; Kimura, Y.; Ito, Y.; Terashima, S.; Sasaki, A.; Sunagawa, M. *T* **1988**, *44*, 2149.
17. Groth, U.; Schmeck, C.; Schöllkopf, U. *LA* **1993**, 321.
18. Voss, G.; Gerlach, H. *HCA* **1983**, *66*, 2294.
19. Hansson, T. G.; Kihlberg, J. O. *JOC* **1986**, *51*, 4490.
20. (a) White, J. D.; Kawasaki, M. *JOC* **1992**, *57*, 5292. (b) Keck, G. E.; Murry, J. A. *JOC* **1991**, *56*, 6606.
21. (a) Imoto, M.; Yoshimura, H.; Yamamoto, M.; Shimamoto, T.; Kusumoto, S.; Shiba, T. *BCJ* **1987**, *60*, 2197. (b) Widmer, U. *S* **1987**, 568.
22. Ito, Y.; Kobayashi, Y.; Kawabata, T.; Takase, M.; Terashima, S. *T* **1989**, *45*, 5767.
23. Keck, G. E.; Andrus, M. B.; Romer, D. R.; *JOC* **1991**, *56*, 417.
24. Barrett, A. G. M.; Pilipauskas, D. *JOC* **1990**, *55*, 5170.
25. Hong, C. Y.; Kishi, Y.; *JACS* **1992**, *114*, 7001.
26. Lawrence, N. J.; Fleming, I. *TL* **1990**, *31*, 3645.
27. Danishefsky, S. J.; Deninno, S.; Lartey, P. *JACS* **1987**, *109*, 2082.

Andrew N. Boa & Paul R. Jenkins
Leicester University, UK

2,4-Bis(4-methoxyphenyl)-1,3,2,4-dithiadiphosphetane 2,4-Disulfide[1]

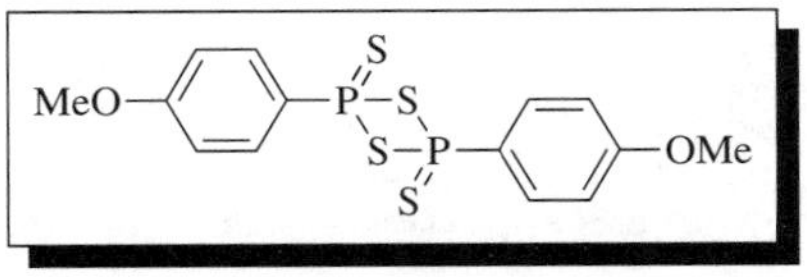

[19172-47-5] $C_{14}H_{14}O_2P_2S_4$ (MW 404.45)

(reagent for the conversion of carbonyl into thiocarbonyl groups[1])

Alternate Name: Lawesson's reagent.
Physical Data: mp 228–229 °C.
Solubility: modestly sol boiling organic solvents such as toluene, chlorobenzene, anisole, dimethoxyethane.
Form Supplied in: yellowish evil-smelling crystals; typical impurity is P_4S_{10}.
Analysis of Reagent Purity: FTIR spectrum.[1c]
Preparative Methods: reaction of P_4S_{10} with anisole in excess refluxing anisole.[2]
Purification: recrystallization from boiling toluene.
Handling, Storage, and Precautions: can be stored for months at rt if moisture is excluded. Prolonged heating in solution causes decomposition (polymerization). It is toxic and should be handled under a fume hood since hazardous H_2S is easily liberated with moisture.

Thionation of Carbonyl Compounds. Lawesson's reagent (LR) is a most powerful reagent for the thionation of a wide variety of carbonyl compounds. Ketones, enones, carboxylic esters, thiolocarboxylic esters, amides, and related substrates are conveniently transformed into the corresponding thiocarbonyl compounds.[1] In some cases, follow-up products are isolated since certain thiones are labile under the reaction conditions. Compared with P_4S_{10}, from which it is easily prepared,[2] LR exhibits several advantages. Its reactivity is significantly higher and it is sufficiently soluble in hot organic solvents, allowing homogeneous reaction conditions to be applied. Therefore many carbonyl compounds can be successfully thionated with LR but not with P_4S_{10}, or at least thionated in higher yields. Several reagents with structures similar to LR, which are even more reactive or more selective than LR, have been developed and are used in particularly difficult cases (see ***2,4-Bis(methylthio)-1,3,2,4-dithiadiphosphetane 2,4-Disulfide***). The workup procedure depends on the reaction conditions applied, in particular on the solvent, and the products formed. If DME is used, the reaction mixture can be poured into water and the product extracted as usual. Low boiling hydrocarbons are best evaporated and the residue, which contains the product together with 2,4,6-tris(4-methoxyphenyl)cyclotriphosphoxane 2,4,6-trisulfide, is subjected to column chromatography. High boiling solvents such as tetralin or trichlorobenzene should be removed by flash chromatography in order to avoid thermal decomposition of the labile thiones.

Thioketones. Thioketones of different types have been obtained from the corresponding ketones. Lawesson, who discovered LR to be a powerful reagent for the conversion of C=O into C=S groups, described the preparation of diaryl thioketones.[3] Substituents such as Me, Br, NO_2, or NMe_2 do not affect the good yields of up to 98% (eq 1).

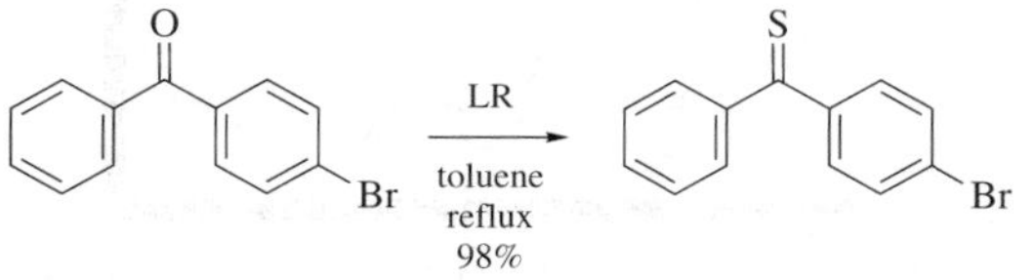

LR, toluene, reflux, 98% (1)

Recently, the diferrocenyl thioketones (**1**) and (**2**) were prepared from the corresponding ketones.[4] Alkyl aryl and dialkyl thioketones (**3**),[3] (**4**),[5] (**5**)[6] are formed in good yields if enethiolization is unfavored or impossible, whereas mainly enethiols are obtained from cycloalkanones (eq 2).[7]

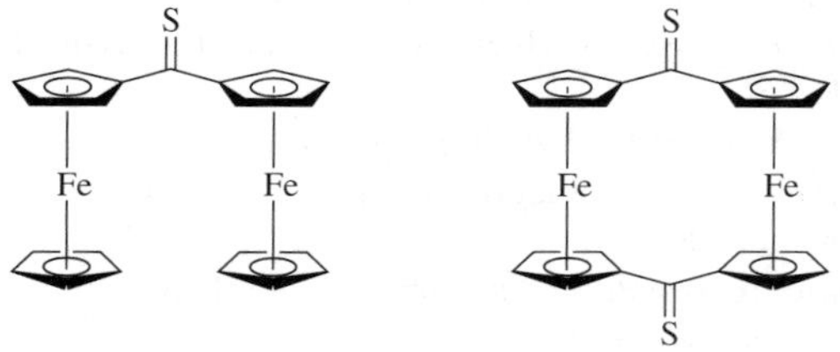

(**1**) 71% (**2**) 90%

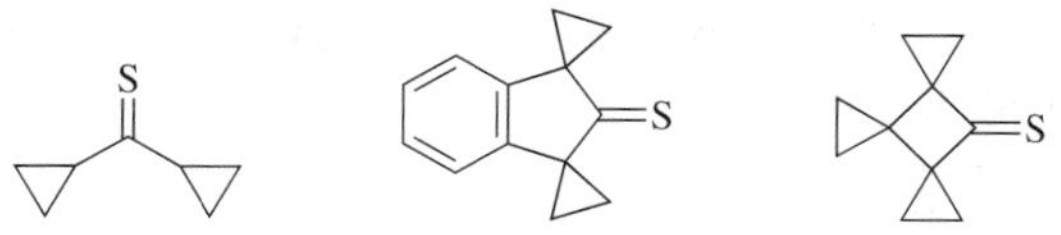

(**3**) 83% (**4**) 85% (**5**) 85%

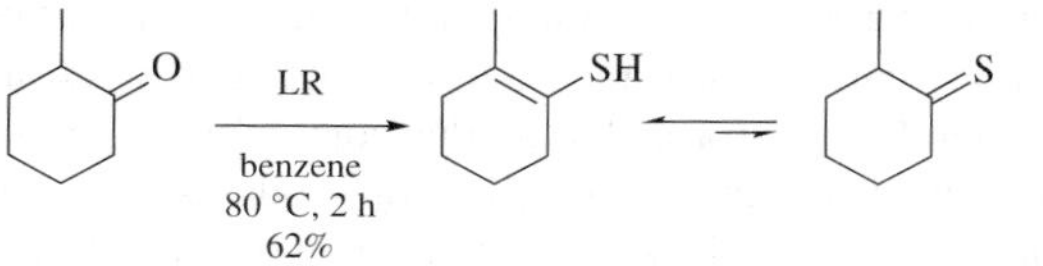

LR, benzene, 80 °C, 2 h, 62% (2)

2-Cyclohexenones are thionated with LR (eq 3),[7] but only the vinylogous dithioester (**6**) is a stable compound. Acyclic enones (chalcones) give on reaction with LR the enethione dimers (eq 4).[8]

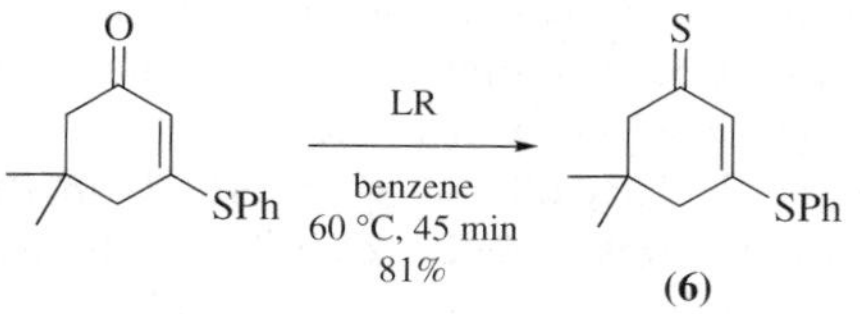

LR, benzene, 60 °C, 45 min, 81% (**6**) (3)

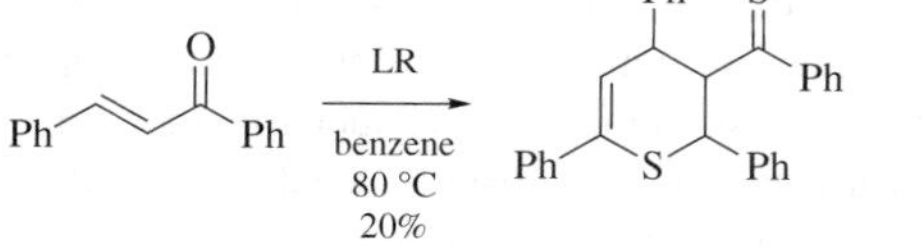

LR, benzene, 80 °C, 20% (4)

Again, stable enethiones are formed if an electron donating substituent (e.g. NR_2) is present in the β-position (eq 5).[9] Certain isolable 2,5-cyclohexadienethiones can be further transformed into bicyclohexadienylidenes (eq 6).[10]

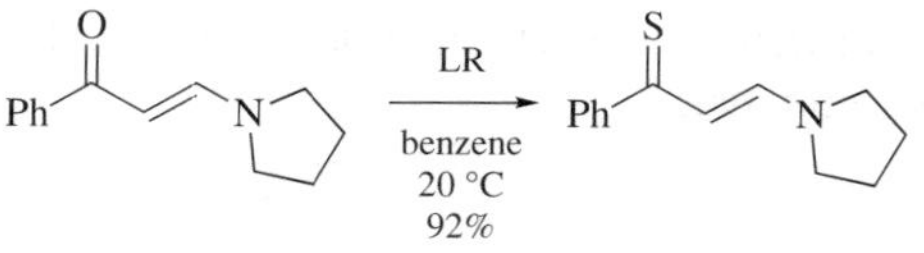

LR, benzene, 20 °C, 92% (5)

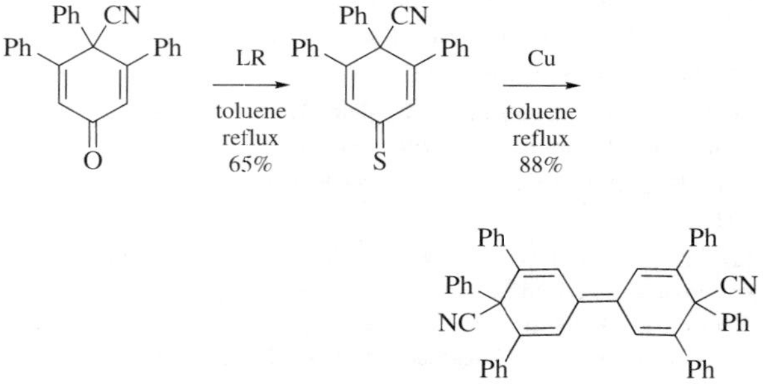

LR, toluene, reflux, 65%; Cu, toluene, reflux, 88% (6)

The thioanalogs of natural products such as the steroid (**7**)[11] or the colchicine alkaloid (**8**)[12] have also been prepared by use of LR.

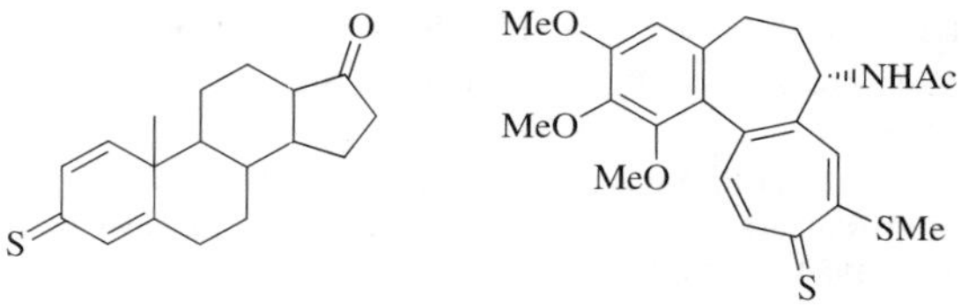

(**7**) 96% (**8**) 33%

Thio Analogs of Dicarbonyl Compounds. In general, the most labile thiono analogs of α-dicarbonyl compounds cannot be prepared by any thionation procedure. However, thioacenaphthoquinone (**9**) as well as thioanthraquinone (**10**) were obtained from the quinones and LR.[13] Even α-thioxo thioamides are available (eq 7),[14] whereas 2,2,5,5-tetramethyl-4-thioxo-3-hexanone, which itself cannot be prepared from the corresponding diketone, reacts with LR to yield 3,4-di-*t*-butyl-1,2-dithiete, the valence isomer of the open-chain dithione (eq 8).[15]

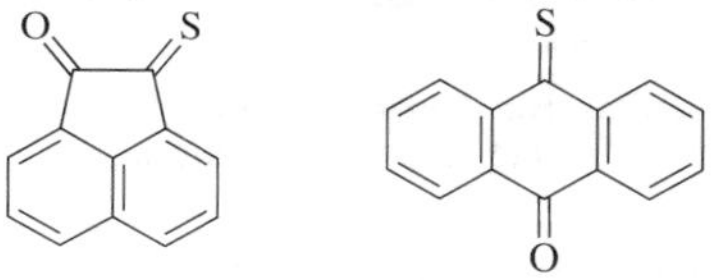

(**9**) 70% (**10**) 65%

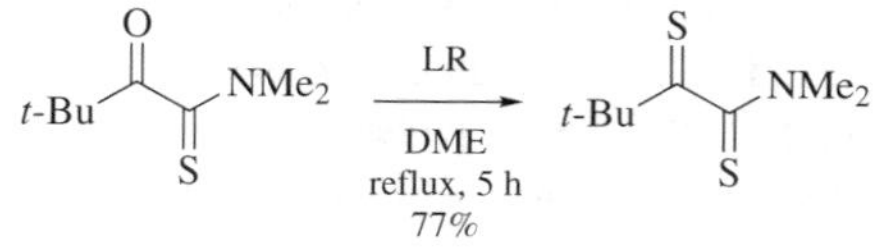

LR, DME, reflux, 5 h, 77% (7)

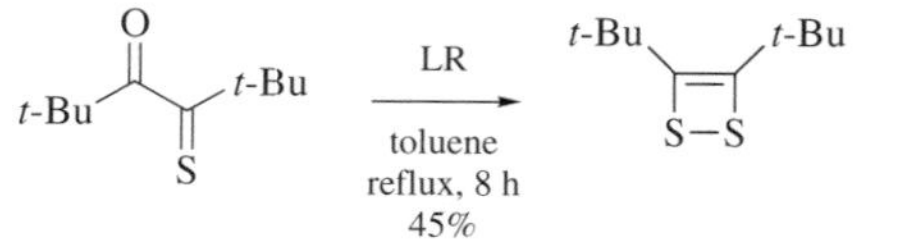

LR, toluene, reflux, 8 h, 45% (8)

Thioketones with pronounced steric hindrance are obtained on reaction of cyclobutane-1,3-diones with LR (eq 9).[16]

(9)

Diketones with remote C=O groups can be transformed by LR into the mono- or dithiones.[17]

Thiono Esters and Lactones. Although thiono esters can be conveniently prepared via imidates,[18] the introduction of LR for the thionation of esters[3] represents a great advance in synthetic methodology since ordinary esters are used as educts and P_4S_{10} only reacts with esters if enforced reaction conditions are applied, under which many thiono esters decompose. Facile conversion of aliphatic and aromatic esters (eq 10),[3,19] α,β-unsaturated esters,[20] and lactones[21] to the corresponding thiono derivatives is effected in high yields (eqs 11 and 12).[22,23]

(10)

(11)

(12)

Thiono esters and lactones are important intermediates in modern organic synthesis since they easily undergo useful follow-up reactions. Ethers are formed on reduction with ***Raney Nickel***[24] or ***Tri-n-butyltin Hydride*** (eq 13).[25]

(13)

Macrocyclic bis-thionolactones have been prepared with LR. These were converted by reduction with ***Sodium Naphthalenide*** into the radical anions, which gave bicyclic systems through radical dimerization and subsequent methylation (eq 14).[26] This method was successfully applied by Nicolaou et al.[27] in the total synthesis of hemibrevetoxin B. One of the crucial steps of the synthesis was the preparation of the bis-thiono ester (**11**), which was achieved by using LR together with tetramethylthiourea in xylene at 175 °C.[27]

(14)

(**11**) 49%

Dithioesters and Related Compounds. Thioesters are transformed into the corresponding dithioesters on reaction with LR.[3,20,28] Again the yields are nearly quantitative even if steric hindrance can be expected (eq 15).[3]

(15)

The reaction is completed within 5–15 min if tetralin at 210 °C is used as solvent.[28] Also, dithio-γ-lactones,[21,28] including dithiophthalides[29] and dithio-α-pyrones,[30] are formed smoothly whereas the hitherto unknown simple β- or δ-dithiolactones cannot be prepared, neither with LR nor by any other method. Interestingly, dithiopilocarpine is formed as a mixture of diastereomers on reaction of pilocarpine with LR (eq 16), i.e. both oxygen atoms of (**12**) are replaced by sulfur.[31] In an interesting sequence of thionation and rearrangement reactions, three different thiono analogs of 1,8-naphthalic anhydride were prepared (eq 17).[32]

(16)

(**12**)

(17)

Thioamides and Related Compounds. Thioamides are the most stable thiocarbonyl compounds and have been prepared, for a century, from amides and P_4S_{10} under rather drastic conditions. However, even for this purpose LR has turned out to be a superior reagent. High yields are obtained for all types of thioamides[1,33] and -lactams[1,33,34] including the elusive unsubstituted acrylothioamide (eq 18)[35] and thioformamides or thioamides bearing sensitive substituents such as NO_2, Z-NH,[36] or OH (eq 19).[33]

(18): 1. LR, HMPA, 80 °C, 90 min; 2. 520 °C, 10^{-3} Torr; 51%

(19): LR, HMPA, 80 °C, 1 h; 70%

β-Lactams are also smoothly transformed[34] (eq 20),[37] which is important for the preparation of thio analogs of β-lactam antibiotics or azetidines. Furthermore, endothio oligopeptides became conveniently available only after LR had been introduced as reagent[1,38,39] (eq 21).[40]

(20): LR, toluene, reflux, 6 h, 47%; Raney Ni, EtOH, reflux, 3 h, 54%

(21): Z-Gly-Gly-Phe-OMe —LR, 66%→ Phe-OMe

Recently the cyclopeptide [D-cysteine][8]cyclosporin has been prepared from [D-serine][8]cyclosporin via selective thionation with LR at the 7-position followed by intramolecular sulfur transfer.[41]

Miscellaneous Reactions of LR. Under particular conditions, certain carbonyl compounds and other substrates react with LR to form thiophosphonates or heterocycles, which fact throws some light on the mechanism of thionation reactions with LR.[1] Carbinols undergo nucleophilic substitution with LR to form the corresponding thiols.[42] The redox properties of LR can be utilized to prepare dithiolactones from dialdehydes (eq 22),[43] α-oxo thioamides from nitro ketones (eq 23),[44] or sulfides from sulfoxides.[45]

(22): LR, $CHCl_3$, rt, 30 min, 96%

(23): LR, toluene, 100 °C

1. (a) Cherkasov, R. A.; Kutyrev, G. A.; Pudovik, A. N. *T* **1985**, *41*, 2567. (b) Cava, M. P.; Levinson, M. I. *T* **1985**, *41*, 5061. (c) Aldrich Library of Infrared Spectra; Aldrich Chemical Co.: Milwaukee, 1985; Vol. 1 (2), p 557B.
2. Thomsen, I.; Clausen, K.; Scheibye, S.; Lawesson, S.-O. *OS* **1984**, *62*, 158.
3. Pedersen, B. S.; Scheibye, S.; Nilsson, N. H.; Lawesson, S.-O. *BSB* **1978**, *87*, 223.
4. Sato, M.; Asai, M. *JOM* **1992**, *430*, 105.
5. Klages, C.-P.; Voss, J. *CB* **1980**, *113*, 2255.
6. Wenck, H.; de Meijere, A.; Gerson, F.; Gleiter, R. *AG(E)* **1986**, *25*, 335.
7. Scheibye, S.; Shabana, R.; Lawesson, S.-O.; Rømming, C. *T* **1982**, *38*, 993.
8. Kametani, S.; Ohmura, H.; Tanaka, H.; Motoki, S. *CL* **1982**, 793.
9. (a) Walter, W.; Proll, T. *S* **1979**, 941. (b) Shabana, R.; Rasmussen, J. B.; Olesen, S. O.; Lawesson, S.-O. *T* **1980**, *36*, 3047. (c) Rasmussen, J. B.; Shabana, R.; Lawesson, S.-O. *T* **1982**, *38*, 1705. (d) Pulst, M.; Greif, D.; Kleinpeter, E. *ZC* **1988**, *28*, 345.
10. Kühn, R.; Otto, H.-H. *AP* **1989**, *322*, 375.
11. Weiss, D.; Gaudig, U.; Beckert, R. *S* **1992**, 751.
12. Muzaffar, A.; Brossi, A. *SC* **1990**, *20*, 713.
13. El-Kateb, A. A.; Hennawy, I. T.; Shabana, R.; Osman, F. H. *PS* **1984**, *20*, 329.
14. Adiwidjaja, G.; Günther, H.; Voss, J. *LA* **1983**, 1116.
15. Köpke, B.; Voss, J. *JCR(S)* **1982**, 314.
16. Strehlow, T.; Voss, J.; Spohnholz, R.; Adiwidjaja, G. *CB* **1991**, *124*, 1397.
17. Ishii, A.; Nakayama, J.; Ding, M.; Kotaka, N.; Hoshino, M. *JOC* **1990**, *55*, 2421.
18. Voss, J. *COS* **1991**, *6*, 435.
19. Jones, B. A.; Bradshaw, J. S. *CRV* **1984**, *84*, 17.
20. Pedersen, B. S.; Scheibye, S.; Clausen, K.; Lawesson, S.-O. *BSB* **1978**, *87*, 293.
21. Scheibye, S.; Kristensen, J.; Lawesson, S.-O. *T* **1979**, *35*, 1339.
22. Takano, S.; Tomita, S.; Takahashi, M.; Ogasawara, K. *S* **1987**, 1116.
23. Nicolaou, K. C.; McGarry, D. G.; Somers, P. K.; Kim, B. H.; Ogilvie, W. W.; Yiannikouros, G.; Prasad, C. V. C.; Veale, C. A.; Hark, R. R. *JACS* **1990**, *112*, 6263.
24. Baxter, S. L.; Bradshaw, J. S. *JOC* **1981**, *46*, 831.
25. Smith, C.; Tunstad, L. M.; Guttierrez, C. G. *PS* **1988**, *37*, 257.
26. Nicolaou, K. C.; Hwang, C.-K.; Marron, B. E.; DeFrees, S. A.; Couladouros, E. A.; Abe, Y.; Carroll, P. J.; Snyder, J. P. *JACS* **1990**, *112*, 3040.
27. Nicolaou, K. C.; Reddy, K. R.; Skokotas, G.; Sato, F.; Xiao, Xiao-Yi; Hwang, C.-K. *JACS* **1993**, *115*, 3558.
28. Ghattas, A.-B. A. G.; El-Khrisy, E.-E. A. M.; Lawesson, S.-O. *Sulfur Lett.* **1982**, *1*, 69.
29. Oparin, D. A.; Kuznetsova, A. S. *Vestsi Akad. Navuk BSSR, Ser. Khim. Navuk* **1990**, (6), 109 (*CA* **1991**, *115*, 29 039y).
30. Hoederath, W.; Hartke, K. *AP* **1984**, *317*, 938.
31. Shapiro, G.; Floersheim, P.; Boelsterli, J.; Amstutz, R.; Bolliger, G.; Gammenthaler, H.; Gmelin, G.; Supavilai, P.; Walkinshaw, M. *JMC* **1992**, *35*, 15.
32. Lakshmikantham, M. V.; Chen, W.; Cava, M. P. *JOC* **1989**, *54*, 4746.
33. Scheibye, S.; Pedersen, B. S.; Lawesson, S.-O. *BSB* **1978**, *87*, 229.
34. Shabana, R.; Scheibye, S.; Clausen, K.; Olesen, S. O.; Lawesson, S.-O. *NJC* **1980**, *4*, 47.
35. Khalid, M.; Vallée, Y.; Ripoll, J.-L. *CI(L)* **1988**, 123.
36. Kohrt, A.; Hartke, K. *LA* **1992**, 595.
37. Verkoyen, C.; Rademacher, P. *CB* **1985**, *118*, 653.
38. Thorsen, M.; Yde, B.; Pedersen, U.; Clausen, K.; Lawesson, S.-O. *T* **1983**, *39*, 3429.
39. Sherman, D. B.; Spatola, A. F. *JACS* **1990**, *112*, 433.
40. Brown, D. W.; Campbell, M. M.; Chambers, M. S.; Walker, C. V. *TL* **1987**, *28*, 2171.
41. Eberle, M. K.; Nuninger, F. *JOC* **1993**, *58*, 673.
42. Nishio, T. *CC* **1989**, 205.

43. Nugara, P. N.; Huang, N.-Z.; Lakshmikantham, M. V.; Cava, M. P. *HC* **1991**, *32*, 1559.
44. Harris, P. A.; Jackson, A.; Joule, J. A. *Sulfur Lett.* **1989**, *10*, 117.
45. Bartsch, H.; Erker, T. *TL* **1992**, *33*, 199.

Jürgen Voss
Universität Hamburg, Germany

Bis(2-oxo-3-oxazolidinyl)phosphinic Chloride[1]

[68641-49-6] $C_6H_8ClN_2O_5P$ (MW 254.57)

(reagent for activating carboxyl groups,[1] converting acids to esters[2] (including thioesters[3] and phosphoesters[4]), amides[5] (including peptides[6–8] and β-lactams[9,10]), and anhydrides;[11] reagent for kinetic resolution of racemic carboxylic acids and alcohols[12])

Alternate Names: *N,N*-bis(2-oxo-3-oxazolidinyl)phosphordiamidic chloride; *N,N*-bis(2-oxo-3-oxazolidinyl)-phosphorodiamidic chloride; phosphoric acid bis(2-oxooxazolidide)-chloride; BOP-Cl; BOPDCl.
Physical Data: white powder, mp 195–198 °C; hygroscopic.
Solubility: slightly sol CH_2Cl_2, MeCN, THF, DMF; dec H_2O.
Form Supplied in: white to off-white powder, typically 97% pure.
Analysis of Reagent Purity: IR (KBr): 1775 (C=O), 765 (P–Cl) cm^{-1}.
Purification: the outcome of reactions can be greatly affected by the purity of the reagent; commercial samples may require to be washed with cold water and recrystallized from MeCN prior to use.[13]
Handling, Storage, and Precautions: store in a cool dry place under nitrogen to preclude contact with moisture and oxygen. Use only in a chemical fume hood. Wear suitable protective clothing and eye/face protection. Causes burns and is harmful if swallowed, inhaled, or absorbed through skin.

Carboxyl Group Activator. BOP-Cl[1] is used for a direct, one-pot conversion of carboxylic acids into esters (eq 1)[2] (including thioesters[3] and phosphoesters[4]), amides (eq 2),[5] and anhydrides (eq 3).[11] Conditions for minimizing side reactions and optimizing yields of both esters and amides are reported.[2,5] Advantages offered over previously used reagents, such as ***N,N′*-Carbonyldiimidazole**, carbodiimides, and unsymmetrical anhydrides, include good yields (typically >80%), mild reaction conditions, and tolerance of both steric bulk and a wide range of functional groups.

(1) BOP-Cl, Et_3N, $MeCONMe_2$, rt, 1 h, 75%

(2) t-$BuCO_2H$ + H_2N–CH(Me)Ph; BOP-Cl, Et_3N, CH_2Cl_2, rt, 75 min, 91%

(3) BOP-Cl, Et_3N, MeCN, 20 °C, 30 min, 94%

The mechanism of these reactions is assumed to involve displacement of chloride by the carboxylate anion to give a phosphorodiamidate (eq 4). Subsequent nucleophilic attack of an alcohol, amine, or a carboxylate anion affords phosphorodiamidic acid and esters, amides, and anhydrides, respectively (eq 5).

(4) RCO_2^-; $-Cl^-$

(5) R^2XH; + R^1COXR^2

X = O, NH, OCO

For lactonization (eq 6),[14] including the synthesis of macrolactones,[15] BOP-Cl is a useful reagent.

(6) BOP-Cl, Et_3N, CH_2Cl_2, 0 °C, 30 min

Chemo- and regioselective acylation with BOP-Cl is illustrated by the formation of 5′-benzoyldeoxyadenosine in good yield (eq 7).[16]

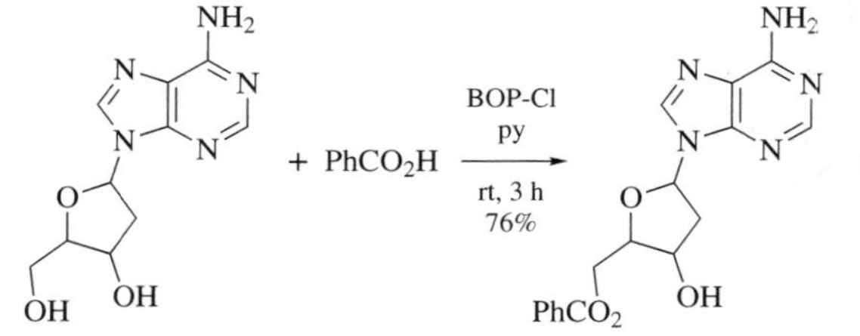

(7)

For the synthesis of oligonucleotides, BOP-Cl is an effective coupling reagent, allowing phosphotriesters to be prepared in high yields without detectable side reactions (eq 8).[17]

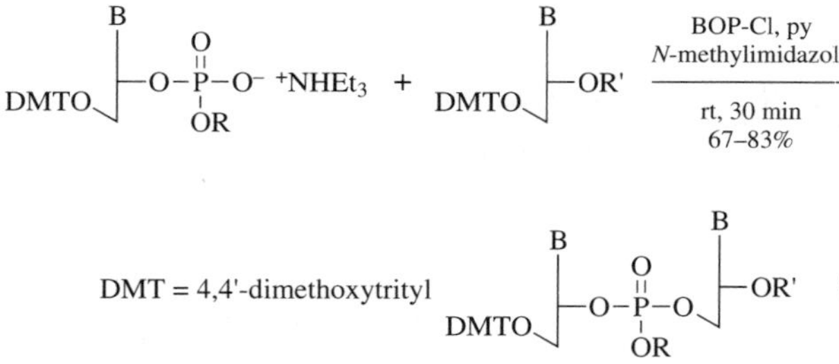

DMT = 4,4'-dimethoxytrityl (8)

β-Lactams may be prepared either by condensation of imines with carboxylic acids (eq 9)[9] or by cyclization of β-amino acids (eq 10).[10]

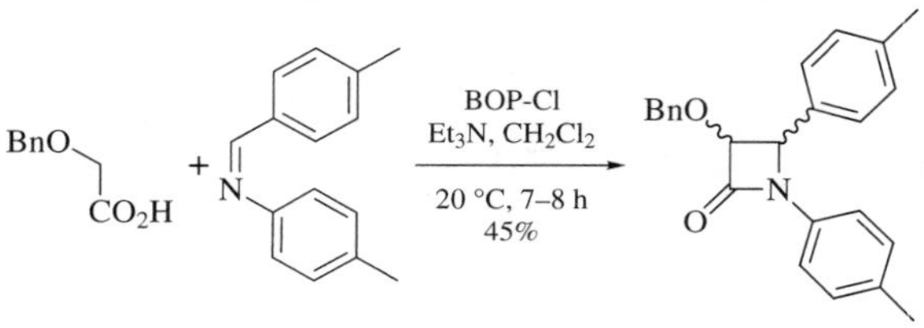

(9)

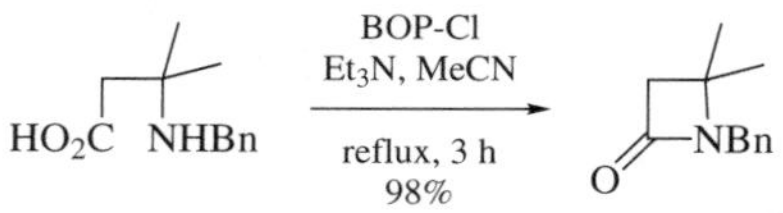

(10)

BOP-Cl has been used quite extensively for the synthesis of peptides.[6–8] It is particularly useful for coupling *N*-alkyl-α-amino acids with minimal racemization (eq 11),[18,19] although the quality of BOP-Cl employed in these reactions can significantly affect yields,[13] and the analogous azide, BOP-N_3, has been proposed as an alternative reagent.[20] Racemization can sometimes be suppressed by inclusion of additives such as ***1-Hydroxybenzotriazole*** (HOBt) or ***Imidazole*** (eq 12),[21] although under certain conditions HOBt can induce epimerization.[22]

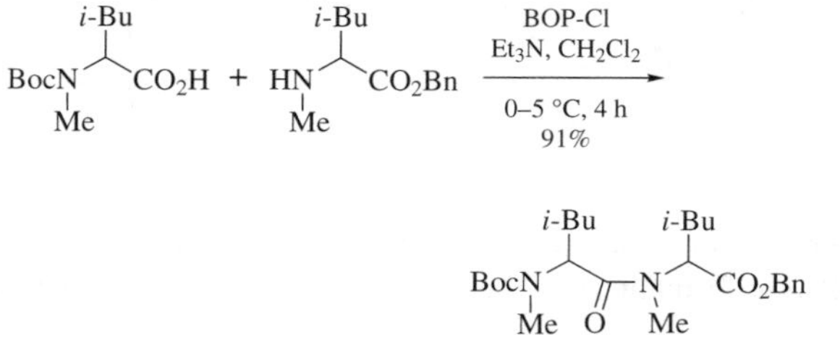

(11)

Cbz-Gly-L-Phe + L-Val-OMe —(BOP-Cl, *N*-methylmorpholine, THF, 0 °C to rt, 20 h)→ Cbz-Gly-D/L-Phe-L-Val-OMe (12)

	Yield (%)	Epimerization (%)
with HO-*t*-Bu	88	0.0
with imidazole	96	0.8
without HO-*t*-Bu or imidazole	78	5.6

Problems, notably racemization and poor yields due to oxazolone formation, are encountered when coupling *N*-acyl amino acids with BOP-Cl;[6] ***1,3-Dicyclohexylcarbodiimide*** (DCC) is a better reagent to use in such cases. Some limitations of BOP-Cl for peptide couplings have been discussed;[7] it is not recommended for β-branched α-amino acids (e.g. Boc-Val) or for couplings in which the nucleophile is a primary amine.[8]

Kinetic Resolution. BOP-Cl is a condensation agent for the kinetic resolution of racemic carboxylic acids and alcohols with chiral alcohols (eq 13) and carboxylic acids, respectively.[12]

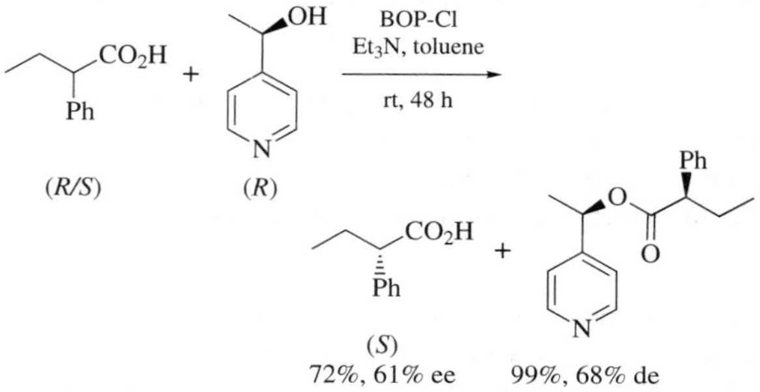

(13)

1. (a) Diago-Meseguer, J.; Palomo-Coll, A. L. *S* **1980**, 547. (b) *FF* **1982**, *10*, 41; **1984**, *11*, 57; **1988**, *13*, 39.
2. Ballester-Rodes, M.; Palomo-Coll, A. L. *SC* **1984**, *14*, 515.
3. Arrieta, A.; Garcia, T.; Lago, J. M.; Palomo, C. *SC* **1983**, *13*, 471.
4. Katti, S. B.; Agarwal, K. *TL* **1986**, *27*, 5327.
5. Cabre, J.; Palomo, A. L. *S* **1984**, 413.
6. Kolodziejczyk, A. M.; Wodecki, Z. J. *Peptides, Proceedings of the 19th European Peptide Symposium, 1986*; Theodoropoulos, D., Ed.; de Gruyter: Berlin, 1987; p 115.
7. Van der Auwera, C.; Anteunis, M. J. O. *Int. J. Pept. Protein Res.* **1987**, *29*, 574.
8. Colucci, W. J.; Tung, R. D.; Petri, J. A.; Rich, D. H. *JOC* **1990**, *55*, 2895.
9. Shridhar, D. R.; Ram, B.; Narayana, V. L. *S* **1982**, 63.
10. Chung, B. Y.; Goh, W.; Nah, C. S. *Bull. Korean Chem. Soc.* **1991**, *12*, 457.
11. Cabre-Castellvi, J.; Palomo-Coll, A.; Palomo-Coll, A. L. *S* **1981**, 616.
12. Mazon, A.; Najera, C.; Yus, M.; Heumann, A. *TA* **1992**, *3*, 1455.
13. Van der Auwera, C.; Anteunis, M. J. O. *BSB* **1986**, *95*, 203.
14. Cruickshank, K. A.; Reese, C. B. *S* **1983**, 199.
15. Corey, E. J.; Hua, D. H.; Pan, B.-C.; Seitz, S. P. *JACS* **1982**, *104*, 6818.
16. Liguori, A.; Perri, E.; Sindona, G.; Uccella, N. *T* **1988**, *44*, 229.
17. Katti, S. B.; Agarwal, K. L. *TL* **1985**, *26*, 2547.
18. Tung, R. D.; Dhaon, M. K.; Rich, D. H. *JOC* **1986**, *51*, 3350.
19. Tung, R. D.; Rich, D. H. *JACS* **1985**, *107*, 4342.
20. Katti, S. B.; Misra, P. K.; Haq, W.; Mathur, K. B. *IJC(B)* **1988**, *27*, 3.

21. Van der Auwera, C.; Van Damme, S.; Anteunis, M. J. O. *Int. J. Pept. Protein Res.* **1987**, *29*, 464.
22. Anteunis, M. J. O.; Sharma, N. K. *BSB* **1988**, *97*, 281.

David C. Rees & Niall M. Hamilton
Organon Laboratories Ltd, Newhouse, Lanarkshire, UK

Bis(tri-*n*-butyltin) Oxide

$(n\text{-}Bu_3Sn)_2O$

[56-35-9] $C_{24}H_{54}OSn_2$ (MW 596.20)

promotes the oxidation of secondary alcohols and sulfides with Br_2; O- and N-activations; dehydrosulfurizations; hydrolysis catalyst)

Physical Data: bp 180 °C/2 mmHg; *d* 1.170 g cm^{-3}.
Solubility: sol ether and hexane.
Form Supplied in: colorless oil.
Handling, Storage, and Precautions: $(Bu_3Sn)_2O$ should be stored in the absence of moisture. Owing to the toxicity of organostannanes, this reagent should be handled in a well-ventilated fume hood. Contact with the eyes and skin should be avoided.

Oxidations. Benzylic, allylic, and secondary alcohols are oxidized to the corresponding carbonyl compounds by using $(Bu_3Sn)_2O$–***Bromine***.[1] This procedure is quite useful for selective oxidation of secondary alcohols in the presence of primary alcohols, which are inert under these conditions (eqs 1–3).[2] $(Bu_3Sn)_2O$–***N-Bromosuccinimide*** can also be applied to the selective oxidation of secondary alcohols (eq 4).[3]

(1) $(Bu_3Sn)_2O$, Br_2, CH_2Cl_2, rt; 76%

(2) $(Bu_3Sn)_2O$, Br_2, CH_2Cl_2, rt; 86%

(3) $(Bu_3Sn)_2O$, Br_2, CH_2Cl_2, rt; 68%

(4) 1. $(Bu_3Sn)_2O$; 2. NBS; 80%

$(Bu_3Sn)_2O$–Br_2 oxidizes sulfides to sulfoxides in CH_2Cl_2 without further oxidation to sulfones, even in the presence of excess reagent (eq 5).[4] This procedure is especially useful for sulfides having long, hydrophobic alkyl chains, for which solubility problems are often encountered in the ***Sodium Periodate*** oxidation in aqueous organic solvents. Oxidation of sulfenamides to sulfinamides can be achieved without formation of sulfonamides using the reagent (eq 6).[4]

$$PhCH_2SCH_2Ph \xrightarrow[92\%]{(Bu_3Sn)_2O,\ Br_2,\ CH_2Cl_2,\ rt} PhCH_2SOCH_2Ph \quad (5)$$

(6) PhS–N(morpholino) → PhSO–N(morpholino); $(Bu_3Sn)_2O$, Br_2, CH_2Cl_2, rt; 92%

$(Bu_3Sn)_2O$–Br_2–***Diphenyl Diselenide*** in refluxing $CHCl_3$ transforms alkenes into α-seleno ketones (eq 7).[5]

(7) Ph–CH=CH$_2$ → PhC(O)CH$_2$SePh; $(Bu_3Sn)_2O$, Br_2, $(PhSe)_2$, $CHCl_3$, reflux; 74%

O- and N-activations. $(Bu_3Sn)_2O$ has been used in the activation of hydroxy groups toward sulfamoylations, acylations, carbamoylations, and alkylations because conversion of alcohols to stannyl ethers enhances the oxygen nucleophilicity. Tributylstannyl ethers are easily prepared by heating the alcohol and $(Bu_3Sn)_2O$, with azeotropic removal of water. Sulfamoylation of alcohols can be achieved via tributyltin derivatives in high yields, whereas direct sulfamoylation gives low yields (eq 8).[6] This activation can be used for selective acylation of vicinal diols (eq 9).[7] In carbohydrate chemistry this approach is extremely useful for the regioselective acylation without the use of a blocking–deblocking technique (eq 10).[8] The order of the activation of hydroxy groups on carbohydrates has been investigated, and is shown in partial structures (**1**), (**2**), and (**3**).[9] Regioselective carbamoylation can also be accomplished by changing experimental conditions (eq 11).[10] On the other hand, alkylations of the tin derivatives are sluggish and less selective than acylations under similar conditions. Regioselective alkylation of sugar compounds, however, can be carried out in high yield by conversion to a tributyltin ether followed by addition of alkylating agent and quaternary ammonium halide catalysts (eq 12).[11]

(8) $(Bu_3Sn)_2O$; then H_2NSO_2Cl, 87%

(9) $(Bu_3Sn)_2O$; then BzCl, 67%

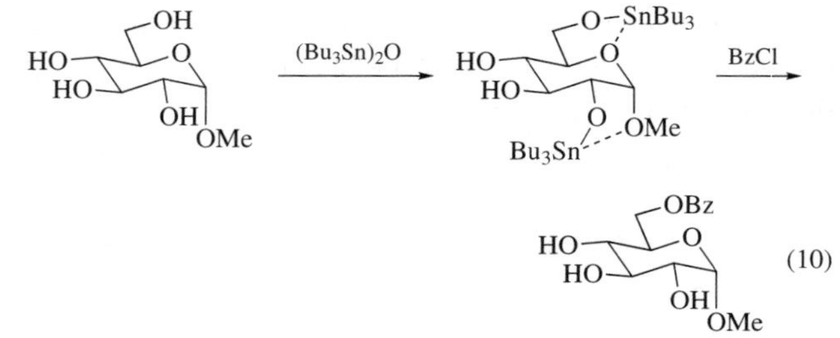

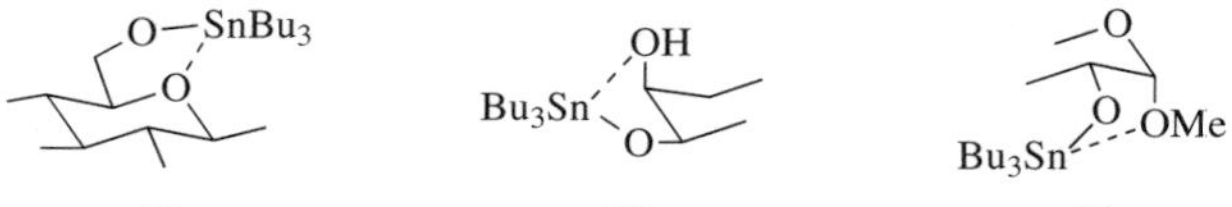

(10)

(1) most reactive

(2) next most reactive

(3) least reactive

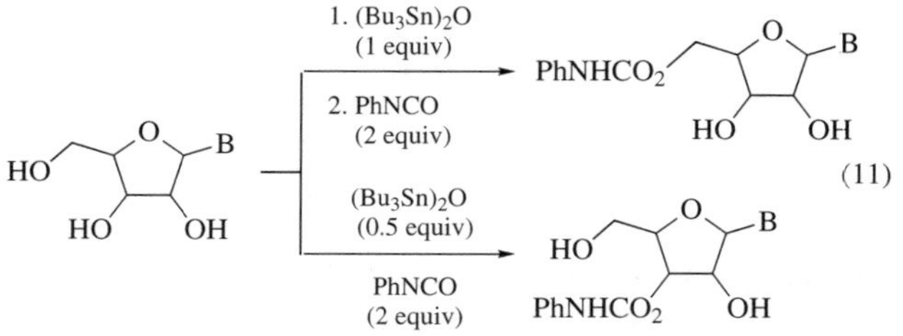

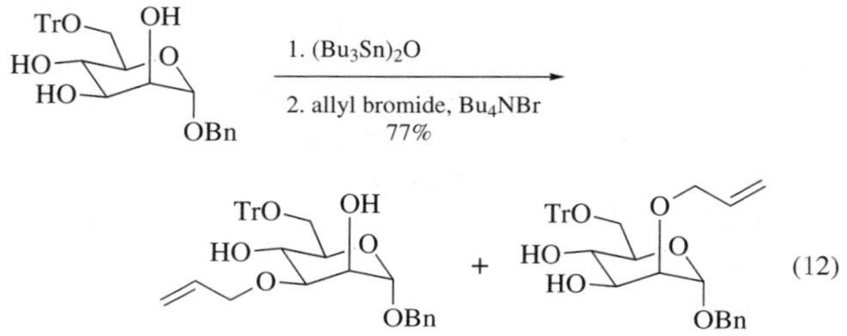

(11)

(12)

80:20

This O-activation is also effective for intramolecular alkylations such as oxetane synthesis (eq 13).[12] Similar N-activation has been used in the synthesis of pyrimidine nucleosides (eq 14).[13]

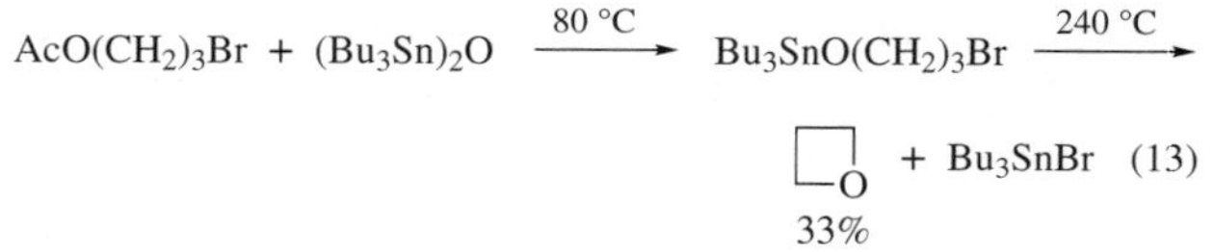

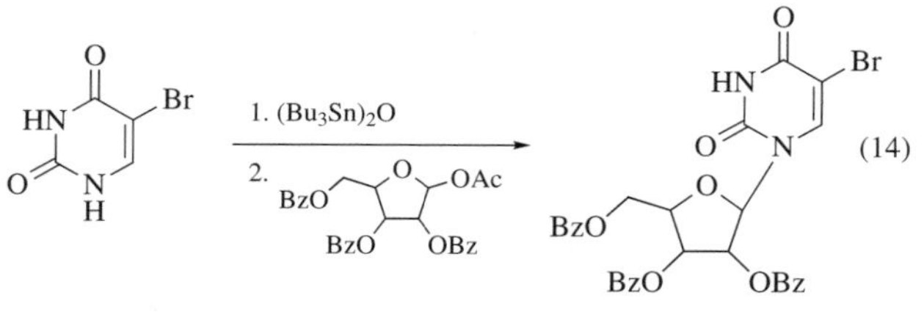

$$AcO(CH_2)_3Br + (Bu_3Sn)_2O \xrightarrow{80\,°C} Bu_3SnO(CH_2)_3Br \xrightarrow{240\,°C} \text{oxetane} + Bu_3SnBr \quad (13)$$

33%

(14)

Dehydrosulfurizations. The thiophilicity of tin compounds is often utilized in functional group transformations. Thus conversion of aromatic and aliphatic thioamides to the corresponding nitriles can be accomplished by using $(Bu_3Sn)_2O$ in boiling benzene under azeotropic conditions (eq 15).[14]

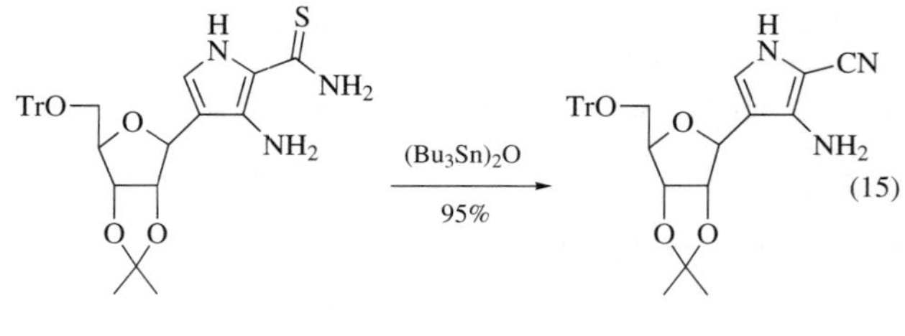

(15)

Hydrolysis. Esters are efficiently hydrolyzed with $(Bu_3Sn)_2O$ under mild conditions (eq 16).[15]

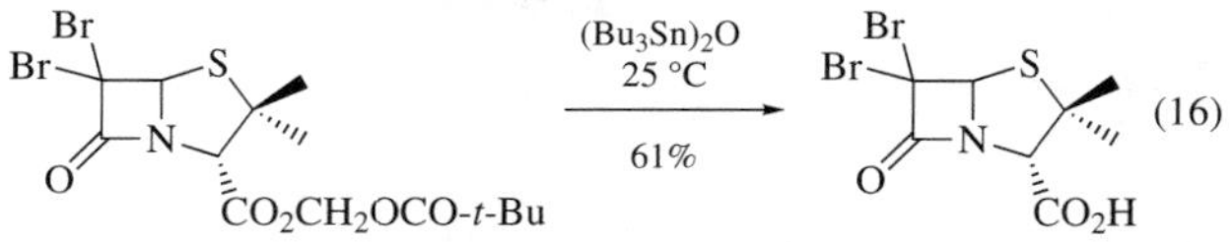

(16)

Transformation of primary alkyl bromides or iodides to the corresponding primary alcohols is achieved in good yield by using $(Bu_3Sn)_2O$–***Silver(I) Nitrate*** (or ***Silver(I) p-Toluenesulfonate***) (eq 17),[16] whereas this method is not applicable to secondary halides due to elimination.

$$MeCO_2(CH_2)_4I \xrightarrow[96\%]{(Bu_3Sn)_2O,\ AgTos,\ DMF,\ 20\,°C} MeCO_2(CH_2)_4OH \quad (17)$$

$(Bu_3Sn)_2O$ is a useful starting material for the preparation of tributyltin hydride, which is a convenient radical reducing reagent in organic synthesis. Thus ***Tri-n-butyltin Hydride*** is easily prepared by using exchange reactions of $(Bu_3Sn)_2O$ with polysiloxanes (eq 18).[17]

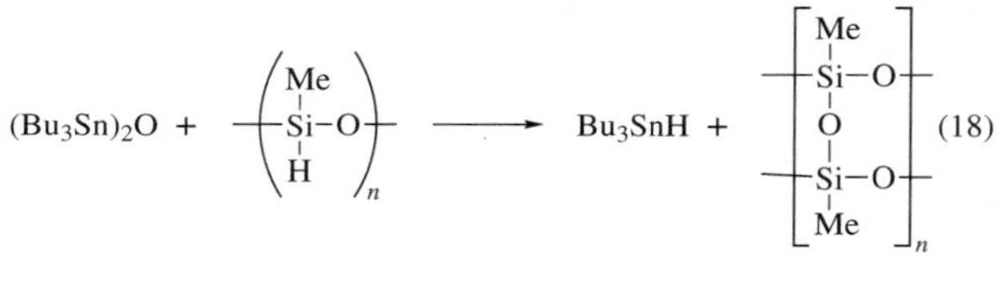

(18)

Related Reagents. Tri-*n*-butyl(methoxy)stannane.

1. Saigo, K.; Morikawa, A.; Mukaiyama, T. *CL* **1975**, 145.
2. Ueno, Y.; Okawara, M. *TL* **1976**, 4597.
3. Hanessian, S.; Roy, R. *CJC* **1985**, *63*, 163.
4. Ueno, Y.; Inoue, T.; Okawara, M. *TL* **1977**, 2413.
5. Kuwajima, I.; Shimizu, M. *TL* **1978**, 1277.
6. Jenkins, I. D.; Verheyden, J. P. H.; Moffatt, J. G. *JACS* **1971**, *93*, 4323.
7. (a) Ogawa, T.; Matsui, M. *T* **1981**, *37*, 2363. (b) David, S.; Hanessian, S. *T* **1985**, *41*, 643.
8. (a) Crowe, A. J.; Smith, P. J. *JOM* **1976**, *110*, C57. (b) Blunden, S. J.; Smith, P. J.; Beynon, P. J.; Gillies, D. G. *Carbohydr. Res.* **1981**, *88*, 9. (c) Ogawa, T.; Matsui, M. *Carbohydr. Res.* **1977**, *56*, C1. (d) Hanessian, S.; Roy, R. *JACS* **1979**, *101*, 5839. (e) Arnarp, J.; Loenngren, J. *CC* **1980**, 1000. (f) Ogawa, T.; Nakabayashi, S.; Sasajima, K. *Carbohydr. Res.* **1981**, *96*, 29.
9. Tsuda, Y.; Haque, M. E.; Yoshimoto, K. *CPB* **1983**, *31*, 1612.

10. (a) Ishido, Y.; Hirao, I.; Sakairi, N.; Araki, Y. *H* **1979**, *13*, 181. (b) Hirao, I.; Itoh, K.; Sakairi, N.; Araki, Y.; Ishido, Y. *Carbohydr. Res.* **1982**, *109*, 181.
11. (a) Alais, J.; Veyrières, A. *JCS(P1)* **1981**, 377. (b) Veyrières, A. *JCS(P1)* **1981**, 1626.
12. Biggs, J. *TL* **1975**, 4285.
13. Ogawa, T.; Matsui, M. *JOM* **1978**, *145*, C37.
14. Lim, M.-I.; Ren, W.-Y.; Klein, R. S. *JOC* **1982**, *47*, 4594.
15. Mata, E. G.; Mascaretti, O. A. *TL* **1988**, *29*, 6893.
16. Gingras, M.; Chan, T. H. *TL* **1989**, *30*, 279.
17. Hayashi, K.; Iyoda, J.; Shiihara, I. *JOM* **1967**, *10*, 81.

Hiroshi Sano
Gunma University, Kiryu, Japan

Bis(trichloromethyl) Carbonate[1]

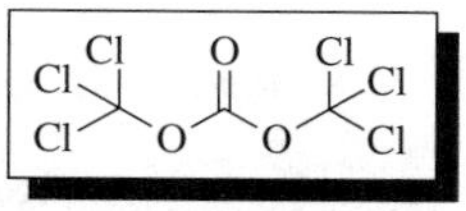

[32315-10-9] $C_3Cl_6O_3$ (MW 296.73)

(a phosgene surrogate)

Alternate Name: triphosgene.
Physical Data: mp 81–83 °C; bp 203–206 °C.
Solubility: sol methanol, ethanol, benzene, diethyl ether, hexane, THF, ethyl acetate; dec slowly in cold water.
Form Supplied in: white crystalline solid.
Handling, Storage, and Precautions: the reagent is somewhat moisture sensitive, but scrupulously anhydrous conditions are not necessary. Rapid handling of the reagent in open air, in the absence of a glove bag or dry box, is usually satisfactory. This reagent should only be handled in a fume hood.

Introduction. Phosgene gas is a versatile reagent for organic synthesis, which has been used in carbonylation, chloroformylation, chlorination, and dehydration reactions, to name a few.[1] However, because of the high toxicity of ***Phosgene***, it has not been used widely. Surrogates for phosgene have long been sought. ***Trichloromethyl Chloroformate*** (diphosgene) was introduced as one such phosgene substitute.[2] To the extent that diphosgene is a highly toxic, hygroscopic, and noxious liquid, it is not a fully satisfactory phosgene substitute. On the other hand, bis(trichloromethyl) carbonate (triphosgene; **1**) is a crystalline solid compound, is not very hygroscopic, and can substitute for phosgene in a plethora of chemical reactions. Triphosgene has been known since 1880,[3] but its synthetic utility had not been exploited until recently.

Carbonylation. The formation of oxazoles,[4] quinazolinediones (eq 1),[5] carbonates (eq 2),[6] urea analogs,[7] and isocyanates (eq 3)[1] have been reported with triphosgene. The reactions give good to excellent yields and often require only 1/3 equiv of triphosgene.

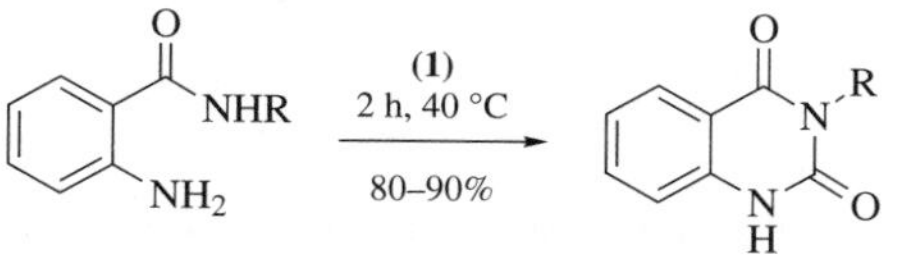

(1): (1), 2 h, 40 °C; 80–90%

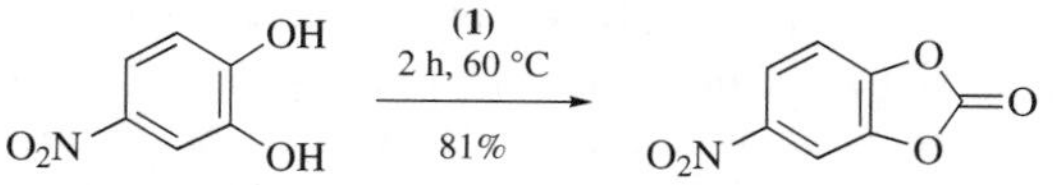

(2): (1), 2 h, 60 °C; 81%

(3): (1), 1–2 h, 70 °C; 68%

Chloroformylation. The use of triphosgene in the formation of chloroformates with both hydroxy compounds (eq 4) and substituted amines (eq 5) has been reported.[1]

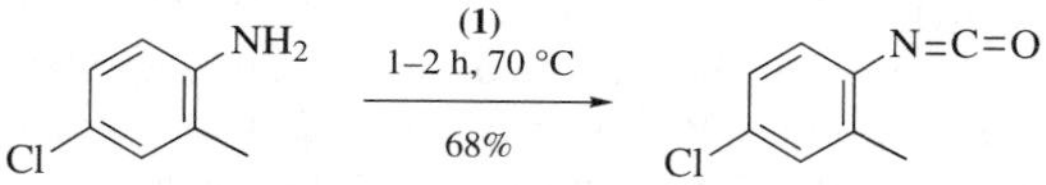

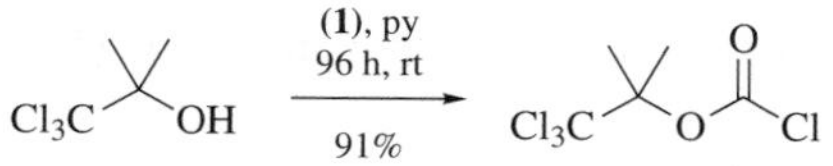

(4): Cl_3C…OH → (1), py, 96 h, rt; 91%

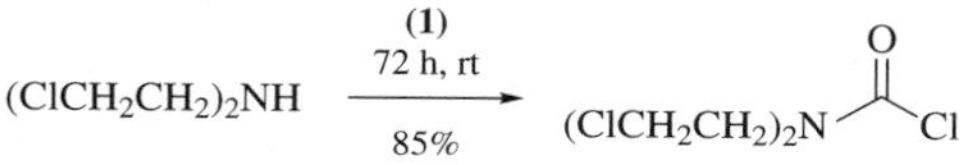

(5): $(ClCH_2CH_2)_2NH$ → $(ClCH_2CH_2)_2NCOCl$; (1), 72 h, rt; 85%

Dehydration.. Synthesis of isocyanides from formamides (eq 6) is the only example of a dehydration reaction reported for triphosgene to date.[1]

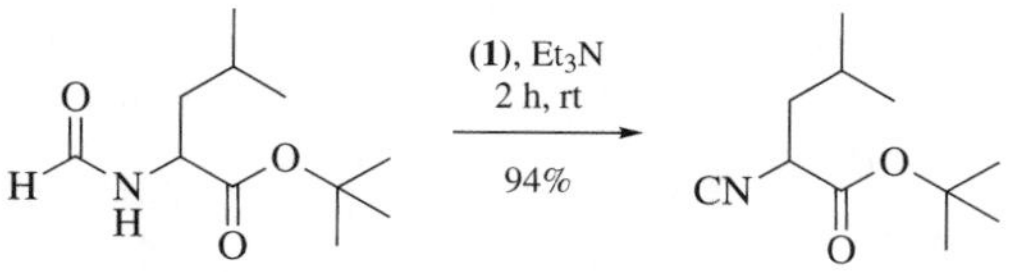

(6): (1), Et_3N, 2 h, rt; 94%

N*-Carboxyamino Acid Anhydrides.** Synthesis of *N*-carboxyamino acid anhydrides (NCAs) is accomplished by the reaction of the zwitterionic unprotected amino acid with triphosgene at elevated temperatures (eq 7).[8] Wilder and Mobashery reported a milder version of this reaction, utilizing *N*-*t*-butoxycarbonyl-α-amino acids; in the presence of a stoichiometric quantity of ***Triethylamine and 1/3 equiv of triphosgene (eq 8) the reaction is believed to proceed via a chloroformic *N*-*t*-butoxycarbonyl-α-amino acid anhydride intermediate to give the corresponding NCA.[9] There are many synthetic routes to NCAs;[10] however, the use of triphosgene in the preparation of NCAs provides a facile, safe, mild, and practical entry to these important molecules.

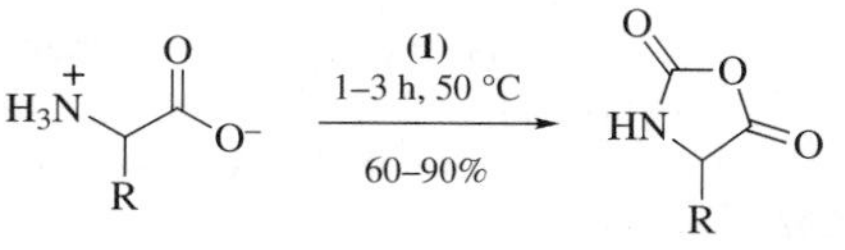

(7): (1), 1–3 h, 50 °C; 60–90%

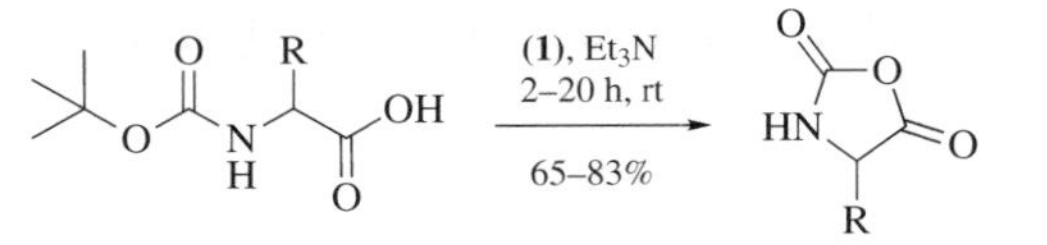

(8)

Chlorination. Triphosgene has been used in the synthesis of acyl chlorides (eq 9).[1] Also, it has been shown to react with a variety of aldehydes to give α-chloro chloroformates (eq 10).[11] Goren et al. have shown that triphosgene mediates chloride substitution reactions on activated alcohols (eq 11);[12] this reaction involves the formation of an intermediary chloroformate species en route to the chlorinated product. The generality of the reaction was shown for benzylic (eq 12), allylic, and propargylic systems. This reaction proceeds primarily via S_N2, with some contribution by S_N1 and/or S_Ni mechanisms.[12] Chlorination with triphosgene is considerably milder than the typical chlorination reactions with ***Thionyl Chloride***, ***Phosphorus(III) Chloride***, ***Phosphorus(V) Chloride***, ***Oxalyl Chloride***, and the like; for example, acid-labile functionalities, such as *t*-butyl carbamate (BOC), benzhydryl, and *p*-methoxybenzyl groups, can be tolerated in the triphosgene-mediated reactions, whereas they do not normally survive reactions with other chlorination reagents.

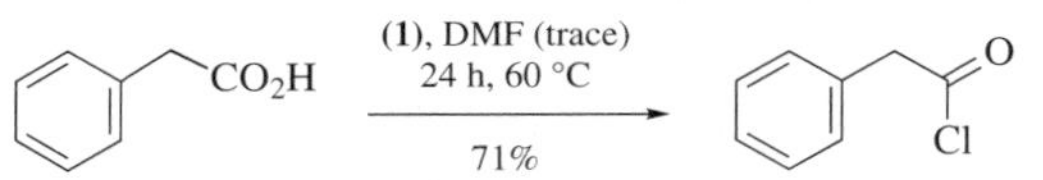

(9)

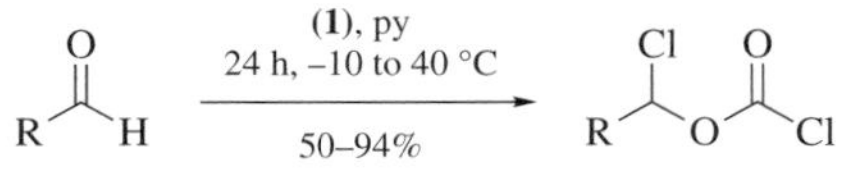

(10)

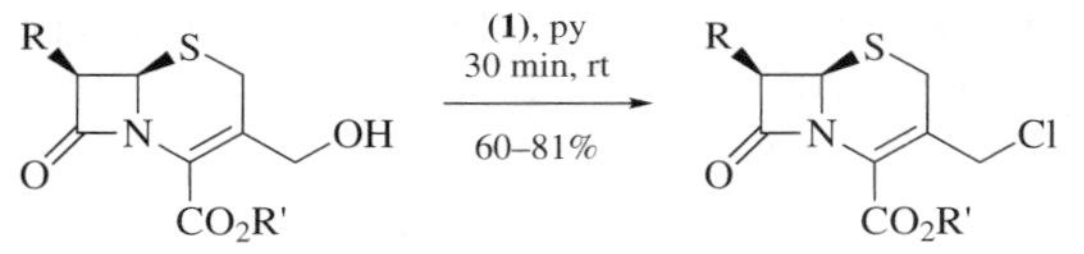

(11)

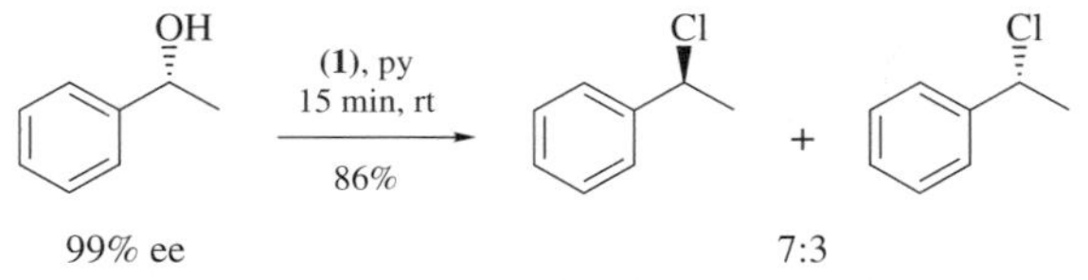

(12)

Oxidation. Palomo et al. have demonstrated that primary and secondary alcohols are readily oxidized by triphosgene in the presence of ***Dimethyl Sulfoxide*** (DMSO) in good to excellent yields;[13a] the reaction is applicable to both activated (eq 13) and unactivated alcohols (eq 14). This reaction is a variation of the one reported by Barton with phosgene gas,[14] and it is an excellent alternative to Swern-type oxidations. Indeed, in some respects, oxidation of alcohols by triphosgene is superior to the widely used Swern oxidation. The triphosgene reaction tolerates acid-labile functionalities and is not as sensitive to residual moisture as oxalyl chloride used in the Swern oxidation. An additional advantage of the method of Palomo et al. is the ease of handling of triphosgene, and the fact that the reaction is amenable to large-scale synthesis. The breadth and scope of this reaction have been investigated with a large variety of alcohol substrates.[13b] See also ***Dimethyl Sulfoxide-Triphosgene***.

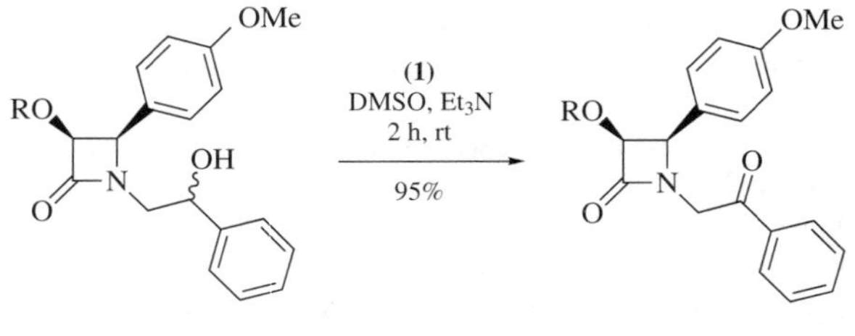

(13)

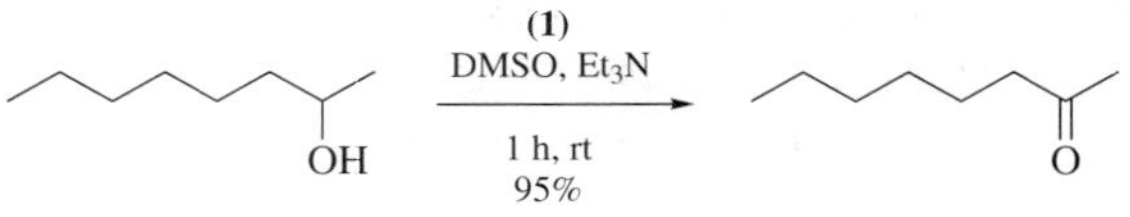

(14)

Related Reagents. Phosgene; Trichloromethyl Chloroformate.

1. Eckert, H.; Foster, B. *AG(E)* **1987**, *26*, 894.
2. Kurita, K.; Iwakura, Y. *OSC* **1988**, *6*, 715.
3. Councler, C. *CB* **1880**, *13*, 1697.
4. Sicker, D. *SC* **1989**, 875. Flouzat, C.; Blanchet, M.; Guillaumet, G. *TL* **1992**, *33*, 4571.
5. Cortez, R.; Rivero, J. A.; Samanathan, R.; Aguire, G.; Ramirez, F. *SC* **1991**, *21*, 285.
6. Laufer, D. A.; Doyle, K.; Zhang, X. *OPP* **1989**, *21*, 771.
7. Zhao, X.; Chang, Y.-L.; Fowler, F. W. *JACS* **1990**, *112*, 6627. Cotarca, L.; Bacaloglu, R.; Csunderlik, C.; Marcu, N.; Tarnaveanu, A. *JPR* **1987**, *329*, 1052.
8. Daly, W. H.; Poché, D. *TL* **1988**, *29*, 5859.
9. Wilder, R.; Mobashery, S. *JOC* **1992**, *57*, 2755.
10. Kricheldorf, H. R. *α-Aminoacid-N-Carboxy-Anhydride and Related Heterocycles*; Springer: Berlin, 1987; pp 3–58. Blacklock, T. J.; Hirschmann, R.; Veber, D. F. *The Peptides*; Academic: New York, 1987; Vol. 9, pp 39–102.
11. Coghlan, M. J.; Caley, B. A. *TL* **1989**, *30*, 2033.
12. Goren, Z.; Heeg, M. J.; Mobashery, S. *JOC* **1991**, *56*, 7188.
13. (a) Palomo, C.; Cossio, F. P.; Ontoria, J. M.; Odrizola, J. M. *JOC* **1991**, *56*, 5948. (b) Rivero, I. A.; Samanathan, R.; Hellberg, L. H. *OPPI Briefs* **1992**, *24*, 363.
14. Barton, D. H. R.; Garner, B. J.; Wightman, R. H. *JCS* **1964**, 1855.

Juliatiek Roestamadji & Shahriar Mobashery
Wayne State University, Detroit, MI, USA

Boron Tribromide[1]

BBr_3

[10294-33-4] BBr_3 (MW 250.52)

(Lewis acid used for deprotection of OH and NH groups; cleaves ethers or esters to alkyl bromides; bromoborates allene and alkynes)

Physical Data: mp −45 °C; bp 91.7 °C; *d* 2.650 g cm^{-3}.

Form Supplied in: colorless, fuming liquid; a 1.0 M solution in dichloromethane and hexane; $BBr_3 \cdot Me_2S$ complex is available as either a white solid or a 1.0 M solution in dichloromethane.

Purification: by distillation.

Handling, Storage, and Precautions: BBr_3 is highly moisture sensitive and decomposes in air with evolution of HBr. Store under a dry inert atmosphere and transfer by syringe or through a Teflon tube. It reacts violently with protic solvents such as water and alcohols. Ether and THF are not appropriate solvents.

Removal of Protecting Groups. BBr_3 is highly Lewis acidic. It coordinates to ethereal oxygens and promotes C–O bond cleavage to an alkyl bromide and an alkoxyborane that is hydrolyzed to an alcohol during workup (eq 1).[2]

$$R^1OR^2 \xrightarrow{BBr_3} R^1Br + Br_2BOR^2 \xrightarrow{H_2O} R^1Br + R^2OH \quad (1)$$

BBr_3 has been widely used to cleave ethers because the reaction proceeds completely under mild conditions. In a special case, BBr_3 has been used to cleave acetals that cannot be deprotected by usual acidic conditions.[3] Because alkyl aryl ethers are cleaved at the alkyl–oxygen bond to give ArOH and alkyl bromides, BBr_3 has been most generally used for the demethylation of methyl aryl ethers,[2,4] for example as the final step of zearalenone synthesis (eq 2).[5] Problems are sometimes encountered in attempts to deprotect more than one nonadjacent methoxy group on one aromatic ring, and when stable chelates are formed.[6] The presence of a carbonyl substituent facilitates the selective deprotection of polymethoxyaryl compounds (eq 3).[7]

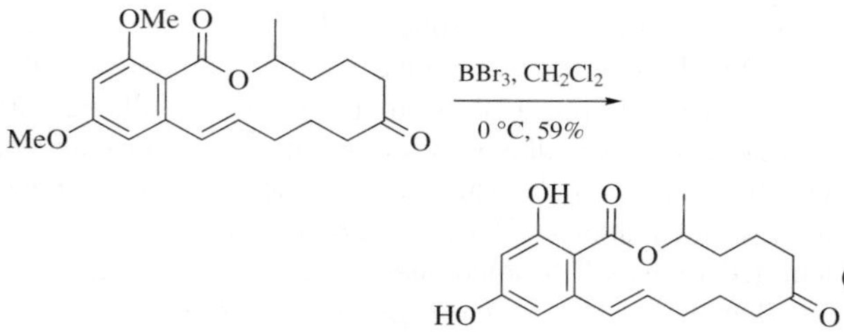

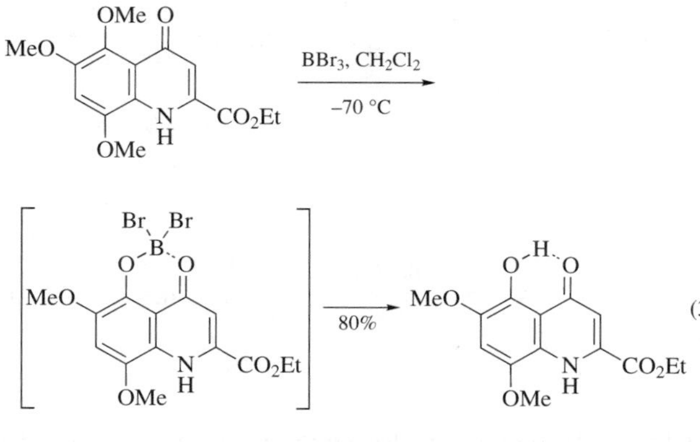

The cleavage of mixed dialkyl ethers occurs at the more substituted carbon–oxygen bond. Methyl ethers of secondary or tertiary alcohols give methanol and secondary or tertiary alkyl bromides selectively by the reaction with BBr_3,[8] although the addition of ***Sodium Iodide*** and ***15-Crown-5*** ether can change this selectivity (eq 4).[9] In contrast, methyl ethers of primary alcohols are generally cleaved at the Me–O bond, as demonstrated in Corey's prostaglandin synthesis (eq 5).[10]

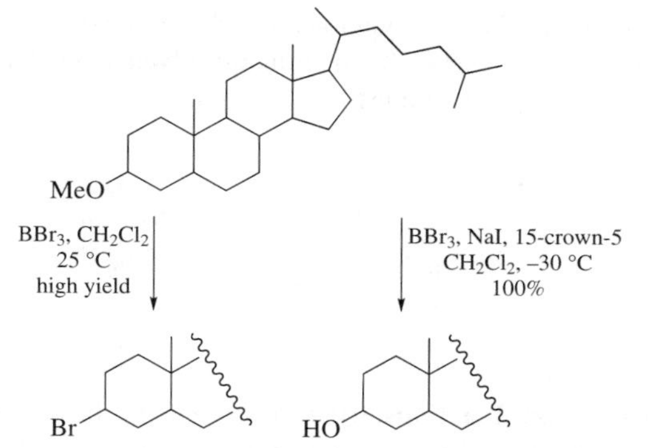

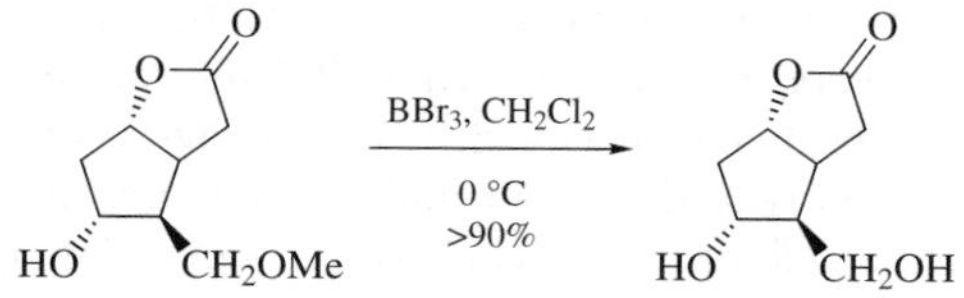

BBr_3 has been also used for the deprotection of carbohydrate derivatives[11] and polyoxygenated intermediates in the synthesis of deoxyvernolepin,[12] vernolepin,[13] and vernomenin.[13] Although one of the model compounds is deprotected cleanly (eq 6),[14] application of BBr_3 to more highly functionalized intermediates leads to cleavage of undesired C–O bonds competitively (eq 7).[12,13]

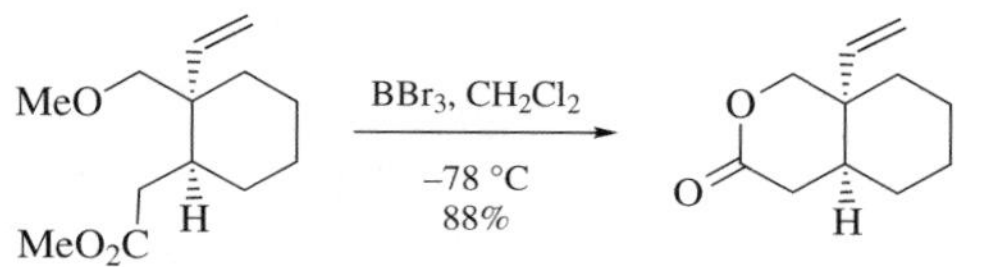

For the complete cleavage, 1 mol of BBr_3 is required for each ether group and other Lewis-basic functional groups. Sometimes it is difficult to find reaction conditions for the selective cleavage of the desired C–O bond. Recently, modified bromoboranes such as ***B-Bromocatecholborane***,[15] dialkylbromoboranes,[16] ***Bromobis(isopropylthio)borane***,[17] and ***9-Bromo-9-borabicyclo[3.3.1]nonane***,[18] have been introduced to cleave C–O bonds more selectively under milder conditions. $BBr_3 \cdot SMe_2$ is also effective for ether cleavage and has the advantage of being more stable than BBr_3. It can be stored for a long time and handled easily. However, a two- to fourfold excess of the reagent is necessary to complete the dealkylation of alkyl aryl ether.[19]

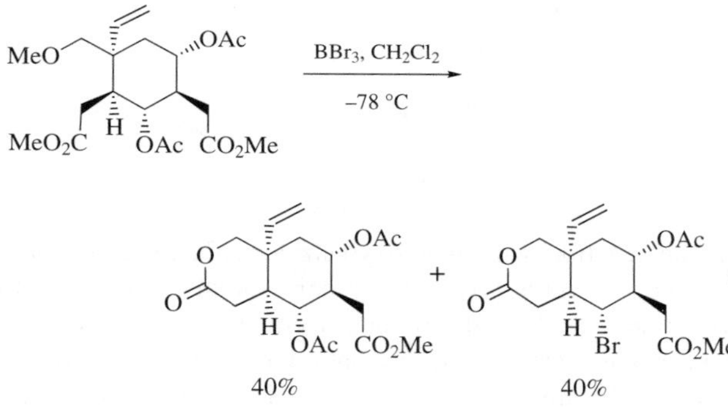

Amino acid protecting groups such as benzyloxycarbonyl and *t*-butoxycarbonyl groups are cleaved by BBr_3. However, the hydrolysis of the ester function also occurs under the same reaction conditions.[20] Debenzylation and debenzyloxymethylation of uracils proceed successfully in aromatic solvents, but demethylation is more sluggish and less facile (eq 8).[21]

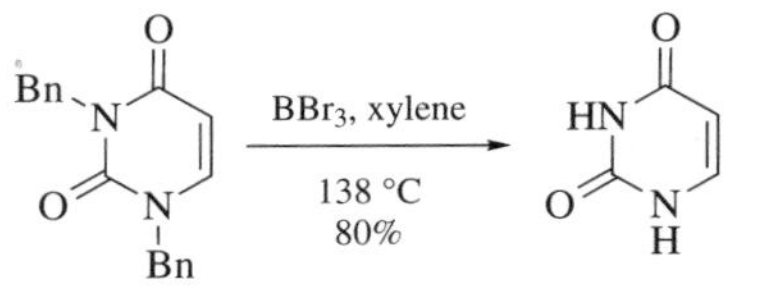

(8)

Substitution Reactions. BBr_3 reacts with cyclic ethers to give tris(ω-bromoalkoxy)boranes which provide ω-bromoalkanols or ω-bromoalkanals when treated with MeOH or ***Pyridinium Chlorochromate***, respectively (eq 9).[22] Unfortunately, unsymmetrically substituted ethers such as 2-methyltetrahydrofuran are cleaved nonregioselectively. Generally, ester groups survive under the reaction conditions for ether cleavage, but the ring opening of lactones occurs under mild conditions to give ω-halocarboxylic acids in good yields (eq 10).[23]

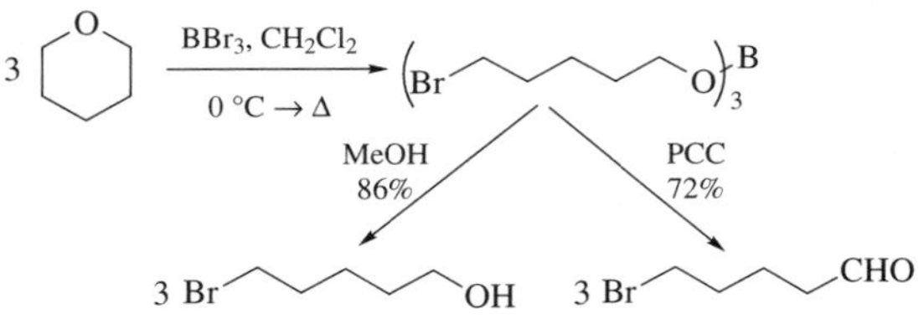

(9)

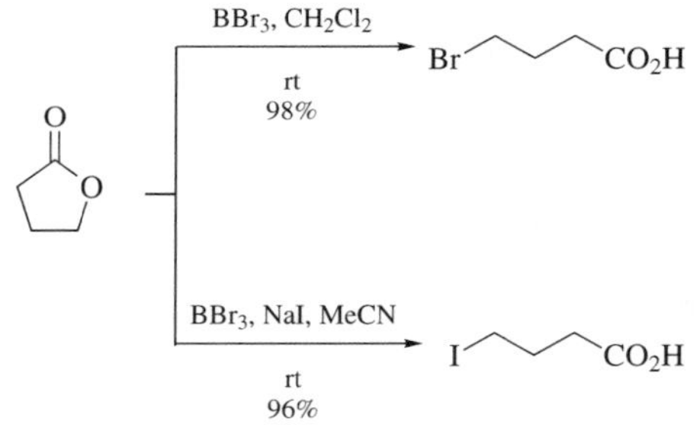

(10)

In the reaction with methoxybenzaldehyde, bromination of the carbonyl group takes place more rapidly than demethylation; therefore benzal bromide formation is generally observed in the reaction with aromatic aldehydes.[24] Cleavage of *t*-butyldimethylsilyl ethers or *t*-butyldiphenylsilyl ethers occurs at the C–O bond to give alkyl bromides.[25] Alcohols can be converted to alkyl bromides by this method.

In a special case, BBr_3 is used for the bromination of hydrocarbons. Adamantane is brominated by a mixture of ***Bromine***, BBr_3, and ***Aluminum Bromide*** to give 1,3-dibromoadamantane selectively.[26] Tetrachlorocyclopropene[27] and hexachlorocyclopentadiene[28] are substituted to the corresponding bromides by BBr_3 and, in the latter case, addition of $AlBr_3$ and Br_2 is effective to improve the result.[29]

Reduction of Sulfur Compounds. Alkyl and aryl sulfoxides are reduced by BBr_3 to the corresponding sulfides in good yields.[30] Addition of ***Potassium Iodide*** and a catalytic amount of ***Tetra-n-butylammonium Iodide*** is necessary for the reduction of sulfonic acids and their derivatives.[31]

Transesterification of Esters or Conversion to Amides. Transesterification reactions of carboxylic esters or conversion into the amides is promoted by a stoichiometric amount of BBr_3.[32]

Removal of Methyl Sulfide from Organoborane–Methyl Sulfide Complexes. Methyl sulfide can be removed from $BrBR_2 \cdot SMe_2$ or $Br_2BR \cdot SMe_2$, which are prepared by the hydroboration reaction of alkenes or alkynes with $BrBH_2 \cdot SMe_2$ or $Br_2BH \cdot SMe_2$, by using BBr_3.[33] The resulting alkenyldibromoboranes are useful for the stereoselective synthesis of bromodienes (eq 11).[34]

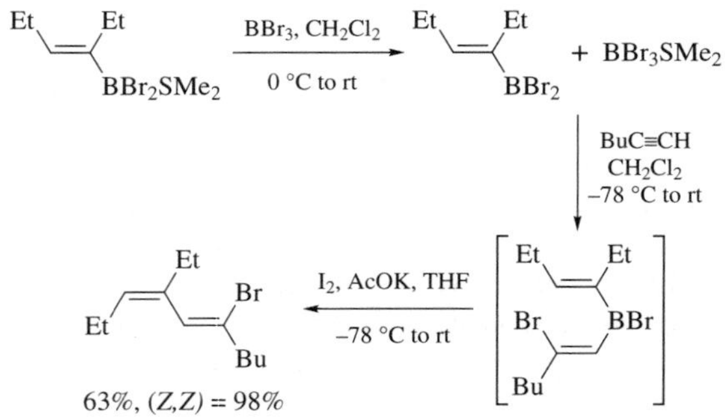

(11)

Bromoboration Reactions. BBr_3 does not add to isolated double bonds, but reacts with allene spontaneously even at low temperature to give (2-bromoallyl)dibromoborane,[35] which provides stable (2-bromoallyl)diphenoxyborane by the addition of anisole.[36] The diphenoxyborane derivative reacts with carbonyl compounds to give 2-bromohomoallylic alcohols in high yields (eq 12). Bromoboration of 1-alkynes provides (*Z*)-(2-bromo-1-alkenyl)dibromoboranes stereo- and regioselectively (eq 13),[37] which are applied for the synthesis of trisubstituted alkenes,[38] α,β-unsaturated esters,[39] and γ,δ-unsaturated ketones,[40] bromodienes,[41] 1,2-dihalo-1-alkenes,[42] 2-bromoalkanals,[43] and β-bromo-α,β-unsaturated amides.[44]

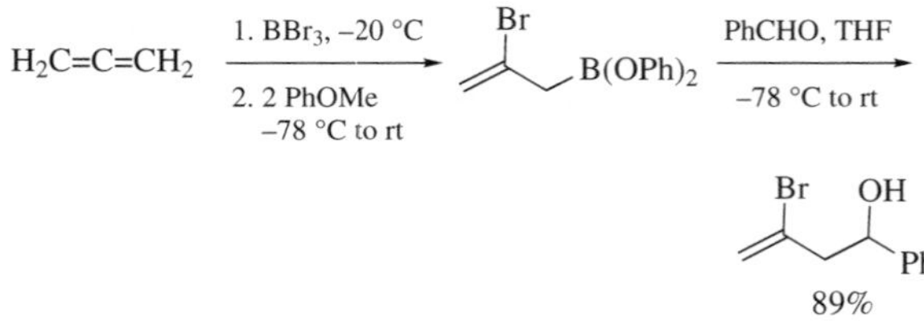

(12)

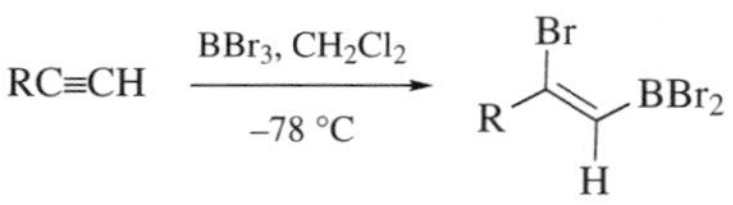

(13)

Chiral Bromoborane Reagents. Complexes made from chiral 1-alkyl-2-(diphenylhydroxymethyl)pyrrolidines and BBr_3 are effective catalysts for asymmetric Diels–Alder reactions.[45] Bromoboranes prepared from chiral 1,2-diphenyl-1,2-bis(arenesulfonamido)ethanes[46,47] are used to prepare chiral allylic boranes,[47,48] allenylic borane,[49] propargylic boranes,[49] and enolates.[46,47,50] The *B*-bromodiazaborolidinene (**1**), prepared from 1,2-diphenyl-1,2-bis(*p*-toluenesulfonamido)ethane, is particularly effective in these applications. The reagents prepared from (**1**) are highly effective for the enantioselective synthesis of homoallylic alcohols (eq 14),[48] homopropargylic alcohols (eq 15),[49] propadienyl carbinols (eq 16),[49] and aldol condensation products (eq 17).[46]

(**1**) 1. 2-chloroallyl-$SnBu_3$, CH_2Cl_2, 0 °C to rt; 2. PhCHO, −78 °C → 84% ee (14)

(**1**) 1. $HC{\equiv}CCH_2SnPh_3$, CH_2Cl_2, 0 °C to rt; 2. PhCHO, −78 °C → 96% ee (15)

(**1**) 1. $H_2C{=}C{=}CHSnPh_3$, CH_2Cl_2, 0 °C to rt; 2. PhCHO, −78 °C → >99% ee (16)

(**1**) 1. 3-pentanone, *i*-Pr_2NEt, CH_2Cl_2, −78 °C; 2. EtCHO, −78 °C → 98% ee, 98% *syn* (17)

Related Reagents. Boron Trichloride; Bromodimethylborane; Bromobis(isopropylthio)borane; 9-Bromo-9-borabicyclo-[3.3.1]nonane; *B*-Bromocatecholborane; Thionyl Bromide; Hydrogen Bromide.

1. Bhatt, M. V.; Kulkarni, S. U. *S* **1983**, 249.
2. McOmie, J. F. W.; Watts, M. L.; West, D. E. *T* **1968**, *24*, 2289.
3. Meyers, A. I.; Nolen, R. L.; Collington, E. W.; Narwid, T. A.; Strickland, R. C. *JOC* **1973**, *38*, 1974.
4. (a) Benton, F. L.; Dillon, T. E. *JACS* **1942**, *64*, 1128. (b) Manson, D. L.; Musgrave, O. C. *JCS* **1963**, 1011. (c) McOmie, J. F. W.; Watts, M. L. *CI(L)* **1963**, 1658. (d) Blatchly, J. M.; Gardner, D. V.; McOmie, J. F. W.; Watts, M. L. *JCS(C)* **1968**, 1545.
5. (a) Vlattas, I.; Harrison, I. T.; Tökés, L.; Fried, J. H.; Cross, A. D. *JOC* **1968**, *33*, 4176. (b) Taub, D.; Girotra, N. N.; Hoffsommer, R. D.; Kuo, C. H.; Slates, H. L.; Weber, S.; Wendler, N. L. *T* **1968**, *24*, 2443.
6. (a) Stetter, H.; Wulff, C. *CB* **1960**, *93*, 1366. (b) Locksley, H. D.; Murray, I. G. *JCS(C)* **1970**, 392. (c) Bachelor, F. W.; Loman, A. A.; Snowdon, L. R. *CJC* **1970**, *48*, 1554.
7. Schäfer, W.; Franck, B. *CB* **1966**, *99*, 160.
8. Youssefyeh, R. D.; Mazur, Y. *CI(L)* **1963**, 609.
9. Niwa, H.; Hida, T.; Yamada, K. *TL* **1981**, *22*, 4239.
10. Corey, E. J.; Weinshenker, N. M.; Schaaf, T. K.; Huber, W. *JACS* **1969**, *91*, 5675.
11. Bonner, T. G.; Bourne, E. J.; McNally, S. *JCS* **1960**, 2929.
12. Grieco, P. A.; Noguez, J. A.; Masaki, Y. *JOC* **1977**, *42*, 495.
13. Grieco, P. A.; Nishizawa, M.; Burke, S. D.; Marinovic, N. *JACS* **1976**, *98*, 1612.
14. (a) Grieco, P. A.; Hiroi, K.; Reap, J. J.; Noguez, J. A. *JOC* **1975**, *40*, 1450. (b) Grieco, P. A.; Reap, J. J.; Noguez, J. A. *SC* **1975**, *5*, 155.
15. (a) Boeckman, Jr., R. K.; Potenza, J. C. *TL* **1985**, *26*, 1411. (b) King, P. F.; Stroud, S. G. *TL* **1985**, *26*, 1415.
16. (a) Guindon, Y.; Morton, H. E.; Yoakim, C. *TL* **1983**, *24*, 3969. (b) Gauthier, J. Y.; Guindon, Y. *TL* **1987**, *28*, 5985. (c) Guindon, Y.; Yoakim, C.; Morton, H. E. *TL* **1983**, *24*, 2969. (d) Guindon, Y.; Yoakim, C.; Morton, H. E. *JOC* **1984**, *49*, 3912.
17. Corey, E. J.; Hua, D. H.; Seitz, S. P. *TL* **1984**, *25*, 3.
18. Bhatt, M. V. *JOM* **1978**, *156*, 221.
19. Williard, P. G.; Fryhle, C. B. *TL* **1980**, *21*, 3731.
20. Felix, A. M. *JOC* **1974**, *39*, 1427.
21. Kundu, N. G.; Hertzberg, R. P.; Hannon, S. J. *TL* **1980**, *21*, 1109.
22. Kulkarni, S. U.; Patil, V. D. *H* **1982**, *18*, 163.
23. Olah, G. A.; Karpeles, R.; Narang, S. C. *S* **1982**, 963.
24. Lansinger, J. M.; Ronald, R. C. *SC* **1979**, *9*, 341.
25. Kim, S.; Park, J. H. *JOC* **1988**, *53*, 3111.
26. (a) Baughman, G. L. *JOC* **1964**, *29*, 238. (b) Talaty, E. R.; Cancienne, A. E.; Dupuy, A. E. *JCS(C)* **1968**, 1902.
27. Tobey, S. W.; West, R. *JACS* **1966**, *88*, 2481.
28. West, R.; Kwitowski, P. T. *JACS* **1968**, *90*, 4697.
29. Ungefug, G. A.; Roberts, C. W. *JOC* **1973**, *38*, 153.
30. Guindon, Y.; Atkinson, J. G.; Morton, H. E. *JOC* **1984**, *49*, 4538.
31. Olah, G. A.; Narang, S. C.; Field, L. D.; Karpeles, R. *JOC* **1981**, *46*, 2408.
32. Yazawa, H.; Tanaka, K.; Kariyone, K. *TL* **1974**, *15*, 3995.
33. (a) Brown, H. C.; Ravindran, N.; Kulkarni, S. U. *JOC* **1979**, *44*, 2417. (b) Brown, H. C.; Ravindran, N.; Kulkarni, S. U. *JOC* **1980**, *45*, 384. (c) Brown, H. C.; Campbell, Jr., J. B. *JOC* **1980**, *45*, 389.
34. Hyuga, S.; Takinami, S.; Hara, S.; Suzuki, A. *TL* **1986**, *27*, 977.
35. Joy, F.; Lappert, M. F.; Prokai, B. *JOM* **1966**, *5*, 506.
36. Hara, S.; Suzuki, A. *TL* **1991**, *32*, 6749.
37. (a) Lappert, M. F.; Prokai, B. *JOM* **1964**, *1*, 384. (b) Blackborow, J. R. *JOM* **1977**, *128*, 161. (c) Suzuki, A.; Hara, S. *Res. Trends Org. Chem.* **1990**, 77. (d) Suzuki, A. *PAC* **1986**, *58*, 629.
38. Satoh, Y.; Serizawa, H.; Miyaura, N.; Hara, S.; Suzuki, A. *TL* **1988**, *29*, 1811.
39. Yamashina, N.; Hyuga, S.; Hara, S.; Suzuki, A. *TL* **1989**, *30*, 6555.
40. (a) Hara, S.; Hyuga, S.; Aoyama, M.; Sato, M.; Suzuki, A. *TL* **1990**, *31*, 247. (b) Aoyama, M.; Hara, S.; Suzuki, A. *SC* **1992**, *22*, 2563.
41. Hyuga, S.; Takinami, S.; Hara, S.; Suzuki, A. *CL* **1986**, 459.
42. Hara, S.; Kato, T.; Shimizu, H.; Suzuki, A. *TL* **1985**, *26*, 1065.
43. Satoh, Y.; Tayano, T.; Koshino, H.; Hara, S.; Suzuki, A. *S* **1985**, 406.
44. Satoh, Y.; Serizawa, H.; Hara, S.; Suzuki, A. *SC* **1984**, *14*, 313.
45. Kobayashi, S.; Murakami, M.; Harada, T.; Mukaiyama, T. *CL* **1991**, 1341.
46. Corey, E. J.; Imwinkelried, R.; Pikul, S.; Xiang, Y. B. *JACS* **1989**, *111*, 5493.
47. Corey, E. J.; Kim, S. S. *TL* **1990**, *31*, 3715.
48. Corey, E. J.; Yu, C.-M.; Kim, S. S. *JACS* **1989**, *111*, 5495.
49. Corey, E. J.; Yu, C.-M.; Lee, D.-H. *JACS* **1990**, *112*, 878.
50. Corey, E. J.; Kim, S. S. *JACS* **1990**, *112*, 4976.

Akira Suzuki & Shoji Hara
Hokkaido University, Sapporo, Japan

Boron Trichloride[1]

BCl_3

[10294-34-5] BCl_3 (MW 117.17)

(Lewis acid capable of selective cleavage of ether and acetal protecting groups; reagent for carbonyl condensations; precursor of organoboron reagents)

Physical Data: bp 12.5 °C; *d* 1.434 g cm^{-3} (0 °C).
Solubility: sol saturated and halogenated hydrocarbon and aromatic solvents; solubility in diethyl ether is approximately 1.5 M at 0 °C; stable for several weeks in ethyl ether at 0 °C, but dec by water or alcohols.
Form Supplied in: colorless gas or fuming liquid in an ampoule; $BCl_3 \cdot SMe_2$ complex (solid) and 1 M solutions in dichloromethane, hexane, heptane, and *p*-xylene are available.
Handling, Storage, and Precautions: a poison by inhalation and an irritant to skin, eyes, and mucous membranes. Reacts exothermically with water and moist air, forming toxic and corrosive fumes. Violent reaction occurs with aniline or phosphine. All operations should be carried out in a well-ventilated fume hood without exposure to the atmosphere. The gas can be collected and measured as a liquid by condensing in a cooled centrifuge tube and then transferred to the reaction system by distillation with a slow stream of nitrogen.

Cleavage of Ethers, Acetals, and Esters. Like many other Lewis acids, BCl_3 has been extensively used as a reagent for the cleavage of a wide variety of ethers, acetals, and certain types of esters.[2] Ether cleavage procedures involve addition of BCl_3, either neat or as a solution in CH_2Cl_2, to the substrate at −80 °C. The vessel is then stoppered and allowed to warm to rt. Whereas the complexes of BCl_3 with dimethyl ether and diethyl ether are rather stable at rt, they decompose to form $ROBCl_2$ or $(RO)_2BCl$ with evolution of alkyl chloride upon heating to 56 °C.[1] Diaryl ethers are unreactive. Mixed dialkyl ethers are cleaved to give the alkyl chloride derived from C–O bond cleavage leading to the more stable carbenium ion. The transition state is predominantly S_N1 in character, as evidenced by partial racemization of chiral ethers[1,2] and the rearrangement of allyl phenyl ethers to *o*-allylphenols.[3] BCl_3 can be used for the deprotection of a variety of methoxybenzenes including hindered polymethoxybenzenes and *peri*-methoxynaphthalene.[1,2,4] When methoxy groups are *ortho* to a carbonyl group, the reaction is accelerated by the formation of a chelate between boron and the carbonyl oxygen atom (Scheme 1).[4a–c]

The reagent is less reactive than ***Boron Tribromide*** for ether cleavage; however, the type and extent of deetherification can be more easily controlled by the ratio of substrate to BCl_3 as well as the reaction temperature and time. The transformation of (−)-β-hydrastine (**1**) to (−)-cordrastine II is efficiently achieved by selective cleavage of the methylenedioxy group in preference to aromatic methoxy groups.[5] The demethylation of (−)-2-*O*-methyl-(−)-inositol in dichloromethane proceeds at −80 °C without cleavage of a tosyl ester group.[6] Methyl glycosides are converted into glycosyl chlorides at −78 °C without effecting benzyl and acetyl protecting groups.[7]

R = Me → H; 81%, rt, 5 min

R = Me → H; 78%, −80 °C

(a) R = Me → H, R' = Me; 80%, rt, 0.5 h
(b) R, R' = Me → H; 97%, rt, 8 h

R = Me → H; 90%, rt, 5 min

(**1**)
R, R' = –CH$_2$– → H; 81%, rt, 6 h

Scheme 1 Demethylation of aromatic ethers by BCl_3 in CH_2Cl_2

One of the difficulties with the use of BCl_3 arises from its tendency to fume profusely in air. The complex of BCl_3 with dimethyl sulfide is solid, stable in air, and handled easily. By using a two- to fourfold excess of the reagent in dichloroethane at 83 °C, aromatic methoxy and methylenedioxy groups can be cleaved in good yields.[8]

Another application of BCl_3 is for the cleavage of highly hindered esters under mild conditions. *O*-Methylpodocarpate (**2**) and methyl adamantane-1-carboxylate are cleaved at 0 °C.[9] The highly selective displacement of the acetoxy group in the presence of other potentially basic groups in 2-cephem ester (**3**) provides the corresponding allylic chloride. On the other hand, treatment of (**3**) with an excess of BCl_3 results in the cleavage of the acetoxy and *t*-butyl ester groups.[10]

(**2**)
R = Me → H; 90%
BCl_3, CH_2Cl_2, 0 °C

(**3**)
(1) R = *t*-Bu, X = OAc → Cl; 64%
BCl_3 (1 equiv), CH_2Cl_2, −5 °C
(2) R = *t*-Bu → H, X = OAc → OH; 68%
BCl_3 (3 equiv), CH_2Cl_2, rt

Tertiary phosphines are cleaved at the P–C bond to give diphenylphosphine oxides. Workup with ***Hydrogen Peroxide*** provides diphenylphosphinic acids (eq 1).[11]

$$Ph_2PCH_2XMe \xrightarrow[2.\ H_2O]{1.\ BCl_3,\ 0\,^\circ C \to rt} Ph_2P(O)H \xrightarrow{H_2O_2} Ph_2P(O)OH \quad (1)$$

X = O, S

Condensation Reactions. Boron trichloride converts ketones into (*Z*)-boron enolates at −95 °C in the presence of ***Diisopropylethylamine***. These enolates react with aldehydes with high *syn* diastereoselectivity (eq 2).[12] A similar condensation of imines with carbonyl compounds also provides crossed aldols in reasonable yields.[13] The reaction was extended to the asymmetric aldol condensation of acetophenone imine and benzaldehyde by using isobornylamine as a chiral auxiliary (48% ee).[14]

BCl_3, $EtN(i\text{-}Pr)_2$, CH_2Cl_2, −95 °C → (**4**); PhCHO, 81%; *syn*:*anti* = 93:7 (2)

(*N*-Alkylanilino)dichloroboranes (**5**), prepared in situ from *N*-alkylanilines and boron trichloride, are versatile intermediates for the synthesis of *ortho*-functionalized aniline derivatives (eqs 3–5).[15] The regioselective *ortho* hydroxyalkylation can be achieved with aromatic aldehydes.[16]

NHMe; BCl_3, Et_3N, benzene, Δ → (**5**)

1. ArCHO; 2. H_2O (3)

1. ArCN, $AlCl_3$; 2. H_2O; R = alkyl, aryl (4)

1. ArCN; 2. AcOH, H_2O; 3. HCl, H_2O; CHO (5)

The reaction of (**5**) with alkyl and aryl nitriles and ***Aluminum Chloride*** catalyst provides *ortho*-acyl anilines.[16] When chloroacetonitrile is used, the products are ideal precursors for indole synthesis.[17] Use of isocyanides instead of nitriles provides *ortho*-formyl *N*-alkylanilines.[18] Although these reactions with BCl_3 are restricted to *N*-alkylanilines, the use of ***Phenylboron Dichloride*** allows the *ortho*-hydroxybenzylation of primary anilines.[19]

Analogously, boron trichloride induces *ortho* selective acylation of phenols at rt with nitriles, isocyanates, or acyl chlorides (eq 6).[20] The efficiency and regioselectivity of these reactions are best with BCl_3 among the representative metal halides that have been examined. In both the aniline and phenol substitutions the boron atom acts as a template to bring the reactants together, leading to cyclic intermediates and exclusively products of *ortho* substitution. A similar *ortho* selective condensation of aromatic azides with BCl_3 provides fused heterocycles containing nitrogen.[21]

1. MeN=C=O, BCl_3, CH_2Cl_2, Δ; 2. H_3O^+; 75% (6)

Aldehydes and ketones condense with ketene in the presence of 1 equiv of boron trichloride to give α,β-unsaturated acyl chlorides.[22] Aryl isocyanates are converted into allophanyl chlorides, which are precursors for industrially important 1,3-diazetidinediones (eq 7).[23]

$$\text{PhN=C=O} \xrightarrow[\text{toluene, 0 °C, 70\%}]{BCl_3} \text{PhNHCON(Ph)COCl} \xrightarrow[\text{rt, 95\%}]{\text{pyridine}} \text{PhN(CO)}_2\text{NPh} \quad (7)$$

Synthesis of Organoboron Reagents. General method of synthesis of organoboranes consists of the transmetallation reaction of organometallic compounds with BX_3.[24] Boronic acid derivatives [$RB(OH)_2$] are most conveniently synthesized by the reaction of $B(OR)_3$ with RLi or RMgX reagents, but boron trihalides are more advantageous for transmetalation reactions with less nucleophilic organometallic reagents based on Pb,[25] Hg,[26] Sn,[27] and Zr[28] (eqs 8 and 9).

$$Ph_4Sn + 2\,BCl_3 \longrightarrow 2\,PhBCl_2 + Ph_2SnCl_2 \quad (8)$$

$$Cp_2Zr(Cl)CH{=}CH(CH_2)_3Me \xrightarrow[0\,°C \to rt,\ >90\%]{BCl_3,\ CH_2Cl_2} Cl_2BCH{=}CH(CH_2)_3Me \quad (9)$$

Redistribution or exchange reactions of R_3B with boron trihalides in the presence of catalytic amounts of hydride provides an efficient synthesis of RBX_2 and R_2BX.[29] Another convenient and general method for the preparation of organodichloroboranes involves treatment of alkyl, 1-alkenyl, and aryl boronates with BCl_3 in the presence of ***Iron(III) Chloride*** (3 mol %).[30] Organodichloroboranes are valuable synthetic reagents because of their high Lewis acidity, and their utility is well demonstrated in the syntheses of piperidine and pyrrolidine derivatives by the intramolecular alkylation of azides (eq 10)[31] or the synthesis of esters by the reaction with ***Ethyl Diazoacetate***.[32] The various organoborane derivatives, R_3B, R_2BCl, and $RBCl_2$, all react with organic azides and diazoacetates. However, especially facile reactions are achieved by using organodichloroboranes ($RBCl_2$).

1. $HBCl_2{\cdot}SMe_2$, BCl_3, pentane; 2. H_2O; 3. NaN_3, EtOH, Δ → $B(OEt)_2$, N_3; BCl_3, CH_2Cl_2, −78 °C → rt → BCl_2, N_3; $-N_2$; H_2O, 40% KOH; 76% (10)

Dichloroborane and monochloroborane etherates or their methyl sulfide complexes have been prepared by the reaction of borane and boron trichloride.[33] However, hydroboration of alkenes with these borane reagents is usually very slow due to the slow dissociation of the complex. Dichloroborane prepared in pentane from boron trichloride and trimethylsilane shows unusually high reactivity with alkenes and alkynes; hydroboration is instantaneous at −78 °C (eq 11).[34]

$$BCl_3 \xrightarrow[\text{pentane, } -78\,°C]{HSiMe_3} HBCl_2 \xrightarrow[95\%]{\alpha\text{-pinene}} \text{(pinanyl)}BCl_2 \quad (11)$$

Direct boronation of benzene derivatives with BCl_3 in the presence of activated aluminum or $AlCl_3$ provides arylboronic acids after hydrolysis (eq 12).[35] Chloroboration of acetylene with boron trichloride produces dichloro(2-chloroethenyl)borane.[36] Similar reaction with phenylacetylene provides (*E*)-2-chloro-2-phenylethenylborane regio- and stereoselectively.[37]

$$PhH + BCl_3 \xrightarrow[I_2 \text{ (trace), MeI (trace), } 150\,°C,\ 60\text{–}79\%]{Al} PhBCl_2 \quad (12)$$

The syntheses of thioaldehydes, thioketones, thiolactones, and thiolactams from carbonyl compounds are readily achieved by in situ preparation of B_2S_3 from bis(tricyclohexyltin) sulfide and boron trichloride (eq 13).[38] The high sulfurating ability of this in situ prepared reagent can be attributed to its solubility in the reaction medium.

$$3\ (Cy_3Sn)=S + 2\ BCl_3 \xrightarrow[\Delta]{\text{toluene}} B_2S_3 \quad (13)$$

94%, Δ, 7 h — 92%, Δ, 3 h — unable to isolate

Related Reagents. Bis(tricyclohexyltin) Sulfide–Boron Trichloride; Catechylphosphorus Trichloride; Boron Tribromide; Hydrogen Chloride.

1. Gerrard, W.; Lappert, M. F. *CRV* **1958**, *58*, 1081.
2. (a) Bhatt, M. V.; Kulkarni, S. U. *S* **1983**, 249. (b) Greene, T. W. *Protective Groups in Organic Synthesis*; Wiley: New York, 1981.
3. (a) Gerrard, W.; Lappert, M. F.; Silver, H. B. *Proc. Chem. Soc.* **1957**, 19. (b) Borgulya, J.; Madeja, R.; Fahrni, P.; Hansen, H.-J.; Schmid, H.; Barner, R. *HCA* **1973**, *56*, 14.
4. (a) Dean, R. B.; Goodchild, J.; Houghton, L. E.; Martin, J. A. *TL* **1966**, 4153. (b) Arkley, V.; Attenburrow, J.; Gregory, G. I.; Walker, T. *JCS* **1962**, 1260. (c) Barton, D. H. R.; Bould, L.; Clive, D. L. J.; Magnus, P. D.; Hase, T. *JCS(C)* **1971**, 2204. (d) Carvalho, C. F.; Seargent, M. V. *CC* **1984**, 227.
5. (a) Teitel, S.; O'Brien, J.; Brossi, A. *JOC* **1972**, *37*, 3368. (b) Teitel, S.; O'Brien, J. P. *JOC* **1976**, *41*, 1657.
6. Gero, S. D. *TL* **1966**, 591.
7. Perdomo, G. R.; Krepinsky, J. J. *TL* **1987**, *28*, 5595.
8. Williard, P. G.; Fryhle, C. B. *TL* **1980**, *21*, 3731.
9. Manchand, P. S. *CC* **1971**, 667.
10. Yazawa, H.; Nakamura, H.; Tanaka, K.; Kariyone, K. *TL* **1974**, 3991.
11. Hansen, K. C.; Solleder, G. B.; Holland, C. L. *JOC* **1974**, *39*, 267.
12. Chow, H-F.; Seebach, D. *HCA* **1986**, *69*, 604.
13. Sugasawa, T.; Toyoda, T. Sasakura, K. *SC* **1979**, *9*, 515.
14. Sugasawa, T.; Toyoda, T.; *TL* **1979**, 1423.
15. Sugasawa, T. *J. Synth. Org. Chem. Jpn.* **1981**, *39*, 39.
16. Sugasawa, T.; Toyoda, T.; Adachi, M.; Sasakura, K. *JACS* **1978**, *100*, 4842.
17. Sugasawa, T.; Adachi, M.; Sasakura, K.; Kitagawa, A. *JOC* **1979**, *44*, 578.
18. Sugasawa, T.; Hamana, H.; Toyoda, T.; Adachi, M. *S* **1979**, 99.
19. Toyoda, T.; Sasakura, K.; Sugasawa, T. *TL* **1980**, *21*, 173.
20. (a) Toyoda, T.; Sasakura, K.; Sugasawa, T. *JOC* **1981**, *46*, 189. (b) Piccolo, O.; Filippini, L.; Tinucci, L.; Valoti, E.; Citterio, A. *T* **1986**, *42*, 885.
21. (a) Zanirato, P. *CC* **1983**, 1065. (b) Spagnolo, P.; Zanirato, P. *JCS(P1)* **1988**, 2615.
22. Paetzold, P. I.; Kosma, S. *CB* **1970**, *103*, 2003.
23. Helfert, H.; Fahr, E. *AG(E)* **1970**, *9*, 372.
24. (a) Nesmeyanov, A. N.; Kocheshkov, K. A. *Methods of Elemento-Organic Chemistry*; North-Holland: Amsterdam, 1967; Vol. 1, pp 20–96. (b) Mikhailov, B. M.; Bubnov, Y. N. *Organoboron Compounds in Organic Synthesis*; Harwood: Amsterdam, 1984.
25. Holliday, A. K.; Jessop, G. N. *JCS(A)* **1967**, 889.
26. Gerrard, W.; Howarth, M.; Mooney, E. F.; Pratt, D. E. *JCS* **1963**, 1582.
27. (a) Niedenzu, K.; Dawson, J. W. *JACS* **1960**, *82*, 4223. (b) Brinkman, F. E.; Stone, F. G. A. *CI(L)* **1959**, 254.
28. Cole, T. E.; Quintanilla, R.; Rodewald, S. *OM* **1991**, *10*, 3777.
29. Brown, H. C.; Levy, A. B. *JOM* **1972**, *44*, 233.
30. (a) Brindley, P. B.; Gerrard, W.; Lappert, M. F. *JCS* **1956**, 824. (b) Brown, H. C.; Salunkhe, A. M.; Argade, A. B. *OM* **1992**, *11*, 3094.
31. (a) Jego, J. M.; Carboni, B.; Vaultier, M.; Carrie', R. *CC* **1989**, 142. (b) Brown, H. C.; Salunkhe, A. M. *TL* **1993**, *34*, 1265.
32. Hooz, J.; Bridson, J. N.; Calzada, J. G.; Brown, H. C.; Midland, M. M.; Levy, A. B. *JOC* **1973**, *38*, 2574.
33. (a) Brown, H. C. *Organic Syntheses via Boranes*; Wiley: New York, 1975; pp 45–47. (b) Brown, H. C.; Kulkarni, S. U. *JOM* **1982**, *239*, 23. (c) Brown, H. C.; Ravindran, N. *IC* **1977**, *16*, 2938.
34. Soundararajan, R.; Matteson, D. S. *JOC* **1990**, *55*, 2274.
35. (a) Muetterties, E. L. *JACS* **1960**, *82*, 4163. (b) Lengyel, B.; Csakvari, B. *Z. Anorg. Allg. Chem.* **1963**, *322*, 103.
36. Lappert, M. F.; Prokai, B. *JOM* **1964**, *1*, 384.
37. Blackborow, J. R. *JCS(P2)* **1973**, 1989.
38. Steliou, K.; Mrani, M. *JACS* **1982**, *104*, 3104.

Norio Miyaura
Hokkaido University, Sapporo, Japan

Boron Trifluoride Etherate

$BF_3 \cdot OEt_2$

($BF_3 \cdot OEt_2$) *[109-63-7]*	$C_4H_{10}BF_3O$	(MW 141.94)
($BF_3 \cdot MeOH$) *[373-57-9]*	CH_4BF_3O	(MW 99.85)

($BF_3 \cdot OEt_2$: easy-to-handle and convenient source of BF_3; Lewis acid catalyst; promotes epoxide cleavage and rearrangement, control of stereoselectivity; $BF_3 \cdot MeOH$: esterification of aliphatic and aromatic acids; cleavage of trityl ethers)

Alternate Names: boron trifluoride diethyl etherate; boron trifluoride ethyl etherate; boron trifluoride ethyl ether complex; trifluoroboron diethyl etherate.

Physical Data: $BF_3 \cdot OEt_2$: bp 126 °C; *d* 1.15 g cm^{-3}; $BF_3 \cdot MeOH$: bp 59 °C/4 mmHg; *d* 1.203 g cm^{-3} for 50 wt % BF_3, 0.868 g cm^{-3} for 12 wt % BF_3.

Solubility: sol benzene, chloromethanes, dioxane, ether, methanol, THF, toluene.

Form Supplied in: $BF_3 \cdot OEt_2$: light yellow liquid, packaged under nitrogen or argon; $BF_3 \cdot MeOH$ is available in solutions of 10–50% BF_3 in MeOH.

Preparative Methods: $BF_3 \cdot OEt_2$ is prepared by passing BF_3 through anhydrous ether;[1a] the $BF_3 \cdot MeOH$ complex is formed from $BF_3 \cdot OET_2$ and methanol.

Purification: oxidation in air darkens commercial boron trifluoride etherate; therefore the reagent should be redistilled prior to use. An excess of the etherate in ether should be distilled in an all-glass apparatus with calcium hydroxide to remove volatile acids and to reduce bumping.[1b]

Handling, Storage, and Precautions: keep away from moisture and oxidants; avoid skin contact and work in a well-ventilated fume hood.

Addition Reactions. $BF_3 \cdot OEt_2$ facilitates the addition of moderately basic nucleophiles like alkyl-, alkenyl-, and aryllithium, imines, Grignard reagents, and enolates to a variety of electrophiles.[2]

Organolithiums undergo addition reactions with 2-isoxazolines to afford *N*-unsubstituted isoxazolidines, and to the carbon–nitrogen double bond of oxime *O*-ethers to give *O*-alkylhydroxylamines.[3] Aliphatic esters react with lithium acetylides in the presence of $BF_3 \cdot OEt_2$ in THF at −78 °C to form alkynyl ketones in 40–80% yields.[4] Alkynylboranes, generated in situ from lithium acetylides and $BF_3 \cdot OEt_2$, were found to react with oxiranes[5] and oxetanes[6] under mild conditions to afford β-hydroxyalkynes and γ-alkoxyalkynes, respectively. (1-Alkenyl)dialkoxyboranes react stereoselectively with α,β-unsaturated ketones[7] and esters[8] in the presence of $BF_3 \cdot OEt_2$ to give γ,δ-unsaturated ketones and α-acyl-γ,δ-unsaturated esters, respectively.

The reaction of imines activated by $BF_3 \cdot OEt_2$ with 4-(phenylsulfonyl)butanoic acid dianion leads to 2-piperidones in high yields.[9] (Perfluoroalkyl)lithiums, generated in situ, add to imines in the presence of $BF_3 \cdot OEt_2$ to give perfluoroalkylated amines.[10] Enolate esters add to 3-thiazolines under mild conditions to form thiazolidines if these imines are first activated with $BF_3 \cdot OEt_2$.[11] The carbon–nitrogen double bond of imines can be alkylated with various organometallic reagents to produce amines.[12] A solution of benzalaniline in acetone treated with $BF_3 \cdot OEt_2$ results in the formation of β-phenyl-β-anilinoethyl methyl ketone.[13] Anilinobenzylphosphonates are synthesized in one pot using aniline, benzaldehyde, dialkyl phosphite, and $BF_3 \cdot OEt_2$;[14] the reagent accelerates imine generation and dialkyl phosphite addition. Similarly, $BF_3 \cdot OEt_2$ activates the nitrile group of cyanocuprates, thereby accelerating Michael reactions.[15]

The reagent activates iodobenzene for the allylation of aromatics, alcohols, and acids.[16] Allylstannanes are likewise activated for the allylation of *p*-benzoquinones, e.g. in the formation of coenzyme Q_n using polyprenylalkylstannane.[17]

Nucleophilic silanes undergo stereospecific addition to electrophilic glycols activated by Lewis acids. The glycosidation is highly stereoselective with respect to the glycosidic linkage in some cases using $BF_3 \cdot OEt_2$. Protected pyranosides undergo stereospecific *C*-glycosidation with C-1-oxygenated allylsilanes to form α-glycosides.[18,19] α-Methoxyglycine esters react with allylsilanes and silyl enol ethers in the presence of $BF_3 \cdot OEt_2$ to give racemic γ,δ-unsaturated α-amino acids and γ-oxo-α-amino acids, respectively.[20] β-Glucopyranosides are synthesized from an aglycon and 2,3,4,6-tetra-*O*-acetyl-β-D-glucopyranose.[21] Alcohols and silyl ethers also undergo stereoselective glycosylation with protected glycosyl fluorides to form β-glycosides.[22]

$BF_3 \cdot OEt_2$ reverses the usual *anti* selectivity observed in the reaction of crotyl organometallic compounds (based on Cu, Cd, Hg, Sn, Tl, Ti, Zr, and V, but not on Mg, Zn, or B) with aldehydes (eq 1a) and imines (eq 1b), so that homoallyl alcohols and homoallylamines are formed, respectively.[23–28] The products show mainly *syn* diastereoselectivity. $BF_3 \cdot OEt_2$ is the only Lewis acid which produces hydroxy- rather than halotetrahydropyrans from the reaction of allylstannanes with pyranosides.[29] The $BF_3 \cdot OEt_2$ mediated condensations of γ-oxygenated allylstannanes with aldehydes (eq 1c) and with 'activated' imines (eq 1d) affords vicinal diol derivatives and 1,2-amino alcohols, respectively, with *syn* diastereoselectivity.[30,31]

The 'activated' imines are obtained from aromatic amines, aliphatic aldehydes, and α-ethoxycarbamates. The reaction of aldehydes with α-(alkoxy)-β-methylallylstannanes with aldehydes in the presence of $BF_3 \cdot OEt_2$ gives almost exclusively *syn*-(*E*)-isomers.[31]

Y~~SnBu$_3$ + RC(=X)H —$BF_3 \cdot OEt_2$→ RCH(X)CH(Y)CH=CH$_2$ (1)

(a) X = O, Y = Me → (a) X = OH, Y = Me
(b) X = NR^2, Y = Me → (b) X = NHR^2, Y = Me
(c) X = O, Y = OMe, OTBDMS → (c) X = OH, Y = OMe, OTBDMS
(d) X = NR^2, Y = OMe, OTBDMS or OCH$_2$OMe → (d) X = NHR^2, Y = OH or derivative

The reaction of α-diketones with allyltrimethylstannane in the presence of $BF_3 \cdot OEt_2$ yields a mixture of homoallylic alcohols, with the less hindered carbonyl group being allylated predominantly.[32] The reaction between aldehydes and allylic silanes with an asymmetric ethereal functionality produces *syn*-homoallyl alcohols when ***Titanium(IV) Chloride*** is coordinated with the allylic silane and *anti* isomers with $BF_3 \cdot OEt_2$.[33]

Chiral oxetanes can be synthesized by the $BF_3 \cdot OEt_2$ catalyzed [2+2] cycloaddition reactions of 2,3-*O*-isopropylidenealdehyde-D-aldose derivatives with allylsilanes, vinyl ethers, or vinyl sulfides.[34] The regiospecificity and stereoselectivity is greater than in the photochemical reaction; *trans*-2-alkoxy- and *trans*-2-phenylthiooxetanes are the resulting products.

2-Alkylthioethyl acetates can be formed from vinyl acetates by the addition of thiols with $BF_3 \cdot OEt_2$ as the catalyst.[35] The yield is 79%, compared to 75% when $BF_3 \cdot OEt_2$ is used in conjunction with ***Mercury(II) Sulfate*** or ***Mercury(II) Oxide***.

α-Alkoxycarbonylallylsilanes react with acetals in the presence of $BF_3 \cdot OEt_2$ (eq 2).[36] The products can be converted into α-methylene-γ-butyrolactones by dealkylation with ***Iodotrimethylsilane***.

$Me_3SiCH_2C(=CH_2)CO_2Et$ + PhCH(OMe)$_2$ —$BF_3 \cdot OEt_2$, 89%→ $CH_2=C(CO_2Et)CH_2CH(Ph)OMe$ (2)

The cuprate 1,4-conjugate addition step in the synthesis of (+)-modhephene is difficult due to the neopentyl environment of C-4 in the enone, but it can occur in the presence of $BF_3·OEt_2$ (eq 3).[37]

(3) [$BF_3·OEt_2$; Me_2CuLi; 70%]

The reagent is used as a Lewis acid catalyst for the intramolecular addition of diazo ketones to alkenes.[38] The direct synthesis of bicyclo[3.2.1]octenones from the appropriate diazo ketones using $BF_3·OEt_2$ (eq 4) is superior to the copper-catalyzed thermal decomposition of the diazo ketone to a cyclopropyl ketone and subsequent acid-catalyzed cleavage.[38]

(4) [$BF_3·OEt_2$, $ClCH_2CH_2Cl$; 0–27 °C; 30–51%; $COCHN_2$]

$BF_3·OEt_2$ reacts with fluorinated amines to form salts which are analogous to Vilsmeier reagents, Arnold reagents, or phosgene–immonium salts (eq 5).[39] These salts are used to acylate electron-rich aromatic compounds, introducing a fluorinated carbonyl group (eq 6).

$$XCHF-CF_2-NR_2 \xrightarrow{BF_3\cdot OEt_2} XCHF-CF{=}\overset{+}{N}R_2\ BF_4^- \quad (5)$$

R = Et; X = Cl, F, CF_3 **(1)**

$$ArH \xrightarrow{(1)} Ar-C(=\overset{+}{N}R_2)CHFX\ BF_4^- \xrightarrow{H_3O^+} Ar-C(=O)CHFX \quad (6)$$

Xenon(II) Fluoride and methanol react to form ***Methyl Hypofluorite***, which reacts as a positive oxygen electrophile in the presence of BF_3 (etherate or methanol complex) to yield anti-Markovnikov fluoromethoxy products from alkenes.[40,41]

Aldol Reactions. Although ***Titanium(IV) Chloride*** is a better Lewis acid in effecting aldol reactions of aldehydes, acetals, and silyl enol ethers, $BF_3·OEt_2$ is more effective for aldol reactions with anions generated from transition metal carbenes and with tetrasubstituted enol ethers such as (*Z*)- and (*E*)-3-methyl-2-(trimethylsilyloxy)-2-pentene.[42,43] One exception involves the preparation of substituted cyclopentanediones from acetals by the aldol condensation of protected four-membered acyloin derivatives with $BF_3·OEt_2$ rather than $TiCl_4$ (eq 7).[44] The latter catalyst causes some loss of the silyl protecting group. The pinacol rearrangement is driven by the release of ring strain in the four-membered ring and controlled by an acyl group adjacent to the diol moiety.

The reagent is the best promoter of the aldol reaction of 2-(trimethylsilyloxy)acrylate esters, prepared by the silylation of pyruvate esters, to afford γ-alkoxy-α-keto esters (eq 8).[45] These esters occur in a variety of important natural products.

(7) [RC(O)R' + Me_3SiO/$OSiMe_3$ cyclobutene; $BF_3·OEt_2$]

(8) [Ph acetal (OMe, OMe) + $CH_2{=}C(OSiMe_3)CO_2Et$; $BF_3·OEt_2$, CH_2Cl_2; –78 to 0 °C; 86%]

$BF_3·OEt_2$ can improve or reverse the aldehyde diastereofacial selectivity in the aldol reaction of silyl enol ethers with aldehydes, forming the *syn* adducts. For example, the reaction of the silyl enol ether of pinacolone with 2-phenylpropanal using $BF_3·OEt_2$ gives enhanced levels of Felkin selectivity relative to the addition of the corresponding lithium enolate.[46,47] In the reaction of silyl enol ethers with 3-formyl-Δ^2-isoxazolines, $BF_3·OEt_2$ gives predominantly *anti* aldol adducts, whereas other Lewis acids give *syn* aldol adducts.[48] The reagent can give high diastereofacial selectivity in the addition of silyl enol ethers or silyl ketones to chiral aldehydes.[49] In the addition of a nonstereogenic silylketene acetal to chiral, racemic α-thioaldehydes, $BF_3·OEt_2$ leads exclusively to the *anti* product.[49]

1,5-Dicarbonyl compounds are formed from the reaction of silyl enol ethers with methyl vinyl ketones in the presence of $BF_3·OEt_2$ and an alcohol (eq 9).[50] α-Methoxy ketones are formed from α-diazo ketones with $BF_3·OEt_2$ and methanol, or directly from silyl enol ethers using iodobenzene/$BF_3·OEt_2$ in methanol.[51]

(9) [$OSiMe_3$; $BF_3·OEt_2$; MeOH]

α-Mercurio ketones condense with aldehydes in the presence of $BF_3·OEt_2$ with predominant *erythro* selectivity (eq 10).[52] Enaminosilanes derived from acylic and cyclic ketones undergo *syn* selective aldol condensations in the presence of $BF_3·OEt_2$.[53]

(10) [HgI ketone + PhCHO; $BF_3·OEt_2$; THF; 60%]

erythro 90:10 *threo*

Cyclizations. Arylamines can undergo photocyclization in the presence of $BF_3 \cdot OEt_2$ to give tricyclic products, e.g. 9-azaphenanthrene derivatives (eq 11).[54]

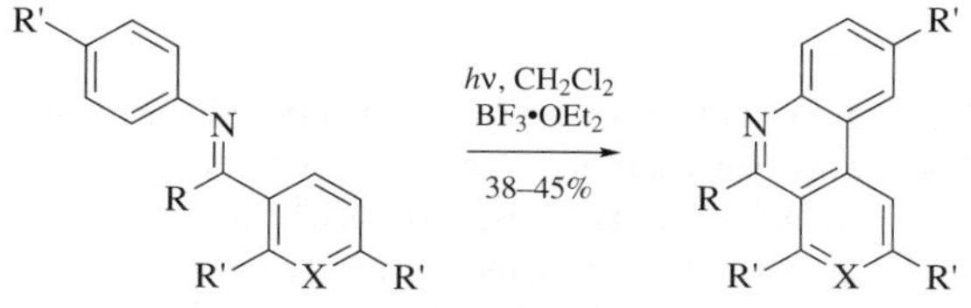

hν, CH_2Cl_2, $BF_3 \cdot OEt_2$, 38–45% (11)

R = H, Me; R' = H, OMe; X = CH, N

Substituted phenethyl isocyanates undergo cyclization to lactams when treated with $BF_3 \cdot OEt_2$.[55] Vinyl ether epoxides (eq 12),[56] vinyl aldehydes,[57] and epoxy β-keto esters[58] all undergo cyclization with $BF_3 \cdot OEt_2$.

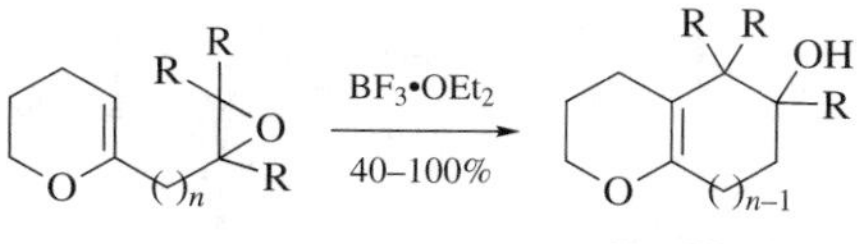

$BF_3 \cdot OEt_2$, 40–100% (12)

R = H, Me

β-Silyl divinyl ketones (Nazarov reagents) in the presence of $BF_3 \cdot OEt_2$ cyclize to give cyclopentenones, generally with retention of the silyl group.[59] $BF_3 \cdot OEt_2$ is used for the key step in the synthesis of the sesquiterpene trichodiene, which has adjacent quaternary centers, by catalyzing the cyclization of the dienone to the tricyclic ketone (eq 13).[60] Trifluoroacetic acid and trifluoroacetic anhydride do not catalyze this cyclization.

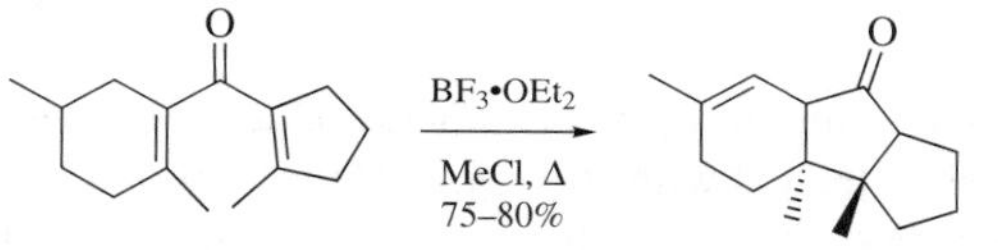

$BF_3 \cdot OEt_2$, MeCl, Δ, 75–80% (13)

Costunolide, treated with $BF_3 \cdot OEt_2$, produces the cyclocostunolide (**2**) and a C-4 oxygenated sesquiterpene lactone (**3**), 4α-hydroxycyclocostunolide (eq 14).[61]

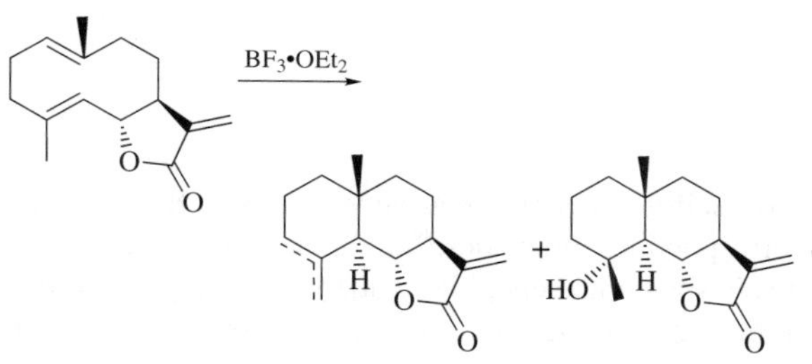

$BF_3 \cdot OEt_2$ (14)

(**2**) (**3**)

Other Condensation Reactions. $BF_3 \cdot MeOH$ and $BF_3 \cdot OEt_2$ with ethanol are widely used in the esterification of various kinds of aliphatic, aromatic, and carboxylic acids;[62] the reaction is mild, and no rearrangement of double bonds occurs. This esterification is used routinely for stable acids prior to GLC analysis. Heterocyclic carboxylic acids,[63] unsaturated organic acids,[64] biphenyl-4,4′-dicarboxylic acid,[65] 4-aminobenzoic acid,[63] and the very sensitive 1,4-dihydrobenzoic acid[65] are esterified directly.

The dianion of acetoacetate undergoes Claisen condensations with tetramethyldiamide derivatives of dicarboxylic acids to produce polyketides in the presence of $BF_3 \cdot OEt_2$ (eq 15).[66] Similarly, 3,5-dioxoalkanoates are synthesized from tertiary amides or esters with the acetoacetate dianion in the presence of $BF_3 \cdot OEt_2$ (eq 16).[66]

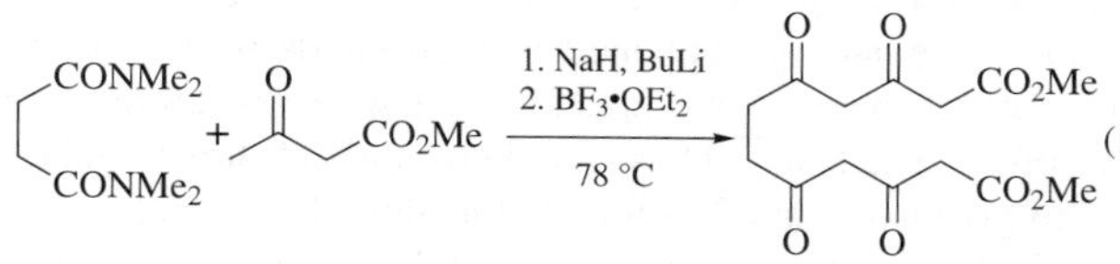

1. NaH, BuLi; 2. $BF_3 \cdot OEt_2$; 78 °C (15)

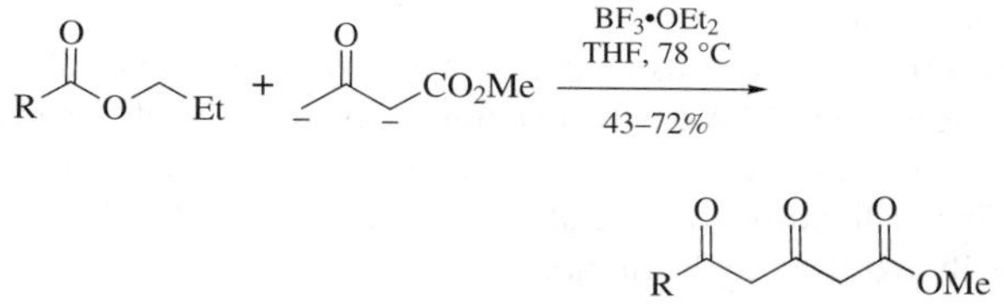

$BF_3 \cdot OEt_2$, THF, 78 °C, 43–72% (16)

R = n-C_9H_{19}, $ClCH_2$, Ph

Aldehydes and siloxydienes undergo cyclocondensation with $BF_3 \cdot OEt_2$ to form pyrones (eq 17).[67] The stereoselectivity is influenced by the solvent.

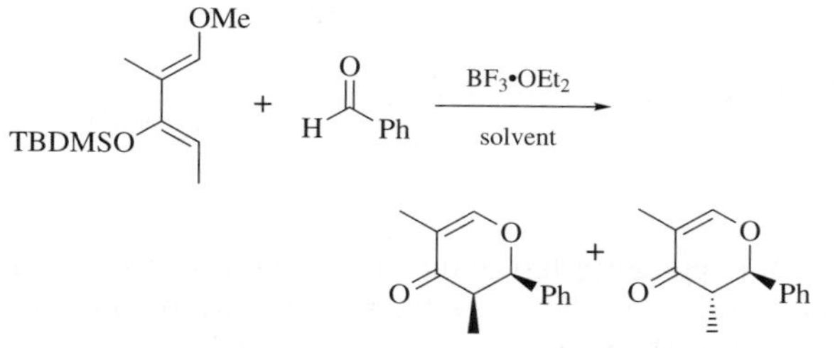

$BF_3 \cdot OEt_2$, solvent (17)

solvent:	
CH_2Cl_2	1:2.3
PhMe	7:1

$BF_3 \cdot OEt_2$ is effective in the direct amidation of carboxylic acids to form carboxamides (eq 18).[68] The reaction is accelerated by bases and by azeotropic removal of water.

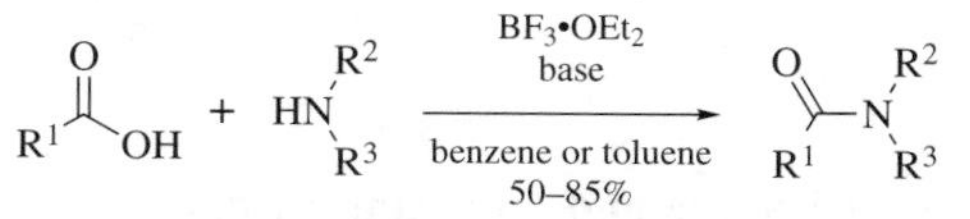

$BF_3 \cdot OEt_2$, base, benzene or toluene, 50–85% (18)

Carbamates of secondary alcohols can be prepared by a condensation reaction with the isocyanate and $BF_3 \cdot OEt_2$ or ***Aluminum Chloride***.[69] These catalysts are superior to basic catalysts such as pyridine and triethylamine. Some phenylsulfonylureas have been prepared from phenylsulfonamides and isocyanates using $BF_3 \cdot OEt_2$ as a catalyst; for example, 1-butyl-3-(p-tolylsulfonyl)urea is prepared from p-toluenesulfonamide and butyl isocyanate.[70] $BF_3 \cdot OEt_2$ is an excellent catalyst for the condensation of amines to form azomethines (eq 19).[71] The temperatures required are much lower than with ***Zinc Chloride***.

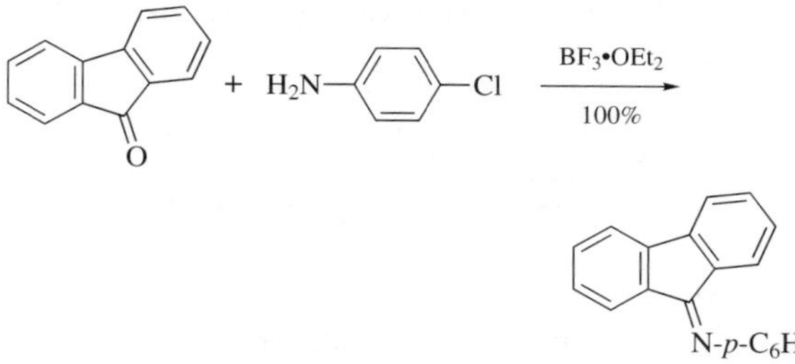

$BF_3 \cdot OEt_2$, 100% (19)

N-p-C_6H_4Cl

Acyltetrahydrofurans can be obtained by $BF_3 \cdot OEt_2$ catalyzed condensation of (*Z*)-4-hydroxy-1-alkenylcarbamates with aldehydes, with high diastereo- and enantioselectivity.[72] Pentasubstituted hydrofurans are obtained by the use of ketones.

Isobornyl ethers are obtained in high yields by the condensation of camphene with phenols at low temperatures using $BF_3 \cdot OEt_2$ as catalyst.[73] Thus camphene and 2,4-dimethylphenol react to give isobornyl 2,4-dimethylphenyl ether, which can undergo further rearrangement with $BF_3 \cdot OEt_2$ to give 2,4-dimethyl-6-isobornylphenol.[73]

The title reagent is also useful for the condensation of allylic alcohols with enols. A classic example is the reaction of phytol in dioxane with 2-methyl-1,4-naphthohydroquinone 1-monoacetate to form the dihydro monoacetate of vitamin K_1 (eq 20), which can be easily oxidized to the quinone.[74]

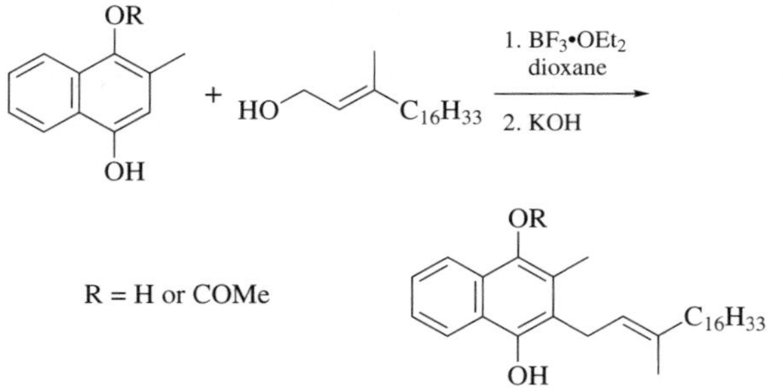

OR OH + HO $C_{16}H_{33}$ 1. $BF_3 \cdot OEt_2$ dioxane 2. KOH

R = H or COMe OR OH $C_{16}H_{33}$ (20)

$BF_3 \cdot OEt_2$ promotes fast, mild, clean regioselective dehydration of tertiary alcohols to the thermodynamically most stable alkenes.[75] 11β-Hydroxysteroids are dehydrated by $BF_3 \cdot OEt_2$ to give $\Delta^{9(11)}$-enes (eq 21).[76,77]

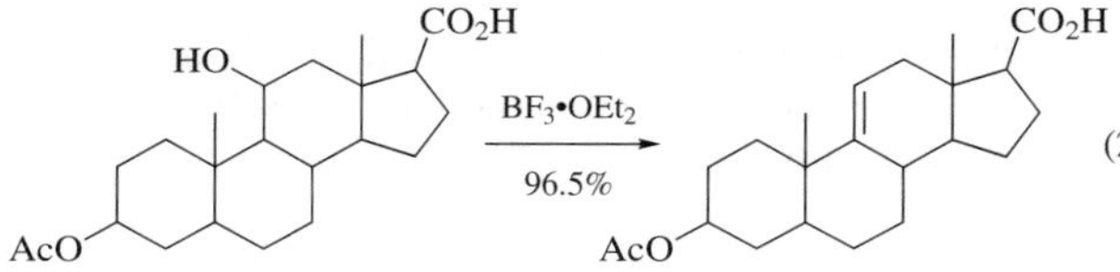

HO CO_2H AcO $BF_3 \cdot OEt_2$ 96.5% CO_2H AcO (21)

Epoxide Cleavage and Rearrangements. The treatment of epoxides with $BF_3 \cdot OEt_2$ results in rearrangements to form aldehydes and ketones (eq 22).[78] The carbon α to the carbonyl group of an epoxy ketone migrates to give the dicarbonyl product.[79] The acyl migration in acyclic α,β-epoxy ketones proceeds through a highly concerted process, with inversion of configuration at the migration terminus.[80] With 5-substituted 2,3-epoxycyclohexanes the stereochemistry of the quaternary carbon center of the cyclopentanecarbaldehyde product is directed by the chirality of the 5-position.[81] Diketones are formed if the β-position of the α,β-epoxy ketone is unsubstituted. The 1,2-carbonyl migration of an α,β-epoxy ketone, 2-cycloheptylidenecyclopentanone oxide, occurs with $BF_3 \cdot OEt_2$ at 25 °C to form the cyclic spiro-1,3-diketone in 1 min (eq 23).[82]

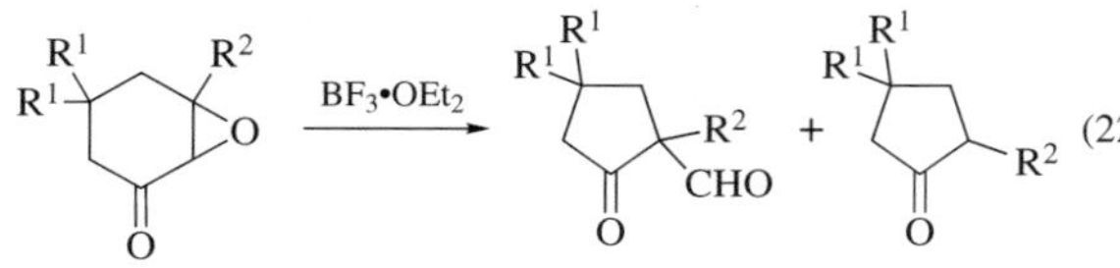

R^1 R^1 R^2 O O $BF_3 \cdot OEt_2$ → R^1 R^1 R^2 CHO O + R^1 R^1 R^2 O (22)

R^1 = Me, H; R^2 = Me, Ph

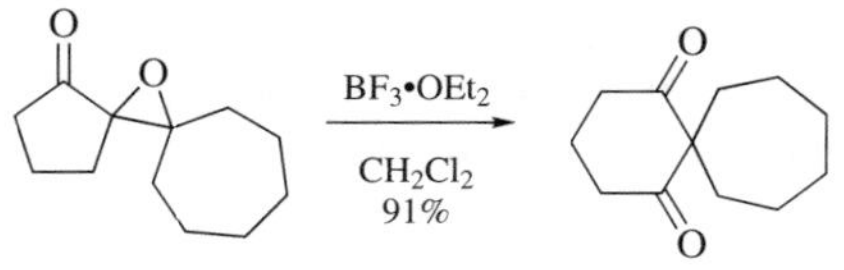

O O $BF_3 \cdot OEt_2$ CH_2Cl_2 91% O O (23)

The migration of the carbonyl during epoxide cleavage is used to produce hydroxy lactones from epoxides of carboxylic acids (eq 24).[83] α-Acyl-2-indanones,[84] furans,[85] and Δ^2-oxazolines[86] (eq 25) can also be synthesized by the cleavage and rearrangement of epoxides with $BF_3 \cdot OEt_2$. The last reaction has been conducted with sulfuric acid and with tin chloride, but the yields were lower. γ,δ-Epoxy tin compounds react with $BF_3 \cdot OEt_2$ to give the corresponding cyclopropylcarbinyl alcohols (eq 26).[87]

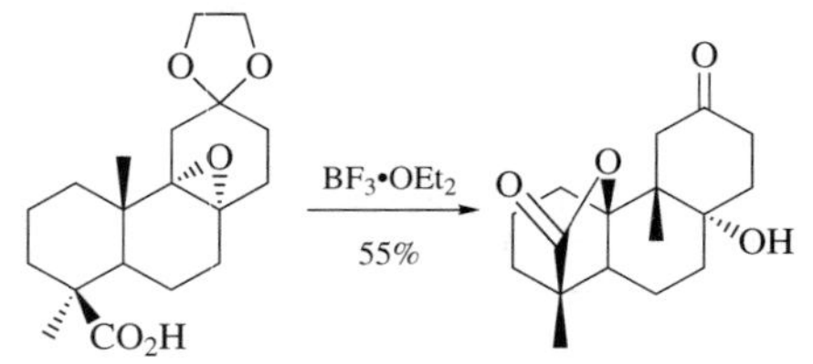

O O O CO_2H $BF_3 \cdot OEt_2$ 55% O O O OH (24)

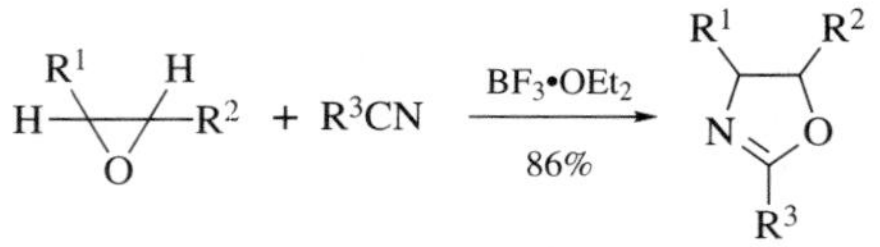

R^1 H H R^2 O + R^3CN $BF_3 \cdot OEt_2$ 86% R^1 R^2 N O R^3 (25)

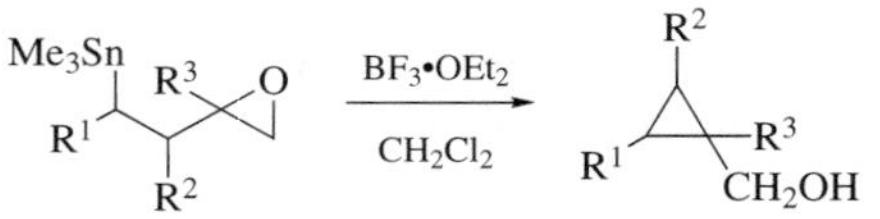

Me_3Sn R^3 O R^1 R^2 $BF_3 \cdot OEt_2$ CH_2Cl_2 R^2 R^1 R^3 CH_2OH (26)

Remotely unsaturated epoxy acids undergo fission rearrangement when treated with $BF_3 \cdot OEt_2$. Hence, *cis* and *trans* ketocyclopropane esters are produced from the unsaturated epoxy ester methyl vernolate (eq 27).[88]

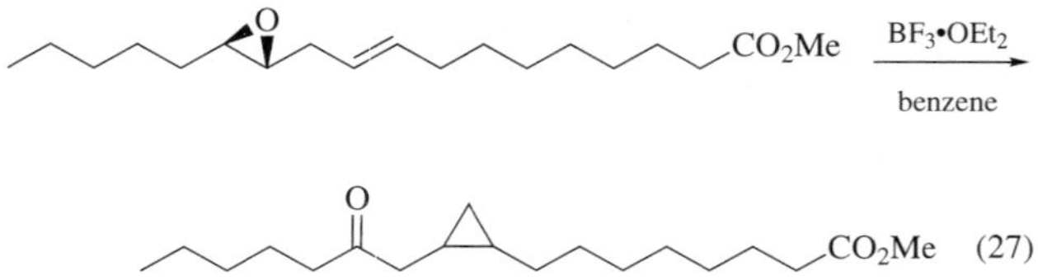

O CO_2Me $BF_3 \cdot OEt_2$ benzene

O CO_2Me (27)

Epoxy sulfones undergo rearrangement with $BF_3 \cdot OEt_2$ to give the corresponding aldehydes.[89] α-Epoxy sulfoxides, like other negatively substituted epoxides, undergo rearrangement in which the sulfinyl group migrates and not the hydrogen, alkyl, or aryl groups (eq 28).[89]

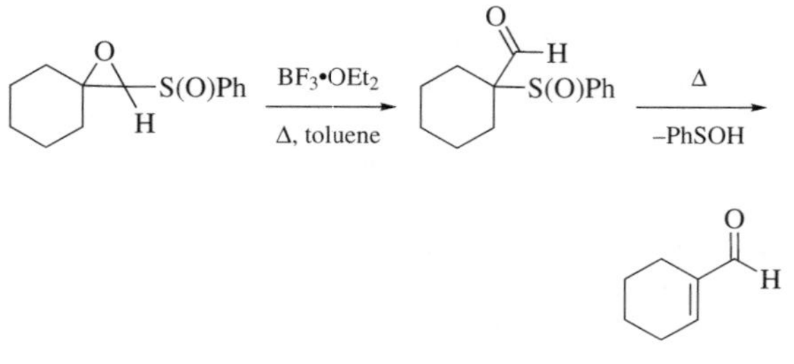

O S(O)Ph H $BF_3 \cdot OEt_2$ Δ, toluene O H S(O)Ph Δ −PhSOH

O H (28)

α,β-Epoxy alcohols undergo cleavage and rearrangement with $BF_3 \cdot OEt_2$ to form β-hydroxy ketones.[90] The rearrangement is stereospecific with respect to the epoxide and generally results in *anti* migration. The rearrangement of epoxy alcohols with β-substituents leads to α,α-disubstituted carbonyl compounds.[91]

The $BF_3 \cdot OEt_2$-induced opening of epoxides with alcohols is regioselective, but the regioselectivity varies with the nature of the substituents on the oxirane ring.[92] If the substituent provides charge stabilization (as with a phenyl ring), the internal position is attacked exclusively. On the other hand, terminal ethers are formed by the regioselective cleavage of the epoxide ring of glycidyl tosylate.[92]

A combination of cyanoborohydride and $BF_3 \cdot OEt_2$ is used for the regio- and stereoselective cleavage of most epoxides to the less substituted alcohols resulting from *anti* ring opening.[93] The reaction rate of organocopper and cuprate reagents with slightly reactive epoxides, e.g. cyclohexene oxide, is dramatically enhanced by $BF_3 \cdot OEt_2$.[94] The Lewis acid and nucleophile work in a concerted manner so that *anti* products are formed.

Azanaphthalene *N*-oxides undergo photochemical deoxygenation reactions in benzene containing $BF_3 \cdot OEt_2$, resulting in amines in 70–80% yield;[95] these amines are important in the synthesis of heterocyclic compounds. ***Azidotrimethylsilane*** reacts with *trans*-1,2-epoxyalkylsilanes in the presence of $BF_3 \cdot OEt_2$ to produce (*Z*)-1-alkenyl azides.[96] The *cis*-1,2-epoxyalkylsilanes undergo rapid polymerization in the presence of Lewis acids.

Other Rearrangements. $BF_3 \cdot OEt_2$ is used for the regioselective rearrangement of polyprenyl aryl ethers to yield polyprenyl substituted phenols, e.g. coenzyme Q_n.[97] The reagent is used in the Fries rearrangement; for example, 5-acetyl-6-hydroxycoumaran is obtained in 96% yield from 6-acetoxycoumaran using this reagent (eq 29).[98]

AcO– (6-acetoxycoumaran) —$BF_3 \bullet OEt_2$→ Ac, HO– (5-acetyl-6-hydroxycoumaran) (29)

Formyl bicyclo[2.2.2]octane undergoes the retro-Claisen rearrangement to a vinyl ether in the presence of $BF_3 \cdot OEt_2$ at 0 °C (eq 30), rather than with HOAc at 110 °C.[99]

MeO_2C–(bicyclooctene)–CHO —$BF_3 \bullet OEt_2$, 0 °C, 72%→ (H, O, H, CO_2Me) (30)

$BF_3 \cdot OEt_2$ is used for a stereospecific 1,3-alkyl migration to form *trans*-2-alkyltetrahydrofuran-3-carbaldehydes from 4,5-dihydrodioxepins (eq 31), which are obtained by the isomerization of 4,7-dihydro-1,3-dioxepins.[100] Similarly, α-alkyl-β-alkoxyaldehydes can be prepared from 1-alkenyl alkyl acetals by a 1,3-migration using $BF_3 \cdot OEt_2$ as catalyst.[101] *Syn* products are obtained from (*E*)-1-alkenyl alkyl acetals and *anti* products from the (*Z*)-acetals.

(R, H, O, O dioxepin) —$BF_3 \bullet OEt_2$, –73 °C, 90–95%→ (O, R, CHO) (31)

The methyl substituent, and not the cyano group, of 4-methyl-4-cyanocyclohexadienone migrates in the presence of $BF_3 \cdot OEt_2$ to give 3-methyl-4-cyanocyclohexadienone.[102] $BF_3 \cdot OEt_2$-promoted regioselective rearrangements of polyprenyl aryl ethers provide a convenient route for the preparation of polyprenyl-substituted hydroquinones (eq 32), which can be oxidized to polyprenylquinones.[103]

(OH, MeO, MeO, O–polyprenyl–H, n) —$BF_3 \bullet OEt_2$, CCl_4→ (OH, MeO, MeO, OH, polyprenyl–H, n) (32)

The (*E*)–(*Z*) photoisomerization of α,β-unsaturated esters,[104] cinnamic esters,[105] butenoic esters,[106] and dienoic esters[106] is catalyzed by $BF_3 \cdot OEt_2$ or ***Ethylaluminum Dichloride***. The latter two reactions also involve the photodeconjugation of α,β-unsaturated esters to β,γ-unsaturated esters. The $BF_3 \cdot MeOH$ complex is used for the isomerization of 1- and 2-butenes to form equal quantities of *cis*- and *trans*-but-2-enes;[107] the $BF_3 \cdot OEt_2$–acetic acid complex is not as effective.

The complex formed with $BF_3 \cdot OEt_2$ and ***Epichlorohydrin*** in DMF acts as a catalyst for the Beckmann rearrangement of oximes.[108] Cyclohexanone, acetaldehyde, and *syn*-benzaldehyde oximes are converted into ε-caprolactam, a mixture of *N*-methylformamide and acetamide, and *N*-phenylacetamide, respectively.

The addition of $BF_3 \cdot OEt_2$ to an α-phosphorylated imine results in the 1,3-transfer of a diphenylphosphinoyl group, with resultant migration of the C–N=C triad.[109] This method is less destructive than the thermal rearrangement. The decomposition of dimethyldioxirane in acetone to methyl acetate is accelerated with $BF_3 \cdot OEt_2$, but acetol is also formed.[110] Propene oxide undergoes polymerization with $BF_3 \cdot OEt_2$ in most solvents, but isomerizes to propionaldehyde and acetone in dioxane.[111]

Hydrolysis. $BF_3 \cdot OEt_2$ is used for stereospecific hydrolysis of methyl ethers, e.g. in the synthesis of (±)-aklavone.[112] The reagent is also used for the mild hydrolysis of dimethylhydrazones.[113] The precipitate formed by the addition of $BF_3 \cdot OEt_2$ to a dimethylhydrazone in ether is readily hydrolyzed by water to the ketone; the reaction is fast and does not affect enol acetate functionality.

Cleavage of Ethers. In aprotic, anhydrous solvents, $BF_3 \cdot MeOH$ is useful for the cleavage of trityl ethers at rt.[114] Under these conditions, *O*- and *N*-acyl groups, *O*-sulfonyl, *N*-alkoxycarbonyl, *O*-methyl, *O*-benzyl, and acetal groups are not cleaved.

$BF_3 \cdot OEt_2$ and iodide ion are extremely useful for the mild and regioselective cleavage of aliphatic ethers and for the removal of the acetal protecting group of carbonyl compounds.[115,116] Aromatic ethers are not cleaved, in contrast to other boron reagents. $BF_3 \cdot OEt_2$, in chloroform or dichloromethane, can be used for the removal of the *t*-butyldimethylsilyl (TBDMS) protecting group of hydroxyls, at 0–25 °C in 85–90% yield.[117] This is an alternative to ether cleavage with ***Tetra-n-butylammonium Fluoride*** or hydrolysis with aqueous ***Acetic Acid***.

In the presence of $BF_3 \cdot OEt_2$, dithio-substituted allylic anions react exclusively at the α-carbons of cyclic ethers, to give high yields of the corresponding alcohol products (eq 33).[118] The dithi-

ITHACA COLLEGE LIBRARY

ane moiety is readily hydrolyzed with ***Mercury(II) Chloride*** to give the keto derivatives.

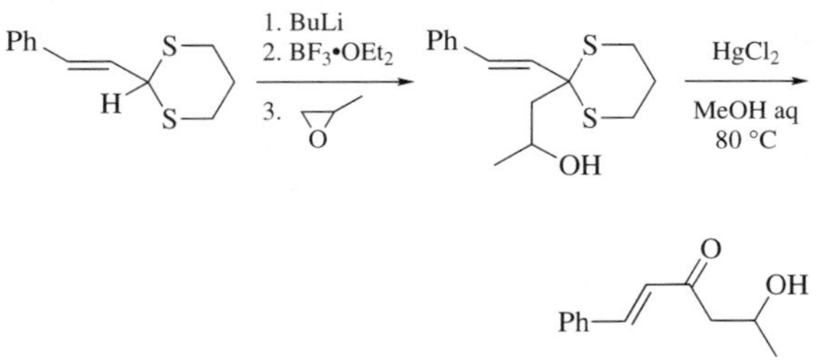

(33)

Inexpensive di-, tri-, and tetramethoxyanthraquinones can be selectively dealkylated to hydroxymethoxyanthraquinones by the formation of difluoroboron chelates with $BF_3 \cdot OEt_2$ in benzene and subsequent hydrolysis with methanol.[119] These unsymmetrically functionalized anthraquinone derivatives are useful intermediates for the synthesis of adriamycin, an antitumor agent. 2,4,6-Trimethoxytoluene reacts with cinnamic acid and $BF_3 \cdot OEt_2$, with selective demethylation, to form a boron heterocycle which can be hydrolyzed to the chalcone aurentiacin (eq 34).[120]

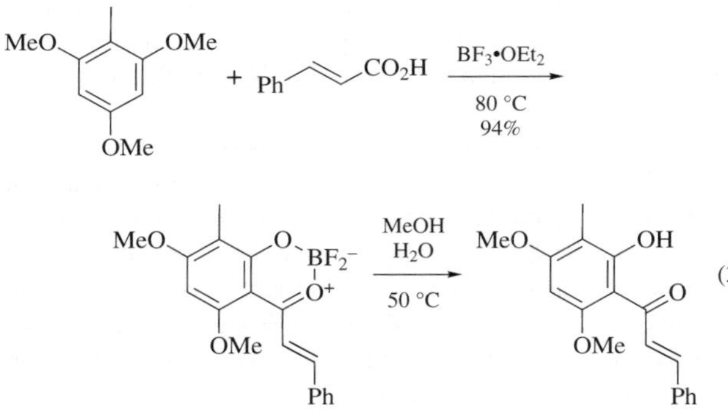

(34)

Reductions. In contrast to hydrosilylation reactions catalyzed by metal chlorides, aldehydes and ketones are rapidly reduced at rt by ***Triethylsilane*** and $BF_3 \cdot OEt_2$, primarily to symmetrical ethers and borate esters, respectively.[121] Aryl ketones like acetophenone and benzophenone are converted to ethylbenzene and diphenylmethane, respectively. Friedel–Crafts acylation–silane reduction reactions can also occur in one step using these reagents; thus ***Benzoyl Chloride*** reacts with benzene, triethylsilane, and $BF_3 \cdot OEt_2$ to give diphenylmethane in 30% yield.[121]

$BF_3 \cdot OEt_2$ followed by ***Diisobutylaluminum Hydride*** is used for the 1,2-reduction of γ-amino-α,β-unsaturated esters to give unsaturated amino alcohols, which are chiral building blocks for α-amino acids.[122] α,β-Unsaturated nitroalkenes can be reduced to hydroxylamines by ***Sodium Borohydride*** and $BF_3 \cdot OEt_2$ in THF;[123,124] extended reaction times result in the reduction of the hydroxylamines to alkylamines. Diphenylamine–borane is prepared from sodium borohydride, $BF_3 \cdot OEt_2$, and diphenylamine in THF at 0 °C.[125] This solid is more stable in air than $BF_3 \cdot THF$ and is almost as reactive in the reduction of aldehydes, ketones, carboxylic acids, esters, and anhydrides, as well as in the hydroboration of alkenes.

Bromination. $BF_3 \cdot OEt_2$ can catalyze the bromination of steroids that cannot be brominated in the presence of HBr or sodium acetate. Hence, 11α-bromoketones are obtained in high yields from methyl 3α,7α-diacetoxy-12-ketocholanate.[126] Bromination (at the 6α-position) and dibromination (at the 6α- and 11α-positions) of methyl 3α-acetoxy-7,12-dioxocholanate can occur, depending on the concentration of bromine.[127]

A combination of $BF_3 \cdot OEt_2$ and a halide ion (tetraethylammonium bromide or iodide in dichloromethane or chloroform, or sodium bromide or iodide in acetonitrile) is useful for the conversion of allyl, benzyl, and tertiary alcohols to the corresponding halides.[128,129]

Diels–Alder Reactions. $BF_3 \cdot OEt_2$ is used to catalyze and reverse the regiospecificity of some Diels–Alder reactions, e.g. with *peri*-hydroxylated naphthoquinones,[130] sulfur-containing compounds,[131] the reaction of 1-substituted *trans*-1,3-dienes with 2,6-dimethylbenzoquinones,[132] and the reaction of 6-methoxy-1-vinyl-3,4-dihydronaphthalene with *p*-quinones.[133] $BF_3 \cdot OEt_2$ has a drastic effect on the regioselectivity of the Diels–Alder reaction of quinoline- and isoquinoline-5,8-dione with piperylene, which produces substituted azaanthraquinones.[134] This Lewis acid is the most effective catalyst for the Diels–Alder reaction of furan with methyl acrylate, giving high *endo* selectivity in the 7-oxabicyclo[2.2.1]heptene product (eq 35).[135]

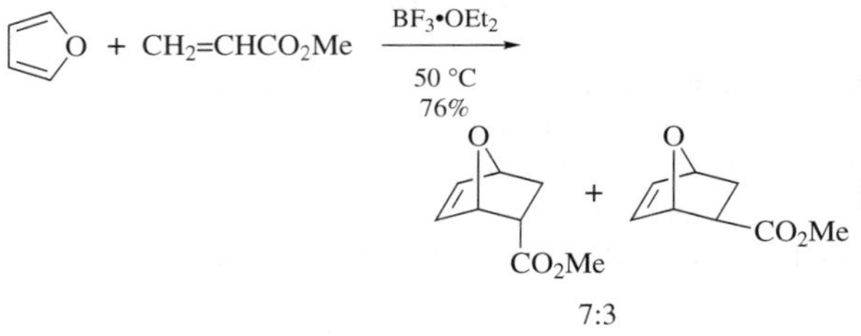

(35)

α-Vinylidenecycloalkanones, obtained by the reaction of ***Lithium Acetylide*** with epoxides and subsequent oxidation, undergo a Diels–Alder reaction at low temperature with $BF_3 \cdot OEt_2$ to form spirocyclic dienones (eq 36).[136]

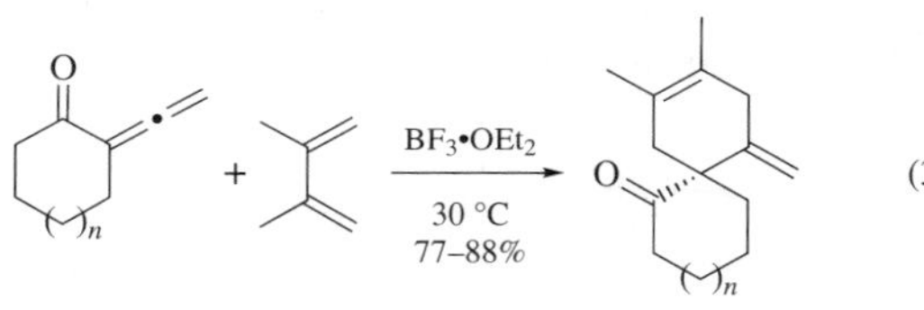

(36)

Other Reactions. The 17-hydroxy group of steroids can be protected by forming the THP (*O*-tetrahydropyran-2-yl) derivative with 2,3-dihydropyran, using $BF_3 \cdot OEt_2$ as catalyst;[137] the yields are higher and the reaction times shorter than with *p*-toluenesulfonic acid monohydrate.

$BF_3 \cdot OEt_2$ catalyzes the decomposition of β,γ-unsaturated diazomethyl ketones to cyclopentenone derivatives (eq 37).[138,139] Similarly, γ,δ-unsaturated diazo ketones are decomposed to β,γ-unsaturated cyclohexenones, but in lower yields.[140]

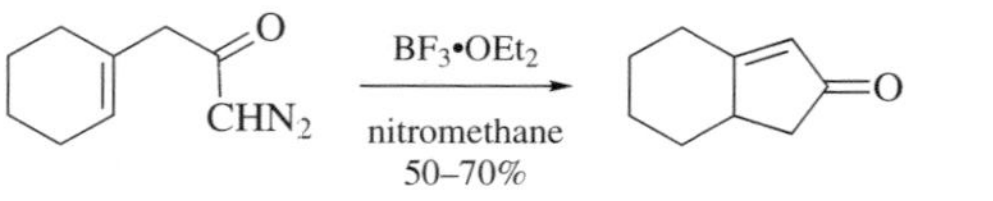

(37)

$BF_3 \cdot OEt_2$ is an effective reagent for debenzyloxycarbonylations of methionine-containing peptides.[141] Substituted 6*H*-1,3-thiazines can be prepared in high yields from $BF_3 \cdot OEt_2$-catalyzed reactions between α,β-unsaturated aldehydes, ketones, or acetals with thioamides, thioureas, and dithiocarbamates (eq 38).[142]

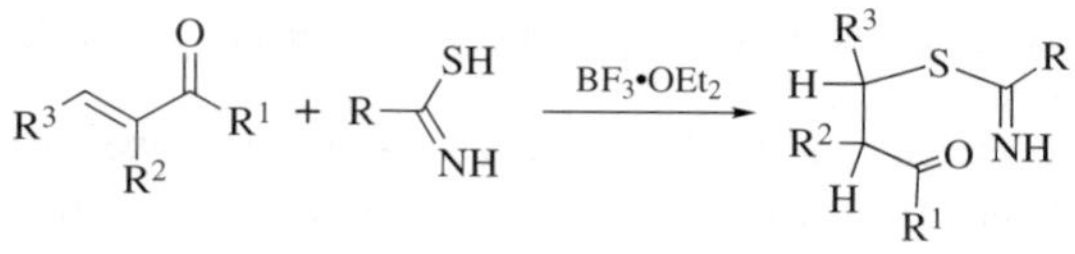

(38)

α-Alkoxy ketones can be prepared from α-diazo ketones and primary, secondary, and tertiary alcohols using $BF_3 \cdot OEt_2$ in ethanol.[143] Nitrogen is released from a solution of α-***Diazoacetophenone*** and $BF_3 \cdot OEt_2$ in ethanol to give α-ethoxyacetophenone.[143]

Anti-diols can be formed from β-hydroxy ketones using ***Tin(IV) Chloride*** or $BF_3 \cdot OEt_2$.[144] The hydroxy ketones are silylated, treated with the Lewis acid, and then desilylated with ***Hydrogen Fluoride***. *Syn*-diols are formed if ***Zinc Chloride*** is used as the catalyst.

$BF_3 \cdot OEt_2$ activates the formal substitution reaction of the hydroxyl group of γ- or δ-lactols with some organometallic reagents (M = Al, Zn, Sn), so that 2,5-disubstituted tetrahydrofurans or 2,6-disubstituted tetrahydropyrans are formed.[145]

A new method of nitrile synthesis from aldehydes has been discovered using *O*-(2-aminobenzoyl)hydroxylamine and $BF_3 \cdot OEt_2$, achieving 78–94% yields (eq 39).[146]

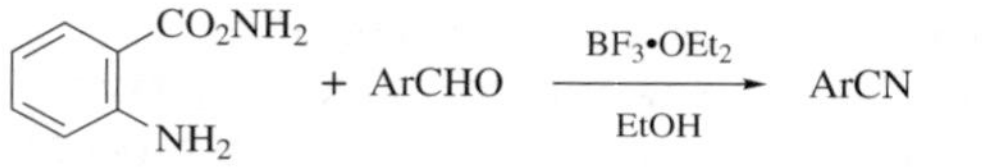

(39)

Carbonyl compounds react predominantly at the α site of dithiocinnamyllithium if $BF_3 \cdot OEt_2$ is present, as the hardness of the carbonyl compound is increased (eq 40).[147] The products can be hydrolyzed to α-hydroxyenones.

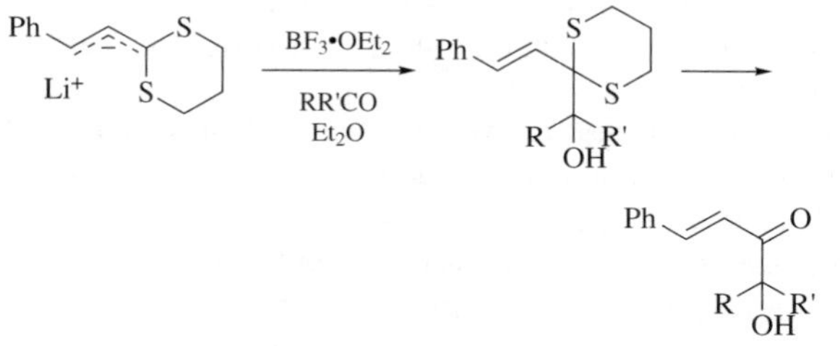

(40)

Optically active sulfinates can be synthesized from sulfinamides and alcohols using $BF_3 \cdot OEt_2$.[148] The reaction proceeds stereospecifically with inversion of sulfinyl configuration; the mild conditions ensure that the reaction will proceed even with alcohols with acid-labile functionality.

Related Reagents. See entries for other Lewis acids, e.g. Zinc Chloride, Aluminum Chloride, Titanium(IV) Chloride; also see entries for Boron Trifluoride (and combination reagents), and combination reagents employing boron trifluoride etherate, e.g. *n*-Butyllithium–Boron Trifluoride Etherate; Cerium(III) Acetate–Boron Trifluoride Etherate; Lithium Aluminum Hydride–Boron Trifluoride Etherate; Methylcopper–Boron Trifluoride Etherate.

1. (a) Hennion, G. F.; Hinton, H. D.; Nieuwland, J. A. *JACS* **1933**, *55*, 2857. (b) Zweifel, G.; Brown, H. C. *OR* **1963**, *13*, 28.
2. Eis, M. J.; Wrobel, J. E.; Ganem, B. *JACS* **1984**, *106*, 3693.
3. (a) Uno, H.; Terakawa, T.; Suzuki, H. *CL* **1989**, 1079. (b) *SL* **1991**, 559.
4. Yamaguchi, M.; Shibato, K.; Fujiwara, S.; Hirao, I. *S* **1986**, 421.
5. Yamaguchi, M.; Hirao, I. *TL* **1983**, *24*, 391.
6. Yamaguchi, M.; Nobayashi, N.; Hirao, I. *T* **1984**, *40*, 4261.
7. Hara, S.; Hyuga, S.; Aoyama, M.; Sato, M.; Suzuki, A. *TL* **1990**, *31*, 247.
8. Aoyama, M.; Hara, S.; Suzuki, A. *SC* **1992**, *22*, 2563.
9. Thompson, C. M.; Green, D. L. C.; Kubas, R. *JOC* **1988**, *53*, 5389.
10. Uno, H.; Okada, S.; Ono, T.; Shiraishi, Y.; Suzuki, H. *JOC* **1992**, *57*, 1504.
11. Volkmann, R. A.; Davies, J. T.; Meltz, C. N. *JACS* **1983**, *105*, 5946.
12. Kawate, T.; Nakagawa, M.; Yamazaki, H.; Hirayama, M.; Hino, T. *CPB* **1993**, *41*, 287.
13. Snyder, H. R.; Kornberg, H. A.; Romig, J. R. *JACS* **1939**, *61*, 3556.
14. Ha, H. J.; Nam, G. S. *SC* **1992**, *22*, 1143.
15. (a) Lipshutz, B. H.; Ellsworth, E. L.; Siahaan, T. J. *JACS* **1988**, *110*, 4834. (b) *JACS* **1989**, *111*, 1351.
16. Ochiai, M.; Fujita, E.; Arimoto, M.; Yamaguchi, H. *CPB* **1985**, *33*, 41.
17. Maruyama, K.; Naruta, Y. *JOC* **1978**, *43*, 3796.
18. Panek, J. S.; Sparks, M. A. *JOC* **1989**, *54*, 2034.
19. Giannis, A.; Sandhoff, K. *TL* **1985**, *26*, 1479.
20. Roos, E. C.; Hiemstra, H.; Speckamp, W. N.; Kaptein, B.; Kamphuis, J.; Schoemaker, H. E. *RTC* **1992**, *111*, 360.
21. Kuhn, M.; von Wartburg, A. *HCA* **1968**, *51*, 1631.
22. Kunz, H.; Sager, W. *HCA* **1985**, *68*, 283.
23. Yamamoto, Y.; Schmid, M. *CC* **1989**, 1310.
24. Yamamoto, Y.; Maruyama, K. *JOM* **1985**, *284*, C45.
25. (a) Keck, G. E.; Abbott, D. E. *TL* **1984**, *25*, 1883. (b) Keck, G. E.; Boden, E. P. *TL* **1984**, *25*, 265.
26. Keck, G. E.; Enholm, E. J. *JOC* **1985**, *50*, 146.
27. Trost, B. M.; Bonk, P. J. *JACS* **1985**, *107*, 1778.
28. Marshall, J. A.; DeHoff, B. S.; Crooks, S. L. *TL* **1987**, *28*, 527.
29. Marton, D.; Tagliavini, G.; Zordan, M.; Wardell, J. L. *JOM* **1990**, *390*, 127.
30. Ciufolini, M. A.; Spencer, G. O. *JOC* **1989**, *54*, 4739.
31. Gung, B. W.; Smith, D. T.; Wolf, M. A. *TL* **1991**, *32*, 13.
32. Takuwa, A.; Nishigaichi, Y.; Yamashita, K.; Iwamoto, H. *CL* **1990**, 1761.
33. Nishigaichi, Y.; Takuwa, A.; Jodai, A. *TL* **1991**, *32*, 2383.
34. Sugimura, H.; Osumi, K. *TL* **1989**, *30*, 1571.
35. Croxall, W. J.; Glavis, F. J.; Neher, H. T. *JACS* **1948**, *70*, 2805.
36. Hosomi, A.; Hashimoto, H.; Sakurai, H. *TL* **1980**, *21*, 951.
37. Smith, A. B. III; Jerris, P. J. *JACS* **1981**, *103*, 194.
38. Erman, W. F.; Stone, L. C. *JACS* **1971**, *93*, 2821.
39. Wakselman, C.; Tordeux, M. *CC* **1975**, 956.
40. Shellhamer, D. F.; Curtis, C. M.; Hollingsworth, D. R.; Ragains, M. L.; Richardson, R. E.; Heasley, V. L.; Shakelford, S. A.; Heasley, G. E. *JOC* **1985**, *50*, 2751.
41. Shellhamer, D. F.; Curtis, C. M.; Hollingsworth, D. R.; Ragains, M. L.; Richardson, R. E.; Heasley, V. L.; Heasley, G. E. *TL* **1982**, *23*, 2157.
42. Wulff, W. D.; Gilbertson, S. R. *JACS* **1985**, *107*, 503.
43. Yamago, S.; Machii, D.; Nakamura, E. *JOC* **1991**, *56*, 2098.
44. Nakamura, E.; Kuwajima, I. *JACS* **1977**, *99*, 961.
45. Sugimura, H.; Shigekawa, Y.; Uematsu, M. *SL* **1991**, 153.
46. Heathcock, C. H.; Flippin, L. A. *JACS* **1983**, *105*, 1667.
47. Evans, D. A.; Gage, J. R. *TL* **1990**, *31*, 6129.
48. Kamimura, A.; Marumo, S. *TL* **1990**, *31*, 5053.
49. Annunziata, R.; Cinquini, M.; Cozzi, F.; Cozzi, P. G. *TL* **1990**, *31*, 6733.

50. Duhamel, P.; Hennequin, L.; Poirier, N.; Poirier, J.-M. *TL* **1985**, *26*, 6201.
51. Moriarty, R. M.; Prakash, O.; Duncan, M. P.; Vaid, R. K. *JOC* **1987**, *52*, 150.
52. Yamamoto, Y.; Maruyama, K. *JACS* **1982**, *104*, 2323.
53. Ando, W.; Tsumaki, H. *CL* **1983**, 1409.
54. Thompson, C. M.; Docter, S. *TL* **1988**, *29*, 5213.
55. Ohta, S.; Kimoto, S. *TL* **1975**, 2279.
56. Boeckman, R. K. Jr.; Bruza, K. J.; Heinrich, G. R. *JACS* **1978**, *100*, 7101.
57. Rigby, J. H. *TL* **1982**, *23*, 1863.
58. Sum, P.-E.; Weiler, L. *CJC* **1979**, *57*, 1475.
59. Chenard, B. L.; Van Zyl, C. M.; Sanderson, D. R. *TL* **1986**, *27*, 2801.
60. Harding, K. E.; Clement, K. S. *JOC* **1984**, *49*, 3870.
61. Jain, T. C.; McCloskey, J. E. *TL* **1971**, 1415.
62. (a) Hinton, H. D.; Nieuwland, J. A. *JACS* **1932**, *54*, 2017. (b) Sowa, F. J.; Nieuwland, J. A. *JACS* **1936**, *58*, 271. (c) Hallas, G. *JCS* **1965**, 5770.
63. Kadaba, P. K. *S* **1972**, 628.
64. Kadaba, P. K. *S* **1971**, 316.
65. Marshall, J. L.; Erikson, K. C.; Folsom, T. K. *TL* **1970**, 4011.
66. Yamaguchi, M.; Shibato, K.; Nakashima, H.; Minami, T. *T* **1988**, *44*, 4767.
67. Danishefsky, S.; Chao, K.-H.; Schulte, G. *JOC* **1985**, *50*, 4650.
68. Tani, J.; Oine, T.; Inoue, I. *S* **1975**, 714.
69. Ibuka, T.; Chu, G.-N.; Aoyagi, T.; Kitada, K.; Tsukida, T.; Yoneda, F. *CPB* **1985**, *33*, 451.
70. Irie, H.; Nishimura, M.; Yoshida, M.; Ibuka, T. *JCS(P1)* **1989**, 1209.
71. Taylor, M. E.; Fletcher, T. L. *JOC* **1961**, *26*, 940.
72. Hoppe, D.; Krämer, T.; Erdbrügger, C. F.; Egert, E. *TL* **1989**, *30*, 1233.
73. Kitchen, L. J. *JACS* **1948**, *70*, 3608.
74. Hirschmann, R.; Miller, R.; Wendler, N. L. *JACS* **1954**, *76*, 4592.
75. Posner, G. H.; Shulman-Roskes, E. M.; Oh, C. H.; Carry, J.-C.; Green, J. V.; Clark, A. B.; Dai, H.; Anjeh, T. E. N. *TL* **1991**, *32*, 6489.
76. Heymann, H.; Fieser, L. F. *JACS* **1952**, *74*, 5938.
77. Clinton, R. O.; Christiansen, R. G.; Neumann, H. C.; Laskowski, S. C. *JACS* **1957**, *79*, 6475.
78. House, H. O.; Wasson, R. L. *JACS* **1957**, *79*, 1488.
79. Bird, C. W.; Yeong, Y. C.; Hudec, J. *S* **1974**, *27*.
80. Domagala, J. M.; Bach, R. D. *JACS* **1978**, *100*, 1605.
81. Obuchi, K.; Hayashibe, S.; Asaoka, M.; Takei, H. *BCJ* **1992**, *65*, 3206.
82. Bach, R. D.; Klix, R. C. *JOC* **1985**, *50*, 5438.
83. Hancock, W. S.; Mander, L. N.; Massy-Westropp, R. A. *JOC* **1973**, *38*, 4090.
84. French, L. G.; Fenlon, E. E.; Charlton, T. P. *TL* **1991**, *32*, 851.
85. Loubinoux, B.; Viriot-Villaume, M. L.; Chanot, J. J.; Caubere, P. *TL* **1975**, 843.
86. Smith, J. R. L.; Norman, R. O. C.; Stillings, M. R. *JCS(P1)* **1975**, 1200.
87. Sato, T.; Watanabe, M.; Murayama, E. *SC* **1987**, *17*, 781.
88. Conacher, H. B. S.; Gunstone, F. D. *CC* **1967**, 984.
89. Durst, T.; Tin, K.-C. *TL* **1970**, 2369.
90. Maruoka, K.; Hasegawa, M.; Yamamoto, H.; Suzuki, K.; Shimazaki, M.; Tsuchihashi, G. *JACS* **1986**, *108*, 3827.
91. Shimazaki, M.; Hara, H.; Suzuki, K.; Tsuchihashi, G. *TL* **1987**, *28*, 5891.
92. Liu, Y.; Chu, T.; Engel, R. *SC* **1992**, *22*, 2367.
93. Hutchins, R. O.; Taffer, I. M.; Burgoyne, W. *JOC* **1981**, *46*, 5214.
94. Alexakis, A.; Jachiet, D.; Normant, J. F. *T* **1986**, *42*, 5607.
95. Hata, N.; Ono, I.; Kawasaki, M. *CL* **1975**, 25.
96. Tomoda, S.; Matsumoto, Y.; Takeuchi, Y.; Nomura, Y. *BCJ* **1986**, *59*, 3283.
97. Yoshizawa, T.; Toyofuku, H.; Tachibana, K.; Kuroda, T. *CL* **1982**, 1131.
98. Davies, J. S. H.; McCrea, P. A.; Norris, W. L.; Ramage, G. R. *JCS* **1950**, 3206.
99. Boeckman, R. K. Jr.; Flann, C. J.; Poss, K. M. *JACS* **1985**, *107*, 4359.
100. Suzuki, H.; Yashima, H.; Hirose, T.; Takahashi, M., Moro-Oka, Y.; Ikawa, T. *TL* **1980**, *21*, 4927.
101. Takahashi, M.; Suzuki, H.; Moro-Oka, Y.; Ikawa, T. *TL* **1982**, *23*, 4031.
102. Marx, J. N.; Zuerker, J.; Hahn, Y. P. *TL* **1991**, *32*, 1921.
103. Yoshizawa, T.; Toyofuku, H.; Tachibana, K.; Kuroda, T. *CL* **1982**, 1131.
104. Lewis, F. D.; Oxman, J. D. *JACS* **1981**, *103*, 7345.
105. Lewis, F. D.; Oxman, J. D.; Gibson, L. L.; Hampsch, H. L.; Quillen, S. L. *JACS* **1986**, *108*, 3005.
106. Lewis, F. D.; Howard, D. K.; Barancyk, S. V.; Oxman, J. D. *JACS* **1986**, *108*, 3016.
107. Roberts, J. M.; Katovic, Z.; Eastham, A. M. *J. Polym. Sci. A1* **1970**, *8*, 3503.
108. Izumi, Y. *CL* **1990**, 2171.
109. Onys'ko, P. P.; Kim, T. V.; Kiseleva, E. I.; Sinitsa, A. D. *TL* **1992**, *33*, 691.
110. Singh, M.; Murray, R. W. *JOC* **1992**, *57*, 4263.
111. Sugiyama, S.; Ohigashi, S.; Sato, K.; Fukunaga, S.; Hayashi, H. *BCJ* **1989**, *62*, 3757.
112. Pearlman, B. A.; McNamara, J. M.; Hasan, I.; Hatakeyama, S.; Sekizaki, H.; Kishi, Y. *JACS* **1981**, *103*, 4248.
113. Gawley, R. E.; Termine, E. J. *SC* **1982**, *12*, 15.
114. Mandal, A. K.; Soni, N. R.; Ratnam, K. R. *S* **1985**, 274.
115. Mandal, A. K.; Shrotri, P. Y.; Ghogare, A. D. *S* **1986**, 221.
116. Pelter, A.; Ward, R. S.; Venkateswarlu, R.; Kamakshi, C. *T* **1992**, *48*, 7209.
117. Kelly, D. R.; Roberts, S. M.; Newton, R. F. *SC* **1979**, *9*, 295.
118. Fang, J.-M.; Chen, M.-Y. *TL* **1988**, *29*, 5939.
119. Preston, P. N.; Winwick, T.; Morley, J. O. *JCS(P1)* **1983**, 1439.
120. Schiemenz, G. P.; Schmidt, U. *LA* **1982**, 1509.
121. Doyle, M. P.; West, C. T.; Donnelly, S. J.; McOsker, C. C. *JOM* **1976**, *117*, 129.
122. Moriwake, T.; Hamano, S.; Miki, D.; Saito, S.; Torii, S. *CL* **1986**, 815.
123. Varma, R. S.; Kabalka, G. W. *OPP* **1985**, *17*, 254.
124. Varma, R. S.; Kabalka, G. W. *SC* **1985**, *15*, 843.
125. Camacho, C.; Uribe, G.; Contreras, R. *S* **1982**, 1027.
126. Yanuka, Y.; Halperin, G. *JOC* **1973**, *38*, 2587.
127. Takeda, K.; Komeno, T.; Igarashi, K. *CPB* **1956**, *4*, 343.
128. Mandal, A. K.; Mahajan, S. W. *TL* **1985**, *26*, 3863.
129. Vankar, Y. D.; Rao, C. T. *TL* **1985**, *26*, 2717.
130. Trost, B. M.; Ippen, J.; Vladuchick, W. C. *JACS* **1977**, *99*, 8116.
131. Kelly, T. R.; Montury, M. *TL* **1978**, 4311.
132. Stojanác, Z.; Dickinson, R. A.; Stojanác, N.; Woznow, R. J.; Valenta, Z. *CJC* **1975**, *53*, 616.
133. Das, J.; Kubela, R.; MacAlpine, G. A.; Stojanac, Z.; Valenta, Z. *CJC* **1979**, *57*, 3308.
134. Ohgaki, E.; Motoyoshiya, J.; Narita, S.; Kakurai, T.; Hayashi, S.; Hirakawa, K. *JCS(P1)* **1990**, 3109.
135. Kotsuki, H.; Asao, K.; Ohnishi, H. *BCJ* **1984**, *57*, 3339.
136. Gras, J.-L.; Guerin, A. *TL* **1985**, *26*, 1781.
137. Alper, H.; Dinkes, L. *S* **1972**, 81.
138. Smith, A. B. III; Branca, S. J.; Toder, B. H. *TL* **1975**, 4225.
139. Smith, A. B. III *CC* **1975**, 274.
140. Smith, A. B. III, Toder, B. H.; Branca, S. J.; Dieter, R. K. *JACS* **1981**, *103*, 1996.

141. Okamoto, M.; Kimoto, S.; Oshima, T.; Kinomura, Y.; Kawasaki, K.; Yajima, H. *CPB* **1967**, *15*, 1618.
142. Hoff, S.; Blok, A. P. *RTC* **1973**, *92*, 631.
143. Newman, M. S.; Beal, P. F. III *JACS* **1950**, *72*, 5161.
144. Anwar, S.; Davis, A. P. *T* **1988**, *44*, 3761.
145. Tomooka, K.; Matsuzawa, K.; Suzuki, K.; Tsuchihashi, G. *TL* **1987**, *28*, 6339.
146. Reddy, P. S. N.; Reddy, P. P. *SC* **1988**, *18*, 2179.
147. Fang, J.-M.; Chen, M.-Y.; Yang, W.-J. *TL* **1988**, *29*, 5937.
148. Hiroi, K.; Kitayama, R.; Sato, S. *S* **1983**, 1040.

Veronica Cornel
Emory University, Atlanta, GA, USA

Bromodimethylborane[1]

Me_2BBr

[5158-50-9] $C_2H_6BBr_2$ (MW 120.78)

(mild Lewis acid capable of selective cleavage of ethers[2,3] and acetals;[4,5] deoxygenation of sulfoxides[6])

Physical Data: mp −129 °C; bp 31–32 °C; *d* 1.238 g cm^{-3}; fp −37 °C.

Solubility: sol dichloromethane, 1,2-dichloroethane, hexane.

Form Supplied in: colorless liquid.

Preparative Method: can be conveniently prepared by treating ***Tetramethylstannane*** with ***Boron Tribromide***.[7]

Handling, Storage, and Precautions: flammable liquid, moisture sensitive; typically stored and dispensed as a 1.5–2 M solution in dichloromethane or dichloroethane. Solutions of this sort are stable for a period of months if stored at −15 °C and properly protected from moisture.

Cleavage of Ethers. Bromodimethylborane (Me_2BBr) reacts with primary, secondary, and aryl methyl ethers,[2] in addition to trityl,[8] benzyl,[2,8] and 4-methoxybenzyl[9] ethers, to regenerate the parent alcohol in good to excellent yield (e.g. eq 1). The tertiary methyl ethers examined afforded the corresponding tertiary bromides.[2] The reaction is typically carried out in dichloromethane or 1,2-dichloroethane between 0 °C and rt, in the presence of 1.3–4 equiv of Me_2BBr. The reaction is usually complete in a matter of hours. ***Triethylamine*** (0.1–0.15 equiv per equiv of Me_2BBr) is often added as an acid scavenger. 4-Methoxybenzyl ethers are more reactive and are cleaved at −78 °C, whereas aryl methyl ethers require elevated temperatures to react. Other functional groups including acetates, benzoates, alcohols, ethyl esters, and *t*-butyldiphenylsilyl ethers are recovered unchanged under the standard reaction conditions.

Bromodimethylborane is also effective for the cleavage of cyclic ethers.[2,3] Epoxides react at −78 °C while the analogous four- to seven-membered ring heterocycles react between 0 °C and rt. In contrast to other boron-containing Lewis acids, Me_2BBr reacts via a predominantly S_N2 mechanism. Tetrahydrofuran derivatives which are substituted at the 2-position give rise to primary bromides as the major or exclusive products. The nature of the substituent has a quantitative influence on the outcome of the reaction via steric effects and/or complexation to the reagent. It is of interest to note that tetrahydrofurans can be cleaved in the presence of acyclic ethers (eq 2).[3]

Me_2BBr, Et_3N; $ClCH_2CH_2Cl$, 0–25 °C, 16–18 h (1)

R = Me 93%
R = Bu 92%

Me_2BBr, Et_3N; CH_2Cl_2, 0 °C, 2 h, 87%; 4 : 1 (2)

It is also of considerable interest to note that no β-elimination of the hydroxy group was observed in the ring-opening of 2-(ethoxycarbonylmethyl)tetrahydrofurans (eq 3),[3] whereas *C*-glycosides bearing more acidic protons on the aglycon react with Me_2BBr to generate acyclic alkenes (eq 4).[10]

Me_2BBr, Et_3N; CH_2Cl_2, 0 °C, 2 h, 82% (3)

1. Me_2BBr, Et_3N, CH_2Cl_2, 0 °C, 0.5 h; 2. AcBr, 54% (4)

Bromodimethylborane has also been used in conjunction with ***Tetra-n-butylammonium Iodide*** to bring about the fragmentation of iodomethyl ether derivatives (eq 5).[11]

Me_2BBr, Bu_4NI; Et_3N, CH_2Cl_2, 0 °C, 0.5 h, 77% (5)

Cleavage of Acetals.[4,5] Cyclic and acyclic acetals react with Me_2BBr at −78 °C to generate the parent aldehydes and ketones in excellent yield (e.g. eq 6). Primary, secondary, and tertiary (2-methoxyethoxy)methyl (MEM), methoxymethyl (MOM), and (methylthio)methyl (MTM) ethers also react at −78 °C to give, after aqueous workup, the corresponding alcohol. It is interesting to note that even tertiary MEM ethers cleanly regenerate the parent alcohol without formation of the corresponding bromide or elimination products (eq 7). Treatment of an acetonide with Me_2BBr gives the parent diol in high yield (eq 8).

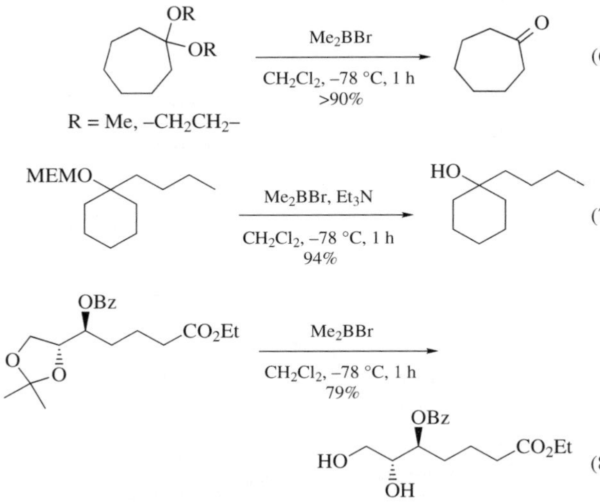

(6)

(7)

(8)

Tetrahydropyranyl (THP) and tetrahydrofuranyl (THF) ethers are converted to the corresponding alcohols by Me_2BBr at rt (eq 9), although the acetals are cleaved at −78 °C (see below).

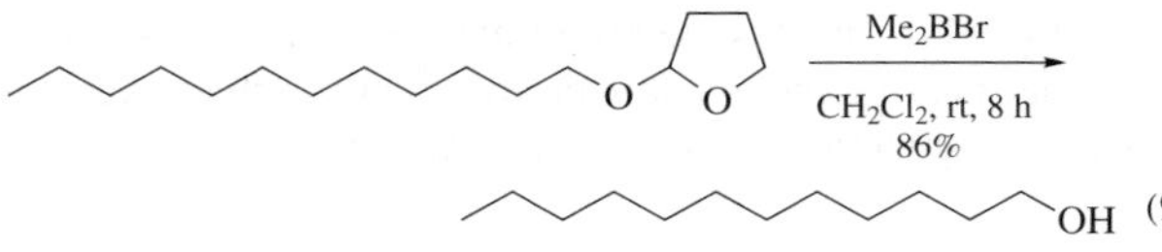

(9)

Bromodiphenylborane (Ph_2BBr) and ***9-Bromo-9-borabicyclo[3.3.1]nonane*** (Br-9-BBN) can often be used in place of Me_2BBr for the cleavage of acetals;[4,5] however, the purification of products from reactions employing Me_2BBr is facilitated by the volatility of Me_2B-containing byproducts, thus making Me_2BBr the reagent of choice in most instances.

Interconversion of Functional Groups. The reaction of Me_2BBr with MEM and MOM ethers is believed to proceed via α-bromo ether intermediates. It is possible to trap these intermediates with nucleophiles such as thiols, alcohols, and cyanide. An example of the utility of this sequence is the conversion of a readily prepared MOM ether into an MTM ether (eq 10).[12]

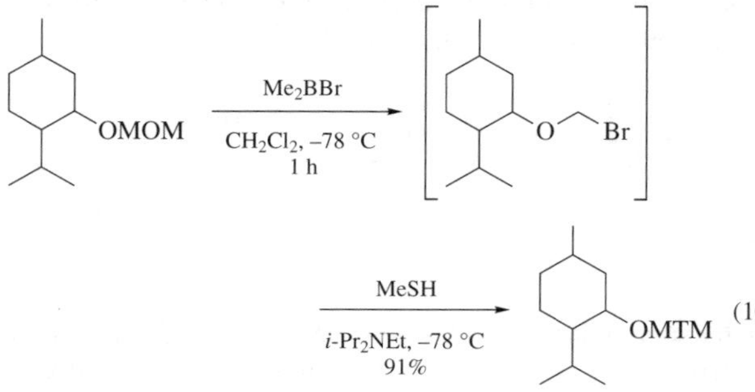

(10)

While THP and THF ethers are converted to the corresponding alcohols by Me_2BBr at rt, the acetal is cleaved at −78 °C.[13] The initial products of the reaction are acyclic α-bromo ethers. These can be trapped with a variety of nucleophiles to generate stable ring-opened products (eq 11).[13] This reaction has been extended to glycosides which, although less reactive, behave in a similar fashion.[14,15]

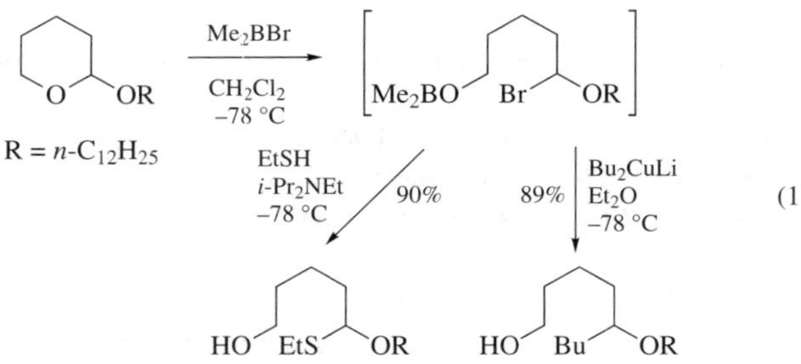

(11)

Benzylidene acetals are recovered unchanged when treated with Me_2BBr under conditions which are used to cleave other acetals.[16] It is, however, possible to cleave benzylidene acetals to generate hydroxy-*O,S*-acetals in excellent yield, by treatment with Me_2BBr at −78 °C followed by ***Thiophenol*** (eq 12).[16] Sterically encumbered bromoboranes optimize regioselective complexation of boron to the least hindered oxygen atom and are, therefore, the reagents of choice for this process (eq 12).[16] These experiments demonstrate that benzylidene acetals do indeed react with Me_2BBr at −78 °C, like other acetals.

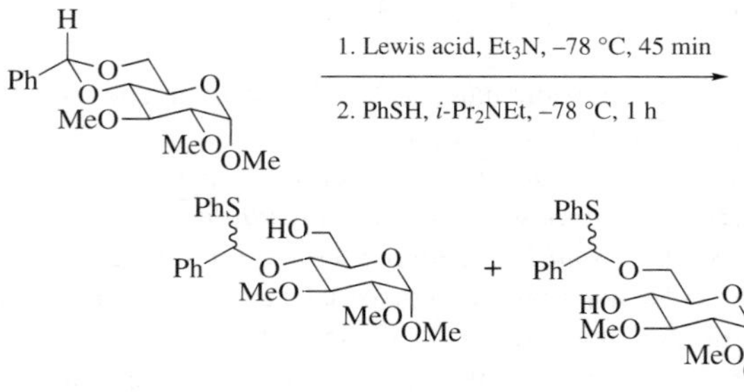

(12)

Me_2BBr	68%	17%
Ph_2BBr	96%	–
9-BrBBN	90%	–

Treatment of glycoside benzylidene acetals with a variety of disubstituted bromoboranes, followed by ***Borane-Tetrahydrofuran***, generates 4-*O*-benzyl-6-hydroxypyranosides in excellent yield (eq 13).[16]

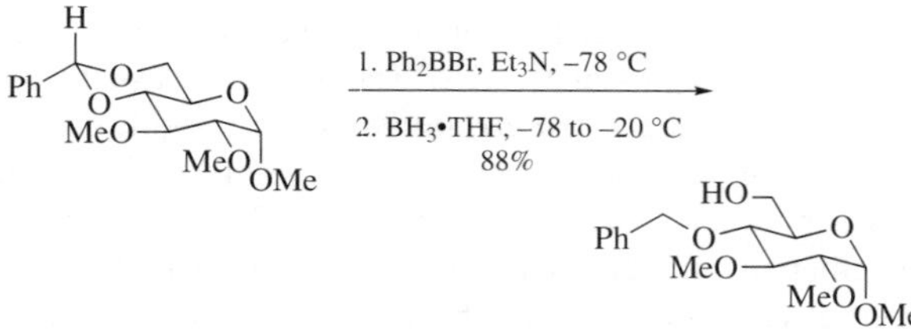

(13)

Acetals derived from ***Dimethyl L-Tartrate*** react with Me_2BBr to generate α-bromo ethers which react further with cuprate reagents to give optically active secondary alcohol derivatives (eq 14).[17] The alcohols may be liberated by treatment with ***Samarium(II) Iodide*** or by a straightforward sequence of reactions (mesylation and elimination to form an enol ether followed by exposure to methoxide in refluxing methanol). Selectivity is enhanced by the use of Ph_2BBr and by careful control of the reaction temperature at each step.

$$\text{MeO}_2\text{C, CO}_2\text{Me dioxolane } (n\text{-}C_9H_{19}) \xrightarrow[\text{2. } Me_2Cu(CN)Li_2,\ -30\ ^\circ C]{\text{1. Lewis acid, } -78 \text{ to rt}} \text{products (two diastereomers)} \quad (14)$$

Me_2BBr, 80% — 34:1
Ph_2BBr, 62% — 82:1

Miscellaneous Reactions. Bromodimethylborane can also be used to convert dialkyl, aryl alkyl, and diaryl sulfoxides to the corresponding sulfides (eq 15).[6] Typically, the sulfoxides are treated with 2.5 equiv of Me_2BBr in dichloromethane at $-23\ ^\circ C$ for 30 min and at $0\ ^\circ C$ for 10 min. Bromine is produced in the reaction and must be removed in order to avoid possible side reactions. This is accomplished by saturating the solution with propene prior to introducing the reagent or by adding cyclohexene. Phosphine oxides and sulfones failed to react under the conditions used to deoxygenate sulfoxides.

$$\xrightarrow[\text{CH}_2\text{Cl}_2,\ \text{cyclohexene};\ -23 \text{ to } 0\ ^\circ C;\ 93\%]{Me_2BBr} \quad (15)$$

Bromodimethylborane has also been used as a catalyst for the Pictet–Spengler reaction (eq 16)[18] and to catalyze the 1,3-transposition of an allylic lactone.[19]

$$\xrightarrow[\text{CH}_2\text{Cl}_2;\ -78 \text{ to rt, } 37\text{ h};\ 87\%]{Me_2BBr} \quad (16)$$

Related Reagents. Boron Tribromide; Boron Trichloride; Bromobis(isopropylthio)borane; 9-Bromo-9-borabicyclo-[3.3.1]borane; *B*-Bromocatechalborane.

1. Guindon, Y.; Anderson, P. C.; Yoakim, C.; Girard, Y.; Berthiaume, S.; Morton, H. E. *PAC* **1988**, *60*, 1705.
2. Guindon, Y.; Yoakim, C.; Morton, H. E. *TL* **1983**, *24*, 2969.
3. Guindon, Y.; Therien, M.; Girard, Y.; Yoakim, C. *JOC* **1987**, *52*, 1680.
4. Guindon, Y.; Morton, H. E.; Yoakim, C. *TL* **1983**, *24*, 3969.
5. Guindon, Y.; Yoakim, C.; Morton, H. E. *JOC* **1984**, *49*, 3912.
6. Guindon, Y.; Atkinson, J. G.; Morton, H. E. *JOC* **1984**, *49*, 4538.
7. Nöth, H.; Vahrenkamp, H. *JOM* **1968**, *11*, 399.
8. Kodali, D. R.; Duclos Jr., R. I. *Chem. Phys. Lipids* **1992**, *61*, 169.
9. Hébert, N.; Beck, A.; Lennox, R. B.; Just, G. *JOC* **1992**, *57*, 1777.
10. Abel, S.; Linker, T.; Giese, B. *SL* **1991**, 171.
11. Gauthier, J. Y.; Guindon, Y. *TL* **1987**, *28*, 5985.
12. Morton, H. E.; Guindon, Y. *JOC* **1985**, *50*, 5379.
13. Guindon, Y.; Bernstein, M. A.; Anderson, P. C. *TL* **1987**, *28*, 2225.
14. Guindon, Y.; Anderson, P. C. *TL* **1987**, *28*, 2485.
15. Hashimoto, H.; Kawanishi, M.; Yuasa, H. *TL*, **1991**, *32*, 7087.
16. Guindon, Y.; Girard, Y.; Berthiaume, S.; Gorys, V.; Lemieux, R.; Yoakim, C. *CJC* **1990**, *68*, 897.
17. Guindon, Y.; Simoneau, B.; Yoakim, C.; Gorys, V.; Lemieux, R.; Ogilvie, W. *TL* **1991**, *32*, 5453.
18. Kawate, T.; Nakagawa, M.; Ogata, K.; Hino, T. *H* **1992**, *33*, 801.
19. Mander, L. N.; Patrick, G. L. *TL* **1990**, *31*, 423.

Yvan Guindon & Paul C. Anderson
Bio-Méga/Boehringer Ingelheim Research, Laval, Québec, Canada

Bromotrimethylsilane[1]

Me_3SiBr

[2857-97-8] C_3H_9BrSi (MW 153.09)

(mild and selective reagent for cleavage of lactones, epoxides, acetals, phosphonate esters and certain ethers; effective reagent for formation of silyl enol ethers; can function as brominating agent)

Alternate Name: TMS-Br.
Physical Data: bp 79 °C; d 1.188 g cm^{-3}; n_D^{20} 1.4240; fp 32 °C.
Solubility: sol CCl_4, $CHCl_3$, CH_2Cl_2, $ClCH_2CH_2Cl$, MeCN, toluene, hexanes; reactive with THF (ethers), alcohols, and somewhat reactive with EtOAc (esters).
Form Supplied in: colorless liquid, packaged in ampules.
Analysis of Reagent Purity: well characterized by ^1H, ^{13}C, and ^{29}Si NMR spectroscopy.
Preparative Methods: although many methods are reported,[1] only a few are provided here: ***Chlorotrimethylsilane*** undergoes halogen exchange with either ***Magnesium Bromide***[2] in Et_2O or ***Sodium Bromide***[3] in MeCN, which allows in situ reagent formation (eq 1); alternatively, ***Hexamethyldisilane*** reacts with ***Bromine*** in benzene solution or neat, to afford only TMS-Br with no byproducts (eq 2).[4] TMS-Br may also be generated by reaction of hexamethyldisiloxane and ***Aluminum Bromide*** (eq 3).[5] However, it should be noted that the reactivity of in situ generated reagent appears to depend upon the method of preparation.

$$Me_3SiCl \xrightarrow[\text{or NaBr, MeCN}]{MgBr_2,\ Et_2O} Me_3SiBr + \text{LiBr or NaBr} \quad (1)$$

$$Me_3Si\text{-}SiMe_3 + Br_2 \longrightarrow Me_3SiBr \quad (2)$$

$$Me_3SiOSiMe_3 + AlBr_3 \longrightarrow Me_3SiBr \quad (3)$$

Purification: by distillation.
Handling, Storage, and Precautions: extremely sensitive to light, air, and moisture; fumes in air due to hydrolysis (HBr), and becomes discolored upon prolonged storage (free Br_2).

Ester Cleavage.[6] Although esters are readily cleaved with ***Iodotrimethylsilane***, reaction of esters with TMS-Br under similar conditions gives somewhat lower yields of silyl esters or acids upon hydrolysis (eq 4). Lactones, however, react with TMS-Br at 100 °C to afford ω-bromocarboxylic acids after hydrolysis of the silyl ester (eq 5).[7]

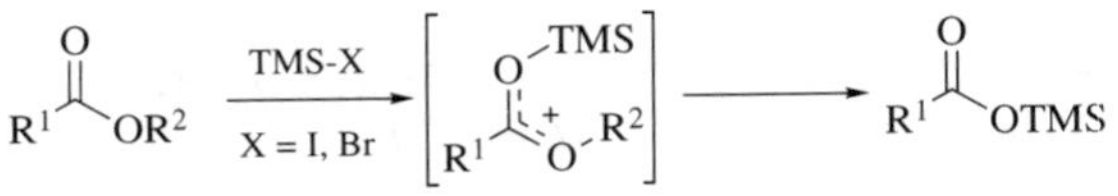

$$R^1C(O)OR^2 \xrightarrow[X = I, Br]{TMS\text{-}X} [\ldots] \longrightarrow R^1C(O)OTMS \quad (4)$$

$$\text{lactone} \xrightarrow[100\,°C]{TMS\text{-}Br} Br(CH_2)_nC(O)OTMS \quad (5)$$

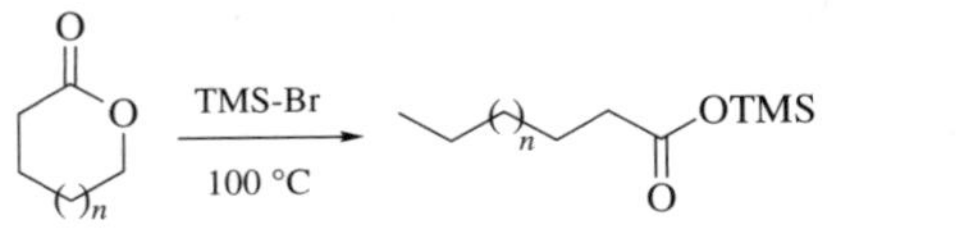

Ether Cleavage. THF[8] reacts with TMS-Br, thereby rendering ethereal solvents incompatible with the reagent. Smooth removal of the methoxymethyl (MOM) protecting group can be accomplished with TMS-Br at 0 °C (eq 6).[9] Whereas acetals, THP and silyl ethers are slowly cleaved with TMS-Br, the reagent generated in situ effects selective MOM ether cleavage in the presence of an acetonide.[10] The majority of published ether cleavages have been accomplished with TMS-I, although limited data show that the more vigorous conditions necessary for ethyl ether cleavage also result in bromide formation.[8]

$$RO\text{-}CH_2OMe \xrightarrow[0\,°C]{TMS\text{-}Br} RO\text{-}H \quad (6)$$

Cleavage of Epoxides. Epoxide opening with TMS-Br occurs to provide the primary alkyl bromide at −60 °C (eq 7).[8]

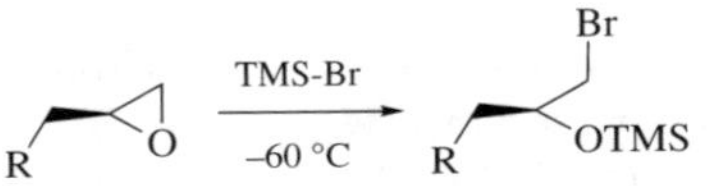

(7) TMS-Br, −60 °C

Cleavage of Acetals. Acetals can be cleaved by analogy to ethers, providing the parent carbonyl species.[3c,6b] Glycosyl bromides have been prepared from the corresponding acetate by reaction with TMS-Br in $CHCl_3$ at rt (eq 8).[11] In conjunction with $CoBr_2$ and ***Tetra-n-butylammonium Bromide***, TMS-Br converts the glucopyranose to the α-D-glucoside in the presence of an alcohol (eq 9).[12]

(8) TMS-Br

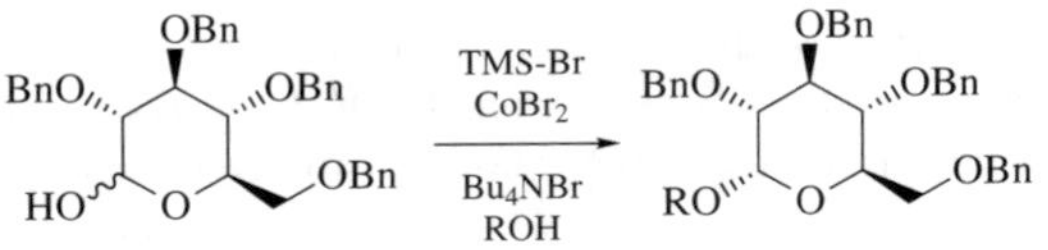

(9) TMS-Br, $CoBr_2$, Bu_4NBr, ROH

An interesting solvent effect was noted in the cleavage of the acetonide moiety in some nucleoside derivatives (eq 10).[13] In CH_2Cl_2, TMS-Br converted the acetonide to the anhydrouridine within 1.5 h, but in MeCN the bromide is formed after 10 min.

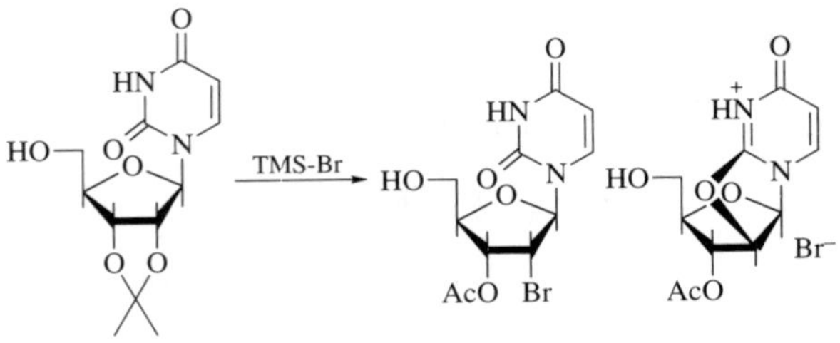

(10) TMS-Br

Formation of Enol Ethers.[14] Bromotrimethylsilane with ***Triethylamine*** in DMF is an effective medium for production of thermodynamic (*Z*) silyl enol ethers (eq 11).

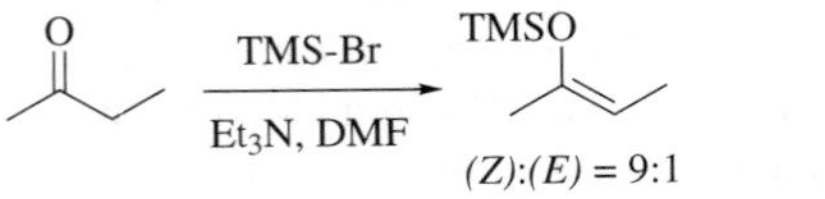

(11) TMS-Br, Et_3N, DMF; (*Z*):(*E*) = 9:1

Formation of Alkyl Bromides.[6b] Alcohols react with excess TMS-Br (1.5–4 equiv) at 25–50 °C to form the alkyl bromide and hexamethyldisiloxane (eq 12). Benzylic and tertiary alcohols react faster than secondary alcohols.

$$R\text{-}CH_2OH \xrightarrow{TMS\text{-}Br} R\text{-}CH_2Br \quad (12)$$

Reaction with Acid Chlorides.[15] Acid bromides may be prepared from acid chlorides by reaction with TMS-Br (eq 13).

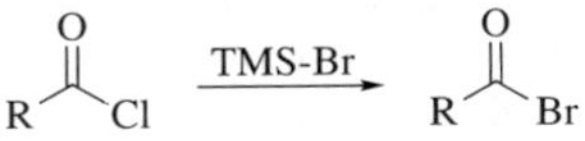

$$RC(O)Cl \xrightarrow{TMS\text{-}Br} RC(O)Br \quad (13)$$

Cleavage of Phosphonate Esters.[16] Compared to the reactivity of TMS-I with phosphonate and phosphate esters, TMS-Br is more selective and will cleave phosphonate esters even in the presence of carboxylic esters and carbamates. Benzyl ester protecting groups on aryl phosphates are selectively removed with TMS-Br.[17] The reaction of phosphonate esters with TMS-Br proceeds through a mechanism similar to ester cleavage, providing a silyl ester which is subsequently hydrolyzed with MeOH or H_2O (eq 14).

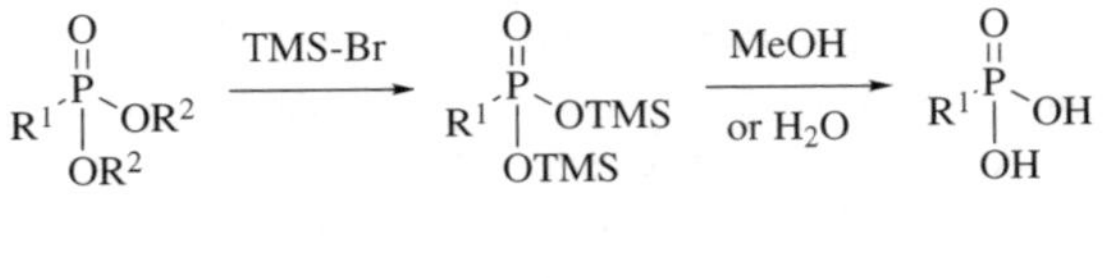

$$R^1P(O)(OR^2)_2 \xrightarrow{TMS\text{-}Br} R^1P(O)(OTMS)_2 \xrightarrow[\text{or } H_2O]{MeOH} R^1P(O)(OH)_2 \quad (14)$$

Reaction with Amines. Amines react with TMS-Br to form isolable adducts, which react readily with ketones to form enamines under mild conditions (eq 15).[18]

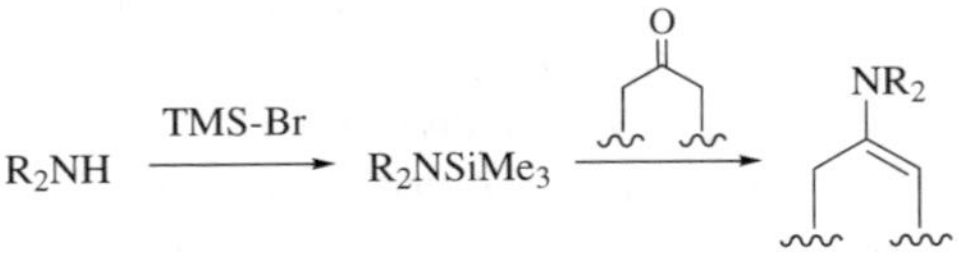

$$R_2NH \xrightarrow{TMS\text{-}Br} R_2NSiMe_3 \longrightarrow \text{enamine } (NR_2) \quad (15)$$

Conjugate Addition. α,β-Unsaturated ketones undergo conjugate addition with TMS-Br. Treatment of the intermediate with ***p-Toluenesulfonic Acid*** and ethylene glycol provides β-bromoethyldioxolanes (eq 16).[19]

1. TMS-Br; 2. p-TsOH, glycol (16)

Bromolactonization. ω-Unsaturated carboxylic acids react with TMS-Br in the presence of a tertiary amine in DMSO yielding bromolactones, resulting from *cis* addition across the double bond (eq 17).[20]

TMS-Br, Et_3N (17)

Ylide Formation. ***Methylenetriphenylphosphorane*** reacts with TMS-Br to provide the corresponding ylide (eq 18).[21]

$$Ph_3P{=}CH_2 \xrightarrow{\text{TMS-Br}} Ph_3\overset{+}{P}CH_2SiMe_3\ Br^- \qquad (18)$$

Related Reagents. Bromodimethylborane; Chlorotrimethylsilane; Hydrogen Bromide; Iodotrimethylsilane.

---

1. Schmidt, A. H. *Aldrichim. Acta* **1981**, *14* (2), 31.
2. Krüerke, U. *CB* **1962**, *95*, 174.
3. (a) Scheibye, S.; Thomsen, I.; Lawesson, S. O. *BSB* **1979**, *88*, 1043. (b) Olah, G. A.; Gupta, B. G. B.; Malhotra, R.; Narang, S. C. *JOC* **1980**, *45*, 1638. (c) Schmidt, A. H.; Russ, M. *CB* **1981**, *114*, 1099.
4. Sakurai, H.; Sasaki, K.; Hosomi, A. *TL* **1980**, *21*, 2329.
5. (a) Voronkov, M. G.; Dolgov, B. N.; Dmitrieva, N. A. *DOK* **1952**, *84*, 959. (b) Gross, H.; Böck, C.; Costisella, B.; Gloede, J. *JPR* **1978**, *320*, 344.
6. (a) Ho, T. L.; Olah, G. A. *S* **1977**, 417. (b) Jung, M. E.; Hatfield, G. L. *TL* **1978**, 4483.
7. Kricheldorf, H. R. *AG(E)* **1979**, *18*, 689.
8. Kricheldorf, H. R.; Mörber, G.; Regel, W. *S* **1981**, 383.
9. Hanessian, S.; Delorme, D.; Dufresne, Y. *TL* **1984**, *25*, 2515.
10. Woodward, R. B.; Logusch, E.; Nambiar, K. P.; Sakan, K.; Ward, D. E.; Au-Yeung, B.; Balaram, P.; Browne, L. J.; Card, P. J.; Chen, C. H.; Chênevert, R. B.; Fliri, A.; Frobel, K.; Gais, H.-J.; Garratt, D. G.; Hayakawa, K.; Heggie, W.; Hesson, D. P.; Hoppe, D.; Hoppe, I.; Hyatt, J. A.; Ikeda, D.; Jacobi, P. A.; Kim, K. S.; Kobuke, Y.; Kojima, K.; Krowicki, K.; Lee, V. J.; Leutert, T.; Malchenko, S.; Martens, J.; Matthews, R. S.; Ong, B. S.; Press, J. B.; Rajan Babu, T. V.; Rousseau, G.; Sauter, H. M.; Suzuki, M.; Tatsuta, K.; Tolbert, L. M.; Truesdale, E. A.; Uchida, I.; Ueda, Y.; Uyehara, T.; Vasella, A. T.; Vladuchick, W. C.; Wade, P. A.; Williams, R. M.; Wong, H. N.-C. *JACS* **1981**, *103*, 3213.
11. Gillard, J. W.; Israel, M. *TL* **1981**, *22*, 513.
12. Morishima, N.; Koto, S.; Kusuhara, C.; Zen, S. *CL* **1981**, 427.
13. Logue, M. W. *Carbohydr. Res.* **1975**, *40*, C9.
14. Ahmad, S.; Khan, M. A.; Iqbal, J. *SC* **1988**, *18*, 1679.
15. Schmidt, A. H.; Russ, M.; Grosse, D. *S* **1981**, 216.
16. (a) McKenna, C. E.; Schmidhauser, J. *CC* **1979**, 739. (b) Breuer, E.; Safadi, M.; Chorev, M.; Gibson, D. *JOC* **1990**, *55*, 6147.
17. Lazar, S.; Guillaumet, G. *SC* **1992**, *22*, 923.
18. Comi, R.; Franck, R. W.; Reitano, M.; Weinreb, S. M. *TL* **1973**, 3107.
19. Hsung, R. P. *SC* **1990**, *20*, 1175.
20. Iwata, C.; Tanaka, A.; Mizuno, H.; Miyashita, K. *H* **1990**, *31*, 987.
21. Seyferth, D.; Grim, S. O. *JACS* **1961**, *83*, 1610.

Michael J. Martinelli
Lilly Research Laboratories, Indianapolis, IN, USA

2-(*t*-Butoxycarbonyloxyimino)-2-phenylacetonitrile

[58632-95-4] $C_{13}H_{14}N_2O_3$ (MW 246.27)

(reagent for protection of amino acids as *t*-butyl carbamates[1–3] and for protection of alcohols as *t*-butyl carbonates[4])

Alternate Name: Boc-ON.
Physical Data: mp 87–89 °C.
Solubility: very sol ethyl acetate, ether, benzene, chloroform, dioxane and acetone; sol methanol, 2-propanol, and *t*-butanol; insol petroleum ether and water.
Form Supplied in: solid.
Handling, Storage, and Precautions: contact with the reagent may cause irritation. The toxicological properties of the reagent have not been thoroughly investigated. The reagent should be stored in a brown bottle at −20 °C to prolong shelf life. After several weeks at rt the reagent undergoes gradual decomposition with evolution of CO_2.

Amine Protection. The reagent, informally referred to as Boc-ON, has been widely employed for the protection of amines as their *t*-butoxycarbonyl (Boc) derivatives. The original description of Boc-ON reports its utility for the N-protection of α-amino acids (eq 1),[1] and this use is still the major application.[5]

Boc-ON + H_2N–CH(R)–CO_2H $\xrightarrow{\text{base}}$ t-BuO–C(O)–NH–CH(R)–CO_2H + HON=C(CN)Ph (1)

The reaction is generally carried out using 1.1 equiv of Boc-ON and 1.5 equiv of ***Triethylamine*** in either 50% aqueous dioxane or 50% aqueous acetone. Boc derivatives are obtained in high yields after 4–5 h at 20 °C or 1 h at 45 °C. The oxime byproduct (eq 1) can be easily and completely removed from the reaction mixture by extraction with ether, ethyl acetate, or benzene. To secure high purity and high yield, distillation of the dioxane or acetone prior to extraction of the oxime byproduct is recommended, though direct extraction may be feasible in most cases. Yields for various amino acids range from good to virtually quantitative (Table 1).

Table 1 Examples of Boc-Amino Acids Prepared with Boc-ON[2]

Boc-amino acid	Reaction time (h)	Oxime extraction solvent	Yield (%)
Ala-OH	4	Ether	80
Asn-OH	20	EtOAc	86
$Glu(OH)_2$	3	EtOAc	78
Gly-OH	2	EtOAc	87
Leu-OH·$0.5H_2O$	3	Ether	99
Met-OH·DCHA[a]	3	EtOAc	82
Phe-OH·DCHA[a]	5	Ether	98
Pro-OH	1.5	Benzene	88
Trp-OH	3	EtOAc	99

[a] DCHA = dicyclohexylamine.

Boc-ON offers a distinct advantage over *t*-Boc azide which can require reaction temperatures of 50–60 °C.[6] In addition, *t*-Boc azide is thermally unstable and decomposes with apparent detonation at temperatures above 80 °C. Boc-ON is comparable to ***Di-t-butyl Dicarbonate*** in terms of reaction times, yields, convenience, availability, and work-up procedure.[7]

Boc-ON has also been employed for the protection of secondary amines (eqs 2 and 3).[8,9]

MeHN–CH(Me)–CO_2H → (Boc-ON, 87%) → BocN(Me)–CH(Me)–CO_2H (2)

MeO_2C–$(CH_2)_4$-pyrrolidine (NH) → (Boc-ON, 93%) → MeO_2C–$(CH_2)_4$-pyrrolidine (N-Boc) (3)

The reagent will selectively protect the amino group in the presence of alcohols and phenols as well as carboxylic acids. Examples include the Boc protection of a kanamycin A derivative (eq 4)[10] and of the glycopeptide antibiotic A41030A (eq 5).[11]

Kanamycin A derivative (CF_3COHN, OH, NH_2, NHZ, ZHN) → (Boc-ON, THF–H_2O, Et_3N, overnight, 83%) → NHBoc derivative (4)

In the case of diamino acids, such as ornithine or lysine, Boc-ON has been used to effect selective protection of N^W at pH 11 (eqs 6 and 7).[12,13]

A41030A (NH_2) → (Boc-ON, Et_3N, acetone–MeOH (1:1), 5 h, ~80%) → NHBoc derivative (5)

H_2N–$(CH_2)_n$–CH(NH_2)–CO_2H, $n = 3, 4$ → (1. Boc-ON (1.1 equiv), pH 11, H_2O–dioxane (1:1), 24 h, rt; 2. ZCl; 68–77% overall) → BocHN–$(CH_2)_n$–CH(NHZ)–CO_2H (6)

Piperazine-2-carboxylic acid → (1. Boc-ON, pH 11, H_2O–dioxane (1:1); 2. ZCl, pH 9.5; 3. CH_2N_2; 53% overall) → 4-Boc-1-Z-piperazine-2-CO_2Me (7)

This selective protection has also been accomplished with copper-chelated diamino acids (eq 8).[14]

Cu chelate of H_2N–$(CH_2)_n$–diamino acid, $n = 1–3$ → (1. Boc-ON (2 equiv), H_2O–dioxane (1:1), NEt_3, 2 d; 2. EDTA; 54–75%) → BocHN–$(CH_2)_n$–CH(NH_2)–CO_2H (8)

Alcohol Protection. Boc-ON has also been employed to protect alcohols as their *t*-butyl carbonates.[4,15,16] For this application the alcohol must first be deprotonated to its alkoxide conjugate base (eqs 9 and 10).

R^1–CH(OH)–CH_2–CH=C(R^2)(R^3) → (1. BuLi; 2. Boc-ON, THF, 4 h, rt; 98%) → R^1–CH(OC(O)O-*t*-Bu)–CH_2–CH=C(R^2)(R^3) (9)

HO, H, OBn, OMe — 1. BuLi (1 equiv); 2. Boc-ON, THF, 4 h, rt, 99% — BocO, H, OBn, OMe (10)

Related Reagents. *N*-(*t*-Butoxycarbonyloxy)phthalimide; *N*-(*t*-Butoxycarbonyloxy)succinimide; *t*-Butyl Azidoformate; *t*-Butyl Chloroformate; Di-*t*-butyl Dicarbonate.

---

1. Itoh, M.; Hagiwara, D.; Kamiya, T. *TL* **1975**, *49*, 4393.
2. Itoh, M.; Hagiwara, D.; Kamiya, T. *BCJ* **1977**, *50*, 718.
3. Itoh, M.; Hagiwara, D.; Kamiya, T. *OS* **1980**, *59*, 95.
4. Barlett, P. A.; Meadows, J. D.; Ottow, E. *JACS* **1984**, *106*, 5304.
5. *FF* **1977**, *6*, 91.
6. *OS* **1977**, *57*, 122.
7. Persio, G.; Piani, S.; De Castiglione, R. *Int. J. Peptide Protein Res.* **1983**, *21*, 227.
8. Galardy, R. E. *B* **1987**, *21*, 5777.
9. Vaurecka, M.; Hesse, M. *HCA* **1989**, *72*, 847.
10. Matsuda, K.; Yasuda, N.; Tsutsumi, H.; Takaya, T. *J. Antibiot.* **1985**, *38*, 1719.
11. Hunt, A. H.; Dorman, D. E.; Debono, M.; Molloy, R. M. *JOC* **1985**, *50*, 2031.
12. Heimer, E. P.; Wang, C.-T.; Lambros, T. J.; Felix, A. M. *OPP* **1983**, *15*, 379.
13. Bigge, C. F.; Hays, S. J.; Novak, P. M.; Drummond, J. T.; Johnson, G.; Bobovski, T. P. *TL* **1989**, *30*, 5193.
14. Rosowsky, A.; Freisham, J. H.; Moran, R. G.; Solan, V. C.; Bader, H.; Wright, J. E.; Radike-Smith, M. *JMC* **1986**, *29*, 655.
15. Bartlett, P. A.; Meadows, J. D.; Brown, E. G.; Morimoto, A.; Jernstedt, K. K. *JOC* **1982**, *47*, 4013.
16. Martin, S. F.; Zinke, P. W. *JOC* **1991**, *56*, 6600.

Michael S. Wolfe
National Institutes of Health, Bethesda, MD, USA

Jeffrey Aubé
University of Kansas, Lawrence, KS, USA

t-Butyl Chloroformate

t-BuO–C(=O)–Cl

[24608-52-4] $C_5H_9ClO_2$ (MW 136.58)

(reagent for the protection of amino groups, including amino acids[2])

Alternate Name: Boc-Cl.
Physical Data: bp 3–4 °C/0.9–1.7 mmHg.[1]
Solubility: sol ether, toluene, and most organic aprotic solvents.
Form Supplied in: not commercially available; should be prepared from *t*-butanol[2] or potassium *t*-butoxide[1,3] and phosgene shortly before use.
Handling, Storage, and Precautions: unstable at rt. *t*-Butyl chloroformate should be kept as a solution in anhydrous solvent at low temperature (deep freezer) and for a period of time not exceeding a few days. Decomposition leads to high pressure in the storage vessel. May contain some residual phosgene. Reacts with DMSO or DMF.

Protection of Amino Groups. *t*-Butyl chloroformate (Boc-Cl), first prepared by Choppin and Rogers in 1948,[1] has been used, despite the reagent's instability, for the *t*-butyloxycarbonyl (Boc) protection of amino compounds and especially of amino acids. Boc protection has gained considerable importance because of the resistance of the protecting group to strong base hydrolysis and catalytic hydrogenation, as well as the ease of deprotection under mildly acidic conditions. Boc protection has become a fundamental tool of modern peptide synthesis and particularly of the Merrifield strategy for solid-phase peptide synthesis. Most amino acids have been protected with Boc-Cl using slightly modified Schotten–Baumann conditions (eq 1).[2]

R–CH(NH_2)–CO_2H + t-BuO–C(=O)–Cl — NaOH, MeOH–H_2O, –5 to 0 °C → R–CH(NH–CO_2-t-Bu)–CO_2H (1)

Reaction of pyridine compounds with Boc-Cl and ***Sodium Borohydride*** in methanol at −65 °C afforded the corresponding *N*-*t*-butyloxycarbonyldihydropyridine derivatives in good yield (eq 2).[4]

t-BuO–C(=O)–Cl, $NaBH_4$ (R¹, R², N–H, N–CO_2-t-Bu) (2)

The main drawback of the reagent remains its very low stability and the difficulties encountered in its preparation. Various preparative procedures are currently available in the literature.[5] It is also necessary to determine the reagent concentration in solution accurately, which can be done by reacting an aliquot with a selected amino compound before use. These liabilities notwithstanding, *t*-butyl chloroformate can be a cheap and valuable reagent for Boc protection and in some cases it is the best reagent available. This is, for instance, the case in the synthesis of Boc-protected *N*-carboxy anhydrides, which are used as building blocks in peptide synthesis (eq 3).[6] ***Di-t-butyl Dicarbonate*** required much longer reaction times in this case. These Boc-UNCAs (urethane-protected *N*-carboxy anhydrides) are, in fact, best prepared from the reaction of Boc-Cl with Leuchs anhydride in the presence of *N*-methylmorpholine or any nonnucleophilic organic base.

t-BuO–C(=O)–Cl + (R, HN, O) — THF, NMM → (R, Boc–N, O) (3)

Preparation of Other Reagents for Boc Protection. Because of its low stability, *t*-butyl chloroformate has mainly been used as a source of various other reagents that are easier to handle under ordinary conditions. ***t*-Butyl Azidoformate**, the most widely employed reagent before the discovery of ***Di-t-butyl Dicarbonate***, is conveniently obtained from Boc-Cl through the formation of *t*-butyl carbazate (eq 4).[7]

$$t\text{-BuO}\text{C(O)Cl} + H_2NNH_2 \xrightarrow[-60\ ^\circ C]{Et_2O} t\text{-BuO}\text{C(O)NHNH}_2 \xrightarrow{NaN_3} t\text{-BuO}\text{C(O)N}_3 \quad (4)$$

t-Butyl azidoformate is also obtained more directly from Boc-Cl and hydrazoic acid (eq 5).[8]

$$t\text{-BuO}\text{C(O)Cl} + HN_3 \xrightarrow[-78\ ^\circ C]{Et_2O} t\text{-BuO}\text{C(O)N}_3 \quad (5)$$

Various active carbonates, which are primarily used for the preparation of stable reagents for the introduction of the Boc protection, have been prepared from Boc-Cl.[9]

Exchange reactions utilizing hydrogen fluoride[10] and various salts such as sodium[11] and silver fluoride do not lead to the more stable *t*-butyl fluoroformate, which has to be prepared instead from carbonyl chlorofluoride or fluorophosgene.[12]

***t*-Butyl Carbonates and Carbamates.** Because Boc-Cl is potentially the cheapest reagent for the synthesis of *t*-butyl carbonates, e.g. (**1**) and (**2**), or carbamates, it has been claimed in numerous patents and may have growing utility in industrial processes, such as for the production of carbonates as components in thermal recording materials.[13]

(**1**) (**2**)

Related Reagents. Allyl Chloroformate; Benzyl Chloroformate; 4-Bromobenzyl Chloroformate; *t*-Butyl Azidoformate; 1-(*t*-Butoxycarbonyl)-1*H*-benzotriazole 3-*N*-oxide; 1-(*t*-Butoxycarbonyl)imidazole; 2-(*t*-Butoxycarbonyloxyimino)-2-phenylacetonitrile; *N*-(*t*-Butoxycarbonyloxy)phthalimide; *N*-(*t*-Butoxycarbonyloxy)succinimide; *N*-(*t*-Butoxycarbonyl)-1,2,4-triazole; Di-*t*-butyl Dicarbonate; Ethyl Chloroformate; 9-Fluorenylmethyl Chloroformate; Isobutyl Chloroformate; Methyl Chloroformate; 2,2,2-Tribromoethyl Chloroformate; 2,2,2-Trichloroethyl Chloroformate; 2,2,2-Trichloro-*t*-butoxycarbonyl Chloride; 2-(Trimethylsilyl)ethyl Chloroformate; Vinyl Chloroformate.

1. Choppin, A. R.; Rogers, J. W. *JACS* **1948**, *70*, 2967.
2. Sakakibara, S.; Shin, M.; Fujino, Y.; Shimonishi, Y.; Inouye, S.; Inukai, N. *BCS* **1969**, *42*, 809.
3. Howard, J. C. *JOC* **1981**, *46*, 1720.
4. Dagnino, L.; Li-Kwong-Ken, M. C.; Wynn, H.; Wolowyk, M. W.; Triggle, C. R.; Knaus, E. E. *JMC* **1987**, *30*, 640.
5. Ger. Patent 2 108 782 (*CA* **1967**, *76*, 24 715).
6. Fuller, W. D.; Cohen, M. P.; Shabankareh, M.; Blair, R. B.; Goodman, M.; Naider, F. *JACS* **1990**, *112*, 7414.
7. Ovchinnikov, Y. A.; Kiryushkin, A. A.; Miroshnikov, A. I. *E* **1965**, *21*, 418.
8. Yajima, H.; Kawatani, H. *CPB* **1968**, *16*, 183.
9. Rzeszotarska, B.; Wiejak, S.; Pawelczak, K. *OPP* **1973**, *5*, 71.
10. Olah, G. A.; Kuhn, S. J. *JOC* **1961**, *26*, 237.
11. Olah, G. A.; Kuhn, S. J. *CB* **1956**, *89*, 862.
12. Schnabel, E.; Herzog, H.; Hoffmann, P.; Klauke, E.; Ugi, I. *AG(E)* **1968**, *7*, 380.
13. Jpn. Patent 04 213 368 (*CA* **1993**, *118*, 14 048).

G. Sennyey
SNPE, Vert-le-Petit, France

t-Butyldimethylchlorosilane

t-$BuMe_2SiCl$

[18162-48-6] $C_6H_{15}ClSi$ (MW 150.72)

(widely used reagent for the protection of alcohols, amines, carboxylic acids, ketones, amides, thiols, and phenols;[1] useful for regioselective silyl enol ether formation[2] and stereoselective silyl keten acetal formation[3])

Alternate Names: *t*-butyldimethylsilyl chloride; TBDMSCl; TBSCl.
Physical Data: mp 86–89 °C; bp 125 °C.
Solubility: very sol nearly all common organic solvents such as THF, methylene chloride, and DMF.
Form Supplied in: moist white crystals, commonly available.
Handling, Storage, and Precautions: hygroscopic, store under N_2; harmful if inhaled, swallowed, or absorbed through skin; should be used and weighed out in a fume hood.

Protecting Group. The reactions of TBDMSCl closely parallel those of ***Chlorotrimethylsilane*** (TMSCl). However, TBDMS ethers are about 10^4 more stable toward hydrolysis than the corresponding TMS ethers.[1,4] The hydrolytic stability of the TBDMS group has made it very valuable for the isolation of many silicon-containing molecules. Since its introduction in 1972, the TBDMS protecting group has undoubtedly become the most widely used silicon protecting group in organic chemistry.[1] Alcohols are most commonly protected as their TBDMS ethers by treatment of the alcohol in DMF (2 mL g^{-1}) at rt with 2.5 equiv of ***Imidazole*** (Im) and 1.2 equiv of TBDMSCl. Alcohol protection in the presence of ***4-Dimethylaminopyridine*** (DMAP) allows a greater range of solvents to be used (protection of alcohols in solvents other than DMF using TBDMSCl alone are sluggish) and distinguishes a kinetic preference for protection of primary alcohols in the presence of secondary alcohols.[5] Table 1 outlines conditions for the selective protection of primary and secondary alcohols using TBDMSCl and DMAP as catalyst.

Table 1 Use of DMAP to Catalyze the Protection of Diols With TBDMSCl

Amine(s)	Solvent	GC ratio of A:B:C
DMAP (0.04 equiv), Et_3N (1.1 equiv)	CH_2Cl_2	95:0:5
Im (2.2 equiv)	DMF	59:11:30
Im (0.04 equiv), Et_3N (1.1 equiv)	DMF	No reaction
DMAP (0.04 equiv), Et_3N (1.1 equiv)	DMF	0:0:96

Alcohol:

OH, Ph, OH

Products:

A: OH, Ph, OTBDMS; B: OTBDMS, Ph, OH; C: OTBDMS, Ph, OTBDMS

Table 2[6] compares the rate of hydrolysis of several bulky silicon ethers with acid, base, and fluoride; TBDMS ethers are less resistant to hydrolysis than the corresponding TIPS (triisopropylsilyl) and TBDPS (*t*-butyldiphenylsilyl) ethers.

Table 2 Comparison of TBDMS, TIPS, and TBDPS rates of hydrolysis by F^-, H^+, and OH^- ($t_{1/2}$).

	R = *n*-Bu[a]		R = Cy[a,b]		
$ROSiR'_3$	H^+	OH^-	H^+	OH^-	F^-
TBDMS	<1 min	1 h	<4 min	26 h	76 min
TIPS	18 min	14 h	100 min	44 h	137 h
TBDPS	244 min	<4 h	360 min	14 h	–

[a] H^+ refers to 1% HCl in 95% EtOH at 22 °C; OH^- refers to 5% NaOH in 95% EtOH at 90 °C. [b] F^- refers to 2 equiv of *n*-Bu_4NF in THF at 22 °C.

The TBDMS group is also suitable for the protection of amines, including heterocycles, carboxylic acids, and phenols. Other more reactive reagents are also available for introduction of the TBDMS group including ***t-Butyldimethylsilyl Trifluoromethanesulfonate***, MTBSA, and TBDMS-imidazole.[1,7]

Anion Trap. TBDMSCl is useful as an anion trapping reagent. For example, TBDMSCl was found to be an efficient trap of the lithio α-phenylthiocyclopropane anion.[8] When dichlorothiophene was treated with 2 equiv of ***n-Butyllithium*** followed by 2 equiv of TBDMSCl the di-TBDMS-thiophene was isolated.[9] The lithium anions of primary (eq 1) and secondary (eq 2) nitriles were trapped with TBDMSCl to give *C,N*-disilyl- and *N*-silylketenimines in excellent yields.[10]

CN → (1. 2 equiv LDA, THF, –78 °C; 2. 4 equiv TBDMSCl; 100%) → TBDMS, NTBDMS (1)

CN → (1. 1 equiv LDA, THF, –78 °C; 2. 2 equiv TBDMSCl; 100%) → NTBDMS (2)

Silyl stannanes have been prepared by trapping tin anions with TBDMSCl or other silyl chlorides. Alkynes treated with with silyl stannanes and catalytic ***Tetrakis(triphenylphosphine)palladium(0)*** give *cis*-silyl stannylalkenes in good yields.[11]

Silyl Ketene Acetals. The lithium enolates of esters may be trapped with TBDMSCl to prepare the corresponding ketene silyl acetals.[3] The resulting TBDMS ketene acetals are more stable than the corresponding TMS ketene acetals and have a greater preference for *O*- vs. *C*-silylation products. When TMSCl was used to trap the enolate of methyl acetate, a 65:35 ratio of *O*- to *C*-silated products was obtained. In addition, *O*-(TMS) silyl ketene acetals are thermally and hydrolytically unstable. However, similar treatment of lithium enolates with TBDMSCl provided the corresponding *O*-(TBDMS) silyl ketene acetals exclusively (eq 3). The *O*-TBDMS ketene acetals generally survive extraction from cold aqueous acid. ***Lithium Diisopropylamide*** was found to be satisfactory for the preparation of the ester enolates. The lower reactivity of TBDMSCl requires that the enol silation be performed at 0 °C with added HMPA.[12]

O, OEt → (1. LICA, THF, –78 °C; 2. HMPA, TBDMSCl, –78 to 0 °C; >90%) → OTBDMS, OEt (3)

A detailed study of the formation of (*E*)- and (*Z*)-silyl ketene acetals was recently published.[3d] It was found that the formation of silyl enolates does not correspond to simple kinetic vs. thermodynamic formation of the enolates. Formation of the ester enolates occurred under kinetic control and a kinetic resolution accounted for selective formation of (*E*)- and (*Z*)-silyl ketene acetals. Table 3 summarizes some of these results.

Table 3 Effects of Reaction Conditions on (*Z*):(*E*) Ratio of TBDMS Silyl Ketene Acetal Formation of Ethyl Propionate

Ester:base ratio	Base	Solvent	Additive	(*Z*):(*E*) Silyl ketene acetal ratio	Yield (%)
1:1	LDA	THF	–	6:94	90
1.05:1	LDA	THF	45% DMPU	≥98:2	80
1.2:1	LDA	THF	23% HMPA	93:7	65
1:1	LDA	THF	23% HMPA	85:15	80
1:1	LHMDS	THF	23% HMPA	>91:9	85
1.1:1	LHMDS	THF	23% HMPA	>95:5	60

OTBDMS, O (*Z*); OTBDMS, O (*E*)

Claisen Rearrangement. The first silyl ketene acetal Claisen rearrangement was introduced in 1972 using TMS ketene acetals. Since then, the silyl Claisen rearrangement using TBDMS ketene acetals has found widespread use in organic synthesis.[13,14] One advantage of the silyl ketene acetal Claisen rearrangement is that the ketene acetal geometry may be predictably controlled (see above). Two components of the reaction contribute to stereocontrol: the geometry of the silyl ketene acetal and the contribution of boat vs. chair transition state. A useful variant of the Claisen rearrangement involves the use of an enantiomerically pure α-silyl secondary alcohol prepared by Brook rearrangement of a TBDMS-protected primary alcohol. In this reaction the stereochemistry at the silicon-bearing center is transferred to the Claisen product.[15] The (*E*)-enolate was prepared by treatment of the ester with ***Lithium Hexamethyldisilazide*** (eq 4); the (*Z*)-enolate was prepared by treatment of the ester with LDA (eq 5).[15b] As expected, the (*Z*)- and (*E*)-silyl ketene acetals gave the corresponding *syn* and *anti* Claisen products in good selectivity. The vinylsilane was hydrolyzed to the alkene using 50% HBF_4 in acetonitrile at 55 °C.

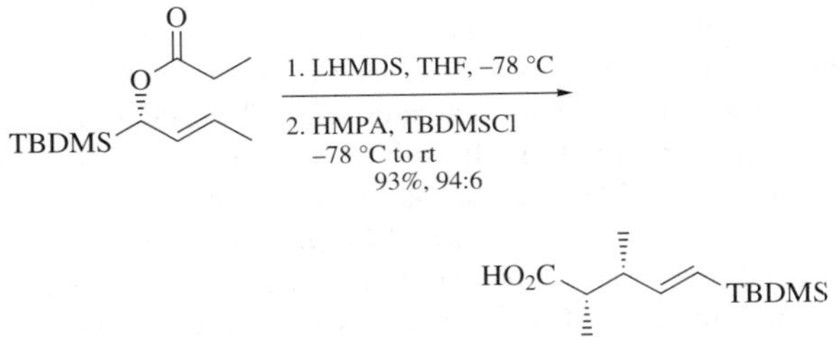

TBDMS … 1. LHMDS, THF, −78 °C; 2. HMPA, TBDMSCl, −78 °C to rt; 93%, 94:6 → HO_2C … TBDMS (4)

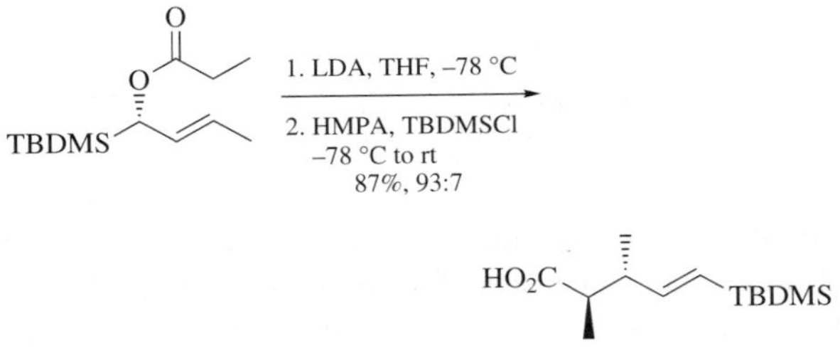

TBDMS … 1. LDA, THF, −78 °C; 2. HMPA, TBDMSCl, −78 °C to rt; 87%, 93:7 → HO_2C … TBDMS (5)

The silyl ketene acetals of methyl α-(allyloxy)acetates were found to undergo [3,3]-sigmatropic rearrangement, whereas the corresponding lithium enolates undergo [2,3]-sigmatropic rearrangement.[16] An interesting ring contraction based on the TBDMS silyl ketene acetal Claisen rearrangement has also been reported.[17]

TBDMS Enol Ethers.[18] Enolates trapped with TBDMSCl to prepare the corresponding enol ethers are more stable than the corresponding TMS enol ethers.[19] The potassium enolate of 2-methylcyclohexanone, prepared by addition of ***Potassium Hydride*** to a solution of the ketone and TBDMSCl in THF at −78 °C followed by warming to rt, gave the thermodynamic enol ether in a 56:44 ratio. In the presence of HMPA, the ratio improves to 98:2 (eq 6). This method works especially well with ketones with a propensity for self-condensation.[20]

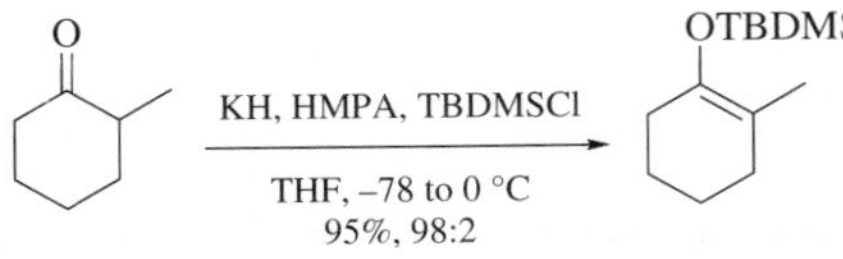

KH, HMPA, TBDMSCl; THF, −78 to 0 °C; 95%, 98:2 → OTBDMS (6)

Potassium enolates derived from acylfulvalenes were trapped with TBDMSCl but not TMSCl or diphenylmethylsilyl chloride.[21] Interestingly, TBDMSCl was found to be compatible with CpK anion at −78 °C. TBDMS enol ethers have also been used as β-acyl anion equivalents.[22] The TBDMS-silyl enol ethers of diketones (eq 7) and β-keto esters (eq 8) may be prepared by mixing them with TBDMSCl in THF with imidazole.[23] Alcohols may be protected under acidic conditions as their TBDMS ethers by treatment with β-silyl enol ethers in polar solvents.

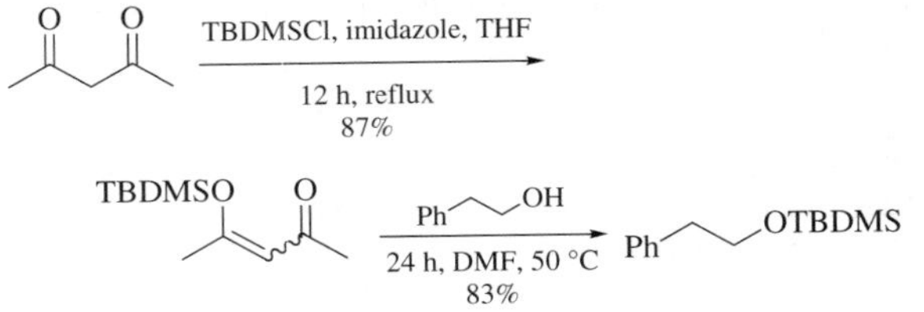

TBDMSCl, imidazole, THF; 12 h, reflux; 87% → TBDMSO … O; Ph⌒OH; 24 h, DMF, 50 °C; 83% → Ph⌒OTBDMS (7)

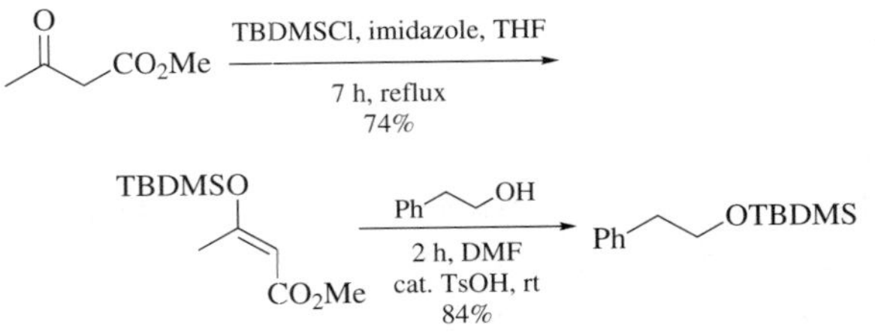

TBDMSCl, imidazole, THF; 7 h, reflux; 74% → TBDMSO … CO_2Me; Ph⌒OH; 2 h, DMF; cat. TsOH, rt; 84% → Ph⌒OTBDMS (8)

Aldol Reaction. The catalyst system $TBDMSCl/InCl_3$ selectively activates aldehydes over acetals for aldol reactions with TBDMS enol ethers.[24] Acetals and aldehydes are activated towards aldol reactions using $TMSCl/InCl_3$ or $Et_3SiCl/InCl_3$ as catalysts (eq 9).

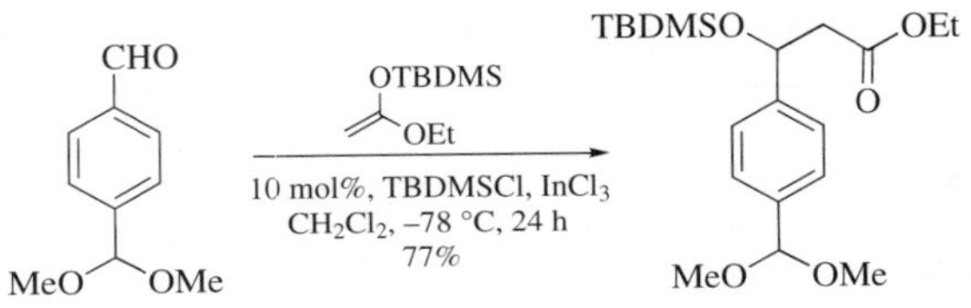

CHO … MeO OMe; OTBDMS, OEt; 10 mol%, TBDMSCl, $InCl_3$; CH_2Cl_2, −78 °C, 24 h; 77% → TBDMSO … OEt, O … MeO OMe (9)

TBDMSCl as Cl^- Source. TBDMSCl was used as a source of chloride ion in the Lewis acid-assisted opening of an epoxide.[25] The epoxide was treated with TBDMSCl and ***Triethylamine*** followed by ***Titanium Tetraisopropoxide*** and additional TBDMSCl to give the *trans* chloride as the major product in 67% yield (eq 10).

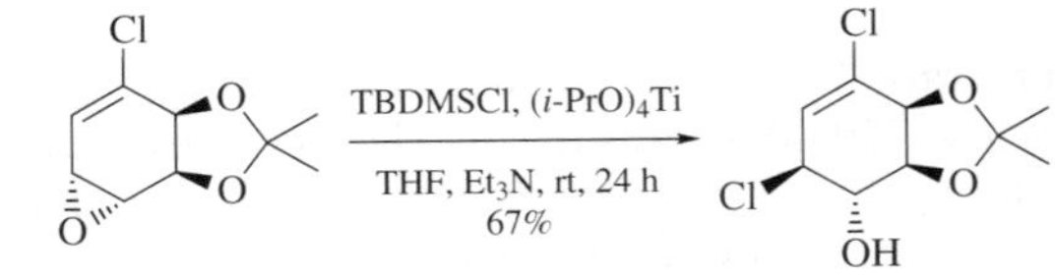

Cl … O; TBDMSCl, (*i*-PrO)$_4$Ti; THF, Et_3N, rt, 24 h; 67% → Cl … Cl, OH (10)

TBDMSCl-Assisted Reactions. Nitro aldol (Henry) reactions have been reported to be promoted by TBDMSCl.[26] To a THF solution of ***Tetra-n-butylammonium Fluoride*** is added sequentially equimolar amounts of the nitro compound, aldehyde, and Et_3N, followed by an excess of TBDMSCl (eq 11). Substitution of TMSCl for TBDMSCl reduces the yield of nitro aldol product. The authors speculate that TBDMSCl is responsible for activation of the aldehyde while n-Bu_4NF activates the nitro compound. In a related method, primary and secondary nitro alkanes were treated with LDA in THF followed by addition of TBDMSCl to give the corresponding silyl nitronates. The silyl nitronates reacted with a variety of aliphatic and aromatic aldehydes which gave vicinal nitro TBDMS aldol products.[27]

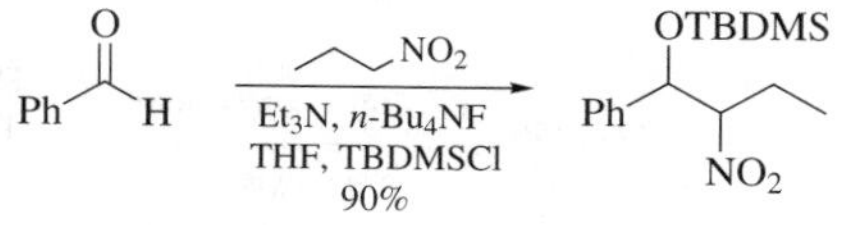

(11)

Reaction with Nucleophiles. TBDMSCl is the reagent of choice for the preparation of other TBDMS-containing reagents. For example, ***t-Butyldimethylsilyl Cyanide*** may be prepared by the reaction of TBDMSCl and ***Potassium Cyanide*** in acetonitrile containing a catalytic amount of ***Zinc Iodide***.[28] TBDMSCN has also been prepared by treatment of TBDMSCl with KCN and ***18-Crown-6*** in CH_2Cl_2 at reflux and by treatment of TBDMSCl with ***Lithium Cyanide*** prepared in situ.[29] ***t-Butyldimethylsilyl Trifluoromethanesulfonate*** is prepared by treatment of TBDMSCl with ***Trifluoromethanesulfonic Acid*** at 60 °C.[30] Other nucleophiles, such as thiolates, also react with TBDMSCl.[31] ***t-Butyldimethylsilyl Iodide*** was prepared by treatment of TBDMSCl with ***Sodium Iodide*** in acetonitrile.[32] In contrast to THF cleavage reactions using TMSI, the more stable TBDMS-protected primary alcohol may be isolated from the reaction in eq 12.

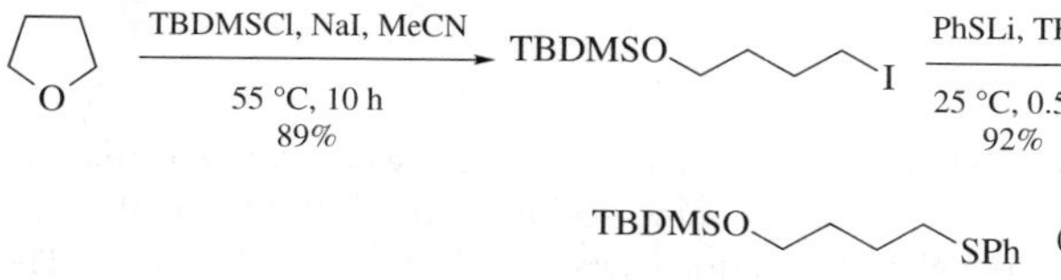

(12)

Mannich Reaction. The Mannich reaction of *N*-methyl-1,3-oxazolidine with 2-methylfuran was shown to proceed smoothly in the presence of TBDMSCl and catalytic ***1,2,4-Triazole*** in 61% yield. Interestingly, this reaction failed with TBDMSOTf due to the destruction of 2-methylfuran. The reaction proceeds with decreased yield (31%) in the absence of triazole. These reaction conditions allowed for the isolation of the TBDMS-protected alcohol (eq 13).[33]

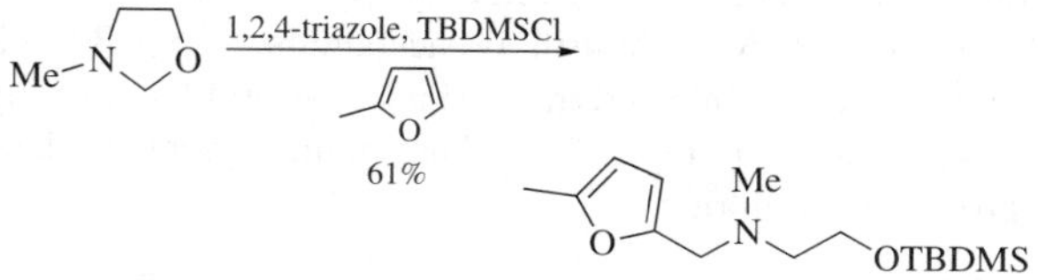

(13)

Acid Chlorides. TBDMS esters, when treated with DMF and ***Oxalyl Chloride*** in methylene chloride at 0 °C, give the corresponding acid chlorides in excellent yields under neutral conditions (eq 14).[34] Similarly, *N*-carboxyamino acid anhydrides were prepared via the intermediacy of an acid chloride prepared from a TBDMS ester (eq 15).[35]

Conjugate Additions. When a mixture of TBDMSCl and a β-aryl enone (2:1) was added to $Bu_2CuCNLi_2$ at −78 °C, the TBDMS group added 1,4 to the enone to give β-silyl carbonyl compounds (eq 16).[36] *N*-TBDMS silyliminocuprates also add to α,β-unsaturated carbonyl compounds.[37]

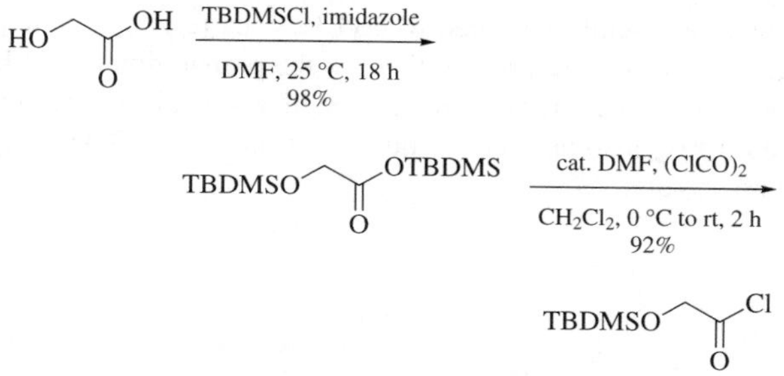

(14)

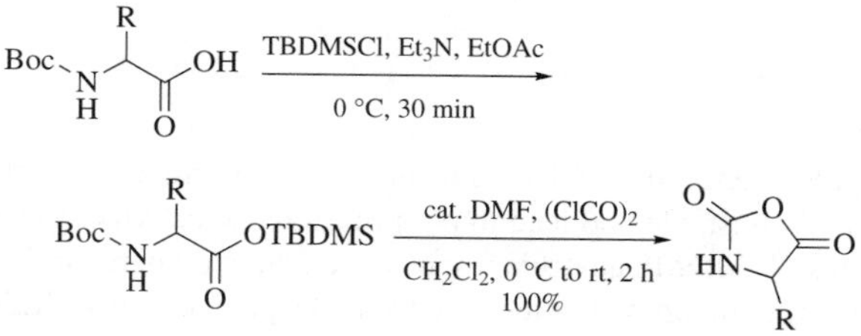

(15)

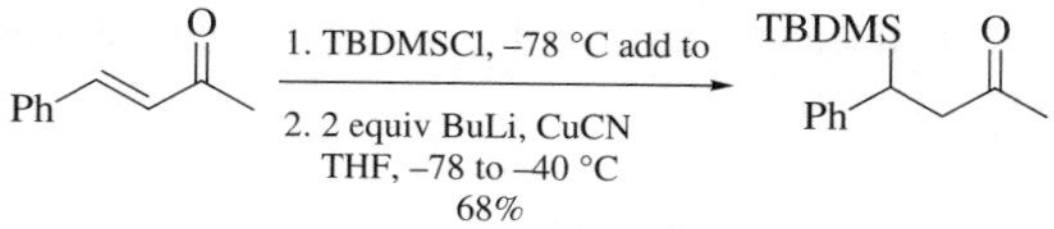

(16)

α-Silyl Aldehydes. Initial attempts to isolate TMS α-silyl aldehydes were unsuccessful due to the lability of the TMS group. However, the α-silyl aldehyde was prepared from the cyclohexyl imine of acetaldehyde by treatment with LDA followed by TBDMSCl (eq 17). Typical of TBDMSCl trapping reactions of imines and hydrazones, *C*-silylation was observed. The imine was hydrolyzed with HOAc in CH_2Cl_2 which gave the α-silyl aldehyde. These compounds, after treatment with organometallic reagents such as ***Ethylmagnesium Bromide*** or ***Ethyllithium***, may be eliminated in a Peterson-like manner to give either *cis* or *trans* alkenes (eq 18).[38]

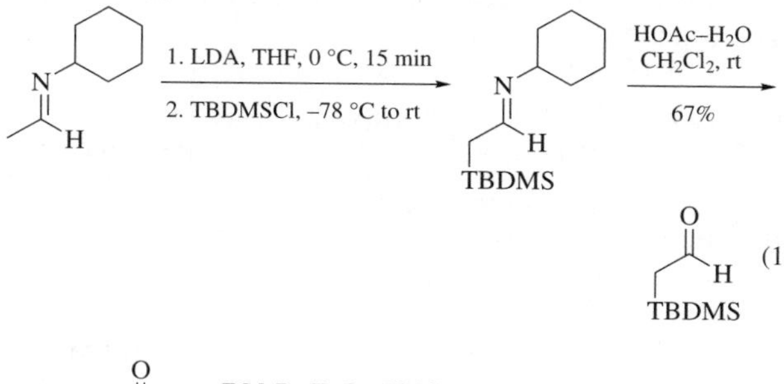

(17)

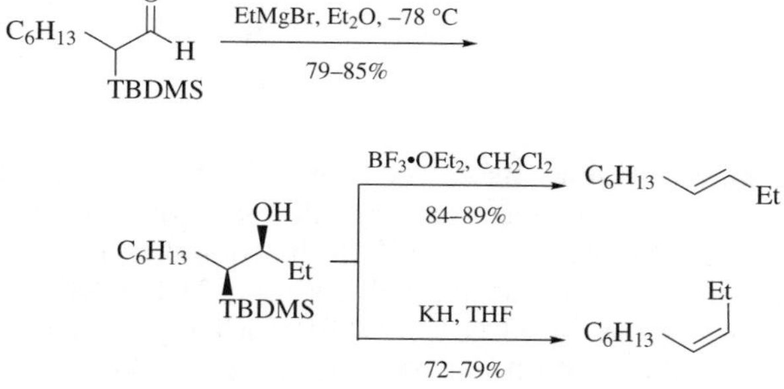

(18)

α-Silyl Ketones. When SAMP or RAMP hydrazones were treated with LDA followed by TBDMSCl, the corresponding α-silyl hydrazones were isolated. Ozonolysis of the hydrazone gave the enantiomerically enriched α-silyl ketones (eq 19).[39] Yields for the overall process are 52–79% for the preparation of α-silyl ketones and 22–42% for the preparation of α-silyl aldehydes. TBDMSOTf may also be used for quench of the SAMP/RAMP hydrazone enolate.

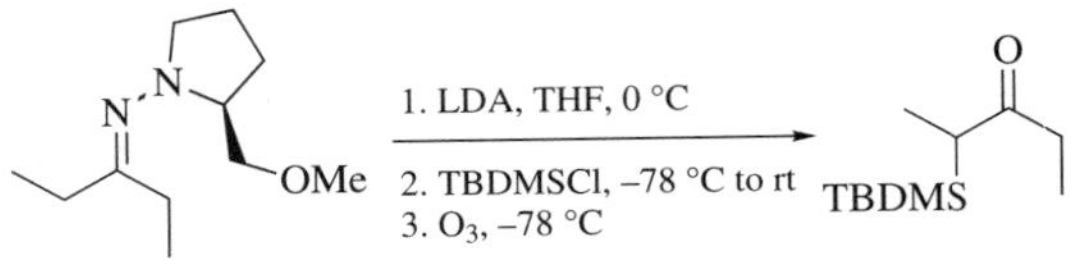

Acyl Silanes. Although acyl trimethylsilanes are known, they are usually unstable and lead to poor diastereoselectivity in aldol reactions.[40] TBDMS acyl silanes, however, were prepared in 50% yield from 1-methoxy-1-lithiopropene in the presence of TMEDA at rt (eq 20). The lithium enolates of TBDMS acyl silanes were treated with aldehydes to give the corresponding aldol products in reasonable yields.

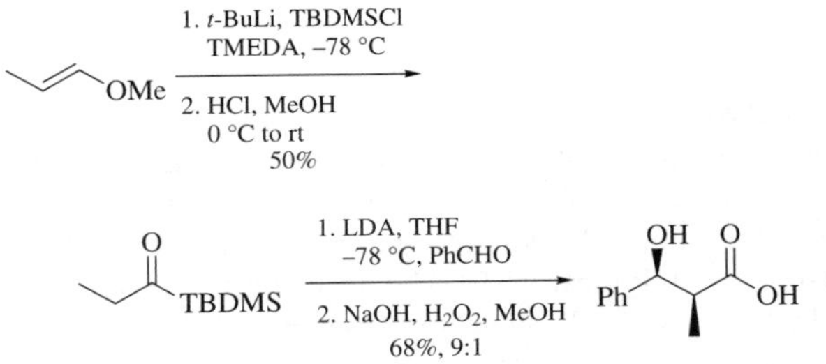

TBDMS acyl enones were prepared by treatment of an ethoxyethyl (EE)-protected alkoxyallene with *n*-BuLi at −85 °C followed by treatment of resulting anion with TBDMSCl (eq 21). Acid hydrolysis of the OEE group led to the TBDMS acyl enones in good yield.[41]

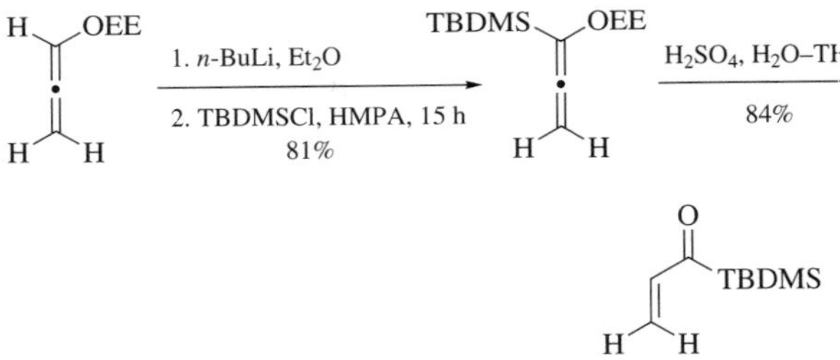

Modified Amine Base. The regioselectivity of ketone deprotonation was improved by the use of lithium *t*-butyldimethylsilylamide as base.[42] The base was prepared by deprotonation of isopropylamine with *n*-BuLi in THF (eq 22). The resulting anion was quenched with TBDMSCl to give the amine in 70% yield after distillation. Deprotonation of various ketones using this amide base was found to be equally or more selective than LDA. For example, the TBDMS-modified base gave a 62:38 ratio of kinetic to thermodynamic enolate, whereas LDA gave a 34:66 ratio with phenyl acetone.

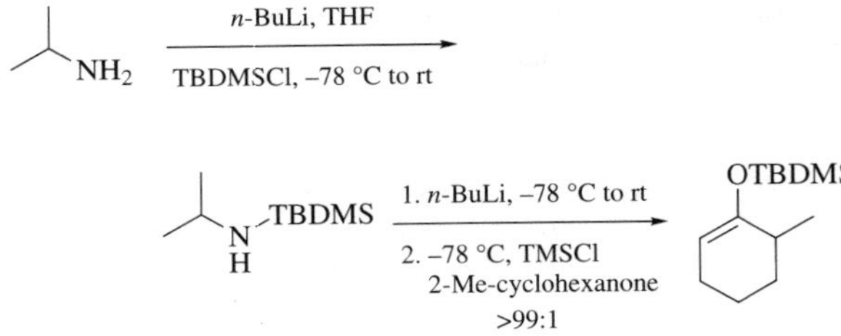

***N*-Formylation.** When secondary amines were treated with TBDMSCl, DMAP, and Et_3N in DMF the corresponding *N*-formyl derivatives were formed (eq 23).[43] It was found that the reaction proceeds through a Vilsmeier type reagent formed by the reaction of TBDMSCl and DMF. It is possible that other TBDMS alkylation reactions, such as protection of alcohols in DMF, may proceed through a similar DMF-derived Vilsmeier reagent.

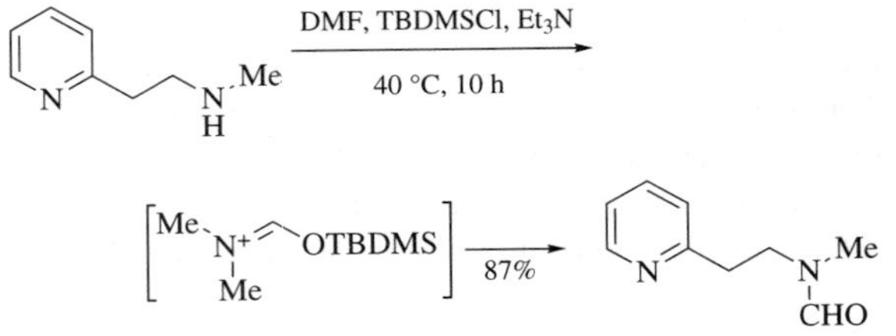

***N*-Silyl Imines.** When aldehydes were treated first with tris(trimethylstannyl)amine followed by TBDMSCl, the corresponding *N*-TBDMS imines were isolated in good yields.[44] These silyl imines reacted with ester enolates to give β-lactams (eq 24).

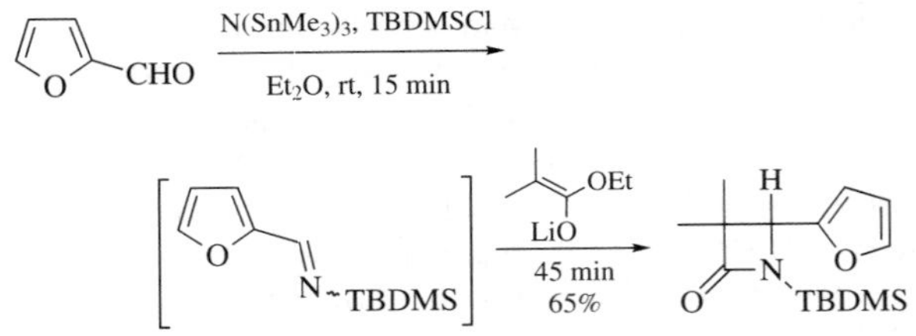

Related Reagents. *t*-Butyldimethylsilyl Cyanide; *N*-(*t*-Butyldimethylsilyl)-*N*-methyltrifluoroacetamide; *t*-Butyldiphenylchlorosilane; Chlorotriphenylsilane; *t*-Butyldimethylsilyl Trifluorosulfonate; Triethylchlorosilane; Trimethylchlorosilane; Triisopropylsilyl Chloride.

1. (a) Corey, E. J.; Venkateswarlu, A. *JACS* **1972**, *94*, 6190. (b) Lalonde, M.; Chan, T. H. *S* **1985**, 817. (c) Greene, T. W.; Wuts, P. G. M. *Protective Groups in Organic Synthesis*; Wiley: New York, 1991. (d) Colvin, E. *Silicon in Organic Synthesis*; Butterworths: London, 1981.
2. Stork, G.; Hudrlik, P. F. *JACS* **1968**, *90*, 4462.
3. (a) Rathke, M. W.; Sullivan, D. F. *SC* **1973**, *3*, 67. (b) Rathke, M. W.; Sullivan, D. F. *TL* **1973**, 1297. (c) An interesting application of TBS silyl ketene acetals as homo-Reformatsky reagents may be found in: Oshino, H.; Nakamura, E.; Kuwajima, I. *JOC* **1985**, *50*, 2802. (d) Ireland, R. E.; Wipf, P.; Armstrong, J. D., III *JOC* **1991**, *56*, 650.
4. For general papers comparing the stability of silanes containing various alkyl groups on Si see: (a) Sommer, L. H.; Tyler, L. J. *JACS* **1954**, *76*, 1030. (b) Ackerman, E. *ACS* **1956**, *10*, 298; *ACS* **1957**, *11*, 373.
5. Hernandez, O.; Chaudhary, S. K. *TL* **1979**, 99.

6. Cunico, R. F.; Bedell, L. *JOC* **1980**, *45*, 4797.
7. Mawhinney, T. P.; Madson, M. A. *JOC* **1982**, *47*, 3336.
8. Wells, G. J.; Yan, T.-H.; Paquette, L. A. *JOC* **1984**, *49*, 3604.
9. Okuda, Y.; Lakshmikantham, M. V.; Cava, M. P. *JOC* **1991**, *56*, 6024.
10. Watt, D. S. *SC* **1974**, *4*, 127.
11. Chenard, B. L.; Van Zyl, C. M. *JOC* **1986**, 3561.
12. Ireland, R. E.; Mueller, R. H. *JACS* **1972**, *94*, 5897.
13. Ireland, R. E.; Mueller, R. H.; Willard, A. K. *JACS* **1976**, *98*, 2868.
14. For other examples see (a) Mohammed, A. Y.; Clive, D. L. J. *CC* **1986**, 588. (b) Kita, Y.; Shibata, N.; Miki, T.; Takemura, Y.; Tamura, O. *CC* **1990**, 727. (c) Metz, P.; Mues, C. *SL* **1990**, 97.
15. (a) Ireland, R. E.; Varney, M. D. *JACS* **1984**, *106*, 3668. (b) Ireland, R. E.; Daub, J. P. *JOC* **1981**, *46*, 479.
16. Raucher, S.; Gustavson, L. M. *TL* **1986**, *27*, 1557.
17. (a) Abelman, M. M.; Funk, R. L.; Munger, J. D., Jr. *JACS* **1982**, *104*, 4030. (b) Funk, R. L.; Munger, J. D., Jr. *JOC* **1984**, *49*, 4320.
18. For a review see: Brownbridge, P. *S* **1983**, 1; *S* **1983**, 29.
19. (a) Ireland, R. E.; Courtney, L.; Fitzsimmons, B. J. *JOC* **1983**, *48*, 5186. (b) Piers, E.; Burmeister, M. S.; Reissig, H.-U. *CJC* **1986**, *64*, 180.
20. (a) Orban, J.; Turner, J. V.; Twitchin, B. *TL* **1984**, *25*, 5099. (b) Orban, J.; Turner, J. V. *TL* **1983**, *24*, 2697. (c) Ireland, R. E.; Thompson, W. J.; Mandel, N. S.; Mandel, G. S. *JOC* **1979**, *44*, 3583.
21. McLoughlin, J. I.; Little, R. D. *JOC* **1988**, *53*, 3624.
22. Trimitsis, G.; Beers, S.; Ridella, J.; Carlon, M.; Cullin, D.; High, J.; Brutts, D. *CC* **1984**, 1088.
23. Veysoglu, T.; Mitscher, L. A. *TL* **1981**, *22*, 1299.
24. Mukaiyama, T.; Ohno, T.; Han, J. S.; Kobayashi, S. *CL* **1991**, 949.
25. Hudlicky, T.; Luna, H.; Olivo, H. F.; Andersen, C.; Nugent, T.; Price, J. D. *JCS(P1)* **1991**, 2907.
26. Fernández, R.; Gasch, C.; Gómez-Sánchez, A.; Vílchez, J. E. *TL* **1991**, *32*, 3225.
27. Colvin, E. W.; Beck, A. K.; Seebach, D. *HCA* **1981**, *64*, 2264.
28. Rawal, V. H.; Rao, J. A.; Cava, M. P. *TL* **1985**, *26*, 4275.
29. (a) Gassman, P. G.; Haberman, L. M. *JOC* **1986**, *51*, 5010. (b) Mai, K.; Patil, G. *JOC* **1986**, *51*, 3545.
30. Corey, E. J.; Cho, H.; Rücker, C.; Hua, D. H. *TL* **1981**, *22*, 3455.
31. Aizpurua, J. M.; Paloma, C. *TL* **1985**, *26*, 475.
32. (a) Nyström, J.-E.; McCanna, T. D.; Helquist, P.; Amouroux, R. *S* **1988**, 56. (b) Detty, M. R.; Seidler, J. D. *JOC* **1981**, *46*, 1283.
33. Fairhurst, R. A.; Heaney, H.; Papageorgiou, G.; Wilkins, R. F.; Eyley, S. C. *TL* **1989**, *30*, 1433.
34. Wissner, A.; Grudzinskas, C. V. *JOC* **1978**, *43*, 3972.
35. Mobashery, S.; Johnston, M. *JOC* **1985**, *50*, 2200.
36. Amberg, W.; Seebach, D. *AG(E)* **1988**, 1718.
37. (a) Murakami, M.; Matsuura, T.; Ito, Y. *TL* **1988**, *29*, 355. (b) Ager, D. J.; Fleming, I.; Patel, S. K. *JCS(P1)* **1981**, 2520.
38. Hudrlik, P. F.; Kulkarni, A. K. *JACS* **1981**, *103*, 6251.
39. (a) Lohray, B. B.; Enders, D. *HCA* **1989**, *72*, 980. (b) Enders, D.; Lohray, B. B. *AG(E)* **1987**, *26*, 351.
40. Schinzer, D. *S* **1989**, 179.
41. Reich, H. J.; Kelly, M. J.; Olsen, R. E.; Holtan, R. C. *T* **1983**, *39*, 949.
42. Prieto, J. A.; Suarez, J.; Larson, G. L. *SC* **1988**, *18*, 253.
43. Djuric, S. W. *JOC* **1984**, *49*, 1311.
44. Busato, S.; Cainelli, G.; Panunzio, M.; Bandini, E.; Martelli, G.; Spunta, G. *SL* **1991**, 243.

Bret E. Huff

Lilly Research Laboratories, Indianapolis, IN, USA

t-Butyldimethylsilyl Trifluoromethanesulfonate[1]

$t\text{-BuMe}_2\text{SiOSO}_2\text{CF}_3$

[69739-34-0] $C_7H_{15}F_3O_3SSi$ (MW 264.33)

(highly reactive silylating agent and Lewis acid capable of converting primary, secondary, and tertiary alcohols[1b] to the corresponding TBDMS ethers, and converting ketones[2] and lactones[2a,3] into their enol silyl ethers; promoting conjugate addition of alkynylzinc compounds[4] and triphenylphosphine[5] to α,β-enones; activation of chromones in [4+2] cycloaddition reactions;[6] rearrangement of allylic tributylstannyl silyl ethers;[7] activation of pyridine rings toward Grignard reagents[8] and transalkylation of tertiary amine *N*-oxides;[9] and transformation of *N*-*t*-butoxycarbonyl groups into *N*-alkoxycarbonyl groups[10])

Alternate Name: TBDMS triflate.
Physical Data: bp 60 °C/7 mmHg; colorless oil, *d* 1.151 g cm^{-3}.
Solubility: sol most organic solvents such as pentane, CH_2Cl_2, etc.
Analysis of Reagent Purity: ^{1}H NMR ($CDCl_3$) δ 1.00 (s, 9H, *t*-Bu), 0.45 (s, 6H, Me).
Form Supplied in: liquid; widely available.
Preparative Method: [1b] to 24 g (0.16 mol) of ***t*-Butyldimethylchlorosilane** at 23 °C under argon is added 14 mL (0.16 mol) of ***Trifluoromethanesulfonic Acid*** dropwise. The solution is heated at 60 °C for 10 h, at which time no further hydrogen chloride evolves (removed through a bubbler). The resulting product is distilled under reduced pressure: 34 g (80% yield) of TBDMS triflate; bp 60 °C/7 mmHg.
Handling, Storage, and Precautions: the material should be stored under argon at 0 °C. The compound has an unpleasant odor and reacts rapidly with water and other protic solvents.

Silylation of Alcohols.[1] Primary, secondary, and tertiary alcohols are silylated by reaction with TBDMS triflate in excellent yields. For instance, treatment of *t*-butanol with 1.5 equiv of TBDMS triflate and 2 equiv of ***2,6-Lutidine*** in CH_2Cl_2 at 25 °C for 10 min gives a 90% yield of (*t*-butoxy)-*t*-butyldimethylsilane.[1b] The following alcohols are similarly silylated in excellent yields (70–90%): 2-phenyl-2-propanol, *endo*-norborneol, *cis*-2,2,4,4-tetramethylcyclobutane-1,3-diol, and 9-*O*-methylmaytansinol (converted to the 3-TBDMS derivative) (eq 1).[1b]

$$\text{R-OH} \xrightarrow[90\%]{\text{TBDMSOTf, 2,6-lutidine, } CH_2Cl_2} \text{R-OTBDMS} \qquad (1)$$

Formation of Enol Silyl Ethers.[2,3] Various sterically hindered ketones have been converted into enol silyl ethers by treatment with 1–2 equiv of TBDMS triflate and 1.5 equiv of ***Triethylamine***

in CH_2Cl_2 or 1,2-dichloroethane at rt. A representative example is depicted in eq 2.[2a]

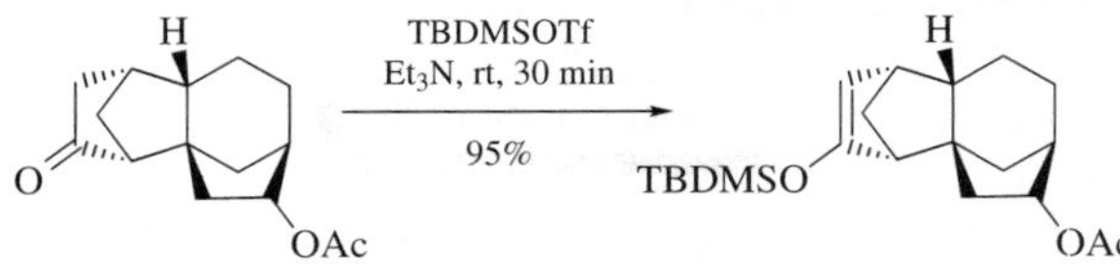

(2)

Reactions of chiral β-keto sulfoxides with 1.1 equiv of ***Lithium Diisopropylamide*** in THF at −78 °C followed by 1.2 equiv of TBDMS triflate at −78 °C produce the corresponding (*Z*)-enol silyl ethers (eq 3).[2b]

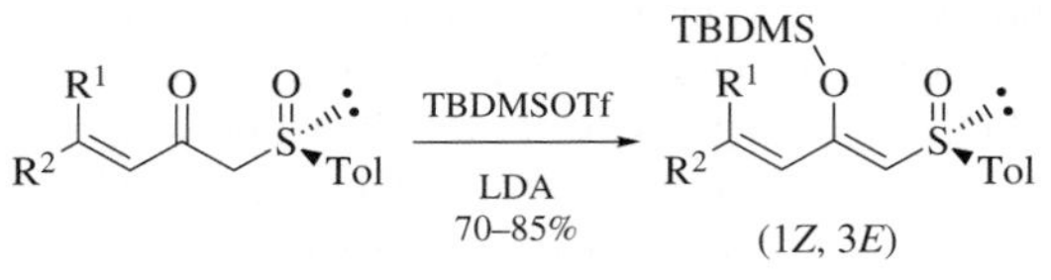

(3)

R^1 = H, R^2 = Me; R^1 = R^2 = Me; R^1 = H, R^2 = Ph

Lactones have also been transformed into silyl ketene acetals upon treatment with TBDMS triflate and triethylamine in CH_2Cl_2 (eqs 4 and 5).[2a,3] In the case of 8a-vinyl-2-oxooctahydro-2*H*-1,4-benzoxazine, the resulting silyl ketene acetal undergoes Claisen rearrangement to provide the octahydroquinoline (eq 6).[3]

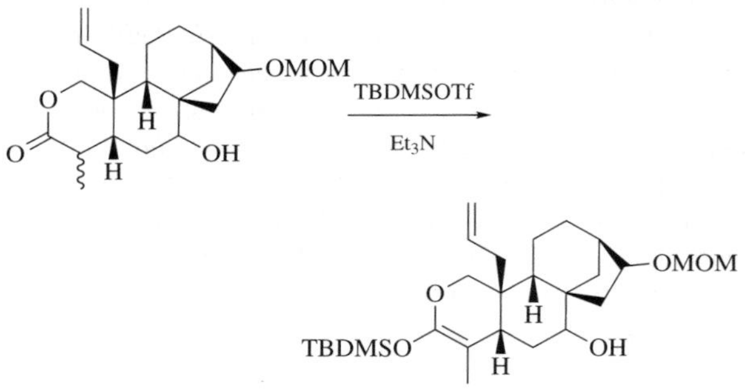

(4)

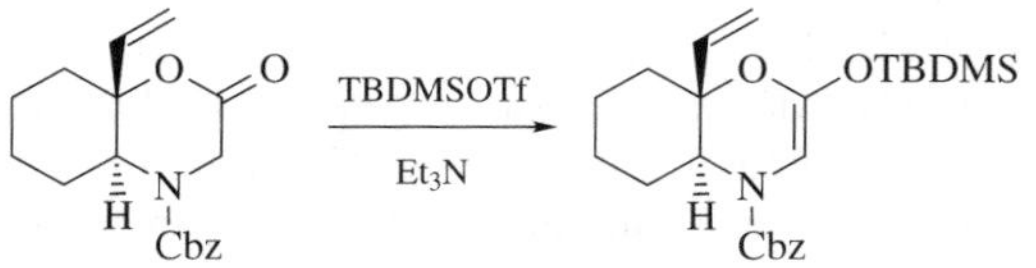

(5)

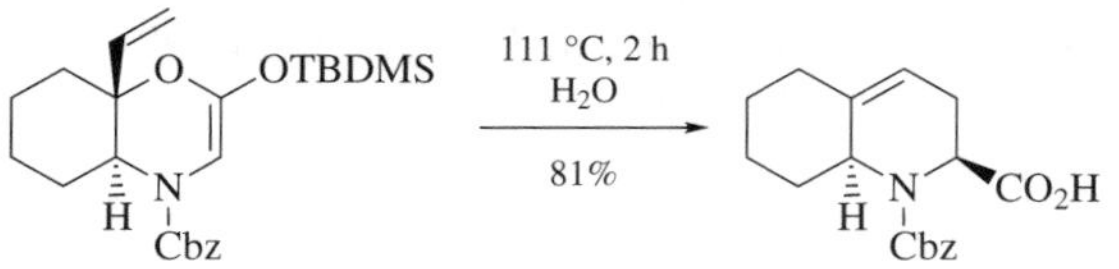

(6)

Conjugate Addition of Alkynylzinc Bromides.[4] Alkynylzinc bromides undergo conjugate addition with α,β-unsaturated ketones in the presence of TBDMS triflate in ether–THF at −40 °C to give the corresponding 1,4-adducts (54–96% yields). A representative example is illustrated in eq 7.[4] Other trialkylsilyl triflates such as ***Triisopropylsilyl Trifluoromethanesulfonate*** or ***Trimethylsilyl Trifluoromethanesulfonate*** can effectively replace TBDMS triflate.

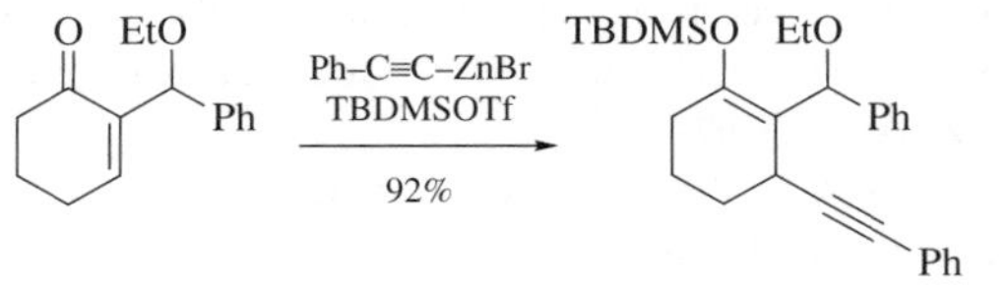

(7)

Phosphoniosilylation.[5] Cyclic enones treated with TBDMS triflate and ***Triphenylphosphine*** in THF at rt provide the corresponding 1-(3-*t*-butyldimethylsilyloxy-2-cycloalkenyl)triphenylphosphonium triflates (eq 8) which, upon lithiation with ***n-Butyllithium*** followed by Wittig reaction with aldehydes, afford various conjuated dienes.[5]

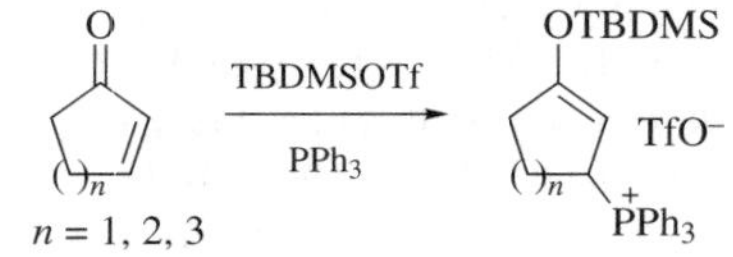

(8)

Silylation of Chromones.[6] The preparation of 4-*t*-butyldimethylsilyloxy-1-benzopyrylium triflate is carried out by heating chromone and TBDMS triflate at 80 °C for 1 h (without solvent) under nitrogen (eq 9).[6b] The silylated chromones undergo addition reaction with enol silyl ethers and 2,6-lutidine,[6b] and [4 + 2]-type cycloaddition reactions with α,β-unsaturated ketones in the presence of TBDMS triflate and 2,6-lutidine. An example of the cycloaddition reaction is shown in eq 10.[6a]

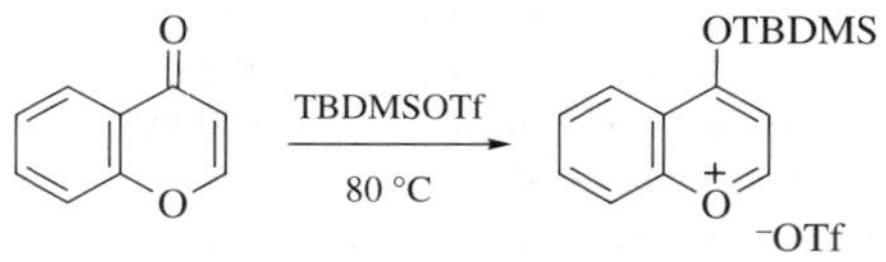

(9)

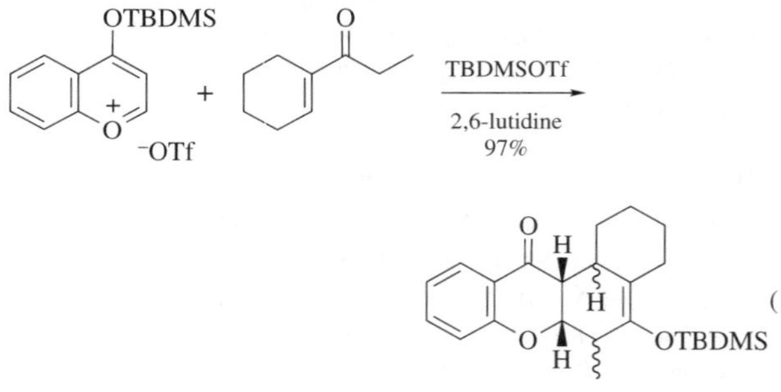

(10)

Rearrangement of Allylic Tributylstannyl Silyl Ethers.[7] α-Silyloxy allylic stannanes are isomerized with TBDMS triflate to (*Z*)-γ-silyloxy allylic stannanes (eq 11).[7] The resulting allylic stannanes undergo addition reactions with aldehydes in the presence of ***Boron Trifluoride Etherate*** to provide the 3-(*t*-butyldimethylsilyloxy)-4-hydroxyalkenes.[7a]

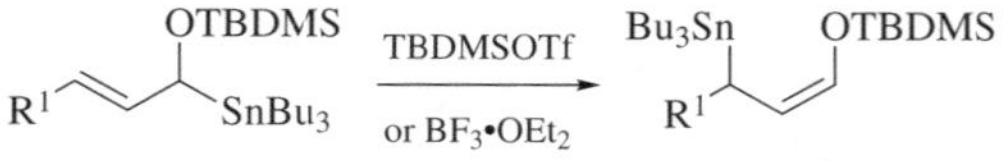

(11)

Activation of Pyridine.[8] *N*-(*t*-Butyldimethylsilyl)pyridinium triflate, prepared from pyridine and TBDMS triflate in CH_2Cl_2

at rt, undergoes addition reactions with alkyl and aryl Grignard reagents to give 4-substituted pyridines after oxidation with oxygen (eq 12).[8] Only about 1% of the 2-substituted pyridines were formed in the cases studied.

(pyridinium N-TBDMS ⁻OTf) —RMgBr, THF, rt, 2–3 h; O_2, 58–68%→ (4-R-pyridine) (12)

Transalkylation of Tertiary Amine *N*-Oxides.[9] *N*-(*t*-Butyldimethylsilyloxy)-*N*-methylpiperidinium triflate is quantitatively formed from the reaction of *N*-methylpiperidine *N*-oxide (eq 13).[9b] The resulting amine salts derived from various trialkylamine *N*-oxides undergo transalkylation by treatment with ***Methyllithium*** in THF at 0 °C followed by alkyl halides and ***Tetra-n-butylammonium Fluoride*** in a sealed tube at 110 °C for 10 h, to afford trisubstituted amines (eq 14).[9a]

(N-methylpiperidine N-oxide) —TBDMSOTf, CH_2Cl_2, ~100%→ (N-Me, N-OTBDMS piperidinium ⁻OTf) (13)

(N-Me, N-OTBDMS piperidinium ⁻OTf) —1. MeLi, THF; 2. RX, Δ, n-Bu_4NF, 33–58%→ (piperidine N–R) (14)

Interconversion of *N*-Boc Group into *N*-Alkoxycarbonyl Group.[10] Treatment of *t*-butyl alkylcarbamates with 1.5 equiv of TBDMS triflate and 2 equiv of 2,6-lutidine in CH_2Cl_2 at rt for about 15 min furnishes the corresponding TBDMS carbamates (eq 15).[10] Desilylation of these silyl carbamates with aqueous fluoride ion gives excellent yields of the corresponding primary amines. The silyl carbamates are also converted into other *N*-alkoxycarbonyl derivatives by treatment with TBAF and alkyl halides in THF at 0 °C (82–88% yields).[10]

RHNC(O)O-*t*-Bu —TBDMSOTf, 2,6-lutidine, rt→ RHNC(O)OTBDMS (15)

Related Reagents. *t*-Butyldimethylchlorosilane; *N*-(*t*-Butyldimethylsilyl)-*N*-methyltrifluoroacetamide; *t*-Butyldimethylsilyl Perchlorate; Triethyl Trifluoromethanesulfonate; Trimethylsilyl Trifluoromethanesulfonate.

---

1. (a) Stewart, R. F.; Miller, L. L. *JACS* **1980**, *102*, 4999. (b) Corey, E. J.; Cho, H.; Rucker, C.; Hua, D. H. *TL* **1981**, *22*, 3455. (c) For a review of trialkylsilyl triflates: Emde, H.; Domsch, D.; Feger, H.; Frick, U.; Gotz, A.; Hergott, H. H.; Hofmann, K.; Kober, W.; Krageloh, K.; Oesterle, T.; Steppan, W.; West, W.; Simchen, G. *S* **1982**, 1.
2. (a) Mander, L. N.; Sethi, S. P. *TL* **1984**, *25*, 5953. (b) Solladie, G.; Mangein, N.; Morreno, I.; Almario, A.; Carreno, M. C.; Garcia-Ruano, J. L. *TL* **1992**, *33*, 4561.
3. Angle, S. R.; Breitenbucker, J. G.; Arnaiz, D. O. *JOC* **1992**, *57*, 5947.
4. Kim, S.; Lee, J. M. *TL* **1990**, *31*, 7627.
5. Kozikowski, A. P.; Jung, S. H. *JOC* **1986**, *51*, 3400.
6. (a) Lee, Y.; Iwasaki, H.; Yamamoto, Y.; Ohkata, K.; Akiba, K. *H* **1989**, *29*, 35. (b) Iwasaki, H.; Kume, T.; Yamamoto, Y.; Akiba, K. *TL* **1987**, *28*, 6355.
7. (a) Marshall, J. A.; Welmaker, G. S. *JOC* **1992**, *57*, 7158. (b) Marshall, J. A.; Welmaker, G. S. *TL* **1991**, *32*, 2101. (c) Marshall, J. A.; Welmaker, G. S. *SL* **1992**, 537.
8. Akiba, K.; Iseki, Y.; Wata, M. *TL* **1982**, *23*, 3935.
9. (a) Tokitoh, N.; Okazaki, R. *CL* **1984**, 1937. (b) Okazaki, R.; Tokitoh, N. *CC* **1984**, 192.
10. Sakaitani, M.; Ohfune, Y. *TL* **1985**, *26*, 5543.

Duy H. Hua & Jinshan Chen
Kansas State University, Manhattan, KS, USA

t-Butyldiphenylchlorosilane[1]

t-$BuPh_2SiCl$

[58479-61-1] $C_{16}H_{19}ClSi$ (MW 274.86)

(reagent for the temporary protection of hydroxy groups as their *t*-butyldiphenylsilyl ethers; selectivity can be obtained for primary vs. secondary hydroxy groups; other protection (carbonyl, amine, etc.) is also possible;[2] deprotection is conveniently effected with fluoride ion, strong acid, or base; the TBDPS ether group is compatible with a large number of organic functional group transformations,[1–3] in nucleoside chemistry,[1] as well as in numerous examples in which polyfunctional molecules are chemically manipulated[4])

Alternate Name: TBDPS-Cl.
Physical Data: colorless liquid, bp 93–95 °C/0.015 mmHg; n_D^{20} 1.5680; *d* 1.057 g cm^{-3}.
Solubility: miscible in most organic solvents.
Form Supplied in: colorless liquid 98%.
Preparative Method: a dry 1 L, three-necked round bottomed flask is equipped with a magnetic stirring bar, a 500 mL equalizing dropping funnel fitted with a rubber septum, a reflux condenser, and nitrogen inlet tube. The flask is flushed with nitrogen, then charged with 127 g (0.5 mol) of diphenyldichlorosilane in 300 mL of redistilled pentane. A solution of ***t-Butyllithium*** in pentane (500 mL, 0.55 mol), is transferred under nitrogen pressure to the dropping funnel using a stainless steel, double-tip transfer needle. This solution is slowly added to the contents of the flask and when the addition is complete, the mixture is refluxed 30 h under nitrogen with stirring. The suspension is allowed to cool to rt, the precipitated lithium chloride is rapidly filtered through a pad of Celite, and the latter is washed with 200 mL of pentane. The solvent is removed by evaporation, and the colorless residue is distilled through a short (10 cm), Vigreux column, to give 125–132 g of the colorless title compound.
Handling, Storage, and Precautions: the reagent is stable when protected from moisture and protic solvents. A standard M solution of the reagent can be prepared in anhyd DMF and

kept at 0 °C under argon in an amber bottle. Use in a fume hood.

Preparation of *t*-Butyldiphenylsilyl Ethers and Related Transformations. A standard protocol[1] involves the addition of the reagent (1.1 equiv) to a solution of the alcohol (1 equiv) and ***Imidazole*** (2 equiv) in DMF at 0 °C or at room temperature. When silylation is complete, the solution is diluted with water and ether or dichloromethane and the organic layer is processed in the usual way. In some instances the addition of ***4-Dimethylaminopyridine*** (DMAP) can enhance the reaction rate.[5,6] Selective protection of the primary hydroxy group in carbohydrate derivatives has been achieved using this reagent and ***Poly(4-vinylpyridine)*** in CH_2Cl_2 or THF in the presence of ***Hexamethylphosphoric Triamide***.[7] Enol *t*-butyldiphenylsilyl ethers are easily formed by trapping of enolates with the chloride.[8] Amines can be selectively converted into the corresponding primary *t*-butyldiphenylsilylamines essentially under the same conditions.[9] *t*-Butyldiphenylsilyl cyanohydrins are prepared in the presence of ***Potassium Cyanide***, ***Zinc Iodide***, and a carbonyl compound.[10]

Selectivity and Compatibility of *O*- and *N*-*t*-Butyldiphenyl Silyl Derivatives. The *O*-*t*-butyldiphenylsilyl ether protective group offers some unique advantages and synthetically useful features compared to other existing counterparts.[1,2] Silylation of a primary hydroxy group takes place in preference to a secondary, giving products that are often crystalline, and detectable on TLC plates under UV light, because of the presence of a strong chromophore. The TBDPS group is compatible with a variety of conditions[1] used in synthetic transformations such as hydrogenolysis ($Pd(OH)_2$, C/H_2, etc.), *O*-alkylation (NaH, DMF, halide), de-*O*-acylation (NaOMe, NH_4OH, K_2CO_3/MeOH), mild chemical reduction with hydride reagents, carbon–carbon bond formation with organometallic reagents, transition metal-mediated reactions, Wittig reactions, etc. These reactions were tested during the total synthesis of thromboxane B_2[11] for which the TBDPS group was originally designed. Subsequently, numerous related organic transformations have been carried out in the presence of TBDPS ethers in the context of natural product synthesis[3,4] and functional group manipulations.[2] It is distinctly more acid stable than the *O*-trityl, *O*-THP, and *O*-TBDMS groups, and is virtually unaffected under conditions that cause complete cleavage of these groups (50% HCO_2H, 2–6 h; 50% aq TFA, 15 min; HBr/AcOH, 0 °C, few min; 80% aq acetic acid, few h).[1] It is also compatible with conditions used for the acid-catalyzed formation and hydrolysis of acetal groups and in the presence of some Lewis acids (e.g. TMSBr,[12] $BCl_3 \cdot SMe_2$,[13] Me_2AlCl,[8] etc.). The mono-TBDPS derivative of a primary amine[9] is stable to base, as well as to alkylating and acylating reagents. Thus a secondary amine can be acylated in the presence of a primary *t*-butyldiphenylsilylamino group.

Deprotection.[1,2] The *O*-TBDPS group can be cleaved under the following conditions: ***Tetra-n-butylammonium Fluoride*** in THF at rt; variations of F^--catalyzed reactions (e.g. F^-/AcOH; HF/pyridine;[14] HF/MeCN,[15] etc.); aqueous acids and bases[1] (1 N HCl, 5 N NaOH, aq methanolic NaOH or KOH); ion exchange resins;[7] ***Potassium Superoxide***, ***Dimethyl Sulfoxide***, ***18-Crown-6***.[16] Selective cleavage of a TBDPS ether in the presence of a TBDMS ether in carbohydrate derivatives involves treatment with ***Sodium Hydride***/HMPA at 0 °C.[17] Although the *O*-TBDPS group is stable to most hydride reagents, ***Lithium Aluminum Hydride*** reduction of an amide group has resulted in the cleavage[18] of an adjacent *O*-TBDPS ether, possibly via internal assistance. TBDPS-amines are cleaved by 80% AcOH or by HF/pyridine.[9]

Related Reagents. *t*-Butyldimethylchlorosilane; *t*-Butyldimethylsilyl Trifluoromethanesulfonate; Chlorotriphenylsilane; Triisopropylsilyl Chloride.

---

1. Hanessian, S.; Lavallée, P. *CJC* **1975**, *53*, 2975; Lavallée, P. Ph.D. Thesis, Université de Montréal, 1977.
2. Greene, T. W.; Wuts, P. G. M. *Protective Groups in Organic Synthesis*, 2nd ed.; Wiley: New York, 1991.
3. Corey, E. J.; Cheng, X.-M. *The Logic of Chemical Synthesis*; Wiley: New York, 1989.
4. Hanessian, S. *The Total Synthesis of Natural Products: The Chiron Approach*; Pergamon: Oxford, 1983.
5. Ireland, R. E.; Obrecht, D. M. *HCA* **1986**, *69*, 1273.
6. Chaudhary, S. K.; Hernandez, O. *TL* **1979**, 99.
7. Cardillo, G.; Orena, M.; Sandri, S.; Tomasini, C. *CI(L)* **1983**, 643.
8. Horiguchi, Y.; Suehiro, I.; Sasaki, A.; Kuwajima, I. *TL* **1992**, *34*, 6077.
9. Overman, L. E.; Okazaki, M. E.; Mishra, P. *TL* **1986**, *27*, 4391.
10. Duboudin, F.; Cazeau, P.; Moulines, F.; Laporte, O. *S* **1982**, 212.
11. Hanessian, S.; Lavallée, P. *CJC* **1981**, *59*, 870.
12. Hanessian, S.; Delorme, D.; Dufresne, Y. *TL* **1984**, *25*, 2515.
13. Congreve, M. S.; Davison, E. C.; Fuhry, M. A. M.; Holmes, A. B.; Payne, A. N.; Robinson, R. A.; Ward, S. E. *SL* **1993**, 663.
14. Nicolaou, K. C.; Seitz, S. P.; Pavia, M. R.; Petasis, N. A. *JOC* **1979**, *44*, 4011.
15. Ogawa, Y.; Nunomoto, M.; Shibasaki, M. *JOC* **1986**, *51*, 1625.
16. Torisawa, Y.; Shibasaki, M.; Ikegami, S. *CPB* **1983**, *31*, 2607.
17. Shekhani, M. S.; Khan, K. M.; Mahmood, K.; Shah, P. M.; Malik, S. *TL* **1990**, *31*, 1669.
18. Rajashekhar, B.; Kaiser, E. T. *JOC* **1985**, *50*, 5480.

Stephen Hanessian
University of Montreal, Quebec, Canada

C

N,N′-Carbonyldiimidazole[1]

[530-62-1] $C_7H_6N_4O$ (MW 162.15)

(reagent for activation of carboxylic acids;[1] synthesis of esters,[2] amides,[3] peptides,[4] aldehydes,[5] ketones,[6] β-keto thioesters,[7] tetronic acids;[8] ureas,[9] isocyanates,[10] carbonates;[9] halides from alcohols;[11] glycosidation;[12] dehydration[13–16])

Physical Data: mp 116–118 °C.
Solubility: no quantitative data available. Inert solvents such as THF, benzene, $CHCl_3$, DMF are commonly used for reactions.
Form Supplied in: commercially available white solid.
Analysis of Reagent Purity: purity can be determined by measuring the amount of CO_2 evolved on hydrolysis.
Preparative Methods: prepared by mixing phosgene with four equivalents of imidazole in benzene/THF.[17]
Purification: may be purified by recrystallization from hot, anhydrous THF with careful exclusion of moisture.[17]
Handling, Storage, and Precautions: moisture sensitive; reacts readily with water with evolution of carbon dioxide. May be kept for long periods either in a sealed tube or in a desiccator over P_2O_5.

Activation of Carboxylic Acids: Synthesis of Acyl Imidazoles. *N,N′*-Carbonyldiimidazole (**1**) converts carboxylic acids into the corresponding acylimidazoles (**2**) (eq 1).[1] The method can be applied to a wide range of aliphatic, aromatic, and heterocyclic carboxylic acids, including some examples (such as formic acid and vitamin A acid) where acid chloride formation is difficult. The reactivity of (**2**) is similar to that of acid chlorides, but the former have the advantage that they are generally crystalline and easily handled. Isolation of (**2**) is simple, but often unnecessary; further reaction with nucleophiles is usually performed in the same reaction vessel. Conversion of (**2**) into acid chlorides (via reaction with HCl),[18] hydrazides,[3] hydroxamic acids,[3] and peroxy esters[19] have all been described. Preparation of the more important carboxylic acid derivatives is described below.

Esters from Carboxylic Acids. Reaction of equimolar amounts of carboxylic acid, alcohol, and (**1**) in an inert solvent (e.g. THF, benzene, or chloroform) results in ester formation (eq 2). Since alcoholysis of the intermediate acylimidazole is relatively slow, the reaction mixture must be heated at 60–70 °C for some time. However, addition of a catalytic amount of a base such as ***Sodium Amide*** to convert the alcohol to the alkoxide, or a catalytic amount of the alkoxide itself, allows rapid and complete formation of the ester at room temperature.[2] The base catalyst must of course be added after formation of (**2**) from the acid is complete, as indicated by cessation of evolution of carbon dioxide.

RCO_2H + (**1**) → (**2**) + CO_2 + imidazole (1)

$$RCO_2H + R^1OH \xrightarrow[\text{2. cat. } R^1ONa \text{ or } NaNH_2]{\text{1. (1)}} RCO_2R^1 \quad (2)$$

Esters of tertiary alcohols may not be prepared from carboxylic acids containing acidic α-protons using this modified procedure, since deprotonation and subsequent condensation, competes. However, the use of stoichiometric ***1,8-Diazabicyclo[5.4.0]undec-7-ene*** as base has been shown to provide good yields of *t*-butyl esters even for acids with acidic α-protons (eq 3).[20] This procedure was unsuccessful for pivalic acid or for *N*-acyl-α-amino acids.

$$Ph(CH_2)_3CO_2H \xrightarrow[\text{2. 1 equiv } t\text{-BuOH, 1 equiv DBU, 40 °C, 24 h}]{\text{1. 1 equiv (1), DMF, 40 °C, 1 h}} Ph(CH_2)_3CO_2\text{-}t\text{-Bu} \quad (3)$$

An alternative approach to increasing the rate of esterification is to activate further the intermediate (**2**). ***N-Bromosuccinimide*** has been used for this purpose,[21] but unsaturation in the carboxylic acid or alcohol is not tolerated. More generally useful is the addition of an activated halide, usually ***Allyl Bromide***, to a chloroform solution of (**1**) and a carboxylic acid, resulting in formation of the acylimidazolium salt (**3**) (eq 4).[22] Addition of the alcohol and stirring for 1–10 h at room temperature or at reflux affords good yields of ester in a one-pot procedure. These conditions work well for the formation of methyl, ethyl, and *t*-butyl esters of aliphatic, aromatic, and α,β-unsaturated acids. Hindered esters such as *t*-butyl pivalate can be prepared cleanly (90% yield). The only limitation is that substrates must not contain functionality that can be alkylated by the excess of the reactive halide.

$$RCO_2H \xrightarrow[\text{4–5 equiv allyl bromide}]{(1),\ CHCl_3} (3) \xrightarrow{R^1OH} RCO_2R^1 \quad (4)$$

Since the purity of commercial (**1**) may be variable due to its water sensitivity, it is common to employ an excess in order to ensure complete conversion of the carboxylic acid to the acylimidazole. It has been suggested that alcohols react faster with residual (**1**) than with the acylimidazole (**2**), thus reducing the yield of ester. A procedure has been developed for removal of excess (**1**) before addition of the alcohol.[23]

Macrolactonization has been accomplished using (**1**).[24] Thiol and selenol esters can also be prepared in one pot from carboxylic acids using (**1**);[25] reaction of the intermediate (**2**) with aromatic thiols or selenols is complete within a few minutes in DMF, while aliphatic thiols require a few hours. Formation of the phenylthiol ester of *N*-Cbz-L-phenylalanine was accompanied by only slight racemization. ***N,N'-Carbonyldi-sym-triazine*** can be used in place of (**1**), with similar results.

Amides from Carboxylic Acids: Peptide Synthesis. Analogous to ester formation, reaction of equimolar amounts of a carboxylic acid and (**1**) in THF, DMF, or chloroform, followed by addition of an amine, allows amide bond formation.[3] The method has been applied to peptide synthesis (eq 5).[4] One equivalent of (**1**) is added to a 1 M solution of an acylamino acid in THF, followed after 1 h by the desired amino acid or peptide ester. The amino acid ester hydrochloride may be used directly instead of the free amino acid ester. An aqueous solution of the amino acid salt can even be used, but yields are lower.

CbzNH–CH(R)–C(=O)OH —[1. (1); 2. EtO_2C–CH(R^1)–NH_2]→ CbzNH–CH(R)–C(=O)–NH–CH(R^1)–CO_2Et (5)

As is the case in esterification reactions, the presence of unreacted (**1**) can cause problems since the amine reacts with this as quickly as it does with the acylimidazole, forming urea byproducts that can be difficult to separate. Use of exactly one equivalent of (**1**) is difficult due to its moisture sensitivity, and also because of the tendency of some peptides or amino acids to form hydrates. Paul and Anderson solved this problem by use of an excess of (**1**) to form the acylimidazole, then cooling to $-5\,^{\circ}C$ and adding a small amount of water to destroy the unreacted (**1**) before addition of the amine.[26]

For the sensitive coupling of Cbz-glycyl-L-phenylalanine and ethyl glycinate (the Anderson test), Paul and Anderson reported the level of racemization as 5% using THF as solvent at $-10\,^{\circ}C$, but as <0.5% in DMF.[4] Performing the same coupling reaction at room temperature, Beyerman and van den Brink later claimed that the degree of racemization in DMF was in fact as high as 17%, and reported better results (no detectable racemization) using the related reagent *N,N'*-carbonyl di-*sym*-triazine in place of (**1**).[27] In a comparative study of several reagents, Weygand and co-workers also observed extensive racemization using (**1**).[28] In the formation of tyrosine esters, Paul reported that the mixed anhydride method is to be preferred to use of (**1**), since *O*-acylation is a major side reaction with the latter.[29]

Aldehydes and Ketones from Carboxylic Acids. Reduction of the derived acylimidazole (**2**) with ***Lithium Aluminum Hydride*** achieves conversion of an aliphatic or aromatic carboxylic acid to an aldehyde (eq 6).[5] ***Diisobutylaluminum Hydride*** has also been used, allowing preparation of α-acylamino aldehydes from *N*-protected amino acids.[30] Similarly, reaction of (**2**) with Grignard reagents affords ketones,[6] with little evidence for formation of tertiary alcohol.

Reaction of acylimidazoles with the appropriate carbon nucleophile has also been used for the preparation of α-nitro ketones[31] and β-keto sulfoxides.[32]

(2) —[$LiAlH_4$]→ RCHO; (2) —[EtMgBr]→ R–C(=O)–Et (6)

***C*-Acylation of Active Methylene Compounds.** Treatment of an acylimidazole, derived from a carboxylic acid and (**1**), with the magnesium salt of a malonic or methylmalonic half thiol ester results in *C*-acylation under neutral conditions (eq 7).[7] The presence of secondary hydroxyl functionality in the carboxylic acid is tolerated, but primary alcohols require protection. Magnesium salts of malonic esters may be used equally effectively. Intramolecular *C*-acylation of ketones has also been reported.[33]

$Ph(CH_2)_2CO_2H$ —[1. (1); 2. $[^-O_2C$–CH(CH_3)–C(=O)SEt$]_2$ Mg^{2+}; 18 h, 25 °C; 100%]→ $Ph(CH_2)_2$–C(=O)–CH(CH_3)–C(=O)–SEt (7)

Tetronic Acids. A synthesis of tetronic acids reported by Smith and co-workers relies on the reaction between (**1**) and the dianion derived from an α-hydroxy ketone (eq 8).[8] The reaction proceeds in moderate yield (31–57%).

$(CH_3)_2C(OH)$–C(=O)–CH_3 —[1. 2 equiv LDA; 2. (1); 48%]→ HO-substituted 5,5-dimethylfuran-2(5H)-one (tetronic acid) (8)

Ureas and Carbonates. Reagent (**1**) may be used as a direct replacement for the highly toxic ***Phosgene*** in reactions with alcohols and amines. Reaction of (**1**) with two equivalents of a primary aliphatic or aromatic amine at room temperature rapidly yields a symmetrical urea (eq 9).[9] If only one equivalent of a primary amine is added to (**1**), then the imidazole-*N*-carboxamide (**4**) is formed (eq 10). These compounds can dissociate into isocyanates and imidazole, even at room temperature, and distillation from the reaction mixture provides a useful synthesis of isocyanates (eq 10).[10] Secondary amines react only at one side of (**1**) at room temperature, again giving the imidazole-*N*-carboxamide of type (**4**).

2 RNH_2 —[(1)]→ RHN–C(=O)–NHR (9)

R = alkyl, aryl

RNH_2 —[(1)]→ RHN–C(=O)–N(imidazolyl) (**4**) ⇌ RNCO + HN(imidazole) (10)

R = alkyl, aryl

Reaction of (**1**) with one equivalent of an alcohol provides the imidazole-*N*-carboxylic ester (**5**) (eq 11).[34] Further treatment with another alcohol or phenol yields an unsymmetrical carbonate;

alternatively, reaction with an amine affords a carbamate (eq 12).[34]

ROH —(1)→ RO(C=O)N-imidazolyl **(5)** (11)

(5) —R'XH→ RO(C=O)XR' (12)

X = O, NH

Heating (**1**) with an excess of an alcohol or phenol gives the symmetrical carbonate.[9] This reaction can be accelerated dramatically by the presence of catalytic base (e.g. ***Sodium Ethoxide***). Reaction under these conditions is so exothermic even at room temperature that only *t*-butanol stops at the imidazole-*N*-carboxylic ester stage.

1,2-Diamines, 1,2-diols, or 1,2-amino alcohols react with (**1**) to form cyclic ureas,[35] carbonates,[36] or oxazolidinones,[35] respectively. In the case of cyclohexane-1,2-diols, the *cis*-diol reacts much more rapidly than the *trans*, as would be expected.[36] Thiazolidinones can also be prepared using (**1**).[37]

Halides from Alcohols. Treatment of an alcohol with (**1**) and an excess (at least three equivalents) of an activated halide results in its conversion to the corresponding halide (eq 13).[11] Any halide more reactive than the product halide may be used, but in practice ***Allyl Bromide*** or ***Iodomethane*** give best results as they are effective and readily removed after the reaction. Acetonitrile is the best solvent and yields are generally high (>80%). Bromide or iodide formation work well, but not chlorination. Optically active alcohols are racemized.

$CH_3CH(OH)CH_2CO_2Et$ —(1), 5 equiv allyl bromide; MeCN, rt, 3 h; 78%→ $CH_3CH(Br)CH_2CO_2Et$ (13)

Glycosidation. A mild glycosidation procedure involving (**1**) has been reported by Ford and Ley.[12] A carbohydrate derivative in ether or dichloromethane reacts with (**1**) through the anomeric C-1-hydroxyl to give the (1-imidazolylcarbonyl) glycoside (IMG) (**6**) (eq 14). Isolation of (**6**) is not usually necessary; treatment with one equivalent of an alcohol and two equivalents of ***Zinc Bromide*** in ether at reflux gives the glycoside. Generally, higher α:β ratios are obtained for more hindered alcohols and when ether is used as solvent rather than the less polar dichloromethane. Along with the fact that the α:β ratio is independent of the configuration of (**6**), this suggests an S_N1-type mechanism. In contrast, treatment of (**6**) with ***Acetyl Chloride*** provides the anomeric chloride with essentially exclusive inversion.

Dehydration. Reagent (**1**) has been used for the dehydration of various substrates, including aldoximes (to give nitriles),[13] β-hydroxy amino acids,[14] and β-hydroxy sulfones.[15] 3-Aryl-2-hydroxyiminopropionic acids undergo dehydration and decarboxylation, to give 2-aryl acetonitriles, upon reaction with (**1**).[16]

Tetra-O-benzyl glucopyranose (BnO, OBn; C-1 OH) —(1), CH_2Cl_2 or Et_2O→ 1-O-(imidazolylcarbonyl) glycoside **(6)** —HO-CH(Ph)CO$_2$Me, 2 equiv $ZnBr_2$, 88%→ glycoside (O-CH(Ph)CO$_2$Me) (14)

α:β = 10:1

Related Reagents. *N*,*N*′-Carbonyldi-syn-triazine.

1. Staab, H. A. *AG(E)* **1962**, *1*, 351.
2. Staab, H. A.; Mannschreck, A. *CB* **1962**, *95*, 1284 (*CA* **1962**, *57*, 5846d).
3. Staab, H. A.; Lüking, M.; Dürr, F. H. *CB* **1962**, *95*, 1275 (*CA* **1962**, *57*, 5908a).
4. Paul, R.; Anderson, G. W. *JACS* **1960**, *82*, 4596.
5. Staab, H. A.; Bräunling, H. *LA* **1962**, *654*, 119 (*CA* **1962**, *57*, 5906c).
6. Staab, H. A.; Jost, E. *LA* **1962**, *655*, 90 (*CA* **1962**, *57*, 15 090g).
7. Brooks, D. W.; Lu, L. D.-L.; Masamune, S. *AG(E)* **1979**, *18*, 72.
8. Jerris, P. J.; Wovkulich, P. M.; Smith, A. B., III. *TL* **1979**, 4517.
9. Staab, H. A. *LA* **1957**, *609*, 75 (*CA* **1958**, *52*, 7332e).
10. Staab, H. A.; Benz, W. *LA* **1961**, *648*, 72 (*CA* **1962**, *57*, 4649g).
11. Kamijo, T.; Harada, H.; Iizuka, K. *CPB* **1983**, *31*, 4189.
12. Ford, M. J.; Ley, S. V. *SL* **1990**, 255.
13. Foley, H. G.; Dalton, D. R. *CC* **1973**, 628.
14. Andruszkiewicz, R.; Czerwinski, A. *S* **1982**, 968.
15. Kang, S.-K.; Park, Y.-W.; Kim, S.-G.; Jeon, J.-H. *JCS(P1)* **1992**, 405.
16. Kitagawa, T.; Kawaguchi, M.; Inoue, S.; Katayama, S. *CPB* **1991**, *39*, 3030.
17. Staab, H. A.; Wendel, K. *OS* **1968**, *48*, 44.
18. Staab, H. A.; Datta, A. P. *AG(E)* **1964**, *3*, 132.
19. Hecht, R.; Rüchardt, C. *CB* **1963**, *96*, 1281 (*CA* **1963**, *59*, 1523h).
20. Ohta, S.; Shimabayashi, A.; Aono, M.; Okamoto, M. *S* **1982**, 833.
21. Katsuki, T. *BCJ* **1976**, *49*, 2019.
22. Kamijo, T.; Harada, H.; Iizuka, K. *CPB* **1984**, *32*, 5044.
23. Morton, R. C.; Mangroo, D.; Gerber, G. E. *CJC* **1988**, *66*, 1701.
24. (a) White, J. D.; Lodwig, S. N.; Trammell, G. L.; Fleming, M. P. *TL* **1974**, 3263. (b) Colvin, E. W.; Purcell, T. A.; Raphael, R. A. *CC* **1972**, 1031.
25. Gais, H.-J. *AG(E)* **1977**, *16*, 244.
26. Paul, R.; Anderson, G. W. *JOC* **1962**, *27*, 2094.
27. Beyerman, H. C.; Van Den Brink, W. M. *RTC* **1961**, *80*, 1372.
28. Weygand, F.; Prox, A.; Schmidhammer, L.; König, W. *AG(E)* **1963**, *2*, 183.
29. Paul, R. *JOC* **1963**, *28*, 236.
30. Khatri, H.; Stammer, C. H. *CC* **1979**, 79.
31. Baker, D. C.; Putt, S. R. *S* **1978**, 478.
32. Ibarra, C. A.; Rodríguez, R. C.; Monreal, M. C. F.; Navarro, F. J. G.; Tesorero, J. M. *JOC* **1989**, *54*, 5620.
33. Garigipati, R. S.; Tschaen, D. M.; Weinreb, S. M. *JACS* **1985**, *107*, 7790.
34. Staab, H. A. *LA* **1957**, *609*, 83 (*CA* **1957**, *52*, 16341e).

35. Wright, W. B., Jr. *JHC* **1965**, *2*, 41.
36. Kutney, J. P.; Ratcliffe, A. H. *SC* **1975**, *5*, 47.
37. D'Ischia, M.; Prota, G.; Rotteveel, R. C.; Westerhof, W. *SC* **1987**, *17*, 1577.

Alan Armstrong
University of Bath, UK

Chloromethyl Methyl Ether[1]

$MeOCH_2Cl$

[107-30-2] C_2H_5ClO (MW 80.51)

(used for the protection of alcohols, phenols, acids, amines, β-keto esters and thiols. A one-carbon synthon for alkylation of aromatics and active methylene derivatives)

Alternate Name: methoxymethyl chloride; MOMCl.
Physical Data: bp 55–57 °C; *d* 1.060 g cm^{-3}.
Solubility: sol nearly all organic solvents.
Form Supplied in: available as a technical grade liquid.
Preparative Methods: besides the original method[2] which uses HCl, formalin, and methanol, MOMCl can be prepared from methoxyacetic acid[3] or methylal (***Dimethoxymethane***).[4]
Handling, Storage, and Precautions: MOMCl is an OSHA-regulated carcinogen and depending on the method of preparation may contain bis(chloromethyl) ether, which is a more potent carcinogen. This reagent should be handled in a fume hood.

Protection of Alcohols. Treatment of an alcohol with MOMCl and i-Pr_2NEt (0 °C, 1 h → 25 °C, 8 h, 86% yield)[5] is the most commonly employed procedure for introduction of the MOM group. Formation of the sodium alkoxide and reaction with MOMCl is also very effective.[6] This method is particularly useful for the protection of the enol of β-keto esters.[7] Because of the carcinogenicity of MOMCl, a number of methods for the introduction of the MOM group have been developed which do not rely on the chloride. The methods are all based on the use of $CH_2(OMe)_2$ with various catalysts such as P_2O_5,[8] Me_3SiI,[9] Nafion H,[10] $CH_2{=}CHCH_2SiMe_3$/TMSOTf/P_2O_5,[11] $FeCl_3$,[12] Montmorillonite clay (H^+),[13] or TsOH/LiBr.[14] In the last case, 1,3-diols give methylene acetals. It is often difficult to achieve monoprotection of 1,3-diols but some success has been recorded, as illustrated in eqs 1–3.[15–17]

MOMCl, NaH; 61% (1)

$(MeO)_3CH$, CSA, CH_2Cl_2, rt, 24 h; DIBAL, −78 °C, 30 min, 0 °C, 10 min (2)

In the orthoester route when unsymmetrical diols are used, the most hindered alcohol is protected, in contrast to the normal methods which are expected to protect the least hindered alcohol. Diols capable of forming a stannylene derivative are efficiently monoprotected (eq 3).[17]

MOMCl, $CHCl_3$; 86% (3)

Cleavage of MOM Ethers. Since the MOM group is an acetal, it can be cleaved using acid hydrolysis. In general, aqueous acid in an organic cosolvent has proven to be effective.[18–21] Use of the weak acid ***Pyridinium p-Toluenesulfonate*** in refluxing t-BuOH or 2-butanone is particularly effective for cleavage of MOM ethers of allylic alcohols.[22] MEM ethers are also cleaved under these conditions. Eq 4 shows that a MOM group can be cleaved with anhydrous ***Trifluoroacetic Acid*** in the presence of an acetonide which is normally cleaved with aqueous acid.[23]

TFA, CH_2Cl_2; >85% (4)

Catecholborane halides, particularly the bromide (***B-Bromocatecholborane***), are effective reagents for the cleavage of MOM ethers. The bromide cleaves the following groups in the order: MOMOR MEMOR > t-Boc > Cbz t-BuOR > BnOR > allylOR > t-BuO_2CR secondary alkylOR > BnO_2CR > primary alkylOR alkylO_2CR. The t-butyldimethylsilyl (TBDMS) group is stable to this reagent. The chloride is less reactive and thus may be more useful for achieving selectivity in multifunctional substrates. Yields are generally >83%.[24] A variety of other boron based reagents have been developed which are useful for MOM ether cleavage. The use of ***Bromobis(isopropylthio)borane*** (eq 5)[25] has the advantage that 1,2- and 1,3-diols do not give formyl acetals as is sometimes the case in cleaving MOM groups with neighboring hydroxyl groups.[26] The reagent also cleaves MEM groups and under basic conditions affords the i-$PrSCH_2OR$ derivatives. With ***Bromodimethylborane***, MEM and MTM ethers are also cleaved, but esters are not affected.[27]

$(i\text{-}PrS)_2BBr$, MeOH; 94% (5)

Bromotrimethylsilane, a reagent similar to but not as powerful as ***Iodotrimethylsilane***, will cleave MOM ethers as well as acetonide, THP, trityl, and t-$BuMe_2Si$ groups, but the reagent will not cleave esters, methyl and benzyl ethers, t-butyl-diphenylsilyl ethers, and amides. One example is shown in eq 6.[28]

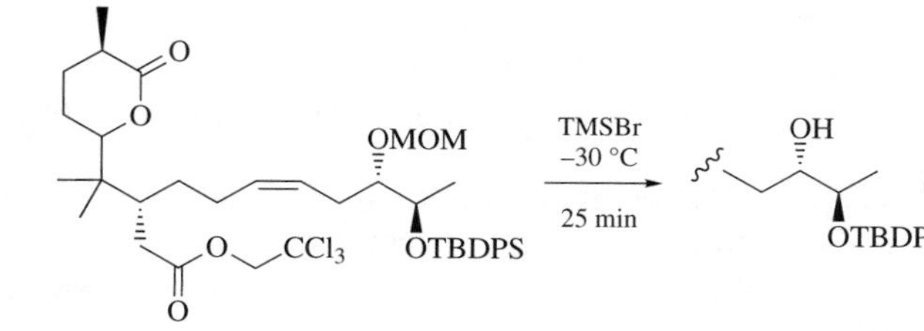

(6)

Boron Trifluoride Etherate in the presence of ***Thioanisole***,[29] ***Magnesium Bromide*** in the presence of butanethiol,[30] ***Triphenylcarbenium Tetrafluoroborate***,[31] and ***Lithium Tetrafluoroborate***[32] are also useful for cleavage of the MOM group. The action of lithium tetrafluoroborate is unique (eq 7) and the mechanism for cleavage is not well understood.

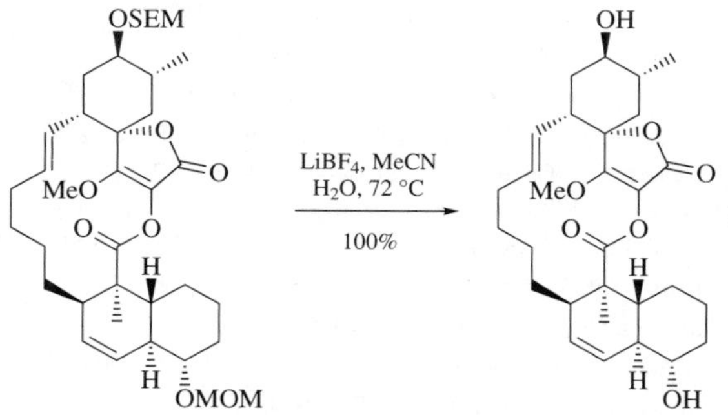

(7)

Protection of Phenols. The reaction of MOMCl with a phenol under phase-transfer conditions works well to give phenolic MOM ethers[33] and will selectivity protect a phenol in the presence of an alcohol.[34] The more classical Williamson ether synthesis[35] also provides excellent results, but may require the addition of a crown ether to enhance the nucleophilicity of the phenolate anion.[36] As in the case of alcohol protection, alternatives using methylal have been developed for phenol protection which do not rely on the carcinogenic MOMCl.[37] Phenolic silyl ethers can be converted directly to MOM ethers by reaction with TASF (***Tris(dimethylamino)sulfonium Difluorotrimethylsilicate***) and MOMCl.[38]

Cleavage of Phenolic MOM Ethers. Again, mild acid[37,39] is used for cleavage, but other alternatives such as ***Sodium Iodide***/acetone/cat. HCl/50 °C,[40] ***Diphosphorus Tetraiodide***/CH_2Cl_2/0 °C → rt,[41] or TMSBr/CH_2Cl_2[42] are also effective.

Protection of Thiols. Lithium[43] and potassium[44] thiolates react with MOMCl to form the thioethers. The use of MOMCl can be avoided by the reaction of the zinc thiolate with methylal.[45]

Protection of Acids. The reaction of MOMCl with an acid in the presence of a base affords the MOM ester.[46–48] As with the thiols, the zinc carboxylate reacts with methylal to afford the MOM ester.[49] MOM esters are quite acid sensitive and are unstable to silica gel chromatography.[50] The MOM ester is easily cleaved with acid, $MgBr_2$,[51] or Me_3SiBr containing a trace of MeOH.[52]

Protection of Amine Derivatives. Heterocyclic amines and amides are protected as their MOM derivatives by reaction of the amine or amide with MOMCl in the presence of a base such as ***Diisopropylethylamine***,[53] ***Sodium Hydride***,[54] LiH, ***Potassium t-Butoxide***,[55] or ***Sodium Hydroxide***.[56] These derivatives are cleaved with ***Hydrogen Chloride***/EtOH[53] or ***Boron Tribromide***.[55] To hydrolyze scopolamine without formation of the rearranged product (**1**) it was necessary to protect the amine nitrogen as the methoxy-methochloride as illustrated in eq 8.[57]

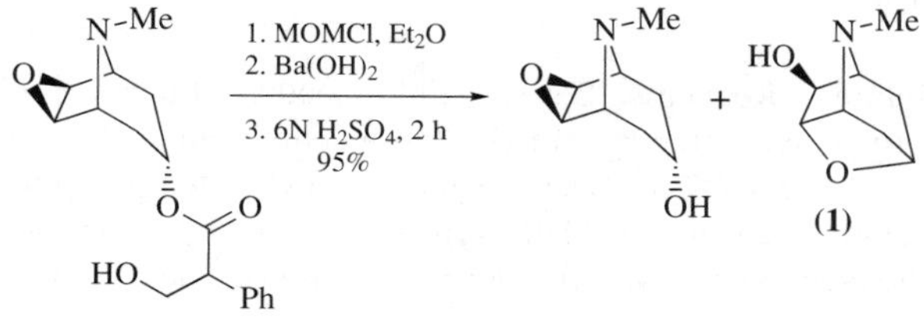

(8)

Use in C–C Bond-Forming Reactions. Since MOMCl is an excellent electrophile it readily reacts with enolates[58] and other carbanions[59] and thus serves as an easily handled one-carbon synthon. Friedel–Crafts alkylation with MOMCl and a Lewis acid such as ***Tin(IV) Chloride***,[60] ***Titanium(IV) Chloride***,[61] ***Zinc Chloride***,[62] and ***Aluminum Chloride***[63] or ***Acetic Acid***[64] is a common method for introduction of a chloromethyl group onto an aromatic nucleus (eqs 9 and 10).[60,64] The reaction is quite general and tolerates a broad range of functionality.

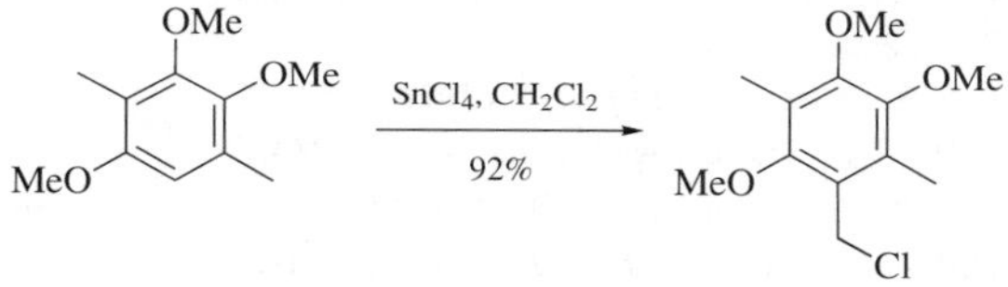

(9)

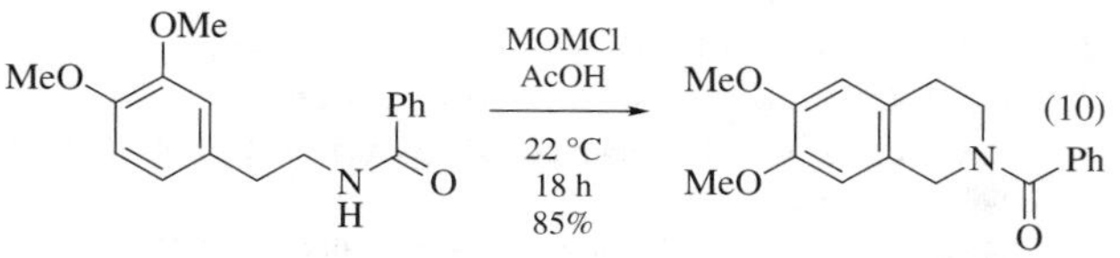

(10)

Alkenes react in a similar fashion giving methoxymethyl derivatives, but in this case the intermediate carbenium ion is trapped with methoxide or another nucleophile such as a nitrile to afford the methyl ether or amide in a Ritter-like reaction (eq 11).[65] In similar fashion, silyl enol ethers give ketones (eq 12),[66,67] allylsilanes afford homologated alkenes (eq 13),[68] and stannylalkynes are converted to propargyl ethers.[69]

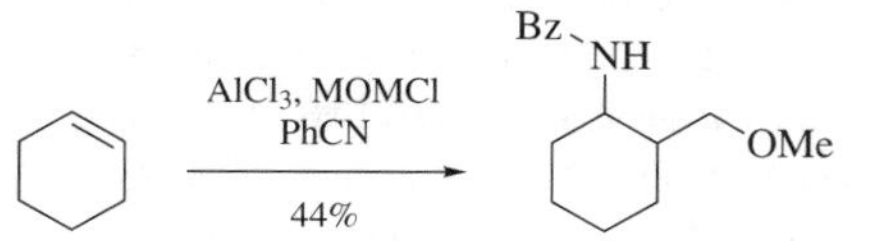

(11)

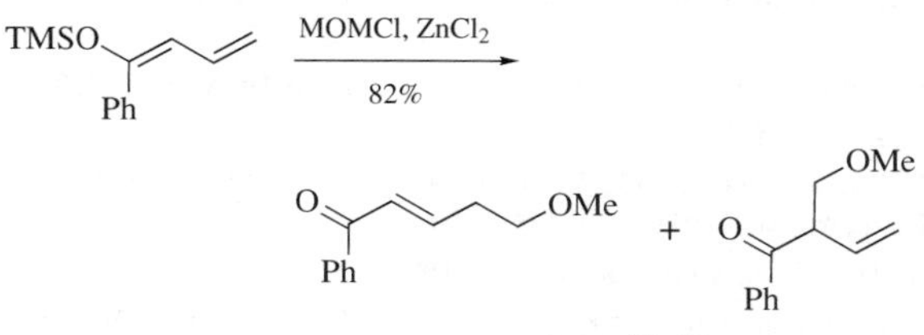

(12)

γ:α = 55:45

$$\text{MeO(O=)C–C(Me)(SiMe}_2\text{Ph)–CH=CH}_2 \xrightarrow[77\%]{\text{MOMCl, TiCl}_4\text{, CH}_2\text{Cl}_2} \text{MeO}\diagup\!\!\diagdown\!\!\diagup\!\!=\!\!\text{C(Me)CO}_2\text{Me} \quad (13)$$

MOMCl can be converted to the Grignard reagent or the lithium reagent and used to introduce one carbon by nucleophilic attack on ketones and esters.[70] These reagents are best prepared in methylal.[71]

Related Reagents. Benzyl Chloromethyl Ether; Benzyl Chloromethyl Sulfide; *t*-Butyl Chloromethyl Ether; 2-Chloroethyl Chloromethyl Ether; Chloromethyl Methyl Sulfide; Dimethoxymethane; (*p*-Methoxybenzyloxy)methyl Chloride; 2-Methoxyethoxymethyl Chloride; 2-(Trimethylsilyl)ethoxymethyl Chloride.

---

1. For a review on the use of this reagent in protective group chemistry, see: Greene, T. W.; Wuts, P. G. M. *Protective Groups in Organic Synthesis*, 2nd ed.; Wiley: New York, 1991.
2. Marvel, C. S.; Porter, P. K., *OSC* **1941**, *1*, 377.
3. (a) Jones, M. *S* **1984**, 727. (b) Stadlwieser, J. *S* **1985**, 490.
4. Amato, J. S.; Karady, S.; Sletzinger, M.; Weinstock, L. M. *S* **1979**, 970.
5. Stork, G.; Takahashi, T. *JACS* **1977**, *99*, 1275.
6. Kluge, A. F.; Untch, K. G.; Fried, J. H. *JACS* **1972**, *94*, 7827.
7. (a) Welch, S. C.; Hagan, C. P. *SC* **1973**, *3*, 29. (b) Roush, W. R.; Brown, B. B. *TL* **1989**, *30*, 7309. (c) Coates, R. M.; Shaw, J. E. *JOC* **1970**, *35*, 2597, 2601.
8. Fuji, K.; Nakano, S.; Fujita, E. *S* **1975**, 276.
9. Olah, G. A.; Husain, A.; Narang, S. C. *S* **1983**, 896.
10. Olah, G. A.; Husain, A.; Gupta, B. G. B.; Narang, S. C. *S* **1981**, 471.
11. Nishino, S.; Ishido, Y. *J. Carbohydr. Chem.* **1986**, *5*, 313.
12. Patney, H. K.; *SL* **1992**, 567.
13. Schaper, U. A. *S* **1981**, 794.
14. Gras, J.-L.; Chang, Y.-Y. K. W.; Guerin, A. *S* **1985**, 74.
15. Ihara, M.; Suzuki, M.; Fukumoto, K.; Kametani, T.; Kabuto, C. *JACS* **1988**, *110*, 1963.
16. Takasu, M.; Naruse, Y.; Yamamoto, H. *TL* **1988**, *29*, 1947.
17. Nobuo, N.; Masaji, O. *CL* **1987**, 141.
18. Auerbach, J.; Weinreb, S. M. *CC* **1974**, 298.
19. Meyers, A. I.; Durandetta, J. L.; Munavu, R. *JOC* **1975**, *40*, 2025.
20. Bremmer, M. L.; Khatri, N. A.; Weinreb, S. M. *JOC* **1983**, *48*, 3661.
21. LaForge, F. B. *JACS* **1933**, *55*, 3040.
22. Monti, H.; Léandri, G.; Klos-Ringquet, M.; Corriol, C. *SC* **1983**, *13*, 1021.
23. Woodward, R. B. et al. *JACS* **1981**, *103*, 3210.
24. Boeckman, R. K., Jr.; Potenza, J. C. *TL* **1985**, *26*, 1411.
25. Corey, E. J.; Hua, D. H.; Seitz, S. P. *TL* **1984**, *25*, 3.
26. Barot, B. C.; Pinnick, H. W. *JOC* **1981**, *46*, 2981.
27. Guindon, Y.; Morton, H. E.; Yoakim, C. *TL* **1983**, *24*, 3969.
28. Hanessian, S.; Delorme, D.; Dufresne, Y. *TL* **1984**, *25*, 2515. For in situ prepared TMSBr see: Woodward, R. B. et al. *JACS* **1981**, *103*, 3213 (note 2).
29. Kieczykowski, G. R.; Schlessinger, R. H. *JACS* **1978**, *100*, 1938.
30. Kim, S. *SL* **1991**, 183.
31. Nakata, T.; Schmid, G.; Vranesic, B.; Okigawa, M.; Smith-Palmer, T.; Kishi, Y. *JACS* **1978**, *100*, 2933.
32. Ireland, R. E.; Varney, M. D. *JOC* **1986**, *51*, 635.
33. van Heerden, F. R.; van Zyl, J. J.; Rall, G. J. H.; Brandt, E. V.; Roux, D. G. *TL* **1978**, 661.
34. Kelly, T. R.; Jagoe, C. T.; Li, Q. *JACS* **1989**, *111*, 4522.
35. Chiarello, J.; Joullié, M. M. *T* **1988**, *44*, 41.
36. Rall, G. J. H.; Oberholzer, M. E.; Ferreira, D.; Roux, D. G. *TL* **1976**, 1033.
37. Yardley, J. P.; Fletcher, H., 3rd, *S* **1976**, 244.
38. Noyori, R.; Nishida, I.; Sakata, J. *TL* **1981**, *22*, 3993.
39. Rahman, M. A. A.; Elliott, H. W.; Binks, R.; Küng, W.; Rapoport, H. *JMC* **1966**, *9*, 1.
40. Williams, D. R.; Barner, B. A.; Nishitani, K.; Phillips, J. G. *JACS* **1982**, *104*, 4708.
41. Saimoto, H.; Kusano, Y.; Hiyama, T. *TL* **1986**, *27*, 1607.
42. Huffman, J. W.; Zheng, X.; Wu, M.-J.; Joyner, H. H.; Pennington, W. T. *JOC* **1991**, *56*, 1481.
43. Brown, H. C.; Imai, T. *JACS* **1983**, *105*, 6285.
44. Fukuyama, T.; Nakatsuka, S.; Kishi, Y. *TL* **1976**, 3393.
45. Dardoize, F.; Gaudemar, M.; Goasdoue, N. *S* **1977**, 567.
46. Jansen, A. B. A.; Russell, T. J. *JCS* **1965**, 2127.
47. Shono, T; Ishige, O.; Uyama, H.; Kashimura, S. *JOC* **1986**, *51*, 546.
48. White, J. D.; Reddy, G.; Nagabhushana, S. G. O. *JACS* **1988**, *110*, 1624.
49. Dardoize, F.; Gaudemar, M.; Goasdoue, N. *S* **1977**, 567.
50. Weinstock, L. M.; Karady, S.; Roberts, F. E.; Hoinowski, A. M.; Brenner, G. S.; Lee, T. B. K.; Lumma, W. C.; Sletzinger, M. *TL* **1975**, 3979.
51. Kim, S.; Park, Y. H.; Kee, I. S. *TL* **1991**, *32*, 3099.
52. Masamune, S. *Aldrichim. Acta* **1978**, *11*, 23.
53. Ozaki, S.; Watanabe, Y.; Fujisawa, H.; Hoshiko, T. *H* **1984**, *22*, 527.
54. Moody C. J.; Ward, J. G. *JCS(P1)* **1984**, 2895.
55. Kirby, G. W.; Robins, D. J.; Stark, W. M. *CC* **1983**, 812.
56. Sundberg, R. J.; Russel, H. F. *JOC* **1973**, *38*, 3324.
57. Meinwald, J.; Chapman, O. L. *JACS* **1957**, *79*, 665.
58. (a) Vila, A. J.; Cravero, R. M.; Gonzalez-Sierra, M. *TL*, **1991**, *32*, 1929. (b) Huet, F.; Pellet, M.; Conia, J. M. *S* **1979**, 33. (c) Greene, A. E.; Coelho, F.; Depres, J.-P.; Brocksom, T. J. *JOC* **1985**, *50*, 1973. (d) Kogen, H.; Tomioka, K.; Hashimoto, S.-I.; Koga, K. *T* **1981**, *37*, 3951.
59. (a) Okazaki, R.; O-oka, M.; Akiyama, T.; Inamoto, N.; Niwa, J. *JACS* **1987**, *109*, 5413. (b) Alexakis, A.; Cahiez, G.; Normant, J. F. *S* **1979**, 826. (c) Micetich, R. G.; Baker, V.; Spevak, P.; Hall, T. W.; Bains, B. K. *H* **1985**, *23*, 943. (d) Corriu, R. J. P.; Huynh, V. Moreau, J. J. E. *TL* **1984**, *25*, 1887.
60. Iyer, S.; Liebeskind, L. S. *JACS* **1987**, *109*, 2759.
61. Tashiro, M.; Yamato, T. *S* **1978**, 435.
62. Steyn, P. S.; Holzapfel, C. W. *T* **1967**, *23*, 4449.
63. Cornforth, J.; Robertson, A. D. *JCS(P1)* **1987**, 867.
64. Jaques, B.; Deeks, R. H. L.; Shah, P. K. J. *CC* **1969**, 1283.
65. Wada Y.; Yamazaki, T.; Nishiura, K.; Tanimoto, S. Okano, M. *BCJ* **1978**, *51*, 1821.
66. Fleming, I.; Lee, T. V. *TL* **1981**, *22*, 705.
67. Shono, T.; Nishiguchi, I.; Komamura, T.; Sasaki, M. *JACS* **1979**, *101*, 984.
68. Uno, H. *BCJ* **1986**, *59*, 2471.
69. Zhai, D.; Zhai, W.; Williams, R. M. *JACS* **1988**, *110*, 2501.
70. Schöllkopf, U. *LA* **1967**, *704*, 120.
71. (a) Runge, F. Taeger, E.; Fiedler, C.; Kahlert, E. *JPR* **1963**, *19*, 37. (b) Schöllkopf, U.; Küppers, H. *TL* **1964**, 1503.

Peter G. M. Wuts
The Upjohn Co., Kalamazoo, MI, USA

2-Chloro-1-methylpyridinium Iodide[1]

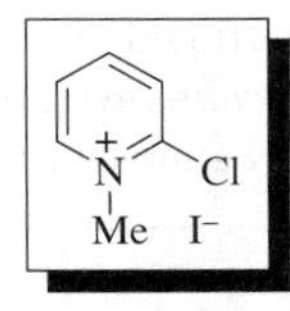

[14338-32-0] C_6H_7ClIN (MW 255.49)

(activation of carboxylic acids;[1] formation of esters[2] and amides;[3] lactonization;[4] ketene formation;[5] β-lactam synthesis;[6,7] carbodiimide synthesis[8])

Alternate Name: Mukaiyama's reagent.
Physical Data: mp 204–206 °C.
Form Supplied in: commercially available yellow solid.
Preparative Method: by reaction between 2-chloropyridine and ***Iodomethane*** in acetone at reflux.[9]
Purification: recrystallization from acetone.
Handling, Storage, and Precautions: hygroscopic.

Activation of Carboxylic Acids: Ester Formation. 2-Chloro-1-methylpyridinium iodide (**1**) reacts with a mixture of a carboxylic acid and an alcohol, in the presence of two equivalents of base, to form an ester (eq 1).[2] The pyridinium salt (**2**) is formed rapidly by displacement of chloride from (**1**) by the carboxylate; subsequent reaction with the alcohol results in formation of the ester, along with 1-methyl-2-pyridone (**3**). A variety of solvents may be employed, but yields are highest in dichloromethane or pyridine. ***Tri-n-butylamine*** or ***Triethylamine*** are often used as base. The co-product (**3**) is insoluble in dichloromethane and so precipitates from this solvent. Good results are obtained even for hindered carboxylic acids and alcohols.

R^1CO_2H + (**1**) —base→ (**2**) —R^2OH, base→ R^2OCOR^1 (62–90%) + (**3**) + base•HI (1)

Amide Synthesis. In a similar way, amides can be prepared efficiently by treatment of an equimolar amount of a carboxylic acid and an amine with (**1**) in the presence of base (eq 2).[3]

$RCO_2H + R^1R^2NH$ —(**1**), Bu_3N, CH_2Cl_2, >82%→ $RCONR^1R^2$ (2)

Lactone Formation. The ester formation method has found widespread use in the synthesis of lactones by cyclization of hydroxy acids (eq 3).[4] The cyclization process is believed to be entropically favored because of the close proximity of all reactants to a central pyridinium salt (**4**). Optimum conditions for the process involve addition over 8 h of a solution of the hydroxy acid (0.0125 M) and triethylamine (8 equiv) in dichloromethane or acetonitrile to a solution of (**1**) (4 equiv; 0.04 M) in the same solvent at reflux. After a further 30 min at reflux, evaporation of the solvent and chromatography affords the lactone. Results of a study of the formation of lactones of various ring sizes are shown in Table 1. Good results are obtained for ring sizes 7 and ≥12, but attempts to form 8- or 9-membered lactones result in substantial dimerization.

$HO(CH_2)_nCO_2H$ —(**1**), Et_3N→ [(**4**)] —Et_3N→ lactone + (**3**) (3)

Table 1 Lactone Formation by Cyclization of Hydroxy Acids with Reagent (**1**)

Ring size	% Yield lactone	% Yield dimer
7	89	0
8	0	93
9	13	34
12	61	24
13	69	14
16	84	3

Cyclization of sensitive *trans*-γ-hydroxy acids to give strained *trans*-bicyclic γ-lactones of the type (**5**) is often problematic due to dehydration of the tertiary alcohol. A study showed (**1**) to be the reagent of choice for this transformation (eq 4).[10] Other reagents, such as ***1,3-Dicyclohexylcarbodiimide/Pyridine*** or ***2,2′-Dipyridyl Disulfide/Triphenylphosphine***, give far lower yields.

—(**1**), Et_3N, CH_2Cl_2, 97%→ (**5**) (4)

Ketene Formation. There is evidence that the above esterification procedure may occur at least partly via formation of the ketene from the carboxylic acid.[5] If a carboxylic acid contains a suitably placed alkene, the resulting ketene can undergo intramolecular [2+2] cycloaddition (eq 5).[5] Benzene is the best solvent for this reaction, giving yields comparable to those obtained by ketene formation from the corresponding acid chloride.

β-Lactam Synthesis. Dehydration of β-amino acids to give β-lactams can be effected in high yield using (**1**) in CH_2Cl_2 or MeCN (eq 6).[6]

β-Lactams can also be prepared by the reaction of carboxylic acids, activated by (**1**), with imines; best results are obtained with ***Tripropylamine*** as base in CH_2Cl_2 at reflux (eq 7).[7]

(5)

(6)

(7)

cis:*trans* = 15:1

Carbodiimides from Thioureas. Conversion of *N*,*N′*-disubstituted thioureas into carbodiimides can be effected by treatment with (**1**) and 2 equiv of base (eq 8).[8]

PhHN–C(=S)–NHC$_6$H$_{11}$ → PhNH=C=NHC$_6$H$_{11}$ (8) [(1), 2 equiv Et$_3$N; MeCN, 5 h; 98%]

1. Mukaiyama, T. *AG(E)* **1979**, *18*, 707.
2. Mukaiyama, T.; Usui, M.; Shimada, E.; Saigo, K. *CL* **1975**, 1045.
3. Bald, E.; Saigo, K.; Mukaiyama, T. *CL* **1975**, 1163.
4. Mukaiyama, T.; Usui, M.; Saigo, K. *CL* **1976**, 49.
5. Funk, R. L.; Abelman, M. M.; Jellison, K. M. *SL* **1989**, 36.
6. Huang, H.; Iwasawa, N.; Mukaiyama, T. *CL* **1984**, 1465.
7. Georg, G. I.; Mashava, P. M.; Guan, X. *TL* **1991**, *32*, 581.
8. Shibanuma, T.; Shiono, M.; Mukaiyama, T. *CL* **1977**, 575.
9. Amin, S. G.; Glazer, R. D.; Manhas, M. S. *S* **1979**, 210.
10. Strekowski, L.; Visnick, M.; Battiste, M. A. *S* **1983**, 493.

Alan Armstrong
University of Bath, UK

Chlorotriethylsilane

Et$_3$SiCl

[994-30-9] C$_6$H$_{15}$ClSi (MW 150.75)

(silylating agent, Lewis acid catalyst)

Alternate Name: triethylsilyl chloride.
Physical Data: bp 145–147 °C; *d* 0.898 g cm^{-3}.
Solubility: readily sol most aprotic organic solvents, although CH_2Cl_2 is most commonly used; reacts with H_2O and other protic solvents.
Form Supplied in: colorless liquid, or as a 1.0 M solution in THF.
Handling, Storage, and Precautions: should be handled and stored under an anhydrous, inert atmosphere.

General Considerations. The triethylsilyl moiety is often used as a protecting group for alcohols because of its greater stability towards hydrolysis than the smaller trimethylsilyl group. Alcohols can normally be silylated with Et$_3$SiCl using either ***Imidazole*** in DMF or ***Triethylamine/4-Dimethylaminopyridine*** in CH_2Cl_2 (eqs 1 and 2).[1,2] For substrates which are more difficult to derivatize, it may be advantageous to carry out the reaction in neat ***Pyridine*** (eq 3)[3] or to generate the alkoxide salt prior to the addition of Et$_3$SiCl (eq 4).[4–6]

(1) Et$_3$SiCl; imidazole, DMF; 95%

(2) Et$_3$SiCl, Et$_3$N; DMAP, CH_2Cl_2; 77%

(3) Et$_3$SiCl; pyridine, 60 °C; 95%

(4) 1. KH, THF; 2. Et$_3$SiCl; 93%; 2 equiv LiNaph; THF, −78 °C; 61%

Silyl ethers have also been obtained by in situ trapping of an intermediate alkoxide that results from a nucleophilic addition onto an ester[7] or epoxide[8] (eqs 5 and 6).

Silyl enol ethers can be prepared from ketones by trapping the kinetically formed lithium enolate with Et$_3$SiCl (eqs 7 and 8),[9,10] or by the conjugate addition of a dialkylcuprate to an α,β-unsaturated ketone with silylation of the enolate intermediate (eq 9).[11–13] Surprisingly, the addition of ***Lithium Di-n-butylcuprate*** to an α,β-unsaturated lactone in the presence of Et$_3$SiCl resulted in the 1,4-addition of the triethylsilyl moiety, apparently via the in situ formation of an intermediate silylcuprate species (eq 10).[14]

Bu–CO–OEt; BuLi / Et_3SiCl → Bu–C(OSiEt$_3$)(OEt)–Bu; H_3O^+ → Bu–CO–Bu, 66% (5)

PhO (epoxide); Et_3SiCl, $Bu_4N^+ Cl^-$, 60 °C, $CHCl_3$, 96% → PhO–CH$_2$–CH(OSiEt$_3$)–CH$_2$Cl (6)

t-Bu–CO–CH$_2$Br; 1. LiN(TMS)$_2$, −78 °C; 2. Et_3SiCl → *t*-Bu–C(OSiEt$_3$)=CHBr; 2 equiv BuLi → *t*-Bu–CO–CH$_2$–SiEt$_3$, 67% (7)

O=C(Me)–CH=CH–CH$_3$; 1. LDA; 2. Et_3SiCl → Et_3SiO–C(=CH$_2$)–CH=CH–CH$_3$ (8)

Et_3SiCl, Bu_2CuLi, 90% → OSiEt$_3$ … Bu + OSiEt$_3$ … Bu, 2:3 mixture (9)

1. 2 equiv Et_3SiCl; 2. add to Bu_2CuLi, THF, −78 °C; 3. NH_3, H_2O; 4. H_3O^+, 70% (*t*-Bu, Ph, O, SiEt$_3$) (10)

Organolithium salts generally react with Et_3SiCl to afford organosilanes. Schlessinger used a deprotonation–silylation procedure to introduce a triethylsilyl group at the α-position of an alkylimine (eq 11).[15] A variety of other stabilized carbanions have been reported to react cleanly with Et_3SiCl (eqs 12–15).[16–19]

N-*t*-Bu; 1. LDA; 2. Et_3SiCl, 73% → N-*t*-Bu, SiEt$_3$ (11)

O_2S–NH; 1. BuLi; 2. Et_3SiCl, 70% → Et_3Si, O_2S–NH (12)

$(EtO)_2P(O)$–CH$_2$CH$_3$; 1. 2 equiv LDA; 2. Et_3SiCl, 90% → $(EtO)_2P(O)$–CH(CH$_3$)SiEt$_3$ (13)

OH; Et_3SiCl, imidazole, DMF → OSiEt$_3$; 1. *s*-BuLi, THF, −78 °C; 2. Et_3SiCl → SiEt$_3$, OSiEt$_3$, 87% (14)

OMe, OMe; *s*-BuLi → Li, OMe, OMe; Et_3SiCl → Et_3Si, OMe, OMe, 70% (15)

In a similar manner, aryl, vinyl-, and alkynylsilanes can be prepared by the reaction of the appropriate carbanion with Et_3SiCl (eqs 16–21).[20–25]

SH; BuLi, TMEDA → SLi, Li; Et_3SiCl → SH, SiEt$_3$, 79% (16)

CHO; 1. LiN(morpholine); 2. *s*-BuLi, THF, −78 °C → [LiO, N, O, Li, O]; Et_3SiCl → Et_3Si, O, CHO, 73% (17)

Br; 1. Mg^0; 2. Et_3SiCl, 65% → SiEt$_3$ (18)

OMe; 1. *t*-BuLi, TMEDA; 2. Et_3SiCl → SiEt$_3$, OMe; HCl, MeOH → O, SiEt$_3$ (19)

F, F, F, Cl; 1. BuLi; 2. Et_3SiCl, −78 °C, 86% → F, F, F, SiEt$_3$ (20)

Hex–≡; 1. EtMgBr; 2. Et_3SiCl, 98% → Hex–≡–SiEt$_3$ (21)

Related Reagents. *t*-Butyldimethylchlorosilane; *t*-Butyldimethylsilyl Trifluoromethanesulfonate; Chlorotrimethylsilane; Chlorotriphenylsilane; Triethylsilyl Trifluoromethanesulfonate; Trimethylsilyl Trifluoromethanesulfonate.

1. Corey, E. J.; Link, J. O. *TL* **1990**, *31*, 601.
2. Anderson, J. C.; Ley, S. V. *TL* **1990**, *31*, 431.
3. Hart, T. W.; Metcalfe, D. A.; Scheinmann, F. *CC* **1979**, 156.
4. Hoffmann, R.; Brückner, R. *CB* **1992**, *125*, 1471.
5. Smith, A. B., III; Richmond, R. E. *JACS* **1983**, *105*, 575.
6. Andrews, D. R.; Barton, D. H. R.; Hesse, R. H.; Pechet, M. M. *JOC* **1986**, *51*, 4819.
7. Cooke, M. P., Jr. *JOC* **1986**, *51*, 951.
8. Andrews, G. C.; Crawford, T. C.; Contillo, L. G., Jr. *TL* **1981**, *22*, 3803.
9. (a) Sampson, P.; Hammond, G. B.; Wiemer, D. F. *JOC* **1986**, *51*, 4342. (b) Sampson, P.; Wiemer, D. F. *CC* **1985**, 1746.

10. (a) Danishefsky, S. J.; Uang, B. J.; Quallich, G. *JACS* **1985**, *107*, 1285. (b) Danishefsky, S. J.; Harvey, D. F.; Quallich, G.; Uang, B. J. *JOC* **1984**, *49*, 392.
11. Horiguchi, Y.; Komatsu, M.; Kuwajima, I. *TL* **1989**, *30*, 7087.
12. Marino, J. P.; Emonds, M. V. M.; Stengel, P. J.; Oliveira, A. R. M.; Simonelli, F.; Ferreira, J. T. B. *TL* **1992**, *33*, 49.
13. (a) Ager, D. J.; Fleming, I.; Patel, S. K. *JCS(P1)* **1981**, 2520; (b) Fleming, I.; Newton, T. W. *JCS(P1)* **1984**, 1805.
14. Amberg, W; Seebach, D. *AG(E)* **1988**, *27*, 1718.
15. Schlessinger, R. H.; Poss, M. A.; Richardson, S.; Lin, P. *TL* **1985**, *26*, 2391.
16. Szymonifka, M. J.; Heck, J. V. *TL* **1989**, *30*, 2873.
17. Savignac, P.; Teulade, M.-P.; Collignon, N. *JOM* **1987**, *323*, 135.
18. Oppolzer, W.; Snowden, R. L.; Simmons, D. P. *HCA* **1981**, *64*, 2002.
19. Seyferth, D.; Mammarella, R. E.; Klein, H. A. *JOM* **1980**, *194*, 1.
20. Block, E.; Eswarakrishnan, V.; Gernon, M.; Ofori-Okai, G.; Saha, C.; Tang, K.; Zubieta, J. *JACS* **1989**, *111*, 658.
21. Lee, G. C. M.; Holmes, J. M.; Harcourt, D. A.; Garst, M. E. *JOC* **1992**, *57*, 3126.
22. Soderquist, J. A.; Lee, S.-J. H. *T* **1988**, *44*, 4033.
23. Schinzer, D. *S* **1989**, 179.
24. Hiyama, T.; Nishide, K.; Obayashi, M. *CL* **1984**, 1765.
25. Uchida, K.; Utimoto, K.; Nozaki, H. *JOC* **1976**, *41*, 2215.

Edward Turos
State University of New York at Buffalo, NY, USA

Chlorotrimethylsilane[1,2]

ClSiMe$_3$

[75-77-4] C_3H_9ClSi (MW 108.64)

(protection of silyl ethers,[3] transients,[5–7] and silylalkynes;[8] synthesis of silyl esters,[4] silyl enol ethers,[9,10] vinylsilanes,[13] and silylvinylallenes;[15] Boc deprotection;[11] TMSI generation;[12] epoxide cleavage;[14] conjugate addition reactions catalyst[16–18])

Alternate Names: trimethylsilyl chloride; TMSCl.
Physical Data: bp 57 °C; *d* 0.856 g cm^{-3}.
Solubility: sol THF, DMF, CH_2Cl_2, HMPA.
Form Supplied in: clear, colorless liquid; 98% purity; commercially available.
Analysis of Reagent Purity: bp, NMR.
Purification: distillation over calcium hydride with exclusion of moisture.
Handling, Storage, and Precautions: moisture sensitive and corrosive; store under an inert atmosphere; use in a fume hood.

Protection of Alcohols as TMS Ethers. The most common method of forming a silyl ether involves the use of TMSCl and a base (eqs 1–3).[3,19–22] Mixtures of TMSCl and ***Hexamethyldisilazane*** (HMDS) have also been used to form TMS ethers. Primary, secondary, and tertiary alcohols can be silylated in this manner, depending on the relative amounts of TMS and HMDS (eqs 4–6).[23]

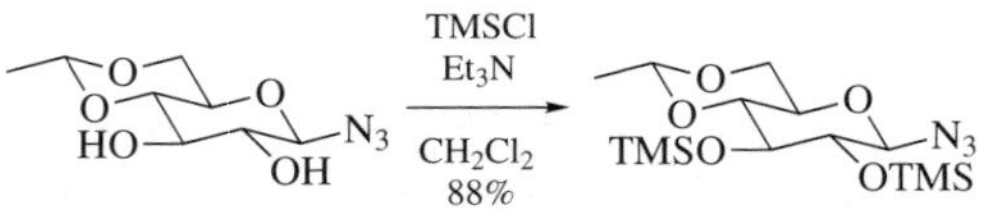

TMSCl, Et_3N, CH_2Cl_2, 88% (1)

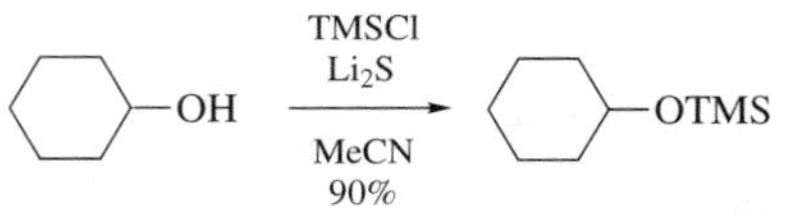

TMSCl, Li_2S, MeCN, 90% (2)

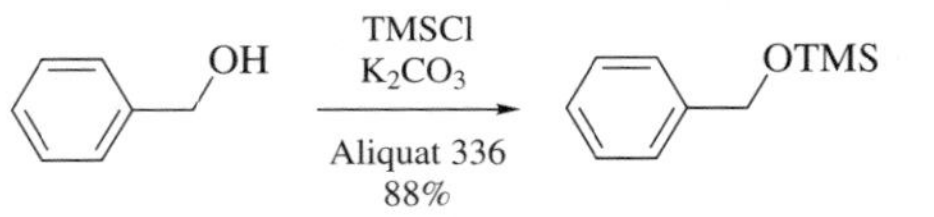

TMSCl, K_2CO_3, Aliquat 336, 88% (3)

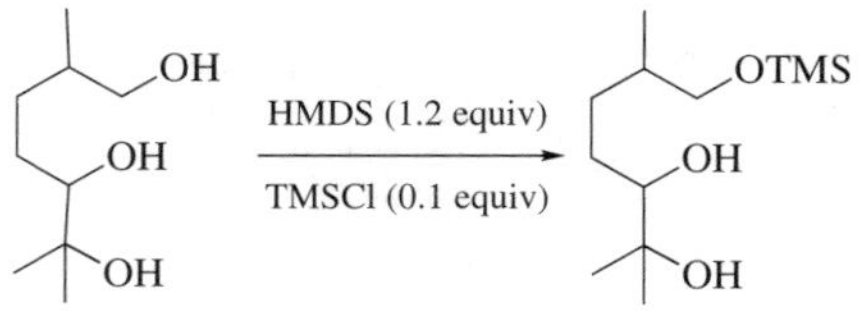

HMDS (1.2 equiv), TMSCl (0.1 equiv) (4)

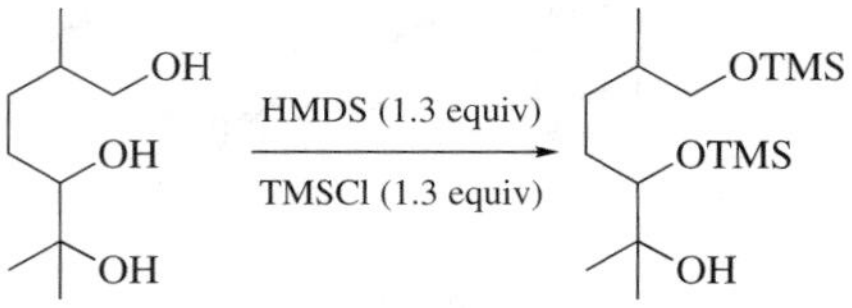

HMDS (1.3 equiv), TMSCl (1.3 equiv) (5)

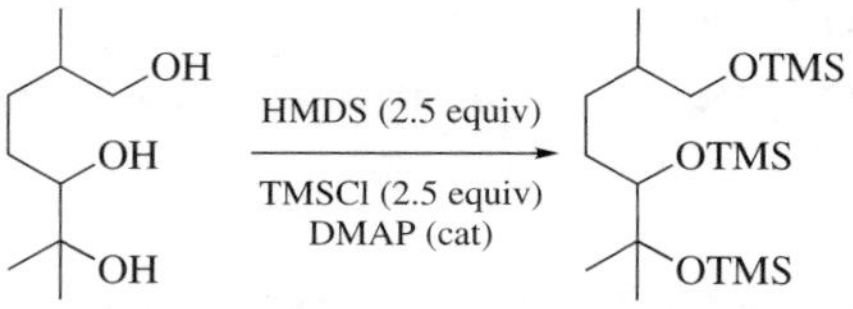

HMDS (2.5 equiv), TMSCl (2.5 equiv), DMAP (cat) (6)

Trimethysilyl ethers can be easily removed under a variety of conditions,[19] including the use of ***Tetra-n-butylammonium Fluoride*** (TBAF) (eq 7),[20] citric acid (eq 8),[24] or ***Potassium Carbonate*** in methanol (eq 9).[25] Recently, resins (OH^- and H^+ form) have been used to remove phenolic or alcoholic TMS ethers selectively (eq 10).[26]

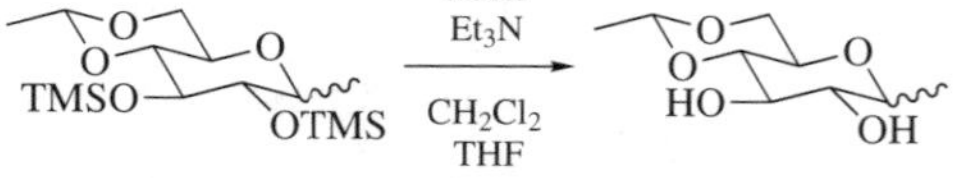

TBAF, Et_3N, CH_2Cl_2, THF (7)

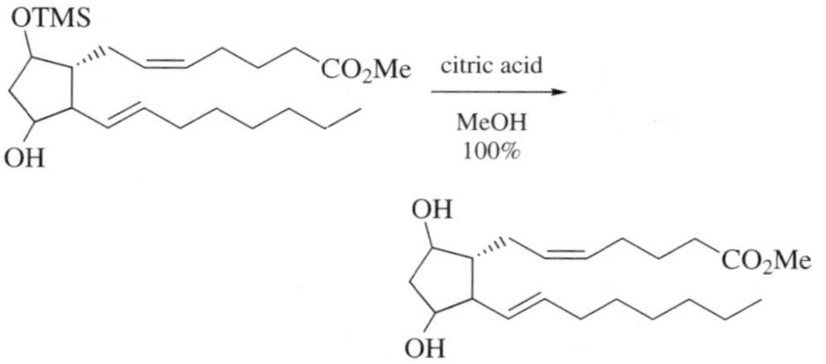

citric acid, MeOH, 100% (8)

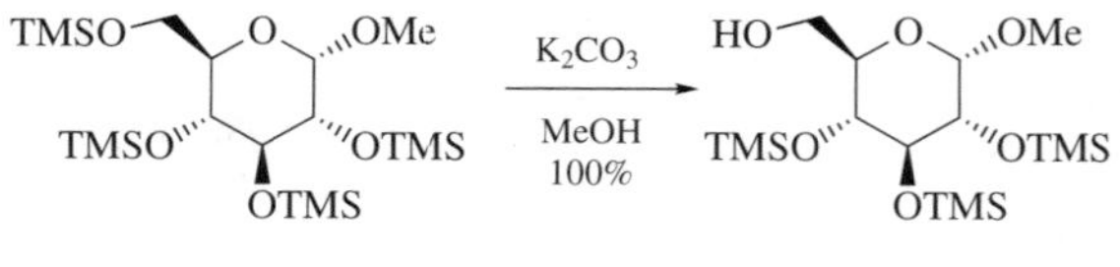

K_2CO_3, MeOH, 100% (9)

Dowex 1-X8 resin (OH⁻ form) 90%; Dowex CCR-2 resin (H⁺ form) 85% (10)

Transient Protection. Silyl ethers can be used for the transient protection of alcohols (eq 11).[27] In this example the hydroxyl groups were silylated to allow tritylation with concomitant desilylation during aqueous workup. The ease of introduction and removal of TMS groups make them well suited for temporary protection.

1. TMSCl, 2,6-lutidine, DMF; 2. TBTrBr; 3. H⁺; 91% (11)

Trimethylsilyl derivatives of amino acids and peptides have been used to improve solubility, protect carboxyl groups, and improve acylation reactions. TMSCl has been used to prepare protected amino acids by forming the *O,N*-bis-trimethylsilylated amino acid, formed in situ, followed by addition of the acylating agent (eq 12).[5] This is a general method which obviates the production of oligomers normally formed using Schotten–Baumann conditions, and which can be applied to a variety of protecting groups.[5]

H_2N–CH(CH_3)–CO_2H → (TMSCl, DIEA, CH_2Cl_2) [TMSHN–CH(CH_3)–CO_2TMS] → (1. FMOCCl; 2. H_2O; 89%) FMOCHN–CH(CH_3)–CO_2H (12)

Transient hydroxylamine oxygen protection has been successfully used for the synthesis of *N*-hydroxamides.[6] Hydroxylamines can be silylated with TMSCl in pyridine to yield the *N*-substituted *O*-TMS derivative. Acylation with a mixed anhydride of a protected amino acid followed by workup affords the *N*-substituted hydroxamide (eq 13).[6]

TMSCl, pyridine; CH_2Cl_2; H^+, 60–86% (13)

Formation of Silyl Esters. TMS esters can be prepared in good yields by reacting the carboxylic acid with TMSCl in 1,2-dichloroethane (eq 14).[4] This method of carboxyl group protection has been used during hydroboration reactions. The organoborane can be transformed into a variety of different carboxylic acid derivatives (eqs 15 and 16).[7] TMS esters can also be reduced with metal hydrides to form alcohols and aldehydes or hydrolyzed to the starting acid, depending on the reducing agent and reaction conditions.[28]

PhCH=CHCO_2H → (TMSCl, $ClCH_2CH_2Cl$, 87%) PhCH=CHCO_2TMS (14)

1. Et_3N, TMSCl; 2. BH_3, THF; 3. H_2O_2, NaOH; 73% (15)

1. Et_3N, TMSCl; 2. BH_3, THF; 3. NaI; 79% (16)

Protection of Terminal Alkynes. Terminal alkynes can be protected as TMS alkynes by reaction with ***n*-Butyllithium** in THF followed by TMSCl (eq 17).[8] A one-pot β-elimination–silylation process (eq 18) can also yield the protected alkyne.

H–C≡C–O-*t*-Bu → (1. *n*-BuLi, THF; 2. TMSCl; 80%) TMS–C≡C–O-*t*-Bu (17)

BrCH=CH–O-*t*-Bu → (1. LDA, THF; 2. TMSCl; 74%) TMS–C≡C–O-*t*-Bu (18)

Silyl Enol Ethers. TMS enol ethers of aldehydes and symmetrical ketones are usually formed by reaction of the carbonyl compound with ***Triethylamine*** and TMSCl in DMF (eq 19), but other bases have been used, including ***Sodium Hydride***[29] and ***Potassium Hydride***.[30]

TMSCl, Et_3N, DMF, 93% (19)

Under the conditions used for the generation of silyl enol ethers of symmetrical ketones, unsymmetrical ketones give mixtures of structurally isomeric enol ethers, with the predominant product being the more substituted enol ether (eq 20).[10] Highly hindered bases, such as ***Lithium Diisopropylamide*** (LDA),[31] favor formation of the kinetic, less substituted silyl enol ether, whereas ***Bromomagnesium Diisopropylamide*** (BMDA)[10] generates the more substituted, thermodynamic silyl enol ether. A combination of TMSCl/***Sodium Iodide*** has also been used to form silyl enol ethers of simple aldehydes and ketones[32] as well as from α,β-unsaturated aldehydes and ketones.[33] Additionally, treatment of α-halo ketones with ***Zinc***, TMSCl, and TMEDA in ether provides a regiospecific method for the preparation of the more substituted enol ether (eq 21).[34]

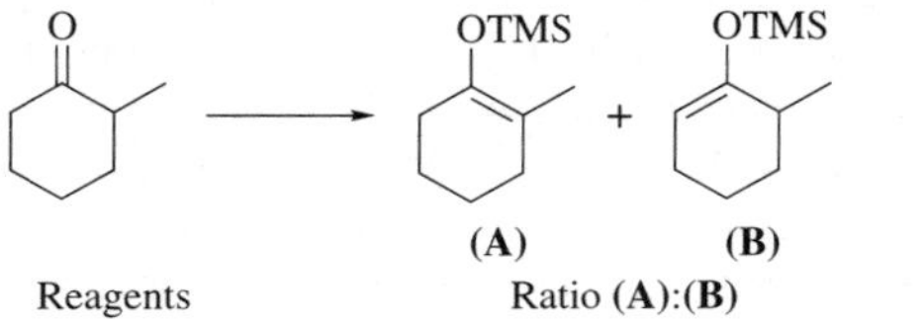

OTMS OTMS

(A) (B) (20)

Reagents	Ratio (A):(B)
LDA, DME; TMSCl	1:99
NaH, DME; TMSCl	73:27
Et_3N, TMSCl, DMF	78:22
KH, THF; TMSCl	67:33
TMSCl, NaI, MeCN, Et_3N	90:10
BMDA, TMSCl, Et_3N	97:3

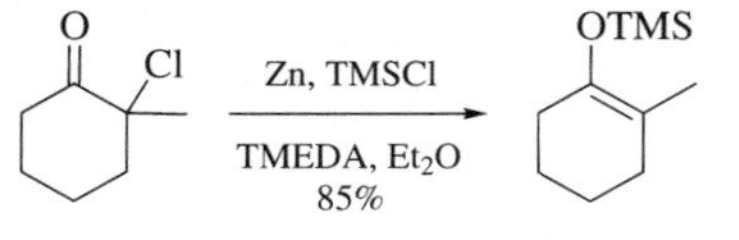

Zn, TMSCl; TMEDA, Et_2O; 85% (21)

Mild Deprotection of Boc Protecting Group. The Boc protecting group is used throughout peptide chemistry. Common ways of removing it include the use of 50% ***Trifluoroacetic Acid*** in CH_2Cl_2, ***Trimethylsilyl Perchlorate***, or ***Iodotrimethylsilane*** (TMSI).[19] A new method has been developed, using TMSCl–phenol, which enables removal of the Boc group in less than one hour (eq 22).[11] The selectivity between Boc and benzyl groups is high enough to allow for selective deprotection.

Boc-Val-OCH_2-resin —(TMSCl, phenol; 20 min; 100%)→ Val-OCH_2-resin (22)

In Situ Generation of Iodotrimethylsilane. Of the published methods used to form TMSI in situ, the most convenient involves the use of TMSCl with NaI in acetonitrile.[12] This method has been used for a variety of synthetic transformations, including cleavage of phosphonate esters (eq 23),[35] conversion of vicinal diols to alkenes (eq 24),[36] and reductive removal of epoxides (eq 25).[37]

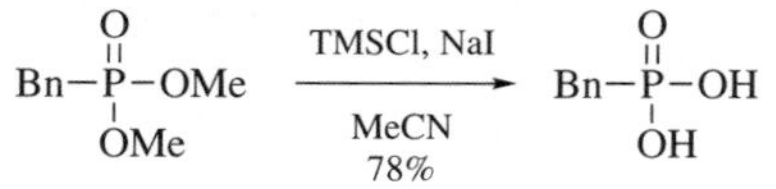

Bn–P(O)(OMe)–OMe —(TMSCl, NaI; MeCN; 78%)→ Bn–P(O)(OH)–OH (23)

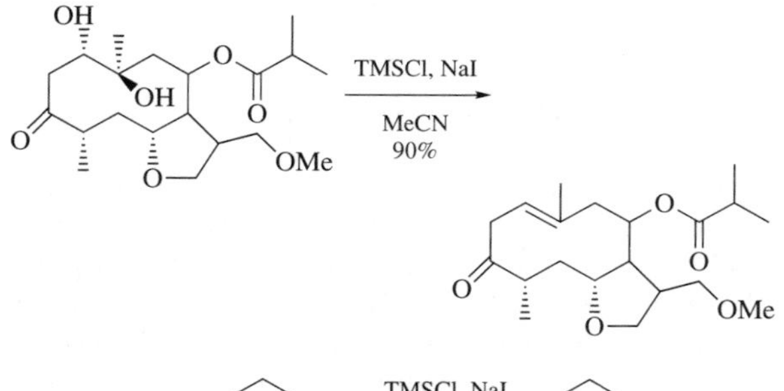

TMSCl, NaI; MeCN; 90% (24)

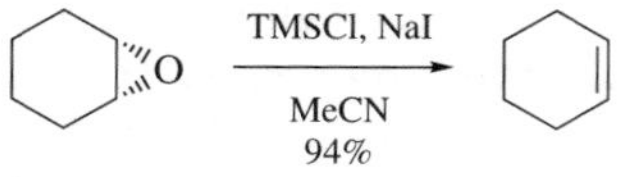

TMSCl, NaI; MeCN; 94% (25)

Conversion of Ketones to Vinylsilanes. Ketones can be transformed into vinylsilanes via intermediate trapping of the vinyl anion from a Shapiro reaction with TMSCl. Formation of either the tosylhydrazone[38] or benzenesulfonylhydrazone (eq 26)[13,39] followed by reaction with *n*-butyllithium in TMEDA and TMSCl gives the desired product.

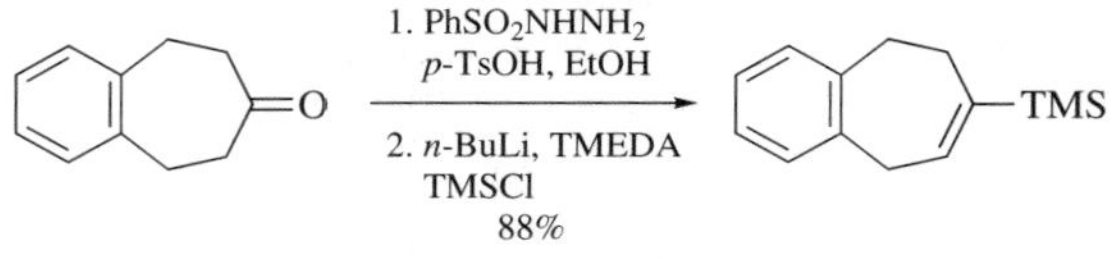

1. $PhSO_2NHNH_2$, *p*-TsOH, EtOH; 2. *n*-BuLi, TMEDA, TMSCl; 88% —TMS (26)

Epoxide Cleavage. Epoxides open by reaction with TMSCl in the presence of ***Triphenylphosphine*** or tetra-*n*-butylammonium chloride to afford *O*-protected vicinal chlorohydrins (eq 27).[14]

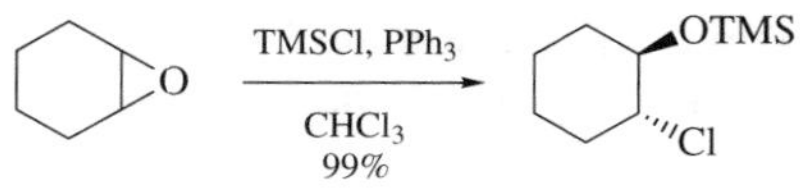

TMSCl, PPh_3; $CHCl_3$; 99% OTMS, Cl (27)

Formation of Silylvinylallenes. Enynes couple with TMSCl in the presence of Li/ether or Mg/***Hexamethylphosphoric Triamide*** to afford silyl-substituted vinylallenes. The vinylallene can be subsequently oxidized to give the silylated cyclopentanone (eq 28).[15]

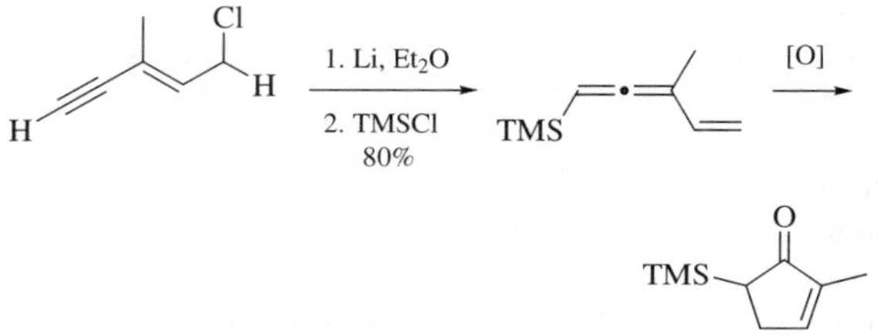

1. Li, Et_2O; 2. TMSCl; 80% TMS; [O] TMS (28)

Conjugate Addition Reactions. In the presence of TMSCl, cuprates undergo 1,2-addition to aldehydes and ketones to afford silyl enol ethers (eq 29).[16] In the case of a chiral aldehyde, addition of TMSCl follows typical Cram diastereofacial selectivity (eq 30).[16,40]

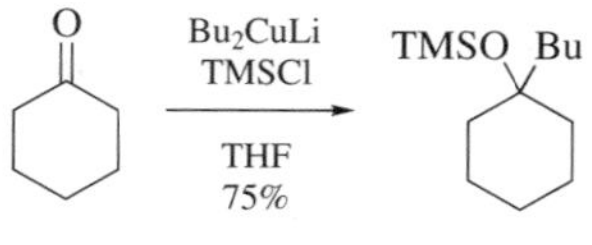

Bu_2CuLi, TMSCl; THF; 75% TMSO Bu (29)

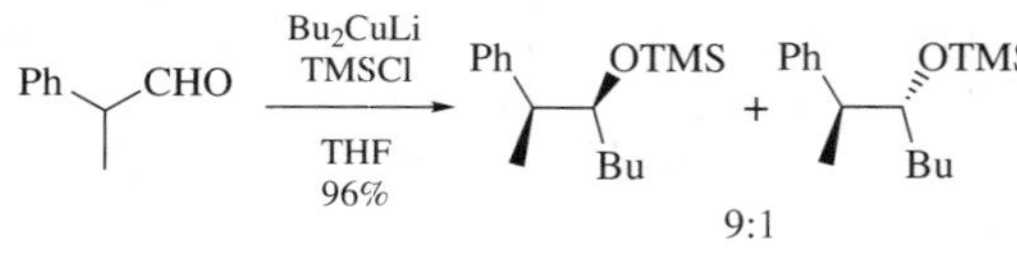

Ph CHO —(Bu_2CuLi, TMSCl; THF; 96%)→ Ph OTMS Bu + Ph OTMS Bu; 9:1 (30)

Conjugate addition of organocuprates to α,β-unsaturated carbonyl compounds, including ketones, esters, and amides, are accelerated by addition of TMSCl to provide good yields of the 1,4-addition products (eq 31).[17,41,42] The effect of additives such as HMPA, DMAP, and TMEDA have also been examined.[18,43] The role of the TMSCl on 1,2- and 1,4-addition has been explored by several groups, and a recent report has been published by Lipshutz.[40] His results appear to provide evidence that there is an interaction between the cuprate and TMSCl which influences the stereochemical outcome of these reactions.

1. $(EtCH{=}CH)_2CuLi$, TMSCl, Et_2O; 2. H_2O; 86% (31)

The addition of TMSCl has made 1,4-conjugate addition reactions to α-(nitroalkyl)enones possible despite the presence of the acidic α-nitro protons (eq 32).[44] Copper-catalyzed conjugate addition of Grignard reagents proceeds in high yield in the presence of TMSCl and HMPA (eq 33).[45] In some instances the reaction gives dramatically improved ratios of 1,4-addition to 1,2-addition.

Bu_2CuLi, TMSCl, THF; 74% (32)

MgBr, 10% CuI, TMSCl, HMPA, THF; 82% (33)

Related Reagents. *N,O*-Bis(trimethylsilyl)acetamide; *N,N'*-Bis(trimethylsilyl)urea; Cyanotrimethylsilane; Hexamethyldisilazane; Hexamethyldisiloxane; Iodotrimethylsilane; *N*-(Trimethylsilyl)imidazole; Trimethylsilyl Trifluoromethanesulfonate.

1. Colvin, E. *Silicon in Organic Synthesis*; Butterworths: Boston, 1981.
2. Weber, W. P., *Silicon Reagents for Organic Synthesis*; Springer: New York, 1983.
3. Langer, S. H.; Connell, S.; Wender, I. *JOC* **1958**, *23*, 50.
4. Hergott, H. H.; Simchen, G. *S* **1980**, 626.
5. Bolin, D. R.; Sytwu, I.-I; Humiec, F.; Meinenhofer, J. *Int. J. Peptide Protein Res.* **1989**, *33*, 353.
6. Nakonieczna, L.; Chimiak, A. *S* **1987**, 418.
7. Kabalka, G. W.; Bierer, D. E. *SC* **1989**, *19*, 2783.
8. Valenti, E.; Pericàs, M. A.; Serratosa, F. *JOC* **1990**, *55*, 395.
9. House, H. O.; Czuba, L. J.; Gall, M.; Olmstead, H. D. *JOC* **1969**, *34*, 2324.
10. Krafft, M. E.; Holton, R. A. *TL* **1983**, *24*, 1345.
11. Kaiser, E.; Tam, J. P.; Kubiak, T. M.; Merrifield, R. B. *TL* **1988**, *29*, 303.
12. Olah, G. A.; Narang, S. C.; Gupta, B. G. B.; Malhotra, R. *JOC* **1979**, *44*, 1247.
13. Paquette, L. A.; Fristad, W. E.; Dime, D. S.; Bailey, T. R. *JOC* **1980**, *45*, 3017.
14. Andrews, G. C.; Crawford, T. C.; Contillo, L. G. *TL* **1981**, *22*, 3803.
15. Dulcere, J.-P; Grimaldi, J.; Santelli, M. *TL* **1981**, *22*, 3179.
16. Matsuzawa, S.; Isaka, M.; Nakamura, E.; Kuwajima, I. *TL* **1989**, *30*, 1975.
17. Alexakis, A.; Berlan, J.; Besace, Y. *TL* **1986**, *27*, 1047.
18. Horiguchi, Y.; Matsuzawa, S.; Nakamura, E.; Kuwajima, I. *TL* **1986**, *27*, 4025.
19. Green, T. W.; Wuts, P. G. M., *Protective Groups in Organic Synthesis*; Wiley: New York, 1991.
20. Allevi, P.; Anastasia, M.; Ciufereda, P. *TL* **1993**, *34*, 7313.
21. Olah, G. A.; Gupta, B. G. B.; Narang, S. C.; Malhotra, R. *JOC* **1979**, *44*, 4272.
22. Lissel, M.; Weiffen, J. *SC* **1981**, *11*, 545.
23. Cossy, J.; Pale, P. *TL* **1987**, *28*, 6039.
24. Bundy, G. L.; Peterson, D. C. *TL* **1978**, 41.
25. Hurst, D. T.; McInnes, A. G. *CJC* **1965**, *43*, 2004.
26. Kawazoe, Y.; Nomura, M.; Kondo, Y.; Kohda, K. *TL* **1987**, *28*, 4307.
27. Sekine, M.; Masuda, N.; Hata, T. *T* **1985**, *41*, 5445.
28. Larson, G. L.; Ortiz, M.; Rodrigues de Roca, M. *SC* **1981**, 583.
29. Stork, G.; Hudrlik, P. F. *JACS* **1968**, *90*, 4462.
30. Negishi, E.; Chatterjee, S. *TL* **1983**, *24*, 1341.
31. Corey, E. J.; Gross, A. W. *TL* **1984**, *25*, 495.
32. Cazeau, P.; Duboudin, F.; Moulines, F.; Babot, O.; Dunogues, J. *T* **1987**, *43*, 2075.
33. Cazeau, P.; Duboudin, F.; Moulines, F.; Babot, O.; Dunogues, J. *T* **1987**, *43*, 2089.
34. Rubottom, G. M.; Mott, R. C.; Krueger, D. S. *SC* **1977**, *7*, 327.
35. Morita, T.; Okamoto, Y.; Sakurai, H. *TL* **1978**, *28*, 2523.
36. Barua, N. C.; Sharma, R. P. *TL* **1982**, *23*, 1365.
37. Caputo, R.; Mangoni, L.; Neri, O.; Palumbo, G. *TL* **1981**, *22*, 3551.
38. Taylor, R. T.; Degenhardt, C. R.; Melega, W. P.; Paquette, L. A. *TL* **1977**, 159.
39. Fristad, W. E.; Bailey, T. R.; Paquette, L. A. *JOC* **1980**, *45*, 3028.
40. Lipschutz, B. H.; Dimock, S. H.; James, B. *JACS* **1993**, *115*, 9283.
41. Nakamura, E.; Matsuzawa, S.; Horiguchi, Y.; Kuwajima, I. *TL* **1986**, *27*, 4029.
42. Corey, E. J.; Boaz, N. W. *TL* **1985**, *26*, 6015.
43. Johnson, C. R.; Marren, T. J. *TL* **1987**, *28*, 27.
44. Tamura, R.; Tamai, S.; Katayama, H.; Suzuki, H. *TL* **1989**, *30*, 3685.
45. Booker-Milburn, K. I.; Thompson, D. F. *TL* **1993**, *34*, 7291.

Ellen M. Leahy
Affymax Research Institute, Palo Alto, CA, USA

Copper(I) Trifluoromethanesulfonate

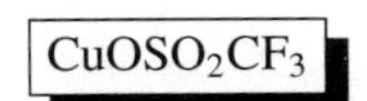

[42152-44-3] $CCuF_3O_3S$ (MW 212.62)
(2:1 benzene complex)
[37234-97-2; 42152-46-5] $C_8H_6Cu_2F_6O_6S_2$ (MW 503.34)

(efficient catalyst for $2\pi + 2\pi$ photocycloadditions and other photoreactions of alkenes,[1] and for alkene cyclopropanation and other reactions of diazo compounds; also a selenophilic and thiophilic Lewis acid that enhances the nucleofugacity of selenide and sulfide leaving groups)

Alternate Name: copper(I) triflate.
Physical Data: moisture-sensitive white crystalline solid.
Solubility: sol MeCN, $MeCO_2H$, 2-butanone, alkenes; slightly sol benzene.
Form Supplied in: must be freshly prepared.
Preparative Methods: copper(I) trifluoromethanesulfonate (CuOTf) was first prepared as a solution in acetonitrile by synproportionation of ***Copper(II) Trifluoromethanesulfonate*** with copper(0).[2] The Cu^I in these solutions is strongly coordinated with acetonitrile, forming complexes analogous to ***Tetrakis(acetonitrile)copper(I) Perchlorate***.[3] A white

crystalline solid benzene complex, $(CuOTf)_2 \cdot C_6H_6$, is prepared by the reaction of a suspension of ***Copper(I) Oxide*** in benzene with ***Trifluoromethanesulfonic Anhydride***.[4] Traces of ***Trifluoromethanesulfonic Acid*** apparently catalyze the reaction.[5] CuOTf is generated in situ by the reduction of $Cu(OTf)_2$ with diazo compounds.[6]

Handling, Storage, and Precautions: moisture sensitive.

Cyclopropanation with Diazo Compounds. Copper(I) triflate is a highly active catalyst for the cyclopropanation of alkenes with diazo compounds.[6] In contrast to other more extensively ligated copper catalysts, e.g. ***Copper(II) Acetylacetonate***, that favor cyclopropanation of the most highly substituted C=C bond, cyclopropanations catalyzed by CuOTf show a unique selectivity for cyclopropanation of the least alkylated C=C bond in both intermolecular (eq 1) and intramolecular (eq 2) competitions. The same selectivity is found with $Cu(OTf)_2$ as *nominal* catalyst. This is because $Cu(OTf)_2$ is reduced by the diazo compound to CuOTf, and CuOTf is the *actual* cyclopropanation catalyst in both cases.[6] Selective cyclopropanation of the least substituted C=C bond is a consequence of the alkene coordinating with the catalyst prior to interaction with the diazo compound, and the increase in stability of Cu^I–alkene complexes with decreasing alkyl substitution on the C=C bond. For catalysts with more strongly ligated Cu^I, an electrophilic carbene or carbenoid intermediate reacts with the free alkene, and the preference for cyclopropanation of the more highly substituted C=C bond arises from the enhancement of alkene nucleophilicity with increasing alkyl substitution (see ***Copper(II) Trifluoromethanesulfonate***).

N_2CHCO_2Et, CuX_n, 80 °C (1)

CuX_n	CO₂Et product	Ratio	EtO₂C product
CuOTf	17%	0.25:1	67%
$Cu(OTf)_2$	21%	0.27:1	77%
$Cu(acac)_2$	61%	1.78:1	35%

CH_2N_2, CuX_2, 0–10 °C (2)

X = acetylacetonate	86%	14%
X = triflate	33%	66%

Cyclopropanecarboxylic esters are conveniently available, even from volatile alkenes, because CuOTf promotes cyclopropanations in good yields at low temperatures. Thus *trans*- and *cis*-2-butenes, boiling under reflux, react stereospecifically with ***Ethyl Diazoacetate*** to produce the corresponding ethyl 2,3-dimethylcyclopropanecarboxylates (eqs 3 and 4),[6] and cyclobutene reacts with ethyl diazoacetate at 0 °C to deliver a mixture of *exo*- and *endo*-5-ethoxycarbonylbicyclo[2.1.0]pentanes (eq 5).[7]

+ N_2CHCO_2Et →(CuOTf, 53%) CO_2Et (3)

+ N_2CHCO_2Et →(CuOTf, 51%) CO_2Et (4)

+ N_2CHCO_2Et →(CuOTf, cyclohexane, 0 °C) EtO_2C 51% + CO_2Et 13% (5)

CuOTf is an outstandingly effective catalyst for the synthesis of cyclopropyl phosphonates by the reaction of ***Diethyl Diazomethylphosphonate*** with alkenes (eq 6).[8] The resulting cyclopropylphosphonates are useful intermediates for the synthesis of alkylidenecyclopropanes by Wadsworth–Emmons alkenation with aromatic carbonyl compounds (eq 7).[8]

→($N_2CHPO(OEt)_2$, CuOTf, CH_2Cl_2, 4–8 °C, 71%) $PO(OEt)_2$ (6)

$PO(OEt)_2$ →(1. BuLi, 2. Ph_2CO, 63%) Ph, Ph (7)

A complex of a chiral, nonracemic bis(oxazoline) with CuOTf is a highly effective catalyst for asymmetric cyclopropanation of alkenes.[9] Copper(II) triflate complexes do not catalyze the reaction unless they are first converted to Cu^I by reduction with a diazo compound or with phenylhydrazine. CuOTf complexes are uniquely effective. Thus the observed enantioselectivity and catalytic activity, if any, are much lower with other Cu^I or Cu^{II} salts including halide, cyanide, acetate, and even perchlorate. Both enantiomers of the bis(oxazoline) ligand are readily available. Spectacularly high levels of asymmetric induction are achieved with both mono- (eq 8) and 1,1-disubstituted alkenes (eq 9).

Ph + N_2CHCO_2Et →(bis(oxazoline) t-Bu, CuOTf (1 mol%), $CHCl_3$, 25 °C) Ph CO_2Et 55%, 98% ee; Ph CO_2Et 20%, 98% ee (8)

+ N_2CHCO_2Et →(bis(oxazoline) t-Bu, CuOTf (1 mol%), $CHCl_3$, 0 °C, 91%) CO_2Et >99% ee (9)

Asymmetric Aziridination. A chiral, nonracemic bis-(oxazoline) complex of copper(I) triflate catalyzes asymmetric aziridination of styrene in good yield (eq 10).[9] However, enantioselectivity is not as high as the corresponding cyclopropanation (eq 8).

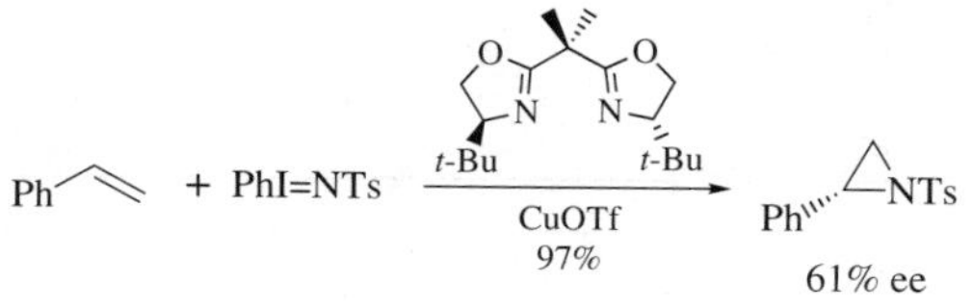

(10)

Photocycloadditions. CuOTf is an exceptionally effective catalyst for $2\pi + 2\pi$ photocycloadditions of alkenes.[1] Thus while CuBr promotes photodimerization of norbornene in only 38% yield,[10] the same reaction affords dimer in 88% yield with CuOTf as catalyst (eq 11).[11] A mechanistic study of this reaction revealed that although both 1:1 and 2:1 alkene Cu^I complexes are in equilibrium with free alkene and both the 1:1 and 2:1 complexes absorb UV light, only light absorbed by the 2:1 complex results in photodimerization. In other words, photodimerization requires precoordination of both C=C bonds with the Cu^I catalyst. Thus the exceptional ability of CuOTf, with its weakly coordinating triflate counter anion, to form π-complexes with as many as four C=C bonds[12] is of paramount importance for its effectiveness as a photodimerization catalyst.

The importance of precoordination is also evident in the CuOTf-promoted $2\pi + 2\pi$ photocycloaddition of *endo*-dicyclopentadiene. This diene forms an isolable 2:1 complex with CuOTf involving *exo*-monodentate coordination with the 8,9-C=C bond of two molecules of diene. Consequently, intermolecular $2\pi + 2\pi$ photocycloaddition involving *exo* addition to the 8,9-C=C bond is strongly favored over intramolecular reaction between the 8,9- and 3,4-C=C bonds (eq 11).[11] This contrasts with the intramolecular photocycloaddition that is promoted by high energy triplet sensitizers.[12]

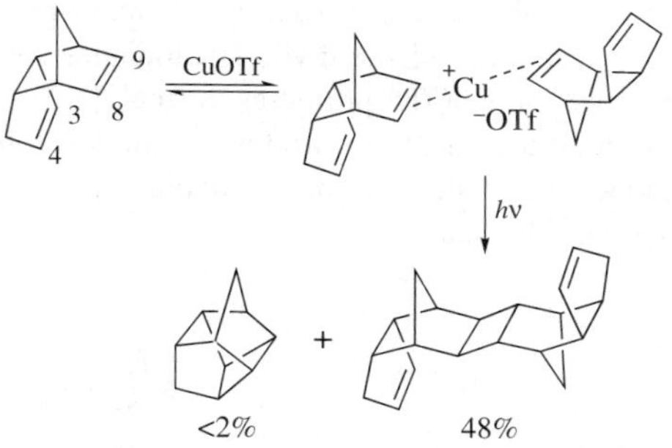

(11)

Especially interesting is the *trans,anti,trans* stereochemistry of the major cyclobutane product generated in the photodimerization of cyclohexene (eq 12).[13] It was noted that the formation of this product may be the result of a preliminary CuOTf promoted *cis–trans* photoisomerization that generates a *trans*-cyclohexene intermediate (eq 13).[13] Since one face of the *trans* C=C bond is shielded by a polymethylene chain, the *trans*-cyclohexene is restricted to suprafacial additions. Although a highly strained *trans*-cyclohexene intermediate could be stabilized by coordination with Cu^I, such a complex has not been isolated.

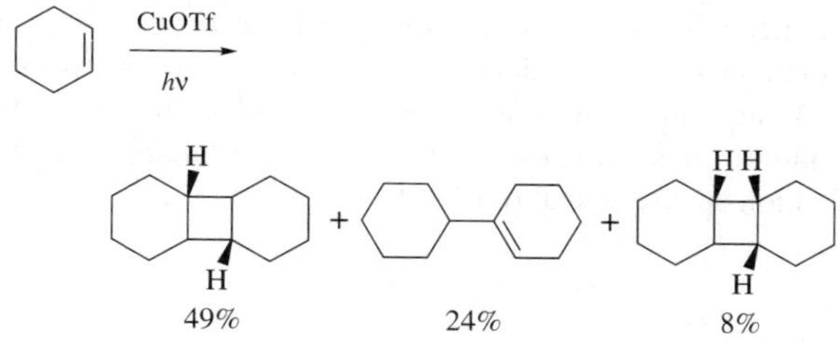

(12)

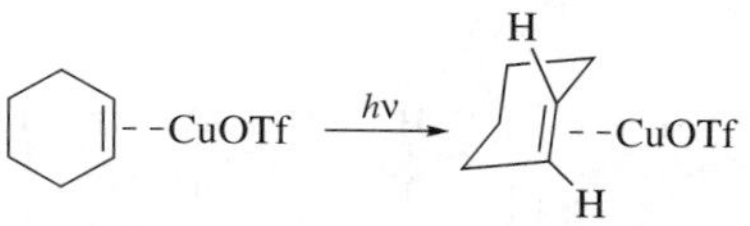

(13)

An isolable CuOTf complex of a highly strained alkene, *trans*-cycloheptene, is produced by UV irradiation of a hexane solution of *cis*-cycloheptene in the presence of CuOTf (eq 14).[14] Photocycloaddition of cycloheptene is also catalyzed by CuOTf. Surprisingly, the major product is not a *trans,anti,trans* dimer analogous to that formed from cyclohexene (eq 12) but rather a *trans,anti,trans,anti,trans* trimer (eq 15).[15]

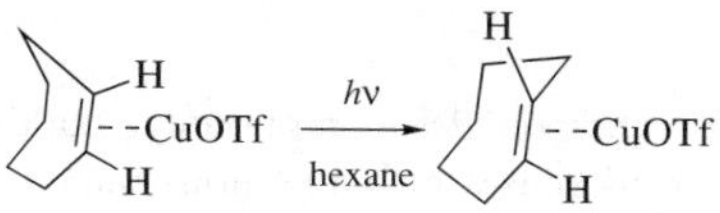

(14)

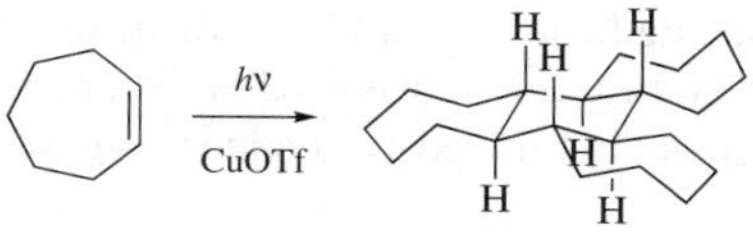

(15)

Dissolution of the *trans*-cycloheptene–CuOTf complex in cycloheptene and evaporation of the solvent delivers a tris alkene complex of CuOTf containing one *trans*-cycloheptene and two *cis*-cycloheptene ligands. Heating *trans*-cycloheptene–CuOTf in neat *cis*-cycloheptene delivers the *trans,anti,trans,anti,trans* trimer (eq 16). Experiments with *cis*-cycloheptene-d_4 show that the cyclotrimerization involves only *trans*-cycloheptene molecules, although the reaction is accelerated by the presence of *cis*-cycloheptene.[16] A likely explanation for these observations is 'concerted "template" cyclotrimerization' of a tris-*trans*-cycloheptene–CuOTf complex formed by ligand redistribution (eq 16).[16]

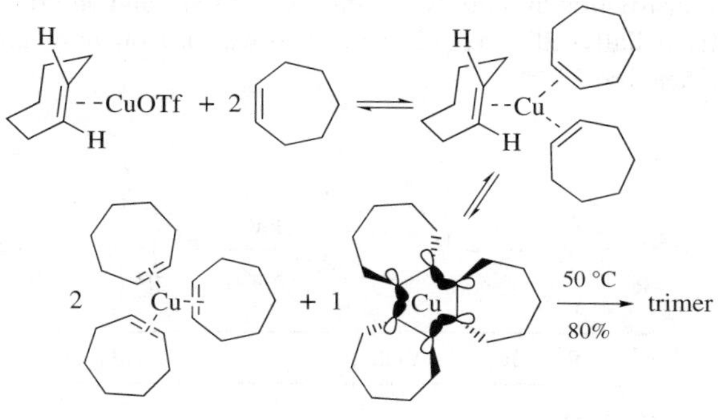

(16)

The involvement of a transient photogenerated *trans*-cyclohexene–CuOTf intermediate was also adduced to explain CuOTf catalysis of photoinduced $2\pi + 4\pi$ cycloaddition between *cis*-cyclohexene and 1,3-butadiene (eq 17).[17] In contrast to thermal

Diels–Alder reactions, this reaction generates *trans*-Δ^2-octalin rather than the *cis* cycloadduct expected for a $2\pi_s + 4\pi_s$ cycloaddition. A mechanism was proposed that involves the $2\pi_s + 4\pi_s$ cycloaddition of a *trans*-cyclohexene with 1,3-butadiene in the coordination sphere of Cu^I (eq 18).[17]

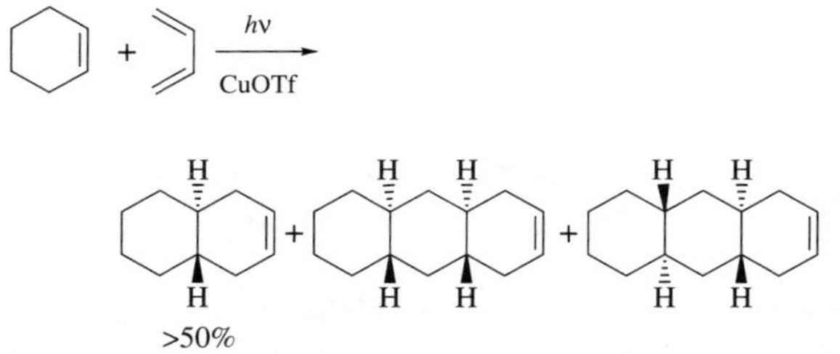

hν, CuOTf (17)

>50%

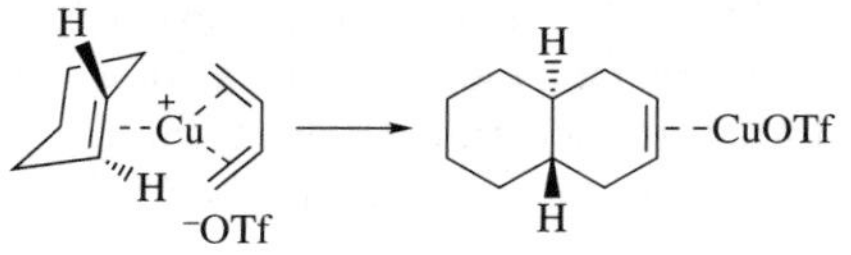

(18)

That CuOTf-catalyzed $2\pi + 2\pi$ photocycloadditions are not restricted to cyclic alkenes was first demonstrated in mixed cycloadditions involving allyl alcohol. To suppress homodimerization of *endo*-dicyclopentadiene (i.e. eq 11) the diene to Cu^I ratio is maintained at $\leq 1{:}1$ and allyl alcohol is used as solvent. Under these conditions, a high yield of mixed cycloadduct is generated (eq 19).[18]

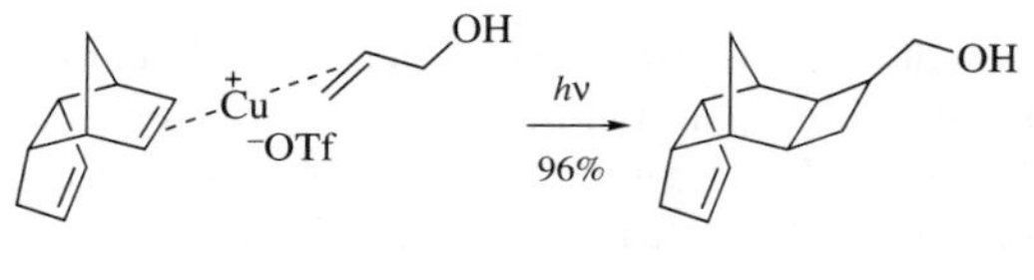

hν, 96% (19)

That both C=C bonds participating in $2\pi + 2\pi$ photocycloadditions can be acyclic is evident from the photobicyclization reactions of simple diallyl ethers that deliver bicyclic tetrahydrofurans (eq 20).[19,20] In conjunction with ***Ruthenium(VIII) Oxide***-catalyzed oxidation by ***Sodium Periodate***, these CuOTf-catalyzed photobicyclizations provide a synthetic route to butyrolactones from diallyl ethers (eq 20).[20] The synthetic method is applicable to the construction of multicyclic tetrahydrofurans and butyrolactones from diallyl ethers (eqs 21 and 22) as well as from homoallyl vinyl ethers (eq 23).[20]

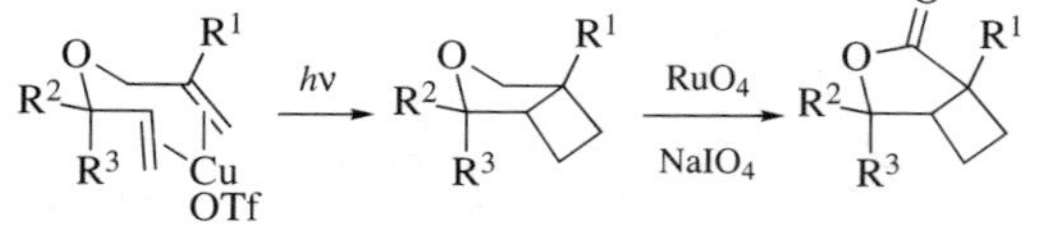

hν; RuO_4, $NaIO_4$ (20)

R^1	R^2	R^3	Yield (%)	Yield (%)
H	H	H	52	91
Me	Me	H	56	94
Me	Me	Me	54	44
Me	H	H	54	56
n-Bu	H	Me	83	83

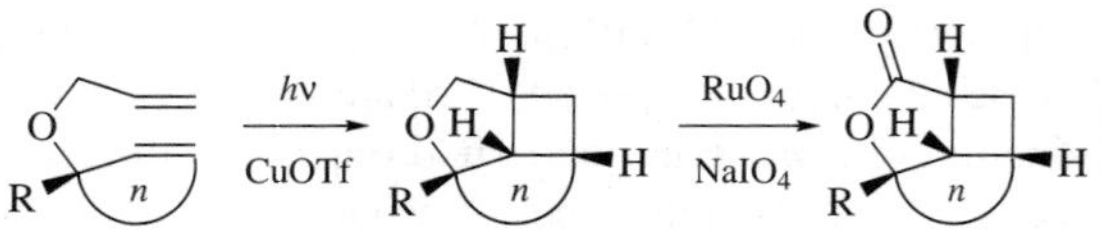

hν, CuOTf; RuO_4, $NaIO_4$ (21)

R	*n*	Yield (%)	Yield (%)
H	5	47	73
H	8	56	56
Me	5	28	87
Me	6	35	82

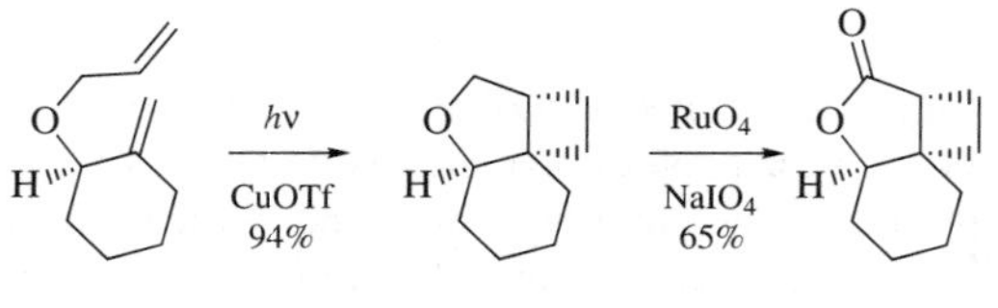

hν, CuOTf, 94%; RuO_4, $NaIO_4$, 65% (22)

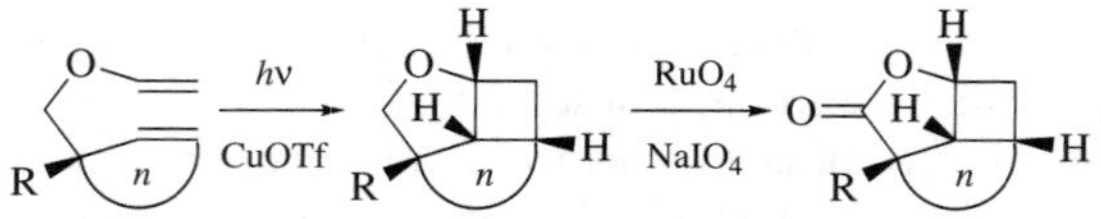

hν, CuOTf; RuO_4, $NaIO_4$ (23)

R	*n*	Yield (%)	Yield (%)
H	5	92	78
Me	5	50	71
H	6	48	85

3-Oxabicyclo[3.2.0]heptanes are also produced in the CuOTf-catalyzed photocycloadditions of allyl 2,4-hexadienyl ethers (eq 24).[21] The CuOTf-catalyzed photocycloadditions of bis-2,4-hexadienyl ethers are more complex. Thus UV irradiation of 5,5′-oxybis[(*E*)-1,3-pentadiene] in THF for 120 h produces vinylcyclohexene and tricyclo[3.3.0.0^{2,6}]octane derivatives (eq 25).[22] However, shorter irradiations reveal that these products arise by secondary CuOTf-catalyzed rearrangements of 6,7-divinyl-3-oxabicyclo[3.2.0]heptanes that are the primary photoproducts (eq 26). UV irradiation of the divinylcyclobutane intermediates in the presence of CuOTf promotes formal [1,3]- and [3,3]-sigmatropic rearrangements to produce a vinylcyclohexene and a 1,5-cyclooctadiene that is the immediate precursor of the tricyclo[3.3.0.0^{2,6}]octane.

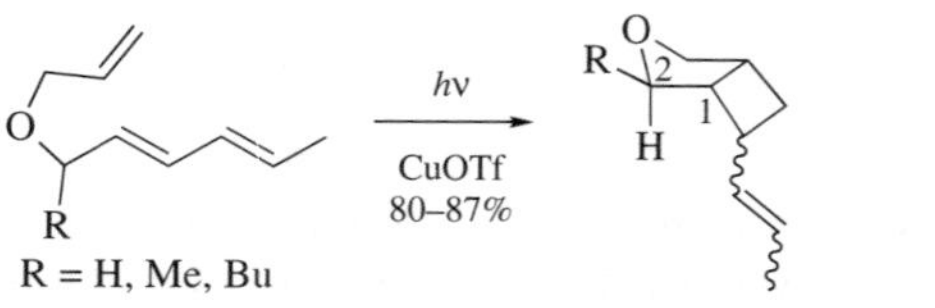

R = H, Me, Bu; hν, CuOTf, 80–87% (24)

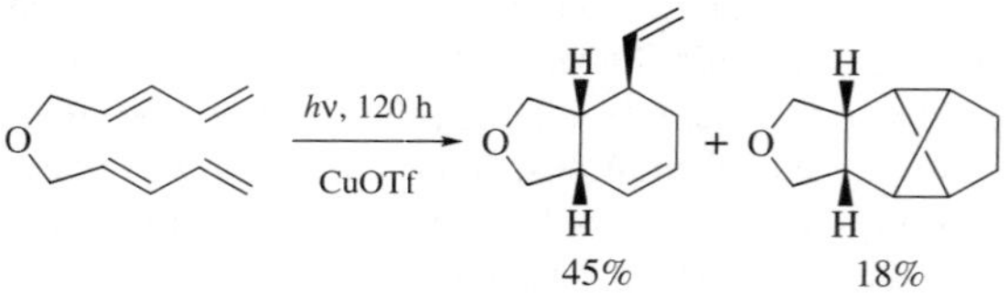

hν, 120 h, CuOTf; 45% + 18% (25)

$h\nu$, 3 h, CuOTf; $h\nu$, CuOTf (26)

CuOTf-catalyzed photobicyclization of 1,6-heptadien-3-ols produces bicyclo[3.2.0]heptan-2-ols (eq 27).[23] In conjunction with pyrolytic fragmentation of the derived ketones, these CuOTf-catalyzed photobicyclizations provide a synthetic route to 2-cyclopenten-1-ones from 1,6-heptadien-3-ols (eq 28).[23] The derived ketones can also be converted into lactones by Baeyer–Villiger oxidation and, in conjunction with pyrolytic fragmentation, CuOTf-catalyzed photobicyclizations provide a synthetic route to enol lactones of glutaraldehydic acid from 1,6-heptadien-3-ols (eq 28).[23]

$h\nu$; CrO_3, H_2SO_4 (27)

R^1	R^2	R^3	Yield (%)	Yield (%)
H	H	H	86	78
Me	H	H	81	67
H	Me	H	91	92
H	H	Me	84	92
Me	Me	H	83	93

580 °C, 54–87%; m-CPBA, 93%; 620 °C, 50% (28)

Copper(I) triflate-catalyzed photobicyclization of β- and γ-(4-pentenyl)allyl alcohols provides a synthetic route to various multicyclic carbon networks in excellent yields (eqs 29–31).[24] The reaction was exploited in a total synthesis of the panasinsene sesquiterpenes (eq 32).[25] It is especially noteworthy in this regard that attempted synthesis of a key tricyclic ketone intermediate for the panasinsenes by the well-known photocycloaddition of ***Isobutene*** to an enone failed to provide any of the requisite cyclobutyl ketone (eq 33).[25]

$h\nu$, CuOTf, 96% (29)

$h\nu$, CuOTf; $n = 5$, 93%; $n = 6$, 94% (30)

$h\nu$, CuOTf, 91% (31)

1. $h\nu$, CuOTf; 2. PCC, NaOAc, 54%; 1. MeLi; 2. $SOCl_2$, py (32)

α-panasinsene, 14% β-panasinsene, 36%

$h\nu$ (33)

In conjunction with carbocationic skeletal rearrangement, photobicyclization of 1,6-heptadien-3-ols provides a synthetic route to 7-hydroxynorbornanes (eq 34).[26] Noteworthy is the stereoselective generation of *exo*-1,2-polymethylenenorbornanes from either the *exo* or *endo* epimer of 2,3-polymethylenebicyclo[3.2.0]heptan-3-ol.

$h\nu$, CuOTf; 1. TFA, H_2O; 2. NaOH, MeOH, H_2O (34)

n	Yield (%)	Yield (%)
5	70–75	64–85
6	51–84	76–77
7	73	73
8	74	66

N,N-Diallylamides are recovered unchanged when irradiated in the presence of CuOTf.[27] This is because the amide chromophore interferes with photoactivation of the Cu^I–alkene complex. Thus CuOTf–alkene complexes containing one, two, three, or even four coordinated C=C bonds exhibit UV absorption at 235 ± 5 nm (ϵ_{max} 2950 ± 450).[12] The CuOTf complex of ethyl *N,N*-diallylcarbamate exhibits $\lambda_{max} = 233.4$ nm (ϵ_{max} 2676) but the free ligand is virtually transparent at this wavelength. Consequently, UV irradiation of ethyl *N,N*-diallylcarbamates in the presence of CuOTf delivers bicyclic (eq 35) or tricyclic (eq 36) pyrrolidines incorporating the 3-azabicyclo[3.2.0]heptane ring system.[27]

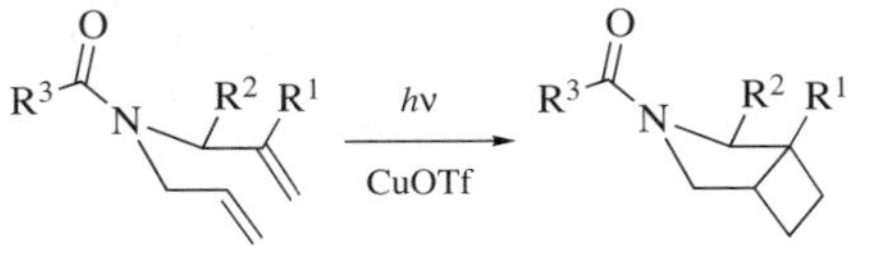

(35)

R^1	R^2	R^3	$\varepsilon_{233\ nm}$	Yield (%)
H	H	Me	192	0
H	H	H	231	0
H	H	OEt	15	74
H	Me	OEt	–	60
Me	H	OEt	–	76

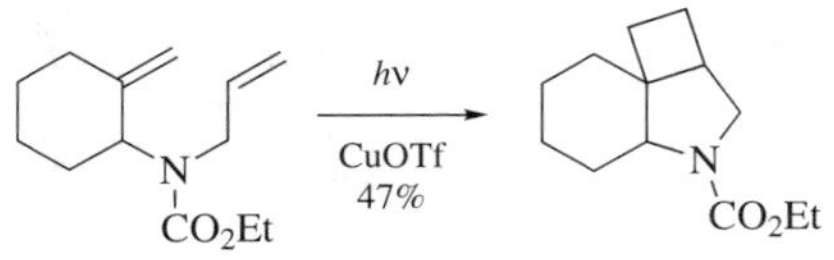

(36)

Catalyzed Diels–Alder Reactions. The uncatalyzed thermal intramolecular Diels–Alder reaction of 5,5′-oxybis[(*E*)-1,3-pentadiene] nonstereoselectively generates four isomeric 4-vinylcyclohexenes (eq 37). The major product has a *trans* ring fusion, in contrast to the single *cis* ring-fused isomer generated in the copper(I) triflate-catalyzed photoreaction of the same tetraene (eq 25). Copper(I) triflate also catalyzes a thermal Diels–Alder reaction of 5,5′-oxybis[(*E*)-1,3-pentadiene] that proceeds under milder conditions than the uncatalyzed reaction. The stereoselectivity is remarkably enhanced, generating mainly the major isomer of the uncatalyzed thermal reaction and a single *cis*-fused isomer (eq 37) that is different than the one favored in the photochemical reaction (eq 25).

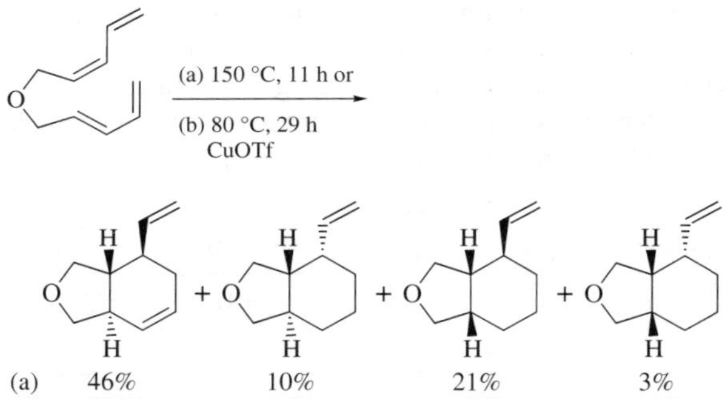

(37)

(a)	46%	10%	21%	3%
(b)	76%	<1%	<1%	11%

C_{sp}–H Bond Activation. Hydrogen–deuterium exchange between terminal alkynes and CD_3CO_2D is catalyzed by CuOTf (eq 38).[28] Proton NMR studies revealed that CuOTf and alkynes form π-complexes that rapidly exchange coordinated with free alkyne. A complex of CuOTf with 1,7-octadiene was isolated (eq 39). The complex rapidly exchanges terminal alkynic hydrogen with deuterium from CD_3CO_2D and undergoes a much slower conversion to a copper alkynide (eq 39).[28] Exchange of alkynic hydrogen and deuterium is also catalyzed by CuOTf (eq 40).[28]

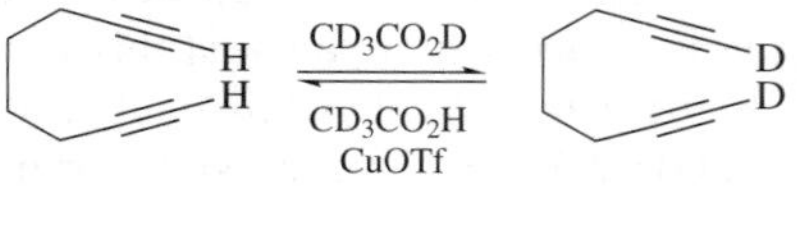

(38)

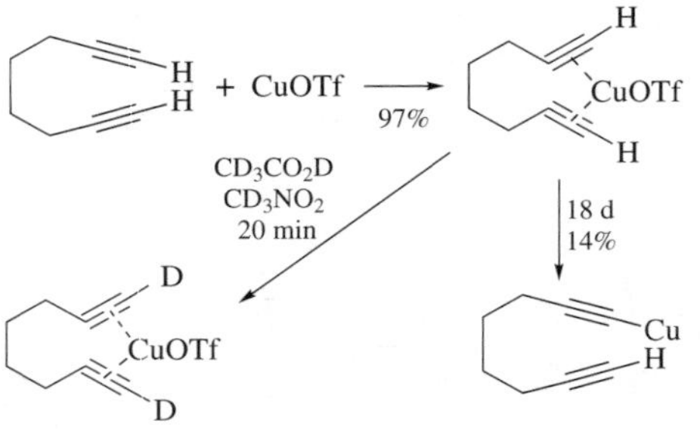

(39)

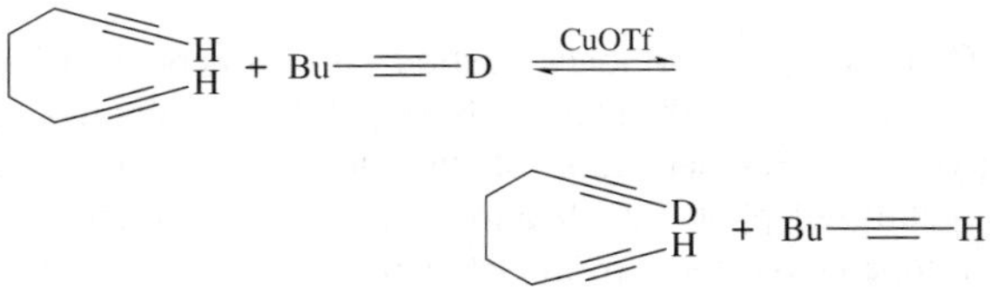

(40)

Activation of Aryl Halides. Ullmann coupling of *o*-bromonitrobenzene is accomplished under exceptionally mild conditions and in homogeneous solution by reaction with copper(I) triflate in the presence of aqueous NH_3 (eq 41).[29] Yields are enhanced by the presence of a small quantity of copper(II) triflate. That the reaction is diverted to reductive dehalogenation by ***Ammonium Tetrafluoroborate*** is presumptive evidence for an organocopper intermediate that can be captured by protonation.

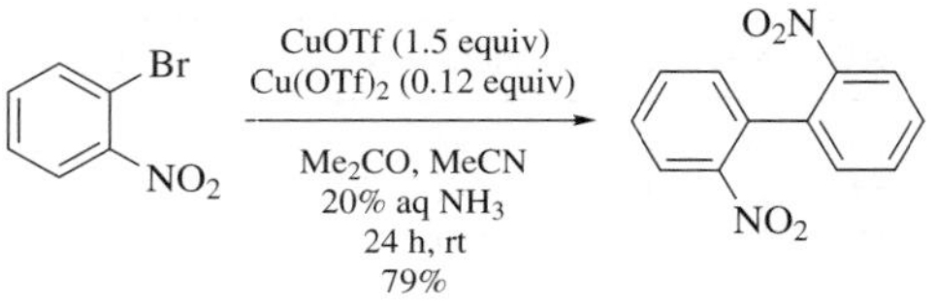

(41)

Biaryl is only a minor product from the reaction of methyl *o*-bromobenzoate with CuOTf (eq 42). The major product can result from replacement of the halide by NH_2, H, or OH, depending on reaction conditions. In the presence of 5% aqueous NH_3, methyl anthranilate is the major product.[30] More concentrated aqueous NH_3 (20%) favors the generation of methyl salicylate, and the yield of this product is enhanced by the presence of a substantial quantity of Cu^{II} ion.[29] Reductive dehalogenation is favored by the presence of ammonium ions, presumably owing to protonolysis of an arylcopper(III) intermediate.[29]

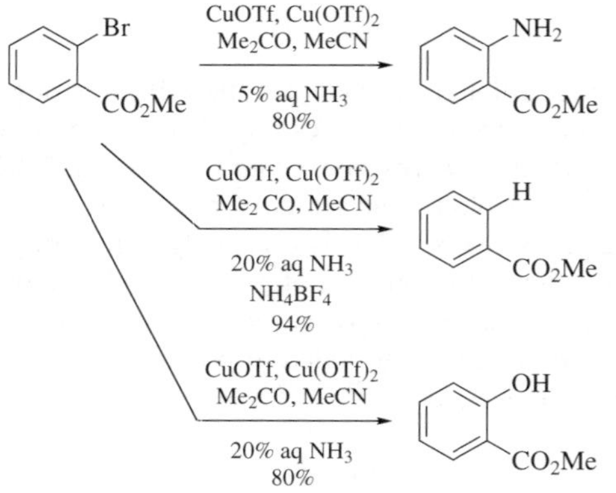

(42)

Activation of Vinyl Halides. Under the optimum conditions for reductive coupling of *o*-bromonitrobenzene (eq 41), diethyl iodofumarate gives very little coupling product; the overwhelming product was diethyl fumarate generated by hydrodehalogenation (eq 43).[29] Reductive coupling delivers *trans,trans*-1,2,3,4-tetraethoxycarbonyl-1,3-butadiene in 95% yield (GLC, or 80% of pure crystalline product) in 2 h if aqueous NH_3 is replaced by anhydrous NH_3.[29] Under the same conditions, diethyl iodomaleate undergoes 45% conversion in 20 h to deliver diethyl maleate, as well as minor amounts of *cis,cis*- and *trans,trans*-tetraethoxycarbonyl-1,3-butadiene (eq 44).[29] The stereospecificity of the reductive dehalogenations in eqs 43 and 44 is presumptive evidence for the noninvolvement of radicals in these reactions.

CuOTf, Cu(OTf)$_2$, Me$_2$CO, MeCN, 20% aq NH$_3$; CuOTf, Cu(OTf)$_2$, Me$_2$CO, MeCN, dry NH$_3$, 2 h, rt, 80% (43)

CuOTf, Cu(OTf)$_2$, Me$_2$CO, MeCN, dry NH$_3$, 20 h, rt; 30%, 8.6%, 2.1% (44)

Elimination of Thiophenol from Thioacetals. Conversion of thioacetals to vinyl sulfides is accomplished under exceptionally mild conditions by treatment with $(CuOTf)_2 \cdot C_6H_6$ (eq 45).[31] The reaction involves an α-phenylthio carbocation intermediate. Three factors contribute to the effectiveness of this synthetic method: the Lewis acidity of a copper(I) cation that is unencumbered by a strongly coordinated counter anion, the solubility of the copper(I) triflate–benzene complex, and the insolubility of CuSPh in the reaction mixture. An analogous elimination reaction provides an effective route to phenylthio enol ethers from ketones (eq 46).[31]

+ CuOTf, benzene, THF, 3 h, rt; + CuSPh; 91% (45)

PhSH; CuOTf, 10 min, rt, 92% (46)

This conversion of thioacetals into vinyl sulfides was applied to a C–C connective synthesis of 2-phenylthio-1,3-butadienes from aldehydes (eq 47).[32] The key elimination step converts cyclobutanone thioacetal intermediates into 1-phenylthiocyclobutenes that undergo electrocyclic ring opening to deliver dienes.

RCHO; 55–80%; HBr, ZnBr$_2$, PhSH (5 equiv), 73–95%; CuOTf, EtN(i-Pr)$_2$, 75–87%; 450 °C, 77–97% (47)

R = Et, i-Pr, c-C$_6$H$_{11}$, p-XC$_6$H$_4$; X = H, Me, OMe

A different synthesis generates 2-phenylthio-1,3-butadienes directly by elimination of two molecules of thiophenol from β-phenylthio thioacetals that are readily available from the corresponding α,β-unsaturated ketones (eq 48).[33]

PhSH, dry HCl, 88%; CuOTf, 0–25 °C, 76% (48)

CuOTf-promoted elimination of thiophenol was exploited in two syntheses of 1-phenylthio-1,3-butadiene, one a C–C connective route from allyl bromide[31] and bis(phenylthio)methyllithium,[34] and another from ***Crotonaldehyde*** (eq 49).[33] A topologically analogous C–C connective strategy provides 2-methoxy-1-phenylthio-1,3-butadiene from acrolein (eq 50).[5,33] That the phenylthio rather than the methoxy substituent in 2-methoxy-1-phenylthio-1,3-butadiene controls the orientation of its Diels–Alder cycloadditions is noteworthy (eq 50).

+ LiCH(SPh)$_2$; CuOTf, 24 h, 46 °C; PhSH, dry HCl, 15 min, 0 °C, 89%; CuOTf, i-Pr$_2$NEt, 25 °C, 14 h, 84% (49)

1. LiCH(SPh)$_2$ 2. Me$_2$SO$_4$, 48–63%; CuOTf, i-Pr$_2$NEt, 70–78 °C, 55–85%; COMe, 25 °C, 71 h, 66% (50)

A synthesis of 4-alkyl-2-methoxy-1-phenylthio-1,3-butadienes by a simple β-elimination of thiophenol from a thioacetal is not possible owing to skeletal rearrangement that is fostered by stabilization of a cyclopropylcarbinyl carbocation intermediate by the alkyl substituent (eq 51).[35] Interconversion of an initial α-phenylthio carbocation to a more stable α-methoxy carbocation intermediate leads to the generation of a 4-alkyl-1-methoxy-2-phenylthio-1,3-butadiene instead.

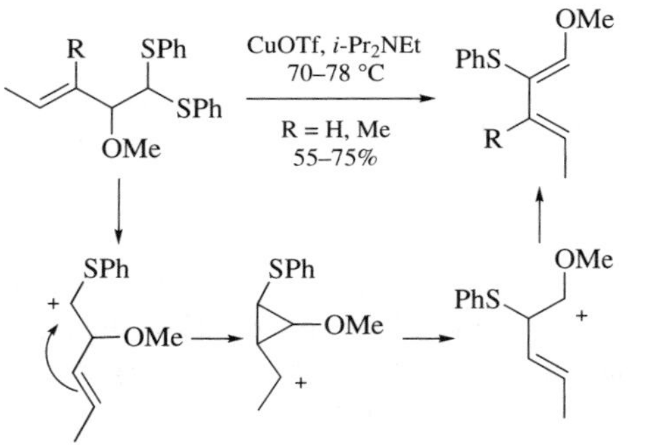

(51)

Syntheses of 1-phenylthio-1,3-butadienes from carboxylic esters (eq 52) and carboxylic acids (eq 53) are achieved by CuOTf-promoted elimination of thiophenol from intermediate thioacetals.[36]

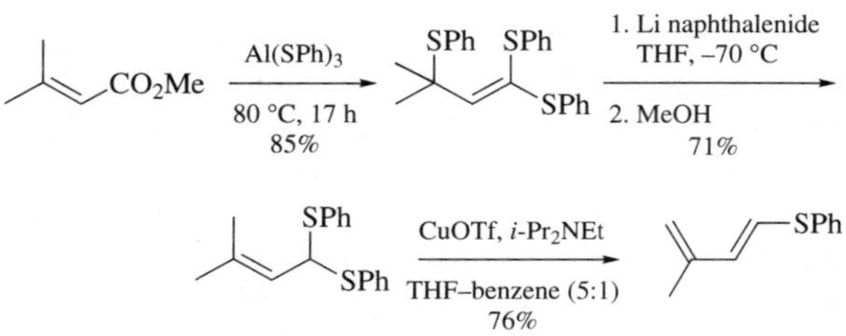

(52)

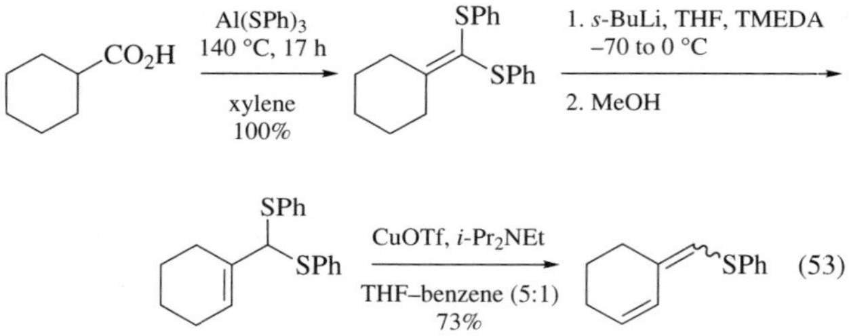

(53)

Heterocyclization of γ-Keto Dithioacetals. A C–C connective synthesis of furans is completed by a CuOTf-promoted heterocyclization of γ-keto thioacetals (eq 54).[31] Rather than simple β-elimination to generate a vinyl sulfide (eq 46), a presumed γ-keto carbocation intermediate is captured intramolecularly by an intimately juxtaposed carbonyl oxygen nucleophile.

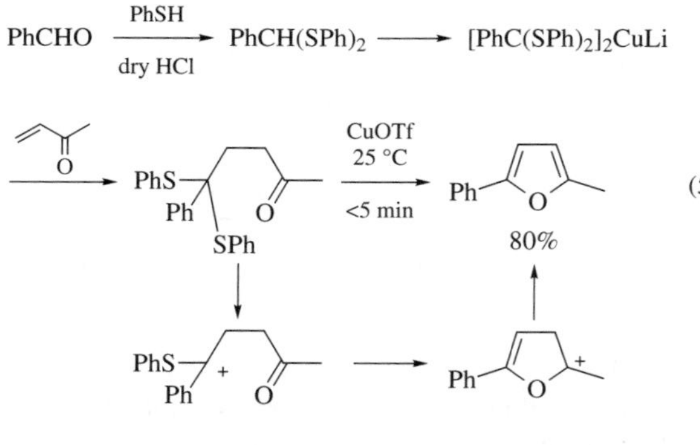

(54)

Friedel–Crafts Alkylation of Arenes with Thioacetals. $(CuOTf)_2 \cdot C_6H_6$ promotes α-thioalkylation of anisole by a dithioacetal under mild conditions (eq 55).[37]

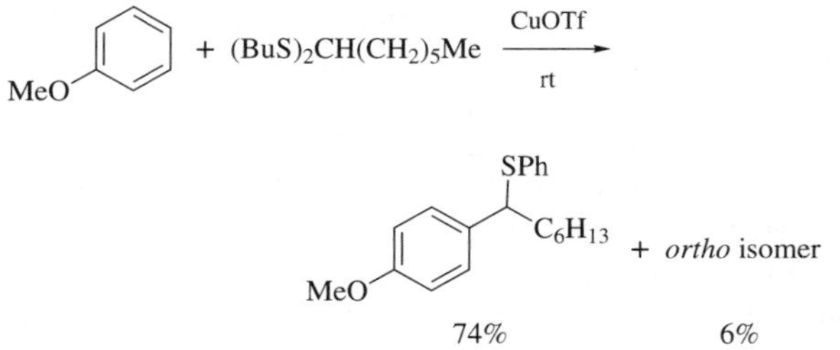

(55)

Elimination of Benzylic Phenyl Thioethers. That C–S bond activation by CuOTf is not limited to substrates that can generate sulfur-stabilized carbocation intermediates is illustrated by a C–C connective synthesis of *trans*-stilbene (eq 56).[31] The elimination of thiophenol under mild conditions is favored by benzylic stabilization of a carbocation intermediate or an E2 transition state with substantial carbocationic character.

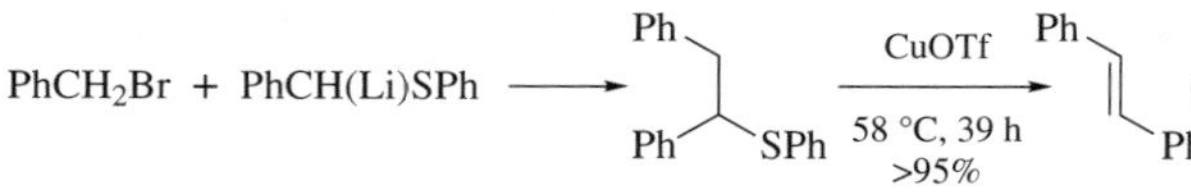

(56)

Hydrolysis of Vinylogous Thioacetals. Carbanions prepared by lithiation of γ-phenylthioallyl phenyl thioethers can serve as synthetic equivalents of β-acyl vinyl anions.[38] Umpölung of the usual electrophilic reactivity of 2-cyclohexenone is achieved by a sequence exploiting electrophilic capture of a lithiated vinylogous thioacetal and subsequent CuOTf-assisted hydrolysis (eq 57).[39] Otherwise unfunctionalized vinylogous thioacetals can be hydrolyzed to enones by ***Mercury(II) Chloride*** in wet acetonitrile.[38] However, the keto-substituted derivative in eq 57 gave only a 25% yield of enone by this method. A superior yield was obtained by CuOTf-assisted hydrolysis.[39]

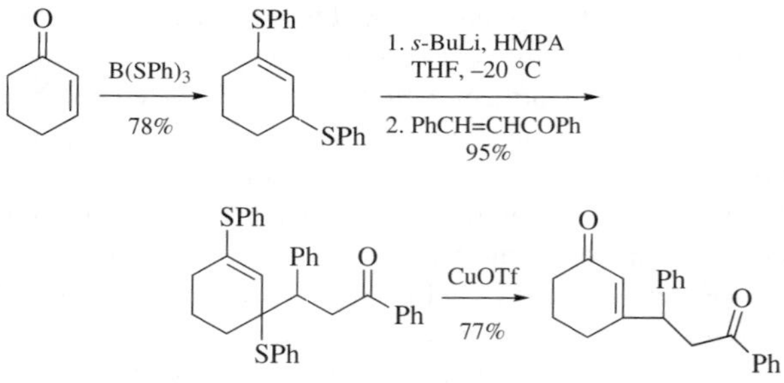

(57)

Grob Fragmentation of β-[Bis(phenylthio)methyl]-alkoxides. A method for achieving Grob-type fragmentation of five- and six-membered rings depends upon the ability of a thiophenyl group to both stabilize a carbanion and serve as an anionic leaving group. For example, reaction of cyclohexene oxide with lithium bis(phenylthio)methide[34] produces a β-[bis(phenylthio)methyl]alkanol that undergoes fragmentation in excellent yield upon treatment with ***n-Butyllithium*** followed by CuOTf (eq 58).[40] Copper(I) trifluoroacetate is equally effective but salts of other thiophilic metals, e.g. mercury or silver, were ineffective. Treatment of the intermediate β-[bis(phenylthio)methyl]alkanol with CuOTf in the absence of added strong base leads primarily to elimination of thiophenol as expected (see eq 45). Fragmentation does not occur with only one equivalent of CuOTf. This suggests a key intermediate with at least one Cu^I ion to coordinate with the alkoxide and another to activate the phenylthio leaving group (eq 58).

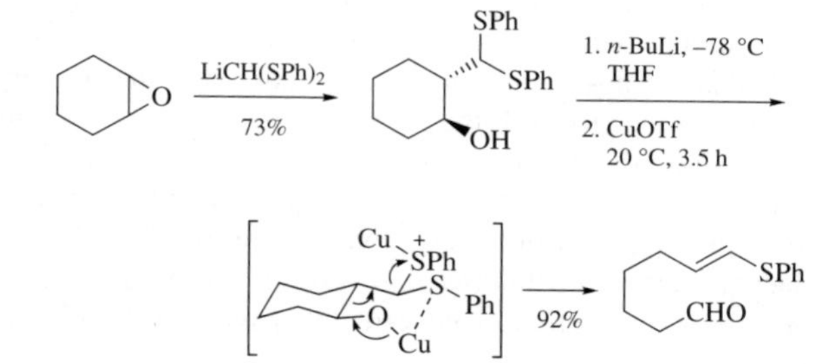

(58)

Ring-Expanding Rearrangements of α-[Bis(phenylthio)methyl]alkanols. A one-carbon ring-expanding synthesis of α-phenylsulfenyl ketones from homologous ketones depends upon the ability of a thiophenyl group to both stabilize a carbanion and serve as an anionic leaving group. For example, reaction of cyclopentanone with lithium bis(phenylthio)methide[34] produces an α-[bis(phenylthio)methyl]alkanol that rearranges to a ring-expanded α-phenylsulfenyl ketone in good yield upon treatment with CuOTf in the presence of ***Diisopropylethylamine*** (eq 59).[41] Epoxy thioether intermediates are generated from the α-[bis(phenylthio)methyl]alkanols by intramolecular nucleophilic displacement of thiophenoxide.

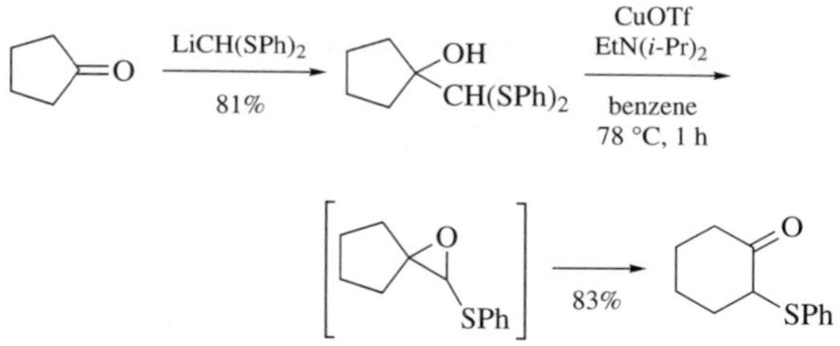

(59)

An analogous synthesis of α,α-bis(methylsulfenyl) ketones from homologous ketones by one-carbon ring expansion depends on copper(I)-promoted rearrangement of an α-tris(methylthio)methyl alkoxide intermediate (eq 60). Both ***Tetrakis(acetonitrile)copper(I) Perchlorate***[42] and ***Tetrakis(acetonitrile)copper(I) Tetrafluoroborate***[43] are effective in promoting the rearrangement but $(CuOTf)_2{\cdot}C_6H_6$, $HgCl_2$, or $Hg(TFA)_2$ are not. Apparently, the MeCN ligand is crucial. Furthermore, treatment of the intermediate α-[tris(methylthio)methyl] alcohol with CuOTf and EtN(i-Pr)$_2$ in toluene followed by aqueous workup delivers an α-hydroxy methylthio ester (eq 61),[43] in contrast to the ring-expanding rearrangement of the analogous α-[bis(phenylthio)methyl]alkanol (eq 59).[41]

The α-[bis(phenylthio)methyl]alkanol derived from cycloheptanone does not undergo ring expansion upon treatment with CuOTf and EtN(i-Pr)$_2$ in benzene. Instead, 1,3-elimination of thiophenol delivers an epoxy thioether intermediate that undergoes a rearrangement involving 1,2-shift of a phenylsulfenyl group to produce an α-phenylsulfenyl aldehyde (eq 62).[41]

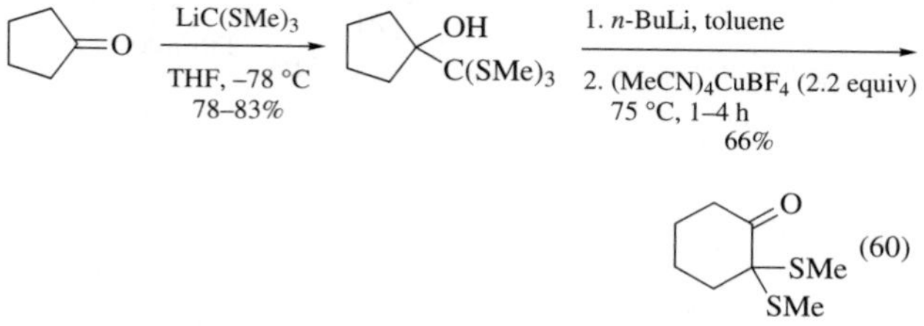

(60)

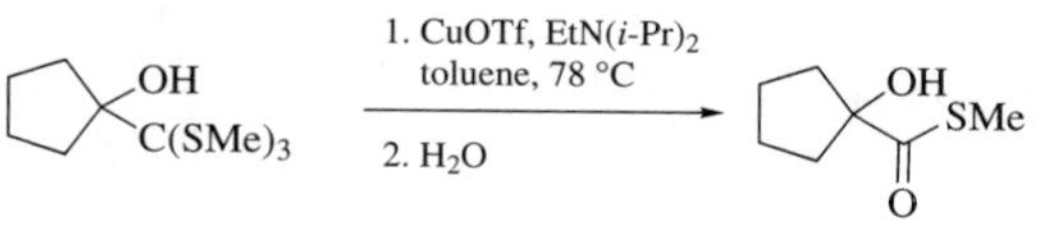

(61)

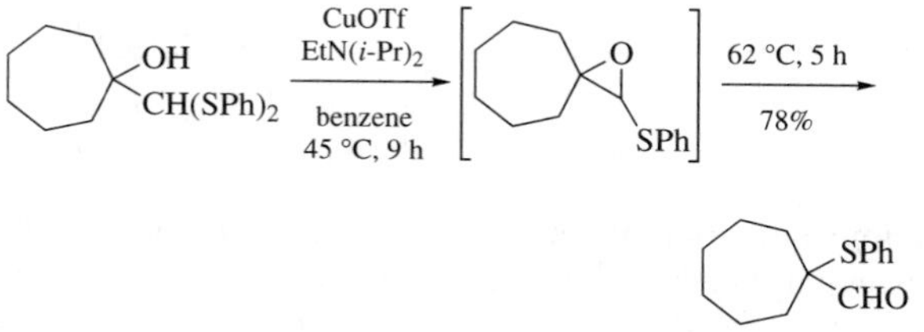

(62)

The α-[bis(phenylthio)methyl]alkanol derived from cyclohexanone, upon treatment with CuOTf and EtN(i-Pr)$_2$ in benzene, undergoes both ring-expanding rearrangement to deliver α-phenylsulfenylcycloheptanone as the major product, as well as rearrangement involving 1,2-shift of a phenylsulfenyl group to produce an α-phenylsulfenylcyclohexanecarbaldehyde (eq 63).[41] In contrast, neither ring expansion nor 1,2-shift of a methylsulfenyl group occurs upon treatment of α-[tris(methylthio)methyl]cyclohexanol with n-butyllithium followed by $(MeCN)_4CuBF_4$. Rather, after aqueous workup, an α-hydroxy methylthio ester is obtained (eq 64).[43]

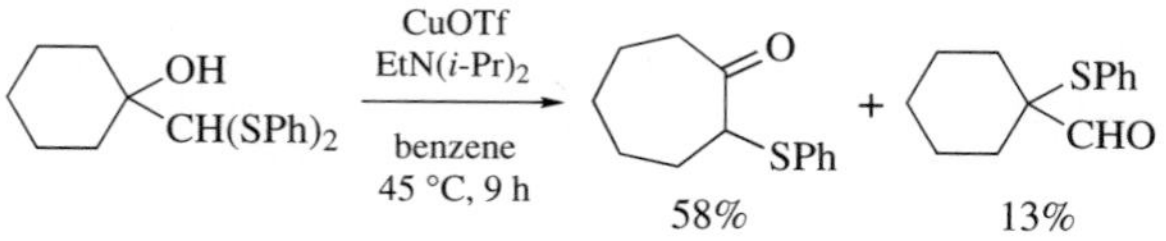

(63)

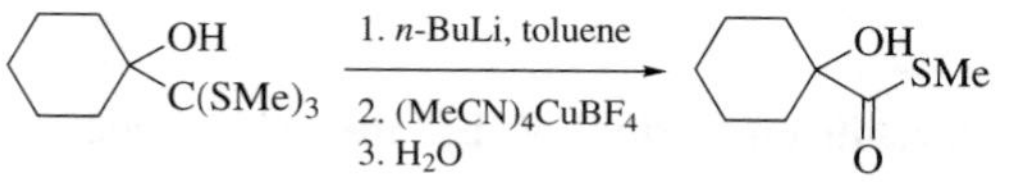

(64)

Chain-Extending Syntheses of α-Phenylsulfenyl Ketones. A C–C connective, chain-extending synthesis of α-phenylsulfenyl ketones from aldehydes (eq 65) or acyclic ketones (eq 66)[41] can be accomplished by a CuOTf-promoted activation of the α-[bis(phenylthio)methyl]alkanols generated by addition of lithium bis(phenylthio)methide.[34] Preferential migration of hydride generates phenylsulfenylmethyl ketones from aldehydes (eq 65). Regioselective insertion of a phenylsulfenylmethylene unit occurs owing to a preference for migration of the more highly substituted alkyl group of dialkyl ketones (eq 66).

Et–CHO + LiCH(SPh)$_2$ —80%→ Et–CH(OH)–CH(SPh)(PhS) —CuOTf, EtN(i-Pr)$_2$, benzene, 78 °C, 1 h, 66%→ Et–CO–CH$_2$SPh (65)

LiCH(SPh)$_2$ + MeCO–i-Pr —85%→ Me–C(OH)(i-Pr)–CH(SPh)$_2$ —CuOTf, EtN(i-Pr)$_2$, benzene, 78 °C, 1.5 h, 81%→ Me–CO–CH(PhS)–i-Pr (66)

Cyclopropanation of Enones. Conjugate addition of lithium tris(phenylthio)methide[44] to α,β-unsaturated ketones produces enolates that cyclize to bis(phenylthio)cyclopropyl ketones at −78 °C upon treatment with nearly one equivalent of $(CuOTf)_2{\cdot}C_6H_6$, i.e. 1.9 equivalents of Cu^I (eq 67).[45] The mild conditions that suffice to bring about nucleophilic displacement of thiophenoxide in the presence of CuOTf are especially noteworthy. In view of the requirement for more than one equivalent of Cu^I to achieve Grob-type fragmentation of β-[bis(phenylthio)methyl]alkoxides (see eq 58), it seems likely, although as yet unproven, that one equivalent of Cu^I coordinates strongly with the enolate oxygen and that a second equivalent of Cu^I is required to activate the thiophenoxide leaving group.

2-cyclohexenone + LiC(SPh)$_3$ —−78 °C, THF→ [lithium enolate, C(SPh)$_3$] —CuOTf, 75%→ 7,7-bis(phenylthio)bicyclo[4.1.0]heptan-2-one (67)

Vinylcyclopropanation of Enones. Conjugate addition of sulfur-stabilized allyl carbanions to α,β-unsaturated ketones produces enolates that cyclize to vinylcyclopropyl ketones upon treatment with nearly one equivalent of $(CuOTf)_2{\cdot}C_6H_6$, i.e. 1.9 equivalents of Cu^I (eq 68).[45]

Friedel–Crafts Acylation with Thio- or Selenoesters. Methylseleno esters are readily available in excellent yields by the reaction of ***Dimethylaluminum Methylselenolate*** with *O*-alkyl esters.[37] These selenoesters will acylate reactive arenes (eq 69) and heterocyclic compounds (eq 70) when activated by CuOTf, a selenophilic Lewis acid.[37] Of the potential activating metal salts tested, $(CuOTf)_2{\cdot}C_6H_6$ is uniquely effective. Mercury(II) or copper(I) trifluoroacetates that are partially organic-soluble, as well as the corresponding chlorides, silver nitrate, and copper(I) oxide that are not organic-soluble, all failed to promote any acylation. The highly reactive CuOTf–benzene complex, in dramatic contrast, was found to readily promote the acylations in benzene solution within minutes at room temperature. The presence of vinyl and keto groups is tolerated by the reaction, and while the alkyl- and vinyl-substituted derivatives afford *para* substitution only, a 2:1 mixture of *para* and *ortho* substitution occurs with the methylseleno ester of levulinic acid. Acylation of toluene is sluggish. Excellent yields of 2-acylfurans, -thiophenes, and -pyrroles are generated by this new variant of the Friedel–Crafts acylation reaction (eq 70). An intramolecular version of this reaction was shown to generate 1-tetralone from the methylseleno ester of γ-phenylbutyric acid (eq 71).[37]

2-cyclohexenone + Li–CH(PhS)–CH=C(X)–SPh —−78 °C, THF→ [lithium enolate adduct] —CuOTf→ 2-oxobicyclo[4.1.0]heptyl–CH=C(X)SPh (68)

X = H, 78%; SPh, 83%

RCOSeMe + MeOC$_6$H$_5$ —CuOTf (1.2 equiv), benzene, 3–40 min, rt→ 4-MeOC$_6$H$_4$COR (69)

R = (CH$_2$)$_5$Me 81%
R = CH$_2$CH$_2$CH=CH$_2$ 63%
R = CH$_2$CH$_2$COMe 60%

Me(CH$_2$)$_5$COSeMe + C$_4$H$_4$X —CuOTf, benzene, 15–20 min, rt→ 2-[CO(CH$_2$)$_5$Me]C$_4$H$_3$X (70)

X = O, 100%; S, 81%
NH, 64%

PhCH$_2$CH$_2$CH$_2$COSeMe —CuOTf (1.2 equiv), benzene, 5 min, rt, 70%→ 1-tetralone (71)

Notwithstanding a prior claim that methylthio esters react only sluggishly under these conditions,[37] such a variant proved effective for a short synthesis of the 4-demethoxy-11-deoxyanthracycline skeleton (eq 72).[46] This is especially significant because methylthio esters are available by an efficient C–C connective process involving *C*-acylation of ketone lithium enolates with ***Carbon Oxysulfide*** (COS) followed by *S*-methylation with ***Iodomethane***.[46] For the deoxyanthracycline synthesis, the requisite enolate was generated by 1,4-addition of a silyl-stabilized benzyllithium derivative to 2-cyclohexenone.

Treatment of the methylseleno ester with $(CuOTf)_2 \cdot C_6H_6$ in benzene, according to the method employed with analogous seleno esters,[37] results in efficient cyclization to deliver a tetracyclic diketone in good yield.

(72)

O*-Acylation with Thioesters.** Activation of a thioester with $(CuOTf)_2 \cdot C_6H_6$ was exploited as a key step in the synthesis of a macrocyclic pyrrolizidine alkaloid ester (eq 73).[47] Since thioesters are relatively unreactive acylating agents, a highly functionalized imidazolide containing acetate and *t*-butyl thioester groups selectively acylated only the primary hydroxyl in the presence of the secondary hydroxyl group in (+)-retronecine. Completion of the synthesis required activation of the *t*-butylthio ester. ***Mercury(II) Trifluoroacetate, that had proven effective for the synthesis of several natural products by lactonization,[48,49] failed to promote any lactonization in the present case.[47] Similarly, ***Mercury(II) Chloride*** and ***Cadmium Chloride***, that have proven effective for promoting lactonizations,[49] had no effect in the present case. Even copper(I) trifluoroacetate failed to induce the crucial lactonization. In contrast, CuOTf was uniquely effective for inducing the requisite macrolactonization by activating the thioester.

(73)

1. (a) Salomon, R. G. *Adv. Chem. Ser.* **1978**, *168*, 174. (b) Salomon, R. G. *T* **1983**, *39*, 485. (c) Salomon, R. G.; Kochi, J. K. *TL* **1973**, 2529.
2. Jenkins, C. L.; Kochi, J. K. *JACS* **1972**, *94*, 843.
3. (a) Hathaway, B. J.; Holah, D. G.; Postlethwaite, J. D. *JCS* **1961**, 3215. (b) Kubota, M.; Johnson, D. L. *J. Inorg. Nucl. Chem.* **1967**, *29*, 769.
4. (a) Salomon, R. G.; Kochi, J. K. *CC* **1972**, 559. (b) Salomon, R. G.; Kochi, J. K. *JACS* **1973**, *95*, 1889. (c) Dines, M. B. *Separ. Sci.* **1973**, *8*, 661.
5. Cohen, T.; Ruffner, R. J.; Shull, D. W.; Fogel, E. R.; Falck, J. R. *OS* **1980**, *59*, 202; *OSC* **1988**, *6*, 737.
6. Salomon, R. G.; Kochi, J. K. *JACS* **1973**, *95*, 3300.
7. Wiberg, K. B.; Kass, S. R.; Bishop, III, K. C. *JACS* **1985**, *107*, 996.
8. Lewis, R. T.; Motherwell, W. B. *TL* **1988**, *29*, 5033.
9. (a) Evans, D. A.; Woerpel, K. A.; Hinman, M. M.; Faul, M. M. *JACS* **1991**, *113*, 726. (b) Evans, D. A.; Woerpel, K. A.; Scott, M. J. *AG(E)* **1992**, *31*, 430.
10. Trecker, D. J.; Foote, R. S. In *Organic Photochemical Synthesis*; Srinivasan, R., Ed.; Wiley: New York, 1971; Vol. 1, p 81.
11. Salomon, R. G.; Kochi, J. K. *JACS* **1974**, *96*, 1137.
12. Salomon, R. G.; Kochi, J. K. *JACS* **1973**, *95*, 1889.
13. Salomon, R. G.; Folting, K.; Streib, W. E.; Kochi, J. K. *JACS* **1974**, *96*, 1145.
14. Evers, J. T. M.; Mackor, A. *RTC* **1979**, *98*, 423.
15. Evers, J. T. M.; Mackor, A. *TL* **1980**, *21*, 415.
16. Spee, T.; Mackor, A. *JACS* **1981**, *103*, 6901.
17. Evers, J. T. M.; Mackor, A. *TL* **1978**, 2317.
18. Salomon, R. G.; Sinha, A. *TL* **1978**, 1367.
19. Evers, J. T. M.; Mackor, A. *TL* **1978**, 821.
20. (a) Raychaudhuri, S. R.; Ghosh, S.; Salomon, R. G. *JACS* **1982**, *104*, 6841. (b) Ghosh, S.; Raychaudhuri, S. R.; Salomon, R. G. *JOC* **1987**, *52*, 83.
21. Avasthi, K.; Raychaudhuri, S. R.; Salomon, R. G. *JOC* **1984**, *49*, 4322.
22. Hertel, R.; Mattay, J.; Runsink, J. *JACS* **1991**, *113*, 657.
23. (a) Salomon, R. G.; Coughlin, D. J.; Easler, E. M. *JACS* **1979**, *101*, 3961. (b) Salomon, R. G.; Ghosh, S. *OS* **1984**, *62*, 125; *OSC* **1990**, *7*, 177. (c) Salomon, R. G.; Coughlin, D. J.; Ghosh, S.; Zagorski, M. G. *JACS* **1982**, *104*, 998.
24. Salomon, R. G.; Ghosh, S.; Zagorski, M. G.; Reitz, M. *JOC* **1982**, *47*, 829.
25. McMurry, J. E.; Choy, W. *TL* **1980**, *21*, 2477.
26. Avasthi, K.; Salomon, R. G. *JOC* **1986**, *51*, 2556.
27. Salomon, R. G.; Ghosh, S.; Raychaudhuri, S.; Miranti, T. S. *TL* **1984**, *25*, 3167.
28. Hefner, J. G.; Zizelman, P. M.; Durfee, L. D.; Lewandos, G. S. *JOM* **1984**, *260*, 369.
29. (a) Cohen, T.; Cristea, I. *JOC* **1975**, *40*, 3649. (b) Cohen, T.; Cristea, I. *JACS* **1976**, *98*, 748.
30. Cohen, T.; Tirpak, J. *TL* **1975**, 143.
31. Cohen, T.; Herman, G.; Falck, J. R.; Mura, Jr., A. J. *JOC* **1975**, *40*, 812.
32. Kwon, T. W.; Smith, M. B. *SC* **1992**, *22*, 2273.
33. Cohen, T.; Mura, A. J.; Shull, D. W.; Fogel, E. R.; Ruffner, R. J.; Falck, J. R. *JOC* **1976**, *41*, 3218.
34. Corey, E. J.; Seebach, D. *JOC* **1966**, *31*, 4097.
35. Cohen, T.; Kosarych, Z. *TL* **1980**, *21*, 3955.
36. Cohen, T.; Gapinski, R. E.; Hutchins, R. R. *JOC* **1979**, *44*, 3599.
37. (a) Kozikowski, A. P.; Ames, A. *JACS* **1980**, *102*, 860. (b) Kozikowski, A. P.; Ames, A. *T* **1985**, *41*, 4821.
38. (a) Corey, E. J.; Noyori, R. *TL* **1970**, 311. (b) Corey, E. J.; Erickson, B. W.; Noyori, R. *JACS* **1971**, *93*, 1724.
39. Cohen, T.; Bennett, D. A.; Mura, A. J. *JOC* **1976**, *41*, 2506.
40. Semmelhack, M. F.; Tomesch, J. C. *JOC* **1977**, *42*, 2657.

41. Cohen, T.; Kuhn, D.; Falck, J. R. *JACS* **1975**, *97*, 4749.
42. Knapp, S.; Trope, A. F.; Ornaf, R. M. *TL* **1980**, *21*, 4301.
43. Knapp, S.; Trope, A. F.; Theodore, M. S.; Hirata, N.; Barchi, J. J. *JOC* **1984**, *49*, 608.
44. Seebach, D. *AG(E)* **1967**, *6*, 442.
45. Cohen, T.; Meyers, M. *JOC* **1988**, *53*, 457.
46. Vedejs, E.; Nader, B. *JOC* **1982**, *47*, 3193.
47. Huang, J.; Meinwald, J. *JACS* **1981**, *103*, 861.
48. (a) Masamune, S. *Aldrichim. Acta* **1978**, *11*, 23. (b) Masamune, S.; Yamamoto, H.; Kamata, S.; Fukuzawa, A. *JACS* **1975**, *97*, 3513.
49. Masamune, S.; Kamata, S.; Schilling, W. *JACS* **1975**, *97*, 3515.

Robert G. Salomon
Case Western Reserve University, Cleveland, OH, USA

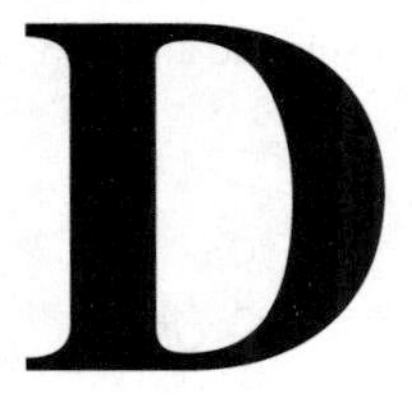

Diazomethane[1]

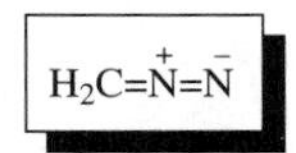

[334-88-3] CH_2N_2 (MW 42.04)

(methylating agent for various functional groups including carboxylic acids, alcohols, phenols, and amides; reagent for the synthesis of α-diazo ketones from acid chlorides, and the cyclopropanation of alkenes[1])

Physical Data: mp −145 °C; bp −23 °C.

Solubility: diazomethane is most often used as prepared in ether, or in ether containing a small amount of ethanol. It is less frequently prepared and used in other solvents such as dichloromethane.

Analysis of Reagent Purity: diazomethane is titrated[2] by adding a known quantity of benzoic acid to an aliquot of the solution such that the solution is colorless and excess benzoic acid remains. Water is then added, and the amount of benzoic acid remaining is back-titrated with NaOH solution. The difference between the amount of acid added and the amount remaining reveals the amount of active diazomethane present in the aliquot.

Preparative Methods: diazomethane is usually prepared by the decomposition of various derivatives of *N*-methyl-*N*-nitrosoamines. Numerous methods of preparation have been described,[3] but the most common and most frequently employed are those which utilize ***N-Methyl-N-nitroso-p-toluenesulfonamide*** (Diazald®; **1**),[4] ***1-Methyl-3-nitro-1-nitrosoguanidine*** (MNNG, **2**),[5] or *N*-methyl-*N*-nitrosourea (**3**).[2]

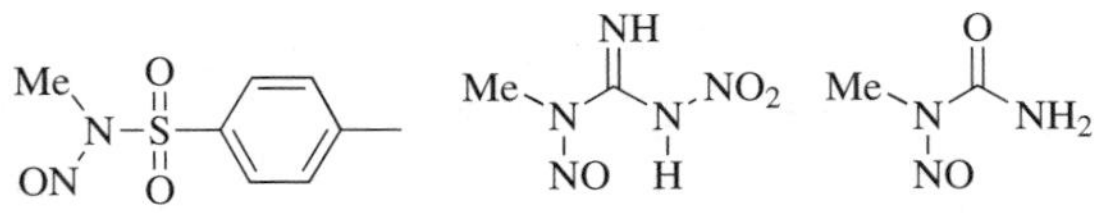

(1) (2) (3)

The various reagents each have their advantages and disadvantages, as discussed below. The original procedure[6] for the synthesis of diazomethane involved the use of *N*-methyl-*N*-nitrosourea, and similar procedures are still in use today. An advantage of using this reagent is that solutions of diazomethane can be prepared without distillation,[7] thus avoiding the most dangerous operation in other preparations of diazomethane. For small scale preparations (1 mmol or less) which do not contain any alcohol, a kit is available utilizing MNNG which produces distilled diazomethane in a closed environment. Furthermore, MNNG is a stable compound and has a shelf life of many years. For larger scale preparations, kits are available for the synthesis of up to 300 mmol of diazomethane using Diazald as the precursor. The shelf life of Diazald (about 1–2 years), however, is shorter than that of MNNG. Furthermore, the common procedure using Diazald produces an ethereal solution of diazomethane which contains ethanol; however, it can be modified to produce an alcohol-free solution. Typical preparations of diazomethane involve the slow addition of base to a heterogeneous aqueous ether mixture containing the precursor. The precursor reacts with the base to liberate diazomethane which partitions into the ether layer and is concomitantly distilled with the ether to provide an ethereal solution of diazomethane. Due to the potentially explosive nature of diazomethane, the chemist is advised to carefully follow the exact procedure given for a particular preparation. Furthermore, since diazomethane has been reported to explode upon contact with ground glass, apparatus which do not contain ground glass should be used. All of the kits previously mentioned avoid the use of ground glass.

Handling, Storage, and Precautions: diazomethane as well as the precursors for its synthesis can present several safety hazards, and must be used with great care.[8] The reagent itself is highly toxic and irritating. It is a sensitizer, and long term exposure can lead to symptoms similar to asthma. It can also detonate unexpectedly, especially when in contact with rough surfaces, or on crystallization. It is therefore essential that any glassware used in handling diazomethane be fire polished and not contain any scratches or ground glass joints. Furthermore, contact with certain metal ions can also cause explosions. Therefore metal salts such as calcium chloride, sodium sulfate, or magnesium sulfate must not be used to dry solutions of the reagent. The recommended drying agent is potassium hydroxide. Strong light is also known to initiate detonation. The reagent is usually generated immediately prior to use and is not stored for extended periods of time. Of course, the reagent must be prepared and used in a well-ventilated hood, preferably behind a blast shield. The precursors used to generate diazomethane are irritants and in some cases mutagens and suspected carcinogens, and care should be exercised in their handling as well.

Methylation of Heteroatoms. The most widely used feature of the chemistry of diazomethane is the methylation of carboxylic acids. Carboxylic acids are good substrates for reaction with diazomethane because the acid is capable of protonating the diazomethane on carbon to form a diazonium carboxylate. The carboxylate can then attack the diazonium salt in what is most likely an S_N2 reaction to provide the ester. Species which are not acidic enough to protonate diazomethane, such as alcohols, require an additional catalyst, such as ***Boron Trifluoride Etherate***, to increase their acidity and facilitate the reaction. The methylation reaction proceeds under mild conditions and is highly reliable and very selective for carboxylic acids. A typical procedure is to add a yellow solution of diazomethane to the carboxylic acid in portions. When the yellow color persists and no more gas is evolved, the reaction is deemed complete. Excess reagent can be destroyed by the addition of a few drops of acetic acid and the entire solution concentrated to provide the methyl ester.

Esterification of Carboxylic Acids and Other Acidic Functional Groups. A variety of functional groups will tolerate the esterification of acids with diazomethane. Thus α,β-unsaturated carboxylic acids and alcohols survive the reaction (eq 1),[9] as do ketones (eq 2),[10] isolated alkenes (eq 3),[11] and amines (eq 4).[12]

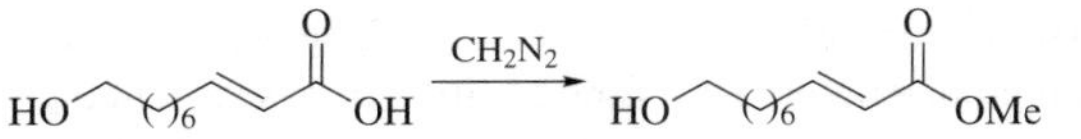

(1)

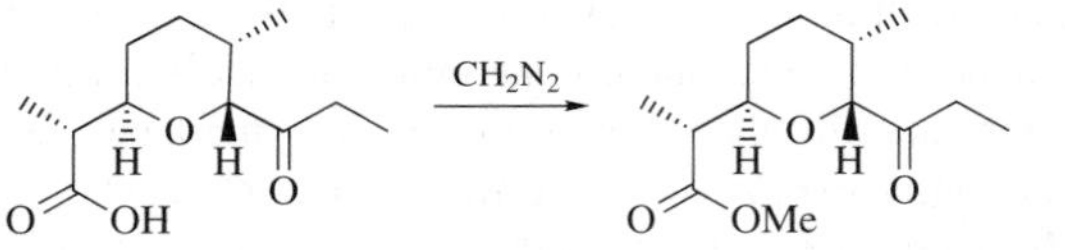

(2)

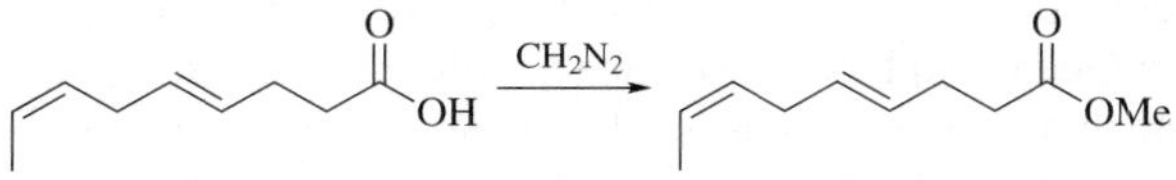

(3)

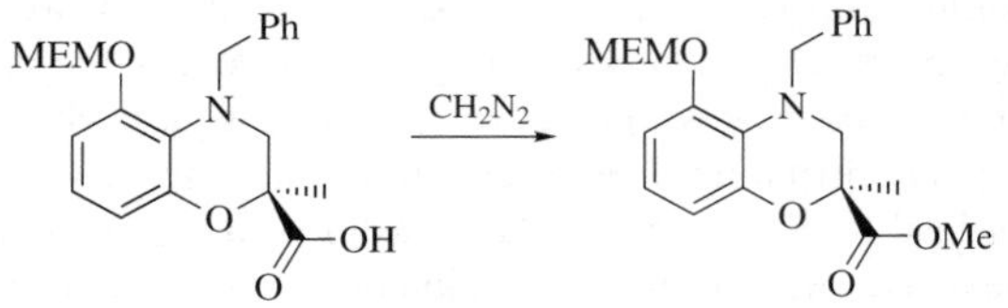

(4)

Other acidic functional groups will also undergo reaction with diazomethane. Thus phosphonic acids (eq 5)[13] and phenols (eq 6)[14] are methylated in high yields, as are hydroxytropolones (eq 7)[15] and vinylogous carboxylic acids (eq 8).[16] The origin of the selectivity in eq 6 is due to the greater acidity of the A-ring phenol.

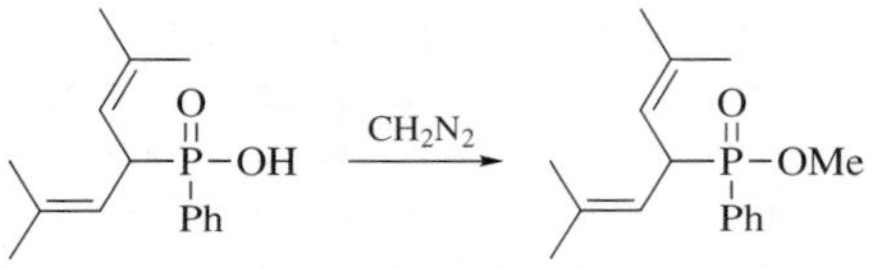

(5)

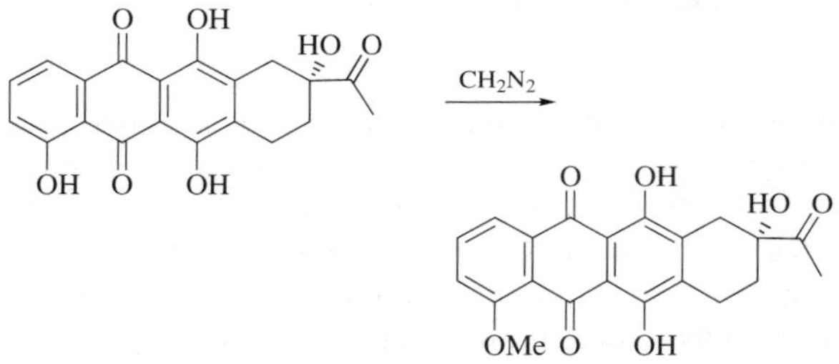

(6)

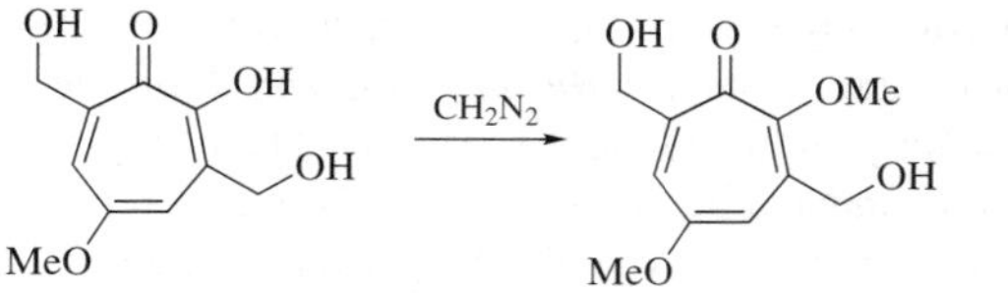

(7)

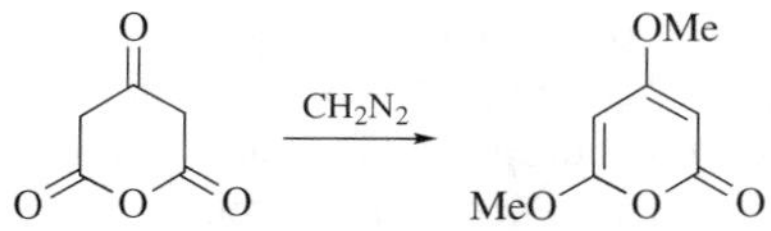

(8)

Selective monomethylation of dicarboxylic acids has been reported using ***Alumina*** as an additive (eq 9).[17] It is thought that one of the two carboxylic acid groups is bound to the surface of the alumina and is therefore not available for reaction. Carboxylic acids that are engaged as lactols will also undergo methylation with diazomethane to provide the methyl ester and aldehyde (eq 10).[18]

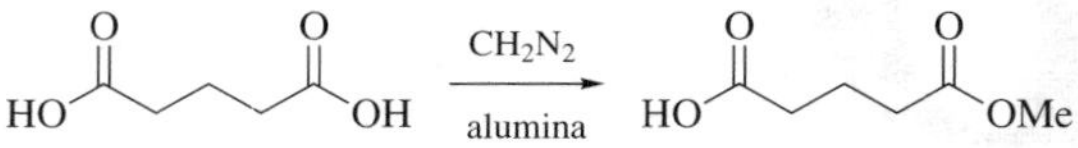

(9)

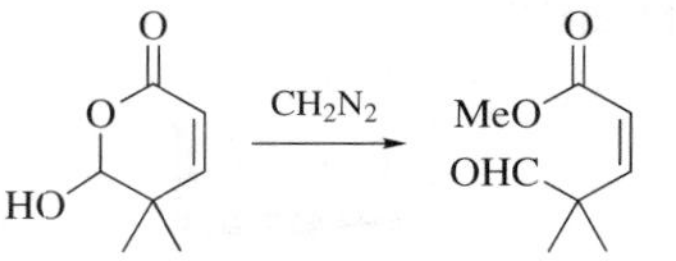

(10)

Methylation of Alcohols and Other Less Acidic Functional Groups. As previously mentioned, alcohols require the addition of a catalyst in order to react with diazomethane. The most commonly used is boron trifluoride etherate (eq 11),[19] but ***Tetrafluoroboric Acid*** has been used as well (eq 12).[20] Mineral acids are not effective since they rapidly react with diazomethane to provide the corresponding methyl halides. Acids as mild as silica gel are also effective (eq 13).[21] Monomethylation of 1,2-diols with diazomethane has been reported using various Lewis acids as promoters, the most effective of which is ***Tin(II) Chloride*** (eq 14).[22]

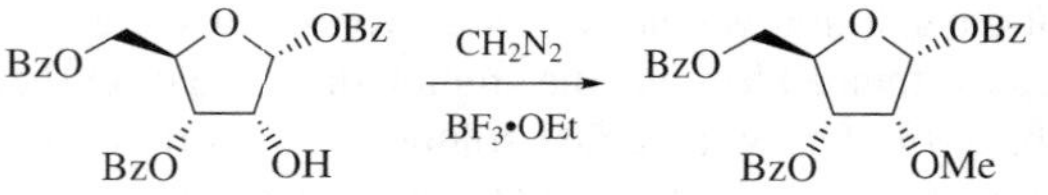

(11)

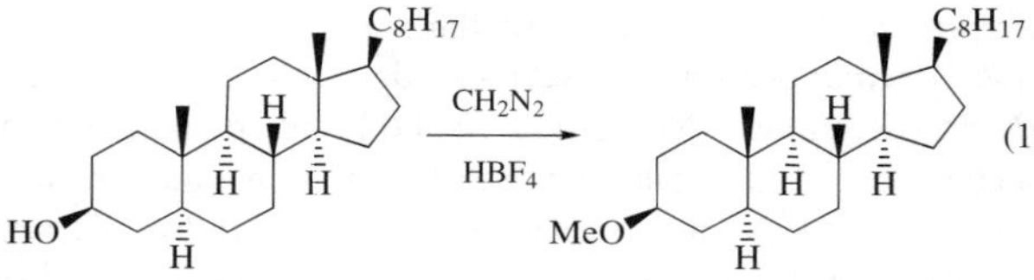

(12)

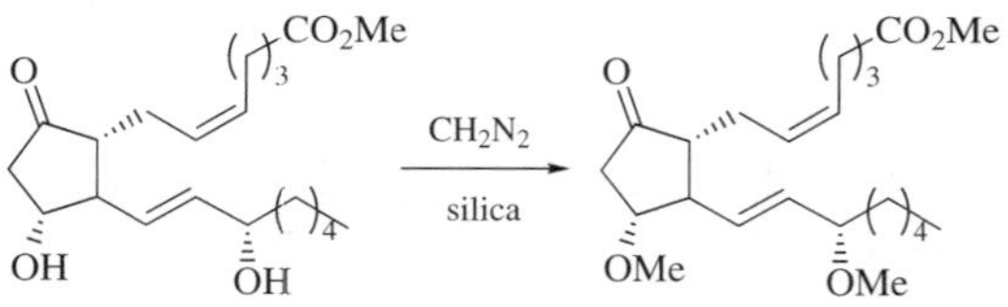

(13)

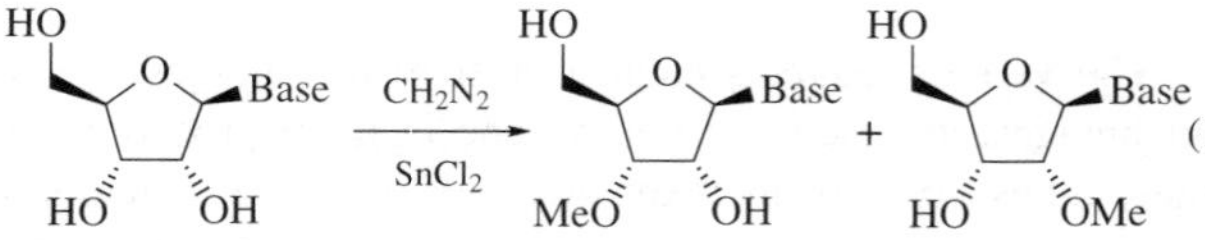

(14)

An interesting case of an alcohol reacting with diazomethane at a rate competitive with a carboxylic acid has been reported (eq 15).[23] In this case, the tertiary structure of the molecule is thought to place the alcohol and the carboxylic acid in proximity to each other. Protonation of the diazomethane by the carboxylic acid leads to a diazonium ion in proximity to the alcohol as well as the carboxylate. These species then attack the diazonium ion at competitive rates to provide the methyl ether and ester. No reaction is observed upon treatment of the corresponding hydroxy ester with diazomethane, indicating that the acid is required to activate the diazomethane.

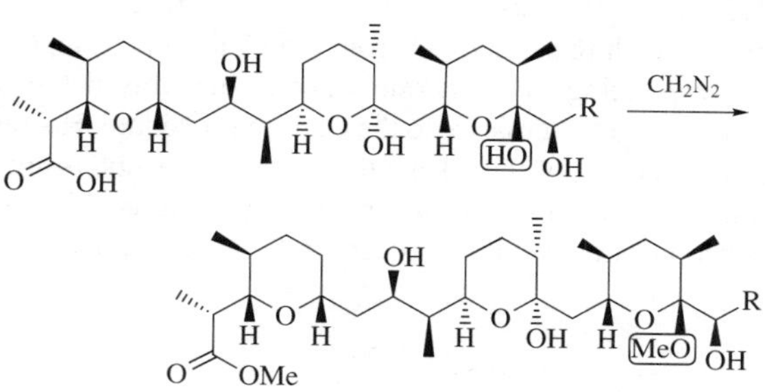

(15)

Amides can also be methylated with diazomethane in the presence of silica gel; however, the reaction requires a large excess of diazomethane (25–60 equiv, eq 16).[24] The reaction primarily provides *O*-methylated material; however, in one case a mixture of *O*- and *N*-methylation was reported. Thioamides are also effectively methylated with this procedure to provide *S*-methylated compounds. Finally, amines have been methylated with diazomethane in the presence of BF_3 etherate, fluoroboric acid,[25] or copper(I) salts;[26] however, the yields are low to moderate, and the method is not widely used.

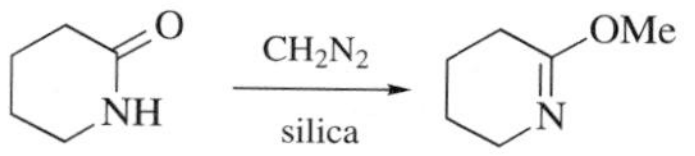

(16)

The Arndt–Eistert Synthesis. Diazomethane is a useful reagent for the one-carbon homologation of acid chlorides via a sequence of reactions known as the Arndt–Eistert synthesis. The first step of this sequence takes advantage of the nucleophilicity of diazomethane in its addition to an active ester, typically an acid chloride,[27] to give an isolable α-diazo ketone and HCl. The HCl that is liberated from this step can react with diazomethane to produce methyl chloride and nitrogen, and therefore at least 2 equiv of diazomethane are typically used. The α-diazo ketone is then induced to undergo loss of the diazo group and insertion into the adjacent carbon–carbon bond of the ketone to provide a ketene. The ketene is finally attacked by water or an alcohol (or some other nucleophile) to provide the homologated carboxylic acid or ester. This insertion step of the sequence is known as the Wolff rearrangement[28] and can be accomplished either thermally (eq 17)[29] or, more commonly, by treatment with a metal ion (usually silver salts, eq 18),[30] or photochemically (eq 19).[31] It has been suggested that the photochemical method is the most efficient of the three.[32] As eqs 18 and 19 illustrate, retention of configuration is observed in the migrating group. The obvious limitations of this reaction are that there must not be functional groups present in the molecule which will react with diazomethane more rapidly than it will attack the acid chloride. Thus carboxylic acids will be methylated under these conditions. Furthermore, electron-deficient alkenes will undergo [2,3] dipolar cycloaddition with diazomethane more rapidly than addition to the acid chloride. Thus when the Arndt–Eistert synthesis is attempted on α,β-unsaturated acid chlorides, cycloaddition to the alkene is observed in the product. In order to prevent this, the alkene must first be protected by addition of HBr and then the reaction carried out in the normal way (eq 20).[33] Cycloaddition to isolated alkenes, however, is not competitive with addition to acid chlorides.

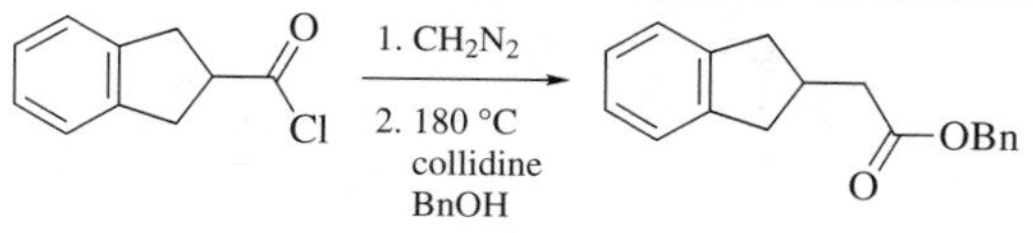

(17)

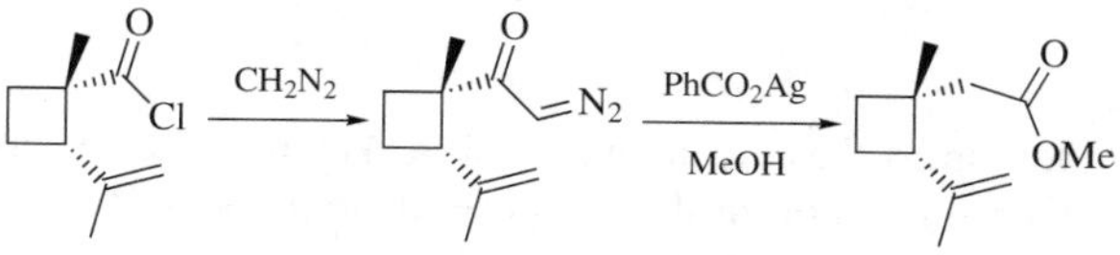

(18)

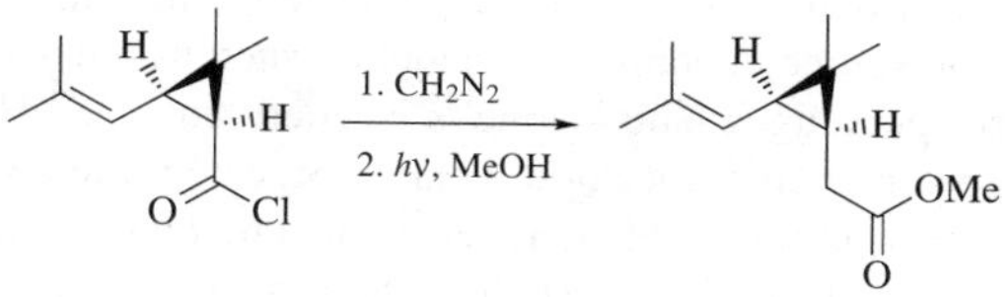

(19)

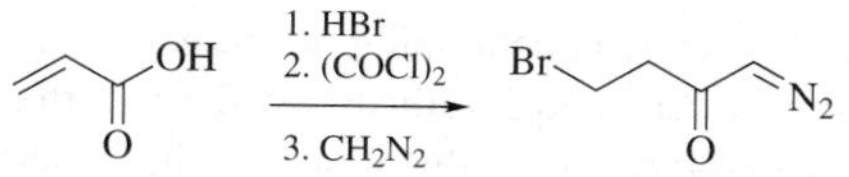

(20)

Other Reactions of α-Diazo Ketones Derived from Diazomethane. Depending on the conditions employed, the Wolff rearrangement may proceed via a carbene or carbenoid intermediate, or it may proceed by a concerted mechanism where the insertion is concomitant with loss of N_2 and no intermediate is formed. In the case where a carbene or carbenoid is involved, other reactions which are characteristic of these species can occur, such as intramolecular cyclopropanation of alkenes. In fact, the reaction conditions can be adjusted to favor cyclopropanation or homologation depending on which is desired. Thus treatment of the dienoic acid chloride shown in eq 21 with diazomethane followed by decomposition of the α-diazo ketone with silver benzoate in the presence of methanol and base provides the homologated methyl ester. However, treatment of the same diazoketone intermediate with Cu^{II} salts provides the cyclopropanation products selectively.[34] This trend is generally observed; that is, silver salts as well as photochemical conditions (eqs 18 and 19) favor the homologation pathway while copper or rhodium salts favor cyclopropanation.[35] Using copper salts to decompose the diazo compounds, hindered alkenes as well as electron-rich aromatics can be cyclopropanated as illustrated in eqs 22 and 23,[36,37] respectively.

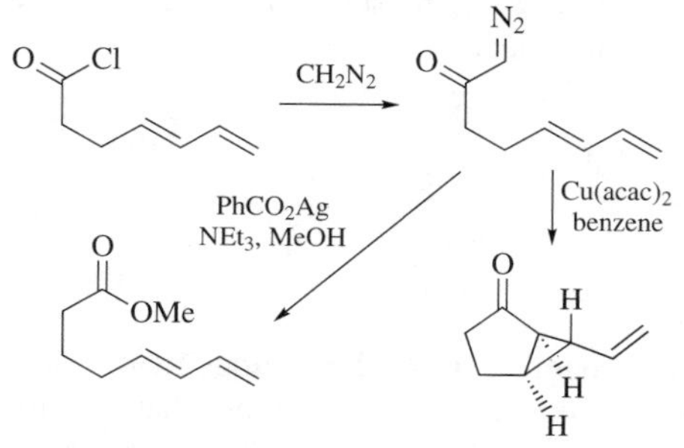

(21)

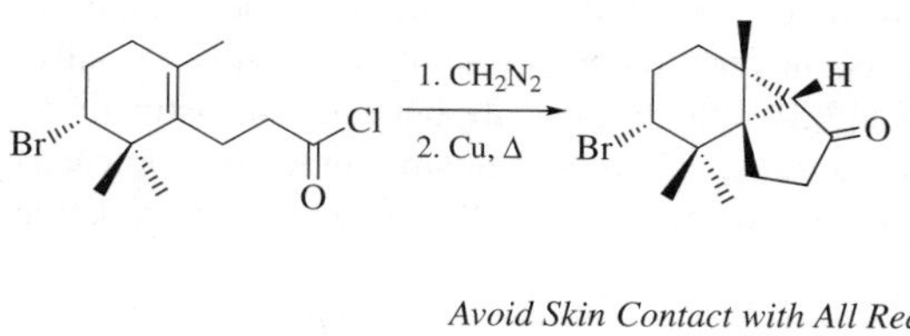

(22)

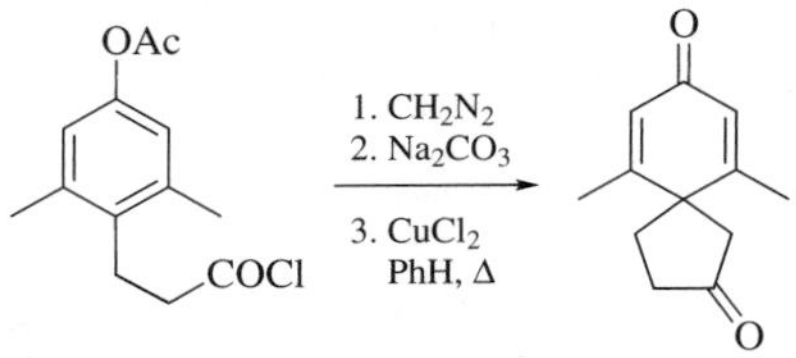

(23)

In addition to these reactions, α-diazo ketones will undergo protonation on carbon in the presence of protic acids[38] to provide the corresponding α-diazonium ketone. These species are highly electrophilic and can undergo nucleophilic attack. Thus if the proton source contains a nucleophile such as a halogen then the corresponding α-halo ketone is isolated (eq 24).[39] However, if the proton source does not contain a nucleophilic counterion then the diazonium species may react with other nucleophiles that are present in the molecule, such as alkenes (eq 25)[40] or aromatic rings (eq 26).[41] Note the similarity between the transformations in eqs 26 and 23 which occur using different catalysts and by different pathways. Also, eq 26 illustrates the fact that other active esters will undergo nucleophilic attack by diazomethane.

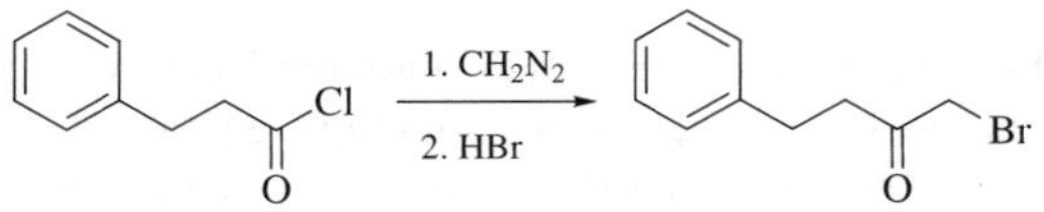

(24)

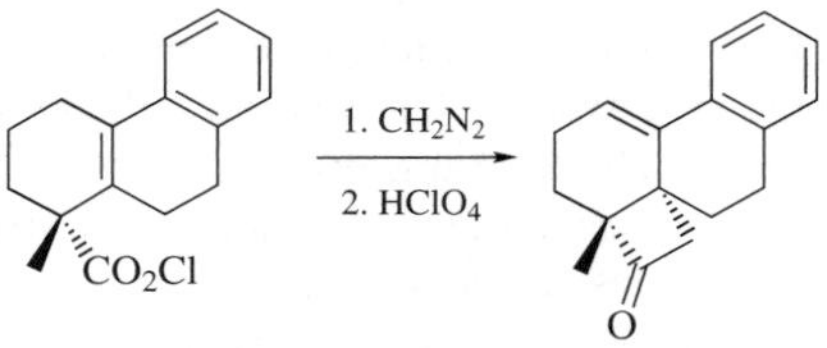

(25)

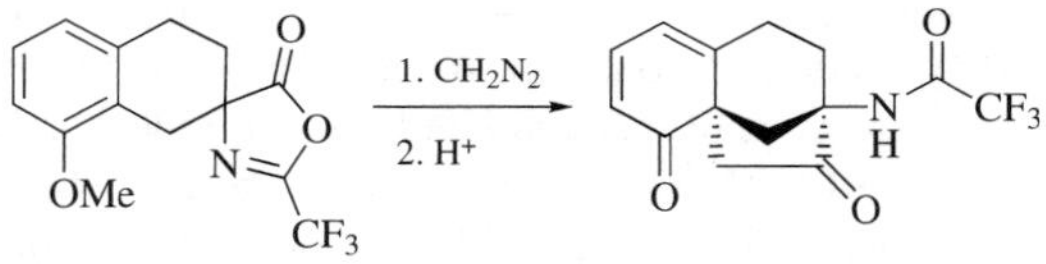

(26)

Lewis acids are also effective in activating α-diazo ketones towards intramolecular nucleophilic attack by alkenes and arenes.[42] The reaction has been used effectively for the synthesis of cyclopentenones (eq 27) starting with β,γ-unsaturated diazo ketones derived from the corresponding acid chloride and diazomethane. It has also been used to initiate polyalkene cyclizations (eq 28). Typically, boron trifluoride etherate is used as the Lewis acid, and electron-rich alkenes are most effective providing the best yields of annulation products.

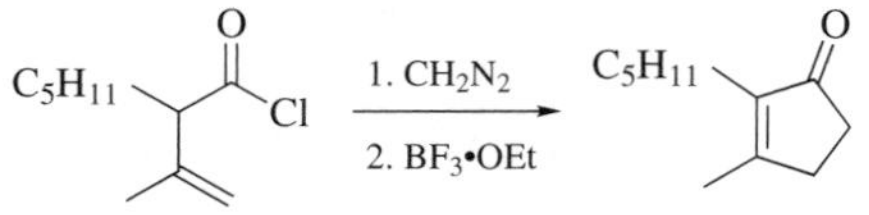

(27)

The Vinylogous Wolff Rearrangement. The vinylogous Wolff rearrangement[43] is a reaction that occurs when the Arndt–Eistert synthesis is attempted on β,γ-unsaturated acid chlorides using copper catalysis. Rather than the usual homologation products, the reaction proceeds to give what is formally the product of a [2,3]-sigmatropic shift, but is mechanistically not derived by this pathway.[44] The mechanism is thought to proceed by an initial cyclopropanation of the alkene by the α-diazo ketone to give a bicyclo[2.1.0]pentanone derivative. This compound then undergoes a fragmentation to a ketene alkene before being trapped by the solvent (eq 29). Inspection of the products reveals that they are identical with those derived from the Claisen rearrangement of the corresponding allylic alcohols, and as such this method can be thought of as an alternative to the Claisen procedure. However, the stereoselectivity of the alkene that is formed is not as high as is typically observed in the Claisen rearrangement (eq 30), and in some substrates the reaction proceeds with no selectivity (eq 31).

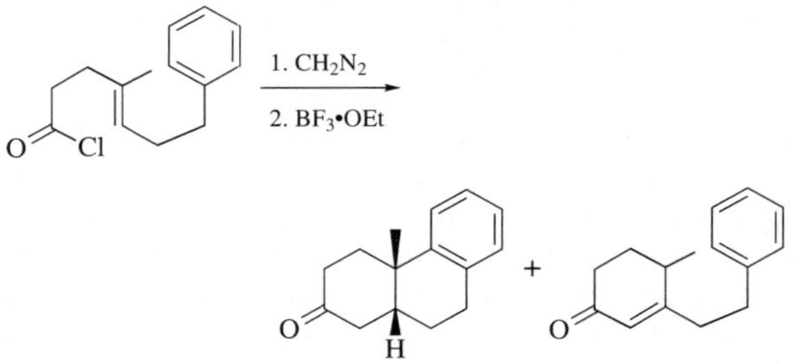

(28)

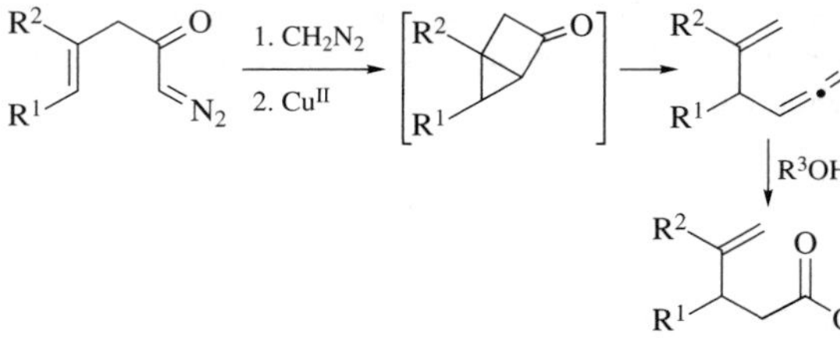

(29)

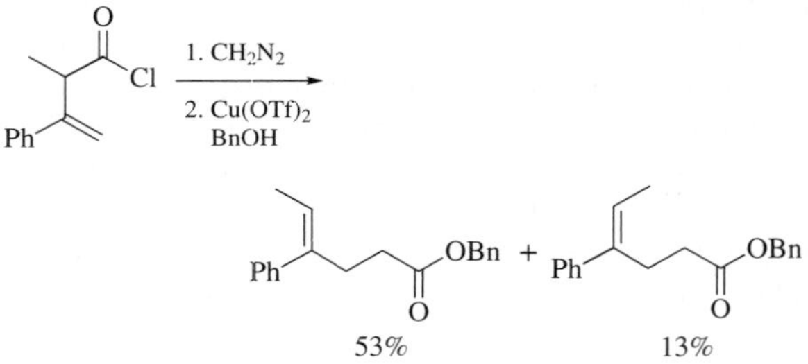

(30)

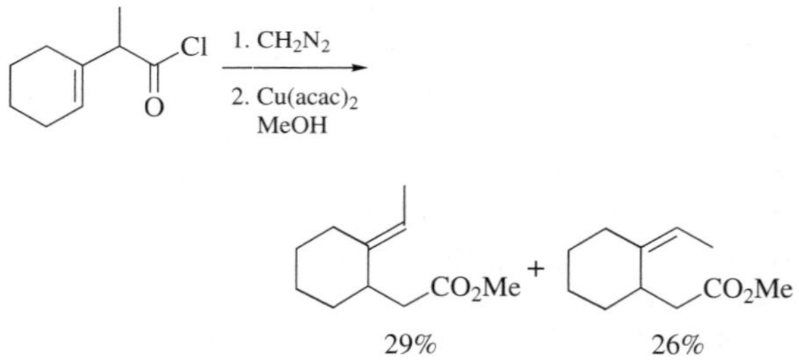

(31)

Insertions into Aldehyde C–H Bonds. The α-diazo ketones (and esters) derived from diazomethane and an acid chloride (or chloroformate) will also insert into the C–H bond of aldehydes to give 1,3-dicarbonyl derivatives.[45] The reaction is catalyzed by $SnCl_2$, but some simple Lewis acids, such as BF_3 etherate, also work. The reaction works well for aliphatic aldehydes, but gives variable results with aromatic aldehydes, at times giving none of the desired diketone (eq 32). Sterically hindered aldehydes will also participate in this reaction, as illustrated in eq 33 with the reaction of ethyl α-diazoacetate and pivaldehyde. In a related reaction, α-diazo phosphonates and sulfonates will react with aldehydes in the presence of $SnCl_2$ to give the corresponding β-keto phosphonates and sulfonates.[46] This reaction is a practical alternative to the Arbuzov reaction for the synthesis of these species.

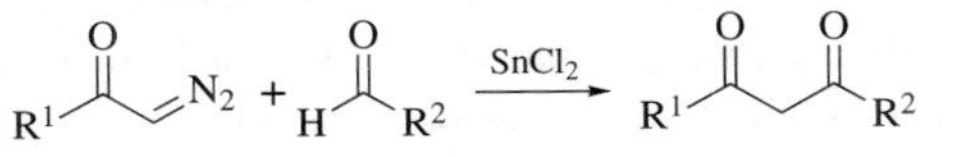

(32)

R^1	R^2	Yield (%)
Ph	H	88
Ph	$PhCH_2$	90
$PhCH_2CH_2$	Ph	0

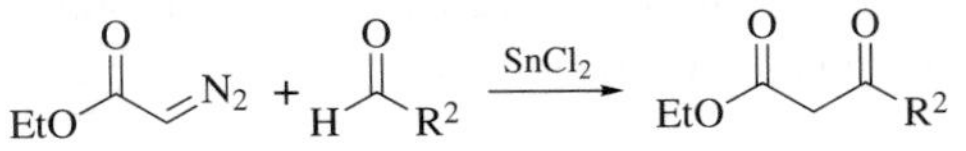

(33)

R^2	Yield (%)
t-Bu	65
Ph	50

Additions to Ketones. The addition of diazomethane to ketones[47] is also a preparatively useful method for one-carbon homologation. This reaction is a one-step alternative to the Tiffeneau–Demjanow rearrangement[48] and proceeds by the mechanism shown in eq 34. It can lead to either homologation or epoxidation depending on the substrate and reaction conditions. The addition of Lewis acids, such as BF_3 etherate, or alcoholic cosolvents tend to favor formation of the homologation products over epoxidation.

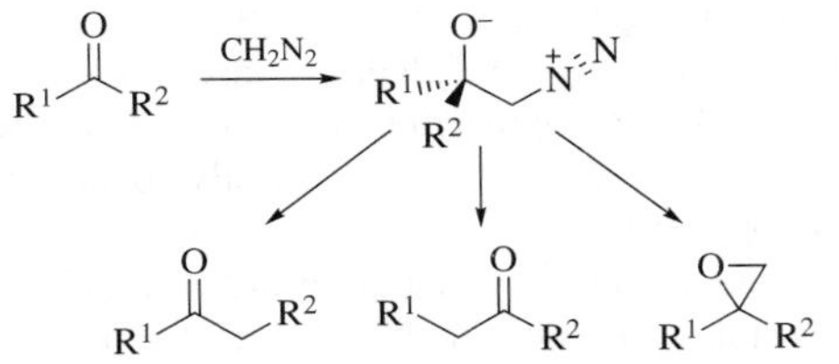

(34)

However, the reaction is limited by the poor regioselectivity observed in the insertion when the groups R^1 and R^2 in the starting ketone are different alkyl groups. What selectivity is observed tends to favor migration of the less substituted carbon,[49] a trend which is opposite to that typically observed in rearrangements of electron-deficient species such as in the Baeyer–Villiger reaction. Furthermore, the product of the reaction is a ketone and is therefore capable of undergoing further reaction with diazomethane. Thus, ideally, the product ketone should be less reactive than the starting ketone. Strained ketones tend to react more rapidly and are therefore good substrates for this reaction (eq 35).[50] This method has also found extensive use in cyclopentane annulation reactions starting with an alkene. The overall process begins with dichloroketene addition to the alkene to produce an α-dichlorocyclobutanone. These species are ideally suited for reaction with diazomethane because the reactivity of the starting ketone is enhanced due to the strain in the cyclobutanone as well as the α-dichloro substitution. Furthermore, the presence of the α-dichloro substituents hinders migration of that group and leads to almost exclusive migration of the methylene group. Thus treatment with diazomethane and methanol leads to a rapid evolution of nitrogen, and produces the corresponding α-dichlorocyclopentanone, which can be readily dehalogenated to the hydrocarbon (eq 36).[51] Aldehydes will also react with diazomethane, but in this case homologation is not observed. Rather, the corresponding methyl ketone derived from migration of the hydrogen is produced (eq 37).

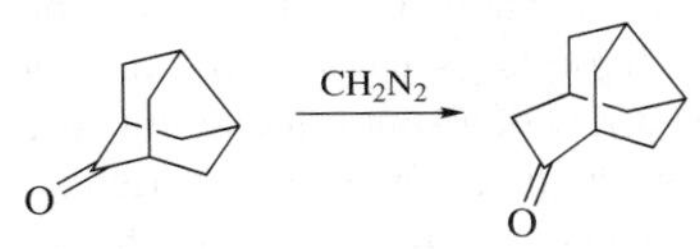

(35)

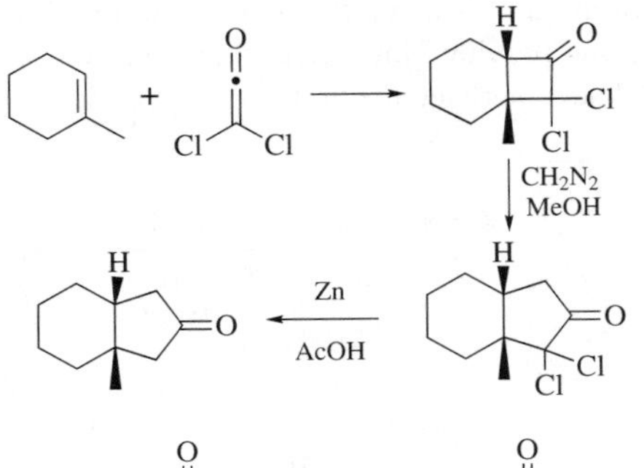

(36)

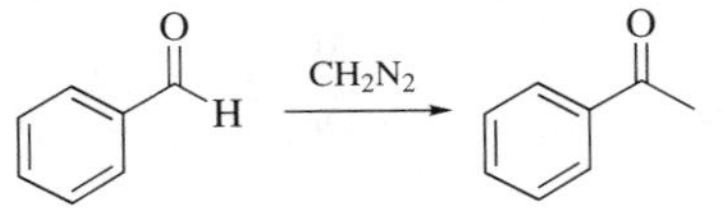

(37)

Cycloadditions with Diazomethane. Diazomethane will undergo [3 + 2] dipolar cycloadditions with alkenes and alkynes to give pyrazolines and pyrazoles, respectively.[52] The reaction proceeds more rapidly with electron-deficient alkenes and strained alkenes and is controlled by FMO considerations with the HOMO of the diazomethane and the LUMO of the alkene serving as the predominant interaction.[53] In the case of additions to electron-deficient alkenes, the carbon atom of the diazomethane behaves as the negatively charged end of the dipole, and therefore the regiochemistry observed is as shown in eq 38. With conjugated alkenes, such as styrene, the terminal carbon has the larger lobe in the LUMO, and as such the reaction proceeds to give the product shown in eq 39. Pyrazolines are most often used as precursors to cyclopropanes by either thermal or photochemical extrusion of N_2. In both cases the reaction may proceed by a stepwise mechanism with loss of stereospecificity. As shown in eq 40, the thermal reaction provides an almost random product distribution, while the photochemical reaction provides variable results ranging from 20:1 to stereospecific extrusion of nitrogen.[54]

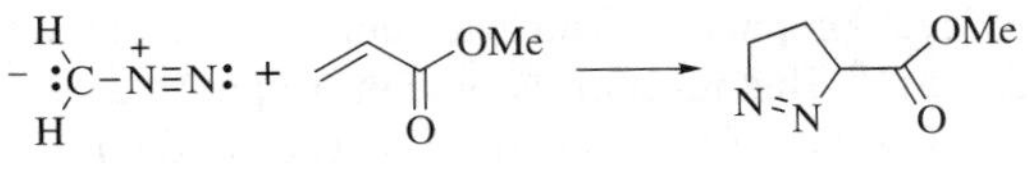

(38)

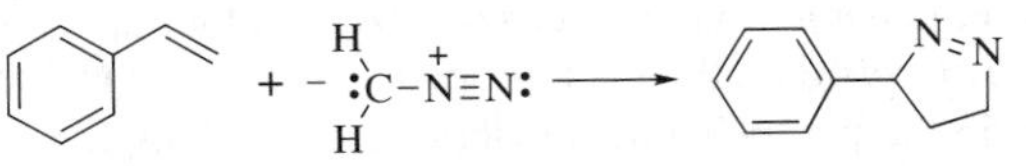

(39)

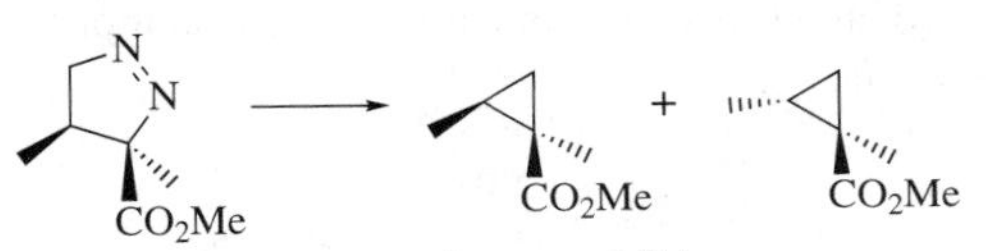

(40)

Δ	1.2:1
hν	20:1 to >100:1

Cyclopropanes can also be directly synthesized from alkenes and diazomethane, either photochemically or by using transition metal salts, usually ***Copper(II) Chloride*** or ***Palladium(II) Acetate***, as promoters. The metal-mediated reactions are more commonly used than the photochemical ones, but they are not as popular as the Simmons–Smith procedure. However, they do occasionally offer advantages. Of the two processes, the Cu-catalyzed reaction produces a more active reagent[55] which will cyclopropanate a variety of alkenes, including enamines as shown in eq 41.[56] These products can then be converted to α-methyl ketones by thermolysis. The cyclopropanation of the norbornenol derivative shown in eq 42 was problematic using the Simmons–Smith procedure and provided low yields, but occurred smoothly using the $CuCl_2$/diazomethane method.[57]

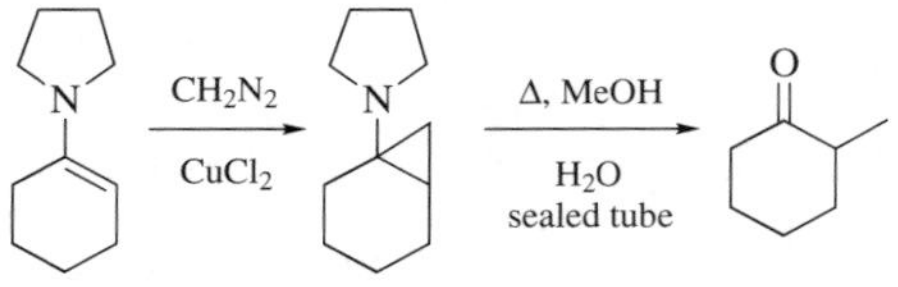

(41)

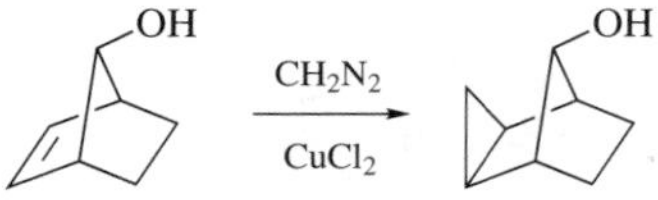

(42)

The $Pd(OAc)_2$-mediated reaction can be used to cyclopropanate electron-deficient alkenes as well as terminal alkenes. Thus selective reaction at a monosubstituted alkene in the presence of others is readily achieved using this method (eq 43).[58] The example shown in eq 44 is one in which the Simmons–Smith procedure failed to provide any of the desired product, whereas the current method provided a 92% yield of cyclopropane.[59]

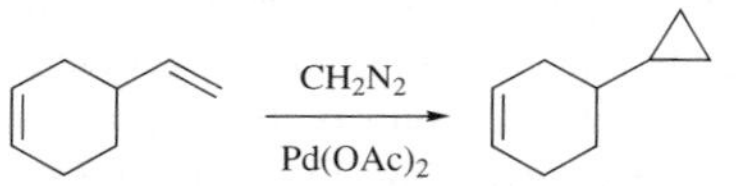

(43)

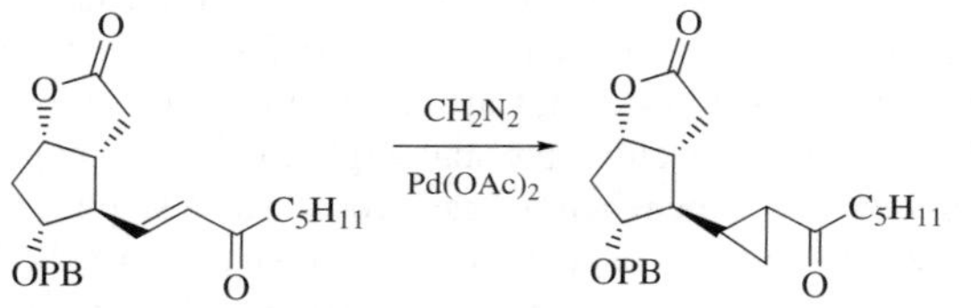

(44)

In the case of the photochemical reaction, irradiation of diazomethane in the presence of *cis*-2-butene provides *cis*-1,2-dimethylcyclopropane with no detectable amount of the *trans* isomer (eq 45).[60] This reaction is thought to proceed via a singlet carbene. However, if the same reaction is carried out via a triplet carbene, generated via triplet sensitization, then a 1.3:1 mixture of *trans* to *cis* dimethylcyclopropane is observed (eq 46).[61] The yields in the photochemical reaction are typically lower than the metal-mediated processes, and are usually accompanied by more side products.

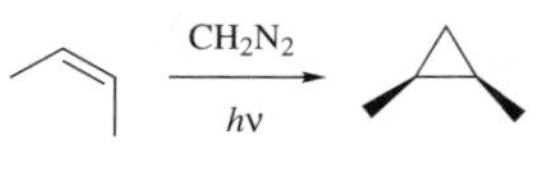

(45)

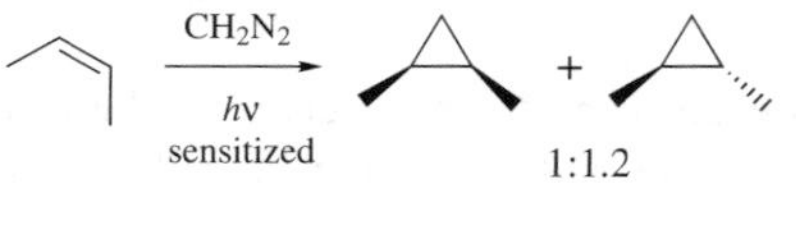

(46)

Additions to Electron-Deficient Species. Diazomethane will also add to highly electrophilic species such as sulfenes or iminium salts to give the corresponding three-membered ring heterocycles. When the reaction is performed on sulfenes, the products are episulfones which are intermediates in the Ramberg–Backlund rearrangement, and are therefore precursors for the synthesis of alkenes via chelotropic extrusion of SO_2. The sulfenes are typically prepared in situ by treatment of a sulfonyl chloride with a mild base, such as ***Triethylamine*** (eq 47).[62] Similarly, the addition of diazomethane to imminium salts has been used to methylenate carbonyls.[63] In this case, the intermediate aziridinium salt is treated with a strong base, such as ***n-Butyllithium***, in order to induce elimination (eq 48).

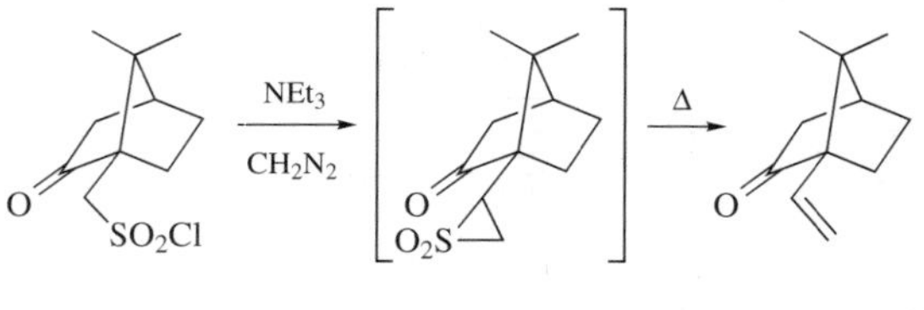

(47)

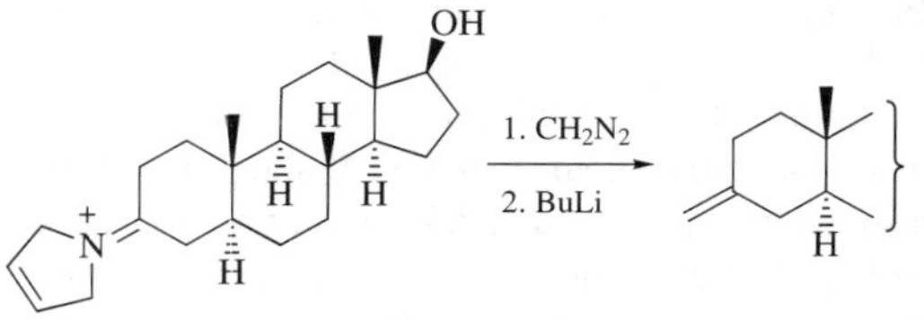

(48)

Miscellaneous Reactions. Diazomethane has been shown to react with vinylsilanes derived from α,β-unsaturated esters to provide the corresponding allylsilane by insertion of CH_2 into the C–Si bond (eq 49).[64] The reaction has been shown to be stereospecific, with *cis*-vinylsilane providing *cis*-allylsilanes; however, the mechanism of the reaction has not been defined. Diazomethane has also been used in the preparation of trimethyloxonium salts. Treatment of a solution of dimethyl ether and trinitrobenzenesulfonic acid with diazomethane provides trimethyloxonium trinitrobenzenesulfonate, which is more stable than the fluoroborate salt.[65]

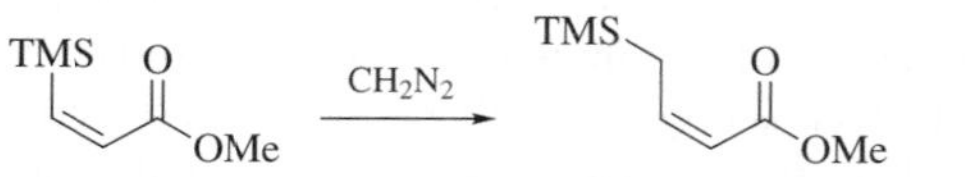

(49)

Related Reagents. 2-Diazopropane; 1-Diazo-2-propene; Dimethyl Sulfate; Diphenyldiazomethane; Methyl Fluorosulfate; Phenyldiazomethane; Trimethyloxonium Tetrafluoroborate; Trimethyl Phosphate; Trimethylsilyldiazomethane.

---

1. (a) Regitz, M.; Maas, G. *Diazo Compounds, Properties and Synthesis*; Academic: Orlando, 1986. (b) Black, T. H. *Aldrichim. acta* **1983**, *16*, 3. (c) Pizey, J. S. *Synthetic Reagents*; Wiley: New York, 1974; Vol. 2, pp 65–142.
2. Arndt, F. *OSC* **1943**, *2*, 165.
3. Moore, J. A.; Reed, D. E. *OSC* **1973**, *5*, 351. Redemann, C. E.; Rice, F. O.; Roberts, R.; Ward, H. P. *OSC* **1955**, *3*, 244. McPhee, W. D.; Klingsberg, E. *OSC* **1955**, *3*, 119.
4. De Boer, Th. J.; Backer, H. J. *OSC* **1963**, *4*, 250. Hudlicky, M. *JOC* **1980**, *45*, 5377. See also Aldrich Chemical Company Technical Bulletins Number AL-121 and AL-131. Note that the preparation described in *FF*,

1967, *1*, 191 is flawed and neglects to mention the addition of ethanol. Failure to add ethanol can result in a buildup of diazomethane and a subsequent explosion.

5. McKay, A. F. *JACS* **1948**, *70*, 1974. See also Aldrich Chemical Company Technical Bulletin Number AL-132.
6. von Pechman, A. *CB* **1894**, *27*, 1888.
7. Ref. 2, note 3.
8. For a description of the safety hazards associated with diazomethane, see: Gutsche, C. D. *OR* **1954**, *8*, 391.
9. Fujisawa, T.; Sato, T.; Itoh, T. *CL* **1982**, 219.
10. Nicolaou, K. C.; Paphatjis, D. P.; Claremon, D. A.; Dole, R. E. *JACS* **1981**, *103*, 6967.
11. Fujisawa, T.; Sato, T.; Kawashima, M.; Naruse, K.; Tamai, K. *TL* **1982**, *23*, 3583.
12. Kozikowski, A. P.; Sugiyama, K.; Springer, J. P. *JOC* **1981**, *46*, 2426.
13. De, B.; Corey, E. J. *TL* **1990**, *31*, 4831. Macomber, R. S. *SC* **1977**, *7*, 405.
14. Blade, R. J.; Hodge, P. *CC* **1979**, 85.
15. Kawamata, A.; Fukuzawa, Y.; Fujise, Y.; Ito, S. *TL* **1982**, *23*, 1083.
16. Ray, J. A.; Harris, T. M. *TL* **1982**, *23*, 1971.
17. Ogawa, H.; Chihara, T.; Taya, K. *JACS* **1985**, *107*, 1365.
18. Frimer, A. A.; Gilinsky-Sharon, P.; Aljadef, G. *TL* **1982**, *23*, 1301.
19. Chavis, C.; Dumont, F.; Wightman, R. H.; Ziegler, J. C.; Imbach, J. L. *JOC* **1982**, *47*, 202.
20. Neeman, M.; Johnson, W. S. *OSC* **1973**, *5*, 245.
21. Ohno, K.; Nishiyama, H.; Nagase, H. *TL* **1979**, *20*, 4405.
22. Robins, M. J.; Lee, A. S. K.; Norris, F. A. *Carbohydr. Res.* **1975**, *41*, 304.
23. Evans, D. S.; Bender, S. L.; Morris, J. *JACS* **1988**, *110*, 2506. For a similar example with the antibiotic lasalocid, see: Westly, J. W.; Oliveto, E. P.; Berger, J.; Evans, R. H.; Glass, R.; Stempel, A.; Toome, V.; Williams, T. *JMC* **1973**, *16*, 397.
24. Nishiyama, H.; Nagase, H.; Ohno, K. *TL* **1979**, *20*, 4671.
25. Muller, v. H.; Huber-Emden, H.; Rundel, W. *LA* **1959**, *623*, 34.
26. Seagusa, T.; Ito, Y.; Kobayashi, S.; Hirota, K.; Shimizu, T. *TL* **1966**, *7*, 6131.
27. In addition to acid chlorides, α-diazo ketones can be synthesized from carboxylic acid anhydrides; however, in this case one equivalent of the carboxylic acid is converted to the corresponding methyl ester. Furthermore, the anhydride can be formed in situ using DCC. See Hodson, D.; Holt, G.; Wall, D. K. *JCS(C)* **1970**, 971.
28. For a review of the Wolff rearrangement, see: Meier, H.; Zeller, K.-P. *AG(E)* **1975**, *14*, 32.
29. Bergmann, E. D.; Hoffmann, E. *JOC* **1961**, *26*, 3555.
30. Clark, R. D. *SC* **1979**, *9*, 325.
31. Smith, A. B.; Dorsey, M.; Visnick, M.; Maeda, T.; Malamas, M. S. *JACS* **1986**, *108*, 3110.
32. Smith, A. B.; Toder, B. H.; Branca, S. J.; Dieter, R. K. *JACS* **1981**, *103*, 1996.
33. Rosenquist, N. R.; Chapman, O. L. *JOC* **1976**, *41*, 3326.
34. Hudliky, T.; Sheth, J. P. *TL* **1979**, *20*, 2667.
35. For a review of intramolecular reactions of α-diazo ketones, see: Burke, S. D.; Grieco, P. A. *OR* **1979**, *26*, 361.
36. Murai, A.; Kato, K.; Masamune, T. *TL* **1982**, *23*, 2887.
37. Iwata, C.; Fusaka, T.; Fujiwara, T.; Tomita, K.; Yamada, M. *CC* **1981**, 463.
38. For a review on the reactions of α-diazo ketones with acid, see: Smith, A. B.; Dieter, R. K. *T* **1981**, *37*, 2407.
39. Ackeral, J.; Franco, F.; Greenhouse, R.; Guzman, A.; Muchowski, J. M. *JHC* **1980**, *17*, 1081.
40. Ghatak, U. R.; Sanyal, B.; Satyanarayana, G.; Ghosh, S. *JCS(P1)* **1981**, 1203.
41. Blair, I. A.; Mander, L. N. *AJC* **1979**, *32*, 1055.
42. Smith, A. B.; Toder, B. H.; Branca, S. J.; Dieter, R. K. *JACS* **1981**, *103*, 1996. Smith, A. B.; Dieter, K. *JACS* **1981**, *103*, 2009. Smith, A. B.; Dieter, K. *JACS* **1981**, *103*, 2017.
43. Smith, A. B.; Toder, B. H.; Branca, S. J. *JACS* **1984**, *106*, 3995.
44. Smith, A. B.; Toder, B. H.; Richmond, R. E.; Branca, S. J. *JACS* **1984**, *106*, 4001.
45. Holmquist, C. R.; Roskamp, E. J. *JOC* **1989**, *54*, 3258. Padwa, A.; Hornbuckle, S. F.; Zhang, Z. Z.; Zhi L. *JOC* **1990**, *55*, 5297.
46. Holmquist, C. R.; Roskamp, E. J. *TL* **1992**, *33*, 1131.
47. For a review of this reaction, see: Gutsche, C. D. *OR* **1954**, *8*, 364.
48. For a review, see: Smith, P. A. S.; Baer, D. R. *OR* **1960**, *11*, 157.
49. House, H. O.; Grubbs, E. J.; Gannon, W. F. *JACS* **1960**, *82*, 4099.
50. Majerski, Z.; Djigas, S.; Vinkovic, V. *JOC* **1979**, *44*, 4064.
51. Greene, A. E.; Depres, J-P. *JACS* **1979**, *101*, 4003.
52. For a review, see: Regitz, M.; Heydt, H. In *1,3-Dipolar Cycloadditions Chemistry*; Padwa, A.; Ed.; Wiley: New York, 1984; p 393.
53. For a discussion of the orbital interactions that control dipolar additions of diazomethane, see: Fleming, I. *Frontier Molecular Orbitals and Organic Chemical Reactions*; Wiley: New York, 1976; pp 148–161.
54. Van Auken, T. V.; Rienhart, K. L. *JACS* **1962**, *84*, 3736.
55. This is a very reactive reagent combination which will cyclopropanate benzene and other aromatic compounds. See: Vogel, E.; Wiedeman, W.; Kiefe, H.; Harrison, W. F. *TL* **1963**, *4*, 673. Muller, E.; Kessler, H.; Kricke, H.; Suhr, H. *TL* **1963**, *4*, 1047.
56. Kuehne, M. E.; King, J. C. *JOC* **1973**, *38*, 304.
57. Pincock, R. E.; Wells, J. I. *JOC* **1964**, *29*, 965.
58. Suda, M. *S* **1981**, 714.
59. Raduchel, B.; Mende, U.; Cleve, G.; Hoyer, G. A.; Vorbruggen, H. *TL* **1975**, *16*, 633.
60. Doering, W. von E.; LaFlamme, P. *JACS* **1956**, *78*, 5447.
61. Duncan, F. J.; Cvetanovic, R. J. *JACS* **1962**, *84*, 3593.
62. Fischer, N.; Opitz, G. *OSC* **1973**, *5*, 877.
63. Hata, Y.; Watanabe, M. *JACS* **1973**, *95*, 8450.
64. Cunico, R. F.; Lee, H. M.; Herbach, J. *JOM* **1973**, *52*, C7.
65. Helmkamp, G. K.; Pettit, D. J. *OSC* **1973**, *5*, 1099.

Tarek Sammakia
University of Colorado, Boulder, CO, USA

Di-*t*-butyl Dicarbonate

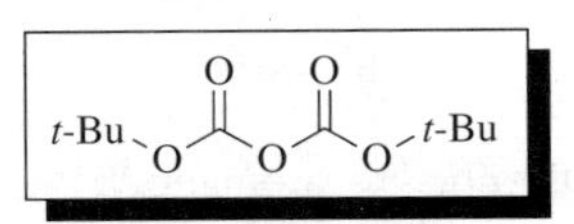

[24424-99-5] $C_{10}H_{18}O_5$ (MW 218.25)

(efficient *t*-butoxycarbonylating agent for amines,[1,2] phenols, and thiols; *t*-butoxycarbonylation of alcohols, amides,[3] lactams, carbamates[4] and NH-pyrroles[5] occurs in the presence of 4-dimethylaminopyridine)

Alternate Names: di-*t*-butyl pyrocarbonate; Boc anhydride; Boc_2O.

Physical Data: mp 22–24 °C; bp 56–57 °C/0.5 mm Hg; fp 37 °C; *d* 0.950 g cm^{-3}; n_D^{20} 1.4103.

Solubility: insol cold H_2O; sol most organic solvents such as decalin, toluene, CCl_4, THF, dioxane, alcohols, acetone, MeCN, DMF.

Form Supplied in: colorless liquid; widely available.

Analysis of Reagent Purity: IR 1810 and 1765 cm^{-1}; NMR 1.50 (CCl_4).

Preparative Method: prepared by the action of ***Phosgene*** on potassium *t*-butyl carbonate, followed by conversion of the obtained di-*t*-butyl tricarbonate under the action of basic catalysts such as ***1,4-Diazabicyclo[2.2.2]octane***.[6]

Handling, Storage, and Precautions: the liquid is flammable and must be stored in a refrigerator in the absence of moisture. Do not heat above 80 °C.

---

t*-Butoxycarbonylation of Amines.** Owing to the instability of the corresponding chloride (t*-Butyl Chloroformate**), the Boc_2O reagent is widely used for the introduction of the *t*-butoxycarbonyl amino protecting group (Boc). This group is easily removed under moderately acidic conditions such as treatment with ***Trifluoroacetic Acid***, or by thermolysis. The Boc–NHR group is not hydrolyzed under basic conditions and is inert to many other nucleophiles.[1] Therefore the use of this amino protecting group is not limited to amino acid and peptide synthesis; it has been extended to the synthesis of amino sugars, alkaloids, etc. The great importance of the *t*-butyl carbamate function in peptide synthesis is not only due to the ease of its cleavage, but also to the fact that the oxazolones formed from the activated *N*-alkoxycarbonyl amino acids do not usually epimerize (racemize) during coupling with amines.[7]

Boc_2O reacts smoothly and rapidly with amino compounds in organic solvents, or aqueous organic solvent mixtures, to form pure derivatives in high yields, the only byproducts being the innocuous and easily removable *t*-butanol and carbon dioxide (eq 1). Innumerable examples are available.[8–11] ***Hydroxylamine*** slightly accelerates the reaction through the in situ formation of the Boc–ONH_2 intermediate.[12] The reaction is also accelerated by sonication of the amine salts.[13]

t-Bu–O–C(=O)–O–C(=O)–O–t-Bu (Boc_2O) —RNH_2→ t-Bu–O–C(=O)–NHR (Boc-NHR) + t-Bu–OH + CO_2 (1)

Aliphatic,[14] alicyclic,[15] aromatic,[16,17] and heterocyclic amines[18,19] have been *t*-butoxycarbonylated under a variety of conditions. Anilines are sometimes heated at reflux in THF to obtain the corresponding Boc derivatives in good yields. However, *t*-butoxycarbonylation also occurs at rt,[17] and 4-aminobenzoic acid reacts in DMF/aq NaOH at rt.[20] The Boc–NHR substituent is an *ortho*-directing group in heteroatom-facilitated lithiations of aromatic and heterocyclic amines.[16,18,19,21] Reaction of the resultant carbanions (eq 2), as well as that of the analogous benzylic carbanions, with electrophiles is a powerful method for the synthesis of substituted anilines and heterocycles.[16,22]

Asymmetric deprotonation of Boc-pyrrolidine with ***s-Butyllithium/(−)-Sparteine*** gives a chiral organolithium reagent, which undergoes reaction with electrophiles to give enantiomerically enriched products.[23] The sequence of equatorial α-lithiation followed by an electrophilic substitution of a series of *N*-Boc cyclic amines provides amines which are substituted adjacent to the nitrogen atom.[15]

A suspension of 4-dimethylaminopyridinium fluoroborate in ethyl acetate reacts with Boc_2O to give the corresponding 1-*t*-butoxycarbonyl-4-dimethylaminopyridinium salt, which is an efficient water-soluble agent for the *t*-butoxycarbonylation of amines (eq 3).[24]

PhNH–C(=O)–O-t-Bu —RLi→ (2-lithiophenyl)N=C(O-t-Bu)–OLi —E^+→ 2-E-C6H4–NH-Boc ⟶ 2-E-C6H4–NH_2 (2)

4-NMe_2-pyridinium(N–H) BF_4^- —Boc_2O, EtOAc, 20 °C, 1 h, 95%→ 1-(t-BuOC(=O))-4-NMe_2-pyridinium BF_4^- + CO_2 + *t*-BuOH (3)

—RNH_2, H_2O→ RNH–Boc

Interesting chemo- and regioselective alkoxycarbonylation of polyfunctional substrates, e.g. diamines, amino alcohols, and amino acids, can be achieved under mild conditions. Treatment of α,ω-alkanediamines (6–10 molar excess) with the reagent in dioxane or acetone as solvent leads to the corresponding mono-Boc diamines in 75–90% yields (eq 4).[25,26]

$H_2N(CH_2)_nNH_2$ ($n = 2, 6$) —6 equiv Boc_2O, dioxane, rt, 22 h→ Boc–NH$(CH_2)_n$$NH_2$ (75–90%) + Boc–NH$(CH_2)_n$NH–Boc (2–9%) (4)

$N^ε$-Boc-L-lysine has been selectively prepared from the copper complex of lysine.[27] This complex, as well as that of $N^δ$-Boc-L-ornithine, has been directly benzyloxycarbonylated to give the corresponding α-Cbz-ε-Boc amino acids (eq 5).[28]

$(H_2N(CH_2)_4CH(NH_2)CO_2^-)_2$ Cu^{2+} —Boc_2O, dioxane, H_2O, rt, 12 h→ $(Boc–NH(CH_2)_4CH(NH_2)CO_2^-)_2$ Cu^{2+} —ion exchange resin, 75–80%→ $N^ε$-Boc-L-Lys; → $N^α$-Cbz-$N^ε$-Boc-L-Lys (5)

In CH_2Cl_2 at low temperature, selective *t*-butoxycarbonylation of a β-amino- in the presence of an α-amino-ester function has been recently realized (eq 6).[29]

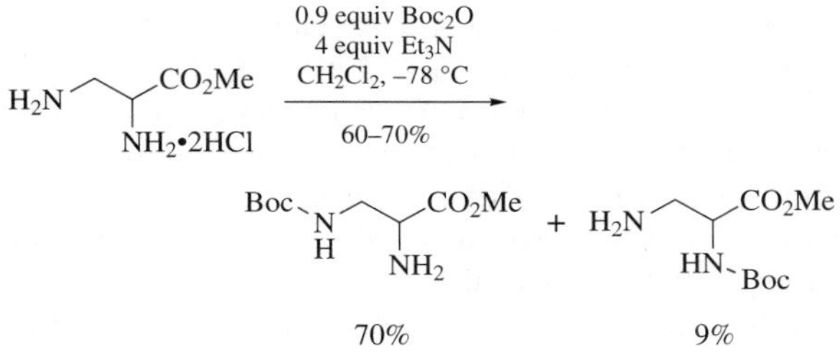

(6)

The substitution of a secondary amino group in the presence of a primary amine has been performed by an indirect route: formation of the labile Schiff base of the primary amine and then *t*-butoxycarbonylation of the secondary nitrogen (eq 7).[30] Selective protection of polyamines may also be achieved by an indirect route in the presence of DMAP (see below).

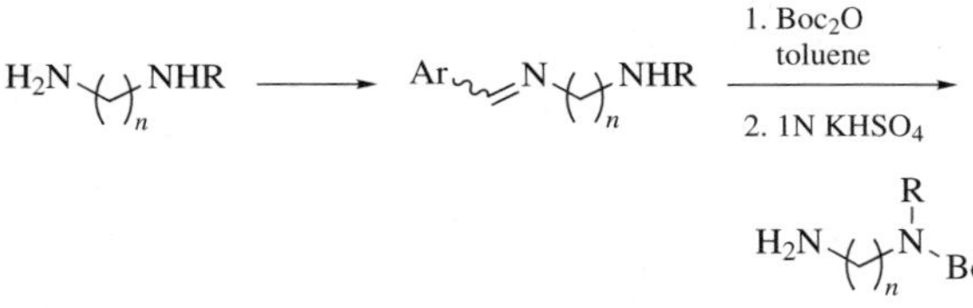

(7)

The Boc_2O reagent has been widely used for the *N*-protection of amino acids and peptides. When a carboxylic acid function is present in the molecule, 1 equiv of base (NaOH, Et_3N, Triton B) is necessary (eq 8).[10,11,31]

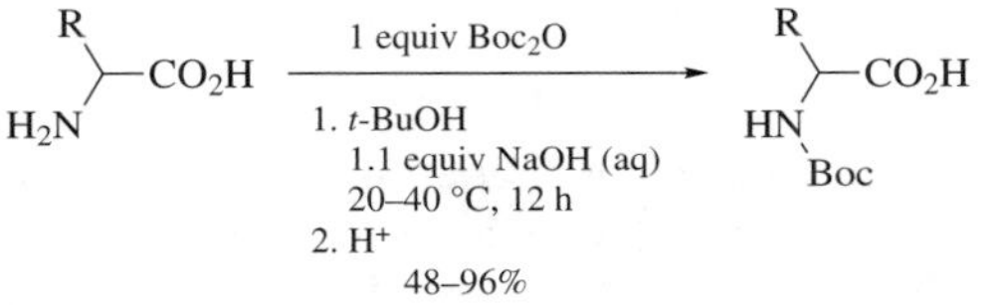

(8)

To avoid loss of the *O*-labeling, trialkylammonium salts of amino acids have been *t*-butoxycarbonylated in dry methanol or DMF.[32] *N-t*-butoxycarbonylation of an alkaline labile and hindered 2,2-dimethylthiazolidine-4-carboxylic acid has been performed in acetonitrile (eq 9).[33]

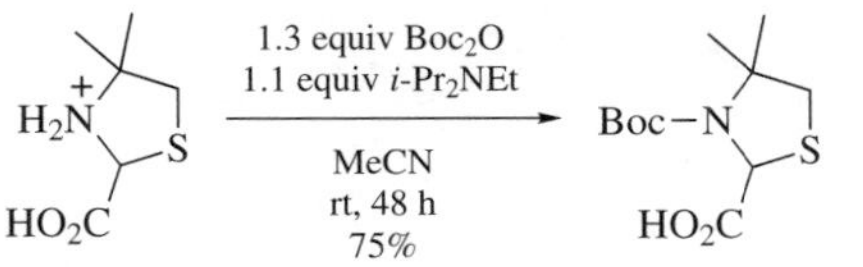

(9)

In the presence of pyridine, a free carboxylic function reacts with the reagent to give a mixed anhydride (see below). This reaction also occurs during the acylation of an amino acid salt with Boc_2O and is responsible for the formation of small amounts of dipeptide byproducts by aminolysis of the mixed anhydride intermediate (eq 10).[34]

t-Butoxycarbonylation of phenolic amino acids in aqueous ***Potassium Hydroxide*** gives the *N,O*-bis-Boc derivatives.[35] Under phase-transfer conditions, selective *N-t*-butoxycarbonylation of tyrosine ethyl ester has been observed (see below). The SH group of cysteine,[36] the imidazole of histidine,[37] and even the guanidine function of arginine[38] have been acylated. Boc_2O has been used for the chemical modification of histidine residues of proteins.[39] For the *t*-butoxycarbonylation of insulin, Boc_2O is much more reactive than the corresponding succinimidyl and *p*-nitrophenyl esters.[40] Bamberger ring-cleavage of the imidazole nucleus slowly occurs in the presence of ***Potassium t-Butoxide***.[41] Selective *N*-protection of amino alcohols is also easily performed (eq 11).[42]

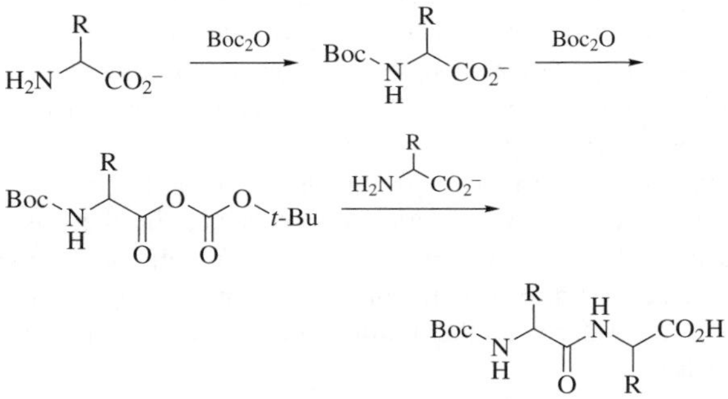

(10)

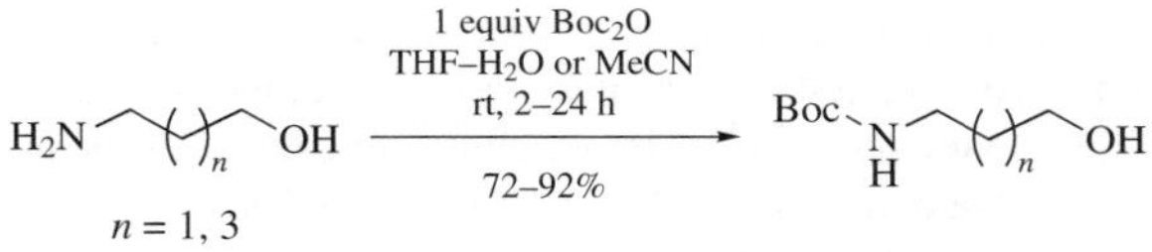

(11)

At pH 10.3, hydroxylamine itself gives the *O*-substituted product,[12] whereas treatment of *N*-alkylhydroxylamines in dioxane furnishes the *N*-Boc derivatives (eq 12).[43] The *O*-Boc hydroxylamine[12,44] as well as *O*-Boc oximes[45] are themselves efficient *t*-butoxycarbonylating agents.

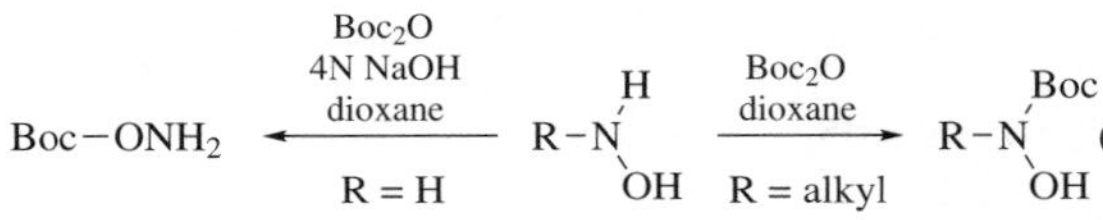

(12)

Hydrazines and hydrazides are acylated by the reagent.[46] The mono-Boc derivatives of hydrazine or phenylhydrazine, obtained in isopropanol, react further with Boc_2O in benzene or dioxane to give the $(Boc–NR)_2$ diacylated derivatives. Regioselective N^{α}-protection of ethyl hydrazinoacetate is performed at 0 °C and the product used for the synthesis of hydrazino peptides (eq 13).[47]

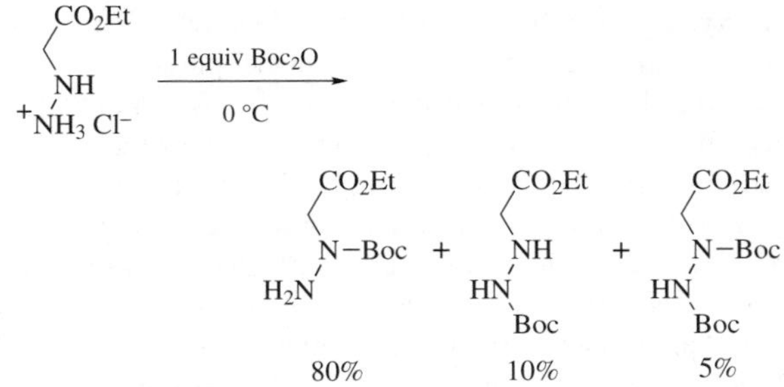

(13)

Boc-amines can also be prepared directly from azides by hydrogenation in the presence of Boc_2O (eq 14).[48]

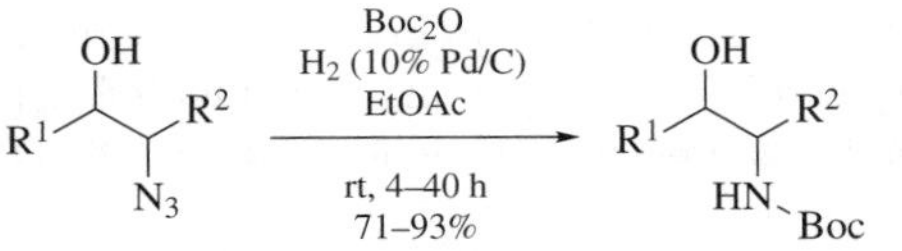

(14)

The alkylation of Boc-amines, particularly that of the more acidic Boc-anilines[17] and the selective alkylation of *N*-Boc amino acids,[49–51] has been done by using an alkylating agent and ***Sodium Hydride*** or ***Silver(II) Oxide*** as a base. For the *N*-perbenzylation of *N*-Boc peptides a P4-phosphazene base (see ***Phosphazene Base*** P_4***-t-Bu***) has been used at −100 °C.[52]

Exchange of Amino-Protecting Groups. *N*-Benzyloxycarbonyl (Cbz) and 9-fluorenylmethyloxycarbonyl (Fmoc) amino protecting groups are also widely used in peptide synthesis.[2] The choice of protecting groups, and hence of the conditions for their cleavage (Cbz: hydrogenolysis or HBr/HOAc; Fmoc: piperidine or F^-)[1] plays a key role in the planning of a synthesis. For instance, orthogonal protection (Boc/Cbz or Fmoc/Boc or Cbz) is of crucial importance in the case of the trifunctional amino acids. For a given peptide segment, chemoselective transformation of an amino protecting group into another one under neutral conditions may be very useful for the synthesis.

In the presence of a slight excess of Boc_2O, conversion of a benzyl carbamate into a *t*-butyl carbamate has been achieved by catalytic hydrogenation,[53] or by the use of cyclohexadiene as an hydrogen donor (eq 15).[54] The catalytic transfer hydrogenation is the more selective: benzyl and BOM ethers are stable under the reaction conditions. The Cbz → Boc transformation can also be performed using triethylsilane and a palladium catalyst.[53] The replacement of a Fmoc group by a Boc group has recently been achieved by the in situ cleavage of the Fmoc group with ***Potassium Fluoride*** in DMF (eq 15).[55]

Ar–CH2–O–C(=O)–NHR → [H_2NR] —(Boc_2O)→ *t*-Bu–O–C(=O)–NHR (15)

Z- or Fmoc-NHR

Ar = Ph
H_2, 5% Pd/C, MeOH, rt, 5–40 h, 93–95%
1,4-cyclohexadiene, 10% Pd/C, EtOH, rt, 0.3–5 h, 86–96%
Et_3SiH, $Pd(OAc)_2$, Et_3N, EtOH, rt, 15–30 h, 91–97%

Ar = 9-fluorenyl: KF, Et_3N, DMF, rt, 5–10 h, 84–93%

***t*-Butoxycarbonylation of Phenols, Alcohols, Enols, and Thiols.** In alkaline media the nucleophilic phenolate salts are easily *t*-butoxycarbonylated.[35,56] Acylation of *p*-dimethylsulfoniophenol gives a sulfonium salt which is a new water-soluble *t*-butoxycarbonylating agent for amino acids.[57] Under anhydrous conditions, thiolate salts are also substituted in good yields using the Boc_2O reagent.[56] Under phase-transfer conditions, *t*-butoxycarbonylation of phenols as well as enols, thiols, or alcohols can be realized. For phenols, enols, and thiols, K_2CO_3 (powder)/18-crown-6/THF are the best experimental conditions. Biphasic conditions (Bu_4NHSO_4/CH_2Cl_2/aq. NaOH) are also effective with sluggish alcoholic groups.[56] Using 1 equiv of Boc_2O, selective *N*-*t*-butoxycarbonylation of tyrosine ethyl ester can be achieved. *N,S*-Bis(Boc)-cysteine ethyl ester has also been prepared.[56] The *t*-butoxycarbonylation of alcohols to give alkyl *t*-butyl carbonates is efficiently catalyzed by DMAP.[58]

t*-Butoxycarbonylation of Amides, Lactams, and Carbamates.** In the presence of ***4-Dimethylaminopyridine, Boc_2O is a strong acylating agent able to react with the weakly nucleophilic amide and carbamate groups under anhydrous conditions. The reaction probably involves a 1-*t*-butoxycarbonyl-4-dimethylaminopyridinium ion intermediate (eq 3).[24] This enhanced reactivity has led to numerous interesting synthetic applications. A wide range of amides, lactams, and urethanes have been converted to the corresponding Boc derivatives using Boc_2O as reagent and DMAP as catalyst.

Secondary amides and lactams react with Boc_2O in the presence of both ***Triethylamine*** and DMAP in CH_2Cl_2 to give stable *N*-*t*-butoxycarbonylated products (eq 16).[3] In the reaction of disubstituted pyrrolidones, the yields are much higher using THF as solvent than with dichloromethane.[59]

R^1C(=O)NHR^2 —(2 equiv Boc_2O, 1 equiv DMAP, 1 equiv Et_3N; CH_2Cl_2, 25 °C, 7 h; 78–96%)→ R^1C(=O)N(Boc)R^2 —(3 equiv LiOH, THF–H_2O; 80–91%)→ R^1COOH (16)

$(CH_2)_n$ lactam NH —(2 equiv Boc_2O, 1 equiv DMAP, 1 equiv Et_3N; CH_2Cl_2, 25 °C, 7 h; 78–96%)→ $(CH_2)_n$ lactam N–Boc —(1.1 equiv NaOMe, MeOH; 85–96%)→ $MeO_2C(CH_2)_n$NHBoc

Exhaustive acylation of a series of secondary amides has also been performed with Boc_2O in dry acetonitrile at ambient temperature using DMAP as catalyst. The method has some steric limitations. For instance, it is retarded in the case of the *ortho*-substituted benzamides. The reaction has been extended to the acylation of carbamates and other amide type functions: sulfonamides, sulfenamides, and phosphinamides.[60] β-Lactams are also *N*-protected in good yields (eq 17).[61]

β-lactam (CO_2-*t*-Bu, NH) —(2 equiv Boc_2O, 0.1 equiv DMAP, MeCN; rt, 24 h; 90%)→ β-lactam (CO_2-*t*-Bu, N-Boc) (17)

Primary carboxamides react with 2 equiv of Boc_2O and 1 equiv of DMAP to give isolable *N*-acylimidodicarbonates which are acylating agents for amines (eq 18).[62] Formamide itself furnishes the unstable di-*t*-butyl *N*-formylimidodicarbonate (eq 19).[63]

RC(=O)NH_2 —(2 equiv Boc_2O, 1 equiv DMAP, MeCN; 25 °C, 2 h; 95%)→ RC(=O)$N(Boc)_2$ —($PhCH_2NH_2$, CH_2Cl_2; 25 °C, 16 h)→ RC(=O)NHCH_2Ph (84%) + $HN(Boc)_2$ (18)

$HCONH_2$ ⟶ ($HCON(Boc)_2$) $\xrightarrow{H_2N(CH_2)_2NEt_2}$ $HN(Boc)_2$ (19)

Grieco has shown that the regioselective hydrolysis or methanolysis of *N-t*-butoxycarbonyl derivatives of amides and lactams affords the corresponding carboxylic acids or methyl esters, respectively (eq 16).[3] ***1,1,3,3-Tetramethylguanidine*** efficiently catalyzes the methanolic cleavage.[64] Grieco's method has been applied to a formal synthesis of gabaculine, in which ***Lithium Hydroxide*** treatment induces both sulfoxide elimination and lactam cleavage (eq 20).[65]

Boc_2O, Et_3N, DMAP (82%); $MeCO_3H$ (79%); LiOH, H_2O–THF, 50 °C (53%) (20)

Other regioselective attacks at the acyl function are known. Aminolysis of substituted pyrrolidones gives the corresponding *N*-Boc γ-amino acid amides.[59] The *N*-Boc five- to eight-membered lactams are cleaved with Grignard reagents at the endocyclic carbonyl group to give ω-Boc-amino ketones,[66–68] which can be deprotected with TFA and cyclized with NaOH to yield cyclic imines.[67] Regioselective ring opening of the *N*-Boc-pyroglutamate ethyl ester with the lithium enolates of carboxylic esters, dithiane anion, alcohols, amines, and thiols occurs without epimerization.[69–71] Selective cleavage of cyclic carbamates gives *N*-protected amino alcohols (eq 21). In this reaction, methyl ester substituents are not transformed into acids by using ***Cesium Carbonate*** as a base.[72]

Boc_2O, Et_3N, DMAP cat.; 2 equiv Cs_2CO_3, MeOH, 3 h; $n = 0, 1$; 70–98% (21)

The selective deacylation of *N*-Boc carboxamides has also been achieved by aminolysis to give *N*-Boc amines. 2-Diethylaminoethylamine in acetonitrile is particularly convenient, but other conditions such as hydrazine in methanol can also be used.[64,73] Aminolysis of di-*t*-butyl *N*-formylimidodicarbonate affords di-*t*-butyl imidodicarbonate, $HN(Boc)_2$ (eq 19).[63] This compound, as well as the unsymmetrical benzyl *t*-butyl imidocarbonate, has been used as a substitute for ***Phthalimide*** in the Gabriel and Mitsunobu reactions to give primary amines. Compared to the classical method, which requires the hydrazinolysis of the phthaloyl protecting group, the main advantage of these novel Gabriel reagents is the much milder and rather specific conditions for the final deprotection step.[74] The lithium salt of $HN(Boc)_2$ has also been used in a palladium-catalyzed allylic amination.[75] Labeled di-*t*-butyl imidodicarbonate leads to ^{13}C and ^{15}N glycine derivatives.[76] Alkylation with triflates prepared from α-hydroxy acid esters gives ^{15}N-labeled chiral Boc-amino acids suitable for the synthesis of labeled peptides.[77]

t-Butoxycarbonylation of benzyl or allyl *N*-Boc or *N*-Cbz amino acid esters using Boc_2O/DMAP gives *N*-bis(alkoxycarbonyl) amino acid esters. Specific cleavage of the allyl protecting group with ***Chlorotris(triphenylphosphine)rhodium(I)***, or hydrogenolysis of the benzyl ester function, respectively, gives N^{α}-benzyloxycarbonyl, N^{α}-*t*-butoxycarbonyl or N^{α}-di-*t*-butoxycarbonyl amino acids (eq 22).[78,79]

1.5 equiv Boc_2O, 0.1 equiv DMAP, MeCN; H_2, Pd/C or $(PPh_3)_3RhCl$ (22)

where H_2, Pd/C is used, R^2 = Bn, Y = Boc
where $(PPh_3)_3RhCl$ is used, R^2 = allyl, Y = Cbz or Boc

The di-*t*-butoxycarbonyl amino acids have been applied to solution peptide synthesis using the ***1,3-Dicyclohexylcarbodiimide*** or *p*-nitrophenyl ester method in DMF for coupling.[79]

Reaction of a bis(alkoxycarbonyl)amino acid pyridinium salt with cyanuric fluoride at $-30\,^{\circ}C$ gives a *N*-bis(alkoxycarbonyl)amino acid fluoride (eq 23). Owing to the high electronegativity and small size of the fluorine atom, this derivative is an efficient acylating agent for amines and pyrrole anions.[80] Use of the Vilsmeier reagent instead of cyanuric fluoride leads to a chemically and optically pure *N*-Boc or *N*-Cbz amino acid *N*-carboxy anhydride (Boc- or Cbz-NCA) without the use of phosgene and the unstable *t*-butyl chlorocarbonate (eq 23).[80] Peptide synthesis with these urethane protected amino acid *N*-carboxy anhydrides (U-NCAs) is a very clean reaction which liberates only CO_2 as a byproduct.[81]

1 equiv $C_3F_3N_3$, CH_2Cl_2, −30 °C, 1.5 h (76–82%); 1 equiv ($SOCl_2$, DMF), MeCN, 20 °C, 2 h; U-NCA 72–92% (23)

R^2 = Bn or *t*-Bu

Treatment with $Boc_2O/DMAP_{cat}$ allows exhaustive *N-t*-butoxycarbonylation of some peptides.[82] The *t*-butoxycarbonylation of the internal amide bonds favors the cyclization of the peptides.[83] Interestingly, one Boc group can be selectively removed from a bis(Boc)-amine[75,76] with a slight excess of trifluoroacetic acid. Use of the $Boc_2O/DMAP_{cat}$/MeCN conditions also allows selective protection of mixed primary–secondary diamines. *t*-Butoxycarbonylation of a

N,N′-bis(benzyloxycarbonyl)diamine followed by hydrogenolysis of the Cbz groups furnishes the *t*-butoxycarbonylated derivative of the primary amino function (eq 24).[84] Analogous strategies have been used for the selective indirect acylation of polyamines such as spermidine.[85,86]

Boc_2O, DMAP, MeCN; H_2 (24)

The *t*-butoxycarbonylation of the sodium salt of thiourea does not require DMAP catalysis. The resulting *N,N′*-bis(*t*-butoxycarbonyl)thiourea reacts with amines to give bis(*t*-butoxycarbonyl)guanidines (eq 25).[87]

2 equiv Boc_2O, NaH, THF; RNH_2, DMF (25)

Sequential diacylation of 1-guanidinylpyrazole gives the *N,N′*-bis(Boc)-guanidinylpyrazole, which is also an efficient reagent for the conversion of amines to protected monosubstituted guanidines and therefore for the synthesis of arginine-containing peptides from their ornithine counterparts (eq 26).[88]

Boc_2O, CH_2Cl_2–DMF, *i*-Pr_2NEt, rt, 2 h; Boc_2O, 3.5 equiv NaH, 65 °C, 4 h; 76%; RNH_2 (26)

N-*t*-Butoxycarbonylation of Pyrroles and Indoles. A series of substituted pyrroles and indoles[89–92] react smoothly with Boc_2O in the presence of catalytic amounts of DMAP in CH_2Cl_2 or acetonitrile to give the corresponding *N*-Boc pyrrole or indole derivatives. The *N*-*t*-butoxycarbonylation allows direct lithiation α to the indole and pyrrole nitrogen at low temperature. Alternatively, lithium–halogen exchange of *N*-Boc-2-bromopyrroles generates the *N*-protected 2-lithiopyrroles (eq 27).[91] After reaction with an electrophile, the Boc protecting group is easily removed either by methoxide ion in methanol or by thermolysis.[90,93]

LiB (27)

X = H, LiB = lithium tetramethylpiperidide
X = Br, LiB = BuLi

In the reaction with Boc_2O/DMAP, N^{α}-Boc-tryptophan methyl ester affords the N^{α},N^{in}-di-Boc derivative in almost quantitative yield. This product has been used in peptide synthesis. Moreover, selective cleavage of the N^{α}-Boc group is possible in the presence of the N^{in}-Boc function by using 2.7 N HCl in dioxane.[94]

Activation of the Carboxyl Group. Formation of Esters and Anhydrides. In the presence of ***Pyridine*** in aprotic solvents, the carboxyl function of a carboxylic acid or a *N*-Boc-amino acid is activated with the Boc_2O reagent. The resulting mixed anhydride can react further with one molecule of the starting acid to give a symmetrical anhydride. At an equimolar ratio of the acid and the Boc_2O reagent a mixture of the mixed and the symmetrical anhydrides is formed. If 2 equiv of acid are used, the symmetrical anhydride becomes the main reaction product. Addition of certain nucleophilic agents to the Boc_2O/carboxylic acid/pyridine mixture leads to the acylation of the nucleophile (eq 28).[58]

Boc_2O, pyridine; RCO_2H; HNu (28)

From a preparative point of view, it is essential that Boc_2O reacts much more rapidly with the carboxylate function than with the nucleophile. The reaction has been applied to the synthesis of aryl and alkyl esters and anilides such as *o*-nitrophenyl, benzyl, or trifluoroethyl esters and *p*-nitroanilides of *N*-protected amino acids. The formation of esters derived from secondary alcohols (benzhydryl, menthyl, or cholesteryl esters) is accelerated by addition of DMAP. Depsipeptides are obtained in good yields. DMAP is a necessary catalyst for the synthesis of *t*-butyl esters and the esterification of *N*-protected amino acids occurs without epimerization.[58]

***t*-Butoxycarbonylation of Carbanions.** Preferential *C*-*t*-butoxycarbonylation of diethyl acetamidomalonate occurs by treatment with $Boc_2O/DMAP_{cat}$ in acetonitrile (eq 29).[60] From the reaction of the potassium salt of diphenylphosphinyl isocyanide and Boc_2O in CH_2Cl_2 at −70 °C, the corresponding Wittig–Horner reagent, *t*-butyl (diphenylphosphinyl)isocyanoacetate, is formed in near quantitative yield.[95] Similarly, treatment of α-lithioalkylphosphonates with Boc_2O gives *C*-*t*-butoxycarbonylated enolates which react with aldehydes to give substituted *t*-butyl acrylates.[96] The lithium salt of the Schiff base derived from the aminomethyldiphenylphosphine oxide reacts with Boc_2O at low temperature to give the imine of *t*-butyl α-(diphenylphosphinoyl)-glycinate.[97]

Boc_2O (29)

A	B	R	Solvent	Temp	Yield
A = B = CO_2Et		R = NHAc	MeCN	rt	93%
A = Ph_2PO	B = N=C	R = H	CH_2Cl_2	−70 °C	92%
A = $(Et_2O)_2PO$	B = alkyl	R = H	THF	−78 °C	quant.
A = Ph_2PO	B = N=CPh	R = H	THF	−60 °C	80%

***t*-Butoxycarbonylation of Imines.** In previous examples, the acylation occurred either at the amino function[30] or at the carbanionic center[97] and not at the imine function. However, the reaction of a nucleophilic imine possessing an electron-releasing substituent (Ar = *p*-$MeOC_6H_4$) with Boc_2O in ethanol proceeds through the intermediacy of an *N*-alkoxycarbonyl iminium ion to give an α-ethoxy carbamate (eq 30). As the trapping of the iminium ion by the alcohol is a reversible process, the α-ethoxy *t*-butyl carbamate is a precursor of the electrophilic *N*-Boc iminium ion in an intermolecular condensation with propargyltrimethylsilane.[98]

Bn–N=CH–Ar → (2 equiv Boc_2O, EtOH, 60–70 °C, 24 h) → [Bn–N^+(=CH–Ar)–C(=O)O-*t*-Bu] → Bn–N(Boc)–CH(OEt)–Ar (80%) (30)

Related Reagents. *t*-Butyl Azidoformate; *t*-Butyl Chloroformate; 1-*N*-(*t*-Butoxycarbonyl)-1*H*-benzotriazole 3-*N*-Oxide; 1-(*t*-Butoxycarbonyl)imidazole; 2-(*t*-Butoxycarbonyloxyimino)-2-phenylacetonitrile; *N*-(*t*-Butoxycarbonyloxy)phthalimide; *N*-(*t*-Butoxycarbonyloxy)succinimide; 1-*t*-Butoxycarbonyl-1,2,4-triazole.

---

1. Greene, T. W.; Wuts, G. M. *Protective Groups in Organic Synthesis* 2nd ed.; Wiley: New York, 1991; p 327.
2. Bodansky, M. *Principles of Peptide Chemistry*; Springer; New York, 1984; p 99.
3. Flynn, D. L.; Zelle, R. E.; Grieco, P. A. *JOC* **1983**, *48*, 2424.
4. Gunnarsson, K.; Grehn, L.; Ragnarsson, U. *AG(E)* **1988**, *27*, 400.
5. Grehn, L.; Ragnarsson, U. *AG(E)* **1984**, *23*, 296.
6. Pope, B. M.; Yamamoto, Y.; Tarbell, D. S. *OS* **1977**, *57*, 45.
7. Benoiton, N. L.; Lee, Y. C.; Steinaur, R.; Chen, F. M. F. *Int. J. Pept. Protein Res.* **1992**, *40*, 559.
8. Pozdnev, V. F. *Khim. Prir. Soedin.* **1971**, 384 (*CA* **1971**, *75*, 110 568x).
9. Pope, B. M.; Yamamoto, Y.; Tarbell, D. S. *PNA* **1972**, *69*, 730.
10. Moroder, L.; Hallett, A.; Wünsch, E.; Keller, O.; Wersin, G. *Z. Physiol. Chem.* **1976**, *357*, 1652.
11. Keller, O.; Keller, W. E.; van Look, G.; Wersin, G. *OS* **1985**, *63*, 160.
12. Harris, R. B.; Wilson, I. B. *TL* **1983**, *24*, 231.
13. Einhorn, J.; Einhorn, C.; Luche, J. L. *SL* **1991**, 37.
14. Guzman, A.; Quintero, C.; Muchowski, J. M. *CJC* **1991**, *69*, 2059.
15. Beak, P.; Lee, W. K. *JOC* **1993**, *58*, 1109.
16. Muchowski, J. M.; Venuti, M. C. *JOC* **1980**, *45*, 4798.
17. Dalla Croce, P.; La Rosa, C.; Ritieni, A. *JCR(S)* **1988**, 346.
18. Fishwick, C. W. G.; Storr, R. C.; Manley, P. W. *CC* **1984**, 1304.
19. Fiakpui, C. Y.; Knaus, E. E. *CJC* **1987**, *65*, 1158.
20. Hofmann, K.; Finn, F.; Kiso, Y. *JACS* **1978**, *100*, 3585.
21. Stanetty, P.; Koller, H.; Mihovilovic, M. *JOC* **1992**, *57*, 6833.
22. Clark, R. D.; Muchowski, J. M.; Fisher, L. E.; Flippin, L. A.; Rebke, D. B.; Souchet, M. *S* **1991**, 871.
23. Kerrick, S. T.; Beak, P. *JACS* **1991**, *113*, 9708.
24. Guibé-Jampel E.; Wakselman, M. *S* **1977**, 772.
25. Shai, Y.; Kirk, K. L.; Channing, M. A.; Dunn, B. B.; Lesniak, M. A.; Eastman, R. C.; Finn, R. D.; Roth, J.; Jakobsen, K. A. *B* **1989**, *28*, 4801.
26. Krapcho, A. P.; Kuell, C. S. *SC* **1990**, *20*, 2559.
27. Scott, J. W.; Parker, D.; Parrish, D. R. *SC* **1981**, *11*, 303.
28. Masukiewicz, E.; Rzeszotarska, B.; Szczerbaniewicz, J. *OPP* **1992**, *24*, 191.
29. Egbertson, M. S.; Homnick, C. F.; Hartman, G. D. *SC* **1993**, *23*, 703.
30. Prugh, J. D.; Birchenough, L. A.; Egberston, M. S. *SC* **1992**, *22*, 2357.
31. Perseo, G.; Piani, S.; De Castiglione, R. *Int. J. Pept. Protein Res.* **1983**, *21*, 227.
32. Ponnusamy, E.; Fotatar, U.; Spisni, A.; Fiat, D. *TL* **1986**, 48.
33. Kemp, D. S.; Carey, R. I. *JOC* **1989**, *54*, 3640.
34. Chen, F. M. F.; Benoiton, N. L. *CJC* **1987**, *65*, 1224.
35. Pozdnev, V. F. *Khim. Prir. Soedin.* **1982**, *129* (*CA* **1983**, *97*, 92 706p).
36. Pozdnev, V. F. *Bioorg. Khim.* **1977**, *3*, 1604 (*CA* **1978**, *88*, 62 595y).
37. Pozdnev, V. F. *ZOB* **1977**, *48*, 476 (*CA* **1978**, *89*, 24 739m).
38. Pozdnev, V. F. *Bioorg. Khim.* **1986**, *12*, 1013.
39. Harris, R. B.; Wilson, I. B. *JBC* **1983**, *258*, 1357.
40. Losse, G.; Naumann, W.; Raddatz, H. *ZC* **1983**, *23*, 22.
41. Altman, Shoef, N.; Wilchek, M.; Warshawsky, A. *CC* **1985**, 1133.
42. Lee, B. H.; Miller, M. J. *JOC* **1983**, *38*, 24.
43. Baldwin, J. E.; Adlington, R. M.; Birch, J. B. *TL* **1985**, *26*, 5931.
44. Harris, R. B.; Wilson, I. B. *Int. J. Pept. Protein Res.* **1984**, *23*, 55.
45. Itoh, H.; Hagiwara, D.; Kamiya, T. *BCJ* **1977**, *50*, 718.
46. Pozdnev, V. F. *ZOR* **1977**, *13*, 2531 (*CA* **1978**, *88*, 89 063k).
47. Lecoq, A.; Marraud, M.; Aubry, A. *TL* **1991**, *32*, 2765.
48. Saito, S.; Nakajima, H.; Inaba, M.; Moriwake, T. *TL* **1989**, *30*, 837.
49. Cheung, S. T.; Benoiton, N. L. *CJC* **1977**, *55*, 906.
50. Hansen, D. W.; Pilipauskas, D. *JOC* **1985**, *50*, 945.
51. Rich, D. H.; Dhaon, M. K.; Dunlap, B.; Miller, S. P. F. *JMC* **1986**, *29*, 978.
52. Pietzonka, T.; Seebach, D. *AG(E)* **1992**, *31*, 1481.
53. Sakatani, M.; Hori, K.; Ohfune, Y. *TL* **1988**, *29*, 2983.
54. Bajwa, J. S. *TL* **1992**, *33*, 2955.
55. Li, W.-R.; Jiang, J.; Joullié, M. M. *TL* **1993**, *34*, 1413.
56. Houlihan, F.; Bouchard, F.; Fréchet, J. M. F.; Wilson, C. G. *CJC* **1985**, *63*, 153.
57. Aruze, I.; Okai, H.; Kouge, K.; Yamamoto, Y.; Koizumi, T. *Chem. Express* **1988**, *3*, 45.
58. Pozdnev, V. F. *Int. J. Pept. Protein Res.* **1992**, *40*, 407.
59. Klutchko, S.; O'Brien, P.; Hodges, J. C. *SC* **1989**, *19*, 2573.
60. Grehn, L.; Gunnarsson, K.; Ragnarsson, U. *ACS* **1986**, *B40*, 745.
61. Baldwin, J. E.; Adlington, R. M.; Godfrey, C. R. A.; Gollins, D. W.; Smith, M; L. Russel, A. *SL* **1993**, 51.
62. Davidsen, S. K.; May, P. D.; Summers, J. B. *JOC* **1991**, *56*, 5482.
63. Grehn, L.; Ragnarsson, U. *S* **1987**, 275.
64. Grehn, L.; Gunnarsson, K.; Ragnarsson, U. *ACS* **1987**, *B41*, 18.
65. Hiemstra, H.; Klaver, W. J.; Speckamp, W. N. *TL* **1986**, *27*, 1411.
66. Ohta, T.; Hosoi, A.; Kimura, T.; Nozoe, S. *CL* **1987**, 2091.
67. Giovannini, A.; Savoia, D.; Umani-Ronchi, A. *JOC* **1989**, *54*, 228.
68. Klutchko, S.; Hamby, J. M.; Reilly, M.; Taylor, M. D.; Hodges, J. C. *SC* **1993**, *23*, 971.
69. Ohta, T.; Kimura, T.; Sato, N.; Nozoe, S. *TL* **1988**, *29*, 4303.
70. Ezquerra, J.; de Mendoza, J.; Pedregal, C.; Ramírez, C. *TL* **1992**, *33*, 5589.
71. Molina, M. T.; del Valle, C.; Escribano, A. M.; Ezquerra, J.; Pedregal C. *T* **1993**, *49*, 3801.
72. Ishizuka, T.; Kunieda, T. *TL* **1987**, *28*, 4185.
73. Grehn, L.; Gunnarsson, K.; Ragnarsson, U. *CC* **1985**, 1317.
74. Ragnarsson, U.; Grehn, L. *ACR* **1991**, *24*, 285.
75. Connell, R. D.; Rein, T.; Åkermark, B.; Helquist, P. *JOC* **1988**, *53*, 3845.
76. Grehn, L.; Bondesson, U.; Pehk, T.; Ragnarsson, U. *CC* **1992**, 1332.
77. Degerbeck, F.; Fransson, B.; Grehn, L.; Ragnarsson, U. *JCS(P1)* **1993**, 11.
78. Gunnarsson, K.; Grehn, L.; Ragnarsson, U. *AG(E)* **1988**, *27*, 400.

79. Gunnarsson, K.; Ragnarsson, U. *ACS* **1990**, *44*, 944.
80. (a) Savrda, J.; Wakselman, M. *CC* **1992**, 812. (b) Carpino, L. A.; Mansour, E.-S. M. E.; El-Faham, A. *JOC* **1993**, *58*, 4162.
81. Fuller, W. D.; Cohen, M. P.; Shabankaresh, M.; Blair, R. K.; Goodman, M.; Naider, F. R. *JACS* **1990**, *112*, 7414.
82. Grehn, L.; Ragnarsson, U. *AG(E)* **1985**, *24*, 400.
83. Cavelier-Frontin, F.; Achmad, S.; Verducci, J.; Pépe, G.; Jacquier, R. *J. Mol. Struct. (Theochem)* **1993**, *105*, 125.
84. Almeida, M. L. S.; Grehn, L.; Ragnarsson, U. *CC* **1987**, 1250.
85. Almeida, M. L. S.; Grehn, L.; Ragnarsson, U. *JCS(P1)* **1988**, 1905.
86. Araújo, M. J. S. M. P.; Ragnarsson, U.; Almeida, M. L. S.; Amaral Trigo, M. J. S. A. *JCR(S)* **1992**, 110.
87. Iwanowicz, E. J.; Poss, M. A.; Lin, J. *SC* **1993**, *23*, 1443.
88. Bernatowicz, M. S.; Wu, Y.; Matsueda, G. R. *TL* **1993**, *34*, 3389.
89. Grehn, L.; Ragnarsson, U. *AG(E)* **1984**, *23*, 296.
90. Rawal, V. H.; Cava, M. P. *TL* **1985**, *26*, 6141.
91. Chen, W.; Cava, M. P. *TL* **1987**, *28*, 6025.
92. Woolridge, E. M.; Rokita, S. E. *TL* **1989**, *30*, 6117.
93. Leroy, J.; Cantacuzene, D.; Wakselman, C. *S* **1982**, 313.
94. Franzén, H.; Grehn, L.; Ragnarsson, U. *CC* **1984**, 1699.
95. Rachoń, J.; Schöllkopf, U. *LA* **1981**, 99.
96. Teulade, M. P.; Savignac, P.; About-Jaudet, E.; Colignon, N. *SC* **1989**, *19*, 71.
97. van Es, J. J. G. S.; Jaarveld, K.; van der Gen, A. *JOC* **1990**, *55*, 4063.
98. Esch, P. M.; Hiemstra, H.; Speckamp, W. N. *T* **1992**, *48*, 3445.

Michel Wakselman
CNRS-CERCOA, Thiais, France

Di-*n*-butyltin Oxide[1,2]

Bu_2SnO

[818-08-6] $C_8H_{18}OSn$ (MW 248.92)

(converts diols to *O*-stannylene acetals[2] which undergo regioselective acylation, alkylation, oxidation, and condensation with activated carboxylic acid derivatives; activates α,ω-hydroxy acids and α,ω-amino acids for lactone and lactam formation, respectively;[3] mediates epoxidation of allylic alcohols by *t*-butyl hydroperoxide;[4] catalyzes $TMSN_3$ addition to nitrile;[5] useful as a source of tin for Otera transesterification[6] and 'organotin phosphate condensate'[7] catalysts)

Alternate Name: DBTO.
Physical Data: mp >300 °C.
Solubility: insol most organic solvents with which it does not react. However solvents like toluene, benzene, and methanol are routinely used for reactions of hydroxylic substrates.
Form Supplied in: white powder.
Handling, Storage, and Precautions: toxic and irritant;[8,9] use in a fume hood.

Introduction. Di-*n*-butyltin oxide, because of its biocidal and antifouling properties, is one of the most widely used industrial chemicals.[9] It is the final product of hydrolysis of dialkyltin halides (eq 1) and exists as an amorphous polymeric solid insoluble in nonreacting solvents. A reticulated network (**1**) consisting of four-membered rings of alternating tin and oxygen has been suggested as a possible structure for this compound.[10] The interaction of diorganotin oxides with various biomolecules like amino acids, peptides, carbohydrates, and nucleic acid components has been reviewed[9] and is beyond the scope of this article.

$$R_2SnX_2 \longrightarrow R_2Sn(X)OSn(X)R_2 \longrightarrow R_2(X)SnOSnR_2(OH) \longrightarrow R_2SnO \quad (1)$$

(**1**)

Formation and Reactions of Dibutylstannylenes. By far the largest application of DBTO in organic chemistry is in the generation and reactions of stannylenes from polyhydroxy compounds. Dibutylstannylenes are prepared in nearly quantitative yields by heating stoichiometric amounts of DBTO and the polyol in a solvent such as benzene or toluene with concomitant removal of water (eq 2).[11] Compounds (**2**)[12] and (**3**)[13] are other prototypical stannylenes that have been prepared by this method. Formation and reactions of 1,6-stannylenes from monosaccharides have also been reported recently.[14] While the stannylene from ethylene glycol (eq 2) is an infinite coordination chain polymer with hexacoordinated Sn,[15] the stannylenes from cyclohexane-1,2-diols and the methyl 4,6-*O*-benzylidene-α-D-glucopyranoside (**3**)[16] are dimeric (e.g. **4**).

HO–CH₂CH₂–OH + DBTO → (benzene, Δ, $-H_2O$, 91%) (2)

mp 223–226 °C

(**2**) (**3**)

Regioselective Alkylation, Acylation, and Sulfonylation. Synthetic applications of stannylenes have followed the elegant studies of Moffatt[17] and Ogawa,[18] who showed that the inherent differences in the nucleophilicities of carbohydrate hydroxys can be amplified by the formation of trialkyltin ethers. Several selective acylations and alkylations could thus be accomplished. Further it was noted that while acylation proceeds without any catalyst, alkylation is a sluggish reaction and needs assistance from tetrabutylammonium halides (eq 3).[19]

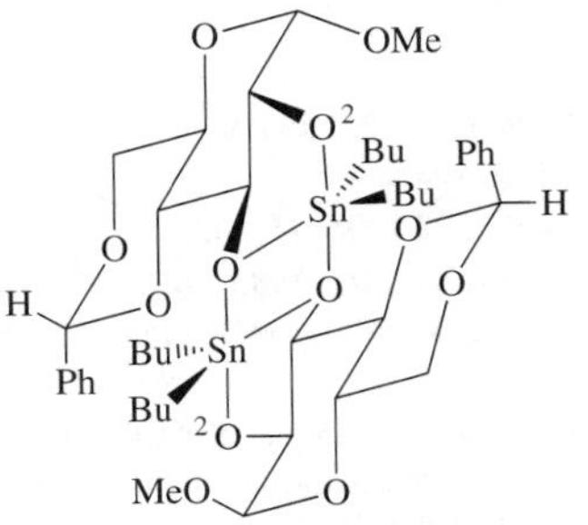

(4)

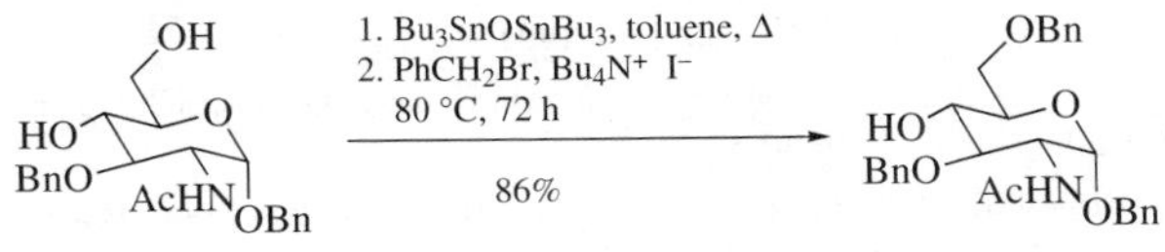

1. $Bu_3SnOSnBu_3$, toluene, Δ
2. $PhCH_2Br$, $Bu_4N^+ I^-$
80 °C, 72 h
86% (3)

Reactions of stannylene (**2**) are typical and are shown in eq 4.[2]

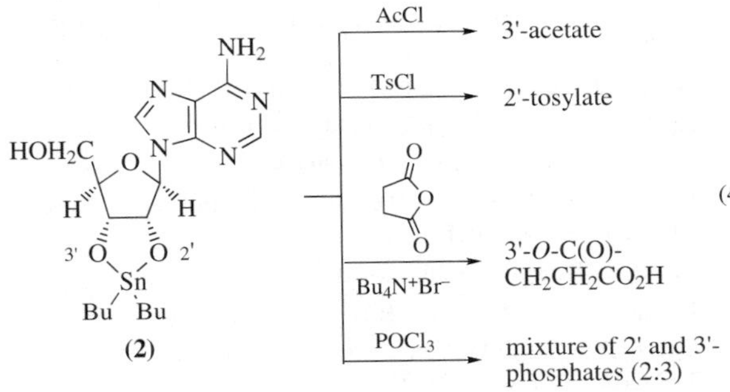

AcCl → 3'-acetate
TsCl → 2'-tosylate
$Bu_4N^+Br^-$ → 3'-*O*-C(O)-$CH_2CH_2CO_2H$
$POCl_3$ → mixture of 2' and 3'-phosphates (2:3) (4)

(2)

The stannylene (**3**) reacts with benzoyl chloride or tosyl chloride to give the corresponding 2-*O*-derivative (eq 5). The alkylation is less selective.[20] Use of nonpolar solvents and tetrabutylammonium halide has been a significant improvement (eq 6) in this alkylation procedure.[21] Several other examples of selective acylation and alkylations are given by David and Hanessian.[2] An intramolecular ether formation using DBTO has been used for the synthesis of octosyl acid.[22] Monoacylation of dimethyl L-tartrate has been carried out using this procedure (eq 7).[23] Two highly readable accounts of how the anomer composition and stoichiometry of reagents affect the regiochemistry of acylation[24] and tosylation[25] have recently appeared.

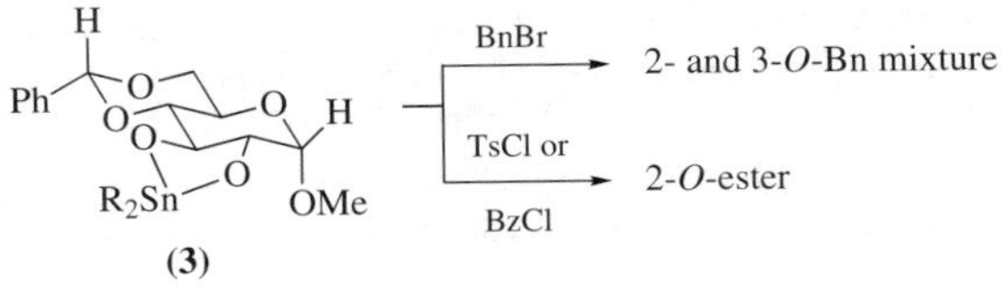

BnBr → 2- and 3-*O*-Bn mixture
TsCl or BzCl → 2-*O*-ester (5)

(3)

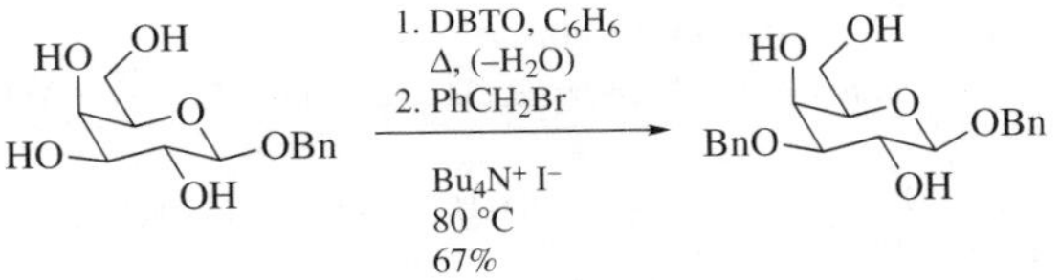

1. DBTO, C_6H_6, Δ, $(-H_2O)$
2. $PhCH_2Br$, $Bu_4N^+ I^-$, 80 °C
67% (6)

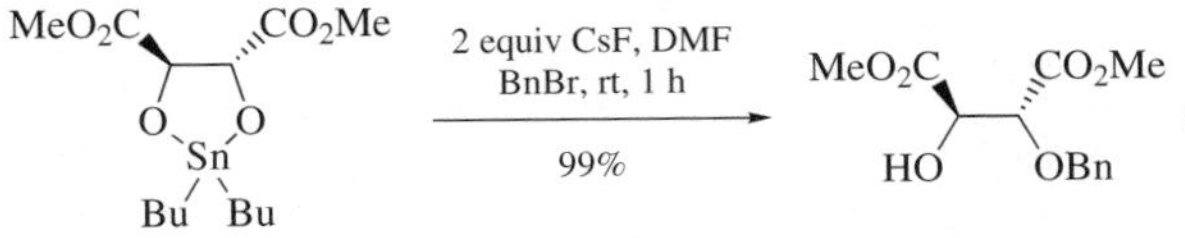

2 equiv CsF, DMF, BnBr, rt, 1 h
99% (7)

(also reactions with *t*-BuCOCl and TBDMS-Cl)

The structure of the dimeric complex (**4**) has been used to explain the observed preferential reactivity of the C-2 hydroxy groups.[2] It is argued that the divalent oxygens that occupy the axial position within the tin cordination sphere (i.e. this case, the C-2 oxygens) are not only more nucleophilic, but also are more accessible sterically. It is believed that the orientational preference of the hydroxy group is related to its acidity, and thus by forming a Sn complex with the more acidic hydroxy group in the axial position, its reactivity towards nucleophiles is enhanced. The stannylene procedure can be used to reverse the stereochemistry of the classical monoacylation regiochemistry observed in some unsymmetrical diols.[26]

Regioselective Oxidation. Following the original discovery by David and Thieffrey,[27] several stannylenes have been oxidized by bromine to acyloins (eq 8). Hanessian employed this reaction for the synthesis of a densely functionalized intermediate which was converted into (+)-spectinomycin.[28] A detailed study of monooxidation of unprotected carbohydrates has also appeared.[29] Stannylenes are oxidized the same way as diols, with ***Sodium Periodate*** and ***Lead(IV) Acetate***.[2]

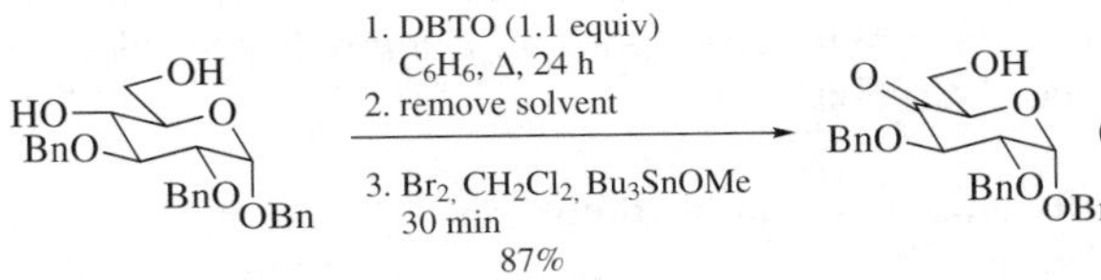

1. DBTO (1.1 equiv), C_6H_6, Δ, 24 h
2. remove solvent
3. Br_2, CH_2Cl_2, Bu_3SnOMe, 30 min
87% (8)

Esterification Catalyst. In the presence of DBTO, alcohols react with carboxylic acids to give esters[30] and the reaction can be extended to make lactones and lactams (eqs 9 and 10).[3] It should be noted that the stereochemistry of the alcohol center is unaffected. A 'template-driven' extrusion process, in which an intermediate where the hydroxy and the carboxyl groups are brought into close proximity by tin oxide, has been proposed for this reaction.

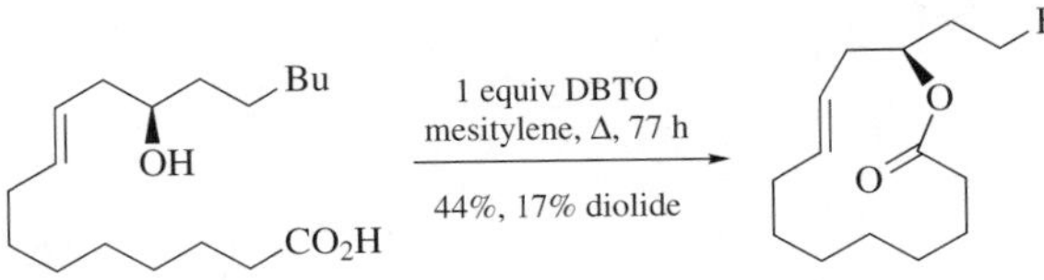

1 equiv DBTO, mesitylene, Δ, 77 h
44%, 17% diolide (9)

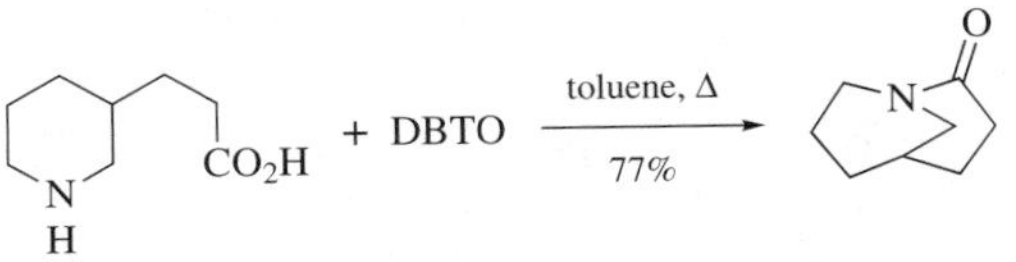

+ DBTO, toluene, Δ
77% (10)

Stannylenes derived from diols react with bifunctional carboxylic acid derivatives to give macrocyclic polyolides of various sizes.[31,32] This approach has been used for the synthesis of (+)-dicrotaline (eq 11).[33]

Bu, Bu Sn–O O H N + O O O xylene −25 °C to Δ 32% O O O O H N (11)

Dibutyltin oxide is the best source of tin for the Otera transesterification catalysts (eq 12),[6] which have found wide use in organic synthesis. DBTO has also been used as a catalyst for hydrolysis of amides in very sensitive molecules, where other procedures have failed.[34]

$$\text{DBTO} + \text{Bu}_2\text{Sn(NCS)}_2 \xrightarrow{\text{EtOH, }\Delta} \quad (12)$$

1 equiv 0.33 equiv

R = Bu, Y = NCS, X = OH

Catalyst in Oxidation Reactions. DBTO has been used as a catalyst in Fe^{III}-mediated oxidation of thiols to disulfides, even though ***Tri-n-butyl(methoxy)stannane*** seems to be better suited for this purpose.[35] Epoxidation of terminal alkenes in a two-phase system (chloroform–water) containing H_2O_2/ammonium molybdate/DBTO has also been reported.[36] A combination of DBTO and ***t-Butyl Hydroperoxide*** oxidizes allylic alcohols with moderate regio- and stereoselectivity.[4] Tri- and tetrasubstituted double bonds are most easily oxidized and the selectivities are comparable to those of the corresponding ***Vanadyl Bis(acetylacetonate)*** mediated reactions.

Miscelleneous Applications. The method of choice for the regioselective opening of benzylic and tertiary epoxides with alcohols (eq 13) appears to be a reaction mediated by 'organotin phosphate condensate (OPC)', which is readily prepared from DBTO and tributyl phosphate.[7] DBTO catalyzes rearrangement of 3-hydroxy-2-oxo carboxylic acid esters (eq 14), a reaction reminiscent of a similar one mediated by the enzyme reductoisomerase.[37] DBTO has been used as a catalyst for the addition of ***Azidotrimethylsilane*** to nitriles[5] for the production of tetrazoles (eq 15).

OMe, 98% ee; MeOH, OPC, Δ; 88%; MeO, OH, OMe (13); 96% ee

OH, CO_2Et, O; DBTO (0.02 equiv), toluene, Δ, 60 h; 79%; O, OH, CO_2Et (14)

MeO_2C, CN; TMS-N_3, DBTO (2:0.1 equiv), toluene, 110 °C, 24 h; 98%; HN–N, N, N, Ar (15)

Related Reagents. Bis(tri-*n*-butyltin) Oxide; 3-Chloro-1-hydroxytetrabutyldistannoxane.

1. Pereyre, M.; Quintard, J. P.; Rahm, A. *Tin in Organic Synthesis*; Butterworth: London, 1987.
2. David, S.; Hanessian, S. *T* **1985**, *41*, 643.
3. Steliou, K.; Poupart, M. A. *JACS* **1983**, *105*, 7130. See also: Steliou, K.; Szczygielska-Nowosielska, A.; Favre, A.; Poupart, M. A.; Hanessian, S. *JACS* **1980**, *102*, 7578.
4. Kanemoto, S.; Nonaka, T.; Oshima, K.; Utimoto, K.; Nozaki, H. *TL* **1986**, *27*, 3387. For the possible structure of the reagent see: Davies, A. G.; Grahan, I. F. *CI(L)* **1963**, 1622.
5. Wittenberger, S. J.; Donner, B. G. *JOC* **1993**, *58*, 4139.
6. Otera, J.; Dan-oh, N.; Nozaki, H. *JOC* **1991**, *56*, 5307.
7. Otera, J.; Niibo, Y.; Nozaki, H. *T* **1991**, *47*, 7625.
8. Selwyn, M. J. In *Chemistry of Tin*; Harrison, P. G., Ed.; Chapman and Hall: New York, 1989; p 359.
9. Molloy, K. C. In *The Chemistry of the Metal–Carbon Bond*; Hartley, F. R., Ed.; Wiley: Chichester, 1989; Vol. 5, p 465.
10. Davies, A. J.; Smith, P. J. In *Comprehensive Organometallic Chemistry*; Wilkinson, G., Ed.; Pergamon: Oxford, 1982; Vol. 2, p 519.
11. Considine, W. J. *JOM* **1966**, *5*, 263. This compound was first described in the patent literature: Ramsden, H. E.; Banks, C. K. *U.S. Patent* 2 789 994, 1957.
12. Wagner, D.; Verheyden, J. P. H.; Moffatt, J. G. *JOC* **1974**, *39*, 24.
13. David, S.; Thieffry, A. *CR(C)* **1974**, *279*, 1045.
14. Köpper, S.; Brandenburg, A. *LA* **1992**, 933.
15. Davies, A. G.; Price, A. J.; Dawes, H. M.; Hursthouse, M. B. *JCS(D)* **1986**, 297. See also Ref. 10.
16. David, S.; Pascard, C.; Cesario, M. *NJC* **1979**, *3*, 63.
17. Jenkins, I. D.; Verheyden, J. P. H.; Moffatt, J. G. *JACS* **1971**, *93*, 4323.
18. Ogawa, T.; Matsui, M. *Carbohydr. Res.* **1977**, *56*, C1. See also: Ogawa, T.; Matsui, M. *T* **1981**, *37*, 2363.
19. Veyrières, A. *JCS(P1)* **1981**, 1626. See also: David, S.; Thieffry, A.; Veyrières, A. *JCS(P1)* **1981**, 1796.
20. Ogawa, T.; Kaburagi, T. *Carbohydr. Res.* **1982**, *53*, 1033. See also: Nashed, M. A.; Anderson, L. *TL* **1976**, 3503. For a recent reference dealing with related chemistry in partially protected carbohydrates see: Tsuda, Y.; Nishimura, M.; Kobayashi, T.; Sato, Y.; Kanemitzu, K. *CPB* **1991**, *39*, 2883.
21. David, S.; Thieffry, A.; Veyrières, A. *JCS(P1)* **1981**, 1796.
22. Danishefsky, S. J.; Hungate, R.; Schulte, G. *JACS* **1988**, *110*, 7434.
23. Nagashima, N.; Ohno, M. *CL* **1987**, 141. For an application in the inositol area see: Yu, K.-L.; Fraser-Reid, B. *TL* **1988**, *29*, 979.
24. Helm, R. F.; Ralph, J.; Anderson, L. *JOC* **1991**, *56*, 7015.
25. Tsuda, Y.; Nishimura, M.; Kobayashi, T.; Sato, Y.; Kanemitzu, K. *CPB* **1991**, *39*, 2883.
26. Ricci, A.; Roelens, S.; Vannuchi, A. *CC* **1985**, 1457.
27. David, S.; Thieffry, A. *JCS(P1)* **1979**, 1568. For such oxidations on unprotected carbohydrates see: Tsuda, Y.; Hanajima, M.; Matsuhira, N.; Okuno, Y.; Kanemitzu, K. *CPB* **1989**, *37*, 2344.
28. Hanessian, S.; Roy, R. *JACS* **1979**, *101*, 5839.
29. Tsuda, Y.; Hanajima, M.; Matsuhira, N.; Okuno, Y.; Kanemitzu, K. *CPB* **1989**, *37*, 2344.
30. Habib, O. M. O.; Malek, J. *CCC* **1976**, *41*, 2724.
31. Shanzer, A. *ACR* **1983**, *16*, 60.
32. Bredenkamp, M. W.; Flowers, H. M.; Holzapfel, C. W. *CB* **1992**, *125*, 1159.
33. Niwa, H.; Okamoto, O.; Ishiwata, H.; Kuroda, A.; Uosaki, Y.; Yamada, K. *BCJ* **1988**, *61*, 3017.

34. Chwang, T. L.; Nemec, J.; Welch, A. D. *J. Carbohydr., Nucleosides Nucleotides* **1980**, *7*, 159.

35. Tsuneo, S.; Otera, J.; Nozaki, H. *TL* **1990**, *31*, 3591.

36. Kamiyama, T.; Inoue, M.; Kashiwagi, H.; Enomoto, S. *BCJ* **1990**, *63*, 1559. and references cited therein.

37. Crout, D. H. G.; Rathone, D. L. *CC* **1987**, 290.

T. V. (Babu) RajanBabu
The Ohio State University, Columbus, OH, USA

1,3-Dicyclohexylcarbodiimide[1]

[538-75-0] $C_{13}H_{22}N_2$ (MW 206.33)

(powerful dehydrating agent commonly used for the preparation of amides,[2] esters,[3] and anhydrides;[4] used with DMSO for the mild oxidation of alcohols to ketones;[5] used in the dehydrative conversion of primary amides to nitriles,[6] β-hydroxy ketones to α,β-unsaturated ketones,[7] and can effect the stereochemical inversion of secondary alcohols[8])

Alternate Name: DCC.
Physical Data: mp 34–35 °C; bp 122–124 °C.
Solubility: highly sol dichloromethane, THF, acetonitrile, DMF.
Form Supplied in: opalescent solid; widely available.
Handling, Storage, and Precautions: is an acute skin irritant in susceptible individuals. Because of its low melting point, it is conveniently handled as a liquid by gentle warming of the reagent container. It should be handled with gloves in a fume hood, and stored under anhydrous conditions.

Amide Formation. Since the initial reports,[9,10] DCC has become the most common reagent in peptide synthesis[2,11,12] and in other amide bond-forming reactions of primary and secondary amines with carboxylic acids. The mechanism is considered to be well understood.[2,5]

Typically, DCC (1.1 equiv) is added to a concentrated solution (0.1–1.0 M) of the carboxylic acid (1.0 equiv), amine (1.0 equiv), and catalyst (when used) in methylene chloride or acetonitrile at 0 °C. The hydrated DCC adduct, dicyclohexylurea (DCU), quickly precipitates and the reaction is generally complete within 1 h at rt. The solvents THF and DMF can be used, but are reported to reduce reaction rates and encourage the formation of the *N*-acylurea side product, as well as increasing racemization in chiral carboxylic acids.[13–16] If the amine is initially present as the salt (i.e. amine hydrochloride), it may be neutralized by adding 1 equiv of ***Diisopropylethylamine*** prior to adding DCC; however, the addition of tertiary amines (particularly ***Triethylamine***) can facilitate *N*-acylurea formation and racemization.[2] Racemization occurs via the formation of an oxazalone intermediate.[2] The addition of coupling agents (acylation catalysts) such as ***1-Hydroxybenzotriazole*** (HOBt),[17] 1-hydroxy-7-azabenzotriazole (HOAt),[18] ***N-Hydroxysuccinimide*** (HOSu),[19] and 3-hydroxy-3,4-dihydro-1,2,3-benzotriazin-4-one[20] can generally ameliorate both problems. These additives are required for the coupling of sterically hindered components, or when the amine is weakly nucleophilic.

Another potential problem with DCC is that at the completion of the reaction some DCU remains in solution with the product, necessitating additional purification. Water-soluble carbodiimide derivatives such as ***1-Cyclohexyl-3-(2-morpholinoethyl)carbodiimide Metho-p-toluenesulfonate***[21] and ***1-Ethyl-3-(3′-dimethylaminopropyl)carbodiimide Hydrochloride*** (EDCI)[22] obviate this problem, as they are removed by a simple extraction. Many newer coupling agents have been developed for peptide synthesis and other acylation reactions. These include ***Benzotriazol-1-yloxytris(dimethylamino)phosphonium Hexafluorophosphate*** (BOP),[23] ***O-Benzotriazol-1-yl-N,N,N′,N′-tetramethyluronium Hexafluorophosphate*** (HBTU),[24] ***Bis(2-oxo-3-oxazolidinyl)phosphinic Chloride***[25] (BOP-Cl), and (1*H*-1,2,3-benzotriazol-1-yloxy)tris(pyrrolidino)phosphonium hexafluorophosphate (PyBOP).[26] In addition to linear and polymeric amides, lactams of various ring sizes have been synthesized using these methods (eq 1).[27]

HO_2C–R–NHTs →(DCC) lactam (N-Ts) (1)

$R = (CH_2)_n$, $n = 0–6$

Ester and Thioester Formation. These reactions occur through the same *O*-acylurea or anhydride active intermediate as in the amide coupling reactions, and the discussion of associated problems applies here as well. In general, alkyl and (particularly) aryl thiols can be efficiently coupled to carboxylic acids using DCC.[28] Reactions of primary and secondary alcohols proceed reliably, but require the presence of an acylation catalyst. This is usually ***4-Dimethylaminopyridine*** (DMAP),[29,30] (see also ***1,3-Dicyclohexylcarbodiimide–4-Dimethylaminopyridine***), but others have been used including 4-pyrrolidinopyridine[31] and pyridine (solvent) with catalytic ***p-Toluenesulfonic Acid***.[32] The acylation of more hindered alcohols often results in reduced yields; however, even *t*-butanol can be acylated, providing a useful route to *t*-butyl esters.[3,31] Various other carbodiimide derivatives have also been used in the preparation of esters.[33,34] As with amides, which are not limited to intermolecular reactions, a wide variety of lactones can also be synthesized.[35,36]

Anhydride Formation. Among other anhydride-forming reagents, including ***Acetic Anhydride***, ***Trifluoroacetic Anhydride***, and ***Phosphorus(V) Oxide***, DCC is one of the simplest, mildest, and most effective reagents for the preparation of symmetrical anhydrides,[4,37] including formic anhydride,[38] which is useful in the preparation of formamides (eq 2).

Anhydride formation is associated with each of the reaction types involving carboxylic acids. The anhydride is often the reactive species, to the extent that it competes with the *O*-acylurea or covalent catalyst adduct (see discussion of amide bond formation reactions).

Sulfinate and Phosphate Esters. Anhydrides of sulfinic and sulfonic acids[39,40] and phosphate monoesters can be prepared using DCC. Further reaction with alcohols or phenols gives the corresponding sulfinates[41] and phosphate diesters.[42,43] Pyrophosphates can be prepared from mono- and disubstituted phosphate esters.[44] Reaction of activated sulfinates with amines gives the corresponding sulfinamides.[45]

$$HCO_2H + DCC \longrightarrow [\text{formic anhydride}] \xrightarrow{\text{serine benzyl ester hydrochloride}} \text{HCO-NH-CH}(CH_2OH)\text{-CO}_2CH_2Ph \quad (2)$$

Oxidation of Alcohols to Aldehydes and Ketones: Moffatt Oxidation. The oxidation of primary and secondary alcohols[46] to aldehydes and ketones, respectively, can be carried out by reaction with DMSO activated by DCC (see ***Dimethyl Sulfoxide-Dicyclohexylcarbodiimide***).[5,47,48] In comparison to the many metal-mediated oxidative reactions, the Moffatt oxidation is carried out under very mild conditions, and has found widespread use for reactions in the presence of sensitive functional groups (eq 3).[49] In addition, over-oxidation of aldehydes to form carboxylic acids is not observed. Under typical reaction conditions, a solution of the alcohol (1 equiv), DCC (3 equiv), and a proton source (pyridinium trifluoroacetate) (0.5 equiv) is stirred in DMSO or DMSO/benzene overnight at rt. After quenching the reaction with aqueous acetic acid and removing DCU by filtration, the product can be isolated by extraction. In modifications of this reaction,[46] DCC can be replaced by other DMSO-activating agents including acetic anhydride (see ***Dimethyl Sulfoxide–Acetic Anhydride***),[50] trifluoroacetic anhydride (see ***Dimethyl Sulfoxide–Trifluoroacetic Anhydride***),[51] and oxalyl chloride (see ***Dimethyl Sulfoxide–Oxalyl Chloride***).[52,53]

(3) [scheme: indole alkaloid secondary alcohol (MeO₂C, OH) → ketone; DCC, DMSO]

Dehydrative-Type Couplings. Because of the power of DCC as a dehydrating agent, it has found many uses in a variety of other dehydrative coupling reactions. These include the reaction of primary amines with DCC and ***Carbon Dioxide*** or ***Carbon Disulfide*** to form ureas (eq 4)[54] or isothiocyanates (eq 5),[55] respectively.

$$2\ R{-}NH_2 + CO_2 \xrightarrow{DCC} RNH{-}CO{-}NHR \quad (4)$$

$$2\ R{-}NH_2 + CS_2 \xrightarrow{DCC} R{-}N{=}C{=}S \quad (5)$$

Symmetrical peroxides can be synthesized from benzoic acid derivatives and hydrogen peroxide (eq 6).[56] Unsymmetrical peroxides can be synthesized from the corresponding carboxylic acid and peroxy acid.[56]

$$2\ Ar{-}CO_2H + H_2O_2 \xrightarrow{DCC} ArCO{-}O{-}O{-}COAr \quad (6)$$

Aromatic and aliphatic alcohols and thiophenols can be coupled to form ethers (eq 7)[57,58] and thioethers (eq 8),[59,60] respectively, using DCC.

$$Ar{-}OH + MeOH \xrightarrow{DCC} Ar{-}O{-}Me \quad (7)$$

$$Ar{-}SH + R{-}OH \xrightarrow{DCC} Ar{-}S{-}Me \quad (8)$$

In cases where sensitivity of the compound precludes the formation of an acid chloride, α-diazo ketones can be synthesized by the DCC-mediated coupling of diazomethane to carboxylic acids (eq 9).[61,62]

$$R{-}CO_2H + CH_2N_2 \xrightarrow{DCC} R{-}CO{-}CHN_2 \quad (9)$$

Dehydration to Alkenes, Epoxides, Nitriles, and Ketenes. β-Hydroxy ketones and β-hydroxy esters can be dehydrated, using DCC, to α,β-unsaturated ketones (eq 10)[7,63] and esters (eq 11),[64] respectively. Cyclopropanes can be synthesized by the 'dehydration' of γ-hydroxy ketones (eq 12).[65]

(10) [scheme: HCON, R¹O, R² cyclopentane with β-hydroxy ketone side chain (C₅H₁₁) → enone; DCC]

(11) [scheme: 1-hydroxycyclohexylacetic acid ethyl ester (HO, OEt) → cyclohexylideneacetate (OEt); DCC]

(12) [scheme: 5-hydroxy-2-pentanone (OH) → cyclopropyl methyl ketone; DCC]

Other dehydration reactions include the conversion of primary amides to nitriles,[6] aldehydes to nitriles (via the hydroxylamine) (eq 13),[66,67] and carboxylic acids to ketenes (eq 14).[68]

$$R{-}CHO + HONH_2{\cdot}HCl \xrightarrow[CuSO_4]{DCC} R{-}C{\equiv}N \quad (13)$$

$$R^1R^2CH{-}CO_2H \xrightarrow{DCC} R^1R^2C{=}C{=}O \quad (14)$$

Dehydroxylation of Alcohols. Alcohols and phenols can be dehydroxylated to the corresponding alkane by hydrogenation of the *O*-acylurea adduct formed from the reaction of the alcohol with DCC (eq 15).[69,70]

$$\text{R–OH} \xrightarrow{\text{DCC}} \left[\begin{array}{c} O\text{-acyl} \\ \text{urea} \end{array}\right] \xrightarrow[\text{Pd/C}]{H_2} \text{R–H} \quad (15)$$

Inversion of Secondary Alcohols. Secondary alcohols can be stereochemically inverted by formylation (or esterification) with DCC, followed by saponification (eq 16).[8]

1. DCC; 2. HCO_2H; 3. NaOMe; 90%; >99% de (16)

Heterocyclization Reactions. DCC has frequently been used both as a reagent and as a reactant in the synthesis of heterocycles.[1] For example, DCC-mediated cyclodesulfurative annulation reactions[71] have been used to synthesize guanosine-type nucleotide analogs (eq 17).[72]

1. MeO_2CNCS; 2. DCC; 3. TMSX (17)

1. (a) Mikolajczyk, M.; Kielbasinski, P. *T* **1981**, *37*, 233. (b) Williams, A.; Ibrahim, I. T. *CRV* **1981**, *81*, 589. (c) Kurzer, F.; Douraghi-Zadeh, K. *CRV* **1967**, *67*, 107.
2. Bodanszky, M. *Peptide Chemistry: A Practical Textbook*; Springer: New York, 1988.
3. Neises, B.; Steglich, W. *OS* **1985**, *63*, 183.
4. Chen, F. M. F.; Kuroda, K.; Benoiton, N. L. *S* **1978**, 928.
5. Moffatt, J. G. *JOC* **1971**, *36*, 1909.
6. Ressler, C.; Ratzkin, H. *JOC* **1961**, *26*, 3356.
7. Corey, E. J.; Andersen, N. H.; Carlson, R., M.; Paust, J.; Vedejs, E.; Vlattas, I.; Winter, R. E. K. *JACS* **1968**, *90*, 3245.
8. Kaulen, J. *AG(E)* **1987**, *26*, 773.
9. Sheehan, J. C.; Hess, G. P. *JACS* **1955**, *77*, 1067.
10. Khorana, H. G. *CI(L)* **1955**, 1087.
11. Hudson, D. *JOC* **1988**, *53*, 617.
12. Wang, S. S.; Tam, J. P.; Wang, B. S. H.; Merrifield, R. B. *Int. J. Pept. Protein Res.* **1981**, *19*, 459.
13. Balcom, B. J.; Petersen, N. O. *JOC* **1989**, *54*, 1922.
14. DeTar, D. F.; Silverstein, R. *JACS* **1966**, *88*, 1013.
15. DeTar, D. F.; Silverstein, R. *JACS* **1966**, *88*, 1020.
16. Mironova, D. F.; Dvorko, G. F.; Skuratovskaya, T. N. *UKZ* **1969**, *35*, 726.
17. König, W.; Geiger, R. *CB* **1970**, *103*, 788.
18. Carpino, L. A. *JACS* **1993**, *115*, 4397.
19. Wünsch, E.; Drees, F. *CB* **1966**, *99*, 110.
20. König, W.; Geiger, R. *CB* **1970**, *103*, 2034.
21. Sheehan, J. C.; Hlavka, J. J. *JOC* **1956**, *21*, 439.
22. Sheehan, J. C.; Cruickshank, P. A.; Boshart, G. L. *JOC* **1961**, *26*, 2525.
23. Castro, B.; Dormoy, J. R.; Evin, G.; Selve, C. *TL* **1975**, 1219.
24. Knorr, R.; Trzeciak, A.; Bannwarth, W.; Gillessen, D. *TL* **1989**, *30*, 1927.
25. Cabré, J.; Palomo, A. L. *S* **1984**, 413.
26. Frérot, E.; Coste, J.; Pantaloni, A.; Dufour, M.; Jouin, P. *TL* **1991**, *47*, 259.
27. Tanner, D.; Somfai, P. *T* **1988**, *44*, 613.
28. Grunwell, J. R.; Foerst, D. L. *SC* **1976**, *6*, 453.
29. Smith, R. J.; Capaldi, R. A.; Muchmore, D.; Dahlquist, F. *B* **1978**, *17*, 3719.
30. Neises, B.; Steglich, W. *AG(E)* **1978**, *17*, 522.
31. Hassner, A.; Alexanian, V. *TL* **1978**, 4475.
32. Holmberg, K.; Hansen, B. *ACS* **1979**, *33*, 410.
33. Dhaon, M. K.; Olsen, R. K.; Ramasamy, K. *JOC* **1982**, *47*, 1962.
34. Ohta, S.; Shimabayashi, A.; Aono, M.; Okamoto, M. *S* **1982**, 833.
35. Johnson, W. S.; Bauer, V. J.; Margrave, J. L.; Frisch, M. A.; Dreger, L. H.; Hubbard, W. N. *JACS* **1961**, *83*, 606.
36. Woodward, R. B.; Bader, F. E.; Bickel, H.; Frey, A. J.; Kierstead, R. W. *T* **1958**, *2*, 1.
37. Rammler, D. H.; Khorana, H. G. *JACS* **1963**, *85*, 1997.
38. Chen, F. M. F.; Benoiton, N. L. *S* **1979**, 709.
39. Samuel, D.; Silver, B. L. *JACS* **1963**, *85*, 1197.
40. Khorana, H. G. *CJC* **1953**, *31*, 585.
41. Mijaji, Y.; Minoto, H.; Kobayashi, M. *BCJ* **1971**, *44*, 862.
42. Kampe, W. *CB* **1965**, *98*, 1031.
43. Pal, B. C.; Schmidt, D. G.; Farrelly, J. G. *Nucleic Acid Chem.* **1978**, *2*, 963.
44. Khorana, H. G.; Todd, A. R. *JCS* **1953**, 2257.
45. Furukawa, M.; Okawara, T. *S* **1976**, 339.
46. Tidwell, T. T. *OR* **1990**, *39*, Chapter 3.
47. Pfitzner, K. E.; Moffatt, J. G. *JACS* **1965**, *87*, 5661.
48. Moffatt, J. G. In *Oxidation*; R. L. Augustine and D. J. Trecker, Ed.; Dekker: New York, 1971; Vol. 2, Chapter 1.
49. Albright, J. D.; Goldman, L. *JOC* **1965**, *30*, 1107.
50. Albright, J. D.; Goldman, L. *JACS* **1967**, *89*, 2416.
51. Huang, S. L.; Omura, K.; Swern, D. *S* **1978**, 297.
52. Omura, K.; Swern, D. *T* **1978**, *34*, 1651.
53. Marx, M.; Tidwell, T. T. *JOC* **1984**, *49*, 788.
54. Ogura, H.; Takeda, K.; Tokue, R.; Kobayashi, T. *S* **1978**, 394.
55. Jochims, J. C.; Seeliger, A. *AG(E)* **1967**, *6*, 174.
56. Greene, F. D.; Kazan, J. *JOC* **1963**, *28*, 2168.
57. Vowinkel, E. *CB* **1966**, *99*, 42.
58. Vowinkel, E. *CB* **1966**, *99*, 1499.
59. Vowinkel, E.; Wolff, C. *CB* **1974**, *107*, 496.
60. Vowinkel, E. *S* **1974**, 430.
61. Penke, B.; Czombos, J.; Baláspiri, L.; Petres, J.; Kovács, K. *HCA* **1970**, *53*, 1057.
62. Hodson, D.; Holt, G.; Wall, D. K. *JCS(C)* **1970**, 971.
63. Alexandre, C.; Rouessac, F. *BSF(2)* **1971**, 1837.
64. Alexandre, C.; Rouessac, F. *CR(C)* **1972**, *274*, 1585.
65. Alexandre, C.; Rouessac, F. *TL* **1970**, 1011.
66. Vowinkel, E.; Bartel, J. *CB* **1974**, *107*, 1221.
67. Vowinkel, E. *AG(E)* **1974**, *13*, 351.
68. Olah, G. A.; Wu, A.; Farooq, O. *S* **1989**, 568.
69. Vowinkel, E.; Wolff, C. *CB* **1974**, *107*, 907.
70. Vowinkel, E.; Büthe, I. *CB* **1974**, *107*, 1353.

71. Mohsen, A.; Omar, M. E. M. E. O.; Habib, N. S.; Aboulwafa, O. M. *S* **1977**, 864.
72. Groziak, M. P.; Chern, J.; Townsend, L. B. *JOC* **1986**, *51*, 1065.

Jeffrey S. Albert & Andrew D. Hamilton
University of Pittsburgh, PA, USA

Diethylaluminum Chloride[1]

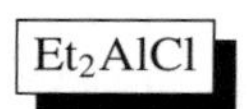

[96-10-6] $C_4H_{10}AlCl$ (MW 120.56)

(strong Lewis acid that can also act as a proton scavenger; reacts with HX to give ethane and EtAlClX)

Alternate Names: chlorodiethylaluminum; diethylchloroalane.
Physical Data: mp −50 °C; bp 125 °C/50 mmHg; *d* 0.961 g cm^{-3}.
Solubility: sol most organic solvents; stable in alkanes or arenes.
Form Supplied in: commercially available neat or as solutions in hexane or toluene.
Analysis of Reagent Purity: solutions are reasonably stable but may be titrated before use by one of the standard methods.[1e]
Handling, Storage, and Precautions: must be transferred under inert gas (Ar or N_2) to exclude oxygen and water. Use in a fume hood.

Introduction. The general properties of alkylaluminum halides as Lewis acids are discussed in the entry for ***Ethylaluminum Dichloride***. Dialkylaluminum halides are less acidic than alkylaluminum dihalides. Et_2AlCl is much cheaper than ***Dimethylaluminum Chloride*** and is used more frequently than Me_2AlCl since comparable results are usually obtained. In some cases, most notably the ene reactions of carbonyl compounds, use of Me_2AlCl is preferable since its methyl groups are less nucleophilic than the ethyl groups of Et_2AlCl, which can act as a reducing agent.

Catalysis of Diels–Alder Reactions. Et_2AlCl has been extensively used as a Lewis acid catalyst for Diels–Alder reactions. *N*-Acyloxazolidinones can form both 1:1 (eq 1) and 1:2 complexes with Et_2AlCl (eq 2).[2] The 1:2 complex is ~100 times as reactive as the 1:1 complex and gives greater *endo* selectivity and higher de. Me_2AlCl gives similar selectivity with fewer byproducts.

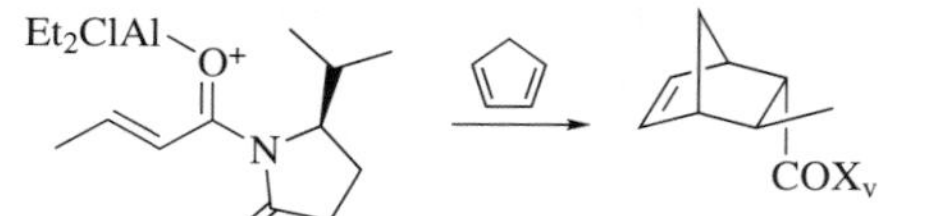

Et_2AlCl has been extensively used as a Lewis acid catalyst for intermolecular[3] and intramolecular[4] Diels–Alder reactions with α,β-unsaturated ketones and esters as dienophiles. It also catalyzes inverse electron demand Diels–Alder reactions of alkenes with quinone methides[5] and Diels–Alder reactions of aldehydes as enophiles.[6]

Catalysis of Ene Reactions. Et_2AlCl has been used as a catalyst for ene reactions with ethyl propiolate as an enophile,[7] for intramolecular ene reactions of aldehydes,[8] and for intramolecular ene reactions with α,β-unsaturated esters as enophiles (eq 3).[3e,9]

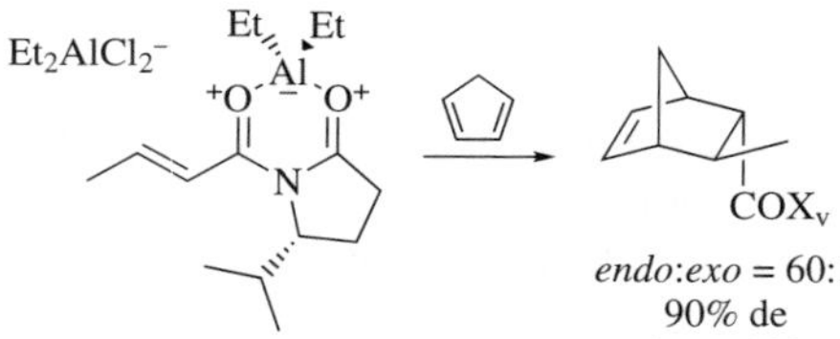

Catalysis of Claisen and Vinylcyclopropane Rearrangements. Et_2AlCl has been used as a catalyst for Claisen rearrangement of aryl allyl ethers.[10] The rearrangement of 2-vinylcyclopropanecarboxylate esters to cyclopentenes is catalyzed by Et_2AlCl (eq 4).[11]

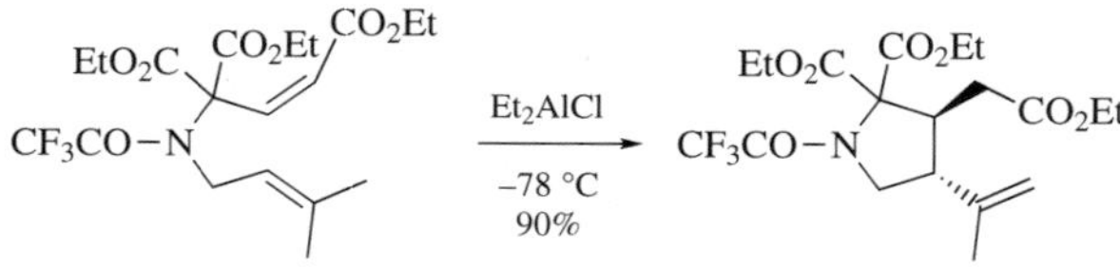

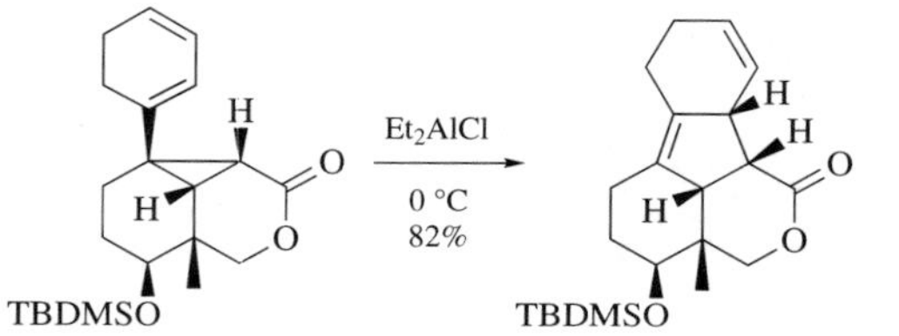

Generation of Electrophilic Cations. Complexation of Et_2AlCl to ketones and aldehydes activates the carbonyl group toward addition of a nucleophilic alkyl- or allylstannane or allylsilane.[12] Et_2AlCl has been used to initiate Beckmann rearrangements of oxime mesylates. The ring-expanded cation can be trapped intermolecularly by enol ethers and cyanide and intramolecularly by alkenes (eq 5).[13]

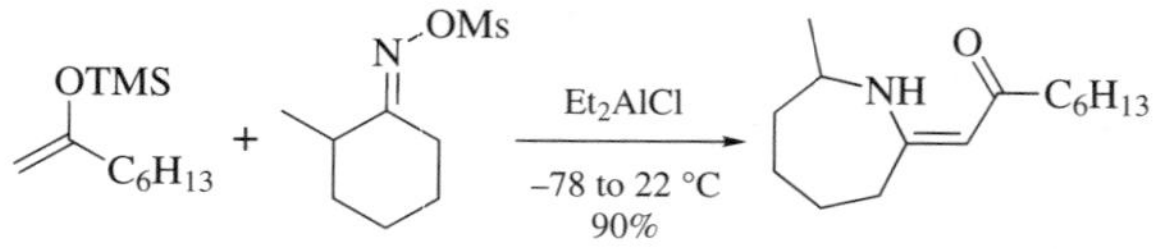

Formation and Reaction of Aluminum Enolates. Et_2AlCl has been used in modified Reformatsky reactions. Aldol adducts are obtained in good yield by reaction of an α-bromo ketone with a ketone in the presence of ***Zinc*** and Et_2AlCl in THF (eq 6).[14] Lithium enolates of esters do not react with epoxides. Reaction of lithium enolates with Et_2AlCl affords aluminum enolates that react with epoxides at the less substituted carbon (eq 7).[15]

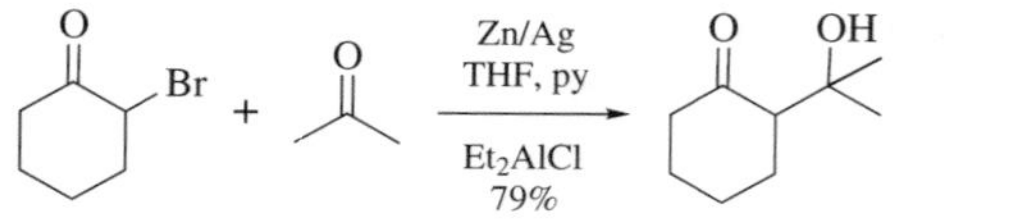

(7) [Scheme: 1. LDA; 2. Et_2AlCl; 3. t-Bu aldehyde, 38%; product syn:anti = 95:5]

Formation and Reaction of Alkynylaluminum Reagents. Lithium acetylides react with Et_2AlCl to form LiCl and diethylaluminum acetylides. The aluminum acetylides are useful reagents for carrying out S_N2 reactions on epoxides (eq 8)[15c,16] and undergo conjugate addition to enones that can adopt an *S-cis* conformation (eq 9).[17]

(8) [Scheme: Et—≡CH; 1. BuLi; 2. Et_2AlCl; 3. OTBDMS epoxide, 77%]

(9) [Scheme: Ph—≡CH; 1. BuLi; 2. Et_2AlCl; 3. PhCH=CHCOPh, 81%]

Reaction as a Nucleophile. Et_2AlCl reacts analogously to ***Ethylmagnesium Bromide*** and transfers an ethyl group to many electrophiles. Since EtMgBr and ***Ethyllithium*** are readily available, use of Et_2AlCl to deliver an ethyl group is needed only when the stereochemistry of addition is an important issue. High levels of asymmetric induction are obtained in the conjugate addition of Et_2AlCl to unsaturated acyloxazolidinones with carbohydrate-derived chiral auxiliaries (eq 10).[18] Et_2AlCl opens epoxides to chlorohydrins.[19]

(10) [Scheme: 1. Et_2AlCl, 84%; 2. LiOOH, 87%; product 92% ee]

Related Reagents. Dimethylaluminum Chloride; Ethylaluminum Dichloride; Methylaluminum Dichloride; Triethylaluminum; Trimethylaluminum.

1. For reviews, see Ref. 1 in ***Ethylaluminum Dichloride***.
2. Evans, D. A.; Chapman, K. T.; Bisaha, J. *JACS* **1988**, *110*, 1238.
3. (a) Schlessinger, R. H.; Schultz, J. A. *JOC* **1983**, *48*, 407. (b) Cohen, T.; Kosarych, Z. *JOC* **1982**, *47*, 4005. (c) Hagiwara, H.; Okano, A.; Uda, H. *CC* **1985**, 1047. (d) Furuta, K.; Iwanaga, K.; Yamamoto, H. *TL* **1986**, *27*, 4507. (e) Oppolzer, W. *AG(E)* **1984**, *23*, 876. (f) Reetz, M. T.; Kayser, F.; Harms, K. *TL* **1992**, *33*, 3453. (g) Midland, M. M.; Koops, R. W. *JOC* **1992**, *57*, 1158.
4. (a) Roush, W. R.; Gillis, H. R. *JOC* **1982**, *47*, 4825. (b) Reich, H. J.; Eisenhart, E. K. *JOC* **1984**, *49*, 5282. (c) Shea, K. J.; Gilman, J. W. *TL* **1983**, *24*, 657. (d) Brown, P. A.; Jenkins, P. R. *JCS(P1)* **1986**, 1303. (e) Reich, H. J.; Eisenhart, E. K.; Olson, R. E.; Kelly, M. J. *JACS* **1986**, *108*, 7791. (f) Funk, R. L.; Bolton, G. L. *JACS* **1986**, *108*, 4655. (g) Taschner, M. J.; Cyr, P. T. *TL* **1990**, *31*, 5297.
5. (a) Tietze, L. F.; Brand, S.; Pfeiffer, T.; Antel, J.; Harms, K.; Sheldrick, G. M. *JACS* **1987**, *109*, 921. (b) Casiraghi, G.; Cornia, M.; Casnati, G.; Fava, G. G.; Belicchi, M. F. *CC* **1986**, 271.
6. Midland, M. M.; Afonso, M. M. *JACS* **1989**, *111*, 4368.
7. Dauben, W. G.; Brookhart, T. *JACS* **1981**, *103*, 237.
8. (a) Kamimura, A.; Yamamoto, A. *CL* **1990**, 1991. (b) Andersen, N. H.; Hadley, S. W.; Kelly, J. D.; Bacon, E. R. *JOC* **1985**, *50*, 4144.
9. (a) Oppolzer, W.; Robbiani, C.; Bättig, K. *T* **1984**, *40*, 1391. (b) Oppolzer, W.; Mirza, S. *HCA* **1984**, *67*, 730.
10. (a) Sonnenberg, F. M. *JOC* **1970**, *35*, 3166. (b) Bender, D. R.; Kanne, D.; Frazier, J. D.; Rapoport, H. *JOC* **1983**, *48*, 2709. (c) Lutz, R. P. *CR* **1984**, *84*, 205.
11. (a) Corey, E. J.; Myers, A. G. *JACS* **1985**, *107*, 5574. (b) Davies, H. M. L.; Hu, B. *TL* **1992**, *33*, 453. (c) Davies, H. M. L.; Hu, B. *JOC* **1992**, *57*, 3186.
12. (a) McDonald, T. L.; Delahunty, C. M.; Mead, K.; O'Dell, D. E. *TL* **1989**, *30*, 1473. (b) Denmark, S. E.; Weber, E. J. *JACS* **1984**, *106*, 7970. (c) Mooiweer, H. H.; Hiemstra, H.; Fortgens, H. P.; Speckamp, N. W. *TL* **1987**, *28*, 3285.
13. (a) Sakane, S.; Matsumura, Y.; Yamamura, Y.; Ishida, Y.; Maruoka, K.; Yamamoto, H. *JACS* **1983**, *105*, 672. (b) Maruoka, K.; Miyazaki, T.; Ando, M.; Matsumura, Y.; Sakane, S.; Hattori, K.; Yamamoto, H. *JACS* **1983**, *105*, 2831. (c) Matsumura, Y.; Fujiwara, J.; Maruoka, K.; Yamamoto, H. *JACS* **1983**, *105*, 6312.
14. (a) Maruoka, K.; Hashimoto, S.; Kitagawa, Y.; Yamamoto, H.; Nozaki, H. *JACS* **1977**, *99*, 7705. (b) Maruoka, K.; Hashimoto, S.; Kitigawa, Y.; Yamamoto, H.; Nozaki, H. *BCJ* **1980**, *53*, 3301. (c) Stokker, G. E.; Hoffmann, W. F.; Alberts, A. W.; Cragoe, E. J., Jr.; Deanna, A. A.; Gilfillan, J. L.; Huff, J. W.; Novello, F. C.; Prugh, J. D.; Smith, R. L.; Willard, A. K. *JMC* **1985**, *28*, 347. (d) Tsuboniwa, N.; Matsubara, S.; Morizawa, Y.; Oshima, K.; Nozaki, H. *TL* **1984**, *25*, 2569. (e) Tsuji, J.; Mandai, T. *TL* **1978**, 1817.
15. (a) Nozaki, H.; Oshima, K.; Takai, K.; Ozawa, S. *CL* **1979**, 379. (b) Sturm, T.-J.; Marolewski, A. E.; Rezenka, D. S.; Taylor, S. K. *JOC* **1989**, *54*, 2039. (c) Danishefsky, S.; Kitahara, T.; Tsai, M.; Dynak, J. *JOC* **1976**, *41*, 1669.
16. (a) Nicolaou, K. C.; Webber, S. E.; Ramphal, J.; Abe, Y. *AG(E)* **1987**, *26*, 1019. (b) Matthews, R. S.; Eickhoff, D. J. *JOC* **1985**, *50*, 3923. (c) Ishiguro, M.; Ikeda, N.; Yamamoto, H. *CL* **1982**, 1029.
17. Hooz, J.; Layton, R. B. *JACS* **1971**, *93*, 7320.
18. (a) Rück, K.; Kunz, H. *AG(E)* **1991**, *30*, 694. (b) Rück, K.; Kunz, H. *SL* **1992**, 343. (c) Rück, K.; Kunz, H. *S* **1993**, 1018.
19. Gao, L.-X.; Saitoh, H.; Feng, F.; Murai, A. *CL* **1991**, 1787.

Barry B. Snider
Brandeis University, Waltham, MA, USA

N,N-Diethylaminosulfur Trifluoride[1]

Et_2NSF_3

[38078-09-0] $C_4H_{10}F_3NS$ (MW 161.18)

(mild reagent for converting primary, secondary, tertiary, allylic, and benzylic alcohols to the corresponding fluorides,[1] aldehydes and ketones to difluorides,[1a,15,19] carboxylic acids to acyl fluorides,[22] and sulfoxides to α-fluoro sulfides[23])

Alternate Name: DAST.
Physical Data: bp 30–32 °C/3 mmHg; *d* 1.220 g cm^{-3}.

Solubility: sol ethereal, chlorinated, and hydrocarbon solvents; reacts violently with water and rapidly with hydroxylic solvents.

Form Supplied in: amber yellow oil; widely available.

Purification: discolored (brown) samples give increasingly lower yields of fluorinated products. Freshly distilled oil is satisfactory for use.

Handling, Storage, and Precautions: can be stored for extended periods of time in the freezer under inert atmosphere. Use in a fume hood.

Fluorodehydroxylation of Alcohols. DAST, like several other variants of dialkylaminotrifluorosulfuranes, converts primary, secondary, tertiary, allylic, and benzylic alcohols to monofluorides.[1] The reaction conditions are mild (temperature as low as −78 to 0 °C for reactive substrates), and a variety of functionalities such as acetonides, isolated double/triple bonds, esters, ethers, amides, and unactivated halides are tolerated (eqs 1–3).[1b,8–10]

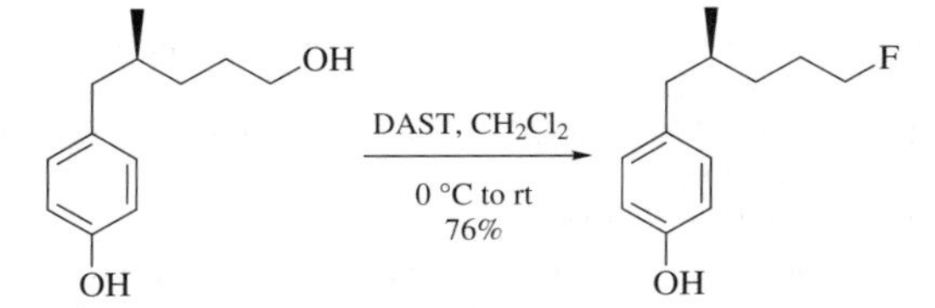

DAST, CH_2Cl_2, 0 °C to rt, 76% (1)

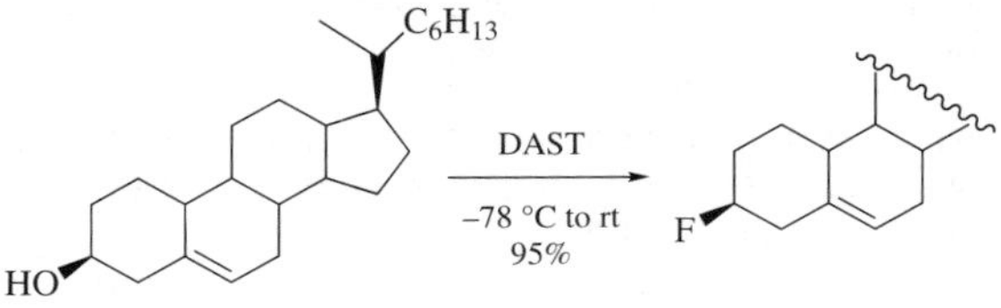

DAST, −78 °C to rt, 95% (2)

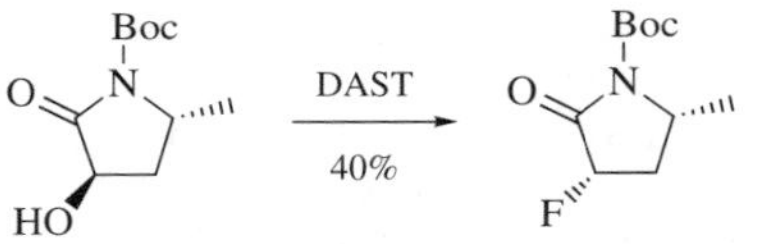

DAST, 40% (3)

The solvents usually employed are dichloromethane, chloroform, carbon tetrachloride, fluorotrichloromethane, ether, THF, benzene, and toluene. Generally, DAST is superior to classical fluorinating agents such as ***Sulfur Tetrafluoride*** in that the latter requires much higher temperatures (typically 100 °C) and gives undesired side products. With DAST, rearrangements are sometimes observed, albeit to a lesser extent (eq 4).[1a,2,7] Thus β-elimination,[1a] ether formation,[3] Friedel–Crafts alkylation,[4] and skeletal rearrangements involving norbornyl cations[5] have been reported. The DAST reaction with isobutyl alcohol yields a mixture of 49% isobutyl fluoride and 21% of *t*-butyl fluoride.[1a] Other functionalities can interfere with the normal course of reaction. A case in point is a pinacol rearrangement with concomitant ring contraction (eq 5).[2a]

S_N2' rearrangement may occur on reacting DAST with allylic alcohols (eq 6).[6]

Highly varied stereochemical outcomes have been obtained in the reactions of DAST with secondary alcohols. Thus, although products of partial or complete racemization through ionic or ion-pair mechanisms[1a] have not been widely observed, the usual steric course is complete inversion or complete retention at the reaction center. An example is the formation of (−)-2-fluorooctane (97.6% optical purity) from (+)-*S*-2-octanol.[30] On the other hand, a mixture of products of inverted and retained stereochemistry are obtained in the reaction of DAST with some protected *myo*-inositol derivatives (eq 7).[11] Interestingly, in unprotected or minimally protected sugars,[12a,b] and inositols,[12c,d] where more than two hydroxy groups are present, only one or two hydroxy groups react regio- and stereoselectively (eq 8).[12d]

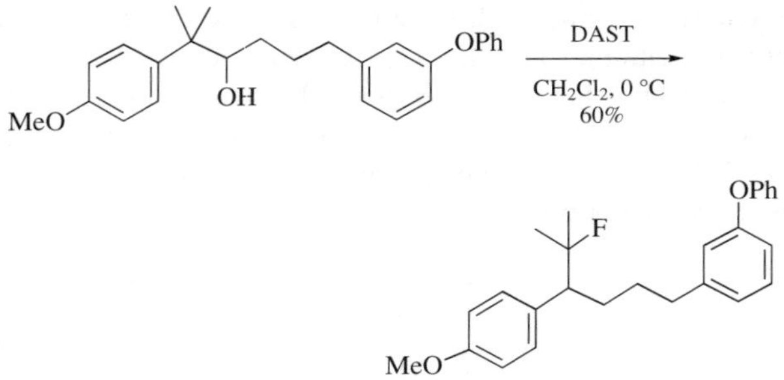

DAST, CH_2Cl_2, 0 °C, 60% (4)

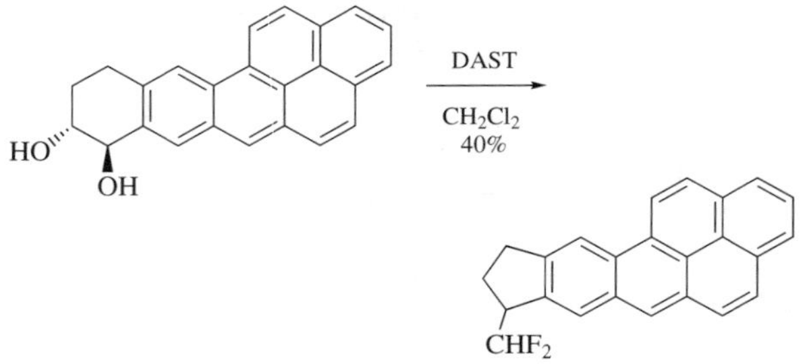

DAST, CH_2Cl_2, 40% (5)

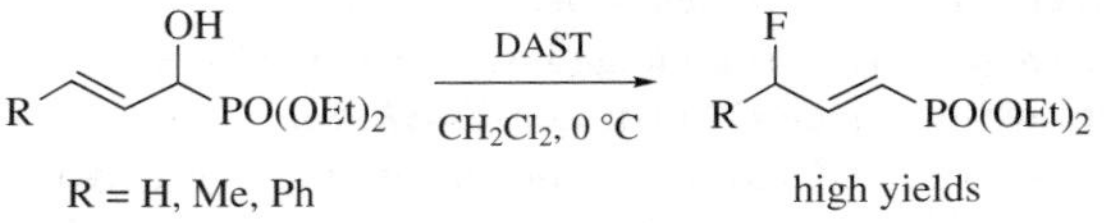

R = H, Me, Ph; DAST, CH_2Cl_2, 0 °C; high yields (6)

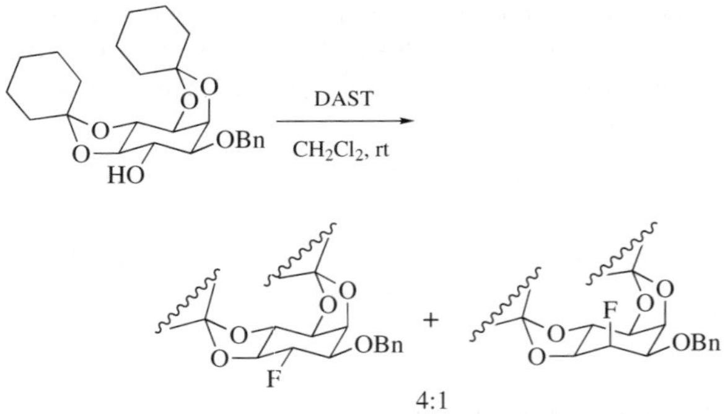

DAST, CH_2Cl_2, rt; 4:1 (7)

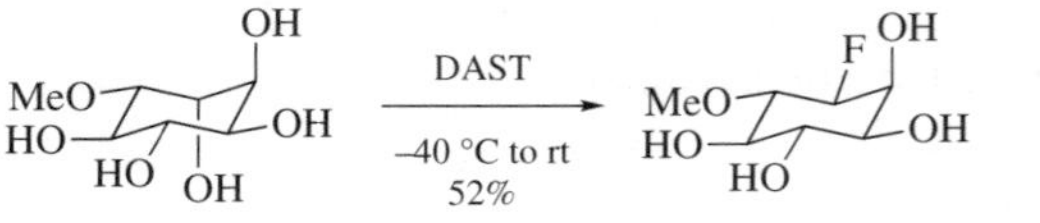

DAST, −40 °C to rt, 52% (8)

Several synthetically useful transformations involving neighboring group participation exist.[13] An example is the reaction of DAST with *N,N*-dibenzyl-L-serine benzyl ester (eq 9).[13b]

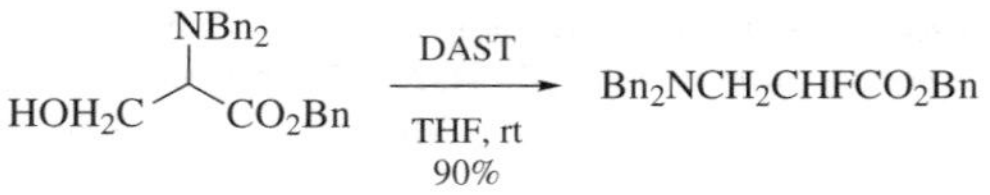

$$HOH_2C\text{–}CH(NBn_2)\text{–}CO_2Bn \xrightarrow[\text{THF, rt, 90\%}]{\text{DAST}} Bn_2NCH_2CHFCO_2Bn \quad (9)$$

Geminal Difluorination of Aldehydes and Ketones. The carbonyl group of aldehydes and ketones can be converted to a

1,1-difluoro group in moderate to high yields.[1a,8,14] The reaction conditions are mild (rt to 80 °C). The solvents generally used are dichloromethane and chloroform. Functional groups compatible with the fluorination of alcohols are also tolerated in the 1,1-difluorination of aldehydes and ketones (see above). An aldehyde carbonyl reacts much faster than a ketone carbonyl and selective difluorination of keto aldehydes at the aldehyde carbonyl has been reported.[14] The general order of reactivity is alcohols > aldehydes > ketones. A variety of aliphatic,[15] aromatic,[1a] and heterocyclic[16] aldehydes have been converted to 1,1-difluorinated compounds (eqs 10–12).[14b,18,19]

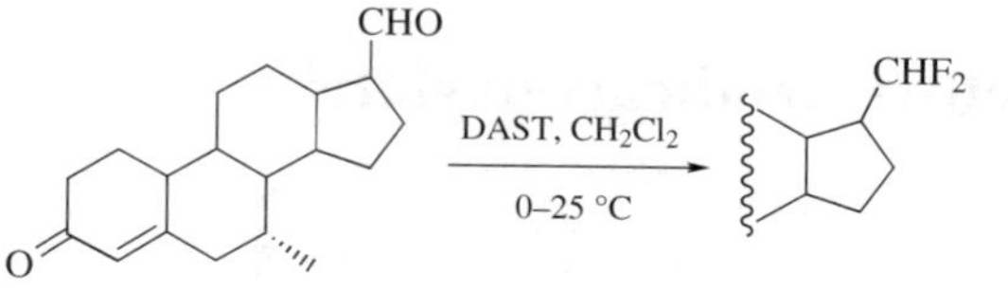

DAST, CH_2Cl_2, 0–25 °C (10)

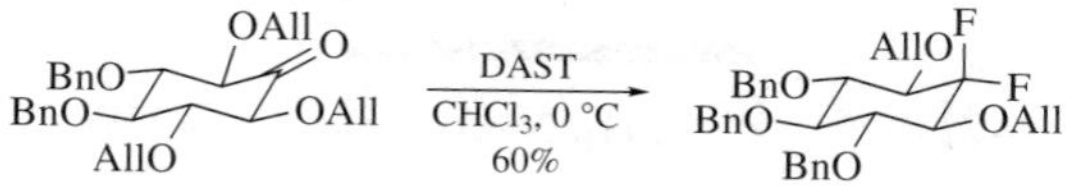

DAST, $CHCl_3$, 0 °C, 60% (11)

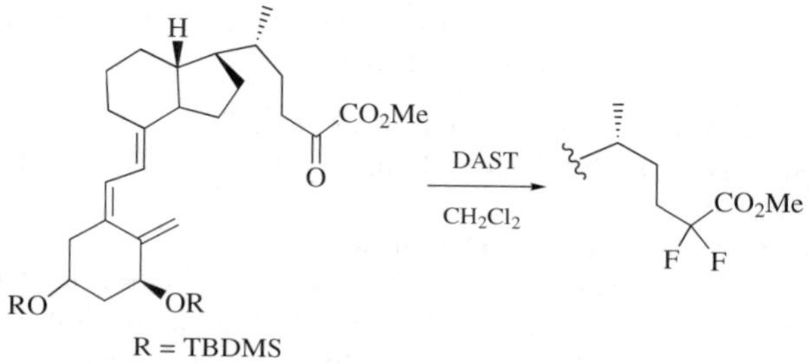

DAST, CH_2Cl_2 (12)

R = TBDMS

While the reactivity of ketones is similar to aldehydes in scope, vinyl fluoride formation is a complication and sometimes DAST, in conjunction with fuming ***Sulfuric Acid*** in glyme,[17a,b] or a mixture of ***Lithium Chloride*** and ***Copper(II) Chloride***,[17b] is used for the preparation of vinyl fluorides (eq 13).

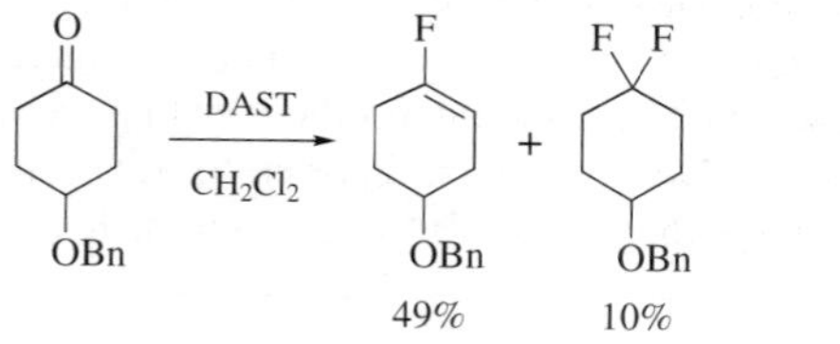

DAST, CH_2Cl_2 (13) 49% 10%

β-Diketones and β-keto esters are oxidatively fluorinated with DAST to furnish α,β-difluoro-α,β-unsaturated ketones and esters, respectively (eq 14).[20]

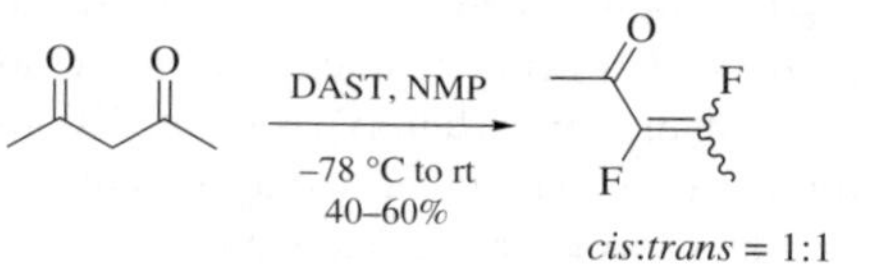

DAST, NMP, –78 °C to rt, 40–60% (14)

cis:*trans* = 1:1

Reaction with Epoxides. The reactivity of DAST with epoxides varies with structure. Thus cyclopentene oxide and cyclohexene oxide give a mixture of 1,1-difluoro and bis(α-fluoro) ethers (eq 15).[21] A stereospecific synthesis of *meso*- or (±)-difluorides has been accomplished in two steps using, respectively, *cis*- or *trans*-epoxides as starting materials (eq 16).[22]

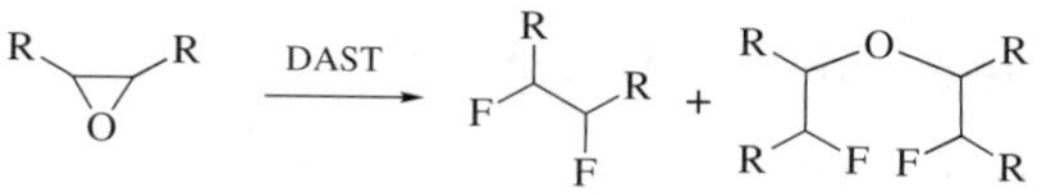

DAST (15)

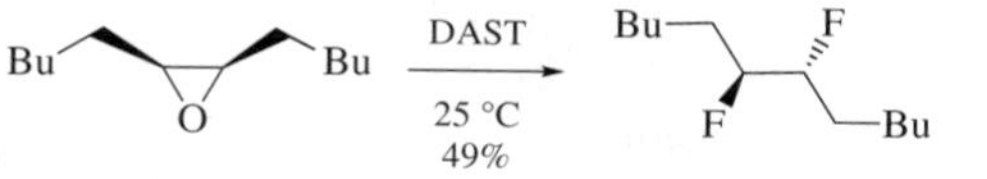

DAST, 25 °C, 49% (16)

Reaction with Organic Acids. Acyl fluorides can be prepared by the reaction of DAST with carboxylic acids in good to excellent yields.[23] α-Hydroxy acids give α-fluoroacyl fluorides which hydrolyze on workup to form α-fluoro acids.[24] Sulfur tetrafluoride, on the other hand, converts the carboxylic group into a trifluoromethyl group.[21]

Reaction with Halides and Sulfonates. Reactive halides such as iodides, allylic, and benzylic halides, and chlorides of organic acids (sulfinic, sulfonic, and phosphonic) react with DAST to form the corresponding fluorides.[1a]

Reaction with Sulfoxides. DAST reacts with α-hydrogen containing dialkyl and aralkyl sulfoxides to form α-fluoroalkyl sulfides in high yields (eq 17).[25] This reaction can be extended to less reactive sulfoxides by catalysis with certain Lewis acids, e.g. ***Antimony(III) Chloride*** (eq 18)[26] and ***Zinc Iodide***.[27]

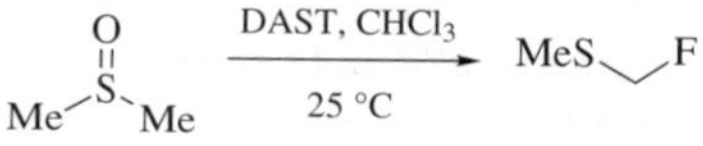

DAST, $CHCl_3$, 25 °C (17)

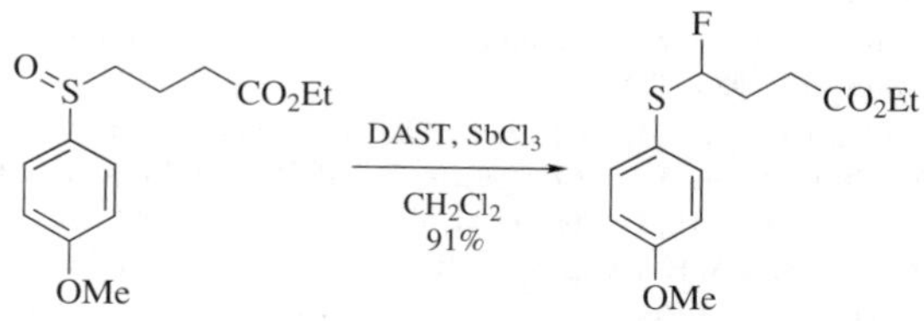

DAST, $SbCl_3$, CH_2Cl_2, 91% (18)

Miscellaneous Reactions. Lactones containing α-hydrogen or fluorine do not react with DAST. However, α-hydroxy lactones undergo normal fluorodehydroxylation along with geminal difluorination at the lactone carbonyl.[28] Glycosyl fluorides can be obtained in high yield, and in a stereospecific manner, either by reacting DAST with hemithioacetals in the presence of ***N-Bromosuccinimide*** or, more simply, hemiacetals.[29b] An interesting fluorination of a phenyl ring of *N*-benzylphenylhydroxylamine has been reported (eq 19).[30]

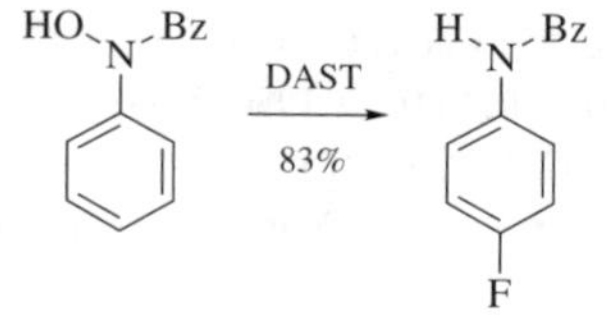

DAST, 83% (19)

An attempt to convert the C-7 hydroxy group of 7-*epi*-taxol to a fluoride failed. Instead, a high yield of a cyclopropanated product with A-ring contraction was obtained (eq 20). A cyclopropane intermediate corner-protonated at C-19 was postulated to explain this anomalous transformation.[31]

7-*epi*-taxol

DAST, CH_2Cl_2, 80–90% (20)

Related Reagents. Pyridinium Poly(hydrogen fluoride); Sulfur Tetrafluoride.

1. (a) Middleton, W. J. *JOC* **1975**, *40*, 574. (b) Hudlický, M. *OR* **1988**, 513.
2. (a) Newman, M. S.; Khanna, V. K.; Kanakarajan, K. *JACS* **1979**, *101*, 6788. (b) Pankiewicz, K. W.; Krzeminski, J.; Ciszewski, L. A.; Ren, W. Y.; Watanabe, K. A. *JOC* **1992**, *57*, 553.
3. Gai, S.; Hakomori, S.; Toyokuni, T. *JOC* **1992**, *57*, 3431.
4. Napolitano, E.; Fiaschi, R.; Hanson, R. *CC* **1989**, 1330.
5. MacLeod, A. M.; Herbert, R.; Hoogsteen, K. *CC* **1990**, 100.
6. (a) Blackburn, G. M.; Kent, D. E. *CC* **1981**, 511. (b) Tellier, F.; Sauvêtre, R. *TL* **1992**, *33*, 3643. (c) Tellier, F.; Sauvêtre, R. *TL* **1991**, *32*, 5963.
7. Uneme, H.; Okada, Y. *BCJ* **1992**, *65*, 2401.
8. Goswami, R.; Harsy, S. G.; Heiman, D. F.; Katzenellenbogen, J. A. *JMC* **1980**, *23*, 1002.
9. Rozen, S.; Faust, Y.; Ben-Yakov, H. *TL* **1979**, *20*, 1823.
10. Avent, A. G.; Bowler, A. N.; Doyle, P. M.; Marchand, C. M.; Young, D. W. *TL* **1992**, *33*, 1509.
11. Moyer, J. D.; Reizes, O.; Malinowski, N.; Jiang, C.; Baker, D. C. *ACS Symp. Ser.* **1988**, *374*, 43.
12. (a) Card, P. J.; Reddy, G. S. *JOC* **1983**, *48*, 4734. (b) Somawardhana, C. W.; Brunngraber, E. G. *Carbohydr. Res.* **1981**, *94*, C14. (c) Kozikowski, A. P.; Fauq, A. H.; Powis, G.; Melder, D. C. *JACS* **1990**, *112*, 4528. (d) Kozikowski, A. P.; Fauq, A. H.; Rusnak, J. M. *TL* **1989**, *30*, 3365.
13. (a) Hasegawa, A.; Goto, M.; Kiso, M. *J. Carbohydr. Chem.* **1985**, *4*, 627. (b) Somekh, L.; Shanzer, A. *JACS* **1982**, *104*, 5836. (c) Castillon, S.; Dessinges, A.; Faghih, R.; Lukacs, G.; Olesker, A.; Thang, T. T. *JOC* **1985**, *50*, 4913.
14. (a) Biollaz, M.; Kalvoda, J. Swiss Patent 616 433, 1980 (*CA* **1980**, *93*, 168 491e). (b) Campbell, J. A. U.S. Patent 4 416 822, 1983.
15. Markovskij, L. N.; Pashinnik, V. E.; Kirsanov, A. V. *S* **1973**, 787.
16. (a) Kotick, M. P.; Polazzi, J. O. *JHC* **1981**, *18*, 1029. (b) Boswell, G. A., Jr.; Brittelli, D. R. U.S. Patent 3 919 204, 1975.
17. (a) Boswell, G. A., Jr. U.S. Patent 4 212 815, 1980 (*CA* **1980**, *93*, 239 789w). (b) Daub, W.; Zuckermann, R. N.; Johnson, W. S. *JOC* **1985**, *50*, 1599.
18. Boehm, M. F.; Prestwich, G. D. *TL* **1988**, *29*, 5217.
19. Ando, K.; Kondo, F.; Koike, F.; Takayama, H. *CPB* **1992**, *40*, 1662.
20. Asato, A. E.; Lieu, R. S. H. *TL* **1986**, *27*, 3337.
21. Hudlicky, M. *JFC* **1987**, *36*, 373.
22. Hamatani, T.; Matsubara, S.; Matsuda, H.; Schlosser, M. *T* **1988**, *44*, 2875.
23. Middleton, W. J. U.S. Patent 3 914 265, 1975.
24. Cantrell, G. L.; Filler, R. *JFC* **1985**, *27*, 35.
25. Robins, M. J.; Wnuk, S. F. *TL* **1988**, *29*, 5729.
26. Wnuk, S. F.; Robins, M. J. *JOC* **1990**, *55*, 4757.
27. McCarthy, J. R.; Peet, N. P. *JACS* **1985**, *107*, 735.
28. Albert, R.; Dax, K.; Katzenbeisser, U.; Sterk, H.; Stuetz, A. E. *J. Carbohydr. Chem.* **1985**, *4*, 521.
29. (a) Nicolaou, K. C.; Dolle, R. E.; Papahatjis, D. P.; Randall, J. L. *JACS* **1984**, *106*, 4189. (b) Posner, G. H.; Haines, S. R. *TL* **1985**, *26*, 5.
30. Leroy, J.; Hebert, E.; Wakselman, C. *JOC* **1979**, *44*, 3406.
31. Chen, S. H.; Huang, S.; Wei, J.; Farina, V. *JOC* **1993**, *58*, 4520.

Abdul H. Fauq
Mayo Foundation, Jacksonville, FL, USA

Diethyl Azodicarboxylate[1]

EtO–C(=O)–N=N–C(=O)–OEt

[1972-28-7] $C_6H_{10}N_2O_4$ (MW 174.16)

(functional group oxidations;[5] dealkylation of amines;[16] enophile;[24] dienophile[33])

Alternate Names: diethyl azidoformate; DEAD; DAD.[2]
Physical Data: bp 108–110 °C/15 mmHg; bp 211 °C/760 mmHg; fp 110 sp;°C, *d* 1.11 g cm^{-3}.
Solubility: sol CH_2Cl_2, Et_2O, toluene.
Form Supplied in: orange liquid; 90% (technical grade) and 95% purity are commonly available, as is a 40% solution in toluene.
Preparative Method: although widely available, diethyl azodicarboxylate can be readily prepared from ethyl chloroformate and hydrazine, followed by oxidation of the resulting diethyl hydrazodicarboxylate.[3]
Purification: in most cases commercial samples are used without purification. Distillation is possible, but *not recommended.*
Handling, Storage, and Precautions: flammable; may produce toxic combustion byproducts. Heat and light sensitive; should be stored in dark containers under refrigerated conditions. These containers should also be vented periodically to reduce pressure. DAD has been reported to occasionally decompose violently when heated.[3a,4] Use in a fume hood.

Introduction. See ***Triphenylphosphine–Diethyl Azodicarboxylate*** for reactions involving the use of the combination of PPh_3 and DAD.

Functional Group Oxidations by Dehydrogenation. The strong electron withdrawing character of diethyl azodicarboxylate makes it suitable for certain types of oxidations. In general, diethyl hydrazocarboxylate is formed as a product of any oxidation. Primary and secondary alcohols are oxidized to aldehydes and ketones, respectively, although application of this method has been limited.[5] Thiols, similarly, are oxidized to disulfides. Unsymmetrical disulfides are also available by a variation of this method.[6] The oxidation of formamides to isocyanates has been accomplished at high temperatures with DAD, although the yields are frequently low and the method does not appear to be general.[7]

Arylhydroxylamines are readily converted into nitroso compounds with DAD at 0 °C (eq 1).[8] One example of the conversion of an *N,N*-dimethylhydrazone to a nitrile has been reported.[9] Sulfur-containing amino acids like methionine and *S*-ethylcysteine can be oxidized to their sulfoxides in virtually quantitative yields, although another reaction pathway occurs with most other thioethers.[10] Thioethers and ethers usually react with DAD to yield α-hydrazo derivatives by hydrogen abstraction.[11] The initial ether/DAD adducts can be formed thermally at 100 °C[12] or photochemically at much lower temperatures.[13]

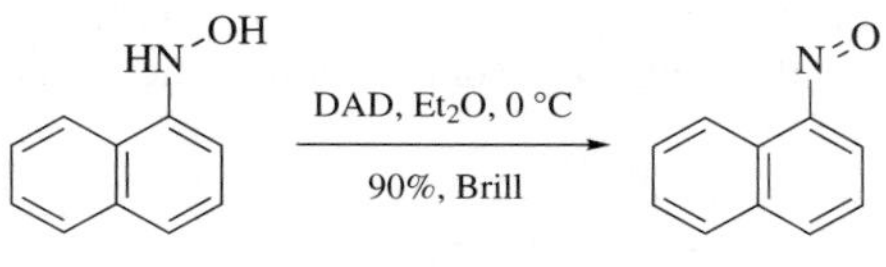

DAD, Et_2O, 0 °C; 90%, Brill (1)

The oxidation of propargylic hydrazine derivatives with DAD to the corresponding propargylic diazenes (with subsequent spontaneous loss of N_2) forms the basis of a powerful, yet mild, allene synthesis (eq 2).[14]

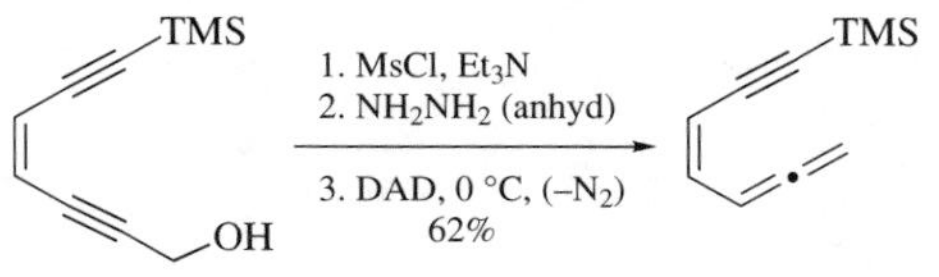

1. MsCl, Et_3N; 2. NH_2NH_2 (anhyd); 3. DAD, 0 °C, ($-N_2$); 62% (2)

Dealkylation of Amines. Treatment of a secondary or tertiary amine with diethyl azodicarboxylate in nonpolar solvents followed by acidic hydrolysis leads to the formation of monodealkylated amines.[15] The mechanism of this reaction is believed to involve the formation of a triaza adduct (by Michael addition) followed by a two-step ylide rearrangement yielding an alkyl-substituted hydrazocarboxylate.[12a] Research on unsymmetrically substituted amines suggests that benzyl groups are more easily removed than alkyl groups; methyl groups are the hardest to remove except in cyclic amines like *N*-methylpiperidine (eq 3).[16] The *N*-dealkylation of imines has also been reported.[17]

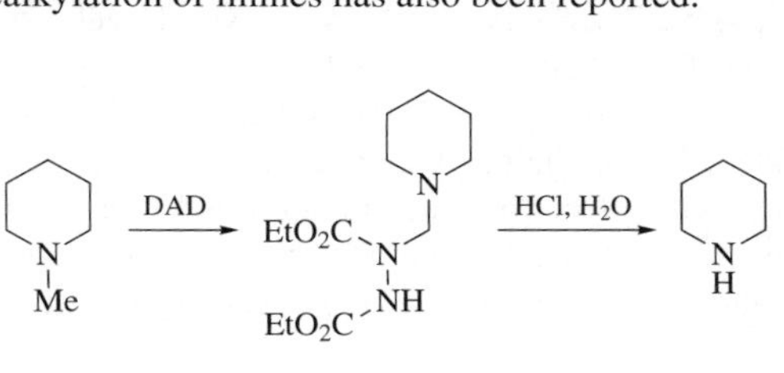

DAD; HCl, H_2O (3)

Purine Synthesis. Aminopyrimidines and aminouracil derivatives can be converted into purines with diethyl azodicarboxylate in a number of closely related synthetic methods. Treatment of compound (**1**) with an aldehyde leads to the formation of imine (**2**), which is converted to purine (**3**) with DAD (eq 4).[18] A number of researchers have applied variations of this approach.[19]

Treatment of 6-aminouracils[20] or 6-aminopyrimidines[21] (unsubstituted at the 5-position) with DAD leads to the initial formation of hydrazino Michael adducts. Treatment of these adducts with excess DAD leads to cyclization (eq 5).

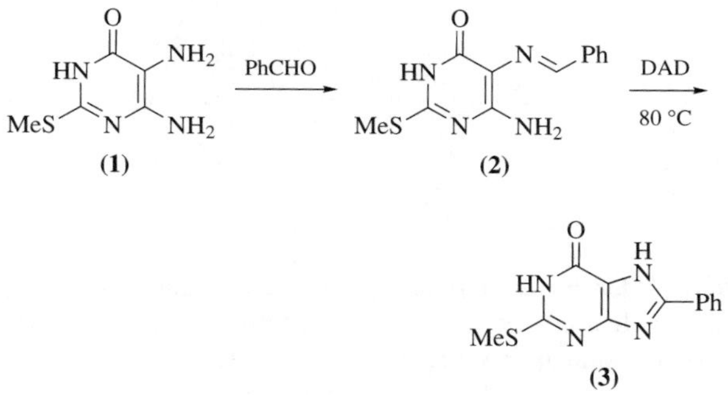

PhCHO; DAD, 80 °C (4)

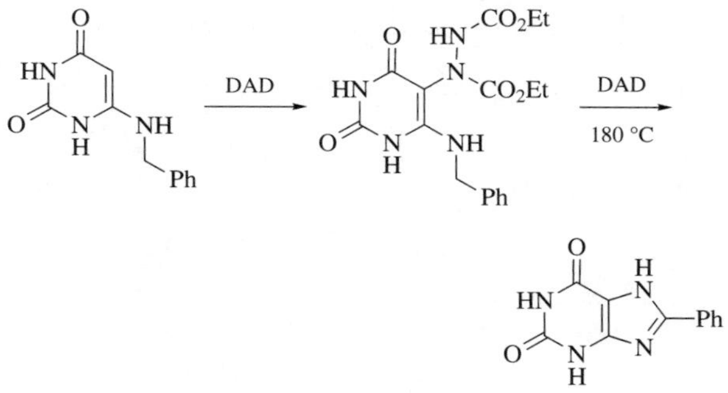

DAD; DAD, 180 °C (5)

This methodology has been applied to the synthesis of the antibiotic fervenulin[22] and 8-dimethylaminotheophylline.[23] In general, high temperatures (>100 °C) are required to complete the cyclization.

Pericyclic Reactions. In general, α-carboxyazo compounds participate in a number of pericyclic processes as noted below.

Ene Reactions. Diethyl azodicarboxylate reacts with most simple alkenes possessing an allyl hydrogen to yield the corresponding ene products, allylic hydrazocarboxylates.[24] The uncatalyzed reaction takes place at moderate temperatures (80 °C) and bis-adducts can be formed if excess DAD is used.[25] The use of ***Tin(IV) Chloride*** catalyzes the reaction, allowing for rapid reaction at −60 °C.[26] This reaction proceeds with a surprising degree of selectivity for (*E*)-alkene geometry (eq 6). A particularly useful application of the ene reaction with DAD is in the synthesis of allyl amines, which are readily available by reduction of the initial adducts with Li/NH_3.[27]

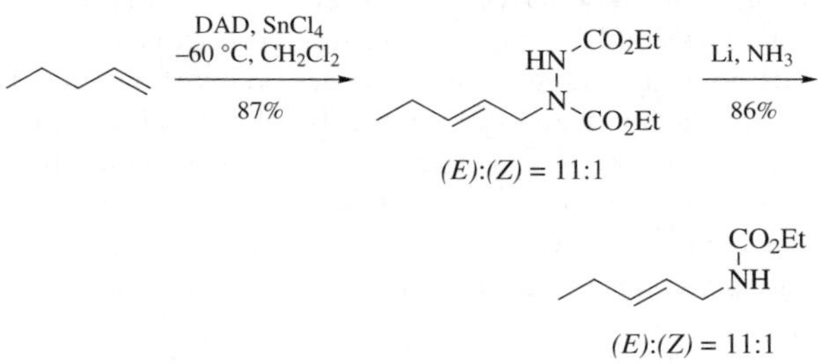

DAD, $SnCl_4$, −60 °C, CH_2Cl_2; 87%; (*E*):(*Z*) = 11:1; Li, NH_3; 86%; (*E*):(*Z*) = 11:1 (6)

Cycloheptatriene reacts with DAD to afford exclusively ene products,[28] as do the enols of some triazene derivatives.[29] Allenes with alkyl substituents react with DAD to yield ene products in most cases (eq 7).[30] Ene reactions with DAD have also been found to be a useful way of cleaving allyl ethers.[31]

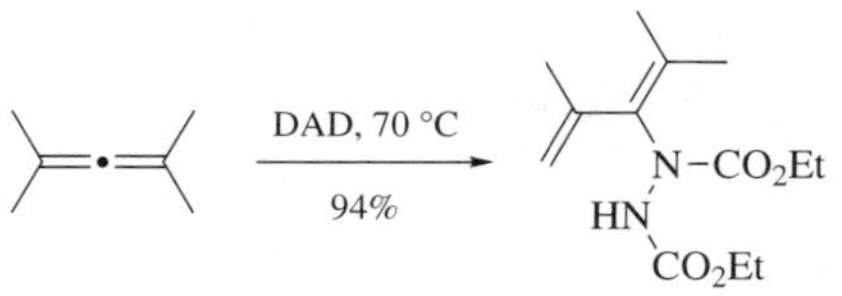

(7)

Diels–Alder [4 + 2] Cycloadditions. Except as noted below, diethyl azodicarboxylate reacts with conjugated dienes to yield [4 + 2] cycloadducts.[32] When the diene moiety is a vinyl aromatic, cycloaddition with DAD is a powerful route for the preparation of annulated tetrahydropyridazine derivatives.[33] Thus the indole derivative (**4**) reacts with DAD at rt to afford cycloadduct (**5**) (eq 8).[34]

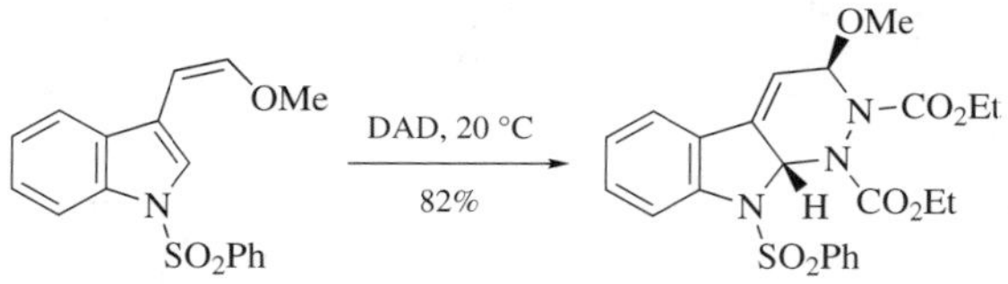

(8)

The reaction of vinylfurans with DAD usually affords intractable mixtures;[33b,e] however, furan itself undergoes cycloaddition readily, provided suitable reaction conditions[35] are employed to avoid decomposition of the reactive products.[36] Oxazole derivatives readily participate in [4 + 2] cycloadditions with DAD.[37] An asymmetric Diels–Alder reaction between DAD and pyridazin-3-ones facilitated by ***Baker's Yeast*** has been reported with ee's in the range of 9.1–62.7%.[38] The reactions of a number of acyclic heterodienes with DAD have also been reported recently, including 2-aza-1,3-dienes,[39] α,β-unsaturated thioketones,[40] and 1-thia-3-aza-1,3-dienes.[41] The quinodimethane derivatives of a number of heterocycles have been used in cycloadditions with DAD (eq 9).[42]

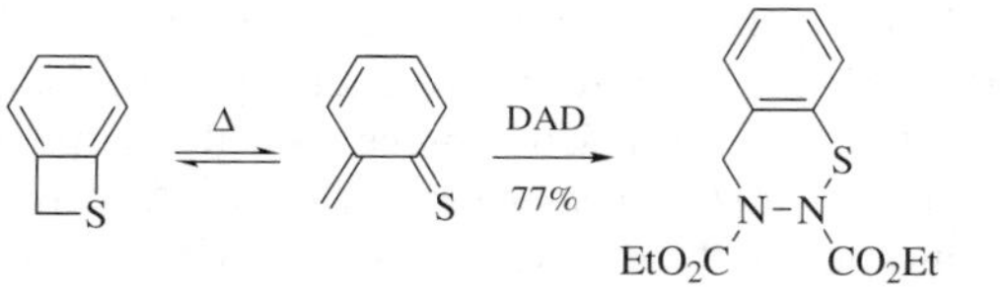

(9)

Competition Between Ene and Diels–Alder Reactions. One area of considerable research centers on the reactivity of diethyl azodicarboxylate with conjugated dienes, systems in which two pericyclic reaction pathways are possible: Diels–Alder and ene. Although a number of researchers have investigated this area, no clear explanation for the preference of one pathway over another has emerged.[43] In most cases, one reaction course seems to be strongly predominant. At this time, only generalizations are possible.

Acyclic 1,3-dienes, for example, in which alkyl groups are present at the terminus of the diene system, tend to give ene products with DAD, while those with internal alkyl substituents yield [4 + 2] adducts. 2,3-Dimethylbutadiene therefore reacts with DAD to give the cycloadduct in 94% yield, while 2,5-dimethyl-2,4-hexadiene gives only ene products.[44,55] In cases where mixed substitution is present, predictive methods fail completely.[46]

Reactivities among cyclic dienes are equally complicated. Cyclopentadiene gives exclusively [4 + 2] adducts.[47,48] 1,3-Cyclohexadiene, however, has been reported to give solely ene products,[49] but careful reexamination of this reaction reveals that 5–15% yields of [4 + 2] cycloadducts can be obtained.[50] The reaction can be optimized to yield exclusively cycloadducts by altering experimental conditions.[43,51] In steroidal systems, dienes in internal rings tend to undergo ene reactions,[52] while dienes located in the terminal ring give mixtures where cycloadditions are the predominant pathways (eq 10).[53] In cholesterol derivatives, reduction of the ene adducts with $Li/EtNH_2$ is a convenient way to isomerize the diene system (eq 11).[54] It has been reported that treatment with DAD will aromatize certain cyclohexadiene systems.[44,55]

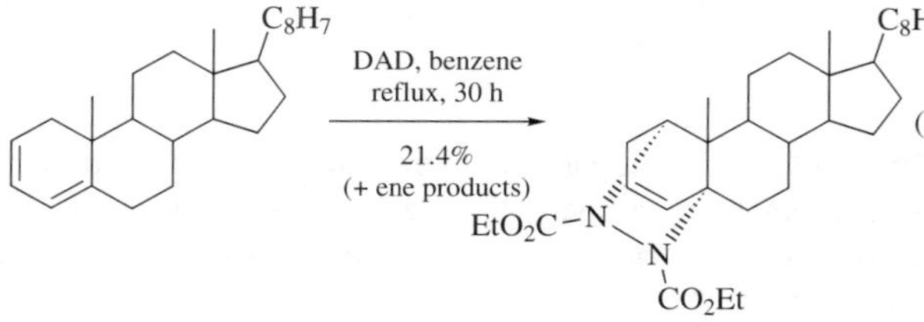

(10)

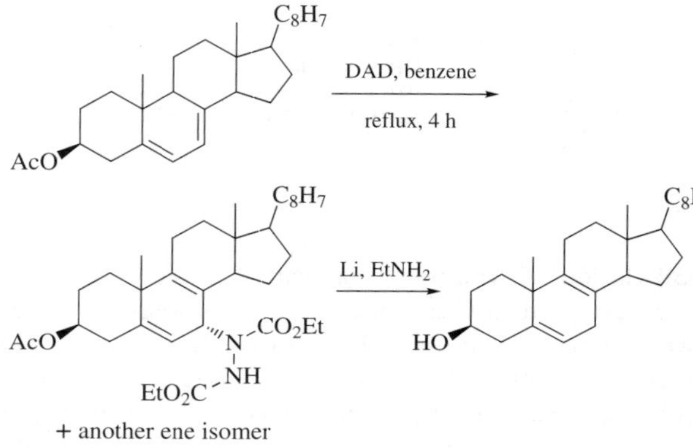

(11)

Other Cycloadditions. Diethyl azodicarboxylate has seen little application to other types of pericyclic reactions. DAD has been reported to undergo [2 + 2][56] cycloadditions with tetramethoxyallene.[57] Two reports of [8 + 2] cycloadditions involving DAD have emerged recently, one involving 7-alkylidene-1,3,5-cyclooctatrienes[58] and the other 3-methoxy-3a-methyl-3a*H*-indene.[59] Indolizine, which is known to undergo [8 + 2] reactions with electron deficient alkenes,[60] reacts with DAD exclusively in Michael fashion.[61] A photochemically induced 1,3-dipolar cycloaddition involving DAD has been reported.[62]

Similar Reagents. In most of the reactions involving diethyl azodicarboxylate, two closely related compounds can be employed instead. In many cases, 4-methyl-1,2,4-triazoline-3,5-dione (MTAD) (**6**) and ***4-Phenyl-1,2,4-triazoline-3,5-dione*** (PTAD) (**7**) exhibit greater reactivity than DAD.[15,30a,37b,63]

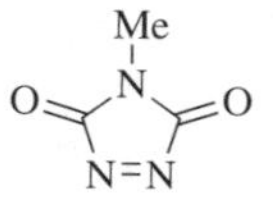

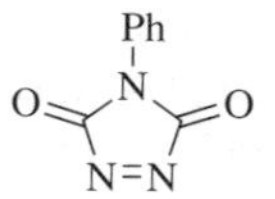

1. Fahr, E.; Lind, H. *AG(E)* **1966**, *5*, 372.
2. The abbreviations DAD and DEAD have been frequently used to represent diethyl acetylenedicarboxylate as well. To avoid confusion, the abbreviation DAD will be used throughout to represent diethyl azodicarboxylate.
3. (a) Kauer; J. C. *OSC* **1963**, *4*, 411. (b) Moriarty, R. M.; Prakash, I.; Penmasta, R. *SC* **1987**, *17*, 409. (c) Rabjohn, N. *OSC* **1955**, *3*, 375. (d) Kenner, G. W.; Stedman, R. J. *JCS* **1952**, 2089. (e) Curtius, T.; Heidenreich, K. *B* **1894**, *27*, 773. (f) Stollé, R.; Mampel, J.; Holzapfel, J.; Leverkus, K. C. *B* **1912**, *45*, 273. (g) Picard, J. P., Boivin, J. L. *CJC* **1951**, *29*, 223.
4. Fieser, L. F.; Fieser, M. F. *FF* **1967**, *1*, 245.
5. (a) Yoneda, F.; Suzuki, K.; Nitta, Y. *JACS* **1966**, *88*, 2328. (b) Yoneda, F.; Suzuki, K.; Nitta, Y. *JOC* **1967**, *32*, 727.
6. Mukaiyama, T.; Takahashi, K. *TL* **1968**, 5907.
7. Fu, P. P.; Boyer, J. H. *JCS(P1)* **1974**, 2246.
8. (a) Taylor, E. C.; Yoneda, F. *CC* **1967**, 199. (b) Brill, E. *E* **1969**, *25*, 680.
9. Borras-Almenar, C.; Sepulveda-Arques, J.; Medio-Simon, M.; Pindur, U. *H* **1990**, *31*, 1927.
10. Axen, R.; Chaykovsky, M.; Witkop, B. *JOC* **1967**, *32*, 4117.
11. Woodward, R. B.; Huesler, J.; Gosteli, J.; Naegeli, P.; Oppolzer, W.; Ramage, R.; Ranganathan, S.; Vorbrügen, U. *JACS* **1966**, *88*, 852.
12. (a) Huisgen, R.; Jakob, F. *LA* **1954**, *37*, 590. (b) Diels, O.; Paquin, M. *B* **1913**, *46*, 2000.
13. Cookson, R. C.; Stevens, I. D. R.; Watts, C. T. *CC* **1965**, 259.
14. Myers, A. G.; Finney, N. S.; Kuo, E. Y. *TL* **1989**, *30*, 5747.
15. Kenner, G. W.; Stedman, R. J. *JCS* **1952**, 2089.
16. Smissman, E. E.; Makriyannis, A. *JOC* **1973**, *38*, 1652.
17. Doleschall, G.; Tóth, G. *T* **1980**, *36*, 1649.
18. Nagamatsu, T.; Yamasaki, H. *H* **1992**, *33*, 775.
19. (a) Kaplita, P. V.; Abreu, M. E.; Connor, J. R.; Erickson, R. H.; Ferkany, J. W.; Hicks, R. P.; Schenden, J. A.; Noronha-Blob, L.; Hanson, R. C. *Drug. Dev. Res.* **1990**, *20*, 429. (b) Yoneda, F.; Higuchi, M. *H* **1976**, *4*, 1759.
20. Yoneda, F.; Matsumoto, S.; Higuchi, M. *CC* **1975**, 146.
21. Taylor, E. C.; Sowinski, F. *JOC* **1974**, *39*, 907.
22. Taylor, E. C.; Sowinski, F. *JACS* **1968**, *90*, 1374.
23. Walsh, E. B.; Nai-Jue, Z.; Fang, G.; Wamhoff, H. *TL* **1988**, *29*, 4401.
24. For reviews of the ene reaction, see: (a) Hoffmann, H. M. R. *AG(E)* **1969**, *8*, 556. (b) Boyd, G. V. In *The Chemistry of Double-Bonded Functional Groups*; Patai, S., Ed; Wiley: New York, 1989; Vol 2, Part 1, pp 477–526. (c) Oppolzer, W.; Snieckus, V. *AG(E)* **1978**, *17*, 476. (d) Mikami, K.; Shimizu, M. *CR* **1992**, *92*, 1021.
25. Thaler, W. A.; Franzus, B. *JOC* **1964**, *29*, 2226.
26. Brimble, M. A.; Heathcock, C. H. *JOC* **1993**, *58*, 5261.
27. Denmark, S. E.; Nicaise, O.; Edwards, J. P. *JOC* **1990**, *55*, 6219.
28. Cinnamon, J. M.; Weiss, K. *JOC* **1961**, *26*, 2644.
29. Bessiére-Chrétien, Y.; Serne, H. *JHC* **1974**, *11*, 317.
30. (a) Lee, C. B.; Taylor, D. R. *JCS(P1)* **1977**, 1463. (b) Lee, C. B.; Newman, J. J.; Taylor, D. R. *JCS(P1)* **1978**, 1161.
31. Ho, T.-L.; Wong, C. M. *SC* **1974**, *4*, 109.
32. For pertinent reviews of the Diels–Alder reaction see: (a) Gillis, B. T. In *1,4-Cycloaddition Reactions*; Hamer, J., Ed.; Academic: New York, 1967; pp 143–177. (b) Needleman, S. B.; Changkuo, M. C. *CR* **1962**, *62*, 405. (c) Weinreb, S. M.; Staib, R. R. *T* **1982**, *36*, 3087.
33. Cycloadducts with (a) 1-phenyl-5-vinylpyrazole: Medio-Simón, M.; Alvarez de Laviada, M. J.; Seqúlveda-Arques, J. *JCS(P1)* **1990**, 2749. (b) 2-(1-Trimethylsilyloxyvinyl)thiophene: Sasaki, T.; Ishibashi, Y.; Ohno, M. *H* **1983**, *20*, 1933. (c) 3-Vinylcoumarins and chromenes: Minami, T.; Matsumoto, Y.; Nakamura, S.; Koyanagi, S.; Yamaguchi, M. *JOC* **1992**, *57*, 167. (d) Cyanovinylthiophene: Abarca, B.; Ballesteros, R.; Soriano, C. *T* **1987**, *43*, 991. (e) Vinylpyridines: Jones, G.; Rafferty, P. *T* **1979**, *35*, 2027. (f) Jones, G.; Rafferty, P. *TL* **1978**, 2731.
34. Pindur, U.; Kim, M.-H.; Rogge, M.; Massa, W.; Molinier, M. *JOC* **1992**, *57*, 910.
35. Yur'ev, Y. K.; Zefirov, N. S. *JGU* **1959**, *29*, 2916.
36. Barenger, P.; Levisalles, J. *BSF(2)* **1957**, 704.
37. (a) Ibata, T.; Nakano, S.; Nakawa, H.; Toyoda, J.; Isogami, Y. *BCJ* **1986**, *59*, 433. (b) Shi, X.; Ibata, T.; Suga, H.; Matsumoto, K. *BCJ* **1992**, *65*, 3315. (c) Ibata, T.; Suga, H.; Isogami, Y.; Tamura, H.; Shi, X. *BCJ* **1992**, *65*, 2998.
38. Kakulapati, R. R.; Nanduri, B.; Yadavalli, V. D. N.; Trichinapally, N. S. *CL* **1992**, 2059.
39. Barluenga, J.; González, F. J.; Fustero, S. *TL* **1990**, *31*, 397.
40. Motoki, S.; Matsuo, Y.; Terauchi, Y. *BCJ* **1990**, *63*, 284.
41. Barluenga, J.; Tomás, M.; Ballesteros, A.; López, L. A. *TL* **1989**, *30*, 6923.
42. (a) *N*-Benzoylindole-2,3-quinodimethane: Haber, M.; Pindur, U. *T* **1991**, *47*, 1925. (b) 4,5-Dihydro-4,5-dimethylene-1-phenyl-1,2,3-triazole: Mertzanos, G. E.; Stephanidou-Stephanatou, J.; Tsoleridis, C. A.; Alexandrou, N. E. *TL* **1992**, *33*, 4499. (c) 4,5-Dihydro-4,5-dimethylene-3-phenylisoxazole: Mitkidou, S.; Stephanidou-Stephanatou, J. *TL* **1991**, *32*, 4603. (d) Benzotheite: Jacob, D.; Peter-Neidermann, H.; Meier, H. *TL* **1986**, *27*, 5703.
43. Jenner, G.; Salem, R. B. *JCS(P2)* **1990**, 1961.
44. Gillis, B. T.; Beck, P. E. *JOC* **1962**, *27*, 1947.
45. Measurement of the rate of reaction with 2,3-dimethylbutadiene in many solvents: Desimoni, G.; Faita, G.; Righetti, P. P.; Toma, L. *T* **1990**, *46*, 7951.
46. (a) Gillis, B. T.; Beck, P. E. *JOC* **1963**, *28*, 3177. (b) Jacobson, B. M.; Feldstein, A. C.; Smallwood, J. I. *JOC* **1977**, *42*, 2849.
47. This reaction is one of the oldest examples of the Diels–Alder reaction: Diels, O.; Blom, J. H.; Koll, W. *LA* **1925**, *443*, 242.
48. For representative applications of the [4+2] reaction of DAD and cyclopentadiene: (a) Kam, S.-T.; Portoghese, P. S.; Gerrard, J. M.; Dunham, E. W. *JMC* **1979**, *22*, 1402. (b) Lyons, B. A.; Pfeifer, J.; Peterson, T. H.; Carpenter, B. K. *JACS* **1993**, *115*, 2427. (c) Adam, W.; Finzel, R. *JACS* **1992**, *114*, 4563. (d) Wyvratt, M. J.; Paquette, L. A. *TL* **1974**, 2433.
49. Pirsh, J.; Jörgl, J. *B* **1935**, *68*, 1324.
50. (a) Franzus, B.; Surridge, J. H. *JOC* **1962**, *27*, 1951. (b) Franzus, B. *JOC* **1963**, *28*, 2954.
51. Askani, R. *B* **1965**, *98*, 2551.
52. van der Gen, A.; Lakeman, J.; Gras, M. A. M. P.; Huisman, H. O. *T* **1964**, *20*, 2521.
53. Tomoeda, M.; Kikuchi, R.; Urata, M.; Futamura, T. *CPB* **1970**, *18*, 542.
54. Anastasia, M.; Fiecchi, A.; Galli, G. *JOC* **1981**, *46*, 3421.
55. (a) Mehta, G.; Kapoor, S. K. *OPP* **1972**, *4*, 257. (b) Medion-Simon, M.; Pindur, U. *HCA* **1991**, *74*, 430.
56. Review: Muller, L. L.; Hamer, J. *1,2-Cycloadditions: The Formation of Three- and Four-Membered Heterocycles*; Wiley: New York, 1967.
57. Hoffmann, R. W. *AG(E)* **1972**, *11*, 324.
58. Ferber, P. H.; Gream, G. E.; Kirkbride, P. K. *TL* **1980**, *21*, 2447.
59. Gilchrist, T. L.; Rees, C. W.; Tuddenham, D. *JSC(P1)* **1981**, 3214.
60. Galbraith, A.; Small, T.; Barnes, R. A.; Bockelheide, V. *JACS* **1961**, *83*, 453.
61. Masamura, M.; Yamashita, Y. *H* **1979**, *12*, 787.
62. Gilgren, P.; Heimgartner, H.; Schmid, H. *HCA* **1974**, *57*, 1382.
63. (a) Stickler, J. C.; Pirkle, W. H. *JOC* **1966**, *31*, 3444. (b) Cookson, R. C.; Galani, S. S. H.; Stevens, I. D. R. *TL* **1962**, 615. (c) McLellan, J. F.; Mortier, R. M.; Orszulik, S. T.; Paton, R. M. *CI(L)* **1963**, 94.

Eric J. Stoner
Abbott Laboratories, North Chicago, IL, USA

Diethyl Phosphorochloridate[1]

[814-49-3] $C_4H_{10}ClO_3P$ (MW 172.55)

(highly electrophilic phosphorylating reagent easily installed on anionic carbon,[2] oxygen,[3] nitrogen,[4] and sulfur;[5] ketones are converted to enol phosphates which can be reduced to alkenes[6] or alkanes,[7] or coupled with organometallic reagents to form substituted alkenes;[8] enol phosphates can be tranformed into β-keto phosphonates,[9] useful for Horner–Emmons homologation, or into terminal alkynes;[10] used to convert carboxylic acids to other carboxylic derivatives[11])

Alternate Name: diethyl chlorophosphate.
Physical Data: bp 60 °C/2 mmHg; *d* 1.194 g cm^{-3}.
Handling, Storage, and Precautions: highly toxic, corrosive. Use in a fume hood.

Phosphorylation. Diethyl phosphorochloridate is highly electrophilic and can be cleanly reacted at an anionic center provided that prior metalation is regio- and chemoselective. For example, phenols,[12] thiophenols,[5] and anilines[4] can be phosphorylated under basic conditions (eq 1). The phosphorylated compound can be isolated, or treated further with bases, resulting in orthometalation followed by a facile 1,3-phosphorus migration from the heteroatom to carbon. If carbon phosphorylation is desired, simply treating thiophenol with 2 equiv of ***n-Butyllithium*** followed by diethyl phosphorochloridate gave a moderate yield (58%).[5]

(1)

X	Step 1 (% yield (2))	Step 2 (% yield (3))
OH	NaH, THF; $(EtO)_2P(O)Cl$ (50–85%)	LDA, THF −78 °C to rt (90%)
SH	NaH, THF, rt; $(EtO)_2P(O)Cl$ (63%)	LDA, THF −78 to 0 °C (16%)
HNMe	$(EtO)_2P(O)Cl$, Et_3N, THF (97%)	LDA, THF −78 to rt (63%)

Synthetic preparation of oligonucleotides in a cost-efficient manner can be complicated by low chemoselectivity during phosphorylation, suggesting the need for prior nitrogen protection. It has been determined that despite the higher acidity of guanosine and thymidine over a hydroxy, addition of 1 equiv of the phosphorylating agent provided only the phosphates (eq 2).[3] Adenosine and cytidine bases were compatible as well.

Phosphorylation of substituted 1,4-dihydropyridine dianions occurred selectively (eq 3).[2b] Other reports documented selective phosphorylation at C-5 of thiophenes[2a] and the *ortho* position of various substituted benzenes.[13] It has been noted that certain heterocycles demonstrated alternative pathways, such as nucleophilic attack at carbon rather than phosphorus (eq 4).[14]

(2)

(4) R¹	(4) R²	B	Conditions	(5) R¹	(5) R²	% Yield
TBS	H	guanosine	t-BuMgCl (2 equiv) THF, rt $(EtO)_2P(O)Cl$	TBS	$P(O)(OEt)_2$	94
H	TBS	thymidine	t-BuLi, (2 equiv) THF, −50 °C $(EtO)_2P(O)Cl$	$P(O)(OEt)_2$	H	85

(3)

(4)

95:5

Trapping of ester, lactone, or ketone enolates results in rapid *O*-phosphorylation, thus providing the enol phosphate (eq 5).[9] Further treatment with base resulted in facile rearrangement to the β-ketophosphonate (eq 5).[15] However, regioisomeric β-ketophosphonates can often be observed from a regiochemically pure enol phosphate (eq 6).[9a] Since the classical Arbuzov reaction is limited to primary alkyl iodides (competing Perkov reaction is known with secondary halides), an alternative method was sought. Installation of an alkene blocking group sufficed (eq 7).[9a] Dienol phosphates have also been prepared and the authors note that ***Lithium 2,2,6,6-Tetramethylpiperidide*** was the preferred base for regiochemical control.[16]

(5)

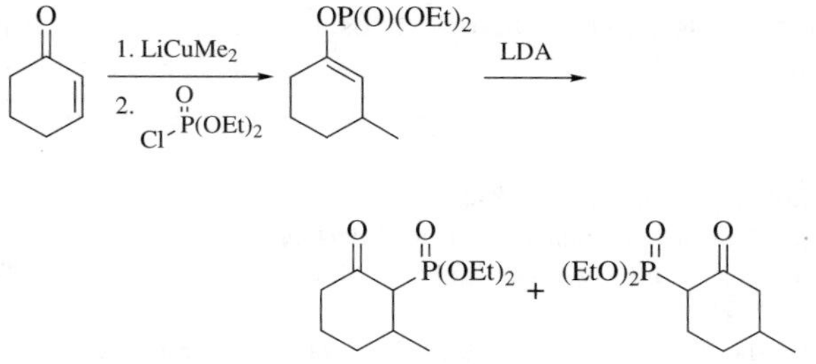

(6)

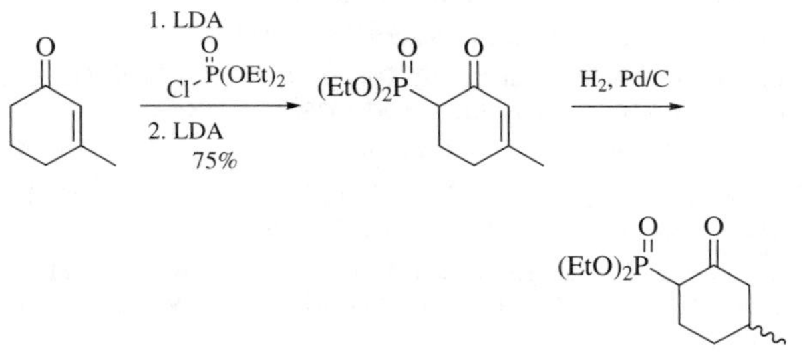

(7)

β-Ketophosphonates are valuable intermediates in the realm of Horner–Emmons alkenation methodology. Acyclic variants are difficult to obtain from enol phosphates due to competing alkyne and allene formation. One solution utilized the dianion derived from α-bromo ketones and trapping with diethyl phosphorochloridate;[17] however, only moderate yields of β-keto phosphonates were reported. The most efficient procedure utilizes the anion derived from dialkyl methylphosphonate, addition to an aldehyde, followed by oxidation (eq 8).[18]

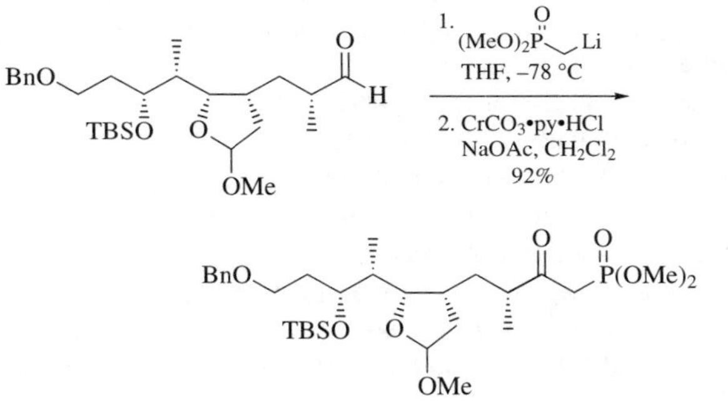

(8)

Interconversion of Carboxylic Acid Derivatives. Diethyl phosphorochloridate is useful for activation of carboxylic acids toward nucleophilic attack. Subsequent treatment of the phosphate ester with thallium sulfides produced thiol esters.[19] Variations[11] on this theme included prior formation of a heterocyclic phosphonate followed by treatment with alcohols, amines, or thiols, thus providing a racemization-free method to prepare esters, amides, and thiol esters, respectively (eq 9).[11b]

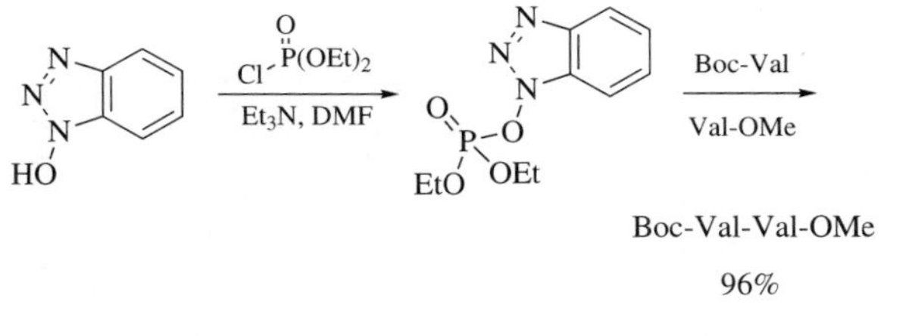

(9)

Amines can be readily transformed to an alcohol surrogate (eq 10)[20] and phenols to anilines.[21]

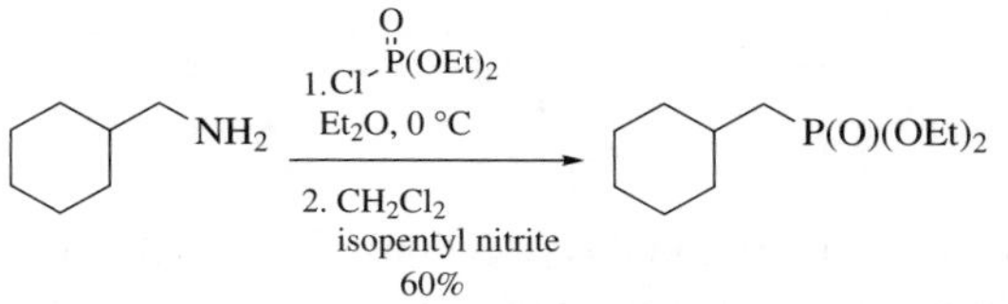

(10)

Deoxygenation of Phenols and Ketones. Excision of oxygen from a molecule is often encountered in a synthetic sequence. Phenols are readily deoxygenated by formation of the phosphate followed by reduction under dissolving metal conditions.[22] It has been noted that the Birch conditions result in low yields, whereas an alternative method utilizing activated ***Titanium*** metal is superior (eq 11).[23] Under this protocol, enol phosphates are efficiently reduced as well.[6b] Conversion of an enone to a regiochemically defined alkene was accomplished via 1,4-reduction, enolate trapping, and reduction of the trapped enol phosphate (eq 12).[6a] Enol phosphates can be fully reduced to alkanes by hydrogenation of palladium catalysts[7] or converted to vinyl iodides when treated with ***Iodotrimethylsilane***.[24]

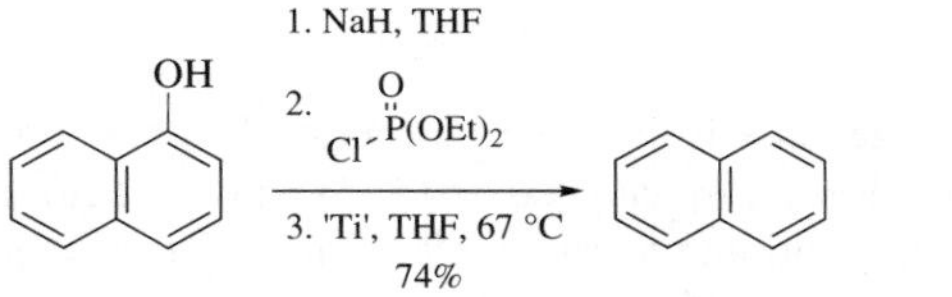

(11)

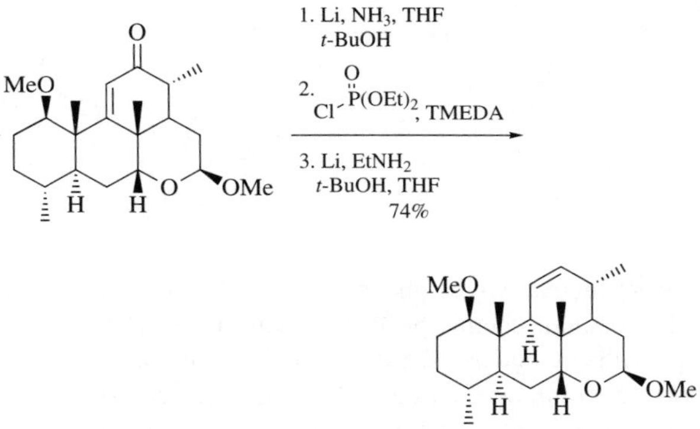

(12)

Synthesis of Alkenes and Alkynes. Enol phosphates are smoothly transformed into substituted alkenes when treated with organometallic reagents. If the enol phosphate was derived from a β-keto ester, cuprate reagents are generally reactive enough to encourage conjugate addition–phosphate elimination (eq 13).[25] In the event that this coupling fails, the combination of Pd^0 and ***Trimethylaluminum*** results in regio- and stereospecific methylation.[8b] Substrates lacking an ester moiety on the enol phosphate can be alkylated with Grignard reagents under nickel catalysis (eq 14).[8a]

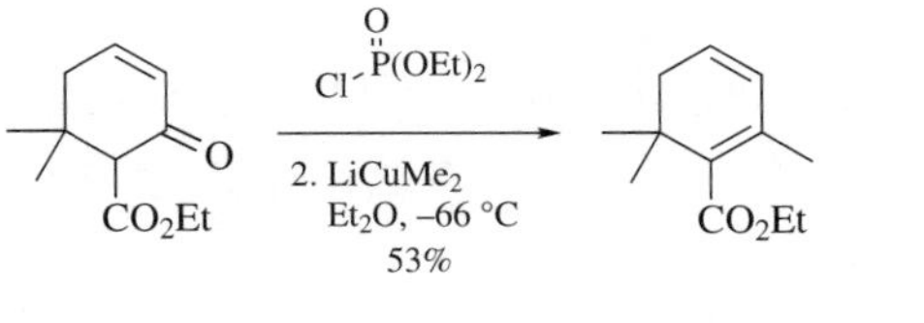

(13)

O; MeO$_2$C; O; O → 1. LDA, THF, −78 °C; O, Cl–P(OEt)$_2$; 2. MeMgI, Ni(acac)$_2$, THF, 0 °C → MeO$_2$C; O; O (14)

One of the most useful applications of β-keto phosphonates is the Horner–Emmons alkenation procedure (eq 15).[18] Variations of this theme have employed β-imino[26] and β-sulfonyl phosphates.[27]

O; O; (MeO)$_2$P; CHO; RO; O; O; O; O; OTBS → K_2CO_3, 18-crown-6; PhMe, 70 °C → O; RO; O; O; O; O; OTBS (15)

Alkynes are available from methyl ketones[10] by elimination of the enol phosphate (eq 16).[28] When the ketone contains α-branching, lithium tetramethylpiperidide has been recommended to circumvent allene formation.

O → 1. LDA, THF, −78 °C; 2. O, Cl–P(OEt)$_2$; 3. LDA; 72% → H (16)

Miscellaneous. Allylic phosphates have found applications in π-allyl palladium chemistry. It has been demonstrated that allylic phosphates undergo oxidative addition more readily than the corresponding acetate, such that chemoselectivity could be achieved when these functionalities were present in the same molecule (eq 17).[29] Finally, β-keto phosphonates were coupled with epoxides to provide useful yields of spirocyclopropanes (eq 18).[30]

(EtO)$_2$P(O)–O; OAc → Pd$_2$(dba)$_3$•CHCl$_3$; NaN$_3$; 92% → N$_3$; OAc + OAc; N$_3$ (17)

80:20

O; O; P(OEt)$_2$ → 1. NaH, C$_6$H$_6$; 2. O (propylene oxide); 76% → O (18)

Related Reagents. Diethyl Phosphorobromidate; Diethyl Phosphorocyanidate.

1. Koh, Y. J.; Oh, D. Y. *SC* **1993**, *23*, 1771.
2. (a) Graham, S. L.; Scholz, T. H. *JOC* **1991**, *56*, 4260. (b) Poindexter, G. S.; Licause, J. F.; Dolan, P. L.; Foley, M. A.; Combs, C. M. *JOC* **1993**, *58*, 3811.
3. (a) Hayakawa, Y.; Aso, Y. *TL* **1983**, *24*, 1165. (b) Uchiyama, M.; Aso, Y.; Noyori, R.; Hayakawa, Y. *JOC* **1993**, *58*, 373.
4. Jardine, A. M.; Vather, S. M. *JOC* **1988**, *53*, 3983.
5. Masson, S.; Saint-Clair, J.-F.; Saquet, M. *S* **1993**, 485.
6. (a) Grieco, P. A.; Nargund, R. P.; Parker, D. T. *JACS* **1989**, *111*, 6287. (b) Welch, S. C.; Walters, M. E. *JOC* **1978**, *43*, 2715.
7. Jung, A.; Engel, R. *JOC* **1975**, *40*, 3652.
8. (a) Iwashima, M.; Nagaoka, H.; Kobayashi, K.; Yamada, Y. *TL* **1992**, *33*, 81. (b) Asao, K.; Iio, H.; Tokoroyama, T. *S* **1990**, 382.
9. (a) Gloer, K. B.; Calogeropoulou, T.; Jackson, J. A.; Wiemer, D. F. *JOC* **1990**, *55*, 2842. (b) Jackson, J. A.; Hammond, G. B.; Wiemer, D. F. *JOC* **1989**, *54*, 4750.
10. Negishi, E.; King, A. O.; Tour, J. M. *OS* **1986**, *64*, 44.
11. (a) Mikaye, M.; Kirisawa, M.; Tokutake, N. *CL* **1985**, 123. (b) Kim, S.; Chang, H.; Ko, Y. K. *TL* **1985**, *26*, 1341.
12. Casteel, D. A.; Peri, S. P. *S* **1991**, 691.
13. Takenaka, H.; Hayase, Y. *H* **1989**, *29*, 1185.
14. Rani, B. R.; Bhalerao, U. T.; Rahman, M. F. *SC* **1990**, *20*, 3045.
15. Hammond, G. B.; Calogeropoulou, T.; Wiemer, D. F. *TL* **1986**, *27*, 4265.
16. Blotny, G.; Pollack, R. M. *S* **1988**, 109.
17. Sampson, P.; Hammond, G. B.; Wiemer, D. F. *JOC* **1986**, *51*, 4342.
18. Nicolaou, K. C.; Pavia, M. R.; Seitz, S. P. *JACS* **1982**, *104*, 2027.
19. Masamune, S.; Kamata, S.; Diakur, J.; Sugihara, Y.; Bates, G. S. *CJC* **1975**, *53*, 3693.
20. Nikolaides, N.; Ganem, B. *TL* **1990**, *31*, 1113.
21. Rossi, R. A.; Bunnett, J. F. *JOC* **1972**, *37*, 3570.
22. Kenner, G. W.; Williams, N. J. *JCS* **1955**, 522.
23. Welch, S. C.; Walters, M. E. *JOC* **1978**, *43*, 4797.
24. Lee, K.; Wiemer, D. F. *TL* **1993**, *34*, 2433.
25. Moorhoff, C. M.; Schneider, D. F. *TL* **1987**, *28*, 4721.
26. (a) Molin, H.; Pring, B. G. *TL* **1985**, *26*, 677. (b) Meyers, A. I.; Shipman, M. *JOC* **1991**, *56*, 7098. (c) Highet, R. J.; Jones, T. H. *JOC* **1992**, *57*, 4038.
27. Musicki, B.; Widlanski, T. S. *TL* **1991**, *32*, 1267.
28. Okuda, Y.; Morizawa, Y.; Oshima, K.; Nokaki, H. *TL* **1984**, *25*, 2483.
29. Murahashi, S.-I.; Taniguchi, Y.; Imada, Y.; Tanigawa, Y. *JOC* **1989**, *54*, 3292.
30. Jacks, T. E.; Nibbe, H.; Wiemer, D. F. *JOC* **1993**, *58*, 4584.

Jonathan R. Young
University of Wisconsin, Madison, WI, USA

3,4-Dihydro-2*H*-pyran[1]

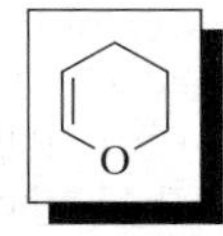

[110-87-2] C_5H_8O (MW 84.12)

(widely used OH-protecting reagent;[1] applicable to alkanols, phenols, thiols; stable at high pH; labile at low pH; generally resistant to nucleophiles and organometallic reagents; resistant to hydride reductions; labile under Lewis acidic conditions)

Physical Data: mp −70 °C; bp 86 °C; fp −15 °C; *d* 0.922 g cm^{-3}.
Solubility: sol water, ethanol.
Handling, Storage, and Precautions: it has been reported that the tetrahydropyranyl (THP) ethers form sensitive organic peroxides when in contact with peroxy reagents. Violent explosions have occurred during purification of these compounds. Precautions normally sufficient in isolation of such products have failed to destroy the sensitive components.[2]

Tetrahydropyranylation of Alcohols. Protection of alcohol functionality as the THP ether is an often-utilized tool in organic synthesis. It must be noted that the reaction of a chiral alcohol with dihydropyran introduces an additional asymmetric center and hence a diastereomeric mixture is obtained (eq 1). This can lead to difficulties with purification, assignment of spectral features, etc., but does not prevent successful implementation.[3]

(1)

A variety of reaction conditions for the synthesis of THP derivatives, the majority acidic, have been proposed in the literature (eq 2). The techniques have been highly optimized and increasingly mild conditions have been developed. Mineral acids have classically been used to effect the reaction.[4] Products have been obtained in 20–93% yields using these methods. However, the use of mineral acids is clearly limiting with respect to substrates containing sensitive functionality. Nevertheless, many applications in steroid[5] and saccharide[6] chemistry are possible using this technology. Significantly, tetrahydropyranyl ethers are not known to migrate in 1,2-diol systems.

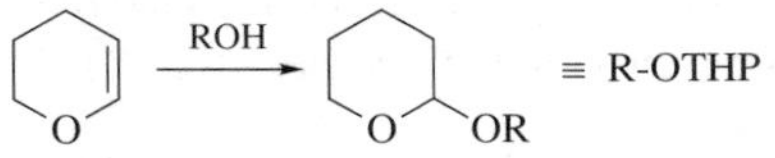

≡ R-OTHP (2)

The standard acid catalyst for effecting tetrahydropyranylation of alcohols has become ***p*-Toluenesulfonic Acid**. Reaction of even tertiary alcohols with excess dihydropyran is usually complete in 0.5–1 h at rt, although sometimes reflux temperatures are required. Yields range from 60 to 84% for tertiary alcohols.[7] Catalysis by *p*-TsOH can give very high yields, depending on conditions and choice of solvent: *p*-TsOH (dioxane, 20 °C, 5 min, 71–97%).[8]

Tetrahydropyranylation of primary and secondary alcohols can give high yields under mild and neutral conditions using ***Iodotrimethylsilane*** (25 °C, 30 min, 80–95%). Workup is particularly convenient, consisting of evaporation of volatiles and chromatography when required. Tertiary alcohols, however, yield the corresponding iodide rather than the THP ether.[9]

Reaction using ***Boron Trifluoride Etherate*** (Et_2O–petroleum ether, 25 °C, 47–76%) gives moderate yields with evaporation of volatiles being the only workup required.[10]

Pyridinium p-Toluenesulfonate (25 °C, 4 h, 94–100%) is a mild and efficient catalyst, particularly applicable to the protection of highly acid sensitive alcohols. Yields were markedly superior to procedures employing $BF_3 \cdot Et_2O$, *p*-TsOH, and ***Hydrogen Chloride***.[11]

Tetrahydropyranylation of tertiary alcohols, while more difficult owing to the hindered nature of the substrates or competitive elimination, can be carried out using dihydropyran and ***Triphenylphosphine Hydrobromide*** (25 °C, 6–24 h, 80–96%).[12] The method is applicable to even very sensitive substrates such as mevalonolactone (eq 3).

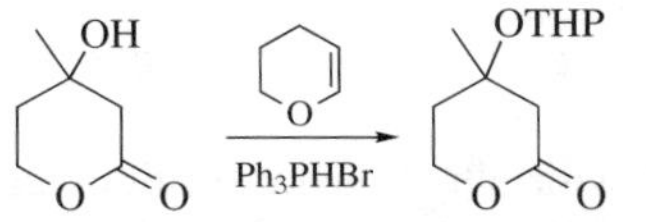

Ph_3PHBr (3)

Tetrahydropyranylation of alcohols can be carried out under very mild conditions in the presence of bis(trimethylsilyl) sulfate (0 °C, 1 h, 89–100%). No rearrangement is observed even with tertiary allylic alcohols.[13]

Protocols have been developed which utilize an insoluble solid catalyst in combination with dihydropyran to effect the protection of alcohols as their corresponding THP ethers. These procedures are advantageous in that the catalyst may be recovered by simple filtration and the products isolated by evaporation of volatiles. In many cases the catalyst can be reused without regeneration. Reaction of alcohols with dihydropyran in the presence of Amberlyst H-15 (25 °C, 1 h, 90–98%) yields THP derivatives. Alternatively, a solution of dihydropyran and the alcohol may be passed slowly through a column of silica overlaid with Amberlyst H-15 to yield the THP ethers directly (73–97%).[14] The acidic clay ***Montmorillonite K10*** (25 °C, 15–30 min, 63–95%) is similarly applicable.[15] Reillex 425 resin (86 °C, 1.5 h, 84–98%)[16] is applicable with the advantage that it does not promote the sometimes troublesome polymerization of dihydropyran.[17] Polymeric derivatives of pyridinium *p*-toluenesulfonate are also effective. Poly(4-vinylpyridinium *p*-toluenesulfonate) and poly(2-vinylpyridinium *p*-toluenesulfonate) catalysts yield tetrahydropyranyl derivatives of primary, secondary, and tertiary alcohols (24 °C, 3–8 h, 72–95%).[18]

Cleavage of Tetrahydropyranyl Derivatives. Deprotection of THP ethers has been carried out using ***Acetic Acid*** (H_2O–THF (3:3:2), 50 °C, 5 h, 97%),[19] HOAc (H_2O–THF (2:1:as required), 47 °C, 4 h, 82%),[20] HOAc (THF–H_2O (4:2:1), 45 °C, 3.5 h),[21] aqueous ***Oxalic Acid*** (MeOH, 50–90 °C, 1–2 h),[22] *p*-TsOH (MeOH, 25 °C, 1 h, 94%),[23] pyridinium *p*-toluenesulfonate (EtOH, 55 °C, 3 h, 98–100%),[24] bis(trimethylsilyl) sulfate (MeOH, 25 °C, 10–90 min, 93–100%).[13] Cleavage with Amberlyst H-15 (MeOH, 45 °C, 0.5–2 h, 93–98%)[25] is advantageous in that the acid catalyst can be recovered by filtration and no aqueous workup is required. Organotin phosphate condensates are effective in the selective cleavage of THP ethers (MeOH, reflux, 2 h, 80–90%) in the presence of MEM ethers, MOM ethers, and 1,3-dioxolanes.[26] The catalyst is insoluble in the reaction medium and can be reused repeatedly.

Selective cleavage of THP ethers in the presence of *t*-butyldimethylsilyl ethers can be accomplished by treatment with mild Lewis acids: ***Dimethylaluminum Chloride*** (−25 °C to 25 °C, 1 h, 98%) or ***Methylaluminum Dichloride*** (−25 °C, 0.5 h, 90%).[27] Selective cleavage of THP ethers of primary, secondary, and tertiary alcohols can be carried out in the presence of *t*-butyldimethylsilyl, acetyl, mesyl, and methoxymethyl ethers, in the presence of thiostannane catalysts: $Me_2Sn(SMe)_2$–$BF_3 \cdot Et_2O$ (−20 °C or 0 °C, 3–25 h, 80–97%).[28]

Tetrahydropyranylation of Thiols. Tetrahydropyranyl derivatives of thiols have been utilized for the masking of this functional group (eq 4).

RSH ≡ R-STHP (4)

Thiols react with dihydropyran in the presence of $BF_3 \cdot Et_2O$ (0 °C, 0.5 h, 25 °C, 1 h) to yield (*S*)-2-tetrahydropyranyl hemithioacetals in satisfactory yields.[29] In contrast to *O*-tetrahydropyranyl ethers, an *S*-tetrahydropyranyl ether is stable to 4 N HCl–MeOH.[30] Deprotection is conveniently accomplished with ***Silver(I) Nitrate*** (0 °C, 10 min)[31] or ***Hydrogen Bromide–Trifluoroacetic Acid*** (90 min)[32] in quantitative yields. Oxidation to disulfides can be carried out with ***Iodine***[33] or ***Thiocyanogen***.[34]

Tetrahydropyranylation of Amides. Aromatic amides, alkyl amides, ureas, sulfonamides, and imides undergo reaction with dihydropyran–***Hydrogen Chloride*** (benzene, reflux, 2 h, 22–73%), yielding the expected adducts.[35]

Tetrahydropyranylation of Amines. Purines react with dihydropyran in the presence of a catalytic amount of *p*-TsOH to give the 9-tetrahydro-2-pyranyl derivatives (50–87%) (eq 5).[36]

p-TsOH (5)

Functional Group Manipulation of Tetrahydropyranyl Derivatives. While tetrahydropyranyl derivatives are often utilized as masked equivalents of alcohols, etc., it is possible to perform functional group manipulation on such species. In these cases the tetrahydropyranyl derivatization serves to activate the alcohol rather than protect it.

Conversion of alcohols to their corresponding alkyl halides can be accomplished in two steps by conversion to the THP ethers followed by treatment with ***Triphenylphosphine–Carbon Tetrabromide*** (23 °C, 24 h, 62–87%).[37] In the case of tertiary substrates, 15–25% elimination is observed. Similarly rapid conversion of tetrahydropyranyl derivatives to their corresponding bromides or iodides can be carried out with 1,2-bis(diphenylphosphino)ethane tetrahalides (0.1–4 h, 0 °C to 23 °C, 81–96%).[38] Methyl esters and *t*-butyldimethylsilyl ethers and alkenes are unreactive under these reaction conditions. Direct conversion of alcohol tetrahydropyranyl ethers to bromides, chlorides, trifluoroacetates, methyl ethers, or nitriles can be effected with ***Triphenylphosphine Dibromide*** (25 °C, 30 min, 43–89%).[39] Tetrahydropyranyl ethers may be converted to benzyl ethers, MEM ethers, benzoates, or tosylates by reaction with the appropriate electrophile in the presence of Bu_3SnSMe–$BF_3 \cdot EtO_2$ (0 °C, 7–70 h, 75–78%). Oxidation of the intermediate alkoxystannane with ***Pyridinium Chlorochromate*** yields aldehydes (83%).[40]

Ring Opening Reactions of Dihydropyran. Treatment of dihydropyran with *n*-pentylsodium,[41] or ***n-Butyllithium***[42] yields *trans*-1-hydroxy-4-nonene (73%) (eq 6).

BuLi OH (6)

Cleavage of dihydropyran with HCl(aq) (30 min, 25 °C) constitutes a convenient preparation of 5-hydroxypentanal (eq 7).[43]

aq HCl (7)

Tetrahydropyranyl ethers are unstable to reduction with ***Lithium Aluminium Hydride–Aluminium Chloride*** (eq 8). Cleavage occurs in the tetrahydropyranyl ring or exocyclicly.[44] The reaction is seldom useful synthetically since the two modes are competitive. In marked contrast, tetrahydropyranyl derivatives of thiols are cleaved selectively at the ring carbon–oxygen bond, giving hydroxyalkyl thioethers (58–82%) (eq 9).[45]

$LiAlH_4$, $AlCl_3$: + ROH + RO⌒OH (8)

$LiAlH_4$, $AlCl_3$: RS⌒OH (9)

Tetrahydropyranyl derivatives of amines are hydrogenolized by $LiAlH_4$ in the absence of any Lewis acid (eq 10).[46]

$LiAlH_4$: R^2R^1N⌒OH (10)

Dihydropyran is a convenient starting material for the preparation of (*E*)-4-hexen-1-ol, in three steps (eq 11).[47]

Cl_2, MeMgBr, Na (11)

Anion Chemistry. Dihydropyran and derivatives can be quantitatively deprotonated with ***t-Butyllithium*** (−78 °C to 5 °C, 0.5 h, THF) (eq 12). The resulting anion reacts with electrophiles such as ketones and alkyl halides in good yields. Cuprates derived from this anion add smoothly in conjugate fashion to α,β-enones in excellent (91%) yields (eq 13).[48]

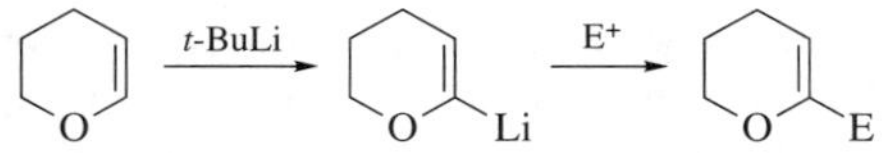

(12)

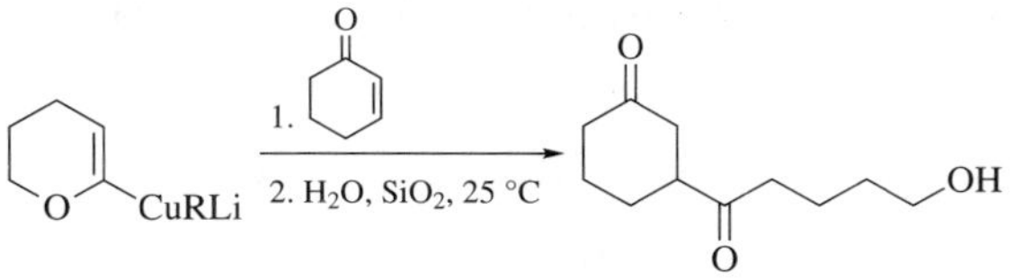

(13)

Electrophilic Addition to Dihydropyran. Electrophilic addition of acetals and orthoesters to dihydropyran occurs in the presence of mild catalyst systems such as ***Chlorotrimethylsilane–Tin(II) Chloride*** (0 °C, 2 h, 55–84%) (eq 14).[49] The synthetic utility of the reaction is however limited by the lack of stereocontrol.

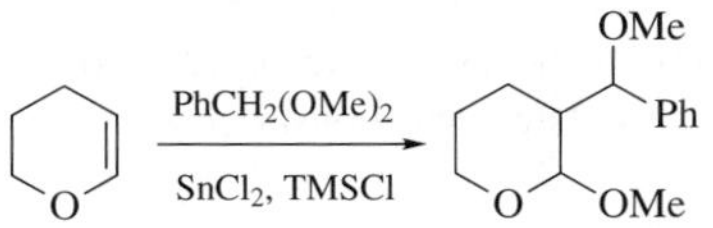

(14)

Specific Applications. The THP protecting group has been widely applied in the field of organic synthesis and no attempt can be made to comprehensively review its use in these pages. However, several specific applications may be of interest.

Introduction of the *trans*-$CH{=}CHCH_2OH$ moiety may be conveniently accomplished using a mixed Gilman cuprate reagent bearing a terminal THP-protected ether (eq 15).[50] The transformation is of considerable utility in several fields of synthesis, including prostaglandin chemistry.

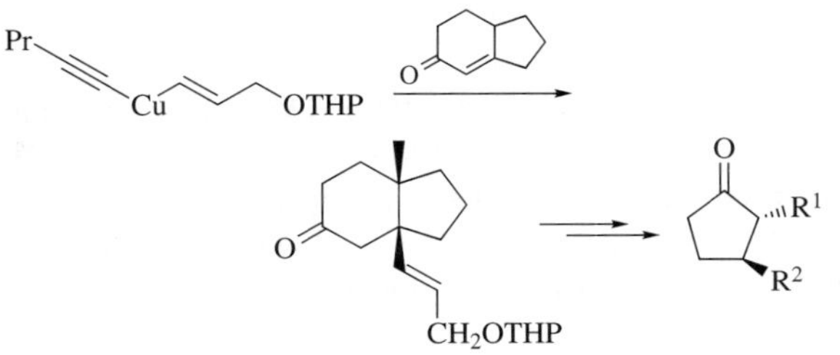

(15)

The tetrahydropyranyl derivative of propargyl alcohol, tetrahydro-2-(2-propynyloxy)-2*H*-pyran, can be converted to methyl 4-hydroxy-2-butynoate in four steps.[51] Diethyl [(2-tetrahydropyranyloxy)methyl]phosphonate is a convenient Wadsworth–Emmons reagent.[52] Tetrahydropyranyl esters of α-bromo acids can be used in the Reformatsky reaction for the preparation of β-hydroxy acids.[53] Elimination of a tetrahydropyranyloxy moiety from butyne-1,4-diols with lithium hydride constitutes an efficient method for the synthesis of allenic alcohols.[54]

Related Reagents. Benzyl Chloromethyl Ether; 2-Chloroethyl Chloromethyl Ether; Chloromethyl Methyl Ether; Dimethoxymethane; Ethyl Vinyl Ether; *p*-Methoxybenzyl Chloride; (*p*-Methoxybenzyloxy)methyl Chloride; 2-Methoxyethoxymethyl Chloride; 2-Methoxypropene; 2-(Trimethylsilyl)ethoxymethyl Chloride.

---

1. Greene, T. W. *Protective Groups in Organic Synthesis*; Wiley: New York, 1981.
2. Meyers, A. I.; Schwartzman, S.; Olson, G. L.; Cheung, H.-C. *TL* **1976**, 2417.
3. Corey, E. J.; Wollenberg, R. H.; Williams, D. R. *TL* **1977**, 2243.
4. (a) Paul, R. *BCF* **1934**, *1*, 971. (b) Woods, G. F.; Kramer, D. N. *JACS* **1947**, *69*, 2246. (c) Parham, W. E.; Anderson, E. L. *JACS* **1948**, *70*, 4187. (d) Jones, R. G.; Mann, M. J. *JACS* **1953**, *75*, 4048.
5. (a) Loewenthal, H. J. E. *T* **1959**, *6*, 269. (b) Dauben, W. G.; Bradlow, H. L. *JACS* **1952**, *74*, 559. (c) Ott, A. C.; Murray, M. F.; Pederson, R. L. *JACS* **1952**, *74*, 1239.
6. (a) Straus, D. B.; Fresco, J. R. *JACS* **1965**, *87*, 1364. (b) Griffin, B. E.; Jarman, M.; Reese, C. B. *T* **1968**, *24*, 639.
7. Robertson, D. N. *JOC* **1960**, *25*, 931.
8. van Boom, J. H.; Herschied, J. D. M.; Reese, C. B. *S* **1973**, 169.
9. Olah, G. A.; Husain, A.; Singh, B. P. *S* **1985**, 703.
10. Alper, H.; Dinkes, L. *S* **1972**, 81.
11. Miyashita, N.; Yoshikoshi, A.; Grieco, P. A. *JOC* **1977**, *42*, 3772.
12. Bolitt, V.; Mioskowski, C.; Shin, D.-S.; Falck, J. R. *TL* **1988**, 4583.
13. Morizawa, Y.; Mori, I.; Hiyama, T.; Nozaki, H. *S* **1981**, 899.
14. Bongini, A.; Cardillo, G.; Orena, M.; Sandri, S. *S* **1979**, 618.
15. Hoyer, S.; Laszlo, P.; Orlovic, M.; Polla, E. *S* **1986**, 655.
16. Johnston, R. D.; Marston, C. R.; Krieger, P. E.; Goe, G. L. *S* **1988**, 393.
17. Olah, G. A.; Husain, A.; Singh, B. P. *S* **1983**, 892.
18. Menger, F. M.; Chu, C. H. *JOC* **1981**, *46*, 5044.
19. Corey, E. J.; Nicolaou, K. C.; Melvin Jr., L. S. *JACS* **1975**, *97*, 654.
20. Corey, E. J.; Schaaf, T. K.; Huber, W., Koelliker, U.; Weinshenker, N. M. *JACS* **1970**, *92*, 397.
21. Bernady, K. F.; Floyd, M. B.; Poletto, J. F.; Weiss, M. J. *JOC* **1979**, *44*, 1438.
22. Grant, H. N.; Prelog, V.; Sneeden, R. P. A. *HCA* **1963**, *46*, 415.
23. Corey, E. J.; Niwa, H.; Knolle, J. *JACS* **1978**, *100*, 1942.
24. Miyashita, N.; Yoshikoshi, A.; Grieco, P. A. *JOC* **1977**, *42*, 3772.
25. Bongini, A.; Cardillo, G.; Orena, M.; Sandri, S. *S* **1979**, 618.
26. Otera, J.; Niibo, Y.; Chikada, S.; Nozaki, H. *S* **1988**, 328.
27. Ogawa, Y.; Shibasaki, M. *TL* **1984**, 663.
28. Sato, T.; Otera, J.; Nozaki, H. *JOC* **1990**, *55*, 4770.
29. (a) Hiskey, R. G.; Tucker, W. P. *JACS* **1962**, *84*, 4789. (b) Holland, G. F.; Cohen, L. A. *JACS* **1958**, *80*, 3765.
30. Griffin, B. E.; Jarman, M.; Reese, C. B. *T* **1968**, *24*, 639.
31. Holland, G. F.; Cohen, L. A. *JACS* **1958**, *80*, 3765.
32. Hammerström, K.; Lunkenheimer, W.; Zahn, H. *Macromol. Chem.*, **1970**, *133*, 41.
33. Holland, G. F.; Cohen, L. A. *JACS* **1958**, *80*, 3765.
34. Hiskey, R. G.; Tucker, W. P. *JACS* **1962**, *84*, 4794.
35. Speziale, A. J.; Ratts, K. W.; Marco, G. J. *JOC* **1961**, *26*, 4311.
36. Robins, R. K.; Godefroi, E. F.; Taylor, E. C.; Lewis, L. R.; Jackson, A. *JACS* **1961**, *83*, 2574.
37. Wagner, A.; Heitz, M.-P.; Mioskowski, C. *TL* **1989**, 557.
38. Schmidt, S. P.; Brooks, D. W. *TL* **1987**, 767.
39. Sonnet, P. E. *SC* **1976**, *6*, 21.
40. Sato, R.; Otera, J.; Nozaki, H. *JOC* **1990**, *55*, 4770.
41. Paul, R.; Tchelitcheff, S. *BCF* **1952**, 808.
42. Pattison, F. L. M.; Dear, R. E. A. *CJC* **1963**, *41*, 2600.
43. (a) Woods, Jr., G. F. *OSC* **1955**, *3*, 470. (b) Paul, R. *BCF* **1934**, *1*, 976.

44. Eliel, E. L.; Nowak, B. E.; Daignault, R. A.; Badding, V. G. *JOC* **1965**, *30*, 2441.
45. Eliel, E. L.; Nowak, B. E.; Daignault, R. A. *JOC* **1965**, *30*, 2448.
46. Eliel, E. L.; Daignault, R. A. *JOC* **1965**, *30*, 2450.
47. Paul, R.; Riobé, O.; Maumy, M. *OSC* **1988**, *6*, 675.
48. Boeckman Jr., R. K.; Bruza, K. J. *TL* **1977**, 4187.
49. Mukaiyama, T.; Wariishi, K.; Saito, Y.; Hayashi, M.; Kobayashi, S. *CL* **1988**, 1101.
50. Corey, E. J.; Wollenberg, R. H. *JOC* **1975**, *40*, 2265.
51. Earl, R. A.; Townsend, L. B. *OS* **1981**, *60*, 81.
52. (a) Kluge, A. F. *OS* **1984**, *64*, 80. (b) Kluge, A. F.; Clousdale, I. S. *JOC* **1979**, *44*, 4847.
53. Bogavac, M.; Arsenijevic, L.; Arsenijevic, V. *BCF* **1980**, 145.
54. (a) Cowie, J. S.; Landor, P. D.; Landor, S. R. *CC* **1969**, 541. (b) Cowie, J. S.; Landor, P. D.; Landor, S. R. *JCS(P1)* **1973**, 720.

Paul Ch. Kierkus
BASF Corporation, Wyandotte, MI, USA

3,4-Dihydro-2*H*-pyrido[1,2-*a*]pyrimidin-2-one

[5439-14-5] $C_8H_8N_2O$ (MW 148.16)

(synthesis of esters and amides from carboxylic acids; glycosylation)

Physical Data: mp 155–156 °C; mp (HCl salt) 284–285.5 °C (dec); mp (HBr salt) 299–300 °C (dec); mp (HI salt) 263–264 °C (dec).

Solubility: sol MeOH, EtOH, H_2O.

Form Supplied in: both the free base and salts are obtained as colorless solids.

Analysis of Reagent Purity: the reagent has a betaine-like structure and behaves as a salt in terms of the reagent's solubility. The literature reports various melting points for both the anhydrous and hydrated forms of the reagent but the conditions used to determine melting point have been shown to be critical and care must be exercised if this property is to be used as a measure of reagent purity.[1]

Preparative Methods: is prepared by reaction of 2-aminopyridine with either ***Acrylic acid*** itself or an equivalent (2-bromo- or 2-iodopropionic acid, ***Methyl Acrylate***, ethyl 2-bromopropionate, ***β-Propiolactone***).[1,2]

Purification: the free base and salts are readily recrystallized from MeOH–Et_2O.[1]

Handling, Storage, and Precautions: the reagent may be dried over P_2O_5 in vacuo (135 °C/17 mmHg).

Introduction. The principal synthetic application of 3,4-dihydro-2*H*-pyrido[1,2-*a*]pyrimidin-2-one (**1**) is an acid scavenger; it is an example of a 'proton sponge' (see ***1,8-Bis(dimethylamino)naphthalene***). The reagent has been used in the esterification and amination of alkyl and aryl carboxylic acids and also in glycosylation of 2-amino-2-deoxy sugars.

Synthesis of Esters from Carboxylic Acids.[3] Esterification (eq 1) proceeds by activation of the carboxyl component using a 2-halopyridinium salt (**2**; X = Cl, Y = I or X = F, Y = OTs) and (**1**) (2.4 equiv) in the presence of an alcohol. The best solvents for this process are CH_2Cl_2 or MeCN, but a range of other solvents may also be used.

$$R^1\text{-}CO_2H \xrightarrow[\text{R}^2\text{OH, CH}_2\text{Cl}_2\text{ or MeCN, 60–97\%}]{\textbf{(2)}, \textbf{(1)}} R^1C(O)OR^2 \quad (1)$$

Synthesis of Amides from Carboxylic Acids.[4] In the corresponding amidation process (eq 2) there is a requirement for the addition of 1 equiv of a tertiary amine (***Tri-n-butylamine***) to ensure efficient utilization of the amine component (R^2R^3NH). Once again, carboxyl activation is achieved using a 2-halopyridinium salt (**2**) but, unlike the esterification reaction, amidation is best carried out as a two-step one-pot process.

$$R^1\text{-}CO_2H \xrightarrow[\text{CH}_2\text{Cl}_2\text{ or MeCN}]{\textbf{(2)}, \text{Bu}_3\text{N}} [\text{N-methyl-2-(OCOR}^1\text{)pyridinium Y}^-] \xrightarrow[\text{HNR}^2\text{R}^3\text{, 70–100\%}]{\textbf{(1)}} R^1C(O)NR^2R^3 \quad (2)$$

Yields for both the esterification (eq 1) and amidation (eq 2) reactions are generally high, but the advantages associated with use of (**1**) over other reagents for these types of transformation have yet to be clearly defined.

Glycosylation.[5] 3, 4-Dihydro-2*H*-pyrido[1, 2-*a*]pyrimidin-2-one has been used in conjunction with ***Silver(I) Trifluoromethanesulfonate*** as a reagent combination to mediate the glycosylation of *N*-protected 2-amino-2-deoxy sugars using a glycosyl bromide as the glycosyl donor (eq 3).

$$\text{glycosyl bromide (OAc)} + \text{HO-acceptor (OBn, BnO, NPhth, OCH}_2\text{CCl}_3) \xrightarrow[\text{CHCl}_3\text{, 61\%}]{\textbf{(1)}, \text{AgOSO}_2\text{CF}_3} \text{disaccharide} \quad (3)$$

---

1. Hurd, C. D.; Hayao, S. *JACS* **1955**, *77*, 117.
2. (a) Magidson, O. Y.; Elina, A. S. *JGU* **1946**, *16*, 1933 (*CA* **1947**, *41*, 6219). (b) Adams, R.; Pachter, I. J. *JACS* **1952**, *74*, 4906, 5491. (c) Krishnan, M.; *PIA(A)* **1955**, *42*, 289. (d) Baltrusis, R.; Maciulis, A.;

Purenas, A. *Lietuvos TSR Moksln Akad. Darbai. Ser. B* **1962**, 125 (*CA* **1963**, *58*, 6827a). (e) Biniecki, S.; Modrzejewska, W. *Acta Pol. Pharm.* **1984**, *41*, 607 (*CA* **1986**, *104*, 50 843p).
3. Mukaiyama, T.; Toda, H.; Kobayashi, S. *CL* **1976**, 13.
4. Mukaiyama, T.; Aikawa, Y.; Kobayashi, S. *CL* **1976**, 57.
5. Ogawa, T.; Nakabayashi, S.; Sasajima, K. *CR*, **1981**, *96*, 29.

Timothy Gallagher
University of Bristol, UK

3,4-Dimethoxybenzyl Bromide

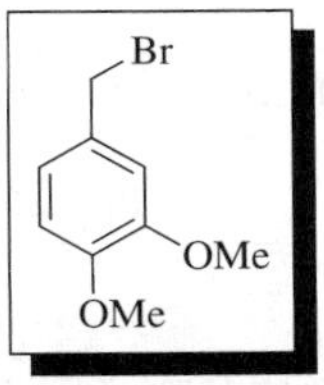

[21852-32-4] $C_9H_{11}BrO_2$ (MW 231.09)

(reagent for introduction of 3,4-dimethoxybenzyl protecting group, which can be cleaved selectively under conditions of benzylic oxidation[1])

Alternate Name: DMBBr.
Physical Data: mp 52–53 °C.
Solubility: reactions are usually carried out in polar solvents such as THF or DMF.
Preparative Methods: synthesized from commercially available 3,4-dimethoxybenzyl alcohol and PBr_3[2] or HBr.[3]
Purification: recrystallization from absolute ethanol.[2]
Handling, Storage, and Precautions: the reagent is quite unstable and is best prepared fresh or stored at low temperature under nitrogen. It is a lachrymator.

3,4-Dimethoxybenzyl Ethers (DMB Ethers). The benzyl group is a frequently used protecting group[4] in organic synthesis due to its acid and base stability, and its facile removal by catalytic hydrogenation or with sodium in liquid ammonia. The group is most commonly used to protect alcohol functions; however, it may not be generally applied to substrates which have other functional groups susceptible to the reductive conditions. To overcome this problem and generally to extend the range of protecting groups available, methoxy-substituted benzyl ethers have been prepared (e.g. the 3,4-dimethoxy[1] and 4-methoxybenzyl[5] (PMB) ethers). DMB ethers are made using the benzyl bromide (or chloride), alcohol, and base,[1] or with base sensitive substrates using the trichloroacetimidate[6] (see ***Benzyl 2,2,2-Trichloroacetimidate***). PMB ethers can also made in similar fashion under either basic[5] or acidic[7] conditions. DMB ethers can be removed oxidatively using ***2,3-Dichloro-5,6-dicyano-1,4-benzoquinone*** (DDQ)[1,5] or trityl tetrafluoroborate,[5b,c,6,7a] and PMB ethers have also been cleaved with a variety of other methods.[4,8] The products obtained are the methoxy substituted benzaldehyde and the parent alcohol (eq 1), although with an allylic alcohol there is a risk of over oxidation to give the enone (eq 2).[9] Methoxybenzyl ethers with free α- or β-hydroxy groups can be converted to acid cleavable acetals with DDQ oxidation under anhydrous conditions, and further oxidized to base cleavable benzoates.[10] Simple benzyl ethers and many other protecting and functional groups remain unaffected under these conditions.[1,5]

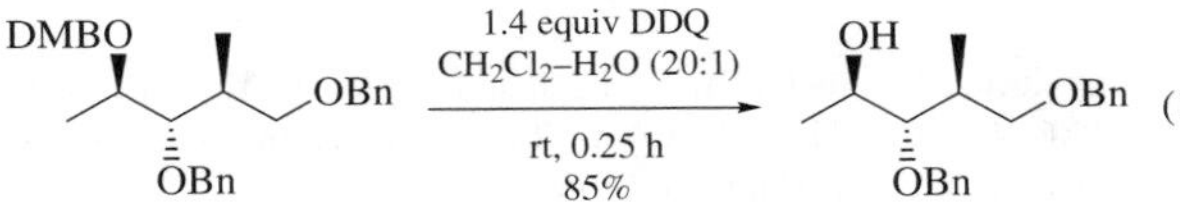

DMBO ... OBn, OBn → (1.4 equiv DDQ, CH_2Cl_2–H_2O (20:1), rt, 0.25 h, 85%) → OH ... OBn, OBn (1)

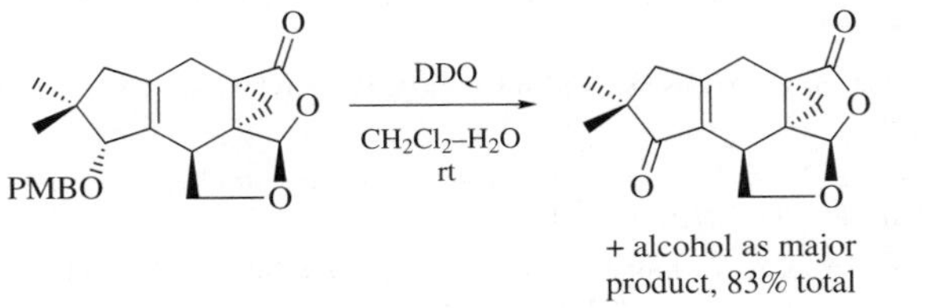

PMBO ... → (DDQ, CH_2Cl_2–H_2O, rt) → enone + alcohol as major product, 83% total (2)

Each substituted benzyl ether has a different oxidation potential (1.45 V for the DMB–OR, and 1.78 V for the PMB–OR), so each methoxy substitution pattern results in different rates of ether cleavage.[1,11] This means that the DMB group can be removed in the presence of a PMB ether with high selectivity (eq 3).[1] It has also been shown that other methoxybenzyl ethers react at widely differing rates.[11] For example, the reaction time for the DMB ether was <20 min (86% yield of alcohol), and the 2,6-dimethoxy isomer required 27.5 h (80% yield) (eq 4).[11] Other substitution patterns give reaction times between these two extremes.

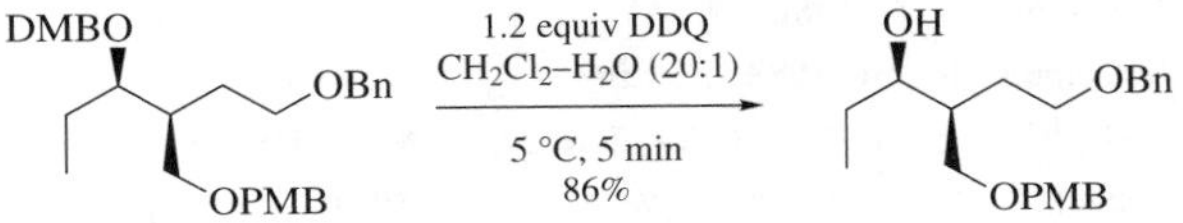

DMBO ... OBn, OPMB → (1.2 equiv DDQ, CH_2Cl_2–H_2O (20:1), 5 °C, 5 min, 86%) → OH ... OBn, OPMB (3)

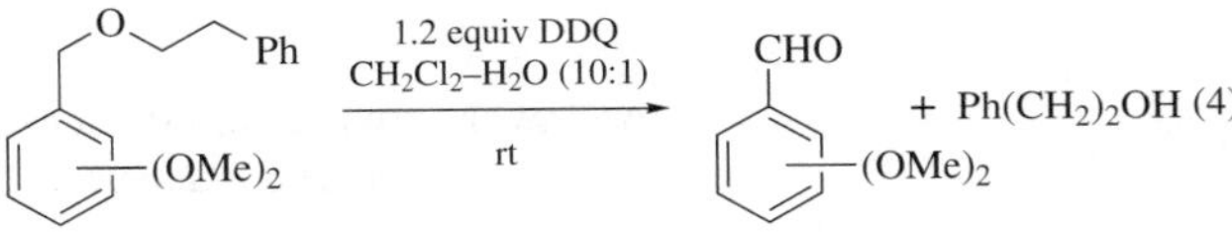

$(OMe)_2C_6H_3CH_2O(CH_2)_2Ph$ → (1.2 equiv DDQ, CH_2Cl_2–H_2O (10:1), rt) → $(OMe)_2C_6H_3CHO$ + $Ph(CH_2)_2OH$ (4)

PMB and DMB ethers are less readily cleaved than simple benzyl ethers using catalytic hydrogenation.[12] W-4 ***Raney Nickel*** gave the best selectivities (eq 5), but deprotection using ***Sodium–Ammonia*** showed no discrimination at all.[12]

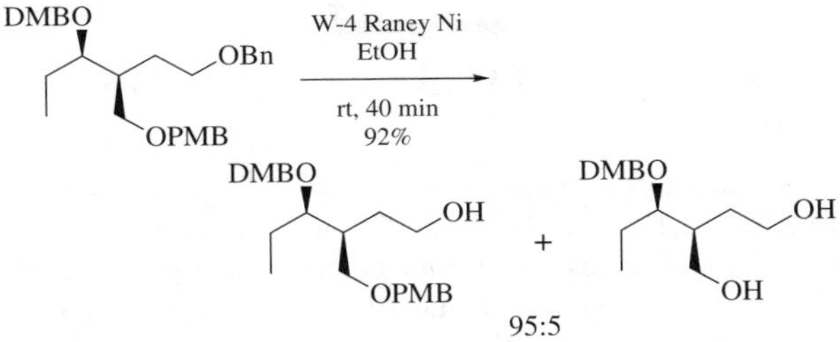

DMBO ... OBn, OPMB → (W-4 Raney Ni, EtOH, rt, 40 min, 92%) → DMBO ... OH, OPMB + DMBO ... OH, OH; 95:5 (5)

The versatility of a combination of the substituted and unsubstituted benzyl ethers as protecting groups has been shown in the multi-step syntheses of 16-membered macrolides[13] and other natural products.[14,15] The DMB group alone has also been used as an alcohol protecting group in molecules ranging from the relatively simple[16] to complex oligoribonucleotides,[6,7a] and occasionally for protection of nitrogen-containing groups.[8,17]

Related Reagents. Benzyl Chloride; Benzyl 2,2,2-Trichloroacetimidate; Benzyl Trifluoromethanesulfonate; *p*-methoxy benzyl bromide; *p*-methoxy 2,2,2-trichloroacetimidate.

---

1. (a) Horita, K.; Yoshioka, T.; Tanaka, T.; Oikawa, Y.; Yonemitsu, O. *T* **1986**, *42*, 3021. (b) Oikawa, Y.; Tanaka, T.; Horita, K.; Yoshioka, T.; Yonemitsu, O. *TL* **1984**, *25*, 5393.
2. Lakhlifi, T.; Sedqui, A.; Laude, B.; Dinh An, N.; Vebrel, J. *CJC* **1991**, *69*, 1156.
3. Coote, S. J.; Davies, S. G.; Middlemiss, D.; Naylor, A. *JCS(P1)* **1989**, 2223.
4. Greene, T. W.; Wuts, P. G. M. *Protective Groups in Organic Synthesis*, 2nd ed.; Wiley: New York, 1991.
5. (a) Oikawa, Y.; Yoshioka, T.; Yonemitsu, O. *TL* **1982**, *23*, 885. (b) Takaku, H.; Kamaike, K. *CL* **1982**, 189. (c) Takaku, H.; Kamaike, K.; Tsuchiya, H. *JOC* **1984**, *49*, 51.
6. Takaku, H.; Ito, T.; Imai, K. *CL* **1986**, 1005.
7. (a) Takaku, H.; Ueda, S.; Ito, T. *TL* **1983**, *24*, 5363. (b) Nakajima, N.; Horita, K.; Abe, R.; Yonemitsu, O. *TL* **1988**, *29*, 4139.
8. Begtrup, M. *BSB* **1988**, *97*, 573.
9. Trost, B. M.; Chung, J. Y. L. *JACS* **1985**, *107*, 4586.
10. (a) Oikawa, Y.; Yoshioka, T.; Yonemitsu, O. *TL* **1982**, *23*, 889. (b) Nozaki, K.; Shirahama, H. *CL* **1988**, 1847.
11. Nakajima, N.; Abe, R.; Yonemitsu, O. *CPB* **1988**, *36*, 4244.
12. Oikawa, Y.; Tanaka, T.; Horita, K.; Yonemitsu, O. *TL* **1984**, *25*, 5397.
13. (a) Nakajima, N.; Hamada, T.; Tanaka, T.; Oikawa, Y.; Yonemitsu, O. *JACS* **1986**, *108*, 4645. (b) Tanaka, T.; Oikawa, Y.; Hamada, T.; Yonemitsu, O. *TL* **1986**, *27*, 3651.
14. Masamune, S. *PAC* **1988**, *60*, 1587.
15. Yadagiri, P.; Shin, D.-S.; Falck, J. R. *TL* **1988**, *29*, 5497.
16. Lebeau, L.; Oudet, P.; Mioskowski, C. *HCA* **1991**, *74*, 1697.
17. Grunder-Klotz, E.; Ehrhardt, J.-D. *TL* **1991**, *32*, 751.

Andrew N. Boa & Paul R. Jenkins
University of Leicester, UK

2,2-Dimethoxypropane

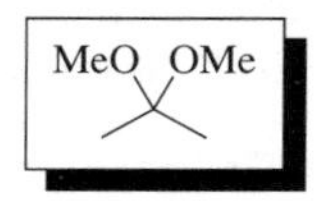

[77-76-9] $C_5H_{12}O_2$ (MW 104.15)

(reagent used in the preparation of acetals,[1,2] acetonides,[3] isopropylidene derivatives of sugars[4–6] and nucleosides,[7] methyl esters of amino acids,[8] and enol ethers;[9] undergoes the aldol condensation with enol silyl ethers;[10] used in the preparation of hypophosphite esters[11] and dimethylboronate esters[12])

Physical Data: bp 83 °C; *d* 0.847 g cm^{-3}.
Form Supplied in: liquid; widely available.
Handling, Storage, and Precautions: flammable liquid; irritant.

Formation of Acetals. The reagent can be used to prepare a variety of acetals via transacetalization. This involves interchanging the alkoxy or ketone groups in an acidic medium. Both symmetrical $[R^1R^2C(OR^3)_2]$ and mixed $[R^1R^2C(OR^3)OR^4]$ acetals can be obtained, depending upon the choice of conditions.[1]

The three equilibria shown in eqs 1–3 are established when 2,2-dimethoxypropane, an alcohol, and a catalytic amount of acid are mixed. Distillation of the methanol formed shifts the equilibrium far in the direction of the new acetals. Methanol and 2,2-dimethoxypropane form a binary azeotrope; however, this can be broken by the addition of a hydrocarbon solvent such as hexane or benzene.

$$Me_2C(OMe)_2 + ROH \rightleftharpoons Me_2C(OMe)(OR) + MeOH \quad (1)$$

$$Me_2C(OMe)(OR) + ROH \rightleftharpoons Me_2C(OR)_2 + MeOH \quad (2)$$

$$2\ Me_2C(OMe)(OR) \rightleftharpoons Me_2C(OR)_2 + Me_2C(OMe)_2 \quad (3)$$

Methyl acetals of other ketones can be obtained by acidifying a mixture of the ketone, methanol, and 2,2-dimethoxypropane and removing the acetone formed by distillation. In this case, the function of the 2,2-dimethoxypropane is to react with the water formed to give methanol and acetone (eqs 4 and 5). This is evidenced by the fact that the reaction is very slow in the absence of alcohol or water, but fast when methanol is present. The rate of the reaction is also directly related to the alcohol concentration.[1]

$$R^1R^2CO + 2\ MeOH \rightleftharpoons R^1R^2C(OMe)_2 + H_2O \quad (4)$$

$$H_2O + Me_2C(OMe)_2 \rightleftharpoons 2\ MeOH + Me_2CO \quad (5)$$

One undesirable side reaction is the formation of enol ethers via the acid-catalyzed loss of one molecule of alcohol from the acetal. This occurs even at relatively low temperatures (50 °C). In order to suppress this reaction, the procedure was modified to include a neutralization with base prior to workup.

A recent example of the use of 2,2-dimethoxypropane in the above capacity involves the reaction of an aldehyde or ketone with ***o*-Nitrobenzyl Alcohol** to afford the bis-*o*-nitrobenzyl acetal derivative.[2] This protecting group has the advantage that it is photoremovable in high yield by irradiation at 350 nm.

Acetonide Formation. The reagent has been used to great advantage as a protecting group for the labile dihydroxyacetone side chain of the corticosteroids. Cyclic acetal formation between the 17α,21-diol grouping of the prednisolone side chain and 2,2-dimethoxypropane in an acid-catalyzed exchange reaction gives 17α,21-isopropylidenedioxypregnane in good yield (eq 6).[3]

$$\text{prednisolone} \xrightarrow[\text{DMF, TsOH}]{Me_2C(OMe)_2} \text{17α,21-isopropylidenedioxy derivative} \quad (6)$$

The base stability of the isopropylidenedioxy function makes it useful for the preparation of C-21 modified cortical hormones whose formation requires strongly basic conditions. The acetonide is easily cleaved with aqueous acetic or formic acid in the presence of heat.

Isopropylidene Derivatives of Sugars and Nucleosides. A mixture of 2,2-dimethoxypropane, DMF, and a catalytic amount of ***p*-*Toluenesulfonic Acid*** constitutes an efficient acetonating agent capable of protecting vicinal, *trans*-diequatorial hydroxy groups and of forming *N*-acetyl-2,2-dimethyloxazolidines from vicinal hydroxy and acetamido groups.[4] Reaction of methyl α-D-glucopyranoside under the above conditions gives recovered starting material, methyl 4,6-*O*-isopropylidene-α-D-glucoside, and a trace amount of the diisopropylidene glucoside (eq 7). Similarly, deoxyinosamine, upon treatment with the same reagent mixture, affords as the major product the diisopropylidene derivative (eq 8).

$Me_2C(OMe)_2$, DMF, TsOH (7)

$Me_2C(OMe)_2$, DMF, TsOH (8)

A further improvement to the existing procedure involves the substitution of ***Tin(II) Chloride*** for the *p*-toluenesulfonic acid and 1,2-dimethoxyethane for DMF.[5a] Studies involving D-mannitol have shown that the tin(II) chloride does not act as a Lewis acid catalyst or as a source of HCl (by reaction with the hydroxy groups) since the use of zinc chloride under identical conditions gives no reaction. In this particular case, it was subsequently demonstrated that no catalyst is required and that isopropylidenation occurs under completely neutral conditions, presumably via acetal exchange.[5b] Tin(II) chloride has been used to catalyze isopropylidene formation in other carbohydrate systems.[6]

The ability of 2,2-dimethoxypropane to act as a water scavenger has resulted in its use (in conjunction with acetone and an acid catalyst) in the protection of the 2′,3′-hydroxy groups of purine and pyrimidine nucleosides.[7] Here again, the formation of acetone and methanol serves to drive the equilibrium towards the product.

Esterification of Amino Acids. Methyl ester hydrochlorides of amino acids have been prepared in which 2,2-dimethoxypropane has been used as the source of the methoxy group, the reaction solvent, and the reagent for the removal of the water formed by virtue of the hydrolysis of the acetal to methanol and acetone.[8a] This procedure has also been used to prepare the methyl esters of fatty acids.[8b]

Enol Ethers. Treatment of testosterone acetate with 2,2-dimethoxypropane/DMF/TsOH converts this material into the enol ether (eq 9).[9] In this case, enol ether formation is attributed to loss of water directly from the unsaturated hemiacetal.

H^+; $Me_2C(OMe)_2$ (9)

$+ MeOH + COMe_2 + H^+$

Aldol Condensation with Enol Silyl Ethers. The first example of an aldol condensation between an unactivated enolate and an electrophilically activated carbonyl substrate was accomplished using enol silyl ethers and acetals or orthoesters in conjunction with a catalytic amount of ***Trimethylsilyl Trifluoromethanesulfonate***. The reaction proceeds very readily at −78 °C and affords the *erythro*-aldol product with high stereoselectivity.[10] The use of 2,2-dimethoxypropane in this reaction affords β-methoxy ketones (eq 10).

Me_3SiOTf, −78 °C, 96% (10)

Hypophosphite Esters. Methyl hypophosphite has been prepared by the reaction between ***Hypophosphorous Acid*** and 2,2-dimethoxypropane (eq 11).[11] However, the product reacts further with the acetone formed to give 1-hydroxy-1,1-dimethylmethylphosphinate (eq 12).

$+ MeOH + Me_2CO$ (11)

$+ Me_2CO$ (12)

Dimethyl Boronate Esters. Dimethyl esters of arylboronic acids have been prepared by heating a mixture of the acid, 2,2-dimethoxypropane, and ***Zinc Chloride*** (catalyst).[12]

Related Reagents. 2-Methoxypropene; Trimethyl orthoformate.

1. (a) Lorette, N. B.; Howard, W. L. *JOC* **1960**, *25*, 521, 525. (b) Brown, B. R.; MacBride, J. A. H. *JCS* **1964**, 3822.
2. Gravel, D; Murray, S.; Ladouceur, G. *CC* **1985**, 1828.
3. (a) Tanabe, M.; Bigley, B. *JACS* **1961**, *83*, 756. (b) Robinson, C. H.; Finckenor, L. E.; Tiberi, R.; Olivetto, E. P. *JOC* **1961**, *26*, 2863.
4. (a) de Belder, A. N. *Adv. Carbohydr. Chem. Biochem* **1965**, *20*, 219. (b) de Belder, A. N., *Adv. Carbohydr. Chem. Biochem.* **1977**, *34*, 179. (c) Evans, M. E.; Parrish, F. W.; Long, L. *Carbohydr. Res* **1967**, *3*, 453. (d) Hasegawa, A.; Fletcher, H. G. *Carbohydr. Res.* **1973**, *29*, 209, 223. (e) Hasegawa, A.; Nakajima, M. *Carbohydr. Res.* **1973**, *29*, 239.
5. (a) Chittenden, G. *Carbohydr. Res.* **1980**, *84*, 350. (b) Chittenden, G. *Carbohydr. Res.* **1980**, *87*, 219.

6. (a) Vekemans, J. A. J. M.; deBruyn, R. G. M.; Caris, R. C. H. M.; Kokx, A. J. P. M.; Konings, J. J. H. G.; Godefroi, E. F.; Chittenden, G. *JOC* **1987**, *52*, 1093. (b) Chittenden, G. *RTC* **1988**, *107*, 455.
7. (a) Hampton, A. *JACS* **1961**, *83*, 3640. (b) Moffatt, J. G. *JACS* **1963**, *85*, 1118. (c) Hampton, A. *JACS* **1965**, *87*, 4654.
8. (a) Rachele, J. R. *JOC* **1963**, *28*, 2898. (b) Radin, N. S.; Hajra, A. K.; Akahori, Y. *J. Lipid Res* **1960**, *1*, 250.
9. Nussbaum, A. L.; Yuan, E.; Dincer, D.; Olivetto, E. P. *JOC* **1961**, *26*, 3925.
10. Murata, S.; Suzuki, M.; Noyori, R. *JACS* **1980**, *102*, 3248.
11. Fitch, S. J. *JACS* **1964**, *86*, 61.
12. Matteson, D. S.; Kramer, E. *JACS* **1968**, *90*, 7261.

Joel Slade
Ciba-Geigy Corporation, Summit, NJ, USA

Dimethylaluminum Chloride[1]

Me_2AlCl

[1184-58-3] C_2H_6AlCl (MW 92.51)

(strong Lewis acid that can also act as a proton scavenger; reacts with HX to give methane and MeAlClX)

Alternate Name: chlorodimethylaluminum.
Physical Data: mp −21 °C; bp 126–127 °C; *d* 0.996 g cm^{-3}.
Solubility: sol most orgatlanic solvents; stable in alkanes or arenes.
Form Supplied in: commercially available neat or as solutions in hexane or toluene.
Analysis of Reagent Purity: solutions are reasonably stable but may be titrated before use by one of the standard methods.[1e]
Handling, Storage, and Precautions: must be transferred under inert gas (Ar or N_2) to exclude oxygen and water. Use in a fume hood.

Introduction. The general properties of alkylaluminum halides as Lewis acids are discussed in the entry for ***Ethylaluminum Dichloride***. Dialkylaluminum halides are less acidic than alkylaluminum dihalides. Me_2AlCl is more expensive than ***Diethylaluminum Chloride***, but the methyl group of Me_2AlCl is much less nucleophilic than the ethyl group of Et_2AlCl. Much higher yields will generally be obtained by use of Me_2AlCl in the ene reactions of carbonyl compounds. In other cases, such as the Diels–Alder reactions of α,β-unsaturated esters, comparable yields will be obtained with either Lewis acid.

Catalysis of Diels–Alder Reactions. Me_2AlCl has been used as a Lewis acid catalyst for inter- and intramolecular Diels–Alder reactions with a wide variety of dienophiles. High diastereoselectivity is obtained from chiral α,β-unsaturated *N*-acyloxazolidinones with more than 1 equiv of Me_2AlCl.[2] Use of Me_2AlCl as a catalyst affords high yields in inter- and intramolecular Diels–Alder reactions of α,β-unsaturated ketones (eq 1),[3,4] and intramolecular Diels–Alder reactions of α,β-unsaturated aldehydes.[4,5] Me_2AlCl-catalyzed Diels–Alder reactions of α,β-unsaturated *N*-acylsultams (eq 2)[6] and 1-mesityl-2,2,2-trifluoroethyl acrylate (eq 3)[7] proceed in high yield with excellent asymmetric induction. Methylaluminum sesquichloride, prepared from Me_2AlCl and ***Methylaluminum Dichloride***, catalyzes an intramolecular Diels–Alder reaction with an aldehyde as the dienophile to afford a dihydropyran.[8]

Me_2AlCl, 69% (1)

Me_2AlCl, −20 °C, 63%; TBDMSO; 100% de (2)

Me_2AlCl, −78 °C, 99%; CF_3; 100% de (3)

Catalysis of Ene Reactions.[9] A wide variety of Lewis acids will catalyze the ene reactions of formaldehyde with electronrich alkenes. Electron-deficient aldehydes, such as chloral and glyoxylate esters, also undergo ene reactions with a variety of Lewis acid catalysts. Ene reactions of aliphatic and aromatic aldehydes with alkenes that can form a tertiary cation, and of formaldehyde with mono- and 1,2-disubstituted alkenes, are best carried out with 1 or more equiv of Me_2AlCl. The alcohol–Me_2AlCl complex produced in the ene reaction decomposes rapidly to give methane and a nonbasic aluminum alkoxide that does not react further (eq 4). This prevents solvolysis of the alcohol–Lewis acid complex or protonation of double bonds. Good to excellent yields of ene adducts are obtained from aliphatic and aromatic aldehydes and 1,1-di- and trisubstituted alkenes. ***Formaldehyde*** is more versatile and gives good yields of ene adducts with all classes of alkenes.[10] When less than 1 equiv of Me_2AlCl is used, γ-chloro alcohols are formed, resulting from the stereospecifically *cis* addition of the hydroxymethyl and chloride groups to the double bond (eq 5). The chloro alcohols are converted to ene-type adducts in the presence of excess Me_2AlCl. Formaldehyde undergoes Me_2AlCl-induced reactions with terminal alkynes to give a 2:3 mixture of the ene adduct allenic alcohol and the (*Z*)-3-chloro allylic alcohol in 50–75% yield (eq 6).[11]

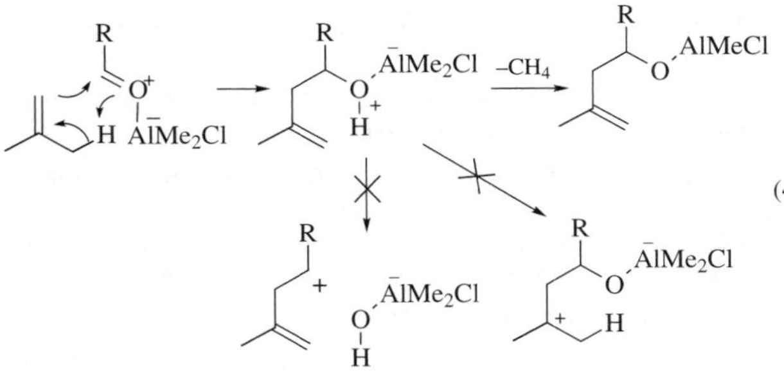

(4)

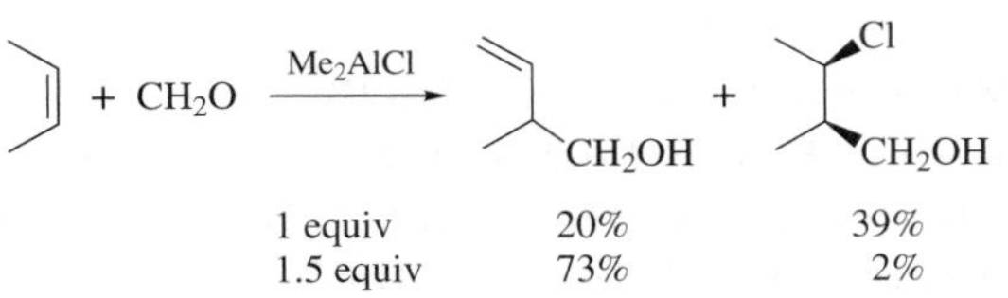

(5)

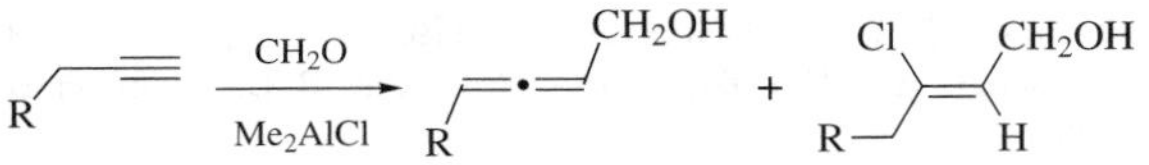

(6)

Me_2AlCl catalyzes the ene reactions of a variety of aldehydes with (Z)-3β-acetoxy-5,17(20)-pregnadiene at −78 °C.[12] The stereoselectivity with aliphatic aldehydes is >10:1 in favor of the 22α-isomer, while aromatic aldehydes produce predominantly the 22β-isomer (eq 7). Me_2AlCl is also the Lewis acid of choice for ene reactions of α-halo aldehydes (eq 8).[13] Ene reactions of vinyl sulfides to produce enol silyl ethers are also catalyzed by Me_2AlCl (eq 9).[14]

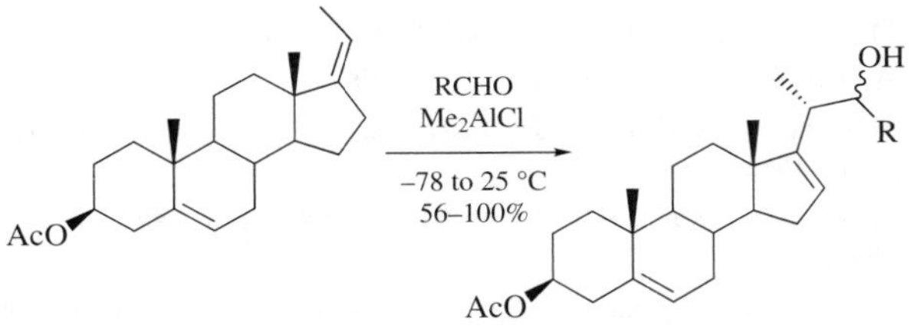

(7)

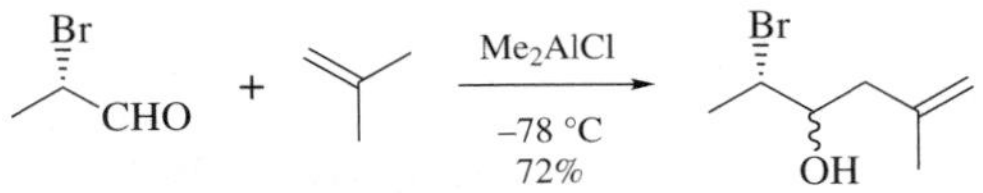

(8)

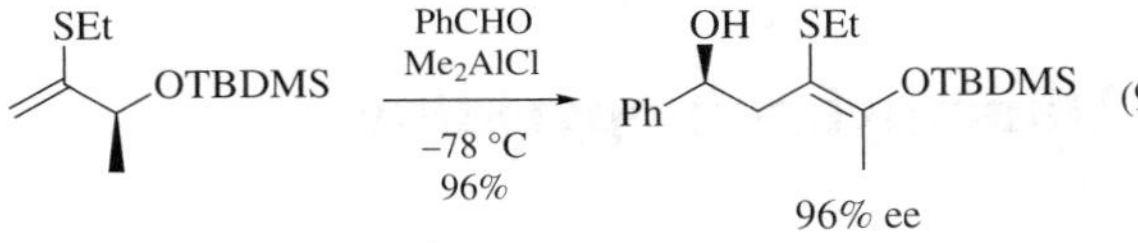

(9)

Type-I intramolecular ene reactions of aldehydes, such as citronellal, that contain electron-rich trisubstituted double bonds proceed readily thermally or with a variety of Lewis acids. Intramolecular ene reactions with less nucleophilic 1,2-disubstituted double bonds proceed efficiently with Me_2AlCl as the Lewis acid catalyst (eqs 10 and 11).[15,16]

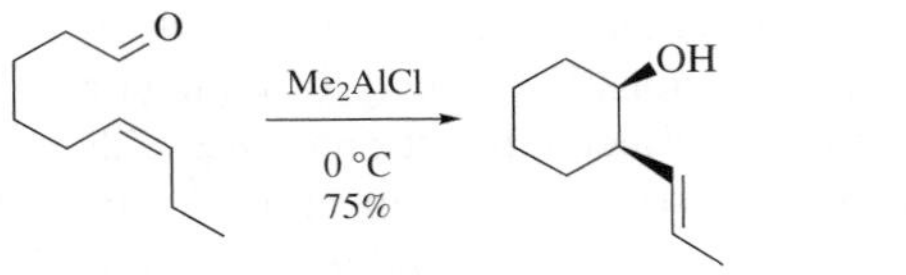

(10)

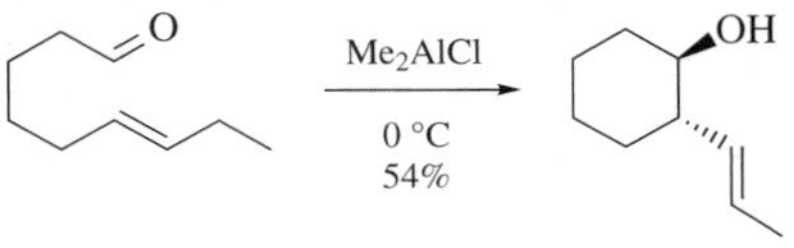

(11)

Type-II intramolecular ene reactions of aldehydes and ketones proceed readily with Me_2AlCl as the Lewis acid.[17–19] Unsaturated aldehydes and ketones can be generated in situ by Me_2AlCl-catalyzed reaction of ***Acrolein*** and ***Methyl Vinyl Ketone*** with alkylidenecycloalkanes at low temperatures (eq 12).[17] The monocyclic aldehyde reacts further under these conditions. The monocyclic ketone can be isolated at low temperature but undergoes a second ene reaction at rt to give the bicyclic alcohol. β-Keto esters form tertiary alcohols in intramolecular ene reactions. The products are stable because they are converted to the aluminum alkoxide (eq 13).[18] Intramolecular Me_2AlCl-catalyzed ene reactions have been used for the preparation of the bicyclic mevinolin ring system (eq 14).[19]

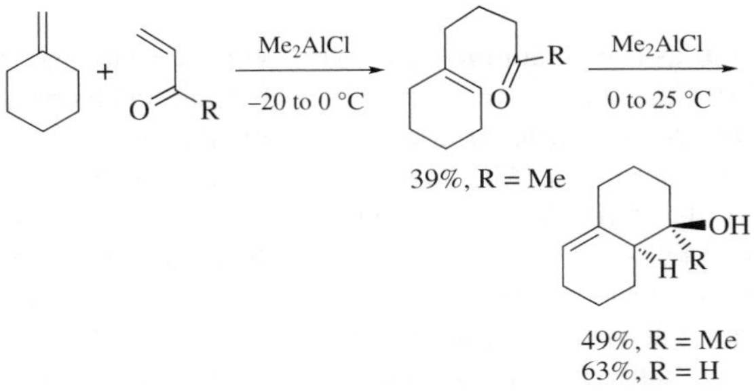

(12)

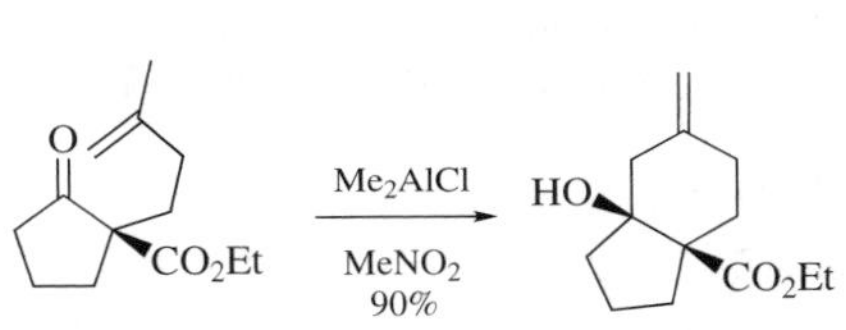

(13)

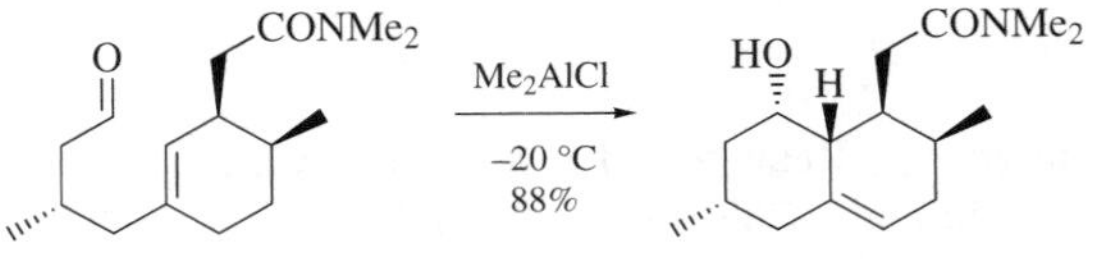

(14)

Generation of Electrophilic Cations. Me_2AlCl in dichloromethane cleaves THP ethers without deprotecting *t*-butyldimethylsilyl ethers.[20] Azido enol silyl ethers undergo Me_2AlCl-catalyzed reactions with ***Allyltributylstannane*** and enol ethers, giving conjugate addition-type products that are isolated as the silyl enol ethers (eq 15).[21] Me_2AlCl will open norbornane epoxide to the rearranged chlorohydrin.[22]

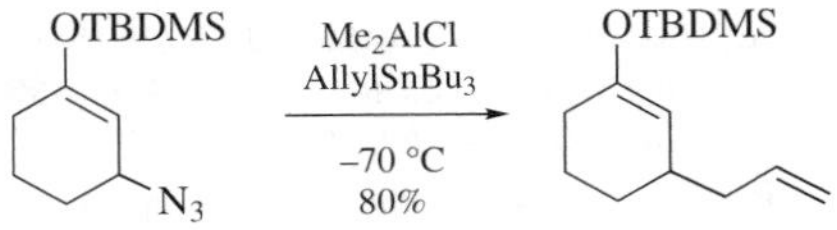

(15)

Formation and Reaction of Aluminum Enolates. Aluminum enolates prepared from esters react with imines to give a β-lactam resulting from aldol-type addition followed by ring closure (eq 16).[23]

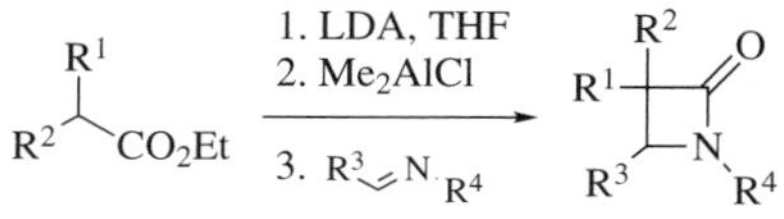

(16)

Formation and Reaction of Alkynylaluminum Reagents. Lithium acetylides react with Me_2AlCl to give dimethylaluminum acetylides that react analogously to the more commonly used diethylaluminum acetylides (see ***Diethylaluminum Ethoxyacetylide***). Addition of the aluminum acetylide to propiolactone results in an S_N2 reaction to give an alkynic acid (eq 17).[24]

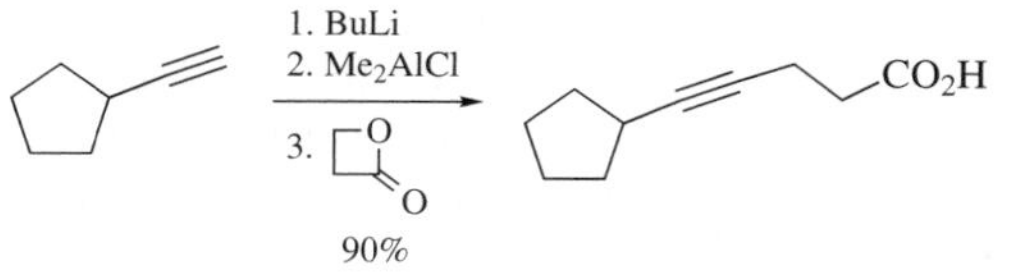

(17)

Reaction as a Nucleophile. Me_2AlCl will react analogously to MeMgBr and transfer a methyl group to many nucleophiles. Since ***Methylmagnesium Bromide*** and ***Methyllithium*** are readily available, use of Me_2AlCl to deliver a methyl group is needed only when the stereochemistry of addition is an important issue. High levels of asymmetric induction are obtained in the conjugate addition of Me_2AlCl to unsaturated acyloxazolidinones with carbohydrate-derived chiral auxiliaries (eq 18).[25] Me_2AlCl differs from higher dialkylaluminum chlorides in that methyl addition is a radical process that requires photochemical or radical initiation. Me_2AlCl will convert acid chlorides to methyl ketones.[26]

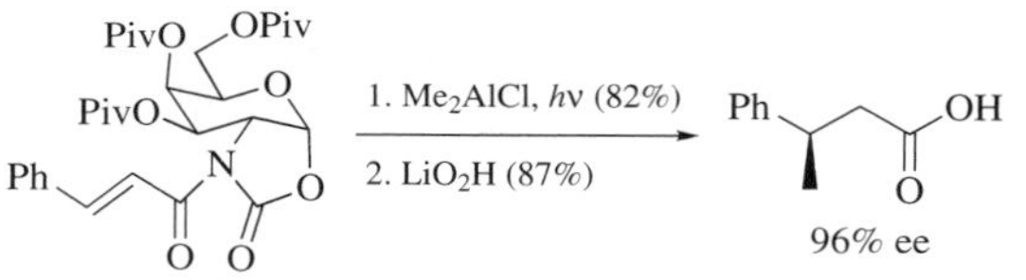

(18)

Related Reagents. 1,8-Bis(dimethylamino)naphthalene; Diethylaluminum Chloride; Dimethylaluminum Iodide; Ethylaluminum Dichloride; Methylaluminum Bis(2,6-di-*t*-butyl-4-methylphenoxide); Methylaluminum Bis(4-bromo-2,6-di-*t*-butylphenoxide); Methylaluminum Bis(2,6-di-*t*-butylphenoxide); Methylaluminum Dichloride; Trimethylsilyl Trifluoromethanesulfonate.

1. For reviews, see Ref. 1 under ***Ethylaluminum Dichloride***.
2. (a) Evans, D. A.; Chapman, K. T.; Bisaha, J. *TL* **1984**, *25*, 4071. (b) Evans, D. A.; Chapman, K. T.; Bisaha, J. *JACS* **1984**, *106*, 4261. (c) Evans, D. A.; Chapman, K. T.; Bisaha, J. *JACS* **1988**, *110*, 1238. (d) Sugahara, T.; Iwata, T.; Yamaoka, M.; Takano, S. *TL* **1989**, *30*, 1821. (e) Hauser, F. M.; Tommasi, R. A. *JOC* **1991**, *56*, 5758.
3. (a) Sakan, K.; Smith, D. A. *TL* **1984**, *25*, 2081. (b) Ireland, R. E.; Dow, W. C.; Godfrey, J. D.; Thaisrivongs, S. *JOC* **1984**, *49*, 1001.
4. Marshall, J. A.; Shearer, B. G.; Crooks, S. L. *JOC* **1987**, *52*, 1236.
5. Takeda, K.; Kobayashi, T.; Saito, K.; Yoshii, E. *JOC* **1988**, *53*, 1092.
6. Oppolzer, W.; Dupuis, D.; Poli, G.; Raynham, T. M.; Bernardinelli, G. *TL* **1988**, *29*, 5885.
7. Corey, E. J.; Cheng, X.-M.; Cimprich, K. A. *TL* **1991**, *32*, 6839.
8. Trost, B. M.; Lautens, M.; Hung, M. H.; Carmichael, C. S. *JACS* **1984**, *106*, 7641.
9. (a) Snider, B. B. *COS* **1991**, *5*, 1. (b) Snider, B. B. *COS* **1991**, *2*, 527. (c) Mikami, K.; Shimizu, M. *CRV* **1992**, *92*, 1021.
10. (a) Snider, B. B.; Rodini, D. J.; Kirk, T. C.; Cordova, R. *JACS* **1982**, *104*, 555. (b) Cartaya-Marin, C. P.; Jackson, A. C.; Snider, B. B. *JOC* **1984**, *49*, 2443. (c) Tietze, L. F.; Beifuss, U.; Antel, J.; Sheldrick, G. M. *AG(E)* **1988**, *27*, 703. (d) Metzger, J. O.; Biermann, U. *S* **1992**, 463.
11. Rodini, D. J.; Snider, B. B. *TL* **1980**, *21*, 3857.
12. (a) Mikami, K.; Loh, T.-P.; Nakai, T. *TL* **1988**, *29*, 6305. (b) Mikami, K.; Loh, T.-P.; Nakai, T. *CC* **1988**, 1430. (c) Houston, T. A.; Tanaka, Y.; Koreda, M. *JOC* **1993**, *58*, 4287.
13. Mikami, K.; Loh, T.-P.; Nakai, T. *CC* **1991**, 77.
14. Tanino, K.; Shoda, H.; Nakamura, T.; Kuwajima, I. *TL* **1992**, *33*, 1337.
15. Snider, B. B.; Karras, M.; Price, R. T.; Rodini, D. J. *JOC* **1982**, *47*, 4538.
16. (a) Smith, A. B., III; Fukui, M. *JACS* **1987**, *109*, 1269. (b) Smith, A. B., III; Fukui, M.; Vaccaro, H. A.; Empfield, J. R. *JACS* **1991**, *113*, 2071.
17. (a) Snider, B. B.; Deutsch, E. A. *JOC* **1982**, *47*, 745. (b) Snider, B. B.; Deutsch, E. A. *JOC* **1983**, *48*, 1822. (c) Snider, B. B.; Goldman, B. E. *T* **1986**, *42*, 2951.
18. Jackson, A. C.; Goldman, B. E.; Snider, B. B. *JOC* **1984**, *49*, 3988.
19. (a) Wovkulich, P. M.; Tang, P. C.; Chadha, N. K.; Batcho, A. D.; Barrish, J. C.; Usković, M. R. *JACS* **1989**, *111*, 2596. (b) Barrish, J. C.; Wovkulich, P. M.; Tang, P. C.; Batcho, A. D.; Usković, M. R. *TL* **1990**, *31*, 2235. (c) Quinkert, G.; Schmalz, H.-G.; Walzer, E.; Kowalczyk-Przewloka, T.; Dürner, G.; Bats, J. W. *AG(E)* **1987**, *26*, 61. (d) Quinkert, G.; Schmalz, H.-G.; Walzer, E.; Gross, S.; Kowalczyk-Przewloka, T.; Schierloh, C.; Dürner, G.; Bats, J. W.; Kessler, H. *LA* **1988**, 283. (e) Cohen, T.; Guo, B.-S. *T* **1986**, *42*, 2803.
20. Ogawa, Y.; Shibasaki, M. *TL* **1984**, *25*, 663.
21. Magnus, P.; Lacour, J. *JACS* **1992**, *114*, 3993.
22. Murray, T. F.; Varma, V.; Norton, J. R. *JOC* **1978**, *43*, 353.
23. Wada, M.; Aiura, H.; Akiba, K.-Y. *TL* **1987**, *28*, 3377.
24. Shinoda, M.; Iseki, K.; Oguri, T.; Hayasi, Y.; Yamada, S.-I.; Shibasaki, M. *TL* **1986**, *27*, 87.
25. (a) Rück, K.; Kunz, H. *AG(E)* **1991**, *30*, 694. (b) *SL* **1992**, 343. (c) *S* **1993**, 1018.
26. Ishibashi, H.; Takamuro, I.; Mizukami, Y.-I.; Irie, M.; Ikeda, M. *SC* **1989**, *19*, 443.

Barry B. Snider
Brandeis University, Waltham, MA, USA

4-Dimethylaminopyridine[1]

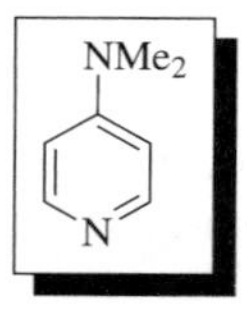

[1122-58-3] $C_7H_{10}N_2$ (MW 122.19)

(catalyst for acylation of alcohols or amines,[1–11] especially for acylations of tertiary or hindered alcohols or phenols[12] and for macrolactonizations;[13–15] catalyst for direct esterification of carboxylic acids and alcohols in the presence of dicyclohexylcarbodiimide (Steglich–Hassner esterification);[5] catalyst for silylation or tritylation of alcohols,[9,10] and for the Dakin–West reaction[20])

Alternate Name: DMAP.
Physical Data: colorless solid; mp 108–110 °C; pK_a 9.7.
Solubility: sol MeOH, $CHCl_3$, CH_2Cl_2, acetone, THF, pyridine, HOAc, EtOAc; partly sol cold hexane or water.
Form Supplied in: colorless solid; commercially available.
Preparative Methods: prepared by heating 4-pyridone with HMPA at 220 °C, or from a number of 4-substituted (Cl, OPh, SO_3H, $OSiMe_3$) pyridines by heating with DMA.[2] Prepared commercially from the 4-pyridylpyridinium salt (obtained from pyridine and $SOCl_2$) by heating with DMF at 155 °C.[1,2]
Purification: can be recrystallized from EtOAc.
Handling, Storage, and Precautions: skin irritant; corrosive, toxic solid.

Acylation of Alcohols. Several 4-aminopyridines speed up esterification of hindered alcohols with acid anhydrides by as much as 10 000 fold; of these, DMAP is the most commonly used but 4-pyrrolidinopyridine (PPY) and 4-tetramethylguanidinopyridine are somewhat more effective.[11] DMAP is usually employed in 0.05–0.2 mol equiv amounts.

DMAP catalyzes the acetylation of hindered 11β- or 12α-hydroxy steroids. The alkynic tertiary alcohol acetal in eq 1 is acetylated at rt within 20 min in the presence of excess DMAP.[3]

HO–C(Me)₂–C≡C–CH(OMe)₂ —[Ac_2O, DMAP; rt; 93%]→ AcO–C(Me)₂–C≡C–CH(OMe)₂ (1)

Esterifications mediated by ***2-Chloro-1-methylpyridinium Iodide*** also benefit from the presence of DMAP.[22]

DMAP acts as an efficient acyl transfer agent, so that alcohols resistant to acetylation by ***Acetic Anhydride–Pyridine*** usually react well in the presence of DMAP.[4a] Sterically hindered phenols can be converted into salicylaldehydes via a benzofurandione prepared by DMAP catalysis (eq 2).[4b]

(2) [scheme: 2,4-di-*t*-Bu-phenol —1. $(ClCO)_2$, DMAP; 2. $AlCl_3$→ benzofurandione → → OHC, OH, *t*-Bu, *t*-Bu salicylaldehyde]

Direct Esterification of Alcohols and Carboxylic Acids. Instead of using acid anhydrides for the esterification of alcohols, it is possible to carry out the reaction in one pot at rt by employing a carboxylic acid, an alcohol, ***1,3-Dicyclohexylcarbodiimide***, and DMAP.[5,6] In this manner, *N*-protected amino acids and even hindered carboxylic acids can be directly esterified at rt using DCC and DMAP or 4-pyrrolidinopyridine (eq 3).[5]

DCC–DMAP has been used in the synthesis of depsipeptides.[6b] Macrocyclic lactones have been prepared by cyclization of hydroxy carboxylic acids with DCC–DMAP. The presence of salts of DMAP,[13a] such as its trifluoroacetate, is beneficial in such cyclizations, as shown for the synthesis of a (9*S*)-dihydroerythronolide.[13b] Other macrolactonizations have been achieved using ***2,4,6-Trichlorobenzoyl Chloride*** and DMAP in ***Triethylamine*** at rt[14] or ***Di-2-pyridyl Carbonate*** (6 equiv) with 2 equiv of DMAP at 73 °C.[15]

Acylation of Amines. Acylation of amines is also faster in the presence of DMAP,[7] as is acylation of indoles,[8a] phosphorylation of amines or hydrazines,[2,8] and conversion of carboxylic acids into anilides by means of ***Phenyl Isocyanate***.[1] β-Lactam formation from β-amino acids has been carried out with DCC–DMAP, but epimerization occurs.[8b]

(3) [scheme: mesitoic acid (CO_2H) + 4-O_2N-phenol (OH) —DCC, rt; PPY, 12 h; 90%→ 4-nitrophenyl mesitoate (NO_2)]

Silylation, Tritylation, and Sulfinylation of Alcohols. Tritylation, including selective tritylation of a primary alcohol in the presence of a secondary one,[9] silylation of tertiary alcohols, selective silylation to *t*-butyldimethylsilyl ethers,[6] and sulfonylation or sulfinylation[10] of alcohols proceed more readily in the presence of DMAP. Silylation of β-hydroxy ketones with ***Chlorodiisopropylsilane*** in the presence of DMAP followed by treatment with a Lewis acid gives diols (eq 4).[16]

(4) [scheme: β-hydroxy ketone (O, OH) —*i*-Pr_2SiHCl, DMAP→ (O, $OSiH(i\text{-}Pr)_2$) —$SnCl_4$→ diol (OH, OH)]

Miscellaneous Reactions. Alcohols, including tertiary ones, can be converted to their acetoacetates by reaction with ***Diketene*** in the presence of DMAP at rt.[17] Decarboxylation of β-keto esters has been carried out at pH 5–7 using 1 equiv of DMAP in refluxing wet toluene (eq 5).[18]

(5) [scheme: 2-CO_2Me-3-methylcyclopentanone —DMAP, toluene, H_2O; reflux; 60%→ 3-methylcyclopentanone]

Elimination of water from a *t*-alcohol in a β-hydroxy aldehyde was carried out using an excess of ***Methanesulfonyl Chloride***–DMAP–H_2O at 25 °C.[23]

Glycosidic or allylic alcohols (even when *s*-) can be converted in a 80–95% yield to alkyl chlorides by means of ***p-Toluenesulfonyl Chloride***–DMAP. Simply primary alcohols react slower and secondary ones are converted to tosylates.[24]

Aldehydes and some ketones can be converted to enol acetates by heating in the presence of TEA, Ac_2O, and DMAP.[2] DMAP catalyzes condensation of malonic acid monoesters with unsaturated aldehydes at 60 °C to afford dienoic esters (eq 6).[19]

(6) [scheme: methacrolein (H, O) + HO_2C–CH_2–CO_2Me —DMAP; py, 60 °C; 86%→ diene–CO_2Me]

The conversion of α-amino acids into α-amino ketones by means of acid anhydrides (Dakin–West reaction)[20] also proceeds faster in the presence of DMAP (eq 7).

MeS–CH₂CH₂–CH(NH₂)–CO₂H → oxazolone; 1. cyclopropanecarbonyl chloride, DMAP; 2. AcOH → MeS–CH₂CH₂–CH(NHCOR)–CO–cyclopropyl (7)

Ketoximes can be converted to nitrimines which react with Ac_2O–DMAP to provide alkynes (eq 8).[21]

cyclododecanone nitrimine ($=NNO_2$) —Ac_2O, DMAP→ cyclododecyne + isomeric allene (8)

For the catalysis by DMAP of the *t*-butoxylcarbonylation of alcohols, amides, carbamates, NH-pyrroles, etc., see ***Di-t-butyl Dicarbonate***.

Related Reagents. 1-Hydroxybenzotriazole; *N*-Hydroxysuccinimide; Imidazole; Pyridine.

1. Hoefle, G.; Steglich, W.; Vorbrueggen, H. *AG(E)* **1978**, *17*, 569.
2. Scriven, E. F. V. *CSR* **1983**, *12*, 129.
3. (a) Steglich, W.; Hoefle, G. *AG(E)* **1969**, *8*, 981. (b) Hoefle, G.; Steglich, W. *S* **1972**, 619.
4. (a) Salomon, R. G., Salomon, M. F.; Zagorski, M. G.; Reuter, J. M.; Coughlin, D. J. *JACS* **1982**, *104*, 1008. (b) Zwanenburg, D. J.; Reynen, W. A. P. *S* **1976**, 624.
5. (a) Neises, B.; Steglich, W. *AG(E)* **1978**, *17*, 522. (b) Hassner, A.; Alexanian, V. *TL* **1978**, 4475.
6. (a) Ziegler, F. E.; Berger, G. D. *SC* **1979**, 539. (b) Gilon, C.; Klausner, Y.; Hassner, A. *TL* **1979**, 3811.
7. Litvinenko, L. M., Kirichenko, A. C.; *DOK* **1967**, *176*, 97. (b) Kirichenko, A. C.; Litvinenko, L. M.; Dotsenko, I. N.; Kotenko, N. G.; Nikkel'sen, E.; Berestetskaya, V. D. *DOK* **1969**, *244*, 1125 (*CA* **1979**, *90*, 157 601).
8. (a) Nickisch, K.; Klose, W.; Bohlmann, F. *CB* **1980**, *113*, 2036. (b) Kametami, T.; Nagahara, T.; Suzuki, Y.; Yokohama, S.; Huang, S.-P.; Ihara, M. *T* **1981**, *37*, 715.
9. (a) Chaudhary, S. K.; Hernandez, O. *TL* **1979**, *95*, 99. (b) Hernandez, O.; Chaudhary, S. K.; Cox, R. H.; Porter, J. *TL* **1981**, *22*, 1491.
10. Guibe-Jampel, E.; Wakselman, M.; Raulais, D. *CC* **1980**, 993.
11. Hassner, A.; Krepski, L. R.; Alexanian, V. *T* **1978**, *34*, 2069.
12. Vorbrueggen, H. (Schering AG) Ger. Offen. 2 517 774, 1976 (*CA* **1977**, *86*, 55 293).
13. (a) Boden, E. P.; Keck, G. E. *JOC* **1985**, *50*, 2394. (b) Stork, G.; Rychnovsky, S. D. *JACS* **1987**, *109*, 1565.
14. Hikota, M.; Tone, H.; Horita, K.; Yonemitsu, O. *JOC* **1990**, *55*, 7.
15. (a) Kim, S.; Lee, J. I.; Ko, Y. K. *TL* **1984**, *25*, 4943. (b) Denis, J.-N.; Greene, A. E.; Guenard, D.; Gueritte-Voegelein, F.; Mangatal, L.; Potier, P. *JACS* **1988**, *110*, 5917.
16. Anwar, S.; Davis, A. P. *T* **1988**, *44*, 3761.
17. Nudelman, A.; Kelner, R.; Broida, N.; Gottlieb, H. E. *S* **1989**, 387.
18. Taber, D. F.; Amedio J. C., Jr.; Gulino, F. *JOC* **1989**, *54*, 3474.
19. Rodriguez, J.; Waegell, B. *S* **1988**, 534.
20. (a) Buchanan, G. L. *CSR* **1988**, *17*, 91. (b) McMurry, J. *JOC* **1985**, *50*, 1112. (c) Hoefle, G., Steglich, W. *CB* **1971**, *104*, 1408.
21. Buechi, G.; Wuest, H. *JOC* **1979**, *44*, 4116.
22. Nicolaou, K. C.; Bunnage, M. E.; Koide, K. *JACS* **1994**, *116*, 8402.
23. Furukawa, J.; Morisaki, N.; Kobayashi, H.; Iwasaki, S.; Nozoe, S.; Okuda, S. *CPB* **1985**, *33*, 440.
24. Hwang, C. K.; Li, W. S.; Nicolaou, K. C. *TL* **1984**, *25*, 2295.

Alfred Hassner
Bar-Ilan University, Ramat Gan, Israel

N,N-Dimethylformamide Diethyl Acetal

$(R^1O)_2CH–NR^2_2$

(**1**; R^1 = Et, R^2 = Me) *[1188-33-6]*	$C_7H_{17}NO_2$	(MW 147.25)
(**2**; R^1 = Me, R^2 = Me) *[4637-24-5]*	$C_5H_{13}NO_2$	(MW 119.19)
(**3**; R^1 = $PhCH_2$, R^2 = Me) *[2016-04-8]*	$C_{17}H_{21}NO_2$	(MW 271.39)
(**4**; R^1 = Et, R^2 = Et) *[22630-13-3]*	$C_9H_{21}NO_2$	(MW 175.31)

(mild and selective reagents for alkylation, formylation, and aminomethylenation[1])

Alternate Name: DMF diethyl acetal.

Physical Data: (**1**) bp 134–136 °C/760 mmHg; pK_b 6.2.[5] (**2**) bp 102–104 °C/720 mmHg; pK_b 6.25.[5] (**3**) bp 138–140 °C/0.5 mmHg; pK_b 6.2.[5] (**4**) bp 57–58 °C/20 mmHg; pK_b 6.4.[5]

Solubility: sol a variety of inert solvents.

Form Supplied in: pure liquid.

Preparative Methods: (1) a solution of sodium alkoxide (1 mol) and the secondary amine (1.1 mol) in 200–300 mL of the respective alcohol is refluxed as chloroform (39.5 g, 0.33 mol) is added. The mixture is refluxed for 2 h and the filtrate obtained is distilled in vacuo. For example, *N,N*-diethylformamide diethyl acetal is obtained in 32% yield.[7] (2) A solution of *N,N*-dialkyl(chloromethylene)ammonium chloride (Vilsmeier reagent, 1 mol) in 640–500 mL of chloroform is stirred while sodium alkoxide (2.1 mol) in 1 L of the respective alcohol is added. After 1 h at 20 °C the sodium chloride is separated and the mixture is distilled in vacuo. For example, the yield of DMF dimethyl acetal is 55%.[2]

Handling, Storage, and Precautions: most of the known orthoamide derivatives are colorless, distillable liquids with an amine-like smell. If moisture and presence of acids and high temperature are avoided, most of the reagents can be stored almost indefinitely. No significant toxicity has been reported.

Reactions and Synthetic Applications: General. The formamide acetals enter into two main categories of reactions, namely alkylation and formylation, mostly via generation of aza-oxo-stabilized carbenium ions (eq 1).[1–7] As alkylation reagents they have been used in the synthesis of esters from acids, in the synthesis of ethers and thioethers from phenols and aromatic and heterocyclic thiols, and in the alkylation of CH-active methines. As formylating agents, formamide acetals are useful in the synthesis of enamines from active methylene compounds and amidines from amines and amides, as well as in the formation and modification of many types of heterocyclic compounds. They can be used for the dehydrative decarboxylation of β-hydroxy carboxylic acids to alkenes, and for the cyclization of *trans* vicinal diols to epoxides.[1] From the numerous literature reports on applications of these reagents,[1–7] some representative reactions are discussed in the following sections.

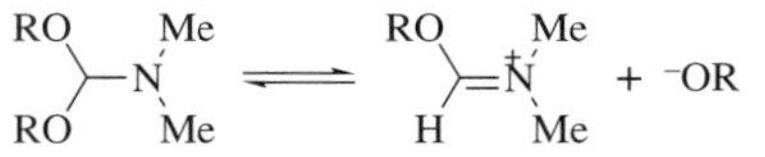

(1)

Alkylation Reactions. DMF dialkyl acetals undergo a variety of reactions with 1,2-diols.[1] For example, the reaction of *trans*-cyclohexane-1,2-diol with DMF dimethyl acetal leads to the formation of cyclohexane epoxide (eq 2)[8] with inversion of configuration. Similarly, *meso*-1,2-diphenyl-1,2-ethanediol gives *trans*-stilbene epoxide stereospecifically (eq 3).[8,9a] This method has also been applied in the synthesis of cholestane epoxide from vicinal diols.[8] If the intermediate 2-dimethylamino-1,3-dioxolane is treated with ***Acetic Anhydride***, reductive elimination to the alkene occurs with retention of stereochemistry (eq 4).[9b]

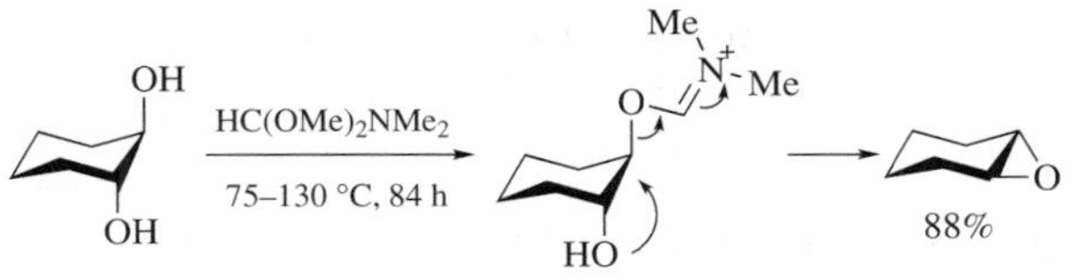

(2)

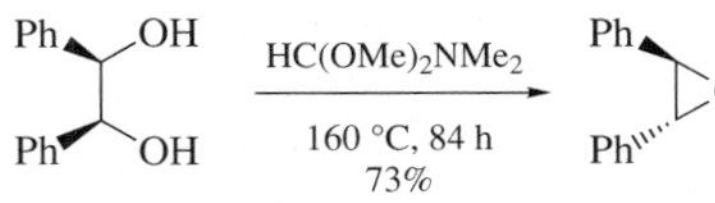

(3)

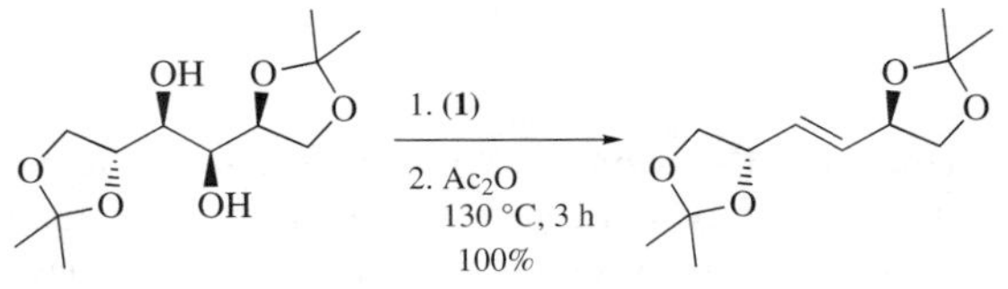

(4)

DMF dimethyl acetal is an effective methylating reagent. For example, heterocyclic thiols are transformed to *S*-methyl heterocycles in high yields (76–86%).[10] DMF dibenzyl acetal is an interesting reagent for selective protection of nucleosides. For example, uridine and guanosine are selectively blocked at the –CONH function (eq 5).[11]

In a very simple procedure, carboxylic acids can be esterified under mild conditions with DMF dialkyl acetals. Some interesting uses are the conversion of carboxylic acids to ethyl and benzyl esters with DMF diethyl and dibenzyl acetals (yield 64–75%).[12] The dibenzyl acetal has been widely used as protecting group reagent for the carboxyl end group in peptides.[13] In several cases, DMF bis(4-dodecylbenzyl) acetal has also been used.[13,14] Some examples are given in Table 1.

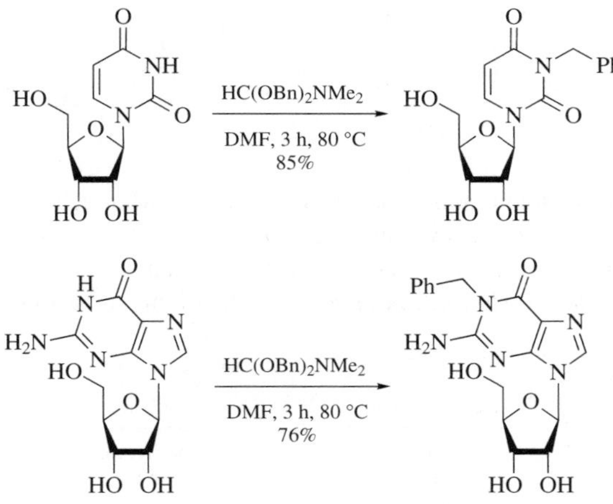

(5)

Table 1 Esterification of Amino Acid and Peptide Derivatives with DMF Dibenzyl Acetal (**3**) and DMF Bis(4-dodecylbenzyl) Acetal (**5**)

RCO_2H[a]	Acetal	Reaction conditions	Yield (%)
N-DOBC-L-Val	(**3**)	C_6H_6, 80 °C, 1.5 h	97[13,14]
N-DOBC-L-Phe	(**3**)	C_6H_6, 80 °C, 48 h	90[14]
N-DOBC-L-Try	(**3**)	C_6H_6, 80 °C, 2 h	78[13,14]
N-DOBC-Gly-L-Leu	(**3**)	C_6H_6, 80 °C, 1 h	73[13,14]
N-DOBC-Gly-L-Tyr	(**3**)	CH_2Cl_2, 25 °C, 48 h	80[14]
N-Boc-Gly-L-Phe	(**3**)	CH_2Cl_2, 25 °C, 90 h	80[14]
N-DOBC-L-Phe	(**5**)	CH_2Cl_2, 20 °C, 18 h	84[13]
N-DOBC-L-Try	(**5**)	CH_2Cl_2, 20 °C, 50 h	68[13]
N-DOBC-L-Phe-L-Ala	(**5**)	CH_2Cl_2, 20 °C, 120 h	75[13]

[a] DOBC = 4-decyloxybenzyloxycarbonyl; Boc = *t*-butoxycarbonyl.

DMF dineopentyl acetal has been used as a reagent for dehydrative decarboxylation in order to avoid the possibility of competing *O*-alkylation of the carboxyl group.[15] This conversion of β-hydroxy carboxylic acids to alkenes by reaction with a DMF acetal involves an *anti* elimination, and it is thus complementary to the *syn* elimination of these hydroxy acids via the β-lactone. These reactions have been used to obtain both (*E*)- and (*Z*)-1-alkoxy-1,3-butadienes (eq 6).[15] For additional examples of alkenes obtained from β-hydroxy acids with DMF acetals, see Scheeren et al.[7]

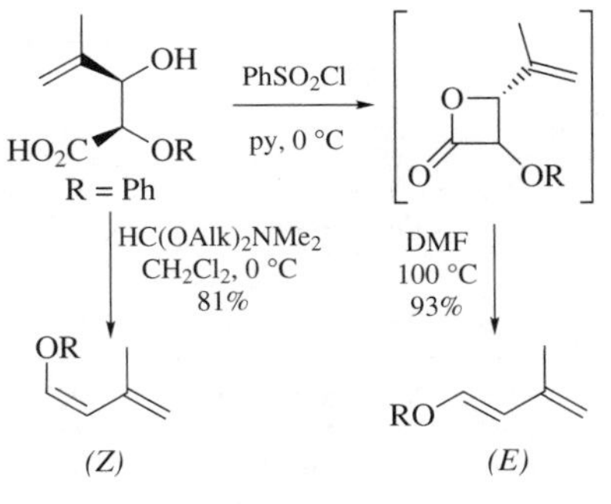

(6)

Formylation Reactions. DMF dialkyl acetals exhibit reactivity as a formyl cation synthon by introducing a C_1 unit at many nucleophilic centers (e.g. at N, S, O, or CH). For example, DMF dimethyl acetal can be used instead of formic acid for conversion of *o*-disubstituted aromatic systems into annulated heterocycles (eq 7).[16]

$HC(OMe)_2NMe_2$, 120–130 °C, 50–70%; X = CH, N (7)

A nonacidic and regioselective route to Mannich bases from ketones and esters involves reaction with DMF acetals at a high temperature to form enamino ketones which are readily reduced by ***Lithium Aluminium Hydride*** to the Mannich bases (eq 8).[17]

$HC(OMe)_2NMe_2$, 110 °C, 44%; $LiAlH_4$, diethyl ether, 81% (8)

CH-acidic groups (including methyl) react with DMF acetals via carbon–carbon bond formation and subsequent elimination of the respective alkanols to form enamines (aminomethylenation). Thus 2,4,7-trimethyl-1,3-dithia-5,6-diazepine reacts with DMF dialkyl acetals to give mono- or bis-aminomethylenated products depending on the amount of reagent and the reaction conditions (eq 9).[18a] The same reagents convert the more nucleophilic 2,5-dimethyl-1,3,4-thiadiazolium salts to the monoenamines (eq 10).[18a] Similarly, 1,2-dimethyl-3-cyano-5-nitroindole condenses with DMF diethyl acetal to give the (*E*)-indol-2-yl enamine (eq 11).[18b]

$HC(OEt)_2NMe_2$ or $HC(OMe)_2NMe_2$, 100 °C, 70%; R = CH=CHNMe$_2$ (9)
2 equiv, 100 °C, 32%; R = CH=CHNMe$_2$

$HC(OEt)_2NMe_2$ or $HC(OMe)_2NMe_2$, 100 °C (10)

R^1	R^2	R^3	X	Yield (%)
Me	Me	Me	$MeSO_4$	93
Bz	Me	Me	Br	85
C_8H_{17}	Me	Me	I	85

$HC(OEt)_2NMe_2$, DMF, reflux, 10 h, 81% (11)

A general indole synthesis involves reaction of an *o*-nitrotoluene derivative with DMF dimethyl acetal in refluxing DMF (eq 12).[19,20] The initially formed *o*-nitroaryl-substituted (*E*)-*N,N*-dimethylenamine is submitted to catalytic hydrogenation to give the indole by spontaneous cyclization. According to a variation of this methodology,[20] 2-arylindoles are readily available by reaction of *o*-nitrotoluene with DMF diethyl acetal and *o*-halobenzoyl chloride. This reaction proceeds via benzoylation of the respective enamine.

$HC(OMe)_2NMe_2$, DMF, reflux, 92%; H_2, Pd/C, 83% (12)

4-Dimethylamino-2-azabutadienes are readily accessible by the reaction of azomethines (imines) with DMF diethyl acetal (eq 13).[21] 1-Dimethylamino-1,3-butadienes can be synthesized in the same manner.[21] Reactions of 2-azavinamidinium salts with DMF diethyl acetal give rise to 2-aza- and 2,4-diazapentamethinium salts (eq 14).[22]

$HC(OEt)_2NMe_2$, reflux (13)

R^1	R^2	R^3	Yield (%)
CO_2Me	H	Ph	83
CO_2Me	H	p-ClC_6H_4	70
CO_2Me	H	p-Tol	82

$HC(OEt)_2NMe_2$, DMF, 80–100 °C, 74–90%; X = CH, N; ClO_4^- (14)

Miscellaneous Reactions. DMF acetals catalyze rearrangement reactions of allylic alcohols to β,γ-unsaturated amides.[23] This reaction, which involves a [2,3]-sigmatropic rearrangement, occurs with complete transfer of chirality. Thus the reaction of the (*R,Z*)-allylic alcohol (eq 15) with DMF dimethyl acetal gives the enantiomerically pure (*R,E*)-β,γ-unsaturated amide as the only product. The (*S,E*)-isomer also rearranges mainly to the (*R,E*)-amide, with only a trace of the (*S,Z*)-isomer. It has been suggested that both rearrangements proceed via a five-membered cyclic transition state with a carbene-like function.[23]

Enol acetates of oxo nucleosides are readily accessible by reaction of the nucleoside with DMF acetal and then with acetic anhydride (eq 16).[24] However, this method is limited to compounds with an oxo group in the 4′-position and with a free CO–NH group in the pyrimidine ring. This reaction is the first synthesis of oxo nucleoside enol acetates by direct enolization. Previous methods of enol formation fail with oxo nucleosides because of their instability in alkaline media.

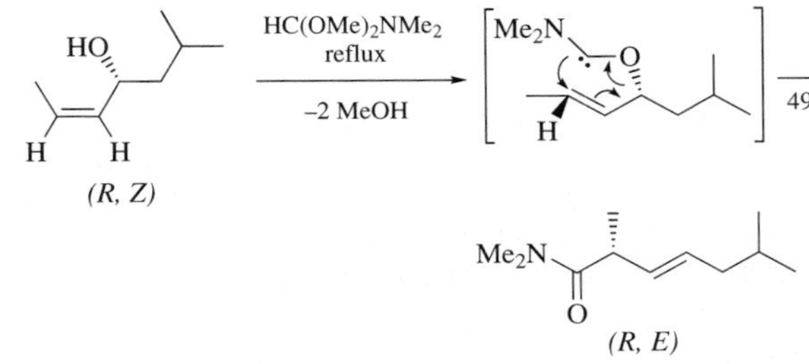

Related Reagents. *t*-Butoxybis(dimethylamino)methane; *N,N*-Dimethylacetamide Dimethyl Acetal; Dimethylchloromethyleneammonium Chloride; *N,N*-Dimethylpropionamide Dimethyl Acetal; Triethyl Orthoformate; Tris(dimethylamino)methane; Tris(formylamino)methane.

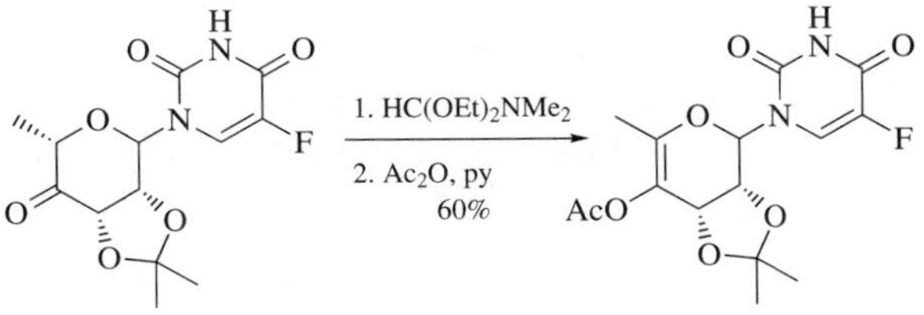

1. Abdulla, R. F.; Brinkmeyer, R. S. *T* **1979**, *35*, 1675.
2. Eilingsfeld, H.; Seefelder, M.; Weidinger, H. *CB* **1963**, *96*, 2671.
3. DeWolfe, R. H. *Carboxylic Ortho Acid Derivatives*, Academic: New York 1970.
4. Kantlehner, W. In *The Chemistry of Acid Derivatives*, Patai, S., Ed.; Wiley: Chichester 1979, Part 1, Suppl. B, p 533.
5. Simchen, G. In *Iminium Salts in Organic Chemistry*; Böhme, H.; Viehe, H. G., Eds.; Wiley: New York, 1979; p 393.
6. Pindur, U. In *The Chemistry of Acid Derivatives*, Patai, S. Ed.; Wiley: Chichester, 1992; Vol. 2, Suppl. B, p 1005.
7. Scheeren, J. W.; Nivard, R. J. F. *RTC* **1969**, *88*, 289.
8. Neumann, H. *C* **1969**, *23*, 267.
9. (a) Harvey, R. G.; Goh, S. H.; Cortez, C. *JACS* **1975**, *97*, 3468. (b) Eastwood, W.; Harrington, K. I.; Josan, J. S.; Pura, I. L. *TL* **1970**, 5223.
10. Holý, A. *TL* **1972**, 585.
11. Philips, K. D.; Horwitz, J. P. *JOC* **1975**, *40*, 1856.
12. (a) Vorbrüggen, H. *AG(E)* **1963**, *2*, 211. (b) Vorbrüggen, H. *LA* **1974**, 821.
13. (a) Brechbühler, H.; Büchi, H.; Hatz, E.; Schreiber, J.; Eschenmoser, A. *AG(E)* **1963**, *2*, 212; (b) Büchi, H.; Steen, K.; Eschenmoser, A. *AG(E)* **1964**, *3*, 62.
14. Brechbühler, H.; Büchi, H.; Hatz, E.; Schreiber, J., Eschenmoser, A. *HCA* **1965**, *48*, 1746.
15. (a) Luengo, J. L.; Koreeda, M. *TL* **1984**, *25*, 4881. (b) Koreeda, M.; Luengo, J. L. *JOC* **1984**, *49*, 2079.
16. Stanovnik, B.; Tisler, M. *S* **1974**, 120.
17. Schuda, P. F.; Ebner, C. B.; Morgan, T. M. *TL* **1986**, *27*, 2567.
18. (a) Kantlehner, W.; Haug, E.; Hagen, H. *LA* **1982**, 298. (b) Krichevski, E. S.; Granik, V. G. *KGS* **1992**, 502.
19. Batcho, A. D.; Leimgruber, W. U.S. Patent 3 976 639, 1973; 3 732 245, 1973.
20. Garcia, E. E.; Fryer, R. I. *JHC* **1974**, *11*, 219.
21. Gompper, R.; Heinemann, U. *AG(E)* **1981**, *20*, 296.
22. Gompper, R.; Heinemann, U. *AG(E)* **1981**, *20*, 297.
23. Yamamoto, H.; Kitatani, K.; Hiyama, T.; Nozaki, H. *JACS* **1977**, *99*, 5816.
24. Bessodes, M.; Ollapally, A.; Antonakis, K. *CC* **1979**, 835.

Ulf Pindur
University of Mainz, Germany

2,2-Dimethyl-1,3-propanediol

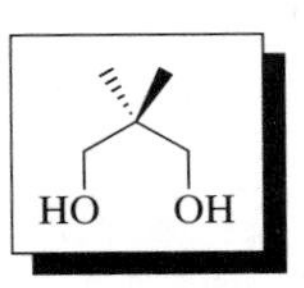

[126-30-7] $C_5H_{12}O_2$ (MW 104.17)

(protecting group for aldehydes and ketones;[9] a ligand used in boron chemistry[12])

Physical Data: mp 123–127 °C.[1]
Solubility: sol H_2O, chloroform, benzene; very sol ethyl alcohol, diethyl ether.[2]
Form Supplied in: white crystals.
Handling, Storage, and Precautions: skin irritant; hygroscopic; store in cool dry place.[1]

Cyclic Acetals. 2,2-Dimethyl-1,3-propanediol (**1**) has been used effectively as a protecting group for ketones and aldehydes. The formation of the cyclic acetal can be accomplished in many ways. The most common is to reflux the ketone and diol (**1**) with a catalytic amount of acid and force the reaction to completion by removing H_2O, usually with a Dean–Stark trap (eq 1).[3] Other methods involve treating the ketone–diol mixture with ***Chlorotrimethylsilane*** (eq 2)[4] or ***Boron Trifluoride Etherate*** (eq 3).[5]

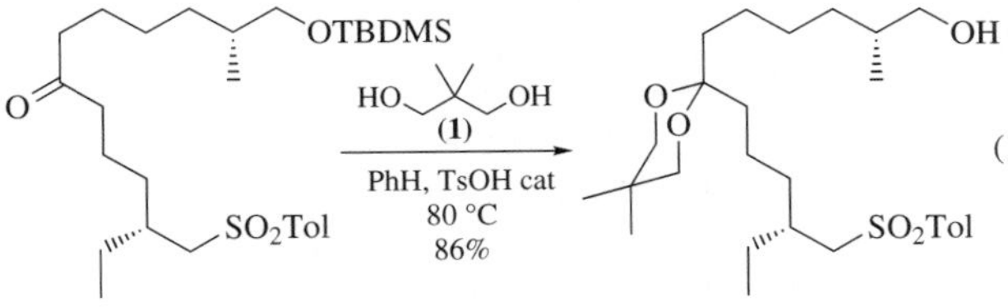

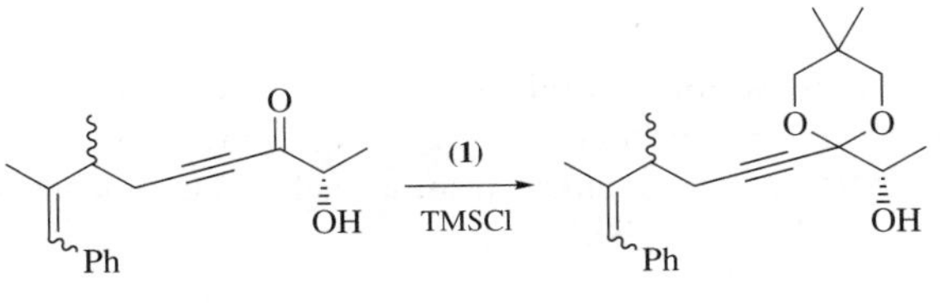

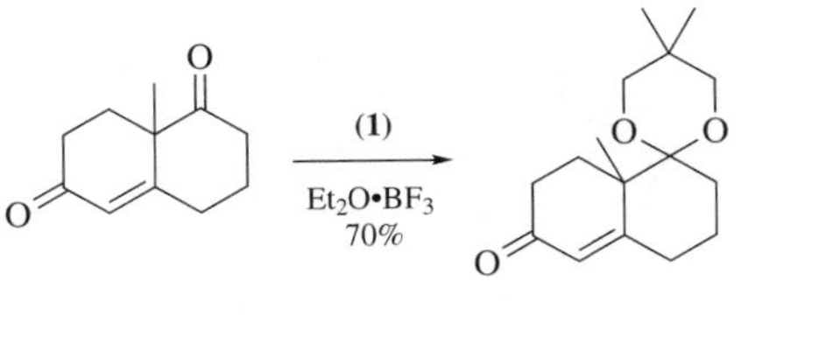

The advantages of diol (**1**) compared to other diols depend on the compound being protected. In the formation of β-halo acetals, for example, reagent (**1**) was found to be more effective than ***Ethylene Glycol*** because of easier separation and a longer shelf life of the product.[6] Diol (**1**) was also considered to be superior to ethylene glycol when Grignard reagents were prepared from β-halo acetals (eq 4).[7]

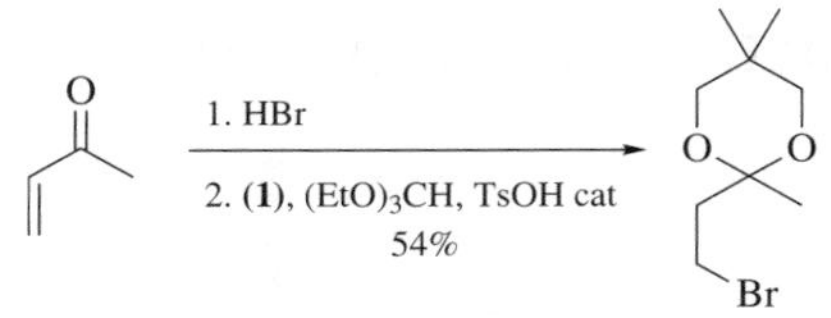

1. HBr
2. (**1**), $(EtO)_3CH$, TsOH cat
54% (4)

In comparing rates of acid hydrolysis of different cyclic acetals formed from a variety of diols, the following order was discovered; 1,3-propanediol ≫ (**1**) > ethylene glycol > 2,2-diethyl-1,3-propanediol > 2,2-diisopropyl-1,3-propanediol.[8] This has been attributed to the *gem*-dialkyl effect which serves to stabilize the cyclic acetal. Cleavage of the protecting group formed from diol (**1**) is easily accomplished via acidic hydrolysis using acids such as H_2SO_4, HCl, or TsOH.

Reagent (**1**) has been used in a variety of syntheses where its merits have not been elucidated. The following reactions show some of these transformations (eqs 5–7).[9–11]

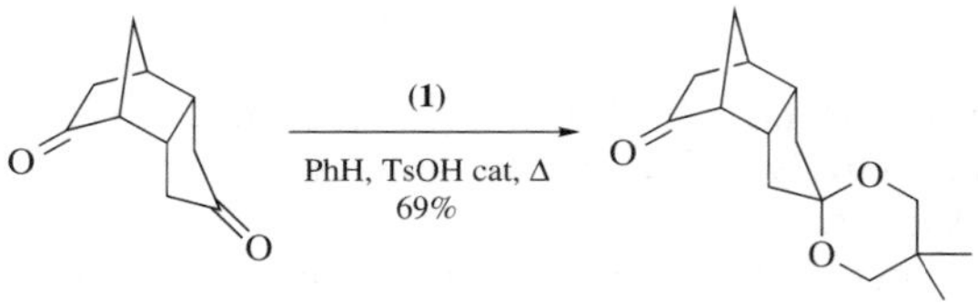

(**1**)
PhH, TsOH cat, Δ
69% (5)

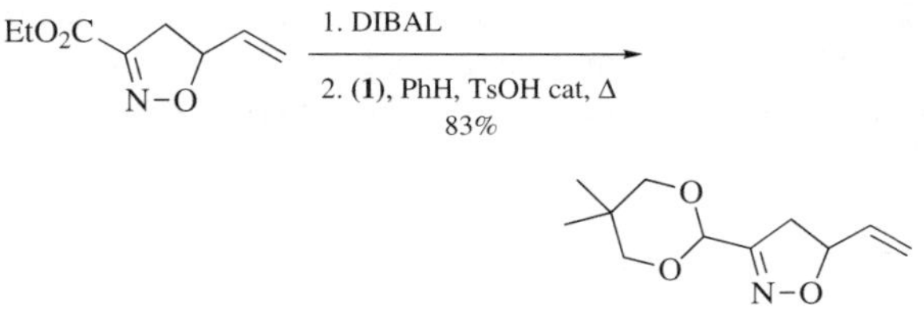

EtO_2C N–O
1. DIBAL
2. (**1**), PhH, TsOH cat, Δ
83% (6)

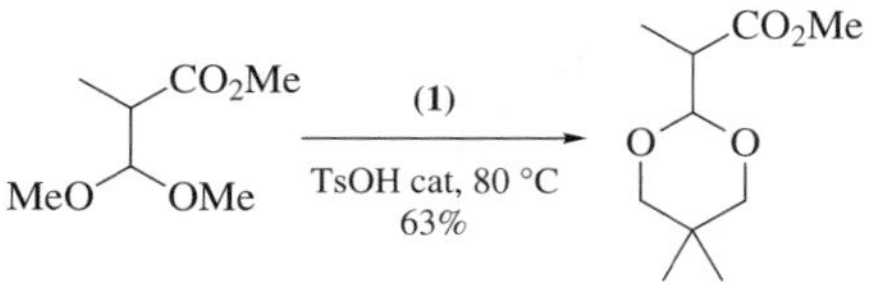

CO_2Me MeO OMe
(**1**)
TsOH cat, 80 °C
63% (7)

Use as a Boron Ligand. This diol has been used to complex boron.[12] For example, diol (**1**) was employed to remove the phenylboronic acid template from a Diels–Alder adduct under neutral conditions (eq 8).[13]

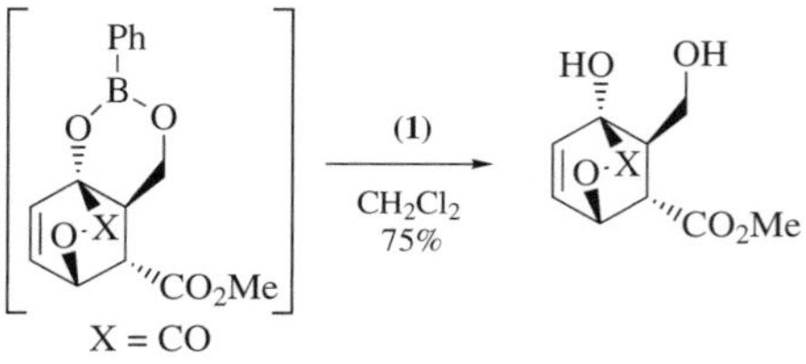

Ph B O O O·X CO_2Me
X = CO
(**1**)
CH_2Cl_2
75% (8)
HO OH O·X CO_2Me

Related Reagents. 3-Bromo-1,2-propanediol; 2,3-Butanediol; Ethylene Glycol; (2*R*,4*R*)-2,4-Pentanediol; 1,3-Propanediol.

1. *Sigma-Aldrich Library of Chemical Safety Data*, 2nd ed.; Lenga, R. E., Ed.; Sigma-Aldrich: Milwaukee, WI, 1988; Vol. 1, p 1411B.
2. *CRC Handbook of Data on Organic Compounds*, 2nd ed.; Weast, R. C.; Grasselli, J. G.; Eds.; CRC: Boca Raton, FL, 1989; Vol. 6, p 3645.
3. Williams, D. R.; McGill, J. M. *JOC* **1990**, *55*, 3457.
4. Honda, Y.; Ori, A.; Tsuchihashi, G.-I. *CL* **1987**, 1259.
5. (a) Gopalakrishnan, G.; Jayaraman, S.; Rajagopalan, K.; Swaminathan, S. *S* **1983**, 797. (b) Brown, E.; Lebreton, J. *TL* **1986**, *27*, 2595.
6. Burnell, D. J.; Wu, Y.-J. *CJC* **1989**, *67*, 816.
7. Stowell, J. C.; Keith, D. R.; King, B. T. *OSC* **1990**, *7*, 59.
8. (a) Newman, M. S.; Harper, R. J., Jr. *JACS* **1958**, *80*, 6350. (b) Smith, S. W.; Newman, M. S. *JACS* **1968**, *90*, 1249. (c) Smith, S. W.; Newman, M. S. *JACS* **1968**, *90*, 1253.
9. (a) Deslongchamps, P.; Cheriyan, U. O.; Lambert, Y.; Mercier, J.-C.; Ruest, L.; Russo, R.; Soucy, P. *CJC* **1978**, *56*, 1687. (b) Carceller, E.; Moyano, A.; Serratosa, F.; Font-Altaba, M.; Solans, X. *JCS(P1)* **1986**, 2055.
10. Annunziata, R.; Cinquini, M.; Cozzi, F.; Raimondi, L.; Restelli, A. *HCA* **1985**, *68*, 1217.
11. Walkup, R. D.; Obeyesekere, N. U. *S* **1987**, 607.
12. Narasaka, K.; Yamamoto, I. *T* **1992**, *48*, 5743.
13. Narasaka, K.; Shimada, S.; Osoda, K.; Iwasawa, N. *S* **1991**, 1171.

Michael A. Walters & John J. Shay
Dartmouth College, Hanover, NH, USA

Dimethyl Sulfate[1]

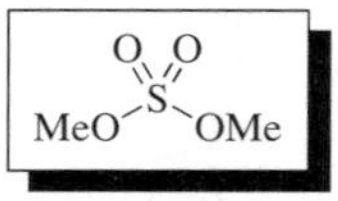

MeO–$S(O)_2$–OMe

[77-78-1] $C_2H_6O_4S$ (MW 126.13)

(effective methylating reagent[1])

Physical Data: bp 188 °C; mp −32 °C; *d* 1.333 g cm^{-3}; fp 83 °C.
Form Supplied in: water-white liquid; widely available.
Handling, Storage, and Precautions: extremely toxic and carcinogenic; use in a fume hood with adequate protection.[2]

***O*-Alkylation.** Dimethyl sulfate is a powerful alkylating agent and has been used for the methylation of almost every imaginable nucleophile over the years.[1] The variety of oxygen nucleophiles include carboxylic acids,[3] alcohols,[4] phenols,[5] lactams,[6] oximes,[7] pyridine *N*-oxides,[8] hydroxylamines,[9] hydroxamic acids,[10] and hydroperoxides.[11]

Carboxylic acids react with dimethyl sulfate in the presence of ***Dicyclohexyl(ethyl)amine*** (DICE) to afford methyl esters in high yield.[3] The method is reported to be facile and particularly useful when ester formation using ***Diazomethane*** or strongly acidic conditions is not possible. For example, β-9,10,12-trihydroxyoctadecanoic acid is readily converted to its methyl ester in 97% yield using this procedure (eq 1). Another example of this type of esterification was demonstrated with the preparation of bile acid methyl esters (eq 2).[12] These steroidal carboxylic acids were methylated using ***Potassium Carbonate*** as a base in refluxing acetone. Both of the aforementioned procedures avoid significant side reactions such as dehydration or ether formation resulting from competing alkylation of free hydroxy groups.

Me_2SO_4, $EtNCy_2$, MeOH, 97% (1)

Me_2SO_4, K_2CO_3, acetone, Δ, 93% (2)

Syntheses of aryl and alkyl methyl ethers using dimethyl sulfate have also been reported. Simple alcohols and phenols may be alkylated using dimethyl sulfate in slightly hydrated solid/liquid heterogeneous media with 1,4-dioxane or triglyme/potassium hydroxide and small amounts of water to give the corresponding methyl ethers in excellent yield.[4] For example, *t*-butyl alcohol is converted to *t*-butyl methyl ether in nearly quantitative yield (eq 3). The procedure calls for the use of only a stoichiometric amount of dimethyl sulfate, which reduces problems associated in the workup and the potential toxicity of remaining dimethyl sulfate. In addition, methylation of alcohols using dimethyl sulfate and ***Alumina*** as a solid adsorbent reaction medium affords excellent yields of the corresponding alkyl alcohols.[13] Phenols and carboxylic acids, however, gave lower yields of methylated product using this procedure. A second example involves the stereospecific formation of highly substituted tetrahydrofurans. Treatment of a 2,5-dihydroxy-4-phenyl sulfide with dimethyl sulfate in dichloromethane at 0 °C cleanly afforded the substituted tetrahydrofuran in excellent yield (eq 4).[14] It has been suggested that the reaction proceeds via methylation of a free hydroxy group followed by solvolysis of methanol assisted by the neighboring thiophenyl group. Subsequent cyclization gave the tetrahydrofuran as a single stereoisomer.

t-BuOH → *t*-BuOMe (Me_2SO_4, KOH, H_2O, dioxane, Δ, 99%) (3)

Me_2SO_4, CH_2Cl_2, 0 °C, 85% (4)

Several methods have been described for preparing aryl methyl ethers from phenols using dimethyl sulfate. For example, defucogilvocarcin M was completed by methylating a sterically hindered C-12 hydroxyl group using dimethyl sulfate and potassium carbonate (eq 5).[5] In addition, the synthesis of a carbazole alkaloid was completed via the selective alkylation of a 5-indolyl alcohol using dimethyl sulfate and ***Sodium Hydride*** to give 4-deoxycarbazomycin B in 96% yield (eq 6).[15]

Me_2SO_4, K_2CO_3, CH_2Cl_2, Δ, 98% (5)

Me_2SO_4, NaH, Et_2O, 96% (6)

The *O*-methylation of lactams to afford *O*-alkylimino ethers has also been reported. For example, the large-scale preparation of *O*-methylcaprolactim proceeds in good yield using caprolactam and 1 equiv of dimethyl sulfate in benzene (eq 7).[6] The treatment of caprolactam with excess dimethyl sulfate gives *N*-methyllactam as the major product.

Me_2SO_4, benzene, Δ, 61–68%; 500 g (7)

Other alkylation processes include the *O*-methylation of oximes, *N*-oxides, hydroxylamines, and hydroxamic acids. The preparation of oxime ethers using dimethyl sulfate has proven an effective methodology for the construction of sidechains designed for new cephalosporin antibiotics. Treatment of α-oxime esters with dimethyl sulfate afforded the α-methoximino esters, which were subsequently hydrolyzed to the acid and used as sidechains via acylation chemistry (eq 8).[7]

1. Me_2SO_4, K_2CO_3; 2. NaOH, EtOH; 46% (8)

The methylation of pyridine, quinoline, and isoquinoline *N*-oxides to afford *N*-methoxy salts has also been investigated. For example, treatment of quinoline *N*-oxide with dimethyl sulfate provides the *N*-methoxy methylsulfate salt in excellent yield (eq 9).[8] The resulting salts were subsequently treated with cyanide ion to afford cyanopyridines in a mild overall process.

Me_2SO_4; NaCN, H_2O; 93%, 2 steps (9)

A new route to 2-substituted indoles derived from 1-methoxyindole was recently described. Dimethyl sulfate proved to be the reagent of choice for the methylation of the unstable 1-hydroxyindole to give 1-methoxyindole in 51% yield (eq 10).[9] The resulting methoxyindole undergoes *O*-lithiation at the 2-position using ***n*-Butyllithium**, and can be trapped with appropriate electrophiles. It should be noted that methylation of 1-hydroxyindole using ***Iodomethane*** afforded only very low yields of the desired methoxyindole.

Na_2WO_4, 30% H_2O_2, MeOH; Me_2SO_4, K_2CO_3, MeOH, 51% (10)

Finally, a convenient synthesis of ***N,O-Dimethylhydroxylamine***, a reagent used to prepare aldehydes from *N*-methoxy-*N*-methylamides, has been reported. It includes a key dimethylation of an intermediate ethyl hydroxycarbamate using dimethyl sulfate and sodium hydroxide (eq 11).[10]

1. $HONH_2$, HCl; 2. Me_2SO_4, NaOH, 70%; HCl, 93%; MeNHOMe•HCl 200 g (11)

N*-Alkylation.** Dimethyl sulfate has also been a useful reagent for the preparation of *N*-methyl alkyl- and aryl-substituted amines, amides, and quaternary ammonium salts. Simple primary amines can be selectively methylated to afford secondary amines by first protecting the amine as a Schiff base or amidine ester,[16] amide,[17] or carbamate,[18] followed by alkylation using dimethyl sulfate and hydrolysis of the resulting amide or iminium ion. For example, the methylation of amidines has been employed in a route to *N*-alkyl amino acids. The amidine of phenylalanine methyl ester was prepared from the amino ester using dimethylformamide dimethyl acetal (see ***N,N-Dimethylformamide Diethyl Acetal), followed by methylation and hydrolysis to afford the corresponding *N*-methylphenylalanine in good yield without racemization (eq 12).[16] Certain substrates such as phenylglycine methyl ester required the use of ***Methyl Trifluoromethanesulfonate*** as the alkylating agent and lower reaction temperatures to retain optical purity throughout the process. The selective synthesis of substituted *N*-monoalkylaromatic amines has also been reported. For example, 2-nitroacetanilide is cleanly converted to 2-nitro-*N*-methylaniline in one pot using phase-transfer conditions (eq 13).[17]

1. Me_2SO_4, toluene; 2. hydrolysis, 58% (12)

Me_2SO_4, NaOH, K_2CO_3, $Bu_4N^+ HSO_4^-$, 98% (13)

Tertiary alkyl and aromatic amines are also conveniently quaternized using dimethyl sulfate. For example, a series of thioquinanthrenediinium salts for use as potential antibiotics were prepared via methylation of their quinoline precursors using dimethyl sulfate (eq 14).[19]

Me_2SO_4 (14)

S*-Alkylation.** The methylation of sulfur-containing compounds using dimethyl sulfate to afford sulfides and sulfonium ions has also been explored. For example, trimethylsulfonium methyl sulfate has been prepared on a large scale via methylation of dimethyl sulfide.[20] This salt has been subsequently used as a precursor of ***Dimethylsulfonium Methylide, a popular reagent for the preparation of epoxides from ketones (eq 15). *O*-Alkyl *S*-methyl dithiocarbamates have also been prepared via condensation of an alkoxide with ***Carbon Disulfide***, followed by methylation using dimethyl sulfate (eq 16).[21] The resulting dithiocarbamates can then be reduced via radical chemistry to afford the alkane.

Me_2SO_4, MeCN; NaOMe, PhCOMe, 85% (15)

1. CS_2, NaH, imidazole; 2. Me_2SO_4, THF, HMPA, 85% (16)

C*-Alkylation.** Several processes involving the formation of C–C bonds via methylation of organometallics using dimethyl sulfate have been reported. For example, a series of 2-methylbutyrolactones and -valerolactones were prepared via asymmetric alkylation using chiral oxazolines and ***Lithium Diisopropylamide followed by treatment with dimethyl sulfate (eq 17).[22] The resulting oxazoline was then hydrolyzed to afford the optically active lactone. Aryllithium species also undergo rapid alkylation using dimethyl sulfate as the electrophile. For example, 3,4-bis(tributylstannyl)furan undergoes a single tin–lithium exchange and upon treatment with dimethyl sulfate and ***N,N′-Dimethylpropyleneurea*** (DMPU) affords the monomethylated product (eq 18).[23] The resulting compounds were then subjected to various palladium-mediated cross-coupling reactions to give 3,4-disubstituted unsymmetrical furans. Dimethyl sulfate was found to be superior to methyl iodide for this particular application.

TMSO … MeO — 1. LDA, THF; 2. Me_2SO_4 → TMSO … MeO — HCl, 58%, 2 steps → 70% ee (17)

Bu_3Sn, $SnBu_3$ — 1. BuLi, THF, −78 °C; 2. Me_2SO_4, DMPU, 65% → Bu_3Sn (18)

Related Reagents. Diazomethane; Iodomethane; Methyl Trifluoromethanesulfonate; Trimethyloxonium Tetrafluoroborate; Trimethyl Phosphate.

1. (a) Suter, C. M. *The Organic Chemistry of Sulfur*; Wiley: New York, 1944; pp 48–74. (b) Kaiser, E. T. *The Organic Chemistry of Sulfur*; Plenum: New York, 1977; p 649. (c) Use as methylating agent: Fieser, L.; Fieser, M. *FF* **1967**, *1*, 293 and further references cited therein.
2. (a) Material Safety Data Sheets (MSDS) on dimethyl sulfate are available from various vendors. (b) *Merck Index* 11th ed.; 1989, p 3247. (c) Review of carcinogenicity studies: *IARC Monographs* **1974**, *4*, 271. (d) Mutagenicity studies: Hoffman, G. R. *Mutat. Res.* **1980**, *75*, 63.
3. Stodola, F. H. *JOC* **1964**, *29*, 2490.
4. Achet, D.; Rocrelle, D.; Murengezi, I.; Delmas, M.; Gaset, A. *S* **1986**, 642.
5. Hart, D. J.; Merriman, G. H. *TL* **1989**, *30*, 5093.
6. Benson, R. E.; Cairns, T. L. *JACS* **1948**, *70*, 2115.
7. Kukolja, S.; Draheim, S. E.; Pfeil, J. L.; Cooper, R. D. G.; Graves, B. J.; Holmes, R. E.; Neel, D. A.; Huffman, G. W.; Webber, J. A.; Kinnick, M. D.; Vasileff, R. T.; Foster, B. J. *JMC* **1985**, *28*, 1886.
8. Feely, W. E.; Beavers, E. M. *JACS* **1959**, *81*, 4004.
9. Kawasaki, T.; Kodama, A.; Nishida, T.; Shimizu, K.; Somei, M. *H* **1991**, *32*, 221.
10. Goel, O. P.; Krolls, U. *OPP* **1987**, 75.
11. Hock, H.; Lang, S.; Duyfjes, W. *CB* **1942**, *75*, 300.
12. Ballini, R.; Carotti, A. *SC* **1983**, 1197.
13. Ogawa, H.; Ichimura, Y.; Chihara, T.; Shousuke, T.; Taya, K. *BCJ* **1986**, *59*, 2481.
14. Williams, D. R.; Phillips, J. G.; Barner, B. A. *JACS* **1981**, *103*, 7398.
15. Knolker, H. J.; Bauermeister, M.; Blaser, D.; Boese, R.; Pannek, J. B. *AG* **1989**, 225.
16. O'Donnel, M. J.; Bruder, W. A.; Daugherty, B. W.; Liu, D.; Wojciechowski, K. *TL* **1984**, 3651.
17. Kalkote, U. R.; Choudhary, A. R.; Natu, A. A.; Lahoti, R. J.; Ayyangar, N. R. *SC* **1991**, 1889.
18. Clark-Lewis, J. W.; Thompson, M. J. *JCS* **1957**, 442.
19. Maslankiewicz, A.; Zieba, A. *H* **1992**, *33*, 247.
20. Kutsuma, T.; Nagayama, I.; Okazaki, T.; Sakamoto, T.; Akaboshhi, S. *H* **1977**, *8*, 397.
21. Barton, D. H. R.; McCombie, S. W. *JCS(P1)* **1975**, 1574.
22. Meyers, A. I.; Yamamoto, Y.; Mihelich, E. D.; Bell, R. A. *JOC* **1980**, *45*, 2792.
23. Yang, Y.; Wong, H. N. C. *CC* **1992**, 1723.

Gregory Merriman
Hoechst-Roussel Pharmaceuticals, Somerville, NJ, USA

Diphenylbis(1,1,1,3,3,3-hexafluoro-2-phenyl-2-propoxy)sulfurane[1]

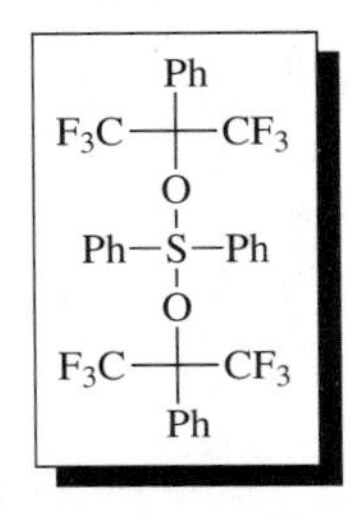

[32133-82-7] $C_{30}H_{20}F_{12}O_2S$ (MW 672.57)

(dehydration of alcohols;[2] synthesis of epoxides and cyclic ethers;[3] cleavage of amides;[5] oxidation of amines[6])

Physical Data: mp 107–109 °C.
Solubility: sol ether, benzene, acetone, alcohols.
Form Supplied in: white crystals.
Analysis of Reagent Purity: NMR, IR.
Preparative Method: by the reaction of the potassium salt of 1,1,1,3,3,3-hexafluoro-2-phenylisopropanol with diphenyl sulfide in the presence of chlorine in ether at −78 °C.[1a]
Handling, Storage, and Precautions: avoid moisture; readily hydrolyzed; stable at rt; decomposes slowly at rt in solution.

Dehydration of Alcohols. The title reagent (**1**) is useful for the dehydration of alcohols. In general, tertiary alcohols are dehydrated instantaneously at rt. Some secondary alcohols are dehydrated. In cyclohexane rings, a *trans*-diaxial orientation of the leaving groups significantly increases the rate of elimination (eq 1). Primary alcohols do not yield products of dehydration unless the β-proton is sufficiently acidic. In most cases, the ether [$(CF_3)_2PhCOR$] is obtained.[2]

Epoxides. Vicinal diols, capable of attaining an antiperiplanar relationship, can be converted to epoxides (eq 2). The reaction requires 1–2 equiv of (**1**) in chloroform, ether, or carbon tetrachloride and takes place at rt. The reaction is postulated to take place via ligand exchange with the sulfone followed by decomposition to the epoxide, diphenyl sulfoxide, and 1,1,1,3,3,3-hexafluoro-2-phenylisopropanol.

(1) rate = 150; (1) rate = 1 (1)

(1)

$+ Ph_2SO + (CF_3)_2PhCOH$ (2)

Other cyclic ethers have been prepared, but yields are highly dependent on product ring size. The following transformations are representative: 2,2-dimethyl-1,3-propanediol to 3,3-dimethyloxetane (86%), 1,4-butanediol to tetrahydrofuran (72%), 1,5-pentanediol to tetrahydropyran (39%), and diethylene glycol to dioxane (40%). Longer chain diols yield ethers $[(CF_3)_2PhCO(CH_2)_n OCPh(CF_3)_2]$.[3]

Eschenmoser used this method to convert (5*R*,6*R*)-5,6-dihydro-β,β-carotene-5,6-diol to its epoxide (eq 3). This reagent is more effective than other reagents due to the unique solubility profile of the dihydrocarotenediol.[4]

(1) (3)

Cleavage of Amides. Secondary amides can be converted to esters with (**1**). The rate is sensitive to steric constraints at the nitrogen and the acyl carbon. In most cases the amine portion is trapped as the sulfilimine and/or the imidate, which are easily converted back to the amine (eq 4). The dual nature of this reaction affords a mild conversion of amides to esters as well as a simple method for deprotection of *N*-acylated amines.[5]

$RCONHBu \xrightarrow{(1)} RCO_2C(CF_3)_2Ph + BuN{=}SPh_2 + BuN{=}COC(CF_3)_2Ph \rightarrow BuNH_2$ (4)

Oxidation of Amines. In a related reaction, (**1**) reacts with primary amines (as well as amides and sulfonamides) to give sulfilimines (eq 5). Secondary amines are converted to imines on reaction with (**1**) whereas benzylamine is converted to benzonitrile (89%) with 2 equiv of (**1**).[6]

$$H_2N\text{-}R \xrightarrow{(1)} Ph_2S{=}N\text{-}R + 2\,(CF_3)_2PhC\text{-}OH \quad (5)$$

1. (a) *FF* **1974**, *4*, 205. (b) *FF* **1975**, *5*, 270. (c) *FF* **1977**, *6*, 239. (d) *FF* **1980**, *8*, 208.
2. Martin, J. C.; Arhart, R. J. *JACS* **1971**, *93*, 4327.
3. Martin, J. C.; Franz, J. A.; Arhart, R. J. *JACS* **1974**, *96*, 4604.
4. Eschenmoser, W.; Engster, C. H. *HCA* **1978**, *61*, 822.
5. (a) Franz, J. A.; Martin, J. C. *JACS* **1973**, *95*, 2017. (b) Franz, J. A.; Martin, J. C. *JACS* **1975**, *97*, 6137.
6. Franz, J. A.; Martin, J. C. *JACS* **1975**, *97*, 583.

Brian A. Roden
Abbott Laboratories, North Chicago, IL, USA

Diphenylphosphinic Chloride

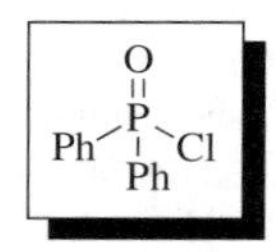

[1499-21-4] $C_{12}H_{10}ClOP$ (MW 236.64)

(agent for acid activation,[1] alkylphosphine oxide formation,[2] and amine protection[3])

Alternate Name: chlorodiphenylphosphine oxide.
Physical Data: bp 222 °C/16 mmHg; *d* 1.240 g cm^{-3}.
Form Supplied in: colorless liquid; widely available.
Handling, Storage, and Precautions: handle only in a chemical fume hood, wear chemical-resistant gloves and safety goggles, do not breathe vapor, and avoid contact with eyes or skin. Store in a 'flammable storage' cabinet under nitrogen. Contact with water generates hydrogen chloride gas.

Acid Activation. Chlorodiphenylphosphine oxide has been utilized for the activation of acids via in situ formation of the diphenylphosphinic (DPP) mixed anhydrides. These anhydrides are superior to carbon-based mixed anhydrides because they do not suffer from disproportionation to symmetrical anhydrides.[1] Also, nucleophiles prefer to attack at carbon rather than phosphorus, solving the regioselectivity problem associated with carbon-based mixed anhydrides.[4] Additionally, diphenylphosphinic mixed anhydrides are more electrophilic. Finally, diphenylphosphinic mixed anhydrides form rapidly allowing shorter activation times.

Diphenylphosphinic mixed anhydrides have been utilized to form peptide bonds.[5] Peptides are easier to isolate by this method than by employing ***1,3-Dicyclohexylcarbodiimide***. These anhydrides are the method of choice for the formation of amides of 2-alkenoic acids (eq 1).[6] Carbodiimide and acyl carbonate methods proved to be inferior. Primary amines result in better yields than secondary amines. This activation protocol can be employed to form thiol esters (eq 2).[7] β-Amino acids are readily converted to β-lactams with chlorodiphenylphosphine oxide (eq 3).[8] Secondary amines work best. This activation protocol has been utilized to convert acids to amines via a Curtius rearrangement.[9] Phenols have been generated from diene acids, presumably via base-induced elimination of diphenylphosphinic acid from the mixed anhydrides to form ketenes which spontaneously cyclize.[10] Acids have been converted to ketones via activation followed by reaction with organometallic reagents (eq 4).[11]

$$\mathrm{ZNHCH(R^1)CO_2H} \xrightarrow[\mathrm{NMM}]{\mathrm{Ph_2POCl}} \left[\mathrm{ZNHCH(R^1)C(O)OP(O)Ph_2}\right] \xrightarrow{\mathrm{H_2NCH(R^2)CO_2H}} \mathrm{ZNHCH(R^1)C(O)NHCH(R^2)CO_2Ph} \quad (1)$$

$$\mathrm{ZNHCH(R)CO_2H} \xrightarrow[\mathrm{Et_3N}]{\mathrm{Ph_2POCl}} \xrightarrow[\mathrm{Et_3N}]{\mathrm{PhSH}} \underset{98\%}{\mathrm{ZNHCH_2COSPh}} \quad (2)$$

$$\mathrm{HO_2C\ \ NHCH_2Ph} \xrightarrow[\mathrm{MeCN,\ 80\ ^\circ C}\ \ 94\%]{\mathrm{Ph_2POCl,\ Et_3N}} \beta\text{-lactam (N-CH}_2\text{Ph)} \quad (3)$$

$$\mathrm{C_5H_{11}C(O)OP(O)Ph_2} \begin{cases} \xrightarrow[67\%]{\mathrm{EtMgBr}} \mathrm{C_5H_{11}C(O)Et} \\ \xrightarrow[\mathrm{CH_2(CO_2Et)_2}\ \ 83\%]{\mathrm{NaH}} \mathrm{C_5H_{11}C(O)CH(CO_2Et)_2} \end{cases} \quad (4)$$

Phosphine Oxides. This reagent reacts with organometallic reagents to form phosphine oxides.[2] Alkyl phosphonates have been converted to diphenylvinylphosphine oxides via trapping of the α-lithioalkylphosphonate with diphenylchlorophosphine oxide, deprotonation, and reaction with aldehydes (eq 5).[12] Organometallic reagents have been converted to 1,3-dienes utilizing Ph2 P(O)Cl and carbonyl compounds).[13]

Thiol Esters and Ketones. The Dpp mixed anhydrides react with thiols to yield thiol esters (eq 2)[7] and undergo carbon-acylation reactions with both Grignard reagents and diethyl sodiomalonate in good yields (eq 6).[14]

$$\mathrm{(RO)_2POCH_2R'} \xrightarrow[\text{2. (EtO)}_2\text{POCl; 3. RCHO}]{\text{1. LDA, }-78^\circ\text{C}} \mathrm{(EtO)_2P(O){-}CR'{=}CHR} \quad (5)$$

$$\mathrm{C_5H_{11}C(O)OP(O)Ph_2} \begin{cases} \xrightarrow[67\%]{\mathrm{EtMgBr}} \mathrm{C_5H_{11}C(O)Et} \\ \xrightarrow[\mathrm{CH_2(CO_2Et)_2}\ \ 83\%]{\mathrm{NaH}} \mathrm{C_5H_{11}C(O)CH(CO_2Et)_2} \end{cases} \quad (6)$$

Protection of Amino Groups. In the presence of *N*-methylmorpholine, Ph_2POCl reacts with amino acid esters in CH_2Cl_2 at 0 °C in 1.5–2 h to give *N*-diphenylphosphinyl (Dpp) amino acid esters (65–80%), which are then hydrolyzed under mild alkaline conditions to give *N*-Dpp amino acids (eq 7).[3,4] Since Ph_2POCl hydrolyzes extremely quickly under aqueous conditions, it is not possible to prepare *N*-Dpp amino acids directly.

$$\mathrm{H_2NCH_2CO_2Me} \xrightarrow[\mathrm{NMM}\ \ 74\%]{\mathrm{Ph_2POCl}} \mathrm{Ph_2P(O)NHCH_2CO_2Me} \quad (7)$$

The Dpp derivatives are slightly more acid-labile than the Boc derivatives and are cleaved by HOAc–HCOOH–H_2O (7:1:2) (24 h), 1 M HCl in H_2O–dioxane (2:1) (3 h), TFA–CH_2Cl_2 (1:1) (40 min), and 95% TFA (10 min). Selective cleavage of the Dpp group in the presence of a *t*-butyl ester and a phenyl ester is possible.

Amination. Ph_2POCl reacts with ***Hydroxylamine*** to yield *O*-(diphenylphosphinyl)hydroxylamine.[15] This reagent aminates stabilized carbanions and Grignard reagents in moderate yields (eq 8).[16] Electrophilic *N*-amination of imide sodium salts is also possible using this reagent (eq 9).[17]

$$\mathrm{Ph_2P(O)ONH_2} \begin{cases} \xrightarrow[51\%]{\mathrm{PhCH_2MgBr}} \mathrm{PhCH_2NH_2} \\ \xrightarrow[45\%]{\mathrm{PhCH{=}C(OLi)OEt}} \mathrm{PhCH(NH_2)CO_2Et} \end{cases} \quad (8)$$

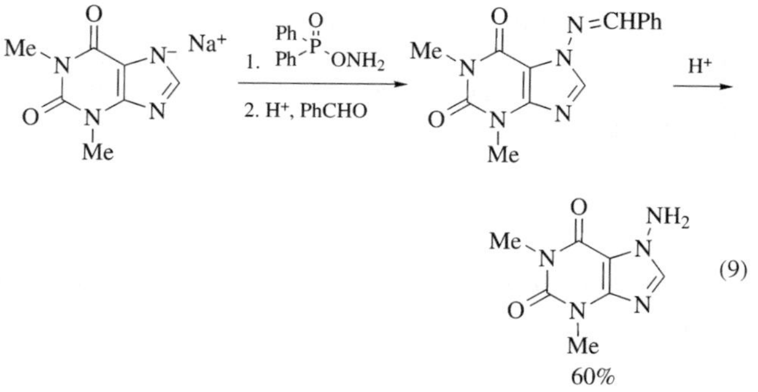

(9)

60%

Diphenylphosphinamide[18] and 1-diphenylphosphinyl-2,2-dimethylaziridine,[19] which are derived from Ph_2POCl, react with Grignard reagents to give high yields of primary amines and moderate yields of α,α-dimethylarylalkylamines, respectively.

1. Jackson, A. G.; Kenner, G. W.; Moore, G. A.; Ramage, R.; Thorpe, W. D. *TL* **1976**, 3627.
2. Takaya, H.; Mashima, K.; Koyano, K.; Yagi, M.; Kumobayashi, H.; Taketomi, T.; Akutagawa, S.; Noyori, R. *JOC* **1986**, *51*, 629.
3. (a) Kenner, G. W.; Moore, G. A.; Ramage, R. *TL* **1976**, 3623. (b) Ramage, R.; Atrash, B.; Hopton, D.; Parrott, M. J. *JCS(P1)* **1985**, 1217.
4. Ramage, R.; Atrash, B.; Hopton, D.; Parrott, M. J. *JCS(P1)* **1985**, 1617.
5. Ramage, R.; Hopton, D.; Parrott, M. J.; Richardson, R. S.; Kenner, G. W.; Moore, G. A. *JCS(P1)* **1985**, 461.
6. Bernasconi, S.; Comini, A.; Corbella, A.; Gariboldi, P.; Sisti, M. *S* **1980**, 385.
7. Horiki, K. *SC* **1977**, *7*, 251.
8. Kim, S.; Lee, P. H.; Lee, T. A. *CC* **1988**, 1242.
9. (a) Armstrong, V. W.; Coulton, S.; Ramage, R. *TL* **1976**, 4311. (b) Ramage, R.; Armstrong, V. W.; Coulton, S. *T* **1981**, *37*, Supplement No. 1, 157.
10. Clinch, K.; Marquez, C. J.; Parrott, M. J.; Ramage, R. *T* **1989**, *45*, 239.
11. Soai, K.; Ookawa, A. *CC* **1986**, 412.
12. (a) Savignac, P.; Teulade, M.-P.; Aboujaoude, E. E.; Collignon, N. *SC* **1987**, *17*, 1559. (b) Aboujaoude, E. E.; Lietje, S.; Collignon, N. *TL* **1985**, *26*, 4435.
13. Davidson, A. H.; Warren, S. *CC* **1975**, 148.
14. Kende, A. S.; Scholz, D.; Schneider, J. *SC* **1978**, 59.
15. (a) Harger, M. J. P. *CC* **1979**, 768. (b) Harger, M. J. P. *JCS(P1)* **1981**, 3284.
16. (a) Colvin, E. W.; Kirby, G. W.; Wilson, A. C. *TL* **1982**, *23*, 3835. (b) Boche, G.; Bernheim, M.; Schrott, W. *TL* **1982**, *23*, 5399.
17. Klotzer, W.; Stadlwieser, J.; Raneburger, J. *OS* **1985**, *64*, 96.
18. Zwierzak, A.; Slusarska, E. *S* **1979**, 691.
19. Buchowiecki, W.; Grosman-Zjawiona, Z.; Zjawiony, J. *TL* **1985**, *26*, 1245.

David N. Deaton
Glaxo Research Institute, Research Triangle Park, NC, USA

Sunggak Kim
Korea Advanced Institute of Science and Technology, Taejon, Korea

Diphenyl Phosphorazidate

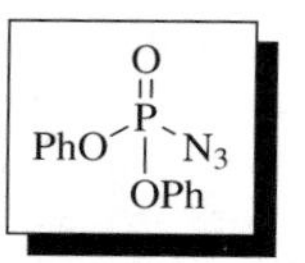

[26386-88-9] $C_{12}H_{10}N_3O_3P$ (MW 275.22)

(peptide synthesis;[1] synthesis of α-aryl carboxylic acids;[4] synthesis of thiol esters;[7,8] stereospecific conversion of alcohols to azides;[9] ring contraction of cycloalkanones;[13] decarbonylation of aldehydes[26])

Alternate Names: DPPA; *O,O*-diphenylphosphoryl azide.
Physical Data: bp 157 °C/0.17 mmHg; *d* 1.277 g cm^{-3}; fp >110 °C.
Solubility: sol toluene, THF, DMF, *t*-butyl alcohol.
Form Supplied in: colorless liquid; commercially available in kilogram quantities.
Preparative Method: by the reaction of ***Diphenyl Phosphorochloridate*** with a slight excess of ***Sodium Azide*** in acetone at rt (85–90% yield).[1]
Purification: vacuum distillation (bath temp <200 °C, dec >200 °C)
Handling, Storage, and Precautions: handle and store under nitrogen. Refrigerate. Decomposes at >200 °C and may produce toxic fumes of phosphorus oxides and/or phosphine. Toxic, irritant. Harmful if swallowed, inhaled, or absorbed through the skin. Use in fume hood. Incompatible with oxidizing agents and acids. Hazardous combustion or decomposition products include carbon monoxide, carbon dioxide, and nitrogen oxides. Prolonged contact with moisture may produce explosive hydrogen azide (HN_3).

General. Diphenyl phosphorazidate is a readily available, nonexplosive, and relatively stable azide widely used as a reagent in peptide synthesis,[1–3] and as a versatile reagent in a wide array of organic transformations. DPPA has been successfully utilized in the synthesis of α-amino acids[4] and α-aryl carboxylic acids;[5,6] direct preparation of thiol esters from carboxylic acids and thiols;[7,8] the stereospecific preparation of alkyl azides;[9] and the phosphorylation of alcohols and amines.[10] The application of DPPA in a modified Curtius[11] reaction permits a simple one-step conversion of carboxylic acids to urethanes under mild reaction conditions. DPPA acts as a nitrene source,[12] and can undergo 1,3-dipolar cycloaddition reactions.[5,13] The Curtius degradation[14] of carboxylic acids in the presence of *t*-butanol gives the Boc-protected amine directly (eq 1).

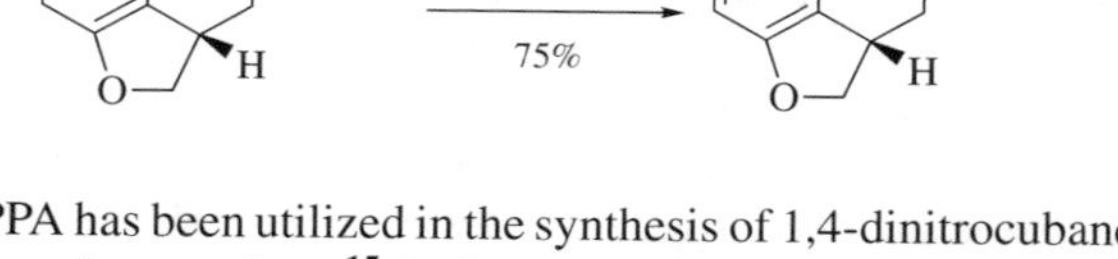

DPPA, *t*-BuOH, Et_3N, 75% (1)

DPPA has been utilized in the synthesis of 1,4-dinitrocubane by Eaton and co-workers.[15] Refluxing cubane-1,4-dicarboxylic acid with DPPA and ***Triethylamine*** in *t*-butanol forms the intermediate *t*-butyl carbamate in nearly quantitative yield. This method avoids the formation of the explosive diacyl azide.

Peptide Synthesis. Urethanes are readily prepared by refluxing equimolar mixtures of DPPA, a carboxylic acid, an alcohol, and triethylamine. The reaction involves transfer of the azide group from DPPA to the carboxylic acid. The resulting acyl azide subsequently undergoes a Curtius rearrangement. This reaction has been successfully applied in the area of peptide synthesis.[1] The coupling of acylamino acids or peptides with amino acid esters or peptide esters in the presence of base proceeds in high yields without racemization, and is compatible with a variety of functional groups. The reaction of malonic acid half-esters results in the corresponding α-amino acid derivatives[4] (eq 2). It should be noted that addition of the alcohol at the beginning of the reaction results in esterification.

$$PhH_2C{-}CH(CO_2Et)(CO_2H) \xrightarrow[2.\ PhCH_2OH]{1.\ DPPA,\ Et_3N} PhH_2C{-}CH(CO_2Et)(NHCO_2Bn) \quad 79\% \qquad (2)$$

α-Aryl Carboxylic Acids. Alkyl aryl ketones are converted into the corresponding α-aryl alkanoic acids[4] via a three-step sequence. Yields of >90% are possible if the synthesis is carried out in a one-flask procedure. The sequence includes a 1,3-dipolar cycloaddition of DPPA to the corresponding enamines (eq 3). Although ***Thallium(III) Nitrate Trihydrate*** has also been used in similar oxidative rearrangements,[16] the DPPA method exhibits higher functional specificity, uses less toxic reagents, and gives preparatively useful yields. Naproxen, a nonsterodial anti-inflammatory therapeutic, has been prepared by this route.[6]

MeCH₂COAr → (BF₃•Et₂O, pyrrolidine; 55–80%) enamine → (DPPA, THF) [triazoline, N–P(O)(OPh)₂] → (–N₂, 82%) amidine N–P(O)(OPh)₂ → (KOH, 83%) Ar–CH(CH₃)–CO₂H (3)

Ar = 6-methoxynaphthyl

Naproxen

Stereospecific Conversion of Alcohols to Azides. Reaction of an alcohol with DPPA, ***Triphenylphosphine***, and ***Diethyl Azodicarboxylate*** forms the corresponding azides in 60–90% yields.[9] The stereospecific nature of this reaction permits the conversion of Δ^5-sterols such as 3β-cholestanol exclusively to the 3α-cholestanyl azide in 75% yield. This synthesis is clearly superior to the alcohol → tosylate → azide route which is longer and also prone to competing elimination reactions.

Ring Contraction Reactions. DPPA undergoes 1,3-dipolar cycloadditions to the enamines of cyclic ketones.[13] The resulting Δ^2-triazolines, which are not isolated, undergo loss of nitrogen to form ring-contracted products, which, on hydrolysis, yield cycloalkanoic acids. In the case of six- to eight-membered cycloalkanones, overall yields as high as 75% have been reported (eq 4).

cyclododecanone → (BF₃•Et₂O, pyrrolidine) enamine → (DPPA, PhMe) [triazoline, N–PO(OPh)₂] → (–N₂) cycloundecyl(pyrrolidinyl)C=NP(O)(OPh)₂ → (KOH) cycloundecane–CO₂H (4)

40% overall

Amination of Aromatic and Heteroaromatic Organometallics. Reaction of organic azides with Grignard and lithium compounds gives 1,3-disubstituted triazenes,[17] which are readily converted to amines by reductive[18,19] or hydrolytic[20,21] workup. These methods are limited to either the aromatic Grignard[18,20] or lithium[21] compounds, and have not been very successful with heteroaromatic organometallics. Aromatic and heteroaromatic organometallics (Grignard and lithium compounds) are aminated in good yields in a one-pot process by treatment with DPPA and reduction of the resulting phosphoryltriazenes with ***Sodium Bis(2-methoxyethoxy)aluminum Hydride*** (eq 5).[22]

$$ArMgBr + DPPA \xrightarrow{Et_2O} [Ar{-}N{=}N{-}N(MgBr){-}P(O)(OPh)_2] \xrightarrow{NaAlH_2(O(CH_2)_2OMe)_2} ArNH_2 \quad 45\text{–}88\% \qquad (5)$$

Metal-Catalyzed Decarbonylation of Primary Aldehydes. The decarbonylation of primary aldehydes under catalytic conditions is difficult, requiring high temperatures or involving radical mechanisms;[23] the rt decarbonylations[24,25] using stoichiometric ***Chlorotris(triphenylphosphine)rhodium(I)*** have had limited practicality due to the high cost of the reagent. Recently, a high-yielding, rt decarbonylation of primary aldehydes using catalytic amounts (5 mol %) of $Rh(PPh_3)_3Cl$ and stoichiometric amounts of DPPA, with minimal formation of alkene byproducts, has been developed[26] (eq 6).

citronellal (CHO) → (DPPA, 5 mol% Rh(PPh₃)₃Cl, 99%) decarbonylated alkene (6)

Synthesis of Macrocyclic Lactams. Macrocyclic lactams are conventionally prepared by the reaction of dicarboxylic acid chlorides with diamines, a method which is effective with simple acyl chlorides. Activation of the carboxylic groups in order to overcome the drawbacks of high dilution and the low yields and purity encountered with larger acyl chlorides has seen only limited success.[27] The superior activating ability of DPPA has recently been demonstrated.[28] The reactions of dicarboxylic acids and diamines in the presence of DPPA form macrobicyclic lactams in yields as high as 82%. By comparison, in a control experiment using conventional high-dilution techniques, the corresponding acyl chlorides cyclized with the diamine to form the lactam in only 24% yield.

DPPA has been used for the direct *C*-acylation of methyl isocyanoacetate with carboxylic acids to give 4-methoxycarbonyloxazoles.[29] L-Daunosamine, the glycone component of anticancer anthracycline antibiotics, has been synthesized from L-lactic acid in 9 steps with a 24% overall yield,[30] where a key step in the sequence is the direct *C*-acylation of methyl isocyanoacetate with the lithium salt of the lactate ester using diphenyl phosphorazidate (eq 7).

$MeOH_2CO$ / CO_2CH_2OMe — 1. LiOH, H_2O–THF; 2. DPPA, DMF; 3. CO_2Me / NC, NaH, DMF →

$MeOH_2CO$ / CO_2Me (oxazole, O, N) ⟹ HO / O / OH / $ClH{\bullet}H_2N$ (7)

Related Reagents. Diphenyl Phosphorochloridate.

1. Shioiri, T.; Ninomiya, K.; Yamada, S. *JACS* **1972**, *94*, 6203.
2. Shioiri, T.; Yamada, S. *CPB* **1974**, *22*, 849.
3. Yamada, S.; Ikota, N.; Shioiri, T.; Tachibana, S. *JACS* **1975**, *97*, 7174.
4. Yamada, S.; Ninomiya, K.; Shioiri, T. *TL* **1973**, 2343.
5. Shioiri, T.; Kawai, N. *JOC* **1978**, *43*, 2936.
6. Riegel, J.; Madox, M. L.; Harrison, I. T. *JMC* **1974**, *17*, 377.
7. Yamada, S.; Yokoyama, Y.; Shioiri, T. *JOC* **1974**, *39*, 3302.
8. Yokoyama, Y.; Shioiri, T.; Yamada, S. *CPB* **1977**, *25*, 2423.
9. Lal, B.; Pramanik, B.; Manhas, M. S.; Bose, A. K. *TL* **1977**, 1977.
10. Cremlyn, R. J. W. *AJC* **1973**, *26*, 1591.
11. (a) Ninomiya, K.; Shioiri, T.; Yamada, S. *T* **1974**, *30*, 2151. (b) Ninomiya, K.; Shioiri, T.; Yamada, S. *CPB* **1974**, *22*, 1398.
12. Breslow, R.; Feiring, A.; Herman, F. *JACS* **1974**, *96*, 5937.
13. Yamada, S.; Hamada, Y.; Ninomiya, K.; Shioiri, T. *TL* **1976**, 4749.
14. Haefliger, W.; Klöppner, E. *HCA* **1982**, *65*, 1837.
15. Eaton, P. E.; Ravi Shanker, B. K. *JOC* **1984**, *49*, 185.
16. Taylor, E. C.; Chiang, C.-S.; McKillop, A.; White, J. F. *JACS* **1976**, *98*, 6750.
17. Dimroth, R. *CB* **1903**, *36*, 909.
18. Smith, P. A. S.; Rowe, C. D.; Bruner, L. B. *JOC* **1969**, *34*, 3430.
19. Reed, J. N.; Snieckus, V. *TL* **1983**, *24*, 3795.
20. Trost, B. M.; Pearson, W. H. *JACS* **1981**, *103*, 2483.
21. Hassner, A.; Munger, P.; Belinka, B. A. *TL* **1982**, *23*, 699.
22. Mori, S.; Aoyama, T.; Shioiri, T. *TL* **1984**, *25*, 429.
23. Domazetis, G.; Tarpey, B.; Dolphin, D.; James, B. R. *CC* **1980**, 939.
24. Tsuji, J.; Ohno, K. *TL* **1965**, 3969.
25. Osborn, J. A.; Jardine, F. H.; Young, J. F.; Wilkinson, G. *JCS(A)* **1966**, 1711.
26. O'Connor J. M.; Ma J. *JOC* **1992**, *57*, 5075.
27. Cazaux, L.; Duriez, M. C.; Picard, C.; Moieties, P. *TL* **1989**, *30*, 1369.
28. Qian, L.; Sun, Z.; Deffo, T.; Mertes, K. B. *TL* **1990**, *31*, 6469.
29. Hamada, Y.; Shioiri, T. *TL* **1982**, *23*, 235, 1226.
30. Hamada, Y.; Kawai, A.; Shioiri, T. *TL* **1984**, *25*, 5409.

Albert V. Thomas
Abbott Laboratories, North Chicago, IL, USA

2,2′-Dipyridyl Disulfide[1]

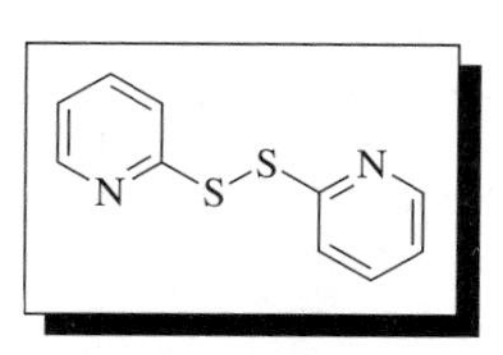

[2127-03-9] $C_{10}H_8N_2S_2$ (MW 220.34)

(macrolactonization of ω-hydroxy acids,[6] acylation of pyrroles,[22] couplings at anomeric centers[27])

Alternate Names: 2,2′-dithiopyridine.
Physical Data: mp 57–58 °C.
Solubility: benzene, THF, CH_2Cl_2, MeCN, DMF.
Form Supplied in: white (colorless) crystalline solid.
Preparative Methods: although 2,2′-dipyridyl disulfide is commercially available, it can be readily prepared from 2-pyridinethione by oxidation with a number of reagents.[2]
Purification: recrystallization from hexane at 30 mL g^{-1} concentration.[3]
Handling, Storage, and Precautions: the solid is irritating to eyes, mucus membranes, and respiratory tract, and physical contact should be avoided. Combustion produces toxic byproducts including nitrogen oxides. Cold storage in air-tight containers is recommended. Use in a fume hood.

Macrolactonization.[4] When a carboxylic acid is treated with 2,2′-dipyridyl disulfide in the presence of ***Triphenylphosphine***, the corresponding 2-pyridinethiol ester is formed.[5] Corey and Nicolaou have developed an efficient method for the synthesis of macrocyclic lactones based on these 2-pyridinethiol esters.[6] When an ω-hydroxy thiolester is heated in refluxing xylene under high dilution conditions (10^{-5} M, typically accomplished with syringe pump techniques), macrolactonization occurs, liberating triphenylphosphine oxide and pyridinethione. The reaction is quite general and is believed to proceed by a 'double activation' mechanism in which the basic 2-pyridinethiol ester simultaneously activates both the hydroxy and the carboxylic acid moieties with a single proton transfer. It has been shown that the cyclization rate is not affected by the presence of acids, bases, or any of the possible reaction contaminants.[7]

This method is mild, highly efficient for the preparation of medium to large rings (7–16,[6] 12–21[7]), and has been applied to the synthesis of a number of important macrocyclic targets including monensin,[8] brefeldin A (eq 1) and erythronolide B,[9] (±)-11-hydroxy-*trans*-8-dodecenoic acid lactone,[10] (±)-vermiculine,[11] enterobactin,[12] and prostaglandins.[8,13]

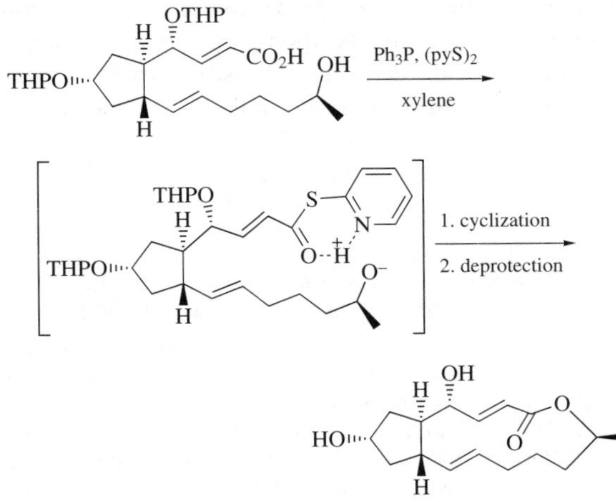

Brefeldin A (1)

In an effort to develop an even milder lactonization protocol, other diaryl disulfides were explored, the most promising being imidazole derivatives (**1**) and (**2**).[14] The formation of dilides (dimers of macrocyclic lactones) is occasionally a problem, necessitating the use of other cyclization methods.[4,15] A variation of Corey's method using silver salts has been developed.[3,16]

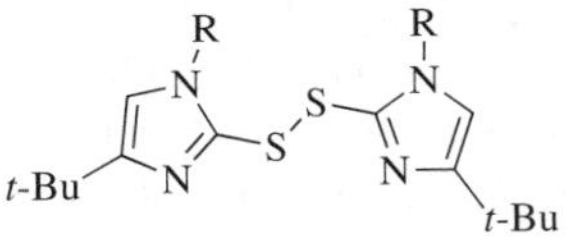

(**1**) R = Me
(**2**) R = *i*-Pr

If an ω-aminopyridinethiol ester is used, macrolactamization occurs.[17] A method for the synthesis of β-lactams from β-amino acids has also been developed where the use of MeCN has been found to be critical (eq 2).[18] The intermolecular version of this reaction is a powerful peptide coupling method.[19,20]

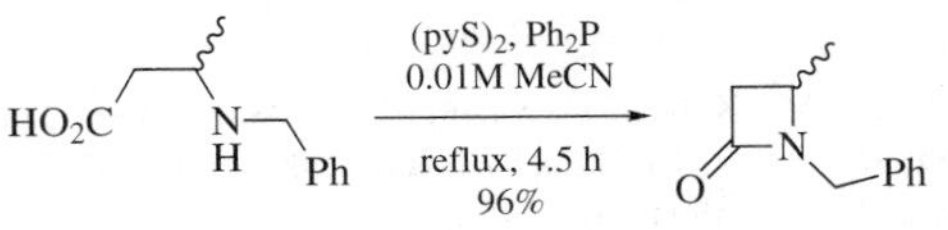

(2)

Related Reactions of 2-Pyridinethiol Esters. In the absence of an internal nucleophile, the thiopyridyl esters generated by the reaction of carboxylic acids with 2,2′-dipyridyl disulfide and triphenylphosphine exhibit other reactivities. Thiolesters are potent acylating agents, reacting with Grignard reagents to yield ketones instead of tertiary alcohols.[21,22] This activity has been exploited for the synthesis of 2-acylpyrroles using pyrrylmagnesium salts as the nucleophilic species. Nicolaou and co-workers have exploited 2-pyridinethiol esters in the synthesis of a number of complex 2-acylpyrroles (eq 3).[23] In certain troublesome cases, the use of stoichiometric ***Copper(I) Iodide*** has been found to significantly improve yields.[24]

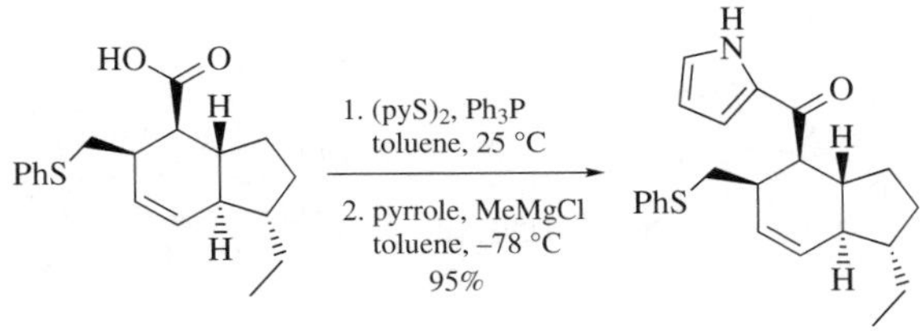

(3)

2-Pyridinethiol esters are readily reduced with ***Sodium Borohydride*** in the presence of isopropanol.[25] Activation of 2-pyridinethiol esters with ***Iodomethane*** allows for mild trapping with alcohols or benzenethiol, yielding esters or thiolesters.[26] In these cases the use of iodomethane avoids the need for thiophilic silver or mercury salts and allows for transthiolesterification.

Disaccharide Formation. Treatment of glycoside (**3**) with 2,2′-dipyridyl disulfide and ***Tri-n-butylphosphine*** in CH_2Cl_2 rapidly yields thiopyridyl derivative (**4**) as a mixture of anomers. Activation of (**4**) with iodomethane, followed by treatment with glycoside acceptor (**5**), affords the disaccharide fragment of avermectin (**6**),[27] exclusively α-linked in 78% yield (eq 4).[28] Although other methods are available, this protocol offers advantages in practicality and stereoselectivity.

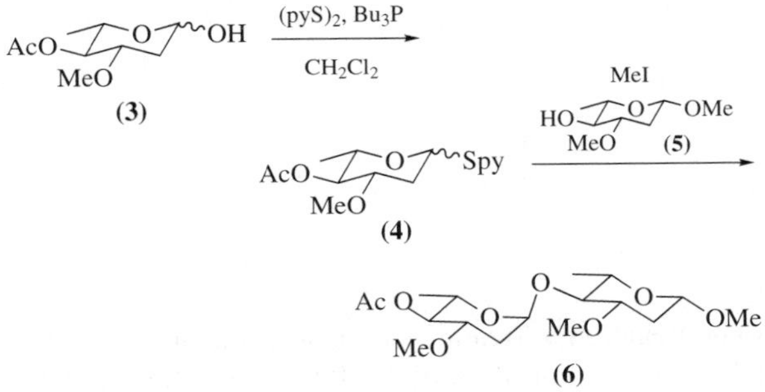

(4)

Carbon–carbon bonds can be similarly formed at the anomeric center of carbohydrates. Oxonium ions, formed by the treatment of 1-(2′-thiopyridyl)glycosides with ***Silver(I) Trifluoromethanesulfonate***, can be trapped with silyl enol ethers, silyl ketene acetals, and reactive aromatic compounds, where the stereoselectivity of the addition is determined by solvent and nucleophile choice.[29] The intramolecular version of this process has also been examined (eq 5).[30] Similarly, bicyclic piperazinediones are available by the intramolecular trapping of iminium ions, generated from the appropriate thiopyridyl derivatives with $PhHgClO_4$ (eq 6).[31]

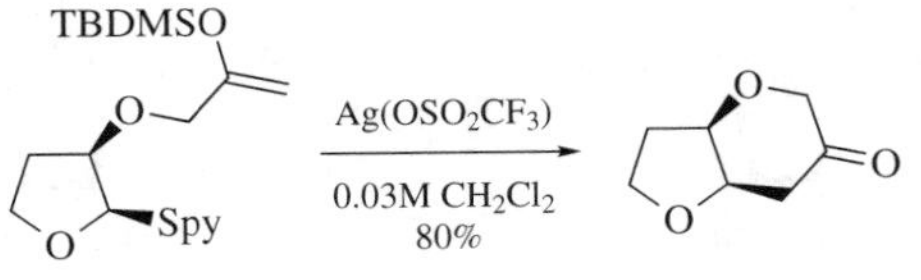

(5)

3 steps

PhHgClO4, THF, 25 °C (6)

(pyS)2, Ph3P, BuOH, 70% (9)

General Reactions of Diaryl Disulfides[1]**.** 2,2′-Dipyridyl disulfide reacts with a variety of carbon nucleophiles, yielding the corresponding thiopyridyl derivatives. These include phosphonate stabilized anions,[39] bromouridine derivatives,[40] indole anions,[41] and heterocyclic stabilized anions.[42]

Alkene Formation. The elimination of pyridinethione or 2-pyridylsulfenic acid forms the basis of a number of alkene syntheses. For example, treatment of dithioacetal anions with 2,2′-dipyridyl disulfide affords the α-thiopyridyl derivative, which undergoes elimination at below room temperature to give ketene dithioacetals.[32] Ester enolates have been similarly treated, although ***m-Chloroperbenzoic Acid*** is first used to generate the sulfoxide which reacts further to yield an α,β-unsaturated ester.[33] This methodology has been applied to the synthesis of methyl dehydrojasmonate (**7**) (eq 7).[34]

(pyS)2, NaOMe, 60% — m-CPBA, toluene, reflux, 73% (7)

(**7**)

Other Reactions. Treatment of an active hydroxy compound with 2,2′-dipyridyl disulfide and n-Bu_3P yields the corresponding thiopyridyl derivative. This methodology has been applied to the preparation of 5-arylthio-5′-deoxyribonucleosides (eq 8).[35] Monophosphate esters [$ROP(O)(OH)_2$] will react similarly to form the activated triphenylphosphonium adduct, which, in the absence of an added external nucleophile, dimerizes yielding a pyrophosphate.[36] *N*-Methylimidazole has been found to catalyze this transformation. The addition of alcohols or amines, however, traps the phosphoryloxyphosphonium salt as the mixed diphosphate ester or mixed ester/amide, respectively (eq 9).[37] ***Chlorotrimethylsilane*** and $(pyS)_2$ have also been reported to facilitate the oxidation of phosphites to phosphates.[38]

(pyS)2, Bu3P, 78% (8)

1. (a) Capozzi, G.; Modena, G. *The Chemistry of the Thiol Group*; Patai, S.; Ed.; Wiley: New York, 1974; Part 2, pp 785–839. (b) Jocelyn, P. C. *The Chemistry of the SH Group*; Academic: New York, 1972.
2. (a) NaOH/KI3: Marckwald, W.; Klemm, W.; Trabert, H. *CB* **1900**, *33*, 1556. (b) Nickel peroxide: Nakagawa, K.; Shiba, S.; Horikawa, M.; Sato, K.; Nakamura, H.; Harada, N.; Harada, F. *SC* **1980**, *10*, 305. (c) I2: McAllan, D. T.; Cullum, T. V.; Dean, R. A.; Fidler, F. A. *JACS* **1951**, *73*, 3627. (d) Zn(BiO3)2: Firouzabadi, H.; Mohammadpour-Baltork, I. *BCJ* **1992**, *65*, 1131. (e) (NH4)2Ce(NO3)6 (CAN): Dhar, D. N.; Bag, A. K. *IJC(B)* **1984**, *23B*, 974. (f) Bromodimethylsulfonium bromide: Olah, G. A.; Arvanaghi, M.; Vankar, Y. D. *S* **1979**, 721. (g) NaBO3: McKillop, A.; Koyunçu, D. *TL* **1990**, *31*, 5007. (h) Diethyl bromomalonate: Kato, E.; Oya, M.; Iso, T.; Iwao, J.-I. *CPB* **1986**, *34*, 486. (i) FeCl3–Bu3SnOMe: Sato, T.; Otera, J.; Nozaki, H. *TL* **1990**, *31*, 3591.
3. Thalmann, A.; Oertle, K.; Gerlach, H. *OSC* **1990**, *7*, 470.
4. Reviews: (a) Nicolaou, K. C. *T* **1977**, *33*, 683. (b) Back, T. G. *T* **1977**, *33*, 3041. (c) Ogliaruso, M. A.; Wolfe, J. A. In *Synthesis of Lactones and Lactams*; Patai, S.; Rappaport, Z.; Eds.; Wiley: New York, 1993.
5. Mukaiyama, T. *SC* **1972**, *2*, 243.
6. Corey, E. J.; Nicolaou, K. C. *JACS* **1974**, *96*, 5614.
7. Corey, E. J.; Brunelle, D. J.; Stork, P. J. *TL* **1976**, 3405.
8. Corey, E. J.; Nicolaou, K. C.; Melvin, L. S. *JACS* **1975**, *97*, 653.
9. Corey, E. J.; Nicolaou, K. C.; Melvin, L. S. *JACS* **1975**, *97*, 654.
10. Corey, E. J.; Ulrich, P.; Fitzpatrick, J. M. *JACS* **1976**, *98*, 222.
11. Corey, E. J.; Nicolaou, K. C.; Toru, T. *JACS* **1975**, *97*, 2287.
12. Corey, E. J.; Bhattacharyya, S. *TL* **1977**, 3919.
13. Bundy, G. L.; Peterson, D. C.; Cornette, J. C.; Miller, W. L.; Spilman, C. H.; Wilks, J. W. *JMC* **1983**, *26*, 1089.
14. Corey, E. J.; Brunelle, D. J. *TL* **1976**, 3409.
15. (a) Karim, M. R.; Sampson, P. *JOC* **1990**, *55*, 598. (b) Justus, K.; Steglich, W. *TL* **1991**, *32*, 5781.
16. Gerlach, H.; Thalmann, A. *HCA* **1974**, *57*, 2661.
17. Bai, D.; Shi, Y. *TL* **1992**, *33*, 943.
18. (a) Ohno, M.; Kobayashi, S.; Iimori, T.; Wang. Y.-F.; Izawa, T. *JACS* **1981**, *103*, 2405. (b) Ohno, M.; Kobayashi, S.; Iimori, T.; Wang. Y.-F.; Izawa, T. *JACS* **1981**, *103*, 2406.
19. Reviews: (a) Klausner, Y. S.; Bodanszky, M. *S* **1972**, 453. (b) Mukaiyama, T. *AG(E)* **1976**, *15*, 94.
20. (a) Mukaiyama, T.; Matsueda, R.; Suzuki, M. *TL* **1970**, 1901. (b) Mukaiyama, T.; Matsueda, R.; Maruyama, H. *BCJ* **1970**, *43*, 1271.
21. (a) Araki, M.; Sakata, S.; Takei, H.; Mukaiyama, T. *BCJ* **1974**, *47*, 1777. (b) Loader, C. E.; Anderson, H. J. *CJC* **1971**, *49*, 45.
22. For other recent methods for the synthesis of 2-acylpyrroles, see: (a) Review: Patterson, J. M. *S* **1976**, 281. (b) Kozikowski, A. P.; Ames, A. *JACS* **1980**, *102*, 860. (c) Martinez, G. R.; Grieco, P. A.; Srinivasan, C. V. *JOC* **1981**, *46*, 3760. (d) Edwards, M. P.; Ley, S. V.; Lister, S. G.; Palmer, B. D. *CC* **1983**, 630.
23. Nicolaou, K. C.; Claremon, D. A.; Papahatjis, D. P. *TL* **1981**, *22*, 4647.
24. Nakahara, Y.; Fujita, A.; Ogawa, T. *ABC* **1985**, *49*, 1491.

25. Nicolaou, K. C.; Maligres, P.; Suzuki, T.; Wendeborn, S. V.; Dai, W.-M.; Chadha, R. K. *JACS* **1992**, *114*, 8890.
26. Ravi, D.; Mereyala, H. B. *TL* **1989**, *30*, 6089.
27. Springer, J. P.; Arison, B. H.; Hirshfield, J. M.; Hoogsteen, K. *JACS* **1981**, *103*, 4221.
28. (a) Mereyala, H. B.; Kulkarni, V. R.; Ravi, D.; Sharma, G. V. M.; Rao, B. V.; Reddy, G. B. *T* **1992**, *48*, 545. (b) Ravi, D.; Kulkarni, V. R.; Mereyala, H. B. *TL* **1989**, *30*, 4287. (c) for mechanistic details see: Mereyala, H. B.; Reddy, G. V. *T* **1991**, *47*, 6435.
29. Williams, R. M.; Stewart, A. O. *TL* **1983**, *24*, 2715.
30. Craig, D.; Munasinghe, V. R. N. *TL* **1992**, *33*, 663.
31. Williams, R. M.; Dung, J.-S.; Josey, J.; Armstrong, R. W.; Meyers, H. *JACS* **1983**, *105*, 3214.
32. Nagao, Y.; Seno, K.; Fujita, E. *TL* **1979**, 4403.
33. Nagao, Y.; Takao, S.; Miyasaka, T.; Fujita, E. *CC* **1981**, 286.
34. Dubs, P.; Stüssi, R. *HCA* **1978**, *61*, 998.
35. (a) Nakagawa, I.; Hata. T. *TL* **1975**, 1409. (b) Nakagawa, I.; Aki, K.; Hata, T. *JCS(P1)* **1983**, 1315.
36. Kanavorioti, A.; Lu, J.; Rosenbach, M. T.; Hurley, T. B. *TL* **1991**, *32*, 6065.
37. Mukaiyama, T.; Hashimoto, M. *BCJ* **1971**, *44*, 196.
38. Hata, T.; Sekine, M. *TL* **1974**, 3943.
39. Ebertino, F. H.; Degenhart, C. R.; Jamieson, L. A.; Burdsall, D. C. *H* **1990**, *30*, 855.
40. Hirota, K.; Tomishi, T.; Maki, Y. *H* **1987**, *26*, 3089.
41. Atkinson, J. G.; Hanel, P.; Girard, Y. *S* **1988**, 480.
42. Mirazaei, Y. R.; Simpson, B. M.; Triggle, D. J.; Natale, N. R. *JOC* **1992**, *57*, 6271.

Eric J. Stoner
Abbott Laboratories, North Chicago, IL, USA

N,N′-Disuccinimidyl Carbonate

[74124-79-1] $C_9H_8N_2O_7$ (MW 256.19)

(reagent for the preparation of (active) esters in peptide chemistry;[1] β-eliminations[2] and carbonyl insertions in heterocyclic and especially peptide chemistry;[3] useful in preparation of some carbamates, ureas, and nitrosoureas[4])

Alternate Name: DSC.
Physical Data: mp 211–215 °C (dec); prisms.[5]
Form Supplied in: white to off-white powder.
Preparative Methods: by treatment of ***N-Hydroxysuccinimide*** (HOSu) with ***Trichloromethyl Chloroformate*** (TCF, $Cl_3COCOCl$) or using trimethylsilylated *N*-hydroxysuccinimide and phosgene, both giving DSC in good yield (eq 1).[1]

N-OH —TCF, Δ→ DSC ←$COCl_2$, 0 °C— N-OTMS (1)

Handling, Storage, and Precautions: moisture sensitive; stable under refrigeration; irritant.

Active Ester Preparation in Peptide Chemistry. Eq 2 shows the method for activating acids for subsequent reaction with amines, to give peptide bonds.[1]

(2)

Di(N-succinimidyl) Oxalate (DSO) has also been used as an alternative to DSC in the preparation of esters.[8]

Carbamate Preparation. DSC reacts with primary and secondary amines to give carbamates (eq 3),[4] in which the carbonyl group is activated by the electron withdrawal of the succinimidyl group. Further attack by the amino groups on the carbonyl of such active carbamates gives ureas. Such carbamates produced by DSC have been used in the preparation of chiral derivatizing agents for amino compounds[6] and in the preparation of nucleoside analogs.[7]

(3)

β-Eliminations. β-Elimination of the hydroxy group of *N*-protected β-hydroxy-α-amino acids has been effected using DSC and ***Triethylamine***.[2] Reaction of Z-threonine with equimolar DSC/Et_3N in acetonitrile gave exclusively the (*Z*)-isomer of Z-DBut-OMe (eq 4). Similarly, use of a 1:2 molar ratio of DSC gave the (*Z*)-isomer of Boc-DBut-OSuc, which on treatment with the methyl ester of alanine gave Z-DBut-Ala-OMe (eq 5).

(4)

(5)

Carbonyl Insertions. DSC has been used as a replacement for ***Phosgene*** in the cyclization of *N*-aralkyl α-amino acids to *N*-carboxyanhydrides (eq 6).[9] This method releases *N*-hydroxysuccinimide as a byproduct, which is only weakly acidic and so does not affect any acid sensitive groups, whereas phosgene releases 2 equiv of hydrogen chloride. Unlike in ester formation using carboxylates, where the central carbonyl group of DSC is released as CO_2, with better nucleophiles, such as secondary α-amino functions, incorporation of the carbonyl group occurs. This is assumed to be by *N*-substitution by R–CO–, followed by intramolecular anhydride formation by tertiary amine. Similar carbonyl insertions by ***N,N′-Carbonyldiimidazole*** (CDI) to give heterocycles[10,11] leads to racemization. Use of DSC as a carbonyl insertion reagent has led to its use in the preparation of ureas and various heterocycles.[3] Reaction with amines (to give ureas), 1,2-diamines (to give cyclic ureas), 1,2-aminothiols (to give cyclic thiocarbamates), and 1,2-hydroxyamines have all been observed. One simple example is the action of DSC on cyclohexylamine to give dicyclohexylurea. A series of (thio)carbamates used as muscle relaxants have also been prepared, e.g. chloroxazone (5-chloro-2-hydroxybenzoxazole) (eq 7). 3-(Benzoxazolyl, benzimidazolyl)pyrazole derivatives have also been made using DSC in MeCN.

(6)

(7)

Related Reagents. Di(*N*-succinimidyl) Oxalate.

1. Ogura, H.; Kobayashi, T.; Shimizu, K.; Kawabe, K.; Takeda, K. *TL* **1979**, *49*, 4745. Ogura, H.; Takeda, K. *NKK* **1981**, *5*, 836.
2. Ogura, H.; Sato, O.; Takeda, K. *TL* **1981**, *22*, 4817.
3. Takeda, K.; Ogura, H. *SC* **1982**, *12*, 213.
4. Takeda, K.; Akagi, Y.; Saika, A.; Tsukahara, T.; Ogura, H. *TL* **1983**, *24*, 4569.
5. *Sigma-Aldrich Handbook of Chemical Safety Data*, 2nd ed.; Lenga, R. E., Ed.; Aldrich: Milwaukee, WI, 1987.
6. Iwaki, K.; Yoshida, S.; Nimura, N.; Kinoshita, T.; Takeda, K.; Ogura, H. *Chromatographia* **1987**, *23*, 899.
7. McCormick, J. E.; McElhinney, R. S.; McMurray, T. B. H.; Maxwell, R. J. *JCS(P1)* **1991**, 877.
8. Yamashina, T.; Higuchi, K.; Hirata, H. *YZ* **1991**, *40*, 155.
9. Halstrom, J.; Kovács, K. *ACS(B)* **1986**, 462.
10. Wright, W. *JHC* **1965**, *2*, 41.
11. Kutney, J.; Ratcliffe, A. *SC* **1975**, *5*, 47.

Edwin C. Davison
University of Cambridge, UK

1,2-Ethanedithiol[1]

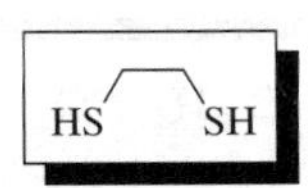

[540-63-6] $C_2H_6S_2$ (MW 94.20)

(1,3-dithiolane formation; ether cleavage; reduction (carbonyl to methylene; sulfoxide to thioether); *gem*-difluoride formation; dithiolenium ion formation)

Physical Data: *d* 1.123 g cm^{-3} (23.5 °C); bp 146 °C/760 mmHg; 63 °C/46 mmHg.
Solubility: slightly sol water; miscible with many organic solvents.
Form Supplied in: widely available.
Handling, Storage, and Precautions: stench! Inhalation can cause chest pain, headache, nausea, pulmonary edema; LD_{50} (oral, mouse) 342 mg kg^{-1}; see ***1,3-Propanedithiol***. Use in a fume hood.

1,3-Dithiolane Formation. 1,2-Ethanedithiol (**1**) condenses with aldehydes, ketones, and acetals to afford 1,3-dithiolanes,[1,2] useful for carbonyl protection (eqs 1 and 2).[3a,b] Stability, condensation selectivity, and conditions for carbonyl regeneration parallel those discussed for ***1,3-Propanedithiol***.[4] Esters and lactones can be protected as ketene dithioacetals and/or dithioortholactones, resistant to nucleophilic attack, using the bis(dimethylalanyl) derivative of (**1**).[5]

(**1**), BF_3•OEt_2, CH_2Cl_2, −10 °C, 98% (1)

(**1**), $ZnCl_2$, HCl, Δ, 82% (2)

Ether Cleavage. An acetonide can be cleaved in the presence of a nearby *t*-butyldiphenylsilyl ether with (**1**).[6a] Under more vigorous conditions, aliphatic methyl ethers are cleaved (eq 3).[6b]

(**1**), BF_3•OEt_2, cat. HCl, 82% (3)

Reduction. ***Raney Nickel*** desulfurization of 1,3-dithiolanes effects overall reduction of C=O to CH_2, as does ***Sodium-Ammonia*** in THF (eqs 4 and 5).[7a,b] Na/hydrazine is an alternative reagent (eq 6).[10a] Peptidic sulfoxides are reduced to thioethers with (**1**) and an electrophilic catalyst.[11a]

Raney Ni, 83% (4)

Na, NH_3, THF, 55% (5)

Na, H_2NNH_2, $(HOCH_2)_2$, 190 °C, 78% (6)

(resolved)

gem*-Difluoride Formation.** 1,3-Dithiolanes yield geminal difluorides on treatment with ***Pyridinium Poly(hydrogen Fluoride) and a mild oxidant (eq 7).[8a,b] 1,3-Dithianes react more slowly and give lower yields.[8a]

py•$(HF)_n$, dibromodimethylhydantoin, 61% (7)

Dithiolenium Ion Formation. Acid chlorides, anhydrides, esters, and orthoesters, upon treatment with (**1**) and a Lewis acid, can produce electrophilic 1,3-dithiolenium cations which react with a variety of carbon nucleophiles.[9] Selective indole formylation in the presence of a primary carboxamide could thus be achieved (eq 8).[9a]

(**1**), HCO_2Et, $TiCl_4$, 58%; SO_2Cl_2, SiO_2, H_2O, $CHCl_3$, 61% (8)

Miscellaneous. Racemic ketones have been resolved by condensation with (*R,R*)- or (*S,S*)-2,3-butanedithiol, then reduced as above to give resolved deoxy compounds (eq 6).[10a,b] Oxidation of prochiral 1,3-dithiolanes under modified Katsuki–Sharpless conditions yields monosulfoxides of high ee.[10c] In peptide synthesis, (**1**) is frequently used as a cation scavenger during deprotection.[11a,b] Transition metal cations are strongly chelated by (**1**), as exemplified by demetalation of Cu^{II} and Ni^{II} metalloporphyrins.[11c]

Related Reagents. Ethanethiol; Methanethiol; 1,3-Propanedithiol.

1. (a) Greene, T. W.; Wuts, P. G. M. *Protective Groups in Organic Synthesis*, 2nd ed.; Wiley: New York, 1991; p 201. (b) Page, P. C. B.; van Niel, M. B.; Prodger, J. C. *T* **1989**, *45*, 7643; see pp 7647–7651.
2. 1,3-Dithiolanes are not generally useful as acyl anion equivalents: (a) Oida, T.; Tanimoto, S.; Terao, H.; Okano, M. *JCS(P1)* **1986**, 1715. (b) An exception: Barton, D. H. R.; Bielska, M. T.; Cardoso, J. M.; Cussans, N. J.; Ley, S. V. *JCS(P1)* **1981**, 1840.
3. (a) Liu, H.-J.; Yeh, W.-L.; Chew, S. Y. *TL* **1993**, *34*, 4435. (b) Wolf, G.; Seligman, A. M. *JACS* **1951**, *73*, 2082.
4. (a) Sato, T.; Otera, J.; Nozaki, H. *JOC* **1993**, *58*, 4971. (b) Stahl, I.; Schramm, B.; Manske, R.; Gosselck, J. *LA* **1982**, 1158. (c) Nakata, T.; Nagao, S.; Takao, S.; Tanaka, T.; Oishi, T. *TL* **1985**, *26*, 73. (d) Cardani, S.; Bernardi, A.; Colombo, L.; Gennari, C.; Scolastico, C.; Venturini, I. *T* **1988**, *44*, 5563. (e) Ni, Z.-J.; Luh, T.-Y. *OS* **1992**, *70*, 240. (f) Bellesia, F.; Boni, M.; Ghelfi, F.; Pagnoni, U. M. *T* **1993**, *49*, 199.
5. Corey, E. J.; Beames, D. J. *JACS* **1973**, *95*, 5829.
6. (a) Williams, D. R.; Sit, S.-Y. *JACS* **1984**, *106*, 2949. (b) Vidari, G.; Ferriño, S.; Grieco, P. A. *JACS* **1984**, *106*, 3539.
7. (a) Maurer, P. J.; Takahata, H.; Rapoport, H. *JACS* **1984**, *106*, 1095. (b) Numazawa, M.; Mutsumi, A. *Biochem. Biophys. Res. Commun.* **1991**, *177*, 401.
8. (a) Sondej, S. C.; Katzenellenbogen, J. A. *JOC* **1986**, *51*, 3508. (b) Jekö, J.; Timár, T.; Jaszberenyi, J. C. *JOC* **1991**, *56*, 6748.
9. (a) Beneš, J.; Semonský, M. *CCC* **1982**, *47*, 1235. (b) Stahl, I. *CB* **1987**, *120*, 135. (c) Okuyama, T.; Fujiwara, W.; Fueno, T. *BCJ* **1986**, *59*, 453. (d) Houghton, R. P.; Dunlop, J. E. *SC* **1990**, *20*, 1.
10. (a) Corey, E. J.; Ohno, M.; Mitra, R. B.: Vatakencherry, P. A. *JACS* **1964**, *86*, 478. (b) Buding, H.; Deppisch, B.; Musso, H.; Snatzke, G. *CB* **1985**, *118*, 4597. (c) Bortolini, O.; Di Furia, F.; Licini, G.; Modena, G.; Rossi, M. *TL* **1986**, *27*, 6257.
11. (a) Futaki, S.; Taike, T.; Akita, T.; Kitagawa, K. *CC* **1990**, 523. (b) Fields, C. G.; Fields, G. B. *TL* **1993**, *34*, 6661. (c) Battersby, A. R.; Jones, K.; Snow, R. J. *AG(E)* **1983**, *22*, 734.

Raymond E. Conrow
Alcon Laboratories, Fort Worth, TX, USA

2-Ethoxy-1-ethoxycarbonyl-1,2-dihydroquinoline

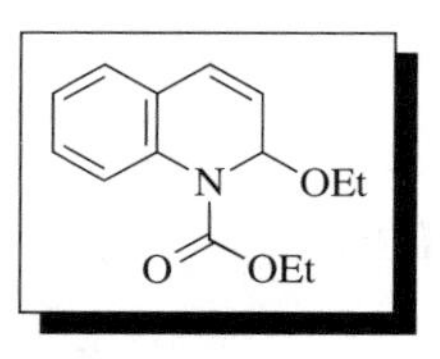

[16357-59-8] $C_{14}H_{17}NO_3$ (MW 247.29)

(activating agent used in peptide synthesis)

Alternate Names: ethyl 1,2-dihydro-2-ethoxy-1-quinolinecarboxylate; EEDQ.
Physical Data: mp 66–67 °C; bp 125–128 °C/0.1 mmHg.
Solubility: sol water, alcohol.
Form Supplied in: white powder; widely available and can be easily prepared.[1]

Peptide Synthesis. In peptide synthesis, peptide bonds are created from the activated carboxylic group of a N^{α}-protected amino acid that is reacted with the nucleophilic free amino group of another amino acid (eq 1).

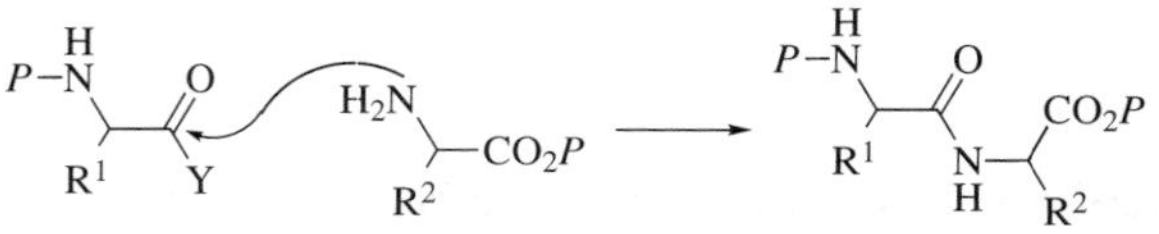

(1)

Activation of the carboxylic acid can be performed according to several strategies, leading mainly to two kinds of intermediates: active esters and acid anhydrides, which can be either symmetrical anhydrides or mixed anhydrides. EEDQ is a reagent used in peptide synthesis for the mixed carbonic anhydride strategy. It allows a slow formation of unsymmetrical anhydrides under very mild conditions (eq 2).[2]

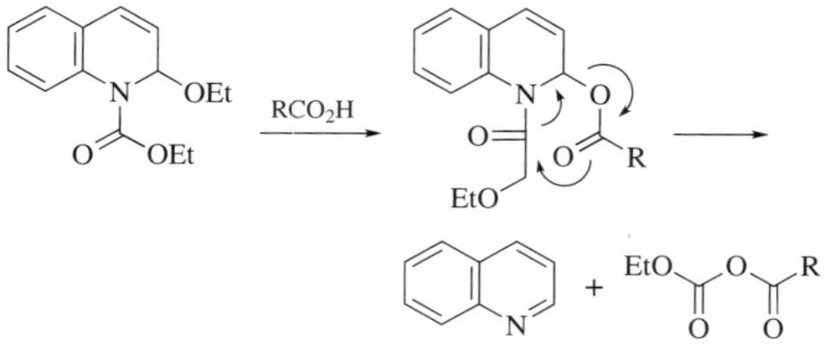

(2)

Faster reaction of the mixed anhydride with nucleophilic species reduces the extent of side reactions that occur with high concentration of the anhydride (eq 3).[3]

Boc–Cys(Bn)–OH + H–Tyr(Bn)–Gln–Asn–Cys(Bn)–OBn —(EEDQ, NEt_3)→ Boc–Cys(Bn)–Tyr(Bn)–Gln–Asn–Cys(Bn)–OBn (3)

The tendency for racemization is limited to a small extent. Nevertheless, wrong-side attack, leading to the formation of urethane, is not suppressed (eq 4).[4]

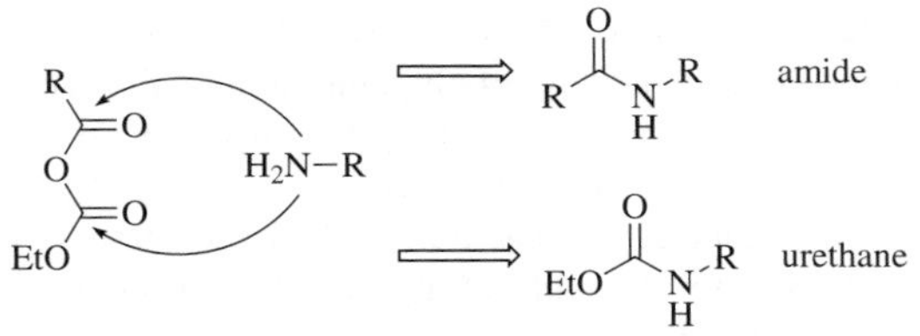

$$\text{(4)}$$

As best results had been obtained with mixed anhydrides of ***Isobutyl Chloroformate***, probably because the isobutyl group decreases the amount of urethane products, IIDQ (**1**), an analog of EEDQ, has been proposed for use in peptide synthesis.[5,6]

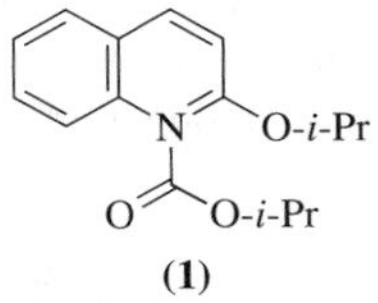

(**1**)

As another alternative, EEDQ has been linked to a polymeric support, providing a supported coupling reagent in high yield (eq 5).[7]

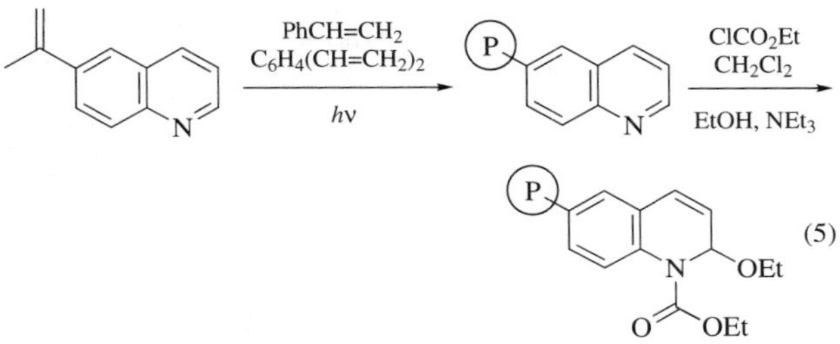

(5)

Coupling reactions mediated by this reagent occur cleanly in high yield and with a low degree of racemization, comparable with that observed with EEDQ itself.[8] Nevertheless, the use of EEDQ in peptide synthesis is rather limited, other coupling reagents being very much more popular. Examples are: the carbodiimides, mainly ***1,3-Dicyclohexylcarbodiimide*** and diisopropylcarbodiimide; the family of BOP reagents such as ***Benzotriazol-1-yloxytris(dimethylamino)phosphonium Hexafluorophosphate***; uroniums such as ***O-Benzotriazol-1-yl-N,N,N′,N′-tetramethyluronium Hexafluorophosphate*** (HBTU), or urethane *N*-protected carbonic anhydrides (UNCA), and active esters.

1. Fieser, M.; Fieser, L. F. *FF* **1969**, *2*, 191.
2. Belleau, B.; Malek, G *JACS* **1968**, *90*, 1651.
3. Mühlemann, M.; Titov, M., I.; Scweyzer, R.; Rudinger, J. *HCA* **1972**, *55*, 2854.
4. Lombardino, J. G.; Anderson, S. L.; Norris, C. P. *JHC* **1978**, *15*, 655.
5. Kiso, Y.; Kai, Y.; Yajima, H. *CPB* **1973**, *90*, 1651.
6. Bodanszky, M.; Bodanszky, A. *The Practice of Peptide Synthesis*; Springer: Berlin, 1984; p 148.
7. Brown, J.; Williams, R. E. *CJC* **1971**, *49*, 3765.
8. Lauren, D. R.; Williams, R. E. *TL* **1972**, 2665.

Jean-Claude Gesquière
Institut Pasteur, Lille, France

Ethylaluminum Dichloride[1]

[563-43-9] $C_2H_5AlCl_2$ (MW 126.95)

(strong Lewis acid that can also act as a proton scavenger; reacts with HX to give ethane and $AlCl_2X$)

Alternate Name: dichloroethylaluminum.
Physical Data: mp 32 °C; bp 115 °C/50 mmHg; *d* 1.207 g cm^{-3}.
Solubility: sol most organic solvents; stable in alkanes or arenes.
Analysis of Reagent Purity: solutions are reasonably stable but may be titrated before use by one of the standard methods.[1e]
Form Supplied in: commercially available neat or as solutions in hexane or toluene.
Handling, Storage, and Precautions: must be transferred under inert gas (Ar or N_2) to exclude oxygen and water. Use in a fume hood.

Alkylaluminum Halides. Since the early 1980s, alkylaluminum halides have come into widespread use as Lewis acid catalysts. These strong Lewis acids offer many advantages over traditional metal halide Lewis acids such as ***Boron Trifluoride***, ***Aluminum Chloride***, ***Titanium(IV) Chloride***, and ***Tin(IV) Chloride***. Most importantly, the alkyl group on the aluminum will react with protons to give an alkane and a new Lewis acid. Alkylaluminum halides are therefore Brønsted bases, as well as Lewis acids. The alkyl groups are also nucleophilic, and this is the major disadvantage in the use of these compounds as Lewis acids.

Pure, anhydrous Lewis acids do not catalyze the polymerization of alkenes or the Friedel–Crafts alkylation of aromatics by alkenes. Cocatalysts, such as water or a protic acid, react with the Lewis acid to produce a very strong Brønsted acid that will protonate a double bond. Therefore, use of strictly anhydrous conditions should minimize side reactions in Lewis acid-catalyzed ene, Diels–Alder, and [2 + 2] cycloaddition reactions. Unfortunately, it is difficult to prepare anhydrous, proton-free $AlCl_3$, BF_3, etc. Alkylaluminum halides are easily prepared and stored in anhydrous form and, more importantly, scavenge any adventitious water, liberating an alkane and generating a new Lewis acid in the process.

Using alkylaluminum halides, Lewis acid-catalyzed reactions can now be carried out under aprotic conditions. This is of value when side reactions can be caused by the presence of adventitious protons; it is of special value when acidic protons are produced by the reaction. In these cases, use of the appropriate alkylaluminum halide in stoichiometric rather than catalytic amounts gives high yields of products not formed at all with other Lewis acids. The Me_2AlCl-catalyzed ene reactions of aliphatic aldehydes with alkenes, which give homoallylic alcohol–Me_2AlCl complexes that react further to give methane and the stable methylaluminum alkoxides, is an example of this type of reaction (eq 1).[1h,2] Loss of methane prevents the alcohol–Lewis acid complex from solvolyzing or protonating double bonds.

The alkylaluminum halides cover a wide range of acidity. Replacing chlorines with alkyl groups decreases Lewis acidity.

$EtAlCl_2$ and ***Methylaluminum Dichloride*** are only slightly less acidic than $AlCl_3$. ***Diethylaluminum Chloride*** and ***Dimethylaluminum Chloride*** are substantially less acidic, and ***Trimethylaluminum***, Me_2AlOR, and $MeAl(OR)_2$ are even weaker Lewis acids. Alkylaluminum halides with fractional ratios of alkyl to chloride are also available. The sesquichlorides are commercially available. Other reagents can be prepared by mixing two reagents in the desired proportion. If no reaction occurs with the alkylaluminum halide, a stronger Lewis acid should be tried. If polymerization or other side reactions compete, a weaker Lewis acid should be used. The sequential ene and Prins reactions shown in eq 2 proceed cleanly with $Me_3Al_2Cl_3$.[3] Complex mixtures are obtained with the stronger Lewis acid $EtAlCl_2$ while the weaker Lewis acid Me_2AlCl reacts with ***Formaldehyde*** to give ethanol.

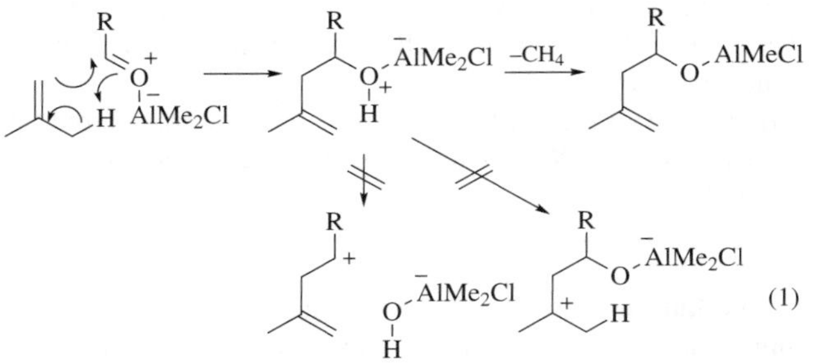

(1)

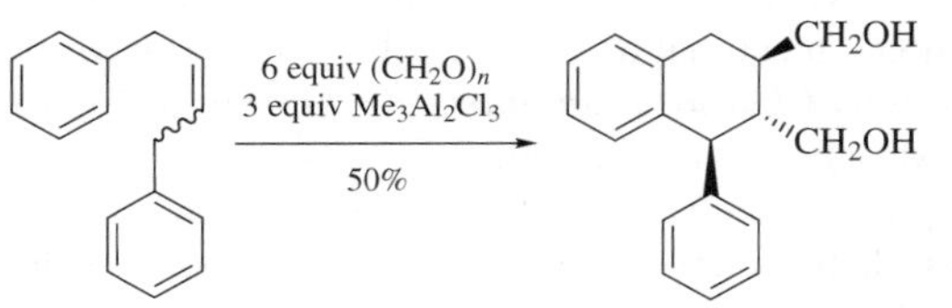

(2)

Use of more than one equivalent of Lewis acid will produce complexes that are formally 1:2 substrate Lewis acid complexes, but are more likely salts with a R_2Al–substrate cation and an aluminate anion.[4–6] This substrate in this salt is much more electrophilic and reactive than that in simple Lewis acid complexes.

These reagents are easier to use than typical inorganic Lewis acids. They are soluble in all organic solvents, including hexane and toluene, in which they are commercially available as standardized solutions. In general, alkane solvents are preferred since toluene can undergo Friedel–Crafts reactions. On a laboratory scale, these reagents are transferred by syringe like alkyllithiums and, unlike anhydrous solid Lewis acids, they do not require a glove bag or dry box for transfer.

While the Brønsted basicity of the alkyl group is advantageous, these alkyl groups are also nucleophilic. The addition of the alkyl group from the aluminum in the Lewis acid–reagent complex to the electrophilic center can be a serious side reaction. The ease of alkyl donation is $R_3Al > R_2AlCl > R_3Al_2Cl_3 > RAlCl_2$. When the nucleophilicity of the alkyl group is a problem, a Lewis acid with fewer alkyl groups should be examined. Addition of diethyl ether or another Lewis base may moderate the reaction if its greater acidity causes problems.

Ethylaluminum compounds are more nucleophilic than methylaluminum compounds and can donate a hydrogen as well as an ethyl group to the electrophilic center. Unfortunately, methylaluminum compounds must be prepared from ***Chloromethane***, while ethylaluminum compounds can be prepared much more cheaply from ***Ethylene***. Therefore ethylaluminum compounds are usually used unless the nucleophilicity of the alkyl group is a problem. Although the predominant use of alkylaluminum halides is as Lewis acids, they are occasionally used for the transfer of an alkyl group or hydride to an electrophilic center.

Eq 3 shows an unusual reaction in which the nature of the reaction depends on the amount, acidity, and alkyl group of the alkylaluminum halide.[6] Use of 1 equiv of Me_2AlCl leads to a concerted ene reaction with the side chains *cis*. Use of 2 equiv of Me_2AlCl produces a more electrophilic aldehyde complex that cyclizes to a zwitterion. Chloride transfer is the major process at $-78\,^{\circ}C$; at $0\,^{\circ}C$, chloride transfer is reversible and a 1,5-proton transfer leads to an ene-type adduct with the side chains *trans*. Use of 2 equiv of $MeAlCl_2$ forms a cyclic zwitterion that undergoes two 1,2-hydride shifts to form the ketone. A similar zwitterion forms with $EtAlCl_2$, but β-hydride transfer leading to the saturated alcohol is faster than 1,2-hydride shifts.

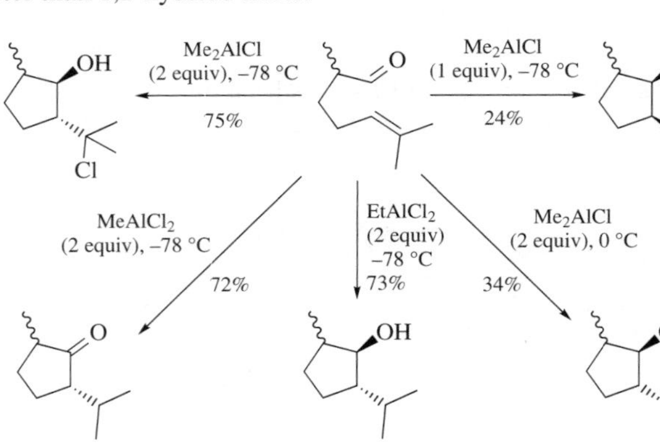

(3)

Catalysis of Diels–Alder Reactions. $EtAlCl_2$ is a useful Lewis acid catalyst for Diels–Alder reactions. It is reported to be more efficacious for the Diels–Alder reaction of ***Acrolein*** and butadiene than either $AlCl_3$ or Et_2AlCl.[7] It is a useful catalyst for intramolecular Diels–Alder reactions with α,β-unsaturated esters (eq 4)[8] and aldehydes[9] as dienophiles. It has also proven to be a very efficient catalyst for the inter-[10] and intramolecular[11] asymmetric Diels–Alder reaction of chiral α,β-unsaturated acyl sultams (eq 5) and has been used to catalyze a wide variety of Diels–Alder reactions.[12]

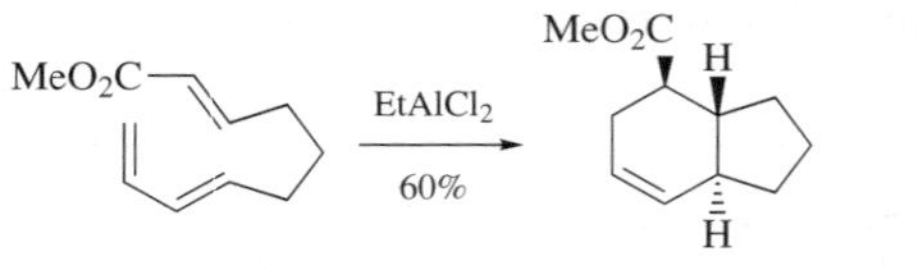

(4)

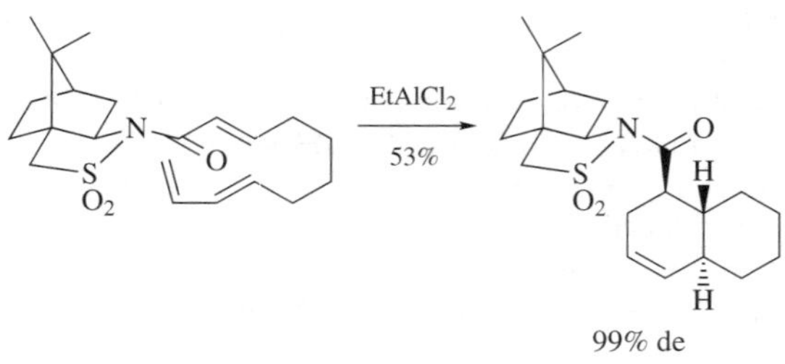

(5)

Catalysis of Ene Reactions. Although $AlCl_3$ can be used as a Lewis acid catalyst for ene reactions of α,β-unsaturated esters,[13a] better results are obtained more reproducibly with $EtAlCl_2$. Ene reactions of ***Methyl Propiolate*** proceed in good yield with 1,1-di-, tri-, and tetrasubstituted alkenes (eq 6).[14] A precursor to 1,25-dihydroxycholesterol can be prepared by an ene reaction with methyl propiolate. Three equiv of $EtAlCl_2$ are needed since the acetate esters are more basic than methyl propiolate (eq 7).[15] *Endo* products are obtained stereospecifically with methyl α-haloacrylates (eq 8).[14] $EtAlCl_2$ has also been used to catalyze intramolecular ene reactions (eq 9).[16]

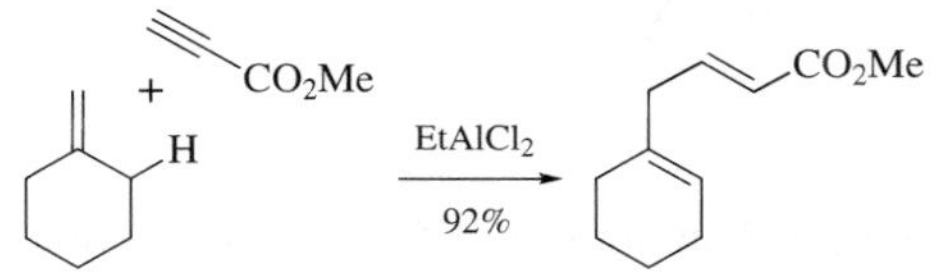

$EtAlCl_2$ 92% (6)

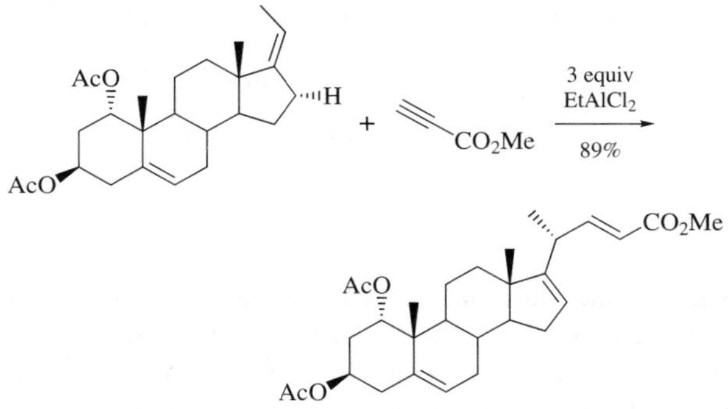

3 equiv $EtAlCl_2$ 89% (7)

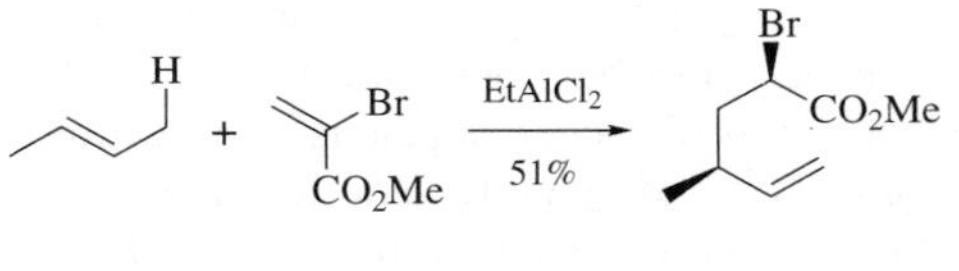

$EtAlCl_2$ 51% (8)

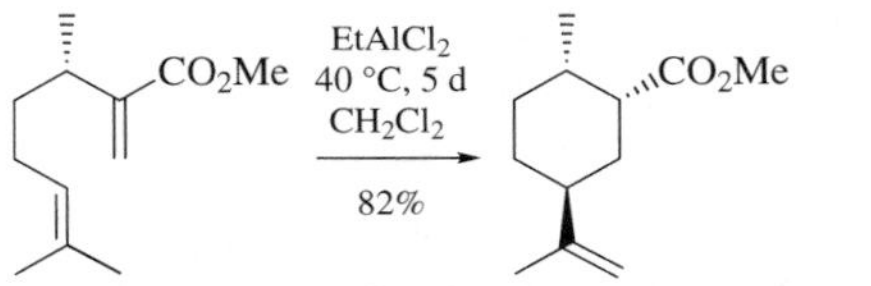

$EtAlCl_2$ 40 °C, 5 d CH_2Cl_2 82% (9)

$EtAlCl_2$ is usually too strong a Lewis acid for ene reactions of carbonyl compounds.[13b,c] However, alkenes that contain basic sites that complex to the Lewis acid do not undergo Me_2AlCl-catalyzed ene reactions with formaldehyde. In these cases, $EtAlCl_2$ is the preferred catalyst.[17,18] The dienyl acetate shown in eq 10 reacts with excess ***Formaldehyde*** and $EtAlCl_2$ to provide the conjugated diene ene adduct that undergoes a quasi-intramolecular Diels–Alder reaction to afford a pseudomonic acid precursor. Me_2AlCl catalyzes the ene reaction of aliphatic aldehydes with 1,1-di-, tri-, and tetrasubstituted alkenes. Terminal alkenes are less nucleophilic, so the only reaction is addition of a methyl group to the aldehyde. $EtAlCl_2$, a stronger Lewis acid with a less nucleophilic alkyl group, catalyzes the reaction of aliphatic aldehydes with terminal alkenes in CH_2Cl_2 at 0 °C to give 50–60% of the ene adduct.[18] Use of $EtAlCl_2$ as a catalyst affords the best diastereoselectivity in the ene reaction of dibenzylleucinal with ***Isobutene*** (eq 11).[19] $EtAlCl_2$ has also been used to catalyze intramolecular ene reactions of trifluoromethyl ketones.[20]

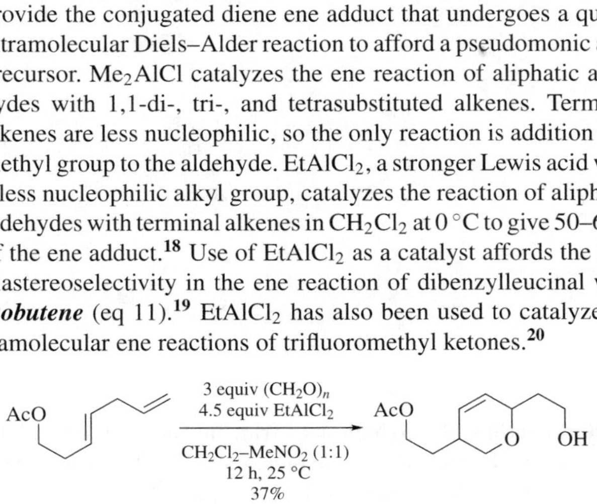

3 equiv $(CH_2O)_n$ 4.5 equiv $EtAlCl_2$ CH_2Cl_2–$MeNO_2$ (1:1) 12 h, 25 °C 37% (10)

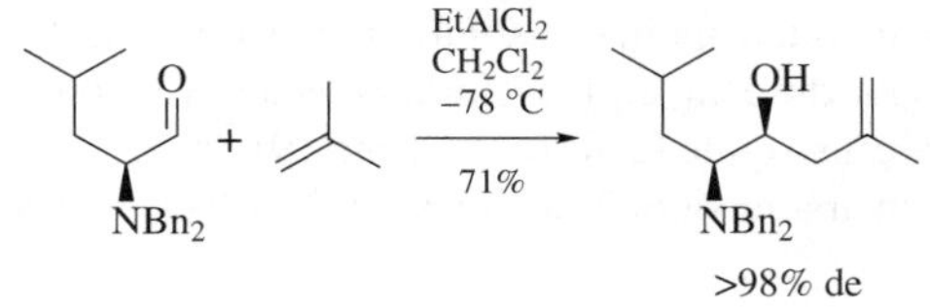

$EtAlCl_2$ CH_2Cl_2 −78 °C 71% >98% de (11)

Catalysis of Intramolecular Sakurai Reactions. $EtAlCl_2$ has been extensively used as a catalyst for intramolecular Sakurai additions. Enones (eqs 12 and 13)[21,22] have been most extensively explored. Different products are often obtained with fluoride or Lewis acid catalysis. $EtAlCl_2$ is the Lewis acid used most often although $TiCl_4$ and BF_3 have also been used. $EtAlCl_2$ also catalyzes intramolecular Sakurai reactions with ketones[23] and other electrophiles.[24] The cyclization of electrophilic centers onto alkylstannanes[25] and Prins-type additions to vinylsilanes[26] are also catalyzed by $EtAlCl_2$.

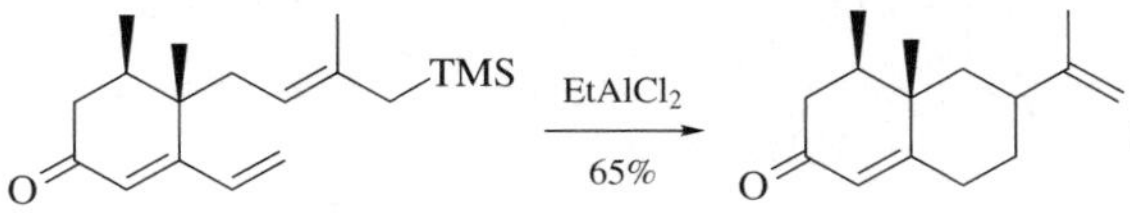

$EtAlCl_2$ 65% nootkatone (12)

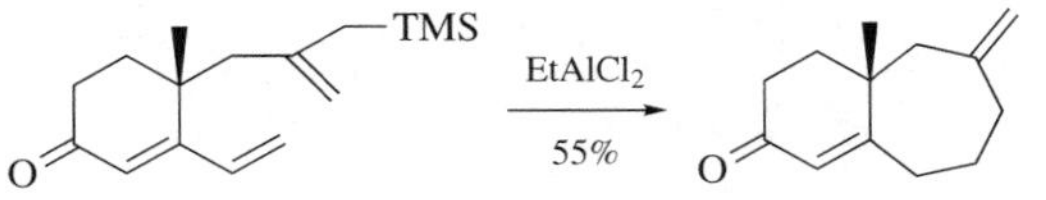

$EtAlCl_2$ 55% (13)

Catalysis of [2 + 2] Cycloadditions. $EtAlCl_2$ catalyzes a wide variety of [2 + 2] cycloadditions. These include the addition of alkynes or allenes to alkenes to give cyclobutenes and alkylidenecyclobutanes (eq 14),[27] the addition of electron-deficient alkenes to allenyl sulfides (eq 15),[28] the addition of propiolate esters to monosubstituted and 1,2-disubstituted alkenes to form cyclobutene carboxylates (eq 16),[14b] and the addition of allenic esters to alkenes to form cyclobutanes.[29]

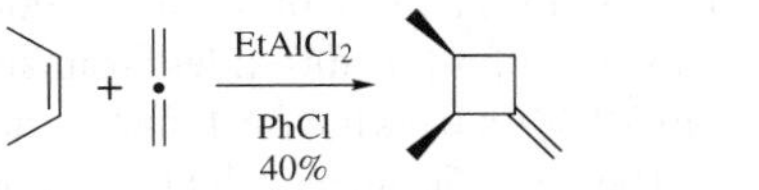

$EtAlCl_2$ PhCl 40% (14)

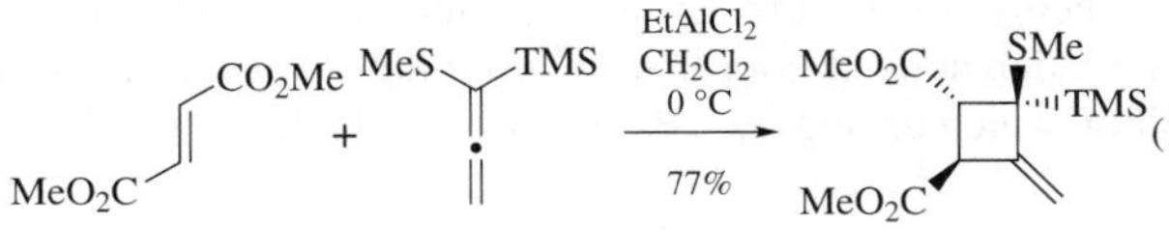

$EtAlCl_2$ CH_2Cl_2 0 °C 77% (15)

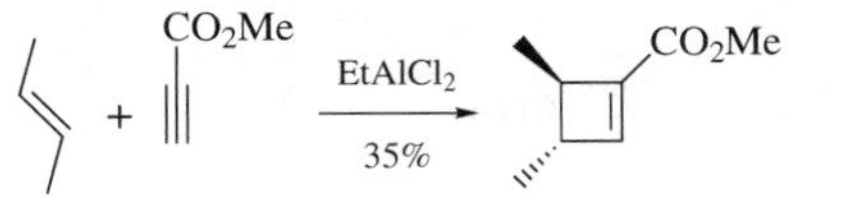

$EtAlCl_2$ 35% (16)

Generation of Electrophilic Cations. $EtAlCl_2$ has proven to be a useful Lewis acid for inducing a wide variety of electrophilic reactions. It is particularly useful since an excess of the reagent can be used so that all nucleophiles are complexed to acid. Under these conditions, intermediates tend to collapse to give cyclobutanes or undergo hydride shifts to give neutral species. Reaction of the 1:2 ***Crotonaldehyde***–$EtAlCl_2$ complex with 2-methyl-2-butene at −80 °C affords a zwitterion that collapses to give mainly the cyclobutane. Closure of the zwitterion is reversible at 0 °C. Since there are no nucleophiles in solution, two 1,2-hydride shifts take place to give the enal (eq 17).[5] Intramolecular versions of these reactions are also quite facile (eqs 18 and 19).[5,21] $EtAlCl_2$ promotes the ring enlargement of 1-acylbicyclo[4.2.0]oct-3-enes (eq 20).[30]

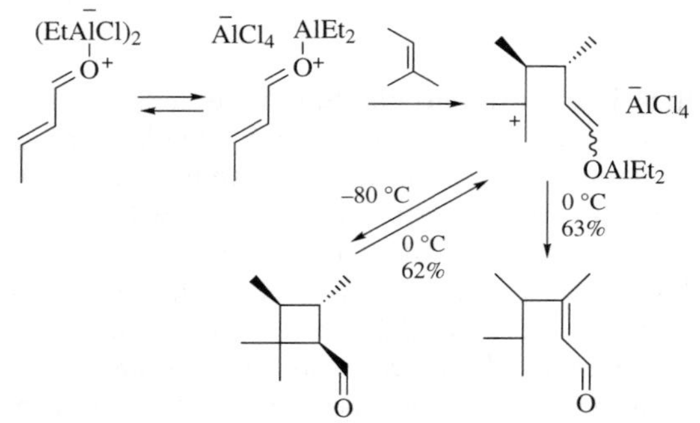

(17)

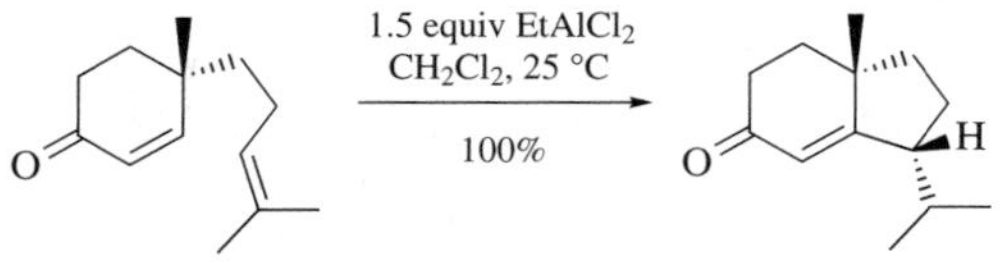

(18)

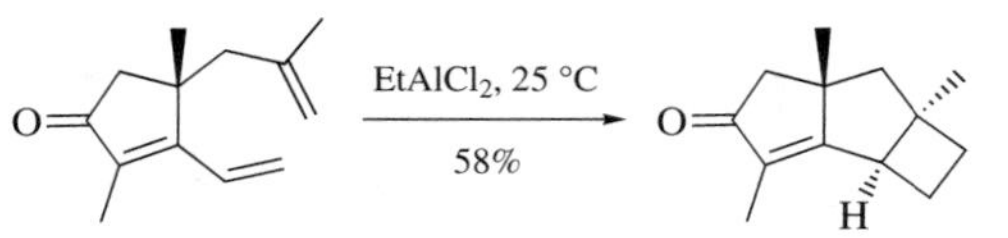

(19)

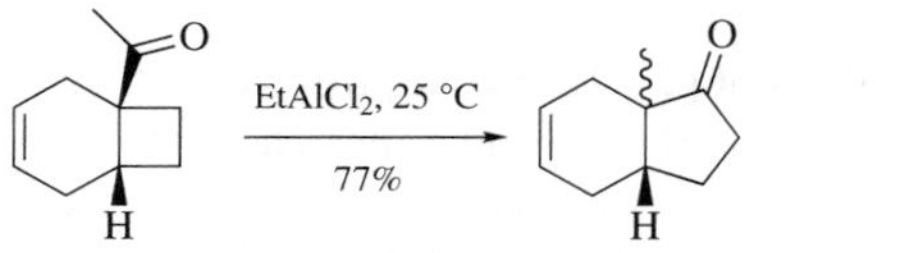

(20)

93:7 *cis:trans*

$EtAlCl_2$ catalyzes the Friedel–Crafts acylation of alkenes with acid chlorides,[31] the formal [3 + 2] cycloaddition of alkenes with cyclopropane-1,1-dicarboxylates (eq 21),[32] the Friedel–Crafts alkylation of anilines and indoles with α-aminoacrylate esters,[33] and the formation of allyl sulfoxides from sulfinyl chlorides and alkenes.[34] $EtAlCl_2$ induces the Beckmann rearrangement of oxime sulfonates. The cationic intermediates can be trapped with enol silyl ethers (eq 22).[35] $EtAlCl_2$ is the preferred catalyst for addition of the cation derived from an α-chloro sulfide to an alkene to give a cation which undergoes a Friedel–Crafts alkylation (eq 23).[36]

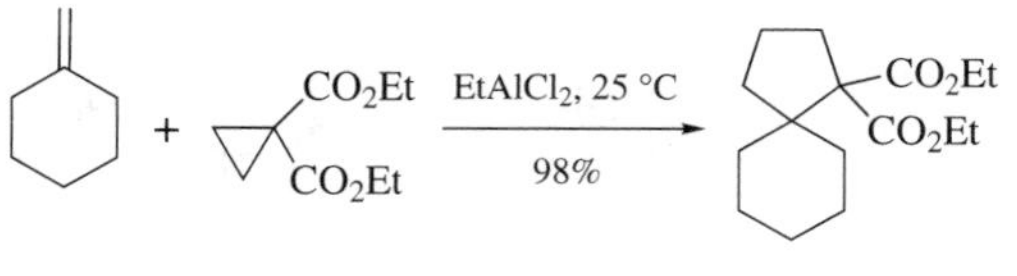

(21)

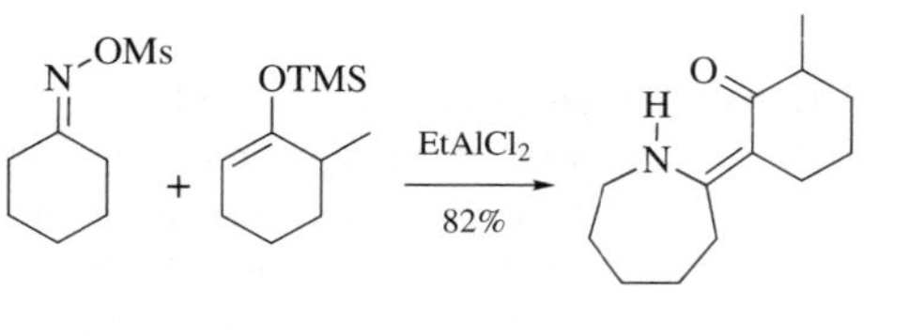

(22)

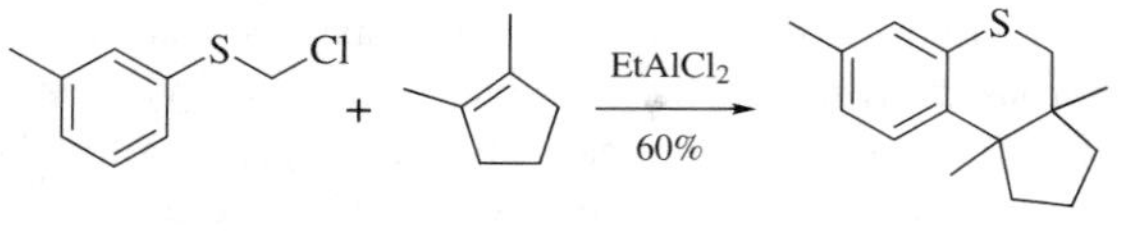

(23)

$EtAlCl_2$ reacts with aliphatic sulfones to generate an aluminum sulfinate and a cation that can be reduced to a hydrocarbon by $EtAlCl_2$,[37] trapped with a nucleophile such as ***Allyltrimethylsilane***,[38] or undergo a pinacol-type rearrangement (eq 24).[39]

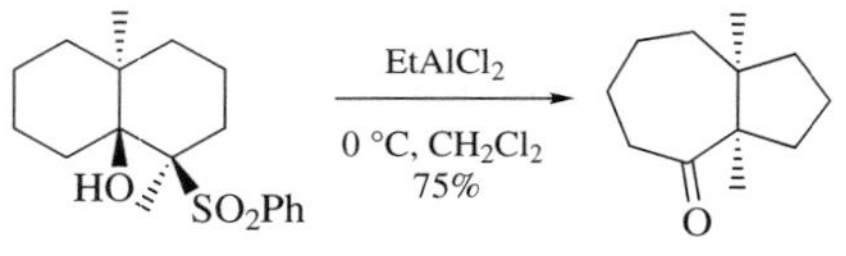

(24)

Elimination Reactions. $EtAlCl_2$ induces elimination of two molecules of HBr from the dibromide to give the dihydropyridine (eq 25).[40] The usual base-catalyzed elimination is ineffective.

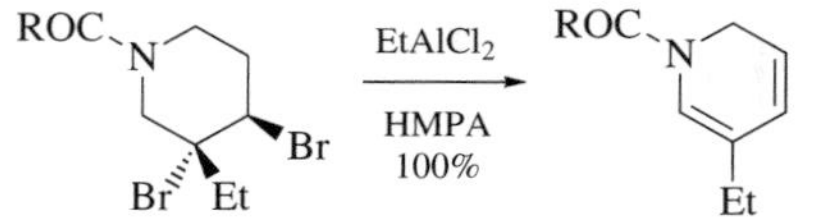

(25)

Nucleophilic Addition. $EtAlCl_2$ has been used to activate conjugated systems toward attack of an external nucleophile or to transfer a hydride or ethyl group as a nucleophile. Addition of a cuprate to the chiral amide in the presence of $EtAlCl_2$ improves the diastereoselectivity, affording a >93:7 mixture of stereoisomers (eq 26).[41] Reaction of butyrolactones with $EtAlCl_2$ reduces the lactone to a carboxylic acid by opening to the cation and hydride delivery (eq 27).[42] Sulfonimidyl chlorides react with $EtAlCl_2$ at −78 °C to provide *S*-ethyl sulfoximines in 65–95% overall yield (eq 28).[43]

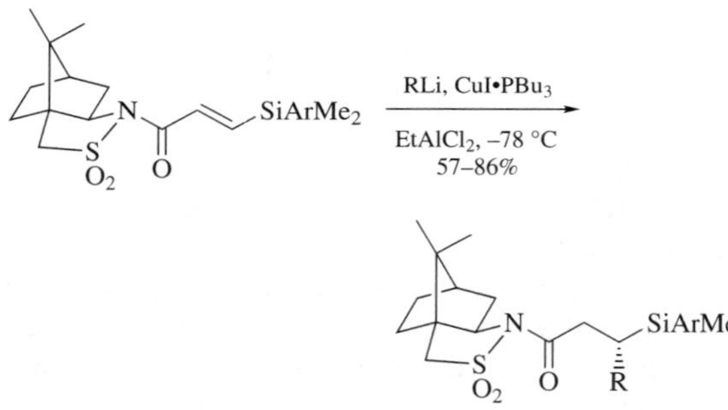

(26)

>93:7 de

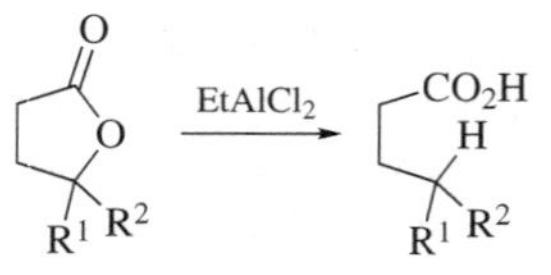

(27)

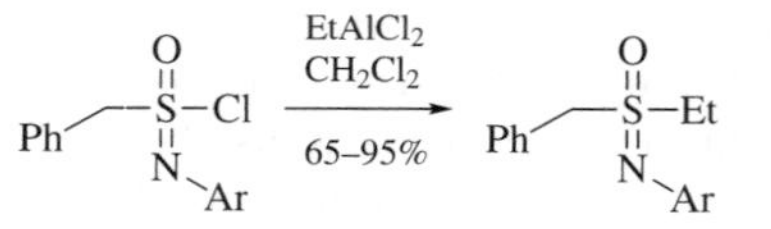

(28)

Modification of Carbanions. Trimethylsilyl allylic carbanions react with aldehydes in the presence of $EtAlCl_2$ exclusively at the α-position to give *threo* adducts (eq 29).[44]

$$\text{TMS}\diagup\!\!\!\sim\!\!\!\sim\text{ Li}^+ + \text{RCHO} \xrightarrow{\text{EtAlCl}_2} \text{R}\text{CH(OH)CH(TMS)CH=CH}_2 \quad (29)$$

Related Reagents. Aluminum Chloride; Boron Trifluoride Etherate; Diethylaluminum Chloride; Diethylaluminum Iodide; Dimethylaluminum Chloride; Dimethylaluminum Iodide; Ethylaluminum Dichloride; Methylaluminum Dichloride; Tin(IV) Chloride; Titanium(IV) Chloride; Triethylaluminum; Trimethylaluminum.

1. (a) Mole, T.; Jeffery, E. A. *Organoaluminum Compounds*; Elsevier: New York, 1972. (b) Bruno, G. *The Use of Aluminum Alkyls in Organic Synthesis*; Ethyl Corporation: Baton Rouge, LA, 1970. (c) Bruno, G. *The Use of Aluminum Alkyls in Organic Synthesis, Supplement, 1969–1972*; Ethyl Corporation: Baton Rouge, LA, 1973. (d) Honeycutt, J. B. *The Use of Aluminum Alkyls in Organic Synthesis, Supplement, 1972–1978*; Ethyl Corporation: Baton Rouge, LA, 1981. (e) *Aluminum Alkyls*; Stauffer Chemical Co.: Westport, CT, 1976. (f) Snider, B. B.; Rodini, D. J.; Karras, M.; Kirk, T. C.; Deutsch, E. A.; Cordova, R.; Price, R. T.; *T* **1981**, *37*, 3927. (g) Yamamoto, H.; Nozaki, H. *AG(E)* **1978**, *17*, 169. (h) Snider, B. B. In *Selectivities in Lewis Acid Promoted Reactions*; Schinzer, D., Ed.; Kluwer: Dordrecht, 1989; Chapter 8. (i) Maruoka, K.; Yamamoto, H. *AG(E)* **1985**, *24*, 668; *T* **1988**, *44*, 5001.
2. (a) Snider, B. B.; Rodini, D. J.; Kirk, T. C.; Cordova, R. *JACS* **1982**, *104*, 555. (b) Cartaya-Marin, C. P.; Jackson, A. C.; Snider, B. B. *JOC* **1984**, *49*, 2443.
3. Snider, B. B.; Jackson, A. C. *JOC* **1983**, *48*, 1471.
4. Evans, D. A.; Chapman, K. T.; Bisaha, J. *JACS* **1988**, *110*, 1238.
5. Snider, B. B.; Rodini, D. J.; van Straten, J. *JACS* **1980**, *102*, 5872.
6. Snider, B. B.; Karras, M.; Price, R. T.; Rodini, D. J. *JOC* **1982**, *47*, 4538.
7. Miyajima, S.; Inukai, T. *BCJ* **1972**, *45*, 1553.
8. (a) Roush, W. R.; Ko, A. I.; Gillis, H. R. *JOC* **1980**, *45*, 4264. (b) Roush, W. R.; Gillis, H. R. *JOC* **1980**, *45*, 4267; *JOC* **1982**, *47*, 4825.
9. (a) Marshall, J. A.; Audia, J. E.; Grote, J. *JOC* **1984**, *49*, 5277. (b) Marshall, J. A.; Grote, J.; Audia, J. E. *JACS* **1987**, *109*, 1186.
10. (a) Oppolzer, W.; Chapuis, C.; Bernardinelli, G. *HCA* **1984**, *67*, 1397. (b) Smith, A. B., III; Hale, K. J.; Laasko, L. M.; Chen, K.; Riéra, A. *TL* **1989**, *30*, 6963.
11. (a) Oppolzer, W.; Dupuis, D. *TL* **1985**, *26*, 5437. (b) Oppolzer, W. *AG(E)* **1984**, *23*, 876.
12. (a) Poll, T.; Metter, J. O.; Helmchen, G. *AG(E)* **1985**, *24*, 112. (b) Herndon, J. W. *JOC* **1986**, *51*, 2853. (c) Metral, J.-L.; Lauterwein, J.; Vogel, P. *HCA* **1986**, *69*, 1287. (d) Waldmann, H.; Braun, M.; Dräger, M. *AG(E)* **1990**, *29*, 1468. (e) Kametani, T.; Takeda, H.; Suzuki, Y.; Honda, T. *SC* **1985**, *15*, 499. (f) Vidari, G.; Ferrino, S.; Grieco, P. A. *JACS* **1984**, *106*, 3539. (g) Funk, R. L.; Zeller, W. E. *JOC* **1982**, *47*, 180. (h) Fringuelli, F.; Pizzo, F.; Taticchi, A.; Wenkert, E. *SC* **1986**, *16*, 245.
13. (a) Snider, B. B. *COS* **1991**, *5*, 1. (b) Snider, B. B. *COS* **1991**, *2*, 527. (c) Mikami, K.; Shimizu, M. *CRV* **1992**, *92*, 1021.
14. (a) Snider, B. B.; Rodini, D. J.; Conn, R. S. E.; Sealfon, S. *JACS* **1979**, *101*, 5283. (b) Snider, B. B.; Roush, D. M.; Rodini, D. J.; Gonzalez, D.; Spindell, D. *JOC* **1980**, *45*, 2773. (c) Snider, B. B.; Duncia, J. V. *JACS* **1980**, *102*, 5926. (d) Duncia, J. V.; Lansbury, P. T., Jr.; Miller, T.; Snider, B. B. *JACS* **1982**, *104*, 1930. (e) Snider, B. B.; Phillips, G. B. *JOC* **1983**, *48*, 3685.
15. (a) Batcho, A. D.; Berger, D. E.; Uskokovic, M. R.; Snider, B. B. *JACS* **1981**, *103*, 1293. (b) Dauben, W. G.; Brookhart, T. *JACS* **1981**, *103*, 237. (c) Batcho, A. D.; Berger, D. E.; Davoust, S. G.; Wovkulich, P. M.; Uskokovic, M. R. *HCA* **1981**, *64*, 1682. (d) Wovkulich, P. M.; Batcho, A. D.; Uskokovic, M. R. *HCA* **1984**, *67*, 612.
16. Snider, B. B.; Phillips, G. B. *JOC* **1984**, *49*, 183.
17. (a) Snider, B. B.; Phillips, G. B. *JACS* **1982**, *104*, 1113. (b) Snider, B. B.; Phillips, G. B.; Cordova, R. *JOC* **1983**, *48*, 3003.
18. Snider, B. B.; Phillips, G. B. *JOC* **1983**, *48*, 464.
19. Mikami, K.; Kaneko, M.; Loh, T.-P.; Tereda, M.; Nakai, T. *TL* **1990**, *31*, 3909.
20. Abouadellah, A.; Aubert, C.; Bégué, J.-P.; Bonnet-Delpon, D.; Guilhem, J. *JCS(P1)* **1991**, 1397.
21. (a) Majetich, G.; Behnke, M.; Hull, K. *JOC* **1985**, *50*, 3615. (b) Majetich, G.; Hull, K.; Lowery, D.; Ringold, C.; Defauw, J. In *Selectivities in Lewis Acid Promoted Reactions*; Schinzer, D., Ed.; Kluwer: Dordrecht, 1989; Chapter 9. (c) Majetich, G.; Khetani, V. *TL* **1990**, *31*, 2243. (d) Majetich, G.; Hull, K.; Casares, A. M.; Khetani, V. *JOC* **1991**, *56*, 3958. (e) Majetich, G.; Song, J.-S.; Ringold, C.; Nemeth, G. A.; Newton, M. G. *JOC* **1991**, *56*, 3973. (f) Majetich, G.; Song, J.-S.; Leigh, A. J.; Condon, S. M. *JOC* **1993**, *58*, 1030.
22. (a) Schinzer, D.; Sólyom, S.; Becker, M. *TL* **1985**, *26*, 1831. (b) Schinzer, D. *S* **1988**, 263.
23. (a) Trost, B. M.; Hiemstra, H. *JACS* **1982**, *104*, 886. (b) Trost, B. M.; Coppola, B. P. *JACS* **1982**, *104*, 6879. (c) Trost, B. M.; Fray, M. J. *TL* **1984**, *25*, 4605.
24. Wada, M.; Shigehisa, T.; Akiba, K. *TL* **1985**, *26*, 5191. (b) Pirrung, M. C.; Thomson, S. A. *TL* **1986**, *27*, 2703.
25. (a) McDonald, T. L.; Delahunty, C. M.; Mead, K.; O'Dell, D. E. *TL* **1989**, *30*, 1473. (b) Plamondon, L.; Wuest, J. D. *JOC* **1991**, *56*, 2066.
26. Casteñeda, A.; Kucera, D. J.; Overman, L. E. *JOC* **1989**, *54*, 5695.
27. (a) Lukas, J. H.; Kouwenhoven, A. P.; Baardman, F. *AG(E)* **1975**, *14*, 709. (b) Lukas, J. H.; Baardman, F.; Kouwenhoven, A. P. *AG(E)* **1976**, *15*, 369.
28. Hayashi, Y.; Niihata, S.; Narasaka, K. *CL* **1990**, 2091.
29. (a) Snider, B. B.; Spindell, D. K. *JOC* **1980**, *45*, 5017. (b) Snider, B. B.; Ron, E. *JOC* **1986**, *51*, 3643.
30. Fujiwara, T.; Tomaru, J.; Suda, A.; Takeda, T. *TL* **1992**, *33*, 2583.
31. Snider, B. B.; Jackson, A. C. *JOC* **1982**, *47*, 5393.
32. (a) Beal, R. B.; Dombroski, M. A.; Snider, B. B. *JOC* **1986**, *51*, 4391. (b) Bambal, R.; Kemmit, R. D. W. *CC* **1988**, 734.
33. Tarzia, G.; Balsamini, C.; Spadoni, G.; Duranti, E. *S* **1988**, 514.
34. Snider, B. B. *JOC* **1981**, *46*, 3155.
35. Matsumura, Y.; Fujiwara, J.; Maruoka, K.; Yamamoto, H. *JACS* **1983**, *105*, 6312.
36. (a) Ishibashi, H.; So, T. S.; Nakatani, H.; Minami, K.; Ikeda, M. *CC* **1988**, 827. (b) Ishibashi, H.; Okoda, M.; Sato, K.; Ikeda, M.; Ishiyama, K.; Tamura, Y. *CPB* **1985**, *33*, 90.
37. Trost, B. M.; Ghadiri, M. R. *JACS* **1984**, *106*, 7260.
38. Barton, D. H. R.; Boivin, J.; Sarma, J.; Da Silva, E.; Zard, S. Z. *TL* **1989**, *30*, 4237.
39. Trost, B. M.; Nielsen, J. B.; Hoogsteen, K. *JACS* **1992**, *114*, 5432.
40. (a) Raucher, S.; Lawrence, R. F. *T* **1983**, *39*, 3731. (b) *TL* **1983**, *24*, 2927.
41. Oppolzer, W.; Mills, R. J.; Pachinger, W.; Stevenson, T. *HCA* **1986**, *69*, 1542.
42. Reinheckel, H.; Sonnek, G.; Falk, F. *JPR* **1974**, *316*, 215.
43. Harmata, M. *TL* **1989**, *30*, 437.
44. Yamamoto, Y.; Saito, Y.; Maruyama, K. *CC* **1982**, 1326.

Barry B. Snider
Brandeis University, Waltham, MA, USA

N-Ethylbenzisoxazolium Tetrafluoroborate

[4611-62-5] $C_9H_{10}BF_4NO$ (MW 234.99)

(peptide-coupling reagent;[1] reacts with a wide range of nucleophiles to give *o*-hydroxy-*N*-ethylbenzamido derivatives[2])

Physical Data: mp 109.5–110.2 °C.
Solubility: sol acetonitrile; slightly sol dichloromethane; sol aqueous acid (unstable above pH 4).
Form Supplied in: crystalline solid; commercially available.
Preparative Methods: ethylation of benzisoxazole with purified ***Triethyloxonium Tetrafluoroborate***.[2]
Handling, Storage, and Precautions: the salt is reported not to be hygroscopic, although it etches glass on long exposure to moist air; should be protected from light; reported to be an irritant.

Peptide Coupling. *N*-Ethylbenzisoxazolium tetrafluoroborate (EBF) is used to couple amines and carboxylic acids in a two-stage procedure. Firstly, EBF reacts with the carboxylic acid to form an active phenolic ester. Secondly, the active ester reacts with the amine to form the peptide linkage (eq 1).[3] The procedure for the formation of the active ester employs a carefully buffered aqueous solution of the Cbz-protected aminoacid (or oligopeptide), to which an equal volume of ethyl acetate or dichloromethane is added, followed by 1.1 equiv of finely powdered EBF. The acylsalicylamide ester is isolated from the organic solvent and may be recrystallized. This product is stable to storage. This procedure does not suffer when used with the sterically demanding valine or proline residues. The active ester cleanly reacts with peptide amines in dipolar aprotic solvents, yielding the amide and *N*-ethylsalicylamide, which can be removed with alkali or carbon tetrachloride. The second step is, however, very slow, taking ca. 48 h at room temperature (roughly 20 times slower than the rate of *p*-nitrophenyl ester couplings[4]), and suffers from unwanted side-reactions (Brenner rearrangements[5]) in the presence of strongly basic amines such as triethylamine. More seriously, significant amounts (up to 1–2%) of epimerization may occur.[6]

$R{-}CO_2^-$ —EBF, H_2O, 0 °C, pH 5, 73–98%→ [aryl ester: NHEt, OCOR] —R'–NH_2, DMF or acetonitrile, rt, 85–90%→ R–CONH–R' (1)

EBF has been used to prepare peptide cyclodimers in yields which compare favorably with other procedures.[1] For this method, intermediate Cbz-protected acylsalicylamide esters were hydrogenolyzed to form the free amines which then yielded the cyclodimers (eqs 2 and 3).

[NHEt, O(Gly)$_3$NHCbz] —H_2, Pd, DMF→ [NHEt, O(Gly)$_3NH_2$] —70%→ *cyclo*-(Gly)$_6$ (2)

[NHEt, OGly-Pro-Gly-NHCbz] —H_2, Pd, DMF, 44%→ *cyclo*-(Gly-Pro-Gly)$_2$ (3)

Peptide cyclotrimers have also been prepared in this manner.[7,8] After deblocking at nitrogen, a highly dilute solution of the resulting free amine in dry pyridine was stirred for 65 h at room temperature to give the cyclotrimer in an impressive 90% yield (eq 4).

[NHEt, O(Gly-Cys(Bn)-Gly)$_3$NHBoc] —1. TFA; 2. 0.001 M in pyridine, rt; 90%→ *cyclo*-(Gly-Cys(Bn)-Gly)$_3$ (4)

Detailed mechanistic studies indicate that the *N*-ethylbenzisoxazolium cation readily undergoes base-catalyzed elimination to form a transitory *N*-ethylbenzoketoketenimine (eq 5).[9]

[N⁺–Et] —base→ [C=NEt, =O] —HX→ [C(NHEt)X, =O] (5)

In the case of reactions with carboxylic acids (X = RCO_2), the initial addition product then rapidly rearranges to the active ester (eq 6).

[C(NHEt)OCOR, =O] ——→ [NHEt, OCOR] (6)

Several hydroxy-,[10] nitro-,[3] and chloro-substituted[11] *N*-ethylbenzisoxazolium cations have been studied in attempts to vary the reactivity of the intermediate active ester (see ***2-Ethyl-7-hydroxybenzisoxazolium Tetrafluoroborate***).

Reactions with Nucleophiles. The *N*-ethylbenzisoxazolium cation reacts rapidly in aqueous solution with a range of nucleophiles, yielding *o*-hydroxy-*N*-ethylbenzimido derivatives. Hydrolysis proceeds smoothly in mild acid, but at pH above 7, polymeric material is obtained (eq 7).[2]

[N⁺–Et] —dilute acid, 90%→ [NHEt, OH] (7)

Treatment of EBF with aqueous solutions of simple nucleophiles (HS^-, F^-, CN^-) or with methanolic methoxide gives high yields of the benzimido derivatives (eq 8).

HX, 66–80% (8)

Reaction of EBF with cyanate, thiocyanate, and thiourea generates initial adducts which can undergo internal rearrangements. For example, the adduct of cyanate to EBF yields a cyclic urethane (eq 9); reaction with thiocyanate yields the analogous thiourethane.[2]

NaCNO, 80% (9)

1. Rajappa, S.; Akerkar, A. S. *TL* **1966**, *25*, 2893.
2. Kemp, D. S.; Woodward, R. B. *T* **1965**, *21*, 3019.
3. Kemp, D. S.; Wang, S.-W.; Mollan, R. C.; Hsia, S.-L.; Confalone, P. N. *T* **1974**, *30*, 3677.
4. Kemp, D. S.; Wang, S.-W. *JACS* **1967**, *89*, 2745.
5. Kemp, D. S.; Duclos, J. M.; Bernstein, Z.; Welch, W. M. *JOC* **1971**, *36*, 157.
6. Kemp, D. S.; Wang, S.-W.; Busby, G., III; Hugel, G. *JACS* **1970**, *92*, 1043.
7. Kemp, D. S.; McNamara, P. *TL* **1981**, *22*, 4571.
8. Kemp, D. S.; McNamara, P. *JOC* **1985**, *50*, 5834.
9. Kemp, D. S. *T* **1967**, *23*, 2001.
10. Kemp, D. S.; Chien, S.-W. *JACS* **1967**, *89*, 2743.
11. Rajappa, S.; Akerkar, A. S. *CC* **1966**, 826.

Ross P. McGeary
University of Cambridge, UK

Ethyl Chloroformate[1]

Cl–C(=O)–OEt

[541-41-3] $C_3H_5ClO_2$ (MW 108.52)

(for ethoxycarbonylation of a wide variety of nucleophiles including anions of carboxylic acids[4] and esters,[5] nitriles,[7] dithianes,[9] ketones, enamines,[10] imines,[11] sulfones,[12] cuprates,[13] stannanes,[14] Grignards,[15] alanes,[18] furans,[17] alkynes;[19] acylation of alcohols,[20,22] carbonyls,[23,25] amines, and aziridines;[27] cleavage of tertiary amines;[32] formation of mixed anhydrides[38])

Physical Data: mp −81 °C; bp 95 °C; *d* 1.135 g cm^{-3}.

Solubility: miscible with benzene, chloroform, diethyl ether, ethanol; insol water.

Form Supplied in: colorless liquid; widely available.

Purification: the liquid is washed with water then distilled at atmospheric pressure using an efficient fractionating column and $CaCl_2$ guard tube.[2]

Handling, Storage, and Precautions: slowly hydrolyzed by water to CO_2 and HCl; store at cool temperatures: pressure may develop within bottle; thermal decomposition to CO_2 and EtCl occurs; highly flammable (fp 2 °C); toxic by inhalation and skin absorption; lachrymatory, irritant; use in a fume hood.

Ethoxycarbonylation of Active Methylene Compounds. Ethyl chloroformate[3] reacts with a wide range of nucleophiles. Enolates derived by deprotonation of aliphatic and aromatic carboxylic acids bearing at least one α-hydrogen with two equivalents of ***Lithium Diisopropylamide*** at low temperature in THF are ethoxycarbonylated to give the monoesters of malonic acids (eq 1).[4] The yields are highest for alkyl carboxylic acids at −78 °C.[5] Although higher yielding than with ***Diethyl Carbonate***, treatment of the α-lithiated esters with ***Carbon Dioxide*** frequently gives better yields of the malonates.[4,6] Mixed malonates, especially those derived from bulky alkyl esters, can be similarly prepared (eq 2).[5]

1. 2 equiv LDA, THF, −78 °C; 2. $ClCO_2Et$; 85% (1)

$MeCO_2$-*t*-Bu → 1. LDA, THF, −78 °C; 2. $ClCO_2Et$, rt; 75% → CO_2Et / CO_2-*t*-Bu (2)

α-Lithionitriles are ethoxycarbonylated in good yield.[7] Acetals[8] and 1,3-dithianes[9] are effectively ethoxycarbonylated (eq 3) if ethyl chloroformate is used as a cosolvent to prevent side reactions. Ethoxycarbonylation of ketones is achieved when enamines are treated with ethyl chloroformate, in some cases giving comparable yields to those obtained using metal enolates.[10]

$ClCO_2Et$, THF, −78 °C to rt; 60% (3)

Enantioselective ethoxycarbonylation has recently been attempted[11] via deprotonation of *N*-benzylidenebenzylamine with two equivalents of a chiral lithium amide base and quenching with ethyl chloroformate; levels of asymmetric induction were low but higher than those obtained using other chloroformates (eq 4).

1. 2 equiv chiral lithium amide; 2. $ClCO_2Et$; 56% → Ph–CH(CO_2Et)–NH_2•HCl, 41% ee (4)

Ethoxycarbonylation of Vinyl Anions. α,β-Unsaturated sulfones are α-ethoxycarbonylated (eq 5) on deprotonation with ***n-Butyllithium*** in THF at −78 °C and quenching with this reagent.[12]

1. $CH_2(OMe)_2$, P_2O_5; 2. BuLi, THF, −78 °C; 3. $ClCO_2Et$, −78 °C; 76% (5)

Vinyl cuprates under Pd[0] catalysis give α,β-alkenic esters (eq 6).[13] Aryl, vinyl, and heterocyclic stannanes[14] react smoothly under Pd^0 catalysis at 100 °C to give ethyl esters (eq 7) in good yields if the catalyst and chloroformate are added to the reaction mixture slowly. A wide variety of functionality on tin is tolerated but allylstannanes are generally low yielding. Use of isobutyl chloroformate improves the yields in these acylcarbonylations.

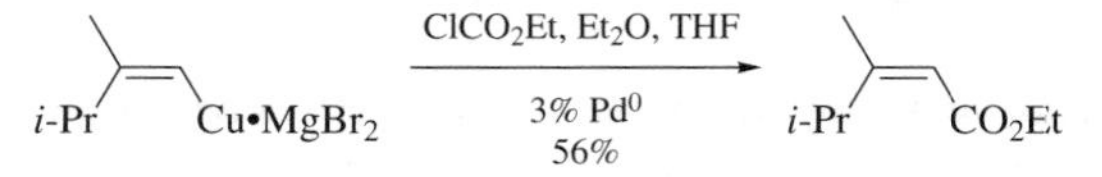

(6)

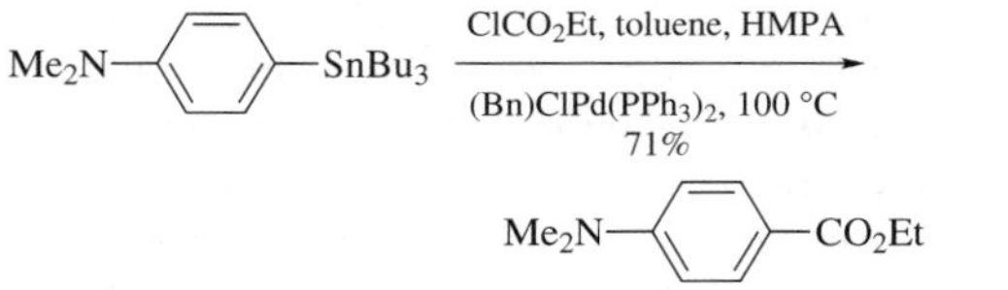

(7)

Ethyl esters are produced from Grignard reagents[15] if the organometallics are not present in excess. This is also the basis of the 5-ethoxycarbonylation of 2,4-dimethylpyrrole.[16] 2-Lithiofurans are ethoxycarbonylated in good yield (eq 8).[17]

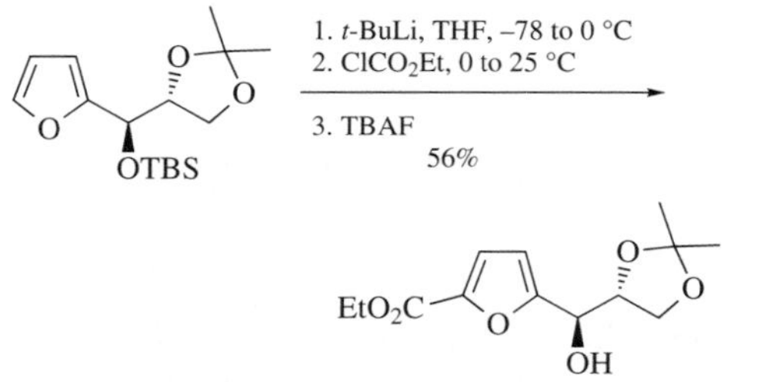

(8)

Trans-α,β-unsaturated esters can be prepared by ethoxycarbonylation of vinylalanes derived from terminal alkynes bearing primary, secondary, or tertiary alkyl substituents without the necessity of forming an intermediate ate complex (eq 9).[18] *Cis*-α,β-unsaturated esters result from ethoxycarbonylation of lithiated alkynes (eq 10)[19] followed by reduction over Lindlar catalyst.

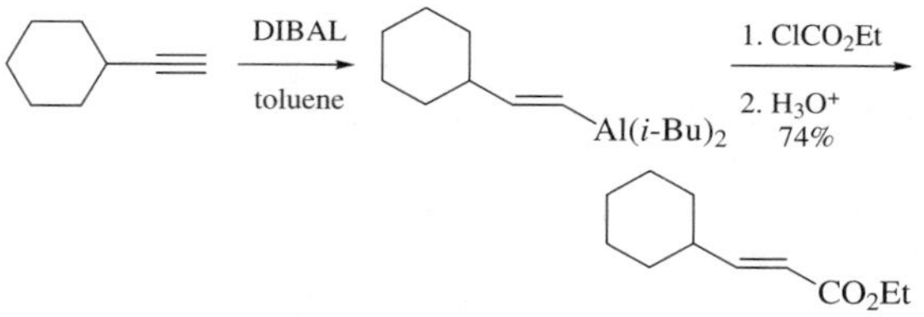

(9)

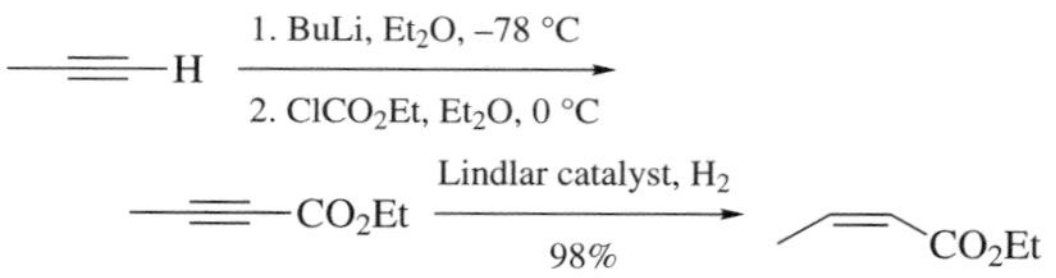

(10)

O-, N-, and S-Acylations. *O*-Ethoxycarbonylation is observed in some instances. Aliphatic alcohols, phenols,[20] and cyanohydrins[21] react in the presence of tertiary amines to give carbonates. Equatorial oriented steroid hydroxyls are selectively acylated in the presence of axial hydroxyl groups.[22] Enol carbonates (eq 11)[23] are readily derived from ketones by formation of the lithium enolate using ***Lithium 2,2,6,6-Tetramethylpiperidide*** in THF/HMPA at −78 °C followed by an ethyl chloroformate quench at room temperature. The use of HMPA prevents competitive *C*-acylation. LiTMP is necessary to avoid side reactions common with other amine-derived bases such as LDA. Sodium and potassium enolates are not useful, generally giving competitive *C*-acylation. This procedure avoids the use of classical methodology relying on the use of Hg^{II} salts.

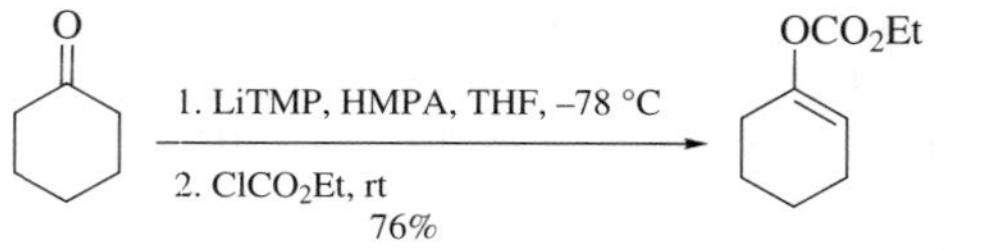

(11)

Trans-1-(1,3-butadienyl) carbamates[24] and carbonates[25] are useful dienes in Diels–Alder chemistry. The carbonates are produced when ***Crotonaldehyde*** (and its congeners) are deprotonated and treated at −78 °C with ethyl chloroformate (eq 12). Use of higher temperatures results in loss of some of the (*Z*) geometry. Similar treatment of crotonaldehyde imines affords the carbamates.

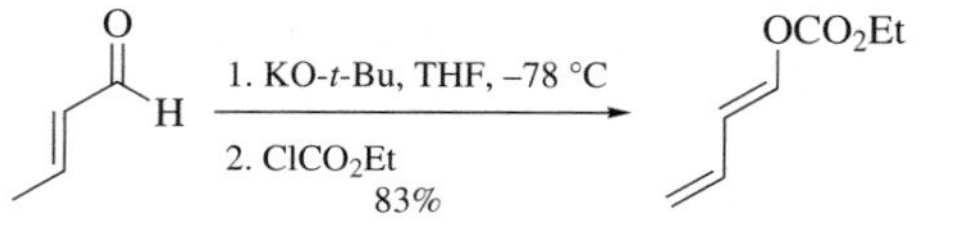

(12)

O-Ethylation occurs when the sodium enolates of α-formyl γ- and δ-lactones are refluxed in THF with ethyl chloroformate.[26]

N-Ethoxycarbonylation forms the basis of a variety of reactions of nitrogenous compounds. Aziridines form *N*-ethoxycarbonylaziridines which are converted, on heating, to allylic amines (eq 13).[27]

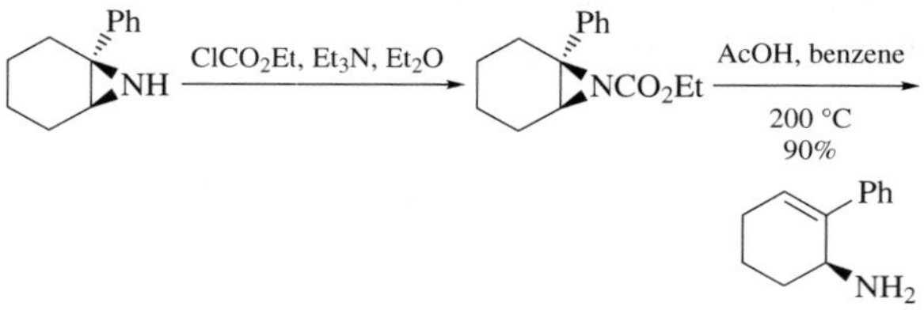

(13)

Tertiary amines may be cleaved by ethyl chloroformate in certain circumstances. *N*-Demethylation has been noted. For example, tropine acetate (eq 14) is readily converted to nortropine hydrochloride[28] on refluxing with ethyl chloroformate and subsequent acid-mediated hydrolysis of the intermediate urethane.

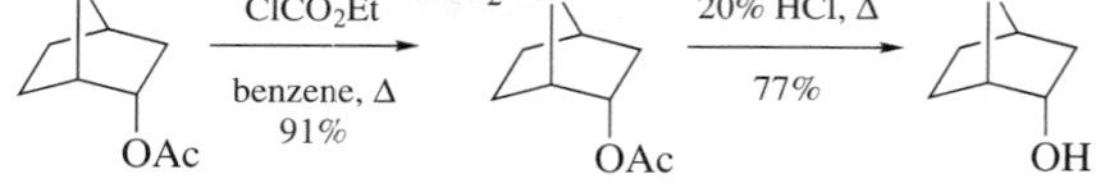

(14)

Reductive C–N bond cleavage has been demonstrated in more complex systems using ethyl chloroformate as a solvent.[29] Chlorination of corticosteroid cyclic ethers has been observed.[30] Tertiary aliphatic and alicyclic amines can be dealkylated.[31] Phenyl chloroformate, however, is usually regarded as the reagent of choice. In summary, deamination, demethylation, debenzylation, and dealkylation of tertiary amines can all occur on treatment with ethyl chloroformate but the regioselectivities of these reactions are difficult to predict.[32]

Imides (e.g. phthalimide) are ethoxycarbonylated if the potassium salt is treated with ethyl chloroformate at 5–10 °C. At elevated temperatures (60–110 °C) the lithium salts are *N*-ethylated in high yield. *N*-Alkylation also occurs with other imide systems such as barbiturates, hydantoins, and succinimides.[33] Thiols give the corresponding monothiocarbonates.[34] Thioamides form ethyl imidates[35] and cyclic iminoethers[36] are obtained from thiolactams on mild warming with ethyl chloroformate. Sulfinic and sulfonic acids yield their ethyl esters.[37]

Mixed Anhydrides. The preparation of mixed anhydrides from ethyl chloroformate and carboxylic acids enables coupling of peptides and amino acids.[38] This coupling procedure has largely been superseded by more effective methods relying on the use of superior reagents; see ***1,3-Dicyclohexylcarbodiimide*** (DCC) and ***1-Hydroxybenzotriazole*** (HOBT).

Ethyl chloroformate enables the one-step cyclization of peptides (eq 15),[39] yields generally being superior to those obtained if ***Bis(2,4-dinitrophenyl) Carbonate*** is used. However, ***Isobutyl Chloroformate***/*N*-methylmorpholine is now the reagent of choice for many solution-phase anhydride couplings.[40]

L-Try–Gly–L-Leu–L-Ala–D-Thr → (1. py•HCl, $ClCO_2Et$, DMF, THF, 15 °C; 2. Et_3N; 30%) cyclo[L-Ala–L-Leu–Gly–COCH–CH_2-β-indolyl / NH–CO–NHCH–CH(OH)Me] (15)

N-Protected amino acids and peptides are rapidly converted to the corresponding amino alcohols in high yields with complete retention of optical purity via reduction of the mixed anhydride by cold ***Sodium Borohydride*** in THF with dropwise addition of methanol (eq 16).[41] The disulfide bridges of cystine, the methyl and benzyl esters of ω-carboxyl-protected glutamic and aspartic acids of peptides, and *N*-Cbz and *N*-Boc protection are compatible with the methodology. The anhydrides derived from ethyl chloroformate are superior to isobutyl and benzyl carbonates both in terms of yield and retention of optical purity.

CbzNH–CH(R)–C(O)OH → (1. NMM, $ClCO_2Et$, THF, −10 °C; 2. $NaBH_4$, THF, dropwise MeOH; 63–90%) CbzNH–CH(R)–CH_2OH (16)

Carboxylic acid triethylamine salts yield mixed anhydrides which, when treated with ethoxymagnesiomalonic esters give the benzoylated malonate.[42] Alternatively, treatment of the mixed anhydrides with sodium tetracarbonylferrate(II) produces the aldehyde derived from the parent aliphatic or aromatic carboxylic acid in good yield.[43]

Formation of mixed anhydrides also provides the basis for the synthesis of acyl azides (eq 17)[44] and hence amines via the Curtius rearrangement. The azide ion is usually delivered to the most electrophilic carbonyl,[45] but steric considerations may be important. The method is applicable even to strained cyclopropyl- and cyclobutylcarboxylic acids.[46] ***Diisopropylethylamine*** (Hünig's base) has been mooted as a superior base for the formation of mixed anhydrides.[47]

1-Ph-cyclopentane-CO_2H → ($ClCO_2Et$, Et_3N, acetone, 0 °C) 1-Ph-cyclopentane-CO_2CO_2Et → (NaN_3, 0 °C) 1-Ph-cyclopentane-CON_3 → (1. toluene, Δ; 2. HCl (aq); 81%) 1-Ph-cyclopentane-NH_2 (17)

(*Z*)-Ethylenic mixed anhydrides are easily prepared without isomerization and react with vinyl cuprates (and other organometallics) under Pd^0 catalysis with >99% stereochemical control to give unsymmetrical divinyl ketones (eq 18).[13] Alternative routes via (*Z*)-ethylenic acyl chlorides produce (*E*/*Z*) mixtures.

Hex–CH=CH–CO_2H → ($ClCO_2Et$, Et_3N, Et_2O; 100%) Hex–CH=CH–CO_2CO_2Et → ([Et–CH=CH]$_2$CuLi($ZnBr_2$), THF, 3% Pd^0; 87%) Et–CH=CH–C(O)–CH=CH–Hex (*Z*, *Z*) (18)

Miscellaneous. β-Cyanoesters (eq 19)[48] are produced from β-amido acids providing that the acid and amide functionality are sufficiently close to enable formation of a cyclic intermediate.

(*Z*)-HO_2C–CH=CH–$CONH_2$ → ($ClCO_2Et$, Et_3N, CH_2Cl_2, 0 °C; 91%) (*Z*)-EtO_2C–CH=CH–CN (19)

Aromatics undergo Friedel–Crafts alkylation in the presence of ***Aluminum Chloride***, probably due to the formation of EtCl in situ.[49]

Ethyl chloroformate has been used to moderate reactivity in polymerizations. For example, PVC with good thermal stability is prepared in the presence of this reagent.[50]

1. Matzner, M.; Kurkjy, R. P.; Cotter, R. J. *CRV* **1964**, *64*, 645.
2. Hamilton, C. S.; Sly, C. S. *JACS* **1925**, *47*, 435; Saunders, J. H.; Slocombe, R. J.; Hardy, E. E. *JACS* **1951**, *73*, 3796.
3. Cappelli, G. *G* **1920**, *50*, 8 (*CA* **1921**, *15*, 524).
4. Krapcho, A. P.; Jahngen, E. G. E.; Kashdan, D. S. *TL* **1974**, *15*, 2721.
5. Brocksom, T. J.; Petragnani, N.; Rodrigues, R. *JOC* **1974**, *39*, 2114.
6. Reiffers, S.; Wynberg, S.; Strating, J. *TL* **1971**, *12*, 3001.
7. Albarella, J. P. *JOC* **1977**, *42*, 2009.
8. Truce, W. E.; Roberts, F. E. *JOC* **1963**, *28*, 961.
9. Seebach, D.; Corey, E. J. *JOC* **1975**, *40*, 231; Corey, E. J.; Erickson, B. W. *JOC* **1971**, *36*, 3553.
10. Stork, G.; Brizzdara, A.; Landesman, H.; Szmuszkovic, J.; Terrell, R. *JACS* **1963**, *85*, 207.
11. Duhamel, L.; Duhamel, P.; Fouguay, S.; Eddine, J. J.; Peschard, O.; Plaguevent, J.-C.; Ravard, A.; Soilard, R.; Valnol, J.-Y.; Vincens, H. *T* **1988**, *44*, 5495.
12. Alcaraz, C.; Carretero, J. C.; Dominguez, E. *TL* **1991**, *32*, 1385.
13. Jabri, N.; Alexakis, A.; Normant, J. F. *T* **1986**, *42*, 1369.

14. Balas, L.; Jousseaume, B.; Shin, H.-A.; Verlhac, J.-B.; Wallian, F. *OM* **1991**, *10*, 366.
15. Gold, J. R.; Sommer, L. H.; Whitmore, F. C. *JACS* **1948**, *70*, 2874; Warrener, R. N.; Cain, E. C. *AJC* **1971**, *24*, 785.
16. Fischer, H. *OSC* **1943**, *2*, 198.
17. Martin, S. F.; Zinke, P. W. *JOC* **1991**, *56*, 6600.
18. Zweifel, G.; Lynd, R. A. *S* **1976**, 625.
19. Taschner, M. J.; Rosen, T.; Heathcock, C. H. *OSC* **1990**, *7*, 226.
20. Claisen, L. *CB* **1894**, *27*, 3182.
21. Davis, O. C. M. *JCS* **1910**, *97*, 949.
22. Fieser, C. F.; Herz, J. E.; Klohs, M. W.; Romero, M. A.; Utne, T. *JACS* **1952**, *74*, 3309.
23. Olofson, R. A.; Cuomo, J.; Bauman, B. A. *JOC* **1978**, *43*, 2073.
24. Oppolzer, W.; Fostl, W. *HCA* **1975**, *58*, 587.
25. DeCusati, P. F.; Olofson, R. A. *TL* **1990**, *31*, 1405.
26. Murray, A. W.; Murray, N. D. *SC* **1986**, *16*, 853.
27. Laurent, A.; Mison, P.; Nafti, A.; Cheikh, R. B.; Chaabouni, R. *JCR(M)* **1984**, 3164.
28. Kraiss, G.; Nador, K. *TL* **1971**, *12*, 57.
29. Hanoaka, M.; Nagami, K.; Imanishi, T. *H* **1979**, *12*, 497.
30. Sugal, S.; Akaboshi, S.; Ikagami, S. *CPB* **1983**, *31*, 12.
31. Hobson, J. D.; McCluskey, J. G. *JCS(C)* **1967**, 2015.
32. Kapnang, H.; Charles, G. *TL* **1983**, *24*, 3233.
33. Vida, J. A. *TL* **1972**, *13*, 3921.
34. Saloman, F. *JPR* **1873**, *6*, 433.
35. Snydam, F. H.; Greth, W. E.; Langerman, N. R. *JOC* **1969**, *34*, 292.
36. Mohacsi, E.; Gordon, E. M. *SC* **1984**, *14*, 1159.
37. Etienne, A.; Vincent, J.; Lonchambon, G. *CR(C)* **1970**, *270*, 841.
38. Greenstein, J. P.; Winitz, M. *The Chemistry of Amino Acids*; Wiley: New York, 1961; Vol. 2, pp 978–981. See also Vaughn, J. R., Jr. *JACS* **1951**, *73*, 3547; Boissonnas, R. A. *HCA* **1951**, *34*, 874; Wieland, T.; Bernhard, H. *LA* **1951**, *572*, 190.
39. Wieland, T.; Faesal, J.; Faustich, H. *LA* **1968**, *713*, 201.
40. Kopple, K. D.; Renick, K. J. *JOC* **1958**, *23*, 1565; Merrifield, R. B.; Mitchell, A. R.; Clarke, J. E. *JOC* **1974**, *39*, 660.
41. Kokotos, G. *S* **1990**, *4*, 299.
42. Price, J. A.; Tarbell, D. S. *OSC* **1963**, *4*, 285.
43. Watanabe, Y.; Yamashita, M.; Mitsudo, T.; Igami, M.; Temi, K.; Takegami, Y. *TL* **1975**, *16*, 1063; Watanabe, Y.; Yamashita, M.; Mitsudo, T.; Igami, M.; Takegami, Y. *BCJ* **1975**, *48*, 2490.
44. Kaiser, C.; Weinstock, J. *OS* **1971**, *51*, 48.
45. Weinstock, J. *JOC* **1961**, *26*, 3511.
46. Finkelstein, J.; Chiang, E.; Vane, F. M.; Lee, J. *JMC* **1966**, *9*, 319.
47. Jessup, R. J.; Petty, C. B.; Roos, J.; Overman, L. E. *OS* **1979**, *59*, 1.
48. Sauers, C. K.; Cotter, R. J. *JOC* **1961**, *26*, 6.
49. Freidel, C.; Crafts, J. A. *LA* **1884**, *1*, 527; see also *The Chemistry of Functional Groups—The Chemistry of Acyl Halides*; S. Patai, Ed.; Interscience: New York, 1972.
50. *CA* **1984**, *100*, 175 541n.

Andrew N. Payne
University of Cambridge, UK

1-Ethyl-3-(3′-dimethylaminopropyl) carbodiimide Hydrochloride[1]

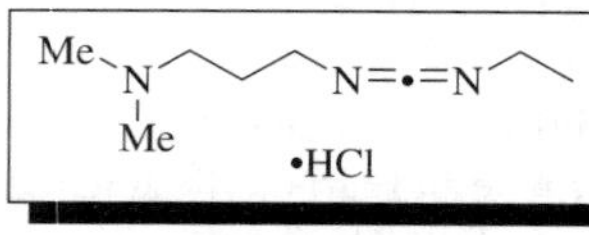

[25952-53-8] $C_8H_{18}ClN_3$ (MW 191.74)
(base)
[1892-57-5]
(·MeI)
[22572-40-3]

(peptide coupling reagent;[1b,2] amide formation;[3] ester formation;[4] protein modification;[5] mild oxidations of primary alcohols[6])

Alternate Name: 1-(3-dimethylaminopropyl)-3-ethylcarbodiimide; water-soluble carbodiimide; EDC; EDCI.
Physical Data: free base is an oil, bp 47–48 °C/0.27 mmHg; HCl salt is a white powder, mp 111–113 °C; MeI salt mp 97–99 °C.
Solubility: sol H_2O, CH_2Cl_2, DMF, THF.
Form Supplied in: commercially available as HCl salt and as methiodide salt that are white solids. Reagents are >98% pure; main impurity is the urea that can form upon exposure to moisture.
Analysis of Reagent Purity: IR: 2150 cm^{-1} (N=C=N stretch); urea has C=O stretch near 1600–1700 cm^{-1}.
Purification: recrystallization from CH_2Cl_2/ether.
Handling, Storage, and Precautions: EDC is moisture-sensitive; store under N_2 in a cool dry place. It is incompatible with strong oxidizers and strong acids. EDC is a skin irritant and a contact allergen; therefore avoid exposure to skin and eyes.

Peptide Coupling Reagent. This carbodiimide (EDC) reacts very similarly to ***1,3-Dicyclohexylcarbodiimide*** and other carbodiimides. The advantage EDC has over DCC is that the urea produced is water soluble and, therefore, is easily extracted. Dicyclohexylurea is only sparingly soluble in many solvents and is removed by filtration which may not be as effective as extraction. A typical example of EDC for peptide coupling is shown in (eq 1).[2]

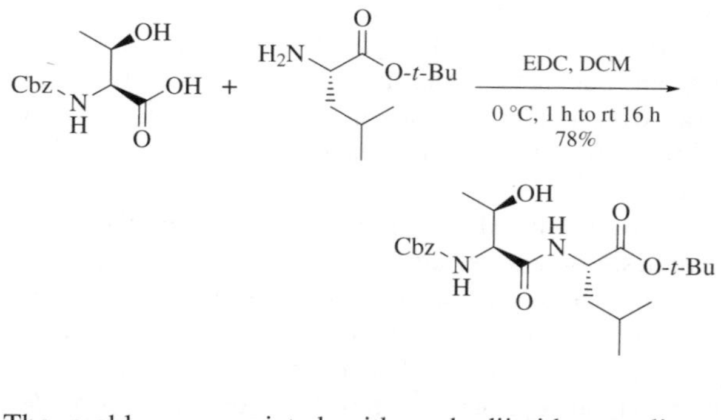

The problems associated with carbodiimide couplings are mainly from the *O*-acylisourea intermediate (**1**) having poor selectivity for specific nucleophiles. This intermediate can rearrange to an *N*-acylurea (**2**), resulting in contamination of the product and low yields. Additionally it can rearrange to 5(4*H*)-oxazolones (**3**) that tautomerize readily, resulting in racemization. Using low dielectric solvents like CH_2Cl_2 or additives to trap (**1**) minimizes these side reactions by favoring intermolecular nucleophilic attack on (**1**).[1b]

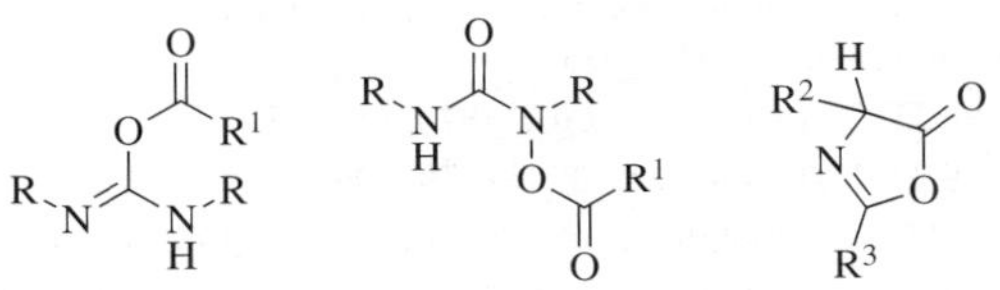

(**1**) (**2**) (**3**)

R = carbodiimide chain; R^1 = amino acid; R^2 = amino acid side chain; R^3 = NH protecting group

When low dielectric constant solvents are used, the carboxylic acid tends to dimerize, thus promoting symmetrical anhydride formation. This intermediate is stable enough to give good yields of the desired product. High dielectric solvents such as DMF retard acylation of amino acids, and *N*-acylurea can be a major byproduct.[1b] Unfortunately, many starting materials require polar solvents for dissolution.

Addition of trapping agents such as ***N-Hydroxysuccinimide***[7] and ***1-Hydroxybenzotriazole*** (HOBt)[8] reduce the extent of many side reactions, especially *N*-acylurea formation. Also, racemization is suppressed when these additives are present. The latter reagent eliminates the intramolecular dehydration of ω-amides of asparagine and glutamine that occurs with carbodiimides (eq 2).[8,9]

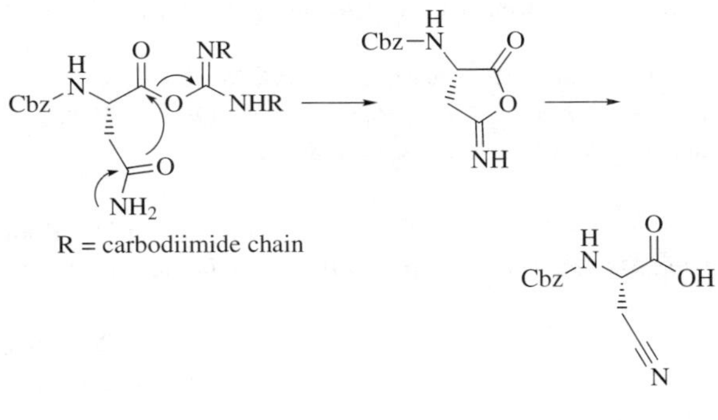

R = carbodiimide chain (2)

With EDC the addition of ***Copper(II) Chloride*** suppresses racemization to <0.1% compared to 0.4% with HOBt under ideal conditions.[10] Combinations of HOBt and $CuCl_2$ are also useful.[10b] The suggested stoichiometry for the highest optical purity and yield is 2 equiv of HOBt and 0.25–0.5 equiv of $CuCl_2$ in DMF.

A practical example of EDC's utility is the solution synthesis of human epidermal growth factor, a 53-residue protein.[11] This included several couplings of fragments ranging from three to six residues in length. All couplings were performed with EDC/HOBt in DMF or NMP. At the completion of the synthesis, no racemized material could be detected by HPLC.

Amide Formation. Formylated amino acids and peptides are prepared in high yields by forming the acid anhydride (eq 3).[3] These products are pure without requiring chromatography or recrystallization.

Another method for formylation of amines is with *p*-nitrophenyl formate, which usually gives products in high yield. However, removing the last traces of the *p*-nitrophenol is difficult.[12]

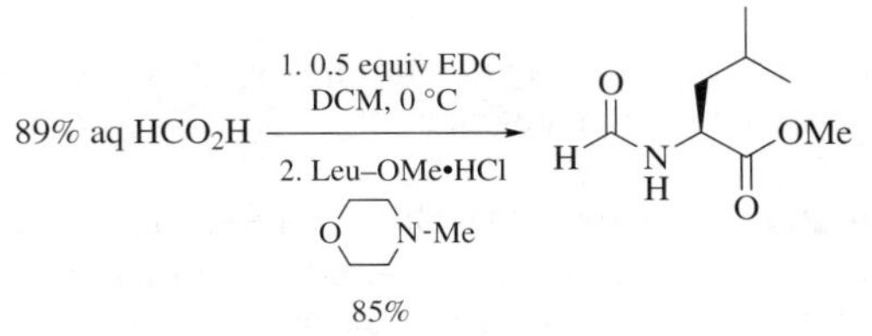

89% aq HCO_2H — 1. 0.5 equiv EDC, DCM, 0 °C; 2. Leu–OMe•HCl, N-methylmorpholine; 85% (3)

Treating carbon dioxide and amines with EDC gives symmetrical ureas (eq 4).[13] DCC with CO_2 at ambient pressure works equally well.

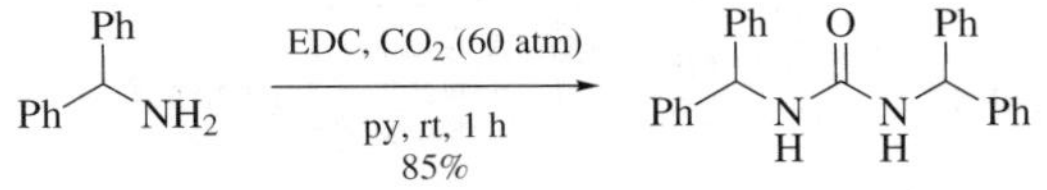

EDC, CO_2 (60 atm); py, rt, 1 h; 85% (4)

EDC facilitates the synthesis of xanthine analogs by condensing a diamine with water-soluble acids (eq 5).[14] The use of water as the solvent precludes the use of DCC in this case.

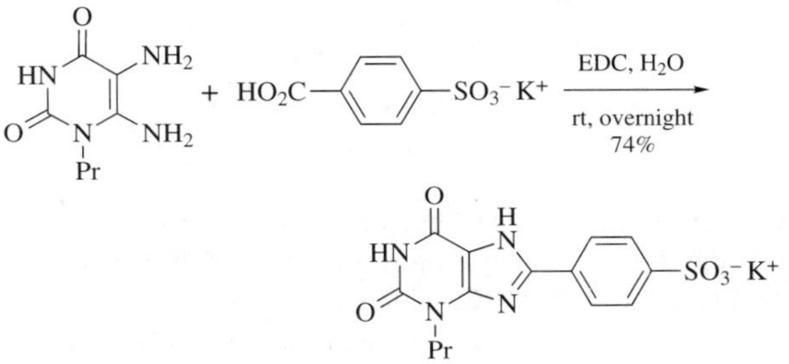

+ HO_2C–C₆H₄–$SO_3^-K^+$ — EDC, H_2O; rt, overnight; 74% (5)

Ester Formation. Esters of *N*-protected amino acids are prepared in high yield with EDC and ***4-Dimethylaminopyridine*** (eq 6).[4] DMAP causes extensive racemization if not used in a catalytic amount.[15] However, when esterifying the α-carboxyl of β- and γ-benzyl esters of aspartyl and glutamyl derivatives, extensive racemization was observed even with DMAP present in catalytic amounts. It is postulated that the side-chain esters contribute in some fashion to the lability of the α-H.

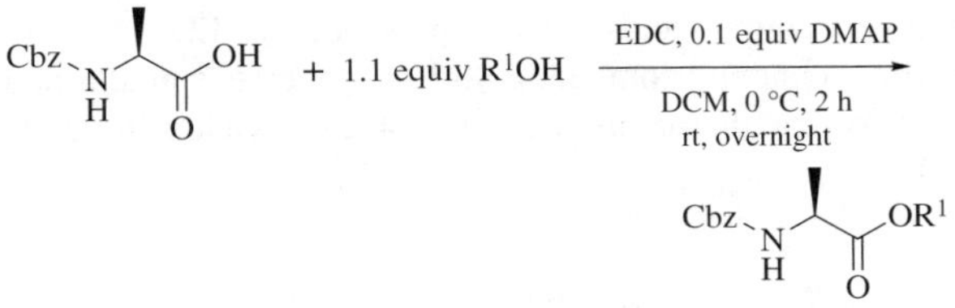

+ 1.1 equiv R^1OH — EDC, 0.1 equiv DMAP; DCM, 0 °C, 2 h; rt, overnight (6)

R^1 = *t*-Bu, 76%; Me, 96%; CH_2CCl_3, 87%

Carbodiimides including EDC are also useful in preparing active esters such as the *p*-nitrophenyl, pentafluorophenyl, and *N*-hydroxysuccinimide esters.[1b] Numerous additional methods have been reported.[16]

Protein Modification. Carboxyl groups in proteins react with EDC, resulting in an activated group that can be trapped with a nucleophile such as glycine methyl ester.[5] The nucleophile can also be amino groups in the protein, causing cross-linking. Applications for carboxylate modification include determining which groups are buried versus exposed, and mechanistic studies for determining which carboxyl groups are essential for an enzyme's activity.

EDC is used extensively in cross-linking proteins to solid supports for affinity chromatography.[17]

Oxidation of Primary Alcohols. One example of EDC substituting for DCC in the Pfitzner–Moffatt oxidation is shown in (eq 7).[6] The use of DCC in this example resulted in a difficult purification of the product and lower yields.

DMSO, 4 equiv EDC; py, TFA, toluene; rt, overnight; 87% (7)

Miscellaneous Reactions. EDC is a useful water scavenger as well. An example of this utility is given in (eq 8).[18] The carbodiimide is required in this reaction, but DCC does not work well in this transformation.

I_2, $AgBF_4$; EDC, CH_2Cl_2; rt, overnight; 75%; (*E*):(*Z*) > 9:1 (8)

Another example of EDC having a marked improvement over DCC is the formation of cyanoguanidines (eq 9).[19] This was originally attempted with DCC, which gives poor yields even after extended reaction times. It is thought the positively charged nitrogen in EDC facilitates the C–S bond cleavage and since DCC lacks this atom the reaction goes slowly.

EDC; DMF, rt, 30 min; 78% (9)

Pyrroles are formed when acetylenes undergo cycloadditions with *N*-acylamino acids in the presence of EDC.[20] An *N*-alkenylmunchnone azomethine ylide is generated in situ and is trapped by the dipolaraphile. Loss of CO_2 yields the pyrrole (eq 10).

EDC, $-H_2O$; *N*-alkenylmunchnone azomethine ylide; E≡E, $-CO_2$ (10)

Related Reagents. Benzotriazol-1-yloxytris(dimethylamino)phosphonium Hexafluorophosphate; *N*,*N*′-Carbonyldiimidazole; 1-Cyclohexyl-3-(2-morpholinoethyl)carbodiimide; 1,3-Dicyclohexylcarbodiimide; Diphenyl Phosphorazidate; Di-*p*-tolylcarbodiimide; *N*-Ethyl-5-phenylisoxazolium-3′-sulfonate; 1-Hydroxybenzotriazole; Isobutyl Chloroformate; 1,1′-Thionylimidazole.

---

1. (a) Kurzer, F.; Douraghi-Zader, K. *CRV* **1967**, *67*, 107. (b) Rich, D. H.; Singh, J. *The Peptides: Analysis, Synthesis, Biology*; Academic: New York, 1979; Vol. 1, pp 241–261.
2. Sheehan, J. C.; Ledis, S. L. *JACS* **1973**, *95*, 875.
3. Chen, F. M. F.; Benoiton, N. L. *S* **1979**, 709.
4. Dhaon, M. K.; Olsen, R. K.; Ramasamy, K. *JOC* **1982**, *47*, 1962.
5. Carraway, K. L.; Koshland, Jr., D. E. *Methods Enzymol.* **1972**, *25*, 616.
6. Ramage, R.; MacLeod, A. M.; Rose, G. W. *T* **1991**, *47*, 5625.
7. Wuensch, E.; Drees, F. *CB* **1966**, *99*, 110.
8. Konig, W.; Geiger, R. *CB* **1970**, *103*, 788.
9. Gish, D. T.; Katsoyannis, P. G.; Hess, G. P.; Stedman, R. J. *JACS* **1956**, *78*, 5954.
10. (a) Miyazawa, T.; Otomatsu, T.; Yamada, T.; Kuwata, S. *TL* **1984**, *25*, 771. (b) Miyazawa, T.; Otomatsu, T.; Fukui, Y.; Yamada, T.; Kuwata, S. *CC* **1988**, 419.
11. Hagiwara, D.; Neya, M.; Miyazaki, Y.; Hemmi, K.; Hashimoto, M. *CC* **1984**, 1676.
12. Okawa, K.; Hase, S. *BCJ* **1963**, *36*, 754.
13. Ogura, H.; Takeda, K.; Tokue, R.; Kobayashi, T. *S* **1978**, 394.
14. Shamim, M. T.; Ukena, D.; Padgett, W. L.; Daly, J. W. *JMC* **1989**, *32*, 1231.
15. Atherton, E.; Benoiton, N. L.; Brown, E.; Sheppard, R. C.; Williams, B. J. *CC* **1981**, 336.
16. Greene, T. W. *Protective Groups in Organic Synthesis*, Wiley: New York, 1981; pp 154–180.
17. Keeton, T. K.; Krutzsch, H.; Lovenberg, W. *Science* **1981**, *211*, 586.
18. Tam, T. F.; Thomas, E.; Kruntz, A. *TL* **1987**, *28*, 1127.
19. Atwal, K. S.; Ahmed, S. Z.; O'Reilly, B. C. *TL* **1989**, *30*, 7313.
20. Anderson, W. A.; Heider, A. R. *SC* **1986**, 357.

Richard S. Pottorf
Marion Merrell Dow Research Institute, Cincinnati, OH, USA

Peter Szeto
University of Bristol, UK

Ethylene Glycol[1]

HO–CH₂CH₂–OH

[107-21-1] $C_2H_6O_2$ (MW 62.08)

(formation of acetals;[1] stereoselective aldol reactions;[2] solvent for Diels–Alder reaction;[3] solvent for Fischer indole reaction;[4] preparation of tertiary alcohols from trialkylboranes[5])

Alternate Name: ethane-1,2-diol.
Physical Data: bp 197 °C; *d* 1.11 g cm^{-3}.

Solubility: completely sol H_2O, low MW alcohols, acetone; insol benzene, chlorinated hydrocarbons.

Form Supplied in: colorless liquid; widely available.

Handling, Storage, and Precautions: harmful or fatal if swallowed. Prolonged or repeated breathing of vapor harmful. Causes eye irritation. May cause kidney and nervous system damage. Causes birth defects in laboratory animals. Use in a fume hood.

Acetal Formation.[1] The dioxolane group is one of the most widely used protecting groups for carbonyl compounds. Dioxolanes are generally stable to bases, Grignard reagents, alkyl lithium reagents, metal hydrides, Na/NH_3, Wittig reagents, hydrogenation conditions, oxidants, and bromination and esterification reagents.[1] This wide range of stability has been especially useful in steroid synthesis.[6] The ease of formation of dioxolanes roughly follows the order: aldehydes > acyclic ketones and cyclohexanones > cyclopentanones > α,β-unsaturated ketones > α-mono- and disubstituted ketones aromatic ketones. Ketones generally require stronger acids than aldehydes, such as sulfuric, hydrochloric, or TsOH.[7–11] A typical protection procedure involves heating a benzene or toluene solution of the ketone and ***p-Toluenesulfonic Acid*** (eq 1).[9] Alternatively, a Lewis acid catalyst, such as ***Boron Trifluoride Etherate***[12–14] may be used (eqs 2 and 3).[13,14]

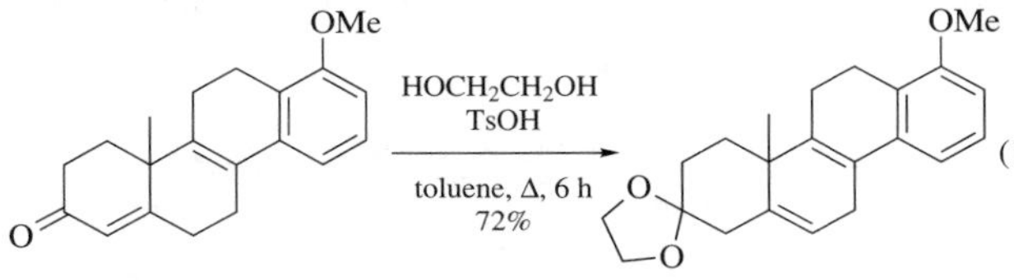

HOCH₂CH₂OH, TsOH; toluene, Δ, 6 h; 72% (1)

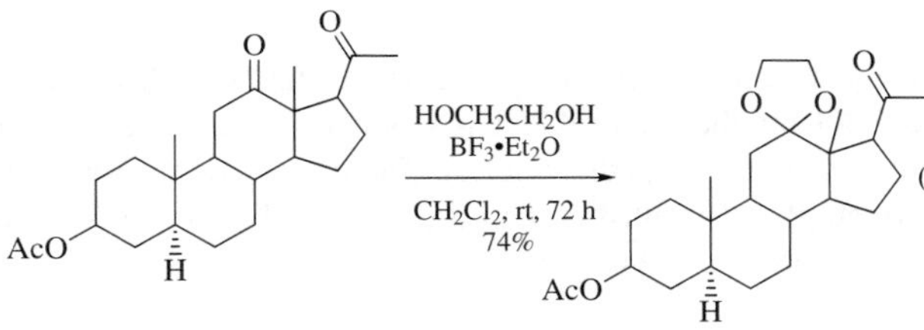

HOCH₂CH₂OH, BF_3•Et_2O; CH_2Cl_2, rt, 72 h; 74% (2)

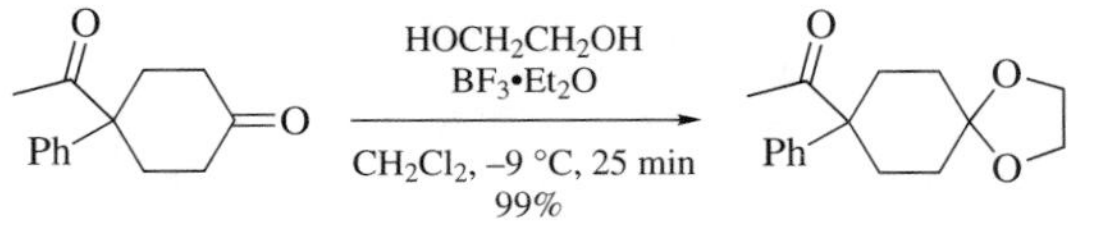

HOCH₂CH₂OH, BF_3•Et_2O; CH_2Cl_2, −9 °C, 25 min; 99% (3)

Water removal can be accomplished by using Dean–Stark distillation,[15] molecular sieves,[16] $CaSO_4$,[17] $CuSO_4$,[18] or water scavengers such as ***Triethyl Orthoformate***[19–21] or dialkyl sulfites (eq 4).[22]

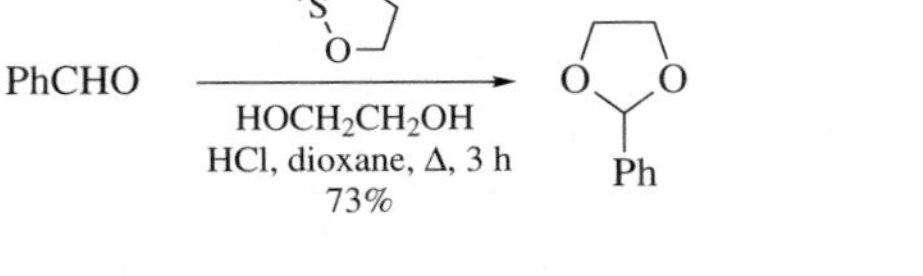

PhCHO → (ethylene sulfite; HOCH₂CH₂OH, HCl, dioxane, Δ, 3 h; 73%) (4)

Compounds that are stable under strongly acidic conditions can be derivatized using a mixture of ethylene glycol and HCl (eq 5).[23,24]

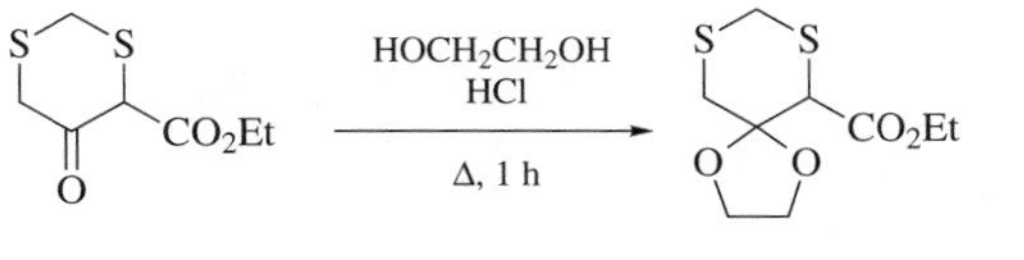

HOCH₂CH₂OH, HCl; Δ, 1 h (5)

Many procedures have also been developed for acid-sensitive compounds. Use of either adipic[25] or ***Oxalic Acid***[26] gives good yields of sensitive steroid Δ^4-3-ethylene acetals. ***Selenium(IV) Oxide*** has also been used effectively for steroid acetalization (eq 6).[27]

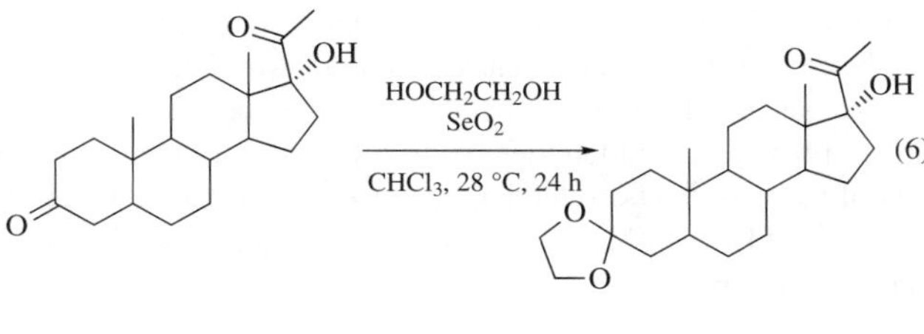

HOCH₂CH₂OH, SeO_2; $CHCl_3$, 28 °C, 24 h (6)

Pyridinium salts are mild catalysts for ethylene acetal formation.[28–31] ***Pyridinium p-Toluenesulfonate*** (PPTS) catalyzes acetal formation with less than 4% epimerization, whereas use of TsOH led to 34% epimerization (eq 7).[30]

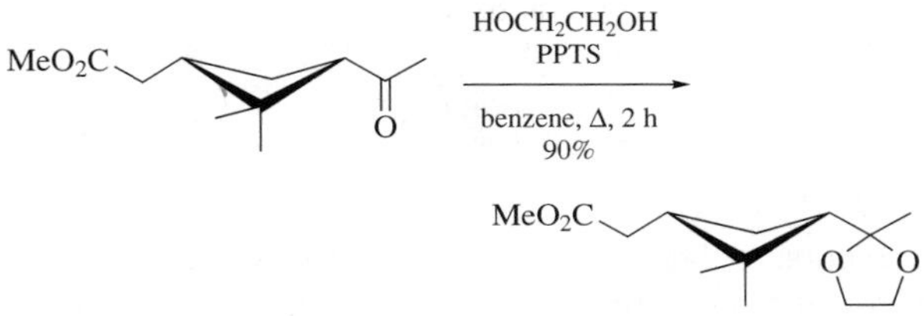

HOCH₂CH₂OH, PPTS; benzene, Δ, 2 h; 90% (7)

The use of a hindered pyridinium salt, such as ***2,4,6-Collidinium p-Toluenesulfonate*** (CTPS), allows acetalization of α,β-unsaturated ketones in the presence of saturated carbonyl systems (eq 8).[31]

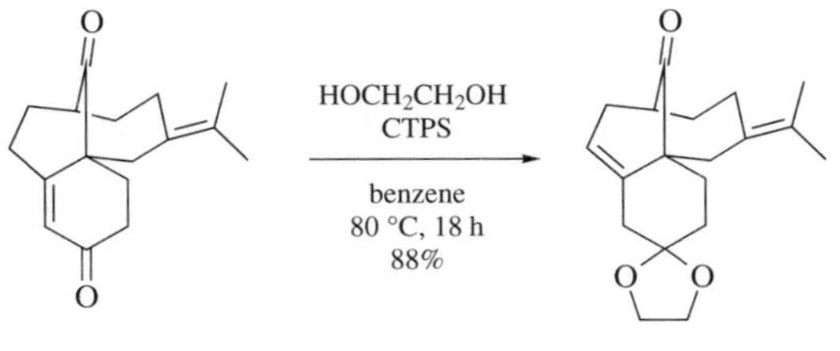

HOCH₂CH₂OH, CTPS; benzene, 80 °C, 18 h; 88% (8)

N,N-Dimethylformamide/Dimethyl Sulfate is another mild acetalization catalyst.[32] Metal catalysts, such as Rh complexes[33] and ***Palladium on Carbon***[1a] have also been used effectively (eqs 9 and 10). Ion-exchange resins offer significant advantages for the preparation of sensitive steroid acetals (eq 11).[34] An especially convenient method involves running a solution of the ketone and ethylene glycol through a column of Amberlyst.[35] Zeolites have also been used to prepare acetals.[36]

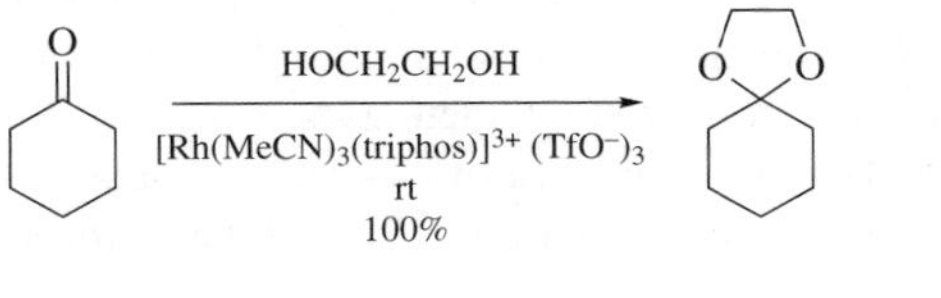

HOCH₂CH₂OH; $[Rh(MeCN)_3(triphos)]^{3+}(TfO^-)_3$, rt; 100% (9)

HOCH$_2$CH$_2$OH; Pd/C, Δ, 3 h; 99% (10)

HOCH$_2$CH$_2$OH, Amberlite; benzene, Δ; 81% (11)

Ketones that react slowly under normal acetalization conditions can be converted into dioxolanes at high pressure (eq 12).[37] A procedure has also been developed for the dioxolanation of ketones in the presence of aldehydes using a bis-silyl ether of ethylene glycol (eq 13).[38] The selective acetalization of aldehydes in the presence of ketones has been accomplished using ethylene glycol and alumina or silica gel.[39]

HOCH$_2$CH$_2$OH, (EtO)$_3$CH, TsOH; benzene, toluene, 15 kbar, 40 °C, 48 h; 82% (12)

1. TMSOTf, Me$_2$S, CH$_2$Cl$_2$, −78 °C, 1 h; 2. TMSOCH$_2$CH$_2$OTMS, −78 °C, 6 h; 80% (13)

In addition to their use as protecting groups, dioxolanes have been employed as intermediates. One example is a ring expansion of cyclopentanones via Grob fragmentation of the acetal (eq 14).[40] A novel synthesis of glycerol from formaldehyde and ethylene glycol involves dioxolane intermediates.[41]

5 equiv HOCH$_2$CH$_2$OH, 7 equiv BF$_3$•Et$_2$O; CH$_2$Cl$_2$, rt, 2–3 h; 98% (14)

Dioxolane formation is not limited to aldehydes and ketones. Amide acetals have been prepared directly and by the alcohol interchange method (eq 15).[42]

Me$_2$N–CH(OMe)$_2$ → HOCH$_2$CH$_2$OH, Δ, 94% → Me$_2$N-dioxolane (15)

The interchange of carbonyl protecting groups can be quite valuable in organic synthesis. For example, replacing a thioacetal with an acetal would allow a subsequent oxidation or hydrogenation which is not possible in the presence of a sensitive sulfur-containing group. Reacting the thioacetal with MeOSO$_2$F, followed by treatment with ethylene glycol, achieves this transformation under mild conditions.[43] ***Ethylene Chloroboronate***, prepared from BCl$_3$ and ethylene glycol, has been used for stereoselective aldol reactions with aliphatic and aromatic aldehydes (eq 16).[2]

ClB(OCH$_2$CH$_2$O), *i*-Pr$_2$NEt, CH$_2$Cl$_2$; 1. PhCHO, 2. H$_2$O; 85%, 98% de (16)

Solvent Effects.[1d] Ethylene glycol is an excellent solvent for many reactions and its high boiling point (197.6 °C) and low freezing point (−13 °C) permit a wide range of operating temperatures. Both NaOH and KOH are very soluble in ethylene glycol. Examples are the hydrolysis of *t*-butylurea,[44] the dehydrochlorination of monovinylacetylene,[45] and the Wolff–Kishner deoxygenation.[46] Dehydroxylations of (*R,R*)-tartrates with ***Samarium(II) Iodide*** are best accomplished in ethylene glycol.[47] The conversion of alkyl bromides to alkyl fluorides has been carried out in ethylene glycol.[48] Removal of the acetonide group can be accelerated by substituting ethylene glycol for H$_2$O.[49] The Fischer indole synthesis can be performed without the addition of an acid catalyst by heating a phenylhydrazone in ethylene glycol.[4,50] Ethylene glycol has also been used as a substrate for the synthesis of indoles and other heterocycles.[51] The Ru complex catalyzed reaction of substituted anilines and ethylene glycol offers an alternative route to the indoles under milder conditions than conventional methods (eq 17).[51a,b]

PhNHMe → HOCH$_2$CH$_2$OH, RuCl$_2$(PPh$_3$)$_3$; 180 °C, 5 h; 46% → N-Me indole (17)

Ethylene glycol is superior to benzene, MeCN, DMSO, MeOH, and H$_2$O for the Diels–Alder reaction involving relatively hydrophobic dienes and dienophiles. The increased reaction rate in ethylene glycol is attributed to the improved aggregation of the diene and dienophile (eq 18).[3]

diene (OCO$_2$Me) + quinone → HOCH$_2$CH$_2$OH; 95 °C, 22 h; 71% (18)

The synthesis of tertiary alcohols from trialkylboranes and CO at 100–150 °C has been improved in the presence of ethylene glycol[5,52–54] (eq 19).[54a] The purpose of the ethylene glycol is to trap the intermediate boronic anhydride, which can form a polymer that is difficult to oxidize.

(norbornyl)$_3$BH → 1. HOCH$_2$CH$_2$OH, CO, 100–150 °C; 2. H$_2$O$_2$, NaOH, 50 °C; 80% → (norbornyl)$_3$COH (19)

Related Reagents. 3-Bromo-1,2-propanediol; 2,3-Butanediol; 2,2-Dimethyl-1,3-propanediol; 2-Methoxy-1,3-dioxolane;

(2*R*,4*R*)-2,4-Pentanediol; 1,3-Propanediol; Triethyl Orthoformate.

1. (a) Meskens, A. J. *S* **1981**, 501. (b) Greene, T. W. *Protective Groups in Organic Synthesis*; Wiley: New York, 1981. (c) McOmie, J. F. W. *Protective Groups in Organic Chemistry*; Plenum: New York, 1973. (d) Fieser, L. F.; Fieser, M. *FF* **1967**, *1*, 375.
2. (a) Gennari, C.; Colombo, L.; Poli, G. *TL* **1984**, *25*, 2279. (b) Gennari, C.; Cardani, S.; Colombo, L.; Scolastico, C. *TL* **1984**, *25*, 2283.
3. Dunams, T.; Hoekstra, W.; Pentaleri, M.; Liotta, D. *TL* **1988**, *29*, 3745.
4. Fitzpatrick, J. T.; Hiser, R. D. *JOC* **1957**, *22*, 1703.
5. Brown, H. C. *ACR* **1969**, *2*, 65.
6. (a) Loewenthal, H. J. E. *T* **1959**, *6*, 269. (b) Keana, J. F. W. *Steroid Reactions*; Djerassi, C., Ed.; Holden-Day: San Francisco, 1963; pp 1–87.
7. Salmi, E. J. *CB* **1938**, *71*, 1803.
8. Sulzbacher, M.; Bergmann, E.; Pariser, E. R. *JACS* **1948**, *70*, 2827.
9. Johnson, W. S.; Rogier, E. R.; Szmuszkovicz, J.; Hadler, H. I.; Ackerman, J.; Bhattacharyya, B. K.; Bloom, B. M.; Stalmann, L.; Clement, R. A.; Bannister, B.; Wynberg, H. *JACS* **1956**, *78*, 6289.
10. Campbell, J. A.; Babcock, J. C.; Hogg, J. A. *JACS* **1958**, *80*, 4717.
11. Daignault, R. A.; Eliel, E. L. *OSC* **1973**, *5*, 303.
12. Fieser, L. F.; Stevenson, R. *JACS* **1954**, *76*, 1728.
13. Engel, C. R.; Rakhit, S. *CJC* **1962**, *40*, 2153.
14. Swenton, J. S.; Blankenship, R. M.; Sanitra, R. *JACS* **1975**, *97*, 4941.
15. Sakane, K.; Otsuji, Y.; Imoto, E. *BCJ* **1974**, *47*, 2410.
16. Roelofsen, D. P.; van Bekkum, H. *S* **1979**, 419.
17. Stenberg, V. I.; Kubik, D. A. *JOC* **1974**, *39*, 2815.
18. Hanzlik, R. P.; Leinwetter, M. *JOC* **1978**, *43*, 438.
19. Caserio, F. F., Jr.; Roberts, J. D. *JACS* **1958**, *80*, 5837.
20. (a) Marquet, A.; Dvolaitzky, M.; Kagan, H. B.; Mamlok, L.; Ouannes, C.; Jacques, J. *BSF* **1961**, 1822. (b) Marquet, A.; Jacques, J. *BSF* **1962**, 90.
21. Doyle, P.; Maclean, I. R.; Murray, R. D. H.; Parker, W.; Raphael, R. A. *JCS* **1965**, 1344.
22. Hesse, G.; Förderreuther, M. *CB* **1960**, *93*, 1249.
23. Vogel, E.; Schinz, H. *HCA* **1950**, *33*, 116.
24. Howard, E. G.; Lindsey, R. V., Jr. *JACS* **1960**, *82*, 158.
25. (a) Brown, J. J.; Lenhard, R. H.; Bernstein, S. *E* **1962**, *18*, 309. (b) Brown, J. J.; Lenhard, R. H.; Bernstein, S. *JACS* **1964**, *86*, 2183.
26. Andersen, N. H.; Uh, H.-S. *SC* **1973**, *3*, 125.
27. Oliveto, E. P.; Smith, H. Q.; Gerold, C.; Weber, L.; Rausser, R.; Hershberg, E. B. *JACS* **1955**, *77*, 2224.
28. Rausser, R.; Lyncheski, A. M.; Harris, H.; Grocela, R.; Murrill, N.; Bellamy, E.; Ferchinger, D.; Gebert, W.; Herzog, H. L.; Hershberg, E. B.; Oliveto, E. P. *JOC* **1966**, *31*, 26.
29. Bond, F. T.; Stemke, J. E.; Powell, D. W. *SC* **1975**, *5*, 427.
30. Sterzycki, R. *S* **1979**, 724.
31. Nitz, T. J.; Paquette, L. A. *TL* **1984**, *25*, 3047.
32. Kantlehner, W.; Gutbrod, H.-D. *LA* **1979**, 1362.
33. Ott, J.; Tombo, G. M. R.; Schmid, B.; Venanzi, L. M.; Wang, G.; Ward, T. R. *TL* **1989**, *30*, 6151.
34. Jones, E. R. H.; Meakins, G. D.; Pragnell, J.; Müller, W. E.; Wilkins, A. L. *JCS(P1)* **1979**, 2376.
35. Dann, A. E.; Davis, J. B.; Nagler, M. J. *JCS(P1)* **1979**, 158.
36. Corma, A.; Climent, M. J.; Hermenegildo, C.; Primo, J. *CA* **1990**, *113*, 61 675x.
37. Dauben, W. G.; Gerdes, J. M.; Look, G. C. *JOC* **1986**, *51*, 4964.
38. Kim, S.; Kim, Y. G.; Kim, D.-i. *TL* **1992**, *33*, 2565.
39. Kamitori, Y.; Hojo, M.; Masuda, R.; Yoshida, T. *TL* **1985**, *26*, 4767.
40. Sakai, K. *CA* **1992**, *117*, 130 946h.
41. Sanderson, J. R.; Lin, J. J.; Duranleau, R. G.; Yeakey, E. L.; Marquis, E. T. *JOC* **1988**, *53*, 2859.
42. Bredereck, H.; Sinchen, G; Rebstat, S.; Kantlehner, W.; Horn, P.; Wohl, R.; Hoffman, H.; Grieshaber; P. *CB* **1968**, *101*, 41.
43. Corey, E. J.; Hase, T. *TL* **1975**, 3267.
44. Pearson, D. E.; Baxter, J. F.; Carter, K. N. *OSC* **1955**, *3*, 154.
45. Hennion, G. F.; Price, C. C.; McKeon, T. F., Jr. *OSC* **1963**, *4*, 683.
46. Georgiadis, M. P.; Tsekouras, A.; Kotretsou, S. I.; Haroutounian, S. A.; Polissiou, M. G. *S* **1991**, 929.
47. Kusuda, K.; Inanaga, J.; Yamaguchi, M. *TL* **1989**, *30*, 2945.
48. Vogel, A. I.; Leicester, J.; Macey, W. A. T. *OSC* **1963**, *4*, 525.
49. Hampton, A.; Fratantoni, J. C.; Carroll, P. M.; Wang, S. *JACS* **1965**, *87*, 5481.
50. Borch, R. F.; Newell, R. G. *JOC* **1973**, *38*, 2729.
51. (a) Tsuji, Y.; Huh, K.-T.; Watanabe, Y. *JOC* **1987**, *52*, 1673. (b) Tsuji, Y.; Huh, K.-T.; Watanabe, Y. *TL* **1986**, *27*, 377. (c) Kondo, T.; Kotachi, S.; Watanabe, Y. *CC* **1992**, 1318.
52. (a) Hillman, M. E. D. *JACS* **1962**, *84*, 4715. (b) Hillman, M. E. D. *JACS* **1963**, *85*, 982.
53. (a) Puzitskii, K. V.; Pirozhkov, S. D.; Ryabova, K. G.; Pastukhova, I. V.; Eidus, Ya. T. *BAU* **1972**, *21*, 1939. (b) Puzitskii, K. V.; Pirozhkov, S. D.; Ryabova, K. G.; Pastukhova, I. V.; Eidus, Ya. T. *BAU* **1973**, *22*, 1760.
54. (a) Brown, H. C.; Rathke, M. W. *JACS* **1967**, *89*, 2737. (b) Negishi, E.; Brown, H. C. *OS* **1983**, *61*, 103.

W. Christopher Hoffman
Union Carbide, South Charleston, WV, USA

N-Ethyl-5-phenylisoxazolium-3′-sulfonate[1]

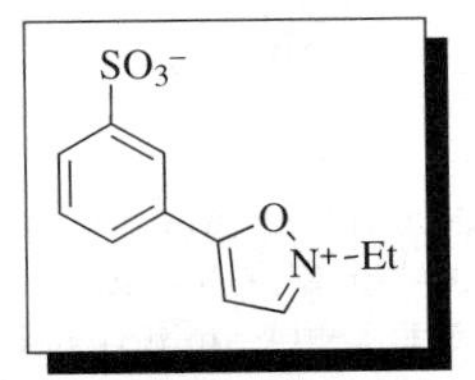

[4156-16-5] $C_{11}H_{11}NO_4S$ (MW 253.30)

(coupling reagent for peptide synthesis;[2] protein modification;[3] reduction of carboxylic acids into alcohols[4])

Alternate Names: Woodward's reagent K; NEPIS.
Physical Data: mp 206–208 °C (dec); UV λ_{max} (ε) 283 nm (22 500) in 0.1 N HCl.
Solubility: insol MeCN, $MeNO_2$, DMF, CH_2Cl_2; sol H_2O.
Form Supplied in: crystalline, anhydrous zwitterion; widely available as 95% pure that may contain some of the *para* sulfonate derivative.
Analysis of Reagent Purity: material that is sufficiently pure has mp 206–208 °C.
Purification: dissolve in aqueous 1 N HCl and reprecipitate by the slow addition of acetone. After filtration and drying the reagent is a fluffy solid.

Handling, Storage, and Precautions: stable to light and temperature and nonhygroscopic. No extra precautions need to be observed.

Peptide Coupling Reagent. Woodward's reagent K reacts with an *N*-protected amino acid, under mild conditions, in the presence of a tertiary amine to form an enol ester intermediate, e.g. (**1**). This enol ester acylates an amino acid ester or peptide to afford the elongated peptide derivative (eq 1). The isoxazolium salt is not soluble in organic solvents, but as it reacts with a carboxyl group in the presence of a tertiary amine the reaction mixture clears.

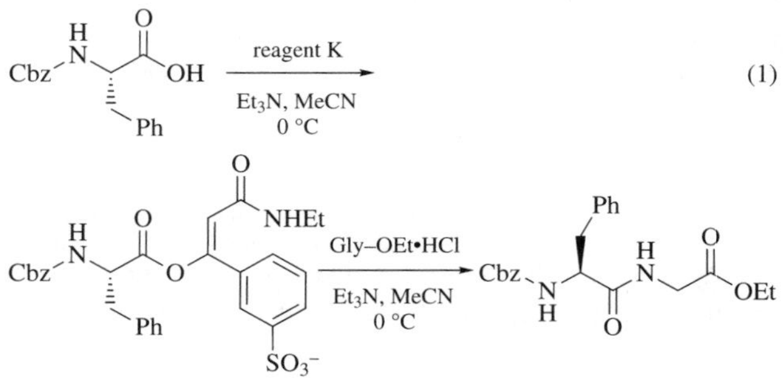

reagent K, Et_3N, MeCN, 0 °C (1)

Gly–OEt•HCl, Et_3N, MeCN, 0 °C

(**1**)

Examples of di- and tripeptide couplings abound with yields typically >80%; minimal purification steps are required since the byproducts are water soluble (eq 2).[2]

Cbz-Gly + Gly-Tyr-OMe → (reagent K, Et_3N, MeCN, 0 °C, 84%) Cbz-Gly-Gly-Tyr-OMe (2)

The recommended coupling procedure is to activate the carboxyl group in $MeNO_2$ or MeCN with vigorous stirring for 10 min at rt or 1 h at 0 °C. Then the amino ester, dissolved in $MeNO_2$ or MeCN, is added and the coupling allowed to go at rt for several hours.

Good yields are obtained in coupling asparagine and glutamine. With some coupling reagents the carboxamide of these residues can undergo an intramolecular dehydration when the α-carboxylate is activated to give a nitrile residue (**2**) in the peptide.[5] However, this byproduct has not been observed with reagent K.

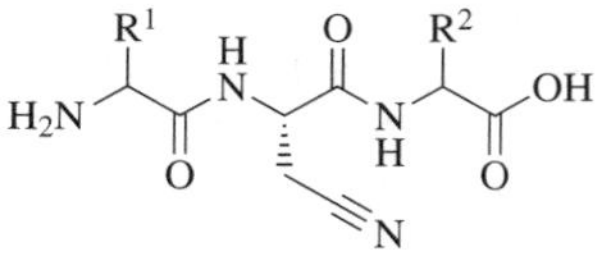

(**2**)

An additional advantage of reagent K is that when serine, threonine, etc., are activated their hydroxyl groups do not require protection.[6] For instance, the hydroxyproline tripeptide in eq 3 was obtained in good yield after crystallization.[2]

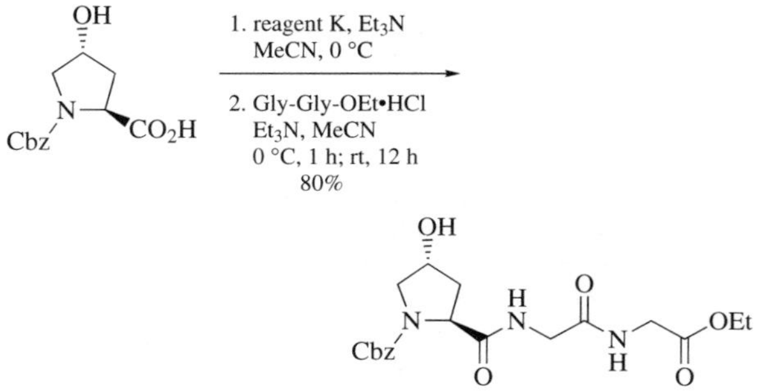

1. reagent K, Et_3N, MeCN, 0 °C
2. Gly-Gly-OEt•HCl, Et_3N, MeCN, 0 °C, 1 h; rt, 12 h, 80% (3)

Several studies[7] on the extent of racemization with peptide coupling reagents have been published. This generally is a problem when performing a fragment condensation, but normally not a problem when the amino group is protected as a carbamate. For best control of the extent of racemization, accurate measurement of the tertiary amine is advised.[7b] The extent of racemization observed with reagent K is just a few percent for acylated amino acids,[7c] which is no worse than most coupling reagents except for the azide coupling procedure. There are some disadvantages of this reagent in peptide couplings. The most serious one is the limitation of solvents. Acetonitrile and nitromethane are the best for high yield, high optical purity products. However, many peptides are not readily soluble in these solvents. The next disadvantage is that the enol ester intermediate can rearrange to the imide side-product (**3**) which is difficult to remove.[1] Another problem with reagent K is its incompatibility with solid-phase peptide synthesis because of its poor solubility in organic solvents and relatively slow activation of carboxyl groups.[8]

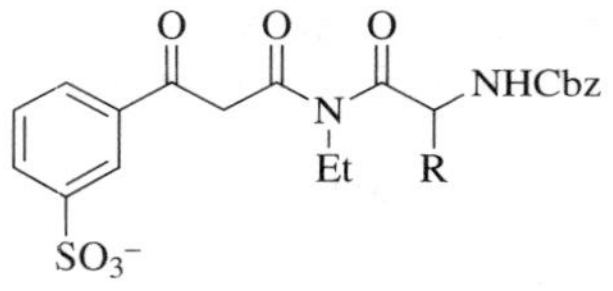

(**3**)

A good comparison of reagent K with other reagents is in the synthesis of a protected derivative of oxytocin (eq 4).[9] This product was further elaborated to give oxytocin. The results are compared with those for ***1,3-Dicyclohexylcarbodiimide*** and ***N,N′-Carbonyldiimidazole*** in Table 1.

Cbz-Cys(Bn)-Tyr-Ile-Gln-Asn + Cys(Bn)-Pro-Leu-Gly-NH_2 → (reagent K, DMP, 0 °C, 85%) Cbz-Cys(Bn)-Tyr-Ile-Gln-Asn-Cys(Bn)-Pro-Leu-Gly-NH_2 (4)

Table 1 Comparison of Coupling Agents for Protected Oxytocin

Coupling method	Yield (%)	Mp (C)	Biological activity[a]
DCC	80	225–230	81
CDI	70.5	225–230	50
Reagent K	85	240–242	134

[a] Units mg^{-1} in avian depressor activity for deprotected and oxidized nonapeptide.

An additional utility for reagent K is peptide cyclization.[10] The activation and cyclization steps are performed separately; therefore each can be carried out under the appropriate dilution conditions. This is advantageous when the carboxyl group activates slowly since it is in concentrated solution.

The isoxazolium salt has been used in the synthesis of other difficult peptide structures. A peptidoglycan has been obtained in good yield by this procedure (eq 5).[11]

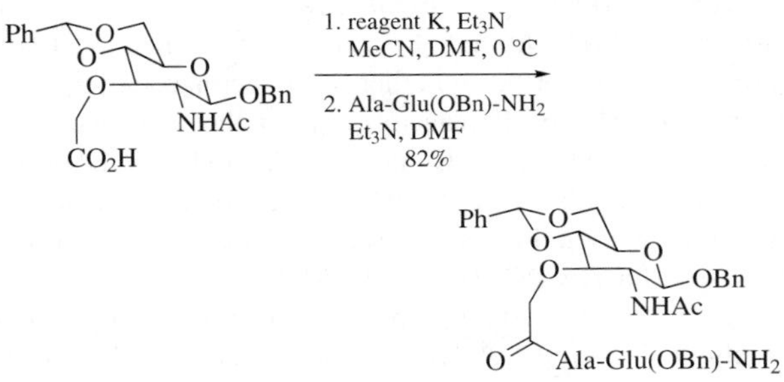

1. reagent K, Et_3N, MeCN, DMF, 0 °C
2. Ala-Glu(OBn)-NH_2, Et_3N, DMF
82% (5)

Additionally, this reagent has been used repetitively in the synthesis of a steroid–peptide adduct with a free hydroxyl group (eq 6).[12] Deprotection and another coupling with Cbz–Arg(N^G-nitro) gave the dipeptide–steroid adduct in good yield. In both couplings the steroid hydroxyl group was not acylated.

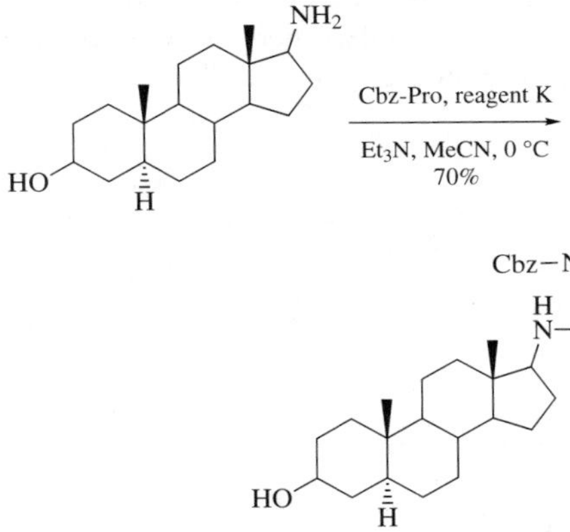

Cbz-Pro, reagent K
Et_3N, MeCN, 0 °C
70% (6)

Protein Modification. Since this reagent is water soluble and is reactive under aqueous conditions below pH 4.75, it can specifically modify carboxyl groups in proteins.[3,13] This has proven useful in determining essential carboxyl groups in various enzymes. The enol ester is stable enough to isolate the modified protein and 'tag' with a nucleophile for eventual identification of the modified amino acid. Other reagents such as the water-soluble carbodiimides are also used extensively in these mechanistic studies. However, an advantage with reagent K is that the enol ester formed has a UV absorbance at 340 nm ($E_{340} = 7000\ M^{-1}\ cm^{-1}$);[14] thus the number of carboxyl groups modified can be quantified.

Reduction of Carboxylic Acids to Alcohols. The conversion of carboxylic acids to alcohols occurs in a two-step process under mild conditions using reagent K with yields for simple alcohols ranging from 50–100%.[4] The reduction of a dipeptide (eq 7) suggests a potential utility in modifying carboxyl groups in proteins.

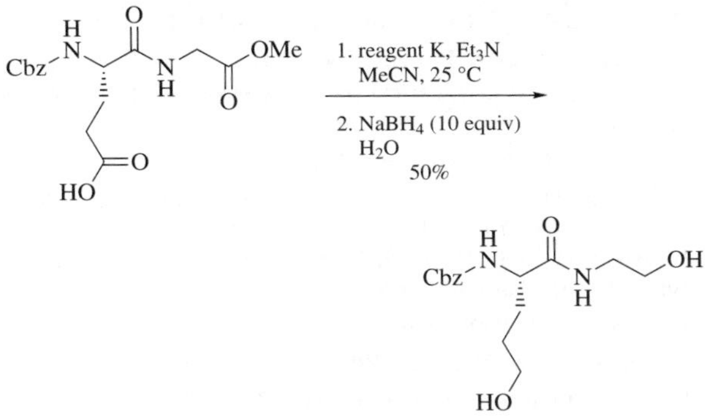

1. reagent K, Et_3N, MeCN, 25 °C
2. $NaBH_4$ (10 equiv), H_2O
50% (7)

Another method[15] to obtain alcohols is to reduce the mixed carbonic acid anhydride prepared with ***Ethyl Chloroformate*** with an excess of ***Sodium Borohydride***. This method uses less costly reagents; thus it is more useful for synthetic transformations.

Miscellaneous Reactions. Pantothenic acid has been coupled to a long chain amine with reagent K (eq 8).[16] This is another example of the nonreactivity of hydroxyl groups with the enol ester intermediate. Attempts to prepare this amide through the mixed anhydride procedure give low yields.

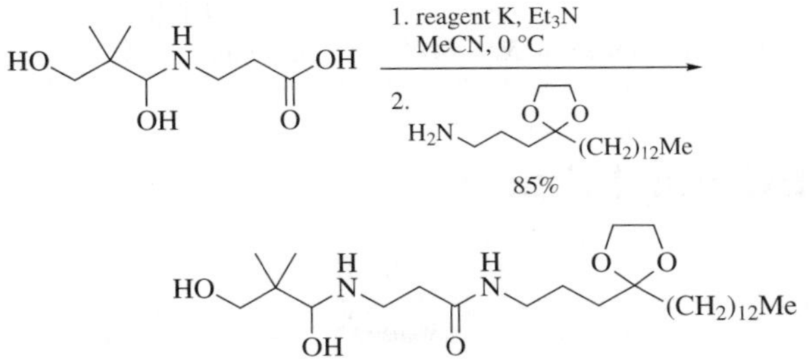

1. reagent K, Et_3N, MeCN, 0 °C
2. H_2N…$(CH_2)_{12}Me$
85% (8)

Reagent K may have utility in sequencing peptides and proteins from the *C*-terminus (eq 9).[17] The 2-thiohydantoin produced is used to analyze a peptide sequence upon cleavage of the thiohydantoin from the peptide. This mild procedure gives good yields with amino acids that have sensitive side chains such as serine and lysine. The original ***Acetic Anhydride*** procedure[18] for *C*-terminal sequencing does not work well with these amino acids; therefore this procedure is an improvement. Another activation reagent, hydroxysuccinimide esters of Fmoc-amino acids, fails to give the thiohydantoin.[17]

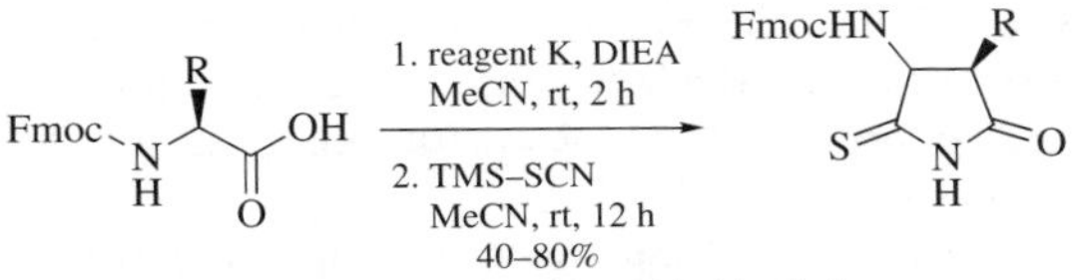

1. reagent K, DIEA, MeCN, rt, 2 h
2. TMS–SCN, MeCN, rt, 12 h
40–80% (9)

R = variety of amino acid side chains

Related Reagents. Benzotriazol-1-yloxytris(dimethylamino)phosphonium Hexafluorophosphate; 1-Cyclohexyl-3-(2-morpholinoethyl)carbodiimide; 1-Ethyl-3-(3′-dimethylaminopropyl)carbodiimide Hydrochloride; Isobutyl Chloroformate; *p*-Nitrophenol; Pentafluorophenol; 1,1′-Thionylimidazole.

1. Woodward, R. B.; Olofson, R. A.; Mayer, H. *T* **1969**, *22(S8)*, 321.
2. Woodward, R. B.; Olofson, R. A. *OS* **1977**, *56*, 88.

3. Bodlaender, P.; Feinstein, G.; Shaw, E. *B* **1969**, *8*, 4941.
4. Hall, P. L.; Perfetti, R. B. *JOC* **1974**, *39*, 111.
5. Gish, D. T.; Katsoyannis, P. G.; Hess, G. P.; Stedman, R. J. *JACS* **1956**, *78*, 5954.
6. Klausner, Y. S.; Bodanszky, M. *S* **1972**, 453.
7. (a) Kemp, D. S. *The Peptides: Analysis, Synthesis and Biology;* Academic: New York, 1979; Vol. 1, pp 315–383. (b) Woodward, R. B.; Woodman, D. J. *JOC* **1969**, *34*, 2742. (c) Kemp, D. S.; Wang, S. W.; Busby, G., III; Hugel, G. *JACS* **1970**, *92*, 1043.
8. Hudson, D. *JOC* **1988**, *53*, 617.
9. Fosker, A. P.; Law, H. D. *JCS* **1965**, 4922.
10. Blaha, K.; Rudinger, J. *CCC* **1965**, *30*, 3325.
11. Merser, C.; Sinay, P.; Adam, A. *Biochem. Biophys. Res. Commun.* **1975**, *66*, 1316.
12. (a) Pettit, G. R.; Gupta, A. K. D.; Smith, R. L. *CJC* **1966**, *44*, 2023. (b) Pettit, G. R.; Smith, R. L.; Klinger, H. *JOC* **1967**, *10*, 145.
13. Baker, A. J.; Weber, B. H. *JBC* **1974**, *249*, 5452.
14. Sinha, U.; Brewer, J. M. *Anal. Biochem.* **1985**, *151*, 327.
15. Ramsamy, K.; Olsen, R. K.; Emery, T. *S* **1982**, 42.
16. Wagner, A. P.; Retey, J. *Eur. J. Biochem.* **1991**, *195*, 699.
17. Boyd, V. L.; Hawke, D. H.; Geiser, T. G. *TL* **1990**, *31*, 3849.
18. Cromwell, L. D.; Stark, G. R. *B* **1969**, *8*, 4735.

Richard S. Pottorf
Marion Merrell Dow Research Institute, Cincinnati, OH, USA

Ethyl Vinyl Ether[1]

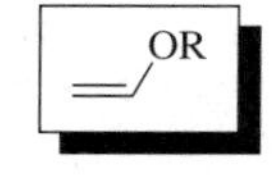

(R = Et)		
[109-92-2]	C_4H_8O	(MW 72.12)
(R = Me)		
[107-25-5]	C_3H_6O	(MW 58.09)
(R = *n*-Bu)		
[111-34-2]	$C_6H_{12}O$	(MW 100.18)

(protection of hydroxyl group; transvinylation, particularly to prepare allyl vinyl ethers for use in the Claisen rearrangement; condensation reactions; cycloaddition reactions)

Physical Data: R = Et, mp −115.8 °C, bp 35–36 °C, *d* 0.7589 g cm^{-3}; R = Me, mp −123 °C, bp 12 °C, *d* 0.7725 g cm^{-3} at 0 °C; R = *n*-Bu, mp −92 °C, bp 94 °C, *d* 0.7888 g cm^{-3}.[2]

Solubility: sol all common organic solvents; slightly sol water.

Form Supplied in: ethyl vinyl ether and *n*-butyl vinyl ether are colorless, extremely flammable liquids, while methyl vinyl ether is a colorless, extremely flammable gas. Ethyl vinyl ether is supplied stabilized with triethanolamine and methyl vinyl ether is supplied in stainless steel cylinders.

Purification: ethyl vinyl ether and *n*-butyl vinyl ether may be distilled from K_2CO_3 or Na if purification of the commercial materials is desirable.

Handling, Storage, and Precautions: all three compounds react *violently* with halogens and with strong oxidizing agents, and undergo rapid polymerization in the presence of acids. These vinyl ethers are lachrymatory and irritate the respiratory system; in particular, methyl vinyl ether can cause rapid suffocation if inhaled. Use in a fume hood.

Protection of the Hydroxyl Group.[3] Acid-catalyzed reaction of primary and secondary alcohols with ethyl vinyl ether gives the α-ethoxyethyl (EE) group. Acids commonly used in the formation of the EE group include ***Trifluoroacetic Acid*** (eq 1),[4] anhyd ***Hydrogen Chloride*** (eq 2),[5] ***p-Toluenesulfonic Acid***,[6] and ***Pyridinium p-Toluenesulfonate*** (PPTS) (eq 3).[7] Whereas attempts to form the methyl or benzyl ether of the hemiacetal (eq 3), and reaction with DHP, result in decomposition, the hemiacetal is successfully converted into the EE group with ethyl vinyl ether in the presence of PPTS (but not *p*-TsOH).[7] For extremely acid-sensitive substrates, ethyl vinyl ether generated in situ from 1-chloroethyl ethyl ether (MeCH(Cl)OEt) with $PhNMe_2$ in CH_2Cl_2 at 0 °C gives EE derivatives of primary, secondary, and some tertiary alcohols in excellent yields (eq 4).[8] Selective protection of a primary alcohol in the presence of a secondary alcohol with ethyl vinyl ether may be achieved at low temperatures (eq 5).[9] The EE group is more readily cleaved by acid hydrolysis than the THP ether, but is more stable than the 1-methyl-1-methoxyethyl ether.[3] In the case of the nucleoside (eq 2), which is subsequently converted into a nucleotide, the ethoxyethyl group is preferred to the THP group because of its ease of removal (5% HOAc, 2 h, 20 °C), with no detectable isomerization of the 3′–5′ nucleotide linkage.[5] Other methods of cleaving the EE group include 0.5N HCl in THF at 0 °C,[6] and PPTS in *n*-PrOH, where cleavage is possible in the presence of an acetonide group (eq 6).[10]

OH, CO_2Et → (OEt; TFA, 0 °C, 20 h, >90%) → O, OEt, CO_2Et (1)

O, NH, N, O, O, OH, O, O–P, OAc, HO, OH (as Ca salt) → (OEt; anhyd. HCl, dioxane, DMF, rt, 24 h, 97%) → O, NH, N, O, O, OEE, O, O–P, OAc, HO, OH (2)

Bu, O, OH, H, H, O → (OEt; PPTS, CH_2Cl_2, rt, 2 h, 88%) → Bu, O, OEE, H, H, O (3)

CHO → (1. *t*-Bu_3SnLi; 2. EtO, Cl; $PhNMe_2$, CH_2Cl_2, 0 °C, 95%) → $SnBu_3$, OEE (4)

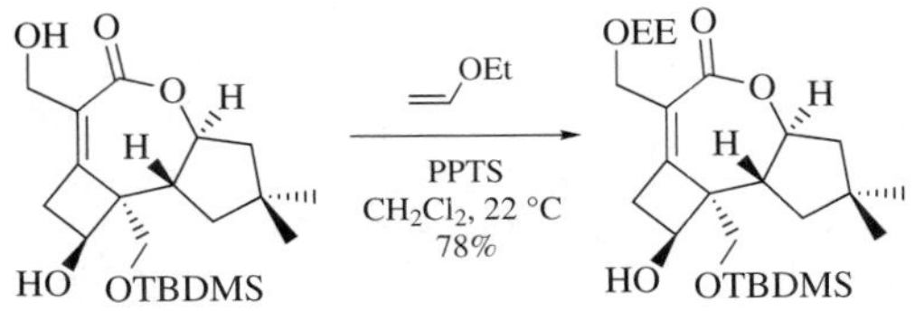

(5)

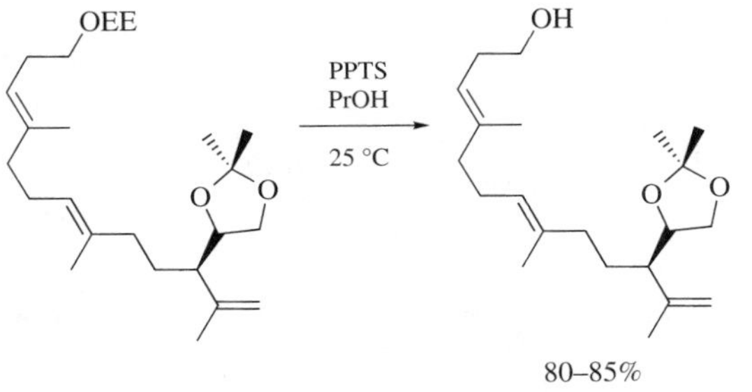

(6)

80–85%

(10)

(11)

(12)

Augmenting its role as a protecting group, the EE group has been used to direct *ortho*-lithiation in suitably substituted aromatic ethers, yielding lithiated species that react rapidly with a variety of electrophiles, such as ***Carbon Dioxide***, to give the phthalide (eq 7).[11]

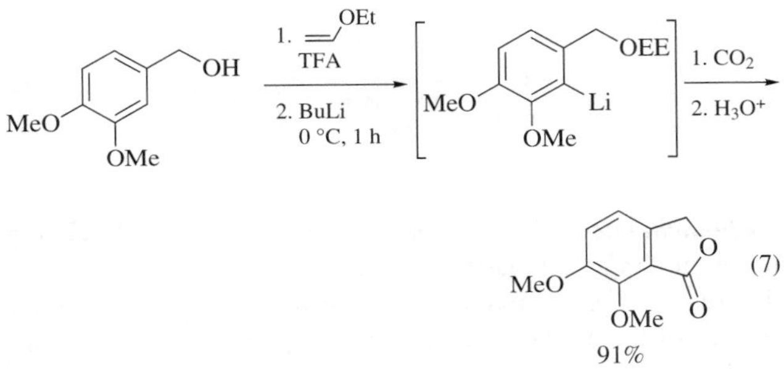

(7)

91%

Transvinylation. Acid-catalyzed transvinylation of allylic alcohols with ethyl vinyl ether gives allyl vinyl ethers, Claisen rearrangement[12] of which affords γ,δ-unsaturated aldehydes of varying complexity (eqs 8–12).[13,14,16–18] ***Mercury(II) Acetate*** (eq 8),[13] ***Phosphoric Acid*** (eq 9),[14] and *p*-TsOH[15] are frequently used to catalyze the transvinylation reaction, and the allyl vinyl ethers produced need not be isolated but may be subjected directly to the Claisen rearrangement, typically at 145–200 °C. $Hg(OAc)_2$, used in the transvinylation reaction, has been reported to effect the rearrangement at lower temperatures (eq 10).[16] Other examples illustrating the stereoselectivity of the reaction are given in eqs 11 and 12.[17,18]

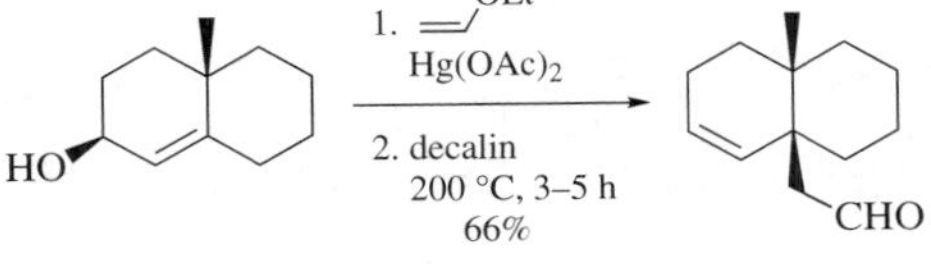

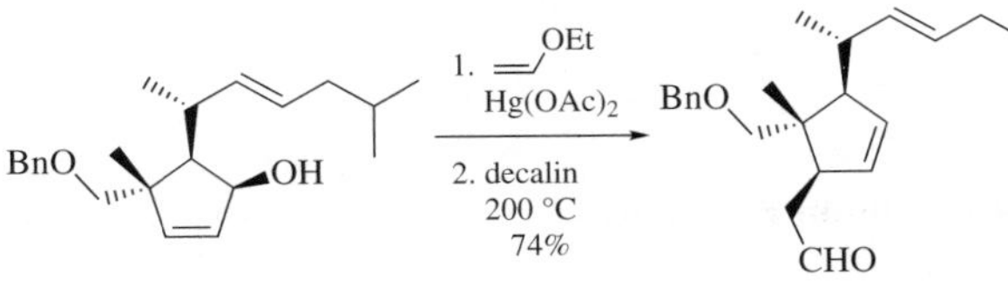

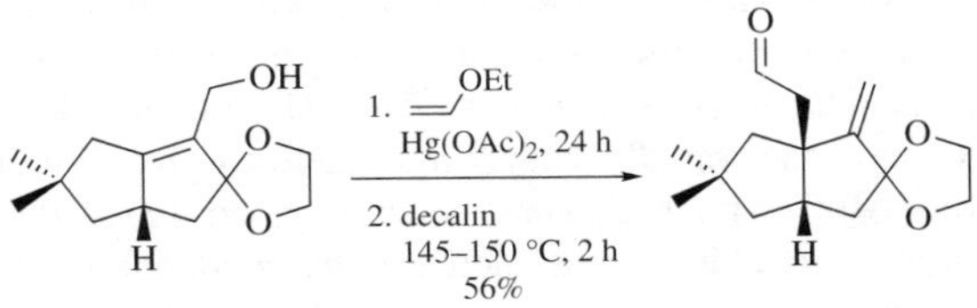

(8)

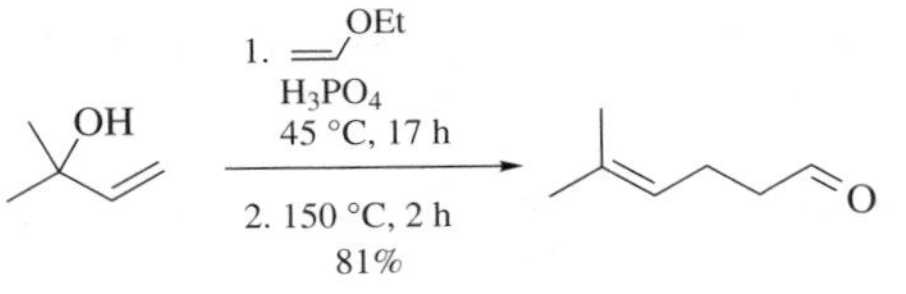

(9)

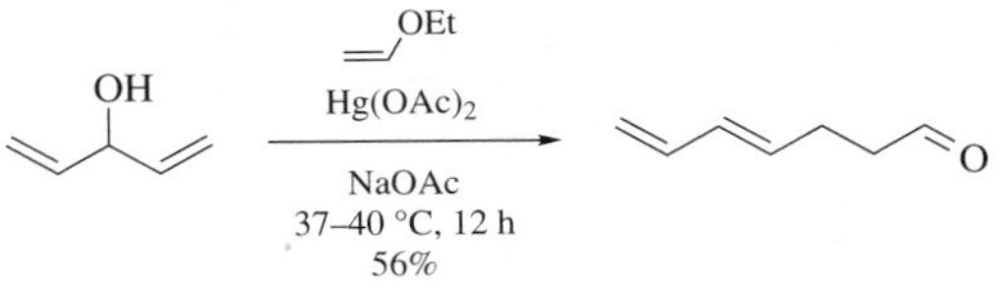

Condensation Reactions.

Acetal Condensations. The acid-catalyzed condensation of ethyl vinyl ether with acetals derived from α,β-unsaturated aldehydes and ketones is an excellent method to lengthen a carbon chain by two carbon atoms, particularly of polyenals. In an industrial process for the manufacture of the C_{16} aldehyde (eq 13), the C_{14} acetal is reacted with ethyl vinyl ether in the presence of ***Zinc Chloride*** to give an alkoxy acetal, which, without isolation, is then hydrolyzed with moist acetic acid.[19] Longer conjugated carotenals have been synthesized by this method,[20] and montmorillonite K-10 clay (see ***Montmorillonite K10***) has been reported to be an excellent replacement for the Lewis acids normally used in the condensation.[21]

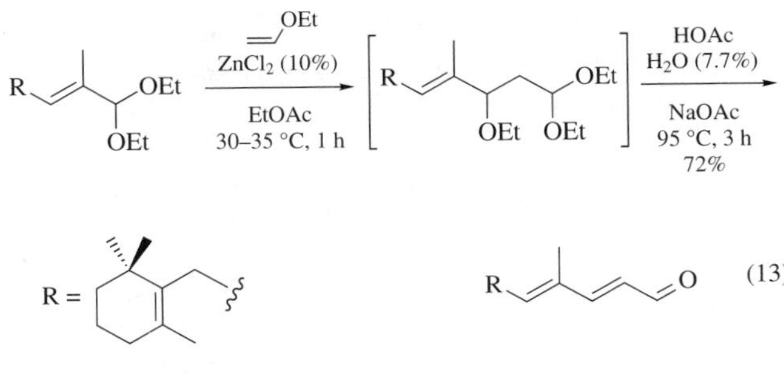

(13)

Acyl Anion Equivalents.[22] The α-H of simple vinyl ethers can be deprotonated with strong bases, such as ***t-Butyllithium***, to give highly reactive α-alkoxyvinyllithium reagents (see also ***1-Ethoxyvinyllithium*** and ***1-Methoxyvinyllithium***). These react rapidly with various electrophiles, including aldehydes and ketones (eq 14),[23] and can undergo conjugate addition to α,β-unsaturated carbonyl compounds via the corresponding cuprates (eq 15).[24]

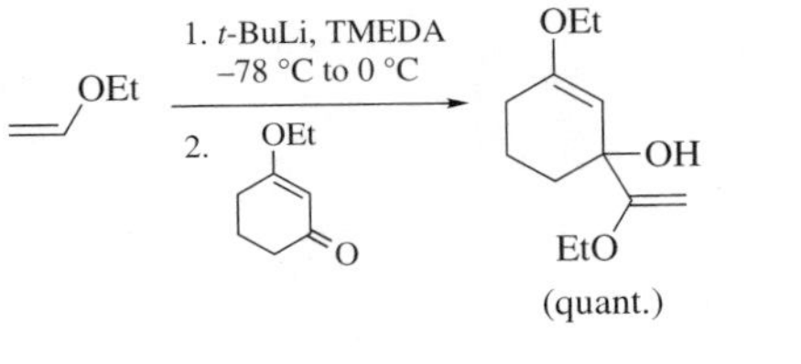

(14)

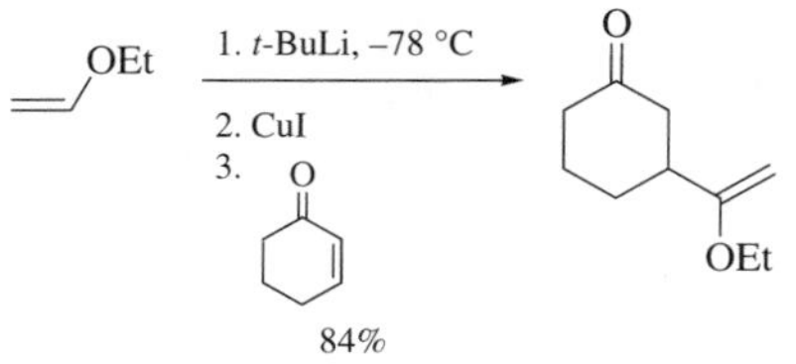

(15)

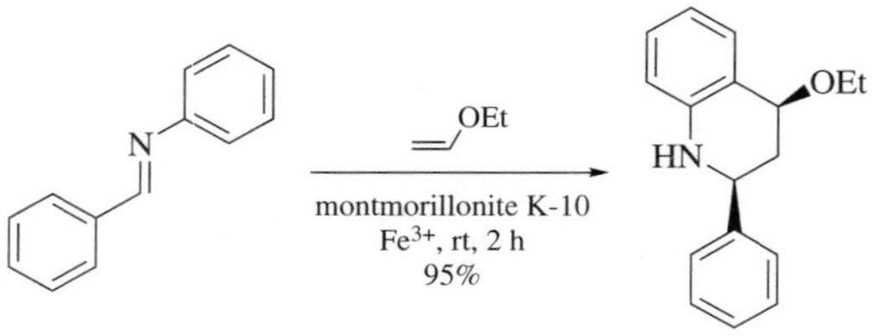

(20)

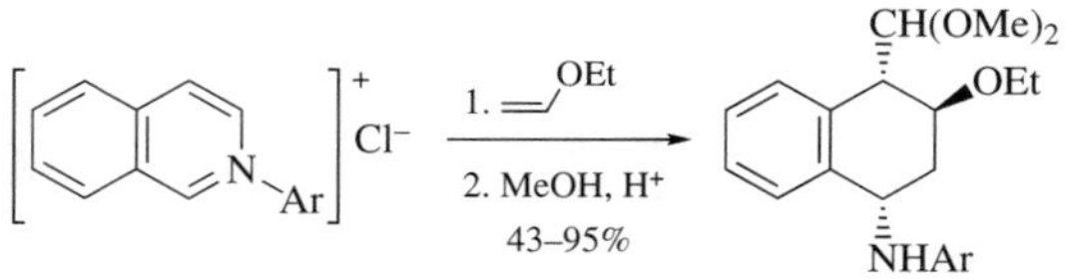

(21)

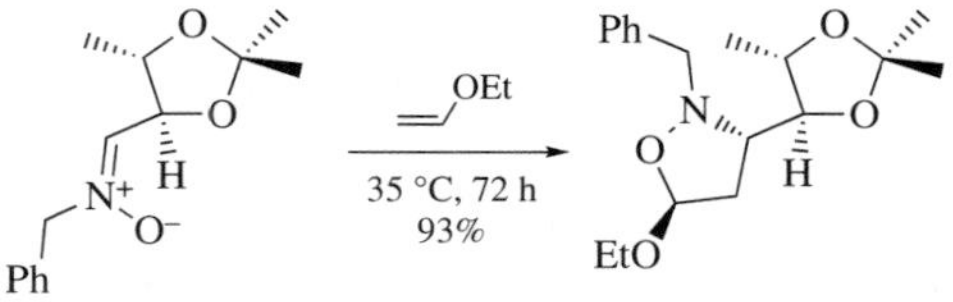

(22)

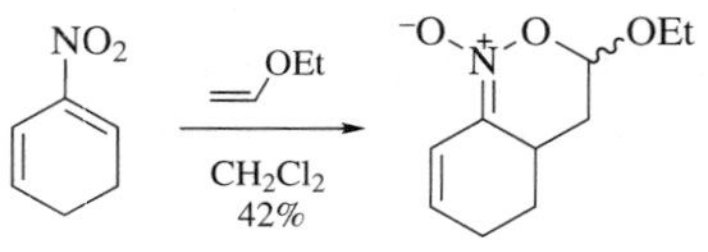

(23)

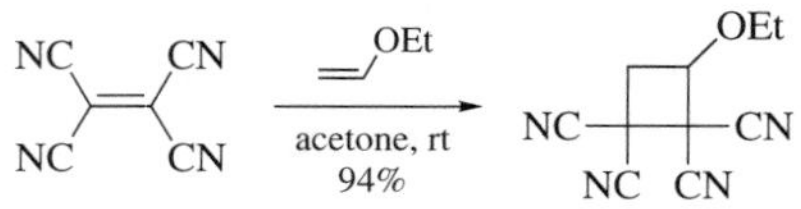

(24)

Cycloaddition Reactions.

Diels–Alder and Related Reactions. Owing to their electron-rich nature, vinyl ethers participate in a number of inverse electron demand Diels–Alder reactions.[25] α,β-Unsaturated aldehydes and ketones react rapidly with vinyl ethers to give dihydropyrans (eq 16),[26] with high *endo* selectivity (eq 17),[27] the proportion of *endo* adduct being increased by conducting the reaction at low temperatures under pressure (eq 18).[28] Lewis acids can also catalyze these reactions and control the stereochemistry of cycloaddition.[25b,29] *o*-Quinone methides yield chromans (eq 19),[30] and imines give tetrahydroquinolines (eq 20).[31] Vinyl ethers react regio- and stereoselectively with isoquinolinium salts giving cycloadducts, which on hydrolysis afford tetralins in excellent yields (eq 21).[32] The cycloaddition of vinyl ethers with nitrones proceeds smoothly, giving, where relevant, isoxazolines with extremely high diastereoselectivity (eq 22).[33] Vinyl ethers react with nitroalkenes to give nitronate esters in moderate yields (eq 23),[34] but excellent yields of cyclobutane derivatives are obtained with tetracyanoethylene (eq 24).[35]

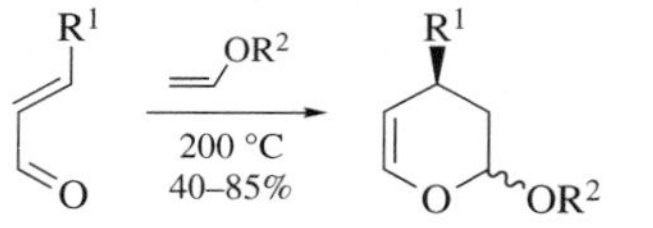

(16)

R^1 = H, Me, Ph; R^2 = Et, Bu, Me

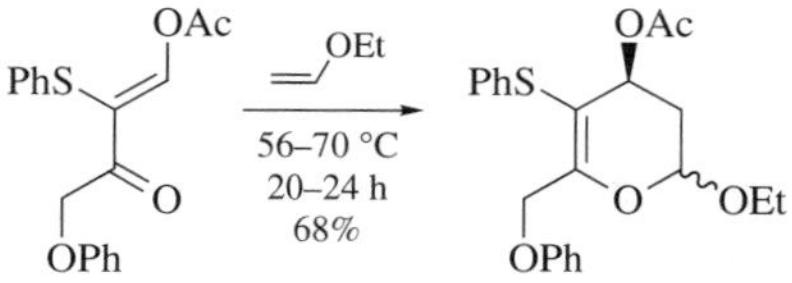

(17)

trans:*cis* = 12.5:1

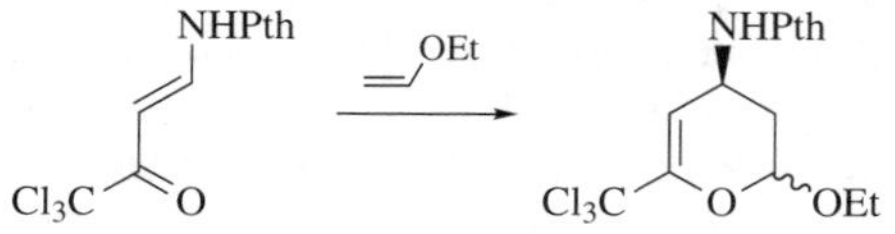

(18)

	β-OEt:α-OEt
90 °C, 1 bar	1.67:1
0.5 °C, 6 kbar	13.6:1

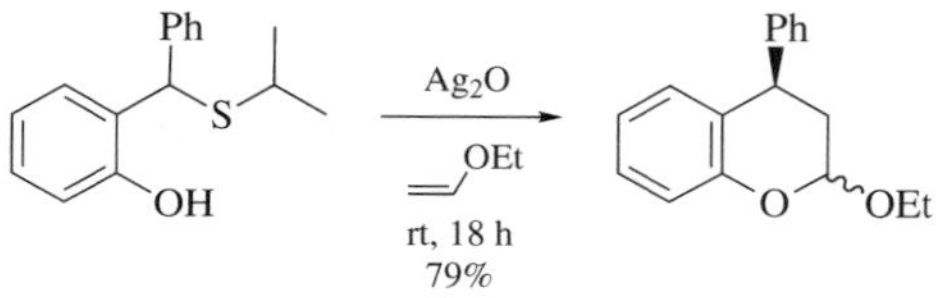

(19)

Photochemical. Vinyl ethers undergo the Paterno–Büchi reaction with simple aliphatic and aromatic carbonyl compounds to give regioisomeric mixtures of alkoxyoxetanes, with a preponderance of the 3-alkoxy regioisomer (eq 25).[36] Photoaddition of vinyl ethers to enones produces alkoxycyclobutanes in excellent yields, with the alkoxy group orientated predominantly away from the carbonyl group (eq 26).[36] Vinyl ethers also undergo photoaddition to arenes, giving, with benzene, a mixture of *ortho*- and *meta*-cycloadducts in low yields; only the *exo* isomer is produced in the *ortho*-cycloaddition, whereas the *meta*-cycloaddition gives a mixture of regio- and stereoisomers (eq 27).[37]

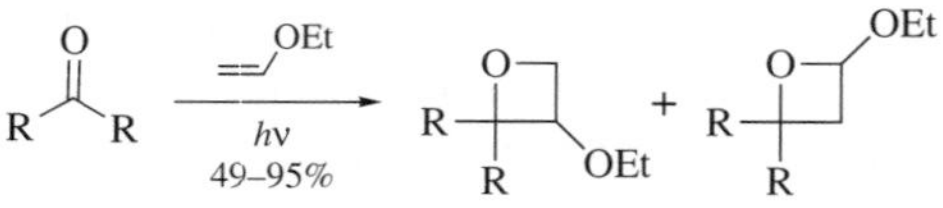

(25)

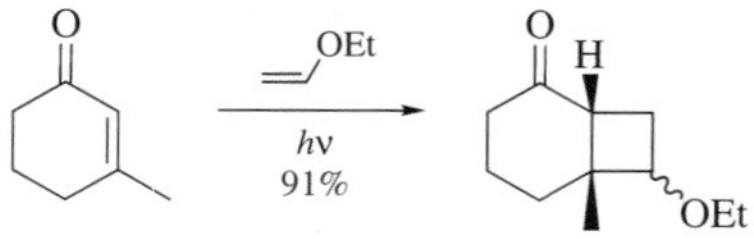

(26)

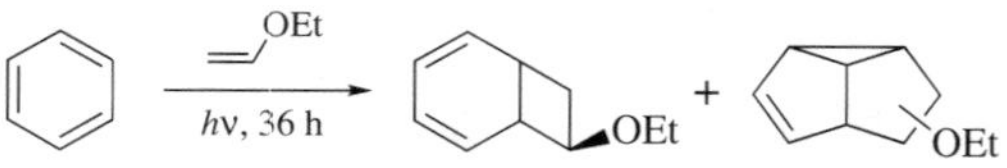

(27)

Related Reagents. 3,4-Dihydro-2*H*-pyran; 1,1-Dimethoxyethane; 1-Ethoxy-1-propene; 2-Methoxypropene; Vinyl Acetate.

1. Mundy, B. P.; Ellerd, M. G. *Name Reactions and Reagents in Organic Synthesis;* Wiley: New York, 1988; p 354.

2. (a) *CRC Handbook of Chemistry and Physics,* 73rd ed.; CRC: Boca Raton, FL, 1992; pp 3–241. (b) *Dictionary of Organic Compounds,* 5th ed.; Chapman & Hall: New York, 1982; Vol. 3, p 2571. (c) *Sax's Dangerous Properties of Industrial Materials,* 8th ed.; Lewis, R. J., Ed.; van Nostrand Rheinhold: New York 1992; Vol. 2, p 1668.
3. Greene, T. W.; Wuts, P. G. M. *Protective Groups in Organic Synthesis*, 2nd ed.; Wiley: New York, 1991; p 38.
4. Manchand, P. S.; Schwartz, A.; Wolff, S.; Belica, P. S.; Madan, P.; Patel, P.; Saposnik, S. J. *H* **1993**, *35*, 1351.
5. (a) Smrt, J.; Chladek, S. *CCC* **1966**, *31*, 2978. (*CA* **1966**, *65*, 10 648c). (b) Young, S. D.; Buse, C. T.; Heathcock, C. H. *OSC* **1990**, *7*, 381.
6. Meyers, A. I.; Comins, D. L.; Roland, D. M.; Henning, R.; Shimizu, K. *JACS* **1979**, *101*, 7104.
7. Anderson, R. C.; Fraser-Reid, B. *TL* **1978**, 3233.
8. Still, W. C. *JACS* **1978**, *100*, 1481.
9. Semmelhack, M. F.; Tomoda, S. *JACS* **1981**, *103*, 2427.
10. Tius, M. A.; Fauq, A. H. *JACS* **1986**, *108*, 1035.
11. Napolitano, E.; Giannone, E.; Fiaschi.; Marsili, A. *JOC* **1983**, *48*, 3653.
12. (a) Rhoads, S. J.; Raulins, N. R. *OR* **1975**, *22*, 1. (b) Ziegler, F. E. *CRV* **1988**, *88*, 1423; (c) Wipf, P. *COS* **1991**, *5*, 827.
13. Dauben, G.; Dietsche, T. J. *JOC* **1972**, *37*, 1212.
14. Marbet, R.; Saucy, G. *HCA* **1967**, *50*, 2095.
15. Saucy, G.; Marbet, R. *HCA* **1967**, *50*, 1158.
16. Reed, S. F. *JOC* **1965**, *30*, 1663.
17. Grieco, P. A.; Takigawa, T.; Moore, D. R. *JACS* **1979**, *101*, 4380.
18. (a) Paquette, L. A.; Annis, G. D.; Schostarez, H. *JACS* **1981**, *103*, 6526. (b) Paquette, L. A.; Annis, G. D.; Schostarez, H.; Blount, J. F. *JOC* **1981**, *46*, 3768.
19. Isler, O.; Lindlar, H.; Montavon, M.; Rüegg, R.; Zeller, P. *HCA* **1956**, *39*, 249.
20. (a) Isler, O.; Schudel, P. *Adv. Org. Chem. Methods Results* **1963**, *4*, 115. (b) Effenberger, F. *AG(E)* **1969**, *8*, 295. (c) *Carotenoids*; Isler, O., Ed.; Birkhauser: Basel, Switzerland, 1971. (d) Makin, S. M. *PAC* **1976**, *47*, 173.
21. Fishman, D.; Klug, J. T.; Shani, A. *S* **1981**, 137.
22. *Umpoled Synthons*; Hase, T. A. Ed.; Wiley: New York, 1987.
23. (a) Kraus, G. A.; Krolski, M. E. *SC* **1982**, 521. (b) Baldwin, J. E.; Höfle, G. A.; Lever, O. W. *JACS* **1974**, *96*, 7125.
24. Boeckman, R. K.; Bruza, K. J. *JOC* **1979**, *44*, 4781.
25. (a) Desimoni, G.; Tacconi, G. *CRV* **1975**, *75*, 651. (b) Boger, D. L. *COS* **1991**, *5* 451. (c) Carruthers, W. *Cycloaddition Reactions in Organic Synthesis*; Pergamon: New York, 1990. (d) Boger, D. L.; Weinreb. S. M. *Hetero Diels–Alder Methodology in Organic Synthesis*; Academic: San Diego, 1987.
26. Longley, R. I.; Emerson, W. J.; Blardinelli, A. J. *OSC* **1963**, *4*, 311.
27. (a) Apparao, S.; Maier, M. E.; Schmidt, R. R. *S* **1987**, 900. (b) Schmidt, R. R. *ACR* **1986**, *19*, 250.
28. Tietze, L. F.; Hübsch, T.; Voss, E.; Buback, M.; Tost, W. *JACS* **1988**, *110*, 4065.
29. Chapleur, Y.; Envrard, M.-N. *CC* **1987**, 884.
30. Inoue, T.; Inoue, S.; Sato, K. *CL* **1989**, 653.
31. Cabral, J.; Laszlo, P.; Montaufier, M. T. *TL* **1988**, *29*, 547.
32. Gupta, R. B.; Franck, R. W. *JACS* **1987**, *109*, 5393.
33. (a) *Advances in Cycloadditions*; Curran, D. P., Ed.; Jai: Greenwich, CT, 1988; Vol. 1. (b) DeShong, P.; Leginus, J. M. *JACS* **1983**, *105*, 1686. (c) Confalone, P. N.; Huie, E. M. *OR* **1988**, *36*, 1.
34. Backvall, J. E.; Karlsson, U.; Chinchilla, R. *TL* **1991**, *32*, 5607.
35. (a) Fatiadi, A. J. *S* **1987**, 749. (b) Baldwin, J. E. *COS* **1991**, *5*, 63.
36. *Synthetic Organic Photochemistry;* Horspool, W. M., Ed.; Plenum: New York, 1984.
37. Gilbert, A.; Taylor, G. N.; binSamsudi, M. W. *JCS(P1)* **1980**, 869.

Percy S. Manchand
Hoffmann-La Roche, Nutley, NJ, USA

9-Fluorenylmethyl Chloroformate[1]

[28920-43-6] $C_{15}H_{11}ClO_2$ (MW 258.71)

(base-labile protecting group for primary and secondary amines;[3] used in peptide synthesis;[5] used for derivatization of amines and amino acids prior to HPLC analysis and fluorescence detection[13])

Alternate Names: 9-fluorenylmethoxycarbonyl chloride; FMOC-Cl.
Physical Data: mp 62–64 °C.
Solubility: sol organics (CH_2Cl_2, THF, dioxane); reacts with alcohols, amines, and water.
Form Supplied in: white crystalline solid; widely available.
Handling, Storage, and Precautions: FMOC-Cl is an acid chloride, and should be protected from moisture and heat. The reagent can evolve CO_2 (pressure!) upon prolonged storage; therefore bottles should be opened carefully. FMOC-Cl is harmful if swallowed, inhaled, or absorbed through skin, is corrosive, and is a strong lachrymator. Inhalation may be fatal. Use in a fume hood.

***N*-Terminus Blocking in Peptide Synthesis.** Solid-phase peptide synthesis based on base-labile FMOC protection[1] is rapidly augmenting and/or replacing older methods based on Boc and Z (Cbz) protection. The major advantage offered by the use of FMOC protection is that repetitive treatment with HF or CF_3CO_2H is no longer required for the deprotection steps. Most FMOC-protected amino acids are commercially available, although the search for orthogonal side-chain functional group protection continues.[2] FMOC protection is currently the method of choice for the synthesis of glycopeptides which carry acid-labile *O*-glycosidic bonds (eq 1).[3,5b,7]

Fmoc-Cl, $NaHCO_3$, H_2O–dioxane (1:1), 0 °C to rt, 92% (1)

FMOC protection of amino acids is accomplished with Shotten–Bauman conditions (e.g. $NaHCO_3$/dioxane/H_2O or $NaHCO_3$/DMF), or with anhydrous conditions (e.g. pyridine/CH_2Cl_2) at room temperature or below (eq 1).[3a,4] A typical synthetic protocol for solid-phase peptide synthesis would involve coupling (eq 2), followed by deprotection with 20–30% ***Piperidine*** in DMF (eq 3), although other amines (e.g. 2% DBU, 10% Et_2NH, or 50% morpholine in DMF) have been used. Many workers no longer use the chloroformate ester FMOC-Cl (**1**), preferring the more shelf-stable carbonate FMOC-OSu (**2**) *[82911-69-1]*,[5] or FMOC-pfp (**3**) *[88744-04-1]*. When used in the standard way,[4] FMOC-Cl (**1**) has been shown to promote the formation of 'FMOC-dipeptides' in 3–7% yield, presumably via the mixed anhydride intermediate.[5a]

peptide–NH_2, –H_2O (2)

:Base; $+ CO_2$ (3)

(**2**) (**3**)

Use of FMOC-OSu (**2**) provides *N*-protected amino acids in comparable yields without the formation of the dipeptide impurities.[5] FMOC-pfp (**3**) has been used to prepare either FMOC-protected amino acids, or the corresponding pfp esters in one pot. Advantages of this reagent include supression of the dipeptide byproducts, as well as the production of the activated pfp esters using the same pentafluorophenol released during the *N*-protection step (eq 4).[6] This same *N*-protection/*C*-activation procedure has been used to produce serine glycosides (eq 5),[3b] and for the solid-phase synthesis of human intestinal mucin fragments (eq 6).[7] The use of FMOC-N_3 (**4**) is also claimed to reduce dipeptide formation during the protection step,[8] but the storage or use of potentially explosive oxycarbonyl azides cannot be recommended by this reviewer. Alternatively, the azide (**4**) may be generated in situ from (**1**).[4]

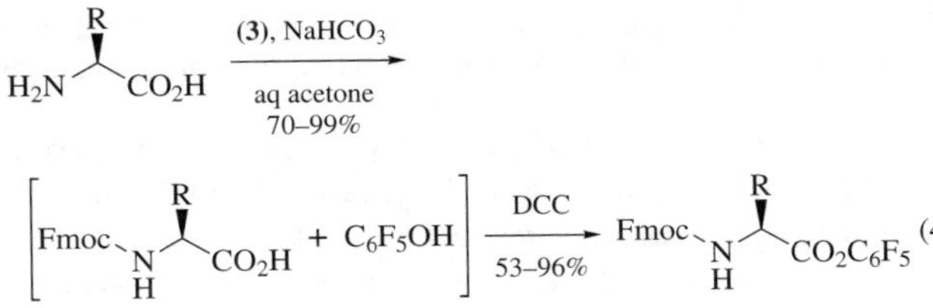

(3), $NaHCO_3$; aq acetone 70–99%; DCC 53–96% (4)

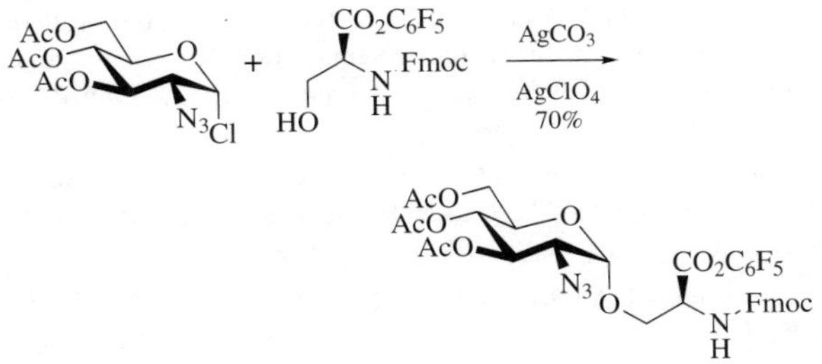

$AgCO_3$; $AgClO_4$ 70% (5)

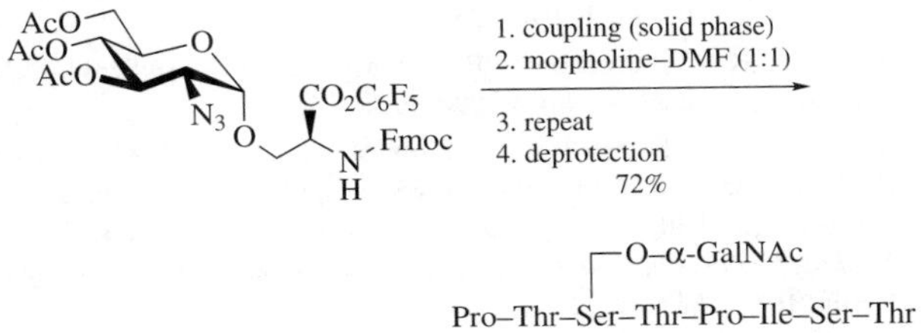

1. coupling (solid phase) 2. morpholine–DMF (1:1) 3. repeat 4. deprotection 72%

O-α-GalNAc
Pro–Thr–Ser–Thr–Pro–Ile–Ser–Thr (6)

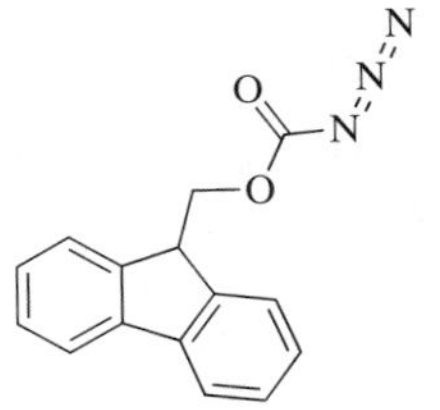

(**4**)

Fluorenylmethyl esters have been used for *C*-terminus protection in combination with FMOC *N*-terminus protection for solution-phase coupling of protected amino acids and peptide fragments. This permits simultaneous deprotection of both termini with Et_2NH/DMF (eq 7).[9]

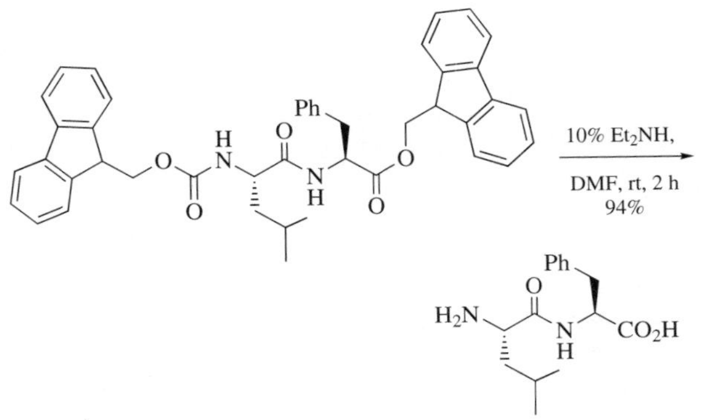

10% Et_2NH, DMF, rt, 2 h 94% (7)

Amine and Hydroxyl Protection for Nucleosides, Carbohydrates, and Natural Products. FMOC has been used to protect the amino groups on various nucleoside bases (adenosine, cytosine, guanine, and their 2-deoxy derivatives) for solid-phase oligonucleotide synthesis in a one-pot procedure.[10] ***Chlorotrimethylsilane*** in dry pyridine was used for transient protection of the 2′,3′,5′- or 3′,5′-hydroxyls, followed by the addition of FMOC-Cl, and hydrolytic workup of the TMS ethers to produce crystalline FMOC-protected nucleosides and 2-deoxynucleosides in good yield (eq 8). The FMOC group was removed from the heterocyclic bases with excess ***Triethylamine*** in dry pyridine.

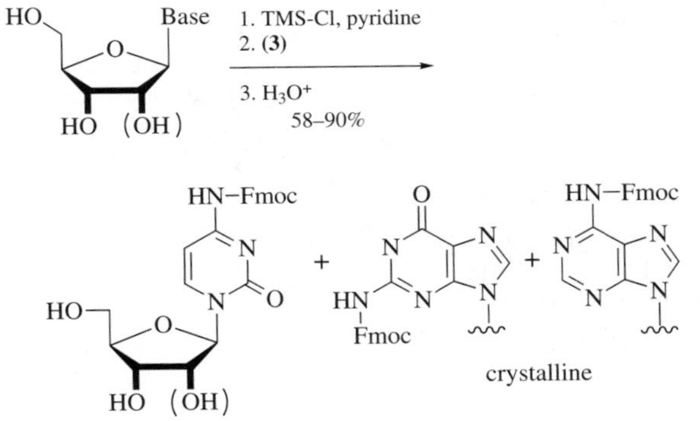

1. TMS-Cl, pyridine 2. (3) 3. H_3O^+ 58–90%

crystalline (8)

FMOC-Cl has been used for 3′,5′-hydroxyl protection of deoxyribose for the introduction of modified bases such as azacytidine using the classical Vorbrüggen reaction.[11] While the carbonate protecting groups do not alter the stereoselectivity of the *N*-glycoside formation, the FMOC groups can be removed under much milder conditions (Et_3N/pyridine) than the typical 4-methylbenzoyl group (eq 9).

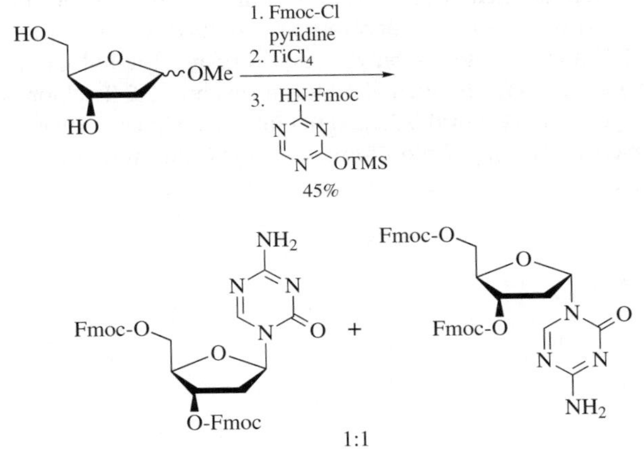

1. Fmoc-Cl pyridine 2. $TiCl_4$ 3. HN-Fmoc triazine OTMS 45%

1:1 (9)

Perhaps the most impressive use of the FMOC group in the synthesis of natural products is provided by the key tetrasaccharide intermediate to calicheamicin.[12] The robust nature of this particular FMOC group is amply demonstrated by the fact that it survives 15 discrete reaction steps, including exposure to cat. $NaH/HOCH_2CH_2OH$, Br_2, EtN_3, K-Selectride, *i*-Bu_2AlH, and NaSMe (eq 10). Clearly, the use of FMOC protection is not limited to peptide synthesis.

FMOC Derivatization for Fluorescence Detection and Analysis. Derivatization of amino acids and other primary and secondary amines with (**1**) for HPLC analysis has been described. The chief advantage of this methodology is the high sensitivity due to the intense fluorescence of the fluorenyl moiety, but the FMOC also reduces polarity for faster elution from the column. Amino acids have been detected in concentrations as low as 26 femtomolar using a precolumn derivatization scheme.[13] FMOC derivatization of *lyso*-gangliosides in a two-phase Et_2O–H_2O system followed by chromatographic purification with fluorescence detection has also been reported.[14] Optically active 1-(9-fluorenyl)ethyl chloroformate (FLEC®) (**5**) is commercially available in both (+) [*107474-79-3*] and (−) forms for the analysis of racemates.

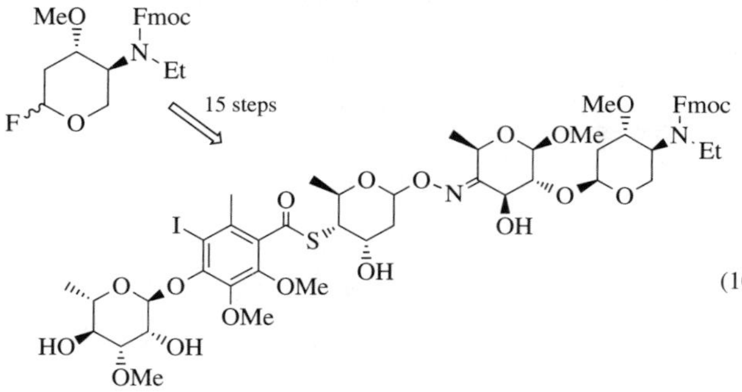

(10)

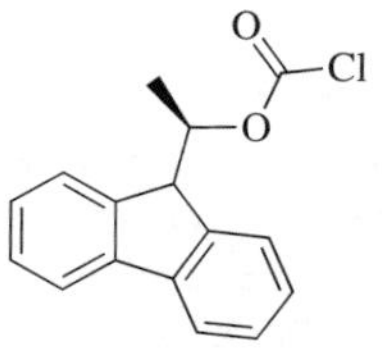

(**5**)

Related Reagents. Allyl Chloroformate; Benzyl Chloroformate; 4-Bromobenzyl Chloroformate; *t*-Butyl Chloroformate; Ethyl Chloroformate; Isobutyl Chloroformate; Methyl Chloroformate; 2,2,2-Tribromoethyl Chloroformate; 2,2,2-Trichloro-*t*-butoxycarbonyl Chloride; 2,2,2-Trichloroethyl Chloroformate; 2-(Trimethylsilyl)ethyl Chloroformate; Vinyl Chloroformate.

1. For illustrative procedures using FMOC and other protection schemes, see *The Practice of Peptide Synthesis*; Bodanszky, M.; Bodanszky, A., Eds.; Springer: Berlin, 1984.
2. (a) Histidine: Fischer, P. M. et al. *Int. J. Pept. Protein Res.* **1992**, *40*, 19. (b) Glutamate: Handa, B. K.; Keech, E. *Int. J. Pept. Protein Res.* **1992**, *40*, 66.
3. (a) Polt, R.; Szabo, L.; Treiberg, J.; Li, Y.; Hruby, V. J. *JACS* **1992**, *114*, 10249. (b) Lüning, B.; Norberg, T.; Tejbrant, J. *CC* **1989**, 1267.
4. Carpino, L. A.; Han, G. Y. *JOC* **1972**, *37*, 3404.
5. (a) Lapatsanis, L.; Milias, G.; Froussios, K.; Kolovos, M. *S* **1983**, 671. (b) Bardaji, E.; Torres, J. L.; Clapes, P.; Albericio, F.; Barany, G.; Rodriguez, R. E.; Sacristin, M. P.; Valencio, G. *JCS(P1)* **1991**, 1755. For an alternate synthesis of FMOC-OSu (**2**) from fluorenylmethyl alcohol and phosgene: (c) Ten Kortenaar, P. B. W. et al. *Int. J. Pept. Protein Res.* **1986**, *27*, 398.
6. Schön, I.; Kisfalady, L. *S* **1986**, 303.
7. (a) Jansson, A. M.; Meldal, M.; Bock, K. *TL* **1990**, *31*, 6991. (b) Bielfeldt, T.; Peters, S.; Meldal, M.; Paulsen, H.; Bock, K. *AG(E)* **1992**, *31*, 857.
8. Tessier, M. et al. *Int. J. Pept. Protein Res.* **1983**, *22*, 125.
9. (a) Bednarek, M. A.; Bodanszky, M. *Int. J. Pept. Protein Res.* **1983**, *21*, 196. (b) Bodanszky, M. et al. *Int. J. Pept. Protein Res.* **1981**, *17*, 444.
10. Heikkila, J.; Chattopadhyaya, J. *ACS* **1983**, *B37*, 263.
11. Ben-Hattar, J.; Jiricny, J. *JOC* **1986**, *51*, 3211.
12. Nicolaou, K. C.; Groneborg, R. D.; Miyazaki, T.; Stylianides, N.; Schulze, T. J.; Stahl, W. *JACS* **1990**, *112*, 8193.
13. (a) Einarsson, S.; Folestad, S.; Josefsson, B. *Anal. Chem.* **1986**, *58*, 1638. (b) Einarsson, S. *J. Chromatogr.* **1985**, *348*, 213. (c) Veuthey, J.-L. *J. Chromatogr.* **1990**, *515*, 385. (d) For a review: Betner, I.; Foeldi, P. In *Modern Methods in Protein Chemistry*; Tschesche, H., Ed.; de Gruyter: Berlin, 1988; Vol. 3, p 227.
14. Neuenhofer, S.; Schwarzmann, G.; Egge, H.; Sandhoff, K. *B* **1985**, *24*, 525.

Robin L. Polt
University of Arizona, Tucson, AZ, USA

Hexacarbonylchromium[1]

$Cr(CO)_6$

[13007-92-6] C_6CrO_6 (MW 220.06)

(catalyst for alkene isomerization, hydrogenation of 1,3-conjugated dienes, oxidation of allylic and benzylic positions; reagent for preparation of tricarbonyl(arene)chromium and Fischer carbene complexes)

Alternate Name: chromium hexacarbonyl.
Physical Data: sublimes at 50–80 °C; mp 154–155 °C (in sealed tube); *d* 1.77 g cm^{-3}.
Solubility: insol water, ethanol; slightly sol ether, $CHCl_3$.
Form Supplied in: white solid; widely commercially available.
Handling, Storage, and Precautions: highly volatile and must be stored and used in a well ventilated fume hood.

Oxidation. Treatment of alkenes with ***t*-Butyl Hydroperoxide** in the presence of 0.5 equiv of $Cr(CO)_6$ or $Cr(CO)_x(MeCN)_y$ species in refluxing MeCN results in the oxidation of allylic methylene groups to give α,β-unsaturated ketones selectively in the presence of certain alcohols (eq 1).[2] Benzylic positions are also oxidized to phenyl ketone derivatives[3] under the same conditions (eq 2).

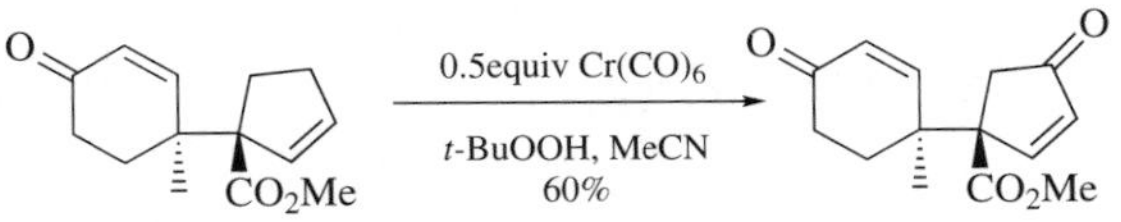

0.5equiv $Cr(CO)_6$; *t*-BuOOH, MeCN; 60% (1)

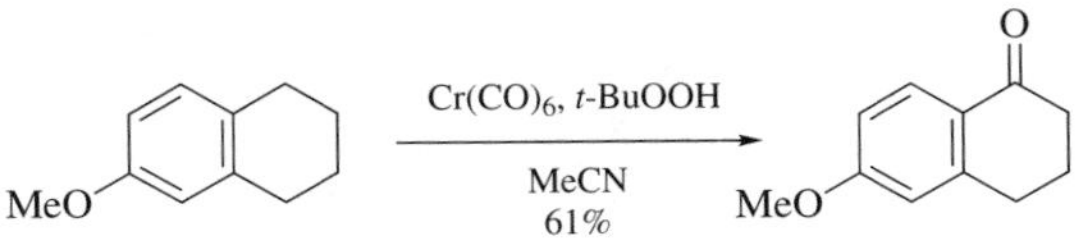

$Cr(CO)_6$, *t*-BuOOH; MeCN; 61% (2)

Hydrogenation and Isomerization of Dienes. The catalytic behavior of (arene)$Cr(CO)_3$ complexes has been exploited in the homogeneous regioselective 1,4-hydrogenation of dienes to *cis*-alkenes.[4] Only 1,3-dienes that can easily achieve the less stable *s-cis* configuration to undergo this catalytic hydrogenation (eq 3). Similar catalytically active species $Cr(CO)_3H_2$(diene) (**1**) can be generated photochemically from $Cr(CO)_6$ or from kinetically thermally labile π-bonded ligated complexes (naphthalene, anthracene, MeCN) in coordinating solvents (THF, acetone, dioxane) under milder conditions (eq 4).[5] This regioselective hydrogenation of dienes can be applied to the total synthesis of complex natural products (eq 5) such as carbacyclin, cyanopcarbacyclin, and deplanchein.[6] 1,4-Hydrosilylation of dienes is also achieved to produce (*Z*)-allyltrimethylsilanes[7] under these conditions. The (arene)chromium complexes are superior catalysts for stereoselective hydrogenation of alkynes to (*Z*)-alkenes, and in addition, cisoid α,β-unsaturated enones and imines can be reduced.[8]

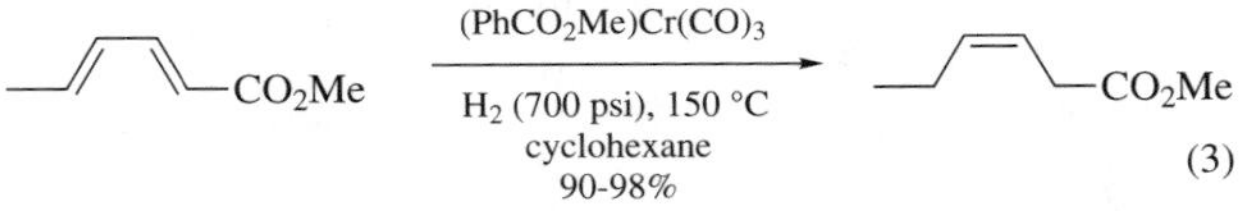

$(PhCO_2Me)Cr(CO)_3$; H_2 (700 psi), 150 °C; cyclohexane; 90-98% (3)

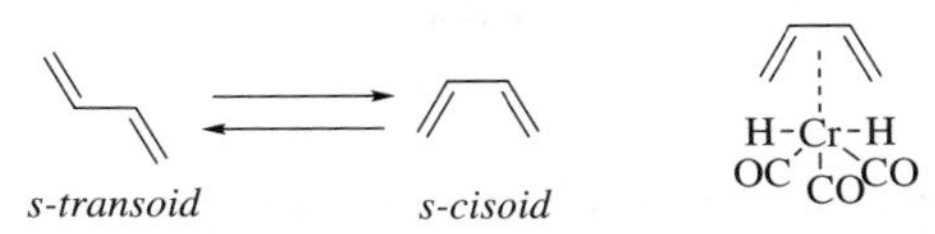

s-transoid ⇌ *s-cisoid* (4) (**1**)

(Naph)$Cr(CO)_3$; THF, H_2 (70 kg cm^{-2}); 45 °C; 97% (5)

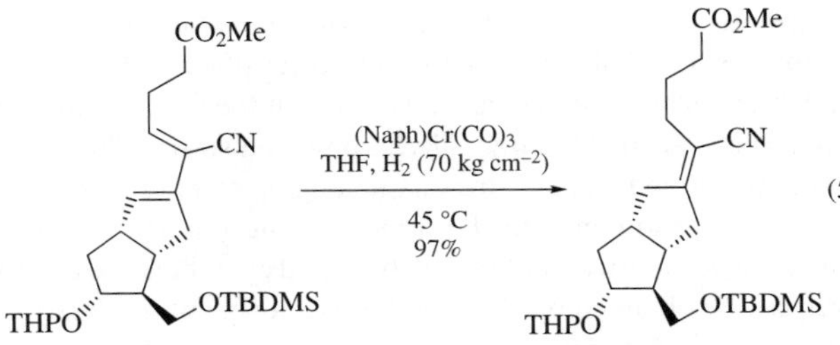

Chromium hexacarbonyl offers a highly regioselective method for the isomerization of cisoid 1,3-dienes to transoid dienes (eq 6).[9] (Naphthalene)chromium tricarbonyl catalyzes isomerization of silyloxymethylbutadiene derivatives to silyl dienol ethers in quantitative yields via a U-shaped pentadienyl intermediate (eq 7).[10]

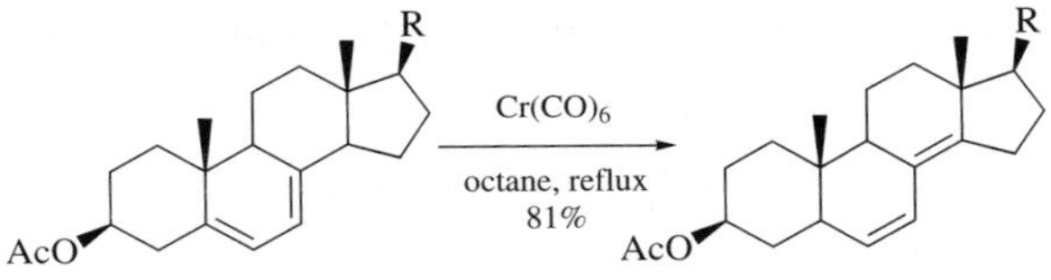

$Cr(CO)_6$; octane, reflux; 81% (6)

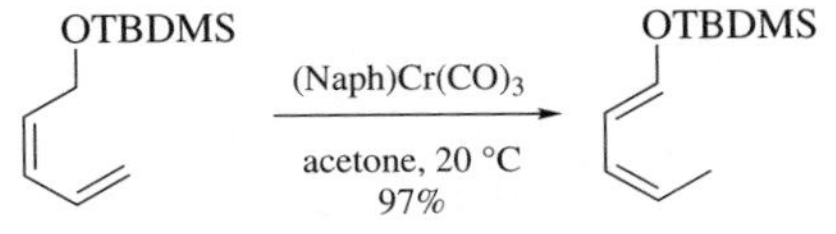

(Naph)$Cr(CO)_3$; acetone, 20 °C; 97% (7)

(Arene)Cr(CO)$_3$ Complexes. Complexation of an arene ring to the chromium tricarbonyl unit is easily accomplished by simple heating with $Cr(CO)_6$ or by ligand transfer with $Cr(CO)_3L_3$.[1,11] The significant properties of (arene)chromium complexes in organic synthesis are as follows: nucleophilic substitution to arene ring; enhancement of acidity of aromatic hydrogen; enhancement of acidity of benzylic hydrogen; enhancement of solvolysis at the benzylic position; and steric hindrance of the $Cr(CO)_3$ group.

Nucleophilic Substitution.[1] Some carbon nucleophiles add to tricarbonyl(arene)chromium complexes to yield anionic η^5-cyclohexadienyl complexes (**2**) (eq 8), which give the substituted arenes via decomplexation by oxidation with iodine. Protolysis of the intermediate cyclohexadienylchromium complexes (**2**) generate cyclohexadienes, and reaction with electrophiles generates either the arene chromium complexes or produces the acylated species.

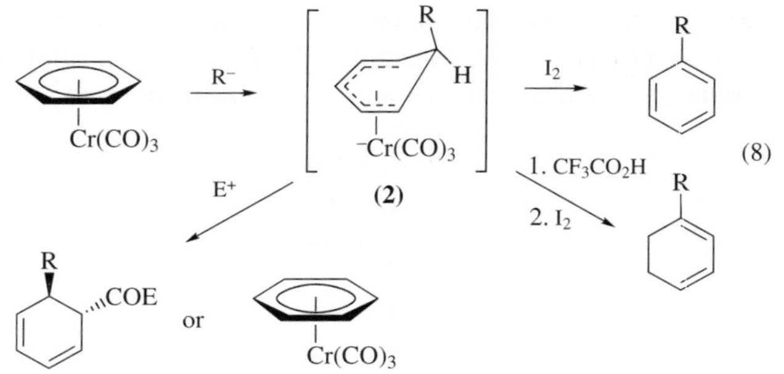

With substituted aromatics, the following regioselectivity for nucleophilic substitution is generally observed (eq 9); alkoxy substituents strongly direct to the *meta* position, and trimethylsilyl groups ensure *para* substitution.[12] Alkylation of the chromium complexes of indoles and benzofurans takes place at the C-4 or C-7 position, depending on the steric effect of the C-3 substituent.[13] Intramolecular nucleophilic alkylation can also be achieved; the products formed depend on chain length, reaction conditions, and other substituents on the aromatic ring (eq 10).[14] With an (anisole)chromium complex, 3-substituted cyclohexenone derivatives can be obtained by the protolytic cleavage of the intermediates (eq 11).[17] A combination of the intramolecular alkylation with protolysis gives an elegant synthesis of the spiro sesquiterpenoids acorenone and acorenone B (eq 12).[15,16]

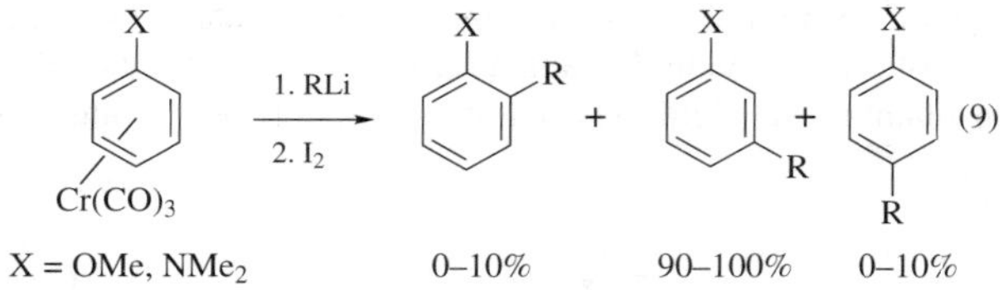

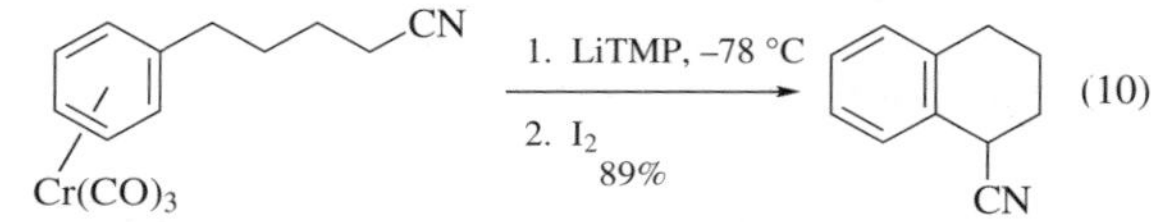

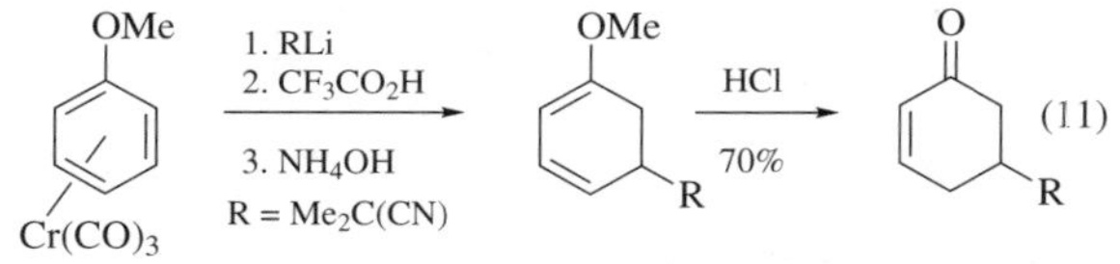

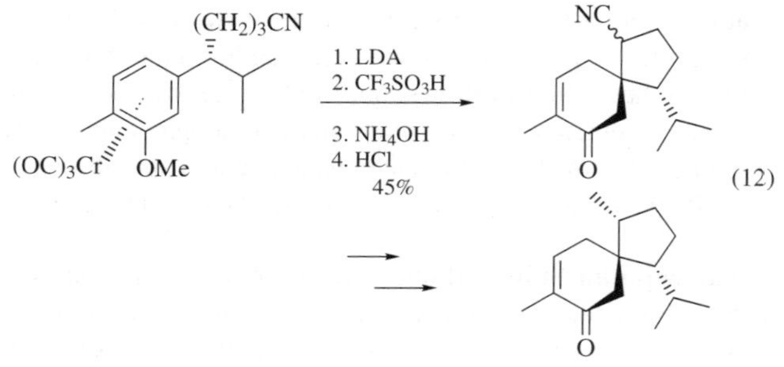

Reaction with a range of electrophiles regenerates the (arene)chromium complexes under reversible conditions, but electrophilic attack upon the cyclohexadienyl complex formed under irreversible conditions from dithiane carbanions produces *trans* addition products (eq 13).[17]

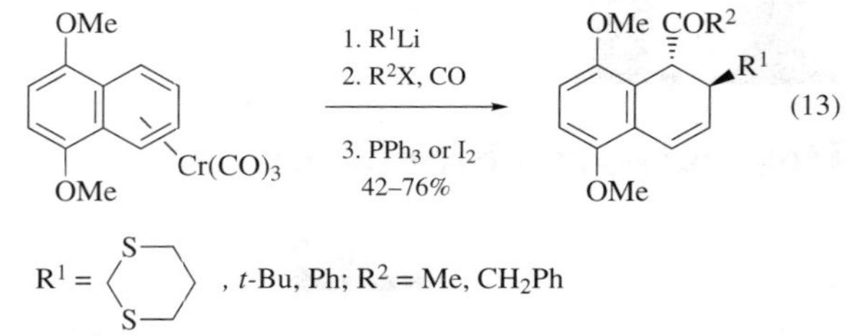

Lithiation. With anisole, fluorobenzene, and chlorobenzene chromium complexes, lithiation always occurs at an *ortho* position of the substituents under mild conditions (eqs 14 and 15).[18] Protected phenol or aniline chromium complexes with sterically bulky substituents produce *meta* lithiation exculsively.[19] The lithiated position of some (arene)chromium complexes (**3**) differs from that of the corresponding chromium-free compounds (**4**).[20]

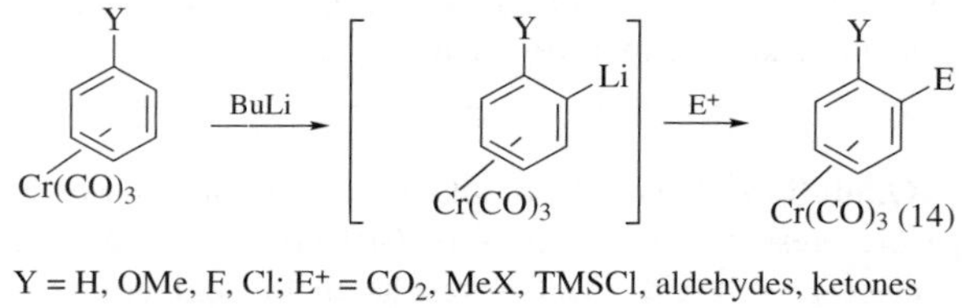

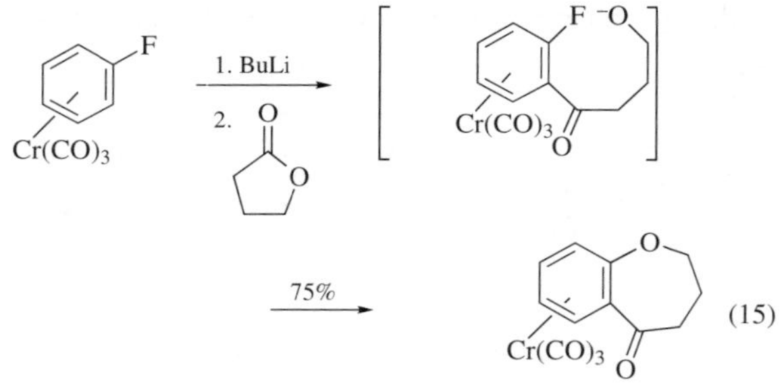

Activation of the Benzylic Position. Both chromium complexed carbanions and carbocations are stabilized at the benzylic position.[21] Dialkylation of alkyl halides at the benzylic position occurs via stabilized carbanions under mild conditions (eq 16).[22] Regio- and stereoselective products are obtained via the benzylic carbanions, depending on the conformation of the tricarbonyl group to the arene (eq 17).[22] (Styrene)chromium complexes stabilize negative charges at the benzylic position by addition of nucleophiles to the β-position, giving bifunctionalization products by trapping with electrophiles (eq 18).[23]

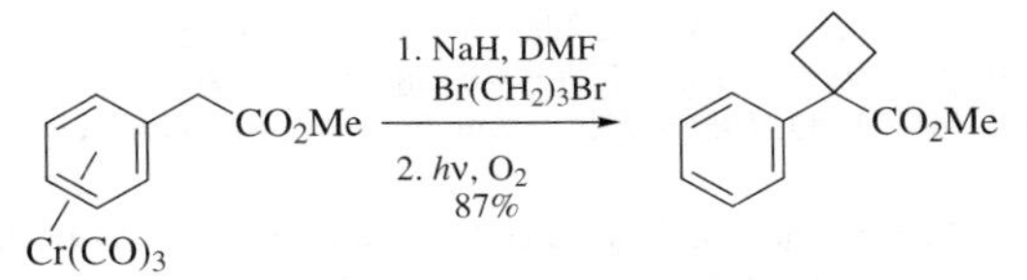

Tricarbonylchromium stabilized benzylic carbocations can be captured by a large variety of nucleophiles,[24] such as alcohols, amines, thiols, nitriles, trimethylsilyl enol ethers, allylsilanes, electron-rich aromatics, dialkylzincs, and trialkylaluminums (eq 19). The relative stereochemistry formed during these reactions via carbocations in acyclic systems proceeds with net retention.[21b,c,25] Friedel–Crafts acylation of (styrene)chromium complexes has been explored via the benzylic cations (eq 20).[26] Tricarbonylchromium-stabilized oxonium ions are also utilized for steroselective carbon–carbon bond forming reactions (eq 21).[27]

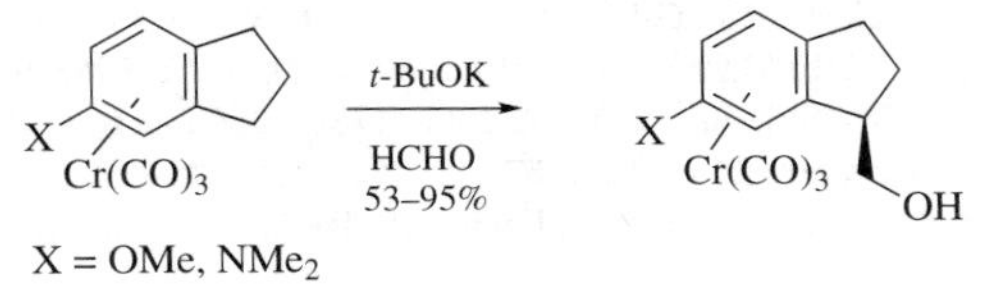

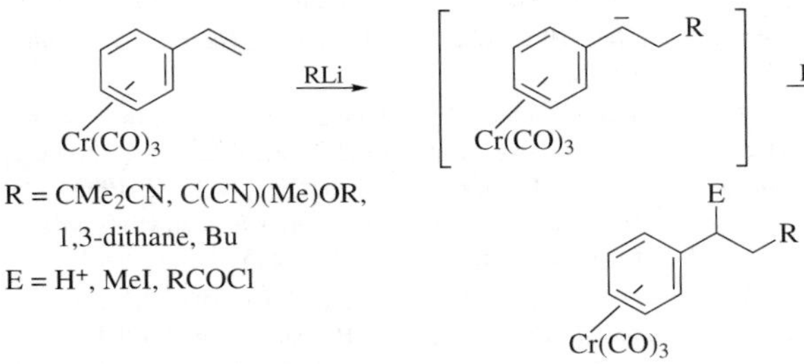

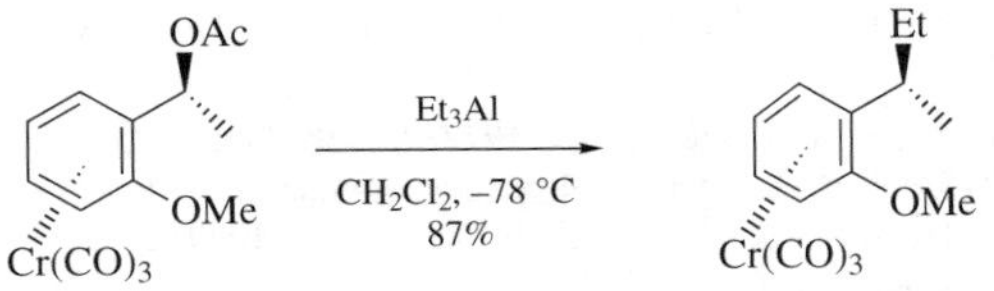

Steric Effect of the $Cr(CO)_3$ Group.[1] A feature of (η^6-arene)chromium complexes is the steric interference offered to the approach of reagents, which occurs exclusively from the *exo*-face at the reactive center in cyclic chromium complexes (eq 22).[28] Even in the (acyclic arene)chromium complexes, such as *o*-substituted benzaldehyde and phenyl alkyl ketones with electron-donating *ortho* substituents (OMe, Me, F), addition of nucleophiles to the chromium complexed benzylic carbonyl group proceeds stereoselectively, giving one diastereomeric complex predominantly (eq 23).[29] A Friedel–Crafts reaction produces predominantly an *exo*-substituted complex (eq 24).[30]

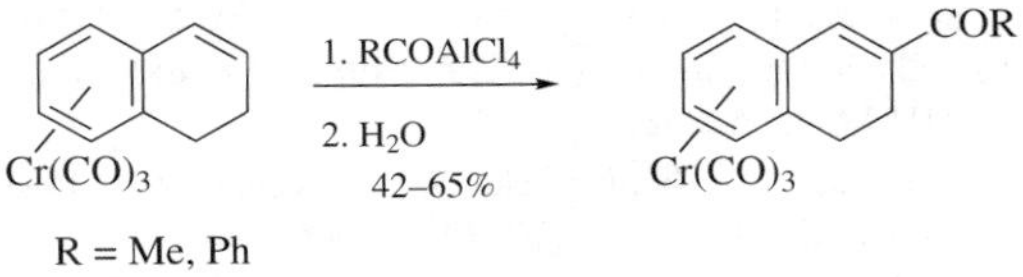

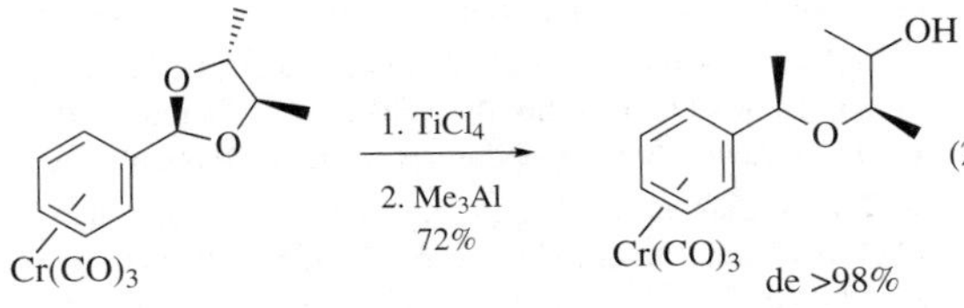

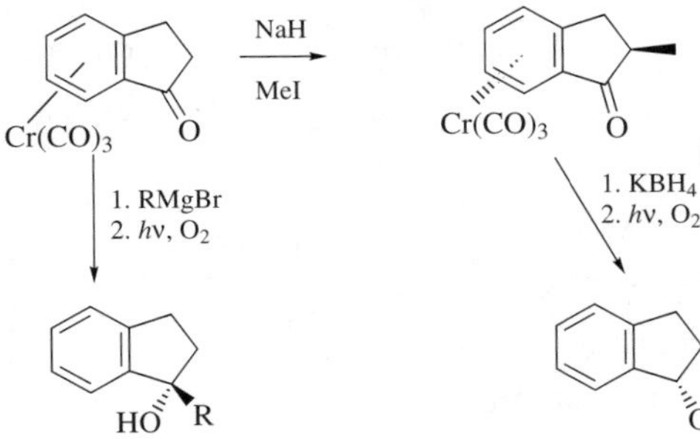

A 1,2 or 1,3 unsymmetrically disubstituted arene is prochiral and therefore the corresponding chromium tricarbonyl compounds are chiral.[31] (Substituted arene) complexes with amine, carboxyl, and formyl groups at the *ortho* position are resolved into optically active chromium complexes through corresponding diastereomeric adducts (eq 25).[32] Biocatalysts also perform the kinetic resolution of racemic chromium complexes (eq 26).[33] The optically active chromium complexes can be prepared by diastereoselective *ortho* lithiation[34] of the chiral benzaldehyde or acetophenone acetal complexes, and diastereoselective chromium complexation[35] of the chiral *ortho*-substituted benzaldehyde aminals (eq 27). Catalytic asymmetric cross-coupling of *meso* (1,2-haloarene)chromium complex produces chiral monosubstituted complexes.[36] The chiral (arene)chromium complexes can be used as ligands in asymmetric reactions.[37]

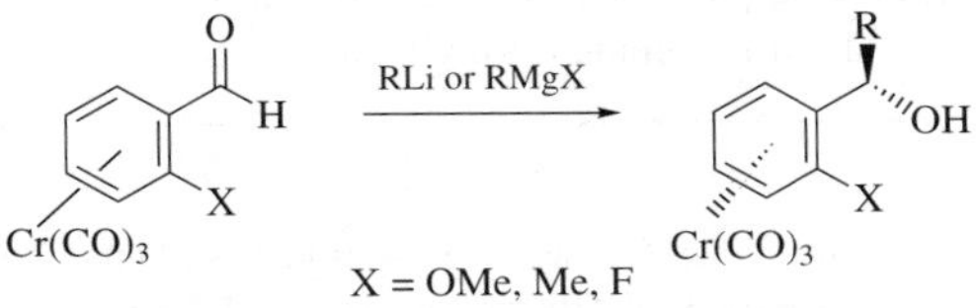

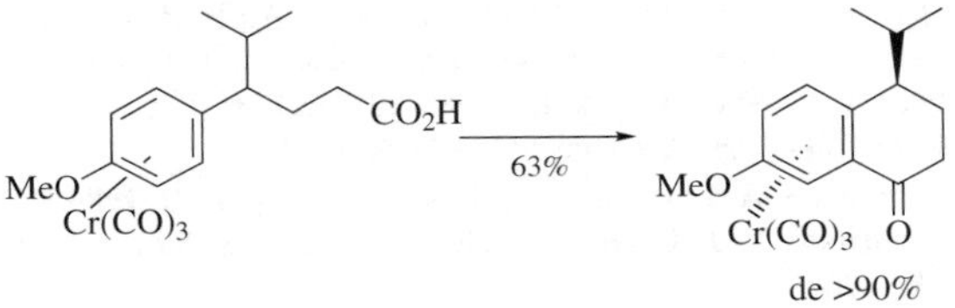

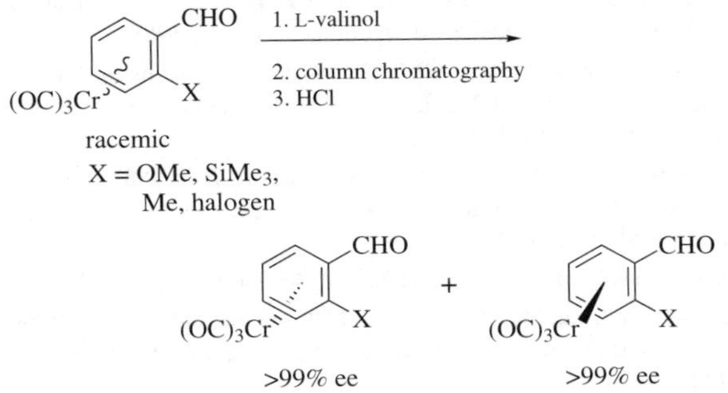

CH_2OH $Cr(CO)_3$ X racemic — lipase, X = Me, OMe, $SiMe_3$ → CH_2OH $Cr(CO)_3$ X 85–99% ee + CH_2OAc $Cr(CO)_3$ X 85–99% ee (26)

MeN N Me X; X = OMe, Me — 1. (Naph)$Cr(CO)_3$, THF, rt; 2. H^+ → CHO $Cr(CO)_3$ X de >94% (27)

Related Reagents. (η^6-Benzene)tricarbonylchromium; Dodecacarbonyltriiron; Pentacarbonyliron.

1. (a) Collman, J. P.; Hegedus, L. S.; Norton, J. R.; Finke, R. G. *Principles and Applications of Organotransition Metal Chemistry*; University Science Books: Mill Valley, CA, 1987. (b) Jaouen, G. In *Transition Metal Organometallics in Organic Synthesis*; Alper, H., Ed.; Academic: New York, 1978; Vol. 2, pp 65–120. (c) Pearson, A. J. *Metallo-organic Chemistry*; Wiley: Chichester, 1985. (d) Heck, R. F. *Organotransition Metal Chemistry*; Academic: New York, 1974.
2. (a) Pearson, A. J.; Chen, Y.-S.; Hus, S.-Y.; Ray, T. *TL* **1984**, *25*, 1235. (b) Pearson, A. J.; Chen, Y.-S.; Han, G. R.; Hsu, S.-Y.; Ray, T. *JCS(P1)* **1985**, 267.
3. Pearson, A. J.; Han, G. R. *JOC* **1985**, *50*, 2791.
4. (a) Cais, M.; Frankel, E. N.; Rejoan, A. *TL* **1968**, 1919. (b) Frankel, E. N.; Selke, E.; Glass, C. A. *JACS* **1968**, *90*, 2446. (c) Frankel, E. N.; Selke, E.; Glass, C. A. *JOC* **1969**, *34*, 3936.
5. (a) Wrighton, M.; Schroeder, M. A. *JACS* **1973**, *95*, 5764. (b) Yagupsky, G.; Cais, M. *ICA* **1975**, *12*, L27.
6. (a) Shibasaki, M.; Sodeoka, M.; Ogawa, Y. *JOC* **1984**, *49*, 4096. (b) Shibasaki, M.; Sodeoka, M. *TL* **1985**, *26*, 3491. (c) Rosenmund, P.; Casutt, M. *TL* **1983**, *24*, 1771.
7. Wrighton, M. S.; Schroeder, M. A. *JACS* **1974**, *96*, 6235.
8. Sodeoka, M.; Shibasaki, M. *JOC* **1985**, *50*, 1147.
9. (a) Barton, D. H. R.; Davies, S. G.; Motherwell, W. B. *S* **1979**, 265. (b) Birch, A. J.; Cross, P. E.; Connor, D. T.; Rao, G. S. R. S. *JCS(C)* **1966**, 54.
10. (a) Sodeoka, M.; Yamada, H.; Shibasaki, M. *JACS* **1990**, *112*, 4906. (b) Sodeoka, M.; Satoh, S.; Shibasaki, M. *JACS* **1988**, *110*, 4823.
11. (a) Mahaffy, C. A. L.; Pauson, P. L. *Inorg. Synth.* **1979**, *19*, 154. (b) Seyferth, D.; Merola, J. S.; Eschbach, C. S. *JACS* **1978**, *100*, 4124. (c) Trahanovsky, W. S.; Wells, D. K. *JACS* **1969**, *91*, 5870. (d) Kündig, E. P.; Perret, C.; Spichiger, S.; Bernardinelli, G. *JOM* **1985**, *286*, 183.
12. (a) Semmelhack, M. F. *PAC* **1981**, *53*, 2379. (b) Semmelhack, M. F.; Clark, G. R.; Garcia, J. L.; Harrison, J. J.; Thebtaranonth, Y.; Wulff, W.; Yamashita, A. *T* **1981**, *37*, 3957.
13. (a) Semmelhack, M. F.; Wulff, W.; Garcia J. L. *JOM* **1982**, *240*, C5. (b) Boutonnet, J–C.; Levisalles, J.; Rose, E.; Precigoux, G.; Courseille, C.; Platzer, N. *JOM* **1983**, *255*, 317.
14. Semmelhack, M. F.; Thebtaranonth, Y.; Keller, L. *JACS* **1977**, *99*, 959.
15. (a) Semmelhack, M. F.; Harrison, J. J.; Thebtaranonth, Y. *JOC* **1979**, *44*, 3275. (b) Boutonnet, J.-C.; Levisalles, J.; Normant, J.-M.; Rose, E. *JOM* **1983**, *255*, C21.
16. Semmelhack, M. F. Yamashita, A. *JACS* **1980**, *102*, 5924.
17. (a) Kündig, E. P. *PAC* **1985**, *57*, 1855. (b) Kündig, E. P.; Thi, N. P. D.; Paglia, P.; Simmons, D. P.; Spichiger, S.; Wenger, E. In *Organometallics in Organic Synthesis*; de Meijere, A.; Tom Dieck, H., Eds; Springer: Berlin, 1987; pp 265–276. (c) Kündig, E. P.; Desobry, V.; Simmons, D. P. *JACS* **1983**, *105*, 6962. (d) Kündig, E. P.; Simmons, D. P. *CC* **1983**, 1320.
18. (a) Semmelhack, M. F.; Bisaha, J.; Czarny, M. *JACS* **1979**, *101*, 768. (b) Semmelhack, M. F.; Zask, A. *JACS* **1983**, *105*, 2034.
19. (a) Dickens, P. J.; Gilday, J. P.; Negri, J. T.; Widdowson, D. A. *PAC* **1990**, *62*, 575. (b) Masters, N. F.; Widdowson, D. A. *CC* **1983**, 955. (c) Gilday, J. P.; Widdowson, D. A. *CC* **1986**, 1235. (d) Clough, J. M.; Mann, I. S.; Widdowson, D. A. *TL* **1987**, *28*, 2645. (e) Fukui, M.; Ikeda, T.; Oishi, T. *TL* **1982**, *23*, 1605. (f) Fukui, M.; Ikeda, T.; Oishi, T. *CPB* **1983**, *31*, 466.
20. (a) Uemura, M.; Nishikawa, N.; Hayashi, Y. *TL* **1980**, *21*, 2069. (b) Uemura, M.; Take, K.; Isobe, K.; Minami, T.; Hayashi, Y. *T* **1985**, *41*, 5771.
21. (a) Davies, S. G.; Coote, S. J.; Goodfellow, C. L. In *Advances in Metal–Organic Chemistry*; Liebeskind, L. S., Ed.; JAI: London, 1991; Vol. 2, pp 1–57. (b) Davies, S. G.; Donohoe, T. J. *SL* **1993**, 323. (c) Uemura, M. In *Advances in Metal–Organic Chemistry*; Liebeskind, L. S., Ed.; JAI: London, 1991; Vol. 2, pp 195–245.
22. (a) Jaouen, G.; Meyer, A.; Simmoneaux, G. *CC* **1975**, 813. (b) Jaouen, G.; Top, S.; Laconi, A.; Couturier, D.; Brocard, J. *JACS* **1984**, *106*, 2207. (c) Caro, B.; Le Bihan, J.-Y.; Guillot, J.-P.; Top, S.; Jaouen, G. *CC* **1984**, 602. (d) Brocard, J.; Laconi, A.; Couturier, D.; Top, S.; Jaouen, G. *CC* **1984**, 475.
23. (a) Semmelhack, M. F.; Seufert, W.; Keller, L. *JACS* **1980**, *102*, 6584. (b) Uemura, M.; Minami, T.; Hayashi, Y. *CC* **1984**, 1193.
24. (a) Top, S.; Caro, B.; Jaouen, G. *TL* **1978**, 787. (b) Top, S.; Jaouen, G. *JOC* **1981**, *46*, 78. (c) Reetz, M. T.; Sauerwald, M. *TL* **1983**, *24*, 2837. (d) Uemura, M.; Isobe, K.; Hayashi, Y. *TL* **1985**, *26*, 767.
25. (a) Uemura, M.; Kobayashi, T.; Isobe, K.; Minami, T.; Hayashi, Y. *JOC* **1986**, *51*, 2859. (b) Uemura, M.; Minami, T.; Hayashi, Y. *JACS* **1987**, *109*, 5277. (c) Top, S.; Jaouen, G.; McGlinchey, M. J. *CC* **1980**, 1110.
26. Senechal-Tocquer, M.-C.; Le Bihan, J.-Y.; Gentic, D.; Senechal, D.; Caro, B. *JOM* **1988**, *356*, C5.
27. (a) Davies, S. G.; Newton, R. F.; Williams, J. M. J. *TL* **1989**, *30*, 2967. (b) Normant, J. F.; Alexakis, A.; Ghribi, A.; Mangeney, P. *T* **1989**, *45*, 507.
28. (a) Meyer, A.; Jaouen, G. *CC* **1974**, 787. (b) Jaouen, G.; Meyer, A. *JACS* **1975**, *97*, 4667. (c) des Abbayes, H.; Boudeville, M. A. *TL* **1976**, 2137. (d) des Abbayes, H.; Boudeville, M. A. *JOC* **1977**, *42*, 4104. (e) Jaouen, G.; Meyer, A. *TL* **1976**, 3547.
29. (a) Besancon, J.; Tirouflet, J.; Card, A.; Dusausoy, Y. *JOM* **1973**, *59*, 267. (b) Solladié-Cavallo, A.; Suffert, J. *S* **1985**, 659. (c) Meyer, A.; Dabard, R. *JOM* **1972**, *36*, C38.
30. (a) Uemura, M.; Isobe, K.; Take, K.; Hayashi, Y. *JOC* **1983**, *48*, 3855. (b) Jaouen, G.; Dabard, R. *BSF* **1974**, 1646.
31. Solladié-Cavallo, A. In *Advances in Metal-Organic Chemistry*; Liebeskind, L. S., Ed.; JAI: London, 1989; Vol. 1, pp 99–133.
32. (a) Solladié-Cavallo, A.; Solladié, G.; Tsamo, E. *Inorg. Synth.* **1985**, *23*, 85. (b) Solladié-Cavallo, A.; Solladié, G.; Tsamo, E. *JOC* **1979**, *44*, 4189. (c) Davies, S. G.; Goodfellow, C. L. *SL* **1989**, 59.
33. (a) Top, S.; Jaouen, G.; Gillois, J.; Baldoli, C.; Maiorana, S. *CC* **1988**, 1284. (b) Yamazaki, Y.; Hosono, K. *TL* **1989**, *30*, 5313. (c) Nakamura, K.; Ishihara, K.; Ohno, A.; Uemura, M.; Nishimura, H.; Hayashi, Y. *TL* **1990**, *31*, 3603.
34. (a) Kondo, Y.; Green, J. R.; Ho, J. *JOC* **1993**, *58*, 6182. (b) Aubé, J.; Heppert, J. A.; Milligan, M. L.; Smith, M. J.; Zenk, P. *JOC* **1992**, *57*, 3563.
35. Alexakis, A.; Mangeney, P.; Marek, I.; Rose-Munch, F.; Rose, E.; Semra, A.; Robert, F. *JACS* **1992**, *114*, 8288.
36. Uemura, M.; Nishimura, H.; Hayashi, T. *TL* **1993**, *34*, 107.

37. (a) Uemura, M.; Miyake, R.; Nakayama, K.; Shiro, M.; Hyashi, Y. *JOC* **1993**, *58*, 1238. (b) Uemura, M.; Miyake, R.; Nishimura, H.; Matsumoto, Y.; Hayashi, T. *TA* **1992**, *3*, 213. (c) Heaton, S. B.; Jones, G. B. *TL* **1992**, *33*, 1693.

Motokazu Uemura
Osaka City University, Japan

Hexamethyldisilazane

$Me_3Si-NH-SiMe_3$

[999-97-3] $C_6H_{19}NSi_2$ (MW 161.44)

(selective silylating reagent;[1] aminating reagent; nonnucleophilic base[2])

Alternate Name: HMDS.
Physical Data: bp 125 °C; *d* 0.765 g cm^{-3}.
Solubility: sol acetone, benzene, ethyl ether, heptane, perchloroethylene.
Form Supplied in: clear colorless liquid; widely available.
Purification: may contain trimethylsilanol or hexamethyldisiloxane; purified by distillation at ambient pressures.
Handling, Storage, and Precautions: may decompose on exposure to moist air or water, otherwise stable under normal temperatures and pressures. Harmful if swallowed, inhaled, or absorbed through skin. Fire hazard when exposed to heat, flames, or oxidizers. Use in a fume hood.

Silylation. Alcohols,[3] amines,[3] and thiols[4] can be trimethylsilylated by reaction with hexamethyldisilazane (HMDS). Ammonia is the only byproduct and is normally removed by distillation over the course of the reaction. Hydrochloride salts, which are typically encountered in silylation reactions employing chlorosilanes, are avoided, thereby obviating the need to handle large amounts of precipitates. Heating alcohols with hexamethyldisilazane to reflux is often sufficient to transfer the trimethylsilyl group (eq 1).[5] Completion of the reaction is indicated by either a change in the reflux temperature (generally a rise) or by the cessation of ammonia evolution.

$$HN(TMS)_2 + 2\,ROH \xrightarrow{\Delta} 2\,ROTMS + NH_3 \quad (1)$$

Silylation with HMDS is most commonly carried out with acid catalysis.[5] The addition of substoichiometric amounts of ***Chlorotrimethylsilane*** (TMSCl) to the reaction mixtures has been found to be a convenient method for catalysis of the silylation reaction.[5,6] The catalytically active species is presumed to be hydrogen chloride, which is liberated upon reaction of the chlorosilane with the substrate. Alternatively, protic salts such as ammonium sulfate can be employed as the catalyst.[7] Addition of catalytic ***Lithium Iodide*** in combination with TMSCl leads to even greater reaction rates.[8] Anilines can be monosilylated by heating with excess HMDS (3 equiv) and catalytic TMSCl and catalytic LiI (eq 2). Silylation occurs without added LiI; however, the reaction is much faster in the presence of iodide, presumably due to the in situ formation of a catalytic amount of the more reactive ***Iodotrimethylsilane***.

(2) [aniline NH_2 + 3 HMDS, cat. TMSCl, cat. LiI, reflux → NH–TMS]

R = H, alkyl, halogen

Hexamethyldisilazane is the reagent of choice for the direct trimethylsilylation of amino acids, for which TMSCl cannot be used due to the amphoteric nature of the substrate.[9] Silylation of glutamic acid with excess hexamethyldisilazane and catalytic TMSCl in either refluxing xylene or acetonitrile followed by dilution with alcohol (methanol or ethanol) yields the derived lactam in good yield (eq 3).[10]

(3) [HO_2C–CH(NH_2)–CH₂CH₂–CO_2H; 1. HMDS, cat. TMSCl; 2. EtOH, 93% → HO_2C-pyrrolidinone (N–H)]

The efficiency of HMDS-mediated silylations can be markedly improved by conducting reactions in polar aprotic solvents. For example, treatment of methylene chloride solutions of primary alcohols or carboxylic acids at ambient temperatures with HMDS (0.5–1 equiv) in the presence of catalytic amounts of TMSCl (0.1 equiv) gives the corresponding silyl ether and the trimethylsilyl ester, respectively (eq 4).[1] *N*-Silylation of secondary amines occurs in preference to primary alcohols when treated with 1 equiv of HMDS and 0.1 equiv TMSCl (eq 5). The silylation of secondary amines cannot be effected in the absence of solvent.[5] Secondary and tertiary alcohols can also be silylated at ambient temperatures in dichloromethane with HMDS and TMSCl mixtures; however, stoichiometric quantities of the silyl chloride are required. Catalysis by ***4-Dimethylaminopyridine*** (DMAP) is necessary for the preparation of tertiary silyl ethers.

(4) [triol (OH, OH, OH); HMDS (1.2 equiv), TMSCl (0.1 equiv), CH_2Cl_2, 0 °C → OTMS, OH, OH]

(5) [Me–NH–CH(Me)CH₂–OH; HMDS (1.2 equiv), TMSCl (0.1 equiv), CH_2Cl_2, 0 °C → Me–N(TMS)–...–OH]

DMF is a useful solvent for HMDS-induced silylation reactions, and reaction rates 10–20 times greater than those carried out in pyridine have been reported.[11] DMSO is also an excellent solvent; however, a cosolvent such as 1,4-dioxane is required to provide miscibility with HMDS.[12]

Imidazole (ImH) catalyzes the silylation reaction of primary, secondary, and tertiary alkanethiols with hexamethyldisilazane.[12] The mechanism is proposed to involve the intermediacy of ***N-(Trimethylsilyl)imidazole*** (ImTMS), since its preparation from hexamethyldisilazane and imidazole to yield 1-(trimethylsilyl)imidazole is rapid.[13] The imidazole-catalyzed reactions of hexamethyldisilazane, however, are more efficient than the silylation reactions effected by ImTMS (eq 6 vs. eq 7) due to reversibility of the latter. Imidazole also catalyzes the reaction of HMDS with ***Hydrogen Sulfide***, which provides a convenient preparation of hexamethyldisilathiane, a reagent which has found utility in sulfur transfer reactions.[14]

SH — HMDS (2 equiv), cat. ImH, reflux, 50% → STMS (6)

SH + ImTMS ⇌ STMS + ImH (7)
15%

Silyl Enol Ethers. Silylation of 1,3-dicarbonyl compounds can be accomplished in excellent yield by heating enolizable 1,3-dicarbonyl compounds with excess HMDS (3 equiv) and catalytic imidazole (eq 8).[15]

O O — HMDS, cat. ImH, reflux, 96% → TMSO O (8)

In combination with TMSI, hexamethyldisilazane is useful in the preparation of thermodynamically favored enol ethers (eq 9).[16] Reactions are carried out at rt or below and are complete within 3 h.

O — HMDS (1.2 equiv), TMSI (1.1 equiv), pentane → OTMS + OTMS (9)
90:10

Related thermodynamic enolization control has been observed using metallated hexamethyldisilazide to give the more substituted bromomagnesium ketone enolates.[17] Metallation reactions of HMDS to yield Li, K, and Na derivatives are well known and the resulting nonnucleophilic bases have found extensive applications in organic synthesis (see ***Lithium Hexamethyldisilazide***, ***Potassium Hexamethyldisilazide***, ***Sodium Hexamethyldisilazide***).[2]

Amination Reactions. Hexamethyldisilazane is a useful synthon for ammonia in amination reactions. Preparation of primary amides by the reaction of acyl chlorides and gaseous Ammonia, for example, is not an efficient process. Treatment of a variety of acyl halides with HMDS in dichloromethane gives, after hydrolysis, the corresponding primary amide (eq 10).[18] Omitting the hydrolysis step allows isolation of the corresponding monosilyl amide.[19]

COCl — 1. HMDS; 2. MeOH, H_3O^+ → O, NH_2 (10)

Reductive aminations of ketones with HMDS to yield α-branched primary amines can be effected in the presence of ***Titanium(IV) Chloride*** (eq 11).[20] The reaction is successful for sterically hindered ketones even though HMDS is a bulky amine and a poor nucleophile. The use of ammonia is precluded in these reactions since it forms an insoluble complex with $TiCl_4$.

O, Ar Ar + HMDS — 1. $TiCl_4$, Et_3N; 2. $NaCNBH_3$ → NH_2, Ar Ar (11)

The reaction of phenols with diphenylseleninic anhydride and hexamethyldisilazane gives the corresponding phenylselenoimines (eq 12).[21] The products thus obtained can be converted to the aminophenol or reductively acetylated using ***Zinc*** and ***Acetic Anhydride***. The use of ammonia or tris(trimethylsilyl)amine in place of HMDS gives only trace amounts of the selenoimines.

OH — $(PhSeO)_2O$ (1.2 equiv), HMDS (1.2 equiv), 65% → O, NSePh (12)

Related Reagents. *N,O*-Bis(trimethylsilyl)acetamide; *N,N'*-Bis(trimethylsilyl)urea; Chlorotrimethylsilane; Cyanotrimethylsilane; Hexamethyldisiloxane; Iodotrimethylsilane; *N*-(Trimethylsilyl)imidazole.

1. Cossy, J.; Pale, P. *TL* **1987**, *28*, 6039.
2. Colvin, E. W. *Silicon in Organic Synthesis*, Butterworths: London, 1981.
3. Speier, J. L. *JACS* **1952**, *74*, 1003.
4. Bassindale, A. R.; Walton, D. R. M. *JOM* **1970**, *25*, 389.
5. Langer, S. H.; Connell, S.; Wender, I. *JOC* **1958**, *23*, 50.
6. Sweeley, C. C.; Bentley, R.; Makita, M.; Wells, W. W. *JACS* **1963**, *85*, 2497.
7. Speier, J. L.; Zimmerman, R.; Webster, J. *JACS* **1956**, *78*, 2278.
8. Smith, A. B., III; Visnick, M.; Haseltine, J. N.; Sprengeler, P. A. *T* **1986**, *42*, 2957.
9. Birkofer, L.; Ritter, A. *AG(E)* **1965**, *4*, 417.
10. Pellegata, R.; Pinza, M.; Pifferi, G. *S* **1978**, 614.
11. Kawai, S.; Tamura, Z. *CPB* **1967**, *15*, 1493.
12. Glass, R. S. *JOM* **1973**, *61*, 83.
13. Birkofer, L.; Richter, P.; Ritter, A. *CB* **1960**, *93*, 2804.
14. Harpp, D. N.; Steliou, K. *S* **1976**, 721.
15. Torkelson, S.; Ainsworth, C. *S* **1976**, 722.
16. (a) Hoeger, C. A.; Okamura, W. H. *JACS* **1985**, *107*, 268. (b) Miller, R. D.; McKean, D. R. *S* **1979**, 730.
17. Kraft, M. E.; Holton, R. A. *TL* **1983**, *24*, 1345.
18. Pellegata R.; Italia, A.; Villa, M. *S* **1985**, 517.
19. Bowser, J. R.; Williams, P. J.; Kuvz, K. *JOC* **1983**, *48*, 4111.
20. Barney, C. L.; Huber, E. W.; McCarthy, J. R. *TL* **1990**, *31*, 5547.
21. Barton, D. H. R.; Brewster, A. G.; Ley, S. V.; Rosenfeld, M. N. *CC* **1977**, 147.

Benjamin A. Anderson
Lilly Research Laboratories, Indianapolis, IN, USA

Hexamethylphosphoric Triamide

[680-31-9] $C_6H_{18}N_3OP$ (MW 179.24)

(high Lewis basicity; dipolar aprotic solvent with superb ability to form cation–ligand complexes; can enhance the rates of a wide variety of main group organometallic reactions and influence regio- or stereochemistry; additive in transition metal chemistry; UV inhibitor in poly(vinyl chloride))

Alternate Names: HMPA; hexamethylphosphoramide; hexametapol; hempa; HMPT.

Physical Data: mp 7.2 °C; bp 230–232 °C/740 mmHg; *d* 1.025 g cm^{-3}; mild amine odor.

Solubility: sol water, polar and nonpolar solvents.

Form Supplied in: water-white liquid.

Drying: distilled from CaH_2 or BaO[1] at reduced pressure and stored under N_2 over molecular sieves.

Handling, Storage, and Precautions: has low to moderate acute toxicity in mammals.[2a] Inhalation exposure to HMPA has been shown to induce nasal tumors in rats,[2b] and has been classified under 'Industrial Substances Suspect of Carcinogenic Potential for Man'.[2c] Adequate precautions must be taken to avoid all forms of exposure to HMPA.

Introduction. Hexamethylphosphoric triamide has been used extensively as an additive in organolithium chemistry.[3] It is among the strongest of electron pair donors and is superior to protic solvents in that it solvates the cation much better than the anion.[4] This coordinating ability gives HMPA its unusual chemical properties.[5,6] For instance, HMPA dramatically enhances the rates of a wide variety of organolithium reactions, as well as significantly influencing regio- or stereochemistry. The reactivity or selectivity effects of HMPA are usually rationalized in terms of changes in either aggregation state or ion pair structure.[7] The breaking up of aggregates to form reactive monomers or solvent-separated ion pairs is often invoked.

Organolithium Reagent Solution Structure. The rate of metalation reactions with ***Lithium Diisopropylamide*** as base are significantly increased through the use of HMPA.[8] Treatment of LDA dimer with HMPA causes sequential solvation of the lithium cation, but no significant deaggregation;[8,9] nor does HMPA promote the break up of tetrameric unhindered phenoxides[10,11] or tetrameric MeLi.[12] A chiral bidentate lithium amide,[13] however, was converted from a dimer to a monomer by HMPA, with an increase in reactivity and enantioselectivity in deprotonation (eq 1). HMPA also converts ***Phenyllithium*** dimer into monomer.[12,14] Other aggregated lithium reagents[12] in THF (MeSLi, LiCl) or ether (Ph_2PLi, PhSeLi) are first deaggregated to monomers, and then solvent-separated ion pair species are formed.[12,15] HMPA may exert its reactivity effects by a combination of one or more of the following:

1. lowering the degree of aggregation[16,17] or forming separated ions,[18]
2. increasing reactivity through cation coordination,[17]
3. activating the aggregate through insertion into the aggregate site normally occupied by the anionic fragment,[5,19]
4. promoting triple ion (ate complex) formation.[16a,20]

HMPA; ether or toluene (1)

HMPA can have large effects on equilibria. Phenyllithium reacts with ***Diphenylmercury***, ***Diphenyl Ditelluride***, and iodobenzene in THF to form ate complexes. In THF/HMPA the ate complex formation constants are dramatically higher than in THF.[14,21]

Enolate Formation. The formation of lithium enolates is one instance where HMPA is sometimes needed.[22] The difficult generation of dimethyl tartrate acetonide enolate[23] and subsequent benzylation (eq 2), as well as the double deprotonation of methyl 3-nitropropanoate,[24] become possible (eq 3) with the addition of HMPA (or ***N,N'-Dimethylpropyleneurea***).[25]

LDA, THF, –78 °C; BnBr (2)

Cosolvent (% vol)	Yield (%)
–	<5
HMPA (17)	54
DMPU (33)	52

2 equiv LDA, THF, –78 °C (3)

Cosolvent (% vol)	Yield (%)
–	<5
HMPA (17)	50–85
DMPU (33)	50–85

Enolate Reactivity. Not only is HMPA necessary to the generation of enolates, it is often needed in the electrophilic trapping of enolates (eqs 4–6).[26,31] Studies of the electrophilic trapping of enolates have demonstrated that substantial increases in reaction rates can be achieved through the use of a polar aprotic solvent such as HMPA.[4]

HMPA (4)

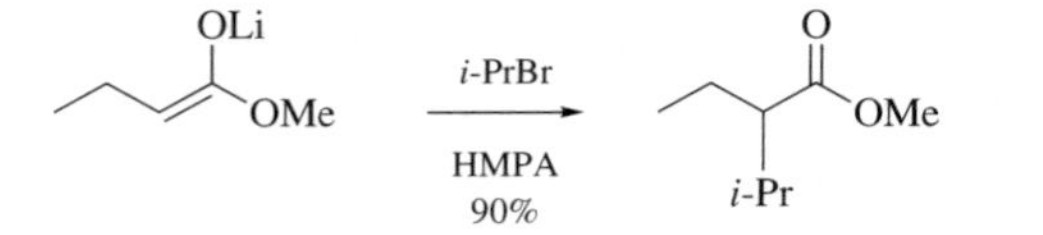

(5)

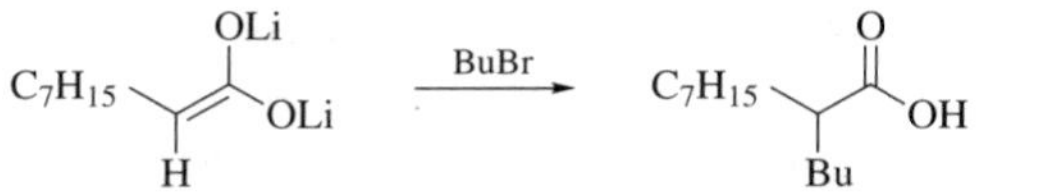

(6)

Solvent	Yield
THF	43%
THF–HMPA	93%

The desired [2,3]-sigmatropic rearrangement of a bis-sulfur cyano-stabilized lithium salt did not proceed (eq 7) without the addition of 25% HMPA.[27] Sometimes higher *O*/*C*-alkylation ratios are obtained in THF–HMPA.[28]

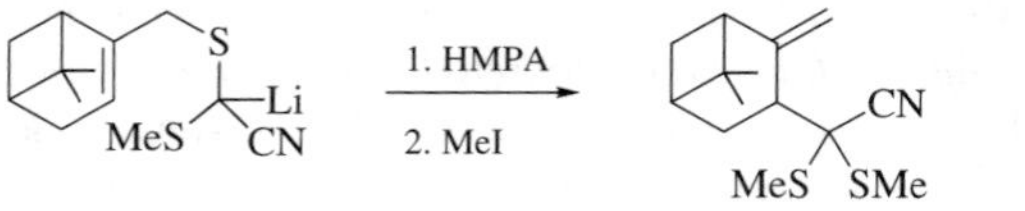

(7)

Enolate Stereochemistry. Stereochemical control of an ester enolate Claisen rearrangement was accomplished through stereoselective enolate formation.[29] The enolization of 3-pentanone with LDA afforded predominantly the (*E*)-enolate in THF and the (*Z*)-enolate in THF–HMPA, as shown by chlorotrialkylsilane trapping experiments (eq 8). Similar stereoselectivity (Z:E = 94:6) was obtained with the dipolar aprotic cosolvent DMPU.[30]

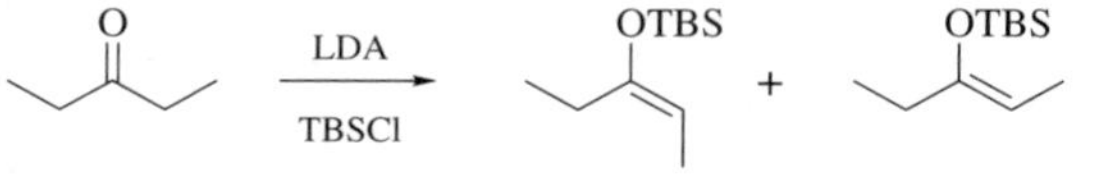

(8)

THF	77	:	23
THF–HMPA	5	:	95

In addition to altering the (*E*/*Z*) isomer ratio of enolates,[17,32] HMPA has a noticeable effect on the metalation of imines and their subsequent alkylation (eq 9).[33] When the metalation (by ***s*-Butyllithium**) of an asymmetric imine is performed in THF, a subsequent alkylation gives about a 1:1 mixture of regioisomers. In the presence of HMPA, however, only the regioisomer due to alkylation at the less-substituted site was observed. A synthetically useful solvent effect for HMPA is also observed in the asymmetric synthesis of trimethylsilyl enol ethers by chiral lithium amide bases.[34] The asymmetric induction in THF can be greatly improved by simply adding HMPA as a cosolvent.

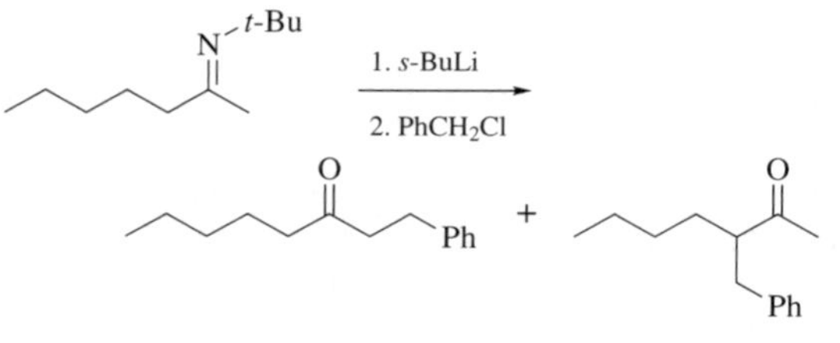

(9)

Carbanion Formation. Often substrates that cannot be metalated by LDA or ***n*-Butyllithium** in THF can be successfully deprotonated by adding HMPA as a cosolvent. Many other weakly acidic C–H acids, e.g. (**1**)–(**6**), can be successfully metalated in the presence of HMPA.[35] HMPA also aids in the formation of dianions[36] and increases the proton abstraction efficiency of ***Sodium Hydride***.[37,38]

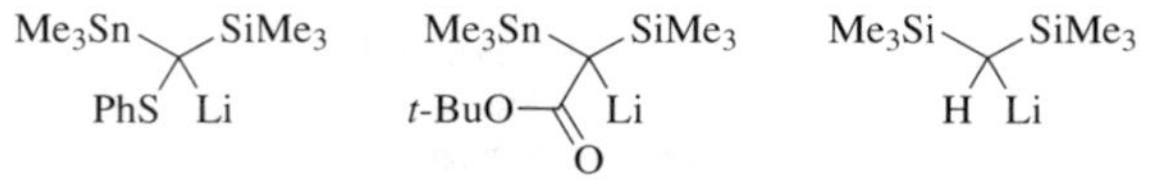

(1) (2) (3)

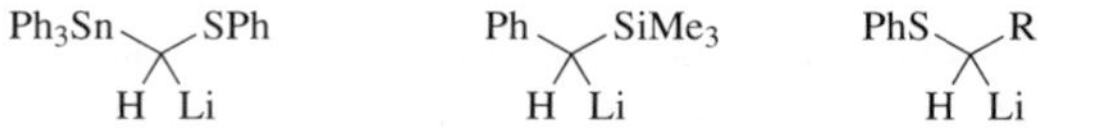

(4) (5) (6)

Carbanion Reactivity. An increase in reaction rate is observed for the reaction of alkynyllithium reagents with alkyl halides[3] and oxiranes[39] (eq 10). The strongly coordinating HMPA probably complexes the lithium cation, thereby increasing the negative charge density on the carbon and creating a much more nucleophilic alkynyl anion. A similar effect is observed for ***(Trimethylstannylmethyl)lithium***, which does not react with oxiranes in THF but in THF–HMPA the reaction proceeds readily.[40]

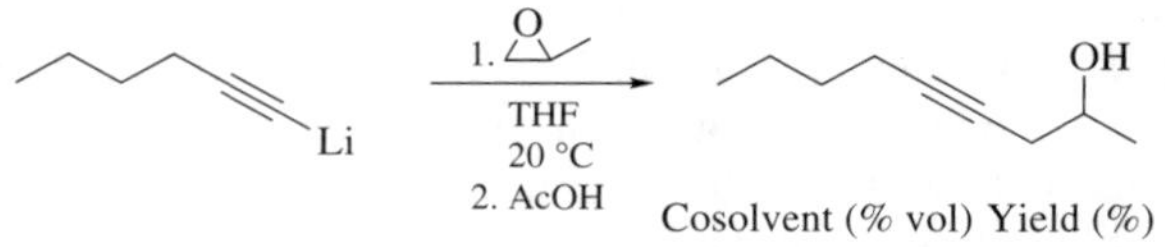

(10)

Cosolvent (% vol)	Yield (%)
–	<5
HMPA (17)	70
DMPU (50)	59

The decarboxylation[41] of 4-*t*-butyl-1-phenyl-1-carboxycyclohexane with ***Methyllithium*** gave a mixture of axial and equatorial products (eq 11), which was highly dependent on the nature of the solvent at the time of aqueous workup. Axial protonation was favored in ether–HMPA.

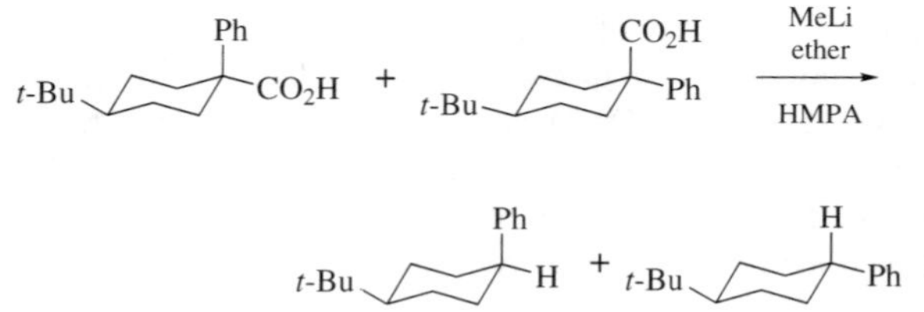

(11)

HMPA is also the only dipolar aprotic solvent to be used extensively with organomagnesium compounds.[42] Large effects are observed when HMPA is used as either a solvent or a cosolvent. As examples, HMPA accelerates addition of an allylic organomagnesium compound to aryl-substituted alkenes,[43] addition of Grignard reagents to ***Carbon Monoxide***,[44] and addition of ***Propargylmagnesium Bromide*** to allylic halides to give allene products.[45]

Carbanion Regioselectivity (1,2- vs. 1,4-Addition). The regioselectivity of addition of certain organolithium reagents to α,β-unsaturated carbonyl compounds is affected by the addition of HMPA. In the addition of 2-lithio-2-substituted-1,3-dithiane to cyclohexenone, there was a complete reversal of regioselectivity from 1,2-addition in THF to 1,4-addition with 2 equiv of HMPA present (eq 12).[46]

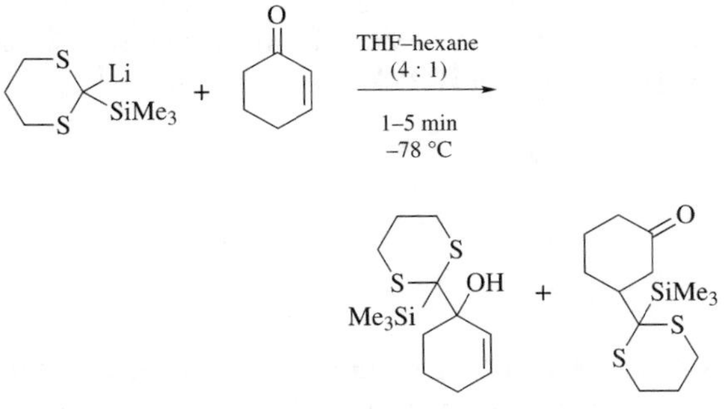

THF–hexane (4 : 1), 1–5 min, –78 °C (12)

Additive	1,2-Addition	1,4-Addition
no HMPA •	>99	<1
HMPA (2 equiv)	5	95

Lithium reagents that exhibit kinetic 1,4-addition in HMPA are shown as (**7**)–(**14**).[47] These include useful acyl anion equivalents[48] like phenylthio(trimethylsilyl)methyllithium (see ***(Phenylthiomethyl)trimethylsilane***).[49]

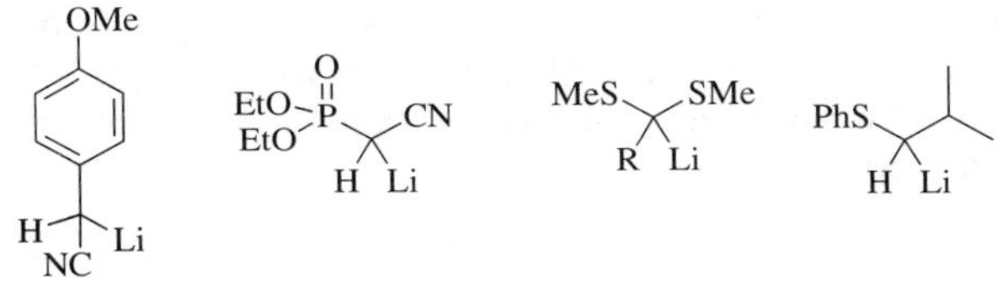

(**7**) (**8**) (**9**) (**10**)

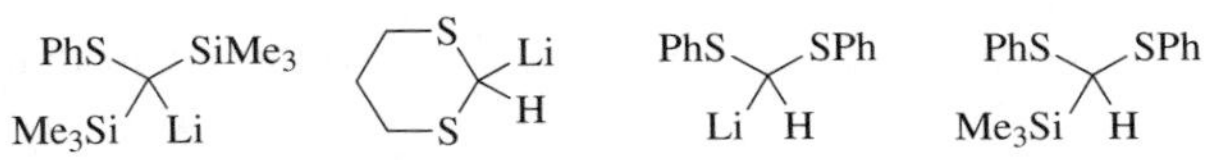

(**11**) (**12**) (**13**) (**14**)

A carboxy anion equivalent was reported to undergo 1,2-addition in the absence of HMPA;[50] however, with 10 equiv of HMPA present only 1,4-addition was observed (eq 13). The addition of 1 equiv of HMPA promotes conjugate addition of alkyl and phenylthioallyl anions to cyclopentanones (eq 14) through the α-position, whereas in THF alone, irreversible 1,2-addition occurs with both α- and γ-attack.[51] The regioselectivities reported for the addition to cyclic enones of ketene dithioacetal anions[47a,52] or ***t-Butyllithium*** (eq 15)[53] are also influenced by HMPA (and counterion).

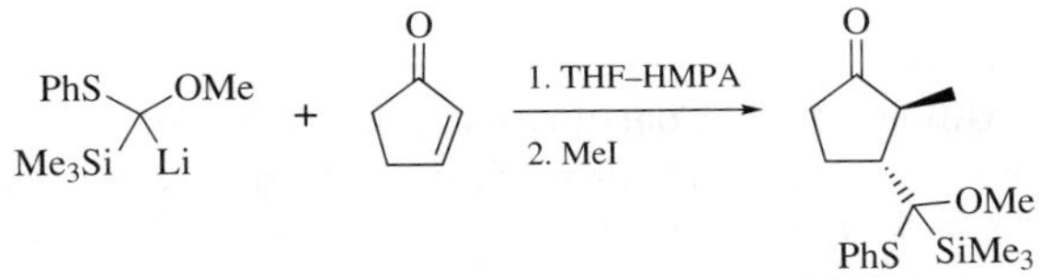

1. THF–HMPA, 2. MeI (13)

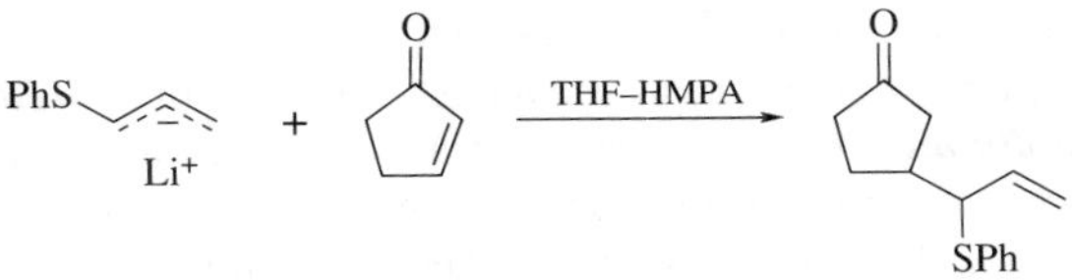

THF–HMPA (14)

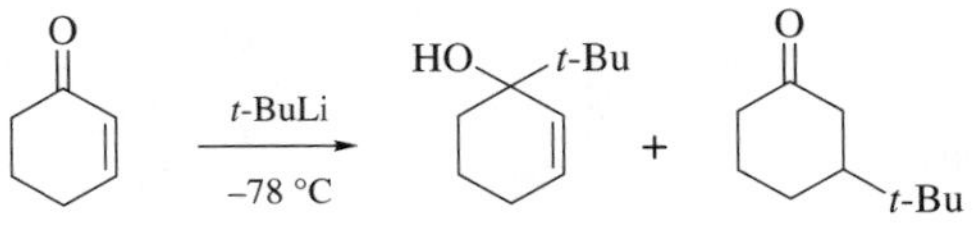

t-BuLi, –78 °C (15)

Solvent		
ether	100	–
THF	95	5
THF–5% HMPA	65	35
THF–20% HMPA	10	90

Ylide Reactivity. HMPA is used as a cosolvent in the Wittig reaction to increase reaction rate, yield, and stereoselectivity. HMPA functions as a lithium cation-complexing agent and removes LiBr salt from ether solution (LiBr/HMPA complexes form precipitates in ether).[54] Such a 'salt-free' Wittig reaction mixture[55] may be responsible for the high level of *cis*-alkene observed in reactions of nonstabilized ylides with aldehydes in THF or ether with added HMPA[25,56] (eq 16). Similarly, increased (*Z*) selectivity is observed in the Wittig alkenation of 2-oxygenated ketones (eq 17) to generate protected (*Z*)-trisubstituted allylic alcohols.[55,57]

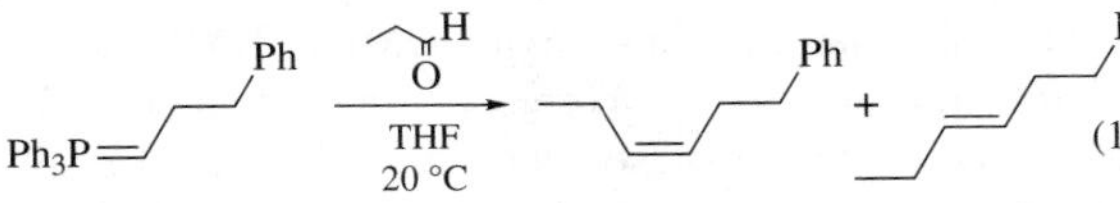

THF, 20 °C (16)

Cosolvent (% vol)	Yield (%)	(*Z*)		(*E*)
–	46	83	:	17
HMPA (35)	44	92	:	8
DMPU (35)	39	93	:	7

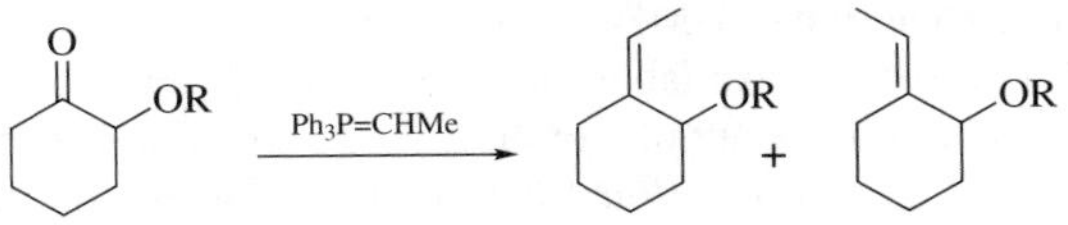

$Ph_3P=CHMe$ (17)

	THF		THF-HMPA	
R	(*E*):(*Z*)	Yield (%)	(*E*):(*Z*)	Yield (%)
Me	2.6:1	50	8:1	76
Ph	5:1	77	10:1	84
TBS	7.6:1	61	26:1	90
Bz	1.1:1	55	36:1	86

With HMPA, Wittig reactions that give (*E*)-alkenes were also observed (eq 18),[58] as was the directed selectivity of a semistabilized arsonium ylide towards carbonyl compounds. The arsenic ylide was generated from LDA in THF or THF/HMPA solution to give exclusively epoxide (eq 19) or diene (eq 20), respectively.[59]

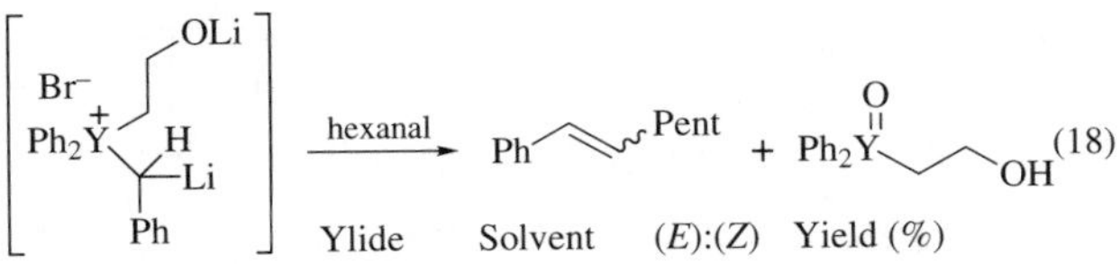

hexanal (18)

Ylide	Solvent	(*E*):(*Z*)	Yield (%)
Y = P	THF	65:35	91
Y = P	THF–HMPA	85:15	87
Y = As	THF	–	–
Y = As	THF–HMPA	100:0	65

(19)

(20)

Nucleophilic Cleavage of Esters and Ethers. The conversion of hindered methyl esters to carboxylic acids, and the demethylation of methyl aryl ethers, can be effectively performed by using HMPA to increase the nucleophilicity of lithium methanethiolate.[60] HMPA (or ***N,N-Dimethylformamide***) will also facilitate the cleavage of methyl aryl ethers and their methylthio analogs by sodium methaneselenolate (see ***Methaneselenol***) to give the phenol or thiophenol, respectively.[60,61] Sodium ethanethiolate (see ***Ethanethiol***) in refluxing DMF or HMPA attacks alkyl aryl selenides to give the corresponding diselenides. The most reactive combination of solvent and halide salt for the decarboxylation of β-keto esters was found to be ***Lithium Chloride***/HMPA, used as part of a stereoselective synthesis of 11-deoxyprostaglandin E_1.[62] In HMPA, the rate of ester cleavage of 2-benzyl-2-methoxycarbonyl-1-cyclopentanone with ***Sodium Cyanide*** is 30 times as fast as the more commonly used DMF.[62a]

Anion Reactivity. HMPA is one of the most potent electron pair donor solvents available for accelerating S_N2 reactions.[3,43a] The formation, for example, of an α-silyl carbanion for use in a Peterson alkenation reaction[63] can be accomplished by the displacement of silicon using ***Sodium Methoxide***[64] or ***Potassium t-Butoxide***[65] in HMPA (eq 21). The increased nucleophilicity of halide ions in the presence of HMPA is seen by the increased rate of silyl-protecting group removal with fluoride ion (***Tetra-n-butylammonium Fluoride***).[66] The substitution of aryl chlorides can be performed using sodium methoxide in HMPA[67] to give anisole derivatives, or by using sodium methanethiolate (MeSNa) (see ***Methanethiol***) in HMPA[61b,68] to generate either aryl methyl sulfides or aryl thiols, depending on the reaction conditions. The increased nucleophilicity of a magnesium alkoxide is demonstrated by the cyclization of a chloro alcohol to a 13-membered cyclic ether (eq 22) upon treatment of the compound with ***Ethylmagnesium Bromide*** in refluxing THF/HMPA.[69]

(21)

(*E*):(*Z*) = 52:1

(22)

The conversion of cyclic alkenes to 1,3-cycloalkadienes can be performed through a bromination/dehydrobromination procedure using LiCl/***Lithium Carbonate***/HMPA (eq 23).[70] The dehydrobromination of 2,3-dibromo-3-methyl-1-butanol to generate a vinyl bromide has been accomplished through the use of 2.3 equiv of LDA and 0.5 equiv of HMPA in THF at −78 °C.[71]

(23)

Low-Valent Metal Coordination (Lanthanoids). HMPA is used extensively as a solvent for dissolving-metal reductions.[72] Used as a cosolvent (5–10%), it remarkably accelerates the one-electron transfer reduction of organic halides by ***Samarium(II) Iodide***.[73] The reductions work on a variety of primary, secondary, or tertiary halides, including chlorides which could not be reduced in pure THF (eq 24). The samarium Barbier reaction, which requires hours in refluxing THF, can be performed as a titration in THF–HMPA at rt (eq 25).[74] The SmI_2/THF/HMPA system has recently been used in the deoxygenation of organoheteroatom oxides,[75] reductive dimerization of conjugated acid derivatives,[76] and selective reduction of α,β-unsaturated carbonyl compounds.[77] In addition, SmI_2 was used in a tandem radical cyclization,[78] and has been a useful reagent in Barbier-type reactions.[73b,79] The dramatic acceleration of electron transfer is also observed for ***Ytterbium(0)***.[80]

R–X → R–H (SmI_2, THF–HMPA, near quant.) (24)

X	THF Time	THF Yield	THF–HMPA Time	THF–HMPA Yield
I	6 h	95%	5 min	>95%
Br	2 days	82%	10 min	>95%
Cl	no reaction		8 h	>95%

(25)

Additive	Time	Yield
none	4 h	82%
HMPA	1 min	95%

Transition Metal Coordination. The ability of HMPA to complex to metals and alter reactivity is also expressed in palladium-catalyzed coupling reactions. For example, ethylbenzene can be formed from the Pd^0 catalyzed cross-coupling of ***Benzyl Bromide*** and ***Tetramethylstannane***, with the formation of almost no bibenzyl.[81] The reaction does not proceed in THF, but requires a highly polar solvent like HMPA or ***1-Methyl-2-pyrrolidinone***. Similarly, HMPA is necessary in the Pd^0-catalyzed coupling of acid chlorides with organotin reagents to give ketones. In highly polar solvents like HMPA the transfer of a chiral group from the tin occurs with preferential inversion of configuration (eq 26).[82] The relative rate of transfer of an alkynic or vinyl group to acid chlorides[82a] or aryl iodides[83] is also greatly accelerated by HMPA. It can increase the rate of alkylation of π-allylpalladium chloride by ester enolates,[84] and alter the chemoselectivity by leading to a cyclopropanation reaction instead of an allylic alkylation.[85]

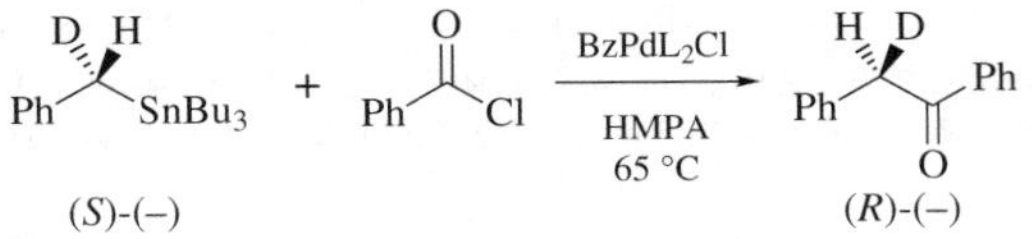

(S)-(−) + PhCOCl —($BzPdL_2Cl$, HMPA, 65 °C)→ (R)-(−) (26)

Hydride Reductions. The reduction of organic compounds by hydride can be influenced by the choice of cosolvent. Cyanoborohydrides (e.g. ***Sodium Cyanoborohydride***) in HMPA provide a mild, effective, and selective reagent system for the reductive displacement of primary and secondary alkyl halides and sulfonate esters in a wide variety of structural types.[86] A ***Sodium Borohydride***/HMPA reagent system was used in the reduction of *N,N*-disulfonamides[87] and in the reduction of dibromides to monobromides.[37a] Similarly, ***Tri-n-butylstannane***/HMPA[88] can be utilized to chemoselectively reduce aldehydes with additional alkene or halide functionality.[89] Finally, the reduction of aldehydes and ketones with hydrosilanes proceeds in the presence of a catalytic amount of Bu_4NF in HMPA.[90] The reaction rate was much lower in DMF than in HMPA, and the reaction yields decreased considerably if less polar solvents like THF or CH_2Cl_2 were used.

Oxidation. The oxidation of acid sensitive alcohols with ***Chromium(VI) Oxide*** in HMPA is one example of the use of HMPA in oxidation reactions.[91] (CAUTION: Do not crush CrO_3 prior to reaction since violent decomposition can occur. The use of DMPU has been reported to have a similar hazardous effect). Recently, a ***Bromine***/$NaHCO_3$/HMPA system[92] was used for the oxidative esterification of alcohols with aldehydes, where the HMPA considerably accelerated the oxidation by bromine and lowered the rate of unwanted halogenation. Epoxidations of alkenes or allylic alcohols have been accomplished using MoO_5·HMPA·pyridine (***Oxodiperoxymolybdenum(pyridine)(hexamethylphosphoric triamide)***; MoOPH).[93]

Effect of HMPA on Protonation. The protonation of (9-anthryl)arylmethyllithium with various oxygen and carbon acids in THF or in THF–HMPA had a significant effect on the product ratio of C-α vs. C-10 protonation (eq 27).[94] Another study[95] found that a nitronate protonation led to mainly one diastereomeric product in a THF solution containing HMPA or DMPU (eq 28). Panek and Rodgers[96] observed stereospecific protonation of 10-*t*-butyl-9-methyl-9-lithio-9,10-dihydroanthracene: >99% *cis* protonation was observed in THF or ether with greater than >99% *trans* protonation observed in HMPA.

Inhibition by HMPA. Finally, it should be added that there are a few cases where HMPA slows the rate of a reaction. Such examples typically involved the inhibition of lithium catalysis by strong coordination of HMPA to lithium.[97] For instance, two-bond ^{13}C–^{13}C NMR coupling in organocuprates is poorly observable in ether or THF at very low temperature.[98] Exchange, however, is slowed in THF/HMPA or THF/12-crown-4, so that coupling is easily observed. The effect of HMPA suggests that Li^+ is involved in the exchange process.

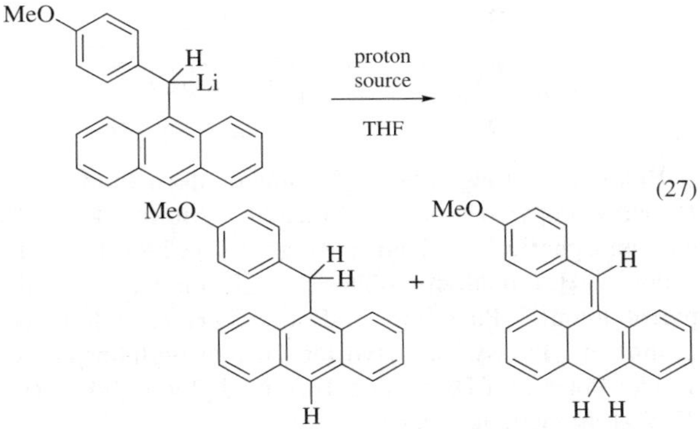

(27) — proton source, THF

Additive	Proton. at C-α	Proton. at C-10
none	57	43
HMPA (12 equiv)	>95	<5

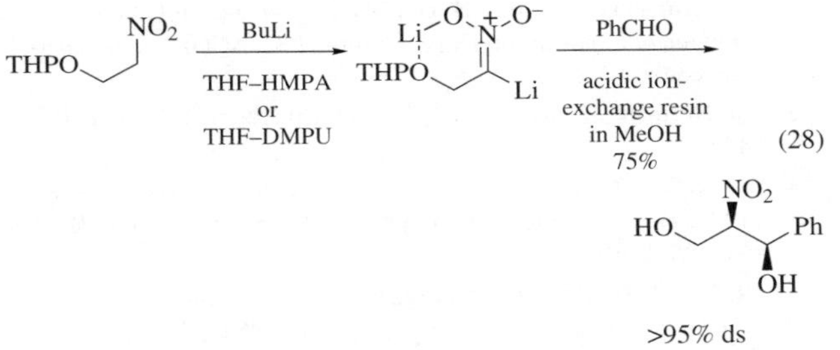

(28) — BuLi, THF–HMPA or THF–DMPU; PhCHO, acidic ion-exchange resin in MeOH, 75%; >95% ds

HMPA Substitutes and Analogs. Researchers have searched for an alternative to HMPA. Such a solvent must be stable to polar organometallic compounds and be comparable to HMPA in its many functions. Replacement solvents typically are useful in some applications but have limited value in others. Examples of some useful alternatives are (**15**)–(**19**).[30,42,99]

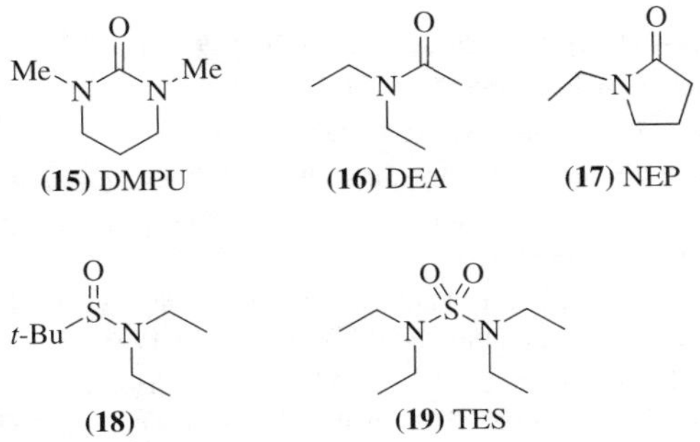

(**15**) DMPU (**16**) DEA (**17**) NEP

(**18**) (**19**) TES

Chiral analogs of HMPA, (**20**)–(**22**), have been used as ligands in transition metal complexes.[93b,100]

(20) (21)

(22)

Related Reagents. *N,N*-Dimethylformamide; *N,N′*-Dimethylpropyleneurea; Dimethyl Sulfoxide; Hexamethylphosphoric Triamide–Thionyl Chloride; Lithium Chloride–Hexamethylphosphoric Triamide; 1-Methyl-2-pyrrolidinone; Potassium *t*-Butoxide–Hexamethylphosphoric Triamide; Potassium Hydride–Hexamethylphosphoric Triamide; Potassium Hydroxide–Hexamethylphosphoric Triamide; *N,N′*-Tetramethylethylenediamine.

---

1. House, H. O.; Lee, T. V. *JOC* **1978**, *43*, 4369.
2. (a) Kimbrough, R. D.; Gaines, T. B. *Nature* **1966**, *211*, 146. Shott, L. D.; Borkovec, A. B.; Knapp, W. A., Jr. *Toxicol. Appl. Pharmacol.* **1971**, *18*, 499. (b) *J. Natl. Cancer Inst.* **1982**, *68*, 157. (c) Mihal, C. P., Jr. *Am. Ind. Hyg. Assoc. J.* **1987**, *48*, 997. American Conference of Governmental Industrial Hygienists: *TLVs-Threshold Limit Values and Biological Exposure Indices for 1986–1987*; ACGIH: Cincinnati, OH; Appendix A2, p 40.
3. (a) Normant, H. *AG(E)* **1967**, *6*, 1046. (b) Normant, H. *RCR* **1970**, *39*, 457.
4. Stowell, J. C. *Carbanions in Organic Synthesis*; Wiley: New York, 1979: House, H. O. *Modern Synthetic Reactions*, 2nd ed.; Benjamin: New York, 1972.
5. Reichardt, C. *Solvent and Solvent Effects in Organic Chemistry*; VCH: Germany, 1988; p. 17. Seebach, D. *AG(E)* **1988**, *27*, 1624.
6. (a) Gutmann, V. *The Donor–Acceptor Approach to Molecular Interactions*; Plenum: New York, 1978. (b) Maria, P.-C.; Gal, J.-F. *JPC* **1985**, *89*, 1296. Maria, P.-C.; Gal, J.-F.; de Franceschi, J.; Fargin, E. *JACS* **1987**, *109*, 483. (c) Persson, I.; Sandström, M.; Goggin, P. L. *ICA* **1987**, *129*, 183.
7. Newcomb, M.; Varick, T. R.; Goh, S.-H. *JACS* **1990**, *112*, 5186. Seebach, D.; Amstutz, R.; Dunitz, J. D. *HCA* **1981**, *64*, 2622.
8. Romesburg, F. E.; Gilchrist, J. H.; Harrison, A. T.; Fuller, D. J.; Collum, D. B. *JACS* **1991**, *113*, 5751 (see also Ref 1). Romesberg, F. E.; Collum, D. B. *JACS* **1992**, *114*, 2112. Romesberg, F. E.; Bernstein, M. P.; Gilchrist, J. H.; Harrison, A. T.; Fuller, D. J.; Collum, D. B. *JACS* **1993**, *115*, 3475.
9. Galiano-Roth, A. S.; Collum, D. B. *JACS* **1989**, *111*, 6772.
10. Seebach, D. *AG(E)* **1988**, *27*, 1624.
11. Jackman, L. M.; Chen, X. *JACS* **1992**, *114*, 403.
12. Reich, H. J.; Borst, J. P.; Dykstra, R. R. *JACS* **1993**, *115*, 8728.
13. Sato, D.; Kawasaki, H.; Shimada, I.; Arata, Y.; Okamura, K.; Date, T.; Koga, K. *JACS* **1992**, *114*, 761.
14. Reich, H. J.; Green, D. P.; Phillips, N. H. *JACS* **1989**, *111*, 3444. Reich, H. J.; Green, D. P.; Phillips, N. H.; Borst, J. P. *PS* **1992**, *67*, 83.
15. Hogen-Esch, T. E.; Smid, J. *JACS* **1966**, *88*, 307. Hogen–Esch, T. E.; Smid, J. *JACS* **1966**, *88*, 318. Smid, J. *Ions and Ion Pairs in Organic Reactions*; Szwarc, M., Ed., Wiley: New York, 1972; Vol. 1, pp 85–151. O'Brien, D. H.; Russell, C. R.; Hart, A. J. *JACS* **1979**, *101*, 633. Grutzner, J. B.; Lawlor, J. M.; Jackman, L. M. *JACS* **1972**, *94*, 2306. Bartlett, P. D.; Goebel, C. V.; Weber, W. P. *JACS* **1969**, *91*, 7425. Seebach, D.; Siegel, H.; Gabriel, J.; Hässig, R. *HCA* **1980**, *63*, 2046. Seebach, D.; Hässig, R.; Gabriel, J. *HCA* **1983**, *66*, 308. Schmitt, B. J.; Schulz, G. V. *Eur. Polym. J.* **1975**, *11*, 119.
16. (a) Jackman, L. M.; Scarmoutzos, L. M.; Porter, W. *JACS* **1987**, *109*, 6524. Fraser, R. R.; Mansour, T. S. *TL* **1986**, *27*, 331. (b) House, H. O.; Prabhu, A. V.; Phillips, W. V. *JOC* **1976**, *41*, 1209.
17. Jackman, L. M.; Lange, B. C. *JACS* **1981**, *103*, 4494.
18. Corset, J.; Froment, F.; Lautie, M.-F.; Ratovelomanana, N.; Seyden-Penne, J.; Strzalko, T.; Roux-Schmitt, M.-C. *JACS* **1993**, *115*, 1684.
19. Raithby, P. R.; Reed, D.; Snaith, R.; Wright, D. S. *AG(E)* **1991**, *30*, 1011. Nudelman, N. S.; Lewkowicz, E.; Furlong, J. J. P. *JOC* **1993**, *58*, 1847.
20. Reich, H. J.; Gudmundsson, B. Ö.; Dykstra, R. R. *JACS* **1992**, *114*, 7937. Jackman, L. M.; Scarmoutzos, L. M.; Smith, B. D.; Williard, P. G. *JACS* **1988**, *110*, 6058. Barr, D.; Doyle, M. J.; Drake, S. R.; Raithby, P. R., Snaith, R.; Wright, D. S. *CC* **1988**, 1415. Fraenkel, G.; Hallden-Abberton, M. P. *JACS* **1981**, *103*, 5657.
21. Reich, H. J.; Green, D. P.; Phillips, N. H. *JACS* **1991**, *113*, 1414.
22. Tsushima, K.; Araki, K.; Murai, A. *CL* **1989**, 1313.
23. Naef, R.; Seebach, D. *AG(E)* **1981**, *20*, 1030.
24. Seebach, D.; Henning, R.; Mukhopadhyay, T. *CB* **1982**, *115*, 1705.
25. Mukhopadhyay, T.; Seebach, D. *HCA* **1982**, *65*, 385.
26. Piers, E.; Tse, H. L. A. *TL* **1984**, 3155. Cregge, R. J.; Herrman, J. L.; Lee, C. S.; Richman, J. E.; Schlessinger, R. H. *TL* **1973**, 2425. Chapdelaine, M. J.; Hulce, M. *OR* **1990**, *38*, 225. Kurth, M. J.; O'Brien, M. J. *JOC* **1985**, *50*, 3846. Odic, Y.; Pereyre, M. *JOM* **1973**, *55*, 273.
27. Snider, B. B.; Hrib, N. J.; Fuzesi, L. *JACS* **1976**, *98*, 7115.
28. Kurts, A. L.; Genkina, N. K.; Macias, I. P.; Beletskaya, I. P.; Reutov, O. A. *T* **1971**, *27*, 4777.
29. Ireland, R. E.; Mueller, R. H.; Williard, A. K. *JACS* **1976**, *98*, 2868.
30. Smrekar, O. *C* **1985**, *39*, 147.
31. Pfeffer, P. E.; Silbert, L. S. *JOC* **1970**, *35*, 262.
32. Corey, E. J.; Gross, A. W. *TL* **1984**, *25*, 495.
33. Hosomi, A.; Araki, Y.; Sakurai, H. *JACS* **1982**, *104*, 2081.
34. Shirai, R.; Tanaka, M.; Koga, K. *JACS* **1986**, *108*, 543.
35. Chan, T. H.; Chang, E. *JOC* **1974**, *39*, 3264. Chan, T. H.; Chang, E.; Vinokur, E. *TL* **1970**, 1137; Dolak, T. M.; Bryson, T. A. *TL* **1977**, 1961. Kauffmann, T. *AG(E)* **1982**, *21*, 410. Kawashima, T.; Iwama, N.; Okazaki, R. *JACS* **1993**, *115*, 2507. Grobel, B.-T.; Seebach, D. *CB* **1977**, *110*, 867. Grobel, B.-T.; Seebach, D. *AG(E)* **1974**, *13*, 83. Ager, D. J.; Cooke, G. E.; East, M. B.; Mole, S. J.; Rampersaud, A.; Webb, V. J. *OM* **1986**, *5*, 1906. Zapata, A.; Fortoul, C. R.; Acuna, C. A. *SC* **1985**, 179. Van Ende, D.; Cravador, A.; Krief, A. *JOM* **1979**, *177*, 1.
36. Bryson, T. A.; Roth, G. A.; Jing-hau, L. *TL* **1986**, *27*, 3685.
37. Caubere, P. *AG(E)* **1983**, *22*, 599.
38. Corey, E. J.; Weigel, L. O.; Chamberlin, A. R.; Lipshutz, B. *JACS* **1980**, *102*, 1439. Smith, A. B., III; Ohta, M.; Clark, W. M.; Leahy, J. W. *TL* **1993**, *34*, 3033.
39. Doolittle, R. E. *OPP* **1980**, *12*, 1. Oehlschlager, A. C.; Czyzewska, E.; Aksela, R.; Pierce, H. D., Jr. *CJC* **1986**, *64*, 1407. Merrer, Y. L.; Gravier-Pelletier, C.; Micas-Languin, D.; Mestre, F.; Dureault, A.; Depezay, J.-C. *JOC* **1989**, *54*, 2409.
40. Murayama, E.; Kikuchi, T.; Sasaki, K.; Sootome, N.; Sato, T. *CL* **1984**, 1897.
41. Gilday, J. P.; Paquette, L. A. *TL* **1988**, *29*, 4505.
42. Richey, H. G., Jr.; Farkas, J., Jr. *JOC* **1987**, *52*, 479. Marczak, S.; Wicha, J. *TL* **1993**, *34*, 6627.
43. (a) Luteri, G. F.; Fork, W. T. *JOC* **1977**, *42*, 820. (b) Luteri, G. F.; Ford, W. T. *JOM* **1976**, *105*, 139.
44. Sprangers, W. J. J. M.; Louw, R. *JCS(P2)* **1976**, 1895.
45. Harding, K. E.; Cooper, J. L.; Puckett, P. M. *JACS* **1978**, *100*, 993.
46. Brown, C. A.; Yamaichi, A. *CC* **1979**, 100.

47. (a) Zieglar, F. E.; Tam, C. C. *TL* **1979**, 4717. (b) Wartski, L.; El-Bouz, M.; Seyden-Penne, J. *JOM* **1979**, *177*, 17. Wartski, L.; El-Bouz, M.; Seyden-Penne, J.; Dumont, W.; Krief, A. *TL* **1979**, 1543. Deschamps, B. *T* **1978**, *34*, 2009. El-Bouz, M.; Nartski, L. *TL* **1980**, 2897. Luccheti, J.; Dumont, W.; Krief, A. *TL* **1979**, 2695.
48. Otera, J.; Niibo, Y.; Aikawa, H. *TL* **1987**, *28*, 2147. Zervos, M.; Wartski, L.; Seydon-Penne, J. *T* **1986**, *42*, 4963. Ager, D. J.; East, M. B. *JOC* **1986**, *51*, 3983.
49. Carey, F. A.; Court, A. S. *JOC* **1972**, *37*, 939. Ager, D. J. *TL* **1981**, *22*, 2803.
50. Hackett, S.; Livinghouse, T. *JOC* **1986**, *51*, 879. Otera, J.; Niibo, Y.; Nozaki, H. *JOC* **1989**, *54*, 5003.
51. Binns, M. R.; Haynes, R. K.; Houston, T. L.; Jackson, W. R. *TL* **1980**, *21*, 573. Binns, M. R.; Haynes, R. K.; Katsifis, A. G.; Schober, P. A.; Vonwiller, S. C. *JACS* **1988**, *110*, 5411. Binns, M. R.; Haynes, R. K. *JOC* **1981**, *46*, 3790.
52. Zieglar, F. E.; Fang, J.-M.; Tam, C. C. *JACS* **1982**, *104*, 7174.
53. Still, W. C. *JACS* **1977**, *99*, 4836; Still, W. C.; Mitra, A. *TL* **1978**, 2659.
54. Magnusson, G. *TL* **1977**, 2713. Barr, D., Doyle, M. J., Mulvey, R. E.; Raithby, P. R.; Reed, D.; Snaith, R.; Wright, D. S. *CC* **1989**, 318. Reich, H. J.; Borst, J. P.; Dykstra, R. R. *OM* **1994**, *13*, 1.
55. (a) Sreekumar, C.; Darst, K. P.; Still, W. C. *JOC* **1980**, *45*, 4260. Koreeda, M.; Patel, P. D.; Brown, L. *JOC* **1985**, *50*, 5910. (b) Vedejs, E.; Peterson, M. J. *Top. Stereochem.* **1993**, *21*, 1. Schlosser, v.-M.; Christmann, K. F. *LA* **1967**, *708*, 1.
56. Waters, R. M.; Voaden, D. J.; Warthen, J. D., Jr. *OPP* **1978**, *10*, 5. Sonnet, P. E. *OPP* **1974**, *6*, 269. Corey, E. J.; Clark, D. A.; Goto, G.; Marfat, A.; Mioskowski, C.; Samuelsson, B.; Hammarström, S. *JACS* **1980**, *102*, 1436. Wernic, D.; DiMaio, J.; Adams, J. *JOC* **1989**, *54*, 4224. Delorme, D.; Girard, Y.; Rokach, J. *JOC* **1989**, *54*, 3635. Yadagiri, P.; Lumin, S.; Falck, J. R.; Karara, A.; Capdevila, J. *TL* **1989**, *30*, 429. Vidal, J. P.; Escale, R.; Niel, G.; Rechencq, E.; Girard, J. P.; Rossi, J. C. *TL* **1989**, *30*, 5129.
57. Cereda, E.; Attolini, M.; Bellora, E.; Donetti, A. *TL* **1982**, *23*, 2219. Inoue, S.; Honda, K.; Iwase, N.; Sato, K. *BCJ* **1990**, *63*, 1629.
58. Corey, E. J.; Marfat, A.; Hoover, D. J. *TL* **1981**, *22*, 1587. Boubia, B.; Mann, A.; Bellamy, F. D.; Mioskowski, C. *AG(E)* **1990**, *29*, 1454.
59. Ousset, J. B.; Mioskowski, C.; Solladie, G. *SC* **1983**, *13*, 1193.
60. Kelly, T. R.; Dali, H. M.; Tsang, W.-G. *TL* **1977**, *44*, 3859. Evers, M. *CS* **1986**, *26*, 585.
61. (a) Evers, M.; Christiaens, L. *TL* **1983**, *24*, 377. Reich, H. J.; Cohen, M. L. *JOC* **1979**, *44*, 3148. (b) Testaferri, L.; Tiecco, M.; Tingoli, M.; Chianelli, D.; Montanucci, M. *S* **1983**, 751.
62. (a) Müller, P.; Siegfried, B. *TL* **1973**, 3565. (b) Kondo, K.; Umemota, T.; Takahatake, Y.; Tunemoto, D. *TL* **1977**, 113.
63. Ager, D. J. *OR* **1990**, *38*, 1.
64. Sakurai, H.; Nishiwaki, K.-i.; Kira, M. *TL* **1973**, 4193.
65. Bassindale, A. R.; Ellis, R. J.; Taylor, P. G. *TL* **1984**, *25*, 2705.
66. Falck, J. R.; Yadagiri, P. *JOC* **1989**, *54*, 5851.
67. Shaw, J. E.; Kunerth, D. C.; Swanson, S. B. *JOC* **1976**, *41*, 732.
68. Ashby, E. C.; Park, W. S.; Goel, A. B.; Su, W.-Y. *JOC* **1985**, *50*, 5184.
69. Marshall, J. A.; Lebreton, J.; DeHoff, B. S.; Jenson, T. M. *JOC* **1987**, *52*, 3883.
70. Normant, J. F.; Deshayes, H. *BSF(2)* **1967**, 2455. Paquette, L. A.; Meisinger, R. H.; Wingard, R. E., Jr. *JACS* **1973**, *95*, 2230. King, P. F.; Paquette, L. A. *S* **1977**, 279. Weisz, A.; Mandelbaum, A. *JOC* **1984**, *49*, 2648.
71. Roush, W. R.; Brown, B. B. *JACS* **1993**, *115*, 2268.
72. Whitesides, G. M.; Ehmann, W. J. *JOC* **1970**, *35*, 3565 and references cited therein.
73. (a) Inanaga, J.; Ishikawa, M.; Yamaguchi, M. *CL* **1987**, 1485. (b) Otsubo, K.; Kawamura, K.; Iwanaga, J.; Yamaguchi, M. *CL* **1987**, 1487.
74. Otsubo, K.; Inanaga, J.; Yamaguchi, M. *TL* **1986**, *27*, 5763.
75. Handa, Y.; Inanaga, J.; Yamaguchi, M. *CC* **1989**, 298.
76. Inanaga, J.; Handa, Y.; Tabuchi, T.; Otsubo, K.; Yamaguchi, M.; Hanamoto, T. *TL* **1991**, *32*, 6557.
77. Cabrera, A.; Alper, H. *TL* **1992**, *33*, 5007.
78. Fevig, T. L.; Elliott, R. L.; Curran, D. P. *JACS* **1988**, *110*, 5064.
79. Curran, D. P.; Wolin, R. L. *SL* **1991**, 317. Curran, D. P.; Yoo, B. *TL* **1992**, *33*, 6931.
80. Hou, Z.; Takamine, K.; Aoki, O.; Shiraishi, H.; Fujiwara, Y.; Taniguchi, H. *JOC* **1988**, *53*, 6077. Hou, Z.; Kobayashi, K.; Yamazaki, H. *CL* **1991**, 265.
81. Milstein, D.; Stille, J. K. *JACS* **1979**, *101*, 4981, 4992.
82. (a) Labadie, J. W.; Stille, J. K. *JACS* **1983**, *105*, 6129. (b) Labadie, J. W.; Stille, J. K. *JACS* **1983**, *105*, 669.
83. Bumagin, N. A.; Bumagina, I. G.; Beletskaya, I. P. *DOK* **1983**, *272*, 1384.
84. Hegedus, L. S.; Williams, R. E.; McGuire, M. A.; Hayashi, T. *JACS* **1980**, *102*, 4973.
85. Hegedus, L. S.; Darlington, W. H.; Russel, C. E. *JOC* **1980**, *45*, 5193.
86. Hutchins, R. O.; Kandasamy, D.; Maryanoff, C. A.; Masilamani, D.; Maryanoff, B. E. *JOC* **1977**, *42*, 82.
87. Hutchins, R. O.; Cistone, F.; Goldsmith, B.; Heuman, P. *JOC* **1975**, *40*, 2018.
88. Shibata, I.; Suzuki, T.; Baba, A.; Matsuda, H. *CC* **1988**, 882.
89. Shibata, I.; Yoshida, T.; Baba, A.; Matsuda, H. *CL* **1989**, 619.
90. Fujita, M.; Hiyama, T. *JOC* **1988**, *53*, 5405.
91. Cardillo, G.; Orena, M.; Sandri, S. *S* **1976**, 394.
92. Al Neirabeyeh, M.; Pujol, M. D. *TL* **1990**, *31*, 2273.
93. (a) Peyronel, J.-F.; Samuel, O.; Fiaud, J.-C. *JOC* **1987**, *52*, 5320 and references cited therein. (b) Arcoria, A.; Ballistreri, F. P.; Tomaselli, G. A.; Di Furia, F.; Modena, G. *JOC* **1986**, *51*, 2374.
94. Takagi, M.; Nojima, M.; Kusabayashi, S. *JACS* **1983**, *105*, 4676.
95. Eyer, M.; Seebach, D. *JACS* **1985**, *107*, 3601.
96. Panek, E. J.; Rodgers, T. J. *JACS* **1974**, *96*, 6921.
97. Lefour, J.-M.; Loupy, A. *T* **1978**, *34*, 2597.
98. Bertz, S. H. *JACS* **1991**, *113*, 5470.
99. Sowinski, A. F.; Whitesides, G. M. *JOC* **1979**, *44*, 2369.
100. (a) Wilson, S. R.; Price, M. F. *SC* **1982**, *12*, 657. (b) Bortolini, O.; Di Furia, F.; Modena, G.; Schionato, A. *J. Mol. Catal.* **1986**, *35*, 47.

Robert R. Dykstra
University of Wisconsin, Madison, WI, USA

Hexamethylphosphorous Triamide[1]

$(Me_2N)_3P$

[1608-26-0] $C_6H_{18}N_3P$ (MW 163.24)

(strong nucleophile;[2] used to synthesize epoxides from aldehydes[2,3] and arene oxides from aryldialdehydes;[4–7] replaces Ph_3P in the Wittig reaction;[8] with CCl_4, converts alcohols to chlorides;[9] with I_2, converts disulfides to sulfides[10] and deoxygenates sulfoxides and azoxyarenes;[11] with dialkyl azodicarboxylate and alcohol, forms mixed carbonates;[12] reduces ozonides[13])

Alternate Name: tris(dimethylamino)phosphine.
Physical Data: mp 26 °C; bp 162–164 °C/760 mmHg, 50 °C/12 mmHg; n_D^{25} 1.4636; *d* 0.911 g cm^{-3}.

Form Supplied in: commercially available; liquid, pure grade >97% (GC).

Preparative Methods: reaction of ***Phosphorus(III) Chloride*** with anhydrous dimethylamine; the same procedure can be used to obtain higher alkyl homologs.[14]

Purification: distillation at reduced pressure; exposure of hot liquid to air should be avoided.

Handling, Storage, and Precautions: very sensitive to air; best stored in nitrogen atmosphere; reacts with carbon dioxide; inhalation should be avoided. Use in a fume hood.

Synthesis of Epoxides. Reaction of $(Me_2N)_3P$ with aromatic aldehydes provides convenient direct synthetic access to symmetrical and unsymmetrical epoxides in generally high yields. A typical example is the reaction of *o*-chlorobenzaldehyde, which provides the corresponding stilbene oxide as a mixture of the *trans* and *cis* isomers (eq 1).[2,3]

2 (o-Cl-C$_6$H$_4$-CHO) + $(Me_2N)_3P$ —(76%)→ bis(o-chlorophenyl)oxirane + $(Me_2N)_3P{=}O$ (1)

trans:*cis* = 1.38:1

The coproduct, ***Hexamethylphosphoric Triamide***, is readily separated by taking advantage of its water solubility. A competing reaction pathway leads to formation of variable amounts of a 1:1 adduct in addition to the epoxide product (eq 2). Originally the adduct was assigned the betaine structure (**1a**). On the basis of more detailed NMR analysis, this was subsequently revised to the phosphonic diamide structure (**1b**).[15]

ArCHO + $(Me_2N)_3P$ ⟶ $(Me_2N)_3\overset{+}{P}-CH(Ar)-O^-$ (**1a**) or $Ar(Me_2N)CH-P(O)(NMe_2)_2$ (**1b**) (2)

The ratio of products depends upon the electronegativity of the aldehyde and the mode of carrying out the reaction. Aromatic aldehydes with electronegative substituents, especially in the *ortho* position, undergo rapid exothermic reaction to yield epoxides exclusively. Conversely, aldehydes bearing electron-releasing substituents react more slowly to afford mainly 1:1 adducts. Slow addition of $(Me_2N)_3P$ to the aldehyde tends to enhance the ratio of the epoxide product. These observations are compatible with a mechanism in which an initially formed 1:1 adduct reacts with a second aldehyde molecule to form a 2:1 adduct which collapses to yield the observed products (eq 3).

$(Me_2N)_3\overset{+}{P}-CH(R)-O^-$ —RCHO→ $(Me_2N)_3P$ (cyclic 2:1 adduct) ⟶ $(Me_2N)_3P{=}O$ + R,R-epoxide (3)

$(Me_2N)_3P$ reacts also with saturated and heterocyclic aldehydes, but 1:1 adducts rather than epoxides are the predominant products. The reaction with ***Chloral*** takes a different course[2] and yields the dichlorovinyloxyphosphonium compound $Cl_2C{=}CH{-}O{-}\overset{+}{P}(NMe_2)_3\ Cl^-$.

The scope of the reaction is considerably extended by its applicability to the synthesis of mixed epoxides.[2] This is accomplished by addition of $(Me_2N)_3P$ to a mixture of aldehydes in which the less reactive aldehyde predominates. For example, addition of $(Me_2N)_3P$ to a mixture of *o*-chlorobenzaldehyde and 2-furaldehyde yields the corresponding mixed epoxide (eq 4).

o-Cl-C$_6$H$_4$-CHO + OHC-(2-furyl) —$(Me_2N)_3P$→ 2-(o-chlorophenyl)-3-(2-furyl)oxirane (4)

An advantage of the method is that it allows the synthesis of epoxides unobtainable by the oxidation of alkene precursors with peroxides or peracids due to the incompatibility of functional groups with these reagents.

Synthesis of Arene Oxides. Reaction of $(Me_2N)_3P$ with aromatic dialdehydes provides arene oxides such as benz[*a*]anthracene 5,6-oxide (**2a**) (eq 5).[4–7] These compounds, also known as oxiranes, are relatively reactive, undergoing thermal and acid-catalyzed rearrangement to phenols and facile hydrolysis to dihydrodiols. Consequently, their preparation and purification requires mild reagents and conditions. The importance of this is underlined by successful synthesis of the reactive arene oxide (**2b**) in 75% yield using appropriate care,[7] despite a previous report of failure of the method.[4] While compound (**2b**) is a relatively potent mutagen, it is rapidly detoxified by mammalian cells.[6] The principal limitation of the method is the unavailability of the dialdehyde precursors, which are obtained through oxidation of the parent hydrocarbons, e.g. by ozonolysis.

dialdehyde + $(Me_2N)_3P$ ⟶ $(Me_2N)_3P{=}O$ + arene oxide (5)

(**2a**) R = H
(**2b**) R = Me

Wittig and Horner–Wittig Reactions. $(Me_2N)_3P$ may be used in place of ***Triphenylphosphine*** in Wittig reactions with aldehydes and ketones (eq 6).[8] It is advantageous because the water solubility of the byproduct, hexamethylphosphoric triamide, renders it readily removable. This method has been used for the preparation of unsaturated esters as well as alkenes (eq 7).

$$PhCH_2Br + (Me_2N)_3P \longrightarrow \underset{\textbf{(3)}}{(Me_2N)_3\overset{+}{P}CH_2Ph\ Br^-} \xrightarrow[Me_2CHCHO]{NaOMe} Me_2CHCH{=}CHPh \quad (6)$$

$$BrCH_2CO_2Et + (Me_2N)_3P \longrightarrow \underset{\textbf{(4)}}{(Me_2N)_3\overset{+}{P}CH_2CO_2Et\ Br^-} \xrightarrow[RCHO]{NaOMe} RCH{=}CHCO_2Et \quad (7)$$

The phosphonic diamide products (**1b**) obtained from the reaction of arylaldehydes having electron-donating groups with $(Me_2N)_3P$ can be deprotonated by ***n-Butyllithium*** in DME at 0 °C.[15] These intermediates participate in Horner–Wittig-type reactions with aromatic aldehydes to give enamines in good yield (eq 8). In the examples studied the enamines have the (*E*) configuration. Mild acid hydrolysis of the reaction mixtures without isolating the intermediate enamines provides the corresponding deoxybenzoins.[15] The overall procedure represents an example of reductive nucleophilic acylation of carbonyl compounds.

$$\textbf{(1b)} \xrightarrow{BuLi} Ar^1(Me_2N)C^-{-}P(O)(NMe_2)_2 \xrightarrow{Ar^2CHO} Ar^1(Me_2N)C{=}CHAr^2 \longrightarrow Ar^1COCH_2Ar^2 \quad (8)$$

Conversion of Alcohols to Alkyl Chlorides and Other Derivatives. The reagent combination $(Me_2N)_3P$ and CCl_4 can be used in place of ***Triphenylphosphine–Carbon Tetrachloride*** for the conversion of alcohols to alkyl chlorides (eq 9).[9] An advantage is the ease of removal of the water-soluble coproduct $(Me_2N)_3P{=}O$. The mechanism entails initial rapid formation of a quasiphosphonium ion, followed by reaction with an alcohol with displacement of chloride, and nucleophilic attack by the chloride ion on the carbon atom of the alcohol in a final rate-determining step to yield an alkyl chloride.

$$(Me_2N)_3P{:}\ Cl{-}CCl_3 \xrightarrow{fast} (Me_2N)_3\overset{+}{P}{-}Cl\ Cl_3C^- \xrightarrow[fast]{ROH}$$

$$Cl_3CH + (Me_2N)_3\overset{+}{P}{-}O{-}R\ Cl^- \xrightarrow{slow} (Me_2N)_3P{=}O + RCl \quad (9)$$

This reagent reacts more rapidly with primary than with secondary alcohols. This property has been made use of to transform the primary hydroxy groups of sugars to salts, which then may be converted to halides (Cl, Br, I), azides, amines, thiols, thiocyanates, etc. by reaction with appropriate nucleophiles (eq 10).[16] Arylalkyl ethers and thioethers may also be prepared by appropriate modification of this method.[17] These reactions generally proceed with high stereoselectivity. Thus reaction of chiral 2-octanol with this reagent afforded 2-chlorooctane with complete inversion of configuration. Also, conversion of the salt prepared from reaction of (*R*)-(−)-2-octanol with $(Me_2N)_3P/CCl_4$ at low temperature to the corresponding hexafluorophosphate salt, followed by reaction of this with potassium phenolate in DMF, gave optically pure (*S*)-(+)-2-phenoxyoctane in 93% yield (eq 11).[17]

$$\text{methyl glycoside }(CH_2OH) \xrightarrow[CCl_4,\ THF]{(Me_2N)_3P} CH_2O\overset{+}{P}(NMe_2)_3\ Cl^- \xrightarrow{Nu^-} CH_2{-}Nu \quad (10)$$

Nu = Cl, Br, I, N_3, NH_2, SCN, H

$$(Me_2N)_3\overset{+}{P}{-}O{-}CH(CH_3)CH_2CH_2Bu\ PF_6^- + PhOK \longrightarrow PhO{-}CH(CH_3)CH_2CH_2Bu + (R_2N)_3P{=}O \quad (11)$$

$(Me_2N)_3P/CCl_4$ may also be employed for the selective functionalization of primary long-chain diols.[17] Reactions of diols of this type with $(Me_2N)_3P$ and CCl_4 in THF followed by addition of KPF_5 gives mono salts in high yield (eq 12).[18] THF serves to precipitate the mono salts as they are formed, thereby blocking their conversion to bis salts. Reactions of the mono salts with various nucleophiles provides the corresponding monosubstituted primary alcohols.

$$HOCH_2(CH_2)_nCH_2OH \xrightarrow[CCl_4,\ -40\ ^\circ C,\ KPF_5]{(Me_2N)_3P} (Me_2N)_3\overset{+}{P}OCH_2(CH_2)_nCH_2OH\ PF_5^- \xrightarrow{Nu^-} NuCH_2(CH_2)_nCH_2OH \quad (12)$$

n = 3–9 Nu = I, N_3, CN, SCN, MeO

Conversion of Disulfides to Sulfides. Alkyl, aralkyl, and alicyclic disulfides undergo facile desulfurization to the corresponding sulfides on treatment with $(Me_2N)_3P$ or $(Et_2N)_3P$ (eq 13).[10] For example, reaction of methyl phenyl disulfide with $(Et_2N)_3P$ in benzene at rt for ≈1 min furnishes methyl phenyl sulfide in 86% yield. The desulfurization process is stereospecific, in that inversion of configuration occurs at one of the carbon atoms α to the disulfide group. Thus desulfurization of *cis*-3,6-dimethoxycarbonyl-1,2-dithiane affords a quantitative yield of *trans*-2,5-dimethoxycarbonylthiolane (eq 14). It is worthy of note that the rates of these reactions are markedly enhanced by solvents of high polarity.

$$PhSSMe + (R_2N)_3P \longrightarrow PhSMe + (R_2N)_3P{=}S \quad (13)$$

$$\textit{cis}\text{-3,6-}(MeO_2C)_2\text{-1,2-dithiane} \xrightarrow{(R_2N)_3P} \textit{trans}\text{-2,5-}(MeO_2C)_2\text{thiolane} + (R_2N)_3P{=}S \quad (14)$$

Deoxygenation of Sulfoxides and Azoxyarenes. Sulfoxides are deoxygenated to sulfides under mild conditions with $(Me_2N)_3P$ activated with ***Iodine*** in acetonitrile (eq 15).[11] Equimolar ratios of the sulfoxide, $(Me_2N)_3P$, and I_2 are generally employed.

Yields are superior to those obtained with either $(Me_2N)_3P/CCl_4$ or Ph_3P/I_2. Reaction time is reduced by addition of ***Sodium Iodide***. Azoxyarenes, such as azoxybenzene, are converted to azoarenes with this reagent combination under similar mild conditions (eq 16).[11]

$$Ph_2S{=}O + (Me_2N)_3P \xrightarrow{I_2} Ph_2S + (Me_2N)_3P{=}O \quad (15)$$

$$PhN(O^-)^{+}{=}NPh + (Me_2N)_3P \xrightarrow{I_2} PhN{=}NPh + (Me_2N)_3P{=}O \quad (16)$$

Preparation of Mixed Carbonates. The reaction of $(Me_2N)_3P$ with alcohols and dialkyl azodicarboxylates proceeds smoothly at rt to provide mixed dialkyl carbonate esters in moderate to good yields (eq 17).[12] An advantage of the method over the chloroformate method is the neutrality of the conditions employed. It should be noted that the related system ***Triphenylphosphine–Diethyl Azodicarboxylate*** converts alcohols into amines.

$$R^1OH + R^2O_2C{-}N{=}N{-}CO_2R^2 \xrightarrow{(R^3{}_2N)_3P} R^1O{-}CO{-}OR^2 + HCO_2R^2 + N_2 \quad (17)$$

Reduction of Ozonides. In the synthesis of ecdysone from ergosterol, the ozonide product produced from (**5**) was reduced with $(Me_2N)_3P$ under mild conditions to the aldehyde (**6**) without isomerization (eq 18).[13] The scope of this method has not been investigated.

(**5**) —— 1. O_3; 2. $(Me_2N)_3P$ ——→ (**6**) (18)

Related Reagents. Diphosphorus Tetraiodide; Tri-*n*-butylphosphine; Triphenylphosphine; Raney Nickel.

1. Wurziger, H. *Kontakte (Darmstadt)* **1990**, 13.
2. Mark, V. *JACS* **1963**, *85*, 1884.
3. Mark, V. *OSC* **1973**, *5*, 358.
4. Newman, M. S.; Blum, S. *JACS* **1964**, *86*, 5598.
5. Harvey, R. G. *S* **1986**, 605.
6. Harvey, R. G. *Polycyclic Aromatic Hydrocarbons: Chemistry and Carcinogenesis*; Cambridge University Press: Cambridge, 1991; Chapter 12.
7. Harvey, R. G.; Goh, S. H.; Cortez, C. *JACS* **1975**, *97*, 3468.
8. Oediger, H.; Eiter, K. *LA* **1965**, *682*, 58.
9. Downie, I. M.; Lee, J. B.; Matough, M. F. S. *CC* **1968**, 1350.
10. Harpp, D. N.; Gleason, J. G. *JACS* **1971**, *93*, 2437.
11. Olah, G. A.; Gupta, B. G. B.; Narang, S. C. *JOC* **1978**, *43*, 4503.
12. Grynkiewicz, G.; Jurczak, J.; Zamojski, A. *T* **1975**, *31*, 1411.
13. Furlenmeier, A.; Fürst, A.; Langemann, A.; Waldvogel, G.; Hocks, P.; Kerb, U.; Wiechert, R. *HCA* **1967**, *50*, 2387.
14. Mark, V. *OSC* **1973**, *5*, 602.
15. Babudri, F.; Fiandanese, V.; Musio, R.; Naso, F.; Sciavovelli, O.; Scilimati, A. *S* **1991**, 225.
16. Castro, B.; Chapleur, Y.; Gross, B. *BSF(2)* **1973**, 3034.
17. Downie, I. M.; Heaney, H.; Kemp, G. *AG(E)* **1975**, *14*, 370.
18. Boigegrain, R.; Castro, B.; Selve, C. *TL* **1975**, 2529.

Ronald G. Harvey
University of Chicago, IL, USA

Hydrazine[1]

N_2H_4

(N_2H_4)		
[302-01-2]	H_4N_2	(MW 32.06)
(hydrate)		
[10217-52-4]		
(monohydrate)		
[7803-57-8]	H_5N_2O	(MW 49.07)
(monohydrochloride)		
[2644-70-4]	ClH_5N_2	(MW 68.52)
(dihydrochloride)		
[5341-61-7]	$Cl_2H_6N_2$	(MW 104.98)
(sulfate)		
[10034-93-2]	$H_6N_2O_4S$	(MW 130.15)

(reducing agent used in the conversion of carbonyls to methylene compounds;[1] reduces alkenes,[9] alkynes,[9] and nitro groups;[14] converts α,β-epoxy ketones to allylic alcohols;[32] synthesis of hydrazides;[35] synthesis of dinitrogen containing heterocycles[42–46])

Physical Data: mp 1.4 °C; bp 113.5 °C; *d* 1.021 g cm^{-3}.

Solubility: sol water, ethanol, methanol, propyl and isobutyl alcohols.

Form Supplied in: anhydrate, colorless oil that fumes in air; hydrate and monohydrate, colorless oils; monohydrochloride, dihydrochloride, sulfate, white solids; all widely available.

Analysis of Reagent Purity: titration.[1]

Purification: anhydrous hydrazine can be prepared by treating hydrazine hydrate with BaO, $Ba(OH)_2$, CaO, NaOH, or Na. Treatment with sodamide has been attempted but this yields diimide, NaOH, and ammonia. An excess of sodamide led to an explosion at 70 °C. The hydrate can be treated with boric acid to give the hydrazinium borate, which is dehydrated by heating. Further heating gives diimide.[1]

Handling, Storage, and Precautions: caution must be taken to avoid prolonged exposure to vapors as this can cause serious damage to the eyes and lungs. In cases of skin contact, wash the affected area immediately as burns similar to alkali contact can occur. Standard protective clothing including an ammonia gas mask are recommended. The vapors of hydrazine are flammable (ignition temperature 270 °C in presence of air). There have been reports of hydrazine, in contact with organic material such as wool or rags, burning spontaneously. Metal oxides can also initiate combustion of hydrazine. Hydrazine and its solutions should be stored in glass containers under nitrogen for extended periods. There are no significant precautions for reaction vessel type with hydrazine; however, there have been reports that stainless steel vessels must be checked for significant oxide formation prior to use. Use in a fume hood.

Reductions. The use of hydrazine in the reduction of carbonyl compounds to their corresponding methylene groups via the Wolff–Kishner reduction has been covered extensively in the literature.[1] The procedure involves the reaction of a carbonyl-containing compound with hydrazine at high temperatures in the presence of a base (usually ***Sodium Hydroxide*** or ***Potassium Hydroxide***). The intermediate hydrazone is converted directly to the fully reduced species. A modification of the original conditions was used by Paquette in the synthesis of (±)-isocomene (eq 1).[2]

$H_2NNH_2{\cdot}H_2O$, K_2CO_3, 200 °C; 80% (1)

Unfortunately, the original procedure suffers from the drawback of high temperatures, which makes large-scale runs impractical. The Huang–Minlon modification[3] of this procedure revolutionized the reaction, making it usable on large scales. This procedure involves direct reduction of the carbonyl compound with hydrazine hydrate in the presence of sodium or potassium hydroxide in diethylene glycol. The procedure is widely applicable to a variety of acid-labile substrates but caution must be taken where base-sensitive functionalities are present. This reaction has seen widespread use in the preparation of a variety of compounds. Other modifications[4] have allowed widespread application of this useful transformation. Barton and co-workers further elaborated the Huang–Minlon modifications by using anhydrous hydrazine and ***Sodium*** metal to ensure totally anhydrous conditions. This protocol allowed the reduction of sterically hindered ketones, such as in the deoxygenation of 11-keto steroids (eq 2).[4a] Cram utilized dry DMSO and ***Potassium t-Butoxide*** in the reduction of hydrazones. This procedure is limited in that the hydrazones must be prepared and isolated prior to reduction.[4b] The Henbest modification[4c] involves the utilization of dry toluene and potassium *t*-butoxide. The advantage of this procedure is the low temperatures needed (110 °C) but it suffers from the drawback that, again, preformed hydrazones must be used. Utilizing modified Wolff–Kishner conditions, 2,4-dehydroadamantanone is converted to 8,9-dehydroadamantane (eq 3).[5]

Hindered aldehydes have been reduced using this procedure.[6] This example is particularly noteworthy in that the aldehyde is sterically hindered and resistant to other methods for conversion to the methyl group.[6a] Note also that the acetal survives the manipulation (eq 4). The reaction is equally useful in the reduction of semicarbazones or azines.

AcO, O, O; H_2NNH_2, RONa, Ac_2O; 69%; AcO (2)

(3)

O, O, OHC; H_2NNH_2, KOH, $(HOCH_2CH_2)_2O$; O, O (4)

In a similar reaction, hydrazine has been shown to desulfurize thioacetals, cyclic and acyclic, to methylene groups (eq 5). The reaction is run in diethylene glycol in the presence of potassium hydroxide, conditions similar to the Huang–Minlon protocol. Yields are generally good (60–95%). In situations where base sensitivity is a concern, the potassium hydroxide may be omitted. Higher temperatures are then required.[7]

R^2, S, R^1, S; H_2NNH_2, KOH, diethylene glycol; R^2, R^1 (5)

Hydrazine, via in situ copper(II)-catalyzed conversion to ***Diimide***, is a useful reagent in the reduction of carbon and nitrogen multiple bonds. The reagent is more reactive to symmetrical rather than polar multiple bonds (C=N, C=O, N=O, S=O, etc.)[8] and reviews of diimide reductions are available.[9] The generation of diimide from hydrazine has been well documented and a wide variety of oxidizing agents can be employed: oxygen (air),[10] ***Hydrogen Peroxide***,[10] ***Potassium Ferricyanide***,[11] ***Mercury(II) Oxide***,[11] ***Sodium Periodate***,[12] and hypervalent ***Iodine***[13] have all been reported. The reductions are stereospecific, with addition occurring *cis* on the less sterically hindered face of the substrate.

Other functional groups have been reduced using hydrazine. Nitroarenes are converted to anilines[14] in the presence of a variety of catalysts such as ***Raney Nickel***,[14a,15] platinum,[14a] ruthenium,[14a] ***Palladium on Carbon***,[16] β-iron(III) oxide,[17] and iron(III) chloride with activated carbon.[18] Graphite/hydrazine reduces aliphatic and aromatic nitro compounds in excellent yields.[19] Halonitrobenzenes generally give excellent yields of haloanilines. In experiments where palladium catalysts are used, significant dehalogenation occurs to an extent that this can be considered a general dehalogenation method.[20] Oximes have also been reduced.[21]

Hydrazones. Reaction of hydrazine with aldehydes and ketones is not generally useful due to competing azine formation or competing Wolff–Kishner reduction. Exceptions have been documented. Recommended conditions for hydrazone preparation are to reflux equimolar amounts of the carbonyl component and hydrazine in *n*-butanol.[22,23] A more useful method for simple hydrazone synthesis involves reaction of the carbonyl compound with dimethylhydrazine followed by an exchange reaction with hydrazine.[24] For substrates where an azine is formed, the hydrazone can be prepared by refluxing the azine with anhydrous hydrazine.[25] *gem*-Dibromo compounds have been converted to hydrazones by reaction with hydrazine (eq 6).[26]

H_2NNH_2, Et_2O, MeOH (6)

Hydrazones are useful synthetic intermediates and have been converted to vinyl iodides[27] and vinyl selenides (eq 7) (see also ***p-Toluenesulfonylhydrazide***).[28]

RX, base (7)

R = I, PhSe; RX = I_2, PhSeBr
base = pentaalkylguanidine, triethylamine

Diazomalonates have been prepared from dialkyl mesoxylates via the ***Silver(I) Oxide***-catalyzed decomposition of the intermediate hydrazones.[29] Monohydrazones of 1,2-diketones yield ketenes after mercury(II) oxide oxidation followed by heating.[30] Dihydrazones of the same compounds give alkynes under similar conditions.[31]

Wharton Reaction. α,β-Epoxy ketones and aldehydes rearrange in the presence of hydrazine, via the epoxy hydrazone, to give the corresponding allylic alcohols. This reaction has been successful in the steroid field but, due to low yields, has seen limited use as a general synthetic tool. Some general reaction conditions have been set. If the intermediate epoxy hydrazone is isolable, treatment with a strong base (potassium *t*-butoxide or ***Potassium Diisopropylamide***) gives good yields, whereas ***Triethylamine*** can be used with nonisolable epoxy hydrazones (eq 8).[32]

(8)

Some deviations from expected Wharton reaction products have been reported in the literature. Investigators found that in some specific cases, treatment of α,β-epoxy ketones under Wharton conditions gives cyclized allylic alcohols (eq 9). No mechanistic interpretation of these observations has been offered. Related compounds have given the expected products, and it therefore appears this phenomenon is case-specific.[33]

H_2NNH_2 (9)

Cyclic α,β-epoxy ketones have been fragmented upon treatment with hydrazine to give alkynic aldehydes.[34]

Hydrazides. Acyl halides,[35] esters, and amides react with hydrazine to form hydrazides which are themselves useful synthetic intermediates. Treatment of the hydrazide with nitrous acid yields the acyl azide which, upon heating, gives isocyanates (Curtius rearrangement).[36] Di- or trichlorides are obtained upon reaction with ***Phosphorus(V) Chloride***.[37] Crotonate and other esters have been cleaved with hydrazine to liberate the free alcohol (eq 10).[38]

H_2NNH_2 (10)

Hydrazine deacylates amides (Gabriel amine synthesis) via the Ing–Manske protocol.[39] This procedure has its limitations, as shown in the synthesis of penicillins and cephalosporins where it was observed that hydrazine reacts with the azetidinone ring. In this case, ***Sodium Sulfide*** was used.[40]

Heterocycle Synthesis. The reaction of hydrazine with α,β-unsaturated ketones yields pyrazoles.[41,42] Although the products can be isolated as such, they are useful intermediates in the synthesis of cyclopropanes upon pyrolysis of cyclopropyl acetates after treatment with ***Lead(IV) Acetate*** (eq 11).

H_2NNH_2; $Pb(OAc)_4$, heat; 60% (11)

3,5-Diaminopyrazoles were prepared by the addition of hydrazine (eq 12), in refluxing ethanol, to benzylmalononitriles (42–73%).[43] Likewise, hydrazine reacted with 1,1-diacetylcyclopropyl ketones to give β-ethyl-1,2-azole derivatives. The reaction mixture must have a nucleophilic component (usually the solvent, i.e. methanol) to facilitate the opening of the cyclopropane ring. Without this, no identifiable products are obtained (eq 13).[44]

H_2NNH_2, ethanol, reflux, 42–73% (12)

Ar = Ph, 4-MeC_6H_4, 3-$NO_2C_6H_4$

R^1, R^2 = Me, Ph
X = Cl, Br, OMe, OEt, OPh, OAc, CN, etc.

In an attempt to reduce the nitro group of nitroimidazoles, an unexpected triazole product was obtained in 66% yield. The suggested mechanism involves addition of the hydrazine to the ring, followed by fragmentation and recombination to give the observed product (eq 14).[45]

Finally, hydrazine dihydrochloride reacted with 2-alkoxynaphthaldehydes to give a product which resulted from an intramolecular $[3^+ + 2]$ criss-cross cycloaddition (42–87%) (eq 15).[46]

Peptide Synthesis. Treatment of acyl hydrazides with nitrous acid leads to the formation of acid azides which react with amines to form amides in good yield. This procedure has been used in peptide synthesis, but is largely superseded by coupling reagents such as ***1,3-Dicyclohexylcarbodiimide***.[47]

---

1. (a) Todd, D. *OR* **1948**, *4*, 378. (b) Szmant, H. H. *AG(E)* **1968**, *7*, 120. (c) Reusch, W. *Reduction*; Dekker: New York, 1968, pp 171–185. (d) Clark, C. *Hydrazine*; Mathieson Chemical Corp.: Baltimore, MD, 1953.
2. Paquette, L. A.; Han, Y. K. *JOC* **1979**, *44*, 4014.
3. (a) Huang-Minlon *JACS* **1946**, *68*, 2487; **1949**, *71*, 3301. (b) Durham, L. J.; McLeod, D. J.; Cason, J. *OSC* **1963**, *4*, 510. (c) Hunig, S.; Lucke, E.; Brenninger, W. *OS* **1963**, *43*, 34.
4. (a) Barton, D. H. R.; Ives, D. A. J.; Thomas, B. R. *JCS* **1955**, 2056. (b) Cram, D. J.; Sahyun, M. R. V.; Knox, G. R. *JACS* **1962**, *90*, 7287. (c) Grundon, M. F.; Henbest, H. B.; Scott, M. D. *JCS* **1963**, 1855. (d) Moffett, R. B.; Hunter, J. H. *JACS* **1951**, *73*, 1973. (e) Nagata, W.; Itazaki, H. *CI(L)* **1964**, 1194.
5. Murray, R. K., Jr.; Babiak, K. A. *JOC* **1973**, *38*, 2556.
6. (a) Zalkow, L. H.; Girotra, N. N. *JOC* **1964**, *29*, 1299. (b) Aquila, H. *Ann. Chim.* **1968**, *721*, 117.
7. van Tamelen, E. E.; Dewey, R. S.; Lease, M. F.; Pirkle, W. H. *JACS* **1961**, *83*, 4302.
8. Georgian, V.; Harrisson, R.; Gubisch, N. *JACS* **1959**, *81*, 5834.
9. (a) Miller, C. E. *J. Chem. Educ.* **1965**, *42*, 254. (b) Hunig, S.; Muller, H. R.; Thier, W. *AG(E)* **1965**, *4*, 271. (c) Hammersma, J. W.; Snyder, E. I. *JOC* **1965**, *30*, 3985.
10. Buyle, R.; Van Overstraeten, A. *CI(L)* **1964**, 839.
11. Ohno, M.; Okamoto, M. *TL* **1964**, 2423.
12. Hoffman, J. M., Jr.; Schlessinger, R. H. *CC* **1971**, 1245.
13. Moriarty, R. M.; Vaid, R. K.; Duncan, M. P. *SC* **1987**, *17*, 703.
14. (a) Furst, A.; Berlo, R. C.; Hooton, S. *CRV* **1965**, *65*, 51. (b) Miyata, T.; Ishino, Y.; Hirashima, T. *S* **1978**, 834.
15. Ayynger, N. R.; Lugada, A. C.; Nikrad, P. V.; Sharma, V. K. *S* **1981**, 640.
16. (a) Pietra, S. *AC(R)* **1955**, *45*, 850. (b) Rondestvedt, C. S., Jr.; Johnson, T. A. *Chem. Eng. News* **1977**, 38. (c) Bavin, P. M. G. *OS* **1960**, *40*, 5.
17. Weiser, H. B.; Milligan, W. O.; Cook, E. L. *Inorg. Synth.* **1946**, 215.
18. Hirashima, T.; Manabe, O. *CL* **1975**, 259.
19. Han, B. H.; Shin, D. H.; Cho, S. Y. *TL* **1985**, *26*, 6233.
20. Mosby, W. L. *CI(L)* **1959**, 1348.
21. Lloyd, D.; McDougall, R. H.; Wasson, F. I. *JCS* **1965**, 822.
22. Schonberg, A.; Fateen, A. E. K.; Sammour, A. E. M. A. *JACS* **1957**, *79*, 6020.
23. Baltzly, R.; Mehta, N. B.; Russell, P. B.; Brooks, R. E.; Grivsky, E. M.; Steinberg, A. M. *JOC* **1961**, *26*, 3669.
24. Newkome, G. R.; Fishel, D. L. *JOC* **1966**, *31*, 677.
25. Day, A. C.; Whiting, M. C. *OS* **1970**, *50*, 3.
26. McBee, E. T.; Sienkowski, K. J. *JOC* **1973**, *38*, 1340.
27. Barton, D. H. R.; Basiardes, G.; Fourrey, J.-L. *TL* **1983**, *24*, 1605.
28. Barton, D. H. R.; Basiardes, G.; Fourrey, J.-L. *TL* **1984**, *25*, 1287.
29. Ciganek, E. *JOC* **1965**, *30*, 4366.
30. (a) Nenitzescu, C. D.; Solomonica, E. *OSC* **1943**, *2*, 496. (b) Smith, L. I.; Hoehn, H. H. *OSC* **1955**, *3*, 356.
31. Cope, A. C.; Smith, D. S.; Cotter, R. J. *OSC* **1963**, *4*, 377.
32. Dupuy, C.; Luche, J. L. *T* **1989**, *45*, 3437.
33. (a) Ohloff, G.; Unde, G. *HCA* **1970**, *53*, 531. (b) Schulte-Elte, K. N.; Rautenstrauch, V.; Ohloff, G. *HCA* **1971**, *54*, 1805. (c) Stork, G.; Williard, P. G. *JACS* **1977**, *99*, 7067.
34. (a) Felix, D.; Wintner, C.; Eschenmoser, A. *OS* **1976**, *55*, 52. (b) Felix, D.; Muller, R. K.; Joos, R.; Schreiber, J.; Eschenmoser, A. *HCA* **1972**, *55*, 1276.
35. ans Stoye, P. In *The Chemistry of Amides (The Chemistry of Functional Groups)*; Zabicky, J., Ed.; Interscience: New York, 1970; pp 515–600.
36. (a) *The Chemistry of the Azido Group*; Interscience: New York, 1971. (b) Pfister, J. R.; Wymann, W. E. *S* **1983**, 38.
37. (a) Mikhailov, Matyushecheva, Derkach, Yagupol'skii *ZOR* **1970**, *6*, 147. (b) Mikhailov, Matyushecheva, Yagupol'skii *ZOR* **1973**, *9*, 1847.
38. Arentzen, R.; Reese, C. B. *CC* **1977**, 270.
39. Ing, H. R.; Manske, R. H. F. *JCS* **1926**, 2348.
40. Kukolja, S.; Lammert, S. R. *JACS* **1975**, *97*, 5582 and 5583.
41. Freeman, J. P. *JOC* **1964**, *29*, 1379.
42. Reimlinger, H.; Vandewalle, J. J. M. *ANY* **1968**, *720*, 117.
43. Vequero, J. J.; Fuentes, L.; Del Castillo, J. C.; Pérez, M. I.; Garcia, J. L.; Soto, J. L. *S* **1987**, 33.

44. Kefirov, N. S.; Kozhushkov, S. I.; Kuzetsova, T. S. *T* **1986**, *42*, 709.
45. Goldman, P.; Ramos, S. M.; Wuest, J. D. *JOC* **1984**, *49*, 932.
46. Shimizu, T.; Hayashi, Y.; Miki, M.; Teramura, K. *JOC* **1987**, *52*, 2277.
47. Bodanszky, M. *The Principles of Peptide Synthesis*; Springer: New York, 1984; p 16.

Brian A. Roden
Abbott Laboratories, North Chicago, IL, USA

1-Hydroxybenzotriazole[1]

[2592-95-2] $C_6H_5N_3O$ (MW 135.14)
(hydrate)
[123333-53-9]

(peptide synthesis; nucleotide synthesis)

Alternate Names: HOBT; HOBt.
Physical Data: mp 155–160 °C; exact melting point depends on the amount of water of hydration present.
Form Supplied in: white solid usually containing 12–17% water of hydration; widely available commercially.
Analysis of Reagent Purity: should be a white solid; if it becomes discolored, it is advisable to purify.
Purification: recrystallize from either water or aqueous ethanol.
Handling, Storage, and Precautions: should be stored in the dark; avoid contact with strong acids, oxidizing agents, and reducing agents; heating above 180 °C causes rapid exothermic decomposition; toxicity not fully investigated so should be treated with caution.

Peptide Coupling. 1-Hydroxybenzotriazole is most widely used in reactions involving the coupling of amino acid units to give peptides.[1] In this context it has been used mainly as an 'additive' to a coupling reaction, although there are also examples of HOBT being incorporated into the coupling reagent itself. The most common use of 1-hydroxybenzotriazole in peptide synthesis is in conjunction with a carbodiimide such as ***1,3-Dicyclohexylcarbodiimide*** (DCC). Although it is quite possible to couple amino acids using DCC alone, it is found that the addition of 1-hydroxybenzotriazole to the reaction system results in improved reaction rates and suppressed epimerization of the chiral centres present in the peptide (eq 1).[2] This reaction appears to proceed via DCC-mediated formation of a hydroxybenzotriazole ester intermediate which then reacts with the amino function of a second amino acid to give the coupled product.

This coupling protocol is not limited to the synthesis of small peptides in solution, but can also be used in the solid-phase preparation of larger peptides, and for the coupling of larger peptide fragments. It should be noted, however, that there are some drawbacks to this method; most notably, it is sometimes difficult to purify the peptide due to contamination with dicyclohexylurea and the method is unsatisfactory for the coupling of *N*-methylated amino acids.[3] The problem of purification can often be overcome by using an alternative carbodiimide, e.g. ones that give water-soluble byproducts can be used, such as EDC·MeI (see ***1-Ethyl-3-(3′-dimethylaminopropyl)carbodiimide Hydrochloride***).[4] Polymer-bound variants of HOBT have also been used to overcome purification problems,[5] since the urea byproduct can be removed from the polymer-bound active ester intermediate (**1**) by washing with chloroform–isopropyl alcohol mixtures. Once the urea has been removed, the active ester can be reacted with the second peptide fragment to give the product (eq 2), regenerating the polymer-bound HOBT. While this method appears quite successful for the preparation of small peptide fragments, it is clearly unsuitable for the preparation of larger peptides by solid-phase synthesis.

Bu$_3$N, HOBT, DCC, CH$_2$Cl$_2$, rt, 18 h (1)

< 0.1% DL
(1.8% DL in absence of HOBT)

DCC, HOBT-polymer

(**1**) (2)

83%

Although the addition of HOBT to the carbodiimide-mediated coupling of amino acids has been shown to suppress epimerization of the chiral centres present, there are still a number of cases where the level of epimerization is unsatisfactory. One method for further suppressing epimerization is to add ***Copper(II) Chloride*** to the reaction system.[6] In this case it is important that the correct stochiometry is determined, since the copper(II) chloride not only suppresses epimerization, but it also slows down the reaction rate and reduces the overall yield (eq 3).

The problem of epimerization during carbodiimide-mediated coupling of amino acids can also be overcome by the use of an alternative coupling agent. Once such system that still employs HOBT as an additive involves the use of bis(2-oxo-3-oxazolidinyl)phosphinic chloride (BOP-Cl) and *N*-

methylmorpholine (NMM) (eq 4).[7] See also ***Bis(2-oxo-3-oxazolidinyl)phosphinic Chloride***.

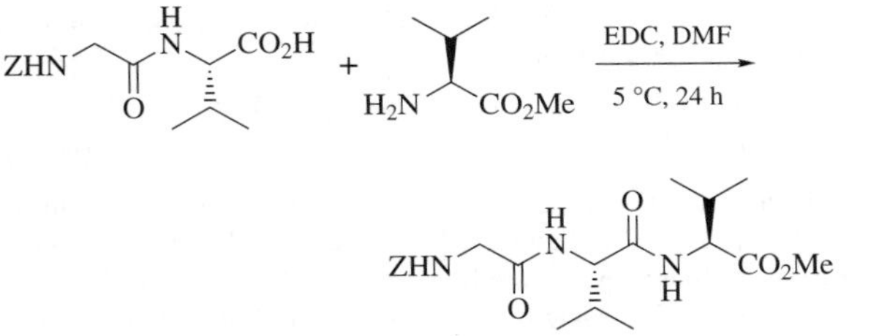

EDC, DMF
5 °C, 24 h
(3)

HOBT	$CuCl_2$	% DL	% Yield
0 equiv	0 equiv	38	22
1 equiv	0 equiv	1.3	96
1 equiv	0.25 equiv	<0.1	79

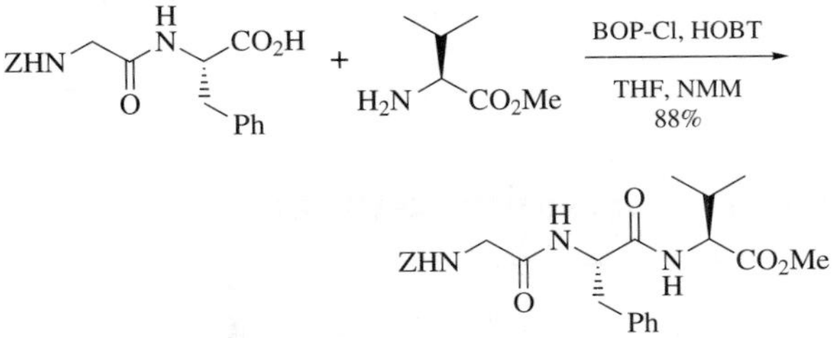

BOP-Cl, HOBT
THF, NMM
88%
(4)
0% DL

HOBT has also been used as an additive in other reactions involving amino acid derivatives.[8] It would appear that most reaction systems that are capable of generating a hydroxybenzotriazole ester intermediate will give successful coupling. One such system involves the use of amino acid derived trichlorophenyl esters.[9] These can be readily reacted with a second amino acid unit in the presence of HOBT to give the coupled product with little or no racemization (eq 5). *p*-Nitrophenyl and pentachlorophenyl esters can also be employed with similar success. This reaction again is thought to proceed via the hydroxybenzotriazole ester intermediate and it has been shown that use of the potassium salt of HOBT in conjunction with a crown ether can lead to substantial rate increase,[10] presumably by enhancing the rate of formation of the active ester intermediate.

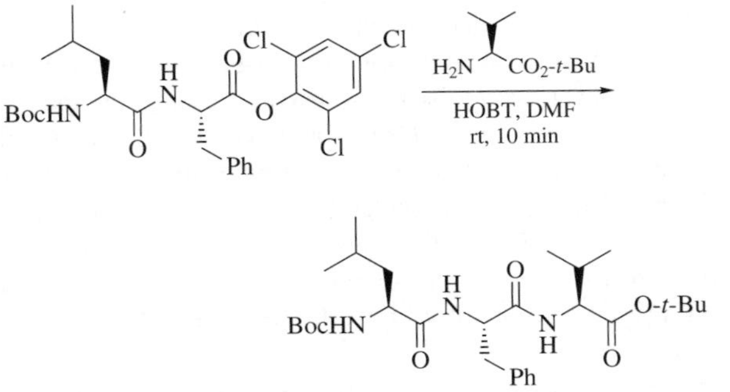

HOBT, DMF
rt, 10 min
(5)
> 2% DL

HOBT has also been used in conjunction with amino acid chlorides to facilitate peptide bond formation. This has proved of particular use in the case of FMOC-protected amino acid chlorides in solid-phase peptide synthesis where direct coupling is rather slow and suffers from competing oxazolone formation.[11]

As mentioned earlier, HOBT has also been incorporated into the peptide coupling reagent itself and one example of such a reagent is benzotriazol-1-yloxytris(dimethylamino)phosphonium chloride (BOP).[12] BOP is prepared by the reaction of HOBT with ***Hexamethylphosphorous Triamide*** in the presence of carbon tetrachloride, and is very effective in the coupling of amino acids (eq 6). See also ***Benzotriazol-1-yloxytris(dimethylamino)phosphonium Hexafluorophosphate*** (also known as BOP).

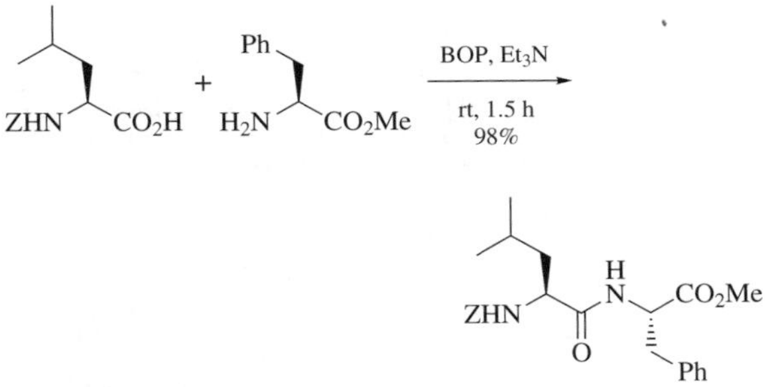

BOP, Et_3N
rt, 1.5 h
98%
(6)

A second peptide coupling reagent that incorporates the HOBT unit is ***O-Benzotriazol-1-yl-N,N,N′,N′-tetramethyluronium Hexafluorophosphate*** (HBTU),[13] which can be prepared by the reaction of tetramethylurea with oxalyl chloride followed by treatment with HOBT and ***Potassium Hexafluorophosphate***. HBTU can then be used as a direct coupling agent as outlined in eq 7.

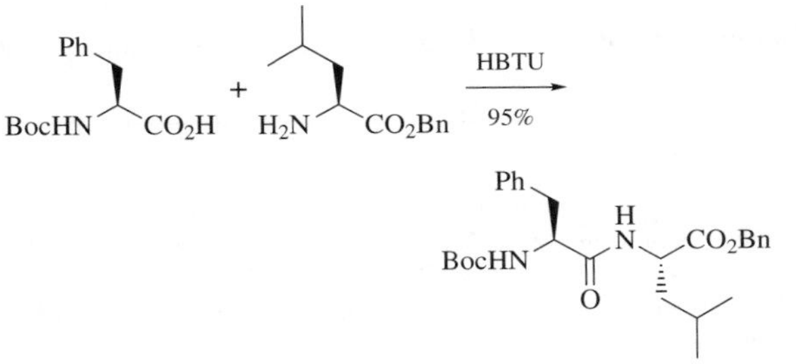

HBTU
95%
(7)

Nucleotide Synthesis. Phosphotriester derivatives of HOBT have been employed in the synthesis of nucleotides.[14] The phosphotriesters are readily formed by reaction of HOBT with the corresponding aryl phosphorodichloridates in the presence of pyridine and can then be used to couple nucleosides as outlined in eq 8. This reaction takes advantage of the fact that a differentially protected nucleoside will react rapidly with the reagent, displacing one molecule of HOBT to form an intermediate phosphotriester (**2**), but at this stage reaction with a second nucleoside molecule is extremely slow due to steric hindrance. Consequently the intermediate can then be reacted with a second nucleoside, the more reactive primary hydroxy displacing the second molecule of HOBT and giving the coupled product. This method for coupling nucleosides has been applied to both solution and solid-phase synthesis of a variety of RNA fragments.

Other Applications. HOBT has also been employed as a catalyst for the conversion of isoamides into maleimides (eq 9).

Ar = 2-chlorophenyl

Related Reagents. 3-Hydroxy-1,2,3-benzotriazine-4(3*H*)-one.

1. For recent general reviews on peptide synthesis involving the use of HOBT, see: Bodanszki, M. *Int. J. Pept. Protein Res.* **1985**, *25*, 449; Kent, S. B. H.; *Annu. Rev. Biochem.* **1988**, *57*, 957; Bodanszky, M. *J. Protein Chem.* **1989**, *8*, 461; Fields, G. B.; Nobel, R. L. *Int. J. Pept. Protein Res.* **1990**, *35*, 161.
2. König, W.; Geiger, R. *CB* **1970**, *103*, 788; König, W.; Geiger, R. *CB* **1970**, *103*, 2024; König, W.; Geiger, R. *CB* **1970**, *103*, 2034; Windridge, G. C.; Jorgensen, E. C. *JACS* **1971**, *93*, 6318; Nagaraj, R.; Balaram, P. *T* **1981**, *37*, 2001; Benoiton, N. L.; Kuroda, K. *Int. J. Pept. Protein Res.* **1981**, *17*, 197; Chen, S. T.; Wu, S. H.; Wang, K. T. *S* **1989**, 37; Dardoize, F.; Goasdoué, C.; Goasdoué, N.; Laborit, H. M.; Topall, G. *T* **1989**, *45*, 7783; Bennoiton, N. L.; Lee, Y. C.; Steinaur, R.; Chen, F. M. F. *Int. J. Pept. Protein Res.* **1992**, *40*, 559; Bennoiton, N. L.; Lee, Y. C.; Chen, F. M. F. *Int. J. Pept. Protein Res.* **1993**, *41*, 587.
3. Coste, J.; Frérot, E.; Jouin, P.; Castro, B. *TL* **1991**, *32*, 1967.
4. Kimura, T.; Takai, M.; Masui, Y.; Morikawa, T.; Sakakibara, S. *Biopolymers* **1981**, *20*, 1823; Hagiwara, D.; Neya, M.; Miyazaki, Y.; Hemmi, K.; Hashimoto, M. *CC* **1984**, 1676.
5. Berrada, A.; Cavelier, F.; Jacquier, R.; Verducci, J. *BSF(1)* **1989**, 511; Grigor'ev, E. I.; Zhil'tsov, O. S. *JOU* **1989**, *25*, 1774; Chen, S. T.; Chang, C. H.; Wang, K. T. *JCR(S)* **1991**, 206.
6. Miyazawa, T.; Otomatsu, T.; Fukui, Y.; Yamada, T.; Kuwata, S. *CC* **1988**, 419; Miyazawa, T.; Otomatsu, T.; Fukui, Y.; Yamada, T.; Kuwata, S. *Int. J. Pept. Protein Res.* **1992**, *39*, 237; Miyazawa, T.; Otomatsu, T.; Fukui, Y.; Yamada, T.; Kuwata, S. *Int. J. Pept. Protein Res.* **1992**, *39*, 308.
7. van der Auwera, C.; van Damme, S.; Anteunis, M. J. O. *Int. J. Pept. Protein Res.* **1987**, *29*, 464.
8. Knorr, R.; Trzeciak, A.; Bannwarth, W.; Gillessen, D. *TL* **1989**, *30*, 1927.
9. König, W.; Geiger, R. *CB* **1973**, *106*, 3626.
10. Horiki, K.; Murakami, A. *H* **1989**, *28*, 615.
11. Carpino, L. A.; Chao, H. G.; Beyermann, M.; Bienert, M. *JOC* **1991**, *56*, 2635; Sivanandaiah, K. M.; Babu, V. V. S.; Renukeshwar, C. *Int. J. Pept. Protein Res.* **1992**, *39*, 201.
12. Castro, B.; Dormoy, J. R.; Evin, G.; Selve, C. *TL* **1975**, 1219.
13. Dourtoglou, V.; Gross, B.; Lambropoulou, V.; Zioudrou, C. *S* **1984**, 572.
14. van der Marel, G.; van Boeckel, C. A. A.; Wille, G.; van Boom, J. H. *TL* **1981**, *22*, 3887; Marugg, J. E.; Tromp, M.; Jhurani, P.; Hoyng, C. F.; van der Marel, G. A.; van Boom, J. H. *T* **1984**, *40*, 73; Gottikh, M.; Ivanovskaya, M.; Shabarova, Z. *Bioorg. Khim.* **1988**, *14*, 500; Hirao, I.; Miura, K. *CL* **1989**, 1799; Colonna, F. P.; Scremin, C. L.; Bonora, G. M. *TL* **1991**, *32*, 3251.

Barry Lygo
Salford University, UK

N-Hydroxypyridine-2-thione

[1121-30-8] C_5H_5NOS (MW 127.18)

(reaction with a carboxylic acid or acid chloride leads to the corresponding *O*-acyl thiohydroxamate; treatment of these intermediates with a radical source leads to alkyl or aryl radicals [R•], the fate of which depends on the precise reactions conditions[1])

Alternate Name: 1-hydroxy-2-(1*H*)-pyridinethione; the tautomeric '*N*-oxide' form *[1121-31-9]*, although the minor component, is often the source of alternate names for this compound, which include 2-pyridinethiol 1-oxide, 2-mercaptopyridine *N*-oxide, 2-mercaptopyridine 1-oxide, and the abbreviated from 'pyrithione'.

Physical Data: mp 70–72 °C.

Form Supplied in: both the pyridinethione and the corresponding sodium salt (sometimes as the hydrate) are commercially available. A 40% aqueous solution of the sodium salt is also available and cheaper (also referred to as sodium omadine). The free thione can be obtained from this by acidification to neutrality using concentrated aq HCl, filtration of the crude product, and crystallization from EtOH. Alternatively, evaporation of the aqueous solution (<50 °C) and crystallization of the residue from ethanol provides the sodium salt, mp 285–290 °C, after slow crystallization from ethanol and drying at 50 °C under vacuum.[3]

Preparative Methods: prepared from 2-bromopyridine by oxidation to the *N*-oxide using either ***Perbenzoic Acid*** or ***Peracetic Acid***, followed by displacement of bromide using either ***Sodium Dithionite*** or ***Sodium Sulfide*** and ***Sodium Hydroxide***.[2] An alternative is to treat the *N*-oxide with ***Thiourea*** and then hydrolyze the resulting thiouronium salt.

Handling, Storage, and Precautions: all operations in this area should be carried out with due regard to both the thermal and photochemical sensitivity of the reagents and intermediate *O*-acyl thiohydroxamates.

Preparation of *O*-Acyl Thiohydroxamates. There are three commonly used options available for the preparation of *O*-acyl thiohydroxamates, outlined in eq 1–;3.[1,4] Firstly, the carboxylic acid can be activated by conversion into a mixed anhydride from ***Isobutyl Chloroformate*** and then coupled directly with the pyridinethione (**1**). Alternatively, *O*-acyl thiohydroxamates (**2**) can be obtained using the DCC (***1,3-Dicyclohexylcarbodiimide***) coupling method (eq 1). If acid chlorides are used, these are best converted into the hydroxamates (**2**) by reaction with the sodium salt (**3**) in the presence of ***4-Dimethylaminopyridine*** (eq 2). A further option is to activate the pyridinethione component by formation of the bicyclic system (**4**) from the parent and ***Phosgene*** and then react this with the triethylammonium salt of the carboxylic acid (eq 3). Compound (**4**) is commercially available as 1-oxa-2-oxo-3-thiaindolizinium chloride (2-oxo-[1.4.2]oxathiazolo[2,3-*a*]pyridinium chloride).[1,4b] The latter two methods appear to be the most favored; the sensitive esters are used promptly with little purification, often just filtration through a short silica column.[3]

R–CO–O–CO–O-*i*-Bu (or RCO_2H + DCC) + (**1**) → (**2**) (1)

R–COCl + (**3**) –[DMAP]→ (**2**) (2)

RCO_2^- $^+HNEt_3$ + (**4**) Cl^- → (**2**) (3)

Decarboxylation. Perhaps the best-known application of this type of pyridinethione derivative in synthesis is in the overall, radical-mediated decarboxylation of carboxylic acids, i.e. $RCO_2H \rightarrow RH$ (eq 4).[1,4b] The transformation is of a wide generality and usually poceeds in good to excellent yields. The chain carrier and hydrogen radical source is either ***Tri-n-butylstannane*** or a tertiary thiol, most often *t*-butanethiol (but see Barton and Crich[6]), and the reactions are generally triggered thermally when tin hydride is employed, or photolytically (tungsten lamp suffices) when a thiol is used.[5] The latter method offers considerable advantages in terms of product purification; for details of this aspect as well as sound advice on how to conduct these and related radical reactions in general, consult Motherwell and Crich.[1]

(**2**) –X–H→ R–H + CO_2 + 2-pyridyl–SX (4)

X = Bu_3Sn, *t*-BuS, $C_{11}H_{23}Me_2CS$

A wide variety of functionality can be incorporated into the substrate acids, such as in the decarboxylation of α-amino acid derivatives which leads to amines (eq 5).[4d] Application of this methodology to the appropriate aspartate and glutamate derivatives provides access to the useful radicals (**5**) which can be trapped in many ways, such as by halides (overall, a Hunsdiecker reaction),[4d] as can many other radical species obtained using this chemistry (see below).

R–CH(NHBoc)–CO–O–N(pyridinethione) –*t*-BuSH, *hv*→ R–CH₂–NHBoc (5)

BocHN–CH(CO₂Bz)–$(CH_2)_n$• (**5**)

Alcohol Deoxygenation. A limitation of the useful Barton–McCombie procedure for the deoxygenation of alcohols is its application to tertiary alcohols because of difficulties in preparing the required xanthate or other derivatives. A solution to this is to first form a monoester with ***Oxalyl Chloride*** and then react the remaining acyl chloride function with the sodium salt (**3**) (eq 6).[6] In these reactions, although *t*-butanethiol is suitable as the chain carrier, in some cases 3-ethyl-3-pentanethiol gives superior yields (50–90%). The intermediate tertiary radicals can be trapped by powerful Michael acceptors such as 1,2-dicyanoethylene.

t-Bu–OH → *t*-BuO–CO–CO–O–N(pyridinethione) → *t*-Bu–H (6)

Decarboxylative Rearrangement. When *O*-acyl thiohydroxamates are simply heated or photolyzed alone, a decarboxylative rearrangement occurs to give 71–92% yields of 2-thiopyridines (eq 7).[4b,7] While the thermal process involves some contribution from a cage recombination mechanism, the photolytic reaction is a purely radical chain process.

R–CO–O–N(pyridinethione) –Δ or *hv*→ R–S–(2-pyridyl) (7)

Formation of Sulfides, Selenides, Tellurides, and Selenocyanates. When *O*-acyl thiohydroxamates are decomposed in the presence of disulfides, diselenides or ditellurides, mixed sulfides, selenides, or tellurides are formed, respectively (eq 8).[8] If the reactions are triggered thermally, a large excess of the disulfide etc. has to be used to avoid competition from decaboxylative rearrangement (see above); however, light-induced reactions proceed at lower temperatures and, in such cases, only a slight excess of the trap is required. When the trapping agent is dicyanogen triselenide then selenocyanates, RSeCN, are produced.

$$R^1CO_2N(\text{pyridine-2-thione}) + R^2X{-}XR^2 \xrightarrow{\Delta \text{ or } h\nu} R^1{-}X{-}R^2 \quad (8)$$

X = S, Se, Te

Decarboxylative Sulfonation. When a solution of an *O*-acyl thiohydroxamate in CH_2Cl_2 containing excess SO_2 is photolyzed at −10 °C, the initially formed radicals are intercepted by the latter before decarboxylative rearrangement can take place. The resulting radicals ($RSO_2\bullet$) then react with a second molecule of the *O*-acyl thiohydroxamate to give an *S*-pyridyl alkylthiosulfonate (38–91%) and the radical R• (eq 9).[9] The products are useful as precursors to unsymmetrical sulfones and sulfonamides, significantly by a nonoxidative procedure.

$$RCO_2N(\text{pyridine-2-thione}) \xrightarrow[CH_2Cl_2,\ -10\,^\circ C]{h\nu,\ SO_2} RSO_2{-}S{-}(2\text{-pyridyl}) \quad (9)$$

The Hunsdiecker Reaction. The classical Hunsdiecker reaction is somewhat restricted due to the relatively harsh conditions required. In the Barton version, alkyl radicals generated from *O*-acyl thiohydroxamates, under either thermal or photolytic conditions, are efficiently trapped either by $Cl_3C{-}X$ (X=Cl or Br; $Cl_3C\bullet$ is the chain carrier) or by ***Iodoform***.[1,4b,10] The method is applicable to sensitive substrates for which the classical methods are unsuitable.[4d,11] thus allowing the preparation of a wide range of alkyl chlorides, bromides, and iodides by the one-carbon degradation of a carboxylic acid. Similar reactions of aromatic acid derivatives tend to require an additional radical initiator (e.g. ***Azobisisobutyronitrile***), if high yields (55–85%) are to be obtained.[12]

Decarboxylative Hydroxylation. A close relative of the foregoing degradation, the transformation $RCO_2H \rightarrow ROH$ can be carried out in two ways. Perhaps the most practical version consists of thermolysis or photolysis of an *O*-acyl thiohydroxamate, typically in oxygen-saturated toluene, followed by reduction of the initial hydroperoxide (eq 10).[4b,13] In similar fashion to decarboxylative sulfonation (see above), the intermediate radicals are trapped much faster ($\sim 10^4$) by oxygen than by H-abstraction from the thiol chain carrier. As an alternative, the initial hydroperoxides can be *O*-tosylated in pyridine to give the corresponding nor-aldehyde or ketone (53–62%). The alternative, mechanistically more convoluted but still efficient and simple procedure (82–93%), has tris(phenylthio)antimony, $(PhS)_3Sb$, as the radical trap in the presence of both oxygen and water (in practice, wet air).[14]

$$RCH_2CO_2N(\text{pyridine-2-thione}) \xrightarrow[O_2]{h\nu \text{ or } \Delta} RCH_2OOH \dashrightarrow RCH_2OH \text{ or } RCHO \quad (10)$$

Decarboxylative Phosphorylation. In similar fashion to the foregoing, radicals generated by decomposition of an *O*-acyl thiohydroxamate can be trapped by tris(phenylthio)phosphine (eq 11).[15] In some cases, the use of alternative thiohydroxamates may be advantageous.

$$RCO_2N(\text{pyridine-2-thione}) \xrightarrow{(PhS)_3P} RP(=O)(SPh)_2 \quad (11)$$

Michael Additions and Other Radical Reactions. As decomposition of *O*-acyl thiohydroxamates provides radicals, these can subsequently participate in a variety of other reactions, the most common of which are Michael additions to electrondeficient alkenes. The most commonly successful pathway (eq 12) involves reaction of the intermediate carbon-centered radical adjacent to the electron-withdrawing group with a second molecule of the thiohydroxamate, rather than hydrogen abstraction, to give an α-*S*-pyridyl derivative,[1,6,7b] which can subsequently be manipulated in a variety of ways.[16] Other thiohydroxamates can lead to higher yields; the mildness and potential of this methodology is illustrated in eq 13.[17]

$$RCO_2N(\text{pyridine-2-thione}) \xrightarrow{CH_2{=}CHCO_2Me} RCH_2CH(SPy)CO_2Me \quad (12)$$

$$\text{(dioxolane }CO_2Me\text{ ester)} \xrightarrow[h\nu \text{ or } \Delta]{CH_2{=}CHCO_2Me} \text{product with } CH_2CH(SPy)CO_2Me \quad (13)$$

Similar reactions, but using nitroalkenes as traps, lead to the corresponding α-pyridylthio nitroalkanes, which are useful as precursors to the one-carbon homologated carboxylic acids, aldehydes, or the related ketones, by starting with an α-substituted nitroalkene (eq 14).[7b,18]

$$R^1{-}CO_2H \dashrightarrow R^1CH(R^2)CH(NO_2)SPy \dashrightarrow R^1CH(R^2)CO_2H \text{ or } R^1C(R^2){=}O \quad (14)$$

Radicals derived from decarboxylation of perfluorocarboxylic acids will undergo addition to ethyl vinyl ether to provide the expected homologs in ~60% yields (eq 15).[19]

$$CF_3(CF_2)_nCO_2N(\text{pyridine-2-thione}) \xrightarrow[h\nu]{CH_2{=}CHOEt} CF_3(CF_2)_nCH_2CH(OEt)SPy \quad (15)$$

Protonated pyridines can also be used as acceptors of carbon radicals generated from thiohydroxamates in general.[20] A useful extension of the Michael addition reacations is to incorporate an additional *t*-butylthio substituent into the acceptor component (eq 16).[18b,21] Tandem reactions are also possible, as exemplified by the conversion of a cyclopentenecarboxylic acid into a bicyclo[2.2.1]heptane (eq 17).[22]

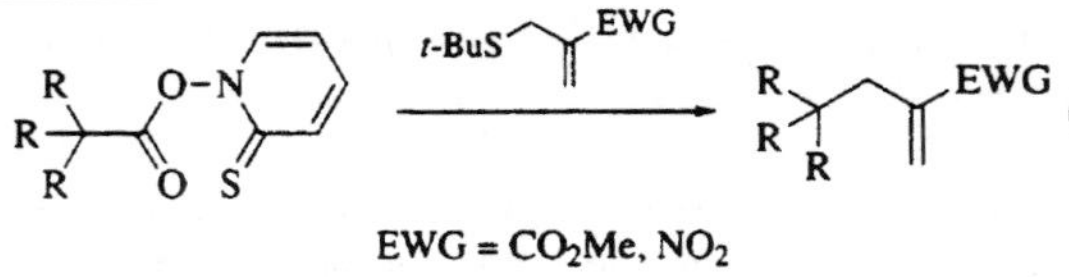

(16)

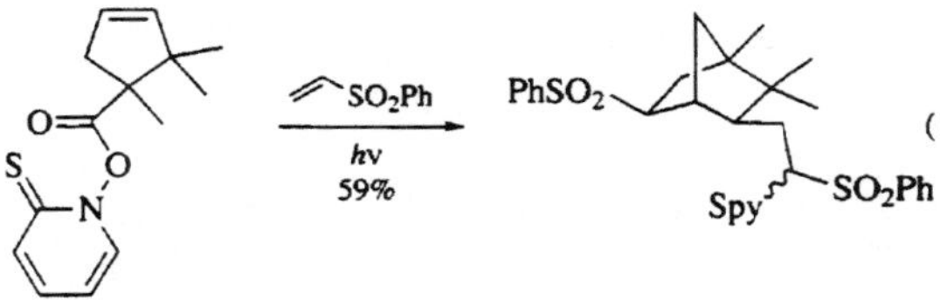

EWG = CO_2Me, NO_2

(17)

Carbon radicals obtained from thiohydroxamates can also be trapped intramolecularly by unactivated alkene functions especially in the 5-*exo*-trig mode,[7b,23] as can aminyl radicals generated in the same way; better yields of cyclic products are obtained in the presence of a weak acid which presumably protonates the *N*-centered radical prior to cyclization (eq 18).[24]

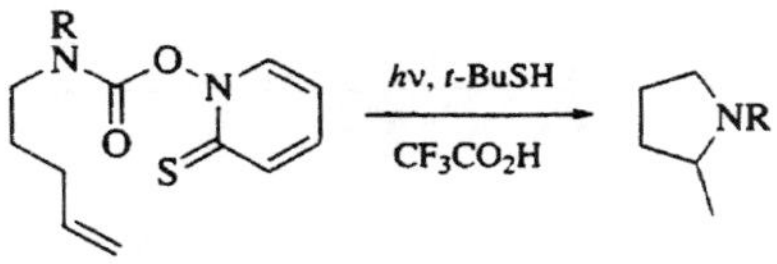

(18)

Related Reagents. Mercury(II) Oxide-Bromine; Thallium(I) Carbonate; Thallium(III) Acetate-Bromine.

1. For an excellent overview of this area, including examples, see Motherwell, W. B.; Crich, D. *Free Radical Chain Reactions in Organic Synthesis*, Academic: London, 1992.
2. Shaw, E.; Bernstein, J.; Losee, K.; Lott, W. A. *JACS* **1950**, *72*, 4362.
3. Barton, D. H. R.; Bridon, D.; Fernandez-Picot, I.; Zard, S. Z. *T* **1987**, *43* 2733.
4. (a) Barton, D. H. R.; Crich, D.; Motherwell, W. B. *CC* **1983**, 939. (b) Barton, D. H. R.; Crich, D.; Motherwell, W. B. *T* **1985**, *41*, 3901. (c) Barton, D. H. R. Herve, Y.; Potier, P.; Thierry, J. *CC* **1984**, 1298. (d) Barton, D. H. R.; Herve, Y.; Potier, P.; Thierry, J. *T* **1988**, *44*, 5479.
5. See, for example: Della, E. W.; Tsanaktsidis, J. *AJC* **1986**, *39*, 2061; Ihara, M.; Suzuki; M.; Fukumoto, K.; Kametani, T.; Kabuto, C. *JACS* **1988**, *110*, 1963; Crich, D.; Ritchie, T. J. *CC* **1988**, 1461; Braeckman, J. C.; Daloze, D.; Kaisin, M.; Moussiaux, B. *T* **1985**, *41*, 4603; Campopiano, O.; Little, R. D.; Petersen, J. L. *JACS* **1985**, *107*, 3721; Otterbach, A.; Musso, H. *AG(E)* **1987**, *26*, 554; Winkler, J. D.; Sridar, V. *JACS* **1986**, **108**, 1708; Winkler, J. D.; Hey, J. P.; Willard, P. G. *JACS*, **1986**, *108*, 6425; Winkler, J. D.; Henegar, K. F.; Williard, P. G. *JACS* **1987**, *109*, 2850.
6. Barton, D. H. R.; Crich, D. *JCS(P1)* **1986**, 1603.
7. (a) Barton, D. H. R.; Crich, D.; Potier, P. *TL* **1985**, *26*, 5943. (b) Barton, D. H. R.; Crich, D.; Kretzschmar, G. *JCS(P1)* **1986**, 39.
8. Barton, D. H. R.; Bridon, D.; Zard, S. Z. *TL* **1984**, *25*, 5777; *H* **1987**, *25*, 449; Barton, D. H. R.; Bridon, D.; Herve, Y.; Potier, P.; Thierry, J.; Zard, S. Z. *T* **1986**, *42*, 4983.
9. Barton, D. H. R.; Lacher, B.; Misterkiewicz, B.; Zard, S. Z. *T* **1988**, *44*, 1153.
10. Barton, D. H. R.; Crich, D.; Motherwell, W. B. *TL* **1983**, *24*, 4979.
11. See, for example: Fleet, G. W. J.; Son, J. C.; Peach, J. M.; Hamor, T. A. *TL* **1988**, *29*, 1449; Rosslein; L.; Tamm, C. *HCA* **1988**, *71*, 47; Kamiyama, K.; Kobayashi, S.; Ohno, M. *CL* **1987**, 29.
12. Vogel, E.; Schieb, T.; Schulz, W. H.; Schmidt, K.; Schmickler, H.; Lex, J. *AG(E)* **1986**, *25*, 723; Barton, D. H. R.; Lacher, B.; Zard, S. Z. *TL* **1985**, *26*, 5939; *T* **1987**, *43*, 4321.
13. Barton, D. H. R.; Crich, D.; Motherwell, W. B. *CC* **1984**, 242.
14. Barton, D. H. R.; Crich, D.; Motherwell, W. B. *CC* **1984**, 242.
15. Barton, D. H. R.; Bridon, D.; Zard, S. Z. *TL* **1986**, *27*, 4309.
16. See for example; Ahmad-Junan, S. A.; Walkington, A. J.; Whiting, D. A. *JCS(P1)* **1992**, 2313.
17. Barton, D. H. R.; Gateau-Olesker, A.; Gero, S. D.; Lacher, B.; Tachdjian, C.; Zard, S. Z. *CC* **1987**, 1790.
18. (a) Barton, D. H. R.; Togo, H.; Zard, S. Z. *TL* **1985**, *26*, 6349; (b) *T* **1985**, *41*, 5507. (c) Barton, D. H. R.; Herve, Y.; Potier, P.; Thierry, J.*T* **1987**, *43*, 4297.
19. Barton, D. H. R.; Lacher, B.; Zard, S. Z. *T* **1986**, *42*, 2325.
20. Barton, D. H. R.; Garcia, B.; Togo, H.; Zard, S. Z. *TL* **1986**, *27*, 1327.
21. Barton, D. H. R.; Crich, D. *JCS(P1)* **1986**, 1613.
22. Barton, D. H. R.; da Silva, E.; Zard, S. Z. *CC* **1988**, 285.
23. See, for example: Green, S. P.; Whiting, D. A. *CC* **1992**, 1754.
24. Newcomb, M.; Deeb, T. M.*JACS* **1987**, *109*, 3163.

David W. Knight
Nottingham University, UK

N-Hydroxysuccinimide

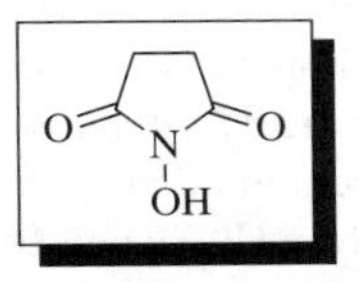

[6066-32-6] $C_4H_5NO_3$ (MW 115.10)

(activating agent for carboxylic acids in amide synthesis and related coupling reactions; acyl transfer reagent)

Alternate Name: HOSu.
Physical Data: mp 96–98 °C (99–100 °C).
Solubility: sol H_2O, DMF, alcohols, EtOAc; insol cold ether.
Form Supplied in: colorless crystalline solid, widely available.
Preparative Methods: by heating ***Succinic Anhydride*** with ***Hydroxylamine*** or, better, hydroxylamine hydrochloride followed by crystallizations from ether, 1-butanol, and finally EtOAc.[1]
Purification: can be crystallized from 1-butanol, EtOAc, or EtOH–EtOAc.

Peptide Bond Formation. The most important use of *N*-hydroxysuccinimide is in the formation of (isolable) activated derivatives of N^{α}-protected α-amino acids, which subsequently undergo generally smooth coupling reactions with amino esters. Typically, ***1,3-Dicyclohexylcarbodiimide*** (DCC) is used as the initial coupling agent (the HOSu–DCC method) (eq 1).[1–3] *N*-Hydroxyphthalimide can be employed in much the same way, but a distinct advantage in the case of HOSu is its high water solubility, which facilitates product purification.

$$R^1COHNCH(R^2)CO_2H \xrightarrow[DCC]{HOSu} R^1COHNCH(R^2)CO_2Su \xrightarrow{H_2NCH(R^3)CO_2R^4} R^1COHNCH(R^2)CONHCH(R^3)CO_2R^4 + HOSu \quad (1)$$

In some cases, up to 3% racemization has been observed when using the HOSu–DCC method, which renders it unsuitable for some types of peptide synthesis.[4,5] Esters derived from HOSu are relatively reactive, as shown by a kinetic study of their rates of hydrolysis in aqueous buffers[6] although, in some examples, *endo-N*-hydroxy-5-norbornene-2,3-dicarboximide has proven superior to HOSu in peptide bond formation.[7] Of course, the HOSu–DCC method is not limited to peptide synthesis and is well suited to the preparation of amides in general, both from ammonia and primary amines (e.g. eq 2).[8]

$$\text{1. HOSu, DCC; 2. } NH_3 \rightarrow \text{RCONH}_2 \quad (2)$$

In order to avoid the need for DCC, a number of activated derivatives of *N*-hydroxysuccinimide itself have been developed which react directly with, for example, an α-amino acid to provide the required activated α-amino acid esters. These include the commercially available carbonate (**1**), a stable crystalline solid obtained from HOSu and ***Trichloromethyl Chloroformate*** or from *O*-trimethylsilyloxysuccinimide and ***Phosgene***,[9] and the related phosphate (**2**) derived from HOSu and ***Diphenyl Phosphorochloridate*** under Schotten–Baumann conditions.[10] Another alternative (also commercially available) is the oxalate (**3**), formed from HOSu and ***Oxalyl Chloride***.[11] In general, these intermediates react rapidly with an N^α-protected α-amino acid, usually in the presence of a mild base such as pyridine, to provide excellent yields of the required hydroxysuccinimide esters, generally with less racemization than is sometimes associated with the HOSu–DCC method. It is also possible to carry out the entire process of peptide synthesis in one pot using these reagents.

(1) (2) (3)

A rather different approach to peptide bond formation and amide synthesis in general is to treat a mixture of a carboxylic acid and an amine with an isocyanide (***2-Morpholinoethyl Isocyanide*** is especially suitable), which effectively acts as a dehydrating agent (eq 3). The procedure can result in extensive racemization of both reactants and products which may be supressed by the addition of HOSu; presumably an HOSu ester is the penultimate intermediate.[12] The addition of HOSu also decreases racemization in polypeptide synthesis when 'Bates reagent' $\{[(Me_2N)_3P^+]_2O\ (BF_4^-)_2\}$ is used as the coupling agent.[13]

$$R^1CO_2H + H_2NR^2 + R^3NC \longrightarrow R^1CONHR^2 + R^3NHCHO \quad (3)$$

A rather unusual coupling reaction between HOSu and glyoxylic acid tosylhydrazone leads to succinimidyl diazoacetate (eq 4); again, the leaving ability of the HOSu residue renders this a useful compound for effecting the direct transfer of a diazoacetyl function to amines, phenols, and peptides.[14]

$$HOSu + HO_2CCH{=}N\text{-}NH\text{-}p\text{-Tol} \longrightarrow SuOCOCHN_2 \quad (4)$$

Modified Barton–McCombie intermediates are best prepared from pentafluorophenyl chlorothionoformate and a catalytic quantity (15–20%) of HOSu in refluxing benzene rather than by using ***4-Dimethylaminopyridine***, the rather more conventional catalyst of acyl group transfer (eq 5).[15]

$$ROH + ClC(S)OC_6F_5 \longrightarrow ROC(S)OC_6F_5 \dashrightarrow RH \quad (5)$$

1. Anderson, G. W.; Zimmerman, J. E.; Callahan, F. M. *JACS* **1964**, *86*, 1839; Wegler, R.; Grewe, F.; Mehlhose, K. U.S. Patent 2 816 111, 1957 (*CA* **1958**, *52*, 6405i).
2. For a review, see: Klausner, Y. S.; Bodansky, M. *S* **1972**, 453.
3. For examples, see: Wunsch, E.; Drees, F. *CB* **1966**, *99*, 110; Manesis, N. J.; Goodman, M. *JOC* **1987**, *52*, 5331; Mukaiyama, T.; Goto, K.; Matsuda, R.; Ueki, M. *TL* **1970**, 1901; Bosshard, H. R.; Schechter, I.; Beger, A. *HCA* **1973**, *56*, 717.
4. Kemp, D. S.; Trangle, M.; Trangle, K. *TL* **1974**, 2695.
5. For a review of side reactions in peptide synthesis, see: Martinez, J. *S* **1981**, 333.
6. Cline, G. W.; Hanna, S. B. *JOC* **1988**, *53*, 3583.
7. Fujino, M.; Kobayashi, S.; Obayashi, M.; Fukuda, T.; Shinagawa, S.; Nishimura, O. *CPB* **1974**, *22*, 1857.
8. Terao, S.; Shiraishi, M.; Kato, K.; Ohkawa, S.; Ashida, Y.; Maki, Y. *JCS(P1)* **1982**, 2909.
9. Ogura, H.; Kobayashi, T.; Shimizu, K.; Kawabe, K.; Takeda, K. *TL* **1979**, 4745.
10. Ogura, H.; Nagai, S.; Takeda, K. *TL* **1980**, *21*, 1467.
11. Takeda, K.; Sawada, I.; Suzuki, A.; Ogura, H. *TL* **1983**, *24*, 4451.
12. Wackerle, L. *S* **1979**, 197; Marquarding, D.; Aignar, H. Ger. Offen. 2 942 606, 1979 (*CA* **1981**, *95*, 62 731j).
13. Bates, A. J.; Galpin, I. J.; Hallett, A.; Hudson, D.; Kenner, G. W.; Ramage, R.; Sheppard, R. C. *HCA* **1975**, *58*, 688.
14. Ouihia, A.; Rene, L.; Guilhem, J.; Pascard, C.; Badet, B. *JOC* **1993**, *58*, 1641.
15. Barton, D. H. R.; Jaszberenyi, J. Cs. *TL* **1989**, *30*, 2619. For an application, see: Gervay, J.; Danishefsky, S. *JOC* **1991**, *56*, 548.

David W. Knight
Nottingham University, UK

I

Imidazole

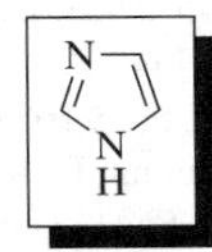

[288-32-4] $C_3H_4N_2$ (MW 68.09)

(nucleophilic catalyst for silylations and acylations; buffer; weak base; iodination methods)

Alternate Names: Im; iminazole; 1,3-diazole; glyoxaline.
Physical Data: mp 90–91 °C; bp 255 °C, 138 °C/12 mmHg.
Solubility: sol water, alcohols, ether, acetone, chloroform.
Form Supplied in: colorless crystalline solid; widely available.
Drying: 40 °C in vacuo over P_2O_5.
Purification: can be crystallized from C_6H_6, CCl_4, CH_2Cl_2, EtOH, petroleum ether, acetone–petroleum ether, or water; can also be purified by vacuum distillation, sublimation, or by zone refining.

Introduction. Imidazole has a pK_a of 7.1 and is thus a stronger base than thiazole (pK_a 2.5), oxazole (pK_a 0.8), and pyridine (pK_a 5.2). It is both a good acceptor and donor of hydrogen bonds. The pK_a for loss of the N–H is ~14.2, i.e. imidazole is a very weak acid.

Silylations. Imidazole is a standard component in silylations of alcohols as well as carboxylic acids, amines, and a variety of other functions, typically in combination with a silyl chloride in DMF (eq 1).[1] A very widely used procedure for alcohol protection is by conversion into the corresponding *t*-butyldimethylsilyl (TBDMS or TBS) ether using the method;[2] in other solvents such as pyridine or THF the reactions are much slower, probably because the primary silylating reagent is $t\text{-BuMe}_2\text{Si–Im}$. In this and many other aspects, imidazole resembles another very useful transfer catalyst, ***4-Dimethylaminopyridine*** (DMAP). Similarly, bulkier and hence more stable silyl groups, such as *t*-butyldiphenylsilyl (TBDPS)[3] and triisopropylsilyl (TIPS),[4] can be introduced. Times for completion of reaction at 20 °C vary (0.5–20 h); these silylating agents, originally developed for nucleoside protection,[5] usually react faster with primary alcohols and with certain secondary alcohols, thus allowing selective protection of polyols to be achieved efficiently. (A faster alternative involves the use of ***1,8-Diazabicyclo[5.4.0]undec-7-ene*** (DBU) in place of imidazole, in a variety of solvents such as CH_2Cl_2, C_6H_6, or MeCN, in combination with R_3SiCl).[6] 1,3-Diones can be efficiently *O*-silylated using TBDMSCl–Im[7] or ***Hexamethyldisilazane*** (HMDS) and imidazole;[8] other reagents are not as suitable, even in cases where the enol content is high. The products are useful as *trans*-silylating agents.[7] The HMDS–Im combination is also useful for the silylation of thiols.[9]

$$R^1\text{CH}_2\text{OH} + R^2_3\text{SiCl} \xrightarrow[\text{DMF}]{\text{imidazole}} R^1\text{CH}_2\text{OSiR}^2_3 \quad (1)$$

Ester Hydrolysis. Inspired by evidence that the imidazole ring of histidine residues present in various hydrolytic enzymes is responsible for their proteolytic activities, imidazole itself has been shown to be an excellent catalyst of ester hydrolysis (e.g. eq 2).[10] In intramolecular transesterifications and hydrolyses of 2-hydroxymethylbenzoic acid derivatives, the accelerating role of imidazole is due to its ability to act as a proton transfer catalyst rather than as a nucleophile.[11]

(eq 2: triacetoxyflavone (AcO, OAc, OAc) → imidazole, 60% aq EtOH, Δ → trihydroxyflavone (HO, OH, OH)) (2)

Peptide Coupling. Peptide couplings involving *p*-nitrophenyl and related esters (see ***p-Nitrophenol***) are dramatically accelerated by the addition of imidazole.[12] However, such reactions, which probably proceed by way of an acylimidazole, can be prone to racemization, in which case ***1,2,4-Triazole*** can be a superior activator.[13] Imidazole also catalyzes peptide coupling using the ***Triphenyl Phosphite*** method, with negligible racemization when the reactions are carried out in dioxane or DMF,[14] and is useful for the activation of phosphomonoester groups in nucleotide coupling, in combination with an arylsulfonyl chloride.[15]

Acylimidazoles and Nucleophiles. Acylimidazoles are readily prepared from the parent carboxylic acids by reaction of the derived acid chloride with imidazole or directly using ***N,N'-Carbonyldiimidazole***. These intermediates react smoothly with a variety of nucleophiles including Grignard reagents (eq 3),[16] ***Lithium Aluminum Hydride*** (eq 4),[16] and nitronates (eq 5).[17] At −20 °C, aroylimidazoles can be reduced to the corresponding aldehydes in the presence of an ester function.[16]

$$\text{RCO–Im} \xrightarrow{\text{PhMgBr}} \text{RCOPh} \quad (3)$$

$$\text{RCO–Im} \xrightarrow{\text{LiAlH}_4} \text{RCHO} \quad (4)$$

$$\text{RCO–Im} \xrightarrow{\text{CH}_2{=}\text{NO}_2^-} \text{RCOCH}_2\text{NO}_2 \quad (5)$$

The activation provided by an imidazole substituent is further illustrated in a route to 1,3-oxathiole-2-thiones from sodium 1-imidazolecarbodithioate, derived from the sodium salt of imidazole and CS_2 (eq 6).[18]

(6)

Other Uses. Imidazole is one of the best catalysts for the preparation of acid chlorides from the corresponding carboxylic acids and phosgene.[19] Aryl triflates can be obtained from phenols, or better phenolates, using ***Trifluoromethanesulfonic Anhydride*** in combination with imidazole; *N*-triflylimidazole is the reactive species.[20] Photochemical deconjugations of enones can be erratic but are promoted by the presence of a weak base such as imidazole or pyridine in polar solvents.[21]

Iodination of Alcohols. Imidazole, ***Triphenylphosphine***, and ***Iodine*** in hot toluene,[22] or preferably toluene–acetonitrile mixtures,[23] is an excellent combination for the conversion of alcohols into iodides, ROH → RI. Secondary alcohols react with inversion (but see below), although the method can be used for the selective iodination of primary hydroxyls. Applications in the area of natural product synthesis[24] emphasize the mildness and generality of the method as well as providing alternative recipes; sometimes, 2,4,5-triiodoimidazole can be a superior reagent.[22] Similarly, $Ph_3P–Cl_2$–Im can be used for the preparation of alkyl chlorides, ROH → RCl, and the addition of imidazole in the ***Triphenylphosphine–Carbon Tetrachloride*** method for alcohol chlorination has a beneficial effect.[25]

Diol Deoxygenation. The Ph_3P–Im–I_2 combination can also be used to convert *vic*-diols into the corresponding alkenes, although 2,4,5-triiodoimidazole is more effective than imidazole itself.[26] Alternative combinations are ***Triphenylphosphine–Iodoform–Imidazole***, which can deoxygenate *cis*-diols but which is better suited to *trans*-isomers,[27] and ***Chlorodiphenylphosphine***–I_2–Im, which can be used to deoxygenate *vic*-diols when both are secondary or when one is secondary and one is primary.[28]

Epoxidation. ***t-Butyl Hydroperoxide*** in combination with $MoO_2(acac)_2$ can be used to oxidatively cleave alkenes, but will epoxidize such functions in the presence of a metalloporphyrin or a simple amine, the choice of which depends upon the substrate structure. Imidazole is the most suitable for 1-phenylpropene, ***Pyridine*** for stilbene, and *N,N*-dimethylethylenediamine for 1-alkenes.[29]

---

1. Lalonde, M.; Chan, T. H. *S* **1985**, 817. Greene, T. W.; Wuts, P. G. M. *Protecting Groups in Organic Synthesis*, 2nd ed.; Wiley: New York, 1991. Kocieński, P. J. *Protecting Groups*; Thieme: Stuttgart, 1994.
2. Corey, E. J.; Venkateswarlu, A. *JACS* **1972**, *94*, 6190.
3. Hanessian, S.; Lavallee, P. *CJC* **1975**, *53*, 2975.
4. Cunico, R. F.; Bedell, L. *JOC* **1980**, *45*, 4797.
5. Ogilvie, K. K. *CJC* **1973**, *51*, 3799. Ogilvie, K. K.; Iwacha, D. J. *TL* **1973**, 317. Ogilvie, K. K.; Sadana, K. L.; Thompson, E. A.; Quilliam, M. A.; Westmore, J. B. *TL* **1974**, 2861. Ogilvie, K. K.; Thompson, E. A.; Quilliam, M. A.; Westmore, J. B. *TL* **1974**, 2865.
6. Aizpurua, J. M.; Palomo, C. *TL* **1985**, *26*, 475.
7. Veysoglu, T.; Mitscher, L. A. *TL* **1981**, *22*, 1299, 1303.
8. Torkelson, S.; Ainsworth, C. *S* **1976**, 722.
9. Glass, R. S. *JOM* **1973**, *61*, 83.
10. Bender, M. L.; Turnquest, B. W. *JACS* **1957**, *79*, 1652. Bruice, T. C.; Schmir, G. L. *JACS* **1957**, *79*, 1663. Bender, M. L. *CRV* **1960**, *60*, 82. Looker, J. H.; Holm, M. J.; Minor, J. L.; Kagal, S. A. *JHC* **1964**, *1*, 253.
11. Fife, T. H.; Benjamin, B. M. *JACS* **1973**, *95*, 2059. Kirby, A. J.; Lloyd, G. J. *JCS(P2)* **1974**, 637. Chiong, K. N. G.; Lewis, S. D.; Shafer, J. A. *JACS* **1975**, *97*, 418. Pollack, R. M.; Dumsha, T. C. *JACS* **1975**, *97*, 377. Belke, C. J.; Su, S. C. K.; Shafer, J. A. *JACS* **1971**, *93*, 4552.
12. Mazur, R. H. *JOC* **1963**, *28*, 2498. Wieland, T.; Vogeler, K. *AG(E)* **1963**, *2*, 42; *LA* **1964**, *680*, 125. McGahren, W. J.; Goodman, M. *T* **1967**, *23*, 2017. Stewart, F. H. C. *CI(L)* **1967**, 1960.
13. Beyerman, H. C.; van der Brink, W. M.; Weygand, F.; Prox, A.; Konig, W.; Schmidhammer, L.; Nintz, E. *RTC* **1965**, *84*, 213.
14. Mitin, Y. V.; Glinskaya, O. V. *TL* **1969**, 5267.
15. Berlin, Yu. A.; Chakhmakhcheva, O. G.; Efimov, V. A.; Kolosov, M. N.; Korobko, V. G. *TL* **1973**, 1353.
16. Staab, H. A.; Braunling, H. *LA* **1962**, *654*, 119. Staab, H. A.; Jost, E. *LA* **1962**, *655*, 90. Staab, H. A. *AG(E)* **1962**, *1*, 351.
17. Baker, D. C.; Putt, S. R. *S* **1978**, 478.
18. Ishida, M.; Sugiura, K.; Takagi, K.; Hiraoka, H.; Kato, S. *CL* **1988**, 1705.
19. Hauser, C. F.; Theiling, L. F. *JOC* **1974**, *39*, 1134.
20. Effenberger, F.; Mack, K. E. *TL* **1970**, 3947.
21. Eng, S. L.; Ricard, R.; Wan, C. S. K.; Weedon, A. C. *CC* **1983**, 236.
22. Garegg, P. J.; Samuelsson, B. *CC* **1979**, 978; *JCS(P1)* **1980**, 2866.
23. Garegg, P. J.; Johansson, R.; Ortega, C.; Samuelsson, B. *JCS(P1)* **1982**, 681.
24. Corey, E. J.; Pyne, S. G.; Su, W. *TL* **1983**, *24*, 4883. Berlage, U.; Schmidt, J.; Peters, U.; Welzel, P. *TL* **1987**, *28*, 3091. Corey, E. J.; Nagata, R. *TL* **1987**, *28*, 5391. Soll, R. M.; Seitz, S. P. *TL* **1987**, *28*, 5457.
25. Garegg, P. J.; Johansson, R.; Samuelsson, B. *S* **1984**, 168.
26. Garegg, P. J.; Samuelsson, B. *S* **1979**, *469*, 813.
27. Bessodes, M.; Abushanab, E.; Panzica, R. P. *CC* **1981**, 26.
28. Liu, Z.; Classon, B.; Samuelsson, B. *JOC* **1990**, *55*, 4273.
29. Kato, J.; Ota, H.; Matsukawa, K.; Endo, T. *TL* **1988**, *29*, 2843.

David W. Knight
Nottingham University, UK

Iodomethane

MeI

[74-88-4] CH_3I (MW 141.94)

(methylating agent for carbon, oxygen, nitrogen, sulfur, and trivalent phosphorus)

Alternate Name: methyl iodide.
Physical Data: bp 41–43 °C; *d* 2.28 g cm^{-3}.
Solubility: sol ether, alcohol, benzene, acetone; moderately sol H_2O.
Form Supplied in: colorless liquid; stabilized by addition of silver wire or copper beads; widely available.

Purification: percolate through silica gel or activated alumina then distill; wash with dilute aqueous $Na_2S_2O_3$, then wash with water, dilute aqueous Na_2CO_3, and water, dry with $CaCl_2$ then distill.

Handling, Storage, and Precautions: toxic, corrosive and a possible carcinogen. Liquid should be stored in brown bottles to prevent liberation of I_2 upon exposure to light. Keep in a cool, dark place. Use only in well ventilated areas.

***C*-Methylation.** Methyl iodide is an 'active' alkylating agent employed in the *C*-methylation of carbanions derived from ketones, esters, carboxylic acids, amides, nitriles, nitroalkanes, sulfones, sulfoxides, imines, and hydrazones.[1] The quantity of methyl iodide utilized in methylations varies from a slight (1.1 equiv) to a large excess (used as solvent).

The monomethylation of carbanions derived from 1,3-cyclohexanedione and acetylacetone has been described (eqs 1 and 2).[2] Selective monomethylation of β-diketones is dependent upon the base employed; variable amounts of *O*-methylation, dimethylation, and carbon–carbon bond cleavage may occur. The tetraethylammonium enolate of β-diketones reportedly provides higher yields of *C*-methylation without competing side reactions (eq 3).[3] Dimethylation is sometimes a desired reaction pathway. In this case, a large excess of both methyl iodide and base favors the dimethylated product (eq 4).[4] Recently, the combination of a potassium base and a catalytic amount of ***18-Crown-6*** (eq 5) has been described to provide a higher yield of dimethylation.[5,6]

MeI; dioxane (aq); 54–56% (1)

MeI; K_2CO_3, acetone; 75–77% (2)

MeI; DMF; Et_4N^+; 92% (3)

t-BuOK, t-BuOH; MeI (xs); 57% (4)

MeI (2.2 equiv); t-BuOK, THF; 18-crown-6; 83% (5)

Methylation of kinetically derived enolates is most readily accomplished via the corresponding silyl enol ether.[7] Lithium enolates, generated by treatment of the silyl enol ether with ***Methyllithium***, may then be alkylated with methyl iodide (eq 6).[8] Alternatively, quarternary ammonium enolates are produced by treatment of the silyl enol ether with the corresponding fluoride salt. For example, monomethylation of a ketone is cleanly effected by treatment of an anhydrous mixture of the trimethylsilyl enol ether of the ketone in methyl iodide with benzyltrimethylammonium fluoride (eq 7).[9] Kinetic enolates produced from the conjugate addition of an organocuprate to an unsaturated ketone or a dissolving metal reduction of an enone may be methylated directly.[7c] However, the choice of solvent is crucial for the success of these reactions. Ether, the solvent typically used in organocopper conjugate additions, is a poor solvent for alkylation reactions.[10] ***N,N,N′,N′-Tetramethylethylenediamine*** (TMEDA), ***Hexamethylphosphoric Triamide*** (HMPA) and liquid ***Ammonia*** have been used as additives to increase the efficiency of alkylation. Alternatively, the solvent used in the conjugate addition can be removed in vacuo and replaced with a more effective medium for alkylation. For example the rate of methylation of the enolate produced in the conjugate addition of ***Lithium Dimethylcuprate*** to an enone is approximately 10^5 times faster in DME than in ether (eq 8).[11]

TMSO; MeLi; MeI; 80% (6)

BnO; OTMS; $BnNMe_3^+F^-$; MeI (xs), rt; 71% (7)

CHSBu; 1. Me_2CuLi; ether; 2. removal of solvent; 3. MeI, DME; <30 s, 0 °C; 93%; 2. MeI, ether; ~5 h, rt (8)

The stereoselectivity of the methylation of ketone enolates is determined by the structure of the substrate.[12] Stereoselective methylation of cyclic ketone enolates has been examined in detail and current models reliably predict the stereochemical outcome (eqs 9–11).[13–15] Diastereoselective methylation of acyclic ketone and ester enolates has been accomplished employing a variety of chiral auxiliaries (eq 12).[12,16] Efficient catalytic enantioselective methylation of 6,7-dichloro-5-methoxy-1-indanone has been accomplished via a chiral phase-transfer catalyst (eq 13).[17] An enantiomeric excess of 92% was observed when employing ***Chloromethane*** as the methylating agent, whereas methyl iodide provided a product of only 36% enantiomeric excess.

$LiNH_2$; MeI (xs), PhH; 97% — 83:17 (9)

LDA; THF, –78 °C; MeI (1.2 equiv); 90% — >97:3 (10)

t-BuOK; MeI (xs); 70% (11)

LDA, THF, –78 °C; MeI (5 equiv); 77% — 91:9 (12)

50% NaOH, PhMe; MeCl; 95% — 92% ee (13)

O-Methylation. Carboxylic acids can be converted to the corresponding methyl ester by stirring a mixture of the carboxylic acid in methanol with an excess of methyl iodide and ***Potassium Carbonate***.[18] A recent report describes the esterification of carboxylic acids using ***Cesium Fluoride*** and methyl iodide in DMF (eq 14).[19] ***Dimethyl Sulfate*** has also been advantageously utilized to effect *O*-methylation of carboxylic acids as well as alcohols (eq 15).[18c,20] These methods often serve as useful alternatives to ***Diazomethane*** for preparative scale esterification of carboxylic acids.

CsF, MeI; DMF; 85% (14)

$(MeO)_2SO_2$; $NaHCO_3$ (aq); 40 °C; 72% (15)

Phenolic hydroxyls are readily methylated by methyl iodide under basic conditions. The most common conditions are methyl iodide and potassium carbonate in acetone (eq 16).[18a,21] The use of ***Lithium Carbonate*** as the base allows for the selective protection of phenols with a $pK_a < 8$ (eq 17).[18c] The *peri*-hydroxy group of an anthraquinone is methylated using methyl iodide and ***Silver(I) Oxide*** in chloroform (eq 18).[18a,22]

K_2CO_3, MeI; acetone; 55–64% (16)

Li_2CO_3, MeI; DMF, 55 °C; 90% (17)

MeI; Ag_2O, $CHCl_3$ (18)

Aliphatic alcohols are also methylated by methyl iodide under basic conditions in dipolar aprotic solvents (eqs 19 and 20).[23] Typical conditions employ ***Sodium Hydride*** as a base, DMF as the solvent and an excess of methyl iodide.[24] Alternatively, dimethyl sulfate or ***Methyl Trifluoromethanesulfonate*** may be used as the methylating agent. Under acidic conditions, diazomethane will also methylate aliphatic hydroxy groups. Finally, methylation of a hydroxy group may be achieved under essentially neutral conditions using silver(1) oxide (eq 21).[23,25]

NaH; MeI (3 equiv), DMF; 90% (19)

t-BuOK; MeI, THF; 100% (20)

MeI; Ag_2O, DMF (21)

S-Alkylation. Methyl iodide alkylates thioalkoxides and sulfides to produce sulfides and sulfonium ions, respectively. For example, thioalkoxides produced from thiocarbonyl compounds are methylated with methyl iodide to generate the corresponding methyl thioether (eq 22).[26] Sulfonium halides, derived from the reaction of methyl iodide with an alkyl sulfide, are sometimes labile in solution and may undergo further reaction (eq 23).[27,28] ***Dimethyl Sulfoxide*** when refluxed with an excess of methyl iodide produces trimethyloxosulfonium iodide, which is collected as a white solid and recrystallized from water. Similarly, methylation of ***Dimethyl Sulfide*** produces trimethylsulfonium iodide.[29] Treatment of trimethyloxosulfonium and trimethylsulfonium salts with a base yields the corresponding ylides, which serve as useful methylene transfer reagents. ***Silver(I) Perchlorate*** promotes the methylation of less reactive sulfides (eq 24).[30]

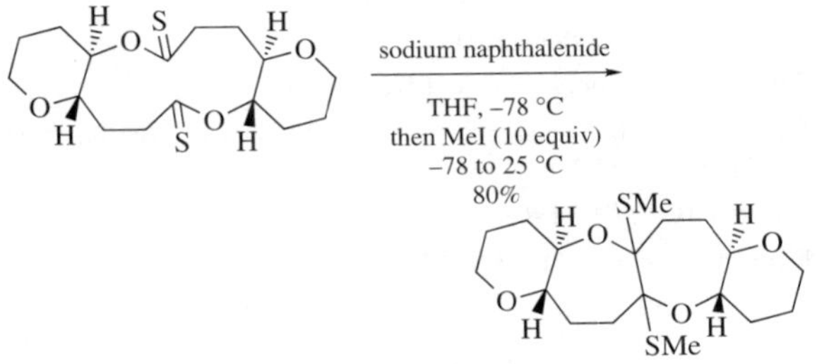

Hydrolysis of sulfonium salts serves as a useful protocol for removal of a protecting group or hydrolysis of a carboxylic acid derivative. For example, thioamides are converted into the corresponding methyl esters by methylation with methanolic methyl iodide followed by treatment with aqueous potassium carbonate (eq 25).[31] Methyl thiomethyl ethers are readily hydrolyzed using an excess of methyl iodide in aqueous acetone (eq 26).[32] Under similar reaction conditions, thioacetals are hydrolyzed to the corresponding carbonyl compounds (eq 27).[33]

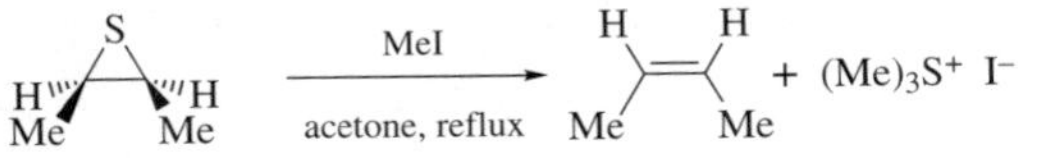

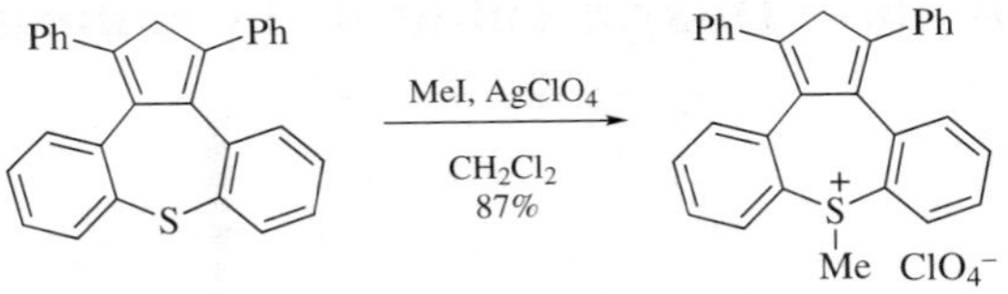

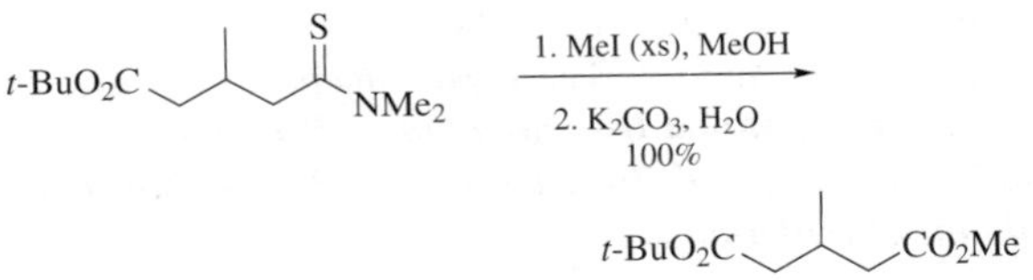

N*-Methylation.** The direct monomethylation of ammonia or a primary amine with methyl iodide is usually not a feasible method for the preparation of primary or secondary amines since further methylation occurs. However, methylation of secondary and tertiary amines leading to the production of tertiary amines and quaternary ammonium salts, respectively, is a useful method. Secondary *N*-methylalkylamines can be prepared from primary amines by a multi-step sequence involving first methylation of the benzylidene of the primary amine followed by hydrolytic removal of the benzylidene group (eq 28).[34] An alternative to the methylation procedure using methyl iodide is the employment of Eschweiler–Clarke conditions.[35] Exhaustive *N*-methylation of amines results in the production of a quaternized amine. The use of ***2,6-Lutidine as base is beneficial to carry out quaternization of amines due to the slow rate of methylation of 2,6-lutidine (eq 29).[36] Quaternized ammonium salts are employed in the Hofmann elimination (eqs 30 and 31).[37,38] As in the case of alcohol methylation, silver(I) salts may be used to facilitate the methylation process (eq 32).[39] Finally, conditions for the methylation of indole have been reported (eq 33).[40]

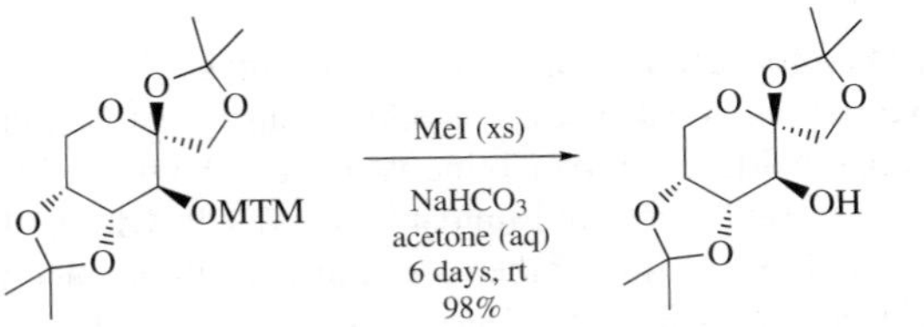

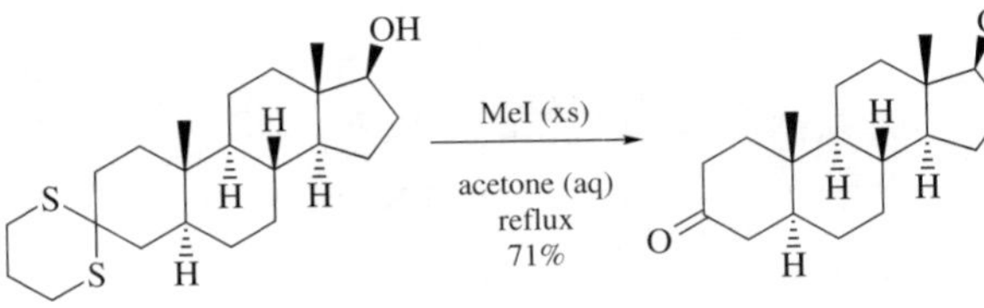

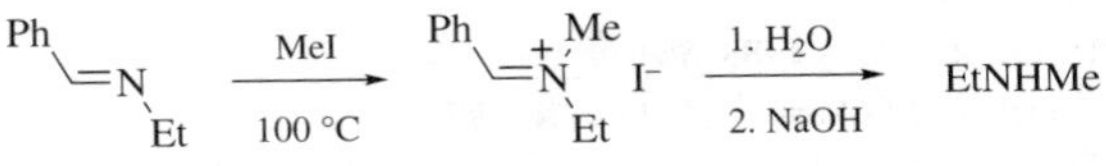

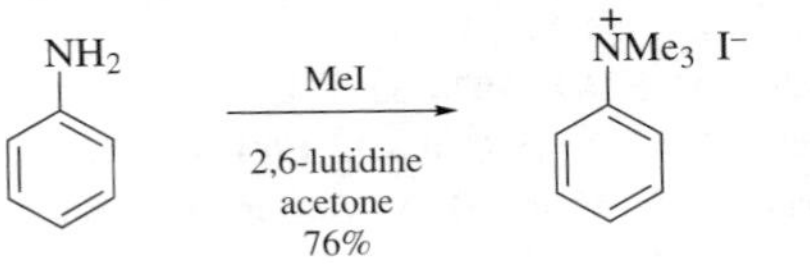

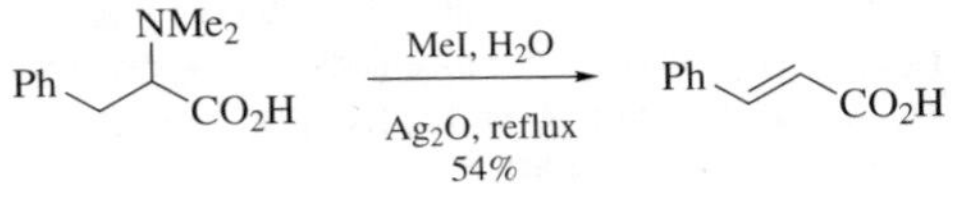

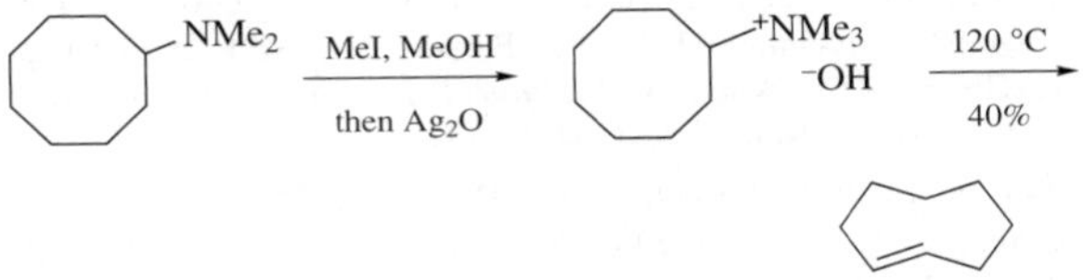

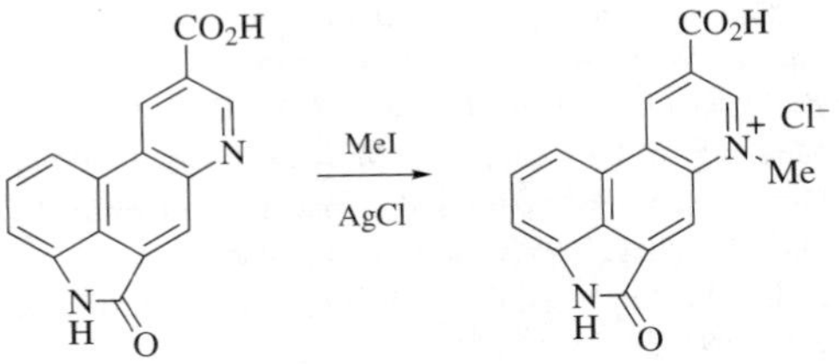

P*-Methylation.** Phosphonium salts are prepared by the quaternization of phosphines with methyl iodide.[41] The displacement reaction is usually conducted in polar solvents such as acetonitrile or DMF. Dialkyl phosphonates are prepared from the reaction of trialkyl phosphites with alkyl halides, commonly known as the Arbuzov reaction.[42] For example, diisopropyl methylphosphonate is prepared by heating a mixture of methyl iodide and ***Triisopropyl Phosphite (eq 34).[43]

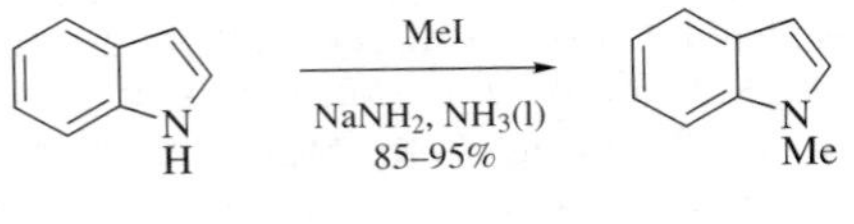

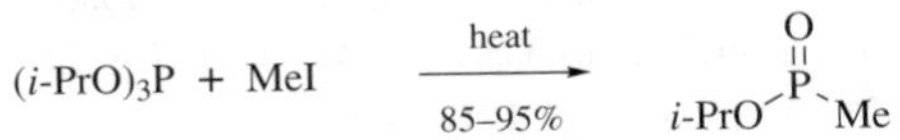

1. (a) Stowell, J. C. *Carbanions in Organic Synthesis*; Wiley: New York, 1979. (b) Caine, D. *COS* **1991**, *3*, Chapter 1. (c) House, H. O. *Modern*

Synthetic Organic Reactions, 2nd ed.; Benjamin: Menlo Park, 1972; Chapter 9.
2. Mekler, A. B.; Ramachandran, S.; Swaminathan, S.; Newman, M. S. *OSC* **1973**, *5*, 742. (b) Johnson, A. W.; Markham, E.; Price, R. *OSC* **1973**, *5*, 785.
3. Shono, T.; Kashiura, S.; Sawamura, M.; Soejima, T. *JOC* **1988**, *53*, 907.
4. Nedelec, L.; Gasc, J. C.; Bucourt, R. *T* **1974**, *30*, 3263.
5. Prasad, G.; Hanna, P. E.; Noland, W. E.; Venkatraman, S. *JOC* **1991**, *56*, 7188.
6. Rubina, K.; Goldverg, Y.; Shymanska, M. *SC* **1989**, *19*, 2489.
7. (a) Rasmussen, J. K. *S* **1979**, 91. (b) Brownbridge, P. *S* **1983**, 1. (c) d'Angelo, J. *T* **1976**, *32*, 2979.
8. Stork, G.; Hudrlik, P. F. *JACS* **1968**, *90*, 4462, 4464.
9. (a) Kuwajima, I.; Nakamura, E. *ACR* **1985**, *18*, 181. (b) Kuwajima, I.; Nakamura, E. *JACS* **1975**, *97*, 3257. (c) Smith, A. B., III; Fukui, M. *JACS* **1987**, *109*, 1269.
10. Taylor, R. J. K. *S* **1985**, 364.
11. Coates, R. M.; Sandfur, L. O. *JOC* **1974**, *39*, 275.
12. Evans, D. A. In *Asymmetric Synthesis*; Morrison, J. D., Ed.; Academic: New York, 1984; Vol. 3, p 1.
13. Kuehne, M. E. *JOC* **1970**, *35*, 171.
14. Bartlett, P. A.; Pizzo, C. F. *JOC* **1981**, *46*, 3896.
15. Ireland, R. E.; Evans, D. A.; Glover, D.; Rubottom, G. M.; Young, H. *JOC* **1969**, *34*, 3717.
16. Evans, D. A.; Ennis, M. D.; Mathre, D. J. *JACS* **1982**, *104*, 1737.
17. (a) Dolling, U.-H.; Davis, P.; Grabowski, E. J. J. *JACS* **1984**, *106*, 446. (b) Hughes, D. L.; Dolling, U.-H.; Ryan, K. M.; Schoenewaldt, E. F.; Grabowski, E. J. J. *JOC* **1987**, *52*, 4745.
18. (a) *FF* **1967**, *1*, 682. (b) Haslam, E. In *Protective Groups in Organic Chemistry*; McOmie, J. F. W., Ed.; Plenum: New York, 1973; Chapter 5. (c) Greene, T. W.; Wuts, P. G. M. *Protective Groups in Organic Synthesis*, 2nd ed.; Wiley: New York, 1991; Chapter 5.
19. Sato, T.; Otera, J.; Nozaki, H. *JOC* **1992**, *57*, 2166.
20. Chung, C. W.; De Bernardo, S.; Tengi, J. P.; Borgese, J.; Weigele, M. *JOC* **1985**, *50*, 3462.
21. (a) Greene, T. W.; Wuts, P. G. M. *Protective Groups in Organic Synthesis*, 2nd ed.; Wiley: New York, 1991; Chapter 3. (b) Wymann, W. E.; Davis, R.; Patterson, Jr., J. W.; Pfister, J. R. *SC* **1988**, *18*, 1379.
22. Manning, W. B.; Kelly, T. R.; Muschik, G. M. *TL* **1980**, *21*, 2629.
23. Greene, T. W.; Wuts, P. G. M. *Protective Groups in Organic Synthesis*, 2nd ed.; Wiley: New York, 1991; Chapter 2.
24. Fisher, M. J.; Myers, C. D.; Joglar, J.; Chen, S.-H.; Danishefsky, S. J. *JOC* **1991**, *56*, 5826.
25. (a) Greene, A. E.; Le Drina, C.; Crabbe, P. *JACS* **1980**, *102*, 7583. (b) Finch, N.; Fitt, J. J.; Hsu, I. H. S. *JOC* **1975**, *40*, 206. (c) Ichikawa, Y.; Tsuboi, K.; Naganawa, A.; Isobe, M. *SL* **1993**, 907.
26. Nicolaou, K. C.; Hwang, C.-K.; Marron, B. E.; DeFrees, S. A.; Coulandouros, E. A.; Abe, Y.; Carroll, P. J.; Snyder, J. P. *JACS* **1990**, *112*, 3040. (b) Nicolaou, K. C.; McGarry, D. G.; Somers, P. K.; Kim, B. H.; Ogilvie, W. W.; Yiannikouros, G.; Prasad, C. V. C.; Veale, C. A.; Hark, R. R. *JACS* **1990**, *112*, 6263.
27. Barrett, G. C. In *Comprehensive Organic Chemistry*; Barton, D. H. R., Ed.; Pergamon: Oxford, 1979; Vol. 3, pp 105–120.
28. Helmkamp, G. K.; Pettitt, D. J. *JOC* **1960**, *25*, 1754.
29. (a) Corey, E. J.; Chaykovsky, M. *JACS* **1965**, *87*, 1353. (b) Kuhn, R.; Trischmann, H. *LA* **1958**, *611*, 117. (c) Emeleus, H. J.; Heal, H. G. *JCS* **1946**, 1126.
30. Hori, M.; Katakoka, T.; Shimizu, H.; Okitsu, M. *TL* **1980**, 4287.
31. Tamaru, Y.; Harada, T.; Yoshida, Z.-I. *JACS* **1979**, *101*, 1316.
32. Pojer, P. M.; Angyal, S. J. *TL* **1976**, 3067.
33. Fetizon, M.; Jurion, M. *CC* **1972**, 382.
34. Wawzonek, S.; McKillip, W.; Peterson, C. J. *OSC* **1973**, *5*, 785.
35. March, J. *Advanced Organic Chemistry*, 3rd. ed.; Wiley: New York, 1985; p 799.
36. Sommer, H. Z.; Jackson, L. L. *JOC* **1970**, *35*, 1558.
37. Cope, A. C.; Trumbull, E. R. *OR* **1960**, *11*, 317.
38. (a) Cope, A. C; Bach, R. D. *OSC* **1973**, *5*, 315. (b) Manitto, P.; Monti, D.; Gramatica, P.; Sabbioni, E. *CC* **1973**, 563.
39. Horwell, D. C. *T* **1980**, *36*, 3123.
40. Potts, K. T.; Saxton, J. E. *OSC* **1973**, *5*, 769.
41. Smith, D. J. H. In *Comprehensive Organic Chemistry*; Barton, D. H. R., Ed.; Pergamon: Oxford, 1979; Vol. 2, p 1160.
42. Arbuzov, B. A. *PAC* **1964**, *9*, 307.
43. Fieser, L. F.; Fieser, M. *FF* **1967**, *1*, 685.

Gary A. Sulikowski & Michelle M. Sulikowski
Texas A&M University, College Station, TX, USA

Iodonium Di-*sym*-collidine Perchlorate

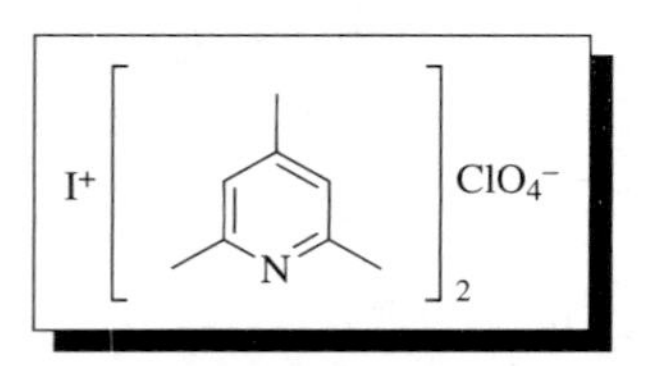

[69417-67-0] $C_{16}H_{22}ClIN_2O_4$ (MW 468.75)

(very reactive electrophile, superior source of I^+,[1] useful in the synthesis of *cis*-β-hydroxy amines,[2] activates glycosides for glycosylation;[5,6] can be used for iodolactonization[12,13] and vicinal *cis*-diol[14] preparation)

Alternate Name: IDCP.
Solubility: sol chloroform; insol ether.
Form Supplied in: fine colorless crystalline powder.
Drying: see ***Bromonium Di-sym-collidine Perchlorate***.
Handling, Storage, and Precautions: see ***Bromonium Di-sym-collidine Perchlorate***.

---

cis*-Oxyamination.**[2] IDCP (**1**) is useful in the synthesis of *cis*-hydroxyamino sugars, e.g. methyl *N*-acetylristosaminide has been obtained from an oxazoline which can be made by the reaction of a trichloromethyl imidate with IDCP. The imidate can be prepared by reaction of the corresponding allylic alcohol with ***Trichloroacetonitrile in presence of ***Sodium Hydride*** (eq 1).

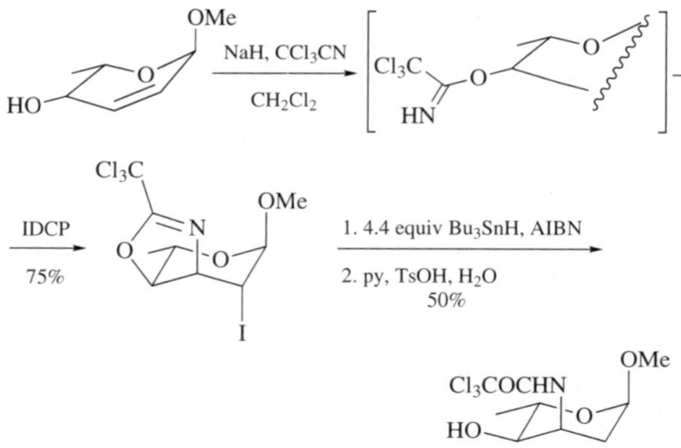

The synthesis of methyl α,L-garosaminide,[3] a key component of aminocyclitol antibiotics, is complicated by the presence of a *cis*-hydroxyamino group and by the tertiary character of the hydroxy group. The problems have been resolved by use of the allylic epoxide as starting material. This epoxide was converted in three steps into an allylic amine. Treatment of iodonium salt gave the iodooxazolidinone in 82% yield. The product was reduced and the ethoxy ethyl group was removed and finally converted to the desired product by hydrolysis (eq 2).

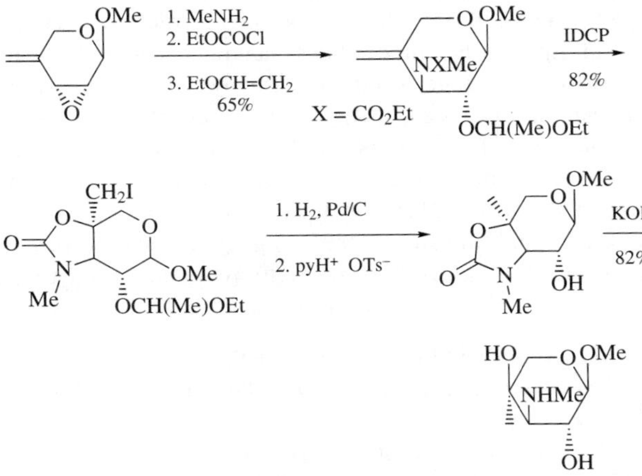

1. $MeNH_2$ 2. $EtOCOCl$ 3. $EtOCH{=}CH_2$ 65%; $X = CO_2Et$; IDCP 82%; 1. H_2, Pd/C 2. pyH^+ OTs^-; KOH 82% (2)

Similar methodology can convert an internal allylic amine into a *cis*-β-hydroxyamine, as illustrated in a synthesis of holacosamin, a component of some glycosteroids (eq 3).[4]

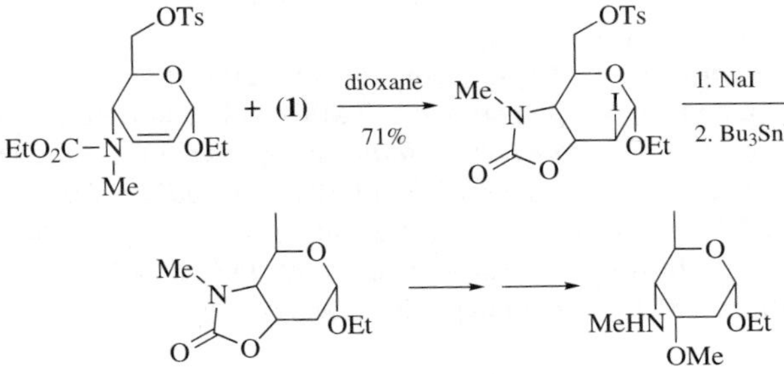

\+ (1) dioxane 71%; 1. NaI 2. Bu_3SnH (3)

α-Linked Disaccharides[5]. The reagent functions as a superior source of I^+, probably because of the nonnucleophilic counterion. Thus a pyranoid diene reacts with IDCP to give a planar ion to which an alcohol adds in a 1,4-sense to give an α-glycoside. Thus tetraacetylfructose reacts to give an α-disaccharide in 45% yield (eq 4). The β-isomer is not detected.

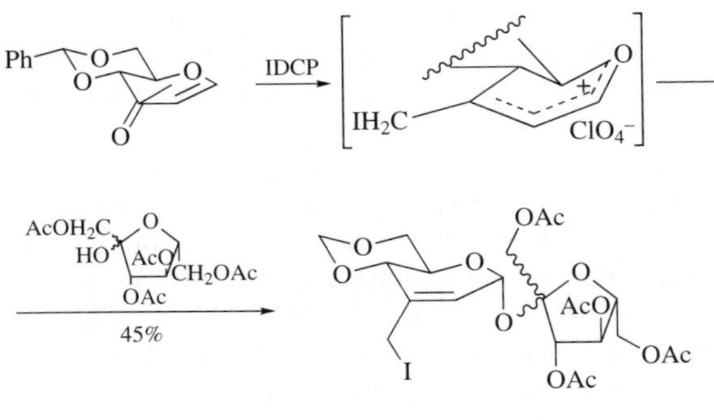

IDCP; 45% (4)

Examples are known where this reagent has activated pent-4-enyl glycosides for glycosylation,[6] but α- and β-glycosides were obtained in various proportions regardless of the nature of donors or acceptors (primary or secondary hydroxyl groups). Glycosylations of 1,2:5,6-di-*O*-isopropylidene-α-D-galactopyranose and methyl 2,3,4-tri-*O*-benzyl-α-D-glucopyranoside were investigated with pent-4-enyl-2,3,4-tri-*O*-benzyl-β-D-glucopyranoside derivatives in which 6-OH was protected by a benzyl, a trityl, or a TBDMS group in order to assess the effect of the bulk of the 6-substituents. The presence of a bulky 6-substituent (a) increases significantly the proportion of the α-product; (b) decreases the yield when a secondary hydroxy group is glycosylated, but the effect is less or opposite when a primary hydroxy group is involved; (c) lowers the increase in yield of the α-product when a primary hydroxy group is glycosylated; (d) gives a much better yield of the α-anomer when there is a higher proportion of ether in the solvent.

Chemospecific glycosidation of partially benzoylated thioglycosides ('disarmed' acceptors) with perbenzylated thioglycosides ('armed' donors) can be realized in the presence of the promotor IDCP (eq 5).[7]

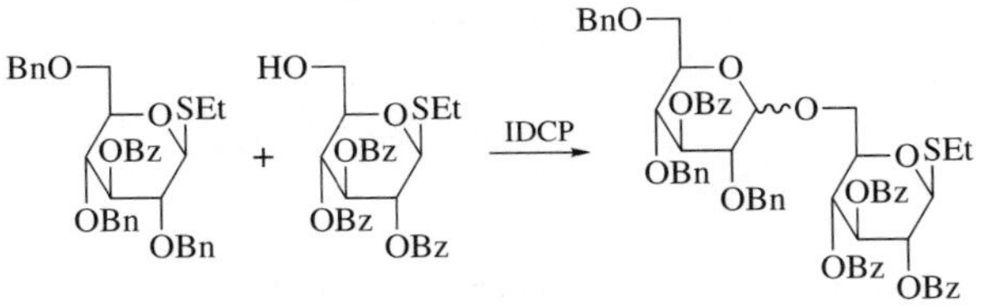

IDCP (5)

Reaction of IDCP with unsaturated alcohols and carboxylic acids in dichloromethane at ambient temperature has afforded three- to seven-membered ring iodoethers and four- to seven-membered ring iodolactones, respectively, in moderate yields and generally with high regioselectivity. The reaction has particular utility for synthesis of 2-(1-iodoalkyl)oxiranes and -oxetanes.[11] Glycosylation has also been achieved by electrophile-induced activation of anomeric *O*-glycosyl *N*-allyl carbamates (eq 6).[8]

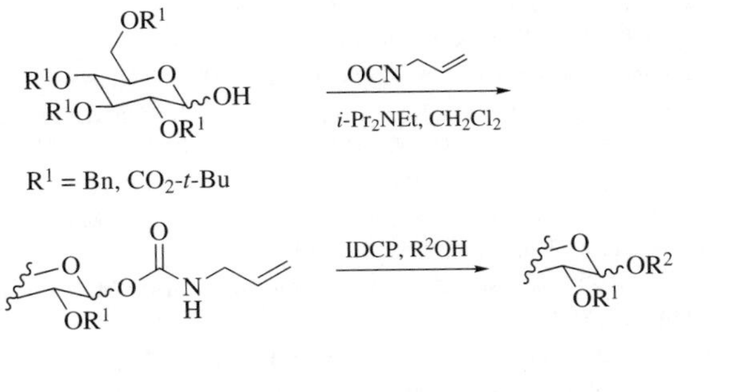

OCN; *i*-Pr_2NEt, CH_2Cl_2; $R^1 = Bn, CO_2$-*t*-Bu; IDCP, R^2OH (6)

Treatment of fully benzylated 1-methylene-D-glucose with IDCP gives easy access to 1-iodoheptuloses or a 1-iodomethylene derivative.[9] The latter compound is, in turn, further amenable to similar IDCP-mediated addition reaction. In the synthesis of ciclamycin O the required trisaccharide glycol was assembled by substituent-directed iodinative coupling of glycals as shown in (eq 7).[10]

IDCP

BnO

OBz OH

OR X

IDCP + OH OBz

IDCP + MeO$_2$C OH OH O OH OH O

OH O MeO$_2$C OH OH O OH O

OR O H H H (7)

OH

Iodolactonization of heptadienoic acid derivatives having oxazolidin-2-ones or a sultam[12] as chiral auxiliary gave the chiral iodolactone with moderate to excellent enantioselectivity.[13]

Vicinal *cis*-Diols.[14] An allylic alcohol is converted into its primary urethane derivative, which is then subjected to iodonium ion induced cyclization to give a single iodocarbonate. The carbonate is then deiodinated reductively and hydrolyzed to afford the vicinal diol.

1. Lemieux, R. U.; Morgan, A. R. *CJC* **1965**, *43*, 2190.
2. Pauls, H. W.; Fraser-Reid, B. *JOC* **1983**, *48*, 1392.
3. Pauls, H. W.; Fraser-Reid, B. *JACS* **1980**, *102*, 3956.
4. George, M.; Fraser-Reid, B. *TL* **1981**, *22*, 4635.
5. Fraser-Reid, B.; Iley, D. E. *CJC* **1979**, *57*, 645.
6. Houdier, S.; Voltero, P. J. A. *Carbohydr. Res.* **1992**, *232*, 349.
7. (a) Veeneman, G. H.; Van Boom, J. H. *TL* **1990**, *31*, 275. (b) Veeneman G. H.; Van Leeuwen, S. H.; Van Boom, J. H. *TL* **1990**, *31*, 1331.
8. Kuns, H.; Simmer, J. *TL* **1993**, *34*, 2907.
9. Noort, D.; Veeneman, G. H.; Boons, G. P. H.; Vander Marel, G. A.; Mulder, G. J.; Van Boom, J. H. *SL* **1990**, 205.
10. Susuki, K.; Sulikowski, G. A.; Friesen, R. W.; Danishefsky, S. J. *JACS* **1990**, *112*, 8895.
11. Evans, R. D.; Magee, J. W.; Schauble, J. H. *S* **1988**, 862.
12. Oppolser, W.; Chapuis, C.; Bernardielli, G. *HCA* **1984**, *67*, 1397.
13. Yokomatsu, T.; Iwasawa, H.; Shibuya, S. *CC* **1992**, 510.
14. Pauls, H. W.; Fraser-Reid, B. *J. Carbohydr. Chem.* **1985**, *4*, 1.

Tapan Ray
Sandoz Research Institute, East Hanover, NJ, USA

Iodotrimethylsilane[1]

Me$_3$SiI

[16029-98-4] C_3H_9ISi (MW 200.11)

(a versatile reagent for the mild dealkylation of ethers, carboxylic esters, lactones, carbamates, acetals, phosphonate and phosphate esters; cleavage of epoxides, cyclopropyl ketones; conversion of vinyl phosphates to vinyl iodides; neutral nucleophilic reagent for halogen exchange reactions, carbonyl and conjugate addition reactions; use as a trimethylsilylating agent for formation of enol ethers, silyl imino esters, and *N*-silylenamines, alkyl, alkenyl and alkynyl silanes; Lewis acid catalyst for acetal formation, α-alkoxymethylation of ketones, for reactions of acetals with silyl enol ethers and allylsilanes; reducing agent for epoxides, enediones, α-ketols, sulfoxides, and sulfonyl halides; dehydrating agent for oximes)

Alternate Names: TMS-I; TMSI; trimethylsilyl iodide.

Physical Data: bp 106–109 °C; *d* 1.406 g cm^{-3}; n_D^{20} 1.4710; fp −31 °C.

Solubility: sol CCl_4, $CHCl_3$, CH_2Cl_2, $ClCH_2CH_2Cl$, MeCN, PhMe, hexanes; reactive with THF (ethers), alcohols, and EtOAc (esters).

Form Supplied in: clear colorless liquid, packaged in ampules, stabilized with copper; widely available.

Analysis of Reagent Purity: easily characterized by ^1H, ^{13}C, or ^{29}Si NMR spectroscopy.

Preparative Methods: although more than 20 methods have been reported[1] for the preparation of TMS-I, only a few are summarized here. ***Chlorotrimethylsilane*** undergoes halogen exchange with either ***Lithium Iodide***[2] in $CHCl_3$ or ***Sodium Iodide***[3] in MeCN, which allows in situ reagent formation (eq 1). Alternatively, ***Hexamethyldisilane*** reacts with ***Iodine*** at 25–61 °C to afford TMS-I with no byproducts (eq 2).[4]

$$\text{TMSCl} \xrightarrow[\text{or NaI, MeCN}]{\text{LiI, CHCl}_3} \text{TMSI} + \text{LiCl or NaCl} \quad (1)$$

$$\text{TMS–TMS} + I_2 \xrightarrow{25\text{–}61\ ^\circ\text{C}} \text{TMSI} \quad (2)$$

Several other methods for in situ generation of the reagent have been described.[5,6] It should be noted, however, that the reactivity of in situ generated reagent appears to depend upon the method of preparation.

Purification: by distillation from copper powder

Handling, Storage, and Precautions: extremely sensitive to light, air, and moisture, it fumes in air due to hydrolysis (HI), and becomes discolored upon prolonged storage due to generation of I_2. It is flammable and should be stored under N_2 with a small piece of copper wire. It should be handled in a well ventilated fume hood and contact with eyes and skin should be avoided.

Use as a Nucleophilic Reagent in Bond Cleavage Reactions.

Ether Cleavage.[5,7] The first broad use of TMS-I was for dealkylation reactions of a wide variety of compounds containing oxygen–carbon bonds, as developed independently by the groups of Jung and Olah. Simple ethers initially afford the trimethylsilyl ether and the alkyl iodide, with further reaction giving the two iodides (eq 3).[7,8] This process occurs under neutral conditions, and is generally very efficient as long as precautions to avoid hydrolysis by adventitious water are taken. Since the silyl ether can be quantitatively hydrolyzed to the alcohol, this reagent permits the use of simple ethers, e.g. methyl ethers, as protective groups in synthesis. The rate of cleavage of alkyl groups is: tertiary ≈ benzylic ≈ allylic ≫ methyl > secondary > primary. Benzyl and *t*-butyl ethers are cleaved nearly instantaneously at low temperature with TMS-I. Cyclic ethers afford the iodo silyl ethers and then the diiodide, e.g. THF gives 4-iodobutyl silyl ether and then 1,4-diiodobutane in excellent yield.[7,8] Alcohols and silyl ethers are rapidly converted into the iodides as well.[8a,9] Alkynic ethers produce the trimethylsilylketene via dealkylative rearrangement.[4b] Phenolic ethers afford the phenols after workup.[5,7,10] In general, ethers are cleaved faster than esters. Selective cleavage of methyl aryl ethers in the presence of other oxygenated functionality has also been accomplished in quinoline.[11] γ-Alkoxyl enones undergo deoxygenation with excess TMS-I (2 equiv), with the first step being conjugate addition of TMS-I.[12]

$$R^1\text{–O–}R^2 \xrightarrow{\text{TMSI}} R^2\text{I} + R^1\text{OTMS} \xrightarrow{\text{TMSI}} R^1\text{I} \quad (3)$$

R^1OTMS $\xrightarrow{H_2O}$ R^1OH $\xrightarrow{\text{TMSI}}$ R^1I

Cleavage of Epoxides. Reaction of epoxides with 1 equiv of TMS-I gives the vicinal silyloxy iodide.[8e] With 2 equiv of TMS-I, however, epoxides are deoxygenated to afford the corresponding alkene (eq 4).[13a,b] However, allylic alcohols are efficiently prepared by reaction of the intermediate iodosilane with base.[13c,d] Furthermore, acyclic 2-ene-1,4-diols react with TMS-I to undergo dehydration, affording the corresponding diene.[13e]

(4)

Ester Dealkylation.[14] Among the widest uses for TMS-I involves the mild cleavage of carboxylic esters under neutral conditions. The ester is treated with TMS-I to form an initial oxonium intermediate which suffers attack by iodide (eq 5). The trimethylsilyl ester is cleaved with H_2O during workup. Although the reaction is general and efficient, it is possible to accomplish selective cleavage according to the reactivity trend: benzyl, *t*-butyl > methyl, ethyl, *i*-propyl. Neutral transesterification is also possible via the silyl ester intermediate.[15] Aryl esters are not cleaved by TMS-I, however, since the mechanism involves displacement of R^2 by I^-. Upon prolonged exposure (75 °C, 3 d) of simple esters to excess TMS-I (2.5 equiv), the corresponding acid iodides are formed.[14b,16] β-Keto esters undergo decarboalkoxylation when treated with TMS-I.[17] An interesting rearrangement reaction provides α-methylene lactones from 1-(dimethylaminomethyl)cyclopropanecarboxylates (eq 6).[18]

(5)

(6) 1. TMSI; 2. heat, 64%

Lactone Cleavage.[14,19] Analogous to esters, lactones are also efficiently cleaved with TMS-I to provide ω-iodocarboxylic acids, which may be further functionalized to afford bifunctional building blocks for organic synthesis (eq 7). Diketene reacts with TMS-I to provide a new reagent for acetoacylation.[20]

(7)

Cleavage of Carbamates.[21] Since strongly acidic conditions are typically required for the deprotection of carbamates, use of TMS-I provides a very mild alternative. Benzyl and *t*-butyl carbamates are readily cleaved at rt,[22] whereas complete cleavage of methyl or ethyl carbamates may require higher temperatures (reflux). The intermediate silyl carbamate is decomposed by the addition of methanol or water (eq 8). Since amides are stable to TMS-I-promoted hydrolysis,[7a] this procedure can be used to deprotect carbamates of amino acids and peptides.[21d]

$$R^1_2N\text{–}CO_2R^2 \xrightarrow{\text{TMSI}} R^1_2N\text{–}CO_2\text{TMS} \xrightarrow[\text{or } H_2O]{\text{MeOH}} R^1_2N\text{–H} \quad (8)$$

A recent example used TMS-I to deprotect three different protecting groups (carbamate, ester, and orthoester) in the same molecule in excellent yield (eq 9).[23]

(9) 1. TMSI; 2. OH⁻, 91%

Cleavage of Acetals.[24] Acetals can be cleaved in analogy to ethers, providing a newly functionalized product (eq 10), or simply the parent ketone (eq 11). Glycals have also been converted to the iodopyrans with TMS-I,[25] and glycosidation reactions have been conducted with this reagent.[26]

(10)

$$R^2O\text{–}C(R^1)_2\text{–}OR^2 \xrightarrow{\text{TMSI}} R^1C(O)R^1 \quad (11)$$

Orthoesters are converted into esters with TMS-I. The dimethyl acetal of formaldehyde, methylal, affords iodomethyl methyl ether in good yield (eq 12)[27a] (in the presence of alcohols, MOM ethers are formed).[27b] α-Acyloxy ethers also furnish the iodo ethers,[28] e.g. the protected β-acetyl ribofuranoside gave the α-iodide which was used in the synthesis of various nucleosides in good yield (eq 13).[28b] Aminals are similarly converted into immonium salts, e.g. Eschenmoser's reagent, ***Dimethyl(methylene)ammonium Iodide***, in good yield.[29]

$$MeO\text{–}CH_2\text{–}OMe \xrightarrow{\text{TMSI}} MeO\text{–}CH_2\text{–}I \quad (12)$$

$$\text{BzO, BzO, OBz ribofuranose-OAc} \xrightarrow{\text{TMSI}} [\text{α-iodide}] \xrightarrow[89\%]{(TMS)_2U} \text{ribofuranosyl-U} \quad (13)$$

Cleavage of Phosphonate and Phosphate Esters.[30] Phosphonate and phosphate esters are cleaved even more readily with TMS-I than carboxylic esters. The reaction of phosphonate esters proceeds via the silyl ester, which is subsequently hydrolyzed with MeOH or H_2O (eq 14).

$$R^1P(O)(OR^2)_2 \xrightarrow{\text{TMSI}} R^1P(O)(OTMS)_2 \xrightarrow{\text{MeOH or } H_2O} R^1P(O)(OH)_2 \quad (14)$$

Conversion of Vinyl Phosphates to Vinyl Iodides.[31] Ketones can be converted to the corresponding vinyl phosphates which react with TMS-I (3 equiv) at rt to afford vinyl iodides (eq 15).

$$\text{cyclohexenyl–O–P(O)(OEt)}_2 \xrightarrow{\text{3 equiv TMSI}} \text{cyclohexenyl–I} \quad (15)$$

Cleavage of Cyclopropyl Ketones.[32] Cyclopropyl ketones undergo ring opening with TMS-I, via the silyl enol ether (eq 16). Cyclobutanones react analogously under these conditions.[33]

$$\text{c-C}_3H_5C(O)R \xrightarrow{\text{TMSI}} I(CH_2)_2CH{=}C(OTMS)R \xrightarrow{H_2O} I(CH_2)_3C(O)R \quad (16)$$

Halogen Exchange Reactions.[34] Halogen exchange can be accomplished with reactive alkyl halides, such as ***Benzyl Chloride*** or ***Benzyl Bromide***, and even with certain alkyl fluorides, by using TMS-I in the presence of $(n\text{-Bu})_4NCl$ as catalyst (eq 17).

$$PhCH_2X \xrightarrow[Bu_4NCl]{\text{TMSI}} PhCH_2I \quad (17)$$

X = Br, Cl

Use of TMS-I in Nucleophilic Addition Reactions.

Carbonyl Addition Reactions.[35] α-Iodo trimethylsilyl ethers are produced in the reaction of aldehydes and TMS-I (eq 18). These compounds may react further to provide the diiodo derivative or may be used in subsequent synthesis.

$$RCHO \xrightarrow{\text{TMSI}} RCH(I)OTMS \quad (18)$$

An example of a reaction of an iodohydrin silyl ether with a cuprate reagent is summarized in eq 19.[36] An interesting reaction of TMS-I with phenylacetaldehydes gives a quantitative yield of the oxygen-bridged dibenzocyclooctadiene, which was then converted in a few steps to the natural product isopavine (eq 20).[35,37]

$$PhCH(CH_3)CHO \xrightarrow[Bu_2Cu(CN)Li_2]{\text{TMSI}} PhCH(CH_3)CH(OH)Bu + PhCH(CH_3)CH(OH)Bu \quad (19)$$

7:1

$$\text{3,4-(MeO)}_2C_6H_3CH_2CHO \xrightarrow[\text{CHCl}_3,\ 100\%]{\text{TMSI}} \text{O-bridged tetramethoxy dibenzocyclooctadiene} \xrightarrow{\text{several steps}} \text{NH-bridged isopavine} \quad (20)$$

β-Iodo ketones have been produced from reactions of TMS-I and ketones with α-hydrogens.[38] This reaction presumably involves a TMS-I catalyzed aldol reaction followed by 1,4-addition of iodide.

Conjugate Addition Reactions.[39] α,β-Unsaturated ketones undergo conjugate addition with TMS-I to afford the β-iodo adducts in high yield (eq 21). The reaction also works well with the corresponding alkynic substrate.[40]

$$R^1CH{=}CHC(O)R^2 \xrightarrow[\text{2. } H_2O]{\text{1. TMSI}} R^1CH(I)CH_2C(O)R^2 \quad (21)$$

TMS-I has also been extensively utilized in conjunction with organocopper reagents to effect highly stereoselective conjugate additions of alkyl nucleophiles.[41]

Use of TMS-I as a Silylating Agent.

Formation of Silyl Enol Ethers.[42] TMS-I in combination with ***Triethylamine*** is a reactive silylating reagent for the formation of silyl enol ethers from ketones (eq 22). TMS-I with ***Hexamethyldisilazane*** has also been used as an effective silylation agent, affording the thermodynamic silyl enol ethers. For example, 2-methylcyclohexanone gives a 90:10 mixture in favor of the tetra-substituted enol ether product.[42a] The reaction of TMS-I with 1,3-diketones is a convenient route to 1,3-bis(trimethylsiloxy)-1,3-dienes.[42c]

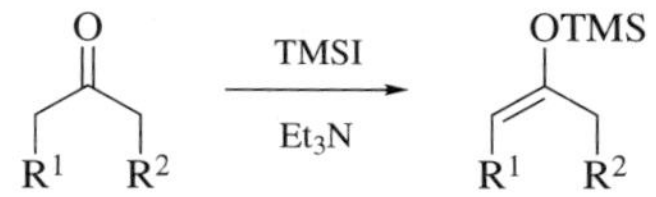

(22)

In an analogous process, TMS-I reacts with lactams in the presence of Et_3N to yield silyl imino ethers (eq 23).[43a]

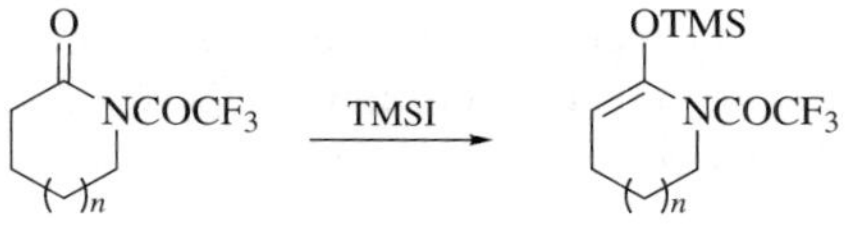

(23)

Halogenation of Lactams.[43b] Selective and high yielding iodination and bromination of lactams occurs with ***Iodine*** or ***Bromine***, respectively, in the presence of TMS-I and a tertiary amine base (eq 24). The proposed reaction mechanism involves intermediacy of the silyl imino ether.

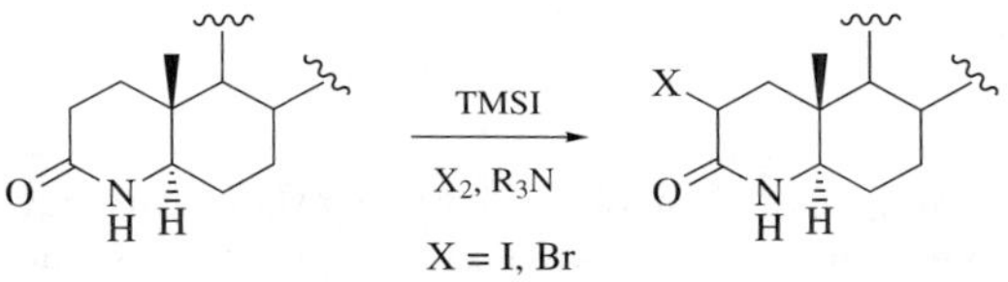

(24)

Reaction with Carbanions.[44] TMS-I has seen limited use in the silylation of carbanions, with different regioselectivity compared to other silylating reagents in the example provided in eq 25.

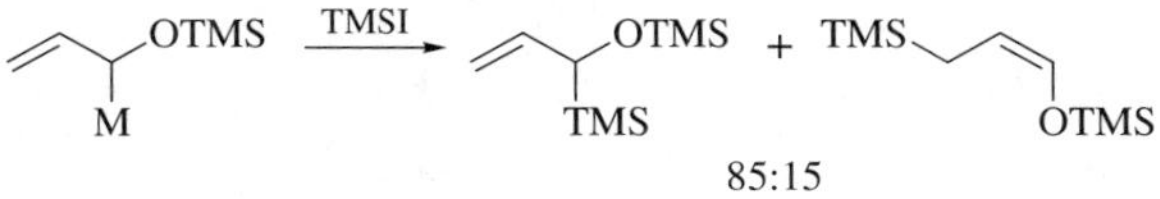

(25)

Silylation of Alkynes and Alkenes.[45] A Heck-type reaction of TMS-I with alkenes in the presence of Pd^0 and Et_3N affords alkenyltrimethylsilanes (eq 26).

$$\text{Ar–CH=CH}_2 \xrightarrow[\text{Pd}^0,\ \text{Et}_3\text{N}]{\text{TMSI}} \text{Ar–CH=CH–TMS} \quad (26)$$

Oxidative addition of TMS-I to alkynes can also be accomplished with a three-component coupling reaction to provide the enyne product (eq 27).

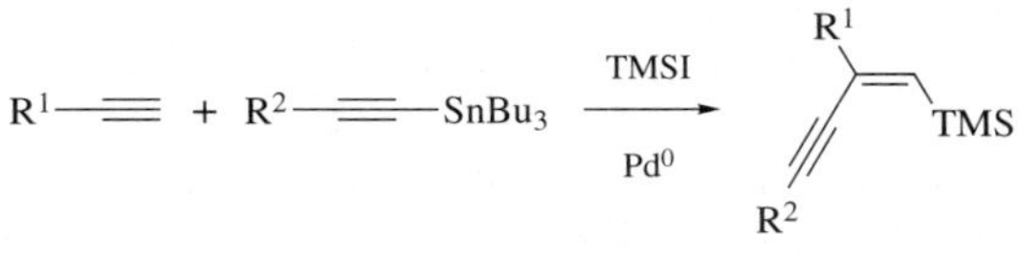

(27)

Use of TMS-I as a Lewis Acid.

Acetalization Catalyst.[46] TMS-I used in conjunction with $(MeO)_4Si$ is an effective catalyst for acetal formation (eq 28).

$$\text{PhCHO} \xrightarrow[(\text{MeO})_4\text{Si}]{\text{TMSI}} \text{PhCH(OMe)}_2 \quad (28)$$

Catalyst for α-Alkoxymethylation of Ketones. Silyl enol ethers react with α-chloro ethers in the presence of TMS-I to afford α-alkoxymethyl ketones (eq 29).[47]

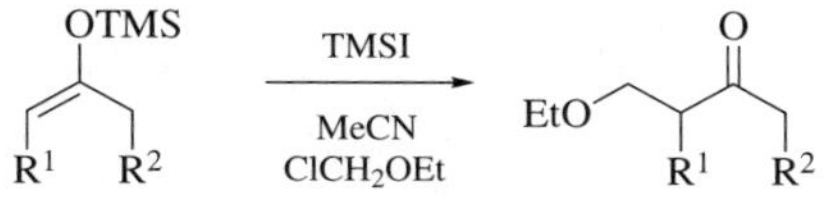

(29)

Catalyst for Reactions of Acetals with Silyl Enol Ethers and Allylsilanes. TMS-I catalyzes the condensation of silyl enol ethers with various acetals (eq 30)[48] and imines,[49] and of allylsilanes with acetals.[50]

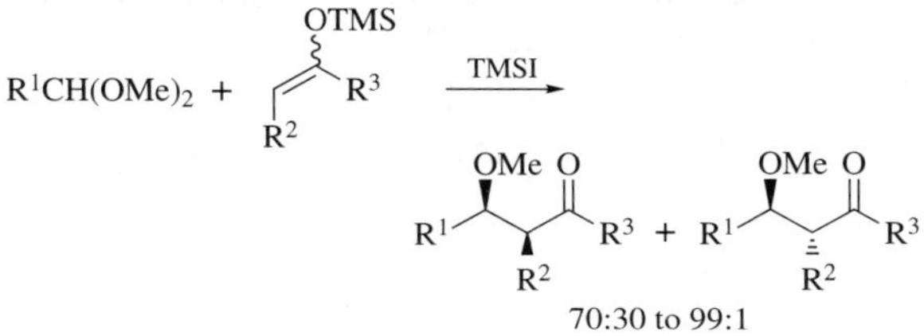

(30)

Use of TMS-I as a Reducing Agent. TMS-I reduces enediones to 1,4-diketones,[51] while both epoxides and 1,2-diols are reduced to the alkenes.[13a,b,52] The Diels–Alder products of benzynes and furans are converted in high yield to the corresponding naphthalene (or higher aromatic derivative) with TMS-I (eq 31).[53]

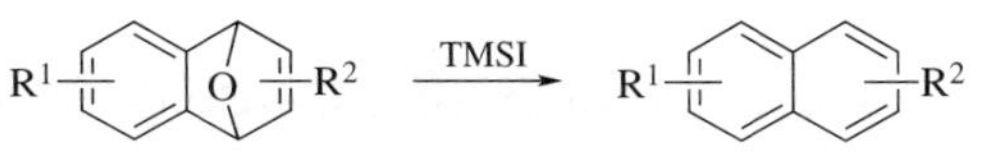

(31)

Styrenes and benzylic alcohols are reduced to the alkanes with TMS-I (presumably via formation of HI).[54] Ketones produce the symmetrical ethers when treated with trimethylsilane as a reducing agent in the presence of catalytic TMS-I.[55]

Reduction of α-Ketols.[56,57] Carbonyl compounds containing α-hydroxy, α-acetoxy, or α-halo groups react with excess TMS-I to give the parent ketone. α-Hydroxy ketone reductions proceed via the iodide, which is then reduced with iodide ion to form the parent ketone (eq 32).

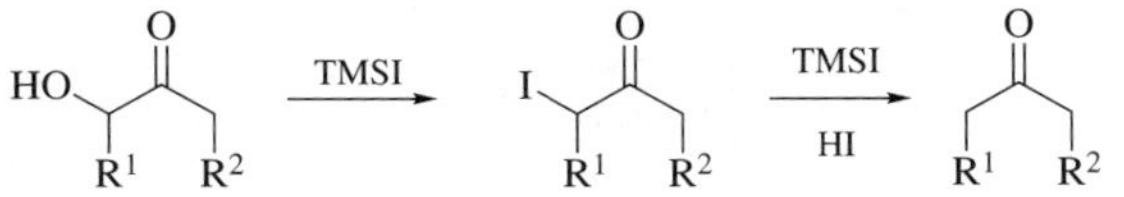

(32)

Sulfoxide Deoxygenation.[58] The reduction of sulfoxides occurs under very mild conditions with TMS-I to afford the corresponding sulfide and iodine (eq 33). Addition of I_2 to the reaction mixture accelerates the second step. The deoxygenation occurs faster in pyridine solution than the reactions with a methyl ester or alcohol.[59]

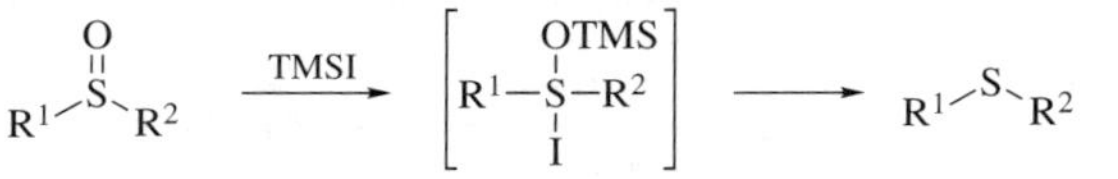

(33)

Pummerer reactions of sulfoxides can be accomplished in the presence of TMS-I and an amine base, leading to vinyl sulfides.[60] An efficient synthesis of dithioles was accomplished with TMS-I and Hünig's base (***Diisopropylethylamine***) (eq 34).[61]

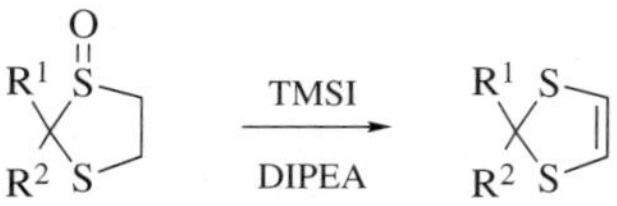

(34)

Reaction with Sulfonyl Halides.[62] Arylsulfonyl halides undergo reductive dimerization to form the corresponding disulfides (eq 35). Alkylsulfonyl halides, however, undergo this process under somewhat more vigorous conditions. Although sulfones generally do not react with TMS-I, certain cyclic sulfones are cleaved in a manner analogous to lactones.[63]

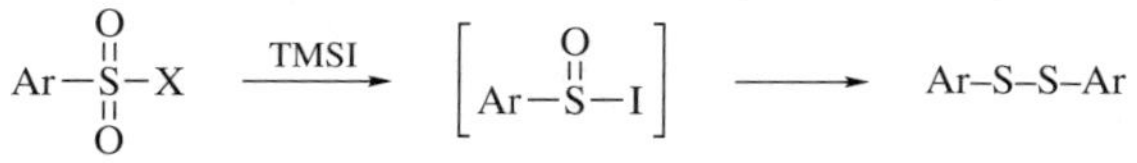

(35)

Other Reactions of TMS-I.

Reaction with Phosphine Oxides.[64] Phosphine oxides react with TMS-I to form stable adducts (eq 36). These *O*-silylated products can undergo further thermolytic reactions such as alkyl group cleavage.

$$R_3P{=}O \xrightarrow{\text{TMSI}} R_3\overset{+}{P}{-}O{-}TMS\ I^- \quad (36)$$

Chlorophosphines undergo halogen exchange reactions with TMS-I.[65]

Reaction with Imines. Imines react with TMS-I to form *N*-silylenamines, in a process analogous to the formation of silyl enol ethers from ketones.[66]

Reaction with Oximes.[67] Oximes are activated for dehydration (aldoximes, with hexamethylsilazane) or Beckmann rearrangement (ketoximes) with TMS-I (eq 37).

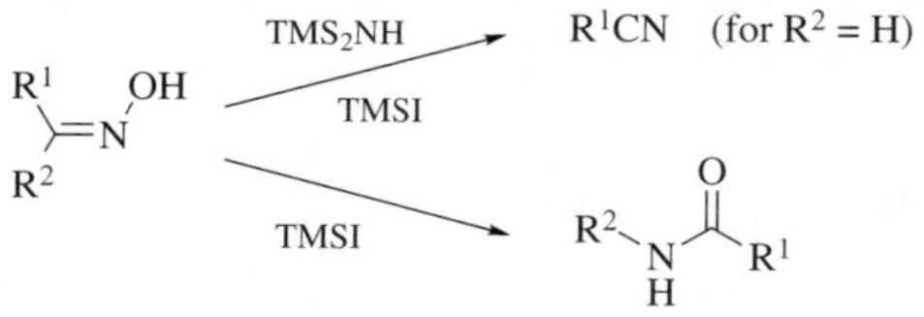

(37)

Reactions with Nitro and Nitroso Compounds.[68] Primary nitro derivatives react with TMS-I to form the oximino intermediate via deoxygenation, which then undergoes dehydration as discussed for the oximes (eq 38). Secondary nitro compounds afford the silyl oxime ethers, and tertiary nitro compounds afford the corresponding iodide. Nitroalkenes, however, react with TMS-I at 0 °C to afford the ketone as the major product (eq 39).[69]

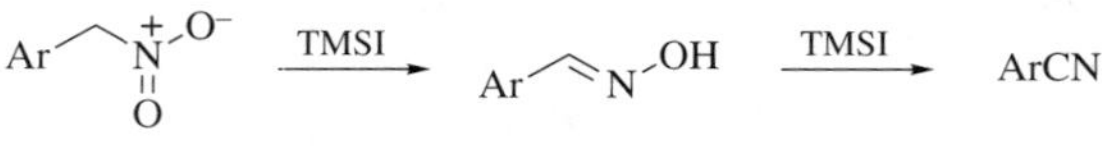

(38)

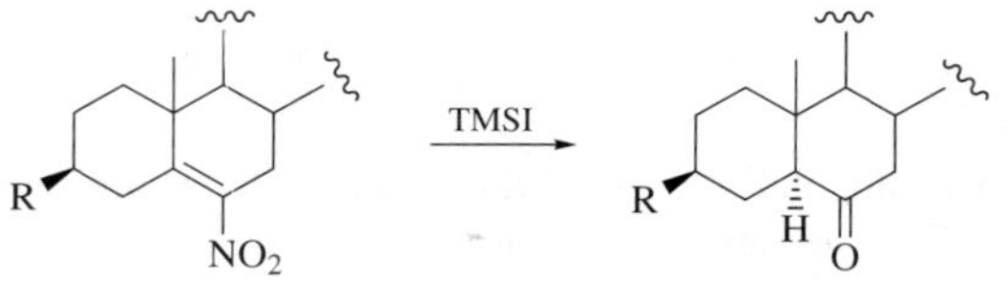

(39)

An interesting analogy to this dehydration process is found in the reductive fragmentation of a bromoisoxazoline with TMS-I, which yields the nitrile (eq 40).[70]

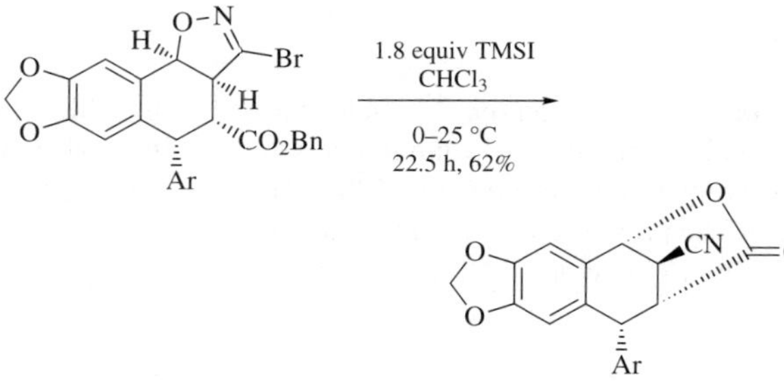

(40)

Rearrangement Reactions. An interesting rearrangement occurs on treatment of a β-alkoxy ketone with TMS-I which effects dealkylation and retro-aldol reaction to give the eight-membered diketone after reductive dehalogenation (eq 41).[71] Tertiary allylic silyl ethers α to epoxides undergo a stereocontrolled rearrangement to give the β-hydroxy ketones on treatment with catalytic TMS-I (eq 42).[72]

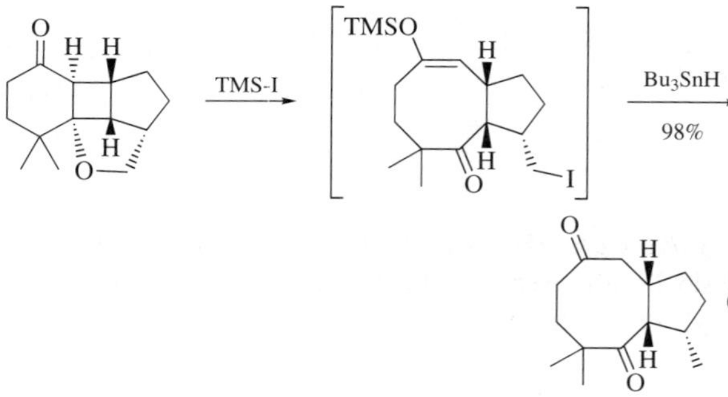

(41)

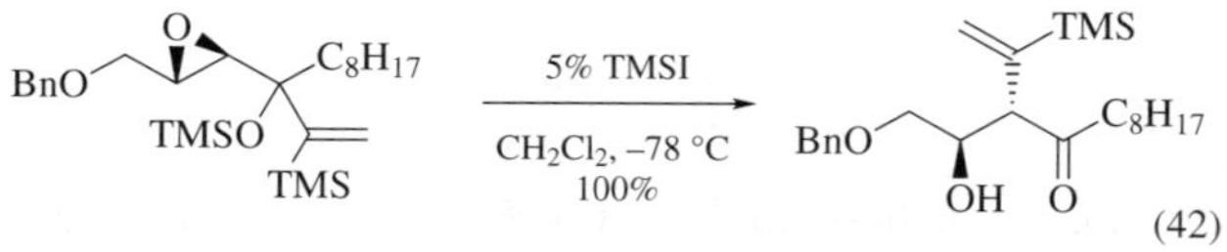

(42)

Related Reagents. Bromotrimethylsilane; Chlorotrimethylsilane; Hydrogen Bromide; Hydrogen Iodide; Trimethylsilyl Trifluoromethanesulfonate.

1. (a) Olah, G. A.; Prakash, G. K.; Krishnamurti, R. *Adv. Silicon Chem.* **1991**, *1*, 1. (b) Lee, S. D.; Chung, I. N. *Hwahak Kwa Kongop Ui Chinbo*

1984, *24*, 735. (c) Olah, G. A.; Narang, S. C. *T* **1982**, *38*, 2225. (d) Hosomi, A. *Yuki Gosei Kagaku Kyokai Shi* **1982**, *40*, 545. (e) Ohnishi, S.; Yamamoto, Y. *Annu. Rep. Tohoku Coll. Pharm.* **1981**, *28*, 1. (f) Schmidt, A. H. *Aldrichim. Acta* **1981**, *14*, 31. (g) Groutas, W. C.; Felker, D. *S* **1980**, *11*, 86. (h) Schmidt, A. H. *CZ* **1980**, *104* (9), 253.

2. (a) Lissel, M.; Drechsler, K. *S* **1983**, 459. (b) Machida, Y.; Nomoto, S.; Saito, I. *SC* **1979**, *9*, 97.
3. (a) Schmidt, A. H.; Russ, M. *CZ* **1978**, *102*, 26, 65. (b) Olah, G. A.; Narang, S. C.; Gupta, B. G. B. *S* **1979**, 61. (c) Morita, T.; Okamoto, Y.; Sakurai, H. *TL* **1978**, 2523; *CC* **1978**, 874.
4. (a) Kumada, M.; Shiiman, K.; Yamaguchi, M. *Kogyo Kagaku Zasshi* **1954**, *57*, 230. (b) Sakurai, H.; Shirahata, A.; Sasaki, K.; Hosomi, A. *S* **1979**, 740.
5. Ho, T. L.; Olah, G. A. *S* **1977**, 417.
6. (a) Jung, M. E.; Lyster, M. A. *OSC* **1988**, *6*, 353. (b) Jung, M. E.; Blumenkopf, T. A. *TL* **1978**, 3657.
7. (a) Jung, M. E.; Lyster, M. A. *JOC* **1977**, *42*, 3761. (b) Voronkov, M. G.; Dubinskaya, E. I.; Pavlov, S. F.; Gorokhova, V. G. *IZV* **1976**, 2355.
8. (a) Olah, G. A.; Narang, S. C.; Gupta, B. G. B.; Malhotra, R. *JOC* **1979**, *44*, 1247. (b) Voronkov, M. G.; Dubinskaya, E. J. *JOM* **1991**, *410*, 13. (c) Voronkov, M. G.; Puzanova, V. E.; Pavlov, S. F.; Dubinskaya, E. J. *BAU* **1975**, *14*, 377. (d) Voronkov, M. G.; Dubinskaya, E. J.; Pavlov, S. F.; Gorokhova, V. G. *BAU* **1976**, *25*, 2198. (e) Voronkov, M. G.; Komarov, V. G.; Albanov, A. I.; Dubinskaya, E. J. *BAU* **1978**, *27*, 2347. (f) Hirst, G. C.; Johnson, T. O., Jr.; Overman, L. E. *JACS* **1993**, *115*, 2992.
9. (a) Jung, M. E.; Ornstein, P. L. *TL* **1977**, 2659. (b) Voronkov, M. G.; Pavlov, S. F.; Dubinskaya, E. J. *DOK* **1976**, *227*, 607 (Eng. p. 218); *BAU* **1975**, *24*, 579.
10. (a) Casnati, A.; Arduini, A.; Ghidini, E.; Pochini, A.; Ungaro, R. *T* **1991**, *47*, 2221. (b) Silverman, R. B.; Radak, R. E.; Hacker, N. P. *JOC* **1979**, *44*, 4970. (c) Vickery, E. H.; Pahler, L. F.; Eisenbraun, E. J. *JOC* **1979**, *44*, 4444. (d) Brasme, B.; Fischer, J. C.; Wartel, M. *CJC* **1977**, *57*, 1720. (e) Rosen, B. J.; Weber, W. P. *JOC* **1977**, *42*, 3463.
11. Minamikawa, J.; Brossi, A. *TL* **1978**, 3085.
12. Hartman, D. A.; Curley, Jr., R. W. *TL* **1989**, *30*, 645.
13. (a) Denis, J. N.; Magnane, R. M.; van Eenoo, M.; Krief, A. *NJC* **1979**, *3*, 705. (b) Detty, M. R.; Seidler, M. D. *TL* **1982**, *23*, 2543. (c) Sakurai, H.; Sasaki, K.; Hosomi, A. *TL* **1980**, *21*, 2329. (d) Kraus, G. A.; Frazier, K. *JOC* **1980**, *45*, 2579. (e) Hill, R. K.; Pendalwar, S. L.; Kielbasinski, K.; Baevsky, M. F.; Nugara, P. N. *SC* **1990**, *20*, 1877.
14. (a) Ho, T. L.; Olah, G. A. *AG(E)* **1976**, *15*, 774. (b) Jung, M. E.; Lyster, M. A. *JACS* **1977**, *99*, 968. (c) Schmidt, A. H.; Russ, M. *CZ* **1979**, *103*, 183, 285. (d) See also refs. 8a, 9a.
15. Olah, G. A.; Narang, S. C.; Salem, G. F.; Gupta, B. G. B. *S* **1981**, 142.
16. Acyl iodides are also available from acid chlorides and TMS-I: Schmidt, A. N.; Russ, M.; Grosse, D. *S* **1981**, 216.
17. (a) Ho, T. L. *SC* **1979**, *9*, 233. (b) Sekiguchi, A.; Kabe, Y.; Ando, W. *TL* **1979**, 871.
18. Hiyama, T.; Saimoto, H.; Nishio, K.; Shinoda, M.; Yamamoto, H.; Nozaki, H. *TL* **1979**, 2043.
19. Kricheldorf, H. R. *AG(E)* **1979**, *18*, 689.
20. Yamamoto, Y.; Ohnishi, S.; Azuma, Y. *CPB* **1982**, *30*, 3505.
21. (a) Jung, M. E.; Lyster, M. A. *CC* **1978**, 315. (b) Rawal, V. H.; Michoud, C.; Monestel, R. F. *JACS* **1993**, *115*, 3030. (c) Wender, P. A.; Schaus, J. M.; White, A. W. *JACS* **1980**, *102*, 6157. (d) Lott, R. S.; Chauhan, V. S.; Stammer, C. H. *CC* **1979**, 495. (e) Vogel, E.; Altenbach, H. J.; Drossard, J. M.; Schmickler, H.; Stegelmeier, H. *AG(E)* **1980**, *19*, 1016.
22. Olah, G. A.; Narang, S. C.; Gupta, B. G. B.; Malhotra, R. *AG(E)* **1979**, *18*, 612.
23. Blaskovich, M. A.; Lajoie, G. A. *JACS* **1993**, *115*, 5021.
24. (a) Jung, M. E.; Andrus, W. A.; Ornstein, P. L. *TL* **1977**, 4175. (b) Bryant, J. D.; Keyser, G. E.; Barrio, J. R. *JOC* **1979**, *44*, 3733. (c) Muchmore, D. C.; Dahlquist, F. W. *Biochem. Biophys. Res. Commun.* **1979**, *86*, 599.
25. Chan, T. H.; Lee, S. D. *TL* **1983**, *24*, 1225.
26. Kobylinskaya, V. I.; Dashevskaya, T. A.; Shalamai, A. S.; Levitskaya, Z. V. *ZOB* **1992**, *62*, 1115.
27. (a) Jung, M. E.; Mazurek, M. A.; Lim, R. M. *S* **1978**, 588. (b) Olah, G. A.; Husain, A.; Narang, S. C. *S* **1983**, 896.
28. (a) Thiem, J.; Meyer, B. *CB* **1980**, *113*, 3075. (b) Tocik, Z.; Earl, R. A.; Beranek, J. *Nucl. Acids Res.* **1980**, *8*, 4755.
29. Bryson, T. A.; Bonitz, G. H.; Reichel, C. J.; Dardis, R. E. *JOC* **1980**, *45*, 524.
30. (a) Zygmunt, J.; Kafarski, P.; Mastalerz, P. *S* **1978**, 609. (b) Blackburn, G. M.; Ingleson, D. *CC* **1978**, 870. (c) Blackburn, G. M.; Ingleson, D. *JCS(P1)* **1980**, 1150.
31. Lee, K.; Wiemer, D. F. *TL* **1993**, *34*, 2433.
32. (a) Miller, R. D.; McKean, D. R. *JOC* **1981**, *46*, 2412. (b) Giacomini, E.; Loreto, M. A.; Pellacani, L.; Tardella, P. A. *JOC* **1980**, *45*, 519. (c) Dieter, R. K.; Pounds, S. *JOC* **1982**, *47*, 3174.
33. (a) Miller, R. D.; McKean, D. R. *TL* **1980**, *21*, 2639. (b) Crimmins, M. T.; Mascarella, S. W. *JACS* **1986**, *108*, 3435.
34. (a) Friedrich, E. C.; Abma, C. B.; Vartanian, P. F. *JOM* **1980**, *187*, 203. (b) Friedrich, E. C.; DeLucca, G. *JOM* **1982**, *226*, 143.
35. Jung, M. E.; Mossman, A. B.; Lyster, M. A. *JOC* **1978**, *43*, 3698.
36. (a) Jung, M. E.; Lewis, P. K. *SC* **1983**, *13*, 213. (b) Lipshutz, B. H.; Ellsworth, E. L.; Siahaan, T. J.; Shirazi, A. *TL* **1988**, *29*, 6677.
37. Jung, M. E.; Miller, S. J. *JACS* **1981**, *103*, 1984.
38. Schmidt, A. H.; Russ, M. *CZ* **1979**, *103*, 183, 285.
39. (a) Miller, R. D.; McKean, D. R. *TL* **1979**, 2305. (b) Larson, G. L.; Klesse, R. *JOC* **1985**, *50*, 3627.
40. Taniguchi, M.; Kobayashi, S.; Nakagawa, M.; Hino, T.; Kishi, Y. *TL* **1986**, *27*, 4763.
41. (a) Corey, E. J.; Boaz, N. W. *TL* **1985**, *26*, 6015, 6019. (b) Bergdahl, M.; Nilsson, M.; Olsson, T.; Stern, K. *T* **1991**, *47*, 9691, and references cited therein.
42. (a) Miller, R. D.; McKean, D. R. *S* **1979**, 730. (b) Hergott, H. H.; Simchen, G. *LA* **1980**, 1718. (c) Babot, O.; Cazeau, P.; Duboudin, F. *JOM* **1987**, *326*, C57.
43. (a) Kramarova, E. P.; Shipov, A. G.; Artamkina, O. B.; Barukov, Y. I. *ZOB* **1984**, *54*, 1921. (b) King, A. O.; Anderson, R. K.; Shuman, R. F.; Karady, S.; Abramson, N. L.; Douglas, A. W. *JOC* **1993**, *58*, 3384.
44. (a) Lau, P. W. K.; Chan, T. H. *JOM* **1979**, *179*, C24. (b) Wilson, S. R.; Phillips, L. R.; Natalie, K. J., Jr. *JACS* **1979**, *101*, 3340.
45. (a) Yamashita, H.; Kobayashi, T.; Hayashi, T.; Tanaka, M. *CL* **1991**, 761. (b) Chatani, N.; Amishiro, N.; Murai, S. *JACS* **1991**, *113*, 7778.
46. Sakurai, H.; Sasaki, K.; Hayashi, J.; Hosomi, A. *JOC* **1984**, *49*, 2808.
47. Hosomi, A.; Sakata, Y.; Sakurai, H. *CL* **1983**, 405.
48. Sakurai, H.; Sasaki, K.; Hosomi, A. *BCJ* **1983**, *56*, 3195.
49. Mukaiyama, T.; Akamatsu, H.; Han, J. S. *CL* **1990**, 889.
50. Sakurai, H.; Sasaki, K.; Hosomi, A. *TL* **1981**, *22*, 745.
51. Vankar, Y. D.; Kumaravel, G.; Mukherjee, N.; Rao, C. T. *SC* **1987**, *17*, 181.
52. Sarma, J. C.; Barua, N. C.; Sharma, R. P.; Barua, J. N. *T* **1983**, *39*, 2843.
53. Jung, K.-Y.; Koreeda, M. *JOC* **1989**, *54*, 5667.
54. Ghera, E.; Maurya, R.; Hassner, A. *TL* **1989**, *30*, 4741.
55. Sassaman, M. B.; Prakash, G. K.; Olah, G. A. *T* **1988**, *44*, 3771.
56. (a) Ho, T.-L. *SC* **1979**, *9*, 665. (b) Sarma, D. N.; Sarma, J. C.; Barua, N. C.; Sharma, R. P. *CC* **1984**, 813. (c) Nagaoka, M.; Kunitama, Y.; Numazawa, M. *JOC* **1991**, *56*, 334. (d) Numazawa, M.; Nagaoka, M.; Kunitama, Y. *CPB* **1986**, *34*, 3722; *CC* **1984**, 31. (e) Hartman, D. A.; Curley, R. W., Jr. *TL* **1989**, *30*, 645. (f) Cherbas, P.; Trainor, D. A.; Stonard, R. J.; Nakanishi, K. *CC* **1982**, 1307.
57. Olah, G. A.; Arvanaghi, M.; Vankar, Y. D. *JOC* **1980**, *45*, 3531.
58. (a) Olah, G. A.; Gupta, B. G. B.; Narang, S. C. *S* **1977**, 583. (b) Pitlik, J.; Sztaricskai, F. *SC* **1991**, *21*, 1769.
59. Nicolaou, K. C.; Barnette, W. E.; Magolda, R. L. *JACS* **1978**, *100*, 2567.

60. Miller, R. D.; McKean, D. R. *TL* **1983**, *24*, 2619.
61. Schaumann, E.; Winter-Extra, S.; Kummert, K.; Scheiblich, S. *S* **1990**, 271.
62. Olah, G. A.; Narang, S. C.; Field, L. D.; Salem, G. F. *JOC* **1980**, *45*, 4792.
63. Shipov, A. G.; Baukov, Y. I. *ZOB* **1984**, *54*, 1842.
64. (a) Beattie, I. R.; Parrett, F. W. *JCS(A)* **1966**, 1784. (b) Livanstov, M. V.; Proskurnina, M. V.; Prischenko, A. A.; Lutsenko, I. F. *ZOB* **1984**, *54*, 2504.
65. Kabachnik, M. M.; Prischenko, A. A.; Novikova, Z. S.; Lutsenko, I. F. *ZOB* **1979**, *49*, 1446.
66. Kibardin, A. M.; Gryaznova, T. V.; Gryaznov, P. I.; Pudovik, A. N. *JGU* **1991**, *61*, 1969.
67. (a) Jung, M. E.; Long-Mei, Z. *TL* **1983**, *24*, 4533. (b) Godleski, S. A.; Heacock, D. J. *JOC* **1982**, *47*, 4820.
68. Olah, G. A.; Narang, S. C.; Field, L. D.; Fung, A. P. *JOC* **1983**, *48*, 2766.
69. Singhal, G. M.; Das, N. B.; Sharma, R. P. *CC* **1989**, 1470.
70. Haber, A. *TL* **1989**, *30*, 5537.
71. Inouye, Y.; Shirai, M.; Michino, T.; Kakisawa, H. *BCJ* **1993**, *66*, 324.
72. Suzuki, K.; Miyazawa, M.; Tsuchihashi, G. *TL* **1987**, *28*, 3515.

Michael E. Jung
University of California, Los Angeles, CA, USA

Michael J. Martinelli
Lilly Research Laboratories, Indianapolis, IN, USA

Isobutene

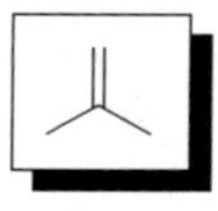

[115-11-7] C_4H_8 (MW 56.12)

(carboxy and hydroxy protection; photochemical cycloaddition with enones; ene reaction with enophiles; Friedel–Crafts and other alkylations; reacts with carbenes; acid-catalyzed cycloadditions)

Alternate Name: isobutylene.
Physical Data: mp −140.4 °C; bp −6.9 °C; vapor density 1.947.
Solubility: insol H_2O; sol THF, pet ether, benzene.
Form Supplied in: gas (lecture bottle); widely available.
Handling, Storage, and Precautions: flammable gas; store and use with adequate ventilation. Do not store with oxidizing materials and compounds that add across double bonds.

Protection for the Carboxy Group. Isobutene has been widely used in synthesis to convert carboxylic acids into corresponding *t*-butyl esters.[1,2] *t*-Butyl esters of aliphatic acids (eq 1),[3] aromatic carboxylic acids (eq 2),[4] and *N*-protected amino acids (eq 3)[5] have been prepared. The hindered *t*-butyl esters are stable to saponification but can be cleaved by acid-catalyzed hydrolysis, with liberation of isobutene.

HO(O)C–CH2–C(O)OH + isobutene → (conc H_2SO_4; Et_2O, 25 °C, 6–12 h, 58–60 %) *t*-BuO(O)C–CH2–C(O)O-*t*-Bu (1)

CO_2H, NO_2, AcO-substituted benzoic acid + isobutene → (H_2SO_4; $CHCl_3$–dioxane (3:1), 40–50 °C, 40 h, sealed tube, 95%) CO_2-*t*-Bu ester (2)

Cbz–NH–CH(R)–C(O)OH + isobutene → (conc H_2SO_4; Et_2O, 25 °C, 6–12 h, 58–60%) Cbz–NH–CH(R)–C(O)O-*t*-Bu (3)

Protection for the Hydroxy Group. In the presence of acid catalysts, isobutene reacts to convert a variety of alcohols and phenols to their corresponding *t*-butyl ethers. *t*-Butyl ethers are stable to most reagents but are readily cleaved by strong acids. Propargyl alcohols (eq 4),[6] steroidal alcohols (eq 5),[7] and phenols (eq 6)[8] have been protected. The *t*-butyl group has also been used to protect the hydroxy group of valinol (eq 7),[9] serine derivatives (eq 8),[10] and tyrosine.[11]

OH + isobutene → (Amberlyst H-15; hexane, 22–45 °C, 3–10 h, 100%) O-*t*-Bu (4)

OH + isobutene → (BF_3•Et_2O, H_3PO_4; CH_2Cl_2, 25 °C, 4.25 h, 86%) O-*t*-Bu (5)

OH + isobutene → (TfOH, CH_2Cl_2; Et_3N, −78 to 22 °C, 4 h, 90%) O-*t*-Bu (6)

OH, HN–CHO + isobutene → (BF_3•Et_2O; dioxane, 22 °C, 3 h, 75–95%) O-*t*-Bu (7)

OH, OBn, NH, OBn + isobutene → (H_2SO_4; CH_2Cl_2, 74%) *t*-BuO (8)

Photochemical Cycloadditions. Isobutene has been widely used in intermolecular [2+2] photocycloaddition reactions with enones.[11] The weakly polarized isobutene is often used to study the regioselectivity of the photocycloaddition. Cyclohexenones (eq 9),[12] cyclopentenones (eq 10),[13] and functionalized enones (eq 11)[14] undergo cycloaddition with isobutene. Paterno–Büchi photoadditions with isobutene have produced various oxetanes (eq 12).[15]

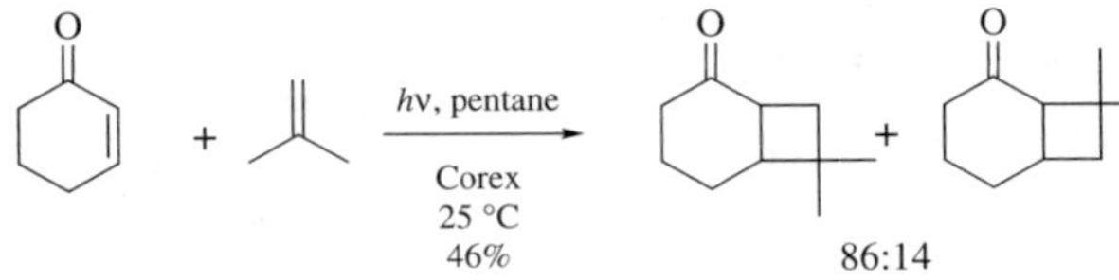

hν, pentane; Corex; 25 °C; 46%; 86:14 (9)

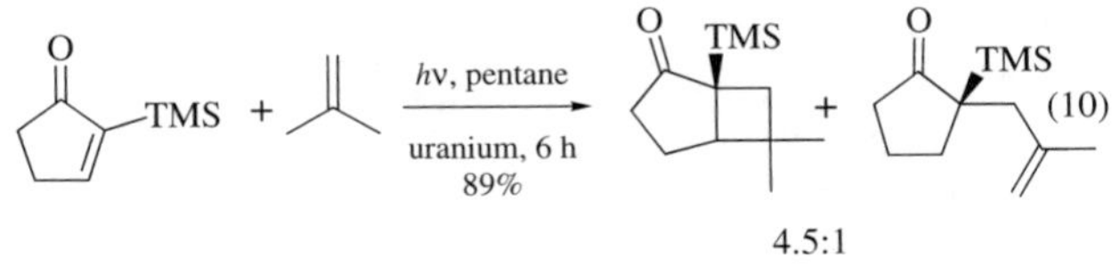

hν, pentane; uranium, 6 h; 89%; 4.5:1 (10)

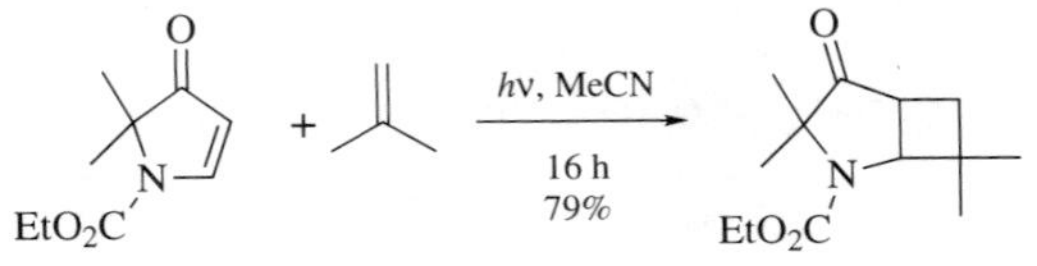

hν, MeCN; 16 h; 79% (11)

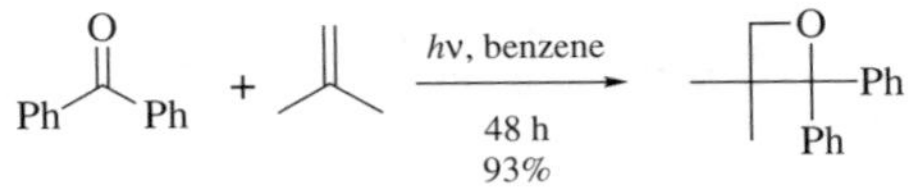

hν, benzene; 48 h; 93% (12)

Ene Reactions. In the presence of Lewis acids, isobutene adds to various enophiles to yield ene adducts.[16] Enophiles such as alkoxyaldehyde (eq 13),[17] dialkyl aminoaldehyde (eq 14),[18] haloaldehyde (eq 15),[19] and vinyl sulfoxides (eq 16)[20] have been utilized. Chiral organoaluminum (eq 17)[21] and organotitanium reagents (eq 18)[22] have been reported to give high levels of asymmetric induction in the ene reaction of isobutene with activated aldehydes.

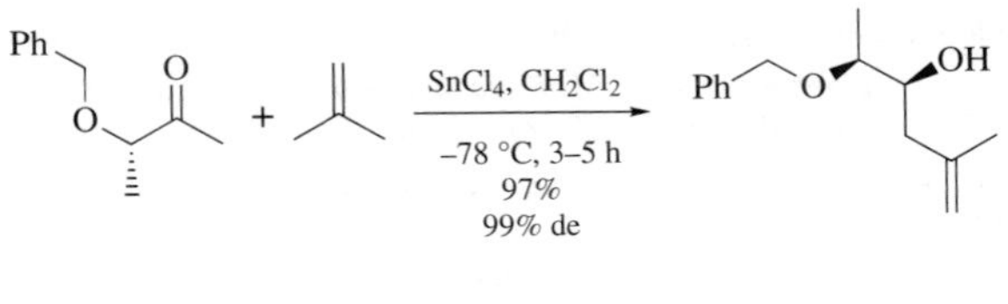

$SnCl_4$, CH_2Cl_2; −78 °C, 3–5 h; 97%; 99% de (13)

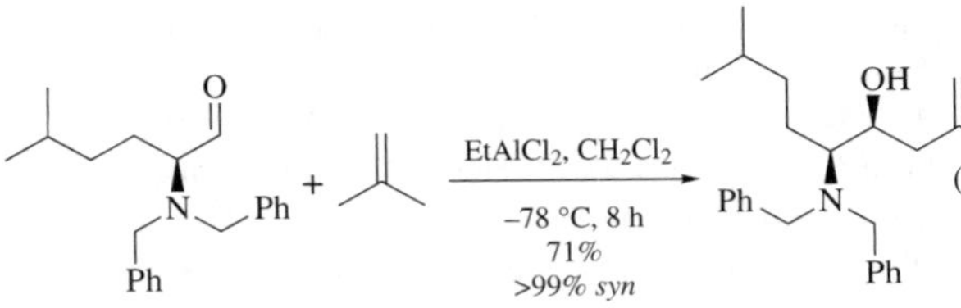

$EtAlCl_2$, CH_2Cl_2; −78 °C, 8 h; 71%; >99% *syn* (14)

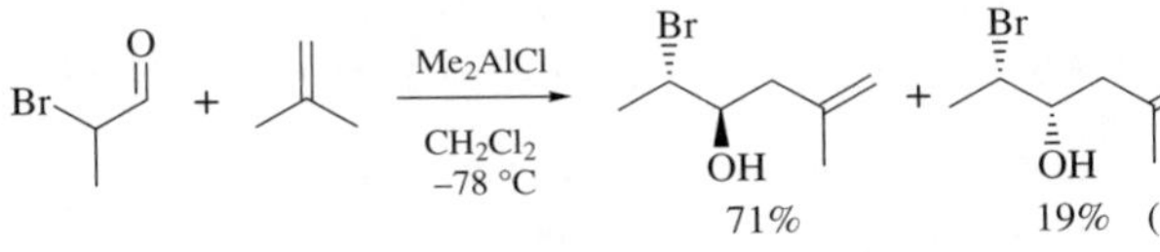

Me_2AlCl; CH_2Cl_2; −78 °C; 71%; 19% (15)

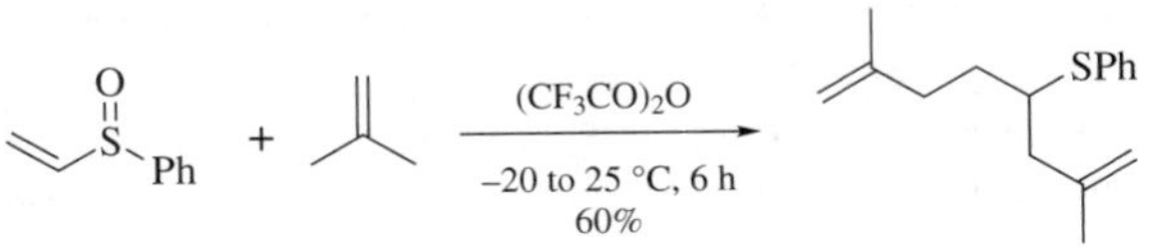

$(CF_3CO)_2O$; −20 to 25 °C, 6 h; 60% (16)

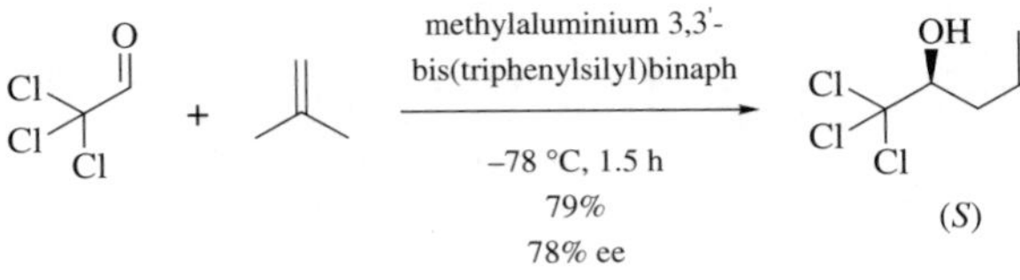

methylaluminium 3,3′-bis(triphenylsilyl)binaph; −78 °C, 1.5 h; 79%; 78% ee; (*S*) (17)

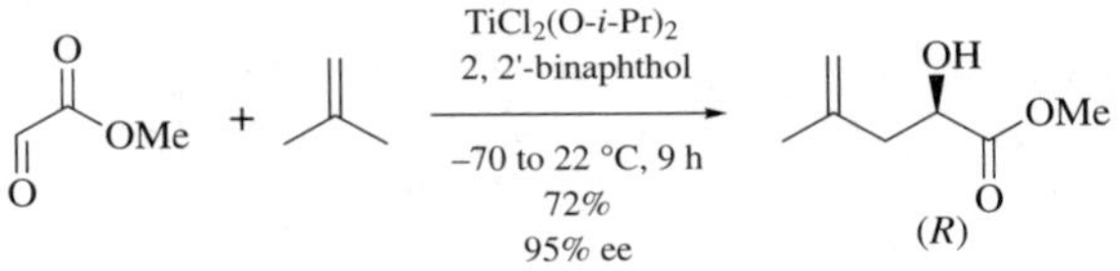

$TiCl_2(O$-*i*-$Pr)_2$; 2, 2′-binaphthol; −70 to 22 °C, 9 h; 72%; 95% ee; (*R*) (18)

Friedel–Crafts and Other Alkylation Reactions. A variety of *t*-butyl-substituted aromatics have been prepared with isobutene in the presence of acids (eqs 19–21).[23–25] Due to steric factors, the regiochemistry is enhanced. Isobutene can be alkylated by benzylic and allylic halides in the presence of ***Zinc Chloride*** (eqs 22–24).[26,27]

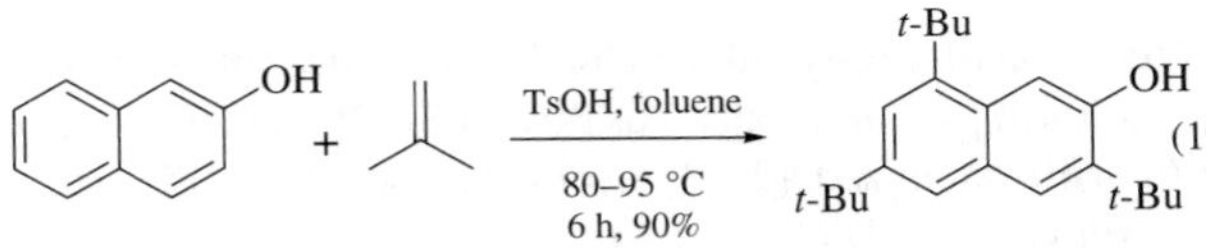

TsOH, toluene; 80–95 °C; 6 h, 90% (19)

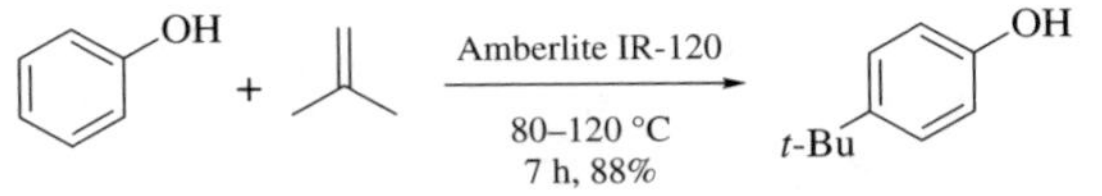

Amberlite IR-120; 80–120 °C; 7 h, 88% (20)

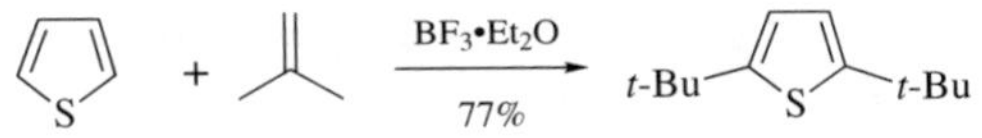

BF_3•Et_2O; 77% (21)

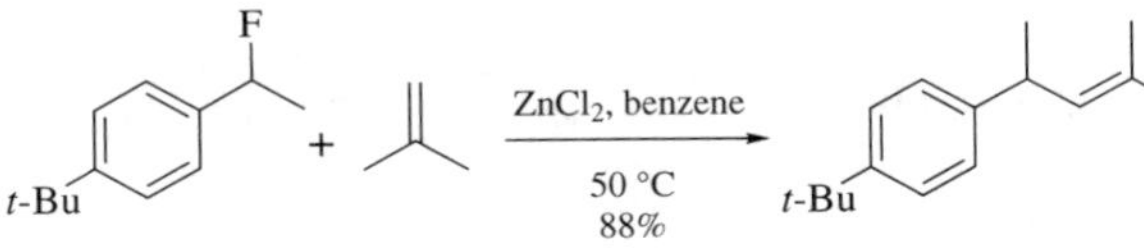

$ZnCl_2$, benzene; 50 °C; 88% (22)

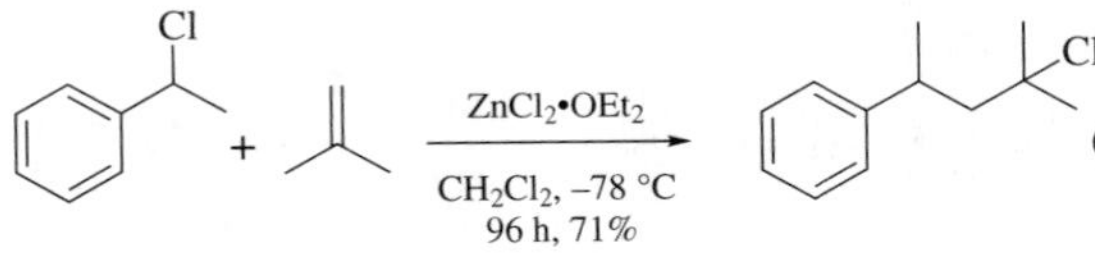

$ZnCl_2$•OEt_2; CH_2Cl_2, −78 °C; 96 h, 71% (23)

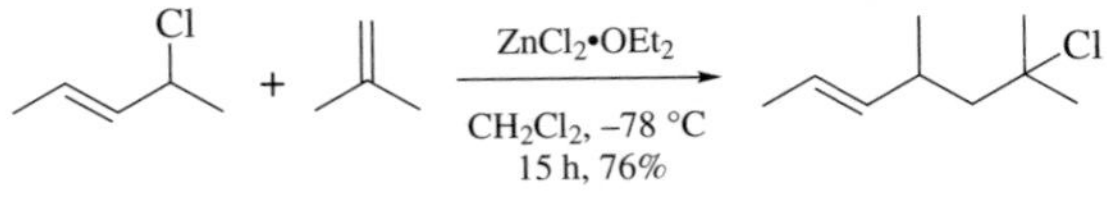

$ZnCl_2$•OEt_2; CH_2Cl_2, −78 °C; 15 h, 76% (24)

Carbene Chemistry. Isobutene reacts readily with carbenes and metallocarbenes to form substituted cyclopropanes (eqs 25–27).[28] Reaction of ***Ethyl Diazoacetate*** with isobutene in the presence of a chiral copper catalyst affords the cyclopropane in high optical purity (eq 28).[29] Isobutene has also been used to form titanocyclobutanes that are more stable forms of the Tebbe reagent (see ***μ-Chlorobis(cyclopentadienyl)(dimethylaluminum)-μ-methylenetitanium***) (eq 29).[30]

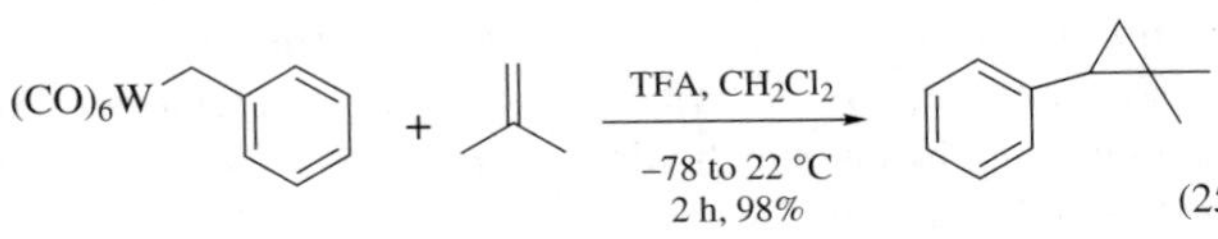

TFA, CH_2Cl_2; −78 to 22 °C; 2 h, 98% (25)

KF, 18-crown-6; diglyme, –78 to 25 °C; 144 h, sealed tube; 97% (26)

Bu_4NCN, DMF; –70 to 22 °C; 4.25 h, 50% (27)

2, 2-bis[2-(4(*S*)-*t*-butyl-1, 3-oxazolinyl)]propane; CuOTf, 0–20 °C; 14 h, 91%, 99% ee (28)

Cp_2TiCl_2 + isobutene —$AlMe_3$→ Cp_2Ti (29)

Other Acid-catalyzed Cycloadditions. Isobutene has been used to form pyrylium salts (eqs 30 and 31),[31] oxetanes by ***Boron Trifluoride***-catalyzed [2+2] cycloaddition (eq 32),[32] and 3,4-dihydropyrans by Diels–Alder reactions (eq 33).[33]

H_2SO_4, Ac_2O; 43–47%; $HO_2CCH_2SO_3^-$ (30)

TfOH; 80%; TfO⁻ (31)

$BF_3 \cdot Et_2O$; CH_2Cl_2, –78 °C; 1 h, 78%; 100% de (32)

$SnCl_4$; CH_2Cl_2, 0 °C; 3 h, 84% (33)

Miscellaneous Reactions. Isobutene, as the carbenium ion source for Ritter reactions, produces *N-t*-butyl amides (eq 34).[34] Isobutene undergoes highly regioselective cycloaddition with isocyanate to form azetidinones and with nitrones to yield isoxazolidines (eqs 35 and 36).[35,36] Allyl thioethers are formed by the electrophilic addition of ***Benzenesulfenyl Chloride*** to isobutene (eq 37).[37] Alkenylphosphonous dichlorides and alkenylthionophosphonic dichlorides are also formed by addition of PCl_5 and P_2S_5 (eqs 38 and 39).[38]

H_2SO_4; AcOH; 71%; NH-*t*-Bu (34)

SO_2, KCl; –20 °C; 65–70% (35)

98% (36)

PhSCl, CH_2Cl_2; 76% (37)

PCl_5, benzene; 25 °C (38)

PCl_5, P_2S_5 (39)

Related Reagents. *t*-Butyl 2,2,2-Trichloroacetimidate.

1. Fieser, L. F.; Fieser, M. *FF* **1967**, *1*, 522.
2. Greene, T. W.; Wuts, P. G. M. *Protective Groups in Organic Synthesis*; Wiley: New York, 1991; p 245.
3. McCloskey, A. L.; Fonken, G. S.; Kluiber, R. W.; Johnson, W. S. *OSC* **1963**, *4*, 261.
4. Barany, G.; Albericio, F. *JACS* **1985**, *107*, 4936.
5. Anderson, G. W.; Callahan, F. M. *JACS* **1960**, *82*, 3359. Valerio, R. M.; Alewood, P. F.; Johns, R. B. *S* **1988**, 786.
6. Alexakis, A.; Gardette, M.; Colin, S. *TL* **1988**, *24*, 2951.
7. Ireland, R. E.; O'Neil, T. H.; Tolman, G. L. *OSC* **1990**, *7*, 66.
8. Holcombe, J. L.; Livinghouse, T. *JOC* **1986**, *51*, 111.
9. Dickman, D. A.; Boes, M.; Meyers, A. I. *OSC* **1993**, *8*, 204.
10. Altmann, E.; Altmann, K.-H.; Mutter, M. *AG(E)* **1988**, *27*, 858.
11. Crimmins, M. T.; Reinhold, T. L. *OR* **1993**, *44*, 297.
12. Corey, E. J.; Bass, J. D.; Le Mahieu, R.; Mitra, R. B. *JACS* **1964**, *86*, 5570.
13. Swenton, J. S.; Fritzen, E. L., Jr. *TL* **1979**, 1951.
14. Patjens, J.; Margaretha, P. *HCA* **1989**, *72*, 1817.
15. Arnold, D. R.; Hinman, R. L.; Glick, A. M. *TL* **1964**, 1425. Jorgenson, M. J. *TL* **1966**, 5811. Kwiatkowski, G. T.; Selley, D. B. *TL* **1968**, 3471.
16. Snider, B. B. *ACR* **1980**, *13*, 426.
17. Mikami, K.; Loh, T.-P.; Nakai, T. *TA* **1990**, *1*, 13.
18. Mikami, K.; Kaneko, M.; Loh, T.-P.; Terada, M.; Nakai, T. *TL* **1990**, *31*, 3909.

19. Mikami, K.; Loh, T.-P.; Nakai, T. *CC* **1991**, 77.
20. Brichard, M.-H.; Janousek, Z.; Merenyi, R.; Viehe, H. G. *TL* **1992**, *33*, 2511.
21. Marouka, K.; Hoshimo, Y.; Shirasaka, T.; Yamamoto, H. *TL* **1988**, *29*, 3967.
22. Mikami, K.; Terada, M.; Nakai, T. *JACS* **1989**, *111*, 1940.
23. Chasar, D. W. *JOC* **1984**, *49*, 4302.
24. Loev, B.; Massengale, J. T. *JOC* **1957**, *22*, 988.
25. Caesar, P. D. *JACS* **1948**, *70*, 3623.
26. Olah, G. A.; Kuhn, S. J.; Barnes, D. G. *JOC* **1964**, *29*, 2685.
27. Mayr, H. *AG(E)* **1981**, *20*, 184. Mayr, H.; Striepe, W. *JOC* **1983**, *48*, 1159. Mayr, H.; Klein, G.; Kolberg, G. *CB* **1984**, *117*, 2555.
28. Casey, C. P.; Polichnowski, S. W. *JACS* **1977**, *99*, 6097. Casey, C. P.; Miles, W. H.; Tukada, H. *JACS* **1985**, *107*, 2924. Curuco, R. F.; Chu, K. S. *SC* **1987**, *17*, 271. Moss, R. A.; Zdrojewski, T.; Krogh-Jespersen, K.; Wlostowski, M.; Matro, A. *TL* **1991**, *32*, 1925.
29. Evans, D. A.; Woerpel, K. A.; Hinman, M. M.; Faul, M. M. *JACS* **1991**, *113*, 726.
30. Dötz, K. H. *AG(E)* **1984**, *23*, 587.
31. Balaban, A. T.; Dinculescu, A. *OPP* **1982**, *14*, 39. Praill, P. F. G.; Whitear, A. L. *JCS* **1961**, 3573. Anderson, A. G.; Stang, P. J. *JOC* **1976**, *41*, 3034.
32. Sugimura, H.; Osumi, K. *TL* **1989**, *30*, 1571.
33. Sera, A.; Ohara, M.; Yamada, H.; Egashira, E.; Ueda, N.; Setsune, J. *CL* **1990**, *11*, 2043.
34. Stephens, C. R.; Beereboom, J. J.; Rennhard, H. H.; Gordon, P. N.; Murai, K.; Blackwood, R. K.; Schach von Wittenau, M. *JACS* **1963**, *85*, 2643. Bishop, R. *COS* **1991**, *6*, 261.
35. Graf, R. *OSC* **1973**, *5*, 673. Arbuzov, B. A.; Zobova, N. N. *S* **1974**, 461.
36. Prosyanik, A. V.; Mischenko, A. I.; Zaichenko, N. L.; Zorin, Y. Z. *KGS* **1979**, 599 (*CA* **1979**, *91*, 175 249).
37. Giese, B.; Mazumdar, P. *CB* **1981**, *114*, 2859.
38. Walsh, E. N.; Beck, T. M.; Woodstock, W. H. *JACS* **1955**, *77*, 929.

Michael T. Crimmins & Agnes S. Kim-Meade
University of North Carolina at Chapel Hill, NC, USA

Isobutyl Chloroformate

[543-27-1] $C_5H_9ClO_2$ (MW 136.59)

(used for making mixed anhydrides which serve as active intermediates for peptide synthesis;[1–3] used to block the 5′-hydroxy function of deoxyribosides in oligonucleotide synthesis;[4] condensing reagent[5])

Physical Data: bp 128.8 °C; *d* 1.04 g cm^{-3}.
Solubility: miscible with benzene, chloroform, ether.
Form Supplied in: clear colorless liquid.
Handling, Storage, and Precautions: gradually decomposed by water and alcohol; highly toxic; vapor irritates eyes and mucous membranes; stable on storage.[6,7] Use in a fume hood.

Peptide Reagent.[1,3] Protected amino acids, such as the *N*-benzyloxycarbonyl (Cbz) derivatives, are brought into solution in toluene or THF by addition of enough ***Triethylamine*** to form the salt, and isobutyl chloroformate is added at 0 °C to produce the mixed anhydride. A solution of an amino acid ester (or peptide ester) to be *N*-acylated is added in an inert solvent and the mixture is allowed to come to rt. Evolution of carbon dioxide begins immediately (eq 1). The isobutyl chloroformate is preferred to lower esters for preparation of peptides of moderate or high molecular weight;[6,7] ethyl chloroformate is preferred for synthesis of dipeptides.

$R^1CH(CO_2H)NHCbz \xrightarrow{Et_3N} R^1CH(CO_2^-\,{}^+NHEt_3)NHCbz \xrightarrow{i\text{-}BuOCOCl} R^1CH(CO_2CO_2\text{-}i\text{-}Bu)NHCbz \xrightarrow{H_2NCHR^2CO_2Me} R^1CH(CONHCHR^2CO_2Me)NHCbz + CO_2 + BuOH$ (1)

Boc-Gly-Gly-OEt and Boc-Gly-Gly-Gly-OEt have been successfully synthesized in reasonable yields (77% and 88%, respectively) via mixed anhydride derivatives from isobutyl chloroformate.[8] Previously it was reported[9] that considerable quantities of diacylated side products are formed in mixed anhydride syntheses with glycine when ethyl chloroformate is used.

5′-Hydroxy Function Blocking Agent for Deoxyribosides. The reagent has been used to block the 5′-hydroxy function of deoxyribosides in oligonucleotide synthesis (eq 2).[4] The reagent shows good selectivity for the 5′-hydroxy group, as is evident in the high yield of (**2**). The blocking group is removed smoothly by alkaline hydrolysis to give (**4**).

(**1**) $\xrightarrow[73\%]{i\text{-}BuOCOCl,\ py}$ (**2**) $\xrightarrow[91\%]{Ph_2C(Cl)C_6H_4\text{-}p\text{-}OMe}$ (**3**) $\xrightarrow[96\%]{dioxane,\ NaOH}$ (**4**) (2)

***N*-Protection.**[10] In the synthesis of enantiomerically pure symmetric vicinal diamines involving chirality transfer from a single chiral source, 1,2-diphenylethane-1,2-diamine, hydrolysis of the bisacetamides to the corresponding vicinal diamines was attempted using a number of amide hydrolysis procedures. However, the various base- and acid-catalyzed procedures did not result in a clean hydrolysis of the acetamide groups. This was achieved by converting amines into biscarbamate derivatives, followed by treatment with 30% HBr in acetic acid (eq 3).

Cyclodehydration.[5] In the synthesis of 1,4-dihydropyran[3,4-*b*]indolones, the reagent was used along with triethylamine for cyclodehydration of hydroxymethylindoleacetic acid to produce the lactone (eq 4).

Regioselective Arylation.[11] Arylation in the 4-position in pyridines results from 1:1 adduct formation between an aryltriisopropoxytitanium reagent and an *N*-isobutyloxycarbonylpyridinium salt, and after successive ***2,3-Dichloro-5,6-dicyano-1,4-benzoquinone*** dehydrogenation and cleavage of the 1-substituent (eq 5).

Isobutyl Sulfonate.[12] With ***Silver(I) p-Toluenesulfonate***, alkyl chloroformate forms sulfonic carbonic anhydride which when heated liberates carbon dioxide, producing alkyl sulfonate. In order to find out the mechanism of the decomposition of the mixed anhydride, reaction of isobutyl chloroformate and silver *p*-toluenesulfonate was investigated. If the free alkyl cation, i.e. isobutyl carbonium ion, is formed, the rearrangement to *s*-butyl and *t*-butyl carbonium ion would be expected, finally resulting in the formation of *s*-butyl *p*-toluenesulfonate. *t-B*utyl *p*-toluenesulfonate is unstable and is not expected to survive under the reaction conditions. In fact, decomposition gave a mixture of isobutyl *p*-toluenesulfonate and *s*-butyl *p*-toluenesulfonate in a 0.8:1 mol ratio, with a quantity of isobutene and free *p*-toluenesulfonic acid. Isobutene was considered to be derived from the *t*-butyl cation.

---

1. Boissonnas, R. A. *HCA* **1951**, *34*, 874; *HCA* **1952**, *35*, 2229 and 2237.
2. Vaughan, J. R. *JACS* **1951**, *73*, 3547; *JACS* **1952**, *74*, 676 and 6137; *JACS* **1953**, *75*, 5556; *JACS* **1954**, *76*, 2474.
3. Wieland, T.; Bernhard, H. *LA* **1951**, *572*, 190.
4. Ogilvie, K. K.; Letsinger, R. L. *JOC* **1967**, *32*, 2365.
5. Fray, E. B.; Moody, C. J.; Shah, P. *T* **1993**, *49*, 439.
6. Vaughan, J. R.; Osato, R. L. *JACS* **1952**, *74*, 676.
7. Anderson, G. W.; Zimmerman, J. E.; Callahan, F. M. *JACS* **1966**, *88*, 1336. ibid **1967**, *89*, 5012.
8. Hoffmann, J. A.; Tilak, M. A. *OPP* **1975**, *7*, 215.
9. Kopple, K. D.; Renick, K. J. *JOC* **1974**, *39*, 660.
10. Nantz, M. H.; Lee, D. A.; Bender, D. M.; Roohi, A. H. *JOC* **1992**, *57*, 6653.
11. Gundersen, L.; Rise, F.; Undheim, K. *T* **1992**, *48*, 5647.
12. Yamamoto, A.; Kobayashi, M. *BCJ* **1966**, *39*, 1283.

Tapan Ray
Sandoz Pharmaceuticals, East Hanover, NJ, USA

Isopropenyl Acetate[1]

[108-22-5] $C_5H_8O_2$ (MW 100.13)

(conversion of carbonyl compounds to their enol acetates;[5] acetylation of oxygen,[12] nitrogen,[14] and carbon;[12a] acetone equivalent in Lewis acid catalyzed aldol reactions;[15] conversion of malonic acids to their isopropylidene derivatives[17])

Physical Data: bp 97 °C;[2] *d* 0.909 g cm^{-3}; fp 18 °C.
Solubility: sol benzene, acetic acid.
Form Supplied in: 99 and 99+% pure liquid widely available.
Analysis of Reagent Purity: the acetone content may be determined by addition of an excess of ***Hydroxylamine*** hydrochloride to an ethanolic solution, followed by back titration against sodium hydroxide.[3]
Handling, Storage, and Precautions: flammable liquid; mild irritant; narcotic in high concentrations. LD_{50} orally in rats: 3.0 g kg^{-1}.[4]

Enol Acetylation. In the presence of catalytic acid, isopropenyl acetate reacts with enolizable ketones to give the enol acetate plus acetone. This reactivity has been exploited to accomplish selective enolization of ketones, and to provide dienol acetates for Diels–Alder reactions. Each of these reactions is described in more detail in the following sections.

Selective Generation of Kinetic and Thermodynamic Enol Acetates. While the use of isopropenyl acetate with catalytic acid is considered to provide conditions for the generation of kinetic enols, the ratio of kinetic to thermodynamic isomers obtained for a given enolizable ketone appears to be highly dependent on the individual system being studied. For example, in steroid chemistry, isopropenyl acetate has been used to provide significant yields of the kinetic Δ^2-enol (**2**) over the thermodynamically favored Δ^3-enol (**3**) (eq 1).[5] For comparison, the results of the corresponding thermodynamically controlled enol acetylation reaction using perchloric acid catalyst and ***Acetic Anhydride*** are also given.

(1)

	Conditions	(**2a**) R = OAc	(**3a**) R = OAc
(**1a**) R = OH	isopropenyl acetate, H_2SO_4, reflux 1 h	47%	53%
	acetic anhydride, $HClO_4$, PhH, CCl_4	25%	75%
		(**2b**) R = H	(**3b**) R = H
(**1b**) R = H	isopropenyl acetate, H_2SO_4, reflux 1 h	29%	71%
	acetic anhydride, $HClO_4$, PhH, CCl_4	6%	94%

As shown in eq 1, introduction of an 11β-hydroxy group causes a 16–18% increase in the amount of the kinetic Δ^2-enol acetate (**2**) regardless of the reaction conditions. Additionally, enol acetylation proceeds with concomitant acetylation of the 11β-hydroxy group under these same reaction conditions. While these conditions provide only a moderate predilection for formation of the kinetic enol acetate in the cases presented above, other related systems have shown excellent selectivity utilizing this method.[6,7]

Selective Generation of (E)- or (Z)-Enol Acetates. The (*E*)- and (*Z*)-enol acetates of β-keto esters can be selectively generated by careful choice of acetylation conditions.[8] Under acidic conditions isopropenyl acetate gives, almost exclusively, the (*Z*) isomer (**6a**) (eq 2). This selectivity is believed to arise because the reactive form of methyl acetoacetate under acidic conditions is the internally hydrogen bound (*Z*)-enol (**5**) (eq 2). Alternatively, the corresponding (*E*)-enol acetate (**6b**) may be selectively generated under basic conditions utilizing ***Acetyl Chloride*** as the acetylating agent (eq 3). The selectivity in this reaction is believed to be due to the solvent-separated ion (**7**) generated in ***Hexamethylphosphoric Triamide*** and ***Triethylamine***. For the anion (**7**), the (*E*) conformation places the negatively charged oxygen atoms at a maximum separation.

(**4**) → p-TsOH → [(**5**)] → 79% → (*Z*)-(**6a**) (2)

(*Z*):(*E*) = 18:1

(**4**) → NEt₃, HMPA → [(**7**)] → 56% → (*E*)-(**6b**) (3)

(*Z*):(*E*) = 1:15

The use of isopropenyl acetate and catalytic acid for the formation of dienol acetates from α,β-unsaturated aldehydes and ketones is known to favor the formation of the (*E*) isomer (**10a**) (eq 4).[9] For the enolization of ***Crotonaldehyde*** (**8**) the reaction gives 1-acetoxy-1-propene (**10**) as an 83:17 (*E*:*Z*) mixture. This selectivity is credited to the intermediacy of cation (**9**) in this process. Cation (**9b**), which would lead to the (*Z*) isomer (**10b**), is disfavored due to an unfavorable steric interaction between the vinyl and acetoxy groups (eq 5).

(**8**) → p-TsOH → [(**9a**)] → (*E*)-(**10a**) (4)

[(**9b**)] → (*Z*)-(**10b**) (5)

Dienol Acetates for the Diels–Alder Reaction. Wolinsky and Login[10] found that the methodology described above allows for the in situ generation of reactive dienes for use in Diels–Alder reactions (eqs 6 and 7). The enol acetate of 2-methyl-2-butenone (**11**) is generated by reaction with isopropenyl acetate and ***p-Toluenesulfonic Acid***. In the presence of ***Dimethyl Acetylenedicarboxylate*** (DMAD), the resulting diene undergoes a cycloaddition reaction, giving, after loss of acetic acid, the phthalic acid derivative (**12**) in 80% yield (eq 6). Similarly, crotonaldehyde (**8**) is converted to 1-acetoxy-1-propene (**10**), as shown in eq 4. Diene (**10**) is then trapped with chloromaleic anhydride (**13**). Loss of acetic acid and HCl gives phthalic anhydride (**14**) in 70% yield (eq 7).

(**11**) → 1. isopropenyl acetate, p-TsOH, DMAD; 2. 200 °C → (**12**) (6)

(**8**) + (**13**) → isopropenyl acetate, p-TsOH, reflux, 16 h → (**14**) (7)

Acetylation of Other O, N, and C Centers. In addition to the preparation of enol acetates from ketones, isopropenyl acetate has also seen use in the acetylation of other O, N, and C centers. Hydroxy acetylation was first noted[11] as a simultaneous reaction occurring during the enol acetylation of steroids bearing unprotected hydroxy groups (eq 1). The reagent has since been used more deliberately for the protection of alcohols.[12] More recently the reagent has been used in conjunction with enzymes to afford chiral acylated material.[13] Kaupp and Matthies[14] were surprised to observe *N*-acetylation of the benzothiazepine derivative (**15**) with no formation of the isomer (**17**) (eq 8). Attempts by the authors to generate the expected isomer (**17**) were unsuccessful.

OAc, *p*-TsOH, reflux, 14 h (8)

(**15**) (**16**)

(**17**)

The 5-position of some *N*-methylpyrrole derivatives (**18**) were successfully acetylated using isopropenyl acetate in the presence of catalytic acid under refluxing conditions (eq 9).[12a]

OAc, H^+, $(CH_2)_2Cl_2$, reflux, 3–5 h (9)

(**18a**) R = CO_2Et (**19a**) 81%
(**18b**) R = H (**19b**) 55%

Aldol Reactions. Isopropenyl acetate participates in some aldol-type reactions. This enol ester, like enol ethers, can react with various acetals (**20**) in the presence of a Lewis acid to afford the aldol-type addition products (**21** and **22**) (eq 10).[15] This same reactivity is observed in the condensation of ***Succinic Anhydride*** with isopropenyl acetate.[16]

OAc, $TiCl_4$, CH_2Cl_2 (10)

(**20a**) R = Ph	(**21a**) 81%	(**22a**) 5%
(**20b**) R = *i*-Pr	(**21b**) 92%	(**22b**) 0%
(**20c**) R = CH=CHPh	(**21c**) 0%	(**22c**) 65%

Isopropenyl Acetate as a Source of Acetone. Under certain reaction conditions, isopropenyl acetate reacts to deliver an equivalent of acetone for acetonide formation or for oxidative addition to an enol. Under acidic conditions, isopropenyl acetate has been used to generate isopropylidene derivatives of malonic acids.[17] Isopropenyl acetate also undergoes an oxidative addition reaction with ketones in the presence of ***Manganese(III) Acetate***.[18] An acetone subunit is added to the α-position of a ketone to give a 1,4-diketone.

Related Reagents. Acetic Anhydride; Acetyl Chloride; 2-Methoxypropene; Vinyl Acetate.

1. Hagemeyer, H. J., Jr.; Hull, D. C. *Ind. Eng. Chem.* **1949**, *41*, 2920.
2. *The Merck Index*, 11th ed.; Budavari, S., Ed.; Merck: Rahway, NJ, 1989; p 5092.
3. Jeffery, E. A.; Satchell, D. P. N. *JCS* **1962**, 1876.
4. Smyth, H. F., Jr.; Carpenter, C. P.; Weil, C. S. *J. Ind. Hyg. Toxicol.* **1949**, *31*, 60.
5. Liston, A. J.; Howarth, M. *JOC* **1967**, *32*, 1034.
6. Nemoto, H.; Kurobe, H.; Fukumoto, K.; Kametani, T. *JOC* **1986**, *51*, 5311.
7. Bold, G; Chao, S.; Bhide, R.; Wu, S.-H.; Patel, D. V.; Sih, C. J.; Chidester, C. *TL* **1987**, *28*, 1973.
8. Casey, C. P.; Marten, D. F. *TL* **1974**, 925.
9. Fukuda, W.; Sato, H.; Kakiuchi, H. *BCJ* **1986**, *59*, 751.
10. Wolinsky, J; Login, R. B. *JOC* **1970**, *35*, 3205.
11. Villotti, R.; Ringold, H. J.; Djerassi, C. *JACS* **1960**, *82*, 5693. The experimental data are cited by Djerassi, C. *Steroid Reactions*; Holden-Day: San Francisco, CA, 1963; p 41.
12. (a) Gizur, T.; Harsanyi, K. *SC* **1990**, *20*, 2365. (b) Shiao, M.-J.; Lin, J. L.; Kuo, Y.-H.; Shih, K.-S. *TL* **1986**, *27*, 4059. (c) Alarcon, P.; Pardo, M.; Soto, J. L. *JHC* **1985**, 273. (d) Pauluth, D.; Hoffmann, H. M. R. *LA* **1985**, 403.
13. Asensio, G.; Andreu, C.; Marco, J. A. *CB* **1992**, 2233.
14. Kaupp, G.; Matthies, D. *CB* **1987**, 1741.
15. Mukaiyama, T.; Izawa, T.; Saigo, K. *CL* **1974**, 323.
16. (a) Merenyi, F.; Nilsson, M. *ACS* **1963**, *17*, 1801; **1964**, *18*, 1368; Nilsson, M. *ACS* **1964**, *18*, 441. (b) Merenyi, F.; Nilsson, M. *OS* **1972**, *52*, 1.
17. (a) Davidson, D.; Bernhard, S. A. *JACS* **1948**, *70*, 3426. (b) Singh, R. K.; Danishefsky, S. *OS* **1981**, *60*, 66. (c) Maier, G.; Wolf, B. *S* **1985**, 871.
18. Dessau, R. M.; Heiba, E. I. *JOC* **1974**, *39*, 3457.

Michael A. Walters & Melissa D. Lee
Dartmouth College, Hanover, NH, USA

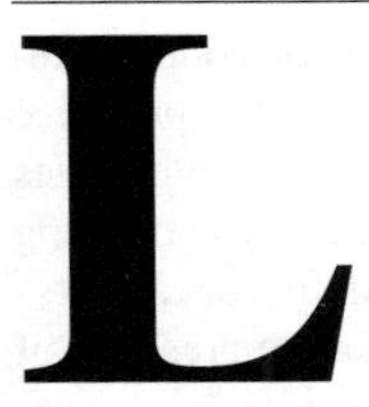

Lithium Bromide[1]

LiBr

[7550-35-8] BrLi (MW 86.85)

(source of nucleophilic bromide;[2] mild Lewis acid;[1] salt effects in organometallic reactions;[1] epoxide opening[1])

Physical Data: mp 550 °C; bp 1265 °C; *d* 3.464 g cm^{-3}.
Solubility: 145 g/100 mL H_2O (4 °C); 254 g/100 mL H_2O (90°C); 73 g/100 mL EtOH (40 °C); 8 g/100 mL MeOH; sol ether, glycol, pentanol, acetone; slightly sol pyridine.
Form Supplied in: anhyd white solid, or as hydrate.
Purification: dry for 1 h at 120 °C/0.1 mmHg before use; or dry by heating in vacuo at 70 °C (oil bath) for 24 h, then store at 110 °C until use.
Handling, Storage, and Precautions: for best results, dry before use in anhyd reactions.

Alkyl and Alkenyl Bromides. LiBr has been extensively used as a source of bromide in nucleophilic substitution and addition reactions. Interconversion of halides[2] and transformation of alcohols to alkyl bromides via the corresponding sulfonate[3] or trifluoroacetate[4] have been widely used in organic synthesis. Primary and secondary alcohols have been directly converted to alkyl bromides upon treatment with a mixture of ***Triphenylphosphine***, ***Diethyl Azodicarboxylate***, and LiBr.[5]

(Z)-3-Bromopropenoates and -propenoic acids have been synthesized stereoselectively by the reaction of LiBr and propiolates or propiolic acid (eq 1).[6]

$\equiv$–CO_2Et → (LiBr, AcOH; 70 °C, 15 h; 91%) Br–CH=CH–CO_2Et (1)

Heterolytic Cleavage of C–X Bonds. In the presence of a Lewis acid, LiBr acts as a nucleophile in the opening of 1,2-oxiranes to produce bromohydrins (eq 2).[7] In the absence of an external Lewis acid or nucleophile, epoxides generally give rise to products resulting from ring-contraction reactions (eq 3).

LiBr, AcOH; THF, rt; 90% (2)

LiBr, alumina; toluene, reflux; 77% (3)

LiBr-mediated decomposition of dioxaphospholanes results in the exclusive formation of the epoxide, whereas the thermal decomposition produces a mixture of products (eq 4).[8]

1. $(EtO)_2PPh_3$; 2. LiBr, rt; 97% (4)

Protection of alcohols as their MOM ethers can be achieved using a mixture of ***Dimethoxymethane***, LiBr, and ***p-Toluenesulfonic Acid***.[9]

Bifunctional Reagents. Activated α-bromo ketones are smoothly converted into the corresponding silyl enol ethers when treated with a mixture of LiBr/R_3N/***Chlorotrimethylsilane***.[10] Aldehydes are converted into the corresponding α,β-unsaturated esters using ***Triethyl Phosphonoacetate*** and ***Triethylamine*** in the presence of LiBr (eq 5).[11,12] Similar conditions were extensively used in the asymmetric cycloaddition and Michael addition reactions of *N*-lithiated azomethine ylides (eq 6).[13]

PhCHO + $(EtO)_2P(O)CH_2CO_2Et$ → (Et_3N, MX, MeCN; 25 °C, 3 h) PhCH=CHCO_2Et (5)

MX = LiCl, 77%; LiBr, 93%; $MgCl_2$, 15%; $MgBr_2$, 71%

1. LiBr, DBU; 2. CH$_2$=CHCO_2Me; 61% (6)

Additive for Organometallic Transformations. The addition of LiBr and ***Lithium Iodide*** was shown to enhance the rate of organozinc formation from primary alkyl chlorides, sulfonates, and phosphonates, and ***Zinc*** dust.[14] Beneficent effects of LiBr addition have also been reported for the Heck-type coupling reactions[15] and for the nickel-catalyzed cross-couplings of alkenyl and α-metalated alkenyl sulfoximines with organozinc reagents.[16] The addition of 2 equiv of LiBr significantly enhances the yield of the conjugate addition products in reactions of certain organocopper reagents (eq 7).[17]

$MeCu(PCy_2)Li$ (7)

LiBr (0 equiv), 61%
LiBr (2 equiv), 96%

Finally, concentrated solutions of LiBr are also known to alter significantly the solubility and the reactivity of amino acids and peptides in organic solvents.[18]

Related Reagents. Tetra-*n*-butylammonium Bromide; Tetramethylammonium Bromide; Phosphorus(III) Bromide; Sodium Bromide; Triphenylphosphine Dibromide.

---

1. Loupy, A.; Tchoubar, B. *Salt effects in Organic and Organometallic Chemistry*; VCH: Weinheim, 1992.

2. Sasson, Y.; Weiss, M.; Loupy, A.; Bram, G.; Pardo, C. *CC* **1986**, 1250.
3. (a) Ingold, K. U.; Walton, J. C. *JACS* **1987**, *109*, 6937. (b) McMurry, J. E.; Erion, M. D. *JACS* **1985**, *107*, 2712.
4. Camps, F.; Gasol, V.; Guerrero, A. *S* **1987**, 511.
5. Manna, S.; Falck, J. R. Mioskowski, C. *SC* **1985**, *15*, 663.
6. (a) Ma, S.; Lu, X. *TL* **1990**, *31*, 7653. (b) Ma, S.; Lu, X. *CC* **1990**, 1643.
7. (a) Bonini, C.; Giuliano, C.; Righi, G.; Rossi, L. *SC* **1992**, *22*, 1863. (b) Shimizu, M.; Yoshida, A.; Fujisawa, T. *SL* **1992**, 204. (c) Bajwa, J. S.; Anderson, R. C. *TL* **1991**, *32*, 3021.
8. (a) Murray, W. T.; Evans Jr., S. A. *NJC* **1989**, *13*, 329. (b) Murray, W. T.; Evans, S. A., Jr. *JOC* **1989**, *54*, 2440.
9. Gras, J.-L.; Chang, Y.-Y. K. W.; Guérin, A. *S* **1985**, 74.
10. Duhamel, L.; Tombret, F.; Poirier, J. M. *OPP* **1985**, *17*, 99.
11. Rathke, M. W.; Nowak, M. *JOC* **1985**, *50*, 2624.
12. Seyden-Penne, J. *BSF* **1988**, 238.
13. (a) Kanemasa, S.; Tatsukawa, A.; Wada, E. *JOC* **1991**, *56*, 2875. (b) Kanemasa, S.; Uchida, O.; Wada, E. *JOC* **1990**, *55*, 4411. (c) Kanemasa, S.; Yoshioka, M.; Tsuge, O. *BCJ* **1989**, *62*, 869. (d) Kanemasa, S.; Yamamoto, H.; Wada, E.; Sakurai, T.; Urushido, K. *BCJ* **1990**, *63*, 2857.
14. Jubert, C.; Knochel, P. *JOC* **1992**, *57*, 5425.
15. (a) Cabri, W.; Candiani, I.; DeBernardinis, S.; Francalanci, F.; Penco, S. *JOC* **1991**, *56*, 5796. (b) Karabelas, K.; Hallberg, A. *JOC* **1989**, *54*, 1773.
16. Erdelmeier, I.; Gais, H.-J. *JACS* **1989**, *111*, 1125.
17. Bertz, S. H.; Dabbagh, G. *JOC* **1984**, *49*, 1119.
18. Seebach, D. *Aldrichim. Acta* **1992**, *25*, 59.

André B. Charette
Université de Montréal, Québec, Canada

Lithium Chloride

[7447-41-8] ClLi (MW 42.39)

(source of Cl^- as nucleophile and ligand; weak Lewis acid that modifies the reactivity of enolates, lithium dialkylamides, and other Lewis bases)

Physical Data: mp 605 °C; bp 1325–1360 °C; *d* 2.068 g cm^{-3}.
Solubility: very sol H_2O; sol methanol, ethanol, acetone, acetonitrile, THF, DMF, DMSO, HMPA.
Form Supplied in: white solid, widely available.
Drying: deliquescent; for most applications, drying at 150 °C for 3 h is sufficient; for higher purity, recrystallization from methanol, followed by drying at 140 °C/0.5 mmHg overnight, is recommended.
Handling, Storage, and Precautions: of low toxicity; take directly from the oven when dryness is required.

Source of Chloride Nucleophile. The solubility of LiCl in many organic dipolar solvents renders it an effective source of nucleophilic chloride anion. Lithium chloride converts alcohols to alkyl chlorides[1] under Mitsunobu conditions,[2] or by way of the corresponding sulfonates[3] or other leaving groups.[4] This salt cleanly and regioselectively opens epoxides to chlorohydrins in the presence of acids and Lewis acids such as ***Acetic Acid***,[5] Amberlyst 15 resin,[6] and ***Titanium(IV) Chloride***.[7] In the presence of acetic acid, LiCl regio- and stereoselectively hydrochlorinates 2-propynoic acid and its derivatives to form the corresponding derivatives of (*Z*)-3-chloropropenoic acid.[8] Oxidative decarboxylation of carboxylic acids by ***Lead(IV) Acetate*** in the presence of 1 equiv of LiCl generates the corresponding chlorides.[9]

In wet DMSO, LiCl dealkoxycarbonylates various activated esters (eq 1).[10,11] If the reaction is performed in anhyd solvent the reaction generates a carbanion intermediate, which can undergo inter- or intramolecular alkylation or elimination. Other inorganic salts (NaCN, NaCl, ***Lithium Iodide***) and other dipolar aprotic solvents (HMPA, DMF) can also be employed. Under similar conditions, lithium chloride cleaves alkyl aryl ethers having electron-withdrawing substituents at the *ortho* or *para* positions.[12]

$$X\text{–}C(R^1)(R^2)\text{–}CO_2R^3 \xrightarrow[\text{wet DMSO}]{\text{LiCl}} X\text{–}C(R^1)(R^2)\text{–}H \quad (1)$$

$X = CO_2R$, COR, CN, SO_2R; R^3 = Me, Et

Source of Chloride Ligand. In palladium-catalyzed reactions, LiCl is often the reagent of choice as a source of chloride ligand. Lithium chloride is a necessary component in palladium-catalyzed coupling and carbonylative coupling reactions of organostannanes and vinyl triflates.[13,14] Lithium chloride has a dramatic effect on the stereochemical course of palladium-catalyzed 1,4-additions to 1,3-dienes.[15] Treatment of 1,3-cyclohexadiene with ***Palladium(II) Acetate*** and LiOAc and the oxidizing agents ***1,4-Benzoquinone*** and ***Manganese Dioxide*** affords 1,4-*trans*-diacetoxy-2-cyclohexene (eq 2). In the presence of a catalytic quantity of LiCl, the *cis* isomer is formed (eq 3). If 2 equiv LiCl are added, the *cis*-acetoxychloro compound forms (eq 4). These methods are general for both cyclic and acyclic dienes, and have recently been extended to the stereospecific formation of fused heterocycles.[16] Lithium chloride is also used in the preparation of ***Dilithium Tetrachloropalladate(II)***[17] and zinc organocuprate reagents.[18]

1,3-cyclohexadiene → (Pd(OAc)$_2$, LiOAc, no added LiCl; 1,4-benzoquinone, MnO_2; 93%) → AcO–cyclohexene–OAc, >91% *trans* (2)

1,3-cyclohexadiene → (Pd(OAc)$_2$, LiOAc, 0.2 equiv LiCl; 1,4-benzoquinone, MnO_2; 85%) → AcO–cyclohexene–OAc, >96% *cis* (3)

1,3-cyclohexadiene → (Pd(OAc)$_2$, LiOAc, 2 equiv LiCl; 1,4-benzoquinone, MnO_2; 89%) → Cl–cyclohexene–OAc, >98% *cis* (4)

Weak Lewis Acid. Lithium chloride is a weak Lewis acid that forms mixed aggregates with lithium dialkylamides, enolates, alkoxides, peptides, and related 'hard' Lewis bases.[19] Thus LiCl often has a dramatic effect on reactions involving these species. In the deprotonation of 3-pentanone by ***Lithium 2,2,6,6-***

Tetramethylpiperidide (LTMP), addition of 0.3 equiv LiCl increases the (*E*)/(*Z*) selectivity from 9:1 to 52:1 (eq 5).[20] Enhancement in the enantioselectivity of deprotonation of prochiral ketones by a chiral lithium amide has also been reported.[21] Lithium chloride stabilizes anions derived from α-phosphonoacetates, permitting amine and amidine bases to be used to perform Horner–Wadsworth–Emmons reactions on base-sensitive aldehydes under exceptionally mild conditions.[22] Lithium chloride and other lithium salts disrupt peptide aggregation and increase the solubilities of peptides in THF and other ethereal solvents, often by 100-fold or greater.[23] These effects render LiCl a useful additive in the chemical modification of peptides (e.g. by the formation and alkylation of peptide enolates).[19,24] Lithium chloride has also shown promise as an additive in solid-phase peptide synthesis, increasing resin swelling and improving the efficiencies of difficult coupling steps.[25]

O; 1. LiTMP (LiCl); 2. TMSCl, THF; OTMS + OTMS (5)

0.0 equiv LiCl 9:1
0.3 equiv LiCl 52:1

Related Reagents. Lithium Chloride–Diisopropylethylamine; Lithium Chloride–Hexamethylphosphoric Triamide; Zinc Chloride.

1. Magid, R. M. *T* **1980**, *36*, 1901.
2. Manna, S.; Falck, J. R.; Mioskowski, C. *SC* **1985**, *15*, 663.
3. (a) Owen, L. N.; Robins, P. A. *JCS* **1949**, 320. (b) Owen, L. N.; Robins, P. A. *JCS* **1949**, 326. (c) Eglinton, G.; Whiting, M. C. *JCS* **1950**, 3650. (d) Collington, E. W.; Meyers, A. I. *JOC* **1971**, *36*, 3044. (e) Stork, G.; Grieco, P. A.; Gregson, M. *TL* **1969**, *18*, 1393.
4. (a) Czernecki, S.; Georgoulis, C. *BSF* **1975**, 405. (b) Camps, F.; Gasol, V.; Guerrero, A. *S* **1987**, 511.
5. Bajwa, J. S.; Anderson, R. C. *TL* **1991**, *32*, 3021.
6. Bonini, C.; Giuliano, C.; Righi, G.; Rossi, L. *SC* **1992**, *22*, 1863.
7. Shimizu, M.; Yoshida, A.; Fujisawa, T. *SL* **1992**, 204.
8. (a) Ma, S.; Lu, X. *TL* **1990**, *31*, 7653. (b) Ma, S.; Lu, X.; Li, Z. *JOC* **1992**, *57*, 709.
9. (a) Kochi, J. K. *JACS* **1965**, *87*, 2500. (b) Review: Sheldon, R. A., Kochi, J. K. *OR* **1972**, *19*, 279.
10. Krapcho, A. P.; Weimaster, J. F.; Eldridge, J. M.; Jahngen, E. G. E., Jr.; Lovey, A. J.; Stephens, W. P. *JOC* **1978**, *43*, 138.
11. Reviews: (a) Krapcho, A. P. *S* **1982**, 805. (b) Krapcho, A. P. *S* **1982**, 893.
12. Bernard, A. M.; Ghiani, M. R.; Piras, P. P.; Rivoldini, A. *S* **1989**, 287.
13. (a) Scott, W. J.; Crisp, G. T.; Stille, J. K. *JACS* **1984**, *106*, 4630. (b) Crisp, G. T.; Scott, W. L.; Stille, J. K. *JACS* **1984**, *106*, 7500.
14. Reviews: (a) Stille, J. K. *AG(E)* **1986**, *25*, 508. (b) Scott, W. J.; McMurry, J. E. *ACR* **1988**, *21*, 47.
15. (a) Bäckvall, J. E.; Byström, S. E.; Nordberg, R. E. *JOC* **1984**, *49*, 4619. (b) Bäckvall, J. E.; Nyström, J. E.; Nordberg, R. E. *JACS* **1985**, *107*, 3676.
16. (a) Bäckvall, J. E.; Andersson, P. G. *JACS* **1992**, *114*, 6374. (b) Review: Bäckvall, J. E. *PAC* **1992**, *64*, 429.
17. Lipshutz, B. H.; Sengupta, S. *OR* **1992**, *41*, 135.
18. (a) Knochel, P.; Yeh, M. C. P.; Berk, S. C.; Talbert, J. *JOC* **1988**, *53*, 2390. (b) Jubert, C.; Knochel, P. *JOC* **1992**, *57*, 5431. (c) Ibuka, T.; Yoshizawa, H.; Habashita, H.; Fujii, N.; Chounan, Y.; Tanaka, M.; Yamamoto, Y. *TL* **1992**, *33*, 3783. (d) Yamamoto, Y.; Chounan, Y.; Tanaka, M.; Ibuka, T. *JOC* **1992**, *57*, 1024. (e) Knochel, P.; Rozema, M. J.; Tucker, C. E.; Retherford, C.; Furlong, M.; AchyuthaRao, S. *PAC* **1992**, *64*, 361.
19. Seebach, D. *AG(E)* **1988**, *27*, 1624.
20. (a) Hall, P. L.; Gilchrist, J. H.; Collum, D. B. *JACS* **1991**, *113*, 9571. (b) Hall, P. L.; Gilchrist, J. H.; Harrison, A. T.; Fuller, D. J.; Collum, D. B. *JACS* **1991**, *113*, 9575.
21. Bunn, B. J.; Simpkins, N. S. *JOC* **1993**, *58*, 533.
22. (a) Blanchette, M. A.; Choy, W.; Davis, J. T.; Essenfeld, A. P.; Masamune, S.; Roush, W. R.; Sakai, T. *TL* **1984**, *25*, 2183. (b) Rathke, M. W.; Nowak, M. *JOC* **1985**, *50*, 2624.
23. Seebach, D.; Thaler, A.; Beck, A. K. *HCA* **1989**, *72*, 857.
24. Seebach, D.; Bossler, H.; Gründler, H.; Shoda, S.-i.; Wenger, R. *HCA* **1991**, *74*, 197.
25. (a) Thaler, A.; Seebach, D.; Cardinaux, F. *HCA* **1991**, *74*, 617. (b) Thaler, A.; Seebach, D.; Cardinaux, F. *HCA* **1991**, *74*, 628.

James S. Nowick
University of California, Irvine, CA, USA

Guido Lutterbach
Johannes Gutenberg University, Mainz, Germany

Lithium Iodide[1]

[10377-51-2] ILi (MW 133.84)

(ester cleavage and decarboxylation;[2] source of nucleophilic iodide;[3] mild Lewis acid;[1] salt effects in organometallic reactions;[1] epoxide opening[4])

Physical Data: mp 449 °C; bp 1180 °C; *d* 4.076 g cm^{-3}.
Solubility: 165 g/100 mL H_2O (20 °C); 433 g/100 mL H_2O (80 °C); 251 g/100 mL EtOH (20 °C); 343 g/100 mL MeOH (20 °C); 43 g/100 mL acetone (18 °C); very sol NH_4OH.
Form Supplied in: anhydrous white solid or as the hydrate.
Preparative Methods: the anhydrous salt of high purity can be prepared from lithium hydride and iodine in ether.[5]
Purification: crystallized from hot H_2O (0.5 mL g^{-1}) by cooling in $CaCl_2$–ice or from acetone. LiI is dried for 2 h at 120 °C (0.1 mmHg, P_2O_5) before use.
Handling, Storage, and Precautions: for best results, LiI should be dried prior to use in anhydrous reactions.

Heterolytic C–X Bond Cleaving Reactions. In the presence of amine bases, LiI has been extensively used as a mild reagent for the chemoselective cleavage of methyl esters (eq 1).[6] Decarboxylation of methyl esters usually occurs when an electron-withdrawing group is present at the α-position of the ester (eq 2).[7] Ester-type glycosyl linkages of acidic tri- and diterpenes can also be selectively cleaved under these conditions.[8] Aryl methyl ethers can be demethylated to afford the corresponding phenols upon heating with LiI and *s*-collidine.[9]

1,2-Oxiranes are readily opened by LiI and a Lewis acid to produce iodohydrins (eq 3).[4] Conversely, 1-oxaspiro[2.2]pentanes and 1-oxaspiro[3.2]hexanes give rise to bond migration products.[10] β-Vinyl-β-propiolactone is efficiently opened by LiI to produce the corresponding substituted allyl iodide (eq 4).[11]

LiI, py, reflux, 91% (1)

LiI•2H₂O, *s*-collidine, reflux, 71% (2)

LiI, AcOH, THF, rt, 100% (3)

LiI, THF, rt, 1 h, 85% (4)

Alkyl and Alkenyl Iodides. LiI has been used as a source of iodide in nucleophilic substitution and addition reactions. Primary alcohols have been directly converted to alkyl iodides upon treatment with a mixture of ***Triphenylphosphine***, ***Diethyl Azodicarboxylate***, and LiI.[3] Tertiary alcohols can be converted into tertiary alkyl iodides upon treatment with ***Hydrogen Iodide*** in the presence of LiI.[12]

(*Z*)-3-Iodopropenoates and -propenoic acids have been synthesized stereoselectively by the reaction of LiI and propiolates or propiolic acid.[13]

C–C Bond Forming Reactions. LiI was shown to efficiently catalyze the Michael addition of β-dicarbonyl compounds,[14] and the intramolecular allylsilane addition to imines to produce 4-methylenepiperidine derivatives (eq 5).[15]

LiI, THF, 40 °C, 12 h, 68% (5)

LiI as an Additive for Organometallic-Mediated Transformations.[16] The *syn*/*anti* selectivity in the reduction of β-alkoxy ketones is markedly increased by the addition of LiI (eq 6).[17]

LiAlH₄, Et₂O, −78 °C, 96% (6)

with LiI *syn*:*anti* = 89:11
without LiI *syn*:*anti* = 79:21

The addition of ***Lithium Bromide*** and LiI was shown to enhance the rate of organozinc formation from primary alkyl chlorides, sulfonates, and phosphonates, and zinc dust.[18] Beneficial effects of LiI addition have also been reported for Heck-type coupling reactions[19] and in conjugate addition to chiral vinyl sulfoximines.[20]

The (*E*)/(*Z*) alkenic ratio in Wittig-type alkenations was shown to be dependent on the amount of Li salt present.[21]

Reduction of α-Alkoxycarbonyl Derivatives. α-Halo ketones are reduced to the corresponding ketones upon treatment with a mixture of LiI and ***Boron Trifluoride Etherate***.[22]

Related Reagents. Aluminum Iodide; Iodotrimethylsilane; Lithium Chloride; Triphenylphosphine–Iodine.

1. Loupy, A.; Tchoubar, B. *Salt Effects in Organic and Organometallic Chemistry*; VCH: Weinheim, 1992.
2. (a) McMurry, J. *OR* **1976**, *24*, 187. (b) Krapcho, A. P. *S* **1982**, 805. (c) Krapcho, A. P. *S* **1982**, 893.
3. Manna, S.; Falck, J. R.; Mioskowski, C. *SC* **1985**, *15*, 663.
4. (a) Bonini, C.; Giuliano, C.; Righi, G.; Rossi, L. *SC* **1992**, *22*, 1863. (b) Shimizu, M.; Yoshida, A.; Fujisawa, T. *SL* **1992**, 204. (c) Bajwa, J. S.; Anderson, R. C. *TL* **1991**, *32*, 3021.
5. Taylor, M. D.; Grant, L. R. *JACS* **1955**, *77*, 1507.
6. Magnus, P.; Gallagher, T. *CC* **1984**, 389.
7. Johnson, F.; Paul, K. G.; Favara, D. *JOC* **1982**, *47*, 4254.
8. Ohtani, K.; Mizutani, K.; Kasai, R.; Tanaka, O. *TL* **1984**, *25*, 4537.
9. (a) Kende, A. S.; Rizzi, J. P. *JACS* **1981**, *103*, 4247. (b) Harrison, I. T. *CC* **1969**, 616.
10. (a) Salaün, J.; Conia, J. M. *CC* **1971**, 1579. (b) Aue, D. H.; Meshishnek, M. J.; Shellhamer, D. F. *TL* **1973**, 4799.
11. Fujisawa, T.; Sato, T.; Takeuchi, M. *CL* **1982**, 71.
12. Masada, H.; Murotani, Y. *BCJ* **1980**, *53*, 1181.
13. (a) Ma, S.; Lu, X. *TL* **1990**, *31*, 7653. (b) Ma, S.; Lu, X. *CC* **1990**, 1643.
14. Antonioletti, R.; Bonadies, F.; Monteagudo, E. S.; Scettri, A. *TL* **1991**, *32*, 5373.
15. Bell, T. W.; Hu, L.-Y. *TL* **1988**, *29*, 4819.
16. For the effect of LiI on organocopper reagents see: Lipshutz, B. H.; Kayser, F.; Siegmann, K. *TL* **1993**, *34*, 6693.
17. (a) Mori, Y.; Kuhara, M.; Takeuchi, A.; Suzuki, M. *TL* **1988**, *29*, 5419. (b) Mori, Y.; Takeuchi, A.; Kageyama, H.; Suzuki, M. *TL* **1988**, *29*, 5423.
18. Jubert, C.; Knochel, P. *JOC* **1992**, *57*, 5425.
19. Cabri, W.; Candiani, I.; DeBernardinis, S.; Francalanci, F.; Penco, S.; Santi, R. *JOC* **1991**, *56*, 5796.
20. (a) Pyne, S. G. *JOC* **1986**, *51*, 81. (b) Pyne, S. G. *TL* **1986**, *27*, 1691.
21. (a) Soderquist, J. A.; Anderson, C. L. *TL* **1988**, *29*, 2425. (b) Soderquist, J. A.; Anderson, C. L. *TL* **1988**, *29*, 2777. (c) Buss, A. D.; Warren, S.; Leake, J. S.; Whitham, G. H. *JCS(P1)* **1983**, 2215. (d) Buss, A. D.; Warren, S. *JCS(P1)* **1985**, 2307.
22. Townsend, J. M.; Spencer, T. A. *TL* **1971**, 137.

André B. Charette
Université de Montréal, Québec, Canada

Lithium Perchlorate[1]

$LiClO_4$

[7791-03-9] $ClLiO_4$ (MW 106.39)

(mild Lewis acid[1] for cycloaddition reactions,[2] conjugate additions,[3] ring opening of epoxides,[4] and ring expansion of cyclopropanes[5])

Physical Data: mp 236 °C; bp 430 °C (dec); *d* 2.428 g cm^{-3}.
Solubility: 60 g/100 mL H_2O (25 °C); 150 g/100 mL H_2O (89 °C); 152 g/100 mL EtOH (25 °C); 137 g/100 mL acetone (25 °C); 182 g/100 mL MeOH (25 °C); 114 g/100 mL ether.
Form Supplied in: white solid; widely available in anhydrous form or as the trihydrate; it is usually used in solution in ether or MeOH.
Purification: anhydrous lithium perchlorate is prepared by heating the commercially available anhydrous material or its trihydrate at 160 °C for 48 h under high vacuum (P_2O_5 trap).
Handling, Storage, and Precautions: the anhydrous material should be used as prepared for best results. The decomposition of lithium perchlorate starts at about 400 °C and becomes rapid at 430 °C, yielding lithium chloride and oxygen. Perchlorates are potentially explosive and should be handled with caution.

Lewis Acid Catalyst for Cycloaddition Reactions and Carbonyl Addition Reactions. This reagent, usually prepared as a 5.0 M solution in diethyl ether, produces a dramatic rate acceleration of Diels–Alder reactions (eq 1). Evidence shows that this rate acceleration, which was initially thought to be a result of high internal solvent pressure, is due to the Lewis acid character of the lithium ion.[6]

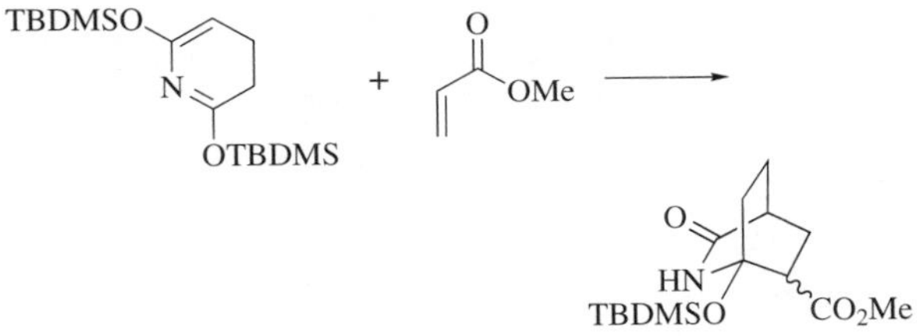

TBDMSO … N … OTBDMS + OMe → O, HN, TBDMSO, CO_2Me (1)

5.0M $LiClO_4$, Et_2O, 5 h, rt, 80% (*endo*:*exo* = 2.9:1)
benzene, 72 h, 60 °C, 74% (*endo*:*exo* = 1:11.5)

Under these conditions, reasonable levels of diastereoselectivity have been observed in the reaction between a chiral diene and ***N*-Phenylmaleimide** (eq 2).[7] An interesting protecting group dependence of diastereoselectivities has also been observed in the hetero-Diels–Alder reaction of *N*-protected α-amino aldehydes with 1-methoxy-3-*t*-butyldimethylsilyloxybutadiene to produce dihydropyrones (eqs 3 and 4).[8]

O-Silylated ketene acetals undergo 1,4-conjugate addition to hindered α,β-unsaturated carbonyl systems[3] and quinones[9] in the presence of $LiClO_4$.

[1,3]-Sigmatropic Rearrangements. In contrast to the [3,3]-sigmatropic rearrangement observed under thermal conditions, allyl vinyl ethers undergo [1,3]-sigmatropic rearrangements at rt when submitted to 1.5–3.0 M $LiClO_4$ in Et_2O (eq 5).[10]

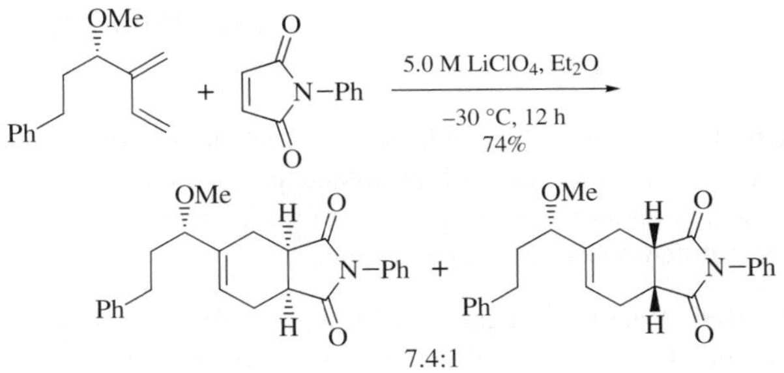

OMe, Ph + N–Ph (5.0 M $LiClO_4$, Et_2O, −30 °C, 12 h, 74%) → OMe, H, O, N–Ph, Ph, H, O + OMe, H, O, N–Ph, Ph, H, O (2)
7.4:1

BnO, O, H, NBn_2 + OMe, OTBDMS (1. 3.0 M $LiClO_4$, Et_2O; 2. TFA, CH_2Cl_2; 66%) → O, BnO, NBn_2, O (3)
single diastereomer

TBDPSO, O, H, NHBn + OMe, OTBDMS (1. 3.0 M $LiClO_4$, Et_2O; 2. TFA, CH_2Cl_2; 79%) → O, TBDPSO, NHBn, O (4)
8:1

O, H, HO, H, HO (3.0M $LiClO_4$, Et_2O, 90%) → O, H, HO, H, HO + O, H, HO, H, HO (5)
5:1

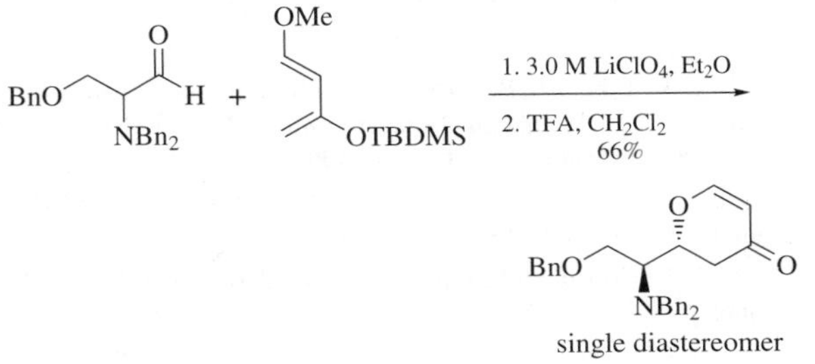

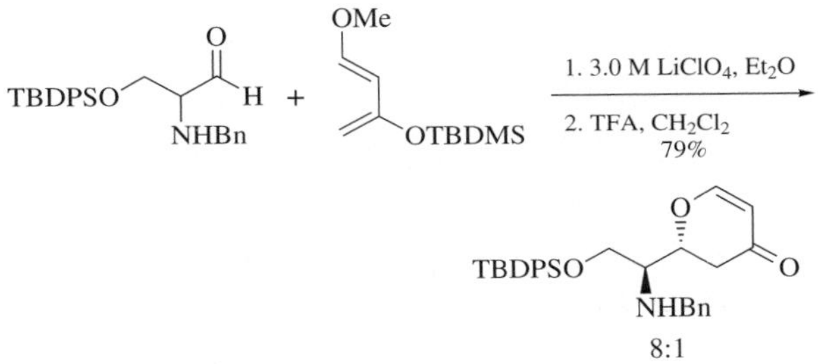

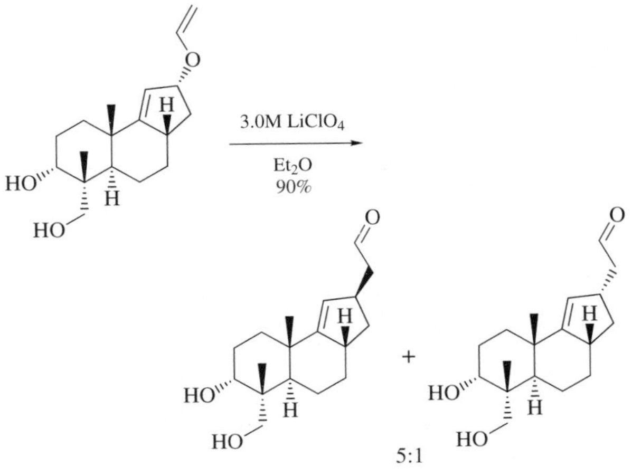

Epoxide Opening. $LiClO_4$ is an efficient promotor for the regioselective nucleophilic opening of oxiranes with amines,[11] cyanide,[12] azide,[13] thiols,[14] halides,[15] and lithium acetylides.[16] The regioselective opening of oxiranes with lithium enolates derived from ketones has also been observed in the presence of $LiClO_4$ (eq 6).[17]

Ph epoxide + OLi / t-Bu enolate —LiClO$_4$, THF, 25 °C, 24 h; 90%→ PhO…OH…C(=O)t-Bu (6)

without LiClO$_4$, 30%

Ring Expansion. The condensation of aldehydes and ketones with diphenylsulfonium cyclopropylide produces oxaspiropentanes which undergo ring expansion to produce cyclobutanones upon treatment with lithium perchlorate.[18]

Related Reagents. Diethylaluminum Chloride; Dimethylaluminum Chloride; Lithium Tetrafluoroborate; Magnesium Bromide.

1. (a) Loupy, A.; Tchoubar, B. *Salts Effects in Organic and Organometallic Chemistry*; VCH: Weinheim, 1992. (b) Grieco, P. A. *Aldrichim. Acta* **1991**, *24*, 59. (c) Waldmann, H. *AG(E)* **1991**, *30*, 1306.
2. (a) Grieco, P. A.; Nunes, J. J.; Gaul, M. D. *JACS* **1990**, *112*, 4595. (b) Braun, R.; Sauer, J. *CB* **1986**, *119*, 1269.
3. Grieco, P. A.; Cooke, R. J.; Henry, K. J.; VanderRoest, J. M. *TL* **1991**, *32*, 4665.
4. Chini, M.; Crotti, P.; Flippin, L. A.; Macchia, F. *JOC* **1990**, *55*, 4265.
5. Rickborn, B.; Gerkin, R. M. *JACS* **1971**, *93*, 1693.
6. (a) Forman, M. A.; Dailey, W. P. *JACS* **1991**, *113*, 2761. (b) Desimoni, G.; Faita, G.; Righetti, P. P.; Tacconi, G. *T* **1991**, *47*, 8399.
7. Hatakeyama, S.; Sugawara, K.; Takano, S. *CC* **1992**, 953.
8. Grieco, P. A.; Moher, E. D. *TL* **1993**, *34*, 5567.
9. Ipaktschi, J.; Heydari, A. *AG(E)* **1992**, *31*, 313.
10. Grieco, P. A.; Clark, J. D.; Jagoe, C. T. *JACS* **1991**, *113*, 5488.
11. Chini, M.; Crotti, P.; Macchia, F. *JOC* **1991**, *56*, 5939.
12. Chini, M.; Crotti, P.; Favero, L.; Macchia, F. *TL* **1991**, *32*, 4775.
13. Chini, M.; Crotti, P.; Macchia, F. *TL* **1990**, *31*, 5641.
14. Chini, M.; Crotti, P.; Giovani, E.; Macchia, F.; Pineschi, M. *SL* **1992**, 303.
15. (a) Chini, M.; Crotti, P.; Gardelli, C.; Macchia, F. *T* **1992**, *48*, 3805. (b) Chini, M.; Crotti, P.; Flippin, L. A.; Macchia, F. *JOC* **1990**, *55*, 4265.
16. Chini, M.; Crotti, P.; Favero, L.; Macchia, F. *TL* **1991**, *32*, 6617.
17. Chini, M.; Crotti, P.; Favero, L.; Pineschi, M. *TL* **1991**, *32*, 7583.
18. Trost, B. M.; Bogdanowicz, M. J. *JACS* **1973**, *95*, 5321.

André B. Charette
Université de Montréal, Québec, Canada

Magnesium Bromide

$MgBr_2$

($MgBr_2$)		
[2923-28-6]	Br_2Mg	(MW 184.11)
($MgBr_2\cdot 6H_2O$)		
[13446-53-2]	$H_{12}Br_2MgO_6$	(MW 292.23)
($MgBr_2\cdot OEt_2$)		
[29858-07-9]	$C_4H_{10}Br_2MgO$	(MW 258.25)

(Lewis acid capable of catalyzing selective nucleophilic additions,[1] cycloadditions,[2] rearrangements,[3] coupling reactions;[4] effective brominating agent[5])

Physical Data: mp 165 °C (dec) (etherate >300 °C; fp 35 °C).

Solubility: sol alcohol, H_2O; etherate sol common organic solvents.

Form Supplied in: white solid, widely available; etherate gray solid.

Preparative Method: the etherate is easily prepared from reacting a slight excess of ***Magnesium*** turnings with ***1,2-Dibromoethane*** in anhydrous diethyl ether.[13]

Handling, Storage, and Precautions: etherate is flammable and moisture sensitive; freshly prepared material is most reactive and anhydrous. Irritant.

Nucleophilic Additions. $MgBr_2$ has been shown to form discrete bidentate chelates with various species,[6] particularly α- and/or α,β-alkoxy carbonyl compounds,[7] and thus functions as a diastereofacial control element in many nucleophilic addition reactions. In many cases, its inclusion completely reverses the nonchelation-controlled stereochemistry observed with nonchelating Lewis acids such as ***Boron Trifluoride Etherate***. Highest diastereoselectivity is observed with α-substituted aldehydes (eq 1).[8] High selectivities are observed for β-alkoxy aldehydes as well, including cases where three contiguous chiral centers are defined during the reaction (eq 2).[9]

(1) $MgBr_2$: *syn:anti* = 250:1; $BF_3\cdot OEt_2$: *syn:anti* = 39:61

(2) 81% diastereofacial selectivity

Nucleophilic additions to α,β-dialkoxy aldehydes via allyl silanes (eq 3)[10] or silyl ketene acetals (Mukaiyama reaction) (eq 4)[11] exhibit similarly high selectivities. α-Thio aldehydes also react under $MgBr_2$-catalyzed Mukaiyama conditions with efficient stereocontrol (eq 5).[1]

(3) $MgBr_2$, CH_2Cl_2, −20 °C, 95%; *syn:anti* >98:2

(4) $MgBr_2$, −78 °C then oxid/elim; *syn:anti* >55:1

(5) $MgBr_2$, 81%; *syn:anti* = 87:13

Rearrangements. The oxophilic nature of $MgBr_2$ renders it effective in mediating many rearrangements wherein polarization of a C–O bond initiates the process. A classic application involves the conversion of an epoxide to an aldehyde (eq 6).[12]

(6) $MgBr_2$, PhH, Δ, 1 h, 95%

β-Lactones, when treated with $MgBr_2$, undergo ionization with further rearrangement to afford either butyrolactones or β,γ-unsaturated carboxylic acids (eq 7).[3] The reaction course is dependent upon whether the cation resulting from lactone ionization can rearrange to a more or equivalently stable cation, in which case ring expansion is observed. If the β-lactone bears an α-chloro substituent, ring expansion is accompanied by elimination of HCl to afford butenolides (eq 8).[13]

(7) $MgBr_2$, Et_2O; R = H, 85%; R = alkyl, 80–90%

(8) $MgBr_2$, CH_2Cl_2, 87%

Cycloadditions. The $MgBr_2$-mediated cyclocondensation of a Danishefsky-type diene with chiral α-alkoxy aldehydes affords a single diastereomer, which reflects a reacting conformer in which the alkoxy group is *syn* to the carbonyl, which is then attacked from its less-hindered face (eq 9).[2] The cycloaddition of ynamines with cycloalkenones occurred selectively at the carbonyl, while in the absence of $MgBr_2$ reaction at the alkene C=C bonds was observed (eq 10).[14]

Organometallic Reactions. $MgBr_2$ often increases the yields of Grignard reactions, as in the synthesis of cyclopropanols from 1,3-dichloroacetone (eq 11).[15] It also serves to form Grignard reagents in situ via first lithiation followed by transmetalation with $MgBr_2$. This technique enables the formation of vinyl Grignards from vinyl sulfones (eq 12),[16] and of α-silyl Grignard reagents from allyl silanes (eq 13).[17] In the latter case, the presence of $MgBr_2$ provided regioselectivity at the α-position; without it, substitution at the γ-position was predominant.

Bromination. Magnesium bromide serves as a source of bromide ion, for displacement of sulfonate esters under mild conditions and in high yield. The reaction is known to proceed with complete inversion at the reacting carbon center when backside attack is possible (eq 14).[5] Even bromination of highly congested bridgehead positions is possible via triflate displacement, although the reaction requires high temperatures, long reaction times, and activation via ultrasound (eq 15).[18]

Epoxides are also converted regiospecifically to bromohydrins by $MgBr_2$.[19]

Carbonyl Condensations. Bis- or tris-TMS ketenimines condense with ketones, mediated by $MgBr_2$, to produce an intermediate that loses hexamethylsiloxane, affording 2-alkenenitriles in high yield with high (*E*) selectivity (eq 16).[20] $MgBr_2$ also enables the use of very mild bases like ***Triethylamine*** in Horner–Wadsworth–Emmons reactions, enabling the synthesis of unsaturated esters from aldehydes or ketones without the need for strongly basic conditions (eq 17).[21]

1. Annunziata, R.; Cinquini, M.; Cozzi, F.; Cozzi, P. G.; Consolandi, E. *JOC* **1992**, *57*, 456.
2. Danishefsky, S.; Pearson, W. H.; Harvey, D. F.; Maring, C. J.; Springer, J. P. *JACS* **1985**, *107*, 1256.
3. (a) Black, T. H.; Hall, J. A.; Sheu, R. G. *JOC* **1988**, *53*, 2371. (b) Black, T. H.; Eisenbeis, S. A.; McDermott, T. S.; Maluleka, S. L. *T* **1990**, *46*, 2307.
4. Cai, D.; Still, W. C. *JOC* **1988**, *53*, 464.
5. Hannesian, S.; Kagotani, M.; Komaglou, K. *H* **1989**, *28*, 1115.
6. Keck, G. E.; Castellino, S. *JACS* **1986**, *108*, 3847.
7. Chen, X.; Hortelano, E. R.; Eliel, E. L.; Frye, S. V. *JACS* **1992**, *114*, 1778.
8. Keck, G. E.; Boden, E. P. *TL* **1984**, *25*, 265.
9. Keck, G. E.; Abbott, D. E. *TL* **1984**, *25*, 1883.
10. Williams, D. R.; Klingler, F. D. *TL* **1987**, *28*, 869.
11. Bernardi, A.; Cardani, S.; Colombo, L.; Poli, G.; Schimperna, G.; Scolastico, C. *JOC* **1987**, *52*, 888.
12. Serramedan, D.; Marc, F.; Pereyre, M.; Filliatre, C.; Chabardes, P.; Delmond, B. *TL* **1992**, *33*, 4457.
13. Black, T. H.; McDermott, T. S.; Brown, G. A. *TL* **1991**, *32*, 6501.
14. Ficini, J.; Krief, A.; Guingant, A.; Desmaele, D. *TL* **1981**, *22*, 725.
15. Barluenga, J.; Florez, J.; Yus, M. *S* **1983**, 647.
16. Eisch, J. J.; Galle, J. E. *JOC* **1979**, *44*, 3279.
17. Lau, P. W. K.; Chan, T. H. *TL* **1978**, 2383.
18. Martinez, A. G.; Vilar, E. T.; Lopez, J. C.; Alonso, J. M.; Hanack, M.; Subramanian, L. R. *S* **1991**, 353.
19. Ueda, Y.; Maynard, S. C. *TL* **1988**, *29*, 5197.
20. Matsuda, I.; Okada, H.; Izumi, Y. *BCJ* **1983**, *56*, 528.
21. Rathke, M. W.; Nowak, M. *JOC* **1985**, *50*, 2624.

T. Howard Black
Eastern Illinois University, Charleston, IL, USA

Mesitylenesulfonyl Chloride

[773-64-8] $C_9H_{11}ClO_2S$ (MW 218.72)

(preparation of sulfonates;[1] condensing agent in nucleotide synthesis[2])

Alternate Name: 2,4,6-trimethylbenzenesulfonyl chloride.
Physical Data: mp 55–57 °C.
Solubility: sol organic solvents.
Form Supplied in: white crystalline solid; widely available.
Purification: can be recrystallized from hexane or pentane.
Handling, Storage, and Precautions: corrosive; moisture sensitive; handle and store under nitrogen.

Sulfonamide Formation. Reaction of mesitylenesulfonyl chloride with amines in the presence of ***Pyridine*** or ***Triethylamine*** yields the corresponding sulfonamides (eq 1),[3] which have been used as intermediates in synthesis.

(1)

The mesityl group has been used as a protecting group for various amines including amino acids (eq 2)[4] and peptides.[5–7]

(2)

The mesitylsulfonyl-protected amino acids can be deblocked by treatment with ***Hydrogen Bromide*** and ***Acetic Acid***. The stability of the sulfonamide towards both acidic and basic conditions makes it a useful protecting group for peptide synthesis. The mesitylenesulfonyl group has been used to protect the indole group of tryptophan[6] and the guanidine group of arginine.[7] These mesityl derivatives can be cleaved with ***Trifluoromethanesulfonic Acid*** or ***Methanesulfonic Acid***. Partial cleavage of the guanidinomesityl group occurs when treated with HBr/HOAc. The mesitylenesulfonyl group has also been used as a blocking group for other indoles. Boteju et al. reported the formation of the indole sulfonamide by treatment with ***n-Butyllithium*** and mesitylenesulfonyl chloride, but the sulfonamide group was not subsequently removed (eq 3).[8]

(3)

Sulfonate Formation. Treatment of alcohols with mesitylenesulfonyl chloride yields the corresponding sulfonates. Mesitylenesulfonyl chloride is particularly useful for the selective sulfonation of polyhydroxylic systems such as carbohydrates (eq 4).[1,9] It is more selective than ***p-Toluenesulfonyl Chloride***, which has been frequently used but gives mixtures of products. Unfortunately, the mesitylenesulfonates are not as reactive as the corresponding tosylates. Numerous 1′-derivatives of sucrose have been synthesized via mesitylenesulfonyl derivatives.[10]

(4)

The selectivity of sulfonation has also been applied to other polyhydroxylic systems such as α- and β-cyclodextrins[11,12] and nucleotides.[13] Monosulfonated products have been reported from the treatment of diols with mesitylenesulfonyl chloride (eq 5).[14,15]

(5)

Mesitylenesulfonyl derivatives which have specific biological activities, such as 2,3-diaziridinyl-1,4-napthoquinone sulfonates[16] and 1-arenesulfonyloxy-2-alkanones,[17] have also been prepared.

Nucleotide Synthesis. Mesitylenesulfonyl chloride has been used as a condensing agent in the diester approach to nucleotide synthesis (eq 6).[2] However, it is not effective in the triester approach due to extensive sulfonation of the 5′-hydroxy group.[18] Other condensing agents include ***1,3-Dicyclohexylcarbodiimide*** (DCC), benzenesulfonyltriazole (BST), (1-mesitylyl-2-sulfonyl)-3-nitro-1,2,4-triazole (MSNT), ***Mesitylsulfonyl-1H-1,2,4-triazole*** (MST), *p*-nitrobenzenesulfonyltriazole (*p*-NBST), *p*-toluenesulfonyltriazole (TST), and triisopropylbenzenesulfonyl chloride (TPS).

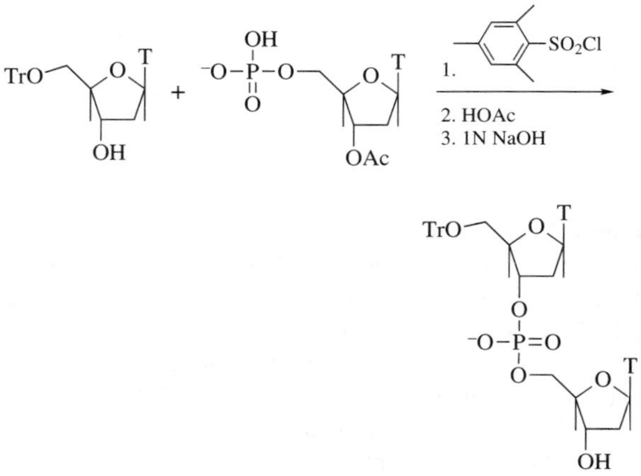

(6)

Condensation Reactions. Mesitylenesulfonyl chloride has also been employed as a condensing agent for the preparation of esters in the synthesis of nonactin (eq 7),[19] *N*-methylmaysenine,[20] and maysine.[21]

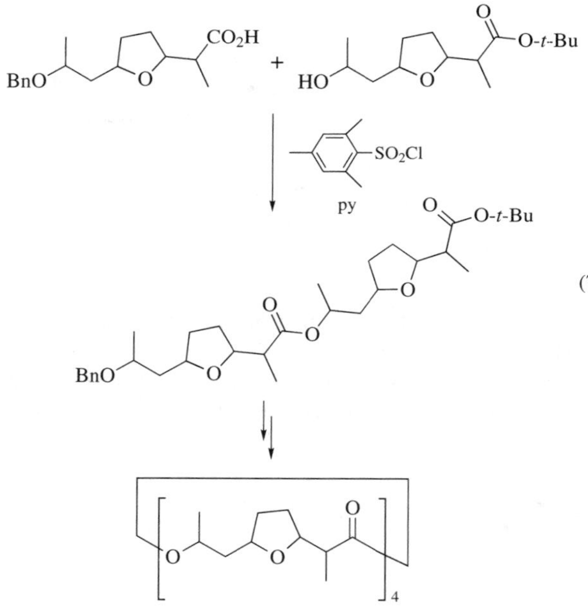

(7)

Sulfinate Ester Synthesis. Sulfinate esters of menthol can be prepared via reaction of mesitylenesulfonyl chloride (as well as other sulfonyl chlorides) with menthol using ***Trimethyl Phosphite*** (eq 8).[22] Sulfinate esters of menthol are used for the preparation of optically active sulfoxides. See also ***10-Camphorsulfonyl Chloride*** and ***p-Toluenesulfonyl Chloride***.

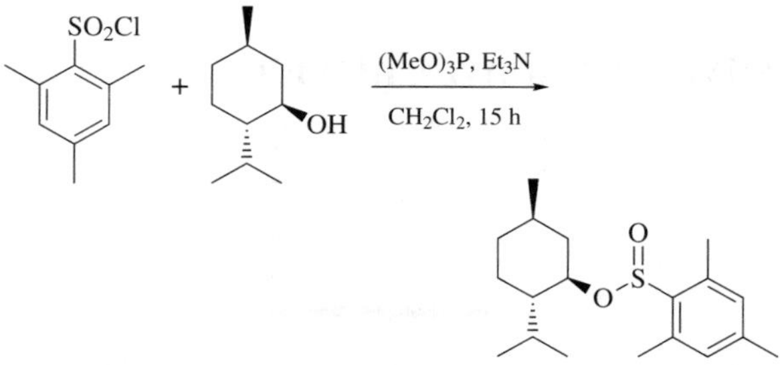

(8)

Sulfonylalkyne Synthesis. Sulfonylalkynes have been synthesized (eq 9) and subsequently reacted with *N*-(1-alkynyl)anilines to give 2-anilino-5-sulfinylfurans.[23]

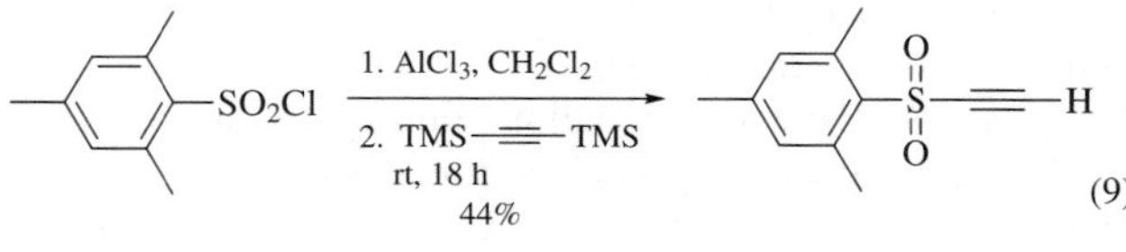

(9)

1. Guthrie, R. D.; Thang, S. *AJC* **1987**, *40*, 2133.
2. (a) Narang, S. A.; Khorana, H. G. *JACS* **1965**, *87*, 2981. (b) *FF* **1967**, *1*, 661.
3. (a) Reetz, M. T.; Kukenhohner, T.; Weinig, P. *TL* **1986**, *27*, 5711. (b) Pavlidis, V. H.; Chan, E. D.; Pennington, L.; McParland, M.; Whitehead, M. *SC* **1988**, *18*, 1615. (c) Hoppe, I.; Hoffmann, H.; Gartner, I.; Krettek, T.; Hoppe, D. *S* **1991**, 1157. (d) Takahashi, H.; Takahashi, K.; Ohno, M.; Yoshioka, M.; Kobayashi, S. *T* **1992**, *48*, 5691.
4. Roemmele, R. C.; Rapoport, H. *JOC* **1988**, 2367.
5. (a) Corey, E. J.; Weigel, L. O.; Floyd, D.; Bock, M. G. *JACS* **1978**, *100*, 2916. (b) *FF* **1980**, *8*, 318.
6. Fuji, N.; Futaki, S.; Yasumura, K.; Yajima, H. *CPB* **1984**, 2660.
7. (a) Yajima, H.; Takeyama, M.; Kanaki, J.; Mitani, K. *CC* **1978**, 482. (b) Yajima, H.; Takeyama, M.; Kanaki, J.; Nishimura, O.; Fujino, M. *CPB* **1978**, *26*, 3752.
8. Boteju, L. W.; Wegner, K.; Hruby, V. J. *TL* **1992**, 7291.
9. Gilchrist, T. L.; Rees, C. W.; Stanton, E. *CC* **1971**, 801.
10. Hough, L.; Phadnis, S. P.; Tarelli, E. *Carbohydr. Res.* **1975**, *44*, C12. (b) Ball, D. H.; Bisett, F. H.; Chalk, R. C. *Carbohydr. Res.* **1977**, *55*, 149. (c) Guthrie, R. D.; Jenkins, I. D.; Watters, J. J. *AJC* **1980**, *33*, 2487.
11. Fujita, K.; Matsunaga, A.; Imoto, T. *JACS* **1984**, *106*, 5740.
12. (a) Fujita, K.; Yamamura, H.; Imoto, T. *JOC* **1985**, *50*, 4393. (b) Fujita, K.; Ishizu, T.; Minamiura, N.; Yamamoto, T. *CL* **1991**, 1889. (c) Fujita, K.; Ishizu, T.; Obe, K.; Minamiura, N.; Yamamoto, T. *JOC* **1992**, *57*, 5606.
13. (a) Ueda, T.; Usui, H.; Shuto, S.; Inoue, H. *CPB* **1984**, *32*, 3410. (b) Tanimura, H.; Sekine, M.; Hata, T. *TL* **1986**, *27*, 4047. (c) Xu, Y. Z.; Zheng, Q.; Swann, P. F. *TL* **1991**, *24*, 2817. (d) Xu, Y. Z.; Zheng, Q.; Swann, P. F. *T* **1992**, 1729.
14. Nakata, T.; Fukui, M.; Ohtsuka, H.; Oishi, T. *TL* **1983**, *24*, 2661.
15. Rama Rao, A. V.; Gaitonde, A.; Prahlada Rao, S. *IJC(B)* **1992**, *31B*, 641.
16. Lin, T. S.; Xu, S. P.; Zhu, L. Y.; Cosby, L. A.; Sartorelli, A. C. *JMC* **1989**, *32*, 1467.
17. Ogawa, K.; Terada, T.; Muranaka, Y.; Hamakawa, T.; Hashimoto, S.; Fujii, S. *CPB* **1986**, *34*, 3252.
18. Katagiri, N.; Itakura, K.; Narang, S. A. *JACS* **1975**, *97*, 7332.
19. (a) Gerlach, H.; Oertle, K.; Thalmann, A.; Servi, S. *HCA* **1975**, *58*, 2036. (b) *FF* **1977**, *6*, 625.

20. Corey, E. J.; Weigel, L. O.; Floyd, D.; Bock, M. G. *JACS* **1978**, *100*, 2916.
21. Kitamura, M.; Isobe, M.; Ichikawa, Y.; Goto, T. *JOC* **1984**, *49*, 3517.
22. Sharpless, K. B.; Klunder, J. M. *JOC* **1987**, *52*, 2598.
23. Kosack, S.; Himbert, G. *CB* **1987**, *120*, 71.

Valerie Vaillancourt & Michele M. Cudahy
The Upjohn Co., Kalamazoo, MI, USA

Methanesulfonyl Chloride

$MeSO_2Cl$

[124-63-0] CH_3ClO_2S (MW 114.56)

(synthesis of methanesulfonates;[1] generation of sulfene)

Alternate Name: mesyl chloride.
Physical Data: bp 60 °C/21 mmHg; *d* 1.480 g cm^{-3}.
Solubility: insol water; sol alcohol, ether.
Form Supplied in: liquid of 98% or 99+% purity.
Handling, Storage, and Precautions: highly toxic; corrosive; lachrymator; moisture sensitive; store in a cool dry place.

Methanesulfonates. The most common use of methanesulfonyl chloride is for the synthesis of sulfonate esters from alcohols. This can be readily accomplished by treatment of an alcohol with mesyl chloride in the presence of a base (usually ***Triethylamine*** or ***Pyridine***).[1] The methanesulfonates formed are functional equivalents of halides. As such they are frequently employed as intermediates for reactions such as displacements, eliminations, reductions,[2] and rearrangements.[3] Selective mesylation of a vicinal diol is a common method of preparation of epoxides.[4] Alkynyl mesylates can be used for the synthesis of trimethylsilyl allenes.[5] Oxime mesylates undergo a Beckmann rearrangement upon treatment with a Lewis acid.[6] Aromatic mesylates have been used as substrates for nucleophilic aromatic substitution.[7] Mesylates are more reactive than tosylates toward nucleophilic substitution, but less reactive toward solvolysis.

Protection of Alcohols. In addition to being used as a halide equivalent, methanesulfonates have also been used as protecting groups for alcohols and phenols. The use of methanesulfonate as a protecting group for alcohols is mainly limited to carbohydrate synthesis due to the lability of the methanesulfonate group toward nucleophilic attack. The sulfonate ester is stable to acidic conditions and can be cleaved by ***Sodium Amalgam***.[8]

Protection of Amines. Amines also react with mesyl chloride to form the corresponding sulfonamide. The sulfonamide group is one of the most stable nitrogen-protecting groups, with good stability to both acidic and basic conditions. The sulfonamide can be cleaved back to the amine by ***Lithium Aluminium Hydride*** or dissolving metal reductions.[9] In addition to treatment of amines with mesyl chloride, other reports for the preparation of sulfonamides include a Reissert-type reaction of mesyl chloride with phthalazine[10] and addition of mesyl chloride to lactim ethers (eq 1).[11]

OMe; MsCl, $Bu_4N^+I^-$, CH_2Cl_2, 79%; O, N, SO_2Me (1)

Chlorination. Methanesulfonyl chloride can effect the direct chlorination of various substrates. The most notable of these reactions is the selective chlorination of the C-6 primary hydroxy group of carbohydrates (eq 2).[12] Under these conditions, 2-furyl alcohol (eq 3)[13] and guanine *N*-oxide (eq 4)[14] can also be chlorinated directly.

HO, O, OMe; glucose or mannose; 2–10 equiv MsCl, DMF, –25 to 65 °C, 38–97%; Cl, O, OMe (2)

OHC, O, OH; MsCl, DMF, 65 °C, 8 h, 84%; OHC, O, Cl (3)

O, HN, H_2N, N, O^-, H, N; MsCl, DMF, 75%; O, HN, H_2N, N, H, N, Cl (4)

Meyers developed a procedure for the direct conversion of allylic alcohols to allylic chlorides that is general for a variety of terpenes (eq 5).[15] Primary alcohols give yields in the 85% range. A single example of a secondary alcohol produced the chloride in 50% yield. Certain allylic alcohols also undergo stereospecific chlorination under the conditions normally employed for mesylate formation (eq 6).[16]

OH; MsCl, LiCl, DMF, collidine, 83%; Cl (5)

O, O, O, OH; MsCl, py, 0 °C, 3 h, 92%; O, O, O, Cl (6)

Under Lewis acid catalysis, mesyl chloride reacts with unactivated benzenes to produce the aromatic chloride and only trace amounts of the corresponding sulfone (eq 7).[17] Mesylates have also been used as intermediates for the transformation of alcohols into halides.

MsCl, $AlCl_3$, xylene, 25 °C, 1 h; SO_2Me ~1% + Cl 70% (7)

Interconversion of Carboxylic Acid Derivatives. Methanesulfonyl chloride is employed in the activation of carboxylic acids for the preparation of anhydrides (eq 8),[18] nitriles (eq 9),[19] amides (eq 10), and esters (eq 11).[20]

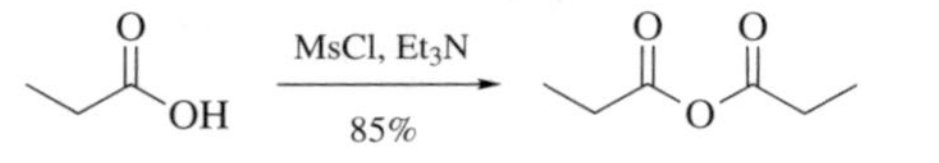

(8)

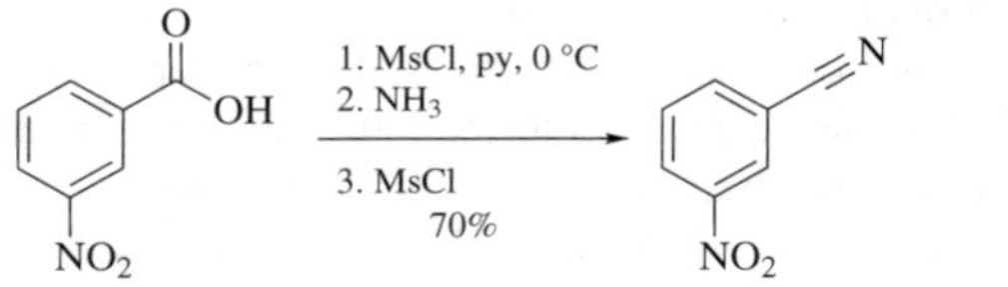

(9)

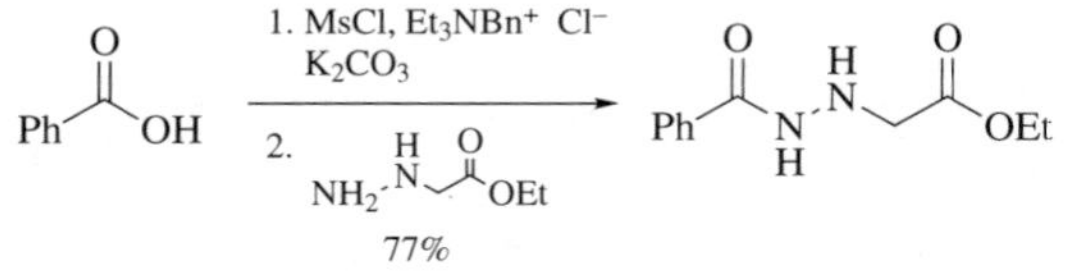

(10)

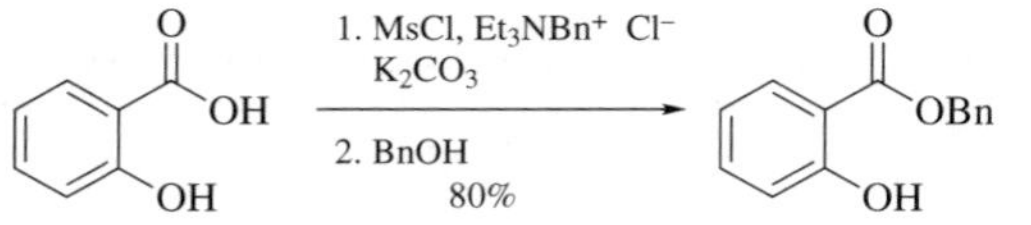

(11)

Eliminations. Alkene formation by elimination can also be achieved with mesyl chloride. This is general for a variety of substrates. Peterson reported the elimination of β-hydroxy silanes to give predominantly *trans*-alkenes upon treatment with mesyl chloride (eq 12).[21] In a similar manner, β-hydroxy phenylselenates can also be eliminated (eq 13).[22]

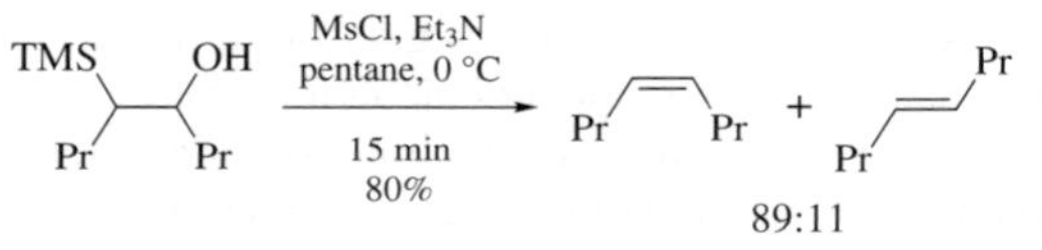

(12)

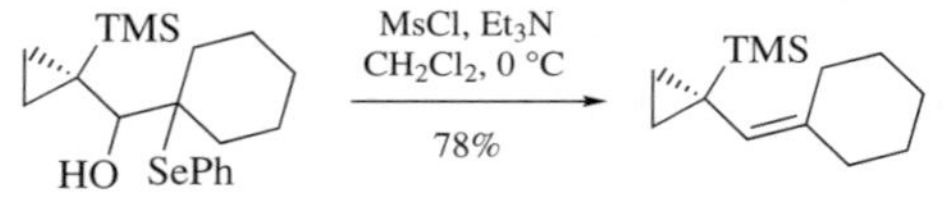

(13)

Corey reported a mild elimination of iodohydrins by activation with mesyl chloride (eq 14).[23] The major advantage of this method over traditional methods is that the use of a strong reducing agent (***Zinc*** or ***Tin(II) Chloride***) is avoided.

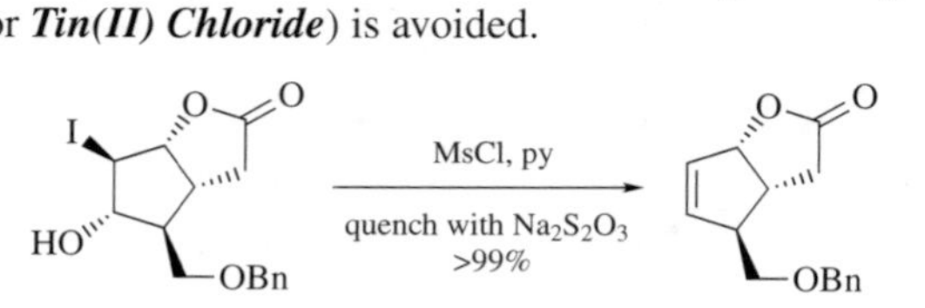

(14)

The direct elimination of tertiary alcohols by treatment with mesyl chloride has also been reported.[24] A variety of acid sensitive functional groups including silyl ethers and acetals are stable to these reaction conditions (eq 15). Mixtures of *cis* and *trans* and internal and terminal alkenes are formed in all cases where possible (eq 16).

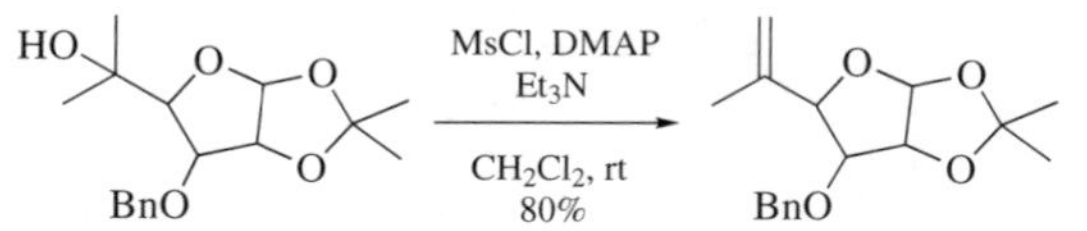

(15)

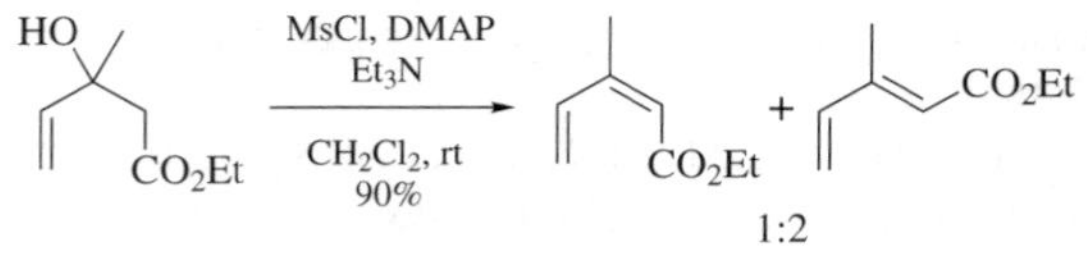

(16)

The synthesis of α-methylene lactones[25] and ketones[26] via the elimination of a mesylate has also been reported (eqs 17 and 18). In these cases, the pyridine or triethylamine is sufficiently basic to induce elimination under standard mesylate forming conditions.

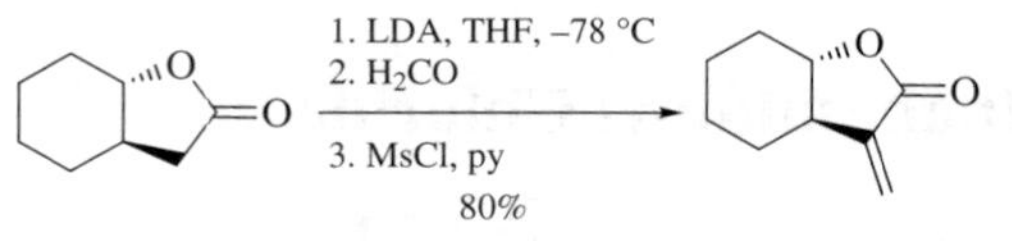

(17)

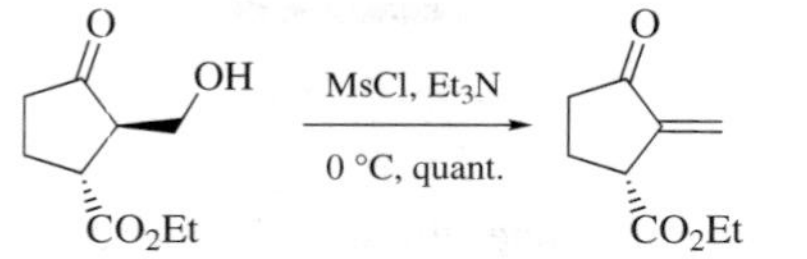

(18)

Heterocycles. Sulfene, generated by the treatment of methanesulfonyl chloride with base, undergoes [2+2], [3+2], and [4+2] cycloadditions to produce four-, five-, and six-membered heterocycles, respectively. Alkynyl amines[27] and enamines[28] both undergo [2+2] cycloadditions with sulfene to produce four-membered cyclic sulfones (eqs 19 and 20).

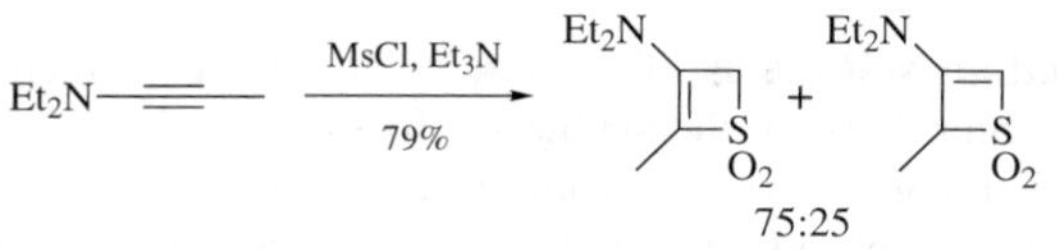

(19)

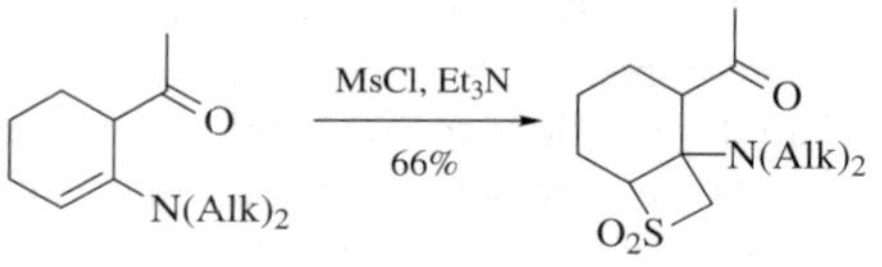

(20)

Five-membered sultones can be formed by reaction of α-hydroxy ketones with sulfene (eq 21).[29] Sulfene adds in a [4+2] manner across azadienes (eq 22)[30] and enones (eq 23).[31]

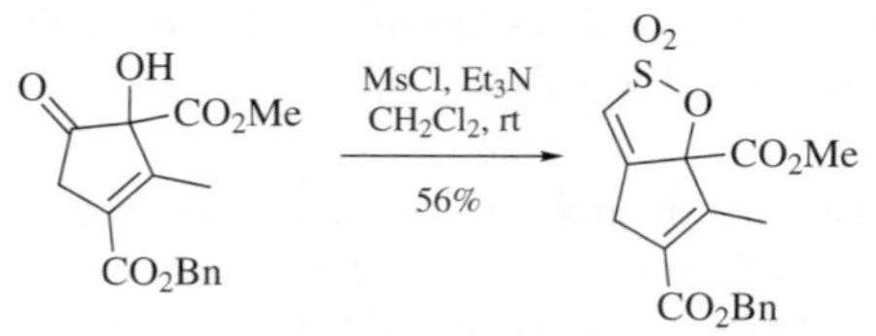

(21)

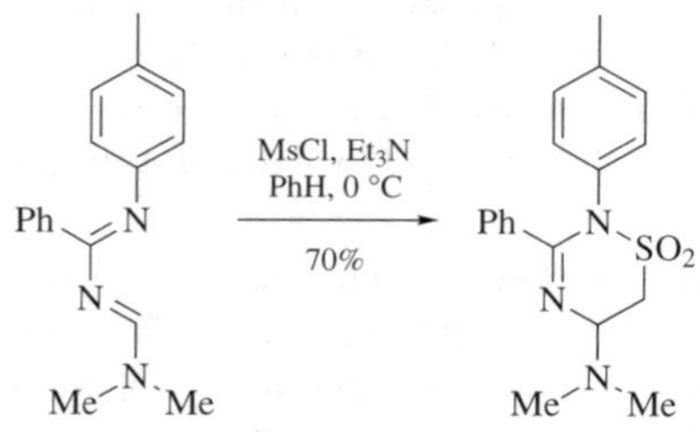

(22)

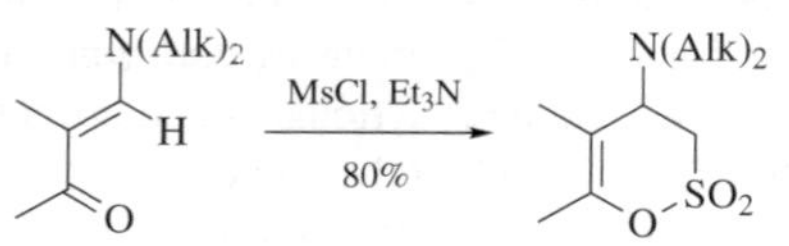

(23)

Addition Across Alkenes and Alkynes. In the presence of a catalyst (usually ***Copper*** or ruthenium), methanesulfonyl chloride adds across alkenes and alkynes to produce β-chloro sulfones.[32] The stereochemistry of the product of addition across phenylacetylene depends on the reaction conditions used. In the absence of an added hydrochloride salt, the *cis* addition product predominates (eq 24), whereas the *trans* product is favored when the salt is added (eq 25).

MsCl, $CuCl_2$, MeCN, 70% — Cl, SO_2Me (24)

MsCl, $CuCl_2$, NEt_3•HCl, MeCN, 70% — Cl, SO_2Me (25)

TMS-substituted alkenes and alkynes are especially useful in these reactions (eqs 26 and 27).[33] Using this methodology, allylic sulfones can be prepared (eq 28). Kamigata reported only minor success in inducing chirality into this reaction by the use of optically active ruthenium catalysts (eq 29).[34]

TMS — MsCl, CuCl, MeCN, CH_2Cl_2, 140 °C, 4 h, 70% — TMS, Cl, SO_2Me (26)

TMS — MsCl, $CuCl_2$, MeCN, CH_2Cl_2, 115 °C, 8 h, 58% — Cl, SO_2Me, TMS; (*Z*):(*E*) = 20:80 (27)

TMS — MsCl, CuCl, MeCN, CH_2Cl_2, 115 °C, 10 h, 35% — TMS, Cl, SO_2Me — −TMSCl — SO_2Me (28)

MsCl, Ru_2Cl_4((−)-DIOP)$_3$, PhH, 100 °C, 24 h, 47% — Cl, SO_2Me; 15% de (29)

Synthesis of Vinyl and Allyl Sulfones. In addition to the methods presented above, vinyl and allyl sulfones have been prepared by the palladium-catalyzed cross coupling of vinyl- and allylstannanes with sulfonyl chlorides (eq 30).[35]

Bu_3Sn, Ph — MsCl, $(Ph_3P)_4Pd$, THF, Δ, 90% — MeO_2S, Ph (30)

Acyliminium Cyclizations. Formation of acyliminium ions by treatment of α-hydroxy amides with methanesulfonyl chloride is a mild way of effecting acyliminium ion cyclizations. Under these conditions, allylstannanes (eq 31)[36] and cyclic vinyl sulfides (eq 32)[37] undergo addition to the iminium ion generated. The allylstannane cyclizes with a high degree of *endo* selectivity.

OH, N, O, $SnBu_3$ — MsCl, Et_3N, CH_2Cl_2, rt, 77% — N, O (31)

OH, N, O, S, S — MsCl, Et_3N, 86% — S, S, N, O (32)

Lactone Inversion. Lansbury demonstrated the use of mesyl chloride for lactone inversion in his synthesis of aromatin (eq 33).[38] This method is superior to the classical methods since the problems of racemization and relactonization are overcome.

O, O, O-*t*-Bu — 1. KOH, MeOH; 2. MsCl, Et_3N, THF, 0 °C; 3. NaOH, Δ; ~85% — O, O, O-*t*-Bu (33)

Rearrangement of Thioacetals to Aldehydes. Sato et al. reported the rearrangement of α-hydroxy thioacetals to α-sulfenyl aldehydes (eq 34).[39] The thioacetals are prepared by the addition of the anion of methoxy(phenylthio)methane to aldehydes and ketones. This reaction is general for alkyl, allyl, and aromatic carbonyl compounds.

OH, OMe, SPh — MsCl, Et_3N, PhH, 20 °C, 79% — SPh, CHO (34)

Related Reagents. Methanesulfonyl Chloride–Dimethylaminopyridine.

1. Furst, A.; Koller, F. *HCA* **1947**, *30*, 1454.
2. Baer, H. H.; Georges, F. F. Z. *CJC* **1977**, *55*, 1348.
3. Ritterskamp, P.; Demuth, M.; Schaffner, K. *JOC* **1984**, *49*, 1155.
4. Graber, R. P.; Meyers, M. B.; Landeryou, V. A. *JOC* **1962**, *27*, 2534.
5. Danheiser, R. L.; Carini, D. J.; Fink, D. M.; Basak, A. *T* **1983**, *39*, 935.
6. Maruoke, K.; Miyazaki, T.; Ando, M.; Matsumura, Y.; Sakane, S.; Hattori, K.; Yamamoto, H. *JACS* **1983**, *105*, 2831.
7. Meltzer, R. I.; Lustgarten, D. M.; Fischman, A. *JOC* **1957**, *22*, 1577.
8. Webster, K. T.; Eby, R.; Schuerch, C. *Carbohydr. Res.* **1983**, *123*, 335.
9. Merlin, P.; Braekman, J. C.; Daloze, D. *TL* **1988**, *29*, 1691.
10. Takeuchi, I.; Hamada, Y.; Hatano, K.; Kurono, Y.; Yashiro, T. *CPB* **1990**, *38*, 1504.

11. Sheu, J.; Smith, M. B.; Oeschger, T. R.; Satchell, J. *OPP* **1992**, *24*, 147.
12. Evans, M. E.; Long, L., Jr.; Parrish, F. W. *JOC* **1968**, *33*, 1074.
13. Sanda, K.; Rigal, L.; Delmas, M.; Gaset, A. *S* **1992**, 541.
14. Wolke, U.; Birdsall, N. J. M.; Brown, G. B. *TL* **1969**, 785.
15. Collington, E. W.; Meyers, A. I. *JOC* **1971**, *56*, 3044.
16. Fujimoto, Y.; Shimizu, T.; Tatsuno, T. *CPB* **1976**, *24*, 365.
17. Hyatt, J. A.; White, A. W. *S* **1984**, 214.
18. Nangia, A.; Chandrasekaran, S. *JCR(S)* **1984**, 100.
19. Dunn, A. D.; Mills, M. J.; Henry, W. *OPP* **1982**, 396.
20. Jaszay, Z. M.; Petnehazy, I.; Toke, L. *S* **1989**, 745.
21. (a) Hudrlik, P. F.; Peterson, D. *TL* **1974**, 1133. (b) Hudrlik, P. F.; Peterson, D. *JACS* **1975**, *97*, 1464.
22. Paquette, L. A.; Yan, T.-H.; Wells, G. J. *JOC* **1984**, *49*, 3610.
23. Corey, E. J.; Grieco, P. A. *TL* **1972**, 107.
24. Yadav, J. S.; Mysorekar, S. V. *SC* **1989**, *19*, 1057.
25. Grieco, P. A.; Hiroi, K. *CC* **1972**, 1317.
26. Kozikowski, A. P.; Stein, P. D. *JACS* **1982**, *104*, 4023.
27. Rosen, M. H. *TL* **1969**, 647.
28. (a) Hasek, R. H.; Gott, P. G.; Meen, R. H.; Martin, J. C. *JOC* **1963**, *28*, 2496. (b) Ziegler, E.; Kappe, T. *AG* **1964**, *76*, 921.
29. (a) Patonay, T.; Batta, G.; Dinya, Z. *JHC* **1988**, *25*, 343. (b) Takada, D.; Suemune, H.; Sakai, K. *CPB* **1990**, *38*, 234.
30. Mazumdar, S. N.; Sharma, M.; Mahajan, M. P. *TL* **1987**, *28*, 2641.
31. (a) Ziegler, E.; Kappe, T. *AG* **1964**, *76*, 921. (b) Menozzi, G.; Bargagna, A.; Mosti, L.; Schenone, P. *JHC* **1987**, *24*, 633.
32. Amiel, Y. *TL* **1971**, 661.
33. Pillot, J. P.; Dunogues, J.; Calas, R. *S* **1977**, 469.
34. Kameyama, M.; Kamigata, N. *BCJ* **1989**, *62*, 648.
35. Labadie, S. S. *JOC* **1989**, *54*, 2496.
36. (a) Keck, G.; Enholm, E. J. *TL* **1985**, *26*, 3311. (b) Keck, G.; Cressman, E. N. K.; Enholm, E. J. *JOC* **1989**, *54*, 4345.
37. Chamgerlin, A. R.; Nguyen, H. D.; Chung, J. Y. L. *JOC* **1984**, *49*, 1682.
38. Lansbury, P. T.; Vacca, J. P. *T* **1982**, *38*, 2797.
39. Sato, T.; Okazaki, H.; Otera, J.; Nozaki, H. *JACS* **1988**, *110*, 5209.

Valerie Vaillancourt & Michele M. Cudahy
The Upjohn Co., Kalamazoo, MI, USA

p-Methoxybenzyl Chloride[1]

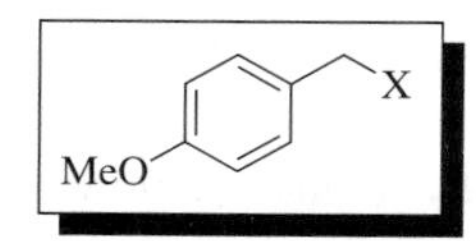

(X = Cl)
[824-98-6] C_8H_9ClO (MW 156.62)
(X = Br)
[2746-25-0] C_8H_9BrO (MW 201.07)

(protection of alcohols,[3] thiols,[42] phenols,[29] amides,[55] amines,[50] and carboxylic acids[31])

Alternate Name: MPMCl.
Physical Data: X = Cl: bp 117–118 °C/14 mmHg; *d* 1.155 g cm^{-3}.
Solubility: sol all organic solvents.
Form Supplied in: the chloride is commercially available. The bromide is not commercially available and is prepared immediately before use because of its instability.[2]
Preparative Methods: to prepare the bromide, an ether solution of the alcohol is stirred with concd HBr, the phases are separated, and the organic phase is washed with saturated aq. NaBr, dried over K_2CO_3, and concentrated.
Handling, Storage, and Precautions: the chloride is stored over K_2CO_3 to stabilize it; the bromide is stored in the freezer to prevent polymerization, which occurs within a few days at rt.[2]

Protection of Alcohols. The inherent stability of the MPM ether, coupled with a large repertoire of methods for its removal under mild conditions that do not normally effect other functional groups, makes it a particularly effective derivative for the protection of alcohols. The most common method for its introduction is by the Williamson ether synthesis. A number of bases can be used to generate the alkoxide, but ***Sodium Hydride***[3] in DMF (eq 1) or THF[4] (eq 2) is the most common. Other bases such as ***n-Butyllithium***,[5] ***Potassium Methylsulfinylmethylide*** (dimsylpotassium)[6] (eq 3), and ***Sodium Hydroxide*** under phase-transfer conditions[7] are also used. From these results, it is clear that protection can be achieved without interference from Payne rearrangement, and considerable selectivity can be obtained. In the ribose case, selectivity is probably achieved because of the increased acidity of the 2′-hydroxy group. The additive ***Tetra-n-butylammonium Iodide***[8] is used for in situ preparation of the highly reactive *p*-methoxybenzyl iodide, thus improving the protection of very hindered alcohols. Selective monoprotection of diols is readily occasioned with *O*-stannylene acetals.[9]

HO, O, A; HO OH → *p*-MeOC$_6$H$_4$CH$_2$Br, DMF; NaH, 1 h, –5 °C; 65% → HO, O, A; HO OMPM (1)

O; OH → *p*-MeOC$_6$H$_4$CH$_2$Br, THF; NaH, 0 °C; 94% → O; OMPM (2)

DMPMO OH OMe O → KCH$_2$S(O)Me, DMSO; *p*-MeOC$_6$H$_4$CH$_2$Cl; >60% → DMPMO OMPM OMe O (3)

Cleavage of the MPM group is usually achieved oxidatively with ***2,3-Dichloro-5,6-dicyano-1,4-benzoquinone***.[10–12] When DDQ is used, overoxidation of the alcohol occasionally occurs,[13,14] but for the most part this is not a problem. In general, selective cleavage of the MPM group is achieved in molecules containing benzyl groups, but more forcing conditions will result in benzyl cleavage.[15] ***Cerium(IV) Ammonium Nitrate***,[16] trityl tetrafluoroborate,[17] photolysis,[18] electrolytic oxidation,[19] ***Bromodimethylborane***,[20] ***Boron Trichloride***,[21] ***Iodotrimethylsilane***,[22] ***Tin(II) Chloride***/***Chlorotrimethylsilane***/anisole,[23] hydrogenolysis,[24] and ozonolysis[25] have also been used to cleave this derivative. Selectivity for the *O*-MPM is achieved over an amide MPM derivative with DDQ.[26]

In the presence of neighboring hydroxyl groups, the DDQ oxidation of MPM ethers leads to cyclic acetals, thus affording a level of selectivity which may not be possible to achieve otherwise.[27] In a synthesis of erythronolide (eq 4), the critical oxidative cyclization had to be performed with freshly recrystallized DDQ supported on molecular sieves in order to prevent acid-catalyzed isomerization of the acetal from its initial kinetic disposition to the more thermodynamically favorable orientation.[28]

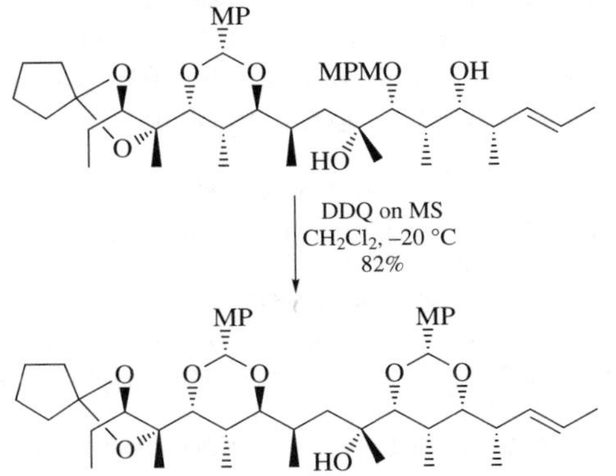

DDQ on MS, CH_2Cl_2, −20 °C, 82% (4)

Protection of Phenols. Protection of phenols is achieved using *p*-methoxybenzyl chloride ($Bu_4N^+I^-$, K_2CO_3) in acetone,[29] or the benzyl bromide (*i*-Pr_2NEt) in CH_2Cl_2.[30] Deprotection is occasioned with ***Trifluoroacetic Acid***[29] or ***10-Camphorsulfonic Acid***/$Me_2C(OMe)_2$.[30]

Protection of Carboxylic Acids. *p*-Methoxybenzyl esters have been prepared from the Ag^I salt of amino acids and the benzyl halide (Et_3N, $CHCl_3$, 25 °C, 24 h, 60% yield),[31] and from cephalosporin precursors and the benzyl alcohol [$Me_2NCH(OCH_2$-*t*-$Bu)_2$, CH_2Cl_2, 90% yield],[32] and from the benzyl halide ($NaHCO_3$, DMF).[33] They are also prepared by activation of the acid with isopropenyl oxychloroformate ($MeOC_6H_4CH_2OH$, DMAP, 0 °C, CH_2Cl_2, 91%).[34] They are cleaved by acidic hydrolysis (CF_3CO_2H/PhOMe, 25 °C, 3 min, 98% yield;[35] HCO_2H, 22 °C, 1 h, 81% yield),[31] by heating with phenol,[36] by treatment with $AlCl_3$ (anisole, CH_2Cl_2 or $MeNO_2$, −50 °C; $NaHCO_3$, −50 °C, 73–95% yield),[37,38] or by $CF_3CO_2H/B(OTf)_3$.[39]

Protection of Thiols. The protection of thiols as *S-p*-methoxybenzyl thioethers is occasioned with MPMCl and a base such as NH_3,[40] KOH (MeOH),[41] and NaH (THF, 60 °C, 1 h).[42] The reduction of disulfides with ***Lithium Aluminium Hydride*** affords an intermediate thiolate that can be trapped with MPMCl to give the thioether.[43] The cleavage of *S-p*-methoxybenzyl thioethers is effected with $Hg(OAc)_2/CF_3CO_2H$ (0 °C, 10–30 min), $Hg(OCOCF_3)_2$/aq AcOH (20 °C, 2–3 h, followed by H_2S or $HSCH_2CH_2OH$, 100% yield),[44,45] $Hg(OCOCF_3)_2/CF_3CO_2H$/anisole,[46] CF_3CO_2H/reflux,[40] anhydrous HF/anisole (25 °C, 1 h, quant),[47] $(4\text{-}BrC_6H_4)_3NH^+ SbCl_6^-$ (75%),[48] HCO_2H (5 °C, 30 min, 45%).[42] During the synthesis of peptides which contain 4-methoxybenzyl-protected cysteine residues, sulfoxide formation may occur (eq 5). These sulfoxides, when treated with HF/anisole, form thiophenyl ethers that cannot be deprotected; therefore the peptides should be subjected to a reduction step prior to deprotection.[49]

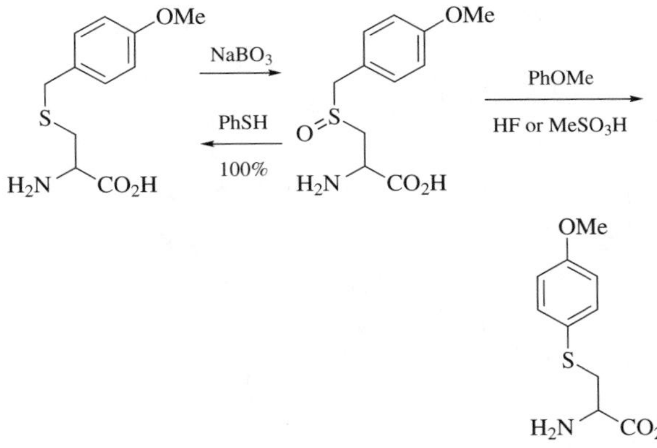

$NaBO_3$; PhSH, 100%; PhOMe, HF or $MeSO_3H$ (5)

Protection of Amines. The use of the MPM group to protect amines has not seen many applications, but a few examples are available in the literature. The MPM group can be introduced on an aromatic amine with MPMBr (DMF, KI, K_2CO_3, 92%),[50] or by reductive amination (anisaldehyde, PhH, H_2O; $NaBH_4$, MeOH, 92%),[51] but for the most part the MPM group acts as a carrier for the nitrogen to be introduced into a molecule. MPM amines can be cleaved by hydrogenolysis,[52] oxidation to the imine followed by hydrolysis,[53] or acylation with MeCHClOCOCl (CH_2Cl_2, −7 °C, 5 min; MeOH, reflux, 89%).[54]

Protection of the Amide Nitrogen. Of the few methods for protection of the amide NH, the MPM group has found success where other methods have not been entirely successful.[55] The group can be introduced using methods similar to those used for alcohols: MPMCl/NaH/DMF,[56] MPMBr/NaH,[55] MPMCl (DBU, MeCN, 60 °C, 5 h, 92%),[57] MPMCl/Ag_2O.[58]

The most common method for removal of the amide MPM group is by oxidation with DDQ[59] or CAN.[60] With CAN, the benzoyl imide is sometimes a byproduct of the reaction (eq 6), but this is easily cleaved by hydrolysis.[61] The amide MPM group is also cleaved by solvolysis in TFA[51] or $AlCl_3$/anisole.[62] Metalation with ***t-Butyllithium*** followed by reaction with oxygen (62%) has been used successfully (eq 7).[63]

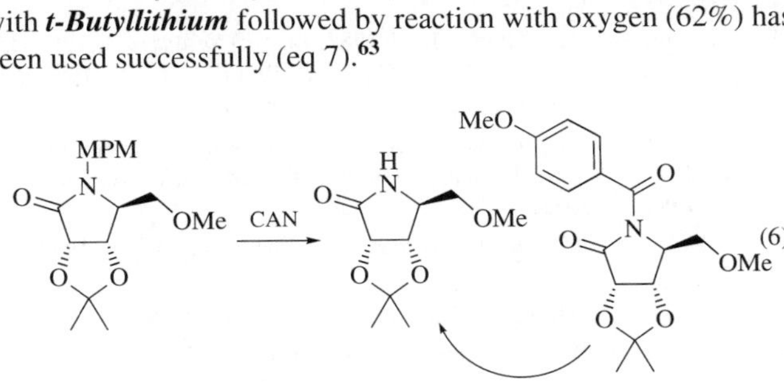

CAN (6)

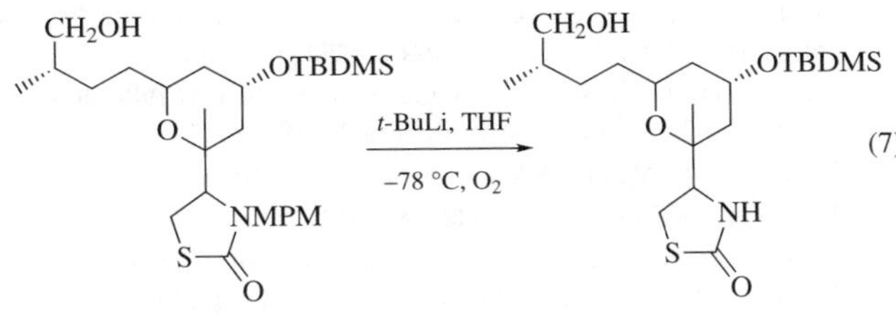

t-BuLi, THF, −78 °C, O_2 (7)

Protection of Phosphates. There is one example where the MPM group is used for phosphate protection. It is cleaved with 48% aq HF/MeCN/H_2O (eq 8).[64]

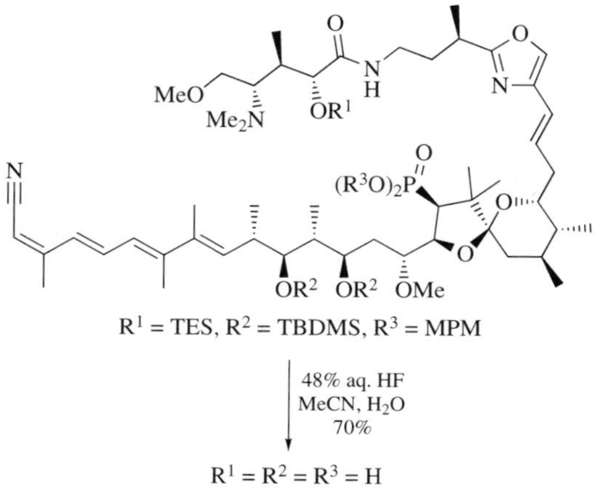

R^1 = TES, R^2 = TBDMS, R^3 = MPM

48% aq. HF
MeCN, H_2O
70%

$R^1 = R^2 = R^3 = H$ (8)

Related Reagents. Benzoyl Trifluoromethanesulfonate; Benzyl Bromide; Benzyl Chloride; Benzyl Iodide; Benzyl 2,2,2-Trichloroacetimidate; 3,4-Dimethoxybenzyl Bromide; (*p*-Methoxybenzyloxy)methyl Chloride; 4-Methoxybenzyl 2,2,2-Trichloroacetimidate.

---

1. For a review of protective groups see: Greene, T. W.; Wuts, P. G. M. *Protective Groups in Organic Synthesis*, 2nd ed.; Wiley: New York, 1991.
2. Ruder, S. M.; Ronald, R. C. *TL* **1987**, *28*, 135.
3. Takaku, H.; Kamaike, K.; Tsuchiya, H. *JOC* **1984**, *49*, 51.
4. Schreiber, S. L.; Smith, D. B. *JOC* **1989**, *54*, 9.
5. Hoshi, H.; Ohnuma, T.; Aburaki, S.; Konishi, M.; Oki, T. *TL* **1993**, *34*, 1047.
6. Nakajima, N.; Hamada, T.; Tanaka, T.; Oikawa, Y.; Yonemitsu, O. *JACS* **1986**, *108*, 4645.
7. Garegg, P. J.; Oscarson, S.; Ritzen, H. *Carbohydr. Res.* **1988**, *181*, 89.
8. (a) Mootoo, D. R.; Fraser Reid, B. *T* **1990**, *46*, 185. (b) Trost, B. M.; Hipskind, P. A.; Chung, J. Y. L.; Chan, C. *AG(E)* **1989**, *28*, 1502.
9. (a) Nagashima, N.; Ohno, M. *CL* **1987**, 141. (b) Verduyn, R.; Elle, C. J. J.; Dreef, C. E.; Van der Marel, G. A.; van Boom, J. H. *RTC* **1990**, *109*, 591. (c) Danishefsky, S.; Lee, J. Y. *JACS* **1989**, *111*, 4829.
10. (a) Horita, K.; Yoshioka, T.; Tanaka, T.; Oikawa, Y.; Yonemitsu, O. *T* **1986**, *42*, 3021. (b) Tanaka, T.; Oikawa, Y.; Hamada, T.; Yonemitsu, O. *TL* **1986**, *27*, 3651.
11. Oikawa, Y.; Tanaka, T.; Horita, K.; Yonemitsu, O. *TL* **1984**, *25*, 5397.
12. Oikawa, Y.; Yoshioka, T.; Yonemitsu, O. *TL* **1982**, *23*, 885.
13. Nagaoka, H.; Baba, A.; Yamada, Y. *TL* **1991**, *32*, 6741.
14. Trost, B. M.; Chung, J. Y. L. *JACS* **1985**, *107*, 4586.
15. (a) Horita, K.; Nagato, S.; Oikawa, Y.; Yonemitsu, O. *TL* **1987**, *28*, 3253. (b) Ikemoto, N.; Schreiber, S. L. *JACS* **1992**, *114*, 2525.
16. (a) Classon, B.; Garegg, P. J.; Samuelsson, B. *ACS(B)* **1984**, *B38*, 419. (b) Johansson, R.; Samuelsson, B. *JCS(P1)* **1984**, 2371.
17. Nakajima, N.; Abe, R.; Yonemitsu, O. *CPB* **1988**, *36*, 1988.
18. Pandey, G.; Krishna, A. *SC* **1988**, *18*, 2309.
19. (a) Schmidt, W.; Steckhan, E. *AG(E)* **1979**, *18*, 801. (b) Weinreb, S. M.; Epling, G. A.; Comi, R.; Reitano, M. *JOC* **1975**, *40*, 1356.
20. Hebert, N.; Beck, A.; Lennox, R. B.; Just, G. *JOC* **1992**, *57*, 1777.
21. Vanhessche, K.; Bello, C. G.; Vandewalle, M. *SL* **1991**, 921.
22. Gordon, D. M.; Danishefsky, S. J. *JACS* **1992**, *114*, 695.
23. Akiyama, T.; Shima, H.; Ozaki, S. *SL* **1992**, 415.
24. Hikota, M.; Tone, H.; Horita, K.; Yonemitsu, O. *JOC* **1990**, *55*, 7.
25. Hirama, M.; Shimizu, M. *SC* **1983**, *13*, 781.
26. Hamada, Y.; Tanada, Y.; Ykokawa, F.; Siori, T. *TL* **1991**, *32*, 5983.
27. Oikawa, Y.; Yoshioka, T.; Yonemitsu, O. *TL* **1982**, *23*, 889.
28. Stürmer, R.; Ritter, K.; Hoffmann, R. W. *AG(E)* **1993**, *32*, 101.
29. White, J. D.; Amedio, J. C., Jr. *JOC* **1989**, *54*, 736.
30. Nagaoka, H.; Schmid, G.; Iio, H.; Kishi, Y. *TL* **1981**, *22*, 899.
31. Stelakatos, G. C.; Argyropoulos, N. *JCS(C)* **1970**, 964.
32. Webber, J. A.; Van Heyningen, E. M.; Vasileff, R. T. *JACS* **1969**, *91*, 5674.
33. Baldwin, J. E.; Flinn, A. *TL* **1987**, *28*, 3605.
34. Jouin, P.; Castro, B.; Zeggaf, C.; Pantaloni, A.; Senet, J. P.; Lecolier, S.; Sennyey, G. *TL* **1987**, *28*, 1661.
35. Stewart, F. H. C. *AJC* **1968**, *21*, 2543.
36. Tanaka, H.; Taniguchi, M.; Kameyama, Y.; Torii, S.; Sasaoka, M.; Shiroi, T.; Kikuchi, R.; Kawahara, I.; Shimabayashi, A.; Nagao, S. *TL* **1990**, *31*, 6661.
37. Ohtani, M.; Watanabe, F.; Narisada, M. *JOC* **1984**, *49*, 5271.
38. Tsuji, T.; Kataoka, T.; Yoshioka, M.; Sendo, Y.; Nishitani, Y.; Hirai, S.; Maeda, T.; Nagata, W. *TL* **1979**, 2793.
39. Young, S. D.; Tamburini, P. P. *JACS* **1989**, *111*, 1933.
40. Akabori, S.; Sakakibara, S.; Shimonishi, Y.; Nobuhara, Y. *BCJ* **1964**, *37*, 433.
41. Chambers, M. S.; Thomas, E. J. *CC* **1989**, 23.
42. Taylor, A. W.; Dean, D. K. *TL* **1988**, *29*, 1845.
43. Baidoo, K. E.; Lever, S. Z. *TL* **1990**, *31*, 5701.
44. Nishimura, O.; Kitada, C.; Fujino, M. *CPB* **1978**, *26*, 1576.
45. Gordon, E. M.; Godfrey, J. D.; Delaney, N. G.; Asaad, M. M.; Von Langen, D.; Cushman, D. W. *JMC* **1988**, *31*, 2199.
46. Holler, T. P.; Spaltenstein, A.; Turner, E.; Klevit, R. E.; Shapiro, B. M.; Hopkins, P. B. *JOC* **1987**, *52*, 4420.
47. Sakakibara, S.; Shimonishi, Y. *BCJ* **1965**, *38*, 1412.
48. Platten, M.; Steckhan, E. *LA* **1984**, 1563.
49. Funakoshi, S.; Fujii, N.; Akaji, K.; Irie, H.; Yajima, H. *CPB* **1979**, *27*, 2151.
50. Tamato, M.; Takeuchi, Y.; Ikeda, Y. *H* **1987**, *26*, 191.
51. Smith, A. B., III; Rano, T. A.; Chida, N.; Sulikowski, G. A. *JOC* **1990**, *55*, 1136.
52. Rowley, M. *T* **1992**, *48*, 3557.
53. Maruyama, K.; Kusukawa, T.; Higuchi, Y.; Nishinaga, A. *CL* **1991**, 1093.
54. Yang, B. V.; O'Rourke, D.; Li, J. *SL* **1993**, 195.
55. Smith, A. B., III; Leahy, J. W.; Noda, I.; Remiszewski, S. W.; Liverton, N. J.; Zibuck, R. *JACS* **1992**, *114*, 2995.
56. Williams, R. M.; Maruyama, L. K. *JOC* **1987**, *52*, 4044.
57. Akiyama, T. *BCJ* **1990**, *63*, 3356.
58. Takahashi, Y.; Yamshita, H.; Kobayashi, S.; Ohno, M. *CPB* **1986**, *34*, 2732.
59. Smith, A. B., III; Rano, T. A.; Chida, N.; Sulikowski, G. A.; Wood, J. L. *JACS* **1992**, *114*, 8008.
60. Williams, R. M.; Sabol, M. R.; Kim, H.-d.; Kwast, A. *JACS* **1991**, *113*, 6621.
61. Smith, A. B., III; Salvatore, B. A.; Hull, K. G.; Duan, J. J.-W. *TL* **1991**, *32*, 4859.
62. Akiyama, T.; Kumegawa, M.; Takesue, Y.; Nishimoto, H.; Ozaki, S. *CL* **1990**, 339.

63. Smith, A. B., III; Noda, I.; Remiszewski, S. W.; Liverton, N. J.; Zibuck, R. *JOC* **1990**, *55*, 3977.

64. Evans, D. A.; Gage, J. R.; Leighton, J. L. *JACS* **1992**, *114*, 9434.

Peter G. M. Wuts
The Upjohn Co., Kalamazoo, MI, USA

(Methoxycarbonylsulfamoyl) triethylammonium Hydroxide[1]

$Et_3N^+-SO_2-N^--CO_2Me$

[29684-56-8] $C_8H_{18}N_2O_4S$ (MW 238.35)

(dehydration agent[1])

Alternate Name: Burgess reagent.
Physical Data: mp 76–79 °C.
Solubility: sol benzene, toluene, THF, triglyme, acetonitrile.
Form Supplied in: colorless solid.
Preparative Method: although the Burgess reagent is commercially available, it can be easily synthesized in two steps:[1a] the reaction of anhydrous methanol with ***Chlorosulfonyl Isocyanate*** followed by the addition of ***Triethylamine*** affords the reagent in good yield.
Handling, Storage, and Precautions: moisture sensitive; the use of anhydrous solvent and atmosphere in reactions with this reagent is recommended.

Formation of Carbamates. The reaction of primary alcohols with the Burgess reagent (**1**) produces alkyl *N*-methoxycarbonylsulfamate salts, which decompose on thermolysis via an S_N2 (or S_Ni) mechanism to give methyl urethanes (eqs 1 and 2).[1a,b] The products are isolated in good yields. Hydrolysis of the carbamate produces primary amines. Transformations of alcohols to amines are often multistep syntheses; therefore this reaction provides a distinct advantage over the existing methodology for such substitution reactions.

OH → (1), 95 °C, 75% → NHCO₂Me (1)

OH → (1), 95 °C, 80% → NHCO₂Me (2)

It was noted that when this reaction was applied to allylic alcohols, either elimination (see below) or an S_Ni' rearrangement ensued.[1c] The fate of this reaction can be determined by the experimental conditions employed. Whereas the reaction in triglyme gave good yields of dienes (eq 3), thermal decomposition of the *N*-methoxycarbonylsulfamate intermediate at 80 °C as a solid provided >90% of the allylurethane product (eq 4).

OH → (1), 80 °C, triglyme, 70% → (3)

OH → (1) → [MeO₂CNS(O)O–] → 80 °C, >90% → HN–CO₂Me (4)

Formation of Alkenes. The products of the reaction of the Burgess reagent with secondary and tertiary alcohols decompose smoothly at temperatures between 30 and 80 °C in a variety of solvents to give alkenes.[1] The stereochemical course of the reaction was established to be stereospecific *cis* elimination. This reaction minimizes the type of rearrangement products that are typical for carbonium ion-mediated elimination reactions. For example, the reaction of the Burgess reagent with 3,3-methyl-2-butanol gave three products in the ratios given in eq 5.[1a] The last two alkenes were generated as a consequence of the Wagner–Meerwein rearrangement, whereas the first product formed by direct β-proton removal. The choice of solvent did not affect the product ratios significantly. Usually, competition between the Wagner–Meerwein rearrangement and β-proton abstraction lies entirely in the direction of the former, which is not seen when the Burgess reagent is used.

OH → (1), 70 °C, 84% → + + (5)
1:3:1.5

The reaction of the Burgess reagent with 2-*endo*-methylbicyclo[2.2.1]heptan-2-ol afforded a mixture of the two cycloalkenes shown in eq 6.[1a] The ratio of the products was unexpectedly 1:1, an observation which was attributed to the steric effects of the *endo* C-5 hydrogen, whereby the rate of proton abstraction at C-3 would be attenuated. No rearrangement products were reported. The reaction in eq 7 proceeded to give the sole product shown; no evidence for the formation of 2-cyclopropylidenepropane was found on grounds that the β-cyclopropyl hydrogen is sterically encumbered and unavailable for elimination.[1a] The reaction in eq 8 afforded a product mixture favoring the (*Z*)-isomer.[2] The selectivity is somewhat less in the case of acid-catalyzed dehydration of the same compound. It is noteworthy that similar CS_2-mediated dehydrations in the presence of base require pyrolysis at temperatures exceeding 200 °C.[2] Some related applications for the Burgess reagent are shown in eqs 9 and 10.[3,4] Dehydration of eq 9 with ***Acetic Anhydride*** in the presence of ***Pyridine*** at 70 °C gave only 56% of the desired product.

OH → (1), 50 °C, 80% → + (6)
1:1

OH → (1), rt, 66% → (7)

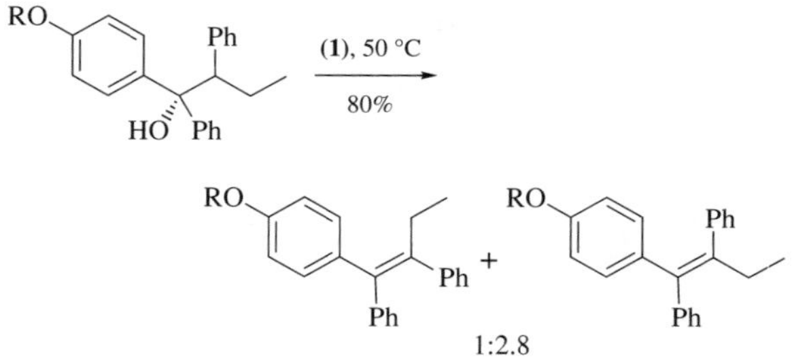

(8)

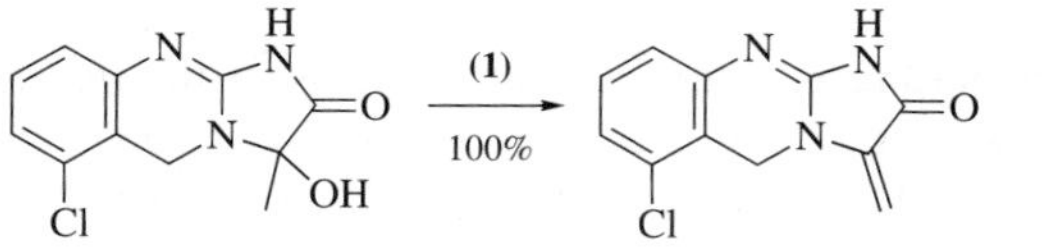

(9)

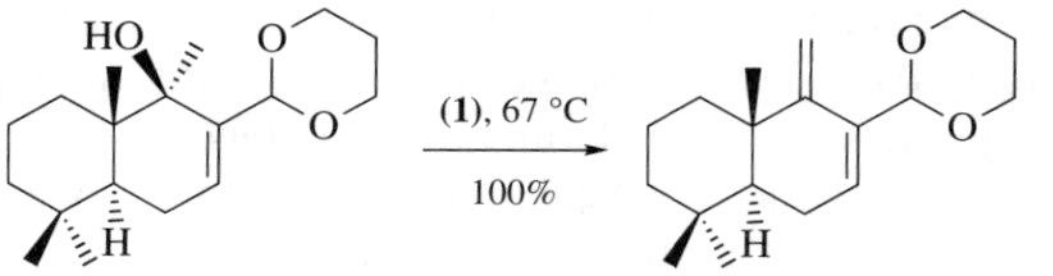

(10)

The requirement for a *cis* elimination chemistry was discussed earlier; however, there are a few exceptional reaction outcomes reported in steroid chemistry.[5] The results depicted in eqs 11 and 12 were unanticipated since the 11α-hydroxyl and 9β-hydrogen (eq 11) are *trans*-diaxial, as are the 3α-hydroxyl and 4β-hydrogen in eq 12. These results are rationalized by an intramolecular 1,2-hydrogen transfer. For example, it was suggested for eq 11 that a C-11 cation is formed first, which subsequently undergoes an intramolecular hydrogen transfer from C-9 to C-11. The resultant C-9 carbocation was postulated to be quenched by the loss of a hydrogen from C-11 and the formation of the Δ^9-double bond.[5]

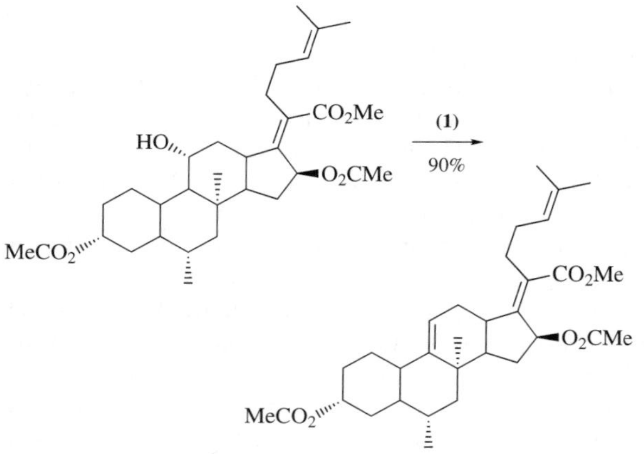

(11)

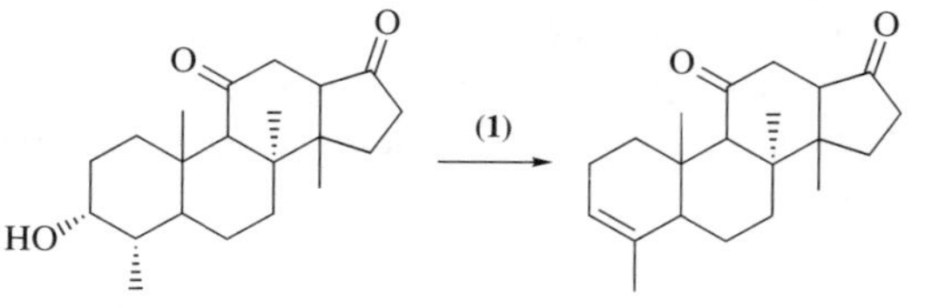

(12)

Formation of Dienones and Furanones. Furanones are formed typically by treatment of α-hydroxy enones with acid; the reaction is presumed to proceed via an acid-labile dienone, although no direct evidence for the existence of such an intermediate has been demonstrated. It has been shown that α-hydroxy enones undergo dehydration in the presence of the Burgess reagent to give isolable dienones, which are cyclized to give the corresponding furanones in a separate acid-catalyzed step (eq 13).[6]

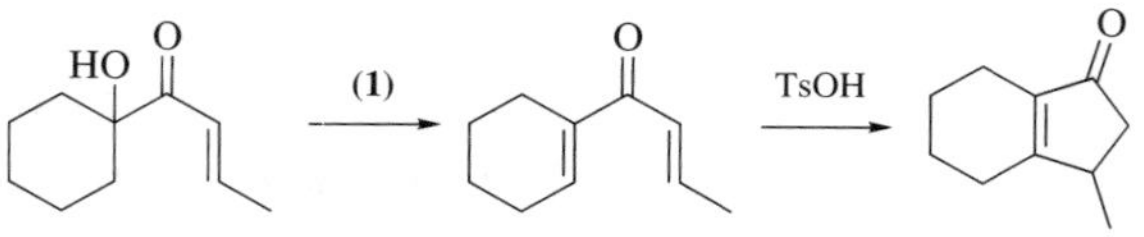

(13)

Formation of Vinyltributyltin Compounds and Tributyltin Isocyanate. The use of the Burgess reagent in the facile formation of vinyltributyltin compounds has been reported.[7] This type of molecule is otherwise typically synthesized under drastic flash pyrolytic conditions. Reaction of the butyltin Grignard reagent with ketones results in tertiary alcohols, which in the presence of the Burgess reagent furnish the vinytributyltin products (eq 14).

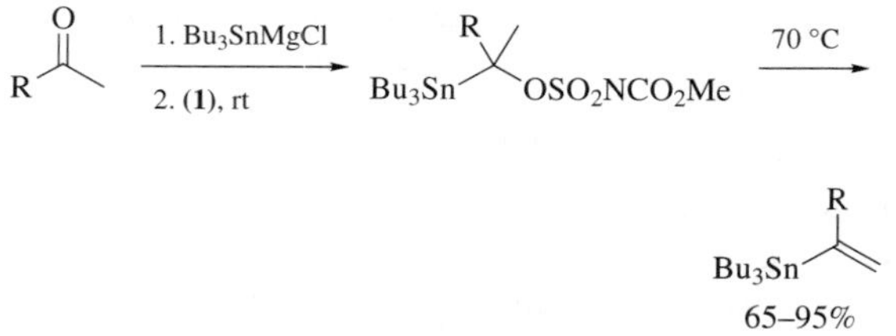

(14)

It was discovered that a side reaction during the vinyltributyltin formation, was the production of tributyltin isocyanate.[8] The formation of tributyltin isocyanate could account for the major portion of the product mixture when cyclic ketones are used, an observation which was attributed to the thermal instability of the intermediary organotin alcohols. These reagents decompose at elevated temperatures to the corresponding ketone and tributyltin hydride, which react with (**1**) to give rise to the isocyanate (eq 15). Correspondingly, it has been shown that tin hydride reduction of (**1**) gives tributyltin isocyanate in quantitative yield (eq 16).

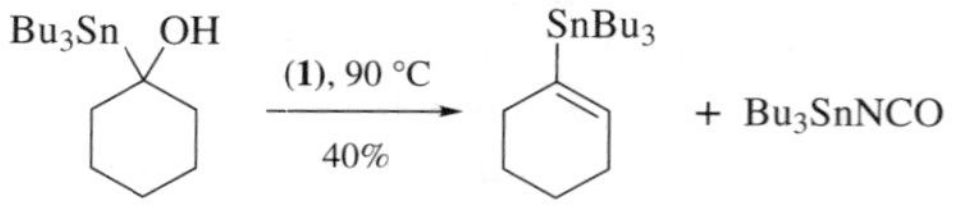

(15)

$$HSnBu_3 \xrightarrow[100\%]{(1),\ 90\ ^\circ C} Bu_3SnNCO + Et_3N + SO_2 + MeOH \quad (16)$$

Formation of Nitriles. Dehydration of the primary amide function by the Burgess reagent takes place under mild conditions (eqs 17 and 18), and there is often no need for protection of many functionalities in more complex molecules; the reagent has been shown to tolerate alcohols, esters epoxides, ketones, carbamates, and secondary amides.[9] Excellent chemoselectivity is observed in this reaction, which takes place at room temperature.

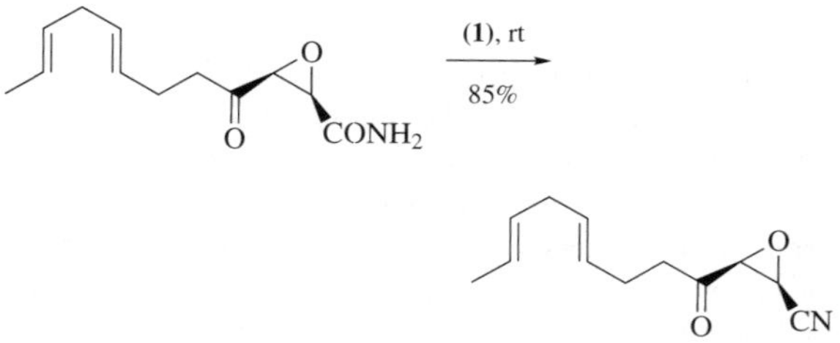

(17)

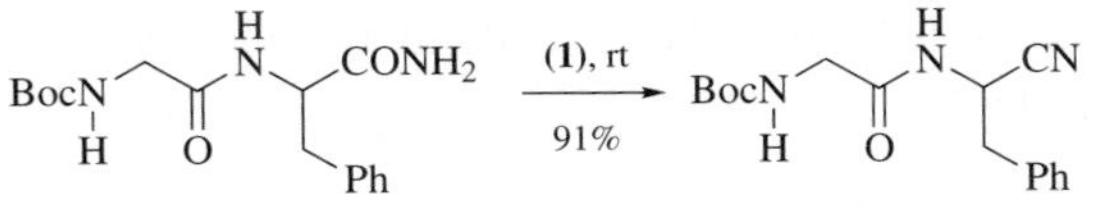

(18)

Electrophilic Addition. The ethyl analog of (**1**), (ethoxycarbonylsulfamoyl)triethylammonium hydroxide (**2**), has also been reported.[10] Compound (**2**) has been shown to undergo electrophilic addition to alkenes in average to good yields (eqs 19 and 20).

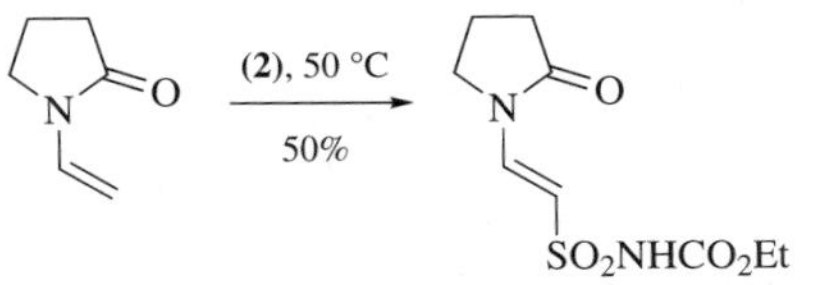

(19)

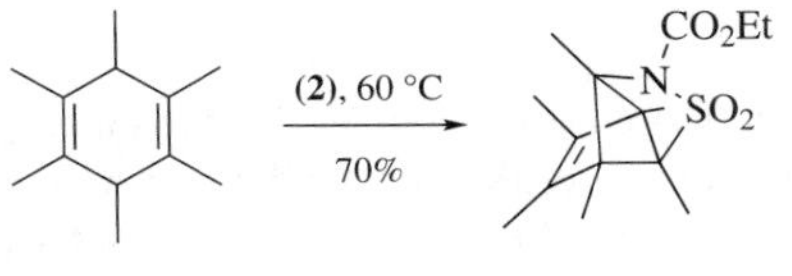

(20)

Related Reagents. Carbon Disulfide; *p*-Toluenesulfonyl Chloride.

1. (a) Burgess, E. M.; Penton, H. R.; Taylor, E. A. *JOC* **1973**, *38*, 26. (b) Burgess, E. M.; Penton, H. R.; Taylor, E. A.; Williams, W. M. *OSC* **1988**, *6*, 788. (c) Burgess, E. M.; Penton, H. R.; Taylor, E. A. *JACS* **1970**, *92*, 5224.
2. McCague, R. *JCS(P1)* **1987**, 1011.
3. Stalder, H. *HCA* **1986**, *69*, 1887.
4. Goldsmith, D. J.; Kezar, H. S. *TL* **1980**, *21*, 3543. Marino, J. P.; Ferro, M. P. *JOC* **1981**, *46*, 1912. Crabbe, P.; Leon, C. *JOC* **1970**, *35*, 2594.
5. O'Grodnick, J. S.; Ebersole, R. C.; Wittstruck, T.; Caspi, E. *JOC* **1974**, *39*, 2124.
6. Jacobson, R. M.; Lahm, G. P. *JOC* **1979**, *44*, 462.
7. Ratier, M.; Khatmi, D.; Duboudin, J. G.; Minh, D. T. *SC* **1989**, *19*, 285.
8. Ratier, M.; Khatmi, D.; Duboudin, J. G.; Minh, D. T. *SC* **1989**, *19*, 1929.
9. Claremon, D. A.; Phillips, B. T. *TL* **1988**, *29*, 2155.
10. Atkins, G. M.; Burgess, E. M. *JACS* **1972**, *94*, 6135. Atkins, G. M.; Burgess, E. M. *JACS* **1968**, *90*, 4744.

Pascale Taibi & Shahriar Mobashery
Wayne State University, Detroit, MI, USA

2-Methoxyethoxymethyl Chloride[1]

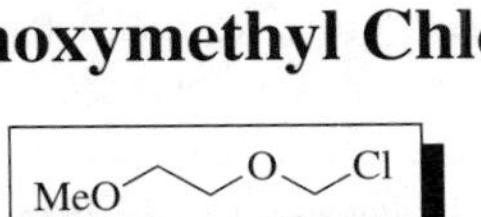

[3970-21-6] $C_4H_9ClO_2$ (MW 124.58)

(protection of alcohols,[2] phenols,[15] and carboxylic acids;[20] functionalized one-carbon synthon[23])

Alternate Name: MEMCl.
Physical Data: bp 50–52 °C/13 mmHg; *d* 1.094 g cm^{-3}.
Handling, Storage, and Precautions: moisture sensitive; a lachrymator and possible carcinogen.

Protection of Alcohols. The 2-methoxyethylmethyl (MEM) group was developed as a protective group that can be introduced with the ease of the methoxymethyl (MOM) group but can be cleaved under milder conditions, thus affording another level of selectivity in deprotection.[2] The most common method for its introduction is occasioned with MEMCl and ***Diisopropylethylamine*** (CH_2Cl_2, 25 °C, 3 h, quant.). Introduction of the MEM group is also faster than the formation of the related MOM derivative, which may be the result of anchimeric assistance. Alternatively the salt, $MEMNEt_3^+ Cl^-$ (MeCN, reflux, 30 min, >90% yield) may be used for its introduction, but this procedure has not seen extensive use. An alkoxide prepared from ***Sodium Hydride*** or ***Potassium Hydride***, when treated with MEMCl (THF or DME, 0 °C, 10–60 min, >95% yield), affords the MEM ether. The reaction of tributyltin ethers in the presence of ***Cesium Fluoride***, DMF, and MEMCl has also been used to prepare MEM ethers, but this method is primarily used as part of a scheme for the interconversion of protective groups.[3] For example, the reaction of THP ethers with $Bu_3SnSMe/BF_3{\cdot}Et_2O$ gives the corresponding tin ethers. Since the MEM group is of low steric demand, little selectivity is expected when protecting two similar alcohols, but electronic effects can be used to advantage to achieve modest selectivity (eq 1).[4]

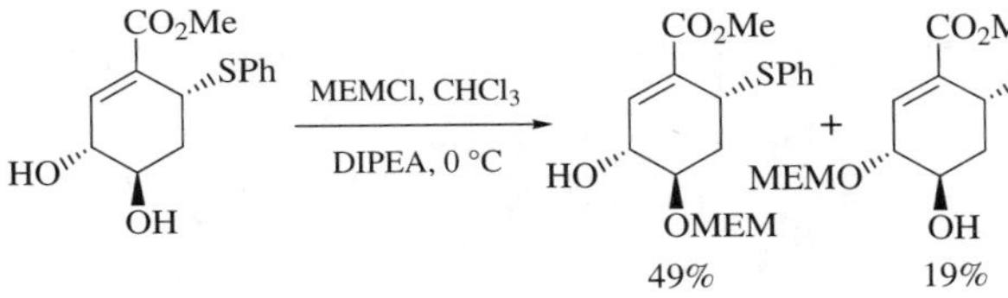

(1)

The original method devised to effect deprotection is to treat the MEM derivative with ***Zinc Bromide*** (CH_2Cl_2, 25 °C, 2–10 h, 90% yield)[1] or ***Titanium(IV) Chloride*** (CH_2Cl_2, 0 °C, 20 min, 95% yield)[1] to form a chelate, which upon further reaction reconstitutes the alcohol. Since then a number of other cleavage methods have been developed which rely on the fact that the MEM group can be viewed as an acetal or ether. Reagents such as ***Bromodimethylborane*** (CH_2Cl_2, −78 °C, $NaHCO_3$, H_2O, 87–95% yield),[5] Ph_2BBr (CH_2Cl_2, −78 °C, 71% yield),[6] ($-SCH_2CH_2S-$)BBr,[7] (*i*-PrS)$_2$BBr (DMAP, K_2CO_3, H_2O),[8] and 2-bromo-1,3,2-benzodioxaborole[9] cleave ethers like BBr_3. Me_2BBr cleaves MTM and MOM ethers and acetals, but (*i*-PrS)$_2$BBr converts the MEM group to the *i*-$PrSCH_2$– ether which can be cleaved using the same conditions used to cleave the MTM ether. The cyclic thio derivative appears to be sterically sensitive (eq 2). It does not cleave benzyl, allyl, methyl, THP, TBDMS, and TBDPS ethers, whereas the related oxygen analog (2-bromo-1,3,2-benzodioxaborole) is a much more powerful reagent and cleaves the following groups in the order: MOMOR MEMOR > *t*-Boc > Cbz > *t*-BuOR > BnOR > allylOR > *t*-BuO_2CR secondary alkylOR > BnO_2CR > primary alkylOR alkylO_2CR. The *t*-butyldimethylsilyl group is stable to this reagent.

Protic acids such as ***Pyridinium p-Toluenesulfonate*** (*t*-BuOH, or 2-butanone, heat, 80–99% yield)[10] and ***Tetrafluoroboric Acid*** (CH_2Cl_2, 0 °C, 3 h, 50–60% yield)[11] can also be used to cleave the MEM ethers. The former also cleaves the MOM ether and has the advantage that it cleanly cleaves allylic ethers which can not be cleaved by Corey's original procedure. The MEM group is reasonably stable to CF_3CO_2H/CH_2Cl_2 (1:1), which is used to cleave Boc groups, and to 0.2 N HCl, but it is not stable to 2.0 N HCl or HBr–AcOH.[12] ***Iodotrimethylsilane*** has long been known as a powerful reagent for the cleavage of ethers, esters, carbamates, etc., and thus it is no surprise that it will cleave the MEM ether as well. It does tend to form iodides when cleaving allylic and benzylic ethers, but this is somewhat suppressed when the reagent is prepared in situ (Me_3SiCl, NaI, MeCN, −20 °C, 79%).[13] In a synthesis of tirandamycin, deprotection of a MEM ether was unsuccessful using conventional methods and an entirely new method was developed which oxidizes the methylene (*n*-BuLi, THF; then $Hg(OAc)_2$, H_2O, THF, 81% yield) (eq 3).[14]

Protection of Phenols. Methoxyethoxymethyl chloride can also be used to protect phenols. The conditions for its introduction are similar to those used for alcohols (NaH, THF, 0 °C; $MeOCH_2CH_2OCH_2Cl$, 0 °C → 25 °C, 2 h, 75% yield).[15,16] In contrast to the alcohol derivatives, phenolic MEM ethers can be cleaved with ***Trifluoroacetic Acid*** (CH_2Cl_2, 23 °C, 1 h, 74% yield).[17] 1 M HCl (THF, 5 h, 60 °C)[18] and HBr/EtOH[19] will also effect cleavage. In general, the cleavage conditions used for alcohols are also effective with the phenolic derivatives. During an examination of the asymmetric reduction of an acetophenone derivative with ***(+)-B-Chlorodiisopinocampheylborane***, it was found that a phenolic MEM ether was slowly cleaved (eq 4).[16]

Protection of Acids. Carboxylic acids are converted to MEM esters using conditions similar to those used for alcohol protection (MEMCl, *i*-Pr_2EtN, CH_2Cl_2, 0 °C, 2 h).[20] Deprotection of this ester is effected with 3 M HCl (THF, 40 °C, 12 h) or with $ZnBr_2$[21] or $MgBr_2$.[22]

Miscellaneous Uses. Outside of the protective group arena, MEMCl has been used to alkylate enolates[23] and aryllithium reagents in the presence of Ph_2TlBr.[24] MEM ethers have also proven to be a good one-carbon source for the preparation of isochromans (eq 5)[25] and other oxygen heterocycles (eq 6).[26]

Guaiacolmethyl chloride (GUMCl) (**1**) is a reagent similar to MEMCl, but it produces derivatives that are somewhat more acid sensitive. In fact the GUM group can be removed in the presence of a MEM group. It is a more sterically demanding reagent, and thus it is possible to introduce this group selectively onto a primary alcohol in the presence of a secondary alcohol. Because of its similarity to a *p*-methoxyphenyl group, it should be possible to remove it oxidatively. Cleavage is effected with $ZnBr_2$ in CH_2Cl_2 just as with the MEM group.[27]

Related Reagents. Benzyl Chloromethyl Ether; Benzyl Chloromethyl Sulfide; *t*-Butyl Chloromethyl Ether; 2-Chloroethyl Chloromethyl Ether; Chloromethyl Methyl Ether; Chloromethyl Methyl Sulfide; (*p*-Methoxybenzyloxy)methyl Chloride; 2-(Trimethylsilyl)ethoxymethyl Chloride.

1. Greene, T. W.; Wuts, P. G. M. *Protective Groups in Organic Synthesis*, 2nd ed.; Wiley: New York, 1991.
2. Corey, E. J.; Gras, J.-L.; Ulrich, P. *TL* **1976**, 809.
3. Sato, T.; Otera, J.; Nozaki, H. *JOC* **1990**, *55*, 4770.
4. Posner, G. H.; Haces, A.; Harrison, W.; Kinter, C. M. *JOC* **1987**, *52*, 4836.
5. Quindon, Y.; Morton, H. E.; Yoakim, C. *TL* **1983**, *24*, 3969.
6. Shibasaki, M.; Ishida, Y.; Okabe, N. *TL* **1985**, *26*, 2217.
7. Williams, D. R.; Sakdarat, S. *TL* **1983**, *24*, 3965.
8. Corey, E. J.; Hua, D. H.; Seitz, S. P. *TL* **1984**, *25*, 3.
9. Boeckman, R. K., Jr.; Potenza, J. C. *TL* **1985**, *26*, 1411.
10. Monti, H.; Léandri, G.; Klos-Ringuet, M.; Corriol, C. *SC* **1983**, *13*, 1021.
11. Ikota, N.; Ganem, B. *CC* **1978**, 869.
12. Vadolas, D.; Germann, H. P.; Thakur, S.; Keller, W.; Heidemann, E. *Int. J. Pept. Protein Res.* **1985**, *25*, 554.
13. Rigby, J. H.; Wilson, J. Z. *TL* **1984**, *25*, 1429.
14. Ireland, R. E.; Wuts, P. G. M.; Ernst, B. *JACS* **1981**, *103*, 3205.
15. Corey, E. J.; Danheiser, R. L.; Chandrasekaran, S.; Siret, P.; Keck, G. E.; Gras, J.-L. *JACS* **1978**, *100*, 8031.

16. Everhart, E. T.; Graig, J. C. *JCS(P1)* **1991**, 1701.
17. Mayrargue, J.; Essamkaoui, M.; Moskowitz, H. *TL* **1989**, *30*, 6867.
18. Brade, W.; Vasella, A. *HCA* **1989**, *72*, 1649.
19. Fujita, V.; Ishiguro, M.; Onishi, T.; Nishida, T. *S* **1981**, 469.
20. Meyers, A. I.; Reider, P. J. *JACS* **1979**, *101*, 2501.
21. Posner, G. H.; Weitzberg, M.; Hemal, T. G.; Asiruatham, E.; Cun-Heng, H. *T* **1986**, *42*, 2919.
22. Kim, S.; Park, Y. H.; Kee, I. S. *TL* **1991**, *32*, 3099.
23. Hlasta, D. J.; Casey, F. B.; Ferguson, E. W.; Gangell, S. J.; Heimann, M. R.; Jaeger, E. P.; Kullnig, R. K.; Gordon, R. J. *JMC* **1991**, *34*, 1560. Topgi, R. S. *JOC* **1989**, *54*, 6125. Laredo, G. C.; Maldonado, L. A. *H* **1987**, *25*, 179. Onaka, M.; Matsuoka, Y.; Mukaiyama, T. *CL* **1981**, 531.
24. Marko, I. E.; Kantam, M. L. *TL* **1991**, *32*, 2255.
25. Mohler, D. L.; Thompson, D. W. *TL* **1987**, *28*, 2567.
26. Blumenkopf, T. A.; Look, G. C.; Overman, L. E. *JACS* **1990**, *112*, 4399.
27. Loubinoux, B.; Coudert, G.; Guillaumet, G. *TL* **1981**, *22*, 1973.

Peter G. M. Wuts
The Upjohn Co., Kalamazoo, MI, USA

2-Methoxypropene[1]

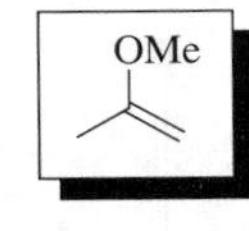

[116-11-0] C_4H_8O (MW 72.12)

(electron-rich alkene used in the monoprotection of aliphatic,[2] allylic,[3] and propargylic[4] alcohols, often followed by Claisen[3b–e] or other[2a,4c] rearrangements; protection of peroxides,[5] cyanohydrins,[6] α-hydroxy ketones,[2b] and phenols;[7] condensation with amines;[8] protection of 1,2-[9,10] and 1,3-diols,[10,11] sometimes followed by Prins[9f,g] rearrangements; protection of 1,2-dithiols[10a] and α-hydroxy carbamates;[12] participant in pericyclic reactions;[13] formation of 2-methoxyallyl halides[14] and substituted furans[4b])

Alternate Names: 2-MP; isopropenyl methyl ether.
Physical Data: bp 34–36 °C; n_D^{20} 1.3820; *d* 0.753 g cm^{-3}; flash point −29 °C.
Form Supplied in: colorless liquid; commercially available.
Preparative Method: prepared in high yield from ***Succinic Anhydride***, ***2,2-Dimethoxypropane***, benzoic acid, and ***Pyridine***.[15]
Handling, Storage, and Precautions: flammable liquid; light-sensitive; should be refrigerated.

Monoprotection of Alcohols. 2-Methoxypropene is used as a protective group for aliphatic,[2,11d] allylic,[3] and propargylic[4] alcohols, masking them as their mixed acetals (eq 1). Deprotection can be accomplished by stirring in MeOH over ion exchange resin,[4a] by reaction in methanol with catalytic ***Acetyl Chloride***,[11d] by ***Potassium Carbonate*** in methanol,[2c] or by 20% ***Acetic Acid***.[3a] A general advantage of this acetal, which undergoes hydrolysis at approximately 10^3 times the rate of THP,[2d,3a] is that it does not confer an additional diastereomeric center to the protected substrate. Access to allyl vinyl ethers for subsequent Claisen rearrangements[3b–e] is illustrated in eq 2.

(1) [OMe-propene; $POCl_3$ (cat.), 85%]

(2) [1. 2-MP; 2. $EtCO_2H$ (cat.); 180–200 °C; 81%; *p*-Tol]

A useful entry into β-keto allenes is provided by the reaction of 2-MP with tertiary propargylic alcohols.[4c] Base-catalyzed rearrangement of the allenes affords conjugated dienones (eq 3).

(3) [2-MP, TsOH (cat.), 95%; NaOH (cat.), 81%; 86% *trans*, 14% *cis*]

Stable allylic peroxyacetals have been prepared by reacting 2-MP with hydroperoxides (eq 4).[5] Organomercury functionality is tolerated in this reaction.[5b] Cyanohydrins,[6] α-hydroxy ketones,[2b] and phenols[7] are similarly protected.

(4) [2-MP, PPTS, 90–100%]

Condensation with Amines. Imine formation with loss of methanol occurs under acidic conditions, as illustrated in eq 5.[8]

(5) [2-MP, PPTS, toluene; 82%]

Protection of 1,2- and 1,3-Diols and 1,3-Dithiols. When another hydroxy group is in close proximity to the initially protected alcohol functionality, acetonide formation occurs.[9–11] The kinetic product is preferentially formed.[10b,c,11c] As anomeric hydroxy groups do not usually[11] participate, this reagent provides complementary selectivity to the more traditional acid/acetone acetonation protocol. Control of the amount of 2-MP can result in varying degrees of protection (eq 6).[10b]

2-MP, H^+; 2-MP, H^+ (6)

Acetonide formation using ***n-Butyltrichlorostannane*** with 2-MP[10a] is quite general for 1,2- and 1,3-diols, as well as for 1,3-dithiols. Monoacetonide formation occurs with 1,2,3- and 1,2,4-triols. Yields of the acetonides (e.g. **1**–**5**) show improvement over conventional procedures.[10a]

(**1**) 83% (**2**) 94% (**3**) 77% (**4**) 82% (**5**) 98%

Protection of α-Hydroxy Carbamates. Aminal formation from α-hydroxy carbamates also occurs using 2-MP, as illustrated in eq 7.[12]

Boc–NH OH; 2-MP, TsOH (cat.), CH_2Cl_2, 50%; BocN (7)

Participation in Pericyclic Reactions. 2-MP is an active participant in pericyclic reactions. A few examples include a substitution-dependent ene reaction with an alkynyl Fischer carbene complex (eq 8),[13c] a [2+3] reaction with 2-amino-1,4-naphthoquinone (eq 9),[13f] a solvent-dependent photochemical cycloaddition with *o*-naphthoquinones,[13j] and reaction with 2-nitro-1,3-dienes to provide α,β-unsaturated 1,4-diones (eq 10).[13g]

$(CO)_5M$, OMe, R + H, OMe ⟶ $M(CO)_5$, MeO, H, R, OMe + $M(CO)_5$, MeO, OMe, R (8)

M = W, Cr	R = TMS	30%	<2%
	R = Me	<2%	63%

NH_2 + OMe; $h\nu$, benzene, 72%; H N (9)

NO_2; DMAP, 2-MP, 38%; ^-O, N^+, O, OMe; HCl, H_2O, 95% (10)

Reaction with *N*-Halosuccinimides. 2-Methoxyallyl and vinyl halides are formed by reaction of 2-MP with *N*-halosuccinimides.[14] Also, treatment of propargylic alcohols with 2-MP and ***N-Bromosuccinimide*** results in the formation of bromo acetals, which are subsequently cyclized to furans in good yield (eq 11).[4b]

Bu, Ar, OH; 1. 2-MP 2. NBS, 90–97%; Bu, Br, Ar, O, OMe; 1. Bu_3SnCl, $NaCNBH_3$, AIBN 2. TsOH (cat.), 65–70%; Bu, Ar, O (11)

Related Reagents. Benzyl Isopropenyl Ether; 3,4-Dihydro-2*H*-pyran; 2,2-Dimethoxypropane; Ethyl Vinyl Ether.

---

1. *FF* **1969**, *2*, 230; **1975**, *5*, 360; **1981**, *9*, 304; **1984**, *11*, 329; **1986**, *12*, 291.
2. (a) Kang, S. H.; Hwang, T. S.; Kim, W. J.; Lim, J. K. *TL* **1991**, *32*, 4015. (b) Wilson, T. M.; Kocienski, P.; Jarowicki, K.; Isaac, K.; Faller, A.; Campbell, S. F.; Bordner, J. *T* **1990**, *46*, 1757. (c) Tietze, L. F.; Lögers, M. *LA* **1990**, 261. (d) Reese, C. B.; Saffhill, R.; Sulston, J. E. *JACS* **1967**, *89*, 3366.
3. (a) Kluge, A. F.; Untch, K. G.; Fried, J. H. *JACS* **1972**, *94*, 7827. (b) Srikrishna, A.; Krishnan, K.; Vankateswarlu, S. *CC* **1993**, 143. (c) Srikrishna, A.; Krishnan, K. *IJC(B)* **1990**, *29B*, 879. (d) Gajewski, J. J.; Gee, K. R.; Jurayj, J. *JOC* **1990**, *55*, 1813. (e) Saucy, G.; Marbet, R. *HCA* **1967**, *50*, 2091.
4. (a) Lampilas, M.; Lett, R. *TL* **1992**, *33*, 773. (b) Srikrishna, A.; Sundarababu, G. *T* **1990**, *46*, 7901. (c) Saucy, G.; Marbet, R. *HCA* **1967**, *50*, 1158.
5. (a) Dussault, P.; Sahli, A.; Westermeyer, T. *JOC* **1993**, *58*, 5469. (b) Bloodworth, A. J.; Cooksey, C. J.; Korkodilos, D. *CC* **1992**, 926. (c) Dussault, P.; Sahli, A. *TL* **1990**, *31*, 5117.
6. Zanderbergen, P.; van den Nieuwendijk, A. M. C. H.; Brussee, J.; van der Gen, A. *T* **1992**, *48*, 3977.
7. Gibson, C. P.; Bem, D. S.; Falloon, S. B.; Hitchens, T. K.; Cortopassi, J. E. *OM* **1992**, *11*, 1742.
8. Freeman, F.; Kim, D. S. H. L. *JOC* **1992**, *57*, 550.

9. (a) Kozikowski, A. P.; Ognyanov, V. I.; Fauq, A. H.; Nahorski, S. R.; Wilcox, R. A. *JACS* **1993**, *115*, 4429. (b) van der Klein, P. A. M.; Filemon, W.; Boons, G. J. P. H.; Veeneman, G. H.; van der Marel, G. A.; van Boom, J. H. *T* **1992**, *48*, 4649. (c) Barbat, J.; Gelas, J.; Horton, D. *Carbohydr. Res.* **1991**, *219*, 115. (d) Shing, T. K. M.; Tang, Y. *T* **1991**, *47*, 4571. (e) Fujisawa, T.; Takemura, I.; Ukaji, Y. *TL* **1990**, *31*, 5479. (f) Hopkins, M. H.; Overman, L. E.; Rishton, G. M. *JACS* **1991**, *113*, 5354. (g) Brown, M. J.; Harrison, T.; Herrington, P. M.; Hopkins, M. H.; Hutchinson, K. D.; Mishra, P.; Overman, L. E. *JACS* **1991**, *113*, 5365.
10. (a) Carofiglio, T.; Marton, D.; Stivanello, D.; Tagliavini, G. *Main Group Chem.* **1992**, *15*, 247. (b) Fanton, E.; Gelas, J.; Horton, D. *JOC* **1981**, *46*, 4057. (c) Gelas, J.; Horton, D. *H* **1981**, *16*, 1587.
11. (a) Guanti, G.; Banfi, L.; Narisano, E.; Thea, S. *SL* **1992**, 311. (b) Srikrishna, A.; Krishnan, K.; Yelamaggad, C. V. *T* **1992**, *48*, 9725. (c) Jung, M. E.; Clevenger, G. L. *TL* **1991**, *32*, 6089. (d) Thomé, M. A.; Giudicelli, M. B.; Picq, D.; Anker, D. *J. Carbohydr. Chem.* **1991**, *10*, 923.
12. Melnick, M. J.; Bisaha, S. N.; Gammill, R. B. *TL* **1990**, *31*, 961.
13. (a) Deaton, M. V.; Ciufolini, M. A. *TL* **1993**, *34*, 2409. (b) Beak, P.; Song, Z.; Resek, J. E. *JOC* **1992**, *57*, 944. (c) Faron, K. L.; Wulff, W. D. *JACS* **1990**, *112*, 6419. (d) Kobayashi, K.; Suzuki, M.; Suginome, H. *JOC* **1992**, *57*, 599. (e) Fannes, C.; Meerpoel, L.; Toppet, S.; Hoornaert, G. *S* **1992**, 705. (f) Kobayashi, K.; Takeuchi, H.; Seko, S.; Suginome, H. *HCA* **1991**, *74*, 1091. (g) Bäckvall, J.-E.; Karlsson, U.; Chinchilla, R. *TL* **1991**, *32*, 5607. (h) Coleman, R. S.; Grant, E. B. *TL* **1990**, *31*, 3677. (i) Gupta, R. B.; Franck, R. W. *SL* **1990**, 355. (j) Takuwa, A.; Sumikawa, M. *CL* **1989**, 9.
14. (a) Jacobson, R. M.; Raths, R. A.; McDonald, J. H., III *JOC* **1977**, *42*, 2545. (b) Greenwood, G.; Hoffmann, H. M. R. *JOC* **1972**, *37*, 611.
15. Newman, M. S.; Vander Zwan, M. C. *JOC* **1973**, *38*, 2910.

K. Sinclair Whitaker
Wayne State University, Detroit, MI, USA

D. Todd Whitaker
Detroit County Day School, Beverly Hills, MI, USA

Methylaluminum Dichloride[1]

$MeAlCl_2$

[917-65-7] CH_3AlCl_2 (MW 112.92)

(strong Lewis acid that can also act as a proton scavenger; reacts with HX to give methane and $AlCl_2X$)

Alternate Name: dichloro(methyl)aluminum.
Physical Data: mp 72.7 °C; bp 94–95 °C/100 mmHg.
Solubility: sol most organic solvents; stable in alkanes or arenes.
Form Supplied in: commercially available as solutions in hexane or toluene.
Analysis of Reagent Purity: solutions are reasonably stable but may be titrated before use by one of the standard methods.[1b]
Handling, Storage, and Precautions: must be transferred under inert gas (Ar or N_2) to exclude oxygen and water. Use in a fume hood.

Introduction. The general properties of alkylaluminum halides as Lewis acids are discussed in the entry for ***Ethylaluminum Dichloride***. $MeAlCl_2$ is used less frequently than $EtAlCl_2$, since $EtAlCl_2$ is much cheaper and comparable results are usually obtained. In some cases, use of $MeAlCl_2$ is preferable, since the methyl group of $MeAlCl_2$ is less nucleophilic than the ethyl group of $EtAlCl_2$ and $EtAlCl_2$ can transfer a hydride.

Catalysis of Diels–Alder Reactions. $MeAlCl_2$ has been used as a catalyst for Diels–Alder[2–4] and retro-Diels–Alder reactions.[5] $MeAlCl_2$ is the catalyst of choice for intramolecular Diels–Alder reactions of furans (eq 1) due to the ease of handling and reaction workup.[2]

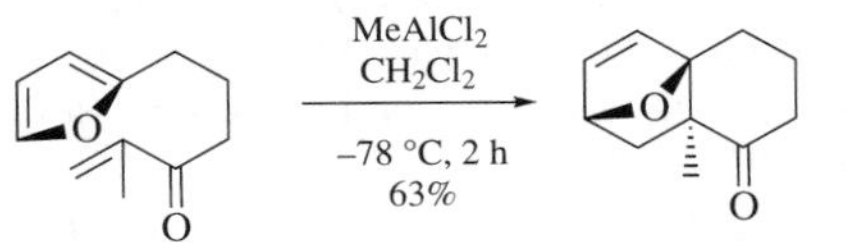

$MeAlCl_2$, CH_2Cl_2, −78 °C, 2 h, 63% (1)

Catalysis of Ene Reactions. $MeAlCl_2$ has been used as a catalyst for ene reactions of trifluoroacetaldehyde[6] and for intramolecular ene reactions of trifluoromethyl ketones (eq 2).[7] Use of $EtAlCl_2$ leads to the ene adduct in lower yield accompanied by 25% of the saturated analog resulting from hydride delivery to the zwitterion.

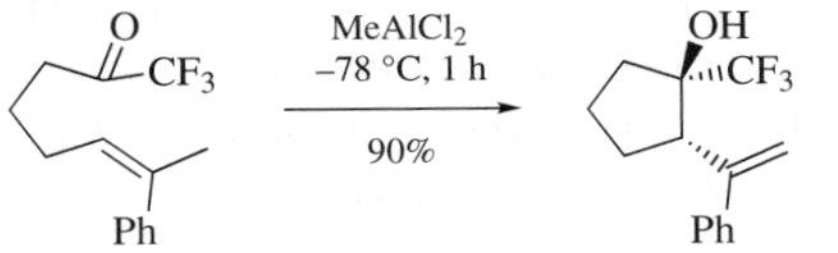

$MeAlCl_2$, −78 °C, 1 h, 90% (2)

Generation of Electrophilic Cations. 1:2 Aldehyde– or ketone–$MeAlCl_2$ complexes add intramolecularly to alkenes to give zwitterions that undergo 1,2-hydride and alkyl shifts to regenerate a ketone (eqs 3 and 4).[8–11] $EtAlCl_2$ can transfer a hydride to the zwitterion to give the reduced product (eq 3).[8]

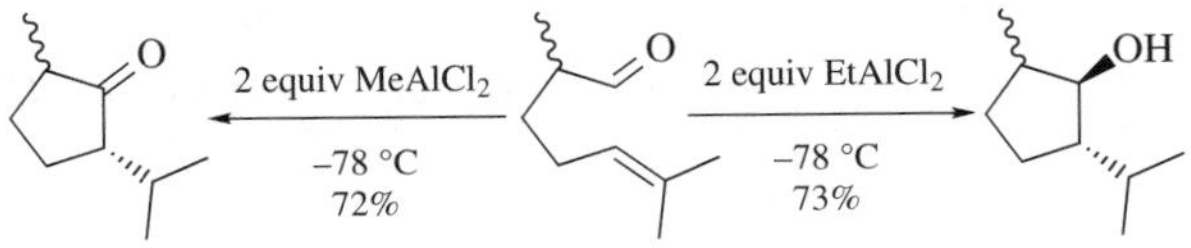

2 equiv $MeAlCl_2$, −78 °C, 72%; 2 equiv $EtAlCl_2$, −78 °C, 73% (3)

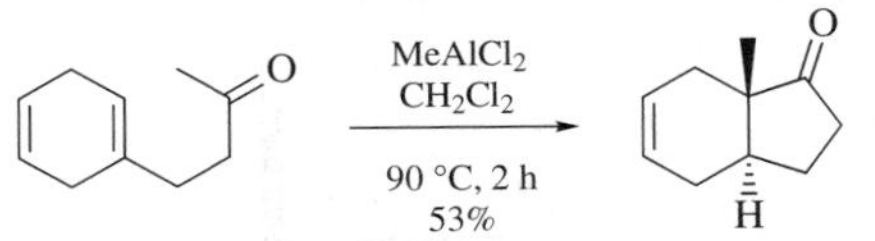

$MeAlCl_2$, CH_2Cl_2, 90 °C, 2 h, 53% (4)

$MeAlCl_2$ has been used as the catalyst for epoxide-initiated cation–alkene cyclizations.[12] A tertiary alcohol has been converted to a hydrazide by reaction with $MeAlCl_2$ in the presence of ***Mesitylenesulfonylhydrazide***.[13] α-Trimethylsilyl enones can be prepared by isomerization of 1-(trimethylsilyl)-2-propynyl trimethylsilyl ethers with $MeAlCl_2$.[14]

Nucleophilic Additions. $MeAlCl_2$ has occasionally been used to introduce a methyl group. γ-Lactones react with $MeAlCl_2$ to give carboxylic acids with a methyl group at the γ-position.[15] The reagent prepared from $MeAlCl_2$ and ***Dichlorobis(cyclopentadienyl)titanium*** adds to trimethylsilylalkynes in a *syn* fashion.[16] Optimal stereoselectivity is obtained using $MeAlCl_2$ to transfer a methyl group to a formyl amide (eq 5).[17]

$$\text{(5)}$$

MeAlCl$_2$, CH$_2$Cl$_2$, 0 °C, 2 h, 66%; ds = 97:3

Related Reagents. Diethylaluminum Chloride; Dimethylaluminum Chloride; Ethylaluminum Dichloride.

1. (a) For reviews, see ref. 1 in ***Ethylaluminum Dichloride***. (b) *Aluminum Alkyls*; Stauffer Chemical Co.: Westport, CT, 1976.
2. (a) Rogers, C.; Keay, B. A. *TL* **1991**, *32*, 6477. (b) Rogers, C.; Keay, B. A. *CJC* **1992**, *70*, 2929.
3. Boeckman, R. K., Jr.; Nelson, S. G.; Gaul, M. D. *JACS* **1992**, *114*, 2258.
4. Trost, B. M.; Lautens, M.; Hung, M. H.; Carmichael, C. S. *JACS* **1984**, *106*, 7641.
5. Grieco, P. A.; Abood, N. *JOC* **1989**, *54*, 6008.
6. Ogawa, K.; Nagai, T.; Nonomura, M.; Takagi, T.; Koyama, M.; Ando, A.; Miki, T.; Kumadaki, I. *CPB* **1991**, *39*, 1707.
7. Abouadellah, A.; Aubert, C.; Bégué, J.-P.; Bonnet-Delpon, D.; Guilhem, J. *JCS(P1)* **1991**, 1397.
8. Snider, B. B.; Karras, M.; Price, R. T.; Rodini, D. J. *JOC* **1982**, *47*, 4538.
9. Snider, B. B.; Rodini, D. J.; van Straten, J. *JACS* **1980**, *102*, 5872.
10. Snider, B. B.; Kirk, T. C. *JACS* **1983**, *105*, 2364.
11. Snider, B. B.; Cartaya-Marin, C. P. *JOC* **1984**, *49*, 153.
12. Corey, E. J.; Sodeoka, M. *TL* **1991**, *32*, 7005.
13. Wood, J. L.; Porco, J. A., Jr.; Taunton, J.; Lee, A. Y.; Clardy, J.; Schreiber, S. L. *JACS* **1992**, *114*, 5898.
14. Enda, J.; Kuwajima, I. *CC* **1984**, 1589.
15. Reinheckel, H.; Sonnek, G.; Falk, F. *JPR* **1974**, *316*, 215.
16. Eisch, J. J.; Piotrowski, A. M.; Brownstein, S. K.; Gabe, E. J.; Lee, F. L. *JACS* **1985**, *107*, 7219.
17. Fujii, H.; Taniguchi, M.; Oshima, K.; Utimoto, K. *TL* **1992**, *33*, 4579.

Barry B. Snider
Brandeis University, Waltham, MA, USA

Methyl Chloroformate

[79-22-1] $C_2H_3ClO_2$ (MW 94.50)

(protecting agent for many functional groups;[1–16] activates *N*-heteroaromatic rings and carboxylic acids toward nucleophilic attack;[17–26] methoxycarbonylating agent for organometallic reagents and enolates[27–39])

Alternate Name: MCF.
Physical Data: bp 70–72 °C; d (20 °C) 1.223 g cm^{-3}; n_D^{20} 1.3870.
Solubility: slightly sol water with slow decomposition; miscible with alcohols, benzene, chloroform, ether, etc.
Form Supplied in: colorless liquid; widely available.
Handling, Storage, and Precautions: flammable liquid; vesicant; its vapors are highly toxic, lachrymatory, and strongly irritating to the eyes. It should be handled with caution in a fume hood using protective gloves.

Functional Group Protection. The protection of alcohols[1] and phenols[2] as their methyl carbonates can be achieved using methyl chloroformate under basic conditions (eq 1). Primary and secondary alkylamines are readily protected as their methyl carbamates using MCF, typically in the presence of Et_3N,[3,4] Na_2CO_3,[5] K_2CO_3,[6] or $NaHCO_3$.[7] Selective *N*-protection of amino alcohols can be achieved under these conditions (eq 2).[5]

MCF, PhNEt$_2$, py, benzene, >98% (1)

MCF, Na$_2$CO$_3$, THF, rt, 2 h, 96% (2)

Methyl carbonates are more resistant to basic hydrolysis than simple esters; however, both alcohol-[1c] and phenol-derived[2] methyl carbonates can be selectively hydrolyzed under mildly basic conditions in the presence of a methyl carbamate derivative, allowing for the overall selective *N*-protection of amino alcohols and amino phenols (eq 3).

1. MCF, K$_2$CO$_3$, acetone, 60 °C, 12 h; 2. 4% NaOH, MeOH, rt, 2 h; 71% (3)

MCF is also useful for the *N*-protection of α-amino acids (Schotten–Baumann conditions),[8] primary arylamines (using pyridine as base),[1b,9] amides (using ***Triethylamine/4-Dimethylaminopyridine*** as base),[10] and pyrrole and indole ring nitrogens (using NaH or KH as base).[11] *N*-Silylpyrroles can be converted to the corresponding methyl carbamates under milder conditions than are required for the direct *N*-protection of pyrroles.[12] *ortho*-Phenylenediamines are readily converted to the corresponding benzimidazole-2-carbamates using MCF in the presence of MeSCH(NH$_2$)=NH.[13] The forcing conditions that are often required for removal of a methyl carbamate protecting group (e.g. ***Hydrazine***, KOH,[14] or aq Ba(OH)$_2$)[15] have limited the utility of this *N*-protecting group when compared with other carbamates; although milder *N*-deprotection conditions have been reported (***Iodotrimethylsilane***, $CHCl_3$, 50 °C; then MeOH),[16] this chemistry has not seen widespread use. ***Lithium Aluminum Hydride*** reduction of methyl carbamates affords the corresponding *N*-methylamines,[3,4] providing an attractive method for the overall *N*-methylation of primary and secondary amines (eq 4).

Activation of *N*-Heteroaromatic Rings Toward Nucleophilic Attack. Pyridine, quinoline, and isoquinoline rings are activated by MCF toward in situ nucleophilic attack, leading to the formation of 1,2- and/or 1,4-dihydropyridines, -quinolines, or -isoquinolines. The addition of alkyl Grignard reagents proceeds with variable and often poor regioselectivity.[17] However, alkenyl- and alkynyl-Grignard reagents (eq 5),[17b] unhindered allylic[18] and allenylic[19] stannanes, cyanide ion,[20] and ***Sodium Borohydride*** (in MeOH at $-70\,^{\circ}C$[21]) all add with high 1,2-selectivity. Highly 1,4-selective addition reactions are observed using alkylcopper reagents (eq 6),[17a,22] dialkylcuprates,[17b,23] silyl enol ethers,[22] and benzylic stannanes.[24] The 1,4-addition of aryllithium reagents,[25] and methyllithium and methylmagnesium bromide,[26] to chiral 3-oxazolinylpyridines proceeds with high regio- and diastereofacial selectivity (eq 7). Removal of the chiral auxiliary affords chiral 4-substituted dihydropyridines in high ee.

Methoxycarbonylation of Organometallic Reagents and Enolates. Deprotonation of terminal alkynes (typically using *n*-BuLi,[27] LDA,[28] or EtMgBr[29]), followed by trapping the resulting alkynyllithium or -Grignard reagent with MCF, offers an excellent route to methyl alkynoates (eq 8) (see ***2,2,2-Trichloroethyl Chloroformate*** if subsequent cleavage to the carboxylic acid is desired). A related approach to methyl alkynoates proceeds via homologation of an aldehyde to the corresponding α,α-dibromoalkene. Dehydrobromination and halogen–metal exchange followed by trapping the resultant alkynyllithium with MCF provides the methyl alkynoate in high overall yield (eq 9).[30] Aryllithium[31] and alkenylaluminium[32] species are also effectively methoxycarbonylated by MCF, but the trapping of simple primary alkyllithiums with MCF appears to be of little synthetic utility.[33] The in situ trapping of a simple phosphonium ylide with MCF followed by deprotonation provides a convenient synthetic approach to α-methoxycarbonylphosphonium ylides (eq 10).[34]

The *C*-methoxycarbonylation of lithium ester[35] and amide[36] enolates can be effectively accomplished using MCF. Addition of *n*-BuLi to a chiral 1-oxazolinylnaphthalene followed by trapping the resulting azaenolate with MCF affords the *C*-acylation product with high diastereoselectivity (eq 11).[37] In contrast, reaction of copper[38] and potassium[39] ketone enolates with MCF typically results in *O*-acylation to afford the corresponding enol carbonates (eq 12). The weak nucleophilicity of enol carbonates (they are stable to peroxy acids and ***Ozone***)[38a] makes them useful masked enolate equivalents.

Activation of Carboxylic Acids as Mixed Anhydrides. Activation of *N*-(alkoxycarbonyl)-α-amino acids as mixed anhydrides using MCF/*N*-methylpiperidine in CH_2Cl_2 minimizes urethane byproduct formation and racemization during peptide bond formation.[40] Reaction of these mixed anhydrides with ***N,O-Dimethylhydroxylamine*** followed by $LiAlH_4$ reduction provides an attractive route to the corresponding *N*-(alkoxycarbonyl)-α-amino aldehydes (eq 13).[41] Attempts to generate mixed anhydrides from *N*-acyl-α-amino acids using MCF result in cyclization to the corresponding 5(4*H*)-oxazolones.[42,43]

$$t\text{-BuO}_2\text{C-NH-CH}(\text{CH}_2\text{-}i\text{-Pr})\text{-CO}_2\text{H} \xrightarrow[\substack{2.\ \text{MeONHMe},\ -12\ °\text{C to rt, 4 h} \\ 3.\ \text{LiAlH}_4,\ \text{ether, rt, 1 h} \\ 81\%}]{1.\ \text{MCF, N-methylmorpholine, CH}_2\text{Cl}_2,\ -12\ °\text{C}} t\text{-BuO}_2\text{C-NH-CH}(\text{CH}_2\text{-}i\text{-Pr})\text{-CHO} \quad (13)$$

α-Diazo ketones may be conveniently prepared by reaction of a mixed anhydride with ***Diazomethane***; the mixed anhydrides are typically prepared either by treatment of the parent carboxylic acid with MCF/***Triethylamine***[44] or by treatment of the corresponding methyl ester with potassium silyloxide and then MCF.[45] Attack by amines[46] or *N*-silylamines[47] on such mixed anhydrides provides a useful approach to amide synthesis (eq 14), while acyl azides are prepared in excellent yield by reaction of the mixed anhydride with ***Sodium Azide***.[48]

(14) [Scheme: 5-bromo-4,8-dimethoxyquinoline-2-carboxylic acid + tryptophan-derived amine ester (H_2N, MeO_2C, indole NH); MCF, Et_3N, 0 °C to rt, 4 h, 97% → corresponding amide]

While the mixed anhydrides derived from the reaction of carboxylic acids with methyl chloroformate using Et_3N as base are relatively stable, the addition of catalytic amounts of ***4-Dimethylaminopyridine*** or *N*-methylmorpholine results in facile conversion to the corresponding methyl esters under very mild reaction conditions that are compatible with sensitive functionality.[49]

Related Reagents. Allyl Chloroformate; Benzyl Chloroformate; 4-Bromobenzyl Chloroformate; *t*-Butyl Chloroacetate; Ethyl Chloroformate; 9-Fluorenylmethyl Chloroformate; Isobutyl Chloroformate; 2,2,2-Tribromoethyl Chloroformate; 2,2,2-Trichloro-*t*-butoxycarbonyl Chloride; 2,2,2-Trichloroethyl Chloroformate; 2-(Trimethylsilyl)ethyl Chloroformate; Vinyl Chloroformate.

---

1. (a) Eren, D.; Keinan, E. *JACS* **1988**, *110*, 4356. (b) Trost, B. M.; Kuo, G. H.; Benneche, T. *JACS* **1988**, *110*, 621. (c) Fang, J.-M.; Cherng, Y.-J. *JCR(M)* **1986**, 1568.
2. Schwartz, M. A.; Pham, P. T. K. *JOC* **1988**, *53*, 2318.
3. Yamada, F.; Hasegawa, T.; Wakita, M.; Sugiyama, M.; Somei, M. *H* **1986**, *24*, 1223.
4. Somei, M.; Yamada, F.; Makita, Y. *H* **1987**, *26*, 895.
5. Knapp, S.; Sebastian, M. J.; Ramanathan, H.; Bharadwaj, P.; Potenza, J. A. *T* **1986**, *42*, 3405.
6. Corey, E. J. et al. *TL* **1978**, 1051.
7. Patjens, J.; Ghaffari-Tabrizi, R.; Margaretha, P. *HCA* **1986**, *69*, 905.
8. Itaya, T.; Mizutani, A.; Watanabe, N. *CPB* **1989**, *37*, 1221.
9. Takai, H. et al. *CPB* **1985**, *33*, 1129.
10. Kubo, A.; Saito, N.; Yamato, H.; Masubuchi, K.; Nakamura, M. *JOC* **1988**, *53*, 4295.
11. Somei, M.; Saida, Y.; Komura, N. *CPB* **1986**, *34*, 4116.
12. Keijsers, J.; Hams, B.; Kruse, C.; Scheeren, H. *H* **1989**, *29*, 79.
13. Akhtar, M. S.; Seth, M.; Bhaduri, A. P. *IJC(B)* **1986**, *25B*, 395.
14. Shono, T.; Matsumura, Y.; Uchida, K.; Tsubata, K.; Makino, A. *JOC* **1984**, *49*, 300.
15. Wovkulich, P. M.; Uskokovic, M. R. *T* **1985**, *41*, 3455.
16. (a) Jung, M. E.; Lyster, M. A. *CC* **1978**, 315. (b) Wender, P. A.; Schaus, J. M.; White, A. W. *JACS* **1980**, *102*, 6157.
17. (a) Gosmini, R.; Mangeney, P.; Alexakis, A.; Commercon, M.; Normant, J. F. *SL* **1991**, 111. (b) Yamaguchi, R.; Nakazono, Y.; Matsuki, T.; Hata, E.-i.; Kawanisi, M. *BCJ* **1987**, *60*, 215.
18. Yamaguchi, R.; Moriyasu, M.; Yoshioka, M.; Kawanisi, M. *JOC* **1985**, *50*, 287; Yamaguchi, R.; Moriyasu, M.; Yoshioka, M.; Kawanisi, M. *JOC* **1988**, *53*, 3507.
19. Yamaguchi, R.; Moriyasu, M.; Takase, I.; Kawanisi, M.; Kozima, S. *CL* **1987**, 1519.
20. Uff, B. C. et al. *JCR(S)* **1986**, 206.
21. Fowler, F. W. *JOC* **1972**, *37*, 1321.
22. Akiba, K.; Ohtani, A.; Yamamoto, Y. *JOC* **1986**, *51*, 5328.
23. Piers, E.; Soucy, M. *CJC* **1974**, *52*, 3563.
24. Yamaguchi, R.; Moriyasu, M.; Kawanisi, M. *TL* **1986**, *27*, 211.
25. Meyers, A. I.; Oppenlaender, T. *CC* **1986**, 920.
26. (a) Meyers, A. I.; Oppenlaender, T. *JACS* **1986**, *108*, 1989. (b) Meyers, A. I.; Natale, N. R.; Wettlaufer, D. G. *TL* **1981**, *22*, 5123.
27. Mori, K.; Fujiwhara, M. *T* **1988**, *44*, 343.
28. Marshall, J. A.; Andrews, R. C.; Lebioda, L. *JOC* **1987**, *52*, 2378.
29. Earl, R. A.; Townsend, L. B. *OS* **1981**, *60*, 81.
30. Boeckman, R. K., Jr.; Perni, R. B. *JOC* **1986**, *51*, 5486.
31. Uemora, M.; Take, K.; Isobe, K.; Minami, T.; Hayashi, Y. *T* **1985**, *41*, 5771.
32. Zwiefel, G.; Lynd, R. A. *S* **1976**, 625.
33. Rucker, C. *JOM* **1986**, *310*, 135.
34. Marshall, J. A.; DeHoff, B. S.; Cleary, D. G. *JOC* **1986**, *51*, 1735.
35. Hersloef, M.; Gronowitz, S. *CS* **1985**, *25*, 257.
36. Ackermann, J.; Matthes, M.; Tamm, C. *HCA* **1990**, *73*, 122.
37. Meyers, A. I.; Roth, G. P.; Hoyer, D.; Barner, B. A.; Laucher, D. *JACS* **1988**, *110*, 4611.
38. (a) Danishefsky, S.; Kahn, M.; Silvestri, M. *TL* **1982**, *23*, 703; however, see also: (b) Trost, B. M.; Hiemstra, H. *T* **1986**, *42*, 3323.
39. Earley, W. G.; Jacobsen, E. J.; Meier, G. P.; Oh, T.; Overman, L. E. *TL* **1988**, *29*, 3781.
40. Chen, F. M. F.; Steinauer, R.; Benoiton, N. L. *JOC* **1983**, *48*, 2939.
41. Goel, O. P.; Krolls, U.; Stier, M.; Kesten, S. *OS* **1989**, *67*, 69.
42. Chen, F. M. F.; Benoiton, N. L. *Int. J. Pept. Protein Res.* **1988**, *31*, 396.
43. Chen, F. M. F.; Slebioda, M.; Benoiton, N. L. *Int. J. Pept. Protein Res.* **1988**, *31*, 339.
44. Padwa, A.; Fryxell, G. E.; Zhi, L. *JACS* **1990**, *112*, 3100.
45. Padwa, A.; Krumpe, K. E.; Kassir, J. M. *JOC* **1992**, *57*, 4940.
46. Rao, A. V. R.; Chavan, S. P.; Sivadasan, L. *T* **1986**, *42*, 5065.
47. Balogh, D. W.; Patterson, L. E.; Wheeler, W. J. *SC* **1988**, *18*, 307.
48. Tius, M. A.; Thurkauf, A. *TL* **1986**, *27*, 4541.
49. (a) Kim, S.; Kim, Y. C.; Lee, J. I. *TL* **1983**, *24*, 3365. (b) Burke, S. D.; Pacofsky, G. J.; Piscopio, A. D. *TL* **1986**, *27*, 3345. (c) Davis, M.; Wu, W.-Y. *AJC* **1987**, *40*, 223.

Paul Sampson
Kent State University, OH, USA

Methyl Cyanoformate[1]

NC–C(=O)–OR

(**1**; R = Me)
[17640-15-2] $C_3H_3NO_2$ (MW 85.07)
(**2**; R = Et)
[623-49-4] $C_4H_5NO_2$ (MW 99.10)
(**3**; R = $PhCH_2$)
[5532-86-5] $C_9H_7NO_2$ (MW 161.17)

(agent for the regioselective methoxycarbonylation of carbanions;[1,2] reacts with organocadmium reagents to form α-keto esters;[3] may function as a dienophile,[4] dipolarophile,[5] or radical cyanating agent[6])

Physical Data: (**1**) mp 26 °C; bp 100–101 °C; *d* 1.072 g cm^{-3}. (**2**) bp 115–116 °C; *d* 1.003 g cm^{-3}. (**3**) bp 66–67 °C/0.6 mmHg; *d* 1.105 g cm^{-3}.
Solubility: sol all common organic solvents; dec by H_2O, alcohols, amines.
Form Supplied in: colorless liquid; methyl cyanoformate, as well as the ethyl and benzyl analogs, is available commercially.
Preparative Methods: small quantities of cyanoformate esters (up to 30 g) may be conveniently prepared from alkyl chloroformates by procedures employing phase-transfer catalysis with either ***18-Crown-6***[7] or ***Tetra-n-butylammonium Bromide***,[8] but several workers have found the products to be unsatisfactory when prepared on a larger scale.
Handling, Storage, and Precautions: store over 4Å molecular sieves; highly toxic; flammable; use in a fume hood.

Regioselective Methoxycarbonylation of Ketones. Methyl cyanoformate gives generally excellent results in the regiocontrolled synthesis of β-keto esters by the *C*-acylation of preformed lithium enolates (eq 1)[1,2] and is normally superior to the more traditional acylating agents such as acyl halides, anhydrides,[9] and CO_2,[10] partly because these reagents afford variable amounts of *O*-acylated products.[11] The enolates may be generated in a variety of ways, including direct enolization of ketones with suitable bases (eq 2),[2] liberation from silyl enol ethers and acetates (eq 3),[12] conjugate additions of cuprates to α,β-unsaturated ketones[13] (eq 4),[14] or by the reduction of enones by lithium in liquid ammonia (eq 5).[1,12]

1. LDA, THF, –78 °C; 2. $NCCO_2Me$, HMPA, –78 °C to 0 °C; 85% (→ CO_2Me) (1)

1. LDA, THF, –78 °C; 2. $NCCO_2Me$, HMPA, –78 °C to 0 °C; 75% (→ CO_2Me) (2)

OTMS → 1. MeLi, THF, –78 °C; 2. $NCCO_2Me$, HMPA, –78 °C to 0 °C; 68% (→ CO_2Me) (3)

1. $R_2Cu(CN)Li_2$, THF, –78 °C; 2. $NCCO_2Me$, HMPA, –78 °C; 82% (→ CO_2Me, OTBDMS) (4)
R = $TBDMSOCH_2CH{=}CH$

1. Li, NH_3, *t*-BuOH, Et_2O, –33 °C; 2. $NCCO_2Me$, –78 °C to 0 °C; 84% (→ H, H, CO_2Me) (5)

Lithium enolates derived from sterically unencumbered cyclohexanones undergo preferential axial acylation (eq 6), whereas equatorial acylation is favored with $\Delta^{1(9)}$-2-octalones (eq 7),[12] even in the absence of an alkyl substituent at C-10.[15]

OTMS, *t*-Bu → 1. MeLi, THF, –20 °C; 2. $NCCO_2Me$, HMPA, –78 to 0 °C → CO_2Me (*t*-Bu) 7% + CO_2Me (*t*-Bu) 75% (6)

R → 1. Li, NH_3, *t*-BuOH, Et_2O, –33 °C; 2. $NCCO_2Me$, –78 to 0 °C → (R, H, CO_2Me) + (R, H, CO_2Me) (7)

R = H	53%	14%
R = Me	78%	0%

For compounds in which the β-carbon of the enolate is sterically hindered, treatment with methyl cyanoformate may result in variable degrees of *O*-acylation, although this problem may be ameliorated by the use of diethyl ether as the solvent. In several cases a switch from predominantly *O*-acylation in THF to predominant *C*-acylation in diethyl ether has been observed (eqs 8–10).[12]

1. Me_2CuLi, –10 °C; 2. $NCCO_2Me$, –78 to 0 °C → OCO_2Me + CO_2Me (8)

	OCO_2Me	CO_2Me
solvent: THF	87%	0%
solvent: Et_2O	6%	78%

OTMS → OCO_2Me + CO_2Me (H) (9)

	OCO_2Me	CO_2Me
MeLi, THF, –20 °C; $NCCO_2Me$, –78 to 0 °C	78%	0%
BuLi, Et_2O, –10 °C; $NCCO_2Me$, –78 to 0 °C	9%	68%

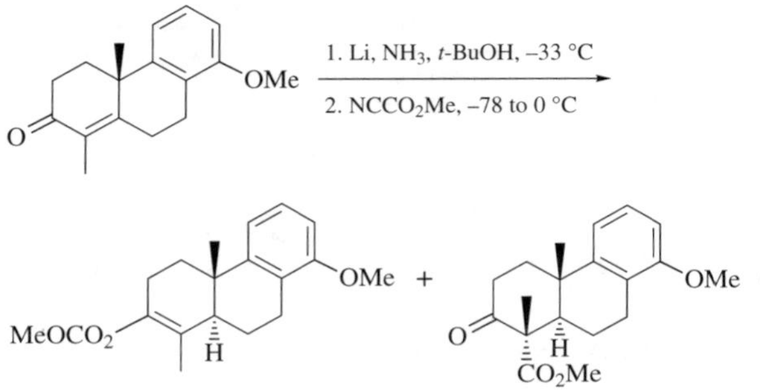

(10)

solvent: THF	67%	8%
solvent: Et_2O	0%	71%

A comparative study of lithium, sodium, and potassium enolates indicated that the lithium derivatives reacted most satisfactorily.[2] There may be substrates for which the thermodynamic enolates are required, however, and the sodium and potassium enolates may therefore be selected. Good results have been reported with these intermediates (eqs 11 and 12), although the latter afford significant amounts of *O*-acylated products.[16,17] Quite apart from the issue of regioselectivity, the cyanoformate based procedure is exceptionally reliable and makes it possible to prepare β-keto esters from ketones under especially mild conditions. It is not only the method of choice with sensitive substrates,[16] but it will often ensure superior results with more robust compounds as well.

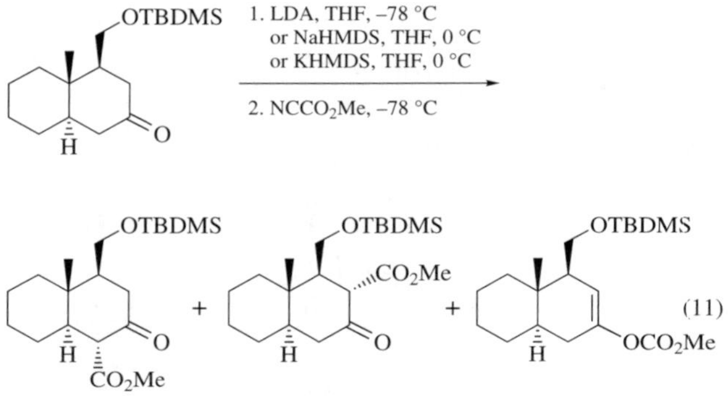

(11)

LDA	4:1:0
NaHMDS	1:7:0
KHMDS	0:1:1.5

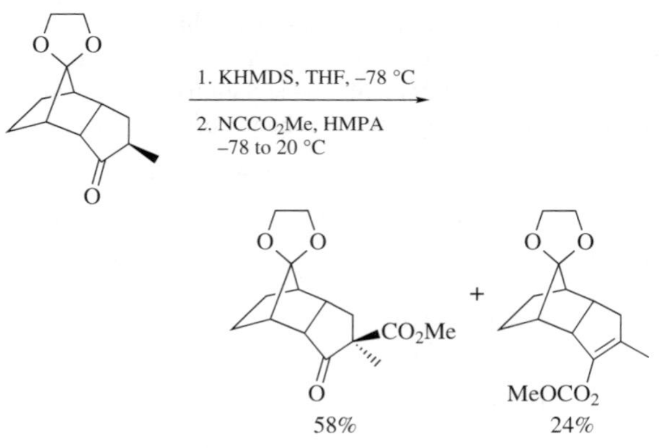

(12)

Methoxycarbonylation of Miscellaneous Carbon Acids. The title reagent has also been applied to the methoxycarbonylation of esters (eq 13),[18] lactones (eq 14),[19] phosphonates (eq 15),[20] imines (eq 16),[21] and the *N*-acylation of lactams (eq 17).[22]

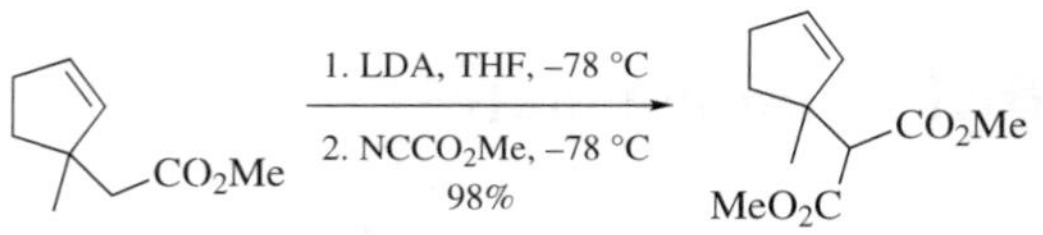

(13)

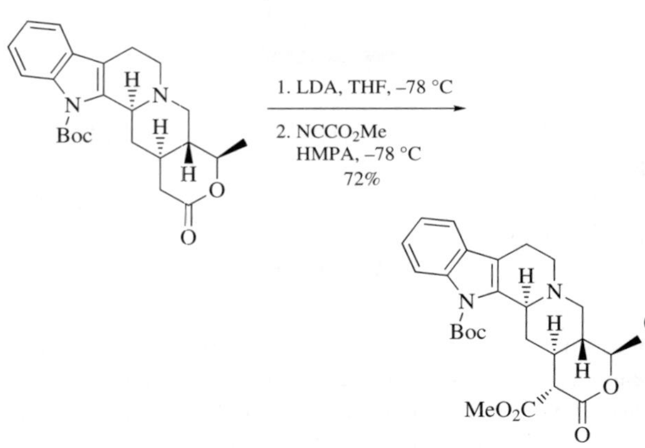

(14)

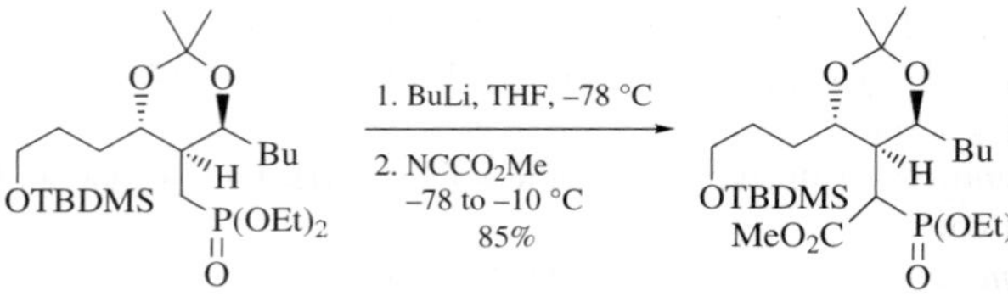

(15)

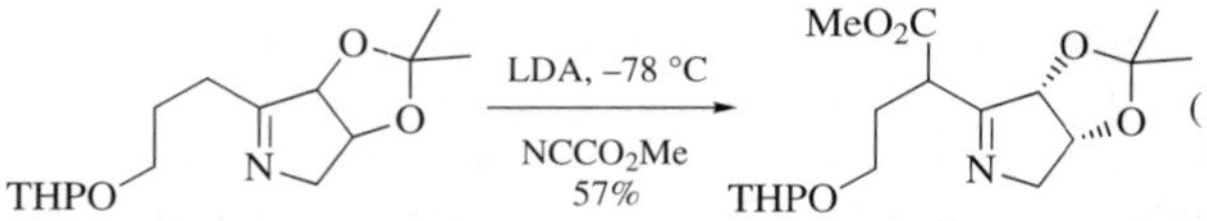

(16)

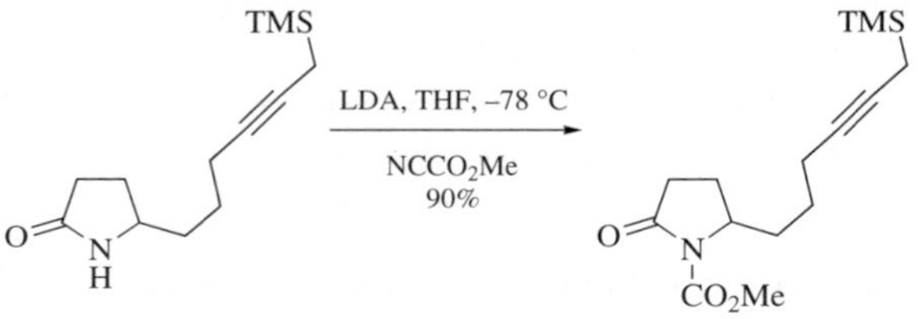

(17)

Higher Alkyl Cyanoformates. A range of other alkyl cyanoformates has been successfully utilized for the acylation of enolate anions, including ethyl,[23] allyl,[24] benzyl,[25] and *p*-methoxybenzyl,[26] but not *t*-butyl cyanoformate, which appears to be insufficiently reactive. Enantiomerically enriched cyanoformates derived from (+)-menthol, (−)-borneol, and the Oppolzer alcohol were reported to furnish good chemical yields, but the level of enantioselectivity was disappointingly low (eq 18).[27]

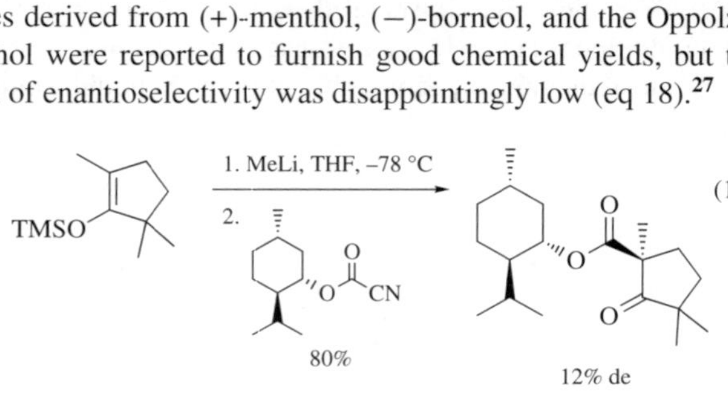

(18)

Additions to the Nitrile Group. Ethyl cyanoformate reacts with organocadmium reagents to afford α-keto esters (eq 19),[3] and with malonate esters, β-keto esters, and other active methylene compounds to give α-aminoacrylates (eq 20).[28] Both processes require catalysis with Lewis acids, of which ***Zinc Chloride*** has proven to be the most effective.

$$\text{RMgBr} \xrightarrow[\text{2. NCCO}_2\text{Et, ZnCl}_2\text{, Et}_2\text{O, 0 °C, 24–66\%}]{\text{1. CdCl}_2} \text{RCOCO}_2\text{Et} \quad (19)$$

R = Ph, *i*-Pr, *s*-Bu, cyclohexyl, cyclopentyl

$$\text{MeOCOCH}_2\text{COCH}_3 \xrightarrow[\text{Et}_3\text{N, CH}_2\text{Cl}_2\text{, 35 °C, 92\%}]{\text{NCCO}_2\text{Et, ZnCl}_2} \text{MeOCOC(COCH}_3\text{)=C(NH}_2\text{)CO}_2\text{Me} \quad (20)$$

Cycloadditions. Methyl and ethyl cyanoformate have been reported to undergo [4+2] cycloadditions, e.g. with cyclopentadienones[4] and 2-alkyl-1-ethoxybuta-1,3-dienes to form pyridines (eq 21),[29] and with cyclobutadienes to form Dewar pyridines (eq 22).[30] Ethyl cyanoformate is also an effective dipolarophile, undergoing 1,3-dipolar addition to azides (eq 23)[31] and cyclic carbonyl ylides (eq 24).[5]

$$\xrightarrow[\text{PhMe, 50 °C, one week, 35–56\%}]{\text{NCCO}_2\text{Et, CpCo(cod)}} \quad \text{ca 2:3} \quad (21)$$

R = Me, Et, Pr, *i*-Pr, Bu, C_5H_{11}

$$\xrightarrow[\text{1.5 h, 57\%}]{\text{NCCO}_2\text{Me, PhH, 80 °C}} \quad (22)$$

$$\xrightarrow[\text{145 °C, 5–6 h, 49–58\%}]{\text{NCCO}_2\text{Et}} \quad (23)$$

R = H, 5-Me. 4-Cl, 5-Cl

$$\xrightarrow[\text{CHCl}_3\text{, 25 °C, 76\%}]{\text{Rh}_2\text{(OAc)}_4\text{, NCCO}_2\text{Et}} \quad (24)$$

Radical Cyanation. The peroxide-initiated radical cyanation of cyclohexane and 2,3-dimethylbutane with methyl cyanoformate has been carried out in 72% and 77% yield, respectively.[6]

Related Reactions. β-Keto ester formation from ketones may be achieved directly with dialkyl carbonates[32] and dialkyl dicarbonates,[33] or indirectly with dialkyl oxalates,[34] methyl magnesium carbonate,[35] and ethyl diethoxyphosphinyl formate.[36] Regiocontrol is problematical, however, and is more reliably effected by trapping enolates with carbon dioxide,[10] carbon disulfide,[37] or carbon oxysulfide[38] followed by methylation. In the latter cases, the dithio and thiol esters are converted into the parent carboxy esters by mercury(II)-catalyzed hydrolysis. The chemistry of acyl cyanides, but excluding cyanoformates, has been the subject of several reviews.[39]

Related Reagents. Acetyl Cyanide; Carbon Dioxide; Carbon Oxysulfide; *N,N'*-Carbonyldiimidazole; Diethyl Carbonate; Methyl Chloroformate; Methyl Magnesium Carbonate.

1. Crabtree, S. R.; Mander, L. N.; Sethi, S. P. *OS* **1991**, *70*, 256.
2. Mander, L. N.; Sethi, S. P. *TL* **1983**, *24*, 5425.
3. Akiyama, Y.; Kawasaki, T.; Sakamoto, M. *CL* **1983**, 1231.
4. Padwa, A.; Akiba, M.; Cohen, L. A.; Gingrich, H. L.; Kamigata, N. *JACS* **1982**, *104*, 286.
5. Padwa, A.; Chinn, R. L.; Hornbuckle, S. F.; Zhang, Z. J. *JOC* **1991**, *56*, 3271.
6. Tanner, D. D.; Rahimi, P. M. *JOC* **1979**, *44*, 1674.
7. Childs, M. E.; Weber, W. P. *JOC* **1976**, *41*, 3486.
8. Nii, Y.; Okano, K.; Kobayashi, S.; Ohno, M. *TL* **1979**, 2517.
9. Caine, D. In *Carbon-Carbon Bond Formation*; Augustine, R. L., Ed.; Dekker: New York, 1979; Vol. 1, pp 250–258.
10. (a) Stork, G.; Rosen, P.; Goldman, N.; Coombs, R. V.; Tsuji, J. *JACS* **1965**, *87*, 275. (b) Caine, D. *OR* **1976**, *23*, 1.
11. (a) House, H. O. *Modern Synthetic Reactions*; Benjamin: Menlo Park, 1972; pp 760–763. (b) Black, T. H. *OPP* **1989**, *21*, 179. (c) Seebach, D.; Weller, T.; Protschuk, G.; Beck, A. K.; Hoekstra, M. S. *HCA* **1981**, *64*, 716. (d) cf. Ref. 5, p 258, footnote 69.
12. Crabtree, S. R.; Chu, W.-L. A.; Mander, L. N. *SL* **1990**, 169.
13. (a) Ihara, M.; Suzuki, T.; Katogi, M.; Taniguchi, N.; Fukumoto, K. *JCS(P1)* **1992**, 865. (b) Haynes, R. K.; Katsifis, A. G. *CC* **1987**, 340.
14. Hashimoto, S.; Kase, S.; Shinoda, T.; Ikegami, S. *CL* **1989**, 1063.
15. cf. Mathews, R. S.; Girgenti, S. J.; Folkers, E. A. *CC* **1970**, 708.
16. Ziegler, F. E.; Klein, S. I.; Pati, U. K.; Wang, T.-F. *JACS* **1985**, *107*, 2730.
17. Schuda, P. F.; Phillips, J. L.; Morgan T. M. *JOC* **1986**, *51*, 2742.
18. Ziegler, F. E.; Sobolov, S. B. *JACS* **1990**, *112*, 2749.
19. (a) Hanessian, S.; Faucher, A.-M. *JOC* **1991**, *56*, 2947. (b) Ziegler, F. E.; Cain, W. T.; Kneisly, A.; Stirchak, E. P.; Wester, R. T. *JACS* **1988**, *110*, 5442. (c) Leonard, J.; Ouali, D.; Rahman, S. K. *TL* **1990**, *31*, 739.
20. McLure, C. K.; Jung, K.-Y. *JOC* **1991**, *56*, 2326.
21. Bennet, R. B.; Cha, J. K. *TL* **1990**, *31*, 5437.
22. (a) Melching, K. H.; Hiemstra, H.; Klaver, W. J.; Speckamp, W. N. *TL* **1986**, *27*, 4799. (b) Esch, P. M.; Hiemstra, H.; Klaver, W. J.; Speckamp, W. N. *H* **1987**, *26*, 75. (c) Pirrung, F. O. H.; Rutjes, F. P. J. T.; Hiemstra, H.; Speckamp, W. N. *TL* **1990**, *31*, 5365.
23. Mori, K.; Ikunaka, M. *T* **1987**, *43*, 45.
24. Barton, D. H. R.; Donnelly, D. M. X.; Finet, J. P.; Guiry, P. J.; Kielty, J. M. *TL* **1990**, *31*, 6637.
25. Hashimoto, S.; Miyazaki, Y.; Shinoda, T.; Ikegami, S. *CC* **1990**, 1100.
26. (a) Winkler, J. D.; Henegar, K. E.; Williard, P. G. *JACS* **1987**, *109*, 2850. (b) Henegar, K. E.; Winkler, J. D. *TL* **1987**, *28*, 1051.
27. Kunisch, F.; Hobert, K.; Welzel, P. *TL* **1985**, *26*, 5433.
28. Iimori, T.; Nii, Y.; Izawa T.; Kobayashi, S.; Ohno, M. *TL* **1979**, 2525.
29. Potthoff, B.; Breitmaier, E. *S* **1986**, 584.
30. (a) Krebs, A.; Franken, E.; Müller, S. *TL* **1981**, *22*, 1675. (b) Fink, J.; Regitz, M. *BSF(2)* **1985**, 239.
31. Klaubert, D. H.; Bell, S. C.; Pattison, T. W. *JHC* **1985**, *22*, 333.
32. Deslongchamps, P.; Ruest, L. *OS* **1974**, *54*, 151.
33. Hellou, J.; Kingston, J. F.; Fallis, A. G. *S* **1984**, 1014.

34. Snyder, H. R.; Brooks, L. A.; Shapiro, S. H. *OSC* **1943**, *2*, 531.
35. (a) Stiles, M. *JACS* **1959**, *81*, 2598. (b) Pelletier, S. W.; Chappell, R. L.; Parthasarathy, P. C.; Lewin, N. *JOC* **1966**, 1747.
36. Shahak, I. *TL* **1966**, 2201.
37. Kende, A. S.; Becker, D. A. *SC* **1982**, *12*, 829.
38. Vedejs, E.; Nader, B. *JOC* **1982**, *47*, 3193.
39. (a) Thesing, J.; Witzel, D.; Brehm, A. *AE* **1956**, *68*, 425. (b) Bayer, O. *MOC* **1977**, *7/2c*, 2487. (c) Hunig, S.; Schaller, R. *AG(E)* **1982**, *21*, 36.

Lewis N. Mander
The Australian National University, Canberra, Australia

N-Methyl-*N*,*N*′-dicyclohexylcarbodiimidium Iodide

[36049-77-1] $C_{14}H_{25}IN_2$ (MW 348.31)

(conversion of alcohols to iodides)

Physical Data: colorless hygroscopic crystals; mp 111–113 °C.

Preparative Method: ***1,3-Dicyclohexylcarbodiimide*** (42 g, 0.204 mol) was dissolved in a large excess of ***Iodomethane*** (90 mL, 1.45 mol) and stirred at 70 °C for 3 days under a nitrogen atmosphere. The excess MeI was removed by distillation under reduced pressure and the residue was dissolved in dry toluene (150 mL) at 40 °C. The resulting crystalline mass was filtered off under an inert atmosphere and the crystals were washed with cold, dry toluene until colorless. The product was dried under vacuum (0.1 mmHg) at rt to provide 35 g (75%) of colorless crystals.

Handling, Storage, and Precautions: the compound should be stored under an inert atmosphere at refrigerator temperatures, where it is reputed to be stable for several months.

Conversion of Alcohols to Iodides. First reported in 1972, the sole reported use of *N*-methyl-*N*,*N*′-dicyclohexylcarbodiimidium idodide (**1**) is in the conversion of alcohols to iodides. The original article[1] contains the first description of the preparation in bulk of a carbodiimidium salt and describes its use in the preparation of some primary and secondary iodides. In three of the examples, inversion of configuration was exclusive, (**2**) → (**4**), hinting at a mechanism such as that shown in eq 1. The reagent was said to be good for the preparation of very hindered secondary iodides, but the yields are generally low (eqs 2 and 3). The reagent is, however, quite successful in preparing iodides from alcohols in multifunctional compounds (**6**),[2] (**7**),[3] and (**8**).[4]

(1) + (2) → (3) → (5) + (4) (1)

Cy-N=•=N⁺(Me)(Cy) I⁻ (2 equiv), THF, 35 °C, 15 h, 84% (2)

(1), THF, 60 °C, 62 h, 18% (3)

(**6**) 54% (**7**) 25% (**8**)

1. R. Scheffold, R.; Saladin, E. *AG(E)* **1972**, *11*, 229.
2. Lischewski, M.; Adam, G. *CA* **1976**, *85*, 108 827u.
3. Gibson, A. R.; Vyas, D. M.; Szarek, W. A. *CI(L)* **1976**, 67.
4. Scartazzini, R.; Bickel, H. *CA* **1974**, *80*, 95 974j.

Kim F. Albizati
University of California, San Diego, CA, USA

N-Methyl-*N*-nitroso-*p*-toluenesulfonamide

[80-11-5] $C_8H_{10}N_2O_3S$ (MW 214.27)

(precursor of diazomethane[1a])

Alternate Names: Diazald; *p*-tolylsulfonylmethylnitrosamide.
Physical Data: mp 61–62 °C.
Solubility: sol petroleum ether, ether, benzene, alcohol, $CHCl_3$, CCl_4; insol H_2O.
Form Supplied in: pale yellow powder; commercially available.
Preparative Methods: prepared by the reaction of ***p*-Toluenesulfonyl Chloride** with methylamine, followed by nitrosation with ***Sodium Nitrite*** in glacial acetic acid.[1c]
Purification: recrystallization is best achieved by dissolving the reagent in hot ether (1 mL g^{-1}), adding an equal volume of petroleum ether (or pentane), and cooling in a refrigerator overnight.[1c]
Handling, Storage, and Precautions: store in a brown bottle. Has a shelf-life of at least 1 year at rt. For longer periods of storage, it is recommended that the reagent be purified by recrystallization[1c] and refrigerated. Toxic; severe skin irritant; handle only in a fume hood.

General Discussion. ***Diazomethane*** is a yellow gas (bp −23 °C, mp −145 °C) with the dipole structure $CH_2{=}N^+{=}N^- \leftrightarrow {}^+CH_2{-}N{=}N^- \leftrightarrow {}^-CH_2{-}N{=}N^+$. This versatile compound behaves as a methylene precursor with release of nitrogen and also functions as a 1,3-dipole in a variety of reactions, as documented in several reviews.[2] Diazomethane itself is both highly toxic and unpredictably explosive, and has been suggested to be a carcinogen.[3] Great care must be taken in handling this substance. However, these risks are minimal when diazomethane is prepared and handled as a dilute solution in an inert solvent such as ether using the proper equipment as discussed below (for further discussion, see ***Diazomethane***). A large number of compounds which previously served as diazomethane precursors can be represented by the formula RN(NO)Me, where R can be sulfonyl, carbonyl, imidoyl, or similar electron-withdrawing groups. Among these the reagent, *N*-methyl-*N*-nitroso-*p*-toluenesulfonamide, introduced by de Boer and Backer in 1954,[1a] is the most common reagent used for the preparation of diazomethane. However, in some cases ***N-[N′-Methyl-N′-nitroso(aminomethyl)]benzamide*** and ***1-Methyl-3-nitro-1-nitrosoguanidine*** have advantages as diazomethane precursors. In addition, *N*-methyl-*N*-nitroso-*p*-toluenesulfonamide does suffer from some drawbacks, including its relatively short shelf life (1–2 years) and its mutagenic activity.[4]

Diazomethane is prepared as a dilute solution in ether by the decomposition (eq 1) of *N*-methyl-*N*-nitroso-*p*-toluenesulfonamide catalyzed by alkali hydroxide.

$$p\text{-}MeC_6H_4SO_2N(Me)NO + OH^- \longrightarrow CH_2N_2 + p\text{-}MeC_6H_4SO_3^- + H_2O \quad (1)$$

This preparation and all reactions involving diazomethane should be carried out in an efficient fume hood and behind a sturdy safety shield. Rough surfaces and strong sunlight are known to initiate detonation. Carefully fire-polished glassware for diazomethane preparation is commercially available.[5] A typical experimental procedure is as follows.[1b] A 125 mL, 3-neck flask is charged with a solution of 6 g of potassium hydroxide in 10 mL of water, 35 mL of 2-(2-ethoxyethoxy)ethanol (carbitol), and 10 mL of ether, and is then fitted with an efficient condenser set downward for distillation. The condenser is connected to two receiving flasks in series, both cooled in an ice-salt bath. The second receiver contains 25 mL of ether and the inlet tube should dip below the surface of the solvent. The generating flask is heated in a water bath at about 70 °C, stirring is started (Teflon-coated magnetic stirring bar), and a solution of 21.4 g (0.1 mol) of *N*-methyl-*N*-nitroso-*p*-toluenesulfonamide in 125 mL of ether is added from the dropping funnel over about 20 min. The rate of addition should about equal the rate of distillation. When the dropping funnel is empty, 50–100 mL of additional ether is added slowly. The distillation is continued until the distilling ether is colorless. The combined ethereal distillate contains 2.7–2.9 g (64–69%) of diazomethane. The content of diazomethane in the codistilled ethereal solution can be determined by titration; excess benzoic acid is added to the solution (to react with diazomethane) and the excess acid is titrated with a standard solution of NaOH. If moisture must be removed from the ethereal diazomethane solution, the drying agent of choice is KOH pellets. It is recommended that diazomethane solutions be used immediately and not stored, even at low temperature.

An alternative popular procedure for the generation of small quantities (1–50 mmol) of diazomethane from Diazald employs a commercial 'mini-Diazald apparatus' with a dry-ice/acetone cold finger in place of the water-jacketed condenser.[5] The generation of diazomethane is carried out by treatment of the Diazald with KOH in aqueous ethanol.[2g]

Related Reagents. Diazomethane; 1-Methyl-3-nitro-1-nitrosoguanidine; *N*-Methyl-*N*-nitrosoacetamide *N*-[*N′*-Methyl-*N′*-nitroso(aminomethyl)]benzamide.

---

1. (a) de Boer, T. J; Backer, H. J. *RTC* **1954**, *73*, 229. (b) de Boer, T. J.; Backer, H. J. *OSC* **1963**, *4*, 250. (c) de Boer, T. J.; Backer, H. J. *OSC* **1963**, *4*, 943.
2. (a) Gutsche, C. D. *OR* **1954**, *8*, 364. (b) Eistert, B.; Regitz, M.; Heck, G.; Schwall, H. *MOC* **1968**, *10/4*, 482. (c) Fieser L. F.; Fieser F. *FF* **1967**, *1*, 191. (d) Pizey, J. S. *Synthetic Reagents*; Wiley: New York, 1974; Vol 2, Chapter 2, p 65. (e) Herrmann, W. A. *AG(E)* **1978**, *17*, 800. (f) Adam, W.; De Lucchi, O. *AG(E)* **1980**, *19*, 762. (g) Black, T. H. *Aldrichim. Acta* **1983**, *16*, 3.
3. Schoental, R. *Nature (London)* **1960**, *188*, 420.

4. (a) Bignami, M.; Carere, A.; Conti, G; Conti, L.; Crebelli, R.; Frabrizi, M. *Mutat. Res.* **1982**, *97*, 293. (b) Druckrey, H.; Preusmann, R. *Nature (London)* **1962**, *195*, 1111.
5. Aldrich Chemical Company, Inc.

Yoshiyasu Terao & Minoru Sekiya
University of Shizuoka, Japan

Methyl Fluorosulfonate and Methyl Trifluoromethanesulfonate

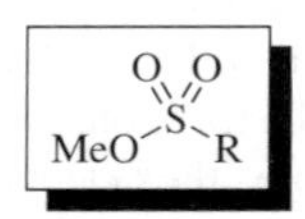

(R = CF_3)
[333-27-7] $C_2H_3F_3O_3S$ (MW 164.12)
(R = F)
[421-20-5] CH_3FO_3S (MW 114.11)

(powerful methylating agents[1,2])

Alternate Name: R = CF_3, methyl triflate; R = F, Magic Methyl.

Physical Data: methyl triflate: bp 99 °C, mp −64 °C, *d* 1.50 g cm^{-3}. Methyl fluorosulfonate: bp 92 °C, mp −92.5 °C, *d* 1.45 g cm^{-3}.

Solubility: both reagents are miscible with all organic solvents, but react with many. They are only sl sol water, but hydrolyze rapidly as they dissolve. Useful inert solvents are CH_2Cl_2, SO_2, sulfolane, nitromethane, Me_2SO_4, and Me_3PO_4.

Form Supplied in: methyl triflate (MeOTf) is available as a colorless liquid. Methyl fluorosulfonate ($MeOSO_2F$) was formerly available as Magic Methyl™ but has been withdrawn (see below).

Analysis of Reagent Purity: MeOTf gives a singlet in ^{1}H NMR at δ 4.18, with ^{13}C absorption at δ 61.60 and 119.32 (q), and a ^{19}F shift of 75.4 ppm. $MeOSO_2F$ absorbs at δ 4.19 in ^{1}H NMR (J_{HF} 0.4 Hz or less), with ^{13}C absorption at δ 62.45, and a ^{19}F shift of −31.2 ppm. ^{33}S and ^{17}O NMR data have been reported for both compounds.[9]

Preparative Methods: both reagents are prepared[3,4] by distilling an equimolar mixture of the corresponding acid with ***Dimethyl Sulfate*** in an all-glass apparatus with a short Vigreux column. They may be dried by standing over fused K_2CO_3 and redistillation. ***Trifluoromethanesulfonic Acid*** and ***Fluorosulfuric Acid*** are both available, and are comparably priced.

Handling, Storage, and Precautions: both reagents are **extremely hazardous**. All possible precautions should be taken to avoid inhalation or absorption through the skin. A fatality has occurred with $MeOSO_2F$ through inhalation of the vapors leading to pulmonary edema.[5] Dexamethasone isonicotinate (Auxiloson® spray) has been recommended as a first aid in the treatment of such pulmonary irritation.[5] The oral LD_{50} of $MeOSO_2F$ is 112 mg/kg in mice, and an LC_{50} for 1 h exposure for rats between 5 and 6 ppm has been reported; severe eye irritation was noted.[6] It is very unlikely[7] that MeOTf is less dangerous, but no data on toxicity have been reported. Both materials are extremely destructive to the tissue of the mucous membranes and upper respiratory tract, eyes, and skin. Inhalation may be fatal as a result of spasm, inflammation, and edema of the larynx and bronchi, chemical pneumonitis, and pulmonary edema. Use in a fume hood.

The reagents are stable in glass when dry, but storage of $MeOSO_2F$ in bottles with ground glass joints should be avoided as these slowly become fused. It is reported[8] that $MeOSO_2F$ after storage over CaH_2 for two weeks contained 17% Me_2SO_4.

General Reactivity. MeOTf and $MeOSO_2F$ are two of the most powerful reagents for methylation and are more reactive by a factor of $\sim 10^4$ than ***Iodomethane*** and Me_2SO_4.[1,2] The only reagents which are substantially more powerful are ***Methyl Fluoride–Antimony(V) Fluoride*** and related reagents[10] and the dimethylhalonium ions; these reagents pose more severe handling problems. The reactivity of MeOTf and $MeOSO_2F$ is not effectively enhanced by addition of Lewis acids. Thus addition of ***Antimony(V) Chloride*** to $MeOSO_2F$ and ***Dimethyl Sulfone*** led to complexation of the sulfone rather than methylation;[11] it is also reported that addition of $SbCl_5$ leads to formation of MeCl.[12]

The qualitative reactivity of some methylating agents toward a range of functional groups is shown in tabular form (Table 1).[1] Few quantitative data are available, but MeOTf and $MeOSO_2F$ are only a little less reactive than the trimethyloxonium ion, with the dimethoxycarbenium ion probably somewhat more reactive again. Substitution rates with various nucleophiles are reported to be in the order: $Me_3O^+ > MeOTf > MeOSO_2F > MeOClO_3$, with rate ratios of 109:23:8:6:1.0 for reaction with acetonitrile at 0 °C.[13,14] Methyl perchlorate, an explosion hazard, therefore never offers a practical advantage. Our experience is that the relative rates for MeOTf and $MeOSO_2F$ rarely differ by more than a factor of 2–5.

It is noteworthy that neither MeOTf nor $MeOSO_2F$ shows any reactivity on their own in Friedel–Crafts methylation reactions even with highly reactive substrates (Me_3O^+ ion is similar). Reaction has been observed in the presence of protic or Lewis acids.[15] Olah has recently suggested that this reactivity is associated with generation of superelectrophiles by further protonation.[16] $MeOSO_2F$ has been reported to react as a methylsulfonylating agent towards phenol and anisole,[17] but this alternative reactivity is usually not significant.

The alkylation reactions of MeOTf and $MeOSO_2F$ are discussed according to the atom alkylated. To avoid unnecessary repetition, reactions can be assumed to use MeOTf unless $MeOSO_2F$ is specified. Historically, $MeOSO_2F$ was mainly used up until about 1980, but MeOTf has since become the reagent of choice. In the great majority of cases these reagents can probably be used interchangeably.

Alkylation at Nitrogen. Most amines react violently with MeOTf or $MeOSO_2F$, and only those with severe steric hindrance or conjugated with strong electron-withdrawing groups really require the use of these reagents. Of derivatives with sp^3 nitrogen, ***Diisopropylamine***, *N*,2,2,6,6-pentamethylpiperidine, and ***1,8-Bis(dimethylamino)naphthalene*** (Proton Sponge™) can all be quaternized by $MeOSO_2F$, and *N,N*,2,6-tetramethylaniline reacts on heating.[1] However, it has been reported that *i*-Pr_3N does not react with $MeOSO_2F$.[19] ***2,6-Lutidine*** reacts exothermically with $MeOSO_2F$ at room temperature, and 2,6-

Table 1 Reactivity[a] of Methylating Agents towards Functional Groups

	Approx. pK_a[18]	MeI	Me_2SO_4	$MeOSO_2F$ (MeOTf)	Me_3O^+	$HC(OMe)_2^+$	Me_2Cl^+
Me_3N	9	+	+	++	++	++	++
Pyridine	5	+	+	++	++	++	++
Me_2S	1	+	+	++	++	++	++
Me_2SO	0	+	+	+	+	+	++
$HCONMe_2$	0	–	+	+	+	+	++
MeCN	–10	–	–	+	+	+	+
Me_2O	–5	–	–	+	(+)	+	+
PhCHO	–6	–	–	+	–	+	+
R_2CO	–6	–	–	–	–	–	+
Lactone	–7	–	–	+	–	+	+
Ester	–7	–	–	(+)	–	+	+
MeI	?	–	–	(+)	–	(+)	+
$MeNO_2$	–10	–	–	–	–	–	+
Me_2SO_2	–10	–	–	–	–	–	+

[a] Key: ++ methylation occurs and may be strongly exothermic; + methylation occurs; – little reaction; (+) equilibrium transfer reaction; – reaction is complex.

dimethoxycarbonylpyridine can be quantitatively quaternized.[1] A range of [*n*](2,6)-pyridinophanes have been methylated with $MeOSO_2F$, with *n* as small as 6 (eq 1).[20]

$$\text{[pyridinophane: N, }(CH_2)_n\text{]} \xrightarrow{MeOSO_2F} \text{[}\overset{+}{N}Me\ (CH_2)_n\text{]}\ FSO_3^- \quad (1)$$

2,6-Di-t-butylpyridine does not react at normal pressure, and this or 2,6-di-*t*-butyl-4-methylpyridine (synthesis[21]), are often used in applications which require base (see below). Note that 2,6-di-*t*-butylpyridine can be alkylated under high pressure with $MeOSO_2F$ to give >90% of the methylation product when water is carefully excluded.[22]

Simple imines, such as benzylideneaniline, react readily. The *N*-methylation of imine and amidine derivatives of amino acid esters, followed by hydrolysis, has been used as a method for the preparation of *N*-alkylated amino acids with minimal racemization.[23] A convenient synthesis of *N*-methyl nitrones has been developed by alkylation of OTMS oximes with MeOTf, followed by treatment with fluoride ion.[24]

Besides the pyridine derivatives already mentioned, most heterocyclic nitrogens can be alkylated. Aspects of the quaternization of heteroaromatics have been recently reviewed.[25] With respect to the limits of reactivity, it is interesting that dimethylation of 2-phenyl-4,6-dimethylpyrimidine could only be achieved with ***Trimethyloxonium Tetrafluoroborate***; $MeOSO_2F$ only gave monoalkylation.[26] 2-Benzoylbenzothiazole can be *N*-alkylated with $MeOSO_2F$ (but not with MeI) and then acts as an active acylating agent.[27] Formation of an *N*-methylindole from an *N*-ethoxycarbonylindole has been achieved with $MeOSO_2F$.[28]

A number of alkylation products from MeOTf and heterocycles have been advocated as useful intermediates. Thus treatment of 2-substituted thiazoles with MeOTf in acetonitrile, followed by reduction of the salt formed with ***Sodium Borohydride***/CuO in CH_2Cl_2, leads to aldehydes.[29] 1-(Benzenesulfonyl)-3-methylimidazolium and 1-(*p*-toluenesulfonyl)-3-methylimidazolium triflates have been proposed as efficient reagents for the preparation of aryl sulfonamides and aryl sulfonates.[30] MeOTf alkylates 2,5-oxazoles to give salts which can be reduced by $PhSiH_3$/CsF to give 4-oxazolines, and these provide a route to stabilized azomethine ylides.[31]

A novel synthesis of 2-aryl-4-piperidones by Mannich cyclization of imino acetals, initiated by methylation of the imine, has been described.[32] MeOTf has been used in the generation of a munchnone for cycloaddition.[33] Finally, methylation of 1-lithio-2-*n*-butyl-1,2-dihydropyridine with MeOTf gives 2-butyl-5-methylpyridine in 42% yield.[34]

Nitriles, with sp hybridized nitrogen, are unreactive to MeI or Me_2SO_4, but are readily methylated by $MeOSO_2F$,[1] MeOTf,[35] or oxonium ions. Nitrilium salts have been shown to have a number of useful applications. The reduction of nitrilium salts by $NaBH_4$ in alcohols leads first to iminoethers and subsequently to amines.[36] Reduction of *N*-alkylnitrilium ions by organosilicon hydrides gives *n*-alkylaldimines, and thus provides a route to aldehydes from nitriles.[37] Nitrilium triflate salts have been shown to be useful reagents for the synthesis of ketones and ketenimines by electrophilic substitution of reactive aromatics, and also provide good routes to amidinium, imidate, and thioimidate salts.[35] Reaction with 2-amino alcohols gives oxazolidines.[38] Synthesis of either 5-substituted 1-methyl-1*H*-tetrazoles or 3,5-disubstituted 1,4-dimethyltriazolium salts from *N*-methylnitrilium triflate salts can be controlled in reactions with $(Me_2N)_2C{=}NH_2^+N_3^-$.[39] Reaction of nitrilium ions with alkyl azides gives 1,2,3-trisubstituted tetrazolium salts.[40] These can be deprotonated to highly reactive 2-methylenetetrazoles.[41]

Amides are alkylated largely on oxygen, as expected (see below), although some *N*-alkylation can be seen by NMR.[1] *N*-Alkylation is more apparent with carbamates (see the section on ambident nucleophiles). *N*,*N*-Dimethylmethanesulfonamides can be *N*-alkylated ($MeOSO_2F$) to provide salts which are effective reagents for mesylation.[42] *N*,*N*-Dimethylsulfamate esters react with $MeOSO_2F$ to give trimethylammoniumsulfate esters, which rapidly give methyl esters unless the O-group is aryl, showing that $Me_3N^+SO_3^-$ is a very powerful leaving group.[43] $Me_3N^+SO_2OPh$ reacts with nucleophiles at either the sulfur or a methyl carbon atom.[44]

Azo compounds, which do not react with methyl iodide, can be *N*-methylated by $MeOSO_2F$.[45]

A steroidal oxaziridine was converted ($MeOSO_2F$) to an oxaziridinium salt which showed oxidizing properties.[46]

Alkylation at Oxygen. Most neutral functionalities with lone pairs on oxygen are not alkylated by MeI or Me_2SO_4 but do react with MeOTf or $MeOSO_2F$, although not all such reactions are preparatively useful.[1,2] Ethers react reversibly, and the ultimate product depends on the conditions. Thus good yields of the oxonium ion can be obtained from reaction of THF with stoichiometric amounts of MeOTf, but the use of catalytic amounts leads to polymerization. Cationic ring opening polymerization, initiated by MeOTf and $MeOSO_2F$ among other reagents, has been extensively investigated and recently reviewed.[47]

Reaction of $MeOSO_2F$ with 2-methoxyethyl carboxylates gives 2-alkyl-1,3-dioxolanium ions.[1] The reaction of these ions with trialkylalkynylborate anions provides versatile and direct routes to (*Z*)-α,β-unsaturated ketones (eq 2). Specifically protected 1,3-diketones and other ketonic species can also be prepared from the intermediates.[48]

$R^1_3B^-$—≡—R^2; 1. Me_2CHCO_2H; 2. aq HCl (2)

Almost all carbonyl functions can be methylated. Enolizable aldehydes and ketones usually lead to complex mixtures, probably because of deprotonation to enol ethers, followed by reaction of these with the electrophilic species in the reaction mixture. Nonenolizable aldehydes and ketones give methoxycarbenium ions cleanly, and the relative thermodynamic stabilities of these have been assessed via pairwise equilibrations.[49] Most esters only generate low equilibrium concentrations of dialkoxycarbenium ions, but lactones are readily alkylated.[1]

Amides, carbamates, and ureas are rapidly alkylated, usually on carbonyl oxygen (see the section on ambident nucleophiles). Alkylation of amides with MeOT5f in CH_2Cl_2 followed by reduction of the salts provides a route for the selective reduction of amides; esters, nitriles, acetals, and double bonds are left unaffected by this procedure.[50] Alkylation of isoindolin-1-ones and subsequent deprotonation can provide routes to methoxyisoindoles.[51]

Alcohols can be converted to methyl ethers by the use of MeOTf + 2,6-di-*t*-butylpyridine or 2,6-di-*t*-butyl-4-methylpyridine. This procedure was initially developed in the carbohydrate field.[52] Me_3PO_4 provides a good polar solvent for this process.[53] A recent application, in the synthesis directed at lonomycin, was to methylation of the complex alcohol (**1**) without causing retro-aldol cleaveage.[54]

(**1**)

MeOTf has been reported to effect complete methylation of inositol polyphosphates, P–OMe groups being formed as well.[55] MeOTf will effect *O*-alkylation at the anomeric center, but the stereochemistry is affected by the presence of a crown ether.[56] Reaction of Meisenheimer complexes with $MeOSO_2F$ can lead to capture of the anion as a nitronate ester.[57]

Alkylation at Sulfur and Selenium. Dialkyl and most arylalkyl sulfides are readily converted into sulfonium salts.[1] Cyclic sulfides, especially dithiolanes, react somewhat faster.[58] Reaction of disulfides with MeOTf gives Me_2SSMe^+ OTf^-; this salt, with ***Triphenylphosphine***, reacts with alkenes in a stereo- and regioselective fashion and the products can be converted into vinylphosphonium salts.[59] Thione groups are also alkylated even when electronegative groups are present. Thus 4,5-bis(trifluoromethyl)-1,3-dithiolane-2-thione was converted into a methylated salt.[60] Sulfoxides are *O*-alkylated,[1] and formation of various oxa- and azasulfonium ions has been reported, up to and including triazasulfonium salts.[61] Alkylation of R_2SO with $MeOSO_2F$, followed by reduction with ***Sodium Cyanoborohydride***, leads to sulfides.[62]

Reaction of MeOTf with the product from P_2S_5/Na_2CO_3 ($Na_2P_4S_{10}O$) gave a useful electrophilic thionation reagent.[63] $MeOSO_2F$ has been used in the conversion of thiols to reactive sulfenating agents (eq 3).[64]

RSH; $MeOSO_2F$; R'SH, *i*-Pr_2NEt; + RSSR' (3)

Methylation of dithioacetals by $MeOSO_2F$, followed by reaction with various nucleophiles, has been used for the removal of this protecting group or its conversion into other protecting groups, e.g. acetals (eq 4).[65–68]

$MeOSO_2F$; Nu^- (4)

The reaction of the sulfonium intermediate with alcohols leads to their protection as hemithioacetals.[69] Treatment of thioglycosides with MeOTf gives an efficient glycosylating agent,[70] and pyruvic acetal formation from a pyruvyl thioacetal has been achieved in a reaction catalyzed by MeOTf (amongst other electrophiles).[71]

Alkenes can be prepared by methylation of selenides by $MeOSO_2F$ and treatment of the selenonium ions formed with ***Potassium t-Butoxide***.[72]

Alkylation at Phosphorus. Phosphines and phosphites undergo easy quaternization. Thus methylation of tris(2,6-dimethylphenoxy)phosphine with MeOTf, followed by treatment of the product with sodium 2,6-dimethylphenoxide, gave methyltetrakis(2,6-dimethylphenoxy)phosphane.[73] Methoxyphosphonium triflates are relatively stable intermediates in Arbuzov reactions.[74] Phosphine oxides and sulfides are alkylated. *S*-Methylation of chiral phosphine sulfides, followed by treatment with ***Hexamethylphosphorous Triamide***, has been advocated as a general synthesis of optically active phosphines.[75]

Ambident Nucleophiles. Amides and related functional groups can be alkylated on oxygen or nitrogen and, as has been noted already, alkylation on carbonyl oxygen normally predominates. In the case of carbamates, *O*-alkylation by $MeOSO_2F$ can be faster, but *N*-alkylation predominates at equilibrium.[76] It has been noted that methylation of secondary amides and thioamides occurs at the protonless heteroatom in the major tautomer.[77] The ionic products of these reactions can be deprotonated to give synthetically useful products, e.g. imidates,[78] but excess $MeOSO_2F$ should be removed before treatment with base.[77]

Reaction of most enolates with MeOTf or $MeOSO_2F$ is always likely to be kinetically controlled. There does not appear to have been a definitive study, but *O*-alkylation is the normal outcome. *O*-Alkylation of a bicyclodecatrienone by $MeOSO_2F$ is enhanced by the use of polar solvents like HMPA.[79] *O*-Alkylation of enolates of appropriate cyclohexadienones by MeOTf has been used to generate various 3a*H*-indenes.[80] A ketene acetal is formed by exclusive *O*-alkylation of the sodium enolate of isopropyl bis(pentachlorophenyl)acetate by MeOTf.[81]

Alkylations at Carbon. In an important recent development, primary α-alkylation of carbonyl compounds under nonbasic conditions has been achieved (eq 5) by alkylation of silyl enol ethers with MeOTf and other primary alkyl triflates, catalyzed by ***Methylaluminum Bis(4-bromo-2,6-di-t-butylphenoxide)*** (MABR).[82]

Bu(OTMS)C=CH–Et —(MeOTf, MABR; CH_2Cl_2, –40 °C; 84%)→ Bu–CO–CH(Me)–Et (5)

Related Reagents. Dimethoxycarbenium Tetrafluoroborate; Dimethyliodonium Hexafluoroantimonate; Dimethyl Sulfate; *O*-Methyldibenzofuranium Tetrafluoroborate; Methyl iodide; Trimethyloxonium Tetrafluoroborate.

1. Alder, R. W. *CI(L)* **1973**, 983.
2. (a) Stang, P. J.; Hanack, M.; Subramanian, L. R. *S* **1982**, 85. (b) For other reviews of trifluoromethanesulfonates, see: Howells, R. D.; McCown, J. D. *CRV* **1977**, *77*, 69 and Stang, P. J.; White, M. R. *Aldrichim. Acta* **1983**, *16*, 15.
3. Ahmed, M. G.; Alder, R. W.; James, G. H.; Sinnott, M. L.; Whiting, M. C. *CC* **1968**, 1533.
4. Beard, C. D.; Baum, K.; Grakauskas, V. *JOC* **1973**, *38*, 3673.
5. (a) van der Ham, D. M. W.; van der Meer, D. *CI(L)* **1976**, 782. (b) van der Ham, D. M. W.; van der Meer, D. *Chem. Eng. News* **1976**, *54* (36), 5.
6. Hite, M.; Rinehart, W.; Braun, W.; Peck, H. *Am. Ind. Hyg. Assoc.* **1979**, *40*, 600 (*CA* **1979**, *91*, 84 580d).
7. Alder, R. W.; Sinnott, M. L.; Whiting, M. C.; Evans, D. A.; *Chem. Eng. News* **1976**, *54* (36), 56.
8. Hase, T.; Kivikari, R. *SC* **1979**, *9*, 107.
9. Barbarella, G.; Chatigilialoglu, C.; Rossini, S.; Tugnoli, V. *JMR* **1986**, *70*, 204.
10. Christie, J. J.; Lewis, E. S.; Casserly, E. F. *JOC* **1983**, *48*, 2531.
11. Minato, H.; Yamaguchi, K.; Miura, T.; Kobayashi, M. *CL* **1976**, 593.
12. Binder, G. E.; Schmidt, A. *Z. Anorg. Allg. Chem.* **1980**, *467*, 197 (*CA* **1980**, *93*, 196 905m).
13. Kevill, D. N.; Lin, G. M. L. *TL* **1978**, 949.
14. (a) Lewis, E. S.; Vanderpool, S. *JACS* **1977**, *99*, 1946; (b) Lewis, E. S.; Vanderpool, S. *JACS* **1978**, *100*, 6421; (c) Lewis, E. S.; Kukes, S.; Slater, C. D. *JACS* **1980**, *102*, 303.
15. (a) Olah, G. A.; Nishimura, J. *JACS* **1974**, *96*, 2214; (b) Booth, B. L.; Haszeldine, R. N.; Laali, K. *JCS(P1)* **1980**, 2887.
16. Olah, G. A. *AG(E)* **1993**, *32*, 767.
17. Kametani, T.; Takanashi, K.; Ogasawara, K. *S* **1972**, 473.
18. Arnett, E. M. *Prog. Phys. Org. Chem.* **1963**, *1*, 223.
19. Wieland, G.; Simchen, G. *LA* **1985**, 2178 (*CA* **1986**, *104*, 33 742n).
20. Weber, H.; Pant, J.; Wunderlich, H. *CB* **1985**, *118*, 4259 (*CA* **1986**, *104*, 129 776c).
21. Anderson, A. G.; Stang, P. J. *OS* **1981**, *60*, 34.
22. Hou, C. J.; Okamoto, Y. *JOC* **1982**, *47*, 1977.
23. O'Donnell, M. J.; Bruder, W. A.; Daugherty, B. W.; Liu, D.; Wojciechowski, K. *TL* **1984**, *25*, 3651.
24. Le Bel, N. A.; Balasubramanian, N. *TL* **1985**, *26*, 4331.
25. Gallo, R.; Roussel, C.; Berg, U *Adv. Heterocycl. Chem.* **1988**, *43*, 173.
26. Douglass, J. E.; Bumgarner, D. L. *JHC* **1981**, *18*, 417.
27. Chikashita, H.; Ishihara, M.; Takigawa, K.; Itoh, K. *BCJ* **1991**, *64*, 3256.
28. Kametani, T.; Suzuki, T.; Ogasawara, K. *CPB* **1972**, *20*, 2057.
29. Dondoni, A.; Marra, A.; Perrone, D. *JOC* **1993**, *58*, 275.
30. O'Connell, J. F.; Rapoport, H. *JOC* **1992**, *57*, 4775.
31. Vedejs, E.; Grissom, J. W. *JACS* **1988**, *110*, 3238.
32. Bosch, J.; Rubiralta, M.; Moral, M.; Valls, M. *JHC* **1983**, *20*, 595.
33. Hershenson, F. M.; Pavia, M. R. *S* **1988**, 999.
34. Knaus, E. E.; Ondrus, T. A.; Giam, C. S. *JHC* **1976**, *13*, 789.
35. Booth, B. L.; Jibodu, K. O.; Proenca, M. F. J. R. P. *CC* **1980**, 1151.
36. Borch, R. F. *CC* **1968**, 442.
37. Fry, J. L.; Ott, R. A. *JOC* **1981**, *46*, 602.
38. Booth, B. L.; Jibodu, K. O.; Proenca, M. F. J. R. P. *JCS(P1)* **1983**, 1067.
39. Amer, M. I. K.; Booth, B. L. *JCR(S)* **1993**, 4.
40. Carboni, B.; Carrie, R. *T* **1984**, *40*, 4115.
41. Quast, H.; Hergenroether, T. *LA* **1992**, 581 (*CA* **1992**, *117*, 48 426e).
42. King, J. F.; du Manior, J. R. *JACS* **1975**, *97*, 2566.
43. King, J. F.; Lee, T. M. *CC* **1978**, 48.
44. King, J. F.; Lee, T. M. L. *CJC* **1981**, *59*, 356.
45. Ferguson, A. N. *TL* **1973**, 2889.
46. Milliet, P.; Picot, A.; Lusinchi, X. *TL* **1976**, 1573, 1577.
47. Penczek, S.; Kubisa, P. In *Ring Opening Polymerization*; Brunelle, D. J., Ed.; Hanser, 1993; Chapter 2.
48. Pelter, A.; Colclough, M. E. *TL* **1986**, *27*, 1935.
49. Quirk, R. P.; Gambill, C. R.; Thyvelikakath, G. X. *JOC* **1981**, *46*, 3181.
50. Tsay, S. C.; Robl, J. A.; Hwu, J. R. *JCS(P1)* **1990**, 757.
51. Kreher, R. P.; Hennige, H.; Konrad, M.; Uhrig, J.; Clemens, A. *ZN(B)* **1991**, *46*, 809 (*CA* **1991**, *115*, 92 001v) and references therein.
52. Arnarp, J.; Kenne, L.; Lindbreg, B.; Lönngren, J. *Carbohydr. Res.* **1975**, *44*, C5; Arnarp, J.; Lönngren, J. *ACS* **1978**, *32B*, 465. Berry, J. M.; Hall, L. D. J. *Carbohydr. Res.* **1976**, *47*, 307. Gilleron, M.; Fournie, J. J.; Pougny, J. R.; Puzo, G. *J. Carbohydr. Chem.* **1988**, *7*, 733.
53. Prehm, P. *Carbohydr. Res.* **1980**, *78*, 372.

54. Evans, D. A.; Sheppard, G. S. *JOC* **1990**, *55*, 5192.
55. Goldman, H. D.; Hsu, F. F.; Sherman, W. R. *Biomed. Environ. Mass Spectrom.* **1990**, *19*, 771.
56. Schmidt, R. R.; Moering, U.; Reichrath, M. *CB* **1982**, *115*, 39 (*CA* **1982**, *96*, 123 134m).
57. Drozd, V. N.; Grandberg, N. V.; Udachin, Y. M. *ZOR* **1982**, *18*, 1249 (*CA* **1982**, *97*, 109 904b).
58. Roberts, R. M. G.; Tillett, J. G.; Ravenscroft, M. *JCS(P2)* **1982**, 1569.
59. Okuma, K.; Koike, T.; Yamamoto, S.; Yonekura, K.; Ohta, H. *CL* **1989**, 1953.
60. Frasch, M.; Mono, S.; Pritzkow, H.; Sundermeyer, W. *CB* **1993**, *126*, 273.
61. Minato, H.; Okuma, K.; Kobayashi, M. *JOC* **1978**, *43*, 652 and references therein.
62. Durst, H. D.; Zubrick, J. W.; Kieczyowski G. R. *TL* **1974**, 1777.
63. Brillon, D. *SC* **1990**, *20*, 3085.
64. Barton, D. H. R.; Hesse, R. H.; O'Sullivan, A. C.; Pechet, M. M. *JOC* **1991**, *56*, 6697.
65. Fetizon, M.; Jurion, M. *CC* **1972**, 382.
66. Ho, T. L.; Wong, C. M. *S* **1972**, 561.
67. Marshall, J. A.; Seitz, D. E. *JOC* **1975**, *40*, 534.
68. Corey, E. J.; Hase, T. *TL* **1975**, 3267.
69. Hase, T. A.; Kivikari, R. *SC* **1979**, *9*, 107.
70. Lonn, H. *Carbohydr. Res.* **1985**, *139*, 105, 115.
71. Liptak, A.; Szabo, L. *Carbohydr. Res.* **1988**, *184*, C5.
72. Halazy, S.; Krief, A. *TL* **1979**, 4233.
73. Szele, I.; Kubisen, S. J.; Westheimer, F. H. *JACS* **1976**, *98*, 3533.
74. (a) Colle, K. S.; Lewis, E. S. *JOC* **1978**, *43*, 571; (b) Lewis, E. S.; Hamp, D. *JOC* **1983**, *48*, 2025.
75. Omelanczuk, J.; Mikolajczyk, M. *TL* **1984**, *25*, 2493.
76. Ahmed, M. G.; Alder, R. W. *CC* **1969**, 1389.
77. Beak, P.; Lee, J.; McKinnie, B. G. *JOC* **1978**, *43*, 1367.
78. Julia, S.; Ryan, R. J. *CR(C)* **1972**, *274*, 1207 (*CA* **1972**, *77*, 5657u).
79. Press, J. B.; Shechter, H. *TL* **1972**, 2677.
80. Foster, S. J.; Rees, C. W.; Williams, D. J. *JCS(P1)* **1985**, 711 and references therein.
81. O'Neill, P.; Hegarty, A. F. *JOC* **1987**, *52*, 2113.
82. Maruoka, K.; Sato, J.; Yamamoto, H. *JACS* **1992**, *114*, 4422.

Roger W. Alder & Justin G. E. Phillips
University of Bristol, UK

Montmorillonite K10

[1318-93-0]

(catalyzes protection reactions of carbonyl and hydroxy groups; promotes ene,[22] condensation,[25] and alkene addition[27] reactions)

Physical Data: the surface acidity of dry K10 corresponds to a Hammett acidity function $H_0 = -6$ to -8.[13]

Form Supplied in: yellowish-grey dusty powder. Formres with water a mud that is difficult to filter, more easily separated by centrifugation; with most organic solvents, forms a well-settling, easy-to-filter suspension.[6]

Handling, Storage, and Precautions: avoid breathing dust; keep in closed containers sheltered from exposure to volatile compounds and moisture.

General. Montmorillonite clays are layered silicates and are among the numerous inorganic supports for reagents used in organic synthesis.[1,2] The interlayer cations are exchangeable, thus allowing alteration of the acidic nature of the material by simple ion-exchange procedures.[3,4] Presently, in fine organic synthesis, the most frequently used montmorillonite is K10, an acidic catalyst, manufactured by alteration of montmorillonite (by calcination and washing with mineral acid; this is probably a proprietary process).

The first part of this article specifically deals with representative laboratory applications to fine chemistry of clearly identified, unaltered K10, excluding its modified forms (cation-exchanged, doped by salt deposition, pillared, etc.) and industrial uses in bulk. This illustrative medley shows the prowess of K10 as a strong Brønsted acidic catalyst. The second part deals with cation-exchanged (mainly Fe^{III}) montmorillonite. Clayfen and claycop, versatile stoichiometric reagents obtained by metal nitrate deposition on K10,[5] are used in oxidation and nitration reactions. They are treated under ***Iron(III) Nitrate–K10 Montmorillonite Clay*** and ***Copper(II) Nitrate–K10 Bentonite Clay***.

K10 is often confused, both in name and in use, with other clay-based acidic catalysts (KSF, K10F, Girdler catalyst, 'acid treated' or 'H^+-exchanged' montmorillonite or clay, etc.) that can be effectively interchanged for K10 in some applications. Between the 1930s and the 1960s, such acid-treated montmorillonites were common industrial catalysts, especially in petroleum processing, but have now been superseded by zeolites.

Activation.[6] K10 clay may be used crude, or after simple thermal activation. Its acidic properties are boosted by cation exchange (i.e. by iron(III)[7] or zinc(II)[8]) or by deposition of Lewis acids, such as zinc(II)[9,10] or iron(III)[11] chloride (i.e. 'clayzic' and 'clayfec'). In addition, K10 is a support of choice for reacting salts, for example nitrates of thallium(III),[12] iron(III) ('clayfen'),[5] or copper(II) ('claycop').[5] Multifarious modifications (with a commensurate number of brand names) result in a surprisingly wide range of applications; coupled with the frequent imprecise identification of the clay (K10 or one of its possible substitutes mentioned above), they turn K10 into a Proteus impossible to grab and to trace exhaustively in the literature.

Preparation of Acetals. Trimethyl orthoformate (see ***Triethyl Orthoformate***) impregnated on K10 affords easy preparation of dimethyl acetals,[14] complete within a few minutes at room temperature in inert solvents such as carbon tetrachloride or hexane (eq 1). The recovered clay can be reused.

O
HC(OMe)3, K10
5 min, 100%
MeO OMe
(1)

Cyclic diacetals of glutaraldehyde are prepared in fair yields by K10-promoted reaction of 2-ethoxy-2,3-dihydro-4*H*-pyran with diols, under benzene azeotropic dehydration (eq 2).[15]

Diastereoisomeric acetal formation catalyzed by K10 has been applied to the resolution of racemic ketones, with diethyl (+)-(*R,R*)-tartrate as an optically active vicinal diol.[16]

1,3-Dioxolanes are also prepared by K10-catalyzed reaction of 1-chloro-2,3-epoxypropane (***Epichlorohydrin***) with aldehydes or ketones, in carbon tetrachloride at reflux (eq 3).[17] In the reaction of acetone with the epichlorohydrin, the efficiency of catalysts varies in the order: K10 (70%) > ***Tin(IV) Chloride*** (65%) > ***Boron Trifluoride*** (60%) = ***Hydrochloric Acid*** (60%) > ***Phosphorus(V) Oxide*** (57%).

Preparation of Enamines. Ketones and amines form enamines in the presence of K10 at reflux in benzene or toluene, with azeotropic elimination of water (eq 4). Typical reactions are over within 3–4 h. With cyclohexanone, the efficiency depends on the nature of the secondary amine: ***Pyrrolidine*** (75%) > ***Morpholine*** (71%) > ***Piperidine*** (55%) > ***Dibutylamine*** (34%).[18] Acetophenone requires longer heating.[19]

The K10-catalyzed reaction of aniline with β-keto esters gives enamines chemoselectively, avoiding the competing formation of anilide observed with other acidic catalysts.[20,21]

Synthesis of γ-Lactones via the Ene Reaction. K10 catalyzes the ene reaction of diethyl oxomalonate and methyl-substituted alkenes at a rather low temperature for this reaction (80 °C), followed by lactonization (eq 5).[22] When alkene isomerization precedes the ene step, it results in a mixture of lactones. Using kaolinite instead of K10 stops the reaction at the ene intermediate, before lactonization.

Synthesis of Enol Thioethers. Using a Dean–Stark water separator, K10 catalyzes formation of alkyl- and arylthioalkenes from cyclic ketones and thiols or thiophenols, in refluxing toluene (eq 6). A similar catalysis is effected by KSF (in a faster reaction) and K10F.[23] The isomer distribution is under thermodynamic control.

Preparation of Monoethers of 3-Chloro-1,2-propanediol. Alcohols react regioselectively with 1-chloro-2,3-epoxypropane to form 1-alkoxy-2-hydroxy-3-chloropropanes. The K10-catalyzed process is carried out in refluxing carbon tetrachloride for 2.5 h (eq 7).[24] Yields are similar to those obtained by ***Sulfuric Acid*** catalysis.

α,β-Unsaturated Aldehydes via Condensation of Acetals with Vinyl Ethers. K10-catalyzed reaction of diethyl acetals with ***Ethyl Vinyl Ether*** leads to 1,1,3-trialkoxyalkanes. Hydrolysis turns these into *trans*-α,β-unsaturated aldehydes.[25] The reaction is performed close to ambient temperatures (eq 8). K10 is superior to previously reported catalysts, such as ***Boron Trifluoride*** or ***Iron(III) Chloride***. The addition is almost instantaneous and needs no solvent. Cyclohexanone diethyl acetal gives an analogous reaction.

Protective Tetrahydropyranylation of Alcohols and Phenols. With an excess of ***3,4-Dihydro-2H-pyran***, in the presence of K10 at room temperature, alcohols are transformed quantitatively into their tetrahydropyranyl derivatives. Run in dichloromethane at room temperature, the reaction is complete within 5–30 min (eq 9). The procedure is applicable to primary, secondary, tertiary, and polyfunctional alcohols as well as to phenols.[26]

Markovnikov Addition of Hydrochloric Acid to Alkenes. 1-Chloro-1-methylcyclohexane, the formal Markovnikov adduct of hydrochloric acid and 1-methylcyclohexene, becomes largely predominant when ***Sulfuryl Chloride*** is the chlorine source and K10 the solid acid.[27] The reaction at 0 °C, in dry methylene chloride, is complete within 2 h (eq 10).

K10, SO_2Cl_2, CH_2Cl_2, 0 °C, 2 h, 98% (10)

1,1:1,2 = 91:9

Porphyrin Synthesis. *Meso*-tetraalkylporphyrins are formed in good yields from condensation of aliphatic aldehydes with pyrrole; thermally activated K10 catalyzes the polymerization–cyclization to porphyrinogen, followed by ***p-Chloranil*** oxidation (eq 11).[28]

1. K10, CH_2Cl_2; 2. *p*-chloranil, 46% (11)

Meso-tetraarylporphyrins, with four identical or with tuneable ratios of different aryl substituents, are made by taking advantage of modified K10 ('clayfen' or Fe^{III}-exchanged) properties.[29,30]

Iron(III)-Doped Montmorillonite.

General Considerations. The acid strength of some cation-exchanged montmorillonites is between ***Methanesulfonic Acid*** (a strong acid) and ***Trifluoromethanesulfonic Acid*** (a superacid) and, in some instances, their catalytic activity is greater than that of a superacid.[31] Iron montmorillonite is prepared by mixing the clay with various Fe^{III} compounds in water.[8,32] The resulting material is filtered and dehydrated to afford the active solid-acid catalyst. These solid-acid catalysts are relatively inexpensive and are generally used in very small quantities to catalyze a wide variety of reactions, including Friedel–Crafts alkylation and acylation, Diels–Alder reactions, and aldol condensations.[1,5]

Diels–Alder Reactions.[33] Stereoselective Diels–Alder reactions involving an oxygen-containing dienophile are accelerated in the presence of Fe^{III}-doped montmorillonite in organic solvents (eq 12).[34] Furans also undergo Diels–Alder reactions with ***Acrolein*** and ***Methyl Vinyl Ketone*** in CH_2Cl_2 to give the corresponding cycloadducts in moderate yield (eq 13).[35] The iron-doped clay also catalyzes the radical ion-initiated self-Diels–Alder cycloaddition of unactivated dienophiles such as 1,3-cyclohexadiene and 2,4-dimethyl-1,3-pentadiene (eq 14).[36]

K10–Fe^{3+}, CH_2Cl_2, −24 °C, 60% (12)

K10–Fe^{3+}, CH_2Cl_2, −24 °C, 98% (13)

endo:exo = 13.5:1

K10–Fe^{3+}, 4-*t*-butylphenol (10%), CH_2Cl_2, 0 °C (14)

endo:exo = 4:1

The role of Fe^{III}-impregnated montmorillonite, and other cation-exchanged montmorillonites, in asymmetric Diels–Alder reactions was found to be limited to the use of small chiral auxiliaries; the results obtained from these reactions are similar to those of homogeneous aluminum catalysts (eq 15).[33]

endo:exo = 98:2, 39% de (15)

Friedel–Crafts Acylation and Alkylation.[37,38] The Friedel–Crafts acylation of aromatic substrates with various acyclic carboxylic acids in the presence of cation-exchanged (H^+, Al^{3+}, Ni^{2+}, Zr^{2+}, Ce^{3+}, Cu^{2+}, La^{3+}) montmorillonites has been reported.[39] Curiously, the use of iron-doped montmorillonite was not included in the report; however, some catalysis is expected. Under these conditions, the yield of the desired ketones was found to be dependent on acid chain length and the nature of the interlayer cation.

The direct arylation of a saturated hydrocarbon, namely adamantane, in benzene using $FeCl_3$-impregnated K10 was recently reported.[11] Additionally, Friedel–Crafts chlorination of adamantane in CCl_4 using the same catalyst was also reported. The alkylation of aromatic substrates with halides under clay catalysis gave much higher yields than conventional Friedel–Crafts reactions employing ***Titanium(IV) Chloride*** or ***Aluminum Chloride*** as catalyst.[8] Higher levels of dialkylation were observed in some cases. The alkylation of aromatic compounds with alcohols and alkenes was also found to be catalyzed with very low levels of cation-exchanged montmorillonites, as compared to standard Lewis acid catalysis; however, iron-doped clays performed poorly compared to other metal-doped clays.

Aldol Condensations. Cation-exchanged montmorillonites accelerate the aldol condensation of silyl enol ethers with acetals and aldehydes.[40] Similarly, the aldol reaction of silyl ketene acetals with electrophiles is catalyzed by solid-acid catalysts. Neither report discussed the use of iron montmorillonite for these reactions; however, some reactivity is anticipated.

Miscellaneous Reactions. The coupling of silyl ketene acetals (enolsilanes) with pyridine derivatives bearing an electron-withdrawing substituent, namely cyano, in the *meta* position is catalyzed by iron montmorillonite and other similar solid-acid catalysts (eq 16).[41]

$$\text{3-NC-pyridine} + R^1O(OTMS)C{=}CR^2R^3 \xrightarrow{\text{K10–}Fe^{3+}} \text{NC-dihydropyridine}(N\text{-TMS})\text{–}C(R^2)(R^3)CO_2R^1 \quad (16)$$

R^1 = Me, Et; R^2 = H, Me; R^3 = H, Me

The resulting *N*-silyldihydropyridines easily undergo desilylation by treatment with ***Cerium(IV) Ammonium Nitrate*** to afford the desired dihydropyridine derivative. The reactivity was found to be dependent on the montmorillonite counterion and to follow the order: $Fe^{3+} > Co^{2+} > Cu^{2+} \approx Zn^{2+} > Al^{3+} \approx Ni^{2+} \approx Sn^{4+}$.

1. Cornélis, A.; Laszlo, P. *SL* **1994**, 155.
2. McKillop, A.; Young, D. W. *S* **1979**, 401.
3. Theng, B. K. G. *The Chemistry of Clay–Organic Reactions*; Hilger: London, 1974.
4. Thomas, J. M. In *Intercalation Chemistry*; Whittingham, M. S.; Jacobson, J. A., Eds.; Academic: New York, 1982; p 55.
5. Cornélis, A.; Laszlo, P. *S* **1985**, 909.
6. Cornélis, A. In *Preparative Chemistry Using Supported Reagents*; Laszlo, P., Ed.; Academic: New York, 1987; pp 99–111.
7. Cornélis, A.; Gerstmans, A.; Laszlo, P.; Mathy, A.; Zieba, I. *Catal. Lett.* **1990**, *6*, 103.
8. Laszlo, P.; Mathy, A. *HCA* **1987**, *70*, 577.
9. Clark, J. A.; Kybett, A. B.; Macquarrie, D. J.; Barlow, S. J.; Landon, P. *CC* **1989**, 1353.
10. Cornélis, A.; Laszlo, P.; Wang, S. *TL* **1993**, *34*, 3849.
11. Chalais, S.; Cornélis, A.; Gerstmans, A.; Kolodziejski, W.; Laszlo, P.; Mathy, A.; Métra, P. *HCA* **1985**, *68*, 1196.
12. Taylor, E. C.; Chiang, C.-S.; McKillop, A.; White, J. F. *JACS* **1976**, *98*, 6750.
13. Pennetreau, P. PhD Thesis, University of Liège (Belgium), 1986.
14. Taylor, E. C.; Chiang, C.-S. *S* **1977**, 467.
15. Vu Moc Thuy; Maitte, P. *BSF(2)* **1979**, 264.
16. Conan, J. Y.; Natat, A.; Guinot, F.; Lamaty, G. *BSF(2)* **1974**, 1400.
17. Vu Moc Thuy; Petit, H.; Maitte, P. *BSB* **1980**, *89*, 759.
18. Hünig, S.; Benzing, E.; Lücke, E. *CB* **1957**, *90*, 2833.
19. Hünig, S.; Hübner, K.; Benzing, E. *CB* **1962**, *95*, 926.
20. Werner, W. *T* **1969**, *25*, 255.
21. Werner, W. *T* **1971**, *27*, 1755.
22. Roudier, J.-F.; Foucaud, A. *TL* **1984**, *25*, 4375.
23. Labiad, B.; Villemin, D. *S* **1989**, 143.
24. Vu Moc Thuy; Petit, H.; Maitte, P. *BSB* **1982**, *91*, 261.
25. Fishman, D.; Klug, J. T.; Shani, A. *S* **1981**, 137.
26. Hoyer, S.; Laszlo, P.; Orlovic, M.; Polla, E. *S* **1986**, 655.
27. Delaude, L.; Laszlo, P. *TL* **1991**, *32*, 3705.
28. Onaka, M.; Shinoda, T.; Izumi, Y.; Nolen, R. *CL* **1993**, 117.
29. Cornélis, A.; Laszlo, P.; Pennetreau, P. *Clay Minerals* **1983**, *18*, 437.
30. Laszlo, P.; Luchetti, J. *CL* **1993**, 449.
31. Kawai, M.; Onaka, M.; Isumi, Y. *BCJ* **1988**, *61*, 1237.
32. Tennakoon, D. T. B.; Thomas, J. M.; Tricker, M. J.; Williams, J. O. *JCS(D)* **1974**, 2207.
33. Cativiela, C.; Figueras, F.; Fraile, J. M.; Garcia, J. I.; Mayoral, J. A. *TA* **1993**, *4*, 223 and references therein.
34. Laszlo, P.; Lucchetti, J. *TL* **1984**, *25*, 2147.
35. Laszlo, P.; Lucchetti, J. *TL* **1984**, *25*, 4387.
36. Laszlo, P.; Lucchetti, J. *TL* **1984**, *25*, 1567.
37. Olah, G. A. *Friedel–Crafts Chemistry*; Wiley: New York, 1973.
38. Olah, G. A.; Reddy, V. P.; Prakash, G. K. S. In *Kirk-Othmer Encyclopedia of Chemical Technology*, 4th ed.; Wiley: New York, 1994; Vol. 11, p 1042.
39. Chiche, B.; Finiels, A.; Gauthier, C.; Geneste, P.; Graille, J.; Piock, D. *J. Mol. Catal.* **1987**, *42*, 229.
40. Onaka, M.; Ohno, R.; Kawai, M.; Isumi, Y. *BCJ* **1987**, *60*, 2689.
41. Onaka, M.; Ohno, R.; Izumi, Y. *TL* **1989**, *30*, 747.

André Cornélis & Pierre Laszlo
Université de Liège, Belgium
Mark W. Zettler
The Dow Chemical Company, Midland, MI, USA

2-Morpholinoethyl Isocyanide

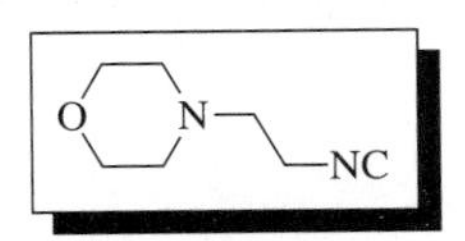

[78375-48-1] $C_7H_{12}N_2O$ (MW 140.21)

(coupling reagent for amino acids; amide synthesis)

Physical Data: bp 72–73 °C/0.7 mmHg; *d* 1.017 g cm^{-3}; n_D 1.469.

Solubility: usually used in CH_2Cl_2.

Form Supplied in: liquid, >98% pure.

Analysis of Reagent Purity: comparison wideth literature boiling point; also ν_{max} (liquid film) at 2150 cm^{-1} is indicative of isocyanide functionality.

Preparative Methods: by the reaction of 2-morpholinoethylamine with ***Formic Acid*** in boiling toluene with a water separator, to yield the formamide which is treated either with ***Phosgene*** and ***Triethylamine*** in CH_2Cl_2 or with ***Phosphorus Oxychloride*** and ***Diisopropylamine***.[1,2] The use of diisopropylamine as base is favored as yields are improved, typically to 68%. ***Trichloroacetic Anhydride*** has also been used as an alternative for $COCl_2$, giving a yield of 74%.[3]

Purification: when phosphorus oxychloride and diisopropylamine are employed, purification is not required.[2]

Introduction. The principal synthetic application of 2-morpholinoethyl isocyanide is as a coupling reagent of amino acids to yield peptides.[2] It also reacts with aldehydes or ketones to yield amides,[4] with acids to yield imidazolinium salts,[5] and has been used as a ligand in spectroscopic studies to examine shielding effects.[6]

Coupling of Amino Acids.[2] The coupling reaction is performed in CH_2Cl_2 with ***N*-Hydroxysuccinimide** or ***1*-Hydroxybenzotriazole** and 2-morpholinoethyl isocyanide (**1**) to yield products in 61–95% yield (eq 1).

$$\text{Cbz–Val–OH} + \text{H–Gly–OMe} \xrightarrow[80\%]{(\mathbf{1}),\ \text{HOSu},\ CH_2Cl_2} \text{Cbz–Val–Gly–OMe} \quad (1)$$

No racemization is observed, as determined by the coupling reaction of CF_3CO–Phe–OH with H–Phe–O-*t*-Bu in the presence of HOSu.

Formation of Amides.[4] The isocyanide moiety of 2-morpholinoethyl isocyanide reacts with aldehydes and ketones to form amides in good yields (eq 2).

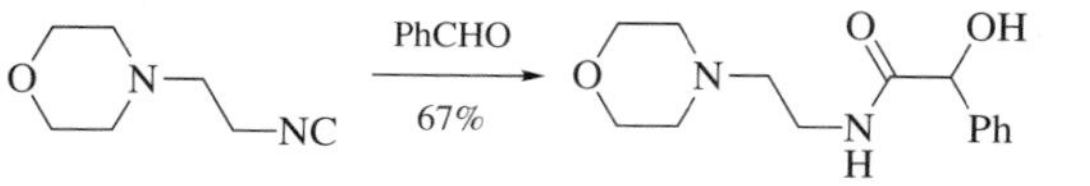

PhCHO, 67% (2)

Formation of Imidazolinium Salts by Cyclization.[5] Treatment of the isocyanide with HZ (Z = Cl, TsO) gives 40–60% of the spiroimidazolium salt which on addition of excess HZ provides the imidazolinium salt in 70–95% yield (eq 3).

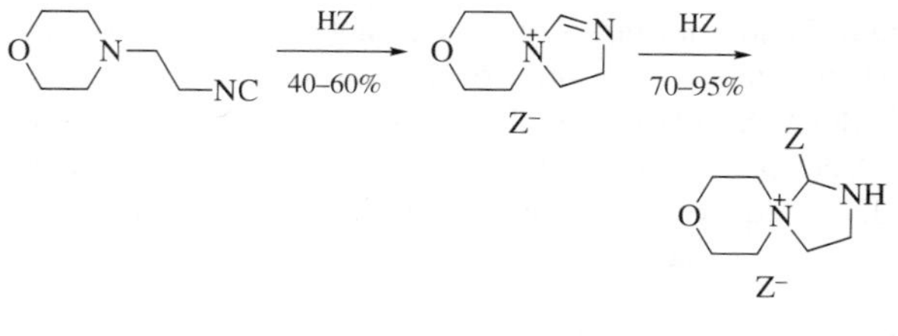

HZ, 40–60%; HZ, 70–95% (3)

Use in Spectroscopic Studies.[6] 2-Morpholinoethyl isocyanide has been used to study the NMR ^{14}N NMR signals of molybdenum (^{95}Mo) and tungsten (^{183}W) carbonyl isocyanide complexes $M(CO)_{6-n}(CNR)_n$ ($n = 0$–6). When isocyanide is coordinated to the metal center, the ^{14}N signal is deshielded but, with further substitution, the N becomes more shielded again. With UV and IR studies, substitution of CO by isocyanide shifts the absorption toward longer wavelength so that the color of the complexes changes from white to orange–red.

---

1. Helmut, A.; Koch, G.; Marquarding, D. *Chem. Pept. Proteins, Proc. USSR–FRG Symp. 3rd*; 1982; p 209.
2. Obrecht, R.; Herrmann, R; Ugi, I. *S* **1985**, *4*, 400.
3. Eckert, H.; Forster, B. *AG* **1987**, *99*, 922.
4. Seebach, D.; Adam, G; Gees, T; Schiess, M; Wiegand, W. *CB* **1988**, *121*, 507.
5. Polyokov, A. I.; Baskakov, Y. A.; Artamonova, O. S.; Baranova, S. S. *KGS* **1983**, *6*, 843.
6. Minelli, M; Masley, W. J. *IC* **1989**, *28*, 2954.

Helen Osborn
University of Bristol, UK

o-Nitrobenzyl Alcohol

[612-25-9] $C_7H_7NO_3$ (MW 153.15)

(photoremovable protective group for ketones and aldehydes,[1] carboxylic acids,[2] anhydrides,[3] acid chlorides,[4] and amines;[5a] reaction with isocyanates to generate protected amines;[5b,c,d] protection of amino acids;[2b,6] protection of the hydroxyl groups of ribonucleotides,[7a,b,c] dinucleotides,[7d] and furanoses;[7e] synthesis of a novel catalyst in the polymerization of epoxides;[8] formation of leaving groups of enhanced ability;[9] reaction with formyl Meldrum's acid to afford the formylacetate;[10] oxidation to the corresponding aldehyde[11] or carboxylic acid[12])

Alternate Name: o-NBA.
Physical Data: mp 70–72 °C; bp 270 °C. The thermal stability of nitrobenzyl alcohols has been examined.[17] Its conformation in solution has been studied using ^{17}O NMR[18a] and ^{13}C NMR[18b] analysis.
Form Supplied in: light tan powder.
Handling, Storage, and Precautions: light-sensitive. Harmful if swallowed or inhaled.

Mechanism of Action. Under photolysis,[19] acidic,[20] and electron bombardment[21] conditions, the transformation of o-nitrobenzyl alcohol or its derivatives involves an internal redox reaction sequence followed by liberation of the deprotected alcohol or amine (eq 1). Analogously, the photorearrangement of esters of o-NBA, obtained through its reaction with acid chlorides[2a,4] or anhydrides,[3] also induces an internal redox reaction (eq 2).

(1)

R = H or protected functionality, X = O or N

Protection of Ketones and Aldehydes. o-NBA serves as a photoremovable protective group for aldehydes and ketones. Bisacetals are prepared in a high-yield exchange reaction, except with hindered ketones for which it is necessary to employ the diol o-$O_2NC_6H_4CH(OH)CH_2OH$. The acetals are removed in excellent yield by irradiation at 350 nm in benzene (eq 3).[1]

(2)

o-NBA, 2,2-dimethoxypropane, β-$C_{10}H_7SO_3H$ (cat.), benzene, 80 °C, 70–95%; hν (350 nm), benzene, 85–90%

Ar = o-$O_2NC_6H_4$ (3)

Generation of Carboxylic Acids. Photolabile phosphatidylcholine (PLPC) has been synthesized by reaction of o-NBA with dodecanedioyl dichloride followed by ***1,3-Dicyclohexylcarbodiimide*** coupling. Subsequent photolysis of the resulting 2-nitrobenzyl esters results in the instantaneous disintegration of the PLPC liposomes, thus opening up the possibility of their use as drug carriers in vivo.[4] Irradiation of copolymers with o-nitrobenzyl ester side groups, synthesized from methacrylyl chloride, generates free carboxylic acids, as illustrated in eq 4.[2a] The positive photosensitive polyiimide biphenyl precursor pictured in eq 5 has been made by treatment of biphenyltetracarboxylic acid dianhydride (BPDA) with o-NBA.[3]

hν, 40% (4)

o-NBA (2 equiv), py (cat.), toluene, 90% (5)

Protection of Amines. Reaction of o-NBA with ***Phosgene*** yields the stable compound o-nitrobenzyloxycarbonyl chloride (eq 6).[5a] Subsequent reaction with 2-amino-2-deoxy-D-glucose cleanly affords the protected amine. The yields of the free amines obtained upon irradiation (eq 7) are lowered by side reactions with the o-nitrobenzaldehyde generated under the reaction conditions.[2a,5a,b] Yields are increased by the addition of polymeric aldehydes or sulfuric acid to the reaction mixture during irradiation,[5a] or by utilizing the related 2,6-dinitrobenzyl alcohol in the first step.[5b] Functional group tolerance for this entire reaction sequence is impressively high.

Reaction with isocyanates also provides a route to carbamates, which serve as masked amines[5b,c] (eq 8) or as diamines[5d] for epoxy resin curing. Carbamates of *o*-NBA possess moderate antitumor activity.[22]

Protection of Amino Acids. Photolabile *N*-protected amino acids and peptides have also been synthesized via their chloroformates.[2b,6] Quantitative deprotection proceeds without racemization[23,24] when an excess of ***Sulfuric Acid*** or ***Semicarbazide*** hydrochloride is added to the reaction mixture (eq 9). Conversely, treatment of serine with *o*-NBA in the presence of catalytic ***p-Toluenesulfonic Acid*** provides the *o*-nitrobenzyl ester of serine (eq 10).[2b]

Protection of Hydroxyl Groups. Photolabile 2′-*O*-(*o*-nitrobenzyl) derivatives of various ribonucleotides[7a–c] as well as dinucleotides[7d] have been synthesized, and their subsequent photodeprotections, which yield the corresponding alcohols, examined (eq 11). Protection of the anomeric hydroxyl functionality of 2-deoxy-D-ribofuranose has been accomplished using *o*-NBA (eq 12).[7e] This protective group is quite stable to acidic conditions, and has proven invaluable in the synthesis of abasic oligonucleotides. Similar conditions are employed in the synthesis of bis(*o*-nitrobenzyl) ethers.[25]

B = purine or pyrimidine

Miscellaneous Reactions. *o*-Nitrobenzyl triphenylsilyl ether, a latent photogenerated source of triphenylsilanol, has been synthesized from *o*-NBA (eq 13). This new catalyst, upon irradiation at 365 nm, serves as a coinitiator in the polymerization of epoxides.[8]

Greatly enhanced leaving group ability has been realized by use of *o*-NBA. Steroidal sulfonates synthesized from the sulfonylating agent formed in eq 14 undergo bimolecular azide displacements more efficiently than traditional sulfonates.[9]

o-Nitrobenzyl formylacetate has been made by the reaction of formyl Meldrum's acid with *o*-NBA in dry toluene (eq 15).[10] α-Hydroxyimino phosphonates can serve as precursors to alkyl metaphosphates when photolabile ester groups are incorporated (eq 16).[26]

1. o-NBA, CH_2Cl_2
2. $NH_2OH{\bullet}HCl$, py
90% (16)

Oxidations and Reductions. Oxidation of o-nitrobenzyl alcohol to its aldehyde has been accomplished using bis(dihydrogentellurato)M^{III} where M = Cu^{III} and Ag^{III},[11a] chromium trioxide and alumina,[11b] biphasic nitric acid,[11c] bromamine-B,[11d] ethyl chlorocarbamate,[11e] potassium bromate in aqueous HCl,[11f] pyridinium fluorochromate,[11g] N-bromoacetamide,[11h] thallium(III),[11i] ammonium molybdate tetrahydrate in 6 M HCl,[11j] 1-chlorobenzotriazole,[11k] pyridine oxodiperoxychromium(VI),[11l] cobalt(III) hydroxide,[11m] cerium(IV) perchlorate,[11n] or silver(II) oxide in acidic media.[11o] Oxidation to o-nitrobenzoic acid is possible using methyltrioctylammonium tetrakis(oxodiperoxotungsto)phosphate,[12a] electrogenerated superoxide,[12b] $(NH_4)VO_3$ and dilute $HClO_4$,[12c] or benzyltrimethylammonium tribromide.[12d] Biological oxidation of o-nitrotoluene to o-NBA by *Pseudomonas* sp. strain JS150 and *Pseudomonas putida* F1 has also been reported.[13]

Reduction of esters of o-nitrobenzyl alcohol to yield o-NBA can be accomplished using sodium borohydride in aqueous media.[14] Reductions of o-nitrobenzaldehyde to o-NBA using sodium borohydride under phase transfer conditions,[15a] dimethoxyborane–$CoCl_2$,[15b] and trimethoxysilane–lithium methoxide[15c] have been reported. Reduction of the nitro group of o-NBA to the amine using catalytic hydrogenation with Pd^0 and polybenzimidazole resin in DMF is likewise known.[16]

Related Reagents. o-Nitrobenzyl Bromide.

1. Gravel, D.; Murray, S.; Ladoucer, G. *CC* **1985**, 1828.
2. (a) Barzynski, H.; Sänger, D. *Angew. Makromol. Chem.* **1981**, *93*, 131. (b) Pirrung, M. C.; Nunn, D. S. *BML* **1992**, *2*, 1489.
3. Kubota, S.; Yamawaki, Y.; Moriwaki, T.; Eto, S. *Polym. Eng. Sci.* **1989**, *29*, 950.
4. Kusumi, A.; Nakahama, S.; Yamaguchi, K. *CL* **1989**, 433.
5. (a) Amit, B.; Zehavi, U.; Patchornik, A. *JOC* **1974**, *39*, 192. (b) Cameron, J. F.; Friéchet, J. M. J. *JACS* **1991**, *113*, 4303. (c) Cummings, R. T.; Krafft, G. A. *TL* **1988**, *29*, 65. (d) Nishikubo, T.; Takehara, E.; Kameyama, A. *Polym. J.* **1993**, *25*, 421.
6. Patchornik, A.; Amit, B.; Woodward, R. B. *JACS* **1970**, *92*, 6333.
7. (a) Ohtsuka, E.; Tanaka, T.; Tanaka, S.; Ikehara, M. *JACS* **1978**, *100*, 4580. (b) Ohtsuka, E.; Tanaka, S.; Ikehara, M. *CPB* **1977**, *25*, 949. (c) Ohtsuka, E.; Tanaka, S.; Ikehara, M. *S* **1977**, 453. (d) Ohtsuka, E.; Tanaka, S.; Ikehara, M. *Nucleic Acids Res.* **1974**, *1*, 1351. (e) Péoc'h, D.; Meyer, A.; Imbach, J.-L.; Rayner, B. *TL* **1991**, *32*, 207.
8. Hayase, S.; Onishi, Y.; Suzuki, S.; Wada, M. *Macromolecules* **1985**, *18*, 1799.
9. Zehavi, U. *JOC* **1975**, *40*, 3870.
10. Sato, M.; Yoneda, N.; Katagiri, N.; Watanabe, H.; Kaneko, C. *S* **1986**, 672.
11. (a) Gupta, K. K. S.; Nandy, B. K.; Gupta, S. S. *JCS(P2)* **1993**, 767. (b) Hirano, M.; Kuroda, H.; Morimoto, T. *BCJ* **1990**, *63*, 2433. (c) Gasparrini, F.; Giovannoli, M.; Misiti, D.; Natile, G.; Palmieri, G. *SC* **1988**, *18*, 69. (d) Mathur, S.; Gupta, A.; Banerji, K. K. *Oxid. Commun.* **1988**, *11*, 137. (e) Jain, S. M.; Banerji, K. K. *BCJ* **1988**, *61*, 1767. (f) Gupta, K. K. S.; Kumar, S. C.; Sen, P. K.; Banerjee, A. *T* **1988**, *44*, 2225. (g) Banerji, K. K. *JOC* **1988**, *53*, 2154. (h) Agarwal, A.; Mathur, S.; Banerji, K. K. *JCR(S)* **1987**, 176. (i) Sur, B.; Adak, M. M.; Pathak, T.; Hazra, B.; Banerjee, A. *S* **1985**, 652. (j) Rangadurai, A.; Thiagarajan, V.; Venkatasubramanian, N. *IJC(B)* **1982**, *21B*, 42. (k) Fleet, G. W. J.; Little, W. *TL* **1977**, 3749. (l) Srinivasan, V. S.; Venkatasubramanian, N. *IJC* **1975**, *13*, 526. (m) Syper, L. *Rocz. Chem.* **1973**, *47*, 43. (n) Balasubramanian, T. R.; Venkatasubramanian, N. *IJC* **1970**, *8*, 305. (o) Syper, L. *TL* **1967**, 4193.
12. (a) Venturello, C.; Gambaro, M. *JOC* **1991**, *56*, 5924. (b) Singh, M.; Singh, K. N.; Dwivedi, S.; Misra, R. A. *S* **1991**, 291. (c) Banerjee, A.; Dutt, S.; Banerjee, G. C.; Hazra, B.; Datta, H.; Banerjee, S. *IJC(B)* **1990**, *29B*, 257. (d) Okamoto, T.; Uesugi, T.; Kakinami, T.; Utsunomiya, T.; Kajigaeshi, S. *BCJ* **1989**, *62*, 3748.
13. Robertson, J. B.; Spain, J. C.; Haddock, J. D.; Gibson, D. T. *Appl. Environ. Microbiol.* **1992**, *58*, 2643.
14. Bianco, A.; Passacantilli, P.; Righi, G. *SC* **1988**, *18*, 1765.
15. (a) Someswara Rao, C.; Deshmukh, A. A.; Patel, B. J. *IJC(B)* **1986**, *25B*, 626. (b) Nose, A.; Kudo, T. *CPB* **1989**, *37*, 808. (c) Hosomi, A.; Hayashida, H.; Kohra, S.; Tominaga, Y. *CC* **1986**, 1411.
16. Li, N.-H.; Fréchet, J. M. J. *CC* **1985**, 1100.
17. Cardillo, P.; Girelli, A. *Chim. Ind. (Milan)* **1986**, *68*, 68.
18. (a) Monti, D.; Orsini, F.; Ricca, G. S. *Spectrosc. Lett.* **1986**, *19*, 505. (b) Abraham, R. J.; Bakke, J. M. *T* **1978**, *34*, 2947.
19. (a) Barltrop, J. A.; Bunce, N. J. *JCS(C)* **1968**, 1467. (b) Wan, P.; Yates, K. *CJC* **1986**, *64*, 2076.
20. Bakke, J. *ACS(B)* **1974**, *28*, 645.
21. McLuckey, S. A.; Glish, G. L. *Org. Mass Spectrom.* **1987**, *22*, 224.
22. Perlman, M. E.; Dunn, J. A.; Piscitelli, T. A.; Earle, J.; Rose, W. C.; Wampler, G. L.; MacDiarmid, J. E.; Bardos, T. J. *JMC* **1991**, *34*, 1400.
23. Barltrop, J. A.; Plant, P. J.; Schofield, P. *CC* **1966**, 822.
24. Manning, J. M.; Moore, S. *JBC* **1968**, *243*, 5591.
25. Smith, M. A.; Weinstein, B.; Greene, F. D. *JOC* **1980**, *45*, 4597.
26. Breuer, E.; Mahajna, M. *JOC* **1991**, *56*, 4791.

D. Todd Whitaker
Detroit Country Day School, Beverly Hills, MI, USA

K. Sinclair Whitaker
Wayne State University, Detroit, MI, USA

o-Nitrophenol

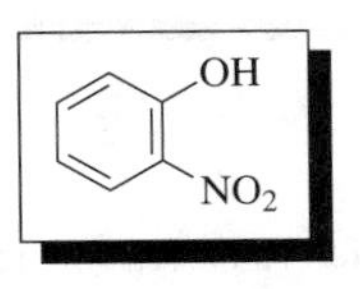

[88-75-5] $C_6H_5NO_3$ (MW 139.12)

(preparation of active esters for peptide synthesis[1])

Physical Data: mp 45 °C; bp 216 °C; pK_a 7.23 (at 25 °C).
Solubility: sol ether, benzene; sparingly sol H_2O.
Form Supplied in: yellow solid, commercially available.
Purification: crystallizes as yellow needles or prisms from ethanol or ether.
Handling, Storage, and Precautions: moderately toxic by inhalation and skin absorption.

Preparation of Active Esters for Peptide Synthesis. Esters of o-nitrophenol, prepared by the ***1,3-Dicyclohexylcarbodiimide***

coupling method, can be used in peptide synthesis (eq 1)[1] in the same way as are the corresponding *p*-nitrophenyl esters (see ***p*-Nitrophenol**). *o*-Nitrophenol is less reactive than *p*-nitrophenol in the DCC coupling step; use of pyridine as solvent instead of ethyl acetate increases the nucleophilicity of the hydroxyl group and minimizes side reactions. In the aminolysis step, the *o*-nitrophenyl esters are more reactive than the *p*-isomers, and the reaction rates are less sensitive to changes in solvent or to steric hindrance. *o*-Nitrophenyl esters have been used in a stepwise synthesis of a nonapeptide in which all synthetic operations are performed in the same reaction vessel, without isolation of intermediates ('in situ' peptide synthesis).[1]

1. Bodanszky, M.; Funk, K. W.; Fink, M. L. *JOC* **1973**, *38*, 3565; Bodanszky, M.; Kondo, M.; Yang Lin, C.; Sigler, G. F. *JOC* **1974**, *39*, 444.

Alan Armstrong
University of Bath, UK

p-Nitrophenol

[*100-02-7*] $C_6H_5NO_3$ (MW 139.12)

(preparation of active esters for peptide synthesis[1])

Physical Data: mp 114 °C; pK_a 7.15 (at 25 °C); dimorphous.
Solubility: moderately sol cold water; sol alcohol, chloroform, ether.
Form Supplied in: generally obtained as a mixture of the two crystalline forms.
Purification: recrystallization from toluene gives the metastable α-form if performed at >63 °C; the β-form is obtained as yellow prisms at <63 °C.
Handling, Storage, and Precautions: the α-form is light stable, whereas the yellow β-form gradually turns red in light. Moderately toxic by inhalation and skin absorption.

Preparation of Active Esters for Peptide Synthesis. *p*-Nitrophenyl esters, prepared by ***1,3-Dicyclohexylcarbodiimide***-mediated coupling between *N*-protected amino acids and *p*-nitrophenol, are among the most useful activated esters for peptide synthesis (eq 1).[1] They are readily purified by recrystallization, often from alcoholic solvents with which they do not react. Aminolysis occurs at a reasonable rate, generally at room temperature without a catalyst. The *p*-nitrophenol generated as a byproduct in this step is easily removed. *o*-Nitrophenyl esters can be used similarly (see ***o*-Nitrophenol**); a study of the rates of aminolysis of Boc-L-leucine nitrophenyl esters showed the *o*-isomer to react 4–9 times faster than the *p*-isomer.[2]

1. Bodanszky, M.; Du Vigneaud, V. *JACS* **1959**, *81*, 5688; Bodanszky, M.; Meienhofer, J.; Du Vigneaud, V. *JACS* **1960**, *82*, 3195.
2. Bodanszky, M.; Bath, R. J. *CC* **1969**, 1259.

Alan Armstrong
University of Bath, UK

Nonacarbonyldiiron[1]

$Fe_2(CO)_9$

[*15321-51-4*] $C_9Fe_2O_9$ (MW 363.79)

(precursor for iron carbonyl complexes, carbonylation, dehalogenation, deoxygenation, and reductionreactions)

Alternate Names: diiron enneacarbonyl; diiron nonacarbonyl; tri-μ-carbonylhexacarbonyldiiron.
Physical Data: mp 100–120 °C (dec); *d* 2.08 g cm^{-3}; X-ray diffraction study.[3]
Solubility: insol organic solvents.
Form Supplied in: orange flaky solid.
Preparative Method: formed by exposing ***Pentacarbonyliron*** to light (eq 1).[1] The complex contains six terminal carbonyl groups and three bridging carbonyl groups.[2]

$$2\,Fe(CO)_5 \xrightarrow[74\%]{h\nu} Fe_2(CO)_9 + CO \quad (1)$$

Handling, Storage, and Precautions: because of its much lower solubility and volatility, $Fe_2(CO)_9$ is less dangerous to handle than $Fe(CO)_5$, though it should be remembered that $Fe(CO)_5$ is often formed in reactions using $Fe_2(CO)_9$; when handling $Fe_2(CO)_9$, avoid dust formed by this light flaky substance; use in a fume hood; store in a freezer; when fully dried, commercial samples can be pyrophoric.[4]

Diene Complexation.

Cyclohexadiene Series. Tricarbonyliron complexes are formed by reaction of $Fe(CO)_4$ with dienes. $Fe_2(CO)_9$ is an important starting material for this reaction because it offers a convenient source of $Fe(CO)_4$ generated under much milder conditions than from pentacarbonyliron (eq 2).

$$Fe_2(CO)_9 \longrightarrow Fe(CO)_4 + Fe(CO)_5 \qquad (2)$$

Starting from a cisoid 1,3-diene, $Fe_2(CO)_9$ can yield, under mild thermal conditions, tricarbonyl(η^4-1,3-diene)iron complexes. Cyclohexa-1,4-dienes (readily available from the Birch reduction of substituted benzenes) often require prior conjugation, but conjugated 1,3-dienes (both cyclic and acylic) can be converted into tricarbonyliron complexes directly. Some of the earliest examples[5–9] of complexation of 1,3-dienes employed $Fe_2(CO)_9$ (eq 3, dates shown in parentheses). (Alternative reagents in the early days were ***Pentacarbonyliron*** or ***Dodecacarbonyltriiron***.)

Ligand —$Fe_2(CO)_9$→ CO_2H / $Fe(CO)_3$ (1971); $Fe(CO)_3$ (1973); $(OC)_3Fe$ (1976); $(OC)_3Fe$, OMe (1977); $Fe(CO)_3$ (1977) (3)

More recently, the use of $Fe_2(CO)_9$ has been increasingly valued because, under mild conditions, improved regio- and stereocontrol are possible, and sensitive functionality is less severely affected by the complexation conditions. Regiocontrol is often a problem in the complexation of 1,4-dienes by high-temperature direct reaction with $Fe(CO)_5$. Preparation of the diene complex (**1**), for example, can be improved by preconjugation of the 1,4-diene with ***p-Toluenesulfonic Acid***[10] and by complexation of the resulting equilibrium concentration of the 1,3-diene with $Fe_2(CO)_9$ under conditions in which the 1,4-isomer does not react (eq 4). Bis-alkene complexes are formed when rearrangement to bring the double bonds into conjugation is not possible.[11]

OMe → 1. TsOH; 2. $Fe_2(CO)_9$, Et_2O, Δ → OMe, $Fe(CO)_3$ (**1**) (4)

Good diastereoselectivity can also be a significant advantage. The dioxygenation of arenes by *Pseudomonas putida* provides optically active *cis*-diol products.[12] In the mild complexation conditions using $Fe_2(CO)_9$, the chirality of the diol is relayed to tricarbonyliron complexes, providing enantiopure intermediates for use in asymmetric synthesis.[13]

Cycloheptadiene Series. The treatment of (**2**) with $Fe_2(CO)_9$ in acetone at 35 °C gives a 1:4 mixture in favor of (**4**). Enriched samples of (**3**) or (**4**) are accessible directly (eq 5), according to the reaction conditions.[14]

CO_2Me (**2**) —$Fe_2(CO)_9$→ $(OC)_3Fe$, CO_2Me (**3**) + $Fe(CO)_3$, CO_2Me (**4**) (5)

Cycloheptatrienone (tropone) reacts efficiently in 60% yield with $Fe_2(CO)_9$ to give a complex (**5**) in which one double bond remains uncomplexed (eq 6).[15] This double bond displays normal reactivity, while the remaining unsaturation is effectively protected. Dinuclear bis-η^3-tricarbonyliron complexes can also be formed.[16]

O= —$Fe_2(CO)_9$→ O=, $Fe(CO)_3$ (**5**) (6)

Complexation of Natural Products. The $Fe(CO)_3$ moiety can provide a method for diene protection. Steroid interconversions employed tricarbonyl(ergosteryl acetate)iron(0)[17] and tricarbonyl(ergosteryl benzoate)iron(0).[18] Thus treatment of ergosteryl acetate or benzoate with tricarbonyl(4-methoxybenzylideneacetone)iron(0) (preformed by complexation using $Fe_2(CO)_9$, or with $Fe_2(CO)_9$ in the presence of 4-methoxybenzylideneacetone[19]), gives the pure desired steroid complex in good yield (60% and 80%, respectively). Nopadiene has been complexed by reaction with $Fe_2(CO)_9$ to form an optically active metal complex.[20]

Exocyclic Diene Complexation. Formation of iron carbonyl complexes of (**6**) illustrates the mild nature of the complexation conditions (eq 7).[21]

Reaction with $Fe_2(CO)_9$ is *exo* face selective, giving a mixture of the corresponding mono complexes (**7**) (*anti-exo*) and (**8**) (*syn-exo*). Further complexation of (**8**) with $Fe_2(CO)_9$ occurs in a nonstereoselective fashion to give mixtures of the diiron complexes (*anti-exo, syn-endo*), (*anti-exo, syn-exo*), and (*anti-endo, syn-exo*).

On treatment of (**9**) with $Fe_2(CO)_9$, the endocyclic double bond is first coordinated to yield the *exo*-$Fe(CO)_4$ complex (**10**), which then further reacts with $Fe_2(CO)_9$ to afford the diiron complex as a major product (eq 8).

(**6**) —$Fe_2(CO)_9$→ (**7**) + (**8**) (7)

(**9**) —$Fe_2(CO)_9$→ (**10**) (8)

Heating promotes deoxygenation to form the substituted tricarbonyl(*o*-quinodimethide)iron complex (**11**). Oxidative removal of the $Fe(CO)_3$ moiety in (**11**) affords the indanone derivative (**12**) by CO insertion (eq 9).[22]

(**11**) —[O]→ (**12**) (9)

Complexes of 7-azanorbornadienes have been examined as sources of nitrenes.[23]

Complexation of Acyclic Dienes. Complexation of *trans*-4-methyl-2,4-pentadienol (**13**) with $Fe_2(CO)_9$ gives the dienol complex (**14**) as a yellow oil. The product is valuable as a precursor for the cation (**15**) by reaction with $HPF_6/Ac_2O/Et_2O$ (eq 10).[24] Substituted dienes with the (*Z*) configuration at a methyl-bearing terminus complex less efficiently.[25]

(**13**) —$Fe_2(CO)_9$, 13%→ (**14**) —HPF_6, Ac_2O, 52%→ (**15**) (10)

$Fe(CO)_3$–diene aldehydes have been formed using $Fe_2(CO)_9$ in work that culminated in an efficient asymmetric induction using chiral allylborane reagents to give access to metal complexes in high optical purity.[26]

1-Methoxybutadiene has been complexed photochemically using $Fe_2(CO)_9$, instead of the more conventional thermal conditions.[27] Complexation of acyclic dienes with $Fe_2(CO)_9$ constituted the first step toward trimethylenemethane complexes.[28] All details concerning the preparation and spectral properties of tricarbonyl(butadiene)iron complexes are gathered in recent reviews.[29]

$Fe_2(CO)_9$/Ultrasound Method of Complexation. The reaction of $Fe_2(CO)_9$ with sensitive dienes can be promoted by ultrasound (eq 11). Syntheses of a variety of tricarbonyl(η^4-diene)iron(0) complexes (**16**) use the convenient high-yielding sonication method.[30]

diene —$Fe_2(CO)_9$,))))), PhH, rt→ (**16**) (11)

R^1 = OAc; R^2, R^3, R^4 = H; 100%

$Fe_2(CO)_9$ Used with a Tertiary Amine. Aliphatic trialkylaminotetracarbonyliron(0) complexes result from direct reaction between R_3N and $Fe_2(CO)_9$.[31] Fair to good yields of (**17**) can be obtained (eq 12).

$$R_3N + Fe_2(CO)_9 \xrightarrow[\text{rt, } N_2]{\text{hexane or THF}} R_3N{\bullet}Fe(CO)_4\ (\mathbf{17}) + Fe(CO)_5 \quad (12)$$

The complex (**17**) transfers $Fe(CO)_4$ to dienes, monoenes (resulting in isomerization), and $Fe(CO)_5$ (to form $Fe_2(CO)_9$ and $Fe_3(CO)_{12}$). The R_3N unit can also be replaced by other amines and by phosphines. The amine/phosphine exchange was performed with $Me_3N{\cdot}Fe(CO)_4$ at 60 °C in quantitative yield after 0.5 h. The amine/diene exchange reaction has been studied with 1,3-cyclohexadiene in hydrocarbon solution at 52 °C (eq 13).

$$\text{cyclohexadiene} + \text{amine}{\bullet}Fe(CO)_4\ (\mathbf{17}) \rightleftharpoons (\text{cyclohexadiene})Fe(CO)_3 + \text{amine} + CO \quad (13)$$

Other Routes to η^4 Complexes. Alternative access to $Fe(CO)_3$ complexes are afforded by a variety of more unusual starting materials: cyclopropenes, dibromides, dihydrothiophene dioxides, alkynes/CO, allenes, and cyclohexadienones are all discussed below.

Enone Complexation. Various α,β-unsaturated ketones form moderately stable tricarbonyliron complexes (**18**),[32,37] and offer access to 1,4-diketones (**19**) by reaction with Grignard reagents, organolithium reagents (eq 14), or organocuprates.[33] Trimethylsilyloxybutadienes require phenyl substituents to form stable complexes by reaction with $Fe_2(CO)_9$.[34] Enone complexes are also useful as transfer reagents to place $Fe(CO)_3$ on diene ligands.[32,35] A related procedure employs complexes of α,β-unsaturated imines in the same way.[36]

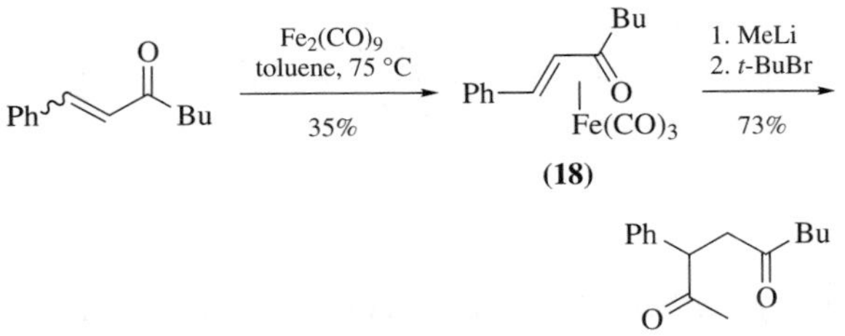

(14)

(19)

Reagents for asymmetric transfer of $Fe(CO)_3$ to dienes have been obtained thermally from (+)-pulegone[37,38] and (−)-3β-acetyloxypregna-5,16-dien-20-one[38] by reaction with $Fe_2(CO)_9$. The chiral enone complexes are used without isolation to provide a direct synthesis of optically active (diene)$Fe(CO)_3$ complexes in up to 40% ee.[38,39]

Complexes from Alkenediols. 2-Butene-1,4-diols react with $Fe_2(CO)_9$ in the presence of Lewis acids under ultrasound conditions to form η^3-allyliron complexes.[40]

Ferralactone Complexes.

From Oxazines. The first ferralactone complex (**21**) was obtained during studies on π-allyliron complexes by Heck. The same compounds are also obtained by treatment of oxazines (**20**) with $Fe_2(CO)_9$ (eq 15).[41]

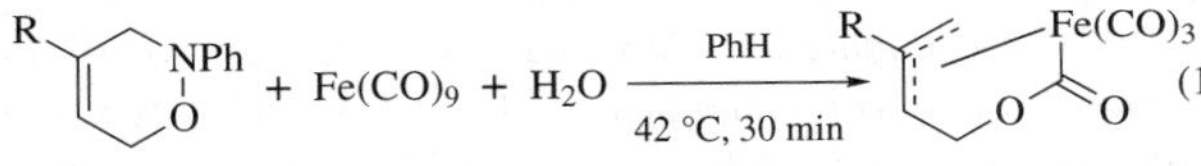

(15)

(20) R = H, Me, D, Cl (21)

Formation of β- and α-Lactones from Vinyloxiranes. An oxidative addition of vinyloxiranes (**22**) to iron(0) complexes provides the most usual synthesis of ferralactone complexes (**21**) (eq 16).[42,43]

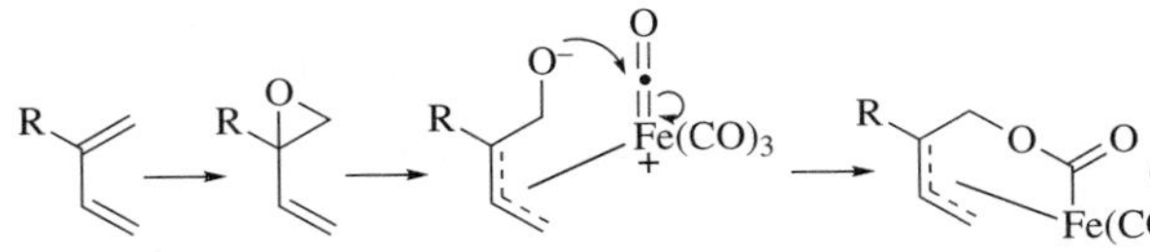

(16)

(22) (21)

Efficient preparations are possible using $Fe_2(CO)_9$ according to two different methods (in THF, or in lower polarity solvents with ultrasound) to produce the ferralactone complexes in moderate to excellent yields.[44] Removal of the metal to form δ-lactones has been successfully applied to total syntheses of parasorbic acid (**23**), the carpenter bee pheromone (**24**), and malyngolide (**25**).[45] Epoxide-derived ferralactone complexes have provided key intermediates in syntheses of routiennocin,[46] avermectin B1a,[47] and (−)-valilactone.[48]

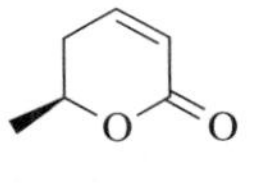

(23) 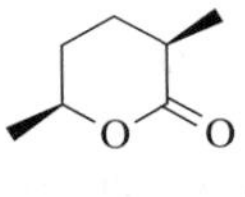 (24) 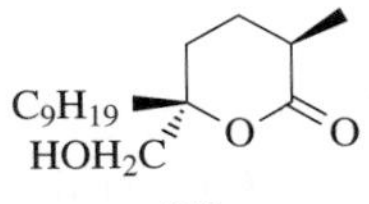 (25)

Demetalation of ferralactone complexes with ***Cerium(IV) Ammonium Nitrate*** produces β-lactones (at low temperature) and/or δ-lactones (at high temperature, under a high pressure of carbon monoxide).

Formation of β-Lactams. The synthesis and oxidative demetallation of ferralactam complexes affords a route to β-lactams.[42] The difficulty of preparation of the monoaziridines limits the scope of direct ring opening, but ferralactam complexes are more readily obtained from ferralactones by reaction with an amine and a mild Lewis acid catalyst.[49] This methodology has been successfully applied to the synthesis of (+)-thienamycin (**28**),[50] via the key intermediates (**26**) and (**27**).

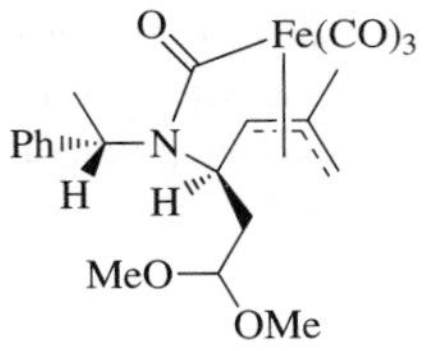

(26) 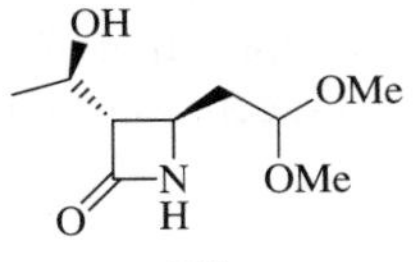 (27)

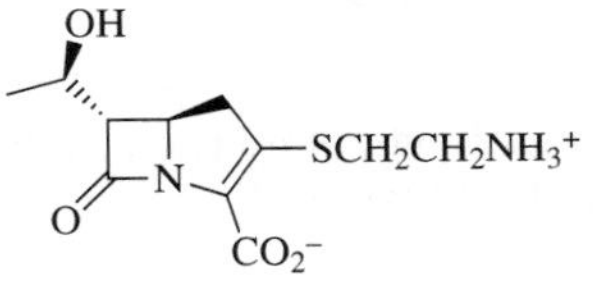

(28)

Complexes from Electrophilic Cyclopropenes. Methyl- and phenyl-substituted cyclopropenes have been shown to react with iron carbonyls by ring opening and carbonylation to yield tricarbonyl(η^3,η^1-allylcarbonyl)iron complexes (eq 17).[51]

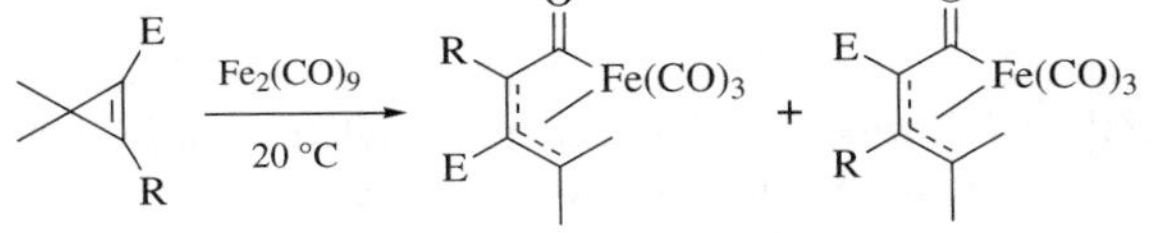

(17)

(29) (30) (31)

Franck-Neumann, starting from the easily accessible electrophilic *gem*-dimethylcyclopropenes (**29**), could obtain the carbonyliron adducts (**30**) and (**31**) in excellent yields.[52,53] The reaction of $Fe_2(CO)_9$ with the isomers of Feist's ester (**32**) leads first to the formation of the corresponding tetracarbonyliron alkene complexes (**33**) (eq 18).[54] Reaction of naphthalenocyclopropene with $Fe_2(CO)_9$ has also been studied.[55]

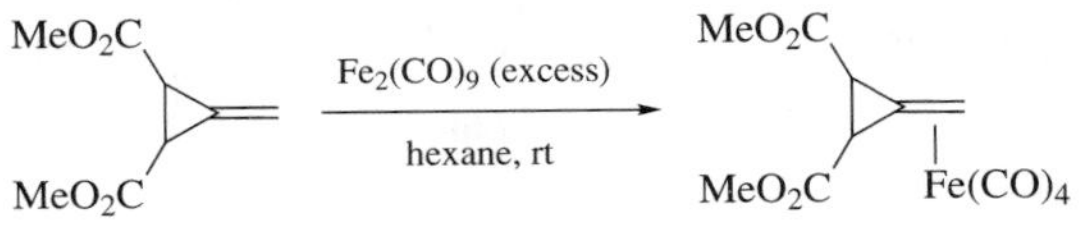

(18)

(32) *cis*	65%	(33) *cis*
(32) *trans*	88%	(33) *trans*

The products from cyclopropene ring opening have the advantage of easy conversion into tricarbonyl(η^4-diene)iron complexes (**34**), its regioisomer and (**35**), by thermal loss of carbon monoxide, or photochemically without CO evolution (eqs 19 and 20).[52,56]

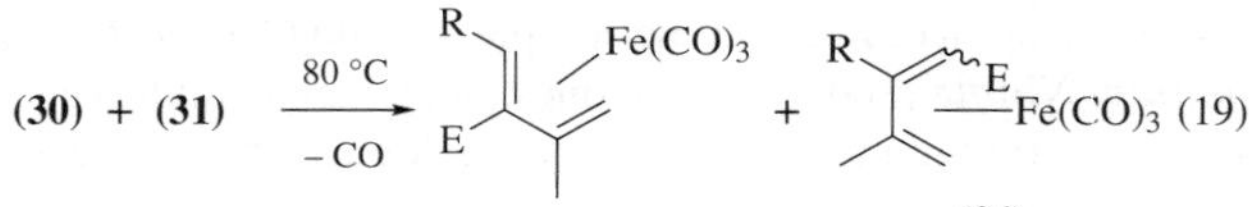

(19)

(20)

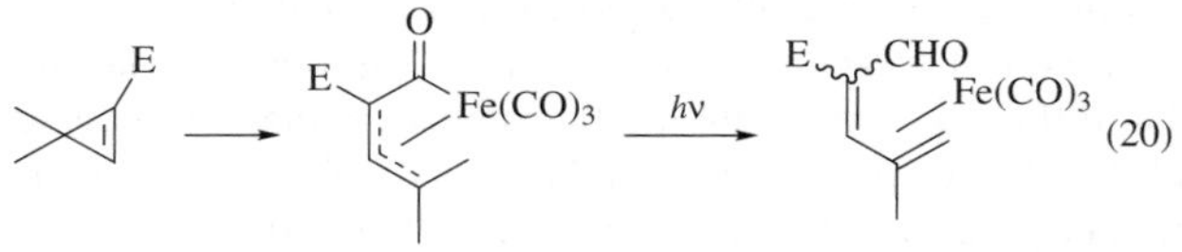

E = CO_2Me

Opening of Cyclopropane Rings. Ring enlargement of compounds such as bicyclo[4.1.0]hept-2-ene (**36**) or its derivatives (+)-2-carene (**37**) or (+)-3-carene (**38**) by carbonylation results in the formation of bicycloheptenones or cycloheptadiene complexes, depending on the reaction conditions.[57]

(**36**) (**37**) (**38**)

Under mild reaction conditions, (+)-2-carene affords the optically active (η^1,η^3) complex (**39**) (eq 21).

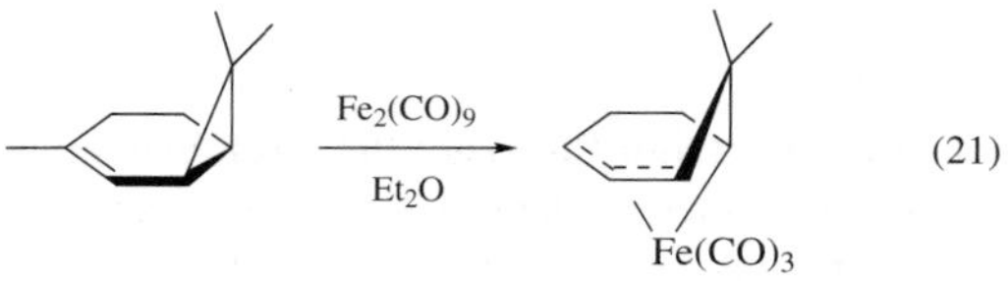

(21)

(**37**) (**39**) $[\alpha]_D = +276°$ >95% ee

Metal and Lewis acid induced carbonylative ring enlargement of chiral and prochiral cyclohexadienes can give access to bicyclo[3.2.1]octanes.[58]

Dehalogenation Reactions.

Formation of Quinonedimethide Complexes. Formation of a stable quinonedimethide complex (**40**) can be achieved by dehalogenation.[59] The product (**40**) decomposes to benzocyclobutene at 500 °C (eq 22).

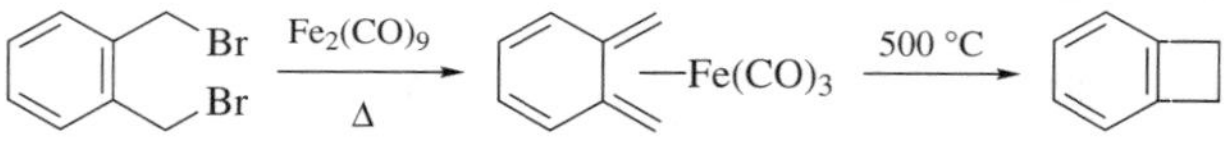

(22)

In the bis(bromomethyl)naphthalene series, a σ-bonded $Fe(CO)_4$ product has been reported.[60]

Formation of Cyclobutadiene Complexes. Cyclobutadiene can be prepared as the stable complex tricarbonyl-(cyclobutadiene)iron(0) (**41**), by reaction of 3,4-dichlorocyclobutene with $Fe_2(CO)_9$ (eq 23).[61]

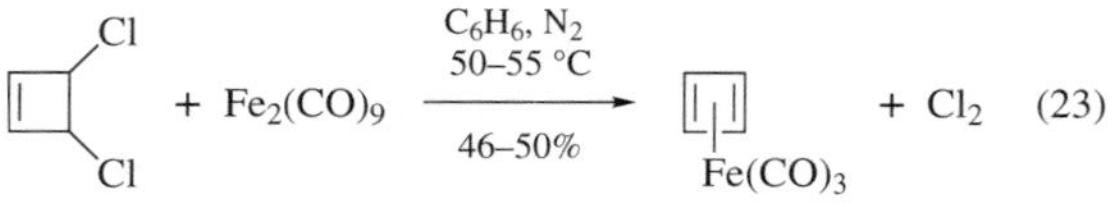

(23)

Dehalogenation of α,α′-dihaloalkenes and Ketones. The preparation (eq 24) of the iron carbonyl complex of trimethylenemethane (**42**)[62] and some $Fe_2(CO)_9$-mediated rearrangements of dibromo ketones[63] constitute the pioneering work in this area. Noyori has extended the dehalogenation reaction of α-halo ketones to α,α′-dibromo ketones and α,α,α′,α′-tetrabromo ketones which can be dehalogenated with $Fe(CO)_5$ or $Fe_2(CO)_9$.[64]

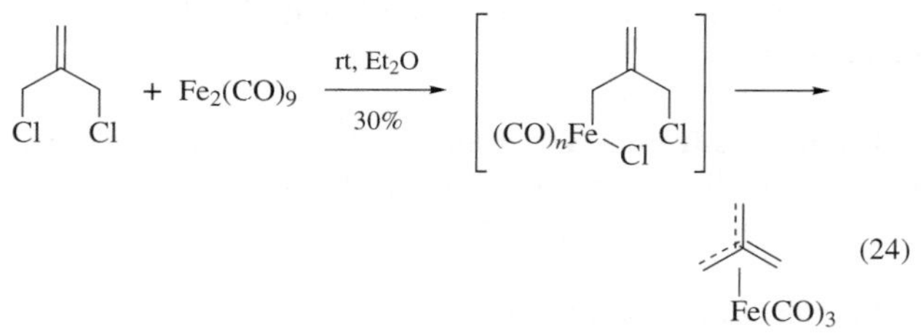

(24)

Iron-stabilized oxallyl cations (generated in situ (eq 25) from α,α′-dibromo ketones and $Fe_2(CO)_9$) react with alkenes. Noyori used this [3 + 2] cycloaddition reaction to produce cyclopentanone or cyclopentanone derivatives, as illustrated by a single-step synthesis of (±)-α-cuparenone (**43**) (eq 26).[65] The reaction of α,α′-dibromo ketones with enamines and $Fe_2(CO)_9$ yields substituted cyclopentenones in 50–100% yield (eq 27), as illustrated by the reaction with the α-morpholinostyrene (**44**).[66]

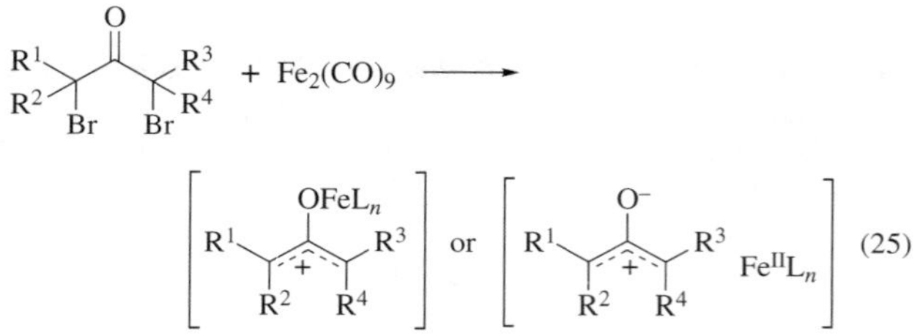

(25)

L = Br, CO, solvent

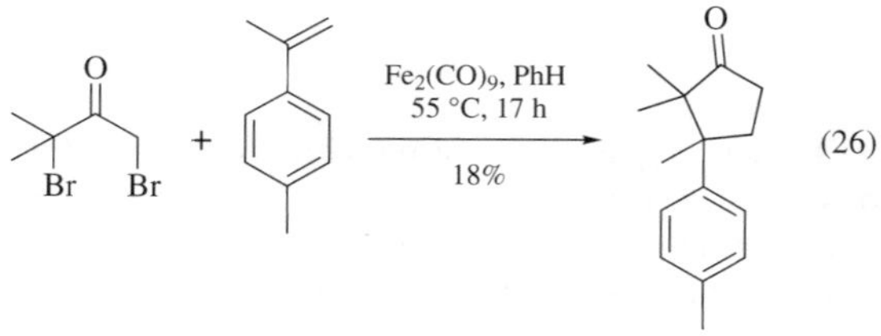

(26)

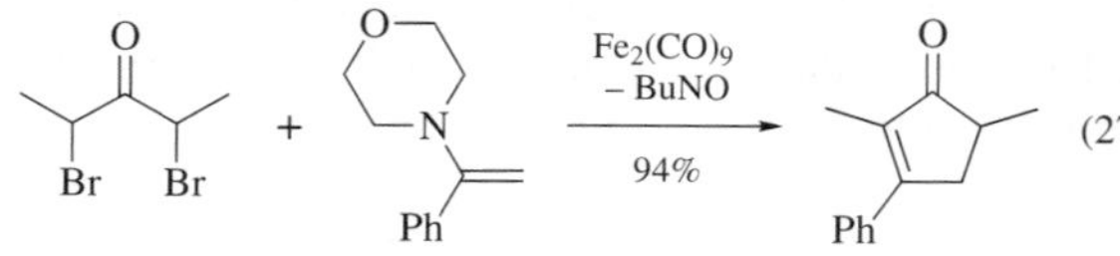

(27)

The reaction with dienes can be carried out either in benzene at 60–80 °C using α,α′-dibromo ketone, diene, and $Fe_2(CO)_9$

(1:large excess:1.2), in benzene at room temperature with irradiation (same ratio of reagents), or in benzene at 80–120 °C using the dibromide and tricarbonyl(η^4-diene)iron(0) complex (1:1.3). Formation of (**45**) (eq 28) provides a typical example.[67]

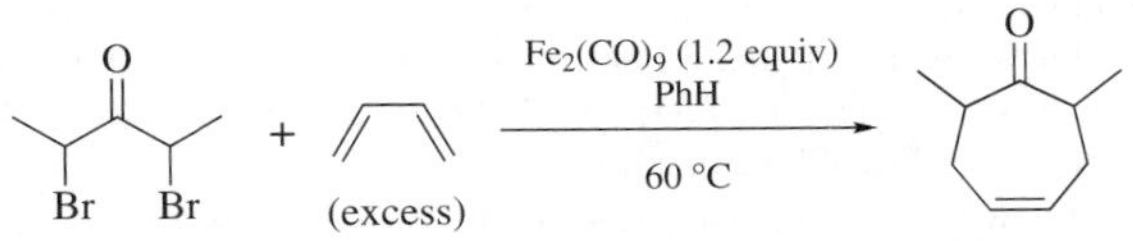

$Fe_2(CO)_9$ (1.2 equiv), PhH, 60 °C (28)

(excess)

(**45**) 44%

Noyori et al. have reported a general synthesis of tropane alkaloids from α,α'-dibromo ketones.[68] Reaction of tetrabromoacetone, *N*-methoxycarbonylpyrrole, and $Fe_2(CO)_9$ (3:1:1.5 mol ratio) in benzene (50 °C, N_2) produces two isomeric cycloadducts in a 2:1 mixture, which can be used in the preparation of the alcohol (**46**), a key intermediate in the synthesis of scopine and other tropane alkaloids.[69] A more recent example gives access to the bicyclo[5.2.0]nonene skeleton.[70]

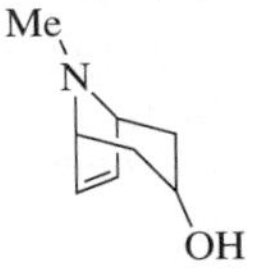

(**46**)

Reactions with Sulfur Compounds.

Desulfurization of Episulfides. The reaction of cyclohexene episulfide with $Fe_3(CO)_{12}$, reported by King[71] to produce cyclohexene, has been improved:[72] 2-butene episulfides, e.g. (**47**) ($R^1 = R^2 = Me$, $R^3 = R^4 = H$), yield alkenes by treatment with $Fe_2(CO)_9$ in refluxing benzene (eq 29). The reaction occurs with retention of the stereochemistry in yields superior to 80%.

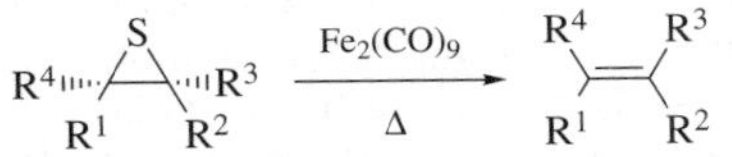

$Fe_2(CO)_9$, Δ (29)

(**47**)

Tetramethylallene episulfide, on the other hand, affords a thioallyl $Fe(CO)_3$ complex.[73]

Reaction with Thioketones. The reaction of thiobenzophenone derivatives (**48**) and similar compounds with $Fe_2(CO)_9$ has been studied by Alper.[74] It has been shown to result in the formation of *ortho*-metalated complexes (**49**) in reasonable to high yields (eq 30). Treatment of the complexes (**49**) with ***Mercury(II) Acetate*** effects *ortho*-mercuration.[75] Thioketene complexes have also been examined.[76] Complexes of type (**49**) offer a new route to lactones.[74]

Reaction with 2,5-Dihydrothiophene 1,1-Dioxide. 2,5-Dihydrothiophene 1,1-dioxides are known to be converted into 1,3-dienes after thermal displacement of sulfur dioxide.[77] Reaction in situ with $Fe_2(CO)_9$ offers a general preparation of highly functionalized tricarbonyl(η^4-buta-1,3-diene)iron(0) complexes (**50**) (eq 31).[78]

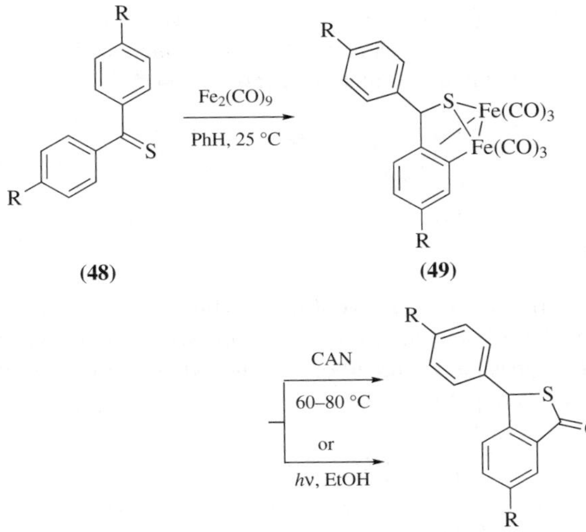

$Fe_2(CO)_9$, PhH, 25 °C

(**48**) (**49**)

CAN, 60–80 °C or $h\nu$, EtOH (30)

R = H, Me, OMe, NMe_2

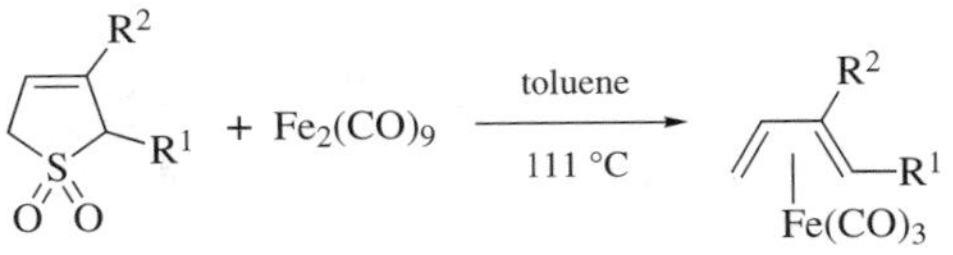

+ $Fe_2(CO)_9$ toluene, 111 °C (31)

(**50**)

Reactions with Alkynes. $Fe_2(CO)_9$ reacts with acetylene by carbonylation/cyclotrimerization leading to the formation of tricarbonyl(tropone)iron(0) (**51**) (eq 32). (For comparison with direct complexation of tropone, see eq 6.) This complex is obtained in low yield (28%) with $Fe_2(CO)_9$, but this represents a significant improvement on results obtained with $Fe(CO)_5$.[79]

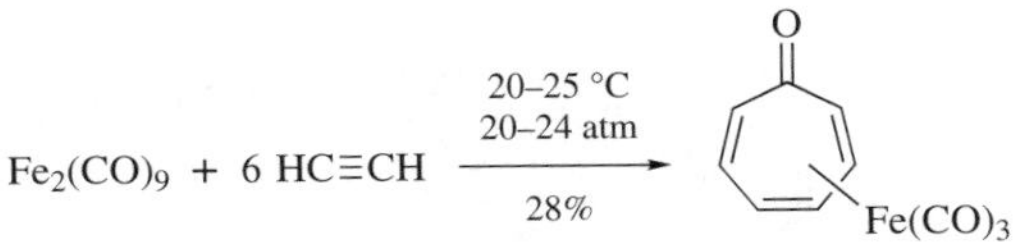

$$Fe_2(CO)_9 + 6\ HC{\equiv}CH \xrightarrow[28\%]{20\text{–}25\ °C,\ 20\text{–}24\ atm} \quad (32)$$

(**51**)

Hexyne is carbonylated by $Fe_2(CO)_9$. An intermediate can be isolated.[80] Alkylthioalkynes are not desulfurized but form iron complexes.[81]

Reaction with Acid Chlorides. The reaction of an acid chloride with one equivalent of $Fe_2(CO)_9$ affords symmetrical ketones.[82]

Formation of 1,2,4-Triazines. 1,2,4-Triazines, e.g. (**54**), are formed regioselectively (eq 33) by cocyclization of adiponitrile (**52**) and a nitrile (**53**) in the presence of $Fe_2(CO)_9$.[83]

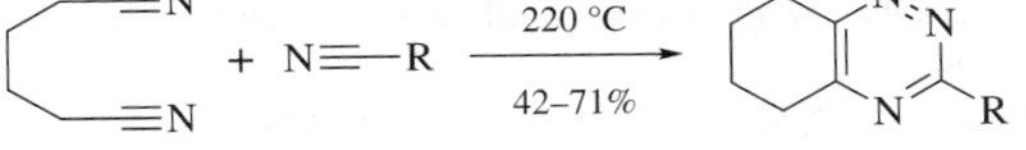

+ N≡R — $Fe_2(CO)_9$, 220 °C, 42–71% (33)

(**52**) (**53**) (**54**)

Deprotection and Deoxygenation Reactions of Oximes and Isobenzofurans. Oximes, oximic ethers, or oxime *O*-acetates can be converted into the corresponding ketones by reaction with $Fe_2(CO)_9$ in methanol at 60 °C.[84] Deoxygenation of 7-oxabicyclo[2.2.1]dienes with $Fe_2(CO)_9$ affords aromatic products.[85] This method is particularly recommended for deoxygenation of (**55**) in the formation of (**56**) (eq 34).[86]

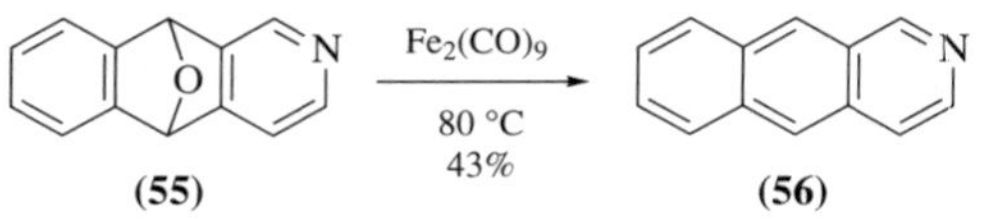

Reaction with Electrophilic Allenes. Formation of tricarbonyl(1,3-butadiene)iron(0) complexes from allenes (e.g. **57**) is possible via tricarbonyl(trimethylenemethane)iron(0) intermediates (eq 35).[87]

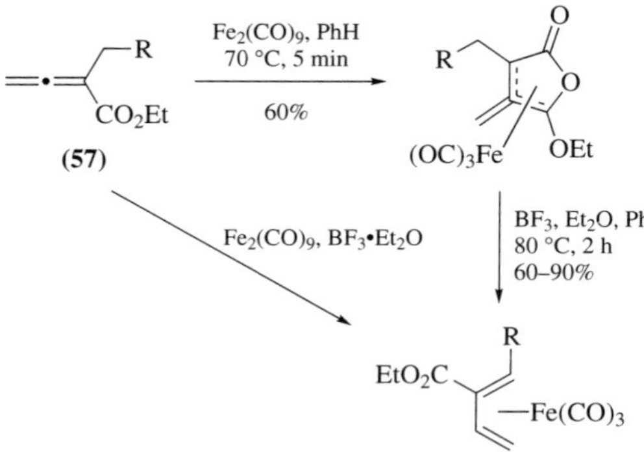

Cyclohexadienone Reductions with $Fe_2(CO)_9$ and Water. Tricarbonyl(cyclohexadienol)iron(0) complexes are formed by selective monohydrogenation of cyclohexadienones (**58**) by reaction with $Fe_2(CO)_9$ and water, under unusually mild conditions (eq 36).[88]

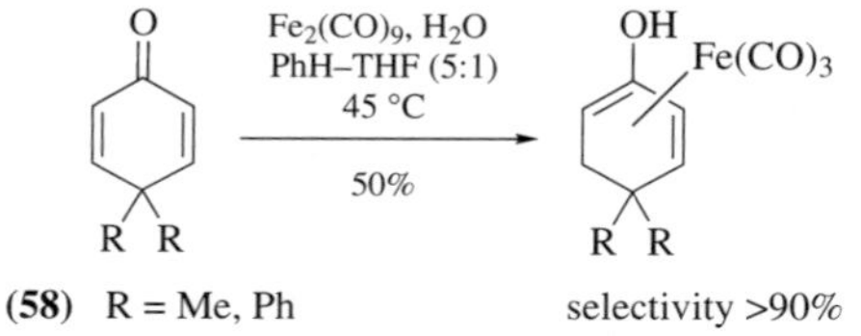

These reactions have been examined with a variety of substrates.[88,89] Regio- and stereoselectivity depends on the pH of the reaction mixture.

1. (a) Braye, E. H.; Hübel, W. *Inorg. Synth.* **1966**, *8*, 178. (b) Speyer, E.; Wolf, H. *CB* **1927**, *60*, 1424.
2. Griffith, W. P.; Wickham, A. J. *JCS(A)* **1969**, 834.
3. Cotton, F. A. *Prog. Inorg. Chem.* **1976**, *21*, 1.
4. (a) Bretherick, L. *Handbook of Reactive Chemical Hazards*, 2nd ed.; Butterworths: Woburn, MA, 1979; p 670 (b) Bretherick, L. *Hazards in the Chemical Laboratory*, 3rd ed.; Royal Society of Chemistry: London, 1981; p 421.
5. Musco, A.; Palumbo, R.; Paiaro, G. *ICA* **1971**, *5*, 157.
6. Banthorpe, D. V.; Fitton, H.; Lewis, J. *JCS(P1)* **1973**, 2051.
7. Nametkine, N. S.; Tyurine, V. D.; Nekhaev, A. I.; Ivanov, V. I.; Bayaouova, F. S. *JOM* **1976**, *107*, 377.
8. Pearson, A. J. *JCS(P1)* **1977**, 2069.
9. Johnson, B. F. G.; Lewis, J.; Parker, D. G.; Postle, S. R. *JCS(D)* **1977**, 794.
10. Birch, A. J.; Dastur, K. P. *JCS(P1)* **1973**, 1650.
11. Sakai, N.; Mashima, K.; Takaya, H.; Yamaguchi, R.; Kozima, S. *JOM* **1991**, *419*, 181.
12. (a) Carless, H. A. J.; Billinge, J. R.; Oak, O. Z. *TL* **1989**, *30*, 3113. (b) Hudlicky, T.; Luna, H.; Price, J. D.; Rulin, F. *TL* **1989**, *30*, 4053. (c) Hudlicky, T.; Luna, H.; Barbieri, G.; Kwart, L. D. *JACS* **1988**, *110*, 4735. (d) For examples of applications of an achiral arene-derived diol, see: Ley, S. V.; Sternfeld, F.; Taylor, S. *TL* **1987**, *28*, 225. (e) Carless, H. A. J.; Oak, O. Z. *TL* **1989**, *30*, 1719.
13. For examples, see: (a) Stephenson, G. R.; Alexander, R. P.; Morley, C.; Howard, P. W. *Philos. Trans. R. Soc. London, Ser. A* **1988**, *326*, 545. (b) Stephenson, G. R.; Howard, P. W.; Taylor, S. C. *CC* **1991**, 127.
14. Pearson, A. J.; Burello, M. P. *OM* **1992**, *11*, 448.
15. Rosenblum, M.; Watkins, J. C. *JACS* **1990**, *112*, 6316.
16. Morita, N.; Kabuto, C.; Asao, T. *BCJ* **1989**, *62*, 1677.
17. Evans, G.; Johnson, B. F. G.; Lewis, J. *JOM* **1975**, *102*, 507.
18. Barton, D. H. R.; Gunatilaka, A. A. L.; Nakanishi, T.; Patin, H.; Widdowson, D. A.; Worth, B. R. *JCS(P1)* **1976**, 821.
19. Howell, J. A. S.; Johnson, B. F. G.; Josty, P. L.; Lewis, J. *JOM* **1972**, *39*, 329.
20. Salzer, A.; Schmalle, H.; Stauber, R.; Streiff, S. *JOM* **1991**, *408*, 403.
21. Rubello, A.; Vogel, P.; Chapuis, G. *HCA* **1987**, *70*, 1638.
22. Bonfantini, E.; Métral, J.-L.; Vogel, P. *HCA* **1987**, *70*, 1791.
23. (a) Sun, C.-H.; Chow, T. J.; Liu, L.-K. *OM* **1990**, *9*, 560. (b) Chow, T. J.; Hwang, J.-J.; Sun, C.-H.; Ding, M.-F. *OM* **1993**, *12*, 3762.
24. Donaldson, W. A. *JOM* **1990**, *395*, 187.
25. Adams, C. M.; Cerioni, G.; Hafner, A.; Kalchhauser, H.; Von Philipsborn, W.; Prewo, R.; Schwenk, A. *HCA* **1988**, *71*, 1116.
26. Roush, W. R.; Park, J. C. *TL* **1990**, *31*, 4707.
27. Yeh, M-C. P.; Chu, C. H.; Sun, M. L.; Kang, K. P. *J. Chin. Chem. Soc. (Taipei)* **1990**, *37*, 547.
28. Kappes, D.; Gerlach, H.; Zbinden, P.; Dobler, M. *HCA* **1990**, *73*, 2136.
29. (a) *The Organic Chemistry of Iron*; Koerner von Gustorf, E. A.; Grevels, F. W.; Fischler, I., Eds.; Academic: London, 1978 (Vol. 1) and 1981 (Vol. 2). (b) Greé, R. *S* **1989**, 341. (c) Greé, R.; Lellouche, J. P. In *Advances in Metal-Organic Chemistry*; Liebeskind, L. S., Ed.; JAI: Greenwich, Conn.; Vol 4, in press.
30. Ley, S. V.; Low, C. M. R.; White, A. D. *JOM* **1986**, *302*, C13.
31. Birencwaig, F.; Shamai, H.; Shvo, Y. *TL* **1979**, 2947.
32. (a) Howell, J. A. S.; Johnson, B. F. G.; Josty, P. L.; Lewis, J. *JOM* **1972**, *39*, 329. (b) Evans, G.; Johnson, B. F. G.; Lewis, J. *JOM* **1975**, *102*, 507. (c) Paquette, L. A.; Photis, J. M.; Ewing, G. D. *JACS* **1975**, *97*, 3538. (d) Domingos, A. J.; Howell, J. A. S.; Johnson, B. F. G.; Lewis, J. *Inorg. Synth.* **1990**, *28*, 52.
33. Danks, T. N.; Rakshit, D.; Thomas, S. E. *JCS(P1)* **1988**, 2091.
34. Thomas, S. E.; Tustin, G. J., Ibbotson, A. *T* **1992**, *48*, 7629.
35. (a) Howell, J. A. S.; Johnson, B. F. G.; Josty, P. L.; Lewis, J. *JOM* **1972**, *39*, 239. (b) Brookhart, M.; Nelson, G. O.; Scholes, G.; Watson, R. A. *CC* **1976**, 195. (c) Barton, D. H. R.; Gunatilaka, A. A. L.; Nakanishi, T.; Patin, H.; Widdowson, D. A.; Worth, B. R. *JCS(P1)* **1976**, 821.
36. Knölker, H.-J.; Gonser, P. *SL* **1992**, 517.
37. Koerner von Gustorf, E.; Grevels, F.- W.; Krüger, C.; Olbrich, G.; Mark, F.; Schulz, D.; Wagner, R. *ZN(B)* **1972**, *27*, 392.
38. Birch, A. J.; Raverty, W. D.; Stephenson, G. R. *TL* **1980**, 197.
39. Birch, A. J.; Raverty, W. D.; Stephenson, G. R. *OM* **1984**, *3*, 1075.
40. Bates, R. W.; Díez-Martín, D.; Kerr, W. J.; Knight, J. G.; Ley, S. V.; Sakellaridis, A. *T* **1990**, *46*, 4063.
41. (a) Heck, R. F.; Boss, C. R. *JACS* **1964**, *86*, 2580. (b) Becker, Y.; Eisenstadt, A.; Shvo, Y., *T* **1974**, *30*, 839. (c) Becker, Y.; Eisenstadt, A.; Shvo, Y. *T* **1976**, *32*, 2123.
42. Aumann, R.; Fröhlich, K.; Ring, H. *AG(E)* **1974**, *13*, 275.
43. (a) Annis, G. D.; Ley, S. V. *CC* **1977**, 581. (b) Annis, G. D.; Ley, S. V.; Self, C. R.; Sivaramakrishnan, R. *JCS(P1)* **1981**, 270.

44. Horton, A. M.; Hollinshead, D. M.; Ley, S. V. *T* **1984**, *40*, 1737.
45. Horton, A. M.; Ley, S. V. *JOM* **1985**, *285*, C17.
46. Kotecha, N. R.; Ley, S. V.; Mantegani, S. *SL* **1992**, 395.
47. Ley, S. V.; Armstrong, A.; Díez-Martín, D.; Ford, M. J.; Grice, P.; Knight, J. G.; Klob, H. C.; Madin, A.; Marby, C. A.; Mukherjee, S.; Shaw, A. N.; Slawin, A. M. Z.; Vile, S.; White, A. D.; Williams, D. J.; Woods, M. *JCS(P1)* **1991**, 667.
48. Bates, R. W.; Fernández-Moro, R.; Ley, S. V. *T* **1991**, *47*, 9929.
49. (a) Annis, G. D.; Hebblethwaite, E. M.; Ley, S. V. *CC* **1980**, 297. (b) Annis, G. D.; Hebblethwaite, E. M.; Hodgson, S. T.; Hollinshead, D. M.; Ley, S. V. *JCS(P1)* **1983**, 2851.
50. (a) Hodgson, S. T.; Hollinshead, D. M.; Ley, S. V. *CC* **1984**, 494. (b) Hodgson, S. T.; Hollinshead, D. M.; Ley, S. V. *T* **1985**, *41*, 5871.
51. (a) Newton, M. G.; Pantaleo, N. S.; King, R. B.; Chu, C. K. *CC* **1979**, 10. (b) Binger, P.; Cetinkaya, B.; Kruger, C. *JOM* **1978**, *159*, 63. (c) Dettlaf, G.; Behrens, U.; Weiss, E. *CB* **1978**, *111*, 3013.
52. Franck-Neumann, M. *PAC* **1983**, *55*, 1715.
53. Franck-Neumann, M.; Dietrich-Buchecker, C.; Khémiss, A. *JOM* **1991**, *220*, 187.
54. Whitesides, T. H.; Slaven, R. W. *JOM* **1974**, *67*, 99.
55. Müller, P.; Bernardinelli, G.; Jacquier, Y. *HCA* **1992**, *75*, 1995.
56. Franck-Neumann, M.; Dietrich-Buchecker, C.; Khémiss, A. *TL* **1981**, *22*, 2307.
57. (a) Aumann, R. *JOM* **1973**, *47*, C29. (b) Aumann, R.; Knecht, J. *CB* **1976**, *109*, 174. (c) Wang, A. H.-J.; Paul, I. C.; Aumann, R. *JOM* **1974**, *69*, 301. (d) Eilbracht, P.; Winkels, I. *CB* **1991**, *124*, 191.
58. (a) Eilbracht, P.; Hittenger, C.; Kufferath, K. *CB* **1990**, *123*, 1071. (b) Eilbracht, P.; Hittenger, C.; Kufferath, K.; Henkel, G. *CB* **1990**, *123*, 1079. (c) Eilbracht, P.; Hittinger, C.; Kufferath, K.; Schmitz, A.; Gilsing, H. D. *CB* **1990**, *123*, 1089.
59. (a) Roth, W. R.; Meier, J. D. *TL* **1967**, 2053. (b) Kerber, R. C.; Ribakove, E. C. *OM* **1991**, *10*, 2848.
60. Azad, S. M.; Azam, K. A.; Hasan, M. K.; Howlader, M. B. H.; Kabir, S. E. *J. Bangladesh Acad. Sci.* **1990**, *14*, 149.
61. (a) Watts, L.; Fitzpatrick, J. D.; Pettit, R. *JACS* **1965**, *87*, 3253. (b) Watts, L.; Fitzpatrick, J. D.; Pettit, R. *JACS* **1966**, *88*, 623. (c) Pettit, R. *JOM* **1975**, *100*, 205.
62. Emerson, G. F.; Ehrlich, K.; Giering, W. P.; Lauterbur, P. C. *JACS* **1966**, *88*, 3172.
63. (a) Noyori, R.; Hayakawa, Y.; Funakura, M.; Takaya, H.; Murai, S.; Kobayashi, R.; Tsutsumi, S. *JACS* **1972**, *94*, 7202. (b) Noyori, R.; Hayakawa, Y.; Takaya, H.; Murai, S.; Kobayashi, R.; Sonoda, N. *JACS* **1978**, *100*, 1759.
64. Review: Noyori, R. *ACR* **1979**, *12*, 61.
65. Hayakawa, Y.; Shimizu, F.; Noyori, R. *TL* **1978**, 993.
66. Noyori, R.; Yokoyama, K.; Makino, S.; Hayakawa, Y. *JACS* **1972**, *94*, 1772.
67. (a) Noyori, R.; Makino, S.; Takaya, H. *JACS* **1971**, *93*, 1272. (b) Noyori, R.; Souchi, T.; Hayakawa, Y. *JOC* **1975**, *40*, 2681. (c) Takaya, H.; Makino, S.; Hayakawa, Y.; Noyori, R. *JACS* **1978**, *100*, 1765.
68. Noyori, R.; Baba, Y.; Hayakawa, Y. *JACS* **1974**, *96*, 3336.
69. (a) Hayakawa, Y.; Baba, Y.; Makino, S.; Noyori, R. *JACS* **1978**, *100*, 1786. (b) Noyori, R.; Baba, Y.; Hayakawa, Y. *JACS* **1974**, *96*, 3336.
70. Hojo, M.; Tomita, K.; Hirohara, Y.; Hosomi, A. *TL* **1993**, *34*, 8123.
71. King, R. B. *IC* **1963**, *2*, 326.
72. Trost, B. M.; Ziman, S. D. *JOC* **1973**, *38*, 932.
73. Choi, N.; Kabe, Y.; Ando, W. *OM* **1992**, *11*, 1506.
74. (a) Alper, H.; Chan, A. S. K. *JACS* **1973**, *95*, 4905. (b) Alper, H.; Root, W. G. *CC* **1974**, 956.
75. Alper, H.; Root, W. G. *TL* **1974**, 1611.
76. Seitz, K.; Benecke, J.; Behrens, U. *JOM* **1989**, *371*, 247.
77. (a) Chou, T. S.; Tso, H. H. *OPP* **1989**, *21*, 257. (b) Chou, T.; Chang, L.-J.; Tso, H.-H. *JCS(P1)* **1986**, 1039. (c) Tso, H.-H.; Chou, T.; Hung, S. C. *CC* **1987**, 1552. (d) Chou, S.-S. P.; Liou, S.-Y.; Tsai, C.-Y.; Wang, A.-J. *JOC* **1987**, *52*, 4468. (e) Chou, T.; Lee, S.-J.; Peng, M.-L.; Sun, D.-J.; Chou, S.-S. P. *JOC* **1988**, *53*, 3027. (f) Tso, H.-H.; Chou, T.; Lai, Y.-L. *JOC* **1989**, *54*, 4138.
78. Yeh, M.-C. P.; Chou, T.; Tso, H.-H.; Tsai, C.-Y. *CC* **1990**, 897.
79. Weiss, E.; Hübel, W. *CB* **1962**, *95*, 1179.
80. Milone, L.; Osella, D.; Ravera, M.; Stanghellini, P. L.; Stein, E. *G* **1992**, *122*, 451.
81. Jeannin, S.; Jeannin, Y.; Robert, F.; Rosenberger, C. *CR(II)* **1992**, *314*, 1165.
82. Flood, T. C.; Sarhangi, A. *TL* **1977**, 3861.
83. Gesing, E. R. F.; Groth, U.; Vollhardt, K. P. C. *S* **1984**, 351.
84. Nitta, M.; Sasaki, I.; Miyano, H.; Kabayashi, T. *BCJ* **1984**, *57*, 3357.
85. Sun, C. H.; Yang, G. Z.; Chow, T. J. *J. Bull. Inst. Chem., Acad. Sin.* **1990**, *37*, 33.
86. Crump, S. L.; Netka, J.; Rickborn, B. *JOC* **1985**, *50*, 2746.
87. Brion, F.; Martina, D. *TL* **1982**, *23*, 861.
88. Eilbracht, P.; Jelitte, R. *CB* **1985**, *118*, 1983.
89. Eilbracht, P.; Jelitte, R. *CB* **1983**, *116*, 243.

Sylvie Samson & G. Richard Stephenson
University of East Anglia, Norwich, UK

Octacarbonyldicobalt

$Co_2(CO)_8$

[10210-68-1] $C_8Co_2O_8$ (MW 341.94)

(catalyst for carbonylation and related reactions; major source of other synthetically useful organocobalt compounds)

Alternate Names: di-μ-carbonylhexacarbonyldicobalt; cobalt carbonyl.
Physical Data: mp 51–52 °C (dec); subl 40 °C/vac.
Solubility: insol H_2O; sol common org solvents.
Form Supplied in: dark orange crystals (commonly stabilized with 5–10% hexane).
Handling, Storage, and Precautions: toxic, moderately air sensitive, and unstable at ambient temperature, slowly releasing carbon monoxide (residual finely divided metal may become pyrophoric); may be stored for long periods at or below 0 °C in tightly closed vessels under inert gas (N_2, Ar) or (best) CO; brief handling (including weighing) in air causes no problems; heating in inert solvents yields the less reactive $Co_4(CO)_{12}$; use in a fume hood.

Introduction. In the crystal, the molecule of this reagent consists of two cobalt atoms linked both directly and by two bridging carbonyls and each carrying three terminal carbonyls (cf. eq 33).[1] In solution this form is in equilibrium with unbridged $(OC)_4Co–Co(CO)_4$.

Halogen oxidation only leads to a halocarbonyl species in the case of iodine; the unstable and incompletely characterized product, probably $ICo(CO)_4$, has found little use. The anion $Co(CO)_4^-$, readily formed by reduction (***Sodium Amalgam***) or by disproportionation on reaction with bases, including not only pyridine and other nitrogen bases but also such donor solvents as methanol and acetonitrile, is treated separately. (See ***Sodium Tetracarbonylcobaltate***.) However, reactions which utilize $Co_2(CO)_8$ as starting material are included here even if the reaction conditions imply that this anion or its conjugate acid, $HCo(CO)_4$ (also formed from $Co_2(CO)_8 + H_2$), may be the 'active' intermediate. The chemistry of the reagent has been reviewed.[2,3]

Catalytic Uses.

Hydroformylation and Hydrogenation. Octacarbonyldicobalt catalyzes a wide range of reactions of carbon monoxide (alone or with hydrogen),[2,4–8] of which the best known is probably the hydroformylation ('oxo reaction') of alkenes. Depending on reaction conditions, notably CO:H_2 ratio and temperature, the products are aldehydes (eq 1) or alcohols ('hydroalkylation') (eq 2).

A valuable feature is the preference for straight chain over branched chain products; since $HCo(CO)_4$ is an efficient alkene isomerization catalyst, this is true even for internal alkenes. Much research has been devoted to optimizing the product ratio for specific uses, e.g. by adding phosphines or other catalyst modifiers. While cobalt carbonyl (probably as $HCo(CO)_4$) is an effective catalyst for these and many related reactions and has found extensive industrial use (albeit commonly generated in situ rather than introduced as such), it has largely been superseded by related rhodium catalysts; the greater cost of the latter is offset by much higher activity and hence efficient conversions using small catalyst concentrations under much milder reaction conditions. Alkyl halides and alcohols can replace alkenes in many such reaction.

$$RCH{=}CH_2 + CO + H_2 \xrightarrow{130–180\,°C} RCH_2CH_2CHO + RCH(Me)CHO \quad (1)$$

$$RCH{=}CH_2 + CO + 2\,H_2 \xrightarrow{160–200\,°C} RCH_2CH_2CH_2OH + RCH(Me)CH_2OH \quad (2)$$

Whereas, in general, reduction of the formyl group (cf. eqs 1 and 2) requires a higher temperature than hydroformylation, reduction of the C=C double bond can be a major side-reaction of cobalt-catalyzed hydroformylation. Thus, in the reaction of eugenol, which is accompanied by cyclization (eq 3), 40% of the starting material is reduced to dihydroeugenol.[9]

MeO, HO (eugenol) + CO + H_2 —[$Co_2(CO)_8$, 120 °C]→ MeO, HO (tetrahydronaphthalene) (3)

The $Co_2(CO)_8$-catalyzed alkene hydrogenation can be useful, as in the smooth reduction of alkylcinnamaldehydes (eq 4),[10] the highly selective hydrogenation of methyl (*E,E*)-octadeca-9,11-dienoate to methyl elaidate (eq 5),[11] and of a range of polycyclic aromatics,[12] e.g. naphthacene (eq 6).

R–C$_6$H$_4$–CH=C(Me)CHO + CO + H_2 (1:12) —[$Co_2(CO)_8$, *i*-Pr_2NH, 107 °C]→ R–C$_6$H$_4$–CH$_2$CH(Me)CHO (4)
e.g. R = *t*-Bu, 96%

Methyl octadeca-9,11-dienoate —[$Co_2(CO)_8$, H_2 (20 bar), 110 °C, 86.5%]→ methyl elaidate (CO_2Me) (5)

Naphthacene + H_2 + CO (1:1, 240 bar) —[$Co_2(CO)_8$, 140–200 °C, 70%]→ 5,12-dihydronaphthacene (6)

Dehydration or loss of alcohol accompanies hydroalkylation (eq 2) when applied to allylic alcohols (giving tetrahydrofurans rather than 1,4-diols) or to acrylic acids or esters, which give butyrolactones (eq 7) in moderate to excellent yields.[13]

$$\text{R-substituted acrylate } (CO_2R') + H_2 + CO \xrightarrow[200\text{–}250\ °C]{Co_2(CO)_8} \text{R-substituted butyrolactone} \quad (7)$$

In more or less related reactions, $Co_2(CO)_8$ is a useful catalyst in carboxylation (viz. alkoxycarbonylation) and a wide range of carbonylation reactions. Both catalysis of the water-gas shift reaction[14] and of Fischer–Tropsch processes[15] have been demonstrated on solid supports and probably involve $Co_2(CO)_8$-derived polynuclear carbonyls.[16]

Carboxylation and Related Reactions. Although nickel, palladium, and other catalysts have been widely studied for carboxylation reactions and are clearly preferable in the case of alkynes, octacarbonyldicobalt (especially in combination with ***Pyridine***) is an effective catalyst for the conversion of alkenes into acids and esters (eq 8) and differs significantly in selectivity from the nickel-based systems. Thus, whereas styrene gives predominantly the branched esters with both types of catalyst, terminal *n*-alkenes give predominantly branched esters with nickel catalysts, but straight-chain esters with $Co_2(CO)_8$, as in eq 9.[4] The related hydrocyanation reaction, on the other hand, gives branched-chain products and, except with the most reactive alkenes, poor yields[17] when using $Co_2(CO)_8$ and better nickel-based catalysts are available.

$$RCH{=}CH_2 + CO + R'OH \xrightarrow{Co_2(CO)_8,\ py} RCH_2CH_2CO_2R' + RCH(Me)CO_2R' \quad (8)$$

R' = H, Me, etc

$$\text{1-octene} + \underset{190\ bar}{CO} + \underset{10\ bar}{H_2} + MeOH \xrightarrow[\substack{(1:12),\ 175\ °C \\ 95\%\ conversion}]{Co_2(CO)_8,\ py}$$

$$\underset{79\%}{C_8H_{17}CO_2Me} + \underset{14\%}{Me(CH_2)_5CHMeCO_2Me} + \underset{4\%}{Me(CH_2)_4CHEtCO_2Me} + \underset{3\%}{Me(CH_2)_3CHPrCO_2Me} \quad (9)$$

The condensation of two alkene molecules with ***Carbon Monoxide*** to yield ketones [e.g. ethylene → 3-pentanone (95%)][4] succeeds well only in special cases, e.g. with methyl acrylate (eq 10).[18]

$$CH_2{=}CHCO_2Me + CO + H_2O \xrightarrow[\substack{135\ °C \\ 94\%}]{\substack{Co_2(CO)_8 \\ Ph_2PCH_2CH_2PPh_2}} O{=}C(CH_2CH_2CO_2Me)_2 \quad (10)$$

Octacarbonyldicobalt-catalyzed homologation of alcohols (eq 11) is difficult to control and of little practical value, while the related carboxylation of methanol to acetic acid requires such extreme conditions that it cannot compete with the technically important rhodium-catalyzed process.

$$ROH + CO + 2\,H_2 \xrightarrow[135\ °C]{Co_2(CO)_8,\ I_2} RCH_2OH \quad (11)$$

Relatively mild conditions suffice for the regioselective carbonylation (eq 12)[19] and carboxylation of epoxides (eq 13).[20] The rather poor yields quoted in these examples have been attributed to extensive side-reactions, but specific cases giving excellent results are also known, e.g. with a carbohydrate epoxide (eq 14).[21]

$$\text{propylene oxide} + CO + H_2 \xrightarrow[\substack{100\ °C \\ 56\%}]{Co_2(CO)_8} CH_3CH(OH)CH_2CHO \quad (12)$$

$$\text{propylene oxide} + \underset{3500\ psi}{CO} + MeOH \xrightarrow[\substack{130\ °C \\ 40\%}]{Co_2(CO)_8} CH_3CH(OH)CH_2CO_2Me \quad (13)$$

$$\text{epoxy furanose (OH, isopropylidene)} + \underset{\substack{1:1 \\ 140\ atm}}{CO + H_2} \xrightarrow[\substack{100\text{–}105\ °C \\ 78\%}]{Co_2(CO)_8} \text{bicyclic product (HO, OH)} \quad (14)$$

Carboxylations of halogen compounds include the technically useful conversion of chloroacetate to malonate (eq 15),[22] and several patent reports of quantitative conversion of ***Benzyl Chloride*** to phenylacetic acid.[23] The incorporation of two molecules of carbon monoxide to give α-keto-acids, as exemplified by the reaction of benzyl chloride (eq 16), is remarkably solvent-dependent[24] and almost certainly involves reaction of the halide with $Co(CO)_4^-$.

$$ClCH_2CO_2Et + \underset{7\ bar}{CO} + NaOEt \xrightarrow[\substack{55\ °C \\ 85\%}]{Co_2(CO)_8} CH_2(CO_2Et)_2 \quad (15)$$

$$PhCH_2Cl + \underset{2\text{–}3\ bar}{CO} + H_2O \xrightarrow[\substack{75\ °C,\ 4.5\ h \\ 75\%}]{\substack{Co_2(CO)_8 \\ aq\ MeCN}} PhCH_2COCO_2H \quad (16)$$

A carboxylation which is followed by a Michael addition occurs with high overall efficiency when ***Acrylonitrile*** reacts in the presence of an alcohol and pyridine (eq 17).[25]

$$CH_2{=}CHCN + ROH + \underset{100\ bar}{CO} \xrightarrow[CH_2Cl_2,\ 100\ °C]{Co_2(CO)_8,\ py} \underset{89\%}{NCCH_2CH_2C(CH_3)(CN)CO_2R} + \underset{3\%}{CH_3CH(CN)CO_2R} + \underset{8\%}{NCCH_2CH_2CO_2R} \quad (17)$$

e.g. R = *i*-Pr

Carbonylation. This term is used here to denote carbon monoxide insertions (with or without cyclization) which do not involve H_2 or ROH.

Unlike epoxides, oxetanes react with carbon monoxide in the presence of $Co_2(CO)_8$ by ring expansion to γ-butyrolactones, rather than ring opening; thietanes (eq 18) as well as azetidines (eqs 19 and 20) behave similarly. Mixtures of cobalt and ruthenium carbonyls (1:1) are reported to give better yields than either catalyst separately.[26] The examples shown illustrate the strong dependence of regioselectivity on substitution pattern in the azetidine case.[27]

$$\text{2-methylthietane} + CO \xrightarrow[Ru_3(CO)_{12}]{Co_2(CO)_8} \text{5-methyl-γ-thiobutyrolactone} \quad (18)$$

(19) 2-phenyl-N-methylazetidine + CO (3.5 atm) → [$Co_2(CO)_8$, 85–90 °C, 90%] 3-phenyl-N-methylpyrrolidin-2-one

(20) 2-(CH_2OMe)-N-t-Bu-azetidine + CO (3.5 atm) → [$Co_2(CO)_8$, 85–90 °C, 91%] pyrrolidinone with CH_2OMe, N–t-Bu

$Co_2(CO)_8$-catalyzed reactions of a variety of other nitrogen compounds, e.g. azo compounds (eq 21),[28] imines (eq 22),[29] and amides (eq 23),[30] involve CO insertion into N–N and N–H bonds and cyclizations which probably proceed via cyclometalation.

(21) PhN=NPh + CO (150 atm) → [190 °C, 53%] 2-phenylindazolone (N-H, N-Ph, C=O); 230 °C → 3-phenylquinazolinedione (H, N, O, N, Ph, O); indazolone → quinazolinedione 80%

(22) PhCR=NR' + CO → isoindolinone (R, N–R', O)

(23) $CH_2=CHCONHBu$ + CO (310 atm) → [$Co_2(CO)_8$, 200 °C, 72%] N-Bu succinimide

The mechanism of the remarkable formation of furans[31] from 3-diethylamino-1-propyne (or -butyne) (eq 24) (R = H or Me) is unknown, but it must involve CO insertion into a C–N bond.

(24) Et_2NCH(R)C≡CH + CO (500–1000 atm) → [$Co_2(CO)_8$] dihydrofuran bearing $Et_2NC(O)$–, R, NEt_2

The carbonylation of ***Diphenyl Diselenide*** (eq 25) illustrates the general behavior of diaryl diselenides and ditellurides,[32] but proceeds more smoothly than most other examples studied.

$$\text{PhSeSePh} + \text{CO} \xrightarrow[\text{MeCN, 200 °C}]{Co_2(CO)_8} \underset{96\%}{\text{PhSeCOPh}} + \underset{2\%}{\text{Ph}_2\text{Se}} \quad (25)$$
(CO at 100 atm)

Reversibility of carbonylation processes is illustrated by decarbonylation of phthalic anhydride by $Co_2(CO)_8$ under H_2 to benzoic acid,[33] and of ***Diphenylketene*** to tetraphenylethylene.[34]

Carbonylation of alkynes is discussed below in connection with the stoichiometric reactions of alkyne–cobalt compounds.

Amidocarbonylation. The reaction of aldehydes with primary amides and carbon monoxide in the presence of cobalt carbonyl (eq 26) provides a valuable alternative to the Strecker reaction as a route to *N*-acyl-α-amino acids.[8,34–36] Its convenience is enhanced by the possibility of generating the aldehyde in situ by isomerization of an allylic alcohol for which a cocatalyst [$PdCl_2(PPh_3)_2$, ***Carbonylhydridotris(triphenylphosphine)rhodium(I)***, or ***Nonacarbonyldiiron***] is advantageous,[37] (e.g. eq 27), or by hydroformylation of an alkene (e.g. by the reactions of eq 28 which can be combined to give the *N*-acetylamino acid in 83% yield[8,38]), or of an alcohol (e.g. eq 29). In the last example[39] the product is isolated (as the methyl ester after ***Diazomethane*** treatment) in 73% total yield as a mixture of *threo* and *erythro* isomers.

$$\text{RCHO} + \text{R'CONH}_2 + \text{CO} \xrightarrow[120\ °C]{Co_2(CO)_8} \text{R–CH(NHCOR')CO}_2\text{H} \quad (26)$$

$$\text{CH}_3\text{CH=CHCH}_2\text{OH} + \text{MeCONH}_2 + \text{CO} + \text{H}_2 \xrightarrow[110\ °C,\ 18\ h]{Co_2(CO)_8,\ PdCl_2(PPh_3)_2} \text{CH}_3\text{CH}_2\text{CH}_2\text{CH(NHAc)CO}_2\text{H} \quad (27)$$
(ca. 50 atm)

$$\text{CF}_3\text{CH=CH}_2 + \text{CO} + \text{H}_2 \xrightarrow[95\%]{Co_2(CO)_8} \text{F}_3\text{CCH}_2\text{CH}_2\text{CHO} \xrightarrow[Co_2(CO)_8,\ 80\%]{AcNH_2,\ CO,\ H_2} \text{F}_3\text{CCH}_2\text{CH}_2\text{CH(NHAc)CO}_2\text{H} \quad (28)$$
(93% straight chain)

(29) 1-cyclopropylethanol (OH) + CO + H_2 + $AcNH_2$ → [$Co_2(CO)_8$] cyclopropyl–CH(CH_3)–CH(AcHN)–CO_2R

Other Octacarbonyldicobalt-Catalyzed Reactions. A retro-Diels–Alder reaction catalyzed by $Co_2(CO)_8$ was observed in which a barrelene derivative loses a C_2H_2 fragment (eq 30).[40] An analogous cleavage accompanies the cyclopentenone synthesis (see below) when norbornadiene reacts in certain solvents with alkynehexacarbonyldicobalts, as shown by the formation of dicarbonylcyclopentadienylcobalt,[41] whereas a Diels–Alder addition catalyzed by a cobalt carbonyl species is involved when the same reaction is applied to cyclohexadiene.[42]

(30) barrelene-fused N-CH_2Ph lactam → [$Co_2(CO)_8$, 80 °C, 64%] 3,3-dimethyl-1-benzyloxindole (N, CH_2Ph, =O)

Efficient and appreciably regioselective ring-opening addition of ***Cyanotrimethylsilane*** to tetrahydrofurans is illustrated by the example shown (eq 31).[43] The same reagent undergoes $Co_2(CO)_8$ (or ***Dicarbonyl(cyclopentadienyl)cobalt(I)***) catalyzed addition to alkynes to give pyrrole derivatives (eq 32).[44]

$$\text{3-methyltetrahydrofuran} + \text{TMSCN} \xrightarrow[\text{C}_6\text{H}_6,\ 150\,^\circ\text{C};\ 88\%]{\text{Co}_2(\text{CO})_8,\ \text{Ar}} \text{TMSO(CH}_2)\text{CH(CH}_3)\text{CH}_2\text{CH}_2\text{CN} + \text{TMSOCH}_2\text{CH}_2\text{CH(CH}_3)\text{CH}_2\text{CN} \quad 77{:}23 \tag{31}$$

$$\text{BuC}{\equiv}\text{CMe} + \text{TMSCN} \xrightarrow[95\%]{\text{Co}_2(\text{CO})_8} \text{pyrroles (NC, Bu, Me, N(TMS)}_2\text{, NH)} \quad 36{:}64 \tag{32}$$

Octacarbonyldicobalt shares with many other metal carbonyls and other organometallics the ability to catalyze the cyclotrimerization of alkynes[45] and related condensations, e.g. of two alkyne and one nitrile molecule to give pyridines. Derived compounds including $Hg[Co(CO)_4]_2$, $Co(CO)_3NO$ (obtained from $Co_2(CO)_8$ with NO or other nitrosating agents) and the alkyne complexes $(RC_2R')Co_2(CO)_6$ (see below) all share this activity, but none of these compare in efficiency with the preferred $CpCoL_2$ catalysts.

The few examples of catalytic cyclopentenone formation from alkyne, alkene, and CO are discussed below, as is the bifurylidenedione synthesis from alkyne and CO. It should also be noted that $Co_4(CO)_{12}$, obtained by heating the dinuclear carbonyl, can probably replace the latter in most of the above catalytic reactions, but without obvious advantage.

Stoichiometric Reactions.

Reactions with Dienes. Conjugated dienes and norbornadiene displace two terminal carbonyl groups from first one and then both metal atoms of the octacarbonyl (eq 33).[46]

$$\text{Co}_2(\text{CO})_8 + \text{R-butadiene} \longrightarrow (\text{diene})\text{Co}_2(\text{CO})_6 \longrightarrow (\text{diene})_2\text{Co}_2(\text{CO})_4 \tag{33}$$

Either of these products can be oxidized to $[(\text{diene})Co(CO)_3]^+$ salts,[47] whose synthetic potential arises from their reactivity towards nucleophiles.[47b,48] Whether 1- or 2-substituted butadiene ligands react highly regioselectively at the 4-position (e.g. eq 34),[47b] or at both C-1 and C-4, depends both on the substituent and on the nucleophile, but attack at C-2 has only been found in the reaction of the butadiene complex with ***Lithium Diisopropylamide***.[48b]

$$\text{TMSCH}_2\text{-butadiene-Co(CO)}_3^+\ \text{BF}_4^- + \text{PhMgBr} \longrightarrow \text{TMSCH}_2\text{-(allyl)-Ph Co(CO)}_3 \tag{34}$$

Greater synthetic interest attaches to the cycloaddition of dienolates of β-diketones or β-keto esters, illustrated for the cyclohexadiene complex (eq 35), which leads, after ***Tetra-n-butylammonium Fluoride*** treatment of an intermediate (not isolated), to dihydrofuran derivatives.

$$(\text{cyclohexadiene})\text{Co(CO)}_3^+\ \text{BF}_4^- + \text{TMSO/OTMS dienol ether} \longrightarrow \longrightarrow \text{acetyl-methyl-hexahydrobenzofuran} \tag{35}$$

Cyclopentadiene does not yield an isolable diene complex when reacting with $Co_2(CO)_8$, yielding instead ***Dicarbonyl(cyclopentadienyl)cobalt(I)***.

Reactions with Alkynes. In contrast to dienes (see above), alkynes replace the two bridging carbonyls of $Co_2(CO)_8$ producing stable, readily isolated complexes in reactions which are normally rapid at room temperature (eq 36).[49] The alkyne moiety lies at right angles to the Co–Co bond, so that the alkyne carbons form a tetrahedron with the two cobalt atoms. If the alkyne is terminal ($R^2 = H$), these products (hereafter drawn in the simplified form of eq 37) react with acids (e.g. ***Sulfuric Acid*** in MeOH) to give the trinuclear compounds $RCH_2C[Co_3(CO)_9]$,[50] which also form directly from the alkyne with $Co_2(CO)_8$ at 85 °C.[49b] (For further elaboration of such trinuclear complexes see ***Sodium Tetracarbonylcobaltate***, from which a wider range of such compounds is most conveniently prepared.)

$$\text{Co}_2(\text{CO})_8 + \text{R}^1\text{C}{\equiv}\text{CR}^2 \longrightarrow (\text{R}^1\text{C}_2\text{R}^2)\text{Co}_2(\text{CO})_6 \tag{36}$$

Reactions of the $(RC_2H)Co_2(CO)_6$ complexes with carbon monoxide under pressure give lactone complexes[51a,b] (reducible to γ-butyrolactones[51c]) by insertion of two CO groups into the Co–C bonds (eq 37); the further stepwise insertion of CO and alkyne provides the mechanism for the efficient $Co_2(CO)_8$-catalyzed synthesis of 2,2′-bifurylidene-5,5′-diones (eq 38).[52] Under relatively mild conditions and with a 1:1 ratio of C_2H_2:CO, the extended bifurandiones of eq 39 also become significant products.[53]

$$\text{R-C}{\equiv}\text{CH}\cdot\text{Co}_2(\text{CO})_6 + \text{CO} \longrightarrow \text{lactone complex (OC)}_3\text{Co}\text{–Co(CO)}_3 \tag{37}$$

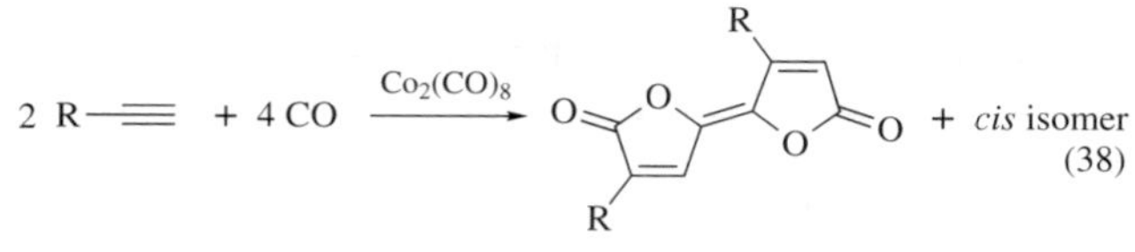

(38)

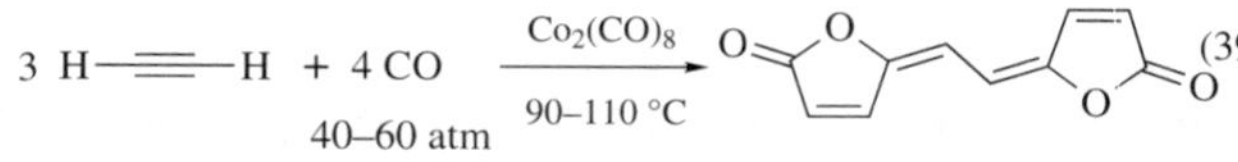

(39)

The lactone complexes react with tertiary propynylamines at room temperature in moderate to good yield to give unsaturated aminolactones (eq 40),[54] while bifurandiones incorporating the propynyl group and a solvent-derived hydrogen atom result when quaternary propynylammonium salts are employed under similar conditions (eq 41).[54]

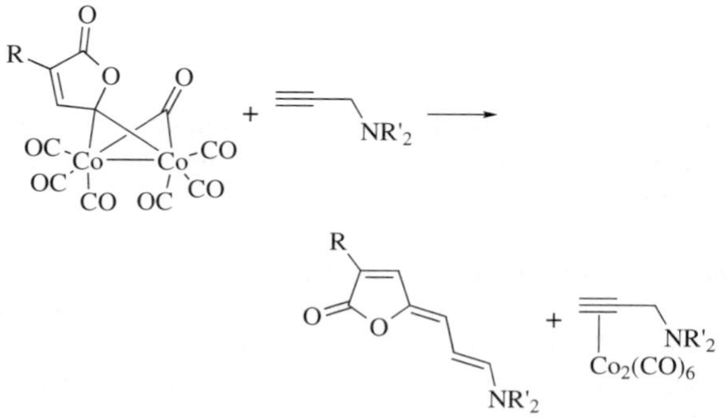

(40)

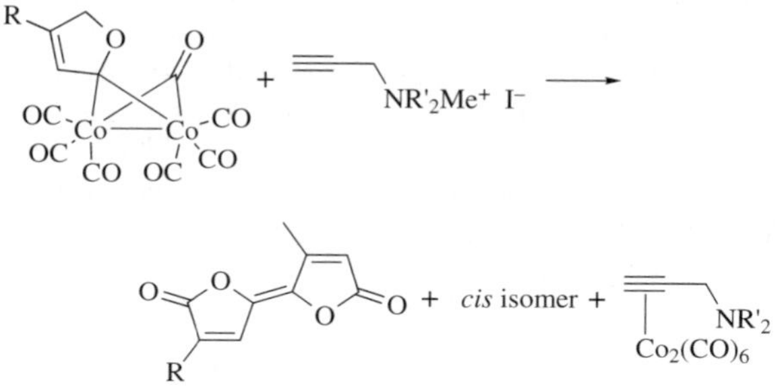

(41)

Without CO pressure, the monoalkyne complexes react on warming with an excess of alkyne to incorporate two more molecules of the latter, giving 'flyover' complexes (eq 42).[55] These cyclize on oxidation (by Br_2) or on heating (typically to 150–170 °C, but in refluxing toluene in the case of the flyover complex from 3 moles of PhC_2COMe[55b]), yielding benzene derivatives. This reaction sequence represents a significantly different mechanism for alkyne cyclotrimerization under $Co_2(CO)_8$ catalysis compared, for example, to the dicarbonylcyclopentadienylcobalt-catalyzed process. The possibility of employing two different alkynes (eq 42) should permit synthesis of a wider range of benzene derivatives and, indeed, the use of alkynes with bulky R^2 groups (e.g. $R^2 = t$-Bu) provides a route to benzene derivatives with two such groups in adjacent positions (e.g. 1,2-di-*t*-butyl- and 1,2,4,5-tetra-*t*-butylbenzenes from the flyovers with $R^1 = H$ or *t*-Bu and $R^2 = t$-Bu).[56] However, since the initial complex, $(R^1C_2H)Co_2(CO)_6$, also undergoes alkyne exchange with the added alkyne, yields are severely limited and product mixtures are inevitably formed.[56]

(42)

Intermediates of the reaction of eq 42 are not normally observed, but some examples of complexes formed from octacarbonyldicobalt and two alkyne moieties are known, e.g. from cyclooctyne, giving a product which promotes the cyclotrimerization of cyclooctyne (eq 43),[57] and also from $MeN(CMe_2C_2H)_2$.[58] The product of the latter reaction has been treated with phenylacetylene,[58] forming an arene system (eq 44). Thus, both of these examples show that this type of bis(alkyne)cobalt complex can also be on the cyclotrimerization pathway.

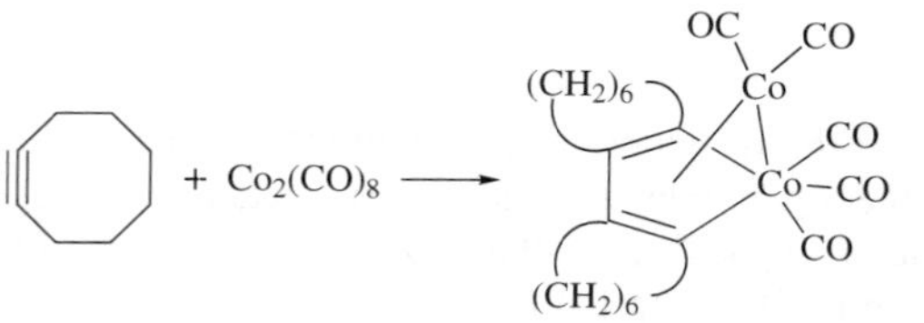

(43)

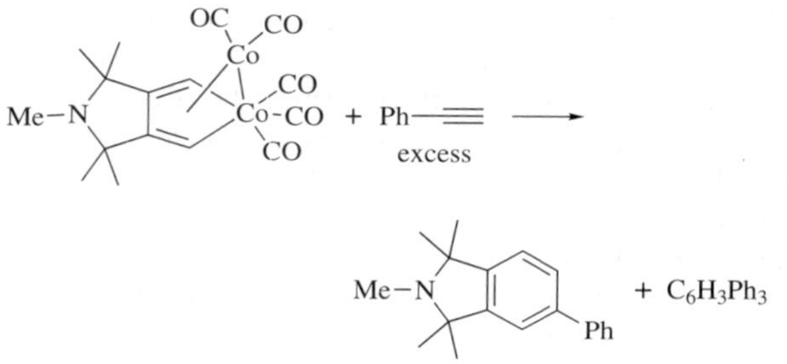

(44)

Alkyne Protection, Distortion, and Altered Substituent Reactivity. The hexacarbonyldicobalt moiety of the bridged alkyne complexes $(R^1C_2R^2)Co_2(CO)_6$ can serve as a valuable protecting group which allows one to perform on other parts of the organic ligand reactions which would not be possible with the free alkyne. At the same time, complexation reduces the bond order from triple to approximately that of a double bond, with corresponding change in bond angles. Hence substituents are in much more favorable locations for many cyclization reactions. Some of the most elegant synthetic applications use both of these features in combination with the α-cation stabilization discussed in the next subsection.

The first demonstration of the stabilizing effect[59] involved Friedel–Crafts acetylation of the tolane complex to give regioselectively the 4-acetyl and 4,4′-diacetyl derivatives and the smooth liberation of the free alkynes by cerium(IV) oxidation. Open-chain enyne complexes (even with the double and triple bonds conjugated) were shown[60] to undergo smooth acid-catalyzed hydration and hydroboration/***Hydrogen Peroxide*** oxidation (e.g. eq 45), as well as reduction of the double bond (by N_2H_2) and again oxidative decomplexation, while the 5,6-double bond of complexed Δ^5-17-ethynylandrostene-3,17-diol was reduced by hydroboration followed by acetic acid treatment.[60] The protective effect has also been utilized to prevent alkyne-to-allene isomerization during oxidation of an allylic alcohol function.[61]

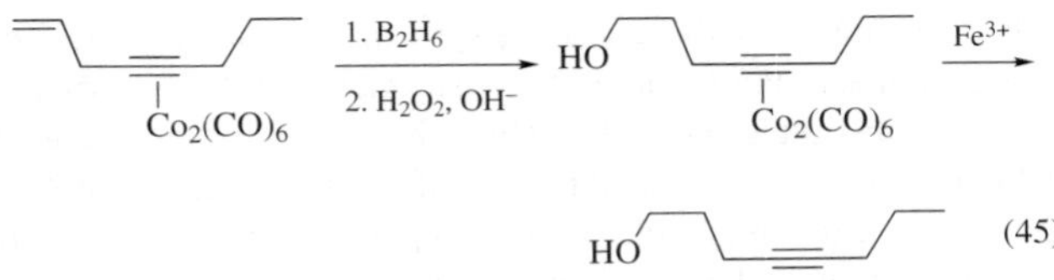

(45)

The effect of distortion is clearly revealed in the macrocyclization of the but-2-yne-1,5-diol complex with diphenyldichlorosilane (eq 46),[62] by elegant syntheses of cycloene-diynes related to calicheamicin (e.g. eq 47),[63] 10- and 11-membered ring lactones (e.g. eq 48),[64] but most strikingly in the construction of the cyclooctadecanonayne system by oxidative coupling of 1,3,5-hexatriyne, complexed to phosphine-substituted cobalt carbonyl (eq 49).[65]

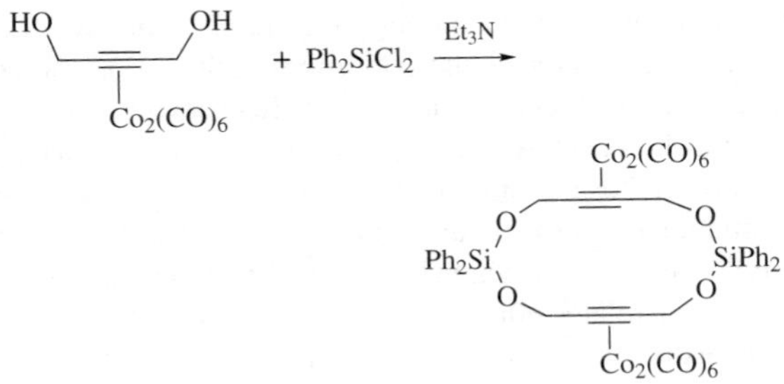

(46)

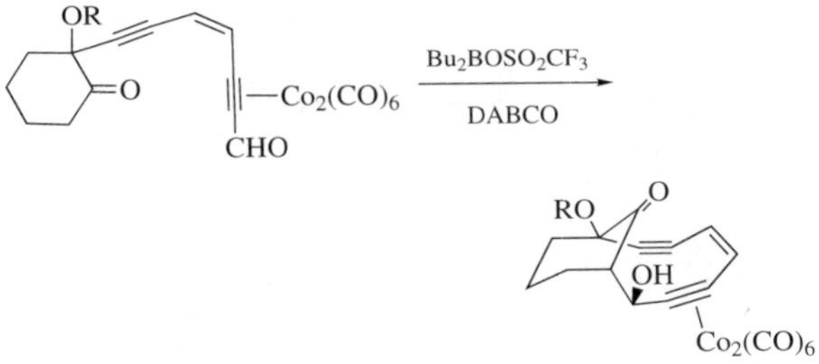

(47)

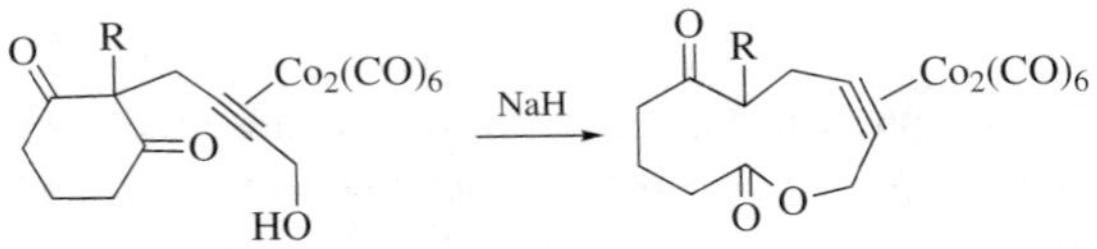

(48)

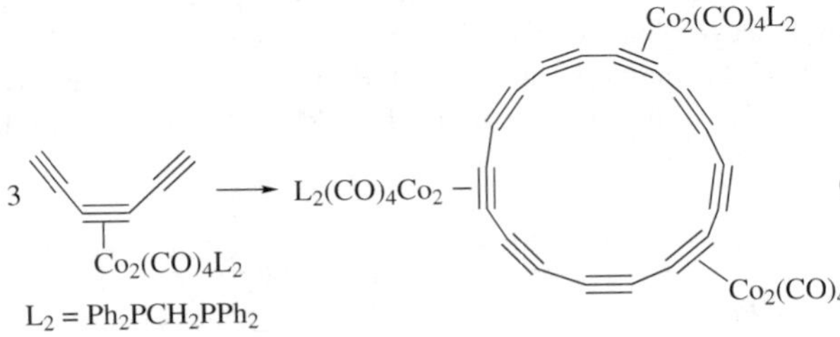

(49)

Steric effects of the hexacarbonyldicobalt moiety may be responsible for altered reactivity, notably enhanced stereoselectivity in reactions of adjacent substituents, e.g. formyl groups in aldol condensations.[66] Thus, condensation of trimethylsilylpropynal with the silyl enol ether of cyclopentanone gives a 90% yield of the aldol product (eq 50) as a 40:60 *erythro:threo* mixture, whereas reaction of the $Co_2(CO)_6$-complexed aldehyde followed by cerium(IV) oxidation gives the same total yield, but in an 87:13 diastereomeric ratio.[66a]

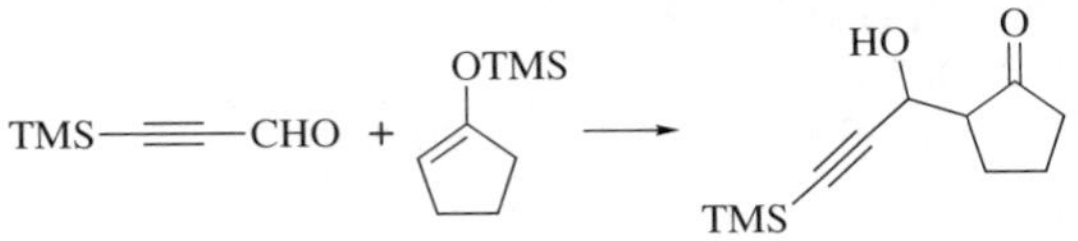

(50)

Cobalt-Complexed Propargyl Cations and Their Reactions. The stability and synthetic utility of hexacarbonyldicobalt complexed cations was first reported in 1971.[60,67] They are readily formed by protonation of propargyl alcohol complexes or addition of electrophiles (protons, or alkyl or acyl cations) to vinylacetylene complexes. They can be isolated in pure form, usually by precipitation as hexafluorophosphates or tetrafluoroborates, but are more commonly generated in situ and used directly. Most of the early work is due to Nicholas, who has comprehensively reviewed the work up to 1986.[68] A detailed description of the procedure for generating the 1-methyl-2-propynyl complex and for its reaction with trimethylsilyloxycyclohexene (eq 51) has been given.[69]

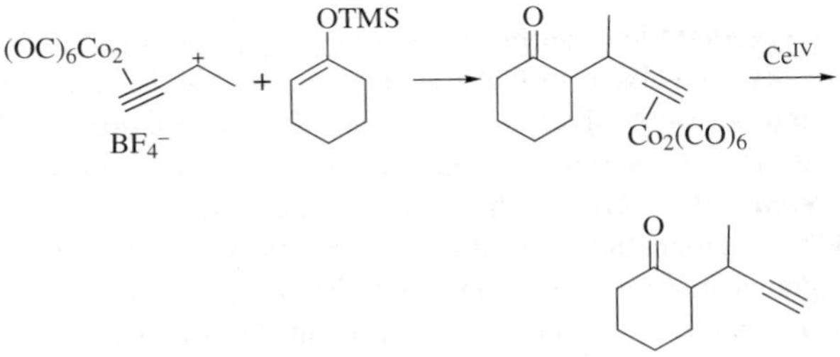

(51)

More recent studies of the utility of the complexed cations in synthesis have paid particular attention to stereoselectivity in their reaction with silyl enolates[70] and enol borates,[71] e.g. eq 52; in this case the same products result, but with only 2.5:1 stereoselection when using the triethylsilyl enol ether and ***Boron Trifluoride Etherate*** catalysis.

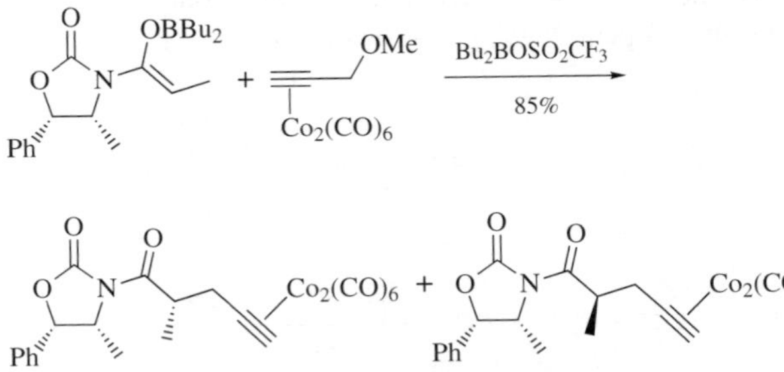

(52)

35:1

Replacement of one carbonyl group by phosphine (thus creating chirality at cobalt) has been shown to permit enantioselective addition of nucleophiles to the cations.[72] Regioselectivity of aromatic substitution by the cations has been studied in resorcinol derivatives[73] and examples of aromatic substitution extended to include indoles.[74]

An alternative approach to the antitumor active enediyne systems (cf. eq 47) using 'Nicholas cations' has been extensively studied[75] and applied to the dynemycin core structure (eq 53).[75c] Another example of the use of such cations in forming strained rings is the preparation of a cobalt complex of a trithia-crown ether (eq 54), which has been shown to complex Cu^I and Ag^I.[76]

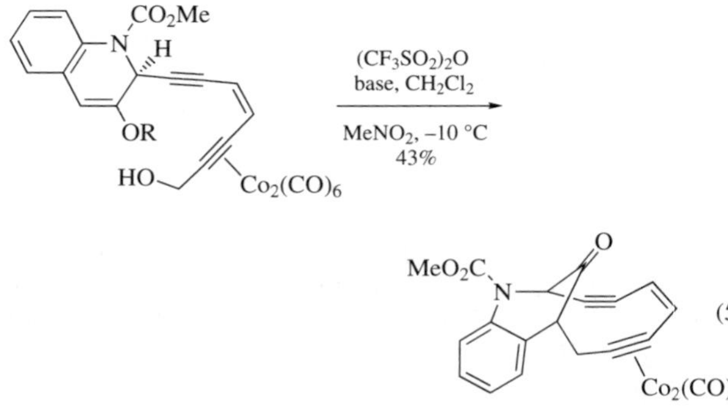

(53)

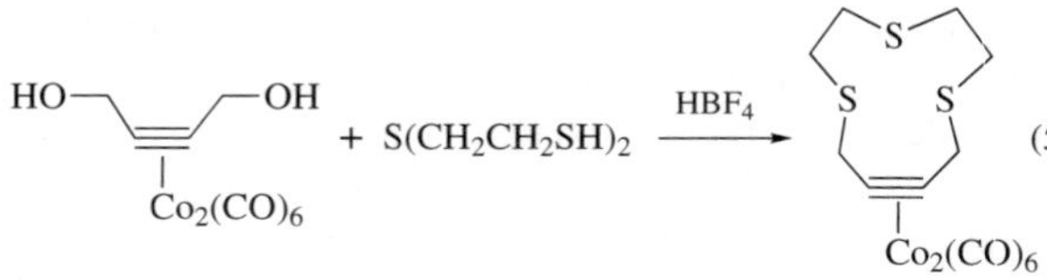

(54)

Cyclopentenone Synthesis. Synthesis of cyclopentenones from alkynes + alkenes + CO has become a widely used process and is variously known as the Khand reaction or the Pauson–Khand reaction. Although inter alia examples promoted by ***Pentacarbonyliron***, ***Hexacarbonylmolybdenum***, tetracarbonylbis(cyclopentadienyl)ditungsten or -molybdenum, and carbene(pentacarbonyl)chromium have been described, octacarbonyldicobalt is the generally used reactant. The method has been extensively reviewed.[52a,77]

The general reaction (eq 55) is most commonly conducted by preparing the intermediate alkyne complex $(R^1C_2R^2)Co_2(CO)_6$, with or without isolation, before adding the alkene component, but this is not necessary. Under a CO atmosphere, regeneration of this complex from added free alkyne occurs and in at least one case[78] efficient catalytic synthesis has been achieved (eq 56). Under conditions of complete conversion of 1-heptyne, the product was obtained in 47–49% yield and contained only 1–2% of the regioisomer, 3-pentylcyclopent-2-en-1-one.

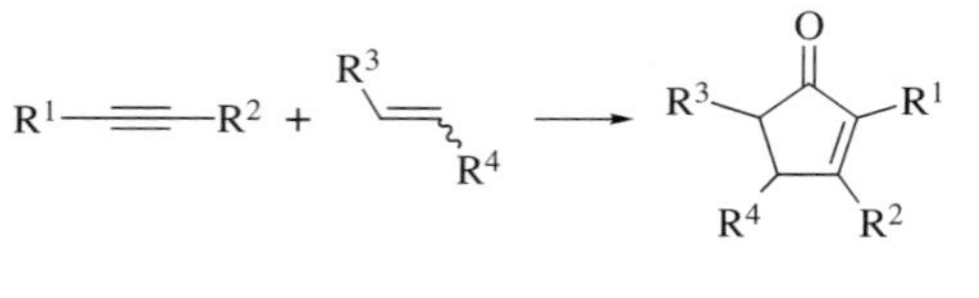

(55)

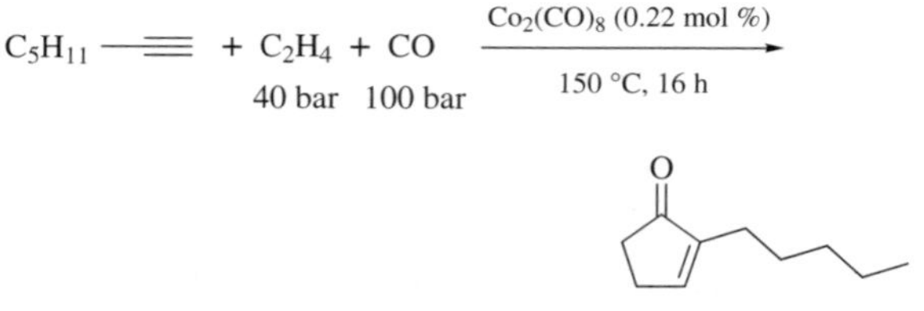

(56)

Such high regioselectivity is typical of terminal alkynes, and the less reactive internal alkynes show only slightly lower selectivity in forming the products with the bulkier group (R^1) in the 2-position. Only strong electronic effects may override this sterically determined preference (e.g. see eq 63 below). Regioselectivity with respect to the alkene is much more variable. Simple terminal alkenes show little or no selectivity when the alkyne is also terminal, but significant preference for 5-substitution (R^3 in eq 55) when the alkyne is internal. More efficient regioselection may be achieved with functional substituents, notably electron-donor groups in the homoallylic position, which favor formation of the 5-substituted product.[79] Thus, the example shown in eq 57 is the key step in a formal synthesis of PGA_2;[80] it also illustrates the good yields obtainable by the simple thermal reaction in the case of alkenes with donor substituents.

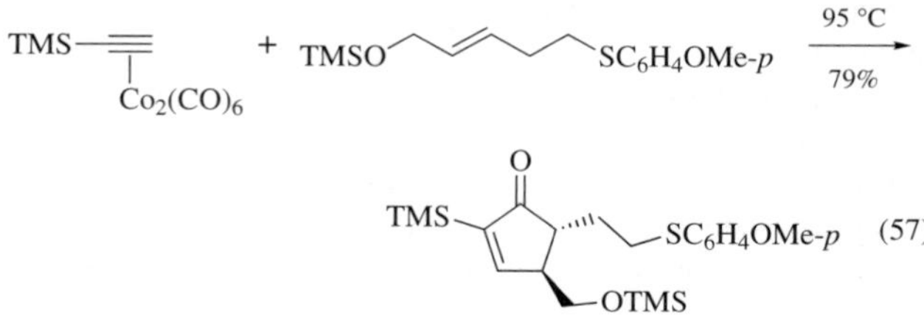

(57)

Three more general techniques have greatly improved the earlier yields. Discussion of the solid-phase adsorption method is included in more recent reviews.[77a–d] It probably remains the method of choice in selected cases and sometimes has results different from those of other methods. For example, the predominant reduction to cyclopentanone in the intramolecular reaction of *N*-acetyl-*N*-allylpropargylamine (eq 58)[81] and the reductive ether cleavage when related ethers react on alumina (eq 59),[82] is avoided on silica, in the presence of oxygen, with formation of the expected bicyclic products.

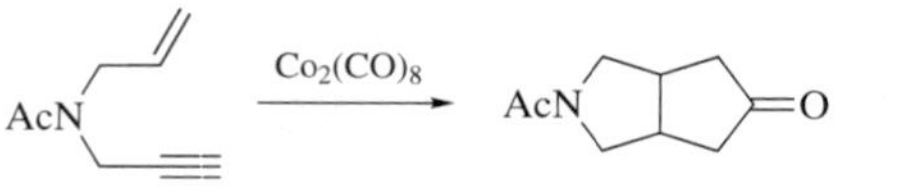

(58)

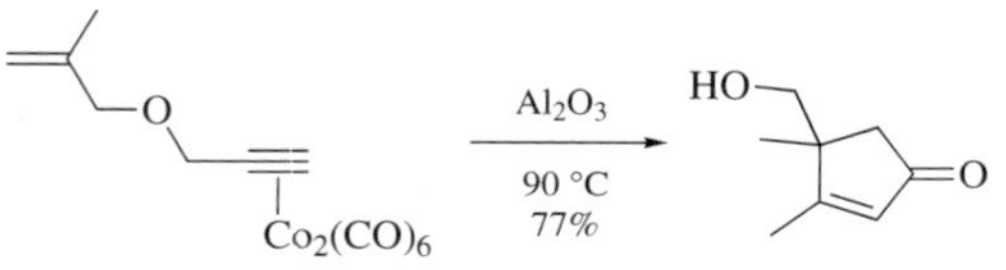

(59)

This solid-phase technique has made possible efficient syntheses of bi- and polycyclic systems from precursors frequently generated via Nicholas cations, (e.g. eq 60)[83] and has been used for intermolecular reactions with methylenecyclopropane (eq 61) and methylenecyclobutane to give spirocyclic cyclopentenone derivatives. (**1**) is the major product when the alkyne is terminal (R^2 = H), but only (**2**) was isolated when EtC_2Et or Me_3SiC_2Me were used.[84]

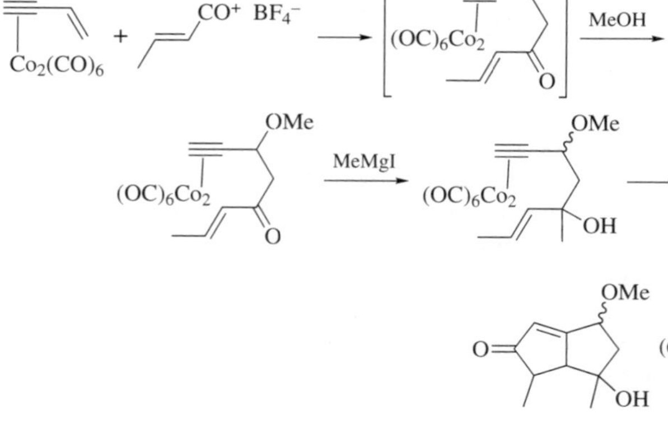

(60)

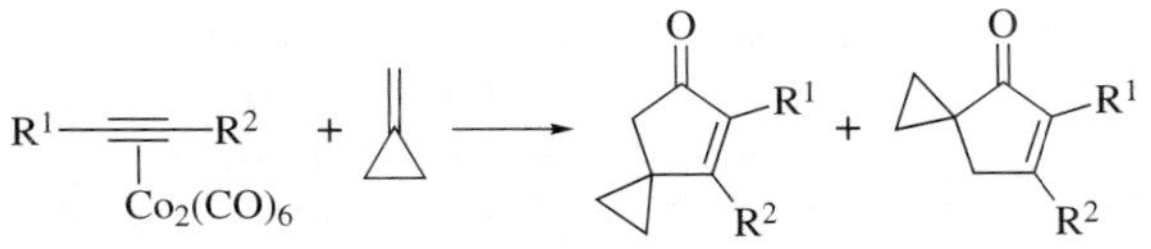

The more recent and now widely used improvement involves the use of amine oxides, usually ***Trimethylamine N-Oxide***[85a] or ***N-Methylmorpholine N-Oxide***.[86] Alternatively, while requiring slightly higher temperature and longer reaction time, DMSO (or other sulfoxides) can give equally good yields. A range of other polar solvents (e.g. MeOH, MeCN, HCO_2Et) also exert a distinct, but smaller, promoting effect.[87]

N-Methylmorpholine *N*-oxide, preferably as monohydrate,[79b] as well as Me_3NO not only induces cyclopentenone formation under very mild conditions (0–20 °C) and in frequently excellent yields, but also removes some earlier limitations. The reductive processes (eqs 58 and 59) do not occur when these promoters or DMSO are used under oxygen.[85,87] Hence two efficient syntheses of (−)-kainic acid[88] have employed cyclizations of allylpropargylamine derivatives as the key step; in one of these (eq 62), as well as in a synthesis of the dendrobine skeleton,[89] use of a chiral precursor was shown to result in enantiospecific bicyclization.[88a]

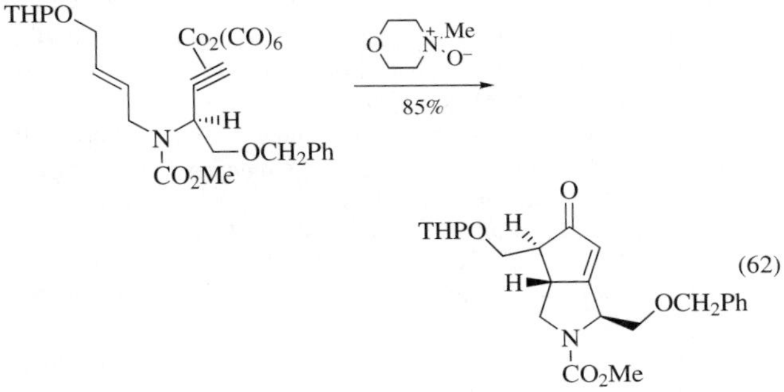

(62)

Intramolecular cyclizations of enynes using catalytic amounts of octacarbonyldicobalt (3 mol %) succeed in the presence of biphenyl phosphite (10 mol %) under 3 atm of CO, typically in DMF at 120 °C.[85b]

Whereas in earlier work the (uncatalyzed) reaction of alkenes bearing electron-withdrawing groups with the alkynecobalt complexes gave conjugated dienes as the only or major products,[77] intramolecular cyclizations involving such groups have now been accomplished by both the solid-state method[90] and using *N*-oxide promotion.[91] Moreover, electron-deficient alkynes have been successfully employed in both inter- and intramolecular cases. The use of ethyl 2-butynoate (eq 63) illustrates the preferred orientation in this case.[92]

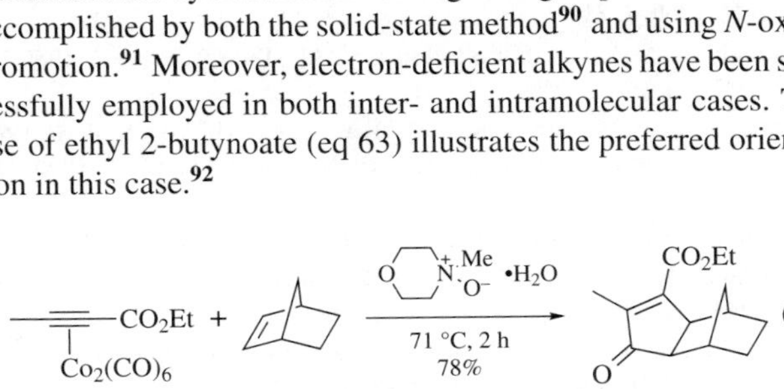

(63)

Further examples of the efficiency of the *N*-oxide method are the formation of bicyclic lactols (e.g. eq 64), utilized in a synthesis of (±)-loganin,[93] and the quantitative cyclization (eq 65) which is the key step in a synthesis of decarboxydemethylquadrone.[94]

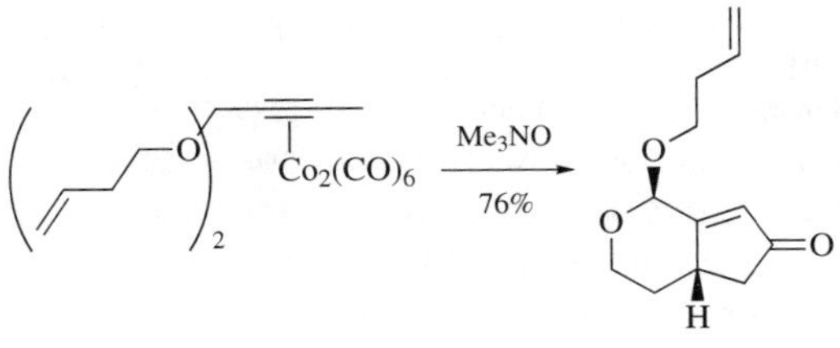

(64)

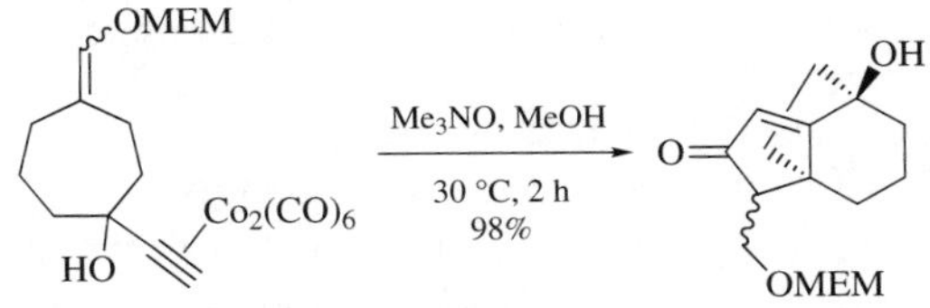

(65)

Finally, reference is made to the remarkably facile cyclization which alkynyl Fischer carbene complexes can undergo[95] under conditions normally leading only to alkyne–cobalt complex formation (eq 66) (M = Cr or W; R = Ph or Et).

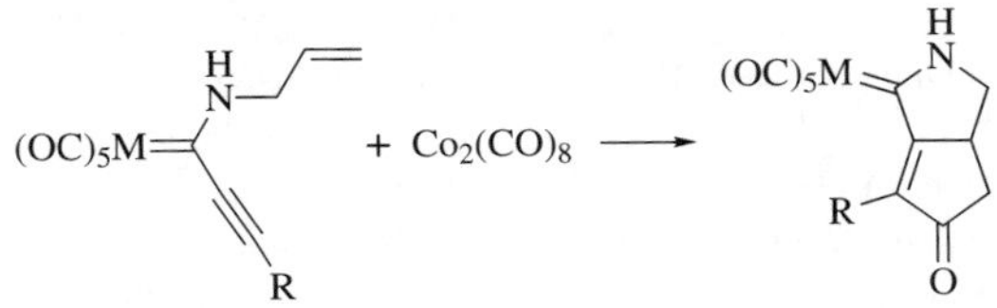

(66)

Related Reagents. Dicarbonyl(cyclopentadienyl)cobalt(I); Octacarbonyldicobalt–Diethyl(methyl)silane–Carbon Monoxide; Sodium Tetracarbonylcobaltate.

1. Leung, P. C.; Coppens, P. *Acta Crystallogr., Sect. B* **1983**, *B39*, 535.
2. Wender, I.; Pino, P. *Organic Synthesis via Metal Carbonyls*; Interscience: New York, 1968.
3. *Gmelin Handbuch der Anorganischen Chemie*, 8th ed.; Springer: Berlin, 1961; Suppl. Vol. 58A, p 677.
4. *MOC* **1986**, *E18*, Part 2.
5. Falbe, J. *Synthesen mit Kohlenoxid*; Springer: Berlin, 1967.
6. Falbe, J. *New Syntheses with Carbon Monoxide*; Springer: Berlin, 1980.
7. Henrici-Olivé, G.; Olivé, S. *Catalyzed Hydrogenation of Carbon Monoxide*; Springer: Berlin 1984.
8. Ojima, I. *CRV* **1988**, *88*, 1011.
9. Gaslini, F.; Nahum, L. Z. *JOC* **1964**, *29*, 1177.
10. Kogami, K.; Kumanotani, J. *BCJ* **1973**, *46*, 3562.
11. Ucciani, E.; Pelloquin, A.; Cecchi, G. *J. Mol. Catal.* **1977/78**, *3*, 363.
12. Friedman, S.; Metlin, S.; Svedi, A.; Wender, I. *JOC* **1959**, *24*, 1287.
13. Falbe, J.; Huppes, N.; Korte, F. *CB* **1964**, *97*, 863.
14. e.g. Haenel, M. W.; Schanne, L.; Woestefeld, E. *Erdöl, Kohle, Erdgas, Petrochem.* **1986**, *39*, 505 (*CA* **1987**, *106*, 20 865).
15. See e.g. Withers, H. P.; Eliezer, K. F.; Mitchell, J. W. *Ind. Eng. Chem., Res.* **1990**, *29*, 1807.
16. Masters, C. *Adv. Organomet. Chem.* **1979**, *17*, 61.
17. Arthur, P., Jr.; England, D. C.; Pratt, B. C.; Whitman, G. M. *JACS* **1954**, *76*, 5364.
18. Murata, K.; Matsuda, M. *BCJ* **1982**, *55*, 2195.
19. Yokokawa, C.; Watanabe, Y.; Takegami, Y. *BCJ* **1964**, *37*, 677, 935.
20. Eisenmann, J. L.; Yamartino, R. L.; Howard, J. F. *JOC* **1961**, *26*, 2102.
21. Rosenthal, A.; Kan, G. *TL* **1967**, 477.
22. El-Chahawi, M.; Prange, U. *CZ* **1978**, *102*, 1.
23. See e.g. *CA* **1990**, *113*, 5940, 171 682.

24. See e.g. (a) Landis, C. R.; Kowaja, H. U.S. Patent 4 948 920, 1989 (*CA* **1991**, *114*, 61 697). (b) Lapidus, A. L.; Krylova, A. Yu.; Kozlova, G. V.; Kondrat'ev, L. T. *IZV* **1989**, 2425; *BAU* **1989**, 2226.
25. Sisak, A.; Ungvary, F.; Marko, L. *JOC* **1990**, *55*, 2508.
26. Wang, M. D.; Calet, S.; Alper, H. *JOC* **1989**, *54*, 20.
27. Roberto, D.; Alper, H. *JACS* **1989**, *111*, 7539.
28. Murahashi, S.; Horiie, S. *JACS* **1956**, *78*, 4816.
29. (a) Murahashi, S.; Horiie, S. *JACS* **1955**, *77*, 6403. (b) Murahashi, S.; Horiie, S.; Jo, T. *Nippon Kagaku Zasshi* **1958**, *79*, 68, 72, 75 (*CA* **1960**, *54*, 5558, 5559).
30. Falbe, J.; Korte, F. *CB* **1962**, *95*, 2680.
31. Sauer, J. C.; Howk, B. W.; Stiehl, R. T. *JACS* **1959**, *81*, 693.
32. Uemura, S.; Takahashi, H.; Che, K.; Sugita, N. *JOM* **1989**, *361*, 63.
33. Friedman, S.; Harris, S. R.; Wender, I. *Ind. Eng. Chem., Prod. Res. Dev.* **1970**, *9*, 347.
34. Hong, P.; Sonogashira, K.; Hagihara, N. *J. Chem. Soc. Jpn.* **1968**, *89*, 74.
35. Wakamatsu, H.; Uda, J.; Yamakami, N. *CC* **1971**, 1540.
36. Parnaud, J. J.; Campari, G.; Pino, P. *J. Mol. Catal.* **1979**, *6*, 341.
37. Hirai, K.; Takahashi, Y.; Ojima, I. *TL* **1982**, *23*, 2491.
38. (a) Ojima, I.; Kato, K.; Okabe, M.; Fuchikami, T. *JACS* **1987**, *109*, 7714. (b) Ojima, I.; Kato, K.; Nakahashi, K.; Fuchikami, T.; Fujita, M. *JOC* **1989**, *54*, 4511.
39. Amino, Y.; Izawa, K. *BCJ* **1991**, *64*, 1040.
40. Trifonov, L.; Orakhovats, A. *HCA* **1989**, *72*, 648.
41. Khand, I. U.; Knox, G. R.; Pauson, P. L.; Watts, W. E.; Foreman, M. I. *JCS(P1)* **1973**, 977.
42. Khand, I. U.; Pauson, P. L.; Habib, M. J. A. *JCR(S)* **1978**, 346; *JCR(M)* **1978**, 4401.
43. Okuda, F.; Watanabe, Y. *BCJ* **1990**, *63*, 1201.
44. Chatani, N.; Hanafusa, T. *JOC* **1991**, *56*, 2166.
45. Hübel, W.; Hoogzand, C. *CB* **1960**, *93*, 103.
46. (a) Winkhaus, G.; Wilkinson, G. *JCS* **1961**, 602. (b) Fischer, E. O.; Kuzel, P.; Fritz, H. P. *ZN(B)* **1961**, *16b*, 138. (c) Fischer, E. O.; Palm, C. *ZN(B)* **1959**, *14b*, 598.
47. (a) Chaudhary, F. M.; Pauson, P. L. *JOM* **1974**, *69*, C31. (b) Pankayatselvan, R.; Nicholas, K. M. *JOM* **1990**, *384*, 361.
48. (a) Barinelli, L. S.; Nicholas, K. M. *JOC* **1988**, *53*, 2114. (b) Miller, M.; Nicholas, K. M. *JOM* **1989**, *362*, C15.
49. (a) Greenfield, H.; Sternberg, H. W.; Friedel, R. A.; Wotiz, J. H.; Markby, R.; Wender, I. *JACS* **1956**, *78*, 120. (b) Dickson, R. S.; Yawney, D. B. W. *AJC* **1969**, *22*, 533.
50. Markby, R.; Wender, I.; Friedel, R. A.; Cotton, F. A.; Sternberg, H. W. *JACS* **1958**, *80*, 6529.
51. (a) Sternberg, H. W.; Shukys, J. G.; Donne, C. D.; Markby, R.; Friedel, R. A.; Wender, I. *JACS* **1959**, *81*, 2339. (b) Varadi, G.; Vecsei, I.; Ötvös, Z.; Pályi, G.; Marko, L. *JOM* **1979**, *182*, 415. (c) Sato, S.; Morishima, A.; Wakamatsu, H. *J. Chem. Soc. Jpn.* **1970**, *91*, 557.
52. (a) Schore, N. E. *CRV* **1988**, *88*, 1081. (b) Rautenstrauch, V.; Mégard, P.; Gamper, B.; Bourdin, B.; Walther, E.; Bernardinelli, G. *HCA* **1989**, *72*, 811 and references therein.
53. Albanesi, G.; Farina, R.; Taccioli, A. *Chim. Ind. (Milan)* **1966**, *48*, 1151.
54. Tasi, M.; Horvath, I. T.; Andreeti, G. D.; Palyi, G. *CC* **1989**, 426.
55. (a) Mills, O. S.; Robinson, G. *Proc. Chem. Soc.* **1964**, 187. (b) Gervasio, G.; Sappa, E.; Markó, L. *JOM* **1993**, *444*, 203.
56. (a) Hoogzand, C.; Hübel, W. *AG* **1961**, *73*, 680. (b) Hoogzand, C.; Hübel, W. *TL* **1961**, 637.
57. Bennett, M. A.; Donaldson, P. B. *IC* **1978**, *17*, 1995.
58. Predieri, G.; Tiripicchio, A.; Camellini, M. T.; Costa, M.; Sappa, E. *JOM* **1992**, *423*, 129.
59. Seyferth, D.; Nestle, M. O.; Wehman, A. T. *JACS* **1975**, *97*, 7417.
60. Nicholas, K. M.; Pettit, R. *TL* **1971**, 3475.
61. Marshall, J. A.; Robinson, E. D.; Lebreton, J. *JOC* **1990**, *55*, 227.
62. Cragg, R. H.; Jeffery, J. C.; Went, M. J. *JCS(D)* **1991**, 137.
63. (a) Magnus, P.; Annoura, H.; Harling, J. *JOC* **1990**, *55*, 1709. (b) Magnus, P.; Davies, M. *CC* **1991**, 1522. (c) Magnus, P.; Pitterna, T. *CC* **1991**, 541.
64. Najdi, S. D.; Olmstead, M. M.; Schore, N. E. *JOM* **1992**, *431*, 335.
65. Rubin, Y.; Knobler, C. B.; Diederich, F. *JACS* **1990**, *112*, 4966.
66. (a) Mukai, C.; Nagami, K.; Hanaoka, M. *TL* **1989**, *30*, 5623, 5627. (b) Mukai, C.; Suzuki, K; Hanaoka, M. *CPB* **1990**, *38*, 567. (c) Ju, J.; Reddy, B. R.; Khan, M.; Nicholas, K. M. *JOC* **1989**, *54*, 5426. (d) Roush, W. R.; Park, J. C. *JOC* **1990**, 55, 1143. (e) Roush, W. R.; Park, J. C. *TL* **1991**, *32*, 6285.
67. Nicholas, K. M.; Pettit, R. *JOM* **1972**, *44*, C21.
68. Nicholas, K. M. *ACR* **1987**, *20*, 207.
69. Varghese, V.; Saha, M.; Nicholas, K. M. *OS* **1988**, *67*, 141.
70. (a) Montana Pedrero, A. M.; Nicholas, K. M. *MRC* **1990**, *28*, 486. (b) Tester, R.; Varghese, V.; Montana, A. M.; Khan, M.; Nicholas, K. M. *JOC* **1990**, *55*, 186.
71. Schreiber, S. L.; Klimas, M. T.; Sammiaka, T. *JACS* **1987**, *109*, 5749.
72. Bradley, D. H.; Khan, M. A.; Nicholas, K. M. *OM* **1992**, *11*, 2598.
73. Gruselle, M.; Rossignol, J.-L.; Vessieres, A.; Jaouen, G. *JOM* **1987**, *328*, C12.
74. Nakagawa, M.; Ma, J.; Hino, T. *H* **1990**, *30*, 451.
75. (a) Magnus, P.; Carter, P. A. *JACS* **1988**, *110*, 1626. (b) Magnus, P.; Lewis, R. T.; Huffman, J. C. *JACS* **1988**, *110*, 6921. (c) Magnus, P.; Fortt, S. M. *CC* **1992**, 544. (d) Magnus, P.; Carter, P.; Elliot, J.; Lewis, R.; Harling, J.; Pitterna, T.; Bauta, W. E.; Fortt, S. *JACS* **1992**, *114*, 2544.
76. (a) Gelling, A.; Jeffery, J. C.; Povey, D. C.; Went, M. J. *CC* **1991**, 349. (b) Demirhan, F.; Irisli, S.; Salek, S. N.; Sentürk, O. S.; Went, M. J.; Jeffery, J. C. *JOM* **1993**, *453*, C30.
77. (a) Schore, N. E. *OR* **1991**, *40*, 1. (b) Schore, N. E. *COS* **1991**, *5*, 1037. (c) Pauson, P. L. In *Organometallics in Organic Synthesis: Aspects of a Modern Interdisciplinary Field*; de Meijere, A.; tom Dieck, H., Eds.; Springer: Berlin, 1988; p 233. (d) Pauson, P. L. *T* **1985**, *41*, 5855. (e) Pauson, P. L.; Khand, I. U. *ANY* **1977**, *295*, 2. (f) Harrington, P. J. *Transition Metals in Total Synthesis*; Wiley: New York, 1990; Chapter 9.
78. Rautenstrauch, V.; Mégard, P.; Conesa, J.; Küster, W. *AG(E)* **1990**, *29*, 1413.
79. (a) Krafft, M. E.; Juliano, C. A. *JOC* **1992**, *57*, 5106. (b) Krafft, M. E.; Scott,, I. L.; Romero, R. H.; Feibelmann, S.; Van Pelt, C. E. *JACS* **1993**, *115*, 7199 and references therein.
80. Krafft, M. E.; Wright, C. *TL* **1992**, *33*, 151.
81. Brown, S. W.; Pauson, P. L. *JCS(P1)* **1990**, 1205.
82. Smit, W. A.; Simonyan, S. O.; Tarasov, V. A.; Mikaelian, G. S.; Gybin, A. S.; Ibragimov, I. I.; Caple, R.; Froen, D.; Kreager, A. *S* **1989**, 472.
83. Gybin, A. S.; Smit, V. A.; Veretenov, A. L. Simonyan, S. O.; Shashkov, A. S.; Struchkov, Yu. T.; Kuz'mina, L. G.; Caple, R. *IZV* **1989**, 2756; *BAU* **1989**, 2521.
84. Smit, W. A.; Kireev, S. L.; Nefedov, O. M.; Tarasov, V. A. *TL* **1989**, *30*, 4021.
85. (a) Jeong, N.; Chung, Y. K.; Lee, B. Y.; Lee, S. H.; Yoo, S.-E. *SL* **1991**, 204. (b) Jeong, N.; Hwang, S. H.; Lee, Y.; Chung, Y. K. *JACS* **1994**, *116*, 3159.
86. Shambayati, S.; Crowe, W. E.; Schreiber, S. L. *TL* **1990**, *31*, 5289.
87. Chung, Y. K.; Lee, B. Y.; Jeong, N.; Hudecek, M.; Pauson, P. L. *OM* **1993**, *12*, 220.
88. (a) Takano, S.; Inomata, K.; Ogasawara, K. *CC* **1992**, 169. (b) Yoo, E.-e.; Lee, S.-H.; Jeong, N.; Cho, I. *TL* **1993**, *34*, 3435.
89. Takano, S.; Inomata, K.; Ogasawara, K. *CL* **1992**, 443.
90. (a) Veretenov, A. L.; Gybin, A. S.; Smit, V. A. *IZV* **1990**, 1908; *BAU* **1990**, 1736. (b) Veretenov, A. L.; Smit, W. A.; Vorontsova, L. G.; Kurella, M. G.; Caple, R.; Gybin, A. S. *TL* **1991**, *32*, 2109.
91. van der Waals, A.; Keese, R. *CC* **1992**, 570.

92. (a) Krafft, M. E.; Romero, R. H.; Scott, I. L. *JOC* **1992**, *57*, 5277. (b) Fonquerna, S.; Moyano, A.; Pericàs, M. A.; Riera, A. *T* **1995**, *51*, 4239.
93. Jeong, N.; Lee, B. Y.; Lee, S. M.; Chung, Y. K.; Lee, S.-G. *TL* **1993**, *34*, 4023.
94. Forsyth, G. S.; Kerr, W. J.; Ladduwahetty, T. Personal communication.
95. Camps, F.; Moretó, J. M.; Ricart, S.; Viñas, J. M. *AG(E)* **1991**, *30*, 1470.

Peter L. Pauson
University of Strathclyde, Glasgow, UK

Oxalyl Chloride

[79-37-8] $C_2Cl_2O_2$ (MW 126.92)

(versatile agent for preparation of carboxylic acid chlorides;[1] phosphonic acid dichlorides;[2] alkyl chlorides;[3] β-chloro enones;[4] acyl isocyanates[5])

Physical Data: mp −12 °C; bp 63–64 °C/763 mmHg; *d* 1.48 g cm^{-3}; n_D^{20} 1.4305.
Solubility: sol hexane, benzene, diethyl ether, halogenated solvents, e.g. dichloromethane and chloroform, acetonitrile.
Form Supplied in: colorless, fuming liquid; widely available; 2M soln in dichloromethane.
Handling, Storage, and Precautions: liquid and solution are toxic, corrosive, and severely irritating to the eyes, skin, and respiratory tract. Use in a fume hood and wear protective gloves, goggles, and clothing. Bottles should be stored in a cool, dry place and kept tightly sealed to preclude contact with moisture. Decomposes violently with water, giving toxic fumes of CO, CO_2, HCl.

Preparation of Carboxylic Acid Chlorides (and Anhydrides). Oxalyl chloride has found general application for the preparation of carboxylic acid chlorides since the reagent was introduced by Adams and Ulich.[1] Acid chlorides produced by this means have subsequently featured in the synthesis of acyl azides,[6] bromoalkenes,[7] carboxamides,[8] cinnolines,[9] diazo ketones,[10] (thio)esters,[11] lactones,[12] ketenes for cycloaddition reactions,[13] intramolecular Friedel–Crafts acylation reactions,[14] and the synthesis of pyridyl thioethers.[11]

Like ***Thionyl Chloride***, oxalyl chloride gives gaseous byproducts with acids and the chlorides can be readily isolated in a pure form by evaporation of the solvent and any excess reagent, or used in situ for further elaboration (eq 1).

Prior formation of an amine or alkali metal salt, with or without pyridine,[1] has been used to advantage with substrates that are sensitive to strong acids or are bases (see also ***Oxalyl Chloride–Dimethylformamide*** for a procedure conducted under neutral conditions using silyl esters). By adjusting the molar proportions of oxalyl chloride to substrate, anhydrides can also be prepared using these methods (eq 2).[15] *N*-Carboxy-α-amino acid anhydrides can also be made this way.[16]

$$RCO_2H + (COCl)_2 \longrightarrow RCOCl + CO_2 + CO + HCl \quad (1)$$

3-Pyridyl-COCl ← $(COCl)_2$, benzene, reflux, 85% — 3-Pyridyl-$CO_2^-K^+$ — 0.5 equiv $(COCl)_2$, benzene, reflux, 85% → (3-Pyridyl-CO)$_2$O (2)

The use of nonpolar solvents such as hexane or toluene allows for the removal of inorganic or amine salts which may otherwise interfere with subsequent reactions.

Under the mild conditions employed (eqs 3 and 4),[17] racemization of stereogenic centers, skeletal rearrangement, or byproduct formation, seen with other reagents such as thionyl chloride/pyridine,[18] are seldom observed.

(3): $(COCl)_2$, CH_2Cl_2, reflux

(4): HO_2C–CH(i-Pr)–$NHCOCF_3$ → $(COCl)_2$, pyridine, 0 °C to rt → ClCO–CH(i-Pr)–$NHCOCF_3$

Conversion of β-bromoacrylic acid to the acid chloride using thionyl chloride/DMF, ***Phosphorus(III) Chloride***, or benzotrichloride/zinc chloride also resulted in bromine for chlorine exchange. Use of oxalyl chloride with the preformed ammonium salt provided a mild, general method to β-bromoacryloyl chlorides (eq 5)[19] without halogen exchange or (*E*/*Z*) equilibration. β-Fluoro- and iodoacrylic acids have been cleanly converted to the acid chlorides without prior salt formation.

$$BrCH{=}CHCO_2^-NH_4^+ \xrightarrow[\text{hexane, reflux, 80\%}]{(COCl)_2} BrCH{=}CHCOCl \quad (5)$$

As well as forming acid chlorides, α-tertiary amino acids can react with oxalyl chloride and undergo an oxidative decarboxylation to give iminium salts, or ring expansion, depending on the substituents and their stereochemistry (eq 6).[20]

Preparation of Phosphonic Acid Chlorides. Phosphonic acid dichlorides have been obtained in high yield (determined by ^{31}P

NMR) at low temperature from the corresponding acids using oxalyl chloride and ***Pyridine*** (eq 7).[2]

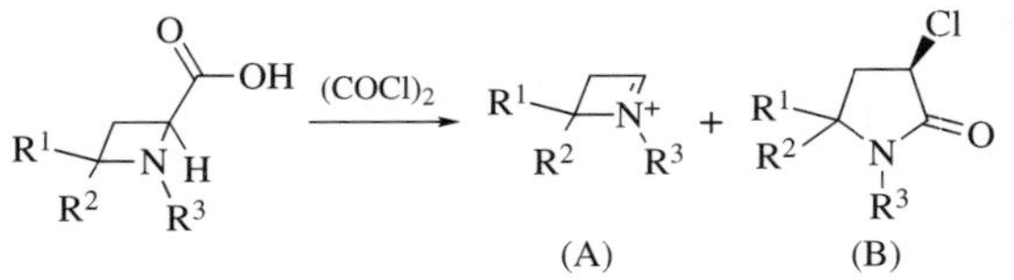

(6)

R^1	R^2	R^3	% Yield (A)	% Yield (B)
Me	H	*t*-Bu	0	69
H	CO_2Me	*c*-Hex	0	59
H	Me	Me	30	0

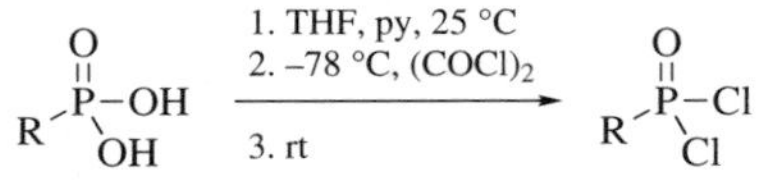

1. THF, py, 25 °C; 2. –78 °C, $(COCl)_2$; 3. rt (7)

R = Et, $PhCH_2$, CF_2H, arabinomethyl, phthalidyl

Similarly, monoalkyl methylphosphonochloridates (eq 8)[21] can be made from dialkyl esters; thionate acid chlorides could not be made by this method. Thionyl chloride and PCl_5 were also used to make this type of compound (see also ***Oxalyl Chloride–Dimethylformamide***).

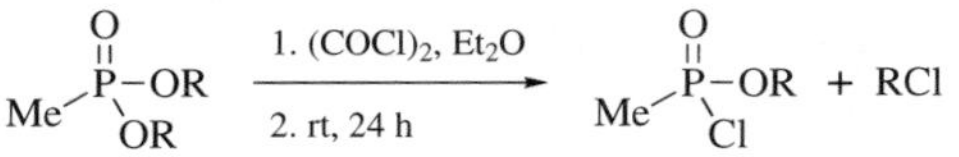

1. $(COCl)_2$, Et_2O; 2. rt, 24 h (8)

R = Me, Et, Pr, *i*-Pr, Bu

Numerous other reagents such as PCl_3, PCl_5, $POCl_3$, and Ph_3P/CCl_4 are available for the preparation of acid chlorides and anhydrides but may not be as convenient as the byproducts are not so easily removed, or the reactions require more vigorous conditions.

Direct Introduction of the Chlorocarbonyl Group (Halocarbonylation). Alkanes or cycloalkanes react with oxalyl chloride under radical conditions; typically, mixtures are produced.[22] However, bicyclo[2.2.1]heptane undergoes regio- and stereospecific chlorocarbonylation, giving the ester on subsequent methanolysis (eq 9).[23]

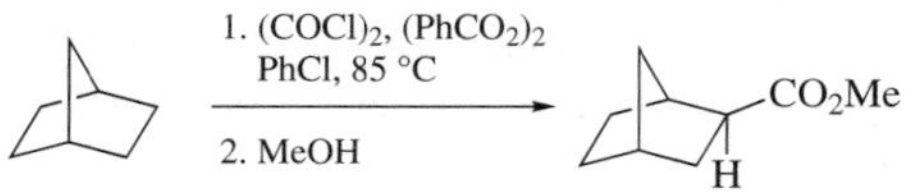

1. $(COCl)_2$, $(PhCO_2)_2$, PhCl, 85 °C; 2. MeOH (9)

Certain alkenes such as 1-methylcyclohexene and styrene react with oxalyl chloride, under ionic conditions without added catalyst, to give alkenoic acid chlorides in variable yields. Alkenes such as octene and stilbene did not react under these conditions.[24]

Reactions of aromatic compounds with oxalyl chloride/Lewis acid catalysts have been reviewed.[25] Anthracene is unusual as it undergoes substitution without added catalyst (eq 10).[26]

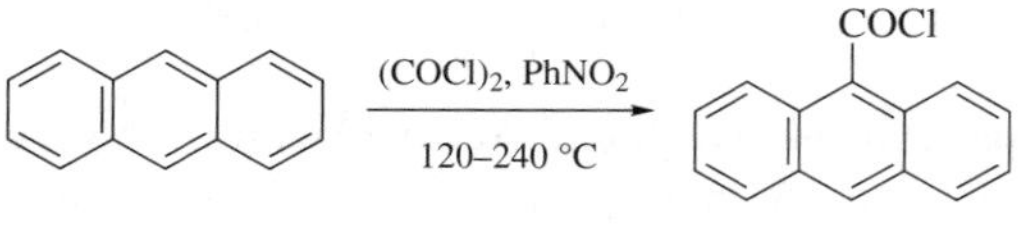

$(COCl)_2$, $PhNO_2$, 120–240 °C (10)

Preparation of Chloroalkanes. Alcohols react with oxalyl chloride to give oxalyl monoalkyl esters, which if heated in the presence of pyridine give the alkyl chloride (eq 11).[3]

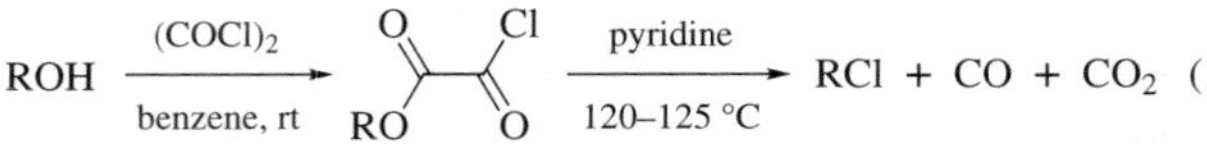

$$ROH \xrightarrow[\text{benzene, rt}]{(COCl)_2} ROCOCOCl \xrightarrow[120\text{–}125\ °C]{\text{pyridine}} RCl + CO + CO_2 \quad (11)$$

Tertiary alcohols have been converted to tertiary chlorides in a Barton–Hunsdiecker type radical process using hydroxamate esters (eq 12).[27]

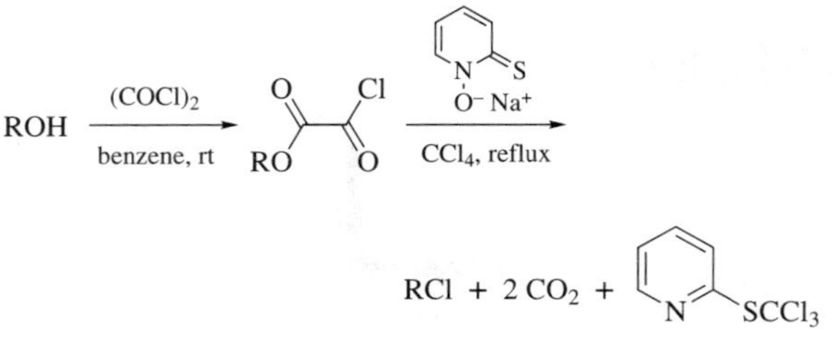

$$ROH \xrightarrow[\text{benzene, rt}]{(COCl)_2} ROCOCOCl \xrightarrow[CCl_4,\ \text{reflux}]{} RCl + 2\,CO_2 + \text{2-pyridyl-}SCCl_3 \quad (12)$$

e.g. R = $Me(CH_2)_{16}CMe_2$–

Chlorination of Alkenes. A novel stereospecific dichlorination of electron rich alkenes has been reported using a manganese reagent generated from ***Benzyltriethylammonium Chloride*** and oxalyl chloride (eqs 13–17).[28] No oxygenation byproducts are observed.

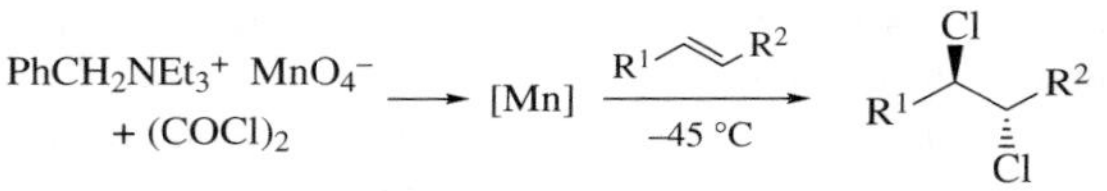

$PhCH_2NEt_3^+\ MnO_4^-$ + $(COCl)_2$ → [Mn]; $R^1CH=CHR^2$, –45 °C (13)

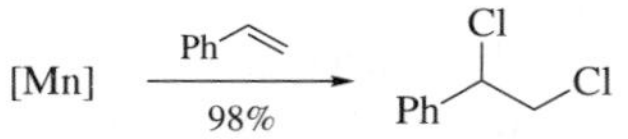

[Mn], PhCH=CH₂, 98% (14)

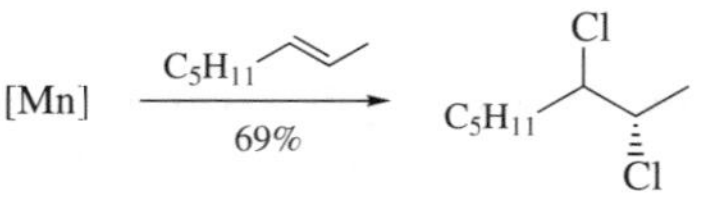

[Mn], $C_5H_{11}CH=CHCH_3$, 69% (15)

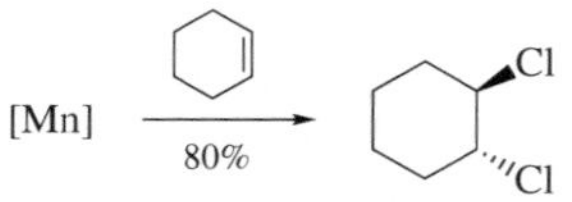

[Mn], cyclohexene, 80% (16)

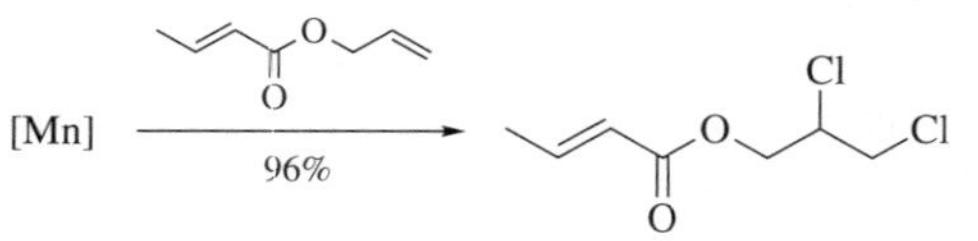

[Mn], allyl crotonate, 96% (17)

Reactions with Carbonyl Groups. Unsaturated 3-keto steroids give the corresponding 3-chloro derivatives with oxalyl chloride (eq 18).[4] Prolonged heating can give rise to aromatization.[4] Tropone gives the chlorotropylium chloride in high yield.[4] In a related reaction, 1,2-dithiol-3-ones and -3-thiones give dithiolium salts when heated in toluene or chloroform with the reagent.[4] A range of β-chloro enones has been prepared from diketones. Dimedone gives the β-chloro enone in high yield (eq 19).[29] Keto esters did not react to give β-chloro esters.

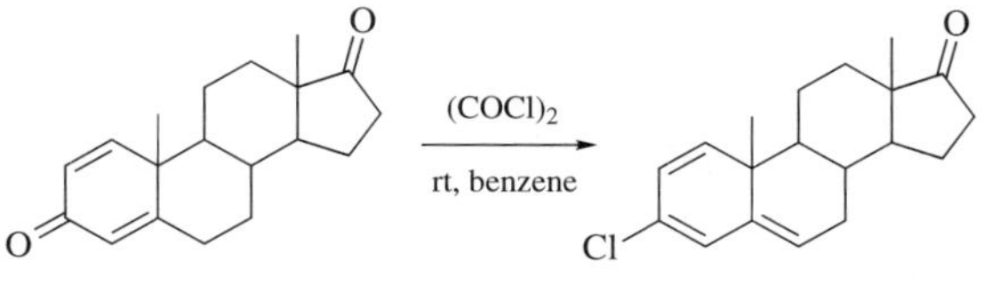

$(COCl)_2$, rt, benzene (18)

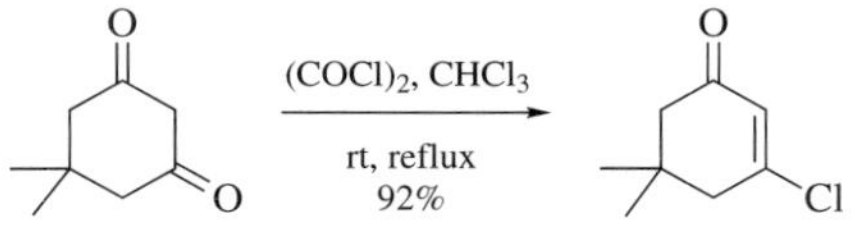

$(COCl)_2$, $CHCl_3$, rt, reflux, 92% (19)

β-Keto aldehydes give a single regio- and stereospecific isomer, the chlorine being *cis* to the carbonyl group (eq 20).

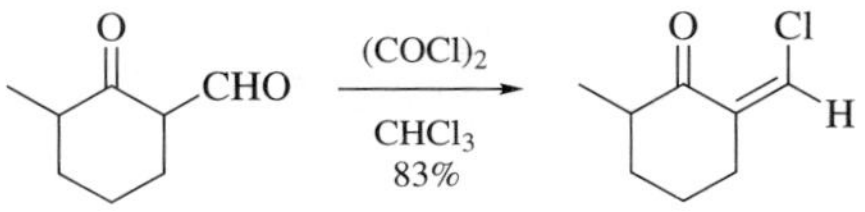

$(COCl)_2$, $CHCl_3$, 83% (20)

Certain triketones give 3-chlorides with excess oxalyl chloride, in good yield (eq 21).[30]

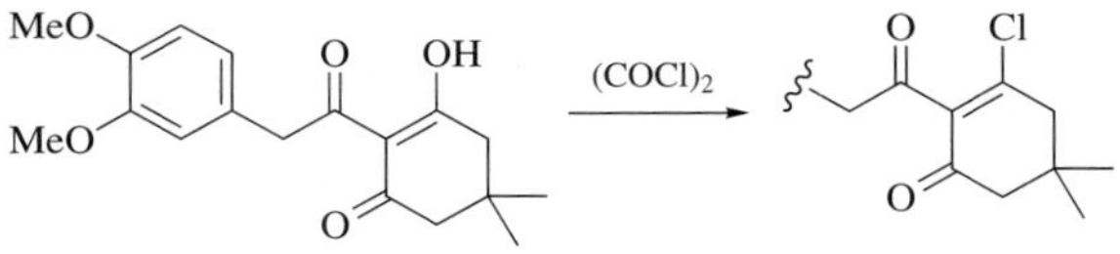

$(COCl)_2$ (21)

Preparation of Acyl Isocyanates and Aryl Isocyanates. Certain primary carboxamides can be converted to acyl isocyanates in yields from 36–97% with the reagent (eq 22);[5] ***Phosgene*** gives nitriles under similar conditions. Oxalyl chloride has found limited application for the preparation of triazine and quinone isocyanates.[5]

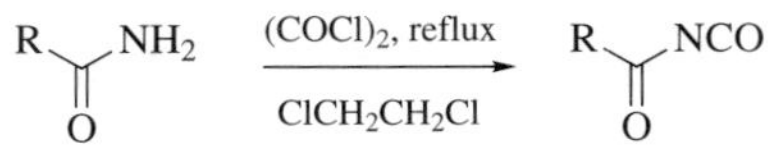

$RC(O)NH_2 \xrightarrow[ClCH_2CH_2Cl]{(COCl)_2,\ reflux} RC(O)NCO$ (22)

R = $ClCH_2$, CCl_3, $PhCH_2$, 3,4-$Cl_2C_6H_3$, Ph_2CH

Miscellaneous Applications. Oxalyl chloride has been used in the preparation of 2,3-furandiones from alkenyloxysilanes,[31] *o*-aminophenols from *N*-aryl nitrones,[32] dihydroquinolines via a modified Bischler–Napieralski ring closure,[33] 2,3-β-furoquinoxalines from quinoxazolones,[34] sterically hindered salicylaldehydes from phenoxyoxalyl chlorides,[35] and in mild cleavage of 7-carboxamido groups in cephalosporin natural products, without cleavage of the lactam ring or disruption of optical centers.[36]

Related Reagents. Dimethyl Sulfoxide–Oxalyl Chloride; Oxalyl Chloride–Aluminum Chloride.

1. Adams, R; Ulich, L. H. *JACS* **1920**, *42*, 599.
2. Stowell, M. H. B.; Ueland, J. M.; McClard, R. W. *TL* **1990**, *31*, 3261.
3. Rhoads, S. J.; Michel, R. E. *JACS* **1963**, *85*, 585.
4. (a) Moersch, G. W.; Neuklis, W. A.; Culbertson, T. P.; Morrow, D. F.; Butler, M. E. *JOC* **1964**, *29*, 2495. (b) Haug, E.; Fohlisch, B. *ZN(B)* **1969**, *24*, 1353 (*CA* **1970**, *72*, 43 079m). (c) Bader, J. *HCA* **1968**, *51*, 1409. (d) Faust, J.; Mayer, R. *LA* **1965**, *688*, 150.
5. (a) Speziale, A. J.; Smith, L. R. *JOC* **1962**, *27*, 3742. (b) von Gizycki, U. *AG(E)* **1971**, *10*, 403.
6. (a) van Reijendam, J. W.; Baardman, F. *S* **1973**, 413. (b) Lemmens, J. M.; Blommerde, W. W. J. M.; Thijs, L.; Zwanenburg, B. *JOC* **1984**, *49*, 2231.
7. Paquette, L. A.; Dahnke, K.; Doyon, J.; He, W.; Wyant, K.; Friendrich, D. *JOC* **1991**, *56*, 6199.
8. Keller-Schierlein, W.; Muller, A.; Hagmann, L.; Schneider, U.; Zähner, H. *HCA* **1985**, *68*, 559.
9. Hutchings, M. G.; Devonald, D. P. *TL* **1989**, *30*, 3715.
10. (a) Wilds, A. L.; Shunk, C. H. *JACS* **1948**, *70*, 2427. (b) Hudlicky, T; Kutchan, T. *TL* **1980**, *21*, 691. (c) Duddeck, H.; Ferguson, G.; Kaitner, B.; Kennedy, M. *JCS(P1)* **1990**, 1055.
11. (a) Szmuszkovicz, J. *JOC* **1964**, *29*, 843. (b) Hatanaka, M.; Yamamoto, Y.; Nitta, H.; Ishimaru, T. *TL* **1981**, *22*, 3883. (c) Cochrane, E. J.; Lazer, S. W.; Pinhey, J. T.; Whitby, J. D. *TL* **1989**, *30*, 7111.
12. Bacardit, R.; Cervelló, J.; de March, P.; Marquet, J.; Moreno-Mañas, M.; Roca, J. L. *JHC* **1989**, *26*, 1205.
13. (a) Brady, W. T.; Giang, Y. F. *JOC* **1985**, *50*, 5177. (b) Snider, B. B.; Allentoff, A. J. *JOC* **1991**, *56*, 321.
14. (a) Burke, S. D.; Murtiashaw, C. W.; Dike, M. S.; Smith-Strickland, S. M.; Saunders, J. O. *JOC* **1981**, *46*, 2400. (b) Satake, K.; Imai, T.; Kimura, M.; Morosawa, S. *H* **1981**, *16*, 1271.
15. (a) Wingfield, H. N.; Harlan, W. R.; Hanmer, H. R. *JACS* **1953**, *75*, 4364. (b) Schrecker, A. W.; Maury, P. B. *JACS* **1954**, *76*, 5803.
16. Konopinska, D.; Siemion, I. Z. *AG(E)* **1967**, *6*, 248.
17. (a) Ihara, M.; Yasui, K.; Takahashi, M.; Taniguchi, N.; Fukumoto, K. *JCS(P1)* **1990**, 1469. (b) Nordlander, J. E.; Njoroge, F. G.; Payne, M. J.; Warman, D. *JOC* **1985**, *50*, 3481.
18. (a) Simon, M. S.; Rogers, J. B.; Saenger, W.; Gououtas, J. Z. *JACS* **1967**, *89*, 5838. (b) Krubsack, A. J.; Higa, T. *TL* **1968**, 5149.
19. (a) Stack, D. P.; Coates, R. M. *S* **1984**, 434. (b) Gillet, J. P.; Sauvêtre, R.; Normant, J. F. *S* **1982**, 297. (c) Wilson, R. M.; Commons, T. J. *JOC* **1975**, *40*, 2891.
20. (a) Wasserman, H. H.; Han, W. T.; Schaus, J. M.; Faller, J. W. *TL* **1984**, *25*, 3111. (b) Clough, S. C.; Deyrup, J. A. *JOC* **1974**, *39*, 902. (c) Sardina, F. J.; Howard, M. H.; Koskinen, A. M. P.; Rapoport, H. J. *JOC* **1989**, *54*, 4654.
21. Pelchowicz, Z. *JCS* **1961**, 238.
22. (a) Kharasch, M. S.; Brown, H. C. *JACS* **1942**, *64*, 329. (b) Tabushi, I.; Hamuro, J.; Oda, R. *JOC* **1968**, *33*, 2108.
23. Tabushi, I.; Okada, T.; Oda, R. *TL* **1969**, 1605.
24. (a) Kharasch, M. S.; Kane, S. S.; Brown, H. C. *JACS* **1942**, *64*, 333. (b) Bergmann, F.; Weizmann, M.; Dimant, E.; Patai, J; Szmuskowicz, J. *JACS* **1948**, *70*, 1612.
25. Olah, G. A.; Olah, J. A. In *The Friedel–Crafts Reaction*; Olah, G. A., Ed.; Interscience: New York, 1964; Vol. 3, p. 1257.
26. Latham, H. G.; May, E. L.; Mosettig, E. *JACS* **1948**, *70*, 1079.
27. (a) Barton, D. H. R.; Crich, D. *CC* **1984**, 774. (b) Crich, D.; Forth, S. M. *S* **1987**, 35.
28. Markó, I. E.; Richardson, P. F. *TL* **1991**, *32*, 1831.

29. (a) Clark, R. D.; Heathcock, C. H. *JOC* **1976**, *41*, 636. (b) Pellicciari, R; Fringuelli, F; Sisani, E.; Curini, M. *JCS(P1)* **1981**, 2566. (c) Büchi, G.; Carlson, J. A. *JACS* **1969**, *91*, 6470.
30. Lakhvich, F. A.; Khlebnicova, T. S.; Ahren, A. A. *S* **1985**, 784.
31. Murai, S.; Hasegawa, K; Sonoda, N. *AG(E)* **1975**, *14*, 636.
32. Liotta, D.; Baker, A. D.; Goldman, N. L.; Engel, R. *JOC* **1974**, *39*, 1975.
33. Larsen, R. D.; Reamer, R. A.; Corley, E. G.; Davis, P.; Grabowski, E. J. J.; Reider, P. J.; Shinkai, I. *JOC* **1991**, *56*, 6034.
34. Kollenz, G. *LA* **1972**, *762*, 13.
35. Zwanenburg, D. J.; Reynen, W. A. P. *S* **1976**, 624.
36. Shiozaki, M; Ishida, N.; Iino, K.; Hiraoka, T. *TL* **1977**, 4059.

Roger Salmon
Zeneca Agrochemicals, Bracknell, UK

Pentacarbonyliron[1]

$Fe(CO)_5$

[13463-40-6] C_5FeO_5 (MW 195.90)

(diene complexation;[4] carbonyl insertion;[62–64] cycloaddition; reduction[96] and deoxygenation)

Physical Data: mp −20 °C; bp 103 °C; fp −15 °C; *d* 1.490 g cm^{-3}; *n* 1.5196; X-ray diffraction studies at −100 °C confirm the trigonal bipyramidal structure of $Fe(CO)_5$.[2]
Solubility: miscible with organic solvents.
Form Supplied in: straw-yellow liquid.
Handling, Storage, and Precautions: $Fe(CO)_5$ is a volatile, flammable, and highly toxic liquid,[3] which does not react with air at room temperature. It has a relatively high vapor pressure (21 mmHg at 20 °C). When $Fe(CO)_5$ is exposed to light, $Fe_2(CO)_9$ impurity slowly accumulates. The monomeric pentacarbonyl complex can be purified by trap-to-trap distillation in a vacuum system, but for routine use this purification step is usually omitted. Use in a fume hood.

Diene Complexation. Tricarbonyl(η^4-butadiene)iron(0), the first iron–diene complex, was prepared as long ago as 1930 by Reihlen and co-workers, by heating butadiene with $Fe(CO)_5$.[4] Subsequently, substituted derivatives were studied by Pettit's group and by others.[5] The first complex of a cyclic diene, tricarbonyl(η^4-cyclohexadiene)iron(0), became available in 1958.[6] The real breakthrough for the development of synthetic methods sprang from pioneering work with substituted complexes in the cyclic (particularly cyclohexadiene) series in the 1960s and 1970s, by Birch,[7–10] Lewis,[11,12] and co-workers, then of the group of Pearson,[13] work which was initiated by the discovery by Fischer and Fischer in 1960[14] of hydride abstraction from tricarbonyl(η^4-cyclohexadiene)iron(0) to yield an η^5 iron-complexed cyclohexadienyl cation.

A wide range of methods are now available for the preparation of η^4-diene tricarbonyliron complexes, but by far the most common technique remains the direct complexation of free dienes. $Fe(CO)_5$ is less reactive than $Fe_2(CO)_9$ or $Fe_3(CO)_{12}$, but under the right conditions it provides an inexpensive and efficient route to iron–alkene complexes, particularly from the 1,4-cyclohexadienes.

1,4-Diene Complexation.

Thermal Complexation. Because 1,4-cyclohexadienes are readily available from the metal–ammonia reduction (Birch reduction) of the corresponding aromatic compounds,[15] they have been extensively studied as substrates for direct 1,4-diene complexation, leading to isomerized 1,3-diene complexes. This methodology suffers, however, from the disadvantage that products are often obtained as mixtures of regioisomers. $Fe(CO)_5$ provides a simple, low-cost source of $Fe(CO)_4$ for initial addition to the π-system of a double bond. By recycling the reagents and the uncomplexed ligand, preparations on a 50–60 g scale can be achieved from a series of overnight thermal reactions. The procedure most frequently used, that of Cais and Maoz,[16] modified by Birch,[8] involves heating a cyclohexadiene with pentacarbonyliron in boiling di-*n*-butyl ether. In the extensive literature on the formation of differently functionalized tricarbonyl(1,3-cyclohexadiene)iron complexes from substituted 1,4-cyclohexadienes[8,9,11,12,17,20] (including the tricarbonyliron complexes of a number of dihydroanisic esters,[18] various cyclohexadiene esters,[9,19] and 1-methoxy-2,4-dimethyl-1,4-cyclohexadiene[21]), at present the most useful and representative intermediates for synthetic applications carry methoxy substituents on the diene ligand. For example, 2,5-dihydroanisole (**1**) or 4-methyl-2,5-dihydroanisole (**2**) upon treatment with pentacarbonyliron at elevated temperature afford a mixture of the 1- and 2-OMe substituted complexes (**3**) and (**5**) or (**4**) and (**6**) (eq 1).[7–9,11,12,22]

$Fe(CO)_5$, Bu_2O, 140 °C (1)
(**1**) R = H; (**2**) R = Me → (**3**) R = H; (**4**) R = Me + (**5**) R = H; (**6**) R = Me

Although 1-methoxy-1,4-cyclohexadiene itself leads to a mixture of the 1- and 2-OMe complexes, bicyclic compounds of substitution type (**7**) do not give the 2-OMe regioisomer, affording instead a mixture of complexes (**8**) and (**9**) in a 1:1 ratio (eq 2).[22,23] In a more unusual case, an amide substituent was present on the 1,4-diene.[24]

(**7**) → (**8**) + (**9**), 1:1 (2)

More extensive rearrangement of the position of unsaturation in α,ω-dienes also leads to η^4-diene complexes.[25] In other cases where rearrangement is blocked, homoconjugated dienes can form stable bis(alkene) complexes by thermal reaction with $Fe(CO)_5$.[26] Isotetralene offers an unusual case where rearrangement to the 1,3-diene is possible, but a 1,4-diene complex is reported.[27]

Photolysis. Photolysis offers an alternative for the direct conversion of 1,4-dienes into 1,3-diene complexes. In eq 3 the deuterium labeling reveals that isomerization during complexation of the 1,4-diene precursor proceeds by a 1,3-hydrogen transfer.[28]

$Fe(CO)_5$, $h\nu$ (3)

The direct complexation of 1,4-dienes can form specific stereoisomers. Initial π-complexation occurs before the conjugation step, and the hydrogen atom moves only on the face of the ligand carrying the metal.[29] Unfortunately, a mixture of regioisomers usually results, which may present separation problems. Loss of OMe can also sometimes occur, particularly from some 1,3-dimethoxy-1,4-cyclohexadienes. In these cases, preconjugation of the diene is essential[30] if a dimethoxydiene complex is required. Mixtures of complexes obtained from 1,4-dienes may also be equilibrated[31] to thermodynamic ratios of isomeric complexes, which do not correspond to the equilibrium ratios of the free dienes.[32,33]

1,3-Diene Complexation.

Thermal Complexation. Low yields and mixed products render some nonconjugated dienes unattractive as precursors for η^4-complexes. For many synthetic applications, cleaner routes are required to form a single defined compound in good yield. A good approach employs a conjugation step prior to complexation.[30] The 1-NR_2 series can usually be conjugated thermally.[33] Equilibration of 1-methoxy-1,4-cyclohexadiene with the 1,3-diene (ca. 75%) using potassium in liquid ammonia as catalyst yields only the 1-OMe substitution pattern on the conjugated isomer.[34] Other conjugation catalysts include bases,[35] Wilkinson's catalyst,[36] Cr derivatives,[29] and charge-transfer complexation reagents such as dichloromaleic anhydride.[37] Frequently, only about 70–80% of the conjugated isomer is present at equilibrium,[38] but the complexation product, if formed under mild conditions, corresponds to the conjugated isomer present. This process of preconjugation followed by complexation, using the usual $Fe(CO)_5$ thermal methodology, has led, for example, to an important key intermediate in the synthesis[39] of limaspermine. The use of azadiene catalysts improves the thermal complexation reaction.[40]

Complexation of dienes with several methyl,[41] alkyl,[42] or methoxy[43] groups is straightforward, but with bulky substituents, rearrangements can lead to the formation of mixtures of products,[44] and care is also needed in the presence of side-chain leaving groups.[45] Trialkylsilyl-substituted dienes complex well. In the reactions of 2-(trimethylsilyl)-1,3-cyclohexadiene derivatives with excess pentacarbonyliron in refluxing di-*n*-butyl ether,[46] (**10**) gave rise to an 8:2 mixture of (**11**) and its epimer (**12**), with coordination to $Fe(CO)_3$ occurring preferentially on the less hindered face of the 1,3-cyclohexadienyl ring (eq 4).

TMS; $Fe(CO)_5$, Δ; $(OC)_3Fe$ TMS + $(OC)_3Fe$ TMS (4)

(**10**) (**11**) (**12**)

Photochemical Complexation. The studies of Birch et al. on the thebaine modification using tricarbonyliron complexes[47,48] exemplify the possibility of complexation of 1,3-dienes by photolysis, and illustrate the potential for iron complexes to ensure control of skeletal rearrangement. Thus, via the initial step of complexation of the methoxycyclohexadiene ring of thebaine (**13**), practically quantitative access to northebaine (**14**) and 14α-substituted thebainone derivatives could be easily obtained with $Fe(CO)_5$ under UV irradiation (eq 5).[47]

R N MeO O OMe; $Fe(CO)_5$, $h\nu$; R N $Fe(CO)_3$ MeO O OMe (5)

(**13**) R = Me
(**14**) R = H

Thermal and Photochemical Reactions in Other Ring Sizes. Similar complexation reactions can be performed in seven- and eight-membered rings,[49] but with cyclopentadiene the reaction takes a different course, producing a dimeric cyclopentadienyl complex.[50] Complexation of silyl-substituted cyclopentadienes afforded dinuclear products.[51] In more unsaturated systems, competing reduction processes are possible. Cycloheptatriene reacts with pentacarbonyliron to give a mixture of tricarbonyliron complexes of cycloheptatriene and 1,3-cycloheptadiene.[52] Substituted cycloheptatrienes[53] and cycloheptadienes[54] can also be complexed. Thermal complexation of divinylpyrrolidines forms exocyclic diene complexes by rearrangement of the location of unsaturation.[55]

Complexation Using $Fe(CO)_5$ and Me_3NO. Reaction of a conjugated diene with $Fe(CO)_5$ in the presence of ***Trimethylamine N-Oxide*** resulted in the formation of tricarbonyl(1,3-diene)iron complexes. The use of amine oxides, preferably Me_3NO, permits fast reactions even at low temperatures. Reduction of amine oxides by $Fe(CO)_5$ has been described by Alper et al.[56] and, in the case of trimethylamine *N*-oxide, the intermediate compound trimethylaminotetracarbonyliron was isolated (eq 6).[57,58]

$$Fe(CO)_5 + Me_3NO \longrightarrow Me_3N{\bullet}Fe(CO)_4 + CO_2 \qquad (6)$$

The reverse reaction, namely the known disengagement of the ligand from the (diene)$Fe(CO)_3$ complex, presents no problem because of the lower reaction temperature and the smaller molar ratio of Me_3NO:$Fe(CO)_5$ employed in the complexation procedure.[57] Trimethylaminotetracarbonyliron is attractive as a mild reagent but further generalization is needed.

Acyclic Series. The first complex in the acyclic series was prepared from butadiene by the thermal method.[4] Heating isoprene and pentacarbonyliron at high temperature, however, is inefficient due to competitive Diels–Alder dimerization. Despite the formation of some bis(diene) iron carbonyl complexes on prolonged irradiation, the photochemical method is superior in this case. Complexation of acyclic dienes by $Fe(CO)_3$ is limited to those that can adopt a cisoid conformation, with the *syn* substitution pattern normally preferred. 2,4-Hexadienoic acid, for example, can be conveniently complexed by a photolytic procedure.[59] Trialkylsilyl-substituted dienes have also been complexed.[60]

Isomerization of Alkenes. In some cases the positional rearrangement of alkenes affords useful products which are not 1,3-diene complexes. The isomerization of allyl ethers in the presence of $Fe(CO)_5$ under UV irradiation has also been reported.[61]

Ring-Opening Reactions with Carbonyl Insertion.

Cyclopropane rings. Synthesis of (diene)$Fe(CO)_3$ complexes, and of cyclohexenones, lactones, ketones, or sulfones, can follow from the opening of vinylcyclopropanes with $Fe(CO)_5$ under either thermal or photochemical conditions. Reactions of the vinylcyclopropane (**15**) with $Fe(CO)_5$ under thermal conditions resulted in ring opening with concomitant hydrogen shift to form the diene complex (**16**) (eq 7).[62,63,64]

(15) + $Fe(CO)_5$ —Δ→ (16) —hν→ (17) → (18) (7)

The reaction of (**16**) under photochemical conditions led to (**17**) by ring opening and carbonyl insertion. Conjugation may occur, forming the cyclohexenone (**18**) (eq 7).[65] The expansion of the carbon skeleton of terpene hydrocarbons via carbonylation offers an interesting means to prepare optically active strained ketones (eq 8).[66]

(19) + $Fe(CO)_5$ → ... (8)

Regiocontrol, however, is a limitation. In the reaction between (+)-2-carene (**19**) and $Fe(CO)_5$, five compounds are isolated. A recent study[67] improved the reaction with (+)-2-carene and extended the method to (+)-3-carene.

Epoxide Rings. The synthesis of alcohols, β-lactones, and bicyclic derivatives occurs via a ferralactone complex. Irradiation of α,β-unsaturated epoxides in the presence of $Fe(CO)_5$ yields ferralactone complexes by ring opening and CO insertion (eq 9). For example, the vinyloxirane (**20**) gives the lactone (**21**).[68]

(20) —$Fe(CO)_5$, PhH, hν, 93%→ (21) (9)

While the photochemical reaction[69] proceeds stereospecifically, the thermal process between $Fe(CO)_5$ and vinyl epoxides gives mixtures of diastereoisomers. Subsequent treatment of ferralactones with ***Cerium(IV) Ammonium Nitrate*** results predominantly in the formation of β-lactones, with a few exceptions where δ-lactones are obtained.[70]

A thermal treatment of the ferralactone displaces one molecule of CO. The α,β-unsaturated epoxide (**22**) forms the lactone (**23**) after irradiation in the presence of $Fe(CO)_5$, but subsequent heating leads to the formation of the (hydroxycyclohexadiene)$Fe(CO)_3$ complex (**24**) (eq 10).[67]

(22) + $Fe(CO)_5$ —60%→ (23) + CO; (23) —−CO, Δ→ (24) (10)

UV irradiation of 9-oxabicyclo[6.1.0]nona-2,4,6-triene (**25**) (the monoepoxide of cyclooctatetraene) in the presence of $Fe(CO)_5$ promotes an overall 1,2 to 1,4 skeletal rearrangement via the lactone (**26**) and the $Fe(CO)_3$ complex (**27**) to form the $Fe(CO)_3$–$Fe(CO)_4$ complex (**28**). Subsequent decomposition of the bis(iron) complex (**28**) gave the previously unknown 9-oxabicyclo[4.2.1]nona-2,4,7-triene (**29**) (eq 11).[71]

(25) + $Fe(CO)_5$ —hν→ [(26) → (27)] —70%→ (28) —[O], Me_3NO, 48%→ (29) (11)

Carbonylative Ring Expansion. When either α- or β-pinene (**30**) or (**31**) is heated neat with an equimolar amount of $Fe(CO)_5$ under an initial pressure of 30 psi CO at 160 °C, two ketones (**32**) and (**33**) are obtained, formed by insertion of CO into the cyclobutane ring (eq 12).[72] Both ketones are optically active.

Dehalogenation Reactions.

Coupling of gem-Dihalides. Pentacarbonyliron effects dehalogenative coupling of dichloro- or dibromodiphenylmethane and benzylic *gem*-dihalides to form alkenic compounds in good yields (eq 13).[73]

α-Halo Ketones. Treatment of α-halo ketones (**34**) with $Fe(CO)_5$ was found by Alper to result in formation of coupled 1,4-diketones (**35**) and reduced monoketones (**36**) (eq 14).[74]

(30) or (31) $\xrightarrow[\Delta]{Fe(CO)_5}$ (30) + (32) 29% + (33) 34% (12)

$(C_6H_5)_2CCl_2 \xrightarrow[PhH,\ \Delta]{Fe(CO)_5} Ph_2C{=}CPh_2$ (13)

(34) $\xrightarrow[2.\ H_2O]{1.\ Fe(CO)_5,\ DME,\ \Delta}$ $(ArCOCR^1R^2)_2$ (35) + $ArCOCHR^1R^2$ (36) (14)

α,α′-Dibromo Ketones and α,α,α′,α′-Tetrabromo Ketones. Although dehalogenation combined with carbon–carbon bond formation is more commonly performed with ***Nonacarbonyldiiron***, $Fe(CO)_5$ may also be used. The use of a tribromo ketone, followed by a separate reductive dehalogenation step, is illustrated for the synthesis of carbocamphenilone (eq 15), a compound of interest as a skewed glyoxal model.[75]

$\xrightarrow[2.\ Zn/Cu]{1.\ Fe(CO)_5}$ (15)

The reaction between the allylic bromide (**37**) and $Fe(CO)_5$ constitutes an alternative method of preparation of tricarbonyl(1,3-cyclohexadiene)iron (**38**) (eq 16).[76] Allylic alcohols react in a similar way (eq 17) under neutral conditions,[77] but in the presence of acid (HBF_4), cationic η^3 iron carbonyl complexes are produced.[78]

(37) $\xrightarrow{Fe(CO)_5}$ (38) $Fe(CO)_3$ (16)

$\xrightarrow[\text{hydroquinone, 72\%}]{Fe(CO)_5,\ Bu_2O}$ $Fe(CO)_3$ (17)

Benzohydroxamoyl Chlorides. Benzohydroxamoyl chlorides (**39**) are converted into nitriles (**40**) when refluxed with $Fe(CO)_5$ in THF (eq 18).[79]

$Ar(Cl)C{=}NOH$ (39) $\xrightarrow[40\text{–}75\%]{Fe(CO)_5,\ THF}$ $Ar{-}C{\equiv}N$ (40) (18)

Dehydration and Dehydrosulfuration Reactions. Certain amides and thioamides, on treatment with $Fe(CO)_5$, yield nitriles or imines depending on the degree of substitution present (eq 19).[80] Dithioesters, on the other hand, form trinuclear iron carbonyl complexes.[81]

$PhC(O)NH_2$ (32%) / $PhC(S)NH_2$ (64%) $\xrightarrow{Fe(CO)_5}$ PhCN (19)

Reaction with Oximes; Deprotection of Ketones. Regeneration of protected ketones from their oximes can be achieved by refluxing the oxime in di-*n*-butyl ether with one equiv of $Fe(CO)_5$ and a catalytic amount of a Lewis acid (eq 20).[82]

HON= $\xrightarrow[Bu_2O,\ \Delta,\ 16\ h;\ 67\%]{Fe(CO)_5,\ BF_3}$ O= (20)

Other Reactions Occurring with Carbonyl Insertion.

Carbonylation of Benzyl Halides. Benzyl halides can be converted into phenylacetic acids by reaction with $Fe(CO)_5$.[83] Phase-transfer catalysis is useful in this reaction.[84]

Diazonium Salts. Carboxylic acids or ketones are obtained from the reaction of diazonium salts with $Fe(CO)_5$, depending on the reaction conditions.[85]

Nucleophiles. Nucleophiles such as amines (eq 21)[86] and organolithium reagents (eq 22)[87] are also found to react with $Fe(CO)_5$ to give products of CO insertion. Cyclopentadienide, on the other hand, undergoes an unusual tricarbonylation reaction to form a cyclopentadienyl complex with an iron–acyl bridge connecting the metal and the cyclopentadienyl ring (eq 23).[88] The product was isolated by protecting the two OH groups. An aryllithium reagent, in which the aromatic ring is bound to a tricarbonylchromium group, also follows a different reaction path, and bimetallic tetracarbonyliron ferrate derivatives are isolated.[89]

$$RNH_2 + Fe(CO)_5 \longrightarrow RNHCHO$$
$$R_2NH + Fe(CO)_5 \longrightarrow R_2NCHO \quad (21)$$

$$\text{2,5-dimethylphenyllithium} \xrightarrow[\text{2. } H^+ \quad 65\%]{\text{1. } Fe(CO)_5,\ -60\ ^\circ C} \text{2,5-dimethylbenzaldehyde} \quad (22)$$

$$\text{cyclopentadienide} \xrightarrow{Fe(CO)_5} (CO)_2Fe\text{-complex} \quad (23)$$

An arylmethylmalonate derivative, with an *ortho* iodine substituent on the aromatic ring, undergoes a carbonylative cyclization by oxidative addition to the aryl iodide and reaction at the nucleophilic center of the malonate group.[90]

Formylation and Acylation of Pyridine. Direct formylation and acylation at the β-position of pyridine is possible by reaction with phenyllithium and then with $Fe(CO)_5$. 2-Phenylpyridine is also formed in these reactions.[91]

Formation of N,N′-Disubstituted Ureas. The substituted ureas are prepared in about 50–95% yield by reaction of aryl or alkyl nitro compounds with the MgBr derivative of an amine in the presence of $Fe(CO)_5$ (eq 24).[92]

$$R^1NHMgBr + R^2NO_2 \xrightarrow[\text{2. } H_2SO_4]{\text{1. } Fe(CO)_5} R^1NHCONHR^2 \quad (24)$$

Formation of Carboxylic Esters[93,95] and Ketones.[94] Grignard reagents can also be converted into esters and ketones by $Fe(CO)_5$ (to supply CO). A cathodic ester synthesis from alcohols and alkyl halides uses carbon monoxide (at 1 atm) and $Fe(CO)_5$ as a catalyst.[95]

Deoxygenation Reactions and Reductions.

Deoxygenation of N–O Bonds. Deoxygenation of amine oxides has been discussed above (see eq 6); pyridine *N*-oxides provide a further illustration (eq 25).[96]

$$\text{pyridine } N\text{-oxide} \xrightarrow{Fe(CO)_5} \text{pyridine} + \text{iron oxides} \quad (25)$$

Nitroso Compounds. Similar treatment of nitrosobenzene resulted in a 75% yield of azobenzene. Aromatic *N*-nitroso amines have been converted into the secondary amines in 85–92% yields.[97] Under photolytic conditions, azoxyarenes are produced.[98] The choice of solvent is important. Dialkylnitrosoamines afford ureas (eq 26).[97]

Aromatic Nitro Compounds. Indoles are formed[99] by $Fe(CO)_5$ catalyzed deoxygenation of *o*-nitrostyrenes with CO (eq 27).

$$\text{N-nitroso heterocycle} \xrightarrow[\Delta]{Fe(CO)_5} \text{urea} \quad (26)$$

X = CH_2, O

$$\text{methyl } o\text{-nitrocinnamate } (trans:cis = 85:15) + 2\ CO \xrightarrow[\text{PhMe, } 220\ ^\circ C \quad 75\%]{Fe(CO)_5} \text{indole-2-}CO_2Me + 2\ CO_2 \quad (27)$$

Deoxygenation of Epoxides. $Fe(CO)_5$ in *N,N*-dimethylacetamide or tetramethylurea deoxygenates epoxides (2 h, 145 °C). The reaction is not stereospecific: *trans*-stilbene oxide is converted into both *trans*- (56%) and *cis*-stilbene (22%).[100] Epoxides of 1-alkenes are converted mainly into mixtures of internal alkenes.

Stabilization of Reactive or Unstable Species. The complex (**42**), formed from the reaction between the dibromoxylene (**41**) and $Fe_2(CO)_9$ or $Na_2Fe(CO)_4$, reacts with $Fe(CO)_5$ to yield the mixture of stereoisomers (**43**) and the complex (**44**) (eq 28).[101] With pentacarbonyliron, however, (**41**) is taken directly on to an indanone by carbonylative cyclization (see above).[102]

$$(\mathbf{41}) \xrightarrow[Bu_2O,\ \Delta]{Fe_2(CO)_9 \text{ or } Na_2Fe(CO)_4} (\mathbf{42})\text{—}Fe(CO)_3 \xrightarrow[h\nu]{Fe(CO)_5} (OC)_3Fe\text{—}(\mathbf{43})\text{—}Fe(CO)_3 + (\mathbf{44}) \quad (28)$$

A pentafulvalene has been trapped as a diiron complex using $Fe(CO)_5$.[103] Thiophene dioxide has also been stabilized as its tricarbonyliron complex.[104] The irradiation of α-pyrone (**45**) in the presence of $Fe(CO)_5$ gives cyclobutadiene stabilized by complexation with the $Fe(CO)_3$ moiety (**46**), together with the (α-pyrone)$Fe(CO)_3$ complex (**47**) (eq 29).[105]

$$(\mathbf{45}) \xrightarrow[h\nu]{Fe(CO)_5} \text{cyclobutadiene—}Fe(CO)_3\ (\mathbf{46}) + (OC)_3Fe\text{—}\alpha\text{-pyrone}\ (\mathbf{47}) \quad (29)$$

Reactions with Alkynes.

Dimerization. Two alkynes can be combined to form a cyclobutadiene ligand. The reaction has been reported for a dialkylamino trimethylsilyl-substituted alkyne, and forms a diaminocyclobutadiene complex.[106]

Dimerization with Insertion of One Molecule of CO. A number of tricarbonyliron complexes of substituted cyclopentadienones have been prepared from alkynes. The reaction was first described by Jones et al.[107] for $Fe(CO)_5$ and Ph–C≡CH in the presence of $Ni(CO)_4$. A number of other alkynes (PhC_2Me, PhC_2SiMe_3, Me_3SiC_2R, $BrC_6H_4C_2H$, $CF_3C_2CF_3$, PhC_2Ph and α,ω-diynes) give cyclopentadienone complexes (eq 30).[108]

$$2\ \text{R}{-}{\equiv} + Fe(CO)_5 \longrightarrow (\text{2,5-R}_2\text{-cyclopentadienone})Fe(CO)_3 \quad (30)$$

Cyclization of an alkyne to an alkene with carbonyl insertion (analogous to the Pauson–Khand reaction) has also been achieved using pentacarbonyliron.[109]

Dimerization with Insertion of Two Molecules of CO. By manipulating reaction conditions, formation of benzoquinones can be achieved (eq 31).[110]

$$2\ \text{HC}{\equiv}\text{CH} + Fe(CO)_5 \xrightarrow[\text{50–80 °C, 40 atm}]{h\nu,\ \text{moist EtOH}} (\text{benzoquinone})Fe(CO)_3 \quad (31)$$

Reaction with Allenes. A diallene, hexatetraene (eq 32), reacts with $Fe(CO)_5$ with insertion of carbon monoxide to form 2,5-dimethylenecyclopent-3-enone.[111] Allenyl ketones, on the other hand, form dinuclear complexes.[112]

$$\text{hexatetraene} \xrightarrow{Fe(CO)_5} \text{2,5-dimethylenecyclopent-3-enone} \quad (32)$$

Formation of Naphthoquinones via Iron Metallocycles. Benzocyclobutenedione (**48**) forms the iron complex (**49**) by irradiation with $Fe(CO)_5$.[113] Complex (**49**) reacts with a wide variety of alkynes to give naphthoquinones, in yields usually >70% (eq 33).

$$(\mathbf{48}) + Fe(CO)_5 \xrightarrow[93\%]{h\nu} (\mathbf{49})\ [\text{phthaloyl}Fe(CO)_4] \xrightarrow[22\text{–}100\%]{\text{R}^1{-}{\equiv}{-}\text{R}^2,\ \text{MeCN, 100 °C}} \text{2-R}^1\text{-3-R}^2\text{-1,4-naphthoquinone} \quad (33)$$

Trimerization of Benzonitriles. Benzonitrile is trimerized in good yield by heating with $Fe(CO)_5$ for several hours (eq 34).[114]

Formation of Carbodiimides. Carbodiimides are obtained in ca. 50% yield from the reaction of azides with isocyanides catalyzed by $Fe(CO)_5$.[115]

$$\text{Ph}{-}{\equiv}\text{N} + Fe(CO)_5 \xrightarrow[\Delta]{N_2} \text{2,4,6-triphenyl-1,3,5-triazine} \quad (34)$$

Anionic Iron Carbonyl Reagents. The reagent $[HFe(CO)_4]^-$ can be generated in situ from pentacarbonyliron, allowing straightforward, regiocontrolled, hydroxycarboxylation of acrylic acid by $Fe(CO)_5$, $Ca(OH)_2$, and H_2O/*i*-PrOH under 1 atm of CO.[116]

Formation of Iron Carbene Complexes. $Fe(CO)_5$ is the starting material for the preparation of the tetracarbonyl(ethoxyphenylmethylidene)iron(0) (**50**) (eq 35).[117]

$$Fe(CO)_5 \xrightarrow[42\%]{\text{1. PhLi, ether; 2. FSO}_3\text{Et, HMPA}} (CO)_4Fe{=}C(OEt)R \quad (35)$$

(**50**) R = Ph

Iron carbene complexes react with alkynes under CO atmosphere to form (α-pyrone)$Fe(CO)_3$ complexes.[118]

Modified Corey–Winter Alkene Synthesis. Daub et al.[119] have obtained satisfactory yields by using $Fe(CO)_5$ in an iron-mediated version (eq 36) of the reaction of thionocarbonates with ***Trimethyl Phosphite***, which gives poor yields in the case of thermally labile alkenes.

$$\text{tetramethyl-1,3-dioxolane-2-thione} \xrightarrow{\text{1 equiv } Fe(CO)_5} \text{tetramethylethylene} \quad (36)$$

1. (a) Pincass, H. *CZ* **1929**, *53*, 525. (b) Shabel'nikov, V. G.; Zolotenina, S. P. *Khim. Neft. Mashinostr.* **1990**, *3*, 1 (*CA* **1990**, *113*, 43 305x).
2. Braga, D.; Grepioni, F.; Orpen, A. G. *OM* **1993**, *12*, 1481.
3. (a) Brief, R. S.; Ajemian, R. S.; Confer, R. G. *Am. Ind. Hyg. Assoc. J.* **1967**, *28*, 21. (b) Permitted levels in air, see: *Fed. Regist.* **1992**, *57*, 26 002 (*CA* **1992**, *118*, 65 829b).
4. (a) Preparation: Reihlen, H.; Gruhl, A.; von Hessling, G.; Pfrengle, O. *LA* **1930**, *482*, 161. (b) Structure: Hallam, B. F.; Pauson, P. L. *JCS* **1958**, 642.
5. (a) Mahler, J. E.; Pettit, R. *JACS* **1963**, *85*, 3955. (b) Emerson, G. F.; Watts, L.; Pettit, R. *JACS* **1965**, *87*, 131. (c) Emerson, G. F.; Pettit, R. *JACS* **1962**, *84*, 4591. (d) Pettit, R.; Barborak, J. C.; Watts, L. *JACS* **1966**, *88*, 1328. (e) Lillya, C. P.; Sahatjian, R. A. *JOM* **1971**, *32*, 371. (f) Graf, R. E.; Lillya, C. P. *CC* **1973**, 271. (g) Whitesides, T. H.; Arhart, R. W. *JACS* **1971**, *93*, 5296. (h) Whitesides, T. H.; Arhart, R. W.; Slaven, R. W. *JACS* **1973**, *95*, 5792. (i) Whitesides, T. H.; Neilan, J. P. *JACS* **1973**, *95*, 5811. (j) Whitesides, T. H.; Neilan, J. P. *JACS* **1976**, *98*, 63. (k) Whitesides, T. H.; Slaven, R. W. *JOM* **1974**, *67*, 99.
6. Hallam, B. F.; Pauson, P. L. *JCS* **1958**, 642.
7. Birch, A. J.; Chamberlain, K. B.; Haas, M. A.; Thompson, D. J. *JCS(P1)* **1973**, 1882.
8. Birch, A. J.; Haas, M. A. *JCS(C)* **1971**, 2465.
9. Birch, A. J.; Williamson, D. H. *JCS(P1)* **1973**, 1892.
10. Birch, A. J.; Bandara, B. M. R.; Chamberlain, K.; Chauncy, B.; Dahler, P.; Day, A. I.; Jenkins, I. D.; Kelly, L. F.; Khor, T.-C.; Kretschmer, G.; Liepa, A. J.; Narula, A. S.; Raverty, W. D.; Rizzardo, E.; Sell, C.; Stephenson, G. R.; Thompson, D. J.; Williamson, D. H. *T* **1981**, *37*, (*Suppl. 1*), 289.

11. Birch, A. J.; Cross, P. E.; Lewis, J.; White, D. A. *CI(L)* **1964**, 838.
12. Birch, A. J.; Cross, P. E.; Lewis, J.; White, D. A.; Wild, S. B. *JCS(A)* **1968**, 332.
13. (a) Pearson, A. J.; Chandler, M. *JCS(P1)* **1980**, 2238. (b) Pearson, A. J.; O'Brien, M. K. *JOC* **1989**, *54*, 4663. (c) Pearson, A. J. *PAC* **1983**, *55*, 1767.
14. Fischer, E. O.; Fischer, R. D. *AG* **1960**, *72*, 919.
15. Birch, A. J.; Subba Rao, G. S. *Adv. Org. Chem.* **1972**, *8*, 1.
16. Cais, M.; Maoz, N. *JOM* **1966**, *5*, 370.
17. (a) Pearson, A. J. *CC* **1977**, 339. (b) Gibson, D. H.; Ong, T.-S.; Khoury, F. G. *JOM* **1978**, *157*, 81. (c) Birch, A. J.; Jenkins, I. D. In *Transition Metal Organometallics in Organic Synthesis: 1*; Academic: New York, 1991.
18. Birch, A. J.; Pearson, A. J. *JCS(P1)* **1978**, 638.
19. Birch, A. J.; Bandara, B. M. R.; Raverty, W. D. *JCS(P1)* **1982**, 1755.
20. Birch, A. J.; Kelly, L. F.; Thompson, D. J. *JCS(P1)* **1981**, 1006.
21. Curtis, H.; Johnson, B. F. G.; Stephenson, G. R. *JCS(D)* **1985**, 1723.
22. Pearson, A. J. *JCS(P1)* **1977**, 2069.
23. Pearson, A. J. *JCS(P1)* **1978**, 495.
24. Ong, C. W.; Hwang, W. S.; Liou, W. T. *J. Chin. Chem. Soc. (Taipei)* **1991**, *38*, 243.
25. Rodriguez, J.; Brun, P.; Waegell, B. *JOM* **1987**, *333*, C25.
26. Sakai, N.; Mashima, K.; Takaya, H.; Yamaguchi, R.; Kozima, S. *JOM* **1991**, *419*, 181.
27. Abser, M. N.; Hashem, M. A.; Kabir, S. E.; Ullah, S. S. *IJC(A)* **1988**, *27A*, 1050.
28. Alper, H.; LePort, P. C.; Wolfe, S. *JACS* **1969**, *91*, 7553.
29. Birch, A. J.; Kelly, L. F. *JOM* **1985**, *285*, 267.
30. Birch, A. J. *ANY* **1980**, *333*, 107.
31. Birch, A. J.; Chauncy, B.; Kelly, L. F.; Thompson, D. J. *JOM* **1985**, *286*, 37.
32. Birch, A. J.; Hutchinson, E. G.; Rao, G. S. *JCS(C)* **1971**, 637.
33. Birch, A. J.; Dyke, S. F. *AJC* **1978**, *31*, 1625.
34. Birch, A. J. *JCS* **1950**, 1551.
35. Birch, A. J.; Shoukry, E. M. A.; Stansfield, F. *JCS* **1961**, 5376.
36. Birch, A. J.; Subba Rao, G. S. R. *TL* **1968**, 3797.
37. Birch, A. J.; Dastur, K. P. *TL* **1972**, 4195.
38. Taskinen, E. *ACS* **1974**, *28B*, 201.
39. Pearson, A. J.; Rees, D. C. *JCS(P1)* **1982**, 2467.
40. Knölker, H.-J.; Gonser, P. *SL* **1992**, 517.
41. Eilbracht, P.; Hittinger, C.; Kufferath, K. *CB* **1990**, *123*, 1071.
42. Eilbracht, P.; Hittinger, C.; Kufferath, K.; Henkel, G. *CB* **1990**, *123*, 1079.
43. Palotai, I. M.; Stephenson, G. R.; Ross, W. J.; Tupper, D. E. *JOM* **1989**, *364*, C11.
44. Ong, C. W.; Liou, W. T.; Hwang, W. S. *JOM* **1990**, *384*, 133.
45. Randall, G. P.; Stephenson, G. R.; Chrystal, E. J. T. *JOM* **1988**, *353*, C47.
46. Paquette, L. A.; Daniels, R. G.; Gleiter, R. *OM* **1984**, *3*, 560.
47. Birch, A. J.; Fitton, H. *AJC* **1969**, *22*, 971.
48. Birch, A. J.; Kelly, L. F.; Liepa, A. J. *TL* **1985**, *26*, 501.
49. Burton, R.; Pratt, L.; Wilkinson, G. *JCS* **1961**, 594.
50. Piper, T. S.; Cotton, F. A.; Wilkinson, G. *J. Inorg. Nucl. Chem.* **1955**, *1*, 165.
51. (a) Siemeling, U.; Jutzi, P.; Neumann, B.; Stammler, H.-G.; Hursthouse, M. B. *OM* **1992**, *11*, 1328. (b) Sun, H.; Xu, S.; Zhou, X.; Wang, H.; Wang, R.; Yao, X. *JOM* **1993**, *444*, C41. (c) Morán, M.; Pascual, M. C.; Cuadrado, I.; Losada, J. *OM* **1993**, *12*, 811.
52. Dauben, Jr., H. J.; Bertelli, D. J. *JACS* **1961**, *83*, 497.
53. Nitta, M.; Nishimura, M.; Miyano, H. *JCS(P1)* **1989**, 1019.
54. Pearson, A. J.; Burello, M. P. *CC* **1989**, 1332.
55. Lassalle, G.; Grée, R. *TL* **1990**, *31*, 655.
56. Alper, H.; Edward, J. T. *CJC* **1970**, *48*, 1543.
57. Shvo, Y.; Hazum, E. *CC* **1975**, 829.
58. Elzinga, J.; Hogeveen, H. *CC* **1977**, 705.
59. Santos, E. H.; Stein, E.; Vichi, E. J. S.; Saitovich, E. B. *JOM* **1989**, *375*, 197.
60. Franck-Neumann, M.; Sedrati, M.; Mokhi, M. *JOM* **1987**, *326*, 389.
61. Iranpoor, N.; Imanieh, H.; Forbes, E. J. *SC* **1989**, 2955.
62. (a) Sarel, S.; Ben-Shoshan, R.; Kirson, B. *JACS* **1965**, *87*, 2517. (b) Sarel, S.; Ben-Shoshan, R.; Kirson, B. *Isr. J. Chem.* **1972**, *10*, 787.
63. Aumann, R. *JACS* **1974**, *96*, 2631.
64. Aumann, R. *JOM* **1974**, *77*, C33.
65. Victor, R.; Ben-Shoshan, R.; Sarel, S. *TL* **1970**, 4253.
66. Santelli-Rouvier, C.; Santelli, M.; Zahra, J.-P. *TL* **1985**, *26*, 1213.
67. Eilbracht, P.; Winkels, I. *CB* **1991**, *124*, 191.
68. Aumann, R.; Fröhlich, K.; Ring, H. *AG(E)* **1974**, *13*, 275.
69. Chen, K.-N.; Moriarty, R. M.; DeBoer, B. G.; Churchill, R. M.; Yeh, H. J. C. *JACS* **1975**, *97*, 5602.
70. Annis, G. D.; Ley, S. V. *CC* **1977**, 581.
71. Aumann, R.; Averbeck, H. *JOM* **1975**, *85*, C4.
72. Stockis, A.; Weissberger, E. *JACS* **1975**, *97*, 4288.
73. Coffey, C. E. *JACS* **1961**, *83*, 1623.
74. Alper, H.; Keung, E. C. H. *JOC* **1972**, *37*, 2566.
75. Noyori, R.; Souchi, T.; Hayakawa, Y. *JOC* **1975**, *40*, 2681.
76. Slupczyński, M.; Wolszczak, I.; Kosztolowicz, P. *ICA* **1979**, *33*, L97.
77. Fărcasiu, D.; Marino, G. *JOM* **1983**, *253*, 243.
78. Krivykh, V. V.; Gusev, O. V.; Rybinskaya, M. I. *JOM* **1989**, *362*, 351.
79. Genco, N. A.; Partis, R. A.; Alper, H. *JOC* **1973**, *38*, 4365.
80. (a) Alper, H.; Edward, J. T. *CJC* **1968**, *46*, 3112. (b) Alper, H. unpublished data cited in: *Organic Syntheses via Metal Carbonyls*; Wender, I.; Pino, P., Eds.; Wiley: New York, 1977; Vol. 2, p 545.
81. Kruger, G. J.; Lombard, A. van A.; Raubenheimer, H. G. *JOM* **1987**, *331*, 247.
82. (a) Alper, H.; Edward, J. T. *JOC* **1967**, *32*, 2938. (b) Frojmovic, M. M.; Just, G. *CJC* **1968**, *46*, 3719.
83. des Abbayes, H.; Clément, J.-C.; Laurent, P.; Tanguy, G.; Thilmont, N. *OM* **1988**, *7*, 2293.
84. Palyi, G.; Sampar Szerencses, E.; Gulamb, V.; Palagyi, J.; Marko, L. Hung. Patent 49 843, 1989 (*CA* **1989**, *113*, 5940t).
85. Schrauzer, G. N. *CB* **1961**, *94*, 1891.
86. (a) Edgell, W. F.; Yang, M. T.; Bulkin, B. J.; Bayer, R.; Koizumi, N. *JACS* **1965**, *87*, 3080. (b) Edgell, W. F.; Bulkin, B. J. *JACS* **1966**, *88*, 4839. (c) Bulkin, B. J.; Lynch, J. A. *IC* **1968**, *7*, 2654.
87. (a) Ryang, M.; Sawa, Y.; Masada, H.; Tsutsumi, S. *Kogyo Kagaku Zasshi* **1963**, *66*, 1086 (*CA* **1965**, *62*, 7670h). (b) Ryang, M.; Rhee, I.; Tsutsumi, S. *BCJ* **1964**, *37*, 341.
88. Nakanishi, S.; Otsuji, Y.; Adachi, T. *Chem. Express* **1992**, *7*, 729.
89. Heppert, J. A.; Thomas-Miller, M. E.; Swepston, P. N.; Extine, M. W. *CC* **1988**, 280.
90. Negishi, E.; Zhang, Y.; Shimoyama, I; Wu, G. *JACS* **1989**, *111*, 8018.
91. Giam, C.-S.; Ueno, K. *JACS* **1977**, *99*, 3166.
92. Yamashita, M.; Mizushima, K.; Watanabe, Y.; Mitsudo, T.-A.; Takegami, Y. *CC* **1976**, 670.
93. Yamashita, M.; Suemitsu, R. *TL* **1978**, 1477.
94. Yamashita, M.; Suemitsu, R. *TL* **1978**, 761.
95. Hashiba, S.; Fuchigami, T.; Nonaka, T. *JOC* **1989**, *54*, 2475.
96. Hieber, W.; Lipp, A. *CB* **1959**, *92*, 2085.
97. Alper, H.; Edward, J. T. *CJC* **1970**, *48*, 1543.
98. Herndon, J. W.; McMullen, L. A. *JOM* **1989**, *368*, 83.

99. Crotti, C.; Cenini, S.; Rindone, B.; Tollari, S.; Demartin, F. *CC* **1986**, 784.
100. Alper, H.; Des Roches, D. *TL* **1977**, 4155.
101. Victor, R.; Ben-Shoshan, R. *JOM* **1974**, *80*, C1.
102. Shim, S. C.; Park, W. H.; Doh, C. H.; Lee, H. K. *Bull. Korean Chem. Soc.* **1988**, *9*, 61.
103. Bister, H.-J.; Butenschön, H. *SL* **1992**, 22.
104. Albrecht, R.; Weiss, E. *JOM* **1990**, *399*, 163.
105. Rosenblum, M.; Gatsonis, C. *JACS* **1967**, *89*, 5074.
106. King, R. B.; Murray, R. M.; Davis, R. E.; Ross, P. K. *JOM* **1987**, *330*, 115.
107. Jones, E. R. H.; Wailes, P. C.; Whiting, M. C. *JCS* **1955**, 4021.
108. (a) Pearson, A. J.; Shively, Jr., R. J. *OM* **1994**, *13*, 578. (b) Pearson, A. J.; Shively, Jr., R. J.; Dubbert, R. A. *OM* **1992**, *11*, 4096. (c) Knölker, H.-J.; Heber, J.; Mahler, C. H. *SL* **1992**, 1002.
109. Pearson, A. J.; Dubbert, R. A. *OM* **1994**, *13*, 1656.
110. Reppe, W.; Vetter, H. *LA* **1953**, *582*, 133.
111. Eaton, B. E.; Rollman, B.; Kaduk, J. A. *JACS* **1992**, *114*, 6245.
112. Trifonov, L. S.; Orahovats, A. S.; Linden, A.; Heimgartner; H. *HCA* **1992**, *75*, 1872.
113. Liebeskind, L. S.; Baysdon, S. L.; South, M. S.; Blount, J. F. *JOM* **1980**, *202*, C73.
114. Kettle, S. F. A.; Orgel, L. E. *Proc. Chem. Soc.* **1959**, 307.
115. Saegusa, T.; Ito, Y.; Shimizu, T. *JOC* **1970**, *35*, 3995.
116. Brunet, J.-J.; Niebecker, D.; Srivastava, R. S. *TL* **1993**, *34*, 2759.
117. (a) Fischer, E. O.; Beck, H.-J.; Kreiter, C. G.; Lynch, J.; Müller, J.; Winkler, E. *CB* **1972**, *105*, 162. (b) Fischer, E. O.; Kreissl, F. R.; Winkler, E.; Kreiter, C. G. *CB* **1972**, *105*, 588. (c) Semmelhack, M. F.; Tamura, R. *JACS* **1983**, *105*, 4099. (d) Chen, J.; Lei, G.; Yin, J.; *Huaxue Xuebao* **1989**, *47*, 1105 (*CA* **1990**, *113*, 40941c). (e) Lotz, S.; Dillen, J. L. M.,; Van Dyk, M. M. *JOM* **1989**, *371*, 371.
118. Semmelhack, M. F.; Tamura, R.; Schnatter, W.; Springer, J. *JACS* **1984**, *106*, 5363.
119. Daub, J.; Trautz, V.; Erhardt, U. *TL* **1972**, 4435.

Sylvie Samson & G. Richard Stephenson
University of East Anglia, Norwich, UK

Pentafluorophenol

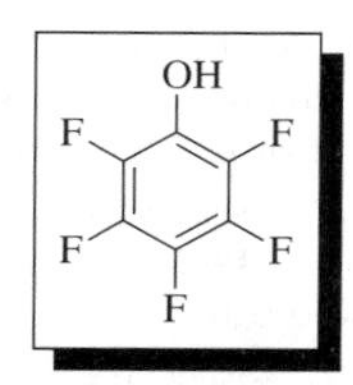

[771-61-9] C_6HF_5O (MW 184.07)

(preparation of pentafluorophenyl active esters for peptide synthesis;[1] formylating agent; preparation of aromatic fluoro derivatives)

Physical Data: white solid, mp 34–36 °C; bp 143 °C; fp >72 °C; n_D 1.4263.
Solubility: sol most organic solvents, particularly aprotic ones.
Form Supplied in: available commercially.
Handling, Storage, and Precautions: toxic and irritant. Harmful if swallowed, inhaled, or absorbed through the skin. Incompatible with strong oxidizing agents, bases, acid chlorides, and anhydrides. Keep away from heat and naked flames: combustible solid/liquid. Store in a cool, dry place.

Use in Peptide Synthesis.

Solution Techniques. Simple alkyl esters of amino acids (which may be protected) undergo aminolysis at a rate too slow for peptide bond synthesis. Phenyl esters are more reactive and, if electronegative substituents are present on the ring, may undergo aminolysis at rates similar to those of anhydrides. The overriding requirements in peptide bond formation are efficiency and freedom from side reactions. Esters of pentafluorophenol have found wide application in the synthesis of peptides (cf. ***Pentachlorophenol***).[1a–c] Pentafluorophenol (**1**) serves as a coupling reagent, i.e. a compound added to a mixture of carboxyl and amino components to promote condensation. However it is very difficult to find efficient coupling reagents which do not give undesirable side reactions. Kinetic studies have shown the superiority of pentafluorophenyl esters compared to other coupling reagents with respect to speed of coupling (and thus lowering or eliminating undesirable reactions, e.g. α-hydrogen abstraction leading to racemization).[2a,b] Relative rates are OPFP ≫ OPCP > ONp, corresponding to 111:3.4:1, where PFP = pentafluorophenyl; PCP = pentachlorophenyl; Np = nitrophenyl.

Pentafluorophenyl esters have been used in both solution and solid phase peptide syntheses, the difference being that solid phase synthesis involves linking the peptide to a solid support (see below), whilst solution techniques involve just one phase. There are many examples of pentafluorophenyl esters being used in peptide synthesis,[1a,3a–e] e.g. the peptide sequence *t*-Boc–Phe–Pro–Pro–Phe–Phe–Val–Pro–Pro–Ala–Phe–OMe[3a] and the alkaloid peptide (**2**)[3b] have been synthesized by these methods.

(2)

More recently, the base labile 9-fluorenylmethoxycarbonyl (Fmoc) group has become well established in peptide synthesis for the protection of amino functionalities.[4,5] The problem associated with this group is cleavage of the Fmoc group by the free amino component to be acylated, which results in double acylations. The principal method of avoiding this now is to use the Fmoc group in association with highly reactive pentafluorophenyl esters to greatly increase the rate of coupling compared with the undesirable reactions. Formation of these esters is shown in eq 1.[4]

For example, this methodology was first used in the synthesis of Fmoc-glycyl-L-tryptophyl-L-leucyl-L-aspartyl-L-phenylalaninamide (Fmoc–Gly–Trp–Leu–Asp–Phe–NH_2) with the isolation of all pentafluorophenyl ester intermediates in excellent yields (75–99%); the coupled amino acids were subsequently prepared stepwise, again in very good yields, with the minimum of undesirable side reactions.[4]

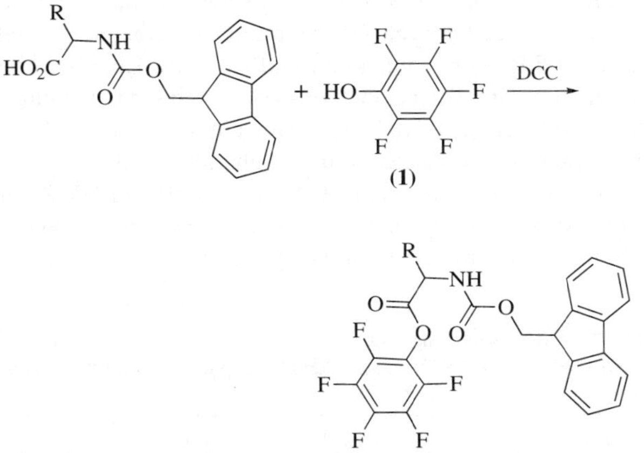

(1)

This method has since been modified and improved further by utilizing 9-fluorenylmethyl pentafluorophenyl carbonate (eq 2).[5a]

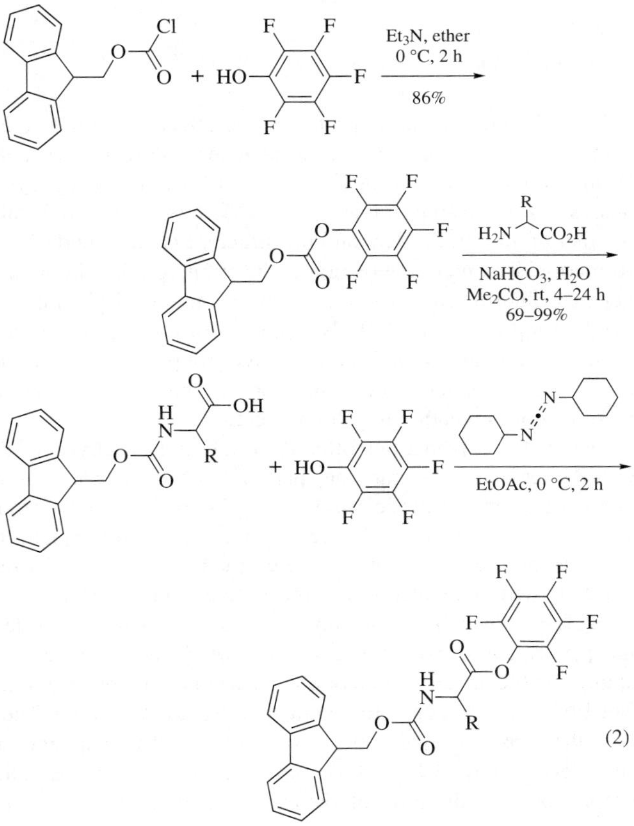

(2)

There is a newer alternative method of synthesizing Fmoc-protected pentafluorophenyl ester amino acids employing pentafluorophenyl trifluoroacetate (**3**).[5b]

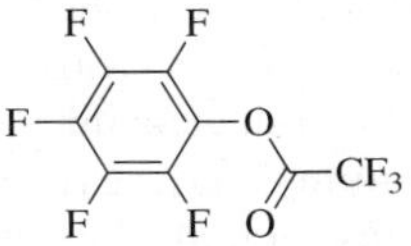

(**3**)

Table 1 gives some typical examples of the *N*-9-Fluorenylmethoxycarbonyl (Fmoc) amino acids and their pentafluorophenyl (PFP) esters which have been prepared by these methods.[4,6,7] The physical data are in agreement with either preparative method, and with early preparations of these compounds.[4,5a,b]

Table 1 Examples of Fmoc-Protected Amino Acids and their Pentafluorophenyl (PFP) Esters

Amino acid	Fmoc derivative mp (°C)	Fmoc derivative yield (%)	Fmoc amino acid PFP ester mp (°C)	Fmoc amino acid PFP ester yield (%)	Ref.
Gly	171–172	84	158–160	72	4,6
Ala	141–142	93	173–175	74	4,6
Leu	151–152	83	113–114	78	4,6
Val	143–144	90	121–123	77	4,7
Pro	114–115	82	125–127	89	4,7
Phe	181–182	86	152–154	90	4,6
Trp	163–165	89	184–185	85	4,6
Tyr	98–107	70	152–156	70	4
Ser	90–94	78	124–126	83	4,7
Ser (*t*-Bu)	–	–	67–71	67	4,7

Solid Phase Peptide Synthesis (SPPS). Pentafluorophenyl esters have been used in both solution and solid phase peptide synthesis. So far, only solution methods have been considered. The principal of SPPS is incorporation of single amino acid residues into peptide chains built on insoluble polymeric supports. The famous Merrifield paper[8] started the tradition of SPPS using *t*-butoxycarbonyl (Boc) amino acids activated with ***1,3-Dicyclohexylcarbodiimide***, although this has now been superseded with the introduction of the Fmoc group in polystyrene- and polyamide-based SPPS. The theory of SPPS will not be discussed here. Needless to say, the use of pentafluorophenyl esters in SPPS was inevitable and was first reported by Atherton in 1985.[9] Fmoc–amino acid pentafluorophenyl active esters were used in the polyamide solid phase series in polar DMF solutions in the presence of ***1-Hydroxybenzotriazole*** as a catalyst.

The preparation of the decapeptide H–Val–Gln–Ala–Ile–Asp–Tyr –Ile–Asn–Gly–OH (**4**) will be used as an example of pentafluorophenyl esters in SPPS.[9,10] This is a good example to use since it contains normal, sterically hindered, and functionalized amino acids. All Fmoc–pentafluorophenyl esters were prepared by the method of Kisfaludy and Schön already described.[4,5] Beaded polymethylacrylamide gel resin was used as the solid phase with fivefold excesses of the pentafluorophenyl esters dissolved in a minimum of DMF. The resin was first functionalized with a reference norleucine residue and then with an acid-labile linkage agent (**5**).

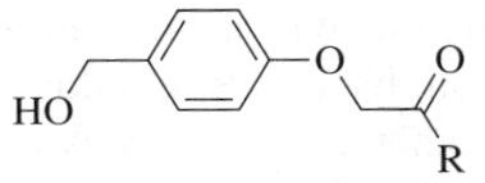

(**5**) R = 2,4,5-$Cl_3C_6H_2O$

Esterification of the first residue, glycine, to the free OH used the symmetrical anhydride/***4-Dimethylaminopyridine*** procedure.[11] The subsequent peptides were all coupled using standard techniques like those already described above and are fully documented. Slightly varying conditions and catalysts were required for some of the amino acids.[10] After addition of the last amino acid, the final Fmoc group was cleaved and the complete decapeptide detached from the support resin using 95% aqueous ***Trifluoroacetic Acid***. Side chain protecting groups were also cleaved simultaneously. Crude peptide analysis was achieved by HPLC and showed the samples obtained by this method to be extremely pure.

More recently, SPPS has also been employed in glycopeptide synthesis.[12] There is an excellent review of active esters in SPPS, including pentafluorophenyl esters.[13]

A New and Efficient Derivative of Pentafluorophenol.[14] Pentafluorophenol has found many applications in both solution and solid phase peptide syntheses. Diphenylphosphinic mixed anhydrides are also well known in peptide chemistry. A new and efficient reagent has been prepared from these precursors: pentafluorophenyl diphenylphosphinate (FDPP; **6**) which can be used directly as a coupling reagent without side reactions. FDPP is prepared from mixing equimolar amounts of ***Diphenylphosphinic Chloride***, pentafluorophenol (**1**), and ***Imidazole*** in CH_2Cl_2 at room temperature (eq 3).[14]

$$Ph_2P(O)Cl + C_6F_5OH\ (\mathbf{1}) \xrightarrow[CH_2Cl_2]{\text{imidazole}} Ph_2P(O)OC_6F_5\ (\mathbf{6}) \quad (3)$$

FDPP may be kept refrigerated for several months and is used as an efficient coupling reagent, by simply mixing the carbonyl and amine components with a tertiary amine and FDPP in an organic solvent, of which DMF was found to be the most efficient (eq 4).

$$R^1NHCH(R^2)CO_2H + H_2NCH(R^3)CO_2R^4 \xrightarrow[\text{tertiary amine}]{\text{FDPP}} R^1NHCH(R^2)C(O)NHCH(R^3)CO_2R^4 + C_6F_5OH + Ph_2P(O)OH\text{•tertiary amine} \quad (4)$$

FDPP has been used in both solid and solution peptide syntheses and analysis has shown it to have the lowest rate of racemization of any of the common coupling reagents.[14]

Comparison of Pentafluorophenyl Active Esters with Pentachlorophenyl Esters.[15] It is interesting to compare these two compounds with respect to active ester formation since they are very closely related and both have been employed in both liquid and solid phase peptide syntheses (see ***Pentachlorophenol***). Pentachlorophenyl esters generally have higher melting points, which can be of advantage if crystallization is difficult or a compound is of low melting point, e.g. for serine and threonine the oily pentafluorophenyl esters are usually replaced by the pentachlorophenyl esters.

As already mentioned, pentafluorophenyl derivatives are much more reactive than pentachlorophenyl ones. The reason for this is still not clear. Many calculations have been performed on the net atomic energies of these esters and show that there is no significant difference between the environment of the electrophilic center in OPFP esters compared to OPCP ones. Analysis of the carbonyl stretching frequencies of these compounds is quite revealing (Table 2), i.e. there is no significant difference in the carbonyl stretching frequencies of these functionalized esters. This, plus additional evidence from MM calculations and NMR studies, suggests that electronic factors do not play a significant role in determining the reactivity of these esters.

Table 2 IR Stretching Frequencies of Carbonyl Groups in Pentachlorophenyl (OPCP) and Pentafluorophenyl (OPFP) Active Esters[a]

Amino acid	OPCP	OPFP	$\Delta\nu$
Z–Gly	1794	1800	6
Boc–Val	1782	1788	6
Boc–Phe	1784	1790	6
Boc–Tyr(*t*-Bu)	1786	1794	8
Z–Cys(Bn)	1784	1794	10

[a] In $CHCl_3$; in cm^{-1}.

There is however, an important steric effect, as would be expected given the differing sizes of the halogens involved (van der Waals atomic radii = 1.75 Å for Cl and 1.47 Åfor F). X-ray crystal analysis is interesting in that it shows there is no intramolecular interaction in OPFP esters and the structure is an extended conformation. The main interaction is the strong dipole–dipole of a negative fluorine to a positive carbon of another PFP ring in a neighboring molecule. CO···NH interactions appear to be only minor in comparison and the carbonyl group and PFP ring are almost perpendicular. Studies have shown that the same conformation applies whether in solid or solution.

One of the explanations offered for the high reactivity of the OPFP ester is neighboring group participation of an *ortho* fluorine atom in accelerating the collapse of a tetrahedral intermediate transition state, i.e. anchimeric acceleration. This would explain the low solvent effect on aminolysis rates, the high second-order rate constant, and unaffected third-order rate constant. Pentahalogen esters fall midway between quinol-8-yl esters, where there is definite participation, and nitrophenol esters where there is none (from studies on the model system of phenyl acetates and piperidine).[15] The PFP ester shows greater interaction than the PCP ester. There are other possible explanations: this could still be explained in steric terms, or by the high solvation of the leaving group (pentafluorophenol is much more soluble and reactive in aprotic solvents than pentachlorophenol).

Formylation Reactions.[16] The problem with ***Acetic Formic Anhydride***, the most commonly used formylating agent, is undesirable side reactions, particularly if there are acid sensitive functionalities in the compound. One of the best alternatives is to use pentafluorophenyl formate (**7**), an easily prepared and stable compound which reacts with N-nucleophiles in minutes at room temperature to yield the *N*-formyl derivative (eq 5) (Table 3). Significantly, no reaction occurs with alcohols, thiols, or sterically hindered amines.

$$RNH_2 + C_6F_5O\text{–}C(=O)H \ (7) \xrightarrow[76\text{–}99\%]{\text{organic solvent, rt, 5–30 min}} R\text{–}NH\text{–}C(=O)H + C_6F_5OH \quad (5)$$

Table 3 Examples of *N*-Formylamines Prepared from Primary Amines and Pentafluorophenyl Esters

Amine	Solvent	Yield (%)	mp (°C)	R_f
Aniline	$CHCl_3$	90	48	0.6[a]
2-Aminobiphenyl	$CHCl_3$	99	67–70	0.6[a]
Benzylamine	$CHCl_3$	90	62–63	0.35[a]
H–Met–Leu–Phe–OMe	$CHCl_3$	93	135–136	0.4[b]

[a] Chloroform–methanol (9:1). [b] Ethyl acetate.

Preparation of Aromatic Fluoro Derivatives. Reaction of (**1**) with VF_2, VF_4, ***Vanadyl Trifluoride***, or ***Xenon(II) Fluoride*** in HF, $CFCl_3$, or MeCN at −30 to 20 °C yields mixtures of perfluoro-2,5-cyclohexadien-1-one together with its 2- and 4-pentafluorophenoxy derivatives and/or perfluoro-6-phenoxy-2,4-cyclohexadien-1-one, along with traces of perfluoro-2-cyclohexen-1-one or perfluoro-*p*-benzoquinone. Using ***Antimony(V) Fluoride*** or NbF_4, however, only gives stable complexes or pentafluorophenolates.[17]

Pentafluorophenol (**1**) reacts with ***Phosphorus(V) Chloride*** in the presence of a base, triethylamine or γ-collidine, to yield perfluoropentaphenoxyphosphorane (**8**) plus perfluorotriphenyl phosphate (**9**) and perfluorodiphenyl ether (**10**) (eq 6).[18]

$$PCl_5 + C_6F_5OH \xrightarrow[\text{or } \gamma\text{-collidine}]{NEt_3} \underset{(8)}{(C_6F_5O)_5P} + \underset{(9)}{(C_6F_5O)_3P{=}O} + \underset{(10)}{(C_6F_5)_2O} \quad (6)$$

This reaction has also been used to prepare pentafluorothiophenol phosphorus derivatives. These compounds have been utilized in ^{31}P, ^{19}F, and ^{1}H variable temperature NMR studies on intramolecular ligand rearrangement processes in phosphorus compounds.[18]

The preparation and reactions of Polyfluoroaromatics, including pentafluorophenol derivatives, have been studied by various Russian workers, who have also evaluated the pentafluorophenoxy group as a leaving group (eq 7).[19a]

The main product of the reaction was the 4-substituted tetrafluoropyridine, although varying yields of the 2-, 2,6- and 2,4,6-pentafluorophenoxypyridines were obtained depending on the ratios of (**11**):(**1**) and also on the levels of ***Potassium Fluoride***, ***18-Crown-6***, the reaction temperature, and solvent. This reaction has been studied in comparison with $4\text{-}NO_2C_6H_4$ (which is of similar basicity) and in competition with F^- anions.

More recently it has been shown that the substitution reaction of pentafluoropyridine with (**1**) in the presence of ***Cesium Fluoride***/graphite gave much better yields of (**12**), but still with traces of the other products. The catalyst activity was found to increase in the order: NaF < KF < RbF < CsF and was similar for the free fluorides and their graphite-supported analogs.[19b]

$$\underset{(11)}{C_5F_5N} + \underset{(1)}{C_6F_5OH} \xrightarrow[\text{18-crown-6, 0 °C (varing ratios attempted)}]{\text{KF, MeCN}} \underset{(12)}{4\text{-}C_6F_5O\text{-}C_5F_4N} \quad (7)$$

+ various amounts of the 2-, 2,6-, and 2,4,6-substituted products depending on ratios of (**11**):(**1**) and of KF and 18-crown-6

Related Reagents. Pentachlorophenol.

1. For examples of early work on pentafluorophenyl esters in peptide synthesis, see: (a) Kisfaludy, L.; Roberts, J. E.; Johnson, R. H.; Mayers, G. L.; Kovacs, J. *JOC* **1970**, *35*, 3563 and references therein. (b) Kisfaludy, L.; Schön, I.; Szirtes, T.; Nyéki, O.; Löw, M. *TL* **1974**, 1785 and references therein. (c) Kovacs, J.; Kisfaludy, L.; Ceprini, M. Q. *JACS* **1967**, *89*, 183.
2. (a) For a kinetic study of coupling rates and competing reactions, see: Gross, E.; Meienhofer, J. In *The Peptides–Analysis, Synthesis and Biology*, 2nd ed.; Academic: New York, 1980; p 519; (b) see also Ref. 1(c).
3. There are many examples of solution peptide synthesis utilizing pentafluorophenyl esters. A few notable examples are cited here: (a) Kisfaludy, L.; Schön, I.; Szirtes, T.; Nyéki, O.; Low, M. *TL* **1974**, 1785. (b) Schmidt, U.; Schanbacher, U. *LA* **1984**, *6*, 1205. See also Ref. 1(a). (c) Schmidt, U.; Lieberknecht, A.; Griesser, H.; Haeusler, J. *LA* **1982**, *12*, 2153. (d) Karel'skii, V. N.; Krysin, E. P.; Antonov, A. A.; Rostovskaya, G. E. *Khim. Prir. Soedin.* **1982**, *1*, 96 (*CA* **1982**, *97*, 92 730s). (e) Polevaya, L. K.; Vegners, R.; Ars, G.; Grinsteine, I.; Cipens, G. *Latv. PSR Zinat. Akad. Vestis, Khim. Ser.* **1981**, *4*, 469 (*CA* **1982**, *96*, 7059s).
4. Kisfaludy, L.; Schön, I. *S* **1983**, 325.
5. (a) Kisfaludy, L.; Schön, I. *S* **1986**, 303. (b) Green, M.; Berman, J. *TL* **1990**, *31*, 5851.
6. Meienhofer, J.; Waki, M.; Heimer, E. P.; Lambros, T. J.; Makofske, R. C.; Chang, C-D. *Int. J. Pept. Protein Res.* **1979**, *13*, 35.
7. Chang C-D.; Waki, M.; Ahmad, M.; Meienhofer, J.; Lundell, E. O.; Haug, J. D. *Int. J. Pept. Protein Res.* **1980**, *15*, 59.
8. Merrifield, R. B. *JACS* **1963**, *85*, 2149.
9. Atherton, E.; Sheppard, R. C. *CC* **1985**, 165.
10. Atherton, E.; Cameron, L. R.; Sheppard, R. C. *T* **1988**, *44*, 843.
11. Atherton, E.; Logan, C. J.; Sheppard, R. C. *JCS(P1)* **1981**, 538.
12. For the use of pentafluorophenyl esters in SPPS of glycopeptides, see: Meldal, M.; Jensen, K. J. *CC* **1990**, 483 and references therein.
13. For a review of active esters in SPPS, see: Bodanszky, M.; Bednarek, M. A. *J. Protein Chem.* **1989**, *8*, 461.
14. Chen, S.; Jiecheng, X. *TL* **1991**, *32*, 6711.
15. Kisfaludy, L.; Low, M.; Argay, G.; Czugler, M.; Komives, T.; Sohar, P.; Darvas, F. In *Peptides (Proceedings 14th European Peptide Symposium)*; Loffet, A., Ed.; Editions de l'Université de Bruxelles: Bruxelles, 1976; p 55 and references therein.
16. Kisfaludy, L.; Ötvos, Jr., L. *S* **1987**, 510.
17. Avramenko, A. A.; Bardin, V. V.; Karelin, A. L.; Krasil'nikov, V. A.; Tushin, P. P.; Furin, G. G.; Yakobson, G. G. *ZOR* **1985**, *21*, 822 (*CA* **1985**, *103*, 141 551n).
18. Denny, D. B.; Denney, D. Z.; Liu L-T. *PS* **1982**, *13*, 1.

19. (a) Aksenov, V. V.; Vlasov, V. M.; Yakobson, G. G. *JFC* **1982**, *20*, 439. (b) Aksenov, V. V.; Vlasov, V. M.; Danilkin, V. I.; Naumova, O. Y.; Rodionov, P. P.; Chertok, V. S.; Shnitko, G. N.; Yakobson, G. G. *IZV* **1984**, *9*, 2158 (*CA* **1985**, *102*, 131 881k).

Keith Jones
King's College London, UK

Phenyl Chlorothionocarbonate

Ph–O–C(=S)–Cl

[1005-56-7] C_7H_5ClOS (MW 172.64)

(forms thionocarbonate ester derivatives of alcohols which can be deoxygenated with tin hydride reagents;[1–5] converts ribonucleosides to 2′-deoxynucleosides;[2] provides allylic thionocarbonates which undergo [3,3]-sigmatropic shifts;[6–9] provides precursors for radical bond-forming reactions;[10,11] reagent for thioacylation[12–16])

Physical Data: bp 81–83 °C/6 mmHg; fp 81 °C; *d* 1.248 g cm^{-3}.
Solubility: sol chloroform, THF.
Form Supplied in: colorless liquid.
Handling, Storage, and Precautions: corrosive; moisture sensitive; should be stored in airtight containers which exclude moisture; incompatible with alcohol solvents.

Deoxygenation of Secondary Alcohols. Reaction of secondary alcohols with the reagent in the presence of ***Pyridine*** or ***4-Dimethylaminopyridine*** (DMAP) provides thiocarbonate ester derivatives which can be reduced to alkanes using ***Tri-n-butylstannane*** (eq 1).[1] The advantage of this method is the ability to deoxygenate alcohols via radical intermediates and thereby avoid problems associated with ionic reaction conditions (i.e. carbonium ion rearrangements, reduction of other functional groups). This method is particularly useful for the conversion of ribonucleosides to 2′-deoxynucleosides. For example, adenosine can be converted to 2′-deoxyadenosine in 78% overall yield by initial protection of the 3′- and 5′-hydroxyl groups as a cyclic disiloxane, thiocarbonylation, reductive cleavage, and then final deprotection using a fluoride source (eq 2).[2] Treatment of 2′-bromo-3′-phenoxythiocarbonyl nucleosides with tributyltin hydride affords 2′,3′-didehydro-2′,3′-dideoxy nucleosides via radical β-elimination.[3]

Synthetic intermediates can be selectively deoxygenated without reduction of other functional groups such as esters, ketones, and oxime ethers (eq 3),[4] as well as epoxides, acetate esters, and alkenes.[5]

ROH + Ph–O–C(=S)–Cl —py→ RO–C(=S)–OPh —Bu_3SnH, AIBN→ RH (1)

Adenosine: 1. $(i\text{-}Pr_2SiCl)_2O$, py; 2. PhOCSCl, DMAP, MeCN → 3′,5′-O-disiloxane 2′-OCSOPh derivative: 1. Bu_3SnH, AIBN, PhH; 2. $Bu_4N^+ F^-$ → 2′-deoxyadenosine (2)

1. PhOCSCl, DMAP, CH_2Cl_2, rt, 26 h, 83%; 2. Bu_3SnH, AIBN, PhH, Δ, 2 h, 65% (3)

Sigmatropic Rearrangements of Allylic Thionocarbonates. The reaction of allyl alcohols with aryl chlorothionocarbonates affords *S*-allyl aryl thiocarbonates by [3,3]-sigmatropic rearrangement via the intermediate thionocarbonate esters.[6–9] For example, treatment of 2-methyl-1-penten-3-ol with the reagent in pyridine at −20 °C affords phenyl 2-methyl-2-pentenyl thionocarbonate in 67% yield (*E*:*Z* = 96.5:3.5).[7] This type of rearrangement, coupled with tin hydride mediated reduction of the phenyl thiocarbonate ester product, was used as a key step in the synthesis of isocarbacyclin (eq 4).[8] Rearrangement of cyclic thionocarbonates contained in eight-membered rings or smaller provides two-atom ring enlarged thiocarbonates having (*Z*) double bond geometry (eq 5).[9] Depending on the system, the cyclic thiocarbonates are obtained by either treatment of the diol monothionocarbonate with base or by reaction of the diol with 1,1′-thiocarbonyldi-2,2′-pyridone. Cyclic thionocarbonates of ring size nine or larger afford ring expanded products with exclusive (*E*) double bond geometry in modest yields.

PhOCSCl, DMAP, MeCN, 0–25 °C, 20 h, 90%; Bu_3SnH, AIBN, PhH, Δ, 3 h, 90% (4)

(5)

Radical Coupling and Cyclization Reactions. Phenyl thionocarbonate esters derived from alcohols serve as efficient precursors for the generation of radical intermediates which can be used for the formation of new carbon–carbon bonds. For example, a 4-thionocarbonate ester derived from L-lyxose undergoes a stereoselective allylation upon photolysis in toluene in the presence of 2.0 equiv of ***Allyltributylstannane*** (eq 6).[10] Photochemical initiation is preferable to chemical initiation using ***Azobisisobutyronitrile*** which results in the formation of side products at the expense of the desired product. The allylation product was used further in a total synthesis of pseudomic acid C.

(6)

Oxime ethers derived from hydroxy aldehydes, upon conversion to their phenyl thionocarbonate esters, undergo radical cyclizations resulting in the formation of carbocycles.[11] For example, an oxime ether obtained from D-glucose is converted into its phenyl thionocarbonate ester at C-5 and, upon heating in benzene in the presence of tributyltin hydride, affords cyclopentanes in 93% yield as a 62:38 mixture of two diastereomers (eq 7). In general, only low to modest stereoselectivity between the newly formed stereocenters is observed in a number of substrates examined.

(7)

Thioacylation Reagent. The regioselective thioacylation of unprotected carbohydrates via the agency of ***Di-n-butyltin Oxide*** and the reagent has been investigated.[12] Glycopyranosides having a *cis*-diol arrangement, e.g. galactose, form cyclic thiocarbonates which can be either converted into dihydropyranosides using the Corey elimination procedure or deoxygenated to a mixture of 3- and 4-deoxyglycopyranosides (eq 8). Methyl D-glucopyranoside is monothioacylated with the regioselectivity dependent upon the configuration at the anomeric carbon; the α-epimer gives 83% of 2-thionocarbonate and the β-epimer gives 85% of 6-thionocarbonate. Further treatment with tributyltin hydride affords the corresponding deoxyglucose derivatives in high yield.

(8)

In a key step leading to a synthesis of saxitoxin, radical fragmentation of a pyrazolidine ring followed by intramolecular thioacylation afforded the ring expanded tetrahydropyrimidine intermediate (eq 9).[13] The thionocarbamate activation of the pyrazolidine *N*-H was found to be necessary to effect the desired transformation.

The intramolecular thioacylation of an ester enolate was used for the synthesis of 2-alkylthiopenem carboxylic acid derivatives.[14] Sequential acylations have led to the synthesis of zwitterionic pyrazole-5-thiones from acyclic precursors,[15] whereas 2-ethoxyoxazolidines react with the reagent to afford the products of *N*-acylation.[16]

(9)

Heating of *O*-phenyl thionocarbonates of pyrrolidine and piperidine-2-ethanols in acetonitrile gives a ring expanded azepine or an octahydroazocine accompanied by the pyrrolidine and piperidine *O*-phenyl ethers (eq 10).[17] These products arise via the internal expulsion of carbonyl sulfide, leading to formation of an azetidinium intermediate followed by nucleophilic ring opening with phenoxide ion.

(10)

Related Reagents. Carbon Disulfide; *N*-Hydroxypyridine-2-thione; 1,1′-Thiocarbonyldiimidazole; *p*-Tolyl Chlorothionocarbonate; Phenyl Chlorothionocarbonate.

1. Barton, D. H. R.; Subramanian, R. *JCS(P1)* **1977**, 1718.
2. Robins, M. J.; Wilson, J. S. *JACS* **1981**, *103*, 932; Robins, M. J.; Wilson, J. S.; Hansske, F. *JACS* **1983**, *105*, 4059.

3. Serafinowski, P. *S* **1990**, 411.
4. Martin, S. F.; Dappen, M. S.; Dupre, B.; Murphy, C. J. *JOC* **1987**, *52*, 3706.
5. Schuda, P. F.; Potlock, S. J.; Wannemacher, R. W., Jr. *J. Nat. Prod.* **1984**, *47*, 514.
6. Garmaise, D. L.; Uchiyama, A.; McKay, A. F. *JOC* **1962**, *27*, 4509.
7. Faulkner, D. J.; Petersen, M. R. *JACS* **1973**, *95*, 553.
8. Torisawa, Y.; Okabe, H.; Ikegami, S. *CC* **1984**, 1602.
9. Harusawa, S.; Osaki, H.; Kurokawa, T.; Fujii, H.; Yoneda, R.; Kurihara, T. *CPB* **1991**, *39*, 1659.
10. Keck, G. E.; Kachensky, D. F.; Enholm, E. J. *JOC* **1985**, *50*, 4317.
11. Bartlett, P. A.; McLaren, K. L.; Ting, P. C. *JACS* **1988**, *110*, 1633.
12. Haque, M. E.; Kikuchi, T.; Kanemitsu, K.; Tsuda, Y. *CPB* **1987**, *35*, 1016.
13. Jacobi, P. A.; Martinelli, M. J.; Polanc, S. *JACS* **1984**, *106*, 5594.
14. Leanza, W. J.; DiNinno, F.; Muthard, D. A.; Wilkering, R. R.; Wildonger, K. J.; Ratcliffe, R. W.: Christensen, B. G. *T* **1983**, *39*, 2505.
15. Grohe, K.; Heitzer, H.; Wendisch, D. *LA* **1982**, 1602.
16. Widera, R.; Muehlstaedt, M. *JPR* **1982**, *324*, 1005.
17. Sakanoue, S.; Harusawa, S.; Yamazaki, N.; Yoneda, R.; Kurihara, T. *CPB* **1990**, *38*, 2981.

Eric D. Edstrom
Utah State University, Logan, UT, USA

Phenyl *N*-Phenylphosphoramidochloridate

Ph–O–P(O)(Cl)–NHPh

[51766-21-3] $C_{12}H_{11}ClNO_2P$ (MW 267.66)

(phosphorylating agent in oligonucleotide synthesis via the phosphotriester approach; for the conversion of acids to their symmetrical anhydrides and for the coupling of acids and amines to give amides; two important derivatives are also considered)

Physical Data: mp 132–134 °C.
Solubility: sol org solvents, e.g. acetone, dichloromethane.
Form Supplied in: white solid; widely available commercially.
Analysis of Reagent Purity: FT-IR;[22] R_f (pure) = 0.63 (chloroform–acetone (5:1) on silica).
Drying: is moisture sensitive; dry in vacuo ($\leq 10^{-2}$ mmHg) over P_2O_5 at rt.
Preparative Methods: obtainable from phenyl phosphorodichloridate (**2**) and aniline (eq 1).[1]

Ph–O–P(O)Cl–Cl (**2**) —$PhNH_2$→ Ph–O–P(O)Cl–NHPh (**1**) (1)

Handling, Storage, and Precautions: irritant and moisture sensitive. May be harmful by inhalation (irritating to mucous membranes and upper respiratory tract), ingestion, or skin absorption. Incompatible with strong oxidizing agents and strong bases.

Phosphorylating Agent. Phenyl *N*-phenylphosphoramidochloridate (**1**) reacts readily at 0 °C in dry pyridine with equimolar amounts of a corresponding alcohol or phenol to give the *O*-alkyl-*O*-phenyl-*N*-phenylphosphoramides (eq 2).[2]

(**1**) —ROH, py, 70–86%→ Ph–O–P(O)(OR)–NHPh (**3**) —1. i-C_5H_{11}ONO; 2. NaOH, 85–95%→ Ph–O–P(O)(OR)–O^- Na^+ (**4**) (2)

R = alkyl, aryl, e.g. *t*-$BuCH_2$, O_2N–C_6H_4–, bornyl

The compound (**3**) is a useful intermediate for the synthesis of mixed diesters of phosphoric acid (**4**). There are several methods of cleaving the N–P bond, of which the method of Ikehara is the most convenient.[3] Reaction of (**3**) with a 40 molar excess of 3-methylbutyl nitrite in acetic acid–pyridine (1:1 v/v) is successful in removing the aniline protecting group from the phosphoryl moiety in almost quantitative yields, giving the *O*-alkyl/aryl-*O*-phenyl hydrogen phosphate.

Given the good yields and simplicity of the process, this procedure has been used to prepare intermediates for oligonucleotide synthesis, e.g. in the synthesis of 5′-*O*-monomethoxytritylthymidine 3′-phenylphosphate (**7**) (eq 3). Phosphorylation of 5′-*O*-monomethoxytritylthymidine (**5**) gave the protected nucleoside phosphate (**6**) in 98% yield which, after removal of the aniline protecting group and workup, gave the pyridinium salt in 85% yield.

(**5**) —(**1**)→ (**6**) —1. i-C_5H_{11}ONO; 2. py→ (**7**) (3)

T = thymidinyl
MMTr = monomethoxytrityl

This chemistry has additionally been utilized in the preparation of nucleoside phosphorothioates.[4] Reaction of (**6**) with NaH to remove the *P*-anilido group followed by CS_2 gave the corresponding 5′-monomethoxytritylthymidine 3′-*O*-phenylphosphorothioate (as a mixture of diastereomers, **8**) which was converted to the *S*-methyl ester (**9**) (eq 4). Owing to the chiral nature of pentavalent phosphorus, the compound (**9**) could exist as a mixture of diastereomers; however, the above reaction was the first example of an asymmetric synthesis of a phosphorothioate. There are now more examples of stereospecific synthesis of nucleoside derivatives utilizing this type of chemistry.[5]

$$(6)\ \xrightarrow[CS_2]{NaH}\ (8)\ \xrightarrow{MeI}\ (9) \quad (4)$$

(**8**) MMTrO-nucleoside (T), 3'-O-P(=O)(OAr)S⁻ Na⁺; (**9**) MMTrO-nucleoside (T), 3'-O-P(=O)(OAr)SMe

There are numerous reviews on phosphorylation methods in biological molecules, some of which include aryl phosphoramidochloridates.[6]

Activation of Carboxylic Acids.

Formation of Symmetrical Anhydrides. Symmetrical carboxylic anhydrides are useful reagents for peptide or amide synthesis.[7,8] Unfortunately, most methods require transformation of the acid first into the acid chloride or the use of condensation agents, e.g. ***1,3-Dicyclohexylcarbodiimide*** (DCC), which can lead to separation problems of the product anhydride from the hydrated derivatives of the condensation reagents. The use of ***Thionyl Chloride/Pyridine*** to form the acid chloride is often complicated by racemization with amino acid derivatives.

However, carboxylic acids are converted almost quantitatively to their corresponding symmetrical anhydride by treatment with 1 equiv of ***Triethylamine*** or ***1-Ethylpiperidine*** and 0.5 equiv of phenyl *N*-phenylphosphoramidochloridate (or ***Diphenyl Phosphorochloridate***) in acetone or dichloromethane, with the product anhydride (**10**) being obtained by filtration or from evaporation of the solvent after washing with water.[9] The resulting crude products are very nearly pure and the phenyl *N*-phenylphosphoramidochloridate can be regenerated as the aqueous solutions yield the phosphoramidic acid on acidification, which can be converted to (**1**) by reaction with ***Phosphorus(V) Chloride***. The proposed pathway for this is through the mixed anhydride (**11**), which then reacts with a second molecule of acid. The only limitation of this method is the reaction of some of the anhydrides with water, e.g. as observed with trichloroacetic acid.

$$RCO_2H + PhO\text{-}P(O)(NHPh)Cl\ (1) \longrightarrow PhO\text{-}P(O)(NHPh)\text{-}O\text{-}C(O)R\ (11) \xrightarrow{RCO_2H} RC(O)\text{-}O\text{-}C(O)R\ (10) + PhO\text{-}P(O)(NHPh)OH \quad (5)$$

The Synthesis of Polyanhydrides by Dehydrative Coupling. The synthesis of polyanhydrides was first reported in 1909[10] and led on to a series of aliphatic polyanhydrides being prepared in the 1930s, as potential materials for the textile industry. One method for the formation of such polyanhydrides, particularly where sensitive monomers are involved, is to find suitably powerful dehydrating agents that function in mild reaction conditions. Organophosphorus compounds appear to be one such group of compounds. Phenyl *N*-phenylphosphoramidochloridate was found to be one of the most successful dehydrating agents.[11] The proposed mechanism of polyanhydride formation is shown in eq 6.

$$ClP(O)R_2 + R'CO_2H \xrightarrow{Et_3N} R'C(O)\text{-}O\text{-}P(O)R_2 + Et_3N{\bullet}HCl \xrightarrow[Et_3N]{R'CO_2H} \left(R'C(O)\text{-}O\text{-}C(O)\right)_n + Et_3N{\bullet}HO\text{-}P(O)R_2 \quad (6)$$

Formation of Carboxamides. There are many reagents for the conversion of carboxylic acids to carboxamides.[8,12–14] Activating agents such as DCC[15] can lead to purification problems, whilst involvement of the acid chloride can lead to racemization; there are also examples using phosphorus containing reagents, e.g. ***Diethyl Phosphorocyanidate***,[13,16] but again these reagents only react well in certain systems. Other methods involve long reaction times, high temperatures, or low yields.

However, the use of phenyl *N*-phenylphosphoramidochloridate as a condensing agent for carboxylic acids and amines yields carboxamides in a one-step method (eq 7).[17] The amide (**12**) is prepared from the carboxylic acid, 2 equiv of triethylamine, 1 equiv of amine, and 1 equiv of reagent (**1**) and isolated by filtration or evaporation of solvent and washing with water. Once again, a mixed anhydride intermediate (**13**) is presumed, which results from nucleophilic attack by the carboxylate anion on the phosphorus atom with elimination of a chloride ion; a further nucleophilic attack by the amine on the carbonyl group of (**13**) and breakdown of the complex yields the amide (**12**) and phenyl *N*-phenylphosphoramidate (**14**) which can be recycled to (**1**) by reaction with phosphorus pentachloride.

$$PhO\text{-}P(O)(Cl)NHPh\ (1) \xrightarrow{R^1CO_2^-} [PhO\text{-}P(O)(N(Ph)H\cdots O{=}C(R^1)\text{-}O\text{-})]\ (13) \xrightarrow{R^2NH_2} PhO\text{-}P(O)(O^-)NHPh\ (14) + R^1C(O)NHR^2\ (12) \quad (7)$$

A distinct order of reactivity is observed in these reactions with (**1**): carboxylate anion ≫ aliphatic amines ≫ aromatic amines. This has led to slight variations in the practical procedures for preparing the amides but these are well documented in the literature.[17] The first route involves a one-step procedure; the second involves reaction of (**1**) with the carboxylate anion and subsequent treatment with the amine and is the preferred method for converting alkanamines to the amides since the first one-step method produces large amounts of phenyl phosphorodiamidates as well as the amide, by nucleophilic attack of (**1**) by the amine. The final method prevents the formation of phenyl phosphorodiamidates as the carboxylic acid first reacts with 0.5 equiv of (**1**) to form its anhydride, which is then converted to the amide in almost quantitative yield (this is largely utilizing the anhydride-formation chemistry described above) (eq 8).

2 R¹CO₂H + PhO–P(O)(Cl)NHPh (1) —NEt₃, –PhO–P(O)(OH)NHPh, –HCl→ [R¹C(O)–O–C(O)R¹] —R^2NH_2, $-R^1CO_2H$→ R¹C(O)NHR² (8)

Generally, phenyl *N*-phenylphosphoramidochloridate and other very similar related organophosphorus reagents e.g. phosphorus oxychloride, phenyl dichlorophosphate, diphenylchlorophosphate and *N,N′*-bis(2-oxo-3-oxazolinyl)phosphorodiamidic chloride have been utilised in recent years not only to form symmetrical carboxylic acid anhydrides but also in esterification and thiol esterification reactions; although phenyl *N*-phenylphosphoramidochloridate is not so efficient for the latter two reactions compared to its analogs but is probably the best choice for forming the anhydride.[18]

Two Important Derivatives of Phenyl *N*-Phenylphosphoramidochloridate.

Phenyl N-Phenylphosphoramidoazidate. This compound (**15**) is a white solid, which is easily prepared from (**1**) (eq 9).[19] It is obtained in excellent yields and is unusual for an organophosphorus compound in that it shows little sensitivity to moisture at room temperature for several months.

PhO–P(O)(Cl)NHPh (1) —NaN_3, MeCN, reflux, 4 h; –NaCl; 80%→ PhO–P(O)(N_3)NHPh (15) (9)

It is an excellent reagent for the preparation of *N,N′*-diarylureas. Reaction of a carboxylic acid, primary amine, and (**15**) in acetonitrile leads to the disubstituted urea (**16**) (eq 10). There is an additional competing reaction to the formation of the diarylurea: formation of phenyl *N*-phenylphosphorodiamidates, involving nucleophilic attack of the carboxylate ion and the alkylamine on the phosphorus atom of phenyl *N*-phenylphosphoramidoazidate. This work has been fully reviewed.[20]

R^1NH_2 + R^2CO_2H —(**15**), MeCN, reflux, 1.5 h→ R¹NHC(O)NHR² (16) + PhO–P(O)(OH)NHPh (10)

Phenyl N-Methyl-N-phenylphosphoramidochloridate. Like (**1**), phenyl *N*-methyl-*N*-phenylphosphoramidochloridate (**17**) reacts under analogous conditions to give anhydrides and carboxamides (see above). However, phenyl *N*-methyl-*N*-phenylphosphoramidochloridate undergoes one additional reaction which is not observed to any significant degree in (**1**). It is used for the conversion of imines to α-substituted β-lactams (eq 11).[21]

The reaction is believed not to work for (**1**) as it requires an excess of triethylamine in the presence of a less nucleophilic imine, which is unfavorable to a mixed anhydride intermediate. The proposed pathway for this reaction is shown in eq 11, and has been performed on a wide variety of imines, proceeding in good yields (44–78%).

PhO–P(O)(Cl)N(Me)Ph (17) + $R^1CH_2CO_2H$ —Et_3N→ [PhO–P(O)(N(Me)Ph)–O–C(O)CH₂R¹] —$R^2CH{=}NR^3$ (e.g. 1-phenyl-3,4-dihydroisoquinoline)→ β-lactam (11)

1. Cremlyn, R. W. J.; Kishore, D. N. *JCS(P1)* **1972**, 585.
2. Zielinski, W. S.; Lesnikowski, Z. J. *S* **1976**, 185.
3. Ikehara, M.; Uesugi, S.; Fukui, T. *CPB* **1967**, *15*, 440.
4. Zielinski, W. S.; Lesnikowski, Z. J.; Stec, W. J. *CC* **1976**, 772.
5. Lesnikowski, Z. J.; Niewiarowski, W.; Zielinski, W. S.; Stec, W. J. *T* **1984**, *40*, 15.
6. Slotin, L. A. *S* **1977**, 737 and references therein.
7. Chen, F. M. F.; Koroda, K.; Benoiton, N. L. *S* **1978**, 928.
8. Jones, J. H. In *The Peptides* Gross, E.; Meienhofer, J., Eds.; Academic: New York, 1979; p 65.
9. Mestres, R.; Palomo, C. *S* **1981**, 219.
10. Bucher, J. E.; Slade, W. C. *JACS* **1909**, *31*, 1319.
11. Leong, K. W.; Simonte, V.; Langer, R. *Macromolecules* **1987**, *20*, 705.
12. Ogliaruso, M. A.; Wolfe, J. F. In *The Chemistry of Acid Derivatives* Wiley: New York, 1979; Part 1, p 474.
13. Haslam, E., *CI(L)* **1979**, 610.
14. Haslam, E. *T* **1980**, *36*, 2409.
15. Mikolajczyk, M.; Kielbasinski, P. *T* **1981**, *37*, 233.
16. Shiori, T.; Yokoyama, Y.; Kasai, Y.; Yamada, S. *T* **1976**, *32*, 2213.
17. Mestres, R.; Palomo, C. *S* **1982**, 288.
18. Arrieta, A.; Garcia, T.; Lago, J. M.; Palomo, C. *SC* **1983**, *13*, 471.
19. Arrieta, A.; Palomo, C. *TL* **1981**, *22*, 1729.
20. Arrieta, A.; Palomo, C. *BSF(2)* **1982**, 7.
21. Shridhar, D. R.; Bhagat, R.; Narayana, V. L.; Awasthi, A. K.; Reddy, G. J. *S* **1984**, 846.
22. Pouchert, C. J. *Aldrich Library of FT-IR Spectra*; Aldrich: Milwaukee, WI, 1989; Vol. 1, p 556C.

Adrian P. Dobbs
King's College London, UK

Phenyl Phosphorodi(1-imidazolate)

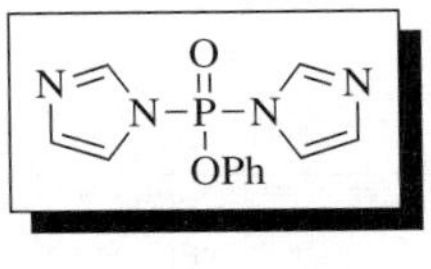

[15706-68-0] $C_{12}H_{11}N_4O_2P$ (MW 274.24)

(used in phosphorylation,[1] as a peptide coupling reagent,[1] and in the dehydration of aldoximes to nitriles[2])

Alternate Name: phenyl diimidazol-1-ylphosphinate.
Physical Data: an exceedingly hygroscopic white solid, mp 96–97 °C[2] (mp 90–92 °C also reported[1]).
Solubility: sol acetonitrile, dioxane, THF; relatively insol benzene, ether.
Form Supplied in: not commercially available.
Preparative Methods: this reagent was first prepared in 81% yield by reaction of 4 equiv of imidazole with 1 equiv of phenyl phosphorodichloridate in benzene at 60 °C.[4] The use of 2 equiv of imidazole and 2 equiv of triethylamine as base has been reported to give a yield of 98% of (**1**) (eq 1).[2]

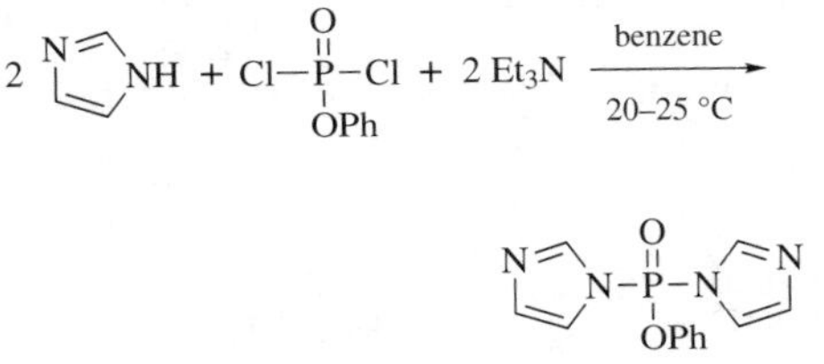

2 imidazole + Cl–P(O)(OPh)–Cl + 2 Et_3N —(benzene, 20–25 °C)→ (**1**) + 2 Et_3NHCl (1)

Purification: recrystallization from benzene, filtration under a water-free atmosphere, and drying under vacuum.
Handling, Storage, and Precautions: owing to its hygroscopic nature, exclusion of moisture in handling and use is essential; should be stored in a sealed tube at −20 °C.[3]

Phosphorylation. The title reagent (**1**) reacts with amines, alcohols, and water to give amides and esters of phenyl phosphate (eqs 2–4).[1] The monoimidazole derivatives are formed rapidly with a much slower reaction to give the disubstitution products.[1]

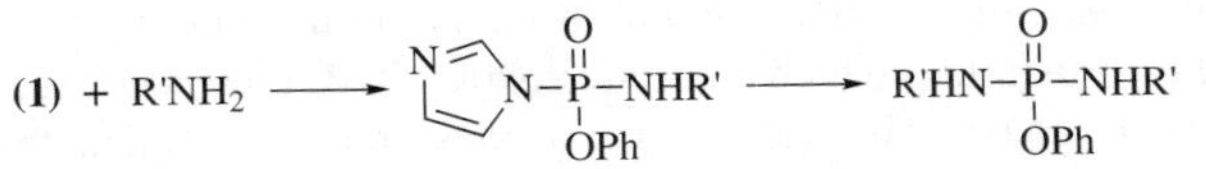

(**1**) + R'NH_2 → Im–P(O)(OPh)–NHR' → R'HN–P(O)(OPh)–NHR' (2)

(**1**) + R'OH → Im–P(O)(OPh)–OR' → R'O–P(O)(OPh)–OR' (3)

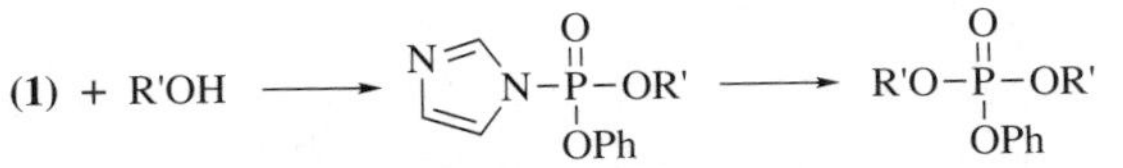

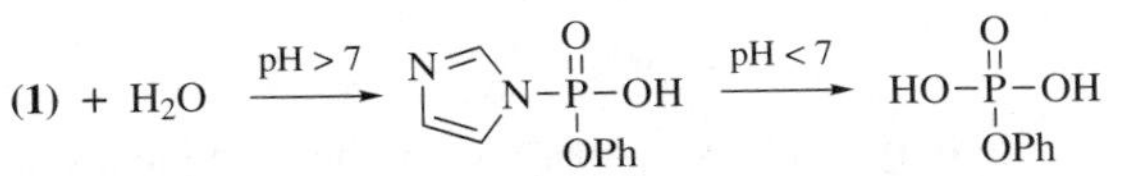

(**1**) + H_2O —(pH > 7)→ Im–P(O)(OPh)–OH —(pH < 7)→ HO–P(O)(OPh)–OH (4)

This behavior can be used to synthesize mixed diphosphates by treatment with a small excess of alcohol followed by hydrolysis of the monoimidazole derivative. In this manner the monophenyl ester of 2′,3′-isopropylideneadenosine 5′-phosphate (**2**) has been synthesized in 53.5% yield (eq 5).[1]

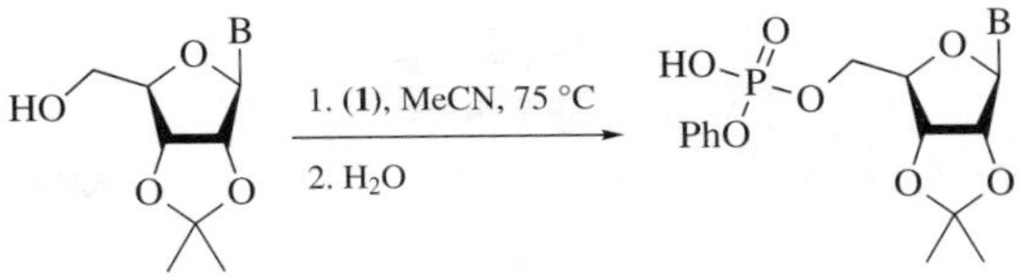

1. (**1**), MeCN, 75 °C; 2. H_2O (5)

B = adenine (**2**)

Peptide Synthesis. Reagent (**1**) has been used as a coupling reagent in the synthesis of benzyloxycarbonylalanylalanine (**3**).[1] Initial treatment of benzyloxycarbonylalanine with (**1**) in acetonitrile at 45 °C for 12 h followed by reaction with an excess of an aqueous solution (pH 8–9) of alanine gave the protected dipeptide in 85% yield (eq 6). No reference to the stereochemistry of the starting materials or product was made.

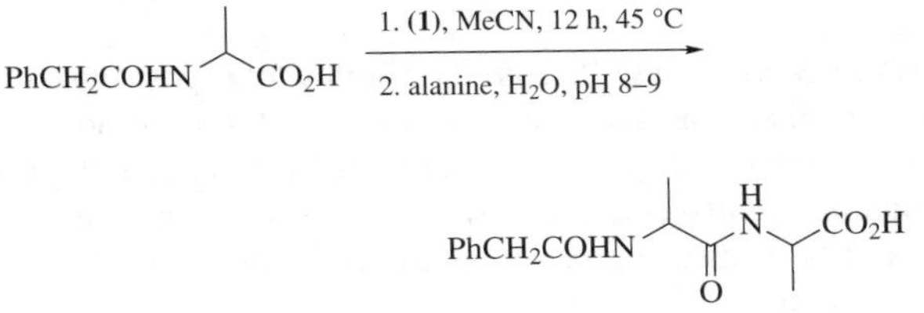

$PhCH_2COHN$–CH(CH_3)–CO_2H —(1. (**1**), MeCN, 12 h, 45 °C; 2. alanine, H_2O, pH 8–9)→ $PhCH_2COHN$–CH(CH_3)–C(O)–NH–CH(CH_3)–CO_2H (**3**) (6)

Acylimidazolides have been isolated from the reaction of (**1**) with carboxylic acids[1] and may be involved in the above reaction.

Dehydration of Aldoximes to Nitriles. Reagent (**1**) has been shown to be excellent for the dehydration of aldoximes of both aromatic and aliphatic aldehydes.[2] This is accomplished by reacting the aldoxime with (**1**) in dioxane at 20 °C for 20 h (eq 7).[2]

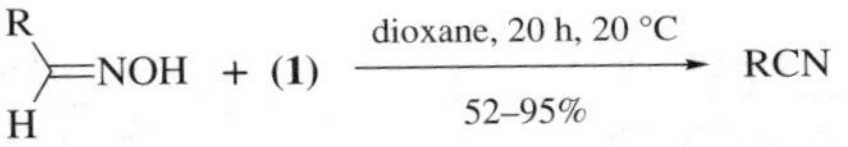

RCH=NOH + (**1**) —(dioxane, 20 h, 20 °C; 52–95%)→ RCN (7)

R = pentyl, hexyl, cyclohexyl, undecyl, cinnamyl, 4-methoxyphenyl, 4-chlorophenyl, 2,4-dichlorophenyl, 4-nitrophenyl, 2-pyridyl

Reagent (**1**) is superior in terms of yield to a variety of phosphorochloridates, including ***Phosphorus Oxychloride***. For example, in the case of 2-pyridyl aldoxime, with (**1**) a yield of 52% was obtained, whereas with phosphorus oxychloride the yield was 33% and with phenyl phosphorodichloridate the yield was 47%.[2] The reaction is thought to proceed via an intermediate formed by *O*-phosphorylation of the aldoxime.[2] The use of spin-labeled versions of (**1**) has been explored, in which the phenyl is replaced with the 2,2,6,6-tetramethylpiperidyl *N*-oxide group.[3,5]

1. Cramer, F.; Schaller, H. *CB* **1961**, *94*, 1634.
2. Konieczny, M.; Sosnovsky, G. *ZN(B)* **1978**, *33b*, 1033.
3. Konieczny, M.; Sosnovsky, G. *S* **1976**, 537.
4. Cramer, F.; Schaller, H.; Staab, H. A. *CB* **1961**, *94*, 1612.
5. Konieczny, M.; Sosnovsky, G. *ZN(B)* **1977**, *32b*, 1179.

Keith Jones
King's College London, UK

Phosgene[1]

$COCl_2$

[75-44-5] CCl_2O (MW 98.91)

(chlorinating agent;[2] carbonylating agent; dehydrating agent;[1] with DMF forms a chloromethyleniminium salt, used in formylation reactions[3])

Alternate Name: carbonyl chloride.
Physical Data: mp −118 °C; bp 8.3 °C; *d* 1.381 g cm^{-3}.
Solubility: slightly sol H_2O (dec); sol toluene, chloroform.
Form Supplied in: liquefied gas in cylinders; solution in toluene.
Purification: chlorine is an impurity which can be detected by bubbling through mercury (gives discoloration) and removed by bubbling through two wash bottles containing cottonseed oil.
Handling, Storage, and Precautions: highly toxic gas; avoid contact or inhalation; there may be a delay of several hours before symptoms of exposure develop (pulmonary edema); keep container tightly closed and well-ventilated; excess phosgene should be vented into a water-fed scrubbing tower. Use in a fume hood.

Reactions with Carboxylic Acids, Alcohols, and Amines. Phosgene reacts with carboxylic acids to give anhydrides, with liberation of carbon dioxide.[4] This has been used to activate acids to attack from nucleophiles, e.g. in esterification,[5] lactonization,[6] and thiolation.[7] For example, acids can be protected as 3-butenyl esters using phosgene and ***Pyridine***,[5] followed by addition of 3-buten-1-ol (eq 1).

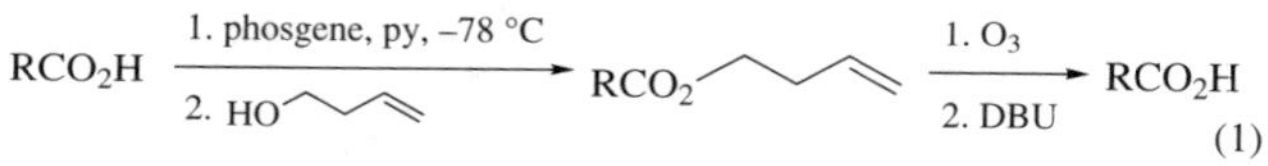

RCO_2H —(1. phosgene, py, −78 °C; 2. $HOCH_2CH_2CH{=}CH_2$)→ $RCO_2CH_2CH_2CH{=}CH_2$ —(1. O_3; 2. DBU)→ RCO_2H (1)

The ester is removed under mild conditions by ozonolysis and β-elimination of the resulting aldehyde. In contrast, reaction with acids in the presence of a variety of catalysts (including ***Imidazole***, ***1,8-Diazabicyclo[5.4.0]undec-7-ene***, and ***1,3-Dicyclohexylcarbodiimide***) cleanly affords the corresponding acid chloride.[2] ***N,N-Dimethylformamide*** has also been used as a catalyst (eq 2).[8]

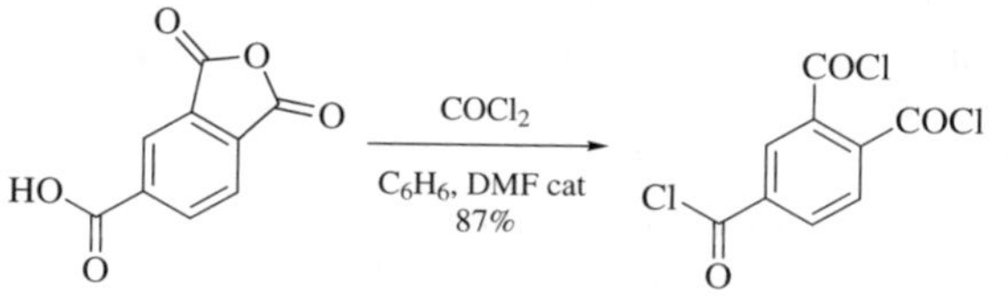

(2) $COCl_2$, C_6H_6, DMF cat, 87%

Phosgene reacts with alcohols to give chloroformates,[9,1b] and with secondary amines to give chloroformamides[10] (it reacts preferentially with the latter).[11] The formation of a chloroformate is the first step in the Barton oxidation of primary alcohols to aldehydes,[12] which is followed by complex formation with ***Dimethyl Sulfoxide*** and elimination by base (eq 3),[12b] a mechanism closely related to that of the Swern and Moffatt oxidations.[13]

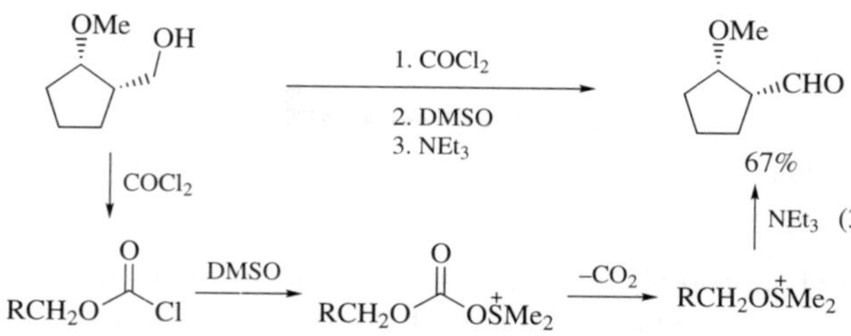

(3) 1. $COCl_2$; 2. DMSO; 3. NEt_3; 67%. RCH_2OCOCl —DMSO→ $RCH_2OCOO\overset{+}{S}Me_2$ —($-CO_2$)→ $RCH_2O\overset{+}{S}Me_2$ —NEt_3→

There are many examples of further displacement of the chloroformate in an intramolecular or intermolecular sense by suitably disposed second nucleophiles (e.g. hydroxy, amino, thio, or acid groups), resulting in carbonates (eq 4),[14] ureas (eq 5),[15] carbamates,[11,16] dithiocarbonates (eq 6),[17] etc.

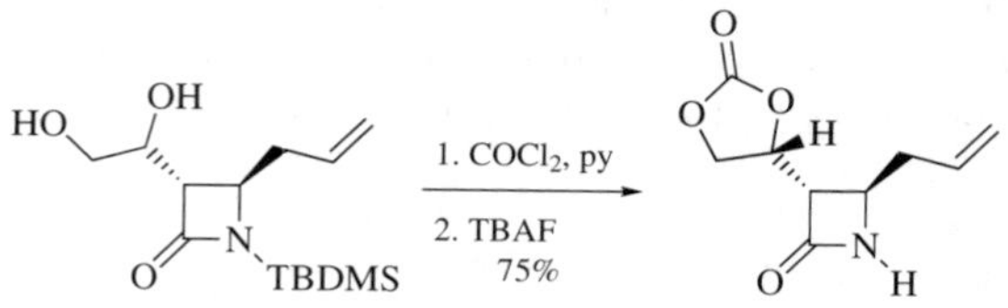

(4) 1. $COCl_2$, py; 2. TBAF; 75%

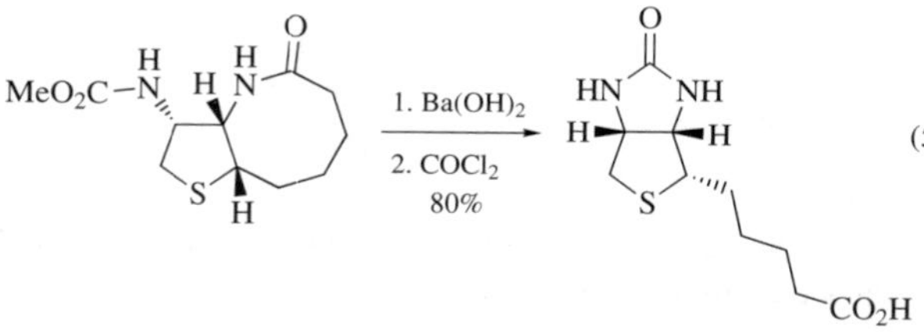

(5) 1. $Ba(OH)_2$; 2. $COCl_2$; 80%

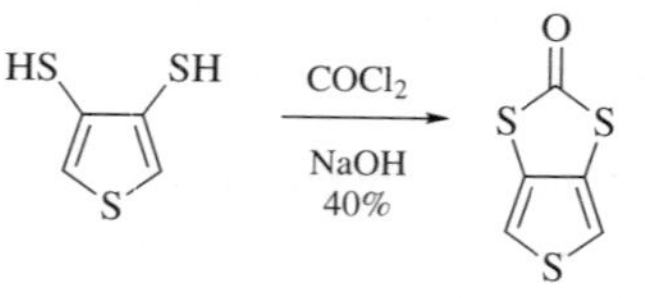

(6) $COCl_2$, NaOH, 40%

The reaction of phosgene with primary amines initially forms chloroformamides which on heating above 50 °C eliminate HCl to give isocyanates. This useful reaction constitutes a general method for isocyanate synthesis (eq 7).[18]

$$RNH_2 + COCl_2 \longrightarrow RNHCOCl \xrightarrow{-HCl} RN{=}C{=}O \quad (7)$$

With tertiary amines, phosgene–amine complexes can be formed.[1a] Warming results in decomposition to an *N,N*-dialkylchloroformamide by elimination of alkyl chloride.[19] This has been utilized for the mild and high yielding deprotection of *N*-methyl amines (eq 8).[20] α-Chloroethyl chloroformate, prepared from phosgene and ***Acetaldehyde***, has also been used for this purpose.[21]

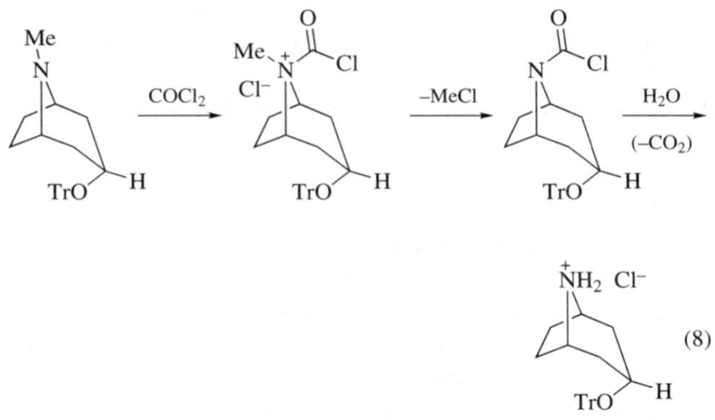

(8)

Reactions with Amides and Thioamides. The action of phosgene on amides (and thioamides) initially gives dehydration to a chloromethyleneiminium chloride (eq 9). Primary amides react further, losing HCl to form nitriles in the presence of a base such as pyridine[22] (this can also be accomplished by the Vilsmeier–Haack reagent generated using phosgene – see below). Secondary formamides are converted to isocyanides in the presence of ***Triethylamine*** or pyridine (eq 10).[23] *N*-Formyl enamines also form isocyanides (eq 11);[24] ***1,4-Diazabicyclo[2.2.2]octane*** is claimed to be a more efficient base due to the avoidance of azeotrope formation with the low boiling vinyl isocyanide products. Other secondary amides lose HCl to give imidoyl chlorides, which can undergo further reaction, e.g. to form pyrroles (eq 12)[25] and oxazolones (eq 13).[26]

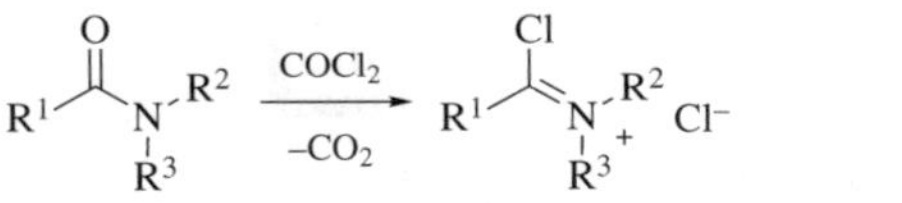

(9)

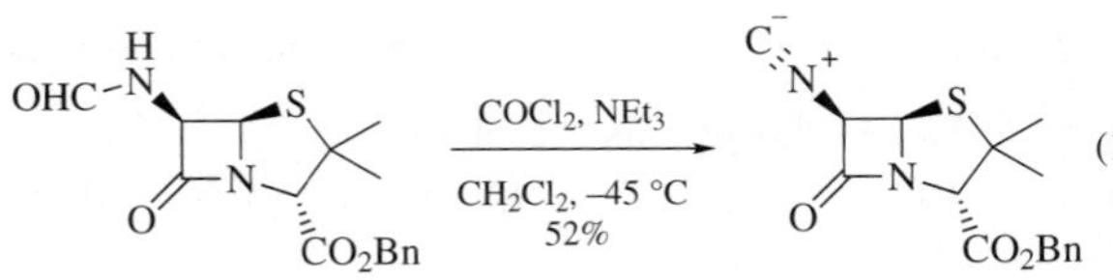

(10)

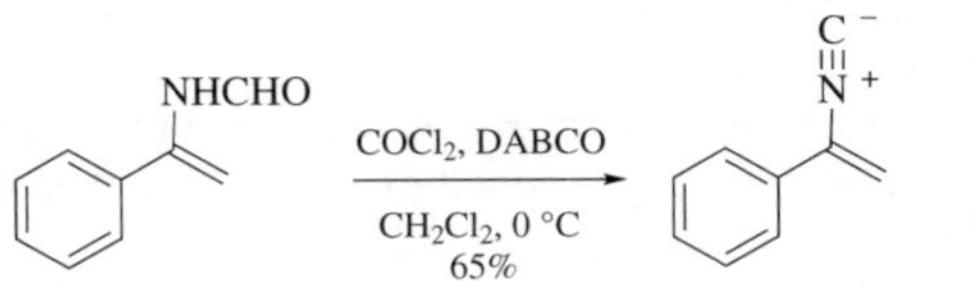

(11)

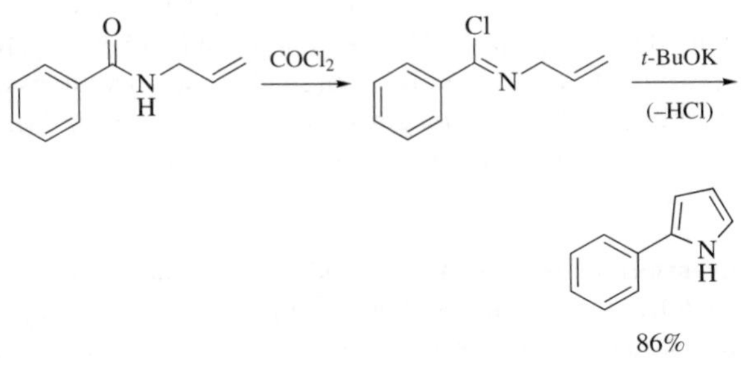

(12)

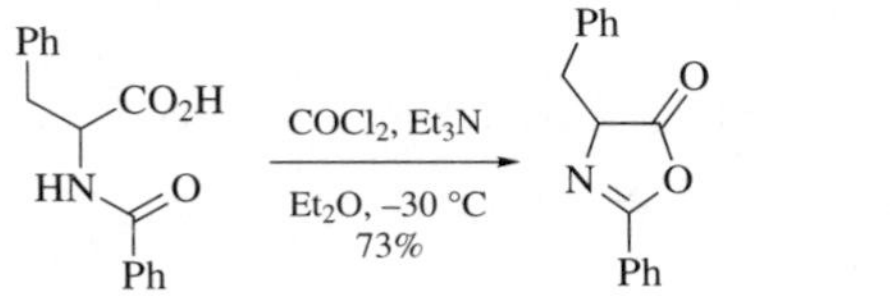

(13)

Secondary thioureas can be dehydrated to carbodiimides (eq 14).[27]

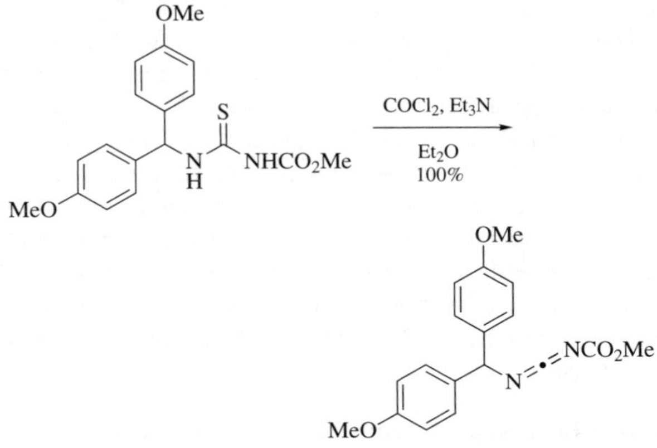

(14)

Tertiary amides give stable chloromethyleneiminium chlorides. A particularly important example is the complex formed between phosgene and DMF (known as the Vilsmeier–Haack reagent), which is a versatile formylating agent.[3] An identical or equivalent species can be generated with ***Oxalyl Chloride***, ***Thionyl Chloride***, or ***Phosphorus Oxychloride***, and these are perhaps preferred in view of phosgene's toxicity. However, the Vilsmeier reagent generated from phosgene is claimed to be a more efficient reagent for the dehydration of primary amides to nitriles.[28]

Tertiary amides (and thioamides)[29] with an α-proton react instead to give chloro enamines.[30] These can undergo nucleophilic substitution (eq 15).[30,31] Chloro enamines are also intermediates in synthetically useful dehydrogenation reactions, using ***Pyridine N-Oxide*** or DMSO as oxidant (eq 16).[32] Tertiary ureas and thioureas form related adducts which have been utilized in the synthesis of hindered guanidine bases by further reaction with amines (eq 17).[33]

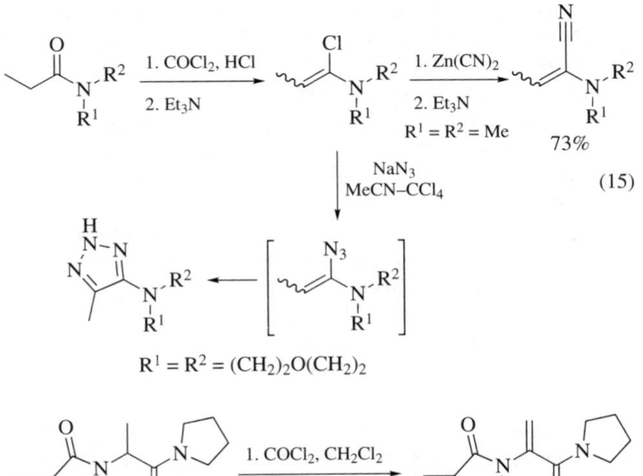

(15)

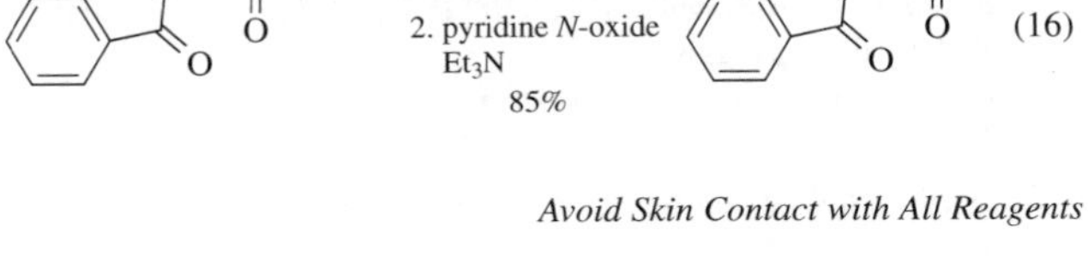

(16)

S; i-Pr, N, N, i-Pr; i-Pr, i-Pr → ($COCl_2$; $EtNH_2$; 85%) → N-Et; i-Pr, N, N, i-Pr; i-Pr, i-Pr (17)

Phosgene Equivalents. In recent years, ***Bis(trichloromethyl) Carbonate*** (triphosgene) has been introduced as a crystalline phosgene source;[34] similarly, ***Trichloromethyl Chloroformate*** (diphosgene) has been used as a liquid equivalent[35] (bp 128 °C, d_{15} 1.65 g mL^{-1}). They have the advantages of much easier manipulation (particularly for small-scale work), safer transport and storage, and they are less susceptible to hydrolysis. Triphosgene has so far been shown to perform several reactions undergone by phosgene itself, e.g. carbonylation[36] and Barton oxidation.[37] Diphosgene generates phosgene slowly at elevated temperatures, or rapidly in the presence of a charcoal catalyst;[38] it is also effective in several of the reactions of phosgene, including *N*-carboxyanhydride formation[38] and the conversion of amino alcohols to oxazolidinones,[39] where it has been recommended for large-scale use.

Related Reagents. Dimethyl Sulfoxide–Phosgene.

1. (a) Babad, H.; Zeiler, A. G. *CRV* **1973**, *73*, 75. (b) Fieser, L. F.; Fieser, M. *FF* **1967**, *1*, 856; **1969**, *2*, 328; **1972**, *3*, 225; **1974**, *4*, 157; **1975**, *5*, 532; **1979**, *7*, 289; **1990**, *15*, 265.
2. Hauser, C. F.; Theiling, L. F. *JOC* **1974**, *39*, 1134.
3. Marson, C. M. *T* **1992**, *48*, 3659.
4. Rinderknecht, H.; Guterstein, M. *OSC* **1973**, *5*, 822.
5. Barrett, A. G. M.; Lebold, S. A.; Zhang, X. *TL* **1989**, *30*, 7317.
6. Wehrmeister, H. L.; Robertson, D. E. *JOC* **1968**, *33*, 4173.
7. Bates, G. S.; Diakur, J; Masumune, S. *TL* **1976**, 4423.
8. Ulrich, H.; Richter, R. *JOC* **1973**, *38*, 2557.
9. (a) Matzner, M.; Kurkjy, R. P.; Cottner, R. J. *CRV* **1964**, *64*, 645. (b) Kevill, D. N. In *The Chemistry of Acyl Halides*; Patai, S, Ed.; Wiley: New York, 1972; p 381. (c) Samsel, E. G.; Norton, J. R. *JACS* **1984**, *106*, 5505.
10. (a) Marley, H.; Wright, S. H. B.; Preston, P. N. *JCS(P1)* **1989**, 1727. (b) Goerdeler, J.; Bartsch, H-J. *CB* **1985**, *118*, 2294.
11. Hall, H. K., Jr; El-Shekeil, A. *JOC* **1980**, *45*, 5325.
12. (a) Barton, D. H. R.; Garner, B. J.; Wightman, R. H. *JCS(P1)* **1964**, 1855. (b) Epstein, W. W.; Sweat, F. W. *CR* **1967**, *67*, 247. (c) Finch, N.; Fitt, J. J.; Hsu, I. H. S. *JOC* **1975**, *40*, 206.
13. Mancuso, A. J.; Swern, D. *S* **1981**, 165.
14. (a) Ona, H.; Uyeo, S. *TL* **1984**, *25*, 2237. (b) Krespan, C. G.; Smart, B. E. *JOC* **1986**, *51*, 320. (c) Das, J.; Vu, T.; Harris, D. N.; Ogletree, M. L. *JMC* **1988**, *31*, 930. (d) Nishiyama, T.; Yamaguchi, H.; Yamada, F. *JHC* **1990**, *27*, 143.
15. (a) Confalone, P. N.; Lollar, E. D.; Pizzolato, G. *JACS* **1978**, *100*, 6291. (b) Henning, R.; Lattrell, R.; Garhards, H. J.; Leven, M. *JMC* **1987**, *30*, 814. (c) Cram, D. J.; Dicker, I. B.; Lauer, M.; Knobler, C. B.; Trueblood, K. N. *JACS* **1984**, *106*, 7150.
16. (a) Gelb. M. H.; Abeles, R. H. *JMC* **1986**, *5*, 585. (b) Uff, B. C.; Joshi, B. L.; Popp, F. D. *JSC(P1)* **1986**, 2295.
17. Rasheed, K.; Warkentin, J. D. *JHC* **1981**, *18*, 1581.
18. (a) Richter, R.; Ulrich, H. In *The Chemistry of Cyanates and their Derivatives*; Patai, S., Ed.; Wiley: New York, 1977; Part 2, p 619. (b) Butler, D. E.; Alexander, S. M. *JHC* **1982**, *19*, 1173. (c) Weisenfeld, R. B. *JOC* **1986**, *51*, 2434.
19. Cooley, J. H.; Evain, E. J. *S* **1989**, 1.
20. Banholzer, R.; Heusner, A.; Schulz, W. *LA* **1975**, 2227.
21. (a) Olofson, R. A.; Martz, J. T.; Senet, J-P.; Piteau, M.; Malfroot, T. *JOC* **1984**, *49*, 2081. (b) Olofson, R. A.; Abbott, D. E. *JOC* **1984**, *49*, 2795.
22. Ficken, G. E.; France, H.; Linstead, R. P. *JCS* **1954**, 3731.
23. Bentley, P. H.; Clayton, J. P.; Boles, M. O.; Girven, R. J. *JCS(P1)* **1979**, 2455.
24. Barton, D. H. R.; Bowles, T.; Husinec, S.; Forbes, J. E.; Llobera, A.; Porter, A. E. A.; Zard, S. Z. *TL* **1988**, *29*, 3343.
25. Engel, N.; Steglich, W. *AG(E)* **1978**, *17*, 676.
26. Miyoshi, M. *BCJ* **1973**, *46*, 212.
27. Groziak, M. P.; Townsend, L. B. *JOC* **1986**, *51*, 1277.
28. Bargar, T.; Riley, C. M. *SC* **1980**, *10*, 479.
29. Dietliker, K.; Heimgartner, H. *HCA* **1983**, *66*, 262.
30. (a) Henriet, M.; Houtekie, M.; Techy, B.; Touillaux, R.; Ghosez, L. *TL* **1980**, *21*, 223. (b) Bernard, C.; Ghosez, L. *CC* **1980**, 940.
31. Toye, J.; Ghosez, L. *JACS* **1975**, *97*, 2276.
32. Da Costa, R.; Gillard, M.; Falmagne, J. B.; Ghosez, L. *JACS* **1979**, *101*, 4381.
33. Barton, D. H. R.; Elliott, J. D.; Gero, S. D. *JCS(P1)* **1982**, 2085.
34. Eckert, H.; Forster, B. *AG(E)* **1987**, *26*, 894.
35. Kurita, K.; Iwakura, Y. *OSC* **1988**, *6*, 715.
36. Burk, R. M.; Roof, M. B. *TL* **1993**, *34*, 395.
37. Palomo, C.; Cossio, F. P.; Ontario, J. M.; Odriozola, J. M. *JOC* **1991**, *56*, 5948.
38. Katakai, R.; Iizuka, Y. *JOC* **1985**, *50*, 715.
39. Pridgen, L. N.; Prol, J.; Alexander, B.; Gillyard, L. *JOC* **1989**, *54*, 3231.

Peter Hamley
Fisons Pharmaceuticals, Loughborough, UK

Phosphorus(III) Bromide

[7789-60-8] Br_3P (MW 270.67)

(brominating agent for conversion of alcohols to bromides)

Alternate Name: phosphorus tribromide.
Physical Data: mp 41.5 °C; bp 168–170 °C/725 mmHg; *d* 2.85 g cm^{-3}.
Solubility: sol acetone, CH_2Cl_2, CS_2.
Form Supplied in: widely available as liquid and 1.0 M solution in CH_2Cl_2.
Preparative Method: from ***Bromine*** and red phosphorus.[1]
Purification: generally used without purification; can be distilled under N_2 at atmospheric pressure.
Handling, Storage, and Precautions: corrosive. The colorless, fuming liquid has a very penetrating odor. It has a vapor pressure of 10 mmHg at 48 °C. The reagent is stable if kept dry, but reacts violently with water. It is extremely destructive to tissue of mucous membranes, upper respiratory tract, eyes, and skin. This reagent should only be used in a fume hood.

Conversion of Alcohols to Bromides. The conversion of alcohols (ROH) into bromides (RBr) using PBr_3 is very general. Reaction conditions for this transformation are quite varied. Each of the bromine atoms in the reagent is available for reaction with an alcohol. The reagent can be used to prepare chiral bromides from chiral alcohols (eq 1).[2] It is generally important to carry out the conversion under 'relatively mild' conditions (between 4 °C and rt). Addition of HBr at the end of workup increases both the optical purity and the isolated yield. An example is provided by eq 2. A polyhydroxylic compound has been converted to a polybrominated product (eq 3).[3]

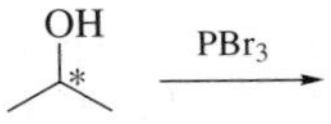

OH * → (PBr_3) → Br * (1)

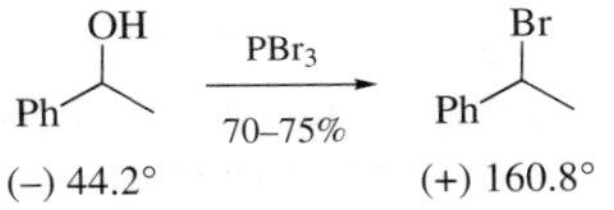

Ph–CH(OH)–CH₃ (–) 44.2° → (PBr_3, 70–75%) → Ph–CH(Br)–CH₃ (+) 160.8° (2)

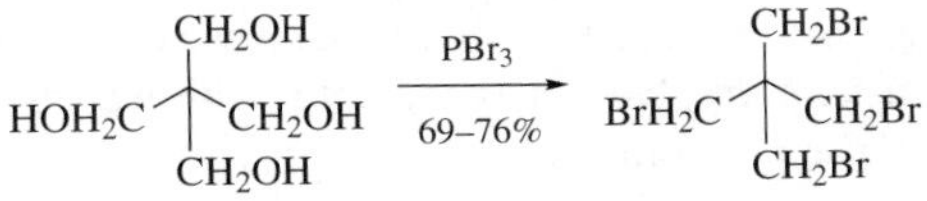

HOH_2C–C(CH_2OH)₃ → (PBr_3, 69–76%) → BrH_2C–C(CH_2Br)₃ (3)

Bromination of Allylic Alcohols. The reaction of an allylic alcohol with PBr_3 in ether at 0 °C leads to both stereoselective and regioselective replacement of the hydroxy group by bromine.[4] Examples of this transformation are given in eqs 4 and 5.[5,6] The latter example shows that the reaction can be run in the presence of significant unsaturation. There are many examples of similar reactions.[7]

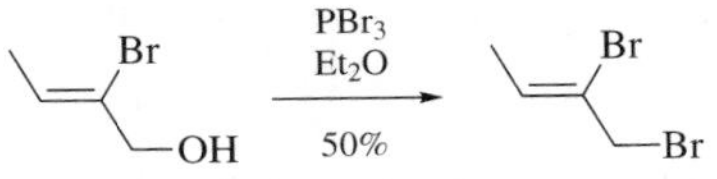

Br, OH → (PBr_3, Et_2O, 50%) → Br, Br (4)

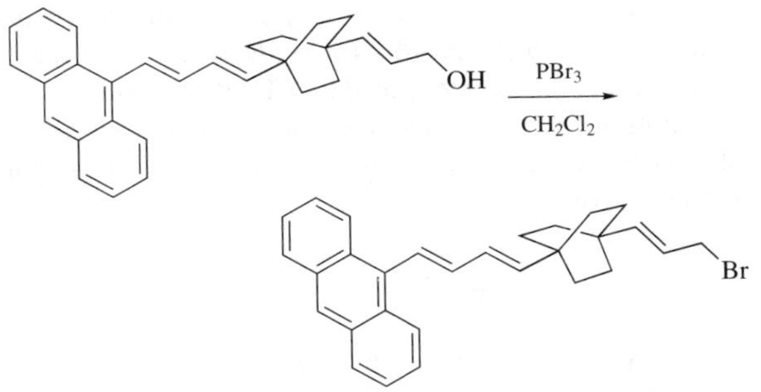

OH → (PBr_3, CH_2Cl_2) → Br (5)

Silverstein[8] followed an older procedure[9] that results in rearrangement of a secondary allylic alcohol to a terminal bromide (eq 6). In a similar fashion, the conversion of propargylic alcohols to allenyl bromides has been noted (eq 7).[10]

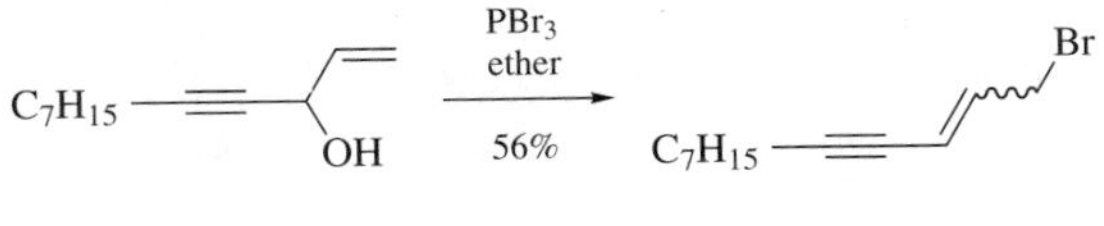

C_7H_{15}, OH → (PBr_3, ether, 56%) → C_7H_{15}, Br (6)

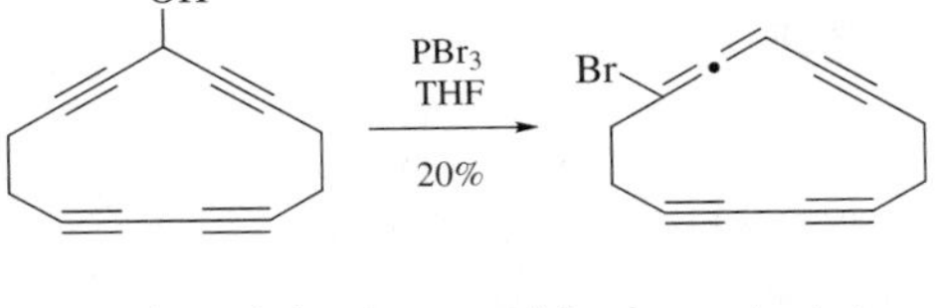

OH → (PBr_3, THF, 20%) → Br (7)

Remote π-bond participation provides important stereochemical control in the reaction. Heathcock (eq 8)[11] has reported retention of stereochemistry due to π-bond participation, in contrast to a similar system without the π-system (eq 9).[12]

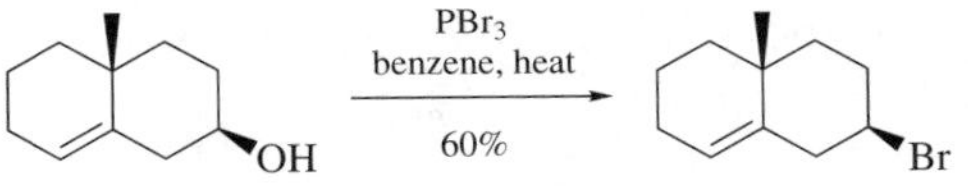

OH → (PBr_3, benzene, heat, 60%) → Br (8)

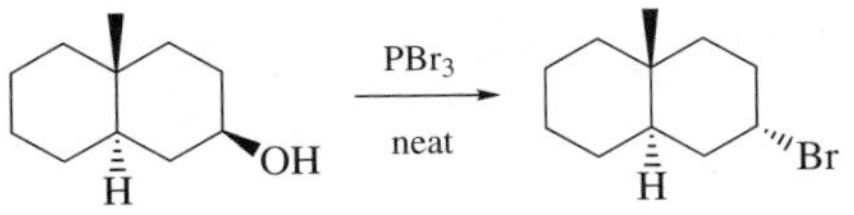

H, OH → (PBr_3, neat) → H, Br (9)

The reagent will convert an alcohol to a bromide in the presence of an ether (eq 10).[13] Acetals are also stable to the bromination procedure,[14] although an interesting reaction of a cyclopropanone acetal has been reported (eq 11).[15]

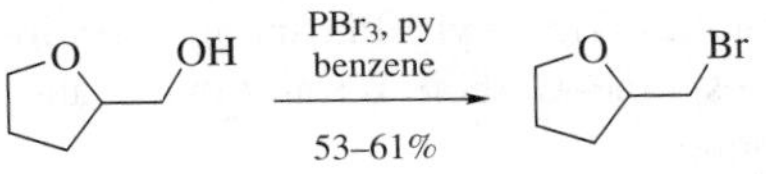

O, OH → (PBr_3, py, benzene, 53–61%) → O, Br (10)

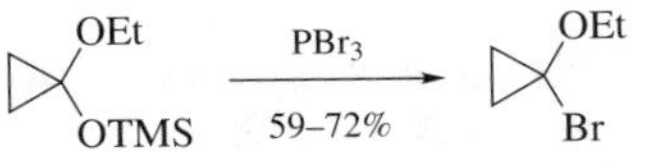

OEt, OTMS → (PBr_3, 59–72%) → OEt, Br (11)

Use for Alkene Preparation. Replacement of two alcohol groups by bromine, followed by zinc dehalogenation, provides an interesting alkene synthesis methodology (eq 12).[16] The preparation of 1,3,5-hexatriene is readily accomplished via a PBr_3 step (eq 13).[17] In the presence of DMF, PBr_3 has been used to eliminate a hindered diallylic tertiary alcohol (eq 14).[18] See also ***Phosphorus(III) Bromide–Copper(I) Bromide–Zinc***.

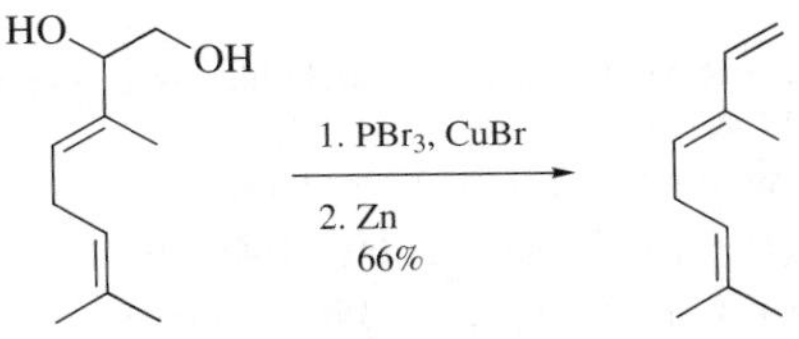

HO, OH → (1. PBr_3, CuBr; 2. Zn, 66%) → (12)

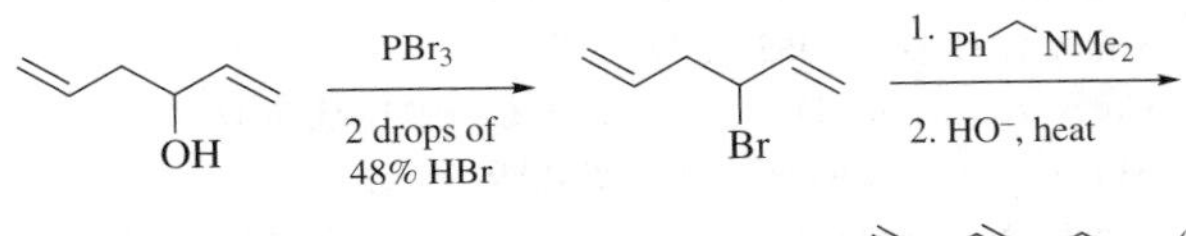

OH → (PBr_3, 2 drops of 48% HBr) → Br → (1. $PhCH_2NMe_2$; 2. HO^-, heat) → (13)

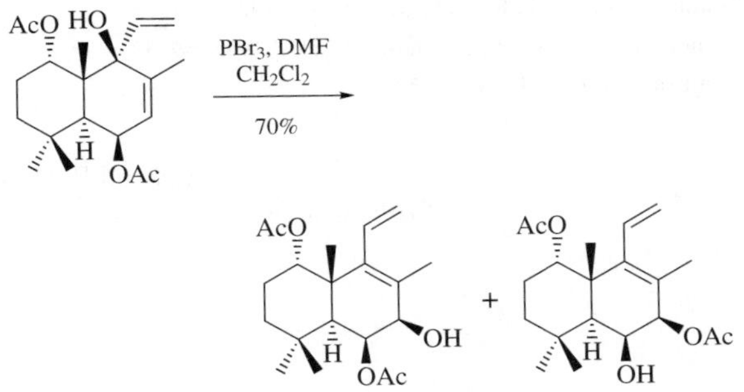

AcO, HO, H, OAc → (PBr_3, DMF, CH_2Cl_2, 70%) → AcO, H, OAc, OH + AcO, H, OH, OAc (14)

Other Reactions. A simple one-step preparation of 2-bromo-2,3-dihydro-1,3,4,2-oxadiazaphospholes has been reported (eq 15).[19] An interesting preparation of 1,5-dibromopentane from piperidine is available (eq 16).[20]

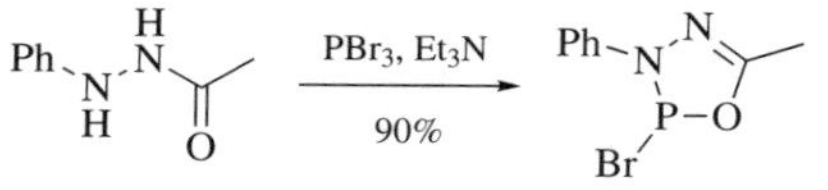

(15)

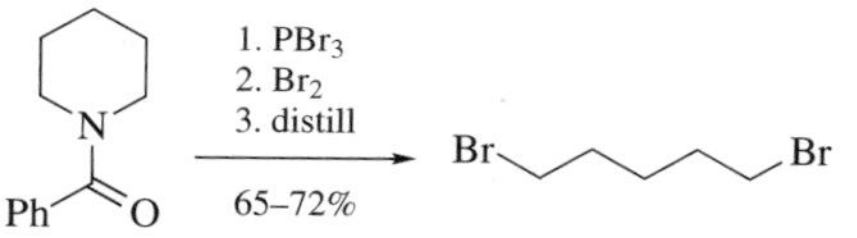

(16)

Related Reagents. A number of related reagents have been used for the conversion of alcohols to bromides. Other reagents for this purpose covered in this encyclopedia include 1,2-Bis(diphenylphosphino)ethane Tetrabromide, Bromine–Triphenyl Phosphite, Hydrogen Bromide, Triphenylphosphine–Carbon Tetrabromide, and Triphenylphosphine Dibromide.

1. (a) Gay, J. F.; Marxson, R. N. *Inorg. Synth* **1946**, *2*, 147. (b) Noller, C. R.; Dinsmore, R. *OSC* **1943**, *2*, 358.
2. Hutchins, R. O.; Masilamani, D.; Maryanoff, C. A. *JOC* **1976**, *41*, 1071.
3. Schurink, H. B. *OSC* **1943**, *2*, 476.
4. Corey, E. J.; Kirst, H. A.; Katzenellenbogen, J. A. *JACS* **1970**, *92*, 6314.
5. Miyaura, N.; Ishikawa, M.; Suzuki, A. *TL* **1992**, *33*, 2571.
6. Effenberger, F.; Kesmarszky, T. *CB* **1992**, *125*, 2103.
7. See, for example: (a) Marshall, J. A.; Faubl, H.; Warne, T. M., Jr. *CC* **1967**, 753. (b) Piers, E.; Britton, R. W.; deWaal, W. *TL* **1969**, 1251. (c) Mori, K. *T* **1974**, *30*, 3807. (d) Bestmann, H. J.; Stransky, W.; Vostrowsky, O. *CB* **1976**, *109*, 1942. (e) Mori, K.; Tominaga, M.; Matsui, M. *T* **1975**, *31*, 1846. (f) Gonzales, A. G.; Martin, J. D.; Melian, M. A. *TL* **1976**, 2279.
8. Silverstein, R. M.; Rodin, J. O.; Burkholder, W. E.; Gorman, J. E. *Science* **1967**, *157*, 85.
9. Celmer, W. D.; Solomons, I. A. *JACS* **1953**, *75*, 3430.
10. Leznoff, C. C.; Sondheimer, F. *JACS* **1968**, *90*, 731.
11. Heathcock, C. H.; Kelly, T. R. *T* **1968**, *24*, 1801.
12. Mathur, R. K.; Rao, A. S. *T* **1967**, *23*, 1259.
13. Smith, L. H. *OSC* **1955**, *3*, 793.
14. Corey, E. J.; Cane, D. E.; Libit, L. *JACS* **1971**, *93*, 7016.
15. Miller, S. A.; Gadwood, R. C. *OS* **1989**, *67*, 210.
16. Tanaka, S.; Yasuda, A.; Yamamoto, H.; Nozaki, H. *JACS* **1975**, *97*, 3252.
17. Hwa, J. C. H.; Sims, H. *OS* **1961**, *41*, 49.
18. Nampalli, S.; Bhide, R. S.; Nakai, H. *SC* **1992**, *22*, 1165.
19. Kimura, H.; Konno, H.; Takahashi, N. *BCJ* **1993**, *66*, 327.
20. von Braun, J. *OSC* **1941**, *1*, 428.

Bradford P. Mundy
Colby College, Waterville, ME, USA

Phosphorus(V) Bromide

[7789-69-7] Br_5P (MW 430.47)

(bromination agent for conversion of alcohols to bromides)

Alternate Name: phosphorus pentabromide.
Physical Data: mp 179–181 °C; *The Merck Index* (11th ed.) reports dec above 100 °C.
Solubility: sol CS_2, CCl_4.
Form Supplied in: yellow solid; commercially available.
Preparative Method: from ***Bromine*** and ***Phosphorus(III) Bromide***,[1] with which it is in equilibrium.
Handling, Storage, and Precautions: corrosive and highly toxic; stable if kept dry, but reacts violently with water; store under N_2 at rt; use in a fume hood.

Conversion of Alcohols to Bromides. The conversion of alcohols (ROH) into bromides (RBr) using PBr_5 would appear to be quite general for alcohols; however, the literature does not report many examples of this reaction. Eliel[2] reported that PBr_5 is better able to convert 4-*t*-butylcyclohexanol to 4-*t*-butylbromocyclohexane than ***Phosphorus(III) Bromide*** (eq 1).

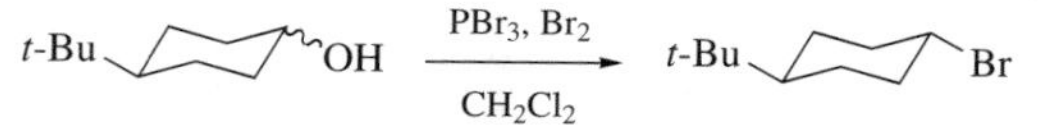

(1)

In the steroid series, D-ring enlargement accompanies reaction of a secondary alcohol with PBr_5 (eq 2).[3] It should be noted that reaction of ***Phosphorus(V) Chloride*** with this same substrate gave direct replacement of the OH with Cl with inversion of configuration.[4] Rearrangement and addition of bromine across a double bond has been noted in the reaction with a propargyl alcohol derivative (eq 3).[5]

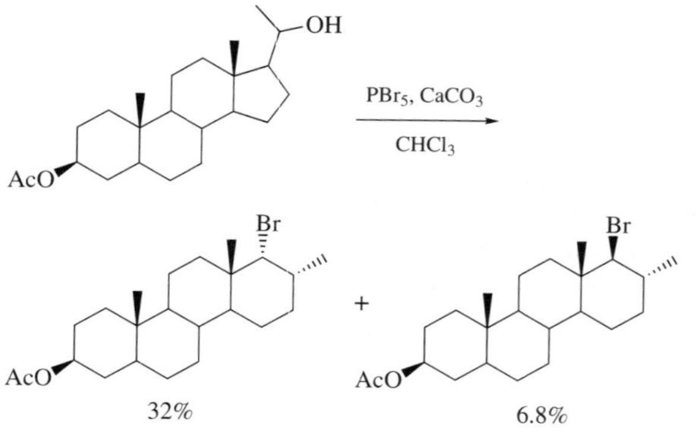

(2)

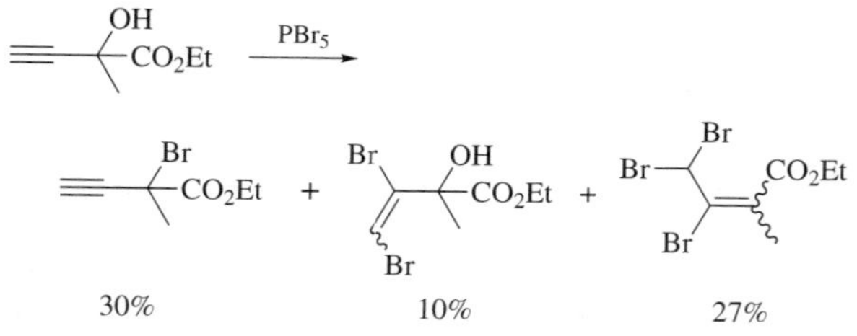

(3)

Cleavage of Ethers. PBr_5 has been used to cleave diisopropyl ether to isopropyl bromide.[5]

Reaction with Ketones. A regioselective dibromination of a diketone has been reported (eq 4).[6]

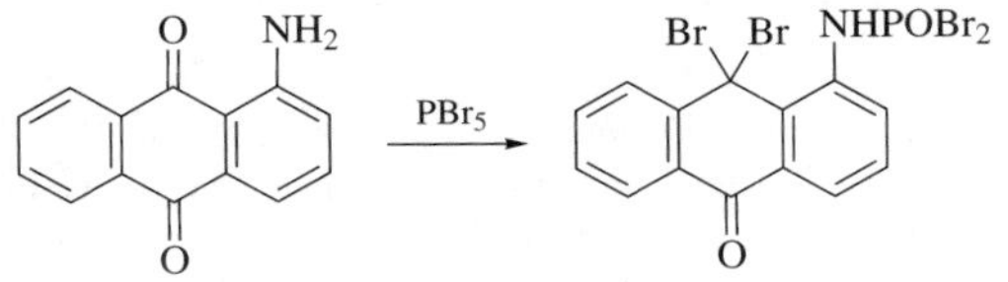

PBr_5 (4)

Related Reagents. Many related reagents are available for the conversion of alcohols to bromides (see ***Phosphorus(III) Bromide*** and related reagents cited therein).

1. (a) Gay, J. F.; Marxson, R. N. *Inorg. Synth.* **1946**, *2*, 147. (b) Noller, C. R.; Dinsmore, R. *OSC* **1943**, *2*, 358. (c) Mellor, J. W. *Comprehensive Treatise on Inorganic and Theoretical Chemistry*; Longmans: London, 1936; Vol. 8, pp 1034–1036 provides an old, but interesting, account of early work with the reagent. (d) Kaslow, C. E.; Marsh, M. M. *JOC* **1947**, *12*, 456.
2. Eliel, E. L.; Haber, R. G. *JOC* **1959**, *24*, 143.
3. Li, R. T.; Sato, Y. *JOC* **1968**, *33*, 3632.
4. Adam, G.; Schreiber, K. *T* **1966**, *22*, 3581.
5. Verny, M.; Vessiere, R. *BSF* **1968**, 2585.
6. Kutyrev, A. A.; Biryukov, V. V.; Litvinov, I. A.; Katayeva, O. N.; Musin, R. Z.; Enikeyev, K. M.; Naumov, V. A.; Ilyasov, A. V.; Moskva, V. V. *T* **1990**, *46*, 4333.

Bradford P. Mundy
Colby College, Waterville, ME, USA

Phosphorus(III) Chloride[1]

PCl_3

[7719-12-2] Cl_3P (MW 137.32)

(reactive chlorinating agent;[1,2] strong oxo, aza, thio, and dienophilicity; reactive with organometallics and metal salts; phosphorus source for metal ion ligand synthesis[28])

Alternate Name: phosphorus trichloride.
Physical Data: mp $-112\,^{\circ}C$; bp $76\,^{\circ}C$; *d* $1.586\,g\,cm^{-3}$.
Solubility: sol benzene, $CHCl_3$, CH_2Cl_2 ether, CS_2; dec in water or alcohols.
Form Supplied in: water-white liquid; widely available.
Purification: generally used as received without further purification. To purify, heat under reflux to expel dissolved hydrogen chloride, then distill at atmospheric pressure under N_2. Further purification is possible by vacuum fractionation several times through a $-45\,^{\circ}C$ trap into a $-78\,^{\circ}C$ receiver.
Handling, Storage, and Precautions: the liquid is highly toxic by ingestion and slightly toxic by single dermal applications. Inhalation can cause delayed, massive, acute pulmonary edema and death. Contact with water will be violent and result in sufficient gas evolution to rupture closed or inadequately vented containers. Acids produced by contact with water can evolve hydrogen on contact with metals. This reagent must be used in a fume hood.

Chlorinations. Alcohols, aldehydes, ketones, and carboxylic acids have been chlorinated with PCl_3. The alternative phosphorochloridites, $SOCl_2$, HCl, PCl_5, Ph_3PCl_2, Ph_3P–hexachloroacetone, $Ph_3P–CCl_4$, and NCS–SEt_2,[1,2] have been used to convert alcohols to chlorides. Aldehydes and ketones are converted to phosphites by PCl_3. Conversion of a carboxylic acid to the acid chloride has been accomplished using PCl_3, PCl_5, $POCl_3$, and $SOCl_2$. In some cases, these reactions tend to be sluggish and substrate, solvent, and temperature dependent. Modified phosphorus halides, i.e. 2,2,2-trichloro-1,3,2-benzodioxaphospholes or Ph_3PCl_2, are superior phosphorus reagents for acid chloride formation.[1] Carboxylic acids are α-brominated with a combination of PCl_3 and Br_2.[3] Imidoyl halides, generated from amides, thioamides, or hydroxamic acid derivatives with PCl_3, are valuable synthetic intermediates and provide access to a wide range of heterocycles, e.g. quinazolinones and oxadiazoles.[4]

Preparation of Phosphites. Alcohols, diols, and phenols are readily converted to phosphites. Analogous reactions are seen with thiols[1] and amines.[5] Diaryl phenols react to form medium-sized heterocyclic ring systems (eq 1).[6]

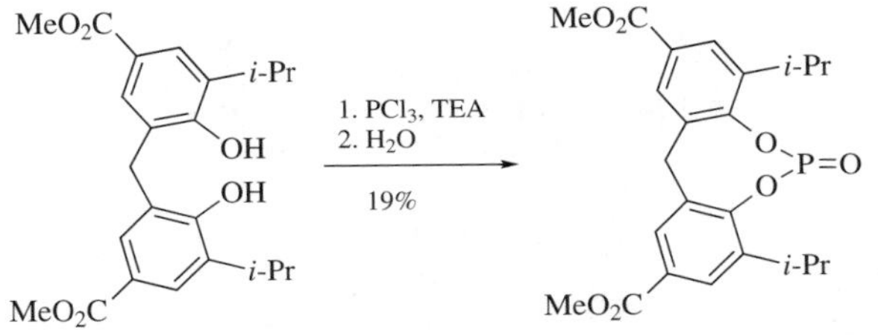

1. PCl_3, TEA
2. H_2O
19% (1)

Dihydroxyarenes yield dioxaphospholes with PCl_3.[7] A special subclass of phosphite reactions are phosphitylations. Dichlorophosphite derived from PCl_3 and an alcohol, epoxide, or $P(OMe)_3$ are common phosphitylation reagents. Replacement of the alcohol by an amine leads to dialkylaminodichlorophosphites and subsequent phosphoramidation of the substrate.[8] Direct phosphitylation with PCl_3 is useful with sterically hindered axial hydroxy functions which fail to phosphorylate with other reagents (eq 2).[9]

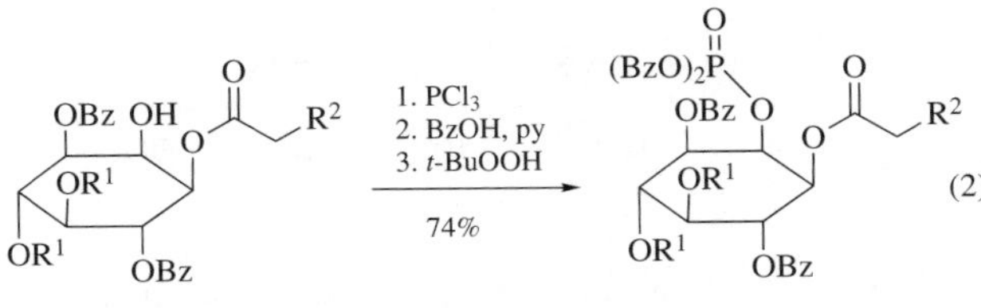

1. PCl_3
2. BzOH, py
3. *t*-BuOOH
74% (2)

R^1 = 4-$MeOC_6H_4CO$, R^2 = 1-menthol

Oxidations. Oxidation of PCl_3 with ozone,[10] diazomethane/chlorine,[11] or oxygen/ethylene,[12] produces $POCl_3$ or, in the latter two cases, phosphate esters.

Deoxygenations. Sulfoxides,[13] amine *N*-oxides,[14] and hydroxy amides[15] are readily reduced with PCl_3 to the corresponding sulfides, amines, and amides. Alternative sulfoxide reducing agents include AcCl, RCOCl, I^-, Sn^{2+}, Fe^{2+}, $SiHCl_3$, $Si_2H_2Cl_2$, $LiAlH_4$, $NaBH_4/CoCl_2$, $BHCl_2$ (aliphatic aryl), Rh/H_2, $Pd/C/H_2$, $Fe(CO)_5$, $TiCl_3$, TMSI, TMSBr, I_2, and $PhSiMe_3$. The presence of active hydrogens will complicate attempts to reduce amine *N*-oxides. Amine *N*-oxides can also be reduced using H_2SO_3, Raney Ni, $Pd/C/H_2$, and R_3P (R = alkyl or aryl).

Amines. Reaction of PCl_3 with quaternary ammonium salts yields phosphaindolizines and thiazolodiazaphospholes.[16] Imines produce phosphonates[17] whereas hydrazones are converted to pyrazoles, indoles, and nitriles.[18] Nitrones undergo rearrangement to secondary or tertiary amides[19] and primary alkylnitro compounds are reduced to nitriles (eq 3).[20] Diazonium salts are transformed to phosphonic acids.[21]

OAc, AcO, AcO, AcO, O, NO_2 — PCl_3, py; 64% → OAc, AcO, AcO, AcO, O, CN (3)

Reaction with Organometallic Compounds. Organomagnesium (eq 4),[22,23] lithium,[24] and other main group metals[25] readily undergo ligand transfer with PCl_3 to form mono-, di-, or trialkyl(aryl) phosphines. A subclass of reactions include ***Aluminum Chloride***-assisted transformations to form phosphetanes (eq 5).[26]

$$2\ i\text{-Pr}{-}\text{MgCl} \xrightarrow[55\text{–}60\%]{PCl_3,\ Et_2O} (i\text{-Pr})_2\text{P}{-}\text{Cl} \quad (4)$$

Cl-benzene + *t*-Bu alkene — $AlCl_3$, PCl_3, CH_2Cl_2, reflux → phosphetane oxide (O, P, Cl) (5)

cis/*trans* = 77%

Enols and Ketene Silyl Acetals. Reaction of a silyl enol ether with PCl_3/***Zinc Chloride*** leads to the formation of dichlorophosphine addition products (see also ***Phosphorus(III) Chloride-Zinc(II) Chloride***). Ketene silyl acetals react directly with PCl_3 under very mild conditions to form similar addition products (eq 6).[27]

TMS, OTMS, OEt — PCl_3, ether, −30 °C; 100% → TMS, O, Cl_2P, OEt (6)

Ligands. Synthesis of some phosphorus-containing ligand molecules have utilized PCl_3 as a key reagent for introduction of the phosphorus atom.[28]

Related Reagents. Acetyl Chloride; Bromotrimethylsilane; *N*-Chlorosuccinimide–Dimethyl Sulfide; Dichloroborane Diethyl Etherate; Hydrogen Chloride; Iodine; Iodotrimethylsilane; Lithium Aluminium Hydride; Pentacarbonyliron; Phosphorus(V) Chloride; Phosphorus Oxychloride; Sodium Borohydride; Thionyl Chloride; Titanium(III) Chloride; Trichlorosilane; Triphenylphosphine Dichloride; Triphenylphosphine–Hexachloroacetone.

1. (a) *Comprehensive Organic Synthesis*; Trost, B. M., Ed.; Pergamon: Oxford, 1991. (b) *Comprehensive Organic Chemistry*; Barton, D.; Ollis, W. D., Eds.; Pergamon: Oxford, 1979. (c) *The Chemistry of Organophosphorus Compounds*; Hartley, F. R., Ed.; Wiley: New York, 1990. (d) *Encyclopedia of Chemical Technology*, 3rd ed.; Grayson, M., Ed.; Wiley: New York, 1982; Vol. 17. (e) Perrin, D. D.; Armarego, W. L. F.; Perrin, D. R. *Purification of Laboratory Chemicals*, 2nd ed.; Pergamon: Oxford, 1980.
2. (a) Boughdady, N. M.; Chynoweth, K. R.; Hewitt, D. G. *AJC* **1987**, *40*, 767. (b) Sanda, K.; Rigal, L.; Gaset, A. *CR* **1989**, *187*, 15. (c) Newkome, G. R.; Theriot, K. J.; Majestic, V. K.; Spruell, P. A.; Baker, G. R. *JOC* **1990**, *55*, 2838. (d) Gazizov, M. B.; Khairullin, R. A.; Moskva, V. V.; Savel' eva, E. I.; Ostanina, L. P.; Nikolaeva, V. G. *ZOB* **1990**, *60*, 1766.
3. Clarke, H. T.; Taylor, E. R. *OSC* **1941**, *1*, 115.
4. *The Chemistry of Amidines and Imidates*; Patai, S.; Rappoport, Z., Eds.; Wiley: New York, 1991.
5. (a) Skolimowski, J. J.; Quin, L. D.; Hughes, A. N. *JOC* **1989**, *54*, 3493. (b) Cabral, J.; Laszlo, P.; Montaufier, M.-T.; Randriamahefa, S. L. *TL* **1990**, *31*, 1705. (c) Bansal, R. K.; Mahnot, R.; Sharma, D. C.; Karaghiosoff, K. *S* **1992**, 267.
6. (a) Nachev, I. A. *PS* **1988**, *37*, 149. (b) Mukmeneva, N. A.; Kadyrova, V. K.; Zharkova, V. M.; Cherkasova O. A.; Voskresenskaya, O. V. *ZOB* **1986**, *56*, 2267.
7. Pastor, S. D.; Spivack, J. D. *JHC* **1991**, *28*, 1561.
8. Yamana, K.; Nishijima, Y.; Oka, A.; Nakano, H.; Sangen, O.; Ozaki, H.; Shimidzu, T. *T* **1989**, *45*, 4135.
9. Watanabe, Y.; Ogasawara, T.; Shiotani, N.; Ozaki, S. *TL* **1987**, *28*, 2607.
10. Caminade, A. M.; El Khatib, F.; Baceiredo, A.; Koenig, M. *PS* **1987**, *29*, 365.
11. Calvo, K. C. *J. Labelled Compd. Radiopharm.* **1989**, *27*, 395.
12. Stringer, O. D.; Charig, A. *J. Labelled Compd. Radiopharm.* **1989**, *27*, 647.
13. *The Chemistry of Sulphones and Sulphoxides*; Patai, S.; Rappoport, Z.; Stirling, C., Eds.; Wiley: New York, 1988.
14. Begtrup, M.; Jonsson, G. *ACS(B)* **1987**, *B41*, 724.
15. Goerlitzer, K.; Heinrici, C. *AP* **1988**, *321*, 477.
16. (a) Bansal, R. K.; Mahnot, R.; Sharma, D. C.; Karaghiosoff, K. *S* **1992**, 267. (b) Bansal, R. K.; Kabra, V.; Gupta, N.; Karaghiosoff, K. *IJC(B)* **1992**, *31B*, 254.
17. Lukanov, L. K.; Venkov, A. P. *S* **1992**, 263.
18. (a) Bartoli, G.; Bosco, M.; Dalpozzo, R.; Marcantoni, E. *TL* **1990**, *31*, 6935. (b) Baccolini, G.; Evangelisti, D.; Rizzoli, C.; Sgarabotto, P. *JCS(P1)* **1992**, 1729.
19. Brever, E.; Aurich, H. G.; Nielson, A. *Nitrones, Nitronates, and Nitroxides*; Wiley: New York: 1989.
20. Koll, P.; Kopf, J.; Wess, D.; Brandenburg, H. *LA* **1988**, 685.
21. Klumpp, E.; Eifert, G.; Boros, P.; Szulágyi, J.; Tamás, J.; Czira, G. *CB* **1989**, *122*, 2021.
22. Voskvil, W.; Arens, J. F. *OSC* **1973**, *5*, 211.
23. Denmark, S. E.; Dorow, R. L. *JOC* **1989**, *54*, 5.
24. Kuhn, N.; Kuhn, A.; Schulten, M. *JOM* **1991**, *402*, C41.
25. Douglas, T.; Theopold, K. H. *AG(E)* **1989**, *28*, 1367.
26. (a) Lochschmidt, S.; Mathey, F.; Schmidpeter, A. *TL* **1986**, *27*, 2635. (b) Yao, E.; Szewczyk, J.; Quin, L. D. *S* **1987**, 265.
27. Pellon, P.; Hamelin, J. *TL* **1986**, *27*, 5611.

28. (a) Cunningham, A. F., Jr.; Kündig, E. P. *JOC* **1988**, *53*, 1823. (b) Meyer, T. G.; Jones, P. G.; Schmutzler, R. *ZN(B)* **1992**, *47*, 517. (c) *Comprehensive Coordination Chemistry*; Wilkinson, G., Ed.; Pergamon: Oxford, 1987.

Kenneth P. Moder
Eli Lilly and Company, Lafayette, IN, USA

Phosphorus(V) Chloride[1]

[10026-13-8] Cl_5P (MW 208.22)

(chlorination of alcohols,[2] carboxylic acids,[3] amides,[4] aldehydes, ketones, and enols; Bischler–Napieralski synthesis of 3,4-dihydroisoquinolines;[5] Beckmann rearrangement of oximes[6])

Alternate Name: phosphorus pentachloride.
Physical Data: mp 179–181 °C (sublimes); *d* 1.6 g cm^{-3}.
Solubility: sol CS_2, CCl_4. Typical solvents for reactions which use PCl_5 are CCl_4, $CHCl_3$, CH_2Cl_2, hexane, decane, benzene, toluene, and diethyl ether.
Form Supplied in: white to pale yellow solid; widely available. In the solid state, PCl_5 exists in the ionic form, $[PCl_4]^+ [PCl_6]^-$.
Purification: by vacuum sublimation.[7]
Handling, Storage, and Precautions: reacts with moisture to liberate hydrochloric acid and phosphoric acids. It is extremely corrosive to the skin, eyes, and mucous membranes. It should be stored in a dry area in containers impervious to moisture and resistant to corrosion. Laboratory quantities may be stored in polyethylene bags placed within glass containers. This reagent should be used in a fume hood.

Chlorides from Alcohols and Phenols. Alcohols are converted to alkyl chlorides with the formation of 1 equiv of HCl and $POCl_3$ by the action of PCl_5.[2] The stability of the substrate and product to acidic reaction conditions and the ease by which the product can be separated from the reaction byproducts greatly influences the suitability of PCl_5 for this functional group transformation. Depending upon the structure of the alcohol, dehydration to form an alkene[8] or rearrangement[9] may be significant competing processes. The conversion of the 3-hydroxymethyl cephalosporin to the 3-chloromethyl derivative[10] shown in eq 1 illustrates the use of PCl_5 for the chlorination of a primary alcohol with a substrate of considerable functional complexity and lability.

PCl_5, CH_2Cl_2, py, −30 °C, 1.5 h, 98% (1)

R = $CO(CH_2)_3CH(NHCOPh)CO_2CHPh_2$

In select cases, secondary alcohols undergo chlorination with inversion of configuration upon reaction with PCl_5 (eq 2).[11]

PCl_5 (2)

Tertiary alcohols are converted to tertiary chlorides under mild conditions with retention of configuration, presumably through an S_N1, tight ion-pair mechanism (eq 3).[12] The reactions are complete within minutes of mixing the tertiary alcohol with PCl_5 in $CHCl_3$ or ether at 0 °C in the presence of calcium carbonate. Chlorination of 1,1,1-trifluoro-2-(trifluoromethyl)-3-butyn-2-ol with PCl_5 leads to 1-chloro-3,3-bis(trifluoromethyl)allene (eq 4).[13] This tertiary alcohol is unreactive towards $SOCl_2$, $POCl_3$, PCl_3, concd. HCl, and $POCl_3$/pyridine. Phenols substituted with strong electron-withdrawing groups can be converted to aryl chlorides with PCl_5.[14]

PCl_5, $CaCO_3$, Et_2O, 0 °C, 3 min (3)

PCl_5, 105 °C, 7 h (4)

A polymer-supported form of PCl_5 has been developed by slurrying PCl_5 with Amberlite IRA 93 resin in CCl_4.[15] The resulting resin is effective for the chlorination of primary and secondary alcohols in yields of 70–98%. Some elimination products are observed with secondary alcohols. The reaction may be conducted in hydrocarbon, ether, or halogenated hydrocarbon solvents. The phosphorus byproducts are retained on the resin and separated from the product by filtration. Regeneration of the resin is possible by washing with aqueous acid and base. Acid-sensitive substrates are susceptible to degradation.

Carboxylic Acid Chlorides from Acids. PCl_5 may be employed for the conversion of carboxylic acids to their acid chlorides.[16] Typically, the acid is combined with PCl_5 either neat[17] or in solvents such as ether[18] or benzene.[19] The use of PCl_5 is most convenient when the product can be isolated by direct crystallization[20] or distillation[21] from the crude reaction mixture. In some cases, aqueous workups may be employed to remove acidic byproducts without significant hydrolysis.[22] Because of the difficulties associated with separating the acid chloride from the byproduct $POCl_3$, the use of PCl_5 for this functional group transformation has largely been supplanted by $SOCl_2$, $COCl_2$, or $(COCl)_2$,[23] particularly in conjunction with catalytic amounts of DMF.[24] The polymer-supported form of PCl_5,[15] described above for the chlorination of alcohols, also converts carboxylic acids to the corresponding acid chlorides in 48–91% yields.

Nitriles from Amides. Primary amides are dehydrated to afford nitriles upon heating with PCl_5.[4a–e] Due to the severity of the reaction conditions, substrates of limited functional complexity are tolerated. The reactions are generally conducted either neat[25] or with $POCl_3$[26] as a solvent. This method is useful for products which can be isolated by distillation from the crude reaction mixture. Secondary amides, most notably *N*-alkylbenzamides, when subjected to PCl_5 may undergo the von Braun degradation to produce a benzonitrile and an alkyl chloride.[27]

Chloromethyleneiminium Chlorides and Imidoyl Chlorides from Amides, Oximes, Hydroxamic Acids, and Ureas. Tertiary amides react with PCl_5 to afford chloromethyleneiminium chlorides,[4d,g–i] whereas secondary amides afford imidoyl chlorides.[4d–f,j] In each case, $POCl_3$ and HCl are generated as reaction byproducts. The amide chlorides and imidoyl chlorides are frequently used as intermediates for subsequent chemical transformation and are not isolated. When desired, isolation has been accomplished by either distillation or crystallization. In choosing PCl_5 as a reagent for this transformation, consideration must be given to the acid lability and reactivity of the substrate. Depending upon the substrate and reaction conditions, the formation of imidoyl chlorides by the action of PCl_5 may be accompanied by further reaction to produce a nitrile and alkyl chloride via the von Braun degradation.[27] α-Chlorination can be a problem if excess PCl_5 is used, and self-condensation can occur at elevated temperatures if an α-CH is present in the imidoyl chloride.[4g] The presence of a strong electron-withdrawing group α to the amide carbonyl can influence the course of the chlorination of an amide with PCl_5, as seen by the formation of the diazadiphosphetidine (eq 5) when *N*-methyltrifluoroacetamide is reacted with PCl_5.[28]

$$F_3CONHMe \xrightarrow[\Delta]{PCl_5} F_3C(Cl)C{=}NMe \;(44\%) + [Cl_3P{-}N(Me)]_2 \;(52\%) \quad (5)$$

A highly practical example of the reaction of PCl_5 with a tertiary amide is the synthesis of ***Dimethylchloromethyleneammonium Chloride***, the Vilsmeier reagent. This salt is obtained in 88% yield by the careful addition of PCl_5 to ***N,N-Dimethylformamide*** while allowing the reaction to exotherm to 100 °C followed by cooling to 0 °C, filtration, and washing.[29] The Vilsmeier reagent is useful for the chlorination of alcohols and acids.

PCl_5 has found considerable application for the removal of amide side chains[30] in the industrial scale manufacture of semisynthetic cephalosporins[31] and penicillins.[32] The intermediate imidoyl chlorides, produced by the action of PCl_5 on a secondary amide, are converted to imino ethers upon the introduction of an alcohol to the reaction mixture. The imino ethers undergo further reaction to afford the deprotected C(7)-amino cephalosporin or C(6)-amino penicillin.[33]

Iminium and imidoyl chlorides have a prominent position as intermediates for the synthesis of heterocyclic compounds. In this regard, PCl_5 has found extensive application. Imidoyl chlorides derived from the treatment of *N*-(2-phenylethyl)carboxamides with PCl_5 can undergo Bischler–Napieralski cyclization[5] to form 3,4-dihydroisoquinolines (eq 6).[34] In addition to PCl_5, this cyclization has been effected with P_2O_5, $POCl_3$, $POCl_3$ in refluxing xylene or decalin, P_2O_5 in pyridine, and polyphosphoric acid. *N,N′*-Disubstituted oxamides (eq 7)[35] and the imines or hydrazones of α-acylamino ketones (eq 8)[36] react with PCl_5 to form imidoyl chlorides which undergo subsequent intramolecular cyclization to produce imidazoles.

(6) [PCl_5, $CHCl_3$]

(7) [PCl_5, $POCl_3$, 75%]

(8) [PCl_5, $CHCl_3$, 81%]

The reaction of oximes with PCl_5 can lead to imidoyl chlorides via Beckmann rearrangement.[6] The reaction is frequently conducted in ether at or below ambient temperature. Aromatic, hydrocarbon, and halogenated hydrocarbon solvents are also used. The oxime of 4-phenyl-3-methyl-3-buten-2-one undergoes Beckmann rearrangement within 5 to 15 min upon treatment with PCl_5 in decalin at 0 °C to give an imidoyl chloride.[37] Subsequent treatment with P_2O_5 at reflux affords 1,3-dimethylisoquinoline (eq 9). 2-(Oximino)indan-1-ones afford 1,3-dichloroisoquinolines when subjected to 1 equiv of PCl_5 in $POCl_3$ followed by treatment with anhydrous HCl and then a second equiv of PCl_5 (eq 10).[38] Treatment of an oxime which contains an α-alkoxy, α-alkylthio, or α-alkylamino group with PCl_5 can lead to 'second-order' Beckmann rearrangement,[39] as exemplified in eq 11.[40]

(9) [1. PCl_5, decalin; 2. P_2O_5; 65%]

(10) [1. PCl_5, $POCl_3$; 2. HCl; 3. PCl_5; 61%]

(11) [1. PCl_5, Et_2O, 0 °C; 2. H_2O; 90%] → $NC(CH_2)_6CHO$

The reaction of alkylbenzohydroxamates with PCl_5 in ether offers a general preparative method for the synthesis of *O*-alkylbenzohydroximoyl chlorides.[41] Ketenimines[42] can be derived by the treatment of secondary amides bearing a single α-CH with PCl_5 followed by dehydrohalogenation with triethylamine.[43] Alkoxycarbonylimidoyl chlorides are obtained by the reaction of the esters of *N*-acylcarbamic acids with PCl_5.[44] *N,N′*-Dialkylureas in which the alkyl groups are primary react with PCl_5 to form 1,3-diaza-2-P^V-phosphetidinones, whereas chloroformamidinium chlorides are produced when the alkyl groups are secondary.[45] *N*-Alkyl-*N*-nitrosamides undergo rearrangement to the corresponding *N*-alkyloxamides via an intermediate *N*-nitrosochloroiminium chloride upon reaction with PCl_5.[46]

***gem*-Dichlorides from Ketones, Aldehydes, and Esters.** Ketones and aldehydes may be converted into *gem*-dichlorides by the action of PCl_5.[47] Formate esters[48] and esters with strong electron-withdrawing groups α to the carbonyl[47] react with PCl_5 to give the corresponding *gem*-dichlorides. Conversions have been effected either neat,[49] or with PCl_3,[50] $POCl_3$,[51] or halogenated hydrocarbons[52] as solvents. The reaction rate and product distribution are influenced by the solvent employed.[52a,b] Polar solvents, such as nitromethane and acetonitrile, greatly retard the reaction. The presence of an α-CH group can lead to the formation of vinyl chlorides as frequent byproducts.[52a,53] Perhaloacetones are resistant to chlorination with PCl_5 under the typical conditions employed for this conversion; however, the *gem*-dichlorides can be obtained in modest yields upon heating at 275–300 °C in an autoclave.[54]

Vinyl Chlorides from Enols. PCl_5 reacts with the enolic form of enolizable cyclic[55] and acyclic[56] carbonyls to form vinyl chlorides. A side-reaction encountered with esters has been formation of the acid chloride, in which case reesterification during the reaction workup affords improved yields (eq 12).[55] Hydroxypyridines,[57] hydroxyquinolines,[58] and hydroxyisoquinolines[38] are converted to chloropyridines, chloroquinolines, and chloroisoquinolines by the action of PCl_5.

O, CO$_2$Me (cyclopentanone) — 1. PCl_5, hexane 0–60 °C; 2. MeOH; 80% → Cl, CO$_2$Me (cyclopentene) (12)

Other Applications. Alkanes,[59] arylalkanes,[59a] heteroaromatics,[60] and alkenes[59,61] have been chlorinated with PCl_5 to afford chloro alkanes, chloro aromatics and *vic*-dichlorides. The use of PCl_5 as a catalyst for the chlorination of alkanes, cycloalkanes, and arylalkanes has been reported.[62] Secondary nitriles are α-chlorinated when subjected to PCl_5 either neat[63] or in an inert solvent such as CCl_4 or $CHCl_3$.[64] Anhydrides are converted to diacid chlorides by the action of PCl_5.[65] Sulfonyl and sulfamoyl chlorides may be prepared by heating the corresponding acid[66] or acid salt[67] either neat or in an inert solvent with PCl_5. PCl_5 has been used to deoxygenate *N*-oxides,[68] hydroxylamines,[69] and sulfoxides.[70] Hydroxylamines in which the α-carbon is tertiary can undergo the Stieglitz rearrangement to afford imines when treated with PCl_5.[71] Highly fluorinated carbinols bearing an α-CH, which prove resistant to the action of $SOCl_2$, SO_2Cl_2, HCl, and $ZnCl_2$, are converted to alkenes by the action of PCl_5.[72] Cyclic acetals of α-keto acids, when reacted with PCl_5 in CH_2Cl_2, afford esters of halohydrins.[73] Glyoxal bis-acetals are converted to 1,2-dichloro-1,2-dialkoxyethanes by the action of PCl_5.[74] This has been developed into a procedure for the synthesis of alkynic diethers.[75] Imines have been converted into amides upon heating with PCl_5, either neat or in a solvent such as xylene, followed by treatment with water.[76] A method has been developed for the synthesis of alkyl chlorides by the reaction of PCl_5 with the alkyl salicylates.[77] The yields are higher in several cases for this procedure when compared to other methods for the chlorination of alcohols; however, synthesis of the alkyl salicylate is required.

Related Reagents. Dimethylchloromethyleneammonium Chloride; Oxalyl Chloride; Phosphorus(III) Chloride; Phosphorus(V) Oxide; Phosphorus Oxychloride; Thionyl Chloride; Triphenylphosphine–Carbon Tetrachloride; Triphenylphosphine Dichloride.

---

1. (a) *The Chemistry of the Hydroxyl Group*; Patai, S., Ed.; Interscience: New York, 1971; Part 1. (b) *The Chemistry of Acyl Halides*; Patai, S., Ed.; Interscience: New York, 1972. (c) *The Chemistry of the Cyano Group*; Rappoport, Z., Ed.; Interscience: New York, 1970. (d) *The Chemistry of Amides*; Zabicky, Z., Ed.; Interscience: New York, 1970. (e) *The Chemistry of the Carbon–Nitrogen Double Bond*; Patai, S., Ed.; Interscience: New York, 1970.
2. (a) Brown, G. W. In *The Chemistry of the Hydroxyl Group*; Patai, S. Ed.; Interscience: New York, 1971; Part 1, pp 603–612. (b) *COS* **1991**, *6*, 204.
3. (a) *Comprehensive Organic Chemistry*; Barton, D. H. R.; Ollis, W. D., Eds.; Pergamon: Oxford, 1979; Vol. 2, pp 633–634. (b) *COS* **1991**, *6*, 301. (c) Ansell, M. F. In *The Chemistry of Acyl Halides*; Patai, S., Ed.; Interscience: New York, 1972; pp 40–44. (d) Buehler, C. A.; Pearson, D. E. *Survey of Organic Syntheses*; Wiley: New York, 1970; pp 859–873.
4. (a) Mowry, D. T. *CRV* **1948**, *42*, 189. (b) Friedrich, K.; Wallenfels, K. In *The Chemistry of the Cyano Group*; Rappoport, Z., Ed.; Interscience: New York, 1970; pp 96–103. (c) Bieron, J. F.; Dinan, F. J. In *The Chemistry of Amides*; Zabicky, Z., Ed.; Interscience: New York, 1970; pp 274–279. (d) Challis, B. C.; Challis, J. A. In *The Chemistry of Amides*; Zabicky, Z., Ed.; Interscience: New York, 1970; pp 801–814. (e) Sustman, R.; Korth, H.-G. *MOC* **1985**, *E5*, 628. (f) Bonnett, R. In *The Chemistry of the Carbon–Nitrogen Double Bond*; Patai, S., Ed.; Interscience: New York, 1970; pp 601–606. (g) *COS* **1991**, *6*, 495. (h) Eilingsfeld, H.; Seefelder, M.; Weidinger, H. *AG* **1960**, *72*, 836. (i) Kantlehner, W. *Adv. Org. Chem.* **1979**, *9*, 65. (j) *COS* **1991**, *6*, 524.
5. (a) Gensler, W. J. In *Heterocyclic Compounds*; Elderfield, R. C., Ed.; Wiley: New York, 1952; p 344. (b) *Comprehensive Organic Chemistry*; Barton, D. H. R.; Ollis, W. D., Eds.; Pergamon: Oxford, 1979; Vol. 4, p 207. (c) Whaley, W. M.; Govindachari, T. R. *OR* **1951**, *6*, 74. (d) Fodor, G.; Nagubandi, S. *T* **1980**, *36*, 1279.
6. (a) Donaruma, L. G.; Heldt, W. Z. *OR* **1960**, *11*, 1. (b) Beckwith, A. L. J. In *The Chemistry of Amides*; Zabicky, J., Ed.; Interscience: New York, 1970; pp 131–137. (c) March, J. *Advanced Organic Chemistry: Reactions, Mechanisms, and Structures*, 3rd ed.; Wiley: New York, 1985; pp 987–989.
7. Suter, R. W.; Knachel, H. C.; Petro, V. P.; Howatson, J. H.; Shore, S. G. *JACS* **1973**, *95*, 1474.
8. Shoppee, C. W.; Lack, R. E.; Sharma, S. C.; Smith, L. R. *JCS(C)* **1967**, 1155.
9. (a) Carman, R. M.; Shaw, I. M. *AJC* **1980**, *33*, 1631. (b) Fry, J. L.; West, J. W. *JOC* **1981**, *46*, 2177.
10. Yamanaka, H.; Chiba, T.; Kawabata, K.; Takasugi, H.; Masugi, T.; Takaya, T. *J. Antibiot.* **1985**, *38*, 1738.
11. Shoppee, C. W.; Coll, J. C. *JCS(C)* **1970**, 1124.
12. Carman, R. M.; Shaw, I. M. *AJC* **1976**, *29*, 133.
13. (a) Abele, H.; Haas, A.; Lieb, M. *CB* **1986**, *119*, 3502. (b) Simmons, H. E.; Wiley, D. W. *JACS* **1960**, *82*, 2288.
14. *Comprehensive Organic Chemistry*; Barton, D. H. R.; Ollis, W. D., Eds.; Pergamon: Oxford, 1979; Vol. 1, p 736.
15. Cainelli, G.; Contento, M.; Manescalchi, F.; Plessi, L.; Panunzio, M. *S* **1983**, 306.
16. (a) *Comprehensive Organic Chemistry*; Barton, D. H. R.; Ollis, W. D., Eds.; Pergamon: Oxford, 1979; Vol. 2, pp 633–634. (b) *COS* **1991**, *6*, 301. (c) Ansell, M. F. In *The Chemistry of Acyl Halides*; Patai, S., Ed.; Interscience: New York, 1972; pp 40–44. (d) Buehler, C. A.; Pearson, D. E. *Survey of Organic Syntheses*; Wiley: New York, 1970; pp 859–873.

17. Adams, R.; Jenkins, R. L. *OSC* **1941**, *1*, 394.
18. Ireland, R. E.; Chaykovsky, M. *OS* **1961**, *41*, 5.
19. Sherk K. W.; Augur, M. V.; Soffer, M. D. *JACS* **1945**, *67*, 2239.
20. Braun, C. E.; Cook, C. D. *OS* **1961**, *41*, 79.
21. Feuer, H.; Pier, S. M. *OSC* **1963**, *4*, 554.
22. Ansell, M. F. In *The Chemistry of Acyl Halides*; Patai, S., Ed.; Interscience: New York, 1972; p 41.
23. Buehler, C. A.; Pearson, D. E. *Survey of Organic Syntheses*; Wiley: New York, 1970; pp 860–861.
24. *COS* **1991**, *6*, 302.
25. Corson, B. B.; Scott, R. W.; Vose, C. E. *OSC* **1943**, *2*, 379.
26. Taylor, E. C., Jr.; Crovetti, A. J. *OSC* **1963**, *4*, 166.
27. (a) Bieron, J. F.; Dinan, F. J. In *The Chemistry of Amides*; Zabicky, Z., Ed.; Interscience: New York, 1970; p 277. (b) Challis, B. C.; Challis, J. A. In *The Chemistry of Amides*; Zabicky, Z., Ed.; Interscience: New York, 1970; p 809. (c) Leonard, N. J.; Nommensen, E. W. *JACS* **1949**, *71*, 2808. (d) Vaughan, W. R.; Carlson, R. D. *JACS* **1962**, *84*, 769.
28. Norris, W. P.; Jonassen, H. B. *JOC* **1962**, *27*, 1449.
29. Hepburn, D. R.; Hudson, H. R. *CI(L)* **1974**, 664.
30. Huber, F. M.; Chauvette, R. R.; Jackson, B. G. In *Cephalosporins and Penicillins: Chemistry and Biology*; Flynn, E. H., Ed.; Academic: New York, 1972; pp 39–70.
31. Fechtig, B.; Peter, H.; Bickel, H.; Vischer, E. *HCA* **1968**, *51*, 1108.
32. Weissenburger, H. W. O.; van der Hoeven, M. G. *RTC* **1970**, *89*, 1081.
33. Hatfield, L. D.; Lunn, W. H. W.; Jackson, B. G.; Peters, L. R.; Blaszczak, L. C.; Fisher, J. W.; Gardner, J. P.; Dunigan, J. M. In *Recent Advances in the Chemistry of β-Lactam Antibiotics*; Gregory, G. I., Ed.; Royal Society of Chemistry: London, 1980; pp 109–124.
34. (a) Lindenmann, A. *HCA* **1949**, *32*, 69. (b) Pfister, J. R. *H* **1986**, *24*, 2099. (c) Badia, D.; Carrillo, L.; Dominguez, E.; Cameno, A. G.; deMarigorta, E. M.; Vicento, T. *JHC* **1990**, *27*, 1287. (d) Fodor, G.; Gal, J.; Phillips, B. A. *AG(E)* **1972**, *11*, 919.
35. Godefroi, E. F.; van der Eycken, C. A. M.; Janssen, P. A. J. *JOC* **1967**, *32*, 1259.
36. Engel, N.; Steglich, W. *LA* **1978**, 1916.
37. Zielinski, W. *S* **1980**, 70.
38. Simchen, G.; Krämer, W. *CB* **1969**, *102*, 3666.
39. *Comprehensive Organic Chemistry*; Barton, D. H. R.; Ollis, W. D., Eds.; Pergamon: Oxford, 1979; Vol. 2, p 533.
40. Ohno, M.; Naruse, N.; Torimitsu, S.; Teresawa, I. *JACS* **1966**, *88*, 3168.
41. (a) Johnson, J. E.; Nalley, E. A.; Weidig, C. *JACS* **1973**, *95*, 2051. (b) Johnson, J. E.; Springfield, J. R.; Hwang, J. S.; Hayes, L. J.; Cunningham, W. C.; McClaugherty, D. L. *JOC* **1971**, *36*, 284. (c) Taylor, E. C.; Kienzle, F. *JOC* **1971**, *36*, 233.
42. Krow, G. R. *AG(E)* **1971**, *10*, 435.
43. Stevens, C. L.; French, J. C. *JACS* **1954**, *76*, 4398.
44. Neidlein, R.; Bottler, R. *TL* **1966**, 1069.
45. Ulrich, H.; Sayigh, A. A. R. *AG(E)* **1966**, *5*, 704.
46. Murakami, M.; Akagi, K.; Takahashi, K. *JACS* **1961**, *83*, 2002.
47. March, J. *Advanced Organic Chemistry: Reactions, Mechanisms, and Structures*, 3rd ed.; Wiley: New York, 1985; pp 807–809.
48. (a) Gross, H.; Rieche, A.; Höft, E.; Beyer, E. *OSC* **1973**, *5*, 365. (b) Rieche, A.; Gross, H.; Höft, E. *CB* **1960**, *93*, 88.
49. Buck, J. S.; Zimmerman, F. J. *OSC* **1943**, *2*, 549.
50. (a) Rieke, R. D.; Bales, S. E. *OSC* **1988**, *6*, 845. (b) Wiberg, K. B.; Lowry, B. R.; Colby, T. H. *JACS* **1961**, *83*, 3998.
51. Looker, J. J. *JOC* **1966**, *31*, 3599.
52. (a) Newman, M. S.; Fraenkel, G.; Kirn, W. N. *JOC* **1963**, *28*, 1851. (b) Newman, M. S.; Kaugars, G. *JOC* **1966**, *31*, 1379. (c) Schoberth, W.; Hanack, M. *S* **1972**, 703.
53. (a) Paukstelis, J. V.; Macharia, B. W. *JOC* **1973**, *38*, 646. (b) Bryson, T. A.; Dolak, T. M. *OSC* **1988**, *6*, 505.
54. Farah, B. S.; Gilbert, E. E. *JOC* **1965**, *30*, 1241.
55. Harding, K. E.; Tseng, C. *JOC* **1978**, *43*, 3974.
56. Friedrich, K.; Thieme, H. K. *S* **1973**, 111.
57. Berrie, A. H.; Newbold, G. T.; Spring, F. S. *JCS* **1952**, 2042.
58. Beisler, J. A. *JMC* **1971**, *14*, 1116.
59. (a) Wyman, D. P.; Wang, J. Y. C.; Freeman, W. R. *JOC* **1963**, *28*, 3173. (b) Fell, B.; Kung, L.-H. *CB* **1965**, *98*, 2871.
60. (a) Chambers, R. D.; MacBride, J. A. H.; Musgrave, W. K. R. *CI(L)* **1966**, 1721. (b) *Comprehensive Organic Chemistry*; Barton, D. H. R.; Ollis, W. D., Eds.; Pergamon: Oxford, 1979; Vol. 1, p 523.
61. (a) Speigler, L.; Tinker, J. M. *JACS* **1939**, *61*, 940. (b) Uemura, S.; Okazaki, H.; Onoe, A.; Okano, M. *BCJ* **1978**, *51*, 3568.
62. (a) Messina, G.; Moretti, M. D.; Ficcadenti, P.; Cancellieri, G. *JOC* **1979**, *44*, 2270. (b) Olah, G. A.; Schilling, P.; Renner, R.; Kerekes, I. *JOC* **1974**, *39*, 3472.
63. Stevens, C. L.; Coffield, T. H. *JACS* **1951**, *73*, 103.
64. (a) Belluš, D.; von Bredow, K.; Sauter, H.; Weis, C. D. *HCA* **1973**, *56*, 3004. (b) Belluš, D.; Sauter, H.; Weis, C. D. *OSC* **1988**, *6*, 427.
65. Ott, E. *OSC* **1943**, *2*, 528.
66. (a) Bartlett, P. D.; Knox, L. H. *OSC* **1973**, *5*, 196. (b) Kloek, J. A.; Leschinsky, K. L. *JOC* **1976**, *41*, 4028.
67. (a) Goldberg, A. A. *JCS* **1945**, 464. (b) Adams, R.; Marvel, C. S. *OSC* **1941**, *1*, 84. (c) Loev, B.; Kormendy, M. *JOC* **1962**, *27*, 1703. (d) Kloek, J. A.; Leschinsky, K. L. *JOC* **1976**, *41*, 4028.
68. Duty, R. C.; Lyons, G. *JOC* **1972**, *37*, 4119.
69. Haszeldine, R. N.; Mattinson, B. J. H. *JCS* **1957**, 1741.
70. Wakisaka, M.; Hatanaka, M.; Nitta, H.; Hatamura, M.; Ishimaru, T. *S* **1980**, *67*.
71. *Comprehensive Organic Chemistry*; Barton, D. H. R.; Ollis, W. D., Eds.; Pergamon: Oxford, 1979; Vol. 2, p 200.
72. Kaufman, M. H.; Braun, J. D. *JOC* **1966**, *31*, 3090.
73. (a) Newman, M. S.; Chen, C. H. *JOC* **1973**, *38*, 1173. (b) Newman, M. S.; Chen, C. H. *JACS* **1972**, *94*, 2149.
74. Fiesselmann, H.; Hörndler, F. *CB* **1954**, *87*, 911.
75. (a) Pericas, M. A.; Serratosa, F. *TL* **1977**, 4433. (b) Pericas, M. A.; Serratosa, F. *TL* **1978**, 2603. (c) Bou, A.; Pericas, M. A.; Serratosa, F. *T* **1981**, *37*, 1441. (d) Pericas, M. A.; Serratosa, F.; Valenti, E. *T* **1987**, *43*, 2311.
76. Singh, H.; Aggarwal, S. K.; Malhotra, N. *S* **1983**, *10*, 791.
77. Pinkus, A. G.; Lin, W. H. *S* **1974**, 279.

John E. Burks, Jr.
Eli Lilly and Company, Lafayette, IN, USA

Phosphorus(III) Iodide

[13455-01-1] I_3P (MW 411.67)

(conversion of alcohols to iodides; deoxygenation of epoxides to alkenes; reduction of sulfoxides, selenoxides, selenones, ozonides, and α-halo ketones; elimination of β-hydroxy sulfides, dithioacetals, and orthothioesters or their selenium counterparts to form alkenes; conversidion of aldoximes and nitro compounds to nitriles; cleavage of dimethyl acetals; water-soluble phosphorus byproducts are usually formed)

Alternate Name: phosphorus triiodide.
Physical Data: mp 61 °C.

Solubility: very sol CS_2; reacts with protic solvents.
Form Supplied in: red solid; widely available commercially.
Handling, Storage, and Precautions: commercial samples are generally suitable for use without purification; store under an inert atmosphere.

Conversion of Alcohols to Iodides. PI_3 is a classic reagent for the conversion of iodides to alcohols. For example, cetyl alcohol is converted to 1-iodohexadecane in 85% yield by treatment with PI_3 (neat).[1] More recently, Krief and co-workers[2] found that a cyclopropyl carbinol (eq 1) is cleanly converted to the iodide with ring opening upon treatment with PI_3 (Et_3N, CH_2Cl_2, 0 °C, 0.5 h, 73%). ***Diphosphorus Tetraiodide***, a commercially available phosphorus halide which is in equilibrium with PI_3 in solution, is a more commonly employed reagent for the conversion of alcohols to iodides and effects many of the same transformations as PI_3.[3] See also ***Triphenylphosphine–Iodine*** for the conversion of alcohols to iodides.

TMS, OH, $C_{10}H_{21}$ — 1 equiv PI_3, CH_2Cl_2, Et_3N, 73% → I–CH₂CH₂–C(TMS)=CH–$C_{10}H_{21}$ (1)

Deoxygenations and Reductions. PI_3 has found use as a deoxygenating and reducing agent. Epoxides are stereospecifically deoxygenated to the corresponding alkenes in good yield and with retention of configuration. For example, the *cis*- and *trans*-epoxides shown in eq 2 are converted (CS_2, 20 °C, 6–8 h) to *cis*- and *trans*-9-octadecene in 83 and 90% yields, respectively.[4] Diphosphorus tetraiodide (P_2I_4) and ***Iodotrimethylsilane*** (TMSI) give similar results. A number of other methods for epoxide deoxygenation have been developed.[5]

R^1, O, R^3, R^2 — PI_3, CS_2, 20 °C → R^1, R^3, R^2 alkene (2)

(a) R^1, R^3 = $(CH_2)_7Me$; R^2 = H; 83%
(b) R^1 = H; R^2, R^3 = $(CH_2)_7Me$; 90%

A variety of methods for the reduction of sulfoxides to sulfides are available.[6] PI_3 rapidly reduces aryl alkyl and dialkyl sulfoxides and selenoxides, usually at −78 °C (eq 3).[7] For example, treatment of ethyl phenyl sulfoxide with 1 equiv of PI_3 (CH_2Cl_2, −78 °C, 15 min) affords ethyl phenyl sulfide in 91% yield. Others have successfully employed this procedure.[8,9] Dialkyl sulfoxides generally react in somewhat lower yield. A phenyl vinyl sulfoxide (eq 3, entry c) requires ambient temperatures to react. Selenoxides behave similarly, and P_2I_4 can usually be used in place of PI_3.[10] Treatment of decyl phenyl selenone with PI_3 (CH_2Cl_2, 0 °C, 30 min) affords a mixture of the reduced product, decyl phenyl selenide (69%), and the substitution product, *n*-decyl iodide.[11]

PI_3 can be used for the efficient reduction of an ozonide, as illustrated by the formation of the cyclopropanecarbaldehyde in eq 4.[12,13] An advantage over the traditional ***Triphenylphosphine*** reduction is the formation of water-soluble phosphorus byproducts from PI_3. It should be noted, however, that PI_3 is reported to react with certain aldehydes.[14]

R^1–X(=O)–R^2 — 1 equiv PI_3, CH_2Cl_2, −78 °C → R^1–X–R^2 (3)

(a) X = S; R^1 = Ph; R^2 = Et; 91%
(b) X = S; R^1, R^2 = Pr; 71%
(c) X = S; R^1 = Ph; R^2 = $CH{=}CHC_5H_{11}$; 75%
(d) X = Se; R^1 = Ph; R^2 = $C_{10}H_{21}$; 93%
(e) X = Se; R^1 = Me; R^2 = $CH(Me)C_6H_{13}$; 87%

$C_{10}H_{21}$, C_6H_{13} — 1. O_3, CH_2Cl_2, −78 °C; 2. PI_3; 87% → $C_{10}H_{21}$, H, =O (4)

PI_3 is reported to dehalogenate α-bromo and α-iodo ketones.[15] For example, as shown in eq 5, 1-iodo-2-dodecanone is reduced to 2-dodecanone by PI_3 in CH_2Cl_2 (25 °C, 1 h, 89%). 1-Bromo-2-octanone affords 2-octanone in 75% yield (25 °C, 4.5 h). This PI_3 reduction can be coupled with the regioselective opening of terminal epoxides by TMSI and alcohol oxidation for a three-step transformation of a terminal epoxide to a methyl ketone.[15,16] P_2I_4,[15] triphenylphosphine,[17] ***1,3-Dimethyl-2-phenylbenzimidazoline***,[18] $PhSiH_3/Mo^0$,[19] ***Samarium(II) Iodide***,[20] ***Zinc–Acetic Acid***,[21] ***Iron–Graphite***,[22] ***Tri-n-butylstannane***,[23] ***Sodium O,O-Diethyl Phosphorotelluroate***,[24] lithium 2-thiophenetellurolates,[25] and a wide variety of other reagents[26,27] can also be used for the reduction of α-halo carbonyl compounds.

R^1C(O)CH(R^2)X — 1.1 equiv PI_3, CH_2Cl_2, 25 °C → R^1C(O)CH₂R^2 (5)

R^1 = $C_{10}H_{21}$, R^2 = H, X = I; 89%
R^1 = C_6H_{13}, R^2 = H, X = Br; 75%
R^1, R^2 = C_8H_{17}, X = I; 97%

Elimination Reactions. PI_3 or P_2I_4 effects the stereospecific *anti*-elimination of β-hydroxy sulfides to alkenes.[28] For example (eq 6), treatment of the *anti*-β-hydroxy sulfide with PI_3 (CH_2Cl_2–Et_3N, 50 °C, 1.5 h) affords (*E*)-9-octadecene (93% yield). Treatment of the *syn*-isomer affords (*Z*)-9-octadecene (93% yield, *Z*:*E* = 94:6). Trisubstituted, but apparently not tetrasubstituted, alkenes can also be formed from the appropriately substituted β-hydroxy sulfides using these reagents. Similarly, β-hydroxy selenides undergo elimination to the corresponding alkenes. With these latter substrates, certain tetrasubstituted alkenes are accessible.[29] ***Methanesulfonyl Chloride***/Et_3N[30] and *N*-ethyl-2-fluoropyridinium tetrafluoroborate/lithium iodide[31] are alternative reagents.

C_8H_{17}, OH, R^1, R^2, SMe — 1 equiv PI_3, CH_2Cl_2, Et_3N → C_8H_{17}, R^1, R^2 alkene (6)

anti; R^1 = C_8H_{17}, R^2 = H (*E*) isomer, 93%
syn; R^1 = H, R^2 = C_8H_{17} (*Z*) isomer, 90%

Vinyl sulfides and ketene dithioacetals[32] are available via treatment of β-hydroxy thioacetals and β-hydroxy orthothioesters with PI_3 or P_2I_4.[33] For example (eq 7), treatment of the β-hydroxy orthothioester (entry a) with PI_3 (CH_2Cl_2–Et_3N, 0 °C, 0.5 h) affords the thioketene acetal (**1**) in 69% yield. With certain substitution patterns the rearranged product (**2**) competes (e.g. entry b) or is the predominant product (e.g. entry c). The rearrangement is less of a problem with the corresponding β-hydroxy orthoselenoester (i.e. entry c vs. entry d). ***Thionyl Chloride*** is also a useful reagent in some cases.

$R^1R^2C(OH)C(XMe)_3$ —(1 equiv PI_3; CH_2Cl_2, Et_3N)→ $R^1R^2C{=}C(XMe)_2$ (**1**) + $R^1(MeX)C{=}C(XMe)_2$ (**2**) (7)

(a) $R^1 = c\text{-}C_3H_5$, $R^2 = Me$, X = S (**1**):(**2**) = 69: 0
(b) $R^1 = C_5H_{11}$, $R^2 = H$, X = S (**1**):(**2**) = 63:13
(c) $R^1 = Ph$, $R^2 = H$, X = S (**1**):(**2**) = 7:68
(d) $R^1 = Ph$, $R^2 = H$, X = Se (**1**):(**2**) = 86: 0

Vinyl sulfides and ketene thioacetals[32] are also available via treatment of thioacetals and orthothioesters with PI_3 or P_2I_4 (eqs 8 and 9).[34] The corresponding seleno analogs are similarly available. For example (eq 8), treatment of the trimethyl orthoselenoester with PI_3 (CH_2Cl_2–Et_3N, 20 °C, 1 h) affords the ketene selenoacetal in 90% yield. ***Tin(IV) Chloride*** or ***Titanium(IV) Chloride*** in combination with ***Diisopropylamine*** is also effective for the transformation of orthothioesters and orthoselenoesters to ketene acetal derivatives.[35] ***Copper(I) Trifluoromethanesulfonate***,[36] ***Mercury(II) Trifluoroacetate/Lithium Carbonate***,[37] and ***Benzenesulfenyl Chloride***[38] are useful reagents for the conversion of dithioacetals to vinyl sulfides.

$EtC(SeMe)_3$ —(1.1 equiv PI_3; CH_2Cl_2, Et_3N; 90%)→ $MeCH{=}C(SeMe)_2$ (8)

1,1-bis(methylseleno)cyclohexane —(80%)→ 1-(methylseleno)cyclohexene (9)

In the presence of ***Triethylamine***, PI_3[7] or P_2I_4[10] effects the conversion of aldehyde oximes or primary nitro compounds to nitriles. For example, treatment of the aldoxime derived from phenylacetaldehyde with 1 equiv of PI_3 (CH_2Cl_2–Et_3N, 25 °C, 15 min) affords benzyl cyanide in 83% yield (eq 10). Similarly, treatment of 1-nitrodecane with 2 equiv of PI_3 (CH_2Cl_2–Et_3N, 25 °C, 15 min) affords 1-cyanononane (82%) (eq 11). Other reagents for the direct conversion of primary nitro compounds to nitriles include $Sn(SPh)_4/Ph_3P$/DEAD,[39] TMSI,[40] ***Sulfur Dioxide***/Et_3N,[41] and ***Phosphorus(III) Chloride***.[42]

$RCH{=}NOH$ —(1 equiv PI_3; CH_2Cl_2, Et_3N)→ $R{-}C{\equiv}N$ (10)

$R = PhCH_2$, 83%
$R = C_{10}H_{21}$, 85%

RCH_2NO_2 —(2 equiv PI_3; CH_2Cl_2, Et_3N)→ $R{-}C{\equiv}N$ (11)

$R = C_9H_{19}$, 82%

Miscellaneous Transformations. PI_3 has been used for the cleavage of dimethyl acetals under nonaqueous conditions.[43] For example (eq 12), treatment of the dimethyl acetal derived from 2-decanone with PI_3 (CH_2Cl_2, 20 °C, 15 min; aq workup) affords the ketone in 85% yield. Alternative nonaqueous acetal cleavage reagents include P_2I_4,[43] TMSI,[44,45] ***Diiodosilane***,[46] $TiCl_4$/***Lithium Iodide***,[47] ***Bromodimethylborane*** or bromodiphenylborane,[48] and ***Samarium(III) Chloride/Chlorotrimethylsilane***.[49]

$R^1R^2C(OMe)_2$ —(1 equiv PI_3, CH_2Cl_2; 20 °C, 15 min)→ $R^1C(O)R^2$ (12)

(a) $R^1 = C_9H_{19}$, $R^2 = Me$, 85%
(b) $R^1 = C_8H_{17}$, $R^2 = H$, 57%

PI_3 is a useful ***Hydrogen Iodide*** precursor in the silica- and alumina-mediated additions of HI to alkenes.[50,51] PI_3 has also found use as a source of electrophilic phosphorus in the synthesis of novel organophosphorus compounds (e.g. $(CF_3)_3P$[52]) and novel phosphorus-containing organometallic compounds.[53]

Related Reagents. Diphosphorus Tetraiodide; Iodotrimethylsilane; Triphenylhosphine–Iodine.

1. Hartman, W. W.; Byers, J. R.; Dickey, J. B. *OS* **1943**, *2*, 322.
2. Halazy, S.; Dumont, W.; Krief, A. *TL* **1981**, *22*, 4737.
3. Lauwers, M.; Regnier, B.; Van Eenoo, M.; Denis, J. N.; Krief, A. *TL* **1979**, 1801.
4. Denis, J. N.; Magnane, R.; Eenoo, M. v.; Krief, A. *NJC* **1979**, *3*, 705.
5. Wong, H. N. C.; Fok, C. C. M.; Wong, T. *H* **1987**, *26*, 1345.
6. Madesclaire, M. *T* **1988**, *44*, 6537.
7. Denis, J. N.; Krief, A. *CC* **1980**, 544.
8. Lown, J. W.; Koganty, R. R.; Joshua, A. V. *JOC* **1982**, *47*, 2027.
9. Binns, M. R.; Haynes, R. K. *JOC* **1981**, *46*, 3790.
10. Denis, J. N.; Krief, A. *TL* **1979**, 3995.
11. Krief, A.; Dumont, W.; Denis, J. N. *CC* **1985**, 571.
12. Denis, J. N.; Krief, A. *CC* **1983**, 229.
13. Halazy, S.; Krief, A. *TL* **1981**, *22*, 4341.
14. Michie, J. K.; Miller, J. A.; Nunn, M. J.; Stewart, D. *JCS(P1)* **1981**, 1744.
15. Denis, J. N.; Krief, A. *TL* **1981**, *22*, 1431.
16. Denis, J. N.; Krief, A. *TL* **1981**, *22*, 1429.
17. Borowitz, I. J.; Grossman, L. I. *TL* **1962**, 471.
18. Chikashita, H.; Ide, H.; Itoh, K. *JOC* **1986**, *51*, 5400.
19. Perez, D.; Greenspoon, N.; Keinan, E. *JOC* **1987**, *52*, 5570.
20. Molander, G. A.; Hahn, G. *JOC* **1986**, *51*, 1135.
21. Zimmerman, H. E.; Mais, A. *JACS* **1959**, *81*, 3644.
22. Savoia, D.; Tagliavini, E.; Trombini, C.; Umani-Ronchi, A. *JOC* **1982**, *47*, 876.
23. Bak, D. A.; Brady, W. T. *JOC* **1979**, *44*, 107.
24. Clive, D. L. J.; Beaulieu, P. L. *JOC* **1982**, *47*, 1124.
25. Engman, L.; Cava, M. P. *JOC* **1982**, *47*, 3946.
26. DeKimpe, N.; Nagy, M.; Boeykens, M.; Schueren, D. v. d. *JOC* **1992**, *57*, 5761.
27. DeKimpe, N.; Verhe, R. In *The Chemistry of α-Haloketones, α-Haloaldehydes and α-Haloimines*; Patai, S.; Rappoprt, Z. Eds.; Wiley: Chichester, 1988.
28. Denis, J. N.; Dumont, W.; Krief, A. *TL* **1979**, 4111.
29. Halazy, S.; Krief, A. *CC* **1979**, 1136.
30. Reich, H. J.; Chow, F. *CC* **1975**, 790.
31. Mukaiyama, T.; Imaoka, M. *CL* **1978**, 413.
32. Kolb, M. *S* **1990**, 171.

33. Denis, J. N.; Desauvage, S.; Hevesi, L.; Krief, A. *TL* **1981**, *22*, 4009.
34. Denis, J. N.; Krief, A. *TL* **1982**, *23*, 3407.
35. Nsunda, K. M.; Hevesi, L. *CC* **1985**, 1000.
36. Cohen, T.; Herman, G.; Falck, J. R.; Mura, A. J. *JOC* **1975**, *40*, 812.
37. Trost, B. M.; Lovoie, A. C. *JACS* **1983**, *105*, 5075.
38. Bartels, B.; Hunter, R.; Simon, C. D.; Tomlinson, G. D. *TL* **1987**, *28*, 2985.
39. Urpi, F.; Vilarrasa, J. *TL* **1990**, *31*, 7497.
40. Olah, G.; Narang, S. C.; Field, L. D.; Fung, A. P. *JOC* **1983**, *48*, 2766.
41. Olah, G. A.; Vankar, Y. D.; Gupta, B. G. B. *S* **1979**, 36.
42. Werli, P. A.; Schaer, B. *JOC* **1977**, *42*, 3956.
43. Denis, J. N.; Krief, A. *AG* **1980**, *92*, 1039.
44. Jung, M. E.; Andrus, W. A.; Orenstein, P. L. *TL* **1977**, 4175.
45. Olah, G. A.; Husain, A.; Singh, B. P.; Mehrotra, A. K. *JOC* **1983**, *48*, 3667.
46. Keinan, E.; Perez, D.; Sahai, M.; Shvily, R. *JOC* **1990**, *55*, 2927.
47. Balme, G.; Gore, J. *JOC* **1983**, *48*, 3336.
48. Guindon, Y.; Yoakim, C.; Morton, H. E. *JOC* **1984**, *49*, 3912.
49. Ukaji, Y.; Koumoto, N.; Fujisawa, T. *CL* **1989**, 1623.
50. Kropp, P. J.; Daus, K. A.; Tubergen, M. W.; Kepler, K. D.; Wilson, V. P.; Craig, S. L.; Baillargeon, M. M.; Breton, G. W. *JACS* **1993**, *115*, 3071.
51. Kropp, P. J.; Daus, K. A.; Crawford, S. D.; Tubergen, M. W.; Kepler, K. D.; Craig, S. L.; Wilson, V. P. *JACS* **1990**, *112*, 7433.
52. Krause, L. J.; Morrison, J. A. *JACS* **1981**, *103*, 2995.
53. Binger, P.; Wettling, T.; Schneider, R.; Zurmuehlen, F.; Bergstraesser, U.; Hoffmann, J.; Maas, G.; Regitz, M. *AG* **1991**, *103*, 208.

James M. Takacs
University of Nebraska-Lincoln, NE, USA

Phosphorus(V) Oxide

(P_2O_5)		
[1314-56-3]	O_5P_2	(MW 141.94)
(P_4O_{10})		
[16752-60-6]	$O_{10}P_4$	(MW 283.88)

(dehydrating agent for acids and amides;[1] cyclization catalyst)

Alternate Names: phosphorus pentoxide; phosphoric anhydride.
Physical Data: mp 569 °C; sublimes at 250 °C under vacuum.
Solubility: sol H_2SO_4, $MeSO_3H$.
Form Supplied in: white powder; commercially available from many sources.
Preparative Method: prepared by combustion of phosphorus in dry air.
Purification: generally used without further purification; can be sublimed under vacuum.
Handling, Storage, and Precautions: corrosive! Reacts violently with water, releasing H_3PO_4 and heat.

Dehydration of Amides to Nitriles. Dehydration of amides with P_2O_5 is one of many methods for the synthesis of nitriles.[2] Distillation of an unsubstituted amide from an excess of P_2O_5 leads to good yields of the nitrile (eq 1).[3,4] In many cases the reaction is carried out in the absence of solvents, although suitably high-boiling solvents can be used. These conditions are not compatible with many common functional groups, and this transformation can be achieved by other reagents,[5] including ***Thionyl Chloride*** and ***Phosphorus Oxychloride***.

P_2O_5, mesitylene, reflux, 70% (1)

Amides can be made by treatment of silyl esters with P_2O_5, followed by treatment with ***Hexamethyldisilazane***.[6] Some *N*-substituted amides afford ketene imines upon treatment with P_2O_5 (eq 2).[7]

P_2O_5, py, reflux, 87% (2)

Formation of Anhydrides. Dehydration of carboxylic acids with P_2O_5 is an effective method for the preparation of carboxylic anhydrides (eq 3). Acids have been converted to anhydrides with silica gel-supported P_2O_5, marketed as 'Sicapent' by Merck.[8] Similar results are obtained by heating a suspension of a carboxylic acid with a powdered mixture of P_2O_5, $CuSO_4$, and Na_2SO_4 in a primary alcohol. This gives the ester in 60–80% yields.[9]

P_2O_5/silica, 69% (3)

Cyclization Catalyst. A 10% by weight (approximately saturated) solution of P_2O_5 in ***Methanesulfonic Acid*** (Eaton's reagent) has been used as an effective replacement for ***Polyphosphoric Acid*** (PPA) as an acid cyclization catalyst. Advantages to this reagent include the ease of stirring $MeSO_3H$ solutions relative to PPA and the greater solubility of many organic compounds in $MeSO_3H$ relative to PPA. Yields are comparable to or better than those obtained with PPA in Beckmann rearrangements (eq 4).[10] The combination of P_2O_5 and methanesulfonic acid is an efficient catalyst for acid/alkene cycloadditions (eq 5).[11] See also ***Phosphorus(V) Oxide–Methanesulfonic Acid***.

P_2O_5, $MeSO_3H$, 100 °C, 79% (4)

P_2O_5, $MeSO_3H$, 75–85% (5)

This mixture is a particularly useful catalyst in the synthesis of 2-substituted indole derivatives via the Fischer indole reaction of unsymmetrical ketones (eq 6).[12] Other catalysts (PPA, H_2SO_4/MeOH, PPSE, Amberlyst 15) give lower regioselectivity. Decomposition of the starting material can be minimized by dilution of the P_2O_5/$MeSO_3H$ mixture with CH_2Cl_2. Treatment of (**1**) with P_2O_5 at high temperature results in cyclization to produce 7-deazahypoxanthines (eq 7).[13]

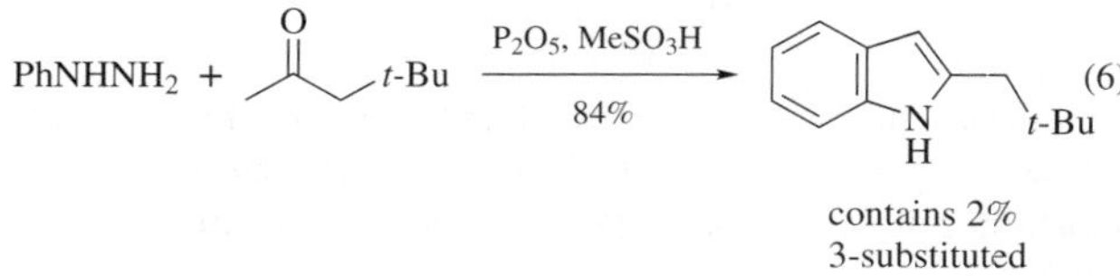

PhNHNH2 + [ketone, t-Bu] —(P_2O_5, $MeSO_3H$; 84%)→ [2-substituted indole, t-Bu] (6)

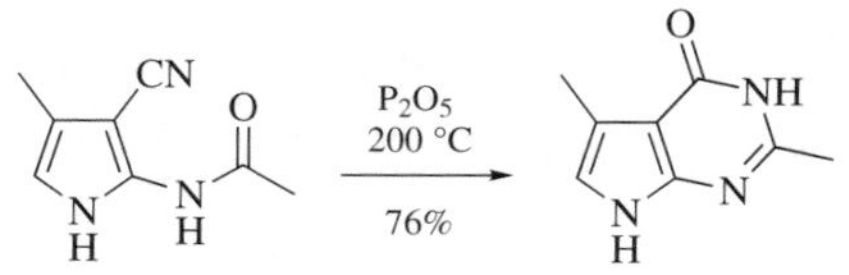

contains 2% 3-substituted

(**1**) —(P_2O_5, 200 °C; 76%)→ (7)

Condensing Agent. Heating a mixture of P_2O_5, a primary or secondary amine (often using the amine as solvent), and a heterocyclic ketone or alcohol is a general route to substituted adenines (eq 8),[14,15] purines,[16,17] and pyrimidines.[18] Cyclization and ketone/amine condensations can often be combined in the one-pot transformations of 4,5-disubstituted imidazoles to substituted adenines (eqs 9 and 10).[19]

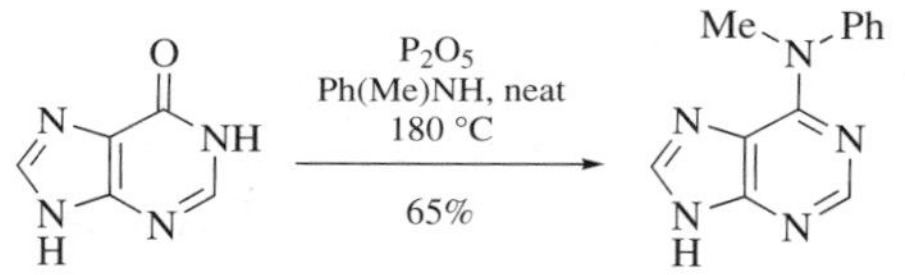

—(P_2O_5, Ph(Me)NH, neat, 180 °C; 65%)→ (8)

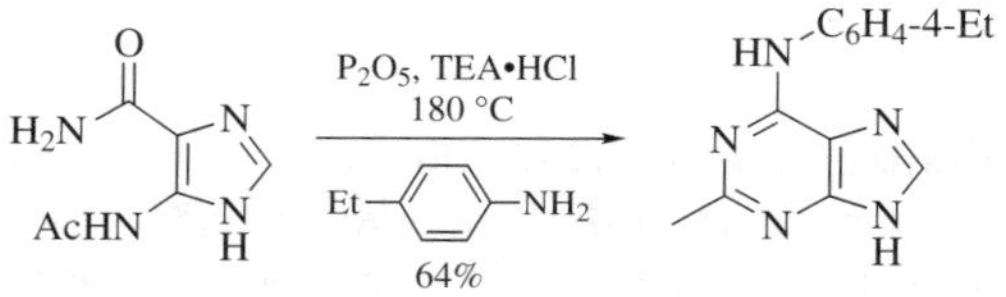

—(P_2O_5, TEA•HCl, 180 °C; Et-C6H4-NH2; 64%)→ (9)

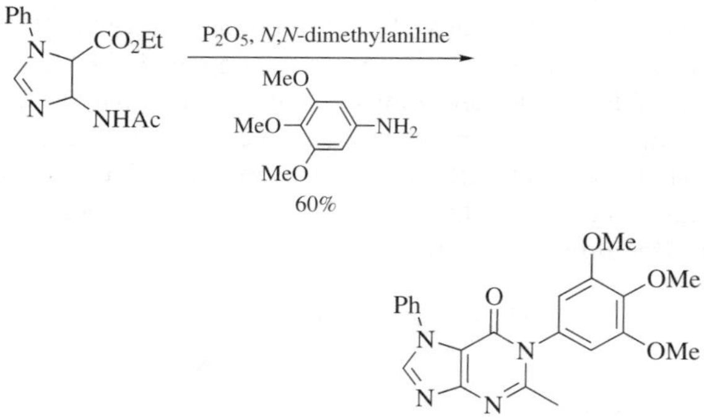

—(P_2O_5, N,N-dimethylaniline; 3,4,5-trimethoxyaniline; 60%)→ (10)

Cyclodehydrations. The Pictet–Gams modification of the Bischler–Napieralski reaction is an important method for the preparation of isoquinolines.[19] This reaction relies on the dehydration of molecules such as (**2**) with P_2O_5 at high temperature (eq 11).[20] Lower temperatures lead to oxazoles, which can be converted to the isoquinolines by heating to 196 °C.

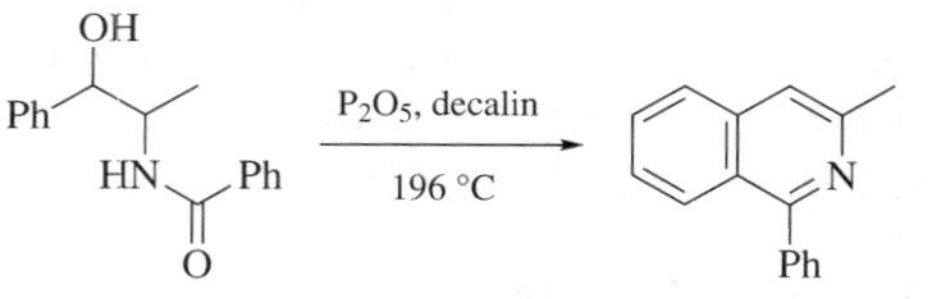

(**2**) —(P_2O_5, decalin, 196 °C)→ (11)

Trimethylsilyl Polyphosphate. Heating a mixture of P_2O_5 with ***Hexamethyldisiloxane*** results in the formation of ***Trimethylsilyl Polyphosphate*** (PPSE), an organic soluble, aprotic alternative to PPE or PPA.[21] This reagent has proven useful for Beckmann rearrangements (eq 12),[21–23] and for generation and cyclization of nitrilium ions.[24]

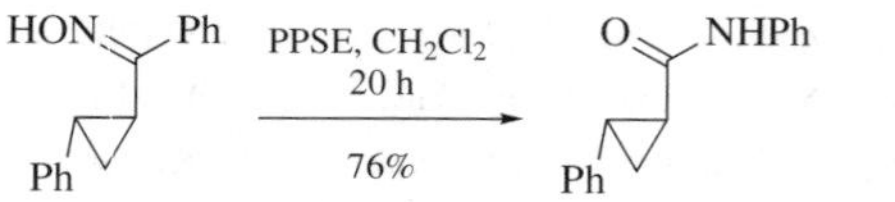

—(PPSE, CH_2Cl_2, 20 h; 76%)→ (12)

Formamidines. Condensation of formanilides with secondary amines in the presence of P_2O_5 produces formamidines in good yield (eq 13).[25] The reaction is compatible with electron-rich and electron-deficient arenes, but is limited to secondary amines.

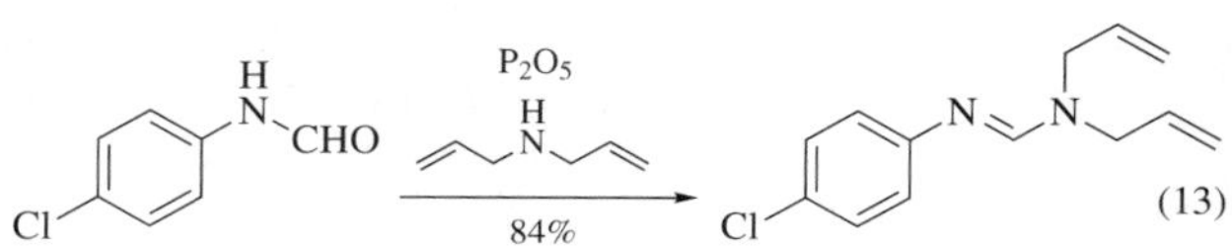

—(P_2O_5, diallylamine; 84%)→ (13)

Related Reagents. Dimethyl Sulfoxide–Phosphorus Pentoxide; Phosphorus(V) Oxide–Methanesulfonic Acid; Phosphorus(V) Oxide-Phosphoric Acid.

---

1. *FF* **1967**, *1*, 871.
2. Mowry, D. T. *CR* **1948**, *42*, 189.
3. Reisner, D. B.; Hornung, E. C. *OSC* **1963**, *4*, 144.
4. Rickborn, B.; Jensen, F. R. *JOC* **1962**, *27*, 4608.
5. Larock, R. C. *Comprehensive Organic Transformations*; VCH: New York, 1989; pp 991–992.
6. Rao, C. S., Rambabu, M.; Srinirasan, P. S. *IJC(B)* **1987**, *26B*, 407.
7. Stevens, C. L.; Singhal, G. H. *JOC* **1964**, *29*, 34.
8. Burton, S. G.; Kaye, P. T. *SC* **1989**, *19*, 3331.
9. Banerjee, A.; Sengupta, S.; Adak, M. M.; Banerjee, G. C. *JOC* **1983**, *48*, 3106.
10. Jeffs, P. W.; Molina, G.; Cortese, N. A.; Hauck, P. R.; Wolfram, J. *JOC* **1982**, *47*, 3876.
11. Eaton, P. E.; Carlson, G. R.; Lee, J. T. *JOC* **1973**, *38*, 4071.
12. Zhao, D.; Hughes, D. L.; Bender, D. R.; DeMarco, A. M.; Reider, P. J. *JOC* **1991**, *56*, 3001.
13. Girgis, N. S.; Jørgensen, A.; Pedersen, E. B. *S* **1985**, 101.
14. Girgis, N. S.; Pedersen, E. B. *S* **1982**, 480.
15. Nielsen, F. E.; Pedersen, E. B. *T* **1982**, *38*, 1435.
16. El-Bayouki, K. A. M.; Nielsen, F. E.; Pedersen, E. B. *S* **1985**, 104.
17. Nielsen, F. E.; Nielsen, K. E.; Pedersen, E. B. *CS* **1984**, *24*, 208.

18. Hilmy, K. M. H.; Mogensen, J.; Jorgensen, A.; Pedersen, E. B. *H* **1990**, *31*, 367.
19. Kametani, T.; Fukumoto, K. In *Isoquinolines*; Grethe, G., Ed.; Wiley: New York, 1981; Vol. 31, Part 1, pp 142–160.
20. Fitton, A. O.; Frost, J. R.; Zakaria, M. M.; Andrew, G. *CC* **1973**, 889.
21. Imamoto, T.; Yokoyama, H.; Yokoyama, M. *TL* **1981**, *22*, 1803.
22. Donaruma, L. G.; Heldt, W. Z. *OR* **1960**, *11*, 1.
23. Gawley, R. E. *OR* **1988**, *35*, 1.
24. Gawley, R. E.; Chemburkar, S. R. *H* **1989**, *29*, 1283.
25. Hansen, B. W.; Pedersen, E. B. *ACS* **1980**, *5*, 369.

Mark S. Meier
University of Kentucky, Lexington, KY, USA

Phosphorus(V) Oxide–Methanesulfonic Acid[1]

P_2O_5–$MeSO_3H$

[39394-84-8]	$CH_4O_8P_2S$	(MW 238.06)
(P_2O_5) *[1314-56-3]*	O_5P_2	(MW 141.94)
($MeSO_3H$) *[75-75]-2*	CH_4O_3S	(MW 96.12)

(acidic dehydrating agent used in cycloalkenone synthesis,[1] Friedel–Crafts reactions,[1] the Fischer indole synthesis,[2] the Beckmann rearrangement,[1] and other dehydrations; an alternative to polyphosphoric acid[1])

Alternate Name: Eaton's reagent.
Physical Data: 7.5 wt% solution: bp 122 °C/1 mmHg; *d* 1.500 g cm^{-3}.
Solubility: sol ether, alcohol, MeCN, CH_2Cl_2; insol toluene, hexane.[2,3]
Form Supplied in: 7.5 wt% solution is commercially available.
Preparative Method: prepared[1] by adding ***Phosphorus(V) Oxide*** (P_2O_5, 36 g) in one portion to ***Methanesulfonic Acid*** (360 g) and stirring at rt[3] until the P_2O_5 dissolves.[4] Although Eaton recommends the use of freshly distilled methanesulfonic acid to allow for a clean workup and good yields,[1] others report using the acid as purchased.[5,6]
Handling, Storage, and Precautions: Eaton's reagent is toxic and corrosive. Direct contact with this reagent should be avoided. The solution begins to yellow upon standing for long periods of time; however, this does not appear to affect the viability of the reagent.[1] Use in a fume hood.

Reagent Description. The reagent was conceived as an alternative to the widely used, but often inconvenient, ***Polyphosphoric Acid*** (PPA) (see also ***Polyphosphate Ester***, PPE).[1] Eaton's reagent successfully addresses the drawbacks of PPA's physical properties. It is much less viscous, and is, therefore, easier to stir. Organic compounds are generally soluble in Eaton's reagent, and the hydrolytic workup is less tedious.[1] Reactions are run at ambient or slightly elevated temperatures. Standard aqueous workup is easy and clean. Eaton recommends quenching the reaction with water; quenching in ice may cause methanesulfonic anhydride to precipitate and be extracted into the organic layer; quenching in aqueous base may cause extensive foaming.[1] In addition to its ease of handling, yields obtained with Eaton's reagent compare favorably with those obtained with PPA.[7] Few modifications of Eaton's original procedure have appeared. A 1:5 by weight ratio has been reported to be as effective as a 1:10 ratio.[8] It has been noted that, to avoid polymer formation, only the minimum amount of reagent needed to effect condensation should be used.[9] The nature of the reagent has not been rigorously determined. It appears that the reactive or catalytic species may vary by reaction. In certain acid-catalyzed reactions, P_2O_5 has been found to be superfluous.[10]

Cycloalkenones. $P_2O_5/MeSO_3H$ is used as a reagent in several reactions leading to cycloalkenones. First described in Eaton's original paper, the lactone-to-cyclopentenone rearrangement (eq 1)[1] has since found wide use.[11]

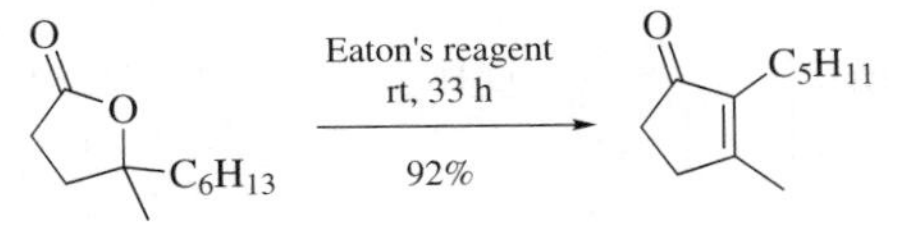

Eaton's reagent, rt, 33 h; 92% (1)

In a related reaction, readily available nitroalkanoic acids cyclize to form cyclopentenones (eq 2).[12] As illustrated in eq 3,[13a] vinylcyclobutanones undergo acyl migration to produce either cyclopentenones or cyclohexenones.[13]

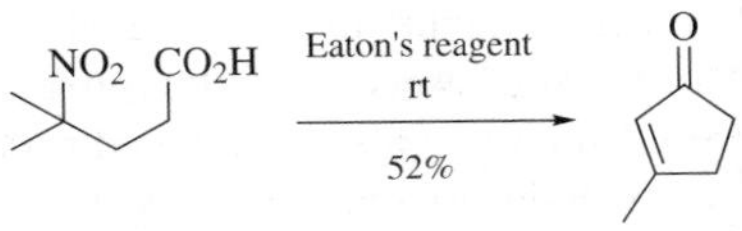

Eaton's reagent, rt; 52% (2)

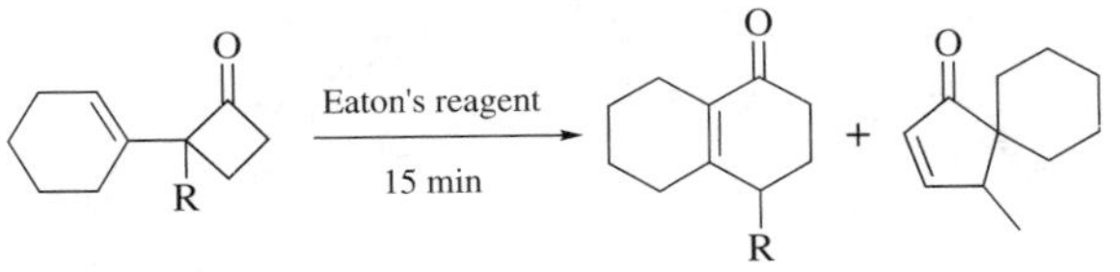

Eaton's reagent, 15 min (3)

R = H	65%	0%
R = Me	13%	51%

Vinylcyclopentenones have undergone the Nazarov cyclization in good yield in the presence of Eaton's reagent (eq 4).[14] However, other reagents may be more generally useful, since there are reports of Eaton's reagent not providing optimal results in this reaction.[15]

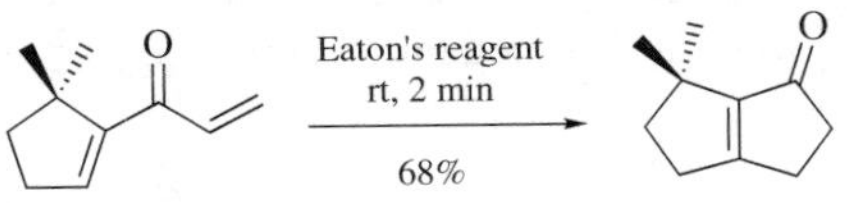

Eaton's reagent, rt, 2 min; 68% (4)

Friedel–Crafts Acylations of Aromatic Rings. Eaton's reagent has been used widely and very effectively[16a] to catalyze Friedel–Crafts acylations.[16] One of the few potential drawbacks is the deprotection of an aryl ether. The examples shown below compare the utility of Eaton's reagent vs. PPA with regard to this deprotection problem. In eq 5, although the cyclization proceeds with an undesired protecting group exchange, the cyclization fails in PPA.[17] Deprotection of an aryl methyl ether is avoided by using Eaton's reagent in place of PPA in one case (eq 6),[18] but not in another.[19] Intramolecular Friedel–Crafts acylations have been observed to occur without the addition of P_2O_5.[10] A comparative study found this observation to be generally applicable to intramolecular acylations, but not intermolecular acylations.[10b,20]

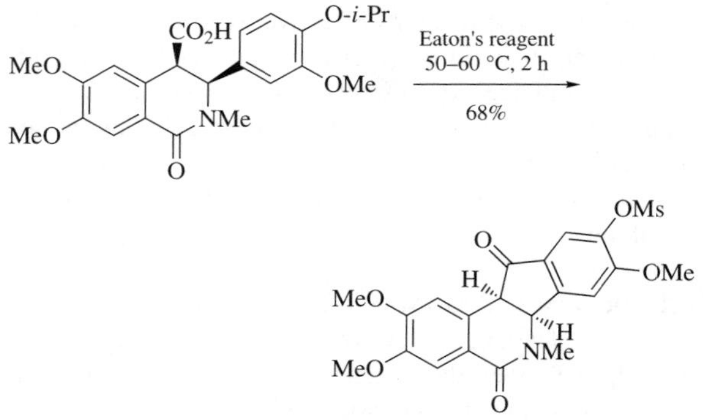

Eaton's reagent
50–60 °C, 2 h
68%

(5)

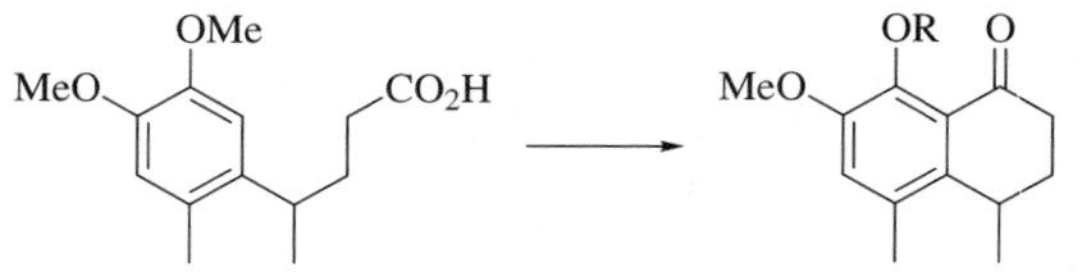

Eaton's reagent R = Me, 70%
PPA R = H, 28%

(6)

Friedel–Crafts Alkylations. $P_2O_5/MeSO_3H$ compares favorably with other reagents in the Friedel–Crafts alkylation reaction.[21] Mechanistic aspects of this reaction have been discussed.[21b] Eq 7 shows an alkene-initiated alkylation that provides (+)-*O*-methylpodocarpate selectively.[22]

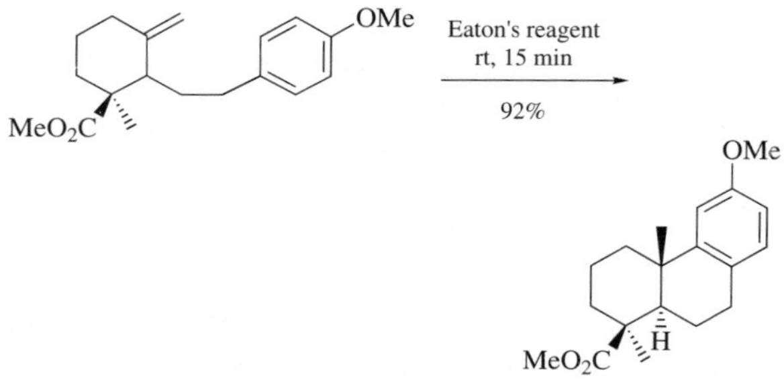

Eaton's reagent
rt, 15 min
92%

(7)

Dehydration. Alcohols have been dehydrated to alkenes with Eaton's reagent (eq 8).[23] In a formal dehydration, a cyclopentenone has been transformed into a diene (eq 9).[24]

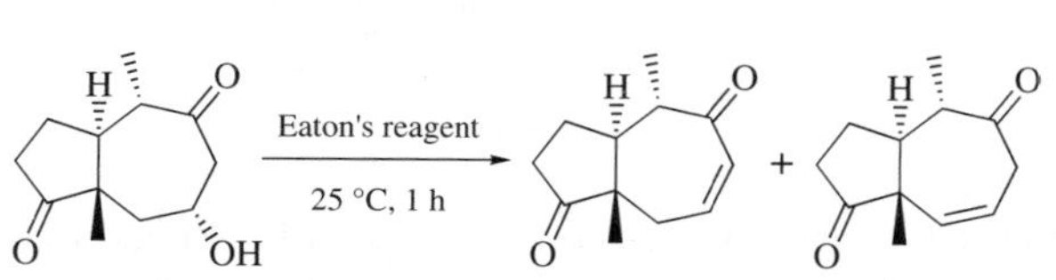

Eaton's reagent
25 °C, 1 h

91:9

(8)

Eaton's reagent
60 °C, 6 h
30–50%

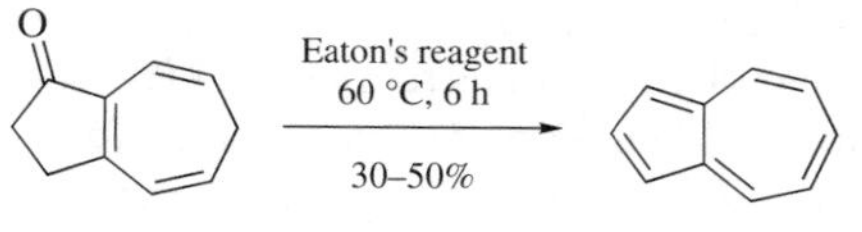

(9)

Fischer Indole Synthesis. The use of Eaton's reagent as the acid catalyst in the Fischer indole reaction results in unprecedented regiocontrol favoring 2-substituted indoles (eq 10).[2] In cases where the harshness of the reagent results in low yields of indoles, dilution of the reaction mixture in sulfolane or CH_2Cl_2 attenuates the problem. Mechanistic studies indicate that the catalytic species, in this reaction, is $MeSO_3H$. The role of P_2O_5 is to act as a drying agent. Further experiments indicate that for the Friedel–Crafts acylation this is not the case; a mixed anhydride is the catalytic species.[2]

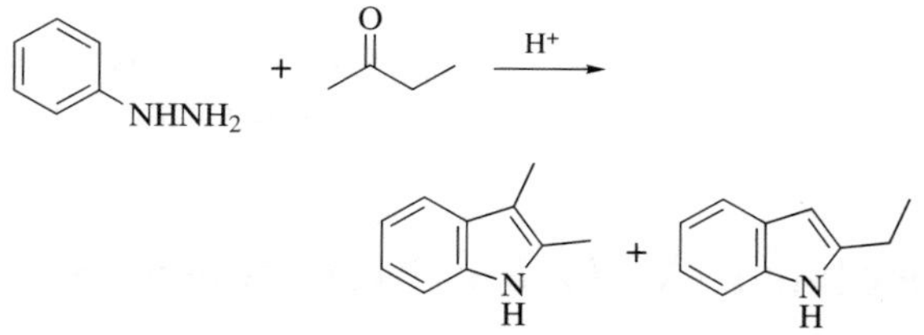

(10)

AcOH	100: 1	
PPA	50:50	
Eaton's reagent	78:22	95%

Heterocycle Preparation. Various heterocycles have been prepared through $P_2O_5/MeSO_3H$-mediated cyclizations. Condensation, and subsequent dehydration, of aminothiophenol and the appropriate acid provides benzothiazoles (eq 11).[25]

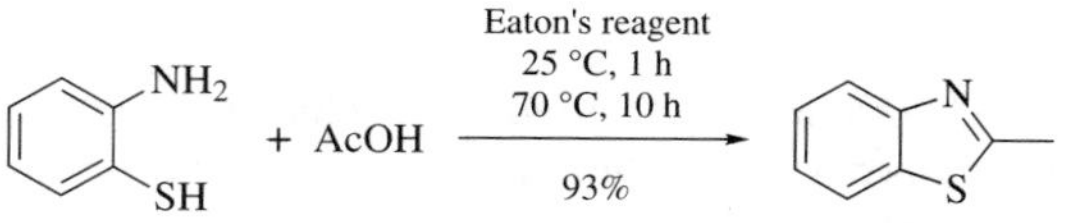

Eaton's reagent
25 °C, 1 h
70 °C, 10 h
93%

(11)

Oxadiazoles can be prepared from diacylhydrazines (eq 12).[26] Furans are formed from the cyclodehydration of a phenolic ketone (eq 13).[27]

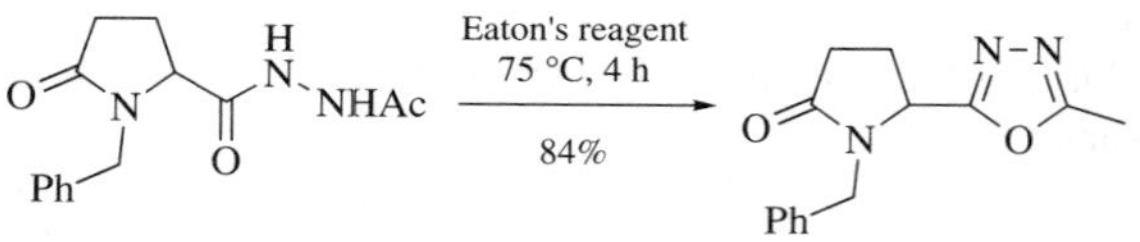

Eaton's reagent
75 °C, 4 h
84%

(12)

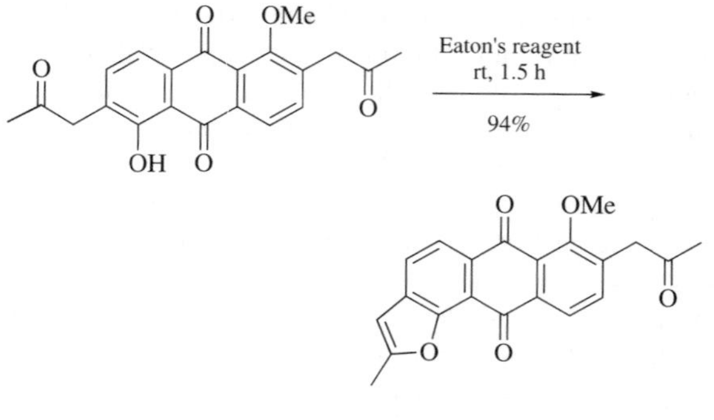

Eaton's reagent
rt, 1.5 h
94%

(13)

Butenolides have been prepared by cyclization of keto esters (eq 14)[28] or by elimination of H_2O from a preformed hydroxy butenolide (eq 15).[29]

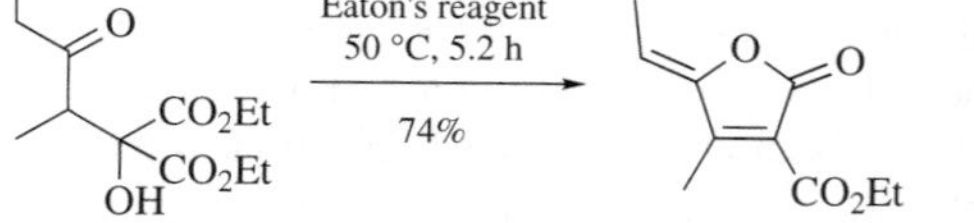

Eaton's reagent
50 °C, 5.2 h
74%

(14)

Eaton's reagent
rt, 1.5 h
95%
(15)

Yields in the synthesis of thiadiazolo[3,2-*a*]pyrimidin-5-ones have been greatly improved by using Eaton's reagent in the place of PPA (eq 16).[8b]

(16)

P_2O_5, $MeSO_3H$; 1:5 87%
PPA 18%

Eaton's reagent is superior to PPA in the addition of an amide across a double bond (eq 17).[30] In another synthesis of lactams, 3-alkenamides reacted stereoselectively with benzaldehyde to provide lactams containing three contiguous stereogenic centers (eq 18).[31]

Eaton's reagent
80–90 °C, 6 h
60%
(17)

PhCHO +
Eaton's reagent
35 °C, 18 h
63%
(18)

Beckmann Rearrangement. Eaton's disclosure of P_2O_5/$MeSO_3H$ as an alternative to PPA compared the two reagents' ability to effect the Beckmann rearrangement.[1] Eaton's reagent has been reported to be superior to other reagents at inducing stereospecific rearrangement of the (*E*)- and (*Z*)-oximes of phenylacetone.[4] However, this is not a general finding. Rearrangement of the oxime in eq 19 does not provide the product expected from an *anti*-migration process.[32]

Eaton's reagent
100 °C, 1 h
72%
(19)

1. Eaton, P. E.; Carlson, G. R.; Lee, J. T. *JOC* **1973**, *38*, 4071.
2. This paper reports that in the preparation of 2-substituted indoles Eaton's reagent is superior to PPA, H_2SO_4, and PPSE: Zhao, D.; Hughes, D. L.; Bender, D. R.; DeMarco, A. M.; Reider, P. J. *JOC* **1991**, *56*, 3001.
3. These authors report the presence of a finely-divided solid after stirring for 6 h. It was removed by filtration under nitrogen.
4. Alternatively, the solution may be heated during dissolution of the P_2O_5. See: Stradling, S. S.; Hornick, D.; Lee, J.; Riley, J. *J. Chem. Educ.* **1983**, *60*, 502.
5. Akhtar, S. R.; Crivello, J. V.; Lee, J. L. *JOC* **1990**, *55*, 4222.
6. Corey, E. J.; Boger, D. L. *TL* **1978**, 5.
7. Examples of exceptions: (a) PPA is superior to either Eaton's reagent or H_2SO_4 in a Friedel–Crafts acylation: Hormi, O. E. O.; Moisio, M. R.; Sund, B. C. *JOC* **1987**, *52*, 5272. (b) PPA is superior to Eaton's reagent, $BF_3{\cdot}OEt_2$, HCO_2H, $ZnCl_2$, CF_3CO_2H, *p*-TsOH, and H_2SO_4 in a Friedel–Crafts alkylation: Maskill, H. *JCS(P1)* **1987**, 1739. (c) PPA is superior to Eaton's reagent or $CF_3CO_2H/(CF_3CO)_2O/BF_3{\cdot}OEt_2$ in a Friedel–Crafts acylation: Hands, D.; Marley, H.; Skittrall, S. J.; Wright, S. H. B.; Verhoeven, T. R. *JHC* **1986**, *23*, 1333. (d) PPA is superior to Eaton's reagent in a Friedel–Crafts acylation: Bosch, J.; Rubiralta, M.; Domingo, A.; Bolos, J.; Linares, A.; Minguillon, C.; Amat, M.; Bonjoch, J. *JOC* **1985**, *50*, 1516. (e) PPA is superior to Eaton's reagent or PPE in a Friedel–Crafts acylation: Jilek, J.; Holubek, J.; Svatek, E.; Schlanger, J.; Pomykacek, J.; Protiva, M. *CCC* **1985**, *50*, 519. (f) In a Friedel–Crafts acylation, where PPA or Eaton's reagent fails to give satisfactory results, the corresponding acid chloride is cyclized using $AlCl_3$: Barco, A.; Benetti, S.; Pollini, G. P. *OPP* **1976**, *8*, 7.
8. (a) Eaton, P. E.; Mueller, R. H.; Carlson, G. R.; Cullison, D. A.; Cooper, G. F.; Chou, T.-C.; Krebs, E.-P. *JACS* **1977**, *99*, 2751. (b) Tsuji, T.; Takenaka, K. *BCJ* **1982**, *55*, 637.
9. Parish, W. W.; Stott, P. E.; McCausland, C. W.; Bradshaw, J. S. *JOC* **1978**, *43*, 4577.
10. (a) Leon, A.; Daub, G.; Silverman, I. R. *JOC* **1984**, *49*, 4544. (b) Premasagar, V.; Palaniswamy, V. A.; Eisenbraun, E. J. *JOC* **1981**, *46*, 2974.
11. (a) Jacobson, R. M.; Lahm, G. P.; Clader, J. W. *JOC* **1980**, *45*, 395. (b) Inouye, Y.; Fukaya, C.; Kakisawa, H. *BCJ* **1981**, *54*, 1117. (c) Murthy, Y. V. S.; Pillai, C. N. *T* **1992**, *48*, 5331. (d) Eaton, P. E.; Srikrishna, A.; Uggeri, F. *JOC* **1984**, *49*, 1728. (e) Pohmakotr, M.; Reutrakul, V.; Phongpradit, T.; Chansri, A. *CL* **1982**, 687. (f) Baldwin, J. E.; Beckwith, P. L. M. *CC* **1983**, 279. (g) Mundy, B. P.; Wilkening, D.; Lipkowitz, K. B. *JOC* **1985**, *50*, 5727. (h) Mehta, G.; Karra, S. R. *TL* **1991**, *32*, 3215. (i) Ho, T.-L.; Yeh, W.-L; Yule, J.; Liu, H.-J. *CJC* **1992**, *70*, 1375.
12. Ho, T.-L. *CC* **1980**, 1149.
13. (a) Matz, J. R.; Cohen, T. *TL* **1981**, *22*, 2459. (b) For a related ring expansion of 1-alkenylcyclopropanols to cyclopentenones, see: Barnier, J.-P.; Karkour, B.; Salaun, J. *CC* **1985**, 1270.
14. Paquette, L. A.; Stevens, K. E. *CJC* **1984**, *62*, 2415.
15. (a) This paper reports obtaining Nazarov cyclization products in 8–10% yield with either Eaton's reagent or $FeCl_3$. A silicon assisted Nazarov was also explored: Cheney, D. L.; Paquette, L. A. *JOC* **1989**, *54*, 3334. (b) PPA is superior to Eaton's reagent or methanesulfonic acid in effecting cyclization of 1,1′-dicyclopentenyl ketone: Eaton, P. E.; Giordano, C.; Schloemer, G.; Vogel, U. *JOC* **1976**, *41*, 2238. (c) Many other reagents including HCO_2H/H_3PO_4, HCl, H_2SO_4, $SnCl_4$, and TsOH have been used in this type of Nazarov cyclization. For a review of the Nazarov cyclization, see: Santelli-Rouvier, C.; Santelli, M. *S* **1983**, 429.
16. Examples: (a) McGarry, L. W.; Detty, M. R. *JOC* **1990**, *55*, 4349. (b) Grunewald, G. L.; Sall, D. J.; Monn, J. A. *JMC* **1988**, *31*, 433. (c) Russell, R. K.; Rampulla, R. A.; van Nievelt, C. E.; Klaubert, D. H. *JHC* **1990**, *27*, 1761. (d) Ye, Q.; Grunewald, G. L. *JMC* **1989**, *32*, 478. (e) Kelly, T. R.; Ghoshal, M. *JACS* **1985**, *107*, 3879. (f) Eck, G.; Julia, M.; Pfeiffer, B.; Rolando, C. *TL* **1985**, *26*, 4723. (g) Kitazawa, S.; Kimura, K.; Yano, H.; Shono, T. *JACS* **1984**, *106*, 6978. (h) Stott, P. E.; Bradshaw, J. S.; Parish, W. W.; Copper, J. W. *JOC* **1980**, *45*, 4716. (i) Cushman, M.; Abbaspour, A.; Gupta, Y. P. *JACS* **1983**, *105*, 2873. (j) Acton, D.; Hill, G.; Tait, B. S. *JMC* **1983**, *26*, 1131. (k) Miller, S. J.; Proctor, G. R.; Scopes, D. I. C. *JCS(P1)* **1982**, 2927.
17. Cushman, M.; Mohan, P. *JMC* **1985**, *28*, 1031.
18. Inouye, Y.; Uchida, Y.; Kakisawa, H. *BCJ* **1977**, *50*, 961.
19. Falling, S. N.; Rapoport, H. *JOC* **1980**, *45*, 1260.
20. For an example of an intermolecular acylation of cyclohexenone, see: Cargill, R. L.; Jackson, T. E. *JOC* **1973**, *38*, 2125.
21. (a) Fox, J. L.; Chen, C. H.; Stenberg, J. F. *OPP* **1985**, *17*, 169. (b) Davis, B. R.; Hinds, M. G.; Johnson, S. J. *AJC* **1985**, *38*, 1815.
22. Hao, X.-J.; Node, M.; Fuji, K. *JCS(P1)* **1992**, 1505.
23. Ziegler, F. E.; Fang, J.-M.; Tam, C. C. *JACS* **1982**, *104*, 7174.

24. Scott, L. T.; Minton, M. A.; Kirms, M. A. *JACS* **1980**, *102*, 6311.
25. Boger, D. L. *JOC* **1978**, *43*, 2296.
26. Rigo, B.; Couturier, D. *JHC* **1986**, *23*, 253.
27. Cambie, R. C.; Howe, T. A.; Pausler, M. G.; Rutledge, P. S.; Woodgate, P. D. *AJC* **1987**, *40*, 1063.
28. Schultz, A. G.; Yee, Y. K. *JOC* **1976**, *41*, 561.
29. Schultz, A. G.; Godfrey, J. D. *JACS* **1980**, *102*, 2414.
30. Tilley, J. W.; Clader, J. W.; Wirkus, M.; Blount, J. F. *JOC* **1985**, *50*, 2220.
31. Marson, C. M.; Grabowska, U.; Walsgrove, T.; Eggleston, D. S.; Baures, P. W. *JOC* **1991**, *56*, 2603.
32. Jeffs, P. W.; Molina, G.; Cortese, N. A.; Hauck, P. R.; Wolfram, J. *JOC* **1982**, *47*, 3876.

Lisa A. Dixon
The R. W. Johnson Pharmaceutical Research Institute, Raritan, NJ, USA

Phosphorus Oxychloride

[10025-87-3] Cl_3OP (MW 153.32)

(formylation of aromatic rings[1,2] (Vilsmeier–Haack reaction); phosphorylating agent;[3] dehydrating agent for amides; halogenation of alcohols, phenols, and heterocycles)

Alternate Name: phosphoryl chloride.
Physical Data: mp 1 °C; bp 106 °C; *d* 1.675 g cm^{-3}; *n* 1.461.
Solubility: sol THF, MeCN, CH_2Cl_2, many other solvents.
Form Supplied in: colorless, fuming liquid; commercially available.
Purification: by distillation in vacuo.[4]
Handling, Storage, and Precautions: toxic and corrosive; reacts vigorously with alcohols and water, liberating HCl, phosphoric acid, and heat. Protect from water. Use in a fume hood.

Chloromethyleneiminium Ions.[5–8] Phosphorus oxychloride (**1**) is a strong Lewis acid that is widely used in synthesis. Of particular importance are reactions of (**1**) with substituted amides, most often ***N,N-Dimethylformamide*** or dimethylacetamide (DMA). These reactions lead to the formation of chloromethyleneiminium salts (**2**) (eq 1). These salts (Vilsmeier reagents, see ***Dimethylchloromethyleneammonium Chloride***) are highly versatile intermediates[6] and are involved in numerous important reactions, including the Vilsmeier–Haack[2,6] and Bischler–Napieralski[9] reactions. Formation of (**2**) can be achieved using ***Thionyl Chloride*** or ***Phosgene***, although there have been reports of differences in reactivity between $POCl_3$/DMF, $SOCl_2$/DMF, and $COCl_2$/DMF systems.[7,10,11]

$$R^1C(O)NR^2R^3 \xrightarrow{POCl_3} R^1C(Cl)=N^+R^2R^3 \; PO_2Cl_2^- \quad (1)$$

(**2**)

Formylation of Aromatic Rings.[2,5,6,8,11] The Vilsmeier reagent attacks electron-rich aromatic systems to form arylmethyleneiminium ions which liberate a formylated aromatic compound upon hydrolysis (eq 2). Thio- and selenoaldehydes can be prepared by hydrolysis in the presence of ***Sodium Hydrogen Sulfide***[6,12] or ***Sodium Hydrogen Selenide***.[13] A wide range of aromatic systems[6,11] can be formylated in this fashion, including benzene derivatives, polyaromatic hydrocarbons (eq 3), and azulene.[11] Substitution occurs at relatively electron-rich positions.

POCl3, *N*-methylformanilide, 50 °C; 86% (2)

POCl3, DMF; 86% (3)

Formylation of Heterocycles.[6,11,14] Many heterocycles are readily formylated by $POCl_3$/DMF, including pyrroles, thiophenes, furans, indoles, quinolines, pteridines, and purines.[6] Reaction at the 2- and 5-positions of pyrroles, furans, and thiophenes is preferred, and indoles undergo attack at the 3-position (eq 4).[6]

POCl3, DMF; 75% (4)

Formylation of Alkenes.[10,11] Both activated and unactivated alkenes undergo electrophilic attack by (**2**).[5] Reaction at terminal positions is preferred.[6] Styrene derivatives give cinnamaldehydes; enamines[6] and aryl polyenes[15,16] react at terminal positions (eq 5).

1. POCl3, DMF; 2. hydrolysis; 91% (5)

Bischler–Napieralski Reaction. A widely used method for cyclization of *N*-β-phenylethyl amides to form dihydroisoquinolines and isoquinolines (eq 6)[17–19] is the Bischler–Napieralski reaction.[9] A nitrilium ion intermediate has been implicated in this reaction.[20,21]

POCl3, MeCN; 85% (6)

Fragmentation of the nitrilium ion interferes with the cyclization[20,21] of 1,2-diphenylethane derivatives, leading to formation of stilbenes rather than cyclization products. This problem has been overcome by using ***Oxalyl Chloride*** instead of (**1**).[22] Fragmentation (von Braun reaction or retro-Ritter reaction) occurs whenever a highly stable cation can be formed, such as benzyl[20] or *t*-butyl (eq 7).[21,23,24]

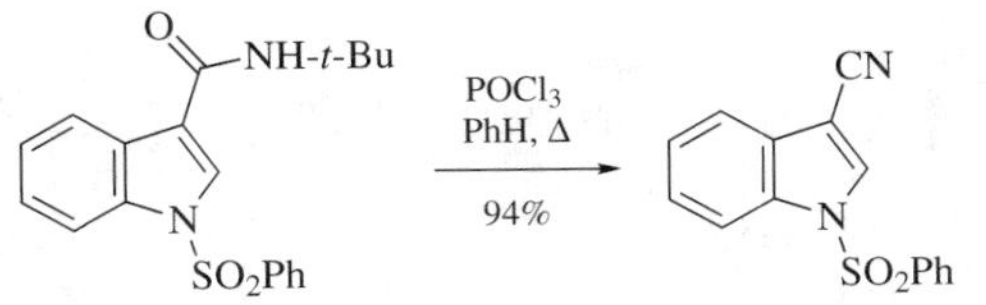

(7)

Treatment of tertiary amides with $POCl_3$/DMF (**2**) results in formation of β-dimethylamino α,β-unsaturated amides. The highly electrophilic iminium ion (**3**) is formed in this reaction, and in the presence of an alkene this undergoes cyclization (eq 8).[14,25]

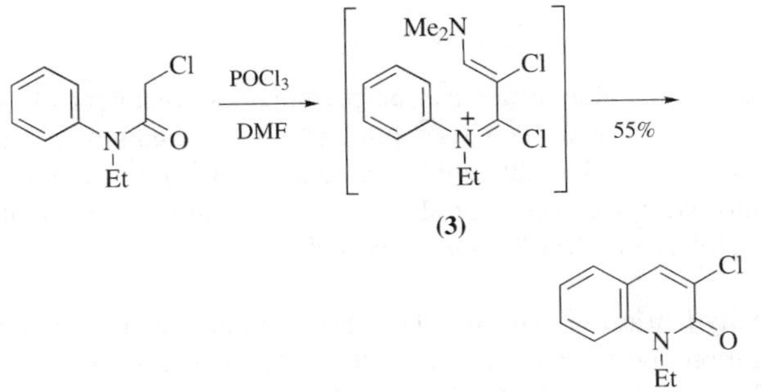

(8)

Excess $POCl_3$ and DMF can lead to formylation of the initial heterocycle (eq 9). A series of α-substituted acrylonitriles has been prepared by dehydration of α,β-unsaturated amides (eq 10). $POCl_3$/DMF and $SOCl_2$/DMF give comparable results.[26]

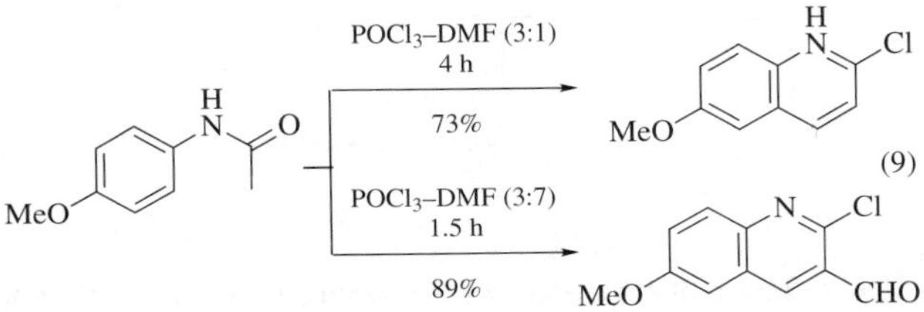

(9)

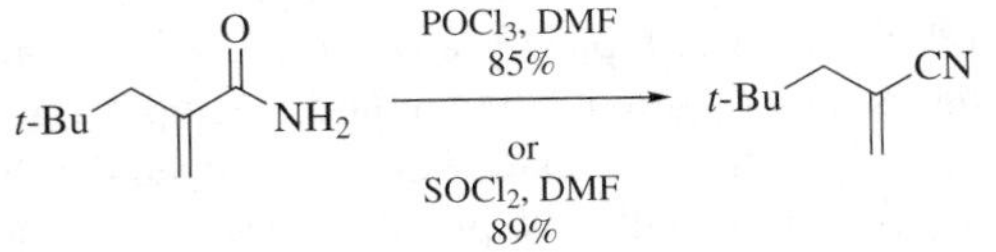

(10)

Chloroaldehydes from Carbonyl Compounds.[7] The enol tautomers of ketones react with $POCl_3$/DMF to form β-chloroacrolein derivatives (eq 11). The regiochemistry is determined both by the stability of the enol as well as by steric factors. Nonsymmetrical ketones often give a mixture of regioisomers, and most ketones produce a mixture of (*E*) and (*Z*) isomers of the product.[27] Aldehydes will undergo this type of reaction, although there are relatively few reports.

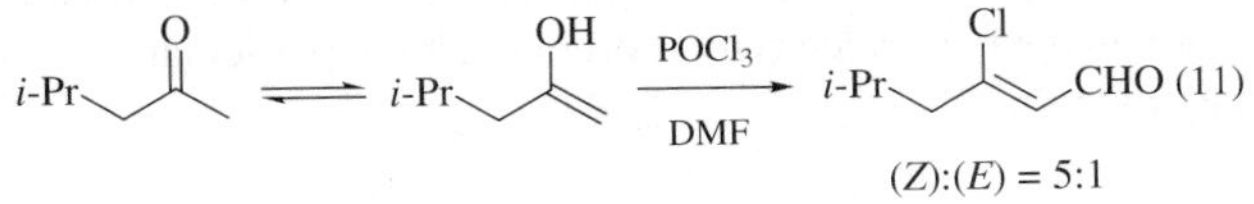

(11)

Reaction of $POCl_3$/DMF with carboxylic acids results in vinamidinium ions (eq 12).[28,29] These ions react with a variety of nucleophiles to produce pyrazoles, oxazoles, pyrimidines, diazepines, quinolines, quinolizines, and vinylogous sulfonamides.[30]

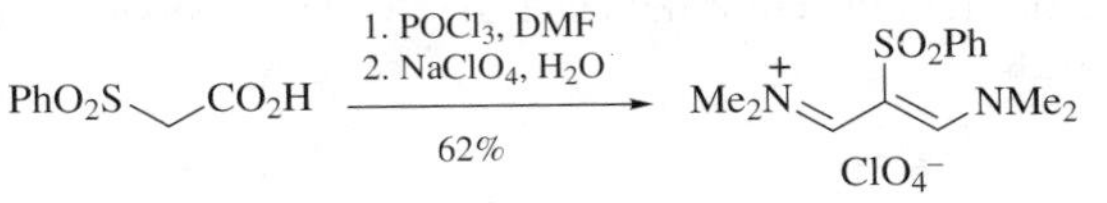

(12)

The Vilsmeier reagent (**2**) reacts with Grignard reagents as well as with alkylzinc and alkylaluminum reagents to form tertiary amines.[31] It also reacts with heteroatom nucleophiles, including thiols,[32] alcohols, and amines,[8] as well as with nucleophilic amines containing other functional groups.[5,33,34] Aromatic alcohols react with $POCl_3$/DMF to form aryl formates in 50–80% yield,[35] but aliphatic alcohols are more efficiently formylated with ***Benzoyl Chloride***/DMF.[36] Homoallylic alcohols such as (**4**) produce biphenyls in 30–98% yield[16] through initial dehydration[6] (eq 13) followed by electrophilic attack on the resulting aryl diene.

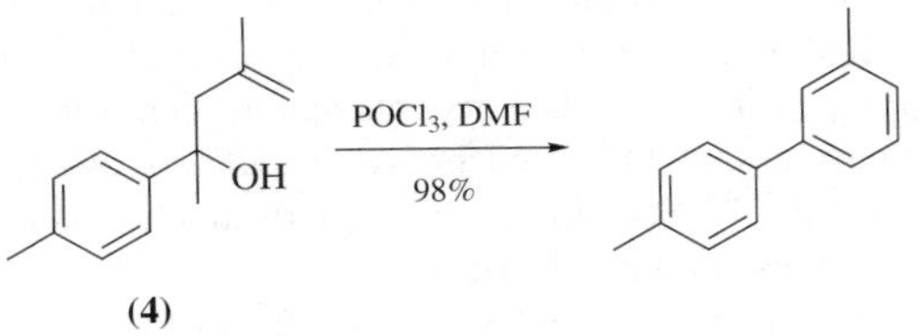

(13)

Halogenation of Alcohols. The combination of $POCl_3$ and DMF can be used to halogenate primary, secondary, and tertiary aliphatic alcohols (eq 14),[37] whereas the reaction of primary alcohols and $POCl_3$ without DMF or DMA will generally lead to the formation of trialkyl phosphates.

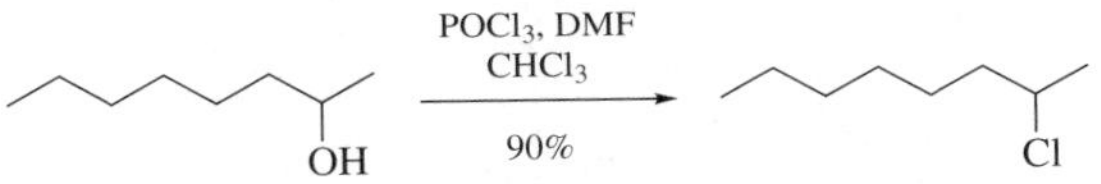

(14)

A comparison of $POCl_3$ with SO_2Cl_2, PCl_3, MsCl, and $SOCl_2$ indicated that optimum conditions for the conversion of (**5**) to (**6**) (eq 15) are 1.1 equiv $POCl_3$ in DMF at 0 °C.[38]

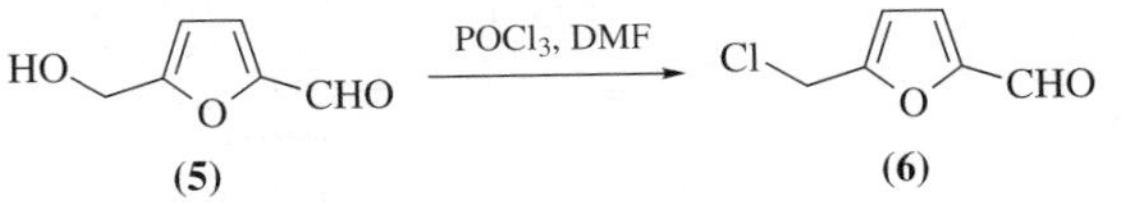

(15)

Halogenation of Heterocycles. Phosphorus oxychloride is widely used in the chlorination of heterocycles. In general, heterocyclic ketones (eq 16) or alcohols (eq 17) react readily with (**1**), including pteridines,[39] purines,[40,41] and others.

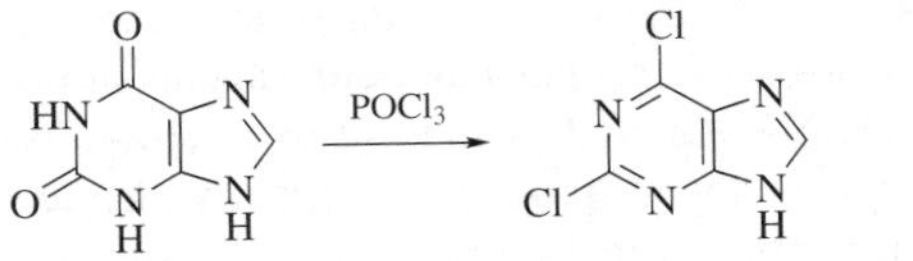

(16)

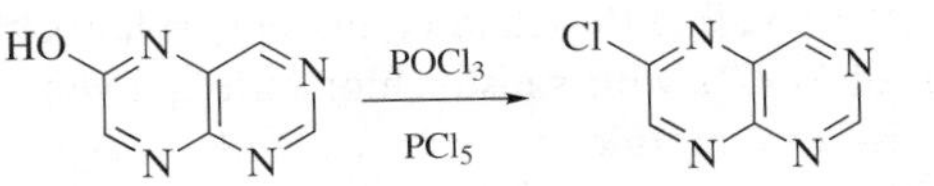

(17)

In the chlorination of the isoxazole (**7**), it was noted that freshly distilled $POCl_3$ is ineffective but that an older sample leads to the formation of the chloride (**8**). Further investigation showed that a mixture of $POCl_3$, acid (H_3PO_4), and an additional chloride source (pyridinium chloride) leads to reliable conversion of (**7**) to (**8**) (eq 18).[42]

$POCl_3$, py•HCl, H_3PO_4; 68% (18)

(**7**) $CO_2CH_2CF_3$ → (**8**) $CO_2CH_2CF_3$

Phosphorylation.[3,43] Phosphorus oxychloride reacts with alcohols, amines, and thiols, resulting in phosphorylation of these functional groups. Trimethyl phosphate is a particularly effective solvent,[44] and tertiary amine bases are generally used as well. Treatment of primary alcohols with $POCl_3$ results in the formation of phosphonyl dichloride intermediates which, in the presence of excess alcohol, convert to symmetrical trialkyl phosphates.[10] It is generally possible to isolate aryl phosphorodichloridates when a two-fold excess of (**1**) is used and $AlCl_3$, KCl, or pyridine is used as a catalyst.[45,46] Secondary and tertiary alcohols tend to form alkyl chlorides and phosphoric acid.

Phosphorus oxychloride is very useful in the preparation of nucleoside 5′-phosphates (eq 19) given appropriate mixtures of $POCl_3$, pyridine, and water.[44,47] The primary hydroxy (5′) is sufficiently reactive for there to be minimal formation of cyclic phosphates at the 2,3-positions of nucleosides. The intermediate phosphorodichloridate can be converted to the corresponding triphosphate by treatment with inorganic phosphate in a convenient one-pot fashion.

$POCl_3$–py–H_2O (2:2:1); 92% (19)

Under anhydrous conditions, $POCl_3$ can be used to produce cyclic dialkyl phosphonyl chlorides from diols, and these can be hydrolyzed with water to the cyclic phosphates or with an alcohol to trialkyl phosphates. Treatment of forskolin (**9**) with $POCl_3$ leads to the formation of the cyclic phosphoryl chloride (**10**) in 55% yield as a mixture of two stereoisomers at phosphorus (eq 20).[48] A comparison of several phosphorylating agents showed that $POCl_3$, $(MeOC_6H_4)_2POCl$, and 2-chloro-2-oxo-1,3,2-dioxaphospholane all give the analogous products in virtually identical yield.[48]

Similar results can be obtained with primary amines,[49] although it is possible to obtain diamidophosphorochloridates by reaction of $POCl_3$ with substoichiometric amounts of amine.[50] Phosphorus oxychloride reacts more readily with primary amines than with primary alcohols, so that it is possible to prepare phosphoramides.[51] Alkoxides, however, react more rapidly than amines so it is possible to phosphorylate hydroxyls in the presence of amines. Again, like the reaction of $POCl_3$ with diols, diamines react with $POCl_3$ to form diazaphospholidines (eq 21).[52]

(**9**) → $POCl_3$, Et_3N, py, Et_2O; 55% → (**10**) (20)

$POCl_3$, Et_3N; 75% (21)

Carboxylic anhydrides can be prepared by treatment of carboxylic acids with $POCl_3$,[32] allowing the formation of both esters[32] and amides, although other methods are more common.[53] Dichloroacetyl chloride produces dichloroketene when treated with $POCl_3$ and ***Zinc/Copper Couple***.[54]

Dehydration of Amides. Unsubstituted amides undergo dehydration upon treatment with $POCl_3$. This reaction can also be performed with P_2O_5, $SOCl_2$, or other reagents, and in a study of racemization at the α-position it was determined that dehydration with $POCl_3$ leads to more (albeit little) epimerization than dehydration with P_2O_5 or $SOCl_2$ (eq 22).[55]

$POCl_3$ (22)

98:2

Dehydration of alkyl and aryl *N*-formyl compounds with $POCl_3$ is one of the more general routes to alkyl and aryl isocyanides (eq 23).[56–59] A base, typically an amine base or ***Potassium t-Butoxide***, is also required. The method is simple and effective, although less useful for small, volatile isocyanides than other techniques.[60] The Bischler–Napieralski reaction can occur preferentially to isocyanide formation.[17]

$POCl_3$, py, pet. ether, 55 °C; 69% (23)

Dehydration of Alcohols. The combination of $POCl_3$ and ***Pyridine*** is an effective dehydrating agent for alcohols (eq 24)[61] and cyanohydrins.[62] The stereochemistry of elimination is *anti*, although the regioselectivity is often not high, particularly for tertiary alcohols.[63]

$POCl_3$, py; 75% (24)

The combination of phosphorus oxychloride and ***Tin(II) Chloride*** reduces halohydrins to alkenes (eq 25). The elimination proceeds in an *anti* fashion.[64]

$POCl_3$, $SnCl_2$, py

(25)

86% from the epoxide precursor

Beckmann Rearrangement. Treatment of ketoximes with $POCl_3$ induces Beckmann rearrangement to form amides (eq 26).[65] Numerous other Lewis acids can be used for this transformation.[66,67]

$POCl_3$, DMF

(26)

Related Reagents. Bis(trichloromethyl) Carbonate; Phosphorus(V) Oxide; Phosphorus Oxychloride–Zinc(II) Chloride; Thionyl Chloride; *p*-Tolyl Vinyl Sulfoxide; Trichloromethyl Chloroformate.

1. *FF* **1967**, *1*, 876.
2. Seshadri, S. *J. Sci. Ind. Res.* **1973**, *32*, 128.
3. Hayakawa, H. *COS* **1991**, *6*, 601.
4. Perrin, D. D.; Armarego, W. L. F. *Purification of Laboratory Chemicals*, 3rd ed.; Pergamon: Oxford, 1988.
5. Meth-Cohn, O.; Stanforth, S. P. *COS* **1991**, *2*, 777.
6. Jutz, C. In *Iminium Salts in Organic Chemistry*; Böhme, H.; Viehe, H. G.; Eds.; Wiley: New York, 1976; Vol. 9, pp 225–342.
7. Marson, C. M. *T* **1992**, *48*, 3659.
8. Kantlenher, W. *COS* **1991**, *6*, 485.
9. Hilger, C. S.; Fugmann, B.; Steglich, W. *TL* **1985**, *26*, 5975.
10. Burn, D. *CI(L)* **1973**, 870.
11. Khimii, U. *RCR* **1960**, *29*, 599.
12. Lin, Y.; Lang, S. A. *JOC* **1980**, *45*, 4857.
13. Reid, D. H.; Webster, R. G.; McKenzie, S. *JCS(P1)* **1979**, 2334.
14. Meth-Cohn, O. *H* **1993**, *35*, 539.
15. Jutz, C.; Heinicke, R. *CB* **1969**, *103*, 623.
16. Suresh Chandler Rao, M. S. C.; Krishna Rao, G. S. K. *S* **1987**, 231.
17. Badia, D.; Carrillo, L.; Dominguez, E.; Cameno, A. G.; Martinez de Marigorta, E.; Vincente, T. *JHC* **1990**, *27*, 1287.
18. Kametani, T.; Fukumoto, K. In *Isoquinolines*; G. Grethe, Ed.; Wiley: New York, 1981; Vol. 31, Part 1, pp 142–160.
19. Knabe, J.; Krause, W.; Powilleit, H.; Sierocks, K. *Pharmazie* **1970**, *25*, 313.
20. Fodor, G.; Gal, J.; Phillips, B. A. *AG(E)* **1972**, *11*, 919.
21. Fodor, G.; Nagubandi, S. *T* **1980**, *36*, 1279.
22. Larsen, R. D.; Reamer, R. A.; Corley, E. G.; Davis, P.; Grabowski, E. J. J.; Reider, P. J.; Shinkai, I. *JOC* **1991**, *56*, 6034.
23. Ketcha, D. M.; Gribble, G. W. *JOC* **1985**, *50*, 5451.
24. Perni, R. B.; Gribble, G. W. *OPP* **1983**, *15*, 297.
25. Meth-Cohn, O.; Tarnowski, B. *Adv. Heterocycl. Chem.* **1982**, *31*, 207.
26. Bargar, T. M.; Riley, C. M. *SC* **1980**, *10*, 479.
27. Schellhorn, H.; Hauptmann, S.; Frischleder, H. *ZC* **1973**, *13*, 97.
28. Gupton, J. T.; Riesinger, S. W.; Shah, A. S.; Bevirt, K. M. *JOC* **1991**, *56*, 976.
29. Gupton, J. T.; Gall, J. E.; Riesinger, S. W.; Smith, S. Q.; Bevirt, K. M.; Sikorski, J. A.; Dahl, M. L.; Arnold, Z. *JHC* **1991**, *28*, 1281.
30. McNab, H.; Lloyd, D. *AG(E)* **1976**, *15*, 459.
31. Mesnard, D.; Miginiac, L. *JOM* **1989**, *373*, 1.
32. Arrieta, A.; Garcia, T.; Lago, J. M.; Palomo, C. *SC* **1983**, *13*, 471.
33. Harris, L. N. *S* **1981**, 907.
34. Zelenin, K. N.; Khrustalev, V. A.; Sergutina, V. P. *CA* **1980**, *93*, 70 910.
35. Morimura, S.; Horiuchi, H.; Muruyama, K. *BCJ* **1977**, *50*, 2189.
36. Barluenga, J.; Campos, P. J.; Gonzalez-Nuñez, E.; Asensio, G. *S* **1985**, 426.
37. Yoshihara, M.; Eda, T.; Sakaki, K.; Maeshima, T. *S* **1980**, 746.
38. Sanda, K.; Rigal, L.; Delmas, M.; Gaset, A. *S* **1992**, *6*, 541.
39. Albert, A.; Clark, J. *JCS* **1964**, 1666.
40. Golovchinskaya, E. S. *RCR* **1974**, *43*, 1089.
41. Robins, R. K.; Revankar, G. R.; O'Brien, D. E.; Springer, R. H.; Novinson, T.; Albert, A.; Senga, K.; Miller, J. P.; Streeter, D. G. *JHC* **1985**, *22*, 601.
42. Andersen, K.; Begtrup, M. *ACS* **1992**, *46*, 1130.
43. Edmunson, R. S. In *Comprehensive Organic Chemistry*; Barton, D. H. R.; Ollis, W. D., Eds.; Pergamon: Oxford, 1979; Vol. 2, pp 1262–1263.
44. Yoshikawa, M.; Kato, T.; Takenishi, T. *BCJ* **1969**, *42*, 3505.
45. Owen, G. R.; Rees, C. B.; Ransom, C. J.; van Boom, J. H.; Herscheid, J. D. H. *S* **1974**, 704.
46. Taguchi, Y.; Mushika, Y. *TL* **1975**, 1913.
47. Sowa, T.; Ouchi, S. *BCJ* **1975**, 2084.
48. Lal, B.; Gangopadhyay, A. K. *JCS(P1)* **1992**, *15*, 1993.
49. Edmunson, R. S. In *Comprehensive Organic Chemistry*; Barton, D. H. R.; Ollis, W. D., Eds.; Pergamon: Oxford, 1979; Vol. 2, pp 1262–1265.
50. Ireland, R. E.; O'Neil, T. H.; Tolman, G. L. *OS* **1983**, *61*, 116.
51. Crans, D. C.; Whitesides, G. M. *JACS* **1985**, *107*, 7008.
52. Alexakis, A.; Mutti, S.; Mangeney, P. *JOC* **1992**, *57*, 1224.
53. Mulzer, J. *COS* **1991**, *6*, 323.
54. Hassner, A.; Dillon, J. *S* **1979**, 689.
55. Rickborn, B.; Jensen, F. R. *JOC* **1962**, *27*, 4608.
56. Sandler, S. R.; Karo, W. In *Organic Functional Group Preparations*, 2nd ed.; Academic: San Diego, 1989; Vol. 3, pp 207–238.
57. Ugi, I.; Meyr, R.; Lipinski, M.; Bodesheim, F.; Rosendahl, F. *OSC* **1973**, *5*, 300.
58. van Leusen, D.; van Leusen, A. M. *RTC* **1992**, *47*, 1249.
59. van Leusen, A. M.; Boerma, G. J. M.; Helmholdt, R. B.; Siderius, H.; Strating, J. *TL* **1972**, 2367.
60. Höfle, G.; Lange, B. *OS* **1983**, *61*, 14.
61. Mehta, G.; Murthy, A. N.; Reddy, D. S.; Reddy, A. V. *JACS* **1986**, *108*, 3443.
62. Oda, M.; Yamauro, A.; Watabe, T. *CL* **1979**, 1427.
63. Giner, J.-L.; Margot, C.; Djerassi, C. *JOC* **1989**, *54*, 369.
64. Cornforth, J. W.; Cornforth, R. H.; Mathew, K. K. *JCS* **1959**, 2539.
65. Fujita, S.; Kotauna, K.; Inagaki, Y. *S* **1982**, 68.
66. Gawley, R. E. *OR* **1988**, *35*, 1.
67. Donaruma, L. G.; Heldt, W. Z. *OR* **1960**, *11*, 1.

Mark S. Meier
University of Kentucky, Lexington, KY, USA

Suzanne M. Ruder
Virginia Commonwealth University, Richmond, VA, USA

4-Picolyl Chloride Hydrochloride

•HCl

[1822-51-1] $C_6H_7Cl_2N$ (MW 164.04)

(protection of alcohols,[1,2] phenols,[3] thiols,[3,4] and carboxylic acids[5])

Alternate Name: 4-(chloromethyl)pyridine hydrochloride.
Physical Data: mp 160–163 °C.
Solubility: sol H_2O, EtOH; insol CH_2Cl_2, acetone.
Form Supplied in: white or off-white crystalline solid; available from several commercial sources.
Handling, Storage, and Precautions: the free base is reported to be unstable,[1] and should be generated shortly before use.

Protection Reagent. Alcohols (eq 1)[1,2] (including hydroxy-containing amino acid derivatives),[2] phenols,[3] carboxylic acids (eq 2),[5] and thiols (eq 3)[3,4] have all been alkylated by this reagent (**1**). Esters (**2**) are also available from the respective acids by ***1,3-Dicyclohexylcarbodiimide*** coupling with 4-pyridylmethanol in CH_2Cl_2[5] in comparable yield.

$C_{10}H_{21}OH$ → (NaH, dioxane, HMPA; (**1**); 70%) → $C_{10}H_{21}O$-CH$_2$-4-pyridyl (1)

RCO_2H → ((**1**), tetramethylguanidine; DMF; 50–60%) → (**2**) (2)

Cystine → (1. Na, NH_3; 2. (**1**); 68%) → S-(4-picolyl)cysteine (3)

Dihydric alcohols such as (**3**) are preferentially alkylated on the primary OH, but (**4**) yields a mixture of dialkylated and both monoalkylated products.[1]

(**3**) (**4**)

The protected materials have been of particular value in peptide chemistry. Cleavage[5] of esters (**2**) takes place not only under the more obvious aqueous alkali or catalytic hydrogenation conditions, but also with ***Sodium-Ammonia*** or by electrolytic reduction at a mercury cathode. The protected alcohols,[1,2] phenols,[3] and thiols[3,4] are similarly deprotectable by electrolysis.

The use of 2-[1] and 3-picolyl[2] ethers in such cases has also been investigated, and an *N*-protecting group, (4-picolyloxy)carbonyl,[6] has also been described. The increased polarity of these picolyl-containing materials, relative to the more typical benzylic protecting groups, can appreciably assist in their purification.[2] Nevertheless, these protecting groups remain outside the mainstream of peptide chemistry, probably due to the greater familiarity and availability of, for example, the benzyl- or *t*-butyl-based groups, and the fact that the Na/NH_3 or electrolytic deprotection of the 4-picolyl group is likely to be incompatible with these. In this regard, the use of 2-(2′- or 4′-pyridyl)ethyl protecting groups[7] ('Pyoc' carbamates,[7a] 'Pet' esters,[7b,7c]), likely to impart similar polarity, but deprotectable cleanly in the presence of these other common functionalities, appears distinctly advantageous.

---

1. Wieditz, S.; Schäfer, H. J. *ACS* **1983**, *B37*, 475.
2. Rizo, J.; Albericio, F.; Romero, G.; Garcia-Echeverria, C.; Claret, J.; Muller, C.; Giralt, E.; Pedroso, E. *JOC* **1988**, *53*, 5386.
3. Gosden, A.; Stevenson, D.; Young, G. T. *CC* **1972**, 1123.
4. Gosden, A.; Macrae, R.; Young, G. T. *JCR(S)* **1977**, 22.
5. Camble, R.; Garner, R.; Young, G. T. *JCS(C)* **1969**, 1911.
6. Veber, D. F.; Paleveda, W. J.; Lee, Y. C.; Hirschmann, R. *JOC* **1977**, *42*, 3286.
7. (a) Kunz, H.; Barthels, R. *AG(E)* **1983**, *22*, 783. (b) Katritzky, A. R.; Khan, G. R.; Schwarz, O. A. *TL* **1984**, *25*, 1223. (c) Kessler, H.; Becker, G.; Kogler, H.; Wolff, M. *TL* **1984**, *25*, 3971.

Peter Ham
SmithKline Beecham Pharmaceuticals, Harlow, UK

1,3-Propanediol[1]

HO OH

[504-63-2] $C_3H_8O_2$ (MW 76.10)

(preparation of acetals,[2,3] quinolines,[4] indoles,[5] boron esters;[6] traps Wacker aldehyde intermediates;[7] reaction solvent for Wolff–Kishner reduction[8])

Alternate Name: trimethylene glycol.
Physical Data: liquid, bp 214 °C; mp −27 °C; *d* 1.053 g cm^{-3}.
Solubility: sol H_2O, alcohol; *v* sol ether; sol (hot) benzene.
Form Supplied in: >97% purity, depending on supplier.
Purification: dried with K_2CO_3 and distilled under reduced pressure.

Reactions with Carbonyls. The chemistry of 1,3-propanediol (**1**) is dominated by acetal[2,3,9,10] formation.[11] Its ketone acetals show differential hydrolytic stability: cyclopentanone acetals hydrolyze faster than cyclohexanone acetals, and both hydrolyze faster than ethylene glycol-derived acetals (Table 1).[3a] The ketone–acetal equilibrium lies far to the left.[3] Its aldehyde acetals, however, show the opposite behavior and are hydrolytically more stable than ethylene glycol acetals.[3b,12] Polyketones[3] and 1,4-diones[13] may be selectively protected by judicious choice of substituted 1,3-propanediols. For protection of steroidal ketones, (**1**) is preferred over ethylene glycol, which is susceptible to attack by alkyllithium reagents.[10]

Table 1 Relative Hydrolysis Rates of 5α-Androstane Acetals

Glycol	3-Acetal	17-Acetal
Ethylene glycol	1.0	1.64
1,3-Propanediol	14.5	40.5

$n = 0, 1$

Conversion to Reactive Intermediates. Acrolein condenses with (**1**) to give 2-(2-bromoethyl)-1,3-dioxane,[2] which finds use as a three-carbon homologating agent (eq 1)[14] and in the preparation of γ-keto aldehydes.[15] Use of (**1**) is essential as the corresponding ethylene glycol dioxolane is thermally unstable[2] and gives Grignard reagents that tend to autodestruct.[15,16] Hydrolysis of the product acetals requires special conditions.[14,15]

HBr (g), TsOH, CH_2Cl_2, H_2O; 0–22 °C, 8 h; 65% (1)

Diol (**1**) is used to prepare cyclopropenone acetal (**2**).[17,18] Its highly nucleophilic double bond forms addition products with alcohols and amines[17] and cycloaddition products with dienes to give norcarenes,[17] ketones to give furanones and oxetanes,[17] aldehydes to give butenolide, furan, and γ-keto ester derivatives (eq 2),[18] electrophilic alkenes to give cyclopropanes[19] and functionalized cyclopentenones,[20] and an α-pyrone to give a cycloheptatrienone.[21]

1. (**1**), pentane, H_2SO_4, H_2O, 22–140 °C, 8 h; 2. K, NH_3 (liq), $FeCl_3$ (cat), −78 to −50 °C, ether, 3 h; 45–62% (**2**) (2)

Formation of Boron Esters. Esterification of boronic acid derivatives occurs readily with (**1**), and has been used in many hydroboration reaction sequences.[6,22] The six-membered ring boronates are more stable than the corresponding five-membered ring or acyclic products made from 1,2-diols or alcohols. The title reagent has been used to regenerate carbohydrates from boronate derivatives.[23]

Heterocycle Synthesis. Alkyl (**3**)[24] and alkoxy (**4**)[25] 1,3,2-dioxophosphorinane 2-oxides are prepared in good yields by base-catalyzed condensation of (**1**) with the appropriate phosphonous dichloride. Iminosulfinyl dichlorides react similarly.[26] With DCC, (**1**) gives 2-imino-1,3-oxazine (**5**) (CuCl, 95%) which is converted into urea derivatives in high yields.[27]

(**3**) (**4**) (**5**)

In an alternative route to the Skraup synthesis, quinolines are prepared from (**1**) and primary anilines (eq 3). Choice of stoichiometry, solvent, and ligand greatly affect yield.[4]

(**1**), $RuCl_3 \cdot 3H_2O$, PBu_3, diglyme; 5 h, reflux, $-H_2$; 59% (3)

Regioselective synthesis of indoles is accomplished by in situ trapping of Wacker aldehydes with (**1**) (eq 4). *o*-Vinylacetanilides give indoles directly under similar reaction conditions.[5] Isocoumarins and 1-isoquinolinones are also prepared by this chemistry.[28] Electron-deficient alkenes give acetals with (**1**) under Wacker conditions (see ***Palladium(II) Chloride-Copper(I) Chloride***).[7]

$PdCl_2$, CuCl, O_2, (**1**), DME, 24 h; 50–60 °C; 90%. 1. 10% Rh/C, H_2, 1 atm, rt, 3 h; 2. 10% HCl, rt, 3 h; 75% (4)

Miscellaneous Transformations. 1,3-Propanediol can be converted to 3-chloro-1- (60%),[29] 3-iodo-1- (68%),[30] and 3-bromo-1-propanol (90%)[31] or to 1,3-dibromopropane (85–95%).[32] With imidoyl chlorides, 3-chloropropyl benzoates are produced.[33] The title reagent has good solvent properties and finds use in the Wolff–Kishner reduction.[8] Cyclohexenones and (**1**) give hydroxypropyl phenols under oxidative conditions (eq 5).[34]

(**1**), CAN, I_2; MeCN, reflux, 10 h; 78–91% (5)

R^1, R^2 = H, Me

Related Reagents. 3-Bromo-1,2-propanediol; 2,3-Butanediol; 2,2-Dimethyl-1,3-propanediol; Ethylene Glycol; 2-Methoxy-1,3-dioxolane; (2*R*,4*R*)-2,4-Pentanediol; Triethyl Orthoformate.

1. (a) Cameron, D. C.; Tong, I. T.; Skraly, F. A. *Prepr.–Am. Chem. Soc., Div. Pet. Chem.*, **1993**, *38*, 294. (b) Avots, A.; Glemite, G.; Dzenitis, J. *Latv. PSR Zinat. Akad. Vestis, Kim. Ser.* **1986**, 398 (*CA* **1987**, *106*, 17 529d).
2. Stowell, J. C.; Keith, D. R.; King, B. T. *OS* **1984**, *62*, 140.

3. (a) Smith, S. W.; Newman, M. S. *JACS* **1968**, *90*, 1249. (b) Newman, M. S., Harper, R. J. *JACS* **1958**, *80*, 6350. (c) Smith, S. W.; Newman, M. S. *JACS* **1968**, *90*, 1253.
4. Tsuji, Y.; Huh, K.-T.; Watanabe, Y. *JOC* **1987**, *52*, 1673.
5. Kasahara, A.; Izumi, T.; Murakami, S.; Miyamoto, K.; Hino, T. *JHC* **1989**, *26*, 1405.
6. Brown, H. C.; Bhat, N. G.; Somayaji, V. *OM* **1983**, *2*, 1311.
7. (a) Hosokawa, T.; Ataka, Y.; Murahashi, S.-I. *BCJ* **1990**, *63*, 166. (b) Hosokawa, T.; Ohta, T.; Kanayama, S.; Murahashi, S.-I. *JOC* **1987**, *52*, 1758.
8. Campbell, T. W.; Ginsig, R.; Schmid, H. *HCA* **1953**, *36*, 1489.
9. Marton, D.; Slaviero, P.; Tagliavini, G. *G* **1989**, *119*, 359.
10. Dann, A. E.; Davis, J. B.; Nagler, M. J. *JCS(P1)* **1979**, 158.
11. Sandler, S. R.; Karo, W. *Organic Functional Group Preparations* 2nd ed.; Academic: New York, 1989; Vol. 3, p 1.
12. Stowell, J. C. *JOC* **1976**, *41*, 560.
13. Cole, J. E.; Johnson, W. S.; Robins, R. A.; Walker, J. *JCS* **1962**, 244.
14. Stowell, J. C.; Keith, D. R. *S* **1979**, 132.
15. Stowell, J. C. *JOC* **1976**, *41*, 560.
16. Ponaras, A. A. *TL* **1976**, *36*, 3105.
17. (a) Butler, G. B.; Herring, K. H.; Lewis, P. L.; Sharpe, V. V.; Veazey, R. L. *JOC* **1977**, *42*, 679. (b) Albert, R. M.; Butler, G. B. *JOC* **1977**, *42*, 674.
18. Boger, D. L.; Brotherton, C. E.; Georg, G. I. *OS* **1986**, *65*, 32.
19. Boger, D. L.; Brotherton, C. E. *TL* **1984**, *25*, 5611.
20. Boger, D. L.; Brotherton, C. E. *JACS* **1984**, *106*, 805.
21. Boger, D. L.; Brotherton, C. E. *JOC* **1985**, *50*, 3425.
22. For alkenyl borinate transformations, see: (a) Brown, H. C.; Bhat, N. G. *TL* **1988**, *29*, 21. (b) Brown, H. C.; Imai, T.; Bhat, N. G. *JOC* **1986**, *51*, 5277. (c) Srebnik, M.; Bhat, N. G.; Brown, H. C. *TL* **1988**, *29*, 2635. (d) Brown, H. C.; Bhat, N. G.; Iyer, R. R. *TL* **1991**, *32*, 3655. For asymmetric hydroboration of prochiral alkenes, see: Brown, H. C.; Imai, T.; Desai, M. C.; Singaram, B. *JACS* **1985**, *107*, 4980.
23. Ferrier, R. J.; Prasad, D.; Rudowski, A.; Sangster, I. *JCS* **1964**, 3330.
24. Yuan, C.; Li, S.; Cheng, Z. *S* **1988**, 186.
25. Boisdon, M.-T.; Munoz, A.; Vives, J.-P. *CR(C)* **1961**, *253*, 1570.
26. Picard, C.; Cazaux, L.; Tisnes, P. *PS* **1981**, *10*, 35.
27. Vowinkel, E.; Gleichenhagen, P. *TL* **1974**, *2*, 143.
28. Izumi, T.; Nishimoto, Y.; Kohei, K.; Kasahara, A. *JHC* **1990**, *27*, 1419.
29. Marvel, C. S.; Calvery, H. O. *OSC* **1941**, *1*, 533.
30. Buijs, W.; van Elburg, P.; van der Gen, A. *SC* **1983**, *13*, 387.
31. Kang, S.-K.; Kim, W.-S.; Moon, B.-H. *S* **1985**, 1161.
32. Kamm, O.; Marvel, C. S. *OSC* **1941**, *1*, 25.
33. Back, T. G.; Barton, D. H. R.; Rao, B. L. *JCS(P1)* **1977**, 1715.
34. Horiuchi, C. A.; Fukunishi, H.; Kajita, M.; Yamaguchi, A.; Kiyomiya, H.; Kiji, S. *CL* **1991**, 1921.

Kenneth C. Caster
Union Carbide Corporation, South Charleston, WV, USA

1,3-Propanedithiol[1]

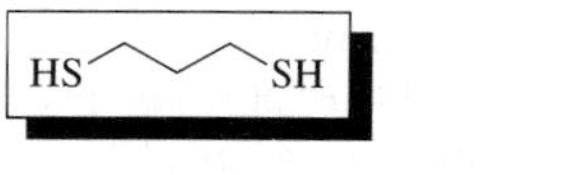

[109-80-8] $C_3H_8S_2$ (MW 108.25)

(1,3-dithiane formation; reduction (carbonyl to methylene, azide to primary amine, peptidic disulfide to dithiol, demercuration); ketene dithioacetal formation)

Physical Data: d^{20} 1.077 g cm^{-3}; bp 170 °C/760 mmHg, 92–98 °C/56 mmHg.
Solubility: slightly sol water; miscible with many organic solvents.
Form Supplied in: liquid; widely available.
Handling, Storage, and Precautions: stench! Use in a fume hood. Can undergo air oxidation to form disulfides. The cyclic disulfide forms a polymeric precipitate in methanol.[17a] Extraction into aqueous NaOH serves to separate thiols from nonacidic impurities.[1f] For toxicity data, see ***1,2-Ethanedithiol***.

1,3-Dithiane Formation. 1,3-Propanedithiol (**1**) condenses under protic or Lewis acid catalysis with aldehydes and ketones to afford 1,3-dithianes (eqs 1–4).[2–5] Useful for carbonyl protection, the 1,3-dithianyl group is compatible with aqueous acid, strong bases, anionic (eq 3)[4] and Pd-catalyzed (eq 5)[6a] C–C bond-forming reactions, catalytic hydrogenation in the presence of the Crabtree catalyst,[6b] and many other synthetic processes.[7] Often, (**1**) shows useful selectivity in reactions with dicarbonyl compounds (eq 3; structures **2**–**5**).[4,8] In contrast to acetalization with diols,[9a] (**1**) reacts selectively with α,β-alkenyl ketones in the presence of moderately hindered saturated ketones;[1c] double bond migration is not observed;[1a] conjugate addition can compete if sterically favored.[9b] In α,β-alkynyl ketones (not aldehydes),[10a] conjugate addition prevails (eq 6).[10b]

OHC, TBDMSO, OTBDMS; (1), $MgBr_2$, Et_2O, 84%; S, S, OTBDMS, OTBDMS (1)

H, N, O, N, H, O, CO_2Et; (1), BF_3•OEt_2, 74%; O, H, N, N, H, S, S, CO_2Et (2)

O, O, Ph; (1), HCl, $ZnCl_2$, Et_2O, 83%; S, S, O, Ph; 1. PhMgBr, 2. HgO, BF_3•OEt_2, aq THF, 59%; O, OH, Ph, Ph (3)

CHO; (1), $Mg(ClO_4)_2$, $CHCl_3$, −10 °C, 95%; S, S; 1. BuLi, THF, −30 °C, 2. BuCHO, −78 °C, 90%; S, S, OH, Bu (4)

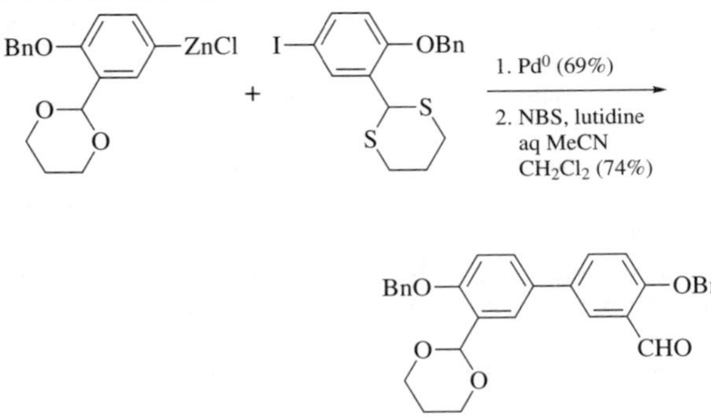

(5)

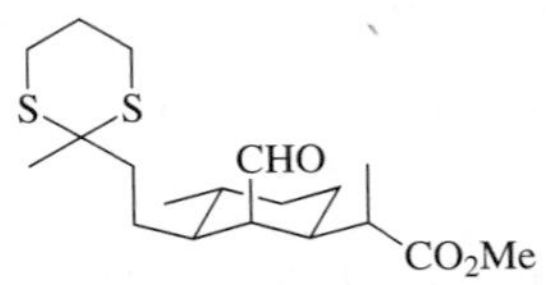

(8)

(9)

R = $SiPh_2$-*t*-Bu

(2) [(1), BF_3•OEt_2, 55–65%] 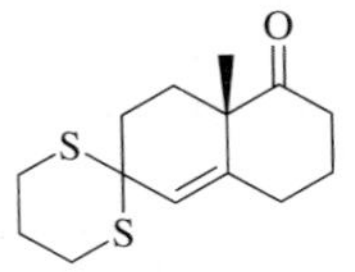 (3) [(1), $TeCl_4$, 68%]

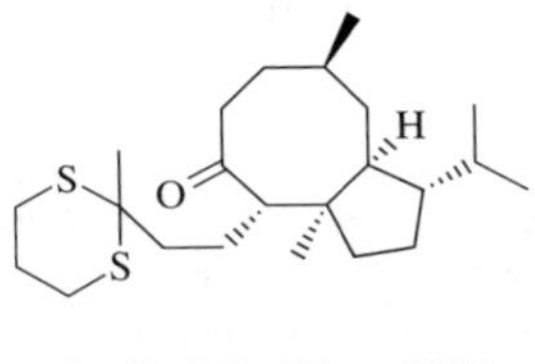

(10)

(11)

(4) [(1), BF_3•OEt_2, >98%] 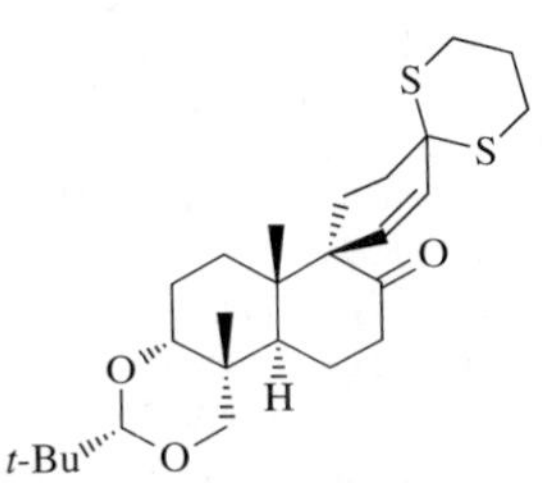 (5) [bis-TMS-(1), ZnI_2, 88%]

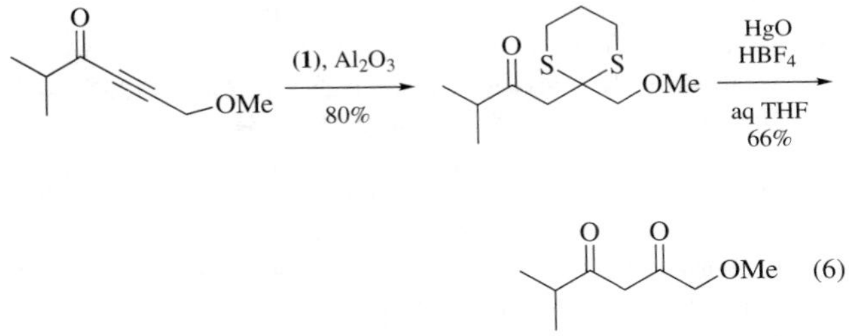

(12)

(6)

Acetals, enol ethers, and oxazolidines also give dithianes with (1) (eqs 7–12).[11a–f,12] This reaction can be useful for opening resistant cyclic structures;[13a,c] ***Titanium(IV) Chloride*** is an especially effective catalyst in such cases.[13b] Reaction of (1) with dihalomethanes,[14a,b] or with carboxylic acids in the presence of ***Tin(II) Chloride***,[14c] also yields 1,3-dithianes.

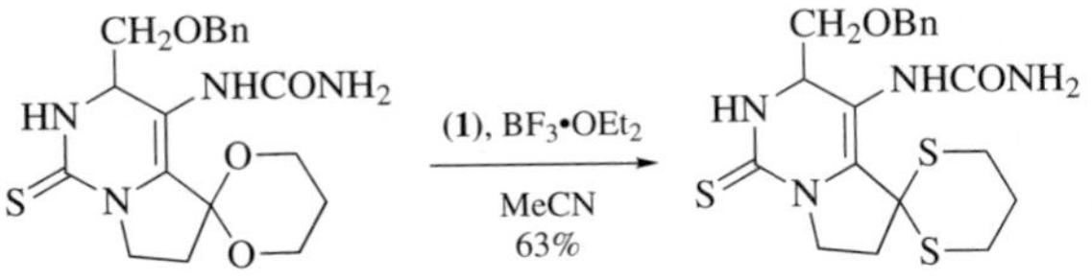

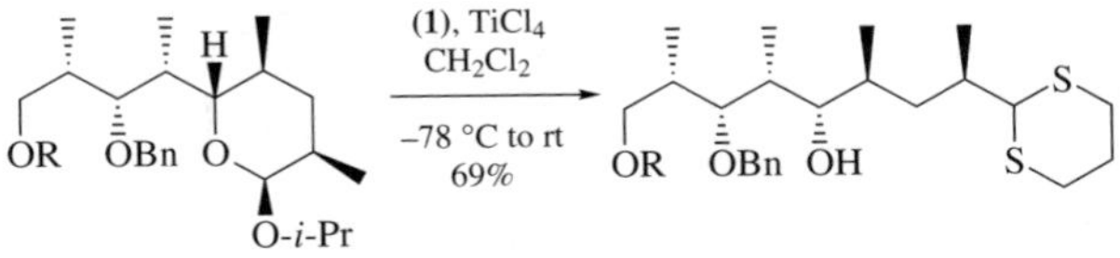

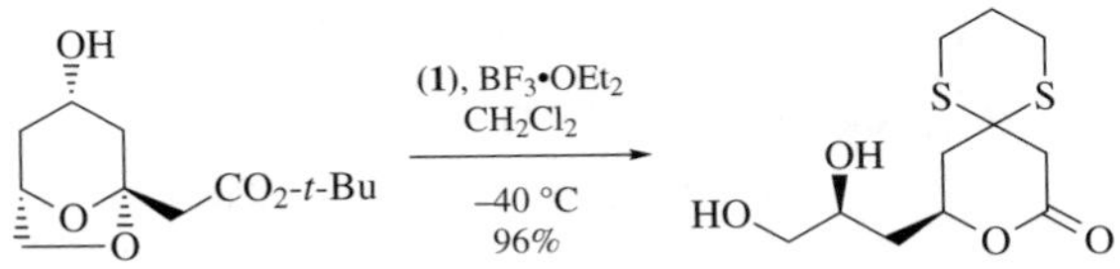

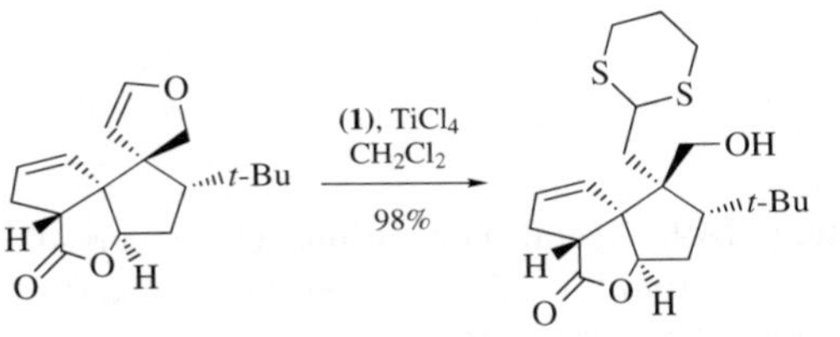

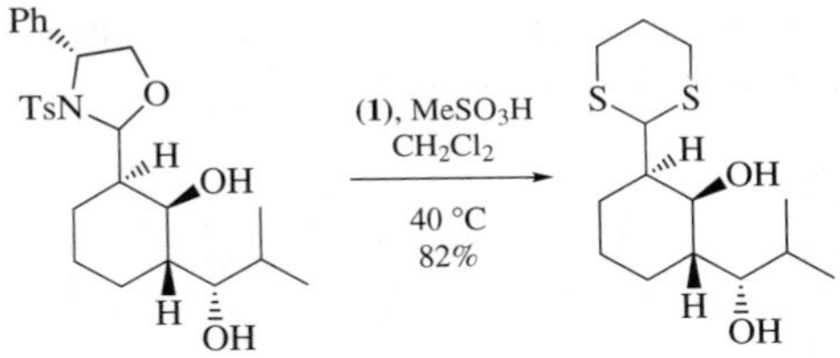

(7)

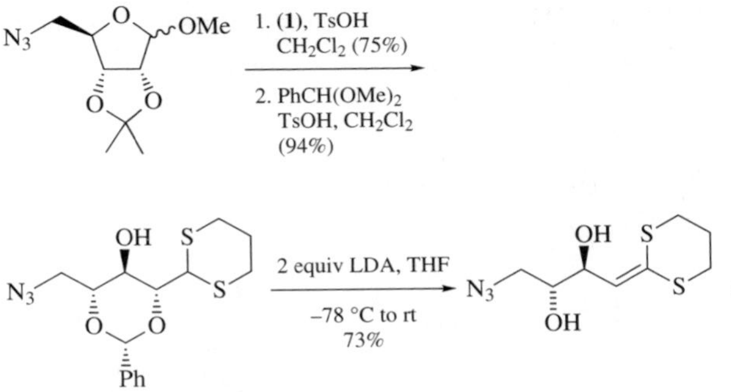

1,3-Dithianes derived from aldehydes are important synthetic intermediates because they undergo deprotonation to 2-lithio-1,3-dithianes which function as carbonyl anion equivalents (umpolung reagents).[1b–e] The vinylogous process (eq 4)[5] is a subject of current study.[15]

Regeneration of the carbonyl group can entail a trial-and-error process with the many procedures available.[1a–e] Hg^{II} salts and *N*-halo amides remain the most frequently employed reagents. Aqueous ***Iodomethane*** is a mild alternative.[8c] Epimerization at an α stereocenter does not normally attend the formation or removal of the dithianyl group (eqs 8, 9, 12, 13).

Reduction. ***Raney Nickel*** treatment of thioacetals is a standard method for carbonyl-to-methylene reduction.[1b] In the presence of ***Triethylamine***, (1) reduces azides to primary amines (eq 14).[16a,b] Under the acidic conditions employed for thioacetal formation, this reduction does not occur (eq 8).[11b,16c] Reduction of peptidic disulfides to dithiols can be conveniently accomplished with (1).[17]

Treatment with (**1**) effected demercuration α to an ester diastereoselectively (eq 15).[18]

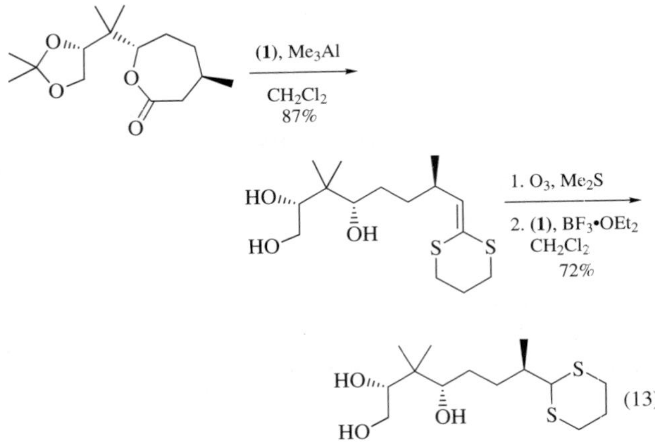

(1), Me_3Al / CH_2Cl_2 / 87%

1. O_3, Me_2S / 2. (1), $BF_3•OEt_2$ / CH_2Cl_2 / 72% (13)

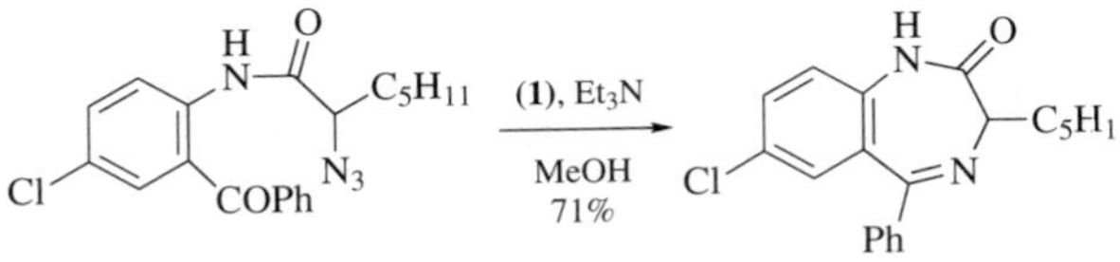

(1), Et_3N / MeOH / 71% (14)

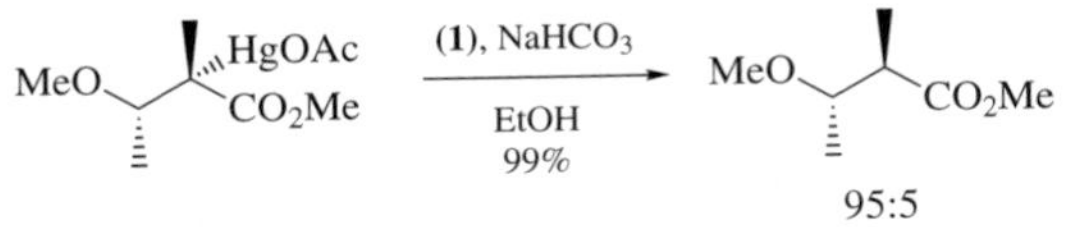

(1), $NaHCO_3$ / EtOH / 99% / 95:5 (15)

Ketene Dithioacetal Formation. These versatile intermediates[1c] arise from 1,3-dithiane anions by elimination (eq 8)[11b,19] or vinylogous alkylation (eq 4),[5] and by condensation of carboxylic acid derivatives with (**1**) (eq 13).[20a] Ketene dithioacetals derived from lactones can cyclize to give dithio orthoesters, which can be selectively deprotected (eq 16).[20b]

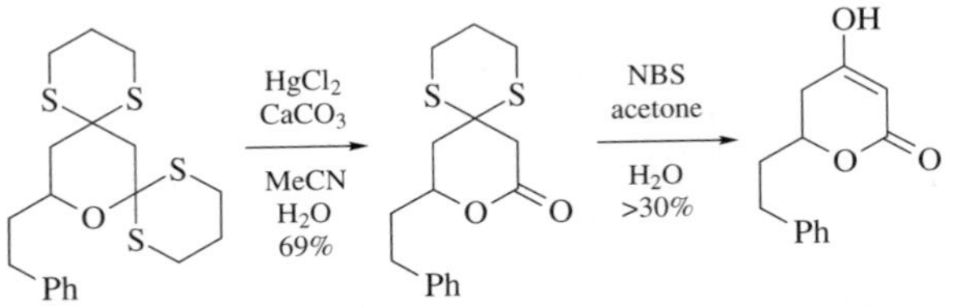

$HgCl_2$ / $CaCO_3$ / MeCN / H_2O / 69% ; NBS / acetone / H_2O / >30% (16)

Related Reagents. 1,2-Ethanedithiol; Ethanethiol; Methanethiol.

1. (a) Greene, T. W.; Wuts, P. G. M. *Protective Groups in Organic Synthesis*, 2nd ed.; Wiley: New York, 1991, p 201. (b) Gröbel, B.-T.; Seebach, D. *S* **1977**, 357. (c) Page, P. C. B.; van Niel, M. B.; Prodger, J. C. *T* **1989**, *45*, 7643. (d) Kolb, M. *S* **1990**, 171. (e) Ogura, K. *COS* **1991**, *1*, Chapter 2.3; Krief, A. *COS* **1991**, *3*, Chapter 1.3. (f) Perrin, D. D.; Armarego, W. L. F.; Perrin, D. R. *Purification of Laboratory Chemicals*, 2nd ed.; Pergamon: Oxford, 1980.
2. Ohmori, K.; Suzuki, T.; Miyazawa, K.; Nishiyama, S.; Yamamura, S. *TL* **1993**, *34*, 4981.
3. Jacobi, P. A.; Brownstein, A.; Martinelli, M.; Grozinger, K. *JACS* **1981**, *103*, 239.
4. (a) Stahl, I.; Manske, R.; Gosselck, J. *CB* **1980**, *113*, 800. (b) Stahl, I.; Gosselck, J. *S* **1980**, 561.
5. Fang, J.-M.; Liao, L.-F.; Hong, B.-C. *JOC* **1986**, *51*, 2828.
6. (a) Schmidt, U.; Meyer, R.; Leitenberger, V.; Griesser, H.; Lieberknecht, A. *S* **1992**, 1025. (b) Schreiber, S. L.; Sommer, T. J. *TL* **1983**, *24*, 4781.
7. (a) Chakraborty, T. K.; Reddy, G. V. *JOC* **1992**, *57*, 5462. (b) Jones, T. K.; Mills, S. G.; Reamer, R. A.; Askin, D.; Desmond, R.; Volante, R. P.; Shinkai, I. *JACS* **1989**, *111*, 1157. (c) Chen, S. H.; Horvath, R. F.; Joglar, J.; Fisher, M. J.; Danishefsky, S. J. *JOC* **1991**, *56*, 5834. (d) Rosen, T.; Taschner, M. J.; Thomas, J. A.; Heathcock, C. H. *JOC* **1985**, *50*, 1190. (e) Golec, J. M. C.; Hedgecock, C. J. R.; Kennewell, P. D. *TL* **1992**, *33*, 547.
8. (a) Xu, X.-X.; Zhu, J.; Huang, D.-Z.; Zhou, W.-S. *T* **1986**, *42*, 819. (b) Tani, H.; Masumoto, K.; Inamasu, T.; Suzuki, H. *TL* **1991**, *32*, 2039. (c) Myers, A. G.; Condroski, K. R. *JACS* **1995**, *117*, 3057. (d) Corey, E. J.; Tius, M. A.; Das, J. *JACS* **1980**, *102*, 1742.
9. (a) Reference 1 (a), p. 188. (b) Hoppmann, A.; Weyerstahl, P.; Zummack, W. *LA* **1977**, 1547.
10. (a) Johnson, W. S.; Frei, B.; Gopalan, A. S. *JOC* **1981**, *46*, 1512. (b) Ranu, B. C.; Bhar, S.; Chakraborti, R. *JOC* **1992**, *57*, 7349.
11. (a) Tanino, H.; Nakata, T.; Kaneko, T.; Kishi, Y. *JACS* **1977**, *99*, 2818. (b) Moss, W. O.; Bradbury, R. H.; Hales, N. J.; Gallagher, T. *JCS(P1)* **1992**, 1901. (c) Myles, D. C.; Danishefsky, S. J.; Schulte, G. *JOC* **1990**, *55*, 1636. (d) Nakata, T.; Nagao, S.; Oishi, T. *TL* **1985**, *26*, 75. (e) Corey, E. J.; Kang, M.-c.; Desai, M. C.; Ghosh, A. K.; Houpis, I. N. *JACS* **1988**, *110*, 649. (f) Hoppe, I.; Hoppe, D.; Herbst-Irmer, R.; Egert, E. *TL* **1990**, *31*, 6859.
12. (a) Sato, T.; Otera, J.; Nozaki, H. *JOC* **1993**, *58*, 4971. (b) Sánchez, I. H.; López, F. J.; Soria, J. J.; Larraza, M. I.; Flores, H. J. *JACS* **1983**, *105*, 7640. (c) Burford, C.; Cooke, F.; Roy, G.; Magnus, P. *T* **1983**, *39*, 867.
13. (a) Alonso, R. A.; Vite, G. D.; McDevitt, R. E.; Fraser-Reid, B. *JOC* **1992**, *57*, 573. (b) Page, P. C. B.; Roberts, R. A.; Paquette, L. A. *TL* **1983**, *24*, 3555. (c) Corey, E. J.; Reichard, G. A. *TL* **1993**, *34*, 6973.
14. (a) Page, P. C. B.; Klair, S. S.; Brown, M. P.; Smith, C. S.; Maginn, S. J.; Mulley, S. *T* **1992**, *48*, 5933. (b) Lissel, M. *LA* **1982**, 1589. (c) Kim, S.; Kim, S. S.; Lim, S. T.; Shim, S. C. *JOC* **1987**, *52*, 2114.
15. (a) Moss, W. O.; Jones, A. C.; Wisedale, R.; Mahon, M. F.; Molloy, K. C.; Bradbury, R. H.; Hales, N. J.; Gallagher, T. *JCS(P1)* **1992**, 2615. (b) Köksal, Y.; Raddatz, P.; Winterfeldt, E. *LA* **1984**, 450.
16. (a) Goldstein, S. W.; McDermott, R. E.; Makowski, M. R.; Eller, C. *TL* **1991**, *32*, 5493. (b) Lim, M.-I.; Marquez, V. E. *TL* **1983**, *24*, 5559. (c) Durette, P. L. *Carbohydr. Res.* **1982**, *100*, C27.
17. (a) Ranganathan, S.; Jayaraman, N. *CC* **1991**, 934. (b) Lees, W. J.; Whitesides, G. M. *JOC* **1993**, *58*, 642.
18. Gouzoules, F. H.; Whitney, R. A. *JOC* **1986**, *51*, 2024.
19. (a) Muzard, M.; Portella, C. *JOC* **1993**, *58*, 29. (b) Barton, D. H. R.; Gateau-Olesker, A.; Anaya-Mateos, J.; Cleophax, J.; Gero, S. D.; Chiaroni, A.; Riche, C. *JCS(P1)* **1990**, 3211.
20. (a) Corey, E. J.; Pan, B.-C.; Hua, D. H.; Deardorff, D. R. *JACS* **1982**, *104*, 6816. (b) Dziadulewicz, E.; Giles, M.; Moss, W. O.; Gallagher, T.; Harman, M.; Hursthouse, M. B. *JCS(P1)* **1989**, 1793.

Raymond E. Conrow
Alcon Laboratories, Fort Worth, TX, USA

Silver(I) Tetrafluoroborate

$AgBF_4$

[14104-20-2] $AgBF_4$ (MW 194.68)

(mild Lewis acid with a high affinity for organic halides)

Physical Data: mp 200 °C (dec).
Solubility: sol benzene, toluene, nitromethane, diethyl ether, water.
Form Supplied in: white solid; widely available.
Analysis of Reagent Purity: contents of Ag can be assayed conveniently by volumetric titration of Ag^I.
Preparative Method: can be prepared by reacting ***Silver(I) Fluoride*** with ***Boron Trifluoride*** in nitromethane.[1]
Handling, Storage, and Precautions: should be protected from light and moisture; very hygroscopic.

Introduction. This reagent has replaced ***Silver(I) Perchlorate*** to a large extent because of the sensitivity of perchlorates.

Activation of Acyl Chlorides. In several cases, $AgBF_4$ has been used to increase the reactivity of acyl chlorides towards nucleophiles.[2] For example, *N*-acylammonium salts were prepared for the first time by the reaction of a tertiary amine and an acyl chloride in the presence of $AgBF_4$ (eq 1).[3]

NMe2 + PhCOCl — AgBF4, MeCN, 4 °C, 85% → Me2N+–COPh BF4− (1)

Nucleophilic Substitution on Alkyl Halides by Heteroatoms. A number of more or less activated alkyl halides, such as benzyl halides[4] and allyl halides,[5] undergo substitution reactions mediated by $AgBF_4$ in the presence of a heteroatom nucleophile. For example, treatment of pentamethylcyclopentadienyl bromide with $AgBF_4$ in the presence of a nucleophile gives the corresponding substituted product (eq 2). Thiols, amines, and alcohols have been used as nucleophiles.[6]

Br — AgBF4, MeOH, CH2Cl2, −10 °C, 77% → OMe (2)

Adenine analogs are prepared stereoselectively from cyclopentene derivatives using a two-step procedure (eq 3). The reaction probably involves a seleniranium salt as an intermediate.[7]

i-PrO, i-PrO, P=O + Cl purine (N, H) — AgBF4, PhSeCl, MeNO2, rt, ultrasound, 46% → SePh (3)

Intramolecular substitutions mediated by Ag^+ do not seem to require activated halides.[8] For example, ω-chloro amides react with $AgBF_4$, giving products from intramolecular attack of the amide oxygen. Depending on the structure of the amide, imino lactones,[9] imino lactonium salts,[10] or lactone hydrazones (eq 4)[11] are obtained as products. Fluorination of α-bromo ketones using $AgBF_4$ has also been reported.[12]

Cl–(CH2)3–C(O)–NH–NMe2 — AgBF4, THF, rt, 75% → Me2N–N= lactone (4)

Nucleophilic Aromatic Substitution. In one case, it has been reported that $AgBF_4$ promotes the nucleophilic substitution of an aromatic chloride (eq 5).[13] This is not due to activation of the halide, but apparently to suppression of halide-promoted decomplexation of the arene–manganese derivative.

MeO2C, NHBoc, OMe + MeO, Cl, Mn(CO)3+ — AgBF4, CH2Cl2, 0 °C, 79% → (CO)3Mn+, MeO, O, OMe, MeO2C, NHBoc (5)

Carbon–Carbon Bond Formation via Cationic Intermediates. In analogy with the heteroatom substitutions described above, certain aliphatic halides undergo substitution reactions with carbon nucleophiles promoted by $AgBF_4$. For example, Eschenmoser and co-workers used $AgBF_4$ in order to transform α-chloro nitrones into 1,3-dipoles which react with ordinary alkenes in a cycloaddition manner (eq 6).[14]

−O–N+, Cl — 1. AgBF4, (ClCH2)2; 2. cyclohexene; 3. KCN, H2O, 85% → O–N, CN (6)

Livinghouse and co-workers have shown that acylnitrilium ions, prepared from isocyanides and an acid chloride followed by treatment with $AgBF_4$, are useful intermediates in the synthesis of nitrogen-containing heterocycles (eq 7).[15]

OTBS; 1. (isobutyryl chloride) Cl; 2. $AgBF_4$, CH_2Cl_2, −78 °C, 78% (7)

In certain cases, allylsilanes[16] and trimethylsilyl enol ethers[17] react with alkyl halides with the formation of a new carbon–carbon bond. α-Bromo imidates[18] (eq 8) and β-chloro imines[19] have been reported to undergo electrophilic aromatic substitution on relatively electron-rich aromatics in the presence of $AgBF_4$.

$AgBF_4$, $(ClCH_2)_2$, rt, 60% (8)

Synthesis via Iminium Ions. α-Cyano amines react with $AgBF_4$ with the formation of an intermediate iminium ion.[20] This has been used synthetically as a method for removal of the cyano group either by a consecutive reduction[21,22] to the amine (eq 9) or by elimination to the imine[23] or enamine.[24]

1. $AgBF_4$, THF, rt; 2. $Zn(BH_4)_2$; 92% (9)

Rearrangements. A number of strained alkyl and/or reactive halides, such as cyclopropyl[25] and bicyclic[26,27] chlorides, rearrange on treatment with $AgBF_4$. For example, β-bromotetrahydropyrans rearrange to tetrahydrofurans stereoselectively on treatment with $AgBF_4$ (eq 10).[28] Other examples include the rearrangement of α-haloalkyl aryl ketones into arylacetic acid derivatives,[29] and the rearrangement of α-haloalkylsilanes upon treatment with $AgBF_4$.[30]

$AgBF_4$, MeOH, rt, 88% (10)

In the presence of strained hydrocarbons, $AgBF_4$ functions as a mild Lewis acid and causes rearrangements.[31–33] For example, the tricyclic hydrocarbon (**1**) rearranges upon treatment with a catalytic amount of $AgBF_4$ to the less strained hydrocarbon (**2**) (eq 11).[34]

$AgBF_4$, $CDCl_3$, rt (11)

(**1**) (**2**)

Numerous examples include the rearrangement of propargyl esters into allenyl esters (see also ***Silver(I) Trifluoromethanesulfonate***)[35] or to dihydrofurans,[36] the Claisen rearrangement of aryl allenylmethyl ethers,[37] and the rearrangement of silyloxycyclopropanes (eq 12, also effected by ***Copper(II) Tetrafluoroborate***).[38]

TMSO; $AgBF_4$, Et_2O, −20 °C, 70% (12)

Activation of Thiol Esters. Pyridyl thiol esters are converted into esters on treatment with $AgBF_4$ and an alcohol.[39] Acylation of alkynylsilanes can also be carried out using thiol esters in the presence of $AgBF_4$.[40]

Alkylation of Thioethers. Thioethers can be methylated by ***Iodomethane*** in the presence of $AgBF_4$.[41] Benzylation of thioethers in the presence of $AgBF_4$ has also been reported.[42]

Electrophilic Aromatic Substitution. Electrophilic nitration using a combination of NO_2Cl and $AgBF_4$ has been reported.[43] Conversion of arylsilanes into iodides and bromides has been achieved using a combination of the halogen and $AgBF_4$ (eq 13).[44]

TMS, TMS, O_2; $AgBF_4$, I_2, CH_2Cl_2, 0 °C, 97%; I, I, O_2 (13)

Catalysis of Cycloadditions. Addition of catalytic amounts of $AgBF_4$ greatly increases the selectivity of [2+4] cycloadditions of benzyne.[45]

Related Reagents. Dimethyl Sulfoxide–Silver Tetrafluoroborate.

1. Olah, G. A.; Quinn, H. W. *J. Inorg. Nucl. Chem.* **1960**, *14*, 295.
2. Schegolev, A. A.; Smit, W. A.; Roitburd, G. V.; Kucherov, V. F. *TL* **1974**, 3373.
3. King, J. A., Jr.; Bryant, G. L., Jr. *JOC* **1992**, *57*, 5136.
4. Zimmerman, H. E.; Paskovich, D. H. *JACS* **1964**, *86*, 2149.
5. Bloodworth, A. J.; Tallant, N. A. *CC* **1992**, 428.
6. Jutzi, P.; Mix, A. *CB* **1992**, *125*, 951.
7. Wolff-Kugel, D.; Halazy, S. *TL* **1991**, *32*, 6341.
8. Lucchini, V.; Modena, G.; Pasquato, L. *CC* **1992**, 293.
9. Peter, H.; Brugger, M.; Schreiber, J.; Eschenmoser, A. *HCA* **1963**, *46*, 577.
10. Nader, R. B.; Kaloustain, M. K. *TL* **1979**, 1477.
11. Enders, D.; Brauer-Scheib, S.; Fey, P. *S* **1985**, 393.
12. Fry, A. J.; Migron, Y. *TL* **1979**, 3357.
13. Pearson, A. J.; Shin, H. *T* **1992**, *48*, 7527.
14. Kempe, H. M.; Das Gupta, T. K.; Blatt, K.; Gygax, P.; Felix, D.; Eschenmoser, A. *HCA* **1972**, *55*, 2187.
15. Lee, C. H.; Westling, M.; Livinghouse, T.; Williams, A. C. *JACS* **1992**, *114*, 4089; Luedtke, G.; Westling, M.; Livinghouse, T. *T* **1992**, *48*, 2209.
16. Nishiyama, H.; Naritomi, T.; Sakuta, T.; Itoh, K. *JOC* **1983**, *48*, 1557.
17. Padwa, A.; Ishida, M. *TL* **1991**, *41*, 5673; Padwa, A.; Austin, D. J.; Ishida, M.; Muller, C. M.; Murphree, S. S.; Yeske, P. E. *JOC* **1992**, *57*, 1161.

18. Shatzmiller, S.; Bercovici, S. *LA* **1992**, 997.
19. Kuehne, M.; Matson, P. A.; Bornmann, W. G. *JOC* **1991**, *56*, 513.
20. Grierson, D. S.; Bettiol, J. L.; Buck, I.; Husson, H. P. *JOC* **1992**, *57*, 6414.
21. Bettiol, J. L.; Buck, I.; Husson, H. P.; Grierson, D. S. *TL* **1991**, *32*, 5413.
22. Theodorakis, E.; Royer, J.; Husson, H. P. *SC* **1991**, 521.
23. Belattar, A.; Saxton, J. E. *JCS(P1)* **1992**, 1583.
24. Agami, C.; Couty, F.; Lin, J. *H* **1993**, *36*, 25.
25. Birch, A. J.; Keeton, R. *JCS(C)* **1968**, 109.
26. Yamada, Y.; Kimura, M.; Nagaoka, H.; Ohnishi, K. *TL* **1977**, 2379.
27. Kraus, G. A.; Zheng, D. *SL* **1993**, 71.
28. Ting, P. C.; Bartlett, P. A. *JACS* **1984**, *106*, 2668.
29. Giordano, C.; Castaldi, G.; Casagrande, F.; Belli, A. *JCS(P1)* **1982**, 2575.
30. Eaborn, C.; Lickiss, P. D.; Najim, S. T.; Stanczyk, W. A. *JCS(P2)* **1993**, 59; Eaborn, C.; Lickiss, P. D.; Najim, S. T. *JCS(P2)* **1993**, 391.
31. Paquette, L. A. *S* **1975**, 349.
32. Paquette, L. A. *JACS* **1970**, *92*, 5765.
33. Fitjer, L.; Justus, K.; Puder, P.; Dittmer, M.; Hassler, C.; Noltemeyer, M. *AG(E)* **1991**, *30*, 436.
34. Paquette, L. A.; Leichter, L. M. *JACS* **1972**, *94*, 3653.
35. Koch-Pomeranz, U.; Hansen, H. J.; Schmid, H. *HCA* **1973**, *56*, 2981.
36. Shigemans, Y.; Yasui, M.; Ohrai, S.; Sasaki, M.; Sashiwa, H.; Saimoto, H. *JOC* **1991**, *56*, 910.
37. Dikshit, D. K.; Singh, S.; Panday, S. K. *JCR(S)* **1991**, 298.
38. Ruy, I.; Ando, M.; Ogawa, A.; Murai, S.; Sonoda, N. *JACS* **1983**, *105*, 7192.
39. Gerlach, H.; Thalmann, A. *HCA* **1974**, *57*, 2661.
40. Kawanami, Y.; Katsuki, T.; Yamaguchi, M. *TL* **1983**, *24*, 5131.
41. Ishibashi, H.; Tabata, T.; Kobayashi, T.; Takamuro, I.; Ikeda, M. *CPB* **1991**, *39*, 2878.
42. Beerli, R.; Borschberg, H. J. *HCA* **1991**, *74*, 110.
43. Olah, G. A.; Pavláth, A.; Kuhn, S. *CI(L)* **1957**, *50*; Kuhn, S. J.; Olah, G. A. *JACS* **1961**, *83*, 4564.
44. Furukawa, N.; Hoshiai, H.; Shibutani, T.; Higaki, M.; Iwasaki, F.; Fujihara, H. *H* **1992**, *34*, 1085.
45. Crews, P.; Beard, J. *JOC* **1973**, *38*, 529.

Lars-G. Wistrand
Nycomed Innovation, Malmö, Sweden

Tetra-*n*-butylammonium Fluoride[1]

n-Bu_4NF

(TBAF)		
[429-41-4]	$C_{16}H_{36}FN$	(MW 261.53)
(TBAF·$3H_2O$)		
[87749-50-6]	$C_{16}H_{42}FNO_3$	(MW 315.59)
(TBAF·xH_2O)		
[22206-57-1]		

(can be used for most fluoride-assisted reactions; deprotection of silyl groups;[1e] desilylation;[2,3] fluorination;[4] used as a base[5,6])

Alternate Name: TBAF.
Physical Data: TBAF·xH_2O: mp 62–63 °C.
Solubility: sol H_2O, THF, MeCN.
Form Supplied in: trihydrate, 1.0 M solution in THF, and 75 wt % solution in water.
Preparative Method: aqueous ***Hydrofluoric Acid*** is passed through an Amberlite IRA 410 OH column, followed by an aqueous solution of ***Tetra-n-butylammonium Bromide***. After the resin is washed with water, the combined water fractions are repeatedly evaporated until no water is present. Tetrabutylammonium fluoride is collected as an oil in quantitative yield.
Handling, Storage, and Precautions: use in a fume hood.

Deprotection of Silyl Groups. Tetrabutylammonium fluoride has been used widely as a reagent for the efficient cleavage of various silyl protecting groups such as *O*-silyls of nucleosides,[7,8] pyrophosphate,[9] *N*-silyls,[10,11] CO_2-silyl, and *S*-silyl derivatives.[1e] These reactions are often carried out under very mild conditions in excellent yields. Thus it has been used in the synthesis of base-sensitive chlorohydrins (eq 1)[12] and β-lactams.[10,13] 2-(Trimethylsilyl)ethoxymethyl groups can also be effectively removed from various substrates (eq 2).[14–18] Silyl ethers can be converted to esters in one pot when they are treated with TBAF, followed by exposure to acyl chlorides[19,20] or anhydride[21] in the presence of base (eq 3). Treatment of triisopropylsilyl enol ethers with ***Iodosylbenzene/Azidotrimethylsilane***, followed by desilylation and elimination with TBAF, gives good yields of the α,β-unsaturated ketones (eq 4).[22,23]

Bu_4NF, THF, 25 °C, 30 min, 50% (1)

Cyclobutanone alkyl silyl acetals, obtained from [2+2] cycloadditions, can be deprotected with 1 equiv of TBAF in THF to give the open-chain cyano esters in excellent yields (eq 5).[24] When 4-chloro-2-cyanocyclobutane alkyl silyl acetals are used as substrates for this reaction, (*E/Z*) mixtures of 2-cyanocyclopropanecarboxylates are obtained by an intramolecular cyclization (eq 6).

TBAF, THF, reflux, 60% (2)

1. TBAF, THF; 2. Et_3N, 80 °C (3)

TBAF, THF, 0 °C (4)

TBAF, THF, 90% (5)

TBAF, THF, 83% (6)

11-Membered pyrrolizidine dilactones have been synthesized by treating a trimethylsilylethyl ester with TBAF in MeCN to form an anion, which then undergoes cyclization by displacement of the mesylate.

Desilylation Reagent. Cleavage of carbon–silicon bonds with fluoride has been studied very extensively. TBAF is a very powerful reagent for desilylation of a wide range of silicon-containing compounds, such as vinylsilanes,[2,25,26] alkynylsilanes,[23,27] arylsilanes,[28,29] acylsilanes,[30] β-silyl sulfones,[31–33] and other silane derivatives.[3,34–37] It appears that cleavage of sp-C–Si bonds is more facile than that of sp^2-C–Si and sp^3-C–Si bonds and that substituted groups, such as phenyl and alkoxyl, can often facilitate cleavage. A dimethylphenylsilyl group can be removed from a vinyl carbon by TBAF with retention of the alkene stereochemistry (eq 7).[2] This method has been applied to the synthesis of terminal conjugated trienes (eqs 8 and 9).[38] The five-membered siloxanes can be desilylated with 3 equiv of TBAF in DMF and this protodesilylation is very sensitive to subtle structure changes (eq 10).[39]

TBAF, 80 °C, THF–DMSO (2:1), 99% (7)

1. AcONa, AcOH, 40 °C, 99%; 2. TBAF, DMSO–THF, 80 °C, 99%; 98% (8)

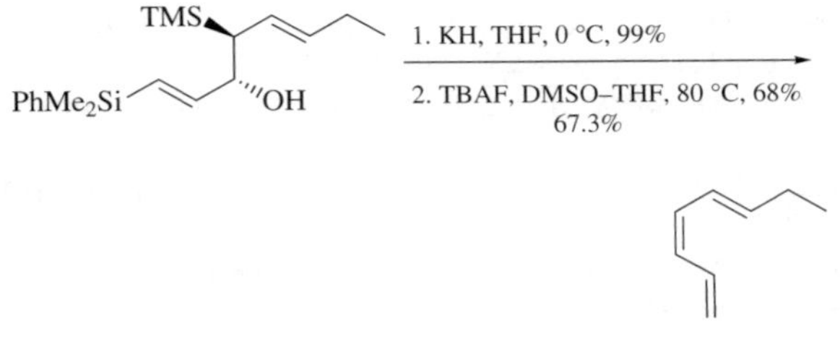

1. KH, THF, 0 °C, 99%
2. TBAF, DMSO–THF, 80 °C, 68%
67.3%

(9)

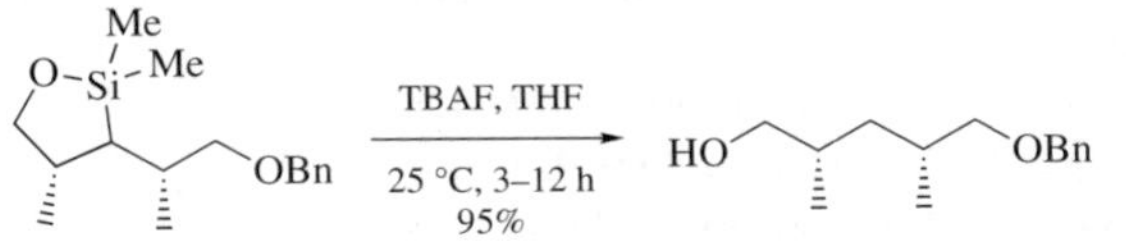

TBAF, THF
25 °C, 3–12 h
95%
(10)

The anions, generated in situ by desilylation of silylacetylenes,[40,41] allylsilanes,[42–44] propargylsilanes,[45] α-silyloxetanones,[46] bis(trimethylsilylmethyl) sulfides,[47] and other silane derivatives,[48–51] can undergo nucleophilic addition to ketones and aldehydes (eq 11).[52] *N*-(*C*,*C*-bis(trimethylsilyl)methyl) amido derivatives can add to aldehydes followed by Peterson alkenation to form acyl enamines.[48,53] Treatment of 2-trimethylsilyl-1,3-dithianes can generate dithianyl anions, which are capable of carbocyclization via direct addition to carbonyl or Michael addition (eq 12). The fluoride-catalyzed Michael additions are more general than Lewis acid-catalyzed reactions and proceed well even for those compounds with enolizable protons and/or severe steric hindrance (eq 13).[54,55]

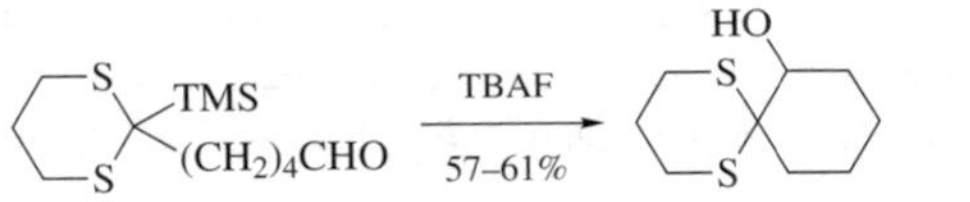

TBAF
57–61%
(11)

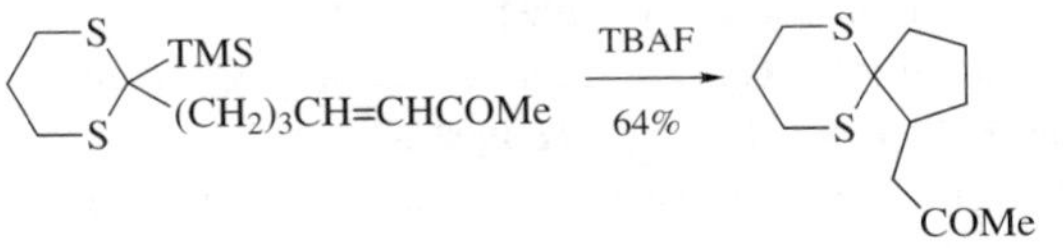

TBAF
64%
(12)

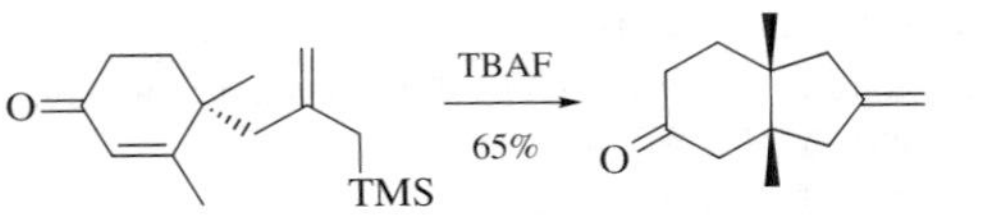

TBAF
65%
(13)

Direct fluoride-induced trifluoromethylation of α-keto esters (eq 14),[56] ketones,[57] aldehydes,[58,59] and sulfoxides[59] have been reported using ***Trifluoromethyltrimethylsilane*** with TBAF in THF.

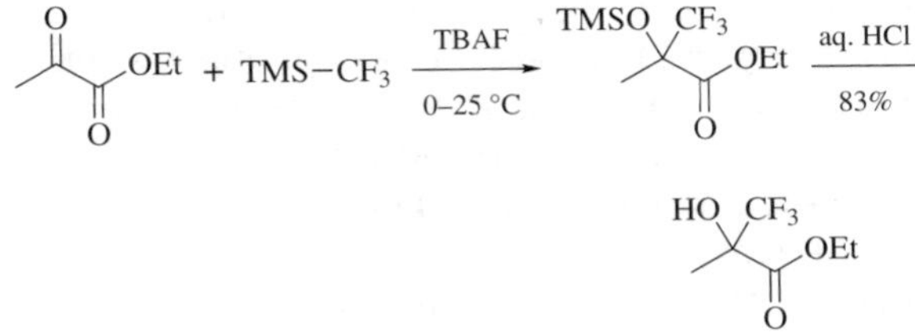

TBAF, 0–25 °C; aq. HCl, 83%
(14)

Desilylation of some compounds can generate very reactive species such as benzynes,[60] pyridynes,[61] xylylenes,[62,63] and benzofuran-2,3-xylylenes.[64] 1,4-Elimination of *o*-(α-trimethylsilylalkyl)benzyltrimethylammonium halides with TBAF in acetonitrile generates *o*-xylylenes, which undergo intermolecular and intramolecular cycloadditions (eq 15).[62–64] Treatment of α-silyl disulfides with ***Cesium Fluoride*** or TBAF forms thioaldehydes, which have been trapped by cycloaddition with cyclopentadiene (eq 16).[65]

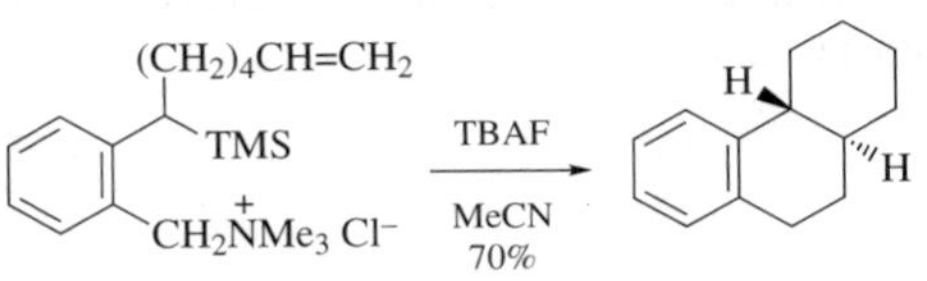

TBAF
MeCN
70%
(15)

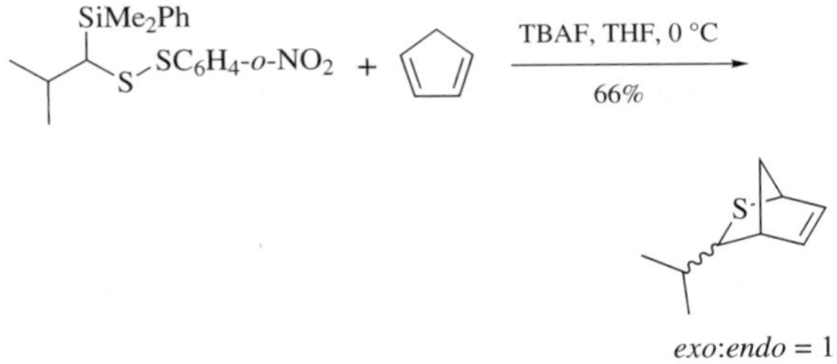

TBAF, THF, 0 °C
66%
(16)

exo:*endo* = 1:7

Use as a Base. TBAF has been widely used for a variety of base-catalyzed reactions such as alkylation,[66] elimination,[67] halogenation,[68] Michael addition,[69–71] aldol condensation, and intramolecular cyclizations.[5,72–74] It is especially useful when other inorganic bases face solubility problems in organic solvents. The reactions are usually carried out below 100 °C due to the low thermal stability of TBAF.[1e]

TBAF is very useful for alkylation of nucleic acid derivatives. Methylation[75] or benzylation[66] of uracil gives almost quantitative yields of alkylated product when using alkyl bromides, dialkyl sulfates (eq 17), trialkyl phosphates, or alkyl chlorides with TBAF. Alkylation of the thiol anions generated from deprotection by 1,2-dibromoethane produces interesting tetrachalcogenofulvalenes.[14] Under phase-transfer conditions, selective mono- and dialkylations of malononitrile have been achieved by using neat TBAF with ***Potassium Carbonate*** or ***Potassium t-Butoxide*** and controlling the amount of alkyl bromides or iodides used (eqs 18 and 19).[76]

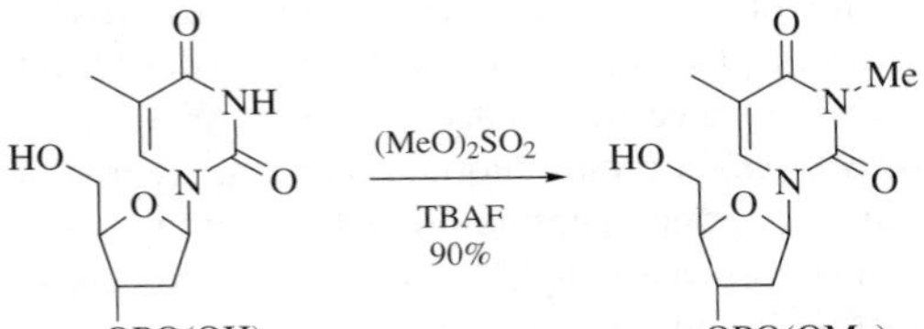

$(MeO)_2SO_2$, TBAF, 90%
(17)

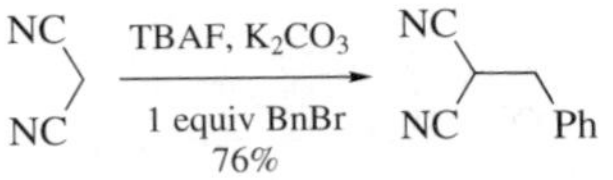

TBAF, K_2CO_3, 1 equiv BnBr, 76%
(18)

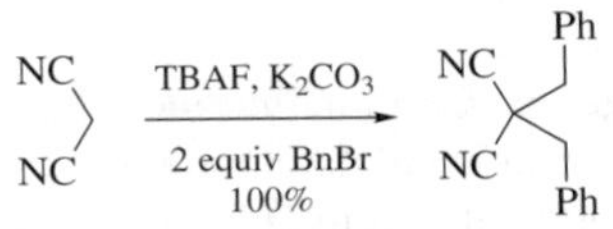

TBAF, K_2CO_3, 2 equiv BnBr, 100%
(19)

Enol silyl ethers react with aldehydes with a catalytic amount of TBAF to give the aldol silyl ethers in good yields. These reactions generally proceed under very mild conditions and within shorter periods of time than conventional strong acidic or basic conditions. The products from 4-*t*-butyl-1-methyl-2-(trimethylsilyloxy)cyclohexene and benzaldehyde show very good axial selectivity and a little *anti–syn* selectivity (eq 20).[77] The aldol condensation of ketones and aldehydes can be achieved in one pot when ethyl (trimethylsilyl)acetate is used as a silylation agent with TBAF (eq 21).

Silyl nitronates undergo aldol condensation with aldehydes in the presence of a catalytic amount of anhydrous TBAF to form highly diastereoselective *erythro* products, which can be elaborated to give synthetically useful 1,2-amino alcohols (eq 22).[6,78] A one-pot procedure has been developed for direct aldol condensation of nitroalkanes with aldehydes by using TBAF trihydrate with ***Triethylamine*** and ***t-Butyldimethylchlorosilane***.[79] It appears that silyl nitronates are not reactive intermediates in this case, and the reactions proceed by a different mechanism.

Miscellaneous. Fluoride ion from anhydrous TBAF undergoes nucleophilic displacement of tosylates,[4,80] halides,[80] and aryl nitro compounds[81] to give fluorinated products. When used with ***N-Bromosuccinimide***, bromofluorination products are obtained.[82]

Several important peptide-protecting groups such as 9-fluorenylmethyloxycarbonyl,[83] benzyl,[84] 4-nitrobenzyl,[85] 2,2,2-trichloroethyl,[85] and acetonyl (eq 23)[86] can be removed by TBAF under mild conditions.

Related Reagents. Cesium Fluoride; Lithium Fluoride; Potassium Fluoride; Tetra-*n*-butylammonium Fluoride (Supported); Tris(dimethylamino)sulfonium Difluorotrimethylsilicate.

1. (a) Corey, E. J.; Snider, B. B. *JACS* **1972**, *94*, 2549. (b) Hudlicky, M. *Chemistry of Organic Fluorine Compounds*, 2nd ed.; Horwood: New York, 1992. (c) Umemoto, T. *Yuki Gosei Kagaku Kyokaishi* **1992**, *50*, 338. (d) Clark, J. H. *CRV* **1980**, *80*, 429. (e) Greene, T. W.; Wuts, P. G. M. *Protective Groups in Organic Synthesis*, 2nd ed.; Wiley: New York, 1991. (f) Sharma, R. K.; Fry, J. L. *JOC* **1983**, *48*, 2112. (g) Cox, D. P.; Terpinski, J.; Lawrynowicz, W. *JOC* **1984**, *49*, 3216.
2. Oda, H.; Sato, M.; Morizawa, Y.; Oshima, K.; Nozaki, H. *T* **1985**, *41*, 3257.
3. Dhar, R. K.; Clawson, D. K.; Fronczek, F. R.; Rabideau, P. W. *JOC* **1992**, *57*, 2917.
4. Gerdes, J. M.; Bishop, J. E.; Mathis, C. A. *JFC* **1991**, *51*, 149.
5. Pless, J. *JOC* **1974**, *39*, 2644.
6. Seebach, D.; Beck, A. K.; Mukhopadhyay, T.; Thomas, E. *HCA* **1982**, *65*, 1101.
7. Krawczyk, S. H.; Townsend, L. B. *TL* **1991**, *32*, 5693.
8. Meier, C.; Tam, H.-D. *SL* **1991**, 227.
9. Valentijn, A. R. P. M.; van der Marel, G. A.; Cohen, L. H.; van Boom, J. *SL* **1991**, 663.
10. Hanessian, S.; Sumi, K.; Vanasse, B. *SL* **1992**, 33.
11. Kita, Y.; Shibata, N.; Tamura, O.; Miki, T. *CPB* **1991**, *39*, 2225.
12. Solladié-Cavallo, A.; Quazzotti, S.; Fischer, J.; DeCian, A. *JOC* **1992**, *57*, 174.
13. Konosu, T.; Oida, S. *CPB* **1991**, *39*, 2212.
14. Zambounis, J. S.; Mayer, C. W. *TL* **1991**, *32*, 2737.
15. Kita, H.; Tohma, H.; Inagaki, M.; Hatanaka, K. *H* **1992**, *33*, 503.
16. Stephenson, G. R.; Owen, D. A.; Finch, H.; Swanson, S. *TL* **1991**, *32*, 1291.
17. Fugina, N.; Holzer, W.; Wasicky, M. *H* **1992**, *34*, 303.
18. Shakya, S.; Durst, T. *H* **1992**, *34*, 67.
19. Beaucage, S. L.; Ogilvie, K. K. *TL* **1977**, 1691.
20. Ma, C.; Miller, M. J. *TL* **1991**, *32*, 2577.
21. Mandai, T.; Murakami, T.; Kawada, M.; Tsuji, J. *TL* **1991**, *32*, 3399.
22. Magnus, P.; Evans, A.; Lacour, J. *TL* **1992**, *33*, 2933.
23. Ihara, M.; Suzuki, S.; Taniguchi, N.; Fukumoto, K.; Kabuto, C. *CC* **1991**, 1168.
24. Rousseau, G.; Quendo, A. *T* **1992**, *48*, 6361.
25. Fleming, I.; Newton, T. W.; Sabin, V.; Zammattio, F. *T* **1992**, *48*, 7793.
26. Ito, T.; Okamoto, S.; Sato, F. *TL* **1990**, *31*, 6399.
27. Lopp, M.; Kanger, T.; Müraus, A.; Pehk, T.; Lille, Ü. *TA* **1991**, *2*, 943.
28. Yu, S.; Keay, B. A. *JCS(P1)* **1991**, 2600.
29. Mukai, C.; Kim, I. J.; Hanaoka, M. *TA* **1992**, *3*, 1007.
30. Degl'Innocenti, A.; Stucchi, E.; Capperucci, A.; Mordini, A.; Reginato, G.; Ricci, A. *SL* **1992**, 329.
31. Kocienski, P. J. *TL* **1979**, 2649.
32. Kocienski, P. J. *JOC* **1980**, *45*, 2037.
33. Hsiao, C. N.; Hannick, S. M. *TL* **1990**, *31*, 6609.
34. Bonini, B. F.; Masiero, S.; Mazzanti, G.; Zani, P. *TL* **1991**, *32*, 2971.
35. Nativi, C.; Palio, G.; Taddei, M. *TL* **1991**, *32*, 1583.
36. Okamoto, S.; Yoshino, T.; Tsujiyama, H.; Sato, F. *TL* **1991**, *32*, 5793.
37. Kobayashi, Y.; Ito, T.; Yamakawa, I.; Urabe, H.; Sato, F. *SL* **1991**, 813.
38. Kishi, N.; Maeda, T.; Mikami, K.; Nakai, T. *T* **1992**, *48*, 4087.
39. Hale, M. R.; Hoveyda, A. H. *JOC* **1992**, *57*, 1643.
40. Nakamura, E.; Kuwajima, I. *AG(E)* **1976**, *15*, 498.
41. Mohr, P. *TL* **1991**, *32*, 2223.
42. Furuta, K.; Mouri, M.; Yamamoto, H. *SL* **1991**, 561.
43. Hosomi, A.; Shirahata, A.; Sakurai, H. *TL* **1978**, 3043.
44. Nakamura, H.; Oya, T.; Murai, A. *BCJ* **1992**, *65*, 929.
45. Pornet, J. *TL* **1981**, *22*, 455.
46. Mead, K. T.; Park, M. *JOC* **1992**, *57*, 2511.
47. Hosomi, A.; Ogata, K.; Ohkuma, M.; Hojo, M. *SL* **1991**, 557.
48. Lasarte, J.; Palomo, C.; Picard, J. P.; Dunogues, J.; Aizpurua, J. M. *CC* **1989**, 72.
49. Watanabe, Y.; Takeda, T.; Anbo, K.; Ueno, Y.; Toru, T. *CL* **1992**, 159.

50. Paquette, L. A.; Blankenship, C.; Wells, G. J. *JACS* **1984**, *106*, 6442.
51. Seitz, D. E.; Milius, R. A.; Quick, J. *TL* **1982**, *23*, 1439.
52. Grotjahn, D. B.; Andersen, N. H. *CC* **1981**, 306.
53. Palomo, C.; Aizpurua, J. M.; Legido, M.; Picard, J. P.; Dunogues, J.; Constantieux, T. *TL* **1992**, *33*, 3903.
54. Majetich, G.; Casares, A.; Chapman, D.; Behnke, M. *JOC* **1986**, *51*, 1745.
55. Majetich, G.; Desmond, R. W.; Soria, J. J. *JOC* **1986**, *51*, 1753.
56. Ramaiah, P.; Prakash, G. K. S. *SL* **1991**, 643.
57. Coombs, M. M.; Zepik, H. H. *CC* **1992**, 1376.
58. Bansal, R. C.; Dean, B.; Hakomori, S.; Toyokuni, T. *CC* **1991**, 796.
59. Patel, N. R.; Kirchmeier, R. L. *IC* **1992**, *31*, 2537.
60. Himeshima, Y.; Sonoda, T.; Kobayashi, H. *CL* **1983**, 1211.
61. Tsukazaki, M.; Snieckus, V. *H* **1992**, *33*, 533.
62. Ito, Y.; Nakatsuka, M.; Saegusa, T. *JACS* **1980**, *102*, 863.
63. Ito, Y.; Miyata, S.; Nakatsuka, M.; Saegusa, T. *JOC* **1981**, *46*, 1043.
64. Bedford, S. B.; Begley, M. J.; Cornwall, P.; Knight, D. W. *SL* **1991**, 627.
65. Krafft, G. A.; Meinke, P. T. *TL* **1985**, *26*, 1947.
66. Botta, M.; Summa, V.; Saladino, R.; Nicoletti, R. *SC* **1991**, *21*, 2181.
67. Ben Ayed, T.; Amri, H.; El Gaied, M. M. *T* **1991**, *47*, 9621.
68. Sasson, Y.; Webster, O. W. *CC* **1992**, 1200.
69. Kuwajima, I.; Murofushi, T.; Nakamura, E. *S* **1976**, 602.
70. Yamamoto, Y.; Okano, H.; Yamada, J. *TL* **1991**, *32*, 4749.
71. Arya, P.; Wayner, D. D. M. *TL* **1991**, *32*, 6265.
72. Taguchi, T.; Suda, Y.; Hamochi, M.; Fujino, Y.; Iitaka, Y. *CL* **1991**, 1425.
73. Ley, S. V.; Smith, S. C.; Woodward, P. R. *T* **1992**, *48*, 1145, 3203.
74. White, J. D.; Ohira, S. *JOC* **1986**, *51*, 5492.
75. Ogilvie, K. K.; Beaucage, S. L.; Gillen, M. F. *TL* **1978**, 1663.
76. Díez-Barra, E.; De La Hoz, A.; Moreno, A.; Sánchez-Verdú, P. *JCS(P1)* **1991**, 2589.
77. Nakamura, E.; Shimizu, M.; Kuwajima, I.; Sakata, J.; Yokoyama, K.; Noyori, R. *JOC* **1983**, *48*, 932.
78. Colvin, E. W.; Seebach, D. *CC* **1978**, 689.
79. Fernández, R.; Gasch, C.; Gómez-Sánchez, A.; Vílchez, J. E. *TL* **1991**, *32*, 3225.
80. Cox, D. P.; Terpinski, J.; Lawrynowicz, W. *JOC* **1984**, *49*, 3216.
81. Clark, J. H.; Smith, D. K. *TL* **1985**, *26*, 2233.
82. Maeda, M.; Abe, M.; Kojima, M. *JFC* **1987**, *34*, 337.
83. Ueki, M.; Amemiya, M. *TL* **1987**, *28*, 6617.
84. Ueki, M.; Aoki, H.; Katoh, T. *TL* **1993**, *34*, 2783.
85. Namikoshi, M.; Kundu, B.; Rinehart, K. L. *JOC* **1991**, *56*, 5464.
86. Kundu, B. *TL* **1992**, *33*, 3193.

Hui-Yin Li
Du Pont Merck Pharmaceutical Company, Wilmington, DE, USA

Tetrafluoroboric Acid

BF_4H

[16872-11-0]	BF_4H	(MW 87.82)
($HBF_4{\cdot}OMe_2$)		
[67969-83-9]	$C_2H_7BF_4O$	(MW 133.90)
($HBF_4{\cdot}OEt_2$)		
[67969-82-8]	$C_4H_{11}BF_4O$	(MW 161.96)

(strong acid with p$K_a = -0.44$[1] and a noncoordinating counterion[2])

Solubility: sol H_2O, alcohols and ethers.

Form Supplied in: 48% aq solution; diethyl ether complex; dimethyl ether complex.

Purification: commercial 50% solutions can be concentrated by evaporation at 50–60 °C/5 mmHg to give a residue of 11 N total acidity. Water can also be removed by slow addition to ice-cold acetic anhydride (dangerously exothermic reaction: caution is advised!).[3]

Handling, Storage, and Precautions: storage in glass containers is not recommended, although dilute solutions can be used in glass containers over short periods. Poisonous: causes burns to eyes, skin, and mucous membranes and may be fatal if ingested. Use in a fume hood with safety goggles and chemically resistant gloves and clothing. Incompatible with cyanides and strong bases: decomposes to form HF.

General Considerations. Tetrafluoroboric acid is a useful strong Brønsted acid with a nonnucleophilic counterion. Its solubility in polar organic solvents makes it a useful strong acid under nonaqueous conditions.

Preparation of Arenediazonium Tetrafluoroborates. Arenediazonium salts can be precipitated as their tetrafluoroborate salts by addition of a cold solution of HBF_4 to a solution of the initial arenediazonium chloride. 4-Methoxybenzenediazonium tetrafluoroborate is thus obtained in 94–98% yield. The arylamine can be diazotized with ***Sodium Nitrite*** in the presence of HBF_4, and the diazonium fluoroborate then precipitates directly. 3-Nitroaniline, when treated in this manner, provides a 90–97% yield of 3-nitrobenzenediazonium tetrafluoroborate (eq 1).[4] Diazotizations of arylamines and heteroarylamines in organic solvents can be conveniently conducted using $HBF_4{\cdot}OEt_2$ and alkyl nitrites.[5]

$$\text{3-}O_2N\text{-}C_6H_4\text{-}NH_2 \xrightarrow[\text{0 °C, 90–97\%}]{NaNO_2,\ \text{aq. } HBF_4} \text{3-}O_2N\text{-}C_6H_4\text{-}N_2BF_4 \quad (1)$$

Protecting-Group Manipulations. A preparation of monoesters of glutamic and aspartic acids using $HBF_4{\cdot}OEt_2$ as a catalyst has been developed. The diesters are generally side products when other acids are used, but $HBF_4{\cdot}OEt_2$ appears to suppress the diesterification. Thus L-glutamic acid γ-benzyl ester was obtained in 94% yield by treatment of the carboxylic acid in BnOH with the $HBF_4{\cdot}OEt_2$ catalyst and a drying agent.[6]

When di-*t*-butyl carbonate is treated with DMAP·HBF_4, water-soluble Boc-pyridinium tetrafluoroborate is formed.[36] This reagent installs the Boc protecting group onto amino acids in aqueous NaOH. Thus L-proline is converted to its *N*-Boc derivative in nearly quantitative yield in 10 min.

Carbohydrate protection and deprotection reactions are amenable to HBF_4 catalysis; aqueous HBF_4 and $HBF_4{\cdot}OEt_2$ are complementary in these applications. Transacetalization with benzaldehyde dimethyl acetal and ethereal HBF_4 gives a monobenzylidene product without disturbing the isopropylidene and trityl groups of the substrate (eq 2). Aqueous HBF_4 is useful in selective deprotection reactions, cleaving a trityl group in the presence of other acid-sensitive functionality (eq 3).[7]

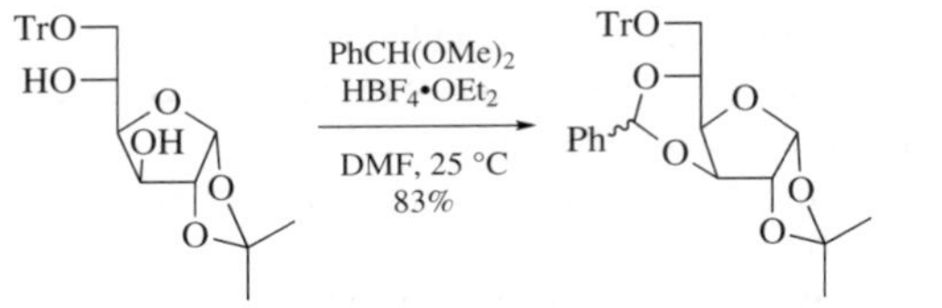

(2)

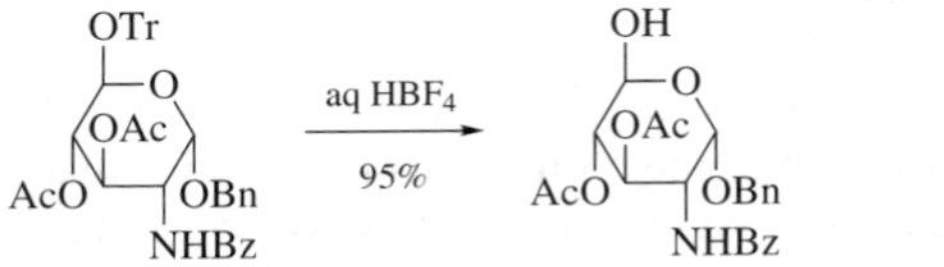

(3)

Recently developed in solid-phase peptide synthesis is the use of HBF_4/thioanisole/*m*-cresol as a general deprotection reagent. Most commonly used groups are cleaved from termini or side chain functionality under mild conditions (1 M HBF_4, 4 °C, 30–60 min) in very high yields. Human glucagon (a 29 residue peptide) has been synthesized as a demonstration of the effectiveness of the method,[8] which has been successful in both solid-phase and solution-phase syntheses.[9]

General Acid Catalysis. HBF_4 is a versatile acid catalyst which is applicable to many typical acid-catalyzed reactions. Some adducts obtained upon reaction of cyclopropyl phenyl sulfide anion with carbonyl compounds can be rearranged to cyclobutanones under the catalytic influence of HBF_4 (eq 4), although different acids have been required for other substrates.[10] Acid-catalyzed rearrangement of 2-cyclopropylphenyl phenyl ether or sulfide with HBF_4 has been reported (eq 5).[11]

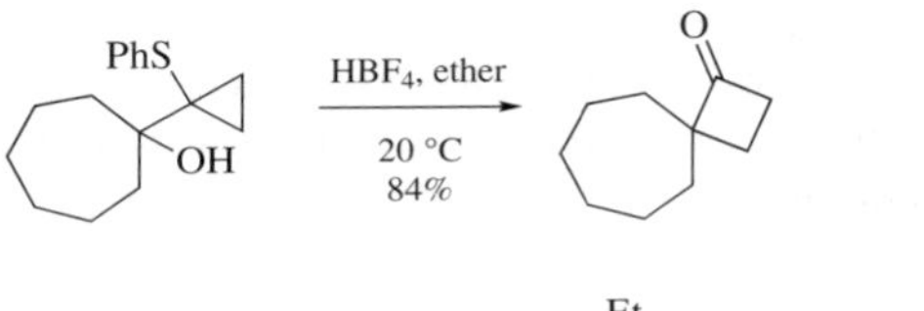

(4)

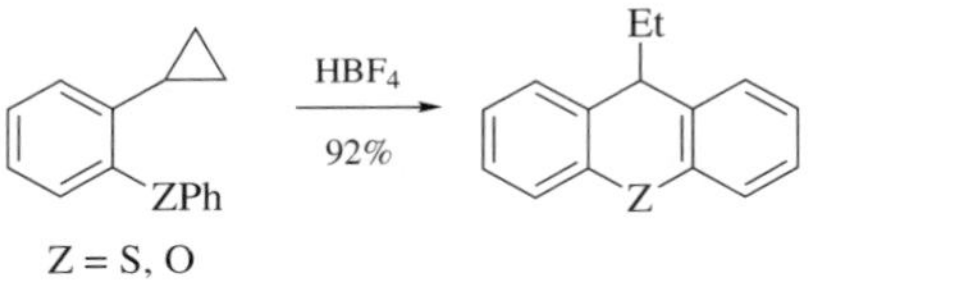

(5)

Acid-catalyzed oxidation of epoxides with $HBF_4 \cdot OMe_2$/DMSO results in the formation of α-hydroxy ketones (eq 6).[12] This procedure in an acidic medium complements the α-hydroxylation of ketone enolates under strongly basic conditions.

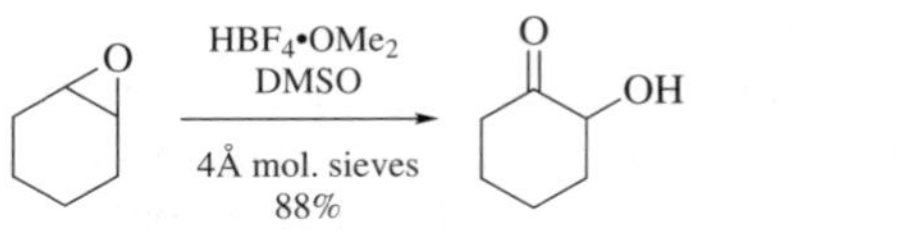

(6)

An intramolecular alkylation of a diazomethyl ketone was achieved with catalysis by HBF_4, providing an angularly fused cyclopentanone hydrofluorene (eq 7).[13]

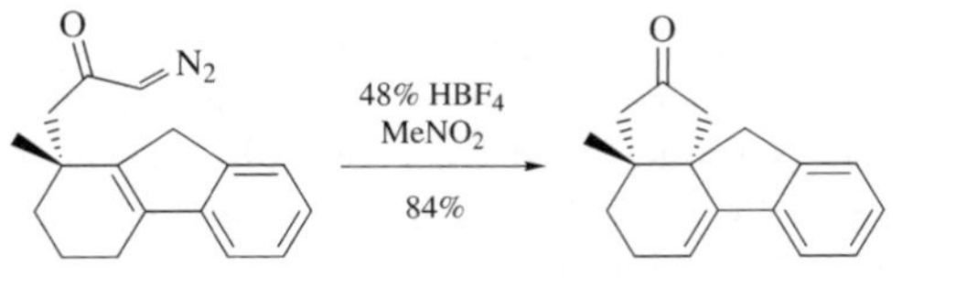

(7)

Other catalytic applications of HBF_4 include alkene isomerizations,[14] alkylation of alcohols with diazoalkanes,[15] preparations of substituted pyridines,[16] hydrolysis of α-hydroxyketene or α-(methylthio)ketene thioacetals to α,β-unsaturated thioesters,[17] and terpene formation from isoprenic precursors.[18]

Preparation of Annulated Triazolones. Cyclization of the isocyanate in eq 8 in the presence of HBF_4 affords an oxotriazolium tetrafluoroborate, which then rearranges to form a 1,5-heteroannulated 1,2-dihydro-2-phenyl-3*H*-1,2,4-triazol-3-one (eq 8).[19]

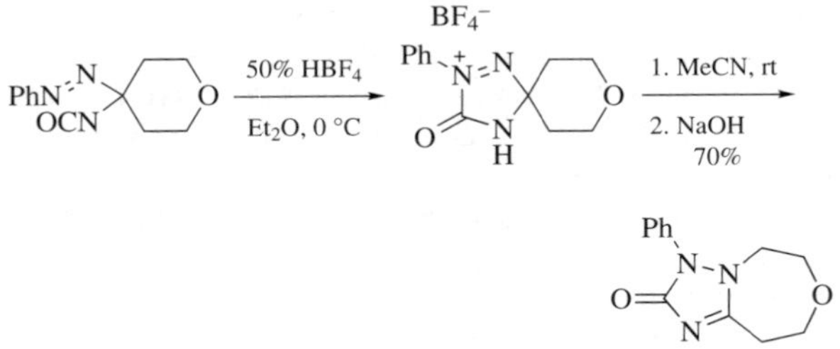

(8)

Carbenium Tetrafluoroborate Preparation. 4,6,8-Trimethylazulene is converted by ethereal HBF_4 into 4,6,8-trimethylazulenium tetrafluoroborate.[20] A convenient preparation of tropylium tetrafluoroborate employs HBF_4 to precipitate the product from a solution of the double salt $[C_7H_7]PCl_6 \cdot [C_7H_7]Cl$ in ethanol (eq 9). An indefinitely stable, nonhygroscopic, and nonexplosive white solid is obtained, a distinct advantage of a fluoroborate salt.[21]

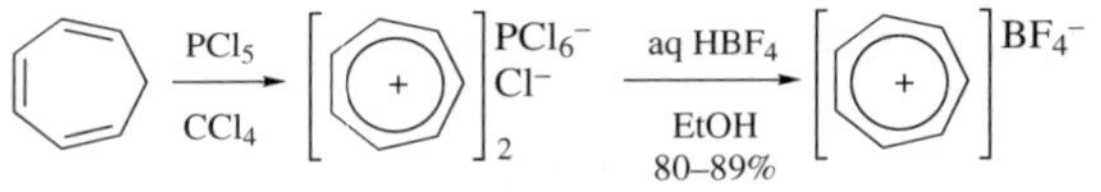

(9)

The formyl cation equivalent 1,3-benzodithiolylium tetrafluoroborate can be made in 94% yield by treatment of a dithioorthoformate with HBF_4/Ac_2O (eq 10).[22] 2-Deuterio-1,3-benzothiolylium tetrafluoroborate prepared by this method has been used to produce 1-deuterioaldehydes by homologation.[23] An analogous preparation of a similar formyl cation equivalent, 1,3-dithiolan-2-ylium tetrafluoroborate, employs $HBF_4 \cdot OEt_2$ (eq 11) to provide the intermediate salt. Subsequent reaction with a silyl enol ether forms a masked 2-formylcyclohexanone in high yield (eq 11).

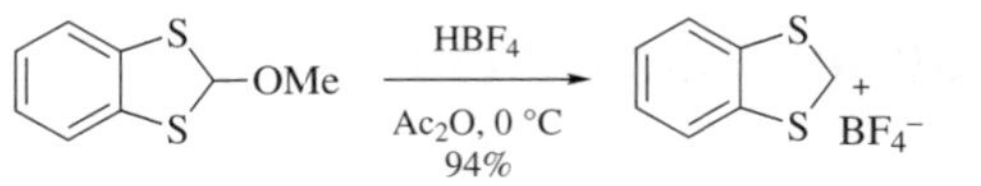

(10)

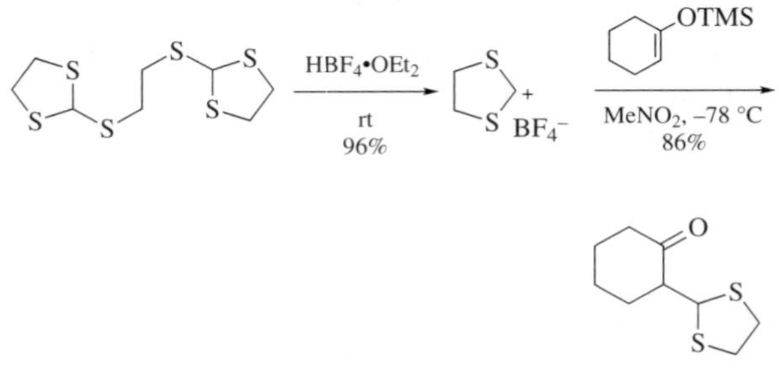

(11)

Ethynylfluorenylium dyes can be obtained by treatment of appropriate tertiary alcohols with HBF_4 (eq 12).[24]

$$\text{(fluorenyl tertiary alcohol, } p\text{-}MeC_6H_4\text{-alkynyl)} \xrightarrow[52\%]{HBF_4,\ MeOH,\ CH_2Cl_2} \text{fluorenylium } BF_4^- \quad (12)$$

In addition to the aforementioned carbenium ions and diazonium ions, numerous other organic cations can be obtained as their fluoroborate salts by treatment of appropriate precursors with HBF_4.[25]

Dimerization of Carbodiimides. Treatment of alkyl carbodiimides with anhydrous $HBF_4{\cdot}OEt_2$ in CH_2Cl_2 results in rapid dimerization to tetrafluoroborate salts in 95% yield (eq 13). Basification converts the salts to diazetidines.[26] In the same work, aryl carbodiimides undergo a similar reaction, but substituted quinazolines are obtained.

$$R{-}N{=}C{=}N{-}R \xrightarrow[CH_2Cl_2]{HBF_4{\cdot}OEt_2} \text{diazetidinium } BF_4^- \xrightarrow{OH^-} RN{=}\text{diazetidine}{=}NR \quad (13)$$

Mercury(II) Oxide/Tetrafluoroboric Acid. Yellow ***Mercury(II) Oxide*** is added to 48% aqueous HBF_4 to yield, upon solvent removal, $HgO{\cdot}2HBF_4$ as a hygroscopic white solid.[27] This reagent is useful in applications involving mercuration of alkenes, including diamination of alkenes and preparation of *trans*-cinnamyl ethers from allylbenzene.[28] Alkylations of carboxylic acids[29] and alcohols[30] with alkyl halides are also facilitated by $HgO{\cdot}2HBF_4$. Mercury(II) oxide and HBF_4 in alcohol effected mild solvolysis of 2-hydroxytrithioorthoesters to yield α-hydroxycarboxylic esters in high yield.[31] See also ***Mercury(II) Oxide–Tetrafluoroboric Acid***.

Preparation of a Useful Hypervalent Iodine Reagent. When treated with $HBF_4{\cdot}OMe_2$ at low temperatures, ***Iodosylbenzene*** reacts with silyl enol ethers to form a hypervalent iodine adduct capable of useful carbon–carbon bond formation reactions with alkenes (eq 14).[32]

Synthesis of Cationic Organometallic Complexes. Tetrafluoroborate is frequently encountered as the counterion in cationic organometallic compounds; its lack of nucleophilic reactivity makes HBF_4 and its etherates ideal reagents for delivery of protons without side reactions. The poorly coordinating conjugate base of HBF_4 allows substrates greater opportunity to bind to metals in organometallic reactions requiring the presence of acids.[33]

Propargylium complexes of cobalt, obtained by treatment of propargylic alcohols with $HBF_4{\cdot}OEt_2$, have been studied with regard to their selectivity as alkylating agents. *N*-Acetyl-3,4-dimethoxyphenethylamine undergoes selective aromatic substitution, whereas the unprotected amine undergoes *N,N*-dialkylation but not aromatic substitution (eq 15).[34]

$$PhIO + HBF_4{\cdot}OMe_2 \xrightarrow[-50\ \text{to}\ -20\ °C]{CH_2Cl_2} PhIO{\cdot}HBF_4 \xrightarrow{\text{OTMS-styrene}} [PhICH_2COPh]^+BF_4^- \quad (14)$$

cyclohexene → 90%; 1-(trimethylsilyloxy)cyclohexene (OTMS) → 50%; $Me_2C{=}CMe_2$ → 80%

$$\text{HO-CH}_2\text{-}[C_2Co_2(CO)_6] \xrightarrow{HBF_4{\cdot}OEt_2} {}^+H_2C{-}C{\equiv}C\ (Co_2(CO)_6)\ BF_4^- \xrightarrow{\text{MeO, MeO-C}_6H_3\text{-CH}_2CH_2NHAc} \text{aryl-}CH_2C{\equiv}C\text{-}(OC)_6Co_2 \quad (15)$$

Oxidation. In a reaction proposed as a model for substrate reactions at metal–sulfur centers of enzymes, HBF_4 apparently functions as an oxidizing agent in a two-electron oxidation of a ruthenium benzenedithiolate complex (eq 16).[35]

$$NBu_4^+\,[Ru(PMe_3)_2(S_2C_6H_4)_2]^- \xrightarrow[20\ °C]{HBF_4,\ MeOH} [Ru(PMe_3)_2(S_2C_6H_4)_2]^+BF_4^- \quad (16)$$

1. (a) Sudakova, T. N.; Krasnoshchekov, V. V. *Zh. Neorg. Khim.* **1978**, *23*, 1506. (b) Acidity relative to other strong acids: Bessiere, J. *BSF* **1969**, *9*, 3356.
2. Ellis, R.; Henderson, R. A.; Hills, A.; Hughes, D. L. *JOM* **1987**, *333*, C6.
3. (a) Lichtenberg, D. W.; Wojcicki, A. *JOM* **1975**, *94*, 311. (b) Wudl, F.; Kaplan, M. L. *Inorg. Synth.* **1979**, *19*, 27.
4. Roe, A. *OR* **1949**, *5*, 193. See also: (a) Starkey, E. B. *OSC* **1943**, *2*, 225. (b) Curtin, D. Y.; Ursprung, J. A. *JOC* **1956**, *21*, 1221. (c) Schiemann, G.; Winkelmuller, W. *OSC* **1943**, *2*, 299.
5. (a) Cohen, T.; Dietz, A. G., Jr.; Miser, J. R. *JOC* **1977**, *42*, 2053. (b) Allmann, R.; Debaerdemaeker, T.; Grehn, W. *CB* **1974**, *107*, 1555.
6. Albert, R.; Danklmaier, J.; Honig, H.; Kandolf, H. *S* **1987**, 635.
7. Albert, R.; Dax, K.; Pleschko, R.; Stutz, A. E. *Carbohydr. Res.* **1985**, *137*, 282.

8. Akaji, K.; Yoshida, M.; Tatsumi, T.; Kimura, T.; Fujiwara, Y.; Kiso, Y. *CC* **1990**, 288.
9. Kiso, Y.; Yoshida, M.; Tatsumi, T.; Kimura, T.; Fujiwara, Y.; Akaji, K. *CPB* **1989**, *37*, 3432.
10. Trost, B. M.; Keeley, D. E.; Arndt, H. C.; Bogdanowicz, M. J. *JACS* **1977**, *99*, 3088.
11. Shabarov, Y. S.; Pisanova, E. V.; Saginova, L. G. *ZOR* **1980**, *16*, 418 (*CA* **1981**, *94*, 3819a).
12. Tsuji, T. *BCJ* **1989**, *62*, 645.
13. Ray, C.; Saha, B.; Ghatak, U. R. *SC* **1991**, *21*, 1223.
14. Powell, J. W.; Whiting, M. C. *Proc. Chem. Soc.* **1960**, 412.
15. (a) Neeman, M.; Johnson, W. S. *OS* **1961**, *41*, 9. (b) Brückner, R.; Peiseler, B. *TL* **1988**, *29*, 5233.
16. (a) Schulz, W.; Pracejus, H.; Oehme, G. *J. Mol. Catal.* **1991**, *66*, 29. (b) Kanemasa, S.; Asai, Y.; Tanaka, J. *BCJ* **1991**, *64*, 375.
17. Dieter, R. K.; Lin, Y. J.; Dieter, J. W. *JOC* **1984**, *49*, 3183.
18. Babin, D.; Fourneron, J.-D.; Julia, M. *BSF(2)* **1980**, 588.
19. Gstasch, H.; Seil, P. *S* **1990**, 1048.
20. (a) Hafner, K.; Pelster, H.; Schneider, J. *LA* **1961**, *650*, 62. (b) Hafner, K.; Pelster, H.; Patzelt, H. *LA* **1961**, *650*, 80.
21. Conrow, K. *OS* **1963**, *43*, 101.
22. Nakayama, J.; Fujiwara, K.; Hoshino, M. *CL* **1975**, 1099.
23. Nakayama, J. *BCJ* **1982**, *55*, 2289.
24. Nakatsuji, S.; Nakazumi, H.; Fukuma, H.; Yahiro, T.; Nakashima, K.; Iyoda, M.; Akiyama, S. *JCS(P1)* **1991**, 1881.
25. A few examples: (a) oxotriazolium: Gstasch, H.; Seil, P. *S* **1990**, 1048. (b) pyridinium: Paley, M. S.; Meehan, E. J.; Smith, C. D.; Rosenberger, F. E.; Howard, S. C.; Harris, J. M. *JOC* **1989**, *54*, 3432. (c) pyridinium: Guibe-Jampel, E.; Wakselman, M. *S* **1977**, 772. (d) tetrameric dication from 2-aminobenzaldehyde: Skuratowicz, J. S.; Madden, I. L.; Busch, D. H. *IC* **1977**, *16*, 1721. (e) tetrathiafulvenium: Wudl, F. *JACS* **1975**, *97*, 1962. Wudl, F.; Kaplan, M. L. *Inorg. Synth.* **1979**, *19*, 27. (f) sulfonium: LaRochelle, R. W.; Trost, B. M. *JACS* **1971**, *93*, 6077. (g) diazetidinium: Hartke, K.; Rossbach, F. *AG(E)* **1968**, *7*, 72.
26. Hartke, K.; Rossbach, F. *AG(E)* **1968**, *7*, 72.
27. Barluenga, J.; Alonso-Cires, L.; Asensio, G. *S* **1979**, 962.
28. Barluenga, J.; Alonso-Cires, L.; Asensio, G. *TL* **1981**, *22*, 2239.
29. Barluenga, J.; Alonso-Cires, L.; Campos, P. J.; Asenio, G. *S* **1983**, 649.
30. Barluenga, J.; Alonso-Cires, L.; Campos, P. J.; Asensio, G. *T* **1984**, *40*, 2563.
31. Scholz, D. *SC* **1982**, *12*, 527.
32. Zhdankin, V. V.; Tykwinski, R.; Caple, R.; Berglund, B.; Koz'min, A. S.; Zefirov, N. S. *TL* **1988**, *29*, 3703.
33. Ellis, R.; Henderson, R. A.; Hills, A.; Hughes, D. L. *JOM* **1987**, *333*, C6. Some other recent examples of the use of $HBF_4 \cdot OEt_2$ in synthesis and/or reactions of organometallics: (a) Field, J. S.; Haines, R. J.; Stewart, M. W.; Sundermeyer, J.; Woollam, S. F. *JCS(D)* **1993**, 947. (b) Dawson, D. M.; Henderson, R. A.; Hills, A.; Hughes, D. L. *JCS(D)* **1992**, 973. (c) Lemos, M. A. N. D. A.; Pombeiro, A. J. L.; Hughes, D. L.; Richards, R. L. *JOM* **1992**, *434*, C6. (d) Arliguie, T.; Chaudret, B.; Jalon, F. A.; Otero, A.; Lopez, J. A.; Lahoz, F. J. *OM* **1991**, *10*, 1888. (e) Bassner, S. L.; Sheridan, J. B.; Kelley, C.; Geoffroy, G. L. *OM* **1989**, *8*, 2121. For use of $HBF_4 \cdot OMe_2$: (a) Schrock, R. R.; Liu, A. H.; O'Regan, M. B.; Finch, W. C.; Payack, J. F. *IC* **1988**, *27*, 3574. (b) Blagg, J.; Davies, S. G.; Goodfellow, C. L.; Sutton, K. H. *JCS(P1)* **1987**, 1805.
34. Gruselle, M.; Philomin, V.; Chaminant, F.; Jaouen, G.; Nicholas, K. M. *JOM* **1990**, *399*, 317.
35. Sellmann, D.; Binker, G.; Knoch, F. *ZN(B)* **1987**, *42*, 1298.
36. Guibe-Jampel, E.; Wakselman, M. *S* **1977**, 772.

Gregory K. Friestad & Bruce P. Branchaud
University of Oregon, Eugene, OR, USA

N,N,N′,N′-Tetramethylethylenediamine[1]

Me_2N–CH_2CH_2–NMe_2

[110-18-9] $C_6H_{16}N_2$ (MW 116.24)

(bidentate tertiary amine Lewis base with good solvating properties; used as an additive to stabilize and activate organometallic reagents and inorganic salts; enhances the rate of metalation of a variety of aromatic and unsaturated systems as well as influencing the regiochemical outcome of these reactions; effective as a neutral amine in base catalyzed reactions)

Alternate Name: TMEDA.
Physical Data: bp 121 °C; d 0.781 g cm^{-3}.
Solubility: very sol water, most organic solvents.
Form Supplied in: colorless liquid, typically of 99% purity as obtained commercially.
Drying: for uses in conjunction with organometallic reagents, moisture exclusion is necessary. Removal of water is best achieved by refluxing over lithium aluminum hydride or calcium hydride for 2 h under nitrogen and distilling immediately prior to use.
Handling, Storage, and Precautions: TMEDA should be used directly after distilling. However, it may be stored under nitrogen and transferred by using a syringe and septum cap as required. For most applications the amine is removed during aqueous workup simply by washing with water owing to its high water solubility. Use in a fume hood.

Lithiation of Difficult Substrates. TMEDA, through an erstwhile perception of enhanced chelating ability,[1b] but more likely through presentation of a more labile environment, activates organolithium reagents.[1j] ***n-Butyllithium*** forms hexamers in hexane but in the presence of TMEDA exists as a solvated tetramer.[2] Thus the use of the *n*-butyllithium/TMEDA complex in hexane[3] effects the dilithiation of ***Furan*** and thiophene,[4] and lithiation of benzene,[5] in high yields. The allylic deprotonation of unactivated alkenes is normally difficult to achieve with BuLi alone. However, in the presence of TMEDA, propene is monolithiated or dilithiated in the allylic position,[6] and limonene is selectively lithiated at C-10.[7] The resulting allylic carbanion and electrophiles such as ***Paraformaldehyde*** give functionalized products (eq 1).[8]

7; BuLi, TMEDA; $(HCHO)_n$; (1); 10; Li^+; HO

However, when vinylic metalation is desired, competing allylic deprotonation may occur. In general, thermodynamic acidity and the kinetic preference for vinylic deprotonation of cyclic alkenes decrease with increasing ring size.[1h] The stable alkane-soluble reagent ***n-Butyllithium–Potassium t-Butoxide***–TMEDA in hexane[9] metalates ***Ethylene*** with potassium[10] and effects selective vinylic deprotonation of cyclopentene (eq 2),[11] cyclobutene,[12] norbornene, and norbornadiene.[10]

BuLi, t-BuOK, TMEDA; hexane, –20 °C; K; K; 9:1 (2)

N-Alkyl- and *N*-arylpyrroles are readily α-lithiated.[1c] For example, *N*-methylpyrrole is deprotonated with ***Ethyllithium***/TMEDA and the lithiated product has been treated with ***Carbon Dioxide*** to give the corresponding carboxylic acid in 70% yield (eq 3).[13]

N Me; 1. EtLi, TMEDA; 2. CO_2; 70%; N Me; CO_2H (3)

Numerous examples exist in which TMEDA not only facilitates the lithiation of aromatic and heteroaromatic substrates but also controls the regioselectivity of lithiation.[1c] While tertiary benzamides are susceptible to nucleophilic attack by *n*-butyllithium to give aryl butyl ketones, the use of ***s-Butyllithium***/TMEDA in THF at −78 °C provides the synthetically useful *ortho* metalated tertiary benzamide which may be treated with a large variety of electrophiles (eq 4).[1d,14] Even with compounds having a second more acidic site the above conditions allow *ortho* lithiation to take place under kinetic control. Thus a *p*-toluamide is *ortho* lithiated with *s*-butyllithium/TMEDA in THF at −78 °C, but when ***Lithium Diisopropylamide*** is used as the base in THF at 0 °C the thermodynamically favored benzyllithium species is obtained (eq 5).[15] The very marked influence of TMEDA on the lithiation of naphthyl methyl ether in hydrocarbon solvents is dramatically illustrated in the example in eq 6.[16]

Y; NEt_2; O; *s*-BuLi, TMEDA; THF, –78 °C; Li; EX; E (4)

EX = D_2O, MeI, DMF, CO_2, $(CO_2Et)_2$, ArCHO, I_2, TMSCl, $B(OMe)_3/H_2O_2$, PhNCO, Ph_2CO

O; N(*i*-Pr)$_2$; Li; *s*-BuLi, TMEDA; THF, –78 °C; LDA; THF, 0 °C; Li (5)

OMe; 1. RLi; 2. CO_2, CH_2N_2; CO_2Me; MeO_2C (6)

R = Bu, hexane/TMEDA >99.3:<0.3 60%
R = *t*-Bu, cyclohexane 1:99 35%

The use of the *s*-butyllithium/TMEDA system in THF at −78 °C is widely employed for the lithiation of amide and thioamide derivatives adjacent to nitrogen.[1f] The resulting lithiated species undergo reaction with a variety of electrophiles such as aldehydes and ketones[17a] and dihalides (eq 7).[17b]

S; *t*-Bu; N; Me; *s*-BuLi, TMEDA; –78 °C; Li; Br; Cl (7)

Ligand for Crystallographic Studies. TMEDA markedly facilitates the isolation of otherwise inaccessible crystalline organolithium reagents suitable for structural determination by X-ray crystallographic means.[18] In most cases, bidentate binding of the TMEDA ligand to the lithium ion is observed.

Control of Regioselectivity and Stereoselectivity. The recognition by Ireland and co-workers[19] that ***Hexamethylphosphoric Triamide*** has a profound effect on the stereochemistry of lithium enolates has led to the examination of the effects of other additives, as the ability to control enolate stereochemistry is of utmost importance for the stereochemical outcome of aldol reactions. Kinetic deprotonation of 3-pentanone with ***Lithium 2,2,6,6-Tetramethylpiperidide*** at 0 °C in THF containing varying amounts of HMPA or TMEDA was found to give predominantly the (*Z*)-enolate at a base:ketone:additive ratio of ca. 1:1:1, whereas with a base:ketone:additive ratio 1:0.25:1, formation of the (*E*)-enolate was favored (Table 1).[20] This remarkable result contrasts with those cases where HMPA:base ratios were varied towards larger amounts of HMPA, which favored formation of the (*Z*)-enolate.[21]

However, TMEDA unlike HMPA, does not cause flow over from a carbonyl to conjugate addition manifold for many lithiated systems. For example, lithiated allylic sulfides undergo conjugate (or '1,4') addition to cyclopent-2-enone in the presence of HMPA (see ***Allyl Phenyl Sulfide***), but in the presence of TMEDA, carbonyl addition only is observed.[22] The perception that TMEDA is unable to form solvent-separated ion pairs required for conjugate addition in this case now requires reevaluation.[1j,23] In the reaction of lithio α-trimethylsilylmethyl phenyl sulfide with cyclohexenone, HMPA promotes predominant conjugate addition, whereas TMEDA has little effect on the normal carbonyl addition pathway taken in THF alone (eq 8).[24]

Table 1 Deprotonation of 3-Pentanone with LiTMP (1.0 mmol) in THF at 0 C in the Presence of TMEDA or HMPA[19]

3-Pentanone (mmol)	TMEDA (mmol)	HMPA (mmol)	(*E*)-Enolate	(*Z*)-Enolate	Total % yield of silyl enol ether derivatives
0.9	1.0	–	17	83	70
0.45	1.0	–	91	9	90
0.25	1.0	–	95	5	70
0.9	–	1.0	8	92	89
0.45	–	1.0	65	35	75
0.25	–	1.0	66	34	80

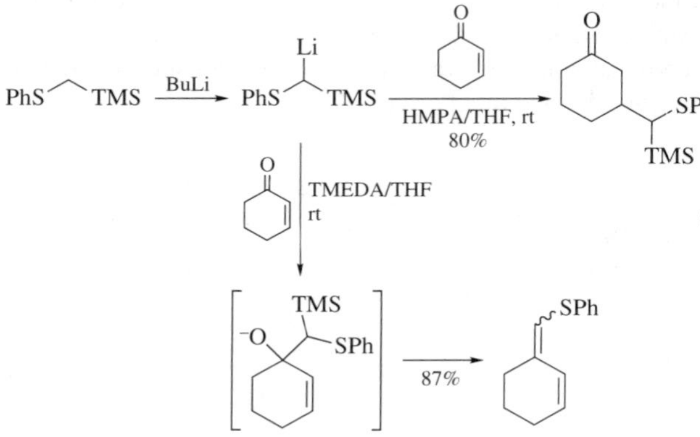

Likewise, TMEDA in THF has little effect on the stereochemical outcome of the Horner–Wittig reaction of lithiated ethyldiphenylphosphine oxide with benzaldehyde in THF at low temperature compared to the reaction in THF alone: a very slight enhancement in favor of the *erythro* (*anti*) hydroxy phosphine oxide intermediate (*erythro*:*threo* from 85:15 to 88:12), thus leading to slightly enhanced (*Z*)-alkene formation, is observed. By contrast, a reaction in ether alone provides less of the *erythro* product (*erythro*:*threo* 60:40).[25] These examples serve to emphasize current thought that TMEDA is not a 'good' chelating agent for lithium in relation to THF itself.[1j] It is noteworthy that the substitution of methyl in TMEDA by chiral binaphthylmethyl ligands generates a reagent which efficiently catalyzes asymmetric addition of butyllithium to benzaldehyde in ether at low temperature (eq 9).[26]

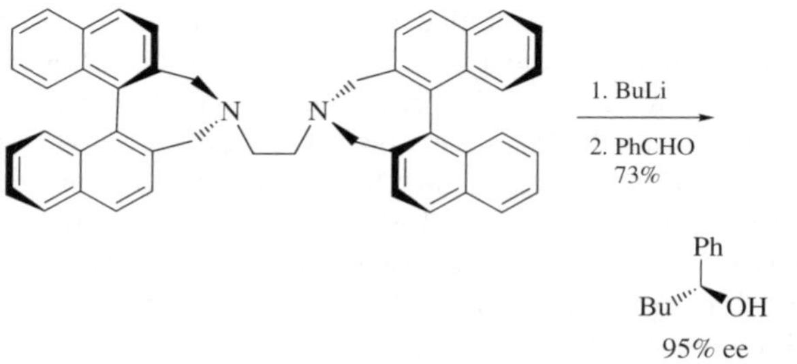

Butylmagnesium bromide in THF in the presence of excess TMEDA undergoes addition to a chiral crotonamide derivative to give the conjugate adduct in modest diastereomeric excess (67%) compared with 16% in the absence of TMEDA.[27] On the other hand, diastereoselection in the alkylation of enolates of chiral diamides derived from piperazines in THF containing TMEDA was minimal, with better results being provided by HMPA.[28]

Stabilization and Activation of Organometallic Reagents. The development and use of organocopper reagents as nucleophiles in the conjugate addition to enones is an important area of organic synthesis. It has been found that the reactivity of aryl- and alkylcopper reagents is dramatically improved through the use of ***Chlorotrimethylsilane*** and TMEDA.[29] The TMEDA not only stabilizes and solubilizes the organocopper reagent but also facilitates the trapping of the resulting enolates, thereby affording silyl enol ethers in excellent yields.[29a] Such a role is also played by HMPA[30] and ***4-Dimethylaminopyridine***.[30b] However, the low toxicity and cost of TMEDA makes it an attractive alternative. Conjugate addition of ***Lithium Di-n-butylcuprate*** to methyl 2-butynoate in diethyl ether provides a 74:26 mixture of (*E*)- and (*Z*)-alkylated enoates, whereas in the presence of TMEDA this ratio increases to 97:3. Stereoselectivity of the conjugate addition of the copper reagent derived from butylmagnesium bromide and ***Copper(I) Iodide*** to ethyl pentynoate in diethyl ether is also enhanced when TMEDA or pyrrolidone are used as additives, to give the (*E*)- and (*Z*)-enoates in a ratio of 99:1. In contrast, the use of HMPA affords a selectivity of only 78:22 for the (*E*)-isomer.[31]

The preparation of a trifluoromethylvinyl anion equivalent has been described in which the vinyl bromide is converted into the zinc reagent with ***Zinc/Silver Couple*** in the presence of TMEDA. The conversion proceeds very cleanly to afford a thermally stable vinylzinc bromide–TMEDA complex which can undergo reactions with electrophiles. TMEDA is essential for the conversion (eq 10).[32]

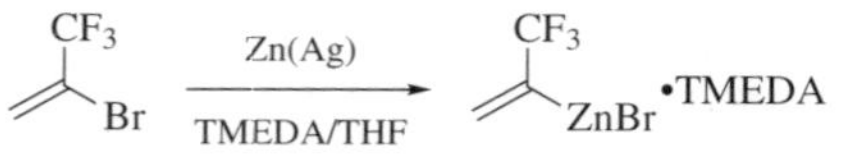

The intramolecular insertion of unactivated alkenes into carbon–lithium bonds to give cycloalkylmethyllithium compounds provides a high-yielding alternative to analogous radical cyclizations.[1g,33] In many cases the addition of Lewis bases such as TMEDA increases the rate of cyclization[1g,33] and dramatically improves those cyclizations which are otherwise sluggish at room temperature (eq 11).[34]

Open chain allylic alcohols add organolithium reagents in the presence of TMEDA. The reactions are regio- and stereoselective; the suggestion is made that the TMEDA complexes the alkoxide lithium counterion, allowing the alkoxide to orientate the incoming organolithium reagent and stabilize the resulting intermediate (eq 12).[35]

1. t-BuLi, pentane–Et_2O (3:2)
2. H^+, H_2O

(11)

TMEDA (2 equiv)	68%	31%
no TMEDA	6%	93%

PrLi, TMEDA, 68% (12)

threo:*erytho* = >50:1

Inorganic Complexes useful in Organometallic Reactions and Organic Synthesis. The complexing properties of TMEDA have made it possible to prepare and handle salts which are otherwise air and moisture sensitive. Thus ***Zinc Chloride*** in the presence of one equivalent of TMEDA forms a crystalline air stable solid, $ZnCl_2$·TMEDA,[36] which with three equivalents of an alkyllithium reagent is converted into trialkylzinclithium. Likewise, CuI and TMEDA react to form the CuI·TMEDA complex,[29] a stable solid which is used for the preparation of stabilized organocopper reagents. The rate of Cu^I-catalyzed oxidative coupling of terminal alkynes in the presence of oxygen to form diynes is considerably increased by using TMEDA as a solubilizing agent for ***Copper(I) Chloride***.[37] Magnesium hydride, rapidly acquiring widespread recognition as a versatile reducing agent (see ***Magnesium Hydride–Copper(I) Iodide***), is prepared from phenylsilane and ***Dibutylmagnesium*** in the presence of TMEDA to give a very active THF-soluble complex free of halide or impurities derived from the usual reducing agents such as ***Lithium Aluminum Hydride***.[38] TMEDA (see also ***1,4-Diazabicyclo[2.2.2]octane***) forms insoluble adducts with boranes and alanes. In particular, the formation of air-stable adducts with monoalkyl boranes is of synthetic usefulness. The free monoalkyl borane may be regenerated by treating the adduct with ***Boron Trifluoride Etherate*** in THF and filtering off the newly formed TMEDA·$2BF_3$ precipitate.[1e]

Base Catalyzed Reactions. TMEDA can be monoprotonated (pK_a 8.97) and diprotonated (pK_a 5.85).[39] Titanium enolate formation from ketones and acid derivatives has been achieved by using ***Titanium(IV) Chloride*** and tertiary amines including TMEDA in dichloromethane at 0 °C.[40] The reactive species, which is likely to be a complex with the tertiary amine, undergoes aldol reaction with aldehydes to form *syn* adducts with high stereoselectivity (eq 13).

In the case of TMEDA, stereoselection in favor of the *syn* product (98:2) is enhanced over that achieved with ***Diisopropylethylamine*** (94:6).[40] Along with bases such as ***Triethylamine*** and ethylisopropylamine, TMEDA facilitates the preparation of cyanohydrin trimethylsilyl ethers from aldehydes and ***Cyanotrimethylsilane***.[41] It has been suggested that coordination by nitrogen induces formation of an active hypervalent cyanation intermediate from cyanotrimethylsilane. The conjugate addition of thiols to enones has been successfully catalyzed by using TMEDA in methanol at room temperature, as exemplified by the reaction of 10-mercaptoisoborneol and 4-t-butoxycyclopentenone (eq 14).[42] In this case the relative mildness of the reaction conditions prevents subsequent elimination of t-butoxide from occurring to give the unwanted enone.

$TiCl_4$, TMEDA, CH_2Cl_2, 0 °C; $OTiCl_3$; CHO; Bn; OH (13)

TMEDA	98:2
i-Pr_2NEt	94:6

OH, SH, t-BuO; TMEDA, MeOH, 93–95% (14)

Related Reagents. Hexamethylphosphoric Triamide; N,N,N',N'',N''-Pentamethyldiethylenetriamine; Potassium Hydride-s-Butyllithium-N,N,N',N'-Tetramethylethylenediamine; (−)-Sparteine.

1. (a) Agami, C. *BSF(2)* **1970**, 1619. (b) Wakefield, B. J. *The Chemistry of Organolithium Compounds*; Pergamon: Oxford, 1974. (c) Gschwend, H. W.; Rodriguez, H. R. *OR* **1979**, *26*, 1. (d) Beak, P.; Snieckus, V. *ACR* **1982**, *15*, 306. (e) Singaram, B.; Pai, G. G. *H* **1982**, *18*, 387. (f) Beak, P.; Zajdel, W. J.; Reitz, D. B. *CRV* **1984**, *84*, 471. (g) Klumpp, G. W. *RTC* **1986**, *105*, 1. (h) Brandsma, L.; Verkruijsse, H. *Preparative Polar Organometallic Chemistry 1*; Springer: Berlin, 1987. (i) *Advances in Carbanion Chemistry*; Snieckus, V., Ed.; JAI: Greenwich, CT, 1992; Vol. 1. (j) For a review and critical analysis of TMEDA complexation to lithium, see: Collum, D. B. *ACR* **1992**, *25*, 448.
2. Lewis, H. L.; Brown, T. L. *JACS* **1970**, *92*, 4664.
3. Peterson, D. J. *JOC* **1967**, *32*, 1717.
4. Chadwick, D. J.; Willbe, C. *JCS(P1)* **1977**, 887.
5. Eberhardt, G. G.; Butte, W. A. *JOC* **1964**, *29*, 2928. Rausch, M. D.; Ciappenelli, D. J. *JOM* **1967**, *10*, 127.
6. Klein, J.; Medlik-Balan, A. *CC* **1975**, 877.
7. Crawford, R. J.; Erman, W. F.; Broaddus, C. D. *JACS* **1972**, *94*, 4298.
8. Crawford, R. J. *JOC* **1972**, *37*, 3543.
9. Schade, C.; Bauer, W.; Schleyer, P. v. R. *JOM* **1985**, *295*, C25.
10. Brandsma, L.; Verkruijsse, H. D.; Schade, C.; Schleyer, P. v. R. *CC* **1986**, 260.
11. Broaddus, C. D.; Muck, D. L. *JACS* **1967**, *89*, 6533.
12. Stähle, M.; Lehmann, R.; Kramar, J.; Schlosser, M. *C* **1985**, *39*, 229.
13. Gjøs, N.; Gronowitz, S. *ACS* **1971**, *25*, 2596.
14. Beak, P.; Brown, R. A. *JOC* **1977**, *42*, 1823.
15. Beak, P.; Brown, R. A. *JOC* **1982**, *47*, 34.
16. Shirley, D. A.; Cheng, C. F. *JOM* **1969**, *20*, 251.
17. (a) Reitz, D. B.; Beak, P.; Tse, A. *JOC* **1981**, *46*, 4316. Beak, P.; Zajdel, W. J. *JACS* **1984**, *106*, 1010. (b) Lubosch, W.; Seebach, D. *HCA* **1980**, *63*, 102.

18. Seebach, D. *AG(E)* **1988**, *27*, 1624. Boche, G. *AG(E)* **1989**, *28*, 277; Zarges, W.; Marsch, M.; Harms, K.; Koch, W.; Frenking, G.; Boche, G. *CB* **1991**, *124*, 543.
19. Ireland, R. E.; Mueller, R. H.; Willard, A. K. *JACS* **1976**, *98*, 2868.
20. Fataftah, Z. A.; Kopka, I. E.; Rathke, M. W. *JACS* **1980**, *102*, 3959.
21. Romesberg, F. E.; Gilchrist, J. H.; Harrison, A. T.; Fuller, D. J.; Collum, D. B. *JACS* **1991**, *113*, 5751.
22. Binns, M. R.; Haynes, R. K.; Houston, T. L.; Jackson, W. R. *TL* **1980**, *21*, 573.
23. Cohen, T.; Abraham, W. D.; Myers, M. *JACS* **1987**, *109*, 7923. Binns, M. R.; Haynes, R. K.; Katsifis, A. G.; Schober, P. A.; Vonwiller, S. C. *JOC* **1989**, *54*, 1960.
24. Ager, D. J.; East, M. B. *JOC* **1986**, *51*, 3983.
25. Buss, A. D.; Warren, S. *TL* **1983**, *24*, 3931.
26. Mazaleyrat, J.-P.; Cram, D. J. *JACS* **1981**, *103*, 4585; Maigrot, N.; Mazaleyrat, J.-P. *CC* **1985**, 508.
27. Soai, K.; Machida, H.; Yokota, N. *JCS(P1)* **1987**, 1909.
28. Soai, K.; Hayashi, H.; Shinozaki, A.; Umebayashi, H.; Yamada, Y. *BCJ* **1987**, *60*, 3450.
29. (a) Johnson, C. R.; Marren, T. J. *TL* **1987**, *28*, 27. (b) Van Heerden, P. S.; Bezuidenhoudt, B. C. B.; Steenkamp, J. A.; Ferreira, D. *TL* **1992**, *33*, 2383 and cited references.
30. (a) Horiguchi, Y.; Matsuzawa, S.; Nakamura, E.; Kuwajima, I. *TL* **1986**, *27*, 4025. (b). Nakamura, E.; Matsuzawa, S.; Horiguchi, Y.; Kuwajima, I. *TL* **1986**, *27*, 4029.
31. Anderson, R. J.; Corbin, V. L.; Cotterrell, G.; Cox, G. R.; Henrick, C. A.; Schaub, F.; Siddall, J. B. *JACS* **1975**, *97*, 1197.
32. Jiang, B.; Xu, Y. *JOC* **1991**, *56*, 7336.
33. Bailey, W. F.; Khanolkar, A. D.; Gavaskar, K.; Ovaska, T. V.; Rossi, K.; Thiel, Y.; Wiberg, K. B. *JACS* **1991**, *113*, 5720.
34. Bailey, W. F.; Nurmi, T. T.; Patricia, J. T.; Wang, W. *JACS* **1987**, *109*, 2442.
35. Felkin, H.; Swierczewski, G.; Tambuté, A. *TL* **1969**, 707.
36. Watson, R. A.; Kjonaas, R. A. *TL* **1986**, *27*, 1437. Isobe, M.; Kondo, S.; Nagasawa, N.; Goto, T. *CL* **1977**, 679.
37. Hay, A. S. *JOC* **1962**, *27*, 3320. Jones, G. E.; Kendrick, D. A.; Holmes, A. B. *OS* **1987**, *65*, 52.
38. Michalczyk, M. J. *OM* **1992**, *11*, 2307.
39. Spialter, L.; Moshier, R. W. *JACS* **1957**, *79*, 5955.
40. Evans, D. A.; Rieger, D. L.; Bilodeau, M. T.; Urpi, F. *JACS* **1991**, *113*, 1047.
41. Kobayashi, S.; Tsuchiya, Y.; Mukaiyama, T. *CL* **1991**, 537.
42. Eschler, B. M.; Haynes, R. K.; Ironside, M. D.; Kremmydas, S.; Ridley, D. D.; Hambley, T. W. *JOC* **1991**, *56*, 4760.

Richard K. Haynes
Hong Kong University of Science and Technology, Hong Kong

Simone C. Vonwiller
The University of Sydney, NSW, Australia

1,1′-Thiocarbonyldiimidazole[1]

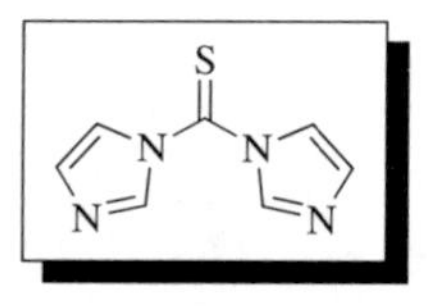

[6160-65-2] $C_7H_6N_4S$ (MW 178.24)

(conversion of vicinal diols to alkenes;[1] deoxygenation of alcohols;[13] thiocarbonyl transfer agent[29])

Alternate Name: TCDI.
Physical Data: mp 101–103 °C.
Solubility: sol many organic solvents including THF, CH_2Cl_2, toluene.
Form Supplied in: yellow solid; 90% pure or >97% pure.
Preparative Methods: prepared by the reaction of ***Thiophosgene*** with 2 equiv of ***Imidazole***.
Purification: can be recrystallized from THF to give yellow crystals; can also be sublimed.
Handling, Storage, and Precautions: very hygroscopic; should be reacted and stored in a dry atmosphere.

Alkene Synthesis. The Corey–Winter alkene synthesis is an effective method for the deoxygenation of vicinal diols.[1,2] The method involves formation of a 1,3-dioxolane-2-thione [cyclic thionocarbonate (or thiocarbonate)] by treatment of a vicinal diol with TCDI. Decomposition of the thionocarbonate, usually with a phosphorus compound, affords the alkene (eq 1).[3] The breakdown of the thionocarbonate occurs in a stereospecific sense; details of investigations into the mechanism have been summarized.[1]

OH OH
PhCO2
TCDI, py
110 °C, 12 h
91%

S
O O
$(EtO)_3P$
reflux, 12 h
90%
(1)

The conditions required for the introduction of the thiocarbonyl group vary greatly, depending on the structure of the diol.[1,4] The thionocarbonate may also be prepared without TCDI. The reaction of vicinal diols with thiophosgene/DMAP/CH_2Cl_2/0 °C,[5] with base/CS_2/MeI/heat,[1] or with 1,1′-thiocarbonyl-2,2′-pyridone/toluene/110 °C[6] also provides thionocarbonates. Along with phosphines and phosphites, other reagents can decompose the thionocarbonates, but not always in a stereospecific manner. The Corey–Hopkins reagent 1,3-dimethyl-2-phenyl-1,3,2-diazaphospholidine,[5] along with ***Raney***

Nickel[7] and ***Bis(1,5-cyclooctadiene)nickel(0)***,[8] affords products of stereospecific *cis*-elimination, while the alkyl iodide/***Zinc*** combination[9] and ***Pentacarbonyliron***[10] yield products of nonstereospecific elimination or unknown stereochemistry.

The Corey–Winter synthesis has proved useful for the generation of a large number of structurally interesting alkenes,[1] including some unstable alkenes that must be captured in situ (eq 2).[4]

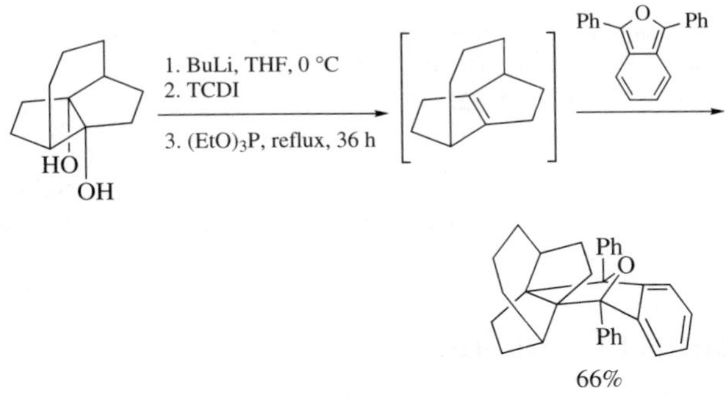

1. BuLi, THF, 0 °C
2. TCDI
3. $(EtO)_3P$, reflux, 36 h

66% (2)

The method has found its niche in the chemistry of sugars. Their polyhydroxylated nature make them excellent substrates provided the by-standing hydroxyls are protected (eq 3).[11] The 2′- and 3′-oxygens of nucleosides are readily are removed under the Corey–Winter conditions.[12]

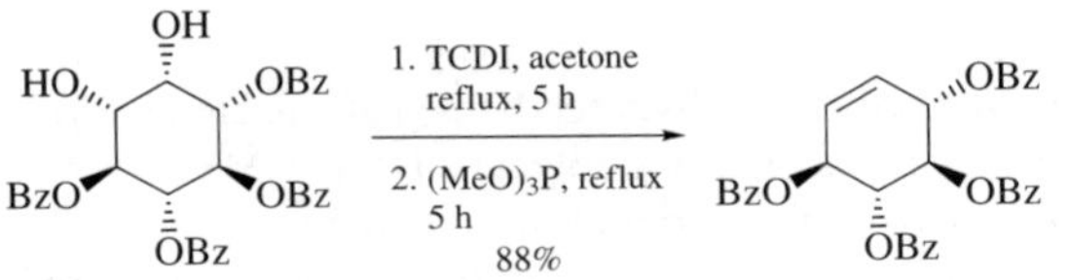

1. TCDI, acetone reflux, 5 h
2. $(MeO)_3P$, reflux 5 h
88% (3)

Radical Chemistry. Treatment of secondary alcohols with 1 equiv of TCDI affords an imidazole-1-thiocarbonyl derivative (imidazolide), which can be reduced to a CH_2 unit under ***Tri-n-butylstannane*** (TBTH) radical chain reaction conditions.[13,14] The deoxygenation of secondary alcohols by way of an imidazolide or other thiocarbonyl derivative is called the Barton–McCombie reaction (eq 4).[6,14] Since imidazolide formation (TCDI, reflux, 65 °C) and the subsequent radical chemistry are done under neutral or near-neutral conditions, the overall reduction is tolerant of the presence of many sorts of functional groups. Furthermore, the low solvation requirements of radical species permits deoxygenation in sterically congested environments (eq 5).[15]

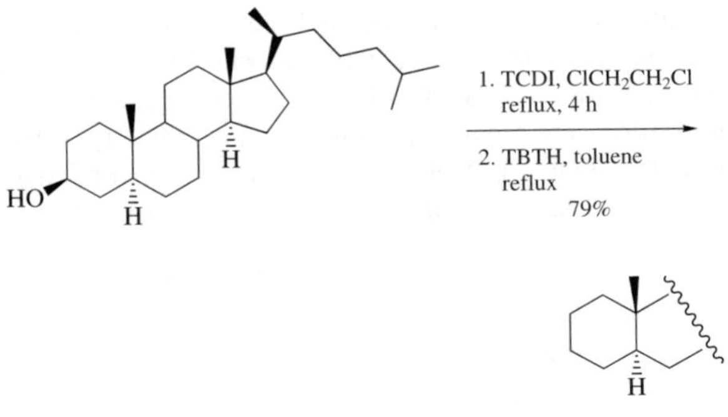

1. TCDI, $ClCH_2CH_2Cl$ reflux, 4 h
2. TBTH, toluene reflux
79% (4)

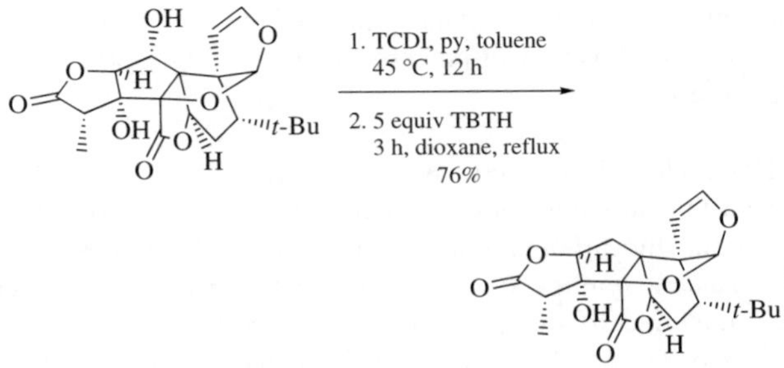

1. TCDI, py, toluene 45 °C, 12 h
2. 5 equiv TBTH 3 h, dioxane, reflux
76% (5)

The mechanism of the radical attack at the thiocarbonyl group and the ensuing breakdown has been outlined elsewhere.[13,14] Some groups have studied the comparative chemistries of the possible thiocarbonyl derivatives, with no particularly obvious trends arising.[13,16,17] There are numerous reports of the deoxygenation of carbohydrates[13,16,18] and related polyhydroxylated species.[14,17,19,20] In one instance of amino glycoside deoxygenation, the C(3′)-OH group can be selectively functionalized with TCDI, while the C(5)-OH group is unaffected and the radical reduction proceeds without protection of the C(5)-OH group (eq 6).[20]

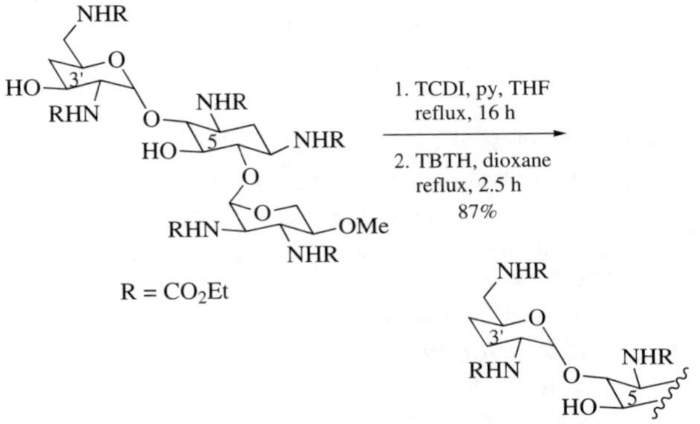

1. TCDI, py, THF reflux, 16 h
2. TBTH, dioxane reflux, 2.5 h
87%

$R = CO_2Et$ (6)

Reduction of tertiary alcohols is not recommended due to the instability of the tertiary imidazolide, although, as an alternative, a thioformaldehyde version has been developed.[21] Imidazolides from primary alcohols do not respond well to the same conditions as the secondary systems, but performing the radical chemistry at 130–150 °C can lead to deoxygenation.[22] As an alternative to the high-temperature radical chemistry, the imidazolide can be converted by methanolysis to a thiocarbonate which can be reduced at room temperature (Et_3B, TBTH, benzene).[23] Tin hydride treatment of thionocarbonates derived from primary/secondary diols affords, after base exposure, the product of selective reduction of the secondary site, leaving the primary hydroxyl (eq 7).[14,24] The approach has also been successfully applied to 1,3-diols.[24,25]

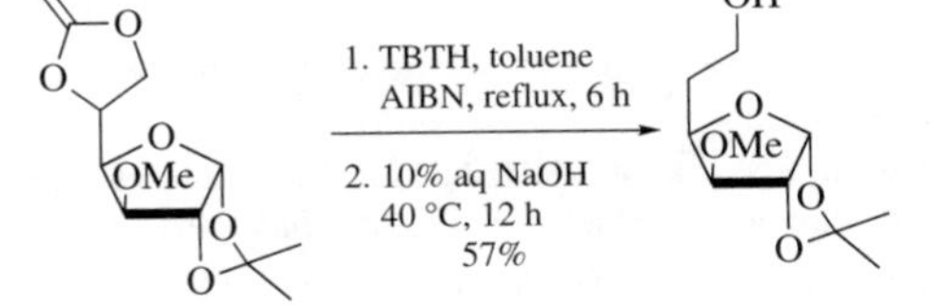

1. TBTH, toluene AIBN, reflux, 6 h
2. 10% aq NaOH 40 °C, 12 h
57% (7)

Since the reduction method proceeds via radical chemistry, the intermediate radicals are prone to bimolecular chemistry and unimolecular reactions such as rearrangements or β-scission, if suitable functionality is proximal.[14] Indeed, the imidazolides have been employed as radical precursors when rearrangement is the desired result.[14]

Thiocarbonyl Transfer. While virtually all uses of TCDI involve a thiocarbonyl transfer reaction, this section covers those uses of the thiocarbonyl transfer that do not lead to radical chemistry or alkene syntheses. TCDI has been used for the simple placement of a thiocarbonyl group between two nucleophilic atoms of one[26] or two molecules.[27] An alcohol which has been converted to an imidazolide is a reactive functionality for coupling reactions,[28] for sigmatropic rearrangements,[29] and for elimination (eq 8).[30] The TCDI alternative 1,1′-thiocarbonyl-2,2′-pyridone[6] seems to be an excellent thiocarbonyl transfer agent.

AcO, OH, H, O, H, CO_2Me — 1. TCDI, C_6H_6, 12 h, rt; 2. C_6H_6, reflux, 4 h; 55% → AcO, O, H, CO_2Me (8)

Related Reagents. Carbon Disulfide; *N*,*N*-Dimethylformamide Diethyl Acetal; *N*-Hydroxypyridine-2-thione; Phenyl Chlorothionocarbonate; Triethyl Orthoformate.

1. Block, E. *OR* **1984**, *30*, 457.
2. Corey, E. J.; and Winter, R. A. E. *JACS* **1963**, *85*, 2677.
3. Koreeda, M.; Koizumi, N.; Teicher, B. A. *CC* **1976**, 1035.
4. Greenhouse, R.; Borden, W. T.; Ravindranathan, T.; Hirotsu, K.; Clardy, J. *JACS* **1977**, *99*, 6955.
5. Corey, E. J.; Hopkins, P. B. *TL* **1982**, *23*, 1979.
6. Kim, S.; Yi, K. Y. *JOC* **1986**, *51*, 2613.
7. Ireland, R. E.; Anderson, R. C.; Badoud, R.; Fitzsimmons, B. J.; McGarvey, G. J.; Thaisrivongs, S.; Wilcox, C. S. *JACS* **1983**, *105*, 1988.
8. Semmelhack, M. F.; Stauffer, R. D. *TL* **1973**, 2667.
9. Vedejs, E.; Wu, E. S. C. *JOC* **1974**, *39*, 3641.
10. Daub, J.; Trautz, V.; Erhardt, U. *TL* **1972**, 4435.
11. (a) Akiyama, T.; Shima, H.; Ozaki, S. *TL* **1991**, *32*, 5593. (b) Horton, D.; Turner, W. N. *TL* **1964**, 2531.
12. Manchand, P. S.; Belica, P. S.; Holman, M. J.; Huang, T.-N.; Maehr, H.; Tam, S. Y.-K.; Yang, R. T. *JOC* **1992**, *57*, 3473.
13. Barton, D. H. R.; McCombie, S. W. *JCS(P1)* **1975**, 1574.
14. (a) McCombie, S. W. *COS* **1991**, *8*, 811. (b) Hartwig, W. *T* **1983**, *39*, 2609. (c) Crich, D.; Quintero, L. *CRV* **1989**, *89*, 1413.
15. Corey, E. J.; Ghosh, A. K. *TL* **1988**, *29*, 3205.
16. Rasmussen, J. R.; Slinger, C. J.; Kordish, R. J.; Newman-Evans, D. D. *JOC* **1981**, *46*, 4843.
17. Robins, M. J.; Wilson, J. S.; Hansske, F. *JACS* **1983**, *105*, 4059.
18. Lin, T.-H.; Kováč, P.; Glaudemans, C. P. J. *Carbohydr. Res.* **1989**, *188*, 228.
19. Piccirilli, J. A.; Krauch, T.; MacPherson, L. J.; Benner, S. A. *HCA* **1991**, *74*, 397.
20. Carney, R. E.; McAlpine, J. B.; Jackson, M.; Stanaszek, R. S.; Washburn, W. H.; Cirovic, M.; Mueller, S. L. *J. Antibiot.* **1978**, *31*, 441.
21. Barton, D. H. R.; Hartwig, W.; Hay Motherwell, R. S.; Motherwell, W. B.; Stange, A. *TL* **1982**, *23*, 2019.
22. Barton, D. H. R.; Motherwell, W. B.; Stange, A. *S* **1981**, 743.
23. Chu, C. K.; Ullas, G. V.; Jeong, L. S.; Ahn, S. K.; Doboszewski, B.; Lin, Z. X.; Beach, J. W.; Schinazi, R. F. *JMC* **1990**, *33*, 1553.
24. Barton, D. H. R.; Subramanian, R. *JCS(P1)* **1977**, 1718.
25. (a) Mubarak, A. M.; Brown, D. M. *TL* **1981**, *22*, 683. (b) Suzuki, M.; Yanagisawa, A.; Noyori, R. *TL* **1984**, *25*, 1383.
26. Chiang, L.-Y., Shu, P.; Holt, D.; Cowan, D. *JOC* **1983**, *48*, 4713.
27. Sugimoto, H.; Makino, I.; Hirai, K. *JOC* **1988**, *53*, 2263.
28. Ley, S. V.; Armstrong, A.; Diez-Martin, D.; Ford, M. J.; Grice, P.; Knight, J. G.; Kolb, H. C.; Madin, A.; Marby, C. A.; Mukherjee, S.; Shaw, A. N.; Slawin, A. M. Z.; Vile, S.; White, A. D.; Williams, D. J.; Woods, M. *JCS(P1)* **1991**, 667.
29. Nicolaou, K. C.; Groneberg, R. D.; Miyazaki, T.; Stylianides, N. A.; Schulze, T. J.; Stahl, W. *JACS* **1990**, *112*, 8193.
30. Ge, Y.; Isoe, S. *CL* **1992**, 139.

Adrian L. Schwan
University of Guelph, Ontario, Canada

Thionyl Chloride[1]

$SOCl_2$

[7719-09-7] Cl_2OS (MW 118.97)

(chlorination of alcohols,[2] carboxylic acids,[3] and sulfonic acids;[4] dehydration of amides;[5] formation of imidoyl chlorides,[6] sulfur-containing heterocycles,[7] and the Vilsmeier reagent)

Physical Data: [8] bp 76 °C; *d* 1.631 g cm^{-3}.
Solubility: sol ethers, hydrocarbons, halogenated hydrocarbons; reacts with water, protic solvents, DMF, DMSO.
Form Supplied in: colorless to yellow liquid.
Purification: impurities cause formation of yellow or red colors. Iron contamination can cause a black color. Fractional distillation, either directly[9] or from triphenyl phosphite[10] or quinoline and linseed oil,[11] has been recommended.
Handling, Storage, and Precautions: very corrosive and reactive; the vapor and liquid are irritating to the eyes, mucous membranes, and skin. Protective gloves should be worn when handling to avoid skin contact. Thionyl chloride should be stored in glass containers at ambient temperature and protected from moisture. It reacts with water to liberate the toxic gases HCl and SO_2. Since most reactions using $SOCl_2$ evolve these gases, they should be performed only in well-ventilated hoods. Above 140 °C, $SOCl_2$ decomposes to Cl_2, SO_2, and S_2Cl_2. Iron and/or zinc contamination may cause catastrophic decomposition.[12]

Chlorides from Alcohols. Thionyl chloride reacts with primary, secondary, and tertiary alcohols to form chlorosulfite esters (eq 1) which can be isolated.[13] The fate of these esters depends on the reaction conditions, especially the stoichiometry, solvent, and base. If 2 equiv of alcohol and pyridine are used relative to thionyl chloride, dialkyl sulfites may be isolated (eq 2).[14]

$$ROH + SOCl_2 \longrightarrow ROSOCl + HCl \quad (1)$$

$$2\,BuOH + SOCl_2 \xrightarrow[77\text{–}84\%]{py} BuOSO_2Bu \quad (2)$$

When thionyl chloride is in excess, or when alkyl chlorosulfites are treated with thionyl chloride,[15] alkyl chlorides are produced and the byproducts are HCl and SO_2. The mechanism of this reaction may be S_N1, S_N2, S_Ni, or S_N2' (with allylic or propargylic alcohols). In the absence of base the S_Ni mechanism operates; retention of configuration is observed in the alkyl chloride (eq 3).[16] Considerable ionic character may be involved in this process and small amounts of elimination or rearrangement are common as side reactions.[17]

R^1, R^2 (O–S(=O)Cl) —40–180 °C→ R^1, R^2 ···Cl + SO_2 (3)

Thus, simple heating of the alcohol with thionyl chloride alone is often the preferred method for retaining the configuration of the alcohol in the chloride. Alternatively, use of amine bases, most commonly ***Pyridine***,[18] results in a shift in the mechanism to S_N2 (eq 4) and consequently overall inversion. Excess pyridine may cause dehydrochlorination or alkylation of the base. If inversion of the substrate is desired or the stereochemical outcome need not be controlled, the use of pyridine or other amine base is recommended since the resulting high chloride concentrations minimize rearrangements and eliminations.[19,17a] Alternatively, catalysis of the chlorination by ***Hexamethylphosphoric Triamide*** results in inversion.[20] With allylic substrates, rearranged products generally predominate, especially in nonpolar solvents.[21] Use of ***Tri-n-butylamine*** as the base can result in unrearranged allylic products.[22]

R^1, R^2, Cl^- (O–S(=O)–N^+ pyridinium) ⟶ R^1, R^2 Cl + SO_2 (4)

Representative procedures are available for the chlorination of tetrahydrofurfuryl alcohol,[23] benzoin,[24] and ethyl mandelate.[25] The use of thionyl chloride for this transformation usually gives less rearrangement and elimination than conc HCl, PCl_3, or PCl_5. For compounds which are acid-sensitive, ***Triphenylphosphine–Carbon Tetrachloride*** may be used.

Carboxylic Acid Chlorides from Acids. Thionyl chloride has been the most widely used reagent for the preparation of acid chlorides. Carboxylic anhydrides also react with $SOCl_2$, giving 2 equiv of the acid chloride.[26] The procedure normally involves heating the carboxylic acid with a slight to large excess of $SOCl_2$ in an inert solvent or with $SOCl_2$ itself as the solvent. Operation at the boiling point of the solvent speeds removal of the byproducts, HCl and SO_2. The ease of removal of the byproducts is the chief advantage of $SOCl_2$ over PCl_5 and PCl_3. This simple procedure gives good yields of the acid chlorides of butyric,[27] cinnamic,[28] and adipic acid.[29] The reactions are first order in each reactant and electron-withdrawing groups on the acid decrease the rate.[30] Several agents have been developed for the catalysis of this reaction. Pyridine was historically the most common; its presumed mode of action is to ensure the presence of soluble chloride (as ***Pyridinium Chloride***) which reacts with the intermediate as shown in eq 5. Catalytic pyridine ensures that the chlorination of certain diacids such as succinic acid proceeds completely to the diacid chloride rather than stopping at the anhydride.[31] The preparation of aromatic acid chlorides also commonly uses pyridine as a catalyst (eq 6).[32] Trichloroacetic acid, which is unreactive toward thionyl chloride alone, is chlorinated by the use of pyridine catalyst.[33]

RC(=O)–O–S(=O)Cl + Cl^- ⟶ RCOCl + SO_2 + Cl^- (5)

1,3,5-$C_6H_3(CO_2H)_3$ —excess $SOCl_2$, cat. pyridine, reflux, 20 h, 97%→ 1,3,5-$C_6H_3(COCl)_3$ (6)

N,N-Dimethylformamide is a particularly effective catalyst for the formation of acid chlorides with $SOCl_2$ and is now commonly the catalyst of choice.[34] The chloromethyleneiminium chloride intermediate is the active chlorinating species (see ***Dimethylchloromethyleneammonium Chloride***).

Esters of amino acids may be produced directly from the amino acid, alcohol, and thionyl chloride. One procedure involves the sequential addition of thionyl chloride and then the acid to chilled methanol.[35] Alternatively, benzyl esters have been prepared by the slow addition of $SOCl_2$ to a suspension of the amino acid in benzyl alcohol at 5 °C.[36]

For the formation of acid chlorides, the competing reactions of concern are α-oxidation and acid-catalyzed degradation. The use of solid ***Sodium Carbonate*** in the reaction mixture can minimize the latter for some compounds.[37] As shown in eq 7, even simple carboxylic acids with enolizable hydrogens α to the carboxylic functionality are subject to oxidation under standard chlorinating conditions.[38] Since the unoxidized acid chloride is the precursor to the sulfenyl chloride, careful attention to stoichiometry and reaction time can effectively minimize this problem.

$PhCH_2CH_2CO_2H$ + excess $SOCl_2$ + 0.08 equiv pyridine —reflux, 14 h, 61%→ $PhCH_2C(Cl)(SCl)COCl$ (7)

Sulfonyl and Sulfinyl Chlorides from Sulfonic and Sulfinic Acids. Alkyl or arylsulfonyl chlorides are prepared by heating the acid with thionyl chloride; DMF catalyzes this reaction. (+)-Camphorsulfonyl chloride is produced in 99% yield without a catalyst.[39] Use of the salts of sulfinic acids minimizes their oxidation; *p*-toluenesulfinyl chloride is produced in about 70% yield from sodium *p*-toluenesulfinate dihydrate with excess thionyl chloride.[40] ***Phosphorus(V) Chloride*** is more commonly used for this transformation.

Nitriles and Isocyanides via Amide Dehydration. Thionyl chloride dehydrates primary amides to form nitriles (eq 8); for example, 2-ethylhexanonitrile is produced in about 90% yield by heating with $SOCl_2$ in benzene.[41] Substituted benzonitriles are readily produced from benzamides.[42] These reactions may also be catalyzed by DMF.[43] *N*-Alkylformamides may be dehydrated to isocyanides.[44]

$$RCONH_2 + SOCl_2 \longrightarrow RCN + SO_2 + 2\,HCl \quad (8)$$

Reactions with Secondary Amides. Treatment of *N*-alkyl or *N*-aryl secondary amides with thionyl chloride in an inert solvent such as methylene chloride results in the formation of imidoyl chlorides (eq 9).[45] Upon heating, the imidoyl chlorides from *N*-alkylamides undergo scission to generate nitriles and alkyl chlorides via the von Braun degradation (eq 10).[46]

$$RCONHPh + SOCl_2 \xrightarrow[100\%]{} Ph(Cl)C{=}NPh \quad (9)$$

$$PhCONHBu \xrightarrow[69\text{–}75\%]{SOCl_2} PhCN + BuCl + SO_2 + HCl \quad (10)$$

Reactions with Aldehydes and Ketones. Aromatic or α,β-unsaturated aldehydes or their bisulfite addition compounds are converted to *gem*-dichlorides by treatment with $SOCl_2$, either neat or in an inert solvent such as nitromethane (eq 11).[47] This process is readily catalyzed by HMPA.[48] Thionyl chloride may be preferred over the more commonly used PCl_5 if removal of byproducts is problematic with the latter reagent.

$$PhCHO + SOCl_2 \xrightarrow[89\%]{\text{cat. DMF}, -10\,°C} PhCHCl_2 \quad (11)$$

Carbonyl compounds or nitriles[49] with α-hydrogens may be oxidized at this position by thionyl chloride. This reaction appears more often as a troublesome side reaction than a useful synthetic procedure. Nitriles with one α-hydrogen produce α-cyanosulfinyl chlorides (eq 12) while those with two α-hydrogens give moderate yields of α-chloro-α-cyanosulfinyl chlorides.[50]

$$Me_2CHCN + SOCl_2 \xrightarrow[0\,°C,\ 7\ d;\ 50\%]{\text{HCl, ether}} NC\text{–}C(Me)_2\text{–}SOCl \quad (12)$$

Oxidation of Activated C–H Bonds. As shown in eq 13, extensive oxidation adjacent to carbonyl groups is possible with $SOCl_2$ under relatively mild conditions.[51] The process often stops after formation of the α-chlorosulfenyl chloride. Remarkably, these may be easily hydrolyzed back to the carbonyl compounds from which they were derived.[52] Alternatively, they may be treated with a secondary amine such as morpholine followed by hydrolysis to yield the α-dicarbonyl compound. Similar oxidation of acid derivatives during acid chloride formation is possible (see above).

$$PhCOMe + \text{excess } SOCl_2 \xrightarrow[20\,°C,\ 2.5\ h;\ 95\%]{\text{cat pyridine}} PhCOCH(Cl)SCl + PhCOC(Cl){=}S \quad (13)$$

Methyl groups attached to benzenoid rings may be oxidized by $SOCl_2$ without added catalysts. The range of reactivity is large and not well understood; the products may be monochlorinated or further oxidation to the trichloromethyl aromatic system is possible.[53] 2-Methylpyrroles are similarly oxidized.[54] In systems with vicinal methyl and carboxylic acid groups, both oxidation of the methyl group and cyclization to a γ-thiolactone can occur.[55] In conjunction with radical initiators, $SOCl_2$ can chlorinate alkanes.[56] Aryl methyl ethers can be oxidized on the aromatic ring to give arylsulfenyl chlorides which may undergo further reactions.[57] This latter process, which presumably occurs via electrophilic aromatic substitution, is another potential serious side reaction for active substrates.

Rearrangement Reactions. Thionyl chloride can act as the dehydrating agent in the Beckman[58] and Lossen rearrangements[59] and promotes the Pummerer rearrangement.[60]

Synthesis of Heterocyclic Compounds. Thionyl chloride and pyridine at elevated temperatures convert diarylalkenes, styrenes, and cinnamic acids to benzo[*b*]thiophenes[61] and adipic acid to 2,5-bis(chlorocarbonyl)thiophene.[62] Additional heterocycles which have been prepared include thiazolo[3,2-*a*]indol-3(2*H*)-ones,[63] oxazolo[5,4-*d*]pyrimidines,[64] and 1,2,3-thiadiazoles.[65] Treatment of 1,2-diamino aromatic compounds with thionyl chloride gives good yields of fused 1,2,5-thiadiazoles.[66]

Other Applications. Thionyl chloride has been used to convert epoxides to vicinal dichlorides[67] and for the preparation of dialkyl sulfides from Grignard reagents.[68] Phenols react with $SOCl_2$ to produce aryl chlorosulfites and diaryl sulfites[69] or nuclear substitution products. As shown in eq 14, ***Aluminum Chloride*** catalysis yields symmetric sulfoxides, while in the absence of Lewis acids, aromatic thiosulfonates are the principal products.[70] Primary amines, especially aromatic ones, react with $SOCl_2$ to produce *N*-sulfinylamines, which are potent enophiles and useful precursors to some heterocyclic compounds.[71]

(14)
2,6-dimethylphenol (HO–Ar) → excess $SOCl_2$, 0–80 °C, 56% → HO–Ar–$S(O_2)$–S–Ar–OH
2,6-dimethylphenol → $SOCl_2$, $AlCl_3$, CCl_4, 0–25 °C, 53% → HO–Ar–S(=O)–Ar–OH

Related Reagents. Dimethylchloromethyleneammonium Chloride; Hexamethylphosphoric Triamide–Thionyl Chloride; Hydrogen Chloride; Oxalyl Chloride; Phosgene; Phosphorus(III) Chloride; Phosphorus(V) Chloride; Phosphorus(V) Oxide; Phosphorus Oxychloride; Triphenylphosphine–Carbon Tetrachloride; Triphenylphosphine Dichloride.

1. (a) Pizey, J. S. *Synthetic Reagents*; Wiley: 1974; Vol. 1, pp 321–357. (b) Davis, M.; Skuta, H.; Krubsack, A. J. *Mech. React. Sulfur Compd.* **1970**, *5*, 1.
2. Brown, G. W. In *The Chemistry of the Hydroxyl Group*; Patai, S., Ed.; Interscience; London, 1971; Part 1, pp 593–622.
3. Ansell, M. F. In *The Chemistry of Acyl Halides*; Patai, S., Ed.; Interscience: London, 1972; pp 35–68.

4. Hoyle, J. In *The Chemistry of Sulfonic Acids, Esters, and Their Derivatives*; Patai, S.; Rappoport, H., Eds.; Wiley: New York, 1991; pp 379–386.
5. Mowry, D. T. *CRV* **1948**, *42*, 257.
6. Ulrich, H. *Chemistry of Imidoyl Halides*; Plenum: New York, 1968; pp 13–54.
7. Davis, M. *Adv. Heterocycl. Chem.* **1982**, *30*, 62.
8. Weil, E. D. In *Kirk-Othmer Encyclopedia of Chemical Technology*, 3rd Ed.; Wiley: New York, 1978; Vol. 22, pp 127–131.
9. Bell, K. H. *AJC* **1985**, *38*, 1209.
10. Fieser, L. F.; Fieser, M. *FF* **1967**, *1*, 1158.
11. Martin, E. L.; Fieser, L. F. *OSC* **1943**, *2*, 569.
12. Spitulnik, M. J. *Chem. Eng. News* 1977, (Aug 1), 31.
13. Gerrard, W. *JCS* **1940**, 218.
14. Suter, C. M.; Gerhart, H. L. *OSC* **1943**, *2*, 112.
15. Frazer, M. J.; Gerrard, W.; Machell, G.; Shepherd, B. D. *CI(L)* **1954**, 931.
16. Lewis, E. S.; Boozer, C. E. *JACS* **1952**, *74*, 308.
17. (a) Hudson, H. R.; de Spinoza, G. R. *JCS(P1)* **1976**, 104. (b) Lee, C. C.; Finlayson, A. J. *CJC* **1961**, *39*, 260.
18. Ward, A. M. *OSC* **1943**, *2*, 159.
19. Stille, J. K.; Sonnenberg, F. M. *JACS* **1966**, *88*, 4915.
20. Normant, J. F.; Deshayes, H. *BSF* **1972**, 2854.
21. Caserio, F. F.; Dennis, G. E.; DeWolfe, R. H.; Young, W. G. *JACS* **1955**, *77*, 4182.
22. Young, W. G.; Caserio, F. F., Jr.; Brandon, D. D., Jr.; *JACS* **1960**, *82*, 6163.
23. Brooks, L. A.; Snyder, H. R. *OSC* **1955**, *3*, 698.
24. Ward, A. M. *OSC* **1943**, *2*, 159.
25. Eliel, E. L.; Fisk, M. T.; Prosser, T. *OSC* **1963**, *4*, 169.
26. Kyrides, L. P. *JACS* **1937**, *59*, 206.
27. Helferich, B.; Schaefer, W. *OSC* **1932**, *1*, 147.
28. Womack, E. B.; McWhirter, J. *OSC* **1955**, *3*, 714.
29. Fuson, R. C.; Walker, J. T. *OSC* **1943**, *2*, 169.
30. Beg, M. A.; Singh, H. N. *Z. Phys. Chem.* **1968**, *237*, 128; **1964**, *227*, 272.
31. Cason, J.; Reist, E. J. *JOC* **1958**, *23*, 1492.
32. Sandler, S. R.; Karo, W. *Organic Functional Group Preparations*, 2nd ed.; Academic: Orlando, 1983; Vol. 1, p 157.
33. Gerrard, W.; Thrush, A. M. *JCS* **1953**, 2117.
34. Bosshard, H. H.; Mory, R.; Schmid, M.; Zollinger, H. *HCA* **1959**, *42*, 1653.
35. Uhle, F. C.; Harris, L. S. *JACS* **1956**, *78*, 381.
36. Patel, R. P.; Price, S. *JOC* **1965**, *30*, 3575.
37. Coleman, G. H.; Nichols, G.; McCloskey, C. M.; Anspon, H. D. *OSC* **1955**, *3*, 712.
38. Krubsack, A. J.; Higa, T. *TL* **1968**, 5149.
39. Sutherland, H.; Shriner, R. L. *JACS* **1936**, *58*, 62.
40. Kurzer, F. *OSC* **1963**, *4*, 937.
41. Krynitsky, J. A.; Carhart, H. W. *OSC* **1963**, *4*, 436.
42. Goldstein, H.; Voegeli, R. *HCA* **1943**, *26*, 1125.
43. Thurman, J. C. *CI(L)* **1964**, 752.
44. (a) Tennant, G. In *Comprehensive Organic Chemistry;* Barton, D. H. R.; Ollis, W. D., Eds.; Pergamon: Oxford, 1979; Vol. 2, p 569. (b) Niznik, G. E.; Morrison, W. H., III; Walborsky, H. M. *OS* **1971**, *51*, 31.
45. von Braun, J.; Pinkernelle, W. *CB* **1934**, *67*, 1218.
46. (a) Challis, B. C.; Challis, J. A. *The Chemistry of Amides*; Zabicky, J., Ed.; Interscience; New York, 1970; pp 809–810. (b) Vaughn, W. R.; Carlson, R. D. *JACS* **1962**, *84*, 769.
47. Newman, M. S.; Sujeeth, P. K. *JOC* **1978**, *43*, 4367.
48. Khurana, J. M.; Mehta, S. *IJC(B)* **1988**, *27*, 1128.
49. Martinetz, D. *ZC* **1980**, *20*, 332.
50. Ohoka, M.; Kojitani, T.; Yanagida, S.; Okahara, M.; Komori, S. *JOC* **1975**, *40*, 3540.
51. (a) Oka, K.; Hara, S. *TL* **1976**, 2783. (b) Oka, K. *S* **1981**, 661.
52. Oka, K.; Hara, S. *TL* **1977**, 695.
53. Davis, M.; Scanlon, D. B. *AJC* **1977**, *30*, 433.
54. Brown, D.; Griffiths, D. *SC* **1983**, *13*, 913.
55. Walser, A.; Flynn, T. *JHC* **1978**, *15*, 687.
56. Krasniewski, J. M., Jr.; Mosher, M. W. *JOC* **1974**, *39*, 1303.
57. Bell, K. H.; McCaffery, L. F. *AJC* **1992**, *45*, 1213.
58. Donaruma, L. G.; Heldt, W. Z. *OR* **1960**, *11*, 1.
59. Yale, H. L. *CRV* **1943**, *33*, 242.
60. (a) Russell, G. A.; Mikol, G. J. In *Mechanisms of Molecular Migrations*; Thyagarajan, B. S., Ed.; Interscience: New York, 1968; Vol. 1, pp 157–207. (b) Durst, T. *Adv. Org. Chem.* **1969**, *6*, 356.
61. (a) Blatt, H.; Brophy, J. J.; Colman, L. J.; Tairych, W. J. *AJC* **1976**, *29*, 883. (b) Campaigne, E. In *Comprehensive Heterocyclic Chemistry*; Katritzky, A. R.; Rees, C. W., Eds.; Pergamon: Oxford, 1984; Vol. 4, p 870, 889.
62. Nakagawa, S.; Okumura, J.; Sakai, F.; Hoshi, H.; Naito, T. *TL* **1970**, 3719.
63. Showalter, H. D. H.; Shipchandler, M. T.; Mitscher, L. A.; Hagaman, E. W. *JOC* **1979**, *44*, 3994.
64. Senga, K.; Sato, J.; Nishigaki, S. *H* **1977**, *6*, 689.
65. Meier, H.; Trickes, G.; Laping, E.; Merkle, U. *CB* **1980**, *113*, 183.
66. (a) Komin, A. P.; Carmack, M. *JHC* **1975**, *12*, 829. (b) Hartman, G. D.; Biffar, S. E.; Weinstock, L. M.; Tull, R. *JOC* **1978**, *43*, 960.
67. Campbell, J. R.; Jones, J. K. N.; Wolfe, S. *CJC* **1966**, *44*, 2339.
68. Drabowicz, J.; Kielbasinski, P.; Mikolajczyk, M. In *The Chemistry of Sulphones and Sulphoxides*; Patai, S.; Rappoport, Z.; Stirling, C. J. M., Eds.; Wiley: New York, 1988; Chapter 8, p 257.
69. Gerrard, W. *JCS* **1940**, 218.
70. (a) Takimoto, H. H.; Denault, G. C. *JOC* **1964**, *29*, 759. (b) Oae, S.; Zalut, C. *JACS* **1960**, *82*, 5359.
71. Kresze, G.; Wucherpfennig, W. *AG(E)* **1967**, *6*, 149.

David D. Wirth
Eli Lilly & Co., Lafayette, IN, USA

1,1′-Thionylimidazole[1]

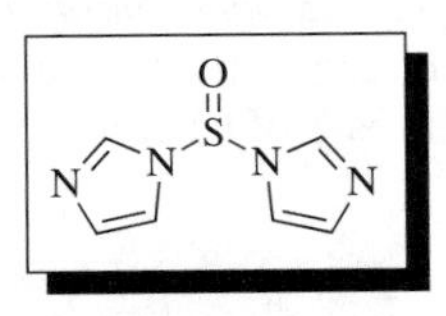

[3005-50-3] $C_6H_6N_4OS$ (MW 182.23)

(forms *N*-acylimidazolides[1] used to synthesize amides, peptides,[2] esters,[3] and ketones;[4] synthesis of symmetrical sulfoxides,[5] inorganic esters,[6] and *N*-alkylimidazoles[7])

Alternate Names: *N,N′*-thionylimitedazole; *N,N′*-thionyldiimidazole; TDI.
Physical Data: mp 78–79 °C.
Solubility: sol THF, CH_2Cl_2, toluene.
Form Supplied in: not commercially available; usually used in THF solution.

Analysis of Reagent Purity: usual preparations are used immediately without analysis.
Preparative Method: the reagent is prepared by reacting ***Thionyl Chloride*** with 4 equiv of ***Imidazole*** in anhydrous THF at 0 °C. The precipitated imidazole hydrochloride is filtered and the solution is pure enough to use.[8]
Purification: filter to remove imidazole hydrochloride. Analytically pure samples have been prepared by reaction of thionyl chloride and ***N-(Trimethylsilyl)imidazole***.[9]
Handling, Storage, and Precautions: extremely hygroscopic; reacts exothermically with water and alcohols. Store as a solution in THF or CH_2Cl_2 under dry N_2.[8b] Use in a fume hood.

N-Acylimidazolides. *N,N′*-Thionylimidazole reacts with carboxylic acids to form *N*-acylimidazolides (eq 1) that readily acylate amines, amino acids, or alcohols.[1] The products are readily purified since the byproducts, imidazole and SO_2, are easily removed. ***N,N′-Carbonyldiimidazole*** (CDI) is another useful reagent for formation of *N*-acylimidazolides. TDI is used less frequently for formation of *N*-acylimidazolides because of its high reactivity; also, CDI is commercially available and easier to handle. Most reported transformations from *N*-acylimidazolides involve CDI; however, TDI should give similar results.

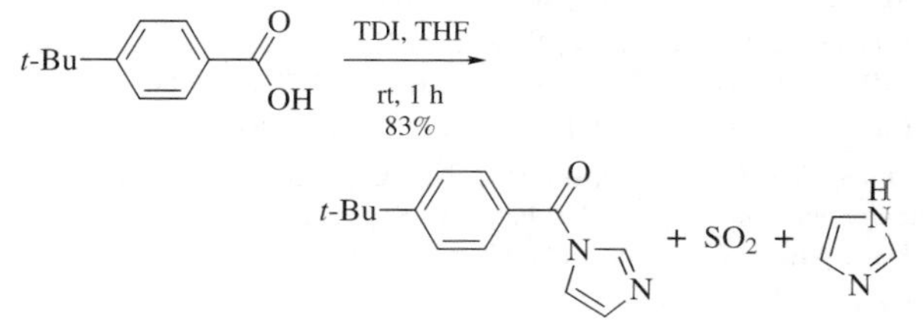

TDI, THF; rt, 1 h; 83% (1)

Peptide Coupling Reagent. *N*-Acylimidazolides are useful acylating intermediates in peptide synthesis (eq 2).[2] These are direct couplings where the amine and acid components are premixed in solution. There are only a few examples of peptides being synthesized with this reagent. Therefore it is not known what limitations exist, such as extent of racemization, ease of handling, etc. Comparisons with many of the well-known peptide coupling reagents (e.g. ***1,3-Dicyclohexylcarbodiimide***) have not been published. CDI is used more frequently if the imidazolide intermediate is desired because of its ease of handling.

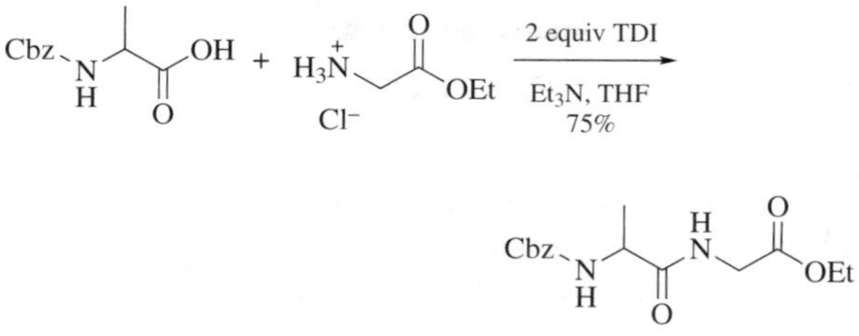

2 equiv TDI; Et_3N, THF; 75% (2)

Ester Formation. *N*-acylimidazolides react with alcohols very fast in the presence of catalytic amounts of the alkoxide (eq 3).[3] With this method the methyl ester of vitamin A has been obtained in 91% yield.[3]

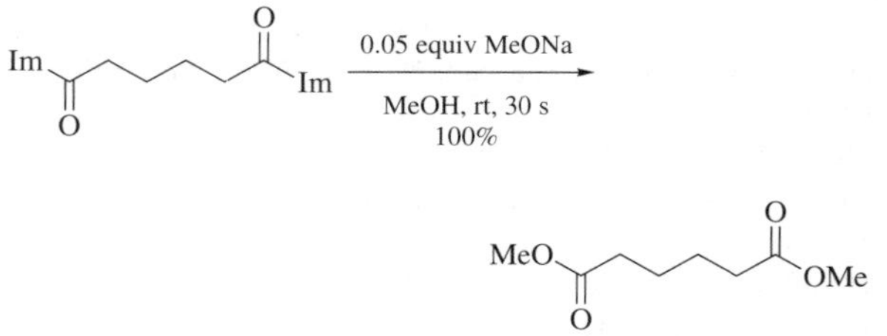

0.05 equiv MeONa; MeOH, rt, 30 s; 100% (3)

Ketones. *N*-Acylimidazolides react with Grignard reagents to give ketones in good yields.[4] A comparison of *N*-acylimidazolides and mixed acid anhydrides in Grignard reactions indicates that the imidazolides result in higher yields (eq 4).[10] There are many procedures for reacting Grignard reagents with 'activated' carboxylic acids to give ketones. Other reagents used with varying degrees of success are acid chlorides (e.g. ***Ethyl Chloroformate***), nitriles, and 2-pyridylthiol esters (e.g. ***2-Thiopyridyl Chloroformate***).[11]

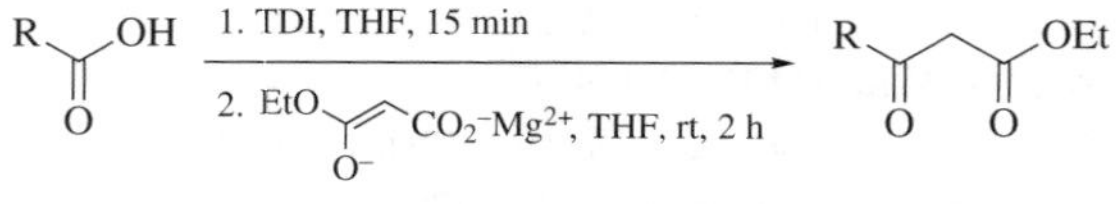

1. TDI, THF, 15 min; 2. EtO(O⁻)C=CH$CO_2^-Mg^{2+}$, THF, rt, 2 h (4)

R	Anhydride	Imidazolide
Et	57	68
Pr	57	74
C_6H_{13}	62	74
Ph	0	79

Symmetrical Sulfoxides. Thionylimidazole reacts directly with Grignard reagents resulting in symmetrical sulfoxides in reasonable yields (eq 5).[5] This method offers an easy workup and purification since the main byproduct is imidazole.

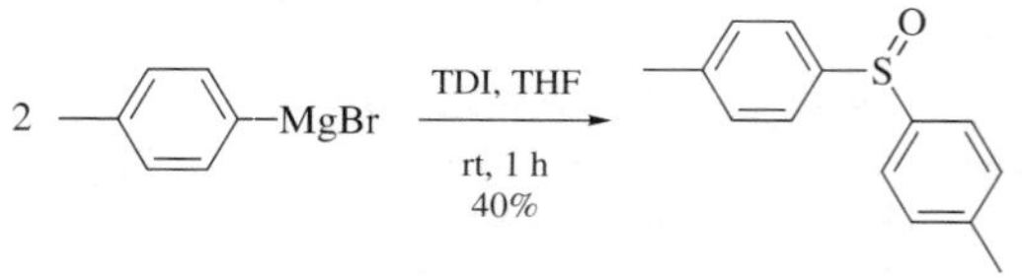

TDI, THF; rt, 1 h; 40% (5)

Inorganic Ester Formation. Phenols and alcohols react exothermically with TDI to give sulfonic acid diesters in high yield (eq 6).[6]

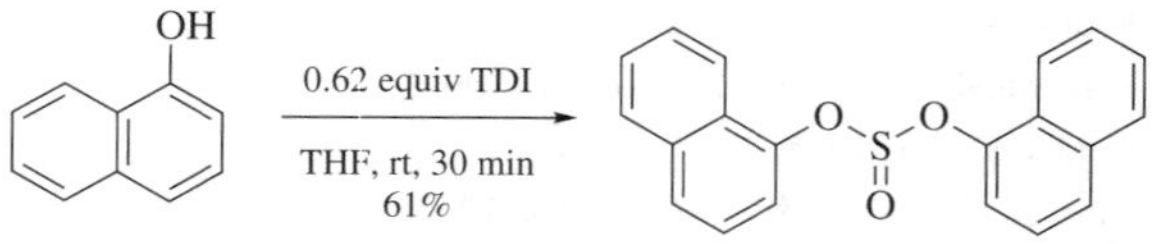

0.62 equiv TDI; THF, rt, 30 min; 61% (6)

N-Alkylimidazoles. Thionylimidazole has found great utility in the synthesis of *N*-alkylimidazoles.[7] Tertiary alcohols react with TDI to form monosulfonic ester imidazolides that rearrange to transfer the imidazole to the tertiary carbon (eq 7).[12] The products are also obtained by treatment of the trityl chloride with imidazole; however, comparison of the two procedures is not possible as yields were not reported.[12]

TDI, MeCN, 12 h, rt (7)

R = H, Me, NH_2; R^1 = H, Me, di-Me; R^2 = H, Cl, Ph

This imidazole transfer reaction occurs readily with *o*-hydroxybenzyl alcohols, giving the imidazole from a primary or secondary alcohol (eq 8).[7] It is thought that the *o*-hydroxy group contributes to the facile transfer of the imidazole. Similarly, *o*-amines promote the transfer of the imidazole. Carbonyl diimidazole also works in this imidazole transfer reaction, but it usually gives lower yields of the desired product.[7]

1.5 equiv TDI, CH_2Cl_2, rt, 30 min, 61% (8)

This imidazole transfer reaction works with ketones as well; however, in this case two products, the mono- and diimidazoles, are obtained (eq 9).[13] Increasing the temperature does not change the product ratio or yield. Some examples contain hydrogen-bonding groups on the ring, but this does not modify the product ratio or yield either. CDI gives similar results in most cases. This imidazole transfer reaction does not work with aldehydes under these conditions.[13]

1.5 equiv TDI, CH_2Cl_2, rt, 96 h; 37% + 12% (9)

1. Staab, H. A. *AG(E)* **1962**, *1*, 351.
2. Wieland, T.; Vogeler, K. *AG* **1961**, *73*, 435.
3. Staab, H. A.; Rohr, W.; Mannschreck, A. *AG* **1961**, *73*, 143 (*CA* **1961**, *55*, 14 437h).
4. Staab, H. A.; Jost, E. *LA* **1962**, *655*, 90 (*CA* **1962**, *57*, 15 090g).
5. Bast, S.; Andersen, K. K. *JOC* **1968**, *33*, 846.
6. Staab, H. A.; Wendel, K. *LA* **1966**, *694*, 86 (*CA* **1966**, *65*, 12 195c).
7. Ogata, M.; Matsumoto, H.; Kida, S.; Shimizu, S. *CI(L)* **1980**, 85.
8. (a) Staab, H. A.; Wendel, K. *AG* **1961**, *73*, 26 (*CA* **1961**, *55*, 12 389e). (b) *FF* **1967**, *1*, 1163.
9. Birkofer, L.; Gilgenberg, W.; Ritter, A. *AG* **1961**, *73*, 143.
10. Bram, G.; Vilkas, M. *BSF(2)* **1964**, 945.
11. (a) Shirley, D. A. *OR* **1954**, *8*, 28. (b) Posner, G.; Witten, C. E. *TL* **1970**, 4647. (c) Mukaiyama, T.; Araki, M.; Takei, H.; *JACS* **1973**, *95*, 4763.
12. Buechel, K. H.; Draber, W.; Regal, E.; Plempel, M. *AF* **1972**, *22*, 1260 (*CA* **1972**, *77*, 152 067e).
13. Ogata, M.; Matsumoto, H.; Kida, S.; Shimizu, S. *TL* **1979**, 5011.

Richard S. Pottorf
Marion Merrell Dow Research Institute, Cincinnati, OH, USA

2-Thiopyridyl Chloroformate[1]

[73371-99-0] C_6H_4ClNOS (MW 173.63)

(convenient preparation of 2-pyridylthiol esters;[2] subsequent transformation into lactones,[3] peptides,[4] and ketones[5])

Physical Data: ^1H NMR ($CDCl_3$) δ 8.64 m (1H), 7.75 m (2H), 7.38 m (1H).
Solubility: sol CH_2Cl_2, ether.
Form Supplied in: colorless oil; main impurity is bis(thiopyridyl) carbonate; not commercially available.
Analysis of Reagent Purity: IR (CH_2Cl_2) 1765 cm^{-1} (C=O); main impurity: IR (CH_2Cl_2) 1715 cm^{-1} (C=O).
Preparative Method: ***Phosgene*** (5 equiv) in toluene and CH_2Cl_2 is cooled to 0 °C. Dropwise addition (5 min) of a CH_2Cl_2 solution of ***Triethylamine*** (slight excess) and ***2-Pyridinethiol*** is followed by stirring for 10 min. After removal of excess phosgene and CH_2Cl_2 in vacuo, hexane is added and the resulting precipitate is filtered. After concentration of the combined filtrates, the colorless oil (96%) is dissolved in CH_2Cl_2 and stored at −25 °C.
Handling, Storage, and Precautions: very unstable to water and silica gel; however, it can be handled in air. It is stable for one month if stored at −25 °C. Since phosgene is used in preparing this reagent, preparation should be in a working fume hood and extreme caution is required.

Preparation of 2-Pyridylthiol Esters. 2-Pyridylthiol esters are formed by treating a carboxylic acid with 2-thiopyridyl chloroformate under extremely mild conditions (eq 1).[2] The $Et_3N{\cdot}HCl$ is removed by filtration or by washing with cold aqueous acid and base. After thorough drying, the thiol esters are generally pure enough to use in many synthetic applications.

pySCOCl, Et_3N, ether, 0 °C, 0.5 h, 98–100%; + CO_2 + Et_3NHCl (1)

There are several other reagents for preparing 2-pyridylthiol esters. ***1,3-Dicyclohexylcarbodiimide***[4] generally gives lower yields of the thiol ester and removal of the dicyclohexylurea can be difficult. Treating an acid chloride with thallium(I) 2-pyridinethiolate[6] has been used, but thallium salts are toxic. The method used most frequently before the introduction of 2-thiopyridyl chloroformate involves reacting a carboxylic acid with ***Triphenylphosphine*** and ***2,2′-Dipyridyl Disulfide***.[7] This procedure suffers from the necessity of removing 2-pyridinethiol and triphenylphosphine oxide by chromatography, which precludes preparing large batches of the thiol esters since there can be a loss of product by reaction with the silica gel. 2-Thiopyridyl chloroformate provides access to many reported synthetic transformations. In some of these reports, the thiol ester has been prepared by other methods. However, use of 2-thiopyridyl chloroformate should make these transformations even more accessible.

Lactone Formation. There are several examples of the synthesis of lactones with 2-pyridylthiol esters.[3] Since this is a facile reaction, reasonable yields of complex macrocycles are obtained (eq 2).[8] The pyridine nitrogen may provide anchimeric assistance for the approaching nucleophile, thus facilitating the acylation.[3]

MeS SMe; MeO; CO_2H; OH — 1. pySCOCl, Et_3N, CH_2Cl_2, 0 °C, 2 h; 2. xylene, Δ; 68% — MeS SMe; MeO; O (2)

1:1 ratio of diastereomers

Ester Formation. 2-Pyridylthiol esters acylate lysophosphatidylcholines rapidly when catalyzed by silver ion, giving mixed-chain phosphatidylcholines in high yields and isomeric purity (eq 3).[9] The main advantages of this procedure over the usual ***4-Dimethylaminopyridine*** catalyzed acylation with acid anhydrides[10] are that less acylating reagent is required to give high yields and rearrangement of the fatty acids is minimized. The main disadvantage is the sensitivity of the 2-pyridylthiol ester to water.

HO; O; R^1; OPC — R^2(C=O)Spy; $AgClO_4$, benzene, rt, 10 min; 95% — R^2; O; R^1; OPC (3)

$R^1 = Me(CH_2)_{16}$
$R^2 = Me(CH_2)_{14}$

1.5% isomeric purity

PC = P(=O)(O−)–$CH_2CH_2\overset{+}{N}Me_3$

Peptide Coupling. These thiol esters have been used in the synthesis of several dipeptides. Many of the examples involve highly sterically hindered amino acids and result in very good yields and high optical purity (eq 4).[4] There are many useful active esters for peptide coupling, such as ***Pentafluorophenol***, ***p-Nitrophenol***, and ***N-Hydroxysuccinimide***. However, these do not work as well with the hindered amino acids shown in eq 4. Additionally, there are many direct coupling procedures; therefore the 2-pyridylthiol esters have not been used frequently.

N; O; Spy — ValOMe•HCl; Et_3N, DMF, 2 h; 80% — N; O; N H; OMe (4)

Ketones. Reaction of 2-pyridylthiol esters with Grignard reagents occurs rapidly to give ketones in almost quantitative yields (eq 5).[5] In all studies, less than 1% of the tertiary alcohol is observed. This procedure works reasonably well with a suitably protected amino acid to give the ketone (eq 6).[11] Phthaloyl protection of the amine is required since esters with amide NH groups decompose when treated with Grignard reagents.[11]

Ph; Spy; O — CyMgBr; THF, 0 °C; 95% — Ph; O (5)

N; Ph; Spy; O — BrMg (dioxolanylethyl); THF, 30–35 °C; 35% — N; Ph; O (6)

Preparation of thiol esters of highly elaborated phosphoranes with thiopyridyl chloroformate has been reported. Treatment of the thiol ester with an aryl Grignard reagent gives the corresponding ketone without epimerization of the chiral centers (eq 7).[12] There are many procedures for reacting Grignard reagents with 'activated' carboxylic acids to give ketones. Other reagents used with varying degrees of success include acid chlorides, nitriles, acid anhydrides, and *N*-acylimidazolides (e.g. ***N,N′-Carbonyldiimidazole***, ***1,1′-Thionylimidazole***).[13] Many of these give higher yields of the tertiary alcohol. However, treating acid chlorides with Grignard reagents at −78 °C gives better results,[14] so this may be an alternative to the thiol esters.

TBDMSO, Spy, O, N, PPh$_3$, CO_2 — BrMg-(2-pyridyl), ether, THF, 0 °C, 1 h, 40% → TBDMSO, O, N, PPh$_3$, CO_2 (7)

1. Haslam, E. *T* **1980**, *36*, 2409.
2. Corey, E. J.; Clark, D. A. *TL* **1979**, *31*, 2875.
3. Corey, E. J.; Nicolaou, K. C. *JACS* **1974**, *96*, 5614.
4. Lloyd, K.; Young, G. T. *JCS(C)* **1971**, 2890.
5. Mukaiyama, T.; Araki, M.; Takei, H. *JACS* **1973**, *95*, 4763.
6. Masamune, S.; Kamata, S.; Diakur, J.; Sugihara, Y.; Bates, G. S. *CJC* **1975**, *53*, 3693.
7. Mukaiyama, T.; Matsueda, R.; Suzuki, M. *TL* **1970**, 1901.
8. Le Drian, C.; Greene, A. E. *JACS* **1982**, *104*, 5473.
9. Nicholas, A. W.; Khouri, L. G.; Ellington, J. C., Jr.; Porter, N. A. *Lipids* **1983**, *18*, 434.
10. Gupta, C. M.; Radhakrishran, R.; Khorana, H. G. *PNA* **1977**, *74*, 4315.
11. Almquist, R. G.; Chao, W.-R.; Ellis, M. E.; Johnson, H. L. *JMC* **1980**, *23*, 1392.
12. Guthikonda, R. N.; Cama, L. D.; Quesada, M.; Woods, M. F.; Salzmann, T. N.; Christensen, B. G. *JMC* **1987**, *30*, 871.
13. (a) Shirley, D. A. *OR* **1954**, *8*, 28. (b) Posner, G.; Witten, C. E. *TL* **1970**, 4647. (c) Staab, H. A.; Jost, E. *LA* **1962**, *655*, 90.
14. Sato, F.; Inoue, M.; Oguro, K.; Sato, M. *TL* **1979**, 4303.

Richard S. Pottorf
Marion Merrell Dow Research Institute, Cincinnati, OH, USA

Tin(IV) Chloride

[7646-78-8] Cl_4Sn (MW 260.51)

(strong Lewis acid used to promote nucleophilic additions, pericyclic reactions, and cationic rearrangements; chlorination reagent)

Alternate Name: stannic chloride.

Physical Data: colorless liquid; mp −33 °C; bp 114.1 °C; *d* 2.226 g cm^{-3}.

Solubility: reacts violently with water; sol cold H_2O; dec hot H_2O; sol alcohol, Et_2O, CCl_4, benzene, toluene, acetone.

Form Supplied in: colorless liquid; 1 M soln in CH_2Cl_2 or heptane; widely available.

Purification: reflux with mercury or P_2O_5 for several hours, then distill under reduced nitrogen pressure into receiver with P_2O_5. Redistill. Typical impurities: hydrates.

Handling, Storage, and Precautions: hygroscopic; should be stored in a glove box or over P_2O_5 to minimize exposure to moisture. Containers should be flushed with N_2 or Ar and tightly sealed. Perform all manipulations under N_2 or Ar. Solvating with H_2O liberates much heat. Use in a fume hood.

Introduction. $SnCl_4$ is used extensively in organic synthesis as a Lewis acid for enhancing a variety of reactions. $SnCl_4$ is classified as a strong Lewis acid according to HSAB theory, and therefore interacts preferentially with hard oxygen and nitrogen bases. Six-coordinate 1:2 species and 1:1 chelates are the most stable coordination complexes, although 1:1 five-coordinate species are also possible.[1] $SnCl_4$ can be used in stoichiometric amounts, in which case it is considered a 'promoter', or in substoichiometric amounts as a catalyst, depending upon the nature of the reaction. $SnCl_4$ is an attractive alternative to boron, aluminum, and titanium Lewis acids because it is monomeric, highly soluble in organic solvents, and relatively easy to handle. $SnCl_4$ and $TiCl_4$ are among the most common Lewis acids employed in 'chelation control' strategies for asymmetric induction. However, $SnCl_4$ is not often the Lewis acid of choice for optimum selectivities and yields.

$SnCl_4$ is also the principal source for alkyltin chlorides, R_nSnCl_{4-n}.[2] Allyltrialkyltin reagents react with $SnCl_4$ to produce allyltrichlorotin species through an S_E2' pathway (eq 1).[3] Silyl enol ethers react with $SnCl_4$ to give α-trichlorotin ketones (eq 2).[4] Transmetalation or metathesis reactions of this type are competing pathways to nucleophilic addition reactions where $SnCl_4$ is present as an external Lewis acid. As a consequence, four important experimental variables must be considered when using $SnCl_4$ as a promoter: (1) the stoichiometry between the substrate and the Lewis acid; (2) the reaction temperature; (3) the nature of the Lewis base site(s) in the substrate; and (4) the order of addition. These variables influence the reaction pathway and product distribution.[5]

allyl-SnR_3 + $SnCl_4$ → allyl-$SnCl_3$ + $SnBu_3Cl$ (1)

R-C(OTMS)=CH_2 + $SnCl_4$ → R-C(=O)-CH_2-$SnCl_3$ + TMSCl (2)

Nucleophilic Additions to Aldehydes. $SnCl_4$ is effective in promoting the addition of nucleophiles to simple aldehydes. Among the most synthetically useful additions are allylstannane and -silane additions. The product distribution in the stannane reactions can be influenced by the order of addition, stoichiometry, and reaction temperature. The *anti* geometry of the tin–aldehyde complex is favored due to steric interactions. Furthermore, the six-coordinate 2:1 complex is most likely the reactive intermediate in these systems. The use of crotylstannanes provides evidence for competing transmetalation reaction pathways (eq 3).[6] Superior selectivities are provided by ***Titanium(IV) Chloride***.

The presence of additional Lewis base sites within the molecule can result in the formation of chelates with $SnCl_4$ or $TiCl_4$, which can lead to 1,2- or 1,3-asymmetric induction with the appropriate substitution at the C-2 or C-3 centers. NMR studies have provided a basis for explaining the levels of diastereofacial selectivity observed in nucleophilic additions to Lewis acid chelates of β-alkoxy aldehydes with substitution at the C-2 or C-3 positions.[7] These studies reveal that $SnCl_4$ chelates are dynamically unstable when substrates are sterically crowded at the alkoxy center, thus enhancing the formation of 2:1 complexes and/or competing metathesis pathways. Furthermore, for β-siloxy aldehydes, the 2:1 $SnCl_4$ complex is formed preferentially over the corresponding chelate.[8]

Conditions	Lewis acid	Product 1	Product 2	Product 3	Product 4
normal addition	1.3 equiv $SnCl_4$	22.8	26.0	36.4	14.8
inverse addition	1.3 equiv $SnCl_4$	21.8	74.9	1.2	2.2
normal addition	1.05 equiv $TiCl_4$	90.5	7.0	2.1	0.5
inverse addition	2.1 equiv $TiCl_4$	4.4	90.8	–	4.9

(3)

Mukaiyama Aldol Additions. Lewis acid-promoted additions of a chiral aldehyde to a silyl enol ether or silyl ketene acetal (the Mukaiyama[9] aldol addition) occurs with good diastereofacial selectivity.[10] The reaction has been investigated with nonheterosubstituted aldehydes, α- and β-alkoxy aldehydes,[11] α- and β-amino aldehydes,[12] and thio-substituted aldehydes.[13] High diastereoselectivity is observed in the $SnCl_4$- or $TiCl_4$-promoted aldol addition of silyl enol ethers to α- and β-alkoxy aldehydes. Prior chelation of the aldehyde before addition of the enol silane is important because certain enol silanes interact with $SnCl_4$ to produce α-trichlorostannyl ketones, which provide lower selectivity.[14] Simple diastereoselectivity is independent of the geometry of the enol silane, and the reaction does not proceed through prior Si–Ti or Si–Sn exchange. Good *anti* selectivities (up to 98:2) are obtained in the $SnCl_4$-promoted reactions of chiral α-thio-substituted aldehydes only with α-phenylthio-substituted aldehydes (eq 4). Stereorandom results are obtained with $SnCl_4$ when other alkylthio-substituted aldehydes, such as α-isopropylthio-substituted aldehydes, are used. ***Boron Trifluoride Etherate*** catalysis gives better *anti* selectivities than $SnCl_4$ for aldehydes with smaller alkylthio substituents. Excellent *syn* selectivities are obtained for α-thio-substituted aldehydes with $TiCl_4$.

(4)

Additions to Nitriles. $SnCl_4$-promoted addition of malonates and bromomalonates to simple nitriles (not electron deficient) gives α,β-dehydro-β-amino acid derivatives (eq 5).[15] $SnCl_4$ is the Lewis acid of choice for the condensation of aroyl chlorides with sodium isocyanate, affording aroyl isocyanates in 70–85% yields.[16] Nonaromatic acyl chlorides react under more variable reaction conditions.

(5)

Hydrochlorination of Allenic Ketones. $SnCl_4$ is also a source for generating chloride anions which form new carbon–chlorine bonds. This occurs through a ligand exchange pathway which has been exploited in the formation of β-chloro enones from conjugated allenic ketones (eq 6).[17] Yields range from 36–82% with complete selectivity for the *trans* geometry. A variety of substituents (R^1, R^2) can be tolerated including aryl, rings, and alkoxymethyl groups (R^1).

(6)

Glycosylations. The reaction of glycofuranosides having a free hydroxyl group at C-2 with functionalized organosilanes, in the presence of $SnCl_4$, provides *C*-glycosyl compounds in high stereoselectivity (eq 7).[18] Organosilanes such as 4-(chlorodimethylsilyl)toluene, chlorodimethylvinylsilane, ***Allyltrimethylsilane***, and allylchlorodimethylsilane are effective reagents. The presence of a leaving group on the silane is essential for good selectivity since the reaction proceeds intramolecularly through a 2-*O*-organosilyl glycoside. The availability of furanosides in the ribo, xylo, and arabino series make this reaction valuable for the stereoselective synthesis of *C*-furanosides. Regioselective glycosylation of nitrogen-containing heterocycles is also effectively promoted by $SnCl_4$, and has been used in the synthesis of pentostatin-like nucleosides, such as (**1**).[19]

(7)

R = H: pentostatin
R = OH: coformycin

(**1**)

Selective De-*O*-benzylation. Regioselective de-*O*-benzylation of polyols and perbenzylated sugars is achieved with organotin reagents or other Lewis acids.[20,21] The equatorial *O*-benzyl group of 1,6-anhydro-2,3,4-tri-*O*-benzyl-β-D-mannopyranose is selectively cleaved by $SnCl_4$ or $TiCl_4$ (eq 8).[2] The equatorial *O*-benzyl group is also selectively cleaved when one of the axial *O*-benzyl groups is replaced by an *O*-methyl group. The 2-*O*-benzyl group of 1,2,3-tris(benzyloxy)propane is selectively cleaved (eq 9), but no debenzylation is observed with 1,2-bis(benzyloxy)ethane.

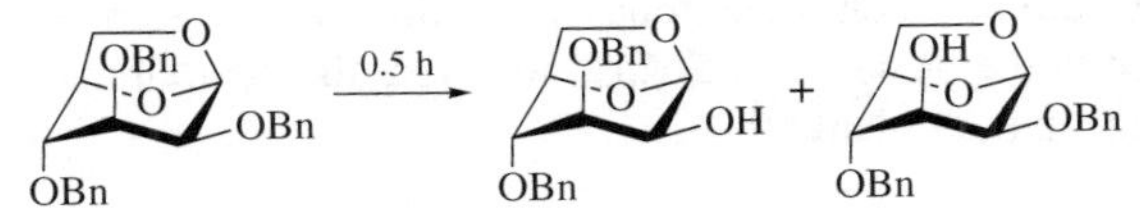

(8) 0.5 h

$SnCl_4$	92%	5%
$TiCl_4$	77%	19%

(9) $SnCl_4$, 86%

Rearrangement of Allylic Acetals. Lewis acid-promoted ($SnCl_4$ or ***Diethylaluminum Chloride***) rearrangements of allylic acetals provide substituted tetrahydrofurans.[22] Upon addition of Lewis acid, (**2**) rearranges to the *all-cis* furan (**3**) (eq 10). No racemization is observed with optically active allylic acetals; however, addition of KOH completely epimerizes the furan–carbonyl bond, as does quenching at room temperature. Acetals successfully undergo similar rearrangement provided the alkene is substituted. Completely substituted tetrahydrofurans are synthesized stereoselectively (>97% ee) by the rearrangement of disubstituted allyl acetals (eq 11). This reaction is related to the acid-catalyzed rearrangements of 5-methyl-5-vinyloxazolidines to 3-acetylpyrrolidines, which involves an aza-Cope rearrangement and Mannich cyclization.[23]

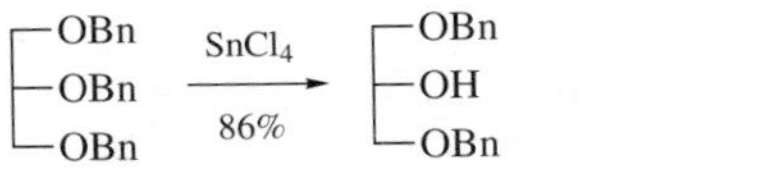

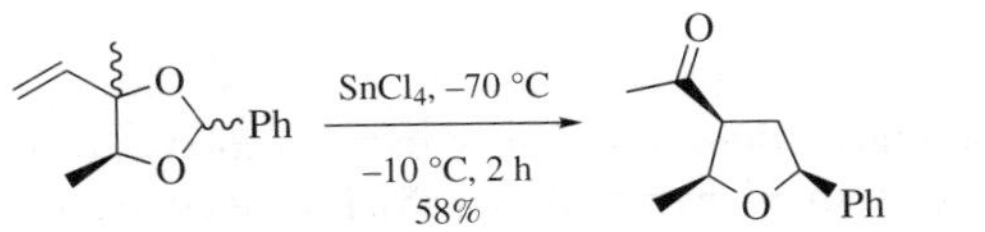

(10) (**2**) → (**3**): $SnCl_4$, –70 °C; –10 °C, 2 h; 58%

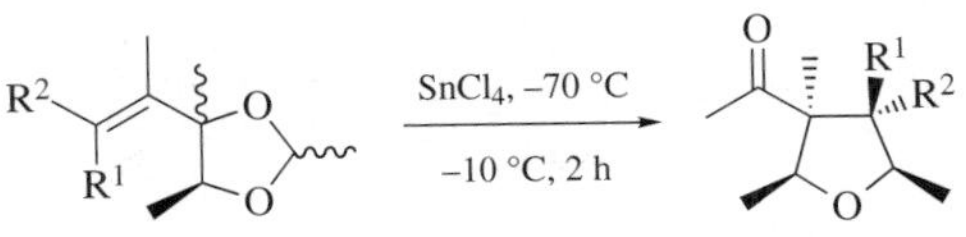

(11) $SnCl_4$, –70 °C; –10 °C, 2 h

R^1 = Me, R^2 = H; 90%
R^1 = H, R^2 = Me; 73%

The rearrangement is also useful for furan annulations, through enlargement of the starting carbocycle.[24] Thus addition of $SnCl_4$ to either diastereomer of the allylic acetal (**4**) produces the *cis*-fused cycloheptatetrahydrofuran (**5**) in 48–76% yield (eq 12). Acetals derived from *trans*-diols rearrange to the same *cis*-fused bicyclics in higher yield. The stereochemistry of a terminal alkene is transmitted to the C-3 carbon of the bicyclic products (eq 13). Rearrangements of acetals require substitution at the internal alkene carbon.

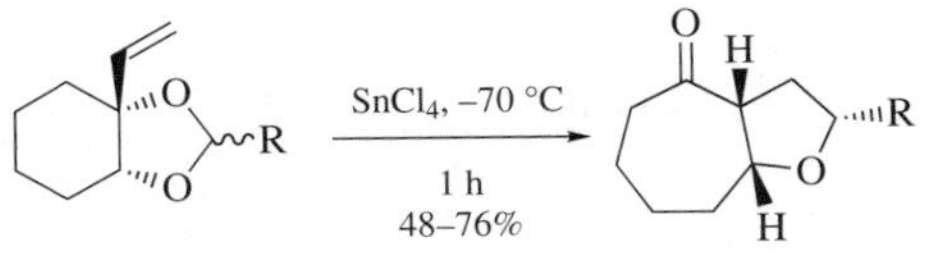

(12) (**4**) → (**5**): $SnCl_4$, –70 °C; 1 h; 48–76%

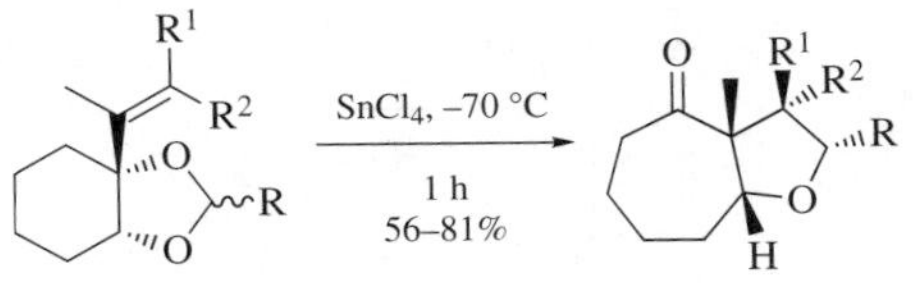

(13) $SnCl_4$, –70 °C; 1 h; 56–81%

α-*t*-Alkylations. $SnCl_4$-promoted α-*t*-alkylations of alkenyl β-dicarbonyl compounds is a particularly useful cyclization reaction.[25] Cyclization occurs through initial formation of a stannyl enol ether, followed by protonation of the alkene to form a carbocation which undergoes subsequent closure (eq 14). The analogous α-*s*-alkylation reactions are best catalyzed by other Lewis acids.

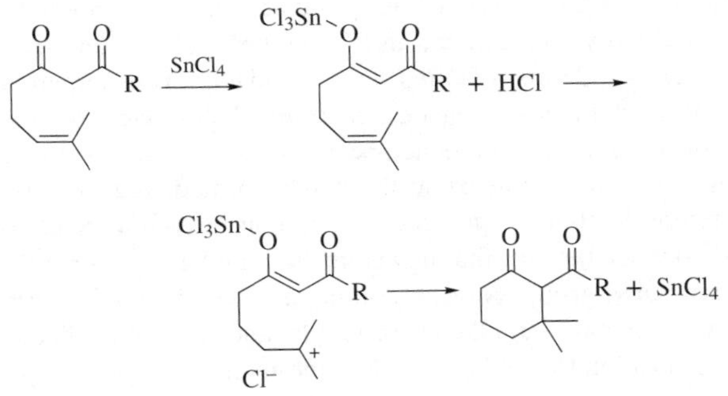

(14) $SnCl_4$; + HCl; + $SnCl_4$

This reaction is useful for cyclizations involving 6-*endo*-trigonal (eq 14) and 'allowed' 7-*endo* trigonal processes (eq 15), but not for those involving 5-*endo* trigonal processes (eq 16). These observations are consistent with the Baldwin rules.

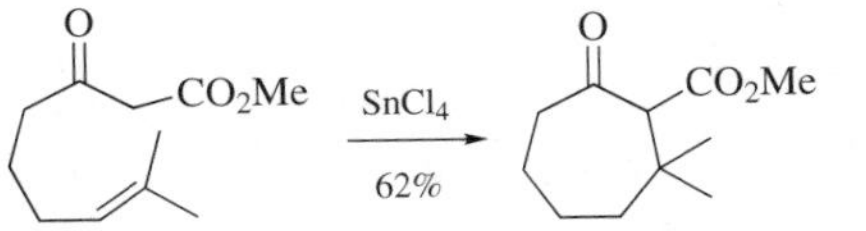

(15) $SnCl_4$, 62%

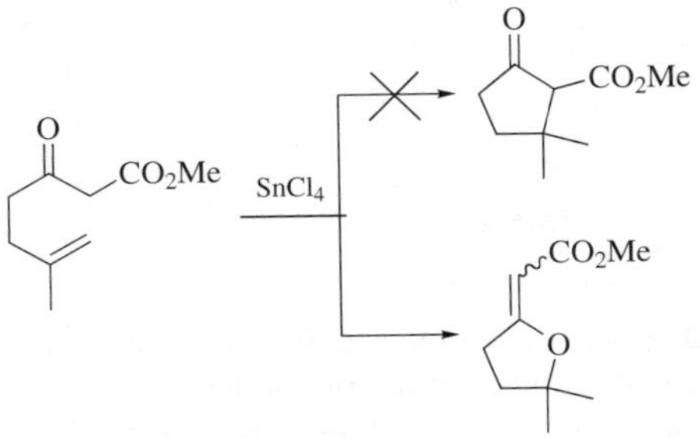

(16) $SnCl_4$

Reactions involving 4- and 6-*exo* trigonal cyclizations result in poor yields or undesired products, while those involving 5-*exo* trigonal cyclizations produce higher yields (eq 17). This synthetic strategy can also be used to form bicyclic and spiro compounds (eqs 18 and 19).

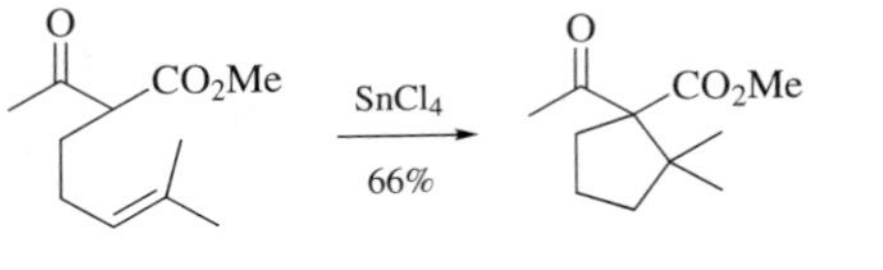

$SnCl_4$, 66% (17)

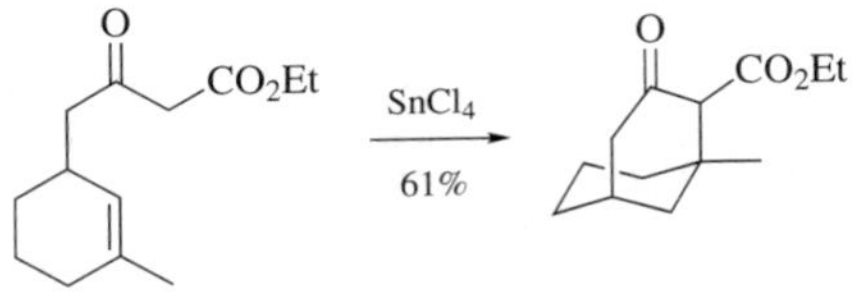

$SnCl_4$, 61% (18)

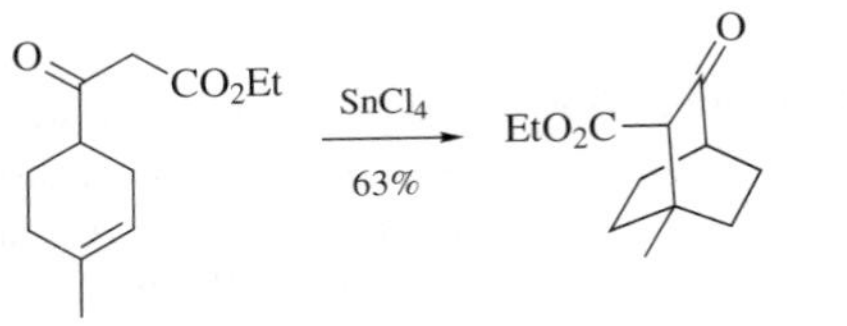

$SnCl_4$, 63% (19)

Alkene Cyclizations. Cationic cyclizations of polyenes, containing initiating groups such as cyclic acetals, are promoted by $SnCl_4$ and have been utilized in the synthesis of *cis*- and *trans*-decalins, *cis*- and *trans*-octalins, and tri- and tetracyclic terpenoids and steroids.[26] In most instances, *all-trans*-alkenes yield products with *trans,anti,trans* stereochemistry (eq 20), while *cis*-alkenes lead to *syn* stereochemistry at the newly formed ring junctions. The stereoselectivity of polyene cyclizations are often greatly diminished when the terminating alkene is a vinyl group rather than an isopropenyl group. Acyclic compounds which contain terminal acyclic acetals and alkenes or vinylsilanes can be cyclized in a similar fashion to yield eight- and nine-membered cyclic ethers (eq 21).[27]

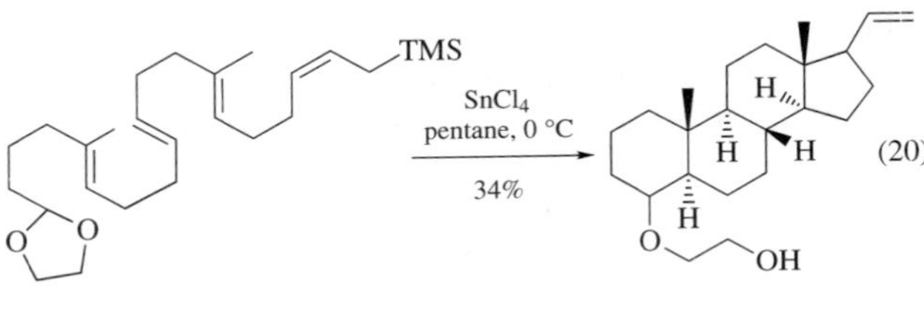

$SnCl_4$, pentane, 0 °C, 34% (20)

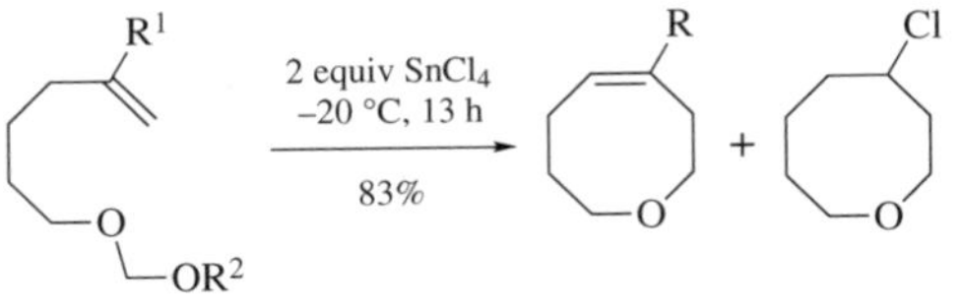

2 equiv $SnCl_4$, −20 °C, 13 h, 83% (21)

R^1 = H, TMS; R^2 = $(CH_2)_2OMe$

The analogous cyclization of chiral imines occurs in high yields (75–85%) with good asymmetric induction (36–65% ee).[28] For example, the cyclization of aldimine (**6**), derived from methyl citronellal, using $SnCl_4$ affords only the *trans*-substituted aminocyclohexane (**7**) in high yield (eq 22). *Exo* products are formed exclusively or preferentially over the thermodynamically favored *endo* products.

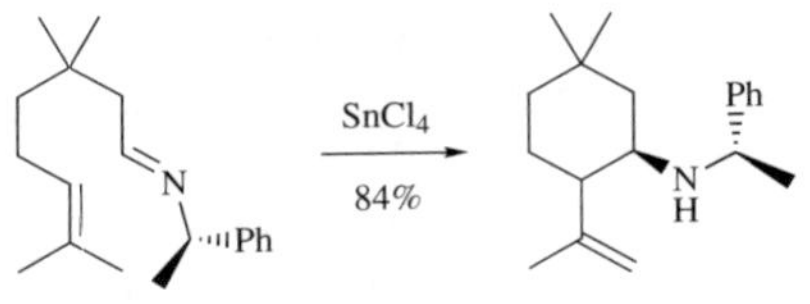

(**6**) → $SnCl_4$, 84% → (**7**) (22)

$SnCl_4$-induced cyclizations between alkenes and enol acetates result in cycloalkanes or bicycloalkanes in high yield (eq 23). It is interesting to note that the TMSOTf-catalyzed reaction can yield fused products rather than bicyclo products. Alkenic carboxylic esters, allylic alcohols, sulfones, and sulfonate esters are also cyclized in the presence of $SnCl_4$; however, alkenic oxiranes often cyclize in poor yield.[26a]

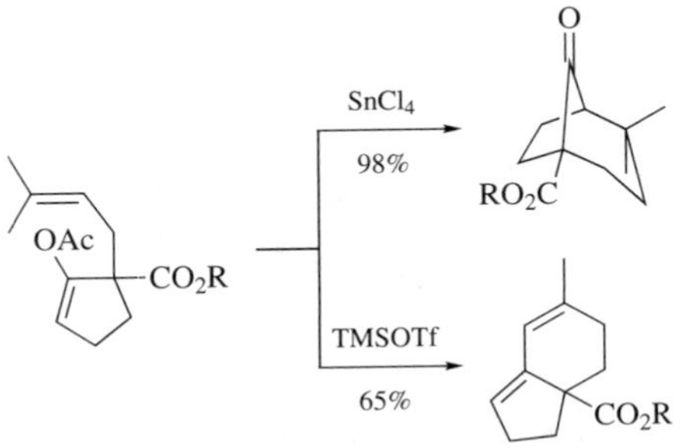

$SnCl_4$, 98%; TMSOTf, 65% (23)

$SnCl_4$ is also effective in the opening of cyclopropane rings to produce cationic intermediates useful in cyclization reactions. For example, the cyclization of aryl cyclopropyl ketones to form aryl tetralones, precursors of aryl lignan lactones and aryl naphthalene lignans, is mediated by $SnCl_4$ (eq 24).[29] The reaction is successful in nitromethane, but not in benzene or methylene chloride. Analogous cyclizations with epoxides result in very low yields (2–5%).

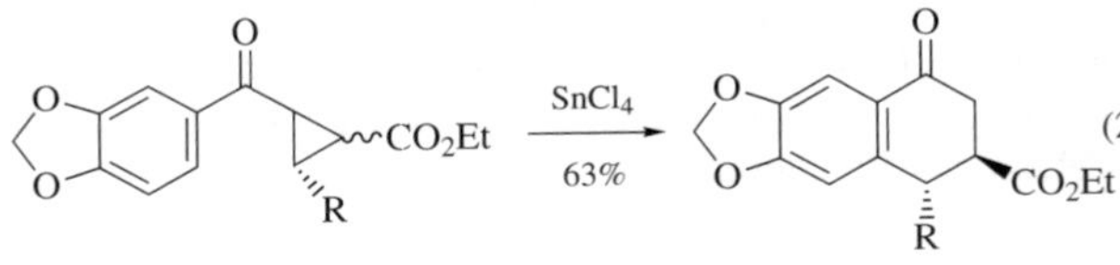

$SnCl_4$, 63% (24)

Polymerization. Cationic polymerizations are catalyzed by $SnCl_4$ and other Lewis acids (eq 25). Propagation is based upon the formation of a cationic species upon complexation with $SnCl_4$.[30] Radical pathways are also possible for polymer propagation.[31]

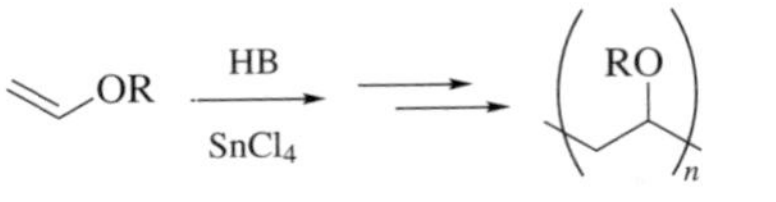

HB, $SnCl_4$ (25)

Diels–Alder Reactions. Diels–Alder reactions are enhanced through the complexation of dienophiles or dienes by Lewis acids.[32] Furthermore, Lewis acids have been successfully employed in asymmetric Diels–Alder additions.[33] Although $SnCl_4$ is a useful Lewis acid in Diels–Alder reactions, in most instances titanium or aluminum Lewis acids provide higher yields and/or selectivities. The stereoselectivity in Lewis acid-promoted Diels–Alder reactions between chiral α,β-unsaturated *N*-acyloxazolidinones shows unexpected selectivities as a function of the Lewis acid (eq 26).[34] Optimum selectivity is expected for chelated intermediates, yet both $SnCl_4$ and $TiCl_4$ perform poorly relative to Et_2AlCl (1.4 equiv). The formation of the $SnCl_4$–*N*-acyloxazolidinone chelate has been confirmed by solution NMR studies.[35] These data suggest that other factors such as the steric bulk associated with complexes may contribute to stereoselectivity.

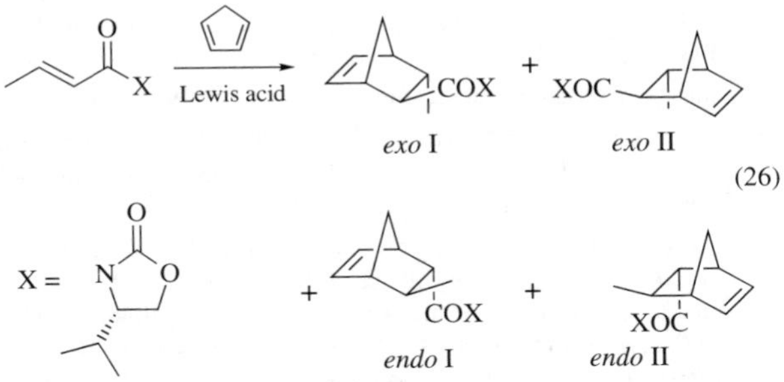

(26)

Lewis acid	conv %	Σ*endo*:Σ*exo*	*endo* I:*endo* II
1.1 equiv $SnCl_4$	70	14.9	3.1
1.1 equiv $TiCl_4$	100	9.9	2.7
1.4 equiv Et_2AlCl	100	50.0	17

In Lewis acid-promoted Diels–Alder reactions of cyclopentadiene with the acrylate of (*S*)-ethyl lactate, good diastereofacial and *endo*/*exo* selectivity are obtained with $SnCl_4$ (84:16; *endo*/*exo* = 18:1) and $TiCl_4$ (85:15; *endo*/*exo* = 16:1).[36] It is interesting to note that boron, aluminum, and zirconium Lewis acids give the opposite diastereofacial selectivity (33:67 to 48:52). Competing polymerization of the diene is observed in methylene chloride, particularly with $TiCl_4$, but not in solvent mixtures containing *n*-hexane.

Cycloalkenones generally perform poorly as dienophiles in Diels–Alder reactions but their reactivity can be enhanced by Lewis acids.[37] $SnCl_4$ is effective in promoting the Diels–Alder reaction between simple 1,3-butadienes, such as isoprene and piperylene, and cyclopentenone esters. For example, the $SnCl_4$-promoted cycloaddition between (**8**) and isoprene is completely regioselective, providing the substituted indene in 86% yield (eq 27).[38] However, cycloaddition does not occur in the presence of $SnCl_4$ when the diene contains an oxygen-bearing substituent such as an alkoxy or siloxy group. In these cases, as is generally true for the Diels–Alder reactions of cycloalkenones, other Lewis acids are more effective. For example, $SnCl_4$-promotion of the cycloaddition between (**8**) and 3-methyl-2-(*t*-butyldimethylsiloxy)butadiene yields 37% of the desired product, while ***Zinc Chloride*** provides a 90% yield. When furan or 2-methyl-1-alkylsiloxybutadiene are utilized as dienes, only decomposition of the starting material is observed with $SnCl_4$.

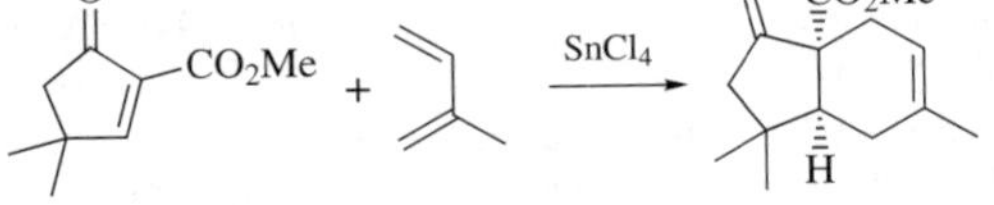

(27)

The Lewis acid-promoted Diels–Alder reaction has been employed in the assembly of steroid skeletons.[39] The cycloaddition reaction between a substituted bicyclic diene and 2,6-dimethylbenzoquinone produces two stereoisomers in a 1:5 ratio with a yield of 83% when $SnCl_4$ is used in acetonitrile. $TiCl_4$ provides slightly higher selectivities (1:8) but lower yield (70%) (eq 28).

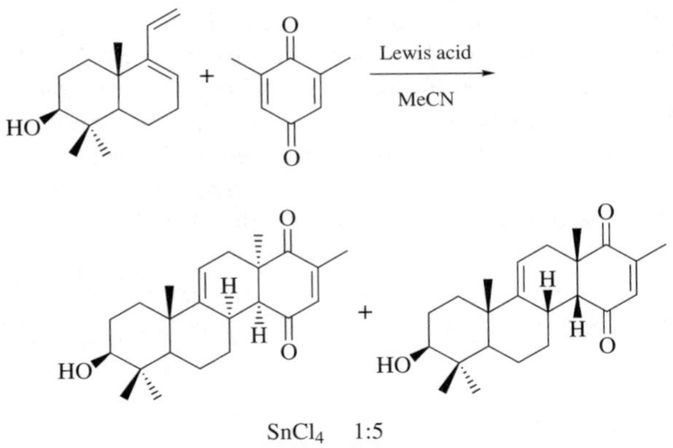

(28)

$SnCl_4$	1:5
$TiCl_4$	1:8

When the dienophile *N*-α-methylbenzylmaleimide (**9**) is reacted with 2-*t*-butyl-1,3-butadiene in the presence of Lewis acids, cycloadducts (**10**) and (**11**) are formed (eq 29).[40] While $SnCl_4$ provides (**10**) and (**11**) in a 5:1 ratio, $TiCl_4$ and $EtAlCl_2$ both provide a 15:1 ratio. Polymerization of the diene competes with adduct formation under all conditions.

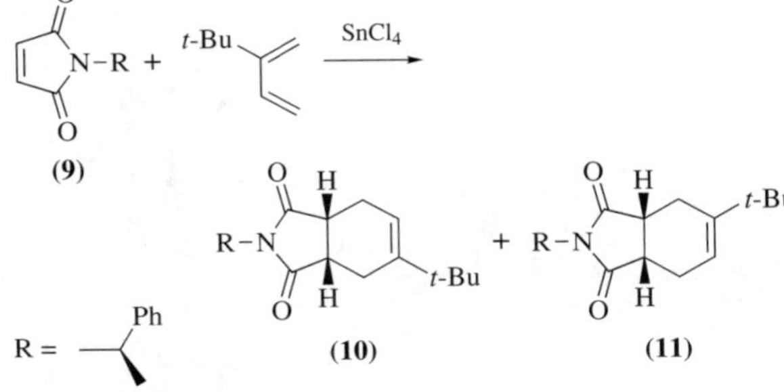

(29)

[4 + 3] Cycloadditions. Oxyallyl cations,[41] which react as C_3 rather than C_2 components in cyclization reactions, are generated by the addition of $SnCl_4$ to substrates which contain silyl enol ethers which are conjugated with a carbonyl moiety. Thus 2-(trimethylsiloxy)propenal undergoes cyclization with cyclopentadiene or furan (eq 30).[42] Substituted 1,1-dimethoxyacetones also form these intermediates and undergo subsequent cyclizations (eq 31).[43] This method complements the usual synthesis of oxyallyl cations involving reductive elimination of halogens from halogenated ketones or electronically equivalent structures.[44]

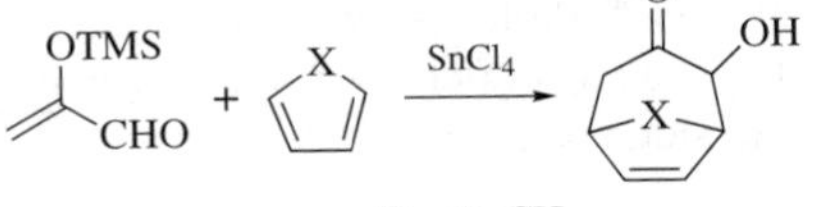

(30)

X = O, CH_2

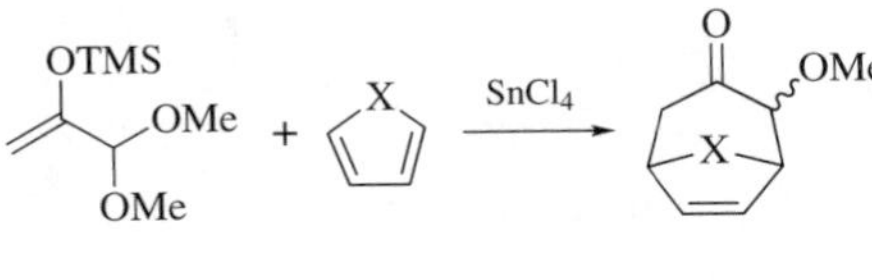

(31)

[3 + 2] Cycloadditions. Lewis acid-mediated [3 + 2] cycloadditions of oxazoles and aldehydes or diethyl ketomalonate have been observed using organoaluminum and Sn^{IV} Lewis acids.[45] The reactions are highly regioselective, with stereoselectivity extremely dependent upon Lewis acid (eq 32). For example, the (BINOL)AlMe-promoted reaction between benzaldehyde and the oxazole (**12**) provides the oxazoline with a *cis/trans* ratio of 98:2. The selectivity is reversed with $SnCl_4$ which provides a *cis/trans* ratio of 15:85. *trans*-5-Substituted 4-alkoxycarbonyl-2-oxazolines are synthesized under thermodynamic conditions in the aldol reaction of isocyanoacetates with aldehydes.[46]

Lewis acid, PhCHO (32)

(**12**)

[2 + 2] Cycloadditions. The regioselectivity in the cycloaddition reactions of 2-alkoxy-5-allyl-1,4-benzoquinones with styrenes is controlled by the choice of Ti^{IV} or $SnCl_4$ Lewis acids (eq 33).[47] The use of an excess of $TiCl_4$ or mixtures of $TiCl_4$ and $Ti(O\text{-}i\text{-}Pr)_4$ produces cyclobutane (**13**) as the major or exclusive product, while $SnCl_4$ promotion with one equivalent of Lewis acid results in the formation of (**14**) only. These reactions represent a classic example of the mechanistic variability often associated with seemingly modest changes in Lewis acid.

Lewis acid, −78 °C (33)

X = H, 3,4-$(OMe)_2$, 3,4-(-OCH_2O-); R = Me, R = Bn

(**13**) (**14**)

Ene Reactions. The Lewis acid-catalyzed ene reaction is synthetically useful methodology for forming new carbon–carbon bonds.[48] Ene reactions utilizing reactive enophiles such as formaldehyde and chloral can be promoted by $SnCl_4$. $SnCl_4$ also enhances intramolecular ene reactions, such as the cyclization of (**15**) which produces the α-hydroxy δ-lactone in 85% yield (eq 34).[49] The ene cyclization of citronellal to give isopulegol has also been reported.[50] Proton scavenging aluminum Lewis acids such as $RAlCl_2$ are most often used in ene reactions to eliminate proton-induced side reactions.

$SnCl_4$, 85% (34)

(**15**)

Related Reagents. Tin(IV) Chloride–Zinc Chloride.

1. (a) Shambayati, S.; Crowe, W. E.; Schreiber, S. L. *AG(E)* **1990**, *29*, 256. (b) Reetz, M. T. In *Selectivities in Lewis Acid Promoted Reactions*; Schinzer, D., Ed.; Kluwer: Dordrecht, 1989; pp 107–125. (c) Denmark, S. E.; Almstead, N. G. *JACS* **1993**, *115*, 3133.
2. Davies, G. A.; Smith, P. J. In *Comprehensive Organometallic Chemistry*; Wilkinson, G.; Stone, F. G. A.; Abel, E. W., Eds.; Pergamon: New York, 1982; Vol. 2, p 519.
3. Naruta, Y.; Nishigaichi, Y.; Maruyama, K. *T* **1989**, *45*, 1067.
4. (a) Nakamura, E.; Kuwajima, I. *CL* **1983**, 59. (b) Yamaguchi, M.; Hayashi, A.; Hirama, M. *JACS* **1993**, *115*, 3362. (c) Yamaguchi, M.; Hayashi, A.; Hirama, M. *CL* **1992**, 2479.
5. (a) Keck, G. E.; Castellino, S.; Andrus, M. B. In *Selectivities in Lewis Acid Promoted Reactions*; Schinzer, D., Ed.; Kluwer: Dordrecht, 1989; pp 73–105. (b) Keck, G. E.; Andrus, M. B.; Castellino, S. *JACS* **1989**, *111*, 8136. (c) Denmark, S. E.; Wilson, T.; Wilson, T. M. *JACS* **1988**, *110*, 984. (d) Boaretto, A.; Marton, D.; Tagliavini, G.; Ganis, P. *JOM* **1987**, *321*, 199. (e) Yamamoto, T.; Maeda, N.; Maruyama, K. *CC* **1983**, 742. (f) Quintard, J. P.; Elissondo, B.; Pereyre, M. *JOC* **1983**, *48*, 1559.
6. Keck, G. E.; Abbott, D. E.; Boden, E. P.; Enholm, E. J. *TL* **1984**, *25*, 3927.
7. (a) Keck, G. E.; Castellino, S. *JACS* **1986**, *108*, 3847. (b) Keck, G. E.; Castellino, S.; Wiley, M. R. *JOC* **1986**, *51*, 5478.
8. Keck, G. E.; Castellino, S. *TL* **1987**, *28*, 281.
9. Mukaiyama, T.; Banno, K.; Narasaka, K. *JACS* **1974**, *96*, 7503.
10. Review of Mukaiyama aldol reaction: Gennan, C. *COS* 1991, Vol. 2.
11. Reetz, M. T.; Kesseler, K.; Jung, A. *T* **1984**, *40*, 4327.
12. (a) Reetz, M. T. *AG(E)* **1984**, *23*, 556. (b) ref 11.
13. (a) Annunziata, R.; Cinquini, M.; Cozzi, F.; Cozzi, P. G.; Consolandi, E. *JOC* **1992**, *57*, 456. (b) Annunziata, R.; Cinquini, M.; Cozzi, F.; Cozzi, P. G. *TL* **1990**, *31*, 6733.
14. Nakamura, E.; Kawajima, I. *TL* **1983**, *24*, 3343.
15. Scavo, F.; Helquist, P. *TL* **1985**, *26*, 2603.
16. Deng, M. Z.; Caubere, P.; Senet, J. P.; Lecolier, S. *T* **1988**, *44*, 6079.
17. Gras, J. L.; Galledou, B. S. *BSF(2)* **1982**, 89.
18. Martin, O. R.; Rao, S. P.; Kurz, K. G.; El-Shenawy, H. A. *JACS* **1988**, *110*, 8698.
19. Showalter, H. D. H.; Putt, S. R. *TL* **1981**, *22*, 3155.
20. Wagner, D.; Verheyden, J. P. H.; Moffat, J. G. *JOC* **1974**, *39*, 24.
21. Hori, H.; Nishida, Y.; Ohrui, H.; Meguro, H. *JOC* **1989**, *54*, 1346.
22. Hopkins, M. H.; Overman, L. E. *JACS* **1987**, *109*, 4748.
23. Overman, L. E.; Kakimoto, M. E.; Okazaki, M. E.; Meier, G. P. *JACS* **1983**, *105*, 6622.
24. Herrington, P. M.; Hopkins, M. H.; Mishra, P.; Brown, M. J.; Overman, L. E. *JOC* **1987**, *52*, 3711.
25. Review of α-alkylations to carbonyl compounds: Reetz, M. T. *AG(E)* **1982**, *21*, 96.
26. (a) Review of asymmetric alkene cyclization: Bartlett, P. A. *Asymmetric Synthesis*; Morrison, J. D., Ed.; Academic: New York, 1984; Vol. 3, Part B, pp 341–409. (b) Review of thermal cycloadditions: Fallis, A. G.; Lu, Y.-F. *Advances in Cycloaddition*; Curran, D. P., Ed.; JAI: Greenwich, CT, 1993; Vol. 3, pp 1–66.
27. (a) Overman, L. E.; Blumenkopf, T. A.; Castaneda, A.; Thompson, A. S. *JACS* **1986**, *108*, 3516. (b) Overman, L. E.; Castaneda, A.; Blumenkopf, T. A. *JACS* **1986**, *108*, 1303.
28. Demailly, G.; Solladie, G. *JOC* **1981**, *46*, 3102.
29. Murphy, W. S.; Waltanansin, S. *JCS(P1)* **1982**, 1029.
30. (a) Kamigaito, M.; Madea, Y.; Sawamota, M.; Higashimura, T. *Macromolecules* **1993**, *26*, 1643. (b) Takahashi, T.; Yokozawa, T.; Endo, T. *Makromol. Chem.* **1991**, *192*, 1207. (c) Ran, R. C.; Mao, G. P.

J. *Macromol. Sci. Chem.* **1990**, *A27*, 125. (d) Kurita, K.; Inoue, S.; Yamamura, K.; Yoshino, H.; Ishii, S.; Nishimura, S. I. *Macromolecules* **1992**, *25*, 3791. (e) Yokozawa, T.; Hayashi, R.; Endo, T. *Macromolecules* **1993**, *26*, 3313.

31. (a) Tanaka, H.; Kato, H.; Sakai, I.; Sato, T.; Ota, T. *Makromol. Chem. Rapid Commun.* **1987**, *8*, 223. (b) Yuan, Y.; Song, H.; Xu, G. *Polym. Int.* **1993**, *31*, 397.
32. Birney, D. M.; Houk, K. N. *JACS* **1990**, *112*, 4127.
33. For leading references on asymmetric Diels–Alder reactions, see: (a) Paquette, L. A. In *Asymmetric Synthesis*; Morrison, J. D., Ed.; Academic: New York, 1984; Vol. 3, pp 455–483. (b) Oppolzer, W. *AG(E)* **1984**, *23*, 876. (c) Carruthers, W. *Cycloaddition Reactions in Organic Synthesis*; Pergamon: New York, 1990; pp 61–72.
34. Evans, D. A.; Chapman, K. T.; Bisaha, J. *JACS* **1988**, *110*, 1238.
35. Castellino, S. *JOC* **1990**, *55*, 5197.
36. Poll, T.; Helmchen, G.; Bauer, B. *TL* **1984**, *25*, 2191.
37. (a) Fringuelli, F.; Pizzo, F.; Taticchi, A.; Wenkert, E. *JOC* **1983**, *48*, 2802. (b) Fringuelli, F.; Pizzo, F.; Taticchi, A.; Halls, T. D. J.; Wenkert, E. *JOC* **1982**, *47*, 5056.
38. Liu, H. J.; Ulibarri, G.; Browne, E. N. C. *CJC* **1992**, *70*, 1545.
39. Arseniyadis, A.; Rodriguez, R.; Spanevello, J. C.; Thompson, A.; Guittet, E.; Ourisson, G. *T* **1992**, *48*, 1255.
40. Baldwin, S. W.; Greenspan, P.; Alaimo, C.; McPhail, A. T. *TL* **1991**, *42*, 5877.
41. For a recent review of oxyallyl cations, see: Mann, J. *T* **1986**, *42*, 4611.
42. Masatomi, O.; Kohki, M.; Tatsuya, H.; Shoji, E. *JOC* **1990**, *55*, 6086.
43. Murray, D. H.; Albizati, K. F. *TL* **1990**, *31*, 4109.
44. Hoffman, H. M. R. *AG(E)* **1973**, *12*, 819; **1984**, *23*, 1.
45. Suga, H.; Shi, X.; Fujieda, H.; Ibata, T. *TL* **1991**, *32*, 6911.
46. For examples of enantioselective synthesis of *trans*-4-alkoxy-2-oxazolines, see: Ito. Y; Sawamura, M.; Shirakawa, E.; Hayashizaki, K.; Hayashi, T. *TL* **1988**, *29*, 235; *T* **1988**, *44*, 5253.
47. Engler, T. A.; Wei, D.; Latavic, M. A. *TL* **1993**, *34*, 1429.
48. Reviews of ene reactions: (a) Hoffman, H. M. R. *AG(E)* **1969**, *8*, 556. (b) Oppolzer, W.; Sniekus, V. *AG(E)* **1978**, *17*, 476. (c) Snyder, B. B. *ACR* **1980**, *13*, 426.
49. Lindner, D. L.; Doherty, J. B.; Shoham, G.; Woodward, R. B. *TL* **1982**, *23*, 5111.
50. Nakatani, Y.; Kawashima, K. *S* **1978**, 147.

Stephen Castellino
Rhône-Poulenc Ag. Co., Research Triangle Park, NC, USA

David E. Volk
North Dakota State University, Fargo, ND, USA

Titanium(IV) Chloride

$TiCl_4$

[7550-45-0] Cl_4Ti (MW 189.68)

(Lewis acid catalyst;[52] affects stereochemical course in cycloaddition[38,39] and aldol reactions;[4–17] electrophilic substitutions;[55] powerful dehydrating agent; when reduced to low-valent state, effects C–C bond formation by reductive coupling;[7–9] reduction of functional groups[81])

Physical Data: mp −24 °C, bp 136.4 °C, *d* 1.726 g cm^{-3}.

Solubility: sol THF, toluene, CH_2Cl_2.

Form Supplied in: neat as colorless liquid; solution in CH_2Cl_2; solution in toluene; $TiCl_4$·2THF *[31011-57-1]*.

Preparative Methods: see Perrin et al.[1]

Purification: reflux with mercury or a small amount of pure copper turnings and distill under N_2 in an all-glass system. Organic material can be removed by adding aluminum chloride hexahydrate as a slurry with an equal amount of water (ca. 2% weight of the amount of $TiCl_4$), refluxing the mixture for 2–6 h while bubbling in Cl_2 which is subsequently removed by a stream of dry air, before the $TiCl_4$ is distilled, refluxed with copper, and distilled again. Volatile impurities can be removed using a technique of freezing, pumping, and melting.[2]

Handling, Storage, and Precautions: moisture sensitive; reacts violently, almost explosively, with water; highly flammable; toxic if inhaled or swallowed; causes burns on contact with skin; use in a fume hood.

General Discussion. $TiCl_4$ is a strong Lewis acid and is used as such in organic reactions. It has high affinity for oxygenated organic molecules and possesses a powerful dehydrating ability. Low-valent titanium, from the reduction of $TiCl_4$ by metal or metal hydrides, is used for reductive coupling reactions and for reduction of functional groups.

Carbon–carbon bond formation by reductive coupling of ketones by low-valent titanium leads to vicinal diols and alkenes, as does coupling using low-valent reagents from ***Titanium(III) Chloride***. Reviews are available on the preparation and reactions of low-valent titanium.[3–6] The reagent from ***Titanium(IV) Chloride-Zinc*** in THF or dioxane in the presence of pyridine transforms ketones into tetrasubstituted alkenes. Unsymmetrical ketones yield (*E*)/(*Z*)-isomer mixtures (eq 1). Strongly hindered ketones react slowly and may preferentially be reduced to alcohols.[7] Reductive coupling of diketo sulfides yields 2,5-dihydrothiophenes (eq 2).[8] The macrocyclic porphycene has been obtained, albeit in low yield, by McMurry coupling of diformylbipyrrole (eq 3).[9] Aldimines are reductively coupled in a reaction analogous to the reaction of carbonyl compounds. The product is a 1:1 mixture of *meso*- and (±)-isomers (eq 4).[10]

$TiCl_4$–Zn, py, THF (1)

$TiCl_4$–Zn, CH_2Cl_2, 0 °C, 70–90% (2)

$TiCl_4$–Zn, THF, py, 2–3% (3)

$TiCl_4$–Mg, 92% (4)

Methylenation of aldehydes and ketones results from reactions with the complex from $TiCl_4$–Zn and ***Diiodomethane*** or

Dibromomethane (eq 5). The reagent can be used for methylenation of enolizable oxo compounds.[6] Recommended modifications to the reagent have been reported.[11] In keto aldehydes, selective methenylation of the keto group results when the aldehyde is precomplexed with ***Titanium Tetrakis(diethylamide)***. Chemoselective methenylation of the aldehyde function is possible by direct use of CH_2I_2/Zn/***Titanium Tetraisopropoxide*** (eq 6).[12] Cyclopropanation to *gem*-dihalocyclopropanes uses ***Lithium Aluminum Hydride-Titanium(IV) Chloride***. The exclusion of a strong base, as frequently used in alternative procedures, is an advantage (eq 7).[13] Allylation of imines has been effected by low-valent titanium species generated from $TiCl_4$ and ***Aluminum*** foil (eq 8).[14]

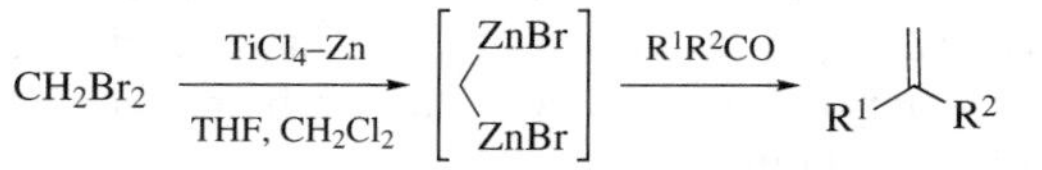

R^1, R^2 = H, aryl, alkyl (5)

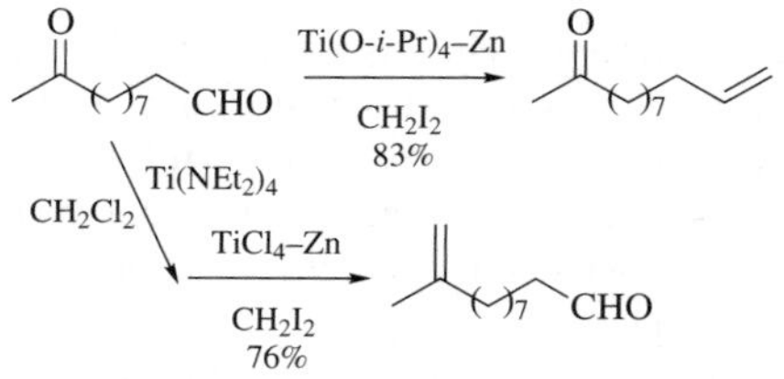

(6)

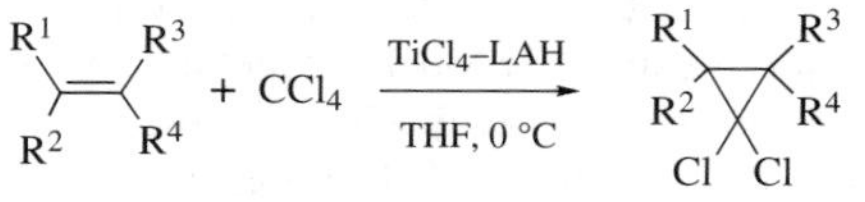

(7)

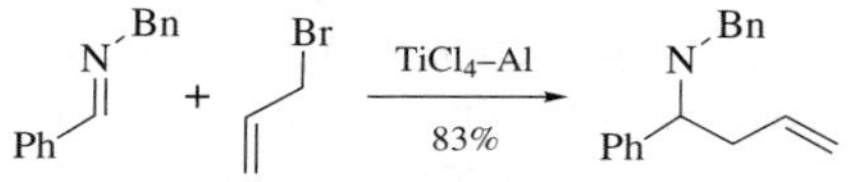

(8)

$TiCl_4$ is a powerful activator of carbonyl groups and promotes nucleophilic attack by a silyl enol ether. The product is a titanium salt of an aldol which, on hydrolysis, yields a β-hydroxy ketone. $TiCl_4$ is generally the best catalyst for this reaction. The temperature range for reactions with ketones is normally 0–20 °C; aldehydes react even at −78 °C, which allows for chemoselectivity (eq 9).[15] In α- or β-alkoxy aldehydes, the aldol reaction can proceed with high 1,2- or 1,3-asymmetric induction. With the nonchelating Lewis acid ***Boron Trifluoride Etherate***, the diastereoselectivity may be opposite to that obtained for the chelating $TiCl_4$ or ***Tin(IV) Chloride*** (eq 10).[16]

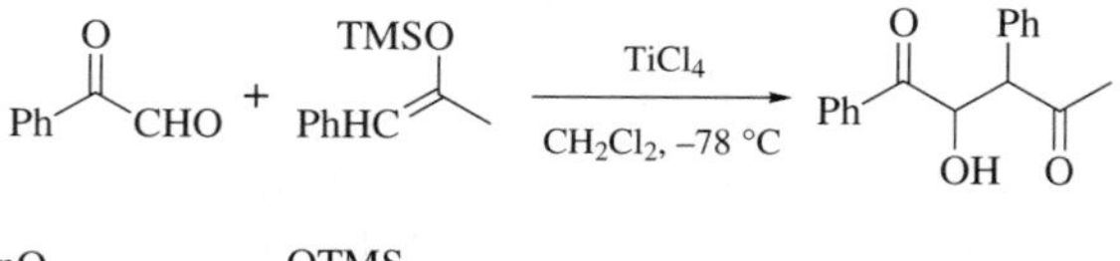

(9)

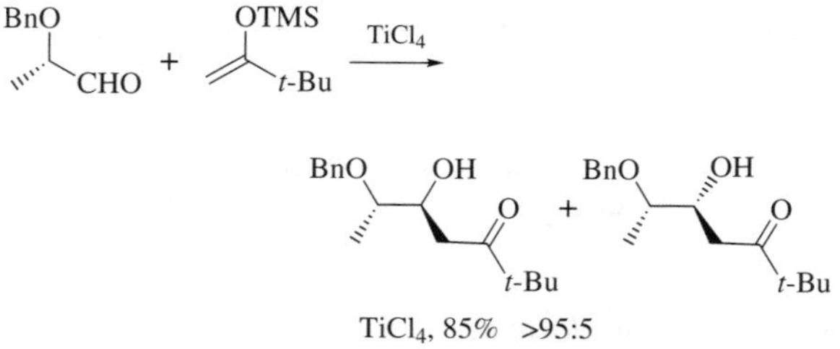

(10)

$TiCl_4$, 85% >95:5
BF_3 10:90

Titanium enolates are generally prepared by transmetalation of alkali-metal enolates but may also arise as structural parts of intermediates in $TiCl_4$-promoted reactions. This is illustrated in a stereoselective alkylation using an oxazolidinone as a chiral auxiliary (eq 11). The enolate, or its ate complex, may be the intermediate in the reaction.[17]

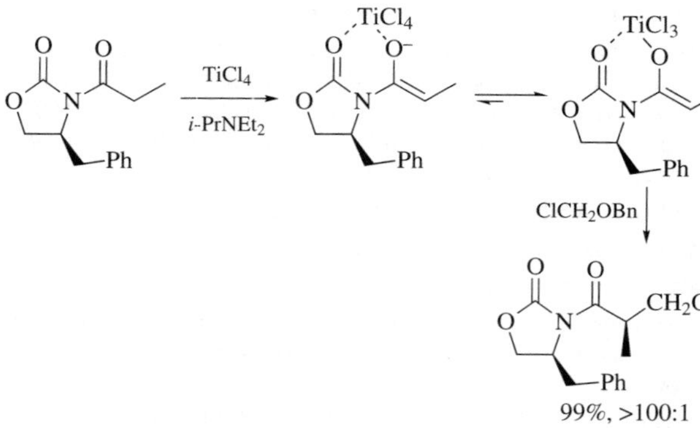

(11)

99%, >100:1

The stereoselectivity in $TiCl_4$-promoted reaction of silyl ketene acetals with aldehydes may be improved by addition of ***Triphenylphosphine*** (eq 12).[18] Enol ethers, as well as enol acetates, can be the nucleophile (eqs 13 and 14).[19] 2-Acetoxyfuran, in analogy to vinyl acetates, reacts with aldehydes to furnish 4-substituted butenolides under the influence of $TiCl_4$ (eq 15).[20]

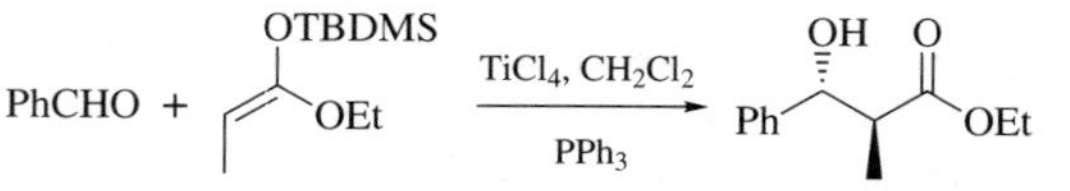

(12)

anti:*syn* = 10.5:1
no PPh_3, 4:1

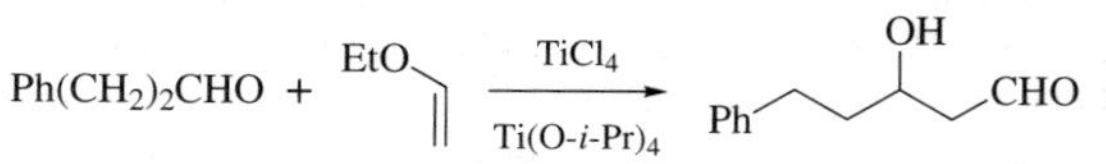

(13)

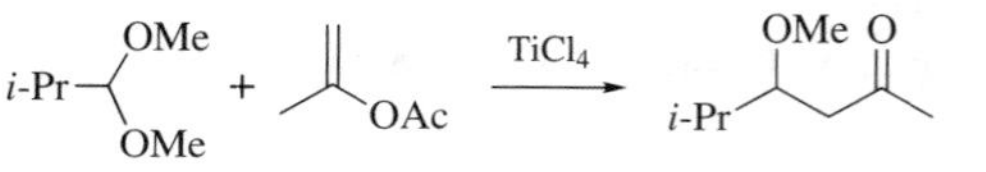

(14)

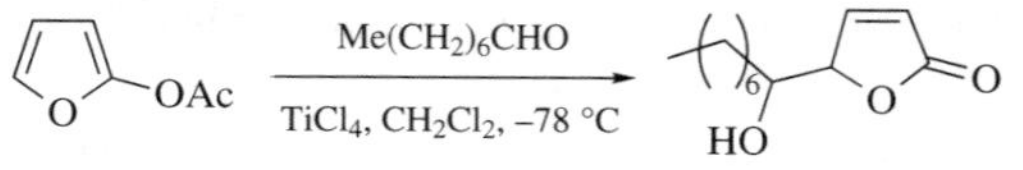

(15)

Silyl enol ethers react with acetals at −78 °C to form β-alkoxy ketones.[21] In intramolecular reactions, six-, seven-, and eight-membered rings are formed.[22] With 1,3-dioxolanes (acetals), 1–2 equiv of $TiCl_4$ leads to pyranone formation (eq 16), whereas no cyclization products are obtained with $SnCl_4$ or $ZnCl_2$.[23]

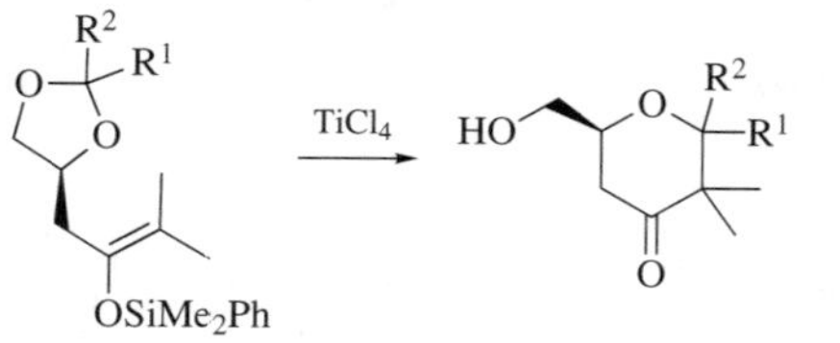

(16)

Alkynyltributyltin compounds react with steroidal aldehydes. In the presence of $TiCl_4$ the reaction gives 9:1 diastereoselectivity (eq 17). Reactions of alkynylmetals with chiral aldehydes generally show only slight diastereoselectivity.[24] With silyloxyacetylenes and aldehydes, α,β-unsaturated carboxylic acid esters are formed with high (*E*) selectivity (eq 18).[25]

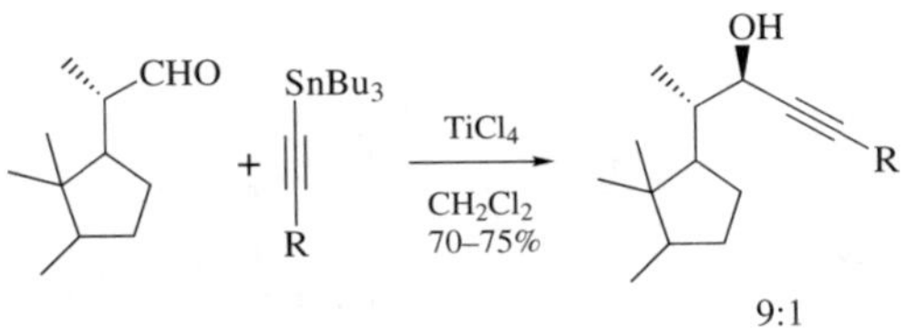

(17)

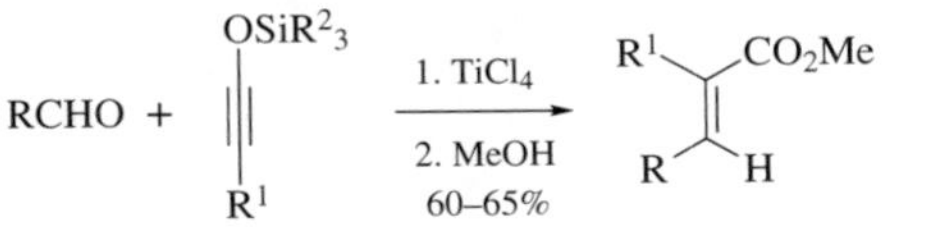

(18)

Allylsilanes are regiospecific in their Lewis acid-catalyzed reactions, the electrophile bonding to the terminus of the allyl unit remote from the silyl group (eq 19).[26,27] Allylstannanes are less reliable in this respect.[26,28] The example in eq 19 illustrates regioselectivity.[29] On extended conjugation the reaction takes place at the terminus of the extended system (eq 20).[30]

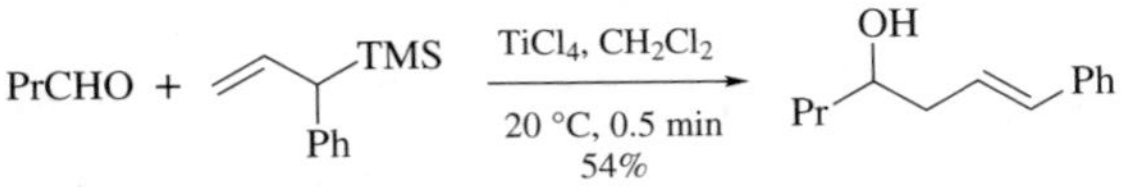

(19)

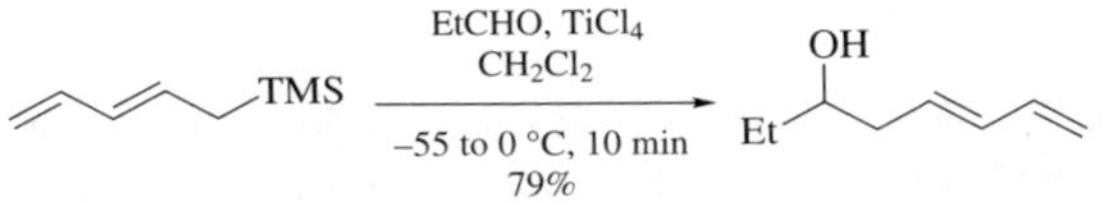

(20)

Intramolecular reactions work well, both for allylstannanes and allylsilanes. An epoxide can be the electrophile. Opening of the epoxide with allylic attack gives the cyclic product (eq 21). $TiCl_4$ is superior to Ti(O-*i*-Pr)$_4$, $SnCl_4$, and ***Aluminum Chloride***, the other Lewis acids tested in the cyclization reaction. The reaction is stereospecific.[31]

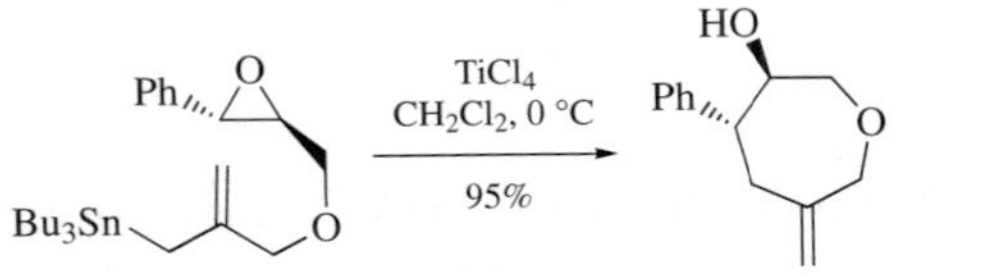

(21)

Stereochemical aspects in these $TiCl_4$-promoted reactions have been covered in reviews.[26,27] Acetals are very good electrophiles for allylsilanes, and may be better than the parent oxo compounds because the products are less prone to further reaction; intramolecular reactions are facile.[26,32] In the unsymmetrical acetal in eq 22, the methoxyethoxy group departs selectively because of chelation to the Lewis acid.[33]

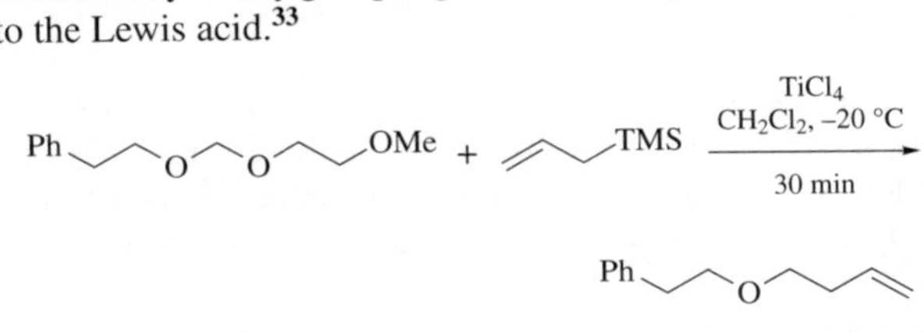

(22)

The $TiCl_4$-mediated addition of allenylsilanes to aldehydes and ketones provides a general, regiocontrolled route to a wide variety of substituted homopropargylic alcohols. With acetals, corresponding ethers are formed (eq 23).[34] Allenylation of acetals results from the reaction of propargylsilanes with acetals. The products are α-allenyl ethers (eq 24).[35]

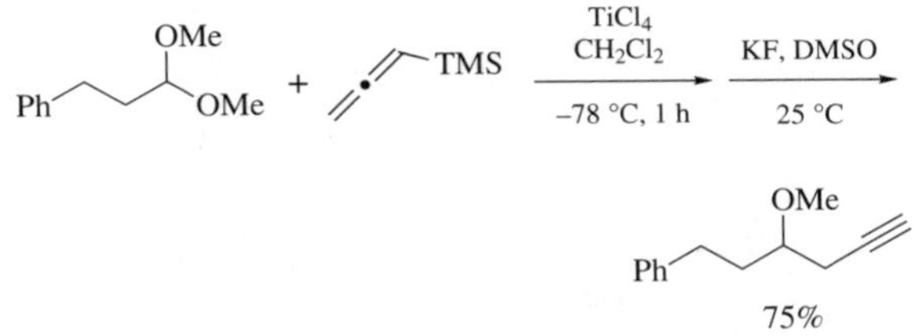

(23)

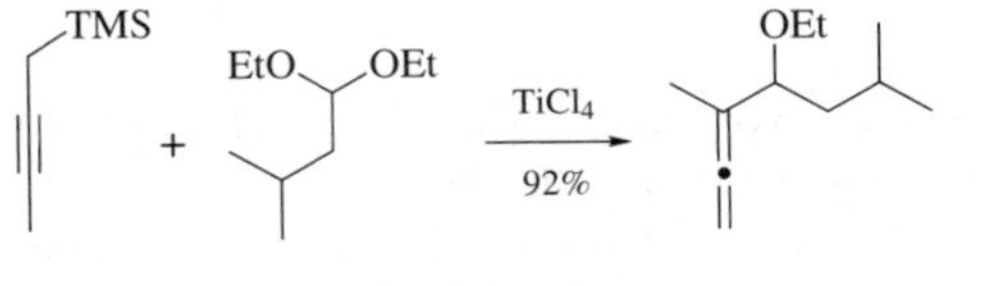

(24)

Homoallylamines are formed in $TiCl_4$- or $BF_3·OEt_2$-mediated reactions between allylstannanes and aldimines. Crotyltributyltin gives mainly *syn*-β-methyl homoallylamines under optimal conditions (eq 25).[36]

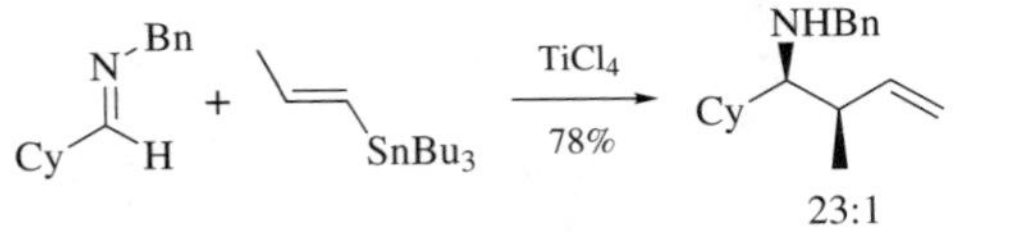

(25)

Diastereoselective Mannich reactions may result with $TiCl_4$ as adjuvant (eq 26); transmetalation of the initial lithium alkoxide adduct and displacement by a lithium enolate gives a diastereoselectivity of 78%.[37]

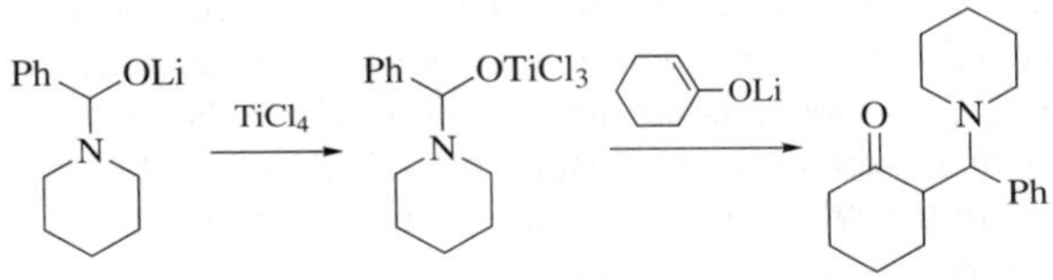

(26)

Lewis acid catalysis increases the reactivity of dienophiles in Diels–Alder reactions by complexing to basic sites on the dienophile. In aldehydes, complexation takes place via the lone pair on the carbonyl oxygen. The stereochemistry is strongly influenced by the Lewis acids.[38] Under chelating conditions, when α-alkoxy aldehydes are used, the prevalent products from $TiCl_4$ catalysis have *cis* configuration (eqs 27 and 28). Stereochemical aspects of pericyclic pathways or passage through aldol type intermediates have been summarized and discussed.[38,39]

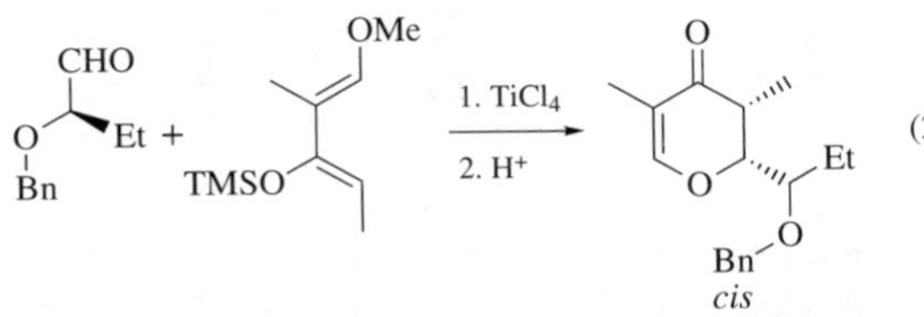

(27)

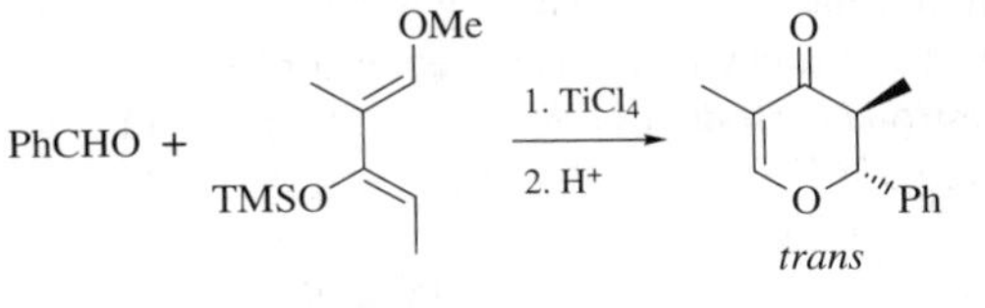

(28)

Ketene silyl acetals can be dimerized to succinates on treatment with $TiCl_4$ in CH_2Cl_2 (eq 29). No reaction occurs with $TiCl_3$, nor are other metal salts efficient.[40] β-Amino esters are formed in the presence of Schiff bases (eq 30).[41]

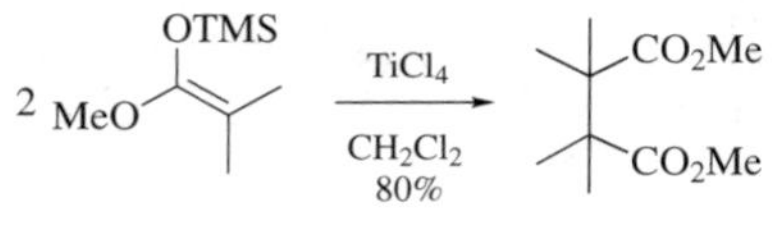

Cycloaddition of alkenes to quinones is effected by Ti^{IV} derivatives. The composition of the Ti^{IV} adjuvant largely controls the type of cycloadduct. $TiCl_4$ favors [3 + 2] cycloaddition; [2 + 2] cycloaddition requires a mixed $TiCl_4$–Ti(O-*i*-Pr)$_4$ catalyst (eq 31).[42]

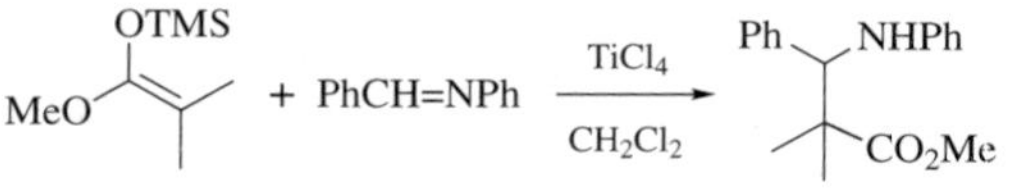

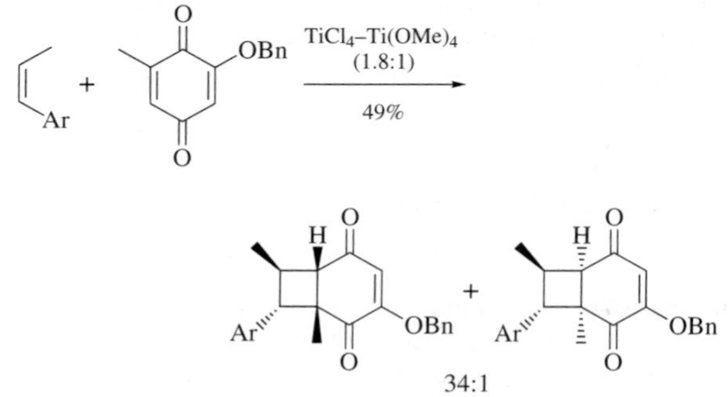

The $TiCl_4$-promoted Michael reaction proceeds under very mild conditions (−78 °C); this suppresses side-reactions and 1,5-dicarbonyl compounds are formed in good yields.[3] For $TiCl_4$-sensitive compounds, a mixture of $TiCl_4$ and Ti(O-*i*-Pr)$_4$ is used. From silyl enol ethers and α,β-unsaturated ketones, 1,5-dicarbonyl compounds are formed (eq 32).[3] The reaction also proceeds for α,β-unsaturated acetals.[43] Silylketene acetals react with α,β-unsaturated ketones or their acetals to form δ-oxo esters (eq 33).[44]

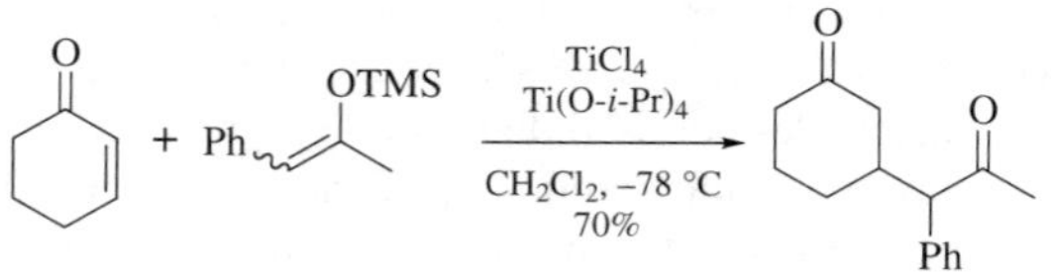

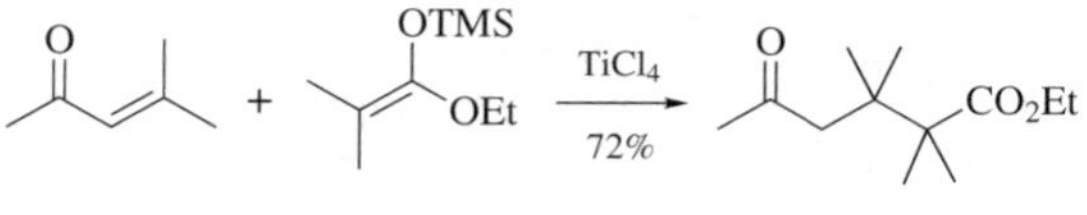

Conjugate addition of allylsilanes to enones results in regiospecific introduction of the allyl group (eq 34).[45] The reaction can be intramolecular (eq 35).[46] $TiCl_4$ or $SnCl_4$ activates nitro alkenes for Michael addition with silyl enol ethers or ketene silyl acetals. The silyl nitronate product is hydrolyzed to a 1,4-diketone or γ-keto ester (eq 36).[47]

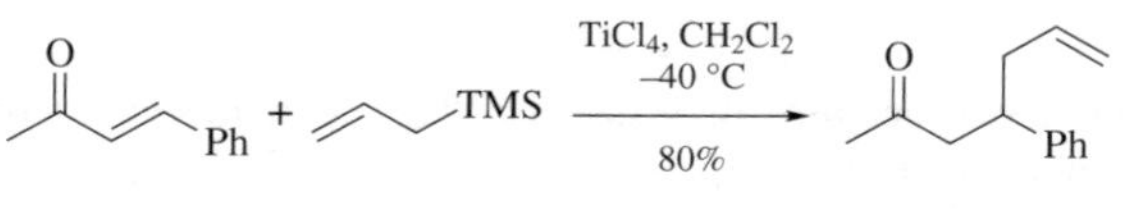

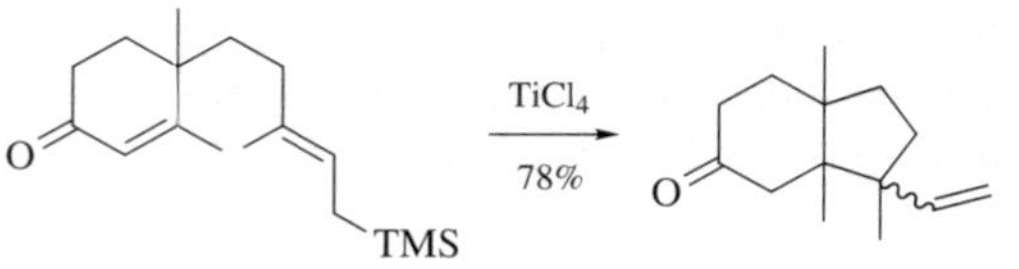

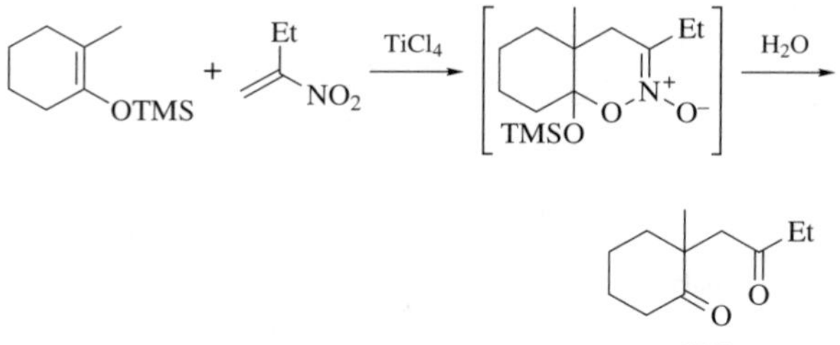

Conjugate propynylation of enones results from $TiCl_4$-mediated addition of allenylstannanes. Other Lewis acids are ineffective in this reaction (eq 37).[48]

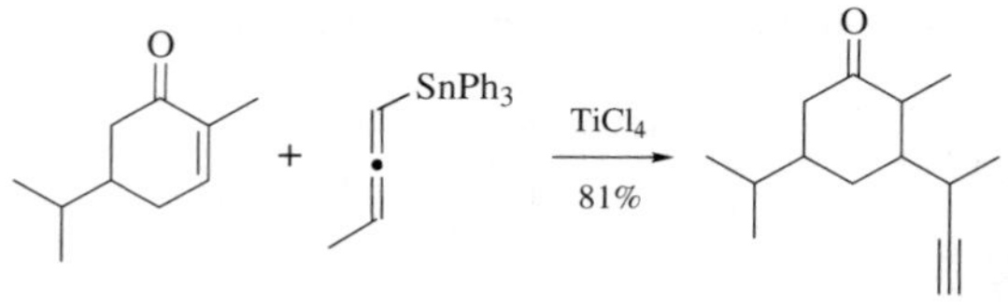

The 3,4-dichloro derivative of squaric acid, on $TiCl_4$-catalysis, reacts with silyl enol ethers or allylsilanes to form 1,2- or 1,4-adducts depending on the substitution. The 3,4-diethoxy derivative, however, adds silyl enol ethers in a 1,4-fashion; the adduct subsequently eliminates the ethoxy group (eq 38).[49]

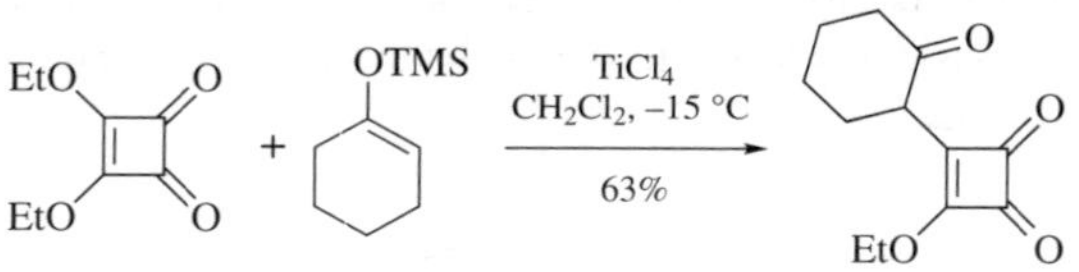

The Knoevenagel reaction with $TiCl_4$ and a tertiary base (eq 39) is recommended over methods which rely on strongly basic conditions.[50] The $TiCl_4$-procedure at low temperature is suitable for base-sensitive substrates.[51]

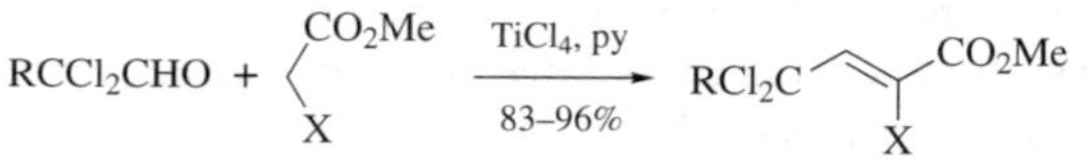

$TiCl_4$, as a Lewis acid, is used as a catalyst in Friedel–Crafts reactions. $AlCl_3$, $SnCl_4$, and $BF_3\cdot OEt_2$ are more commonly used Friedel–Crafts catalysts for reactions with arenes.[52] Use of $TiCl_4$ as catalyst in the preparation of aromatic aldehydes is shown in eq 40.[53]

The formylation reaction can be used to prepare (*E*)-α,β-unsaturated aldehydes from vinylsilanes by *ipso* substitution (eq 41).[54] Regioselectivity in alkylation reactions may depend on the Lewis catalyst. In a fluorene, regioselective 1,8-chloromethylation results with $TiCl_4$ (eq 42).[55] $TiCl_4$ activates nitroalkenes for electrophilic substitution into arenes. The intermediate is hydrolyzed to the oxoalkylated product (eq 43).[56]

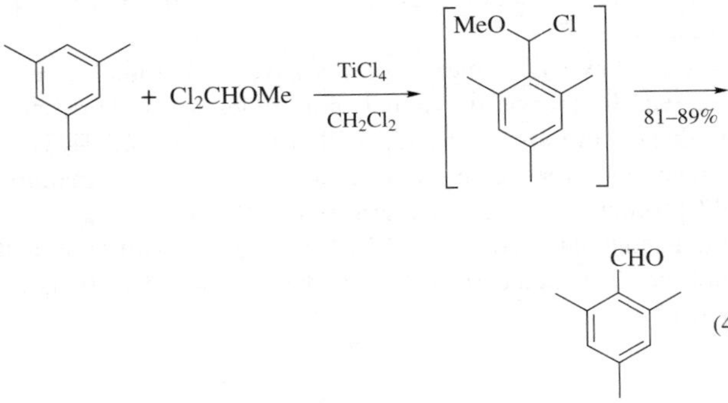

(40)

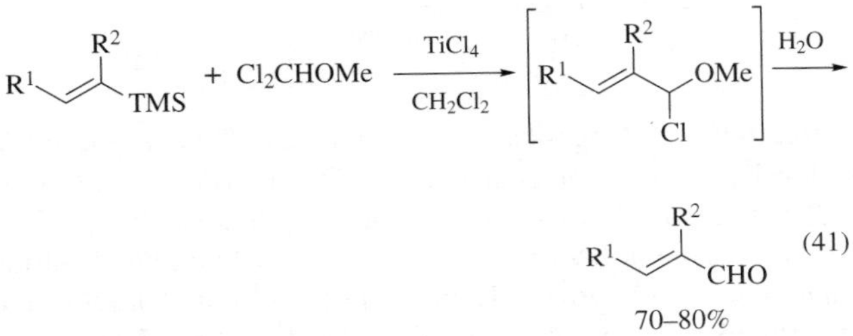

(41)

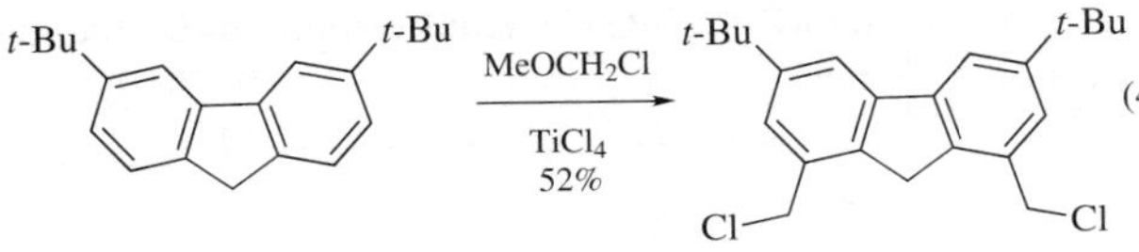

(42)

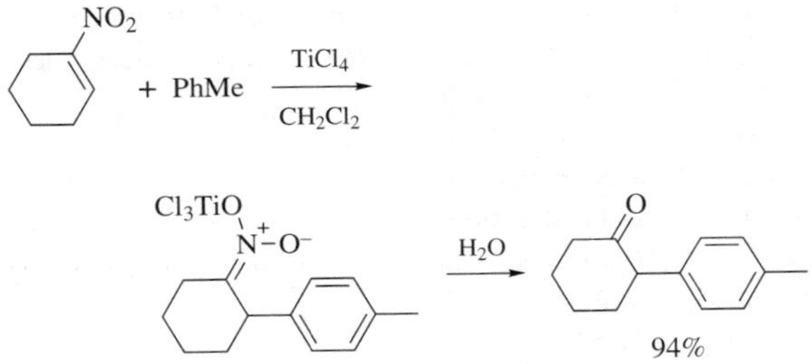

(43)

$TiCl_4$ is generally a good catalyst for Friedel–Crafts acylation of activated alkenes. In the acylation of pyrrolidine-2,4-diones, particularly with unsaturated acyl substrates, $TiCl_4$ is superior to $SnCl_4$ and $BF_3 \cdot OEt_2$ (eq 44).[57]

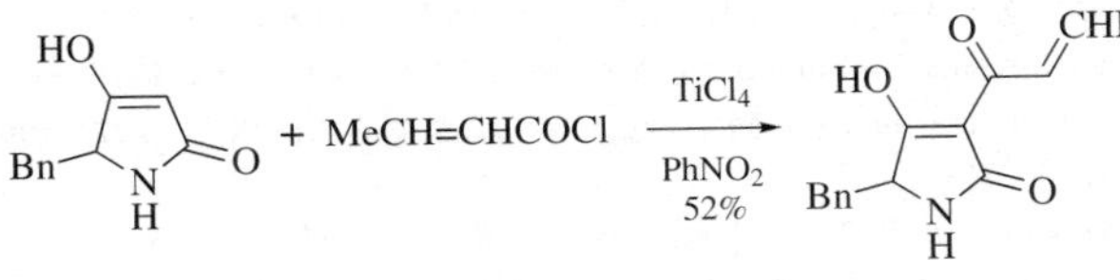

(44)

The ready *ipso* substitution of a silyl group favors substitution rather than addition of electrophiles in alkenylsilanes.[58] Direct acylation of isobutene is not satisfactory, but the acylation is successful on silyl derivatives (eq 45).[59] Intramolecular acylation of alkenylsilanes leads to cyclic products. Even (*E*)-silylalkenes have been cyclized to enones (eq 46).[60]

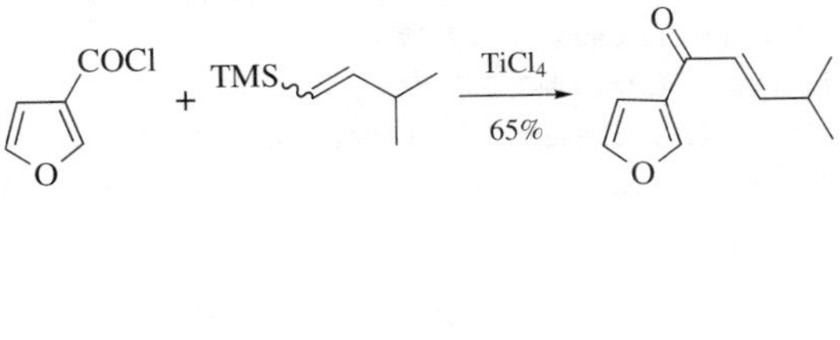

(45)

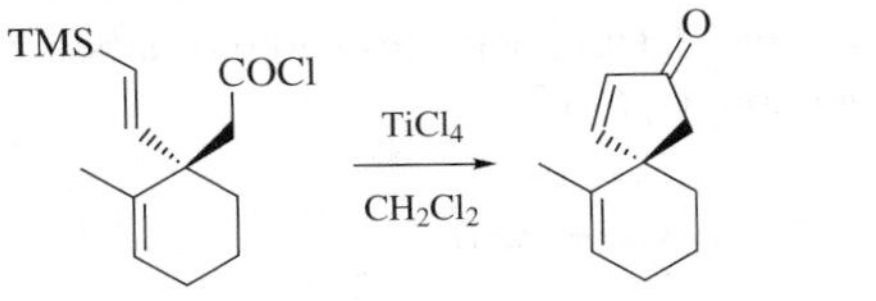

(46)

In analogy to the acylation reactions, vinylsilanes can be alkylated by *ipso* substitution. In the example in eq 47, the MEM group is activated as leaving group by metal complexation[61] The silyl group in allylsilanes directs the incoming electrophile to the allylic γ-carbon (eq 48). $TiCl_4$ is one of several Lewis acids used for catalysis in acylations.[58] The same reaction is seen in allylsilanes with extended conjugation.[62] eq 49 shows the $TiCl_4$-catalyzed alkylation of an allylsilane.[63]

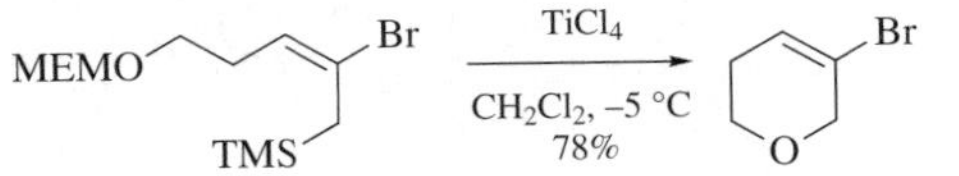

(47)

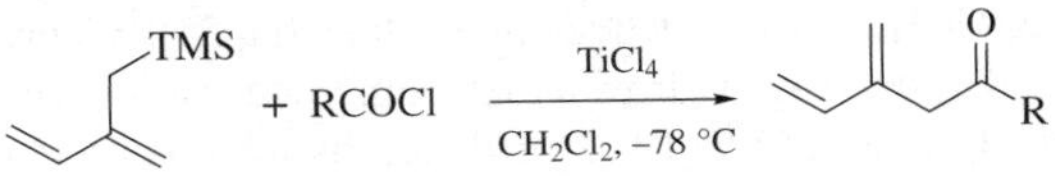

(48)

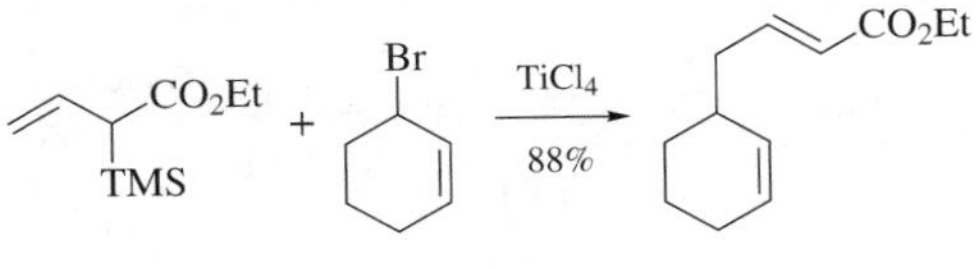

(49)

$TiCl_4$ is a useful catalyst for the Fries rearrangement of phenol esters to *o*- or *p*-hydroxy ketones (eq 50). $TiCl_4$ is a cleaner catalyst than the more frequently used $AlCl_3$, which may cause alkyl migrations.[64]

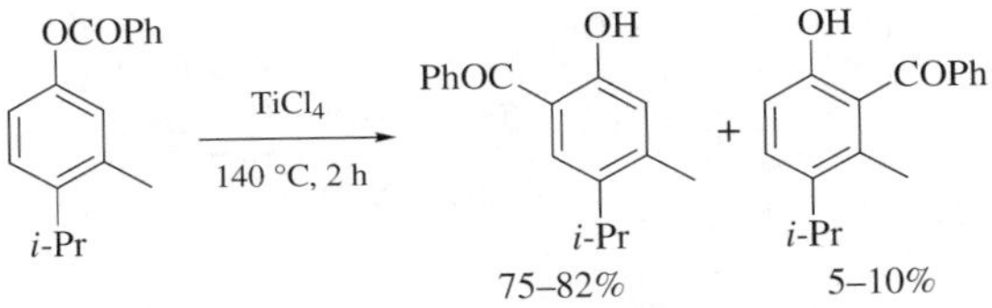

(50)

$TiCl_4$ is both a strong Lewis acid and a powerful dehydrating agent, and hence useful as a water scavenger in the synthesis of enamines. It is particularly useful for the preparation of enamines of acyclic ketones. It is recommended that $TiCl_4$ is complexed with the amine before addition of the ketone.[65] Highly sterically hindered enamines are available by this method (eq 51).[66] Primary amines generally react very slowly with aromatic ketones to form imines. $TiCl_4$ is a good catalyst (eq 52).[67]

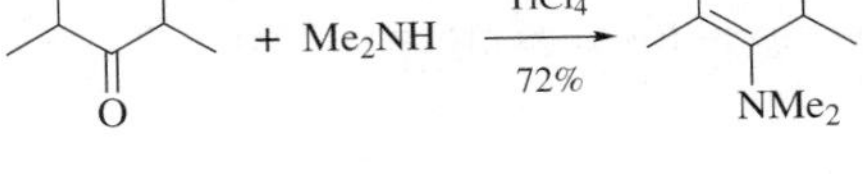

(51)

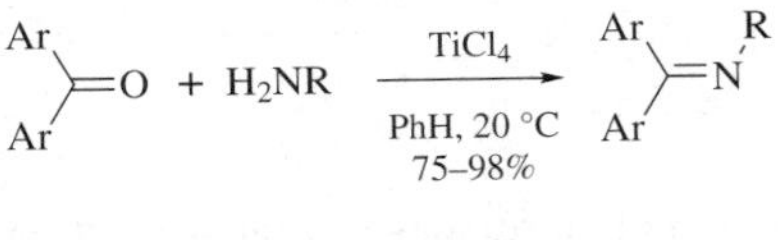

(52)

$TiCl_4$ with a tertiary base provides mild conditions for dehydration of both aldoximes and primary acid amides to form nitriles.[68] Vinyl sulfides are formed from oxo compounds using $TiCl_4$ and a tertiary amine.[3] $TiCl_4$ activates carbon–carbon double bonds for thiol addition (eq 53).[3]

$$CH_2{=}CHCO_2Me + EtSH \xrightarrow[CH_2Cl_2,\ 20\ ^\circ C,\ 94\%]{TiCl_4} EtSCH_2CH_2CO_2Me \quad (53)$$

$TiCl_4$ mediates thioacetalization of aldehydes and ketones with alkanethiols or alkanedithiols in yields >90%. The reaction is satisfactory also for readily enolizable oxo compounds.[69] γ-Lactols, which are generally more stable than acyclic analogs, are amenable to dithiol cleavage (eq 54).[70]

(54) [lactol (i-Pr, H, OH) + HS(CH2)2SH → TiCl4, 85% → dithiolane alcohol]

α-Hydroxy amides are formed in a reaction involving an isocyanide, $TiCl_4$, and an aldehyde or a ketone (eq 55).[71] Vinyl chlorides undergo ready hydrolysis on combined use of $TiCl_4$ and $MeOH–H_2O$ (eq 56).[72] $TiCl_4$-mediated hydrolysis of vinyl sulfides is a good preparative route to ketones.[3,73]

$$MeNC \xrightarrow[CH_2Cl_2]{TiCl_4} Me{-}N{=}C(Cl)TiCl_3 \xrightarrow{R^1COR^2} R^1R^2C(OTiCl_3)C(Cl){=}NMe \xrightarrow{H_2O} R^1R^2C(OH)C(O)NHMe \quad (55)$$

65–95%

(56) [3-chloro-3,5-diene steroid propionate → $TiCl_4$, H_2O–MeOH; CH_2Cl_2, Me_2CO, 20 °C, 96% → 4-en-3-one steroid propionate (OCOEt)]

Low-valent titanium can be used to reduce sulfides and haloarenes to the corresponding hydrocarbons (eq 57).[3,74] Low-valent titanium, prepared from $TiCl_4$ and ***Magnesium Amalgam***, will reduce nitroarenes to amines in THF/*t*-BuOH at 0 °C without affecting halo, cyano, and ester groups.[75] $SnCl_2{\cdot}2H_2O$ is an alternative reagent for nitro group reduction.[76]

$$\text{1-chloronaphthalene} \xrightarrow[THF,\ \Delta,\ 83\%]{TiCl_4–LAH\ (1:2)} \text{naphthalene} \quad (57)$$

Deoxygenation of sulfoxides is a rapid reaction using $[TiCl_2]$ formed in situ by reduction of $TiCl_4$ with Zn dust in CH_2Cl_2 or Et_2O at rt, with yields in the range 85–90%.[77] For deoxygenation of *N*-oxides of pyridine-based heterocycles, a reagent prepared from $TiCl_4–NaBH_4$ (1:2) in DME has been described.[78] Carboxylic acids can be reduced to primary alcohols by $TiCl_4–NaBH_4$ (ratio 1:3). For reduction of amides and lactams the optimum molar ratio is 1:2.[79]

Several low-valent metal species have been found active in reductive elimination of vicinal dibromides to the corresponding alkenes. In most cases, e.g. with $TiCl_4$–LAH and $TiCl_4$–Zn, the reactions proceed through predominant *anti* elimination to yield alkenes with high isomeric purity.[80] The low-valent titanium reagent obtained from $TiCl_4$–LAH (ca. 2:1) will saturate the double bond of enedicarboxylates in the presence of triethylamine (eq 58).[81]

(58) [norbornadiene dicarboxylic acid (CO_2H, CO_2H) → $TiCl_4$–LAH, NEt_3; THF, 100 °C → norbornene dicarboxylic acid]

Related Reagents. Dibromomethane–Zinc-Titanium(IV) Chloride; Diiodomethane–Zinc–Titanium(IV) Chloride; (4*R*, 5*R*)-2,2-Dimethyl-4, 5-bis(hydroxydiphenylmethyl)-1,3-dioxolane–Titanium(IV) Chloride; Lithium Aluminum Hydride-Titanium(IV) Chloride; Titanium(IV) Chloride-Diazabicyclo-[5.4.0]undec-7-ene; Titanium(IV) Chloride-2,2,6,6-Tetramethylpiperidine; Titanium(IV) Chloride-Triethylaluminum; Titanium-(IV) Chloride–Zinc.

---

1. Perrin, D. D.; Armarego, W. L. F.; Perrin, D. R. *Purification of Laboratory Chemicals*, 2nd ed.; Pergamon: Oxford, 1980; p 542.
2. *Gmelin Handbuch der Anorganischen Chemie*; Verlag Chemie: Weinheim, 1951; *Titan*, p 92, 299.
3. Mukaiyama, T. *AG(E)* **1977**, *16*, 817.
4. Betschart, C.; Seebach, D. *C* **1989**, *43*, 39.
5. Pons, J.-M.; Santelli, M. *T* **1988**, *44*, 4295.
6. Reetz, M. T. *Organotitanium Reagents in Organic Synthesis*; Springer: Berlin, 1986; p 223.
7. Lenoir, D. *S* **1977**, 553.
8. (a) Nakayama, J.; Machida, H.; Hoshino, M. *TL* **1985**, *26*, 1981. (b) Nakayama, J.; Machida, H.; Satio, R.; Hoshino, M. *TL* **1985**, *26*, 1983.
9. Vogel, E.; Köcher, M.; Schmickler, H.; Lex, J. *AG(E)* **1986**, *25*, 257.
10. Betschart, C.; Schmidt, B.; Seebach, D. *HCA* **1988**, *71*, 1999.
11. (a) Lombardo, L. *TL* **1982**, *23*, 4293. (b) Lombardo, L. *OS* **1987**, *65*, 81.
12. Hibino, J.; Okazoe, T.; Takai, K.; Nozaki, H. *TL* **1985**, *26*, 5579.
13. Mukaiyama, T.; Shiono, M.; Watanabe, K.; Onaka, M. *CL* **1975**, 711.
14. Tanaka, H.; Inoue, K.; Pokorski, U.; Taniguchi, M.; Torii, S. *TL* **1990**, *31*, 3023.
15. Mukaiyama, T.; Narasaka, K.; Banno, K. *JACS* **1974**, *96*, 7503.
16. (a) Reetz, M. T. *AG(E)* **1984**, *23*, 556. (b) Reetz, M. T.; Kesseler, K. *CC* **1984**, 1079.
17. Evans, D. A.; Urpí, F.: Somers, T. C.; Clark, J. S.; Bilodeau, M. T. *JACS* **1990**, *112*, 8215.
18. Palazzi, C.; Colombo, L.; Gennari, C. *TL* **1986**, *27*, 1735.
19. (a) Kitazawa, E.; Imamura, T.; Saigo, K.; Mukaiyama, T. *CL* **1975**, 569. (b) Mukaiyama, T.; Izawa, T.; Saigo, K. *CL* **1974**, 323.
20. Shono, T.; Matsumura, Y.; Yamane, S. *TL* **1981**, *22*, 3269.
21. Mukaiyama, T.; Banno, K. *CL* **1976**, 279.
22. Mukaiyama, T. *OR* **1982**, *28*, 203.
23. Cockerill, G. S.; Kocienski, P. *CC* **1983**, 705.
24. Yamamoto, Y.; Nishii, S.; Maruyama, K. *CC* **1986**, 102.

25. Kowalski, C. J.; Sakdarat, S. *JOC* **1990**, *55*, 1977.
26. Fleming, I. *COS* **1991**, *2*, 563.
27. Fleming, I.; Dunoguès, J.; Smithers, R. H. *OR* **1989**, *37*, 57.
28. Yamamoto, Y. *ACR* **1987**, *20*, 243.
29. Hosomi, A.; Shirahata, A.; Sakurai, H. *TL* **1978**, 3043.
30. Seyferth, D.; Pornet, J.; Weinstein, R. M. *OM* **1982**, *1*, 1651.
31. (a) Molander, G. A.; Shubert, D. C. *JACS* **1987**, *109*, 576. (b) Molander, G. A.; Andrews, S. W. *JOC* **1989**, *54*, 3114.
32. Mukaiyama, T.; Murakami, M. *S* **1987**, 1043.
33. Nishiyama, H.; Itoh, K. *JOC* **1982**, *47*, 2496.
34. (a) Danheiser, R. L.; Carini, D. J.; Kwasigroch, C. A. *JOC* **1986**, *51*, 3870. (b) Danheiser, R. L.; Carini, D. J. *JOC* **1980**, *45*, 3925.
35. Pornet, J.; Miginiac, L.; Jaworski, K.; Randrianoelina, B. *OM* **1985**, *4*, 333.
36. Keck, G. E.; Enholm, E. J. *JOC* **1985**, *50*, 146.
37. Seebach, D.; Betschart, C.; Schiess, M. *HCA* **1984**, *67*, 1593.
38. Bednarski, M. D.; Lyssikatos, J. P. *COS* **1991**, *2*, 661.
39. Danishefsky, S. J.; Pearson, W. H.; Harvey, D. F. *JACS* **1984**, *106*, 2456.
40. Inaba, S.; Ojima, I. *TL* **1977**, 2009.
41. Ojima, I.; Inaba, S.; Yoshida, K. *TL* **1977**, 3643.
42. Engler, T. A.; Combrink, K. D.; Ray, J. E. *JACS* **1988**, *110*, 7931.
43. Narasaka, K.; Soai, K.; Aikawa, Y.; Mukaiyama, T. *BCJ* **1976**, *49*, 779.
44. Saigo, K.; Osaki, M.; Mukaiyama, T. *CL* **1976**, 163.
45. Hosomi, A.; Sakurai, H. *JACS* **1977**, *99*, 1673.
46. Majetich, G.; Hull, K.; Defauw, J.; Shawe, T. *TL* **1985**, *26*, 2755.
47. Miyashita, M.; Yanami, T.; Kumazawa, T.; Yoshikoshi, A. *JACS* **1984**, *106*, 2149.
48. Haruta, J.; Nishi, K.; Matsuda, S.; Tamura, Y.; Kita, Y. *CC* **1989**, 1065.
49. Ohno, M.; Yamamoto, Y.; Shirasaki, Y.; Eguchi, S. *JCS(P1)* **1993**, 263.
50. (a) Lehnert, W. *S* **1974**, 667. (b) Campaigne, E.; Beckman, J. C. *S* **1978**, 385.
51. Courtheyn, D.; Verhe, R.; De Kimpe, N.; De Buyck, L.; Schamp, N. *JOC* **1981**, *46*, 3226.
52. (a) Olah, G. A.; Krishnamurti, R.; Prakash, G. K. S. *COS* **1991**, *3*, 293. (b) Heaney, H. *COS* **1991**, *2*, 733.
53. (a) Rieche, A.; Gross, H.; Höft, E. *OS* **1967**, *47*, 1. (b) Rieche, A.; Gross, H.; Höft, E.; Beyer, E. *OS* **1967**, *47*, 47.
54. (a) Yamamoto, K.; Nunokawa, O.; Tsuji, J. *S* **1977**, 721. (b) Yamamoto, K.; Yoshitake, J.; Qui, N. T., Tsuji, J. *CL* **1978**, 859.
55. Tsuge, A.; Yamasaki, T.; Moriguchi, T.; Matsuda, T.; Nagano, Y.; Nago, H.; Mataka, S.; Kajigaeshi, S.; Tashiro, M. *S* **1993**, 205.
56. Lee, K.; Oh, D. Y. *TL* **1988**, *29*, 2977.
57. Jones, R. C. F.; Sumaria, S. *TL* **1978**, 3173.
58. Eyley, S. *COS* **1991**, *2*, 707.
59. Pillot, J.-P.; Bennetau, B.; Dunogues, J.; Calas, R. *TL* **1980**, *21*, 4717.
60. (a) Burke, S. D.; Murtiashaw, C. W.; Dike, M. S.; Strickland, S. M. S.; Saunders, J. O. *JOC* **1981**, *46*, 2400. (b) Nakamura, E.; Fukuzaki, K.; Kuwajima, I. *CC* **1983**, 499.
61. Overman, L. E.; Castañeda, A.; Blumenkopf, T. A. *JACS* **1986**, *108*, 1303.
62. Hosomi, A.; Saito, M.; Sakurai, H. *TL* **1979**, 429.
63. (a) Albaugh-Robertson, P.; Katzenellenbogen, J. A. *TL* **1982**, *23*, 723. (b) Morizawa, Y.; Kanemoto, S.; Oshima, K.; Nozaki, H. *TL* **1982**, *23*, 2953.
64. Martin, R.; Demerseman, P. *S* **1992**, 738.
65. Carlson, R.; Nilsson, Å.; Strömqvist, M. *ACS* **1983**, *B37*, 7.
66. (a) White, W. A.; Weingarten, H. *JOC* **1967**, *32*, 213. (b) White, W. A.; Chupp, J. P.; Weingarten, H. *JOC* **1967**, *32*, 3246.
67. Moretti, I.; Torre, G. *S* **1970**, 141.
68. (a) Lehnert, W. *TL* **1971**, 559. (b) Lehnert, W. *TL* **1971**, 1501.
69. Kumar, V.; Dev, S. *TL* **1983**, *24*, 1289.
70. Bulman-Page, P. C.; Roberts, R. A.; Paquette, L. A. *TL* **1983**, *24*, 3555.
71. Schiess, M.; Seebach, D. *HCA* **1983**, *66*, 1618.
72. (a) Mukaiyama, T.; Imamoto, T.; Kobayashi, S. *CL* **1973**, 261. (b) Mukaiyama, T.; Imamoto, T.; Kobayashi, S. *CL* **1973**, 715.
73. Seebach, D.; Neumann, H. *CB* **1974**, *107*, 847.
74. Mukaiyama, T.; Hayashi, M.; Narasaka, K. *CL* **1973**, 291.
75. George, J.; Chandrasekaran, S. *SC* **1983**, *13*, 495.
76. (a) Ballamy, F. D.; Ou, K. *TL* **1984**, *25*, 839. (b) Varma, R. S.; Kabalka, G. W. *CL* **1985**, 243.
77. Drabowicz, J.; Mikolajczyk, M. *S* **1978**, 138.
78. Kano, S.; Tanaka, Y.; Hibino, S. *H* **1980**, *14*, 39.
79. Kano, S.; Tanaka, Y.; Sugino, E.; Hibino, S. *S* **1980**, 695.
80. Imamato, T. *COS* **1991**, *1*, 231.
81. Hung, C. W.; Wong, H. N. C. *TL* **1987**, *28*, 2393.

Lise-Lotte Gundersen
Norwegian College of Pharmacy, Oslo, Norway

Frode Rise & Kjell Undheim
University of Oslo, Norway

Titanium Tetraisopropoxide[1]

$Ti(O\text{-}i\text{-}Pr)_4$

[546-68-9] $C_{12}H_{28}O_4Ti$ (MW 284.28)

(mild Lewis acid used as a catalyst in transesterification reactions, nucleophilic cleavages of 2,3-epoxy alcohols and isomerization reactions; additive in the Sharpless epoxidation reaction and in various reactions involving nucleophilic additions to carbonyl and α,β-unsaturated carbonyl compounds)

Physical Data: mp 18–20 °C; bp 218 °C/10 mmHg; *d* 0.955 g cm^{-3}.
Solubility: sol a wide range of solvents including ethers, organohalides, alcohols, benzene.
Form Supplied in: low-melting solid; widely available.
Handling, Storage, and Precautions: flammable and moisture sensitive; acts as an irritant.

Transesterification and Lactamization Reactions.[2–5] $Ti(O\text{-}i\text{-}Pr)_4$, as well as other titanium(IV) alkoxides, has been recommended as an exceptionally mild and efficient transesterification catalyst which can be used with many acid-sensitive substrates (eq 1).[2,3] Thus the acetonide as well as C=O, OH, OTBDMS, and lactam functional groups (eq 2) are unaffected by these conditions, although acetates are hydrolyzed to the parent alcohol. The alcohol solvent employed in such processes need not be anhydrous, nor need it be identical with the OR group in the titanate, because exchange of these moieties is generally slow compared to the transesterification reaction.

MeO_2C / MeO_2C (acetonide) → [$Ti(O\text{-}i\text{-}Pr)_4$; i-PrOH, 70 °C; 91%] → $i\text{-}PrO_2C$ / $i\text{-}PrO_2C$ (acetonide) (1)

(2)

N-Protected esters of dipeptides can be transesterified using $Ti(O\text{-}i\text{-}Pr)_4$ in the presence of 4 Å molecular sieves and this procedure provides a good way of converting methyl esters into their benzyl counterparts. Such reactions (eq 3) proceed without racemization in 70–85% yield.[4] In a related process,[5] β-, γ-, and δ-amino acids undergo lactamization on treatment with $Ti(O\text{-}i\text{-}Pr)_4$ in refluxing 1,2-dichloroethane (eq 4).

(3)

(4)

R, R¹ = H, Me
n = 1–3

Nucleophilic Cleavage of 2,3-Epoxy Alcohols and Related Compounds.[6–10] Titanium alkoxides are weak Lewis acids which generally have no effect on simple epoxides. However, reaction of 2,3-epoxy alcohols (available, for example, from the Sharpless-type epoxidation of allylic alcohols) with nucleophiles in the presence of $Ti(O\text{-}i\text{-}Pr)_4$ results in highly regioselective ring-cleavage reactions involving preferential nucleophilic attack at C-3 (eq 5).[6] In the absence of the titanium alkoxide, no reaction is observed under otherwise identical conditions, except in the case of PhSNa.

(5)

R = $Me(CH_2)_2$; YH = $CH_2{=}CHCH_2OH$	90%; C-3:C-2 > 100:1
R = $Me(CH_2)_2$; YH = Et_2NH	90%; C-3:C-2 = 20:1
R = $Me(CH_2)_2$; YH = $PhCO_2H$	74%; C-3:C-2 > 100:1
R = $Me(CH_2)_2$; YH = TsOH (with 2,6-lutidine)	64%; C-3:C-2 > 100:1
R = $Me(CH_2)_2$; YH = PhSH	95%; C-3:C-2 = 6.4:1
R = $Me(CH_2)_6$; YH = AcSH	91%; C-3:C-2 > 100:1
R = $Me(CH_2)_6$; Y^- = NO_3^-	86%; C-3:C-2 > 100:1

These types of conversion can be extended[7] to 2,3-epoxy carboxylic acids (glycidic acids) and the related amides. The former compounds are readily available through ***Ruthenium(VIII) Oxide***-mediated oxidation of the appropriate 2,3-epoxy alcohols (eq 6).

(6)

The C-3:C-2 selectivity is, in the cases shown above, greater than 100:1, although lower selectivities (e.g. 20:1) are observed with other nucleophiles such as Et_2NH and PhSH. It should be noted that glycidic acids and amides react preferentially with dialkylamines at C-2, but one equivalent or greater of $Ti(O\text{-}i\text{-}Pr)_4$ ensures reaction occurs with high selectivity at C-3. Combining this type of protocol with the Sharpless asymmetric epoxidation reaction has permitted the development of stereoselective syntheses (eqs 7 and 8)[6,7] of all four stereoisomers of the unusual *N*-terminal amino acid of amastatin, a tripeptide competitive inhibitor of aminopeptidases.[8]

(7)

natural isomer

The reaction of both open-chain and cyclic 2,3-epoxy alcohols with molecular ***Bromine*** or ***Iodine*** in the presence of $Ti(O\text{-}i\text{-}Pr)_4$ at 0 °C leads to the regioselective formation of halo diols (eq 9).[9a] Interestingly, if these reactions are conducted at 25 °C a 1:1 mixture of the C-2 and C-3 cleavage products is obtained, and the same outcome is observed, even at 0 °C, when the acetate derivative of the 2,3-epoxy alcohol is involved as substrate. Dialkylamine hydrochlorides can be used as sources of halide nucleophiles in these types of epoxide ring-cleavage reactions.[9b]

2,3-Epithio alcohols have been obtained by reacting 2,3-epoxy alcohols with ***Thiourea*** at room temperature or 0 °C in the presence of $Ti(O\text{-}i\text{-}Pr)_4$ and using THF as solvent (eq 10).[10] The reactions proceed with high regio- and stereoselectivity, *trans*-substituted 2,3-epoxy alcohols giving only *trans*-2,3-epithio alcohols with complete inversion of configuration at both stereogenic centers. However, when *cis*-2,3-epoxy alcohols are used as starting materials the yields of epithio alcohols were low and thiodiols were also formed. Epithiocinnamyl alcohols could also be prepared from the corresponding epoxycinnamyl alcohols at 0 °C.

However, these products were found to decompose to cinnamyl alcohol and sulfur on standing. Without $Ti(O\text{-}i\text{-}Pr)_4$, thiourea was insoluble in THF and the reaction did not proceed. One equivalent of $Ti(O\text{-}i\text{-}Pr)_4$ was required to achieve complete reaction, and THF was the best solvent (no reaction was observed in ether, CH_2Cl_2, or benzene under similar conditions).

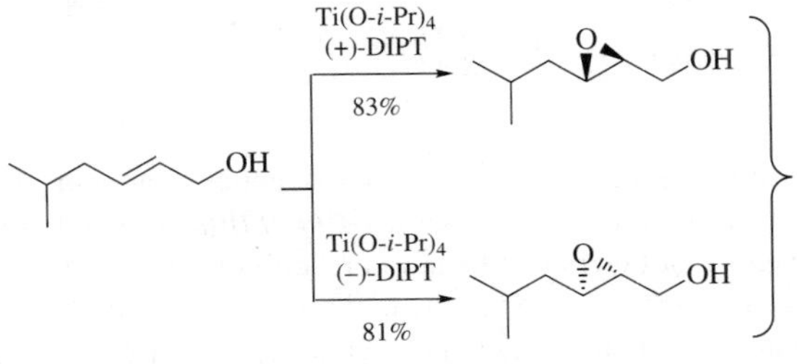

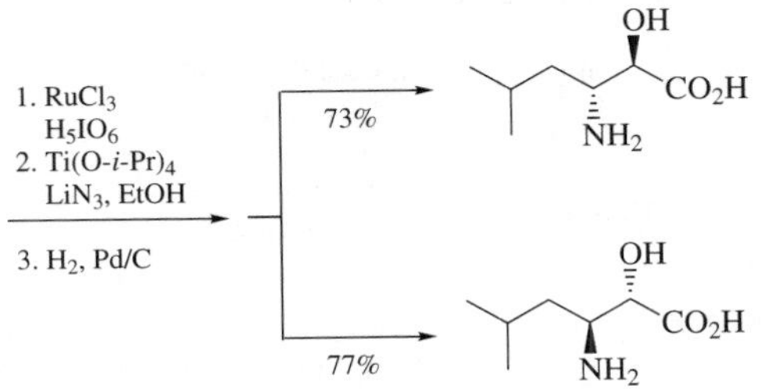

(8)

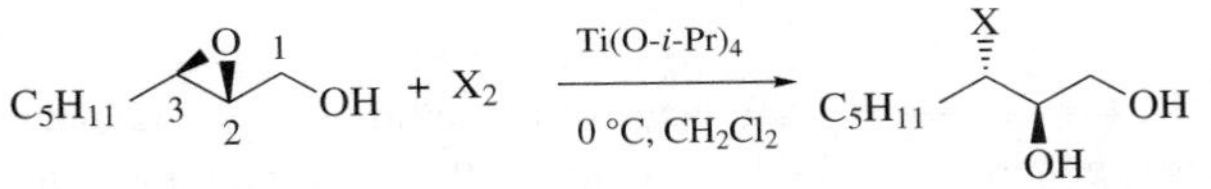

(9)

87% (X = I); C-3:C-2 = 7:1
78% (X = Br); C-3:C-2 = 15:1

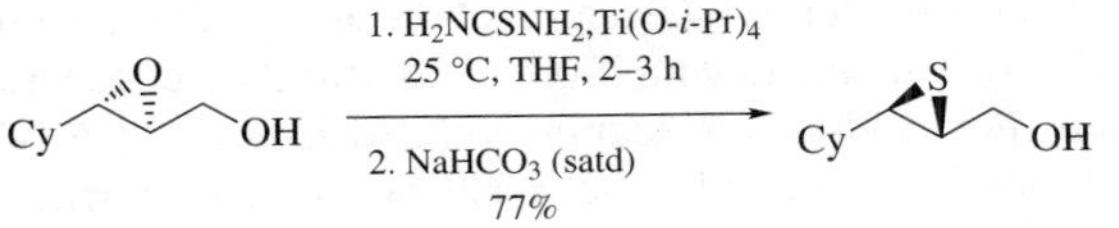

(10)

Isomerization Reactions.[11–14] The reaction of certain 2,3-epoxy alcohols with $Ti(O\text{-}i\text{-}Pr)_4$ can result in isomerization. For example (eq 11), reaction of the illustrated substrate in CH_2Cl_2 results in rearrangement to the isomeric enediol, and this conversion represents a key step in a synthesis of the marine natural product pleraplysillin.[11]

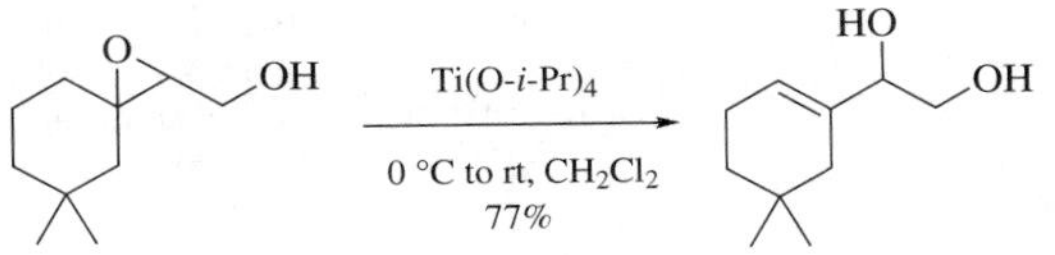

(11)

Under similar conditions, the pendant double bond attached to a 2,3-epoxy alcohol acts as an internal nucleophile attacking at C-3, resulting, after proton loss, in a mixture of cyclized products (eq 12).[12] The cyclopropane-containing products are believed to arise via a retro-homo Prins reaction. Pendant triple bonds can also participate in related cyclization reactions and cyclic allenes result (eq 13).[12] The observation that the *threo* isomer of the substrate shown in eq 12 is stable to $Ti(O\text{-}i\text{-}Pr)_4$ has led to the suggestion that an intramolecular metal alkoxide is the active catalyst in successful cyclization reactions.

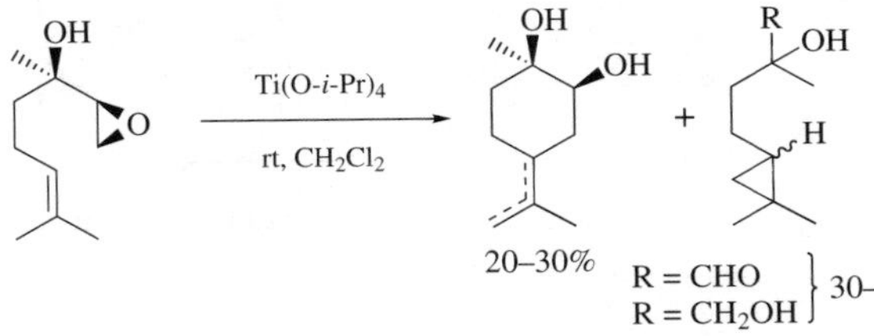

(12)

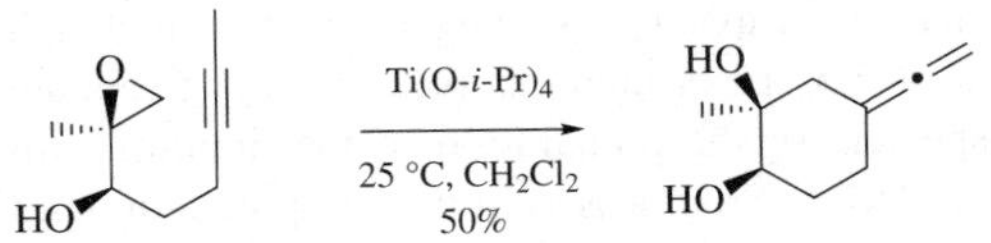

(13)

$Ti(O\text{-}i\text{-}Pr)_4$ has also played a key role in the synthesis of taxanes (eq 14).[12b,c]

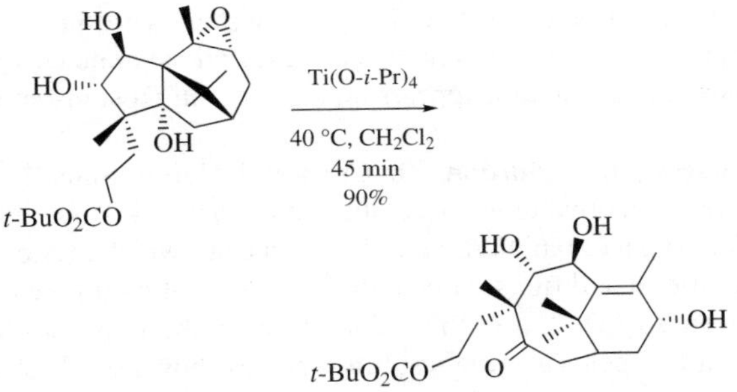

(14)

Allylic hydroperoxides, which are readily obtained by reaction of the corresponding alkene with ***Singlet Oxygen***, have been shown to isomerize to the corresponding 2,3-epoxy alcohol when treated with catalytic amounts of $Ti(O\text{-}i\text{-}Pr)_4$ (eq 15).[13] The title reagent is the one of choice when converting di-, tri-, and tetrasubstituted alkenes (both cyclic and acyclic) into the corresponding 2,3-epoxy alcohols by this protocol. The reactions are generally highly stereoselective and deoxygenation of the allylic hydroperoxide (to give the corresponding allylic alcohol) is not normally a process which competes significantly with the isomerization reaction.

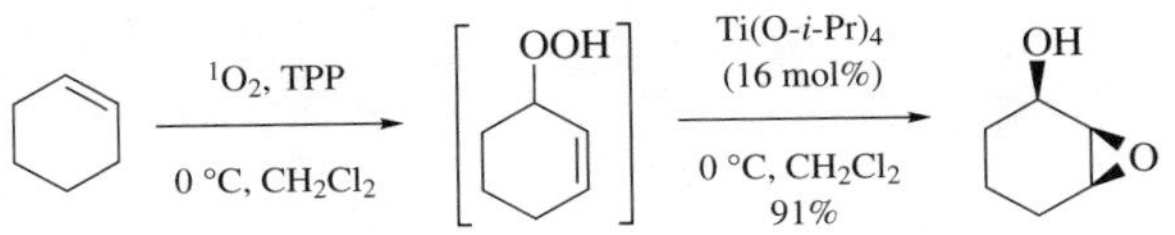

(15)

cis:*trans* = 98:2

This type of chemistry has been extended to the preparation of epoxy diols from chiral allylic alcohols (eq 16).[13,14] The methodology is impressive in that three successive chiral centers are constructed with predictable configuration. Furthermore, the rapid rate of the isomerization process is remarkable, given that α,β-unsaturated diols are generally poor substrates for titanium metal-mediated epoxidations. This rate enhancement is attributed to the tridentate nature of the intermediate hydroperoxides.

$^1O_2$, TPP; $Ti(O\text{-}i\text{-}Pr)_4$ (0.05 equiv), –25 °C, CH_2Cl_2, 1 h

80% of a 86:14 mixture (16)

Asymmetric Epoxidation Reactions.[15] While $Ti(O\text{-}i\text{-}Pr)_4$ clearly has the capacity to bring about the nucleophilic ring-cleavage of 2,3-epoxy alcohols (see above), it remains the preferred species for the preparation of the titanium tartrate complex central to the Sharpless asymmetric epoxidation process (see, for example, eq 7). Since *t*-butoxide-mediated ring-opening of 2-substituted 2,3-epoxy alcohols (a subclass of epoxy alcohols particularly sensitive to nucleophilic ring-cleavage) is much slower than by isopropoxide, the use of $Ti(O\text{-}t\text{-}Bu)_4$ is sometimes recommended in place of $Ti(O\text{-}i\text{-}Pr)_4$. However, with the reduced amount of catalyst that is now needed for all asymmetric epoxidations, this precaution appears unnecessary in most instances.

Nucleophilic Additions to Carbonyl Compounds.[16–28] A wide range of organometallic compounds react with $Ti(O\text{-}i\text{-}Pr)_4$ to produce organotitanium/titanium-'ate' species which may exhibit reactivities that differ significantly from those of their precursor.[1] Thus $Ti(O\text{-}i\text{-}Pr)_4$ forms '-ate' complexes with, amongst others, Grignard reagents and the resulting species show useful selectivities in their reaction with carbonyl compounds. For example, the complex with allylmagnesium chloride is highly selective in its reaction with aldehydes in the presence of ketones, or ketones in the presence of esters (eq 17).[16] Interestingly, the corresponding amino titanium-'ate' complexes react selectively with ketones in the presence of aldehydes.

MgCl; $Ti(O\text{-}i\text{-}Pr)_4$, –78 °C, THF, 81% (17)

Reaction of certain sulfur-substituted allylic anions with Ti(O-*i*-Pr)$_4$ produces a 3-(alkylthio)allyltitanium reagent that condenses, through its α-terminus, with aldehydes to give *anti*-β-hydroxy sulfides in a highly stereo- and regioselective manner (eq 18). These latter compounds can be transformed, stereoselectively, into *trans*-vinyloxiranes or 1,3-dienes.[17]

$Ti(O\text{-}i\text{-}Pr)_4$, –78 °C, THF; R^1CHO, –78 to 0 °C, 70–99%

erythro:*threo* = >6:1

Me_3OBF_4 then NaOH (aq.) 62–96%

1. BuLi, MsCl 2. *t*-BuLi R = Ph (18)

The titanium species derived from sequential treatment of α-alkoxy-substituted allylsilanes with ***s-Butyllithium*** then Ti(O-*i*-Pr)$_4$ engages in a Peterson alkenation reaction with aldehydes to give, via electrophilic attack at the α-terminus of the allyl anion, 2-oxygenated 1,3-butadienes which can be hydrolyzed to the corresponding vinyl ketone (eq 19).[18a]

1. *s*-BuLi, THF, –78 °C 2. $Ti(O\text{-}i\text{-}Pr)_4$ 3. R^1CHO 45–80%

HCl, THF 100% (19)

The titanium-'ate' complexes of α-methoxy allylic phosphine oxides, generated in situ by reaction of the corresponding lithium anion and $Ti(O\text{-}i\text{-}Pr)_4$, condense with aldehydes exclusively at the α-position to produce homoallylic alcohols in a diastereoselective fashion.[18b] The overall result is the three-carbon homologation of the original aldehyde, and this protocol has been used in a synthesis of (−)-aplysin-20 from nerolidol.[19] The titanium-'ate' complex produced by reaction of the chiral lithium anion of an (*E*)-crotyl carbamate with $Ti(O\text{-}i\text{-}Pr)_4$ affords γ-condensation products (homoaldols) on reaction with aldehydes.[18c,d] Allyl anions produced by the reductive metalation of allyl phenyl sulfides condense with α,β-unsaturated aldehydes in a 1,2-manner at the more substituted (α) allyl terminus in the presence of Ti(O-*i*-Pr)$_4$.[20] 1,2-Addition of dialkylzincs to α,β-unsaturated aldehydes can be achieved with useful levels of enantiocontrol when the reaction is conducted using a chiral titanium(IV) catalyst in the presence of $Ti(O\text{-}i\text{-}Pr)_4$ (eq 20).[21] Higher ee values are observed when an α-substituent (e.g. bromine) is attached to the substrate aldehyde, but a β-substituent *cis*-related to the carbonyl group has the opposite effect.

R_2Zn, $Ti(O\text{-}i\text{-}Pr)_4$, –60 to –55 °C, 8–12 h (20)

R = Et, 68%; 93% ee
R = C_5H_{11}, 75%; 85% ee
R = $(CH_2)_5OPiv$, 90%; 92% ee

Highly enantioselective trimethylsilylcyanation of various aldehydes can be achieved by using ***Cyanotrimethylsilane*** in the presence of a modified Sharpless catalyst consisting of $Ti(O\text{-}i\text{-}Pr)_4$

and chiral diisopropyl tartrate.[22] Best results are obtained using dichloromethane as solvent and isopropanol as additive and running the reaction at 0 °C. The same type of catalyst system also effects the asymmetric ring-opening of symmetrical cycloalkene oxides with ***Azidotrimethylsilane***. *trans*-2-Azidocycloalkanols are obtained in up to 63% optical yield.[22] The titanium amide complexes derived from the reaction of lithium dialkylamides with $Ti(O\text{-}i\text{-}Pr)_4$ condense with alkyl and aryl aldehydes and the resulting aminal derivatives undergo C–O bond displacement by benzylmagnesium chloride, thereby generating α-substituted β-phenethylamines (eq 21).[23]

$LiNR^1_2$ → [1. $Ti(O\text{-}i\text{-}Pr)_4$, Et_2O, –20 °C, 20 min; 2. R^2CHO, 20 °C, 3–5 h; 3. $PhCH_2MgCl$, Et_2O, rt, 1 h; 4. H_2O; 73–93%] → $PhCH_2CH(R^2)NR^1_2$ (21)

R^1 = alkyl; R^2 = alkyl, aryl

The aldol-type condensation of aldehydes and ketones with ketenimines[24] and ketones[25] can be catalyzed by titanium alkoxides and, in appropriate cases, useful levels of stereocontrol can be achieved (eq 22).

$Me_2C(TMS)C{=}N{-}TMS$ + $C_8H_{17}CHO$ → [$TiCl_4$, $Ti(O\text{-}i\text{-}Pr)_4$, CH_2Cl_2; –78 °C] → $C_8H_{17}CH(OH)C(Me)(TMS)CN$ → [$BF_3{\bullet}Et_2O$; 86%] → $C_8H_{17}CH{=}C(Me)CN$ (22)

E:*Z* = 8:92

A variety of useful reducing agents have been generated by combining hydrides with $Ti(O\text{-}i\text{-}Pr)_4$.[26–28] For example, the combination of 5 mol % $Ti(O\text{-}i\text{-}Pr)_4$ with 2.5–3.0 equiv of ***Triethoxysilane*** cleanly hydrosilylates esters to silyl ethers at 40–55 °C, and these latter compounds can be converted into the corresponding primary alcohols via aqueous alkaline hydrolysis (eq 23).[26] The actual reducing agent is presumed to be a titanium hydride species which is produced by a σ-bond metathesis process involving $Ti(O\text{-}i\text{-}Pr)_4$ and the silane. The procedure has considerable merit in that no added solvent is required and the active reagent can be generated and used in air. Halides, epoxides, alcohols, and an alkyne all survive the reduction process. ***Lithium Borohydride*** reduction of 2,3-epoxy alcohols yields 1,2-diols highly regioselectively when used in the presence of $Ti(O\text{-}i\text{-}Pr)_4$,[27] while the combination of the reagent with ***Sodium Cyanoborohydride*** is reported[28] to offer superior results in reductive amination processes with difficult carbonyls and those sensitive to acidic conditions.

$RC(O)OR^1$ → [5% $Ti(O\text{-}i\text{-}Pr)_4$, 2.3–3 equiv $(EtO)_3SiH$; 40–55 °C, 4–22 h] → [1 M NaOH, THF; rt, 2–4.5 h] → RCH_2OH (23)

70–95%

Miscellaneous Applications.[29–31] The chemoselective oxidation of alcohols and diols using $Ti(O\text{-}i\text{-}Pr)_4$/***t-Butyl Hydroperoxide*** has been reported.[29] The title reagent has also been employed as a catalyst in Diels–Alder reactions[30] and as an additive in the palladium-catalyzed reaction of aryl-substituted allylic alcohols with zinc enolates of β-dicarbonyl compounds (eq 24).[31] The latter reaction is presumed to generate *C*-allylated β-dicarbonyl compounds as the primary products of reaction, but these compounds suffer deacylation in the presence of $Ti(O\text{-}i\text{-}Pr)_4$.

$PhCH{=}CHCH_2OH$ + $CH_3COCH_2COCH_3$ → [$ZnCl_2$, Et_3N, $PdCl_2(PPh_3)_2$, $Ti(O\text{-}i\text{-}Pr)_4$; $MeO(CH_2)_2OEt$, 120 °C] → $PhCH{=}CHCH_2CH_2COCH_3$ (24)

1. (a) Shiihara, I.; Schwartz, W. T., Jr.; Post, H. W. *CRV* **1961**, *61*, 1. (b) Reetz, M. T. *Top. Curr. Chem.* **1982**, *106*, 3. (c) Seebach, D.; Weidmann, B; Widler, L. *Mod. Synth. Methods* **1983**, *3*, 217. (d) Reetz, M. T. *Organotitanium Reagents in Organic Synthesis*; Springer: Berlin, 1986. (e) Hoppe, D.; Krämer, T.; Schwark, J.-R.; Zschage, O. *PAC* **1990**, *62*, 1999.
2. (a) Imwinkelried, R.; Schiess, M.; Seebach, D. *OS* **1987**, *65*, 230. (b) Seebach, D.; Hungerbühler, E.; Schnurrenberger, P.; Weidmann, B.; Züger, M. *S* **1982**, 138.
3. (a) Schnurrenberger, P.; Züger, M. F.; Seebach, D. *HCA* **1982**, *65*, 1197. (b) Férézou, J. P.; Julia, M.; Liu, L. W.; Pancrazi, A. *SL* **1991**, 618.
4. Rehwinkel, H.; Steglich, W. *S* **1982**, 826.
5. Mader, M.; Helquist, P. *TL* **1988**, *29*, 3049.
6. (a) Caron, M.; Sharpless, K. B. *JOC* **1985**, *50*, 1557. (b) *Aldrichim. Acta* **1985**, *18*, 53.
7. Chong, J. M.; Sharpless, K. B. *JOC* **1985**, *50*, 1560.
8. Tobe, H.; Morishima, H.; Aoyagi, T.; Umezawa, H.; Ishiki, K.; Nakamura, K.; Yoshioka, T.; Shimauchi, Y.; Inui, T. *ABC* **1982**, *46*, 1865.
9. (a) Alvarez, E.; Nuñez, T.; Martin, V. S. *JOC* **1990**, *55*, 3429. (b) Gao, L.; Murai, A. *CL* **1989**, 357.
10. Gao, Y.; Sharpless, K. B. *JOC* **1988**, *53*, 4114.
11. Masaki, Y.; Hashimoto, K.; Serizawa, Y.; Kaji, K. *BCJ* **1984**, *57*, 3476.
12. (a) Morgans, D. J., Jr.; Sharpless, K. B.; Traynor, S. G. *JACS* **1981**, *103*, 462. (b) Holton, R. A.; Juo, R. R.; Kim, H. B.; Williams, A. D.; Harusawa, S.; Lowenthal, R. E.; Yogai, S. *JACS* **1988**, *110*, 6558. (c) Wender, P. A.; Mucciaro, T. P. *JACS* **1992**, *114*, 5878.
13. (a) Mihelich, E. D. US Patent 4 345 984 (*CA* **1983**, *98*, 125 739c). (b) Adam, W.; Braun, M.; Griesbeck, A.; Lucchini, V.; Staab, E.; Will, B. *JACS* **1989**, *111*, 203. (c) Adam, W.; Nestler, B. *JACS* **1993**, *115*, 7226.
14. Adam, W.; Nestler, B. *AG(E)* **1993**, *32*, 733.
15. Johnson, R. A.; Sharpless, K. B. *COS* **1991**, *7*, 389.
16. Reetz, M. T.; Wenderoth, B. *TL* **1982**, *23*, 5259.
17. Furuta, K.; Ikeda, Y.; Meguriya, N.; Ikeda, N.; Yamamoto, H. *BCJ* **1984**, *57*, 2781.
18. (a) Murai, A.; Abiko, A.; Shimada, N.; Masamune, T. *TL* **1984**, *25*, 4951. (b) Birse, E. F.; McKenzie, A.; Murray, A. W. *JCS(P1)* **1988**, 1039. (c) Férézou, J. P.; Julia, M.; Khourzom, R.; Pancrazi, A.; Robert, P. *SL* **1991**, 611. (d) Hoppe, D.; Zschage, O. *AG(E)* **1989**, *28*, 69.
19. Murai, A.; Abiko, A.; Masamune, T. *TL* **1984**, *25*, 4955.
20. Cohen, T.; Guo, B.-S. *T* **1986**, *42*, 2803.
21. Rozema, M. J.; Eisenberg, C.; Lütjens, H.; Ostwald, R.; Belyk, K.; Knochel, P. *TL* **1993**, *34*, 3115.
22. (a) Hayashi, M.; Matsuda, T.; Oguni, N. *JCS(P1)* **1992**, 3135. (b) Hayashi, M.; Kohmura, K.; Oguni, N. *SL* **1991**, 774.
23. Takahashi, H.; Tsubuki, T.; Higashiyama, K. *S* **1988**, 238.
24. Okada, H.; Matsuda, I.; Izumi, Y. *CL* **1983**, 97.
25. Vuitel, L.; Jacot-Guillarmod, A. *HCA* **1974**, *57*, 1703.

26. Berk, S. C.; Buchwald, S. L. *JOC* **1992**, *57*, 3751.
27. Dai, L.; Lou, B.; Zhang, Y.; Guo, G. *TL* **1986**, *27*, 4343.
28. Mattson, R. J.; Pham, K. M.; Leuck, D. J.; Cowen, K. A. *JOC* **1990**, *55*, 2552.
29. Yamawaki, K.; Ishii, Y.; Ogawa, M. *Chem. Express* **1986**, *1*, 95.
30. McFarlane, A. K.; Thomas, G.; Whiting, A. *TL* **1993**, *34*, 2379.
31. Itoh, K.; Hamaguchi, N.; Miura, M.; Nomura, M. *JCS(P1)* **1992**, 2833.

Martin G. Banwell
University of Melbourne, Parkville, Victoria, Australia

p-Toluenesulfonyl Chloride[1]

[98-59-9] $C_7H_7ClO_2S$ (MW 190.66)

(sulfonyl transfer reagent; *O*-sulfonylation of alcohols[2] for conversion to chlorides[3] or intermolecular[4] and intramolecular[5] displacements, vicinal diols for epoxidation,[6] 1,3-diols for oxetane (formation,[7] carboxylic acids for esterification[8] or decarboxylation,[9] oximes for Beckmann rearrangements[10] or fragmentations[11] and Neber rearrangements,[12] hydroxamic acids for Lossen rearrangements,[13] nitrones for rearrangements,[14] conversion of *N*-cyclopropylhydroxylamines to β-lactams;[15] *N*-sulfonylation of aliphatic amines[16] for subsequent deamination[17] or displacement,[18] aromatic amines for protection;[19] *C*-sulfonylation of alkenes[20] and silylalkynes;[21] dehydration of ureas,[22] formamides,[23] and amides[24])

Alternate Names: tosyl chloride.
Physical Data: mp 67–69 °C; bp 146 °C/15 mmHg.
Solubility: insol H_2O; freely sol ethanol, benzene, chloroform, ether.
Form Supplied in: white solid, widely available.
Purification: upon prolonged standing the material develops impurities of *p*-toluenesulfonic acid and HCl. Tosyl chloride is purified by dissolving 10 g in a minimum volume of $CHCl_3$ (ca. 25 mL), filtering, and diluting with five volumes (ca. 125 mL) of petroleum ether (bp 30–60 °C) to precipitate impurities (mostly tosic acid, mp 101–104 °C). The solution is filtered, clarified with charcoal, and concentrated to ca. 40 mL by evaporation. Further evaporation to a very small volume gives 7 g of pure white crystals (mp 67.5–68.5 °C).[25]

Tosyl chloride may also be recrystallized from petroleum ether, from benzene, or from toluene/petroleum ether (bp 40–60 °C) in the cold. Tosyl chloride in diethyl ether can be washed with aqueous 10% NaOH until colorless, then dried with Na_2SO_4 and crystallized by cooling in powdered dry ice. It can also be purified by dissolving in benzene, washing with aqueous 5% NaOH, drying with K_2CO_3 or $MgSO_4$, and distilling under reduced pressure.[26]

Handling, Storage, and Precautions: freshly purified tosyl chloride should be used for best results. Tosyl chloride is a moisture-sensitive, corrosive lachrymator.

General Discussion. The tosylation of alcohols is one of the most prevalent reactions in organic chemistry.[1] Optimized conditions for this reaction include the use of a 1:1.5:2 ratio of alcohol/tosyl chloride/pyridine in chloroform (eq 1).[2a] This procedure avoids formation of unwanted pyridinium salts inherent to reactions where higher relative quantities of pyridine have been employed.[2b]

ROH (1 equiv) + TsCl (2 equiv) → (1.5 equiv py, 0 °C, 87–98%) ROTs (1)

Good yields (81–88%) for tosylation have also been observed under biphasic conditions where ***Benzyltriethylammonium Chloride*** is employed as a phase-transfer catalyst between benzene and aqueous sodium hydroxide solution.[2c]

It has long been known that it is possible to selectively tosylate primary over secondary alcohols (eq 2).[2d]

(TsCl, py, 0 °C, 86%) (2)

It is also possible to regioselectively mono-*O*-tosylate various nonprotected hydroxyl functionalities, as illustrated in Scheme 1.[2e] This method employs a preliminary activation of a glycopyranoside with ***Di-n-butyltin Oxide*** and usually requires the use of a basic catalyst such as ***4-Dimethylaminopyridine*** (DMAP) in conjunction with tosyl chloride. Regioselectivity differs markedly from acylation reactions and is thought to be a function of changes in the kinetics of the reactions with the various tin intermediates which are in equilibrium.

Regioselective mono- and ditosylation of aldonolactones has also been reported.[2f] A few of these products are shown in Scheme 2. In no instances was the β-hydroxy function tosylated.

The regioselective tosylation of various cyclodextrins has been reported. The two major products resulting from the tosylation of β-cyclodextrin are heptakis(6-*O*-(*p*-tosyl))-β-cyclodextrin (**1**) and heptakis(6-*O*-(*p*-tosyl))-2-*O*-(*p*-tosyl))-β-cyclodextrin (**2**).[2g,h] Yamamura has also prepared hexakis(6-*O*-(*p*-tosyl))-α-cyclodextrins[2i] as well as polytosylated γ-cyclodextrins.[2j]

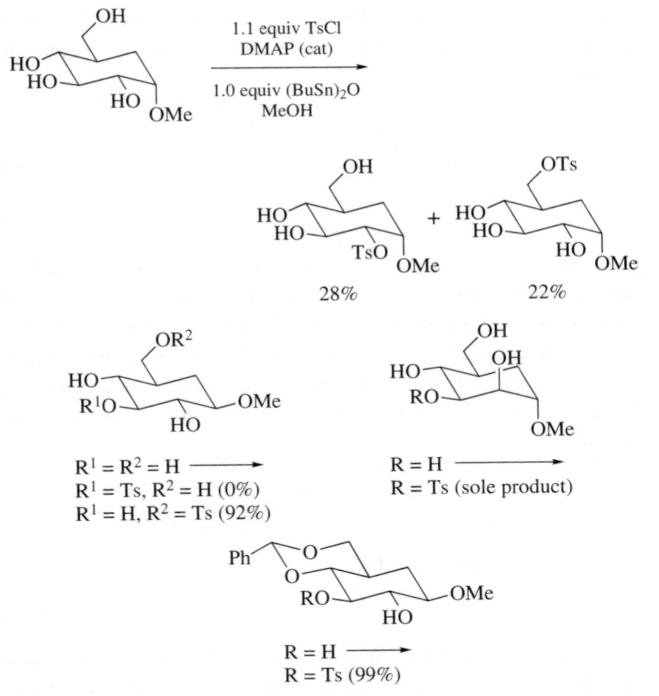

Scheme 1

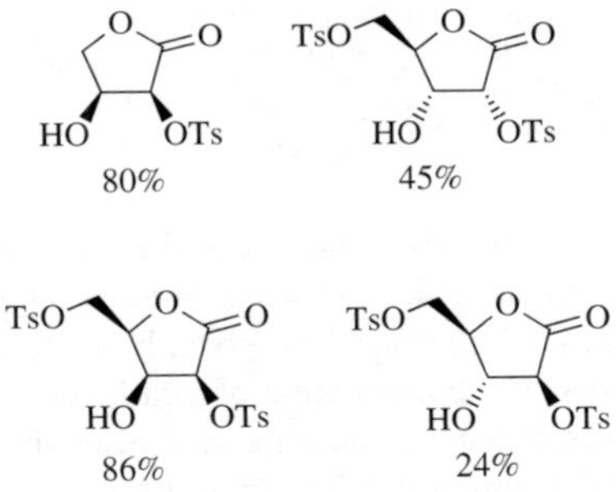

Scheme 2

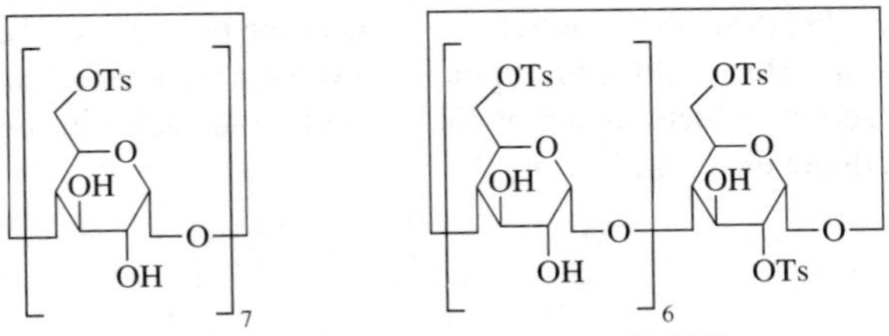

(1) 24%

(2) 15%

By proper choice of base it is possible to selectively *O*-tosylate in the presence of a free amine or *N*-tosylate in the presence of a free hydroxyl, with yields in excess of 94% (eq 3). The probable explanation for selective *O*-tosylation in the presence of the stronger base is formation of an adequate amount of phenoxide to allow the anion to act as the nucleophile.[2k]

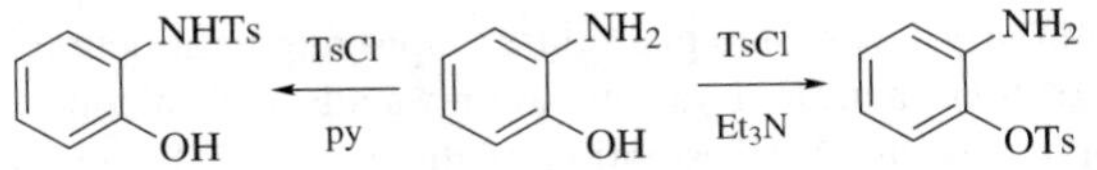

(3)

Tosyl chloride has been used in the preparation of allylic chlorides from their respective alcohols, leaving in place sensitive groups and with no rearrangement of the allylic substrate (eq 4).[21]

(4)

Studies on reactions of various types of alcohols with the related tosyl chloride/dimethylaminopyridine (TsCl/DMAP) system have led to the following conclusions: allylic, propargylic, and glycosidic hydroxyls quickly react to form the corresponding chlorides, 2,3-epoxy and selected primary alcohols yield chlorides but at a slower rate, and reactions of aliphatic secondary alcohols stop at the tosylate stage. Example reactions are illustrated in eqs 5–10.[3a]

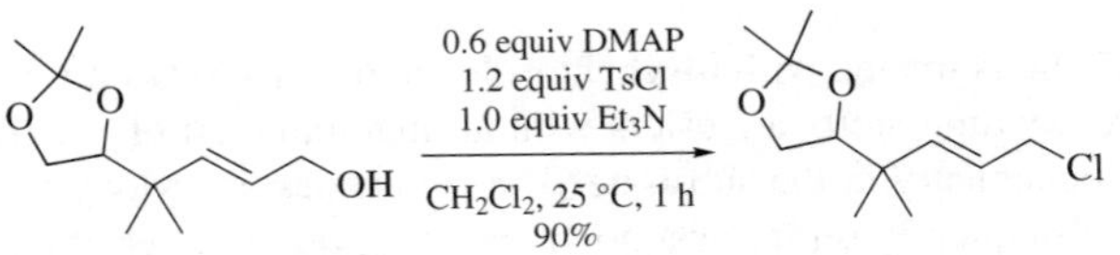

(5)

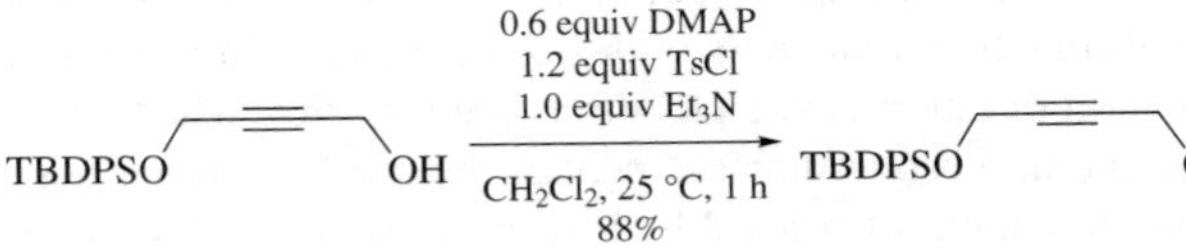

(6)

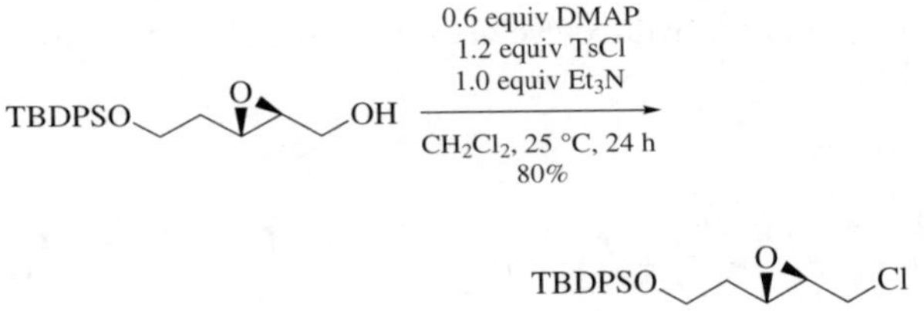

(7)

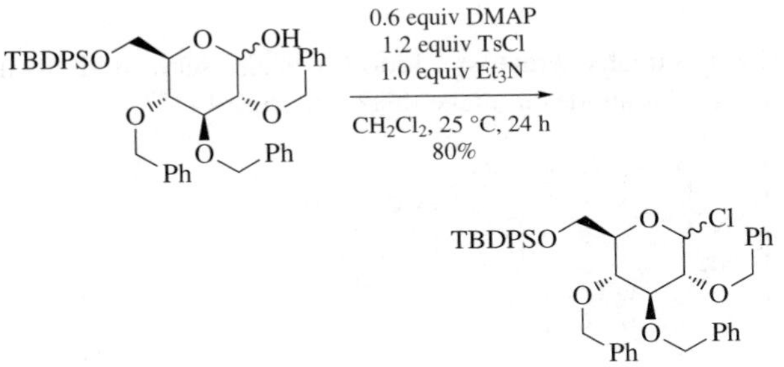

(8)

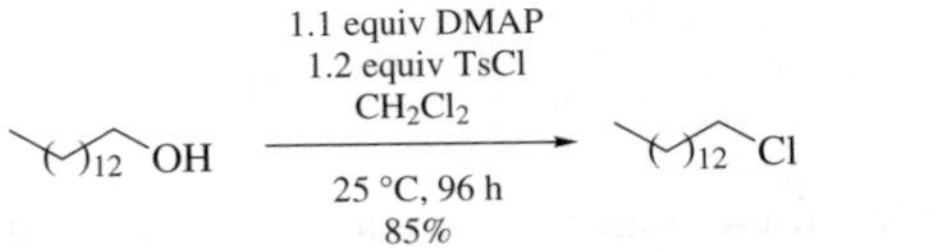

(9)

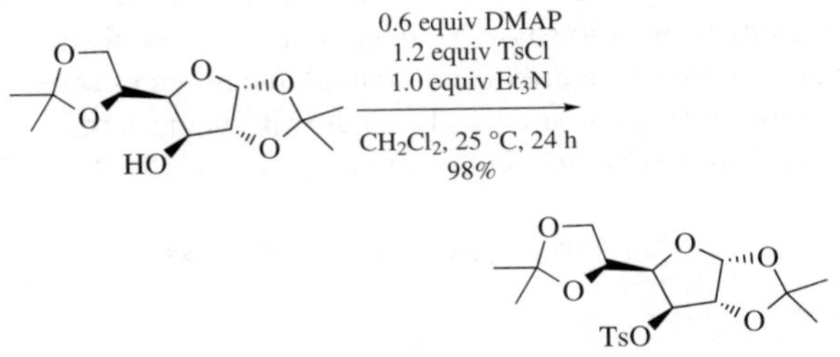

(10)

Attempted tosylation of 3β-methoxy-21-hydroxy-5α-pregnan-20-one afforded its α-chloro derivative.[3b] This reinforces the rule that as hybridization α to the alcohol increases in s character, displacement of the intermediate tosylate is facilitated (eq 11).

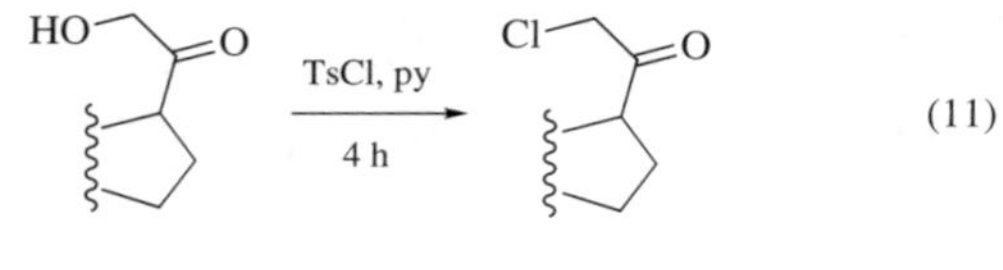

(11)

The α-chlorination of sulfoxides can also be achieved using tosyl chloride and pyridine, albeit in poor yield (32%), as shown in eq 12.[3c]

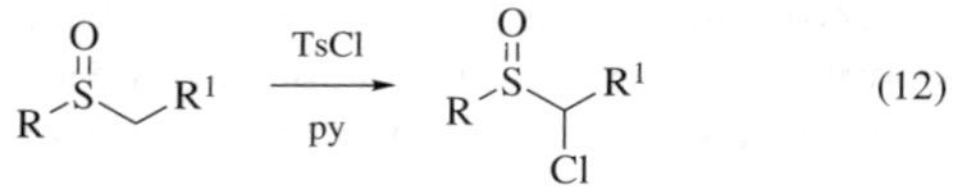

(12)

Alcohols treated with tosyl chloride are transformed into their corresponding sulfonate esters without manipulation of existing stereochemistry in the substrate. The *p*-toluenesulfonyloxy moiety subsequently serves as a good leaving group in intermolecular nucleophilic substitution or elimination reactions.[4] Treatment of multifunctional alcohols with tosyl chloride can result in the formation of intermediate *O*-sulfonylated species, which can undergo intramolecular displacement of the tosylate group. For example, treatment of *trans*-2-hydroxycyclohexaneacetamide with tosyl chloride in pyridine is the key step in the conversion of the lactone of *trans*-2-hydroxycyclohexaneacetic acid to its *cis* isomer (eq 13).[5a]

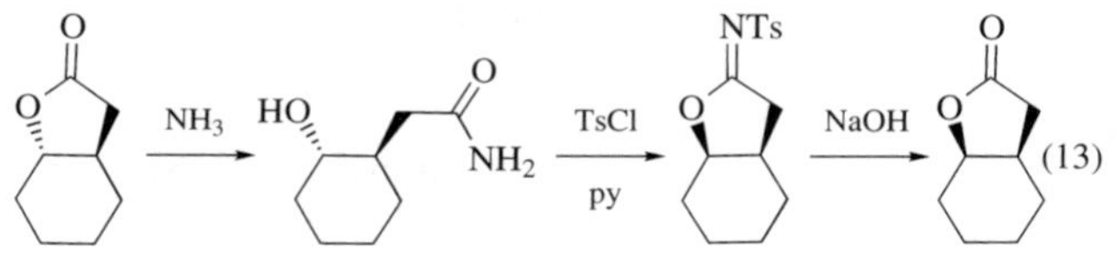

(13)

Alcohols treated with tosyl chloride can also serve as alkylating agents for thioamides to make thiazolines (eq 14).[5b]

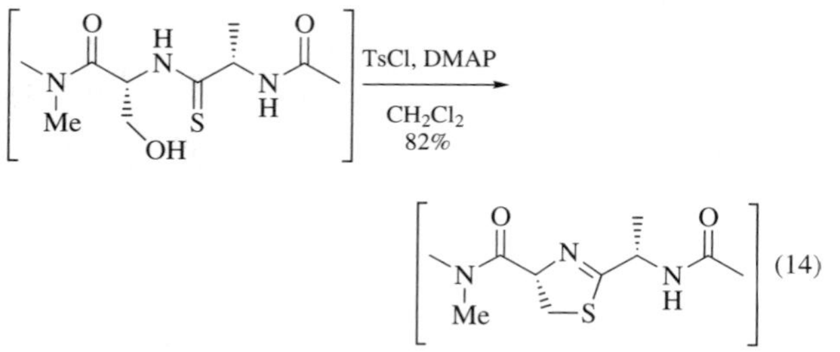

(14)

Several useful one-pot procedures employing the reaction of tosyl chloride with vicinal diols have been used to form epoxides. Treatment of variously substituted diols with 1 equiv of tosyl chloride and sodium hydroxide in monoglyme provides access to a variety of 1-alkynyloxiranes (R^1 and/or R^3 being terminal or substituted alkynes) in moderate to good yield (eq 15).[6a]

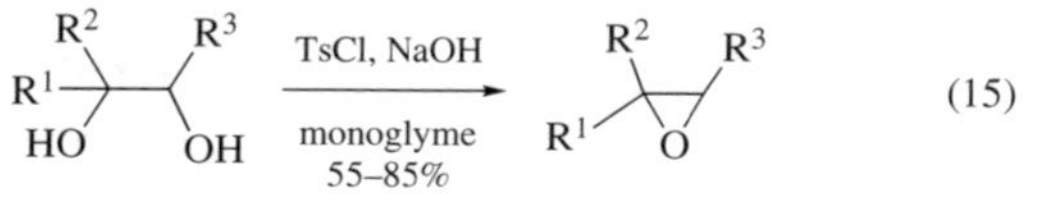

(15)

Another method entails treatment of the diol in THF with 2.2 equiv of ***Sodium Hydride*** followed by reaction with a slight molar excess of tosyl chloride. This fast reaction, illustrated in eq 16, can be used to prepare enantiopure epoxides in good to excellent yields.[6b]

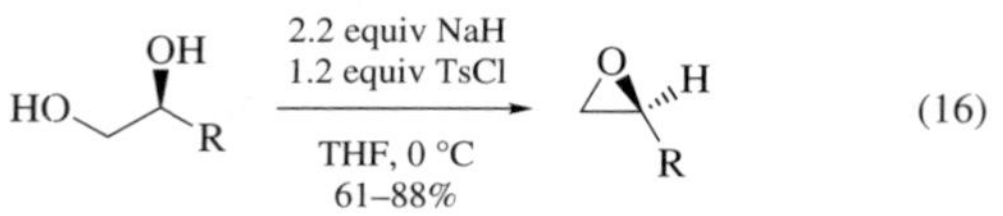

(16)

Phase-transfer catalysis has also been successfully employed to achieve epoxidation. A variety of cyclic *trans*-substituted diols in dichloromethane were treated with a 50% aqueous solution of sodium hydroxide in the presence of the phase-transfer catalyst benzyltriethylammonium bromide. Consistently good yields were achieved for these as well as glycosidic and acyclic substrates (eq 17).[6c]

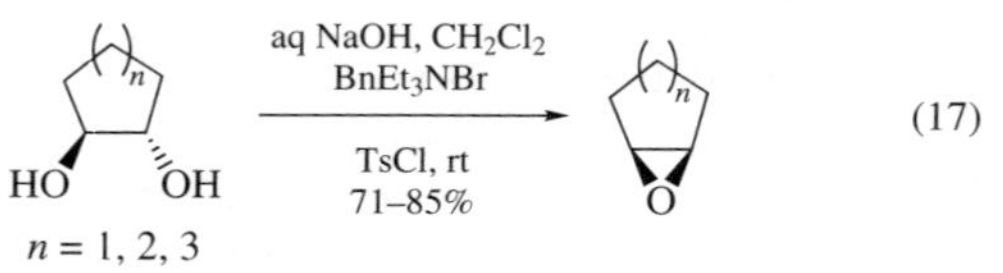

(17)

A one-pot conversion of a variety of 1,3-diols to oxetanes has also been reported.[7] The procedure entailed alcohol deprotonation with one equivalent of ***n-Butyllithium*** followed by treatment with tosyl chloride and then a second equivalent of base (eq 18).

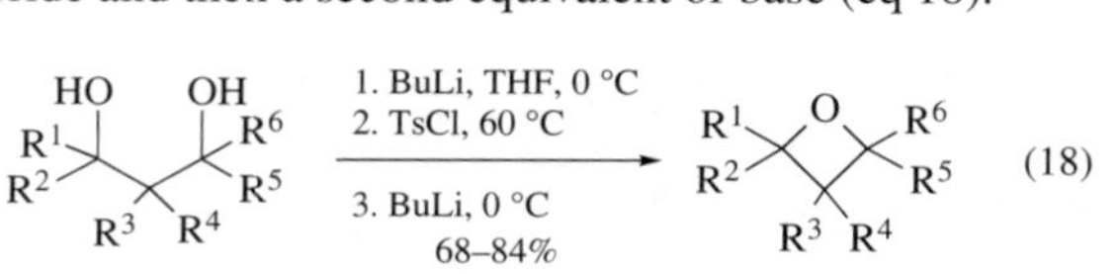

(18)

When a solution of a carboxylic acid and an alcohol in pyridine is treated with tosyl chloride, an ester is formed rapidly in excellent yield. This procedure is useful especially in the esterification of tertiary alcohols. The combination of a carboxylic acid and tosyl chloride serves as a convenient method of in situ preparation of symmetrical acid anhydrides for further formation of esters and amides (eq 19). The novelty of this protocol is that the acid can be recycled through the anhydride stage in the presence of the alcohol, thereby resulting in complete conversion to the ester (eq 20).[8a] Reactivity is determined by the strength of the acid: strong acids facilitate the esterifications.[8b]

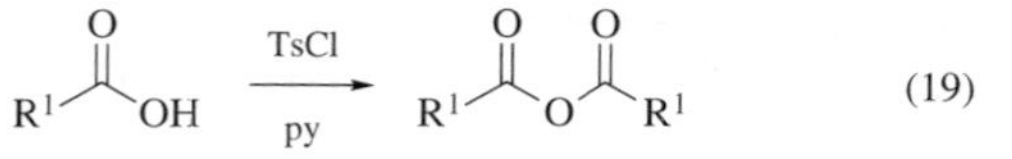

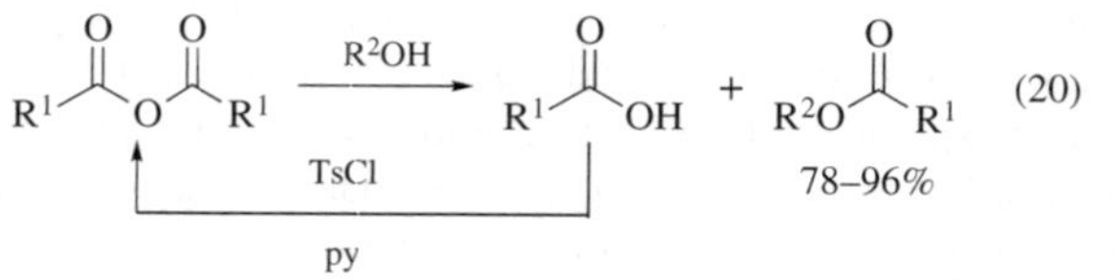

(19)

(20)

Similarly, a two-step procedure employing treatment of a mixture of tosic acid and various amino acids with alcoholic tosyl chloride results in the isolation of the esters of the amino acids as their *p*-toluenesulfonate salts in excellent yield. The tosic acid used in the esterification is added to make the amino acids more soluble and to prevent *N*-tosylation (eq 21).[8c]

$$\text{ROH} \xrightarrow{\text{TsCl}} [\text{ROTs}] \xrightarrow{\text{TsOH}\cdot\text{H}_2\text{NCHR'CO}_2\text{H}} \text{TsOH}\cdot\text{H}_2\text{NCHR'CO}_2\text{R} \quad (21)$$

a. ethyl alcohol; b. benzyl alcohol; amino acids: Gly, Leu, Tyr, Trp; a. 90–97%; b. 81–85%

A novel formal decarboxylation of the amino acid α-anilino-α,α-diphenylacetic acid has been observed upon treatment with tosyl chloride in pyridine.[9] It has been proposed that a mixed anhydride of p-toluenesulfonic acid has undergone an elimination via N-deprotonation and synchronous extrusion of carbon monoxide in this reaction. Interestingly, no N-tosylation occurs (eq 22).

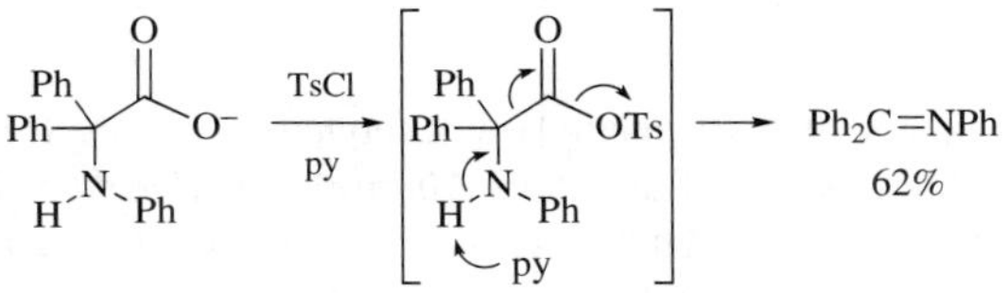

$$\text{Ph}_2\text{C(NHPh)CO}_2^- \xrightarrow[\text{py}]{\text{TsCl}} [\text{Ph}_2\text{C(NHPh)COOTs}] \longrightarrow \text{Ph}_2\text{C=NPh} \quad (22)$$

62%

Spontaneous Beckmann rearrangement has been observed upon O-tosylation of oximes. Lactams can therefore be conveniently prepared from cyclic oximes, as shown in eq 23.[10a] The rearrangement has long been known to proceed with retention of configuration of the migrating group.[10b,c] This is complementary to the reaction of oximes with ***Sulfuric Acid*** (eq 24).[10b]

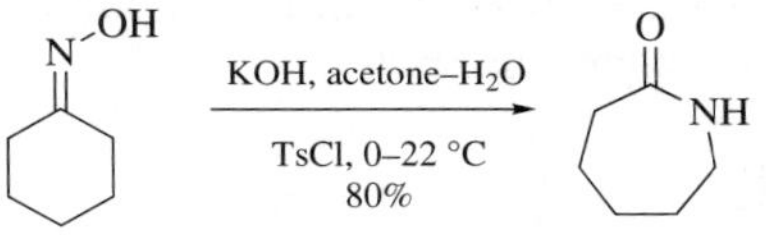

(23) cyclohexanone oxime → caprolactam; KOH, acetone–H_2O; TsCl, 0–22 °C; 80%

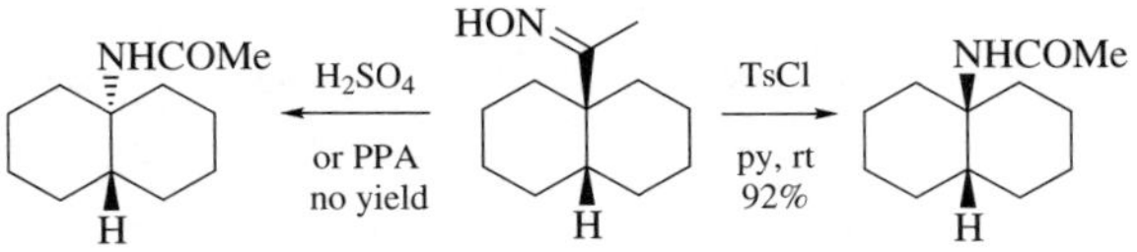

(24) H_2SO_4 or PPA, no yield; TsCl, py, rt, 92%; NHCOMe; HON; H

A Beckmann fragmentation of oximes using tosyl chloride in a basic ethanol–water system has also been observed.[11] Formation of the cyclic product shown in eq 25 was consistent with a base-induced opening of an intermediate lactone followed by rearrangement and incipient extrusion of benzoic acid.

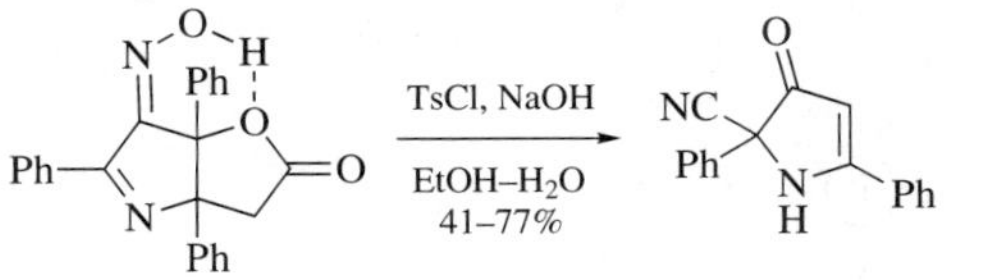

(25) TsCl, NaOH; EtOH–H_2O; 41–77%

Oximes which are treated with tosyl chloride can also be used as substrates in the Neber rearrangement to α-amino ketones.[12a] The mixture shown in eq 26 was further subjected to reductive amination to yield 14% of the CNS-active *cis-N,N*-dimethylated α-amino alcohol.[12b]

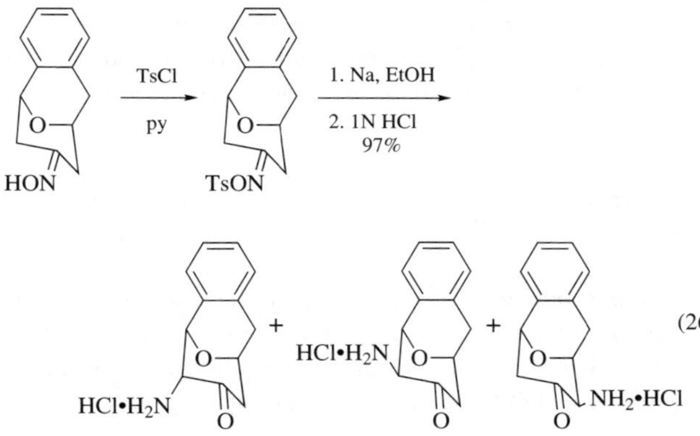

(26) HON → TsON: TsCl, py; 1. Na, EtOH; 2. 1N HCl; 97%

$HCl\cdot H_2N$ + $HCl\cdot H_2N$ + $NH_2\cdot HCl$

2:1:1 inseperable mixture,

Rearrangements of hydroxamic acids using sulfonyl chlorides have been accomplished, albeit without reported yields, the net result being a unique variation of the Lossen rearrangement (eq 27).[13]

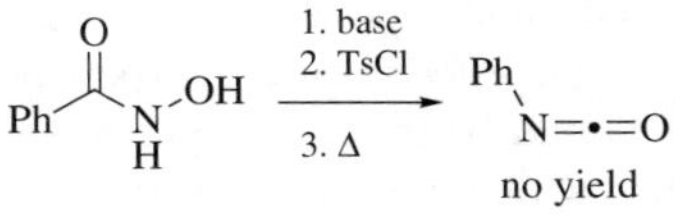

$$\text{PhC(O)NHOH} \xrightarrow[\text{3. }\Delta]{\text{1. base; 2. TsCl}} \text{PhN=C=O} \quad (27)$$

no yield

As an alternative to the Beckmann rearrangement, ketonic nitrones can be treated with tosyl chloride in pyridine in the presence of water, as illustrated in eq 28.[14]

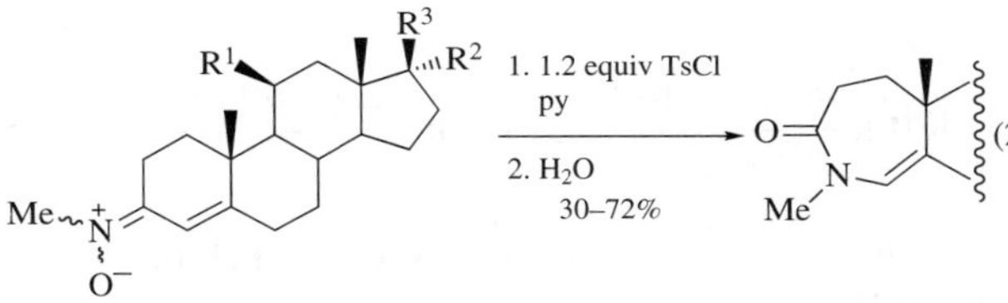

(28) 1. 1.2 equiv TsCl, py; 2. H_2O; 30–72%

Rearrangement reactions which utilize tosyl chloride seem to progress in a two-step process: the displacement of chlorine from the sulfonyl halide resulting in the formation of an oxygen–sulfur bond, followed by the migration, elimination, or intramolecular displacement of the sulfonylate anion.[8a] When the series of carbinolamines depicted in eq 29 were treated with tosyl chloride, they decomposed into the corresponding electron-deficient nitrogen species, which subsequently triggered ring enlargement to the β-lactams in moderate yields.[15]

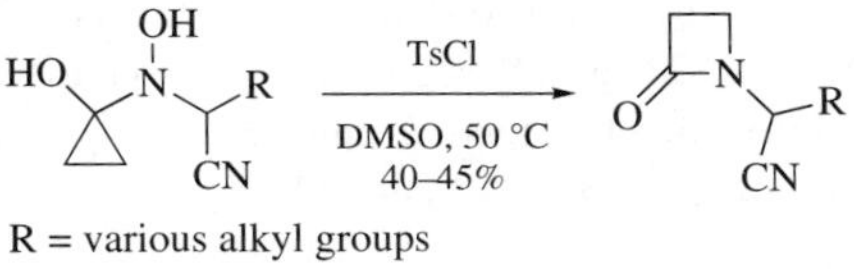

(29) TsCl; DMSO, 50 °C; 40–45%

R = various alkyl groups

N-Tosylation is a facile procedure. For inexpensive amines a useful procedure entails treating 2 equiv of the amine with tosyl chloride. This is the first step in the convenient preparation of Diazald (***N-Methyl-N-nitroso-p-toluenesulfonamide***), a ***Diazomethane*** precursor (eq 30).[16a]

$$2\ \text{MeNH}_2 \xrightarrow{\text{TsCl}} \text{TsNHMe} \xrightarrow{\text{HNO}_3} \text{TsN(NO)Me} \quad (30)$$

Diazald

Protection of 3-amino-3-pyrazoline sulfate, a key intermediate in the preparation of 3(5)-aminopyrazole, was achieved in the presence of an excess of sodium bicarbonate in good yields (eq 31).[16b]

$$\text{3-amino-3-pyrazoline}\cdot H_2SO_4 \xrightarrow[58\text{–}75\%]{\text{TsCl, 2.5 equiv } NaHCO_3} \text{N-Ts product} \quad (31)$$

Sodium Carbonate can also be used as the base in the tosylation of amines, as shown in the reaction of anthranilic acid with tosyl chloride. There was no competing nucleophilic attack at sulfur by the resonance-stabilized carboxylate group (eq 32).[16c]

$$\text{2-}H_2N\text{-}C_6H_4\text{-}CO_2H \xrightarrow[88\text{–}91\%]{\text{TsCl, 2.4 equiv } Na_2CO_3} \text{2-TsHN-}C_6H_4\text{-}CO_2Na \quad (32)$$

Primary, benzyl, and unhindered secondary amines can be ditosylated and deaminated in good to excellent yields, using ***Sodium Borohydride*** as the nucleophile, to afford the corresponding alkanes; only highly congested substrates experience competing attack at nitrogen.[17] Similarly, primary aliphatic amines, when ditosylated and treated with iodide ion in DMF at 90–120 °C yielded, as their major products, the corresponding alkyl iodides, with some competition arising from elimination reactions (eqs 33 and 34).[18]

$$H_2NCH_2R \xrightarrow[\text{NaOH}]{\text{TsCl, DMF–}H_2O} \xrightarrow{\text{TsCl, DMF, NaH}} Ts_2NCH_2R \quad (33)$$

$$ICH_2R \xleftarrow[110\text{–}120\ °C]{\text{2 equiv KI, DMF}} Ts_2NCH_2R \xrightarrow[175\ °C]{\text{2 equiv } NaBH_4} MeR \quad (34)$$

Tosyl chloride has been used in the protection of the imidazole residue of N^{α}-acylhistidine peptides[19a,b] and the guanidino residue of arginine peptides.[19c] The resulting nitrogen-protecting group is stable to most conditions but is easily removable in anhydrous hydrogen fluoride at 0 °C.

The tosylation of carbon can be accomplished using electron transfer conditions. Treatment of styrene[20a] and analogs[20b] with ***Copper(II) Chloride*** and tosyl chloride or ***Benzenesulfonyl Chloride*** results in a formal replacement of the vinyl proton by the sulfonyl moiety (eq 35). The intermediacy of a *trans*-β-chloro sulfone has been demonstrated by ^{1}H NMR. Treatment with base induced the elimination of HCl. A variety of other sulfonyl transfer reagents can be employed in the synthesis of isolated β-chloro sulfones, with good results (60–97% yield) for a variety of alkenes (ethylene, 1-butene, 2-butene, 1-octene, acrylonitrile, methyl acrylate, and 1,3-butadiene).[20a]

$$\text{4-phenyl-1,2-dihydronaphthalene} \xrightarrow[\text{2. } Et_3N,\ C_6H_6;\ 45\%]{\text{1. TsCl, } CuCl_2} \text{3-Ts product} \quad (35)$$

Although it was found that tosyl chloride does not react with 3-sulfolene at temperatures below 110 °C, upon further warming to 140 °C addition to the diene afforded 1-chloro-4-tosyl-2-butene in a convenient procedure that did not require the use of a sealed tube or bomb (eq 36).[20c] Cycloheptatriene, 1,4-norbornadiene, and phenylacetylene are a few examples of a variety of compounds which are reactive substrates in this protocol.

$$\text{3-sulfolene} + \text{TsCl} \xrightarrow[140\ °C,\ 67\%]{CuCl_2} Ts\text{–}CH_2CH{=}CHCH_2\text{–}Cl \quad (36)$$

Radical addition of tosyl chloride to norbornene and aldrin without skeletal rearrangement can be achieved using ***Dibenzoyl Peroxide*** as an initiator (eq 37).[20a,d] Reaction with 1,4-norbornadiene gives rise to a rearranged addition product.

$$\text{norbornene} + \text{TsCl} \xrightarrow[100\ °C,\ 64\%]{\text{dibenzoyl peroxide}} \text{2-}SO_2C_6H_4Me\text{-3-Cl-norbornane} \quad (37)$$

Treatment of bis(trimethylsilyl)acetylene with tosyl chloride and a slight excess of anydrous ***Aluminum Chloride*** affords *p*-tolyl (trimethylsilyl)ethynyl sulfone, a precursor to the reactive Michael acceptor ethynyl *p*-tolyl sulfone (eq 38).[21]

$$\text{TsCl} + \text{TMS–≡–TMS} \xrightarrow[CH_2Cl_2,\ 80\%]{AlCl_3} \text{Ts–≡–TMS} \quad (38)$$

The dehydration of ureas employing tosyl chloride provides access to substituted carbodiimides, as depicted in eq 39. The urea made from treatment of ethyl isocyanate with *N,N*-dimethyl-1,3-propanediamine was treated in situ with 1.1 equiv of tosyl chloride and a large excess of ***Triethylamine*** to afford the corresponding carbodiimide in high yield.[22a,b]

$$EtNHC(O)NH(CH_2)_3NMe_2 \xrightarrow[CH_2Cl_2,\ \text{reflux};\ 88\text{–}97\%]{Et_3N,\ \text{TsCl}} EtN{=}C{=}N(CH_2)_3NMe_2 \quad (39)$$

Various ureas upon treatment with tosyl or benzenesulfonyl chloride in the presence of a phase transfer catalyst, benzyltriethylammonium chloride, results in moderate to excellent yields of carbodiimides (eq 40).[22c] The polymeric carbodiimide in eq 41 offers the advantages of cleaner workup and recyclability if used to prepare aldehydes under Moffatt oxidation conditions.[22d]

$$R^1NHC(O)NHR^2 \xrightarrow[\text{various solvents; } 30\text{–}98\%]{\text{TsCl, } K_2CO_3,\ BnNEt_3Cl} R^1N{=}C{=}NR^2 \quad (40)$$

$$\text{Ⓟ–}C_6H_4CH_2NHC(O)NH\text{-}i\text{-Pr} \xrightarrow[CH_2Cl_2,\ \text{reflux; no yield}]{Et_3N,\ \text{TsCl}} \text{Ⓟ–}C_6H_4CH_2N{=}C{=}N\text{-}i\text{-Pr} \quad (41)$$

Ⓟ = styrene–divinylbenzene copolymer

Isocyanides can be made from treatment of formamides with tosyl chloride in pyridine, as illustrated in eq 42.[23a,b] Similarly, nitriles can be synthesized in moderate to good yields by dehydration of primary amides using tosyl chloride in pyridine[23a,b] or quinoline (eq 43).[23c]

$$\mathrm{RC(O)NH_2} \xrightarrow[\text{30–82\%}]{\text{TsCl, py}} \mathrm{R{-}CN} \qquad (42)$$

$$\mathrm{RNHCHO} \xrightarrow[\text{2 equiv py or 4 equiv quinoline, 50–74\%}]{\text{1.5–2 equiv TsCl}} \mathrm{R{-}NC} \qquad (43)$$

1. *FF* **1967**, *1*, 1179; *FF* **1972**, *3*, 292; *FF* **1974**, *4*, 510; *FF* **1975**, *5*, 676; *FF* **1977**, *6*, 598; *FF* **1980**, *8*, 489; *FF* **1981**, *9*, 472; *FF* **1984**, *11*, 536; *FF* **1988**, *13*, 313.
2. (a) Kabalka, G. W.; Varma, M.; Varma, R. S. *JOC* **1986**, *51*, 2386. (b) Marvel, C. S.; Sekera, V. C. *OSC* **1955**, *3*, 366. (c) Szeja, W. *S* **1979**, 822. (d) Johnson, W. S.; Collins, J. C., Jr.; Pappo, R.; Rubin, M. B.; Kropp, P. J.; Johns, W. F.; Pike, J. E.; Bartmann, W. *JACS* **1963**, *85*, 1409. (e) Tsuda, Y.; Nishimura, M.; Kobayashi, T.; Sato, Y.; Kanemitsu, K. *CPB* **1991**, *39*, 2883. (f) Lundt, I.; Madsen, R. *S* **1992**, 1129. (g) Yamamura, H.; Fujita, K. *CPB* **1991**, *39*, 2505. (h) Ashton, P. R.; Ellwood, P.; Staton, I.; Stoddart, J. F. *JOC* **1991**, *56*, 7274. (i) Fujita, K. E.; Ohta, K.; Masunari, K.; Obe, K.; Yamamura, H. *TL* **1992**, *33*, 5519. (j) Yamamura, H.; Kawase, Y.; Kawai, M.; Butsugan, Y. *BCJ* **1993**, *66*, 585. (k) Kurita, K. *CI(L)* **1974**, 345. (l) Stork, G.; Grieco, P. A.; Gregson, M. *OSC* **1988**, *6*, 638.
3. (a) Hwang, C. K.; Li, W. S.; Nicolaou, K. C. *TL* **1984**, *25*, 2295. (b) Revelli, G. A.; Gros, E. G. *SC* **1993**, *23*, 1111. (c) Hojo, M.; Yoshita, Z.-i. *JACS* **1968**, *90*, 4496.
4. (a) For a general discussion of S_N2 displacement reactions see: Carey, F. A.; Sundberg, R. J. *Advanced Organic Chemistry Part A: Structure and Mechanisms*, 3rd ed.; Plenum: New York, 1990; pp 261–264. (b) For a general discussion of leaving group effects see: *ibid.*, pp 290–293.
5. (a) Brewster, J. H.; Kucera, C. H. *JACS* **1955**, *77*, 4564. (b) Eberle, M. K.; Nuninger, F. *JOC* **1993**, *58*, 673.
6. (a) Holand, S.; Epsztein, R. *S* **1977**, 706. (b) Murthy, V. S.; Gaitonde, A. S.; Rao, S. P. *SC* **1993**, 285. (c) Sjeja, W. *S* **1985**, 983.
7. Picard, P.; Leclercq, D.; Bats, J.-P.; Moulines, J. *S* **1981**, 550.
8. (a) Brewster, J. H.; Ciotti, C. J., Jr. *JACS* **1955**, *77*, 6214. (b) Hennion, G. F.; Barrett, S. O. *JACS* **1957**, *79*, 2146. (c) Arai, I.; Muramatsu, I. *JOC* **1983**, *48*, 121.
9. Sheehan, J. C.; Frankenfeld, J. W. *JOC* **1962**, *27*, 628.
10. (a) Oxley, P.; Short, W. F. *JCS* **1948**, 1514. (b) Hill, R. K.; Chortyk, O. T. *JACS* **1962**, *84*, 1064. (c) Wheland, G. W. *Advanced Organic Chemistry*, 3rd ed., Wiley: New York, 1960; pp 597–610.
11. Mataka, S.; Suzuki, H.; Sawada, T.; Tashiro, M. *BCJ* **1993**, *66*, 1301.
12. (a) O'Brien, C. *CRV* **1964**, *64*, 81. (b) Wünsch, B.; Zott, M.; Höfner, G. *LA* **1992**, 1225.
13. Hurd, C. D.; Bauer, L. *JACS* **1954**, *76*, 2791.
14. (a) Barton, D. H. R.; Day, M. J.; Hesse, R. H.; Pechet, M. M. *JCS(P1)* **1975**, 1764. (b) Barton, D. H. R.; Day, M. J.; Hesse, R. H.; Pechet, M. M. *CC* **1971**, 945.
15. Wasserman, H. H.; Glazer, E. A.; Hearn, M. J. *TL* **1973**, 4855.
16. (a) De Boer, T. J.; Backer, H. J. *OSC* **1963**, *4*, 943. (b) Dorn, H.; Zubek, A. *OSC* **1973**, *5*, 39. (c) Scheifele, H. J., Jr.; De Tar, D. F. *OSC* **1963**, *4*, 35.
17. Hutchins, R. O.; Cistone, F.; Goldsmith, B; Heuman, P. *JOC* **1975**, *40*, 2018.
18. DeChristopher, P. J.; Adamek, J. P.; Lyon, G. D.; Galante, J. J.; Haffner, H. E.; Boggio, R. J.; Baumgarten, R. J. *JACS* **1969**, *91*, 2384.
19. (a) Sakakibari, S.; Fujii, T. *BCJ* **1969**, *42*, 1466. (b) Fujii, T.; Sakakibari, S. *BCJ* **1974**, *47*, 3146. (c) Mazur, R. H.; Plume, G. *E* **1968**, *24*, 661.
20. (a) Asscher, M.; Vofsi, D. *JCS* **1964**, 4962. (b) Truce, W. E.; Goralski, C. T. *JOC* **1970**, *35*, 4220. (c) Truce, W. E.; Goralski, C. T.; Christensen, L. W.; Bavry, R. H. *JOC* **1970**, *35*, 4217. (d) Cristol, S. J.; Reeder, J. A. *JOC* **1961**, *26*, 2182.
21. Waykole, L.; Paquette, L. A. *OS* **1989**, *67*, 149.
22. (a) Sheehan, J. C.; Cruickshank, P. A. *OSC* **1973**, *5*, 555. (b) Sheehan, J. C.; Cruickshank, P. A.; Boshart, G. L. *JOC* **1961**, *26*, 2525. (c) Jászay, Z, M.; Petneházy, I.; Töke, L.; Szajáni, B. *S* **1987**, 520. (d) Weinshenker, N. M.; Shen, C. M.; Wong, J. Y. *OSC* **1988**, *6*, 951.
23. (a) Stephens, C. R.; Bianco, E. J.; Pilgrim, F. J. *JACS* **1955**, *77*, 1701. (b) Stephens, C. R.; Conover, L. H.; Pasternack, R.; Hochstein, F. A.; Moreland, W. T.; Regna, P. P.; Pilgrim, F. J.; Pilgrim, F. J.; Brunings, K. J.; Woodward, R. B. *JACS* **1954**, *76*, 3568.
24. (a) Hertler, W. R.; Corey, E. J. *JOC* **1958**, *23*, 1221. (b) Corey, E. J.; Hertler, W. R. *JACS* **1959**, *81*, 5209. (c) Schuster, R. E.; Scott, J. E.; Casanova, J., Jr. *OSC* **1973**, *5*, 772.
25. Pelletier, S. W. *CI(L)* **1953**, 1034.
26. Perrin, D. D.; Armarego, W. L. F. *Purification of Laboratory Chemicals*, 3rd ed.; Pergamon: Oxford, 1988; p 291.

D. Todd Whitaker, K. Sinclair Whitaker & Carl R. Johnson
Wayne State University, Detroit, MI, USA

1,2,4-Triazole

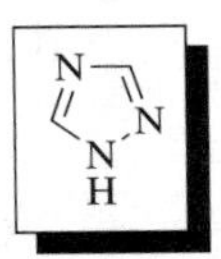

[288-88-0] $C_2H_3N_3$ (MW 69.08)

(transacylating agent used for ester and amide synthesis, especially in peptide synthesis; cyclization reactions; oligonucleotide synthesis)

Physical Data: mp 120–121 °C; bp 260 °C (dec above 187 °C); fp 140 °C.
Solubility: sol water, alcohol.
Form Supplied in: white powder.

Acylating Reagent. The great ability of triazole to accept and transfer acyl groups has made it a catalyst for the synthesis of esters and amides. In particular, triazole has been used in peptide synthesis, for the formation of peptide links as well as for the introduction of amino-protecting groups. It acts as a bifunctional catalyst, accelerating the aminolysis of *p*-nitrophenyl esters[1] and *p*-thiocresyl esters.[2] The catalytic effect is both by a general base mechanism (eq 1) and a specific mechanism (eq 2).

$$\mathrm{R{-}C(=O)OR'} \leftarrow \mathrm{:N(R^1)(R^2){-}H} \leftarrow \mathrm{:N\ (1H\text{-}1,2,4\text{-triazole})} \qquad (1)$$

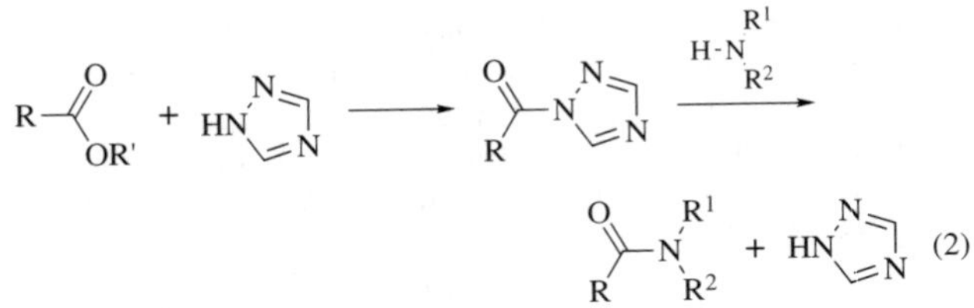

(2)

(1) (2)

It has been used to introduce the Boc group onto an amino acid from *t*-butyl carbonates (eq 3), and it allows the coupling of an *N*-protected amino acid with a fully unprotected amino acid (eq 4). Interestingly, racemization is not observed during coupling of amino acids in the presence of 1,2,4-triazole.[1] Nevertheless, its coupling efficiency depends on the reaction medium and in solvents like DMF, which increase the rate of aminolysis, its usefulness is limited. 1,2,4-Triazole is now seldom used in peptide synthesis, whereas other catalytic additives like ***1-Hydroxybenzotriazole*** are widely used as activated esters as well as in conjunction with coupling reagents like ***1,3-Dicyclohexylcarbodiimide***.

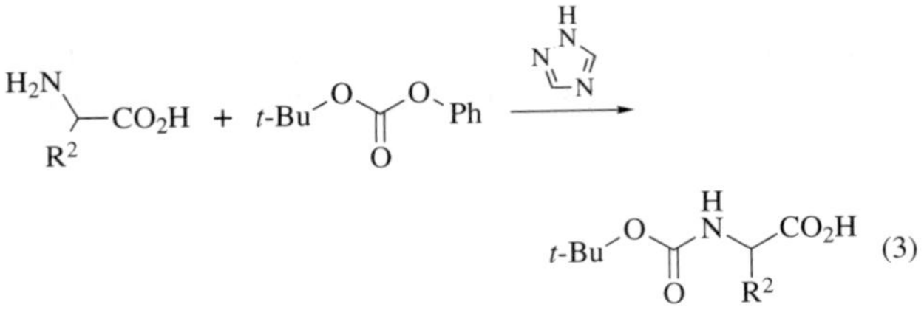

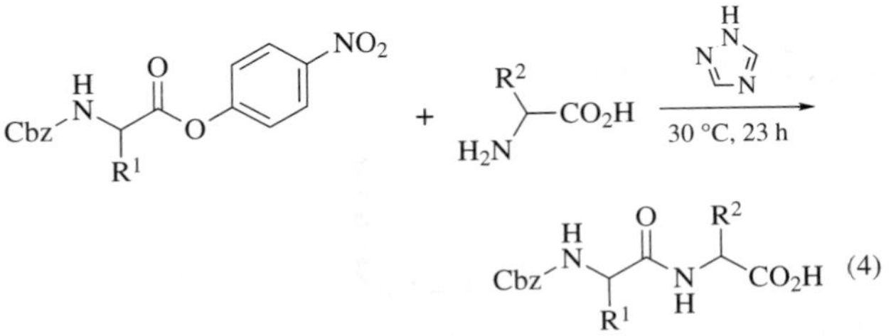

(3)

30 °C, 23 h

(4)

Catalysis of Ring Formation. 1,2,4-Triazole can induce the formation of isoxazole rings from conjugated ketones and hydroxylamine (eq 5).[3]

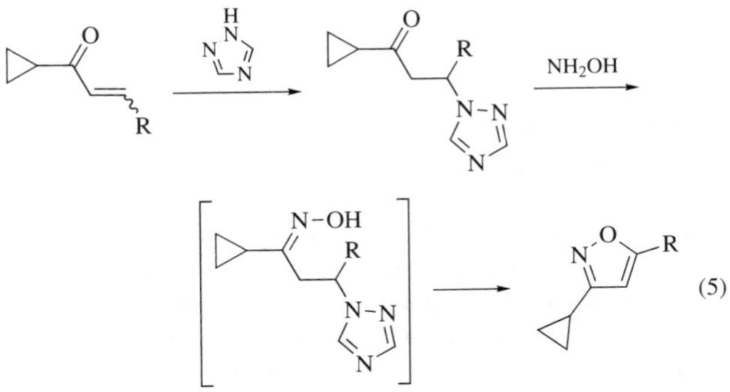

NH_2OH

(5)

Oligonucleotide Synthesis. Sulfonyl derivatives of 1,2,4-triazole, e.g. (**1**), are used as efficient coupling reagents in oligonucleotide synthesis by the phosphotriester method.[4] A nitro derivative (**2**) allows esterification of amino acids onto hydroxymethyl polymers.[5]

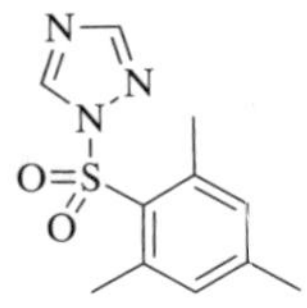

1. Beyerman, H. C.; Massen van den Brink, W.; Weygand, F.; Prox, A.; König, W.; Scmidhammer, L.; Nintz, E. *RTC* **1965**, *84*, 213.
2. Wieland, T.; Kahle, W. *LA* **1966**, *691*, 212.
3. Rentzea, C. N. *AG(E)* **1981**, *20*, 885.
4. Reese, C. B.; Titmas, R. C.; Yau, L. *TL* **1978**, 2727.
5. Blankemeyer-Menge, B.; Nintz, M.; Frank, R. *TL* **1990**, *31*, 1701.

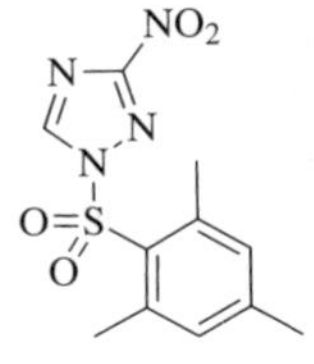

Jean-Claude Gesquière
Institut Pasteur, Lille, France

Trichloroacetonitrile

Cl_3CCN

[545-06-2] C_2Cl_3N (MW 144.38)

(makes trichloroacetimidates from alcohols; trichloroacetimidates are used to introduce nitrogen into molecules via rearrangements[1c] and cyclizations; trichloroacetimidates are useful alkylating agents[1d,e])

Physical Data: mp −42 °C; bp 83–84 °C; *d* 1.440 g cm^{-3}.
Solubility: sol most organic solvents.
Form Supplied in: neat colorless liquid.
Analysis of Reagent Purity: ^{1}H NMR, ^{13}C NMR.
Handling, Storage, and Precautions: toxic lachrymator; use only in a fume hood. Reagent can be absorbed through the skin. Always wear gloves when handling this reagent.

Preparation of Trichloroacetimidates. The imidates derived from the addition of alcohols to trichloroacetonitrile have become important and versatile intermediates in synthetic chemistry.[1] Consequently the principal synthetic use of trichloroacetonitrile has been in the formation of these useful trichloroacetimidate intermediates. The imidates are most often prepared by simple addition of a sodium or potassium alkoxide to the electron-deficient trichloroacetonitrile (eq 1).[1c,d,2] In certain cases, slight modifications of the above procedure are utilized.[2d,3] The product imidates are isolable and, despite their propensity to rearrange,[2b,e,4] can in some instances be purified by distillation[2a,d,e] or chromatography. Typically the addition reaction is sufficiently clean to use the imidates without further purification.[2f]

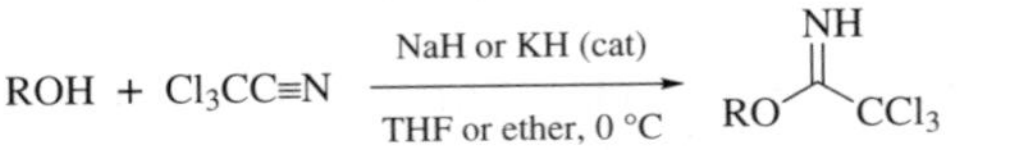

$$ROH + Cl_3CC{\equiv}N \xrightarrow[\text{THF or ether, 0 °C}]{\text{NaH or KH (cat)}} RO\text{–}C(=NH)\text{–}CCl_3 \quad (1)$$

Rearrangement of Trichloroacetimidates. The thermal rearrangement of imidates has been known for many years.[2b,4] In 1974, Overman[2c] described a useful conversion of allylic alcohols to allylic amines utilizing a [3,3]-sigmatropic rearrangement of allylic trichloroacetimidates (eq 2).[1c,2e] In addition to simple thermolysis it was found that the rearrangement could be catalyzed at room temperature with either Hg^{II} or Pd^{II} salts.[1c,5] As with many sigmatropic rearrangements, the reaction is stereoselective and produces double bonds of defined stereochemistry while efficiently transferring chirality.[1c,5,6]

xylene, reflux, 8 h; 74%; $NHCOCCl_3$ (2)

A comparison of reaction conditions[5b,c,d] has shown that palladium catalysis is particularly effective in achieving complete chiral transfer (eq 3).[5b] Propargylic imidates also rearrange when heated in refluxing xylene.[1c,7] The initially formed allene undergoes a series of tautomerizations so that amino 1,3-dienes are the ultimate products (eq 4).[1c] These dienes have found use in the Diels–Alder reaction.[1c]

NaH, ether; then CCl_3CN

$PdCl_2(PhCN)_2$, THF, 25 °C; 84% overall (3)

xylene, reflux, 5 h; heat; $NHCOCCl_3$ (4)

Electrophilic Cyclization of Trichloroacetimidates. Trichloroacetimidates derived from allylic[8a,c–f] and homoallylic[8b,c] alcohols undergo electrocyclic ring closure when treated with a source of I^+ (eq 5).[8c] Cyclization can also be triggered via the Lewis acid-mediated opening of epoxides.[9] In at least one case the imidate proved more reactive than related reactions using carbamates.[9c] These are useful methods for the stereoselective introduction of nitrogen into cyclic and acyclic systems.

$I(collidine)_2ClO_4$; 75% (5)

Alkylation using Trichloroacetimidates. Under mildly acidic conditions, alkyl trichloroacetimidates act as reasonable alkylating agents[2f,10] toward heteroatom nucleophiles, most notably alcohols (eq 6). Alkyl groups which have been transferred include benzyl,[2f,10a,b,d] allyl,[2f] propargyl,[10e] and *t*-butyl.[10c] This methodology has been used to protect alcohols. The use of glycosyl trichloroacetimidates has found widespread use in the synthesis of oligosaccharides and glycoconjugates (eq 7).[1d,e,11] This is a particularly powerful method for controlling stereochemistry at the anomeric position.

$$RO-C(=NH)-CCl_3 + R^1OH \xrightarrow{CF_3SO_3H} R^1OR + Cl_3C-C(=O)-NH_2 \quad (6)$$

$R = ArCH_2$, $CH_2{=}CHCH_2$, $MeC{\equiv}CCH_2$, *t*-Bu

ROH, $BF_3{\bullet}Et_2O$, CH_2Cl_2 (7)

1. (a) Sandler, S. R.; Karo, W. *Organic Functional Group Preparations*; Academic: New York, 1972; Vol. 3, Chapter 8. (b) Patai, S. *The Chemistry of Amidines and Imidates*; Wiley: New York, 1975. (c) Overman, L. E. *ACR* **1980**, *13*, 218. (d) Schmidt, R. R. *AG(E)* **1986**, *25*, 212. (e) Schmidt, R. R. *PAC* **1989**, *61*, 1257.
2. (a) Cramer, F.; Pawelzik, K.; Baldauf, H. J. *CB* **1958**, *91*, 1049. (b) Cramer, F.; Hennrich, N. *CB* **1961**, *94*, 976. (c) Overman, L. E. *JACS* **1974**, *96*, 597. (d) Overman, L. E. *JACS* **1976**, *98*, 2901. (e) Clizbe, L. A.; Overman, L. E. *OS* **1978**, *58*, 4. (f) Wessel, H.-P.; Iverson, T.; Bundle, D. R. *JCS(P1)* **1985**, 2247.
3. (a) Hauser, F. M.; Ellenberger, S. R.; Glusker, J. P.; Smart, C. J.; Carrell, H. L. *JOC* **1986**, *51*, 50. (b) Oehler, E.; Kotzinger, S. *S* **1993**, 497.
4. (a) Mumm, O.; Möller, F. *CB* **1937**, *70*, 2214. (b) McCarty, C. G.; Garner, L. A. In *The Chemistry of Amidines*; Patai, S., Ed.; Wiley: New York, 1975; Chapter 4.
5. (a) Overman, L. E. *AG(E)* **1984**, *23*, 579. (b) Metz, P.; Mues, C.; Schoop, A. *T* **1992**, *48*, 1071. (c) Mehmandoust, M.; Petit, Y.; Larcheveque, M. *TL* **1992**, *33*, 4313. (d) Doherty, A. M.; Kornberg, B. E.; Reily, M. D. *JOC* **1993**, *58*, 795.
6. (a) Yamamoto, Y.; Shimoda, H.; Oda, J.; Inouye, Y. *BCJ* **1976**, *49*, 3247. (b) Isobe, M.; Fukuda, Y.; Nishikawa, T.; Chabert, P.; Kawai, T.; Goto, T. *TL* **1990**, *31*, 3327.
7. Overman, L. E.; Clizbe, L. A. *JACS* **1976**, *98*, 2352.
8. (a) Cardillo, G.; Orena, M.; Porzi, G.; Sandri, S. *JCS(C)* **1982**, 1308. (b) Cardillo, G.; Orena, M.; Porzi, G.; Sandri, S. *JCS(C)* **1982**, 1308. (c) Fraser-Reid, B.; Pauls, H. W. *JOC* **1983**, *48*, 1392. (d) Bongini, A.; Cardillo, G.; Orena, M.; Sandri, S.; Tomasini, C. *JOC* **1986**, *51*, 4905. (e) Cardillo, G.; Orena, M.; Sandri, S.; Tomasini, C. *T* **1986**, *42*, 917. (f) Sammes, P. G.; Thetford, D. *JCS(P1)* **1988**, 111.
9. (a) Jacobsen, S. *ACS* **1988**, *B42*, 605. (b) Schuerrle, K.; Beier, B.; Piepersberg, W. *JCS(P1)* **1991**, 2407. (c) Hart, T. W.; Vacher, B. *TL* **1992**, *33*, 3009.
10. (a) Amouroux, R.; Gerin, B.; Chastrette, M. *T* **1985**, *41*, 5321. (b) Widmer, U. *S* **1987**, 568 (c) Armstrong, A.; Brackenridge, I.; Jackson, R. F. W.; Kirk, J. M. *TL* **1988**, *29*, 2483. (d) Audia, J. E.; Boisvert, L.; Patten, A. D.; Villalobos, A.; Danishefsky, S. J. *JOC* **1989**, *54*, 3738, (e) Wei, S. Y.; Tomooka, K.; Nakai, T. *T* **1993**, *49*, 1025. (f) Bourgeois, M. J.; Montaudon, E.; Maillard, B. *T* **1993**, *49*, 2477.
11. (a) Nicolaou, K. C.; Daines, R. A.; Ogawa, Y.; Chakraborty, T. K. *JACS* **1988**, *110*, 4696. (b) Barrett, A. G. M.; Pilipauskas, D. *JOC* **1991**, *56*, 2787.

Patrick G. McDougal
Reed College, Portland, OR, USA

2,4,6-Trichlorobenzoyl Chloride[1]

[4136-95-2] $C_7H_2Cl_4O$ (MW 243.89)

(strong acylating agent; used with DMAP in the formation of esters,[1] thioesters,[2] amides,[3] and, particularly, small- and medium-ring lactones[1,4])

Physical Data: bp 107–108 °C/6 mmHg; *d* 1.561 g cm^{-3}.
Solubility: sol most organic solvents.
Form Supplied in: liquid; commercially available.
Purification: redistill under reduced pressure.[5]
Preparative Methods: first prepared by Yamaguchi from 2,4,6-trichloroaniline,[7] but Seebach has reported a simpler synthesis from 1,3,5-trichlorobenzene (**1**) (eq 1).[8]

(1) — BuLi, CO_2 → SO_2Cl_2 → (1)

Handling, Storage, and Precautions: irritant and moisture sensitive. Incompatible with strong bases (see Aldrich safety index for ***Benzoyl Chloride***).[6]

Carboxylic Acid Derivatives. 2,4,6-Trichlorobenzoyl chloride was designed to react rapidly with carboxylic acids to form mixed anhydrides (**2**),[9] which would then quickly react with alcohols, amines, and thiols to give esters (**3**), amides (**4**), and thioesters (**5**) in high yields (eq 2).

+ RCO_2H —Et_3N, THF→ (**2**) + $Et_3N{\bullet}HCl$; (**2**) —R^1OH or R^2R^3NH or R^4SH, DMAP, PhH→ (**6**) + RCO_2R^1 or $RCONR^2R^3$ or $RCOSR^4$ (2)
(**3**) 95–100% (**4**) ~90% (**5**) 80–90%

The trichloro-substituted phenyl ring serves two purposes: to prevent any side reaction occurring, due to attack of the alcohol at the carbonyl adjacent to the aromatic ring, and to make the carboxylate anion (**6**) a good leaving group, enhancing the rate of reaction.

Lactone Formation. Although many methods exist for lactonizing α,ω-hydroxy acids,[10] 2,4,6-trichlorobenzoyl chloride remains one of the most powerful and has been widely used in the synthesis of naturally occurring macrolides.[11] Yamaguchi first showed how 9-, 12-, and 13-membered lactones could be synthesized using high dilution techniques (Table 1)[1] and went on to demonstrate the potential of his reagents by using them to synthesize methynolide and (±)-brefeldin A (**7**) (eq 3), 12- and 13-membered lactones, respectively.[7,12]

Table 1 Synthesis of Lactones[1]

Ring size of lactone formed	DMAP (equiv)	Time of addition (h)	Yield of monomer (%)	Yield of dimer (%)
9	3	8	36	23
10[13]	6	20	33	0
12	6	5	48	20
13	6	1.5	67	10

MEMO, OTHP, CO_2H, OH —$Cl_3C_6H_2COCl$, Et_3N, DMAP→ MEMO, HO —$TiCl_4$→ HO, HO (**7**) (3)

Other groups have also used the Yamaguchi lactonization procedure to synthesize natural products, most notably Seebach in the synthesis of (+)-myscovirescine M_2, a 27-membered lactone, in which the seco acid was converted to the lactone in 83% yield.[14] Symmetrical diolides (**8**) have been prepared from unprotected seco acids in fair yields (eq 4)[15] when other lactonization procedures failed.[9] Unsymmetrical diolides require protected seco acids.[16]

Ph_3CO, OH, O, OH —$Cl_3C_6H_2COCl$, Et_3N, DMAP, 49%→ (**8**) $OCPh_3$, Ph_3CO (4)

More recently, treatment of 3-hydroxybutanoic acid (**9**) under the Yamaguchi lactonization conditions of high dilution gave, in equal proportions, macropentolides, macrohexolides, and macroheptolides in good yields (eq 5).[17]

The Shanzer lactonization method produces a lower overall yield of macrolides, but an increased ratio of the higher homologs.[18]

Effect on Stereochemistry. The Yamaguchi esterification is also often used in preference to other methods because of its lack of racemization of stereogenic centers. Kunz used it to esterify an *N*-Boc protected aspartate (**10**)[19] when other methods have been known to racemize the chiral center (eq 6).[20]

The Yamaguchi lactonization also leaves intact any stereochemistry at the carbon bearing the hydroxyl group. This is in contrast to the Mitsunobu lactonization, which inverts that stereochemistry (eq 7).[21]

Furthermore, when no base is present during esterification, double bonds are also little affected, as Greene showed when converting a variety of alcohols to angelate esters (**12**) in quantitative yield; standard esterification procedures gave a mixture of angelate and tiglate esters (eq 8).[22]

Limitations of the Yamaguchi Lactonization Conditions. The tendency of the Yamaguchi lactonization conditions to form diolides from some seco acids, as shown by Seebach,[15] can also pose a problem. Steglich has shown that some α,ω-hydroxy acids (**13**) tend to dimerize under the Yamaguchi lactonization conditions. However, by using a modified Mitsunobu lactonization, the monolide was obtained in 59% yield (eq 9). All other lactonization procedures also gave varying amounts of diolide, including the standard Mitsunobu lactonization procedure.[23]

Related Reagents. 2-Chloro-1-methylpyridinium Iodide; 1,3-Dicyclohexylcarbodiimide; 2-Thiopyridyl Chloroformate; Trimethylacetyl Chloride.

1. Inanaga, J.; Hirata, K.; Saeki, H.; Katsuki, T.; Yamaguchi, M. *BCJ* **1979**, *52*, 1989.
2. (a) *FF* **1984**, *11*, 552. (b) Kawanami, Y.; Dainobu, Y.; Inanga, J.; Katsuki, T.; Yamaguchi, M. *BCJ* **1981**, *54*, 943.
3. Tanaka, T.; Sato, T.; Imura, T. Jpn. Patent 04 99 757 [92 99 757] (*CA* **1992**, *117*, 111 155h)
4. *FF* **1981**, *9*, 478.
5. Harwood, L. M.; Moody, C. J. *Experimental Organic Chemistry*; Blackwell: Oxford, 1989; p 147.
6. *Sigma-Aldrich Library of Chemical Safety Data*, 2nd ed.; Lenga, R. E., Ed.; Sigma-Aldrich: Milwaukee, 1987; Vol. 2, p 371B.
7. Inanaga, J.; Katsuki, T.; Takimoto, S.; Ouchida, S.; Inoue, K.; Nakano, A.; Okukado, N.; Yamaguchi, M. *CL* **1979**, 1021.
8. Sutter, M. A.; Seebach, D. *LA* **1983**, 939.
9. Albertson, N. F. *OR* **1962**, *12*, 157.
10. (a) Corey, E. J.; Nicolaou, K. C. *JACS* **1974**, *96*, 5614. (b) Corey, E. J.; Brunelle, D. J. *TL* **1976**, 3409. (c) Mukiyama, T.; Usui, M.; Saigo, K. *CL* **1976**, 49. (d) Boden, E. P.; Keck, G. E. *JOC* **1985**, *50*, 2394. (e) Mitsunobu, O. *S* **1981**, 1.
11. See, for example: (a) Seebach, D.; Brändli, U.; Müller, H.-M.; Dobler, M.; Egli, M.; Przybylski, M.; Schneider, K. *HCA* **1989**, *72*, 1704. (b) Mulzer, J.; Mareski, P. A.; Buschmann, J.; Luger, P. *S* **1992**, 215.

12. Honda, M.; Hirata, K.; Sueoka, H.; Katsuki, T.; Yamaguchi, M. *TL* **1981**, *22*, 2679.
13. Inanaga, J.; Kawanami, Y.; Yamaguchi, M. *BCJ* **1986**, *59*, 1521.
14. Seebach, D.; Maestro, M. A.; Sefkow, M.; Neidlein, A.; Sternfeld, F.; Adam, G.; Sommerfeld, T. *HCA* **1991**, *74*, 2112.
15. Seebach, D.; Chow, H.-F.; Jackson, R. F. W.; Sutter, M. A.; Thaisrivongs, S.; Zimmerman, J. *LA* **1986**, 1281 and references cited therein.
16. Schregenberger, C.; Seebach, D. *LA* **1986**, 2081.
17. (a) Seebach, D.; Brändli, U.; Schnurrenberger, P. *HCA* **1988**, *71*, 155. (b) For a review of poly(hydroxyalkanates), see: Müller, H.-M.; Seebach, D. *AG(E)* **1993**, *32*, 477.
18. Seebach, D.; Brändli, U.; Müller, H.-M.; Dobler, M.; Egli, M.; Przybylski, M.; Schneider, K. *HCA* **1989**, *72*, 1704.
19. Waldmann, H.; Kunz, H. *JOC* **1988**, *53*, 4172.
20. Dhaon, M. K.; Olsen, R. K.; Ramasamy, K. *JOC* **1982**, *47*, 1962.
21. Tsutsui, H.; Mitsunobu, O. *TL* **1984**, *25*, 2163.
22. Hartmann, B.; Kanazawa, A. M.; Deprés, J.-P.; Greene, A. E. *TL* **1991**, *32*, 5077.
23. Justus, K.; Steglich, W. *TL* **1991**, *32*, 5781.

Richard A. Ewin
King's College London, UK

Triethylaluminum[1]

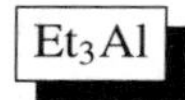

[97-93-8] $C_6H_{15}Al$ (MW 114.19)

(Lewis acid and source of nucleophilic ethyl groups;[2–5] couples with alkenyl halides in the presence of a transition metal catalyst;[8–11] selective hydrocyanation agent in combination with HCN[12])

Physical Data: mp −58 °C; bp 62 °C 0.8 mmHg; *d* 0.835 g cm^{-3} (25 °C).

Solubility: freely miscible with saturated and aromatic hydrocarbons; reacts violently with H_2O and protic solvents.

Form Supplied in: as a neat liquid in a stainless container or as a solution in hydrocarbon solvents (hexane, heptane, toluene).

Analysis of Reagent Purity: brochures from manufacturers, describe an apparatus and method for assay.

Handling, Storage, and Precautions: indefinitely stable under an inert atmosphere. The neat liquid or dense solutions are highly pyrophoric. Solutions more dilute than a certain concentration are not pyrophoric and are safer to handle. The nonpyrophoric limits are 13 wt % in isopentane, 12 wt % in hexane, and 12 wt % in heptane, respectively. Use of halogenated hydrocarbons as solvents should be avoided because of possible explosive reactions sometimes observed for mixtures of CCl_4 and organoaluminums.

Ethylation. Et_3Al, like other organoaluminums, can act as a Lewis acid to activate Lewis basic functionalities and also as a captor of electrophilic species by ethylation, as illustrated in eq 1.[2] Substitution reactions of glycosyl fluorides (eq 2)[3] and bromides[4] can be effected with Et_3Al. γ-Lactols react with Et_3Al in the presence of ***Boron Trifluoride Etherate*** to deliver 2,5-disubstituted tetrahydrofurans stereoselectively (eq 3).[5]

OAc → (Et_3Al, hexane, 0 °C, 75%) → Et + Et (70:30) (1)

BnO, OBn, F, OBn → (Et_3Al, toluene, 0 °C, 93%) → BnO, OBn, Et, OBn; α:β = 6:1 (2)

OH, OTBDPS → (BF_3•OEt_2, Et_3Al, CH_2Cl_2, −78 °C to rt, 61%) → Et, OTBDPS + Et, OTBDPS; *trans*:*cis* = 13:1 (3)

Conjugate addition of Et_3Al to 2-nitrofurans provides, after hydrolysis, dihydro-2(3*H*)-furanones (eq 4).[6] Regioselective addition of Et_3Al to unsymmetrical 1,1′-azodicarbonyl compounds has been reported.[7]

NO_2 → (Et_3Al, hexane–benzene, −95 to 0 °C, 84%) → Et, O (4)

Cross coupling of Et_3Al with aryl phosphates[8] or alkynyl bromides[9] proceeds with a Ni catalyst to provide alkylated arenes or alkynes (eqs 5 and 6). While Pd or Cu complexes are used as the catalyst for the coupling of Et_3Al with carboxylic acid chlorides or thioesters (eq 7),[10] ***Iron(III) Chloride*** is used for reactions of propargyl acetates to give substituted allenes (eq 8).[11]

OP(O)(OEt)$_2$ → (Et_3Al, cat $NiCl_2$(dppp), Et_2O, rt, 95%) → Et (5)

s-Bu—≡—Br → (Et_3Al, cat Ni(mesal)$_2$, pentane, 25 °C, 84%) → *s*-Bu—≡—Et (6)

PhC(O)Cl → (Et_3Al, Cu(acac)$_2$, PPh_3, THF, 88%) → PhC(O)Et (7)

AcO → (Et_3Al, cat $FeCl_3$, Et_2O, 72%) → Et (8)

Lewis Acid. Et_3Al–***Hydrogen Cyanide*** and ***Diethylaluminum Cyanide*** are two optional reagents for conjugate hydrocyanation of α,β-unsaturated ketones.[12,13] The results from these two reagents often differ. Due to a rather slow reaction rate between Et_3Al and HCN, the former reagent contains a proton source (HCN) which can quench aluminum enolate intermediates. An impressive example is shown in eq 9. Preformed Et_2AlCN gives *cis*-isomer, whereas HCN–Et_3Al leads to *trans*-isomer.[14] A variant involves trapping of the enolates as TMS ethers by use of ***Cyanotrimethylsilane***–Et_3Al (eq 10).[14]

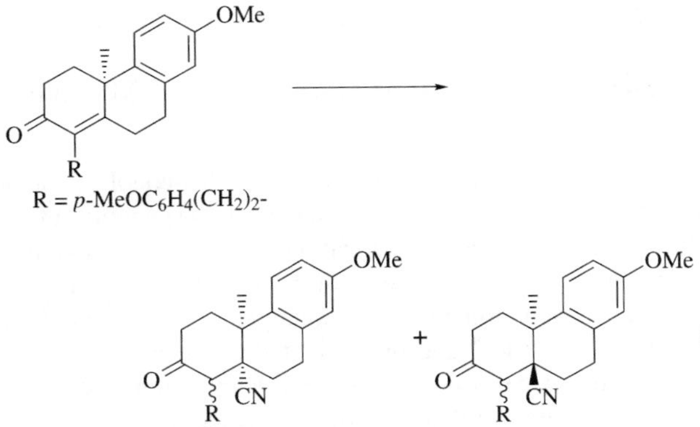

R = *p*-$MeOC_6H_4(CH_2)_2$-

(9)

Et_2AlCN, Ph, 99%	15:1
Et_3Al–HCN, THF, 89%	1:5

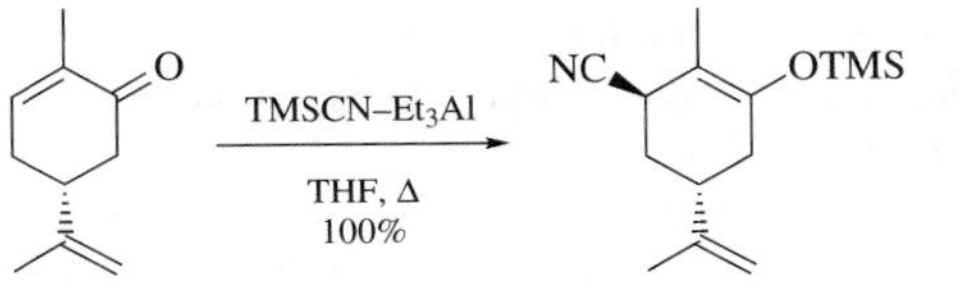

TMSCN–Et_3Al, THF, Δ, 100% (10)

The HCN–Et_3Al system has been used to cleave oxiranes in steroids[15a] or carbohydrates (eq 11) to give β-cyano alcohols.[15b]

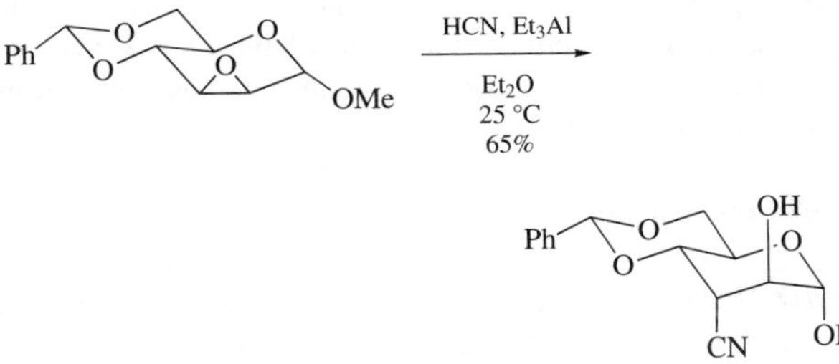

HCN, Et_3Al, Et_2O, 25 °C, 65% (11)

The ate complex, formed on metalation of allyl *i*-propyl ether with ***s-Butyllithium*** followed by addition of Et_3Al, reacts with carbonyl compounds at the α-position in *syn*-selective manner (eq 12).[16]

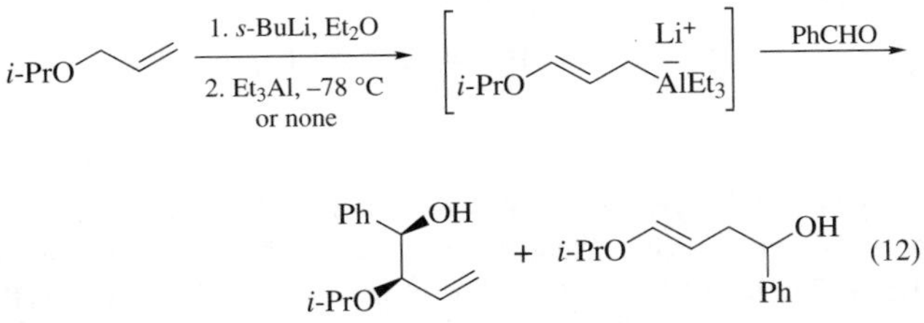

1. *s*-BuLi, Et_2O; 2. Et_3Al, –78 °C or none; PhCHO (12)

syn:*anti* = 92:8

none, 95%	28:72
Et_3Al, 81%	>99:<1

Rearrangements. Allyl vinyl ethers undergo [3,3]-sigmatropic rearrangements promoted by Et_3Al, which also effects subsequent ethylation of the resulting aldehydes (eq 13). Use of ***Triisobutylaluminum*** leads to primary alcohols by β-hydride reduction.[17]

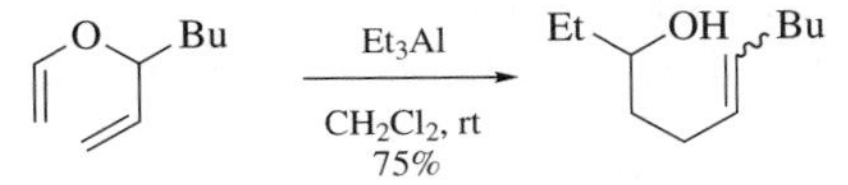

Et_3Al, CH_2Cl_2, rt, 75% (13)

Alkylative Beckmann rearrangements of oxime sulfonates are promoted by trialkylaluminums. The rearrangements give the imines, which are reduced with ***Diisobutylaluminum Hydride*** to the corresponding amines (eq 14).[18] Related alkylative Beckmann fragmentations have also been reported.[19]

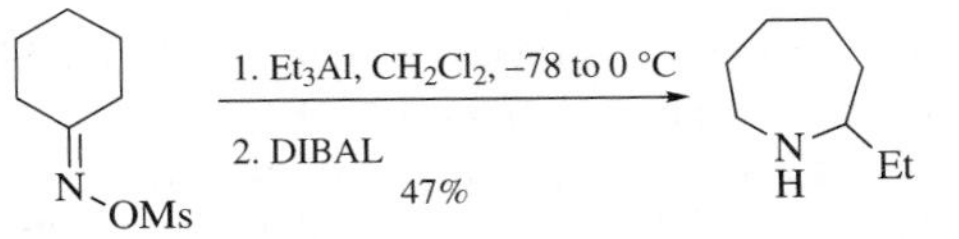

1. Et_3Al, CH_2Cl_2, –78 to 0 °C; 2. DIBAL, 47% (14)

Et_3Al promotes stereospecific pinacol-type rearrangements of chiral β-methanesulfonyloxy alcohols. Aryl[20a] or alkenyl groups[20b] cleanly take part in the 1,2-migration to provide a range of α-chiral ketones (eq 15). The 1,2-migration of alkyl groups is effected by the more Lewis acidic ***Diethylaluminum Chloride***.[20c] The reagent combination of DIBAL and Et_3Al effects the reductive 1,2-rearrangement of α-mesyloxy ketones (eq 16).[20d,e]

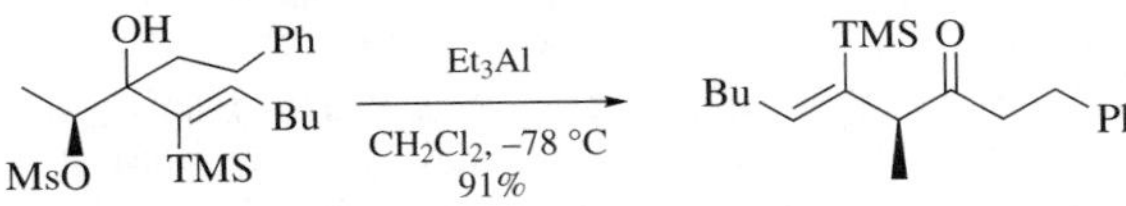

Et_3Al, CH_2Cl_2, –78 °C, 91% (15)

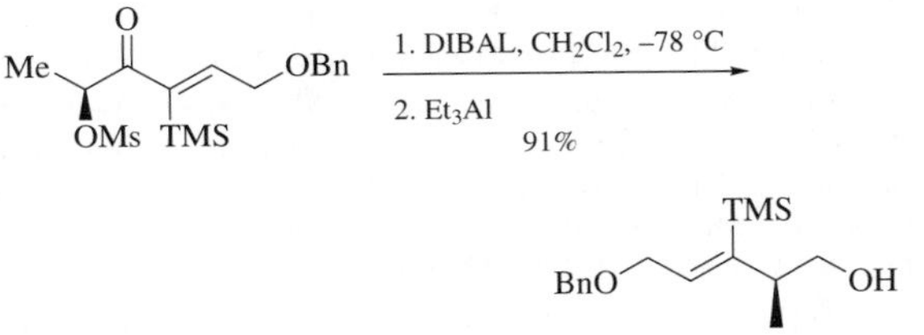

1. DIBAL, CH_2Cl_2, –78 °C; 2. Et_3Al, 91% (16)

Cyclopropanation. The reagent combination of ***Diiodomethane*** and Et_3Al (or other organoaluminums) leads to cyclopropanation of alkenes (eq 17).[21]

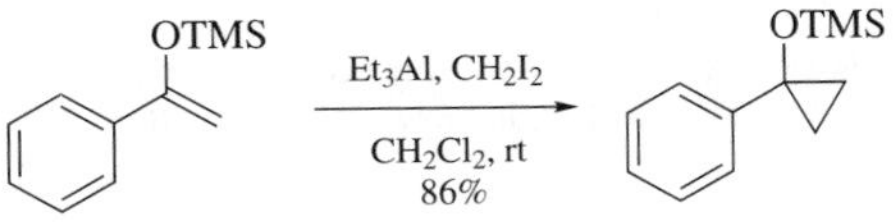

Et_3Al, CH_2I_2, CH_2Cl_2, rt, 86% (17)

Related Reagents. Titanium(IV) Chloride–Triethylaluminum.

---

1. (a) Mole, T.; Jeffery, E. A. *Organoaluminum Compounds*; Elsevier: Amsterdam, 1972. (b) Reinheckel, H.; Haage, K.; Jahnke, D. *Organomet. Chem. Rev. A* **1969**, *4*, 47. (c) Lehmkuhl, H.; Ziegler, K.; Gellert, H. G. *MOC* **1970**, *8/4*. (d) Negishi, E. *JOM Libr.* **1976**, *1*, 93. (e) Yamamoto, H.; Nozaki, H. *AG(E)* **1985**, *17*, 169. (f) Negishi, E. *Organometallics in Organic Synthesis*; Wiley: New York, 1980; Vol. 1, pp 286–393, (g)

Eisch, J. J. In *Comprehensive Organometallic Chemistry*, Wilkinson, G.; Stone, F. G. A.; Abel, E. W., Eds.; Pergamon: Oxford, 1982; Vol. 1, pp 555–682. (h) Zietz, J. R. Jr.; Robinson, G. C.; Lindsay, K. L. In *Comprehensive Organometallic Chemistry*, Wilkinson, G.; Stone, F. G. A.; Abel, E. W., Eds.; Pergamon: Oxford, 1982; Vol. 7, pp 365–464. (i) Maruoka, K.; Yamamoto, H. *AG(E)* **1985**, *24*, 668. (j) Maruoka, K. Yamamoto, H. *T* **1988**, *44*, 5001.

2. Hashimoto, S.; Kitagawa, Y.; Iemura, S.; Yamamoto, H.; Nozaki, H. *TL* **1976**, 2615.
3. (a) Posner, G. H.; Haines, S. R. *TL* **1985**, *26*, 1823. (b) Nicolaou, K. C.; Dolle, R. E.; Chucholowski, A.; Randall, J. L. *CC* **1984**, 1153.
4. Tolstikov, G. A.; Prokhorova, N. A.; Spivak, A. Yu.; Khalilov, L. M.; Sultanmuratova, V. R. *ZOK* **1991**, *27*, 2101.
5. Tomooka, K.; Matsuzawa, K.; Suzuki, K.; Tsuchihashi, G. *TL* **1987**, *28*, 6339.
6. Pecunioso, A.; Menicagli, R. *JCR(S)* **1988**, 228.
7. Yamamoto, Y.; Yumoto, M.; Yamada, J. *TL* **1991**, *32*, 3079.
8. Hayashi, T.; Katsuro, Y.; Okamoto, Y.; Kumada, M. *TL* **1981**, *22*, 4449.
9. Giacomelli, G.; Lardicci, L. *TL* **1978**, 2831.
10. (a) Takai, K.; Oshima, K.; Nozaki, H. *BCJ* **1981**, *54*, 1281. (b) Wakamatsu, K.; Okuda, Y.; Oshima, K.; Nozaki, H. *BCJ* **1985**, *58*, 2425.
11. Tolstikov, G. A.; Romanova, T. Yu.; Kuchin, A. V. *JOM* **1985**, *285*, 71.
12. (a) Nagata, W.; Yoshioka, M. *OS* **1972**, *52*, 100. (b) Nagata, W.; Yoshioka, M. *OR* **1977**, *25*, 255.
13. Ireland, R. E.; Dawson, M. I.; Welch, S. C.; Hagenbach, A.; Bordner, J.; Trus, B. *JACS* **1973**, *95*, 7829.
14. (a) Utimoto, K.; Obayashi, M.; Shishiyama, Y.; Inoue, M.; Nozaki, H. *TL* **1980**, *21*, 3389. (b) Utimoto, K.; Wakabayashi, Y.; Horiie, T.; Inoue, M.; Shishiyama, Y.; Obayashi, M.; Nozaki, H. *T* **1983**, *39*, 967.
15. (a) Nagata, W.; Yoshioka, M.; Okumura, T. *TL* **1966**, 847. (b) Davidson, B. E.; Guthrie, R. D.; McPhail, A. T. *CC* **1968**, 1273.
16. (a) Yamamoto, Y.; Yatagai, H.; Maruyama, K. *JOC* **1980**, *45*, 195. (b) Yamamoto, Y.; Yatagai, H.; Saito, Y.; Maruyama, K. *JOC* **1984**, *49*, 1096. (c) Yamamoto, Y.; Saito, Y.; Maruyama, K. *JOM* **1985**, *292*, 311.
17. (a) Takai, K.; Mori, I.; Oshima, K.; Nozaki, H. *TL* **1981**, *22*, 3985. (b) Takai, K.; Mori, I.; Oshima, K.; Nozaki, H. *BCJ* **1984**, *57*, 446.
18. (a) Hattori, K.; Matsumura, Y.; Miyazaki, T.; Maruoka, K.; Yamamoto, H. *JACS* **1981**, *103*, 7368. (b) Sakane, S.; Matsumura, Y.; Yamamura, Y.; Ishida, Y.; Maruoka, K.; Yamamoto, H. *JACS* **1983**, *105*, 672. (c) Maruoka, K.; Miyazaki, T.; Ando, M.; Matsumura, Y.; Sakane, S.; Hattori, K.; Yamamoto, H. *JACS* **1983**, *105*, 2831.
19. Fujioka, H.; Yamanaka, T.; Takuma, K.; Miyazaki, M.; Kita, Y. *CC* **1991**, 533.
20. (a) Suzuki, K.; Katayama, E.; Tsuchihashi, G. *TL* **1983**, *24*, 4997. (b) Suzuki, K.; Katayama, E.; Tsuchihashi, G. *TL* **1984**, *25*, 1817. (c) Suzuki, K.; Tomooka, K.; Tsuchihashi, G. *TL* **1984**, *25*, 4253. (d) Suzuki, K.; Tomooka, K.; Katayama, E.; Matsumoto, T.; Tsuchihashi, G. *JACS* **1986**, *108*, 5221. (e) Suzuki, K.; Katayama, E.; Matsumoto, T.; Tsuchihashi, G. *TL* **1984**, *25*, 3715.
21. (a) Maruoka, K.; Fukutani, Y.; Yamamoto, H. *JOC* **1985**, *50*, 4412. (b) Maruoka, K.; Sakane, S.; Yamamoto, H. *OS* **1988**, *67*, 176.

Keisuke Suzuki & Tetsuya Nagasawa
Keio University, Yokohama, Japan

Triethyl Orthoformate[1]

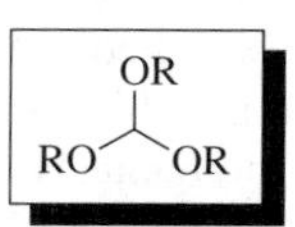

(**1**; R = Et)		
[122-51-0]	$C_7H_{16}O_3$	(MW 148.23)
(**2**; R = Me)		
[149-73-5]	$C_4H_{10}O_3$	(MW 106.14)

(precursor for higher analogs by reaction with alcohols,[1c] including cyclic orthoesters from polyols;[2] for deoxygenation of 1,2-diols, affording alkenes;[3] acetalization of carbonyl compounds;[4] a dehydrating agent[5] for enol ether formation;[6] esterification of acids;[7] formylation of active methylene compounds,[8] heteroatom nucleophiles[9] and organometallic reagents;[10] formylation of electron-rich species; dialkoxycarbenium ion precursor;[11] solvent for thallium trinitrate reactions[12])

Alternate Name: triethoxymethane.
Physical Data: (**1**) bp 146 °C; $d = 0.891\ g\,cm^{-3}$; (**2**) bp 102 °C; $d = 0.970\ g\,cm^{-3}$.
Solubility: sol most organic solvents.
Form Supplied in: clear liquid; widely available.
Purification: distillation.
Handling, Storage, and Precautions: highly moisture sensitive; flammable; irritant with high volatility. Use in a fume hood.

Introduction. The orthoformates are a remarkably useful group of reagents. They are shelf-stable, yet highly reactive. As alkylating agents, they readily transfer the associated alkyl group, a large variety of which are easily available. As formylation reagents, they are reactive under both acidic and basic conditions. The choice of ester is often arbitrary in this context.

Transesterification.

Higher Orthoformates. While there are many ways to obtain esters of orthoformic acid,[1c] an easy method takes advantage of the rapid equilibrium among orthoesters (eq 1). By starting with the lowest analog, trimethyl orthoformate, essentially complete conversion to the higher esters is possible by carrying out the reaction at such a temperature as to distill away the evolving methanol. It is important that reactions be carried out under anhydrous conditions to avoid formation of the formate through hydrolysis.

$$HC(OR)_3 + 3\,R'OH \rightleftharpoons HC(OR')_3 + 3\,ROH \qquad (1)$$

Cyclic Orthoformates.[2] When the above strategy is applied to polyols, cyclic orthoformates can be isolated. Most common are the cyclization of 1,3-diols (eq 2)[2,13] and 1,2-diols,[14] as well as the formation of caged structures from the use of polyols (eq 3).[15,16]

Those orthoformates obtained from 1,2-diols (eq 4)[3] can undergo cycloelimination upon pyrolysis to afford alkenes in high yield.[17] There are a variety of methods for carrying out this overall process,[18] but the orthoester route is competitive if the alkene is thermally stable.

$(MeO)_3CH$, p-TsOH; 12 h, reflux; 43% (2)

$(MeO)_3CH$; p-TsOH, DMF (3)

$(EtO)_3CH$, 95.3%; 175 °C, 70% (4)

Acetals and Enol Ethers. The conversion of the orthoformate to formate is energetically favored. As a result, the acetalization of ketones by orthoformate is a highly favored process and allows the formation of acetals under exceedingly mild conditions. A wide variety of ketones can be converted into dimethyl acetals by the action of trimethyl orthoformate and ***p-Toluenesulfonic Acid***. The methyl formate thus evolved is distilled away.[4]

The process is general and allows isolation of quite sensitive acetals (eq 5).[19] The technique accommodates protection of α,β-unsaturated carbonyl compounds.[20,21,22] While some form of acid catalysis is usually needed, there is great flexibility in the choice of acid, including Amberlyst 15[23] and, in particular, ***Montmorillonite K10***.[24] Cyclic acetals are also accessible (eq 6).[25]

1. O_3, MeOH, −70 °C; 2. Me_2S, $(MeO)_3CH$, MeOH, HCl; 90% (5)

1. CH_2=$CHCH_2MgBr$, $CeCl_3$; 2. $(MeO)_3CH$ (6)

Upon distillation of some acetals,[26] or upon attempted acetalization of highly conjugated species,[26,27] the enol ether can also be observed. The choice of acid often determines whether the acetal or the enol ether is isolated.

Orthoformates are also useful in promoting the formation of other acetals[5] by functioning as dehydrating agents. This function is useful for β-lactone formation as well (eq 7).[28,29]

$(MeO)_3CH$, benzene, reflux; Δ, 83% overall (7)

Esterification. In a related process, orthoformates are good esterification agents; they operate on carboxylic acids (eq 8),[7] sulfonic acids,[30] and carboxyboranes,[31] often without the need for acid catalysis.

CO_2H + $(EtO)_3CH$ →(Δ, no acid) CO_2Et (8)

Formylation. The orthoformate carbon is highly reactive in a number of bond-forming reactions. It is capable of reaction under both electrophilic and nucleophilic conditions and serves as a formylation reagent.

Active Methylene Compounds. Triethyl orthoformate can formylate diethyl malonate under slightly acidic conditions.[8] With less activated compounds it can be induced to undergo a Mannich reaction,[32] and can also formylate a cyclohexanone enolate anion (eq 9).[33]

$(MeO)_3CH$; BF_3•Et_2O; 79% (9)

This reaction is noteworthy in its propensity for *C*-alkylation and the fact that the protected acetal raises the pK_a of the product relative to the unprotected β-dicarbonyl compound.

Heteroatom Nucleophiles. The formylation of anilines is well known[34] and provides entry to a large array of functional groups.[35] Of greater interest is the ability of the product to be trapped in a subsequent reaction to afford heterocycles (eq 10).[9,36–38] Orthoformates also react readily with phosphorus nucleophiles.[39]

HCl, $(EtO)_3CH$; DMF, Δ; 63% (10)

Organometallic Reactions. In addition to the enolate reactions described above, orthoformates can also carry out the formal formylation of Grignard reagents.[10]

Electrophilic Formylation. It has been shown that the dialkoxycarbenium ion can be readily formed by acid treatment of orthoformates. While the reactive ion can be isolated and used directly,[11] the typical practice is to generate it in situ.

While not as universal as the Gatterman–Koch reaction, the cation works well for the formylation of activated aromatic compounds (eq 11).[40,41] Of greater synthetic utility is the effective formylation of alkynes (eq 12)[42] and alkenes (see below).

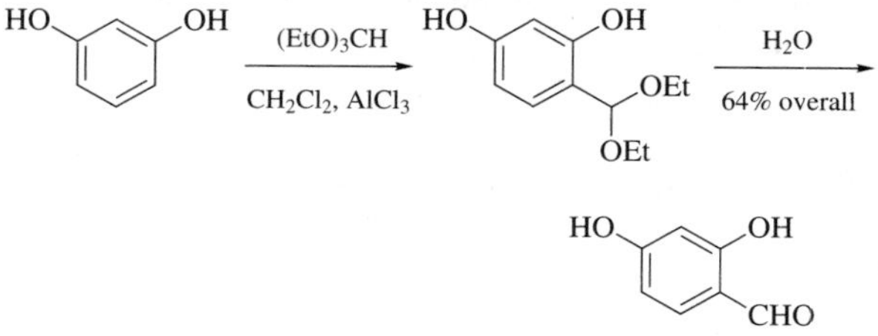

(11)

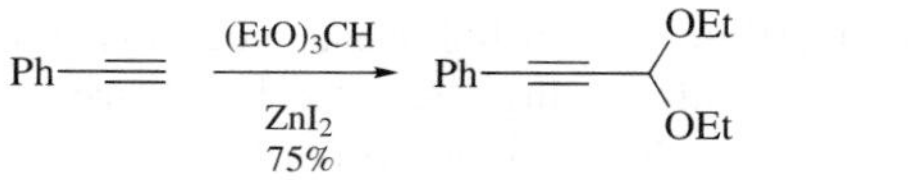

(12)

Reactions of silyl enol ethers with dialkoxycarbenium ions result in α-formyl ketones (eq 13),[43] much like those achieved above through the use of enolate anions. With a dienol silane (eq 14), regioselective γ-formylation is achieved.[44] In extended alkenic systems, cationic cyclization (eq 15) can be realized.[45]

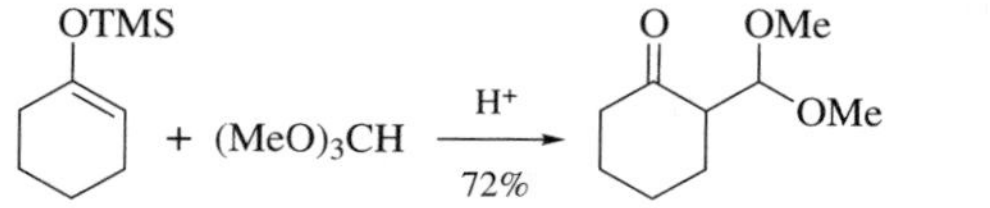

(13)

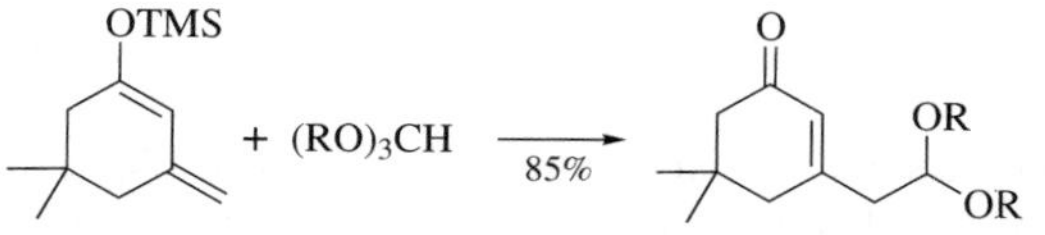

(14)

R = CH_2=$CHCH_2$

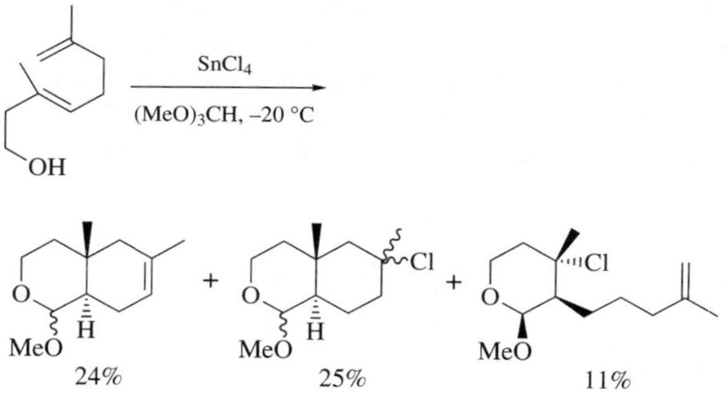

(15)

Solvent for Thallium Trinitrate Oxidations. While ***Thallium(III) Nitrate*** oxidations of aromatic ketones and chalcones often gives rise to mixtures of products, the use of trimethyl orthoformate as solvent gives substantially cleaner reactions and higher yields (eq 16).[12]

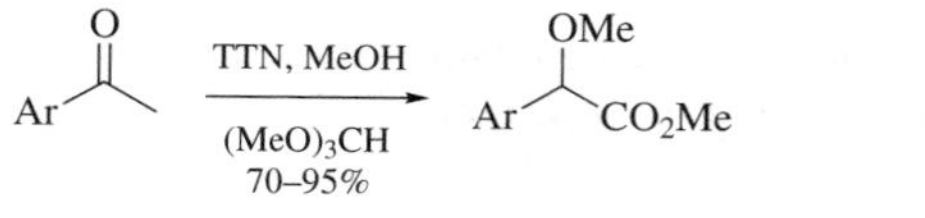

(16)

Related Reagents. Diethyl Phenyl Orthoformate; Dimethoxycarbenium Tetrafluoroborate; Dimethylchloromethyleneammonium Chloride; *N*,*N*-Dimethylformamide; *N*,*N*-Dimethylformamide Diethyl Acetal; Methyl Formate.

1. (a) DeWolfe, R. H. *Carboxylic Ortho Acid Derivatives*; Academic: New York, 1970. (b) Ghosh, S.; Ghatak, U. R. *Proc. Ind. Acad. Sci.* **1988**, *100*, 235. (c) DeWolfe, R. H. *S* **1974**, 153.
2. Denmark, S. E.; Almstead, N. G. *JOC* **1991**, *56*, 6458.
3. Camps, P.; Cardellach, J.; Font, J.; Ortuno, R. M.; Ponsati, O. *T* **1982**, *38*, 2395.
4. Napolitano, E.; Fiaschi, R.; Mastrorilli, E. *S* **1986**, 122.
5. Marquet, A.; Dvolaitzky, M.; Kagan, H. B.; Mamlok, L.; Ouannes, C.; Jacques, J. *BSF* **1961**, 1822.
6. Wohl, R. A. *S* **1974**, 38.
7. Cohen, H.; Mier, J. D. *CI(L)* **1965**, 349.
8. Parham, W. E.; Reed, L. J. *OSC* **1955**, *3*, 395.
9. Harden, M. R.; Jarvest, R. L.; Parratt, M. J. *JCS(P1)* **1992**, 2259.
10. Bachman, G. B. *OSC* **1943**, *2*, 323.
11. Pindur, U.; Flo, C. *SC* **1989**, *19*, 2307.
12. Taylor, E. C.; Robey, R. L.; Liu, K.-T.; Favre, B.; Bozimo, H. T.; Conley, R. A.; Chiang, C.-S.; McKillop, A.; Ford, M. E. *JACS* **1976**, *98*, 3037.
13. Gardi, R.; Vitali, R.; Ercoli, A. *TL* **1961**, 448.
14. Takasu, M.; Naruse, Y.; Yamamoto, H. *TL* **1988**, *29*, 1947.
15. Stetter, H.; Steinacker, K. H. *CB* **1953**, *86*, 790.
16. Yu, K.-L.; Fraser-Reid, B. *TL* **1988**, *29*, 979.
17. Burgstahler, A. W.; Boger, D. L.; Naik, N. C. *T* **1976**, *32*, 309.
18. Block, E. *OR* **1984**, *30*, 457.
19. Frickel, F. *S* **1974**, 507.
20. Taylor, E. C.; Conley, R. A.; Johnson, D. K.; McKillop, A. *JOC* **1977**, *42*, 4167.
21. van Allen, J. A. *OSC* **1963**, *4*, 21.
22. Wengel, J.; Lau, J.; Pedersen, E. B.; Nielson, C. M. *JOC* **1991**, *56*, 3591.
23. Patwardhan, S. A.; Dev, S. *S* **1974**, 348.
24. Taylor, E. C.; Chiang, C.-S. *S* **1977**, 467.
25. Rychnovsky, S. D.; Griesgraber, G. *CC* **1993**, 291.
26. Meek, E. G.; Turnbull, J. H.; Wilson, W. *JCS* **1953**, 811.
27. van Hulle, F.; Sipido, V.; Vandewalle, M. *TL* **1973**, 2213.
28. Rogic, M. M.; van Peppe, J. F.; Klein, K. P.; Demmin, T. R. *JOC* **1974**, *39*, 3424.
29. Blume, R. C. *TL* **1969**, 1047.
30. Padmapriya, A. A.; Just, G.; Lewis, N. G. *SC* **1985**, *15*, 1057.
31. Mittakanti, M.; Feakes, D. A.; Morse, K. W. *S* **1992**, 380.
32. El Cherif, S.; Rene, L. *S* **1988**, 138.
33. Suzuki, S.; Yanagisawa, A.; Noyori, R. *TL* **1982**, *23*, 3595.
34. Roberts, R. M.; Vogt, P. J. *JACS* **1956**, *78*, 4778.
35. Crochet, R. A.; Blanton, C. D., Jr. *S* **1974**, 55.
36. Jenkins, G. L.; Knevel, A. M.; Davis, C. S. *JOC* **1961**, *26*, 274.
37. Patridge, M. W.; Slorach, S. A.; Vipond, H. J. *JCS* **1964**, 3670.
38. Lee, K.-J.; Kim, S. H.; Kim, S.; Cho, Y. R. *S* **1992**, 929.
39. Baille, A. C.; Cornell, C. L.; Wright, B. J.; Wright, K. *TL* **1992**, *33*, 5133.
40. Treibs, W. *TL* **1967**, 4707.
41. Gross, H.; Rieche, A.; Matthey, G. *CB* **1963**, *96*, 308.
42. Howk, B. W.; Sauer, J. C. *OSC* **1963**, *4*, 801.
43. Mukaiyama, T.; Iwakiri, H. *CL* **1985**, 1363.
44. Pirrung, M. C.; Thomson, S. A. *TL* **1986**, *27*, 2703.
45. Perron-Sierra, F.; Promo, M. A.; Martin, V. A.; Albizati, K. F. *JOC* **1991**, *56*, 6188.

Richard T. Taylor
Miami University, Oxford, OH, USA

Trifluoroacetic Anhydride[1]

$F_3C-C(=O)-O-C(=O)-CF_3$

[407-25-0] $C_4F_6O_3$ (MW 210.04)

(activating agent; oxidation)

Alternate Name: TFAA.
Physical Data: mp −65 °C; bp 39–40 °C; *d* 1.487 g cm^{-3}.
Solubility: sol C_6H_6, CH_2Cl_2, Et_2O, DMF, THF, MeCN.
Form Supplied in: colorless liquid; commercially available.
Analysis of Reagent Purity: by standard analytical techniques.
Preparative Methods: by distilling ***Trifluoroacetic Acid*** from ***Phosphorus(V) Oxide***.[1]
Handling, Storage, and Precautions: corrosive and moisture sensitive; toxic by inhalation; should be freshly distilled prior to reaction. Use in a fume hood.

Activated Esters. Mixed anhydrides may be prepared by reaction of carboxylic acids with trifluoroacetic anhydride.[2] The method has been used widely as a means of activating carboxyl groups to nucleophilic attack. The method has also been highly useful in Friedel–Crafts acylation of arenes (eqs 1 and 2).[3,4] Sulfonic acids are also activated to nucleophilic attack; substituted sulfones result. Intramolecular Friedel–Crafts reactions are also facilitated, but only for formation of six-membered rings.[5]

$$PhCO_2H \xrightarrow[50\text{–}60\ ^\circ C]{TFAA,\ anisole} 4\text{-}MeOC_6H_4COPh \quad (1)$$

$$RCO_2H \xrightarrow[2.\ ArX]{1.\ 3\ equiv\ TFAA,\ 85\%\ H_3PO_4,\ MeCN} X\text{-}C_6H_4COR \quad (2)$$

Anhydrides of diacids are prepared in (about) 90% yield by reaction with TFAA in ether.[6] The method involves formation of the monotrifluoroacetyl mixed anhydride, which may be converted to the cyclic anhydride by heating under vacuum (eq 3).[7]

$$HO_2C\text{-(tetrahydrofuran)-}CO_2H \xrightarrow[2.\ 67\ ^\circ C,\ 0.1\ mmHg;\ >90\%]{1.\ TFAA} \text{bicyclic anhydride} \quad (3)$$

An inversion of alkene geometry is made possible by the reaction of their derived epoxides with lithium halides and TFAA (eq 4);[8] the products of this reaction are the corresponding halohydrin trifluoroacetates, which undergo a *syn* elimination upon reaction with ***Lithium Iodide***.

$$(Z)\text{-}RCH{=}CHR' \xrightarrow[3.\ NaI,\ DMF]{1.\ m\text{-}CPBA;\ 2.\ LiCl,\ TFAA} (E)\text{-}RCH{=}CHR' \quad (4)$$

Primary alkanols and 2-alkenols are converted into the corresponding halides in high yield by a one-pot, two-step reaction via transformation into intermediate trifluoroacetates, followed by nucleophilic substitution with lithium halides (eq 5).[9]

$$ROH \xrightarrow[2.\ LiX,\ THF/HMPA;\ 70\text{–}98\%]{1.\ TFAA,\ THF,\ rt} RX \quad (5)$$

Sulfoxides are reduced to sulfides under mild conditions with ***Trifluoroacetic Anhydride–Sodium Iodide*** (eq 6).[10]

$$RR'S{=}O \xrightarrow[NaI,\ 0\ ^\circ C]{TFAA} RR'S \quad (6)$$

Primary amides are converted under mild conditions to the corresponding nitriles using TFAA in pyridine (eq 7).[11] Similarly, aldoximes are converted to nitriles by TFAA in pyridine (eq 8).[12]

$$RC(O)NH_2 \xrightarrow[py]{TFAA} RCN \quad (7)$$

$$RCH{=}NOH \xrightarrow[py]{TFAA} RCN \quad (8)$$

TFAA in triethylamine can be used for the dehydration of aldols to enones where other methods are less successful (eq 9).[13] When ***Acetic Anhydride*** is employed instead of TFAA, the reaction proceeds slowly.

$$PhC(O)CH_2C(OH)Me_2 \xrightarrow[Et_3N;\ 88\text{–}97\%]{TFAA} PhC(O)CH{=}CMe_2 \quad (9)$$

The Pummerer rearrangement of sulfoxides to α-acyloxy sulfides, induced by TFAA, has been used as a means of converting sulfoxides to aldehydes (eq 10).[14]

$$RCH_2S(O)Ph \xrightarrow[2.\ NaHCO_3,\ H_2O,\ MeCN;\ 53\text{–}72\%]{1.\ TFAA,\ 0\ ^\circ C} RCHO \quad (10)$$

e.g. R = C_7H_{15}, $PhCH_2OCH_2$

Methyl aryl sulfides are converted in a mild, one-pot, three-step procedure via Pummerer rearrangement of the corresponding sulfoxides, and without purification of intermediates, to provide arylthiols in excellent yields (eq 11).[15]

$$MeS\text{-}C_6H_4\text{-}X \xrightarrow[4.\ MeOH,\ Et_3N;\ 5.\ NH_4Cl]{1.\ m\text{-}CPBA;\ 2.\ Ca(OH)_2;\ 3.\ TFAA,\ 40\ ^\circ C} HS\text{-}C_6H_4\text{-}X \quad (11)$$

Oxidation. Trifluoroacetoxydimethylsulfonium trifluoroacetate is prepared in situ from ***Dimethyl Sulfoxide*** and TFAA below −50 °C and reacts rapidly with alcohols in the presence of ***Triethylamine*** to give the corresponding carbonyl compounds (eq 12).[16]

Me, Me, S+, O, CF3, O — 1. RR'CHOH; 2. Et_3N → R, R', =O (12)

TFAA/DMSO also efficiently converts vicinal diols to the corresponding α-dicarbonyl compounds or derivatives thereof (eq 13).[17] Unlike the Swern oxidant (***Dimethyl Sulfoxide–Oxalyl Chloride***), this reagent mixture gives good yields for halogenated substrates. The preparation of previously inaccessible compounds (such as σ-homo-*ortho*-benzoquinones) is thus facilitated.

Br, Br, OH, OH — TFAA, DMSO → Br, Br, O, O (13)

Undesired electrophilic chlorination as a side-reaction in Swern oxidations is avoided by use of TFAA in place of oxalyl chloride.[18] See also ***Dimethyl Sulfoxide–Trifluoroacetic Anhydride***.

Miscellaneous. Enamines of α-amino acids react rapidly with TFAA to give pyrrole derivatives (eq 14).[19]

Ph, H, N, CO_2Et, CO_2Na — TFAA, 0 °C; 30% → HO, CO_2Et, Ph, N (14)

The reaction of TFAA with triethylamine *N*-oxide leads to formation of the trifluoroacetate salt of *N,N*-dimethylformaldimmonium ion. This ion is a superior reagent in the Mannich reaction.[20]

TFAA is used to prepare trifluoroacetamides from amines; these amides may be used in Gabriel-type reactions. The use of TFAA delivers an amide of enhanced acidity and which is easily hydrolyzed in situ, thereby allowing a one-pot alkylation process (eq 15).[21]

RNH_2 — 1. TFAA; 2. KH, THF; 3. 18-crown-6; 4. R'X; 5. KOH, MeOH → RNHR' (15)

Reaction of TFAA with ammonium nitrate provides an in situ source of trifluoroacetyl nitrate, which has been used for the nitration of enol acetates[22] and as an *N*-nitrating agent.[23]

TFAA efficiently cleaves *N*-tosyl protecting groups from histidine (eq 16).[24]

R, N, N–SO_2Tol — 1. TFAA, py; 2. H_2O → R, N, NH (16)

Related Reagents. Acetic Anhydride; Trifluoroacetyl Chloride.

1. Tedder, J. M. *CR* **1955**, *55*, 787.
2. Emmons, W. D.; McCallum, K. S.; Ferris, A. F. *JACS* **1953**, *75*, 6047.
3. Bourne, E. J.; Stacey, M.; Tatlow, J. C.; Tedder, J. M. *JCS* **1951**, 718.
4. Galli, C. *S* **1979**, 703.
5. Ferrier, R. J.; Tedder, J. M. *JCS* **1957**, 1435.
6. Duckworth, A. C. *JOC* **1962**, *27*, 3146.
7. Moore, J. A.; Kelly, J. E. *Org. Prep. Proc. Int.* **1974**, *6*, 255.
8. Sonnet, P. E. *JOC* **1980**, *45*, 154.
9. Camps, F.; Gasol, V.; Guerrero, A. *S* **1987**, 511.
10. Drabowitz, J.; Oae, S. *S* **1977**, 404.
11. Campagna, F.; Carotti, A.; Casini, G. *TL* **1977**, 1813.
12. Carotti, A.; Canpagna, F. *S* **1979**, 56.
13. Narasaka, K. *OS* **1987**, *65*, 12.
14. Sugiharo, H.; Tanikoga, R.; Kaji, A. *S* **1978**, 881.
15. Young, R. N.; Gauthier, J. Y.; Coombs, W. *TL* **1984**, *25*, 1753.
16. Omuro, K.; Sharma, A. K.; Swern, D. *JOC* **1976**, *41*, 957.
17. Amon, C. M.; Banwell, N. G.; Gravatt, G. L. *JOC* **1987**, *52*, 4851.
18. Smith, A. B.; Leenay, T. L.; Liu, H. J.; Nelson, L. A. K.; Ball, R. G. *TL* **1988**, *29*, 49.
19. Gupta, S. K. *S* **1975**, 726.
20. Ahond, A.; Cavé, A.; Kan-Fan, C.; Potier, P. *BCF* **1970**, 2707.
21. Nordlander, J. E.; Catalare, D. B.; Eberlein, T. H.; Farkas, L. V.; Howe, R. S.; Stevens, R. M.; Tripoulas, N. A. *TL* **1978**, *19*, 4987.
22. Dampawan, P.; Zajac, W. W. *S* **1983**, 545.
23. Suri, S. C.; Chapman, R. D. *S* **1988**, 743.
24. van der Eijk, J. M.; Nolte, R. J. M.; Zwikker, J. W. *JOC* **1980**, *45*, 547.

Joseph Sweeney & Gemma Perkins
University of Bristol, UK

Trifluoromethanesulfonic Anhydride[1]

$(CF_3SO_2)_2O$

[358-23-6] $C_2F_6O_5S_2$ (MW 282.16)

(preparation of triflates;[1] mild dehydrating reagent; promoter for coupling reactions in carbohydrates[2])

Alternate Name: triflic anhydride.
Physical Data: bp 81–83 °C/745 mmHg; *d* 1.677 g cm^{-3}; n_D^{20} 1.3210.
Solubility: sol dichloromethane; insol hydrocarbons.
Form Supplied in: colorless liquid in ampules. Once opened it should be immediately used.
Analysis of Reagent Purity: IR, NMR.
Preparative Methods: by distillation of ***Trifluoromethanesulfonic Acid*** with an excess of ***Phosphorus(V) Oxide***.[1]
Purification: by redistillation with a small amount of P_2O_5. It is advisable to freshly distill the reagent from a small quantity of P_2O_5 before use.
Handling, Storage, and Precautions: the pure reagent is a colorless liquid that does not fume in air and is stable for a long period. It is not soluble in water and hydrolyzes only very slowly to triflic acid over several days at room temperature. Preferably stored under N_2 in a stoppered flask.

Reaction with Alcohols and Phenols. The reaction of alcohols and phenols with triflic anhydride (Tf_2O) at ~0 °C in the presence of a base (usually ***Pyridine***) in an inert solvent (usually dichloromethane) for 2–24 h affords the corresponding reactive trifluoromethanesulfonate esters (triflates).[1] When triflic anhydride and pyridine are combined, the pyridinium salt forms immediately and normally precipitates out from the reaction mixture. Nevertheless, the salt is an effective esterifying agent, reacting with the added alcohol to give triflates in high yields (eq 1).[3]

$$\text{pyridine} \xrightarrow{Tf_2O} \text{N-Tf pyridinium } OTf^- \xrightarrow{ROH} ROTf + \text{pyridinium (N-H) } OTf^- \quad (1)$$

R = alkyl, aryl

Pyridine can become involved in nucleophilic substitution when very reactive triflates are being synthesized.[2,3] One approach to minimize this disadvantage is to replace it with sterically hindered bases, such as 2,6-di-*t*-butyl-4-methylpyridine,[3,4] 2,4,6-trisubstituted pyrimidines,[5] or nonnucleophilic aliphatic amines (usually *N,N*-diisobutyl-2,4-dimethyl-3-pentylamine). No salt formation appears to take place under these conditions. The triflic anhydride seems to be the direct triflating agent and the base only neutralizes the triflic acid formed. Numerous alkyl triflates have been prepared in the literature[1b] by the above method. Some recent examples of triflates prepared from alcohols are illustrated in eqs 2 and 3.[6,7] As an exception, 2,6-dinitrobenzyl alcohol does not react with Tf_2O although similar sulfonyl esters could be prepared.[8]

$$\text{HO-ketone} \xrightarrow[\text{CH}_2\text{Cl}_2,\ 0\ ^\circ\text{C},\ 87\%]{Tf_2O,\ py} \text{TfO-ketone} \quad (2)$$

$$\text{Cl-bicyclic-OH} \xrightarrow[\text{CH}_2\text{Cl}_2,\ -10\ ^\circ\text{C},\ 85\%]{Tf_2O,\ py} \text{Cl-bicyclic-OTf} \quad (3)$$

Alkyl triflates have come to be recognized as useful intermediates for the functionalization of organic substrates by nucleophilic substitution, e.g. in carbohydrate chemistry.[9] Triflate is the best leaving group known[1b] next to the nonaflate and hence a large number of triflates, obtained in good yields by reaction of the corresponding alcohols (or alkoxides) with Tf_2O, have been used to generate unstable or destabilized carbocations under solvolytic conditions.[1b] Some new typical examples are shown in (**1**)–(**4**).[10–13]

OTf **(1)**

TMS, OTf **(2)**

OTf, D **(3)**

Br, OTf **(4)**

Alkyl triflates are known to be powerful reagents for the alkylation of aromatic compounds.[1b,14] However, the reaction of alkyl triflates with heterocycles affords *N*-alkylation products.[15]

In an improved modification of the Ritter reaction, primary and secondary alcohols react with Tf_2O in CH_2Cl_2 in the presence of a 2:1 excess of nitriles to give the corresponding amides in good yields (eq 4).[16]

$$R^1(H(R^2))CH{-}OH \xrightarrow[CH_2Cl_2]{Tf_2O} \left[R^1(H(R^2))CH{-}OTf \right] \xrightarrow{R_3(Ar){-}CN} \left[R^1(H(R^2))CH{-}\overset{+}{N}{\equiv}{-}R^3(Ar)\ OTf^- \right] \xrightarrow[50\text{–}98\%]{NaHCO_3} R^1(H(R^2))CH{-}NHCOR^3(Ar) \quad (4)$$

Aryl triflates are prepared from phenols at 0 °C using pyridine as solvent.[1b] Sometimes it is useful to conduct the reaction in CH_2Cl_2 at −77 °C, as in the preparation of 3,5-di-*t*-butyl-4-hydroxyphenyl triflate (eq 5).[17] Aryl triflates are synthetically transformed into several products of interest and applications in organic chemistry, by cross-coupling reactions with organometallics (eqs 5 and 6).[18]

$$\text{2,6-di-}t\text{-Bu-hydroquinone} \xrightarrow[\text{CH}_2\text{Cl}_2,\ -77\ ^\circ\text{C to rt},\ 88\%]{Tf_2O,\ py} \text{3,5-di-}t\text{-Bu-4-HO-C}_6\text{H}_2\text{-OTf} \quad (5)$$

$$Ar{-}OH \xrightarrow[0\ ^\circ C]{Tf_2O,\ py} Ar{-}OTf \xrightarrow{Pd^0,\ RX} Ar{-}R \quad (6)$$

Reaction of Tf_2O with Amines. The reaction of 1 equiv of Tf_2O in CH_2Cl_2 and Et_3N with amines (or their salts) affords trifluoromethanesulfonamides (triflamides) in good yields.[19,20] If 2 equiv of Tf_2O are used, triflimides are formed. The triflamides are soluble in alkali and readily alkylated to triflimides (eq 7).[19,20]

$$R^1{-}NH_2 \xrightarrow[-78\ ^\circ C]{Tf_2O,\ Et_3N,\ CH_2Cl_2} R^1{-}NHTf \xrightarrow[R^2X]{K_2CO_3} R^1R^2NTf$$

$$R^1{-}NHTf \xrightarrow{Tf_2O} R^1NTf_2 \xrightarrow[-33\ ^\circ C]{Na/NH_3,\ t\text{-BuOH}} R^1NH_2 \quad (7)$$

$R^1 = R^2$ = alkyl

Triflamides can be deprotected reductively (***Sodium-Ammonia***) to yield the corresponding amines.[21] This protocol has been employed in the facile two-step synthesis of aza macrocycles starting from trifluoromethanesulfonyl derivatives of linear tetramines (eq 8).[22]

Several triflamides (**5**–**8**)[23] and *O*-triflylammonium salts[24] have been used for the formation of vinyl triflates from regiospecifically generated metalloenolates or for preparing triflates from alcohols.

$$\text{H}_2\text{N}\frown\text{N(H)}\frown\text{N(H)}\frown\text{NH}_2 \xrightarrow[\text{CH}_2\text{Cl}_2]{\text{Tf}_2\text{O, Et}_3\text{N}}$$

$$\text{TfHN}\frown\text{N(Tf)}\frown\text{N(Tf)}\frown\text{NHTf} \xrightarrow[\text{DMF, 110 °C}]{\text{K}_2\text{CO}_3\text{, Br(CH}_2)_4\text{Br}} \text{tetraaza macrocycle (N–X)}_4 \quad (8)$$

X = Tf → X = H (Na, NH$_3$, −33 °C)

NTf$_2$ (5) (6) (7) (8)

Reaction of Tf$_2$O with Carbonyl Compounds. The reaction of Tf$_2$O with carbonyl compounds consists of the electrophilic attack of the anhydride on the carboxylic oxygen, resulting in the formation of triflyloxycarbenium ions as intermediates (eq 9). According to the nature of the carbonyl compound, the triflyloxycarbenium cations can eliminate a proton giving a vinyl triflate, undergo a rearrangement, or be trapped by the gegenion yielding *gem*-bistriflates (eq 9).

$$\text{R}^1\text{C(O)CH}_2\text{R}^2 \xrightarrow[\text{CH}_2\text{Cl}_2\text{, 0 °C to rt}]{\text{Tf}_2\text{O, base}} \text{R}^1\text{C}^+(\text{OTf})\text{CH}_2\text{R}^2 \xrightarrow{-\text{H}^+} \text{R}^1\text{C(OTf)=CHR}^2 \quad (9)$$

→ TfO OTf R^1 R^2 ; → rearranged products

In the case of acyclic and monocyclic ketones, the reaction with Tf$_2$O affords vinyl triflates in good yields. Several methods exist to realize this reaction.[1b,25] For example, the reaction is carried out at room temperature in CH$_2$Cl$_2$ (or pentane) in the presence of 2,4-di-*t*-butyl-4-methylpyridine (DTBMP) (eq 10).[4]

$$\text{R}^1\text{CH}_2\text{C(O)CH}_2\text{R}^2 \xrightarrow[\text{CH}_2\text{Cl}_2]{\text{Tf}_2\text{O, DTBMP}} \text{R}^1\text{CH=C(OTf)CH}_2\text{R}^2 + \text{R}^1\text{CH}_2\text{C(OTf)=CHR}^2 \quad (10)$$

Other bases such as pyridine,[1b] lutidine,[1b] Et$_3$N,[1b] polymer-bound 2,6-di-*t*-butyl-4-methylpyridine,[26] and 2,4,6-trialkyl-substituted pyrimidines[27] were also used. The commercially available *N,N*-diisobutyl-2,4-dimethyl-3-pentylamine is a very convenient base to prepare the vinyl triflates.[28] In the case of non-functionalized ketones, anhydrous Na$_2$CO$_3$ has been proved to be very successful.[1b,25]

The reaction of ketones with Tf$_2$O is governed by Markovnikov's rule and results in the formation of the more substituted triflate as the major product. When the reaction of ketones[27] and α-halo ketones[29] with Tf$_2$O is carried out in the presence of a nitrile, the intermediate trifloxy cation (eq 9) can be trapped, forming pyrimidines in good yields (eq 11).

$$\text{R}^1\text{C(O)CH}_2\text{X} \xrightarrow[\text{CH}_2\text{Cl}_2\text{, 1–5 d, rt; 55–91\%}]{\text{Tf}_2\text{O, R}^2\text{CN}} \text{pyrimidine (R}^1\text{, R}^2\text{, R}^2\text{, X)} \quad (11)$$

X = Cl, Br, I
R^1 = R^2 = alkyl, aryl

The reaction of Tf$_2$O with strained bicyclic ketones such as 2-norbornanone and nopinone takes place with Wagner–Meerwein rearrangement of the corresponding triflyloxy cations, forming bridgehead triflates in good yields (eq 12).[30] These triflates are key compounds in the preparation of other bridgehead derivatives by substitution[31] and of substituted cyclopentanes by fragmentation.[32]

Tf$_2$O, DTBMP, CH$_2$Cl$_2$, 24 h, rt → OTf 93% + OTf 5% (12)

In the reaction of Tf$_2$O with norcaranones and spiro[2.5]-octan-4-one, the cyclopropane ring undergoes fragmentation to give vinyl triflates (eqs 13 and 14).[33]

Tf$_2$O, Na$_2$CO$_3$, CH$_2$Cl$_2$, 30 min, rt → OTf 36% + OTf 12% (13)

Tf$_2$O, Na$_2$CO$_3$, CH$_2$Cl$_2$, 30 min, rt → OTf 37% + OTf, OTf 37% (14)

However, a cyclopropane ring is formed in the reaction of 5-methylnorborn-5-en-2-one with Tf$_2$O under the same conditions (eq 15).[34]

Tf$_2$O, Na$_2$CO$_3$, CH$_2$Cl$_2$, 30 min, rt → OTf 50% (15)

When the ketone can accomplish neither the stereoelectronic conditions for the elimination of TfOH nor for a rearrangement, the reaction of ketones with Tf_2O results in the formation of a *gem*-bistriflate (eqs 16 and 17).[35]

(16)

(17)

Sensitive ketones such as 3-pentyn-2-one also afford the corresponding vinyl triflate on treatment with Tf_2O in the presence of a base (eq 18).[36]

(18)

Substituted cyclopropenones and tropones react with Tf_2O with the formation of the corresponding dication ether salts (eq 19).[37]

(19)

Treatment of trifluoroacetyl ylides with Tf_2O results in the formation of *gem*-bistriflates (eq 20).[38]

(20)

Reaction of Tf_2O with Aldehydes. The reaction of aliphatic aldehydes with Tf_2O in the presence of 2,6-di-*t*-butyl-4-methylpyridine (DTBMP) in refluxing CH_2Cl_2 or $ClCH_2CH_2Cl$ for 2 h affords the corresponding vinyl triflates as a mixture of (*Z*)- and (*E*)-isomers.[4,39] When the reaction is carried out at 0 °C, *gem*-bistriflates are formed as products (eq 21).[40] The *gem*-bistriflates result due to the trapping of the intermediate triflyloxycarbenium ion by the triflate anion. Primary vinyl triflates have been used extensively in the generation of alkylidene carbenes,[41] and *gem*-bistriflates are interesting precursors for *gem*-dihaloalkanes[42,43] and (*E*)-iodoalkenes.[44]

Reaction with Dicarbonyl Compounds. 1,3-Diketones can be reacted with an equimolar amount of Tf_2O or in excess to furnish the corresponding vinyl triflates or dienyl triflates (eq 22).[45] These triflates are transferred into monoketones, monoalcohols, alkanes, and unsaturated ketones by means of various reducing reagents.[45]

(21)

(22)

The reaction of 3-methylcyclopentane-1,2-dione with Tf_2O/Et_3N affords the vinyl triflate in 53% yield (eq 23).[46] The reaction takes place probably through the enol form. The product was coupled with alkenylzinc compounds in the presence of a palladium catalyst.[46]

(23)

The reaction of β-keto esters[47] with Tf_2O in the presence of a base results in the formation of 2-carboxyvinyl triflates (eq 24). These substrates undergo nucleophilic substitution of the TfO-group (eq 24)[47] and also coupling reactions.[48]

(24)

PNB = *p*-nitrobenzyl carbamate

Reaction with Carboxylic Acids and Esters. The reaction of carboxylic acids and esters with Tf_2O takes place according to the scheme shown in eq 25.[49]

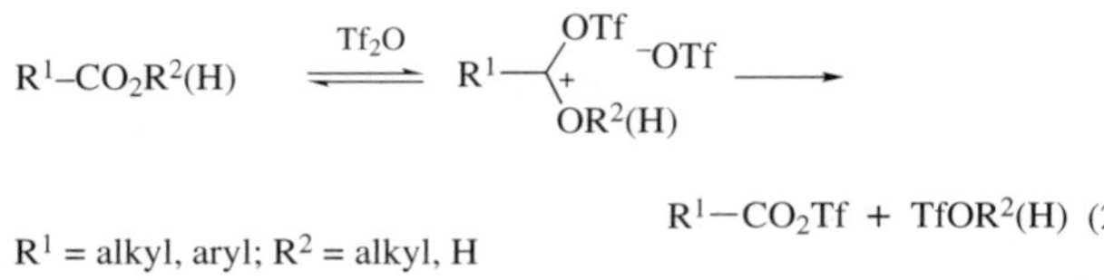

$$R^1{-}CO_2R^2(H) \underset{}{\overset{Tf_2O}{\rightleftharpoons}} R^1{-}C^+(OTf)(OR^2(H))\ {}^-OTf \longrightarrow R^1{-}CO_2Tf + TfOR^2(H) \quad (25)$$

R^1 = alkyl, aryl; R^2 = alkyl, H

The trifluoromethanesulfonic carboxylic anhydrides are highly effective acylation agents, which react without catalysts even with deactivated aromatics to yield aryl ketones (eq 26).[50]

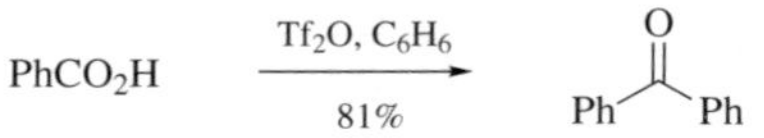

$$PhCO_2H \xrightarrow[81\%]{Tf_2O,\ C_6H_6} PhCOPh \quad (26)$$

Alkyl arylacetates react with Tf_2O to give a cation which in the presence of a nitrile affords isoquinoline derivatives via cyclization of the intermediate nitrilium cation (eq 27).[50]

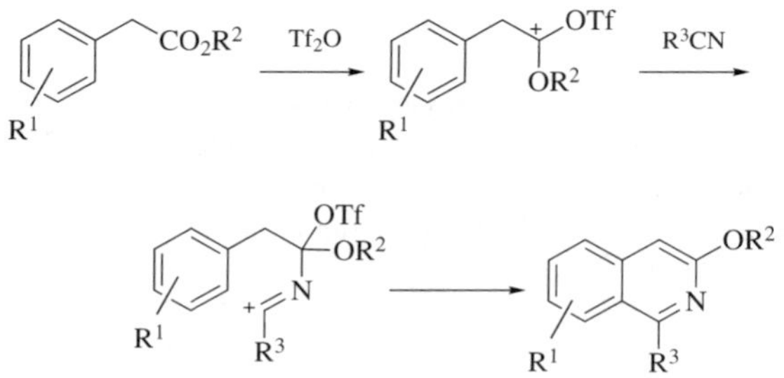

(27)

R^1 = H, 6-Me, 7–Cl, 5–NO_2, 6,7-$(OMe)_2$; R^2 = Et, Me

Reaction of Tf_2O with Amides. The reaction of a 2-oxo-1,2-dihydroquinoline with Tf_2O in the presence of pyridine affords the corresponding 2-quinoline triflate (eq 28).[51]

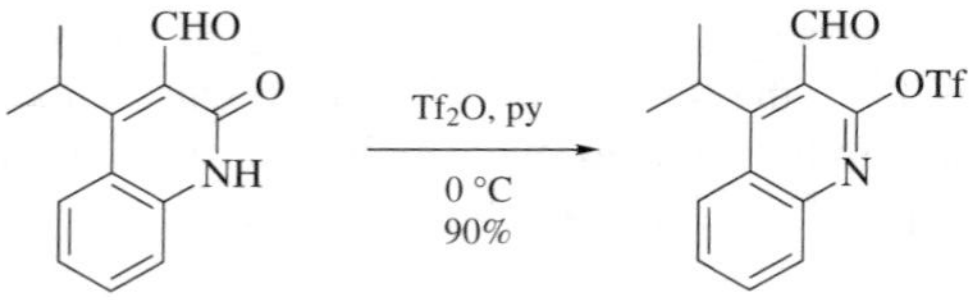

$$\xrightarrow[0\ °C,\ 90\%]{Tf_2O,\ py} \quad (28)$$

The reaction of tertiary amides with Tf_2O gives a mixture of *O*-sulfonylated (major) and *N*-sulfonylated (minor) products. In the presence of collidine and an alkene, [2 + 2] cycloadducts are formed which hydrolyze to give cyclobutanones (eq 29).[52]

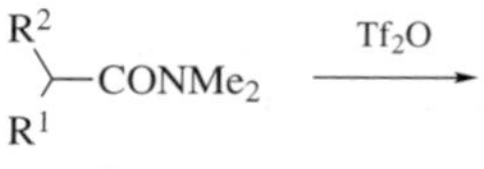

R^1, R^2 = H, Me, Ph

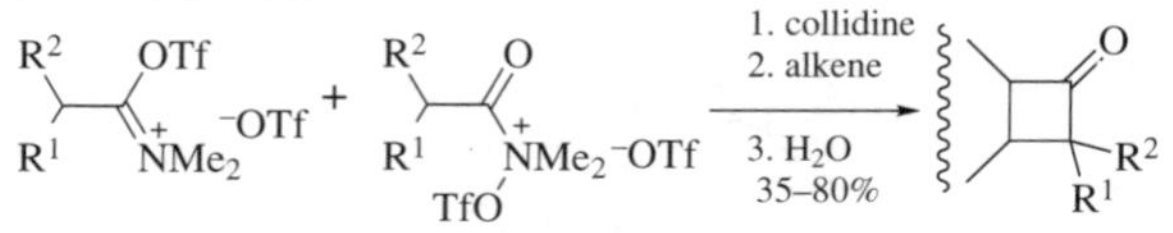

$$\xrightarrow[3.\ H_2O,\ 35\text{–}80\%]{1.\ collidine,\ 2.\ alkene} \quad (29)$$

Treatment of DMF with Tf_2O results in the formation of an imminium triflate, which formylates less active aromatics. It is a convenient variation of the Vilsmeier–Haack reaction (eq 30).[53]

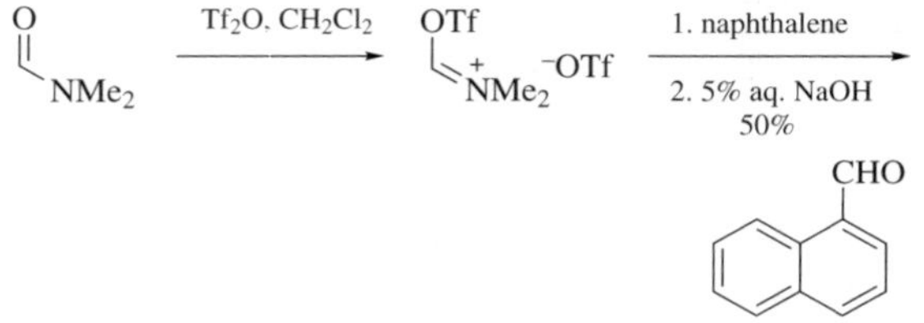

$$\xrightarrow{Tf_2O,\ CH_2Cl_2} \xrightarrow[2.\ 5\%\ aq.\ NaOH,\ 50\%]{1.\ naphthalene} \quad (30)$$

The reaction of *N*-methylpyridone and substituted urea systems with Tf_2O gives heteroatom-stabilized dicarbonium salts (eqs 31 and 32).[37,54]

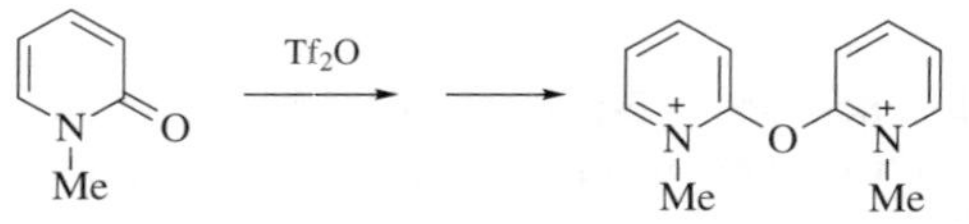

(31)

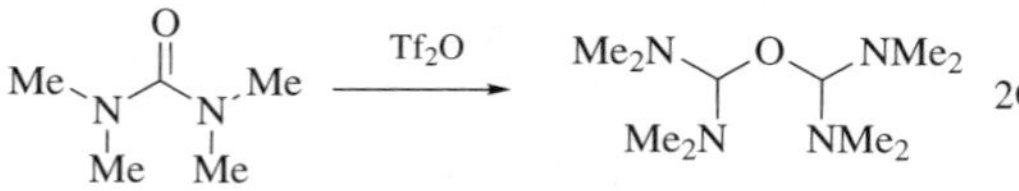

(32)

Secondary amides can be converted to tetrazoles with Tf_2O in the presence of ***Sodium Azide*** (eq 33).[55]

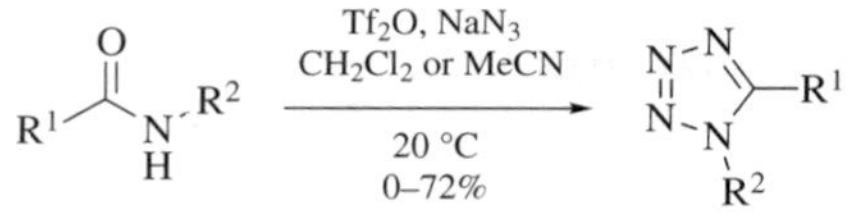

$$\xrightarrow[20\ °C,\ 0\text{–}72\%]{Tf_2O,\ NaN_3,\ CH_2Cl_2\ or\ MeCN} \quad (33)$$

R^1 = alkyl, Ph; R^2 = alkyl, $(CH_2)_2OAc$, $(CH_2)_2OTBDMS$

Other Applications. Activated arenes can be converted to aryl triflones by Friedel–Crafts reaction with Tf_2O using ***Aluminum Chloride*** as catalyst (eq 34).[56]

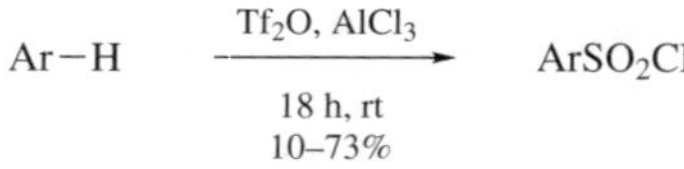

$$Ar{-}H \xrightarrow[18\ h,\ rt,\ 10\text{–}73\%]{Tf_2O,\ AlCl_3} ArSO_2CF_3 \quad (34)$$

The reaction of Tf_2O with Ph_3PO in CH_2Cl_2 at 0 °C affords triphenylphosphine ditriflate, which can be used as an oxygen activator, and then to a diphosphonium salt (eq 35).[57]

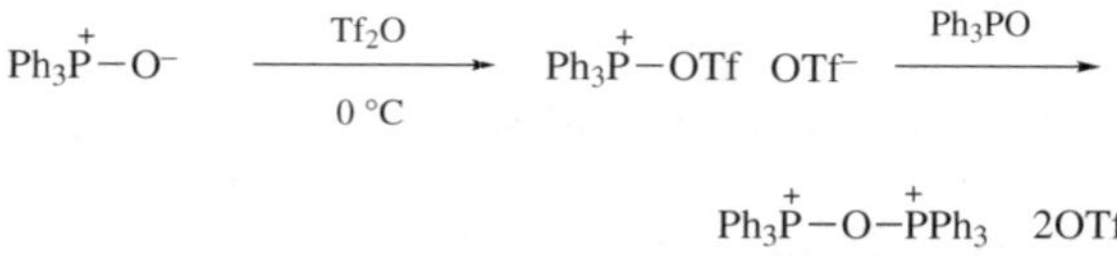

$$Ph_3\overset{+}{P}{-}O^- \xrightarrow[0\ °C]{Tf_2O} Ph_3\overset{+}{P}{-}OTf\ OTf^- \xrightarrow{Ph_3PO} Ph_3\overset{+}{P}{-}O{-}\overset{+}{P}Ph_3\ 2OTf^- \quad (35)$$

The less stable dimethyl sulfide ditriflate, obtained from Tf_2O and DMSO, has been used to oxidize alcohols (eq 36).[58]

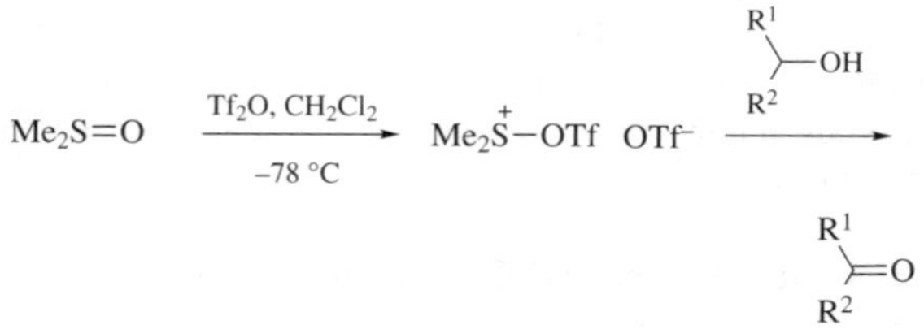

$$Me_2S{=}O \xrightarrow[-78\ °C]{Tf_2O,\ CH_2Cl_2} Me_2\overset{+}{S}{-}OTf\ OTf^- \xrightarrow{R^1R^2CHOH} R^1R^2C{=}O \quad (36)$$

Tetrahydropyran is not a suitable solvent in reactions involving Tf_2O because it is cleaved, affording 1,5-bistrifloxypentane (eq 37).[59]

Tf_2O, 20 °C → TfO(CH$_2$)$_5$OTf (37)

Diols react with Tf_2O to yield the corresponding ditriflates; however, the reaction of 1,1,2,2-tetraphenyl-1,2-ethanediol with Tf_2O takes place with rearrangement (eq 38).[59]

Tf_2O, py, 20 °C (38)

Vinylene 1,2-bistriflates are formed by the reaction of azobenzils with Tf_2O (eq 39).[60]

Tf_2O (39)

83% 3%

The reaction of enolates, prepared from silyl enol ethers and ***Methyllithium***, with Tf_2O affords vinyl triflates (eq 40).[61]

TMSCl, DMF, Et_3N; 1. MeLi, 2. Tf_2O (40)

The combination of equimolecular quantities of ***Iodosylbenzene*** and Tf_2O generates $PhI(OTf)_2$, a compound also formed by treatment of Zefiro's reagent with Tf_2O. As shown in eq 41, this compound can be used to prepare *para*-disubstituted benzene derivatives in good yields.[62]

PhIO →(Tf_2O) $PhI(OTf)_2$ →(ArH); PPh_3 → $Ph_3\overset{+}{P}$–C$_6$H$_4$–$\overset{+}{P}Ph_3$ 2OTf$^-$ (41)

Tf_2O is a suitable promoter for the stereoselective glucosidation of glycosyl acceptors using sulfoxides as donors.[63]

The reaction of Tf_2O with a catalytic amount of ***Antimony(V) Fluoride*** at 25 °C produces trifluoromethyl triflate in 94% yield (eq 42).[64]

$$(CF_3SO_2)_2O \xrightarrow[80\%]{SbF_5} CF_3OSO_2CF_3 \qquad (42)$$

Useful application of Tf_2O as dehydrating reagent is accounted by the synthesis of isocyanides from formamides and vinylformamides (eq 43).[65]

Tf_2O, i-Pr_2NEt, CH_2Cl_2, −78 °C, 86% (43)

Reaction of enaminones with Tf_2O in a 1:1 molar ratio affords 3-trifloxypropeniminium triflates by *O*-sulfonylation. From a cyclic enaminone, by using a 2:1 molar ratio, the corresponding bis(3-amino-2-propenylio) bistriflate is obtained (eq 44).[66]

Tf_2O, 81% (44)

Related Reagents. Methanesulfonyl Chloride; Mesitylenesulfonyl Chloride; *N*-Phenyltrifluoromethanesulfonimide; *p*-Toluenesulfonyl Chloride; Trifluoromethanesulfonyl Chloride.

1. (a) Gramstad, T.; Haszeldine, R. N. *JCS* **1957**, 4069. (b) Stang, P. J.; Hanack, M.; Subramanian, L. R. *S* **1982**, 85. (c) Stang, P. J.; White, M. R. *Aldrichim. Acta* **1983**, *16*, 15.
2. Binkley, R. W.; Ambrose, M. G. *J. Carbohydr. Chem.* **1984**, *3*, 1.
3. Ambrose, M. G.; Binkley, R. W. *JOC* **1983**, *48*, 674.
4. Stang, P. J.; Treptow, W. *S* **1980**, 283.
5. García Martínez, A.; Herrera Fernandez, A.; Martínez Alvarez, R.; Silva Losada, M. C.; Molero Vilchez, D.; Subramanian, L. R.; Hanack, M. *S* **1990**, 881.
6. Yoshida, M.; Takeuchi, K. *JOC* **1993**, *58*, 2566.
7. Jeanneret, V.; Gasparini, F.; Péchy, P.; Vogel, P. *T* **1992**, *48*, 10637.
8. Neenan, T. X.; Houlihan, F. M.; Reichmanis, E.; Kometani, J. M.; Bachman, B. J.; Thompson, L. F. *Macromolecules* **1990**, *23*, 145.
9. (a) Sato, K.; Hoshi, T.; Kajihara, Y. *CL* **1992**, 1469. (b) Izawa, T.; Nakayama, K.; Nishiyama, S.; Yamamura, S.; Kato, K.; Takita, T. *JCS(P1)* **1992**, 3003. (c) Knapps, S.; Naughton, A. B. J.; Jaramillo, C.; Pipik, B. *JOC* **1992**, *57*, 7328.
10. Takeuchi, K.; Kitagawa, T.; Ohga, Y.; Yoshida, M.; Akiyama, F.; Tsugeno, A. *JOC* **1992**, *57*, 280.
11. Eaton, P. E.; Zhou, J. P. *JACS* **1992**, *114*, 3118.
12. Spitz, U. P. *JACS* **1993**, *115*, 10174.
13. Zheng, C. Y.; Slebocka-Tilk, H.; Nagorski, R. W.; Alvarado, L.; Brown, R. S. *JOC* **1993**, *58*, 2122.
14. Effenberger, F.; Weber, Th. *CB* **1988**, *121*, 421.
15. (a) Rubinsztajn, S.; Fife, W. K.; Zeldin, M. *TL* **1992**, *33*, 1821. (b) Dodd, R. H.; Poissonnet, G.; Potier, P. *H* **1989**, *29*, 365.
16. García Martínez, A.; Martínez Alvarez, R.; Teso Vilar, E.; García Fraile, A.; Hanack, M. Subramanian, L. R. *TL* **1989**, *30*, 581.
17. Sonoda, T.; García Martínez, A.; Hanack, M.; Subramanian, L. R. *Croat. Chim. Acta* **1992**, *65*, 585 (*CA* **1993**, *118*, 168 491).
18. Ritter, K. *S* **1993**, 735.
19. Hendrickson, J. B.; Bergeron, R. *TL* **1973**, 4607.
20. Hendrickson, J. B.; Bergeron, R.; Giga, A.; Sternbach, D. D. *JACS* **1973**, *95*, 3412.
21. Edwards, M. L.; Stemerick, D.; McCarthy, J. R. *TL* **1990**, *31*, 3417.
22. Panetta, V.; Yaouanc, J. J.; Handel, H. *TL* **1992**, *33*, 5505.
23. (a) McMurry, J. E.; Scott, W. J. *TL* **1983**, *24*, 979. (b) Comins, J. E.; Dehghani, A. *TL* **1992**, *33*, 6299. (c) Crisp, G. T.; Flynn, B. L. *T* **1993**, *49*, 5873.
24. Anders, E.; Stankowiak, A. *S* **1984**, 1039.
25. García Martínez, A.; Herrera Fernandez, A.; Alvarez, R. M.; Sánchez García, J. M. *An. Quim., Ser. C* **1981**, *77c*, 28 (*CA* **1982**, *97*, 5840).

26. (a) Wright, M. E.; Pulley, S. R. *JOC* **1987**, *52*, 5036. (b) Dolle, R. E.; Schmidt, S. J.; Erhard, K. F.; Kruse, L. I. *JACS* **1989**, *111*, 278.
27. García Martínez, A.; Herrera Fernandez, A.; Moreno Jiménez, F.; García Fraile, A.; Subramanian, L. R.; Hanack, M. *JOC* **1992**, *57*, 1627.
28. Stang, P. J.; Kowalski, M. H.; Schiavelli, M. D.; Longford, R. *JACS* **1989**, *111*, 3347.
29. García Martínez, A.; Herrera Fernández, A.; Molero Vilchez, D.; Hanack, M.; Subramanian, L. R. *S* **1992**, 1053.
30. (a) Bentz, H.; Subramanian, L. R.; Hanack, M.; García Martínez, A.; Gómez Marin, M.; Perez-Ossorio, R. *TL* **1977**, 9. (b) Kraus, W.; Zartner, G. *TL* **1977**, 13. (c) García Martínez, A.; García Fraile, A.; Sánchez García, J. M. *CB* **1983**, *116*, 815. (d) García Martínez, A.; Teso Vilar, E.; Gómez Marin, M.; Ruano Franco, C. *CB* **1985**, *118*, 1282.
31. (a) García Martínez, A.; Teso Vilar, E.; López, J. C.; Manrique Alonso, J.; Hanack, M.; Subramanian, L. R. *S* **1991**, 353. (b) García Martínez, A.; Teso Vilar, E.; García Fraile, A.; Ruano Franco, C.; Soto Salvador, J.; Subramanian, L. R.; Hanack, M. *S* **1987**, 321.
32. (a) García Martínez, A.; Teso Vilar, E.; García Fraile, A.; Osío Barcina, J.; Hanack, M.; Subramanian, L. R. *TL* **1989**, *30*, 1503. (b) García Martínez, A.; Teso Vilar, E.; Osío Barcina, J.; Manrique Alonso, J.; Rodríguez Herrero, E.; Hanack, M.; Subramanian, L. R. *TL* **1992**, *33*, 607.
33. García Martínez, A.; Herrera Fernandez, A.; Sánchez García, J. M. *An. Quim.* **1979**, *75*, 723 (*CA* **1980**, *93*, 45 725).
34. García Martínez, A.; Espada Rios, I.; Osío Barcina, J.; Teso Vilar, E. *An. Quim., Ser. C* **1982**, *78c*, 299 (*CA* **1983**, *98*, 159 896).
35. (a) García Martínez, A.; Espada Rios, I.; Teso Vilar, E. *S* **1979**, 382. (b) García Martínez, A.; Espada Rios, I.; Osío Barcina, J.; Montero Hernando, M. *CB* **1984**, *117*, 982.
36. Hanack, M.; Hadenteufel, J. R. *CB* **1982**, *115*, 764.
37. Stang, P. J.; Maas, G.; Smith, D. L.; McCloskey, J. A. *JACS* **1981**, *103*, 4837.
38. Wittmann, H.; Ziegler, E.; Sterk, H. *M* **1987**, *118*, 531.
39. Wright, M. E.; Pulley, S. R. *JOC* **1989**, *54*, 2886.
40. García Martínez, A.; Martínez Alvarez, R.; García Fraile, A.; Subramanian, L. R.; Hanack, M. *S* **1987**, 49.
41. Stang, P. J. *CRV* **1978**, *78*, 383.
42. García Martínez, A.; Herrera Fernandez, A.; Martínez Alvarez, R.; García Fraile, A.; Calderón Bueno, J.; Osío Barcina, J.; Hanack, M.; Subramanian, L. R. *S* **1986**, 1076.
43. García Martínez, A.; Osío Barcina, J.; Rys, A. Z.; Subramanian, L. R. *TL* **1992**, *33*, 7787.
44. García Martínez, A.; Martínez Alvarez, R.; Martínez Gonzalez, S.; Subramanian, L. R.; Conrad, M. *TL* **1992**, *33*, 2043.
45. García Martínez, A.; Martínez Alvarez, R.; Madueño Casado, M.; Subramanian, L. R.; Hanack, M. *T* **1987**, *43*, 275.
46. Negishi, E.; Owczarczyk, Z.; Swanson, D. R. *TL* **1991**, *32*, 4453.
47. Evans, D. A.; Sjogren, E. B. *TL* **1985**, *26*, 3787.
48. (a) Houpis, I. N. *TL* **1991**, *32*, 6675. (b) Cook, G. K.; Hornback, W. J.; Jordan, C. L.; McDonald III, J. H.; Munroe, J. E. *JOC* **1989**, *54*, 5828.
49. (a) Effenberger, F.; Sohn, E.; Epple, G. *CB* **1983**, *116*, 1195. (b) Effenberger, F. *AG* **1980**, *92*, 147; *AG(E)* **1980**, *19*, 151.
50. García Martínez, A.; Herrera Fernández, A.; Molero Vilchez, D.; Laorden Gutiérrez, L.; Subramanian, L. R. *SL* **1993**, 229.
51. Robl, J. A. *S* **1991**, 56.
52. Falmagne, J.-B.; Escudero, J.; Taleb-Sahraoui, S.; Ghosez, L. *AG* **1981**, *93*, 926; *AG(E)* **1981**, *20*, 879.
53. García Martínez, A.; Martínez Alvarez, R.; Osío Barcina, J.; de la Moya Cerero, S.; Teso Vilar, E.; García Fraile, A.; Hanack, M.; Subramanian, L. R. *CC* **1990**, 1571.
54. Gramstad, T.; Husebye, S.; Saebo, J. *TL* **1983**, *24*, 3919.
55. Thomas, E. W. *S* **1993**, 767.
56. Hendrickson, J. B.; Bair, K. W. *JOC* **1977**, *42*, 3875.
57. Hendrickson, J. B.; Schwartzman, S. M. *TL* **1975**, 277.
58. Hendrickson, J. B.; Schwartzman, S. M. *TL* **1975**, 273.
59. Lindner, E.; v. Au, G.; Eberle, H.-J. *CB* **1981**, *114*, 810.
60. Maas, G.; Lorenz, W. *JOC* **1984**, *49*, 2273.
61. (a) Stang, P. J.; Magnum, M. G.; Fox, D. P.; Haak, P. *JACS* **1974**, *96*, 4562. (b) Hanack, M.; Märkl, R.; García Martínez, A. *CB* **1982**, *115*, 772.
62. (a) Kitamura, T.; Furuki, R.; Nagata, K.; Taniguchi, H.; Stang, P. J. *JOC* **1992**, *57*, 6810. (b) Stang, P. J.; Zhdankin, V. V.; Tykwinski, R.; Zefirov, N. *TL* **1992**, *33*, 1419.
63. Raghavan, S.; Kahne, D. *JACS* **1993**, *115*, 1580.
64. Taylor, S. L.; Martin, J. C. *JOC* **1987**, *52*, 4147.
65. Baldwin, J. E.; O'Neil, I. A. *SL* **1990**, 603.
66. Singer, B.; Maas, G. *CB* **1987**, *120*, 485.

Antonio García Martínez, Lakshminarayanapuram R. Subramanian & Michael Hanack
Universität Tübingen, Germany

Triisopropylsilyl Chloride

$i\text{-}Pr_3SiCl$

[131154-24-0] $C_9H_{21}ClSi$ (MW 192.84)

(hydroxy protecting group;[1,2] formation of triisopropylsilyl ynol ethers;[3] *N*-protection of pyrroles;[4,5] prevents chelation with Grignard reagents[6])

Alternate Names: TIPSCl; chlorotriisopropylsilane.
Physical Data: bp 198 °C/739 mmHg; *d* 0.901 g mL^{-3}.
Solubility: sol THF, DMF, CH_2Cl_2.
Form Supplied in: clear, colorless liquid; commercially available (99% purity).
Analysis of Reagent Purity: bp; NMR.
Purification: distillation under reduced pressure.
Handling, Storage, and Precautions: moisture sensitive; therefore should be stored under an inert atmosphere; corrosive; use in a fume hood.

Hydroxy Protecting Group. Several hindered triorganosilyl protecting groups have been developed to mask the hydroxy functionality. Although the *t*-butyldiphenylsilyl (TBDPS) and *t*-butyldimethylsilyl (TBDMS) groups are the most widely used, the triisopropylsilyl group (TIPS) has several properties which make it particularly attractive for use in a multi-step synthesis.[1,2]

Introduction of the TIPS group is most frequently accomplished using TIPSCl and ***Imidazole*** in DMF,[1] although several other methods exist, including using TIPSCl and ***4-Dimethylaminopyridine*** in CH_2Cl_2. It is possible to silylate hydroxy groups selectively in different steric environments. For example, primary alcohols can be silylated in the presence of secondary alcohols (eq 1)[7] and less hindered secondary alcohols can be protected in the presence of more hindered ones (eq 2).[8]

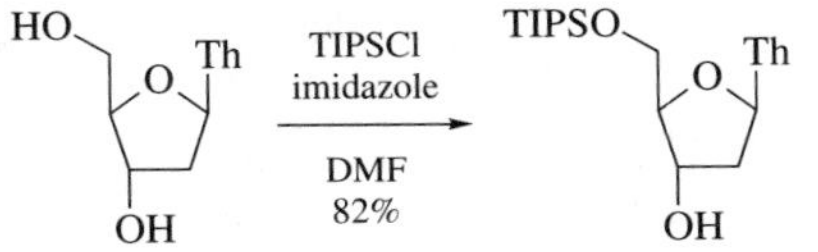

A comparison of the stability of different trialkylsilyl groups has shown that TIPS ethers are more stable than TBDMS ethers but less stable than TBDPS ethers toward acid hydrolysis.[2] The rate difference is large enough that a TBDMS group can be removed in the presence of a TIPS group (eq 3).[7] Under basic hydrolysis, TIPS ethers are more stable than TBDMS or TBDPS ethers.[2] The cleavage of TIPS ethers can also be accomplished by using ***Tetra-n-butylammonium Fluoride*** (TBAF) in THF at rt.[1] This is most convenient if only one silyl protecting group is present or if all of those present can be removed in one synthetic step.

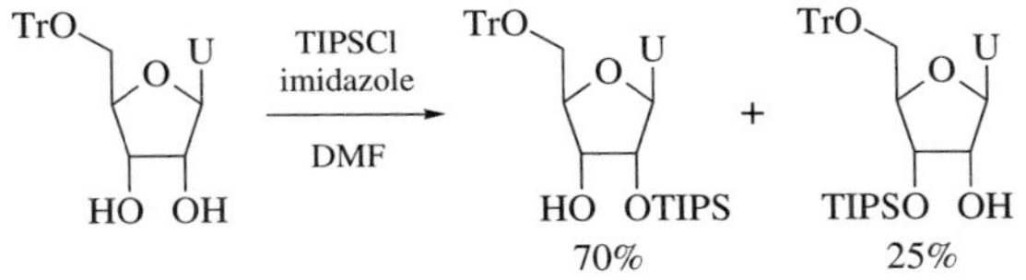

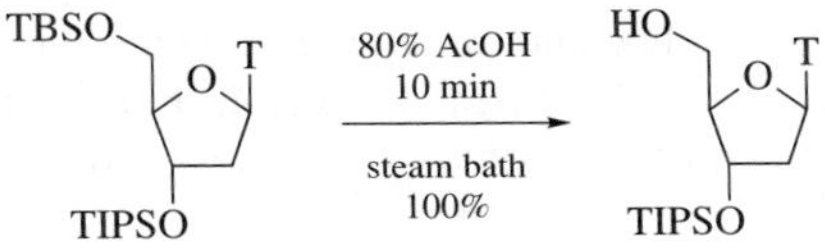

An additional feature of TIPS ethers is that they are volatile enough to make them amenable to GC and MS analysis. In fact, MS fragmentation patterns have been used to discern the structures of isomeric nucleosides.[9]

Formation of Silyl Ynol Ethers. Esters can also be transformed into triisopropylsilyl ynol ethers.[3] The ester is first converted to the ynolate anion, followed by treatment with TIPSCl to furnish the TIPS ynol ether (eq 4). This method has even proven successful with lactones as starting materials (eq 5).

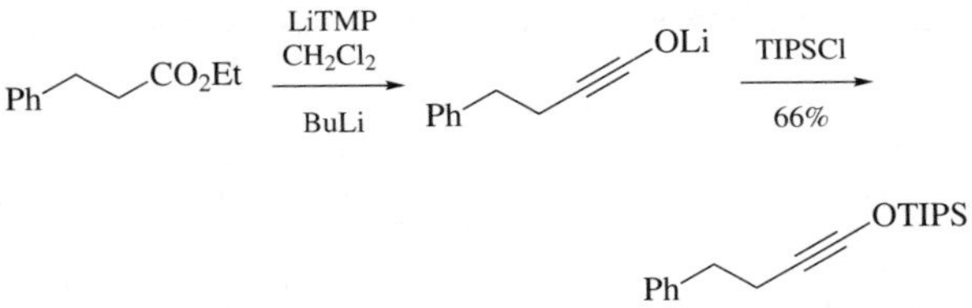

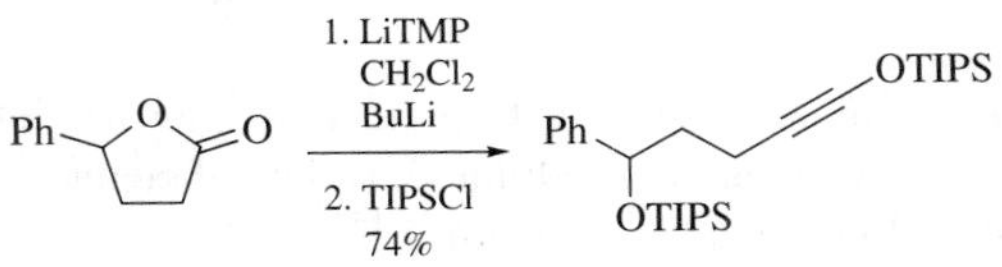

N*-Protection of Pyrroles.** Pyrroles typically undergo electrophilic substitution at the α-(2)-position, but when protected as *N*-triisopropylsilylpyrroles, substitution occurs exclusively at the β-(3)-position (eq 6).[4] It has been shown that 3-bromo-1-(triisopropylsilyl)pyrrole undergoes rapid halogen–metal exchange with ***n-Butyllithium to generate the 3-lithiopyrrole, which can be trapped by electrophiles to provide the silylated 3-substituted pyrrole (eq 7).[5] As expected, the silyl group can be removed with TBAF to furnish the 3-substituted pyrrole. It should also be mentioned that the *N*-TIPS group also enhances the stability of some pyrroles. For example, 3-bromopyrrole is very unstable, but the *N*-silylated derivative is stable for an indefinite period of time.[5]

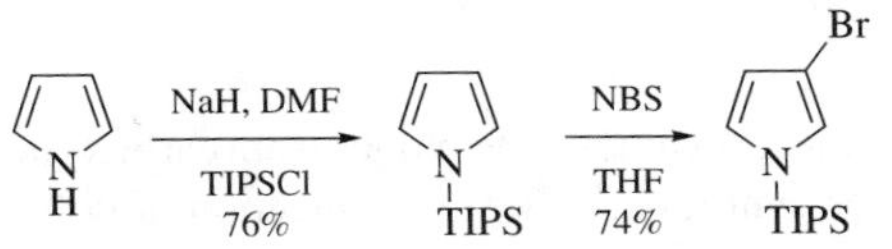

Prevention of Chelation in Grignard Reactions. The bulky TIPS protecting group has proven extremely effective in preventing competing chelation of α- and β-oxygen functionalities during Grignard reactions (eq 8).[6] It was determined that this effect is steric and not electronic since TMS and TBDMS ethers did not affect selectivity.

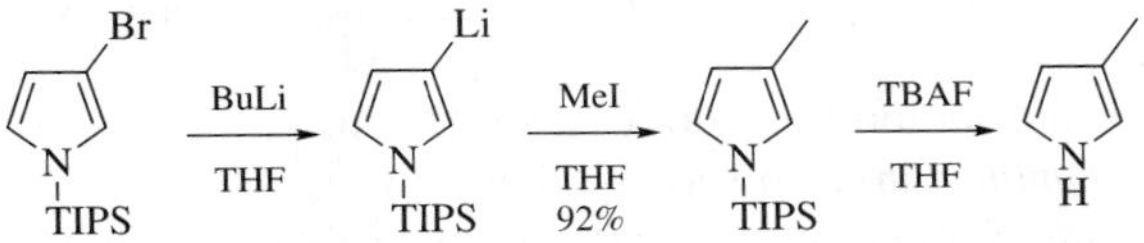

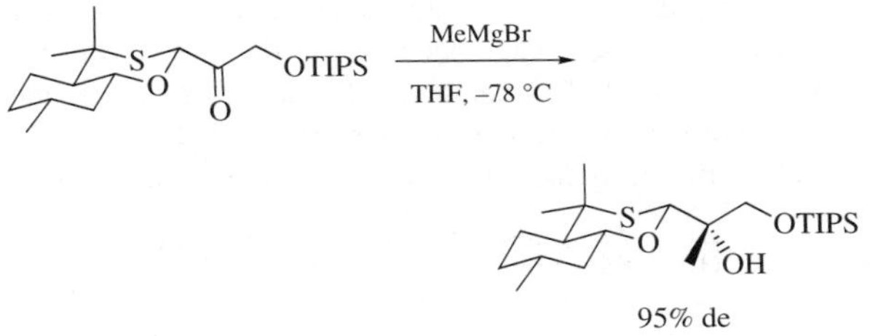

Related Reagents. *t*-Butyldimethylchlorosilane; *t*-Butyldimethylsilyl Trifluoromethanesulfonate; 3*t*-Butyldiphenylchlorosilane; Chlorotriphenylsilane; Triisopropylsilyl Trifluoromethanesulfonate.

1. Green, T. W.; Wuts, P. G. M. *Protective Groups in Organic Synthesis*, 2nd ed.; Wiley: New York, 1991; p 74.
2. Cunico, R. F.; Bedell, L. *JOC* **1980**, *45*, 4797.
3. Kowalski, C. J.; Lal, G. S.; Haque, M. S. *JACS* **1986**, *108*, 7127.
4. Muchowski, J. M.; Solas, D. R. *TL* **1983**, *24*, 3455.
5. Kozikowski, A. P.; Cheng, X.-M. *JOC* **1984**, *49*, 3239.
6. Frye, S. V.; Eliel, E. L. *TL* **1986**, *27*, 3223.
7. Ogilvie, K. K.; Thompson, E. A.; Quilliam, M. A.; Westmore, J. B. *TL* **1974**, 2865.
8. Ogilvie, K. K.; Sadana, K. L.; Thompson, E. A.; Quilliam, M. A.; Westmore, J. B. *TL* **1974**, 2861.
9. Ogilvie, K. K.; Beaucage, S. L.; Entwistle, D. W.; Thompson, E. A.; Quilliam, M. A.; Westmore, J. B. *J. Carbohydr., Nucleosides, Nucleotides* **1976**, 197.

Ellen M . Leahy
Affymax Research Institute, Palo Alto, CA, USA

Trimethylacetyl Chloride

[3282-30-2] C_5H_9ClO (MW 120.59)

(ether cleavage;[1] synthesis of ketones from carboxylic acids;[2] peptide synthesis;[3] selective protection of polyols[4])

Alternate Names: pivaloyl chloride; 2,2-dimethylpropanoyl chloride.
Physical Data: bp 105–106 °C; *d* 0.979 g cm^{-3}.
Form Supplied in: colorless liquid; widely available.
Handling, Storage, and Precautions: flammable; can cause severe burns and is irritating to the eyes and respiratory system; toxic by contact with the skin, by inhalation, and if swallowed. It should be used in a fume hood. During use, avoid sources of ignition and any contact. Avoid exposure to moisture.[5]

Ether Cleavage. Ethers can be cleaved by the action of trimethylacetyl chloride (*t*-BuCOCl) and ***Sodium Iodide*** to give the corresponding trimethylacetate and alkyl iodide.[1] The reaction is regioselective, the cleavage occurring at the less substituted α-carbon–oxygen bond.[1,6] Other acyl chlorides can be used, but the reaction is more regioselective with trimethylacetyl chloride.[1] The reaction is particularly useful for cleaving methyl ethers (eq 1).[1]

Oct—OMe → (*t*-BuCOCl, NaI; MeCN, rt; 90%) Oct–O–C(O)–*t*-Bu (1)

Ketone Synthesis. Carboxylic acids can be converted to ketones by first forming the mixed anhydride with trimethylacetyl chloride and then treating it with a Grignard reagent (eq 2).[2] In addition to trimethylacetyl chloride, *ortho*-substituted benzoyl chlorides such as *o*-anisoyl chloride can be used to form the mixed anhydride. Yields were found to be slightly higher than when trimethylacetyl chloride was used.[2]

Ph–CH$_2$CH$_2$–COOH → (*t*-BuCOCl, Et$_3$N; THF, –30 °C) mixed anhydride (O–C(O)–*t*-Bu) → (PhMgBr; THF, –78 °C) Ph–CH$_2$CH$_2$–C(O)–Ph, 73% (2)

The carboxylic acid (**1**) has been converted to the ketone (**2**) via the mixed anhydride with trimethylacetyl chloride (eq 3).[7]

HO$_2$C(CH$_2$)$_7$CO$_2$Me (**1**) → (1. *t*-BuCOCl, Et$_3$N, THF, –15 °C; 2. BrMg(CH$_2$)$_3$SPh, THF, –78 °C to rt; 70%) PhS(CH$_2$)$_3$C(O)(CH$_2$)$_7$CO$_2$Me (**2**) (3)

A γ-keto aldehyde synthesis has been developed using similar methodology. The ethoxy lactam (**3**) was converted to its *N*-trimethylacetyl derivative (**4**), which was then treated with a Grignard reagent. Acid workup gave the γ-keto aldehyde (**5**) (eq 4).[8]

(**3**) → (1. BuLi, THF, –78 °C; 2. *t*-BuCOCl, THF, –78 °C) (**4**) → (3. HexMgX, THF, –78 °C; 4. 6N HCl, reflux) Hex–C(O)–CH$_2$CH$_2$–CHO (**5**) 74% (4)

Peptide Synthesis. Trimethylacetyl chloride has been used to synthesize peptidic bonds by the mixed anhydride method (eq 5).

R^1–C(O)–OH → (*t*-BuCOCl, base) R^1–C(O)–O–C(O)–*t*-Bu → (H$_2$NR^2) R^1–C(O)–NH–R^2 (5)

A comparison of a range of carboxylic acid chlorides showed trimethylacetyl chloride to be one of the most effective for peptide synthesis, though slightly inferior to diethylacetyl chloride.[3] The extent of racemization that occurred using various acyl chlorides has been studied and trimethylacetyl chloride was shown to give only a very small amount of racemization.[9] Trimethylacetyl chloride has been used in the synthesis of adrenocorticotrophic hormone.[10,11]

Selective Protection of Polyols. The protection (as the trimethylacetate) of a primary alcohol in the presence of two secondary alcohols has been achieved in high yield using 1 equiv of trimethylacetyl chloride (eq 6).[4]

Triol → (*t*-BuCOCl, pyridine; CH$_2$Cl$_2$, 0 °C to rt; 90%) primary O$_2$C-*t*-Bu ester (6)

The selective protection of a less hindered primary alcohol in the presence of another more hindered one has also been achieved (eq 7).[12]

Diol → (*t*-BuCOCl, pyridine; CH$_2$Cl$_2$, –20 °C; 48%) mono-*t*-BuCO$_2$ ester (7)

Sucrose has been selectively protected using trimethylacetyl chloride. By varying the conditions a variety of penta-, hexa-, and heptatrimethylacetates was obtained.[13]

Related Reagents. Acetic Anhydride; Isobutyl Chloroformate; 2,4,6-Trichlorobenzoyl Chloride.

---

1. Oku, A.; Harada, T.; Kita, K. *TL* **1982**, *23*, 681.
2. Araki, M.; Mukaiyama, T. *CL* **1974**, 663.
3. Vaughan, J. R.; Osato, R. L. *JACS* **1951**, *73*, 5553.

4. Nicolaou, K. C.; Webber, S. E. *S* **1986**, 453.
5. For further information, see: *The Sigma-Aldrich Library of Chemical Safety Data*, 2nd ed.; Leng, R. E., Ed.; Sigma-Aldrich: Milwaukee, 1987; Vol. 2, p 3479.
6. Rodriguez, J.; Dulcere, J.-P.; Bertrand, M. *TL* **1984**, *25*, 527.
7. Bakuzis, P.; Bakuzis, M. L. F. *JOC* **1977**, *42*, 2362.
8. Savoia, D.; Concialini, V.; Roffia, S.; Tarsi, L. *JOC* **1991**, *56*, 1822.
9. Taschner, E.; Smulkowski, M.; Lubiewska-Nakonieczna, L. *LA* **1970**, *739*, 228.
10. Schwyzer, R.; Sieber, P. *Nature* **1963**, *199*, 172.
11. Schwyzer, R.; Sieber, P. *HCA* **1965**, *49*, 134.
12. Schuda, P. F.; Heimann, M. R. *TL* **1983**, *24*, 4267.
13. Hough, L.; Chowdhary, M. S.; Richardson, A. C. *CC* **1978**, 664.

Christopher J. Urch
Zeneca Agrochemicals, Bracknell, UK

Trimethyloxonium Tetrafluoroborate[1]

$Me_3O^+BF_4^-$

[420-37-1] $C_3H_9BF_4O$ (MW 147.93)

(methylating agent;[1] activates C–X multiple bonds;[2] esterifies polyfunctional carboxylic acids;[3] catalyst for polymerization of cyclic sulfides and ethers; Beckmann rearrangement of oximes[4])

Physical Data: mp 179.6–180.0 °C (sealed tube, with dec).
Solubility: sol nitrobenzene, nitromethane, $CHCl_3$, acetone (hot), SO_2 (liq); slightly sol CH_2Cl_2; insol common organic solvents.
Form Supplied in: white crystalline solid; commercially available; is contaminated by ethyldimethyloxonium tetrafluoroborate.
Analysis of Reagent Purity: ^1H NMR (CD_2Cl_2/SO_2) δ 4.68 (s, CH_3); ^{13}C NMR (CD_2Cl_2/SO_2) δ 78.8 (CH_3).
Purification: highly pure oxonium salt is obtained from dimethoxycarbenium tetrafluoroborate and dimethyl ether.[5] The resulting solid is vacuum dried at 50 °C/1 mmHg for 30 min.
Handling, Storage, and Precautions: when prepared according to Curphey,[6] the oxonium salt is stable, nonhygroscopic, and may be readily handled in the air for short periods of time. The dry oxonium salt should be stored under argon at −15 °C or as a suspension in CH_2Cl_2 at −20 °C for prolonged periods. Batches stored in this manner for over a year have been successfully used for alkylations. Highly pure samples[5] showed no change in spectral data after this time. Reactions of the reagent should be performed under an argon or a nitrogen atmosphere; dry solvents are also necessary. Because of its caustic nature and potent properties as an alkylating agent, direct contact with the skin must be avoided.

Functional Group Methylations. The powerful alkylating property of trimethyloxonium tetrafluoroborate allows methylation of sensitive or weakly nucleophilic functional groups. Smooth alkylation of a variety of anions or uncharged molecules has been reported. Examples include carboxylic acids,[7] ketones,[2,8] lactones,[4,9] nitriles,[4,10] *O*-acetals,[4] *S*-acetals,[11] sulfoxides,[12] sulfides,[13] thioamides,[14] thiophenes,[13g] and sulfonium ylides.[14] As should be expected for strongly electrophilic agents,[1c] the hard oxonium salt yields considerable amounts of products derived from methylation at the site of the highest electron density (HSAB principle) when it reacts with ambident nucleophiles. *O*-Methylation of sulfoxides,[15] *N*-methylation of sulfoximides,[16] sulfinyl amines,[17] sulfodiimides, and nitriles[10] and *S*-methylation of thioamides[14b] have been observed. The very reactive cationic intermediates (e.g. carboxonium, carbosulfonium, or nitrilium ions) as primary products in those reactions are versatile intermediates for synthetic transformations. Ketones can be converted into α-acetoxy ketones via *N*-methylation of the oxime acetates (eq 1)[18] and thioamides to amides (eq 2).[14a,b]

$Pr_2C{=}O$ —(1. H_2NOH; 2. Ac_2O)→ $Pr_2C{=}NOAc$ —($Me_3O^+ BF_4^-$, $MeNO_2$, 25 °C)→ [$Pr_2C{=}N^+(Me)OAc$] BF_4^- —(Et_3N)→ → → PrC(=O)CH(OAc)Et (1)
48%

PhC(=S)N(*i*-Pr)$_2$ —($Me_3O^+ BF_4^-$, C_6H_6, 25 °C)→ PhC(SMe)=N^+(*i*-Pr)$_2$ BF_4^- —(OH^-)→ PhC(=O)N(*i*-Pr)$_2$ (2)
85%

The reactions of thioacetals with $Me_3O^+BF_4^-$ constitute a high-yielding method for deprotection of thioacetals to give ketones (eq 3).[11a,19]

$R^1R^2C(S)_2(CH_2)_n$ —(1. 6 equiv $Me_3O^+ BF_4^-$, CH_2Cl_2; 2. H_2O; 90–95%)→ $R^1R^2C{=}O$ (3)

The reagent has been employed as a quaternizing agent for a variety of N-heterocycles.[20] Regioselective *N*-methylation has been observed (eq 4).[20c] Thiazoles are exclusively *N*-methylated,[20a,b] whereas tetrazoles yield regioisomeric products.[20d,h] The dication of squaric acid bis(amidine) is prepared by regioselective *N*-methylation (eq 5).[21]

(eq 4: O_2N-aryl bicyclic aziridine–imidazoline —($Me_3O^+ BF_4^-$, CH_2Cl_2)→ *N*-Me iminium BF_4^- salt) (4)

8 equiv $Me_3O^+BF_4^-$, $ClCH_2CH_2Cl$, i-Pr_2NEt, N_2, Δ, 4 h, 79% (5)

1,3-Benzothiazolines react preferentially at nitrogen,[22] whereas heterocycles containing dipolar carbonyl groups usually yield products resulting from *O*-methylation.[2,23] Furthermore, $Me_3O^+BF_4^-$ is the reagent of choice for the synthesis of higher alkylated products which are otherwise not obtainable. Examples are the preparation of vinylidene disulfonium salts (eq 6),[13c] dionium salts of *N,S*-[13b] and *S,S*-acetals,[11a–c,24] or 1,4-dithianium ditetrafluoroborates (eq 7),[19] and a heterocyclophane-type tetrasulfonium salt (eq 8). The latter introduces a hydrophobic cavity into the aqueous phase, thus serving as an inclusion catalyst.[25]

$Me_3O^+BF_4^-$, CH_2Cl_2, 20 °C; $Me_3O^+BF_4^-$; $2BF_4^-$ + $2BF_4^-$; 70% (6)

2.1 equiv $Me_3O^+BF_4^-$, CH_2Cl_2, rt, 14 d, 88%; $2BF_4^-$ (7)

mixture of two stereoisomers

$Me_3O^+BF_4^-$; $4 BF_4^-$ (8)

Trialkyloxonium salts such as $Me_3O^+BF_4^-$ are excellent reagents for the generation of cyclopropenylium cations by *O*-methylation of suitable cyclopropenones (eq 9).[26]

$Me_3O^+BF_4^-$, CH_2Cl_2; OMe BF_4^- (9)

The reaction of trialkyloxonium salts with carboxylic acids is a mild, general esterification procedure that does not require the use of more hazardous reagents such as ***Hexamethylphosphoric Triamide–Thionyl Chloride***, ***Iodomethane***, ***Dimethyl Sulfate***, ***3-Methyl-1-p-tolyltriazene***, or ***Diazomethane***. The reaction proceeds smoothly with sterically hindered acids, as well as with acids containing various functional groups such as amides or nitriles (eq 10).[3,7]

$Me_3O^+BF_4^-$, i-Pr_2NH, CH_2Cl_2, 25 °C, 86% (10)

One of the major advantages of the use of oxonium salts is that alkylations can be effected under reaction conditions that are generally much milder than those necessary with conventional alkyl halides and sulfonates. $Me_3O^+BF_4^-$ has been used as a suspension in CH_2Cl_2 or dichloroethane, or as a solution in nitromethane or liquid SO_2. Alkylations in water[27] and trifluoroacetic acid[28] have been described, and direct fusion has been used in cases where other conditions failed.[29] An S_N2 mechanism must be assumed in the reaction of the reagent with nucleophiles. The rates of the complete hydrolysis decrease in the sequence $Me_3O^+ \gg Et_3O^+ > Pr_3O^+ > c\text{-}C_5H_{10}\overset{+}{O}Et$.[30] Comparative studies on the alkylating ability of common methylating agents[31] show that Me_3O^+ ions are more reactive than the trifluoromethanesulfonate ester (factor 5 to 12), following the order $Me_3O^+ > MeOSO_2CF_3 > MeOSO_2F > MeOClO_3$. In terms of availability, stability, and freedom from hazards, however, oxonium salts often appear the reagents of choice (including comparisons with other powerful alkylating agents such as dialkoxycarbenium ions,[32] dialkylhalonium ions,[33] and haloalkanes in the presence of silver salts[34]). While the carcinogenic properties of other potent alkylating agents are well documented,[35] any such dangers associated with trialkyloxonium salts are presumably minimized by the fact that these compounds are water-soluble, nonvolatile, crystalline solids which are rapidly solvolyzed in aqueous solution.[1b,36] Although ***Triethyloxonium Tetrafluoroborate*** is the cheaper and much more used oxonium salt in synthetic chemistry, $Me_3O^+BF_4^-$ effects alkylations which the triethyl analog does not.[11,37]

Use in Transition Metal Chemistry. $Me_3O^+BF_4^-$ has been successfully used in the generation of transition metal carbene complexes by direct methylation of lithium acylcarbonylmetalates (eq 11).[38] Oxidative addition processes by which metal–carbon σ-bonds are formed have also been observed.[39] $Me_3O^+BF_4^-$ also serves as a halide acceptor in the reaction with square planar platinum(II) complexes.[40]

$Cr(CO)_6$ — 1. MeLi, Et_2O; 2. $Me_3O^+BF_4^-$; 87% (11)

Catalytic Properties. $Me_3O^+BF_4^-$ has been used as a catalyst for the polymerization of cyclic sulfides and in the polymerization of THF to macrocyclic ethers.[4] A valuable modern application is the catalytic Beckmann rearrangement of oximes in homogeneous liquid phase,[41] the active species being a formamidinium salt.

Other Reactions. The reaction of γ-alkenyl-γ-butyrolactones with allylsilanes in the presence of $Me_3O^+BF_4^-$ as the lactone-activating agent is reported to proceed with high regio- and stereoselectivity to afford methyl (*E*)-4,8-alkadienoates in high yields (eq 12).[42]

TMS; $Me_3O^+BF_4^-$; CH_2Cl_2, rt, 94 h; 100%; O; OMe (12)

$Me_3O^+BF_4^-$ has also been used as a methylating agent of the lithium enolate of a β-keto sulfoxide, leading to a mixture of isomeric dienol ethers, which are of potential interest as substrates for asymmetric Diels–Alder cycloadditions (eq 13).[43]

1. 1.1 equiv LDA THF, –30 °C; 2. $Me_3O^+BF_4^-$ 75%; *p*-Tol; OMe; (*EZ*):(*EE*) = 84:16 (13)

Related Reagents. Diazomethane; Dimethyl Sulfate; Iodomethane; Methyl Flurosulfonate; Methyl Trifluoromethanesulfonate; Triethyloxonium Tetrafluoroborate; Trimethyl Phosphate.

1. (a) Meerwein, H. *MOC* **1965**, *6/3*, 335. (b) Perst, H. *Oxonium Ions in Organic Chemistry*; Verlag Chemie: Weinheim, 1971. (c) Granik, V. G.; Pyatin, B. M.; Glushkov, R. G. *RCR* **1971**, *40*, 747. (d) Perst, H. In *Carbonium Ions*; Olah, G. A.; Schleyer, P. v. R., Eds.; Wiley: New York, 1976; Vol. 5, pp 1961–2047.
2. Meerwein, H.; Hinz, G.; Hofmann, P.; Kroning, E.; Pfeil, E. *JPR* **1937**, *147*, 257.
3. (a) Raber, D. J.; Gariano, P. *TL* **1971**, 4741. (b) Raber, D. J.; Gariano, P., Jr.; Brod, A. O.; Gariano, A. L.; Guida, W. C. *OSC* **1988**, *6*, 576.
4. Meerwein, H.; Borner, P.; Fuchs, O.; Sasse, H. J.; Schrodt, H.; Spille, J. *CB* **1956**, *89*, 2060.
5. Earle, M. J.; Fairhurst, R. A.; Giles, R. G.; Heaney, H. *SL* **1991**, 728.
6. Curphey, T. J. *OSC* **1988**, *6*, 1019.
7. Raber, D. J.; Gariano P., Jr.; Brod, A. O.; Gariano, A.; Guida, W. C.; Guida, A. R.; Herbst, M. D. *JOC* **1979**, *44*, 1149.
8. Mock, W. L.; Hartmann, M. E. *JOC* **1977**, *42*, 466.
9. (a) Pirkle, W. H.; Dines, M. *JHC* **1969**, *6*, 313. (b) Deslongchamps, P.; Chenevert, R.; Taillefer, R. J.; Moreau, C.; Saunders, J. K. *CJC* **1975**, *53*, 1601. (c) Kaloustian, M. K.; Khouri, F. *TL* **1981**, *22*, 413.
10. Eyley, S. C.; Giles, R. G.; Heaney, H. *TL* **1985**, *26*, 4649.
11. (a) Stahl, I.; Hetschko, M.; Gosselck, J. *TL* **1971**, 4077. (b) Böhme, H.; Krack, W. *LA* **1972**, *758*, 143. (c) Wolfe, S.; Chamberlain, P.; Garrard, T. F. *CJC* **1976**, *54*, 2847.
12. (a) La Rochelle, R. W.; Trost, B. M. *JACS* **1971**, *93*, 6077. (b) Tsumori, K.; Minato, H.; Kobayashi, M. *BCJ* **1973**, *46*, 3503. (c) Andersen, K. K.; Caret, R. L.; Karup-Nielsen, I. *JACS* **1974**, *96*, 8026. (d) Andersen, K. K.; Caret, R. L.; Ladd, D. L. *JOC* **1976**, *41*, 3096. (e) Lucchini, V.; Modena, G.; Pasquato, L. *CC* **1992**, 293. (f) Solladiė, G.; Maugein, N.; Morreno, I.; Almario, A.; Carreno Carmen, M.; Garcia-Ruano, J. L. *TL* **1992**, *33*, 4561.
13. (a) Shanklin, J. R.; Johnson, C. R.; Ollinger, J.; Coates, R. M. *JACS* **1973**, *95*, 3429. (b) Boehme, H.; Daehler, G.; Krack, W. *LA* **1973**, 1686. (c) Braun, H.; Amann, A. *AG* **1975**, *87*, 773. (d) Minato, H.; Miura, T.; Kobayashi, M. *CL* **1975**, 1055. (e) Watanabe, Y.; Shiono, M.; Mukaiyama, T. *CL* **1975**, 871. (f) Boehme, H.; Krack, W. *LA* **1977**, 51. (g) Hoffmann, R. W.; Ladner, W. *CB* **1983**, *116*, 1631. (h) Furuta, K.; Ikeda, Y.; Meguriya, N.; Ikeda, N.; Yamamoto, H. *BCJ* **1984**, *57*, 2781. (i) Bodwell, J. R.; Patwardhan, B. H.; Dittmer, D. C. *JOC* **1984**, *49*, 4192.
14. (a) Mukherjee, R. *JCS(D)* **1971**, 1113. (b) Mukherjee, R. *IJC(B)* **1977**, *15B*, 502. (c) Kosbahn, W.; Schaefer, H. *AG* **1977**, *89*, 826. (d) Casadei, M. A.; Di Rienzo, B.; Moracci, F. M. *SC* **1983**, *13*, 753.
15. (a) Hayashi, Y.; Nozaki, H. *BCJ* **1972**, *45*, 198. (b) Johnson, C. R.; Rogers, P. E. *JOC* **1973**, *38*, 1798.
16. Johnson, C. R. *ACR* **1973**, *6*, 341.
17. (a) Kresze, G.; Rössert, M. *AG* **1978**, *90*, 61. (b) Kresze, G.; Rössert, M. *LA* **1981**, 58.
18. House, H. O.; Richey, F. A., Jr. *JOC* **1969**, *34*, 1430.
19. Gundermann, K. D.; Hoenig, W.; Berrada, M.; Giesecke, H.; Paul, H. G. *LA* **1974**, 809.
20. (a) Altman, L. J.; Richheimer, S. L. *TL* **1971**, 4709. (b) Meyers, A. I.; Munavu, R.; Durandetta, J. *TL* **1972**, 3929. (c) Heine, H. W.; Newton, T. A.; Blosick, G. J.; Irving, K. C.; Meyer, C.; Corcoran G. B., III *JOC* **1973**, *38*, 651. (d) Quast, H.; Bieber, L. *CB* **1981**, *114*, 3253. (e) Hünig, S.; Prockschy, F. *CB* **1984**, *117*, 2099. (f) Meyers, A. I.; Hoyer, D. *TL* **1984**, *25*, 3667. (g) Reichardt, C.; Kaufmann, N. *CB* **1985**, *118*, 3424. (h) Quast, H.; Bieber, L.; Meichsner, G. *LA* **1987**, 469. (i) Hanquet, G.; Lusinchi, X.; Milliet, P. *TL* **1987**, *28*, 6061.
21. Hünig, S.; Pütter, H. *CB* **1977**, *110*, 2532.
22. Akiba, K.; Ohara, Y.; Inamoto, N. *BCJ* **1982**, *55*, 2976.
23. Kosbahn, W.; Schäfer, H. *AG* **1977**, *89*, 826.
24. Iwamura, H.; Fukunaga, M. *CL* **1974**, 1211.
25. Tabushi, I.; Sasaki, H.; Kuroda, Y. *JACS* **1976**, *98*, 5727.
26. Dehmlow, E. V. *AG* **1974**, *86*, 203.
27. Aumann, R.; Fischer, E. O. *CB* **1968**, *101*, 954.
28. Hesse, G.; Broll, H.; Rupp, W. *LA* **1966**, *697*, 22.
29. Rapko, J. N.; Feistel, G. *IC* **1970**, *9*, 1401.
30. Meerwein, H.; Battenberg, E.; Gold, H.; Pfeil, E.; Willfang, G. *JPR* **1939**, *154*, 83.
31. (a) Lewis, E. S.; Vanderpool, S. *JACS* **1977**, *99*, 1946. (b) Kevill, D. N.; Lin, G. M. L. *TL* **1978**, 949.
32. Kabuss, S. *AG(E)* **1966**, *5*, 675.
33. Olah, G. A.; DeMember, J. R. *JACS* **1970**, *92*, 2562.
34. (a) Meerwein, H.; Hederich, V.; Wunderlich, K. *AP* **1958**, *291*, 541. (b) Boulton, A. J.; Gray, A. C. G.; Katritzky, A. R. *JCS(B)* **1967**, 911. (c) Acheson, R. M.; Harrison, D. R. *JCS(C)* **1970**, 1764.
35. (a) Poirier, L. A.; Stoner, G. D.; Shimkin, M. B. *Cancer Res.* **1975**, *35*, 1411. (b) McCann, J.; Choi, E.; Yamasaki, E.; Ames, B. N. *PNA* **1975**, *72*, 5135.
36. Diem, M. J.; Burow, D. F.; Fry, J. L. *JOC* **1977**, *42*, 1801.
37. Baldwin, J. E.; Hackler, R. E.; Kelly, D. P. *JACS* **1968**, *90*, 4758.
38. (a) Fischer, E. O. *AG* **1974**, *86*, 651. (b) Fischer, E. O.; Fischer, H. *CB* **1974**, *107*, 657. (c) Rausch, M. D.; Moser, G. A.; Meade, C. F. *JOM* **1973**, *51*, 1.
39. (a) Strope, D.; Shriver, D. F. *JACS* **1973**, *95*, 8197. (b) Olgemöller, B.; Bauer, H.; Löbermann, H.; Nagel, U.; Beck, W. *CB* **1982**, *115*, 2271.
40. Treichel, P. M.; Wagner, K. P.; Knebel, W. J. *ICA* **1972**, 674.
41. Izumi, Y. *CL* **1990**, 2171.
42. Fujisawa, T.; Kawashima, M.; Ando, S. *TL* **1984**, *25*, 3213.

43. Solladié, G.; Maugein, N.; Moreno, I.; Almario, A.; Carreno, M. C.; Garcia-Ruano, J. L. *TL* **1992**, *33*, 4561.

Ingfried Stahl
Universität Kassel, Germany

Trimethylsilyldiazomethane[1]

Me_3Si–CH=N_2

[18107-18-1] $C_4H_{10}N_2Si$ (MW 114.25)

(one-carbon homologation reagent; stable, safe substitute for diazomethane; [C–N–N] 1,3-dipole for the preparation of azoles[1])

Physical Data: bp 96 °C/775 mmHg; n_D^{20} 1.4362.[2]
Solubility: sol most organic solvents; insol H_2O.
Form Supplied in: commercially available as 2 M and 10 w/w% solutions in hexane, and 10 w/w% solution in CH_2Cl_2.
Analysis of Reagent Purity: concentration in hexane is determined by ^{1}H NMR.[3]
Preparative Method: prepared by the diazo-transfer reaction of ***Trimethylsilylmethylmagnesium Chloride*** with ***Diphenyl Phosphorazidate*** (DPPA) (eq 1).[3]

TMS–CH$_2$Cl —(Mg)→ TMS–CH$_2$MgCl —($(PhO)_2P(O)N_3$)→ TMS–CH=N_2 (1)

Handling, Storage, and Precautions: should be protected from light.

One-Carbon Homologation. Along with its lithium salt, which is easily prepared by lithiation of trimethylsilyldiazomethane ($TMSCHN_2$) with ***n-Butyllithium***, $TMSCHN_2$ behaves in a similar way to ***Diazomethane*** as a one-carbon homologation reagent. $TMSCHN_2$ is acylated with aromatic acid chlorides in the presence of ***Triethylamine*** to give α-trimethylsilyl diazo ketones. In the acylation with aliphatic acid chlorides, the use of 2 equiv of $TMSCHN_2$ without triethylamine is recommended. The crude diazo ketones undergo thermal Wolff rearrangement to give the homologated carboxylic acid derivatives (eqs 2 and 3).[4]

1-naphthyl-COCl —(1. $TMSCHN_2$, Et_3N; 2. $PhNH_2$, 180 °C, 2,4,6-trimethylpyridine; 80%)→ 1-naphthyl-CH$_2$CONHPh (2)

N-CO_2Bn-prolyl-COCl —(1. 2 equiv $TMSCHN_2$; 2. $PhCH_2OH$, 180 °C, 2,4,6-trimethylpyridine; 77%)→ N-CO_2Bn-pyrrolidinyl-CH_2CO_2Bn (3)

Various ketones react with $TMSCHN_2$ in the presence of ***Boron Trifluoride Etherate*** to give the chain or ring homologated ketones (eqs 4–6).[5] The bulky trimethylsilyl group of $TMSCHN_2$ allows for regioselective methylene insertion (eq 5). Homologation of aliphatic and alicyclic aldehydes with $TMSCHN_2$ in the presence of ***Magnesium Bromide*** smoothly gives methyl ketones after acidic hydrolysis of the initially formed β-keto silanes (eq 7).[6]

$PhCOCH_2Ph$ —($TMSCHN_2$, BF_3•Et_2O; CH_2Cl_2, –15 °C, 1 h; 74%)→ $PhCOCH_2CH_2Ph$ (4)

2-methylcyclohexanone —($TMSCHN_2$, BF_3•Et_2O; CH_2Cl_2, –15 °C, 4 h; 69%)→ 2-methylcycloheptanone (5)

fluorenone —($TMSCHN_2$, BF_3•Et_2O; CH_2Cl_2, –15 to –10 °C, 3 h; 80%)→ 9-phenanthrol (OH) (6)

t-BuCHO —(1. $TMSCHN_2$, $MgBr_2$; 2. 10% aq HCl; 89%)→ t-BuCOMe (7)

O-Methylation of carboxylic acids, phenols, enols, and alcohols can be accomplished with $TMSCHN_2$ under different reaction conditions. $TMSCHN_2$ instantaneously reacts with carboxylic acids in benzene in the presence of methanol at room temperature to give methyl esters in nearly quantitative yields (eq 8).[7] This method is useful for quantitative gas chromatographic analysis of fatty acids. Similarly, *O*-methylation of phenols and enols with $TMSCHN_2$ can be accomplished, but requires the use of ***Diisopropylethylamine*** (eqs 9 and 10).[8] Although methanol is recommended in these *O*-methylation reactions, methanol is not the methylating agent. Various alcohols also undergo *O*-methylation with $TMSCHN_2$ in the presence of 42% aq. ***Tetrafluoroboric Acid***, smoothly giving methyl ethers (eq 11).[9]

HO–C_6H_4–CO_2H —($TMSCHN_2$, MeOH–benzene; rt, 30 min)→ HO–C_6H_4–CO_2Me (8)
quantitative

2-HO-C_6H_4-CO_2Me —($TMSCHN_2$, i-Pr_2NEt, MeOH–MeCN; rt, 15 h; 78%)→ 2-MeO-C_6H_4-CO_2Me (9)

$PhCOCH_2CO_2Et$ —($TMSCHN_2$, i-Pr_2NEt, MeOH–MeCN; rt, 15 h; 89%)→ Ph(MeO)C=CHCO_2Et (10)

$CH_3(CH_2)_8C(CH_3)_2OH$ —($TMSCHN_2$, 42% aq HBF_4; CH_2Cl_2, 0 °C, 2 h; 74%)→ $CH_3(CH_2)_8C(CH_3)_2OMe$ (11)

Alkylation of the lithium salt of $TMSCHN_2$ ($TMSC(Li)N_2$) gives α-trimethylsilyl diazoalkanes which are useful for the preparation of vinylsilanes and acylsilanes. Decomposition of α-trimethylsilyl diazoalkanes in the presence of a catalytic amount of ***Copper(I) Chloride*** gives mainly (*E*)-vinylsilanes (eq 12),[10] while replacement of CuCl with rhodium(II) pivalate affords (*Z*)-vinylsilanes as the major products (eq 12).[11] Oxidation of α-trimethylsilyl diazoalkanes with ***m*-Chloroperbenzoic Acid** in a two-phase system of benzene and phosphate buffer (pH 7.6) affords acylsilanes (α-keto silanes) (eq 12).[12]

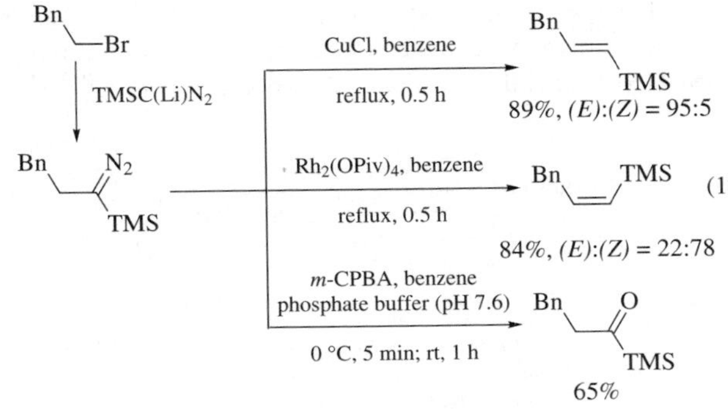

(12)

(*E*)-β-Trimethylsilylstyrenes are formed by reaction of alkanesulfonyl chlorides with $TMSCHN_2$ in the presence of triethylamine (eq 13).[13] $TMSC(Li)N_2$ reacts with carbonyl compounds to give α-diazo-β-hydroxy silanes which readily decompose to give α,β-epoxy silanes (eq 14).[14] However, benzophenone gives diphenylacetylene under similar reaction conditions (eq 15).[15]

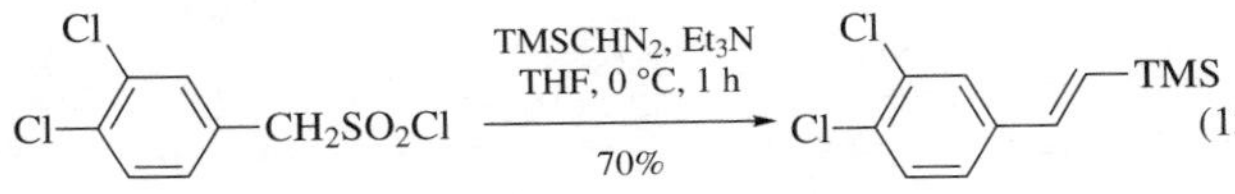

(13)

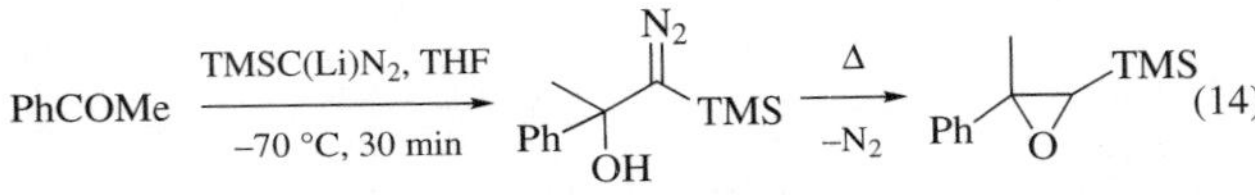

(14)

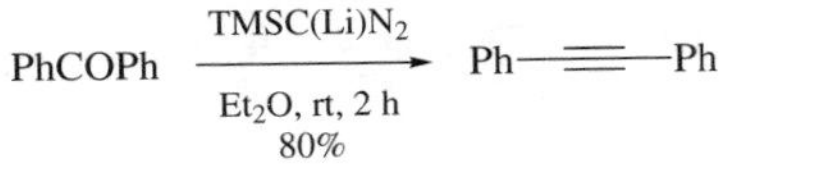

(15)

Silylcyclopropanes are formed by reaction of alkenes with $TMSCHN_2$ in the presence of either ***Palladium(II) Chloride*** or CuCl depending upon the substrate (eqs 16 and 17).[16] Silylcyclopropanones are also formed by reaction with trialkylsilyl and germyl ketenes (eq 18).[17]

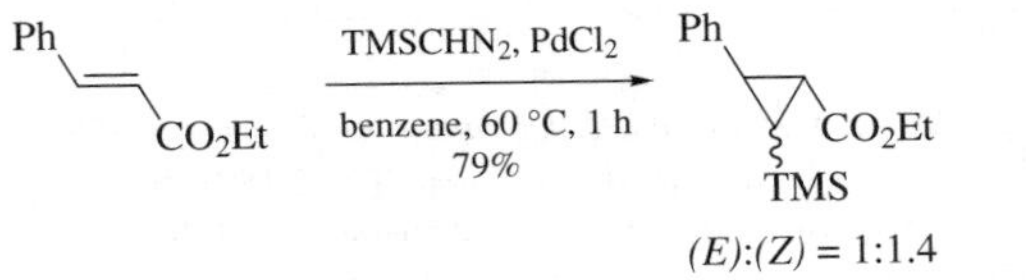

(16)

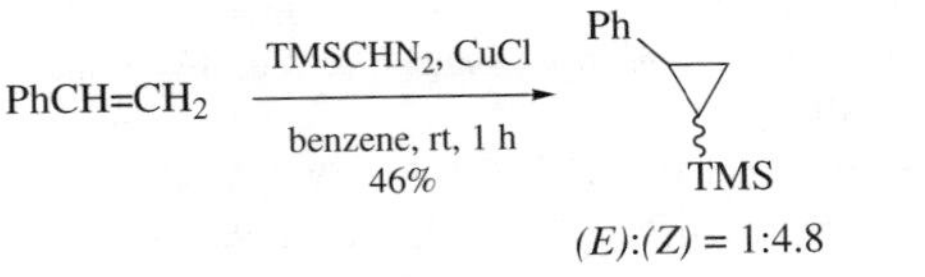

(17)

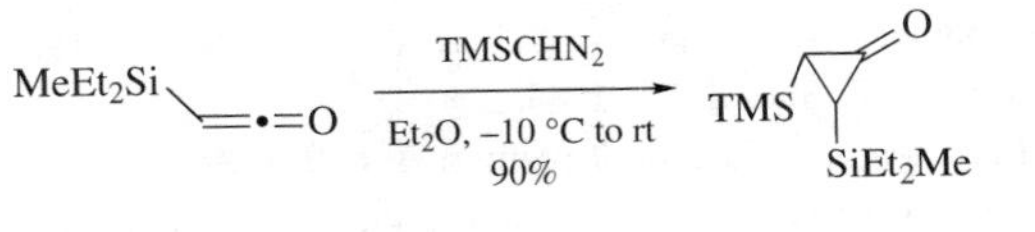

(18)

[C–N–N] Azole Synthon. $TMSCHN_2$, mainly as its lithium salt, $TMSC(Li)N_2$, behaves like a 1,3-dipole for the preparation of [C–N–N] azoles. The reaction mode is similar to that of diazomethane but not in the same fashion. $TMSC(Li)N_2$ (2 equiv) reacts with carboxylic esters to give 2-substituted 5-trimethylsilyltetrazoles (eq 19).[18] Treatment of thiono and dithio esters with $TMSC(Li)N_2$ followed by direct workup with aqueous methanol gives 5-substituted 1,2,3-thiadiazoles (eq 20).[19] While reaction of di-*t*-butyl thioketone with $TMSCHN_2$ produces the episulfide with evolution of nitrogen (eq 21),[20] its reaction with $TMSC(Li)N_2$ leads to removal of one *t*-butyl group to give the 1,2,3-thiadiazole (eq 21).[20]

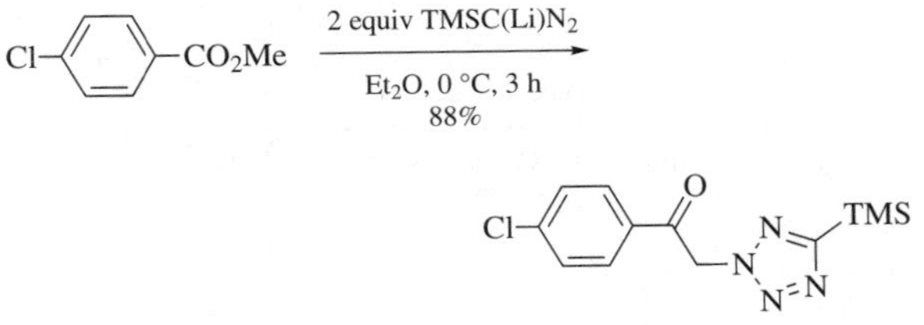

(19)

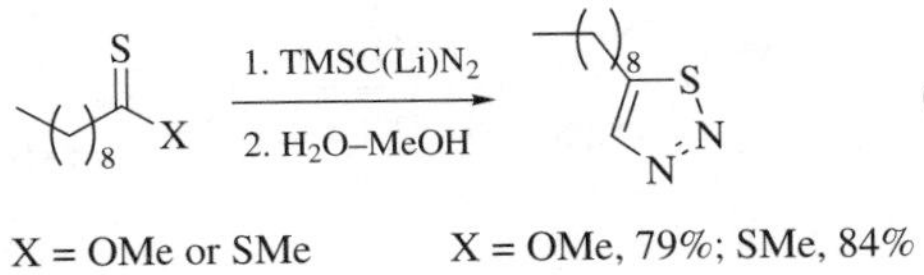

(20)

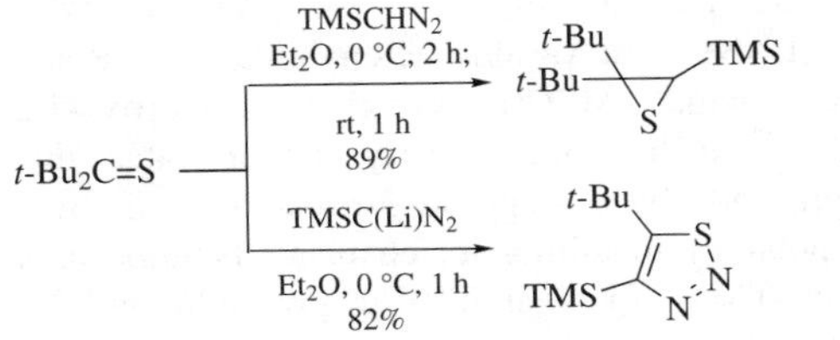

(21)

$TMSCHN_2$ reacts with activated nitriles only, such as cyanogen halides, to give 1,2,3-triazoles.[21] In contrast with this, $TMSC(Li)N_2$ smoothly reacts with various nitriles including aromatic, heteroaromatic, and aliphatic nitriles, giving 4-substituted 5-trimethylsilyl-1,2,3-triazoles (eq 22).[22] However, reaction of α,β-unsaturated nitriles with $TMSC(Li)N_2$ in Et_2O affords 3(or 5)-trimethylsilylpyrazoles, in which the nitrile group acts as a leaving group (eq 23).[23] Although α,β-unsaturated nitriles bearing bulky substituents at the α- and/or β-positions of the nitrile group undergo reaction with $TMSC(Li)N_2$ to give pyrazoles, significant amounts of 1,2,3-triazoles are also formed. Changing the reaction solvent from Et_2O to THF allows for predominant formation of pyrazoles (eq 24).[23] Complete exclusion of the formation of 1,2,3-triazoles can be achieved when the nitrile group is replaced by a phenylsulfonyl species.[24] Thus reaction of α,β-unsaturated sulfones with $TMSC(Li)N_2$ affords pyrazoles in excellent yields (eq 25). The geometry of the double bond of α,β-unsaturated sulfones is not critical in the reaction. When both a cyano and a sulfonyl group are present as a leaving group, elimination of the sulfonyl group occurs preferentially (eq 26).[24] The trimethylsilyl group attached to the heteroaromatic products is easily removed with 10% aq. KOH in EtOH or HCl–KF.

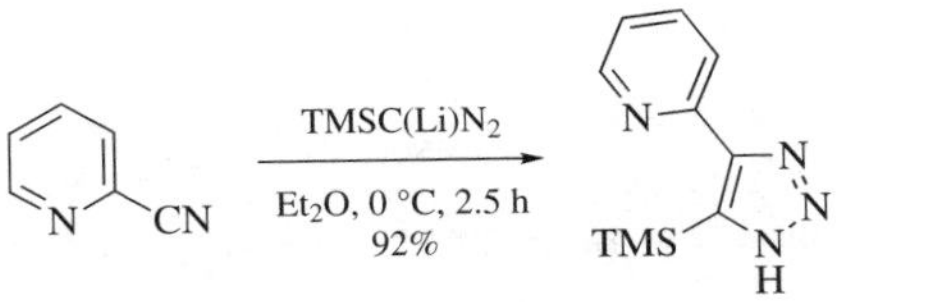

(22)

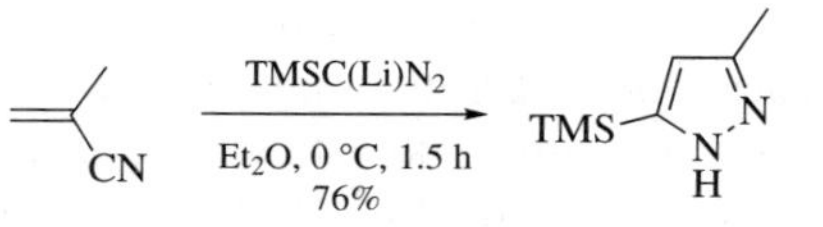

Various 1,2,3-triazoles can be prepared by reaction of $TMSC(Li)N_2$ with various heterocumulenes. Reaction of isocyanates with $TMSC(Li)N_2$ gives 5-hydroxy-1,2,3-triazoles (eq 27).[25] It has been clearly demonstrated that the reaction proceeds by a stepwise process and not by a concerted 1,3-dipolar cycloaddition mechanism. Isothiocyanates also react with $TMSC(Li)N_2$ in THF to give lithium 1,2,3-triazole-5-thiolates which are treated in situ with alkyl halides to furnish 1-substituted 4-trimethylsilyl-5-alkylthio-1,2,3-triazoles in excellent yields (eq 28).[26] However, changing the reaction solvent from THF to Et_2O causes a dramatic solvent effect. Thus treatment of isothiocyanates with $TMSC(Li)N_2$ in Et_2O affords 2-amino-1,3,4-thiadiazoles in good yields (eq 28).[27] Reaction of ketenimines with $TMSC(Li)N_2$ smoothly proceeds to give 1,5-disubstituted 4-trimethylsilyl-1,2,3-triazoles in high yields (eq 29).[28] Ketenimines bearing an electron-withdrawing group at one position of the carbon–carbon double bond react with $TMSC(Li)N_2$ to give 4-aminopyrazoles as the major products (eq 30).[29]

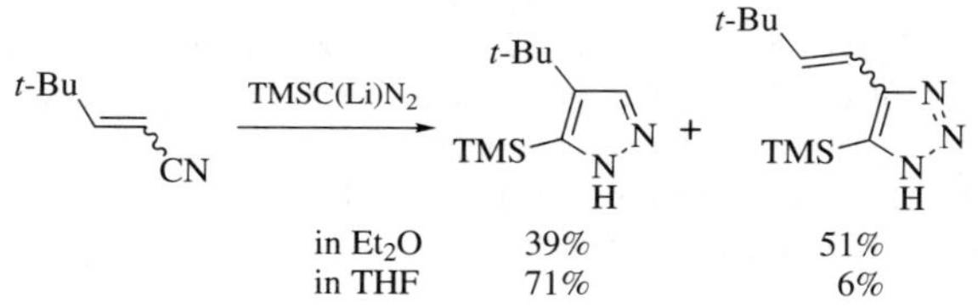

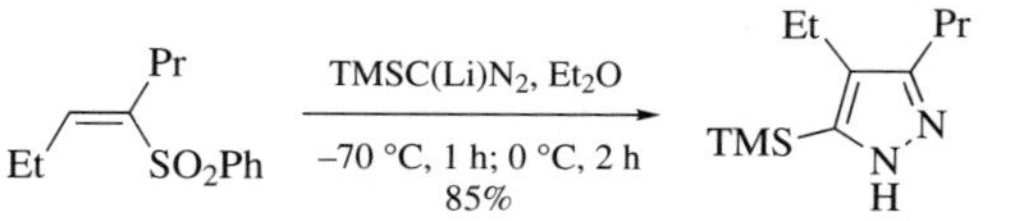

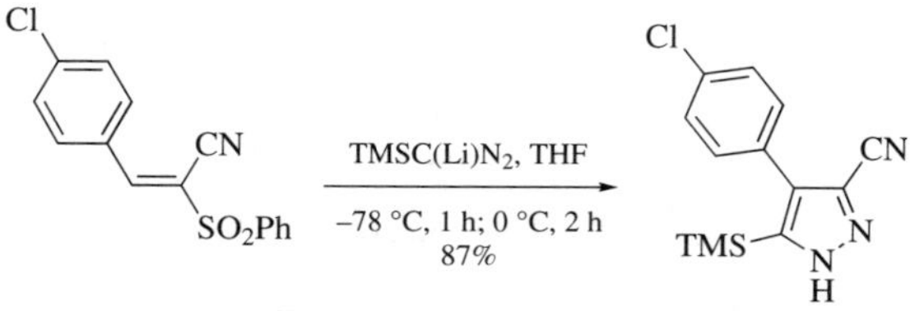

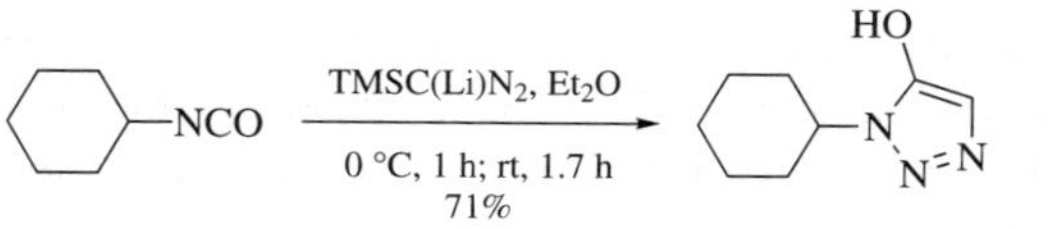

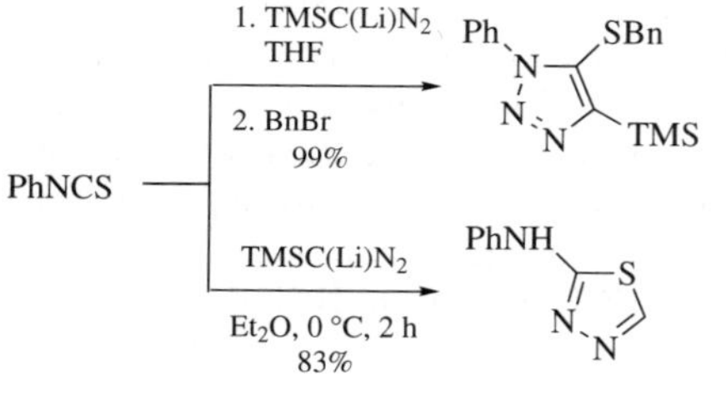

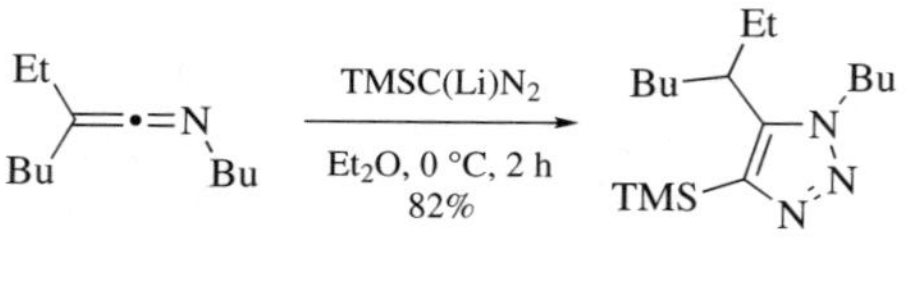

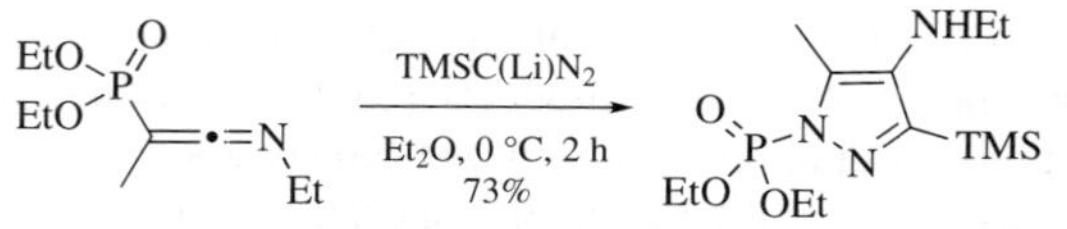

Pyrazoles are formed by reaction of $TMSCHN_2$ or $TMSC(Li)N_2$ with some alkynes (eqs 31 and 32)[24,30] and quinones (eq 33).[31] Some miscellaneous examples of the reactivity of $TMSCHN_2$ or its lithium salt are shown in eqs 34–36.[20,31,32]

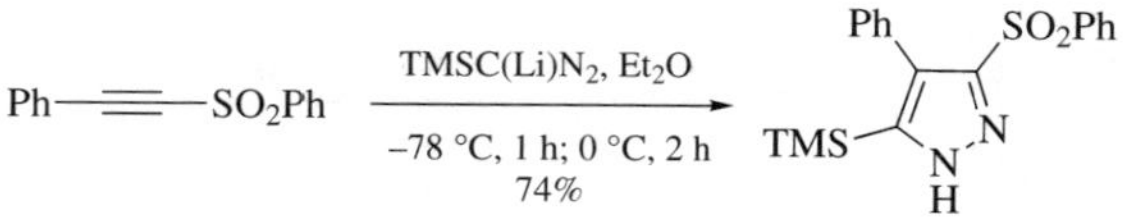

Related Reagents. Diazomethane.

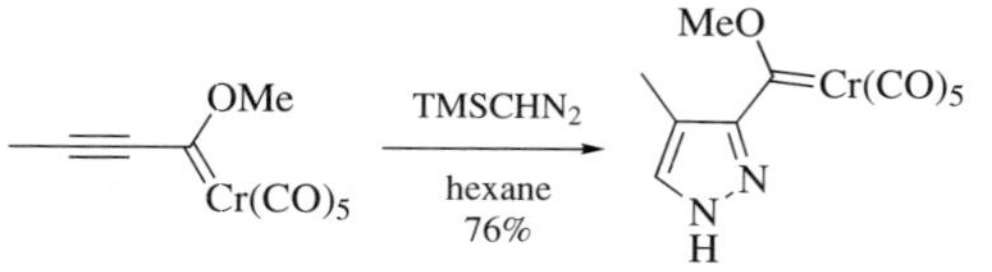

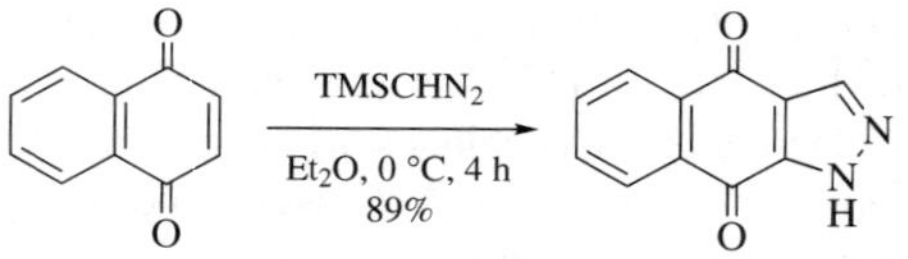

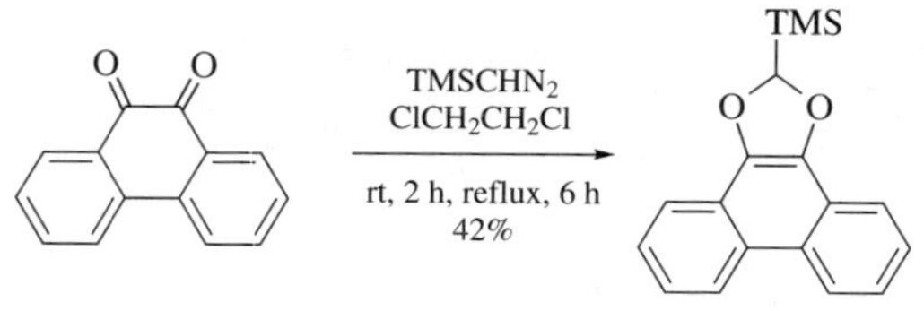

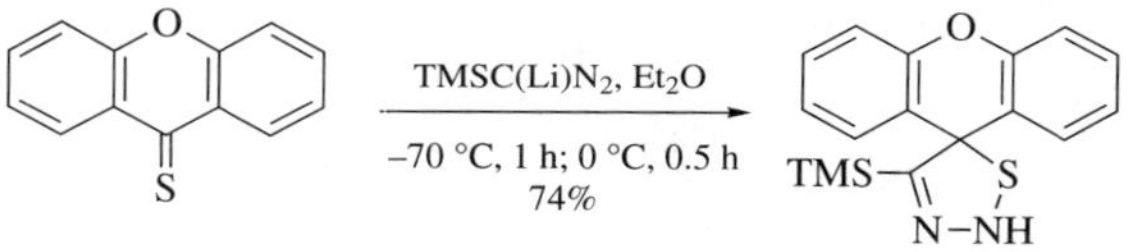

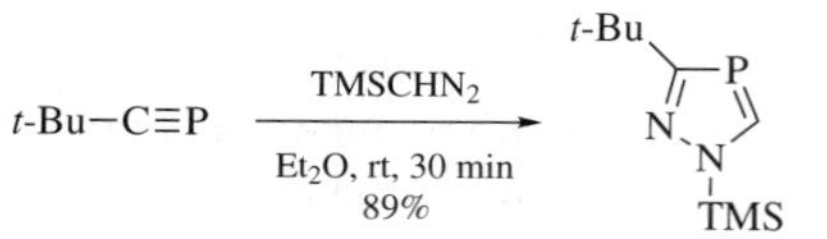

1. (a) Shioiri, T.; Aoyama, T. *J. Synth. Org. Chem. Jpn* **1986**, *44*, 149 (*CA* **1986**, *104*, 168 525q). (b) Aoyama, T. *YZ* **1991**, *111*, 570 (*CA* **1992**, *116*, 58 332q). (c) Anderson, R.; Anderson, S. B. In *Advances in Silicon Chemistry*; Larson, G. L., Ed.; JAI: Greenwich, CT, 1991; Vol. 1, pp 303–325. (d) Shioiri, T.; Aoyama, T. In *Advances in the Use of Synthons in Organic Chemistry*; Dondoni, A., Ed.; JAI: London, 1993; Vol. 1, pp 51–101.
2. Seyferth, D.; Menzel, H.; Dow, A. W.; Flood, T. C. *JOM* **1972**, *44*, 279.
3. Shioiri, T.; Aoyama, T.; Mori, S. *OS* **1990**, *68*, 1.
4. Aoyama, T.; Shioiri, T. *CPB* **1981**, *29*, 3249.
5. Hashimoto, N.; Aoyama, T.; Shioiri, T. *CPB* **1982**, *30*, 119.
6. Aoyama, T.; Shioiri, T. *S* **1988**, 228.
7. Hashimoto, N.; Aoyama, T.; Shioiri, T. *CPB* **1981**, *29*, 1475.
8. Aoyama, T.; Terasawa, S.; Sudo, K.; Shioiri, T. *CPB* **1984**, *32*, 3759.
9. Aoyama, T.; Shioiri, T. *TL* **1990**, *31*, 5507.
10. Aoyama, T.; Shioiri, T. *TL* **1988**, *29*, 6295.
11. Aoyama, T.; Shioiri, T. *CPB* **1989**, *37*, 2261.

12. Aoyama, T.; Shioiri, T. *TL* **1986**, *27*, 2005.
13. Aoyama, T.; Toyama, S.; Tamaki, N.; Shioiri, T. *CPB* **1983**, *31*, 2957.
14. Schöllkopf, U.; Scholz, H.-U. *S* **1976**, 271.
15. Colvin, E. W.; Hamill, B. J. *JCS(P1)* **1977**, 869.
16. Aoyama, T.; Iwamoto, Y.; Nishigaki, S.; Shioiri, T. *CPB* **1989**, *37*, 253.
17. Zaitseva, G. S.; Lutsenko, I. F.; Kisin, A. V.; Baukov, Y. I.; Lorberth, J. *JOM* **1988**, *345*, 253.
18. Aoyama, T.; Shioiri, T. *CPB* **1982**, *30*, 3450.
19. Aoyama, T.; Iwamoto, Y.; Shioiri, T. *H* **1986**, *24*, 589.
20. Shioiri, T.; Iwamoto, Y.; Aoyama, T. *H* **1987**, *26*, 1467.
21. Crossman, J. M.; Haszeldine, R. N.; Tipping, A. E. *JCS(D)* **1973**, 483.
22. Aoyama, T.; Sudo, K.; Shioiri, T. *CPB* **1982**, *30*, 3849.
23. Aoyama, T,; Inoue, S.; Shioiri, T. *TL* **1984**, *25*, 433.
24. Asaki, T.; Aoyama, T.; Shioiri, T. *H* **1988**, *27*, 343.
25. Aoyama, T.; Kabeya, M.; Fukushima, A.; Shioiri, T. *H* **1985**, *23*, 2363.
26. Aoyama, T.; Kabeya, M.; Shioiri, T. *H* **1985**, *23*, 2371.
27. Aoyama, T.; Kabeya, M.; Fukushima, A.; Shioiri, T. *H* **1985**, *23*, 2367.
28. Aoyama, T.; Katsuta, S.; Shioiri, T. *H* **1989**, *28*, 133.
29. Aoyama, T.; Nakano, T.; Marumo, K.; Uno, Y.; Shioiri, T. *S* **1991**, 1163.
30. Chan, K. S.; Wulff, W. D. *JACS* **1986**, *108*, 5229.
31. Aoyama, T.; Nakano, T.; Nishigaki, S.; Shioiri, T. *H* **1990**, *30*, 375.
32. Rösch, W.; Hees, U.; Regitz, M. *CB* **1987**, *120*, 1645.

Takayuki Shioiri & Toyohiko Aoyama
Nagoya City University, Japan

β-Trimethylsilylethanesulfonyl Chloride

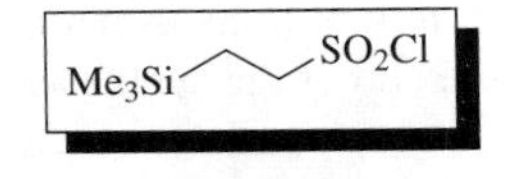

[97203-62-8] $C_5H_{13}ClO_2SSi$ (MW 200.79)

(protection of primary and secondary amines as their sulfonamides, which are cleaved by fluoride ion[1])

Alternate Name: SESCl.
Physical Data: bp 60 °C/0.1 mmHg; yellow oil.
Solubility: sol most common organic solvents.
Preparative Methods: can be most conveniently synthesized from commercially available ***Vinyltrimethylsilane*** (**1**) (eq 1).[1] Radical addition of sodium bisulfite to the vinyl group catalyzed by *t*-butyl perbenzoate yields the sulfonate salt (**2**) which can be directly converted to SESCl (**3**) with ***Phosphorus(V) Chloride***. The chloride (**3**) can then be purified by distillation. The intermediate sulfonate salt (**2**) is commercially available. The chloride (**3**) can also be prepared in 62% yield from the salt (**2**) using ***Sulfuryl Chloride*** and ***Triphenylphosphine*** (eq 2).[2] A less convenient procedure to synthesize SESCl (**3**) using β-trimethylsilylethylmagnesium chloride (**4**) and sulfuryl chloride has also been developed (eq 3).[1]

TMS–CH=CH$_2$ (**1**) —[$NaHSO_3$, $PhCO_3$-*t*-Bu; MeOH, Δ; 70%]→ TMS–CH$_2$CH$_2$–SO_3Na (**2**) —[PCl_5, CCl_4; 64%]→ TMS–CH$_2$CH$_2$–SO_2Cl (**3**) (1)

(**2**) —[SO_2Cl_2, PPh_3; CH_2Cl_2; 62%]→ (**3**) (2)

TMS–CH$_2$CH$_2$–MgCl (**4**) —[SO_2Cl_2; CH_2Cl_2; 50%]→ (**3**) (3)

Handling, Storage, and Precautions: stable liquid that can be stored at room temperature for weeks. Prone to hydrolysis.

Protection of Amines. Sulfonamides are among the most stable of amine protecting groups and it is this stability that detracts from their utility, since harsh reaction conditions are often needed for their removal. The advantage of using the β-trimethylsilylethanesulfonyl (SES) protecting group is that the sulfonamide (**5**) can be easily cleaved to regenerate the parent amine (eq 4)[1] in good yields with either ***Cesium Fluoride*** (2–3 equiv) in DMF at 95 °C for 9–40 h, or ***Tetra-n-butylammonium Fluoride*** trihydrate (3 equiv) in refluxing MeCN. The main disadvantage of the latter procedure is an occasional difficulty in separating tetrabutylammonium salts from some amines.

TMS–CH$_2$CH$_2$–SO_2NRR' (**5**) —[CsF, DMF, 95 °C, 9–40 h; or $TBAF{\cdot}3H_2O$, MeCN, reflux]→ TMSF + $H_2C{=}CH_2$ + SO_2 + RR'NH (4)

The sulfonamides can be prepared from a wide variety of primary and secondary amines using sulfonyl chloride (**3**) in DMF containing ***Triethylamine***. For aromatic and heterocyclic amines, ***Sodium Hydride*** is the preferred base. The sulfonamides are generally quite stable and are untouched by refluxing TFA, 6 M HCl in refluxing THF, 1 M TBAF in refluxing THF, $LiBF_4$ in refluxing MeCN, $BF_3{\cdot}OEt_2$, and 40% HF in ethanol. Table 1 lists a few examples of various amines and their protection as SES sulfonamides and cleavage with CsF in DMF at 95 °C.[1]

Table 1 Synthesis and Cleavage of SES Sulfonamides with CsF, DMF, 95 °C

Amine	% Yield sulfonamide	% Yield of cleavage (time)
NH	92	82 (29 h)
NH$_2$	92	86 (16 h)
NH$_2$, Cl	83	93 (40 h)
N, NH	93	91 (9 h)
NH$_2$, O, O, Ph	88	89 (24 h)

SESCl in Synthesis.[3] The SES group has been used successfully in the synthesis of glycosides (eq 5).[4] Reaction of (**6**) with SES-sulfonamide and ***Iodonium Di-sym-collidine Perchlorate*** provides the iodo sulfonamide (**7**) in 82% yield. Treatment of (**7**) with (benzyloxy)tributylstannane in the presence of ***Silver(I) Trifluoromethanesulfonate*** provides the β-benzyl glycoside (**8**). Fluoride treatment of (**8**) removes both the silyl ether and the SES group, giving the amino alcohol (**9**).

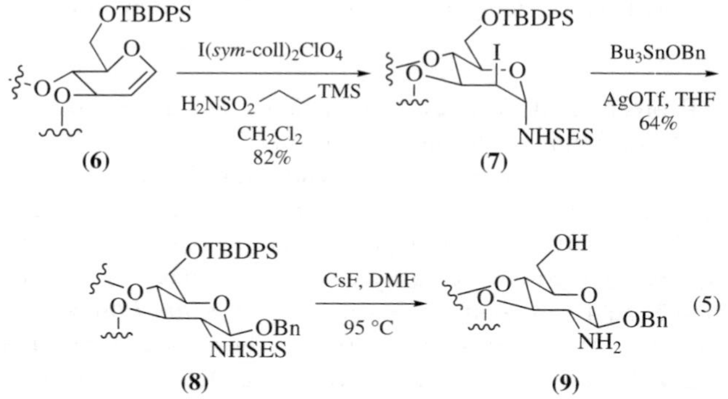

(5)

The smooth removal of the SES protecting group from pyrrole and pyrrole-containing peptides demonstrates the synthetic potential of this protecting group in heterocyclic chemistry (eq 6).[5] The SES group of (**10**) is removed with TBAF·$3H_2O$ in DMF at room temperature to yield (**11**). Other protecting groups (i.e. mesylate, triflate) cannot be removed without destruction of the substrates.

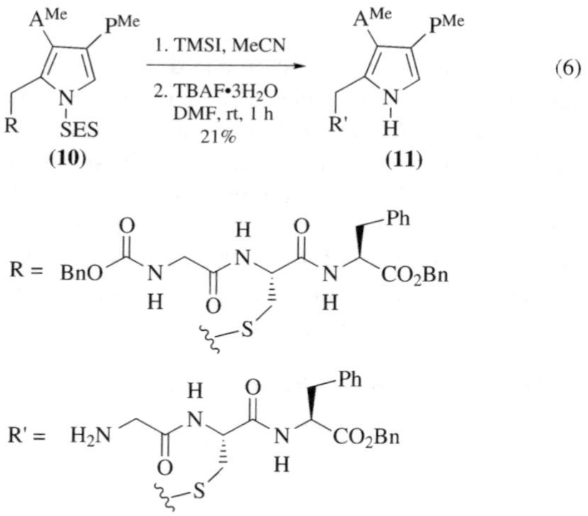

(6)

The *N*-SES group can be incorporated by treating an aldehyde with *N*-sulfinyl-β-trimethylsilylethanesulfonamide (SESNSO) (**13**), which can be made by treating the sulfonamide (**12**) with ***Thionyl Chloride*** and a catalytic amount of *N,N*-dichloro-*p*-toluenesulfonamide (eq 7) (see also ***N-Sulfinyl-p-toluenesulfonamide***).[6] The *N*-sulfonyl imine can be used in situ in a number of reactions. For example, the *N*-sulfonyl imine from aldehyde (**14**) reacts with 2,3-dimethylbutadiene (eq 8)[7] to give the Diels–Alder adduct (**15**). Treatment of (**15**) with fluoride ion affords the bicyclic lactam (**16**). Also, the *N*-sulfonyl imine derived from isobutyraldehyde and (**13**) reacts with ***Vinylmagnesium Bromide*** to provide the allylic SES-sulfonamide (**17**) in 65% yield (eq 9).[8]

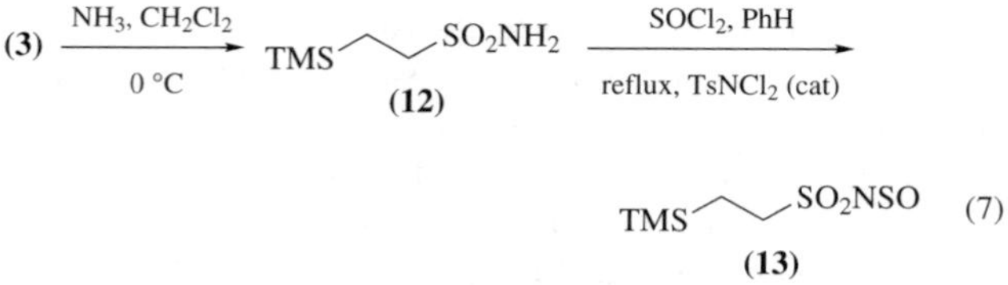

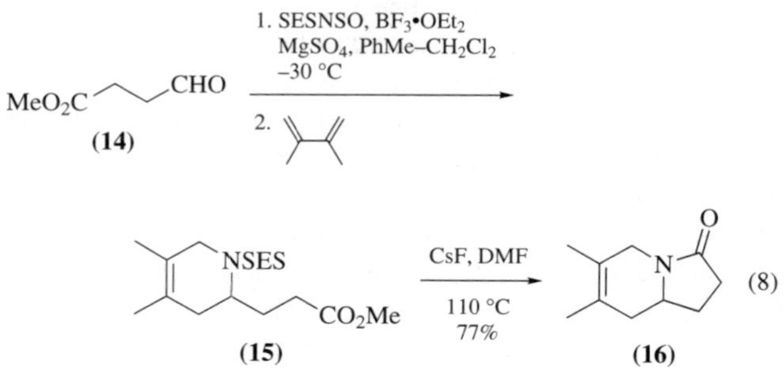

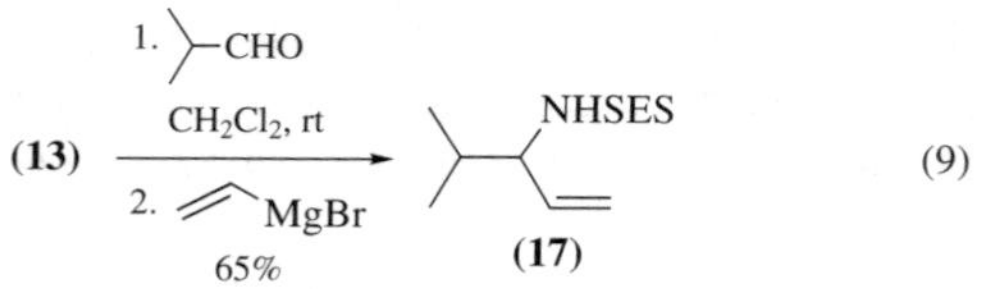

(7)

(8)

(9)

In a total synthesis of the antitumor antibiotic (−)-bactobolin (**18**), the choice of protecting group on the nitrogen was crucial (eq 10).[6] Unlike other protecting groups, the SES group is compatible with a wide variety of transformations and reagents, and is easily removed at the end of the synthesis.

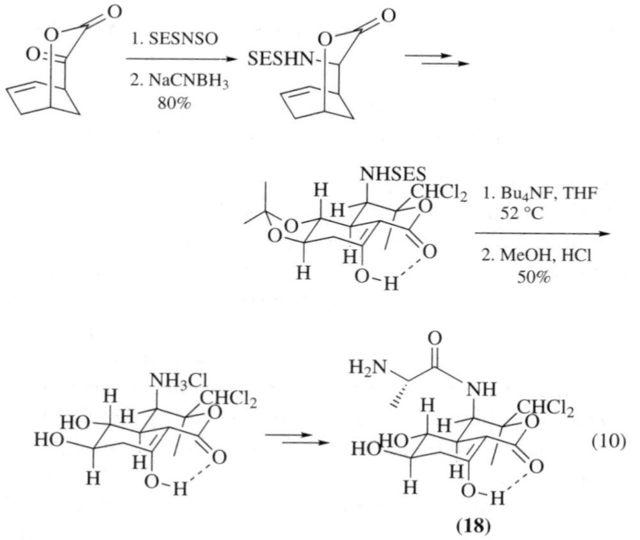

(10)

Related Reagents. Benzenesulfonyl Chloride; 4-Bromobenzenesulfonyl Chloride; Mesitylenesulfonyl Chloride; Methanesulfonyl Chloride; *p*-Toluenesulfonyl Chloride; Trifluoromethanesulfonic Anhydride.

1. Weinreb, S. M.; Demko, D. M.; Lessen, T. A. *TL* **1986**, *27*, 2099.
2. Huang, J.; Widlanski, T. S. *TL* **1992**, *33*, 2657.

3. For the use of SESCl in the synthesis of *t*-butyl [[2-(trimethylsilyl)ethyl]sulfonyl]carbamate, a useful reagent in Mitsunobu reactions, see ***N-(t-Butoxycarbonyl)-p-toluenesulfonamide***; Campbell, J. A.; Hart, D. J. *JOC* **1993**, *58*, 2900.
4. Danishefsky, S. J.; Koseki, K.; Griffith, D. A.; Gervay, J.; Peterson, J. M.; McDonald, F. E.; Oriyama, T. *JACS* **1992**, *114*, 8331.
5. Miller, A. D.; Leeper, F. J.; Battersby, A. R. *JCS(P1)* **1989**, 1943.
6. Garigipati, R. S.; Tschaen, D. M.; Weinreb, S. M. *JACS* **1990**, *112*, 3475.
7. Sisko, J.; Weinreb, S. M. *TL* **1989**, *30*, 3037.
8. Sisko, J.; Weinreb, S. M. *JOC* **1990**, *55*, 393.

Steven M. Weinreb & Janet L. Ralbovsky
The Pennsylvania State University, University Park, PA, USA

2-(Trimethylsilyl)ethanol

Me_3Si–CH_2CH_2–OH

[2916-68-9] $C_5H_{14}OSi$ (MW 118.28)

(protecting reagent for carboxyl, phosphoryl, hydroxyl, and amino groups)

Physical Data: bp 50–52 °C/10 mmHg, 71–73 °C/35 mmHg; *d* 0.825 g cm^{-3}.

Form Supplied in: colorless liquid; 99% purity; widely available.

Preparative Methods: three methods of preparation have been reported: (a) from the treatment of ethyl bromoacetate with ***Zinc*** followed by the reaction with ***Chlorotrimethylsilane***[1] and subsequent reduction of the resultant ***Ethyl Trimethylsilylacetate*** with ***Lithium Aluminum Hydride***[2,3] or ***Borane–Tetrahydrofuran***[4] (eq 1); (b) from the hydroboration/oxidation[5,6] or oxymercuration/demercuration[7] of ***Vinyltrimethylsilane*** (eq 2); and (c) most conveniently, by the reaction of the Grignard reagent formed from ***(Chloromethyl)trimethylsilane*** with ***Paraformaldehyde*** (eq 3).[8]

(1) Zn, TMSCl, PhH, Et_2O, Δ, 55–72%; then $LiAlH_4$, Et_2O or BH_3•THF, 78%

(2) 1. 9-BBN, THF; 2. aq NaOH, H_2O_2, 97%; or 1. $Hg(OAc)_2$, THF, H_2O; 2. $NaBH_4$, 40–60%

(3) 1. Mg, Et_2O; 2. $(CH_2O)_3$, Δ, 96%

Handling, Storage, and Precautions: corrosive; irritant.

A variety of methods are available for the protection of carboxylic acids as 2-(trimethylsilyl)ethyl esters.[9] 2-(Trimethylsilyl)ethyl esters are stable to conditions used in peptide synthesis. Deprotection of the ester is readily accomplished by treatment with ***Tetra-n-butylammonium Fluoride*** (TBAF).[9] Hemisuccinates have been prepared indirectly under nonacidic conditions by monoprotection of succinic anhydride as 2-(trimethylsilyl)ethyl esters followed by esterification (eq 4) and then selective deprotection of the resultant diester (eq 5).[10]

(4) TMS–CH_2CH_2–OH, Et_3N, EtOAc, Δ; R = H → R = R': R'OH, DCC, DMAP, 83–93%

(5) Bu_4NF, THF, rt, 70–81%

Transesterification of methyl esters to 2-(trimethylsilyl)ethyl esters under mild and neutral conditions takes place in the presence of ***Titanium Tetraisopropoxide*** (eqs 6–8).[11] Deprotection of the 2-(trimethylsilyl)ethyl ester in the presence of an *O*-TBDMS protected secondary hydroxyl group has been achieved (eq 9).[11b] An alternative method for the transesterification uses ***1,8-Diazabicyclo[5.4.0]undec-7-ene/Lithium Bromide*** and 2-(trimethylsilyl)ethanol.[12]

(6) TMS–CH_2CH_2–OH, Ti(O-*i*-Pr)$_4$, Δ, 71%; R = CH_2CH_2TMS

(7) TMS–CH_2CH_2–OH, Ti(O-*i*-Pr)$_4$, Δ, 74%; R = CH_2CH_2TMS

(8) TMS–CH_2CH_2–OH, Ti(O-*i*-Pr)$_4$, Δ, 85%

(9) $Bu_4NF•H_2O$ + *p*-TsOH•H_2O (3:1), THF

Protection of hydroxyl groups as 2-(trimethylsilyl)ethyl ethers[13a] and 2-(trimethylsilyl)ethyl carbonates[13b] (eq 10) has been utilized as illustrated below. The 2-(trimethylsilyl)ethyl carbonate group can be cleaved under mild conditions using TBAF in dry THF (eq 11).[13b]

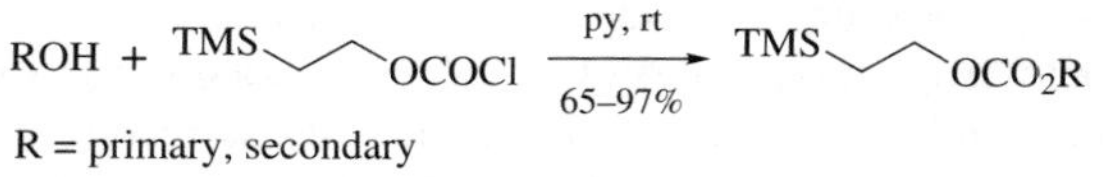

ROH + TMS–CH₂CH₂–OCOCl —(py, rt; 65–97%)→ TMS–CH₂CH₂–OCO₂R (10)

R = primary, secondary

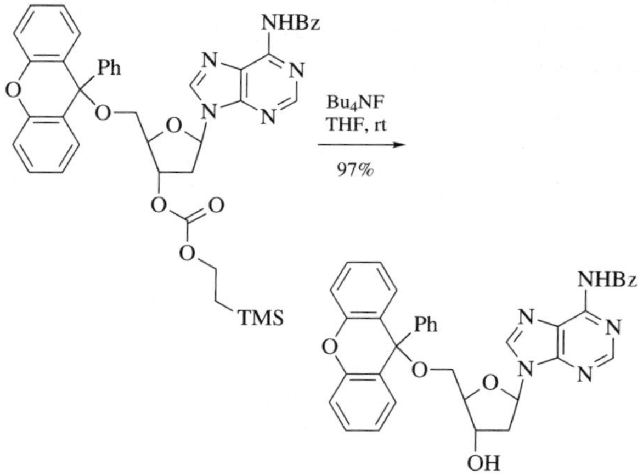

Bu₄NF, THF, rt, 97% (11)

Methods for the protection of pyranosides and furanosides as 2-(trimethylsilyl)ethyl glycosides (eq 12) and deprotection using dry ***Lithium Tetrafluoroborate*** in MeCN have been developed (eq 13).[14]

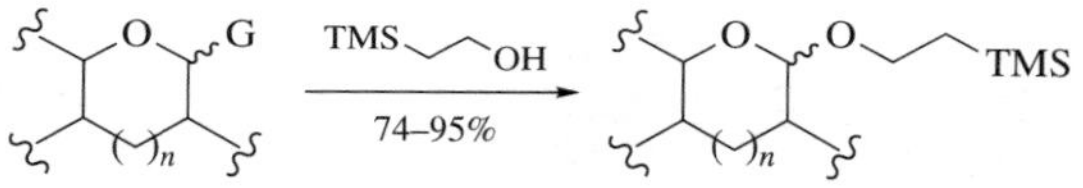

TMS–CH₂CH₂–OH, 74–95% (12)

G = OH, halogen; $n = 0, 1$

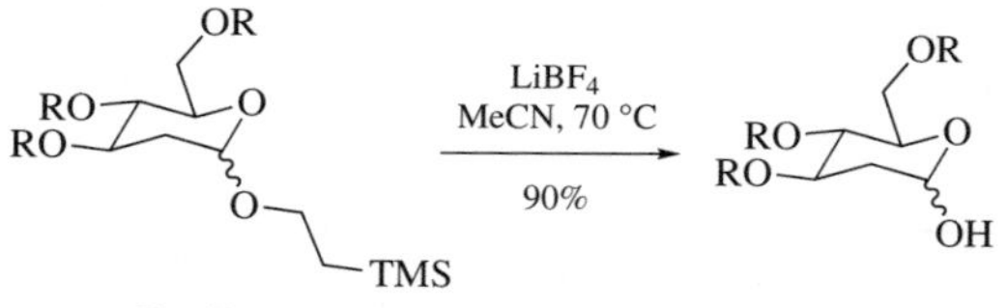

LiBF₄, MeCN, 70 °C, 90% (13)

R = Bn

2-(Trimethylsilyl)ethanol has been used as a protecting group in phosphate monoester synthesis and involves the use of 2-(trimethylsilyl)ethyl dichlorophosphite (eq 14).[15] Bis[2-(trimethylsilyl)ethyl] *N,N*-diisopropylphosphoramidite has been prepared from dichloro(diisopropylamino)phosphine and 2-(trimethylsilyl)ethanol and used as a phosphitylating agent in the synthesis of phosphotyrosine containing peptides (eq 15).[16]

2-(Trimethylsilyl)ethoxycarbonyl (Teoc) groups have been used to protect amine functionalities (eq 16).[17] Using a mixture of tetra-*n*-butylammonium chloride and KF·2H_2O deprotects the Teoc group.[18] *N*-Debenzylation and concurrent protection as *N*-Teoc results when tertiary *N*-benzylamines are treated with ***2-(Trimethylsilyl)ethyl Chloroformate***.[19]

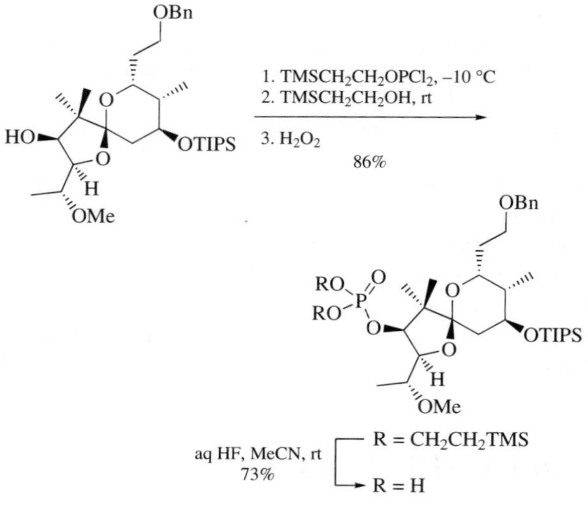

1. $TMSCH_2CH_2OPCl_2$, −10 °C
2. $TMSCH_2CH_2OH$, rt
3. H_2O_2
86% (14)

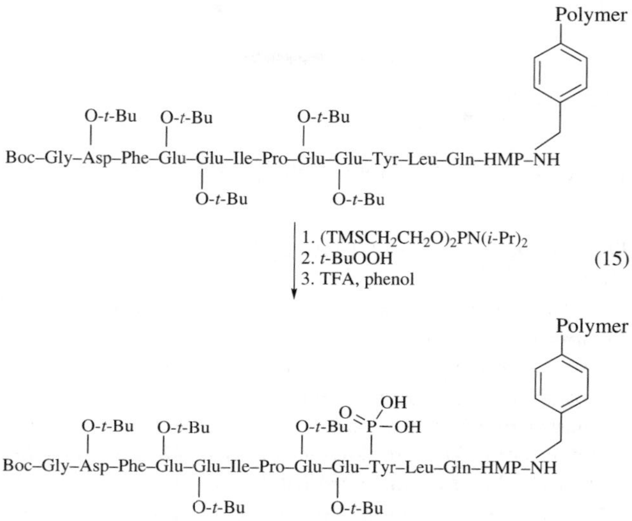

aq HF, MeCN, rt, 73%: R = CH_2CH_2TMS → R = H

Boc–Gly–Asp–Phe–Glu–Glu–Ile–Pro–Glu–Glu–Tyr–Leu–Gln–HMP–NH–Polymer

1. $(TMSCH_2CH_2O)_2PN(i\text{-}Pr)_2$
2. *t*-BuOOH
3. TFA, phenol
(15)

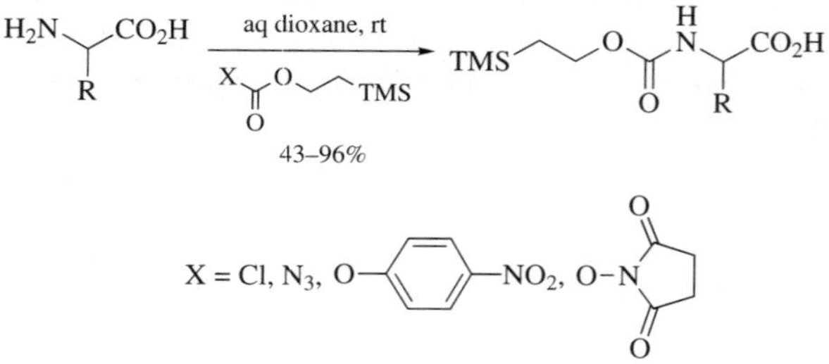

H_2N–CH(R)–CO_2H —(aq dioxane, rt; X–C(O)–O–CH₂CH₂–TMS; 43–96%)→ TMS–CH₂CH₂–O–C(O)–NH–CH(R)–CO_2H (16)

X = Cl, N_3, O–C₆H₄–NO_2, O–N(succinimide)

A new method for the synthesis of imidazolones involves the replacement of the C-2 nitro group of *N*-protected dinitroimidazoles by nucleophilic addition of the sodium salt of 2-(trimethylsilyl)ethanol (eq 17).[20]

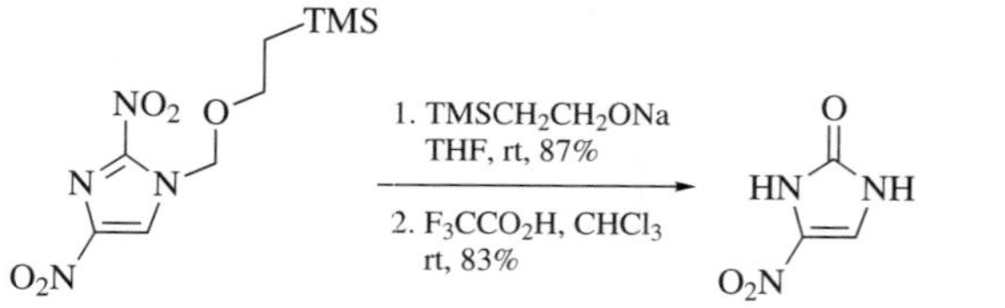

1. $TMSCH_2CH_2ONa$, THF, rt, 87%
2. F_3CCO_2H, $CHCl_3$, rt, 83%
(17)

Reaction of 2-(trimethylsilyl)ethyl benzenesulfenate with halides in the presence of TBAF yields phenyl sulfoxides (eq 18).[21] 2-(Trimethylsilyl)ethyl benzenesulfenate is prepared by the reaction of ***Benzenesulfenyl Chloride*** and the lithium salt of 2-(trimethylsilyl)ethanol.

$$\text{PhSCl} + \text{LiO}\frown\text{TMS} \xrightarrow[81\%]{Et_2O,\ -78\,°C} \text{PhS-O}\frown\text{TMS} \xrightarrow[RCH_2I,\ 21\text{–}66\%]{Bu_4NF,\ THF} \text{Ph-S(=O)-CH}_2\text{R} \quad (18)$$

Related Reagents. 2-(Trimethylsilyl)ethoxycarbonylimidazole; 2-(Trimethylsilyl)ethoxymethyl Chloride; 2-(Trimethylsilyl)-ethyl Chloroformate.

---

1. Fessenden, R. J.; Fessenden, J. S. *JOC* **1967**, *32*, 3535.
2. Gerlach, H. *HCA* **1977**, *60*, 3039.
3. Jansson, K.; Ahlfors, S.; Frejd, T.; Kihlberg, J.; Magnusson, G.; Dahmen, J.; Noori, G.; Stenvall, K. *JOC* **1988**, *53*, 5629.
4. Rosowsky, A.; Wright, J. E. *JOC* **1983**, *48*, 1539.
5. Soderquist, J. A.; Hassner, A. *JOM* **1978**, *156*, C12.
6. Soderquist, J. A.; Brown, H. C. *JOC* **1980**, *45*, 3571.
7. Soderquist, J. A.; Thompson, K. L. *JOM* **1978**, *159*, 237.
8. Mancini, M. L.; Honek, J. F. *TL* **1982**, *23*, 3249.
9. From acids: (a) Sieber, P. *HCA* **1977**, *60*, 2711. (b) Brook, M. A.; Chan, T. H. *S* **1983**, 201. (c) White, J. D.; Jayasinghe, L. R. *TL* **1988**, *29*, 2139. From an acid chloride: see Ref. 2. From an anhydride: Vedejs, E.; Larsen, S. D. *JACS* **1984**, *106*, 3030. For cleavage: see Ref. 9a, and Forsch, R. A.; Rosowsky, A. *JOC* **1984**, *49*, 1305.
10. Pouzar, V.; Drasar, P.; Cerny, I.; Havel, M. *SC* **1984**, *14*, 501.
11. (a) Seebach, D.; Hungerbühler, E.; Naef, R.; Schnurrenberger, P.; Weidmann, B.; Züger, M. *S* **1982**, 138. (b) Férézou, J. P.; Julia, M.; Liu, L. W.; Pancrazi, A. *SL* **1991**, 618.
12. Seebach, D.; Thaler, A.; Blaser, D.; Ko, S. Y. *HCA* **1991**, *74*, 1102.
13. (a) Burke, S. D.; Pacofsky, G. J.; Piscopio, A. D. *TL* **1986**, *27*, 3345. (b) Gioeli, C.; Balgobin, N.; Josephson, S.; Chattopadhyaya, J. B. *TL* **1981**, *22*, 969.
14. Lipshutz, B. H.; Pegram, J. J.; Morey, M. C. *TL* **1981**, *22*, 4603; also see Ref. 3.
15. Sawabe, A.; Filla, S. A.; Masamune, S. *TL* **1992**, *33*, 7685.
16. Chao, H.-G.; Bernatowicz, M. S.; Klimas, C. E.; Matsueda, G. R. *TL* **1993**, *34*, 3377.
17. (a) Shute, R. E.; Rich, D. H. *S* **1987**, 346. (b) Ref. 4. (c) Carpino, L. A.; Tsao, J.-H. *CC* **1978**, 358.
18. Carpino, L. A.; Sau, A. C. *CC* **1979**, 514.
19. Campbell, A. L.; Pilipauskas, D. R.; Khanna, I. K.; Rhodes, R. A. *TL* **1987**, *28*, 2331.
20. Marlin, J. E.; Killpack, M. O. *H* **1992**, *34*, 1385.
21. Oida, T.; Ohnishi, A.; Shimamaki, T.; Hayashi, Y.; Tanimoto, S. *BCJ* **1991**, *64*, 702.

Jayachandra P. Reddy
Indiana University, Bloomington, IN, USA

2-(Trimethylsilyl)ethoxymethyl Chloride[1]

$Me_3Si\frown O \frown Cl$

[76513-69-4] $C_6H_{15}ClOSi$ (MW 166.75)

(protection of alcohols,[1,2] secondary aryl amines,[3] and imidazole, indole, and pyrrole nitrogens;[4–6] electrophilic formaldehyde equivalent;[7] acyl anion equivalent[8])

Alternate Name: SEM-Cl.
Physical Data: bp 57–59 °C/8 mmHg; *d* 0.942 g cm^{-3}.
Solubility: sol most organic solvents (pentane, CH_2Cl_2, Et_2O, THF, DMF, DMPU, HMPA).
Form Supplied in: liquid; commercially available 90–95% pure; HCl is typical impurity.
Analysis of Reagent Purity: GC.
Preparative Methods: several syntheses have been reported.[2,9–12]
Handling, Storage, and Precautions: water sensitive; corrosive; should be stored in a glass container under an inert atmosphere; a lachrymator; flammable (fp 46 °C).

Reagent for the Protection of Alcohols. Lipshutz and coworkers introduced the use of SEM-Cl for the protection of primary, secondary, and tertiary alcohols (eq 1).[2] SEM-Cl is now widely employed in organic synthesis for the protection of hydroxyl functionalities (eqs 2 and 3).[13,14]

(1) Diacetonide sugar with free OH $\xrightarrow[2M\ CH_2Cl_2,\ 25\,°C,\ 1\,h,\ 98\%]{4\text{–}5\ equiv\ i\text{-}Pr_2NEt,\ 3\ equiv\ SEMCl}$ OSEM product

(2) OBn/OH/CO_2Me arene $\xrightarrow[then\ SEMCl,\ 89\%]{NaH,\ Et_2O}$ OBn/OSEM/CO_2Me arene

(3) H, OTBDMS, dithiane, OH, acetonide, OBn substrate $\xrightarrow[1.1\ equiv\ Bu_4NI,\ i\text{-}Pr_2NEt,\ 97\%]{SEMCl,\ CH_2Cl_2}$ SEMO product

The resulting SEM ethers are stable under a variety of conditions. Most SEM ethers are cleaved with a fluoride anion source (eqs 4 and 5),[2,15–17] although this fragmentation is generally much less facile than fluoride-induced cleavage of silyloxy bonds. Therefore the selective deprotection of other silyl ethers in the presence of SEM ethers is possible (eq 6).[15] Vigorous conditions with anhydrous fluoride ion are required for the cleavage of some tertiary SEM ethers (eq 7).[18,19]

2 M TBAF, 5 equiv THF, 45 °C, 4 h, 85% (4)

CsF, DMF, reflux, 80%, no alkene scrambling (5)

1.5 equiv TBAF, THF, rt, 4 h, 98% (6)

TBAF, HMPA, mol. sieves 4Å, 100 °C, 98%; R = SEM → R = H (7)

The stability of SEM ethers has necessitated the development of a variety of other deprotection methods. SEM protective groups are stable to the acidic conditions used to hydrolyze THP, TBDMS, and MOM ethers (AcOH, H_2O, THF, 45 °C),[2] but can be removed under strongly acidic conditions with ***Trifluoroacetic Acid***.[20] SEM, MTM, and MOM ethers are selectively cleaved in the presence of MEM, TBDMS, and benzyl ethers with ***Magnesium Bromide*** (eq 8).[21] SEM, MOM, and MEM phenolic protective groups are removed with ***Diphosphorus Tetraiodide*** (eq 9).[22]

5 equiv $MgBr_2$, 2.5 equiv BuSH, Et_2O, rt, 7 h, 95% (8)

0.75 equiv P_2I_4, CH_2Cl_2, 0 °C to rt, 30 min, 86% (9)

SEM-Cl has also proven useful in carbohydrate synthesis where the resulting SEM ethers can be cleaved under less acidic conditions than those required for the cleavage of MOM and MEM ethers (eq 10).[23]

i-Pr_2NEt, SEMCl, CH_2Cl_2, 80%; 1.5% HCl in MeOH, 16 h, 80% (10)

Reagent for the Protection of Acids. Carboxylic acids can be protected as SEM esters (eq 11).[24,25] Cleavage of the SEM esters occurs with refluxing methanol[24] or magnesium bromide (eq 12).[25,26]

Et_3N, SEMCl, THF, 0 °C, 80% (11)

2 equiv $MgBr_2$, Et_2O, 1 h, rt, 94% (12)

Reagent for the Protection of Secondary Amines. SEM-Cl can be used to protect secondary aromatic amines (eq 13),[3] and is an ideal reagent for the protection of imidazoles, indoles, and pyrroles (eqs 14–16).[4–6,27] Many functional groups are compatible with the introduction and cleavage of SEM amines, and the SEM substituent is unusually stable to further functionalization of the molecule.[4–6,27] Recently, a one-pot method for protection and alkylation of imidazoles employing ***n-Butyllithium*** and SEM-Cl has been developed (eq 17).[28–30]

NaH, DMF then SEMCl at 0 °C 100%

1. THF, –70 °C, 2. 2N HCl, 50 °C, 5 h 88% (13)

NaH, DMF then SEMCl, 85%; 1M TBAF, THF, reflux or 0.5M aq. HCl in EtOH (14)

1. 1.6 equiv NaH, DMF, 0 °C 2. SEMCl, DMF, DMSO, rt 95% (15)

NaH, DMF, then SEMCl at 0 °C, 1 h, 91%; $BF_3 \cdot Et_2O$, Triton B, MeCN, 86% (16)

1. 1.1 equiv BuLi, THF, –20 °C 2. SEMCl, –20 °C to rt 3. 1.1 equiv BuLi, –78 °C 4. TMSCl (17)

One-Carbon Homologations. SEM-Cl has functioned as an alternative to the use of ***Formaldehyde*** (eq 18).[7] The enolate of the lactone (**1**) undergoes *C*-alkylation with SEM-Cl, and the resulting SEM substituent can be subsequently treated with trifluoroacetic acid to afford the alcohol (**2**).

SEM-Cl has been transformed to a formaldehyde carbanion equivalent.[8] SEM-Cl is converted to the ylide upon treatment with ***Triphenylphosphine*** and ***Sodium Hydride***. This ylide reacts with a variety of aldehydes and ketones affording enol ethers (eq 19).[8] Hydrolysis with 5% aqueous HF gives the corresponding aldehyde.

1. $KN(TMS)_2$, THF, –75 °C then SEMCl, 76% 2. $NaN(TMS)_2$, crown ether THF, –78 °C, then $ICH_2CH_2CH_2Cl$, 70% (**1**)

1. several steps 2. CF_3CO_2H (18) (**2**)

1. Ph_3P, PhH 45 °C, 48 h, 75% 2. DMSO, NaH then cyclohexanone 80%

5% HF, MeCN (19)

Related Reagents. BenzylChloromethyl Ether; *t*-Butyl Chloromethyl Ether; 2-(Chloroethyl) chloromethyl Ether; Chloromethyl Methyl Ether; Chloromethyl Methyl Sulfide; (*p*-Methoxybenzyloxy)methyl Chloride; 2-Methoxyethoxymethyl Chloride; 2-(Trimethylsilyl)ethanol.

1. Greene, T. W.; Wuts, P. G. M. *Protective Groups in Organic Synthesis*, 2nd ed.; Wiley: New York, 1991.
2. Lipshutz, B. H.; Pegram, J. J. *TL* **1980**, *21*, 3343.
3. Zeng, Z.; Zimmerman, S. C. *TL* **1988**, *29*, 5123.
4. Whitten, J. P.; Matthews, D. P.; McCarthy, J. R. *JOC* **1986**, *51*, 1891.
5. Ley, S. V.; Smith, S. C.; Woodward, P. R. *T* **1992**, *48*, 1145.
6. Muchowski, J. M.; Solas, D. R. *JOC* **1984**, *49*, 203.
7. Baldwin, J. E.; Lee, V.; Schofield, C. J. *SL* **1992**, 249.
8. Schönauer, K.; Zbiral, E. *TL* **1983**, *24*, 573.
9. Sonderquist, J. A.; Hassner, A. *JOM* **1978**, *156*, C12.
10. Sonderquist, J. A.; Thompson, K. L. *JOM* **1978**, *159*, 237.
11. Gerlach, H. *HCA* **1977**, *60*, 3039.
12. Fessenden, R. J.; Fessenden, J. S. *JOC* **1967**, *32*, 3535.
13. Kotecha, N. R.; Ley, S. V.; Mantegani, S. *SL* **1992**, 395.
14. Lipshutz, B. H.; Moretti, R.; Crow, R. *TL* **1989**, *30*, 15.
15. Williams, D. R.; Jass, P. A.; Tse, H.-L. A.; Gaston, R. D. *JACS* **1990**, *112*, 4552.
16. Shull, B. K.; Koreeda, M. *JOC* **1990**, *55*, 99.
17. Ireland, R. E.; Varney, M. D. *JOC* **1986**, *51*, 635.
18. Kan, T.; Hashimoto, M.; Yanagiya, M.; Shirahama, H. *TL* **1988**, *29*, 5417.
19. Lipshutz, B. H.; Miller, T. A. *TL* **1989**, *30*, 7149.
20. Schlessinger, R. H.; Poss, M. A.; Richardson, S. *JACS* **1986**, *108*, 3112.
21. Kim, S.; Kee, I. S.; Park, Y. H.; Park, J. H. *SL* **1991**, 183.
22. Saimoto, H.; Kusano, Y.; Hiyama, T. *TL* **1986**, *27*, 1607.
23. Pinto, B. M.; Buiting, M. M. W.; Reimer, K. B. *JOC* **1990**, *55*, 2177.
24. Logusch, E. W. *TL* **1984**, *25*, 4195.

25. Kim, S; Park, Y. H.; Kee, I. S. *TL* **1991**, *32*, 3099.
26. Salomon, C. J.; Mata, E. G.; Mascaretti, O. A. *T* **1993**, *49*, 3691.
27. Lipshutz, B. H.; Vaccaro, W.; Huff, B. *TL* **1986**, *27*, 4095.
28. Lipshutz, B. H.; Huff, B.; Hagen, W. *TL* **1988**, *29*, 3411.
29. Demuth, T. P., Jr.; Lever, D. C.; Gorgos, L. M.; Hogan, C. M.; Chu, J. *JOC* **1992**, *57*, 2963.
30. For a similar method with pyrroles, see Ref. 13.

Jill Earley
Indiana University, Bloomington, IN, USA

Trimethylsilyl Trifluoromethanesulfonate[1]

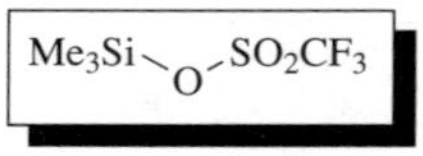

$Me_3Si-O-SO_2CF_3$

[88248-68-4; 27607-77-8] $C_4H_9F_3O_3SSi$ (MW 222.29)

Alternate Name: TMSOTf.
Physical Data: bp 45–47 °C/17 mmHg, 39–40 °C/12mmHg; *d* 1.225 g cm^{-3}.
Solubility: sol aliphatic and aromatic hydrocarbons, haloalkanes, ethers.
Form Supplied in: colorless liquid; commercially available.
Preparative Methods: may be prepared by a variety of methods.[2]
Handling, Storage, and Precautions: flammable; corrosive; very hygroscopic.

Silylation. TMSOTf is widely used in the conversion of carbonyl compounds to their enol ethers. The conversion is some 10^9 faster with TMSOTf/***Triethylamine*** than with ***Chlorotrimethylsilane*** (eqs 1–3).[3–5]

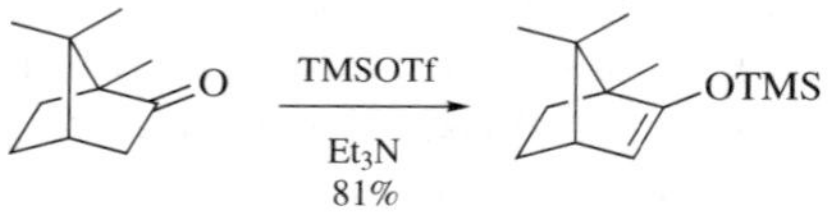

TMSOTf, Et3N, 81% (1)

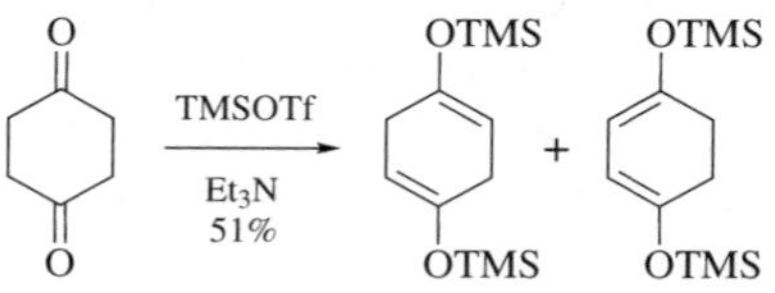

TMSOTf, Et3N, 51% (2)

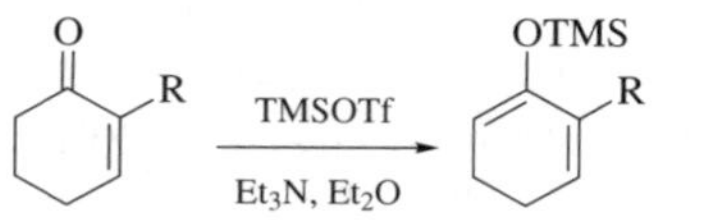

TMSOTf, Et3N, Et2O (3)

Dicarbonyl compounds are converted to the corresponding bis-enol ethers; this method is an improvement over the previous two-step method (eq 4).[6]

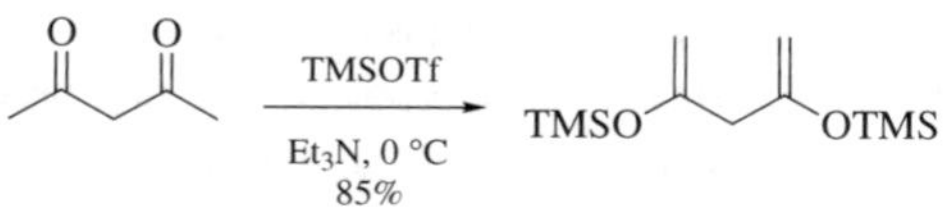

TMSOTf, Et3N, 0 °C, 85% (4)

In general, TMSOTf has a tendency to *C*-silylation which is seen most clearly in the reaction of esters, where *C*-silylation dominates over *O*-silylation. The exact ratio of products obtained depends on the ester structure[7] (eq 5).[8] Nitriles undergo *C*-silylation; primary nitriles may undergo *C,C*-disilylation.[9]

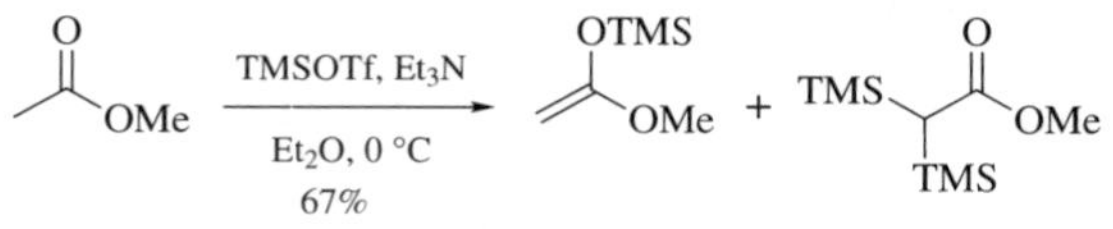

TMSOTf, Et3N, Et2O, 0 °C, 67%; 84:16 (5)

TMS enol ethers may be prepared by rearrangement of α-ketosilanes in the presence of catalytic TMSOTf (eq 6).[10,11]

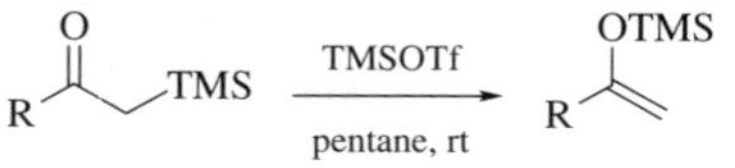

TMSOTf, pentane, rt (6)

Enhanced regioselectivity is obtained when trimethylsilyl enol ethers are prepared by treatment of α-trimethylsilyl ketones with catalytic TMSOTf (eq 7).[12]

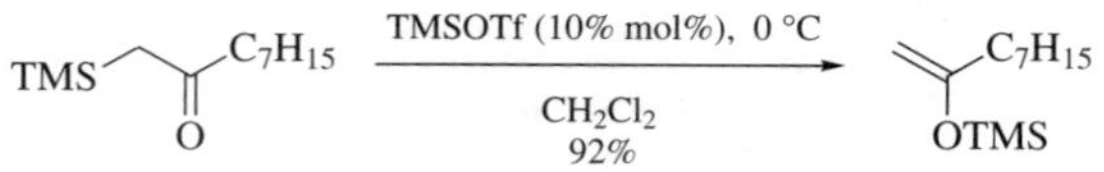

TMSOTf (10% mol%), 0 °C, CH2Cl2, 92% (7)

The reaction of imines with TMSOTf in the presence of Et_3N gives *N*-silylenamines.[13]

Ethers do not react, but epoxides are cleaved to give silyl ethers of allylic alcohols in the presence of TMSOTf and ***1,8-Diazabicyclo[5.4.0]undec-7-ene***; The regiochemistry of the reaction is dependent on the structure of the epoxide (eq 8).[14]

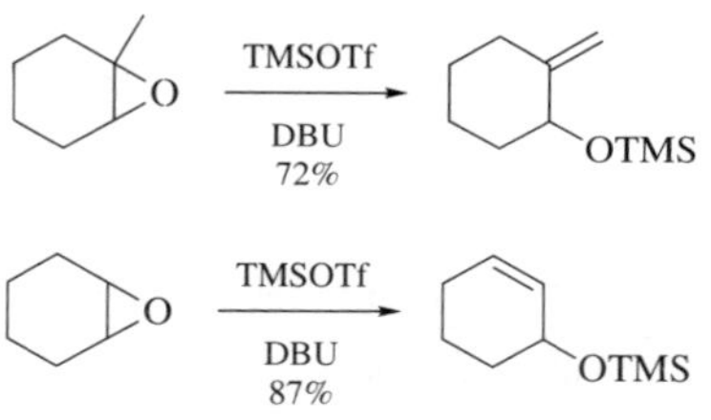

TMSOTf, DBU, 72%; TMSOTf, DBU, 87% (8)

Indoles and pyrroles undergo efficient *C*-silylation with TMSOTf (eq 9).[15]

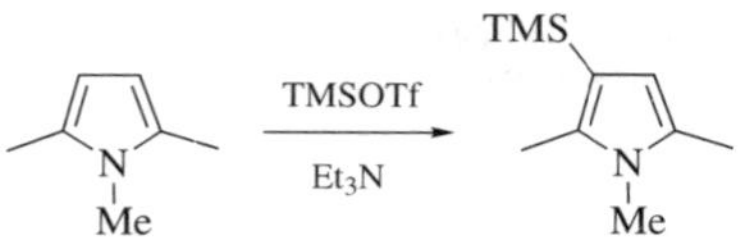

TMSOTf, Et3N (9)

t-Butyl esters are dealkylatively silylated to give TMS esters by TMSOTf; benzyl esters are inert under the same conditions.[16]

Imines formed from unsaturated amines and α-carbonyl esters undergo ene reactions in the presence of TMSOTf to form cyclic amino acids.[17]

Carbonyl Activation. 1,3-Dioxolanation of conjugated enals is facilitated by TMSOTf in the presence of 1,2-bis(trimethylsilyloxy)ethane. In particular, highly selective protection of sterically differentiated ketones is possible (eq 10).[18] Selective protection of ketones in the presence of enals is also facilitated (eq 11).[19]

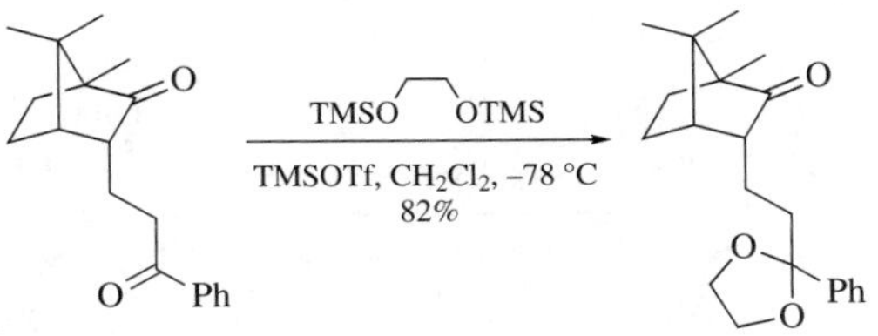

O; O Ph; TMSO OTMS; TMSOTf, CH_2Cl_2, –78 °C; 82%; O; O O Ph (10)

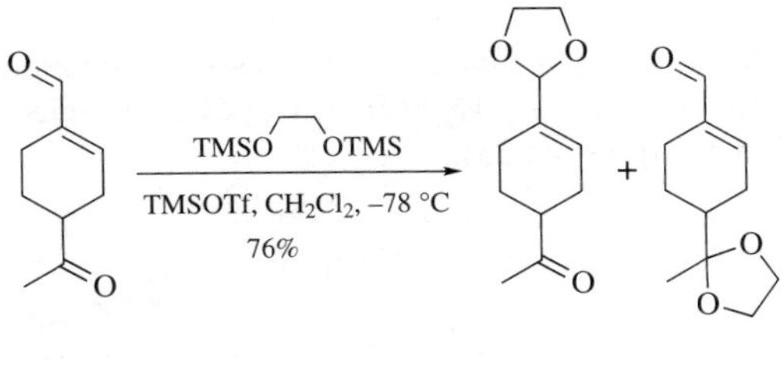

O; O; TMSO OTMS; TMSOTf, CH_2Cl_2, –78 °C; 76%; O O; O; + O; O O (11)

1:27

The similar reaction of 2-alkyl-1,3-disilyloxypropanes with chiral ketones is highly selective and has been used to prepare spiroacetal starting materials for an asymmetric synthesis of α-tocopherol subunits (eq 12).[20]

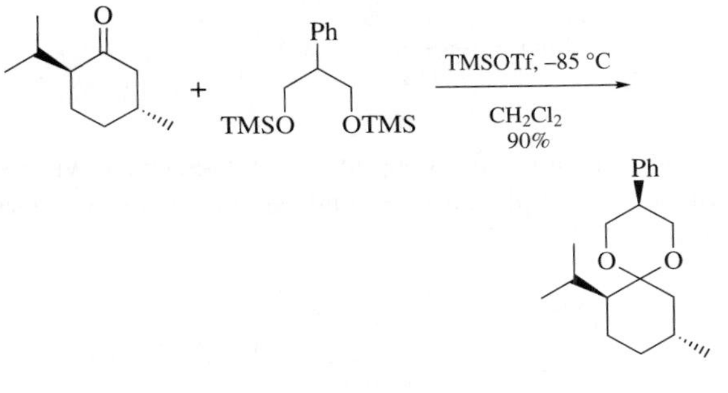

O; + Ph; TMSO OTMS; TMSOTf, –85 °C; CH_2Cl_2; 90%; Ph; O O (12)

The preparation of spiro-fused dioxolanes (useful as chiral glycolic enolate equivalents) also employs TMSOTf (eq 13).[21]

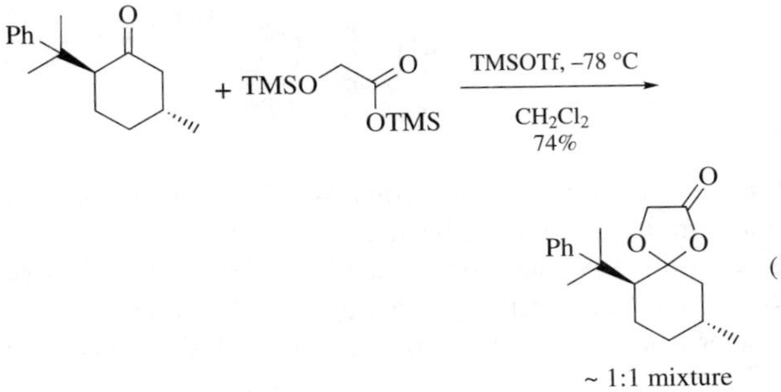

Ph; O; + TMSO O; OTMS; TMSOTf, –78 °C; CH_2Cl_2; 74%; O; Ph O O (13)

~ 1:1 mixture

TMSOTf mediates a stereoselective aldol-type condensation of silyl enol ethers and acetals (or orthoesters). The nonbasic reaction conditions are extremely mild. TMSOTf catalyzes many aldol-type reactions; in particular, the reaction of relatively nonnucleophilic enol derivatives with carbonyl compounds is facile in the presence of the silyl triflate. The activation of acetals was first reported by Noyori and has since been widely employed (eq 14).[22,23]

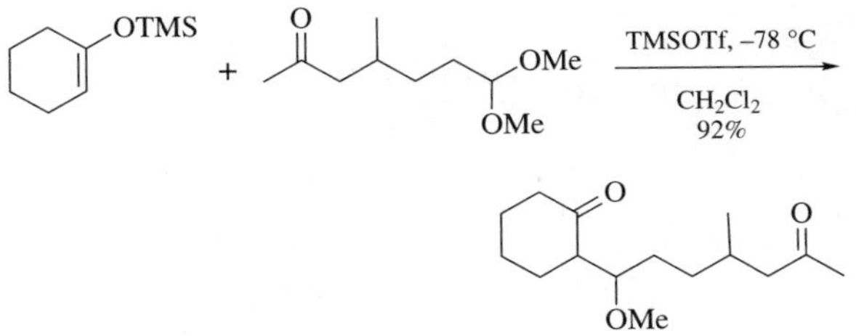

OTMS; + O; OMe; OMe; TMSOTf, –78 °C; CH_2Cl_2; 92%; O; O; OMe (14)

In an extension to this work, TMSOTf catalyzes the first step of a [3 + 2] annulation sequence which allows facile synthesis of fused cyclopentanes possessing bridgehead hydroxy groups (eq 15).[24]

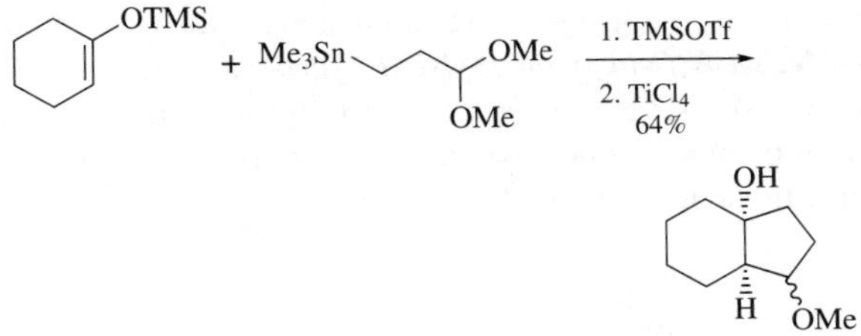

OTMS; + Me_3Sn; OMe; OMe; 1. TMSOTf; 2. $TiCl_4$; 64%; OH; H OMe (15)

The use of TMSOTf in aldol reactions of silyl enol ethers and ketene acetals with aldehydes is ubiquitous. Many refinements of the basic reaction have appeared. An example is shown in eq 16.[25]

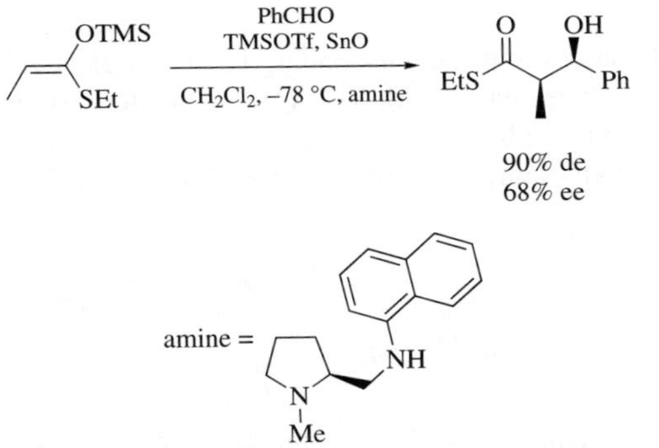

OTMS; SEt; PhCHO; TMSOTf, SnO; CH_2Cl_2, –78 °C, amine; O OH; EtS Ph (16)

90% de
68% ee

amine = NH; N; Me

The use of TMSOTf in the reaction of silyl ketene acetals with imines offers an improvement over other methods (such as Ti^{IV}- or Zn^{II}-mediated processes) in that truly catalytic amounts of activator may be used (eq 17);[26] this reaction may be used as the crucial step in a general synthesis of 3-(1′-hydroxyethyl)-2-azetidinones (eq 18).[27]

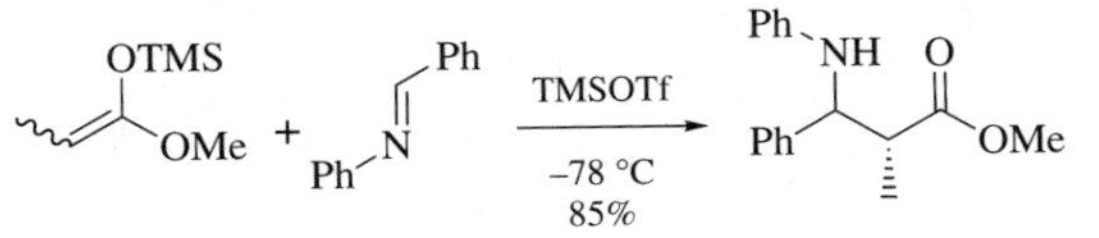

OTMS; OMe; + Ph; Ph N; TMSOTf; –78 °C; 85%; Ph NH O; Ph OMe (17)

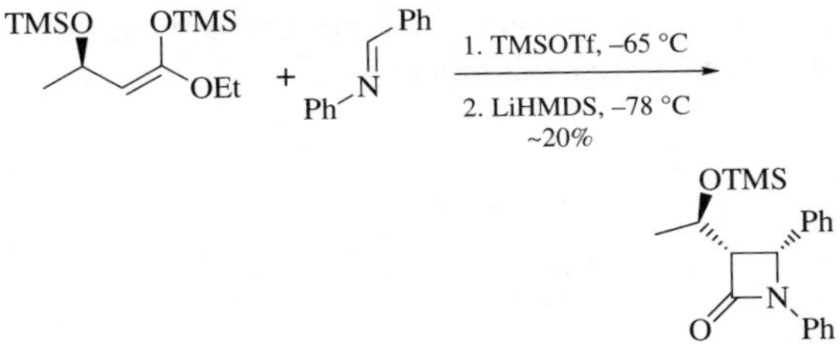

TMSO OTMS; OEt; + Ph; Ph N; 1. TMSOTf, –65 °C; 2. LiHMDS, –78 °C; ~20%; OTMS; Ph; O N Ph (18)

Stereoselective cyclization of α,β-unsaturated enamide esters is induced by TMSOTf and has been used as a route to quinolizidines and indolizidines (eq 19).[28]

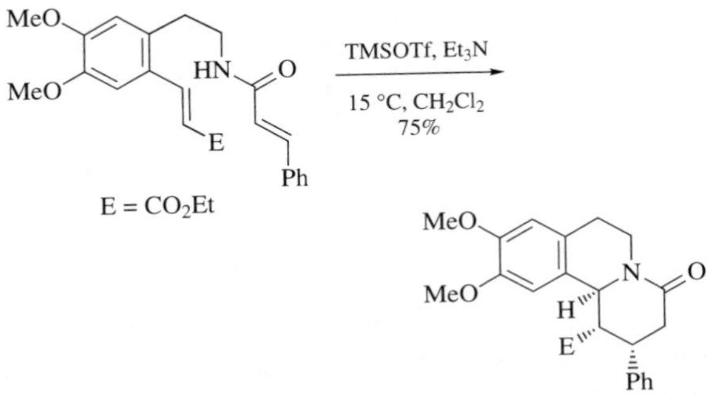

TMSOTf, Et_3N; 15 °C, CH_2Cl_2; 75%; $E = CO_2Et$ (19)

The formation of nitrones by reaction of aldehydes and ketones with ***N-Methyl-N,O-bis(trimethylsilyl)hydroxylamine*** is accelerated when TMSOTf is used as a catalyst; the acceleration is particularly pronounced when the carbonyl group is under a strong electronic influence (eq 20).[29]

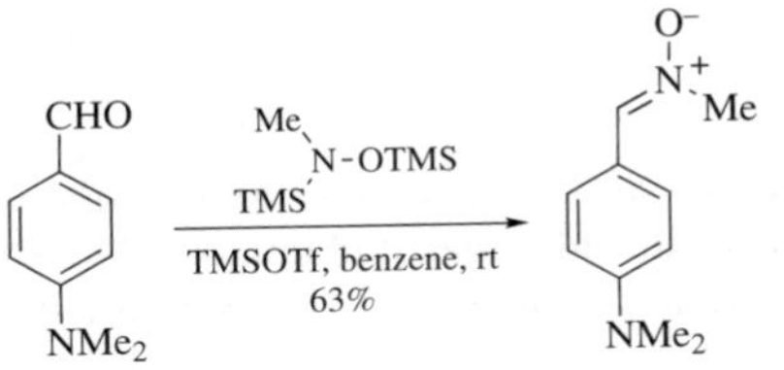

Me(TMS)N-OTMS; TMSOTf, benzene, rt; 63% (20)

β-Stannylcyclohexanones undergo a stereoselective ring contraction when treated with TMSOTf at low temperature. When other Lewis acids were employed, a mixture of ring-contracted and protiodestannylated products was obtained (eq 21).[30]

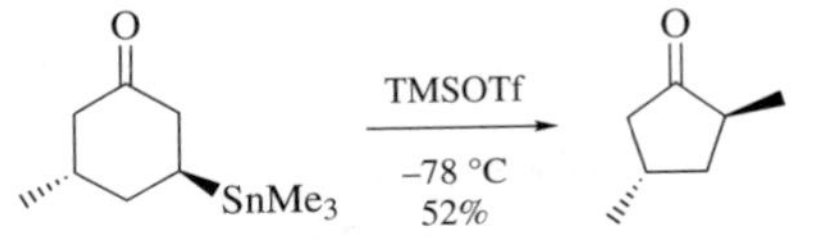

TMSOTf; −78 °C; 52% (21)

The often difficult conjugate addition of alkynyl organometallic reagents to enones is greatly facilitated by TMSOTf. In particular, alkynyl zinc reagents (normally unreactive with α,β-unsaturated carbonyl compounds) add in good yield (eq 22).[31] The proportion of 1,4-addition depends on the substitution pattern of the substrate.

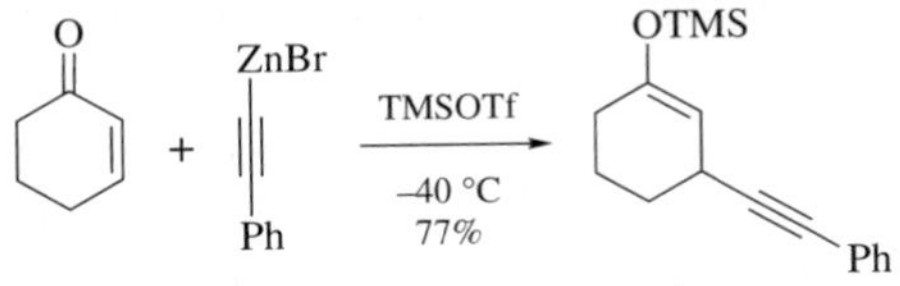

TMSOTf; −40 °C; 77% (22)

The 1,4-addition of phosphines to enones in the presence of TMSOTf gives β-phosphonium silyl enol ethers, which may be deprotonated and alkylated in situ (eq 23).[32]

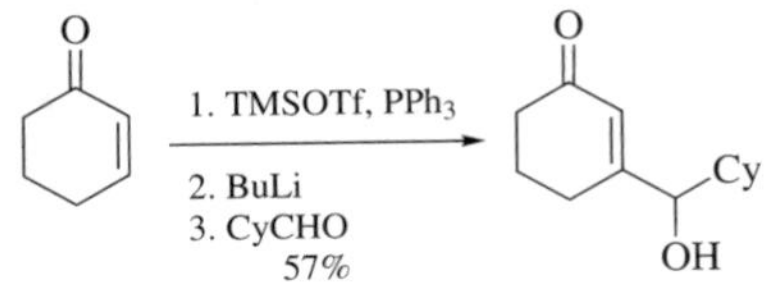

1. TMSOTf, PPh_3; 2. BuLi; 3. CyCHO; 57% (23)

Miscellaneous. Methyl glucopyranosides and glycopyranosyl chlorides undergo allylation with allylsilanes under TMSOTf catalysis to give predominantly α-allylated carbohydrate analogs (eq 24).[33]

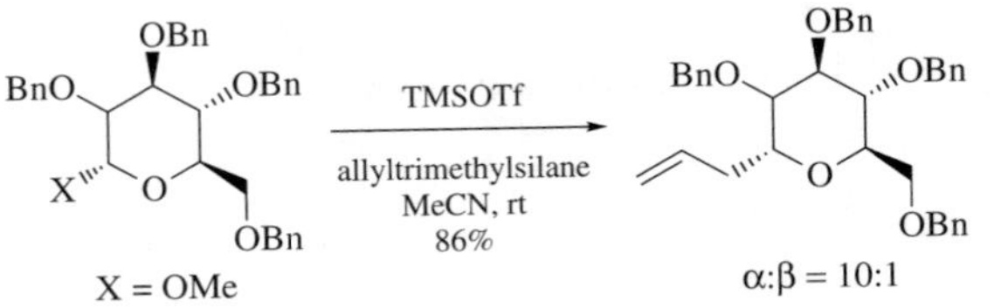

TMSOTf; allyltrimethylsilane, MeCN, rt; 86%; X = OMe; α:β = 10:1 (24)

Glycosidation is a reaction of massive importance and widespread employment. TMSOTf activates many selective glycosidation reactions (eq 25).[34]

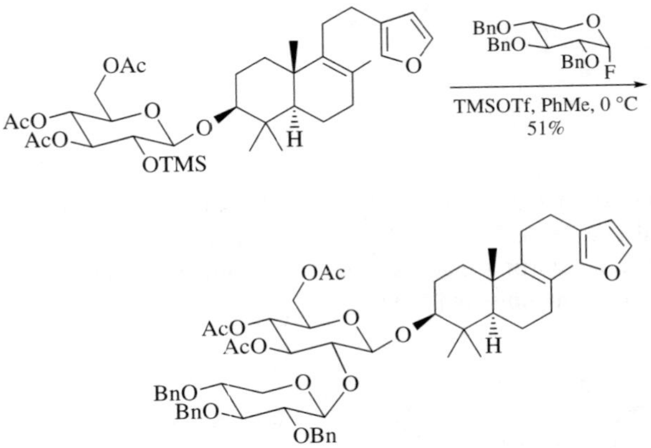

TMSOTf, PhMe, 0 °C; 51% (25)

TMSOTf activation for coupling of 1-*O*-acylated glycosyl donors has been employed in a synthesis of avermectin disaccharides (eq 26).[35]

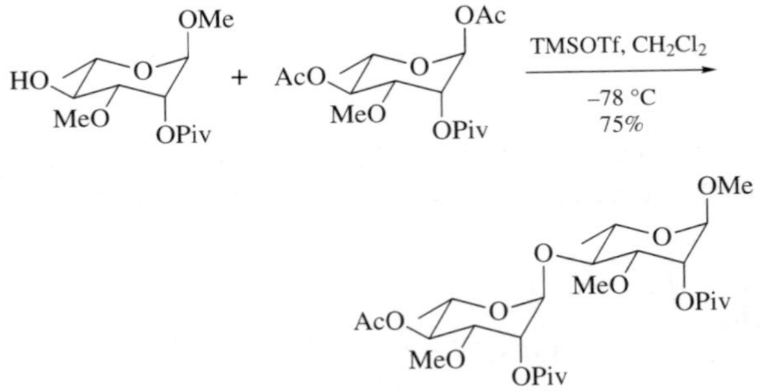

TMSOTf, CH_2Cl_2; −78 °C; 75% (26)

Similar activation is efficient in couplings with trichloroimidates[36] and *O*-silylated sugars.[37,38]

2-Substituted Δ^3-piperidines may be prepared by the reaction of 4-hydroxy-1,2,3,4-tetrahydropyridines with a variety of carbon and heteronucleophiles in the presence of TMSOTf (eqs 27 and 28).[39]

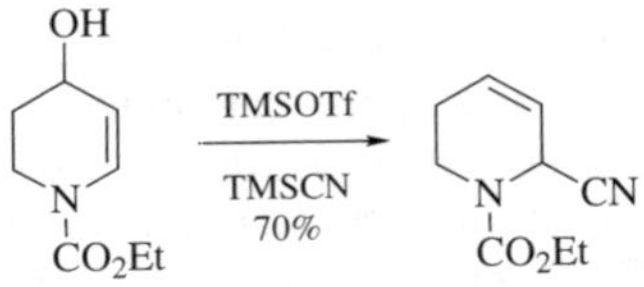

TMSOTf; TMSCN; 70% (27)

TMSOTf, allyltrimethylsilane, 51% (28)

Iodolactamization is facilitated by the sequential reaction of unsaturated amides with TMSOTf and ***Iodine*** (eq 29).[40]

1. TMSOTf, CH_2Cl_2, Et_3N; 2. I_2; 64% (29)

By use of a silicon-directed Beckmann fragmentation, cyclic (*E*)-β-trimethylsilylketoxime acetates are cleaved in high yield in the presence of catalytic TMSOTf to give the corresponding unsaturated nitriles. Regio- and stereocontrol are complete (eq 30).[41]

1. TMSOTf, CH_2Cl_2, 0 °C; 2. Et_3N; 94% (30)

A general route to enol ethers is provided by the reaction of acetals with TMSOTf in the presence of a hindered base (eq 31).[42] The method is efficient for dioxolanes and noncyclic acetals.

TMSOTf, *i*-Pr_2NEt, −20 °C, CH_2Cl_2, 98% (31)

α-Halo sulfoxides are converted to α-halovinyl sulfides by reaction with excess TMSOTf (eq 32),[43] while α-cyano- and α-alkoxycarbonyl sulfoxides undergo a similar reaction (eq 33).[44] TMSOTf is reported as much superior to ***Iodotrimethylsilane*** in these reactions.

2.2 equiv TMSOTf, Et_3N, 0 °C (32)

2.2 equiv TMSOTf, HMDS, 25 °C (33)

X = CN or CO_2R'

Related Reagents. *t*-Butyldimethylsilyl Trifluoromethanesulfonate; Chlorotrimethylsilane; Iodotrimethylsilane; Triethylsilyl Trifluoromethanesulfonate.

1. Reviews: (a) Emde, H.; Domsch, D.; Feger, H.; Frick, U.; Götz, H. H.; Hofmann, K.; Kober, W.; Krägeloh, K.; Oesterle, T.; Steppan, W.; West, W.; Simchen, G. *S* **1982**, 1. (b) Noyori, R.; Murata, S.; Suzuki, M. *T* **1981**, *37*, 3899. (c) Stang, P. J.; White, M. R. *Aldrichim. Acta* **1983**, *16*, 15. Preparation: (d) Olah, G. H.; Husain, A.; Gupta, B. G. B.; Salem, G. F.; Narang, S. C. *JOC* **1981**, *46*, 5212. (e) Morita, T.; Okamoto, Y.; Sakurai, H. *S* **1981**, 745. (f) Demuth, M.; Mikhail, G. *S* **1982**, 827. (g) Ballester, M.; Palomo, A. L. *S* **1983**, 571. (h) Demuth, M.; Mikhail, G. *T* **1983**, *39*, 991. (i) Aizpurua, J. M.; Palomo, C. *S* **1985**, 206.
2. Simchen, G.; Kober, W. *S* **1976**, 259.
3. Hergott, H. H.; Simchen, G. *LA* **1980**, 1718.
4. Simchen, G.; Kober, W. *S* **1976**, 259.
5. Emde, H.; Götz, A.; Hofmann, K.; Simchen, G. *LA* **1981**, 1643.
6. Krägeloh, K.; Simchen, G. *S* **1981**, 30.
7. Emde, H.; Simchen, G. *LA* **1983**, 816.
8. Emde, H.; Simchen, G. *S* **1977**, 636.
9. Emde, H.; Simchen, G. *S* **1977**, 867.
10. Yamamoto, Y.; Ohdoi, K.; Nakatani, M.; Akiba, K. *CL* **1984**, 1967.
11. Emde, H.; Götz, A.; Hofmann, K.; Simchen, G. *LA* **1981**, 1643.
12. Matsuda, I.; Sato, S.; Hattori, M.; Izumi, Y. *TL* **1985**, *26*, 3215.
13. Ahlbrecht, H.; Düber, E. O. *S* **1980**, 630.
14. Murata, S.; Suzuki, M.; Noyori, R. *JACS* **1980**, *102*, 2738.
15. Frick, U.; Simchen, G. *S* **1984**, 929.
16. Borgulya, J.; Bernauer, K. *S* **1980**, 545.
17. Tietze, L. F.; Bratz, M. *S* **1989**, 439.
18. Hwu, J. R.; Wetzel, J. M. *JOC* **1985**, *50*, 3946.
19. Hwu, J. R.; Robl, J. A. *JOC* **1987**, *52*, 188.
20. Harada, T.; Hayashiya, T.; Wada, I.; Iwa-ake, N.; Oku, A. *JACS* **1987**, *109*, 527.
21. Pearson, W. H.; Cheng, M-C. *JACS* **1986**, *51*, 3746.
22. Murata, S.; Suzuki, M.; Noyori, R. *JACS* **1980**, *102*, 3248.
23. Murata, S.; Suzuki, M.; Noyori, R. *T* **1988**, *44*, 4259.
24. Lee, T. V.; Richardson, K. A. *TL* **1985**, *26*, 3629.
25. Mukaiyama, T.; Uchiro, H.; Kobayashi, S. *CL* **1990**, 1147.
26. Guanti, G.; Narisano, E.; Banfi, L. *TL* **1987**, *28*, 4331.
27. Guanti, G.; Narisano, E.; Banfi, L. *TL* **1987**, *28*, 4335.
28. Ihara, M.; Tsuruta, M.; Fukumoto, K.; Kametani, T. *CC* **1985**, 1159.
29. Robl, J. A.; Hwu, J. R. *JOC* **1985**, *50*, 5913.
30. Sato, T.; Watanabe, T.; Hayata, T.; Tsukui, T. *CC* **1989**, 153.
31. Kim, S.; Lee, J. M. *TL* **1990**, *31*, 7627.
32. Kim, S.; Lee, P. H. *TL* **1988**, *29*, 5413.
33. Hosomi, A.; Sakata, Y.; Sakurai, H. *TL* **1984**, *25*, 2383.
34. Yamada, H.; Nishizawa, M *T* **1992**, 3021.
35. Rainer, H.; Scharf, H.-D.; Runsink, J. *LA* **1992**, 103.
36. Schmidt, R. R. *AG(E)* **1986**, *25*, 212.
37. Tietze, L.-F.; Fischer, R.; Guder, H.-J. *TL* **1982**, *23*, 4661.
38. Mukaiyama, T.; Matsubara, K. *CL* **1992**, 1041.
39. Kozikowski, A. P.; Park, P. *JOC* **1984**, *49*, 1674.
40. Knapp, S.; Rodriques, K. E. *TL* **1985**, *26*, 1803.
41. Nishiyama, H.; Sakuta, K.; Osaka, N.; Itoh, K. *TL* **1983**, *24*, 4021.
42. Gassman, P. G.; Burns, S. J. *JOC* **1988**, *53*, 5574.
43. Miller, R. D.; Hässig, R., *SC* **1984**, *14*, 1285.
44. Miller, R. D.; Hässig, R., *TL* **1985**, *26*, 2395.

Joseph Sweeney & Gemma Perkins
University of Bristol, UK

Triphenylcarbenium Tetrafluoroborate

$Ph_3C^+ BF_4^-$

[341-02-6] $C_{19}H_{15}BF_4$ (MW 330.15)

(easily prepared[1] hydride abstractor used for conversion of dihydroaromatics to aromatics,[2–4] and the preparation of aromatic and benzylic cations;[5–8] oxidative hydrolysis of ketals[9] and thioketals;[10] conversion of acetonides to α-hydroxy ketones;[9] oxidation of acetals[11] and thioacetals;[12] selective oxidation of alcohols and ethers to ketones;[9,13–15] oxidation of silyl enol ethers to enones;[16] hydrolysis of TBS and MTM ethers;[17] oxidation of amines and amides to iminium salts;[18–20] oxidation of organometallics to give alkenes;[21–23] sensitizer for photooxidation using molecular oxygen;[24] Lewis acid catalyst for various reactions;[25] polymerization catalyst;[26] other reactions[27–30])

Alternate Name: trityl fluoroborate.
Physical Data: mp ~200 °C (dec).
Solubility: sol most standard organic solvents; reacts with some nucleophilic solvents.
Form Supplied in: yellow solid; commercially available.
Preparative Methods: the most convenient procedure involves the reaction of Ph_3CCl with ***Silver(I) Tetrafluoroborate*** in ethanol.[1b] The most economical route employs the reaction of Ph_3CCl with the anhydrous ***Tetrafluoroboric Acid***–Et_2O complex,[1c] or Ph_3COH with HBF_4 in acetic anhydride.[1d]
Purification: recrystallization of commercial samples from a minimal amount of dry MeCN provides material of improved purity, but the recovery is poor.[1a]
Handling, Storage, and Precautions: moisture-sensitive and corrosive. Recrystallized reagent can be stored at rt for several months in a desiccator without significant decomposition. This compound is much less light-sensitive than other trityl salts such as the perchlorate.[1a]

Preparation of Aromatic Compounds via Dehydrogenation. Dihydroaromatic compounds are easily converted into the corresponding aromatic compound by treatment with triphenylcarbenium tetrafluoroborate followed by base.[2] Certain α,α-disubstituted dihydroaromatics are converted to the 1,4-dialkylaromatic compounds with rearrangement (eq 1).[3] Nonbenzenoid aromatic systems, e.g. benzazulene[4a] or dibenzosesquifulvalene,[4b] are readily prepared from their dihydro counterparts. Aromatic cations are also easily prepared by hydride abstraction, for example, tropylium ion (e.g. in the synthesis of heptalene (eq 2)),[5] cyclopropenyl cation,[6] and others, including heterocyclic systems.[7] Some benzylic cations, especially ferrocenyl cations,[8] can also be formed by either hydride abstraction or trityl addition.

Oxidation by Hydride Abstraction. In the early 1970s, Barton developed a method for the oxidative hydrolysis of ketals to ketones, e.g. in the tetracycline series (eq 3).[9] The same conditions can also be used to hydrolyze thioketals.[10] Acetonides of 1,2-diols are oxidized to the α-hydroxy ketones in good yield by this reagent (eq 4).[9] The hydrogen of acetals is easily abstracted (eq 5), providing a method for the conversion of benzylidene units in sugars to the hydroxy benzoates.[11] The hydrogen of dithioacetals is also abstracted to give the salts.[12] Since benzylic hydrogens are readily abstracted, this is also a method for deprotection of benzyl ethers.[9,13] Trimethylsilyl, *t*-butyl, and trityl ethers of simple alcohols are oxidized to the corresponding ketones and aldehydes in good yield. Primary–secondary diols are selectively oxidized at the secondary center to give hydroxy ketones by this method (eq 6).[14] 2,2-Disubstituted 1,4-diols are oxidized only at the 4-position to give the corresponding lactones.[15] Trimethylsilyl enol ethers are oxidized to α,β-unsaturated ketones, thereby providing a method for ketone to enone conversion (eq 7).[16] *t*-Butyldimethylsilyl (TBDMS) ethers are not oxidized but rather hydrolyzed to the alcohols, as are methylthiomethyl (MTM) ethers.[17] Benzylic amines and amides can be oxidized to the iminium salts,[18] allylic amines and enamines afford eniminium salts,[19] and orthoamides give triaminocarbocations.[20]

Bn Bn → ($Ph_3C^+BF_4^-$, AcOH, Δ, 0.5 h, 67%) Bn, Bn (1)

(1. $Ph_3C^+ BF_4^-$, 88%; 2. Me_3N, 41%) (2)

O, H, O, Ph, HO, O → ($Ph_3C^+ BF_4^-$, 65%) O, H, O, H, O, Ph (3)

H, O, O, H, H → ($Ph_3C^+ BF_4^-$, 79%) O, HO, H (4)

OMe, BnO, O, O, O, Ph → (1. $Ph_3C^+ BF_4^-$; 2. $NaHCO_3$, 75%) OMe, BnO, O, HO, OBz (5)

OH, OH → (1. Ph_3CCl; 2. $Ph_3C^+ BF_4^-$, 91%) O, OH (6)

OTMS → ($Ph_3C^+ BF_4^-$, collidine; CH_2Cl_2, 25 °C, 85%) O (7)

Generation of Alkenes from Organometallics. Various β-metalloalkanes can be oxidized by trityl fluoroborate to the corresponding alkenes.[21–23] The highest yields are obtained for the β-iron derivatives (eq 8), which are easily prepared from the corresponding halides or tosylates.[21] Grignard reagents and organolithiums also undergo this reaction (eq 9),[22] as do Group 14 organometallics (silanes, stannanes, etc.).[23]

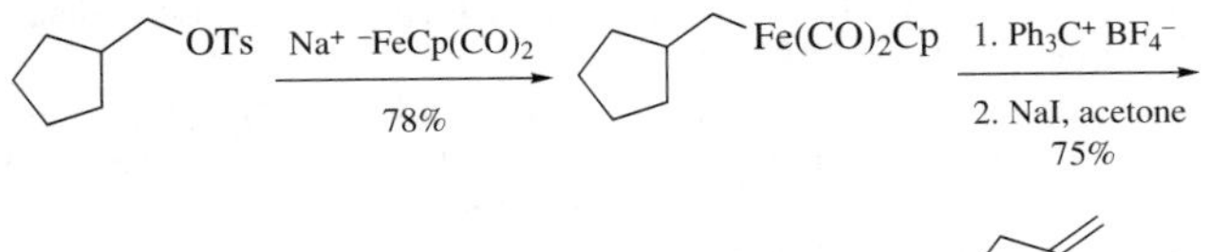

(8)

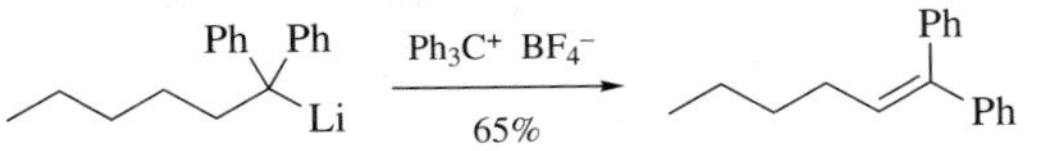

(9)

Sensitizer of Photooxygenation. Barton showed that oxygen, in the presence of trityl fluoroborate and ordinary light, adds to cisoid dienes at −78 °C in very high yields.[24] For example, the peroxide of ergosterol acetate is formed in quantitative yields under these conditions (eq 10),[24a,b] which have been used also for photocycloreversions of cyclobutanes.[24c]

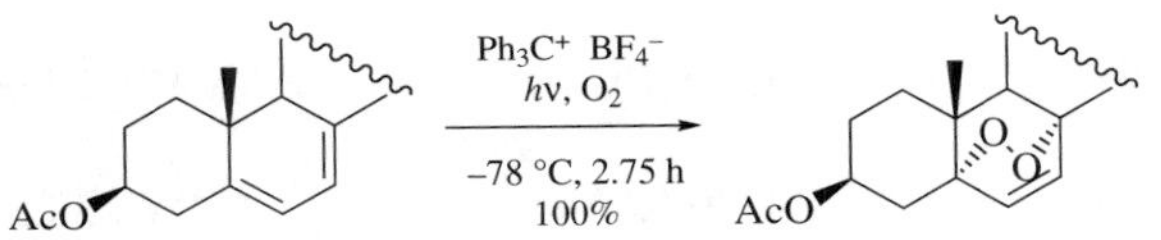

(10)

Lewis Acid Catalysis. Trityl fluoroborate is a good Lewis acid for various transformations,[25] e.g. the Mukaiyama-type aldol reaction using a dithioacetal and silyl enol ether (eq 11).[25a] It has also been used as the catalyst for the formation of glycosides from alcohols and sugar dimethylthiophosphinates (eq 12)[25b] and for the formation of disaccharides from a protected α-cyanoacetal of glucose and a 6-*O*-trityl hexose.[25c] Michael additions of various silyl nucleophiles to conjugated dithiolenium cations also proceed well (eq 13).[25d,e] Finally, the [4 + 2] cycloaddition of cyclic dienes and oxygenated allyl cations has been effected with trityl fluoroborate.[25f]

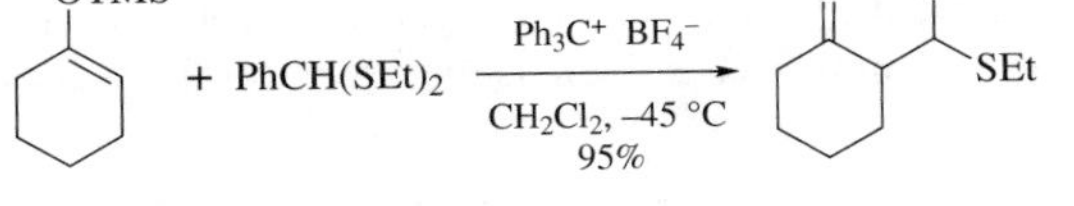

(11)

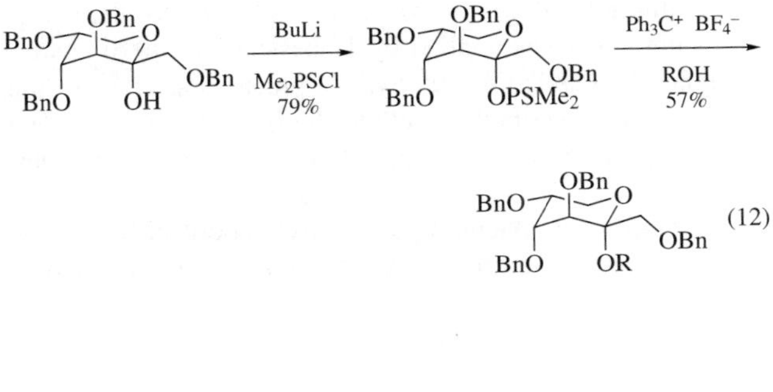

(12)

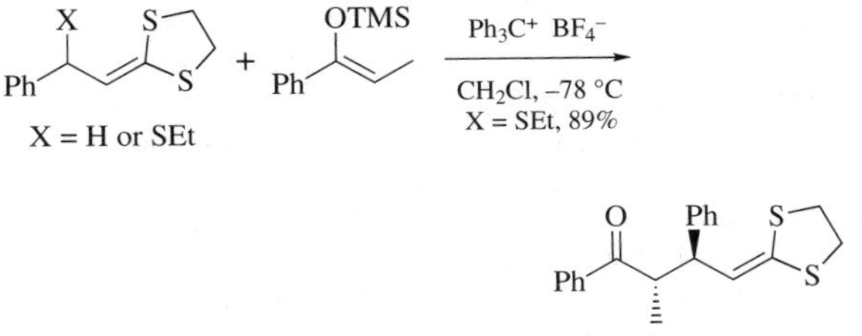

(13)

Polymerization Catalyst. Several types of polymerization[26] have been promoted by trityl fluoroborate, including reactions of orthocarbonates[26a] and orthoesters,[26b–d] vinyl ethers,[26e–g] epoxides,[26h,i] and lactones.[26j,k]

Other Reactions. Trityl fluoroborate has been used often to prepare cationic organometallic complexes, as in the conversion of dienyl complexes of iron, ruthenium, and osmium into their cationic derivatives.[27] It alkylates pyridines on the nitrogen atom in a preparation of dihydropyridines[28a] and acts as a tritylating agent.[28b] It has also been used in attempts to form silyl cations and silyl fluorides from silanes.[29] Finally, it has been reported to be a useful desiccant.[30]

1. (a) Dauben, H. J., Jr.; Honnen, L. R.; Harmon, K. M. *JOC* **1960**, *25*, 1442. (b) Fukui, K.; Ohkubo, K.; Yamabe, T. *BCJ* **1969**, *42*, 312. (c) Olah, G. A.; Svoboda, J. J.; Olah, J. A. *S* **1972**, 544. (d) Pearson, A. J. Iron Compounds in Organic Synthesis, Academic Press, 1994, p. 155.
2. (a) Müller, P. *HCA* **1973**, *56*, 1243. (b) Giese, G.; Heesing, A. *CB* **1990**, *123*, 2373.
3. (a) Karger, M. H.; Mazur, Y. *JOC* **1971**, *36*, 540. (b) Acheson, R. M.; Flowerday, R. F. *JCS(P1)* **1975**, 2065.
4. (a) O'Leary, M. A.; Richardson, G. W.; Wege, D. *T* **1981**, *37*, 813. (b) Prinzbach, H.; Seip, D.; Knothe, L.; Faisst, W. *LA* **1966**, *698*, 34.
5. (a) Dauben, H. J., Jr.; Gadecki, F. A.; Harmon, K. M.; Pearson, D. L. *JACS* **1957**, *79*, 4557. (b) Dauben, H. J., Jr.; Bertelli, D. J. *JACS* **1961**, *83*, 4657, 4659. (c) Peter-Katalinic, J.; Zsindely, J.; Schmid, H. *HCA* **1973**, *56*, 2796. (d) Vogel, E.; Ippen, J. *AG(E)* **1974**, *13*, 734. (e) Beeby, J.; Garratt, P. J. *JOC* **1973**, *38*, 3051. (f) Murata, I.; Yamamoto, K.; Kayane, Y. *AG(E)* **1974**, *13*, 807, 808. (g) Kuroda, S.; Asao, T. *TL* **1977**, 285. (h) Komatsu, K.; Takeuchi, K.; Arima, M.; Waki, Y.; Shirai, S.; Okamoto, K. *BCJ* **1982**, *55*, 3257. (i) Müller, J.; Mertschenk, B. *CB* **1972**, *105*, 3346. (j) Schweikert, O.; Netscher, T.; Knothe, L.; Prinzbach, H. *CB* **1984**, *117*, 2045. (k) Bindl, J.; Seitz, P.; Seitz, U.; Salbeck, E.; Salbeck, J.; Daub, J. *CB* **1987**, *120*, 1747.
6. (a) Zimmerman, H. E.; Aasen, S. M. *JOC* **1978**, *43*, 1493. (b) Komatsu, K.; Tomioka, I.; Okamoto, K. *BCJ* **1979**, *52*, 856.
7. (a) Yamamura, K.; Miyake, H.; Murata, I. *JOC* **1986**, *51*, 251. (b) Matsumoto, S.; Masuda, H.; Iwata, K.; Mitsunobu, O. *TL* **1973**, 1733. (c) Yano, S.; Nishino, K.; Nakasuji, K.; Murata, I. *CL* **1978**, 723. (d) Kedik, L. M.; Freger, A. A.; Viktorova, E. A. *KGS* **1976**, *12*, 328 (*Chem. Heterocycl. Compd. (Engl. Transl.)* **1976**, *12*, 279). (e) Reichardt, C.; Schäfer, G.; Milart, P. *CCC* **1990**, *55*, 97.
8. (a) Müller, P. *HCA* **1973**, *56*, 500. (b) Boev, V. I.; Dombrovskii, A. V. *ZOB* **1987**, *57*, 938, 633. (c) Klimova, E. I.; Pushin, A. N.; Sazonova, V. A. *ZOB* **1987**, *57*, 2336. (d) Abram, T. S.; Watts, W. E. *JCS(P1)* **1975**, 113; *JOM* **1975**, *87*, C39. (e) Barua, P.; Barua, N. C.; Sharma, R. P. *TL* **1983**, *24*, 5801. (f) Akgun, E.; Tunali, M. *AP* **1988**, *321*, 921.
9. (a) Barton, D. H. R.; Magnus, P. D.; Smith, G.; Strecker, G.; Zurr, D. *JCS(P1)* **1972**, 542. (b) Barton, D. H. R.; Magnus, P. D.; Smith, G.; Zurr, D. *CC* **1971**, 861.
10. Ohshima, M.; Murakami, M.; Mukaiyama, T. *CL* **1986**, 1593.

11. (a) Hanessian, S.; Staub, A. P. A. *TL* **1973**, 3551. (b) Jacobsen, S.; Pedersen, C. *ACS* **1974**, *28B*, 1024, 866. (c) Wessel, H.-P.; Bundle, D. R. *JCS(P1)* **1985**, 2251.
12. (a) Nakayama, J.; Fujiwara, K.; Hoshino, M. *CL* **1975**, 1099; *BCJ* **1976**, *49*, 3567. (b) Nakayama, J.; Imura, M.; Hoshino, M. *BCJ* **1980**, *53*, 1661. (c) Nakayama, J. *BCJ* **1982**, *55*, 2289. (d) Bock, H.; Brähler, G.; Henkel, U.; Schlecker, R.; Seebach, D. *CB* **1980**, *113*, 289. (e) Neidlein, R.; Droste-Tran-Viet, D.; Gieren, A.; Kokkinidis, M.; Wilckens, R.; Geserich, H.-P.; Ruppel, W. *HCA* **1984**, *67*, 574. (f) However, azide abstraction is seen with azidodithioacetals: Nakayama, J.; Fujiwara, K.; Hoshino, M. *JOC* **1980**, *45*, 2024.
13. (a) Barton, D. H. R.; Magnus, P. D.; Streckert, G.; Zurr, D. *CC* **1971**, 1109. (b) Doyle, M. P.; Siegfried, B. *JACS* **1976**, *98*, 163. (c) Hoye, T. R.; Kurth, M. J. *JACS* **1979**, *101*, 5065. (d) For simple ethers, see: Deno, N. C.; Potter, N. H. *JACS* **1967**, *89*, 3550.
14. (a) Jung, M. E. *JOC* **1976**, *41*, 1479. (b) Jung, M. E.; Speltz, L. M. *JACS* **1976**, *98*, 7882. (c) Jung, M. E.; Brown, R. W. *TL* **1978**, 2771.
15. Doyle, M. P.; Dow, R. L.; Bagheri, V.; Patrie, W. J. *JOC* **1983**, *48*, 476; *TL* **1980**, *21*, 2795.
16. (a) Jung, M. E.; Pan, Y.-G.; Rathke, M. W.; Sullivan, D. F.; Woodbury, R. P. *JOC* **1977**, *42*, 3961. (b) Reetz, M. T.; Stephan, W. *LA* **1980**, 533.
17. (a) Metcalf, B. W.; Burkhardt, J. P.; Jund, K. *TL* **1980**, *21*, 35. (b) Chowdhury, P. K.; Sharma, R. P.; Baruah, J. N. *TL* **1983**, *24*, 4485. (c) Niwa, H.; Miyachi, Y. *BCJ* **1991**, *64*, 716.
18. (a) Damico, R.; Broaddus, C. D. *JOC* **1966**, *31*, 1607. (b) Barton, D. H. R.; Bracho, R. D.; Gunatilaka, A. A. L.; Widdowson, D. A. *JCS(P1)* **1975**, 579. (c) Wanner, K. T.; Praschak, I.; Nagel, U. *AP* **1990**, *322*, 335; *H* **1989**, *29*, 29.
19. Reetz, M. T.; Stephan, W.; Maier, W. F. *SC* **1980**, *10*, 867.
20. Erhardt, J. M.; Grover, E. R.; Wuest, J. D. *JACS* **1980**, *102*, 6365.
21. (a) Laycock, D. E.; Hartgerink, J.; Baird, M. C. *JOC* **1980**, *45*, 291. (b) Laycock, D. E.; Baird, M. C. *TL* **1978**, 3307. (c) Slack, D.; Baird, M. C. *CC* **1974**, 701. (d) Bly, R. S.; Bly, R. K.; Hossain, M. M.; Silverman, G. S.; Wallace, E. *T* **1986**, *42*, 1093. (e) Bly, R. S.; Silverman, G. S.; Bly, R. K. *OM* **1985**, *4*, 374.
22. Reetz, M. T.; Schinzer, D. *AG(E)* **1977**, *16*, 44.
23. (a) Traylor, T. G.; Berwin, H. J.; Jerkunica, J.; Hall, M. L. *PAC* **1972**, *30*, 597. (b) Jerkunica, J. M.; Traylor, T. G. *JACS* **1971**, *93*, 6278. (c) Washburne, S. S.; Szendroi, R. *JOC* **1981**, *46*, 691. (d) Washburne, S. S.; Simolike, J. B. *JOM* **1974**, *81*, 41. (e) However, organostannanes lacking a β-hydrogen afford alkyltriphenylmethanes in good yield. Kashin, A. N.; Bumagin, N. A.; Beletskaya, I. P.; Reutov, O. A. *JOM* **1979**, *171*, 321.
24. (a) Barton, D. H. R.; Haynes, R. K.; Leclerc, G.; Magnus, P. D.; Menzies, I. D. *JCS(P1)* **1975**, 2055. (b) Barton, D. H. R.; Leclerc, G.; Magnus, P. D.; Menzies, I. D. *CC* **1972**, 447. (c) Okada, K.; Hisamitsu, K.; Mukai, T. *TL* **1981**, *22*, 1251. (d) Futamura, S.; Kamiya, Y. *CL* **1989**, 1703.
25. (a) Ohshima, M.; Murakami, M.; Mukaiyama, T. *CL* **1985**, 1871. (b) Inazu, T.; Yamanoi, T. Jpn. Patent 02 240 093, 02 255 693 (*CA* **1991**, *114*, 143 907j, 143 908k); Jpn. Patent 01 233 295 (*CA* **1990**, *112*, 198 972r). (c) Bochkov, A. F.; Kochetkov, N. K. *Carbohydr. Res.* **1975**, *39*, 355; for polymerizations of carbohydrate cyclic orthoesters, see: Bochkov, A. F.; Chernetskii, V. N.; Kochetkov, N. K. *Carbohydr. Res.* **1975**, *43*, 35; *BAU* **1975**, *24*, 396. (d) Hashimoto, Y.; Mukaiyama, T. *CL* **1986**, 1623, 755. (e) Hashimoto, Y.; Sugumi, H.; Okauchi, T.; Mukaiyama, T. *CL* **1987**, 1691. (f) Murray, D. H.; Albizati, K. F. *TL* **1990**, *31*, 4109.
26. (a) Endo, T.; Sato, H.; Takata, T. *Macromolecules* **1987**, *20*, 1416. (b) Uno, H.; Endo, T.; Okawara, M. *J. Polym. Sci., Polym. Chem. Ed.* **1985**, *23*, 63. (c) Nishida, H.; Ogata, T. Jpn. Patent 62 295 920 (*CA* **1988**, *109*, 57 030h). (d) See also Ref. 25c. (e) Kunitake, T. *J. Macromol. Sci., Chem.* **1975**, *A9*, 797. (f) Kunitake, T.; Takarabe, K.; Tsugawa, S. *Polym. J.* **1976**, *8*, 363. (g) Spange, S.; Dreier, R.; Opitz, G.; Heublein, G. *Acta Polym.* **1989**, *40*, 55. (h) Mijangos, F.; León, L. M. *J. Polym. Sci., Polym. Lett. Ed.* **1983**, *21*, 885; *Eur. Polym. J.* **1983**, *19*, 29. (i) Bruzga, P.; Grazulevicius, J.; Kavaliunas, R.; Kublickas, R. *Polym. Bull. (Berlin)* **1991**, *26*, 193. (j) Khomyakov, A. K.; Gorelikov, A. T.; Shapet'ko, N. N.; Lyudvig, E. B. *Vysokomol. Soedin., Ser. A* **1976**, *18*, 1699, 1053; *DOK* **1975**, *222*, 1111.
27. (a) For a review, see any basic organometallic text, e.g. Coates, G. E.; Green, M. L. H.; Wade, K. *Organometallic Compounds*; Methuen: London, 1968; Vol. 2, pp 136ff. (b) Birch, A. J.; Cross, P. E.; Lewis, J.; White, D. A. *CI(L)* **1964**, 838. (c) Cotton, F. A.; Deeming, A. J.; Josty, P. L.; Ullah, S. S.; Domingos, A. J. P.; Johnson, B. F. G.; Lewis, J. *JACS* **1971**, *93*, 4624.
28. (a) Lyle, R. E.; Boyce, C. B. *JOC* **1974**, *39*, 3708. (b) Hanessian, S.; Staub, A. P. A. *TL* **1973**, 3555.
29. (a) Sommer, L. H.; Bauman, D. L. *JACS* **1969**, *91*, 7076. (b) Bulkowski, J. E.; Stacy, R.; Van Dyke, C. H. *JOM* **1975**, *87*, 137. (c) Chojnowski, J.; Fortuniak, W.; Stanczyk, W. *JACS* **1987**, *109*, 7776.
30. Burfield, D. R.; Lee, K.-H.; Smithers, R. H. *JOC* **1977**, *42*, 3060.

Michael E. Jung
University of California, Los Angeles, CA, USA

Triphenylphosphine–*N*-Bromosuccinimide[1]

Ph_3P + O N O X

(Ph_3P)		
[603-35-0]	$C_{18}H_{15}P$	(MW 262.30)
(NBS)		
[128-08-5]	$C_4H_4BrNO_2$	(MW 177.99)
(NCS)		
[128-09-6]	$C_4H_4ClNO_2$	(MW 133.54)
(NIS)		
[516-12-1]	$C_4H_4INO_2$	(MW 224.99)

(conversion of primary alcohols to alkyl halides)

Physical Data: See ***Triphenylphosphine***, ***N-Bromosuccinimide***, ***N-Chlorosuccinimide***, and ***N-Iodosuccinimide***.

Preparative Methods: the mixed reagent is prepared in situ as needed by combining triphenylphosphine, the alcohol, and *N*-halosuccinimide in a suitable solvent (usually DMF) and heating.

Handling, Storage, and Precautions: precautions in handling should be made as described for the individual reagents. This mixed reagent is not appropriate for storage. Recrystallization of the triphenylphosphine and *N*-halosuccinimide is necessary to achieve the most efficient halogenation reaction. In addition, the use of anhydrous solvent is important for the success of the reaction.

Introduction. The combination of triphenylphosphine and *N*-halosuccinimide was applied by Hanessian[2] to the preparation of primary halides from carbohydrate precursors. This method is related to a number of other methods for hydroxyl substitution via oxyphosphonium salts as described in a comprehensive review.[1]

Conversion of Alcohols to Alkyl Halides. The typical procedure[2] as applied to the preparation of methyl 5-deoxy-5-

iodo-2,3-*O*-isopropylidene-β-D-ribofuranose (eq 1) involves the addition of 2 equiv of *N*-iodosuccinimide followed by 2 equiv of triphenylphosphine to a cooled solution of the alcohol in anhydrous DMF and heating the mixture to 50 °C for 20 min. Methyl alcohol is added to destroy excess reagent, followed by *n*-butanol to aid in the removal of DMF under reduced pressure. After concentration, the product is purified from the byproducts (succinimide and triphenylphosphine oxide) by addition of ether and aqueous extraction followed by column chromatography. Isolated yields of 70–85% are regularly obtained using this procedure for the preparation of chlorides, bromides, and iodides. In addition, the reaction conditions allow the preparation of primary halides in the presence of many functional groups used in carbohydrate chemistry, including the ester, amide, lactone, benzylidene acetal, and acetonide groups (eq 2). In fact, moderate yields (49–82%) of a number of primary halides are obtained using this procedure on sugars containing one or more free secondary hydroxyl groups (eq 3).[2–4]

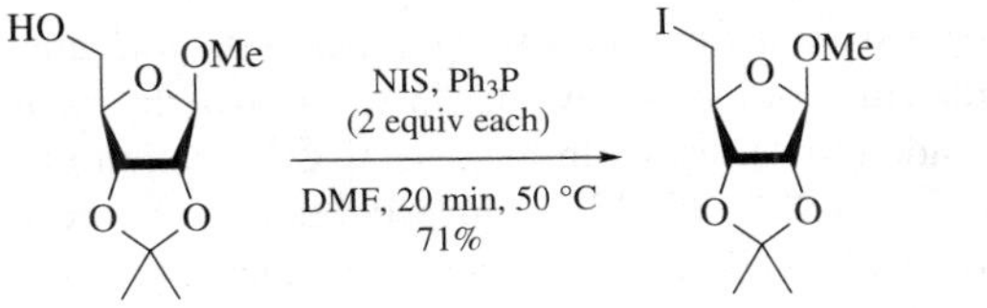

NIS, Ph_3P (2 equiv each); DMF, 20 min, 50 °C; 71% (1)

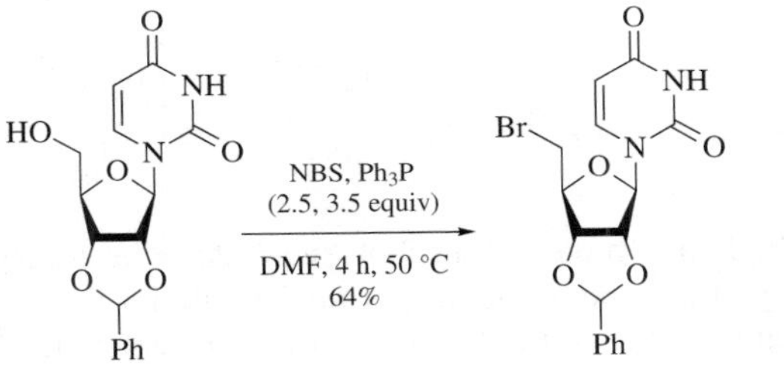

NBS, Ph_3P (2.5, 3.5 equiv); DMF, 4 h, 50 °C; 64% (2)

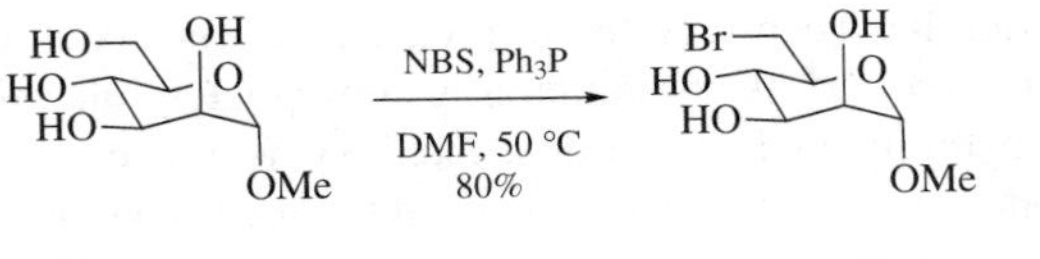

NBS, Ph_3P; DMF, 50 °C; 80% (3)

Other examples of the use of this method include the preparation of tetrahydrofurfuryl bromide,[5] 3α-cholestanyl bromide (using THF as solvent, 85% yield),[6] and intermediates toward the syntheses of alkaloids (eqs 4 and 5).[7] The reaction of the reagent with DMF alone generates the formamidinium salt which can ultimately result in the isolation of the formate ester derivative of the alcohol substrate (eq 6).[1,8] It is therefore important that the *N*-halosuccinimide is added last for the preparation of alkyl halides. Other solvents used for this reaction include CH_2Cl_2 (which can facilitate the workup of the reaction products) and HMPA.[9]

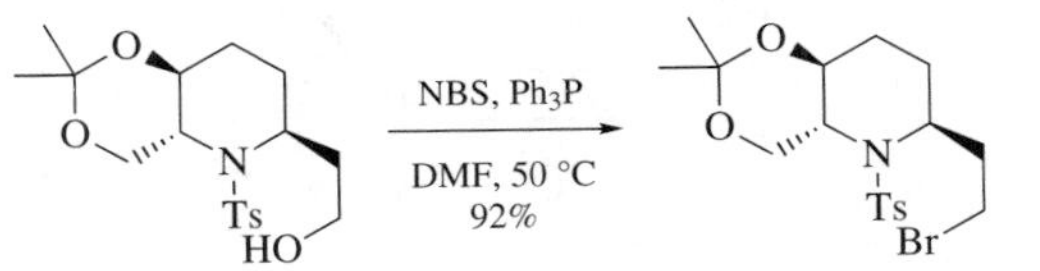

NBS, Ph_3P; DMF, 50 °C; 92% (4)

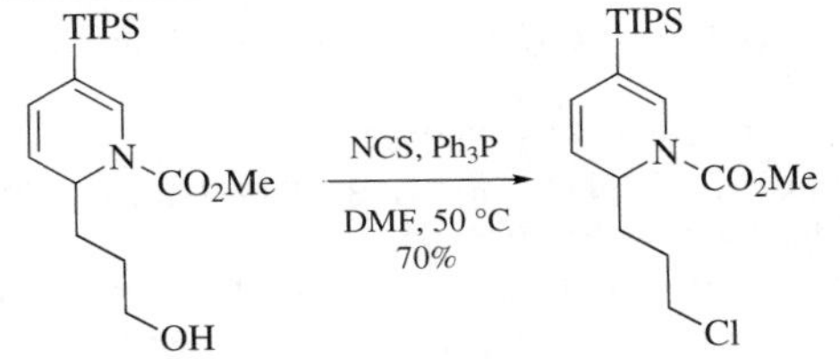

NCS, Ph_3P; DMF, 50 °C; 70% (5)

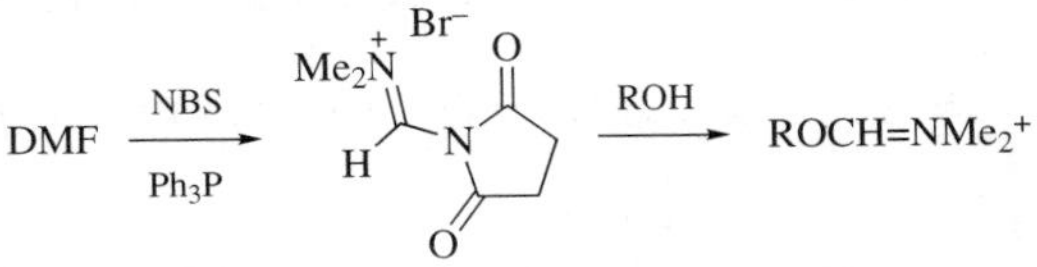

DMF —(NBS, Ph_3P)→ formamidinium salt —(ROH)→ $ROCH{=}NMe_2^+$ (6)

The standard procedure has often been found to be unsatisfactory for substrates which are prone to acid-catalyzed rearrangements (including migration of acetonide protecting groups) and cleavage and is also not suitable for substrates containing labile protecting groups such as THP ethers.[10,11] In some cases this problem has been circumvented by the addition of various bases such as $BaCO_3$ (eq 7),[12] ***Pyridine***, and ***Imidazole***. A recent study has achieved the preparation of 3-bromo-3-deoxy-1,2:5,6-di-*O*-isopropylidene-α-D-allofuranose by addition of 2 equiv of *N*-halosuccinimide to a refluxing solution of the secondary alcohol and 2 equiv each of PPh_3 and imidazole in toluene or chlorobenzene (eq 8). Previous applications of PPh_3 and *N*-halosuccinimide alone afforded only the primary halide products resulting from acetonide migration. The use of pyridine and imidazole have also been found to play an essential role in the conversion of carbohydrates to halides using the reagents ***Triphenylphosphine–Iodine*** and ***Triphenylphosphine–Carbon Tetrachloride***, respectively.[13]

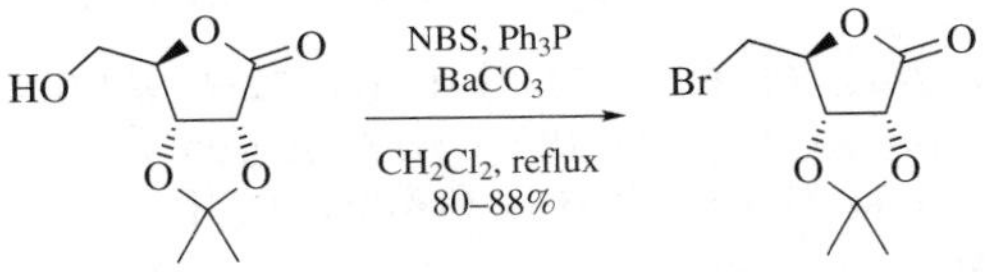

NBS, Ph_3P, $BaCO_3$; CH_2Cl_2, reflux; 80–88% (7)

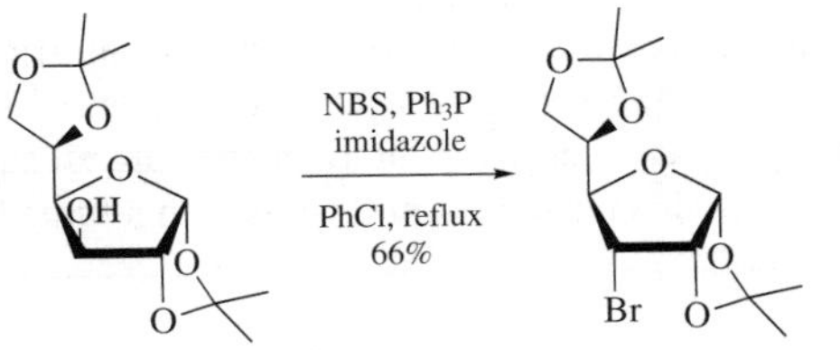

NBS, Ph_3P, imidazole; PhCl, reflux; 66% (8)

Related Reagents. Triphenylphosphine-*N*-Chlorosuccinimide; Triphenylphosphine-*N*-iodosuccinimide.

---

1. Castro, B. R. *OR* **1983**, *29*, 1.
2. (a) Ponpipom, M. M.; Hanessian, S. *Carbohydr. Res.* **1971**, *18*, 342. (b) Hanessian, S.; Ponpipom, M. M.; Lavallee, P. *Carbohydr. Res.* **1972**, *24*, 45.
3. (a) Nakane, M.; Hutchinson, C. R.; Gollman, H. *TL* **1980**, *21*, 1213. (b) Aspinall, G. O.; Carpenter, R. C.; Khondo, L. *Carbohydr. Res.* **1987**, *165*, 281.
4. Hasegawa, A.; Morita, M.; Ishida, H.; Kiso, M. *J. Carbohydr. Chem.* **1989**, *8*, 579.

5. Schweizer, E. E.; Creasy, W. S.; Light, K. K.; Shaffer, E. T. *JOC* **1969**, *34*, 212.
6. Bose, A. K.; Lal, B. *TL* **1973**, 3937.
7. (a) Birkinshaw, T. N.; Holmes, A. B. *TL* **1987**, *28*, 813. (b) Comins, D. L.; Myoung, Y. C. *JOC* **1990**, *55*, 292.
8. Hodosi, G.; Podányi, B.; Kuszmann, J. *Carbohydr. Res.* **1992**, *230*, 327.
9. Baker, C. W.; Whistler, R. L. *Carbohydr. Res.* **1975**, *45*, 237.
10. Yunker, M. B.; Tam, S. Y.-K.; Hicks, D. R.; Fraser-Reid, B. *CJC* **1976**, *54*, 2411.
11. Gensler, W. J.; Marshall, J. P.; Langone, J. J.; Chen, J. C. *JOC* **1977**, *42*, 118.
12. Jäger, V.; Häfele, B. *S* **1987**, 801.
13. (a) Garegg, P. J.; Johansson, R.; Samuelsson, B. *S* **1984**, 168. (b) Whistler, R. L.; Anisuzzaman, A. K. M. *Methods Carbohydr. Chem.* **1980**, *8*, 227.

Scott C. Virgil
Massachusetts Institute of Technology, Cambridge, MA, USA

Triphenylphosphine–Carbon Tetrabromide[1]

$Ph_3P–CBr_4$

(Ph_3P)		
[603-35-0]	$C_{18}H_{15}P$	(MW 262.30)
(CBr_4)		
[558-13-4]	CBr_4	(MW 331.61)

(reagent combination for the conversion of alcohols to bromides, aldehydes and ketones to dibromoalkenes, and terminal alkynes to 1-bromoalkynes; carboxyl activation)

Physical Data: Ph_3P: mp 79–81 °C; bp 377 °C; *d* 1.0749 g cm^{-3}. CBr_4: mp 90–91 °C; bp 190 °C; *d* 3.273 g cm^{-3}.
Solubility: sol MeCN, CH_2Cl_2, pyridine, DMF.
Preparative Method: the reactive species is generated in situ by reaction of Ph_3P with CBr_4.
Handling, Storage, and Precautions: Ph_3P is an irritant; CBr_4 is toxic and a cancer suspect agent. All solvents used must be carefully dried because the intermediates are all susceptible to hydrolysis. This reagent should be used in a fume hood.

Conversion of Alcohols to Alkyl Bromides. The reaction of alcohols with ***Triphenylphosphine*** and ***Carbon Tetrabromide*** results in the formation of alkyl bromides. The conditions are sufficiently mild to allow for the efficient conversion of alcohols into the corresponding bromides. The uridine derivative (eq 1) is transformed into its bromide with Ph_3P and CBr_4.[2]

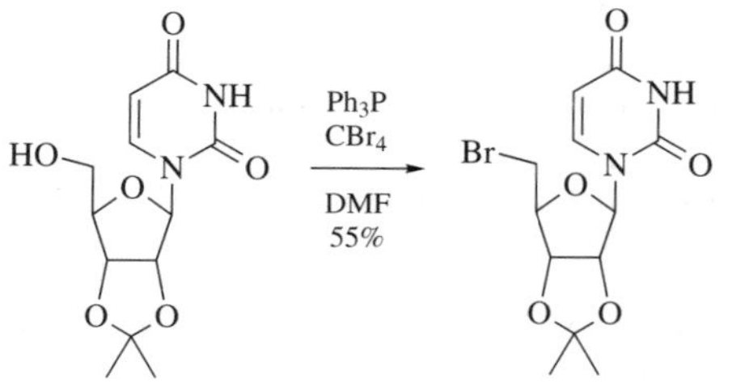

(1)

The geometry about the double bond of an allylic alcohol is usually not compromised (eq 2).[3] Allylic rearrangement is also not commonly observed. Although ketones are known to react with this reagent combination, the reaction of an allylic alcohol has been selectively achieved in the presence of a ketone (eq 3).[4]

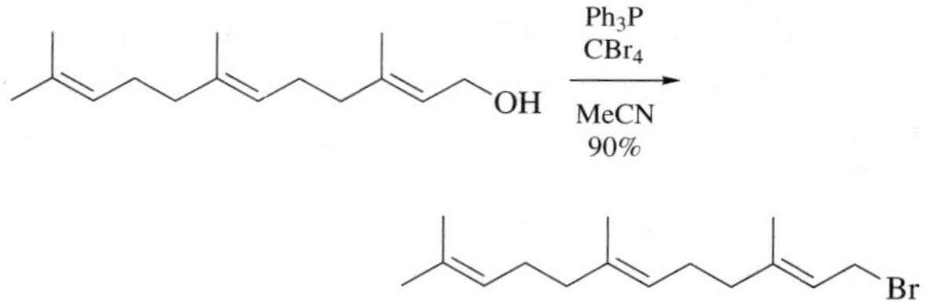

(2)

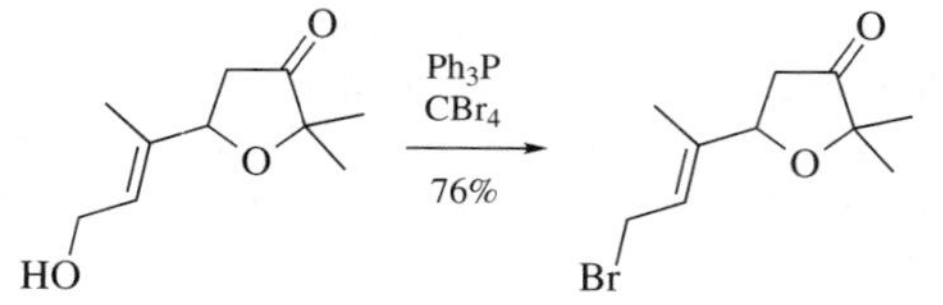

(3)

Regioselective bromination of primary alcohols in the presence of secondary alcohols is possible. These reactions are usually performed in pyridine as solvent. The reaction of the methyl glucopyranoside (eq 4) results in the selective formation of the primary bromide in 98% yield.[5] An investigation of this reaction with a chiral deuterated neopentyl alcohol yielded a partially racemized bromide.[6]

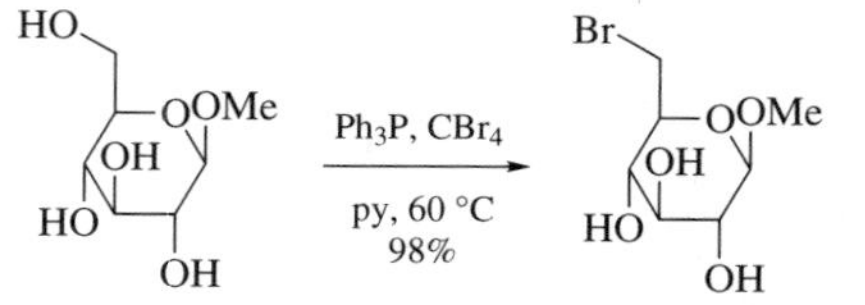

(4)

Silyl ether protected alcohols (eq 5) have been converted directly into the corresponding bromides with Ph_3P and CBr_4. The reaction works best if 1.5 equiv of acetone are added.[7] Tetrahydropyranyl ether protected alcohols have also been directly transformed into the bromides using this reagent combination. The reaction has been reported to proceed with inversion of configuration (eq 6).[8] If unsaturation is appropriately placed within a tetrahydropyranyl (eq 7) or a methoxymethyl (eq 8) protected alcohol, cyclization occurs to afford tetrahydropyrans.[9] The conversion of an alcohol to the bromide without complications with a methoxymethyl protected alcohol in the molecule is possible (eq 9).[10]

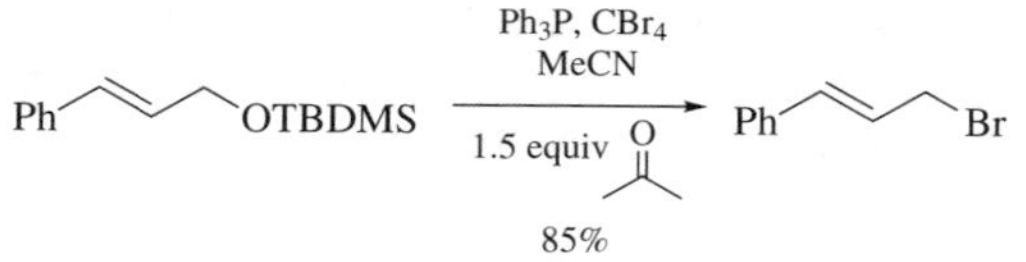

(5)

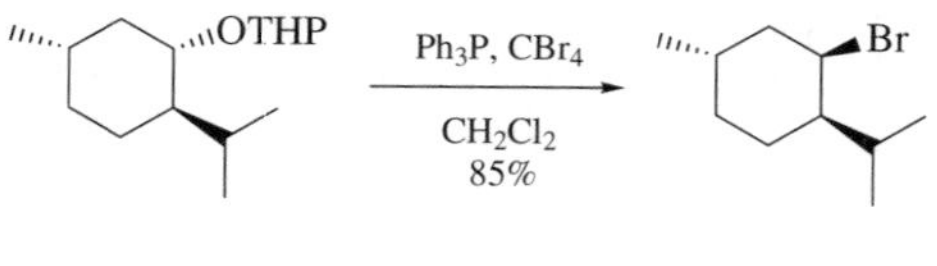

(6)

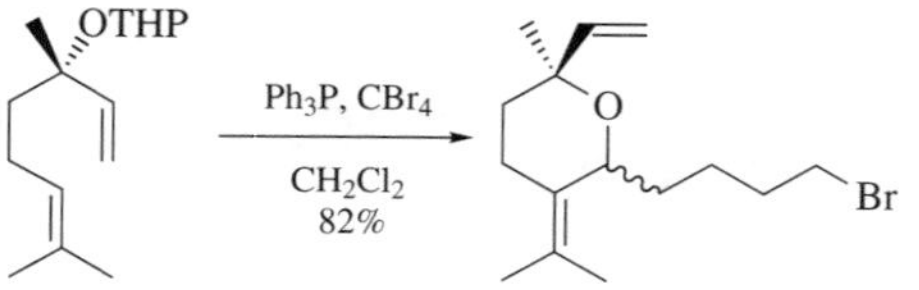

(7)

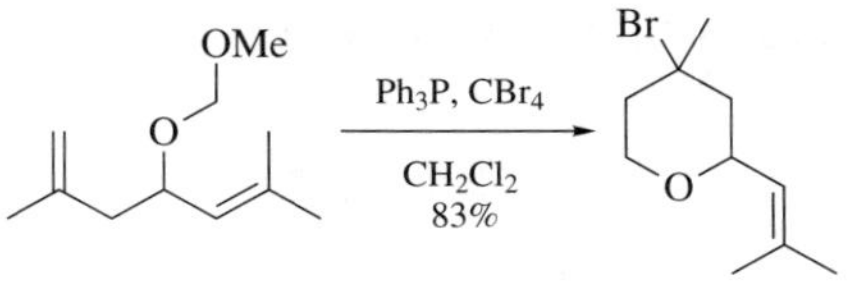

(8)

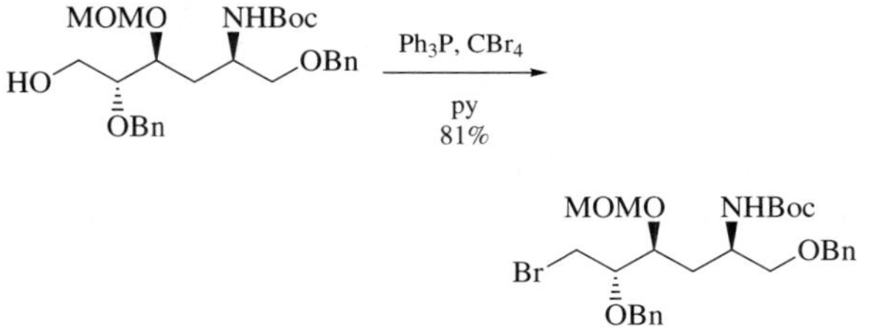

(9)

Amides from Carboxylic Acids. *N*-Methoxy-*N*-methyl amides can be prepared from carboxylic acids and the amine hydrochlorides. In the case of an α-phenyl carboxylic acid (eq 10), the amide is formed in 71% yield and no racemization is detected.[11]

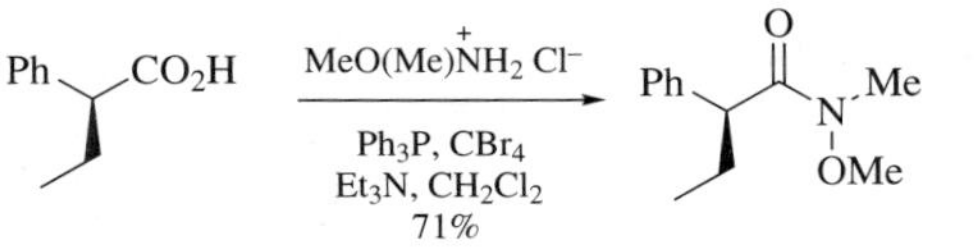

(10)

Dibromoalkenes from Aldehydes and Ketones. Benzaldehyde is transformed into the dibromoalkene in 84% yield when treated with Ph_3P and CBr_4 (eq 11).[12] An alternative procedure for the conversion of an aldehyde to the dibromoalkene uses ***Zinc*** dust in place of an excess of the phosphine. This allows the amount of Ph_3P and CBr_4 to be reduced to 2 equiv each as opposed to 4 equiv. This procedure gives comparable results to the original procedure. The dibromoalkenes can be reacted with ***n-Butyllithium*** to form the intermediate lithium acetylide. The acetylides can then be reacted with electrophiles such as H_2O (eq 12) and CO_2 (eq 13). This offers a convenient method for the formyl to ethynyl conversion.[13]

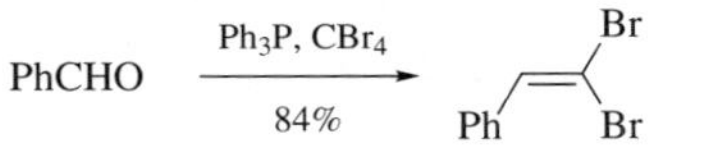

(11)

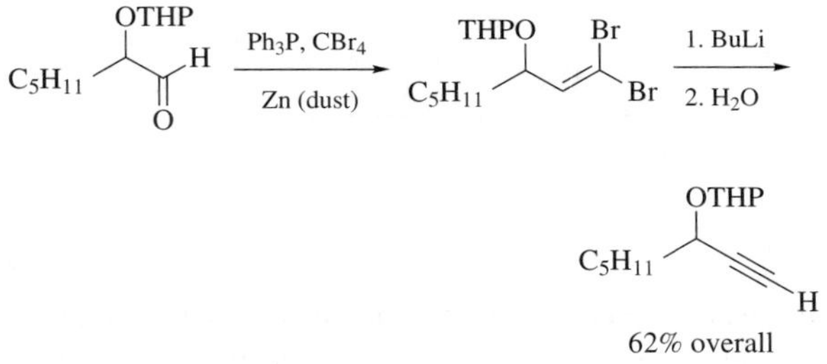

(12)

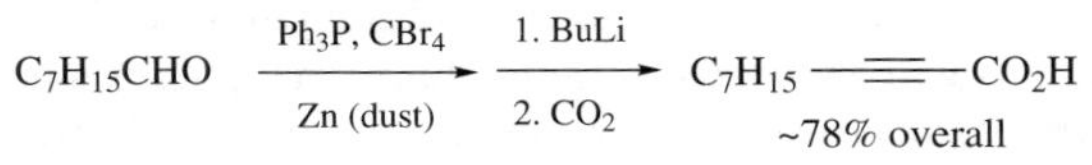

(13)

Ketones are converted to dibromomethylene derivatives. These intermediates can be transformed to isopropylidene compounds by reaction with ***Lithium Dimethylcuprate*** and ***Iodomethane*** (eq 14).[14] No racemization was reported for the chain extension of the aldehyde derived from (*S*)-ethyl lactate under the reaction conditions (eq 15).[15]

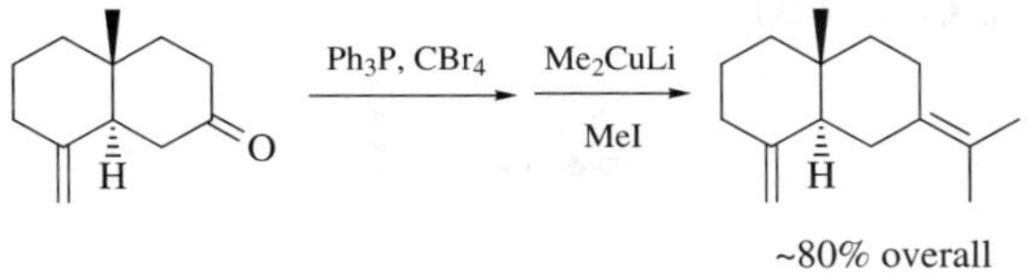

(14)

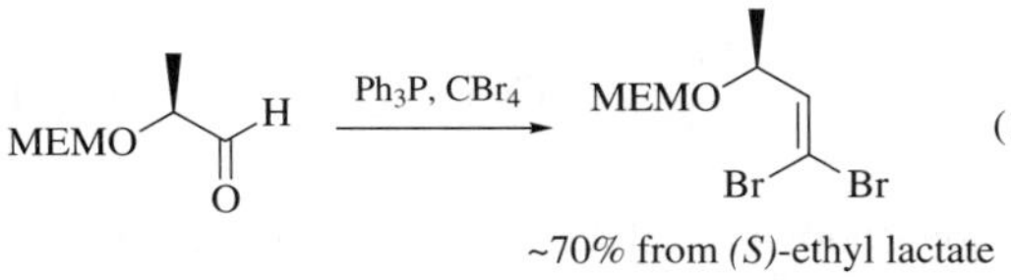

(15)

β-Bromo Enones from 1,3-Diketones. The reaction of Ph_3P and CBr_4 with a 1,3-diketone efficiently converts it to the β-bromo enone (eq 16).[16]

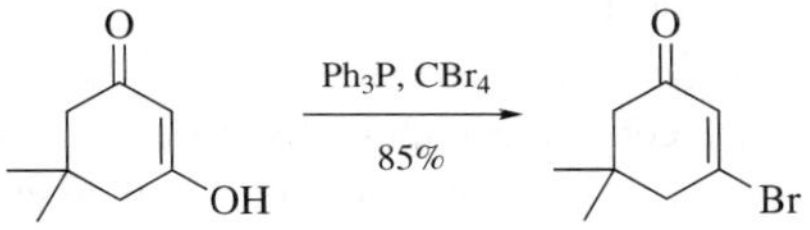

(16)

1-Bromoalkynes from Terminal Alkynes. Terminal alkynes on reaction with Ph_3P and CBr_4 afford 1-bromoalkynes in high yield (eq 17).[17]

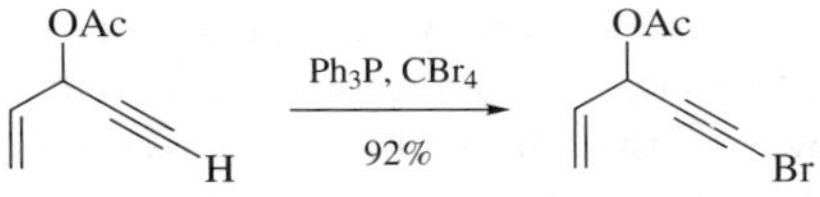

(17)

1. Castro, B. R. *OR* **1983**, *29*, 1.
2. Verheyden, J. P. H.; Moffatt, J. G. *JOC* **1972**, *37*, 2289.
3. Axelrod, E. H.; Milne, G. M.; van Tamelen, E. E. *JACS* **1970**, *92*, 2139.
4. Kang, S. H.; Hong, C. Y. *TL* **1987**, *28*, 675.
5. Kashem, A.; Anisuzzaman, M.; Whistler, R. L. *CR* **1978**, *61*, 511.
6. Weiss, R. G.; Snyder, E. I. *JOC* **1971**, *36*, 403.
7. Mattes, H.; Benezra, C. *TL* **1987**, *28*, 1697.
8. Wagner, A.; Heitz, M.-P.; Mioskowski, C. *TL* **1989**, *30*, 557.
9. Wagner, A.; Heitz, M.-P.; Mioskowski, C. *TL* **1989**, *30*, 1971.
10. Clinch, K.; Vasella, A.; Schauer, R. *TL* **1987**, *28*, 6425.
11. Einhorn, J.; Einhorn, C.; Luche, J.-L. *SC* **1990**, *20*, 1105.
12. Ramirez, F.; Desai, N. B.; McKelvie, N. *JACS* **1962**, *84*, 1745.
13. Corey, E. J.; Fuchs, P. L. *TL* **1972**, 3769.
14. Posner, G. H.; Loomis, G. L.; Sawaya, H. S. *TL* **1975**, 1373.
15. Mahler, H.; Braun, M. *TL* **1987**, *28*, 5145.
16. Gruber, L.; Tömösközi, I.; Radics, L. *S* **1975**, 708.
17. Wagner, A.; Heitz, M.-P.; Mioskowski, C. *TL* **1990**, *31*, 3141.

Michael J. Taschner
The University of Akron, OH, USA

Triphenylphosphine–Carbon Tetrachloride[1]

$Ph_3P–CCl_4$

(Ph_3P)		
[605-35-0]	$C_{18}H_{15}P$	(MW 262.30)
(CCl_4)		
[56-23-5]	CCl_4	(MW 153.81)

(reagent combination for the conversion of a number of functional groups into their corresponding chlorides and for dehydrations)

Physical Data: Ph_3P: mp 79–81 °C; bp 377 °C; *d* 1.0749 g cm^{-3}. CCl_4: mp −23 °C; bp 77 °C; *d* 1.594 g cm^{-3}.
Solubility: sol CCl_4, MeCN, CH_2Cl_2, 1,2-dichloroethane.
Preparative Method: reactive intermediates are generated in situ by reaction of Ph_3P and CCl_4.
Handling, Storage, and Precautions: Ph_3P is an irritant; CCl_4 is toxic and a cancer suspect agent; use in a fume hood. Solvents must be carefully dried because the intermediates are all susceptible to hydrolysis.

Combination of Triphenylphosphine and Carbon Tetrachloride. This reagent combination is capable of performing a range of chlorinations and dehydrations. The reactions are typically run using the so-called two-component or three-component systems. Carbon tetrachloride can function as both the reagent and solvent. However, the rates of the reactions are highly solvent-dependent, with MeCN providing the fastest rates.[1]

Conversion of Alcohols to Alkyl Chlorides. The reaction of alcohols with ***Triphenylphosphine*** and carbon tetrachloride results in the formation of alkyl chlorides.[2] The mild, neutral conditions allow for the efficient conversion of even sensitive alcohols into the corresponding chlorides (eqs 1 and 2).[3,4] The reaction typically proceeds with inversion of configuration.[5] In eq 3, it is interesting to note that the reaction not only proceeds with inversion of configuration but also no acyloxy migration is observed.[6]

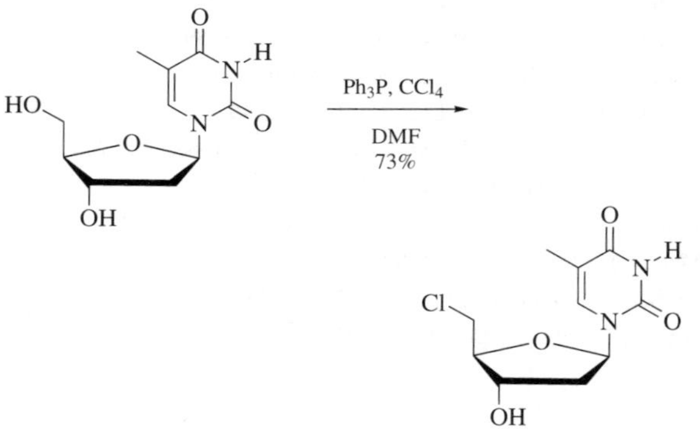

Ph3P, CCl4; DMF; 73% (1)

The conversion of an allylic alcohol to an allylic chloride occurs with no or minimal allylic rearrangement (eqs 4 and 5).[7] For the synthesis of low boiling allylic alcohols, it is advantageous to substitute hexachloroacetone (HCA) for CCl_4 (eq 6). The stereochemical integrity of the double bond also remains intact under these conditions.[8]

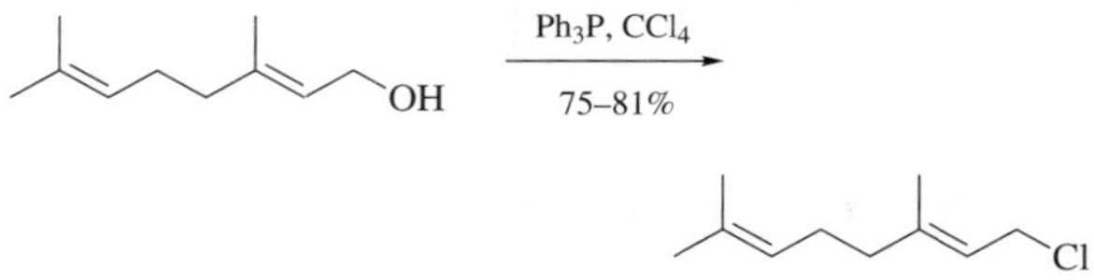

Ph3P, CCl4; 75–81% (2)

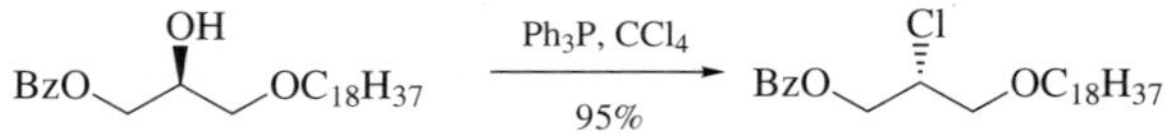

Ph3P, CCl4; 95% (3)

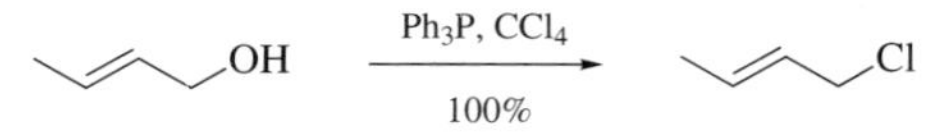

Ph3P, CCl4; 100% (4)

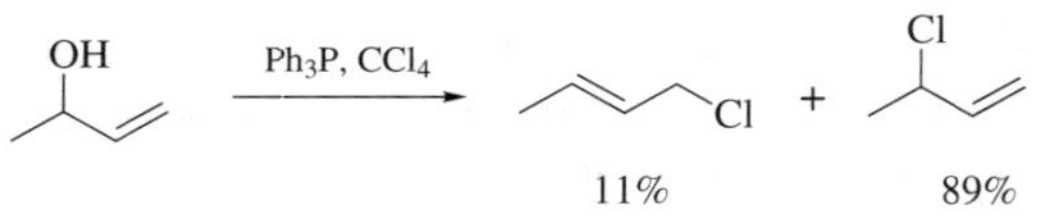

Ph3P, CCl4; 11% + 89% (5)

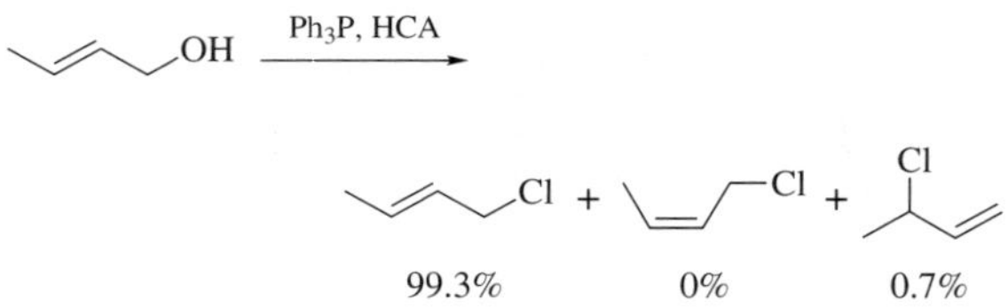

Ph3P, HCA; 99.3% + 0% + 0.7% (6)

If the conversion of the alcohol to the chloride is attempted in refluxing MeCN, dehydration to form the alkene occurs (eqs 7 and 8).[9] Occasionally the separation of the product from the triphenylphosphine oxide produced in the reaction can be problematic. This can be overcome by using a polymer-supported phosphine.[10] Simple filtration and evaporation of the solvent are all that is required under these conditions. Not only is the workup facilitated, but the rate of the reaction is also increased by employing the supported reagent.[10c]

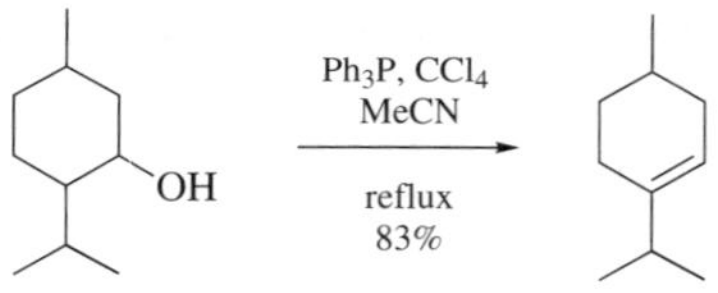

Ph3P, CCl4; MeCN; reflux; 83% (7)

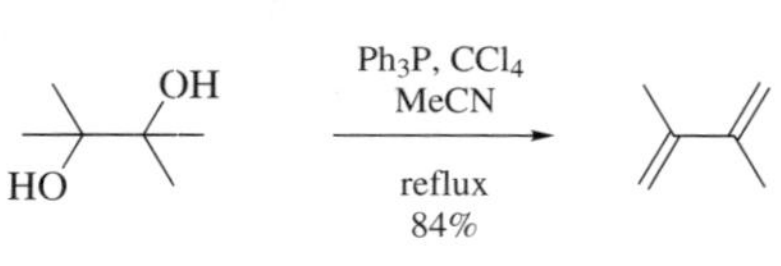

Ph3P, CCl4; MeCN; reflux; 84% (8)

Conversion of Acids to Acid Chlorides. The reaction of carboxylic acids with triphenylphosphine–CCl_4 reportedly produces acid chlorides in good yield under mild conditions (eq 9).[11] These conditions will allow acid sensitive functional groups to survive. Phosphoric mono- and diesters are successfully converted into the phosphoric monoester dichlorides and diester chlorides, respectively. The reaction of the diethyl ester does not produce the acid chloride. Instead, the anhydride is formed. Phosphinic acid chlorides can also be prepared from the corresponding phosphinic acid under these conditions (eq 10).[12]

CO_2H — Ph_3P, CCl_4 — Cl (9)

O, P, R, OH, R — Ph_3P, CCl_4; MeCN; 50–60 °C — O, P, R, Cl, R (10)

R = Me, Ph

Epoxides to *cis*-1,2-Dichloroalkanes. Epoxides are converted into *cis*-1,2-dichlorides by the action of triphenylphosphine in refluxing CCl_4.[13] Cyclohexene oxide forms *cis*-1,2-dichlorocyclohexane (eq 11) contaminated by a trace of the *trans*-isomer. Cyclopentene oxide gives only the *cis*-isomer.

O — Ph_3P, CCl_4; reflux; 80% — Cl, Cl (11)

Dehydrations. Diols may be cyclodehydrated to the corresponding cyclic ethers. The reaction is most effective for 1,4-diols (eq 12). Dehydration of 1,3- and 1,5-diols is not as successful, except in the case of the configurationally constrained 1,5-diol shown in eq 13.[14] The reaction of *trans*-1,2-cyclohexanediol with the reagent affords *trans*-2-chlorocyclohexanol with none of the *cis*-isomer or the *trans*-dichloride being detected. If the reaction is performed in the presence of K_2CO_3 as an HCl scavenger, the epoxide is formed (eq 14).[14] The yields are not as good with substituted acyclic diols.[15]

HO, OH — Ph_3P, CCl_4; 78% — O (12)

OH, OH — Ph_3P, CCl_4; 80% — O (13)

O — Ph_3P, CCl_4; K_2CO_3; 86% — OH, OH — Ph_3P, CCl_4; 88% — OH, Cl (14)

Dehydration of *N*-substituted β-amino alcohols with triphenylphosphine–CCl_4 and ***Triethylamine*** produces aziridines in good yield (eq 15).[16] This reaction has been successfully employed in the preparation of stable arene imines (eq 16).[17] Azetidines can be obtained from the corresponding 3-aminoalkanols (eq 17).[18] Additionally, reaction of 2-(3-hydroxypropyl)piperidine under these conditions yields octahydroindolizine (eq 18).

OH, Et, NHBn — Ph_3P, CCl_4, Et_3N; MeCN; 73% — Bn, N, Et (15)

OH, NHBn — Ph_3P, CCl_4, Et_3N; MeCN; 42% — NHBn (16)

Ph, OH, NHMe — Ph_3P, CCl_4; 41% — Ph, N, Me (17)

NH, OH — Ph_3P, CCl_4, Et_3N; MeCN; 42% — N (18)

Substituted hydroxamic acids successfully cyclize to form β-lactams as long as Et_3N is present (eq 19). In the absence of the base, complex mixtures are formed.[19] Unsubstituted amides can be converted into nitriles via dehydration (eq 20).[20] This is the reagent of choice for the transformation of the amide to the nitrile in eq 21.[21]

NHBoc, BnONH, OH, O — Ph_3P, CCl_4, Et_3N; 78% — BocNH, NOBn, O (19)

$CONH_2$ — Ph_3P, CCl_4, $ClCH_2CH_2Cl$; 90% — CN (20)

H, N, O, $CONH_2$, Cl, N, Ph — Ph_3P, CCl_4; dioxane; 76% — H, N, O, CN, Cl, N, Ph (21)

Nitriles can also be obtained from aldoximes using this reagent (eq 22).[22] Ketoximes produce imidoyl chlorides via a Beckmann rearrangement under these conditions (eq 23).[23] Imidoyl chlorides are also available by the reaction of monosubstituted amides with $Ph_3P–CCl_4$ in acetonitrile (eq 24).[24]

N, OH — Ph_3P, CCl_4, Et_3N; $ClCH_2CH_2Cl$; 88% — CN (22)

N, OH — Ph_3P, CCl_4; MeCN; 67% — Cl, N, Ph (23)

H, N, O — Ph_3P, CCl_4; MeCN; 71% — N, Cl (24)

N,N,N′-Trisubstituted ureas afford chloroformamidine derivatives (eq 25),[25] while *N,N′*-disubstituted ureas and thioureas produce carbodiimides (eq 26).[26] When carbamoyl chlorides are treated with the reagent in MeCN, they are converted into isocyanates (eq 27).[27] Dehydration of *N*-substituted formamides provides access to isocyanides (eq 28).[28]

PhHN–C(=O)–N(Me)Ph →(Ph₃P, CCl₄; MeCN; 83%) PhN=C(Cl)–N(Me)Ph (25)

PhHN–C(=X)–NH–cyclohexyl →(Ph₃P, CCl₄; Et₃N) Ph–N=•=N–cyclohexyl (26)

X = O, S; X = O, 84%; X = S, 92%

PhNH–C(=O)Cl →(Ph₃P, CCl₄; MeCN; 87%) PhN=C=O (27)

PhCH₂NH–CHO →(Ph₃P, CCl₄; Et₃N; 89%) PhCH₂NC (28)

Amide Formation. The synthesis of an amide can be accomplished by initial reaction of an acid with $Ph_3P–CCl_4$ and then reaction of the intermediate with 2 equiv of the appropriate amine. A tertiary amine, such as ***Diisopropylethylamine***, can be employed as the HCl scavenger in cases where one would not want to waste any of a potentially valuable amine.[29] This method has been used in the construction of an amide in the synthesis of the skeleton of the lycorine alkaloids (eq 29).[30]

Ph₃P, CCl₄; MeCN; 0 °C to rt; 93% (29)

The method is effective for peptide coupling in that the yields are typically good; however, the reaction is often accompanied by racemization of the product.[31] The racemization problem can be suppressed, but this involves a change in the phosphine employed[32] or slightly modified reaction conditions. These modified conditions have been used successfully in the construction of a hexapeptide without racemization (eq 30).[33]

Cbz–Leu–Ala–Val–Phe–Gly–OH + Pro–OBn →(Ph₃P, CCl₄; Et₃N; 68%) Cbz–Leu–Ala–Val–Phe–Gly–Pro–OBn (30)

If amino alcohols, amino thiols, or diamines are used, the intermediate amides cyclodehydrate under the reaction conditions to form Δ^2-oxazolines (eq 31), Δ^2-oxazines, Δ^2-thiazolines, or Δ^2-imidazolines.[33] The reaction requires the use of 3 equiv of the $Ph_3P–CCl_4$ reagent. The reaction is reported to fail if a commercial sample of the polymer-supported phosphine is used instead of triphenylphosphine.

PhCH₂CO₂H + H₂NC(CH₃)₂CH₂OH →(Ph₃P, CCl₄; Et₃N; 67%) 2-benzyl-4,4-dimethyl-Δ²-oxazoline (31)

1,1-Dichloroalkenes and Vinyl Chlorides. The reaction of aldehydes produces a mixture of the 1,1-dichloroalkene and the dichloromethylene derivatives. These are the result of reaction of the in situ generated ***Dichloromethylenetriphenylphosphorane*** and ***Triphenylphosphine Dichloride***, respectively. Benzaldehyde (eq 32) produces a 1:1 mixture of the dichloroalkene and benzal chloride in 72% yield.[34]

PhCHO →(Ph₃P, CCl₄; 72%) PhCH=CCl₂ + PhCHCl₂ (32)

1:1

Ketones can also be used in this transformation, but sometimes enolizable ketones lead to the formation of the vinyl chloride derived from the enol. This is usually more of a problem for six-membered ring ketones than for five-membered ring ketones. Cyclopentanone yields predominantly the 1,1-dichloroalkene (eq 33), while cyclohexanone provides mainly the vinyl chloride (eq 34).[35]

cyclopentanone →(Ph₃P, CCl₄; 65%) (dichloromethylene)cyclopentane + 1-chlorocyclopentene (33)

95:5

cyclohexanone →(Ph₃P, CCl₄; 65%) (dichloromethylene)cyclohexane + 1-chlorocyclohexene (34)

7:93

1. (a) Appel, R. *AG(E)* **1975**, *14*, 801. (b) Appel, R.; Halstenberg, M. *Organophosphorus Reagents in Organic Synthesis*; Cadogan, J. I. G., Ed.; Academic: New York, 1979; pp 387–431.
2. (a) Lee, J. B.; Nolan, T. J. *CJC* **1966**, *44*, 1331. (b) Lee, J. B.; Downie, I. W. *T* **1967**, *23*, 359.
3. Verheyden, J. P. H.; Moffat, J. G. *JOC* **1972**, *37*, 2289.
4. Calzada, J. G.; Hooz, J. *OS* **1974**, *54*, 63.
5. Weiss, R. G.; Snyder, E. I. *JOC* **1970**, *35*, 1627.
6. Aneja, R.; Davies, A. P.; Knaggs, J. A. *CC* **1973**, 110.
7. Snyder, E. I. *JOC* **1972**, *37*, 1466.
8. Majid, R. M.; Fruchey, O. S.; Johnson, W. L. *TL* **1977**, 2999.
9. Appel, R.; Wihler, H.-D. *CB* **1976**, *109*, 3446.
10. (a) Harrison, C. R.; Hodge, P. *CC* **1975**, 622. (b) Regen, S. L.; Lee, D. P. *JOC* **1975**, *40*, 1669. (c) Harrison, C. R.; Hodge, P. *CC* **1978**, 813.
11. Lee, J. B. *JACS* **1966**, *88*, 3440.
12. Appel, R.; Einig, H. *Z. Anorg. Allg. Chem.* **1975**, *414*, 236.

13. Isaacs, N. S.; Kirkpatrick, D. *TL* **1972**, 3869.
14. Barry, C. N.; Evans, S. A. *JOC* **1981**, *46*, 3361.
15. Barry, C. N.; Evans, S. A. *TL* **1983**, *24*, 661.
16. Appel, R.; Kleinstück, R. *CB* **1974**, *107*, 5.
17. Ittah, Y.; Shahak, I.; Blum, J. *JOC* **1978**, *43*, 397.
18. Stoilova, V.; Trifonov, L. S.; Orahovats, A. S. *S* **1979**, 105.
19. Miller, M. J.; Mattingly, P. G.; Morrison, M. A.; Kerwin, J. F. *JACS* **1980**, *102*, 7026.
20. (a) Yamato, E.; Sugasawa, S. *TL* **1970**, 4383. (b) Appel, R.; Kleinstück, R.; Ziehn, K.-D. *CB* **1971**, *104*, 1030.
21. Juanin, R.; Arnold, W. *HCA* **1973**, *56*, 2569.
22. Appel, R.; Kohnke, J. *CB* **1971**, *104*, 2023.
23. Appel, R.; Warning, K. *CB* **1975**, *108*, 1437.
24. Appel, R.; Warning, K.; Ziehn, K.-D. *CB* **1973**, *106*, 3450.
25. (a) Appel, R.; Warning, K.; Ziehn, K.-D. *CB* **1974**, *107*, 698. (b) Appel, R.; Warning, K.; Ziehn, K.-D. *CB* **1973**, *106*, 2093.
26. Appel, R.; Warning, K.; Ziehn, K.-D. *CB* **1971**, *104*, 1335.
27. Appel, R.; Warning, K.; Ziehn, K.-D.; Gilak, A. *CB* **1974**, *107*, 2671.
28. Appel, R.; Kleinstück, R.; Ziehn, K.-D. *AG(E)* **1971**, *10*, 132.
29. Barstow, L. E.; Hruby, V. J. *JOC* **1971**, *36*, 1305.
30. Stork, G.; Morgans, D. J. *JACS* **1979**, *101*, 7110.
31. Wieland, T.; Seeliger, A. *CB* **1971**, *104*, 3992.
32. Takeuchi, Y.; Yamada, S. *CPB* **1974**, *22*, 832.
33. Appel, R.; Bäumer, G.; Strüver, W. *CB* **1975**, *108*, 2680.
34. Rabinowitz, R.; Marcus, R. *JACS* **1962**, *84*, 1312.
35. Isaacs, N.; Kirkpatrick, D. *CC* **1972**, 443.

Michael J. Taschner
The University of Akron, OH, USA

Triphenylphosphine Dibromide[1]

Ph_3PBr_2

[1034-39-5] $C_{15}H_{18}Br_2P$ (MW 389.10)

(bromination of alcohols,[3,4] phenols,[22] and enols;[3] cleavage of ethers[23,32,33] and acetals[25] to alkyl bromides; cyclization of amino alcohols[34,35] to cyclic amines; conversion of carboneoxylic acid derivatives into acyl bromides;[36,37] bromination[39] or dehydration[38] of carboxamide groups)

Alternate Names: bromotriphenylphosphonium bromide; dibromotriphenylphosphorane.

Physical Data: adduct:[2] colorless crystalline solid; mp 235 °C (dec). Ph_3P: mp 80.5 °C; bp 377 °C (in inert gas); d_4^{25} 1.194 g cm^{-3}; n_D^{80} 1.6358. Br_2: mp −7.2 °C; bp 59.5 °C; d_4^{25} 3.12 g cm^{-3}.

Solubility: adduct: sol CH_2Cl_2, MeCN, PhCN, DMF; less sol PhH, PhCl. Ph_3P: sol ether, PhH, $CHCl_3$, AcOH; less sol alcohol; pract. insol. H_2O. Br_2: sol H_2O (3.58 g/100 mL); very sol. alcohol, ether, $CHCl_3$, CCl_4, CS_2.

Form Supplied in: adduct: hygroscopic solid; commercially available; purity 95–96%. Both parent compounds, Ph_3P and Br_2, are widely available. Ph_3P: odorless platelets or prisms; purity ≈99%; typical impurity Ph_3PO ≈1%. Br_2: dark reddish liquid, volatile, with suffocating odor; vaporizes rapidly at rt; purity >99.5%.

Analysis of Reagent Purity: adduct: ^{31}P NMR (CH_2Cl_2) +49 ppm (ionic form);[2] Raman (solid phase) 239 cm^{-1} (P–Br). Typical impurities, Ph_3P and Ph_3PO: ^{31}P NMR (various solvents) −5 to −8 ppm and +25 to +29 ppm, respectively.

Preparative Methods: several preparations are described,[2,23b,32b] but in most cases the hygroscopic reagent is prepared just before use by addition of ***Bromine*** to ***Triphenylphosphine*** in a dry solvent with cooling.

Handling, Storage, and Precautions: adduct: corrosive; very hygroscopic; moisture sensitive; incompatible with strong oxidizing agents and strong bases; may decompose on exposure to moist air or water; do not get in eyes, on skin, or on clothing; keep containers tightly closed and store in a cool dry place; these reagents should be handled in a fume hood.

Introduction. The Ph_3PBr_2 adduct was first used in synthesis in 1959[3] for the preparation of alkyl and acyl bromides from alcohols and carboxylic acids, and for the dehydration of amides and oximes to nitriles. Since then it has been widely employed as a versatile reagent for a number of synthetic reactions.

Conversion of Alcohols into Alkyl Bromides. It has been established that Ph_3PX_2 reagents present considerable advantages over phosphorus halides (i.e. PX_3, PX_5) for which only two replaceable groups would be 'necessary and desirable'.[4a] The mechanism[4b] of the reaction of Ph_3PBr_2 with an alcohol involves initial rapid formation of an alkyloxyphosphonium bromide,[4,5,7a] which then collapses slowly to the phosphine oxide and an alkyl bromide (eq 1). S_N2 substitution and high yields are generally observed with primary and secondary alcohols.[3,4a] This reagent shows little tendency to induce either carbonium ion rearrangement in the carbon skeleton or elimination reactions; this is illustrated by the preparation of neopentyl bromide from neopentyl alcohol (eq 2)[4a] and in the steroid field,[6] in which cholesterol and 3β-cholestanol give respectively 3β-bromocholest-5-ene and 3α-bromocholestane in yields of about 80%. Inversion of configuration is found in the S_N2 reactions of (+)-*endo*-norbornanol (eq 3)[7a,8] and 7-norbornanol. However, a similar reaction with (−)-*endo*-norbornanol leads mainly to racemic bromide via a nonclassical norbornyl cation intermediate.[7b] Likewise, little rearrangement is observed in the bromination of 3-methyl-2-butanol (eq 4).[9] Greater amounts of rearranged product are formed in the same reaction with Ph_3PCNBr (3%), ***Bromine–Triphenyl Phosphite*** (55.6%), or ***Triphenylphosphine–Carbon Tetrabromide*** (6–10%).

$$\text{ROH} \xrightarrow[\text{fast}]{Ph_3PBr_2} \text{R–O–}\overset{+}{P}Ph_3\ Br^- \xrightarrow[\text{slow}]{} \text{RBr} + \text{OPPh}_3 \quad (1)$$

$$t\text{-Bu}\diagup\!\!\diagdown\text{OH} \xrightarrow[\text{DMF, <55 °C, 79\%}]{Ph_3PBr_2} t\text{-Bu}\diagup\!\!\diagdown\text{Br} \quad (2)$$

HO–(norbornane) $\xrightarrow[\text{triglyme, heat, 62\%}]{Ph_3PBr_2}$ Br–(norbornane) (3)

HO — Ph_3PBr_2, PhH, pyridine, 40–45 °C → Br (43.8%) + Br, Et (1.4%) (4)

Ph_3PBr_2 can be used with sensitive substrates which incorporate cyclopropyl rings (e.g. cyclopropyl carbinol[11,12]) or unsaturation[10,13] (e.g. cinnamyl alcohol and alkynediols) with little if any side reactions. Cyclopropyl carbinols are transformed into bromides in good yield without ring opening or ring expansion, with the formation of only trace amounts of homoallylic bromides; no bromocyclobutane derivatives are detected (eq 5). Ph_3PBr_2 also proves superior to ***Phosphorus(III) Bromide*** for the bromination of alkynediols to dibromoalkynes (eq 6).[13]

R, OH — Ph_3PBr_2, DMF, –5 °C then rt, 65–80% → Br, R (5)

R = H, Me, Et, Pr, Bu, *i*-Pr, *t*-Bu

$HO–CR_2–(C{\equiv}C)_n–CR_2–OH$ — Ph_3PBr_2, MeCN, 40–50 °C, 78–92% → $Br–CR_2–(C{\equiv}C)_n–CR_2–Br$ (6)

R = H, Me; *n* = 1, 2

Ph_3PBr_2 provides comparable yields to Ph_3P/CBr_4 in the bromination of diethyl 1-hydroxyalkylphosphonates.[14] A number of other primary,[15] secondary,[11–14] and even tertiary alcohols are brominated with Ph_3PBr_2, but with preponderant elimination[4a,6,16] in the latter case. A polymer-supported Ph_3PBr_2 reagent[17] has been successfully applied to alcohol bromination in refluxing $CHCl_3$; product isolation is facilited by the ready separation of supported from the nonsupported species. However, note that a polymer-supported Ph_3P/CBr_4 reagent leads to better bromination results than the Ph_3PBr_2 system in a number of cases.

Similar mono- and dibrominations have also been achieved in the carbohydrate field (eq 7).[18] Configurational inversion at chiral centers is observed throughout. Ph_3PBr_2/***Imidazole*** (ImH) or Ph_3P/2,4,5-tribromoimidazole/ImH combinations in refluxing PhMe or in PhMe–pyridine mixtures are preferred in some cases. In the nucleoside field,[19] double activation is achieved with Ph_3PBr_2 in pyridine, leading both to bromination of the 5′-position of the sugar moiety and to substitution on the C-6 position of the adenine part (eq 8).[19b]

OH, HO, HO, HO, OMe — Ph_3PBr_2, ImH, PhMe, reflux, 74% → Br, Br, HO, HO, OMe (7)

O, HN, N, N, HO — Ph_3PBr_2, then ArSH, pyridine, 25 °C, 57–71% → SAr, 6, N, N, N, Br, 5′ (8)

Ar = Ph, 1-naphthyl

A variety of solvents such as CCl_4, PhH, triglyme, $CHCl_3$, PhMe, pyridine, NMP, MeCN, DMF, and mixtures of these have been employed in these bromination reactions, depending on substrate solubility. Both *O*- and *C*-formylations are encountered in some cases when DMF is the solvent. Ph_3PBr_2 and both Ph_3PX_2 analogs (X = Cl, I) can react with DMF according to the sequence in eq 9.[20] The intermediate alkoxyimonium bromide (**3**) affords either the expected bromide by heating, or the formyl ester by hydrolytic workup with secondary[20] or tertiary[16] alcohols (e.g. eq 10). Vilsmeier-type formylation reactions on nucleophilic carbon atoms can also take place, as developed early[38a] and then encountered later as a side reaction in the steroid field.[21] Such *O*-formylation of secondary alcohols is interestingly used in the differentiation of diols, leading to a new synthesis of bromo formates (eq 10).[20b] Under mild conditions, primary alcohols still afford bromides whereas secondary ones are converted almost exclusively to the corresponding formate esters.

$$Ph_3PBr_2 + Me_2NCHO \longrightarrow \underset{(\mathbf{1})}{Me_2\overset{+}{N}{=}CH–O–PPh_3Br\ Br^-} \xrightarrow{-Ph_3PO}$$

$$\underset{(\mathbf{2})}{Me_2\overset{+}{N}{=}CH–Br\ Br^-} + ROH \longrightarrow \underset{(\mathbf{3})}{ROCH{=}\overset{+}{N}Me_2\ Br^-} \xrightarrow{heat} RBr + DMF \quad (9)$$

$$(\mathbf{3}) \xrightarrow{H_2O} RO–CHO + HBr{\bullet}HNMe_2$$

$$\underset{(\mathbf{4})}{Ph_3\overset{+}{P}OR\ Br^-} + Me_2NCHO \xrightarrow{-Ph_3PO} (\mathbf{3})$$

R^3, CH_2OH, $(CH_x)_n$, R^2, R^1, OH — Ph_3PBr_2, DMF, –25 to –5 °C, 50–90% → R^3, CH_2Br, $(CH_x)_n$, R^2, R^1, OCHO (10)

$x = 1, 2$; $R^1 \neq H$; R^2, R^3 = H, Me, $(CH_x)_n$

Synthesis of Aryl Bromides from Phenols. Already developed in the pioneering work,[4a] the preparation of aryl bromides from phenols with Ph_3PBr_2 was later extended to a number of substrates (eq 11).[22]

R^1, R^2, X, Y, OH, R^3 — Ph_3PBr_2, MeCN then neat, 200–250 °C, 48–92% → R^1, R^2, X, Y, Br, R^3 (11)

X, Y = C, N; R^2 = H; R^3 = H, Cl, NO_2, MeO;
$R^1 + R^2$ = Ph; 2,3-pyridyl; $R^2 + R^3$ = Ph, 3,4-phenol

Alkyl Bromides from Ethers and Acetals. Ph_3PBr_2 cleaves dialkyl ethers to give the two alkyl bromides under essentially neutral conditions.[23] This reaction offers obvious advantages as it avoids both the strongly acidic and basic media which are usually employed for ether cleavage. Brominations are generally conducted at reflux in high boiling solvents such as PhCl, PhCN, DMF, or NMP. Primary and secondary alkyl groups provide good yields of bromides without rearrangement in PhCl or PhCN, at

60 to 120 °C,[23a,b] as illustrated in eq 12.[23c] Phenyl alkyl ethers initially afford aryloxyphosphonium bromides which collapse to bromobenzene only by heating at higher temperature (>230 °C as seen before). Alkyl *t*-butyl ethers are cleaved more easily in DMF[23b] or in MeCN,[24] usually between 60 and 110 °C; the *t*-butyl group is converted into isobutene in this process. In a related reaction, endoxides are transformed into arenes[28] by Ph_3PBr_2 treatment in PhCl. Aromatization takes place via HBr elimination from the initially formed dibromide.

Ph_3Br_2; PhCl, 120 °C; 91% (12)

Direct conversion of tetrahydropyranyl (THP)[25] and tetrahydrofuranyl ethers[26] (THF) to bromides can be achieved by Ph_3PBr_2 treatment under milder conditions (eq 13).[25a] THP and THF ethers afford good yields of acyclic, saturated, and unsaturated,[25c] primary, secondary, and even tertiary alkyl bromides; under the same conditions, cyclohexyl THF ethers provide mainly cyclohexene derivatives. In view of the fact that the preceding reaction of THP ethers can be stopped (at low temperature) at the stage of either the alkoxyphosphonium (see eq 1) or the pentacovalent $ROPPh_3Br$ intermediate, hydrolysis of the reaction mixture at −50 °C, leads to the corresponding alcohols in good yield (74–97%) (eq 14).[29] The method is efficient for the deprotection of secondary and tertiary THP ethers, as well as for acyclic acetal ether, dialkyl acetals, and *O*-glucosides. Acetal functions can be removed with retention of stereochemistry, as illustrated in the conversion of (−)-menthol THP ether into (−)-menthol with full recovery of the optical activity. This procedure is not applicable to primary THP ethers, where the corresponding bromides are formed under these conditions.

Ph_3PBr_2; CH_2Cl_2, rt; 91% (13)

Ph_3PBr_2; CH_2Cl_2, rt; 92% (14)

Reaction of hindered trialkylsilyl ethers with Ph_3PBr_2 in CH_2Cl_2 at rt affords primary and secondary alkyl bromides in excellent yield (70–94%); the reaction is valuable in the β-lactam field.[27a] The reaction rate is increased by addition of a catalytic amount of ***Zinc Bromide***. Silylated enol ethers, such as trimethylsilyl(1-phenylvinyloxy)silane, provide vinyl bromides such as α-bromostyrene with Ph_3PBr_2 in refluxing CCl_4.[27b]

α-Alkynyl Ketones and β-Bromo-α-Vinyl Ketones from β-Diketones. The previously described bromination of alcohols can be applied to the enol form of β-diketones. The first attempts in this field were conducted with the Ph_3PBr_2/***Triethylamine*** system on dibenzoylmethane, leading to phenylethynyl phenyl ketone.[30a] Elimination at the oxyphosphonium stage or HBr elimination from the initially generated vinyl bromide is probably involved in this reaction. Analogously, reaction of unsymmetrical β-diketones[30b] with excess Ph_3PBr_2/Et_3N proceeds with the formation of 3:1 mixtures of α,β-ynones in good overall yield (eq 15). Ethynyltriphenylphosphonium bromides are similarly obtained[30c] starting from (benzoylmethylene)- and (carbamoylmethylene)triphenylphosphoranes.

Ph_3PBr_2, Et_3N; CH_2Cl_2, rt; 80–95%

R = Ph, CF_3, C_3F_7

3:1 (15)

Similar treatment of unconjugated β-diketones[31] in PhH or MeCN as solvent leads to β-bromo-α,β-unsaturated ketones with superior results relative to the same transformation with PBr_3 as the brominating agent (eq 16).[31c] The reaction of acyclic 2,4-pentanedione with Ph_3PBr_2/Et_3N in PhH–MeCN[31c] or in CH_2Cl_2,[31d] but in the absence of Et_3N, appears not to be totally stereoselective, since a mixture of the geometrically isomeric bromo enones is produced, with 88% ($E/Z \approx 87/13$) and 85% ($E/Z \approx 93/7$) yields, respectively.

Ph_3PBr_2, Et_3N; PhH or MeCN, rt; 80–97% (16)

R^1 = H, Me, allyl, 2-cyanoethyl
n = 0, 1; R^2 = H, Me; R^3 = H, Me

Epoxide Opening to Vicinal Dibromides or Bromohydrins. Epoxide ring opening with Ph_3PBr_2[32a] takes place in MeCN or PhH at 20–50 °C, affording vicinal dibromides (32–74%); *cis–trans* isomer mixtures are obtained in the case of cycloalkene epoxides. This reaction involves initial cleavage of the epoxide C–O bond at the most substituted epoxide carbon; bromoalkoxyphosphonium salts are thus formed, and these undergo subsequent substitution to give dibromides and Ph_3PO. In further studies,[32b] reaction of *cis*-epoxides in PhH produced *erythro*-dibromides exclusively (eq 17); *trans*-epoxides exhibit less specificity, leading to a mixture of *threo*- and *erythro*-dibromides. Use of a more polar solvent such as CH_2Cl_2 or MeCN, instead of PhH, with *cis*-epoxides provides *erythro–threo* mixtures (60–40 to 50–50). By reacting epoxides first with HCl and then with Ph_3PBr_2, it is possible to obtain vicinal bromochlorides stereospecifically that are also products of two S_N2 displacements.

Ph_3Br_2; PhH, rt; 85% (17)

R^1 = nonyl; R^2 = Bu

The reaction can also be carried out so as to give bromohydrins. In the steroid field[33a] with conformationally rigid epoxides, oxirane cleavage appears to be quite stereoselective and leads only to the bromohydrin resulting from the usual *anti* opening of the ring in high yield (90–97%). Less hindered and rigid substrates afford regioisomeric mixtures of cyclic *trans*-bromohydrins. Equivalent results[33b] are obtained, in the steroid series, by use of polymeric Ph_3PBr_2 for the transformation of epoxides to bromohydrins under mild nonacidic conditions; oxirane ring opening remains regio- and stereoselective.

Cyclization of β- and γ-Amino Alcohols to Aziridines and Azetidines. β-Amino alcohols undergo cyclization upon Ph_3PBr_2/Et_3N treatment,[34] providing the corresponding aziridines along with substituted piperazines as dimeric byproducts (eq 18). Walden inversion is observed in the ring closure of both *threo*- and *erythro*-ephedrine and this supports the postulated 1,2-*trans*-elimination. Application of this procedure to the synthesis of 1-monosubstituted aziridines is unsuccessful, giving only the piperazine byproduct. *N*-Aryl γ-amino alcohols are cyclized to azetidines by Ph_3PBr_2/Et_3N in MeCN (eq 19).[35a] Under these conditions, some starting material is recovered unchanged along with mixed tetrahydroquinolines as side products and the expected *N*-arylazetidine. Comparable results are obtained (45–65%)[35b] with some other *N*-protecting groups.

Ph_3PBr_2, Et_3N; MeCN, rt; 50–74% (18)

R^1 = Bu, *t*-Bu, cyclohexyl, benzyl; R^2 = Me, Ph

Ph_3PBr_2, Et_3N; MeCN, –5 °C; 20–60% → 3-$MeOC_6H_4$–N (19)

R^1 = Me, Ph; R^2 = H, Me; R^3 = H, Me, Ph

Acyl Bromides from Carboxylic Acids, Anhydrides, and Esters. Carboxylic acid bromides are prepared by reaction of Ph_3PBr_2 with various carboxylic acids and anhydrides in boiling PhCl.[36a] Milder conditions and better yields were later obtained in a comparative study[36b] by using CH_2Cl_2 at rt (eq 20). Further improvements are observed through the use of the corresponding trimethylsilyl esters; acyl bromides are thus obtained under mild and neutral conditions, allowing reactions with sensitive acid substrates without side reactions (eq 20). This transformation is also applicable to generate the acyl bromides of more hindered trialkylsilyl esters such as the TBDMS, TIPS, and TBDPS carboxylates; in CH_2Cl_2, the reaction rate increases in the presence of a catalytic amount of $ZnBr_2$.[27a]

$$RCO_2Z \xrightarrow{Ph_3PBr_2} RCOBr \quad (20)$$

Z = H; PhCl, reflux	50–80%
Z = H; CH_2Cl_2, rt	64–90%
Z = TMS; CH_2Cl_2, rt	70–90%

R = Ph, 2-ClC_6H_4, 4-ClC_6H_4, Et, $PhCH_2CH_2$, PhCH=CH, $PhCH(CO_2Z)$, MeCH=CH

Direct reaction of less reactive alkyl esters and lactones[37] with Ph_3PBr_2 affords both acyl and alkyl bromides. This reaction is achieved at MeCN reflux[37a–c] for α-halogenated esters such as CF_3CO_2R, but only at higher temperatures and with longer reaction times for unhalogenated ones. The reaction of an ester with Ph_3PBr_2 presumably forms an oxonium salt, which then undergoes cleavage through reaction with bromide ion. Prolonged heating (80–110 °C) of lactones with Ph_3PBr_2 gives the expected acyl alkyl dibromides, but in low yield (≤50%) due to a concomitant degradation.[37d]

Dehydration of Ureas and Amides to Carbodiimides, Nitriles, Isocyanides, and Ketenimines. Dehydration of *N,N′*-disubstituted ureas by heating with Ph_3PBr_2/Et_3N in PhH or PhCl at 70–80 °C affords carbodiimides[38a] in good yields. Subsequent improvements[38d] involving milder conditions provides access to certain unstable derivatives from *N,N′*-disubstituted ureas; this procedure compares favorably with previous ones and with another related approach using the ***Triphenylphosphine–Carbon Tetrachloride*** reagent system (eq 21). Reaction of *N,N′*-dialkylidene ureas with Ph_3PBr_2 in PhH leads to α-haloalkyl carbodiimides.[38e]

$$R^1HN{-}C(O){-}NHR^2 \xrightarrow{Ph_3PBr_2,\ Et_3N} R^1N{=}C{=}NR^2 \quad (21)$$

PhH or PhCl, reflux	57–75%
CH_2Cl_2, 0 °C	70–90%
Ph_3P/CCl_4, Et_3N, CH_2Cl_2, rt	30–40%

R^1 = Me, Et, Bu, Cy, Ph

R^2 = Et, $ClCH_2CH_2$, $BrCH_2CH_2$, Bu, Hex, Cy, Ph

By the same Ph_3PBr_2/Et_3N procedure carried out in refluxing PhH, disubstituted cyanamides such as Me_2NCN are obtained from the *N,N*-disubstituted urea Me_2NCONH_2 (67%); similarly, isocyanides result from monosubstituted formamides (56–73%),[38a,40b] and nitriles from primary amides and oximes (58–68%).[3] Via the same route, ketenimines ($R^1R^2C{=}C{=}NR^3$) are prepared[38a,40b] in CH_2Cl_2 at reflux by dehydration of secondary amides having a C–H bond adjacent to the carbonyl function (45–93%). Application of this procedure to sulfimides gives highly reactive ketenimines, which are used for further reactions without isolation; [2 + 2] cycloaddition of such species with Schiff bases provides, in good yield, *N*-(tosyl)azetidin-2-imines related to β-lactam derivatives (eq 22).[38b] The diphosphorylated ketenimine $[(EtO)_2(O)P]_2C{=}C{=}NPh$ is obtained analogously in 87% yield with Ph_3PBr_2/Et_3N from the α,α-diphosphorylated acetanilide.[38c]

i-PrCONHTs → (Ph_3PBr_2, Et_3N; CH_2Cl_2, reflux) → [$Me_2C{=}C{=}N$-Ts] → (PhCH=NHPh; 86%) → TsN= (azetidine, N-Ph, –Ph) (22)

Imidoyl Bromides from Secondary Amides. In the case of the secondary diarylamide benzanilide, reaction with Ph_3PBr_2/Et_3N in refluxing PhH provides *N*-phenylbenzimidoyl bromide in 65% yield, whereas a 70% yield is attained with the Ph_3P/CBr_4 reagent system.[39] The closely related *N*-methylbenzamide affords a dimeric compound[40b] with Ph_3PBr_2 in boiling CH_2Cl_2 by intermolecular *N*-imidoylation (see below for a similar intramolecular process) between generated imidoyl bromide or oxyphosphonium species and the starting carboxamide. Cyclodehydration of secondary carboxylic diamides with Ph_3PBr_2/Et_3N affords intramolecular *O*-imidoylation products with a benzoxadiazepine structure.[40] Without Et_3N, intramolecular *N*-imidoylation products with an ω-imino lactam structure are obtained; *O*-imidoylated products rearrange to *N*-analogs under acidic conditions.

Iminophosphoranes from Amines, Hydrazines, and Related Derivatives. Compounds containing a P=N bond, such as iminophosphoranes, are widely used as synthetic intermediates, especially in the heterocyclic field. Aromatic primary amines are generally transformed into iminotriphenylphosphoranes, and aliphatic amines into aminophosphonium bromides with Ph_3PBr_2/Et_3N in PhH, PhCl, CCl_4, or CH_2Cl_2 as solvent;[41a,b] more basic conditions are usually required to convert an alkylaminophosphonium salt into the corresponding iminophosphorane.[41b] As protected primary amines, these phosphinimines (iminophosphoranes) can then be subjected to monoalkylation, the Ph_3P protecting group being cleaved in a subsequent step to provide secondary amines (eq 23).[41a–c]

$$RNH_2 \xrightarrow[\text{PhH, 0 °C to reflux}]{Ph_3PBr_2,\ Et_3N} RN{=}PPh_3 \xrightarrow[\text{2. OH}^-\text{, 45–80\%}]{\text{1. R}^1\text{X}} RNHR^1 \quad (23)$$

$$RN{=}PPh_3 \xrightarrow{CO_2} RN{=}C{=}O; \quad RN{=}PPh_3 \xrightarrow{CS_2} RN{=}C{=}S \quad (70\text{–}90\%)$$

R = alkyl, aryl
R^1 = Me, Et, *n*-Bu
X = Cl, Br, I

Some aliphatic and aromatic phosphinimines (prepared as above) undergo a Wittig-like condensation with CO_2 or CS_2, giving rise to isocyanates and isothiocyanates, respectively (eq 23).[41d] Dehydration reactions promoted by Ph_3PBr_2/Et_3N are used for intramolecular cyclizations involving an imide carbonyl group and an aromatic amine;[42] an aza-Wittig reaction of a phosphinimide intermediate is assumed to occur. Iminophosphoranes derived from hydrazines[43a] and acylhydrazines[43b,c] are also used as intermediates in heterocyclic synthesis. Reactions of metalated amines RNHMgX with Ph_3PBr_2 in PhH–ether can also provide aminophosphonium salts $Ph_3\overset{+}{P}NHR\ Br^-$ in fair to good yields (33–74%).[43d]

Other Applications. Reaction of Ph_3PBr_2 with benzoins in MeCN at rt provides diaryl α-diketones in excellent yields (75–98%).[44] Ph_3PBr_2 is used as a precursor in the synthesis of dialkoxytriphenylphosphoranes such as $(CF_3CH_2O)_2PPh_3$[45] and $(t\text{-}BuCH_2O)_2PPh_3$,[46] which are used as alkylating or acylating and cyclodehydration agents, respectively. Direct synthesis of β-lactams by [2+2] cycloaddition between carboxylic acids and imides, thus avoiding the use of acid halides, is achieved with Ph_3PBr_2 (40–55%).[47] Ph_3PBr_2, as a dehydrating agent, effects esterification reactions with tertiary alcohols,[48a] such as *t*-BuOH in PhH–HMPA, and with aromatic or aryl allylic acids and primary or secondary aliphatic alcohols in petroleum ether (27–85%).[48b]

Preparation of ***Bromotrimethylsilane*** in quantitative yield occurs by deoxygenation of $(Me_3Si)_2O$ with Ph_3PBr_2 (in 1,2-dichlorobenzene at reflux) in the presence of a catalytic amount of Zn; a one-pot preparation of pseudohalogenosilanes, in excellent yield (80–95%), can follow by subsequent addition of the appropriate salt XM in DMF ($X = N_3$, NCO, CN; M = K, Na,).[49] Treatment of Ph_3PBr_2 with ***Iodotrimethylsilane*** in CH_2Cl_2 at 0 °C yields Ph_3PI_2 (86%).[50] Reaction of Ph_3PBr_2/Et_3N with $EtO_2CCH_2CH{=}CHCO_2Et$ in PhH leads to the corresponding alkylidenephosphonium ylide in 80% yield.[51] Beckmann rearrangement of cycloalkanone oximes into lactams (74–81% yields) is effected by Ph_3PBr_2 in dry PhH at 50–60 °C.[52] The Ph_3PBr_2–Et_3N reagent system compares favorably with a number of other related phosphorus reagents for the rearrangement of *N*-allylamides into nitriles under mild conditions via 3-aza-Claisen reaction.[53]

---

1. Castro, B. R. *OR* **1983**, *29*, 1.
2. Physical studies on the Ph_3P–dihalide adducts indicates that the extent of ionic (and covalent) behavior is a function of solvent polarity; see especially: Bricklebank, N.; Godfrey, S. M.; Mackie, A. G.; McAuliffe, C. A.; Pritchard, R. G. *CC* **1992**, 355, and references cited therein.
3. Horner, L.; Oediger, H.; Hoffmann, H. *LA* **1959**, *626*, 26.
4. (a) Wiley, G. A.; Hershkowitz, R. L.; Rein, B. M.; Chung, B. C. *JACS* **1964**, *86*, 964. (b) Wiley, G. A.; Rein, B. M.; Hershkowitz, R. L. *TL* **1964**, 2509.
5. Kaplan, L. *JOC* **1966**, *31*, 3454.
6. (a) Levy, D.; Stevenson, R. *TL* **1965**, 341. (b) Levy, D.; Stevenson, R. *JOC* **1965**, *30*, 3469.
7. (a) Schaefer, J. P.; Weinberg, D. S. *JOC* **1965**, *30*, 2635. (b) Schaefer, J. P.; Weinberg, D. S. *JOC* **1965**, *30*, 2639.
8. Brett, D.; Downie, I. M.; Lee, J. B.; Matough, M. F. S. *CI(L)* **1969**, 1017.
9. Arain, R. A.; Hargreaves, M. K. *JCS(C)* **1970**, 67.
10. (a) Schaefer, J. P.; Higgins, J. G.; Shenov, P. K. *OS* **1968**, *48*, 51. (b) Sandri, J.; Viala, J. *SC* **1992**, *22*, 2945.
11. Hrubiec, R. T.; Smith, M. B. *JOC* **1984**, *49*, 431.
12. Hanack, M.; Auchter, G. *JACS* **1985**, *107*, 5238.
13. Machinek, R.; Lüttke, W. *S* **1975**, 255.
14. Gajda, T. *PS* **1990**, *53*, 327.
15. (a) Piper, J. R.; Rose, L. M.; Johnston, T. P.; Grenan, M. M. *JMC* **1979**, *22*, 631. (b) Zakharkin, L. I.; Kovredov, A. I.; Kazantsev, A. V.; Meiramov, M. G. *JGU* **1981**, *51*, 289. (c) Janda, K. D.; Weinhouse, M. I.; Danon, T.; Pacelli, K. A.; Schloeder, D. M. *JACS* **1991**, *113*, 5427. (d) Skvarchenko, V. R.; Lapteva, V. L.; Gorbunova, M. A. *JOU* **1990**, *26*, 2244.
16. Caubere, P.; Mourad, M. S. *T* **1974**, *30*, 3439.
17. Hodge, P.; Khoshdel, E. *JCS(P1)* **1984**, 195.
18. (a) Garegg, P. J. *PAC* **1984**, *56*, 845. (b) Classon, B.; Garegg, P. J.; Samuelsson, B. *CJC* **1981**, *59*, 339.
19. (a) Haga, K.; Yoshikawa, M.; Kato, T. *BCJ* **1970**, *43*, 3922. (b) Bridges, A. J. *Nucleosides Nucleotides* **1988**, *7*, 375.
20. (a) Herr, M. E.; Johnson, R. A. *JOC* **1972**, *37*, 310. (b) Boeckman, R. K., Jr.; Ganem, B. *TL* **1974**, 913.
21. Dahl, T.; Stevenson, R.; Bhacca, N. S. *JOC* **1971**, *36*, 3243.
22. (a) Schaefer, J. P.; Higgins, J. *JOC* **1967**, *32*, 1607. (b) Schaefer, J. P.; Higgins, J.; Shenoy, P. K. *OS* **1969**, *49*, 6. (c) Porzi, G.; Concilio, C. *JOM* **1977**, *128*, 95.
23. (a) Anderson, A. G., Jr.; Freenor, F. J. *JACS* **1964**, *86*, 5037. (b) Anderson, A. G., Jr.; Freenor, F. J. *JOC* **1972**, *37*, 626. (c) Kato, M.; Nomura, S.; Kobayashi, H.; Miwa, T. *CL* **1986**, 281.
24. Marchand, A. P.; Weimar, W. R., Jr. *CI(L)* **1969**, 200.
25. (a) Schwartz, M.; Oliver, J. E.; Sonnet, P. E. *JOC* **1975**, *40*, 2410. (b) Sonnet, P. E. *SC* **1976**, *6*, 21. (c) Buser, H. R.; Guerin, P. M.; Toth, M.; Szöcs, G.; Schmid, A.; Francke, W.; Arn, H. *TL* **1985**, *26*, 403.
26. Kruse, C. G.; Jonkers, F. L.; Dert, V.; van der Gen, A. *RTC* **1979**, *98*, 371.

27. (a) Aizpurua, J. M.; Cossío, F. P.; Palomo, C. *JOC* **1986**, *51*, 4941. (b) Lazukina, L. A.; Kolodyazhnyi, O. I.; Pesotskaya, G. V.; Kukhar, V. P. *JGU* **1976**, *46*, 1931.
28. De Wit, J.; Wynberg, H. *RTC* **1973**, *92*, 281.
29. Wagner, A.; Heitz, M.-P.; Mioskowski, C. *CC* **1989**, 1619.
30. (a) Hoffmann, H.; Diehr, H. J. *TL* **1962**, 583. (b) Chechulin, P. I.; Filyakova, V. I.; Pashkevich, K. I. *BAU* **1989**, 189. (c) Bestmann, H. J.; Kisielowski, L. *CB* **1983**, *116*, 1320.
31. (a) Carnduff, J.; Miller, J. A.; Stockdale, B. R.; Larkin, J.; Nonhebel, D. C.; Wood, H. C. S. *JCS(P1)* **1972**, 692. (b) Piers, E.; Nagakura, I. *SC* **1975**, *5*, 193. (c) Piers, E.; Grierson, J. R.; Lau, C. K.; Nagakura, I. *CJC* **1982**, *60*, 210. (d) Buono, G. *S* **1981**, 872.
32. (a) Thakore, A. N.; Pope, P.; Oehlschlager, A. C. *T* **1971**, *27*, 2617. (b) Sonnet, P. E.; Oliver, J. E. *JOC* **1976**, *41*, 3279.
33. (a) Palumbo, G.; Ferreri, C.; Caputo, R. *TL* **1983**, *24*, 1307. (b) Caputo, R.; Ferreri, C.; Noviello, S.; Palumbo, G. *S* **1986**, 499.
34. Okada, I.; Ichimura, K.; Sudo, R. *BCJ* **1970**, *43*, 1185.
35. (a) Gogte, V. N.; Kulkarni, S. B.; Tilak, B. D. *TL* **1973**, 1867. (b) Freeman, J. P.; Mondron, P. J. *S* **1974**, 894.
36. (a) Bestmann, H.-J.; Mott, L. *LA* **1966**, *693*, 132. (b) Aizpurua, J. M.; Palomo, C. *S* **1982**, 684.
37. (a) Burton, D. J.; Koppes, W. M. *CC* **1973**, 425. (b) Burton, D. J.; Koppes, W. M. *JOC* **1975**, *40*, 3026. (c) Anderson, A. G., Jr.; Kono, D. H. *TL* **1973**, 5121. (d) Smissman, E. E.; Alkaysi, H. N.; Creese, M. W. *JOC* **1975**, *40*, 1640.
38. (a) Bestmann, H. J.; Lienert, J.; Mott, L. *LA* **1968**, *718*, 24. (b) Van Camp, A.; Goossens, D.; Moya-Portuguez, M.; Marchand-Brynaert, J.; Ghosez, L. *TL* **1980**, *21*, 3081. (c) Bestmann, H. J.; Lehnen, H. *TL* **1991**, *32*, 4279. (d) Palomo, C.; Mestres, R. *S* **1981**, 373. (e) Fetyukhin, V. N.; Vovk, M. V.; Samarai, L. I. *JOU* **1981**, *17*, 1263.
39. Appel, R.; Warning, K.; Ziehn, K.-D. *CB* **1973**, *106*, 3450.
40. (a) Mazurkiewicz, R. *M* **1988**, *119*, 1279. (b) Mazurkiewicz, R. *Acta Chim. Hung.* **1990**, *127*, 439 (*CA* **1990**, *114*, 206 693).
41. (a) Horner, L.; Oediger, H. *LA* **1959**, *627*, 142. (b) Zimmer, H.; Jayawant, M.; Gutsch, P. *JOC* **1970**, *35*, 2826. (c) Briggs, E. M.; Brown, G. W.; Jiricny, J.; Meidine, M. F. *S* **1980**, 295. (d) Molina, P.; Alajarin, M.; Arques, A. *S* **1982**, 596.
42. (a) Al-Khathlan, H.; Zimmer, H. *JHC* **1988**, *25*, 1047. (b) Al-Khathlan, H. Z.; Al-Lohedan, H. A. *PS* **1991**, *61*, 367.
43. (a) Cullen, E. R.; Guziec, F. S., Jr.; Hollander, M. I.; Murphy, C. J. *TL* **1981**, *22*, 4563. (b) Molina, P.; Alajarin, M.; Saez, J. R. *S* **1984**, 983. (c) Froyen, P. *PS* **1991**, *57*, 11. (d) Zbiral E.; Berner-Fenz, L. *M* **1967**, *98*, 666.
44. Ho, T.-L. *S* **1972**, 697.
45. Kubota, T.; Miyashita, S.; Kitazume, T.; Ishikawa, N. *JOC* **1980**, *45*, 5052.
46. Kelly, J. W.; Evans, S. A., Jr. *JOC* **1986**, *51*, 5490.
47. Cossío, F. P.; Ganboa, I.; Palomo, C. *TL* **1985**, *26*, 3041.
48. (a) Haynes, R. K.; Katsifis, A.; Vonwiller, S. C. *AJC* **1984**, *37*, 1571. (b) Lajis, N. *Pertanika* **1985**, *8*, 67 (*CA* **1986**, *105*, 6277d).
49. Aizpurua, J. M.; Palomo, C. *NJC* **1984**, *8*, 51.
50. Romanenko, V. D.; Tovstenko, V. I.; Markovski, L. N. *S* **1980**, 823.
51. Labuschagne, A. J. H.; Schneider, D. F. *TL* **1982**, *23*, 4135.
52. Sakai, I.; Kawabe, N.; Ohno, M. *BCJ* **1979**, *52*, 3381.
53. Walters, M. A.; Hoem, A. B.; Arcand, H. R.; Hegeman, A. D.; McDonough, C. S. *TL* **1993**, *34*, 1453.

Jean-Robert Dormoy & Bertrand Castro
SANOFI Chimie, Gentilly, France

Triphenylphosphine Dichloride[1]

[2526-64-9] $C_{18}H_{15}Cl_2P$ (MW 333.20)

(conversion of alcohols,[6–10] phenols,[13] and enols[15] into alkyl chlorides; synthesis of *vic*-dichlorides[16] or chlorohydrins[17] from epoxides; conversion of carboxylic acids[11] and derivatives[20] into acyl chlorides; chlorination[23,24] or dehydration[11] of the CONH group; production of iminophosphoranes from amines and related compounds[27–29])

Alternate Names: dichlorotriphenylphosphorane; triphenyldichlorophosphorane; chlorotriphenylphosphonium chloride.

Physical Data: adduct:[2] white crystalline solid; mp 85–100 °C;[4] fp 20 °C. Ph_3P: mp 80.5 °C; bp 377 °C (in inert gas); d_4^{25} 1.194 g mL^{-3}; n_D^{80} 1.6358. Cl_2: mp −101 °C; bp −34.6 °C; d^0 3.214 g L^{-1}.

Solubility: adduct: slightly sol ether, THF, PhMe; sol CCl_4, CH_2Cl_2, DMF, MeCN, pyridine; insol petroleum ether. Ph_3P: sol ether, PhH, $CHCl_3$, AcOH; less sol alcohol; pract insol H_2O. Cl_2: sol CCl_4, alcohol.

Form Supplied in: very hygroscopic solid; commercially available; purity ~80%; the remainder is 1,2-dichloroethane. The precursors Ph_3P and Cl_2 are both widely available. Ph_3P: odorless platelets or prisms; purity ~99%; typical impurity is Ph_3PO ~1%. Cl_2: greenish-yellow gas, with suffocating odor; purity ≥99.3%; typical impurities are Br_2, C_2Cl_6, C_6Cl_6, and H_2O.

Analysis of Reagent Purity: adduct:[2,3] ^{31}P NMR (various solvents and solid state) +47 to +66 (ionic form), −6.5 to +8 (covalent form); Raman (solid state) ν(P–Cl) 593 cm^{-1} (ionic form), 274 cm^{-1} (covalent form). Typical impurities, Ph_3P and Ph_3PO: ^{31}P NMR (various solvents) −5 to −9 and +25 to +42, respectively.

Preparative Methods: various preparations have been described;[5] the adduct is usually prepared just before use by addition of a stoichiometric amount of ***Chlorine*** to ***Triphenylphosphine*** in a dry solvent.[2]

Handling, Storage, and Precautions: adduct: exceedingly sensitive to moisture; incompatible with strong oxidizing agents and strong bases; may decompose on exposure to moist air or water; do not get in eyes, on skin, or on clothing; keep containers closed and store in a cool dry place; these reagents should be handled in a fume hood.

Alkyl Chlorides from Alcohols and Ethers. The reaction of Ph_3PCl_2 with alcohols provides an excellent synthetic method for the preparation of alkyl chlorides.[6] Mechanistic studies[6c] suggest the rapid initial formation of an alkyloxyphosphonium intermediate which then undergoes slow conversion into Ph_3PO and alkyl chloride (eq 1).[6b,c] It is assumed that chlorination takes place by an S_N2 reaction in most cases; thus, inversion of configuration is observed in the transformation of (−)-menthol to (+)-neomenthyl chloride (eq 2).[6a] As illustrated in eq 1, primary,[6b,7] secondary,[6] and even tertiary[6b] alcohols are chlorinated with Ph_3PCl_2, although reactions of tertiary alcohols are often accompanied by elimination (10%).

$$\text{ROH} \xrightarrow[\text{fast}]{\text{Ph}_3\text{PCl}_2} \text{RO}\overset{+}{\text{P}}\text{Ph}_3\ \text{Cl}^- \xrightarrow[\text{slow}]{\text{S}_\text{N}2} \underset{90\text{–}99\%}{\text{OPPh}_3 + \text{RCl}} \quad (1)$$

R = Bu, neopentyl, Cy, *t*-Bu; DMF, reflux

Ph$_3$PCl$_2$, PhH, reflux, 93% (2)

'Ph$_3$PCl$_2$', generated in situ from Ph$_3$P and ***Hexachloroacetone*** (HCA) has proven to be a very efficient reagent for the regio- and stereoselective chlorination of allylic alcohols (eq 3),[8] and for the regioselective conversion of sterically hindered cyclopropylcarbinyl alcohols into cyclopropylcarbinyl chlorides (eq 4).[9] Chlorination of allylic alcohols occurs in less than 20 min, with total preservation of the double bond geometry and with >99% inversion of configuration for optically active alcohols. Primary and secondary alcohols give predominantly the unrearranged chlorides, while tertiary alcohols provide mostly rearranged products, with elimination to dienes becoming an important side reaction with more highly substituted systems. Similarly, cyclopropylcarbinyl alcohols yield the corresponding chlorides with no trace of homoallylic chlorides or cyclobutane derivatives.

Ph$_3$PCl$_2$, HCA, 0 °C then rt, 90–100% → α-isomer + γ-isomer (3)

R^1, R^2, R^3, R^4 = H, Me α:γ = 100:0 to 82:18

Ph$_3$PCl$_2$, HCA, 20 °C, 79–93% (4)

R = H, Me, Et, Pr, Bu, *i*-Pr, *t*-Bu

In the carbohydrate field, primary and secondary alcohols are chlorinated in excellent yield (80–95%) with the Ph$_3$PCl$_2$/ImH (***Imidazole***) reagent system in PhMe, MeCN, or a MeCN/pyridine mixture at rt to reflux temperature.[10] Polymer-supported Ph$_3$PCl$_2$,[11] prepared by the Ph$_3$PO/COCl$_2$ (***Phosgene***) procedure,[5b] has been used to transform PhCH$_2$OH to PhCH$_2$Cl (88%) in MeCN as a solvent; the simple workup consists of filtration of polymeric phosphine oxide and solvent removal. Several examples have been reported of the direct conversion of ethers into chlorides, as in eq 5.[7] Enol ethers such as acetophenone trimethylsilyl ether give α-chlorostyrene (30%) by Ph$_3$PCl$_2$ treatment at CCl$_4$ reflux.[12]

Ph$_3$PCl$_2$, CCl$_4$, reflux, 91% → CH$_2$Cl, CH$_2$Cl (5)

Aryl Chlorides from Phenols and Arenes. Heating phenols with Ph$_3$PCl$_2$ at 120–140 °C leads to the corresponding aryl chlorides in good yield (eq 6).[13] A related chlorination reaction employs the Ph$_3$PCl$_2$/BSPO (***Bis(trimethylsilyl) Peroxide***) reagent system as the electrophilic chlorine source. With this reagent, in MeCN at rt, aromatic hydrocarbons bearing electron-donating substituents, such as 2,4,6-tri-*t*-butylbenzene and mesitylene, afford 1-monochloroarenes, while anisole gives a mixture of 2- and 4-chloro derivatives, in moderate to good yields (44–86%).[14a] Similar aromatic *para* chlorination has also been observed by heating anisole with Ph$_3\overset{+}{\text{P}}$Cl PCl$_5^-$, albeit in low yield (33%).[14b]

Ph$_3$PCl$_2$, CCl$_4$ then neat, 120–140 °C, 57–79% (6)

R^1 = H, NO$_2$, Me; R^2 = H; R^3 = H, Me, NO$_2$, Ph
R^1 + R^2, R^2 + R^3 = Ph

Vinyl Chlorides, Alkynyl Ketones, and β-Chloro-α-vinyl Ketones from Ketones and β-Diketones. 1-Chlorocyclohexene (45%) and α,α-dichlorotoluene (59%) are produced by the reaction of cyclohexanone and benzaldehyde with Ph$_3$PCl$_2$/***Triethylamine*** and Ph$_3$PCl$_2$ alone, respectively, in refluxing PhH.[6a] In MeCN as solvent, polymer-supported Ph$_3$PCl$_2$ converts acetophenone into α-chlorostyrene (75%).[11] In a similar fashion, unsymmetrical fluorinated β-diketones give 3:1 mixtures of α,β-ynones (eq 7) in good overall yields (slightly lower than those obtained with ***Triphenylphosphine Dibromide***).[15a] β-Chloro-α,β-unsaturated ketones are prepared in high yield from cyclic β-diketones (eq 8).[15b,c]

Ph$_3$PCl$_2$, Et$_3$N, CH$_2$Cl$_2$, rt, 84–85% (7)

R = CF$_3$, C$_3$F$_7$ 3:1

Ph$_3$PCl$_2$, Et$_3$N, PhH, CCl$_4$, rt, 90–97% (8)

R^1, R^2, R^3 = H, Me; *n* = 0, 1

Epoxide Cleavage to Vicinal Dichlorides and Chlorohydrins. Early reports of work in this area described the ring opening of ethylene oxide with Ph$_3$PCl$_2$ in CCl$_4$ at rt, leading to 1,2-dichloroethane.[16a] Subsequently,[16b,c] excellent yields were reported in the reaction of Ph$_3$PCl$_2$ with aliphatic epoxides to produce the corresponding vicinal dichlorides. The ring opening takes place stereospecifically with both *cis* and *trans* epoxides in PhH or CH$_2$Cl$_2$ at reflux, in each case providing the dichloride derived from S$_\text{N}$2 displacement on each C–O bond (eq 9).[16c] Alkoxyphosphonium chloride intermediates have even been isolated and characterized in a study involving ethylene oxide derivatives.[16d]

Ph$_3$PCl$_2$, PhH, 0 °C then reflux, 71–73% (9)

The reaction of epoxides with Ph$_3$PCl$_2$ in anhydrous CH$_2$Cl$_2$ at rt[17a] results in chlorohydrins in generally high yields (90–96%).

With conformationally rigid epoxides the oxirane ring cleavage appears to be quite stereoselective, leading only to the products resulting from the usual *anti* opening of the ring. Less hindered and rigid cyclic substrates provide regioisomeric mixtures of cyclic *trans*-chlorohydrins. The reaction of the polymer-supported Ph_3PCl_2 reagent proceeds in a similar fashion with even higher yields and easier workup; simple filtration and evaporation provides the product (eq 10).[17b]

$$\text{epoxide (R}^1\text{, R}^2\text{, R}^3\text{, H)} \xrightarrow[\text{CH}_2\text{Cl}_2\text{, rt; 96–98\%}]{\text{polymer-C}_6\text{H}_4\text{PPh}_2\text{Cl}_2} \text{R}^1\text{R}^2\text{C(Cl)–C(R}^3\text{)(H)OH} \quad (10)$$

Acid Chlorides from Acids and Esters. Mono- and dicarboxylic acids give acyl chlorides on reaction with Ph_3PCl_2[6a] or polymer-supported Ph_3PCl_2[11] in PhH, CH_2Cl_2, or MeCN (eq 11).[11] On similar Ph_3PCl_2 treatment in PhH at $-10\,°C$ to rt, sulfamic acid (H_2NSO_3H) is transformed into $Ph_3P{=}NSO_2Cl$ in 95% yield.[18] Analogously, Ph_3PCl_2, generated in situ from Ph_3P and $(EtO)_2P({=}O)SCl$, reacts with Et_3N and $(EtO)_2P(O)SH$ at $-78\,°C$ to provide the corresponding acid chloride $(EtO)_2P(S)Cl$.[19]

$$\text{R(CO}_2\text{H)}_n \xrightarrow[\text{CH}_2\text{Cl}_2\text{ or MeCN, rt or reflux; 87–100\%}]{\text{polymer-C}_6\text{H}_4\text{PPh}_2\text{Cl}_2} \text{R(COCl)}_n \quad (11)$$

R = $PhCH_2$, 4-MeC_6H_4 (n = 1); 1,4-C_6H_4, 1,3-C_6H_4 (n = 2)

Direct cleavage of esters or lactones[20] to both acid and alkyl chlorides is achieved with Ph_3PCl_2; halogenated esters (RCO_2Me; R = CF_3, CCl_3, CH_2Cl) are readily cleaved in refluxing MeCN, while esters of nonhalogenated acids and lactones (eq 12) require higher temperatures or/and the use of additives such as ***Boron Trifluoride***. Cleavage is considerably retarded by steric hindrance in the alkoxy fragment. A mechanism involving initial nucleophilic cleavage of the O–alkyl bond with Cl^- is proposed for halogenated esters, whereas an initial electrophilic attack by $Ph_3\overset{+}{P}Cl$ on the carbonyl oxygen is assumed for the cleavage of nonhalogenated esters.

$$\text{phthalide} \xrightarrow[\text{97\%}]{\text{Ph}_3\text{PCl}_2\text{, neat, 180 °C}} \text{2-(chloromethyl)benzoyl chloride} \quad (12)$$

Similar transformations of esters to acid chlorides have also been achieved in the phosphonate diester field[21] and in the conversion of trialkyl phosphites into dialkyl chlorophosphites.[22] Ph_3PCl_2 acts as a mild reagent for the replacement of a single ester linkage by a chloride in phosphonate diesters (eq 13).

$$\text{RP(O)(OMe)OMe} \xrightarrow[\text{73–97\%}]{\text{Ph}_3\text{PCl}_2\text{, CHCl}_3\text{, rt}} \text{RP(O)(OMe)Cl} \quad (13)$$

R = Me, Pr, allyl, MeOCO, *i*-PrOCO, PhOCO

Chlorination and Dehydration of Substituted Carboxamide Groups. Secondary amides and *N,N,N'*-trisubstituted ureas possessing an N–H bond are converted to imidoyl chlorides[11,23] and chloroformamidines,[5a,24] respectively, by reaction with Ph_3PCl_2 (eqs 14 and 15). Unsupported or polymer-supported reagent has been used, with or without Et_3N, in a variety of solvents. Under very similar conditions, with Ph_3PCl_2 in CH_2Cl_2 at reflux, the primary amide $PhCONH_2$ is dehydrated to the nitrile PhCN (78%),[11] while aryl-substituted arylhydroxamic acids are dehydrated to the corresponding aryl isocyanates.[25] In the latter case, dehydration occurs via Lossen-type rearrangement of a phosphorane intermediate. In a related reaction, chlorination–dehydrochlorination of *N*-acylated hydrazines with the Ph_3PCl_2/Et_3N system is a smooth one-pot procedure to generate nitrilimines.[26] Thus, by such treatment, PhCONHNHPh affords $[PhC{\equiv}\overset{+}{N}{-}\overset{-}{N}{-}Ph]$ in situ; this reacts with alkenic or alkynic dipolarophiles to give pyrazolines and pyrazoles.

$$\text{R}^1\text{CONHR}^2 \xrightarrow[\text{MeCN, rt or reflux; 35–93\%}]{\text{Ph}_3\text{PCl}_2\text{, Et}_3\text{N or polymeric reagent}} \text{R}^1\text{C(Cl)=NR}^2 \quad (14)$$

R^1 = Ph, Me; R^2 = Ph, Cy

$$\text{R}^1\text{NHCONR}^2\text{R}^3 \xrightarrow[\text{rt or reflux; 61–75\%}]{\text{Ph}_3\text{PCl}_2\text{, Et}_3\text{N, MeCN or PhH}} \text{R}^1\text{N=C(Cl)NR}^2\text{R}^3 \quad (15)$$

R^1 = Et, Pr, Ph; R^2 = Et, Bu, Ph; R^3 = Me, Et, Bu

Iminophosphoranes from Amines, Hydrazines, and Related Derivatives. Iminophosphoranes (or phosphinimines) are commonly used as intermediates, especially in heterocyclic synthesis. Phosphinimines can be obtained via phosphorylation of primary amines with Ph_3PCl_2 alone,[27] or in the presence of Et_3N,[28] if necessary, to ensure the last dehydrochlorination step. In the heterocyclic β-enamino ester field, iminophosphoranes are submitted to vinylogous alkylation[28a] or cycloaddition (eq 16)[28a,b] with ring enlargement. The $=PPh_3$ moiety, which serves as a temporary amino protecting group, is then cleaved hydrolytically. Reaction of phosphinimines with aryl isocyanates affords carbodiimides,[28c] whereas iminophosphoranes of acyl hydrazines undergo dimerization to tetrazines.[28d] Other phosphinimines and phosphonium salts have been prepared from phosphorylation of amino derivatives, such as hydrazines,[29a] urethanes,[29b] and *N*-silylated imines,[29c] with Ph_3PCl_2.

$$\text{N-Ts-2-amino-3-(CO}_2\text{Me)pyrroline} \xrightarrow[\text{MeCN, rt then reflux; 85\%}]{\text{Ph}_3\text{PCl}_2\text{, Et}_3\text{N}} \text{N-Ts-2-(N=PPh}_3\text{)-3-(CO}_2\text{Me)pyrroline} \xrightarrow[\text{2. H}^+\text{, EtOH; 85\%}]{\text{1. HC}\equiv\text{CCO}_2\text{Me}} \text{N-Ts-azepine (MeO}_2\text{C, CO}_2\text{Me, NH}_2\text{)} \quad (16)$$

Other Applications. Ph_3PCl_2 is a good condensation reagent for the synthesis of ketones (48–90%) from carboxylic acids and Grignard reagents. The versatility of the method is illustrated by the chemoselective reaction of carboxylic acids possessing such functional groups as halogen, cyano, and carbonyl (eq 17).[30]

$$\mathrm{EtC(O)(CH_2)_4CO_2Li} \xrightarrow[\substack{2.\ \mathrm{Ph(CH_2)_2MgBr} \\ \mathrm{THF},\ -30\,^{\circ}\mathrm{C} \\ 72\%}]{1.\ \mathrm{Ph_3PCl_2}} \mathrm{EtC(O)(CH_2)_4C(O)(CH_2)_2Ph} \quad (17)$$

Beckmann rearrangement of benzophenone oxime to PhC(Cl)=NPh (82%) is promoted by Ph_3PCl_2/Et_3N in CH_2Cl_2 at rt.[31a] Cyclopentanone and cyclohexanone oxime are converted to δ-valero- and ε-caprolactam (76 and 86%) with Ph_3PCl_2 in PhH at 50–60 °C.[31b] The action of heat (120–130 °C) on a mixture of Ph_3PCl_2 with the highly fluorinated propanol $(CF_3)_2C(OH)CH_2SO_2CF_3$ leads to elimination of the CF_3SO_2 group and formation of the vinylic chloride $(CF_3)_2C{=}CHCl$.[32] Ph_3PCl_2 reacts with $Pb(SCN)_2$ to form $Ph_3\overset{+}{P}{-}N{=}C{=}S\ SCN^-$, another reagent of the pseudohalophosphonium type, which is used for converting hydroxy groups into thiocyanate and isothiocyanate functions.[33] Ylides such as triphenylphosphonio(vinylsulfonylphenylsulfonylmethanide), $CH_2{=}CHSO_2\ \overset{-}{C}{-}(\overset{+}{P}Ph_3)SO_2Ph$, can be prepared by reaction of Ph_3PCl_2 with the very activated methylene derivative $CH_2{=}CHSO_2CH_2SO_2Ph$.[34] Heating Ph_3PCl_2 with $(Me_3Si)_2S$ at 60–70 °C leads to the formation of Ph_3PS (85%) after distillation of the Me_3SiCl byproduct.[35] Ph_3PCl_2 is reduced to Ph_3P with formation of alkyl or aryl chlorides by reaction with organometallic reagents (Mg, Li).[36]

---

1. Castro, B. R. *OR* **1983**, *29*, 1.
2. The Ph_3PCl_2 adduct can be prepared by different routes; the most common involves the reaction of Ph_3P and Cl_2, both being generally used in solution, in order to ensure a correct 1/1 stoichiometry. As illustrated by physical studies,[3] the resultant product is often a mixture of several compounds depending on starting materials (Ph_3P, Ph_3PO, Cl_2, $COCl_2$, CCl_3CCl_3, CCl_3COCCl_3), on their stoichiometries, and on the polarity of the solvent used for the preparation. ^{31}P NMR studies under various conditions, and solid state Raman measurements, lead to different values for ionic P^{IV} and molecular P^{V} species or to average values corresponding to equilibrated mixtures of the both structures according to the polarity of the solvent.
3. (a) Al-Juboori, M. A. H. A.; Gates, P. N.; Muir, A. S. *CC* **1991**, 1270. (b) Dillon, K. B.; Lynch, R. J.; Reeve, R. N.; Waddington, T. C. *JCS(D)* **1976**, 1243, and literature cited therein.
4. Another preparation from Ph_3P and CCl_3CCl_3 in MeCN provides material of higher mp: 207–210 °C (from MeCN–petroleum ether); see Appel, R.; Schöler, H. *CB* **1977**, *110*, 2382.
5. Apart from the classical $Ph_3P + Cl_2$ method, the following routes have been reported: (a) $Ph_3P + COCl_2$: Appel, R.; Ziehn, K. D.; Warning, K. *CB* **1973**, *106*, 2093, and literature cited therein. (b) $Ph_3PO + COCl_2$: Masaki, M.; Kakeya, N. *AG(E)* **1977**, *16*, 552, and literature cited therein. (c) $Ph_3P + CCl_4$ in the presence of RNHCOCl: Appel, R.; Warning, K.; Ziehn, K. D.; Gilak, A. *CB* **1974**, *107*, 2671. (d) With $Ph_3P + CCl_4$ alone, a mixture of equal amounts of Ph_3PCl_2 and $Ph_3P{=}CCl_2$ is obtained: Rabinowitz, R.; Marcus, R. *JACS* **1962**, *84*, 1312; Appel, R.; Knoll, F.; Michel, W.; Morbach, W.; Wihler, H.-D.; Veltmann, H. *CB* **1976**, *109*, 58. (e) $Ph_3P + CCl_3CCl_3$, e.g. Ref. 4 and: Appel, R.; Halstenberg, M. In *Organophosphorus Reagent in Organic Synthesis*, Cadogan, J. I. G., Ed.; Academic: New York, 1979; Chapter 9 and literature cited therein. (f) $Ph_3P + CCl_3COCCl_3$ for in situ preparation of Ph_3PCl_2: see Ref. 8 and 9. (g) Ph_3PO or $Ph_3P + PCl_5$: Dillon, K. B.; Reeve, R. N.; Waddington, T. C. *J. Inorg. Nucl. Chem.* **1976**, *38*, 1439, and literature cited therein.
6. (a) Horner, L.; Oediger, H.; Hoffmann, H. *LA* **1959**, *626*, 26. (b) Wiley, G. A.; Hershkowitz, R. L.; Rein, B. M.; Chung, B. C. *JACS* **1964**, *86*, 964. (c) Wiley, G. A.; Rein, B. M.; Hershkowitz, R. L. *TL* **1964**, 2509.
7. Skvarchenko, V. R.; Lapteva, V. L.; Gorbunova, M. A. *JOU* **1990**, *26*, 2244.
8. (a) Magid, R. M.; Fruchey, O. S.; Johnson, W. L. *TL* **1977**, 2999. (b) Magid, R. M.; Fruchey, O. S.; Johnson, W. L.; Allen, T. G. *JOC* **1979**, *44*, 359.
9. (a) Hrubiec, R. T.; Smith, M. B. *SC* **1983**, *13*, 593. (b) Hrubiec, R. T.; Smith, M. B. *JOC* **1984**, *49*, 431.
10. (a) Garegg, P. J.; Johansson, R.; Samuelsson, B. *S* **1984**, 168. (b) Garegg, P. J. *PAC* **1984**, *56*, 845.
11. Relles, H. M.; Schluenz, R. W. *JACS* **1974**, *96*, 6469.
12. Lazukina, L. A.; Kolodyazhnyi, O. I.; Pesotskaya, G. V.; Kukhar', V. P. *JGU* **1976**, *46*, 1931.
13. Hoffmann, H.; Horner, L.; Wippel, H. G.; Michael, D. *CB* **1962**, *95*, 523.
14. (a) Shibata, K.; Itoh, Y.; Tokitoh, N.; Okazaki, R.; Inamoto, N. *BCJ* **1991**, *64*, 3749. (b) Timokhin, B. V.; Dudnikova, V. N.; Kron, V. A.; Glukhikh, V. I. *JOU* **1979**, *15*, 337.
15. (a) Chechulin, P. I.; Filyakova, V. I.; Pashkevich, K. I. *BAU* **1989**, *38*, 189. (b) Piers, E.; Nagakura, I. *SC* **1975**, *5*, 193. (c) Piers, E.; Grierson, J. R.; Lau, C. K.; Nagakura, I. *CJC* **1982**, *60*, 210.
16. (a) Gloede, J.; Keitel, I.; Gross, H. *JPR* **1976**, *318*, 607. (b) Sonnet, P. E.; Oliver, J. E. *JOC* **1976**, *41*, 3279. (c) Oliver, J. E.; Sonnet, P. E. *OS* **1978**, *58*, 64. (d) Appel, R.; Gläsel, V. I. *ZN(B)* **1981**, *36*, 447.
17. (a) Palumbo, G.; Ferreri, C.; Caputo, R. *TL* **1983**, *24*, 1307. (b) Caputo, R.; Ferreri, C.; Noviello, S.; Palumbo, G. *S* **1986**, 499.
18. Arrington, D. E.; Norman, A. D. *Inorg. Synth.* **1992**, *29*, 27.
19. Krawczyk, E.; Mikolajczak, J.; Skowrońska, A.; Michalski, J. *JOC* **1992**, *57*, 4963.
20. (a) Burton, D. J.; Koppes, W. M. *CC* **1973**, 425. (b) Burton, D. J.; Koppes, W. M. *JOC* **1975**, *40*, 3026.
21. Ylagan, L.; Benjamin, A.; Gupta, A.; Engel, R. *SC* **1988**, *18*, 285.
22. Gazizov, M. B.; Zakharov, V. M.; Khairullin, R. A.; Moskva, V. V. *JGU* **1986**, *56*, 1471.
23. Appel, R.; Warning, K.; Ziehn, K.-D. *CB* **1973**, *106*, 3450.
24. Appel, R.; Warning, K.; Ziehn, K.-D. *CB* **1974**, *107*, 698.
25. von Hinrichs, E.; Ugi, I. *JCR(S)* **1978**, 338; *JCR(M)* **1978**, 3973.
26. Wamhoff, H.; Zahran, M. *S* **1987**, 876.
27. (a) Roesky, H. W.; Giere, H. H. *CB* **1969**, *102*, 2330. (b) Gotsmann, G.; Schwarzmann, M. *LA* **1969**, *729*, 106.
28. (a) Wamhoff, H.; Haffmanns, G.; Schmidt, H. *CB* **1983**, *116*, 1691. (b) Wamhoff, H.; Hendrikx, G. *CB* **1985**, *118*, 863. (c) Wamhoff, H.; Haffmanns, G. *CB* **1984**, *117*, 585. (d) Farkas, L.; Keuler, J.; Wamhoff, H. *CB* **1980**, *113*, 2566.
29. (a) Zhmurova, I. N.; Yurchenko, V. G.; Pinchuk, A. M. *JGU* **1983**, *53*, 1360. (b) Shevchenko, V. I.; Shtepanek, A. S.; Kirsanov, A. V. *JGU* **1962**, *32*, 2557. (c) Lazukina, L. A.; Kristhal', V. S.; Sinitsa, A. D.; Kukhar', V. P. *JGU* **1980**, *50*, 1761.
30. Fujisawa, T.; Iida, S.; Uehara, H.; Sato, T. *CL* **1983**, 1267.
31. (a) Appel, R.; Warning, K. *CB* **1975**, *108*, 1437. (b) Sakai, I.; Kawabe, N.; Ohno, M. *BCJ* **1979**, *52*, 3381.
32. Samusenko, Y. V.; Aleksandrov, A. M.; Yagupol'skii, L. M. *JOU* **1975**, *11*, 622.
33. Burski, J.; Kieszkowski, J.; Michalski, J.; Pakulski, M.; Skowrońska, A. *CC* **1978**, 940.
34. Diefenbach, H.; Ringsdorf, H.; Wilhelms, R. E. *CB* **1970**, *103*, 183.
35. Markovskii, L. N.; Dubinina, T. N.; Levchenko, E. S.; Kukhar', V. P.; Kirsanov, A. V. *JOU* **1972**, *8*, 1869.
36. (a) Denney, D. B.; Gross, F. J. *JOC* **1967**, *32*, 3710. (b) Dmitriev, V. I.; Timokhin, B. V.; Kalabina, A. V. *JGU* **1979**, *49*, 1936.

Jean-Robert Dormoy & Bertrand Castro
SANOFI Chimie, Gentilly, France

Triphenylphosphine–Diethyl Azodicarboxylate[1]

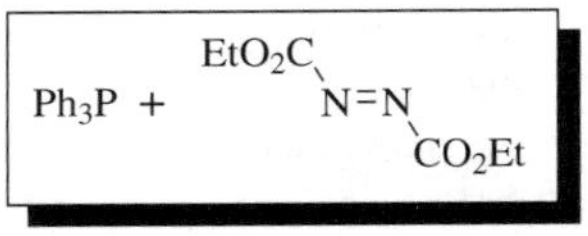

$$Ph_3P + EtO_2C{-}N{=}N{-}CO_2Et$$

(Ph_3P)
[603-35-0] $C_{18}H_{15}P$ (MW 262.30)
(DEAD)
[1972-28-7] $C_6H_{10}N_2O_4$ (MW 174.18)

(reagent in the Mitsunobu reaction, which is a versatile, mild dehydration reaction, widely used for the synthesis of esters and ethers, for the formation of C–N, C–S, and C–halogen bonds, and for inverting the configuration of a stereogenic carbon containing an OH group[1])

Physical Data: see ***Triphenylphosphine*** and ***Diethyl Azodicarboxylate***.

Handling, Storage, and Precautions: all reagents and solvents must be anhydrous. In general, the Mitsunobu betaine (**1**) is generated in situ from the phosphine and the azodicarboxylate; however, in some cases it is essential to preform the betaine. The betaine is an unstable, colorless, crystalline solid, rapidly hydrolyzed on contact with moisture. It can be crystallized from dry THF or $CHCl_3$/hexane.

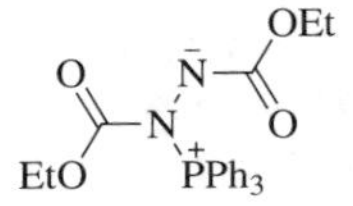

(**1**)

The Mitsunobu Reaction. A mixture of triphenylphosphine (TPP) and diethyl azodicarboxylate (DEAD) is generally used; however, diisopropyl azodicarboxylate (DIAD) is cheaper and works just as well. The overall reaction enables the replacement of the hydroxyl group of an alcohol by a nucleophile X (eq 1).

$$ROH + HX \xrightarrow[\text{DIAD}]{\text{TPP}} RX \quad (1)$$

The phosphine accepts the oxygen (to give triphenylphosphine oxide) while the azodicarboxylate accepts the two hydrogens (to give the corresponding hydrazine derivative). The reaction occurs under mild (0–25 °C), essentially neutral conditions, can be carried out in the presence of a wide range of functional groups, and usually gives good yields (60–90%). In general, the reaction proceeds well with primary and secondary alcohols, and inversion of configuration is observed.[1] Retention of configuration may result from neighboring group participation,[2] or a change in mechanism if the alcohol is very hindered.[3a] Racemization may result if the C–O bond is prone to S_N1 cleavage.[4] Examples of H–X in the reaction (eq 1) include carboxylic acids, thioacids, phenols, thiols, imides, sulfonamides, hydrazoic acid, heterocyclic compounds, hydrogen halides, phosphate diesters, phosphinic acids, and certain active methylene compounds. The pK_a of H–X should generally be less than 13. LiX or ZnX_2 sometimes give better results than HX.[1c] THF is the most commonly used solvent, but other solvents such as dichloromethane, chloroform, benzene, toluene, ethyl acetate, DMF, and HMPA can also be used. The reaction is faster in nonpolar solvents; for example, formation of ethyl benzoate is about 30 times faster in $CHCl_3$ than in MeCN.[5a] A problem commonly encountered with the Mitsunobu reaction is the separation of the product from triphenylphosphine oxide. This can be overcome by the use of a phosphine containing a basic group such as diphenyl(2-pyridyl)phosphine[6] or (4-dimethylaminophenyl)diphenylphosphine,[7] where an aqueous acid wash removes the corresponding phosphine oxide, or by using a polymer-bound phosphine.[8] Similarly, dimethyl azodicarboxylate (the corresponding hydrazine is water soluble)[9] and a polymer-supported azodicarboxylate[10] can be used to facilitate product isolation. The reaction is best explained by assuming successive formation of the betaine (**1**), the protonated *N*-phosphonium salt (**2**), and the alkoxyphosphonium salt (**3**), which collapses in S_N2 fashion (eq 2).

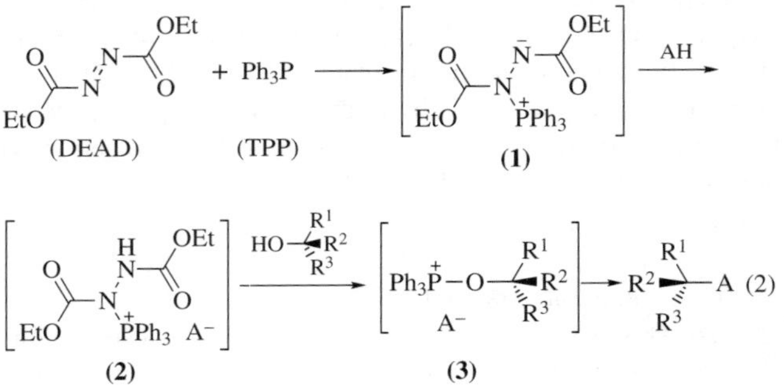

(DEAD) + Ph_3P (TPP) → [(**1**)] —AH→ [(**2**) A^-] —$HO{-}CR^1R^2R^3$→ [$Ph_3\overset{+}{P}{-}O{-}CR^1R^2R^3$ A^- (**3**)] → $R^2R^1R^3C{-}A$ (2)

Without acidic components, *N*-alkylhydrazinedicarboxylates (**4**) can be formed.[5b,11] NMR studies of the reaction of alcohols with DEAD and TPP in the absence of acidic components have revealed that the key intermediate is the pentavalent dialkoxyphosphorane (**5**).[1c,5] In the presence of an acidic component, phosphorane intermediates are present in equilibrium with the oxyphosphonium salt (**3**).[3]

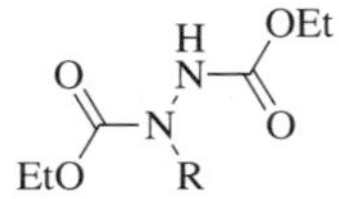

(**4**) 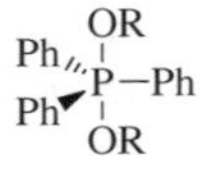 (**5**)

The combination of DEAD and TPP has been utilized most often as the condensation reaction system, but dimethyl, diisopropyl, and di-*t*-butyl azodicarboxylates (**6**) can also be used instead of DEAD; the azodicarboxamide (**7**) has been used as well.[1,12] Triphenylphosphine can be substituted by a variety of trivalent phosphorus compounds, including substituted triarylphosphines and trialkylphosphines.[1c]

Order of Addition of the Reagents. The order of addition can be crucial. The most commonly used procedure is to dissolve the alcohol (ROH), the acid (HX), and triphenylphosphine in THF, cool to 0 °C, add the DEAD (or DIAD) slowly with stirring, and then stir at room temperature for several hours. If this procedure fails to give the desired product, the betaine (TPP–DEAD adduct) should be preformed (addition of DEAD to TPP in THF at 0 °C) before addition of a mixture of the ROH and the HX,[13] or addition of ROH followed by slow addition of HX or HX and a buffering salt (strong acids),[14,15] or addition of HX followed by ROH (weak acids).[16] The advantage of preforming the betaine is that X^- is not generated in the presence of a powerful oxidant (the azodicarboxylate). Preformation of the betaine should minimize radical-induced side-reactions (radicals are produced when TPP and DEAD are mixed[17]).

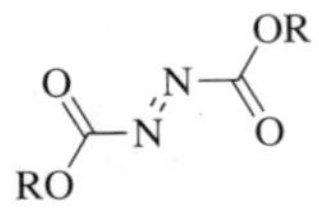

(6)
R = Me: DMAD
R = Et: DEAD
R = *i*-Pr: DIAD
R = *t*-Bu: DBAD

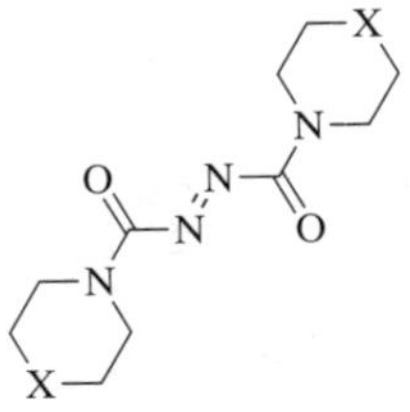

(7)
X = CH_2: ADDP
X = O, NMe

Ester Formation. A variety of alcohols react at room temperature with carboxylic acids in the presence of DEAD and TPP to produce the corresponding esters.[1] When polyols are used, the reaction generally takes place at the less hindered hydroxyl group, as exemplified in the reaction of 1,3-butanediol (eq 3).[18,19] When 1,2-propanediol or styrene glycol is used, however, the more sterically encumbered C-2 benzoate is predominantly obtained with complete inversion of the stereochemistry.[19] This result has been explained by the formation of a dioxaphospholane (**8**) as the key intermediate (eq 4).[3b,19,20a] With acyclic 1,4-diols such as isomaltitol,[20b] five-membered cyclic ether formation takes place in preference to esterification.

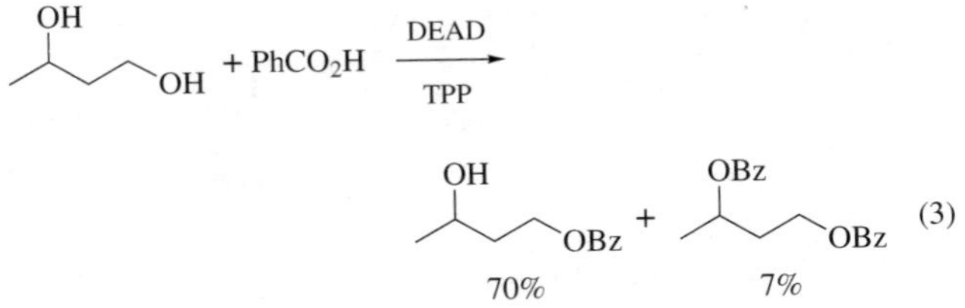

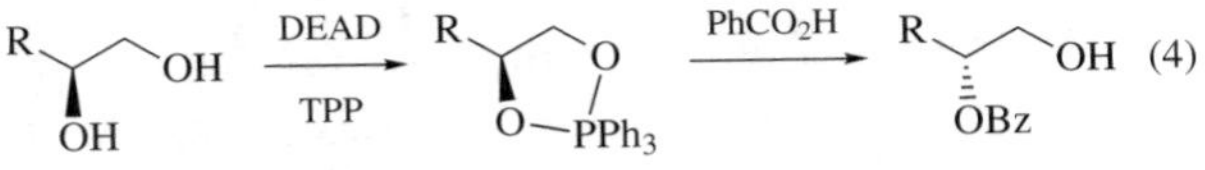

+ $PhCO_2H$ —DEAD, TPP→ (3)
70% 7%

—DEAD, TPP→ (**8**) —$PhCO_2H$→ (4)

Thymidine (**9**) reacts with aromatic carboxylic acids to give 5′-*O*-aroylthymidines (**10**) (eq 5). The yield of esters increases with the acidity of the carboxylic acid, indicating the importance of the protonated *N*-phosphonium salt (**2**).[21,22]

The reaction of chiral secondary carbinols with carboxylic acids, DEAD, and TPP gives the corresponding esters with inverted configuration. Removal of the carboxyl group affords the enantiomer of the parent alcohol (eqs 6–9).[23–27] For the purpose of inverting the stereochemistry of a secondary carbinol center, 3,5-dinitrobenzoic acid,[28] *p*-nitrobenzoic acid,[22b] and chloroacetic acid[29] have been recommended, because of increased yield and ease of purification and removal of the carboxyl group from the resulting esters.

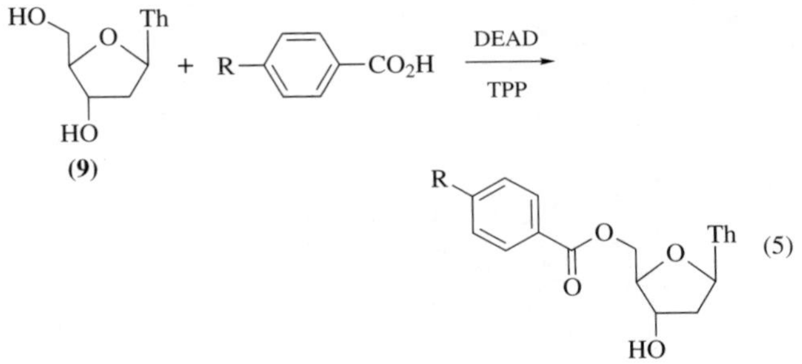

(**9**) + R–C_6H_4–CO_2H —DEAD, TPP→ (**10**) (5)

R = Me, 65%; OMe, 66%; H, 74%; NC, 80%; O_2N, 85%

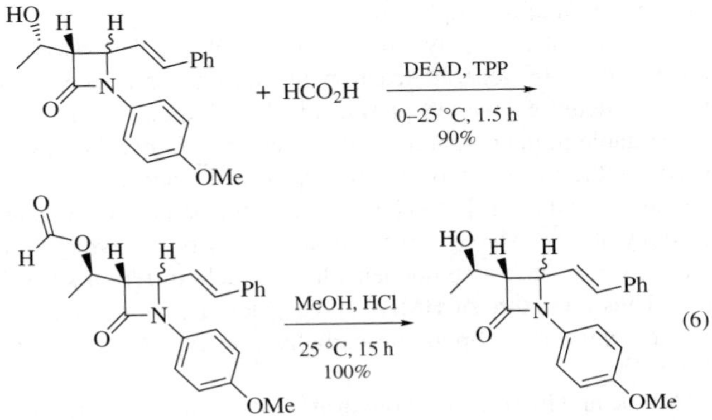

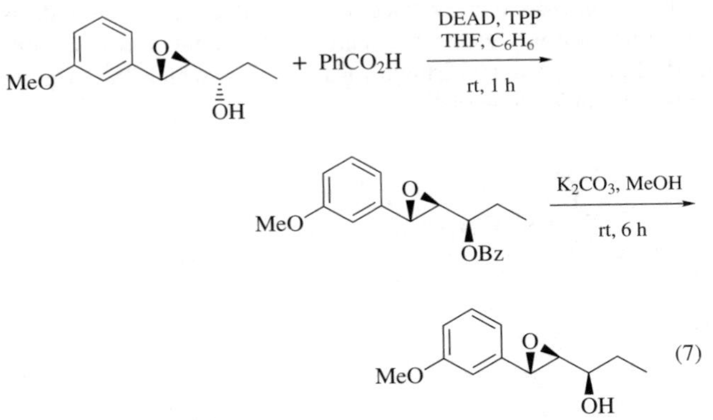

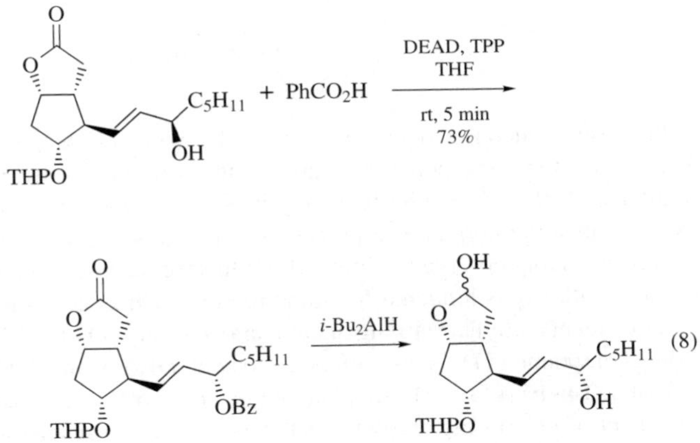

+ HCO_2H —DEAD, TPP, 0–25 °C, 1.5 h, 90%→ —MeOH, HCl, 25 °C, 15 h, 100%→ (6)

+ $PhCO_2H$ —DEAD, TPP, THF, C_6H_6, rt, 1 h→ —K_2CO_3, MeOH, rt, 6 h→ (7)

+ $PhCO_2H$ —DEAD, TPP, THF, rt, 5 min, 73%→ —*i*-Bu_2AlH→ (8)

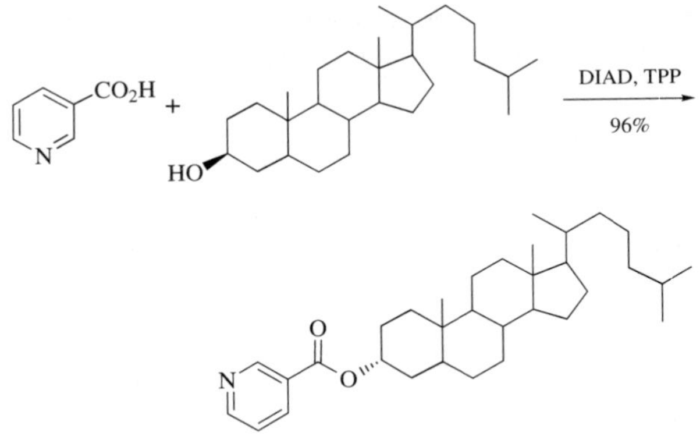

DIAD, TPP
96%

(9)

With some exceptions, a primary hydroxyl group is generally more reactive than a secondary one.[22a,30] Steric congestion and electronic effects sometimes retard the reaction and/or result in the formation of a complex mixture.[31]

The solvent also plays an important role in the success of the reaction. In general, reaction in benzene or toluene gives higher yields of inverted products.[1,22b,32] Although pyridine is not suitable in the preparation of nucleotides,[33a] pyridine can be used for the synthesis of sucrose epoxide,[33b] and a mixture of dioxane–pyridine (9:1) can be utilized in the preparation of sugar carboxylates.[22a] Mixed solvent systems may be necessary when the acid and alcohol components have widely differing solubilities. Thus a mixture of HMPA and dichloromethane works well in the synthesis of lipophilic carbohydrate esters such as cord factor.[33c]

Allylic and benzylic cyanohydrins react with carboxylic acids, DEAD, and TPP to give inverted esters with 92–99% ee, while extensive racemization takes place with benzaldehyde cyanohydrins containing strongly electron-donating substituents (eq 10). Saturated aliphatic cyanohydrins afford esters in which the original configuration is retained.[34]

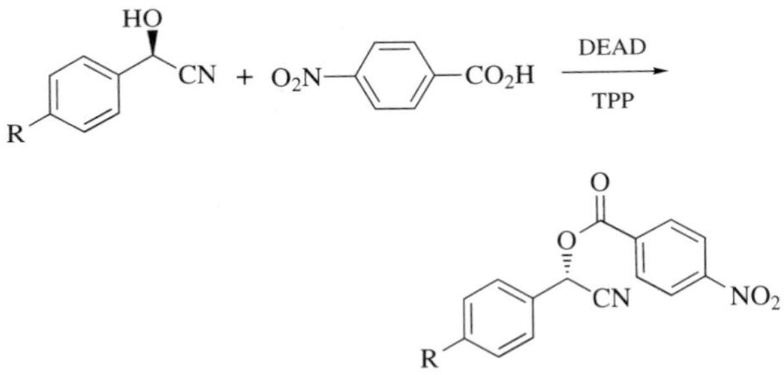

DEAD
TPP

(10)

R = H, 76% (96% ee)
R = MeO, 80% (6% ee)

Propargylic alcohols react with carboxylic acids, DEAD, and TPP to give the corresponding inverted esters without formation of allenic alcohols.[35] Similarly, in allylic alcohols where no bias exists against the normal S_N2 process, clean regiospecific inversion is invariably observed or implied.[1,36] In specific cases where the S_N2 pathway is hindered for some reason, such as eclipsing of substituents, allylic rearrangement can occur (eqs 11–13).[1,37] When compound (**11**) reacts with benzoic acid, DEAD, and TPP, (**13**) and (**14a,b**) (**a/b** = 4:1) are obtained in a ratio of 1:1. Similar treatment of (**11**) in the presence of $PdCl_2$ (0.1 equiv) results in the nearly exclusive production of (**14a,b**) (73%; **a/b** = 10:1) via (**12**).[37d]

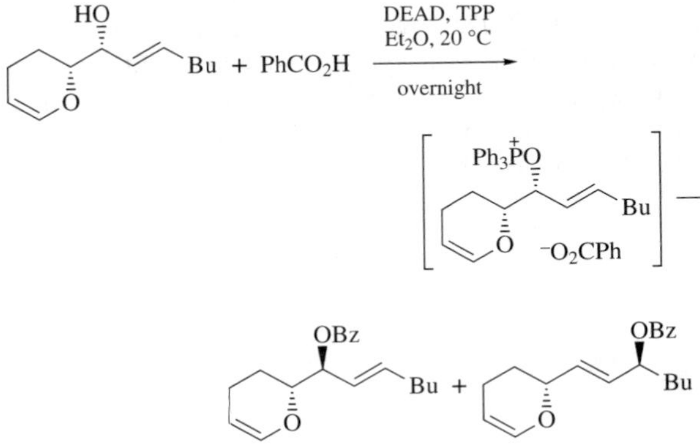

DEAD, TPP
Et_2O, 20 °C
overnight

(11)

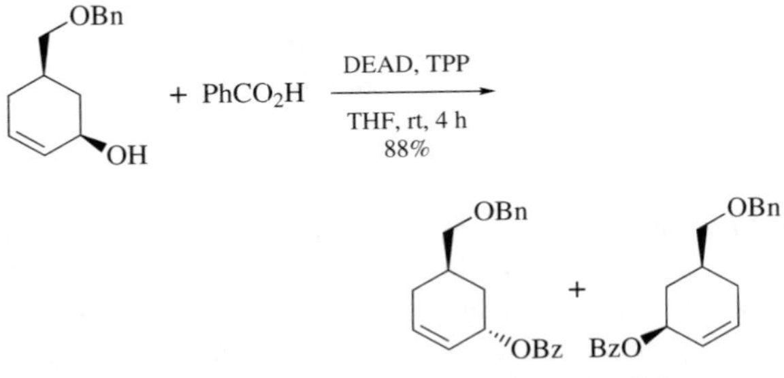

DEAD, TPP
THF, rt, 4 h
88%

(12)

trans:*cis* = 9:1

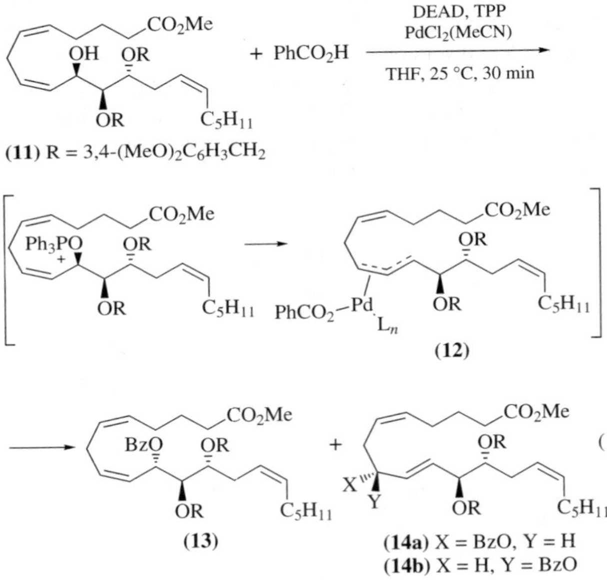

DEAD, TPP
$PdCl_2(MeCN)$
THF, 25 °C, 30 min

(**11**) R = $3,4\text{-}(MeO)_2C_6H_3CH_2$

(**12**)

(13)

(**13**)

(**14a**) X = BzO, Y = H
(**14b**) X = H, Y = BzO

Reaction with Thiocarboxylic Acids, Phosphoric Acids, Sulfonic Acids, and their Derivatives.[1] Thiocarboxylic acids,[38,13] dithiocarboxylic acids,[39] and dimethyldithiocarbamic acid zinc salt,[40] as well as various phosphorus oxyacids[41] and phosphorus thioacids,[42] can also be utilized. α,ω-Mercapto alcohols form cyclic thioethers whereas thiols react with both DEAD and TPP–DEAD to form disulfides.[43a] 2-Mercaptoazoles also react with alcohols in the presence of DEAD and TPP.[43b] Although arenesulfonic acids do not enter into the reaction, a combination of DEAD–TPP with methyl *p*-toluenesulfonate as a nucleophile carrier gives the corresponding alkyl sulfonates (eq 14).[44] Alternatively, ***Zinc p-Toluenesulfonate*** reacts with a variety of secondary alcohols to produce the desired tosylates with clean inversion and in good yields, with some exceptions (eq 15).[45]

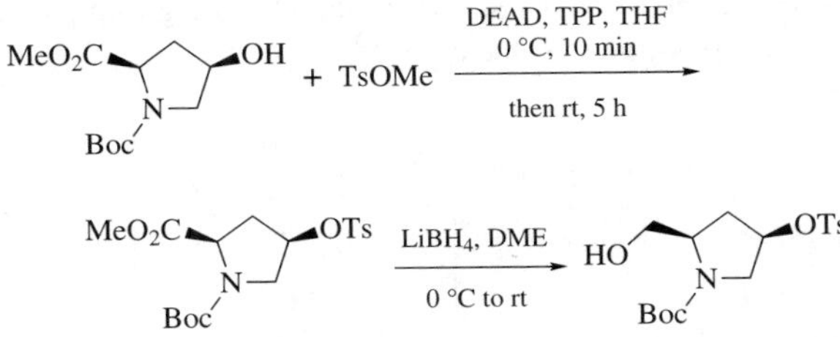

(14)

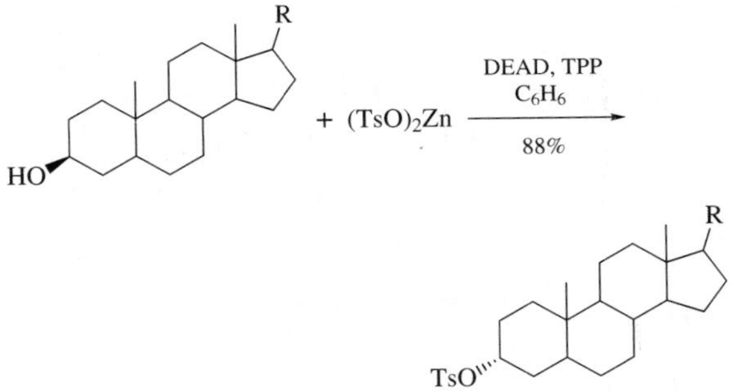

(15)

Reaction with Hydroxy Acids. A study of reactions of *threo*-3-hydroxycarboxylic acids (**15**) with DEAD and TPP revealed that both hydroxyl group activation (HGA) and carboxyl group activation processes (CGA) are involved. With small R^1 and R^2, the zwitterion (**16**) is more stable than (**17**), so that HGA is exclusively observed, resulting in the formation of alkenes. On the other hand, the CGA process via (**17**) is progressively preferred as the size of R^1 and R^2 increases (eq 16).[46a,b] When the reaction is carried out in acetonitrile, CGA predominates.[46c]

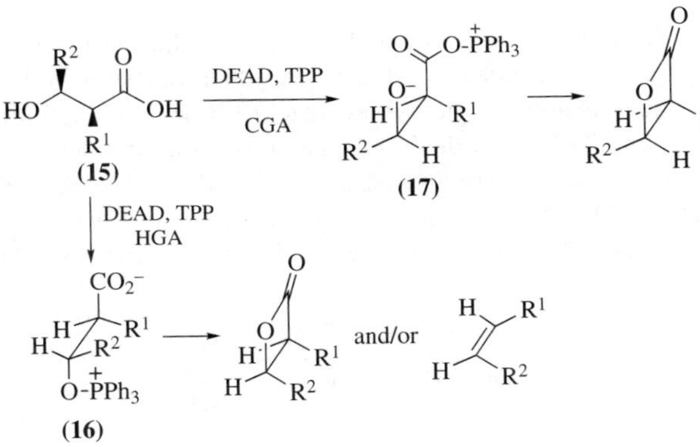

(16)

Hydroxy acids $HO–(CH_2)_n–CO_2H$ with $n \geq 3$ react with DEAD and TPP to afford the corresponding lactones.[1] This procedure can be utilized in the preparation of macrolide antibiotics and related compounds. Macrolactonization by the use of DEAD–TPP depends on the reaction conditions and the structure of the seco acids. Thus dropwise addition of the hydroxy acid (**18**) over a period of 10 h to a mixture of DEAD (7.7 equiv) and TPP (7.5 equiv) affords the lactone (**19**) in 59% yield as well as the unwanted dilactide (**20**) in a yield of <1%. On the other hand, the reaction of (**18**) with DEAD (5.0 equiv) and TPP (5.0 equiv) at 25 °C for 18 h gives (**19**) and (**20**) in 2% and 40% yields, respectively (eq 17).[47]

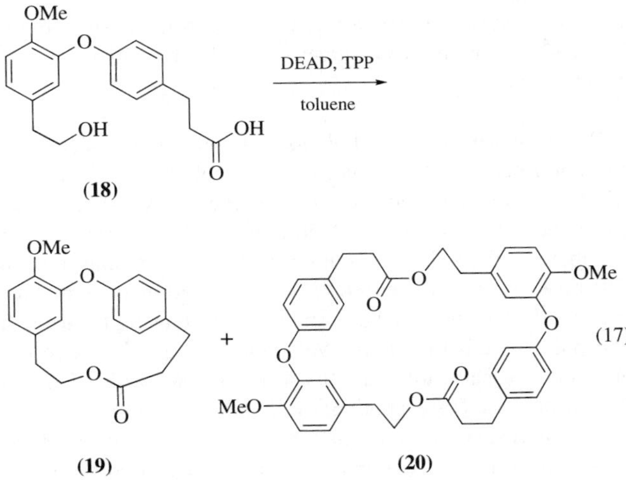

(17)

The reaction of the seco acid (**21**) of colletodiol with DEAD and TPP gives the lactone (**22**) in 45% yield after recrystallization (eq 18),[48a] while the seco acid (**23**), which has a closely related structure to (**21**), affords the corresponding lactone (**24**) in only 4% yield (eq 19).[48b,49]

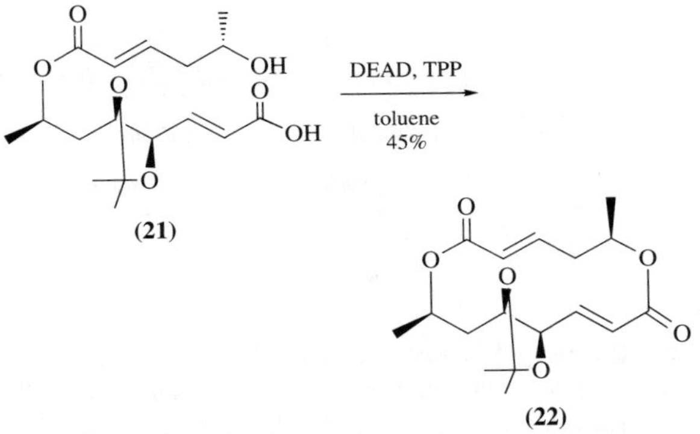

(18)

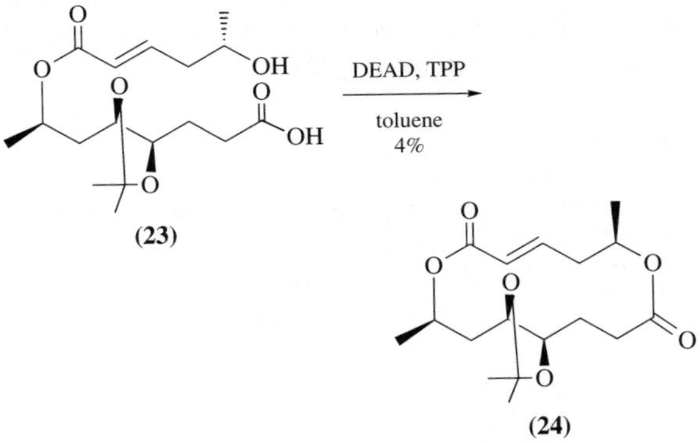

(19)

Reaction with Phenols and Other Oxygen Nucleophiles. When alcohols react with phenols, DEAD, and TPP, the corresponding aryl alkyl ethers are produced. A tertiary amine may facilitate the reaction.[50a] In general, the reaction proceeds with clean inversion of chiral secondary carbinol centers (eq 2; AH = a phenol).[50b] Depending on the structure of the substrate, allylic rearrangement[51,52] and neighboring group participation can be observed.[2,53] Phenols having a hydroxyalkyl chain in a suitable position for cyclization afford the corresponding cyclic aryl ether.[54] Silanols react with phenols and with alcohols in the presence of DEAD and TPP to afford the corresponding silyl ethers.[55] Oximes[56] and *N*-hydroxy imides[57] can also be utilized as acidic components.

Reaction with Di- and Polyols. Although intermolecular dehydration between two molecules of alcohols to afford acyclic ethers usually does not occur with the DEAD–TPP system, intramolecular cyclization of diols to produce three to seven-membered ethers is a common and high yielding reaction. Contrary to an early report,[58a] 1,3-propanediol does not form oxetane.[5b] Oxetanes can be formed, however, using the trimethyl phosphite modification of the Mitsunobu reaction.[1c] The reaction of (*S*)-1,2-propanediol and (*R*)-1,4-pentanediol with DEAD and TPP affords the corresponding cyclic ethers with 80–87% retention of stereochemistry at the chiral carbon, while (*S*)-phenyl-1,2-ethanediol affords racemic styrene oxide. In contrast to the reaction of the same 1,2-diols with benzoic acid (eq 4), oxyphosphonium salts (**25a**) and (**25b**) have been postulated as key intermediates in the present reaction (eq 20).[58b,e]

DEAD, TPP; (**25a**) retention; (**25b**) inversion (20)

The reaction of cyclic *trans*-1,2-diols with DEAD and TPP generally gives the corresponding oxiranes via phosphoranes.[1,33b,59] Cyclic *cis*-1,2-diols also react with DEAD/TPP to afford primarily phosphoranes (**26**) which decompose into final product(s) such as oxiranes (with retention of configuration) (eq 21)[60] and dehydrated alcohols and/or ketones (eq 22).[61]

DEAD, TPP; DMF, 12 h; (**26a**) (21)

Carbon–Nitrogen Bond Formation. ***Phthalimide*** reacts with alcohols, DEAD, and TPP to give the corresponding *N*-alkylphthalimides with inversion of configuration (eq 2; HA = phthalimide).[1] Nitrophthalimide is more reactive than phthalimide.[22a] If the reaction site is hindered and an S_N1 process favored, some stereochemical scrambling is observed. For example, the reaction of (**27**) with phthalimide in the presence of DEAD and TPP results in complete epimerization, giving (**28a**) and (**28b**) in a ratio of 1:1 (eq 23).[62] Neighboring group participation is also possible in some cases.[63]

Cholestane-5α,6α-diol; DEAD, TPP; benzene, 80 °C; (**26b**); 24%; 28%; 4% (22)

(**27**); PhtNH, DEAD; TPP, THF; rt, 12 h; (**28a**) 32%; (**28b**) 32% (23)

PhtNH =

Alcohols having an acidic hydrogen atom adjacent to the hydroxyl group may undergo dehydration rather than substitution.[1] Thus the reaction of diethyl maleate (**29**) with phthalimide, DEAD, and TPP gives diethyl fumarate without any detectable formation of the substituted product.[64a] With ***Hydrazoic Acid***, however, (**29**) gives the expected azido succinate (**30**) in 74% yield (eq 24).[44a] Dehydration is regioselective in some cases.[64b,c]

(**29**); PhtNH, DEAD; TPP; HN_3, DEAD; TPP; (**30**) (24)

Since azides are very useful in organic synthesis (especially for the introduction of primary amino groups and the construction of heterocyclic systems), the direct conversion of alcohols into azides has been extensively studied (eq 2; HA = HN_3).[1,65] ***Diphenyl Phosphorazidate***[66] and the zinc azide/bis-pyridine complex[67] can be used instead of HN_3. Treatment of alcohols with hydrazoic acid, DIAD, and excess TPP in THF, followed by addition of water or aqueous acid, affords amines in moderate to good yields (eq 25).[68]

$$\text{ROH} + \text{HN}_3 \text{ (in benzene)} \xrightarrow[\text{then 50 °C, 3 h}]{\text{DIAD, TPP, THF, rt, 1 h}} \text{RN=PPh}_3 \xrightarrow[\text{1M HCl}]{\text{H}_2\text{O or}} \text{RNH}_2 \text{ or } \text{R}\overset{+}{\text{N}}\text{H}_3\text{Cl}^- \quad (25)$$

35–85%

Acyclic amides and imides with $pK_a < 13$ can be expected to function as acidic components in the present reaction system. Thus the reaction of *N*-benzyloxycarbonylbenzamide with ethyl (*S*)-(−)-lactate gives the inverted *N*-alkylated product (**31a**) and the *O*-alkylated product (**31b**) in 19% and 48% yields, respectively (eq 26).[69] This result suggests that the alkoxyphosphonium salt (**3**) is a hard alkylating reagent.

PhC(O)NHCO₂Bn + HOCH(CH₃)CO₂Et —DEAD, TPP→ (**31a**) 19% + (**31b**) 48% (26)

(**31a**) —HBr, AcOH→ 97%; (**31b**) —HBr, AcOH→ PhC(O)NH₂

In the alkylation of imidodicarbonates and tosylcarbamates, the yield of alkylated product increases as the acidity of the NH acids increases. Thus imidodicarbonates and tosylcarbamates with $pK_a < 13$ give the corresponding *N*-alkylated products in 83–93% yields, while lower yields are obtained with less acidic substrates.[70a] The combination of ***1,1′-(Azodicarbonyl)dipiperidine*** (ADDP) with ***Tri-n-butylphosphine*** appears to be superior to that of DEAD with TPP in the alkylation of amides.[12] In the presence of DEAD and TPP, dibenzyl imidodicarbonate does not react with alcohols, but it does react with α-hydroxy stannanes to give the corresponding *N*-alkylated products (eq 27).[71]

EtC(O)SnBu₃ —1. (*S*)-BINAL-H; 2. (BnO₂C)₂NH, DEAD, TPP→ (BnO₂C)₂N–CH(Et)SnBu₃ (27)

The *O*-acylhydroxamate (**32a**) (pK_a 6–7) reacts with benzyl alcohol, DEAD, and TPP to give *N*- and *O*-alkylated products, while the less acidic *O*-alkylhydroxamate (**32b**) (pK_a 9–10) gives only the *N*-alkyl product (eq 28).[72a]

PhC(O)NHOR + PhCH₂OH —DEAD, TPP→ PhC(O)N(Bn)OR + PhC(OBn)=NOR (28)

	Yield	Ratio
(**32a**) R = PhCO	82%	65:35
(**32b**) R = Bn	91%	100:0

Intramolecular dehydration of 3-hydroxy carboxamides affords the corresponding β-lactams. Side reactions include elimination and formation of aziridines and oxazolidines (eq 29).[72,73] The efficiency of β-lactam formation is dependent on substrate substituents (including protecting groups for the side-chain amino group) as well as on the choice of azodicarboxylate and P^{III} compound.[1c,74–76] For example, the dipeptide (**33**) reacts with DEAD and TPP to give a 2:1 mixture of the β-lactams (**34a**) and (**34b**). Control of the labile C-5 stereocenter can be achieved by use of ***Triethyl Phosphite*** instead of TPP; (**34a**) and (**34b**) are obtained in a >50:1 ratio (eq 30).[75] 3-Hydroxy *O*-alkylhydroxamates can also be converted into the corresponding β-lactams (eq 31).[1c,77]

(29)

(**33**) → (**34a**) + (**34b**) (30)

	(**34a**):(**34b**)
DEAD–TPP	2:1
DEAD–(EtO)₃P	50:1

—DIAD, TPP, THF→ (31)

Ox = 4,5-diphenyl-2-oxo-oxazol-3-yl

Purines and pyrimidines react with various alcohols in the presence of DEAD and TPP (eq 32).[78] The alcohol (**35**) gives a carbocyclic nucleoside (**36**) in 65% yield along with a small amount (9%) of the 7-substituted purine derivative (eq 33).[79–81]

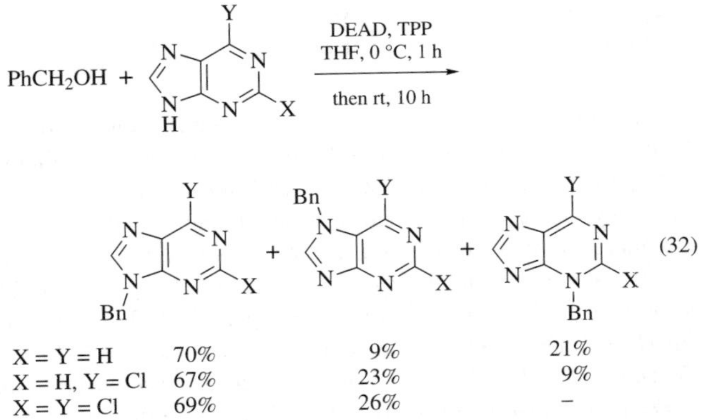

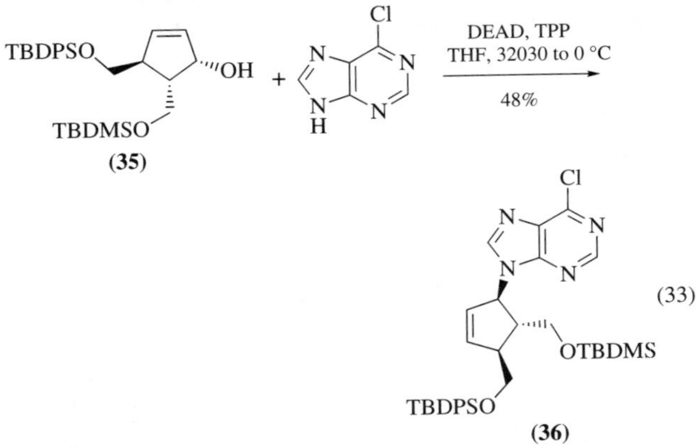

(32)

X = Y = H	70%	9%	21%
X = H, Y = Cl	67%	23%	9%
X = Y = Cl	69%	26%	–

(33)

Reaction with Amino Alcohols and Related Compounds. Both secondary and tertiary carbamates bearing a vicinal hydroxyl group react with carboxylic acids in the presence of DEAD and TPP to afford the corresponding esters with inversion of configuration (eq 34).[82] Reaction of *N*-*t*-butoxycarbonylserine with methanol, DEAD, and TPP gives the corresponding methyl ester in >98% yield (eq 35).[83] Without methanol, *N*-benzyloxycarbonyl- or *N*-*t*-butoxycarbonylserine reacts with the preformed adduct of DEAD and TPP to afford the β-lactone (eq 35).[84] The *N*-protected serine methyl ester (**37a**) and threonine methyl ester (**37b**) react with DEAD and TPP to give the dehydrated products (**38a**) and (**38b**) (eq 36).[85]

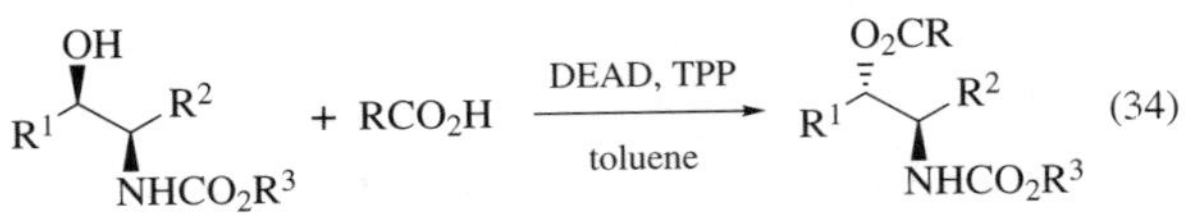

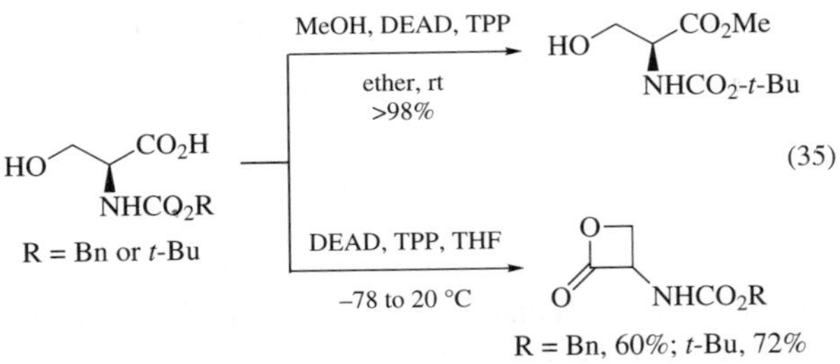

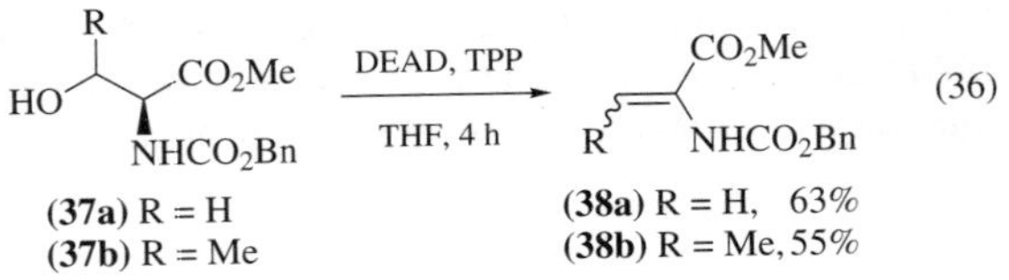

(34)

(35)

R = Bn, 60%; *t*-Bu, 72%

(36)

(**37a**) R = H
(**37b**) R = Me

(**38a**) R = H, 63%
(**38b**) R = Me, 55%

In the absence of an acidic component, β-(*N*-carbonylamino) alcohols afford aziridines or oxazolines, depending on the structure of the substrate. Contrary to the case of (**37**), the *N*-benzyloxycarbonylamino alcohol (**39**), with no acidic hydrogen present at the position adjacent to the hydroxyl group, affords the aziridine (**40**) in 85% yield (eq 37),[86,87] while the *N*-acylamino alcohols (**41**) give the oxazolines (**42**) in 31–68% yields (eq 38).[88]

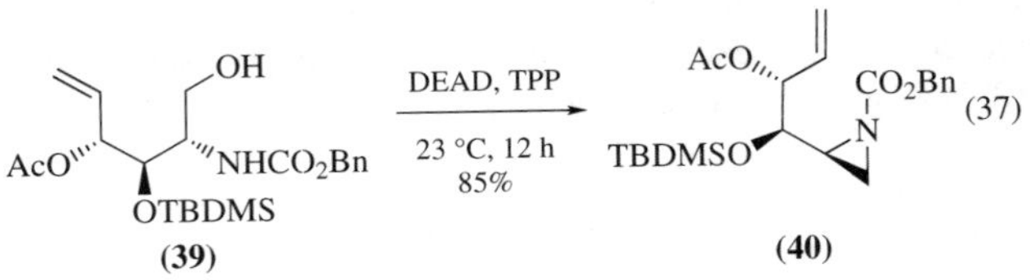

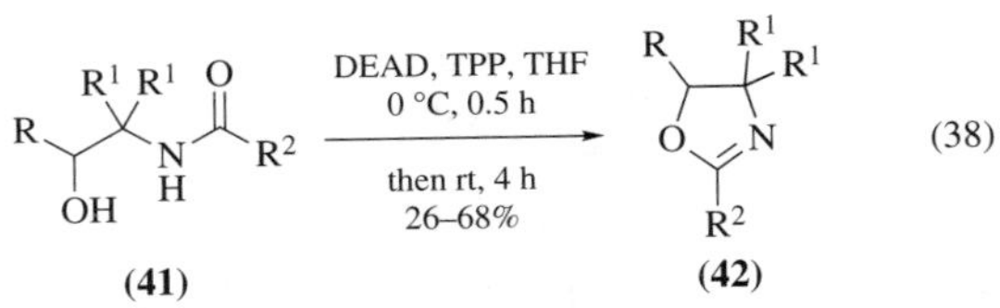

(37)

(38)

Protection of the amino group by an electron-withdrawing group is not necessarily required for the cyclization. Amino alcohols of general structure HO–(C)$_n$–NHR react with DEAD and TPP to form the corresponding cyclic amines. Thus 2-aminoethanols give aziridines in 18–90% yields.[89] The reaction of the tetrafluoroborate salt of 3-(benzylamino)-1-propanol with DEAD and TPP gives the corresponding azetidine in 50% yield (eq 39).[90] Five- and six-membered cyclic amines are also prepared by the reaction of amino alcohols with DEAD and TPP (eqs 40 and 41).[91] Although azepines are not formed from straight-chain amino alcohols, compound (**43**) gives the azepines (**45a**) and (**45b**) by S_N2' substitution of the intermediate phosphorane (**44**) (eq 42).[92]

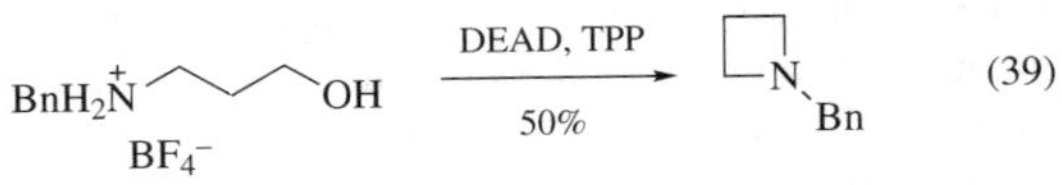

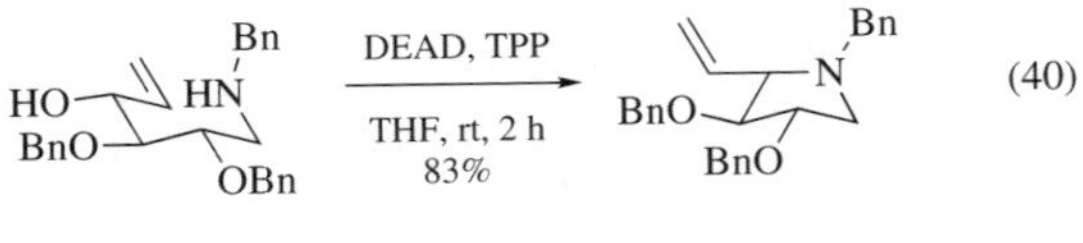

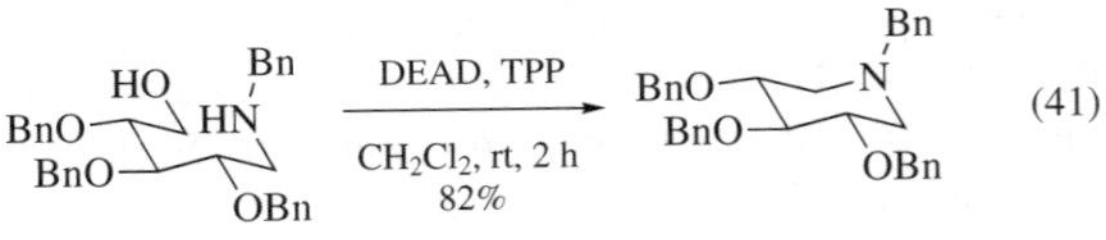

(39)

(40)

(41)

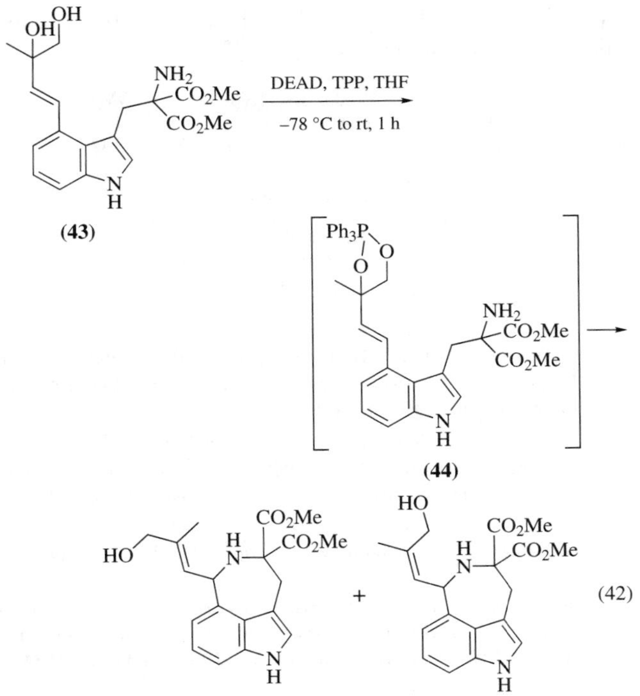

(45a) 85% **(45b)** 3% (42)

O-**Glycosidation with Phenols.** Monosaccharides possessing a free anomeric hydroxyl group react with phenol, DEAD, and trivalent phosphorus compounds to give phenyl glycosides. The yield is higher with trialkylphosphines (Bu_3P, Pr_3P).[93] Coupling of phenols and 2-α-(phenylthio)- or 2-α-(phenylseleno)-α-D-pyranoses by the use of DEAD and TPP gives aryl 2-deoxy-β-D-glycosides in 70–85% yields with high stereoselectivity (inversion of anomeric center) (eq 43).[94] Intramolecular phenolic glycosidation is also possible.[95]

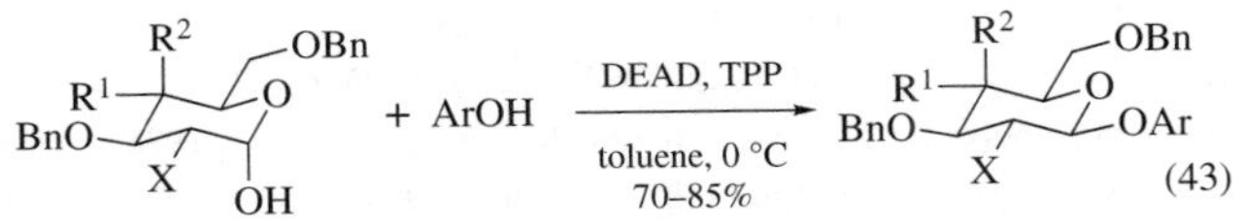

(43)

selectivity = 87:13 to 95:5

R^1 = H, R^2 = BnO or R^1 = BnO, R^2 = H
X = PhS or PhSe; ArOH = 2-naphthol, phenol, or 2-cresol

O-**Glycosidation with Carboxylic Acids.** Reaction of carbohydrates having a free anomeric hydroxyl group with carboxylic acids gives the corresponding glycosyl esters (eq 44).[96] Intramolecular glycosidation of (**46**) can be accomplished by the use of the preformed Bu_3P–DEAD betaine at low temperature (eq 45).[97]

Glycosidation with Nitrogen Nucleophiles and *N*-Hydroxy Compounds. Reaction of an anomeric hydroxyl group with nitrogen nucleophiles and with *N*-hydroxy heterocycles affords the corresponding glycosides.[1,98,99] The α-lactol (**48**) (α:β = ca. 8:1) reacts with *N*-hydroxyphthalimide, DIAD, and TPP to give the β-glycoside (**49**) (eq 46).[100]

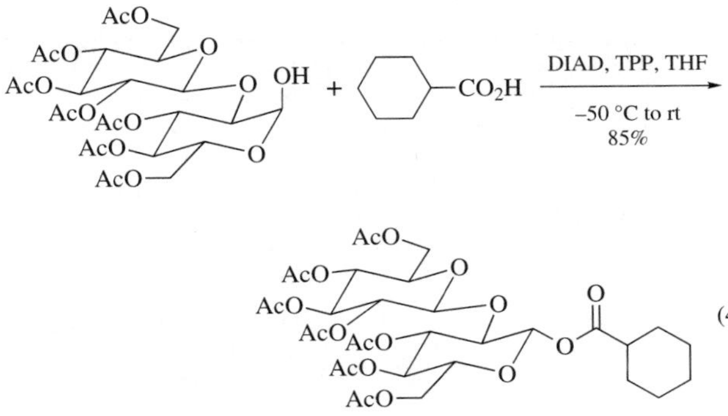

(44)

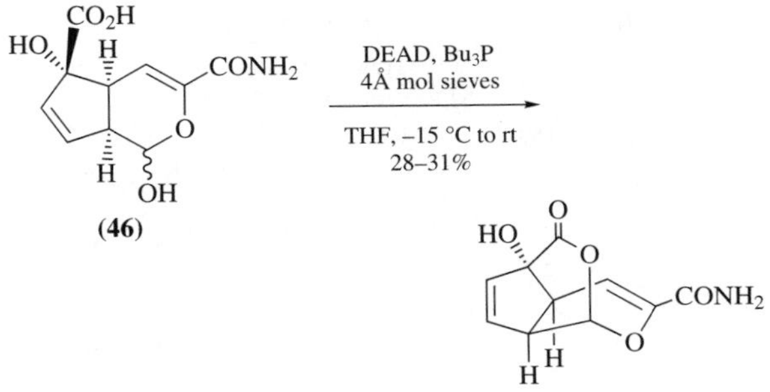

(45)

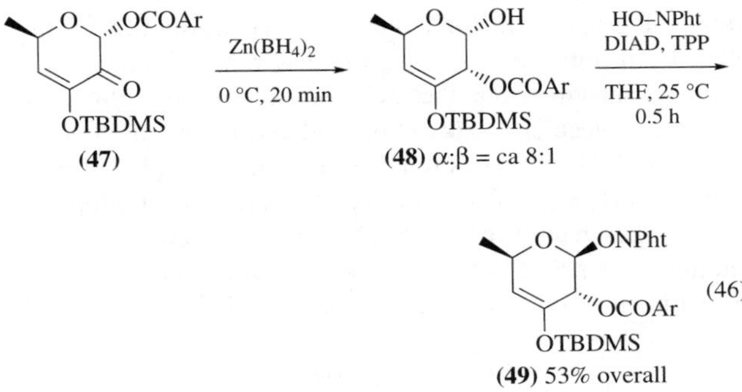

(46)

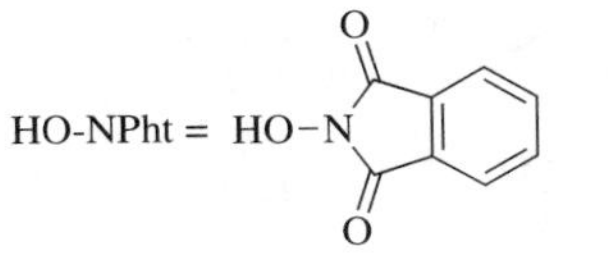

Carbon–Carbon Bond Formation. Direct coupling of an alcohol with a carbon acid by the use of DEAD/TPP has been limited to only a few examples because of the lack of carbon acids with $pK_a < 13$. If enolizable carbonyl compounds are used, alkylation generally takes place on oxygen.[69a,101] Thus the reaction of 1,3-cyclohexanedione with isopropanol, DEAD, and TPP affords exclusively the *O*-alkylated product (eq 47).[101a] Cyanoacetate reacts at the carbon atom, while diethyl malonate is unreactive.[101a] However, the reaction of diethyl malonate with alcohols in the presence of ADDP and Bu_3P affords *C*-alkylated products in 3–56% yields (eq 48).[12] Ethyl nitroacetate and nitromalonate oxidize the alcohol via an *aci*-nitro ester (eq 49).[102] *o*-Nitroarylacetonitriles of general formula (**50**) undergo alkylation on carbon (eq 50).[16,103]

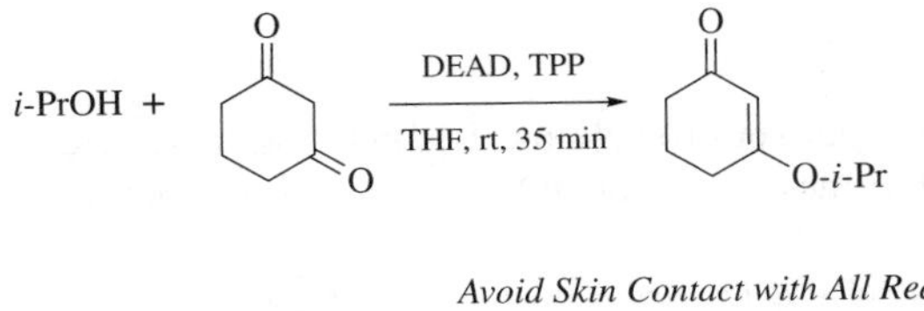

(47)

$$CH_2(CO_2Et)_2 + ROH \xrightarrow[\text{benzene, rt, 24 h; 3–56\%}]{\text{ADDP, Bu}_3\text{P}} RCH(CO_2Et)_2 \quad (48)$$

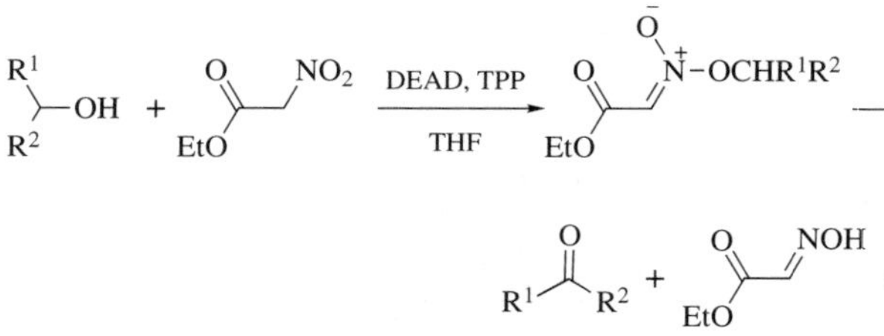

DEAD, TPP; THF (49)

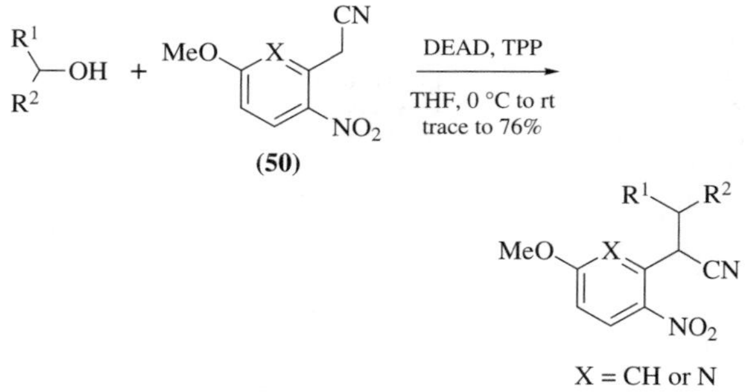

DEAD, TPP; THF, 0 °C to rt; trace to 76% (50)

X = CH or N

γ-Nitro alcohols react with DEAD and TPP to afford α-nitrocyclopropanes in good to excellent yields (eq 51).[104a] On the other hand, nitro alcohols bearing electron-withdrawing or unsaturated substituents at the α-carbon experience exclusive intra- and intermolecular *O*-alkylation and furnish good to excellent yields of alkyl nitronates (eq 52). In contrast, 1-phenylsulfonyl-1-nitro-3-propanol affords the corresponding cyclopropane.[104b] Bis-sulfones readily undergo alkylation by alcohols.[104c] Novel intramolecular S_N2 and S_N2' type arylation occurs with certain aromatic and allylic alcohols (eqs 53 and 54).[105]

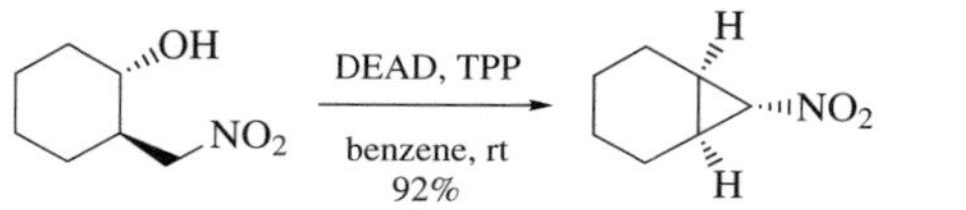

DEAD, TPP; benzene, rt; 92% (51)

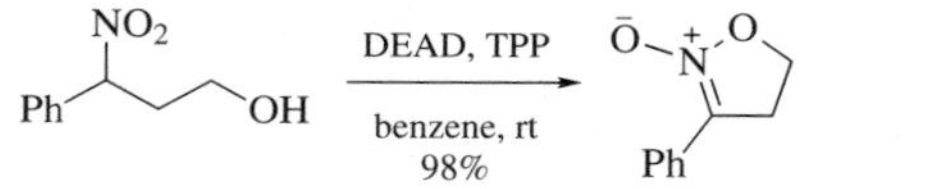

DEAD, TPP; benzene, rt; 98% (52)

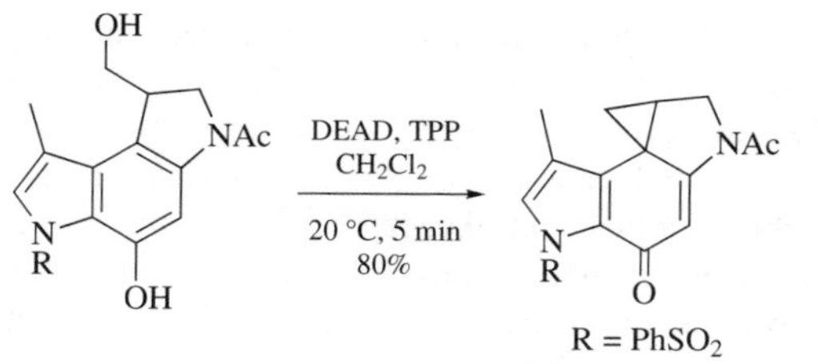

DEAD, TPP, CH_2Cl_2; 20 °C, 5 min; 80% (53)

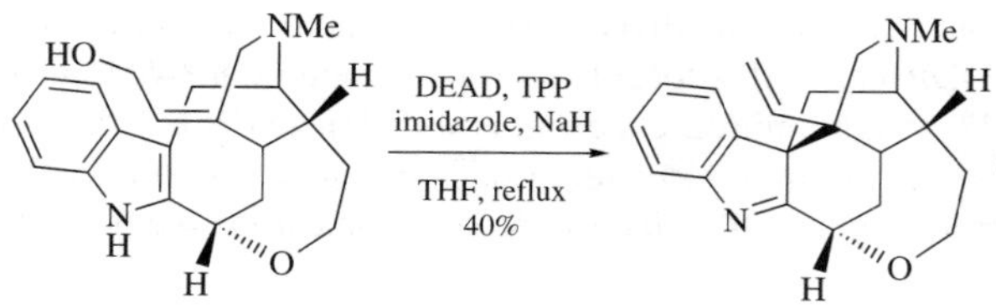

R = $PhSO_2$

DEAD, TPP, imidazole, NaH; THF, reflux; 40% (54)

Miscellaneous Reactions. Dialkyl phosphonates react with alcohols in the presence of DIAD and TPP to give trialkyl phosphites (eq 55).[5a] TPP–DIAD reacts with KHF_2 or HF–pyridine to form difluorotriphenylphosphorane (eq 56).[106]

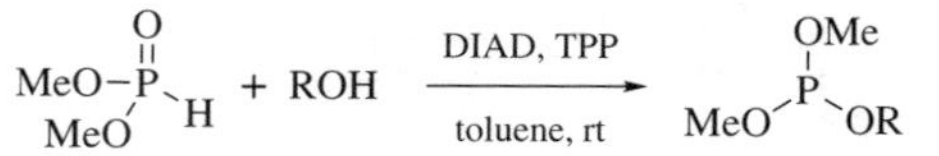

$$(MeO)_2P(O)H + ROH \xrightarrow[\text{toluene, rt}]{\text{DIAD, TPP}} (MeO)_2P\text{–}OR \quad (55)$$

$$Ph_3P \xrightarrow[\text{rt, sonication; 60\%}]{\text{KHF}_2\text{, MeCN, DIAD}} Ph_3PF_2 \quad (56)$$

1. (a) Mitsunobu, O. *S* **1981**, 1. (b) Castro, B. R. *OR* **1983**, *29*, 1. (c) Hughes, D. L. *OR* **1992**, *42*, 335. (d) Mitsunobu, O. *COS* **1991**, *6*, 1. (e) Mitsunobu, O. *COS* **1991**, *6*, 65.
2. Freedman, J.; Vaal, M. J.; Huber, E. W. *JOC* **1991**, *56*, 670.
3. (a) Camp, D.; Jenkins, I. D. *JOC* **1989**, *54*, 3049. (b) Camp, D.; Jenkins, I. D. *JOC* **1989**, *54*, 3045. (c) Camp, D.; Jenkins, I. D. *AJC* **1992**, *45*, 47.
4. Brown, R. F. C.; Jackson, W. R.; McCarthy, T. D. *TL* **1993**, *34*, 1195.
5. (a) Harvey, P.; Jenkins, I. D., unpublished results. (b) Von Itzstein, M.; Jenkins, I. D. *JCS(P1)* **1986**, 437. (c) Von Itzstein, M.; Jenkins, I. D. *JCS(P1)* **1987**, 2057. (d) Von Itzstein, M.; Jenkins, I. D. *AJC* **1984**, *37*, 2447.
6. Camp, D.; Jenkins, I. D. *AJC* **1988**, *41*, 1835.
7. Von Itzstein, M.; Mocerino, M. *SC* **1990**, *20*, 2049; Von Itzstein, M.; Jenkins, M. J.; Mocerino, M. *Carbohydr. Res.* **1990**, *208*, 287.
8. Amos, R. A.; Emblidge, R. W.; Havens, N. *JOC* **1983**, *48*, 3598.
9. Spry, D. O.; Bhala, A. R. *H* **1986**, *24*, 1653.
10. Arnold, L. D.; Assil, H. I.; Vederas, J. C. *JACS* **1989**, *111*, 3973.
11. Dialkyl phosphonates have been reported to promote this reaction: Osikawa, T.; Yamashita, M. *Shizuoka Daigaku Kogakubu Kenkyu Hokoku* **1984**, 37 (*CA* **1985**, *103*, 142095d).
12. Tsunoda, T.; Yamamiya, Y.; Ito, S. *TL* **1993**, *34*, 1639.
13. Volante, R. P. *TL* **1981**, *22*, 3119.
14. Varassi, M.; Walker, K. A. M.; Maddox, M. L. *JOC* **1987**, *52*, 4235.
15. Hughes, D. L.; Reamer, R. A.; Bergan, J. J.; Grabowski, E. J. J. *JACS* **1988**, *110*, 64.
16. Macor, J. E.; Wehner, J. M. *H* **1993**, *35*, 349.
17. Camp, D.; Hanson, G.; Jenkins, I. D. *JOC* **1995**, *60*, 2977.
18. Mitsunobu, O.; Kimura, J.; Iizumi, K.; Yanagida, N. *BCJ* **1976**, *49*, 510.
19. (a) Pautard, A. M.; Evans, S. A., Jr. *JOC* **1988**, *53*, 2300. (b) Pautard-Cooper, A.; Evans, S. A., Jr. *JOC* **1989**, *54*, 2485. (c) Pautard-Cooper, A.; Evans, S. A., Jr. *JOC* **1989**, *54*, 4974. (d) Pautard-Cooper, A.; Evans, S. A., Jr. *T* **1991**, *47*, 1603.
20. (a) Pawlak, J. L.; Padykura, R. E.; Kronis, J. D.; Aleksejczyk, R. A.; Berchtold, G. A. *JACS* **1989**, *111*, 3374. (b) Jenkins, I. D.; Richards, G., unpublished results.
21. Mitsunobu, O.; Kimura, K.; Fujisawa, Y. *BCJ* **1972**, *45*, 245.
22. (a) Grynkiewicz, G. *Pol. J. Chem.* **1979**, *53*, 2501. (b) Martin, S. P.; Dodge, J. A. *TL* **1991**, *32*, 3017.; Dodge, J. A.; Trujillo, J. I.; Presnell, M. *JOC* **1994**, *59*, 234.
23. Georg, G. I.; Kant, J.; Gill, H. S. *JACS* **1987**, *109*, 1129.
24. Ito, Y.; Kobayashi, Y.; Kawabata, T.; Takase, M.; Terashima, S. *T* **1989**, *45*, 5767.
25. Evans, D. A.; Gauchet-Prunet, J. A.; Carreira, E. M.; Charette, A. B. *JOC* **1991**, *56*, 741.
26. Danishefsky, S. L.; Cabal, M. P.; Chow, K. *JACS* **1989**, *111*, 3456.
27. Breslow, R.; Link, T. *TL* **1992**, *33*, 4145.
28. (a) Mori, K.; Otsuka, T.; Oda, M. *T* **1984**, *40*, 2929. (b) Mori, K.; Ikunaka, M. *T* **1984**, *40*, 3471. (c) Mori, K.; Ishikura, M. *LA* **1989**, 1263. (d) Mori, K.; Watanabe, H. *TL* **1984**, *25*, 6025.

29. Saïah, M.; Bessodes, M.; Antonakis, K. *TL* **1992**, *33*, 4317.

30. Weinges, K.; Haremsa, S.; Maurer, W. *Carbohydr. Res.* **1987**, *164*, 453.

31. See, For example: (a) Lindhorst, T. K.; Thiem, J. *LA* **1990**, 1237. (b) Sakamoto, S.; Tsuchiya, T.; Umezawa, S.; Umezawa, H. *BCJ* **1987**, *60*, 1481. (c) Palmer, C. F.; Parry, K. P.; Roberts, S. M.; Ski, V. *JCS(P1)* **1991**, 2051. (d) Bartlett, P. A.; Meadows, J. D.; Ottow, E. *JACS* **1984**, *106*, 5304. (e) Jeker, N.; Tamm, C. *HCA* **1988**, *71*, 1904. See also references 1a and 1c.

32. See, for example: (a) Loibner, H.; Zbiral, E. *HCA* **1977**, *60*, 417. (b) Goya, S.; Takadate, A.; Fujino, H.; Irikura, M. *YZ* **1981**, *101*, 1064.

33. (a) Kimura, J.; Fujisawa, Y.; Yoshizawa, T.; Fukuda, K.; Mitsunobu, O. *BCJ* **1979**, *52*, 1191. (b) Guthrie, R. D.; Jenkins, I. D.; Thang, S.; Yamasaki, R. *Carbohydr. Res.* **1983**, *121*, 109; **1988**, *176*, 306. (c) Jenkins, I. D.; Goren, M. B. *Chem. Phys. Lipids* **1986**, *41*, 225.

34. Warmerdam, E. G. J. C.; Brussee, J.; Kruse, C. G.; van der Gen, A. *T* **1993**, *49*, 1063.

35. (a) Kang, S. H.; Kim, W. J. *TL* **1989**, *30*, 5915. (b) Jarosz, S.; Glodek, J.; Zamojski, A. *Carbohydr. Res.* **1987**, *163*, 289.

36. See, for example: (a) Grossé-Kobo, B.; Mosset, P.; Grée, R. *TL* **1989**, *30*, 4235. (b) Balan, A.; Ziffer, H. *CC* **1990**, 175. (c) Clive, D. L. J.; Daigneault, S. *JOC* **1991**, *56*, 3801.

37. (a) Farina, V. *TL* **1989**, *30*, 6645, and references cited therein. (b) Jefford, C. W.; Moulin, M.-C. *HCA* **1991**, *74*, 336. (c) Carda, M.; Marco, J. A. *T* **1992**, *48*, 9789. (d) Lumin, S.; Falck, J. R.; Capdevila, J.; Karara, A. *TL* **1992**, *33*, 2091. (e) Charette, A. B.; Cote, B. *TL* **1993**, *34*, 6833.

38. (a) Kanematsu, K.; Yoshiyasu, T.; Yoshida, M. *CPB* **1990**, *38*, 1441. (b) Moree, W. J.; van der Marel, G. A.; Liskamp, R. M. J. *TL* **1992**, *33*, 6389. (c) Strijtveen, B.; Kellogg, R. M. *RTC* **1987**, *106*, 539. (d) Jenkins, I. D.; Thang, S. *AJC* **1984**, *37*, 1925.

39. Kpegba, K.; Metzner, P. *S* **1989**, 137.

40. Rollin, P. *TL* **1986**, *35*, 4169.

41. (a) Mlotkowska, B.; Zwierzak, A. *Pol. J. Chem.* **1979**, *53*, 359. (b) Campbell, D. A. *JOC* **1992**, *57*, 6331.; Campbell, D. A.; Bermak, J. C. *JOC* **1994**, *59*, 658.

42. Mlotkowska, B.; Wartalowska-Graczyk, M. *JPR* **1987**, *329*, 735.

43. (a) Camp, D.; Jenkins, I. D. *AJC* **1990**, *43*, 161. (b) Dancy, I.; Laupichler, L.; Rollin, P.; Thiem, J. *SL* **1992**, 283, and references cited therein.

44. (a) Loibner, H.; Zbiral, E. *HCA* **1976**, *59*, 2100. (b) Peterson, M. L.; Vince, R. *JMC* **1991**, *34*, 2787.

45. (a) Still, W. C.; Galynker, I. *JACS* **1982**, *104*, 1774. (b) Galynker, I.; Still, W. C. *TL* **1982**, *23*, 4461.

46. (a) Mulzer, J.; Brüntrup, G.; Chucholowski, A. *AG(E)* **1979**, *18*, 622. (b) Mulzer, J.; Lammer, O. *AG* **1983**, *Suppl.*, 887. (c) Adam, W.; Narita, N.; Nishizawa, Y. *JACS* **1984**, *106*, 1843.

47. Justus, K.; Steglich, W. *TL* **1991**, *32*, 5781.

48. (a) Tsutsui, H.; Mitsunobu, O. *TL* **1984**, *25*, 2163. (b) Ohta, K.; Miyagawa, O.; Tsutsui, H.; Mitsunobu, O. *BCJ* **1993**, *66*, 523.

49. For recent reports of macrolactonization, see, for example: (a) Barrett, A. G. M.; Carr, R. A. E.; Attwood, S. V.; Richardson, G.; Walshe, N. D. A. *JOC* **1986**, *51*, 4840. (b) Smith A. B., III; Noda, I.; Remiszewski, S. W.; Liverton, N. J.; Zibuck, R. *JOC* **1990**, *55*, 3977. (c) White, J. D.; Kawasaki, M. *JACS* **1990**, *112*, 4991. (d) Hanessian, S.; Chemla, P. *TL* **1991**, *32*, 2719.

50. (a) Riehter, L. S.; Gadek, T. R. *TL* **1994**, *35*, 4705. (b) See, for example: Heumann, A.; Faure, R. *JOC* **1993**, *58*, 1276, and references cited therein.

51. For alkyl aryl ether formation without allylic rearrangement, see, for example: Pirrung, M. C.; Brown, W. L.; Rege, S.; Laughton, P. *JACS* **1991**, *113*, 8561.

52. For alkyl aryl ether formation with allylic rearrangement, see, for example: Danishefsky, S.; Berman, E. M.; Ciufolini, M.; Etheredge, S. J.; Segmuller, B. E. *JACS* **1985**, *107*, 3891.

53. Walba, D. M.; Eidman, K. F.; Haltiwanger, R. C. *JOC* **1989**, *54*, 4939.

54. (a) Sugihara, H.; Mabuchi, H.; Hirata, M.; Imamoto, T.; Kawamatsu, Y. *CPB* **1987**, *35*, 1930. (b) Macor, J. E.; Ryan, K.; Newman, M. E. *T* **1992**, *48*, 1039.

55. Clive, D. L. J.; Kellner, D. *TL* **1991**, *32*, 7159.

56. See, for example: Stachulski, A. V. *JCS(P1)* **1991**, 3065.

57. See, for example: Brown, P.; Calvert, S. H.; Chapman, P. C. A.; Cosham, S. C.; Eglington, A. J.; Elliot, R. L.; Harris, M. A.; Hinks, J. D.; Lowther, J.; Merrikin, D. J.; Pearson, M. J.; Ponsford, R. J.; Syms, J. V. *JCS(P1)* **1991**, 881.

58. (a) Carlock, J. T.; Mack, M. P. *TL* **1978**, 5153. (b) Robinson, P. L.; Barry, C. N.; Bass, S. W.; Jarvis, S. E.; Evans, S. A., Jr. *JOC* **1983**, *48*, 5396. (c) Bestmann, H. J.; Pecher, B.; Riemer, C. *S* **1991**, 731.

59. (a) Mitsunobu, O.; Kudo, T.; Nishida, M.; Tsuda, N. *CL* **1980**, 1613. (b) Brandstetter, H. H.; Zbiral, E. *HCA* **1980**, *63*, 327. (c) Mark, E.; Zbiral, E.; Brandstetter, H. H. *M* **1980**, *111*, 289. (d) Guthrie, R. D.; Jenkins, I. D.; Yamasaki, R.; Skelton, B. W.; White, A. H. *JCS(P1)* **1981**, 2328. (e) Guthrie, R. D.; Jenkins, I. D. *AJC* **1981**, *34*, 1997. (f) Guthrie, R. D.; Jenkins, I. D.; Yamasaki, R. *Carbohydr. Res.* **1980**, *85*, C5. (g) Guthrie, R. D.; Jenkins, I. D.; Yamasaki, R.; Skelton, B. W.; White, A. H. *JCS(P1)* **1981**, 2328. (h) McGowan, D. A.; Berchtold, G. A. *JOC* **1981**, *46*, 2381.

60. Palomino, E.; Schaap, A. P.; Heeg, M. J. *TL* **1989**, *30*, 6797.

61. Penz, G.; Zbiral, E. *M* **1981**, *112*, 1045.

62. (a) Iida, H.; Yamazaki, N.; Kibayashi, C. *JOC* **1987**, *52*, 1956. (b) Yamazaki, N.; Kibayashi, C. *TL* **1988**, *29*, 5767.

63. Audia, J. E.; Colocci, N. *TL* **1991**, *32*, 3779.

64. (a) Wada, M.; Sano, T.; Mitsunobu, O. *BCJ* **1973**, *46*, 2833. (b) Akita, H.; Yamada, H.; Matsukura, H.; Nakata, T.; Oishi, T. *CPB* **1990**, *38*, 2377. (c) Iimori, T.; Ohtsuka, Y.; Oishi, T. *TL* **1991**, *32*, 1209.

65. For an unsuccessful result, see: Roemmele, R. C.; Rapoport, H. *JOC* **1989**, *54*, 1866.

66. (a) Lal, B.; Pramanik, B. N.; Manhas, M. S.; Bose, A. K. *TL* **1977**, 1977. (b) Pearson, W. H.; Hines, J. V. *JOC* **1989**, *54*, 4235.

67. (a) Viaud, M. C.; Rollin, P. *S* **1990**, 130. (b) Gajda, T.; Matusiak, M. *S* **1992**, 367. (c) Duclos, O.; Duréault, A.; Depezay, J. C. *TL* **1992**, *33*, 1059. (d) Duclos, O.; Mondange, M.; Duréault, A.; Depezay, J. C. *TL* **1992**, *33*, 8061.

68. Fabiano, E.; Golding, B. T.; Sadeghi, M. M. *S* **1987**, 190.

69. (a) Morimoto, H.; Furukawa, T.; Miyazima, K.; Mitsunobu, O. *CL* **1973**, 821. For *C*- and *N*-alkylation of imides and related compounds, see also: (b) Sammes, P. G.; Thetford, D. *JCS(P1)* **1989**, 655. (c) Kim, T. H.; Rapoport, H. *JOC* **1990**, *55*, 3699.

70. (a) Koppel, I.; Koppel, J.; Degerbeck, F; Grehn, L.; Ragnarsson, U. *JOC* **1991**, *56*, 7172. See also, for example: (b) Slusarska, E.; Zwierzak, A. *LA* **1986**, 402. (c) Tanner, D.; Birgersson, C.; Dhaliwal, H. K. *TL* **1990**, *31*, 1903. (d) Arnould, J. C.; Landier, F.; Pasquet, M. J. *TL* **1992**, *33*, 7133. (e) Macor, J. E.; Blank, D. H.; Post, R. J.; Ryan, K. *TL* **1992**, *33*, 8011. (f) Campbell, J. A.; Hart, D. J. *JOC* **1993**, *58*, 2900.

71. Chong, J. M.; Park, S. B. *JOC* **1992**, *57*, 2220.

72. (a) Miller, M. J.; Mattingly, P. G.; Morrison, M. A.; Kerwin, J. F., Jr. *JACS* **1980**, *102*, 7026. (b) Teng, M.; Miller, M. J. *JACS* **1993**, *115*, 548.

73. (a) Bose, A. K.; Sahu, D. P.; Manhas, M. S. *JOC* **1981**, *46*, 1229. (b) Bose, A. K.; Manhas, M. S.; Sahu, D. P.; Hegde, V. R. *CJC* **1984**, *62*, 2498.

74. Huang, N. Z.; Kalish, V. J.; Miller, M. J. *T* **1990**, *46*, 8067.

75. Salituro, G. M.; Townsend, C. A. *JACS* **1990**, *112*, 760, and references cited therein.

76. For the reaction of hydroxyamino acid derivatives with DEAD–TPP and related systems, see, for example: (a) Nakajima, K.; Tanaka, T.; Neya, M.; Okawa, K. *BCJ* **1982**, *55*, 3237. (b) Nakajima, K.; Sasaki, H.; Neya, M.; Morishita, M.; Sakai, S.; Okawa, K. In *Peptide Chemistry 1982*; Sakakibara, S., Ed.; Protein Research Foundation: Osaka, 1983; pp 19–24.

77. Lotz, B. T.; Miller, M. J. *JOC* **1993**, *58*, 618, and references cited therein.
78. Toyota, A.; Katagiri, N.; Kaneko, C. *SC* **1993**, *23*, 1295.
79. Katagiri, N.; Toyota, A.; Shiraishi, T.; Sato, H.; Kaneko, C. *TL* **1992**, *33*, 3507.
80. For related reports, see, for example: (a) Toyota, A.; Katagiri, N.; Kaneko, C. *CPB* **1992**, *40*, 1039. (b) Katagiri, N.; Sato, H.; Arai, S.; Toyota, A.; Kaneko, C. *H* **1992**, *34*, 1097. (c) Toyota, A.; Katagiri, N.; Kaneko, C. *H* **1993**, *36*, 1625. (d) Jenny, T. F.; Previsani, N.; Benner, S. A. *TL* **1991**, *32*, 7029. (e) Jenny, T. F.; Schneider, K. C.; Benner, S. A. *Nucleosides Nucleosides* **1992**, *11*, 1257. (f) Overberger, C. G.; Chang, J. Y. *TL* **1989**, *30*, 51. (g) Bestmann, H. J.; Roth, D. *AG(E)* **1990**, *29*, 99. (h) See ref. 31(b).
81. For *O*-alkylation of purines, see, for example: Schulz, B. S.; Pfleiderer, W. *HCA* **1987**, *70*, 210, and references cited therein.
82. (a) Lipshutz, B. H.; Miller, T. A. *TL* **1990**, *31*, 5253. For neighboring group participation in related reactions, see: (b) Campbell, J. A.; Hart, D. J. *TL* **1992**, *33*, 6247.
83. Branquet, E.; Durand, P.; Vo-Quang, L.; Le Goffic, F. *SC* **1993**, *23*, 153.
84. (a) Arnold, L. D.; Kalantar, T. H.; Vederas, J. C. *JACS* **1985**, *107*, 7105. (b) Ramer, S. E.; Moore, R. N.; Vederas, J. C. *CJC* **1986**, *64*, 706. See also: (c) Parker, W. L.; Rathnum, M. L.; Liu, W.-C. *J. Antibiot.* **1982**, *35*, 900.
85. (a) Wojciechowska, H.; Pawlowicz, R.; Andruszkiewicz, R.; Grzybowska, J. *TL* **1978**, 4063. (b) Andruszkiewicz, R.; Grzybowska, J.; Wojciechowska, H. *Pol. J. Chem.* **1981**, *55*, 67.
86. Coleman, R. S.; Carpenter, A. J. *JOC* **1992**, *57*, 5813.
87. (a) Wipf, P.; Miller, C. P. *TL* **1992**, *33*, 6267. See also: (b) Galéotti, N.; Montagne, C.; Poncet, J.; Jouin, P. *TL* **1992**, *33*, 2807.
88. Roush, D. M.; Patel, M. M. *SC* **1985**, *15*, 675.
89. Pfister, J. R. *S* **1984**, 969.
90. Sammes, P. G.; Smith, S. *CC* **1983**, 682.
91. (a) Bernotas, R. C.; Cube, R. V. *TL* **1991**, *32*, 161. (b) Bernotas, R. C. *TL* **1990**, *31*, 469. (c) Holladay, M. W.; Nadzan, A. M. *JOC* **1991**, *56*, 3900.
92. Kozikowski, A. P.; Okita, M. *TL* **1985**, *26*, 4043.
93. Grynkiewicz, G. *Pol. J. Chem.* **1979**, *53*, 1571.
94. Roush, W. R.; Lin, X.-F. *JOC* **1991**, *56*, 5740.
95. Delgado, A.; Clardy, J. *JOC* **1993**, *58*, 2862.
96. (a) Smith, A. B., III; Hale, K. J.; Rivero, R. A. *TL* **1986**, *27*, 5813. (b) de Masmaeker, A.; Hoffmann, P.; Ernst, B. *TL* **1989**, *30*, 3773. (c) Smith, A. B., III; Rivero, R. A.; Hale, K. J.; Vaccaro, H. A. *JACS* **1991**, *113*, 2092. (d) Smith, A. B., III; Hale, K. J.; Vaccaro, H. A.; Rivero, R. A. *JACS* **1991**, *113*, 2112.
97. Smith, A. B., III; Sulikowski, G. A.; Sulikowski, M. M.; Fujimoto, K. *JACS* **1992**, *114*, 2567.
98. For glycosidation with HN_3, see: Schörkhuber, W.; Zbiral, E. *LA* **1980**, 1455.
99. For glycosidation with *N*-hydroxy heterocycles, see: Grochowski, E.; Stepowska, H. *S* **1988**, 795, and references cited therein.
100. Nicolaou, K. C.; Groneberg, R. D. *JACS* **1990**, *112*, 4085.
101. (a) Kurihara, T.; Sugizaki, M.; Kime, I.; Wada, M.; Mitsunobu, O. *BCJ* **1981**, *54*, 2107. (b) Bajwa, J. S.; Anderson, R. C. *TL* **1990**, *31*, 6973. (c) Nakayama, E.; Watanabe, K.; Miyauchi, M.; Fujimoto, K.; Ide, J. *J. Antibiot.* **1990**, *43*, 1122. (d) Czernecki, S.; Valéry, J.-M. *S* **1991**, 239.
102. Mitsunobu, O.; Yoshida, N. *TL* **1981**, *22*, 2295.
103. Macor, J. E.; Wehner, J. M. *TL* **1991**, *32*, 7195.
104. (a) Yu, J.; Falck, J. R.; Mioskowski, C. *JOC* **1992**, *57*, 3757. (b) Falck, J. R.; Yu, J. *TL* **1992**, *33*, 6723. (c) Yu, J.; Cho, H.-S.; Falck, J. R. *JOC* **1993**, *58*, 5892.
105. (a) Magnus, P.; Gallagher, T.; Schulz, J.; Or, Y.-S.; Ananthanarayan, T. P. *JACS* **1987**, *109*, 2706. (b) Boger, D. L.; Wysocki, R. J., Jr; Ishizaki, T. *JACS* **1990**, *112*, 5230. (c) Magnus, P.; Mugrage, B.; DeLuca, M. R.; Cain, G. A. *JACS* **1990**, *112*, 5220.
106. Harvey, P.; Jenkins, I. D. *TL* **1994**, *35*, 9775.

Ian D. Jenkins
Griffith University, Brisbane, Australia
Oyo Mitsunobu
Aoyama Gakuin University, Tokyo, Japan

Tris(dimethylamino)sulfonium Difluorotrimethylsilicate[1]

$(R_2N)_3S^+Me_3SiF_2^-$

(R = Me)		
[59218-87-0]	$C_9H_{27}F_2N_3SSi$	(MW 275.55)
(R = Et)		
[59201-86-4]	$C_{15}H_{39}F_2N_3SSi$	(MW 359.73)

(anhydrous fluoride ion source; synthesis of C–F compounds by nucleophilic displacement of sulfonates;[3] promoter for electrophilic reactions of silyl enolates of ketones and esters;[4–6] source of sulfonium cation capable of stabilizing or imparting high nucleophilic reactivity to other anions;[46,8a,10a] activator of vinylsilanes in Pd-catalyzed cross-coupling reactions;[7] also used for generation[8] and reactions[10] of α- and β-halo carbanions; hydrosilylation[11] and cyanomethylation[14] of ketones)

Alternate Name: TASF.
Physical Data: R = Me, mp 98–101 °C; R = Et, mp 90–95 °C.
Solubility: R = Me, sol MeCN, pyridine, benzonitrile; partially sol THF. R = Et, sol THF, MeCN. Both react slowly with MeCN.
Form Supplied in: R = Me, white crystalline solid, ~90% pure; major impurity is tris(dimethylamino)sulfonium bifluoride $[(Me_2N)_3S^+\ HF_2^-]$.
Analysis of Reagent Purity: mp; ^{19}F NMR δ (at 200 MHz, $CFCl_3$ standard) $TASMe_3SiF_2$ (CD_3CN) δ −60.3; $TASHF_2$ −145.8 (d, J_{HF} = 120 Hz).
Preparative Methods: the methyl derivative is prepared by the reaction of dimethylaminotrimethylsilane and ***Sulfur Tetrafluoride*** at −70 °C to rt in ether; the precipitated solid is filtered off.[1a] The ethyl derivative is best prepared by the reaction of ***N,N-Diethylaminosulfur Trifluoride*** (DAST) and diethylaminotrimethylsilane.[1b,11b]
Handling, Storage, and Precautions: because of the extreme hygroscopic nature of this compound, it is best handled in a dry box or a polyethylene glove bag filled with high purity nitrogen. Use in a fume hood.

Introduction. The acronym TASF has been used to refer to both $(Me_2N)_3S^+F_2Me_3Si^-$ and $(Et_2N)_3S^+F_2Me_3Si^-$. To eliminate confusion, the sulfonium salt containing dimethylamino groups is referred to as TASF(Me), and the sulfonium salt containing diethylamino groups is referred to as TASF(Et). Reactions of both reagents are similar. Since both of these salts can be prepared in a rigorously anhydrous state, they have an advantage over quaternary ammonium fluorides which usually contain some water. TASF(Me) has a slight advantage over TASF(Et) in that it is highly crystalline and easier to prepare in a high state of purity, whereas TASF(Et) has an advantage over TASF(Me) in that it has greater solubility in organic solvents. The tris(dialkylamino)sulfonium cation is often referred to by the acronym TAS.

TASF is a source of organic soluble fluoride ion[2] with a bulky noncoordinating counter ion (eq 1).[9]

$$TAS^+ \, Me_3SiF_2^- \rightleftharpoons Me_3SiF + TAS^+ + F^- \quad (1)$$

Fluoride Ion Source in Nucleophilic Displacements. TASF(Me) can be used to prepare fluorides from halides[1b] and sulfonates[3] under relatively mild conditions (eq 2).

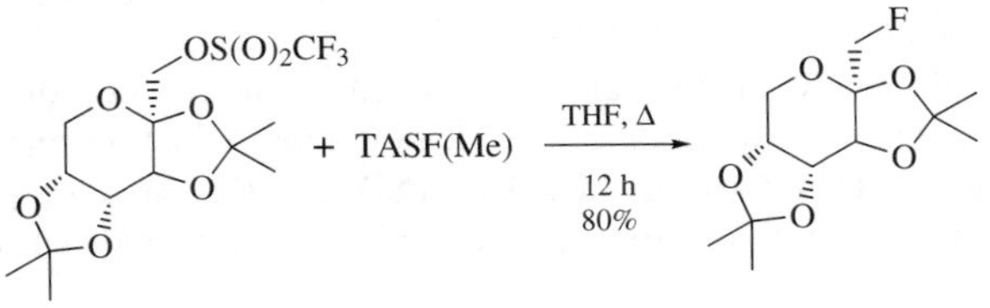

Generation of Enolates and Enolate Surrogates from Enol Silanes. Enol silanes react with TASF(Et) to give highly reactive 'naked enolates' which have been characterized by NMR and electrochemical measurements.[4] These enolates, generated in situ, can be regioselectively alkylated without complications from polyalkylation or rearrangements of the alkylating agent (eq 3).

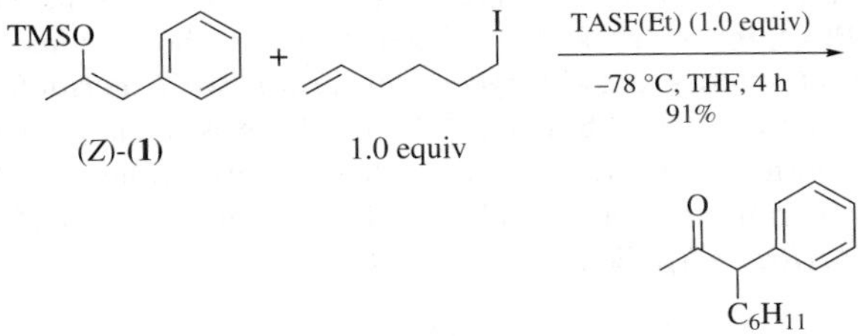

In the presence of excess ***Fluorotrimethylsilane***, TASF(Et) catalyzes aldol reactions of silyl enol ethers and aldehydes.[4] The stereochemical course of the reaction (*syn* selectivity, independent of the enol geometry, (*Z*)- or (*E*)-(**1**), eq 4), has been interpreted as arising from an extended transition state in which steric and charge repulsions are minimized.

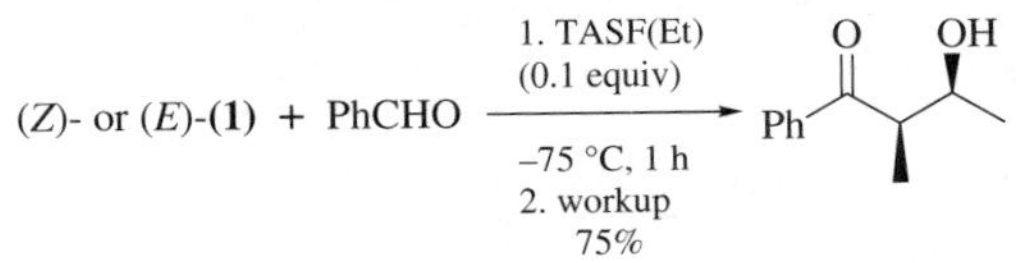

Very potent carbon nucleophiles formally equivalent to ester enolates are generated by the interaction of TASF(Me) with unhindered trialkylsilyl ketene acetals. In contrast to lithium enolates, these TAS enolates add 1,4 (nonstereoselectively) to α,β-unsaturated ketones. These adducts can be alkylated in situ to form two new C–C bonds in one pot, or they can be hydrolyzed to give 1,5-dicarbonyl compounds (eq 5).[5a]

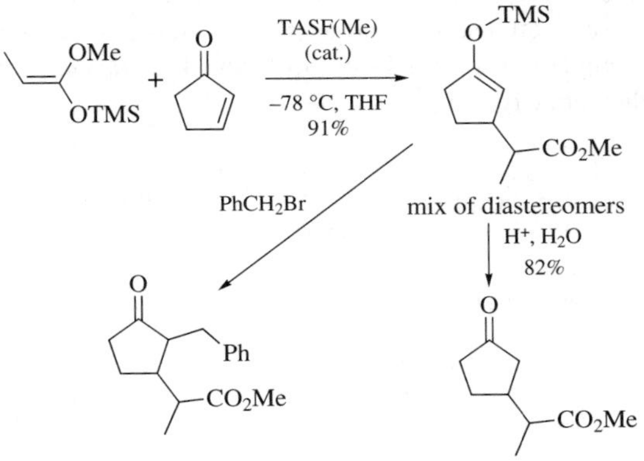

Conjugated esters undergo sequential additions to form polymers (group transfer polymerization).[5b] The molecular weight and the end group functionality of the polymer can be controlled by this method. Mechanistic studies indicate an associative intramolecular silicon transfer process via (**2**), with concomitant C–C bond formation during the polymer growth (eq 6).

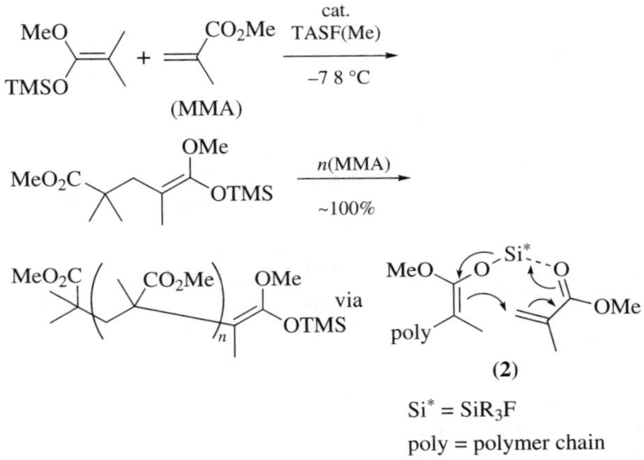

Silyl enol ethers and ketene silyl acetals add to aromatic nitro compounds in the presence of TASF(Me) to give intermediate dihydro aromatic nitronates which can be oxidized with ***Bromine*** or ***2,3-Dichloro-5,6-dicyano-1,4-benzoquinone*** to give α-nitroaryl carbonyl compounds;[6a] the latter are precursors for indoles and oxindoles.[6b] The reaction is widely applicable to alkyl-, halo-, and alkoxy-substituted aromatic nitro compounds, including heterocyclic and polynuclear derivatives (eq 7).

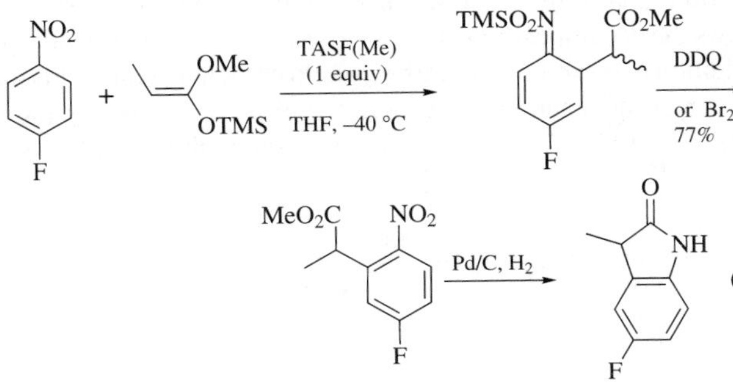

Cross-Coupling Reactions. TASF(Et) activates vinyl-, alkynyl-, and allylsilanes in the Pd-mediated cross-coupling with vinyl and aryl iodides and bromides.[7] As illustrated in eqs 8–10, the reaction is stereospecific and chemoselective. This cross-coupling protocol is remarkably tolerant towards a variety of other functional groups such as carbonyl, amino, hydroxy, and nitro. Vinylsilanes can be synthesized from ***Hexamethyldisilane*** and vinyl iodides in the presence of TASF(Et) (eq 10) via cleavage of a Si–Si bond.[7a] Aryl iodides can also be synthesized by this method. TASF is superior to ***Tetra-n-butylammonium Fluoride*** for these reactions. In the absence of a vinylsilane reagent, one of the methyl groups from the difluorotrimethylsilicate is substituted for the halide (eq 11).[7d]

(8) 4-iodonitrobenzene + allyl-TMS (1.3 equiv) + TASF(Et) (1.3 equiv) → [$(\eta^3\text{-}C_3H_5PdCl)_2$ (0.025 equiv), HMPA, 50 °C, 83%] 4-vinylnitrobenzene derivative

(9) TMS-alkynol + Br-vinyl-Ph → [TASF(Et), $(\eta^3\text{-}C_3H_5PdCl)_2$ (0.025 equiv), 84%] enynol-Ph

(10) C_6H_{13}-CH=CH-I + TMS–TMS → [TASF(Et), $Pd(Ph_3P)_4$, HMPA, 0 °C, 74%] C_6H_{13}-CH=CH-TMS

(11) 4-bromoacetophenone + TASF(Et) → [$(RPdCl)_2$, THF, 20 °C, 82%] 4-methylacetophenone

Generation and Reactions of Unusual Carbanions. Both α- and β-halo carbanions are generally labile species and their generation and reactions require extremely low temperatures. TASF(Me) has been used to prepare several stable and isolable perfluorinated carbanions (eq 12)[8a] or alkoxides.[8b] As compared to the corresponding metal salts, the TAS^+ counterion has little coordination to the fluorines of the anion,[9] and this presumably slows the decomposition of the TAS salts to carbenoids or alkenes. Addition of α- and β-halo carbanions to carbonyl compounds may be achieved by the in situ generation of these species by the reaction of TASF(Et) with the corresponding silylated derivatives (eq 13).[10]

(12) TASF(Me) + $(F_3C)_2C{=}CF_2$ → [THF, 0 °C to rt, 94%] $(CF_3)_3C^-TAS^+$ + TMSF; mp 170 °C!

(13) PhCH(Me)CHO + $TMSCCl_3$ → [1. 0.1 equiv TASF(Et), THF, 0 °C, 4 h; 2. H^+; 75%] PhCH(Me)CH(OH)CCl_3, 87% *syn*

Other Applications. TASF(Et) catalyzes the addition of ***Dimethyl(phenyl)silane*** to α-alkoxy, -acyl or -amido ketones to give the corresponding *anti* aldols (eq 14).[11] This complements the acid-catalyzed reduction which gives the *syn* isomer.

(14) PhCOCH(Me)CO-piperidide + TASF(Et) (0.08 equiv) + $PhMe_2SiH$ (1.70 equiv) → [THF, HMPA, 0 °C, 12 h, 77%] PhCH(OH)CH(Me)CO-piperidide; *anti*:*syn* = 99:1

Dimethyl(phenyl)silane will reduce aldehydes and ketones to hydroxyl compounds under very mild conditions in the presence of a catalytic amount of TASF(Et).[11c]

Aryl or vinyl anions can be generated by the reaction of the corresponding vinyl iodide with Bu_3Sn anion, which in turn is produced from TASF(Et) and $Bu_3SnSiMe_3$. With an appropriately placed carbonyl group, an intramolecular cyclization ensues (eq 15).[12]

(15) 2-(2-bromobenzyl)-2-methylcyclohexane-1,3-dione + TMS–$SnBu_3$ → [TASF(Et), DMF, 1 h, −30 °C, 93%] tricyclic hydroxy ketone

A useful variation of the Peterson alkenation relies on the generation of α-silyl carbanions from geminal disilyl compounds containing an additional stabilizing group (CO_2R, SPh, SO_2Ph, OMe, CN, Ph) at the α-carbon.[13] A related reaction is the cyanomethylation of ketones and aldehydes with ***Trimethylsilylacetonitrile*** in the presence of TASF(Me).[14] TASF(Me) was found to be the best fluoride ion source for the synthesis of aryl trifluoromethyl sulfones from the corresponding sulfonyl fluorides and ***Trifluoromethyltrimethylsilane*** or Me_3SnCF_3 (eq 16).[15]

$$PhSO_2F + TMSCF_3 \xrightarrow[\text{THF, 20 °C, 96\%}]{\text{TASF(Me)}} PhSO_2CF_3 \quad (16)$$

The carbanions formed by scission of a C–Si bond with TASF can also be oxygenated. Benzylic trimethylsilyl groups can be converted to hydroxyl groups in 20–95% yield by reaction with TASF(Me) in the presence of oxygen and ***Trimethyl Phosphite*** (eq 17).[16] No other source of fluoride ion was found that could replace TASF.

(17) ArCH(R)TMS → [TASF(Me)] $ArCH(R)^-\ TAS^+$ → [O_2] $ArCH(R)OO^-\ TAS^+$ → [$P(OMe)_3$, 20–95%] ArCH(R)OH

Related Reagents. Cesium Fluoride; Potassium Fluoride; Tetra-*n*-butylammonium Fluoride.

1. (a) Middleton, W. J. *OS* **1985**, *64*, 221. (b) Middleton, W. J. U.S. Patent 3 940 402, 1976 (*CA* **1976**, *85*, 6388j). See also Ref. 11(b).
2. For a review of applications of fluoride ion in organic synthesis, see: Clark, J. H. *CRV* **1980**, *80*, 429.
3. (a) Card, P. J.; Hitz, W. D. *JACS* **1984**, *106*, 5348. (b) Doboszewski, B.; Hay, G. W.; Szarek, W. A. *CJC* **1987**, *65*, 412.
4. (a) Noyori, R.; Nishida, I.; Sakata, J. *JACS* **1983**, *105*, 1598. (b) Noyori, R.; Nishida, I.; Sakata, J. *TL* **1981**, *22*, 3993.
5. (a) RajanBabu, T. V. *JOC* **1984**, *49*, 2083. (b) Webster, O. W.; Hertler, W. R.; Sogah, D. Y.; Farnham, W. B.; RajanBabu, T. V. *JACS* **1983**, *105*, 5706.
6. (a) RajanBabu, T. V.; Reddy, G. S.; Fukunaga, T. *JACS* **1985**, *107*, 5473. (b) RajanBabu, T. V.; Chenard, B. L.; Petti, M. A. *JOC* **1986**, *51*, 1704.
7. (a) Hatanaka, Y.; Hiyama, T. *TL* **1987**, *28*, 4715. (b) Hatanaka, Y.; Hiyama, T. *JOC* **1988**, *53*, 918. (c) Hatanaka, Y.; Fukushima, S.; Hiyama, T. *H* **1990**, *30*, 303. (d) Hatanaka, Y.; Hiyama, T. *TL* **1988**, *29*, 97.
8. (a) Smart, B. E.; Middleton, W. J.; Farnham, W. B. *JACS* **1986**, *108*, 4905. (b) Farnham, W. B.; Smart, B. E.; Middleton, W. J.; Calabrese, J. C.; Dixon, D. A. *JACS* **1985**, *107*, 4565.
9. For a discussion of structural aspects of TAS salts, see: (a) Farnham, W. B.; Dixon, D. A.; Middleton, W. J.; Calabrese, J. C.; Harlow, R. L.; Whitney, J. F.; Jones, G. A.; Guggenberger, L. J. *JACS* **1987**, *109*, 476. (b) Dixon, D. A.; Farnham, W. B.; Heilemann, W.; Mews, R.; Noltemeyer, M. *HC* **1993**, *4*, 287.
10. (a) Fujita, M.; Hiyama, T. *JACS* **1985**, *107*, 4085. (b) Hiyama, T.; Obayashi, M.; Sawahata, M. *TL* **1983**, *24*, 4113. See also: de Jesus, M. A.; Prieto, J. A.; del Valle, L.; Larson, G. L. *SC* **1987**, *17*, 1047.
11. (a) Fujita, M.; Hiyama, T. *JOC* **1988**, *53*, 5405. (b) Fujita, M.; Hiyama, T. *OS* **1990**, *69*, 44. (c) Fujita, M; Hiyama, T. *TL* **1987**, *28*, 2263.
12. Mori, M.; Isono, N.; Kaneta, N.; Shibasaki, M. *JOC* **1993**, *58*, 2972.
13. Palomo, C.; Aizpurua, J. M.; García, J. M.; Ganboa, I.; Cossio, F. P.; Lecea, B.; López, C. *JOC* **1990**, *55*, 2498. Also see: Padwa, A.; Chen. Y.-Y.; Dent, W.; Nimmesgern, H. *JOC* **1985**, *50*, 4006.
14. Palomo, C.; Aizpurua, J. M.; López, M. C.; Lecea, B. *JCS(P1)* **1989**, 1692.
15. Kolomeitsev, A. A.; Movchun, V. N.; Kondratenko, N. V.; Yagupolski, Yu. L. *S* **1990**, 1151.
16. Vedejs, E.; Pribish, J. R. *JOC* **1988**, *53*, 1593.

T. V. (Babu) RajanBabu
The Ohio State University, Columbus, OH, USA

William J. Middleton & Victor J. Tortorelli
Ursinus College, Collegeville, PA, USA

Zinc Bromide[1]

ZnBr₂

[7699-45-8] Br_2Zn (MW 225.19)

(used in the preparation of organozinc reagents via transmetalation;[1] a mild Lewis acid useful for promoting addition[2] and substitution reactions[3])

Physical Data: mp 394 °C; bp 697 °C (dec); *d* 4.201 g cm^{-3}.
Solubility: sol Et_2O, H_2O (1 g/25 mL), 90% EtOH (1 g/0.5 mL).
Form Supplied in: granular white powder; principal impurity is H_2O.
Analysis of Reagent Purity: melting point.
Purification: heat to 300 °C under vacuum (2×10^{-2} mmHg) for 1 h, then sublime.
Handling, Storage, and Precautions: very hygroscopic; store under anhydrous conditions. Irritant.

Organozinc Reagents. The transmetalation of organomagnesium, organolithium, and organocopper reagents by anhydrous $ZnBr_2$ in ethereal solvents offers a convenient method of preparing organozinc bromides and diorganozinc reagents.[1a] Alternatively, anhydrous $ZnBr_2$ may be reduced by potassium metal to result in highly activated Zn^0, which is useful for the preparation of zinc reagents through oxidative addition to organic halides.[4] Alkyl, allylic, and propargylic zinc reagents derived by these methods have shown considerable value in their stereoselective and regioselective addition reactions with aldehydes, ketones, imines, and iminium salts.[1a,5] Zinc enolates used in the Reformatsky reaction may also be prepared through transmetalation using $ZnBr_2$.[1b] Organozinc species are especially useful in palladium- and nickel-catalyzed coupling reactions of sp^2 carbon centers. In this fashion, sp^2–sp^3 (eq 1)[6] and sp^2–sp^2 (eqs 2 and 3)[7,8] carbon–carbon bonds are formed selectively in high yields. The enantioselective cross coupling of secondary Grignard reagents with vinyl bromide is strongly affected by the presence of $ZnBr_2$, which accelerates the reaction and inverts its enantioselectivity (eq 4).[9]

PhCH₂Li → (1. ZnBr₂, THF; 2. I–C₆H₄–CO₂Me, Ni(PPh₃)₄, THF–Et₂O, 25 °C; 88%) → PhCH₂–C₆H₄–CO₂Me (1)

CH₂=C(Li)OCONEt₂ → (1. ZnBr₂, THF; 2. BrCH=CHPh, Pd(PPh₃)₄; 80%) → CH₂=C(OCONEt₂)CH=CHPh (2)

Organozinc intermediates formed via transmetalation using $ZnBr_2$ have been used to effect carbozincation of alkenes and alkynes through metallo-ene and metallo-Claisen reactions. Both intermolecular and intramolecular variants of these reactions have been described, often proceeding with high levels of stereoselectivity and affording organometallic products that may be used in subsequent transformations (eqs 5 and 6),[10] including alkenation (eq 6).[10b,c] Bimetallic zinc–zirconium reagents have also been developed that offer a method for the alkenation of carbonyl compounds (eq 7).[11]

(Et–CH=CH)₂CuLi → (ZnBr₂, Pd(PPh₃)₄, THF–Et₂O; ICH=CHC₅H₁₁; 96%) → Et–CH=CH–CH=CH–C₅H₁₁
(Et–CH=CH)₂CuLi → (ZnBr₂, THF, Pd(PPh₃)₄; I–C₆H₄–CO₂Me; 81%) → Et–CH=CH–C₆H₄–CO₂Me (3)

ArCH(Me)MgCl (Ar = 4-isobutylphenyl) → (CH₂=CHBr, NiCl₂, Et₂O, 0 °C; MeS–CH₂CH₂–CH(NMe₂)–PPh₂) → ArC(R¹)(R²)CH=CH₂ (4)

without $ZnBr_2$: R¹ = Me, R² = H; >95% (52% ee)
with $ZnBr_2$: R¹ = H, R² = Me; >95% (49% ee)

Enyne (OMe, TMS) → (1. BuLi, THF; 2. ZnBr₂; 80%) → cyclopentane (CH₂X, OMe, C≡C–TMS) (5)

X = ZnBr → (H₂O) → X = H

t-BuO(Pr)CH–CH=CHI → (1. t-BuLi; 2. CH₃CH=CHCH₂MgBr; 3. ZnBr₂) → [zinc chelate intermediate] →

M¹, M² adduct (Pr, O-t-Bu) → (1. PhCHO; 2. H₃O⁺; 68%) → diene (Ph, Pr, O-t-Bu), 95% de (6)

M¹, M² = metal → (H₃O⁺) → M¹, M² = H, 75%, 95% de

(7)

83%, 100% (E)

(12)

$ZnBr_2$, THF, 20 °C, 6 h	100:0
CsF, CH_2Cl_2. 20 °C, 14 h	5:95

Concerted Ring-Forming Reactions. The mild Lewis acid character of $ZnBr_2$ sometime imparts a catalytic effect on thermally allowed pericyclic reactions. The rate and stereoselectivity of cycloaddition reactions (eq 8),[12] including dipolar cycloadditions (eq 9),[13] are significantly improved by the presence of this zinc salt.

(8)

no catalyst, 5 h	70%, 78% *endo*
with $ZnBr_2$, 3 h	80%, 95% *endo*

(9)

single isomer

Some intramolecular ene reactions benefit from $ZnBr_2$ catalysis to afford the cyclic products under milder conditions, in higher yields and selectivities (eqs 10 and 11).[14,15] Generally, the use of $ZnBr_2$ is preferred over ***Zinc Chloride*** or ***Zinc Iodide*** in this type of reaction.[15]

(10)

180 °C, *o*-dichlorobenzene	66% (83% de)
$ZnBr_2$, CH_2Cl_2, 25 °C	79% (95% de)

(11)

(R^1 = CN, R^2 = H):(R^1 = H, R^2 = CN) = 88:12

Activation of C=X Bonds. Lewis acid activation of carbonyl compounds by $ZnBr_2$ promotes the addition of allylsilanes and silyl ketene acetals.[16] Addition to imines has also been reported.[17] In general, other Lewis acids have been found to be more useful, though in some instances $ZnBr_2$ has proven to be advantageous (eq 12).[2]

Activation of C–X Bonds. Even more important than carbonyl activation, $ZnBr_2$ promotes substitution reactions with suitably active organic halides with a variety of nucleophiles. Alkylation of silyl enol ethers and silyl ketene acetals using benzyl and allyl halides proceeds smoothly (eq 13).[3] Especially useful electrophiles are α-thio halides which afford products that may be desulfurized or oxidatively eliminated to result in α,β-unsaturated ketones, esters, and lactones (eq 14).[18] Other electrophiles that have been used with these alkenic nucleophiles include ***Chloromethyl Methyl Ether***, $HC(OMe)_3$, and ***Acetyl Chloride***.[3,19]

(13)

(14)

Enol ethers and allylic silanes and stannanes will engage cyclic α-seleno sulfoxides,[20] ω-acetoxy lactams,[21] and acyl glycosides (eq 15)[22] in the presence of $ZnBr_2$ catalysis. Along these lines, it has been found that $ZnBr_2$ is superior to ***Boron Trifluoride Etherate*** in promoting glycoside bond formation using trichloroimidate-activated glycosides (eq 16).[23] Imidazole carbamates are also effective activating groups for $ZnBr_2$-mediated glycosylation (eq 17).[24]

(15)

α:β = 1:1

Cyclic acetals also undergo highly selective, Lewis acid-dependent ring opening substitution with ***Cyanotrimethylsilane*** (eq 18).[25]

Reduction. Complexation with $ZnBr_2$ has been shown to markedly improve stereoselectivity in the reduction of certain heteroatom-substituted ketones (eqs 19 and 20).[26,27] Furthermore, the *anti* selectivity observed in $BF_3 \cdot OEt_2$-mediated intramolecular hydrosilylation of ketones is reversed when $ZnBr_2$ is used instead (eq 21).[28]

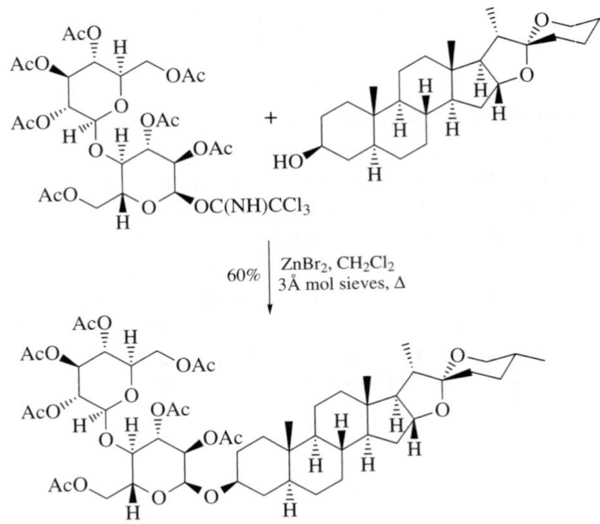

60% $ZnBr_2$, CH_2Cl_2, 3Å mol sieves, Δ (16)

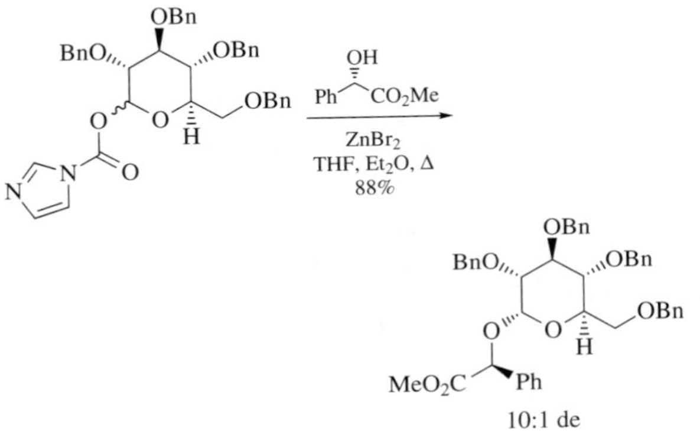

$ZnBr_2$, THF, Et_2O, Δ, 88% (17)

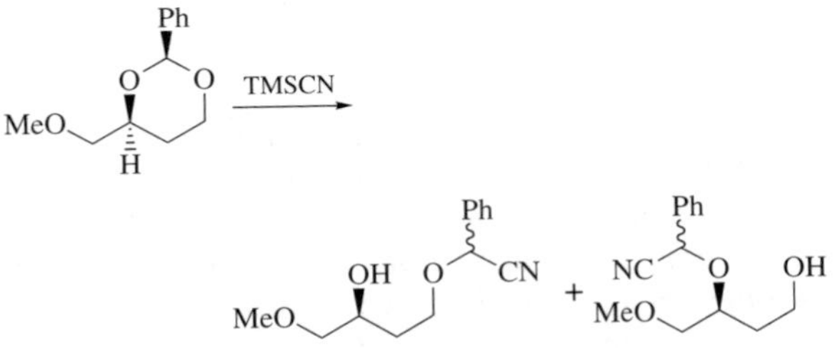

10:1 de

TMSCN (18)

Conditions	Ratio
$ZnBr_2$, CH_2Cl_2, 25 °C, 20 h	1:250
$TiCl_4$, CH_2Cl_2, 25 °C, 20 h	250:1

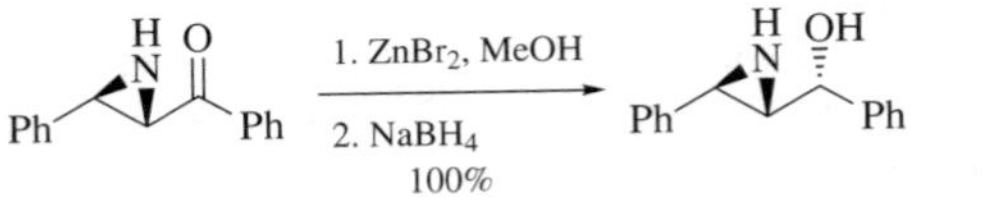

1. $ZnBr_2$, MeOH; 2. $NaBH_4$, 100% (19)

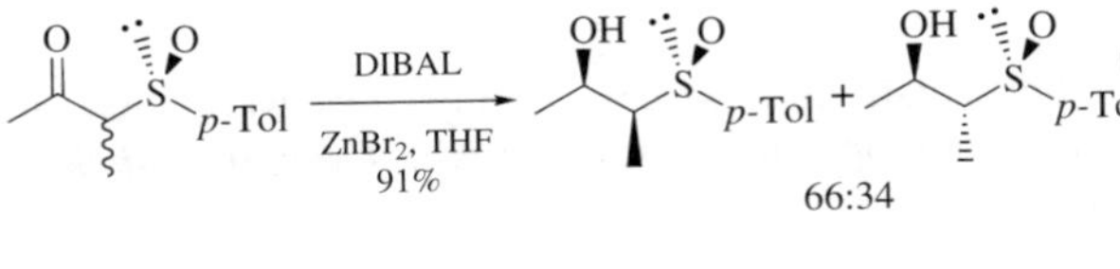

DIBAL, $ZnBr_2$, THF, 91%; 66:34 (20)

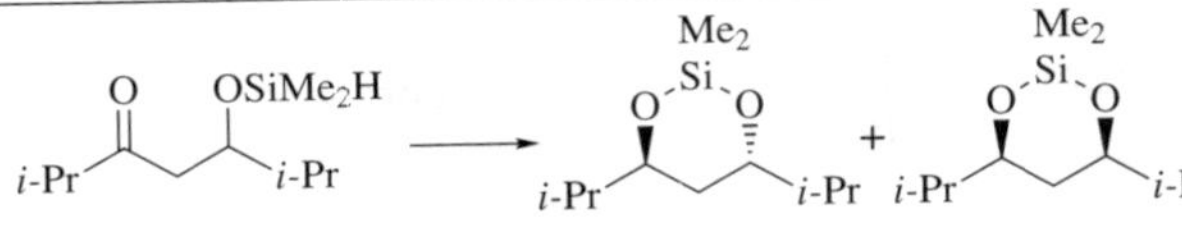

(21)

Conditions	Ratio
$BF_3 \cdot OEt_2$, CH_2Cl_2, –80 °C, 2 h	23:1
$ZnBr_2$, CH_2Cl_2, –80 °C, 8 h	1:6

Deprotection. $ZnBr_2$ is a very mild reagent for several deprotection protocols, including the detritylation of nucleotides[29] and deoxynucleotides,[30] *N*-deacylation of *N,O*-peracylated nucleotides,[31] and the selective removal of Boc groups from secondary amines in the presence of Boc-protected primary amines.[32] Perhaps the most widespread use of $ZnBr_2$ for deprotection is in the mild removal of MEM ethers to afford free alcohols (eq 22).[33]

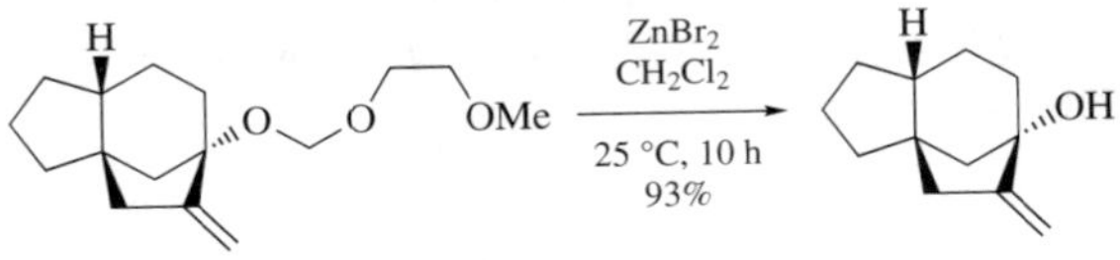

$ZnBr_2$, CH_2Cl_2, 25 °C, 10 h, 93% (22)

Miscellaneous. An important method for the synthesis of stereodefined trisubstituted double bonds involves the treatment of cyclopropyl bromides with $ZnBr_2$. The (*E*) isomer is obtained almost exclusively by this method (eq 23).[34]

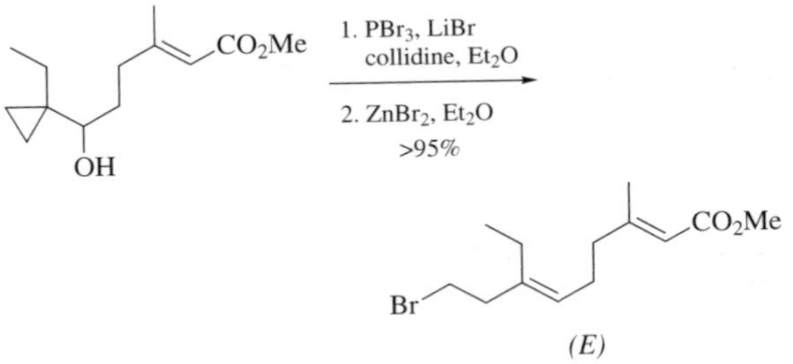

1. PBr_3, LiBr, collidine, Et_2O; 2. $ZnBr_2$, Et_2O, >95% (23)

(*E*)

The rearrangement of a variety of terpene oxides has been examined (eq 24).[35] While $ZnBr_2$ is generally a satisfactory catalyst for this purpose, other Lewis acids, including $ZnCl_2$[36] and ***Magnesium Bromide***,[37] are advantageous in some instances.

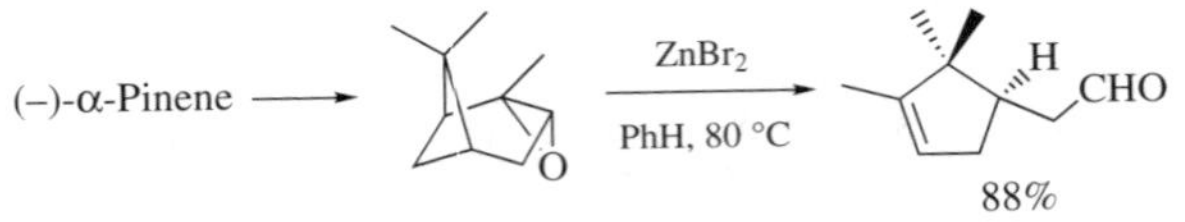

(–)-α-Pinene → ; $ZnBr_2$, PhH, 80 °C; 88% (24)

In the presence of $ZnBr_2$/48% ***Hydrobromic Acid***, suitably functionalized cyclopropanes undergo ring expansion to afford cyclobutane (eq 25)[38] and α-methylene butyrolactone products (eq 26).[39] One-carbon ring expansion has been reported when certain trimethylsilyl dimethyl acetals are exposed to $ZnBr_2$ with warming (eq 27).[40]

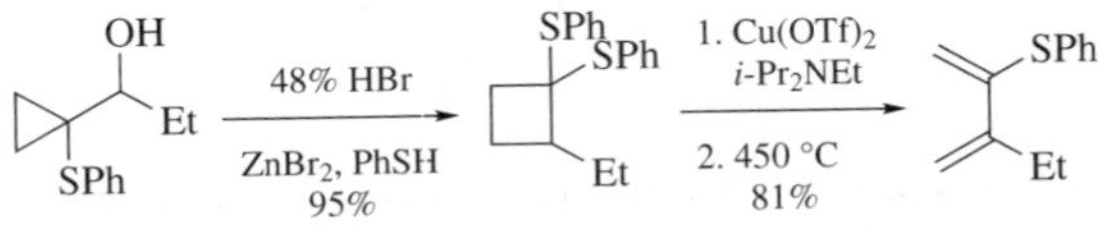

48% HBr, $ZnBr_2$, PhSH, 95%; 1. $Cu(OTf)_2$, *i*-Pr_2NEt; 2. 450 °C, 81% (25)

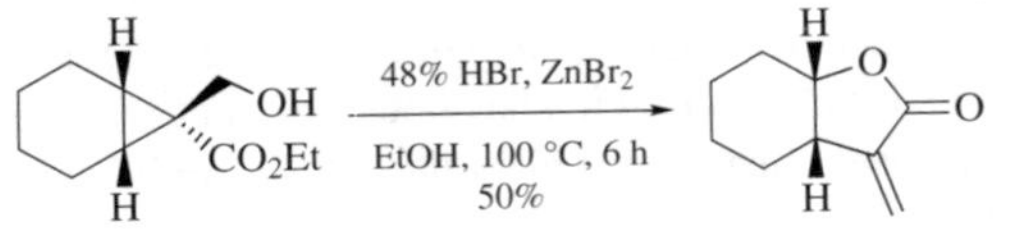

48% HBr, $ZnBr_2$, EtOH, 100 °C, 6 h, 50% (26)

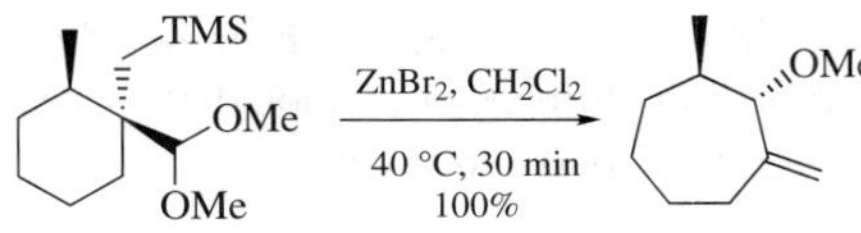

$\xrightarrow[\text{40 °C, 30 min; 100\%}]{ZnBr_2,\ CH_2Cl_2}$ (27)

Related Reagents. Magnesium Bromide; Zinc Chloride; Zinc Iodide.

---

1. (a) Knochel, P. *COS* **1991**, *1*, Chapter 1.7. (b) Rathke, M. W.; Weipert, P. *COS* **1991**, *2*, Chapter 1.8.
2. For an example: Bellassoued, M.; Ennigrou, R.; Gaudemar, M. *JOM* **1988**, *338*, 149.
3. For examples: (a) Reetz, M. T.; Maier, W. F. *AG(E)* **1978**, *17*, 48. (b) Reetz, M. T.; Chatziiosifidis, I.; Löwe, W. F.; Maier, W. F. *TL* **1979**, 1427. (c) Paterson, I. *TL* **1979**, 1519.
4. Riecke, R. D.; Uhm, S. J.; Hudnall, P. M. *CC* **1973**, 269.
5. For representative examples of allylic and propargylic zinc reagents: (a) Yamamoto, Y.; Nishii, S.; Maruyama, K.; Komatsu, T.; Ito, W. *JACS* **1986**, *108*, 7778. (b) Yamamoto, Y.; Ito, W. *T* **1988**, *44*, 5414. (c) Yamamoto, Y.; Ito, W.; Maruyama, K. *CC* **1985**, 1131. (d) Yamanoto, Y.; Komatsu, T.; Maruyama, K. *CC* **1985**, 814. (e) Fronza, G.; Fuganti, C.; Grasselli, P.; Pedrocchi-Fantoni, G.; Zirotti, C. *TL* **1982**, *23*, 4143. (f) Fujisawa, T.; Kojima, E.; Itoh, T.; Sato, T. *TL* **1985**, *26*, 6089. (g) Pornet, J.; Miginiac, L. *BSF* **1975**, 841. (h) Yamamoto, Y.; Komatsu, T.; Maruyama, K. *JOM* **1985**, *285*, 31. (i) Bouchoule, C.; Miginiac, P. *CR(C)* **1968**, *266*, 1614. (j) Miginiac, L.; Mauzé, B. *BSF* **1968**, 3832. (k) Arous-Chtara, R.; Gaudemar, M.; Moreau, J.-L. *CR(C)* **1976**, *282*, 687. (l) Moreau, J.-L.; Gaudemar, M. *BSF* **1971**, 3071. (m) Miginiac, L.; Mauzé, B. *BSF* **1968**, 2544.
6. Negishi, E.; King, A. O.; Okudado, N. *JOC* **1977**, *42*, 1821.
7. Sengupta, S.; Snieckus, V. *JOC* **1990**, *55*, 5680. See also: Gilchrist, T. L.; Summersell, R. J. *TL* **1987**, *28*, 1469.
8. (a) Jabri, N.; Alexakis, A.; Normant, J. F. *BSF(2)* **1983**, 321. (b) Jabri, N.; Alexakis, A.; Normant, J. F. *TL* **1982**, *23*, 1589. (c) Jabri, N.; Alexakis, A.; Normant, J. F. *TL* **1981**, *22*, 959. (d) Jabri, N.; Alexakis, A.; Normant, J. F. *TL* **1981**, *22*, 3851.
9. Cross, G.; Vriesema, B. K.; Boven, G.; Kellogg, R. M.; van Bolhuis, F. *JOM* **1989**, *370*, 357.
10. (a) Courtemanche, G.; Normant, J.-F. *TL* **1991**, *32*, 5317. (b) Marek, I.; Normant, J.-F. *TL* **1991**, *32*, 5973. (c) Marek, I.; Lefrançois, J.-M.; Normant, J.-F. *SL* **1992**, 633.
11. Tucker, C. E.; Knochel, P. *JACS* **1991**, *113*, 9888.
12. Narayana Murthy, Y. V. S.; Pillai, C. N. *SC* **1991**, *21*, 783. See also: López, R.; Carretero, J. C. *TA* **1991**, *2*, 93.
13. Kanemasa, S.; Tsuruoka, T.; Wada, E. *TL* **1993**, *34*, 87.
14. Tietze, L. F.; Biefuss, U.; Ruther, M. *JOC* **1989**, *54*, 3120. See also: (a) Tietze, L. F.; Ruther, M. *CB* **1990**, *123*, 1387. (b) Nakatani, Y.; Kawashima, K. *S* **1978**, 147.
15. Hiroi, K.; Umemura, M. *TL* **1992**, *33*, 3343.
16. (a) Mikami, K.; Kawamoto, K.; Loh, T.-P.; Nakai, T. *CC* **1990**, 1161. (b) Bellassoued, M.; Gaudemar, M. *TL* **1988**, *29*, 4551.
17. Gaudemar, M.; Bellassoued, M. *TL* **1990**, *31*, 349.
18. Khan, H. A.; Paterson, I. *TL* **1982**, *23*, 5083. See also: (a) Paterson, I. *T* **1988**, *44*, 4207. (b) Khan, H. A.; Paterson, I. *TL* **1982**, *23*, 4811. (c) Paterson, I.; Fleming, I. *TL* **1979**, *20*, 993, 995, 2179.
19. Fleming, I.; Goldhill, J.; Paterson, I. *TL* **1979**, 3209.
20. Ren, P.; Ribezzo, M. *JACS* **1991**, *113*, 7803.
21. Ohta, T.; Shiokawa, S.; Iwashita, E.; Nozoe, S. *H* **1992**, *34*, 895.
22. Kozikowski, A. P.; Sorgi, K. L. *TL* **1982**, *23*, 2281.
23. Urban, F. J.; Moore, B. S.; Breitenbach, R. *TL* **1990**, *31*, 4421.
24. Ford, M. J.; Ley, S. V. *SL* **1990**, 255.
25. Corcoran, R. C. *TL* **1990**, *31*, 2101.
26. Bartnik, R.; Lesniak, S.; Laurent, A. *TL* **1981**, *22*, 4811.
27. Barros, D.; Carreño, M. C.; Ruano, J. L. G.; Maestro, M. C. *TL* **1992**, *33*, 2733.
28. Anwar, S.; Davis, A. P. *T* **1988**, *44*, 3761.
29. Waldemeier, F.; De Bernardini, S.; Leach, C. A.; Tamm, C. *HCA* **1982**, *65*, 2472.
30. (a) Kohli, V.; Blöcker, H.; Köster, H. *TL* **1980**, *21*, 2683. (b) Matteuci, M. D.; Caruthers, M. H. *TL* **1980**, *21*, 3243.
31. Kierzek, R.; Ito, H.; Bhatt, R.; Itakura, K. *TL* **1981**, *22*, 3761.
32. Nigam, S. C.; Mann, A.; Taddei, M.; Wermuth, C.-G. *SC* **1989**, *19*, 3139.
33. Corey, E. J.; Gras, J.-L.; Ulrich, P. *TL* **1976**, 809.
34. Johnson, W. S.; Li, T.; Faulkner, D. J.; Campbell, S. F. *JACS* **1968**, *90*, 6225. See also: (a) Brady, S. F.; Ilton, M. A.; Johnson, W. S. *JACS* **1968**, *90*, 2882. (b) Nakamura, H.; Yamamoto, H.; Nozaki, H. *TL* **1973**, 111.
35. Lewis, J. B.; Hendrick, G. W. *JOC* **1965**, *30*, 4271. See also: (a) Settine, R. L.; Parks, G. L.; Hunter, G. L. K. *JOC* **1964**, *29*, 616. (b) Bessière-Chréieu, Y.; Bras, J. P. *CR(C)* **1970**, *271*, 200. (c) Clark, Jr., B. C.; Chafin, T. C.; Lee, P. L.; Hunter, G. L. K. *JOC* **1978**, *43*, 519. (d) Watanabe, H.; Katsuhara, J.; Yamamoto, N. *BCJ* **1971**, *44*, 1328.
36. Kaminski, J.; Schwegler, M. A.; Hoefnagel, A. J.; van Bekkum, H. *RTC* **1992**, *111*, 432.
37. Serramedan, D.; Marc, F.; Pereyre, M.; Filliatre, C.; Chabardès, P.; Delmond, B. *TL* **1992**, *33*, 4457.
38. Kwan, T. W.; Smith, M. B. *SC* **1992**, *22*, 2273.
39. Hudrlik, P. F.; Rudnick, L. R.; Korzeniowski, S. H. *JACS* **1973**, *95*, 6848.
40. (a) Tanino, K.; Katoh, T.; Kuwajima, I. *TL* **1988**, *29*, 1815. (b) Tanino, K.; Katoh, T.; Kuwajima, I. *TL* **1988**, *29*, 1819. See also: Tanino, K.; Sato, K.; Kuwajima, I. *TL* **1989**, *30*, 6551.

Glenn J. McGarvey
University of Virginia, Charlottesville, VA, USA

Zinc Chloride[1]

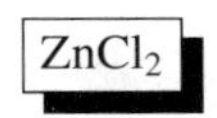

[7646-85-7] Cl_2Zn (MW 136.29)

(used in the preparation of organozinc reagents via transmetalation;[1] a mild Lewis acid useful for promoting cycloaddition,[2] substitution,[3] and addition reactions,[4] including electrophilic aromatic additions;[5] has found use in selective reductions[6])

Physical Data: mp 293 °C; bp 732 °C; *d* 2.907 g cm^{-3}.
Solubility: sol H_2O (432 g/100 g at 25 °C), EtOH (1 g/1.3 mL), glycerol (1 g/2 mL).
Form Supplied in: white, odorless, very deliquescent granules; principal impurities are H_2O and zinc oxychloride.
Analysis of Reagent Purity: melting point.
Purification: reflux (50 g) in dioxane (400 mL) in the presence of Zn^0 dust, then filter hot and allow to cool to precipitate purified $ZnCl_2$. Also, anhydrous material may be sublimed under a stream of dry HCl, followed by heating to 400 °C in a stream of dry N_2.
Handling, Storage, and Precautions: very hygroscopic; store under anhydrous conditions; moderately irritating to skin and mucous membranes.

Organozinc Reagents. The transmetalation of organomagnesium, organolithium, and organocopper reagents is an important and versatile method of preparing useful zinc reagents.[1a] Alternatively, $ZnCl_2$ may be employed for direct insertion of Zn^0 into carbon–halogen bonds using Mg^0/ultrasound[7] or prior reduction with K^0.[8] The addition of the resulting allylic and propargylic or allenic zinc reagents to carbonyl compounds, imines, and iminium salts represents an important method of selective carbon–carbon bond formation.[9,11] These reactions are generally guided by chelation and the Zimmerman–Traxler transition state[10] to control the relative stereochemistry about the new carbon–carbon bond (eq 1),[12] as well as with respect to preexisting stereocenters (eq 2).[13] Anions derived from propargylic deprotonation add to carbonyl compounds with a high degree of regiochemical integrity in the presence of $ZnCl_2$, i.e. allenic organozinc intermediates cleanly afford the propargylic product (eq 3).[11b]

Lewis acid		
none	63%	53:47
0.5 equiv $ZnCl_2$	68%	89:11

Alkylzinc reagents, often in the presence of copper salts, effectively participate in conjugate addition reactions[14] and clean S_N2' reactions (eq 4).[15]

Organozinc reagents are superior species for palladium- and nickel-catalyzed coupling reactions.[16] This offers an exceptional method for the selective formation of sp^2–sp^3 (eq 5),[17] sp^2–sp^2 (eqs 6 and 7),[18,19] and sp^2–sp (eq 8)[20] carbon–carbon bonds. In addition, the palladium-catalyzed coupling reactions of sp and sp^2 halides with vinylalanes (eq 9),[21] vinylcuprates,[22] vinylzirconium (eq 10)[23] and acyliron species[24] often proceed more effectively in the presence of $ZnCl_2$. This effect has been noted in other metal-mediated carbon bond formations,[25] though the role of this $ZnCl_2$ catalysis is not understood at present.

Organozinc reagents have been successfully exploited in asymmetric carbon bond formation.[26] Chiral glyoxylate esters engage organozinc species derived from Grignard reagents in selective addition reactions to afford enantiomerically enriched α-hydroxy acids (eq 11).[27] Enantioselective addition reactions of Grignard reagents have been achieved by sequential addition of $ZnCl_2$ and a chiral catalyst to afford secondary alcohols (eq 12).[28]

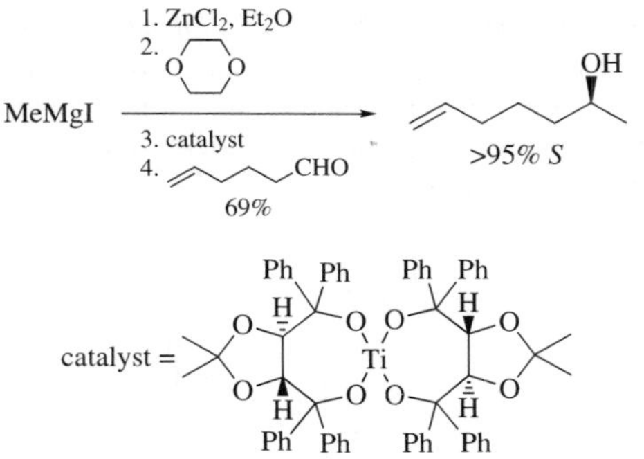

(12)

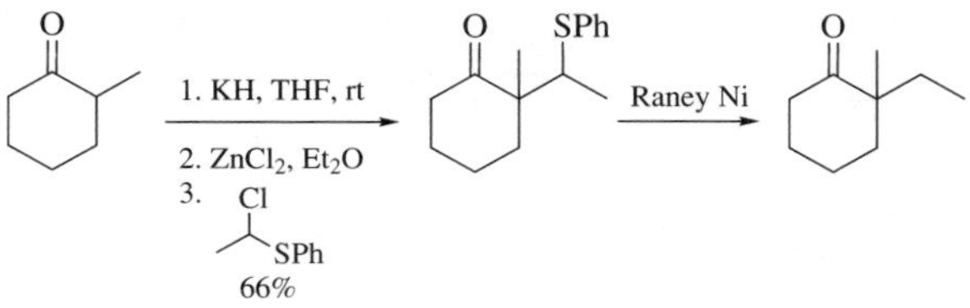

(15)

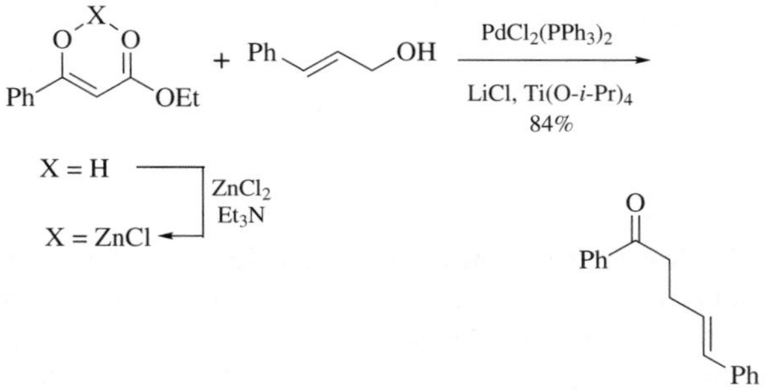

(16)

Zinc Enolates. Zinc enolates can be prepared by deprotonation of carbonyl compounds using standard bases, followed by transmetalation with $ZnCl_2$.[1b] These enolates offer important opportunities for stereoselective aldol condensations, including higher yields and stereochemical selection for *threo* crossed-aldol products (eq 13).[29] Both of these consequences are the result of the reversible formation of a six-membered zinc chelate intermediate (**1**) which favors the *anti* disposition of its substituents. Though their use has been mostly replaced by kinetically controlled aldol methodology, zinc enolates sometimes present advantages in specific cases.[30] For example, chelated zinc enolates derived from α-amino acids have been shown to be of value in the stereoselective synthesis of β-lactams (eq 14).[31] Controlled monoalkylation of unsymmetrical ketones has been accomplished via treatment of zinc enolates with α-chlorothio ethers (eq 15),[32] and mild allylation of β-dicarbonyl compounds may be realized through exposure of the corresponding zinc enolate to an allyl alcohol in the presence of a palladium catalyst (eq 16).[33]

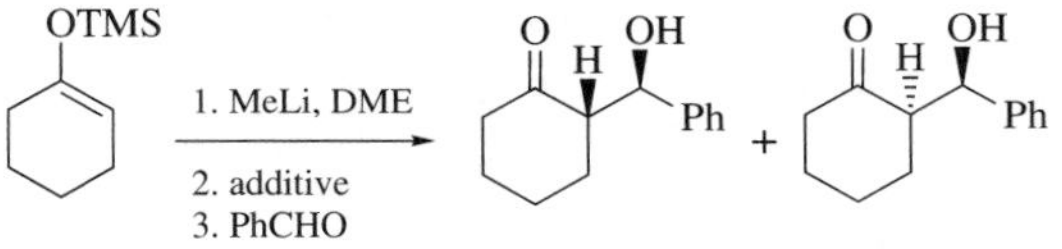

(13)

additive	
none, –72 °C	52:48
$ZnCl_2$, DME, 14 °C	83:17

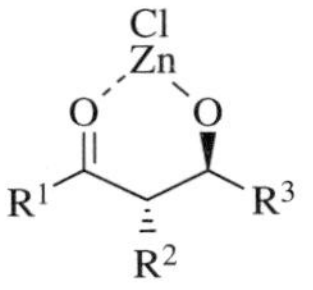

(1)

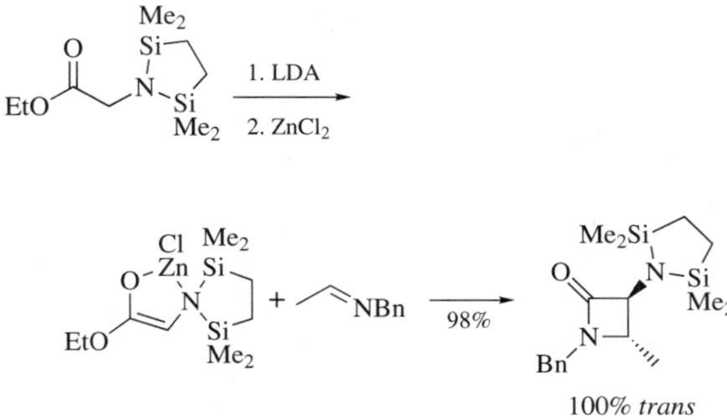

(14)

Homoenolates. Readily available mixed Me_3Si/alkyl acetals are converted into zinc homoenolates in high yield through exposure to $ZnCl_2$ in Et_2O (eq 17).[34] These mildly reactive species, which tolerate asymmetry α to the carbonyl, are useful for a variety of bond-forming reactions. In the presence of Me_3Si activation they undergo addition to aldehydes and, when admixed with CuBr·DMS (see ***Copper(I) Bromide***), can be allylated or conjugatively introduced to α,β-unsaturated ketones (eq 18).[35] Copper salts are not required for conjugate addition to propargylic esters (eq 19).[36] These zinc species also participate in useful palladium-catalyzed bond-forming processes (eq 20).[37]

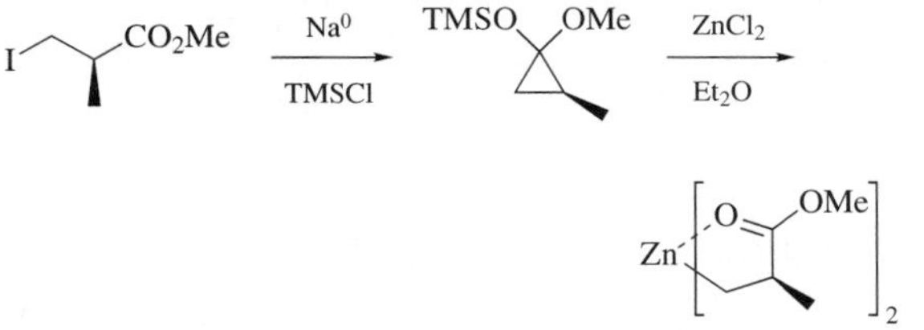

(17)

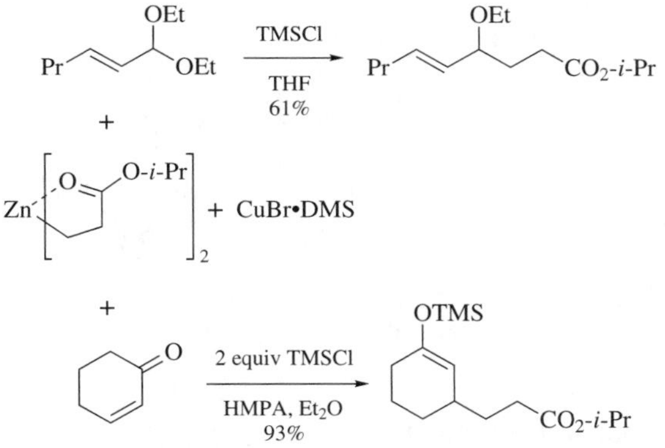

(18)

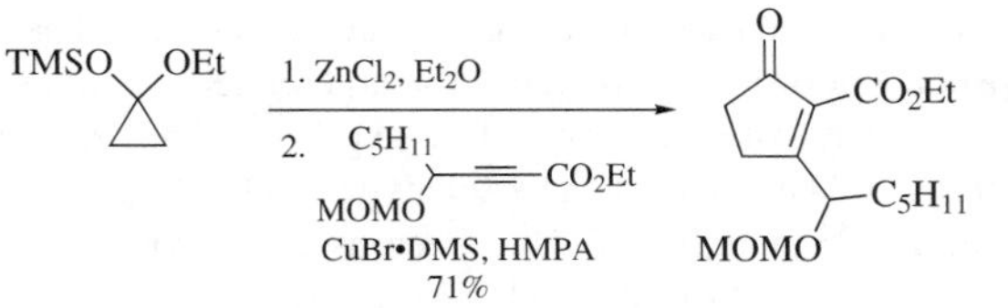

(19)

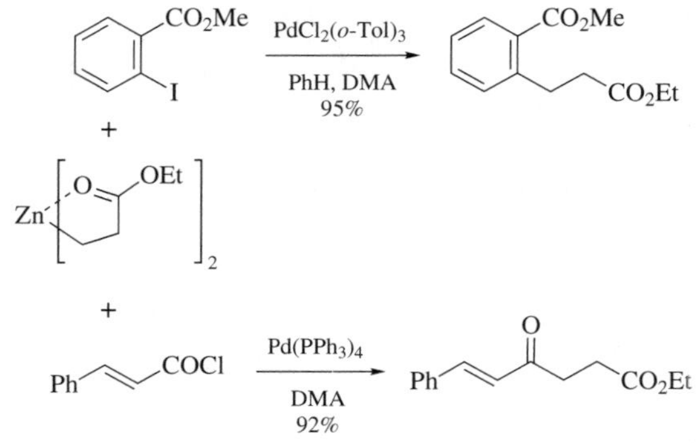

(20)

Cycloaddition Reactions. Catalysis by $ZnCl_2$ is often a powerful influence on cycloaddition reactions.[2] In addition to improving the rate of Diels–Alder reactions, enhanced control over regioselectivity (eq 21)[2c] and stereochemistry (eq 22)[2e] may be observed. An exceedingly important application of $ZnCl_2$ catalysis is found in the reaction of electron-rich dienes with carbonyl compounds.[2a,b] Intensive mechanistic studies on these reactions has uncovered two mechanistic pathways that afford stereochemically contrasting products, depending upon which Lewis acid is employed (eq 23).[38] It was concluded that cycloaddition reactions catalyzed by $ZnCl_2$ proceed via a classical [4 + 2] concerted process, whereas Lewis acids such as ***Boron Trifluoride Etherate*** and ***Titanium(IV) Chloride*** afford products through sequential aldol/cyclization processes. Examples abound wherein advantage has been taken of the predictable Cram–Felkin selectivity of this $ZnCl_2$ catalysis to exploit asymmetric variants of this reaction to synthesize a variety of natural products (eq 24).[39] The selective cycloaddition of electron-rich dienes with imines has also been catalyzed by $ZnCl_2$ (eq 25).[40]

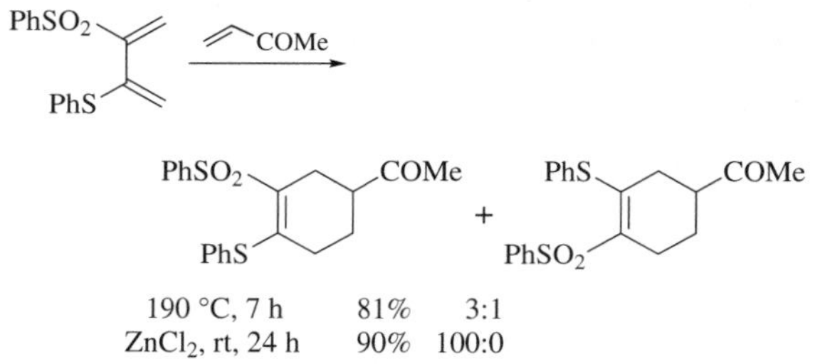

(21)

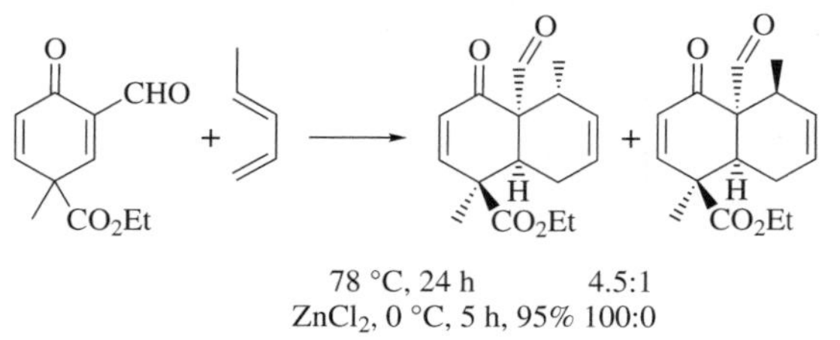

190 °C, 7 h	81%	3:1
$ZnCl_2$, rt, 24 h	90%	100:0

(22)

78 °C, 24 h 4.5:1
$ZnCl_2$, 0 °C, 5 h, 95% 100:0

Activation of C=X Bonds. The mild Lewis acid character of $ZnCl_2$ is frequently exploited to promote the addition of various nucleophiles to carbon–heteroatom double bonds.[4] The well-established Knoevenagel condensation and related reactions have been effectively catalyzed by $ZnCl_2$ (eq 26).[41] The addition of enol ethers and ketene acetals to aldehydes and ketones has been noted (eq 27),[42] though $ZnCl_2$ has been used less widely in these aldol-type condensations than other Lewis acids, including $TiCl_4$, ***Magnesium Bromide***, and ***Tin(IV) Chloride***, to name a few.[4] In some instances, excellent levels of stereocontrol have been observed (eq 28).[43] Analogous additions to imines have been noted as well.[4e–g,44] Some conjugate addition reactions may also benefit from $ZnCl_2$ catalysis (eq 29).[45]

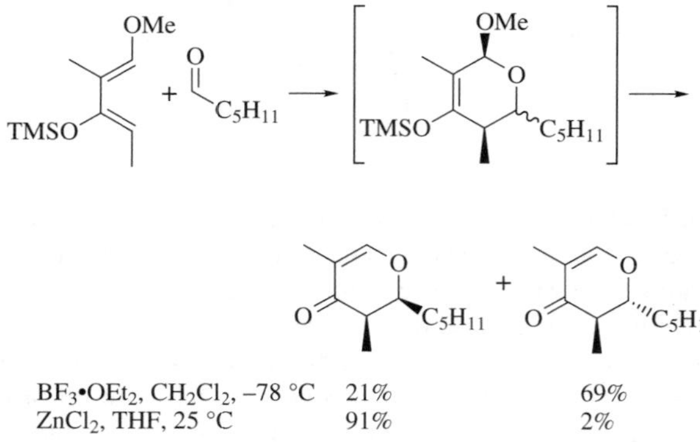

(23)

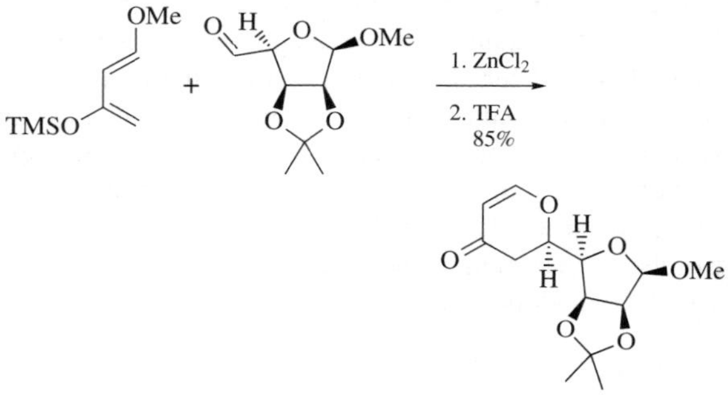

$BF_3 \cdot OEt_2$, CH_2Cl_2, –78 °C	21%	69%
$ZnCl_2$, THF, 25 °C	91%	2%

(24)

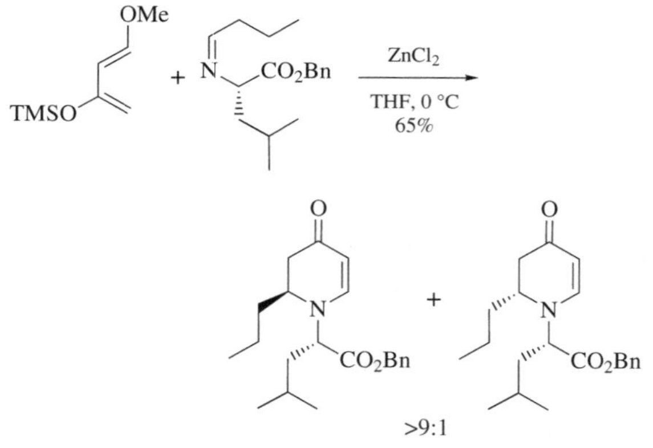

(25)

>9:1

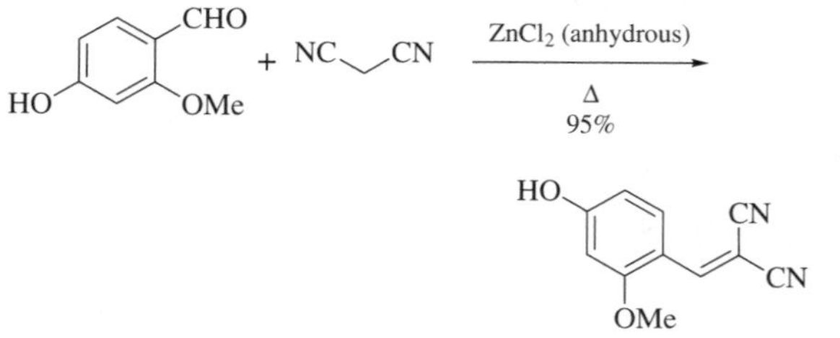

(26)

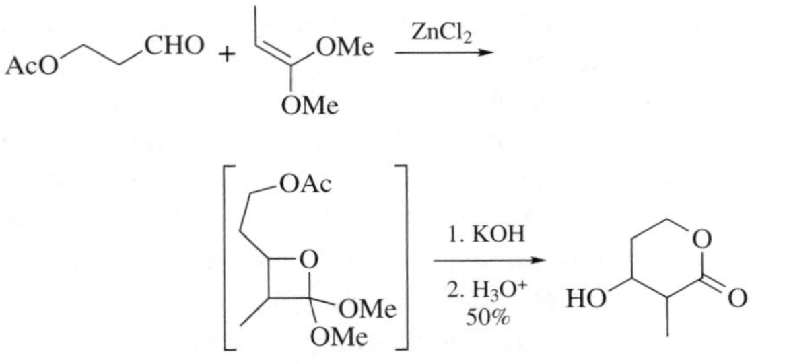

(27)

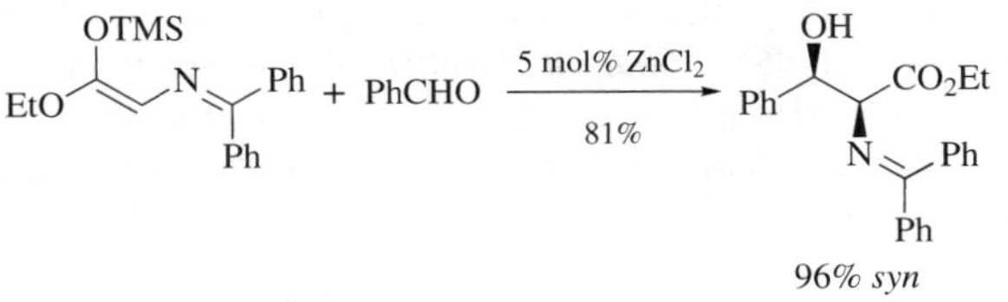

(28)

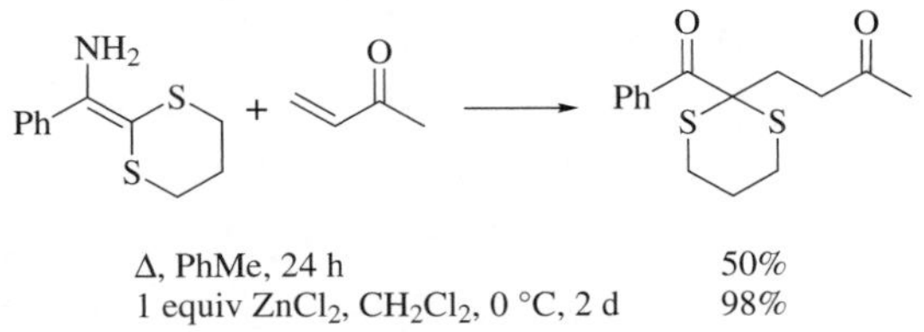

96% *syn*

(29)

Δ, PhMe, 24 h 50%
1 equiv $ZnCl_2$, CH_2Cl_2, 0 °C, 2 d 98%

The formation of cyanohydrins using ***Cyanotrimethylsilane*** and ***Isoselenocyanatotrimethylsilane*** has been effectively catalyzed by $ZnCl_2$ (eq 30),[46] as has Strecker amino acid synthesis via the treatment of imines with $Me_3SiCN/ZnCl_2$.[47] The combination of carbonyl compounds with ***Acetyl Chloride*** or ***Acetyl Bromide*** may be promoted by $ZnCl_2$ to afford protected vicinal halohydrins (eq 31).[48]

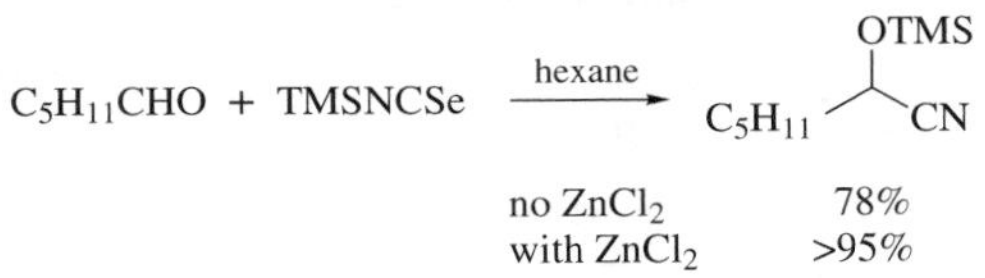

(30)

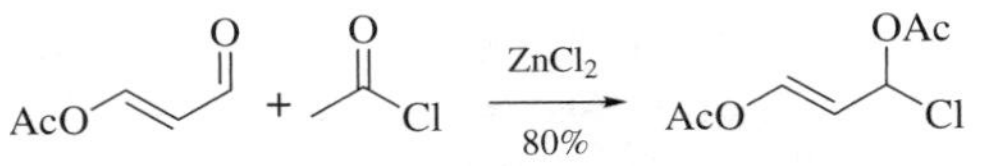

no $ZnCl_2$ 78%
with $ZnCl_2$ >95%

(31)

It is noteworthy that the treatment of carbonyl compounds with ***Chlorotrimethylsilane***/$ZnCl_2$ results in a useful synthesis of Me_3Si enol ethers which, in turn, may be useful for other carbon bond-forming processes (eq 32).[49]

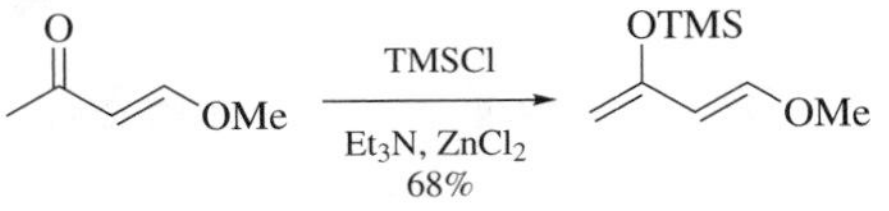

(32)

Activation of C–X Bonds. The activation of C–X single bonds toward nucleophilic substitution is also mediated by the Lewis acidic character of $ZnCl_2$.[3] Benzylic (eq 33),[50] allylic (eq 34),[3d,51] propargylic,[52] and tertiary halides (eq 35)[53] undergo substitution with mild carbon and heteroatom[3e] nucleophiles.

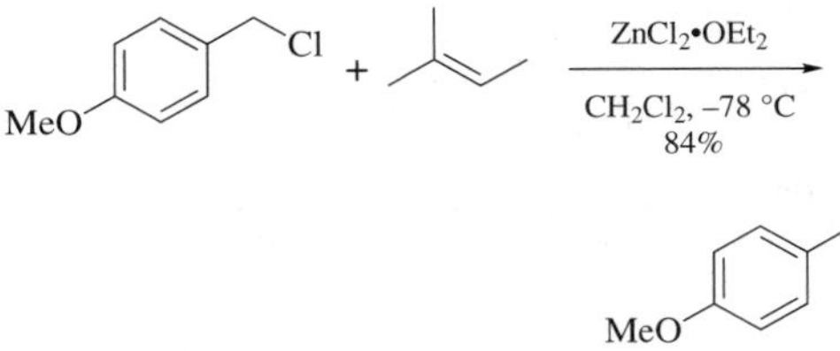

(33)

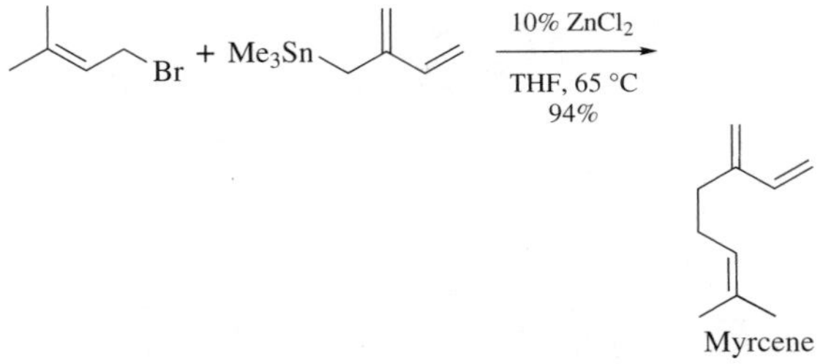

(34)

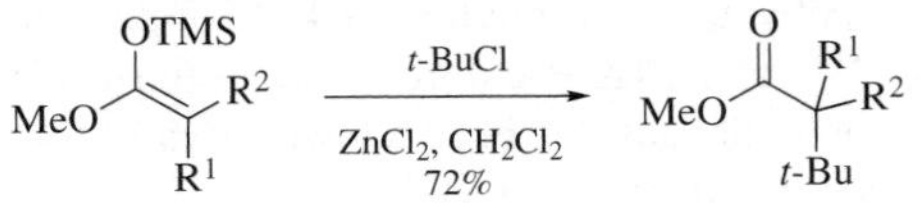

(35)

In a similar fashion, acetals (eq 36)[54] and orthoesters (eq 37)[55] may be used as electrophiles in substitution reactions with electron-rich alkenic nucleophiles. The combination of $ZnCl_2$ with co-catalysts has sometimes proven advantageous in these reactions (eq 38).[56]

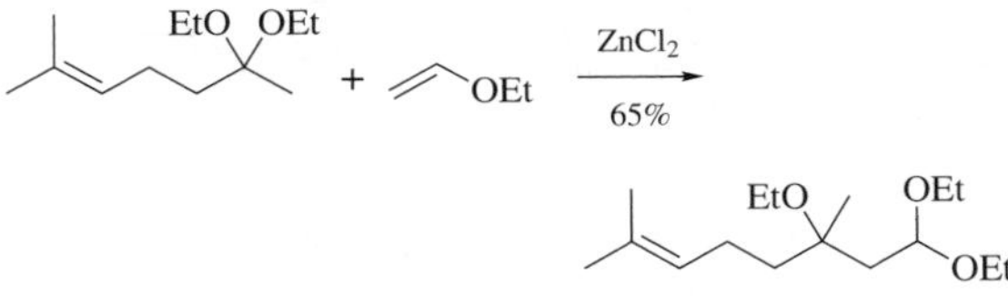

(36)

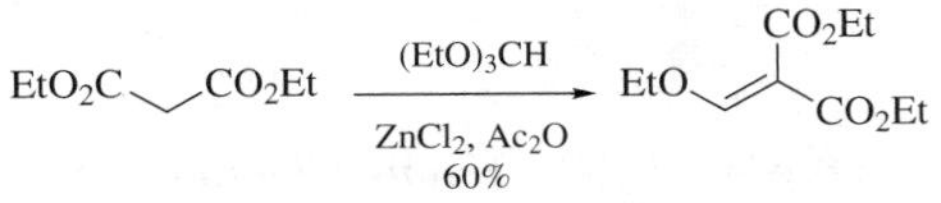

(37)

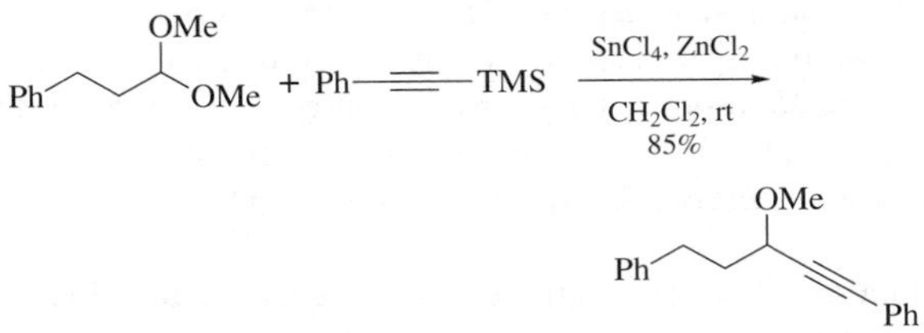

(38)

The regioselective ring opening reactions of epoxides (eq 39),[57,58] oxetanes,[58] and tetrahydrofurans (eq 40)[59] has been promoted by $ZnCl_2$ to afford adducts with suitable nucleophiles.

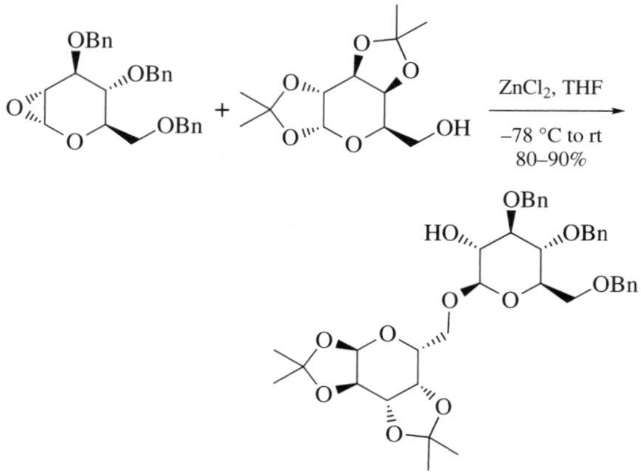

(39)

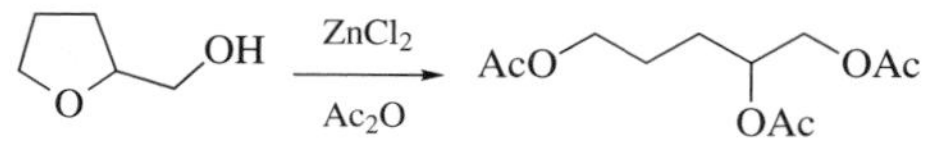

(40)

Activation by $ZnCl_2$ of allylic (eq 41)[60] and propargylic chlorides (eq 42),[61] as well as α-chloroenamines (eq 43),[62] in the presence of simple alkenes has been shown to yield four-membered and five-membered cycloadducts.

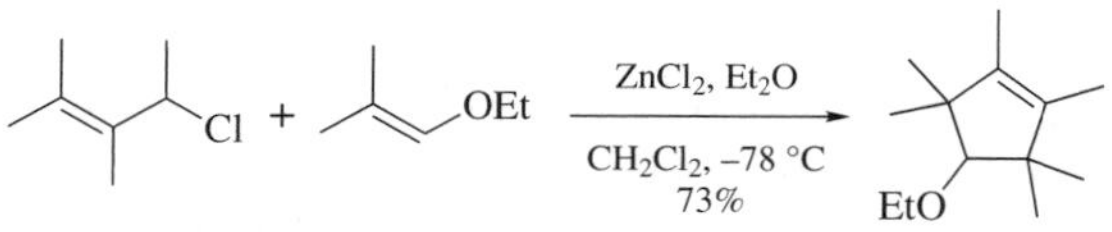

(41)

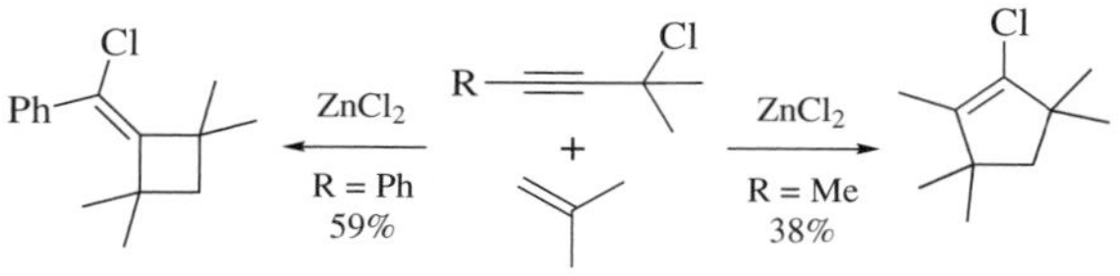

(42)

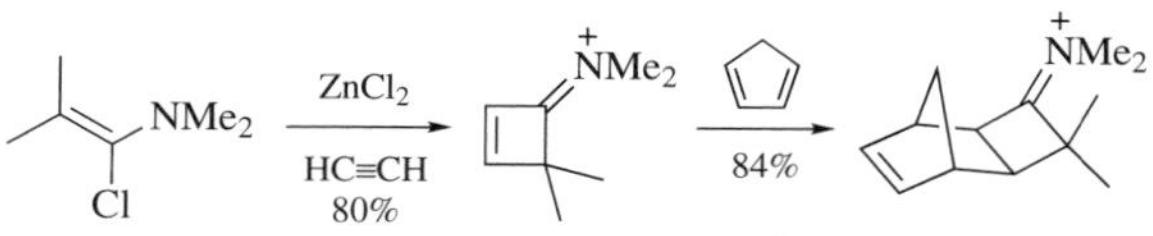

(43)

Chlorination of alcohols by ***Thionyl Chloride***,[63] the preparation of acyl chlorides from lactones and anhydrides,[64] and the bromination and iodination of aromatic rings by ***Benzyltrimethylammonium Tribromide*** and ***Benzyltrimethylammonium Dichloroiodate***, respectively,[65] are all effectively catalyzed by the presence of $ZnCl_2$. In addition, $ZnCl_2$ acts as a source of chloride for the halogenation of primary, secondary, and allylic alcohols using ***Triphenylphosphine–Diethyl Azodicarboxylate***.[66]

Reduction. The very useful reducing agent ***Zinc Borohydride*** is prepared by exposure of $NaBH_4$ to $ZnCl_2$ in ether solvents.[67] Its use in selective reductions is described elsewhere.[6a] A complex reducing agent resulting from the mixture of NaH/*t*-pentyl-OH/$ZnCl_2$, given the acronym ZnCRA (***Zinc Complex Reducing Agents***), is found to open epoxides in a highly regioselective fashion, favoring hydride delivery at the least hindered position.[68] A related reagent, ZnCRASi, which includes Me_3SiCl in the mixture, selectively reduces ketones in high yield, though the levels of selectivity do not compete with other selective reagents.[69]

The reduction of ketones and aldehydes by silicon and tin hydrides in the presence of $ZnCl_2$ has been documented.[70] In the presence of Pd^0 catalysts and $ZnCl_2$, these hydrides selectively reduce α,β-unsaturated aldehydes and ketones to the corresponding saturated products (eq 44).[71] The presence of $ZnCl_2$ has been shown to modify the reactivity of several common reducing agents. For example, the mixture of ***Sodium Cyanoborohydride*** and $ZnCl_2$ selectively reduces tertiary, allylic, and benzylic halides (eq 45),[72] ***Sodium Borohydride***, in the presence of $ZnCl_2$ and $PhNMe_2$ will reduce aryl esters to primary alcohols,[73] and ***Lithium Aluminum Hydride*** with $ZnCl_2/CuCl_2$ desulfurizes dithianes (eq 46).[74]

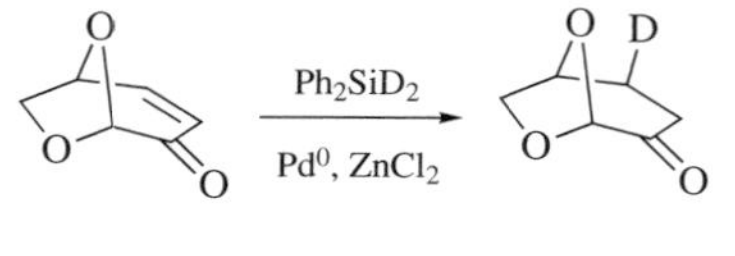

(44)

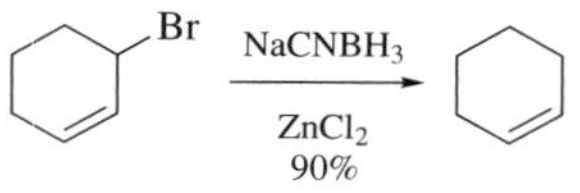

(45)

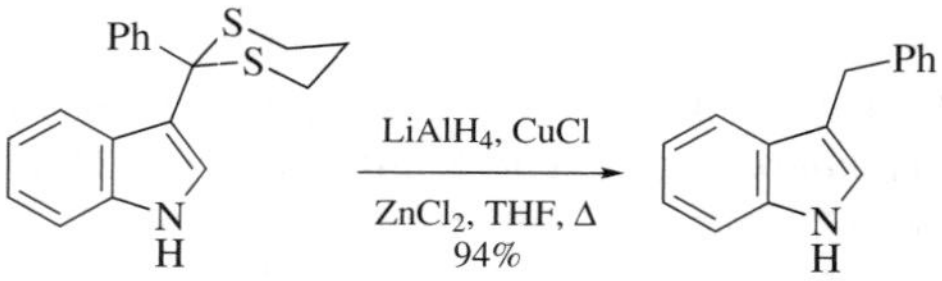

(46)

Stereoselectivity is also modified by the presence of $ZnCl_2$ (eq 47).[75] Enantioselective reduction of aryl ketones has been observed with ***Diisobutylaluminum Hydride*** (DIBAL) modified by $ZnCl_2$ and chiral diamine ligands (eq 48).[76]

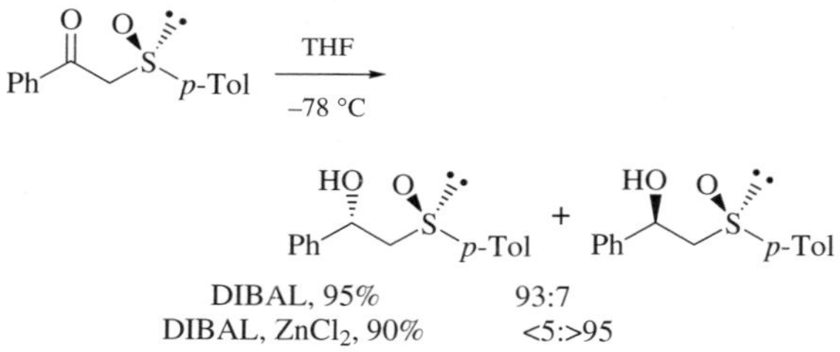

(47)

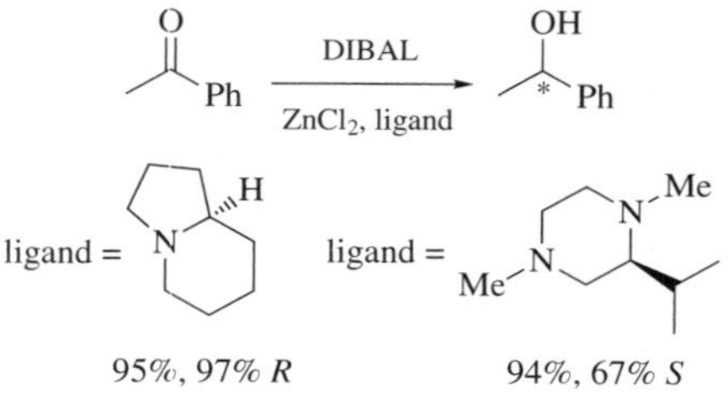

DIBAL, 95% 93:7
DIBAL, $ZnCl_2$, 90% <5:>95

(48)

95%, 97% *R* 94%, 67% *S*

Protection/Deprotection. The acetylation of carbohydrates and other alcohols has been realized using ***Acetic Anhydride***/$ZnCl_2$ (eq 49).[77] It is found that $ZnCl_2$ imparts selectivity to both acetylation[78] and acetonide formation[79] of polyols, which is useful in the synthetic manipulation of carbohydrates. Synthetically useful selective deprotection of acetates (eq 50)[80] and dimethyl acetals (eq 51)[81] has been reported to be mediated by $ZnCl_2$.

Ac_2O, $ZnCl_2$; H_2O, Δ; 79% (49)

(50)

$ZnCl_2$, EtOH	R^1 = H, R^2 = Ac, 100%
$SnCl_2$	R^1 = Ac, R^2 = H, 100%

AcCl; $ZnCl_2$; 89% (51)

Acylation. Unsaturated esters are obtained through acylation of alkenes by anhydrides using activation by $ZnCl_2$ (eq 52).[82] The mixture of $RCOCl/ZnCl_2$ is effective in the acylation of silyl enol ethers to afford β-dicarbonyl products (eq 53).[83] Friedel–Crafts acylation is catalyzed by $ZnCl_2$, using anhydrides or acyl halides as the electrophiles (eqs 54 and 55).[84,85]

$(EtCO)_2O$; $ZnCl_2$, H_2O; 79% (52)

t-BuCOCl; $ZnCl_2$; 90% (53)

Ac_2O; $ZnCl_2$, Δ; 65% (54)

$ZnCl_2$; $POCl_3$; 68% (55)

Aromatic Substitution. Several important classes of aromatic substitutions are mediated by $ZnCl_2$, including the Hoesch reaction (eq 56)[86] and the Fischer indole synthesis (eq 57).[87] Haloalkylation of aromatic rings using ***Formaldehyde*** or ***Chloromethyl Methyl Ether*** is readily accomplished through the agency of $ZnCl_2$ and warming (eq 58).[88]

MeCN, $ZnCl_2$; HCl, Et_2O; H_2O; 87% (56)

$ZnCl_2$; cumene, Δ; 86% (57)

Cl⌒OMe; $ZnCl_2$, Δ; 86% (58)

Related Reagents. Aluminum Chloride; Diphenylsilane–Tetrakis(triphenylphosphine)palladium(0)–Zinc Chloride; Phosphorus(III) Chloride–Zinc(II) Chloride; Phosphorus Oxychloride–Zinc(II) Chloride; Tin(IV) Chloride; Tin(IV) Chloride–Zinc Chloride; Titanium(IV) Chloride; Zinc Bromide.

1. (a) Knochel, P. *COS* **1991**, *1*, Chapter 1.7. (b) Rathke, M. W.; Weipert, P. *COS* **1991**, 2, Chapter 1.8.
2. For examples: (a) Danishefsky, S.; DeNinno, M. P. *AG(E)* **1987**, *26*, 15. (b) Danishefsky, S. *Aldrichim. Acta* **1986**, *19*, 59. (c) Chou, S.-S.; Sun, D.-J. *CC* **1988**, 1176. (d) Liu, H.-J.; Ulibarri, G.; Browne, E. N. C. *CJC* **1992**, *70*, 1545. (e) Liu, H.-J.; Han, Y. *TL* **1993**, *34*, 423.
3. For examples: (a) Zhai, D.; Zhai, W.; Williams, R. M. *JACS* **1988**, *110*, 2501. (b) Williams, R. M.; Sinclair, P. J.; Zhai, D.; Chen, D. *JACS* **1988**, *110*, 1547. (c) Paterson, I.; Fleming, I. *TL* **1979**, *20*, 993, 995, 2175. (d) Godschalz, J. P.; Stille, J. K. *TL* **1983**, *24*, 1905. (e) Miller, J. A. *TL* **1975**, 2050. (f) Shikhmamedbekova, A. Z.; Sultanov, R. A. *JGU* **1970**, *40*, 72. (g) Ishibashi, H.; Nakatani, H.; Umei, Y.; Yamamoto, W.; Ikeda, M. *JCS(P1)* **1987**, 589. (h) Mori, I.; Bartlett, P. A.; Heathcock, C. H. *JACS* **1987**, *109*, 7199.
4. For examples: (a) Eliel, E. L.; Hutchins, Sr., R. O.; Knoeber, M. *OS* **1970**, *50*, 38. (b) Angle, S. R.; Turnbull, K. D. *JACS* **1989**, *111*, 1136. (c) Takai, K.; Heathcock, C. H. *JOC* **1985**, *50*, 3247. (d) Chiba, T.; Nakai, T. *CL* **1987**, 2187. (e) Taguchi, T.; Kitagawa, O.; Suda, Y.; Ohkawa, W.; Hashimoto, A.; Iitaka, Y.; Kobayashi, Y. *TL* **1988**, *29*, 5291. (f) Kunz, H.; Pfrengle, W. *AG(E)* **1989**, *28*, 1067. (g) Kunz, H.; Pfrengle, W. *JOC* **1989**, *54*, 4261.
5. For examples: (a) Gulati, K. C.; Seth, S. R.; Venkataraman, K. *OSC* **1943**, 2, 522. (b) Böhmer, V.; Deveaux, J. *OPP* **1972**, *4*, 283. (c) Chapman, N. B.; Clarke, K.; Hughes, H. *JCS* **1965**, 1424.
6. For examples: (a) Oishi, T.; Nakata, T. *ACR* **1983**, *24*, 2653. (b) Fort, Y.; Feghouli, A.; Vanderesse, R.; Caubère, P. *JOC* **1990**, *55*, 5911. (c) Solladié, G.; Demailly, G.; Greck, C. *OS* **1977**, *56*, 8.
7. Boerma, J. In *Comprehensive Organometallic Chemistry*; Wilkinson, G., Ed.; Pergamon: Oxford, 1982, 2, Chapter 16.
8. Rieke, R. D.; Uhm, S. J. *S* **1975**, 452.
9. Allylic organozinc chlorides: (a) Masuyama, Y.; Kinugawa, N.; Kurusu, Y. *JOC* **1987**, *52*, 3702. (b) Tamao, K.; Nakajo, E.; Ito, Y. *JOC* **1987**, *52*, 957. (c) Pétrier, C.; Luche, J.-L. *JOC* **1985**, *50*, 910. (d) Tamao, K.; Nakajo, E.; Ito, Y. *T* **1988**, *44*, 3997. (e) Jacobson, R. N.; Clader, J. W. *TL* **1980**, *21*, 1205. (f) Hua, D. H.; Chon-Yu-King, R.; McKie, J. A.; Myer, L. *JACS* **1987**, *109*, 5026. (g) Courtois, G.; Harama, M.; Miginiac, P. *JOM* **1981**, *218*, 275. (h) Evans, D. A.; Sjogren, E. B. *TL* **1986**, *27*, 4961. (i)

Fujisawa, T.; Kofima, E.; Itoh, T.; Sato, T. *TL* **1985**, *26*, 6089. (j) Fang, J.-M.; Hong, B.-C. *JOC* **1987**, *52*, 3162. (k) Auvray, P.; Knochel, P.; Normant, J.-F. *TL* **1986**, *27*, 5091. For general reviews on the reactions of allylic organometallic reagents, see ref. 10.

10. (a) Yamamoto, Y. *ACR* **1987**, *20*, 243. (b) Hofmann, R. W. *AG(E)* **1982**, *21*, 555. (c) Yamamoto, Y.; Maruyama, K. *H* **1982**, *18*, 357. (d) Courtois, G.; Miginiac, L. *JOM* **1974**, *69*, 1.
11. (a) Zweifel, G.; Hahn, G. *JOC* **1984**, *49*, 4565. (b) Evans, D. A.; Nelson, J. V. *JACS* **1980**, *102*, 774.
12. Verlhac, J.-B.; Pereyre, M. *JOM* **1990**, *391*, 283.
13. (a) Fuganti, C.; Grasselli, P.; Pedrocchi-Fantoni, G. *JOC* **1983**, *48*, 909. (b) Fronza, G.; Fuganti, C.; Grasselli, P.; Pedrocchi-Fantoni, G. *J. Carbohydr. Chem.* **1983**, *2*, 225.
14. Watson, R. A.; Kjonaas, R. A. *TL* **1986**, *27*, 1437.
15. Yamamoto, Y.; Chounan, Y.; Tanaka, M.; Ibuka, T. *JOC* **1992**, *57*, 1024. See also: Arai, M.; Kawasuji, T.; Nakamura, E. *CL* **1993**, 357.
16. For examples: (a) Negishi, E.; Takahashi, T.; King, A. O. *OS* **1988**, *66*, 67. (b) Murahashi, S.-I.; Yamamura, M.; Yanagisawa, K.; Mita, N.; Kondo, K.; Negishi, E. *JOC* **1983**, *48*, 1560. (c) Tius, M. A.; Trehan, S. *JOC* **1986**, *51*, 765. (d) Rossi, R.; Carpita, A.; Cossi, P. *T* **1992**, *48*, 8801. (e) Shiragami, H.; Kawamoto, T.; Imi, K.; Matsubara, S.; Utimoto, K.; Nozaki, H. *T* **1988**, *44*, 4009. (f) Matsushita, H.; Negishi, E. *JACS* **1981**, *103*, 2882. (g) Pelter, A.; Rowlands, M.; Jenkins, I. H. *TL* **1987**, *28*, 5213. (h) Andreini, B. P.; Carpita, A.; Rossi, R. *TL* **1988**, *29*, 2239. (i) Negishi, E.; Okukado, N.; Lovich, S. F.; Luo, T.-T. *JOC* **1984**, *49*, 2629.
17. Tamao, K.; Ishida, M.; Kumada, M. *JOC* **1983**, *48*, 2120.
18. Russell, C. E.; Hegedus, L. S. *JACS* **1983**, *105*, 943.
19. Takahashi, K.; Ogiyama, M. *CC* **1990**, 1196.
20. (a) King, A. O.; Okukado, N.; Negishi, E. *CC* **1977**, 683. (b) King, A. O.; Negishi, E.; Villiani, Jr., F. J.; Silveira, A. *JOC* **1978**, *43*, 358.
21. Negishi, E.; Takahashi, T.; Baba, S. *OS* **1988**, *66*, 60. See also: Zweifel, G.; Miller, J. A. *OR* **1984**, *32*, 375.
22. (a) Jabri, N.; Alexakis, A.; Normant, J.-F. *BSF(2)* **1983**, *321*, 332. (b) Jabri, N.; Alexakis, A.; Normant, J.-F. *TL* **1981**, *22*, 959, 3851. (c) Nunomoto, S.; Kawakami, Y.; Yamashita, Y. *JOC* **1983**, *48*, 1912.
23. Negishi, E.; Okukado, N.; King, A. O.; Van Horn, D. E.; Spiegel, B. I. *JACS* **1978**, *100*, 2254. See also: Van Horn, D. E.; Valente, L. F.; Idacavage, M. J.; Negishi, E. *JOM* **1978**, *156*, C20.
24. Koga, T.; Makinouchi, S.; Okukado, N. *CL* **1988**, 1141.
25. (a) Godschalx, J.; Stille, J. K. *TL* **1980**, *21*, 2599. (b) Erdelmeier, I.; Gais, H.-J. *JACS* **1989**, *111*, 1125. (c) Cahiez, G.; Chavant, P.-Y. *TL* **1989**, *30*, 7373.
26. For examples: (a) Jansen, J. F. G. A.; Feringa, B. L. *JOC* **1990**, *55*, 4168. (b) Soai, K.; Kawase, Y.; Oshio, A. *JCS(P1)* **1991**, 1613.
27. Boireau, G.; Deberly, A.; Abeuhaim, D. *T* **1989**, *45*, 5837. See also: Fujisawa, T.; Ukaji, Y.; Funabora, M.; Yamashita, M.; Sato, T. *BCJ* **1990**, *63*, 1894.
28. Seebach, D.; Behrendt, L.; Felix, D. *AG(E)* **1991**, *30*, 1008.
29. House, H. O.; Crumrine, D. S.; Teranishi, A. Y.; Olmstead, H. D. *JACS* **1973**, *95*, 3310.
30. For a general review on stereoselective aldol condensations, see: Evans, D. A.; Nelson, J. V.; Taber, T. R. *Top. Stereochem.* **1982**, *13*, 1.
31. van der Steen, F. H.; Jastrzebski, J. T. B. H.; van Koten, G. *TL* **1988**, *29*, 765, 2467. See also: van der Steen, F. H.; Boersma, J.; Spek, A. L.; van Koten, G. *OM* **1991**, *10*, 2467.
32. Groth, U.; Huhn, T.; Richter, N. *LA* **1993**, 49.
33. Itoh, K.; Hamaguchi, N.; Miura, M.; Nomura, M. *JCS(P1)* **1992**, 2833.
34. Nakamura, E.; Sekiya, K.; Kuwajima, I. *TL* **1987**, *28*, 337.
35. Nakamura, E.; Aoki, S.; Sekiya, K.; Oshino, H.; Kuwajima, I. *JACS* **1987**, *109*, 8056.
36. Crimmins, M. T.; Nantermet, P. G. *JOC* **1990**, *55*, 4235.
37. (a) Tamaru, Y.; Ochiai, H.; Nakamura, T.; Yoshida, Z. *TL* **1986**, *27*, 955. (b) Tamaru, Y.; Ochiai, H.; Nakamura, T.; Tsubaki, K.; Yoshida, Z.-i. *TL* **1985**, *26*, 5559.
38. (a) Danishefsky, S. J.; Kerwin, Jr., J. F.; Kobayashi, S. *JACS* **1982**, *104*, 358. (b) Danishefsky, S. J.; Larson, E. R.; Askin, D. *JACS* **1982**, *104*, 6437. (c) Larson, E. R.; Danishefsky, S. J. *JACS* **1982**, *104*, 6458. (d) Danishefsky, S. J.; Maring, C. J. *JACS* **1985**, *107*, 1269. (e) Danishefsky, S. J.; Larson, E. R.; Askin, D.; Kato, N. *JACS* **1985**, *107*, 1246. (f) Danishefsky, S. J.; Kato, N.; Askin, D.; Kerwin, Jr., J. F. *JACS* **1982**, *104*, 360. (g) Midland, M. M.; Graham, R. S. *JACS* **1984**, *106*, 4294. (h) Garner, P. *TL* **1984**, *25*, 5855.
39. Danishefsky, S. J.; Hungate, R. *JACS* **1986**, *108*, 2486.
40. Waldmann, H.; Braun, M. *JOC* **1992**, *57*, 4444.
41. Rao, P. S.; Venkataratnam, R. V. *TL* **1991**, *32*, 5821. See also Ref. 4a.
42. Hofstraat, R. G.; Lange, J.; Scheeren, H. W.; Nivard, R. J. F. *JCS(P1)* **1988**, 2315.
43. van der Werf, A. W.; Kellogg, R. M.; van Bolhuis, F. *CC* **1991**, 682.
44. Beslin, P.; Marion, P. *TL* **1992**, *33*, 5339.
45. Page, P. C. B.; Harkin, S. A.; Marchington, A. P. *SC* **1989**, *19*, 1655. See also: Yamauchi, M.; Shirota, M.; Watanabe, T. *H* **1990**, *31*, 1699.
46. Sukata, K. *JOC* **1989**, *54*, 2015. See also: (a) Deuchert, K.; Hertenstein, W.; Wehner, G. *CB* **1979**, *112*, 2045. (b) Ruano, J. L. G.; Castro, A. M. M.; Rodriguez, J. H. *TL* **1991**, *32*, 3195.
47. Kunz, H.; Sager, W.; Pfrengle, W.; Schanzenback, D. *TL* **1988**, *29*, 4397.
48. (a) Neuenschwander, M.; Bigler, P.; Christen, K.; Iseli, R.; Kyburz, R.; Mühle, H. *HCA* **1978**, *61*, 2047. (b) Bigler, P.; Schonholzer, S.; Neuenschwander, M. *HCA* **1978**, *61*, 2059. (c) Bigler, P.; Neuenschwander, M. *HCA* **1978**, *61*, 2165.
49. Danishefsky, S. J.; Kitahara, T. *JACS* **1974**, *96*, 7807.
50. Bäuml, E.; Tschemschlok, K.; Pock, R.; Mayr, H. *TL* **1988**, *29*, 6925. See also: (a) Clark, J. H.; Kybett, A. P.; Macquarrie, D. C.; Barlow, S. J.; Landon, P. *CC* **1989**, 1353. (b) Reetz, M. T.; Sauerwald, M. *TL* **1983**, *24*, 2837.
51. Other examples: (a) Koschinsky, R.; Köhli, T.-P.; Mayr, H. *TL* **1988**, *29*, 5641. (b) Alonso, F.; Yus, M. *T* **1991**, *47*, 9119.
52. Mayr, H.; Klein, H. *JOC* **1981**, *46*, 4097.
53. Reetz, M. T.; Schwellus, K. *TL* **1978**, 1455.
54. Isler, O.; Schudel, P. *Adv. Org. Chem.* **1963**, *4*, 128. See also: Oriyama, T.; Iwanami, K.; Miyauchi, Y.; Koga, G. *BCJ* **1990**, 3716.
55. Parham, W. E.; Reed, L. J. *OSC* **1955**, *3*, 395. See also: (a) Grégoire, de Bellemont, E. *BSF* **1901**, *25*, 18. (b) Hatanaka, K.; Tanimoto, S.; Sugimoto, T.; Okano, M. *TL* **1981**, *22*, 3243.
56. Hayashi, M.; Inubushi, A.; Mukaiyama, T. *BCJ* **1988**, *61*, 4037.
57. Halcomb, R. L.; Danishefsky, S. J. *JACS* **1989**, *111*, 6661.
58. Scheeren, H. W.; Dahman, F. J. M.; Bakker, C. G. *TL* **1979**, 2925. See also: (a) Sukata, K. *BCJ* **1990**, *63*, 825. (b) Hug, E.; Mellor, M.; Scovell, E. G.; Sutherland, J. K. *CC* **1978**, 526.
59. Grummett, O.; Stearns, J. A.; Arters, A. A. *OSC* **1955**, *3*, 833.
60. Klein, H.; Mayr, H. *AG(E)* **1981**, *20*, 1027.
61. Mayr, H.; Seitz, B.; Halberstat-Kausch, I.-K. *JOC* **1981**, *46*, 1041.
62. Hoornaert, C.; Hesbain-Frisque, A. M.; Ghosez, L. *AG(E)* **1975**, *14*, 569. See also: Sidani, A.; Marchand-Brynaert, J.; Ghosez, L. *AG(E)* **1974**, *13*, 267.
63. Squires, T. G.; Schmidt, W. W.; McCandlish, Jr., C. S. *JOC* **1975**, *40*, 134.
64. (a) Goel, O. P.; Seamans, R. E. *S* **1973**, 538. (b) Kyrides, L. P. *OSC* **1943**, *2*, 528.
65. (a) Kajigaeshi, S.; Kakinami, T.; Moriwaki, M.; Tanaka, T.; Fujisaki, S.; Okamoto, T. *BCJ* **1989**, *62*, 439. (b) Kajigaeshi, S.; Kakinami, T.; Watanabe, F.; Odanoto, T. *BCJ* **1989**, *62*, 1349.
66. Ho, P.-T.; Davies, N. *JOC* **1984**, *49*, 3027.
67. Gensler, W. J.; Johnson, F.; Sloan, A. D. B. *JACS* **1960**, *82*, 6074.

68. Fort, Y.; Vanderesse, R.; Caubère, P. *TL* **1985**, *26*, 3111.
69. See Ref. 6b and Caubère, P.; Vanderesse, R.; Fort, Y. *ACS* **1991**, *45*, 742.
70. (a) Anwar, S.; Davies, A. P. *T* **1988**, *44*, 3761. (b) Laurent, A. J.; Lesniak, S. *TL* **1992**, *33*, 3311.
71. Keinan, E.; Greenspoon, N. *JACS* **1986**, *108*, 7314.
72. Kim, S.; Kim, Y. J.; Ahn, K. H. *TL* **1983**, *24*, 3369.
73. Yamakawa, T.; Masaki, M.; Nohira, H. *BCJ* **1991**, *64*, 2730.
74. (a) Mukaiyama, T. *IJS(B)* **1972**, *7*, 173. (b) Stütz, P.; Stadler, P. A. *OS* **1977**, *56*, 8.
75. Solladié, G.; Demailly, G.; Greck, C. *TL* **1985**, *26*, 435. See also: Hanamoto, T.; Fuchikama, T. *JOC* **1990**, *55*, 4969 and Ref. 6c.
76. Falorni, M.; Giacomelli, G.; Lardicci, L. *G* **1990**, *120*, 765.
77. Braun, C. E.; Cook, C. D. *OS* **1961**, *41*, 79.
78. Hanessian, S.; Kagotani, M. *Carbohydr. Res.* **1990**, *202*, 67.
79. Angyal, S. J.; Gilham, P. T.; Macdonald, C. G. *JCS* **1957**, 1417.
80. Griffen, R. J.; Lowe, P. R. *JSC(P1)* **1992**, 1811.
81. Chang, C.; Chu, K. C.; Yue, S. *SC* **1992**, *22*, 1217.
82. Groves, J. K.; Jones, N. *JCS(C)* **1968**, 2898. See also: (a) Marshall, J. A.; Andersen, N. H.; Schlicher, J. W. *JOC* **1979**, *35*, 858. (b) Groves, J. K.; Jones, N. *JCS(C)* **1969**, 2350. (c) House, H. O.; Gilmore, W. F. *JACS* **1961**, *83*, 3980.
83. Tirpak, R. E.; Rathke, M. W. *JOC* **1982**, *47*, 5099. See also: Reetz, M. T.; Kyung, S.-H. *TL* **1985**, *26*, 6333.
84. Cooper, S. R. *OSC* **1955**, *3*, 761. See also: Baddeley, G.; Williamson, R. *JCS* **1956**, 4647. See also: Zani, C. L.; de Oliveira, A. B.; Snieckus, V. *TL* **1987**, *28*, 6561.
85. Dike, S. Y.; Merchant, J. R.; Sapre, N. Y. *T* **1991**, *47*, 4775. See also: (a) Shah, V. R.; Bose, J. L.; Shah, R. C. *JOC* **1960**, *25*, 677. (b) Dallacker, F.; Kratzer, P.; Lipp, M. *LA* **1961**, *643*, 97.
86. Ref. 5a and Ruske, W. In *Friedel–Crafts and Related Reactions*; Olah, G. A., Ed.; Interscience: New York, 1964, Vol. 3, p 383.
87. Ref. 5c and (a) Shriner, R. L.; Ashley, W. C.; Welch, E. *OSC* **1955**, *3*, 725. (b) Prochazki, M. P.; Cartson, R. *ACS* **1989**, *43*, 651.
88. Ref. 5b and Olah, G. A.; Tolgyesi, W. S. In *Friedel–Crafts and Related Reactions*; Olah, G. A., Ed.; Interscience: New York, 1964, Vol. 2, p 659.

Glenn J. McGarvey
University of Virginia, Charlottesville, VA, USA

Zinc Iodide[1]

ZnI_2

[10139-47-6] I_2Zn (MW 319.19)

(used in the preparation of organozinc reagents via transmetalation;[1] a mild Lewis acid useful for promoting addition[2] and substitution reactions[3])

Physical Data: mp 446 °C; bp 625 °C (dec); *d* 4.740 g cm^{-3}.
Solubility: sol H_2O (1 g/0.3 mL), glycerol (1 g/2 mL); freely sol EtOH, Et_2O.
Form Supplied in: white, odorless, granular solid; principal impurities are H_2O and iodine.
Analysis of Reagent Purity: mp.
Purification: heat to 300 °C under vacuum for 1 h, then sublime.
Handling, Storage, and Precautions: very hygroscopic and light sensitive; store under anhydrous conditions in the absence of light.

Organozinc Reagents. Organozinc reagents may be prepared through transmetalation of organolithium, organomagnesium, and organocopper species with ZnI_2, although ***Zinc Chloride*** and ***Zinc Bromide*** are used far more frequently.[1a] It is more usually the case that organozinc iodides are prepared by zinc insertion into alkyl iodides using ***Zinc/Copper Couple***.[1a] These zinc reagents possess reactivity analogous to the other organozinc halides, finding particular use in palladium-catalyzed coupling reactions (eq 1).[4] An alternative method for the preparation of the Simmons–Smith reagent (***Iodomethylzinc Iodide***) involving the treatment of ***Diazomethane*** with ZnI_2 has been reported (eq 2).[5]

+ IZn⁀⁀CO2Et —[PdCl2(PPh3)2, PhH, DMA, rt, 78%]→ (1)

$$CH_2N_2 + ZnI_2 \longrightarrow ICH_2ZnI + N_2 \quad (2)$$

Cycloaddition Reactions. The catalytic effect of ZnI_2 on the Diels–Alder reaction has been noted (eq 3),[6] but its use in such cycloaddition reactions is rare compared with $ZnCl_2$ and $ZnBr_2$.

(3) ZnI_2, 40 °C, 55%; *endo:exo* = 67:33

Activation of C=X Bonds. Catalysis of aldol condensation reactions using silyl ketene acetals and ZnI_2 has been the subject of several studies.[7] It is observed that ZnI_2 favors the activation of functionalized carbonyl compounds via β-chelates to impart useful stereoselectivity (eq 4).[7a] In analogous fashion, diastereoselective additions to imines and nitrones have been reported which offer useful access to β-lactams (eq 5).[8] α,β-Unsaturated esters are subject to conjugate addition reactions by silyl ketene acetals, also through the agency of ZnI_2 activation (eq 6).[9]

(4) ZnI_2, MeCN, 67%; 96% *anti*

An important role for ZnI_2 has been found in the catalysis of R_3SiCN addition to ketones and aldehydes to afford silyl protected cyanohydrins.[10] This is a very general reaction that is effective even with very hindered carbonyl compounds (eq 7).[11] Diastereoselective cyanohydrin formation has been reported when these reaction conditions are applied to asymmetric carbonyl substrates (eq 8).[2a]

Activation of C–X Bonds. The Lewis acidity of ZnI_2 may be exploited to activate various carbon–heteroatom bonds to nucleophilic substitution. Treatment of epoxides and oxetanes with ***Cyanotrimethylsilane***/ZnI_2 results in selective C–N bond formation with ring opening (eq 9).[12] Similarly, C–S and C–Se bonds may be formed by treating cyclic ethers with ZnI_2 and $RSSiMe_3$ and $RSeSiMe_3$, respectively (eq 10).[13]

Treatment of 4-acetoxy- and 4-sulfoxyazetidin-2-ones with ZnI_2 results in the formation of the corresponding imine or iminium species which subsequently suffers silyl-mediated addition. These reactions result in the formation of *trans* substitution (eqs 11 and 12).[14,15]

(R^1 = Me, R^2 = H):(R^1 = H, R^2 = Me) = 3:1

Substitution reactions of orthoesters[3b] and acetals,[3c] including anomeric bond formation in carbohydrates (eq 13),[16] have been catalyzed by ZnI_2.

Reduction. Allyl and aryl ketones, aldehydes, and alcohols are reduced to the corresponding hydrocarbon by ***Sodium Cyanoborohydride***/ZnI_2 (eq 14).[17] This is a reasonably reactive reducing mixture which will also attack nitro and ester groups.

Deprotection. Methyl and benzyl ethers may be cleaved by the combination of ***(Phenylthio)trimethylsilane***/ZnI_2 (eq 15).[18] Similar cleavage of alkyl and benzyl ethers takes place using ***Acetic Anhydride***/ZnI_2 to afford the corresponding acetate (eq 16).[19]

Related Reagents. Zinc Bromide; Zinc Chloride.

1. (a) Knochel, P. *COS* **1991**, *1*, Chapter 1.7. (b) Rathke, M. W.; Weipert, P. *COS* **1991**, *2*, Chapter 1.8.
2. For examples: (a) Effenberger, T.; Hopf, M.; Ziegler, T.; Hudelmayer, J. *CB* **1991**, *124*, 1651. (b) Klimba, P. G.; Singleton, D. A. *JOC* **1992**, *57*, 1733. (c) Colvin, E. W.; McGarry, D. G. *CC* **1985**, 539.
3. For examples: (a) Paquette, L. A.; Lagerwall, D. R.; King, J. L.; Niwayama, S.; Skerlj, R. *TL* **1991**, *32*, 5259. (b) Howk, B. W.; Sauer, J. C. *OSC* **1963**, *4*, 801. (c) Kubota, T.; Iijima, M.; Tanaka, T. *TL* **1992**, *33*, 1351.
4. (a) Sakamoto, T.; Nishimura, S.; Kondo, Y.; Yamanaka, H. *S* **1988**, 485. (b) See also: Yamanaka, H.; An-naka, M.; Kondo, Y.; Sakamoto, T. *CPB* **1985**, *38*, 4309.
5. (a) Wittig, G.; Schwarzenback, K. *LA* **1961**, *650*, 1. (b) Wittig, G.; Wingher, F. *LA* **1962**, *656*, 18.
6. Brion, F. *TL* **1982**, *23*, 5239.
7. (a) Kita, Y.; Yasuda, H.; Tamura, O.; Itoh, F.; Yuan Ke, Y.; Tamura, Y. *TL* **1985**, *26*, 5777. (b) Kita, Y.; Tamura, O.; Itoh, F.; Yasuda, H.; Kishino, H.; Yuan Ke, Y.; Tamura, Y. *JOC* **1988**, *53*, 554. (c) Annunziata, R.; Cinquini, M.; Cozzi, P. G.; Consolandi, E. *JOC* **1992**, *57*, 456.
8. (a) Colvin, E. W.; McGarry, D.; Nugent, M. J. *T* **1988**, *44*, 4157, and reference 2c. See also: (b) Kita, Y.; Itoh, F.; Tamura, O.; Yuan Ke, Y.; Tamura, Y. *TL* **1987**, *28*, 1431. (c) Reider, P. J.; Grabowski, E. J. J. *TL* **1982**, *23*, 2293. (d) Chiba, T.; Nakai, T. *CL* **1987**, 2187. (e) Chiba, T.; Nagatsuma, M.; Nakai, T. *CL* **1985**, 1343.
9. Quendo, A.; Rousseau, G. *TL* **1988**, *29*, 6443.
10. (a) Rassmussen, J. K.; Heihmann, S. M. *S* **1978**, 219. (b) Gassman, P.; Talley, J. J. *TL* **1978**, 3773. (c) Foley, L. H. *SC* **1984**, *14*, 1291. (d) Higuchi, K.; Onaka, M.; Izumi, Y. *CC* **1991**, 1035. (e) Quast, H.; Carlson, J.; Klaubert, C. A.; Peters, E.-M.; Peters, K.; von Schnering, H. G. *LA* **1992**, 759. (f) Batra, M. S.; Brunet, E. *TL* **1993**, *34*, 711.
11. Golinski, M.; Brock, C. P.; Watt, D. S. *JOC* **1993**, *58*, 159.
12. (a) Gassman, P. G.; Gremban, R. S. *TL* **1984**, *25*, 3259. (b) See also: Gassman, P. G.; Haberman, L. M. *TL* **1985**, *26*, 4971.
13. (a) Miyoshi, N.; Hatayama, Y.; Ryu, I.; Kambe, N.; Murai, T.; Murai, S.; Sonoda *S* **1988**, 175. (b) See also: Guidon, Y.; Young, R. N.; Frenette, R. *SC* **1981**, *11*, 391.
14. (a) Kita, Y.; Shibata, N.; Yoshida, N.; Tohjo, T. *TL* **1991**, *32*, 2375. (b) See also: Kita, Y.; Shibata, N.; Miki, T.; Takemura, Y.; Tamura, O. *CC* **1990**, 727.
15. Fuentes, L. M.; Shinkai, I.; Salzmann, T. N. *JACS* **1986**, *108*, 4675.
16. (a) Hanessian, S.; Guidon, Y. *Carbohydr. Res.* **1980**, *86*, C3. (b) Chu, S.-H. L.; Anderson, L. *Carbohydr. Res.* **1976**, *50*, 227.
17. Lau, C. K.; Dufresne, C.; Bélanger, P. C.; Piétré, S.; Scheigetz, J. *JOC* **1986**, *51*, 3038.
18. Hanessian, S.; Guidon, Y. *TL* **1980**, *21*, 2305.
19. Benedetti, M. O. V.; Monteagudo, E. S.; Burton, G. *JCS(S)* **1990**, 248.

Glenn J. McGarvey
University of Virginia, Charlottesville, VA, USA

Zinc *p*-Toluenesulfonate

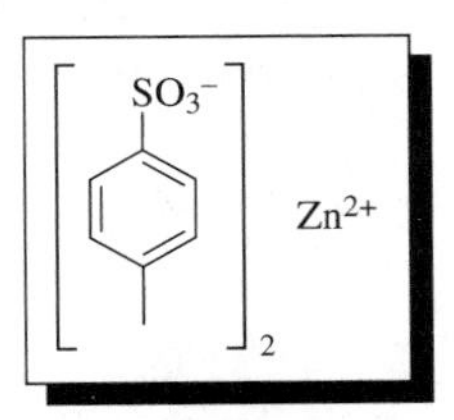

[13438-45-4] $C_{14}H_{14}O_6S_2Zn$ (MW 407.81)
($\cdot 6H_2O$)
[60553-56-2]
($\cdot xH_2O$; 'hydrate')
[123334-05-4]

(agent for tosylation[1] under mild Mitsunobu conditions, usually with inversion)

Alternate Name: zinc tosylate.
Solubility: insol H_2O, benzene.
Form Supplied in: white solid; the 'hydrate' is commercially available.
Preparative Method: anhydrous material[1] is prepared when the precipitate formed from mixture of ***p-Toluenesulfonic Acid*** and ***Zinc Chloride*** in water is filtered off and dried in vacuo at rt for 1 h. Purification is not usually required.
Handling, Storage, and Precautions: no unusual precautions are noted, but storage in a well stoppered container is advisable.

Tosylate Formation with Inversion of Configuration. Alkyl tosylates can be formed directly from secondary alcohol functionality with retention of carbon stereochemistry by treatment with ***p-Toluenesulfonyl Chloride*** and ***Pyridine***. However, conversion of an alcohol to the corresponding tosylate of opposite stereochemistry typically requires a minimum of three steps. For example, inversion of the stereocenter with benzoic acid under Mitsunobu reaction conditions, hydrolysis of the resulting ester, and finally conventional tosylation of the alcohol, provides an attractive route for this transformation.[2] A similar route, the inversion of a secondary alcohol directly with *p*-TsOH, ***Diethyl Azodicarboxylate*** (DEAD), and ***Triphenylphosphine***, does not produce the desired tosylate product.[1]

Modification of this procedure, through the use of bis(*p*-toluenesulfonato)zinc ((*p*-TsO)$_2$Zn) with Mitsunobu conditions, allows this transformation to proceed in a single step.[1] Although somewhat slower than the analogous inversion with benzoic acid, treatment of secondary alcohols with (*p*-TsO)$_2$Zn, DEAD, and Ph_3P gives the desired tosylate product in high yield. This reaction is effective for tosylate formation from either acyclic or cyclic secondary alcohols with clean inversion of stereochemistry (eq 1). As expected, this reaction is very sensitive to the steric hindrance at the reacting sp^3 center; moderately hindered alcohols, e.g. (**1**), require longer reaction times and greater quantities of DEAD and Ph_3P to undergo the transformation in high yield (eq 2). Similarly, the use of *p*-TsOLi is effective for this transformation, but yields are slightly lower than those obtained with (*p*-TsO)$_2$Zn.

$(p\text{-TsO})_nM$, DEAD, PPh_3, benzene, rt, 3 h (1)

$(p\text{-TsO})_2Zn$	88%
p-TsOLi	75%

$(p\text{-TsO})_nM$, DEAD, PPh_3, benzene, rt, 48 h (2)

(1)

$(p\text{-TsO})_2Zn$	90%
p-TsOLi	68%

Tosylation with Retention of Stereochemistry[1]. Although dihydrocholesterol (**2**) undergoes clean tosylation with inversion in 88% yield using the procedure described above, cholesterol (**3a**) itself gives an 85% yield of the 3β-tosylate (**3b**).

(2)

(**3a**) X = H
(**3b**) X = Ts

It was suggested that this anomaly could be explained by the intermediacy of a cyclopropylmethyl/homoallylic species, trapped by addition of tosylate ion. Indeed, the more vigorous Mitsunobu reaction[2] of (**3a**) with benzoic acid leads to a mixture of four products, confirming involvement of this type of homoallylic rearrangement.

Although anomalous, it is conceivable that this Mitsunobu procedure could be of use in effecting tosylation with retention in less common systems in which neighboring group effects, for example, override the normal stereochemical outcome.

1. Galynker, I.; Still, W. C. *TL* **1982**, *23*, 4461.
2. (a) Mitsunobu, O.; Eguchi, M. *BCJ* **1971**, *44*, 3427. (b) Mitsunobu, O. *S* **1981**, 1.
3. Keinan, E.; Sinha, S. C.; Sinha-Bagchi, A. *JOC* **1992**, *57*, 3631.
4. Ager, D. J.; East, M. B. *T* **1992**, *48*, 2803.

Peter Ham
SmithKline Beecham Pharmaceuticals, Harlow, UK

John R. Stille
Michigan State University, East Lansing, MI, USA

List of Contributors

Fernando Albericio — *Millipore Corporation, Bedford, MA, USA*
- *O*-Benzotriazol-1-yl-*N*,*N*,*N*′,*N*′-tetramethyluronium Hexafluorophosphate 40
- Azidotris(dimethylamino)phosphonium Hexafluorophosphate 34

Jeffrey S. Albert — *University of Pittsburgh, PA, USA*
- 1,3-Dicyclohexylcarbodiimide 133

Kim F. Albizati — *University of California, San Diego, CA, USA*
- *N*-Methyl-*N*,*N*′-dicyclohexylcarbodiimidium Iodide 276

Roger W. Alder — *University of Bristol, UK*
- Methyl fluorosulfonate and Methyl Trifluoromethanesulfonate 278

Benjamin A. Anderson — *Lilly Research Laboratories, Indianapolis, IN, USA*
- Hexamethyldisilazane 205

Paul C. Anderson — *Bio-Méga/Boehringer Ingelheim Research, Laval, Québec, Canada*
- Bromodimethylborane 77

Toyohiko Aoyama — *Nagoya City University, Japan*
- Trimethylsilyldiazomethane 422

Alan Armstrong — *University of Bath, UK*
- *N*,*N*′-Carbonyldiimidazole 93
- *o*-Nitrophenol 289
- *p*-Nitrophenol 290
- 2-Chloro-1-methylpyridinium Iodide 99

Jeffrey Aubé — *University of Kansas, Lawrence, KS, USA*
- 2-(*t*-Butoxycarbonyloxyimino)-2-phenylacetonitrile 81

Martin G. Banwell — *University of Melbourne, Parkville, Victoria, Australia*
- Titanium Tetraisopropoxide 389

William E. Bauta — *Sandoz Research Institute, East Hanover, NJ, USA*
- Benzyl Bromide 45

T. Howard Black — *Eastern Illinois University, Charleston, IL, USA*
- Magnesium Bromide 253

Andrew N. Boa — *Leicester University, UK*
- Benzyl 2,2,2-Trichloroacetimidate 51
- 3,4-Dimethoxybenzyl Bromide 151

Bruce P. Branchaud — *University of Oregon, Eugene, OR, USA*
- Tetrafluoroboric Acid 361

John E. Burks, Jr. — *Eli Lilly and Company, Lafayette, IN, USA*
- Phosphorus(V) Chloride 335

Barry Lygo	*Salford University, UK*	
	• 1-Hydroxybenzotriazole	220
Percy S. Manchand	*Hoffmann-La Roche, Nutley, NJ, USA*	
	• Ethyl Vinyl Ether	194
Lewis N. Mander	*The Australian National University, Canberra, Australia*	
	• Methyl Cyanoformate	273
Michael J. Martinelli	*Lilly Research Laboratories, Indianapolis, IN, USA*	
	• Bromotrimethylsilane	79
	• Iodotrimethylsilane	234
Antonio García Martínez	*Universität Tübingen, Germany*	
	• Trifluoromethanesulfonic Anhydride	410
Patrick G. McDougal	*Reed College, Portland, OR, USA*	
	• Trichloroacetonitrile	400
Glenn J. McGarvey	*University of Virginia, Charlottesville, VA, USA*	
	• Zinc Bromide	468
	• Zinc Chloride	471
	• Zinc Iodide	479
Ross P. McGeary	*University of Cambridge, UK*	
	• *N*-Ethylbenzisoxazolium Tetrafluoroborate	182
Mark S. Meier	*University of Kentucky, Lexington, KY, USA*	
	• Phosphorus Oxychloride	346
	• Phosphorus(V) Oxide	341
Gregory Merriman	*Hoechst-Roussel Pharmaceuticals, Somerville, NJ, USA*	
	• Dimethyl Sulfate	162
William J. Middleton	*Ursinus College, Collegeville, PA, USA*	
	• Tris(dimethylamino)sulfonium Difluorotrimethylsilicate	464
Oyo Mitsunobu	*Aoyama Gakuin University, Tokyo, Japan*	
	• Triphenylphosphine–Diethyl Azodicarboxylate	454
Norio Miyaura	*Hokkaido University, Sapporo, Japan*	
	• Boron Trichloride	66
Shahriar Mobashery	*Wayne State University, Detroit, MI, USA*	
	• (Methoxycarbonylsulfamoyl)triethylammonium Hydroxide	263
	• Bis(trichloromethyl) Carbonate	61
Kenneth P. Moder	*Eli Lilly and Company, Lafayette, IN, USA*	
	• Phosphorus(III) Chloride	333
Bradford P. Mundy	*Colby College, Waterville, ME, USA*	
	• Phosphorus(III) Bromide	330
	• Phosphorus(V) Bromide	332
Tetsuya Nagasawa	*Keio University, Yokohama, Japan*	
	• Triethylaluminum	404
James S. Nowick	*University of California, Irvine, CA, USA*	
	• Lithium Chloride	248
George A. Olah	*University of Southern California, Los Angeles, CA, USA*	
	• Antimony(V) Fluoride	29

Helen Osborn — *University of Bristol, UK*
- 2-Morpholinoethyl Isocyanide 285

Peter L. Pauson — *University of Strathclyde, Glasgow, UK*
- Octacarbonyldicobalt 298

Andrew N. Payne — *University of Cambridge, UK*
- Ethyl Chloroformate 183

Bruce A. Pearlman — *The Upjohn Company, Kalamazoo, MI, USA*
- Acetyl Chloride 16

Gemma Perkins — *University of Bristol, UK*
- Trifluoroacetic Anhydride 409
- Trimethylsilyl Trifluoromethanesulfonate 432

Justin G. E. Phillips — *University of Bristol, UK*
- Methyl fluorosulfonate and Methyl Trifluoromethanesulfonate 278

Ulf Pindur — *University of Mainz, Germany*
- *N,N*-Dimethylformamide Diethyl Acetal 158

Harold W. Pinnick — *Bucknell University, Lewisburg, PA, USA*
- Benzyl Chloromethyl Ether 50

Robin L. Polt — *University of Arizona, Tucson, AZ, USA*
- 9-Fluorenylmethyl Chloroformate 198

Richard S. Pottorf — *Marion Merrell Dow Research Institute, Cincinnati, OH, USA*
- *N*-Ethyl-5-phenylisoxazolium-3′-sulfonate 191
- 1,1′-Thionylimidazole 373
- 2-Thiopyridyl Chloroformate 375

G. K. Surya Prakash — *University of Southern California, Los Angeles, CA, USA*
- Antimony(V) Fluoride 29

T. V. (Babu) RajanBabu — *The Ohio State University, Columbus, OH, USA*
- Di-*n*-butyltin Oxide 130
- Tris(dimethylamino)sulfonium Difluorotrimethylsilicate 464

Janet L. Ralbovsky — *The Pennsylvania State University, University Park, PA, USA*
- β-Trimethylsilylethanesulfonyl Chloride 425

Tapan Ray — *Sandoz Pharmaceuticals, East Hanover, NJ, USA*
- Isobutyl Chloroformate 243
- Iodonium Di-*sym*-collidine Perchlorate 232

Jayachandra P. Reddy — *Indiana University, Bloomington, IN, USA*
- 2-(Trimethylsilyl)ethanol 427

David C. Rees — *Organon Laboratories Ltd, Newhouse, Lanarkshire, UK*
- Bis(2-oxo-3-oxazolidinyl)phosphinic Chloride 57

Frode Rise — *University of Oslo, Norway*
- Titanium(IV) Chloride 383

Brian A. Roden — *Abbott Laboratories, North Chicago, IL, USA*
- Diphenylbis(1,1,1,3,3,3-hexafluoro-2-phenyl-2-propoxy)sulfurane 165
- Hydrazine 216

Juliatiek Roestamadji — *Wayne State University, Detroit, MI, USA*
- Bis(trichloromethyl) Carbonate 61

Lakshminarayanapuram R. Subramanian — *Universität Tübingen, Germany*
- Trifluoromethanesulfonic Anhydride 410

Gary A. Sulikowski — *Texas A&M University, College Station, TX, USA*
- Iodomethane 228

Michelle M. Sulikowski — *Texas A&M University, College Station, TX, USA*
- Iodomethane 228

Akira Suzuki — *Hokkaido University, Sapporo, Japan*
- Boron Tribromide 62

Keisuke Suzuki — *Keio University, Yokohama, Japan*
- Triethylaluminum 404

Joseph Sweeney — *University of Bristol, UK*
- Trifluoroacetic Anhydride 409
- Trimethylsilyl Trifluoromethanesulfonate 432

Peter Szeto — *University of Bristol, UK*
- 1-Ethyl-3-(3′-dimethylaminopropyl)carbodiimide Hydrochloride 186

Pascale Taibi — *Wayne State University, Detroit, MI, USA*
- (Methoxycarbonylsulfamoyl)triethylammonium Hydroxide 263

James M. Takacs — *University of Nebraska-Lincoln, NE, USA*
- Phosphorus(III) Iodide 338

Michael J. Taschner — *The University of Akron, OH, USA*
- Triphenylphosphine–Carbon Tetrabromide 440
- Triphenylphosphine–Carbon Tetrachloride 442

Richard T. Taylor — *Miami University, Oxford, OH, USA*
- Triethyl Orthoformate 406

Yoshiyasu Terao — *University of Shizuoka, Japan*
- *N*-Methyl-*N*-nitroso-*p*-toluenesulfonamide 277

Albert V. Thomas — *Abbott Laboratories, North Chicago, IL, USA*
- Diphenyl Phosphorazidate 168

Victor J. Tortorelli — *Ursinus College, Collegeville, PA, USA*
- Tris(dimethylamino)sulfonium Difluorotrimethylsilicate 464

Edward Turos — *State University of New York at Buffalo, NY, USA*
- Chlorotriethylsilane 100

Motokazu Uemura — *Osaka City University, Japan*
- Hexacarbonylchromium 201

Kjell Undheim — *University of Oslo, Norway*
- Titanium(IV) Chloride 383

Christopher J. Urch — *Zeneca Agrochemicals, Bracknell, UK*
- Trimethylacetyl Chloride 418

Valerie Vaillancourt — *The Upjohn Co., Kalamazoo, MI, USA*
- Mesitylenesulfonyl Chloride 255
- Methanesulfonyl Chloride 257

Scott C. Virgil — *Massachusetts Institute of Technology, Cambridge, MA, USA*
- Triphenylphosphine–*N*-Bromosuccinimide 438

David E. Volk — *North Dakota State University, Fargo, ND, USA*
- Tin(IV) Chloride 377

Simone C. Vonwiller — *The University of Sydney, NSW, Australia*
- *N*,*N*,*N*′,*N*′-Tetramethylethylenediamine 364

Jürgen Voss — *Universität Hamburg, Germany*
- 2,4-Bis(4-methoxyphenyl)-1,3,2,4-dithiadiphosphetane 2,4-Disulfide 53

Michel Wakselman — *CNRS-CERCOA, Thiais, France*
- Di-*t*-butyl Dicarbonate 123

Michael A. Walters — *Dartmouth College, Hanover, NH, USA*
- 2,2-Dimethyl-1,3-propanediol 161
- Isopropenyl Acetate 244

Qi Wang — *University of Southern California, Los Angeles, CA, USA*
- Antimony(V) Fluoride 29

Steven M. Weinreb — *The Pennsylvania State University, University Park, PA, USA*
- β-Trimethylsilylethanesulfonyl Chloride 425

D. Todd Whitaker — *Detroit Country Day School, Beverly Hills, MI, USA*
- *o*-Nitrobenzyl Alcohol 287
- 2-Methoxypropene 267

Wayne State University, Detroit, MI, USA
- *p*-Toluenesulfonyl Chloride 394

K. Sinclair Whitaker — *Wayne State University, Detroit, MI, USA*
- *o*-Nitrobenzyl Alcohol 287
- *p*-Toluenesulfonyl Chloride 394
- 2-Methoxypropene 267

David D. Wirth — *Eli Lilly & Co., Lafayette, IN, USA*
- Thionyl Chloride 370

Lars-G. Wistrand — *Nycomed Innovation, Malmö, Sweden*
- Silver(I) Tetrafluoroborate 355

Michael S. Wolfe — *National Institutes of Health, Bethesda, MD, USA*
- 2-(*t*-Butoxycarbonyloxyimino)-2-phenylacetonitrile 81

Peter G. M. Wuts — *The Upjohn Co., Kalamazoo, MI, USA*
- *p*-Methoxybenzyl Chloride 260
- 2-Methoxyethoxymethyl Chloride 265
- Chloromethyl Methyl Ether 96

Jonathan R. Young — *University of Wisconsin, Madison, WI, USA*
- Diethyl Phosphorochloridate 144

Mark W. Zettler — *The Dow Chemical Company, Midland, MI, USA*
- Montmorillonite K10 282

Regina Zibuck — *Wayne State University, Detroit, MI, USA*
- Acetic Anhydride 9

Reagent Formula Index

Subject Index

Reference Abbreviations

ABC	Agric. Biol. Chem.
AC(R)	Ann. Chim. (Rome)
ACR	Acc. Chem. Res.
ACS	Acta Chem. Scand.
AF	Arzneim.-Forsch.
AG	Angew. Chem.
AG(E)	Angew. Chem., Int. Ed. Engl.
AJC	Aust. J. Chem.
AK	Ark. Kemi
ANY	Ann. N. Y. Acad. Sci.
AP	Arch. Pharm. (Weinheim, Ger.)
B	Biochemistry
BAU	Bull. Acad. Sci. USSR, Div. Chem. Sci.
BBA	Biochim. Biophys. Acta
BCJ	Bull. Chem. Soc. Jpn.
BJ	Biochem. J.
BML	Bioorg. Med. Chem. Lett.
BSB	Bull. Soc. Chim. Belg.
BSF(2)	Bull. Soc. Chem. Fr. Part 2
C	Chimia
CA	Chem. Abstr.
CB	Ber. Dtsch. Chem. Ges./Chem. Ber.
CC	Chem. Commun./J. Chem. Soc., Chem. Commun.
CCC	Collect. Czech. Chem. Commun.
CED	J. Chem. Eng. Data
CI(L)	Chem. Ind. (London)
CJC	Can. J. Chem.
CL	Chem. Lett.
COS	Comprehensive Organic Synthesis
CPB	Chem. Pharm. Bull.
CR(C)	C. R. Hebd. Seances Acad. Sci., Ser. C
CRV	Chem. Rev.
CS	Chem. Ser.
CSR	Chem. Soc. Rev.
CZ	Chem.-Ztg.
DOK	Dokl. Akad. Nauk SSSR
E	Experientia
FES	Farmaco Ed. Sci.
FF	Fieser & Fieser
G	Gazz. Chim. Ital.
H	Heterocycles
HC	Heteroatom Chem.
HCA	Helv. Chim. Acta
IC	Inorg. Chem.
ICA	Inorg. Chim. Acta
IJC(B)	Indian J. Chem., Sect. B
IJS(B)	Int. J. Sulfur Chem., Part B
IZV	Izv. Akad. Nauk SSSR, Ser. Khim.
JACS	J. Am. Chem. Soc.
JBC	J. Biol. Chem.
JCP	J. Chem. Phys.
JCR(M)	J. Chem. Res. (M)
JCR(S)	J. Chem. Res. (S)
JCS	J. Chem. Soc.
JCS(C)	J. Chem. Soc. (C)
JCS(D)	J. Chem. Soc., Dalton Trans.
JCS(F)	J. Chem. Soc., Faraday Trans.
JCS(P1)	J. Chem. Soc., Perkin Trans. 1
JCS(P2)	J. Chem. Soc., Perkin Trans. 2
JFC	J. Fluorine Chem.
JGU	J. Gen. Chem. USSR (Engl. Transl.)
JHC	J. Heterocycl. Chem.
JIC	J. Indian Chem. Soc.
JMC	J. Med. Chem.
JMR	J. Magn. Reson.
JOC	J. Org. Chem.
JOM	J. Organomet. Chem.
JOU	J. Org. Chem. USSR (Engl. Transl.)
JPOC	J. Phys. Org. Chem.
JPP	J. Photochem. Photobiol.
JPR	J. Prakt. Chem.
JPS	J. Pharm. Sci.
KGS	Khim. Geterotsikl. Soedin.
LA	Justus Liebigs Ann. Chem./Liebigs Ann. Chem.
M	Monatsh. Chem.
MOC	Methoden Org. Chem. (Houben-Weyl)
MRC	Magn. Reson. Chem.
N	Naturwissenschaften
NJC	Nouv. J. Chim.
NKK	Nippon Kagaku Kaishi
OM	Organometallics
OMR	Org. Magn. Reson.
OPP	Org. Prep. Proced. Int.
OR	Org. React.
OS	Org. Synth.
OSC	Org. Synth., Coll. Vol.
P	Phytochemistry
PAC	Pure Appl. Chem.
PIA(A)	Proc. Indian Acad. Sci., Sect. A